Win the complete library of F

EFFICIENT

ACCURATE

CUSTOMIZABLE

Register your book below to receive your **free** quarterly updates. **Plus** you will be entered into a quarterly drawing to win the **complete** RSMeans Online library of 2016 data!

Be sure to keep up-to-date in 2016!

Fill out the card below for RSMeans' free quarterly updates,
as well as a chance to win the complete RSMeans Online library.
Please provide your name, address, and email below and return
this card by mail, or register online:

info.thegordiangroup.com/2016updates.html

Name _____

Email_____ Title_____

Company _____

Street _____

City/Town _____ State/Prov. _____ Zip/Postal Code _____

Win the complete library of RSMeans Online data!

EFFICIENT

ACCURATE

CUSTOMIZABLE

Register your book below to receive your **free** quarterly updates.
Plus you will be entered into a quarterly drawing to win the **complete** RSMeans Online library of 2016 data!

Mechanical Cost Data

2016
39th annual edition

Melville J. Mossman, PE, Senior Editor

RSMeans dedicates this book to Mel Mossman: its creator and keeper for 40 years. A respected and loved member of our team who is a master of his craft.

40 years
Congratulations, Mel!

Engineering Director
Bob Mewis, CCP (*1, 2, 4, 13, 14, 31, 32, 33, 34, 35, 41, 44, 46*)

Contributing Editors
Christopher Babbitt
Adrian C. Charest, PE
Cheryl Elsmore
Wafaa Hamitou
Joseph Kelble
Charles Kibbee
Robert J. Kuchta (*8*)
Michael Landry
Thomas Lane (*6, 7*)
Genevieve Medeiros
Elisa Mello
Chris Morris (*26, 27, 28, 48*)

Melville J. Mossman, PE (*21, 22, 23*)
Marilyn Phelan, AIA (*9, 10, 11, 12*)
Stephen C. Plotner (*3, 5*)
Stephen Rosenberg
Kevin Souza
Keegan Spraker
Tim Tonello
David Yazbek

Vice President Data & Engineering
Chris Anderson

Product Manager
Andrea Sillah

Production Manager
Debbie Panarelli

Production
Sharon Larsen
Jonathan Forgit
Sheryl Rose
Mary Lou Geary

Technical Support
Gary L. Hoitt
Kathryn S. Rodriguez

Cover Design
Blaire Gaddis

RSMeans
FROM THE GORDIAN GROUP®

THE GORDIAN GROUP®

Numbers in italics are the divisional responsibilities for each editor. Please contact the designated editor directly with any questions.

RSMeans
Construction Publishers & Consultants
1099 Hingham Street, Suite 201
Rockland, MA 02370
United States of America
1-877-739-3689
www.RSMeans.com

Printed in the United States of America
ISSN 1524-3702
ISBN 978-1-943215-13-3

0026　$219.99 per copy (in United States)
Price is subject to change without prior notice.

Related RSMeans Products and Services

The engineers at RSMeans suggest the following products and services as companion information resources to *RSMeans Mechanical Cost Data:*

Construction Cost Data
Assemblies Cost Data 2016
Square Foot Costs 2016

Reference Books
Estimating Building Costs
RSMeans Estimating Handbook
Green Building: Project Planning & Estimating
How to Estimate with RSMeans Data
Plan Reading & Material Takeoff
Project Scheduling & Management for Construction
Universal Design Ideas for Style, Comfort & Safety

Seminars and In-House Training
Unit Price Estimating
RSMeans Online® Training
Practical Project Management for Construction Professionals
Scheduling with MSProject for Construction Professionals
Mechanical & Electrical Estimating

RSMeans Online Store
Visit RSMeans at www.RSMeans.com for the most reliable and current resources available on the market. Learn more about our more than 20 data sets available in Online, Book, eBook, and CD formats. Our library of reference books is also available, along with professional development seminars aimed at improving cost estimating, project management, administration, and facilities management skills.

RSMeans Electronic Data
Receive the most up-to-date cost information with RSMeans Online. This web-based service is quick, intuitive, and easy to use, giving you instant access to RSMeans' comprehensive database. Learn more at: **www.RSMeans.com/Online**.

RSMeans Custom Solutions
Building owners, facility managers, building product manufacturers, attorneys, and even insurance firms across the public and private sectors have engaged in RSMeans' custom solutions to solve their estimating needs including:

Custom Cost Engineering Solutions
Knowing the lifetime maintenance cost of your building, how much it will cost to build a specific building type in different locations across the country and globe, and if the estimate from which you are basing a major decision is accurate are all imperative for building owners and their facility managers.

Market & Custom Analytics
Taking a construction product to market requires accurate market research in order to make critical business development decisions. Once the product goes to market, one way to differentiate your product from the competition is to make any cost saving claims you have to offer.

Third-Party Legal Resources
Natural disasters can lead to reconstruction and repairs just as new developments and expansions can lead to issues of eminent domain. In these cases where one party must pay, it is plausible for legal disputes over costs and estimates to arise.

Construction Costs for Software Applications
More than 25 unit price and assemblies cost databases are available through a number of leading estimating and facilities management software partners. For more information, see "Other RSMeans Products and Services" at the back of this publication.

RSMeans data is also available to federal, state, and local government agencies as multi-year, multi-seat licenses.

For procurement construction cost data such as Job Order Contracting and Change Order Management please refer to The Gordian Group's procurement solutions.

For information on our current partners, call 1-877-739-3689.

Table of Contents

Foreword

Who We Are

Since 1942, RSMeans has delivered construction cost estimating information and consulting throughout North America. In 2014, RSMeans was acquired by The Gordian Group, combining two industry-leading construction cost databases. Through the RSMeans line of products and services, The Gordian Group provides innovative construction cost estimating data to organizations pursuing efficient and effective construction planning, estimating, procurement, and information solutions.

Our Offerings

With RSMeans' construction cost estimating data, contractors, architects, engineers, facility owners, and managers can utilize the most up-to-date data available in many formats to meet their specific planning and estimating needs.

When you purchase information from RSMeans, you are, in effect, hiring the services of a full-time staff of construction and engineering professionals.

Our thoroughly-experienced and highly-qualified staff works daily to collect, analyze, and disseminate comprehensive cost information to meet your needs.

These staff members have years of practical construction experience and engineering training prior to joining the firm. As a result, you can count on them not only for accurate cost figures, but also for additional background reference information that will help you create a realistic estimate.

The RSMeans organization is equipped to help you solve construction problems through its variety of data solutions.

Access our comprehensive database electronically with RSMeans Online. Quick, intuitive, easy to use, and updated continuously throughout the year, RSMeans Online is the most accurate and up-to-date cost estimating data.

This up-to-date, accurate, and localized data can be leveraged for a broad range of applications, such as Custom Cost Engineering solutions, Market and Custom Analytics, and Third-Party Legal Resources.

To ensure you are getting the most from your cost estimating data, we also offer a myriad of reference guides, training, and professional seminars (learn more at **www.RSMeans.com/Learn**).

In short, RSMeans can provide you with the tools and expertise for developing accurate and dependable construction estimates and budgets in a variety of ways.

Our Commitment

Today at RSMeans, we do more than talk about the quality of our data and the usefulness of the information. We stand behind all of our data—from historical cost indexes to construction materials and techniques—to craft current costs and predict future trends.

If you have any questions about our products or services, please call us toll-free at 1-877-739-3689. You can also visit our website at: **www.RSMeans.com**.

How the Cost Data Is Built: An Overview

A Powerful Construction Tool

You now have one of the most powerful construction tools available today. A successful project is built on the foundation of an accurate and dependable estimate. This tool will enable you to construct such an estimate.

For the casual user the information is designed to be:

- quickly and easily understood so you can get right to your estimate.
- filled with valuable information so you can understand the necessary factors that go into a cost estimate.

For the professional user, the information is designed to be:

- a handy reference that can be quickly referred to for key costs.
- a comprehensive, fully reliable source of current construction costs and productivity rates so you'll be prepared to estimate any project.
- a source for preliminary project costs, product selections, and alternate materials and methods.

To meet all of these requirements, we have organized the information into the following clearly defined sections.

Estimating with RSMeans Unit Price Cost Data*

Please refer to these steps for guidance on completing an estimate using RSMeans unit price cost data.

How to Use the Information: The Details

This section contains an in-depth explanation of how the information is arranged and how you can use it to determine a reliable construction cost estimate. It includes how we develop our cost figures and how to prepare your estimate.

Unit Prices*

All cost data has been divided into 50 divisions according to the MasterFormat system of classification and numbering.

Assemblies*

The cost data in this section has been organized in an "Assemblies" format. These assemblies are the functional elements of a building and are arranged according to the 7 elements of the UNIFORMAT II classification system. For a complete explanation of a typical "Assembly", see "How RSMeans Assemblies Data Works."

Residential Models*

Model buildings for four classes of construction—economy, average, custom, and luxury—are developed and shown with complete costs per square foot.

Commercial/Industrial/ Institutional Models*

This section contains complete costs for 77 typical model buildings expressed as costs per square foot.

Green Commercial/Industrial/ Institutional Models*

This section contains complete costs for 25 green model buildings expressed as costs per square foot.

References*

This section includes information on Equipment Rental Costs, Crew Listings, Historical Cost Indexes, City Cost Indexes, Location Factors, Reference Tables, and Change Orders, as well as a listing of abbreviations.

- **Equipment Rental Costs:** Included are the average costs to rent and operate hundreds of pieces of construction equipment.
- **Crew Listings:** This section lists all the crews referenced in the cost data. A crew is composed of more than one trade classification and/or the addition of power equipment to any trade classification. Power equipment is included in the cost of the crew. Costs are shown both with bare labor rates and with the installing contractor's overhead and profit added. For each, the total crew cost per eight-hour day and the composite cost per labor-hour are listed.

- **Historical Cost Indexes:** These indexes provide you with data to adjust construction costs over time.
- **City Cost Indexes:** All costs in this data set are U.S. national averages. Costs vary by region. You can adjust for this by CSI Division to over 700 locations throughout the U.S. and Canada by using this data.
- **Location Factors:** You can adjust total project costs to over 900 locations throughout the U.S. and Canada by using the weighted number, which applies across all divisions.
- **Reference Tables:** At the beginning of selected major classifications in the Unit Prices are reference numbers indicators. These numbers refer you to related information in the Reference Section. In this section, you'll find reference tables, explanations, and estimating information that support how we develop the unit price data, technical data, and estimating procedures.
- **Change Orders:** This section includes information on the factors that influence the pricing of change orders.
- **Abbreviations:** A listing of abbreviations used throughout this information, along with the terms they represent, is included.

Index (printed versions only)

A comprehensive listing of all terms and subjects will help you quickly find what you need when you are not sure where it occurs in MasterFormat.

Conclusion

This information is designed to be as comprehensive and easy to use as possible.

The Construction Specifications Institute (CSI) and Construction Specifications Canada (CSC) have produced the 2014 edition of MasterFormat®, a system of titles and numbers used extensively to organize construction information.

All unit prices in the RSMeans cost data are now arranged in the 50-division MasterFormat® 2014 system.

---

* Not all information is available in all data sets

Note: The material prices in RSMeans cost data are "contractor's prices." They are the prices that contractors can expect to pay at the lumberyards, suppliers'/distributors' warehouses, etc. Small orders of specialty items would be higher than the costs shown, while very large orders, such as truckload lots, would be less. The variation would depend on the size, timing, and negotiating power of the contractor. The labor costs are primarily for new construction or major renovation rather than repairs or minor alterations. With reasonable exercise of judgment, the figures can be used for any building work.

Estimating with RSMeans Unit Prices

Following these steps will allow you to complete an accurate estimate using RSMeans Unit Price cost data.

1. Scope Out the Project

- Think through the project and identify the CSI Divisions needed in your estimate.
- Identify the individual work tasks that will need to be covered in your estimate.
- The Unit Price data has been divided into 50 Divisions according to the CSI MasterFormat® 2014.
- In printed versions, the Unit Price Section Table of Contents on page 1 may also be helpful when scoping out your project.
- Experienced estimators find it helpful to begin with Division 2 and continue through completion. Division 1 can be estimated after the full project scope is known.

2. Quantify

- Determine the number of units required for each work task that you identified.
- Experienced estimators include an allowance for waste in their quantities. (Waste is not included in RSMeans Unit Price line items unless otherwise stated.)

3. Price the Quantities

- Use the search tools available to locate individual Unit Price line items for your estimate.
- Reference Numbers indicated within a Unit Price section refer to additional information that you may find useful.
- The crew indicates who is performing the work for that task. Crew codes are expanded in the Crew Listings in the Reference Section to include all trades and equipment that comprise the crew.
- The Daily Output is the amount of work the crew is expected to do in one day.
- The Labor-Hours value is the amount of time it will take for the crew to install one unit of work.
- The abbreviated Unit designation indicates the unit of measure upon which the crew, productivity, and prices are based.
- Bare Costs are shown for materials, labor, and equipment needed to complete the Unit Price line item. Bare costs do not include waste, project overhead, payroll insurance, payroll taxes, main office overhead, or profit.
- The Total Incl O&P cost is the billing rate or invoice amount of the installing contractor or subcontractor who performs the work for the Unit Price line item.

4. Multiply

- Multiply the total number of units needed for your project by the Total Incl O&P cost for each Unit Price line item.
- Be careful that your take off unit of measure matches the unit of measure in the Unit column.
- The price you calculate is an estimate for a completed item of work.
- Keep scoping individual tasks, determining the number of units required for those tasks, matching each task with individual Unit Price line items, and multiplying quantities by Total Incl O&P costs.
- An estimate completed in this manner is priced as if a subcontractor, or set of subcontractors, is performing the work. The estimate does not yet include Project Overhead or Estimate Summary components such as general contractor markups on subcontracted work, general contractor office overhead and profit, contingency, and location factors.

5. Project Overhead

- Include project overhead items from Division 1–General Requirements.
- These items are needed to make the job run. They are typically, but not always, provided by the general contractor. Items include, but are not limited to, field personnel, insurance, performance bond, permits, testing, temporary utilities, field office and storage facilities, temporary scaffolding and platforms, equipment mobilization and demobilization, temporary roads and sidewalks, winter protection, temporary barricades and fencing, temporary security, temporary signs, field engineering and layout, final cleaning, and commissioning.
- Each item should be quantified and matched to individual Unit Price line items in Division 1, then priced and added to your estimate.
- An alternate method of estimating project overhead costs is to apply a percentage of the total project cost, usually 5% to 15% with an average of 10% (see General Conditions).
- Include other project related expenses in your estimate such as:
  - Rented equipment not itemized in the Crew Listings
  - Rubbish handling throughout the project (see section 02 41 19.19)

6. Estimate Summary

- Include sales tax as required by laws of your state or county.
- Include the general contractor's markup on self-performed work, usually 5% to 15% with an average of 10%.
- Include the general contractor's markup on subcontracted work, usually 5% to 15% with an average of 10%.
- Include the general contractor's main office overhead and profit:
  - RSMeans gives general guidelines on the general contractor's main office overhead (see section 01 31 13.60 and Reference Number R013113-50).
  - RSMeans gives no guidance on the general contractor's profit.
  - Markups will depend on the size of the general contractor's operations, projected annual revenue, the level of risk, and the level of competition in the local area and for this project in particular.
- Include a contingency, usually 3% to 5%, if appropriate.
- Adjust your estimate to the project's location by using the City Cost Indexes or the Location Factors in the Reference Section:
  - Look at the rules in "How to Use the City Cost Indexes" to see how to apply the Indexes for your location.
  - When the proper Index or Factor has been identified for the project's location, convert it to a multiplier by dividing it by 100, then multiply that multiplier by your estimated total cost. The original estimated total cost will now be adjusted up or down from the national average to a total that is appropriate for your location.

Editors' Note:
We urge you to spend time reading and
understanding the supporting material.
An accurate estimate requires experience,
knowledge, and careful calculation. The more
you know about how we at RSMeans developed
the data, the more accurate your estimate will
be. In addition, it is important to take into
consideration the reference material such as
Equipment Listings, Crew Listings, City Cost
Indexes, Location Factors, and Reference
Tables.

How to Use the Cost Data: The Details

What's Behind the Numbers? The Development of Cost Data

The staff at RSMeans continually monitors developments in the construction industry in order to ensure reliable, thorough, and up-to-date cost information. While overall construction costs may vary relative to general economic conditions, price fluctuations within the industry are dependent upon many factors. Individual price variations may, in fact, be opposite to overall economic trends. Therefore, costs are constantly tracked, and complete updates are performed yearly. Also, new items are frequently added in response to changes in materials and methods.

Costs—$ (U.S.)

All costs represent U.S. national averages and are given in U.S. dollars. The RSMeans City Cost Indexes (CCI) can be used to adjust costs to a particular location. The CCI for Canada can be used to adjust U.S. national averages to local costs in Canadian dollars. No exchange rate conversion is necessary because it has already been factored in.

G The processes or products identified by the green symbol in our publications have been determined to be environmentally responsible and/or resource-efficient solely by the RSMeans engineering staff. The inclusion of the green symbol does not represent compliance with any specific industry association or standard.

Material Costs

The RSMeans staff contacts manufacturers, dealers, distributors, and contractors all across the U.S. and Canada to determine national average material costs. If you have access to current material costs for your specific location, you may wish to make adjustments to reflect differences from the national average. Included within material costs are fasteners for a normal installation. RSMeans engineers use manufacturers' recommendations, written specifications, and/or standard construction practices for size and spacing of fasteners. Adjustments to material costs may be required for your specific application or location. The manufacturer's warranty is assumed. Extended warranties are not included in the material costs. **Material costs do not include sales tax.**

Labor Costs

Labor costs are based upon a mathematical average of trade-specific wages in 30 major U.S. cities. The type of wage (union, open shop, or residential) is identified on the inside back cover of printed publications or is selected by the estimator when using the electronic products. Markups for the wages can also be found on the inside back cover of printed publications and/or under the labor references found in the electronic products.

- If wage rates in your area vary from those used, or if rate increases are expected within a given year, labor costs should be adjusted accordingly.

Labor costs reflect productivity based on actual working conditions. In addition to actual installation, these figures include time spent during a normal weekday on tasks, such as material receiving and handling, mobilization at site, site movement, breaks, and cleanup.

Productivity data is developed over an extended period so as not to be influenced by abnormal variations and reflects a typical average.

Equipment Costs

Equipment costs include not only rental, but also operating costs for equipment under normal use. The operating costs include parts and labor for routine servicing, such as repair and replacement of pumps, filters, and worn lines. Normal operating expendables, such as fuel, lubricants, tires, and electricity (where applicable), are also included. Extraordinary operating expendables with highly variable wear patterns, such as diamond bits and blades, are excluded. These costs are included under materials. Equipment rental rates are obtained from industry sources throughout North America—contractors, suppliers, dealers, manufacturers, and distributors.

Rental rates can also be treated as reimbursement costs for contractor-owned equipment. Owned equipment costs include depreciation, loan payments, interest, taxes, insurance, storage, and major repairs.

Equipment costs do not include operators' wages.

Equipment Cost/Day—The cost of equipment required for each crew is included in the Crew Listings in the Reference Section (small tools that are considered essential everyday tools are not listed out separately). The Crew Listings itemize specialized tools and heavy equipment along with labor trades. The daily cost of itemized equipment included in a crew is based on dividing the weekly bare rental rate by 5 (number of working days per week), then adding the hourly operating cost times 8 (the number of hours per day). This Equipment Cost/Day is shown in the last column of the Equipment Rental Costs in the Reference Section.

Mobilization/Demobilization—The cost to move construction equipment from an equipment yard or rental company to the job site and back again is not included in equipment costs. Mobilization (to the site) and demobilization (from the site) costs can be found in the Unit Price Section. If a piece of equipment is already at the job site, it is not appropriate to utilize mob./demob. costs again in an estimate.

Overhead and Profit

Total Cost including O&P for the installing contractor is shown in the last column of the Unit Price and/or Assemblies. This figure is the sum of the bare material cost plus 10% for profit, the bare labor cost plus total overhead and profit, and the bare equipment cost plus 10% for profit. Details for the calculation of overhead and profit on labor are shown on the inside back cover of the printed product and in the Reference Section of the electronic product.

General Conditions

Cost data in this data set is presented in two ways: Bare Costs and Total Cost including O&P (Overhead and Profit). General Conditions, or General Requirements, of the contract should also be added to the Total Cost including O&P when applicable. Costs for General Conditions are listed in Division 1 of the Unit Price Section and in the Reference Section.

General Conditions for the installing contractor may range from 0% to 10% of the Total Cost including O&P. For the general or prime contractor, costs for General Conditions may range from 5% to 15% of the Total Cost including O&P, with a figure of 10% as the most typical allowance. If applicable, the Assemblies and Models sections use costs that include the installing contractor's overhead and profit (O&P).

Factors Affecting Costs

Costs can vary depending upon a number of variables. Here's a listing of some factors that affect costs and points to consider.

Quality—The prices for materials and the workmanship upon which productivity is based represent sound construction work. They are also in line with industry standard and manufacturer specifications and are frequently used by federal, state, and local governments.

Overtime—We have made no allowance for overtime. If you anticipate premium time or work beyond normal working hours, be sure to make an appropriate adjustment to your labor costs.

Productivity—The productivity, daily output, and labor-hour figures for each line item are based on working an eight-hour day in daylight hours in moderate temperatures. For work that extends beyond normal work hours or is performed under adverse conditions, productivity may decrease.

Size of Project—The size, scope of work, and type of construction project will have a significant impact on cost. Economies of scale can reduce costs for large projects. Unit costs can often run higher for small projects.

Location—Material prices are for metropolitan areas. However, in dense urban areas, traffic and site storage limitations may increase costs. Beyond a 20-mile radius of metropolitan areas, extra trucking or transportation charges may also increase the material costs slightly. On the other hand, lower wage rates may be in effect. Be sure to consider both of these factors when preparing an estimate, particularly if the job site is located in a central city or remote rural location. In addition, highly specialized subcontract items may require travel and per-diem expenses for mechanics.

Other Factors—

- season of year
- contractor management
- weather conditions
- local union restrictions
- building code requirements
- availability of:
  - adequate energy
  - skilled labor
  - building materials
- owner's special requirements/restrictions
- safety requirements
- environmental considerations
- access

Unpredictable Factors—General business conditions influence "in-place" costs of all items. Substitute materials and construction methods may have to be employed. These may affect the installed cost and/or life cycle costs. Such factors may be difficult to evaluate and cannot necessarily be predicted on the basis of the job's location in a particular section of the country. Thus, where these factors apply, you may find significant but unavoidable cost variations for which you will have to apply a measure of judgment to your estimate.

Rounding of Costs

In printed publications only, all unit prices in excess of $5.00 have been rounded to make them easier to use and still maintain adequate precision of the results.

How Subcontracted Items Affect Costs

A considerable portion of all large construction jobs is usually subcontracted. In fact, the percentage done by subcontractors is constantly increasing and may run over 90%. Since the workers employed by these companies do nothing else but install their particular product, they soon become experts in that line. The result is, installation by these firms is accomplished so efficiently that the total in-place cost, even adding the general contractor's overhead and profit, is no more, and often less, than if the principal contractor had handled the installation. Companies that deal with construction specialties are anxious to have their product perform well and, consequently, the installation will be the best possible.

Contingencies

The allowance for contingencies generally provides for unforeseen construction difficulties. On alterations or repair jobs, 20% is not too much. If drawings are final and only field contingencies are being considered, 2% or 3% is probably sufficient, and often nothing needs to be added. Contractually, changes in plans will be covered by extras. The contractor should consider inflationary price trends and possible material shortages during the course of the job. These escalation factors are dependent upon both economic conditions and the anticipated time between the estimate and actual construction. If drawings are not complete or approved, or a budget cost is wanted, it is wise to add 5% to 10%. Contingencies, then, are a matter of judgment.

Important Estimating Considerations

The productivity, or daily output, of each craftsman or crew assumes a well-managed job where tradesmen with the proper tools and equipment, along with the appropriate construction materials, are present. Included are daily set-up and cleanup time, break time, and plan layout time. Unless otherwise indicated, time for material movement on site (for items that can be transported by hand) of up to 200' into the building and to the first or second floor is also included. If material has to be transported by other means, over greater distances, or to higher floors, an additional allowance should be considered by the estimator.

While horizontal movement is typically a sole function of distances, vertical transport introduces other variables that can significantly impact productivity. In an occupied building, the use of elevators (assuming access, size, and required protective measures are acceptable) must be understood at the time of the estimate. For new construction, hoist wait and cycle times can easily be 15 minutes and may result in scheduled access extending beyond the normal work day. Finally, all vertical transport will impose strict weight limits likely to preclude the use of any motorized material handling.

The productivity, or daily output, also assumes installation that meets manufacturer/designer/standard specifications. A time allowance for quality control checks, minor adjustments, and any task required to ensure the proper function or operation is also included. For items that require connections to services, time is included for positioning, leveling, securing the unit, and for making all the necessary connections (and start up where applicable), ensuring a complete installation. Estimating of the services themselves (electrical, plumbing, water, steam, hydraulics, dust collection, etc.) is separate.

In some cases, the estimator must consider the use of a crane and an appropriate crew for the installation of large or heavy items.

For those situations where a crane is not included in the assigned crew and as part of the line item cost, equipment rental costs, mobilization and demobilization costs, and operator and support personnel costs must be considered.

Labor-Hours

The labor-hours expressed in this publication are derived by dividing the total daily labor-hours for the crew by the daily output. Based on average installation time and the assumptions listed above, the labor-hours include: direct labor, indirect labor, and nonproductive time. A typical day for a craftsman might include, but is not limited to:

- Direct Work
  - ☐ Measuring and layout
  - ☐ Preparing materials
  - ☐ Actual installation
  - ☐ Quality assurance/quality control
- Indirect Work
  - ☐ Reading plans or specifications
  - ☐ Preparing space
  - ☐ Receiving materials
  - ☐ Material movement
  - ☐ Giving or receiving instruction
  - ☐ Miscellaneous
- Non-Work
  - ☐ Chatting
  - ☐ Personal issues
  - ☐ Breaks
  - ☐ Interruptions (i.e., sickness, weather, material or equipment shortages, etc.)

If any of the items for a typical day do not apply to the particular work or project situation, the estimator should make any necessary adjustments.

Final Checklist

Estimating can be a straightforward process provided you remember the basics. Here's a checklist of some of the steps you should remember to complete before finalizing your estimate.

Did you remember to:

- factor in the City Cost Index for your locale?
- take into consideration which items have been marked up and by how much?
- mark up the entire estimate sufficiently for your purposes?
- read the background information on techniques and technical matters that could impact your project time span and cost?
- include all components of your project in the final estimate?
- double check your figures for accuracy?
- call RSMeans if you have any questions about your estimate or the data you've used? Remember, RSMeans stands behind its information. If you have any questions about your estimate, about the costs you've used from our data, or even about the technical aspects of the job that may affect your estimate, feel free to call the RSMeans editors at 1-877-739-3689.

Free Quarterly Updates

Stay up-to-date throughout the year with RSMeans' free cost data updates four times a year. Sign up online to make sure you have access to the newest data. Every quarter we provide the city cost adjustment factors for hundreds of cities and key materials.

Register at: **http://info.TheGordianGroup.com/2016Updates.html**.

Unit Price Section

Table of Contents

1

How RSMeans Unit Price Data Works

All RSMeans unit price data is organized in the same way.

It is important to understand the structure, so that you can find information easily and use it correctly.

03 30 Cast-In-Place Concrete

03 30 53 – Miscellaneous Cast-In-Place Concrete

03 30 53.40 Concrete In Place	Crew	Daily Output	Labor-Hours	Unit	Material	2016 Bare Costs Labor	Equipment	Total	Total Incl O&P
0010 **CONCRETE IN PLACE**									
0020 Including forms (4 uses), Grade 60 rebar, concrete (Portland cement									
0050 Type I), placement and finishing unless otherwise indicated									
0500 Chimney foundations (5000 psi), over 5 C.Y.	C-14C	32.22	3.476	C.Y.	155	160	.97	315.97	415
0510 (3500 psi), under 5 C.Y.	"	23.71	4.724	"	182	217	1.32	400.32	535
3540 Equipment pad (3000 psi), 3' x 3' x 6" thick	C-14H	45	1.067	Ea.	47.50	50.50	.70	98.70	130
3550 4' x 4' x 6" thick		30	1.600		70.50	75.50	1.06	147.06	195
3560 5' x 5' x 8" thick		18	2.667		124	126	1.76	251.76	330
3570 6' x 6' x 8" thick		14	3.429		167	162	2.26	331.26	435
3580 8' x 8' x 10" thick		8	6		360	284	3.96	647.96	835
3590 10' x 10' x 12" thick		5	9.600		615	455	6.35	1,076.35	1,375
3800 Footings (3000 psi), spread under 1 C.Y.	C-14C	28	4	C.Y.	171	184	1.12	356.12	470
3825 1 C.Y. to 5 C.Y.		43	2.605		207	120	.73	327.73	415
3850 Over 5 C.Y.		75	1.493		190	68.50	.42	258.92	315

❶ Line Numbers

RSMeans **Line Numbers** consist of 12 characters, which identify a unique location in the database for each task. The first 6 or 8 digits conform to the Construction Specifications Institute MasterFormat® 2014. The remainder of the digits are a further breakdown by RSMeans in order to arrange items in understandable groups of similar tasks. Line numbers are consistent across all RSMeans publications, so a line number in any RSMeans product will always refer to the same unit of work.

❷ Descriptions

RSMeans **Descriptions** are shown in a hierarchical structure to make them readable. In order to read a complete description, read up through the indents to the top of the section. Include everything that is above and to the left that is not contradicted by information below. For instance, the complete description for line 03 30 53.40 3550 is "Concrete in place, including forms (4 uses), Grade 60 rebar, concrete (Portland cement Type 1), placement and finishing unless otherwise indicated; Equipment pad (3000 psi), 4' x 4' x 6" thick."

❸ RSMeans Data

When using **RSMeans data**, it is important to read through an entire section to ensure that you use the data that most closely matches your work. Note that sometimes there is additional information shown in the section that may improve your price. There are frequently lines that further describe, add to, or adjust data for specific situations.

❹ Reference Information

RSMeans engineers have created **reference** information to assist you in your estimate. **If** there is information that applies to a section, it will be indicated at the start of the section. The Reference Section is located in the back of the data set.

❺ Crews

Crews include labor and/or equipment necessary to accomplish each task. In this case, Crew C-14H is used.

RSMeans selects a crew to represent the workers and equipment that are typically used for that task. In this case, Crew C-14H consists of one carpenter foreman (outside), two carpenters, one rodman, one laborer, one cement finisher, and one gas engine vibrator. Details of all crews can be found in the reference section.

Crews - Standard

Crew No.	Bare Costs		Incl. Subs O&P		Cost Per Labor-Hour	
Crew C-14H	Hr.	Daily	Hr.	Daily	Bare Costs	Incl. O&P
1 Carpenter Foreman (outside)	$50.45	$403.60	$77.40	$619.20	$47.32	$72.33
2 Carpenters	48.45	775.20	74.30	1188.80		
1 Rodman (reinf.)	53.00	424.00	82.10	656.80		
1 Laborer	37.90	303.20	58.15	465.20		
1 Cement Finisher	45.65	365.20	67.70	541.60		
1 Gas Engine Vibrator		31.60		34.76	.66	.72
48 L.H., Daily Totals		$2302.80		$3506.36	$47.98	$73.05

6 Daily Output

The **Daily Output** is the amount of work that the crew can do in a normal 8-hour workday, including mobilization, layout, movement of materials, and cleanup. In this case, crew C-14H can install thirty 4' x 4' x 6" thick concrete pads in a day. Daily output is variable and based on many factors, including the size of the job, location, and environmental conditions. RSMeans data represents work done in daylight (or adequate lighting) and temperate conditions.

7 Labor-Hours

The figure in the **Labor-Hours** column is the amount of labor required to perform one unit of work–in this case the amount of labor required to construct one 4' x 4' equipment pad. This figure is calculated by dividing the number of hours of labor in the crew by the daily output (48 labor-hours divided by 30 pads = 1.6 hours of labor per pad). Multiply 1.600 times 60 to see the value in minutes: 60 x

1.6 = 96 minutes. Note: the labor-hour figure is not dependent on the crew size. A change in crew size will result in a corresponding change in daily output, but the labor-hours per unit of work will not change.

8 Unit of Measure

All RSMeans unit cost data includes the typical **Unit of Measure** used for estimating that item. For concrete-in-place the typical unit is cubic yards (C.Y.) or each (Ea.). For installing broadloom carpet it is square yard, and for gypsum board it is square foot. The estimator needs to take special care that the unit in the data matches the unit in the take-off. Unit conversions may be found in the Reference Section.

9 Bare Costs

Bare Costs are the costs of materials, labor, and equipment that the installing contractor pays. They represent the cost, in U.S. dollars, for one unit of work. They do not include any markups for profit or labor burden.

10 Bare Total

The **Total column** represents the total bare cost for the installing contractor, in U.S. dollars. In this case, the sum of $70.50 for material + $75.50 for labor + $1.06 for equipment is $147.06.

11 Total Incl O&P

The **Total Incl O&P column** is the total cost, including overhead and profit, that the installing contractor will charge the customer. This represents the cost of materials plus 10% profit, the cost of labor plus labor burden and 10% profit, and the cost of equipment plus 10% profit. It does not include the general contractor's overhead and profit. Note: See the inside back cover of the printed product or the reference section of the electronic product for details of how RSMeans calculates labor burden.

National Average

*The data in RSMeans print publications represents a "national average" cost. This data should be modified to the project location using the **City Cost Indexes** or **Location Factors** tables found in the Reference Section. Use the location factors to adjust estimate totals if the project covers multiple trades. Use the city cost indexes (CCI) for single trade projects or projects where a more detailed analysis is required. All figures in the two tables are derived from the same research. The last row of data in the CCI, the weighted average, is the same as the numbers reported for each location in the location factor table.*

How RSMeans Unit Price Data Works (Continued)

①

Project Name: Pre-Engineered Steel Building				Architect: As Shown				
Location:	Anywhere, USA						**01/01/16**	**STD**
Line Number	**Description**	**Qty**	**Unit**	**Material**	**Labor**	**Equipment**	**SubContract**	**Estimate Total**
03 30 53.40 3940	Strip footing, 12" x 24", reinforced	34	C.Y.	$4,896.00	$3,638.00	$22.10	$0.00	
03 30 53.40 3950	Strip footing, 12" x 36", reinforced	15	C.Y.	$2,070.00	$1,290.00	$7.80	$0.00	
03 11 13.65 3000	Concrete slab edge forms	500	L.F.	$165.00	$1,220.00	$0.00	$0.00	
03 22 11.10 0200	Welded wire fabric reinforcing	150	C.S.F.	$2,632.50	$4,125.00	$0.00	$0.00	
03 31 13.35 0300	Ready mix concrete, 4000 psi for slab on grade	278	C.Y.	$31,414.00	$0.00	$0.00	$0.00	
03 31 13.70 4300	Place, strike off & consolidate concrete slab	278	C.Y.	$0.00	$4,795.50	$161.24	$0.00	
03 35 13.30 0250	Machine float & trowel concrete slab	15,000	S.F.	$0.00	$9,000.00	$450.00	$0.00	
03 15 16.20 0140	Cut control joints in concrete slab	950	L.F.	$47.50	$389.50	$85.50	$0.00	
03 39 23.13 0300	Sprayed concrete curing membrane	150	C.S.F.	$1,905.00	$960.00	$0.00	$0.00	
Division 03	**Subtotal**			**$43,130.00**	**$25,418.00**	**$726.64**	**$0.00**	**$69,274.64**
08 36 13.10 2650	Manual 10' x 10' steel sectional overhead door	8	Ea.	$9,800.00	$3,440.00	$0.00	$0.00	
08 36 13.10 2860	Insulation and steel back panel for OH door	800	S.F.	$3,800.00	$0.00	$0.00	$0.00	
Division 08	**Subtotal**			**$13,600.00**	**$3,440.00**	**$0.00**	**$0.00**	**$17,040.00**
13 34 19.50 1100	Pre-Engineered Steel Building, 100' x 150' x 24'	15,000	SF Flr.	$0.00	$0.00	$0.00	$352,500.00	
13 34 19.50 6050	Framing for PESB door opening, 3' x 7'	4	Opng.	$0.00	$0.00	$0.00	$2,260.00	
13 34 19.50 6100	Framing for PESB door opening, 10' x 10'	8	Opng.	$0.00	$0.00	$0.00	$9,200.00	
13 34 19.50 6200	Framing for PESB window opening, 4' x 3'	6	Opng.	$0.00	$0.00	$0.00	$3,300.00	
13 34 19.50 5750	PESB door, 3' x 7', single leaf	4	Opng.	$2,380.00	$680.00	$0.00	$0.00	
13 34 19.50 7750	PESB sliding window, 4' x 3' with screen	6	Opng.	$2,430.00	$582.00	$67.80	$0.00	
13 34 19.50 6550	PESB gutter, eave type, 26 ga., painted	300	L.F.	$2,040.00	$798.00	$0.00	$0.00	
13 34 19.50 8650	PESB roof vent, 12" wide x 10' long	15	Ea.	$540.00	$3,195.00	$0.00	$0.00	
13 34 19.50 6900	PESB insulation, vinyl faced, 4" thick	27,400	S.F.	$12,330.00	$9,316.00	$0.00	$0.00	
Division 13	**Subtotal**			**$19,720.00**	**$14,571.00**	**$67.80**	**$367,260.00**	**$401,618.80**
			Subtotal	$76,450.00	$43,429.00	$794.44	$367,260.00	$487,933.44
Division 01 **②**	**General Requirements @ 7%**			5,351.50	3,040.03	55.61	25,708.20	
			Estimate Subtotal	$81,801.50	$46,469.03	$850.05	$392,968.20	$487,933.44
		③	Sales Tax @ 5%	4,090.08		42.50	9,824.21	
			Subtotal A	85,891.58	46,469.03	892.55	402,792.41	
		④	GC O & P	8,589.16	25,093.28	89.26	40,279.24	
			Subtotal B	94,480.73	71,562.31	981.81	443,071.65	$610,096.49
		⑤	Contingency @ 5%					30,504.82
			Subtotal C					$640,601.32
		⑥	Bond @ $12/1000 +10% O&P					8,455.94
			Subtotal D					$649,057.25
		⑦	Location Adjustment Factor		102.30			14,928.32
			Grand Total					**$663,985.57**

This estimate is based on an interactive spreadsheet. A copy of this spreadsheet is located on the RSMeans website at **www.RSMeans.com/2016extras.** You are free to download it and adjust it to your methodology.

4

Sample Estimate

This sample demonstrates the elements of an estimate, including a tally of the RSMeans data lines and a summary of the markups on a contractor's work to arrive at a total cost to the owner. The RSMeans Location Factor is added at the bottom of the estimate to adjust the cost of the work to a specific location.

1 Work Performed

The body of the estimate shows the RSMeans data selected, including the line number, a brief description of each item, its take-off unit and quantity, and the bare costs of materials, labor, and equipment. This estimate also includes a column titled "SubContract." This data is taken from the RSMeans column "Total Incl O&P" and represents the total that a subcontractor would charge a general contractor for the work, including the sub's markup for overhead and profit.

2 Division 1, General Requirements

This is the first division numerically but the last division estimated. Division 1 includes project-wide needs provided by the general contractor. These requirements vary by project but may include temporary facilities and utilities, security, testing, project cleanup, etc. For small projects a percentage can be used, typically between 5% and 15% of project cost. For large projects the costs may be itemized and priced individually.

3 Sales Tax

If the work is subject to state or local sales taxes, the amount must be added to the estimate. Sales tax may be added to material costs, equipment costs, and subcontracted work. In this case, sales tax was added in all three categories. It was assumed that approximately half the subcontracted work would be material cost, so the tax was applied to 50% of the subcontract total.

4 GC O&P

This entry represents the general contractor's markup on material, labor, equipment, and subcontractor costs. RSMeans' standard markup on materials, equipment, and subcontracted work is 10%. In this estimate, the markup on the labor performed by the GC's workers uses "Skilled Workers Average" shown in Column F on the table "Installing Contractor's Overhead & Profit," which can be found on the inside-back cover of the printed product or in the Reference Section of the electronic product.

5 Contingency

A factor for contingency may be added to any estimate to represent the cost of unknowns that may occur between the time that the estimate is performed and the time the project is constructed. The amount of the allowance will depend on the stage of design at which the estimate is done and the contractor's assessment of the risk involved. Refer to section 01 21 16.50 for contingency allowances.

6 Bonds

Bond costs should be added to the estimate. The figures here represent a typical performance bond, ensuring the owner that if the general contractor does not complete the obligations in the construction contract the bonding company will pay the cost for completion of the work.

7 Location Adjustment

RSMeans published data is based on national average costs. If necessary, adjust the total cost of the project using a location factor from the "Location Factor" table or the "City Cost Index" table. Use location factors if the work is general, covering multiple trades. If the work is by a single trade (e.g., masonry) use the more specific data found in the "City Cost Indexes."

Estimating Tips
01 20 00 Price and Payment Procedures
- Allowances that should be added to estimates to cover contingencies and job conditions that are not included in the national average material and labor costs are shown in section 01 21.
- When estimating historic preservation projects (depending on the condition of the existing structure and the owner's requirements), a 15%–20% contingency or allowance is recommended, regardless of the stage of the drawings.

01 30 00 Administrative Requirements
- Before determining a final cost estimate, it is a good practice to review all the items listed in Subdivisions 01 31 and 01 32 to make final adjustments for items that may need customizing to specific job conditions.
- Requirements for initial and periodic submittals can represent a significant cost to the General Requirements of a job. Thoroughly check the submittal specifications when estimating a project to determine any costs that should be included.

01 40 00 Quality Requirements
- All projects will require some degree of Quality Control. This cost is not included in the unit cost of construction listed in each division. Depending upon the terms of the contract, the various costs of inspection and testing can be the responsibility of either the owner or the contractor. Be sure to include the required costs in your estimate.

01 50 00 Temporary Facilities and Controls
- Barricades, access roads, safety nets, scaffolding, security, and many more requirements for the execution of a safe project are elements of direct cost. These costs can easily be overlooked when preparing an estimate. When looking through the major classifications of this subdivision, determine which items apply to each division in your estimate.
- Construction Equipment Rental Costs can be found in the Reference Section in section 01 54 33. Operators' wages are not included in equipment rental costs.
- Equipment mobilization and demobilization costs are not included in equipment rental costs and must be considered separately.

- The cost of small tools provided by the installing contractor for his workers is covered in the "Overhead" column on the "Installing Contractor's Overhead and Profit" table that lists labor trades, base rates and markups and, therefore, is included in the "Total Incl. O&P" cost of any Unit Price line item.

01 70 00 Execution and Closeout Requirements
- When preparing an estimate, thoroughly read the specifications to determine the requirements for Contract Closeout. Final cleaning, record documentation, operation and maintenance data, warranties and bonds, and spare parts and maintenance materials can all be elements of cost for the completion of a contract. Do not overlook these in your estimate.

Reference Numbers
Reference numbers are shown at the beginning of some major classifications. These numbers refer to related items in the Reference Section. The reference information may be an estimating procedure, an alternate pricing method, or technical information.

Note: Not all subdivisions listed here necessarily appear. ■

Did you know?
RSMeans Online gives you the same access to RSMeans' data with 24/7 access:
- Quickly locate costs in the searchable database.
- Build cost lists, estimates, and reports in minutes.
- Adjust costs to any location in the U.S. and Canada with the click of a button.

Start your free trial today at **www.RSMeansOnline.com**

RSMeans Online
FROM THE GORDIAN GROUP®

01 11 Summary of Work

01 11 31 – Professional Consultants

01 11 31.10 Architectural Fees		Crew	Daily Output	Labor-Hours	Unit	Material	2016 Bare Costs Labor	Equipment	Total	Total Incl O&P
0010	**ARCHITECTURAL FEES** R011110-10									
2000	For "Greening" of building [G]				Project				3%	3%

01 11 31.30 Engineering Fees		Crew	Daily Output	Labor-Hours	Unit	Material	2016 Bare Costs Labor	Equipment	Total	Total Incl O&P
0010	**ENGINEERING FEES** R011110-30									
0020	Educational planning consultant, minimum				Project				.50%	.50%
0100	Maximum				"				2.50%	2.50%
0200	Electrical, minimum				Contrct				4.10%	4.10%
0300	Maximum								10.10%	10.10%
0400	Elevator & conveying systems, minimum								2.50%	2.50%
0500	Maximum								5%	5%
0600	Food service & kitchen equipment, minimum								8%	8%
0700	Maximum								12%	12%
1000	Mechanical (plumbing & HVAC), minimum								4.10%	4.10%
1100	Maximum								10.10%	10.10%

01 21 Allowances

01 21 16 – Contingency Allowances

01 21 16.50 Contingencies

		Crew	Daily Output	Labor-Hours	Unit	Material	Labor	Equipment	Total	Total Incl O&P
0010	**CONTINGENCIES**, Add to estimate									
0020	Conceptual stage				Project				20%	20%
0050	Schematic stage								15%	15%
0100	Preliminary working drawing stage (Design Dev.)								10%	10%
0150	Final working drawing stage								3%	3%

01 21 53 – Factors Allowance

01 21 53.50 Factors

		Crew	Daily Output	Labor-Hours	Unit	Material	Labor	Equipment	Total	Total Incl O&P
0010	**FACTORS** Cost adjustments R012153-10									
0100	Add to construction costs for particular job requirements									
0500	Cut & patch to match existing construction, add, minimum				Costs	2%	3%			
0550	Maximum					5%	9%			
0800	Dust protection, add, minimum					1%	2%			
0850	Maximum					4%	11%			
1100	Equipment usage curtailment, add, minimum					1%	1%			
1150	Maximum					3%	10%			
1400	Material handling & storage limitation, add, minimum					1%	1%			
1450	Maximum					6%	7%			
1700	Protection of existing work, add, minimum					2%	2%			
1750	Maximum					5%	7%			
2000	Shift work requirements, add, minimum						5%			
2050	Maximum						30%			
2300	Temporary shoring and bracing, add, minimum					2%	5%			
2350	Maximum					5%	12%			
2400	Work inside prisons and high security areas, add, minimum						30%			
2450	Maximum						50%			

01 21 55 – Job Conditions Allowance

01 21 55.50 Job Conditions

		Crew	Daily Output	Labor-Hours	Unit	Material	Labor	Equipment	Total	Total Incl O&P
0010	**JOB CONDITIONS** Modifications to applicable									
0020	cost summaries									
0100	Economic conditions, favorable, deduct				Project				2%	2%
0200	Unfavorable, add								5%	5%
0300	Hoisting conditions, favorable, deduct								2%	2%

01 21 Allowances

01 21 55 – Job Conditions Allowance

01 21 55.50 Job Conditions		Crew	Daily Output	Labor-Hours	Unit	Material	2016 Bare Costs Labor	Equipment	Total	Total Incl O&P
0400	Unfavorable, add				Project				5%	5%
0700	Labor availability, surplus, deduct								1%	1%
0800	Shortage, add								10%	10%
0900	Material storage area, available, deduct								1%	1%
1000	Not available, add								2%	2%
1100	Subcontractor availability, surplus, deduct								5%	5%
1200	Shortage, add								12%	12%
1300	Work space, available, deduct								2%	2%
1400	Not available, add								5%	5%

01 21 57 – Overtime Allowance

01 21 57.50 Overtime

0010	**OVERTIME** for early completion of projects or where	R012909-90								
0020	labor shortages exist, add to usual labor, up to					Costs		100%			

01 21 63 – Taxes

01 21 63.10 Taxes

0010	**TAXES**	R012909-80								
0020	Sales tax, State, average					%	5.08%				
0050	Maximum	R012909-85					7.50%				
0200	Social Security, on first $118,500 of wages							7.65%			
0300	Unemployment, combined Federal and State, minimum							.60%			
0350	Average							9.60%			
0400	Maximum							12%			

01 31 Project Management and Coordination

01 31 13 – Project Coordination

01 31 13.20 Field Personnel

0010	**FIELD PERSONNEL**									
0020	Clerk, average					Week		460		460	710
0100	Field engineer, minimum							1,100		1,100	1,675
0120	Average							1,425		1,425	2,200
0140	Maximum							1,625		1,625	2,500
0160	General purpose laborer, average							1,525		1,525	2,325
0180	Project manager, minimum							2,025		2,025	3,125
0200	Average							2,325		2,325	3,600
0220	Maximum							2,675		2,675	4,100
0240	Superintendent, minimum							1,975		1,975	3,050
0260	Average							2,175		2,175	3,350
0280	Maximum							2,475		2,475	3,800
0290	Timekeeper, average							1,275		1,275	1,950

01 31 13.30 Insurance

0010	**INSURANCE**	R013113-40								
0020	Builders risk, standard, minimum					Job				.24%	.24%
0050	Maximum	R013113-50								.64%	.64%
0200	All-risk type, minimum									.25%	.25%
0250	Maximum	R013113-60								.62%	.62%
0400	Contractor's equipment floater, minimum					Value				.50%	.50%
0450	Maximum					"				1.50%	1.50%
0600	Public liability, average					Job				2.02%	2.02%
0800	Workers' compensation & employer's liability, average										
0850	by trade, carpentry, general					Payroll		14.37%			

01 31 Project Management and Coordination

01 31 13 – Project Coordination

01 31 13.30 Insurance

		Crew	Daily Output	Labor-Hours	Unit	Material	2016 Bare Costs Labor	Equipment	Total	Total Incl O&P
1000	Electrical				Payroll		5.64%			
1150	Insulation						11.56%			
1450	Plumbing						6.94%			
1550	Sheet metal work (HVAC)						8.84%			

01 31 13.50 General Contractor's Mark-Up

		Crew	Daily Output	Labor-Hours	Unit	Material	Labor	Equipment	Total	Total Incl O&P
0010	**GENERAL CONTRACTOR'S MARK-UP** on Change Orders									
0200	Extra work, by subcontractors, add				%				10%	10%
0250	By General Contractor, add								15%	15%
0400	Omitted work, by subcontractors, deduct all but								5%	5%
0450	By General Contractor, deduct all but								7.50%	7.50%
0600	Overtime work, by subcontractors, add								15%	15%
0650	By General Contractor, add								10%	10%

01 31 13.60 General Contractor's Main Office Overhead

		Crew	Daily Output	Labor-Hours	Unit	Material	Labor	Equipment	Total	Total Incl O&P
0010	**GENERAL CONTRACTOR'S MAIN OFFICE OVERHEAD**									
0020	As percent of direct costs, minimum				%				5%	
0050	Average								13%	
0100	Maximum								30%	

01 31 13.80 Overhead and Profit

		Crew	Daily Output	Labor-Hours	Unit	Material	Labor	Equipment	Total	Total Incl O&P
0010	**OVERHEAD & PROFIT** Allowance to add to items in this									
0020	data set that do not include Subs O&P, average				%				25%	
0100	Allowance to add to items in this data set that									
0110	do include Subs O&P, minimum				%				5%	5%
0150	Average								10%	10%
0200	Maximum								15%	15%
0300	Typical, by size of project, under $100,000								30%	
0350	$500,000 project								25%	
0400	$2,000,000 project								20%	
0450	Over $10,000,000 project								15%	

01 31 13.90 Performance Bond

		Crew	Daily Output	Labor-Hours	Unit	Material	Labor	Equipment	Total	Total Incl O&P
0010	**PERFORMANCE BOND**	R013113-80								
0020	For buildings, minimum				Job				.60%	.60%
0100	Maximum				"				2.50%	2.50%

01 32 Construction Progress Documentation

01 32 33 – Photographic Documentation

01 32 33.50 Photographs

		Crew	Daily Output	Labor-Hours	Unit	Material	Labor	Equipment	Total	Total Incl O&P
0010	**PHOTOGRAPHS**									
0020	8" x 10", 4 shots, 2 prints ea., std. mounting				Set	535			535	585
0100	Hinged linen mounts					530			530	580
0200	8" x 10", 4 shots, 2 prints each, in color					475			475	525
0300	For I.D. slugs, add to all above					5			5	5.50
1500	Time lapse equipment, camera and projector, buy				Ea.	2,700			2,700	2,975
1550	Rent per month				"	1,275			1,275	1,400
1700	Cameraman and film, including processing, B.&W.				Day	1,225			1,225	1,350
1720	Color				"	1,425			1,425	1,550

01 41 Regulatory Requirements

01 41 26 – Permit Requirements

01 41 26.50 Permits		Crew	Daily Output	Labor-Hours	Unit	Material	2016 Bare Costs Labor	Equipment	Total	Total Incl O&P
0010	**PERMITS**									
0020	Rule of thumb, most cities, minimum				Job				.50%	.50%
0100	Maximum				"				2%	2%

01 45 Quality Control

01 45 23 – Testing and Inspecting Services

01 45 23.50 Testing		Crew	Daily Output	Labor-Hours	Unit	Material	2016 Bare Costs Labor	Equipment	Total	Total Incl O&P
0010	**TESTING** and Inspecting Services									
5820	Non-destructive metal testing, dye penetrant				Day				310	340
5840	Magnetic particle								310	340
5860	Radiography								450	495
5880	Ultrasonic								310	340
9000	Thermographic testing, for bldg envelope heat loss, average 2,000 S.F. [G]				Ea.				500	500

01 51 Temporary Utilities

01 51 13 – Temporary Electricity

01 51 13.80 Temporary Utilities		Crew	Daily Output	Labor-Hours	Unit	Material	2016 Bare Costs Labor	Equipment	Total	Total Incl O&P
0010	**TEMPORARY UTILITIES**									
0350	Lighting, lamps, wiring, outlets, 40,000 S.F. building, 8 strings	1 Elec	34	.235	CSF Flr	5	12.95		17.95	25
0360	16 strings	"	17	.471		10	26		36	50
0400	Power for temp lighting only, 6.6 KWH, per month								.92	1.01
0430	11.8 KWH, per month								1.65	1.82
0450	23.6 KWH, per month								3.30	3.63
0600	Power for job duration incl. elevator, etc., minimum								47	51.50
0650	Maximum								110	121
1000	Toilet, portable, see Equip. Rental 01 54 33 in Reference Section									

01 52 Construction Facilities

01 52 13 – Field Offices and Sheds

01 52 13.20 Office and Storage Space		Crew	Daily Output	Labor-Hours	Unit	Material	2016 Bare Costs Labor	Equipment	Total	Total Incl O&P
0010	**OFFICE AND STORAGE SPACE**									
0020	Office trailer, furnished, no hookups, 20' x 8', buy	2 Skwk	1	16	Ea.	10,700	800		11,500	13,000
0250	Rent per month					193			193	213
0300	32' x 8', buy	2 Skwk	.70	22.857		16,000	1,150		17,150	19,400
0350	Rent per month					246			246	270
0400	50' x 10', buy	2 Skwk	.60	26.667		24,600	1,325		25,925	29,200
0450	Rent per month					350			350	385
0500	50' x 12', buy	2 Skwk	.50	32		30,100	1,600		31,700	35,600
0550	Rent per month					440			440	485
0700	For air conditioning, rent per month, add					48.50			48.50	53.50
0800	For delivery, add per mile				Mile	11			11	12.10
0890	Delivery each way				Ea.	1,525			1,525	1,700
0900	Bunk house trailer, 8' x 40' duplex dorm with kitchen, no hookups, buy	2 Carp	1	16		85,000	775		85,775	94,500
0910	9 man with kitchen and bath, no hookups, buy		1	16		90,000	775		90,775	100,000
0920	18 man sleeper with bath, no hookups, buy		1	16		95,000	775		95,775	105,500
1000	Portable buildings, prefab, on skids, economy, 8' x 8'		265	.060	S.F.	23.50	2.93		26.43	30.50
1100	Deluxe, 8' x 12'		150	.107	"	19.35	5.15		24.50	29.50

For customer support on your Mechanical Cost Data, call 877.739.3689.

11

01 52 Construction Facilities

01 52 13 – Field Offices and Sheds

01 52 13.20 Office and Storage Space		Crew	Daily Output	Labor-Hours	Unit	Material	2016 Bare Costs Labor	Equipment	Total	Total Incl O&P
1200	Storage boxes, 20' x 8', buy	2 Skwk	1.80	8.889	Ea.	2,975	445		3,420	3,950
1250	Rent per month					78			78	85.50
1300	40' x 8', buy	2 Skwk	1.40	11.429		3,825	570		4,395	5,075
1350	Rent per month				↓	108			108	118

01 52 13.40 Field Office Expense

		Crew	Daily Output	Labor-Hours	Unit	Material	Labor	Equipment	Total	Total Incl O&P
0010	**FIELD OFFICE EXPENSE**									
0100	Office equipment rental average				Month	200			200	220
0120	Office supplies, average				"	80			80	88
0125	Office trailer rental, see Section 01 52 13.20									
0140	Telephone bill; avg. bill/month incl. long dist.				Month	85			85	93.50
0160	Lights & HVAC				"	160			160	176

01 54 Construction Aids

01 54 16 – Temporary Hoists

01 54 16.50 Weekly Forklift Crew

		Crew	Daily Output	Labor-Hours	Unit	Material	Labor	Equipment	Total	Total Incl O&P
0010	**WEEKLY FORKLIFT CREW**									
0100	All-terrain forklift, 45' lift, 35' reach, 9000 lb. capacity	A-3P	.20	40	Week		1,975	2,675	4,650	5,925

01 54 19 – Temporary Cranes

01 54 19.50 Daily Crane Crews

		Crew	Daily Output	Labor-Hours	Unit	Material	Labor	Equipment	Total	Total Incl O&P
0010	**DAILY CRANE CREWS** for small jobs, portal to portal									
0100	12-ton truck-mounted hydraulic crane	A-3H	1	8	Day		420	875	1,295	1,600
0200	25-ton	A-3I	1	8			420	1,025	1,445	1,750
0300	40-ton	A-3J	1	8			420	1,250	1,670	2,000
0400	55-ton	A-3K	1	16			785	1,650	2,435	3,000
0500	80-ton	A-3L	1	16			785	2,375	3,160	3,800
0600	100-ton	A-3M	1	16	↓		785	2,375	3,160	3,825
0900	If crane is needed on a Saturday, Sunday or Holiday									
0910	At time-and-a-half, add				Day		50%			
0920	At double time, add				"		100%			

01 54 19.60 Monthly Tower Crane Crew

		Crew	Daily Output	Labor-Hours	Unit	Material	Labor	Equipment	Total	Total Incl O&P
0010	**MONTHLY TOWER CRANE CREW**, excludes concrete footing									
0100	Static tower crane, 130' high, 106' jib, 6200 lb. capacity	A-3N	.05	176	Month		9,200	25,200	34,400	41,600

01 54 23 – Temporary Scaffolding and Platforms

01 54 23.70 Scaffolding

		Crew	Daily Output	Labor-Hours	Unit	Material	Labor	Equipment	Total	Total Incl O&P
0010	**SCAFFOLDING** R015423-10									
0015	Steel tube, regular, no plank, labor only to erect & dismantle									
0090	Building exterior, wall face, 1 to 5 stories, 6'-4" x 5' frames	3 Carp	8	3	C.S.F.		145		145	223
0200	6 to 12 stories	4 Carp	8	4			194		194	297
0301	13 to 20 stories	5 Clab	8	5			190		190	291
0460	Building interior, wall face area, up to 16' high	3 Carp	12	2			97		97	149
0560	16' to 40' high		10	2.400	↓		116		116	178
0800	Building interior floor area, up to 30' high	↓	150	.160	C.C.F.		7.75		7.75	11.90
0900	Over 30' high	4 Carp	160	.200	"		9.70		9.70	14.85
0906	Complete system for face of walls, no plank, material only rent/mo				C.S.F.	35.50			35.50	39
0908	Interior spaces, no plank, material only rent/mo				C.C.F.	4.30			4.30	4.73
0910	Steel tubular, heavy duty shoring, buy									
0920	Frames 5' high 2' wide				Ea.	81.50			81.50	90
0925	5' high 4' wide					93			93	102
0930	6' high 2' wide				↓	93.50			93.50	103

12

01 54 23.70 Scaffolding	Crew	Daily Output	Labor-Hours	Unit	Material	2016 Bare Costs Labor	Equipment	Total	Total Incl O&P
0935 6' high 4' wide				Ea.	109			109	120
0940 Accessories									
0945 Cross braces				Ea.	15.50			15.50	17.05
0950 U-head, 8" x 8"					19.10			19.10	21
0955 J-head, 4" x 8"					13.90			13.90	15.30
0960 Base plate, 8" x 8"					15.50			15.50	17.05
0965 Leveling jack					33.50			33.50	36.50
1000 Steel tubular, regular, buy									
1100 Frames 3' high 5' wide				Ea.	89			89	97.50
1150 5' high 5' wide					104			104	114
1200 6'-4" high 5' wide					147			147	162
1350 7'-6" high 6' wide					158			158	173
1500 Accessories cross braces					18.10			18.10	19.90
1550 Guardrail post					16.40			16.40	18.05
1600 Guardrail 7' section					8.70			8.70	9.60
1650 Screw jacks & plates					21.50			21.50	24
1700 Sidearm brackets					30.50			30.50	33.50
1750 8" casters					31			31	34
1800 Plank 2" x 10" x 16'-0"					59.50			59.50	65.50
1900 Stairway section					286			286	315
1910 Stairway starter bar					32			32	35
1920 Stairway inside handrail					58.50			58.50	64
1930 Stairway outside handrail					86.50			86.50	95
1940 Walk-thru frame guardrail					41.50			41.50	46
2000 Steel tubular, regular, rent/mo.									
2100 Frames 3' high 5' wide				Ea.	5			5	5.50
2150 5' high 5' wide					5			5	5.50
2200 6'-4" high 5' wide					5.35			5.35	5.90
2250 7'-6" high 6' wide					10			10	11
2500 Accessories, cross braces					1			1	1.10
2550 Guardrail post					1			1	1.10
2600 Guardrail 7' section					1			1	1.10
2650 Screw jacks & plates					2			2	2.20
2700 Sidearm brackets					2			2	2.20
2750 8" casters					8			8	8.80
2800 Outrigger for rolling tower					3			3	3.30
2850 Plank 2" x 10" x 16'-0"					10			10	11
2900 Stairway section					35			35	38.50
2940 Walk-thru frame guardrail					2.50			2.50	2.75
3000 Steel tubular, heavy duty shoring, rent/mo.									
3250 5' high 2' & 4' wide				Ea.	8.50			8.50	9.35
3300 6' high 2' & 4' wide					8.50			8.50	9.35
3500 Accessories, cross braces					1			1	1.10
3600 U - head, 8" x 8"					2.50			2.50	2.75
3650 J - head, 4" x 8"					2.50			2.50	2.75
3700 Base plate, 8" x 8"					1			1	1.10
3750 Leveling jack					2.50			2.50	2.75
5700 Planks, 2" x 10" x 16'-0", labor only to erect & remove to 50' H	3 Carp	72	.333			16.15		16.15	25
5800 Over 50' high	4 Carp	80	.400			19.40		19.40	29.50
6000 Heavy duty shoring for elevated slab forms to 8'-2" high, floor area									
6100 Labor only to erect & dismantle	4 Carp	16	2	C.S.F.		97		97	149
6110 Materials only, rent/mo.				"	44.50			44.50	49
6500 To 14'-8" high									

01 54 23 – Temporary Scaffolding and Platforms

01 54 23.70 Scaffolding		Crew	Daily Output	Labor-Hours	Unit	Material	2016 Bare Costs Labor	Equipment	Total	Total Incl O&P
6600	Labor only to erect & dismantle	4 Carp	10	3.200	C.S.F.		155		155	238
6610	Materials only, rent/mo	"			"	65			65	71.50

01 54 23.75 Scaffolding Specialties

	01 54 23.75 Scaffolding Specialties	Crew	Daily Output	Labor-Hours	Unit	Material	2016 Bare Costs Labor	Equipment	Total	Total Incl O&P
0010	**SCAFFOLDING SPECIALTIES**									
1200	Sidewalk bridge, heavy duty steel posts & beams, including									
1210	parapet protection & waterproofing (material cost is rent/month)									
1220	8' to 10' wide, 2 posts	3 Carp	15	1.600	L.F.	43.50	77.50		121	167
1230	3 posts	"	10	2.400	"	67.50	116		183.50	253
1500	Sidewalk bridge using tubular steel scaffold frames including									
1510	planking (material cost is rent/month)	3 Carp	45	.533	L.F.	8.55	26		34.55	49
1600	For 2 uses per month, deduct from all above					50%				
1700	For 1 use every 2 months, add to all above					100%				
1900	Catwalks, 20" wide, no guardrails, 7' span, buy				Ea.	166			166	183
2000	10' span, buy					233			233	256
3720	Putlog, standard, 8' span, with hangers, buy					66.50			66.50	73
3730	Rent per month					20			20	22
3750	12' span, buy					99			99	109
3755	Rent per month					25			25	27.50
3760	Trussed type, 16' span, buy					229			229	252
3770	Rent per month					30			30	33
3790	22' span, buy					274			274	300
3795	Rent per month					40			40	44
3800	Rolling ladders with handrails, 30" wide, buy, 2 step					291			291	320
4000	7 step					815			815	900
4050	10 step					1,175			1,175	1,275
4100	Rolling towers, buy, 5' wide, 7' long, 10' high					1,300			1,300	1,450
4200	For 5' high added sections, to buy, add					244			244	269
4300	Complete incl. wheels, railings, outriggers,									
4350	21' high, to buy				Ea.	2,225			2,225	2,425
4400	Rent/month = 5% of purchase cost				"	200			200	220

01 54 36 – Equipment Mobilization

01 54 36.50 Mobilization

	01 54 36.50 Mobilization		Crew	Daily Output	Labor-Hours	Unit	Material	2016 Bare Costs Labor	Equipment	Total	Total Incl O&P
0010	**MOBILIZATION** (Use line item again for demobilization)	R015436-50									
0015	Up to 25 mi. haul dist. (50 mi. RT for mob/demob crew)										
1200	Small equipment, placed in rear of, or towed by pickup truck		A-3A	4	2	Ea.		98.50	39	137.50	192
1300	Equipment hauled on 3-ton capacity towed trailer		A-3Q	2.67	3			147	67.50	214.50	298
1400	20-ton capacity		B-34U	2	8			370	236	606	820
1500	40-ton capacity		B-34N	2	8			375	375	750	985
1600	50-ton capacity		B-34V	1	24			1,150	1,050	2,200	2,900
1700	Crane, truck-mounted, up to 75 ton (driver only)		1 Eqhv	4	2			105		105	158
1800	Over 75 ton (with chase vehicle)		A-3E	2.50	6.400			305	62.50	367.50	530
2400	Crane, large lattice boom, requiring assembly		B-34W	.50	144			6,575	7,525	14,100	18,200
2500	For each additional 5 miles haul distance, add							10%	10%		
3000	For large pieces of equipment, allow for assembly/knockdown										
3100	For mob/demob of micro-tunneling equip, see Section 33 05 23.19										

14

For customer support on your Mechanical Cost Data, call 877.739.3689.

01 55 Vehicular Access and Parking

01 55 23 – Temporary Roads

01 55 23.50 Roads and Sidewalks	Crew	Daily Output	Labor-Hours	Unit	Material	2016 Bare Costs Labor	Equipment	Total	Total Incl O&P
0010 **ROADS AND SIDEWALKS** Temporary									
0050 Roads, gravel fill, no surfacing, 4" gravel depth	B-14	715	.067	S.Y.	3.70	2.69	.51	6.90	8.75
0100 8" gravel depth	"	615	.078	"	7.40	3.13	.59	11.12	13.60
1000 Ramp, 3/4" plywood on 2" x 6" joists, 16" O.C.	2 Carp	300	.053	S.F.	1.51	2.58		4.09	5.60
1100 On 2" x 10" joists, 16" O.C.	"	275	.058	"	2.22	2.82		5.04	6.75

01 56 Temporary Barriers and Enclosures

01 56 13 – Temporary Air Barriers

01 56 13.60 Tarpaulins

01 56 13.60 Tarpaulins	Crew	Daily Output	Labor-Hours	Unit	Material	2016 Bare Costs Labor	Equipment	Total	Total Incl O&P
0010 **TARPAULINS**									
0020 Cotton duck, 10 oz. to 13.13 oz. per S.Y., 6' x 8'				S.F.	.88			.88	.97
0050 30' x 30'					.64			.64	.70
0100 Polyvinyl coated nylon, 14 oz. to 18 oz., minimum					1.42			1.42	1.56
0150 Maximum					1.43			1.43	1.57
0200 Reinforced polyethylene 3 mils thick, white					.04			.04	.04
0300 4 mils thick, white, clear or black					.11			.11	.12
0400 5.5 mils thick, clear					.19			.19	.21
0500 White, fire retardant					.56			.56	.62
0600 12 mils, oil resistant, fire retardant					.43			.43	.47
0700 8.5 mils, black					.23			.23	.25
0710 Woven polyethylene, 6 mils thick					.19			.19	.21
0730 Polyester reinforced w/integral fastening system 11 mils thick					.22			.22	.24
0740 Mylar polyester, non-reinforced, 7 mils thick					1.17			1.17	1.29

01 56 13.90 Winter Protection

01 56 13.90 Winter Protection	Crew	Daily Output	Labor-Hours	Unit	Material	2016 Bare Costs Labor	Equipment	Total	Total Incl O&P
0010 **WINTER PROTECTION**									
0100 Framing to close openings	2 Clab	500	.032	S.F.	.43	1.21		1.64	2.33
0200 Tarpaulins hung over scaffolding, 8 uses, not incl. scaffolding		1500	.011		.24	.40		.64	.88
0250 Tarpaulin polyester reinf. w/integral fastening system 11 mils thick		1600	.010		.22	.38		.60	.82
0300 Prefab fiberglass panels, steel frame, 8 uses		1200	.013		2.40	.51		2.91	3.42

01 56 16 – Temporary Dust Barriers

01 56 16.10 Dust Barriers, Temporary

01 56 16.10 Dust Barriers, Temporary	Crew	Daily Output	Labor-Hours	Unit	Material	2016 Bare Costs Labor	Equipment	Total	Total Incl O&P
0010 **DUST BARRIERS, TEMPORARY**									
0020 Spring loaded telescoping pole & head, to 12', erect and dismantle	1 Clab	240	.033	Ea.		1.26		1.26	1.94
0025 Cost per day (based upon 250 days)				Day	.26			.26	.29
0030 To 21', erect and dismantle	1 Clab	240	.033	Ea.		1.26		1.26	1.94
0035 Cost per day (based upon 250 days)				Day	.54			.54	.60
0040 Accessories, caution tape reel, erect and dismantle	1 Clab	480	.017	Ea.		.63		.63	.97
0045 Cost per day (based upon 250 days)				Day	.36			.36	.40
0060 Foam rail and connector, erect and dismantle	1 Clab	240	.033	Ea.		1.26		1.26	1.94
0065 Cost per day (based upon 250 days)				Day	.12			.12	.13
0070 Caution tape	1 Clab	384	.021	C.L.F.	3.13	.79		3.92	4.65
0080 Zipper, standard duty		60	.133	Ea.	8	5.05		13.05	16.55
0090 Heavy duty		48	.167	"	9.75	6.30		16.05	20.50
0100 Polyethylene sheet, 4 mil		37	.216	Sq.	2.55	8.20		10.75	15.35
0110 6 mil		37	.216	"	3.70	8.20		11.90	16.60
1000 Dust partition, 6 mil polyethylene, 1" x 3" frame	2 Carp	2000	.008	S.F.	.30	.39		.69	.91
1080 2" x 4" frame	"	2000	.008	"	.33	.39		.72	.95

01 56 Temporary Barriers and Enclosures

01 56 23 – Temporary Barricades

01 56 23.10 Barricades	Crew	Daily Output	Labor-Hours	Unit	Material	2016 Bare Costs Labor	Equipment	Total	Total Incl O&P
0010 **BARRICADES**									
0020 5' high, 3 rail @ 2" x 8", fixed	2 Carp	20	.800	L.F.	5.70	39		44.70	66
0150 Movable		30	.533		4.74	26		30.74	44.50
1000 Guardrail, wooden, 3' high, 1" x 6", on 2" x 4" posts		200	.080		1.22	3.88		5.10	7.30
1100 2" x 6", on 4" x 4" posts	↓	165	.097		2.41	4.70		7.11	9.85
1200 Portable metal with base pads, buy					15.55			15.55	17.10
1250 Typical installation, assume 10 reuses	2 Carp	600	.027	↓	2.55	1.29		3.84	4.79
1300 Barricade tape, polyethylene, 7 mil, 3" wide x 500' long roll				Ea.	25			25	27.50
3000 Detour signs, set up and remove									
3010 Reflective aluminum, MUTCD, 24" x 24", post mounted	1 Clab	20	.400	Ea.	2.41	15.15		17.56	26
4000 Roof edge portable barrier stands and warning flags, 50 uses	1 Rohe	9100	.001	L.F.	.07	.03		.10	.12
4010 100 uses	"	9100	.001	"	.03	.03		.06	.09

01 56 26 – Temporary Fencing

01 56 26.50 Temporary Fencing	Crew	Daily Output	Labor-Hours	Unit	Material	2016 Bare Costs Labor	Equipment	Total	Total Incl O&P
0010 **TEMPORARY FENCING**									
0020 Chain link, 11 ga., 4' high	2 Clab	400	.040	L.F.	1.59	1.52		3.11	4.08
0100 6' high		300	.053		3.75	2.02		5.77	7.25
0200 Rented chain link, 6' high, to 1000' (up to 12 mo.)		400	.040		2.89	1.52		4.41	5.50
0250 Over 1000' (up to 12 mo.)	↓	300	.053		2.89	2.02		4.91	6.30
0350 Plywood, painted, 2" x 4" frame, 4' high	A-4	135	.178		6.25	8.15		14.40	19.30
0400 4" x 4" frame, 8' high	"	110	.218		12.20	10		22.20	28.50
0500 Wire mesh on 4" x 4" posts, 4' high	2 Carp	100	.160		9.95	7.75		17.70	23
0550 8' high	"	80	.200	↓	15.10	9.70		24.80	31.50

01 56 29 – Temporary Protective Walkways

01 56 29.50 Protection	Crew	Daily Output	Labor-Hours	Unit	Material	2016 Bare Costs Labor	Equipment	Total	Total Incl O&P
0010 **PROTECTION**									
0020 Stair tread, 2" x 12" planks, 1 use	1 Carp	75	.107	Tread	5.05	5.15		10.20	13.50
0100 Exterior plywood, 1/2" thick, 1 use		65	.123		2.04	5.95		7.99	11.40
0200 3/4" thick, 1 use		60	.133	↓	2.73	6.45		9.18	12.90
2200 Sidewalks, 2" x 12" planks, 2 uses		350	.023	S.F.	.84	1.11		1.95	2.62
2300 Exterior plywood, 2 uses, 1/2" thick		750	.011		.34	.52		.86	1.16
2400 5/8" thick		650	.012		.38	.60		.98	1.33
2500 3/4" thick	↓	600	.013	↓	.46	.65		1.11	1.49

01 58 Project Identification

01 58 13 – Temporary Project Signage

01 58 13.50 Signs	Crew	Daily Output	Labor-Hours	Unit	Material	2016 Bare Costs Labor	Equipment	Total	Total Incl O&P
0010 **SIGNS**									
0020 High intensity reflectorized, no posts, buy				Ea.	25			25	27.50

01 74 Cleaning and Waste Management

01 74 13 – Progress Cleaning

01 74 13.20 Cleaning Up		Crew	Daily Output	Labor-Hours	Unit	Material	2016 Bare Costs Labor	Equipment	Total	Total Incl O&P
0010	**CLEANING UP**									
0020	After job completion, allow, minimum				Job				.30%	.30%
0040	Maximum				"				1%	1%
0050	Cleanup of floor area, continuous, per day, during const.	A-5	24	.750	M.S.F.	2.23	29	2.59	33.82	49.50
0100	Final by GC at end of job	"	11.50	1.565	"	2.36	60	5.40	67.76	101

01 91 Commissioning

01 91 13 – General Commissioning Requirements

01 91 13.50 Building Commissioning		Crew	Daily Output	Labor-Hours	Unit	Material	2016 Bare Costs Labor	Equipment	Total	Total Incl O&P
0010	**BUILDING COMMISSIONING**									
0100	Systems operation and verification during turnover				%				.25%	.25%
0150	Including all systems subcontractors								.50%	.50%
0200	Systems design assistance, operation, verification and training								.50%	.50%
0250	Including all systems subcontractors				▼				1%	1%

01 93 Facility Maintenance

01 93 13 – Facility Maintenance Procedures

01 93 13.15 Mechanical Facilities Maintenance			Crew	Daily Output	Labor-Hours	Unit	Material	2016 Bare Costs Labor	Equipment	Total	Total Incl O&P
0010	**MECHANICAL FACILITIES MAINTENANCE**										
0100	Air conditioning system maintenance										
0130	Belt, replace		1 Stpi	15	.533	Ea.		32.50		32.50	49
0170	Fan, clean			16	.500			30.50		30.50	46
0180	Filter, remove, clean, replace			12	.667			40.50		40.50	61
0190	Flexible coupling alignment, inspect			40	.200			12.15		12.15	18.30
0200	Gas leak locate and repair			4	2			121		121	183
0250	Pump packing gland, remove and replace			11	.727			44		44	66.50
0270	Tighten			32	.250			15.20		15.20	23
0290	Pump disassemble and assemble			4	2			121		121	183
0300	Air pressure regulator, disassemble, clean, assemble		1 Skwk	4	2			100		100	154
0310	Repair or replace part		"	6	1.333			66.50		66.50	102
0320	Purging system		1 Stpi	16	.500			30.50		30.50	46
0400	Compressor, air, remove or install fan wheel		1 Skwk	20	.400			19.95		19.95	30.50
0410	Disassemble or assemble 2 cylinder, 2 stage			4	2			100		100	154
0420	4 cylinder, 4 stage			1	8			400		400	615
0430	Repair or replace part			2	4	▼		200		200	305
0700	Demolition, for mech. demolition see Section 23 05 05.10 or 22 05 05.10										
0800	Ductwork, clean										
0810	Rectangular										
0820	6"	G	1 Shee	187.50	.043	L.F.		2.44		2.44	3.73
0830	8"	G		140.63	.057			3.26		3.26	4.98
0840	10"	G		112.50	.071			4.07		4.07	6.20
0850	12"	G		93.75	.085			4.89		4.89	7.45
0860	14"	G		80.36	.100			5.70		5.70	8.70
0870	16"	G		70.31	.114	▼		6.50		6.50	9.95
0900	Round										
0910	4"	G	1 Shee	358.10	.022	L.F.		1.28		1.28	1.95
0920	6"	G		238.73	.034			1.92		1.92	2.93
0930	8"	G		179.05	.045			2.56		2.56	3.91
0940	10"	G		143.24	.056	▼		3.20		3.20	4.89

For customer support on your Mechanical Cost Data, call 877.739.3689.

17

01 93 13 – Facility Maintenance Procedures

01 93 13.15 Mechanical Facilities Maintenance		Crew	Daily Output	Labor-Hours	Unit	Material	2016 Bare Costs Labor	Equipment	Total	Total Incl O&P
0950	12"	**G** 1 Shee	119.37	.067	L.F.		3.84		3.84	5.85
0960	16"	**G** ↓	89.52	.089	↓		5.10		5.10	7.80
1000	Expansion joint, not screwed, install or remove	1 Stpi	3	2.667	Ea.		162		162	244
1010	Repack	"	6	1.333	"		81		81	122
1200	Fire protection equipment									
1220	Fire hydrant, replace	Q-1	3	5.333	Ea.		284		284	430
1230	Service, lubricate, inspect, flush, clean	1 Plum	7	1.143			67.50		67.50	102
1240	Test	"	11	.727	↓		43		43	65
1310	Inspect valves, pressure, nozzle,	1 Spri	4	2	System		115		115	175
1500	Instrumentation preventive maintenance, rule of thumb									
1510	Annual cost per instrument for calibration & trouble shooting									
1520	Includes flow meters, analytical transmitters, flow control									
1530	chemical feed systems, recorders and distrib. control sys.				Ea.				325	325
1800	Plumbing fixtures, for installation see Section 22 41 00									
1801	Plumbing fixtures									
1840	Open drain with snake	1 Plum	13	.615	Ea.		36.50		36.50	55
1850	Open drain with toilet auger		16	.500			29.50		29.50	44.50
1860	Grease trap, clean		2	4			237		237	355
1870	Plaster trap, clean	↓	6	1.333			79		79	119
1881	Clean commode	1 Clab	20	.400		.37	15.15		15.52	24
1882	Clean commode seat		48	.167		.19	6.30		6.49	9.90
1883	Clean single sink		28	.286		.24	10.85		11.09	16.85
1884	Clean double sink		20	.400		.37	15.15		15.52	24
1885	Clean lavatory		28	.286		.24	10.85		11.09	16.85
1886	Clean bathtub		12	.667		.74	25.50		26.24	40
1887	Clean fiberglass shower		12	.667		.74	25.50		26.24	40
1888	Clean fiberglass tub/shower		8	1		.93	38		38.93	59
1889	Clean faucet set		32	.250		.15	9.50		9.65	14.70
1891	Clean shower head		80	.100		.02	3.79		3.81	5.80
1893	Clean water heater	↓	16	.500		1.30	18.95		20.25	30.50
1900	Relief valve, test and adjust	1 Stpi	20	.400			24.50		24.50	36.50
1910	Clean pump, heater, or motor for whirlpool	1 Clab	16	.500		.19	18.95		19.14	29
1920	Clean thermal cover for whirlpool	"	16	.500		1.85	18.95		20.80	31
2000	Repair or replace-steam trap	1 Stpi	8	1			60.50		60.50	91.50
2020	Y-type or bell strainer		6	1.333			81		81	122
2040	Water trap or vacuum breaker, screwed joints	↓	13	.615	↓		37.50		37.50	56.50
2100	Steam specialties, clean									
2120	Air separator with automatic trap, 1" fittings	1 Stpi	12	.667	Ea.		40.50		40.50	61
2130	Bucket trap, 2" pipe		7	1.143			69.50		69.50	105
2150	Drip leg, 2" fitting		45	.178			10.80		10.80	16.30
2200	Thermodynamic trap, 1" fittings		50	.160			9.70		9.70	14.65
2210	Thermostatic		65	.123			7.45		7.45	11.25
2240	Screen and seat in y-type strainer, plug type.		25	.320			19.40		19.40	29.50
2242	Screen and seat in y-type strainer, flange type.		12	.667			40.50		40.50	61
2500	Valve, replace broken handwheel	↓	24	.333	↓		20		20	30.50
3000	Valve, overhaul, regulator, relief, flushometer, mixing									
3040	Cold water, gas	1 Stpi	5	1.600	Ea.		97		97	147
3050	Hot water, steam		3	2.667			162		162	244
3080	Globe, gate, check up to 4" cold water, gas		10	.800			48.50		48.50	73.50
3090	Hot water, steam		5	1.600			97		97	147
3100	Over 4" ID hot or cold line	↓	1.40	5.714			345		345	525
3120	Remove and replace, gate, globe or check up to 4"	Q-5	6	2.667			146		146	220
3130	Over 4"	"	2	8			435		435	660

18

01 93 13.15 Mechanical Facilities Maintenance		Crew	Daily Output	Labor-Hours	Unit	Material	2016 Bare Costs Labor	Equipment	Total	Total Incl O&P
3150	Repack up to 4"	1 Stpi	13	.615	Ea.		37.50		37.50	56.50
3160	Over 4"	"	4	2	↓		121		121	183

For customer support on your Mechanical Cost Data, call 877.739.3689.

19

Division Notes

	CREW	DAILY OUTPUT	LABOR-HOURS	UNIT	BARE COSTS				TOTAL INCL O&P
					MAT.	LABOR	EQUIP.	TOTAL	

Estimating Tips

02 30 00 Subsurface Investigation

In preparing estimates on structures involving earthwork or foundations, all information concerning soil characteristics should be obtained. Look particularly for hazardous waste, evidence of prior dumping of debris, and previous stream beds.

02 40 00 Demolition and Structure Moving

The costs shown for selective demolition do not include rubbish handling or disposal. These items should be estimated separately using RSMeans data or other sources.

- Historic preservation often requires that the contractor remove materials from the existing structure, rehab them, and replace them. The estimator must be aware of any related measures and precautions that must be taken when doing selective demolition and cutting and patching. Requirements may include special handling and storage, as well as security.

- In addition to Subdivision 02 41 00, you can find selective demolition items in each division. Example: Roofing demolition is in Division 7.

- Absent of any other specific reference, an approximate demolish-in-place cost can be obtained by halving the new-install labor cost. To remove for reuse, allow the entire new-install labor figure.

02 40 00 Building Deconstruction

This section provides costs for the careful dismantling and recycling of most of low-rise building materials.

02 50 00 Containment of Hazardous Waste

This section addresses on-site hazardous waste disposal costs.

02 80 00 Hazardous Material Disposal/ Remediation

This subdivision includes information on hazardous waste handling, asbestos remediation, lead remediation, and mold remediation. See reference R028213-20 and R028319-60 for further guidance in using these unit price lines.

02 90 00 Monitoring Chemical Sampling, Testing Analysis

This section provides costs for on-site sampling and testing hazardous waste.

Reference Numbers

Reference numbers are shown at the beginning of some major classifications. These numbers refer to related items in the Reference Section. The reference information may be an estimating procedure, an alternate pricing method, or technical information.

Note: Not all subdivisions listed here necessarily appear. ■

02 41 Demolition

02 41 13 – Selective Site Demolition

02 41 13.17 Demolish, Remove Pavement and Curb	Crew	Daily Output	Labor-Hours	Unit	Material	2016 Bare Costs Labor	Equipment	Total	Total Incl O&P
0010 **DEMOLISH, REMOVE PAVEMENT AND CURB** R024119-10									
5010 Pavement removal, bituminous roads, up to 3" thick	B-38	690	.058	S.Y.		2.50	1.83	4.33	5.85
5050 4" to 6" thick		420	.095			4.11	3.01	7.12	9.55
5100 Bituminous driveways		640	.063			2.70	1.98	4.68	6.30
5200 Concrete to 6" thick, hydraulic hammer, mesh reinforced		255	.157			6.75	4.96	11.71	15.80
5300 Rod reinforced		200	.200			8.65	6.30	14.95	20
5400 Concrete, 7" to 24" thick, plain		33	1.212	C.Y.		52.50	38.50	91	122
5500 Reinforced		24	1.667	"		72	52.50	124.50	168
5600 With hand held air equipment, bituminous, to 6" thick	B-39	1900	.025	S.F.		1.01	.12	1.13	1.68
5700 Concrete to 6" thick, no reinforcing		1600	.030			1.20	.14	1.34	2
5800 Mesh reinforced		1400	.034			1.38	.16	1.54	2.28
5900 Rod reinforced		765	.063			2.52	.30	2.82	4.18

02 41 13.23 Utility Line Removal

	Crew	Daily Output	Labor-Hours	Unit	Material	Labor	Equipment	Total	Total Incl O&P
0010 **UTILITY LINE REMOVAL**									
0015 No hauling, abandon catch basin or manhole	B-6	7	3.429	Ea.		143	52	195	276
0020 Remove existing catch basin or manhole, masonry		4	6			250	91.50	341.50	480
0030 Catch basin or manhole frames and covers, stored		13	1.846			77	28	105	148
0040 Remove and reset		7	3.429			143	52	195	276
0900 Hydrants, fire, remove only	B-21A	5	8			380	96	476	680
0950 Remove and reset	"	2	20			945	240	1,185	1,700

02 41 13.30 Minor Site Demolition

	Crew	Daily Output	Labor-Hours	Unit	Material	Labor	Equipment	Total	Total Incl O&P
0010 **MINOR SITE DEMOLITION** R024119-10									
4000 Sidewalk removal, bituminous, 2" thick	B-6	350	.069	S.Y.		2.86	1.04	3.90	5.50
4010 2-1/2" thick		325	.074			3.08	1.12	4.20	5.95
4050 Brick, set in mortar		185	.130			5.40	1.98	7.38	10.40
4100 Concrete, plain, 4"		160	.150			6.25	2.28	8.53	12.05
4110 Plain, 5"		140	.171			7.15	2.61	9.76	13.75
4120 Plain, 6"		120	.200			8.35	3.05	11.40	16.05
4200 Mesh reinforced, concrete, 4"		150	.160			6.65	2.44	9.09	12.85
4210 5" thick		131	.183			7.65	2.79	10.44	14.70
4220 6" thick		112	.214			8.95	3.26	12.21	17.20

02 41 19 – Selective Demolition

02 41 19.13 Selective Building Demolition

0010 **SELECTIVE BUILDING DEMOLITION**									
0020 Costs related to selective demolition of specific building components									
0025 are included under Common Work Results (XX 05)									
0030 in the component's appropriate division.									

02 41 19.19 Selective Demolition

	Crew	Daily Output	Labor-Hours	Unit	Material	Labor	Equipment	Total	Total Incl O&P
0010 **SELECTIVE DEMOLITION**, Rubbish Handling R024119-10									
0020 The following are to be added to the demolition prices									
0600 Dumpster, weekly rental, 1 dump/week, 6 C.Y. capacity (2 Tons)				Week	415			415	455
0700 10 C.Y. capacity (3 Tons)					480			480	530
0725 20 C.Y. capacity (5 Tons) R024119-20					565			565	625
0800 30 C.Y. capacity (7 Tons)					730			730	800
0840 40 C.Y. capacity (10 Tons)					775			775	850
2000 Load, haul, dump and return, 0 – 50' haul, hand carried	2 Clab	24	.667	C.Y.		25.50		25.50	39
2005 Wheeled		37	.432			16.40		16.40	25
2040 0 – 100' haul, hand carried		16.50	.970			37		37	56.50
2045 Wheeled		25	.640			24.50		24.50	37
2080 Haul and return, add per each extra 100' haul, hand carried		35.50	.451			17.10		17.10	26
2085 Wheeled		54	.296			11.25		11.25	17.25

22

For customer support on your Mechanical Cost Data, call 877.739.3689.

02 41 Demolition

02 41 19 – Selective Demolition

02 41 19.19 Selective Demolition

02 41 19.19 Selective Demolition		Crew	Daily Output	Labor-Hours	Unit	Material	2016 Bare Costs		Total	Total Incl O&P
							Labor	Equipment		
2120	For travel in elevators, up to 10 floors, add	2 Clab	140	.114	C.Y.		4.33		4.33	6.65
2130	0 – 50' haul, incl. up to 5 riser stair, hand carried		23	.696			26.50		26.50	40.50
2135	Wheeled		35	.457			17.35		17.35	26.50
2140	6 – 10 riser stairs, hand carried		22	.727			27.50		27.50	42.50
2145	Wheeled		34	.471			17.85		17.85	27.50
2150	11 – 20 riser stairs, hand carried		20	.800			30.50		30.50	46.50
2155	Wheeled		31	.516			19.55		19.55	30
2160	21 – 40 riser stairs, hand carried		16	1			38		38	58
2165	Wheeled		24	.667			25.50		25.50	39
2170	0 – 100' haul, incl. 5 riser stair, hand carried		15	1.067			40.50		40.50	62
2175	Wheeled		23	.696			26.50		26.50	40.50
2180	6 – 10 riser stair, hand carried		14	1.143			43.50		43.50	66.50
2185	Wheeled		21	.762			29		29	44.50
2190	11 – 20 riser stair, hand carried		12	1.333			50.50		50.50	77.50
2195	Wheeled		18	.889			33.50		33.50	51.50
2200	21 – 40 riser stair, hand carried		8	2			76		76	116
2205	Wheeled		12	1.333			50.50		50.50	77.50
2210	Haul and return, add per each extra 100' haul, hand carried		35.50	.451			17.10		17.10	26
2215	Wheeled		54	.296			11.25		11.25	17.25
2220	For each additional flight of stairs, up to 5 risers, add		550	.029	Flight		1.10		1.10	1.69
2225	6 – 10 risers, add		275	.058			2.21		2.21	3.38
2230	11 – 20 risers, add		138	.116			4.39		4.39	6.75
2235	21 – 40 risers, add		69	.232			8.80		8.80	13.50
3000	Loading & trucking, including 2 mile haul, chute loaded	B-16	45	.711	C.Y.		28.50	15.35	43.85	60
3040	Hand loading truck, 50' haul	"	48	.667			26.50	14.40	40.90	56.50
3080	Machine loading truck	B-17	120	.267			11.20	6.50	17.70	24.50
5000	Haul, per mile, up to 8 C.Y. truck	B-34B	1165	.007			.30	.59	.89	1.10
5100	Over 8 C.Y. truck	"	1550	.005			.22	.45	.67	.83

02 41 19.20 Selective Demolition, Dump Charges

		Crew	Daily Output	Labor-Hours	Unit	Material	Labor	Equipment	Total	Total Incl O&P
0010	**SELECTIVE DEMOLITION, DUMP CHARGES** R024119-10									
0020	Dump charges, typical urban city, tipping fees only									
0100	Building construction materials				Ton	74			74	81
0200	Trees, brush, lumber					63			63	69.50
0300	Rubbish only					63			63	69.50
0500	Reclamation station, usual charge					74			74	81

02 41 19.27 Selective Demolition, Torch Cutting

		Crew	Daily Output	Labor-Hours	Unit	Material	Labor	Equipment	Total	Total Incl O&P
0010	**SELECTIVE DEMOLITION, TORCH CUTTING** R024119-10									
0020	Steel, 1" thick plate	E-25	333	.024	L.F.	.84	1.33	.03	2.20	3.21
0040	1" diameter bar	"	600	.013	Ea.	.14	.74	.02	.90	1.42
1000	Oxygen lance cutting, reinforced concrete walls									
1040	12" to 16" thick walls	1 Clab	10	.800	L.F.		30.50		30.50	46.50
1080	24" thick walls	"	6	1.333	"		50.50		50.50	77.50

02 65 Underground Storage Tank Removal

02 65 10 – Underground Tank and Contaminated Soil Removal

02 65 10.30 Removal of Underground Storage Tanks		Crew	Daily Output	Labor-Hours	Unit	Material	2016 Bare Costs Labor	Equipment	Total	Total Incl O&P
0010	**REMOVAL OF UNDERGROUND STORAGE TANKS** R026510-20									
0011	Petroleum storage tanks, non-leaking									
0100	Excavate & load onto trailer									
0110	3000 gal. to 5000 gal. tank	G B-14	4	12	Ea.		480	91.50	571.50	835
0120	6000 gal. to 8000 gal. tank	G B-3A	3	13.333			540	345	885	1,200
0130	9000 gal. to 12000 gal. tank	"	2	20			810	515	1,325	1,825
0190	Known leaking tank, add				%				100%	100%
0200	Remove sludge, water and remaining product from tank bottom									
0201	of tank with vacuum truck									
0300	3000 gal. to 5000 gal. tank	G A-13	5	1.600	Ea.		78.50	152	230.50	286
0310	6000 gal. to 8000 gal. tank	G	4	2			98.50	190	288.50	355
0320	9000 gal. to 12000 gal. tank	G	3	2.667			131	253	384	475
0390	Dispose of sludge off-site, average				Gal.				6.25	6.80
0400	Insert inert solid CO_2 "dry ice" into tank									
0401	For cleaning/transporting tanks (1.5 lb./100 gal. cap)	G 1 Clab	500	.016	Lb.	1.17	.61		1.78	2.22
1020	Haul tank to certified salvage dump, 100 miles round trip									
1023	3000 gal. to 5000 gal. tank				Ea.				760	830
1026	6000 gal. to 8000 gal. tank								880	960
1029	9000 gal. to 12000 gal. tank								1,050	1,150
1100	Disposal of contaminated soil to landfill									
1110	Minimum				C.Y.				145	160
1111	Maximum				"				400	440
1120	Disposal of contaminated soil to									
1121	bituminous concrete batch plant									
1130	Minimum				C.Y.				80	88
1131	Maximum				"				115	125
2010	Decontamination of soil on site incl poly tarp on top/bottom									
2011	Soil containment berm, and chemical treatment									
2020	Minimum	G B-11C	100	.160	C.Y.	7.80	7.10	3.65	18.55	23.50
2021	Maximum	G "	100	.160		10.10	7.10	3.65	20.85	26
2050	Disposal of decontaminated soil, minimum								135	150
2055	Maximum								400	440

02 81 Transportation and Disposal of Hazardous Materials

02 81 20 – Hazardous Waste Handling

02 81 20.10 Hazardous Waste Cleanup/Pickup/Disposal

		Crew	Daily Output	Labor-Hours	Unit	Material	2016 Bare Costs Labor	Equipment	Total	Total Incl O&P
0010	**HAZARDOUS WASTE CLEANUP/PICKUP/DISPOSAL**									
0100	For contractor rental equipment, i.e., Dozer,									
0110	Front end loader, Dump truck, etc., see 01 54 33 Reference Section									
1000	Solid pickup									
1100	55 gal. drums				Ea.				240	265
1120	Bulk material, minimum				Ton				190	210
1130	Maximum				"				595	655
1200	Transportation to disposal site									
1220	Truckload = 80 drums or 25 C.Y. or 18 tons									
1260	Minimum				Mile				3.95	4.45
1270	Maximum				"				7.25	7.35
3000	Liquid pickup, vacuum truck, stainless steel tank									
3100	Minimum charge, 4 hours									
3110	1 compartment, 2200 gallon				Hr.				140	155
3120	2 compartment, 5000 gallon				"				200	225

02 81 Transportation and Disposal of Hazardous Materials

02 81 20 – Hazardous Waste Handling

02 81 20.10 Hazardous Waste Cleanup/Pickup/Disposal		Crew	Daily Output	Labor-Hours	Unit	Material	2016 Bare Costs Labor	Equipment	Total	Total Incl O&P
3400	Transportation in 6900 gallon bulk truck				Mile				7.95	8.75
3410	In teflon lined truck				"				10.20	11.25
5000	Heavy sludge or dry vacuumable material				Hr.				140	160
6000	Dumpsite disposal charge, minimum				Ton				140	155
6020	Maximum				"				415	455

02 82 Asbestos Remediation

02 82 13 – Asbestos Abatement

02 82 13.39 Asbestos Remediation Plans and Methods

		Crew	Daily Output	Labor-Hours	Unit	Material	2016 Bare Costs Labor	Equipment	Total	Total Incl O&P
0010	**ASBESTOS REMEDIATION PLANS AND METHODS**									
0100	Building Survey-Commercial Building				Ea.				2,200	2,400
0200	Asbestos Abatement Remediation Plan				"				1,350	1,475

02 82 13.41 Asbestos Abatement Equipment

		Crew	Daily Output	Labor-Hours	Unit	Material	2016 Bare Costs Labor	Equipment	Total	Total Incl O&P
0010	**ASBESTOS ABATEMENT EQUIPMENT** R028213-20									
0011	Equipment and supplies, buy									
0200	Air filtration device, 2000 CFM				Ea.	975			975	1,075
0250	Large volume air sampling pump, minimum					355			355	390
0260	Maximum					340			340	375
0300	Airless sprayer unit, 2 gun					3,275			3,275	3,600
0350	Light stand, 500 watt					51			51	56
0400	Personal respirators									
0410	Negative pressure, 1/2 face, dual operation, min.				Ea.	27			27	29.50
0420	Maximum					30			30	33
0450	P.A.P.R., full face, minimum					127			127	140
0460	Maximum					169			169	185
0470	Supplied air, full face, incl. air line, minimum					172			172	189
0480	Maximum					410			410	450
0500	Personnel sampling pump					230			230	254
1500	Power panel, 20 unit, incl. GFI					615			615	675
1600	Shower unit, including pump and filters					1,300			1,300	1,450
1700	Supplied air system (type C)					3,475			3,475	3,825
1750	Vacuum cleaner, HEPA, 16 gal., stainless steel, wet/dry					440			440	485
1760	55 gallon					570			570	630
1800	Vacuum loader, 9 – 18 ton/hr.					96,000			96,000	106,000
1900	Water atomizer unit, including 55 gal. drum					286			286	315
2000	Worker protection, whole body, foot, head cover & gloves, plastic					9			9	9.90
2500	Respirator, single use					25.50			25.50	28
2550	Cartridge for respirator					5.85			5.85	6.45
2570	Glove bag, 7 mil, 50" x 64"					9.35			9.35	10.30
2580	10 mil, 44" x 60"					5.90			5.90	6.45
2590	6 mil, 44" x 60"					5.90			5.90	6.45
3000	HEPA vacuum for work area, minimum					310			310	345
3050	Maximum					830			830	915
6000	Disposable polyethylene bags, 6 mil, 3 C.F.					.84			.84	.92
6300	Disposable fiber drums, 3 C.F.					18.55			18.55	20.50
6400	Pressure sensitive caution labels, 3" x 5"					3.54			3.54	3.89
6450	11" x 17"					7.50			7.50	8.25
6500	Negative air machine, 1800 CFM					865			865	950

For customer support on your Mechanical Cost Data, call 877.739.3689.

25

02 82 13 – Asbestos Abatement

02 82 13.42 Preparation of Asbestos Containment Area	Crew	Daily Output	Labor-Hours	Unit	Material	2016 Bare Costs Labor	2016 Bare Costs Equipment	Total	Total Incl O&P
0010 **PREPARATION OF ASBESTOS CONTAINMENT AREA** R028213-20									
0100 Pre-cleaning, HEPA vacuum and wet wipe, flat surfaces	A-9	12000	.005	S.F.	.02	.28		.30	.46
0200 Protect carpeted area, 2 layers 6 mil poly on 3/4" plywood	"	1000	.064		2	3.42		5.42	7.50
0300 Separation barrier, 2" x 4" @ 16", 1/2" plywood ea. side, 8' high	2 Carp	400	.040		3.40	1.94		5.34	6.70
0310 12' high		320	.050		3.40	2.42		5.82	7.45
0320 16' high		200	.080		2.30	3.88		6.18	8.50
0400 Personnel decontam. chamber, 2" x 4" @ 16", 3/4" ply ea. side		280	.057		4.40	2.77		7.17	9.10
0450 Waste decontam. chamber, 2" x 4" studs @ 16", 3/4" ply ea. side		360	.044		4.40	2.15		6.55	8.15
0500 Cover surfaces with polyethylene sheeting									
0501 Including glue and tape									
0550 Floors, each layer, 6 mil	A-9	8000	.008	S.F.	.04	.43		.47	.71
0551 4 mil		9000	.007		.03	.38		.41	.62
0560 Walls, each layer, 6 mil		6000	.011		.04	.57		.61	.93
0561 4 mil		7000	.009		.03	.49		.52	.79
0570 For heights above 12', add						20%			
0575 For heights above 20', add						30%			
0580 For fire retardant poly, add					100%				
0590 For large open areas, deduct					10%	20%			
0600 Seal floor penetrations with foam firestop to 36 sq. in.	2 Carp	200	.080	Ea.	9.50	3.88		13.38	16.45
0610 36 sq. in. to 72 sq. in.		125	.128		19.05	6.20		25.25	30.50
0615 72 sq. in. to 144 sq. in.		80	.200		38	9.70		47.70	57
0620 Wall penetrations, to 36 square inches		180	.089		9.50	4.31		13.81	17.10
0630 36 sq. in. to 72 sq. in.		100	.160		19.05	7.75		26.80	33
0640 72 sq. in. to 144 sq. in.		60	.267		38	12.90		50.90	62
0800 Caulk seams with latex	1 Carp	230	.035	L.F.	.17	1.69		1.86	2.77
0900 Set up neg. air machine, 1-2k CFM/25 M.C.F. volume	1 Asbe	4.30	1.860	Ea.		99.50		99.50	155
0950 Set up and remove portable shower unit	2 Asbe	4	4	"		214		214	330

02 82 13.43 Bulk Asbestos Removal

	Crew	Daily Output	Labor-Hours	Unit	Material	2016 Bare Costs Labor	2016 Bare Costs Equipment	Total	Total Incl O&P
0010 **BULK ASBESTOS REMOVAL**									
0020 Includes disposable tools and 2 suits and 1 respirator filter/day/worker									
0200 Boiler insulation	A-9	480	.133	S.F.	.47	7.15		7.62	11.60
0210 With metal lath, add				%				50%	50%
0300 Boiler breeching or flue insulation	A-9	520	.123	S.F.	.37	6.60		6.97	10.65
0310 For active boiler, add				%				100%	100%
0400 Duct or AHU insulation	A-10B	440	.073	S.F.	.22	3.89		4.11	6.30
0500 Duct vibration isolation joints, up to 24 sq. in. duct	A-9	56	1.143	Ea.	3.41	61		64.41	99
0520 25 sq. in. to 48 sq. in. duct		48	1.333		3.98	71.50		75.48	115
0530 49 sq. in. to 76 sq. in. duct		40	1.600		4.77	85.50		90.27	138
0600 Pipe insulation, air cell type, up to 4" diameter pipe		900	.071	L.F.	.21	3.80		4.01	6.15
0610 4" to 8" diameter pipe		800	.080		.24	4.28		4.52	6.90
0620 10" to 12" diameter pipe		700	.091		.27	4.89		5.16	7.90
0630 14" to 16" diameter pipe		550	.116		.35	6.20		6.55	10.10
0650 Over 16" diameter pipe		650	.098	S.F.	.29	5.25		5.54	8.50
0700 With glove bag up to 3" diameter pipe		200	.320	L.F.	9.30	17.10		26.40	37
1000 Pipe fitting insulation up to 4" diameter pipe		320	.200	Ea.	.60	10.70		11.30	17.30
1100 6" to 8" diameter pipe		304	.211		.63	11.25		11.88	18.20
1110 10" to 12" diameter pipe		192	.333		.99	17.80		18.79	28.50
1120 14" to 16" diameter pipe		128	.500		1.49	26.50		27.99	43
1130 Over 16" diameter pipe		176	.364	S.F.	1.08	19.45		20.53	31.50
1200 With glove bag, up to 8" diameter pipe		75	.853	L.F.	6.40	45.50		51.90	78
2000 Scrape foam fireproofing from flat surface		2400	.027	S.F.	.08	1.43		1.51	2.31
2100 Irregular surfaces		1200	.053		.16	2.85		3.01	4.61

02 82 13 – Asbestos Abatement

02 82 13.43 Bulk Asbestos Removal

		Crew	Daily Output	Labor-Hours	Unit	Material	2016 Bare Costs Labor	Equipment	Total	Total Incl O&P
3000	Remove cementitious material from flat surface	A-9	1800	.036	S.F.	.11	1.90		2.01	3.08
3100	Irregular surface		1000	.064		.14	3.42		3.56	5.45
6000	Remove contaminated soil from crawl space by hand		400	.160	C.F.	.48	8.55		9.03	13.80
6100	With large production vacuum loader	A-12	700	.091	"	.27	4.89	1.08	6.24	9.10
7000	Radiator backing, not including radiator removal	A-9	1200	.053	S.F.	.16	2.85		3.01	4.61
9000	For type B (supplied air) respirator equipment, add				%				10%	10%

02 82 13.44 Demolition In Asbestos Contaminated Area

		Crew	Daily Output	Labor-Hours	Unit	Material	2016 Bare Costs Labor	Equipment	Total	Total Incl O&P
0010	**DEMOLITION IN ASBESTOS CONTAMINATED AREA**									
0200	Ceiling, including suspension system, plaster and lath	A-9	2100	.030	S.F.	.09	1.63		1.72	2.64
0210	Finished plaster, leaving wire lath		585	.109		.33	5.85		6.18	9.45
0220	Suspended acoustical tile		3500	.018		.05	.98		1.03	1.58
0230	Concealed tile grid system		3000	.021		.06	1.14		1.20	1.84
0240	Metal pan grid system		1500	.043		.13	2.28		2.41	3.69
0250	Gypsum board		2500	.026		.08	1.37		1.45	2.21
0260	Lighting fixtures up to 2' x 4'		72	.889	Ea.	2.65	47.50		50.15	77
0400	Partitions, non load bearing									
0410	Plaster, lath, and studs	A-9	690	.093	S.F.	.91	4.96		5.87	8.70
0450	Gypsum board and studs	"	1390	.046	"	.14	2.46		2.60	3.98
9000	For type B (supplied air) respirator equipment, add				%				10%	10%

02 82 13.45 OSHA Testing

		Crew	Daily Output	Labor-Hours	Unit	Material	2016 Bare Costs Labor	Equipment	Total	Total Incl O&P
0010	**OSHA TESTING**									
0100	Certified technician, minimum				Day				200	220
0110	Maximum								300	330
0120	Industrial hygienist, minimum								250	250
0130	Maximum								400	440
0200	Asbestos sampling and PCM analysis, NIOSH 7400, minimum	1 Asbe	8	1	Ea.	14	53.50		67.50	98.50
0210	Maximum		4	2		22	107		129	190
1000	Cleaned area samples		8	1		200	53.50		253.50	305
1100	PCM air sample analysis, NIOSH 7400, minimum		8	1		14	53.50		67.50	98.50
1110	Maximum		4	2		2.20	107		109.20	168
1200	TEM air sample analysis, NIOSH 7402, minimum								80	106
1210	Maximum								360	450

02 82 13.46 Decontamination of Asbestos Containment Area

		Crew	Daily Output	Labor-Hours	Unit	Material	2016 Bare Costs Labor	Equipment	Total	Total Incl O&P
0010	**DECONTAMINATION OF ASBESTOS CONTAINMENT AREA**									
0100	Spray exposed substrate with surfactant (bridging)									
0200	Flat surfaces	A-9	6000	.011	S.F.	.40	.57		.97	1.33
0250	Irregular surfaces		4000	.016	"	.40	.86		1.26	1.77
0300	Pipes, beams, and columns		2000	.032	L.F.	.60	1.71		2.31	3.32
1000	Spray encapsulate polyethylene sheeting		8000	.008	S.F.	.40	.43		.83	1.11
1100	Roll down polyethylene sheeting		8000	.008	"		.43		.43	.67
1500	Bag polyethylene sheeting		400	.160	Ea.	1	8.55		9.55	14.40
2000	Fine clean exposed substrate, with nylon brush		2400	.027	S.F.		1.43		1.43	2.22
2500	Wet wipe substrate		4800	.013			.71		.71	1.11
2600	Vacuum surfaces, fine brush		6400	.010			.53		.53	.83
3000	Structural demolition									
3100	Wood stud walls	A-9	2800	.023	S.F.		1.22		1.22	1.90
3500	Window manifolds, not incl. window replacement		4200	.015			.81		.81	1.27
3600	Plywood carpet protection		2000	.032			1.71		1.71	2.66
4000	Remove custom decontamination facility	A-10A	8	3	Ea.	15	161		176	267
4100	Remove portable decontamination facility	3 Asbe	12	2	"	13.35	107		120.35	181
5000	HEPA vacuum, shampoo carpeting	A-9	4800	.013	S.F.	.10	.71		.81	1.22
9000	Final cleaning of protected surfaces	A-10A	8000	.003	"		.16		.16	.25

02 82 13.47 Asbestos Waste Pkg., Handling, and Disp.

		Crew	Daily Output	Labor-Hours	Unit	Material	2016 Bare Costs Labor	2016 Bare Costs Equipment	Total	Total Incl O&P
0010	**ASBESTOS WASTE PACKAGING, HANDLING, AND DISPOSAL**									
0100	Collect and bag bulk material, 3 C.F. bags, by hand	A-9	400	.160	Ea.	.84	8.55		9.39	14.20
0200	Large production vacuum loader	A-12	880	.073		1	3.89	.86	5.75	8.10
1000	Double bag and decontaminate	A-9	960	.067		.84	3.56		4.40	6.45
2000	Containerize bagged material in drums, per 3 C.F. drum	"	800	.080		18.55	4.28		22.83	27
3000	Cart bags 50' to dumpster	2 Asbe	400	.040			2.14		2.14	3.32
5000	Disposal charges, not including haul, minimum				C.Y.				61	67
5020	Maximum				"				355	395
9000	For type B (supplied air) respirator equipment, add				%				10%	10%

02 82 13.48 Asbestos Encapsulation With Sealants

		Crew	Daily Output	Labor-Hours	Unit	Material	2016 Bare Costs Labor	2016 Bare Costs Equipment	Total	Total Incl O&P
0010	**ASBESTOS ENCAPSULATION WITH SEALANTS**									
0100	Ceilings and walls, minimum	A-9	21000	.003	S.F.	.40	.16		.56	.69
0110	Maximum		10600	.006	"	.50	.32		.82	1.05
0300	Pipes to 12" diameter including minor repairs, minimum		800	.080	L.F.	.50	4.28		4.78	7.20
0310	Maximum		400	.160	"	1.20	8.55		9.75	14.60

02 85 Mold Remediation

02 85 16 – Mold Remediation Preparation and Containment

02 85 16.50 Preparation of Mold Containment Area

		Crew	Daily Output	Labor-Hours	Unit	Material	2016 Bare Costs Labor	2016 Bare Costs Equipment	Total	Total Incl O&P
0010	**PREPARATION OF MOLD CONTAINMENT AREA** R028213-20									
0100	Pre-cleaning, HEPA vacuum and wet wipe, flat surfaces	A-9	12000	.005	S.F.	.02	.28		.30	.46
0300	Separation barrier, 2" x 4" @ 16", 1/2" plywood ea. side, 8' high	2 Carp	400	.040		3.40	1.94		5.34	6.70
0310	12' high		320	.050		3.40	2.42		5.82	7.45
0320	16' high		200	.080		2.30	3.88		6.18	8.50
0400	Personnel decontam. chamber, 2" x 4" @ 16", 3/4" ply ea. side		280	.057		4.40	2.77		7.17	9.10
0450	Waste decontam. chamber, 2" x 4" studs @ 16", 3/4" ply each side		360	.044		4.40	2.15		6.55	8.15
0500	Cover surfaces with polyethylene sheeting									
0501	Including glue and tape									
0550	Floors, each layer, 6 mil	A-9	8000	.008	S.F.	.04	.43		.47	.71
0551	4 mil		9000	.007		.03	.38		.41	.62
0560	Walls, each layer, 6 mil		6000	.011		.04	.57		.61	.93
0561	4 mil		7000	.009		.03	.49		.52	.79
0570	For heights above 12', add						20%			
0575	For heights above 20', add						30%			
0580	For fire retardant poly, add					100%				
0590	For large open areas, deduct					10%	20%			
0600	Seal floor penetrations with foam firestop to 36 sq. in.	2 Carp	200	.080	Ea.	9.50	3.88		13.38	16.45
0610	36 sq. in. to 72 sq. in.		125	.128		19.05	6.20		25.25	30.50
0615	72 sq. in. to 144 sq. in.		80	.200		38	9.70		47.70	57
0620	Wall penetrations, to 36 square inches		180	.089		9.50	4.31		13.81	17.10
0630	36 sq. in. to 72 sq. in.		100	.160		19.05	7.75		26.80	33
0640	72 sq. in. to 144 sq. in.		60	.267		38	12.90		50.90	62
0800	Caulk seams with latex caulk	1 Carp	230	.035	L.F.	.17	1.69		1.86	2.77
0900	Set up neg. air machine, 1-2k CFM/25 M.C.F. volume	1 Asbe	4.30	1.860	Ea.		99.50		99.50	155

02 85 33 – Removal and Disposal of Materials with Mold

02 85 33.50 Demolition in Mold Contaminated Area

		Crew	Daily Output	Labor-Hours	Unit	Material	2016 Bare Costs Labor	2016 Bare Costs Equipment	Total	Total Incl O&P
0010	**DEMOLITION IN MOLD CONTAMINATED AREA**									
0200	Ceiling, including suspension system, plaster and lath	A-9	2100	.030	S.F.	.09	1.63		1.72	2.64
0210	Finished plaster, leaving wire lath		585	.109		.33	5.85		6.18	9.45
0220	Suspended acoustical tile		3500	.018		.05	.98		1.03	1.58

02 85 Mold Remediation

02 85 33 – Removal and Disposal of Materials with Mold

02 85 33.50 Demolition in Mold Contaminated Area	Crew	Daily Output	Labor-Hours	Unit	Material	2016 Bare Costs Labor	Equipment	Total	Total Incl O&P
0230	Concealed tile grid system	A-9	3000	.021	S.F.	.06	1.14		1.20	1.84
0240	Metal pan grid system		1500	.043		.13	2.28		2.41	3.69
0250	Gypsum board		2500	.026		.08	1.37		1.45	2.21
0255	Plywood		2500	.026		.08	1.37		1.45	2.21
0260	Lighting fixtures up to 2' x 4'		72	.889	Ea.	2.65	47.50		50.15	77
0400	Partitions, non load bearing									
0410	Plaster, lath, and studs	A-9	690	.093	S.F.	.91	4.96		5.87	8.70
0450	Gypsum board and studs		1390	.046		.14	2.46		2.60	3.98
0465	Carpet & pad		1390	.046		.14	2.46		2.60	3.98
0600	Pipe insulation, air cell type, up to 4" diameter pipe		900	.071	L.F.	.21	3.80		4.01	6.15
0610	4" to 8" diameter pipe		800	.080		.24	4.28		4.52	6.90
0620	10" to 12" diameter pipe		700	.091		.27	4.89		5.16	7.90
0630	14" to 16" diameter pipe		550	.116		.35	6.20		6.55	10.10
0650	Over 16" diameter pipe		650	.098	S.F.	.29	5.25		5.54	8.50
9000	For type B (supplied air) respirator equipment, add				%				10%	10%

For customer support on your Mechanical Cost Data, call 877.739.3689.

29

Division Notes

		CREW	DAILY OUTPUT	LABOR-HOURS	UNIT	BARE COSTS				TOTAL INCL O&P
						MAT.	LABOR	EQUIP.	TOTAL	

Estimating Tips

General

- Carefully check all the plans and specifications. Concrete often appears on drawings other than structural drawings, including mechanical and electrical drawings for equipment pads. The cost of cutting and patching is often difficult to estimate. See Subdivision 03 81 for Concrete Cutting, Subdivision 02 41 19.16 for Cutout Demolition, Subdivision 03 05 05.10 for Concrete Demolition, and Subdivision 02 41 19.19 for Rubbish Handling (handling, loading and hauling of debris).

- Always obtain concrete prices from suppliers near the job site. A volume discount can often be negotiated, depending upon competition in the area. Remember to add for waste, particularly for slabs and footings on grade.

03 10 00 Concrete Forming and Accessories

- A primary cost for concrete construction is forming. Most jobs today are constructed with prefabricated forms. The selection of the forms best suited for the job and the total square feet of forms required for efficient concrete forming and placing are key elements in estimating concrete construction. Enough forms must be available for erection to make efficient use of the concrete placing equipment and crew.

- Concrete accessories for forming and placing depend upon the systems used. Study the plans and specifications to ensure that all special accessory requirements have been included in the cost estimate, such as anchor bolts, inserts, and hangers.

- Included within costs for forms-in-place are all necessary bracing and shoring.

03 20 00 Concrete Reinforcing

- Ascertain that the reinforcing steel supplier has included all accessories, cutting, bending, and an allowance for lapping, splicing, and waste. A good rule of thumb is 10% for lapping, splicing, and waste. Also, 10% waste should be allowed for welded wire fabric.

- The unit price items in the subdivisions for Reinforcing In Place, Glass Fiber Reinforcing, and Welded Wire Fabric include the labor to install accessories such as beam and slab bolsters, high chairs, and bar ties and tie wire. The material cost for these accessories is not included; they may be obtained from the Accessories Subdivisions.

03 30 00 Cast-In-Place Concrete

- When estimating structural concrete, pay particular attention to requirements for concrete additives, curing methods, and surface treatments. Special consideration for climate, hot or cold, must be included in your estimate. Be sure to include requirements for concrete placing equipment, and concrete finishing.

- For accurate concrete estimating, the estimator must consider each of the following major components individually: forms, reinforcing steel, ready-mix concrete, placement of the concrete, and finishing of the top surface. For faster estimating, Subdivision 03 30 53.40 for Concrete-In-Place can be used; here, various items of concrete work are presented that include the costs of all five major components (unless specifically stated otherwise).

03 40 00 Precast Concrete
03 50 00 Cast Decks and Underlayment

- The cost of hauling precast concrete structural members is often an important factor. For this reason, it is important to get a quote from the nearest supplier. It may become economically feasible to set up precasting beds on the site if the hauling costs are prohibitive.

Reference Numbers

Reference numbers are shown at the beginning of some major classifications. These numbers refer to related items in the Reference Section. The reference information may be an estimating procedure, an alternate pricing method, or technical information.

Note: Not all subdivisions listed here necessarily appear. ■

Did you know?

RSMeans Online gives you the same access to RSMeans' data with 24/7 access:

- Quickly locate costs in the searchable database.
- Build cost lists, estimates, and reports in minutes.
- Adjust costs to any location in the U.S. and Canada with the click of a button.

Start your free trial today at **www.RSMeansOnline.com**

03 01 Maintenance of Concrete

03 01 30 – Maintenance of Cast-In-Place Concrete

03 01 30.62 Concrete Patching		Crew	Daily Output	Labor-Hours	Unit	Material	2016 Bare Costs Labor	Equipment	Total	Total Incl O&P
0010	**CONCRETE PATCHING**									
0100	Floors, 1/4" thick, small areas, regular grout	1 Cefi	170	.047	S.F.	1.49	2.15		3.64	4.83
0150	Epoxy grout	"	100	.080	"	8.25	3.65		11.90	14.45
2000	Walls, including chipping, cleaning and epoxy grout									
2100	1/4" deep	1 Cefi	65	.123	S.F.	7.25	5.60		12.85	16.30
2150	1/2" deep		50	.160		14.45	7.30		21.75	27
2200	3/4" deep	↓	40	.200	↓	21.50	9.15		30.65	37.50

03 11 Concrete Forming

03 11 13 – Structural Cast-In-Place Concrete Forming

03 11 13.40 Forms In Place, Equipment Foundations

		Crew	Daily Output	Labor-Hours	Unit	Material	2016 Bare Costs Labor	Equipment	Total	Total Incl O&P
0010	**FORMS IN PLACE, EQUIPMENT FOUNDATIONS**									
0020	1 use	C-2	160	.300	SFCA	3.35	14.10		17.45	25
0050	2 use		190	.253		1.84	11.90		13.74	20
0100	3 use		200	.240		1.34	11.30		12.64	18.75
0150	4 use	↓	205	.234	↓	1.09	11		12.09	18.10

03 11 13.45 Forms In Place, Footings

		Crew	Daily Output	Labor-Hours	Unit	Material	2016 Bare Costs Labor	Equipment	Total	Total Incl O&P
0010	**FORMS IN PLACE, FOOTINGS**									
0020	Continuous wall, plywood, 1 use	C-1	375	.085	SFCA	6.80	3.91		10.71	13.50
0050	2 use		440	.073		3.74	3.33		7.07	9.20
0100	3 use		470	.068		2.72	3.12		5.84	7.75
0150	4 use		485	.066		2.21	3.02		5.23	7.10
5000	Spread footings, job-built lumber, 1 use		305	.105		2.15	4.81		6.96	9.70
5050	2 use		371	.086		1.20	3.95		5.15	7.35
5100	3 use		401	.080		.86	3.66		4.52	6.55
5150	4 use	↓	414	.077	↓	.70	3.54		4.24	6.20

03 11 13.65 Forms In Place, Slab On Grade

		Crew	Daily Output	Labor-Hours	Unit	Material	2016 Bare Costs Labor	Equipment	Total	Total Incl O&P
0010	**FORMS IN PLACE, SLAB ON GRADE**									
3000	Edge forms, wood, 4 use, on grade, to 6" high	C-1	600	.053	L.F.	.33	2.44		2.77	4.12
6000	Trench forms in floor, wood, 1 use		160	.200	SFCA	1.79	9.15		10.94	16
6050	2 use		175	.183		.98	8.40		9.38	13.95
6100	3 use		180	.178		.71	8.15		8.86	13.30
6150	4 use	↓	185	.173	↓	.58	7.90		8.48	12.80

03 15 Concrete Accessories

03 15 05 – Concrete Forming Accessories

03 15 05.75 Sleeves and Chases

			Crew	Daily Output	Labor-Hours	Unit	Material	2016 Bare Costs Labor	Equipment	Total	Total Incl O&P
0010	**SLEEVES AND CHASES**										
0100	Plastic, 1 use, 12" long, 2" diameter		1 Carp	100	.080	Ea.	2.23	3.88		6.11	8.40
0150	4" diameter			90	.089		6.25	4.31		10.56	13.50
0200	6" diameter			75	.107		11.05	5.15		16.20	20
0250	12" diameter			60	.133		32.50	6.45		38.95	46
5000	Sheet metal, 2" diameter	**G**		100	.080		1.49	3.88		5.37	7.60
5100	4" diameter	**G**		90	.089		1.86	4.31		6.17	8.65
5150	6" diameter	**G**		75	.107		1.86	5.15		7.01	10
5200	12" diameter	**G**		60	.133		3.72	6.45		10.17	14
6000	Steel pipe, 2" diameter	**G**		100	.080		3.27	3.88		7.15	9.55
6100	4" diameter	**G**		90	.089		12.10	4.31		16.41	19.90
6150	6" diameter	**G**		75	.107		39	5.15		44.15	51

For customer support on your Mechanical Cost Data, call 877.739.3689.

03 15 Concrete Accessories

03 15 05 – Concrete Forming Accessories

03 15 05.75 Sleeves and Chases		Crew	Daily Output	Labor-Hours	Unit	Material	2016 Bare Costs Labor	Equipment	Total	Total Incl O&P
6200	12" diameter	**G** 1 Carp	60	.133	Ea.	93.50	6.45		99.95	113

03 15 16 – Concrete Construction Joints

03 15 16.20 Control Joints, Saw Cut

		Crew	Daily Output	Labor-Hours	Unit	Material	Labor	Equipment	Total	Total Incl O&P
0010	**CONTROL JOINTS, SAW CUT**									
0100	Sawcut control joints in green concrete									
0120	1" depth	C-27	2000	.008	L.F.	.04	.37	.08	.49	.67
0140	1-1/2" depth		1800	.009		.05	.41	.09	.55	.76
0160	2" depth	↓	1600	.010	↓	.07	.46	.11	.64	.88
0180	Sawcut joint reservoir in cured concrete									
0182	3/8" wide x 3/4" deep, with single saw blade	C-27	1000	.016	L.F.	.05	.73	.17	.95	1.33
0184	1/2" wide x 1" deep, with double saw blades		900	.018		.10	.81	.19	1.10	1.52
0186	3/4" wide x 1-1/2" deep, with double saw blades	↓	800	.020		.21	.91	.21	1.33	1.81
0190	Water blast joint to wash away laitance, 2 passes	C-29	2500	.003			.12	.03	.15	.22
0200	Air blast joint to blow out debris and air dry, 2 passes	C-28	2000	.004	↓		.18	.01	.19	.28
0300	For backer rod, see Section 07 91 23.10									
0342	For joint sealant, see Section 07 92 13.20									

03 15 19 – Cast-In Concrete Anchors

03 15 19.10 Anchor Bolts

			Crew	Daily Output	Labor-Hours	Unit	Material	Labor	Equipment	Total	Total Incl O&P
0010	**ANCHOR BOLTS**										
0015	Made from recycled materials										
0025	Single bolts installed in fresh concrete, no templates										
0030	Hooked w/nut and washer, 1/2" diameter, 8" long	**G**	1 Carp	132	.061	Ea.	1.32	2.94		4.26	5.95
0040	12" long	**G**		131	.061		1.47	2.96		4.43	6.15
0050	5/8" diameter, 8" long	**G**		129	.062		3.09	3		6.09	8
0060	12" long	**G**		127	.063		3.81	3.05		6.86	8.85
0070	3/4" diameter, 8" long	**G**		127	.063		3.81	3.05		6.86	8.85
0080	12" long	**G**	↓	125	.064	↓	4.76	3.10		7.86	10
0090	2-bolt pattern, including job-built 2-hole template, per set										
0100	J-type, incl. hex nut & washer, 1/2" diameter x 6" long	**G**	1 Carp	21	.381	Set	5.55	18.45		24	34.50
0110	12" long	**G**		21	.381		6.15	18.45		24.60	35.50
0120	18" long	**G**		21	.381		7	18.45		25.45	36
0130	3/4" diameter x 8" long	**G**		20	.400		10.80	19.40		30.20	41.50
0140	12" long	**G**		20	.400		12.70	19.40		32.10	43.50
0150	18" long	**G**		20	.400		15.55	19.40		34.95	46.50
0160	1" diameter x 12" long	**G**		19	.421		22	20.50		42.50	55.50
0170	18" long	**G**		19	.421		25.50	20.50		46	59.50
0180	24" long	**G**		19	.421		30.50	20.50		51	65
0190	36" long	**G**		18	.444		40.50	21.50		62	77.50
0200	1-1/2" diameter x 18" long	**G**		17	.471		42	23		65	81
0210	24" long	**G**		16	.500		49	24		73	91
0300	L-type, incl. hex nut & washer, 3/4" diameter x 12" long	**G**		20	.400		13.75	19.40		33.15	44.50
0310	18" long	**G**		20	.400		16.70	19.40		36.10	48
0320	24" long	**G**		20	.400		19.65	19.40		39.05	51
0330	30" long	**G**		20	.400		24	19.40		43.40	56
0340	36" long	**G**		20	.400		27	19.40		46.40	59.50
0350	1" diameter x 12" long	**G**		19	.421		21	20.50		41.50	54.50
0360	18" long	**G**		19	.421		25.50	20.50		46	59.50
0370	24" long	**G**		19	.421		30.50	20.50		51	65
0380	30" long	**G**		19	.421		35.50	20.50		56	70.50
0390	36" long	**G**		18	.444		40	21.50		61.50	77
0400	42" long	**G**		18	.444		48	21.50		69.50	86
0410	48" long	**G**		18	.444		53.50	21.50		75	92

For customer support on your Mechanical Cost Data, call 877.739.3689.

33

03 15 19.10 Anchor Bolts		Crew	Daily Output	Labor-Hours	Unit	Material	2016 Bare Costs Labor	Equipment	Total	Total Incl O&P
0420	1-1/4" diameter x 18" long	G	1 Carp	18	.444	Set	30	21.50		51.50	66
0430	24" long	G		18	.444		35	21.50		56.50	71.50
0440	30" long	G		17	.471		40	23		63	79
0450	36" long	G		17	.471		45	23		68	84.50
0460	42" long	G	2 Carp	32	.500		50.50	24		74.50	92.50
0470	48" long	G		32	.500		57	24		81	100
0480	54" long	G		31	.516		67	25		92	112
0490	60" long	G		31	.516		73	25		98	119
0500	1-1/2" diameter x 18" long	G		33	.485		46	23.50		69.50	86.50
0510	24" long	G		32	.500		53	24		77	95.50
0520	30" long	G		31	.516		59.50	25		84.50	104
0530	36" long	G		30	.533		68	26		94	115
0540	42" long	G		30	.533		77.50	26		103.50	125
0550	48" long	G		29	.552		86.50	26.50		113	136
0560	54" long	G		28	.571		105	27.50		132.50	158
0570	60" long	G		28	.571		114	27.50		141.50	169
0580	1-3/4" diameter x 18" long	G		31	.516		75	25		100	121
0590	24" long	G		30	.533		87.50	26		113.50	136
0600	30" long	G		29	.552		101	26.50		127.50	152
0610	36" long	G		28	.571		115	27.50		142.50	169
0620	42" long	G		27	.593		129	28.50		157.50	185
0630	48" long	G		26	.615		141	30		171	201
0640	54" long	G		26	.615		174	30		204	237
0650	60" long	G		25	.640		188	31		219	255
0660	2" diameter x 24" long	G		27	.593		112	28.50		140.50	167
0670	30" long	G		27	.593		126	28.50		154.50	182
0680	36" long	G		26	.615		138	30		168	197
0690	42" long	G		25	.640		153	31		184	216
0700	48" long	G		24	.667		175	32.50		207.50	243
0710	54" long	G		23	.696		208	33.50		241.50	281
0720	60" long	G		23	.696		224	33.50		257.50	298
0730	66" long	G		22	.727		239	35		274	315
0740	72" long	G		21	.762		262	37		299	345
1000	4-bolt pattern, including job-built 4-hole template, per set										
1100	J-type, incl. hex nut & washer, 1/2" diameter x 6" long	G	1 Carp	19	.421	Set	7.90	20.50		28.40	40
1110	12" long	G		19	.421		9.05	20.50		29.55	41.50
1120	18" long	G		18	.444		10.85	21.50		32.35	45
1130	3/4" diameter x 8" long	G		17	.471		18.40	23		41.40	55.50
1140	12" long	G		17	.471		22	23		45	59.50
1150	18" long	G		17	.471		28	23		51	65.50
1160	1" diameter x 12" long	G		16	.500		40.50	24		64.50	81.50
1170	18" long	G		15	.533		48	26		74	92.50
1180	24" long	G		15	.533		58	26		84	103
1190	36" long	G		15	.533		78	26		104	126
1200	1-1/2" diameter x 18" long	G		13	.615		80.50	30		110.50	134
1210	24" long	G		12	.667		95	32.50		127.50	155
1300	L-type, incl. hex nut & washer, 3/4" diameter x 12" long	G		17	.471		24.50	23		47.50	61.50
1310	18" long	G		17	.471		30	23		53	68
1320	24" long	G		17	.471		36	23		59	75
1330	30" long	G		16	.500		45	24		69	86.50
1340	36" long	G		16	.500		51	24		75	93
1350	1" diameter x 12" long	G		16	.500		38.50	24		62.50	79.50
1360	18" long	G		15	.533		47.50	26		73.50	91.50

03 15 19.10 Anchor Bolts		Crew	Daily Output	Labor-Hours	Unit	Material	2016 Bare Costs Labor	Equipment	Total	Total Incl O&P
1370	24" long	G 1 Carp	15	.533	Set	58	26		84	103
1380	30" long	G	15	.533		67.50	26		93.50	114
1390	36" long	G	15	.533		77	26		103	124
1400	42" long	G	14	.571		93	27.50		120.50	145
1410	48" long	G	14	.571		104	27.50		131.50	157
1420	1-1/4" diameter x 18" long	G	14	.571		57	27.50		84.50	105
1430	24" long	G	14	.571		67	27.50		94.50	116
1440	30" long	G	13	.615		77	30		107	130
1450	36" long	G	13	.615		87	30		117	141
1460	42" long	G 2 Carp	25	.640		98	31		129	156
1470	48" long	G	24	.667		111	32.50		143.50	172
1480	54" long	G	23	.696		131	33.50		164.50	196
1490	60" long	G	23	.696		143	33.50		176.50	209
1500	1-1/2" diameter x 18" long	G	25	.640		89	31		120	146
1510	24" long	G	24	.667		103	32.50		135.50	164
1520	30" long	G	23	.696		116	33.50		149.50	180
1530	36" long	G	22	.727		133	35		168	200
1540	42" long	G	22	.727		151	35		186	220
1550	48" long	G	21	.762		170	37		207	243
1560	54" long	G	20	.800		206	39		245	287
1570	60" long	G	20	.800		225	39		264	310
1580	1-3/4" diameter x 18" long	G	22	.727		147	35		182	215
1590	24" long	G	21	.762		171	37		208	245
1600	30" long	G	21	.762		199	37		236	276
1610	36" long	G	20	.800		226	39		265	310
1620	42" long	G	19	.842		254	41		295	340
1630	48" long	G	18	.889		279	43		322	370
1640	54" long	G	18	.889		345	43		388	445
1650	60" long	G	17	.941		375	45.50		420.50	480
1660	2" diameter x 24" long	G	19	.842		220	41		261	305
1670	30" long	G	18	.889		248	43		291	340
1680	36" long	G	18	.889		272	43		315	365
1690	42" long	G	17	.941		305	45.50		350.50	405
1700	48" long	G	16	1		350	48.50		398.50	455
1710	54" long	G	15	1.067		415	51.50		466.50	535
1720	60" long	G	15	1.067		445	51.50		496.50	570
1730	66" long	G	14	1.143		475	55.50		530.50	610
1740	72" long	G	14	1.143		520	55.50		575.50	655
1990	For galvanized, add				Ea.	75%				

03 15 19.45 Machinery Anchors

0010	**MACHINERY ANCHORS**, heavy duty, incl. sleeve, floating base nut,									
0020	lower stud & coupling nut, fiber plug, connecting stud, washer & nut.									
0030	For flush mounted embedment in poured concrete heavy equip. pads.									
0200	Stud & bolt, 1/2" diameter	G E-16	40	.400	Ea.	52.50	21.50	3.68	77.68	98.50
0300	5/8" diameter	G	35	.457		61	25	4.21	90.21	114
0500	3/4" diameter	G	30	.533		72.50	29	4.91	106.41	134
0600	7/8" diameter	G	25	.640		82.50	34.50	5.90	122.90	156
0800	1" diameter	G	20	.800		88.50	43.50	7.35	139.35	179
0900	1-1/4" diameter	G	15	1.067		118	58	9.80	185.80	239

03 21 Reinforcement Bars

03 21 11 – Plain Steel Reinforcement Bars

03 21 11.60 Reinforcing In Place		Crew	Daily Output	Labor-Hours	Unit	Material	2016 Bare Costs Labor	Equipment	Total	Total Incl O&P
0010	**REINFORCING IN PLACE**, 50-60 ton lots, A615 Grade 60									
0020	Includes labor, but not material cost, to install accessories									
0030	Made from recycled materials									
0502	Footings, #4 to #7	G 4 Rodm	4200	.008	Lb.	.48	.40		.88	1.16
0552	#8 to #18	G	7200	.004		.48	.24		.72	.89
0602	Slab on grade, #3 to #7	G ↓	4200	.008	↓	.48	.40		.88	1.16
0900	For other than 50 – 60 ton lots									
1000	Under 10 ton job, #3 to #7, add						25%	10%		
1010	#8 to #18, add						20%	10%		
1050	10 – 50 ton job, #3 to #7, add						10%			
1060	#8 to #18, add						5%			
1100	60 – 100 ton job, #3 to #7, deduct						5%			
1110	#8 to #18, deduct						10%			
1150	Over 100 ton job, #3 to #7, deduct						10%			
1160	#8 to #18, deduct						15%			

03 22 Fabric and Grid Reinforcing

03 22 11 – Plain Welded Wire Fabric Reinforcing

03 22 11.10 Plain Welded Wire Fabric

		Crew	Daily Output	Labor-Hours	Unit	Material	Labor	Equipment	Total	Total Incl O&P
0010	**PLAIN WELDED WIRE FABRIC** ASTM A185									
0020	Includes labor, but not material cost, to install accessories									
0030	Made from recycled materials									
0050	Sheets									
0100	6 x 6 - W1.4 x W1.4 (10 x 10) 21 lb. per C.S.F.	G 2 Rodm	35	.457	C.S.F.	13.50	24		37.50	52.50

03 22 13 – Galvanized Welded Wire Fabric Reinforcing

03 22 13.10 Galvanized Welded Wire Fabric

		Crew	Daily Output	Labor-Hours	Unit	Material	Labor	Equipment	Total	Total Incl O&P
0010	**GALVANIZED WELDED WIRE FABRIC**									
0100	Add to plain welded wire pricing for galvanized welded wire				Lb.	.23			.23	.25

03 22 16 – Epoxy-Coated Welded Wire Fabric Reinforcing

03 22 16.10 Epoxy-Coated Welded Wire Fabric

		Crew	Daily Output	Labor-Hours	Unit	Material	Labor	Equipment	Total	Total Incl O&P
0010	**EPOXY-COATED WELDED WIRE FABRIC**									
0100	Add to plain welded wire pricing for epoxy-coated welded wire				Lb.	.22			.22	.24

03 30 Cast-In-Place Concrete

03 30 53 – Miscellaneous Cast-In-Place Concrete

03 30 53.40 Concrete In Place

		Crew	Daily Output	Labor-Hours	Unit	Material	Labor	Equipment	Total	Total Incl O&P
0010	**CONCRETE IN PLACE**									
0020	Including forms (4 uses), Grade 60 rebar, concrete (Portland cement									
0050	Type I), placement and finishing unless otherwise indicated									
0500	Chimney foundations (5000 psi), over 5 C.Y.	C-14C	32.22	3.476	C.Y.	155	160	.97	315.97	415
0510	(3500 psi), under 5 C.Y.	"	23.71	4.724	"	182	217	1.32	400.32	535
3540	Equipment pad (3000 psi), 3' x 3' x 6" thick	C-14H	45	1.067	Ea.	47.50	50.50	.70	98.70	130
3550	4' x 4' x 6" thick	↓	30	1.600	↓	70.50	75.50	1.06	147.06	195
3560	5' x 5' x 8" thick		18	2.667		124	126	1.76	251.76	330
3570	6' x 6' x 8" thick		14	3.429		167	162	2.26	331.26	435
3580	8' x 8' x 10" thick		8	6		360	284	3.96	647.96	835
3590	10' x 10' x 12" thick	↓	5	9.600	↓	615	455	6.35	1,076.35	1,375
3800	Footings (3000 psi), spread under 1 C.Y.	C-14C	28	4	C.Y.	171	184	1.12	356.12	470

03 30 Cast-In-Place Concrete

03 30 53 – Miscellaneous Cast-In-Place Concrete

03 30 53.40 Concrete In Place	Crew	Daily Output	Labor-Hours	Unit	Material	2016 Bare Costs Labor	Equipment	Total	Total Incl O&P
3825	1 C.Y. to 5 C.Y.	C-14C	43	2.605	C.Y.	207	120	.73	327.73	415
3850	Over 5 C.Y.	↓	75	1.493		190	68.50	.42	258.92	315
3900	Footings, strip (3000 psi), 18" x 9", unreinforced	C-14L	40	2.400		131	108	.79	239.79	310
3920	18" x 9", reinforced	C-14C	35	3.200		154	147	.90	301.90	395
3925	20" x 10", unreinforced	C-14L	45	2.133		127	95.50	.70	223.20	287
3930	20" x 10", reinforced	C-14C	40	2.800		146	129	.78	275.78	360
3935	24" x 12", unreinforced	C-14L	55	1.745		125	78.50	.58	204.08	259
3940	24" x 12", reinforced	C-14C	48	2.333		144	107	.65	251.65	325
3945	36" x 12", unreinforced	C-14L	70	1.371		121	61.50	.45	182.95	227
3950	36" x 12", reinforced	C-14C	60	1.867		138	86	.52	224.52	285
4000	Foundation mat (3000 psi), under 10 C.Y.	↓	38.67	2.896		209	133	.81	342.81	435
4050	Over 20 C.Y.	↓	56.40	1.986		184	91.50	.56	276.06	345
4650	Slab on grade (3500 psi), not including finish, 4" thick	C-14E	60.75	1.449		129	68.50	.52	198.02	248
4700	6" thick	"	92	.957	↓	124	45	.34	169.34	207
4701	Thickened slab edge (3500 psi), for slab on grade poured									
4702	monolithically with slab; depth is in addition to slab thickness;									
4703	formed vertical outside edge, earthen bottom and inside slope									
4705	8" deep x 8" wide bottom, unreinforced	C-14L	2190	.044	L.F.	3.57	1.97	.01	5.55	6.95
4710	8" x 8", reinforced	C-14C	1670	.067		5.85	3.09	.02	8.96	11.15
4715	12" deep x 12" wide bottom, unreinforced	C-14L	1800	.053		7.30	2.39	.02	9.71	11.70
4720	12" x 12", reinforced	C-14C	1310	.086		11.40	3.94	.02	15.36	18.65
4725	16" deep x 16" wide bottom, unreinforced	C-14L	1440	.067		12.30	2.99	.02	15.31	18.15
4730	16" x 16", reinforced	C-14C	1120	.100		17.25	4.60	.03	21.88	26
4735	20" deep x 20" wide bottom, unreinforced	C-14L	1150	.083		18.70	3.75	.03	22.48	26.50
4740	20" x 20", reinforced	C-14C	920	.122		25	5.60	.03	30.63	36
4745	24" deep x 24" wide bottom, unreinforced	C-14L	930	.103		26.50	4.63	.03	31.16	36
4750	24" x 24", reinforced	C-14C	740	.151	↓	34.50	6.95	.04	41.49	49
4751	Slab on grade (3500 psi), incl. troweled finish, not incl. forms									
4760	or reinforcing, over 10,000 S.F., 4" thick	C-14F	3425	.021	S.F.	1.43	.91	.01	2.35	2.94
4820	6" thick		3350	.021		2.08	.93	.01	3.02	3.69
4840	8" thick		3184	.023		2.85	.98	.01	3.84	4.62
4900	12" thick		2734	.026		4.28	1.14	.01	5.43	6.40
4950	15" thick	↓	2505	.029	↓	5.35	1.24	.01	6.60	7.75

03 31 Structural Concrete

03 31 13 – Heavyweight Structural Concrete

03 31 13.35 Heavyweight Concrete, Ready Mix

		Crew	Daily Output	Labor-Hours	Unit	Material	2016 Bare Costs Labor	Equipment	Total	Total Incl O&P
0010	**HEAVYWEIGHT CONCRETE, READY MIX**, delivered									
0012	Includes local aggregate, sand, Portland cement (Type I) and water									
0015	Excludes all additives and treatments									
0020	2000 psi				C.Y.	102			102	113
0100	2500 psi					105			105	115
0150	3000 psi					107			107	118
0200	3500 psi					110			110	121
0300	4000 psi					113			113	124
1000	For high early strength (Portland cement Type III), add					10%				
1300	For winter concrete (hot water), add					4.45			4.45	4.90
1410	For mid-range water reducer, add					3.33			3.33	3.66
1420	For high-range water reducer/superplasticizer, add					6			6	6.60
1430	For retarder, add					3.23			3.23	3.55
1440	For non-Chloride accelerator, add	-			↓	5.70			5.70	6.25

For customer support on your Mechanical Cost Data, call 877.739.3689.

37

03 31 Structural Concrete

03 31 13 – Heavyweight Structural Concrete

03 31 13.35 Heavyweight Concrete, Ready Mix

		Crew	Daily Output	Labor-Hours	Unit	Material	2016 Bare Costs Labor	Equipment	Total	Total Incl O&P
1450	For Chloride accelerator, per 1%, add				C.Y.	3			3	3.30
1460	For fiber reinforcing, synthetic (1 lb./C.Y.), add					6.70			6.70	7.35
1500	For Saturday delivery, add				↓	7.50			7.50	8.25
1510	For truck holding/waiting time past 1st hour per load, add				Hr.	96.50			96.50	106
1520	For short load (less than 4 C.Y.), add per load				Ea.	74.50			74.50	82
2000	For all lightweight aggregate, add				C.Y.	45%				

03 31 13.70 Placing Concrete

		Crew	Daily Output	Labor-Hours	Unit	Material	2016 Bare Costs Labor	Equipment	Total	Total Incl O&P
0010	**PLACING CONCRETE**									
0020	Includes labor and equipment to place, level (strike off) and consolidate									
1900	Footings, continuous, shallow, direct chute	C-6	120	.400	C.Y.		15.80	.53	16.33	24.50
1950	Pumped	C-20	150	.427			17.40	5.25	22.65	32.50
2000	With crane and bucket	C-7	90	.800			33	13.45	46.45	65.50
2100	Footings, continuous, deep, direct chute	C-6	140	.343			13.55	.45	14	21
2150	Pumped	C-20	160	.400			16.30	4.94	21.24	30.50
2200	With crane and bucket	C-7	110	.655			27	11	38	53
2400	Footings, spread, under 1 C.Y., direct chute	C-6	55	.873			34.50	1.15	35.65	54
2450	Pumped	C-20	65	.985			40	12.15	52.15	74.50
2500	With crane and bucket	C-7	45	1.600			66	27	93	131
2600	Over 5 C.Y., direct chute	C-6	120	.400			15.80	.53	16.33	24.50
2650	Pumped	C-20	150	.427			17.40	5.25	22.65	32.50
2700	With crane and bucket	C-7	100	.720			30	12.10	42.10	59
2900	Foundation mats, over 20 C.Y., direct chute	C-6	350	.137			5.40	.18	5.58	8.45
2950	Pumped	C-20	400	.160			6.50	1.98	8.48	12.10
3000	With crane and bucket	C-7	300	.240	↓		9.95	4.03	13.98	19.55

03 35 Concrete Finishing

03 35 13 – High-Tolerance Concrete Floor Finishing

03 35 13.30 Finishing Floors, High Tolerance

		Crew	Daily Output	Labor-Hours	Unit	Material	2016 Bare Costs Labor	Equipment	Total	Total Incl O&P
0010	**FINISHING FLOORS, HIGH TOLERANCE**									
0012	Finishing of fresh concrete flatwork requires that concrete									
0013	first be placed, struck off & consolidated									
0015	Basic finishing for various unspecified flatwork									
0100	Bull float only	C-10	4000	.006	S.F.		.26		.26	.39
0125	Bull float & manual float		2000	.012			.52		.52	.77
0150	Bull float, manual float, & broom finish, w/edging & joints		1850	.013			.56		.56	.84
0200	Bull float, manual float & manual steel trowel	↓	1265	.019	↓		.82		.82	1.22
0210	For specified Random Access Floors in ACI Classes 1, 2, 3 and 4 to achieve									
0215	Composite Overall Floor Flatness and Levelness values up to FF35/FL25									
0250	Bull float, machine float & machine trowel (walk-behind)	C-10C	1715	.014	S.F.		.60	.03	.63	.93
0300	Power screed, bull float, machine float & trowel (walk-behind)	C-10D	2400	.010			.43	.05	.48	.70
0350	Power screed, bull float, machine float & trowel (ride-on)	C-10E	4000	.006	↓		.26	.06	.32	.46
0352	For specified Random Access Floors in ACI Classes 5, 6, 7 and 8 to achieve									
0354	Composite Overall Floor Flatness and Levelness values up to FF50/FL50									
0356	Add for two-dimensional restraightening after power float	C-10	6000	.004	S.F.		.17		.17	.26
0358	For specified Random or Defined Access Floors in ACI Class 9 to achieve									
0360	Composite Overall Floor Flatness and Levelness values up to FF100/FL100									
0362	Add for two-dimensional restraightening after bull float & power float	C-10	3000	.008	S.F.		.34		.34	.52
0364	For specified Superflat Defined Access Floors in ACI Class 9 to achieve									
0366	Minimum Floor Flatness and Levelness values of FF100/FL100									
0368	Add for 2-dim'l restraightening after bull float, power float, power trowel	C-10	2000	.012	S.F.		.52		.52	.77

03 54 Cast Underlayment

03 54 16 – Hydraulic Cement Underlayment

03 54 16.50 Cement Underlayment	Crew	Daily Output	Labor-Hours	Unit	Material	2016 Bare Costs Labor	Equipment	Total	Total Incl O&P
0010 **CEMENT UNDERLAYMENT**									
2510 Underlayment, P.C. based, self-leveling, 4100 psi, pumped, 1/4" thick	C-8	20000	.003	S.F.	1.58	.12	.04	1.74	1.95
2520 1/2" thick		19000	.003		3.15	.12	.04	3.31	3.70
2530 3/4" thick		18000	.003		4.73	.13	.04	4.90	5.45
2540 1" thick		17000	.003		6.30	.14	.04	6.48	7.20
2550 1-1/2" thick		15000	.004		9.45	.16	.05	9.66	10.70
2560 Hand placed, 1/2" thick	C-18	450	.020		3.15	.76	.13	4.04	4.79
2610 Topping, P.C. based, self-leveling, 6100 psi, pumped, 1/4" thick	C-8	20000	.003		2.37	.12	.04	2.53	2.83
2620 1/2" thick		19000	.003		4.75	.12	.04	4.91	5.45
2630 3/4" thick		18000	.003		7.10	.13	.04	7.27	8.10
2660 1" thick		17000	.003		9.50	.14	.04	9.68	10.70
2670 1-1/2" thick		15000	.004		14.25	.16	.05	14.46	15.95
2680 Hand placed, 1/2" thick	C-18	450	.020		4.75	.76	.13	5.64	6.50

03 63 Epoxy Grouting

03 63 05 – Grouting of Dowels and Fasteners

03 63 05.10 Epoxy Only

	Crew	Daily Output	Labor-Hours	Unit	Material	2016 Bare Costs Labor	Equipment	Total	Total Incl O&P
0010 **EPOXY ONLY**									
1500 Chemical anchoring, epoxy cartridge, excludes layout, drilling, fastener									
1530 For fastener 3/4" diam. x 6" embedment	2 Skwk	72	.222	Ea.	4.84	11.10		15.94	22.50
1535 1" diam. x 8" embedment		66	.242		7.25	12.10		19.35	26.50
1540 1-1/4" diam. x 10" embedment		60	.267		14.50	13.30		27.80	36.50
1545 1-3/4" diam. x 12" embedment		54	.296		24	14.80		38.80	49.50
1550 14" embedment		48	.333		29	16.65		45.65	57.50
1555 2" diam. x 12" embedment		42	.381		38.50	19		57.50	72
1560 18" embedment		32	.500		48.50	25		73.50	91.50

03 82 Concrete Boring

03 82 13 – Concrete Core Drilling

03 82 13.10 Core Drilling

	Crew	Daily Output	Labor-Hours	Unit	Material	2016 Bare Costs Labor	Equipment	Total	Total Incl O&P
0010 **CORE DRILLING**									
0015 Includes bit cost, layout and set-up time									
0020 Reinforced concrete slab, up to 6" thick									
0100 1" diameter core	B-89A	17	.941	Ea.	.18	41.50	6.90	48.58	71.50
0150 For each additional inch of slab thickness in same hole, add		1440	.011		.03	.49	.08	.60	.87
0200 2" diameter core		16.50	.970		.28	42.50	7.10	49.88	73.50
0250 For each additional inch of slab thickness in same hole, add		1080	.015		.05	.65	.11	.81	1.17
0300 3" diameter core		16	1		.38	44	7.30	51.68	76
0350 For each additional inch of slab thickness in same hole, add		720	.022		.06	.98	.16	1.20	1.75
0500 4" diameter core		15	1.067		.50	47	7.80	55.30	81
0550 For each additional inch of slab thickness in same hole, add		480	.033		.08	1.46	.24	1.78	2.61
0700 6" diameter core		14	1.143		.76	50	8.35	59.11	87
0750 For each additional inch of slab thickness in same hole, add		360	.044		.13	1.95	.32	2.40	3.50
0900 8" diameter core		13	1.231		1.04	54	9	64.04	94
0950 For each additional inch of slab thickness in same hole, add		288	.056		.17	2.44	.41	3.02	4.39
1100 10" diameter core		12	1.333		1.48	58.50	9.75	69.73	102
1150 For each additional inch of slab thickness in same hole, add		240	.067		.25	2.93	.49	3.67	5.30
1300 12" diameter core		11	1.455		1.77	64	10.65	76.42	112
1350 For each additional inch of slab thickness in same hole, add		206	.078		.29	3.41	.57	4.27	6.20

03 82 Concrete Boring

03 82 13 – Concrete Core Drilling

03 82 13.10 Core Drilling		Crew	Daily Output	Labor-Hours	Unit	Material	2016 Bare Costs Labor	Equipment	Total	Total Incl O&P
1500	14" diameter core	B-89A	10	1.600	Ea.	2.07	70	11.70	83.77	123
1550	For each additional inch of slab thickness in same hole, add		180	.089		.34	3.90	.65	4.89	7.10
1700	18" diameter core		9	1.778		2.78	78	13	93.78	137
1750	For each additional inch of slab thickness in same hole, add		144	.111		.46	4.88	.81	6.15	8.90
1754	24" diameter core		8	2		3.97	88	14.60	106.57	155
1756	For each additional inch of slab thickness in same hole, add	▼	120	.133		.66	5.85	.97	7.48	10.80
1760	For horizontal holes, add to above				▼		20%	20%		
1770	Prestressed hollow core plank, 8" thick									
1780	1" diameter core	B-89A	17.50	.914	Ea.	.24	40	6.70	46.94	69
1790	For each additional inch of plank thickness in same hole, add		3840	.004		.03	.18	.03	.24	.34
1794	2" diameter core		17.25	.928		.38	40.50	6.80	47.68	70.50
1796	For each additional inch of plank thickness in same hole, add		2880	.006		.05	.24	.04	.33	.47
1800	3" diameter core		17	.941		.51	41.50	6.90	48.91	71.50
1810	For each additional inch of plank thickness in same hole, add		1920	.008		.06	.37	.06	.49	.70
1820	4" diameter core		16.50	.970		.66	42.50	7.10	50.26	74
1830	For each additional inch of plank thickness in same hole, add		1280	.013		.08	.55	.09	.72	1.03
1840	6" diameter core		15.50	1.032		1.02	45.50	7.55	54.07	79
1850	For each additional inch of plank thickness in same hole, add		960	.017		.13	.73	.12	.98	1.40
1860	8" diameter core		15	1.067		1.39	47	7.80	56.19	82
1870	For each additional inch of plank thickness in same hole, add		768	.021		.17	.91	.15	1.23	1.77
1880	10" diameter core		14	1.143		1.97	50	8.35	60.32	88.50
1890	For each additional inch of plank thickness in same hole, add		640	.025		.25	1.10	.18	1.53	2.16
1900	12" diameter core		13.50	1.185		2.35	52	8.65	63	92
1910	For each additional inch of plank thickness in same hole, add	▼	548	.029	▼	.29	1.28	.21	1.78	2.52

03 82 16 – Concrete Drilling

03 82 16.10 Concrete Impact Drilling

		Crew	Daily Output	Labor-Hours	Unit	Material	2016 Bare Costs Labor	Equipment	Total	Total Incl O&P
0010	**CONCRETE IMPACT DRILLING**									
0020	Includes bit cost, layout and set-up time, no anchors									
0050	Up to 4" deep in concrete/brick floors/walls									
0100	Holes, 1/4" diameter	1 Carp	75	.107	Ea.	.07	5.15		5.22	8.05
0150	For each additional inch of depth in same hole, add		430	.019		.02	.90		.92	1.40
0200	3/8" diameter		63	.127		.06	6.15		6.21	9.50
0250	For each additional inch of depth in same hole, add		340	.024		.01	1.14		1.15	1.77
0300	1/2" diameter		50	.160		.06	7.75		7.81	11.95
0350	For each additional inch of depth in same hole, add		250	.032		.02	1.55		1.57	2.40
0400	5/8" diameter		48	.167		.10	8.10		8.20	12.50
0450	For each additional inch of depth in same hole, add		240	.033		.03	1.61		1.64	2.51
0500	3/4" diameter		45	.178		.13	8.60		8.73	13.35
0550	For each additional inch of depth in same hole, add		220	.036		.03	1.76		1.79	2.74
0600	7/8" diameter		43	.186		.18	9		9.18	14
0650	For each additional inch of depth in same hole, add		210	.038		.04	1.85		1.89	2.88
0700	1" diameter		40	.200		.16	9.70		9.86	15
0750	For each additional inch of depth in same hole, add		190	.042		.04	2.04		2.08	3.17
0800	1-1/4" diameter		38	.211		.27	10.20		10.47	15.95
0850	For each additional inch of depth in same hole, add		180	.044		.07	2.15		2.22	3.37
0900	1-1/2" diameter		35	.229		.41	11.05		11.46	17.45
0950	For each additional inch of depth in same hole, add	▼	165	.048	▼	.10	2.35		2.45	3.71
1000	For ceiling installations, add						40%			

For customer support on your Mechanical Cost Data, call 877.739.3689.

Estimating Tips

04 05 00 Common Work Results for Masonry

- The terms mortar and grout are often used interchangeably, and incorrectly. Mortar is used to bed masonry units, seal the entry of air and moisture, provide architectural appearance, and allow for size variations in the units. Grout is used primarily in reinforced masonry construction and is used to bond the masonry to the reinforcing steel. Common mortar types are M(2500 psi), S(1800 psi), N(750 psi), and O(350 psi), and conform to ASTM C270. Grout is either fine or coarse and conforms to ASTM C476, and in-place strengths generally exceed 2500 psi. Mortar and grout are different components of masonry construction and are placed by entirely different methods. An estimator should be aware of their unique uses and costs.

- Mortar is included in all assembled masonry line items. The mortar cost, part of the assembled masonry material cost, includes all ingredients, all labor, and all equipment required. Please see reference number R040513-10.

- Waste, specifically the loss/ droppings of mortar and the breakage of brick and block, is included in all unit cost lines that include mortar and masonry units

in this division. A factor of 25% is added for mortar and 3% for brick and concrete masonry units.

- Scaffolding or staging is not included in any of the Division 4 costs. Refer to Subdivision 01 54 23 for scaffolding and staging costs.

04 20 00 Unit Masonry

- The most common types of unit masonry are brick and concrete masonry. The major classifications of brick are building brick (ASTM C62), facing brick (ASTM C216), glazed brick, fire brick, and pavers. Many varieties of texture and appearance can exist within these classifications, and the estimator would be wise to check local custom and availability within the project area. For repair and remodeling jobs, matching the existing brick may be the most important criteria.

- Brick and concrete block are priced by the piece and then converted into a price per square foot of wall. Openings less than two square feet are generally ignored by the estimator because any savings in units used is offset by the cutting and trimming required.

- It is often difficult and expensive to find and purchase small lots of historic brick. Costs can vary widely. Many design issues affect costs, selection of mortar mix, and repairs or replacement of masonry

materials. Cleaning techniques must be reflected in the estimate.

- All masonry walls, whether interior or exterior, require bracing. The cost of bracing walls during construction should be included by the estimator, and this bracing must remain in place until permanent bracing is complete. Permanent bracing of masonry walls is accomplished by masonry itself, in the form of pilasters or abutting wall corners, or by anchoring the walls to the structural frame. Accessories in the form of anchors, anchor slots, and ties are used, but their supply and installation can be by different trades. For instance, anchor slots on spandrel beams and columns are supplied and welded in place by the steel fabricator, but the ties from the slots into the masonry are installed by the bricklayer. Regardless of the installation method, the estimator must be certain that these accessories are accounted for in pricing.

Reference Numbers

Reference numbers are shown at the beginning of some major classifications. These numbers refer to related items in the Reference Section. The reference information may be an estimating procedure, an alternate pricing method, or technical information.

Note: Not all subdivisions listed here necessarily appear. ■

04 05 23 – Masonry Accessories

04 05 23.19 Masonry Cavity Drainage, Weepholes, and Vents	Crew	Daily Output	Labor-Hours	Unit	Material	2016 Bare Costs Labor	Equipment	Total	Total Incl O&P
0010 **MASONRY CAVITY DRAINAGE, WEEPHOLES, AND VENTS**									
0020 Extruded aluminum, 4" deep, 2-3/8" x 8-1/8"	1 Bric	30	.267	Ea.	35	12.35		47.35	57
0050 5" x 8-1/8"		25	.320		45.50	14.80		60.30	72.50
0100 2-1/4" x 25"		25	.320		79.50	14.80		94.30	110
0150 5" x 16-1/2"		22	.364		65.50	16.80		82.30	97.50
0175 5" x 24"		22	.364		85.50	16.80		102.30	120
0200 6" x 16-1/2"		22	.364		92.50	16.80		109.30	128
0250 7-3/4" x 16-1/2"		20	.400		79	18.50		97.50	115
0400 For baked enamel finish, add					35%				
0500 For cast aluminum, painted, add					60%				
1000 Stainless steel ventilators, 6" x 6"	1 Bric	25	.320		210	14.80		224.80	255
1050 8" x 8"		24	.333		232	15.40		247.40	280
1100 12" x 12"		23	.348		266	16.10		282.10	320
1150 12" x 6"		24	.333		254	15.40		269.40	305
1200 Foundation block vent, galv., 1-1/4" thk, 8" high, 16" long, no damper		30	.267		16.20	12.35		28.55	36.50
1250 For damper, add					3.11			3.11	3.42

04 21 13 – Brick Masonry

04 21 13.15 Chimney

	Crew	Daily Output	Labor-Hours	Unit	Material	2016 Bare Costs Labor	Equipment	Total	Total Incl O&P
0010 **CHIMNEY**, excludes foundation, scaffolding, grout and reinforcing									
0100 Brick, 16" x 16", 8" flue	D-1	18.20	.879	V.L.F.	25	37		62	84
0150 16" x 20" with one 8" x 12" flue		16	1		39	42		81	108
0200 16" x 24" with two 8" x 8" flues		14	1.143		57	48		105	137
0250 20" x 20" with one 12" x 12" flue		13.70	1.168		46.50	49		95.50	126
0300 20" x 24" with two 8" x 12" flues		12	1.333		65	56		121	157
0350 20" x 32" with two 12" x 12" flues		10	1.600		81.50	67.50		149	193

Estimating Tips

05 05 00 Common Work Results for Metals

- Nuts, bolts, washers, connection angles, and plates can add a significant amount to both the tonnage of a structural steel job and the estimated cost. As a rule of thumb, add 10% to the total weight to account for these accessories.

- Type 2 steel construction, commonly referred to as "simple construction," consists generally of field-bolted connections with lateral bracing supplied by other elements of the building, such as masonry walls or x-bracing. The estimator should be aware, however, that shop connections may be accomplished by welding or bolting. The method may be particular to the fabrication shop and may have an impact on the estimated cost.

05 10 00 Structural Steel

- Steel items can be obtained from two sources: a fabrication shop or a metals service center. Fabrication shops can fabricate items under more controlled conditions than can crews in the field. They are also more efficient and can produce items more economically. Metal service centers serve as a source of long mill shapes to both fabrication shops and contractors.

- Most line items in this structural steel subdivision, and most items in 05 50 00 Metal Fabrications, are indicated as being shop fabricated. The bare material cost for these shop fabricated items is the "Invoice Cost" from the shop and includes the mill base price of steel plus mill extras, transportation to the shop, shop drawings and detailing where warranted, shop fabrication and handling, sandblasting and a shop coat of primer paint, all necessary structural bolts, and delivery to the job site. The bare labor cost and bare equipment cost for these shop fabricated items is for field installation or erection.

- Line items in Subdivision 05 12 23.40 Lightweight Framing, and other items scattered in Division 5, are indicated as being field fabricated. The bare material cost for these field fabricated items is the "Invoice Cost" from the metals service center and includes the mill base price of steel plus mill extras, transportation to the metals service center, material handling, and delivery of long lengths of mill shapes to the job site. Material costs for structural bolts and welding rods should be added to the estimate. The bare labor cost and bare equipment cost for these items is for both field fabrication and field installation or erection, and include time for cutting, welding and drilling in the fabricated metal items. Drilling into concrete and fasteners to fasten field fabricated items to other work are not included and should be added to the estimate.

05 20 00 Steel Joist Framing

- In any given project the total weight of open web steel joists is determined by the loads to be supported and the design. However, economies can be realized in minimizing the amount of labor used to place the joists. This is done by maximizing the joist spacing, and therefore minimizing the number of joists required to be installed on the job. Certain spacings and locations may be required by the design, but in other cases maximizing the spacing and keeping it as uniform as possible will keep the costs down.

05 30 00 Steel Decking

- The takeoff and estimating of metal deck involves more than simply the area of the floor or roof and the type of deck specified or shown on the drawings. Many different sizes and types of openings may exist. Small openings for individual pipes or conduits may be drilled after the floor/roof is installed, but larger openings may require special deck lengths as well as reinforcing or structural support. The estimator should determine who will be supplying this reinforcing. Additionally, some deck terminations are part of the deck package, such as screed angles and pour stops, and others will be part of the steel contract, such as angles attached to structural members and cast-in-place angles and plates. The estimator must ensure that all pieces are accounted for in the complete estimate.

05 50 00 Metal Fabrications

- The most economical steel stairs are those that use common materials, standard details, and most importantly, a uniform and relatively simple method of field assembly. Commonly available A36/A992 channels and plates are very good choices for the main stringers of the stairs, as are angles and tees for the carrier members. Risers and treads are usually made by specialty shops, and it is most economical to use a typical detail in as many places as possible. The stairs should be pre-assembled and shipped directly to the site. The field connections should be simple and straightforward to be accomplished efficiently, and with minimum equipment and labor.

Reference Numbers

Reference numbers are shown at the beginning of some major classifications. These numbers refer to related items in the Reference Section. The reference information may be an estimating procedure, an alternate pricing method, or technical information.

Note: Not all subdivisions listed here necessarily appear. ■

05 05 19 – Post-Installed Concrete Anchors

05 05 19.10 Chemical Anchors		Crew	Daily Output	Labor-Hours	Unit	Material	2016 Bare Costs Labor	Equipment	Total	Total Incl O&P
0010	**CHEMICAL ANCHORS**									
0020	Includes layout & drilling									
1430	Chemical anchor, w/rod & epoxy cartridge, 3/4" diam. x 9-1/2" long	B-89A	27	.593	Ea.	9.25	26	4.33	39.58	55
1435	1" diameter x 11-3/4" long		24	.667		16.95	29.50	4.87	51.32	69
1440	1-1/4" diameter x 14" long		21	.762		34	33.50	5.55	73.05	95
1445	1-3/4" diameter x 15" long		20	.800		63.50	35	5.85	104.35	130
1450	18" long		17	.941		76.50	41.50	6.90	124.90	155
1455	2" diameter x 18" long		16	1		101	44	7.30	152.30	187
1460	24" long		15	1.067		132	47	7.80	186.80	226

05 05 19.20 Expansion Anchors			Crew	Daily Output	Labor-Hours	Unit	Material	2016 Bare Costs Labor	Equipment	Total	Total Incl O&P
0010	**EXPANSION ANCHORS**										
0100	Anchors for concrete, brick or stone, no layout and drilling										
0200	Expansion shields, zinc, 1/4" diameter, 1-5/16" long, single	G	1 Carp	90	.089	Ea.	.44	4.31		4.75	7.10
0300	1-3/8" long, double	G		85	.094		.55	4.56		5.11	7.60
0400	3/8" diameter, 1-1/2" long, single	G		85	.094		.66	4.56		5.22	7.75
0500	2" long, double	G		80	.100		1.20	4.85		6.05	8.75
0600	1/2" diameter, 2-1/16" long, single	G		80	.100		1.21	4.85		6.06	8.80
0700	2-1/2" long, double	G		75	.107		1.94	5.15		7.09	10.10
0800	5/8" diameter, 2-5/8" long, single	G		75	.107		2.01	5.15		7.16	10.15
0900	2-3/4" long, double	G		70	.114		2.74	5.55		8.29	11.50
1000	3/4" diameter, 2-3/4" long, single	G		70	.114		3.18	5.55		8.73	12
1100	3-15/16" long, double	G		65	.123		5.10	5.95		11.05	14.75
2100	Hollow wall anchors for gypsum wall board, plaster or tile										
2300	1/8" diameter, short	G	1 Carp	160	.050	Ea.	.25	2.42		2.67	4
2400	Long	G		150	.053		.25	2.58		2.83	4.24
2500	3/16" diameter, short	G		150	.053		.46	2.58		3.04	4.47
2600	Long	G		140	.057		.64	2.77		3.41	4.95
2700	1/4" diameter, short	G		140	.057		.64	2.77		3.41	4.95
2800	Long	G		130	.062		.70	2.98		3.68	5.35
3000	Toggle bolts, bright steel, 1/8" diameter, 2" long	G		85	.094		.21	4.56		4.77	7.25
3100	4" long	G		80	.100		.27	4.85		5.12	7.75
3200	3/16" diameter, 3" long	G		80	.100		.28	4.85		5.13	7.75
3300	6" long	G		75	.107		.39	5.15		5.54	8.40
3400	1/4" diameter, 3" long	G		75	.107		.38	5.15		5.53	8.35
3500	6" long	G		70	.114		.55	5.55		6.10	9.10
3600	3/8" diameter, 3" long	G		70	.114		.93	5.55		6.48	9.50
3700	6" long	G		60	.133		1.43	6.45		7.88	11.45
3800	1/2" diameter, 4" long	G		60	.133		2	6.45		8.45	12.10
3900	6" long	G		50	.160		2.59	7.75		10.34	14.75
5000	Screw anchors for concrete, masonry,										
5100	stone & tile, no layout or drilling included										
5700	Lag screw shields, 1/4" diameter, short	G	1 Carp	90	.089	Ea.	.37	4.31		4.68	7
5800	Long	G		85	.094		.41	4.56		4.97	7.45
5900	3/8" diameter, short	G		85	.094		.71	4.56		5.27	7.80
6000	Long	G		80	.100		.94	4.85		5.79	8.50
6100	1/2" diameter, short	G		80	.100		1.01	4.85		5.86	8.55
6200	Long	G		75	.107		1.31	5.15		6.46	9.40
6300	5/8" diameter, short	G		70	.114		1.57	5.55		7.12	10.25
6400	Long	G		65	.123		2	5.95		7.95	11.35
6600	Lead, #6 & #8, 3/4" long	G		260	.031		.19	1.49		1.68	2.50
6700	#10 - #14, 1-1/2" long	G		200	.040		.39	1.94		2.33	3.40
6800	#16 & #18, 1-1/2" long	G		160	.050		.42	2.42		2.84	4.18

05 05 19 – Post-Installed Concrete Anchors

05 05 19.20 Expansion Anchors

		Crew	Daily Output	Labor-Hours	Unit	Material	2016 Bare Costs Labor	Equipment	Total	Total Incl O&P
6900	Plastic, #6 & #8, 3/4" long	1 Carp	260	.031	Ea.	.04	1.49		1.53	2.33
7000	#8 & #10, 7/8" long		240	.033		.04	1.61		1.65	2.52
7100	#10 & #12, 1" long		220	.036		.05	1.76		1.81	2.76
7200	#14 & #16, 1-1/2" long		160	.050		.08	2.42		2.50	3.81
8000	Wedge anchors, not including layout or drilling									
8050	Carbon steel, 1/4" diameter, 1-3/4" long	G 1 Carp	150	.053	Ea.	.43	2.58		3.01	4.43
8100	3-1/4" long	G	140	.057		.56	2.77		3.33	4.87
8150	3/8" diameter, 2-1/4" long	G	145	.055		.52	2.67		3.19	4.67
8200	5" long	G	140	.057		.91	2.77		3.68	5.25
8250	1/2" diameter, 2-3/4" long	G	140	.057		1.02	2.77		3.79	5.35
8300	7" long	G	125	.064		1.75	3.10		4.85	6.70
8350	5/8" diameter, 3-1/2" long	G	130	.062		1.92	2.98		4.90	6.70
8400	8-1/2" long	G	115	.070		4.08	3.37		7.45	9.65
8450	3/4" diameter, 4-1/4" long	G	115	.070		3.01	3.37		6.38	8.45
8500	10" long	G	95	.084		6.85	4.08		10.93	13.75
8550	1" diameter, 6" long	G	100	.080		9.55	3.88		13.43	16.45
8575	9" long	G	85	.094		12.40	4.56		16.96	20.50
8600	12" long	G	75	.107		13.35	5.15		18.50	22.50
8650	1-1/4" diameter, 9" long	G	70	.114		25	5.55		30.55	36.50
8700	12" long	G	60	.133		32.50	6.45		38.95	45.50
8750	For type 303 stainless steel, add					350%				
8800	For type 316 stainless steel, add					450%				
8950	Self-drilling concrete screw, hex washer head, 3/16" diam. x 1-3/4" long	G 1 Carp	300	.027	Ea.	.21	1.29		1.50	2.21
8960	2-1/4" long	G	250	.032		.24	1.55		1.79	2.64
8970	Phillips flat head, 3/16" diam. x 1-3/4" long	G	300	.027		.21	1.29		1.50	2.21
8980	2-1/4" long	G	250	.032		.24	1.55		1.79	2.64

05 05 21 – Fastening Methods for Metal

05 05 21.15 Drilling Steel

		Crew	Daily Output	Labor-Hours	Unit	Material	2016 Bare Costs Labor	Equipment	Total	Total Incl O&P
0010	**DRILLING STEEL**									
1910	Drilling & layout for steel, up to 1/4" deep, no anchor									
1920	Holes, 1/4" diameter	1 Sswk	112	.071	Ea.	.10	3.80		3.90	6.55
1925	For each additional 1/4" depth, add		336	.024		.10	1.27		1.37	2.26
1930	3/8" diameter		104	.077		.09	4.09		4.18	7.05
1935	For each additional 1/4" depth, add		312	.026		.09	1.36		1.45	2.41
1940	1/2" diameter		96	.083		.11	4.43		4.54	7.65
1945	For each additional 1/4" depth, add		288	.028		.11	1.48		1.59	2.63
1950	5/8" diameter		88	.091		.16	4.84		5	8.40
1955	For each additional 1/4" depth, add		264	.030		.16	1.61		1.77	2.91
1960	3/4" diameter		80	.100		.20	5.30		5.50	9.20
1965	For each additional 1/4" depth, add		240	.033		.20	1.77		1.97	3.22
1970	7/8" diameter		72	.111		.26	5.90		6.16	10.30
1975	For each additional 1/4" depth, add		216	.037		.26	1.97		2.23	3.63
1980	1" diameter		64	.125		.23	6.65		6.88	11.50
1985	For each additional 1/4" depth, add		192	.042		.23	2.22		2.45	4
1990	For drilling up, add						40%			

05 05 21.90 Welding Steel

		Crew	Daily Output	Labor-Hours	Unit	Material	2016 Bare Costs Labor	Equipment	Total	Total Incl O&P
0010	**WELDING STEEL**, Structural									
0020	Field welding, 1/8" E6011, cost per welder, no operating engineer	E-14	8	1	Hr.	4.46	55	18.40	77.86	118
0200	With 1/2 operating engineer	E-13	8	1.500		4.46	80	18.40	102.86	156
0300	With 1 operating engineer	E-12	8	2		4.46	104	18.40	126.86	193
0500	With no operating engineer, 2# weld rod per ton	E-14	8	1	Ton	4.46	55	18.40	77.86	118
0600	8# E6011 per ton	"	2	4		17.85	221	73.50	312.35	475

For customer support on your Mechanical Cost Data, call 877.739.3689.

45

05 05 21 – Fastening Methods for Metal

05 05 21.90 Welding Steel	Crew	Daily Output	Labor-Hours	Unit	Material	2016 Bare Costs Labor	Equipment	Total	Total Incl O&P
0800	With one operating engineer per welder, 2# E6011 per ton	E-12	8	2	Ton	4.46	104	18.40	126.86	193
0900	8# E6011 per ton	"	2	8	↓	17.85	415	73.50	506.35	770
1200	Continuous fillet, down welding									
1300	Single pass, 1/8" thick, 0.1#/L.F.	E-14	150	.053	L.F.	.22	2.94	.98	4.14	6.30
1400	3/16" thick, 0.2#/L.F.		75	.107		.45	5.90	1.96	8.31	12.60
1500	1/4" thick, 0.3#/L.F.		50	.160		.67	8.85	2.94	12.46	18.95
1610	5/16" thick, 0.4#/L.F.		38	.211		.89	11.60	3.87	16.36	25
1800	3 passes, 3/8" thick, 0.5#/L.F.		30	.267		1.12	14.70	4.91	20.73	31.50
2010	4 passes, 1/2" thick, 0.7#/L.F.		22	.364		1.56	20	6.70	28.26	43
2200	5 to 6 passes, 3/4" thick, 1.3#/L.F.		12	.667		2.90	37	12.25	52.15	79
2400	8 to 11 passes, 1" thick, 2.4#/L.F.	↓	6	1.333		5.35	73.50	24.50	103.35	158
2600	For vertical joint welding, add						20%			
2700	Overhead joint welding, add						300%			
2900	For semi-automatic welding, obstructed joints, deduct						5%			
3000	Exposed joints, deduct				↓		15%			
4000	Cleaning and welding plates, bars, or rods									
4010	to existing beams, columns, or trusses	E-14	12	.667	L.F.	1.12	37	12.25	50.37	77

05 05 23 – Metal Fastenings

05 05 23.30 Lag Screws

			Crew	Daily Output	Labor-Hours	Unit	Material	Labor	Equipment	Total	Total Incl O&P
0010	**LAG SCREWS**										
0020	Steel, 1/4" diameter, 2" long	G	1 Carp	200	.040	Ea.	.09	1.94		2.03	3.07
0100	3/8" diameter, 3" long	G		150	.053		.32	2.58		2.90	4.31
0200	1/2" diameter, 3" long	G		130	.062		.68	2.98		3.66	5.30
0300	5/8" diameter, 3" long	G	↓	120	.067	↓	1.22	3.23		4.45	6.30

05 05 23.35 Machine Screws

			Crew	Daily Output	Labor-Hours	Unit	Material	Labor	Equipment	Total	Total Incl O&P
0010	**MACHINE SCREWS**										
0020	Steel, round head, #8 x 1" long	G	1 Carp	4.80	1.667	C	4.27	81		85.27	129
0110	#8 x 2" long	G		2.40	3.333		6.65	162		168.65	255
0200	#10 x 1" long	G		4	2		5.40	97		102.40	155
0300	#10 x 2" long	G	↓	2	4	↓	10.80	194		204.80	310

05 05 23.50 Powder Actuated Tools and Fasteners

			Crew	Daily Output	Labor-Hours	Unit	Material	Labor	Equipment	Total	Total Incl O&P
0010	**POWDER ACTUATED TOOLS & FASTENERS**										
0020	Stud driver, .22 caliber, single shot					Ea.	149			149	164
0100	.27 caliber, semi automatic, strip					"	450			450	495
0300	Powder load, single shot, .22 cal, power level 2, brown					C	5.50			5.50	6.05
0400	Strip, .27 cal, power level 4, red						7.90			7.90	8.70
0600	Drive pin, .300 x 3/4" long	G	1 Carp	4.80	1.667		4.23	81		85.23	129
0700	.300 x 3" long with washer	G	"	4	2	↓	13	97		110	163

05 05 23.55 Rivets

			Crew	Daily Output	Labor-Hours	Unit	Material	Labor	Equipment	Total	Total Incl O&P
0010	**RIVETS**										
0100	Aluminum rivet & mandrel, 1/2" grip length x 1/8" diameter	G	1 Carp	4.80	1.667	C	8.05	81		89.05	133
0200	3/16" diameter	G		4	2		11.90	97		108.90	162
0300	Aluminum rivet, steel mandrel, 1/8" diameter	G		4.80	1.667		10.30	81		91.30	135
0400	3/16" diameter	G		4	2		16.50	97		113.50	167
0500	Copper rivet, steel mandrel, 1/8" diameter	G		4.80	1.667		9.50	81		90.50	134
1200	Steel rivet and mandrel, 1/8" diameter	G		4.80	1.667		7.75	81		88.75	133
1300	3/16" diameter	G	↓	4	2	↓	12.55	97		109.55	163
1400	Hand riveting tool, standard					Ea.	76.50			76.50	84
1500	Deluxe						385			385	425
1600	Power riveting tool, standard						500			500	555
1700	Deluxe					↓	1,650			1,650	1,825

05 05 Common Work Results for Metals

05 05 23 – Metal Fastenings

05 05 23.70 Structural Blind Bolts		Crew	Daily Output	Labor-Hours	Unit	Material	2016 Bare Costs		Total	Total Incl O&P
							Labor	Equipment		
0010	**STRUCTURAL BLIND BOLTS**									
0100	1/4" diameter x 1/4" grip	G	1 Sswk	240	.033	Ea.	1.44	1.77		3.21	4.58
0150	1/2" grip	G		216	.037		1.55	1.97		3.52	5.05
0200	3/8" diameter x 1/2" grip	G		232	.034		2.04	1.83		3.87	5.35
0250	3/4" grip	G		208	.038		2.14	2.05		4.19	5.80
0300	1/2" diameter x 1/2" grip	G		224	.036		4.66	1.90		6.56	8.35
0350	3/4" grip	G		200	.040		6.50	2.13		8.63	10.75
0400	5/8" diameter x 3/4" grip	G		216	.037		9.60	1.97		11.57	13.95
0450	1" grip	G		192	.042		11.05	2.22		13.27	15.90

05 12 Structural Steel Framing

05 12 23 – Structural Steel for Buildings

05 12 23.40 Lightweight Framing

		Crew	Daily Output	Labor-Hours	Unit	Material	Labor	Equipment	Total	Total Incl O&P
0010	**LIGHTWEIGHT FRAMING**									
0015	Made from recycled materials									
0400	Angle framing, field fabricated, 4" and larger	G	E-3	440	.055	Lb.	.77	2.94	.33	4.04	6.20
0450	Less than 4" angles	G		265	.091		.80	4.88	.56	6.24	9.75
0600	Channel framing, field fabricated, 8" and larger	G		500	.048		.80	2.59	.29	3.68	5.55
0650	Less than 8" channels	G		335	.072		.80	3.86	.44	5.10	7.90
1000	Continuous slotted channel framing system, shop fab, simple framing	G	2 Sswk	2400	.007		4.11	.35		4.46	5.10
1200	Complex framing	G	"	1600	.010		4.64	.53		5.17	6
1250	Plate & bar stock for reinforcing beams and trusses	G					1.46			1.46	1.60
1300	Cross bracing, rods, shop fabricated, 3/4" diameter	G	E-3	700	.034		1.59	1.85	.21	3.65	5.10
1310	7/8" diameter	G		850	.028		1.59	1.52	.17	3.28	4.52
1320	1" diameter	G		1000	.024		1.59	1.29	.15	3.03	4.10
1330	Angle, 5" x 5" x 3/8"	G		2800	.009		1.59	.46	.05	2.10	2.59
1350	Hanging lintels, shop fabricated	G		850	.028		1.59	1.52	.17	3.28	4.52
1380	Roof frames, shop fabricated, 3'-0" square, 5' span	G	E-2	4200	.013		1.59	.70	.36	2.65	3.30
1400	Tie rod, not upset, 1-1/2" to 4" diameter, with turnbuckle	G	2 Sswk	800	.020		1.72	1.06		2.78	3.69
1420	No turnbuckle	G		700	.023		1.66	1.22		2.88	3.88
1500	Upset, 1-3/4" to 4" diameter, with turnbuckle	G		800	.020		1.72	1.06		2.78	3.69
1520	No turnbuckle	G		700	.023		1.66	1.22		2.88	3.88

05 12 23.60 Pipe Support Framing

		Crew	Daily Output	Labor-Hours	Unit	Material	Labor	Equipment	Total	Total Incl O&P
0010	**PIPE SUPPORT FRAMING**									
0020	Under 10#/L.F., shop fabricated	G	E-4	3900	.008	Lb.	1.78	.44	.04	2.26	2.74
0200	10.1 to 15#/L.F.	G		4300	.007		1.75	.40	.03	2.18	2.64
0400	15.1 to 20#/L.F.	G		4800	.007		1.72	.36	.03	2.11	2.53
0600	Over 20#/L.F.	G		5400	.006		1.70	.32	.03	2.05	2.44

For customer support on your Mechanical Cost Data, call 877.739.3689.

47

Division Notes

	CREW	DAILY OUTPUT	LABOR-HOURS	UNIT	BARE COSTS				TOTAL INCL O&P
					MAT.	LABOR	EQUIP.	TOTAL	

Division Notes

Estimating Tips
06 05 00 Common Work Results for Wood, Plastics, and Composites

- Common to any wood-framed structure are the accessory connector items such as screws, nails, adhesives, hangers, connector plates, straps, angles, and hold-downs. For typical wood-framed buildings, such as residential projects, the aggregate total for these items can be significant, especially in areas where seismic loading is a concern. For floor and wall framing, the material cost is based on 10 to 25 lbs. of accessory connectors per MBF. Hold-downs, hangers, and other connectors should be taken off by the piece.

  Included with material costs are fasteners for a normal installation. RSMeans engineers use manufacturer's recommendations, written specifications, and/or standard construction practice for size and spacing of fasteners. Prices for various fasteners are shown for informational purposes only. Adjustments should be made if unusual fastening conditions exist.

06 10 00 Carpentry

- Lumber is a traded commodity and therefore sensitive to supply and demand in the marketplace. Even in "budgetary" estimating of wood-framed projects, it is advisable to call local suppliers for the latest market pricing.

- Common quantity units for wood-framed projects are "thousand board feet" (MBF). A board foot is a volume of wood, 1" x 1' x 1', or 144 cubic inches. Board-foot quantities are generally calculated using nominal material dimensions—dressed sizes are ignored. Board foot per lineal foot of any stick of lumber can be calculated by dividing the nominal cross-sectional area by 12. As an example, 2,000 lineal feet of 2 x 12 equates to 4 MBF by dividing the nominal area, 2 x 12, by 12, which equals 2, and multiplying by 2,000 to give 4,000 board feet. This simple rule applies to all nominal dimensioned lumber.

- Waste is an issue of concern at the quantity takeoff for any area of construction. Framing lumber is sold in even foot lengths, i.e., 8', 10', 12', 14', 16' and, depending on spans, wall heights, and the grade of lumber, waste is inevitable. A rule of thumb for lumber waste is 5%–10% depending on material quality and the complexity of the framing.

- Wood in various forms and shapes is used in many projects, even where the main structural framing is steel, concrete, or masonry. Plywood as a back-up partition material and 2x boards used as blocking and cant strips around roof edges are two common examples. The estimator should ensure that the costs of all wood materials are included in the final estimate.

06 20 00 Finish Carpentry

- It is necessary to consider the grade of workmanship when estimating labor costs for erecting millwork and interior finish. In practice, there are three grades: premium, custom, and economy. The RSMeans daily output for base and case moldings is in the range of 200 to 250 L.F. per carpenter per day. This is appropriate for most average custom-grade projects. For premium projects, an adjustment to productivity of 25%–50% should be made, depending on the complexity of the job.

Reference Numbers

Reference numbers are shown at the beginning of some major classifications. These numbers refer to related items in the Reference Section. The reference information may be an estimating procedure, an alternate pricing method, or technical information.

Note: Not all subdivisions listed here necessarily appear. ∎

06 16 36 – Wood Panel Product Sheathing

06 16 36.10 Sheathing	Crew	Daily Output	Labor-Hours	Unit	Material	2016 Bare Costs Labor	Equipment	Total	Total Incl O&P
0010 **SHEATHING**									
0012 Plywood on roofs, CDX									
0030 5/16" thick	2 Carp	1600	.010	S.F.	.62	.48		1.10	1.43
0035 Pneumatic nailed		1952	.008		.62	.40		1.02	1.30
0050 3/8" thick		1525	.010		.62	.51		1.13	1.46
0055 Pneumatic nailed		1860	.009		.62	.42		1.04	1.32
0100 1/2" thick		1400	.011		.68	.55		1.23	1.60
0105 Pneumatic nailed		1708	.009		.68	.45		1.13	1.45
0200 5/8" thick		1300	.012		.76	.60		1.36	1.75
0205 Pneumatic nailed		1586	.010		.76	.49		1.25	1.59
0300 3/4" thick		1200	.013		.91	.65		1.56	1.99
0305 Pneumatic nailed		1464	.011		.91	.53		1.44	1.81
0500 Plywood on walls, with exterior CDX, 3/8" thick		1200	.013		.62	.65		1.27	1.67
0505 Pneumatic nailed		1488	.011		.62	.52		1.14	1.48
0600 1/2" thick		1125	.014		.68	.69		1.37	1.81
0605 Pneumatic nailed		1395	.011		.68	.56		1.24	1.60
0700 5/8" thick		1050	.015		.76	.74		1.50	1.97
0705 Pneumatic nailed		1302	.012		.76	.60		1.36	1.75
0800 3/4" thick		975	.016		.91	.80		1.71	2.22
0805 Pneumatic nailed		1209	.013		.91	.64		1.55	1.98

Estimating Tips

07 10 00 Dampproofing and Waterproofing

- Be sure of the job specifications before pricing this subdivision. The difference in cost between waterproofing and dampproofing can be great. Waterproofing will hold back standing water. Dampproofing prevents the transmission of water vapor. Also included in this section are vapor retarding membranes.

07 20 00 Thermal Protection

- Insulation and fireproofing products are measured by area, thickness, volume or R-value. Specifications may give only what the specific R-value should be in a certain situation. The estimator may need to choose the type of insulation to meet that R-value.

07 30 00 Steep Slope Roofing
07 40 00 Roofing and Siding Panels

- Many roofing and siding products are bought and sold by the square. One square is equal to an area that measures 100 square feet.

  This simple change in unit of measure could create a large error if the estimator is not observant. Accessories necessary for a complete installation must be figured into any calculations for both material and labor.

07 50 00 Membrane Roofing
07 60 00 Flashing and Sheet Metal
07 70 00 Roofing and Wall Specialties and Accessories

- The items in these subdivisions compose a roofing system. No one component completes the installation, and all must be estimated. Built-up or single-ply membrane roofing systems are made up of many products and installation trades. Wood blocking at roof perimeters or penetrations, parapet coverings, reglets, roof drains, gutters, downspouts, sheet metal flashing, skylights, smoke vents, and roof hatches all need to be considered along with the roofing material. Several different installation trades will need to work together on the roofing system. Inherent difficulties in the scheduling and coordination of various trades must be accounted for when estimating labor costs.

07 90 00 Joint Protection

- To complete the weather-tight shell, the sealants and caulkings must be estimated. Where different materials meet—at expansion joints, at flashing penetrations, and at hundreds of other locations throughout a construction project—caulking and sealants provide another line of defense against water penetration. Often, an entire system is based on the proper location and placement of caulking or sealants. The detailed drawings that are included as part of a set of architectural plans show typical locations for these materials. When caulking or sealants are shown at typical locations, this means the estimator must include them for all the locations where this detail is applicable. Be careful to keep different types of sealants separate, and remember to consider backer rods and primers if necessary.

Reference Numbers

Reference numbers are shown at the beginning of some major classifications. These numbers refer to related items in the Reference Section. The reference information may be an estimating procedure, an alternate pricing method, or technical information.

Note: Not all subdivisions listed here necessarily appear. ■

Did you know?

RSMeans Online gives you the same access to RSMeans' data with 24/7 access:

- Quickly locate costs in the searchable database.
- Build cost lists, estimates, and reports in minutes.
- Adjust costs to any location in the U.S. and Canada with the click of a button.

Start your free trial today at **www.RSMeansOnline.com**

RSMeans Online
FROM THE G⊕RDIAN GROUP®

07 65 Flexible Flashing

07 65 10 – Sheet Metal Flashing

07 65 10.10 Sheet Metal Flashing and Counter Flashing	Crew	Daily Output	Labor-Hours	Unit	Material	2016 Bare Costs Labor	Equipment	Total	Total Incl O&P
0010 **SHEET METAL FLASHING AND COUNTER FLASHING**									
0011 Including up to 4 bends									
0020 Aluminum, mill finish, .013" thick	1 Rofc	145	.055	S.F.	.83	2.30		3.13	4.83
0030 .016" thick		145	.055		.95	2.30		3.25	4.97
0060 .019" thick		145	.055		1.25	2.30		3.55	5.30
0100 .032" thick		145	.055		1.35	2.30		3.65	5.40
0200 .040" thick		145	.055		2.31	2.30		4.61	6.45
0300 .050" thick		145	.055		2.64	2.30		4.94	6.80
0400 Painted finish, add					.32			.32	.35
1600 Copper, 16 oz., sheets, under 1000 lb.	1 Rofc	115	.070		7.85	2.90		10.75	13.60
1700 Over 4000 lb.		155	.052		7.85	2.15		10	12.30
1900 20 oz. sheets, under 1000 lb.		110	.073		10.30	3.03		13.33	16.50
2000 Over 4000 lb.		145	.055		9.80	2.30		12.10	14.65
2200 24 oz. sheets, under 1000 lb.		105	.076		14.60	3.18		17.78	21.50
2300 Over 4000 lb.		135	.059		13.85	2.47		16.32	19.45
2500 32 oz. sheets, under 1000 lb.		100	.080		18.50	3.34		21.84	26
2600 Over 4000 lb.		130	.062		17.60	2.57		20.17	23.50
5800 Lead, 2.5 lb. per S.F., up to 12" wide		135	.059		5.40	2.47		7.87	10.15
5900 Over 12" wide		135	.059		3.99	2.47		6.46	8.60
8900 Stainless steel sheets, 32 ga.,		155	.052		3.25	2.15		5.40	7.25
9000 28 ga.		155	.052		4.20	2.15		6.35	8.30
9100 26 ga.		155	.052		4.45	2.15		6.60	8.55
9200 24 ga.		155	.052		4.95	2.15		7.10	9.10
9400 Terne coated stainless steel, .015" thick, 28 ga		155	.052		8.05	2.15		10.20	12.50
9500 .018" thick, 26 ga		155	.052		9	2.15		11.15	13.55
9600 Zinc and copper alloy (brass), .020" thick		155	.052		10.20	2.15		12.35	14.85
9700 .027" thick		155	.052		12	2.15		14.15	16.85
9800 .032" thick		155	.052		15.40	2.15		17.55	20.50
9900 .040" thick		155	.052		20	2.15		22.15	25.50

07 65 13 – Laminated Sheet Flashing

07 65 13.10 Laminated Sheet Flashing

	Crew	Daily Output	Labor-Hours	Unit	Material	Labor	Equipment	Total	Total Incl O&P
0010 **LAMINATED SHEET FLASHING**, Including up to 4 bends									
4300 Copper-clad stainless steel, .015" thick, under 500 lb.	1 Rofc	115	.070	S.F.	6.40	2.90		9.30	12
4400 Over 2000 lb.		155	.052		6.70	2.15		8.85	11
4600 .018" thick, under 500 lb.		100	.080		7.85	3.34		11.19	14.35
4700 Over 2000 lb.		145	.055		7.50	2.30		9.80	12.15
8550 Shower pan, 3 ply copper and fabric, 3 oz.		155	.052		4	2.15		6.15	8.05
8600 7 oz.		155	.052		4.20	2.15		6.35	8.30
9300 Stainless steel, paperbacked 2 sides, .005" thick		330	.024		3.88	1.01		4.89	6

07 65 19 – Plastic Sheet Flashing

07 65 19.10 Plastic Sheet Flashing and Counter Flashing

	Crew	Daily Output	Labor-Hours	Unit	Material	Labor	Equipment	Total	Total Incl O&P
0010 **PLASTIC SHEET FLASHING AND COUNTER FLASHING**									
7300 Polyvinyl chloride, black, 10 mil	1 Rofc	285	.028	S.F.	.21	1.17		1.38	2.22
7400 20 mil		285	.028		.26	1.17		1.43	2.28
7600 30 mil		285	.028		.34	1.17		1.51	2.36
7700 60 mil		285	.028		.82	1.17		1.99	2.89
7900 Black or white for exposed roofs, 60 mil		285	.028		1.80	1.17		2.97	3.97
8060 PVC tape, 5" x 45 mils, for joint covers, 100 L.F./roll				Ea.	172			172	189
8850 Polyvinyl chloride, 30 mil	1 Rofc	160	.050	S.F.	1.45	2.09		3.54	5.15

52

07 71 Roof Specialties

07 71 23 – Manufactured Gutters and Downspouts

07 71 23.10 Downspouts	Crew	Daily Output	Labor-Hours	Unit	Material	2016 Bare Costs Labor	Equipment	Total	Total Incl O&P
0010 **DOWNSPOUTS**									
0020 Aluminum, embossed, .020" thick, 2" x 3"	1 Shee	190	.042	L.F.	1.05	2.41		3.46	4.84
0100 Enameled		190	.042		1.39	2.41		3.80	5.20
4800 Steel, galvanized, round, corrugated, 2" or 3" diameter, 28 ga.		190	.042		2.10	2.41		4.51	6
4900 4" diameter, 28 ga.		145	.055		2.52	3.16		5.68	7.60
5100 5" diameter, 26 ga.		130	.062		3.73	3.52		7.25	9.50
5400 6" diameter, 28 ga.		105	.076		4.15	4.36		8.51	11.20
5500 26 ga.		105	.076		3.50	4.36		7.86	10.50

07 72 Roof Accessories

07 72 26 – Ridge Vents

07 72 26.10 Ridge Vents and Accessories

	Crew	Daily Output	Labor-Hours	Unit	Material	Labor	Equipment	Total	Total Incl O&P
0010 **RIDGE VENTS AND ACCESSORIES**									
2300 Ridge vent strip, mill finish	1 Shee	155	.052	L.F.	3.88	2.95		6.83	8.80

07 72 53 – Snow Guards

07 72 53.10 Snow Guard Options

	Crew	Daily Output	Labor-Hours	Unit	Material	Labor	Equipment	Total	Total Incl O&P
0010 **SNOW GUARD OPTIONS**									
0100 Slate & asphalt shingle roofs, fastened with nails	1 Rofc	160	.050	Ea.	12.55	2.09		14.64	17.35
0200 Standing seam metal roofs, fastened with set screws		48	.167		16.95	6.95		23.90	30.50
0300 Surface mount for metal roofs, fastened with solder		48	.167		6.05	6.95		13	18.50
0400 Double rail pipe type, including pipe		130	.062	L.F.	34	2.57		36.57	42

07 72 73 – Pitch Pockets

07 72 73.10 Pitch Pockets, Variable Sizes

	Crew	Daily Output	Labor-Hours	Unit	Material	Labor	Equipment	Total	Total Incl O&P
0010 **PITCH POCKETS, VARIABLE SIZES**									
0100 Adjustable, 4" to 7", welded corners, 4" deep	1 Rofc	48	.167	Ea.	16.45	6.95		23.40	30
0200 Side extenders, 6"	"	240	.033	"	2.75	1.39		4.14	5.40

07 76 Roof Pavers

07 76 16 – Roof Decking Pavers

07 76 16.10 Roof Pavers and Supports

	Crew	Daily Output	Labor-Hours	Unit	Material	Labor	Equipment	Total	Total Incl O&P
0010 **ROOF PAVERS AND SUPPORTS**									
1000 Roof decking pavers, concrete blocks, 2" thick, natural	1 Clab	115	.070	S.F.	3.37	2.64		6.01	7.75
1100 Colors		115	.070	"	3.73	2.64		6.37	8.15
1200 Support pedestal, bottom cap		960	.008	Ea.	3	.32		3.32	3.78
1300 Top cap		960	.008		4.80	.32		5.12	5.80
1400 Leveling shims, 1/16"		1920	.004		1.20	.16		1.36	1.56
1500 1/8"		1920	.004		1.20	.16		1.36	1.56
1600 Buffer pad		960	.008		2.50	.32		2.82	3.23
1700 PVC legs (4" SDR 35)		2880	.003	Inch	.12	.11		.23	.30
2000 Alternate pricing method, system in place		101	.079	S.F.	7	3		10	12.30

For customer support on your Mechanical Cost Data, call 877.739.3689.

53

07 84 13 – Penetration Firestopping

07 84 13.10 Firestopping	Crew	Daily Output	Labor-Hours	Unit	Material	2016 Bare Costs Labor	Equipment	Total	Total Incl O&P
0010 **FIRESTOPPING** R078413-30									
0100 Metallic piping, non insulated									
0110 Through walls, 2" diameter	1 Carp	16	.500	Ea.	3.60	24		27.60	41
0120 4" diameter		14	.571		6.30	27.50		33.80	49.50
0130 6" diameter		12	.667		9.25	32.50		41.75	59.50
0140 12" diameter		10	.800		18.45	39		57.45	80
0150 Through floors, 2" diameter		32	.250		1.80	12.10		13.90	20.50
0160 4" diameter		28	.286		3.15	13.85		17	24.50
0170 6" diameter		24	.333		4.60	16.15		20.75	30
0180 12" diameter		20	.400		9.25	19.40		28.65	39.50
0190 Metallic piping, insulated									
0200 Through walls, 2" diameter	1 Carp	16	.500	Ea.	6.30	24		30.30	44
0210 4" diameter		14	.571		9.25	27.50		36.75	52.50
0220 6" diameter		12	.667		12.40	32.50		44.90	63
0230 12" diameter		10	.800		21	39		60	83
0240 Through floors, 2" diameter		32	.250		3.15	12.10		15.25	22
0250 4" diameter		28	.286		4.60	13.85		18.45	26
0260 6" diameter		24	.333		6.20	16.15		22.35	32
0270 12" diameter		20	.400		10.60	19.40		30	41
0280 Non metallic piping, non insulated									
0290 Through walls, 2" diameter	1 Carp	12	.667	Ea.	45.50	32.50		78	99.50
0300 4" diameter		10	.800		87	39		126	155
0310 6" diameter		8	1		170	48.50		218.50	262
0330 Through floors, 2" diameter		16	.500		30	24		54	70
0340 4" diameter		6	1.333		55.50	64.50		120	160
0350 6" diameter		6	1.333		103	64.50		167.50	212
0370 Ductwork, insulated & non insulated, round									
0380 Through walls, 6" diameter	1 Carp	12	.667	Ea.	12.40	32.50		44.90	63
0390 12" diameter		10	.800		21	39		60	83
0400 18" diameter		8	1		25.50	48.50		74	103
0410 Through floors, 6" diameter		16	.500		6.20	24		30.20	44
0420 12" diameter		14	.571		10.60	27.50		38.10	54
0430 18" diameter		12	.667		12.85	32.50		45.35	63.50
0440 Ductwork, insulated & non insulated, rectangular									
0450 With stiffener/closure angle, through walls, 6" x 12"	1 Carp	8	1	Ea.	16.20	48.50		64.70	92.50
0460 12" x 24"		6	1.333		32.50	64.50		97	135
0470 24" x 48"		4	2		65	97		162	221
0480 With stiffener/closure angle, through floors, 6" x 12"		10	.800		8.10	39		47.10	68.50
0490 12" x 24"		8	1		16.20	48.50		64.70	92.50
0500 24" x 48"		6	1.333		32.50	64.50		97	135
0510 Multi trade openings									
0520 Through walls, 6" x 12"	1 Carp	2	4	Ea.	45	194		239	345
0530 12" x 24"	"	1	8		147	390		537	755
0540 24" x 48"	2 Carp	1	16		525	775		1,300	1,775
0550 48" x 96"	"	.75	21.333		1,975	1,025		3,000	3,750
0560 Through floors, 6" x 12"	1 Carp	2	4		49.50	194		243.50	350
0570 12" x 24"	"	1	8		68	390		458	670
0580 24" x 48"	2 Carp	.75	21.333		138	1,025		1,163	1,725
0590 48" x 96"	"	.50	32		390	1,550		1,940	2,800
0600 Structural penetrations, through walls									
0610 Steel beams, W8 x 10	1 Carp	8	1	Ea.	68.50	48.50		117	150
0620 W12 x 14		6	1.333		99	64.50		163.50	208
0630 W21 x 44		5	1.600		138	77.50		215.50	271

07 84 Firestopping

07 84 13 – Penetration Firestopping

07 84 13.10 Firestopping		Crew	Daily Output	Labor-Hours	Unit	Material	2016 Bare Costs Labor	Equipment	Total	Total Incl O&P
0640	W36 x 135	1 Carp	3	2.667	Ea.	219	129		348	440
0650	Bar joists, 18" deep		6	1.333		40.50	64.50		105	144
0660	24" deep		6	1.333		54	64.50		118.50	158
0670	36" deep		5	1.600		81	77.50		158.50	208
0680	48" deep	↓	4	2	↓	108	97		205	267
0690	Construction joints, floor slab at exterior wall									
0700	Precast, brick, block or drywall exterior									
0710	2" wide joint	1 Carp	125	.064	L.F.	9.50	3.10		12.60	15.20
0720	4" wide joint	"	75	.107	"	14.85	5.15		20	24.50
0730	Metal panel, glass or curtain wall exterior									
0740	2" wide joint	1 Carp	40	.200	L.F.	9.50	9.70		19.20	25.50
0750	4" wide joint	"	25	.320	"	14.85	15.50		30.35	40.50
0760	Floor slab to drywall partition									
0770	Flat joint	1 Carp	100	.080	L.F.	13.20	3.88		17.08	20.50
0780	Fluted joint		50	.160		15.05	7.75		22.80	28.50
0790	Etched fluted joint	↓	75	.107	↓	18.50	5.15		23.65	28.50
0800	Floor slab to concrete/masonry partition									
0810	Flat joint	1 Carp	75	.107	L.F.	13.20	5.15		18.35	22.50
0820	Fluted joint	"	50	.160	"	15.05	7.75		22.80	28.50
0830	Concrete/CMU wall joints									
0840	1" wide	1 Carp	100	.080	L.F.	18.85	3.88		22.73	26.50
0850	2" wide		75	.107		23.50	5.15		28.65	33.50
0860	4" wide	↓	50	.160	↓	33.50	7.75		41.25	48.50
0870	Concrete/CMU floor joints									
0880	1" wide	1 Carp	200	.040	L.F.	18.85	1.94		20.79	23.50
0890	2" wide		150	.053		23.50	2.58		26.08	29.50
0900	4" wide	↓	100	.080	↓	33.50	3.88		37.38	42.50

07 91 Preformed Joint Seals

07 91 13 – Compression Seals

07 91 13.10 Compression Seals

		Crew	Daily Output	Labor-Hours	Unit	Material	Labor	Equipment	Total	Total Incl O&P
0010	**COMPRESSION SEALS**									
4900	O-ring type cord, 1/4"	1 Bric	472	.017	L.F.	.43	.78		1.21	1.67
4910	1/2"		440	.018		1.20	.84		2.04	2.60
4920	3/4"		424	.019		2.40	.87		3.27	3.97
4930	1"		408	.020		4.08	.91		4.99	5.85
4940	1-1/4"		384	.021		7.40	.96		8.36	9.60
4950	1-1/2"		368	.022		9.20	1.01		10.21	11.65
4960	1-3/4"		352	.023		18.55	1.05		19.60	22
4970	2"	↓	344	.023	↓	22	1.08		23.08	26

07 91 16 – Joint Gaskets

07 91 16.10 Joint Gaskets

		Crew	Daily Output	Labor-Hours	Unit	Material	Labor	Equipment	Total	Total Incl O&P
0010	**JOINT GASKETS**									
4400	Joint gaskets, neoprene, closed cell w/adh, 1/8" x 3/8"	1 Bric	240	.033	L.F.	.32	1.54		1.86	2.70
4500	1/4" x 3/4"		215	.037		.62	1.72		2.34	3.31
4700	1/2" x 1"		200	.040		1.45	1.85		3.30	4.42
4800	3/4" x 1-1/2"	↓	165	.048	↓	1.60	2.24		3.84	5.20

For customer support on your Mechanical Cost Data, call 877.739.3689.

55

07 91 Preformed Joint Seals

07 91 23 – Backer Rods

07 91 23.10 Backer Rods	Crew	Daily Output	Labor-Hours	Unit	Material	2016 Bare Costs Labor	Equipment	Total	Total Incl O&P
0010 **BACKER RODS**									
0030 Backer rod, polyethylene, 1/4" diameter	1 Bric	4.60	1.739	C.L.F.	2.28	80.50		82.78	126
0050 1/2" diameter		4.60	1.739		4	80.50		84.50	127
0070 3/4" diameter		4.60	1.739		6.50	80.50		87	130
0090 1" diameter		4.60	1.739		11.50	80.50		92	136

07 91 26 – Joint Fillers

07 91 26.10 Joint Fillers	Crew	Daily Output	Labor-Hours	Unit	Material	2016 Bare Costs Labor	Equipment	Total	Total Incl O&P
0010 **JOINT FILLERS**									
4360 Butyl rubber filler, 1/4" x 1/4"	1 Bric	290	.028	L.F.	.22	1.28		1.50	2.20
4365 1/2" x 1/2"		250	.032		.90	1.48		2.38	3.25
4370 1/2" x 3/4"		210	.038		1.35	1.76		3.11	4.17
4375 3/4" x 3/4"		230	.035		2.02	1.61		3.63	4.68
4380 1" x 1"		180	.044		2.70	2.06		4.76	6.10
4390 For coloring, add					12%				
4980 Polyethylene joint backing, 1/4" x 2"	1 Bric	2.08	3.846	C.L.F.	12.90	178		190.90	286
4990 1/4" x 6"		1.28	6.250	"	28.50	289		317.50	470
5600 Silicone, room temp vulcanizing foam seal, 1/4" x 1/2"		1312	.006	L.F.	.46	.28		.74	.93
5610 1/2" x 1/2"		656	.012		.91	.56		1.47	1.86
5620 1/2" x 3/4"		442	.018		1.37	.84		2.21	2.78
5630 3/4" x 3/4"		328	.024		2.05	1.13		3.18	3.97
5640 1/8" x 1"		1312	.006		.46	.28		.74	.93
5650 1/8" x 3"		442	.018		1.37	.84		2.21	2.78
5670 1/4" x 3"		295	.027		2.73	1.25		3.98	4.91
5680 1/4" x 6"		148	.054		5.45	2.50		7.95	9.80
5690 1/2" x 6"		82	.098		10.90	4.51		15.41	18.90
5700 1/2" x 9"		52.50	.152		16.40	7.05		23.45	29
5710 1/2" x 12"		33	.242		22	11.20		33.20	41

07 92 Joint Sealants

07 92 13 – Elastomeric Joint Sealants

07 92 13.20 Caulking and Sealant Options	Crew	Daily Output	Labor-Hours	Unit	Material	2016 Bare Costs Labor	Equipment	Total	Total Incl O&P
0010 **CAULKING AND SEALANT OPTIONS**									
0050 Latex acrylic based, bulk				Gal.	27			27	29.50
0055 Bulk in place 1/4" x 1/4" bead	1 Bric	300	.027	L.F.	.08	1.23		1.31	1.97
0060 1/4" x 3/8"		294	.027		.14	1.26		1.40	2.07
0065 1/4" x 1/2"		288	.028		.19	1.28		1.47	2.17
0075 3/8" x 3/8"		284	.028		.21	1.30		1.51	2.22
0080 3/8" x 1/2"		280	.029		.28	1.32		1.60	2.33
0085 3/8" x 5/8"		276	.029		.35	1.34		1.69	2.44
0095 3/8" x 3/4"		272	.029		.42	1.36		1.78	2.54
0100 1/2" x 1/2"		275	.029		.38	1.35		1.73	2.46
0105 1/2" x 5/8"		269	.030		.47	1.38		1.85	2.62
0110 1/2" x 3/4"		263	.030		.56	1.41		1.97	2.77
0115 1/2" x 7/8"		256	.031		.66	1.45		2.11	2.93
0120 1/2" x 1"		250	.032		.75	1.48		2.23	3.09
0125 3/4" x 3/4"		244	.033		.85	1.52		2.37	3.25
0130 3/4" x 1"		225	.036		1.13	1.64		2.77	3.75
0135 1" x 1"		200	.040		1.50	1.85		3.35	4.47
0190 Cartridges				Gal.	32.50			32.50	36
0200 11 fl. oz. cartridge				Ea.	2.80			2.80	3.08

07 92 Joint Sealants

07 92 13 – Elastomeric Joint Sealants

07 92 13.20 Caulking and Sealant Options		Crew	Daily Output	Labor-Hours	Unit	Material	2016 Bare Costs Labor	Equipment	Total	Total Incl O&P
0500	1/4" x 1/2"	1 Bric	288	.028	L.F.	.23	1.28		1.51	2.21
0600	1/2" x 1/2"		275	.029		.46	1.35		1.81	2.55
0800	3/4" x 3/4"		244	.033		1.03	1.52		2.55	3.45
0900	3/4" x 1"		225	.036		1.37	1.64		3.01	4.02
1000	1" x 1"		200	.040		1.72	1.85		3.57	4.71
1400	Butyl based, bulk				Gal.	35			35	38.50
1500	Cartridges				"	38.50			38.50	42
1700	1/4" x 1/2", 154 L.F./gal.	1 Bric	288	.028	L.F.	.23	1.28		1.51	2.21
1800	1/2" x 1/2", 77 L.F./gal.	"	275	.029	"	.45	1.35		1.80	2.55
2300	Polysulfide compounds, 1 component, bulk				Gal.	82			82	90
2600	1 or 2 component, in place, 1/4" x 1/4", 308 L.F./gal.	1 Bric	300	.027	L.F.	.27	1.23		1.50	2.17
2700	1/2" x 1/4", 154 L.F./gal.		288	.028		.53	1.28		1.81	2.55
2900	3/4" x 3/8", 68 L.F./gal.		272	.029		1.21	1.36		2.57	3.41
3000	1" x 1/2", 38 L.F./gal.		250	.032		2.16	1.48		3.64	4.63
3200	Polyurethane, 1 or 2 component				Gal.	53			53	58.50
3300	Cartridges				"	65			65	71.50
3500	Bulk, in place, 1/4" x 1/4"	1 Bric	300	.027	L.F.	.17	1.23		1.40	2.07
3655	1/2" x 1/4"		288	.028		.34	1.28		1.62	2.34
3800	3/4" x 3/8"		272	.029		.78	1.36		2.14	2.94
3900	1" x 1/2"		250	.032		1.38	1.48		2.86	3.78
4100	Silicone rubber, bulk				Gal.	54			54	59.50
4200	Cartridges				"	53			53	58.50

07 92 16 – Rigid Joint Sealants

07 92 16.10 Rigid Joint Sealants

		Crew	Daily Output	Labor-Hours	Unit	Material	Labor	Equipment	Total	Total Incl O&P
0010	**RIGID JOINT SEALANTS**									
5800	Tapes, sealant, PVC foam adhesive, 1/16" x 1/4"				C.L.F.	10			10	11
5900	1/16" x 1/2"					10			10	11
5950	1/16" x 1"					16			16	17.60
6000	1/8" x 1/2"					9.50			9.50	10.45

07 92 19 – Acoustical Joint Sealants

07 92 19.10 Acoustical Sealant

		Crew	Daily Output	Labor-Hours	Unit	Material	Labor	Equipment	Total	Total Incl O&P
0010	**ACOUSTICAL SEALANT**									
0020	Acoustical sealant, elastomeric, cartridges				Ea.	8.50			8.50	9.35
0025	In place, 1/4" x 1/4"	1 Bric	300	.027	L.F.	.35	1.23		1.58	2.26
0030	1/4" x 1/2"		288	.028		.69	1.28		1.97	2.72
0035	1/2" x 1/2"		275	.029		1.39	1.35		2.74	3.57
0040	1/2" x 3/4"		263	.030		2.08	1.41		3.49	4.44
0045	3/4" x 3/4"		244	.033		3.12	1.52		4.64	5.75
0050	1" x 1"		200	.040		5.55	1.85		7.40	8.90

	CREW	DAILY OUTPUT	LABOR-HOURS	UNIT	BARE COSTS				TOTAL INCL O&P
					MAT.	LABOR	EQUIP.	TOTAL	

Estimating Tips
08 10 00 Doors and Frames

All exterior doors should be addressed for their energy conservation (insulation and seals).

- Most metal doors and frames look alike, but there may be significant differences among them. When estimating these items, be sure to choose the line item that most closely compares to the specification or door schedule requirements regarding:
  - □ type of metal
  - □ metal gauge
  - □ door core material
  - □ fire rating
  - □ finish
- Wood and plastic doors vary considerably in price. The primary determinant is the veneer material. Lauan, birch, and oak are the most common veneers. Other variables include the following:
  - □ hollow or solid core
  - □ fire rating
  - □ flush or raised panel
  - □ finish
- Door pricing includes bore for cylindrical lockset and mortise for hinges.

08 30 00 Specialty Doors and Frames

- There are many varieties of special doors, and they are usually priced per each. Add frames, hardware, or operators required for a complete installation.

08 40 00 Entrances, Storefronts, and Curtain Walls

- Glazed curtain walls consist of the metal tube framing and the glazing material. The cost data in this subdivision is presented for the metal tube framing alone or the composite wall. If your estimate requires a detailed takeoff of the framing, be sure to add the glazing cost and any tints.

08 50 00 Windows

- Most metal windows are delivered preglazed. However, some metal windows are priced without glass. Refer to 08 80 00 Glazing for glass pricing. The grade C indicates commercial grade windows, usually ASTM C-35.
- All wood windows and vinyl are priced preglazed. The glazing is insulating glass. Add the cost of screens and grills if required, and not already included.

08 70 00 Hardware

- Hardware costs add considerably to the cost of a door. The most efficient method to determine the hardware requirements for a project is to review the door and hardware schedule together. One type of door may have different hardware, depending on the door usage.
- Door hinges are priced by the pair, with most doors requiring 1-1/2 pairs per door. The hinge prices do not include installation labor, because it is included in door installation.

Hinges are classified according to the frequency of use, base material, and finish.

08 80 00 Glazing

- Different openings require different types of glass. The most common types are:
  - □ float
  - □ tempered
  - □ insulating
  - □ impact-resistant
  - □ ballistic-resistant
- Most exterior windows are glazed with insulating glass. Entrance doors and window walls, where the glass is less than 18" from the floor, are generally glazed with tempered glass. Interior windows and some residential windows are glazed with float glass.
- Coastal communities require the use of impact-resistant glass, dependant on wind speed.
- The insulation or 'u' value is a strong consideration, along with solar heat gain, to determine total energy efficiency.

Reference Numbers

Reference numbers are shown at the beginning of some major classifications. These numbers refer to related items in the Reference Section. The reference information may be an estimating procedure, an alternate pricing method, or technical information.

Note: Not all subdivisions listed here necessarily appear. ■

08 34 13 – Cold Storage Doors

08 34 13.10 Doors for Cold Area Storage	Crew	Daily Output	Labor-Hours	Unit	Material	2016 Bare Costs Labor	Equipment	Total	Total Incl O&P
0010 **DOORS FOR COLD AREA STORAGE**									
0020 Single, 20 ga. galvanized steel									
0300 Horizontal sliding, 5' x 7', manual operation, 3.5" thick	2 Carp	2	8	Ea.	3,325	390		3,715	4,250
0400 4" thick		2	8		3,300	390		3,690	4,250
0500 6" thick		2	8		3,500	390		3,890	4,450
0800 5' x 7', power operation, 2" thick		1.90	8.421		5,700	410		6,110	6,900
0900 4" thick		1.90	8.421		5,800	410		6,210	7,000
1000 6" thick		1.90	8.421		6,600	410		7,010	7,875
1300 9' x 10', manual operation, 2" insulation		1.70	9.412		4,550	455		5,005	5,700
1400 4" insulation		1.70	9.412		4,775	455		5,230	5,950
1500 6" insulation		1.70	9.412		5,700	455		6,155	6,950
1800 Power operation, 2" insulation		1.60	10		7,825	485		8,310	9,375
1900 4" insulation		1.60	10		8,000	485		8,485	9,550
2000 6" insulation	↓	1.70	9.412	↓	9,000	455		9,455	10,600
2300 For stainless steel face, add					25%				
3000 Hinged, lightweight, 3' x 7'-0", 2" thick	2 Carp	2	8	Ea.	1,450	390		1,840	2,175
3050 4" thick		1.90	8.421		1,800	410		2,210	2,600
3300 Polymer doors, 3' x 7'-0"		1.90	8.421		1,375	410		1,785	2,125
3350 6" thick		1.40	11.429		2,375	555		2,930	3,475
3600 Stainless steel, 3' x 7'-0", 4" thick		1.90	8.421		1,775	410		2,185	2,575
3650 6" thick		1.40	11.429		3,000	555		3,555	4,150
3900 Painted, 3' x 7'-0", 4" thick		1.90	8.421		1,300	410		1,710	2,050
3950 6" thick	↓	1.40	11.429	↓	2,375	555		2,930	3,475
5000 Bi-parting, electric operated									
5010 6' x 8' opening, galv. faces, 4" thick for cooler	2 Carp	.80	20	Opng.	7,350	970		8,320	9,575
5050 For freezer, 4" thick		.80	20		8,100	970		9,070	10,400
5300 For door buck framing and door protection, add		2.50	6.400		655	310		965	1,200
6000 Galvanized batten door, galvanized hinges, 4' x 7'		2	8		1,875	390		2,265	2,650
6050 6' x 8'		1.80	8.889		2,525	430		2,955	3,425
6500 Fire door, 3 hr., 6' x 8', single slide		.80	20		8,400	970		9,370	10,700
6550 Double, bi-parting	↓	.70	22.857	↓	12,400	1,100		13,500	15,300

08 91 16 – Operable Wall Louvers

08 91 16.10 Movable Blade Louvers	Crew	Daily Output	Labor-Hours	Unit	Material	2016 Bare Costs Labor	Equipment	Total	Total Incl O&P
0010 **MOVABLE BLADE LOUVERS**									
0100 PVC, commercial grade, 12" x 12"	1 Shee	20	.400	Ea.	82	23		105	125
0110 16" x 16"		14	.571		96	32.50		128.50	156
0120 18" x 18"		14	.571		99	32.50		131.50	159
0130 24" x 24"		14	.571		140	32.50		172.50	204
0140 30" x 30"		12	.667		168	38		206	244
0150 36" x 36"		12	.667		197	38		235	276
0300 Stainless steel, commercial grade, 12" x 12"		20	.400		219	23		242	276
0310 16" x 16"		14	.571		265	32.50		297.50	340
0320 18" x 18"		14	.571		282	32.50		314.50	360
0330 20" x 20"		14	.571		325	32.50		357.50	405
0340 24" x 24"		14	.571		385	32.50		417.50	475
0350 30" x 30"		12	.667		430	38		468	535
0360 36" x 36"	↓	12	.667	↓	520	38		558	635

08 91 Louvers

08 91 19 – Fixed Louvers

08 91 19.10 Aluminum Louvers

		Crew	Daily Output	Labor-Hours	Unit	Material	2016 Bare Costs Labor	Equipment	Total	Total Incl O&P
0010	**ALUMINUM LOUVERS**									
0020	Aluminum with screen, residential, 8" x 8"	1 Carp	38	.211	Ea.	20	10.20		30.20	37.50
0100	12" x 12"		38	.211		16	10.20		26.20	33.50
0200	12" x 18"		35	.229		20	11.05		31.05	39
0250	14" x 24"		30	.267		29	12.90		41.90	52
0300	18" x 24"		27	.296		32	14.35		46.35	57
0500	24" x 30"		24	.333		60	16.15		76.15	91
0700	Triangle, adjustable, small		20	.400		58	19.40		77.40	93.50
0800	Large		15	.533		79.50	26		105.50	127
1200	Extruded aluminum, see Section 23 37 15.40									
2100	Midget, aluminum, 3/4" deep, 1" diameter	1 Carp	85	.094	Ea.	.77	4.56		5.33	7.85
2150	3" diameter		60	.133		2.48	6.45		8.93	12.65
2200	4" diameter		50	.160		4.95	7.75		12.70	17.35
2250	6" diameter		30	.267		4.10	12.90		17	24.50
3000	PVC, commercial grade, 12" x 12"	1 Shee	20	.400		97	23		120	142
3010	12" x 18"		20	.400		109	23		132	155
3020	12" x 24"		20	.400		146	23		169	196
3030	14" x 24"		20	.400		152	23		175	202
3100	Aluminum, commercial grade, 12" x 12"		20	.400		171	23		194	223
3110	24" x 24"		16	.500		249	28.50		277.50	320
3120	24" x 36"		16	.500		300	28.50		328.50	375
3130	36" x 24"		16	.500		310	28.50		338.50	385
3140	36" x 36"		14	.571		390	32.50		422.50	480
3150	36" x 48"		10	.800		400	46		446	510
3160	48" x 36"		10	.800		390	46		436	500
3170	48" x 48"		10	.800		435	46		481	545
3180	60" x 48"		8	1		485	57.50		542.50	625
3190	60" x 60"		8	1		600	57.50		657.50	750

08 91 19.20 Steel Louvers

		Crew	Daily Output	Labor-Hours	Unit	Material	2016 Bare Costs Labor	Equipment	Total	Total Incl O&P
0010	**STEEL LOUVERS**									
3300	Galvanized Steel, fixed blades, commercial grade, 18" x 18"	1 Shee	20	.400	Ea.	197	23		220	252
3310	24" x 24"		20	.400		232	23		255	290
3320	24" x 36"		16	.500		300	28.50		328.50	375
3330	36" x 24"		16	.500		298	28.50		326.50	375
3340	36" x 36"		14	.571		380	32.50		412.50	470
3350	36" x 48"		10	.800		385	46		431	495
3360	48" x 36"		10	.800		385	46		431	495
3370	48" x 48"		10	.800		425	46		471	535
3380	60" x 48"		10	.800		490	46		536	610
3390	60" x 60"		10	.800		590	46		636	715

For customer support on your Mechanical Cost Data, call 877.739.3689.

61

08 95 Vents

08 95 13 – Soffit Vents

08 95 13.10 Wall Louvers	Crew	Daily Output	Labor-Hours	Unit	Material	2016 Bare Costs Labor	Equipment	Total	Total Incl O&P
0010 **WALL LOUVERS**									
2330 Soffit vent, continuous, 3" wide, aluminum, mill finish	1 Carp	200	.040	L.F.	.76	1.94		2.70	3.81
2340 Baked enamel finish		200	.040	"	6.05	1.94		7.99	9.60
2400 Under eaves vent, aluminum, mill finish, 16" x 4"		48	.167	Ea.	1.93	8.10		10.03	14.50
2500 16" x 8"	↓	48	.167	"	2.23	8.10		10.33	14.85

Estimating Tips

General

- Room Finish Schedule: A complete set of plans should contain a room finish schedule. If one is not available, it would be well worth the time and effort to obtain one.

09 20 00 Plaster and Gypsum Board

- Lath is estimated by the square yard plus a 5% allowance for waste. Furring, channels, and accessories are measured by the linear foot. An extra foot should be allowed for each accessory miter or stop.

- Plaster is also estimated by the square yard. Deductions for openings vary by preference, from zero deduction to 50% of all openings over 2 feet in width. The estimator should allow one extra square foot for each linear foot of horizontal interior or exterior angle located below the ceiling level. Also, double the areas of small radius work.

- Drywall accessories, studs, track, and acoustical caulking are all measured by the linear foot. Drywall taping is figured by the square foot. Gypsum wallboard is estimated by the square foot. No material deductions should be made for door or window openings under 32 S.F.

09 60 00 Flooring

- Tile and terrazzo areas are taken off on a square foot basis. Trim and base materials are measured by the linear foot. Accent tiles are listed per each. Two basic methods of installation are used. Mud set is approximately 30% more expensive than thin set. In terrazzo work, be sure to include the linear footage of embedded decorative strips, grounds, machine rubbing, and power cleanup.

- Wood flooring is available in strip, parquet, or block configuration. The latter two types are set in adhesives with quantities estimated by the square foot. The laying pattern will influence labor costs and material waste. In addition to the material and labor for laying wood floors, the estimator must make allowances for sanding and finishing these areas, unless the flooring is prefinished.

- Sheet flooring is measured by the square yard. Roll widths vary, so consideration should be given to use the most economical width, as waste must be figured into the total quantity. Consider also the installation methods available, direct glue down or stretched.

09 70 00 Wall Finishes

- Wall coverings are estimated by the square foot. The area to be covered is measured, length by height of wall above baseboards, to calculate the square footage of each wall. This figure is divided by the number of square feet in the single roll which is being used. Deduct, in full, the areas of openings such as doors and windows. Where a pattern match is required allow 25%–30% waste.

09 80 00 Acoustic Treatment

- Acoustical systems fall into several categories. The takeoff of these materials should be by the square foot of area with a 5% allowance for waste. Do not forget about scaffolding, if applicable, when estimating these systems.

09 90 00 Painting and Coating

- A major portion of the work in painting involves surface preparation. Be sure to include cleaning, sanding, filling, and masking costs in the estimate.

- Protection of adjacent surfaces is not included in painting costs. When considering the method of paint application, an important factor is the amount of protection and masking required. These must be estimated separately and may be the determining factor in choosing the method of application.

Reference Numbers

Reference numbers are shown at the beginning of some major classifications. These numbers refer to related items in the Reference Section. The reference information may be an estimating procedure, an alternate pricing method, or technical information.

Note: Not all subdivisions listed here necessarily appear. ■

Did you know?

RSMeans Online gives you the same access to RSMeans' data with 24/7 access:

- Quickly locate costs in the searchable database.
- Build cost lists, estimates, and reports in minutes.
- Adjust costs to any location in the U.S. and Canada with the click of a button.

Start your free trial today at **www.RSMeansOnline.com**

RSMeans Online
FROM THE GORDIAN GROUP®

09 22 Supports for Plaster and Gypsum Board

09 22 03 – Fastening Methods for Finishes

09 22 03.20 Drilling Plaster/Drywall		Crew	Daily Output	Labor-Hours	Unit	Material	2016 Bare Costs Labor	Equipment	Total	Total Incl O&P
0010	**DRILLING PLASTER/DRYWALL**									
1100	Drilling & layout for drywall/plaster walls, up to 1" deep, no anchor									
1200	Holes, 1/4" diameter	1 Carp	150	.053	Ea.	.01	2.58		2.59	3.97
1300	3/8" diameter		140	.057		.01	2.77		2.78	4.26
1400	1/2" diameter		130	.062		.01	2.98		2.99	4.58
1500	3/4" diameter		120	.067		.02	3.23		3.25	4.97
1600	1" diameter		110	.073		.02	3.52		3.54	5.40
1700	1-1/4" diameter		100	.080		.03	3.88		3.91	6
1800	1-1/2" diameter	▼	90	.089	▼	.05	4.31		4.36	6.65
1900	For ceiling installations, add						40%			

09 69 Access Flooring

09 69 13 – Rigid-Grid Access Flooring

09 69 13.10 Access Floors		Crew	Daily Output	Labor-Hours	Unit	Material	2016 Bare Costs Labor	Equipment	Total	Total Incl O&P
0010	**ACCESS FLOORS**									
0015	Access floor pkg. including panel, pedestal, & stringers 1500 lbs load									
0100	Package pricing, conc. fill panels, no fin., 6" ht.	4 Carp	750	.043	S.F.	15.85	2.07		17.92	20.50
0105	12" ht.		750	.043		16.40	2.07		18.47	21
0110	18" ht.		750	.043		16.05	2.07		18.12	21
0115	24" ht.		750	.043		17.05	2.07		19.12	22
0120	Package pricing, steel panels, no fin., 6" ht.		750	.043		12.55	2.07		14.62	16.95
0125	12" ht.		750	.043		15.20	2.07		17.27	19.85
0130	18" ht.		750	.043		15.25	2.07		17.32	19.95
0135	24" ht.		750	.043		15.35	2.07		17.42	20
0140	Package pricing, wood core panels, no fin., 6" ht.		750	.043		11.50	2.07		13.57	15.80
0145	12" ht.		750	.043		11.50	2.07		13.57	15.80
0150	18" ht.		750	.043		11.65	2.07		13.72	15.95
0155	24" ht.		750	.043		11.75	2.07		13.82	16.10
0160	Pkg. pricing, conc. fill pnls, no stringers, no fin., 6" ht, 1250 lbs load		700	.046		10.20	2.21		12.41	14.65
0165	12" ht.		700	.046		10	2.21		12.21	14.40
0170	18" ht.		700	.046		10.55	2.21		12.76	15
0175	24" ht.	▼	700	.046		11.10	2.21		13.31	15.60
0250	Panels, 2' x 2' conc. fill, no fin.	2 Carp	500	.032		20.50	1.55		22.05	25
0255	With 1/8" high pressure laminate		500	.032		61	1.55		62.55	70
0260	Metal panels with 1/8" high pressure laminate		500	.032		72.50	1.55		74.05	82
0265	Wood core panels with 1/8" high pressure laminate		500	.032		47.50	1.55		49.05	55
0400	Aluminum panels, no fin.	▼	500	.032		33	1.55		34.55	38.50
0600	For carpet covering, add					8.80			8.80	9.70
0700	For vinyl floor covering, add					9.20			9.20	10.15
0900	For high pressure laminate covering, add					7.50			7.50	8.25
0910	For snap on stringer system, add	2 Carp	1000	.016	▼	1.56	.78		2.34	2.91
1000	Machine cutouts after initial installation	1 Carp	50	.160	Ea.	20	7.75		27.75	34
1050	Pedestals, 6" to 12"	2 Carp	85	.188		8.40	9.10		17.50	23.50
1100	Air conditioning grilles, 4" x 12"	1 Carp	17	.471		71	23		94	114
1150	4" x 18"	"	14	.571	▼	97.50	27.50		125	150
1200	Approach ramps, steel	2 Carp	60	.267	S.F.	25	12.90		37.90	48
1300	Aluminum	"	40	.400	"	34	19.40		53.40	67
1500	Handrail, 2 rail, aluminum	1 Carp	15	.533	L.F.	109	26		135	159

09 91 23.52 Miscellaneous, Interior	Crew	Daily Output	Labor-Hours	Unit	Material	2016 Bare Costs Labor	Equipment	Total	Total Incl O&P
0010 **MISCELLANEOUS, INTERIOR** R099100-10									
3800 Grilles, per side, oil base, primer coat, brushwork	1 Pord	520	.015	S.F.	.13	.62		.75	1.08
3850 Spray		1140	.007		.13	.28		.41	.58
3880 Paint 1 coat, brushwork		520	.015		.22	.62		.84	1.18
3900 Spray		1140	.007		.24	.28		.52	.70
3920 Paint 2 coats, brushwork		325	.025		.43	.99		1.42	1.97
3940 Spray		650	.012		.49	.50		.99	1.29
3950 Prime & paint 1 coat		325	.025		.35	.99		1.34	1.88
3960 Prime & paint 2 coats		270	.030		.34	1.20		1.54	2.17
4500 Louvers, one side, primer, brushwork		524	.015		.10	.62		.72	1.03
4520 Paint one coat, brushwork		520	.015		.10	.62		.72	1.04
4530 Spray		1140	.007		.11	.28		.39	.55
4540 Paint two coats, brushwork		325	.025		.18	.99		1.17	1.70
4550 Spray		650	.012		.20	.50		.70	.97
4560 Paint three coats, brushwork		270	.030		.27	1.20		1.47	2.10
4570 Spray		500	.016		.30	.65		.95	1.30
5000 Pipe, 1" - 4" diameter, primer or sealer coat, oil base, brushwork	2 Pord	1250	.013	L.F.	.10	.52		.62	.89
5100 Spray		2165	.007		.09	.30		.39	.55
5200 Paint 1 coat, brushwork		1250	.013		.11	.52		.63	.90
5300 Spray		2165	.007		.10	.30		.40	.56
5350 Paint 2 coats, brushwork		775	.021		.19	.83		1.02	1.47
5400 Spray		1240	.013		.22	.52		.74	1.02
5420 Paint 3 coats, brushwork		775	.021		.29	.83		1.12	1.58
5450 5" - 8" diameter, primer or sealer coat, brushwork		620	.026		.20	1.04		1.24	1.79
5500 Spray		1085	.015		.33	.60		.93	1.26
5550 Paint 1 coat, brushwork		620	.026		.29	1.04		1.33	1.89
5600 Spray		1085	.015		.33	.60		.93	1.26
5650 Paint 2 coats, brushwork		385	.042		.39	1.68		2.07	2.96
5700 Spray		620	.026		.43	1.04		1.47	2.04
5720 Paint 3 coats, brushwork		385	.042		.58	1.68		2.26	3.16
5750 9" - 12" diameter, primer or sealer coat, brushwork		415	.039		.30	1.56		1.86	2.67
5800 Spray		725	.022		.35	.89		1.24	1.73
5850 Paint 1 coat, brushwork		415	.039		.30	1.56		1.86	2.67
6000 Spray		725	.022		.33	.89		1.22	1.70
6200 Paint 2 coats, brushwork		260	.062		.58	2.48		3.06	4.38
6250 Spray		415	.039		.64	1.56		2.20	3.05
6270 Paint 3 coats, brushwork		260	.062		.86	2.48		3.34	4.69
6300 13" - 16" diameter, primer or sealer coat, brushwork		310	.052		.40	2.08		2.48	3.58
6350 Spray		540	.030		.44	1.20		1.64	2.29
6400 Paint 1 coat, brushwork		310	.052		.40	2.08		2.48	3.58
6450 Spray		540	.030		.44	1.20		1.64	2.29
6500 Paint 2 coats, brushwork		195	.082		.78	3.31		4.09	5.85
6550 Spray		310	.052		.86	2.08		2.94	4.09
6600 Radiators, per side, primer, brushwork	1 Pord	520	.015	S.F.	.10	.62		.72	1.04
6620 Paint, one coat		520	.015		.09	.62		.71	1.04
6640 Two coats		340	.024		.18	.95		1.13	1.63
6660 Three coats		283	.028		.27	1.14		1.41	2.02

For customer support on your Mechanical Cost Data, call 877.739.3689.

65

Division Notes

	CREW	DAILY OUTPUT	LABOR-HOURS	UNIT	BARE COSTS				TOTAL INCL O&P
					MAT.	LABOR	EQUIP.	TOTAL	

Estimating Tips
General

- The items in this division are usually priced per square foot or each.
- Many items in Division 10 require some type of support system or special anchors that are not usually furnished with the item. The required anchors must be added to the estimate in the appropriate division.
- Some items in Division 10, such as lockers, may require assembly before installation. Verify the amount of assembly required. Assembly can often exceed installation time.

10 20 00 Interior Specialties

- Support angles and blocking are not included in the installation of toilet compartments, shower/ dressing compartments, or cubicles. Appropriate line items from Divisions 5 or 6 may need to be added to support the installations.
- Toilet partitions are priced by the stall. A stall consists of a side wall, pilaster, and door with hardware. Toilet tissue holders and grab bars are extra.
- The required acoustical rating of a folding partition can have a significant impact on costs. Verify the sound transmission coefficient rating of the panel priced to the specification requirements.

- Grab bar installation does not include supplemental blocking or backing to support the required load. When grab bars are installed at an existing facility, provisions must be made to attach the grab bars to solid structure.

Reference Numbers

Reference numbers are shown at the beginning of some major classifications. These numbers refer to related items in the Reference Section. The reference information may be an estimating procedure, an alternate pricing method, or technical information.

Note: Not all subdivisions listed here necessarily appear. ■

---

10 31 Manufactured Fireplaces

10 31 13 – Manufactured Fireplace Chimneys

10 31 13.10 Fireplace Chimneys	Crew	Daily Output	Labor-Hours	Unit	Material	2016 Bare Costs Labor	Equipment	Total	Total Incl O&P
0010 **FIREPLACE CHIMNEYS**									
0500 Chimney dbl. wall, all stainless, over 8'-6", 7" diam., add to fireplace	1 Carp	33	.242	V.L.F.	84	11.75		95.75	111
0600 10" diameter, add to fireplace		32	.250		112	12.10		124.10	143
0700 12" diameter, add to fireplace		31	.258		167	12.50		179.50	202
0800 14" diameter, add to fireplace		30	.267	↓	227	12.90		239.90	270
1000 Simulated brick chimney top, 4' high, 16" x 16"		10	.800	Ea.	435	39		474	540
1100 24" x 24"	↓	7	1.143	"	555	55.50		610.50	695

10 31 23 – Prefabricated Fireplaces

10 31 23.10 Fireplace, Prefabricated	Crew	Daily Output	Labor-Hours	Unit	Material	2016 Bare Costs Labor	Equipment	Total	Total Incl O&P
0010 **FIREPLACE, PREFABRICATED**, free standing or wall hung									
0100 With hood & screen, painted	1 Carp	1.30	6.154	Ea.	1,475	298		1,773	2,075
0150 Average		1	8		1,700	390		2,090	2,475
0200 Stainless steel		.90	8.889	↓	3,050	430		3,480	4,000
1500 Simulated logs, gas fired, 40,000 BTU, 2' long, manual safety pilot		7	1.143	Set	440	55.50		495.50	570
1600 Adjustable flame remote pilot		6	1.333		1,200	64.50		1,264.50	1,425
1700 Electric, 1,500 BTU, 1'-6" long, incandescent flame		7	1.143		205	55.50		260.50	310
1800 1,500 BTU, LED flame	↓	6	1.333	↓	320	64.50		384.50	450

10 35 Stoves

10 35 13 – Heating Stoves

10 35 13.10 Woodburning Stoves	Crew	Daily Output	Labor-Hours	Unit	Material	2016 Bare Costs Labor	Equipment	Total	Total Incl O&P
0010 **WOODBURNING STOVES**									
0015 Cast iron, °1500 S.F.	2 Carp	1.30	12.308	Ea.	1,300	595		1,895	2,350
0020 1500-2000 S.F.		1	16		2,050	775		2,825	3,450
0030 2,000 S.F.	↓	.80	20		2,800	970		3,770	4,550
0050 For gas log lighter, add				↓	46.50			46.50	51

Estimating Tips

General

- The items in this division are usually priced per square foot or each. Many of these items are purchased by the owner for installation by the contractor. Check the specifications for responsibilities and include time for receiving, storage, installation, and mechanical and electrical hookups in the appropriate divisions.

- Many items in Division 11 require some type of support system that is not usually furnished with the item. Examples of these systems include blocking for the attachment of casework and support angles for ceiling-hung projection screens. The required blocking or supports must be added to the estimate in the appropriate division.

- Some items in Division 11 may require assembly or electrical hookups. Verify the amount of assembly required or the need for a hard electrical connection and add the appropriate costs.

Reference Numbers

Reference numbers are shown at the beginning of some major classifications. These numbers refer to related items in the Reference Section. The reference information may be an estimating procedure, an alternate pricing method, or technical information.

Note: Not all subdivisions listed here necessarily appear. ■

11 11 Vehicle Service Equipment

11 11 13 – Compressed-Air Vehicle Service Equipment

11 11 13.10 Compressed Air Equipment	Crew	Daily Output	Labor-Hours	Unit	Material	2016 Bare Costs Labor	Equipment	Total	Total Incl O&P
0010 **COMPRESSED AIR EQUIPMENT**									
0030 Compressors, electric, 1-1/2 H.P., standard controls	L-4	1.50	16	Ea.	470	725		1,195	1,650
0550 Dual controls		1.50	16		765	725		1,490	1,975
0600 5 H.P., 115/230 volt, standard controls		1	24		1,850	1,075		2,925	3,700
0650 Dual controls		1	24		2,825	1,075		3,900	4,775

11 11 19 – Vehicle Lubrication Equipment

11 11 19.10 Lubrication Equipment

	Crew	Daily Output	Labor-Hours	Unit	Material	Labor	Equipment	Total	Total Incl O&P
0010 **LUBRICATION EQUIPMENT**									
3000 Lube equipment, 3 reel type, with pumps, not including piping	L-4	.50	48	Set	8,800	2,175		10,975	13,000
3700 Pump lubrication, pneumatic, not incl. air compressor									
3710 Oil/gear lube	Q-1	9.60	1.667	Ea.	815	89		904	1,025
3720 Grease	"	9.60	1.667	"	1,100	89		1,189	1,325

11 11 33 – Vehicle Spray Painting Equipment

11 11 33.10 Spray Painting Equipment

	Crew	Daily Output	Labor-Hours	Unit	Material	Labor	Equipment	Total	Total Incl O&P
0010 **SPRAY PAINTING EQUIPMENT**									
4000 Spray painting booth, 26' long, complete	L-4	.40	60	Ea.	10,400	2,725		13,125	15,700

11 21 Retail and Service Equipment

11 21 73 – Commercial Laundry and Dry Cleaning Equipment

11 21 73.16 Drying and Conditioning Equipment

	Crew	Daily Output	Labor-Hours	Unit	Material	Labor	Equipment	Total	Total Incl O&P
0010 **DRYING AND CONDITIONING EQUIPMENT**									
0100 Dryers, not including rough-in									
1500 Industrial, 30 lb. capacity	1 Plum	2	4	Ea.	3,225	237		3,462	3,900
1600 50 lb. capacity	"	1.70	4.706	"	3,650	279		3,929	4,450

11 21 73.26 Commercial Washers and Extractors

	Crew	Daily Output	Labor-Hours	Unit	Material	Labor	Equipment	Total	Total Incl O&P
0010 **COMMERCIAL WASHERS AND EXTRACTORS**, not including rough-in									
6000 Combination washer/extractor, 20 lb. capacity	L-6	1.50	8	Ea.	6,025	465		6,490	7,325
6100 30 lb. capacity		.80	15		9,375	865		10,240	11,600
6200 50 lb. capacity		.68	17.647		11,000	1,025		12,025	13,600
6300 75 lb. capacity		.30	40		20,600	2,325		22,925	26,200
6350 125 lb. capacity		.16	75		27,700	4,325		32,025	37,000

11 21 73.33 Coin-Operated Laundry Equipment

	Crew	Daily Output	Labor-Hours	Unit	Material	Labor	Equipment	Total	Total Incl O&P
0010 **COIN-OPERATED LAUNDRY EQUIPMENT**									
0990 Dryer, gas fired									
1000 Commercial, 30 lb. capacity, coin operated, single	1 Plum	3	2.667	Ea.	3,350	158		3,508	3,950
1100 Double stacked	"	2	4	"	7,500	237		7,737	8,600
5290 Clothes washer									
5300 Commercial, coin operated, average	1 Plum	3	2.667	Ea.	1,250	158		1,408	1,625

11 21 83 – Photo Processing Equipment

11 21 83.13 Darkroom Equipment

	Crew	Daily Output	Labor-Hours	Unit	Material	Labor	Equipment	Total	Total Incl O&P
0010 **DARKROOM EQUIPMENT**									
0020 Developing sink, 5" deep, 24" x 48"	Q-1	2	8	Ea.	4,150	425		4,575	5,225
0050 48" x 52"		1.70	9.412		4,425	500		4,925	5,625
0200 10" deep, 24" x 48"		1.70	9.412		1,575	500		2,075	2,475
0250 24" x 108"		1.50	10.667		3,725	570		4,295	4,950
3500 Washers, round, minimum sheet 11" x 14"		2	8		3,250	425		3,675	4,225
3550 Maximum sheet 20" x 24"		1	16		3,575	850		4,425	5,200
3800 Square, minimum sheet 20" x 24"		1	16		3,575	850		4,425	5,200
3900 Maximum sheet 50" x 56"		.80	20		4,900	1,075		5,975	7,000

11 21 Retail and Service Equipment

11 21 83 – Photo Processing Equipment

11 21 83.13 Darkroom Equipment	Crew	Daily Output	Labor-Hours	Unit	Material	2016 Bare Costs Labor	Equipment	Total	Total Incl O&P
4500 Combination tank sink, tray sink, washers, with									
4510 Dry side tables, average	Q-1	.45	35.556	Ea.	10,800	1,900		12,700	14,700

11 30 Residential Equipment

11 30 13 – Residential Appliances

11 30 13.15 Cooking Equipment

		Crew	Daily Output	Labor-Hours	Unit	Material	Labor	Equipment	Total	Total Incl O&P
0010	COOKING EQUIPMENT									
0020	Cooking range, 30" free standing, 1 oven, minimum	2 Clab	10	1.600	Ea.	470	60.50		530.50	610
0050	Maximum		4	4		1,900	152		2,052	2,300
0150	2 oven, minimum		10	1.600		1,250	60.50		1,310.50	1,475
0200	Maximum		10	1.600		2,800	60.50		2,860.50	3,175

11 30 13.16 Refrigeration Equipment

		Crew	Daily Output	Labor-Hours	Unit	Material	Labor	Equipment	Total	Total Incl O&P
0010	REFRIGERATION EQUIPMENT									
5200	Icemaker, automatic, 20 lb. per day	1 Plum	7	1.143	Ea.	1,100	67.50		1,167.50	1,300
5350	51 lb. per day	"	2	4	"	1,600	237		1,837	2,100

11 30 13.17 Kitchen Cleaning Equipment

		Crew	Daily Output	Labor-Hours	Unit	Material	Labor	Equipment	Total	Total Incl O&P
0010	KITCHEN CLEANING EQUIPMENT									
2750	Dishwasher, built-in, 2 cycles, minimum	L-1	4	4	Ea.	299	229		528	675
2800	Maximum		2	8		450	455		905	1,175
2950	4 or more cycles, minimum		4	4		425	229		654	815
2960	Average		4	4		565	229		794	970
3000	Maximum		2	8		1,225	455		1,680	2,000

11 30 13.18 Waste Disposal Equipment

		Crew	Daily Output	Labor-Hours	Unit	Material	Labor	Equipment	Total	Total Incl O&P
0010	WASTE DISPOSAL EQUIPMENT									
3300	Garbage disposal, sink type, minimum	L-1	10	1.600	Ea.	93	91.50		184.50	239
3350	Maximum	"	10	1.600	"	243	91.50		334.50	405

11 30 13.19 Kitchen Ventilation Equipment

		Crew	Daily Output	Labor-Hours	Unit	Material	Labor	Equipment	Total	Total Incl O&P
0010	KITCHEN VENTILATION EQUIPMENT									
4150	Hood for range, 2 speed, vented, 30" wide, minimum	L-3	5	3.200	Ea.	83	167		250	345
4200	Maximum		3	5.333		855	279		1,134	1,375
4300	42" wide, minimum		5	3.200		166	167		333	435
4330	Custom		5	3.200		1,625	167		1,792	2,050
4350	Maximum		3	5.333		1,975	279		2,254	2,600
4500	For ventless hood, 2 speed, add					20			20	22
4650	For vented 1 speed, deduct from maximum					55			55	60.50

11 30 13.24 Washers

		Crew	Daily Output	Labor-Hours	Unit	Material	Labor	Equipment	Total	Total Incl O&P
0010	WASHERS									
5000	Residential, 4 cycle, average	1 Plum	3	2.667	Ea.	890	158		1,048	1,225
6650	Washing machine, automatic, minimum		3	2.667		545	158		703	840
6700	Maximum		1	8		1,650	475		2,125	2,550

11 30 13.25 Dryers

		Crew	Daily Output	Labor-Hours	Unit	Material	Labor	Equipment	Total	Total Incl O&P
0010	DRYERS									
0500	Gas fired residential, 16 lb. capacity, average	1 Plum	3	2.667	Ea.	685	158		843	995
7450	Vent kits for dryers	1 Carp	10	.800	"	40	39		79	104

11 30 15 – Miscellaneous Residential Appliances

11 30 15.13 Sump Pumps

		Crew	Daily Output	Labor-Hours	Unit	Material	Labor	Equipment	Total	Total Incl O&P
0010	SUMP PUMPS									
6400	Cellar drainer, pedestal, 1/3 H.P., molded PVC base	1 Plum	3	2.667	Ea.	140	158		298	390
6450	Solid brass	"	2	4	"	297	237		534	680

For customer support on your Mechanical Cost Data, call 877.739.3689.

71

11 30 Residential Equipment

11 30 15 – Miscellaneous Residential Appliances

11 30 15.13 Sump Pumps

		Crew	Daily Output	Labor-Hours	Unit	Material	2016 Bare Costs Labor	Equipment	Total	Total Incl O&P
6460	Sump pump, see also Section 22 14 29.16									

11 30 15.23 Water Heaters

		Crew	Daily Output	Labor-Hours	Unit	Material	2016 Bare Costs Labor	Equipment	Total	Total Incl O&P
0010	**WATER HEATERS**									
6900	Electric, glass lined, 30 gallon, minimum	L-1	5	3.200	Ea.	815	183		998	1,175
6950	Maximum		3	5.333		1,125	305		1,430	1,700
7100	80 gallon, minimum		2	8		1,475	455		1,930	2,300
7150	Maximum		1	16		2,050	915		2,965	3,625
7180	Gas, glass lined, 30 gallon, minimum	2 Plum	5	3.200		1,450	189		1,639	1,875
7220	Maximum		3	5.333		2,025	315		2,340	2,700
7260	50 gallon, minimum		2.50	6.400		1,525	380		1,905	2,250
7300	Maximum		1.50	10.667		2,125	630		2,755	3,300

11 30 15.43 Air Quality

		Crew	Daily Output	Labor-Hours	Unit	Material	2016 Bare Costs Labor	Equipment	Total	Total Incl O&P
0010	**AIR QUALITY**									
2450	Dehumidifier, portable, automatic, 15 pint	1 Elec	4	2	Ea.	200	110		310	385
2550	40 pint		3.75	2.133		214	118		332	410
3550	Heater, electric, built-in, 1250 watt, ceiling type, minimum		4	2		117	110		227	293
3600	Maximum		3	2.667		204	147		351	445
3700	Wall type, minimum		4	2		191	110		301	375
3750	Maximum		3	2.667		195	147		342	435
3900	1500 watt wall type, with blower		4	2		173	110		283	355
3950	3000 watt		3	2.667		390	147		537	650
4850	Humidifier, portable, 8 gallons per day					152			152	167
5000	15 gallons per day					222			222	244

11 32 Unit Kitchens

11 32 13 – Metal Unit Kitchens

11 32 13.10 Commercial Unit Kitchens

		Crew	Daily Output	Labor-Hours	Unit	Material	2016 Bare Costs Labor	Equipment	Total	Total Incl O&P
0010	**COMMERCIAL UNIT KITCHENS**									
1500	Combination range, refrigerator and sink, 30" wide, minimum	L-1	2	8	Ea.	1,100	455		1,555	1,900
1550	Maximum		1	16		1,250	915		2,165	2,750
1570	60" wide, average		1.40	11.429		1,425	655		2,080	2,550
1590	72" wide, average		1.20	13.333		1,500	760		2,260	2,800
1600	Office model, 48" wide		2	8		2,050	455		2,505	2,925
1620	Refrigerator and sink only		2.40	6.667		2,525	380		2,905	3,350
1640	Combination range, refrigerator, sink, microwave									
1660	Oven and ice maker	L-1	.80	20	Ea.	4,550	1,150		5,700	6,725

11 41 Foodservice Storage Equipment

11 41 13 – Refrigerated Food Storage Cases

11 41 13.10 Refrigerated Food Cases

		Crew	Daily Output	Labor-Hours	Unit	Material	2016 Bare Costs Labor	Equipment	Total	Total Incl O&P
0010	**REFRIGERATED FOOD CASES**									
0030	Dairy, multi-deck, 12' long	Q-5	3	5.333	Ea.	11,300	291		11,591	12,800
0100	For rear sliding doors, add					1,850			1,850	2,025
0200	Delicatessen case, service deli, 12' long, single deck	Q-5	3.90	4.103		8,400	224		8,624	9,600
0300	Multi-deck, 18 S.F. shelf display		3	5.333		7,650	291		7,941	8,875
0400	Freezer, self-contained, chest-type, 30 C.F.		3.90	4.103		9,125	224		9,349	10,300
0500	Glass door, upright, 78 C.F.		3.30	4.848		12,000	265		12,265	13,600
0600	Frozen food, chest type, 12' long		3.30	4.848		8,225	265		8,490	9,450
0700	Glass door, reach-in, 5 door		3	5.333		14,100	291		14,391	15,900

11 41 Foodservice Storage Equipment

11 41 13 – Refrigerated Food Storage Cases

11 41 13.10 Refrigerated Food Cases

		Crew	Daily Output	Labor-Hours	Unit	Material	2016 Bare Costs Labor	2016 Bare Costs Equipment	Total	Total Incl O&P
0800	Island case, 12' long, single deck	Q-5	3.30	4.848	Ea.	7,275	265		7,540	8,425
0900	Multi-deck		3	5.333		8,500	291		8,791	9,800
1000	Meat case, 12' long, single deck		3.30	4.848		7,275	265		7,540	8,425
1050	Multi-deck		3.10	5.161		10,400	282		10,682	11,800
1100	Produce, 12' long, single deck		3.30	4.848		6,875	265		7,140	7,950
1200	Multi-deck		3.10	5.161		8,725	282		9,007	10,000

11 41 13.20 Refrigerated Food Storage Equipment

		Crew	Daily Output	Labor-Hours	Unit	Material	2016 Bare Costs Labor	2016 Bare Costs Equipment	Total	Total Incl O&P
0010	**REFRIGERATED FOOD STORAGE EQUIPMENT**									
2350	Cooler, reach-in, beverage, 6' long	Q-1	6	2.667	Ea.	3,375	142		3,517	3,925
4300	Freezers, reach-in, 44 C.F.		4	4		5,325	213		5,538	6,175
4500	68 C.F.		3	5.333		4,550	284		4,834	5,450

11 44 Food Cooking Equipment

11 44 13 – Commercial Ranges

11 44 13.10 Cooking Equipment

		Crew	Daily Output	Labor-Hours	Unit	Material	2016 Bare Costs Labor	2016 Bare Costs Equipment	Total	Total Incl O&P
0010	**COOKING EQUIPMENT**									
0020	Bake oven, gas, one section	Q-1	8	2	Ea.	5,525	107		5,632	6,225
0300	Two sections		7	2.286		9,150	122		9,272	10,300
0600	Three sections		6	2.667		10,700	142		10,842	12,000
0900	Electric convection, single deck	L-7	4	7		6,225	325		6,550	7,350
6350	Kettle, w/steam jacket, tilting, w/positive lock, SS, 20 gallons		7	4		8,700	186		8,886	9,850
6600	60 gallons		6	4.667		11,800	216		12,016	13,300

11 46 Food Dispensing Equipment

11 46 83 – Ice Machines

11 46 83.10 Commercial Ice Equipment

		Crew	Daily Output	Labor-Hours	Unit	Material	2016 Bare Costs Labor	2016 Bare Costs Equipment	Total	Total Incl O&P
0010	**COMMERCIAL ICE EQUIPMENT**									
5800	Ice cube maker, 50 pounds per day	Q-1	6	2.667	Ea.	1,625	142		1,767	2,000
6050	500 pounds per day	"	4	4	"	2,950	213		3,163	3,550

11 48 Foodservice Cleaning and Disposal Equipment

11 48 13 – Commercial Dishwashers

11 48 13.10 Dishwashers

			Crew	Daily Output	Labor-Hours	Unit	Material	2016 Bare Costs Labor	2016 Bare Costs Equipment	Total	Total Incl O&P
0010	**DISHWASHERS**										
2700	Dishwasher, commercial, rack type										
2720	10 to 12 racks per hour		Q-1	3.20	5	Ea.	3,525	266		3,791	4,275
2730	Energy star rated, 35 to 40 racks/hour	G		1.30	12.308		4,700	655		5,355	6,150
2740	50 to 60 racks/hour	G		1.30	12.308		11,600	655		12,255	13,800
2800	Automatic, 190 to 230 racks per hour		L-6	.35	34.286		14,000	1,975		15,975	18,400
2820	235 to 275 racks per hour			.25	48		31,500	2,775		34,275	38,800
2840	8,750 to 12,500 dishes per hour			.10	120		55,500	6,950		62,450	71,500

For customer support on your Mechanical Cost Data, call 877.739.3689.

73

11 53 Laboratory Equipment

11 53 13 – Laboratory Fume Hoods

11 53 13.13 Recirculating Laboratory Fume Hoods	Crew	Daily Output	Labor-Hours	Unit	Material	2016 Bare Costs Labor	Equipment	Total	Total Incl O&P
0010 **RECIRCULATING LABORATORY FUME HOODS**									
0600 Fume hood, with countertop & base, not including HVAC									
0610 Simple, minimum	2 Carp	5.40	2.963	L.F.	510	144		654	780
0620 Complex, including fixtures		2.40	6.667		980	325		1,305	1,575
0630 Special, maximum	↓	1.70	9.412	↓	1,100	455		1,555	1,900
0670 Service fixtures, average				Ea.	275			275	300
0680 For sink assembly with hot and cold water, add	1 Plum	1.40	5.714	"	735	340		1,075	1,325

11 53 19 – Laboratory Sterilizers

11 53 19.13 Sterilizers

	Crew	Daily Output	Labor-Hours	Unit	Material	Labor	Equipment	Total	Total Incl O&P
0010 **STERILIZERS**									
0700 Glassware washer, undercounter, minimum	L-1	1.80	8.889	Ea.	6,225	510		6,735	7,625
0710 Maximum	"	1	16		13,900	915		14,815	16,700
1850 Utensil washer-sanitizer	1 Plum	2	4	↓	8,450	237		8,687	9,650

11 53 33 – Emergency Safety Appliances

11 53 33.13 Emergency Equipment

	Crew	Daily Output	Labor-Hours	Unit	Material	Labor	Equipment	Total	Total Incl O&P
0010 **EMERGENCY EQUIPMENT**									
1400 Safety equipment, eye wash, hand held				Ea.	405			405	445
1450 Deluge shower				"	775			775	850

11 53 43 – Service Fittings and Accessories

11 53 43.13 Fittings

	Crew	Daily Output	Labor-Hours	Unit	Material	Labor	Equipment	Total	Total Incl O&P
0010 **FITTINGS**									
1600 Sink, one piece plastic, flask wash, hose, free standing	1 Plum	1.60	5	Ea.	1,950	296		2,246	2,600
1610 Epoxy resin sink, 25" x 16" x 10"	"	2	4	"	221	237		458	600
8000 Alternate pricing method: as percent of lab furniture									
8050 Installation, not incl. plumbing & duct work				% Furn.				22%	22%
8100 Plumbing, final connections, simple system								10%	10%
8110 Moderately complex system								15%	15%
8120 Complex system								20%	20%
8150 Electrical, simple system								10%	10%
8160 Moderately complex system								20%	20%
8170 Complex system				↓				35%	35%

11 71 Medical Sterilizing Equipment

11 71 10 – Medical Sterilizers & Distillers

11 71 10.10 Sterilizers and Distillers

	Crew	Daily Output	Labor-Hours	Unit	Material	Labor	Equipment	Total	Total Incl O&P
0010 **STERILIZERS AND DISTILLERS**									
0700 Distiller, water, steam heated, 50 gal. capacity	1 Plum	1.40	5.714	Ea.	19,200	340		19,540	21,700
5600 Sterilizers, floor loading, 26" x 62" x 42", single door, steam					122,000			122,000	134,500
5650 Double door, steam					206,000			206,000	227,000
5800 General purpose, 20" x 20" x 38", single door					12,900			12,900	14,200
6000 Portable, counter top, steam, minimum					3,000			3,000	3,300
6020 Maximum					4,750			4,750	5,225
6050 Portable, counter top, gas, 17" x 15" x 32-1/2"					39,800			39,800	43,700
6150 Manual washer/sterilizer, 16" x 16" x 26"	1 Plum	2	4	↓	54,500	237		54,737	60,500
6200 Steam generators, electric 10 kW to 180 kW, freestanding									
6250 Minimum	1 Elec	3	2.667	Ea.	10,500	147		10,647	11,700
6300 Maximum	"	.70	11.429	↓	22,000	630		22,630	25,100
8200 Bed pan washer-sanitizer	1 Plum	2	4	↓	7,500	237		7,737	8,600

11 73 Patient Care Equipment

11 73 10 – Patient Treatment Equipment

11 73 10.10 Treatment Equipment	Crew	Daily Output	Labor-Hours	Unit	Material	2016 Bare Costs Labor	2016 Bare Costs Equipment	Total	Total Incl O&P
0010 **TREATMENT EQUIPMENT**									
1800 Heat therapy unit, humidified, 26" x 78" x 28"				Ea.	3,125			3,125	3,450
8400 Whirlpool bath, mobile, sst, 18" x 24" x 60"					5,775			5,775	6,350
8450 Fixed, incl. mixing valves	1 Plum	2	4	↓	4,525	237		4,762	5,325

11 74 Dental Equipment

11 74 10 – Dental Office Equipment

11 74 10.10 Diagnostic and Treatment Equipment

11 74 10.10 Diagnostic and Treatment Equipment	Crew	Daily Output	Labor-Hours	Unit	Material	2016 Bare Costs Labor	2016 Bare Costs Equipment	Total	Total Incl O&P
0010 **DIAGNOSTIC AND TREATMENT EQUIPMENT**									
0020 Central suction system, minimum	1 Plum	1.20	6.667	Ea.	1,600	395		1,995	2,350
0100 Maximum	"	.90	8.889		4,575	525		5,100	5,850
0600 Chair, electric or hydraulic, minimum	1 Skwk	.50	16		2,225	800		3,025	3,675
0700 Maximum		.25	32		7,800	1,600		9,400	11,100
2000 Light, ceiling mounted, minimum		8	1		1,125	50		1,175	1,325
2100 Maximum		8	1		2,225	50		2,275	2,525
2200 Unit light, minimum	2 Skwk	5.33	3.002		710	150		860	1,000
2210 Maximum		5.33	3.002		1,600	150		1,750	2,000
2220 Track light, minimum		3.20	5		1,725	250		1,975	2,275
2230 Maximum		3.20	5		2,650	250		2,900	3,300
2300 Sterilizers, steam portable, minimum					1,375			1,375	1,500
2350 Maximum					9,650			9,650	10,600
2600 Steam, institutional					3,175			3,175	3,475
2650 Dry heat, electric, portable, 3 trays				↓	1,225			1,225	1,350

11 78 Mortuary Equipment

11 78 13 – Mortuary Refrigerators

11 78 13.10 Mortuary and Autopsy Equipment

11 78 13.10 Mortuary and Autopsy Equipment	Crew	Daily Output	Labor-Hours	Unit	Material	2016 Bare Costs Labor	2016 Bare Costs Equipment	Total	Total Incl O&P
0010 **MORTUARY AND AUTOPSY EQUIPMENT**									
0015 Autopsy table, standard	1 Plum	1	8	Ea.	9,750	475		10,225	11,400
0020 Deluxe	"	.60	13.333		16,300	790		17,090	19,100
3200 Mortuary refrigerator, end operated, 2 capacity					9,850			9,850	10,800
3300 6 capacity				↓	14,800			14,800	16,300

11 78 16 – Crematorium Equipment

11 78 16.10 Crematory

11 78 16.10 Crematory	Crew	Daily Output	Labor-Hours	Unit	Material	2016 Bare Costs Labor	2016 Bare Costs Equipment	Total	Total Incl O&P
0010 **CREMATORY**									
1500 Crematory, not including building, 1 place	Q-3	.20	160	Ea.	74,000	9,025		83,025	95,000
1750 2 place	"	.10	320	"	105,500	18,000		123,500	143,000

For customer support on your Mechanical Cost Data, call 877.739.3689.

75

11 81 Facility Maintenance Equipment

11 81 19 – Vacuum Cleaning Systems

11 81 19.10 Vacuum Cleaning	Crew	Daily Output	Labor-Hours	Unit	Material	2016 Bare Costs Labor	Equipment	Total	Total Incl O&P
0010	**VACUUM CLEANING**									
0020	Central, 3 inlet, residential	1 Skwk	.90	8.889	Total	1,100	445		1,545	1,875
0200	Commercial		.70	11.429		1,275	570		1,845	2,275
0400	5 inlet system, residential		.50	16		1,550	800		2,350	2,925
0600	7 inlet system, commercial		.40	20		1,750	1,000		2,750	3,450
0800	9 inlet system, residential	↓	.30	26.667		3,875	1,325		5,200	6,300
4010	Rule of thumb: First 1200 S.F., installed								1,425	1,575
4020	For each additional S.F., add				S.F.				.26	.26

11 82 Facility Solid Waste Handling Equipment

11 82 19 – Packaged Incinerators

11 82 19.10 Packaged Gas Fired Incinerators

		Crew	Daily Output	Labor-Hours	Unit	Material	2016 Bare Costs Labor	Equipment	Total	Total Incl O&P
0010	**PACKAGED GAS FIRED INCINERATORS**									
4400	Incinerator, gas, not incl. chimney, elec. or pipe, 50#/hr., minimum	Q-3	.80	40	Ea.	39,200	2,250		41,450	46,500
4420	Maximum		.70	45.714		41,300	2,575		43,875	49,300
4440	200 lb. per hr., minimum (batch type)		.60	53.333		69,000	3,000		72,000	80,500
4460	Maximum (with feeder)		.50	64		76,500	3,600		80,100	89,500
4480	400 lb. per hr., minimum (batch type)		.30	106		75,500	6,000		81,500	92,000
4500	Maximum (with feeder)		.25	128		90,000	7,225		97,225	110,500
4520	800 lb. per hr., with feeder, minimum		.20	160		115,000	9,025		124,025	140,000
4540	Maximum		.17	188		193,000	10,600		203,600	228,500
4560	1,200 lb. per hr., with feeder, minimum		.15	213		141,000	12,000		153,000	173,000
4580	Maximum		.11	290		176,500	16,400		192,900	218,500
4600	2,000 lb. per hr., with feeder, minimum		.10	320		381,500	18,000		399,500	446,500
4620	Maximum		.05	640		616,000	36,100		652,100	732,500
4700	For heat recovery system, add, minimum		.25	128		81,500	7,225		88,725	100,500
4710	Add, maximum		.11	290		254,500	16,400		270,900	304,000
4720	For automatic ash conveyer, add		.50	64	↓	33,900	3,600		37,500	42,800
4750	Large municipal incinerators, incl. stack, minimum		.25	128	Ton/day	20,600	7,225		27,825	33,500
4850	Maximum	↓	.10	320	"	54,500	18,000		72,500	87,000

11 82 26 – Facility Waste Compactors

11 82 26.10 Compactors

		Crew	Daily Output	Labor-Hours	Unit	Material	2016 Bare Costs Labor	Equipment	Total	Total Incl O&P
0010	**COMPACTORS**									
0020	Compactors, 115 volt, 250#/hr., chute fed	L-4	1	24	Ea.	12,200	1,075		13,275	15,100
0100	Hand fed		2.40	10		15,400	455		15,855	17,600
0300	Multi-bag, 230 volt, 600#/hr., chute fed		1	24		15,400	1,075		16,475	18,700
0400	Hand fed		1	24		15,600	1,075		16,675	18,800
0500	Containerized, hand fed, 2 to 6 C.Y. containers, 250#/hr.		1	24		15,400	1,075		16,475	18,600
0550	For chute fed, add per floor		1	24		1,425	1,075		2,500	3,250
1000	Heavy duty industrial compactor, 0.5 C.Y. capacity		1	24		10,300	1,075		11,375	13,100
1050	1.0 C.Y. capacity		1	24		15,800	1,075		16,875	19,100
1100	3.0 C.Y. capacity		.50	48		26,500	2,175		28,675	32,500
1150	5.0 C.Y. capacity		.50	48		33,200	2,175		35,375	39,900
1200	Combination shredder/compactor (5,000 lb./hr.)	↓	.50	48		64,500	2,175		66,675	74,500
1400	For handling hazardous waste materials, 55 gallon drum packer, std.					20,100			20,100	22,100
1410	55 gallon drum packer w/HEPA filter					25,100			25,100	27,600
1420	55 gallon drum packer w/charcoal & HEPA filter					33,400			33,400	36,800
1430	All of the above made explosion proof, add					1,475			1,475	1,625
5800	Shredder, industrial, minimum					24,200			24,200	26,600
5850	Maximum					129,000			129,000	142,000

11 82 Facility Solid Waste Handling Equipment

11 82 26 – Facility Waste Compactors

11 82 26.10 Compactors	Crew	Daily Output	Labor-Hours	Unit	Material	2016 Bare Costs Labor	Equipment	Total	Total Incl O&P
5900	Baler, industrial, minimum				Ea.	9,650			9,650	10,600
5950	Maximum				↓	563,500			563,500	620,000

For customer support on your Mechanical Cost Data, call 877.739.3689.

77

Division Notes

	CREW	DAILY OUTPUT	LABOR-HOURS	UNIT	BARE COSTS				TOTAL INCL O&P
					MAT.	LABOR	EQUIP.	TOTAL	

Estimating Tips
General

- The items and systems in this division are usually estimated, purchased, supplied, and installed as a unit by one or more subcontractors. The estimator must ensure that all parties are operating from the same set of specifications and assumptions, and that all necessary items are estimated and will be provided. Many times the complex items and systems are covered, but the more common ones, such as excavation or a crane, are overlooked for the very reason that everyone assumes nobody could miss them. The estimator should be the central focus and be able to ensure that all systems are complete.

- Another area where problems can develop in this division is at the interface between systems. The estimator must ensure, for instance, that anchor bolts, nuts, and washers are estimated and included for the air-supported structures and pre-engineered buildings to be bolted to their foundations. Utility supply is a common area where essential items or pieces of equipment can be missed or overlooked, because each subcontractor may feel it is another's responsibility. The estimator should also be aware of certain items which may be supplied as part of a package but installed by others, and ensure that the installing contractor's estimate includes the cost of installation. Conversely, the estimator must also ensure that items are not costed by two different subcontractors, resulting in an inflated overall estimate.

13 30 00 Special Structures

- The foundations and floor slab, as well as rough mechanical and electrical, should be estimated, as this work is required for the assembly and erection of the structure. Generally, as noted in the data set, the pre-engineered building comes as a shell. Pricing is based on the size and structural design parameters stated in the reference section. Additional features, such as windows and doors with their related structural framing, must also be included by the estimator. Here again, the estimator must have a clear understanding of the scope of each portion of the work and all the necessary interfaces.

Reference Numbers

Reference numbers are shown at the beginning of some major classifications. These numbers refer to related items in the Reference Section. The reference information may be an estimating procedure, an alternate pricing method, or technical information.

Note: Not all subdivisions listed here necessarily appear. ■

13 11 Swimming Pools

13 11 13 – Below-Grade Swimming Pools

13 11 13.50 Swimming Pools

		Crew	Daily Output	Labor-Hours	Unit	Material	2016 Bare Costs Labor	Equipment	Total	Total Incl O&P
0010	**SWIMMING POOLS** Residential in-ground, vinyl lined, concrete									
0020	Swimming pools, resi in-ground, vyl lined, conc sides, W/equip, sand bot	B-52	300	.187	SF Surf	25.50	8.30	2	35.80	43.50
0100	Metal or polystyrene sides	B-14	410	.117		21.50	4.70	.89	27.09	31.50
0200	Add for vermiculite bottom				↓	1.64			1.64	1.80
0500	Gunite bottom and sides, white plaster finish									
0600	12' x 30' pool	B-52	145	.386	SF Surf	48	17.10	4.13	69.23	83
0720	16' x 32' pool		155	.361		43	16	3.86	62.86	76.50
0750	20' x 40' pool	↓	250	.224	↓	38.50	9.95	2.39	50.84	60.50
0810	Concrete bottom and sides, tile finish									
0820	12' x 30' pool	B-52	80	.700	SF Surf	48.50	31	7.50	87	109
0830	16' x 32' pool		95	.589		40	26	6.30	72.30	91
0840	20' x 40' pool	↓	130	.431	↓	32	19.10	4.60	55.70	69
1100	Motel, gunite with plaster finish, incl. medium									
1150	capacity filtration & chlorination	B-52	115	.487	SF Surf	59	21.50	5.20	85.70	104
1200	Municipal, gunite with plaster finish, incl. high									
1250	capacity filtration & chlorination	B-52	100	.560	SF Surf	76	25	6	107	129
1350	Add for formed gutters				L.F.	112			112	123
1360	Add for stainless steel gutters				"	330			330	365
1700	Filtration and deck equipment only, as % of total				Total				20%	20%
1800	Deck equipment, rule of thumb, 20' x 40' pool				SF Pool				1.18	1.30
1900	5000 S.F. pool				"				1.73	1.90
3000	Painting pools, preparation + 3 coats, 20' x 40' pool, epoxy	2 Pord	.33	48.485	Total	1,725	1,950		3,675	4,850
3100	Rubber base paint, 18 gallons	"	.33	48.485		1,250	1,950		3,200	4,325
3500	42' x 82' pool, 75 gallons, epoxy paint	3 Pord	.14	171		7,300	6,925		14,225	18,500
3600	Rubber base paint	"	.14	171	↓	5,225	6,925		12,150	16,100

13 11 23 – On-Grade Swimming Pools

13 11 23.50 Swimming Pools

		Crew	Daily Output	Labor-Hours	Unit	Material	2016 Bare Costs Labor	Equipment	Total	Total Incl O&P
0010	**SWIMMING POOLS** Residential above ground, steel construction									
0100	Round, 15' dia.	B-80A	3	8	Ea.	845	305	102	1,252	1,500
0120	18' dia.		2.50	9.600		915	365	123	1,403	1,700
0140	21' dia.		2	12		1,050	455	154	1,659	2,050
0160	24' dia.		1.80	13.333		1,150	505	171	1,826	2,250
0180	27' dia.		1.50	16		1,350	605	205	2,160	2,625
0200	30' dia.		1	24		1,500	910	305	2,715	3,400
0220	Oval, 12' x 24'		2.30	10.435		1,575	395	134	2,104	2,475
0240	15' x 30'		1.80	13.333		1,750	505	171	2,426	2,900
0260	18' x 33'	↓	1	24		1,975	910	305	3,190	3,925

13 11 46 – Swimming Pool Accessories

13 11 46.50 Swimming Pool Equipment

		Crew	Daily Output	Labor-Hours	Unit	Material	2016 Bare Costs Labor	Equipment	Total	Total Incl O&P
0010	**SWIMMING POOL EQUIPMENT**									
0020	Diving stand, stainless steel, 3 meter	2 Carp	.40	40	Ea.	15,600	1,950		17,550	20,200
0300	1 meter	"	2.70	5.926		9,500	287		9,787	10,900
2100	Lights, underwater, 12 volt, with transformer, 300 watt	1 Elec	1	8		305	440		745	1,000
2200	110 volt, 500 watt, standard		1	8		293	440		733	980
2400	Low water cutoff type	↓	1	8	↓	293	440		733	980
2800	Heaters, see Section 23 52 28.10									

13 18 Ice Rinks

13 18 13 – Ice Rink Floor Systems

13 18 13.50 Ice Skating

	Crew	Daily Output	Labor-Hours	Unit	Material	2016 Bare Costs Labor	Equipment	Total	Total Incl O&P
0010 **ICE SKATING** Equipment incl. refrigeration, plumbing & cooling									
0020 coils & concrete slab, 85' x 200' rink									
0300 55° system, 5 mos., 100 ton				Total	604,000			604,000	664,000
0700 90° system, 12 mos., 135 ton				"	682,500			682,500	751,000
1200 Subsoil heating system (recycled from compressor), 85' x 200'	Q-7	.27	118	Ea.	42,000	6,850		48,850	56,500
1300 Subsoil insulation, 2 lb. polystyrene with vapor barrier, 85' x 200'	2 Carp	.14	114	"	31,500	5,525		37,025	43,200

13 18 16 – Ice Rink Dasher Boards

13 18 16.50 Ice Rink Dasher Boards

	Crew	Daily Output	Labor-Hours	Unit	Material	2016 Bare Costs Labor	Equipment	Total	Total Incl O&P
0010 **ICE RINK DASHER BOARDS**									
1000 Dasher boards, 1/2" H.D. polyethylene faced steel frame, 3' acrylic									
1020 screen at sides, 5' acrylic ends, 85' x 200'	F-5	.06	533	Ea.	142,000	26,100		168,100	196,000
1100 Fiberglass & aluminum construction, same sides and ends	"	.06	533	"	163,000	26,100		189,100	219,000

13 21 Controlled Environment Rooms

13 21 13 – Clean Rooms

13 21 13.50 Clean Room Components

	Crew	Daily Output	Labor-Hours	Unit	Material	2016 Bare Costs Labor	Equipment	Total	Total Incl O&P
0010 **CLEAN ROOM COMPONENTS**									
1100 Clean room, soft wall, 12' x 12', Class 100	1 Carp	.18	44.444	Ea.	18,000	2,150		20,150	23,100
1110 Class 1,000		.18	44.444		16,000	2,150		18,150	21,000
1120 Class 10,000		.21	38.095		13,200	1,850		15,050	17,300
1130 Class 100,000		.21	38.095		12,100	1,850		13,950	16,100
2800 Ceiling grid support, slotted channel struts 4'-0" O.C., ea. way				S.F.				5.90	6.50
3000 Ceiling panel, vinyl coated foil on mineral substrate									
3020 Sealed, non-perforated				S.F.				1.27	1.40
4000 Ceiling panel seal, silicone sealant, 150 L.F./gal.	1 Carp	150	.053	L.F.	.35	2.58		2.93	4.35
4100 Two sided adhesive tape	"	240	.033	"	.12	1.61		1.73	2.61
4200 Clips, one per panel				Ea.	1.02			1.02	1.12
6000 HEPA filter, 2' x 4', 99.97% eff., 3" dp beveled frame (silicone seal)					405			405	445
6040 6" deep skirted frame (channel seal)					465			465	510
6100 99.99% efficient, 3" deep beveled frame (silicone seal)					415			415	455
6140 6" deep skirted frame (channel seal)					470			470	520
6200 99.999% efficient, 3" deep beveled frame (silicone seal)					470			470	515
6240 6" deep skirted frame (channel seal)					505			505	555
7000 Wall panel systems, including channel strut framing									
7020 Polyester coated aluminum, particle board				S.F.				18.20	18.20
7100 Porcelain coated aluminum, particle board								32	32
7400 Wall panel support, slotted channel struts, to 12' high								16.35	16.35

13 21 26 – Cold Storage Rooms

13 21 26.50 Refrigeration

	Crew	Daily Output	Labor-Hours	Unit	Material	2016 Bare Costs Labor	Equipment	Total	Total Incl O&P
0010 **REFRIGERATION**									
0020 Curbs, 12" high, 4" thick, concrete	2 Carp	58	.276	L.F.	5.15	13.35		18.50	26
1000 Doors, see Section 08 34 13.10									
2400 Finishes, 2 coat Portland cement plaster, 1/2" thick	1 Plas	48	.167	S.F.	1.40	7.50		8.90	12.75
2500 For galvanized reinforcing mesh, add	1 Lath	335	.024		.95	1.13		2.08	2.72
2700 3/16" thick latex cement	1 Plas	88	.091		2.40	4.08		6.48	8.75
2900 For glass cloth reinforced ceilings, add	"	450	.018		.60	.80		1.40	1.85
3100 Fiberglass panels, 1/8" thick	1 Carp	149.45	.054		3.20	2.59		5.79	7.50
3200 Polystyrene, plastic finish ceiling, 1" thick		274	.029		2.95	1.41		4.36	5.40
3400 2" thick		274	.029		3.35	1.41		4.76	5.85
3500 4" thick		219	.037		3.70	1.77		5.47	6.80

For customer support on your Mechanical Cost Data, call 877.739.3689.

81

13 21 Controlled Environment Rooms

13 21 26 – Cold Storage Rooms

	13 21 26.50 Refrigeration	Crew	Daily Output	Labor-Hours	Unit	Material	2016 Bare Costs Labor	Equipment	Total	Total Incl O&P
3800	Floors, concrete, 4" thick	1 Cefi	93	.086	S.F.	1.39	3.93		5.32	7.35
3900	6" thick	"	85	.094		2.18	4.30		6.48	8.75
4000	Insulation, 1" to 6" thick, cork				B.F.	1.35			1.35	1.49
4100	Urethane					.51			.51	.56
4300	Polystyrene, regular					.56			.56	.62
4400	Bead board					.26			.26	.29
4600	Installation of above, add per layer	2 Carp	657.60	.024	S.F.	.45	1.18		1.63	2.31
4700	Wall and ceiling juncture		298.90	.054	L.F.	2.20	2.59		4.79	6.40
4900	Partitions, galvanized sandwich panels, 4" thick, stock		219.20	.073	S.F.	9.20	3.54		12.74	15.50
5000	Aluminum or fiberglass		219.20	.073	"	10.10	3.54		13.64	16.50
5200	Prefab walk-in, 7'-6" high, aluminum, incl. refrigeration, door & floor									
5210	not incl. partitions, 6' x 6'	2 Carp	54.80	.292	SF Flr.	164	14.15		178.15	203
5500	10' x 10'		82.20	.195		132	9.45		141.45	159
5700	12' x 14'		109.60	.146		119	7.05		126.05	142
5800	12' x 20'		109.60	.146		103	7.05		110.05	125
6100	For 8'-6" high, add					5%				
6300	Rule of thumb for complete units, w/o doors & refrigeration, cooler	2 Carp	146	.110		149	5.30		154.30	172
6400	Freezer		109.60	.146		176	7.05		183.05	205
6600	Shelving, plated or galvanized, steel wire type		360	.044	SF Hor.	13.05	2.15		15.20	17.65
6700	Slat shelf type		375	.043		16.10	2.07		18.17	21
6900	For stainless steel shelving, add					300%				
7000	Vapor barrier, on wood walls	2 Carp	1644	.010	S.F.	.20	.47		.67	.94
7200	On masonry walls	"	1315	.012	"	.50	.59		1.09	1.45
7500	For air curtain doors, see Section 23 34 33.10									

13 24 Special Activity Rooms

13 24 16 – Saunas

	13 24 16.50 Saunas and Heaters	Crew	Daily Output	Labor-Hours	Unit	Material	2016 Bare Costs Labor	Equipment	Total	Total Incl O&P
0010	**SAUNAS AND HEATERS**									
0020	Prefabricated, incl. heater & controls, 7' high, 6' x 4', C/C	L-7	2.20	12.727	Ea.	5,325	590		5,915	6,750
0050	6' x 4', C/P		2	14		4,800	650		5,450	6,275
0400	6' x 5', C/C		2	14		5,975	650		6,625	7,575
0450	6' x 5', C/P		2	14		5,425	650		6,075	6,950
0600	6' x 6', C/C		1.80	15.556		6,325	720		7,045	8,075
0650	6' x 6', C/P		1.80	15.556		5,775	720		6,495	7,450
0800	6' x 9', C/C		1.60	17.500		8,400	810		9,210	10,500
0850	6' x 9', C/P		1.60	17.500		7,650	810		8,460	9,650
1000	8' x 12', C/C		1.10	25.455		12,800	1,175		13,975	15,800
1050	8' x 12', C/P		1.10	25.455		11,600	1,175		12,775	14,500
1200	8' x 8', C/C		1.40	20		9,625	930		10,555	12,000
1250	8' x 8', C/P		1.40	20		8,875	930		9,805	11,200
1400	8' x 10', C/C		1.20	23.333		10,600	1,075		11,675	13,400
1450	8' x 10', C/P		1.20	23.333		9,700	1,075		10,775	12,400
1600	10' x 12', C/C		1	28		12,800	1,300		14,100	16,100
1650	10' x 12', C/P		1	28		11,600	1,300		12,900	14,700
2500	Heaters only (incl. above), wall mounted, to 200 C.F.					730			730	805
2750	To 300 C.F.					945			945	1,050
3000	Floor standing, to 720 C.F., 10,000 watts, w/controls	1 Elec	3	2.667		2,950	147		3,097	3,475
3250	To 1,000 C.F., 16,000 watts	"	3	2.667		3,900	147		4,047	4,500

For customer support on your Mechanical Cost Data, call 877.739.3689.

13 24 Special Activity Rooms

13 24 26 – Steam Baths

	13 24 26.50 Steam Baths and Components	Crew	Daily Output	Labor-Hours	Unit	Material	2016 Bare Costs Labor	Equipment	Total	Total Incl O&P
0010	**STEAM BATHS AND COMPONENTS**									
0020	Heater, timer & head, single, to 140 C.F.	1 Plum	1.20	6.667	Ea.	2,200	395		2,595	3,000
0500	To 300 C.F.		1.10	7.273		2,425	430		2,855	3,325
1000	Commercial size, with blow-down assembly, to 800 C.F.		.90	8.889		5,725	525		6,250	7,100
1500	To 2500 C.F.	↓	.80	10		7,650	590		8,240	9,325
2000	Multiple, motels, apts., 2 baths, w/blow-down assm., 500 C.F.	Q-1	1.30	12.308		6,325	655		6,980	7,975
2500	4 baths	"	.70	22.857	↓	10,200	1,225		11,425	13,100
2700	Conversion unit for residential tub, including door					3,550			3,550	3,925

13 34 Fabricated Engineered Structures

13 34 23 – Fabricated Structures

13 34 23.10 Comfort Stations

		Crew	Daily Output	Labor-Hours	Unit	Material	2016 Bare Costs Labor	Equipment	Total	Total Incl O&P
0010	**COMFORT STATIONS** Prefab., stock, w/doors, windows & fixt.									
0100	Not incl. interior finish or electrical									
0300	Mobile, on steel frame, 2 unit				S.F.	178			178	196
0350	7 unit					295			295	325
0400	Permanent, including concrete slab, 2 unit	B-12J	50	.320		249	14.45	17.65	281.10	315
0500	6 unit	"	43	.372	↓	191	16.75	20.50	228.25	259
0600	Alternate pricing method, mobile, 2 fixture				Fixture	6,175			6,175	6,775
0650	7 fixture					10,600			10,600	11,700
0700	Permanent, 2 unit	B-12J	.70	22.857		20,500	1,025	1,250	22,775	25,600
0750	6 unit	"	.50	32	↓	17,400	1,450	1,775	20,625	23,400

13 34 23.16 Fabricated Control Booths

		Crew	Daily Output	Labor-Hours	Unit	Material	2016 Bare Costs Labor	Equipment	Total	Total Incl O&P
0010	**FABRICATED CONTROL BOOTHS**									
0100	Guard House, prefab conc. w/bullet resistant doors & windows, roof & wiring									
0110	8' x 8', Level III	L-10	1	24	Ea.	51,000	1,275	660	52,935	59,000
0120	8' x 8', Level IV	"	1	24	"	58,000	1,275	660	59,935	67,000

13 47 Facility Protection

13 47 13 – Cathodic Protection

13 47 13.16 Cathodic Prot. for Underground Storage Tanks

		Crew	Daily Output	Labor-Hours	Unit	Material	2016 Bare Costs Labor	Equipment	Total	Total Incl O&P
0010	**CATHODIC PROTECTION FOR UNDERGROUND STORAGE TANKS**									
1000	Anodes, magnesium type, 9 #	R-15	18.50	2.595	Ea.	37.50	141	16.70	195.20	270
1010	17 #		13	3.692		76.50	200	24	300.50	410
1020	32 #		10	4.800		121	260	31	412	555
1030	48 #	↓	7.20	6.667		166	360	43	569	770
1100	Graphite type w/epoxy cap, 3" x 60" (32 #)	R-22	8.40	4.438		128	224		352	475
1110	4" x 80" (68 #)		6	6.213		241	315		556	735
1120	6" x 72" (80 #)		5.20	7.169		1,500	360		1,860	2,200
1130	6" x 36" (45 #)	↓	9.60	3.883		760	196		956	1,125
2000	Rectifiers, silicon type, air cooled, 28 V/10 A	R-19	3.50	5.714		2,300	315		2,615	3,025
2010	20 V/20 A		3.50	5.714		2,375	315		2,690	3,075
2100	Oil immersed, 28 V/10 A		3	6.667		3,175	370		3,545	4,025
2110	20 V/20 A	↓	3	6.667	↓	3,175	370		3,545	4,050
3000	Anode backfill, coke breeze	R-22	3850	.010	Lb.	.22	.49		.71	.97
4000	Cable, HMWPE, No. 8		2.40	15.533	M.L.F.	500	785		1,285	1,725
4010	No. 6		2.40	15.533		745	785		1,530	2,000
4020	No. 4		2.40	15.533		1,125	785		1,910	2,425

For customer support on your Mechanical Cost Data, call 877.739.3689.

83

13 47 Facility Protection

13 47 13 – Cathodic Protection

13 47 13.16 Cathodic Prot. for Underground Storage Tanks	Crew	Daily Output	Labor-Hours	Unit	Material	2016 Bare Costs Labor	Equipment	Total	Total Incl O&P
4030	No. 2	R-22	2.40	15.533	M.L.F.	1,750	785		2,535	3,100
4040	No. 1		2.20	16.945		2,400	855		3,255	3,925
4050	No. 1/0		2.20	16.945		2,975	855		3,830	4,550
4060	No. 2/0		2.20	16.945		4,625	855		5,480	6,375
4070	No. 4/0		2	18.640		6,000	940		6,940	8,000
5000	Test station, 7 terminal box, flush curb type w/lockable cover	R-19	12	1.667	Ea.	71.50	92		163.50	217
5010	Reference cell, 2" dia PVC conduit, cplg., plug, set flush	"	4.80	4.167	"	151	230		381	510

Estimating Tips

General

- Many products in Division 14 will require some type of support or blocking for installation not included with the item itself. Examples are supports for conveyors or tube systems, attachment points for lifts, and footings for hoists or cranes. Add these supports in the appropriate division.

14 10 00 Dumbwaiters
14 20 00 Elevators

- Dumbwaiters and elevators are estimated and purchased in a method similar to buying a car. The manufacturer has a base unit with standard features. Added to this base unit price will be whatever options the owner or specifications require. Increased load capacity, additional vertical travel, additional stops, higher speed, and cab finish options are items to be considered. When developing an estimate for dumbwaiters and elevators, remember that some items needed by the installers may have to be included as part of the general contract.

Examples are:

- ☐ shaftway
- ☐ rail support brackets
- ☐ machine room
- ☐ electrical supply
- ☐ sill angles
- ☐ electrical connections
- ☐ pits
- ☐ roof penthouses
- ☐ pit ladders

Check the job specifications and drawings before pricing.

- Installation of elevators and handicapped lifts in historic structures can require significant additional costs. The associated structural requirements may involve cutting into and repairing finishes, moldings, flooring, etc. The estimator must account for these special conditions.

14 30 00 Escalators and Moving Walks

- Escalators and moving walks are specialty items installed by specialty contractors. There are numerous options associated with these items. For specific options, contact a manufacturer or contractor. In a method similar to estimating dumbwaiters and elevators, you should verify the extent of general contract work and add items as necessary.

14 40 00 Lifts
14 90 00 Other Conveying Equipment

- Products such as correspondence lifts, chutes, and pneumatic tube systems, as well as other items specified in this subdivision, may require trained installers. The general contractor might not have any choice as to who will perform the installation, or when it will be performed. Long lead times are often required for these products, making early decisions in scheduling necessary.

Reference Numbers

Reference numbers are shown at the beginning of some major classifications. These numbers refer to related items in the Reference Section. The reference information may be an estimating procedure, an alternate pricing method, or technical information.

Note: Not all subdivisions listed here necessarily appear. ■

---

14 91 Facility Chutes

14 91 33 - Laundry and Linen Chutes

14 91 33.10 Chutes	Crew	Daily Output	Labor-Hours	Unit	Material	2016 Bare Costs Labor	Equipment	Total	Total Incl O&P
0011	**CHUTES**, linen, trash or refuse									
0050	Aluminized steel, 16 ga., 18" diameter	2 Shee	3.50	4.571	Floor	1,700	262		1,962	2,275
0100	24" diameter		3.20	5		1,775	286		2,061	2,400
0200	30" diameter		3	5.333		2,125	305		2,430	2,800
0300	36" diameter		2.80	5.714		2,625	325		2,950	3,400
0400	Galvanized steel, 16 ga., 18" diameter		3.50	4.571		1,000	262		1,262	1,500
0500	24" diameter		3.20	5		1,125	286		1,411	1,700
0600	30" diameter		3	5.333		1,275	305		1,580	1,875
0700	36" diameter		2.80	5.714		1,500	325		1,825	2,150
0800	Stainless steel, 18" diameter		3.50	4.571		3,000	262		3,262	3,700
0900	24" diameter		3.20	5		3,150	286		3,436	3,900
1000	30" diameter		3	5.333		3,750	305		4,055	4,600
1005	36" diameter		2.80	5.714		3,950	325		4,275	4,825
1200	Linen chute bottom collector, aluminized steel		4	4	Ea.	1,400	229		1,629	1,900
1300	Stainless steel		4	4		1,800	229		2,029	2,325
1500	Refuse, bottom hopper, aluminized steel, 18" diameter		3	5.333		1,025	305		1,330	1,600
1600	24" diameter		3	5.333		1,250	305		1,555	1,850
1800	36" diameter		3	5.333		2,500	305		2,805	3,225

14 91 82 - Trash Chutes

14 91 82.10 Trash Chutes and Accessories

		Crew	Daily Output	Labor-Hours	Unit	Material	Labor	Equipment	Total	Total Incl O&P
0010	**TRASH CHUTES AND ACCESSORIES**									
2900	Package chutes, spiral type, minimum	2 Shee	4.50	3.556	Floor	2,425	204		2,629	2,975
3000	Maximum	"	1.50	10.667	"	6,325	610		6,935	7,900

14 92 Pneumatic Tube Systems

14 92 10 - Conventional, Automatic and Computer Controlled Pneumatic Tube Systems

14 92 10.10 Pneumatic Tube Systems

		Crew	Daily Output	Labor-Hours	Unit	Material	Labor	Equipment	Total	Total Incl O&P
0010	**PNEUMATIC TUBE SYSTEMS**									
0020	100' long, single tube, 2 stations, stock									
0100	3" diameter	2 Stpi	.12	133	Total	3,375	8,100		11,475	15,900
0300	4" diameter	"	.09	177	"	4,275	10,800		15,075	21,000
0400	Twin tube, two stations or more, conventional system									
0600	2-1/2" round	2 Stpi	62.50	.256	L.F.	38	15.55		53.55	65.50
0700	3" round		46	.348		38	21		59	74
0900	4" round		49.60	.323		48	19.60		67.60	82.50
1000	4" x 7" oval		37.60	.426		89	26		115	137
1050	Add for blower		2	8	System	5,200	485		5,685	6,450
1110	Plus for each round station, add		7.50	2.133	Ea.	1,350	129		1,479	1,700
1150	Plus for each oval station, add		7.50	2.133	"	1,350	129		1,479	1,700
1200	Alternate pricing method: base cost, economy model		.75	21.333	Total	5,750	1,300		7,050	8,275
1300	Custom model		.25	64	"	11,500	3,875		15,375	18,600
1500	Plus total system length, add, for economy model		93.40	.171	L.F.	8.25	10.40		18.65	25
1600	For custom model		37.60	.426	"	25	26		51	66.50
1800	Completely automatic system, 4" round, 15 to 50 stations		.29	55.172	Station	20,100	3,350		23,450	27,200
2200	51 to 144 stations		.32	50		15,600	3,025		18,625	21,800
2400	6" round or 4" x 7" oval, 15 to 50 stations		.24	66.667		25,200	4,050		29,250	33,800
2800	51 to 144 stations		.23	69.565		21,100	4,225		25,325	29,600

Estimating Tips

22 10 00 Plumbing
Piping and Pumps

This subdivision is primarily basic pipe and related materials. The pipe may be used by any of the mechanical disciplines, i.e., plumbing, fire protection, heating, and air conditioning.

Note: CPVC plastic piping approved for fire protection is located in 21 11 13.

- The labor adjustment factors listed in Subdivision 22 01 02.20 apply throughout Divisions 21, 22, and 23. CAUTION: the correct percentage may vary for the same items. For example, the percentage add for the basic pipe installation should be based on the maximum height that the craftsman must install for that particular section. If the pipe is to be located 14' above the floor but it is suspended on threaded rod from beams, the bottom flange of which is 18' high (4' rods), then the height is actually 18' and the add is 20%. The pipe coverer, however, does not have to go above the 14', and so the add should be 10%.

- Most pipe is priced first as straight pipe with a joint (coupling, weld, etc.) every 10' and a hanger usually every 10'. There are exceptions with hanger spacing such as for cast iron pipe (5') and plastic pipe (3 per 10'). Following each type of pipe there are several lines listing sizes and the amount to be subtracted to delete couplings and hangers. This is for pipe that is to be buried or supported together on trapeze hangers. The reason that the couplings are deleted is that these runs are usually long, and frequently longer lengths of pipe are used. By deleting the couplings, the estimator is expected to look up and add back the correct reduced number of couplings.

- When preparing an estimate, it may be necessary to approximate the fittings. Fittings usually run between 25% and 50% of the cost of the pipe. The lower percentage is for simpler runs, and the higher number is for complex areas, such as mechanical rooms.

- For historic restoration projects, the systems must be as invisible as possible, and pathways must be sought for pipes, conduit, and ductwork. While installations in accessible spaces (such as basements and attics) are relatively straightforward to estimate, labor costs may be more difficult to determine when delivery systems must be concealed.

22 40 00 Plumbing Fixtures

- Plumbing fixture costs usually require two lines: the fixture itself and its "rough-in, supply, and waste."

- In the Assemblies Section (Plumbing D2010) for the desired fixture, the System Components Group at the center of the page shows the fixture on the first line. The rest of the list (fittings, pipe, tubing, etc.) will total up to what we refer to in the Unit Price section as "Rough-in, supply, waste, and vent." Note that for most fixtures we allow a nominal 5' of tubing to reach from the fixture to a main or riser.

- Remember that gas- and oil-fired units need venting.

Reference Numbers

Reference numbers are shown at the beginning of some major classifications. These numbers refer to related items in the Reference Section. The reference information may be an estimating procedure, an alternate pricing method, or technical information.

Note: Not all subdivisions listed here necessarily appear. ∎

22 01 Operation and Maintenance of Plumbing

22 01 02 – Labor Adjustments

22 01 02.10 Boilers, General		Crew	Daily Output	Labor-Hours	Unit	Material	2016 Bare Costs Labor	Equipment	Total	Total Incl O&P
0010	**BOILERS, GENERAL**, Prices do not include flue piping, elec. wiring,									
0020	gas or oil piping, boiler base, pad, or tankless unless noted									
0100	Boiler H.P.: 10 KW = 34 lb./steam/hr. = 33,475 BTU/hr.									
0150	To convert SFR to BTU rating: Hot water, 150 x SFR;									
0160	Forced hot water, 180 x SFR; steam, 240 x SFR									

22 01 02.20 Labor Adjustment Factors		Crew	Daily Output	Labor-Hours	Unit	Material	2016 Bare Costs Labor	Equipment	Total	Total Incl O&P
0010	**LABOR ADJUSTMENT FACTORS** (For Div. 21, 22 and 23) R220102-20									
0100	Labor factors, The below are reasonable suggestions, however									
0110	each project must be evaluated for its own peculiarities, and									
0120	the adjustments be increased or decreased depending on the									
0130	severity of the special conditions.									
1000	Add to labor for elevated installation (Above floor level)									
1080	10' to 14.5' high R221113-70						10%			
1100	15' to 19.5' high						20%			
1120	20' to 24.5' high						25%			
1140	25' to 29.5' high						35%			
1160	30' to 34.5' high						40%			
1180	35' to 39.5' high						50%			
1200	40' and higher						55%			
2000	Add to labor for crawl space									
2100	3' high						40%			
2140	4' high						30%			
3000	Add to labor for multi-story building									
3010	For new construction (No elevator available)									
3100	Add for floors 3 thru 10						5%			
3110	Add for floors 11 thru 15						10%			
3120	Add for floors 16 thru 20						15%			
3130	Add for floors 21 thru 30						20%			
3140	Add for floors 31 and up						30%			
3170	For existing structure (Elevator available)									
3180	Add for work on floor 3 and above						2%			
4000	Add to labor for working in existing occupied buildings									
4100	Hospital						35%			
4140	Office building						25%			
4180	School						20%			
4220	Factory or warehouse						15%			
4260	Multi dwelling						15%			
5000	Add to labor, miscellaneous									
5100	Cramped shaft						35%			
5140	Congested area						15%			
5180	Excessive heat or cold						30%			
9000	Labor factors, The above are reasonable suggestions, however									
9010	each project should be evaluated for its own peculiarities.									
9100	Other factors to be considered are:									
9140	Movement of material and equipment through finished areas									
9180	Equipment room									
9220	Attic space									
9260	No service road									
9300	Poor unloading/storage area									
9340	Congested site area/heavy traffic									

22 05 Common Work Results for Plumbing

22 05 05 – Selective Demolition for Plumbing

22 05 05.10 Plumbing Demolition		Crew	Daily Output	Labor-Hours	Unit	Material	2016 Bare Costs Labor	Equipment	Total	Total Incl O&P
0010	**PLUMBING DEMOLITION** R220105-10									
0400	Air compressor, up thru 2 H.P.	Q-1	10	1.600	Ea.		85.50		85.50	129
0410	3 H.P. thru 7-1/2 H.P. R024119-10		5.60	2.857			152		152	230
0420	10 H.P. thru 15 H.P.	↓	1.40	11.429			610		610	920
0430	20 H.P. thru 30 H.P.	Q-2	1.30	18.462			1,025		1,025	1,550
0500	Backflow preventer, up thru 2" diameter	1 Plum	17	.471			28		28	42
0510	2-1/2" thru 3" diameter	Q-1	10	1.600			85.50		85.50	129
0520	4" thru 6" diameter	"	5	3.200			171		171	257
0530	8" thru 10" diameter	Q-2	3	8			440		440	665
1900	Piping fittings, single connection, up thru 1-1/2" diameter	1 Plum	30	.267			15.80		15.80	24
1910	2" thru 4" diameter		14	.571			34		34	51
1980	Pipe hanger/support removal		80	.100	↓		5.90		5.90	8.95
2000	Piping, metal, up thru 1-1/2" diameter		200	.040	L.F.		2.37		2.37	3.57
2050	2" thru 3-1/2" diameter	↓	150	.053			3.16		3.16	4.77
2100	4" thru 6" diameter	2 Plum	100	.160			9.45		9.45	14.30
2150	8" thru 14" diameter	"	60	.267			15.80		15.80	24
2153	16" thru 20" diameter	Q-18	70	.343			19.40	.83	20.23	30.50
2155	24" thru 26" diameter		55	.436			24.50	1.06	25.56	38.50
2156	30" thru 36" diameter	↓	40	.600			34	1.45	35.45	53
2160	Plastic pipe with fittings, up thru 1-1/2" diameter	1 Plum	250	.032			1.89		1.89	2.86
2162	2" thru 3" diameter	"	200	.040			2.37		2.37	3.57
2164	4" thru 6" diameter	Q-1	200	.080			4.26		4.26	6.45
2166	8" thru 14" diameter		150	.107			5.70		5.70	8.60
2168	16" diameter		100	.160	↓		8.50		8.50	12.85
2180	Pumps, all fractional horse-power		12	1.333	Ea.		71		71	107
2184	1 H.P. thru 5 H.P.		6	2.667			142		142	214
2186	7-1/2 H.P. thru 15 H.P.	↓	2.50	6.400			340		340	515
2188	20 H.P. thru 25 H.P.	Q-2	4	6			330		330	500
2190	30 H.P. thru 60 H.P.		.80	30			1,650		1,650	2,500
2192	75 H.P. thru 100 H.P.		.60	40			2,200		2,200	3,325
2194	150 H.P.	↓	.50	48	↓		2,650		2,650	4,000
2230	Temperature maintenance cable	1 Plum	1200	.007	L.F.		.39		.39	.60
3100	Tanks, water heaters and liquid containers									
3110	Up thru 45 gallons	Q-1	22	.727	Ea.		39		39	58.50
3120	50 thru 120 gallons		14	1.143			61		61	92
3130	130 thru 240 gallons		7.60	2.105			112		112	169
3140	250 thru 500 gallons	↓	5.40	2.963			158		158	238
3150	600 thru 1000 gallons	Q-2	1.60	15			830		830	1,250
3160	1100 thru 2000 gallons		.70	34.286			1,900		1,900	2,850
3170	2100 thru 4000 gallons	↓	.50	48			2,650		2,650	4,000
9100	Valve, metal valves or strainers and similar, up thru 1-1/2" diameter	1 Stpi	28	.286			17.35		17.35	26
9110	2" thru 3" diameter	Q-1	11	1.455			77.50		77.50	117
9120	4" thru 6" diameter	"	8	2			107		107	161
9130	8" thru 14" diameter	Q-2	8	3			166		166	250
9140	16" thru 20" diameter		2	12			665		665	1,000
9150	24" diameter	↓	1.20	20			1,100		1,100	1,675
9200	Valve, plastic, up thru 1-1/2" diameter	1 Plum	42	.190			11.30		11.30	17
9210	2" thru 3" diameter		15	.533			31.50		31.50	47.50
9220	4" thru 6" diameter		12	.667			39.50		39.50	59.50
9300	Vent flashing and caps	↓	55	.145			8.60		8.60	13
9350	Water filter, commercial, 1" thru 1-1/2"	Q-1	2	8			425		425	645
9360	2" thru 2-1/2"	"	1.60	10	↓		535		535	805
9400	Water heaters									

22 05 Common Work Results for Plumbing

22 05 05 – Selective Demolition for Plumbing

22 05 05.10 Plumbing Demolition

		Crew	Daily Output	Labor-Hours	Unit	Material	2016 Bare Costs Labor	2016 Bare Costs Equipment	Total	Total Incl O&P
9410	Up thru 245 GPH	Q-1	2.40	6.667	Ea.		355		355	535
9420	250 thru 756 GPH		1.60	10			535		535	805
9430	775 thru 1640 GPH	↓	.80	20			1,075		1,075	1,600
9440	1650 thru 4000 GPH	Q-2	.50	48			2,650		2,650	4,000
9470	Water softener	Q-1	2	8	↓		425		425	645

22 05 23 – General-Duty Valves for Plumbing Piping

22 05 23.10 Valves, Brass

		Crew	Daily Output	Labor-Hours	Unit	Material	2016 Bare Costs Labor	2016 Bare Costs Equipment	Total	Total Incl O&P
0010	**VALVES, BRASS**									
0032	For motorized valves, see Section 23 09 53.10									
0500	Gas cocks, threaded									
0510	1/4"	1 Plum	26	.308	Ea.	14.70	18.20		32.90	43.50
0520	3/8"		24	.333		14.70	19.75		34.45	46
0530	1/2"		24	.333		13.15	19.75		32.90	44.50
0540	3/4"		22	.364		15.50	21.50		37	49.50
0550	1"		19	.421		31	25		56	71.50
0560	1-1/4"		15	.533		44.50	31.50		76	96.50
0570	1-1/2"		13	.615		61	36.50		97.50	122
0580	2"	↓	11	.727	↓	103	43		146	179
0672	For larger sizes use lubricated plug valve, Section 23 05 23.70									

22 05 23.20 Valves, Bronze

		Crew	Daily Output	Labor-Hours	Unit	Material	2016 Bare Costs Labor	2016 Bare Costs Equipment	Total	Total Incl O&P
0010	**VALVES, BRONZE** R220523-90									
1020	Angle, 150 lb., rising stem, threaded									
1030	1/8"	1 Plum	24	.333	Ea.	134	19.75		153.75	178
1040	1/4"		24	.333		134	19.75		153.75	178
1050	3/8"		24	.333		149	19.75		168.75	194
1060	1/2"		22	.364		149	21.50		170.50	197
1070	3/4"		20	.400		203	23.50		226.50	260
1080	1"		19	.421		293	25		318	365
1090	1-1/4"		15	.533		375	31.50		406.50	465
1100	1-1/2"	↓	13	.615	↓	495	36.50		531.50	595
1102	Soldered same price as threaded									
1110	2"	1 Plum	11	.727	Ea.	795	43		838	940
1300	Ball									
1304	Soldered									
1312	3/8"	1 Plum	21	.381	Ea.	14.95	22.50		37.45	50.50
1316	1/2"		18	.444		14.95	26.50		41.45	56
1320	3/4"		17	.471		25	28		53	69
1324	1"		15	.533		31	31.50		62.50	82
1328	1-1/4"		13	.615		52.50	36.50		89	113
1332	1-1/2"		11	.727		66.50	43		109.50	139
1336	2"		9	.889		83.50	52.50		136	172
1340	2-1/2"		7	1.143		395	67.50		462.50	535
1344	3"	↓	5	1.600	↓	455	94.50		549.50	645
1350	Single union end									
1358	3/8"	1 Plum	21	.381	Ea.	21.50	22.50		44	57.50
1362	1/2"		18	.444		22.50	26.50		49	64.50
1366	3/4"		17	.471		43	28		71	89.50
1370	1"		15	.533		52	31.50		83.50	105
1374	1-1/4"		13	.615		88.50	36.50		125	153
1378	1-1/2"		11	.727		111	43		154	187
1382	2"	↓	9	.889	↓	173	52.50		225.50	270
1398	Threaded, 150 psi									

22 05 23 – General-Duty Valves for Plumbing Piping

22 05 23.20 Valves, Bronze		Crew	Daily Output	Labor-Hours	Unit	Material	2016 Bare Costs Labor	Equipment	Total	Total Incl O&P
1400	1/4"	1 Plum	24	.333	Ea.	14.15	19.75		33.90	45.50
1430	3/8"		24	.333		14.15	19.75		33.90	45.50
1450	1/2"		22	.364		14.15	21.50		35.65	48
1460	3/4"		20	.400		23.50	23.50		47	61
1470	1"		19	.421		35	25		60	76
1480	1-1/4"		15	.533		61	31.50		92.50	115
1490	1-1/2"		13	.615		79.50	36.50		116	143
1500	2"		11	.727		96.50	43		139.50	171
1510	2-1/2"		9	.889		325	52.50		377.50	435
1520	3"	▼	8	1	▼	490	59		549	630
1522	Solder the same price as threaded									
1600	Butterfly, 175 psi, full port, solder or threaded ends									
1610	Stainless steel disc and stem									
1620	1/4"	1 Plum	24	.333	Ea.	18.05	19.75		37.80	50
1630	3/8"		24	.333		18.05	19.75		37.80	50
1640	1/2"		22	.364		19.40	21.50		40.90	54
1650	3/4"		20	.400		31	23.50		54.50	69.50
1660	1"		19	.421		37.50	25		62.50	79
1670	1-1/4"		15	.533		60.50	31.50		92	114
1680	1-1/2"		13	.615		78	36.50		114.50	141
1690	2"	▼	11	.727	▼	97.50	43		140.50	172
1750	Check, swing, class 150, regrinding disc, threaded									
1800	1/8"	1 Plum	24	.333	Ea.	70	19.75		89.75	107
1830	1/4"		24	.333		70	19.75		89.75	107
1840	3/8"		24	.333		74.50	19.75		94.25	112
1850	1/2"		24	.333		79.50	19.75		99.25	118
1860	3/4"		20	.400		105	23.50		128.50	152
1870	1"		19	.421		152	25		177	205
1880	1-1/4"		15	.533		219	31.50		250.50	289
1890	1-1/2"		13	.615		255	36.50		291.50	335
1900	2"	▼	11	.727		375	43		418	480
1910	2-1/2"	Q-1	15	1.067		840	57		897	1,000
1920	3"	"	13	1.231	▼	1,125	65.50		1,190.50	1,325
2000	For 200 lb., add					5%	10%			
2040	For 300 lb., add					15%	15%			
2060	Check swing, 300#, lead free unless noted, sweat, 3/8" size	1 Plum	24	.333	Ea.	123	19.75		142.75	166
2070	1/2"		24	.333		123	19.75		142.75	166
2080	3/4"		20	.400		166	23.50		189.50	219
2090	1"		19	.421		244	25		269	305
2100	1-1/4"		15	.533		345	31.50		376.50	430
2110	1-1/2"		13	.615		405	36.50		441.50	500
2120	2"		11	.727		595	43		638	720
2130	2-1/2", not lead free	Q-1	15	1.067		865	57		922	1,025
2140	3", not lead free	"	13	1.231	▼	1,200	65.50		1,265.50	1,425
2350	Check, lift, class 150, horizontal composition disc, threaded									
2430	1/4"	1 Plum	24	.333	Ea.	158	19.75		177.75	204
2440	3/8"		24	.333		200	19.75		219.75	250
2450	1/2"		24	.333		174	19.75		193.75	221
2460	3/4"		20	.400		213	23.50		236.50	270
2470	1"		19	.421		305	25		330	375
2480	1-1/4"		15	.533		405	31.50		436.50	495
2490	1-1/2"		13	.615		485	36.50		521.50	590
2500	2"	▼	11	.727	▼	810	43		853	955

For customer support on your Mechanical Cost Data, call 877.739.3689.

91

22 05 23 – General-Duty Valves for Plumbing Piping

22 05 23.20 Valves, Bronze		Crew	Daily Output	Labor-Hours	Unit	Material	2016 Bare Costs Labor	Equipment	Total	Total Incl O&P
2850	Gate, N.R.S., soldered, 125 psi									
2900	3/8"	1 Plum	24	.333	Ea.	62.50	19.75		82.25	99
2920	1/2"		24	.333		56.50	19.75		76.25	92
2940	3/4"		20	.400		66	23.50		89.50	108
2950	1"		19	.421		81.50	25		106.50	128
2960	1-1/4"		15	.533		135	31.50		166.50	196
2970	1-1/2"		13	.615		152	36.50		188.50	222
2980	2"		11	.727		198	43		241	282
2990	2-1/2"	Q-1	15	1.067		490	57		547	620
3000	3"	"	13	1.231		605	65.50		670.50	765
3350	Threaded, class 150									
3410	1/4"	1 Plum	24	.333	Ea.	79	19.75		98.75	117
3420	3/8"		24	.333		79	19.75		98.75	117
3430	1/2"		24	.333		76.50	19.75		96.25	115
3440	3/4"		20	.400		86	23.50		109.50	130
3450	1"		19	.421		108	25		133	156
3460	1-1/4" size		15	.533		146	31.50		177.50	208
3470	1-1/2"		13	.615		209	36.50		245.50	284
3480	2"		11	.727		249	43		292	340
3490	2-1/2"	Q-1	15	1.067		635	57		692	785
3500	3" size	"	13	1.231		900	65.50		965.50	1,100
3600	Gate, flanged, 150 lb.									
3610	1"	1 Plum	7	1.143	Ea.	1,150	67.50		1,217.50	1,350
3620	1-1/2"		6	1.333		1,425	79		1,504	1,700
3630	2"		5	1.600		2,100	94.50		2,194.50	2,450
3634	2-1/2"	Q-1	5	3.200		3,250	171		3,421	3,825
3640	3"	"	4.50	3.556		3,775	189		3,964	4,425
3850	Rising stem, soldered, 300 psi									
3900	3/8"	1 Plum	24	.333	Ea.	141	19.75		160.75	185
3920	1/2"		24	.333		116	19.75		135.75	157
3940	3/4"		20	.400		132	23.50		155.50	181
3950	1"		19	.421		179	25		204	235
3960	1-1/4"		15	.533		247	31.50		278.50	320
3970	1-1/2"		13	.615		300	36.50		336.50	385
3980	2"		11	.727		480	43		523	595
3990	2-1/2"	Q-1	15	1.067		1,025	57		1,082	1,200
4000	3"	"	13	1.231		1,575	65.50		1,640.50	1,850
4250	Threaded, class 150									
4310	1/4"	1 Plum	24	.333	Ea.	72.50	19.75		92.25	110
4320	3/8"		24	.333		72.50	19.75		92.25	110
4330	1/2"		24	.333		66	19.75		85.75	103
4340	3/4"		20	.400		77	23.50		100.50	121
4350	1"		19	.421		104	25		129	152
4360	1-1/4"		15	.533		141	31.50		172.50	203
4370	1-1/2"		13	.615		178	36.50		214.50	250
4380	2"		11	.727		239	43		282	330
4390	2-1/2"	Q-1	15	1.067		560	57		617	700
4400	3"	"	13	1.231		775	65.50		840.50	955
4500	For 300 psi, threaded, add					100%	15%			
4540	For chain operated type, add					15%				
4850	Globe, class 150, rising stem, threaded									
4920	1/4"	1 Plum	24	.333	Ea.	102	19.75		121.75	142
4940	3/8"		24	.333		101	19.75		120.75	141

22 05 Common Work Results for Plumbing

22 05 23 – General-Duty Valves for Plumbing Piping

22 05 23.20 Valves, Bronze		Crew	Daily Output	Labor-Hours	Unit	Material	2016 Bare Costs Labor	Equipment	Total	Total Incl O&P
4950	1/2"	1 Plum	24	.333	Ea.	101	19.75		120.75	141
4960	3/4"		20	.400		103	23.50		126.50	150
4970	1"		19	.421		161	25		186	215
4980	1-1/4"		15	.533		256	31.50		287.50	330
4990	1-1/2"		13	.615		310	36.50		346.50	395
5000	2"		11	.727		465	43		508	580
5010	2-1/2"	Q-1	15	1.067		1,225	57		1,282	1,425
5020	3"	"	13	1.231		1,750	65.50		1,815.50	2,025
5120	For 300 lb. threaded, add					50%	15%			
5130	Globe, 300 lb., sweat, 3/8" size	1 Plum	24	.333	Ea.	127	19.75		146.75	170
5140	1/2"		24	.333		131	19.75		150.75	174
5150	3/4"		20	.400		178	23.50		201.50	231
5160	1"		19	.421		246	25		271	310
5170	1-1/4"		15	.533		390	31.50		421.50	480
5180	1-1/2"		13	.615		470	36.50		506.50	570
5190	2"		11	.727		695	43		738	830
5200	2-1/2"	Q-1	15	1.067		1,350	57		1,407	1,575
5210	3"	"	13	1.231		1,750	65.50		1,815.50	2,025
5600	Relief, pressure & temperature, self-closing, ASME, threaded									
5640	3/4"	1 Plum	28	.286	Ea.	206	16.90		222.90	253
5650	1"		24	.333		330	19.75		349.75	395
5660	1-1/4"		20	.400		655	23.50		678.50	755
5670	1-1/2"		18	.444		1,300	26.50		1,326.50	1,475
5680	2"		16	.500		1,325	29.50		1,354.50	1,500
5950	Pressure, poppet type, threaded									
6000	1/2"	1 Plum	30	.267	Ea.	72	15.80		87.80	103
6040	3/4"	"	28	.286	"	79	16.90		95.90	112
6400	Pressure, water, ASME, threaded									
6440	3/4"	1 Plum	28	.286	Ea.	181	16.90		197.90	225
6450	1"		24	.333		278	19.75		297.75	335
6460	1-1/4"		20	.400		430	23.50		453.50	505
6470	1-1/2"		18	.444		610	26.50		636.50	710
6480	2"		16	.500		880	29.50		909.50	1,025
6490	2-1/2"		15	.533		3,450	31.50		3,481.50	3,850
6900	Reducing, water pressure									
6920	300 psi to 25-75 psi, threaded or sweat									
6940	1/2"	1 Plum	24	.333	Ea.	425	19.75		444.75	495
6950	3/4"		20	.400		435	23.50		458.50	510
6960	1"		19	.421		670	25		695	780
6970	1-1/4"		15	.533		1,175	31.50		1,206.50	1,350
6980	1-1/2"		13	.615		1,775	36.50		1,811.50	2,000
6990	2"		11	.727		2,625	43		2,668	2,950
7100	For built-in by-pass or 10-35 psi, add					45.50			45.50	50
7700	High capacity, 250 psi to 25-75 psi, threaded									
7740	1/2"	1 Plum	24	.333	Ea.	640	19.75		659.75	735
7780	3/4"		20	.400		645	23.50		668.50	745
7790	1"		19	.421		850	25		875	975
7800	1-1/4"		15	.533		1,475	31.50		1,506.50	1,675
7810	1-1/2"		13	.615		2,150	36.50		2,186.50	2,425
7820	2"		11	.727		3,125	43		3,168	3,525
7830	2-1/2"		9	.889		4,475	52.50		4,527.50	5,000
7840	3"		8	1		5,375	59		5,434	6,000
7850	3" flanged (iron body)	Q-1	10	1.600		5,300	85.50		5,385.50	5,950

For customer support on your Mechanical Cost Data, call 877.739.3689.

93

22 05 23.20 Valves, Bronze		Crew	Daily Output	Labor-Hours	Unit	Material	2016 Bare Costs Labor	Equipment	Total	Total Incl O&P
7860	4" flanged (iron body)	Q-1	8	2	Ea.	6,875	107		6,982	7,725
7920	For higher pressure, add					25%				
8000	Silent check, bronze trim									
8010	Compact wafer type, for 125 or 150 lb. flanges									
8020	1-1/2"	1 Plum	11	.727	Ea.	335	43		378	430
8021	2"	"	9	.889		355	52.50		407.50	470
8022	2-1/2"	Q-1	9	1.778		370	94.50		464.50	555
8023	3"		8	2		415	107		522	615
8024	4"	↓	5	3.200		710	171		881	1,025
8025	5"	Q-2	6	4		940	221		1,161	1,350
8026	6"		5	4.800		1,100	265		1,365	1,625
8027	8"		4.50	5.333		2,250	295		2,545	2,900
8028	10"		4	6		4,325	330		4,655	5,250
8029	12"	↓	3	8	↓	9,800	440		10,240	11,500
8050	For 250 or 300 lb. flanges, thru 6" no change									
8051	For 8" and 10", add				Ea.	40%	10%			
8060	Full flange wafer type, 150 lb.									
8061	1"	1 Plum	14	.571	Ea.	310	34		344	395
8062	1-1/4"		12	.667		355	39.50		394.50	450
8063	1-1/2"		11	.727		395	43		438	500
8064	2"	↓	9	.889		530	52.50		582.50	660
8065	2-1/2"	Q-1	9	1.778		595	94.50		689.50	800
8066	3"		8	2		660	107		767	885
8067	4"	↓	5	3.200		975	171		1,146	1,325
8068	5"	Q-2	6	4		1,250	221		1,471	1,700
8069	6"	"	5	4.800		1,650	265		1,915	2,225
8080	For 300 lb., add				↓	40%	10%			
8100	Globe type, 150 lb.									
8110	2"	1 Plum	9	.889	Ea.	690	52.50		742.50	840
8111	2-1/2"	Q-1	9	1.778		855	94.50		949.50	1,075
8112	3"	↓	8	2		1,025	107		1,132	1,275
8113	4"	↓	5	3.200		1,425	171		1,596	1,800
8114	5"	Q-2	6	4		1,725	221		1,946	2,225
8115	6"	"	5	4.800	↓	2,400	265		2,665	3,050
8130	For 300 lb., add					20%	10%			
8140	Screwed end type, 250 lb.									
8141	1/2"	1 Plum	24	.333	Ea.	62.50	19.75		82.25	98.50
8142	3/4"		20	.400		62.50	23.50		86	104
8143	1"		19	.421		71	25		96	116
8144	1-1/4"		15	.533		98.50	31.50		130	156
8145	1-1/2"		13	.615		110	36.50		146.50	176
8146	2"	↓	11	.727	↓	149	43		192	229
8350	Tempering, water, sweat connections									
8400	1/2"	1 Plum	24	.333	Ea.	102	19.75		121.75	142
8440	3/4"	"	20	.400	"	136	23.50		159.50	186
8650	Threaded connections									
8700	1/2"	1 Plum	24	.333	Ea.	151	19.75		170.75	196
8740	3/4"		20	.400		825	23.50		848.50	945
8750	1"		19	.421		930	25		955	1,075
8760	1-1/4"		15	.533		1,425	31.50		1,456.50	1,625
8770	1-1/2"		13	.615		1,575	36.50		1,611.50	1,775
8780	2"	↓	11	.727	↓	2,350	43		2,393	2,675

22 05 23 – General-Duty Valves for Plumbing Piping

22 05 23.60 Valves, Plastic		Crew	Daily Output	Labor-Hours	Unit	Material	2016 Bare Costs Labor	Equipment	Total	Total Incl O&P
0010	**VALVES, PLASTIC** R220523-90									
1100	Angle, PVC, threaded									
1110	1/4"	1 Plum	26	.308	Ea.	41	18.20		59.20	72.50
1120	1/2"		26	.308		59	18.20		77.20	92
1130	3/4"		25	.320		70	18.95		88.95	106
1140	1"		23	.348		84.50	20.50		105	124
1150	Ball, PVC, socket or threaded, true union									
1230	1/2"	1 Plum	26	.308	Ea.	42.50	18.20		60.70	74.50
1240	3/4"		25	.320		52	18.95		70.95	85.50
1250	1"		23	.348		62	20.50		82.50	99
1260	1-1/4"		21	.381		88	22.50		110.50	131
1270	1-1/2"		20	.400		98.50	23.50		122	144
1280	2"		17	.471		133	28		161	189
1290	2-1/2"	Q-1	26	.615		186	33		219	255
1300	3"		24	.667		256	35.50		291.50	335
1310	4"		20	.800		435	42.50		477.50	540
1360	For PVC, flanged, add					100%	15%			
1450	Double union 1/2"	1 Plum	26	.308		34	18.20		52.20	65
1460	3/4"		25	.320		44.50	18.95		63.45	77.50
1470	1"		23	.348		53	20.50		73.50	89
1480	1-1/4"		21	.381		73	22.50		95.50	115
1490	1-1/2"		20	.400		88	23.50		111.50	133
1500	2"		17	.471		116	28		144	170
1650	CPVC, socket or threaded, single union									
1700	1/2"	1 Plum	26	.308	Ea.	61	18.20		79.20	95
1720	3/4"		25	.320		77	18.95		95.95	113
1730	1"		23	.348		92.50	20.50		113	133
1750	1-1/4"		21	.381		147	22.50		169.50	196
1760	1-1/2"		20	.400		147	23.50		170.50	198
1770	2"		17	.471		203	28		231	265
1780	3"	Q-1	24	.667		800	35.50		835.50	935
1840	For CPVC, flanged, add					65%	15%			
1880	For true union, socket or threaded, add					50%	5%			
2050	Polypropylene, threaded									
2100	1/4"	1 Plum	26	.308	Ea.	36	18.20		54.20	67
2120	3/8"		26	.308		36	18.20		54.20	67
2130	1/2"		26	.308		36	18.20		54.20	67
2140	3/4"		25	.320		43	18.95		61.95	76
2150	1"		23	.348		49.50	20.50		70	85.50
2160	1-1/4"		21	.381		66.50	22.50		89	108
2170	1-1/2"		20	.400		82	23.50		105.50	126
2180	2"		17	.471		110	28		138	163
2190	3"	Q-1	24	.667		251	35.50		286.50	330
2200	4"	"	20	.800		435	42.50		477.50	545
2550	PVC, three way, socket or threaded									
2600	1/2"	1 Plum	26	.308	Ea.	51.50	18.20		69.70	84.50
2640	3/4"		25	.320		61.50	18.95		80.45	96
2650	1"		23	.348		74.50	20.50		95	113
2660	1-1/2"		20	.400		105	23.50		128.50	151
2670	2"		17	.471		133	28		161	189
2680	3"	Q-1	24	.667		425	35.50		460.50	520
2740	For flanged, add					60%	15%			

For customer support on your Mechanical Cost Data, call 877.739.3689.

95

22 05 Common Work Results for Plumbing

22 05 23 – General-Duty Valves for Plumbing Piping

22 05 23.60 Valves, Plastic		Crew	Daily Output	Labor-Hours	Unit	Material	2016 Bare Costs Labor	Equipment	Total	Total Incl O&P
3150	Ball check, PVC, socket or threaded									
3200	1/4"	1 Plum	26	.308	Ea.	48	18.20		66.20	80.50
3220	3/8"		26	.308		48	18.20		66.20	80.50
3240	1/2"		26	.308		48	18.20		66.20	80.50
3250	3/4"		25	.320		54	18.95		72.95	87.50
3260	1"		23	.348		67.50	20.50		88	105
3270	1-1/4"		21	.381		113	22.50		135.50	158
3280	1-1/2"		20	.400		113	23.50		136.50	160
3290	2"		17	.471		154	28		182	211
3310	3"	Q-1	24	.667		425	35.50		460.50	525
3320	4"	"	20	.800		605	42.50		647.50	730
3360	For PVC, flanged, add					50%	15%			
3750	CPVC, socket or threaded									
3800	1/2"	1 Plum	26	.308	Ea.	74	18.20		92.20	109
3840	3/4"		25	.320		88	18.95		106.95	126
3850	1"		23	.348		105	20.50		125.50	146
3860	1-1/2"		20	.400		180	23.50		203.50	234
3870	2"		17	.471		242	28		270	310
3880	3"	Q-1	24	.667		655	35.50		690.50	775
3920	4"	"	20	.800		885	42.50		927.50	1,050
3930	For CPVC, flanged, add					40%	15%			
4340	Polypropylene, threaded									
4360	1/2"	1 Plum	26	.308	Ea.	46	18.20		64.20	78
4400	3/4"		25	.320		63	18.95		81.95	98
4440	1"		23	.348		67	20.50		87.50	105
4450	1-1/2"		20	.400		129	23.50		152.50	178
4460	2"		17	.471		164	28		192	222
4500	For polypropylene flanged, add					200%	15%			
4850	Foot valve, PVC, socket or threaded									
4900	1/2"	1 Plum	34	.235	Ea.	56.50	13.95		70.45	83.50
4930	3/4"		32	.250		64.50	14.80		79.30	93
4940	1"		28	.286		83	16.90		99.90	117
4950	1-1/4"		27	.296		160	17.55		177.55	203
4960	1-1/2"		26	.308		160	18.20		178.20	204
4970	2"		24	.333		186	19.75		205.75	234
4980	3"		20	.400		445	23.50		468.50	525
4990	4"		18	.444		780	26.50		806.50	900
5000	For flanged, add					25%	10%			
5050	CPVC, socket or threaded									
5060	1/2"	1 Plum	34	.235	Ea.	74	13.95		87.95	103
5070	3/4"		32	.250		92.50	14.80		107.30	125
5080	1"		28	.286		113	16.90		129.90	151
5090	1-1/4"		27	.296		180	17.55		197.55	225
5100	1-1/2"		26	.308		180	18.20		198.20	226
5110	2"		24	.333		225	19.75		244.75	277
5120	3"		20	.400		455	23.50		478.50	540
5130	4"		18	.444		830	26.50		856.50	955
5140	For flanged, add					25%	10%			
5280	Needle valve, PVC, threaded									
5300	1/4"	1 Plum	26	.308	Ea.	58.50	18.20		76.70	91.50
5340	3/8"		26	.308		67.50	18.20		85.70	102
5360	1/2"		26	.308		67.50	18.20		85.70	102
5380	For polypropylene, add					10%				

96

For customer support on your Mechanical Cost Data, call 877.739.3689.

22 05 Common Work Results for Plumbing

22 05 23 – General-Duty Valves for Plumbing Piping

22 05 23.60 Valves, Plastic		Crew	Daily Output	Labor-Hours	Unit	Material	2016 Bare Costs Labor	Equipment	Total	Total Incl O&P
5800	Y check, PVC, socket or threaded									
5820	1/2"	1 Plum	26	.308	Ea.	64	18.20		82.20	97.50
5840	3/4"		25	.320		69	18.95		87.95	105
5850	1"		23	.348		75	20.50		95.50	114
5860	1-1/4"		21	.381		118	22.50		140.50	164
5870	1-1/2"		20	.400		129	23.50		152.50	177
5880	2"		17	.471		160	28		188	218
5890	2-1/2"	↓	15	.533		340	31.50		371.50	425
5900	3"	Q-1	24	.667		340	35.50		375.50	430
5910	4"	"	20	.800		555	42.50		597.50	675
5960	For PVC flanged, add				↓	45%	15%			
6350	Y sediment strainer, PVC, socket or threaded									
6400	1/2"	1 Plum	26	.308	Ea.	51	18.20		69.20	83.50
6440	3/4"		24	.333		54.50	19.75		74.25	90
6450	1"		23	.348		64.50	20.50		85	102
6460	1-1/4"		21	.381		109	22.50		131.50	153
6470	1-1/2"		20	.400		109	23.50		132.50	155
6480	2"		17	.471		134	28		162	190
6490	2-1/2"	↓	15	.533		325	31.50		356.50	410
6500	3"	Q-1	24	.667		325	35.50		360.50	415
6510	4"	"	20	.800		610	42.50		652.50	740
6560	For PVC, flanged, add				↓	55%	15%			

22 05 29 – Hangers and Supports for Plumbing Piping and Equipment

22 05 29.10 Hangers & Supp. for Plumb'g/HVAC Pipe/Equip.		Crew	Daily Output	Labor-Hours	Unit	Material	2016 Bare Costs Labor	Equipment	Total	Total Incl O&P
0010	**HANGERS AND SUPPORTS FOR PLUMB'G/HVAC PIPE/EQUIP.**									
0011	TYPE numbers per MSS-SP58									
0050	Brackets									
0060	Beam side or wall, malleable iron, TYPE 34									
0070	3/8" threaded rod size	1 Plum	48	.167	Ea.	3.14	9.85		12.99	18.35
0080	1/2" threaded rod size		48	.167		4.43	9.85		14.28	19.75
0090	5/8" threaded rod size		48	.167		8.30	9.85		18.15	24
0100	3/4" threaded rod size		48	.167		11.40	9.85		21.25	27.50
0110	7/8" threaded rod size	↓	48	.167		11.80	9.85		21.65	28
0120	For concrete installation, add				↓		30%			
0150	Wall, welded steel, medium, TYPE 32									
0160	0 size, 12" wide, 18" deep	1 Plum	34	.235	Ea.	173	13.95		186.95	212
0170	1 size, 18" wide, 24" deep		34	.235		213	13.95		226.95	255
0180	2 size, 24" wide, 30" deep	↓	34	.235	↓	293	13.95		306.95	340
0200	Beam attachment, welded, TYPE 22									
0202	3/8"	Q-15	80	.200	Ea.	7.20	10.65	.73	18.58	25
0203	1/2"		76	.211		7.35	11.20	.76	19.31	26
0204	5/8"		72	.222		8.55	11.85	.81	21.21	28
0205	3/4"		68	.235		12.40	12.55	.85	25.80	33.50
0206	7/8"		64	.250		15.95	13.30	.91	30.16	38.50
0207	1"	↓	56	.286	↓	32	15.20	1.04	48.24	59
0300	Clamps									
0310	C-clamp, for mounting on steel beam flange, w/locknut, TYPE 23									
0320	3/8" threaded rod size	1 Plum	160	.050	Ea.	2.07	2.96		5.03	6.75
0330	1/2" threaded rod size		160	.050		2.22	2.96		5.18	6.90
0340	5/8" threaded rod size		160	.050		3.74	2.96		6.70	8.60
0350	3/4" threaded rod size		160	.050		4.60	2.96		7.56	9.50
0352	7/8" threaded rod size		140	.057	↓	15.15	3.38		18.53	22

22 05 29 – Hangers and Supports for Plumbing Piping and Equipment

22 05 29.10 Hangers & Supp. for Plumb'g/HVAC Pipe/Equip.	Crew	Daily Output	Labor-Hours	Unit	Material	2016 Bare Costs Labor	Equipment	Total	Total Incl O&P
0400	High temperature to 1050°F, alloy steel									
0410	4" pipe size	Q-1	106	.151	Ea.	27	8.05		35.05	41.50
0420	6" pipe size		106	.151		46	8.05		54.05	62.50
0430	8" pipe size		97	.165		51	8.80		59.80	69.50
0440	10" pipe size		84	.190		75	10.15		85.15	97.50
0450	12" pipe size		72	.222		84	11.85		95.85	110
0460	14" pipe size		64	.250		211	13.30		224.30	252
0470	16" pipe size		56	.286		232	15.20		247.20	278
0480	Beam clamp, flange type, TYPE 25									
0482	For 3/8" bolt	1 Plum	48	.167	Ea.	3.45	9.85		13.30	18.70
0483	For 1/2" bolt		44	.182		4.44	10.75		15.19	21
0484	For 5/8" bolt		40	.200		5.05	11.85		16.90	23.50
0485	For 3/4" bolt		36	.222		5.05	13.15		18.20	25.50
0486	For 1" bolt		32	.250		5.05	14.80		19.85	28
0500	I-beam, for mounting on bottom flange, strap iron, TYPE 21									
0510	2" flange size	1 Plum	96	.083	Ea.	18.50	4.93		23.43	28
0520	3" flange size		95	.084		22	4.99		26.99	31.50
0530	4" flange size		93	.086		5.80	5.10		10.90	14.05
0540	5" flange size		92	.087		6.35	5.15		11.50	14.75
0550	6" flange size		90	.089		7.20	5.25		12.45	15.90
0560	7" flange size		88	.091		8.05	5.40		13.45	16.95
0570	8" flange size		86	.093		8.65	5.50		14.15	17.80
0600	One hole, vertical mounting, malleable iron									
0610	1/2" pipe size	1 Plum	160	.050	Ea.	1.09	2.96		4.05	5.65
0620	3/4" pipe size		145	.055		1.18	3.27		4.45	6.25
0630	1" pipe size		136	.059		1.26	3.48		4.74	6.65
0640	1-1/4" pipe size		128	.063		2.22	3.70		5.92	8.05
0650	1-1/2" pipe size		120	.067		2.55	3.95		6.50	8.75
0660	2" pipe size		112	.071		3.67	4.23		7.90	10.45
0670	2-1/2" pipe size		104	.077		7.15	4.55		11.70	14.70
0680	3" pipe size		96	.083		8.65	4.93		13.58	16.95
0690	3-1/2" pipe size		90	.089		10.55	5.25		15.80	19.55
0700	4" pipe size		84	.095		12	5.65		17.65	21.50
0750	Riser or extension pipe, carbon steel, TYPE 8									
0756	1/2" pipe size	1 Plum	52	.154	Ea.	2.55	9.10		11.65	16.55
0760	3/4" pipe size		48	.167		2.55	9.85		12.40	17.70
0770	1" pipe size		47	.170		2.67	10.10		12.77	18.15
0780	1-1/4" pipe size		46	.174		3.24	10.30		13.54	19.10
0790	1-1/2" pipe size		45	.178		3.46	10.50		13.96	19.70
0800	2" pipe size		43	.186		3.58	11		14.58	20.50
0810	2-1/2" pipe size		41	.195		3.72	11.55		15.27	21.50
0820	3" pipe size		40	.200		4.04	11.85		15.89	22.50
0830	3-1/2" pipe size		39	.205		4.54	12.15		16.69	23.50
0840	4" pipe size		38	.211		5.50	12.45		17.95	25
0850	5" pipe size		37	.216		6.85	12.80		19.65	27
0860	6" pipe size		36	.222		8.45	13.15		21.60	29
0870	8" pipe size		34	.235		14.75	13.95		28.70	37
0880	10" pipe size		32	.250		23	14.80		37.80	48
0890	12" pipe size		28	.286		30.50	16.90		47.40	59
0900	For plastic coating 3/4" to 4", add					190%				
0910	For copper plating 3/4" to 4", add					58%				
0950	Two piece, complete, carbon steel, medium weight, TYPE 4									
0960	1/2" pipe size	Q-1	137	.117	Ea.	2.31	6.20		8.51	11.95

22 05 29.10 Hangers & Supp. for Plumb'g/HVAC Pipe/Equip.		Crew	Daily Output	Labor-Hours	Unit	Material	2016 Bare Costs Labor	Equipment	Total	Total Incl O&P
0970	3/4" pipe size	Q-1	134	.119	Ea.	2.31	6.35		8.66	12.15
0980	1" pipe size		132	.121		2.61	6.45		9.06	12.60
0990	1-1/4" pipe size		130	.123		2.83	6.55		9.38	13
1000	1-1/2" pipe size		126	.127		3.67	6.75		10.42	14.25
1010	2" pipe size		124	.129		4.52	6.85		11.37	15.30
1020	2-1/2" pipe size		120	.133		4.89	7.10		11.99	16.10
1030	3" pipe size		117	.137		5.40	7.30		12.70	16.95
1040	3-1/2" pipe size		114	.140		10.10	7.50		17.60	22.50
1050	4" pipe size		110	.145		7.15	7.75		14.90	19.55
1060	5" pipe size		106	.151		17.55	8.05		25.60	31.50
1070	6" pipe size		104	.154		19.40	8.20		27.60	34
1080	8" pipe size		100	.160		23.50	8.50		32	39
1090	10" pipe size		96	.167		46.50	8.90		55.40	64.50
1100	12" pipe size		89	.180		56	9.60		65.60	76.50
1110	14" pipe size		82	.195		74.50	10.40		84.90	97.50
1120	16" pipe size	▼	68	.235		96	12.55		108.55	125
1130	For galvanized, add				▼	45%				
1150	Insert, concrete									
1160	Wedge type, carbon steel body, malleable iron nut, galvanized									
1170	1/4" threaded rod size	1 Plum	96	.083	Ea.	9.55	4.93		14.48	17.95
1180	3/8" threaded rod size		96	.083		10.30	4.93		15.23	18.80
1190	1/2" threaded rod size		96	.083		10.65	4.93		15.58	19.20
1200	5/8" threaded rod size		96	.083		11.25	4.93		16.18	19.85
1210	3/4" threaded rod size		96	.083		11.95	4.93		16.88	20.50
1220	7/8" threaded rod size	▼	96	.083	▼	12.35	4.93		17.28	21
1250	Pipe guide sized for insulation									
1260	No. 1, 1" pipe size, 1" thick insulation	1 Stpi	26	.308	Ea.	125	18.70		143.70	165
1270	No. 2, 1-1/4"-2" pipe size, 1" thick insulation		23	.348		153	21		174	200
1280	No. 3, 1-1/4"-2" pipe size, 1-1/2" thick insulation		21	.381		153	23		176	203
1290	No. 4, 2-1/2"-3-1/2" pipe size, 1-1/2" thick insulation		18	.444		153	27		180	209
1300	No. 5, 4"-5" pipe size, 1-1/2" thick insulation	▼	16	.500		173	30.50		203.50	236
1310	No. 6, 5"-6" pipe size, 2" thick insulation	Q-5	21	.762		201	41.50		242.50	284
1320	No. 7, 8" pipe size, 2" thick insulation		16	1		254	54.50		308.50	360
1330	No. 8, 10" pipe size, 2" thick insulation	▼	12	1.333		375	73		448	525
1340	No. 9, 12" pipe size, 2" thick insulation	Q-6	17	1.412		375	80		455	535
1350	No. 10, 12"-14" pipe size, 2-1/2" thick insulation		16	1.500		440	85		525	615
1360	No. 11, 16" pipe size, 2-1/2" thick insulation		10.50	2.286		440	129		569	680
1370	No. 12, 16"-18" pipe size, 3" thick insulation		9	2.667		650	151		801	945
1380	No. 13, 20" pipe size, 3" thick insulation		7.50	3.200		650	181		831	990
1390	No. 14, 24" pipe size, 3" thick insulation	▼	7	3.429	▼	855	194		1,049	1,225
1400	Bands									
1410	Adjustable band, carbon steel, for non-insulated pipe, TYPE 7									
1420	1/2" pipe size	Q-1	142	.113	Ea.	.31	6		6.31	9.40
1430	3/4" pipe size		140	.114		.31	6.10		6.41	9.55
1440	1" pipe size		137	.117		.31	6.20		6.51	9.75
1450	1-1/4" pipe size		134	.119		.33	6.35		6.68	9.95
1460	1-1/2" pipe size		131	.122		.33	6.50		6.83	10.15
1470	2" pipe size		129	.124		.33	6.60		6.93	10.30
1480	2-1/2" pipe size		125	.128		.60	6.80		7.40	10.95
1490	3" pipe size		122	.131		.65	7		7.65	11.25
1500	3-1/2" pipe size		119	.134		1.05	7.15		8.20	11.95
1510	4" pipe size		114	.140		1.05	7.50		8.55	12.45
1520	5" pipe size	▼	110	.145	▼	1.67	7.75		9.42	13.55

22 05 29.10 Hangers & Supp. for Plumb'g/HVAC Pipe/Equip.	Crew	Daily Output	Labor-Hours	Unit	Material	2016 Bare Costs Labor	Equipment	Total	Total Incl O&P
1530	6" pipe size	Q-1	108	.148	Ea.	1.96	7.90		9.86	14.05
1540	8" pipe size	↓	104	.154	↓	2.96	8.20		11.16	15.60
1550	For copper plated, add					50%				
1560	For galvanized, add					30%				
1570	For plastic coating, add					30%				
1600	Adjusting nut malleable iron, steel band, TYPE 9									
1610	1/2" pipe size, galvanized band	Q-1	137	.117	Ea.	5	6.20		11.20	14.90
1620	3/4" pipe size, galvanized band		135	.119		5	6.30		11.30	15.05
1630	1" pipe size, galvanized band		132	.121		5.15	6.45		11.60	15.45
1640	1-1/4" pipe size, galvanized band		129	.124		5.20	6.60		11.80	15.65
1650	1-1/2" pipe size, galvanized band		126	.127		5.35	6.75		12.10	16.10
1660	2" pipe size, galvanized band		124	.129		5.60	6.85		12.45	16.50
1670	2-1/2" pipe size, galvanized band		120	.133		8.85	7.10		15.95	20.50
1680	3" pipe size, galvanized band		117	.137		9.25	7.30		16.55	21
1690	3-1/2" pipe size, galvanized band		114	.140		30	7.50		37.50	44.50
1700	4" pipe size, cadmium plated band	↓	110	.145		23.50	7.75		31.25	37.50
1740	For plastic coated band, add					35%				
1750	For completely copper coated, add				↓	45%				
1800	Clevis, adjustable, carbon steel, for non-insulated pipe, TYPE 1									
1810	1/2" pipe size	Q-1	137	.117	Ea.	.94	6.20		7.14	10.45
1820	3/4" pipe size		135	.119		.94	6.30		7.24	10.60
1830	1" pipe size		132	.121		.99	6.45		7.44	10.85
1840	1-1/4" pipe size		129	.124		1.12	6.60		7.72	11.20
1850	1-1/2" pipe size		126	.127		1.23	6.75		7.98	11.55
1860	2" pipe size		124	.129		1.45	6.85		8.30	11.95
1870	2-1/2" pipe size		120	.133		2.41	7.10		9.51	13.35
1880	3" pipe size		117	.137		2.61	7.30		9.91	13.85
1890	3-1/2" pipe size		114	.140		3.03	7.50		10.53	14.65
1900	4" pipe size		110	.145		3.27	7.75		11.02	15.30
1910	5" pipe size		106	.151		4.53	8.05		12.58	17.15
1920	6" pipe size		104	.154		5.50	8.20		13.70	18.40
1930	8" pipe size		100	.160		8.05	8.50		16.55	21.50
1940	10" pipe size		96	.167		14.35	8.90		23.25	29
1950	12" pipe size		89	.180		18.55	9.60		28.15	35
1960	14" pipe size		82	.195		31	10.40		41.40	50
1970	16" pipe size		68	.235		50	12.55		62.55	74
1971	18" pipe size		54	.296		60	15.80		75.80	90
1972	20" pipe size	↓	38	.421	↓	125	22.50		147.50	171
1980	For galvanized, add					66%				
1990	For copper plated 1/2" to 4", add					77%				
2000	For light weight 1/2" to 4", deduct					13%				
2010	Insulated pipe type, 3/4" to 12" pipe, add					180%				
2020	Insulated pipe type, chrome-moly U-strap, add					530%				
2250	Split ring, malleable iron, for non-insulated pipe, TYPE 11									
2260	1/2" pipe size	Q-1	137	.117	Ea.	4.43	6.20		10.63	14.25
2270	3/4" pipe size		135	.119		4.47	6.30		10.77	14.45
2280	1" pipe size		132	.121		4.74	6.45		11.19	14.95
2290	1-1/4" pipe size		129	.124		5.90	6.60		12.50	16.40
2300	1-1/2" pipe size		126	.127		7.15	6.75		13.90	18.05
2310	2" pipe size		124	.129		8.10	6.85		14.95	19.25
2320	2-1/2" pipe size		120	.133		11.65	7.10		18.75	23.50
2330	3" pipe size		117	.137		14.60	7.30		21.90	27
2340	3-1/2" pipe size		114	.140		15.20	7.50		22.70	28

22 05 Common Work Results for Plumbing

22 05 29 - Hangers and Supports for Plumbing Piping and Equipment

22 05 29.10 Hangers & Supp. for Plumb'g/HVAC Pipe/Equip.	Crew	Daily Output	Labor-Hours	Unit	Material	2016 Bare Costs Labor	Equipment	Total	Total Incl O&P
2350	4" pipe size	Q-1	110	.145	Ea.	15.20	7.75		22.95	28.50
2360	5" pipe size		106	.151		20	8.05		28.05	34
2370	6" pipe size		104	.154		45	8.20		53.20	62
2380	8" pipe size		100	.160		66.50	8.50		75	86
2390	For copper plated, add					8%				
2400	Channels, steel, 3/4" x 1-1/2"	1 Plum	80	.100	L.F.	2.47	5.90		8.37	11.65
2404	1-1/2" x 1-1/2"		70	.114		3.16	6.75		9.91	13.70
2408	1-7/8" x 1-1/2"		60	.133		23	7.90		30.90	37.50
2412	3" x 1-1/2"		50	.160		39.50	9.45		48.95	58
2416	Hangers, trapeze channel support, 12 ga. 1-1/2" x 1/1/2", 12" wide, steel		8.80	.909	Ea.	17.95	54		71.95	101
2418	18" wide, steel		8.30	.964		19.55	57		76.55	108
2430	Spring nuts, long, 1/4"		120	.067		.90	3.95		4.85	6.95
2432	3/8"		100	.080		.99	4.74		5.73	8.25
2434	1/2"		80	.100		1.03	5.90		6.93	10.10
2436	5/8"		80	.100		3.62	5.90		9.52	12.95
2438	3/4"		75	.107		6.25	6.30		12.55	16.40
2440	Spring nuts, short, 1/4"		120	.067		1.41	3.95		5.36	7.50
2442	3/8"		100	.080		1.49	4.74		6.23	8.80
2444	1/2"		80	.100		1.69	5.90		7.59	10.80
2500	Washer, flat steel									
2502	3/8"	1 Plum	240	.033	Ea.	.08	1.97		2.05	3.07
2503	1/2"		220	.036		.18	2.15		2.33	3.45
2504	5/8"		200	.040		.35	2.37		2.72	3.96
2505	3/4"		180	.044		.56	2.63		3.19	4.59
2506	7/8"		160	.050		.66	2.96		3.62	5.20
2507	1"		140	.057		.99	3.38		4.37	6.20
2508	1-1/4"		120	.067		1.06	3.95		5.01	7.10
2520	Nut, steel, hex									
2522	3/8"	1 Plum	200	.040	Ea.	.15	2.37		2.52	3.74
2523	1/2"		180	.044		.37	2.63		3	4.38
2524	5/8"		160	.050		.59	2.96		3.55	5.10
2525	3/4"		140	.057		.99	3.38		4.37	6.20
2526	7/8"		120	.067		1.58	3.95		5.53	7.70
2527	1"		100	.080		2.18	4.74		6.92	9.55
2528	1-1/4"		80	.100		5.45	5.90		11.35	14.95
2532	Turnbuckle, TYPE 13									
2534	3/8"	1 Plum	80	.100	Ea.	4.82	5.90		10.72	14.25
2535	1/2"		72	.111		5.45	6.60		12.05	15.95
2536	5/8"		64	.125		9.75	7.40		17.15	22
2537	3/4"		56	.143		12.40	8.45		20.85	26.50
2538	7/8"		48	.167		29.50	9.85		39.35	47.50
2539	1"		40	.200		42	11.85		53.85	64
2540	1-1/4"		32	.250		78	14.80		92.80	109
2650	Rods, carbon steel									
2660	Continuous thread									
2670	1/4" thread size	1 Plum	144	.056	L.F.	1.78	3.29		5.07	6.90
2680	3/8" thread size		144	.056		1.90	3.29		5.19	7.05
2690	1/2" thread size		144	.056		3	3.29		6.29	8.25
2700	5/8" thread size		144	.056		4.26	3.29		7.55	9.65
2710	3/4" thread size		144	.056		7.50	3.29		10.79	13.20
2720	7/8" thread size		144	.056		9.40	3.29		12.69	15.30
2721	1" thread size	Q-1	160	.100		16.20	5.35		21.55	26
2722	1-1/8" thread size	"	120	.133		23.50	7.10		30.60	36

For customer support on your Mechanical Cost Data, call 877.739.3689.

101

22 05 29.10 Hangers & Supp. for Plumb'g/HVAC Pipe/Equip.		Crew	Daily Output	Labor-Hours	Unit	Material	2016 Bare Costs Labor	Equipment	Total	Total Incl O&P
2725	1/4" thread size, bright finish	1 Plum	144	.056	L.F.	1.24	3.29		4.53	6.30
2726	1/2" thread size, bright finish	"	144	.056	↓	4.23	3.29		7.52	9.60
2730	For galvanized, add					40%				
2820	Rod couplings									
2821	3/8"	1 Plum	60	.133	Ea.	1.44	7.90		9.34	13.50
2822	1/2"		54	.148		2.01	8.75		10.76	15.45
2823	5/8"		48	.167		2.53	9.85		12.38	17.70
2824	3/4"		44	.182		3.55	10.75		14.30	20
2825	7/8"		40	.200		6	11.85		17.85	24.50
2826	1"		34	.235		7.95	13.95		21.90	29.50
2827	1-1/8"	↓	30	.267	↓	15.85	15.80		31.65	41.50
2860	Pipe hanger assy, adj. clevis, saddle, rod, clamp, insul. allowance									
2864	1/2" pipe size	Q-5	35	.457	Ea.	35	25		60	75.50
2866	3/4" pipe size		34.80	.460		35.50	25		60.50	77.50
2868	1" pipe size		34.60	.462		22	25.50		47.50	62.50
2869	1-1/4" pipe size		34.30	.466		22.50	25.50		48	63.50
2870	1-1/2" pipe size		33.90	.472		23	26		49	64.50
2872	2" pipe size		33.30	.480		25	26.50		51.50	67
2874	2-1/2" pipe size		32.30	.495		29	27		56	72.50
2876	3" pipe size		31.20	.513		30.50	28		58.50	76
2880	4" pipe size		30.70	.521		33	28.50		61.50	79
2884	6" pipe size		29.80	.537		42	29.50		71.50	90.50
2888	8" pipe size		28	.571		53.50	31		84.50	106
2892	10" pipe size		25.20	.635		60.50	34.50		95	120
2896	12" pipe size	↓	23.20	.690	↓	89.50	37.50		127	156
2900	Rolls									
2910	Adjustable yoke, carbon steel with CI roll, TYPE 43									
2918	2" pipe size	Q-1	140	.114	Ea.	10.70	6.10		16.80	21
2920	2-1/2" pipe size		137	.117		10.70	6.20		16.90	21
2930	3" pipe size		131	.122		11.35	6.50		17.85	22.50
2940	3-1/2" pipe size		124	.129		15.95	6.85		22.80	28
2950	4" pipe size		117	.137		15.95	7.30		23.25	28.50
2960	5" pipe size		110	.145		23	7.75		30.75	37
2970	6" pipe size		104	.154		31	8.20		39.20	46.50
2980	8" pipe size		96	.167		44	8.90		52.90	62
2990	10" pipe size		80	.200		54	10.65		64.65	75.50
3000	12" pipe size		68	.235		86.50	12.55		99.05	114
3010	14" pipe size		56	.286		164	15.20		179.20	203
3020	16" pipe size	↓	48	.333	↓	197	17.75		214.75	243
3050	Chair, carbon steel with CI roll									
3060	2" pipe size	1 Plum	68	.118	Ea.	11.90	6.95		18.85	23.50
3070	2-1/2" pipe size		65	.123		12.95	7.30		20.25	25
3080	3" pipe size		62	.129		13.90	7.65		21.55	27
3090	3-1/2" pipe size		60	.133		17.95	7.90		25.85	31.50
3100	4" pipe size		58	.138		16.40	8.15		24.55	30.50
3110	5" pipe size		56	.143		19.25	8.45		27.70	34
3120	6" pipe size		53	.151		25.50	8.95		34.45	41.50
3130	8" pipe size		50	.160		35	9.45		44.45	53
3140	10" pipe size		48	.167		47.50	9.85		57.35	67.50
3150	12" pipe size		46	.174		73	10.30		83.30	95.50
3170	Single pipe roll, (see line 2650 for rods), TYPE 41, 1" pipe size	Q-1	137	.117		7.95	6.20		14.15	18.15
3180	1-1/4" pipe size		131	.122		8.25	6.50		14.75	18.85
3190	1-1/2" pipe size	↓	129	.124	↓	8.50	6.60		15.10	19.30

For customer support on your Mechanical Cost Data, call 877.739.3689.

22 05 29.10 Hangers & Supp. for Plumb'g/HVAC Pipe/Equip.	Crew	Daily Output	Labor-Hours	Unit	Material	2016 Bare Costs Labor	2016 Bare Costs Equipment	Total	Total Incl O&P
3200	2" pipe size	Q-1	124	.129	Ea.	8.85	6.85		15.70	20
3210	2-1/2" pipe size		118	.136		9.15	7.20		16.35	21
3220	3" pipe size		115	.139		9.50	7.40		16.90	21.50
3230	3-1/2" pipe size		113	.142		10.40	7.55		17.95	23
3240	4" pipe size		112	.143		10.70	7.60		18.30	23.50
3250	5" pipe size		110	.145		12.25	7.75		20	25
3260	6" pipe size		101	.158		26	8.45		34.45	41.50
3270	8" pipe size		90	.178		27	9.45		36.45	44
3280	10" pipe size		80	.200		35	10.65		45.65	54.50
3290	12" pipe size		68	.235		58	12.55		70.55	83
3291	14" pipe size		56	.286		88	15.20		103.20	120
3292	16" pipe size		48	.333		113	17.75		130.75	151
3293	18" pipe size		40	.400		133	21.50		154.50	179
3294	20" pipe size		35	.457		158	24.50		182.50	210
3296	24" pipe size		30	.533		254	28.50		282.50	320
3297	30" pipe size		25	.640		475	34		509	575
3298	36" pipe size	▼	20	.800	▼	585	42.50		627.50	710
3300	Saddles (add vertical pipe riser, usually 3" diameter)									
3310	Pipe support, complete, adjust., CI saddle, TYPE 36									
3320	2-1/2" pipe size	1 Plum	96	.083	Ea.	93.50	4.93		98.43	110
3330	3" pipe size		88	.091		95	5.40		100.40	113
3340	3-1/2" pipe size		79	.101		96.50	6		102.50	115
3350	4" pipe size		68	.118		140	6.95		146.95	165
3360	5" pipe size		64	.125		143	7.40		150.40	168
3370	6" pipe size		59	.136		143	8.05		151.05	169
3380	8" pipe size		53	.151		148	8.95		156.95	177
3390	10" pipe size		50	.160		172	9.45		181.45	204
3400	12" pipe size	▼	48	.167	▼	182	9.85		191.85	216
3450	For standard pipe support, one piece, CI deduct					34%				
3460	For stanchion support, CI with steel yoke, deduct					60%				
3550	Insulation shield 1" thick, 1/2" pipe size, TYPE 40	1 Asbe	100	.080	Ea.	2.76	4.27		7.03	9.70
3560	3/4" pipe size		100	.080		3.09	4.27		7.36	10.05
3570	1" pipe size		98	.082		3.33	4.36		7.69	10.45
3580	1-1/4" pipe size		98	.082		3.55	4.36		7.91	10.70
3590	1-1/2" pipe size		96	.083		3.48	4.45		7.93	10.75
3600	2" pipe size		96	.083		3.69	4.45		8.14	10.95
3610	2-1/2" pipe size		94	.085		3.81	4.54		8.35	11.25
3620	3" pipe size		94	.085		4.58	4.54		9.12	12.10
3630	2" thick, 3-1/2" pipe size		92	.087		4.89	4.64		9.53	12.65
3640	4" pipe size		92	.087		6.80	4.64		11.44	14.75
3650	5" pipe size		90	.089		9.25	4.75		14	17.55
3660	6" pipe size		90	.089		10.55	4.75		15.30	19
3670	8" pipe size		88	.091		17.15	4.85		22	26.50
3680	10" pipe size		88	.091		21.50	4.85		26.35	31
3690	12" pipe size		86	.093		24.50	4.97		29.47	35
3700	14" pipe size		86	.093		41	4.97		45.97	53
3710	16" pipe size		84	.095		43.50	5.10		48.60	56
3720	18" pipe size		84	.095		48.50	5.10		53.60	61.50
3730	20" pipe size		82	.098		53.50	5.20		58.70	67
3732	24" pipe size	▼	80	.100	▼	62.50	5.35		67.85	77.50
3750	Covering protection saddle, TYPE 39									
3760	1" covering size									
3770	3/4" pipe size	1 Plum	68	.118	Ea.	4.65	6.95		11.60	15.60

22 05 29.10 Hangers & Supp. for Plumb'g/HVAC Pipe/Equip.		Crew	Daily Output	Labor-Hours	Unit	Material	2016 Bare Costs Labor	Equipment	Total	Total Incl O&P
3780	1" pipe size	1 Plum	68	.118	Ea.	4.65	6.95		11.60	15.60
3790	1-1/4" pipe size		68	.118		4.65	6.95		11.60	15.60
3800	1-1/2" pipe size		66	.121		4.97	7.20		12.17	16.30
3810	2" pipe size		66	.121		4.97	7.20		12.17	16.30
3820	2-1/2" pipe size		64	.125		4.97	7.40		12.37	16.60
3830	3" pipe size		64	.125		6.65	7.40		14.05	18.45
3840	3-1/2" pipe size		62	.129		7.40	7.65		15.05	19.65
3850	4" pipe size		62	.129		7.40	7.65		15.05	19.65
3860	5" pipe size		60	.133		7.40	7.90		15.30	20
3870	6" pipe size	▼	60	.133	▼	8.85	7.90		16.75	21.50
3900	1-1/2" covering size									
3910	3/4" pipe size	1 Plum	68	.118	Ea.	5.55	6.95		12.50	16.60
3920	1" pipe size		68	.118		5.55	6.95		12.50	16.60
3930	1-1/4" pipe size		68	.118		5.55	6.95		12.50	16.60
3940	1-1/2" pipe size		66	.121		5.55	7.20		12.75	16.95
3950	2" pipe size		66	.121		6	7.20		13.20	17.45
3960	2-1/2" pipe size		64	.125		7.25	7.40		14.65	19.15
3970	3" pipe size		64	.125		7.25	7.40		14.65	19.15
3980	3-1/2" pipe size		62	.129		7.65	7.65		15.30	20
3990	4" pipe size		62	.129		7.65	7.65		15.30	20
4000	5" pipe size		60	.133		7.65	7.90		15.55	20.50
4010	6" pipe size		60	.133		10.35	7.90		18.25	23.50
4020	8" pipe size		58	.138		13	8.15		21.15	26.50
4022	10" pipe size		56	.143		13	8.45		21.45	27
4024	12" pipe size	▼	54	.148	▼	31.50	8.75		40.25	48
4028	2" covering size									
4029	2-1/2" pipe size	1 Plum	62	.129	Ea.	7.35	7.65		15	19.60
4032	3" pipe size		60	.133		8.60	7.90		16.50	21.50
4033	4" pipe size		58	.138		8.60	8.15		16.75	22
4034	6" pipe size		56	.143		12.35	8.45		20.80	26.50
4035	8" pipe size		54	.148		14.55	8.75		23.30	29.50
4080	10" pipe size		58	.138		15.50	8.15		23.65	29.50
4090	12" pipe size		56	.143		35.50	8.45		43.95	52
4100	14" pipe size		56	.143		35.50	8.45		43.95	52
4110	16" pipe size		54	.148		46	8.75		54.75	64
4120	18" pipe size		54	.148		46	8.75		54.75	64
4130	20" pipe size		52	.154		51	9.10		60.10	70.50
4150	24" pipe size		50	.160		60	9.45		69.45	80.50
4160	30" pipe size		48	.167		66	9.85		75.85	87.50
4180	36" pipe size	▼	45	.178	▼	76	10.50		86.50	99.50
4186	2-1/2" covering size									
4187	3" pipe size	1 Plum	58	.138	Ea.	8.75	8.15		16.90	22
4188	4" pipe size		56	.143		9.50	8.45		17.95	23
4189	6" pipe size		52	.154		12.95	9.10		22.05	28
4190	8" pipe size		48	.167		16	9.85		25.85	32.50
4191	10" pipe size		44	.182		17.60	10.75		28.35	35.50
4192	12" pipe size		40	.200		38	11.85		49.85	59.50
4193	14" pipe size		36	.222		38	13.15		51.15	61.50
4194	16" pipe size		32	.250		48.50	14.80		63.30	75.50
4195	18" pipe size	▼	28	.286		54	16.90		70.90	85
4200	Sockets									
4210	Rod end, malleable iron, TYPE 16									
4220	1/4" thread size	1 Plum	240	.033	Ea.	1.43	1.97		3.40	4.55

22 05 29.10 Hangers & Supp. for Plumb'g/HVAC Pipe/Equip.	Crew	Daily Output	Labor-Hours	Unit	Material	2016 Bare Costs Labor	Equipment	Total	Total Incl O&P
4230	3/8" thread size	1 Plum	240	.033	Ea.	1.43	1.97		3.40	4.55
4240	1/2" thread size		230	.035		1.89	2.06		3.95	5.20
4250	5/8" thread size		225	.036		3.53	2.11		5.64	7.05
4260	3/4" thread size		220	.036		4.96	2.15		7.11	8.70
4270	7/8" thread size		210	.038		6.95	2.26		9.21	11.05
4290	Strap, 1/2" pipe size, TYPE 26	Q-1	142	.113		1.56	6		7.56	10.75
4300	3/4" pipe size		140	.114		1.63	6.10		7.73	11
4310	1" pipe size		137	.117		2.40	6.20		8.60	12.05
4320	1-1/4" pipe size		134	.119		2.44	6.35		8.79	12.30
4330	1-1/2" pipe size		131	.122		2.59	6.50		9.09	12.65
4340	2" pipe size		129	.124		2.62	6.60		9.22	12.85
4350	2-1/2" pipe size		125	.128		4.02	6.80		10.82	14.70
4360	3" pipe size		122	.131		5.30	7		12.30	16.35
4370	3-1/2" pipe size		119	.134		5.45	7.15		12.60	16.80
4380	4" pipe size		114	.140		5.75	7.50		13.25	17.60
4400	U-bolt, carbon steel									
4410	Standard, with nuts, TYPE 42									
4420	1/2" pipe size	1 Plum	160	.050	Ea.	.98	2.96		3.94	5.55
4430	3/4" pipe size		158	.051		1.04	3		4.04	5.65
4450	1" pipe size		152	.053		1.08	3.12		4.20	5.90
4460	1-1/4" pipe size		148	.054		1.37	3.20		4.57	6.35
4470	1-1/2" pipe size		143	.056		1.45	3.31		4.76	6.60
4480	2" pipe size		139	.058		1.57	3.41		4.98	6.90
4490	2-1/2" pipe size		134	.060		2.56	3.53		6.09	8.15
4500	3" pipe size		128	.063		2.88	3.70		6.58	8.75
4510	3-1/2" pipe size		122	.066		3.11	3.88		6.99	9.25
4520	4" pipe size		117	.068		3.15	4.05		7.20	9.55
4530	5" pipe size		114	.070		3.70	4.15		7.85	10.30
4540	6" pipe size		111	.072		7.05	4.27		11.32	14.20
4550	8" pipe size		109	.073		7.70	4.34		12.04	15
4560	10" pipe size		107	.075		13.35	4.43		17.78	21.50
4570	12" pipe size		104	.077		19.35	4.55		23.90	28.50
4580	For plastic coating on 1/2" thru 6" size, add					150%				
4700	U-hook, carbon steel, requires mounting screws or bolts									
4710	3/4" thru 2" pipe size									
4720	6" long	1 Plum	96	.083	Ea.	.86	4.93		5.79	8.40
4730	8" long		96	.083		1.05	4.93		5.98	8.60
4740	10" long		96	.083		1.17	4.93		6.10	8.75
4750	12" long		96	.083		1.29	4.93		6.22	8.85
4760	For copper plated, add					50%				
7000	Roof supports									
7006	Duct									
7010	Rectangular, open, 12" off roof									
7020	To 18" wide	Q-9	26	.615	Ea.	150	31.50		181.50	214
7030	To 24" wide	"	22	.727		175	37.50		212.50	251
7040	To 36" wide	Q-10	30	.800		200	42.50		242.50	286
7050	To 48" wide		28	.857		225	46		271	320
7060	To 60" wide		24	1		250	53.50		303.50	355
7100	Equipment									
7120	Equipment support	Q-5	20	.800	Ea.	65	43.50		108.50	138
7300	Pipe									
7310	Roller type									
7320	Up to 2-1/2" diam. pipe									

For customer support on your Mechanical Cost Data, call 877.739.3689.

105

22 05 29 – Hangers and Supports for Plumbing Piping and Equipment

22 05 29.10 Hangers & Supp. for Plumb'g/HVAC Pipe/Equip.	Crew	Daily Output	Labor-Hours	Unit	Material	2016 Bare Costs Labor	Equipment	Total	Total Incl O&P
7324	3-1/2" off roof	Q-5	24	.667	Ea.	20	36.50		56.50	77
7326	Up to 10" off roof	"	20	.800	"	25	43.50		68.50	93.50
7340	2-1/2" to 3-1/2" diam. pipe									
7342	Up to 16" off roof	Q-5	18	.889	Ea.	45.50	48.50		94	124
7360	4" to 5" diam. pipe									
7362	Up to 12" off roof	Q-5	16	1	Ea.	65	54.50		119.50	154
7400	Strut/channel type									
7410	Up to 2-1/2" diam. pipe									
7424	3-1/2" off roof	Q-5	24	.667	Ea.	18	36.50		54.50	75
7426	Up to 10" off roof	"	20	.800	"	23	43.50		66.50	91.50
7440	Strut and roller type									
7452	2-1/2" to 3-1/2" diam. pipe									
7454	Up to 16" off roof	Q-5	18	.889	Ea.	65.50	48.50		114	146
7460	Strut and hanger type									
7470	Up to 3" diam. pipe									
7474	Up to 8" off roof	Q-5	19	.842	Ea.	41.50	46		87.50	116
8000	Pipe clamp, plastic, 1/2" CTS	1 Plum	80	.100		.24	5.90		6.14	9.20
8010	3/4" CTS		73	.110		.25	6.50		6.75	10.10
8020	1" CTS		68	.118		.56	6.95		7.51	11.10
8080	Economy clamp, 1/4" CTS		175	.046		.04	2.71		2.75	4.12
8090	3/8" CTS		168	.048		.04	2.82		2.86	4.29
8100	1/2" CTS		160	.050		.04	2.96		3	4.51
8110	3/4" CTS		145	.055		.05	3.27		3.32	4.99
8200	Half clamp, 1/2" CTS		80	.100		.08	5.90		5.98	9.05
8210	3/4" CTS		73	.110		.12	6.50		6.62	9.95
8300	Suspension clamp, 1/2" CTS		80	.100		.24	5.90		6.14	9.20
8310	3/4" CTS		73	.110		.25	6.50		6.75	10.10
8320	1" CTS		68	.118		.56	6.95		7.51	11.10
8400	Insulator, 1/2" CTS		80	.100		.39	5.90		6.29	9.40
8410	3/4" CTS		73	.110		.40	6.50		6.90	10.25
8420	1" CTS		68	.118		.41	6.95		7.36	10.95
8500	J hook clamp with nail 1/2" CTS		240	.033		.13	1.97		2.10	3.12
8501	3/4" CTS	↓	240	.033	↓	.13	1.97		2.10	3.12
8800	Wire cable support system									
8810	Cable with hook terminal and locking device									
8830	2 mm, (.079") dia cable, (100 lb. cap.)									
8840	1 m, (3.3') length, with hook	1 Shee	96	.083	Ea.	3.03	4.77		7.80	10.65
8850	2 m, (6.6') length, with hook		84	.095		3.50	5.45		8.95	12.20
8860	3 m, (9.9') length, with hook	↓	72	.111		3.97	6.35		10.32	14.05
8870	5 m, (16.4') length, with hook	Q-9	60	.267		4.97	13.75		18.72	26.50
8880	10 m, (32.8') length, with hook	"	30	.533	↓	7.20	27.50		34.70	50
8900	3mm, (.118") dia cable, (200 lb. cap.)									
8910	1 m, (3.3') length, with hook	1 Shee	96	.083	Ea.	3.88	4.77		8.65	11.55
8920	2 m, (6.6') length, with hook		84	.095		4.36	5.45		9.81	13.15
8930	3 m, (9.9') length, with hook	↓	72	.111		4.81	6.35		11.16	15
8940	5 m, (16.4') length, with hook	Q-9	60	.267		5.95	13.75		19.70	27.50
8950	10 m, (32.8') length, with hook	"	30	.533	↓	8.55	27.50		36.05	51.50
9000	Cable system accessories									
9010	Anchor bolt, 3/8", with nut	1 Shee	140	.057	Ea.	1.21	3.27		4.48	6.35
9020	Air duct corner protector		160	.050		.54	2.86		3.40	4.97
9030	Air duct support attachment	↓	140	.057	↓	1.03	3.27		4.30	6.15
9040	Flange clip, hammer-on style									
9044	For flange thickness 3/32" - 9/64", 160 lb. cap.	1 Shee	180	.044	Ea.	.25	2.54		2.79	4.17

For customer support on your Mechanical Cost Data, call 877.739.3689.

22 05 Common Work Results for Plumbing

22 05 29 – Hangers and Supports for Plumbing Piping and Equipment

22 05 29.10 Hangers & Supp. for Plumb'g/HVAC Pipe/Equip.	Crew	Daily Output	Labor-Hours	Unit	Material	2016 Bare Costs Labor	Equipment	Total	Total Incl O&P
9048	For flange thickness 1/8" - 1/4", 200 lb. cap.	1 Shee	160	.050	Ea.	.28	2.86		3.14	4.69
9052	For flange thickness 5/16" - 1/2", 200 lb. cap.		150	.053		.59	3.05		3.64	5.30
9056	For flange thickness 9/16" - 3/4", 200 lb. cap.		140	.057		.78	3.27		4.05	5.85
9060	Wire insulation protection tube		180	.044	L.F.	.34	2.54		2.88	4.26
9070	Wire cutter				Ea.	40.50			40.50	44.50

22 05 33 – Heat Tracing for Plumbing Piping

22 05 33.20 Temperature Maintenance Cable

		Crew	Daily Output	Labor-Hours	Unit	Material	2016 Bare Costs Labor	Equipment	Total	Total Incl O&P
0010	**TEMPERATURE MAINTENANCE CABLE**									
0040	Components									
0080	Heating cable									
0100	208 V									
0150	140°F	Q-1	1060.80	.015	L.F.	9.75	.80		10.55	11.90
0200	120 V									
0220	125°F	Q-1	1060.80	.015	L.F.	8.55	.80		9.35	10.60
0300	Power kit w/1 end seal	1 Elec	48.80	.164	Ea.	79.50	9.05		88.55	101
0310	Splice kit		35.50	.225		94	12.40		106.40	123
0320	End seal		160	.050		9.70	2.76		12.46	14.80
0330	Tee kit w/1 end seal		26.80	.299		99.50	16.45		115.95	135
0340	Powered splice w/2 end seals		20	.400		106	22		128	150
0350	Powered tee kit w/3 end seals		18	.444		136	24.50		160.50	187
0360	Cross kit w/2 end seals		18.60	.430		140	23.50		163.50	190
0500	Recommended thickness of fiberglass insulation									
0510	Pipe size									
0520	1/2" - 1" use 1" insulation									
0530	1-1/4" - 2" use 1-1/2" insulation									
0540	2-1/2" - 6" use 2" insulation									
0560	NOTE: For pipe sizes 1-1/4" and smaller use 1/4" larger diameter									
0570	insulation to allow room for installation over cable.									

22 05 48 – Vibration and Seismic Controls for Plumbing Piping and Equipment

22 05 48.40 Vibration Absorbers

		Crew	Daily Output	Labor-Hours	Unit	Material	2016 Bare Costs Labor	Equipment	Total	Total Incl O&P
0010	**VIBRATION ABSORBERS**									
0100	Hangers, neoprene flex									
0200	10 – 120 lb. capacity				Ea.	27			27	29.50
0220	75 – 550 lb. capacity					38			38	42
0240	250 – 1100 lb. capacity					77.50			77.50	85
0260	1000 – 4000 lb. capacity					145			145	160
0500	Spring flex, 60 lb. capacity					37			37	40.50
0520	450 lb. capacity					53			53	58.50
0540	900 lb. capacity					77.50			77.50	85.50
0560	1100 – 1300 lb. capacity					77.50			77.50	85
0600	Rubber in shear									
0610	45 – 340 lb., up to 1/2" rod size	1 Stpi	22	.364	Ea.	23	22		45	58.50
0620	130 – 700 lb., up to 3/4" rod size		20	.400		49.50	24.50		74	90.50
0630	50 – 1000 lb., up to 3/4" rod size		18	.444		65.50	27		92.50	113
1000	Mounts, neoprene, 45 – 380 lb. capacity					17.60			17.60	19.35
1020	250 – 1100 lb. capacity					50.50			50.50	55.50
1040	1000 – 4000 lb. capacity					113			113	124
1100	Spring flex, 60 lb. capacity					72.50			72.50	79.50
1120	165 lb. capacity					73.50			73.50	81
1140	260 lb. capacity					76			76	84
1160	450 lb. capacity					97.50			97.50	107
1180	600 lb. capacity					102			102	112

For customer support on your Mechanical Cost Data, call 877.739.3689.

107

22 05 48 – Vibration and Seismic Controls for Plumbing Piping and Equipment

22 05 48.40 Vibration Absorbers		Crew	Daily Output	Labor-Hours	Unit	Material	2016 Bare Costs Labor	Equipment	Total	Total Incl O&P
1200	750 lb. capacity				Ea.	112			112	123
1220	900 lb. capacity					110			110	121
1240	1100 lb. capacity					121			121	133
1260	1300 lb. capacity					121			121	133
1280	1500 lb. capacity					181			181	199
1300	1800 lb. capacity					193			193	213
1320	2200 lb. capacity					204			204	224
1340	2600 lb. capacity				↓	204			204	224
1399	Spring type									
1400	2 Piece									
1410	50 – 1000 lb.	1 Stpi	12	.667	Ea.	208	40.50		248.50	289
1420	1100 – 1600 lb.	"	12	.667	"	225	40.50		265.50	310
1500	Double spring open									
1510	150 – 450 lb.	1 Stpi	24	.333	Ea.	143	20		163	188
1520	500 – 1000 lb.		24	.333		143	20		163	188
1530	1100 – 1600 lb.		24	.333		143	20		163	188
1540	1700 – 2400 lb.		24	.333		155	20		175	201
1550	2500 – 3400 lb.	↓	24	.333		264	20		284	320
2000	Pads, cork rib, 18" x 18" x 1", 10-50 psi					184			184	203
2020	18" x 36" x 1", 10-50 psi					370			370	405
2100	Shear flexible pads, 18" x 18" x 3/8", 20-70 psi					80.50			80.50	88.50
2120	18" x 36" x 3/8", 20-70 psi				↓	184			184	203
2150	Laminated neoprene and cork									
2160	1" thick	1 Stpi	16	.500	S.F.	91.50	30.50		122	147
2200	Neoprene elastomer isolation bearing pad									
2230	464 PSI, 5/8" x 19-11/16" x 39-3/8"	1 Stpi	16	.500	Ea.	305	30.50		335.50	380
3000	Note overlap in capacities due to deflections									

22 05 53 – Identification for Plumbing Piping and Equipment

22 05 53.10 Piping System Identification Labels

		Crew	Daily Output	Labor-Hours	Unit	Material	2016 Bare Costs Labor	Equipment	Total	Total Incl O&P
0010	**PIPING SYSTEM IDENTIFICATION LABELS**									
0100	Indicate contents and flow direction									
0106	Pipe markers									
0110	Plastic snap around									
0114	1/2" pipe	1 Plum	80	.100	Ea.	6.70	5.90		12.60	16.30
0116	3/4" pipe		80	.100		7	5.90		12.90	16.65
0118	1" pipe		80	.100		7	5.90		12.90	16.65
0120	2" pipe		75	.107		8.30	6.30		14.60	18.70
0122	3" pipe		70	.114		12.80	6.75		19.55	24.50
0124	4" pipe		60	.133		12.80	7.90		20.70	26
0126	6" pipe		60	.133		13.60	7.90		21.50	27
0128	8" pipe		56	.143		17.90	8.45		26.35	32.50
0130	10" pipe		56	.143		17.90	8.45		26.35	32.50
0200	Over 10" pipe size	↓	50	.160	↓	21.50	9.45		30.95	38.50
1110	Self adhesive									
1114	1" pipe	1 Plum	80	.100	Ea.	3.40	5.90		9.30	12.70
1116	2" pipe		75	.107		3.40	6.30		9.70	13.30
1118	3" pipe		70	.114		5.20	6.75		11.95	15.90
1120	4" pipe		60	.133		5.20	7.90		13.10	17.60
1122	6" pipe		60	.133		5.20	7.90		13.10	17.60
1124	8" pipe		56	.143		7.40	8.45		15.85	21
1126	10" pipe		56	.143		7.40	8.45		15.85	21
1200	Over 10" pipe size	↓	50	.160	↓	10.50	9.45		19.95	26

22 05 Common Work Results for Plumbing

22 05 53 – Identification for Plumbing Piping and Equipment

22 05 53.10 Piping System Identification Labels		Crew	Daily Output	Labor-Hours	Unit	Material	2016 Bare Costs Labor	Equipment	Total	Total Incl O&P
2000	Valve tags									
2010	Numbered plus identifying legend									
2100	Brass, 2" diameter	1 Plum	40	.200	Ea.	3	11.85		14.85	21
2200	Plastic, 1-1/2" diam.	"	40	.200	"	2.70	11.85		14.55	21

22 07 Plumbing Insulation

22 07 16 – Plumbing Equipment Insulation

22 07 16.10 Insulation for Plumbing Equipment

			Crew	Daily Output	Labor-Hours	Unit	Material	2016 Bare Costs Labor	Equipment	Total	Total Incl O&P
0010	**INSULATION FOR PLUMBING EQUIPMENT**										
2900	Domestic water heater wrap kit										
2920	1-1/2" with vinyl jacket, 20-60 gal.	G	1 Plum	8	1	Ea.	15.10	59		74.10	106

22 07 19 – Plumbing Piping Insulation

22 07 19.10 Piping Insulation

			Crew	Daily Output	Labor-Hours	Unit	Material	2016 Bare Costs Labor	Equipment	Total	Total Incl O&P
0010	**PIPING INSULATION**										
0100	Rule of thumb, as a percentage of total mechanical costs					Job				10%	10%
0110	Insulation req'd. is based on the surface size/area to be covered										
0230	Insulated protectors, (ADA)										
0235	For exposed piping under sinks or lavatories										
0240	Vinyl coated foam, velcro tabs										
0245	P Trap, 1-1/4" or 1-1/2"		1 Plum	32	.250	Ea.	16.85	14.80		31.65	41
0260	Valve and supply cover										
0265	1/2", 3/8", and 7/16" pipe size		1 Plum	32	.250	Ea.	16.45	14.80		31.25	40.50
0280	Tailpiece offset (wheelchair)										
0285	1-1/4" pipe size		1 Plum	32	.250	Ea.	13.65	14.80		28.45	37.50
0600	Pipe covering (price copper tube one size less than IPS)										
1000	Mineral wool										
1010	Preformed, 1200°F, plain										
1014	1" wall										
1016	1/2" iron pipe size	G	Q-14	230	.070	L.F.	1.65	3.34		4.99	7
1018	3/4" iron pipe size	G		220	.073		1.72	3.49		5.21	7.35
1022	1" iron pipe size	G		210	.076		1.77	3.66		5.43	7.65
1024	1-1/4" iron pipe size	G		205	.078		1.82	3.75		5.57	7.85
1026	1-1/2" iron pipe size	G		205	.078		1.90	3.75		5.65	7.95
1028	2" iron pipe size	G		200	.080		2.32	3.84		6.16	8.55
1030	2-1/2" iron pipe size	G		190	.084		2.46	4.05		6.51	9
1032	3" iron pipe size	G		180	.089		2.55	4.27		6.82	9.45
1034	4" iron pipe size	G		150	.107		3.14	5.15		8.29	11.45
1036	5" iron pipe size	G		140	.114		3.55	5.50		9.05	12.45
1038	6" iron pipe size	G		120	.133		3.79	6.40		10.19	14.10
1040	7" iron pipe size	G		110	.145		4.43	7		11.43	15.75
1042	8" iron pipe size	G		100	.160		6.10	7.70		13.80	18.65
1044	9" iron pipe size	G		90	.178		6.65	8.55		15.20	20.50
1046	10" iron pipe size	G		90	.178		7.50	8.55		16.05	21.50
1050	1-1/2" wall										
1052	1/2" iron pipe size	G	Q-14	225	.071	L.F.	2.82	3.42		6.24	8.40
1054	3/4" iron pipe size	G		215	.074		2.89	3.58		6.47	8.75
1056	1" iron pipe size	G		205	.078		2.93	3.75		6.68	9.05
1058	1-1/4" iron pipe size	G		200	.080		2.99	3.84		6.83	9.30
1060	1-1/2" iron pipe size	G		200	.080		3.14	3.84		6.98	9.45
1062	2" iron pipe size	G		190	.084		3.32	4.05		7.37	9.95

For customer support on your Mechanical Cost Data, call 877.739.3689.

109

22 07 19.10 Piping Insulation		Crew	Daily Output	Labor-Hours	Unit	Material	2016 Bare Costs		Total	Total Incl O&P
							Labor	Equipment		
1064	2-1/2" iron pipe size	G Q-14	180	.089	L.F.	3.62	4.27		7.89	10.65
1066	3" iron pipe size	G	165	.097		3.77	4.66		8.43	11.40
1068	4" iron pipe size	G	140	.114		4.53	5.50		10.03	13.55
1070	5" iron pipe size	G	130	.123		4.84	5.90		10.74	14.50
1072	6" iron pipe size	G	110	.145		4.99	7		11.99	16.40
1074	7" iron pipe size	G	100	.160		5.90	7.70		13.60	18.45
1076	8" iron pipe size	G	90	.178		6.55	8.55		15.10	20.50
1078	9" iron pipe size	G	85	.188		7.55	9.05		16.60	22.50
1080	10" iron pipe size	G	80	.200		8.30	9.60		17.90	24
1082	12" iron pipe size	G	75	.213		9.45	10.25		19.70	26.50
1084	14" iron pipe size	G	70	.229		10.80	11		21.80	29
1086	16" iron pipe size	G	65	.246		12.30	11.85		24.15	32
1088	18" iron pipe size	G	60	.267		13.85	12.80		26.65	35
1090	20" iron pipe size	G	55	.291		15.25	14		29.25	39
1092	22" iron pipe size	G	50	.320		17.45	15.40		32.85	43
1094	24" iron pipe size	▼	45	.356	▼	18.65	17.10		35.75	47
1100	2" wall									
1102	1/2" iron pipe size	G Q-14	220	.073	L.F.	3.65	3.49		7.14	9.45
1104	3/4" iron pipe size	G	210	.076		3.78	3.66		7.44	9.85
1106	1" iron pipe size	G	200	.080		4.01	3.84		7.85	10.40
1108	1-1/4" iron pipe size	G	190	.084		4.30	4.05		8.35	11.05
1110	1-1/2" iron pipe size	G	190	.084		4.48	4.05		8.53	11.25
1112	2" iron pipe size	G	180	.089		4.70	4.27		8.97	11.80
1114	2-1/2" iron pipe size	G	170	.094		5.30	4.52		9.82	12.90
1116	3" iron pipe size	G	160	.100		5.70	4.81		10.51	13.75
1118	4" iron pipe size	G	130	.123		6.60	5.90		12.50	16.50
1120	5" iron pipe size	G	120	.133		7.35	6.40		13.75	18.05
1122	6" iron pipe size	G	100	.160		7.65	7.70		15.35	20.50
1124	8" iron pipe size	G	80	.200		9.25	9.60		18.85	25
1126	10" iron pipe size	G	70	.229		11.35	11		22.35	29.50
1128	12" iron pipe size	G	65	.246		12.55	11.85		24.40	32.50
1130	14" iron pipe size	G	60	.267		14.05	12.80		26.85	35.50
1132	16" iron pipe size	G	55	.291		15.90	14		29.90	39.50
1134	18" iron pipe size	G	50	.320		17.20	15.40		32.60	43
1136	20" iron pipe size	G	45	.356		19.80	17.10		36.90	48.50
1138	22" iron pipe size	G	45	.356		21.50	17.10		38.60	50
1140	24" iron pipe size	▼	40	.400	▼	22.50	19.20		41.70	55
1150	4" wall									
1152	1-1/4" iron pipe size	G Q-14	170	.094	L.F.	11.65	4.52		16.17	19.85
1154	1-1/2" iron pipe size	G	165	.097		12.10	4.66		16.76	20.50
1156	2" iron pipe size	G	155	.103		12.35	4.96		17.31	21.50
1158	4" iron pipe size	G	105	.152		15.60	7.30		22.90	28.50
1160	6" iron pipe size	G	75	.213		18.40	10.25		28.65	36
1162	8" iron pipe size	G	60	.267		21	12.80		33.80	43
1164	10" iron pipe size	G	50	.320		25	15.40		40.40	51.50
1166	12" iron pipe size	G	45	.356		27.50	17.10		44.60	57
1168	14" iron pipe size	G	40	.400		29.50	19.20		48.70	62.50
1170	16" iron pipe size	G	35	.457		33.50	22		55.50	70.50
1172	18" iron pipe size	G	32	.500		36	24		60	77
1174	20" iron pipe size	G	30	.533		38.50	25.50		64	82.50
1176	22" iron pipe size	G	28	.571		43.50	27.50		71	90
1178	24" iron pipe size	▼	26	.615	▼	46	29.50		75.50	96.50
4280	Cellular glass, closed cell foam, all service jacket, sealant,									

22 07 Plumbing Insulation

22 07 19 – Plumbing Piping Insulation

22 07 19.10 Piping Insulation		Crew	Daily Output	Labor-Hours	Unit	Material	2016 Bare Costs Labor	Equipment	Total	Total Incl O&P
4281	working temp. (-450°F to +900°F), 0 water vapor transmission									
4284	1" wall									
4286	1/2" iron pipe size	**G** Q-14	120	.133	L.F.	6.80	6.40		13.20	17.40
4300	1-1/2" wall,									
4301	1" iron pipe size	**G** Q-14	105	.152	L.F.	10.40	7.30		17.70	23
4304	2-1/2" iron pipe size	**G**	90	.178		14.20	8.55		22.75	29
4306	3" iron pipe size	**G**	85	.188		17.60	9.05		26.65	33.50
4308	4" iron pipe size	**G**	70	.229		21.50	11		32.50	40.50
4310	5" iron pipe size	**G**	65	.246		22	11.85		33.85	42.50
4320	2" wall,									
4322	1" iron pipe size	**G** Q-14	100	.160	L.F.	13.85	7.70		21.55	27
4324	2-1/2" iron pipe size	**G**	85	.188		21.50	9.05		30.55	37.50
4326	3" iron pipe size	**G**	80	.200		21.50	9.60		31.10	38.50
4328	4" iron pipe size	**G**	65	.246		26	11.85		37.85	47
4330	5" iron pipe size	**G**	60	.267		27	12.80		39.80	49.50
4332	6" iron pipe size	**G**	50	.320		31	15.40		46.40	58.50
4336	8" iron pipe size	**G**	40	.400		36.50	19.20		55.70	70.50
4338	10" iron pipe size	**G**	35	.457		45.50	22		67.50	84
4350	2-1/2" wall,									
4360	12" iron pipe size	**G** Q-14	32	.500	L.F.	60.50	24		84.50	104
4362	14" iron pipe size	**G** "	28	.571	"	67	27.50		94.50	116
4370	3" wall,									
4378	6" iron pipe size	**G** Q-14	48	.333	L.F.	41.50	16		57.50	70.50
4380	8" iron pipe size	**G**	38	.421		52	20		72	88.50
4382	10" iron pipe size	**G**	33	.485		60	23.50		83.50	103
4384	16" iron pipe size	**G**	25	.640		84.50	31		115.50	141
4386	18" iron pipe size	**G**	22	.727		108	35		143	173
4388	20" iron pipe size	**G**	20	.800		117	38.50		155.50	189
4400	3-1/2" wall,									
4412	12" iron pipe size	**G** Q-14	27	.593	L.F.	71.50	28.50		100	123
4414	14" iron pipe size	**G** "	25	.640	"	74	31		105	130
4430	4" wall,									
4446	16" iron pipe size	**G** Q-14	22	.727	L.F.	96	35		131	161
4448	18" iron pipe size	**G**	20	.800		105	38.50		143.50	175
4450	20" iron pipe size	**G**	18	.889		118	42.50		160.50	197
4480	Fittings, average with fabric and mastic									
4484	1" wall,									
4486	1/2" iron pipe size	**G** 1 Asbe	40	.200	Ea.	5.85	10.70		16.55	23
4500	1-1/2" wall,									
4502	1" iron pipe size	**G** 1 Asbe	38	.211	Ea.	7.80	11.25		19.05	26
4504	2-1/2" iron pipe size	**G**	32	.250		14.15	13.35		27.50	36.50
4506	3" iron pipe size	**G**	30	.267		14.15	14.25		28.40	37.50
4508	4" iron pipe size	**G**	28	.286		17.85	15.25		33.10	43
4510	5" iron pipe size	**G**	24	.333		22	17.80		39.80	51.50
4520	2" wall,									
4522	1" iron pipe size	**G** 1 Asbe	36	.222	Ea.	11.30	11.85		23.15	31
4524	2-1/2" iron pipe size	**G**	30	.267		17.85	14.25		32.10	41.50
4526	3" iron pipe size	**G**	28	.286		17.85	15.25		33.10	43
4528	4" iron pipe size	**G**	24	.333		22	17.80		39.80	51.50
4530	5" iron pipe size	**G**	22	.364		32.50	19.40		51.90	65.50
4532	6" iron pipe size	**G**	20	.400		45.50	21.50		67	83
4536	8" iron pipe size	**G**	12	.667		76	35.50		111.50	140
4538	10" iron pipe size	**G**	8	1		88.50	53.50		142	181

For customer support on your Mechanical Cost Data, call 877.739.3689.

111

22 07 19 – Plumbing Piping Insulation

22 07 19.10 Piping Insulation		Crew	Daily Output	Labor-Hours	Unit	Material	2016 Bare Costs Labor	Equipment	Total	Total Incl O&P
4550	2-1/2" wall,									
4560	12" iron pipe size	G 1 Asbe	6	1.333	Ea.	162	71		233	289
4562	14" iron pipe size	G "	4	2	"	214	107		321	400
4570	3" wall,									
4578	6" iron pipe size	G 1 Asbe	16	.500	Ea.	65.50	26.50		92	114
4580	8" iron pipe size	G	10	.800		98	42.50		140.50	175
4582	10" iron pipe size	G	6	1.333		133	71		204	257
4900	Calcium silicate, with 8 oz. canvas cover									
5100	1" wall, 1/2" iron pipe size	G Q-14	170	.094	L.F.	4.07	4.52		8.59	11.55
5130	3/4" iron pipe size	G	170	.094		4.10	4.52		8.62	11.55
5140	1" iron pipe size	G	170	.094		3.98	4.52		8.50	11.45
5150	1-1/4" iron pipe size	G	165	.097		4.07	4.66		8.73	11.75
5160	1-1/2" iron pipe size	G	165	.097		4.13	4.66		8.79	11.80
5170	2" iron pipe size	G	160	.100		4.74	4.81		9.55	12.70
5180	2-1/2" iron pipe size	G	160	.100		5.05	4.81		9.86	13.05
5190	3" iron pipe size	G	150	.107		5.45	5.15		10.60	14
5200	4" iron pipe size	G	140	.114		6.75	5.50		12.25	16
5210	5" iron pipe size	G	135	.119		7.10	5.70		12.80	16.70
5220	6" iron pipe size	G	130	.123		7.55	5.90		13.45	17.55
5280	1-1/2" wall, 1/2" iron pipe size	G	150	.107		4.40	5.15		9.55	12.85
5310	3/4" iron pipe size	G	150	.107		4.48	5.15		9.63	12.95
5320	1" iron pipe size	G	150	.107		4.88	5.15		10.03	13.35
5330	1-1/4" iron pipe size	G	145	.110		5.25	5.30		10.55	14.05
5340	1-1/2" iron pipe size	G	145	.110		5.65	5.30		10.95	14.45
5350	2" iron pipe size	G	140	.114		6.20	5.50		11.70	15.35
5360	2-1/2" iron pipe size	G	140	.114		6.75	5.50		12.25	16
5370	3" iron pipe size	G	135	.119		7.10	5.70		12.80	16.65
5380	4" iron pipe size	G	125	.128		8.20	6.15		14.35	18.60
5390	5" iron pipe size	G	120	.133		9.25	6.40		15.65	20
5400	6" iron pipe size	G	110	.145		9.55	7		16.55	21.50
5402	8" iron pipe size	G	95	.168		12.55	8.10		20.65	26.50
5404	10" iron pipe size	G	85	.188		16.40	9.05		25.45	32
5406	12" iron pipe size	G	80	.200		19.40	9.60		29	36.50
5408	14" iron pipe size	G	75	.213		22	10.25		32.25	40
5410	16" iron pipe size	G	70	.229		24.50	11		35.50	44
5412	18" iron pipe size	G	65	.246		27	11.85		38.85	48
5460	2" wall, 1/2" iron pipe size	G	135	.119		6.80	5.70		12.50	16.30
5490	3/4" iron pipe size	G	135	.119		7.15	5.70		12.85	16.70
5500	1" iron pipe size	G	135	.119		7.50	5.70		13.20	17.10
5510	1-1/4" iron pipe size	G	130	.123		8.05	5.90		13.95	18.05
5520	1-1/2" iron pipe size	G	130	.123		8.40	5.90		14.30	18.45
5530	2" iron pipe size	G	125	.128		8.90	6.15		15.05	19.35
5540	2-1/2" iron pipe size	G	125	.128		10.55	6.15		16.70	21
5550	3" iron pipe size	G	120	.133		10.65	6.40		17.05	21.50
5560	4" iron pipe size	G	115	.139		12.30	6.70		19	24
5570	5" iron pipe size	G	110	.145		14	7		21	26.50
5580	6" iron pipe size	G	105	.152		15.30	7.30		22.60	28.50
5581	8" iron pipe size	G	95	.168		18.25	8.10		26.35	32.50
5582	10" iron pipe size	G	85	.188		23	9.05		32.05	39
5583	12" iron pipe size	G	80	.200		25.50	9.60		35.10	43
5584	14" iron pipe size	G	75	.213		28	10.25		38.25	46.50
5585	16" iron pipe size	G	70	.229		30.50	11		41.50	50.50
5586	18" iron pipe size	G	65	.246		33.50	11.85		45.35	55.50

For customer support on your Mechanical Cost Data, call 877.739.3689.

22 07 Plumbing Insulation

22 07 19 – Plumbing Piping Insulation

22 07 19.10 Piping Insulation		Crew	Daily Output	Labor-Hours	Unit	Material	2016 Bare Costs Labor	Equipment	Total	Total Incl O&P
5587	3" wall, 1-1/4" iron pipe size	G	Q-14	100	.160	L.F.	12.45	7.70		20.15	25.50
5588	1-1/2" iron pipe size	G		100	.160		12.60	7.70		20.30	26
5589	2" iron pipe size	G		95	.168		13	8.10		21.10	27
5590	2-1/2" iron pipe size	G		95	.168		15.15	8.10		23.25	29.50
5591	3" iron pipe size	G		90	.178		15.35	8.55		23.90	30
5592	4" iron pipe size	G		85	.188		19.40	9.05		28.45	35.50
5593	6" iron pipe size	G		80	.200		23.50	9.60		33.10	41
5594	8" iron pipe size	G		75	.213		28	10.25		38.25	47
5595	10" iron pipe size	G		65	.246		34	11.85		45.85	55.50
5596	12" iron pipe size	G		60	.267		37.50	12.80		50.30	61
5597	14" iron pipe size	G		55	.291		42	14		56	68
5598	16" iron pipe size	G		50	.320		46.50	15.40		61.90	75
5599	18" iron pipe size	G	▼	45	.356	▼	51	17.10		68.10	82.50
5600	Calcium silicate, no cover										
5720	1" wall, 1/2" iron pipe size	G	Q-14	180	.089	L.F.	3.70	4.27		7.97	10.70
5740	3/4" iron pipe size	G		180	.089		3.70	4.27		7.97	10.70
5750	1" iron pipe size	G		180	.089		3.55	4.27		7.82	10.55
5760	1-1/4" iron pipe size	G		175	.091		3.61	4.39		8	10.80
5770	1-1/2" iron pipe size	G		175	.091		3.64	4.39		8.03	10.85
5780	2" iron pipe size	G		170	.094		4.20	4.52		8.72	11.65
5790	2-1/2" iron pipe size	G		170	.094		4.44	4.52		8.96	11.95
5800	3" iron pipe size	G		160	.100		4.77	4.81		9.58	12.75
5810	4" iron pipe size	G		150	.107		5.95	5.15		11.10	14.55
5820	5" iron pipe size	G		145	.110		6.20	5.30		11.50	15.05
5830	6" iron pipe size	G		140	.114		6.50	5.50		12	15.70
5900	1-1/2" wall, 1/2" iron pipe size	G		160	.100		3.91	4.81		8.72	11.80
5920	3/4" iron pipe size	G		160	.100		3.98	4.81		8.79	11.90
5930	1" iron pipe size	G		160	.100		4.34	4.81		9.15	12.25
5940	1-1/4" iron pipe size	G		155	.103		4.67	4.96		9.63	12.85
5950	1-1/2" iron pipe size	G		155	.103		5	4.96		9.96	13.20
5960	2" iron pipe size	G		150	.107		5.55	5.15		10.70	14.10
5970	2-1/2" iron pipe size	G		150	.107		6.05	5.15		11.20	14.65
5980	3" iron pipe size	G		145	.110		6.30	5.30		11.60	15.20
5990	4" iron pipe size	G		135	.119		7.30	5.70		13	16.90
6000	5" iron pipe size	G		130	.123		8.25	5.90		14.15	18.25
6010	6" iron pipe size	G		120	.133		8.40	6.40		14.80	19.20
6020	7" iron pipe size	G		115	.139		9.85	6.70		16.55	21.50
6030	8" iron pipe size	G		105	.152		11.15	7.30		18.45	23.50
6040	9" iron pipe size	G		100	.160		13.35	7.70		21.05	26.50
6050	10" iron pipe size	G		95	.168		14.75	8.10		22.85	29
6060	12" iron pipe size	G		90	.178		17.50	8.55		26.05	32.50
6070	14" iron pipe size	G		85	.188		19.80	9.05		28.85	36
6080	16" iron pipe size	G		80	.200		22	9.60		31.60	39.50
6090	18" iron pipe size	G		75	.213		24.50	10.25		34.75	43
6120	2" wall, 1/2" iron pipe size	G		145	.110		6.20	5.30		11.50	15.05
6140	3/4" iron pipe size	G		145	.110		6.50	5.30		11.80	15.40
6150	1" iron pipe size	G		145	.110		6.85	5.30		12.15	15.80
6160	1-1/4" iron pipe size	G		140	.114		7.35	5.50		12.85	16.65
6170	1-1/2" iron pipe size	G		140	.114		7.70	5.50		13.20	17
6180	2" iron pipe size	G		135	.119		8.05	5.70		13.75	17.70
6190	2-1/2" iron pipe size	G		135	.119		9.70	5.70		15.40	19.50
6200	3" iron pipe size	G		130	.123		9.75	5.90		15.65	19.90
6210	4" iron pipe size	G		125	.128		11.25	6.15		17.40	22

For customer support on your Mechanical Cost Data, call 877.739.3689.

113

22 07 19.10 Piping Insulation		Crew	Daily Output	Labor-Hours	Unit	Material	2016 Bare Costs Labor	Equipment	Total	Total Incl O&P
6220	5" iron pipe size	G	Q-14	120	.133	L.F.	12.85	6.40		19.25	24
6230	6" iron pipe size	G		115	.139		14.05	6.70		20.75	26
6240	7" iron pipe size	G		110	.145		15.20	7		22.20	27.50
6250	8" iron pipe size	G		105	.152		16.75	7.30		24.05	30
6260	9" iron pipe size	G		100	.160		18.90	7.70		26.60	33
6270	10" iron pipe size	G		95	.168		21	8.10		29.10	35.50
6280	12" iron pipe size	G		90	.178		23.50	8.55		32.05	39
6290	14" iron pipe size	G		85	.188		25.50	9.05		34.55	42
6300	16" iron pipe size	G		80	.200		28.50	9.60		38.10	46
6310	18" iron pipe size	G		75	.213		31	10.25		41.25	50
6320	20" iron pipe size	G		65	.246		38	11.85		49.85	60.50
6330	22" iron pipe size	G		60	.267		42	12.80		54.80	66.50
6340	24" iron pipe size	G		55	.291		43	14		57	69.50
6360	3" wall, 1/2" iron pipe size	G		115	.139		11.20	6.70		17.90	23
6380	3/4" iron pipe size	G		115	.139		11.30	6.70		18	23
6390	1" iron pipe size	G		115	.139		11.35	6.70		18.05	23
6400	1-1/4" iron pipe size	G		110	.145		11.55	7		18.55	23.50
6410	1-1/2" iron pipe size	G		110	.145		11.60	7		18.60	23.50
6420	2" iron pipe size	G		105	.152		12	7.30		19.30	24.50
6430	2-1/2" iron pipe size	G		105	.152		14.10	7.30		21.40	27
6440	3" iron pipe size	G		100	.160		14.20	7.70		21.90	27.50
6450	4" iron pipe size	G		95	.168		18.15	8.10		26.25	32.50
6460	5" iron pipe size	G		90	.178		19.70	8.55		28.25	35
6470	6" iron pipe size	G		90	.178		22	8.55		30.55	38
6480	7" iron pipe size	G		85	.188		24.50	9.05		33.55	41
6490	8" iron pipe size	G		85	.188		26.50	9.05		35.55	43
6500	9" iron pipe size	G		80	.200		30	9.60		39.60	48
6510	10" iron pipe size	G		75	.213		32	10.25		42.25	51
6520	12" iron pipe size	G		70	.229		35	11		46	56
6530	14" iron pipe size	G		65	.246		39.50	11.85		51.35	62
6540	16" iron pipe size	G		60	.267		43.50	12.80		56.30	68
6550	18" iron pipe size	G		55	.291		48	14		62	74.50
6560	20" iron pipe size	G		50	.320		57	15.40		72.40	86.50
6570	22" iron pipe size	G		45	.356		62	17.10		79.10	94.50
6580	24" iron pipe size	G		40	.400		66.50	19.20		85.70	103
6600	Fiberglass, with all service jacket										
6640	1/2" wall, 1/2" iron pipe size	G	Q-14	250	.064	L.F.	.68	3.08		3.76	5.55
6660	3/4" iron pipe size	G		240	.067		.77	3.20		3.97	5.85
6670	1" iron pipe size	G		230	.070		.80	3.34		4.14	6.10
6680	1-1/4" iron pipe size	G		220	.073		.85	3.49		4.34	6.40
6690	1-1/2" iron pipe size	G		220	.073		.98	3.49		4.47	6.55
6700	2" iron pipe size	G		210	.076		1.05	3.66		4.71	6.85
6710	2-1/2" iron pipe size	G		200	.080		1.09	3.84		4.93	7.20
6840	1" wall, 1/2" iron pipe size	G		240	.067		.83	3.20		4.03	5.90
6860	3/4" iron pipe size	G		230	.070		.90	3.34		4.24	6.20
6870	1" iron pipe size	G		220	.073		.97	3.49		4.46	6.50
6880	1-1/4" iron pipe size	G		210	.076		1.05	3.66		4.71	6.85
6890	1-1/2" iron pipe size	G		210	.076		1.13	3.66		4.79	6.95
6900	2" iron pipe size	G		200	.080		1.22	3.84		5.06	7.35
6910	2-1/2" iron pipe size	G		190	.084		1.39	4.05		5.44	7.85
6920	3" iron pipe size	G		180	.089		1.49	4.27		5.76	8.30
6930	3-1/2" iron pipe size	G		170	.094		1.61	4.52		6.13	8.80
6940	4" iron pipe size	G		150	.107		1.97	5.15		7.12	10.15

22 07 Plumbing Insulation

22 07 19 – Plumbing Piping Insulation

22 07 19.10 Piping Insulation		Crew	Daily Output	Labor-Hours	Unit	Material	2016 Bare Costs Labor	Equipment	Total	Total Incl O&P
6950	5" iron pipe size	G	Q-14	140	.114	L.F.	2.22	5.50		7.72	11
6960	6" iron pipe size	G		120	.133		2.36	6.40		8.76	12.55
6970	7" iron pipe size	G		110	.145		2.81	7		9.81	14
6980	8" iron pipe size	G		100	.160		3.85	7.70		11.55	16.20
6990	9" iron pipe size	G		90	.178		4.06	8.55		12.61	17.75
7000	10" iron pipe size	G		90	.178		4.09	8.55		12.64	17.80
7010	12" iron pipe size	G		80	.200		4.45	9.60		14.05	19.85
7020	14" iron pipe size	G		80	.200		5.35	9.60		14.95	21
7030	16" iron pipe size	G		70	.229		6.80	11		17.80	24.50
7040	18" iron pipe size	G		70	.229		7.55	11		18.55	25.50
7050	20" iron pipe size	G		60	.267		8.40	12.80		21.20	29
7060	24" iron pipe size	G		60	.267		10.25	12.80		23.05	31
7080	1-1/2" wall, 1/2" iron pipe size	G		230	.070		1.57	3.34		4.91	6.95
7100	3/4" iron pipe size	G		220	.073		1.57	3.49		5.06	7.20
7110	1" iron pipe size	G		210	.076		1.68	3.66		5.34	7.55
7120	1-1/4" iron pipe size	G		200	.080		1.83	3.84		5.67	8
7130	1-1/2" iron pipe size	G		200	.080		1.92	3.84		5.76	8.10
7140	2" iron pipe size	G		190	.084		2.11	4.05		6.16	8.60
7150	2-1/2" iron pipe size	G		180	.089		2.27	4.27		6.54	9.15
7160	3" iron pipe size	G		170	.094		2.37	4.52		6.89	9.65
7170	3-1/2" iron pipe size	G		160	.100		2.60	4.81		7.41	10.35
7180	4" iron pipe size	G		140	.114		2.70	5.50		8.20	11.50
7190	5" iron pipe size	G		130	.123		3.02	5.90		8.92	12.50
7200	6" iron pipe size	G		110	.145		3.19	7		10.19	14.40
7210	7" iron pipe size	G		100	.160		3.54	7.70		11.24	15.85
7220	8" iron pipe size	G		90	.178		4.47	8.55		13.02	18.20
7230	9" iron pipe size	G		85	.188		4.66	9.05		13.71	19.25
7240	10" iron pipe size	G		80	.200		4.82	9.60		14.42	20.50
7250	12" iron pipe size	G		75	.213		5.45	10.25		15.70	22
7260	14" iron pipe size	G		70	.229		6.55	11		17.55	24.50
7270	16" iron pipe size	G		65	.246		8.50	11.85		20.35	28
7280	18" iron pipe size	G		60	.267		9.55	12.80		22.35	30.50
7290	20" iron pipe size	G		55	.291		9.75	14		23.75	33
7300	24" iron pipe size	G		50	.320		11.95	15.40		27.35	37
7320	2" wall, 1/2" iron pipe size	G		220	.073		2.43	3.49		5.92	8.10
7340	3/4" iron pipe size	G		210	.076		2.51	3.66		6.17	8.45
7350	1" iron pipe size	G		200	.080		2.67	3.84		6.51	8.95
7360	1-1/4" iron pipe size	G		190	.084		2.82	4.05		6.87	9.40
7370	1-1/2" iron pipe size	G		190	.084		2.94	4.05		6.99	9.55
7380	2" iron pipe size	G		180	.089		3.10	4.27		7.37	10.05
7390	2-1/2" iron pipe size	G		170	.094		3.34	4.52		7.86	10.70
7400	3" iron pipe size	G		160	.100		3.54	4.81		8.35	11.40
7410	3-1/2" iron pipe size	G		150	.107		3.84	5.15		8.99	12.20
7420	4" iron pipe size	G		130	.123		4.13	5.90		10.03	13.75
7430	5" iron pipe size	G		120	.133		4.73	6.40		11.13	15.15
7440	6" iron pipe size	G		100	.160		4.88	7.70		12.58	17.30
7450	7" iron pipe size	G		90	.178		5.55	8.55		14.10	19.40
7460	8" iron pipe size	G		80	.200		5.95	9.60		15.55	21.50
7470	9" iron pipe size	G		75	.213		6.50	10.25		16.75	23
7480	10" iron pipe size	G		70	.229		7.10	11		18.10	25
7490	12" iron pipe size	G		65	.246		7.95	11.85		19.80	27
7500	14" iron pipe size	G		60	.267		10.25	12.80		23.05	31
7510	16" iron pipe size	G		55	.291		11.25	14		25.25	34.50

22 07 19.10 Piping Insulation		Crew	Daily Output	Labor-Hours	Unit	Material	2016 Bare Costs Labor	Equipment	Total	Total Incl O&P
7520	18" iron pipe size	G Q-14	50	.320	L.F.	12.55	15.40		27.95	38
7530	20" iron pipe size	G	45	.356		14.25	17.10		31.35	42
7540	24" iron pipe size	G	40	.400		15.30	19.20		34.50	47
7560	2-1/2" wall, 1/2" iron pipe size	G	210	.076		2.88	3.66		6.54	8.85
7562	3/4" iron pipe size	G	200	.080		3.01	3.84		6.85	9.30
7564	1" iron pipe size	G	190	.084		3.14	4.05		7.19	9.75
7566	1-1/4" iron pipe size	G	185	.086		3.25	4.16		7.41	10.05
7568	1-1/2" iron pipe size	G	180	.089		3.42	4.27		7.69	10.40
7570	2" iron pipe size	G	170	.094		3.59	4.52		8.11	11
7572	2-1/2" iron pipe size	G	160	.100		4.13	4.81		8.94	12.05
7574	3" iron pipe size	G	150	.107		4.35	5.15		9.50	12.80
7576	3-1/2" iron pipe size	G	140	.114		4.73	5.50		10.23	13.75
7578	4" iron pipe size	G	120	.133		4.99	6.40		11.39	15.45
7580	5" iron pipe size	G	110	.145		5.95	7		12.95	17.45
7582	6" iron pipe size	G	90	.178		7.15	8.55		15.70	21
7584	7" iron pipe size	G	80	.200		7.20	9.60		16.80	23
7586	8" iron pipe size	G	70	.229		7.65	11		18.65	25.50
7588	9" iron pipe size	G	65	.246		8.35	11.85		20.20	27.50
7590	10" iron pipe size	G	60	.267		9.10	12.80		21.90	30
7592	12" iron pipe size	G	55	.291		11.05	14		25.05	34
7594	14" iron pipe size	G	50	.320		12.90	15.40		28.30	38
7596	16" iron pipe size	G	45	.356		14.80	17.10		31.90	43
7598	18" iron pipe size	G	40	.400		16.05	19.20		35.25	47.50
7602	24" iron pipe size	G	30	.533		21	25.50		46.50	63.50
7620	3" wall, 1/2" iron pipe size	G	200	.080		3.69	3.84		7.53	10.05
7622	3/4" iron pipe size	G	190	.084		3.90	4.05		7.95	10.60
7624	1" iron pipe size	G	180	.089		4.13	4.27		8.40	11.20
7626	1-1/4" iron pipe size	G	175	.091		4.21	4.39		8.60	11.50
7628	1-1/2" iron pipe size	G	170	.094		4.40	4.52		8.92	11.90
7630	2" iron pipe size	G	160	.100		4.76	4.81		9.57	12.75
7632	2-1/2" iron pipe size	G	150	.107		4.95	5.15		10.10	13.45
7634	3" iron pipe size	G	140	.114		5.25	5.50		10.75	14.35
7636	3-1/2" iron pipe size	G	130	.123		5.80	5.90		11.70	15.60
7638	4" iron pipe size	G	110	.145		6.25	7		13.25	17.80
7640	5" iron pipe size	G	100	.160		7.05	7.70		14.75	19.75
7642	6" iron pipe size	G	80	.200		7.55	9.60		17.15	23.50
7644	7" iron pipe size	G	70	.229		8.65	11		19.65	26.50
7646	8" iron pipe size	G	60	.267		9.40	12.80		22.20	30.50
7648	9" iron pipe size	G	55	.291		10.15	14		24.15	33
7650	10" iron pipe size	G	50	.320		10.90	15.40		26.30	36
7652	12" iron pipe size	G	45	.356		13.55	17.10		30.65	41.50
7654	14" iron pipe size	G	40	.400		15.65	19.20		34.85	47
7656	16" iron pipe size	G	35	.457		17.75	22		39.75	53.50
7658	18" iron pipe size	G	32	.500		18.95	24		42.95	58.50
7660	20" iron pipe size	G	30	.533		20.50	25.50		46	62.50
7662	24" iron pipe size	G	28	.571		26	27.50		53.50	71.50
7664	26" iron pipe size	G	26	.615		29	29.50		58.50	77.50
7666	30" iron pipe size	G	24	.667		34	32		66	87.50
7800	For fiberglass with standard canvas jacket, deduct					5%				
7802	For fittings, add 3 L.F. for each fitting									
7804	plus 4 L.F. for each flange of the fitting									
7810	Finishes									
7814	For single layer of felt, add					10%	10%			

22 07 19 – Plumbing Piping Insulation

22 07 19.10 Piping Insulation		Crew	Daily Output	Labor-Hours	Unit	Material	2016 Bare Costs Labor	Equipment	Total	Total Incl O&P
7816	For roofing paper, 45 lb. to 55 lb., add					25%	10%			
7820	Polyethylene tubing flexible closed cell foam, UV resistant									
7828	Standard temperature (-90°F to +212°F)									
7830	3/8" wall, 1/8" iron pipe size	G	1 Asbe	130	.062	L.F.	.22	3.29		3.51	5.35
7831	1/4" iron pipe size	G		130	.062		.23	3.29		3.52	5.35
7832	3/8" iron pipe size	G		130	.062		.26	3.29		3.55	5.40
7833	1/2" iron pipe size	G		126	.063		.29	3.39		3.68	5.60
7834	3/4" iron pipe size	G		122	.066		.34	3.50		3.84	5.80
7835	1" iron pipe size	G		120	.067		.39	3.56		3.95	6
7836	1-1/4" iron pipe size	G		118	.068		.47	3.62		4.09	6.15
7837	1-1/2" iron pipe size	G		118	.068		.60	3.62		4.22	6.30
7838	2" iron pipe size	G		116	.069		.79	3.68		4.47	6.60
7839	2-1/2" iron pipe size	G		114	.070		1.01	3.75		4.76	6.95
7840	3" iron pipe size	G		112	.071		1.56	3.81		5.37	7.65
7842	1/2" wall, 1/8" iron pipe size	G		120	.067		.33	3.56		3.89	5.90
7843	1/4" iron pipe size	G		120	.067		.35	3.56		3.91	5.95
7844	3/8" iron pipe size	G		120	.067		.39	3.56		3.95	6
7845	1/2" iron pipe size	G		118	.068		.43	3.62		4.05	6.10
7846	3/4" iron pipe size	G		116	.069		.49	3.68		4.17	6.30
7847	1" iron pipe size	G		114	.070		.55	3.75		4.30	6.45
7848	1-1/4" iron pipe size	G		112	.071		.65	3.81		4.46	6.65
7849	1-1/2" iron pipe size	G		110	.073		.77	3.88		4.65	6.90
7850	2" iron pipe size	G		108	.074		1.11	3.96		5.07	7.35
7851	2-1/2" iron pipe size	G		106	.075		1.37	4.03		5.40	7.75
7852	3" iron pipe size	G		104	.077		1.89	4.11		6	8.50
7853	3-1/2" iron pipe size	G		102	.078		2.14	4.19		6.33	8.85
7854	4" iron pipe size	G		100	.080		2.44	4.27		6.71	9.35
7855	3/4" wall, 1/8" iron pipe size	G		110	.073		.50	3.88		4.38	6.60
7856	1/4" iron pipe size	G		110	.073		.53	3.88		4.41	6.65
7857	3/8" iron pipe size	G		108	.074		.61	3.96		4.57	6.80
7858	1/2" iron pipe size	G		106	.075		.73	4.03		4.76	7.05
7859	3/4" iron pipe size	G		104	.077		.88	4.11		4.99	7.35
7860	1" iron pipe size	G		102	.078		1.01	4.19		5.20	7.60
7861	1-1/4" iron pipe size	G		100	.080		1.38	4.27		5.65	8.15
7862	1-1/2" iron pipe size	G		100	.080		1.56	4.27		5.83	8.35
7863	2" iron pipe size	G		98	.082		1.98	4.36		6.34	9
7864	2-1/2" iron pipe size	G		96	.083		2.39	4.45		6.84	9.55
7865	3" iron pipe size	G		94	.085		3.15	4.54		7.69	10.50
7866	3-1/2" iron pipe size	G		92	.087		3.51	4.64		8.15	11.10
7867	4" iron pipe size	G		90	.089		3.92	4.75		8.67	11.70
7868	1" wall, 1/4" iron pipe size	G		100	.080		1.06	4.27		5.33	7.80
7869	3/8" iron pipe size	G		98	.082		1.08	4.36		5.44	8
7870	1/2" iron pipe size	G		96	.083		1.35	4.45		5.80	8.40
7871	3/4" iron pipe size	G		94	.085		1.57	4.54		6.11	8.80
7872	1" iron pipe size	G		92	.087		1.91	4.64		6.55	9.35
7873	1-1/4" iron pipe size	G		90	.089		2.17	4.75		6.92	9.80
7874	1-1/2" iron pipe size	G		90	.089		2.48	4.75		7.23	10.15
7875	2" iron pipe size	G		88	.091		3.23	4.85		8.08	11.10
7876	2-1/2" iron pipe size	G		86	.093		4.22	4.97		9.19	12.40
7877	3" iron pipe size	G		84	.095		4.74	5.10		9.84	13.10
7878	Contact cement, quart can	G				Ea.	10.50			10.50	11.55
7879	Rubber tubing, flexible closed cell foam									
7880	3/8" wall, 1/4" iron pipe size	G	1 Asbe	120	.067	L.F.	.33	3.56		3.89	5.90

22 07 19.10 Piping Insulation		Crew	Daily Output	Labor-Hours	Unit	Material	2016 Bare Costs Labor	Equipment	Total	Total Incl O&P
7900	3/8" iron pipe size	G	1 Asbe	120	.067	L.F.	.36	3.56		3.92	5.95
7910	1/2" iron pipe size	G		115	.070		.41	3.72		4.13	6.25
7920	3/4" iron pipe size	G		115	.070		.46	3.72		4.18	6.30
7930	1" iron pipe size	G		110	.073		.52	3.88		4.40	6.60
7940	1-1/4" iron pipe size	G		110	.073		.59	3.88		4.47	6.70
7950	1-1/2" iron pipe size	G		110	.073		.73	3.88		4.61	6.85
8100	1/2" wall, 1/4" iron pipe size	G		90	.089		.54	4.75		5.29	8
8120	3/8" iron pipe size	G		90	.089		.60	4.75		5.35	8.05
8130	1/2" iron pipe size	G		89	.090		.67	4.80		5.47	8.20
8140	3/4" iron pipe size	G		89	.090		.75	4.80		5.55	8.30
8150	1" iron pipe size	G		88	.091		.82	4.85		5.67	8.45
8160	1-1/4" iron pipe size	G		87	.092		.95	4.91		5.86	8.70
8170	1-1/2" iron pipe size	G		87	.092		1.15	4.91		6.06	8.90
8180	2" iron pipe size	G		86	.093		1.47	4.97		6.44	9.35
8190	2-1/2" iron pipe size	G		86	.093		1.85	4.97		6.82	9.80
8200	3" iron pipe size	G		85	.094		2.06	5.05		7.11	10.05
8210	3-1/2" iron pipe size	G		85	.094		2.86	5.05		7.91	10.95
8220	4" iron pipe size	G		80	.100		3.06	5.35		8.41	11.65
8230	5" iron pipe size	G		80	.100		4.17	5.35		9.52	12.90
8240	6" iron pipe size	G		75	.107		4.17	5.70		9.87	13.45
8300	3/4" wall, 1/4" iron pipe size	G		90	.089		.85	4.75		5.60	8.35
8320	3/8" iron pipe size	G		90	.089		.92	4.75		5.67	8.40
8330	1/2" iron pipe size	G		89	.090		1.10	4.80		5.90	8.65
8340	3/4" iron pipe size	G		89	.090		1.35	4.80		6.15	8.95
8350	1" iron pipe size	G		88	.091		1.54	4.85		6.39	9.25
8360	1-1/4" iron pipe size	G		87	.092		2.05	4.91		6.96	9.90
8370	1-1/2" iron pipe size	G		87	.092		2.32	4.91		7.23	10.20
8380	2" iron pipe size	G		86	.093		2.68	4.97		7.65	10.70
8390	2-1/2" iron pipe size	G		86	.093		3.58	4.97		8.55	11.70
8400	3" iron pipe size	G		85	.094		4.08	5.05		9.13	12.30
8410	3-1/2" iron pipe size	G		85	.094		4.81	5.05		9.86	13.10
8420	4" iron pipe size	G		80	.100		5.10	5.35		10.45	13.95
8430	5" iron pipe size	G		80	.100		6.15	5.35		11.50	15.10
8440	6" iron pipe size	G		80	.100		7.45	5.35		12.80	16.50
8444	1" wall, 1/2" iron pipe size	G		86	.093		2.05	4.97		7.02	10
8445	3/4" iron pipe size	G		84	.095		2.48	5.10		7.58	10.65
8446	1" iron pipe size	G		84	.095		2.89	5.10		7.99	11.10
8447	1-1/4" iron pipe size	G		82	.098		3.23	5.20		8.43	11.65
8448	1-1/2" iron pipe size	G		82	.098		3.76	5.20		8.96	12.25
8449	2" iron pipe size	G		80	.100		4.95	5.35		10.30	13.75
8450	2-1/2" iron pipe size	G		80	.100		6.45	5.35		11.80	15.40
8456	Rubber insulation tape, 1/8" x 2" x 30'	G				Ea.	21			21	23
8460	Polyolefin tubing, flexible closed cell foam, UV stabilized, work										
8462	temp. -165°F to +210°F, 0 water vapor transmission										
8464	3/8" wall, 1/8" iron pipe size	G	1 Asbe	140	.057	L.F.	.48	3.05		3.53	5.30
8466	1/4" iron pipe size	G		140	.057		.50	3.05		3.55	5.30
8468	3/8" iron pipe size	G		140	.057		.55	3.05		3.60	5.35
8470	1/2" iron pipe size	G		136	.059		.61	3.14		3.75	5.55
8472	3/4" iron pipe size	G		132	.061		.62	3.24		3.86	5.75
8474	1" iron pipe size	G		130	.062		.83	3.29		4.12	6
8476	1-1/4" iron pipe size	G		128	.063		1.02	3.34		4.36	6.30
8478	1-1/2" iron pipe size	G		128	.063		1.25	3.34		4.59	6.60
8480	2" iron pipe size	G		126	.063		1.50	3.39		4.89	6.95

22 07 19 – Plumbing Piping Insulation

22 07 19.10 Piping Insulation		Crew	Daily Output	Labor-Hours	Unit	Material	2016 Bare Costs Labor	Equipment	Total	Total Incl O&P
8482	2-1/2" iron pipe size [G]	1 Asbe	123	.065	L.F.	2.17	3.47		5.64	7.80
8484	3" iron pipe size [G]		121	.066		2.42	3.53		5.95	8.15
8486	4" iron pipe size [G]		118	.068		4.43	3.62		8.05	10.50
8500	1/2" wall, 1/8" iron pipe size [G]		130	.062		.71	3.29		4	5.90
8502	1/4" iron pipe size [G]		130	.062		.76	3.29		4.05	5.95
8504	3/8" iron pipe size [G]		130	.062		.81	3.29		4.10	6
8506	1/2" iron pipe size [G]		128	.063		.89	3.34		4.23	6.20
8508	3/4" iron pipe size [G]		126	.063		1.02	3.39		4.41	6.40
8510	1" iron pipe size [G]		123	.065		1.18	3.47		4.65	6.70
8512	1-1/4" iron pipe size [G]		121	.066		1.39	3.53		4.92	7.05
8514	1-1/2" iron pipe size [G]		119	.067		1.67	3.59		5.26	7.45
8516	2" iron pipe size [G]		117	.068		2.11	3.65		5.76	8
8518	2-1/2" iron pipe size [G]		114	.070		2.85	3.75		6.60	9
8520	3" iron pipe size [G]		112	.071		3.71	3.81		7.52	10.05
8522	4" iron pipe size [G]		110	.073		5.15	3.88		9.03	11.70
8534	3/4" wall, 1/8" iron pipe size [G]		120	.067		1.08	3.56		4.64	6.75
8536	1/4" iron pipe size [G]		120	.067		1.14	3.56		4.70	6.80
8538	3/8" iron pipe size [G]		117	.068		1.30	3.65		4.95	7.15
8540	1/2" iron pipe size [G]		114	.070		1.45	3.75		5.20	7.45
8542	3/4" iron pipe size [G]		112	.071		1.56	3.81		5.37	7.65
8544	1" iron pipe size [G]		110	.073		2.20	3.88		6.08	8.45
8546	1-1/4" iron pipe size [G]		108	.074		2.94	3.96		6.90	9.40
8548	1-1/2" iron pipe size [G]		108	.074		3.43	3.96		7.39	9.90
8550	2" iron pipe size [G]		106	.075		4.34	4.03		8.37	11
8552	2-1/2" iron pipe size [G]		104	.077		5.45	4.11		9.56	12.40
8554	3" iron pipe size [G]		102	.078		6.80	4.19		10.99	14
8556	4" iron pipe size [G]		100	.080		9	4.27		13.27	16.55
8570	1" wall, 1/8" iron pipe size [G]		110	.073		1.95	3.88		5.83	8.20
8572	1/4" iron pipe size [G]		108	.074		2	3.96		5.96	8.35
8574	3/8" iron pipe size [G]		106	.075		2.04	4.03		6.07	8.50
8576	1/2" iron pipe size [G]		104	.077		2.16	4.11		6.27	8.80
8578	3/4" iron pipe size [G]		102	.078		2.83	4.19		7.02	9.60
8580	1" iron pipe size [G]		100	.080		3.27	4.27		7.54	10.25
8582	1-1/4" iron pipe size [G]		97	.082		3.72	4.40		8.12	10.95
8584	1-1/2" iron pipe size [G]		97	.082		4.25	4.40		8.65	11.55
8586	2" iron pipe size [G]		95	.084		6	4.50		10.50	13.60
8588	2-1/2" iron pipe size [G]		93	.086		7.85	4.59		12.44	15.75
8590	3" iron pipe size [G]		91	.088		9.55	4.69		14.24	17.80
8606	Contact adhesive (R-320) [G]				Qt.	16.85			16.85	18.55
8608	Contact adhesive (R-320) [G]				Gal.	67			67	74
8610	NOTE: Preslit/preglued vs unslit, same price									

22 07 19.30 Piping Insulation Protective Jacketing, PVC

		Crew	Daily Output	Labor-Hours	Unit	Material	2016 Bare Costs Labor	Equipment	Total	Total Incl O&P
0010	**PIPING INSULATION PROTECTIVE JACKETING, PVC**									
0100	PVC, white, 48" lengths cut from roll goods									
0120	20 mil thick									
0140	Size based on OD of insulation									
0150	1-1/2" ID	Q-14	270	.059	L.F.	.20	2.85		3.05	4.65
0152	2" ID		260	.062		.27	2.96		3.23	4.90
0154	2-1/2" ID		250	.064		.32	3.08		3.40	5.15
0156	3" ID		240	.067		.39	3.20		3.59	5.40
0158	3-1/2" ID		230	.070		.44	3.34		3.78	5.70
0160	4" ID		220	.073		.50	3.49		3.99	6

For customer support on your Mechanical Cost Data, call 877.739.3689.

119

22 07 19.30 Piping Insulation Protective Jacketing, PVC

		Crew	Daily Output	Labor-Hours	Unit	Material	2016 Bare Costs Labor	Equipment	Total	Total Incl O&P
0162	4-1/2" ID	Q-14	210	.076	L.F.	.56	3.66		4.22	6.30
0164	5" ID		200	.080		.61	3.84		4.45	6.65
0166	5-1/2" ID		190	.084		.67	4.05		4.72	7.05
0168	6" ID		180	.089		.73	4.27		5	7.45
0170	6-1/2" ID		175	.091		.79	4.39		5.18	7.70
0172	7" ID		170	.094		.84	4.52		5.36	7.95
0174	7-1/2" ID		164	.098		.91	4.69		5.60	8.30
0176	8" ID		161	.099		.96	4.78		5.74	8.50
0178	8-1/2" ID		158	.101		1.02	4.87		5.89	8.65
0180	9" ID		155	.103		1.08	4.96		6.04	8.90
0182	9-1/2" ID		152	.105		1.14	5.05		6.19	9.10
0184	10" ID		149	.107		1.19	5.15		6.34	9.35
0186	10-1/2" ID		146	.110		1.25	5.25		6.50	9.60
0188	11" ID		143	.112		1.31	5.40		6.71	9.80
0190	11-1/2" ID		140	.114		1.37	5.50		6.87	10.05
0192	12" ID		137	.117		1.42	5.60		7.02	10.30
0194	12-1/2" ID		134	.119		1.49	5.75		7.24	10.60
0195	13" ID		132	.121		1.54	5.80		7.34	10.75
0196	13-1/2" ID		132	.121		1.60	5.80		7.40	10.80
0198	14" ID		130	.123		1.66	5.90		7.56	11.05
0200	15" ID		128	.125		1.78	6		7.78	11.30
0202	16" ID		126	.127		1.89	6.10		7.99	11.60
0204	17" ID		124	.129		2.01	6.20		8.21	11.85
0206	18" ID		122	.131		2.13	6.30		8.43	12.15
0208	19" ID		120	.133		2.24	6.40		8.64	12.40
0210	20" ID		118	.136		2.35	6.50		8.85	12.75
0212	21" ID		116	.138		2.47	6.65		9.12	13
0214	22" ID		114	.140		2.58	6.75		9.33	13.35
0216	23" ID		112	.143		2.69	6.85		9.54	13.65
0218	24" ID		110	.145		2.81	7		9.81	14
0220	25" ID		108	.148		2.93	7.10		10.03	14.30
0222	26" ID		106	.151		3.04	7.25		10.29	14.65
0224	27" ID		104	.154		3.16	7.40		10.56	15
0226	28" ID		102	.157		3.27	7.55		10.82	15.35
0228	29" ID		100	.160		3.39	7.70		11.09	15.70
0230	30" ID		98	.163		3.51	7.85		11.36	16.05
0300	For colors, add				Ea.	10%				
1000	30 mil thick									
1010	Size based on OD of insulation									
1020	2" ID	Q-14	260	.062	L.F.	.40	2.96		3.36	5.05
1022	2-1/2" ID		250	.064		.48	3.08		3.56	5.30
1024	3" ID		240	.067		.58	3.20		3.78	5.65
1026	3-1/2" ID		230	.070		.65	3.34		3.99	5.90
1028	4" ID		220	.073		.74	3.49		4.23	6.25
1030	4-1/2" ID		210	.076		.82	3.66		4.48	6.60
1032	5" ID		200	.080		.91	3.84		4.75	7
1034	5-1/2" ID		190	.084		.99	4.05		5.04	7.40
1036	6" ID		180	.089		1.08	4.27		5.35	7.85
1038	6-1/2" ID		175	.091		1.16	4.39		5.55	8.15
1040	7" ID		170	.094		1.25	4.52		5.77	8.45
1042	7-1/2" ID		164	.098		1.34	4.69		6.03	8.75
1044	8" ID		161	.099		1.43	4.78		6.21	9
1046	8-1/2" ID		158	.101		1.51	4.87		6.38	9.20

22 07 Plumbing Insulation

22 07 19 – Plumbing Piping Insulation

22 07 19.30 Piping Insulation Protective Jacketing, PVC		Crew	Daily Output	Labor-Hours	Unit	Material	2016 Bare Costs Labor	Equipment	Total	Total Incl O&P
1048	9" ID	Q-14	155	.103	L.F.	1.60	4.96		6.56	9.45
1050	9-1/2" ID		152	.105		1.68	5.05		6.73	9.70
1052	10" ID		149	.107		1.77	5.15		6.92	10
1054	10-1/2" ID		146	.110		1.85	5.25		7.10	10.25
1056	11" ID		143	.112		1.94	5.40		7.34	10.50
1058	11-1/2" ID		140	.114		2.02	5.50		7.52	10.75
1060	12" ID		137	.117		2.11	5.60		7.71	11.05
1062	12-1/2" ID		134	.119		2.20	5.75		7.95	11.35
1063	13" ID		132	.121		2.29	5.80		8.09	11.55
1064	13-1/2" ID		132	.121		2.37	5.80		8.17	11.65
1066	14" ID		130	.123		2.46	5.90		8.36	11.90
1068	15" ID		128	.125		2.63	6		8.63	12.25
1070	16" ID		126	.127		2.80	6.10		8.90	12.60
1072	17" ID		124	.129		2.97	6.20		9.17	12.90
1074	18" ID		122	.131		3.15	6.30		9.45	13.25
1076	19" ID		120	.133		3.32	6.40		9.72	13.60
1078	20" ID		118	.136		3.49	6.50		9.99	14
1080	21" ID		116	.138		3.66	6.65		10.31	14.35
1082	22" ID		114	.140		3.83	6.75		10.58	14.70
1084	23" ID		112	.143		4	6.85		10.85	15.10
1086	24" ID		110	.145		4.17	7		11.17	15.50
1088	25" ID		108	.148		4.34	7.10		11.44	15.85
1090	26" ID		106	.151		4.51	7.25		11.76	16.25
1092	27" ID		104	.154		4.68	7.40		12.08	16.65
1094	28" ID		102	.157		4.85	7.55		12.40	17.10
1096	29" ID		100	.160		5	7.70		12.70	17.45
1098	30" ID	▼	98	.163	▼	5.20	7.85		13.05	17.90
1300	For colors, add				Ea.	10%				
2000	PVC, white, fitting covers									
2020	Fiberglass insulation inserts included with sizes 1-3/4" thru 9-3/4"									
2030	Size is based on OD of insulation									
2040	90° Elbow fitting									
2060	1-3/4"	Q-14	135	.119	Ea.	.47	5.70		6.17	9.35
2062	2"		130	.123		.58	5.90		6.48	9.85
2064	2-1/4"		128	.125		.65	6		6.65	10.05
2068	2-1/2"		126	.127		.71	6.10		6.81	10.30
2070	2-3/4"		123	.130		.77	6.25		7.02	10.60
2072	3"		120	.133		.82	6.40		7.22	10.85
2074	3-3/8"		116	.138		.93	6.65		7.58	11.30
2076	3-3/4"		113	.142		1.04	6.80		7.84	11.75
2078	4-1/8"		110	.145		1.31	7		8.31	12.35
2080	4-3/4"		105	.152		1.58	7.30		8.88	13.15
2082	5-1/4"		100	.160		1.84	7.70		9.54	13.95
2084	5-3/4"		95	.168		2.43	8.10		10.53	15.25
2086	6-1/4"		90	.178		4.05	8.55		12.60	17.75
2088	6-3/4"		87	.184		4.27	8.85		13.12	18.45
2090	7-1/4"		85	.188		5.35	9.05		14.40	20
2092	7-3/4"		83	.193		5.60	9.25		14.85	20.50
2094	8-3/4"		80	.200		7.20	9.60		16.80	23
2096	9-3/4"		77	.208		9.65	10		19.65	26
2098	10-7/8"		74	.216		10.75	10.40		21.15	28
2100	11-7/8"		71	.225		12.35	10.85		23.20	30.50
2102	12-7/8"	▼	68	.235	▼	15.55	11.30		26.85	35

For customer support on your Mechanical Cost Data, call 877.739.3689.

121

22 07 19.30 Piping Insulation Protective Jacketing, PVC		Crew	Daily Output	Labor-Hours	Unit	Material	2016 Bare Costs Labor	Equipment	Total	Total Incl O&P
2104	14-1/8"	Q-14	66	.242	Ea.	17.80	11.65		29.45	38
2106	15-1/8"		64	.250		19.15	12		31.15	39.50
2108	16-1/8"		63	.254		21	12.20		33.20	42
2110	17-1/8"		62	.258		23	12.40		35.40	45
2112	18-1/8"		61	.262		30	12.60		42.60	52.50
2114	19-1/8"		60	.267		40.50	12.80		53.30	64.50
2116	20-1/8"		59	.271		52	13.05		65.05	78
2200	45° Elbow fitting									
2220	1-3/4" thru 9-3/4" same price as 90° Elbow fitting									
2320	10-7/8"	Q-14	74	.216	Ea.	10.75	10.40		21.15	28
2322	11-7/8"		71	.225		11.85	10.85		22.70	30
2324	12-7/8"		68	.235		13.25	11.30		24.55	32
2326	14-1/8"		66	.242		15.10	11.65		26.75	35
2328	15-1/8"		64	.250		16.25	12		28.25	36.50
2330	16-1/8"		63	.254		18.55	12.20		30.75	39.50
2332	17-1/8"		62	.258		21	12.40		33.40	42.50
2334	18-1/8"		61	.262		25.50	12.60		38.10	47.50
2336	19-1/8"		60	.267		35	12.80		47.80	58.50
2338	20-1/8"		59	.271		39.50	13.05		52.55	64
2400	Tee fitting									
2410	1-3/4"	Q-14	96	.167	Ea.	.88	8		8.88	13.40
2412	2"		94	.170		1	8.20		9.20	13.85
2414	2-1/4"		91	.176		1.09	8.45		9.54	14.35
2416	2-1/2"		88	.182		1.19	8.75		9.94	14.90
2418	2-3/4"		85	.188		1.31	9.05		10.36	15.55
2420	3"		82	.195		1.42	9.40		10.82	16.15
2422	3-3/8"		79	.203		1.63	9.75		11.38	16.95
2424	3-3/4"		76	.211		1.85	10.10		11.95	17.80
2426	4-1/8"		73	.219		2.19	10.55		12.74	18.80
2428	4-3/4"		70	.229		2.73	11		13.73	20
2430	5-1/4"		67	.239		3.28	11.45		14.73	21.50
2432	5-3/4"		63	.254		4.36	12.20		16.56	24
2434	6-1/4"		60	.267		5.75	12.80		18.55	26.50
2436	6-3/4"		59	.271		7.10	13.05		20.15	28.50
2438	7-1/4"		57	.281		11.45	13.50		24.95	33.50
2440	7-3/4"		54	.296		12.55	14.25		26.80	36
2442	8-3/4"		52	.308		15.30	14.80		30.10	40
2444	9-3/4"		50	.320		18	15.40		33.40	44
2446	10-7/8"		48	.333		18.25	16		34.25	45
2448	11-7/8"		47	.340		20.50	16.35		36.85	48
2450	12-7/8"		46	.348		22.50	16.70		39.20	50.50
2452	14-1/8"		45	.356		24.50	17.10		41.60	53.50
2454	15-1/8"		44	.364		26.50	17.45		43.95	56
2456	16-1/8"		43	.372		28.50	17.90		46.40	59
2458	17-1/8"		42	.381		30.50	18.30		48.80	62
2460	18-1/8"		41	.390		33.50	18.75		52.25	66
2462	19-1/8"		40	.400		36.50	19.20		55.70	70
2464	20-1/8"		39	.410		40	19.70		59.70	74.50
4000	Mechanical grooved fitting cover, including insert									
4020	90° Elbow fitting									
4030	3/4" & 1"	Q-14	140	.114	Ea.	3.70	5.50		9.20	12.60
4040	1-1/4" & 1-1/2"		135	.119		4.68	5.70		10.38	14
4042	2"		130	.123		7.40	5.90		13.30	17.35

22 07 19.30 Piping Insulation Protective Jacketing, PVC

		Crew	Daily Output	Labor-Hours	Unit	Material	2016 Bare Costs Labor	Equipment	Total	Total Incl O&P
4044	2-1/2"	Q-14	125	.128	Ea.	8.20	6.15		14.35	18.55
4046	3"		120	.133		9.20	6.40		15.60	20
4048	3-1/2"		115	.139		10.65	6.70		17.35	22
4050	4"		110	.145		11.85	7		18.85	24
4052	5"		100	.160		14.75	7.70		22.45	28
4054	6"		90	.178		22	8.55		30.55	37.50
4056	8"		80	.200		23.50	9.60		33.10	41
4058	10"		75	.213		30.50	10.25		40.75	49.50
4060	12"		68	.235		43.50	11.30		54.80	65.50
4062	14"		65	.246		43.50	11.85		55.35	66.50
4064	16"		63	.254		59.50	12.20		71.70	84.50
4066	18"	↓	61	.262	↓	81.50	12.60		94.10	110
4100	45° Elbow fitting									
4120	3/4" & 1"	Q-14	140	.114	Ea.	3.29	5.50		8.79	12.15
4130	1-1/4" & 1-1/2"		135	.119		4.19	5.70		9.89	13.45
4140	2"		130	.123		6.60	5.90		12.50	16.45
4142	2-1/2"		125	.128		7.30	6.15		13.45	17.60
4144	3"		120	.133		8.20	6.40		14.60	18.95
4146	3-1/2"		115	.139		8.65	6.70		15.35	19.90
4148	4"		110	.145		10.55	7		17.55	22.50
4150	5"		100	.160		13.10	7.70		20.80	26.50
4152	6"		90	.178		19.35	8.55		27.90	35
4154	8"		80	.200		21	9.60		30.60	38
4156	10"		75	.213		27	10.25		37.25	45.50
4158	12"		68	.235		39	11.30		50.30	60
4160	14"		65	.246		50.50	11.85		62.35	74
4162	16"		63	.254		61	12.20		73.20	86.50
4164	18"	↓	61	.262	↓	72	12.60		84.60	99
4200	Tee fitting									
4220	3/4" & 1"	Q-14	93	.172	Ea.	4.82	8.25		13.07	18.15
4230	1-1/4" & 1-1/2"		90	.178		6.10	8.55		14.65	20
4240	2"		87	.184		9.60	8.85		18.45	24.50
4242	2-1/2"		84	.190		10.70	9.15		19.85	26
4244	3"		80	.200		11.95	9.60		21.55	28
4246	3-1/2"		77	.208		12.65	10		22.65	29.50
4248	4"		73	.219		15.45	10.55		26	33.50
4250	5"		67	.239		17.05	11.45		28.50	36.50
4252	6"		60	.267		25	12.80		37.80	47.50
4254	8"		54	.296		27.50	14.25		41.75	52
4256	10"		50	.320		39.50	15.40		54.90	67.50
4258	12"		46	.348		56.50	16.70		73.20	88.50
4260	14"		43	.372		68.50	17.90		86.40	104
4262	16"		42	.381		82	18.30		100.30	119
4264	18"	↓	41	.390	↓	90.50	18.75		109.25	129

22 07 19.40 Pipe Insulation Protective Jacketing, Aluminum

		Crew	Daily Output	Labor-Hours	Unit	Material	2016 Bare Costs Labor	Equipment	Total	Total Incl O&P
0010	**PIPE INSULATION PROTECTIVE JACKETING, ALUMINUM**									
0100	Metal roll jacketing									
0120	Aluminum with polykraft moisture barrier									
0140	Smooth, based on OD of insulation, .016" thick									
0180	1/2" ID	Q-14	220	.073	L.F.	.20	3.49		3.69	5.65
0190	3/4" ID		215	.074		.26	3.58		3.84	5.85
0200	1" ID	↓	210	.076	↓	.32	3.66		3.98	6.05

For customer support on your Mechanical Cost Data, call 877.739.3689.

123

22 07 19 – Plumbing Piping Insulation

22 07 19.40 Pipe Insulation Protective Jacketing, Aluminum	Crew	Daily Output	Labor-Hours	Unit	Material	2016 Bare Costs Labor	Equipment	Total	Total Incl O&P
0210	1-1/4" ID	Q-14	205	.078	L.F.	.38	3.75		4.13	6.25
0220	1-1/2" ID		202	.079		.44	3.81		4.25	6.40
0230	1-3/4" ID		199	.080		.50	3.86		4.36	6.55
0240	2" ID		195	.082		.56	3.94		4.50	6.75
0250	2-1/4" ID		191	.084		.62	4.03		4.65	6.95
0260	2-1/2" ID		187	.086		.69	4.11		4.80	7.15
0270	2-3/4" ID		184	.087		.75	4.18		4.93	7.35
0280	3" ID		180	.089		.82	4.27		5.09	7.55
0290	3-1/4" ID		176	.091		.88	4.37		5.25	7.75
0300	3-1/2" ID		172	.093		.94	4.47		5.41	8
0310	3-3/4" ID		169	.095		1	4.55		5.55	8.20
0320	4" ID		165	.097		1.06	4.66		5.72	8.40
0330	4-1/4" ID		161	.099		1.12	4.78		5.90	8.70
0340	4-1/2" ID		157	.102		1.19	4.90		6.09	8.90
0350	4-3/4" ID		154	.104		1.25	4.99		6.24	9.15
0360	5" ID		150	.107		1.31	5.15		6.46	9.45
0370	5-1/4" ID		146	.110		1.37	5.25		6.62	9.70
0380	5-1/2" ID		143	.112		1.43	5.40		6.83	9.90
0390	5-3/4" ID		139	.115		1.49	5.55		7.04	10.25
0400	6" ID		135	.119		1.55	5.70		7.25	10.55
0410	6-1/4" ID		133	.120		1.61	5.80		7.41	10.75
0420	6-1/2" ID		131	.122		1.67	5.85		7.52	11
0430	7" ID		128	.125		1.79	6		7.79	11.30
0440	7-1/4" ID		125	.128		1.85	6.15		8	11.60
0450	7-1/2" ID		123	.130		1.91	6.25		8.16	11.85
0460	8" ID		121	.132		2.03	6.35		8.38	12.15
0470	8-1/2" ID		119	.134		2.15	6.45		8.60	12.40
0480	9" ID		116	.138		2.27	6.65		8.92	12.80
0490	9-1/2" ID		114	.140		2.39	6.75		9.14	13.15
0500	10" ID		112	.143		2.51	6.85		9.36	13.45
0510	10-1/2" ID		110	.145		2.63	7		9.63	13.80
0520	11" ID		107	.150		2.75	7.20		9.95	14.25
0530	11-1/2" ID		105	.152		2.88	7.30		10.18	14.55
0540	12" ID		103	.155		3	7.45		10.45	14.90
0550	12-1/2" ID		100	.160		3.12	7.70		10.82	15.40
0560	13" ID		99	.162		3.24	7.75		10.99	15.65
0570	14" ID		98	.163		3.49	7.85		11.34	16.05
0580	15" ID		96	.167		3.73	8		11.73	16.55
0590	16" ID		95	.168		3.98	8.10		12.08	17
0600	17" ID		93	.172		4.22	8.25		12.47	17.50
0610	18" ID		92	.174		4.38	8.35		12.73	17.80
0620	19" ID		90	.178		4.70	8.55		13.25	18.45
0630	20" ID		89	.180		4.95	8.65		13.60	18.90
0640	21" ID		87	.184		5.20	8.85		14.05	19.45
0650	22" ID		86	.186		5.45	8.95		14.40	19.85
0660	23" ID		84	.190		5.70	9.15		14.85	20.50
0670	24" ID		83	.193		5.90	9.25		15.15	21
0710	For smooth .020" thick, add					27%	10%			
0720	For smooth .024" thick, add					52%	20%			
0730	For smooth .032" thick, add					104%	33%			
0800	For stucco embossed, add					1%				
0820	For corrugated, add					2.50%				
0900	White aluminum with polysurlyn moisture barrier									

22 07 19 – Plumbing Piping Insulation

22 07 19.40 Pipe Insulation Protective Jacketing, Aluminum	Crew	Daily Output	Labor-Hours	Unit	Material	2016 Bare Costs Labor	Equipment	Total	Total Incl O&P
0910	Smooth, % is an add to polykraft lines of same thickness									
0940	For smooth .016" thick, add				L.F.	35%				
0960	For smooth .024" thick, add				"	22%				
1000	Aluminum fitting covers									
1010	Size is based on OD of insulation									
1020	90° LR elbow, 2 piece									
1100	1-1/2"	Q-14	140	.114	Ea.	3.45	5.50		8.95	12.35
1110	1-3/4"		135	.119		3.45	5.70		9.15	12.65
1120	2"		130	.123		4.34	5.90		10.24	13.95
1130	2-1/4"		128	.125		4.34	6		10.34	14.10
1140	2-1/2"		126	.127		4.34	6.10		10.44	14.25
1150	2-3/4"		123	.130		4.34	6.25		10.59	14.50
1160	3"		120	.133		4.71	6.40		11.11	15.15
1170	3-1/4"		117	.137		4.71	6.55		11.26	15.45
1180	3-1/2"		115	.139		5.55	6.70		12.25	16.55
1190	3-3/4"		113	.142		5.70	6.80		12.50	16.90
1200	4"		110	.145		6	7		13	17.50
1210	4-1/4"		108	.148		6.15	7.10		13.25	17.85
1220	4-1/2"		106	.151		6.15	7.25		13.40	18.05
1230	4-3/4"		104	.154		6.15	7.40		13.55	18.25
1240	5"		102	.157		6.95	7.55		14.50	19.35
1250	5-1/4"		100	.160		8	7.70		15.70	21
1260	5-1/2"		97	.165		9.95	7.95		17.90	23.50
1270	5-3/4"		95	.168		9.95	8.10		18.05	23.50
1280	6"		92	.174		8.50	8.35		16.85	22.50
1290	6-1/4"		90	.178		8.50	8.55		17.05	22.50
1300	6-1/2"		87	.184		10.25	8.85		19.10	25
1310	7"		85	.188		12.40	9.05		21.45	28
1320	7-1/4"		84	.190		12.40	9.15		21.55	28
1330	7-1/2"		83	.193		18.25	9.25		27.50	34.50
1340	8"		82	.195		13.60	9.40		23	29.50
1350	8-1/2"		80	.200		25	9.60		34.60	42.50
1360	9"		78	.205		25	9.85		34.85	43
1370	9-1/2"		77	.208		18.50	10		28.50	36
1380	10"		76	.211		18.50	10.10		28.60	36.50
1390	10-1/2"		75	.213		18.90	10.25		29.15	37
1400	11"		74	.216		18.90	10.40		29.30	37
1410	11-1/2"		72	.222		21.50	10.70		32.20	40
1420	12"		71	.225		21.50	10.85		32.35	40.50
1430	12-1/2"		69	.232		38	11.15		49.15	59.50
1440	13"		68	.235		38	11.30		49.30	59.50
1450	14"		66	.242		51.50	11.65		63.15	74.50
1460	15"		64	.250		53.50	12		65.50	77.50
1470	16"		63	.254		58.50	12.20		70.70	83.50
2000	45° Elbow, 2 piece									
2010	2-1/2"	Q-14	126	.127	Ea.	3.59	6.10		9.69	13.45
2020	2-3/4"		123	.130		3.59	6.25		9.84	13.70
2030	3"		120	.133		4.05	6.40		10.45	14.40
2040	3-1/4"		117	.137		4.05	6.55		10.60	14.70
2050	3-1/2"		115	.139		4.62	6.70		11.32	15.50
2060	3-3/4"		113	.142		4.62	6.80		11.42	15.70
2070	4"		110	.145		4.70	7		11.70	16.05
2080	4-1/4"		108	.148		5.35	7.10		12.45	17

For customer support on your Mechanical Cost Data, call 877.739.3689.

125

22 07 19.40 Pipe Insulation Protective Jacketing, Aluminum		Crew	Daily Output	Labor-Hours	Unit	Material	2016 Bare Costs Labor	Equipment	Total	Total Incl O&P
2090	4-1/2"	Q-14	106	.151	Ea.	5.35	7.25		12.60	17.20
2100	4-3/4"		104	.154		5.35	7.40		12.75	17.40
2110	5"		102	.157		6.30	7.55		13.85	18.70
2120	5-1/4"		100	.160		6.30	7.70		14	18.90
2130	5-1/2"		97	.165		6.60	7.95		14.55	19.60
2140	6"		92	.174		6.60	8.35		14.95	20.50
2150	6-1/2"		87	.184		9.10	8.85		17.95	24
2160	7"		85	.188		9.10	9.05		18.15	24
2170	7-1/2"		83	.193		9.20	9.25		18.45	24.50
2180	8"		82	.195		9.20	9.40		18.60	25
2190	8-1/2"		80	.200		11.45	9.60		21.05	27.50
2200	9"		78	.205		11.45	9.85		21.30	28
2210	9-1/2"		77	.208		15.90	10		25.90	33
2220	10"		76	.211		15.90	10.10		26	33.50
2230	10-1/2"		75	.213		14.85	10.25		25.10	32.50
2240	11"		74	.216		14.85	10.40		25.25	32.50
2250	11-1/2"		72	.222		17.85	10.70		28.55	36.50
2260	12"		71	.225		17.85	10.85		28.70	36.50
2270	13"		68	.235		21	11.30		32.30	40.50
2280	14"		66	.242		25.50	11.65		37.15	46.50
2290	15"		64	.250		39.50	12		51.50	62
2300	16"		63	.254		43	12.20		55.20	66.50
2310	17"		62	.258		42	12.40		54.40	65.50
2320	18"		61	.262		48	12.60		60.60	72
2330	19"		60	.267		58.50	12.80		71.30	84.50
2340	20"		59	.271		56.50	13.05		69.55	82.50
2350	21"		58	.276		61	13.25		74.25	87.50
3000	Tee, 4 piece									
3010	2-1/2"	Q-14	88	.182	Ea.	29	8.75		37.75	45.50
3020	2-3/4"		86	.186		29	8.95		37.95	46
3030	3"		84	.190		32.50	9.15		41.65	50.50
3040	3-1/4"		82	.195		32.50	9.40		41.90	50.50
3050	3-1/2"		80	.200		34.50	9.60		44.10	52.50
3060	4"		78	.205		34.50	9.85		44.35	53.50
3070	4-1/4"		76	.211		36.50	10.10		46.60	56
3080	4-1/2"		74	.216		36.50	10.40		46.90	56
3090	4-3/4"		72	.222		36.50	10.70		47.20	56.50
3100	5"		70	.229		37.50	11		48.50	58.50
3110	5-1/4"		68	.235		37.50	11.30		48.80	59
3120	5-1/2"		66	.242		39.50	11.65		51.15	61.50
3130	6"		64	.250		39.50	12		51.50	62
3140	6-1/2"		60	.267		43.50	12.80		56.30	67.50
3150	7"		58	.276		43.50	13.25		56.75	68
3160	7-1/2"		56	.286		48	13.75		61.75	74.50
3170	8"		54	.296		48	14.25		62.25	75
3180	8-1/2"		52	.308		49.50	14.80		64.30	77.50
3190	9"		50	.320		49.50	15.40		64.90	78.50
3200	9-1/2"		49	.327		36.50	15.70		52.20	65
3210	10"		48	.333		36.50	16		52.50	65.50
3220	10-1/2"		47	.340		39	16.35		55.35	68.50
3230	11"		46	.348		39	16.70		55.70	69
3240	11-1/2"		45	.356		41	17.10		58.10	72
3250	12"		44	.364		41	17.45		58.45	72.50

22 07 19.40 Pipe Insulation Protective Jacketing, Aluminum

		Crew	Daily Output	Labor-Hours	Unit	Material	2016 Bare Costs Labor	Equipment	Total	Total Incl O&P
3260	13"	Q-14	43	.372	Ea.	43	17.90		60.90	75.50
3270	14"		42	.381		45.50	18.30		63.80	78.50
3280	15"		41	.390		49	18.75		67.75	83
3290	16"		40	.400		50.50	19.20		69.70	85.50
3300	17"		39	.410		58	19.70		77.70	94
3310	18"		38	.421		60	20		80	97.50
3320	19"		37	.432		66.50	21		87.50	106
3330	20"		36	.444		68	21.50		89.50	108
3340	22"		35	.457		86.50	22		108.50	129
3350	23"		34	.471		89	22.50		111.50	133
3360	24"	▼	31	.516	▼	91	25		116	139

22 07 19.50 Pipe Insulation Protective Jacketing, St. Stl.

		Crew	Daily Output	Labor-Hours	Unit	Material	2016 Bare Costs Labor	Equipment	Total	Total Incl O&P
0010	**PIPE INSULATION PROTECTIVE JACKETING, STAINLESS STEEL**									
0100	Metal roll jacketing									
0120	Type 304 with moisture barrier									
0140	Smooth, based on OD of insulation, .010" thick									
0260	2-1/2" ID	Q-14	250	.064	L.F.	2.12	3.08		5.20	7.10
0270	2-3/4" ID		245	.065		2.31	3.14		5.45	7.40
0280	3" ID		240	.067		2.50	3.20		5.70	7.75
0290	3-1/4" ID		235	.068		2.69	3.27		5.96	8.05
0300	3-1/2" ID		230	.070		2.88	3.34		6.22	8.35
0310	3-3/4" ID		225	.071		3.07	3.42		6.49	8.70
0320	4" ID		220	.073		3.26	3.49		6.75	9.05
0330	4-1/4" ID		215	.074		3.44	3.58		7.02	9.35
0340	4-1/2" ID		210	.076		3.63	3.66		7.29	9.70
0350	5" ID		200	.080		4.01	3.84		7.85	10.40
0360	5-1/2" ID		190	.084		4.39	4.05		8.44	11.15
0370	6" ID		180	.089		4.76	4.27		9.03	11.90
0380	6-1/2" ID		175	.091		5.15	4.39		9.54	12.50
0390	7" ID		170	.094		5.50	4.52		10.02	13.10
0400	7-1/2" ID		164	.098		5.90	4.69		10.59	13.80
0410	8" ID		161	.099		6.25	4.78		11.03	14.35
0420	8-1/2" ID		158	.101		6.65	4.87		11.52	14.85
0430	9" ID		155	.103		7.05	4.96		12.01	15.45
0440	9-1/2" ID		152	.105		7.40	5.05		12.45	16
0450	10" ID		149	.107		7.80	5.15		12.95	16.60
0460	10-1/2" ID		146	.110		8.15	5.25		13.40	17.20
0470	11" ID		143	.112		8.55	5.40		13.95	17.75
0480	12" ID		137	.117		9.30	5.60		14.90	18.95
0490	13" ID		132	.121		10.05	5.80		15.85	20
0500	14" ID	▼	130	.123		10.80	5.90		16.70	21
0700	For smooth .016" thick, add				▼	45%	33%			
1000	Stainless steel, Type 316, fitting covers									
1010	Size is based on OD of insulation									
1020	90° LR Elbow, 2 piece									
1100	1-1/2"	Q-14	126	.127	Ea.	12.50	6.10		18.60	23.50
1110	2-3/4"		123	.130		12.50	6.25		18.75	23.50
1120	3"		120	.133		13.05	6.40		19.45	24.50
1130	3-1/4"		117	.137		13.05	6.55		19.60	24.50
1140	3-1/2"		115	.139		13.75	6.70		20.45	25.50
1150	3-3/4"		113	.142		14.50	6.80		21.30	26.50
1160	4"	▼	110	.145	▼	15.95	7		22.95	28.50

For customer support on your Mechanical Cost Data, call 877.739.3689.

127

22 07 19.50 Pipe Insulation Protective Jacketing, St. Stl.	Crew	Daily Output	Labor-Hours	Unit	Material	2016 Bare Costs Labor	Equipment	Total	Total Incl O&P
1170	4-1/4"	Q-14	108	.148	Ea.	21	7.10		28.10	34.50
1180	4-1/2"		106	.151		21	7.25		28.25	35
1190	5"		102	.157		21.50	7.55		29.05	36
1200	5-1/2"		97	.165		32.50	7.95		40.45	48
1210	6"		92	.174		35.50	8.35		43.85	52
1220	6-1/2"		87	.184		49	8.85		57.85	67.50
1230	7"		85	.188		49	9.05		58.05	67.50
1240	7-1/2"		83	.193		57	9.25		66.25	77
1250	8"		80	.200		57	9.60		66.60	77.50
1260	8-1/2"		80	.200		59	9.60		68.60	80
1270	9"		78	.205		89.50	9.85		99.35	114
1280	9-1/2"		77	.208		88	10		98	113
1290	10"		76	.211		88	10.10		98.10	113
1300	10-1/2"		75	.213		105	10.25		115.25	132
1310	11"		74	.216		101	10.40		111.40	127
1320	12"		71	.225		114	10.85		124.85	143
1330	13"		68	.235		158	11.30		169.30	192
1340	14"		66	.242		159	11.65		170.65	193
2000	45° Elbow, 2 piece									
2010	2-1/2"	Q-14	126	.127	Ea.	10.40	6.10		16.50	21
2020	2-3/4"		123	.130		10.40	6.25		16.65	21
2030	3"		120	.133		11.20	6.40		17.60	22.50
2040	3-1/4"		117	.137		11.20	6.55		17.75	22.50
2050	3-1/2"		115	.139		11.35	6.70		18.05	23
2060	3-3/4"		113	.142		11.35	6.80		18.15	23
2070	4"		110	.145		14.60	7		21.60	27
2080	4-1/4"		108	.148		20.50	7.10		27.60	33.50
2090	4-1/2"		106	.151		20.50	7.25		27.75	34
2100	4-3/4"		104	.154		20.50	7.40		27.90	34
2110	5"		102	.157		21	7.55		28.55	35
2120	5-1/2"		97	.165		21	7.95		28.95	35.50
2130	6"		92	.174		24.50	8.35		32.85	40
2140	6-1/2"		87	.184		24.50	8.85		33.35	41
2150	7"		85	.188		42	9.05		51.05	60
2160	7-1/2"		83	.193		42.50	9.25		51.75	61.50
2170	8"		82	.195		42.50	9.40		51.90	61.50
2180	8-1/2"		80	.200		50	9.60		59.60	70
2190	9"		78	.205		50	9.85		59.85	70.50
2200	9-1/2"		77	.208		58.50	10		68.50	80
2210	10"		76	.211		58.50	10.10		68.60	80.50
2220	10-1/2"		75	.213		70.50	10.25		80.75	93.50
2230	11"		74	.216		70.50	10.40		80.90	93.50
2240	12"		71	.225		76.50	10.85		87.35	101
2250	13"		68	.235		90	11.30		101.30	117

22 11 Facility Water Distribution

22 11 13 – Facility Water Distribution Piping

22 11 13.14 Pipe, Brass

		Crew	Daily Output	Labor-Hours	Unit	Material	2016 Bare Costs Labor	Equipment	Total	Total Incl O&P
0010	**PIPE, BRASS**, Plain end									
0900	Field threaded, coupling & clevis hanger assembly 10' O.C.									
0920	Regular weight									
1120	1/2" diameter	1 Plum	48	.167	L.F.	6.90	9.85		16.75	22.50
1140	3/4" diameter		46	.174		9.55	10.30		19.85	26
1160	1" diameter		43	.186		13.65	11		24.65	31.50
1180	1-1/4" diameter	Q-1	72	.222		20.50	11.85		32.35	40.50
1200	1-1/2" diameter		65	.246		24.50	13.10		37.60	46.50
1220	2" diameter		53	.302		33.50	16.10		49.60	61.50
1240	2-1/2" diameter		41	.390		52	21		73	89
1260	3" diameter		31	.516		73	27.50		100.50	122
1300	4" diameter	Q-2	37	.649		127	36		163	193
1930	To delete coupling & hanger, subtract									
1940	1/2" diam.					40%	46%			
1950	3/4" diam. to 1-1/2" diam.					39%	39%			
1960	2" diam. to 4" diam.					48%	35%			

22 11 13.16 Pipe Fittings, Brass

		Crew	Daily Output	Labor-Hours	Unit	Material	2016 Bare Costs Labor	Equipment	Total	Total Incl O&P
0010	**PIPE FITTINGS, BRASS**, Rough bronze, threaded.									
1000	Standard wt., 90° Elbow									
1040	1/8"	1 Plum	13	.615	Ea.	18	36.50		54.50	75
1060	1/4"		13	.615		18	36.50		54.50	75
1080	3/8"		13	.615		18	36.50		54.50	75
1100	1/2"		12	.667		18	39.50		57.50	79.50
1120	3/4"		11	.727		24	43		67	91.50
1140	1"		10	.800		39	47.50		86.50	115
1160	1-1/4"	Q-1	17	.941		63.50	50		113.50	145
1180	1-1/2"		16	1		78.50	53.50		132	167
1200	2"		14	1.143		126	61		187	231
1220	2-1/2"		11	1.455		305	77.50		382.50	450
1240	3"		8	2		465	107		572	675
1260	4"	Q-2	11	2.182		945	121		1,066	1,225
1280	5"		8	3		2,625	166		2,791	3,125
1300	6"		7	3.429		3,850	189		4,039	4,525
1500	45° Elbow, 1/8"	1 Plum	13	.615		22	36.50		58.50	79.50
1540	1/4"		13	.615		22	36.50		58.50	79.50
1560	3/8"		13	.615		22	36.50		58.50	79.50
1580	1/2"		12	.667		22	39.50		61.50	84
1600	3/4"		11	.727		31.50	43		74.50	99.50
1620	1"		10	.800		53.50	47.50		101	131
1640	1-1/4"	Q-1	17	.941		85	50		135	169
1660	1-1/2"		16	1		107	53.50		160.50	199
1680	2"		14	1.143		173	61		234	283
1700	2-1/2"		11	1.455		330	77.50		407.50	480
1720	3"		8	2		510	107		617	720
1740	4"	Q-2	11	2.182		1,125	121		1,246	1,425
1760	5"		8	3		2,125	166		2,291	2,600
1780	6"		7	3.429		2,950	189		3,139	3,525
2000	Tee, 1/8"	1 Plum	9	.889		21	52.50		73.50	103
2040	1/4"		9	.889		21	52.50		73.50	103
2060	3/8"		9	.889		21	52.50		73.50	103
2080	1/2"		8	1		21	59		80	113
2100	3/4"		7	1.143		30	67.50		97.50	135

For customer support on your Mechanical Cost Data, call 877.739.3689.

129

22 11 Facility Water Distribution

22 11 13 – Facility Water Distribution Piping

22 11 13.16 Pipe Fittings, Brass		Crew	Daily Output	Labor-Hours	Unit	Material	2016 Bare Costs Labor	Equipment	Total	Total Incl O&P
2120	1"	1 Plum	6	1.333	Ea.	54.50	79		133.50	179
2140	1-1/4"	Q-1	10	1.600		93.50	85.50		179	232
2160	1-1/2"		9	1.778		105	94.50		199.50	259
2180	2"		8	2		175	107		282	355
2200	2-1/2"		7	2.286		415	122		537	645
2220	3"		5	3.200		635	171		806	955
2240	4"	Q-2	7	3.429		1,575	189		1,764	2,000
2260	5"		5	4.800		3,275	265		3,540	4,000
2280	6"		4	6		5,150	330		5,480	6,150
2500	Coupling, 1/8"	1 Plum	26	.308		15.05	18.20		33.25	44
2540	1/4"		22	.364		15.05	21.50		36.55	49
2560	3/8"		18	.444		15.05	26.50		41.55	56
2580	1/2"		15	.533		11.85	31.50		43.35	60.50
2600	3/4"		14	.571		21	34		55	74.50
2620	1"		13	.615		36	36.50		72.50	94.50
2640	1-1/4"	Q-1	22	.727		60	39		99	125
2660	1-1/2"		20	.800		78.50	42.50		121	151
2680	2"		18	.889		130	47.50		177.50	214
2700	2-1/2"		14	1.143		221	61		282	335
2720	3"		10	1.600		305	85.50		390.50	465
2740	4"	Q-2	12	2		630	111		741	860
2760	5"		10	2.400		1,175	133		1,308	1,500
2780	6"		9	2.667		1,675	147		1,822	2,075
3000	Union, 125 lb.									
3020	1/8"	1 Plum	12	.667	Ea.	51.50	39.50		91	116
3040	1/4"		12	.667		51.50	39.50		91	116
3060	3/8"		12	.667		51.50	39.50		91	116
3080	1/2"		11	.727		51.50	43		94.50	122
3100	3/4"		10	.800		70.50	47.50		118	149
3120	1"		9	.889		106	52.50		158.50	197
3140	1-1/4"	Q-1	16	1		154	53.50		207.50	250
3160	1-1/2"		15	1.067		183	57		240	287
3180	2"		13	1.231		247	65.50		312.50	370
3200	2-1/2"		10	1.600		675	85.50		760.50	870
3220	3"		7	2.286		1,050	122		1,172	1,325
3240	4"	Q-2	10	2.400		3,250	133		3,383	3,775
3320	For 250 lb. (navy pattern), add					100%				

22 11 13.23 Pipe/Tube, Copper

22 11 13.23 Pipe/Tube, Copper		Crew	Daily Output	Labor-Hours	Unit	Material	Labor	Equipment	Total	Total Incl O&P
0010	**PIPE/TUBE, COPPER**, Solder joints R221113-50									
0100	Solder									
0120	Solder, lead free, roll				Lb.	10.10			10.10	11.15
1000	Type K tubing, couplings & clevis hanger assemblies 10' O.C.									
1100	1/4" diameter	1 Plum	84	.095	L.F.	3.83	5.65		9.48	12.70
1120	3/8" diameter		82	.098		4.46	5.80		10.26	13.60
1140	1/2" diameter		78	.103		4.98	6.05		11.03	14.65
1160	5/8" diameter		77	.104		6.10	6.15		12.25	16.05
1180	3/4" diameter		74	.108		8.15	6.40		14.55	18.60
1200	1" diameter		66	.121		10.60	7.20		17.80	22.50
1220	1-1/4" diameter		56	.143		13	8.45		21.45	27
1240	1-1/2" diameter		50	.160		16.30	9.45		25.75	32.50
1260	2" diameter		40	.200		24.50	11.85		36.35	45
1280	2-1/2" diameter	Q-1	60	.267		36.50	14.20		50.70	61.50

For customer support on your Mechanical Cost Data, call 877.739.3689.

22 11 13 – Facility Water Distribution Piping

22 11 13.23 Pipe/Tube, Copper		Crew	Daily Output	Labor-Hours	Unit	Material	2016 Bare Costs Labor	Equipment	Total	Total Incl O&P
1300	3" diameter	Q-1	54	.296	L.F.	49.50	15.80		65.30	78.50
1320	3-1/2" diameter		42	.381		66	20.50		86.50	103
1330	4" diameter		38	.421		77	22.50		99.50	119
1340	5" diameter		32	.500		147	26.50		173.50	201
1360	6" diameter	Q-2	38	.632		214	35		249	288
1380	8" diameter	"	34	.706		375	39		414	470
1390	For other than full hard temper, add					13%				
1440	For silver solder, add						15%			
1800	For medical clean, (oxygen class), add					12%				
1950	To delete cplgs. & hngrs., 1/4"-1" pipe, subtract					27%	60%			
1960	1-1/4"-3" pipe, subtract					14%	52%			
1970	3-1/2"-5" pipe, subtract					10%	60%			
1980	6"-8" pipe, subtract					19%	53%			
2000	Type L tubing, couplings & clevis hanger assemblies 10' O.C.									
2100	1/4" diameter	1 Plum	88	.091	L.F.	2.80	5.40		8.20	11.20
2120	3/8" diameter		84	.095		3.50	5.65		9.15	12.35
2140	1/2" diameter		81	.099		3.68	5.85		9.53	12.90
2160	5/8" diameter		79	.101		5.30	6		11.30	14.90
2180	3/4" diameter		76	.105		5.25	6.25		11.50	15.15
2200	1" diameter		68	.118		7.40	6.95		14.35	18.65
2220	1-1/4" diameter		58	.138		10.05	8.15		18.20	23.50
2240	1-1/2" diameter		52	.154		12.60	9.10		21.70	27.50
2260	2" diameter		42	.190		18.40	11.30		29.70	37
2280	2-1/2" diameter	Q-1	62	.258		28	13.75		41.75	51.50
2300	3" diameter		56	.286		37	15.20		52.20	63.50
2320	3-1/2" diameter		43	.372		50.50	19.85		70.35	85.50
2340	4" diameter		39	.410		57.50	22		79.50	96.50
2360	5" diameter		34	.471		106	25		131	155
2380	6" diameter	Q-2	40	.600		146	33		179	210
2400	8" diameter	"	36	.667		248	37		285	330
2410	For other than full hard temper, add					21%				
2590	For silver solder, add						15%			
2900	For medical clean, (oxygen class), add					12%				
2940	To delete cplgs. & hngrs., 1/4"-1" pipe, subtract					37%	63%			
2960	1-1/4"-3" pipe, subtract					12%	53%			
2970	3-1/2"-5" pipe, subtract					12%	63%			
2980	6"-8" pipe, subtract					24%	55%			
3000	Type M tubing, couplings & clevis hanger assemblies 10' O.C.									
3100	1/4" diameter	1 Plum	90	.089	L.F.	3.10	5.25		8.35	11.35
3120	3/8" diameter		87	.092		3.15	5.45		8.60	11.65
3140	1/2" diameter		84	.095		3.21	5.65		8.86	12.05
3160	5/8" diameter		81	.099		4.59	5.85		10.44	13.90
3180	3/4" diameter		78	.103		4.39	6.05		10.44	13.95
3200	1" diameter		70	.114		6.60	6.75		13.35	17.45
3220	1-1/4" diameter		60	.133		9.15	7.90		17.05	22
3240	1-1/2" diameter		54	.148		12.10	8.75		20.85	26.50
3260	2" diameter		44	.182		18.15	10.75		28.90	36
3280	2-1/2" diameter	Q-1	64	.250		26.50	13.30		39.80	49.50
3300	3" diameter		58	.276		35	14.70		49.70	61
3320	3-1/2" diameter		45	.356		49	18.95		67.95	82
3340	4" diameter		40	.400		57.50	21.50		79	95.50
3360	5" diameter		36	.444		115	23.50		138.50	163
3370	6" diameter	Q-2	42	.571		159	31.50		190.50	223

22 11 13.23 Pipe/Tube, Copper		Crew	Daily Output	Labor-Hours	Unit	Material	2016 Bare Costs Labor	Equipment	Total	Total Incl O&P
3380	8" diameter	Q-2	38	.632	L.F.	275	35		310	355
3440	For silver solder, add						15%			
3960	To delete cplgs. & hngrs., 1/4"-1" pipe, subtract					35%	65%			
3970	1-1/4"-3" pipe, subtract					19%	56%			
3980	3-1/2"-5" pipe, subtract					13%	65%			
3990	6"-8" pipe, subtract					28%	58%			
4000	Type DWV tubing, couplings & clevis hanger assemblies 10' O.C.									
4100	1-1/4" diameter	1 Plum	60	.133	L.F.	9.20	7.90		17.10	22
4120	1-1/2" diameter		54	.148		11.30	8.75		20.05	25.50
4140	2" diameter	↓	44	.182		14.75	10.75		25.50	32.50
4160	3" diameter	Q-1	58	.276		26.50	14.70		41.20	51
4180	4" diameter		40	.400		42.50	21.50		64	78.50
4200	5" diameter	↓	36	.444		103	23.50		126.50	150
4220	6" diameter	Q-2	42	.571		148	31.50		179.50	210
4240	8" diameter	"	38	.632	↓	310	35		345	395
4730	To delete cplgs. & hngrs., 1-1/4"-2" pipe, subtract					16%	53%			
4740	3"-4" pipe, subtract					13%	60%			
4750	5"-8" pipe, subtract					23%	58%			
5200	ACR tubing, type L, hard temper, cleaned and									
5220	capped, no couplings or hangers									
5240	3/8" OD				L.F.	1.56			1.56	1.72
5250	1/2" OD					2.38			2.38	2.62
5260	5/8" OD					2.89			2.89	3.18
5270	3/4" OD					4.09			4.09	4.50
5280	7/8" OD					4.55			4.55	5
5290	1-1/8" OD					6.60			6.60	7.25
5300	1-3/8" OD					8.85			8.85	9.75
5310	1-5/8" OD					11.30			11.30	12.45
5320	2-1/8" OD					17.85			17.85	19.60
5330	2-5/8" OD					26.50			26.50	29
5340	3-1/8" OD					35.50			35.50	39
5350	3-5/8" OD					46			46	50.50
5360	4-1/8" OD					56			56	61.50
5380	ACR tubing, type L, hard, cleaned and capped									
5381	No couplings or hangers									
5384	3/8"	1 Stpi	160	.050	L.F.	1.56	3.04		4.60	6.30
5385	1/2"		160	.050		2.38	3.04		5.42	7.20
5386	5/8"		160	.050		2.89	3.04		5.93	7.75
5387	3/4"		130	.062		4.09	3.74		7.83	10.15
5388	7/8"		130	.062		4.55	3.74		8.29	10.65
5389	1-1/8"		115	.070		6.60	4.22		10.82	13.60
5390	1-3/8"		100	.080		8.85	4.86		13.71	17.10
5391	1-5/8"		90	.089		11.30	5.40		16.70	20.50
5392	2-1/8"	↓	80	.100		17.85	6.05		23.90	29
5393	2-5/8"	Q-5	125	.128		26.50	7		33.50	39.50
5394	3-1/8"		105	.152		35.50	8.30		43.80	51.50
5395	4-1/8"	↓	95	.168	↓	56	9.20		65.20	75.50
5800	Refrigeration tubing, dryseal, 50' coils									
5840	1/8" OD				Coil	31.50			31.50	34.50
5850	3/16" OD					37			37	40.50
5860	1/4" OD					42.50			42.50	46.50
5870	5/16" OD					57			57	62.50
5880	3/8" OD				↓	60			60	66

22 11 13 – Facility Water Distribution Piping

22 11 13.23 Pipe/Tube, Copper

	Crew	Daily Output	Labor-Hours	Unit	Material	2016 Bare Costs Labor	2016 Bare Costs Equipment	Total	Total Incl O&P
5890	1/2" OD				Coil	85			85	93.50
5900	5/8" OD					114			114	125
5910	3/4" OD					133			133	147
5920	7/8" OD					199			199	219
5930	1-1/8" OD					330			330	365
5940	1-3/8" OD					570			570	630
5950	1-5/8" OD					725			725	795
9400	Sub assemblies used in assembly systems									
9410	Chilled water unit, coil connections per unit under 10 ton	Q-5	.80	20	System	1,150	1,100		2,250	2,925
9420	Chilled water unit, coil connections per unit 10 ton and up		1	16		1,875	875		2,750	3,400
9430	Chilled water dist. piping per ton, less than 61 ton systems		26	.615		17.40	33.50		50.90	69.50
9440	Chilled water dist. piping per ton, 61 through 120 ton systems	Q-6	31	.774		51.50	44		95.50	123
9450	Chilled water dist. piping/ton, 135 ton systems and up	Q-8	25.40	1.260		68	73	2.29	143.29	188
9510	Refrigerant piping/ton of cooling for remote condensers	Q-5	2	8		370	435		805	1,075
9520	Refrigerant piping per ton up to 10 ton w/remote condensing unit		2.40	6.667		172	365		537	740
9530	Refrigerant piping per ton, 20 ton w/remote condensing unit		2	8		249	435		684	935
9540	Refrigerant piping per ton, 40 ton w/remote condensing unit		1.90	8.421		320	460		780	1,050
9550	Refrigerant piping per ton, 75-80 ton w/remote condensing unit	Q-6	2.40	10		465	565		1,030	1,375
9560	Refrigerant piping per ton, 100 ton w/remote condensing unit	"	2.20	10.909		605	620		1,225	1,600

22 11 13.25 Pipe/Tube Fittings, Copper

	Crew	Daily Output	Labor-Hours	Unit	Material	2016 Bare Costs Labor	2016 Bare Costs Equipment	Total	Total Incl O&P
0010	**PIPE/TUBE FITTINGS, COPPER**, Wrought unless otherwise noted									
0020	For silver solder, add						15%			
0040	Solder joints, copper x copper									
0070	90° elbow, 1/4"	1 Plum	22	.364	Ea.	3.28	21.50		24.78	36
0090	3/8"		22	.364		3.11	21.50		24.61	36
0100	1/2"		20	.400		1.13	23.50		24.63	36.50
0110	5/8"		19	.421		2.53	25		27.53	40.50
0120	3/4"		19	.421		2.53	25		27.53	40.50
0130	1"		16	.500		6.20	29.50		35.70	51.50
0140	1-1/4"		15	.533		9.20	31.50		40.70	57.50
0150	1-1/2"		13	.615		14.35	36.50		50.85	71
0160	2"		11	.727		26	43		69	94
0170	2-1/2"	Q-1	13	1.231		47.50	65.50		113	152
0180	3"		11	1.455		67.50	77.50		145	192
0190	3-1/2"		10	1.600		210	85.50		295.50	360
0200	4"		9	1.778		148	94.50		242.50	305
0210	5"		6	2.667		580	142		722	855
0220	6"	Q-2	9	2.667		775	147		922	1,075
0230	8"	"	8	3		2,875	166		3,041	3,400
0250	45° elbow, 1/4"	1 Plum	22	.364		5.85	21.50		27.35	39
0270	3/8"		22	.364		5	21.50		26.50	38
0280	1/2"		20	.400		2.06	23.50		25.56	38
0290	5/8"		19	.421		8.75	25		33.75	47
0300	3/4"		19	.421		3.52	25		28.52	41.50
0310	1"		16	.500		8.85	29.50		38.35	54
0320	1-1/4"		15	.533		11.85	31.50		43.35	60.50
0330	1-1/2"		13	.615		14.55	36.50		51.05	71
0340	2"		11	.727		24.50	43		67.50	91.50
0350	2-1/2"	Q-1	13	1.231		47	65.50		112.50	151
0360	3"		13	1.231		66	65.50		131.50	172
0370	3-1/2"		10	1.600		113	85.50		198.50	253
0380	4"		9	1.778		136	94.50		230.50	292

For customer support on your Mechanical Cost Data, call 877.739.3689.

133

22 11 13 – Facility Water Distribution Piping

22 11 13.25 Pipe/Tube Fittings, Copper		Crew	Daily Output	Labor-Hours	Unit	Material	2016 Bare Costs Labor	Equipment	Total	Total Incl O&P
0390	5"	Q-1	6	2.667	Ea.	490	142		632	755
0400	6"	Q-2	9	2.667		775	147		922	1,075
0410	8"	"	8	3		3,400	166		3,566	3,975
0450	Tee, 1/4"	1 Plum	14	.571		6.55	34		40.55	58
0470	3/8"		14	.571		5.25	34		39.25	57
0480	1/2"		13	.615		1.92	36.50		38.42	57
0490	5/8"		12	.667		12.05	39.50		51.55	73
0500	3/4"		12	.667		4.64	39.50		44.14	64.50
0510	1"		10	.800		13.95	47.50		61.45	87
0520	1-1/4"		9	.889		19.25	52.50		71.75	101
0530	1-1/2"		8	1		29	59		88	122
0540	2"		7	1.143		46	67.50		113.50	153
0550	2-1/2"	Q-1	8	2		85	107		192	255
0560	3"		7	2.286		122	122		244	320
0570	3-1/2"		6	2.667		355	142		497	605
0580	4"		5	3.200		286	171		457	570
0590	5"		4	4		875	213		1,088	1,275
0600	6"	Q-2	6	4		1,200	221		1,421	1,650
0610	8"	"	5	4.800		4,625	265		4,890	5,475
0612	Tee, reducing on the outlet, 1/4"	1 Plum	15	.533		10.75	31.50		42.25	59.50
0613	3/8"		15	.533		10.40	31.50		41.90	59
0614	1/2"		14	.571		9.65	34		43.65	61.50
0615	5/8"		13	.615		19.40	36.50		55.90	76.50
0616	3/4"		12	.667		6.45	39.50		45.95	66.50
0617	1"		11	.727		21	43		64	88
0618	1-1/4"		10	.800		24	47.50		71.50	98
0619	1-1/2"		9	.889		25	52.50		77.50	107
0620	2"		8	1		40	59		99	134
0621	2-1/2"	Q-1	9	1.778		99	94.50		193.50	252
0622	3"		8	2		103	107		210	274
0623	4"		6	2.667		220	142		362	455
0624	5"		5	3.200		1,125	171		1,296	1,500
0625	6"	Q-2	7	3.429		1,550	189		1,739	1,975
0626	8"	"	6	4		5,950	221		6,171	6,875
0630	Tee, reducing on the run, 1/4"	1 Plum	15	.533		13.90	31.50		45.40	63
0631	3/8"		15	.533		19.60	31.50		51.10	69
0632	1/2"		14	.571		13.25	34		47.25	65.50
0633	5/8"		13	.615		19.55	36.50		56.05	76.50
0634	3/4"		12	.667		14.30	39.50		53.80	75
0635	1"		11	.727		16.95	43		59.95	83.50
0636	1-1/4"		10	.800		27.50	47.50		75	102
0637	1-1/2"		9	.889		47	52.50		99.50	131
0638	2"		8	1		60	59		119	156
0639	2-1/2"	Q-1	9	1.778		128	94.50		222.50	284
0640	3"		8	2		187	107		294	365
0641	4"		6	2.667		380	142		522	635
0642	5"		5	3.200		1,075	171		1,246	1,425
0643	6"	Q-2	7	3.429		1,625	189		1,814	2,075
0644	8"	"	6	4		5,950	221		6,171	6,875
0650	Coupling, 1/4"	1 Plum	24	.333		.77	19.75		20.52	31
0670	3/8"		24	.333		1	19.75		20.75	31
0680	1/2"		22	.364		.85	21.50		22.35	33.50
0690	5/8"		21	.381		2.50	22.50		25	37

22 11 13 – Facility Water Distribution Piping

22 11 13.25 Pipe/Tube Fittings, Copper		Crew	Daily Output	Labor-Hours	Unit	Material	2016 Bare Costs Labor	Equipment	Total	Total Incl O&P
0700	3/4"	1 Plum	21	.381	Ea.	1.72	22.50		24.22	36
0710	1"		18	.444		3.39	26.50		29.89	43
0715	1-1/4"		17	.471		5.85	28		33.85	48.50
0716	1-1/2"		15	.533		7.85	31.50		39.35	56
0718	2"		13	.615		13.15	36.50		49.65	69.50
0721	2-1/2"	Q-1	15	1.067		25.50	57		82.50	114
0722	3"		13	1.231		36	65.50		101.50	139
0724	3-1/2"		8	2		69	107		176	237
0726	4"		7	2.286		73	122		195	265
0728	5"		6	2.667		169	142		311	400
0731	6"	Q-2	8	3		279	166		445	555
0732	8"	"	7	3.429		905	189		1,094	1,275
0741	Coupling, reducing, concentric									
0743	1/2"	1 Plum	23	.348	Ea.	2.22	20.50		22.72	33.50
0745	3/4"		21.50	.372		4.17	22		26.17	38
0747	1"		19.50	.410		5.95	24.50		30.45	43
0748	1-1/4"		18	.444		8.45	26.50		34.95	49
0749	1-1/2"		16	.500		11.05	29.50		40.55	56.50
0751	2"		14	.571		23	34		57	76.50
0752	2-1/2"		13	.615		44	36.50		80.50	104
0753	3"	Q-1	14	1.143		49.50	61		110.50	147
0755	4"	"	8	2		101	107		208	273
0757	5"	Q-2	7.50	3.200		585	177		762	905
0759	6"		7	3.429		1,075	189		1,264	1,475
0761	8"		6.50	3.692		2,150	204		2,354	2,650
0771	Cap, sweat									
0773	1/2"	1 Plum	40	.200	Ea.	.82	11.85		12.67	18.75
0775	3/4"		38	.211		1.52	12.45		13.97	20.50
0777	1"		32	.250		3.59	14.80		18.39	26.50
0778	1-1/4"		29	.276		4.75	16.35		21.10	30
0779	1-1/2"		26	.308		6.95	18.20		25.15	35
0781	2"		22	.364		12.80	21.50		34.30	46.50
0791	Flange, sweat									
0793	3"	Q-1	22	.727	Ea.	174	39		213	250
0795	4"		18	.889		243	47.50		290.50	340
0797	5"		12	1.333		455	71		526	605
0799	6"	Q-2	18	1.333		475	73.50		548.50	630
0801	8"	"	16	1.500		905	83		988	1,125
0850	Unions, 1/4"	1 Plum	21	.381		26	22.50		48.50	63
0870	3/8"		21	.381		26.50	22.50		49	63
0880	1/2"		19	.421		14.05	25		39.05	53
0890	5/8"		18	.444		60	26.50		86.50	106
0900	3/4"		18	.444		17.60	26.50		44.10	59
0910	1"		15	.533		30.50	31.50		62	81
0920	1-1/4"		14	.571		50	34		84	106
0930	1-1/2"		12	.667		65.50	39.50		105	132
0940	2"		10	.800		112	47.50		159.50	195
0950	2-1/2"	Q-1	12	1.333		245	71		316	375
0960	3"	"	10	1.600		635	85.50		720.50	825
0980	Adapter, copper x male IPS, 1/4"	1 Plum	20	.400		11.05	23.50		34.55	47.50
0990	3/8"		20	.400		5.50	23.50		29	41.50
1000	1/2"		18	.444		2.36	26.50		28.86	42
1010	3/4"		17	.471		3.96	28		31.96	46.50

For customer support on your Mechanical Cost Data, call 877.739.3689.

135

22 11 13.25 Pipe/Tube Fittings, Copper		Crew	Daily Output	Labor-Hours	Unit	Material	2016 Bare Costs Labor	Equipment	Total	Total Incl O&P
1020	1"	1 Plum	15	.533	Ea.	10.15	31.50		41.65	58.50
1030	1-1/4"		13	.615		14.55	36.50		51.05	71
1040	1-1/2"		12	.667		17.05	39.50		56.55	78.50
1050	2"		11	.727		29	43		72	97
1060	2-1/2"	Q-1	10.50	1.524		101	81		182	234
1070	3"		10	1.600		125	85.50		210.50	267
1080	3-1/2"		9	1.778		157	94.50		251.50	315
1090	4"		8	2		164	107		271	340
1200	5", cast		6	2.667		1,800	142		1,942	2,200
1210	6", cast	Q-2	8.50	2.824		2,000	156		2,156	2,425
1214	Adapter, copper x female IPS									
1216	1/2"	1 Plum	18	.444	Ea.	3.75	26.50		30.25	43.50
1218	3/4"		17	.471		5.15	28		33.15	47.50
1220	1"		15	.533		11.75	31.50		43.25	60.50
1221	1-1/4"		13	.615		17.05	36.50		53.55	74
1222	1-1/2"		12	.667		26.50	39.50		66	89
1224	2"		11	.727		36	43		79	105
1250	Cross, 1/2"		10	.800		18	47.50		65.50	91.50
1260	3/4"		9.50	.842		35	50		85	114
1270	1"		8	1		59.50	59		118.50	155
1280	1-1/4"		7.50	1.067		85.50	63		148.50	190
1290	1-1/2"		6.50	1.231		122	73		195	244
1300	2"		5.50	1.455		230	86		316	385
1310	2-1/2"	Q-1	6.50	2.462		530	131		661	785
1320	3"	"	5.50	2.909		620	155		775	920
1500	Tee fitting, mechanically formed, (Type 1, 'branch sizes up to 2 in.')									
1520	1/2" run size, 3/8" to 1/2" branch size	1 Plum	80	.100	Ea.		5.90		5.90	8.95
1530	3/4" run size, 3/8" to 3/4" branch size		60	.133			7.90		7.90	11.90
1540	1" run size, 3/8" to 1" branch size		54	.148			8.75		8.75	13.25
1550	1-1/4" run size, 3/8" to 1-1/4" branch size		48	.167			9.85		9.85	14.90
1560	1-1/2" run size, 3/8" to 1-1/2" branch size		40	.200			11.85		11.85	17.85
1570	2" run size, 3/8" to 2" branch size		35	.229			13.55		13.55	20.50
1580	2-1/2" run size, 1/2" to 2" branch size		32	.250			14.80		14.80	22.50
1590	3" run size, 1" to 2" branch size		26	.308			18.20		18.20	27.50
1600	4" run size, 1" to 2" branch size		24	.333			19.75		19.75	30
1640	Tee fitting, mechanically formed, (Type 2, branches 2-1/2" thru 4")									
1650	2-1/2" run size, 2-1/2" branch size	1 Plum	12.50	.640	Ea.		38		38	57
1660	3" run size, 2-1/2" to 3" branch size		12	.667			39.50		39.50	59.50
1670	3-1/2" run size, 2-1/2" to 3-1/2" branch size		11	.727			43		43	65
1680	4" run size, 2-1/2" to 4" branch size		10.50	.762			45		45	68
1698	5" run size, 2" to 4" branch size		9.50	.842			50		50	75
1700	6" run size, 2" to 4" branch size		8.50	.941			55.50		55.50	84
1710	8" run size, 2" to 4" branch size		7	1.143			67.50		67.50	102
1800	ACR fittings, OD size									
1802	Tee, straight									
1808	5/8"	1 Stpi	12	.667	Ea.	1.92	40.50		42.42	63
1810	3/4"		12	.667		12.05	40.50		52.55	74.50
1812	7/8"		10	.800		4.64	48.50		53.14	78.50
1813	1"		10	.800		33	48.50		81.50	110
1814	1-1/8"		10	.800		13.95	48.50		62.45	89
1816	1-3/8"		9	.889		19.25	54		73.25	103
1818	1-5/8"		8	1		29	60.50		89.50	124
1820	2-1/8"		7	1.143		46	69.50		115.50	156

22 11 13.25 Pipe/Tube Fittings, Copper		Crew	Daily Output	Labor-Hours	Unit	Material	2016 Bare Costs Labor	Equipment	Total	Total Incl O&P
1822	2-5/8"	Q-5	8	2	Ea.	85	109		194	259
1824	3-1/8"		7	2.286		122	125		247	320
1826	4-1/8"		5	3.200		286	175		461	580
1830	90° elbow									
1836	5/8"	1 Stpi	19	.421	Ea.	3.67	25.50		29.17	42.50
1838	3/4"		19	.421		6.65	25.50		32.15	46
1840	7/8"		16	.500		6.60	30.50		37.10	53.50
1842	1-1/8"		16	.500		8.85	30.50		39.35	56
1844	1-3/8"		15	.533		9.20	32.50		41.70	59
1846	1-5/8"		13	.615		14.35	37.50		51.85	72.50
1848	2-1/8"		11	.727		26	44		70	95.50
1850	2-5/8"	Q-5	13	1.231		47.50	67		114.50	154
1852	3-1/8"		11	1.455		67.50	79.50		147	195
1854	4-1/8"		9	1.778		148	97		245	310
1860	Coupling									
1866	5/8"	1 Stpi	21	.381	Ea.	.84	23		23.84	36
1868	3/4"		21	.381		2.23	23		25.23	37.50
1870	7/8"		18	.444		1.70	27		28.70	42.50
1871	1"		18	.444		5.85	27		32.85	47
1872	1-1/8"		18	.444		3.39	27		30.39	44
1874	1-3/8"		17	.471		5.95	28.50		34.45	49.50
1876	1-5/8"		15	.533		7.85	32.50		40.35	57.50
1878	2-1/8"		13	.615		13.15	37.50		50.65	71
1880	2-5/8"	Q-5	15	1.067		27	58.50		85.50	118
1882	3-1/8"		13	1.231		37	67		104	142
1884	4-1/8"		7	2.286		81	125		206	277
2000	DWV, solder joints, copper x copper									
2030	90° Elbow, 1-1/4"	1 Plum	13	.615	Ea.	12.05	36.50		48.55	68.50
2050	1-1/2"		12	.667		16	39.50		55.50	77
2070	2"		10	.800		25.50	47.50		73	99.50
2090	3"	Q-1	10	1.600		62	85.50		147.50	197
2100	4"	"	9	1.778		300	94.50		394.50	475
2150	45° Elbow, 1-1/4"	1 Plum	13	.615		9.85	36.50		46.35	66
2170	1-1/2"		12	.667		9.20	39.50		48.70	69.50
2180	2"		10	.800		18.75	47.50		66.25	92
2190	3"	Q-1	10	1.600		42	85.50		127.50	176
2200	4"	"	9	1.778		201	94.50		295.50	365
2250	Tee, Sanitary, 1-1/4"	1 Plum	9	.889		21	52.50		73.50	103
2270	1-1/2"		8	1		26.50	59		85.50	119
2290	2"		7	1.143		36	67.50		103.50	142
2310	3"	Q-1	7	2.286		138	122		260	335
2330	4"	"	6	2.667		330	142		472	580
2400	Coupling, 1-1/4"	1 Plum	14	.571		5	34		39	56.50
2420	1-1/2"		13	.615		6.20	36.50		42.70	62
2440	2"		11	.727		8.65	43		51.65	74.50
2460	3"	Q-1	11	1.455		20	77.50		97.50	139
2480	4"	"	10	1.600		44	85.50		129.50	178
2602	Traps, see Section 22 13 16.60									
3500	Compression joint fittings									
3510	As used for plumbing and oil burner work									
3520	Fitting price includes nuts and sleeves									
3540	Sleeve, 1/8"				Ea.	.13			.13	.14
3550	3/16"					.13			.13	.14

For customer support on your Mechanical Cost Data, call 877.739.3689.

137

22 11 13 – Facility Water Distribution Piping

22 11 13.25 Pipe/Tube Fittings, Copper		Crew	Daily Output	Labor-Hours	Unit	Material	2016 Bare Costs Labor	Equipment	Total	Total Incl O&P
3560	1/4"				Ea.	.05			.05	.06
3570	5/16"					.14			.14	.15
3580	3/8"					.27			.27	.30
3600	1/2"					.36			.36	.40
3620	Nut, 1/8"					.19			.19	.21
3630	3/16"					.20			.20	.22
3640	1/4"					.20			.20	.22
3650	5/16"					.29			.29	.32
3660	3/8"					.32			.32	.35
3670	1/2"					.56			.56	.62
3710	Union, 1/8"	1 Plum	26	.308		1.18	18.20		19.38	29
3720	3/16"		24	.333		1.13	19.75		20.88	31
3730	1/4"		24	.333		1.19	19.75		20.94	31.50
3740	5/16"		23	.348		1.52	20.50		22.02	32.50
3750	3/8"		22	.364		1.53	21.50		23.03	34
3760	1/2"		22	.364		2.36	21.50		23.86	35
3780	5/8"		21	.381		2.61	22.50		25.11	37
3820	Union tee, 1/8"		17	.471		4.49	28		32.49	47
3830	3/16"		16	.500		2.60	29.50		32.10	47.50
3840	1/4"		15	.533		2.36	31.50		33.86	50
3850	5/16"		15	.533		3.31	31.50		34.81	51
3860	3/8"		15	.533		3.37	31.50		34.87	51
3870	1/2"		15	.533		5.70	31.50		37.20	54
3910	Union elbow, 1/4"		24	.333		1.45	19.75		21.20	31.50
3920	5/16"		23	.348		2.56	20.50		23.06	34
3930	3/8"		22	.364		2.47	21.50		23.97	35
3940	1/2"		22	.364		4.03	21.50		25.53	37
3980	Female connector, 1/8"		26	.308		.91	18.20		19.11	28.50
4000	3/16" x 1/8"		24	.333		1.17	19.75		20.92	31.50
4010	1/4" x 1/8"		24	.333		1.03	19.75		20.78	31
4020	1/4"		24	.333		1.39	19.75		21.14	31.50
4030	3/8" x 1/4"		22	.364		1.39	21.50		22.89	34
4040	1/2" x 3/8"		22	.364		2.21	21.50		23.71	35
4050	5/8" x 1/2"		21	.381		3.11	22.50		25.61	37.50
4090	Male connector, 1/8"		26	.308		.74	18.20		18.94	28.50
4100	3/16" x 1/8"		24	.333		.76	19.75		20.51	31
4110	1/4" x 1/8"		24	.333		.79	19.75		20.54	31
4120	1/4"		24	.333		.79	19.75		20.54	31
4130	5/16" x 1/8"		23	.348		.93	20.50		21.43	32
4140	5/16" x 1/4"		23	.348		1.05	20.50		21.55	32
4150	3/8" x 1/8"		22	.364		1.10	21.50		22.60	33.50
4160	3/8" x 1/4"		22	.364		1.27	21.50		22.77	34
4170	3/8"		22	.364		1.27	21.50		22.77	34
4180	3/8" x 1/2"		22	.364		1.74	21.50		23.24	34.50
4190	1/2" x 3/8"		22	.364		1.69	21.50		23.19	34.50
4200	1/2"		22	.364		1.89	21.50		23.39	34.50
4210	5/8" x 1/2"		21	.381		2.18	22.50		24.68	36.50
4240	Male elbow, 1/8"		26	.308		1.52	18.20		19.72	29
4250	3/16" x 1/8"		24	.333		1.67	19.75		21.42	32
4260	1/4" x 1/8"		24	.333		1.27	19.75		21.02	31.50
4270	1/4"		24	.333		1.44	19.75		21.19	31.50
4280	3/8" x 1/4"		22	.364		1.83	21.50		23.33	34.50
4290	3/8"		22	.364		3.14	21.50		24.64	36

22 11 13.25 Pipe/Tube Fittings, Copper	Crew	Daily Output	Labor-Hours	Unit	Material	2016 Bare Costs Labor	Equipment	Total	Total Incl O&P
4300	1/2" x 1/4"	1 Plum	22	.364	Ea.	3.03	21.50		24.53	36
4310	1/2" x 3/8"		22	.364		4.77	21.50		26.27	38
4340	Female elbow, 1/8"		26	.308		3.26	18.20		21.46	31
4350	1/4" x 1/8"		24	.333		1.85	19.75		21.60	32
4360	1/4"		24	.333		2.43	19.75		22.18	32.50
4370	3/8" x 1/4"		22	.364		2.40	21.50		23.90	35
4380	1/2" x 3/8"		22	.364		4.09	21.50		25.59	37
4390	1/2"		22	.364		5.50	21.50		27	38.50
4420	Male run tee, 1/4" x 1/8"		15	.533		2.27	31.50		33.77	50
4430	5/16" x 1/8"		15	.533		3.93	31.50		35.43	52
4440	3/8" x 1/4"		15	.533		3.51	31.50		35.01	51.50
4480	Male branch tee, 1/4" x 1/8"		15	.533		2.63	31.50		34.13	50.50
4490	1/4"		15	.533		2.83	31.50		34.33	50.50
4500	3/8" x 1/4"		15	.533		3.30	31.50		34.80	51
4510	1/2" x 3/8"		15	.533		5.10	31.50		36.60	53
4520	1/2"	▼	15	.533	▼	8.20	31.50		39.70	56.50
4800	Flare joint fittings									
4810	Refrigeration fittings									
4820	Flare joint nuts and labor not incl. in price. Add 1 nut per jnt.									
4830	90° Elbow, 1/4"				Ea.	1.37			1.37	1.51
4840	3/8"					1.95			1.95	2.15
4850	1/2"					2.51			2.51	2.76
4860	5/8"					3.45			3.45	3.80
4870	3/4"					9.85			9.85	10.85
5030	Tee, 1/4"					1.76			1.76	1.94
5040	5/16"					2.11			2.11	2.32
5050	3/8"					2.29			2.29	2.52
5060	1/2"					3.46			3.46	3.81
5070	5/8"					5.05			5.05	5.55
5080	3/4"					5.50			5.50	6.05
5140	Union, 3/16"					.81			.81	.89
5150	1/4"					.56			.56	.62
5160	5/16"					1			1	1.10
5170	3/8"					.93			.93	1.02
5180	1/2"					1.39			1.39	1.53
5190	5/8"					2.23			2.23	2.45
5200	3/4"					6.20			6.20	6.80
5260	Long flare nut, 3/16"	1 Stpi	42	.190		1.74	11.55		13.29	19.35
5270	1/4"		41	.195		.57	11.85		12.42	18.50
5280	5/16"		40	.200		1.03	12.15		13.18	19.45
5290	3/8"		39	.205		1.15	12.45		13.60	20
5300	1/2"		38	.211		1.63	12.80		14.43	21
5310	5/8"		37	.216		3.43	13.10		16.53	23.50
5320	3/4"		34	.235		11.25	14.30		25.55	34
5380	Short flare nut, 3/16"		42	.190		2.88	11.55		14.43	20.50
5390	1/4"		41	.195		.40	11.85		12.25	18.30
5400	5/16"		40	.200		.48	12.15		12.63	18.85
5410	3/8"		39	.205		.63	12.45		13.08	19.50
5420	1/2"		38	.211		.93	12.80		13.73	20.50
5430	5/8"		36	.222		1.30	13.50		14.80	22
5440	3/4"		34	.235		4.33	14.30		18.63	26.50
5500	90° Elbow flare by MIPS, 1/4"					1.66			1.66	1.83
5510	3/8"				▼	1.83			1.83	2.01

22 11 13.25 Pipe/Tube Fittings, Copper		Crew	Daily Output	Labor-Hours	Unit	Material	2016 Bare Costs Labor	Equipment	Total	Total Incl O&P
5520	1/2"				Ea.	2.93			2.93	3.22
5530	5/8"					7.35			7.35	8.10
5540	3/4"					11.95			11.95	13.15
5600	Flare by FIPS, 1/4"					2.78			2.78	3.06
5610	3/8"					3.02			3.02	3.32
5620	1/2"					4.33			4.33	4.76
5670	Flare by sweat, 1/4"					1.96			1.96	2.16
5680	3/8"					5.35			5.35	5.90
5690	1/2"					5.15			5.15	5.70
5700	5/8"					16.25			16.25	17.90
5760	Tee flare by IPS, 1/4"					2.78			2.78	3.06
5770	3/8"					3.90			3.90	4.29
5780	1/2"					6.55			6.55	7.20
5790	5/8"					5.50			5.50	6.05
5850	Connector, 1/4"					.60			.60	.66
5860	3/8"					.90			.90	.99
5870	1/2"					1.47			1.47	1.62
5880	5/8"					3.21			3.21	3.53
5890	3/4"					4.82			4.82	5.30
5950	Seal cap, 1/4"					.39			.39	.43
5960	3/8"					.60			.60	.66
5970	1/2"					.74			.74	.81
5980	5/8"					1.54			1.54	1.69
5990	3/4"					4.50			4.50	4.95
6000	Water service fittings									
6010	Flare joints nut and labor are included in the fitting price.									
6020	90° Elbow, C x C, 3/8"	1 Plum	19	.421	Ea.	44.50	25		69.50	86.50
6030	1/2"		18	.444		44.50	26.50		71	88.50
6040	3/4"		16	.500		57	29.50		86.50	107
6050	1"		15	.533		108	31.50		139.50	166
6080	2"		10	.800		390	47.50		437.50	500
6090	90° Elbow, C x MPT, 3/8"		19	.421		34	25		59	75
6100	1/2"		18	.444		34	26.50		60.50	77
6110	3/4"		16	.500		39.50	29.50		69	87.50
6120	1"		15	.533		107	31.50		138.50	166
6130	1-1/4"		13	.615		220	36.50		256.50	298
6140	1-1/2"		12	.667		220	39.50		259.50	305
6150	2"		10	.800		268	47.50		315.50	365
6160	90° Elbow, C x FPT, 3/8"		19	.421		36.50	25		61.50	77.50
6170	1/2"		18	.444		36.50	26.50		63	79.50
6180	3/4"		16	.500		45	29.50		74.50	94
6190	1"		15	.533		104	31.50		135.50	162
6200	1-1/4"		13	.615		108	36.50		144.50	174
6210	1-1/2"		12	.667		415	39.50		454.50	520
6220	2"		10	.800		600	47.50		647.50	730
6230	Tee, C x C x C, 3/8"		13	.615		65	36.50		101.50	127
6240	1/2"		12	.667		65	39.50		104.50	131
6250	3/4"		11	.727		76.50	43		119.50	149
6260	1"		10	.800		135	47.50		182.50	221
6330	Tube nut, C x nut seat, 3/8"		40	.200		13.90	11.85		25.75	33
6340	1/2"		38	.211		13.90	12.45		26.35	34
6350	3/4"		34	.235		13.20	13.95		27.15	35.50
6360	1"		32	.250		24	14.80		38.80	49

22 11 13 – Facility Water Distribution Piping

22 11 13.25 Pipe/Tube Fittings, Copper		Crew	Daily Output	Labor-Hours	Unit	Material	2016 Bare Costs Labor	Equipment	Total	Total Incl O&P
6380	Coupling, C x C, 3/8"	1 Plum	19	.421	Ea.	40	25		65	81.50
6390	1/2"		18	.444		40	26.50		66.50	83.50
6400	3/4"		16	.500		50.50	29.50		80	100
6410	1"		15	.533		93.50	31.50		125	151
6420	1-1/4"		12	.667		151	39.50		190.50	227
6430	1-1/2"		12	.667		229	39.50		268.50	310
6440	2"		10	.800		345	47.50		392.50	450
6450	Adapter, C x FPT, 3/8"		19	.421		30	25		55	70.50
6460	1/2"		18	.444		30	26.50		56.50	72.50
6470	3/4"		16	.500		36.50	29.50		66	84.50
6480	1"		15	.533		78	31.50		109.50	133
6490	1-1/4"		13	.615		101	36.50		137.50	167
6500	1-1/2"		12	.667		166	39.50		205.50	242
6510	2"		10	.800		215	47.50		262.50	310
6520	Adapter, C x MPT, 3/8"		19	.421		27	25		52	67
6530	1/2"		18	.444		27	26.50		53.50	69
6540	3/4"		16	.500		37.50	29.50		67	85.50
6550	1"		15	.533		66	31.50		97.50	120
6560	1-1/4"		13	.615		164	36.50		200.50	236
6570	1-1/2"		12	.667		164	39.50		203.50	241
6580	2"	▼	10	.800	▼	223	47.50		270.50	315
6992	Tube connector fittings, See Section 22 11 13.76 for plastic ftng.									
7000	Insert type Brass/copper, 100 psi @ 180°F, CTS									
7010	Adapter MPT 3/8" x 3/8" CTS	1 Plum	29	.276	Ea.	3.07	16.35		19.42	28
7020	1/2" x 1/2"		26	.308		3.10	18.20		21.30	31
7030	3/4" x 1/2"		26	.308		3.96	18.20		22.16	32
7040	3/4" x 3/4"		25	.320		4.58	18.95		23.53	33.50
7050	Adapter CTS 1/2" x 1/2" sweat		24	.333		4.16	19.75		23.91	34.50
7060	3/4" x 3/4" sweat		22	.364		1.46	21.50		22.96	34
7070	Coupler center set 3/8" CTS		25	.320		1.51	18.95		20.46	30
7080	1/2" CTS		23	.348		4.16	20.50		24.66	35.50
7090	3/4" CTS		22	.364		1.46	21.50		22.96	34
7100	Elbow 90°, copper 3/8"		25	.320		3.32	18.95		22.27	32
7110	1/2" CTS		23	.348		2.12	20.50		22.62	33.50
7120	3/4" CTS		22	.364		2.58	21.50		24.08	35.50
7130	Tee copper 3/8" CTS		17	.471		3.94	28		31.94	46.50
7140	1/2" CTS		15	.533		2.71	31.50		34.21	50.50
7150	3/4" CTS		14	.571		4.16	34		38.16	55.50
7160	3/8" x 3/8" x 1/2"		16	.500		3.14	29.50		32.64	48
7170	1/2" x 3/8" x 1/2"		15	.533		2.02	31.50		33.52	49.50
7180	3/4" x 1/2" x 3/4"	▼	14	.571	▼	3.14	34		37.14	54.50

22 11 13.27 Pipe/Tube, Grooved Joint for Copper

		Crew	Daily Output	Labor-Hours	Unit	Material	2016 Bare Costs Labor	Equipment	Total	Total Incl O&P
0010	**PIPE/TUBE, GROOVED JOINT FOR COPPER**									
4000	Fittings: coupling material required at joints not incl. in fitting price.									
4001	Add 1 selected coupling, material only, per joint for installed price.									
4010	Coupling, rigid style									
4018	2" diameter	1 Plum	50	.160	Ea.	23.50	9.45		32.95	40
4020	2-1/2" diameter	Q-1	80	.200		26.50	10.65		37.15	45
4022	3" diameter		67	.239		29.50	12.70		42.20	51.50
4024	4" diameter		50	.320		44.50	17.05		61.55	74.50
4026	5" diameter	▼	40	.400		74.50	21.50		96	114
4028	6" diameter	Q-2	50	.480	▼	98.50	26.50		125	148

For customer support on your Mechanical Cost Data, call 877.739.3689.

141

22 11 Facility Water Distribution

22 11 13 – Facility Water Distribution Piping

22 11 13.27 Pipe/Tube, Grooved Joint for Copper		Crew	Daily Output	Labor-Hours	Unit	Material	2016 Bare Costs Labor	Equipment	Total	Total Incl O&P
4100	Elbow, 90° or 45°									
4108	2" diameter	1 Plum	25	.320	Ea.	41	18.95		59.95	74
4110	2-1/2" diameter	Q-1	40	.400		44.50	21.50		66	81
4112	3" diameter		33	.485		62.50	26		88.50	108
4114	4" diameter		25	.640		144	34		178	210
4116	5" diameter		20	.800		405	42.50		447.50	515
4118	6" diameter	Q-2	25	.960		650	53		703	795
4200	Tee									
4208	2" diameter	1 Plum	17	.471	Ea.	67.50	28		95.50	117
4210	2-1/2" diameter	Q-1	27	.593		72	31.50		103.50	127
4212	3" diameter		22	.727		107	39		146	176
4214	4" diameter		17	.941		234	50		284	335
4216	5" diameter		13	1.231		655	65.50		720.50	820
4218	6" diameter	Q-2	17	1.412		810	78		888	1,000
4300	Reducer, concentric									
4310	3" x 2-1/2" diameter	Q-1	35	.457	Ea.	59	24.50		83.50	102
4312	4" x 2-1/2" diameter		32	.500		121	26.50		147.50	173
4314	4" x 3" diameter		29	.552		121	29.50		150.50	178
4316	5" x 3" diameter		25	.640		340	34		374	425
4318	5" x 4" diameter		22	.727		340	39		379	435
4320	6" x 3" diameter	Q-2	28	.857		370	47.50		417.50	475
4322	6" x 4" diameter		26	.923		370	51		421	480
4324	6" x 5" diameter		24	1		370	55.50		425.50	490
4350	Flange, w/groove gasket									
4351	ANSI class 125 and 150									
4355	2" diameter	1 Plum	23	.348	Ea.	178	20.50		198.50	227
4356	2-1/2" diameter	Q-1	37	.432		186	23		209	239
4358	3" diameter		31	.516		194	27.50		221.50	256
4360	4" diameter		23	.696		213	37		250	291
4362	5" diameter		19	.842		287	45		332	385
4364	6" diameter	Q-2	23	1.043		310	57.50		367.50	430

22 11 13.29 Pipe, Fittings and Valves, Copper, Pressed-Joint

		Crew	Daily Output	Labor-Hours	Unit	Material	Labor	Equipment	Total	Total Incl O&P
0010	**PIPE, FITTINGS AND VALVES, COPPER, PRESSED-JOINT**									
0040	Pipe/tube includes coupling & clevis type hanger assy's, 10' O.C.									
0120	Type K									
0130	1/2" diameter	1 Plum	78	.103	L.F.	5.20	6.05		11.25	14.85
0134	3/4" diameter		74	.108		8.40	6.40		14.80	18.90
0138	1" diameter		66	.121		11.15	7.20		18.35	23
0142	1-1/4" diameter		56	.143		13.50	8.45		21.95	27.50
0146	1-1/2" diameter		50	.160		17.55	9.45		27	33.50
0150	2" diameter		40	.200		25.50	11.85		37.35	46.50
0154	2-1/2" diameter	Q-1	60	.267		42.50	14.20		56.70	68
0158	3" diameter		54	.296		56.50	15.80		72.30	86.50
0162	4" diameter		38	.421		84.50	22.50		107	127
0180	To delete cplgs. & hngrs., 1/2" pipe, subtract					19%	48%			
0184	3/4" -2" pipe, subtract					14%	46%			
0186	2-1/2"-4" pipe, subtract					24%	34%			
0220	Type L									
0230	1/2" diameter	1 Plum	81	.099	L.F.	3.88	5.85		9.73	13.10
0234	3/4" diameter		76	.105		5.50	6.25		11.75	15.45
0238	1" diameter		68	.118		7.95	6.95		14.90	19.25
0242	1-1/4" diameter		58	.138		10.55	8.15		18.70	24

142

For customer support on your Mechanical Cost Data, call 877.739.3689.

22 11 13 – Facility Water Distribution Piping

22 11 13.29 Pipe, Fittings and Valves, Copper, Pressed-Joint		Crew	Daily Output	Labor-Hours	Unit	Material	2016 Bare Costs Labor	Equipment	Total	Total Incl O&P
0246	1-1/2" diameter	1 Plum	52	.154	L.F.	13.80	9.10		22.90	29
0250	2" diameter	↓	42	.190		19.65	11.30		30.95	38.50
0254	2-1/2" diameter	Q-1	62	.258		33.50	13.75		47.25	58
0258	3" diameter		56	.286		44	15.20		59.20	71
0262	4" diameter	↓	39	.410	↓	65	22		87	105
0280	To delete cplgs. & hngrs., 1/2" - 1"pipe, subtract					21%	52%			
0284	1-1/4"-2" pipe, subtract					17%	46%			
0286	2-1/2"-4" pipe, subtract					23%	35%			
0320	Type M									
0330	1/2" diameter	1 Plum	84	.095	L.F.	3.41	5.65		9.06	12.25
0334	3/4" diameter		78	.103		4.65	6.05		10.70	14.25
0338	1" diameter		70	.114		7.15	6.75		13.90	18.05
0342	1-1/4" diameter		60	.133		9.65	7.90		17.55	22.50
0346	1-1/2" diameter		54	.148		13.30	8.75		22.05	28
0350	2" diameter	↓	44	.182		19.40	10.75		30.15	38
0354	2-1/2" diameter	Q-1	64	.250		32.50	13.30		45.80	55.50
0358	3" diameter		58	.276		42	14.70		56.70	68.50
0362	4" diameter	↓	40	.400	↓	65	21.50		86.50	104
0380	To delete cplgs. & hngrs., 1/2" pipe, subtract					32%	49%			
0384	3/4"-2" pipe, subtract					21%	46%			
0386	2-1/2"-4" pipe, subtract					25%	36%			
1600	Fittings									
1610	Press joints, copper x copper									
1620	Note: Reducing fittings show most expensive size combination.									
1800	90° elbow, 1/2"	1 Plum	36.60	.219	Ea.	3.19	12.95		16.14	23
1810	3/4"		27.50	.291		5.35	17.20		22.55	32
1820	1"		25.90	.309		10.75	18.30		29.05	39.50
1830	1-1/4"		20.90	.383		21.50	22.50		44	58
1840	1-1/2"		18.30	.437		41	26		67	84
1850	2"	↓	15.70	.510		57	30		87	108
1860	2-1/2"	Q-1	25.90	.618		165	33		198	232
1870	3"		22	.727		208	39		247	288
1880	4"	↓	16.30	.982		258	52.50		310.50	360
2000	45° elbow, 1/2"	1 Plum	36.60	.219		3.70	12.95		16.65	23.50
2010	3/4"		27.50	.291		4.55	17.20		21.75	31
2020	1"		25.90	.309		14.55	18.30		32.85	43.50
2030	1-1/4"		20.90	.383		21	22.50		43.50	57
2040	1-1/2"		18.30	.437		34	26		60	76
2050	2"	↓	15.70	.510		47	30		77	97.50
2060	2-1/2"	Q-1	25.90	.618		109	33		142	170
2070	3"		22	.727		156	39		195	231
2080	4"	↓	16.30	.982		217	52.50		269.50	320
2200	Tee, 1/2"	1 Plum	27.50	.291		4.87	17.20		22.07	31.50
2210	3/4"		20.70	.386		8.55	23		31.55	44
2220	1"		19.40	.412		15.35	24.50		39.85	54
2230	1-1/4"		15.70	.510		26	30		56	74
2240	1-1/2"		13.80	.580		50.50	34.50		85	108
2250	2"	↓	11.80	.678		62.50	40		102.50	129
2260	2-1/2"	Q-1	19.40	.825		204	44		248	292
2270	3"		16.50	.970		256	51.50		307.50	360
2280	4"	↓	12.20	1.311	↓	360	70		430	500
2400	Tee, reducing on the outlet									
2410	3/4"	1 Plum	20.70	.386	Ea.	7.20	23		30.20	42.50

For customer support on your Mechanical Cost Data, call 877.739.3689.

143

22 11 13 – Facility Water Distribution Piping

22 11 13.29 Pipe, Fittings and Valves, Copper, Pressed-Joint	Crew	Daily Output	Labor-Hours	Unit	Material	2016 Bare Costs Labor	Equipment	Total	Total Incl O&P
2420	1"	1 Plum	19.40	.412	Ea.	17.60	24.50		42.10	56.50
2430	1-1/4"		15.70	.510		24.50	30		54.50	72.50
2440	1-1/2"		13.80	.580		53.50	34.50		88	111
2450	2"	↓	11.80	.678		86.50	40		126.50	156
2460	2-1/2"	Q-1	19.40	.825		285	44		329	380
2470	3"		16.50	.970		310	51.50		361.50	420
2480	4"	↓	12.20	1.311	↓	430	70		500	575
2600	Tee, reducing on the run									
2610	3/4"	1 Plum	20.70	.386	Ea.	13.70	23		36.70	49.50
2620	1"		19.40	.412		26.50	24.50		51	66
2630	1-1/4"		15.70	.510		51	30		81	102
2640	1-1/2"		13.80	.580		78.50	34.50		113	138
2650	2"	↓	11.80	.678		86.50	40		126.50	156
2660	2-1/2"	Q-1	19.40	.825		291	44		335	385
2670	3"		16.50	.970		350	51.50		401.50	465
2680	4"	↓	12.20	1.311		500	70		570	655
2800	Coupling, 1/2"	1 Plum	36.60	.219		2.87	12.95		15.82	22.50
2810	3/4"		27.50	.291		4.35	17.20		21.55	31
2820	1"		25.90	.309		8.75	18.30		27.05	37
2830	1-1/4"		20.90	.383		10.85	22.50		33.35	46
2840	1-1/2"		18.30	.437		20	26		46	61
2850	2"	↓	15.70	.510		26	30		56	74
2860	2-1/2"	Q-1	25.90	.618		82.50	33		115.50	140
2870	3"		22	.727		105	39		144	174
2880	4"	↓	16.30	.982		146	52.50		198.50	239
3000	Union, 1/2"	1 Plum	36.60	.219		24	12.95		36.95	45.50
3010	3/4"		27.50	.291		31	17.20		48.20	60.50
3020	1"		25.90	.309		50	18.30		68.30	82.50
3030	1-1/4"		20.90	.383		71	22.50		93.50	113
3040	1-1/2"		18.30	.437		96.50	26		122.50	145
3050	2"	↓	15.70	.510	↓	158	30		188	220
3200	Adapter, tube to MPT									
3210	1/2"	1 Plum	15.10	.530	Ea.	3.75	31.50		35.25	51.50
3220	3/4"		13.60	.588		6.75	35		41.75	60
3230	1"		11.60	.690		12.40	41		53.40	75
3240	1-1/4"		9.90	.808		27	48		75	102
3250	1-1/2"		9	.889		38	52.50		90.50	121
3260	2"	↓	7.90	1.013		73	60		133	171
3270	2-1/2"	Q-1	13.40	1.194		165	63.50		228.50	278
3280	3"		10.60	1.509		208	80.50		288.50	350
3290	4"	↓	7.70	2.078	↓	241	111		352	430
3400	Adapter, tube to FPT									
3410	1/2"	1 Plum	15.10	.530	Ea.	4.51	31.50		36.01	52.50
3420	3/4"		13.60	.588		7.35	35		42.35	60.50
3430	1"		11.60	.690		13.50	41		54.50	76.50
3440	1-1/4"		9.90	.808		30.50	48		78.50	106
3450	1-1/2"		9	.889		44.50	52.50		97	128
3460	2"	↓	7.90	1.013		75	60		135	173
3470	2-1/2"	Q-1	13.40	1.194		221	63.50		284.50	340
3480	3"		10.60	1.509		290	80.50		370.50	440
3490	4"	↓	7.70	2.078	↓	375	111		486	575
3600	Flange									
3620	1"	1 Plum	36.20	.221	Ea.	126	13.10		139.10	158

22 11 13 – Facility Water Distribution Piping

22 11 13.29 Pipe, Fittings and Valves, Copper, Pressed-Joint		Crew	Daily Output	Labor-Hours	Unit	Material	2016 Bare Costs Labor	Equipment	Total	Total Incl O&P
3630	1-1/4"	1 Plum	29.30	.273	Ea.	183	16.15		199.15	226
3640	1-1/2"		25.60	.313		202	18.50		220.50	250
3650	2"	▼	22	.364		227	21.50		248.50	283
3660	2-1/2"	Q-1	36.20	.442		253	23.50		276.50	315
3670	3"		30.80	.519		305	27.50		332.50	375
3680	4"	▼	22.80	.702		345	37.50		382.50	435
3800	Cap, 1/2"	1 Plum	53.10	.151		6.10	8.90		15	20
3810	3/4"		39.80	.201		10.30	11.90		22.20	29.50
3820	1"		37.50	.213		15.90	12.65		28.55	36.50
3830	1-1/4"		30.40	.263		18	15.60		33.60	43.50
3840	1-1/2"		26.60	.301		28	17.80		45.80	57.50
3850	2"	▼	22.80	.351		34	21		55	69
3860	2-1/2"	Q-1	37.50	.427		108	22.50		130.50	154
3870	3"		31.90	.502		137	26.50		163.50	192
3880	4"	▼	23.60	.678	▼	182	36		218	256
4000	Reducer									
4010	3/4"	1 Plum	27.50	.291	Ea.	12.95	17.20		30.15	40.50
4020	1"		25.90	.309		22.50	18.30		40.80	52.50
4030	1-1/4"		20.90	.383		33.50	22.50		56	71
4040	1-1/2"		18.30	.437		40	26		66	83
4050	2"	▼	15.70	.510		54.50	30		84.50	106
4060	2-1/2"	Q-1	25.90	.618		163	33		196	229
4070	3"		22	.727		211	39		250	291
4080	4"	▼	16.30	.982		272	52.50		324.50	380
4100	Stub out, 1/2"	1 Plum	50	.160		7.85	9.45		17.30	23
4110	3/4"		35	.229		13.90	13.55		27.45	36
4120	1"	▼	30	.267	▼	19.05	15.80		34.85	45
6000	Valves									
6200	Ball valve									
6210	1/2"	1 Plum	25.60	.313	Ea.	43.50	18.50		62	75.50
6220	3/4"		19.20	.417		61	24.50		85.50	104
6230	1"		18.10	.442		75	26		101	122
6240	1-1/4"		14.70	.544		122	32		154	183
6250	1-1/2"		12.80	.625		169	37		206	241
6260	2"	▼	11	.727	▼	269	43		312	360
6400	Check valve									
6410	1/2"	1 Plum	30	.267	Ea.	33.50	15.80		49.30	61
6420	3/4"		22.50	.356		43	21		64	79
6430	1"		21.20	.377		49.50	22.50		72	87.50
6440	1-1/4"		17.20	.465		71.50	27.50		99	120
6450	1-1/2"		15	.533		101	31.50		132.50	159
6460	2"	▼	12.90	.620	▼	184	36.50		220.50	259
6600	Butterfly valve, lug type									
6660	2-1/2"	Q-1	9	1.778	Ea.	157	94.50		251.50	315
6670	3"		8	2		193	107		300	375
6680	4"	▼	5	3.200	▼	241	171		412	520

For customer support on your Mechanical Cost Data, call 877.739.3689.

145

22 11 13 – Facility Water Distribution Piping

22 11 13.44 Pipe, Steel	Crew	Daily Output	Labor-Hours	Unit	Material	2016 Bare Costs Labor	Equipment	Total	Total Incl O&P
0010 **PIPE, STEEL** R221113-50									
0020 All pipe sizes are to Spec. A-53 unless noted otherwise R221113-70									
0032 Schedule 10, see Line 22 11 13.48 0500									
0050 Schedule 40, threaded, with couplings, and clevis hanger									
0060 assemblies sized for covering, 10' O.C.									
0540 Black, 1/4" diameter	1 Plum	66	.121	L.F.	4.96	7.20		12.16	16.30
0550 3/8" diameter		65	.123		5.50	7.30		12.80	17.05
0560 1/2" diameter		63	.127		2.90	7.50		10.40	14.55
0570 3/4" diameter		61	.131		3.35	7.75		11.10	15.40
0580 1" diameter		53	.151		4.30	8.95		13.25	18.25
0590 1-1/4" diameter	Q-1	89	.180		5.20	9.60		14.80	20
0600 1-1/2" diameter		80	.200		5.90	10.65		16.55	22.50
0610 2" diameter		64	.250		7.45	13.30		20.75	28
0620 2-1/2" diameter		50	.320		11.55	17.05		28.60	38
0630 3" diameter		43	.372		14.60	19.85		34.45	46
0640 3-1/2" diameter		40	.400		20.50	21.50		42	54.50
0650 4" diameter		36	.444		22	23.50		45.50	60
0809 A-106, gr. A/B, seamless w/cplgs. & clevis hanger assemblies									
0811 1/4" diameter	1 Plum	66	.121	L.F.	6.60	7.20		13.80	18.15
0812 3/8" diameter		65	.123		6.60	7.30		13.90	18.25
0813 1/2" diameter		63	.127		7.25	7.50		14.75	19.30
0814 3/4" diameter		61	.131		8.20	7.75		15.95	20.50
0815 1" diameter		53	.151		9.40	8.95		18.35	24
0816 1-1/4" diameter	Q-1	89	.180		11.15	9.60		20.75	27
0817 1-1/2" diameter		80	.200		15.70	10.65		26.35	33.50
0819 2" diameter		64	.250		14.25	13.30		27.55	35.50
0821 2-1/2" diameter		50	.320		19.55	17.05		36.60	47
0822 3" diameter		43	.372		25.50	19.85		45.35	58
0823 4" diameter		36	.444		40	23.50		63.50	79.50
1220 To delete coupling & hanger, subtract									
1230 1/4" diam. to 3/4" diam.					31%	56%			
1240 1" diam. to 1-1/2" diam.					23%	51%			
1250 2" diam. to 4" diam.					23%	41%			
1280 All pipe sizes are to Spec. A-53 unless noted otherwise									
1281 Schedule 40, threaded, with couplings and clevis hanger									
1282 assemblies sized for covering, 10' O. C.									
1290 Galvanized, 1/4" diameter	1 Plum	66	.121	L.F.	6.70	7.20		13.90	18.20
1300 3/8" diameter		65	.123		7.25	7.30		14.55	19
1310 1/2" diameter		63	.127		3.18	7.50		10.68	14.85
1320 3/4" diameter		61	.131		3.60	7.75		11.35	15.65
1330 1" diameter		53	.151		4.83	8.95		13.78	18.80
1340 1-1/4" diameter	Q-1	89	.180		5.80	9.60		15.40	21
1350 1-1/2" diameter		80	.200		6.60	10.65		17.25	23.50
1360 2" diameter		64	.250		8.45	13.30		21.75	29.50
1370 2-1/2" diameter		50	.320		13.40	17.05		30.45	40
1380 3" diameter		43	.372		16.85	19.85		36.70	48.50
1390 3-1/2" diameter		40	.400		21.50	21.50		43	56
1400 4" diameter		36	.444		25	23.50		48.50	62.50
1750 To delete coupling & hanger, subtract									
1760 1/4" diam. to 3/4" diam.					31%	56%			
1770 1" diam. to 1-1/2" diam.					23%	51%			
1780 2" diam. to 4" diam.					23%	41%			

22 11 13 – Facility Water Distribution Piping

22 11 13.44 Pipe, Steel	Crew	Daily Output	Labor-Hours	Unit	Material	2016 Bare Costs Labor	Equipment	Total	Total Incl O&P
2000	Welded, sch. 40, on yoke & roll hanger assy's, sized for covering, 10' O.C.									
2040	Black, 1" diameter	Q-15	93	.172	L.F.	4.14	9.15	.62	13.91	19.10
2050	1-1/4" diameter		84	.190		5.15	10.15	.69	15.99	21.50
2060	1-1/2" diameter		76	.211		5.55	11.20	.76	17.51	24
2070	2" diameter		61	.262		6.65	14	.95	21.60	29.50
2080	2-1/2" diameter		47	.340		10.55	18.15	1.24	29.94	40.50
2090	3" diameter		43	.372		12.55	19.85	1.35	33.75	45.50
2100	3-1/2" diameter		39	.410		15.10	22	1.49	38.59	51
2110	4" diameter		37	.432		17.50	23	1.57	42.07	56
2120	5" diameter		32	.500		37	26.50	1.82	65.32	82.50
2130	6" diameter	Q-16	36	.667		46	37	1.61	84.61	108
2140	8" diameter		29	.828		73.50	45.50	2	121	152
2150	10" diameter		24	1		95.50	55.50	2.42	153.42	191
2160	12" diameter		19	1.263		112	70	3.06	185.06	231
2170	14" diameter, (two rod roll type hanger for 14" diam. and up)		15	1.600		105	88.50	3.87	197.37	252
2180	16" diameter, (two rod roll type hanger)		13	1.846		158	102	4.47	264.47	335
2190	18" diameter, (two rod roll type hanger)		11	2.182		150	121	5.30	276.30	355
2200	20" diameter, (two rod roll type hanger)		9	2.667		151	147	6.45	304.45	395
2220	24" diameter, (two rod roll type hanger)		8	3		194	166	7.25	367.25	470
2345	Sch. 40, A-53, gr. A/B, ERW, welded w/hngrs.									
2346	2" diameter	Q-15	61	.262	L.F.	7.15	14	.95	22.10	30
2347	2-1/2" diameter		47	.340		9.20	18.15	1.24	28.59	39
2348	3" diameter		43	.372		11.50	19.85	1.35	32.70	44
2349	4" diameter		38	.421		15.55	22.50	1.53	39.58	53
2350	6" diameter	Q-16	37	.649		30	36	1.57	67.57	88.50
2351	8" diameter		29	.828		72.50	45.50	2	120	151
2352	10" diameter		24	1		94.50	55.50	2.42	152.42	190
2560	To delete hanger, subtract									
2570	1" diam. to 1-1/2" diam.					15%	34%			
2580	2" diam. to 3-1/2" diam.					9%	21%			
2590	4" diam. to 12" diam.					5%	12%			
2596	14" diam. to 24" diam.					3%	10%			
2640	Galvanized, 1" diameter	Q-15	93	.172	L.F.	4.86	9.15	.62	14.63	19.90
2650	1-1/4" diameter		84	.190		6.10	10.15	.69	16.94	23
2660	1-1/2" diameter		76	.211		6.70	11.20	.76	18.66	25
2670	2" diameter		61	.262		8.15	14	.95	23.10	31
2680	2-1/2" diameter		47	.340		11.60	18.15	1.24	30.99	41.50
2690	3" diameter		43	.372		14.85	19.85	1.35	36.05	48
2700	3-1/2" diameter		39	.410		17.35	22	1.49	40.84	53.50
2710	4" diameter		37	.432		20.50	23	1.57	45.07	59.50
2720	5" diameter		32	.500		23.50	26.50	1.82	51.82	68
2730	6" diameter	Q-16	36	.667		28.50	37	1.61	67.11	89
2740	8" diameter		29	.828		41	45.50	2	88.50	116
2750	10" diameter		24	1		89.50	55.50	2.42	147.42	184
2760	12" diameter		19	1.263		116	70	3.06	189.06	236
3160	To delete hanger, subtract									
3170	1" diam. to 1-1/2" diam.					30%	34%			
3180	2" diam. to 3-1/2" diam.					19%	21%			
3190	4" diam. to 12" diam.					10%	12%			
3250	Flanged, 150 lb. weld neck, on yoke & roll hangers									
3260	sized for covering, 10' O.C.									
3290	Black, 1" diameter	Q-15	70	.229	L.F.	9.35	12.20	.83	22.38	29.50
3300	1-1/4" diameter		64	.250		10.35	13.30	.91	24.56	32.50

22 11 13 – Facility Water Distribution Piping

22 11 13.44 Pipe, Steel		Crew	Daily Output	Labor-Hours	Unit	Material	2016 Bare Costs Labor	Equipment	Total	Total Incl O&P
3310	1-1/2" diameter	Q-15	58	.276	L.F.	10.75	14.70	1	26.45	35
3320	2" diameter		45	.356		12.65	18.95	1.29	32.89	44
3330	2-1/2" diameter		36	.444		17.30	23.50	1.61	42.41	56.50
3340	3" diameter		32	.500		19.90	26.50	1.82	48.22	64
3350	3-1/2" diameter		29	.552		24.50	29.50	2	56	73.50
3360	4" diameter		26	.615		27	33	2.23	62.23	81.50
3370	5" diameter		21	.762		51.50	40.50	2.77	94.77	122
3380	6" diameter	Q-16	25	.960		61.50	53	2.32	116.82	151
3390	8" diameter		19	1.263		101	70	3.06	174.06	219
3400	10" diameter		16	1.500		143	83	3.63	229.63	286
3410	12" diameter		14	1.714		172	94.50	4.15	270.65	335
3470	For 300 lb. flanges, add					63%				
3480	For 600 lb. flanges, add					310%				
3960	To delete flanges & hanger, subtract									
3970	1" diam. to 2" diam.					76%	65%			
3980	2-1/2" diam. to 4" diam.					62%	59%			
3990	5" diam. to 12" diam.					60%	46%			
4040	Galvanized, 1" diameter	Q-15	70	.229	L.F.	10.05	12.20	.83	23.08	30.50
4050	1-1/4" diameter		64	.250		11.30	13.30	.91	25.51	33.50
4060	1-1/2" diameter		58	.276		11.90	14.70	1	27.60	36
4070	2" diameter		45	.356		13.45	18.95	1.29	33.69	44.50
4080	2-1/2" diameter		36	.444		18.50	23.50	1.61	43.61	58
4090	3" diameter		32	.500		21.50	26.50	1.82	49.82	66
4100	3-1/2" diameter		29	.552		25.50	29.50	2	57	74.50
4110	4" diameter		26	.615		30	33	2.23	65.23	85
4120	5" diameter		21	.762		38	40.50	2.77	81.27	107
4130	6" diameter	Q-16	25	.960		45	53	2.32	100.32	132
4140	8" diameter		19	1.263		58.50	70	3.06	131.56	173
4150	10" diameter		16	1.500		123	83	3.63	209.63	265
4160	12" diameter		14	1.714		146	94.50	4.15	244.65	310
4220	For 300 lb. flanges, add					65%				
4240	For 600 lb. flanges, add					350%				
4660	To delete flanges & hanger, subtract									
4670	1" diam. to 2" diam.					73%	65%			
4680	2-1/2" diam. to 4" diam.					58%	59%			
4690	5" diam. to 12" diam.					43%	46%			
4750	Schedule 80, threaded, with couplings, and clevis hanger assemblies									
4760	sized for covering, 10' O.C.									
4790	Black, 1/4" diameter	1 Plum	54	.148	L.F.	8.65	8.75		17.40	23
4800	3/8" diameter		53	.151		10.90	8.95		19.85	25.50
4810	1/2" diameter		52	.154		4.10	9.10		13.20	18.25
4820	3/4" diameter		50	.160		4.62	9.45		14.07	19.40
4830	1" diameter		45	.178		6.05	10.50		16.55	22.50
4840	1-1/4" diameter	Q-1	75	.213		7.80	11.35		19.15	26
4850	1-1/2" diameter		69	.232		8.75	12.35		21.10	28.50
4860	2" diameter		56	.286		11.50	15.20		26.70	35.50
4870	2-1/2" diameter		44	.364		16.95	19.35		36.30	47.50
4880	3" diameter		38	.421		21.50	22.50		44	57.50
4890	3-1/2" diameter		35	.457		26.50	24.50		51	66
4900	4" diameter		32	.500		31.50	26.50		58	74.50
5061	A-106, gr. A/B seamless with cplgs. & clevis hanger assemblies, 1/4" diam.	1 Plum	63	.127		7.75	7.50		15.25	19.85
5062	3/8" diameter		62	.129		8.90	7.65		16.55	21.50
5063	1/2" diameter		61	.131		7.35	7.75		15.10	19.80

For customer support on your Mechanical Cost Data, call 877.739.3689.

22 11 13.44 Pipe, Steel	Crew	Daily Output	Labor-Hours	Unit	Material	2016 Bare Costs Labor	Equipment	Total	Total Incl O&P
5064	3/4" diameter	1 Plum	57	.140	L.F.	8.60	8.30		16.90	22
5065	1" diameter	↓	51	.157		10	9.30		19.30	25
5066	1-1/4" diameter	Q-1	85	.188		13.15	10.05		23.20	29.50
5067	1-1/2" diameter		77	.208		17.30	11.05		28.35	36
5071	2" diameter		61	.262		18.10	14		32.10	41
5072	2-1/2" diameter		48	.333		19.75	17.75		37.50	48.50
5073	3" diameter	↓	41	.390		25.50	21		46.50	59.50
5074	4" diameter	↓	35	.457	↓	37	24.50		61.50	77.50
5430	To delete coupling & hanger, subtract									
5440	1/4" diam. to 1/2" diam.					31%	54%			
5450	3/4" diam. to 1-1/2" diam.					28%	49%			
5460	2" diam. to 4" diam.					21%	40%			
5510	Galvanized, 1/4" diameter	1 Plum	54	.148	L.F.	9.45	8.75		18.20	23.50
5520	3/8" diameter		53	.151		10.40	8.95		19.35	25
5530	1/2" diameter		52	.154		4.50	9.10		13.60	18.70
5540	3/4" diameter		50	.160		5.35	9.45		14.80	20
5550	1" diameter	↓	45	.178		7	10.50		17.50	23.50
5560	1-1/4" diameter	Q-1	75	.213		9.25	11.35		20.60	27.50
5570	1-1/2" diameter		69	.232		10.60	12.35		22.95	30.50
5580	2" diameter		56	.286		14	15.20		29.20	38.50
5590	2-1/2" diameter		44	.364		21	19.35		40.35	52
5600	3" diameter		38	.421		26.50	22.50		49	63.50
5610	3-1/2" diameter		35	.457		38.50	24.50		63	79
5620	4" diameter	↓	32	.500	↓	35	26.50		61.50	78.50
5930	To delete coupling & hanger, subtract									
5940	1/4" diam. to 1/2" diam.					31%	54%			
5950	3/4" diam. to 1-1/2" diam.					28%	49%			
5960	2" diam. to 4" diam.					21%	40%			
6000	Welded, on yoke & roller hangers									
6010	sized for covering, 10' O.C.									
6040	Black, 1" diameter	Q-15	85	.188	L.F.	6.10	10.05	.68	16.83	22.50
6050	1-1/4" diameter		79	.203		7.85	10.80	.74	19.39	25.50
6060	1-1/2" diameter		72	.222		8.75	11.85	.81	21.41	28.50
6070	2" diameter		57	.281		10.80	14.95	1.02	26.77	35.50
6080	2-1/2" diameter		44	.364		15.90	19.35	1.32	36.57	48
6090	3" diameter		40	.400		19.95	21.50	1.45	42.90	55.50
6100	3-1/2" diameter		34	.471		24.50	25	1.71	51.21	66.50
6110	4" diameter		33	.485		28.50	26	1.76	56.26	72.50
6120	5" diameter, A-106B	↓	26	.615		48	33	2.23	83.23	104
6130	6" diameter, A-106B	Q-16	30	.800		65.50	44	1.94	111.44	141
6140	8" diameter, A-106B		25	.960		98.50	53	2.32	153.82	191
6150	10" diameter, A-106B		20	1.200		159	66.50	2.90	228.40	278
6160	12" diameter, A-106B	↓	15	1.600	↓	340	88.50	3.87	432.37	505
6540	To delete hanger, subtract									
6550	1" diam. to 1-1/2" diam.					30%	14%			
6560	2" diam. to 3" diam.					23%	9%			
6570	3-1/2" diam. to 5" diam.					12%	6%			
6580	6" diam. to 12" diam.					10%	4%			
6610	Galvanized, 1" diameter	Q-15	85	.188	L.F.	7.05	10.05	.68	17.78	23.50
6650	1-1/4" diameter		79	.203		9.30	10.80	.74	20.84	27.50
6660	1-1/2" diameter		72	.222		10.60	11.85	.81	23.26	30.50
6670	2" diameter		57	.281		13.35	14.95	1.02	29.32	38.50
6680	2-1/2" diameter	↓	44	.364		19.70	19.35	1.32	40.37	52

For customer support on your Mechanical Cost Data, call 877.739.3689.

149

22 11 13 – Facility Water Distribution Piping

22 11 13.44 Pipe, Steel		Crew	Daily Output	Labor-Hours	Unit	Material	2016 Bare Costs Labor	Equipment	Total	Total Incl O&P
6690	3" diameter	Q-15	40	.400	L.F.	25	21.50	1.45	47.95	61
6700	3-1/2" diameter		34	.471		36.50	25	1.71	63.21	80
6710	4" diameter		33	.485		32	26	1.76	59.76	76
6720	5" diameter, A-106B	↓	26	.615		46.50	33	2.23	81.73	103
6730	6" diameter, A-106B	Q-16	30	.800		63.50	44	1.94	109.44	139
6740	8" diameter, A-106B		25	.960		143	53	2.32	198.32	241
6750	10" diameter, A-106B		20	1.200		256	66.50	2.90	325.40	385
6760	12" diameter, A-106B	↓	15	1.600	↓	515	88.50	3.87	607.37	700
7150	To delete hanger, subtract									
7160	1" diam. to 1-1/2" diam.					26%	14%			
7170	2" diam. to 3" diam.					16%	9%			
7180	3-1/2" diam. to 5" diam.					10%	6%			
7190	6" diam. to 12" diam.					7%	4%			
7250	Flanged, 300 lb. weld neck, on yoke & roll hangers									
7260	sized for covering, 10' O.C.									
7290	Black, 1" diameter	Q-15	66	.242	L.F.	12.55	12.90	.88	26.33	34.50
7300	1-1/4" diameter		61	.262		14.25	14	.95	29.20	38
7310	1-1/2" diameter		54	.296		15.20	15.80	1.08	32.08	42
7320	2" diameter		42	.381		19.10	20.50	1.38	40.98	53
7330	2-1/2" diameter		33	.485		25.50	26	1.76	53.26	69
7340	3" diameter		29	.552		29.50	29.50	2	61	79
7350	3-1/2" diameter		24	.667		40	35.50	2.42	77.92	100
7360	4" diameter		23	.696		44.50	37	2.53	84.03	108
7370	5" diameter	↓	19	.842		72	45	3.06	120.06	150
7380	6" diameter	Q-16	23	1.043		89.50	57.50	2.53	149.53	188
7390	8" diameter		17	1.412		139	78	3.42	220.42	275
7400	10" diameter		14	1.714		234	94.50	4.15	332.65	405
7410	12" diameter	↓	12	2		435	111	4.84	550.84	645
7470	For 600 lb. flanges, add				↓	100%				
7940	To delete flanges & hanger, subtract									
7950	1" diam. to 1-1/2" diam.					75%	66%			
7960	2" diam. to 3" diam.					62%	60%			
7970	3-1/2" diam. to 5" diam.					54%	66%			
7980	6" diam. to 12" diam.					55%	62%			
8040	Galvanized, 1" diameter	Q-15	66	.242	L.F.	13.50	12.90	.88	27.28	35.50
8050	1-1/4" diameter		61	.262		15.70	14	.95	30.65	39.50
8060	1-1/2" diameter		54	.296		17.05	15.80	1.08	33.93	44
8070	2" diameter		42	.381		21.50	20.50	1.38	43.38	56
8080	2-1/2" diameter		33	.485		29	26	1.76	56.76	73
8090	3" diameter		29	.552		34.50	29.50	2	66	84.50
8100	3-1/2" diameter		24	.667		52	35.50	2.42	89.92	114
8110	4" diameter		23	.696		48	37	2.53	87.53	112
8120	5" diameter, A-106B	↓	19	.842		70.50	45	3.06	118.56	148
8130	6" diameter, A-106B	Q-16	23	1.043		87.50	57.50	2.53	147.53	186
8140	8" diameter, A-106B		17	1.412		184	78	3.42	265.42	325
8150	10" diameter, A-106B		14	1.714		330	94.50	4.15	428.65	515
8160	12" diameter, A-106B	↓	12	2		610	111	4.84	725.84	840
8240	For 600 lb. flanges, add				↓	100%				
8300	To delete flanges & hangers, subtract									
8310	1" diam. to 1-1/2" diam.					72%	66%			
8320	2" diam. to 3" diam.					59%	60%			
8330	3-1/2" diam. to 5" diam.					51%	66%			
8340	6" diam. to 12" diam.					49%	62%			

22 11 13 – Facility Water Distribution Piping

22 11 13.44 Pipe, Steel		Crew	Daily Output	Labor-Hours	Unit	Material	2016 Bare Costs Labor	Equipment	Total	Total Incl O&P
9000	Threading pipe labor, one end, all schedules through 80									
9010	1/4" through 3/4" pipe size	1 Plum	80	.100	Ea.		5.90		5.90	8.95
9020	1" through 2" pipe size		73	.110			6.50		6.50	9.80
9030	2-1/2" pipe size		53	.151			8.95		8.95	13.50
9040	3" pipe size		50	.160			9.45		9.45	14.30
9050	3-1/2" pipe size	Q-1	89	.180			9.60		9.60	14.45
9060	4" pipe size		73	.219			11.70		11.70	17.60
9070	5" pipe size		53	.302			16.10		16.10	24.50
9080	6" pipe size		46	.348			18.55		18.55	28
9090	8" pipe size		29	.552			29.50		29.50	44.50
9100	10" pipe size		21	.762			40.50		40.50	61.50
9110	12" pipe size		13	1.231			65.50		65.50	99
9120	Cutting pipe labor, one cut									
9124	Shop fabrication, machine cut									
9126	Schedule 40, straight pipe									
9128	2" pipe size or less	1 Stpi	62	.129	Ea.		7.85		7.85	11.80
9130	2-1/2" pipe size		56	.143			8.65		8.65	13.10
9132	3" pipe size		42	.190			11.55		11.55	17.45
9134	4" pipe size		31	.258			15.65		15.65	23.50
9136	5" pipe size		26	.308			18.70		18.70	28
9138	6" pipe size		19	.421			25.50		25.50	38.50
9140	8" pipe size		14	.571			34.50		34.50	52.50
9142	10" pipe size		10	.800			48.50		48.50	73.50
9144	12" pipe size		7	1.143			69.50		69.50	105
9146	14" pipe size	Q-5	10.50	1.524			83.50		83.50	126
9148	16" pipe size		8.60	1.860			102		102	153
9150	18" pipe size		7	2.286			125		125	188
9152	20" pipe size		5.80	2.759			151		151	227
9154	24" pipe size		4	4			219		219	330
9160	Schedule 80, straight pipe									
9164	2" pipe size or less	1 Stpi	42	.190	Ea.		11.55		11.55	17.45
9166	2-1/2" pipe size		37	.216			13.10		13.10	19.80
9168	3" pipe size		31	.258			15.65		15.65	23.50
9170	4" pipe size		23	.348			21		21	32
9172	5" pipe size		18	.444			27		27	40.50
9174	6" pipe size		14.60	.548			33.50		33.50	50
9176	8" pipe size		10	.800			48.50		48.50	73.50
9178	10" pipe size	Q-5	14	1.143			62.50		62.50	94
9180	12" pipe size	"	10	1.600			87.50		87.50	132
9200	Welding labor per joint									
9210	Schedule 40,									
9230	1/2" pipe size	Q-15	32	.500	Ea.		26.50	1.82	28.32	42
9240	3/4" pipe size		27	.593			31.50	2.15	33.65	50
9250	1" pipe size		23	.696			37	2.53	39.53	59
9260	1-1/4" pipe size		20	.800			42.50	2.90	45.40	67.50
9270	1-1/2" pipe size		19	.842			45	3.06	48.06	71
9280	2" pipe size		16	1			53.50	3.63	57.13	84.50
9290	2-1/2" pipe size		13	1.231			65.50	4.47	69.97	104
9300	3" pipe size		12	1.333			71	4.84	75.84	112
9310	4" pipe size		10	1.600			85.50	5.80	91.30	135
9320	5" pipe size		9	1.778			94.50	6.45	100.95	150
9330	6" pipe size		8	2			107	7.25	114.25	169
9340	8" pipe size		5	3.200			171	11.60	182.60	270

For customer support on your Mechanical Cost Data, call 877.739.3689.

151

22 11 13.44 Pipe, Steel

		Crew	Daily Output	Labor-Hours	Unit	Material	2016 Bare Costs Labor	2016 Bare Costs Equipment	Total	Total Incl O&P
9350	10" pipe size	Q-15	4	4	Ea.		213	14.50	227.50	335
9360	12" pipe size		3	5.333			284	19.35	303.35	450
9370	14" pipe size		2.60	6.154			330	22.50	352.50	520
9380	16" pipe size		2.20	7.273			385	26.50	411.50	615
9390	18" pipe size		2	8			425	29	454	675
9400	20" pipe size		1.80	8.889			475	32.50	507.50	750
9410	22" pipe size		1.70	9.412			500	34	534	795
9420	24" pipe size	↓	1.50	10.667	↓		570	38.50	608.50	905
9450	Schedule 80,									
9460	1/2" pipe size	Q-15	27	.593	Ea.		31.50	2.15	33.65	50
9470	3/4" pipe size		23	.696			37	2.53	39.53	59
9480	1" pipe size		20	.800			42.50	2.90	45.40	67.50
9490	1-1/4" pipe size		19	.842			45	3.06	48.06	71
9500	1-1/2" pipe size		18	.889			47.50	3.23	50.73	75
9510	2" pipe size		15	1.067			57	3.87	60.87	90.50
9520	2-1/2" pipe size		12	1.333			71	4.84	75.84	112
9530	3" pipe size		11	1.455			77.50	5.30	82.80	123
9540	4" pipe size		8	2			107	7.25	114.25	169
9550	5" pipe size		6	2.667			142	9.70	151.70	225
9560	6" pipe size		5	3.200			171	11.60	182.60	270
9570	8" pipe size		4	4			213	14.50	227.50	335
9580	10" pipe size		3	5.333			284	19.35	303.35	450
9590	12" pipe size	↓	2	8			425	29	454	675
9600	14" pipe size	Q-16	2.60	9.231			510	22.50	532.50	795
9610	16" pipe size		2.30	10.435			575	25.50	600.50	900
9620	18" pipe size		2	12			665	29	694	1,025
9630	20" pipe size		1.80	13.333			735	32.50	767.50	1,125
9640	22" pipe size		1.60	15			830	36.50	866.50	1,300
9650	24" pipe size	↓	1.50	16	↓		885	38.50	923.50	1,375

22 11 13.45 Pipe Fittings, Steel, Threaded

		Crew	Daily Output	Labor-Hours	Unit	Material	2016 Bare Costs Labor	2016 Bare Costs Equipment	Total	Total Incl O&P
0010	**PIPE FITTINGS, STEEL, THREADED**									
0020	Cast Iron									
0040	Standard weight, black									
0060	90° Elbow, straight									
0070	1/4"	1 Plum	16	.500	Ea.	10.15	29.50		39.65	55.50
0080	3/8"		16	.500		14.65	29.50		44.15	60.50
0090	1/2"		15	.533		6.40	31.50		37.90	54.50
0100	3/4"		14	.571		6.70	34		40.70	58.50
0110	1"	↓	13	.615		7.90	36.50		44.40	63.50
0120	1-1/4"	Q-1	22	.727		11.25	39		50.25	71
0130	1-1/2"		20	.800		15.55	42.50		58.05	81.50
0140	2"		18	.889		24.50	47.50		72	98
0150	2-1/2"		14	1.143		58.50	61		119.50	156
0160	3"		10	1.600		95.50	85.50		181	234
0170	3-1/2"		8	2		259	107		366	445
0180	4"	↓	6	2.667	↓	177	142		319	410
0250	45° Elbow, straight									
0260	1/4"	1 Plum	16	.500	Ea.	12.50	29.50		42	58.50
0270	3/8"		16	.500		13.40	29.50		42.90	59.50
0280	1/2"		15	.533		9.85	31.50		41.35	58.50
0300	3/4"		14	.571		9.90	34		43.90	62
0320	1"	↓	13	.615		11.60	36.50		48.10	68

22 11 13.45 Pipe Fittings, Steel, Threaded	Crew	Daily Output	Labor-Hours	Unit	Material	2016 Bare Costs Labor	2016 Bare Costs Equipment	Total	Total Incl O&P
0330	1-1/4"	Q-1	22	.727	Ea.	15.60	39		54.60	75.50
0340	1-1/2"		20	.800		26	42.50		68.50	93
0350	2"		18	.889		30	47.50		77.50	105
0360	2-1/2"		14	1.143		77.50	61		138.50	178
0370	3"		10	1.600		123	85.50		208.50	264
0380	3-1/2"		8	2		288	107		395	475
0400	4"	↓	6	2.667	↓	255	142		397	495
0500	Tee, straight									
0510	1/4"	1 Plum	10	.800	Ea.	15.85	47.50		63.35	89
0520	3/8"		10	.800		15.40	47.50		62.90	88.50
0530	1/2"		9	.889		10	52.50		62.50	90.50
0540	3/4"		9	.889		11.70	52.50		64.20	92.50
0550	1"	↓	8	1		10.40	59		69.40	101
0560	1-1/4"	Q-1	14	1.143		18.95	61		79.95	113
0570	1-1/2"		13	1.231		24.50	65.50		90	126
0580	2"		11	1.455		34.50	77.50		112	155
0590	2-1/2"		9	1.778		89	94.50		183.50	241
0600	3"		6	2.667		137	142		279	365
0610	3-1/2"		5	3.200		276	171		447	560
0620	4"	↓	4	4	↓	276	213		489	625
0660	Tee, reducing, run or outlet									
0661	1/2"	1 Plum	9	.889	Ea.	24.50	52.50		77	107
0662	3/4"		9	.889		21.50	52.50		74	103
0663	1"	↓	8	1		24.50	59		83.50	117
0664	1-1/4"	Q-1	14	1.143		35	61		96	130
0665	1-1/2"		13	1.231		35	65.50		100.50	137
0666	2"		11	1.455		48.50	77.50		126	171
0667	2-1/2"		9	1.778		101	94.50		195.50	254
0668	3"		6	2.667		176	142		318	410
0669	3-1/2"		5	3.200		425	171		596	725
0670	4"	↓	4	4	↓	465	213		678	830
0674	Reducer, concentric									
0675	3/4"	1 Plum	18	.444	Ea.	17.95	26.50		44.45	59.50
0676	1"	"	15	.533		12.15	31.50		43.65	61
0677	1-1/4"	Q-1	26	.615		31.50	33		64.50	84
0678	1-1/2"		24	.667		57	35.50		92.50	116
0679	2"		21	.762		65	40.50		105.50	133
0680	2-1/2"		18	.889		92.50	47.50		140	173
0681	3"		14	1.143		148	61		209	255
0682	3-1/2"		12	1.333		251	71		322	385
0683	4"	↓	10	1.600	↓	270	85.50		355.50	425
0687	Reducer, eccentric									
0688	3/4"	1 Plum	16	.500	Ea.	38	29.50		67.50	86
0689	1"	"	14	.571		41	34		75	96
0690	1-1/4"	Q-1	25	.640		65	34		99	123
0691	1-1/2"		22	.727		86	39		125	154
0692	2"		20	.800		124	42.50		166.50	202
0693	2-1/2"		16	1		171	53.50		224.50	269
0694	3"		12	1.333		270	71		341	405
0695	3-1/2"		10	1.600		385	85.50		470.50	555
0696	4"	↓	9	1.778	↓	485	94.50		579.50	680
0700	Standard weight, galvanized cast iron									
0720	90° Elbow, straight									

22 11 13.45 Pipe Fittings, Steel, Threaded	Crew	Daily Output	Labor-Hours	Unit	Material	2016 Bare Costs Labor	Equipment	Total	Total Incl O&P
0730	1/4"	1 Plum	16	.500	Ea.	16.95	29.50		46.45	63
0740	3/8"		16	.500		16.95	29.50		46.45	63
0750	1/2"		15	.533		19.35	31.50		50.85	69
0760	3/4"		14	.571		19	34		53	72
0770	1"		13	.615		22	36.50		58.50	79
0780	1-1/4"	Q-1	22	.727		34	39		73	96
0790	1-1/2"		20	.800		47	42.50		89.50	116
0800	2"		18	.889		69.50	47.50		117	148
0810	2-1/2"		14	1.143		142	61		203	248
0820	3"		10	1.600		216	85.50		301.50	365
0830	3-1/2"		8	2		345	107		452	540
0840	4"		6	2.667		395	142		537	650
0900	45° Elbow, straight									
0910	1/4"	1 Plum	16	.500	Ea.	19.35	29.50		48.85	66
0920	3/8"		16	.500		19.35	29.50		48.85	66
0930	1/2"		15	.533		19.35	31.50		50.85	69
0940	3/4"		14	.571		22	34		56	75.50
0950	1"		13	.615		27.50	36.50		64	85
0960	1-1/4"	Q-1	22	.727		40.50	39		79.50	104
0970	1-1/2"		20	.800		57.50	42.50		100	128
0980	2"		18	.889		80.50	47.50		128	161
0990	2-1/2"		14	1.143		130	61		191	235
1000	3"		10	1.600		254	85.50		339.50	410
1010	3-1/2"		8	2		465	107		572	670
1020	4"		6	2.667		465	142		607	725
1100	Tee, straight									
1110	1/4"	1 Plum	10	.800	Ea.	17.80	47.50		65.30	91
1120	3/8"		10	.800		20.50	47.50		68	94
1130	1/2"		9	.889		19.75	52.50		72.25	101
1140	3/4"		9	.889		27.50	52.50		80	110
1150	1"		8	1		29.50	59		88.50	122
1160	1-1/4"	Q-1	14	1.143		52	61		113	149
1170	1-1/2"		13	1.231		68.50	65.50		134	175
1180	2"		11	1.455		85.50	77.50		163	211
1190	2-1/2"		9	1.778		174	94.50		268.50	335
1200	3"		6	2.667		390	142		532	645
1210	3-1/2"		5	3.200		490	171		661	795
1220	4"		4	4		550	213		763	925
1300	Extra heavy weight, black									
1310	Couplings, steel straight									
1320	1/4"	1 Plum	19	.421	Ea.	3.99	25		28.99	42
1330	3/8"		19	.421		4.35	25		29.35	42.50
1340	1/2"		19	.421		5.90	25		30.90	44
1350	3/4"		18	.444		6.30	26.50		32.80	46.50
1360	1"		15	.533		8	31.50		39.50	56.50
1370	1-1/4"	Q-1	26	.615		12.85	33		45.85	63.50
1380	1-1/2"		24	.667		12.85	35.50		48.35	67.50
1390	2"		21	.762		19.55	40.50		60.05	83
1400	2-1/2"		18	.889		29	47.50		76.50	104
1410	3"		14	1.143		34.50	61		95.50	130
1420	3-1/2"		12	1.333		46.50	71		117.50	158
1430	4"		10	1.600		54.50	85.50		140	189
1510	90° Elbow, straight									

154

22 11 13.45 Pipe Fittings, Steel, Threaded	Crew	Daily Output	Labor-Hours	Unit	Material	2016 Bare Costs Labor	Equipment	Total	Total Incl O&P
1520	1/2"	1 Plum	15	.533	Ea.	33	31.50		64.50	84
1530	3/4"		14	.571		33.50	34		67.50	88
1540	1"	▼	13	.615		40.50	36.50		77	100
1550	1-1/4"	Q-1	22	.727		60.50	39		99.50	125
1560	1-1/2"		20	.800		75	42.50		117.50	147
1580	2"		18	.889		93	47.50		140.50	174
1590	2-1/2"		14	1.143		225	61		286	340
1600	3"		10	1.600		305	85.50		390.50	465
1610	4"	▼	6	2.667	▼	620	142		762	895
1650	45° Elbow, straight									
1660	1/2"	1 Plum	15	.533	Ea.	47	31.50		78.50	99.50
1670	3/4"		14	.571		45.50	34		79.50	101
1680	1"	▼	13	.615		54.50	36.50		91	115
1690	1-1/4"	Q-1	22	.727		90	39		129	158
1700	1-1/2"		20	.800		99	42.50		141.50	174
1710	2"		18	.889		141	47.50		188.50	227
1720	2-1/2"	▼	14	1.143	▼	243	61		304	360
1800	Tee, straight									
1810	1/2"	1 Plum	9	.889	Ea.	51.50	52.50		104	137
1820	3/4"		9	.889		51.50	52.50		104	137
1830	1"	▼	8	1		62.50	59		121.50	158
1840	1-1/4"	Q-1	14	1.143		93	61		154	194
1850	1-1/2"		13	1.231		120	65.50		185.50	231
1860	2"		11	1.455		148	77.50		225.50	280
1870	2-1/2"		9	1.778		315	94.50		409.50	495
1880	3"		6	2.667		445	142		587	705
1890	4"	▼	4	4	▼	865	213		1,078	1,275
4000	Standard weight, black									
4010	Couplings, steel straight, merchants									
4030	1/4"	1 Plum	19	.421	Ea.	1.08	25		26.08	38.50
4040	3/8"		19	.421		1.29	25		26.29	39
4050	1/2"		19	.421		1.37	25		26.37	39
4060	3/4"		18	.444		1.75	26.50		28.25	41.50
4070	1"	▼	15	.533		2.45	31.50		33.95	50
4080	1-1/4"	Q-1	26	.615		3.13	33		36.13	53
4090	1-1/2"		24	.667		3.96	35.50		39.46	58
4100	2"		21	.762		5.65	40.50		46.15	68
4110	2-1/2"		18	.889		17.75	47.50		65.25	91
4120	3"		14	1.143		25	61		86	120
4130	3-1/2"		12	1.333		44	71		115	156
4140	4"	▼	10	1.600		44	85.50		129.50	178
4166	Plug, 1/4"	1 Plum	38	.211		2.86	12.45		15.31	22
4167	3/8"		38	.211		2.86	12.45		15.31	22
4168	1/2"		38	.211		2.86	12.45		15.31	22
4169	3/4"		32	.250		8.60	14.80		23.40	32
4170	1"	▼	30	.267		9.20	15.80		25	34
4171	1-1/4"	Q-1	52	.308		10.55	16.40		26.95	36
4172	1-1/2"		48	.333		15	17.75		32.75	43.50
4173	2"		42	.381		19.35	20.50		39.85	52
4176	2-1/2"		36	.444		28.50	23.50		52	67
4180	4"	▼	20	.800	▼	49.50	42.50		92	119
4200	Standard weight, galvanized									
4210	Couplings, steel straight, merchants									

22 11 13.45 Pipe Fittings, Steel, Threaded		Crew	Daily Output	Labor-Hours	Unit	Material	2016 Bare Costs Labor	Equipment	Total	Total Incl O&P
4230	1/4"	1 Plum	19	.421	Ea.	1.23	25		26.23	39
4240	3/8"		19	.421		1.58	25		26.58	39
4250	1/2"		19	.421		1.68	25		26.68	39.50
4260	3/4"		18	.444		2.11	26.50		28.61	42
4270	1"		15	.533		2.94	31.50		34.44	50.50
4280	1-1/4"	Q-1	26	.615		3.77	33		36.77	53.50
4290	1-1/2"		24	.667		4.67	35.50		40.17	58.50
4300	2"		21	.762		7	40.50		47.50	69
4310	2-1/2"		18	.889		22	47.50		69.50	95.50
4320	3"		14	1.143		29	61		90	124
4330	3-1/2"		12	1.333		51	71		122	163
4340	4"		10	1.600		51	85.50		136.50	185
4370	Plug, galvanized, square head									
4374	1/2"	1 Plum	38	.211	Ea.	6.90	12.45		19.35	26.50
4375	3/4"		32	.250		6.90	14.80		21.70	30
4376	1"		30	.267		6.90	15.80		22.70	31.50
4377	1-1/4"	Q-1	52	.308		11.50	16.40		27.90	37
4378	1-1/2"		48	.333		15.55	17.75		33.30	44
4379	2"		42	.381		19.30	20.50		39.80	51.50
4380	2-1/2"		36	.444		41	23.50		64.50	80.50
4381	3"		28	.571		52	30.50		82.50	104
4382	4"		20	.800		143	42.50		185.50	223
4700	Nipple, black									
4710	1/2" x 4" long	1 Plum	19	.421	Ea.	2.35	25		27.35	40
4712	3/4" x 4" long		18	.444		2.84	26.50		29.34	42.50
4714	1" x 4" long		15	.533		3.94	31.50		35.44	52
4716	1-1/4" x 4" long	Q-1	26	.615		4.92	33		37.92	55
4718	1-1/2" x 4" long		24	.667		5.80	35.50		41.30	60
4720	2" x 4" long		21	.762		8.10	40.50		48.60	70.50
4722	2-1/2" x 4" long		18	.889		22	47.50		69.50	95.50
4724	3" x 4" long		14	1.143		28	61		89	123
4726	4" x 4" long		10	1.600		39	85.50		124.50	172
4800	Nipple, galvanized									
4810	1/2" x 4" long	1 Plum	19	.421	Ea.	2.88	25		27.88	40.50
4812	3/4" x 4" long		18	.444		3.56	26.50		30.06	43.50
4814	1" x 4" long		15	.533		4.77	31.50		36.27	53
4816	1-1/4" x 4" long	Q-1	26	.615		5.85	33		38.85	56
4818	1-1/2" x 4" long		24	.667		7.45	35.50		42.95	61.50
4820	2" x 4" long		21	.762		9.45	40.50		49.95	72
4822	2-1/2" x 4" long		18	.889		25	47.50		72.50	99
4824	3" x 4" long		14	1.143		33	61		94	128
4826	4" x 4" long		10	1.600		44	85.50		129.50	178
5000	Malleable iron, 150 lb.									
5020	Black									
5040	90° elbow, straight									
5060	1/4"	1 Plum	16	.500	Ea.	4.98	29.50		34.48	50
5070	3/8"		16	.500		4.98	29.50		34.48	50
5080	1/2"		15	.533		3.45	31.50		34.95	51.50
5090	3/4"		14	.571		4.17	34		38.17	55.50
5100	1"		13	.615		7.25	36.50		43.75	63
5110	1-1/4"	Q-1	22	.727		11.95	39		50.95	71.50
5120	1-1/2"		20	.800		15.70	42.50		58.20	82
5130	2"		18	.889		28	47.50		75.50	102

22 11 Facility Water Distribution

22 11 13 – Facility Water Distribution Piping

22 11 13.45 Pipe Fittings, Steel, Threaded	Crew	Daily Output	Labor-Hours	Unit	Material	2016 Bare Costs Labor	Equipment	Total	Total Incl O&P
5140 2-1/2"	Q-1	14	1.143	Ea.	60.50	61		121.50	159
5150 3"		10	1.600		88.50	85.50		174	226
5160 3-1/2"		8	2		243	107		350	430
5170 4"		6	2.667		190	142		332	425
5250 45° elbow, straight									
5270 1/4"	1 Plum	16	.500	Ea.	7.50	29.50		37	53
5280 3/8"		16	.500		7.50	29.50		37	53
5290 1/2"		15	.533		5.70	31.50		37.20	54
5300 3/4"		14	.571		7.05	34		41.05	59
5310 1"		13	.615		8.90	36.50		45.40	65
5320 1-1/4"	Q-1	22	.727		15.70	39		54.70	76
5330 1-1/2"		20	.800		19.45	42.50		61.95	86
5340 2"		18	.889		29.50	47.50		77	104
5350 2-1/2"		14	1.143		85	61		146	186
5360 3"		10	1.600		111	85.50		196.50	251
5370 3-1/2"		8	2		217	107		324	400
5380 4"		6	2.667		217	142		359	450
5450 Tee, straight									
5470 1/4"	1 Plum	10	.800	Ea.	7.25	47.50		54.75	79.50
5480 3/8"		10	.800		7.25	47.50		54.75	79.50
5490 1/2"		9	.889		4.62	52.50		57.12	84.50
5500 3/4"		9	.889		6.65	52.50		59.15	87
5510 1"		8	1		11.35	59		70.35	102
5520 1-1/4"	Q-1	14	1.143		18.40	61		79.40	112
5530 1-1/2"		13	1.231		23	65.50		88.50	124
5540 2"		11	1.455		39	77.50		116.50	160
5550 2-1/2"		9	1.778		84	94.50		178.50	236
5560 3"		6	2.667		124	142		266	350
5570 3-1/2"		5	3.200		287	171		458	570
5580 4"		4	4		299	213		512	650
5601 Tee, reducing, on outlet									
5602 1/2"	1 Plum	9	.889	Ea.	11.85	52.50		64.35	92.50
5603 3/4"		9	.889		12.60	52.50		65.10	93.50
5604 1"		8	1		18.20	59		77.20	110
5605 1-1/4"	Q-1	14	1.143		32	61		93	128
5606 1-1/2"		13	1.231		32.50	65.50		98	135
5607 2"		11	1.455		44	77.50		121.50	166
5608 2-1/2"		9	1.778		127	94.50		221.50	283
5609 3"		6	2.667		173	142		315	405
5610 3-1/2"		5	3.200		405	171		576	700
5611 4"		4	4		355	213		568	710
5650 Coupling									
5670 1/4"	1 Plum	19	.421	Ea.	6.20	25		31.20	44.50
5680 3/8"		19	.421		6.20	25		31.20	44.50
5690 1/2"		19	.421		4.78	25		29.78	43
5700 3/4"		18	.444		5.60	26.50		32.10	45.50
5710 1"		15	.533		8.40	31.50		39.90	56.50
5720 1-1/4"	Q-1	26	.615		10.85	33		43.85	61.50
5730 1-1/2"		24	.667		14.65	35.50		50.15	69.50
5740 2"		21	.762		21.50	40.50		62	85.50
5750 2-1/2"		18	.889		60	47.50		107.50	138
5760 3"		14	1.143		81	61		142	181
5770 3-1/2"		12	1.333		163	71		234	287

For customer support on your Mechanical Cost Data, call 877.739.3689.

157

22 11 13.45 Pipe Fittings, Steel, Threaded		Crew	Daily Output	Labor-Hours	Unit	Material	2016 Bare Costs Labor	Equipment	Total	Total Incl O&P
5780	4"	Q-1	10	1.600	Ea.	163	85.50		248.50	310
5840	Reducer, concentric, 1/4"	1 Plum	19	.421		6.55	25		31.55	44.50
5850	3/8"		19	.421		8.35	25		33.35	46.50
5860	1/2"		19	.421		6.60	25		31.60	45
5870	3/4"		16	.500		7.80	29.50		37.30	53
5880	1"		15	.533		13.15	31.50		44.65	62
5890	1-1/4"	Q-1	26	.615		14.75	33		47.75	65.50
5900	1-1/2"		24	.667		21	35.50		56.50	76.50
5910	2"		21	.762		30.50	40.50		71	95
5911	2-1/2"		18	.889		68	47.50		115.50	147
5912	3"		14	1.143		82.50	61		143.50	183
5913	3-1/2"		12	1.333		261	71		332	395
5914	4"		10	1.600		199	85.50		284.50	345
5981	Bushing, 1/4"	1 Plum	19	.421		.95	25		25.95	38.50
5982	3/8"		19	.421		5	25		30	43
5983	1/2"		19	.421		5	25		30	43
5984	3/4"		16	.500		6.40	29.50		35.90	51.50
5985	1"		15	.533		9.10	31.50		40.60	57.50
5986	1-1/4"	Q-1	26	.615		11.30	33		44.30	62
5987	1-1/2"		24	.667		9.75	35.50		45.25	64.50
5988	2"		21	.762		12.30	40.50		52.80	75
5989	Cap, 1/4"	1 Plum	38	.211		4.79	12.45		17.24	24
5991	3/8"		38	.211		5.85	12.45		18.30	25.50
5992	1/2"		38	.211		3.49	12.45		15.94	22.50
5993	3/4"		32	.250		4.73	14.80		19.53	27.50
5994	1"		30	.267		5.70	15.80		21.50	30.50
5995	1-1/4"	Q-1	52	.308		7.55	16.40		23.95	33
5996	1-1/2"		48	.333		10.35	17.75		28.10	38.50
5997	2"		42	.381		15.10	20.50		35.60	47
6000	For galvanized elbows, tees, and couplings add					20%				
6058	For galvanized reducers, caps and bushings add					20%				
6100	90° Elbow, galvanized, 150 lb., reducing									
6110	3/4" x 1/2"	1 Plum	15.40	.519	Ea.	9.95	31		40.95	57.50
6112	1" x 3/4"		14	.571		12.75	34		46.75	65
6114	1" x 1/2"		14.50	.552		13.50	32.50		46	64.50
6116	1-1/4" x 1"	Q-1	24.20	.661		21.50	35		56.50	76.50
6118	1-1/4" x 3/4"		25.40	.630		25.50	33.50		59	78.50
6120	1-1/4" x 1/2"		26.20	.611		27	32.50		59.50	79
6122	1-1/2" x 1-1/4"		21.60	.741		34	39.50		73.50	97
6124	1-1/2" x 1"		23.50	.681		34	36.50		70.50	92
6126	1-1/2" x 3/4"		24.60	.650		34	34.50		68.50	90
6128	2" x 1-1/2"		20.50	.780		40	41.50		81.50	107
6130	2" x 1-1/4"		21	.762		45.50	40.50		86	112
6132	2" x 1"		22.80	.702		47.50	37.50		85	109
6134	2" x 3/4"		23.90	.669		48.50	35.50		84	108
6136	2-1/2" x 2"		12.30	1.301		135	69.50		204.50	254
6138	2-1/2" x 1-1/2"		12.50	1.280		151	68		219	269
6140	3" x 2-1/2"		8.60	1.860		246	99		345	420
6142	3" x 2"		11.80	1.356		213	72		285	345
6144	4" x 3"		8.20	1.951		555	104		659	765
6160	90° Elbow, black, 150 lb., reducing									
6170	1" x 3/4"	1 Plum	14	.571	Ea.	8.70	34		42.70	60.50
6174	1-1/2" x 1"	Q-1	23.50	.681		20.50	36.50		57	77

22 11 13.45 Pipe Fittings, Steel, Threaded		Crew	Daily Output	Labor-Hours	Unit	Material	2016 Bare Costs Labor	Equipment	Total	Total Incl O&P
6178	1-1/2" x 3/4"	Q-1	24.60	.650	Ea.	23.50	34.50		58	78.50
6182	2" x 1-1/2"		20.50	.780		30	41.50		71.50	96
6186	2" x 1"		22.80	.702		34	37.50		71.50	94
6190	2" x 3/4"		23.90	.669		36	35.50		71.50	93.50
6194	2-1/2" x 2"		12.30	1.301		82	69.50		151.50	195
7000	Union, with brass seat									
7010	1/4"	1 Plum	15	.533	Ea.	26	31.50		57.50	76
7020	3/8"		15	.533		17.80	31.50		49.30	67
7030	1/2"		14	.571		16.15	34		50.15	69
7040	3/4"		13	.615		18.55	36.50		55.05	75.50
7050	1"		12	.667		24	39.50		63.50	86
7060	1-1/4"	Q-1	21	.762		34.50	40.50		75	99.50
7070	1-1/2"		19	.842		43	45		88	115
7080	2"		17	.941		50.50	50		100.50	131
7090	2-1/2"		13	1.231		150	65.50		215.50	264
7100	3"		9	1.778		180	94.50		274.50	340
7120	Union, galvanized									
7124	1/2"	1 Plum	14	.571	Ea.	21.50	34		55.50	74.50
7125	3/4"		13	.615		24.50	36.50		61	82
7126	1"		12	.667		32	39.50		71.50	95
7127	1-1/4"	Q-1	21	.762		46.50	40.50		87	113
7128	1-1/2"		19	.842		56.50	45		101.50	130
7129	2"		17	.941		65	50		115	147
7130	2-1/2"		13	1.231		225	65.50		290.50	345
7131	3"		9	1.778		315	94.50		409.50	490
7500	Malleable iron, 300 lb.									
7520	Black									
7540	90° Elbow, straight, 1/4"	1 Plum	16	.500	Ea.	17.50	29.50		47	64
7560	3/8"		16	.500		15.55	29.50		45.05	61.50
7570	1/2"		15	.533		20	31.50		51.50	69.50
7580	3/4"		14	.571		23	34		57	76
7590	1"		13	.615		29	36.50		65.50	87
7600	1-1/4"	Q-1	22	.727		42.50	39		81.50	105
7610	1-1/2"		20	.800		50	42.50		92.50	120
7620	2"		18	.889		72	47.50		119.50	151
7630	2-1/2"		14	1.143		187	61		248	297
7640	3"		10	1.600		214	85.50		299.50	365
7650	4"		6	2.667		505	142		647	770
7700	45° Elbow, straight, 1/4"	1 Plum	16	.500		26	29.50		55.50	73
7720	3/8"		16	.500		26	29.50		55.50	73
7730	1/2"		15	.533		29	31.50		60.50	79.50
7740	3/4"		14	.571		32	34		66	86.50
7750	1"		13	.615		35.50	36.50		72	94
7760	1-1/4"	Q-1	22	.727		57	39		96	122
7770	1-1/2"		20	.800		74.50	42.50		117	147
7780	2"		18	.889		113	47.50		160.50	196
7790	2-1/2"		14	1.143		251	61		312	370
7800	3"		10	1.600		330	85.50		415.50	495
7810	4"		6	2.667		750	142		892	1,050
7850	Tee, straight, 1/4"	1 Plum	10	.800		22	47.50		69.50	95.50
7870	3/8"		10	.800		23.50	47.50		71	97
7880	1/2"		9	.889		29.50	52.50		82	112
7890	3/4"		9	.889		32	52.50		84.50	115

For customer support on your Mechanical Cost Data, call 877.739.3689.

159

22 11 13 - Facility Water Distribution Piping

22 11 13.45 Pipe Fittings, Steel, Threaded		Crew	Daily Output	Labor-Hours	Unit	Material	2016 Bare Costs Labor	Equipment	Total	Total Incl O&P
7900	1"	1 Plum	8	1	Ea.	38.50	59		97.50	132
7910	1-1/4"	Q-1	14	1.143		55.50	61		116.50	153
7920	1-1/2"		13	1.231		64.50	65.50		130	170
7930	2"		11	1.455		95	77.50		172.50	222
7940	2-1/2"		9	1.778		231	94.50		325.50	395
7950	3"		6	2.667		330	142		472	580
7960	4"		4	4		940	213		1,153	1,350
8050	Couplings, straight, 1/4"	1 Plum	19	.421		15.80	25		40.80	55
8070	3/8"		19	.421		15.80	25		40.80	55
8080	1/2"		19	.421		17.60	25		42.60	57
8090	3/4"		18	.444		20.50	26.50		47	62
8100	1"		15	.533		23	31.50		54.50	73
8110	1-1/4"	Q-1	26	.615		27.50	33		60.50	80
8120	1-1/2"		24	.667		41.50	35.50		77	99
8130	2"		21	.762		59.50	40.50		100	127
8140	2-1/2"		18	.889		109	47.50		156.50	192
8150	3"		14	1.143		158	61		219	265
8160	4"		10	1.600		286	85.50		371.50	445
8200	Galvanized									
8220	90° Elbow, straight, 1/4"	1 Plum	16	.500	Ea.	31.50	29.50		61	79
8222	3/8"		16	.500		32	29.50		61.50	79.50
8224	1/2"		15	.533		38	31.50		69.50	89.50
8226	3/4"		14	.571		42.50	34		76.50	98
8228	1"		13	.615		55	36.50		91.50	116
8230	1-1/4"	Q-1	22	.727		86	39		125	153
8232	1-1/2"		20	.800		93.50	42.50		136	168
8234	2"		18	.889		157	47.50		204.50	244
8236	2-1/2"		14	1.143		350	61		411	475
8238	3"		10	1.600		410	85.50		495.50	585
8240	4"		6	2.667		1,225	142		1,367	1,575
8280	45° Elbow, straight									
8282	1/2"	1 Plum	15	.533	Ea.	56.50	31.50		88	110
8284	3/4"		14	.571		65.50	34		99.50	123
8286	1"		13	.615		71.50	36.50		108	134
8288	1-1/4"	Q-1	22	.727		111	39		150	181
8290	1-1/2"		20	.800		143	42.50		185.50	223
8292	2"		18	.889		199	47.50		246.50	290
8310	Tee, straight, 1/4"	1 Plum	10	.800		44.50	47.50		92	121
8312	3/8"		10	.800		45.50	47.50		93	122
8314	1/2"		9	.889		57	52.50		109.50	142
8316	3/4"		9	.889		62.50	52.50		115	149
8318	1"		8	1		77	59		136	175
8320	1-1/4"	Q-1	14	1.143		110	61		171	213
8322	1-1/2"		13	1.231		115	65.50		180.50	225
8324	2"		11	1.455		183	77.50		260.50	320
8326	2-1/2"		9	1.778		580	94.50		674.50	780
8328	3"		6	2.667		655	142		797	935
8330	4"		4	4		1,675	213		1,888	2,175
8380	Couplings, straight, 1/4"	1 Plum	19	.421		20.50	25		45.50	60
8382	3/8"		19	.421		29.50	25		54.50	70
8384	1/2"		19	.421		30.50	25		55.50	71
8386	3/4"		18	.444		34.50	26.50		61	77
8388	1"		15	.533		45.50	31.50		77	97.50

22 11 13.45 Pipe Fittings, Steel, Threaded		Crew	Daily Output	Labor-Hours	Unit	Material	2016 Bare Costs Labor	Equipment	Total	Total Incl O&P
8390	1-1/4"	Q-1	26	.615	Ea.	59	33		92	115
8392	1-1/2"		24	.667		83.50	35.50		119	145
8394	2"		21	.762		101	40.50		141.50	173
8396	2-1/2"		18	.889		266	47.50		313.50	365
8398	3"		14	1.143		315	61		376	440
8399	4"		10	1.600		465	85.50		550.50	640
8529	Black									
8530	Reducer, concentric, 1/4"	1 Plum	19	.421	Ea.	20	25		45	59.50
8531	3/8"		19	.421		20	25		45	59.50
8532	1/2"		17	.471		24	28		52	68.50
8533	3/4"		16	.500		34.50	29.50		64	82
8534	1"		15	.533		41	31.50		72.50	92.50
8535	1-1/4"	Q-1	26	.615		61.50	33		94.50	117
8536	1-1/2"		24	.667		63.50	35.50		99	124
8537	2"		21	.762		92	40.50		132.50	163
8550	Cap, 1/4"	1 Plum	38	.211		15.50	12.45		27.95	36
8551	3/8"		38	.211		15.50	12.45		27.95	36
8552	1/2"		34	.235		15.35	13.95		29.30	38
8553	3/4"		32	.250		19.40	14.80		34.20	44
8554	1"		30	.267		25.50	15.80		41.30	52
8555	1-1/4"	Q-1	52	.308		28	16.40		44.40	55.50
8556	1-1/2"		48	.333		41	17.75		58.75	72.50
8557	2"		42	.381		57	20.50		77.50	93.50
8570	Plug, 1/4"	1 Plum	38	.211		3.47	12.45		15.92	22.50
8571	3/8"		38	.211		3.64	12.45		16.09	23
8572	1/2"		34	.235		3.82	13.95		17.77	25
8573	3/4"		32	.250		4.76	14.80		19.56	28
8574	1"		30	.267		7.30	15.80		23.10	32
8575	1-1/4"	Q-1	52	.308		15	16.40		31.40	41
8576	1-1/2"		48	.333		17.15	17.75		34.90	46
8577	2"		42	.381		27	20.50		47.50	60.50
9500	Union with brass seat, 1/4"	1 Plum	15	.533		34	31.50		65.50	85
9530	3/8"		15	.533		34.50	31.50		66	85.50
9540	1/2"		14	.571		28	34		62	82
9550	3/4"		13	.615		31	36.50		67.50	89
9560	1"		12	.667		40.50	39.50		80	104
9570	1-1/4"	Q-1	21	.762		66	40.50		106.50	134
9580	1-1/2"		19	.842		69	45		114	144
9590	2"		17	.941		86	50		136	170
9600	2-1/2"		13	1.231		276	65.50		341.50	405
9610	3"		9	1.778		355	94.50		449.50	540
9620	4"		5	3.200		1,125	171		1,296	1,500
9630	Union, all iron, 1/4"	1 Plum	15	.533		53	31.50		84.50	106
9650	3/8"		15	.533		53	31.50		84.50	106
9660	1/2"		14	.571		58.50	34		92.50	116
9670	3/4"		13	.615		60	36.50		96.50	121
9680	1"		12	.667		58	39.50		97.50	124
9690	1-1/4"	Q-1	21	.762		113	40.50		153.50	186
9700	1-1/2"		19	.842		138	45		183	220
9710	2"		17	.941		174	50		224	268
9720	2-1/2"		13	1.231		310	65.50		375.50	440
9730	3"		9	1.778		585	94.50		679.50	790
9750	For galvanized unions, add					15%				

For customer support on your Mechanical Cost Data, call 877.739.3689.

161

22 11 13 – Facility Water Distribution Piping

22 11 13.45 Pipe Fittings, Steel, Threaded	Crew	Daily Output	Labor-Hours	Unit	Material	2016 Bare Costs Labor	Equipment	Total	Total Incl O&P
9757	Forged steel, 3000 lb.									
9758	Black									
9760	90° Elbow, 1/4"	1 Plum	16	.500	Ea.	20.50	29.50		50	67
9761	3/8"		16	.500		20.50	29.50		50	67
9762	1/2"		15	.533		15.70	31.50		47.20	65
9763	3/4"		14	.571		19.70	34		53.70	72.50
9764	1"		13	.615		29.50	36.50		66	87
9765	1-1/4"	Q-1	22	.727		55.50	39		94.50	120
9766	1-1/2"		20	.800		72	42.50		114.50	144
9767	2"		18	.889		87	47.50		134.50	168
9780	45° Elbow 1/4"	1 Plum	16	.500		26	29.50		55.50	73
9781	3/8"		16	.500		26	29.50		55.50	73
9782	1/2"		15	.533		25.50	31.50		57	75.50
9783	3/4"		14	.571		29.50	34		63.50	83.50
9784	1"		13	.615		40.50	36.50		77	99.50
9785	1-1/4"	Q-1	22	.727		56	39		95	120
9786	1-1/2"		20	.800		80	42.50		122.50	153
9787	2"		18	.889		110	47.50		157.50	194
9800	Tee, 1/4"	1 Plum	10	.800		25	47.50		72.50	99
9801	3/8"		10	.800		25	47.50		72.50	99
9802	1/2"		9	.889		22.50	52.50		75	104
9803	3/4"		9	.889		30.50	52.50		83	113
9804	1"		8	1		40	59		99	134
9805	1-1/4"	Q-1	14	1.143		77.50	61		138.50	177
9806	1-1/2"		13	1.231		90.50	65.50		156	199
9807	2"		11	1.455		116	77.50		193.50	245
9820	Reducer, concentric, 1/4"	1 Plum	19	.421		12.55	25		37.55	51.50
9821	3/8"		19	.421		13.05	25		38.05	52
9822	1/2"		17	.471		13.05	28		41.05	56.50
9823	3/4"		16	.500		15.50	29.50		45	61.50
9824	1"		15	.533		20	31.50		51.50	69.50
9825	1-1/4"	Q-1	26	.615		35.50	33		68.50	88.50
9826	1-1/2"		24	.667		38.50	35.50		74	95.50
9827	2"		21	.762		55.50	40.50		96	123
9840	Cap, 1/4"	1 Plum	38	.211		7.30	12.45		19.75	27
9841	3/8"		38	.211		7.70	12.45		20.15	27.50
9842	1/2"		34	.235		7.50	13.95		21.45	29.50
9843	3/4"		32	.250		10.45	14.80		25.25	34
9844	1"		30	.267		16	15.80		31.80	41.50
9845	1-1/4"	Q-1	52	.308		26	16.40		42.40	53
9846	1-1/2"		48	.333		31	17.75		48.75	61
9847	2"		42	.381		44.50	20.50		65	79.50
9860	Plug, 1/4"	1 Plum	38	.211		3.47	12.45		15.92	22.50
9861	3/8"		38	.211		3.64	12.45		16.09	23
9862	1/2"		34	.235		3.82	13.95		17.77	25
9863	3/4"		32	.250		4.76	14.80		19.56	28
9864	1"		30	.267		7.30	15.80		23.10	32
9865	1-1/4"	Q-1	52	.308		15	16.40		31.40	41
9866	1-1/2"		48	.333		17.15	17.75		34.90	46
9867	2"		42	.381		27	20.50		47.50	60.50
9880	Union, bronze seat, 1/4"	1 Plum	15	.533		57	31.50		88.50	110
9881	3/8"		15	.533		57	31.50		88.50	110
9882	1/2"		14	.571		54.50	34		88.50	111

162

For customer support on your Mechanical Cost Data, call 877.739.3689.

22 11 13.45 Pipe Fittings, Steel, Threaded

22 11 13.45 Pipe Fittings, Steel, Threaded		Crew	Daily Output	Labor-Hours	Unit	Material	2016 Bare Costs Labor	Equipment	Total	Total Incl O&P
9883	3/4"	1 Plum	13	.615	Ea.	73	36.50		109.50	135
9884	1"		12	.667		82.50	39.50		122	150
9885	1-1/4"	Q-1	21	.762		160	40.50		200.50	238
9886	1-1/2"		19	.842		165	45		210	249
9887	2"		17	.941		195	50		245	291
9900	Coupling, 1/4"	1 Plum	19	.421		8	25		33	46.50
9901	3/8"		19	.421		8	25		33	46.50
9902	1/2"		17	.471		6.55	28		34.55	49
9903	3/4"		16	.500		8.55	29.50		38.05	54
9904	1"		15	.533		14.85	31.50		46.35	64
9905	1-1/4"	Q-1	26	.615		25	33		58	77
9906	1-1/2"		24	.667		32	35.50		67.50	88.50
9907	2"		21	.762		40	40.50		80.50	106

22 11 13.47 Pipe Fittings, Steel

22 11 13.47 Pipe Fittings, Steel		Crew	Daily Output	Labor-Hours	Unit	Material	2016 Bare Costs Labor	Equipment	Total	Total Incl O&P
0010	**PIPE FITTINGS, STEEL**, Flanged, Welded & Special									
0020	Flanged joints, C.I., standard weight, black. One gasket & bolt									
0040	set, mat'l only, required at each joint, not included (see line 0620)									
0060	90° Elbow, straight, 1-1/2" pipe size	Q-1	14	1.143	Ea.	585	61		646	735
0080	2" pipe size		13	1.231		340	65.50		405.50	470
0090	2-1/2" pipe size		12	1.333		365	71		436	505
0100	3" pipe size		11	1.455		305	77.50		382.50	450
0110	4" pipe size		8	2		375	107		482	575
0120	5" pipe size		7	2.286		895	122		1,017	1,175
0130	6" pipe size	Q-2	9	2.667		590	147		737	870
0140	8" pipe size		8	3		1,025	166		1,191	1,375
0150	10" pipe size		7	3.429		2,225	189		2,414	2,725
0160	12" pipe size		6	4		4,500	221		4,721	5,275
0171	90° Elbow, reducing									
0172	2-1/2" by 2" pipe size	Q-1	12	1.333	Ea.	1,125	71		1,196	1,350
0173	3" by 2-1/2" pipe size		11	1.455		1,150	77.50		1,227.50	1,375
0174	4" by 3" pipe size		8	2		835	107		942	1,075
0175	5" by 3" pipe size		7	2.286		1,825	122		1,947	2,175
0176	6" by 4" pipe size	Q-2	9	2.667		1,050	147		1,197	1,375
0177	8" by 6" pipe size		8	3		1,525	166		1,691	1,925
0178	10" by 8" pipe size		7	3.429		2,875	189		3,064	3,450
0179	12" by 10" pipe size		6	4		5,525	221		5,746	6,400
0200	45° Elbow, straight, 1-1/2" pipe size	Q-1	14	1.143		705	61		766	870
0220	2" pipe size		13	1.231		490	65.50		555.50	640
0230	2-1/2" pipe size		12	1.333		520	71		591	680
0240	3" pipe size		11	1.455		510	77.50		587.50	675
0250	4" pipe size		8	2		570	107		677	790
0260	5" pipe size		7	2.286		1,375	122		1,497	1,675
0270	6" pipe size	Q-2	9	2.667		935	147		1,082	1,250
0280	8" pipe size		8	3		1,375	166		1,541	1,750
0290	10" pipe size		7	3.429		2,875	189		3,064	3,450
0300	12" pipe size		6	4		4,400	221		4,621	5,175
0310	Cross, straight									
0311	2-1/2" pipe size	Q-1	6	2.667	Ea.	1,075	142		1,217	1,400
0312	3" pipe size		5	3.200		1,125	171		1,296	1,500
0313	4" pipe size		4	4		1,475	213		1,688	1,950
0314	5" pipe size		3	5.333		3,075	284		3,359	3,825
0315	6" pipe size	Q-2	5	4.800		3,075	265		3,340	3,800

22 11 Facility Water Distribution

22 11 13 – Facility Water Distribution Piping

22 11 13.47 Pipe Fittings, Steel		Crew	Daily Output	Labor-Hours	Unit	Material	2016 Bare Costs Labor	Equipment	Total	Total Incl O&P
0316	8" pipe size	Q-2	4	6	Ea.	4,975	330		5,305	5,950
0317	10" pipe size		3	8		5,900	440		6,340	7,150
0318	12" pipe size		2	12		8,975	665		9,640	10,900
0350	Tee, straight, 1-1/2" pipe size	Q-1	10	1.600		675	85.50		760.50	870
0370	2" pipe size		9	1.778		370	94.50		464.50	550
0380	2-1/2" pipe size		8	2		540	107		647	755
0390	3" pipe size		7	2.286		380	122		502	600
0400	4" pipe size		5	3.200		575	171		746	885
0410	5" pipe size		4	4		1,550	213		1,763	2,025
0420	6" pipe size	Q-2	6	4		835	221		1,056	1,250
0430	8" pipe size		5	4.800		1,425	265		1,690	1,950
0440	10" pipe size		4	6		3,825	330		4,155	4,725
0450	12" pipe size		3	8		6,025	440		6,465	7,300
0459	Tee, reducing on outlet									
0460	2-1/2" by 2" pipe size	Q-1	8	2	Ea.	1,125	107		1,232	1,375
0461	3" by 2-1/2" pipe size		7	2.286		1,125	122		1,247	1,400
0462	4" by 3" pipe size		5	3.200		1,175	171		1,346	1,550
0463	5" by 4" pipe size		4	4		2,625	213		2,838	3,225
0464	6" by 4" pipe size	Q-2	6	4		1,150	221		1,371	1,575
0465	8" by 6" pipe size		5	4.800		1,775	265		2,040	2,350
0466	10" by 8" pipe size		4	6		4,075	330		4,405	4,975
0467	12" by 10" pipe size		3	8		7,025	440		7,465	8,400
0476	Reducer, concentric									
0477	3" by 2-1/2"	Q-1	12	1.333	Ea.	675	71		746	850
0478	4" by 3"		9	1.778		760	94.50		854.50	980
0479	5" by 4"		8	2		1,175	107		1,282	1,450
0480	6" by 4"	Q-2	10	2.400		1,025	133		1,158	1,325
0481	8" by 6"		9	2.667		1,275	147		1,422	1,625
0482	10" by 8"		8	3		2,650	166		2,816	3,150
0483	12" by 10"		7	3.429		4,550	189		4,739	5,300
0492	Reducer, eccentric									
0493	4" by 3"	Q-1	8	2	Ea.	1,225	107		1,332	1,500
0494	5" by 4"	"	7	2.286		1,925	122		2,047	2,300
0495	6" by 4"	Q-2	9	2.667		1,175	147		1,322	1,500
0496	8" by 6"		8	3		1,475	166		1,641	1,875
0497	10" by 8"		7	3.429		3,750	189		3,939	4,400
0498	12" by 10"		6	4		5,100	221		5,321	5,925
0500	For galvanized elbows and tees, add					100%				
0520	For extra heavy weight elbows and tees, add					140%				
0620	Gasket and bolt set, 150#, 1/2" pipe size	1 Plum	20	.400		2.75	23.50		26.25	38.50
0622	3/4" pipe size		19	.421		2.90	25		27.90	40.50
0624	1" pipe size		18	.444		2.94	26.50		29.44	42.50
0626	1-1/4" pipe size		17	.471		3.08	28		31.08	45.50
0628	1-1/2" pipe size		15	.533		3.25	31.50		34.75	51
0630	2" pipe size		13	.615		5.95	36.50		42.45	61.50
0640	2-1/2" pipe size		12	.667		6.15	39.50		45.65	66.50
0650	3" pipe size		11	.727		6.35	43		49.35	72
0660	3-1/2" pipe size		9	.889		13.05	52.50		65.55	94
0670	4" pipe size		8	1		14.45	59		73.45	105
0680	5" pipe size		7	1.143		18.85	67.50		86.35	123
0690	6" pipe size		6	1.333		19.90	79		98.90	141
0700	8" pipe size		5	1.600		22	94.50		116.50	167
0710	10" pipe size		4.50	1.778		39	105		144	202

For customer support on your Mechanical Cost Data, call 877.739.3689.

22 11 13.47 Pipe Fittings, Steel	Crew	Daily Output	Labor-Hours	Unit	Material	2016 Bare Costs Labor	Equipment	Total	Total Incl O&P
0720	12" pipe size	1 Plum	4.20	1.905	Ea.	41.50	113		154.50	216
0730	14" pipe size		4	2		40.50	118		158.50	224
0740	16" pipe size		3	2.667		46	158		204	289
0750	18" pipe size		2.70	2.963		88.50	175		263.50	365
0760	20" pipe size		2.30	3.478		144	206		350	470
0780	24" pipe size		1.90	4.211		180	249		429	575
0790	26" pipe size		1.60	5		244	296		540	715
0810	30" pipe size		1.40	5.714		475	340		815	1,025
0830	36" pipe size		1.10	7.273		890	430		1,320	1,625
0850	For 300 lb. gasket set, add					40%				
2000	Flanged unions, 125 lb., black, 1/2" pipe size	1 Plum	17	.471	Ea.	82.50	28		110.50	133
2040	3/4" pipe size		17	.471		109	28		137	162
2050	1" pipe size		16	.500		105	29.50		134.50	161
2060	1-1/4" pipe size	Q-1	28	.571		125	30.50		155.50	183
2070	1-1/2" pipe size		27	.593		115	31.50		146.50	175
2080	2" pipe size		26	.615		135	33		168	199
2090	2-1/2" pipe size		24	.667		183	35.50		218.50	256
2100	3" pipe size		22	.727		207	39		246	287
2110	3-1/2" pipe size		18	.889		350	47.50		397.50	455
2120	4" pipe size		16	1		282	53.50		335.50	390
2130	5" pipe size		14	1.143		660	61		721	815
2140	6" pipe size	Q-2	19	1.263		615	70		685	785
2150	8" pipe size	"	16	1.500		1,425	83		1,508	1,700
2200	For galvanized unions, add					150%				
2290	Threaded flange									
2300	Cast iron									
2310	Black, 125 lb., per flange									
2320	1" pipe size	1 Plum	27	.296	Ea.	49.50	17.55		67.05	80.50
2330	1-1/4" pipe size	Q-1	44	.364		59	19.35		78.35	94
2340	1-1/2" pipe size		40	.400		54.50	21.50		76	92
2350	2" pipe size		36	.444		54.50	23.50		78	95.50
2360	2-1/2" pipe size		28	.571		63.50	30.50		94	116
2370	3" pipe size		20	.800		82	42.50		124.50	155
2380	3-1/2" pipe size		16	1		117	53.50		170.50	210
2390	4" pipe size		12	1.333		110	71		181	228
2400	5" pipe size		10	1.600		156	85.50		241.50	300
2410	6" pipe size	Q-2	14	1.714		177	94.50		271.50	335
2420	8" pipe size		12	2		279	111		390	470
2430	10" pipe size		10	2.400		495	133		628	745
2440	12" pipe size		8	3		1,125	166		1,291	1,475
2460	For galvanized flanges, add					95%				
2490	Blind flange									
2492	Cast iron									
2494	Black, 125 lb., per flange									
2496	1" pipe size	1 Plum	27	.296	Ea.	78.50	17.55		96.05	113
2500	1-1/2" pipe size	Q-1	40	.400		89.50	21.50		111	131
2502	2" pipe size		36	.444		99.50	23.50		123	145
2504	2-1/2" pipe size		28	.571		109	30.50		139.50	166
2506	3" pipe size		20	.800		131	42.50		173.50	209
2508	4" pipe size		12	1.333		168	71		239	291
2510	5" pipe size		10	1.600		272	85.50		357.50	430
2512	6" pipe size	Q-2	14	1.714		296	94.50		390.50	470
2514	8" pipe size		12	2		465	111		576	675

For customer support on your Mechanical Cost Data, call 877.739.3689.

165

22 11 13.47 Pipe Fittings, Steel		Crew	Daily Output	Labor-Hours	Unit	Material	2016 Bare Costs Labor	Equipment	Total	Total Incl O&P
2516	10" pipe size	Q-2	10	2.400	Ea.	690	133		823	955
2518	12" pipe size	↓	8	3	↓	1,325	166		1,491	1,700
2520	For galvanized flanges, add					80%				
2570	Threaded flange									
2580	Forged steel,									
2590	Black 150 lb., per flange									
2600	1/2" pipe size	1 Plum	30	.267	Ea.	26.50	15.80		42.30	53
2610	3/4" pipe size		28	.286		26.50	16.90		43.40	54.50
2620	1" pipe size	↓	27	.296		26.50	17.55		44.05	55.50
2630	1-1/4" pipe size	Q-1	44	.364		26.50	19.35		45.85	58
2640	1-1/2" pipe size		40	.400		26.50	21.50		48	61
2650	2" pipe size		36	.444		29.50	23.50		53	68
2660	2-1/2" pipe size		28	.571		36	30.50		66.50	86
2670	3" pipe size		20	.800		36.50	42.50		79	105
2690	4" pipe size		12	1.333		42.50	71		113.50	154
2700	5" pipe size	↓	10	1.600		66.50	85.50		152	203
2710	6" pipe size	Q-2	14	1.714		73	94.50		167.50	224
2720	8" pipe size		12	2		126	111		237	305
2730	10" pipe size	↓	10	2.400	↓	228	133		361	450
2860	Black 300 lb., per flange									
2870	1/2" pipe size	1 Plum	30	.267	Ea.	29	15.80		44.80	56
2880	3/4" pipe size		28	.286		29	16.90		45.90	57.50
2890	1" pipe size	↓	27	.296		29	17.55		46.55	58.50
2900	1-1/4" pipe size	Q-1	44	.364		29	19.35		48.35	61
2910	1-1/2" pipe size		40	.400		29	21.50		50.50	64
2920	2" pipe size		36	.444		34	23.50		57.50	73
2930	2-1/2" pipe size		28	.571		47.50	30.50		78	98
2940	3" pipe size		20	.800		50.50	42.50		93	120
2960	4" pipe size	↓	12	1.333		73	71		144	187
2970	6" pipe size	Q-2	14	1.714	↓	136	94.50		230.50	293
3000	Weld joint, butt, carbon steel, standard weight									
3040	90° elbow, long radius									
3050	1/2" pipe size	Q-15	16	1	Ea.	45	53.50	3.63	102.13	134
3060	3/4" pipe size		16	1		45	53.50	3.63	102.13	134
3070	1" pipe size		16	1		21	53.50	3.63	78.13	107
3080	1-1/4" pipe size		14	1.143		21	61	4.15	86.15	120
3090	1-1/2" pipe size		13	1.231		21	65.50	4.47	90.97	127
3100	2" pipe size		10	1.600		22.50	85.50	5.80	113.80	160
3110	2-1/2" pipe size		8	2		27.50	107	7.25	141.75	200
3120	3" pipe size		7	2.286		29	122	8.30	159.30	225
3130	4" pipe size		5	3.200		48	171	11.60	230.60	325
3136	5" pipe size		4	4		103	213	14.50	330.50	450
3140	6" pipe size	Q-16	5	4.800		106	265	11.60	382.60	530
3150	8" pipe size		3.75	6.400		200	355	15.50	570.50	770
3160	10" pipe size		3	8		400	440	19.35	859.35	1,125
3170	12" pipe size		2.50	9.600		590	530	23	1,143	1,475
3180	14" pipe size		2	12		970	665	29	1,664	2,100
3190	16" pipe size		1.50	16		1,325	885	38.50	2,248.50	2,850
3191	18" pipe size		1.25	19.200		1,525	1,050	46.50	2,621.50	3,325
3192	20" pipe size		1.15	20.870		2,450	1,150	50.50	3,650.50	4,500
3194	24" pipe size	↓	1.02	23.529		3,450	1,300	57	4,807	5,825
3200	45° Elbow, long									
3210	1/2" pipe size	Q-15	16	1	Ea.	62.50	53.50	3.63	119.63	153

22 11 13 – Facility Water Distribution Piping

22 11 13.47 Pipe Fittings, Steel		Crew	Daily Output	Labor-Hours	Unit	Material	2016 Bare Costs Labor	Equipment	Total	Total Incl O&P
3220	3/4" pipe size	Q-15	16	1	Ea.	62.50	53.50	3.63	119.63	153
3230	1" pipe size		16	1		22	53.50	3.63	79.13	109
3240	1-1/4" pipe size		14	1.143		22	61	4.15	87.15	121
3250	1-1/2" pipe size		13	1.231		22	65.50	4.47	91.97	128
3260	2" pipe size		10	1.600		22	85.50	5.80	113.30	160
3270	2-1/2" pipe size		8	2		26.50	107	7.25	140.75	198
3280	3" pipe size		7	2.286		27.50	122	8.30	157.80	224
3290	4" pipe size		5	3.200		49.50	171	11.60	232.10	325
3296	5" pipe size		4	4		75	213	14.50	302.50	420
3300	6" pipe size	Q-16	5	4.800		97.50	265	11.60	374.10	520
3310	8" pipe size		3.75	6.400		163	355	15.50	533.50	730
3320	10" pipe size		3	8		325	440	19.35	784.35	1,050
3330	12" pipe size		2.50	9.600		460	530	23	1,013	1,325
3340	14" pipe size		2	12		615	665	29	1,309	1,700
3341	16" pipe size		1.50	16		1,100	885	38.50	2,023.50	2,600
3342	18" pipe size		1.25	19.200		1,575	1,050	46.50	2,671.50	3,375
3343	20" pipe size		1.15	20.870		1,625	1,150	50.50	2,825.50	3,575
3345	24" pipe size		1.05	22.857		2,450	1,275	55.50	3,780.50	4,650
3346	26" pipe size		.85	28.235		2,700	1,550	68.50	4,318.50	5,400
3347	30" pipe size		.45	53.333		2,975	2,950	129	6,054	7,875
3349	36" pipe size		.38	63.158		3,275	3,500	153	6,928	9,075
3350	Tee, straight									
3352	For reducing tees and concentrics see starting line 4600									
3360	1/2" pipe size	Q-15	10	1.600	Ea.	110	85.50	5.80	201.30	256
3370	3/4" pipe size		10	1.600		110	85.50	5.80	201.30	256
3380	1" pipe size		10	1.600		54.50	85.50	5.80	145.80	195
3390	1-1/4" pipe size		9	1.778		68.50	94.50	6.45	169.45	226
3400	1-1/2" pipe size		8	2		68.50	107	7.25	182.75	245
3410	2" pipe size		6	2.667		54.50	142	9.70	206.20	285
3420	2-1/2" pipe size		5	3.200		75.50	171	11.60	258.10	355
3430	3" pipe size		4	4		84	213	14.50	311.50	430
3440	4" pipe size		3	5.333		118	284	19.35	421.35	580
3446	5" pipe size		2.50	6.400		195	340	23	558	755
3450	6" pipe size	Q-16	3	8		203	440	19.35	662.35	910
3460	8" pipe size		2.50	9.600		355	530	23	908	1,225
3470	10" pipe size		2	12		695	665	29	1,389	1,800
3480	12" pipe size		1.60	15		975	830	36.50	1,841.50	2,375
3481	14" pipe size		1.30	18.462		1,700	1,025	44.50	2,769.50	3,475
3482	16" pipe size		1	24		1,900	1,325	58	3,283	4,175
3483	18" pipe size		.80	30		3,025	1,650	72.50	4,747.50	5,900
3484	20" pipe size		.75	32		4,750	1,775	77.50	6,602.50	7,975
3486	24" pipe size		.70	34.286		6,150	1,900	83	8,133	9,700
3487	26" pipe size		.55	43.636		6,775	2,400	106	9,281	11,200
3488	30" pipe size		.30	80		7,425	4,425	194	12,044	15,000
3490	36" pipe size		.25	96		8,200	5,300	232	13,732	17,300
3491	Eccentric reducer, 1-1/2" pipe size	Q-15	14	1.143		63.50	61	4.15	128.65	167
3492	2" pipe size		11	1.455		43	77.50	5.30	125.80	170
3493	2-1/2" pipe size		9	1.778		50	94.50	6.45	150.95	205
3494	3" pipe size		8	2		59	107	7.25	173.25	234
3495	4" pipe size		6	2.667		75.50	142	9.70	227.20	310
3496	6" pipe size	Q-16	5	4.800		232	265	11.60	508.60	670
3497	8" pipe size		4	6		345	330	14.50	689.50	895
3498	10" pipe size		3	8		435	440	19.35	894.35	1,150

22 11 13 – Facility Water Distribution Piping

22 11 13.47 Pipe Fittings, Steel		Crew	Daily Output	Labor-Hours	Unit	Material	2016 Bare Costs Labor	Equipment	Total	Total Incl O&P
3499	12" pipe size	Q-16	2.50	9.600	Ea.	605	530	23	1,158	1,500
3501	Cap, 1-1/2" pipe size	Q-15	28	.571		24.50	30.50	2.07	57.07	75.50
3502	2" pipe size		22	.727		22	39	2.64	63.64	85.50
3503	2-1/2" pipe size		18	.889		30	47.50	3.23	80.73	108
3504	3" pipe size		16	1		33.50	53.50	3.63	90.63	121
3505	4" pipe size		12	1.333		45	71	4.84	120.84	162
3506	6" pipe size	Q-16	10	2.400		92	133	5.80	230.80	305
3507	8" pipe size		8	3		138	166	7.25	311.25	410
3508	10" pipe size		6	4		212	221	9.70	442.70	580
3509	12" pipe size		5	4.800		283	265	11.60	559.60	725
3511	14" pipe size		4	6		410	330	14.50	754.50	965
3512	16" pipe size		4	6		520	330	14.50	864.50	1,075
3513	18" pipe size		3	8		665	440	19.35	1,124.35	1,425
3517	Weld joint, butt, carbon steel, extra strong									
3519	90° elbow, long									
3520	1/2" pipe size	Q-15	13	1.231	Ea.	55.50	65.50	4.47	125.47	165
3530	3/4" pipe size		12	1.333		55.50	71	4.84	131.34	174
3540	1" pipe size		11	1.455		27	77.50	5.30	109.80	153
3550	1-1/4" pipe size		10	1.600		27	85.50	5.80	118.30	165
3560	1-1/2" pipe size		9	1.778		27	94.50	6.45	127.95	180
3570	2" pipe size		8	2		27.50	107	7.25	141.75	200
3580	2-1/2" pipe size		7	2.286		38.50	122	8.30	168.80	236
3590	3" pipe size		6	2.667		49.50	142	9.70	201.20	279
3600	4" pipe size		4	4		81.50	213	14.50	309	425
3606	5" pipe size		3.50	4.571		195	244	16.60	455.60	600
3610	6" pipe size	Q-16	4.50	5.333		206	295	12.90	513.90	685
3620	8" pipe size		3.50	6.857		390	380	16.60	786.60	1,025
3630	10" pipe size		2.50	9.600		830	530	23	1,383	1,725
3640	12" pipe size		2.25	10.667		1,025	590	26	1,641	2,050
3650	45° Elbow, long									
3660	1/2" pipe size	Q-15	13	1.231	Ea.	62	65.50	4.47	131.97	172
3670	3/4" pipe size		12	1.333		62	71	4.84	137.84	180
3680	1" pipe size		11	1.455		29	77.50	5.30	111.80	154
3690	1-1/4" pipe size		10	1.600		29	85.50	5.80	120.30	167
3700	1-1/2" pipe size		9	1.778		29	94.50	6.45	129.95	182
3710	2" pipe size		8	2		29	107	7.25	143.25	201
3720	2-1/2" pipe size		7	2.286		64	122	8.30	194.30	264
3730	3" pipe size		6	2.667		37	142	9.70	188.70	266
3740	4" pipe size		4	4		58.50	213	14.50	286	400
3746	5" pipe size		3.50	4.571		139	244	16.60	399.60	540
3750	6" pipe size	Q-16	4.50	5.333		160	295	12.90	467.90	635
3760	8" pipe size		3.50	6.857		277	380	16.60	673.60	895
3770	10" pipe size		2.50	9.600		535	530	23	1,088	1,400
3780	12" pipe size		2.25	10.667		790	590	26	1,406	1,800
3800	Tee, straight									
3810	1/2" pipe size	Q-15	9	1.778	Ea.	136	94.50	6.45	236.95	299
3820	3/4" pipe size		8.50	1.882		131	100	6.85	237.85	305
3830	1" pipe size		8	2		52	107	7.25	166.25	227
3840	1-1/4" pipe size		7	2.286		52	122	8.30	182.30	251
3850	1-1/2" pipe size		6	2.667		52	142	9.70	203.70	282
3860	2" pipe size		5	3.200		59.50	171	11.60	242.10	335
3870	2-1/2" pipe size		4	4		99	213	14.50	326.50	445
3880	3" pipe size		3.50	4.571		124	244	16.60	384.60	525

22 11 13 – Facility Water Distribution Piping

22 11 13.47 Pipe Fittings, Steel		Crew	Daily Output	Labor-Hours	Unit	Material	2016 Bare Costs Labor	Equipment	Total	Total Incl O&P
3890	4" pipe size	Q-15	2.50	6.400	Ea.	149	340	23	512	705
3896	5" pipe size	▼	2.25	7.111		375	380	26	781	1,000
3900	6" pipe size	Q-16	2.25	10.667		286	590	26	902	1,225
3910	8" pipe size		2	12		545	665	29	1,239	1,625
3920	10" pipe size		1.75	13.714		845	760	33	1,638	2,125
3930	12" pipe size	▼	1.50	16		1,225	885	38.50	2,148.50	2,725
4000	Eccentric reducer, 1-1/2" pipe size	Q-15	10	1.600		17.30	85.50	5.80	108.60	154
4010	2" pipe size		9	1.778		47.50	94.50	6.45	148.45	203
4020	2-1/2" pipe size		8	2		71.50	107	7.25	185.75	248
4030	3" pipe size		7	2.286		59	122	8.30	189.30	258
4040	4" pipe size		5	3.200		95.50	171	11.60	278.10	375
4046	5" pipe size	▼	4.70	3.404		266	181	12.35	459.35	580
4050	6" pipe size	Q-16	4.50	5.333		259	295	12.90	566.90	745
4060	8" pipe size		3.50	6.857		390	380	16.60	786.60	1,025
4070	10" pipe size		2.50	9.600		665	530	23	1,218	1,550
4080	12" pipe size		2.25	10.667		845	590	26	1,461	1,850
4090	14" pipe size		2.10	11.429		1,575	630	27.50	2,232.50	2,700
4100	16" pipe size	▼	1.90	12.632		2,025	700	30.50	2,755.50	3,300
4151	Cap, 1-1/2" pipe size	Q-15	24	.667		24.50	35.50	2.42	62.42	83
4152	2" pipe size		18	.889		22	47.50	3.23	72.73	99
4153	2-1/2" pipe size		16	1		30	53.50	3.63	87.13	117
4154	3" pipe size		14	1.143		33.50	61	4.15	98.65	133
4155	4" pipe size	▼	10	1.600		45	85.50	5.80	136.30	185
4156	6" pipe size	Q-16	9	2.667		92	147	6.45	245.45	330
4157	8" pipe size		7	3.429		138	189	8.30	335.30	445
4158	10" pipe size		5	4.800		212	265	11.60	488.60	645
4159	12" pipe size	▼	4	6	▼	283	330	14.50	627.50	825
4190	Weld fittings, reducing, standard weight									
4200	Welding ring w/spacer pins, 2" pipe size				Ea.	1.85			1.85	2.04
4210	2-1/2" pipe size					2.01			2.01	2.21
4220	3" pipe size					2.09			2.09	2.30
4230	4" pipe size					2.34			2.34	2.57
4236	5" pipe size					2.92			2.92	3.21
4240	6" pipe size					3.06			3.06	3.37
4250	8" pipe size					3.37			3.37	3.71
4260	10" pipe size					3.92			3.92	4.31
4270	12" pipe size					4.59			4.59	5.05
4280	14" pipe size					5.25			5.25	5.80
4290	16" pipe size					6.10			6.10	6.75
4300	18" pipe size					6.90			6.90	7.55
4310	20" pipe size					7.80			7.80	8.60
4330	24" pipe size					11			11	12.10
4340	26" pipe size					13.55			13.55	14.90
4350	30" pipe size					15.95			15.95	17.55
4370	36" pipe size				▼	20.50			20.50	22.50
4600	Tee, reducing on outlet									
4601	2-1/2" x 2" pipe size	Q-15	5	3.200	Ea.	100	171	11.60	282.60	380
4602	3" x 2-1/2" pipe size		4	4		118	213	14.50	345.50	465
4604	4" x 3" pipe size		3	5.333		124	284	19.35	427.35	590
4605	5" x 4" pipe size	▼	2.50	6.400		268	340	23	631	835
4606	6" x 5" pipe size	Q-16	3	8		370	440	19.35	829.35	1,100
4607	8" x 6" pipe size		2.50	9.600		460	530	23	1,013	1,325
4608	10" x 8" pipe size	▼	2	12		850	665	29	1,544	1,975

22 11 13 – Facility Water Distribution Piping

22 11 13.47 Pipe Fittings, Steel		Crew	Daily Output	Labor-Hours	Unit	Material	2016 Bare Costs Labor	Equipment	Total	Total Incl O&P
4609	12" x 10" pipe size	Q-16	1.60	15	Ea.	1,250	830	36.50	2,116.50	2,675
4610	16" x 12" pipe size		1.50	16		2,000	885	38.50	2,923.50	3,575
4611	14" x 12" pipe size		1.52	15.789		1,900	870	38	2,808	3,450
4618	Reducer, concentric									
4619	2-1/2" by 2" pipe size	Q-15	10	1.600	Ea.	42.50	85.50	5.80	133.80	182
4620	3" by 2-1/2" pipe size		9	1.778		35	94.50	6.45	135.95	189
4621	3-1/2" by 3" pipe size		8	2		89.50	107	7.25	203.75	267
4622	4" by 2-1/2" pipe size		7	2.286		54.50	122	8.30	184.80	253
4623	5" by 3" pipe size		7	2.286		114	122	8.30	244.30	320
4624	6" by 4" pipe size	Q-16	6	4		94	221	9.70	324.70	450
4625	8" by 6" pipe size		5	4.800		124	265	11.60	400.60	550
4626	10" by 8" pipe size		4	6		240	330	14.50	584.50	780
4627	12" by 10" pipe size		3	8		296	440	19.35	755.35	1,000
4660	Reducer, eccentric									
4662	3" x 2" pipe size	Q-15	8	2	Ea.	59	107	7.25	173.25	234
4664	4" x 3" pipe size		6	2.667		71.50	142	9.70	223.20	305
4666	4" x 2" pipe size		6	2.667		87.50	142	9.70	239.20	320
4670	6" x 4" pipe size	Q-16	5	4.800		148	265	11.60	424.60	575
4672	6" x 3" pipe size		5	4.800		232	265	11.60	508.60	670
4676	8" x 6" pipe size		4	6		188	330	14.50	532.50	725
4678	8" x 4" pipe size		4	6		345	330	14.50	689.50	895
4682	10" x 8" pipe size		3	8		247	440	19.35	706.35	960
4684	10" x 6" pipe size		3	8		435	440	19.35	894.35	1,150
4688	12" x 10" pipe size		2.50	9.600		385	530	23	938	1,250
4690	12" x 8" pipe size		2.50	9.600		605	530	23	1,158	1,500
4691	14" x 12" pipe size		2.20	10.909		675	605	26.50	1,306.50	1,675
4693	16" x 14" pipe size		1.80	13.333		1,025	735	32.50	1,792.50	2,250
4694	16" x 12" pipe size		2	12		1,375	665	29	2,069	2,525
4696	18" x 16" pipe size		1.60	15		1,100	830	36.50	1,966.50	2,525
5000	Weld joint, socket, forged steel, 3000 lb., schedule 40 pipe									
5010	90° elbow, straight									
5020	1/4" pipe size	Q-15	22	.727	Ea.	25	39	2.64	66.64	89
5030	3/8" pipe size		22	.727		25	39	2.64	66.64	89
5040	1/2" pipe size		20	.800		13.95	42.50	2.90	59.35	83
5050	3/4" pipe size		20	.800		14.55	42.50	2.90	59.95	83.50
5060	1" pipe size		20	.800		18.70	42.50	2.90	64.10	88
5070	1-1/4" pipe size		18	.889		36	47.50	3.23	86.73	115
5080	1-1/2" pipe size		16	1		41.50	53.50	3.63	98.63	130
5090	2" pipe size		12	1.333		61.50	71	4.84	137.34	180
5100	2-1/2" pipe size		10	1.600		174	85.50	5.80	265.30	325
5110	3" pipe size		8	2		300	107	7.25	414.25	500
5120	4" pipe size		6	2.667		795	142	9.70	946.70	1,100
5130	45° Elbow, straight									
5134	1/4" pipe size	Q-15	22	.727	Ea.	25	39	2.64	66.64	89
5135	3/8" pipe size		22	.727		25	39	2.64	66.64	89
5136	1/2" pipe size		20	.800		18.60	42.50	2.90	64	88
5137	3/4" pipe size		20	.800		21.50	42.50	2.90	66.90	91.50
5140	1" pipe size		20	.800		28.50	42.50	2.90	73.90	98.50
5150	1-1/4" pipe size		18	.889		39	47.50	3.23	89.73	118
5160	1-1/2" pipe size		16	1		47.50	53.50	3.63	104.63	137
5170	2" pipe size		12	1.333		76.50	71	4.84	152.34	196
5180	2-1/2" pipe size		10	1.600		201	85.50	5.80	292.30	355
5190	3" pipe size		8	2		335	107	7.25	449.25	535

22 11 13.47 Pipe Fittings, Steel

		Crew	Daily Output	Labor-Hours	Unit	Material	Labor	Equipment	Total	Total Incl O&P
5200	4" pipe size	Q-15	6	2.667	Ea.	660	142	9.70	811.70	950
5250	Tee, straight									
5254	1/4" pipe size	Q-15	15	1.067	Ea.	27.50	57	3.87	88.37	121
5255	3/8" pipe size		15	1.067		27.50	57	3.87	88.37	121
5256	1/2" pipe size		13	1.231		17.30	65.50	4.47	87.27	123
5257	3/4" pipe size		13	1.231		21	65.50	4.47	90.97	127
5260	1" pipe size		13	1.231		29	65.50	4.47	98.97	135
5270	1-1/4" pipe size		12	1.333		44	71	4.84	119.84	160
5280	1-1/2" pipe size		11	1.455		58.50	77.50	5.30	141.30	187
5290	2" pipe size		8	2		84.50	107	7.25	198.75	262
5300	2-1/2" pipe size		6	2.667		243	142	9.70	394.70	490
5310	3" pipe size		5	3.200		580	171	11.60	762.60	910
5320	4" pipe size		4	4		935	213	14.50	1,162.50	1,350
5350	For reducing sizes, add					60%				
5450	Couplings									
5451	1/4" pipe size	Q-15	23	.696	Ea.	15.90	37	2.53	55.43	76.50
5452	3/8" pipe size		23	.696		15.90	37	2.53	55.43	76.50
5453	1/2" pipe size		21	.762		7.35	40.50	2.77	50.62	72.50
5454	3/4" pipe size		21	.762		9.50	40.50	2.77	52.77	75
5460	1" pipe size		20	.800		10.50	42.50	2.90	55.90	79
5470	1-1/4" pipe size		20	.800		18.60	42.50	2.90	64	88
5480	1-1/2" pipe size		18	.889		20.50	47.50	3.23	71.23	97.50
5490	2" pipe size		14	1.143		33	61	4.15	98.15	133
5500	2-1/2" pipe size		12	1.333		73.50	71	4.84	149.34	193
5510	3" pipe size		9	1.778		165	94.50	6.45	265.95	330
5520	4" pipe size		7	2.286		248	122	8.30	378.30	465
5570	Union, 1/4" pipe size		21	.762		34	40.50	2.77	77.27	102
5571	3/8" pipe size		21	.762		34	40.50	2.77	77.27	102
5572	1/2" pipe size		19	.842		28.50	45	3.06	76.56	102
5573	3/4" pipe size		19	.842		33.50	45	3.06	81.56	107
5574	1" pipe size		19	.842		43	45	3.06	91.06	118
5575	1-1/4" pipe size		17	.941		69.50	50	3.42	122.92	156
5576	1-1/2" pipe size		15	1.067		75	57	3.87	135.87	173
5577	2" pipe size		11	1.455		108	77.50	5.30	190.80	242
5600	Reducer, 1/4" pipe size		23	.696		40.50	37	2.53	80.03	103
5601	3/8" pipe size		23	.696		43	37	2.53	82.53	106
5602	1/2" pipe size		21	.762		26	40.50	2.77	69.27	93
5603	3/4" pipe size		21	.762		26	40.50	2.77	69.27	93
5604	1" pipe size		21	.762		36.50	40.50	2.77	79.77	105
5605	1-1/4" pipe size		19	.842		46.50	45	3.06	94.56	122
5607	1-1/2" pipe size		17	.941		50	50	3.42	103.42	134
5608	2" pipe size		13	1.231		55	65.50	4.47	124.97	164
5612	Cap, 1/4" pipe size		46	.348		15.15	18.55	1.26	34.96	46
5613	3/8" pipe size		46	.348		15.15	18.55	1.26	34.96	46
5614	1/2" pipe size		42	.381		9.25	20.50	1.38	31.13	42
5615	3/4" pipe size		42	.381		10.80	20.50	1.38	32.68	44
5616	1" pipe size		42	.381		16.35	20.50	1.38	38.23	50
5617	1-1/4" pipe size		38	.421		19.30	22.50	1.53	43.33	56.50
5618	1-1/2" pipe size		34	.471		27	25	1.71	53.71	69.50
5619	2" pipe size		26	.615		41.50	33	2.23	76.73	97.50
5630	T-O-L, 1/4" pipe size, nozzle		23	.696		7.10	37	2.53	46.63	66.50
5631	3/8" pipe size, nozzle		23	.696		7.20	37	2.53	46.73	66.50
5632	1/2" pipe size, nozzle		22	.727		7.10	39	2.64	48.74	69

For customer support on your Mechanical Cost Data, call 877.739.3689.

171

22 11 13 – Facility Water Distribution Piping

22 11 13.47 Pipe Fittings, Steel		Crew	Daily Output	Labor-Hours	Unit	Material	2016 Bare Costs Labor	Equipment	Total	Total Incl O&P
5633	3/4" pipe size, nozzle	Q-15	21	.762	Ea.	8.20	40.50	2.77	51.47	73.50
5634	1" pipe size, nozzle		20	.800		9.55	42.50	2.90	54.95	78
5635	1-1/4" pipe size, nozzle		18	.889		15.10	47.50	3.23	65.83	91.50
5636	1-1/2" pipe size, nozzle		16	1		15.10	53.50	3.63	72.23	101
5637	2" pipe size, nozzle		12	1.333		17.20	71	4.84	93.04	131
5638	2-1/2" pipe size, nozzle		10	1.600		57.50	85.50	5.80	148.80	199
5639	4" pipe size, nozzle		6	2.667		130	142	9.70	281.70	370
5640	W-O-L, 1/4" pipe size, nozzle		23	.696		16.80	37	2.53	56.33	77
5641	3/8" pipe size, nozzle		23	.696		16.15	37	2.53	55.68	76.50
5642	1/2" pipe size, nozzle		22	.727		14.70	39	2.64	56.34	77.50
5643	3/4" pipe size, nozzle		21	.762		15.50	40.50	2.77	58.77	81.50
5644	1" pipe size, nozzle		20	.800		16.20	42.50	2.90	61.60	85.50
5645	1-1/4" pipe size, nozzle		18	.889		18.65	47.50	3.23	69.38	95.50
5646	1-1/2" pipe size, nozzle		16	1		19.25	53.50	3.63	76.38	105
5647	2" pipe size, nozzle		12	1.333		19.45	71	4.84	95.29	134
5648	2-1/2" pipe size, nozzle		10	1.600		44.50	85.50	5.80	135.80	184
5649	3" pipe size, nozzle		8	2		48.50	107	7.25	162.75	223
5650	4" pipe size, nozzle		6	2.667		61.50	142	9.70	213.20	292
5651	5" pipe size, nozzle		5	3.200		152	171	11.60	334.60	435
5652	6" pipe size, nozzle		4	4		177	213	14.50	404.50	530
5653	8" pipe size, nozzle		3	5.333		340	284	19.35	643.35	825
5654	10" pipe size, nozzle		2.60	6.154		490	330	22.50	842.50	1,050
5655	12" pipe size, nozzle		2.20	7.273		930	385	26.50	1,341.50	1,650
5674	S-O-L, 1/4" pipe size, outlet		23	.696		9.45	37	2.53	48.98	69
5675	3/8" pipe size, outlet		23	.696		9.45	37	2.53	48.98	69
5676	1/2" pipe size, outlet		22	.727		8.80	39	2.64	50.44	71
5677	3/4" pipe size, outlet		21	.762		8.95	40.50	2.77	52.22	74.50
5678	1" pipe size, outlet		20	.800		9.90	42.50	2.90	55.30	78.50
5679	1-1/4" pipe size, outlet		18	.889		16.55	47.50	3.23	67.28	93.50
5680	1-1/2" pipe size, outlet		16	1		16.55	53.50	3.63	73.68	103
5681	2" pipe size, outlet	▼	12	1.333	▼	18.80	71	4.84	94.64	133
6000	Weld-on flange, forged steel									
6020	Slip-on, 150 lb. flange (welded front and back)									
6050	1/2" pipe size	Q-15	18	.889	Ea.	17.50	47.50	3.23	68.23	94.50
6060	3/4" pipe size		18	.889		17.50	47.50	3.23	68.23	94.50
6070	1" pipe size		17	.941		17.50	50	3.42	70.92	98.50
6080	1-1/4" pipe size		16	1		17.50	53.50	3.63	74.63	104
6090	1-1/2" pipe size		15	1.067		17.50	57	3.87	78.37	110
6100	2" pipe size		12	1.333		19.45	71	4.84	95.29	134
6110	2-1/2" pipe size		10	1.600		25	85.50	5.80	116.30	163
6120	3" pipe size		9	1.778		26.50	94.50	6.45	127.45	179
6130	3-1/2" pipe size		7	2.286		33	122	8.30	163.30	230
6140	4" pipe size		6	2.667		33	142	9.70	184.70	261
6150	5" pipe size	▼	5	3.200		57.50	171	11.60	240.10	335
6160	6" pipe size	Q-16	6	4		54.50	221	9.70	285.20	405
6170	8" pipe size		5	4.800		82.50	265	11.60	359.10	505
6180	10" pipe size		4	6		143	330	14.50	487.50	675
6190	12" pipe size		3	8		212	440	19.35	671.35	920
6191	14" pipe size		2.50	9.600		281	530	23	834	1,125
6192	16" pipe size	▼	1.80	13.333	▼	440	735	32.50	1,207.50	1,625
6200	300 lb. flange									
6210	1/2" pipe size	Q-15	17	.941	Ea.	23.50	50	3.42	76.92	105
6220	3/4" pipe size	▼	17	.941	▼	23.50	50	3.42	76.92	105

For customer support on your Mechanical Cost Data, call 877.739.3689.

22 11 13 – Facility Water Distribution Piping

22 11 13.47 Pipe Fittings, Steel		Crew	Daily Output	Labor-Hours	Unit	Material	2016 Bare Costs Labor	Equipment	Total	Total Incl O&P
6230	1" pipe size	Q-15	16	1	Ea.	23.50	53.50	3.63	80.63	110
6240	1-1/4" pipe size		13	1.231		23.50	65.50	4.47	93.47	129
6250	1-1/2" pipe size		12	1.333		23.50	71	4.84	99.34	138
6260	2" pipe size		11	1.455		31	77.50	5.30	113.80	157
6270	2-1/2" pipe size		9	1.778		35	94.50	6.45	135.95	189
6280	3" pipe size		7	2.286		38	122	8.30	168.30	235
6290	4" pipe size		6	2.667		55.50	142	9.70	207.20	286
6300	5" pipe size		4	4		95	213	14.50	322.50	440
6310	6" pipe size	Q-16	5	4.800		95.50	265	11.60	372.10	520
6320	8" pipe size		4	6		163	330	14.50	507.50	695
6330	10" pipe size		3.40	7.059		278	390	17.10	685.10	915
6340	12" pipe size		2.80	8.571		340	475	20.50	835.50	1,125
6400	Welding neck, 150 lb. flange									
6410	1/2" pipe size	Q-15	40	.400	Ea.	25.50	21.50	1.45	48.45	61.50
6420	3/4" pipe size		36	.444		25.50	23.50	1.61	50.61	65.50
6430	1" pipe size		32	.500		25.50	26.50	1.82	53.82	70
6440	1-1/4" pipe size		29	.552		25.50	29.50	2	57	74.50
6450	1-1/2" pipe size		26	.615		25.50	33	2.23	60.73	80
6460	2" pipe size		20	.800		29.50	42.50	2.90	74.90	99.50
6470	2-1/2" pipe size		16	1		32	53.50	3.63	89.13	119
6480	3" pipe size		14	1.143		35	61	4.15	100.15	135
6500	4" pipe size		10	1.600		42.50	85.50	5.80	133.80	182
6510	5" pipe size		8	2		66.50	107	7.25	180.75	243
6520	6" pipe size	Q-16	10	2.400		64	133	5.80	202.80	276
6530	8" pipe size		7	3.429		113	189	8.30	310.30	420
6540	10" pipe size		6	4		180	221	9.70	410.70	545
6550	12" pipe size		5	4.800		262	265	11.60	538.60	700
6551	14" pipe size		4.50	5.333		380	295	12.90	687.90	875
6552	16" pipe size		3	8		570	440	19.35	1,029.35	1,300
6553	18" pipe size		2.50	9.600		800	530	23	1,353	1,700
6554	20" pipe size		2.30	10.435		975	575	25.50	1,575.50	1,975
6556	24" pipe size		2	12		1,300	665	29	1,994	2,475
6557	26" pipe size		1.70	14.118		1,425	780	34	2,239	2,800
6558	30" pipe size		.90	26.667		1,650	1,475	64.50	3,189.50	4,100
6559	36" pipe size		.75	32		1,900	1,775	77.50	3,752.50	4,825
6560	300 lb. flange									
6570	1/2" pipe size	Q-15	36	.444	Ea.	31	23.50	1.61	56.11	71.50
6580	3/4" pipe size		34	.471		31	25	1.71	57.71	74
6590	1" pipe size		30	.533		31	28.50	1.94	61.44	79
6600	1-1/4" pipe size		28	.571		31	30.50	2.07	63.57	82.50
6610	1-1/2" pipe size		24	.667		31	35.50	2.42	68.92	90
6620	2" pipe size		18	.889		38.50	47.50	3.23	89.23	117
6630	2-1/2" pipe size		14	1.143		44	61	4.15	109.15	145
6640	3" pipe size		12	1.333		44.50	71	4.84	120.34	161
6650	4" pipe size		8	2		73	107	7.25	187.25	249
6660	5" pipe size		7	2.286		111	122	8.30	241.30	315
6670	6" pipe size	Q-16	9	2.667		113	147	6.45	266.45	355
6680	8" pipe size		6	4		195	221	9.70	425.70	560
6690	10" pipe size		5	4.800		355	265	11.60	631.60	805
6700	12" pipe size		4	6		455	330	14.50	799.50	1,025
6710	14" pipe size		3.50	6.857		980	380	16.60	1,376.60	1,675
6720	16" pipe size		2	12		1,150	665	29	1,844	2,300
7740	Plain ends for plain end pipe, mechanically coupled									

22 11 13.47 Pipe Fittings, Steel	Crew	Daily Output	Labor-Hours	Unit	Material	2016 Bare Costs Labor	Equipment	Total	Total Incl O&P
7750	Cplg. & labor required at joints not included, add 1 per								
7760	joint for installed price, see line 9180								
7770	Malleable iron, painted, unless noted otherwise								
7800	90° Elbow 1"				Ea.	119		119	131
7810	1-1/2"					141		141	155
7820	2"					210		210	231
7830	2-1/2"					246		246	270
7840	3"					255		255	281
7860	4"					289		289	320
7870	5" welded steel					330		330	365
7880	6"					420		420	460
7890	8" welded steel					790		790	865
7900	10" welded steel					1,000		1,000	1,100
7910	12" welded steel					1,125		1,125	1,225
7970	45° Elbow 1"					71		71	78.50
7980	1-1/2"					106		106	117
7990	2"					223		223	245
8000	2-1/2"					223		223	245
8010	3"					255		255	281
8030	4"					266		266	292
8040	5" welded steel					330		330	365
8050	6"					380		380	420
8060	8"					445		445	490
8070	10" welded steel					460		460	505
8080	12" welded steel					830		830	915
8140	Tee, straight 1"					133		133	147
8150	1-1/2"					171		171	189
8160	2"					171		171	189
8170	2-1/2"					223		223	246
8180	3"					345		345	380
8200	4"					490		490	540
8210	5" welded steel					670		670	740
8220	6"					585		585	645
8230	8" welded steel					845		845	925
8240	10" welded steel					1,325		1,325	1,450
8250	12" welded steel					1,550		1,550	1,700
8340	Segmentally welded steel, painted								
8390	Wye 2"				Ea.	213		213	234
8400	2-1/2"					213		213	234
8410	3"					238		238	262
8430	4"					355		355	390
8440	5"					435		435	475
8450	6"					595		595	655
8460	8"					770		770	850
8470	10"					1,150		1,150	1,250
8480	12"					1,775		1,775	1,950
8540	Wye, lateral 2"					231		231	255
8550	2-1/2"					266		266	293
8560	3"					315		315	350
8580	4"					435		435	480
8590	5"					740		740	810
8600	6"					760		760	840
8610	8"					1,275		1,275	1,400

22 11 13.47 Pipe Fittings, Steel		Crew	Daily Output	Labor-Hours	Unit	Material	2016 Bare Costs Labor	Equipment	Total	Total Incl O&P
8620	10"				Ea.	1,875			1,875	2,050
8630	12"					2,350			2,350	2,575
8690	Cross, 2"					225			225	248
8700	2-1/2"					225			225	248
8710	3"					269			269	295
8730	4"					365			365	405
8740	5"					520			520	575
8750	6"					685			685	755
8760	8"					880			880	970
8770	10"					1,300			1,300	1,425
8780	12"					1,850			1,850	2,050
8800	Tees, reducing 2" x 1"					146			146	161
8810	2" x 1-1/2"					146			146	161
8820	3" x 1"					146			146	161
8830	3" x 1-1/2"					146			146	161
8840	3" x 2"					148			148	163
8850	4" x 1"					215			215	236
8860	4" x 1-1/2"					215			215	236
8870	4" x 2"					212			212	233
8880	4" x 2-1/2"					219			219	240
8890	4" x 3"					219			219	240
8900	6" x 2"					320			320	350
8910	6" x 3"					345			345	375
8920	6" x 4"					375			375	415
8930	8" x 2"					390			390	425
8940	8" x 3"					395			395	435
8950	8" x 4"					425			425	465
8960	8" x 5"					460			460	505
8970	8" x 6"					495			495	545
8980	10" x 4"					655			655	720
8990	10" x 6"					655			655	720
9000	10" x 8"					695			695	765
9010	12" x 6"					1,000			1,000	1,100
9020	12" x 8"					1,025			1,025	1,125
9030	12" x 10"					1,050			1,050	1,150
9080	Adapter nipples 3" long									
9090	1"				Ea.	20.50			20.50	22.50
9100	1-1/2"					20.50			20.50	22.50
9110	2"					20.50			20.50	22.50
9120	2-1/2"					23.50			23.50	26
9130	3"					29			29	31.50
9140	4"					47			47	52
9150	6"					124			124	137
9180	Coupling, mechanical, plain end pipe to plain end pipe or fitting									
9190	1"	Q-1	29	.552	Ea.	72	29.50		101.50	124
9200	1-1/2"		28	.571		72	30.50		102.50	125
9210	2"		27	.593		72	31.50		103.50	127
9220	2-1/2"		26	.615		72	33		105	129
9230	3"		25	.640		105	34		139	168
9240	3-1/2"		24	.667		121	35.50		156.50	187
9250	4"		22	.727		121	39		160	192
9260	5"	Q-2	28	.857		173	47.50		220.50	262
9270	6"		24	1		211	55.50		266.50	315

22 11 13.47 Pipe Fittings, Steel

		Crew	Daily Output	Labor-Hours	Unit	Material	2016 Bare Costs Labor	Equipment	Total	Total Incl O&P
9280	8"	Q-2	19	1.263	Ea.	370	70		440	515
9290	10"		16	1.500		485	83		568	655
9300	12"	↓	12	2	↓	605	111		716	830
9310	Outlets for precut holes through pipe wall									
9331	Strapless type, with gasket									
9332	4" to 8" pipe x 1/2"	1 Plum	13	.615	Ea.	78.50	36.50		115	142
9333	4" to 8" pipe x 3/4"		13	.615		84	36.50		120.50	148
9334	10" pipe and larger x 1/2"		11	.727		78.50	43		121.50	152
9335	10" pipe and larger x 3/4"	↓	11	.727	↓	84	43		127	158
9341	Thermometer wells with gasket									
9342	4" to 8" pipe, 6" stem	1 Plum	14	.571	Ea.	118	34		152	181
9343	8" pipe and larger, 6" stem	"	13	.615	"	118	36.50		154.50	185
9940	For galvanized fittings for plain end pipe, add					20%				

22 11 13.48 Pipe, Fittings and Valves, Steel, Grooved-Joint

		Crew	Daily Output	Labor-Hours	Unit	Material	2016 Bare Costs Labor	Equipment	Total	Total Incl O&P
0010	**PIPE, FITTINGS AND VALVES, STEEL, GROOVED-JOINT**									
0012	Fittings are ductile iron. Steel fittings noted.									
0020	Pipe includes coupling & clevis type hanger assemblies, 10' O.C.									
0500	Schedule 10, black									
0550	2" diameter	1 Plum	43	.186	L.F.	7.30	11		18.30	24.50
0560	2-1/2" diameter	Q-1	61	.262		9.10	14		23.10	31
0570	3" diameter		55	.291		10.40	15.50		25.90	35
0580	3-1/2" diameter		53	.302		13.15	16.10		29.25	39
0590	4" diameter		49	.327		13.75	17.40		31.15	41.50
0600	5" diameter		40	.400		16.65	21.50		38.15	50.50
0610	6" diameter	Q-2	46	.522		21	29		50	67
0620	8" diameter	"	41	.585	↓	29	32.50		61.50	81
0700	To delete couplings & hangers, subtract									
0710	2" diam. to 5" diam.					25%	20%			
0720	6" diam. to 8" diam.					27%	15%			
1000	Schedule 40, black									
1040	3/4" diameter	1 Plum	71	.113	L.F.	5	6.65		11.65	15.55
1050	1" diameter		63	.127		4.81	7.50		12.31	16.65
1060	1-1/4" diameter		58	.138		5.95	8.15		14.10	18.85
1070	1-1/2" diameter		51	.157		6.60	9.30		15.90	21.50
1080	2" diameter	↓	40	.200		7.65	11.85		19.50	26.50
1090	2-1/2" diameter	Q-1	57	.281		11.60	14.95		26.55	35.50
1100	3" diameter		50	.320		14	17.05		31.05	41
1110	4" diameter		45	.356		19.55	18.95		38.50	50
1120	5" diameter		37	.432		40.50	23		63.50	79.50
1130	6" diameter	Q-2	42	.571		51.50	31.50		83	105
1140	8" diameter		37	.649		84	36		120	147
1150	10" diameter		31	.774		114	43		157	190
1160	12" diameter		27	.889		127	49		176	213
1170	14" diameter		20	1.200		131	66.50		197.50	244
1180	16" diameter		17	1.412		196	78		274	335
1190	18" diameter		14	1.714		196	94.50		290.50	360
1200	20" diameter		12	2		228	111		339	420
1210	24" diameter	↓	10	2.400	↓	257	133		390	485
1740	To delete coupling & hanger, subtract									
1750	3/4" diam. to 2" diam.					65%	27%			
1760	2-1/2" diam. to 5" diam.					41%	18%			
1770	6" diam. to 12" diam.					31%	13%			

22 11 13 – Facility Water Distribution Piping

22 11 13.48 Pipe, Fittings and Valves, Steel, Grooved-Joint	Crew	Daily Output	Labor-Hours	Unit	Material	2016 Bare Costs Labor	Equipment	Total	Total Incl O&P
1780	14" diam. to 24" diam.					35%	10%			
1800	Galvanized									
1840	3/4" diameter	1 Plum	71	.113	L.F.	5.25	6.65		11.90	15.80
1850	1" diameter		63	.127		5.55	7.50		13.05	17.45
1860	1-1/4" diameter		58	.138		6.90	8.15		15.05	19.90
1870	1-1/2" diameter		51	.157		7.70	9.30		17	22.50
1880	2" diameter		40	.200		9.15	11.85		21	28
1890	2-1/2" diameter	Q-1	57	.281		13.35	14.95		28.30	37
1900	3" diameter		50	.320		16.30	17.05		33.35	43.50
1910	4" diameter		45	.356		23	18.95		41.95	53.50
1920	5" diameter		37	.432		27	23		50	65
1930	6" diameter	Q-2	42	.571		34.50	31.50		66	85
1940	8" diameter		37	.649		51.50	36		87.50	111
1950	10" diameter		31	.774		107	43		150	183
1960	12" diameter		27	.889		131	49		180	218
2540	To delete coupling & hanger, subtract									
2550	3/4" diam. to 2" diam.					36%	27%			
2560	2-1/2" diam. to 5" diam.					19%	18%			
2570	6" diam. to 12" diam.					14%	13%			
2600	Schedule 80, black									
2610	3/4" diameter	1 Plum	65	.123	L.F.	5.85	7.30		13.15	17.40
2650	1" diameter		61	.131		6.95	7.75		14.70	19.35
2660	1-1/4" diameter		55	.145		8.85	8.60		17.45	23
2670	1-1/2" diameter		49	.163		10	9.65		19.65	25.50
2680	2" diameter		38	.211		12.25	12.45		24.70	32.50
2690	2-1/2" diameter	Q-1	54	.296		17.20	15.80		33	43
2700	3" diameter		48	.333		21.50	17.75		39.25	51
2710	4" diameter		44	.364		31	19.35		50.35	63
2720	5" diameter		35	.457		52	24.50		76.50	94
2730	6" diameter	Q-2	40	.600		70	33		103	127
2740	8" diameter		35	.686		106	38		144	173
2750	10" diameter		29	.828		170	45.50		215.50	256
2760	12" diameter		24	1		350	55.50		405.50	470
3240	To delete coupling & hanger, subtract									
3250	3/4" diam. to 2" diam.					30%	25%			
3260	2-1/2" diam. to 5" diam.					14%	17%			
3270	6" diam. to 12" diam.					12%	12%			
3300	Galvanized									
3310	3/4" diameter	1 Plum	65	.123	L.F.	6.55	7.30		13.85	18.25
3350	1" diameter		61	.131		7.95	7.75		15.70	20.50
3360	1-1/4" diameter		55	.145		10.30	8.60		18.90	24.50
3370	1-1/2" diameter		46	.174		11.85	10.30		22.15	28.50
3380	2" diameter		38	.211		15.20	12.45		27.65	35.50
3390	2-1/2" diameter	Q-1	54	.296		21.50	15.80		37.30	47.50
3400	3" diameter		48	.333		27.50	17.75		45.25	57
3410	4" diameter		44	.364		35.50	19.35		54.85	68
3420	5" diameter		35	.457		50.50	24.50		75	92.50
3430	6" diameter	Q-2	40	.600		68.50	33		101.50	126
3440	8" diameter		35	.686		151	38		189	223
3450	10" diameter		29	.828		273	45.50		318.50	370
3460	12" diameter		24	1		530	55.50		585.50	670
3920	To delete coupling & hanger, subtract									
3930	3/4" diam. to 2" diam.					30%	25%			

For customer support on your Mechanical Cost Data, call 877.739.3689.

177

22 11 13 – Facility Water Distribution Piping

22 11 13.48 Pipe, Fittings and Valves, Steel, Grooved-Joint		Crew	Daily Output	Labor-Hours	Unit	Material	2016 Bare Costs Labor	Equipment	Total	Total Incl O&P
3940	2-1/2" diam. to 5" diam.					15%	17%			
3950	6" diam. to 12" diam.					11%	12%			
3990	Fittings: coupling material required at joints not incl. in fitting price.									
3994	Add 1 selected coupling, material only, per joint for installed price.									
4000	Elbow, 90° or 45°, painted									
4030	3/4" diameter	1 Plum	50	.160	Ea.	63.50	9.45		72.95	84.50
4040	1" diameter		50	.160		34	9.45		43.45	52
4050	1-1/4" diameter		40	.200		34	11.85		45.85	55.50
4060	1-1/2" diameter		33	.242		34	14.35		48.35	59
4070	2" diameter		25	.320		34	18.95		52.95	66
4080	2-1/2" diameter	Q-1	40	.400		34	21.50		55.50	69.50
4090	3" diameter		33	.485		60	26		86	105
4100	4" diameter		25	.640		65	34		99	123
4110	5" diameter		20	.800		155	42.50		197.50	236
4120	6" diameter	Q-2	25	.960		182	53		235	281
4130	8" diameter		21	1.143		380	63		443	510
4140	10" diameter		18	1.333		700	73.50		773.50	880
4150	12" diameter		15	1.600		770	88.50		858.50	985
4170	14" diameter		12	2		860	111		971	1,100
4180	16" diameter		11	2.182		1,125	121		1,246	1,400
4190	18" diameter	Q-3	14	2.286		1,425	129		1,554	1,750
4200	20" diameter		12	2.667		1,875	150		2,025	2,275
4210	24" diameter		10	3.200		2,700	180		2,880	3,250
4250	For galvanized elbows, add					26%				
4690	Tee, painted									
4700	3/4" diameter	1 Plum	38	.211	Ea.	68.50	12.45		80.95	94
4740	1" diameter		33	.242		53	14.35		67.35	79.50
4750	1-1/4" diameter		27	.296		53	17.55		70.55	84.50
4760	1-1/2" diameter		22	.364		53	21.50		74.50	90.50
4770	2" diameter		17	.471		53	28		81	100
4780	2-1/2" diameter	Q-1	27	.593		53	31.50		84.50	106
4790	3" diameter		22	.727		72	39		111	138
4800	4" diameter		17	.941		110	50		160	197
4810	5" diameter		13	1.231		256	65.50		321.50	380
4820	6" diameter	Q-2	17	1.412		295	78		373	445
4830	8" diam.		14	1.714		645	94.50		739.50	855
4840	10" diameter		12	2		840	111		951	1,100
4850	12" diameter		10	2.400		1,100	133		1,233	1,400
4851	14" diameter		9	2.667		1,150	147		1,297	1,500
4852	16" diameter		8	3		1,300	166		1,466	1,675
4853	18" diameter	Q-3	10	3.200		1,625	180		1,805	2,075
4854	20" diameter		9	3.556		2,325	200		2,525	2,875
4855	24" diameter		8	4		3,550	225		3,775	4,250
4900	For galvanized tees, add					24%				
4906	Couplings, rigid style, painted									
4908	1" diameter	1 Plum	100	.080	Ea.	26.50	4.74		31.24	36
4909	1-1/4" diameter		100	.080		26.50	4.74		31.24	36
4910	1-1/2" diameter		67	.119		26.50	7.05		33.55	39.50
4912	2" diameter		50	.160		33.50	9.45		42.95	51
4914	2-1/2" diameter	Q-1	80	.200		38	10.65		48.65	58
4916	3" diameter		67	.239		44	12.70		56.70	67.50
4918	4" diameter		50	.320		61.50	17.05		78.55	93.50
4920	5" diameter		40	.400		79.50	21.50		101	120

22 11 Facility Water Distribution

22 11 13 – Facility Water Distribution Piping

22 11 13.48 Pipe, Fittings and Valves, Steel, Grooved-Joint	Crew	Daily Output	Labor-Hours	Unit	Material	2016 Bare Costs Labor	Equipment	Total	Total Incl O&P
4922	6" diameter	Q-2	50	.480	Ea.	105	26.50		131.50	156
4924	8" diameter		42	.571		165	31.50		196.50	230
4926	10" diameter		35	.686		320	38		358	410
4928	12" diameter		32	.750		335	41.50		376.50	430
4930	14" diameter		24	1		350	55.50		405.50	470
4931	16" diameter		20	1.200		395	66.50		461.50	535
4932	18" diameter		18	1.333		410	73.50		483.50	565
4933	20" diameter		16	1.500		565	83		648	745
4934	24" diameter		13	1.846		720	102		822	950
4940	Flexible, standard, painted									
4950	3/4" diameter	1 Plum	100	.080	Ea.	19	4.74		23.74	28
4960	1" diameter		100	.080		19	4.74		23.74	28
4970	1-1/4" diameter		80	.100		25	5.90		30.90	36.50
4980	1-1/2" diameter		67	.119		27	7.05		34.05	40
4990	2" diameter		50	.160		29	9.45		38.45	46.50
5000	2-1/2" diameter	Q-1	80	.200		33.50	10.65		44.15	53
5010	3" diameter		67	.239		37.50	12.70		50.20	60
5020	3-1/2" diameter		57	.281		53.50	14.95		68.45	81
5030	4" diameter		50	.320		54	17.05		71.05	84.50
5040	5" diameter		40	.400		81	21.50		102.50	122
5050	6" diameter	Q-2	50	.480		96	26.50		122.50	146
5070	8" diameter		42	.571		156	31.50		187.50	220
5090	10" diameter		35	.686		254	38		292	335
5110	12" diameter		32	.750		289	41.50		330.50	385
5120	14" diameter		24	1		415	55.50		470.50	540
5130	16" diameter		20	1.200		540	66.50		606.50	695
5140	18" diameter		18	1.333		635	73.50		708.50	805
5150	20" diameter		16	1.500		995	83		1,078	1,225
5160	24" diameter		13	1.846		1,100	102		1,202	1,350
5176	Lightweight style, painted									
5178	1-1/2" diameter	1 Plum	67	.119	Ea.	24	7.05		31.05	36.50
5180	2" diameter	"	50	.160		29	9.45		38.45	46.50
5182	2-1/2" diameter	Q-1	80	.200		33.50	10.65		44.15	53
5184	3" diameter		67	.239		37.50	12.70		50.20	60
5186	3-1/2" diameter		57	.281		45.50	14.95		60.45	72.50
5188	4" diameter		50	.320		54	17.05		71.05	84.50
5190	5" diameter		40	.400		81	21.50		102.50	122
5192	6" diameter	Q-2	50	.480		96	26.50		122.50	146
5194	8" diameter		42	.571		156	31.50		187.50	220
5196	10" diameter		35	.686		315	38		353	400
5198	12" diameter		32	.750		355	41.50		396.50	455
5200	For galvanized couplings, add					33%				
5220	Tee, reducing, painted									
5225	2" x 1-1/2" diameter	Q-1	38	.421	Ea.	111	22.50		133.50	156
5226	2-1/2" x 2" diameter		28	.571		111	30.50		141.50	168
5227	3" x 2-1/2" diameter		23	.696		98	37		135	164
5228	4" x 3" diameter		18	.889		132	47.50		179.50	217
5229	5" x 4" diameter		15	1.067		282	57		339	395
5230	6" x 4" diameter	Q-2	18	1.333		310	73.50		383.50	450
5231	8" x 6" diameter		15	1.600		645	88.50		733.50	845
5232	10" x 8" diameter		13	1.846		720	102		822	945
5233	12" x 10" diameter		11	2.182		900	121		1,021	1,175
5234	14" x 12" diameter		10	2.400		1,050	133		1,183	1,350

For customer support on your Mechanical Cost Data, call 877.739.3689.

179

22 11 13.48 Pipe, Fittings and Valves, Steel, Grooved-Joint		Crew	Daily Output	Labor-Hours	Unit	Material	2016 Bare Costs Labor	Equipment	Total	Total Incl O&P
5235	16" x 12" diameter	Q-2	9	2.667	Ea.	1,200	147		1,347	1,550
5236	18" x 12" diameter	Q-3	12	2.667		1,250	150		1,400	1,600
5237	18" x 16" diameter		11	2.909		1,575	164		1,739	1,975
5238	20" x 16" diameter		10	3.200		2,050	180		2,230	2,550
5239	24" x 20" diameter		9	3.556		3,250	200		3,450	3,875
5240	Reducer, concentric, painted									
5241	2-1/2" x 2" diameter	Q-1	43	.372	Ea.	40	19.85		59.85	74
5242	3" x 2-1/2" diameter		35	.457		48	24.50		72.50	89.50
5243	4" x 3" diameter		29	.552		58	29.50		87.50	108
5244	5" x 4" diameter		22	.727		79.50	39		118.50	146
5245	6" x 4" diameter	Q-2	26	.923		92.50	51		143.50	179
5246	8" x 6" diameter		23	1.043		238	57.50		295.50	350
5247	10" x 8" diameter		20	1.200		465	66.50		531.50	615
5248	12" x 10" diameter		16	1.500		835	83		918	1,050
5255	Eccentric, painted									
5256	2-1/2" x 2" diameter	Q-1	42	.381	Ea.	83	20.50		103.50	122
5257	3" x 2-1/2" diameter		34	.471		94.50	25		119.50	142
5258	4" x 3" diameter		28	.571		115	30.50		145.50	173
5259	5" x 4" diameter		21	.762		156	40.50		196.50	234
5260	6" x 4" diameter	Q-2	25	.960		182	53		235	280
5261	8" x 6" diameter		22	1.091		365	60.50		425.50	495
5262	10" x 8" diameter		19	1.263		965	70		1,035	1,175
5263	12" x 10" diameter		15	1.600		1,325	88.50		1,413.50	1,600
5270	Coupling, reducing, painted									
5272	2" x 1-1/2" diameter	1 Plum	52	.154	Ea.	38	9.10		47.10	56
5274	2-1/2" x 2" diameter	Q-1	82	.195		50	10.40		60.40	70.50
5276	3" x 2" diameter		69	.232		56.50	12.35		68.85	80.50
5278	4" x 2" diameter		52	.308		89	16.40		105.40	123
5280	5" x 4" diameter		42	.381		100	20.50		120.50	141
5282	6" x 4" diameter	Q-2	52	.462		151	25.50		176.50	205
5284	8" x 6" diameter	"	44	.545		226	30		256	294
5290	Outlet coupling, painted									
5294	1-1/2" x 1" pipe size	1 Plum	65	.123	Ea.	50.50	7.30		57.80	66.50
5296	2" x 1" pipe size	"	48	.167		51.50	9.85		61.35	72
5298	2-1/2" x 1" pipe size	Q-1	78	.205		79.50	10.95		90.45	104
5300	2-1/2" x 1-1/4" pipe size		70	.229		89.50	12.20		101.70	117
5302	3" x 1" pipe size		65	.246		101	13.10		114.10	131
5304	4" x 3/4" pipe size		48	.333		112	17.75		129.75	150
5306	4" x 1-1/2" pipe size		46	.348		159	18.55		177.55	203
5308	6" x 1-1/2" pipe size	Q-2	44	.545		224	30		254	293
5320	Outlet, strap-on T, painted									
5324	Threaded female branch									
5330	2" pipe x 1-1/2" branch size	1 Plum	50	.160	Ea.	37	9.45		46.45	55
5334	2-1/2" pipe x 1-1/2" branch size	Q-1	80	.200		45	10.65		55.65	65.50
5338	3" pipe x 1-1/2" branch size		67	.239		46.50	12.70		59.20	70.50
5342	3" pipe x 2" branch size		59	.271		53.50	14.45		67.95	81
5346	4" pipe x 2" branch size		50	.320		56.50	17.05		73.55	87.50
5350	4" pipe x 2-1/2" branch size		47	.340		59	18.15		77.15	92
5354	4" pipe x 3" branch size		43	.372		62.50	19.85		82.35	99
5358	5" pipe x 3" branch size		40	.400		77	21.50		98.50	117
5362	6" pipe x 2" branch size	Q-2	50	.480		70.50	26.50		97	118
5366	6" pipe x 3" branch size		47	.511		80.50	28		108.50	131
5370	6" pipe x 4" branch size		44	.545		89.50	30		119.50	144

22 11 13 – Facility Water Distribution Piping

22 11 13.48 Pipe, Fittings and Valves, Steel, Grooved-Joint		Crew	Daily Output	Labor-Hours	Unit	Material	2016 Bare Costs Labor	Equipment	Total	Total Incl O&P
5374	8" pipe x 2-1/2" branch size	Q-2	42	.571	Ea.	123	31.50		154.50	184
5378	8" pipe x 4" branch size	↓	38	.632	↓	134	35		169	200
5390	Grooved branch									
5394	2" pipe x 1-1/4" branch size	1 Plum	78	.103	Ea.	37	6.05		43.05	49.50
5398	2" pipe x 1-1/2" branch size	"	50	.160		37	9.45		46.45	55
5402	2-1/2" pipe x 1-1/2" branch size	Q-1	80	.200		45	10.65		55.65	65.50
5406	3" pipe x 1-1/2" branch size		67	.239		46.50	12.70		59.20	70.50
5410	3" pipe x 2" branch size		59	.271		53.50	14.45		67.95	81
5414	4" pipe x 2" branch size		50	.320		56.50	17.05		73.55	87.50
5418	4" pipe x 2-1/2" branch size		47	.340		59	18.15		77.15	92
5422	4" pipe x 3" branch size		43	.372		62.50	19.85		82.35	99
5426	5" pipe x 3" branch size	↓	40	.400		77	21.50		98.50	117
5430	6" pipe x 2" branch size	Q-2	50	.480		70.50	26.50		97	118
5434	6" pipe x 3" branch size		47	.511		80.50	28		108.50	131
5438	6" pipe x 4" branch size		44	.545		89.50	30		119.50	144
5442	8" pipe x 2-1/2" branch size		42	.571		123	31.50		154.50	184
5446	8" pipe x 4" branch size	↓	38	.632	↓	134	35		169	200
5750	Flange, w/groove gasket, black steel									
5754	See Line 22 11 13.47 0620 for gasket & bolt set									
5760	ANSI class 125 and 150, painted									
5780	2" pipe size	1 Plum	23	.348	Ea.	112	20.50		132.50	155
5790	2-1/2" pipe size	Q-1	37	.432		140	23		163	189
5800	3" pipe size		31	.516		151	27.50		178.50	208
5820	4" pipe size		23	.696		201	37		238	277
5830	5" pipe size		19	.842		233	45		278	325
5840	6" pipe size	Q-2	23	1.043		254	57.50		311.50	365
5850	8" pipe size		17	1.412		287	78		365	435
5860	10" pipe size		14	1.714		455	94.50		549.50	645
5870	12" pipe size		12	2		595	111		706	820
5880	14" pipe size		10	2.400		1,100	133		1,233	1,400
5890	16" pipe size		9	2.667		1,275	147		1,422	1,625
5900	18" pipe size		6	4		1,575	221		1,796	2,050
5910	20" pipe size		5	4.800		1,875	265		2,140	2,475
5920	24" pipe size	↓	4.50	5.333	↓	2,400	295		2,695	3,100
5940	ANSI class 350, painted									
5946	2" pipe size	1 Plum	23	.348	Ea.	140	20.50		160.50	185
5948	2-1/2" pipe size	Q-1	37	.432		162	23		185	214
5950	3" pipe size		31	.516		222	27.50		249.50	286
5952	4" pipe size		23	.696		295	37		332	380
5954	5" pipe size		19	.842		335	45		380	440
5956	6" pipe size	Q-2	23	1.043		390	57.50		447.50	515
5958	8" pipe size		17	1.412		450	78		528	615
5960	10" pipe size		14	1.714		715	94.50		809.50	930
5962	12" pipe size	↓	12	2	↓	760	111		871	1,000
6100	Cross, painted									
6110	2" diameter	1 Plum	12.50	.640	Ea.	94	38		132	160
6112	2-1/2" diameter	Q-1	20	.800		94	42.50		136.50	168
6114	3" diameter		15.50	1.032		167	55		222	266
6116	4" diameter	↓	12.50	1.280		277	68		345	410
6118	6" diameter	Q-2	12.50	1.920		725	106		831	955
6120	8" diameter		10.50	2.286		955	126		1,081	1,250
6122	10" diameter		9	2.667		1,575	147		1,722	1,950
6124	12" diameter	↓	7.50	3.200	↓	2,250	177		2,427	2,750

For customer support on your Mechanical Cost Data, call 877.739.3689.

181

22 11 13.48 Pipe, Fittings and Valves, Steel, Grooved-Joint		Crew	Daily Output	Labor-Hours	Unit	Material	2016 Bare Costs Labor	Equipment	Total	Total Incl O&P
7400	Suction diffuser									
7402	Grooved end inlet x flanged outlet									
7410	3" x 3"	Q-1	27	.593	Ea.	910	31.50		941.50	1,050
7412	4" x 4"		19	.842		1,225	45		1,270	1,425
7414	5" x 5"		14	1.143		1,450	61		1,511	1,675
7416	6" x 6"	Q-2	20	1.200		1,825	66.50		1,891.50	2,100
7418	8" x 8"		15	1.600		3,375	88.50		3,463.50	3,850
7420	10" x 10"		12	2		4,600	111		4,711	5,225
7422	12" x 12"		9	2.667		7,575	147		7,722	8,550
7424	14" x 14"		7	3.429		7,575	189		7,764	8,625
7426	16" x 14"		6	4		7,800	221		8,021	8,900
7500	Strainer, tee type, painted									
7506	2" pipe size	1 Plum	21	.381	Ea.	605	22.50		627.50	705
7508	2-1/2" pipe size	Q-1	30	.533		640	28.50		668.50	745
7510	3" pipe size		28	.571		715	30.50		745.50	835
7512	4" pipe size		20	.800		810	42.50		852.50	955
7514	5" pipe size		15	1.067		1,175	57		1,232	1,375
7516	6" pipe size	Q-2	23	1.043		1,275	57.50		1,332.50	1,475
7518	8" pipe size		16	1.500		1,950	83		2,033	2,275
7520	10" pipe size		13	1.846		2,875	102		2,977	3,300
7522	12" pipe size		10	2.400		3,700	133		3,833	4,250
7524	14" pipe size		8	3		10,200	166		10,366	11,600
7526	16" pipe size		7	3.429		16,700	189		16,889	18,700
7570	Expansion joint, max. 3" travel									
7572	2" diameter	1 Plum	24	.333	Ea.	765	19.75		784.75	870
7574	3" diameter	Q-1	31	.516		880	27.50		907.50	1,000
7576	4" diameter	"	23	.696		1,150	37		1,187	1,325
7578	6" diameter	Q-2	38	.632		1,800	35		1,835	2,025
7790	Valves: coupling material required at joints not incl. in valve price.									
7794	Add 1 selected coupling, material only, per joint for installed price.									
7800	Ball valve w/handle, carbon steel trim									
7810	1-1/2" pipe size	1 Plum	31	.258	Ea.	152	15.30		167.30	191
7812	2" pipe size	"	24	.333		169	19.75		188.75	216
7814	2-1/2" pipe size	Q-1	38	.421		370	22.50		392.50	440
7816	3" pipe size		31	.516		595	27.50		622.50	695
7818	4" pipe size		23	.696		920	37		957	1,050
7820	6" pipe size	Q-2	24	1		2,825	55.50		2,880.50	3,175
7830	With gear operator									
7834	2-1/2" pipe size	Q-1	38	.421	Ea.	745	22.50		767.50	855
7836	3" pipe size		31	.516		1,050	27.50		1,077.50	1,200
7838	4" pipe size		23	.696		1,350	37		1,387	1,525
7840	6" pipe size	Q-2	24	1		3,175	55.50		3,230.50	3,550
7870	Check valve									
7874	2-1/2" pipe size	Q-1	38	.421	Ea.	290	22.50		312.50	355
7876	3" pipe size		31	.516		345	27.50		372.50	415
7878	4" pipe size		23	.696		360	37		397	455
7880	5" pipe size		18	.889		600	47.50		647.50	730
7882	6" pipe size	Q-2	24	1		715	55.50		770.50	870
7884	8" pipe size		19	1.263		975	70		1,045	1,175
7886	10" pipe size		16	1.500		2,600	83		2,683	3,000
7888	12" pipe size		13	1.846		3,100	102		3,202	3,550
7900	Plug valve, balancing, w/lever operator									
7906	3" pipe size	Q-1	31	.516	Ea.	720	27.50		747.50	830

For customer support on your Mechanical Cost Data, call 877.739.3689.

22 11 13 – Facility Water Distribution Piping

22 11 13.48 Pipe, Fittings and Valves, Steel, Grooved-Joint	Crew	Daily Output	Labor-Hours	Unit	Material	2016 Bare Costs Labor	Equipment	Total	Total Incl O&P
7908	4" pipe size	Q-1	23	.696	Ea.	885	37		922	1,025
7909	6" pipe size	Q-2	24	1	↓	1,350	55.50		1,405.50	1,575
7916	With gear operator									
7920	3" pipe size	Q-1	31	.516	Ea.	1,375	27.50		1,402.50	1,550
7922	4" pipe size	"	23	.696		1,425	37		1,462	1,600
7924	6" pipe size	Q-2	24	1		1,900	55.50		1,955.50	2,175
7926	8" pipe size		19	1.263		2,925	70		2,995	3,325
7928	10" pipe size		16	1.500		3,700	83		3,783	4,200
7930	12" pipe size	↓	13	1.846	↓	5,625	102		5,727	6,350
8000	Butterfly valve, 2 position handle, with standard trim									
8010	1-1/2" pipe size	1 Plum	30	.267	Ea.	273	15.80		288.80	325
8020	2" pipe size	"	23	.348		273	20.50		293.50	330
8030	3" pipe size	Q-1	30	.533		390	28.50		418.50	475
8050	4" pipe size	"	22	.727		430	39		469	535
8070	6" pipe size	Q-2	23	1.043		870	57.50		927.50	1,050
8080	8" pipe size		18	1.333		1,175	73.50		1,248.50	1,375
8090	10" pipe size	↓	15	1.600	↓	1,925	88.50		2,013.50	2,225
8200	With stainless steel trim									
8240	1-1/2" pipe size	1 Plum	30	.267	Ea.	345	15.80		360.80	405
8250	2" pipe size	"	23	.348		345	20.50		365.50	410
8270	3" pipe size	Q-1	30	.533		465	28.50		493.50	555
8280	4" pipe size	"	22	.727		505	39		544	615
8300	6" pipe size	Q-2	23	1.043		940	57.50		997.50	1,100
8310	8" pipe size		18	1.333		2,325	73.50		2,398.50	2,650
8320	10" pipe size		15	1.600		3,900	88.50		3,988.50	4,400
8322	12" pipe size		12	2		4,475	111		4,586	5,075
8324	14" pipe size		10	2.400		5,725	133		5,858	6,500
8326	16" pipe size	↓	9	2.667		8,275	147		8,422	9,325
8328	18" pipe size	Q-3	12	2.667		10,200	150		10,350	11,400
8330	20" pipe size		10	3.200		12,500	180		12,680	14,100
8332	24" pipe size	↓	9	3.556	↓	16,800	200		17,000	18,700
8336	Note: sizes 8" up w/manual gear operator									
9000	Cut one groove, labor									
9010	3/4" pipe size	Q-1	152	.105	Ea.		5.60		5.60	8.45
9020	1" pipe size		140	.114			6.10		6.10	9.20
9030	1-1/4" pipe size		124	.129			6.85		6.85	10.35
9040	1-1/2" pipe size		114	.140			7.50		7.50	11.30
9050	2" pipe size		104	.154			8.20		8.20	12.35
9060	2-1/2" pipe size		96	.167			8.90		8.90	13.40
9070	3" pipe size		88	.182			9.70		9.70	14.60
9080	3-1/2" pipe size		83	.193			10.25		10.25	15.50
9090	4" pipe size		78	.205			10.95		10.95	16.50
9100	5" pipe size		72	.222			11.85		11.85	17.85
9110	6" pipe size		70	.229			12.20		12.20	18.40
9120	8" pipe size		54	.296			15.80		15.80	24
9130	10" pipe size		38	.421			22.50		22.50	34
9140	12" pipe size		30	.533			28.50		28.50	43
9150	14" pipe size		20	.800			42.50		42.50	64.50
9160	16" pipe size		19	.842			45		45	67.50
9170	18" pipe size		18	.889			47.50		47.50	71.50
9180	20" pipe size		17	.941			50		50	75.50
9190	24" pipe size	↓	15	1.067	↓		57		57	86
9210	Roll one groove									

For customer support on your Mechanical Cost Data, call 877.739.3689.

183

22 11 Facility Water Distribution

22 11 13 – Facility Water Distribution Piping

22 11 13.48 Pipe, Fittings and Valves, Steel, Grooved-Joint		Crew	Daily Output	Labor-Hours	Unit	Material	2016 Bare Costs Labor	Equipment	Total	Total Incl O&P
9220	3/4" pipe size	Q-1	266	.060	Ea.		3.20		3.20	4.84
9230	1" pipe size		228	.070			3.74		3.74	5.65
9240	1-1/4" pipe size		200	.080			4.26		4.26	6.45
9250	1-1/2" pipe size		178	.090			4.79		4.79	7.25
9260	2" pipe size		116	.138			7.35		7.35	11.10
9270	2-1/2" pipe size		110	.145			7.75		7.75	11.70
9280	3" pipe size		100	.160			8.50		8.50	12.85
9290	3-1/2" pipe size		94	.170			9.05		9.05	13.70
9300	4" pipe size		86	.186			9.90		9.90	14.95
9310	5" pipe size		84	.190			10.15		10.15	15.30
9320	6" pipe size		80	.200			10.65		10.65	16.10
9330	8" pipe size		66	.242			12.90		12.90	19.50
9340	10" pipe size		58	.276			14.70		14.70	22
9350	12" pipe size		46	.348			18.55		18.55	28
9360	14" pipe size		30	.533			28.50		28.50	43
9370	16" pipe size		28	.571			30.50		30.50	46
9380	18" pipe size		27	.593			31.50		31.50	47.50
9390	20" pipe size		25	.640			34		34	51.50
9400	24" pipe size		23	.696			37		37	56

22 11 13.60 Tubing, Stainless Steel

		Crew	Daily Output	Labor-Hours	Unit	Material	2016 Bare Costs Labor	Equipment	Total	Total Incl O&P
0010	**TUBING, STAINLESS STEEL**									
5000	Tubing									
5010	Type 304, no joints, no hangers									
5020	.035 wall									
5021	1/4"	1 Plum	160	.050	L.F.	3.12	2.96		6.08	7.90
5022	3/8"		160	.050		3.94	2.96		6.90	8.80
5023	1/2"		160	.050		5.05	2.96		8.01	10.05
5024	5/8"		160	.050		6.20	2.96		9.16	11.30
5025	3/4"		133	.060		7.50	3.56		11.06	13.60
5026	7/8"		133	.060		7.90	3.56		11.46	14.05
5027	1"		114	.070		8.70	4.15		12.85	15.80
5040	.049 wall									
5041	1/4"	1 Plum	160	.050	L.F.	3.67	2.96		6.63	8.50
5042	3/8"		160	.050		5.05	2.96		8.01	10.05
5043	1/2"		160	.050		5.60	2.96		8.56	10.65
5044	5/8"		160	.050		7.60	2.96		10.56	12.80
5045	3/4"		133	.060		8.10	3.56		11.66	14.25
5046	7/8"		133	.060		8.45	3.56		12.01	14.60
5047	1"		114	.070		8.95	4.15		13.10	16.10
5060	.065 wall									
5061	1/4"	1 Plum	160	.050	L.F.	4.11	2.96		7.07	9
5062	3/8"		160	.050		5.65	2.96		8.61	10.70
5063	1/2"		160	.050		6.85	2.96		9.81	12
5064	5/8"		160	.050		8.30	2.96		11.26	13.55
5065	3/4"		133	.060		8.85	3.56		12.41	15.05
5066	7/8"		133	.060		10.15	3.56		13.71	16.50
5067	1"		114	.070		9.95	4.15		14.10	17.15
5210	Type 316									
5220	.035 wall									
5221	1/4"	1 Plum	160	.050	L.F.	3.96	2.96		6.92	8.85
5222	3/8"		160	.050		5	2.96		7.96	9.95
5223	1/2"		160	.050		7	2.96		9.96	12.15

22 11 13 – Facility Water Distribution Piping

22 11 13.60 Tubing, Stainless Steel

		Crew	Daily Output	Labor-Hours	Unit	Material	2016 Bare Costs Labor	Equipment	Total	Total Incl O&P
5224	5/8"	1 Plum	160	.050	L.F.	9.10	2.96		12.06	14.45
5225	3/4"		133	.060		10.50	3.56		14.06	16.90
5226	7/8"		133	.060		15	3.56		18.56	22
5227	1"		114	.070		11.90	4.15		16.05	19.35
5240	.049 wall									
5241	1/4"	1 Plum	160	.050	L.F.	5.45	2.96		8.41	10.45
5242	3/8"		160	.050		7.15	2.96		10.11	12.30
5243	1/2"		160	.050		7.85	2.96		10.81	13.10
5244	5/8"		160	.050		10.90	2.96		13.86	16.45
5245	3/4"		133	.060		10.50	3.56		14.06	16.90
5246	7/8"		133	.060		13.50	3.56		17.06	20
5247	1"		114	.070		12.30	4.15		16.45	19.80
5260	.065 wall									
5261	1/4"	1 Plum	160	.050	L.F.	5.20	2.96		8.16	10.15
5262	3/8"		160	.050		8.30	2.96		11.26	13.55
5263	1/2"		160	.050		8.90	2.96		11.86	14.25
5264	5/8"		160	.050		12.40	2.96		15.36	18.10
5265	3/4"		133	.060		13.35	3.56		16.91	20
5266	7/8"		133	.060		15.80	3.56		19.36	23
5267	1"		114	.070		15.05	4.15		19.20	23

22 11 13.61 Tubing Fittings, Stainless Steel

		Crew	Daily Output	Labor-Hours	Unit	Material	2016 Bare Costs Labor	Equipment	Total	Total Incl O&P
0010	**TUBING FITTINGS, STAINLESS STEEL**									
8200	Tube fittings, compression type									
8202	Type 316									
8204	90° elbow									
8206	1/4"	1 Plum	24	.333	Ea.	17.60	19.75		37.35	49.50
8207	3/8"		22	.364		21.50	21.50		43	56.50
8208	1/2"		22	.364		35.50	21.50		57	71.50
8209	5/8"		21	.381		40	22.50		62.50	78
8210	3/4"		21	.381		63	22.50		85.50	103
8211	7/8"		20	.400		96.50	23.50		120	142
8212	1"		20	.400		120	23.50		143.50	168
8220	Union tee									
8222	1/4"	1 Plum	15	.533	Ea.	25	31.50		56.50	75
8224	3/8"		15	.533		32	31.50		63.50	83
8225	1/2"		15	.533		49.50	31.50		81	102
8226	5/8"		14	.571		55.50	34		89.50	112
8227	3/4"		14	.571		74	34		108	133
8228	7/8"		13	.615		148	36.50		184.50	218
8229	1"		13	.615		159	36.50		195.50	230
8234	Union									
8236	1/4"	1 Plum	24	.333	Ea.	12.20	19.75		31.95	43.50
8237	3/8"		22	.364		17.40	21.50		38.90	51.50
8238	1/2"		22	.364		26	21.50		47.50	61
8239	5/8"		21	.381		34	22.50		56.50	71
8240	3/4"		21	.381		42	22.50		64.50	80
8241	7/8"		20	.400		68.50	23.50		92	111
8242	1"		20	.400		72.50	23.50		96	116
8250	Male connector									
8252	1/4" x 1/4"	1 Plum	24	.333	Ea.	7.80	19.75		27.55	38.50
8253	3/8" x 3/8"		22	.364		12.20	21.50		33.70	46
8254	1/2" x 1/2"		22	.364		18.05	21.50		39.55	52.50

For customer support on your Mechanical Cost Data, call 877.739.3689.

185

22 11 13.61 Tubing Fittings, Stainless Steel		Crew	Daily Output	Labor-Hours	Unit	Material	2016 Bare Costs Labor	Equipment	Total	Total Incl O&P
8256	3/4" x 3/4"	1 Plum	21	.381	Ea.	27.50	22.50		50	64.50
8258	1" x 1"		20	.400		48.50	23.50		72	88.50

22 11 13.64 Pipe, Stainless Steel

		Crew	Daily Output	Labor-Hours	Unit	Material	Labor	Equipment	Total	Total Incl O&P
0010	**PIPE, STAINLESS STEEL**									
0020	Welded, with clevis type hanger assemblies, 10' O.C.									
0500	Schedule 5, type 304									
0540	1/2" diameter	Q-15	128	.125	L.F.	8.25	6.65	.45	15.35	19.65
0550	3/4" diameter		116	.138		10.25	7.35	.50	18.10	23
0560	1" diameter		103	.155		11.35	8.30	.56	20.21	25.50
0570	1-1/4" diameter		93	.172		15.45	9.15	.62	25.22	31.50
0580	1-1/2" diameter		85	.188		18.60	10.05	.68	29.33	36.50
0590	2" diameter		69	.232		31	12.35	.84	44.19	53.50
0600	2-1/2" diameter		53	.302		33.50	16.10	1.10	50.70	62.50
0610	3" diameter		48	.333		40	17.75	1.21	58.96	73
0620	4" diameter		44	.364		51	19.35	1.32	71.67	86.50
0630	5" diameter		36	.444		103	23.50	1.61	128.11	151
0640	6" diameter	Q-16	42	.571		95.50	31.50	1.38	128.38	154
0650	8" diameter		34	.706		146	39	1.71	186.71	222
0660	10" diameter		26	.923		202	51	2.23	255.23	300
0670	12" diameter		21	1.143		265	63	2.77	330.77	390
0700	To delete hangers, subtract									
0710	1/2" diam. to 1-1/2" diam.					8%	19%			
0720	2" diam. to 5" diam.					4%	9%			
0730	6" diam. to 12" diam.					3%	4%			
0750	For small quantities, add				L.F.	10%				
1250	Schedule 5, type 316									
1290	1/2" diameter	Q-15	128	.125	L.F.	10.30	6.65	.45	17.40	22
1300	3/4" diameter		116	.138		14.75	7.35	.50	22.60	28
1310	1" diameter		103	.155		16.45	8.30	.56	25.31	31
1320	1-1/4" diameter		93	.172		20.50	9.15	.62	30.27	37
1330	1-1/2" diameter		85	.188		27.50	10.05	.68	38.23	46.50
1340	2" diameter		69	.232		42.50	12.35	.84	55.69	66
1350	2-1/2" diameter		53	.302		54.50	16.10	1.10	71.70	85.50
1360	3" diameter		48	.333		69	17.75	1.21	87.96	104
1370	4" diameter		44	.364		85.50	19.35	1.32	106.17	124
1380	5" diameter		36	.444		163	23.50	1.61	188.11	216
1390	6" diameter	Q-16	42	.571		159	31.50	1.38	191.88	224
1400	8" diameter		34	.706		239	39	1.71	279.71	325
1410	10" diameter		26	.923		340	51	2.23	393.23	455
1420	12" diameter		21	1.143		440	63	2.77	505.77	585
1490	For small quantities, add					10%				
1940	To delete hanger, subtract									
1950	1/2" diam. to 1-1/2" diam.					5%	19%			
1960	2" diam. to 5" diam.					3%	9%			
1970	6" diam. to 12" diam.					2%	4%			
2000	Schedule 10, type 304									
2040	1/4" diameter	Q-15	131	.122	L.F.	5.40	6.50	.44	12.34	16.25
2050	3/8" diameter		128	.125		6.50	6.65	.45	13.60	17.70
2060	1/2" diameter		125	.128		6.75	6.80	.46	14.01	18.25
2070	3/4" diameter		113	.142		7.80	7.55	.51	15.86	20.50
2080	1" diameter		100	.160		11.20	8.50	.58	20.28	26
2090	1-1/4" diameter		91	.176		12.70	9.35	.64	22.69	29

22 11 13.64 Pipe, Stainless Steel	Crew	Daily Output	Labor-Hours	Unit	Material	2016 Bare Costs Labor	Equipment	Total	Total Incl O&P
2100	1-1/2" diameter	Q-15	83	.193	L.F.	14.15	10.25	.70	25.10	32
2110	2" diameter		67	.239		17.10	12.70	.87	30.67	39
2120	2-1/2" diameter		51	.314		22	16.70	1.14	39.84	50.50
2130	3" diameter		46	.348		26.50	18.55	1.26	46.31	58.50
2140	4" diameter		42	.381		32.50	20.50	1.38	54.38	67.50
2150	5" diameter		35	.457		43.50	24.50	1.66	69.66	86.50
2160	6" diameter	Q-16	40	.600		51	33	1.45	85.45	108
2170	8" diameter		33	.727		88.50	40	1.76	130.26	160
2180	10" diameter		25	.960		122	53	2.32	177.32	217
2190	12" diameter		21	1.143		152	63	2.77	217.77	266
2250	For small quantities, add					10%				
2650	To delete hanger, subtract									
2660	1/4" diam. to 3/4" diam.					9%	22%			
2670	1" diam. to 2" diam.					4%	15%			
2680	2-1/2" diam. to 5" diam.					3%	8%			
2690	6" diam. to 12" diam.					3%	4%			
2750	Schedule 10, type 316									
2790	1/4" diameter	Q-15	131	.122	L.F.	6.50	6.50	.44	13.44	17.45
2800	3/8" diameter		128	.125		7.60	6.65	.45	14.70	18.90
2810	1/2" diameter		125	.128		8.50	6.80	.46	15.76	20
2820	3/4" diameter		113	.142		9.95	7.55	.51	18.01	23
2830	1" diameter		100	.160		14.15	8.50	.58	23.23	29
2840	1-1/4" diameter		91	.176		17.90	9.35	.64	27.89	34.50
2850	1-1/2" diameter		83	.193		19.80	10.25	.70	30.75	38.50
2860	2" diameter		67	.239		24	12.70	.87	37.57	46.50
2870	2-1/2" diameter		51	.314		31	16.70	1.14	48.84	60.50
2880	3" diameter		46	.348		36.50	18.55	1.26	56.31	70
2890	4" diameter		42	.381		46.50	20.50	1.38	68.38	83
2900	5" diameter		35	.457		56.50	24.50	1.66	82.66	101
2910	6" diameter	Q-16	40	.600		69	33	1.45	103.45	127
2920	8" diameter		33	.727		130	40	1.76	171.76	205
2930	10" diameter		25	.960		171	53	2.32	226.32	271
2940	12" diameter		21	1.143		214	63	2.77	279.77	335
2990	For small quantities, add					10%				
3430	To delete hanger, subtract									
3440	1/4" diam. to 3/4" diam.					6%	22%			
3450	1" diam. to 2" diam.					3%	15%			
3460	2-1/2" diam. to 5" diam.					2%	8%			
3470	6" diam. to 12" diam.					2%	4%			
3500	Threaded, couplings and clevis hanger assemblies, 10' O.C.									
3520	Schedule 40, type 304									
3540	1/4" diameter	1 Plum	54	.148	L.F.	10.45	8.75		19.20	25
3550	3/8" diameter		53	.151		11.65	8.95		20.60	26.50
3560	1/2" diameter		52	.154		13.55	9.10		22.65	28.50
3570	3/4" diameter		51	.157		14.05	9.30		23.35	29.50
3580	1" diameter		45	.178		20.50	10.50		31	38.50
3590	1-1/4" diameter	Q-1	76	.211		28.50	11.20		39.70	48.50
3600	1-1/2" diameter		69	.232		31	12.35		43.35	52.50
3610	2" diameter		57	.281		47.50	14.95		62.45	74.50
3620	2-1/2" diameter		44	.364		81	19.35		100.35	118
3630	3" diameter		38	.421		105	22.50		127.50	149
3640	4" diameter	Q-2	51	.471		141	26		167	194
3740	For small quantities, add					10%				

For customer support on your Mechanical Cost Data, call 877.739.3689.

187

22 11 13.64 Pipe, Stainless Steel		Crew	Daily Output	Labor-Hours	Unit	Material	2016 Bare Costs Labor	Equipment	Total	Total Incl O&P
4200	To delete couplings & hangers, subtract									
4210	1/4" diam. to 3/4" diam.					15%	56%			
4220	1" diam. to 2" diam.					18%	49%			
4230	2-1/2" diam. to 4" diam.					34%	40%			
4250	Schedule 40, type 316									
4290	1/4" diameter	1 Plum	54	.148	L.F.	11.70	8.75		20.45	26
4300	3/8" diameter		53	.151		12.75	8.95		21.70	27.50
4310	1/2" diameter		52	.154		15.55	9.10		24.65	31
4320	3/4" diameter		51	.157		18.05	9.30		27.35	34
4330	1" diameter		45	.178		25.50	10.50		36	44
4340	1-1/4" diameter	Q-1	76	.211		34	11.20		45.20	54.50
4350	1-1/2" diameter		69	.232		39	12.35		51.35	61
4360	2" diameter		57	.281		53.50	14.95		68.45	81.50
4370	2-1/2" diameter		44	.364		91.50	19.35		110.85	130
4380	3" diameter		38	.421		117	22.50		139.50	163
4390	4" diameter	Q-2	51	.471		155	26		181	209
4490	For small quantities, add					10%				
4900	To delete couplings & hangers, subtract									
4910	1/4" diam. to 3/4" diam.					12%	56%			
4920	1" diam. to 2" diam.					14%	49%			
4930	2-1/2" diam. to 4" diam.					27%	40%			
5000	Schedule 80, type 304									
5040	1/4" diameter	1 Plum	53	.151	L.F.	18.15	8.95		27.10	33.50
5050	3/8" diameter		52	.154		20.50	9.10		29.60	37
5060	1/2" diameter		51	.157		25	9.30		34.30	41.50
5070	3/4" diameter		48	.167		26.50	9.85		36.35	44
5080	1" diameter		43	.186		33	11		44	53
5090	1-1/4" diameter	Q-1	73	.219		44.50	11.70		56.20	66.50
5100	1-1/2" diameter		67	.239		56	12.70		68.70	81
5110	2" diameter		54	.296		69.50	15.80		85.30	101
5190	For small quantities, add					10%				
5700	To delete couplings & hangers, subtract									
5710	1/4" diam. to 3/4" diam.					10%	53%			
5720	1" diam. to 2" diam.					14%	47%			
5750	Schedule 80, type 316									
5790	1/4" diameter	1 Plum	53	.151	L.F.	21.50	8.95		30.45	37
5800	3/8" diameter		52	.154		26.50	9.10		35.60	43
5810	1/2" diameter		51	.157		31	9.30		40.30	48.50
5820	3/4" diameter		48	.167		34.50	9.85		44.35	52.50
5830	1" diameter		43	.186		44.50	11		55.50	65.50
5840	1-1/4" diameter	Q-1	73	.219		64	11.70		75.70	88
5850	1-1/2" diameter		67	.239		68.50	12.70		81.20	94.50
5860	2" diameter		54	.296		86	15.80		101.80	119
5950	For small quantities, add					10%				
7000	To delete couplings & hangers, subtract									
7010	1/4" diam. to 3/4" diam.					9%	53%			
7020	1" diam. to 2" diam.					14%	47%			
8000	Weld joints with clevis type hanger assemblies, 10' O.C.									
8010	Schedule 40, type 304									
8050	1/8" pipe size	Q-15	126	.127	L.F.	8.40	6.75	.46	15.61	19.95
8060	1/4" pipe size		125	.128		9.20	6.80	.46	16.46	21
8070	3/8" pipe size		122	.131		10.15	7	.48	17.63	22.50
8080	1/2" pipe size		118	.136		11.55	7.20	.49	19.24	24

For customer support on your Mechanical Cost Data, call 877.739.3689.

22 11 13 – Facility Water Distribution Piping

22 11 13.64 Pipe, Stainless Steel	Crew	Daily Output	Labor-Hours	Unit	Material	2016 Bare Costs Labor	Equipment	Total	Total Incl O&P
8090	3/4" pipe size	Q-15	109	.147	L.F.	11.40	7.80	.53	19.73	25
8100	1" pipe size		95	.168		16.05	8.95	.61	25.61	32
8110	1-1/4" pipe size		86	.186		21.50	9.90	.68	32.08	39
8120	1-1/2" pipe size		78	.205		23	10.95	.74	34.69	43
8130	2" pipe size		62	.258		34.50	13.75	.94	49.19	60
8140	2-1/2" pipe size		49	.327		51.50	17.40	1.19	70.09	84.50
8150	3" pipe size		44	.364		64.50	19.35	1.32	85.17	101
8160	3-1/2" pipe size		44	.364		74	19.35	1.32	94.67	112
8170	4" pipe size		39	.410		84.50	22	1.49	107.99	128
8180	5" pipe size		32	.500		114	26.50	1.82	142.32	168
8190	6" pipe size	Q-16	37	.649		136	36	1.57	173.57	206
8200	8" pipe size		29	.828		185	45.50	2	232.50	274
8210	10" pipe size		24	1		310	55.50	2.42	367.92	425
8220	12" pipe size		20	1.200		500	66.50	2.90	569.40	655
8300	Schedule 40, type 316									
8310	1/8" pipe size	Q-15	126	.127	L.F.	9.25	6.75	.46	16.46	21
8320	1/4" pipe size		125	.128		10.20	6.80	.46	17.46	22
8330	3/8" pipe size		122	.131		10.95	7	.48	18.43	23
8340	1/2" pipe size		118	.136		13.20	7.20	.49	20.89	26
8350	3/4" pipe size		109	.147		14.85	7.80	.53	23.18	28.50
8360	1" pipe size		95	.168		20.50	8.95	.61	30.06	36.50
8370	1-1/4" pipe size		86	.186		26	9.90	.68	36.58	44
8380	1-1/2" pipe size		78	.205		29.50	10.95	.74	41.19	50
8390	2" pipe size		62	.258		38.50	13.75	.94	53.19	64
8400	2-1/2" pipe size		49	.327		56	17.40	1.19	74.59	90
8410	3" pipe size		44	.364		69	19.35	1.32	89.67	106
8420	3-1/2" pipe size		44	.364		80.50	19.35	1.32	101.17	119
8430	4" pipe size		39	.410		87	22	1.49	110.49	130
8440	5" pipe size		32	.500		120	26.50	1.82	148.32	174
8450	6" pipe size	Q-16	37	.649		138	36	1.57	175.57	208
8460	8" pipe size		29	.828		194	45.50	2	241.50	285
8470	10" pipe size		24	1		245	55.50	2.42	302.92	355
8480	12" pipe size		20	1.200		330	66.50	2.90	399.40	470
8500	Schedule 80, type 304									
8510	1/4" pipe size	Q-15	110	.145	L.F.	16.75	7.75	.53	25.03	30.50
8520	3/8" pipe size		109	.147		19.05	7.80	.53	27.38	33.50
8530	1/2" pipe size		106	.151		23	8.05	.55	31.60	38.50
8540	3/4" pipe size		96	.167		24	8.90	.61	33.51	40.50
8550	1" pipe size		87	.184		28.50	9.80	.67	38.97	47
8560	1-1/4" pipe size		81	.198		34	10.50	.72	45.22	53.50
8570	1-1/2" pipe size		74	.216		43.50	11.50	.78	55.78	66.50
8580	2" pipe size		58	.276		53.50	14.70	1	69.20	81.50
8590	2-1/2" pipe size		46	.348		68	18.55	1.26	87.81	104
8600	3" pipe size		41	.390		78.50	21	1.42	100.92	120
8610	4" pipe size		33	.485		96.50	26	1.76	124.26	147
8630	6" pipe size	Q-16	30	.800		171	44	1.94	216.94	257
8640	Schedule 80, type 316									
8650	1/4" pipe size	Q-15	110	.145	L.F.	19.95	7.75	.53	28.23	34.50
8660	3/8" pipe size		109	.147		24.50	7.80	.53	32.83	39.50
8670	1/2" pipe size		106	.151		29	8.05	.55	37.60	44.50
8680	3/4" pipe size		96	.167		30.50	8.90	.61	40.01	47.50
8690	1" pipe size		87	.184		38.50	9.80	.67	48.97	58
8700	1-1/4" pipe size		81	.198		50.50	10.50	.72	61.72	72

For customer support on your Mechanical Cost Data, call 877.739.3689.

189

22 11 13 – Facility Water Distribution Piping

22 11 13.64 Pipe, Stainless Steel		Crew	Daily Output	Labor-Hours	Unit	Material	2016 Bare Costs Labor	Equipment	Total	Total Incl O&P
8710	1-1/2" pipe size	Q-15	74	.216	L.F.	53	11.50	.78	65.28	76.50
8720	2" pipe size		58	.276		64.50	14.70	1	80.20	93.50
8730	2-1/2" pipe size		46	.348		80	18.55	1.26	99.81	117
8740	3" pipe size		41	.390		92.50	21	1.42	114.92	135
8760	4" pipe size	↓	33	.485		122	26	1.76	149.76	175
8770	6" pipe size	Q-16	30	.800	↓	219	44	1.94	264.94	310
9100	Threading pipe labor, sst, one end, schedules 40 & 80									
9110	1/4" through 3/4" pipe size	1 Plum	61.50	.130	Ea.		7.70		7.70	11.60
9120	1" through 2" pipe size		55.90	.143			8.45		8.45	12.80
9130	2-1/2" pipe size		41.50	.193			11.40		11.40	17.20
9140	3" pipe size	↓	38.50	.208			12.30		12.30	18.55
9150	3-1/2" pipe size	Q-1	68.40	.234			12.45		12.45	18.80
9160	4" pipe size		73	.219			11.70		11.70	17.60
9170	5" pipe size		40.70	.393			21		21	31.50
9180	6" pipe size		35.40	.452			24		24	36.50
9190	8" pipe size		22.30	.717			38		38	57.50
9200	10" pipe size		16.10	.994			53		53	80
9210	12" pipe size	↓	12.30	1.301	↓		69.50		69.50	105
9250	Welding labor per joint for stainless steel									
9260	Schedule 5 and 10									
9270	1/4" pipe size	Q-15	36	.444	Ea.		23.50	1.61	25.11	37.50
9280	3/8" pipe size		35	.457			24.50	1.66	26.16	39
9290	1/2" pipe size		35	.457			24.50	1.66	26.16	39
9300	3/4" pipe size		28	.571			30.50	2.07	32.57	48.50
9310	1" pipe size		25	.640			34	2.32	36.32	54
9320	1-1/4" pipe size		22	.727			39	2.64	41.64	61.50
9330	1-1/2" pipe size		21	.762			40.50	2.77	43.27	64.50
9340	2" pipe size		18	.889			47.50	3.23	50.73	75
9350	2-1/2" pipe size		12	1.333			71	4.84	75.84	112
9360	3" pipe size		9.73	1.644			87.50	5.95	93.45	139
9370	4" pipe size		7.37	2.171			116	7.90	123.90	184
9380	5" pipe size		6.15	2.602			139	9.45	148.45	219
9390	6" pipe size		5.71	2.802			149	10.15	159.15	236
9400	8" pipe size		3.69	4.336			231	15.75	246.75	365
9410	10" pipe size		2.91	5.498			293	19.95	312.95	460
9420	12" pipe size	↓	2.31	6.926	↓		370	25	395	585
9500	Schedule 40									
9510	1/4" pipe size	Q-15	28	.571	Ea.		30.50	2.07	32.57	48.50
9520	3/8" pipe size		27	.593			31.50	2.15	33.65	50
9530	1/2" pipe size		25.40	.630			33.50	2.29	35.79	53
9540	3/4" pipe size		22.22	.720			38.50	2.61	41.11	61
9550	1" pipe size		20.25	.790			42	2.87	44.87	66.50
9560	1-1/4" pipe size		18.82	.850			45.50	3.09	48.59	72
9570	1-1/2" pipe size		17.78	.900			48	3.27	51.27	76
9580	2" pipe size		15.09	1.060			56.50	3.85	60.35	89.50
9590	2-1/2" pipe size		7.96	2.010			107	7.30	114.30	170
9600	3" pipe size		6.43	2.488			133	9.05	142.05	210
9610	4" pipe size		4.88	3.279			175	11.90	186.90	277
9620	5" pipe size		4.26	3.756			200	13.65	213.65	315
9630	6" pipe size		3.77	4.244			226	15.40	241.40	355
9640	8" pipe size		2.44	6.557			350	24	374	550
9650	10" pipe size		1.92	8.333			445	30.50	475.50	705
9660	12" pipe size	↓	1.52	10.526	↓		560	38	598	885

22 11 13.64 Pipe, Stainless Steel

		Crew	Daily Output	Labor-Hours	Unit	Material	2016 Bare Costs Labor	Equipment	Total	Total Incl O&P
9750	Schedule 80									
9760	1/4" pipe size	Q-15	21.55	.742	Ea.		39.50	2.70	42.20	62.50
9770	3/8" pipe size		20.75	.771			41	2.80	43.80	65
9780	1/2" pipe size		19.54	.819			43.50	2.97	46.47	69.50
9790	3/4" pipe size		17.09	.936			50	3.40	53.40	79
9800	1" pipe size		15.58	1.027			54.50	3.73	58.23	86.50
9810	1-1/4" pipe size		14.48	1.105			59	4.01	63.01	93.50
9820	1-1/2" pipe size		13.68	1.170			62.50	4.25	66.75	98.50
9830	2" pipe size		11.61	1.378			73.50	5	78.50	117
9840	2-1/2" pipe size		6.12	2.614			139	9.50	148.50	220
9850	3" pipe size		4.94	3.239			173	11.75	184.75	273
9860	4" pipe size		3.75	4.267			227	15.50	242.50	360
9870	5" pipe size		3.27	4.893			261	17.75	278.75	415
9880	6" pipe size		2.90	5.517			294	20	314	465
9890	8" pipe size		1.87	8.556			455	31	486	725
9900	10" pipe size		1.48	10.811			575	39	614	915
9910	12" pipe size		1.17	13.675			730	49.50	779.50	1,150
9920	Schedule 160, 1/2" pipe size		17	.941			50	3.42	53.42	79.50
9930	3/4" pipe size		14.81	1.080			57.50	3.92	61.42	91.50
9940	1" pipe size		13.50	1.185			63	4.30	67.30	100
9950	1-1/4" pipe size		12.55	1.275			68	4.63	72.63	108
9960	1-1/2" pipe size		11.85	1.350			72	4.90	76.90	114
9970	2" pipe size		10	1.600			85.50	5.80	91.30	135
9980	3" pipe size		4.28	3.738			199	13.55	212.55	315
9990	4" pipe size		3.25	4.923			262	17.85	279.85	415

22 11 13.66 Pipe Fittings, Stainless Steel

		Crew	Daily Output	Labor-Hours	Unit	Material	2016 Bare Costs Labor	Equipment	Total	Total Incl O&P
0010	**PIPE FITTINGS, STAINLESS STEEL**									
0100	Butt weld joint, schedule 5, type 304									
0120	90° Elbow, long									
0140	1/2"	Q-15	17.50	.914	Ea.	19.80	48.50	3.32	71.62	99
0150	3/4"		14	1.143		19.80	61	4.15	84.95	119
0160	1"		12.50	1.280		21	68	4.65	93.65	131
0170	1-1/4"		11	1.455		28.50	77.50	5.30	111.30	154
0180	1-1/2"		10.50	1.524		24	81	5.55	110.55	156
0190	2"		9	1.778		28.50	94.50	6.45	129.45	182
0200	2-1/2"		6	2.667		66	142	9.70	217.70	297
0210	3"		4.86	3.292		72.50	175	11.95	259.45	360
0220	3-1/2"		4.27	3.747		182	200	13.60	395.60	515
0230	4"		3.69	4.336		99	231	15.75	345.75	475
0240	5"		3.08	5.195		385	277	18.85	680.85	865
0250	6"	Q-16	4.29	5.594		297	310	13.55	620.55	805
0260	8"		2.76	8.696		625	480	21	1,126	1,450
0270	10"		2.18	11.009		970	610	26.50	1,606.50	2,025
0280	12"		1.73	13.873		1,375	765	33.50	2,173.50	2,700
0320	For schedule 5, type 316, add					30%				
0600	45° Elbow, long									
0620	1/2"	Q-15	17.50	.914	Ea.	19.80	48.50	3.32	71.62	99
0630	3/4"		14	1.143		19.80	61	4.15	84.95	119
0640	1"		12.50	1.280		21	68	4.65	93.65	131
0650	1-1/4"		11	1.455		28.50	77.50	5.30	111.30	154
0660	1-1/2"		10.50	1.524		24	81	5.55	110.55	156
0670	2"		9	1.778		28.50	94.50	6.45	129.45	182

22 11 13 – Facility Water Distribution Piping

22 11 13.66 Pipe Fittings, Stainless Steel		Crew	Daily Output	Labor-Hours	Unit	Material	2016 Bare Costs Labor	Equipment	Total	Total Incl O&P
0680	2-1/2"	Q-15	6	2.667	Ea.	66	142	9.70	217.70	297
0690	3"		4.86	3.292		72.50	175	11.95	259.45	360
0700	3-1/2"		4.27	3.747		182	200	13.60	395.60	515
0710	4"		3.69	4.336		83.50	231	15.75	330.25	460
0720	5"	↓	3.08	5.195		310	277	18.85	605.85	780
0730	6"	Q-16	4.29	5.594		209	310	13.55	532.55	710
0740	8"		2.76	8.696		440	480	21	941	1,225
0750	10"		2.18	11.009		770	610	26.50	1,406.50	1,800
0760	12"	↓	1.73	13.873		965	765	33.50	1,763.50	2,225
0800	For schedule 5, type 316, add				↓	25%				
1100	Tee, straight									
1130	1/2"	Q-15	11.66	1.372	Ea.	59.50	73	4.98	137.48	181
1140	3/4"		9.33	1.715		59.50	91.50	6.25	157.25	210
1150	1"		8.33	1.921		62.50	102	6.95	171.45	231
1160	1-1/4"		7.33	2.183		50.50	116	7.90	174.40	240
1170	1-1/2"		7	2.286		49.50	122	8.30	179.80	248
1180	2"		6	2.667		51.50	142	9.70	203.20	282
1190	2-1/2"		4	4		124	213	14.50	351.50	475
1200	3"		3.24	4.938		198	263	17.95	478.95	635
1210	3-1/2"		2.85	5.614		260	299	20.50	579.50	760
1220	4"		2.46	6.504		141	345	23.50	509.50	705
1230	5"	↓	2	8		445	425	29	899	1,175
1240	6"	Q-16	2.85	8.421		350	465	20.50	835.50	1,100
1250	8"		1.84	13.043		750	720	31.50	1,501.50	1,950
1260	10"		1.45	16.552		1,200	915	40	2,155	2,750
1270	12"	↓	1.15	20.870		1,675	1,150	50.50	2,875.50	3,650
1320	For schedule 5, type 316, add				↓	25%				
2000	Butt weld joint, schedule 10, type 304									
2020	90° elbow, long									
2040	1/2"	Q-15	17	.941	Ea.	13.50	50	3.42	66.92	94
2050	3/4"		14	1.143		13.50	61	4.15	78.65	111
2060	1"		12.50	1.280		14.25	68	4.65	86.90	124
2070	1-1/4"		11	1.455		19.50	77.50	5.30	102.30	144
2080	1-1/2"		10.50	1.524		16.50	81	5.55	103.05	147
2090	2"		9	1.778		19.50	94.50	6.45	120.45	172
2100	2-1/2"		6	2.667		45	142	9.70	196.70	274
2110	3"		4.86	3.292		42	175	11.95	228.95	325
2120	3-1/2"		4.27	3.747		124	200	13.60	337.60	450
2130	4"		3.69	4.336		70.50	231	15.75	317.25	445
2140	5"	↓	3.08	5.195		263	277	18.85	558.85	730
2150	6"	Q-16	4.29	5.594		203	310	13.55	526.55	705
2160	8"		2.76	8.696		430	480	21	931	1,225
2170	10"		2.18	11.009		660	610	26.50	1,296.50	1,675
2180	12"	↓	1.73	13.873	↓	940	765	33.50	1,738.50	2,200
2500	45° elbow, long									
2520	1/2"	Q-15	17.50	.914	Ea.	13.50	48.50	3.32	65.32	92
2530	3/4"		14	1.143		13.50	61	4.15	78.65	111
2540	1"		12.50	1.280		14.25	68	4.65	86.90	124
2550	1-1/4"		11	1.455		19.50	77.50	5.30	102.30	144
2560	1-1/2"		10.50	1.524		16.50	81	5.55	103.05	147
2570	2"		9	1.778		19.50	94.50	6.45	120.45	172
2580	2-1/2"		6	2.667		45	142	9.70	196.70	274
2590	3"		4.86	3.292		34	175	11.95	220.95	315

22 11 13.66 Pipe Fittings, Stainless Steel	Crew	Daily Output	Labor-Hours	Unit	Material	2016 Bare Costs Labor	Equipment	Total	Total Incl O&P
2600	3-1/2"	Q-15	4.27	3.747	Ea.	124	200	13.60	337.60	450
2610	4"		3.69	4.336		57	231	15.75	303.75	430
2620	5"		3.08	5.195		210	277	18.85	505.85	670
2630	6"	Q-16	4.29	5.594		143	310	13.55	466.55	635
2640	8"		2.76	8.696		300	480	21	801	1,075
2650	10"		2.18	11.009		525	610	26.50	1,161.50	1,525
2660	12"		1.73	13.873		655	765	33.50	1,453.50	1,900
2670	Reducer, concentric									
2674	1" x 3/4"	Q-15	13.25	1.208	Ea.	29.50	64.50	4.38	98.38	134
2676	2" x 1-1/2"	"	9.75	1.641		24	87.50	5.95	117.45	165
2678	6" x 4"	Q-16	4.91	4.888		72	270	11.85	353.85	500
2680	Caps									
2682	1"	Q-15	25	.640	Ea.	30	34	2.32	66.32	87
2684	1-1/2"		21	.762		36	40.50	2.77	79.27	104
2685	2"		18	.889		34	47.50	3.23	84.73	112
2686	4"		7.38	2.168		54	116	7.85	177.85	242
2687	6"	Q-16	8.58	2.797		86.50	155	6.75	248.25	335
3000	Tee, straight									
3030	1/2"	Q-15	11.66	1.372	Ea.	40.50	73	4.98	118.48	160
3040	3/4"		9.33	1.715		40.50	91.50	6.25	138.25	189
3050	1"		8.33	1.921		43	102	6.95	151.95	209
3060	1-1/4"		7.33	2.183		49	116	7.90	172.90	238
3070	1-1/2"		7	2.286		49	122	8.30	179.30	247
3080	2"		6	2.667		51	142	9.70	202.70	281
3090	2-1/2"		4	4		35.50	213	14.50	263	375
3100	3"		3.24	4.938		67	263	17.95	347.95	490
3110	3-1/2"		2.85	5.614		98.50	299	20.50	418	580
3120	4"		2.46	6.504		96	345	23.50	464.50	655
3130	5"		2	8		305	425	29	759	1,000
3140	6"	Q-16	2.85	8.421		240	465	20.50	725.50	985
3150	8"		1.84	13.043		510	720	31.50	1,261.50	1,700
3151	10"		1.45	16.552		825	915	40	1,780	2,325
3152	12"		1.15	20.870		1,150	1,150	50.50	2,350.50	3,050
3154	For schedule 10, type 316, add					25%				
3281	Butt weld joint, schedule 40, type 304									
3284	90° Elbow, long, 1/2"	Q-15	12.70	1.260	Ea.	15.75	67	4.57	87.32	123
3288	3/4"		11.10	1.441		15.75	77	5.25	98	139
3289	1"		10.13	1.579		16.50	84	5.75	106.25	151
3290	1-1/4"		9.40	1.702		22	90.50	6.20	118.70	168
3300	1-1/2"		8.89	1.800		17.25	96	6.55	119.80	171
3310	2"		7.55	2.119		25	113	7.70	145.70	205
3320	2-1/2"		3.98	4.020		47.50	214	14.60	276.10	395
3330	3"		3.21	4.984		61	266	18.10	345.10	485
3340	3-1/2"		2.83	5.654		225	300	20.50	545.50	725
3350	4"		2.44	6.557		105	350	24	479	665
3360	5"		2.13	7.512		355	400	27.50	782.50	1,025
3370	6"	Q-16	2.83	8.481		310	470	20.50	800.50	1,075
3380	8"		1.83	13.115		615	725	31.50	1,371.50	1,800
3390	10"		1.44	16.667		1,300	920	40.50	2,260.50	2,875
3400	12"		1.14	21.053		1,650	1,175	51	2,876	3,625
3410	For schedule 40, type 316, add					25%				
3460	45° Elbow, long, 1/2"	Q-15	12.70	1.260	Ea.	15.75	67	4.57	87.32	123
3470	3/4"		11.10	1.441		15.75	77	5.25	98	139

22 11 13.66 Pipe Fittings, Stainless Steel		Crew	Daily Output	Labor-Hours	Unit	Material	2016 Bare Costs Labor	Equipment	Total	Total Incl O&P
3480	1"	Q-15	10.13	1.579	Ea.	16.50	84	5.75	106.25	151
3490	1-1/4"		9.40	1.702		22	90.50	6.20	118.70	168
3500	1-1/2"		8.89	1.800		17.25	96	6.55	119.80	171
3510	2"		7.55	2.119		25	113	7.70	145.70	205
3520	2-1/2"		3.98	4.020		47.50	214	14.60	276.10	395
3530	3"		3.21	4.984		48	266	18.10	332.10	475
3540	3-1/2"		2.83	5.654		225	300	20.50	545.50	725
3550	4"		2.44	6.557		75	350	24	449	635
3560	5"		2.13	7.512		248	400	27.50	675.50	905
3570	6"	Q-16	2.83	8.481		218	470	20.50	708.50	965
3580	8"		1.83	13.115		430	725	31.50	1,186.50	1,600
3590	10"		1.44	16.667		910	920	40.50	1,870.50	2,450
3600	12"		1.14	21.053		1,150	1,175	51	2,376	3,075
3610	For schedule 40, type 316, add					25%				
3660	Tee, straight 1/2"	Q-15	8.46	1.891	Ea.	40.50	101	6.85	148.35	204
3670	3/4"		7.40	2.162		40.50	115	7.85	163.35	227
3680	1"		6.74	2.374		43	126	8.60	177.60	247
3690	1-1/4"		6.27	2.552		100	136	9.25	245.25	325
3700	1-1/2"		5.92	2.703		46	144	9.80	199.80	278
3710	2"		5.03	3.181		68.50	169	11.55	249.05	345
3720	2-1/2"		2.65	6.038		86.50	320	22	428.50	605
3730	3"		2.14	7.477		87	400	27	514	725
3740	3-1/2"		1.88	8.511		164	455	31	650	900
3750	4"		1.62	9.877		164	525	36	725	1,025
3760	5"		1.42	11.268		380	600	41	1,021	1,375
3770	6"	Q-16	1.88	12.766		365	705	31	1,101	1,500
3780	8"		1.22	19.672		735	1,075	47.50	1,857.50	2,525
3790	10"		.96	25		1,425	1,375	60.50	2,860.50	3,725
3800	12"		.76	31.579		1,900	1,750	76.50	3,726.50	4,800
3810	For schedule 40, type 316, add					25%				
3820	Tee, reducing on outlet, 3/4" x 1/2"	Q-15	7.73	2.070	Ea.	59.50	110	7.50	177	240
3822	1" x 1/2"		7.24	2.210		58.50	118	8	184.50	251
3824	1" x 3/4"		6.96	2.299		54	122	8.35	184.35	254
3826	1-1/4" x 1"		6.43	2.488		132	133	9.05	274.05	355
3828	1-1/2" x 1/2"		6.58	2.432		79.50	130	8.85	218.35	293
3830	1-1/2" x 3/4"		6.35	2.520		74.50	134	9.15	217.65	295
3832	1-1/2" x 1"		6.18	2.589		57	138	9.40	204.40	281
3834	2" x 1"		5.50	2.909		102	155	10.55	267.55	360
3836	2" x 1-1/2"		5.30	3.019		85.50	161	10.95	257.45	350
3838	2-1/2" x 2"		3.15	5.079		129	271	18.45	418.45	575
3840	3" x 1-1/2"		2.72	5.882		131	315	21.50	467.50	645
3842	3" x 2"		2.65	6.038		109	320	22	451	630
3844	4" x 2"		2.10	7.619		272	405	27.50	704.50	945
3846	4" x 3"		1.77	9.040		197	480	33	710	980
3848	5" x 4"		1.48	10.811		455	575	39	1,069	1,425
3850	6" x 3"	Q-16	2.19	10.959		505	605	26.50	1,136.50	1,500
3852	6" x 4"		2.04	11.765		435	650	28.50	1,113.50	1,500
3854	8" x 4"		1.46	16.438		1,025	910	40	1,975	2,550
3856	10" x 8"		.69	34.783		1,725	1,925	84	3,734	4,875
3858	12" x 10"		.55	43.636		2,275	2,400	106	4,781	6,300
3950	Reducer, concentric, 3/4" x 1/2"	Q-15	11.85	1.350		30	72	4.90	106.90	147
3952	1" x 3/4"		10.60	1.509		31.50	80.50	5.50	117.50	162
3954	1-1/4" x 3/4"		10.19	1.570		77.50	83.50	5.70	166.70	217

22 11 13.66 Pipe Fittings, Stainless Steel	Crew	Daily Output	Labor-Hours	Unit	Material	2016 Bare Costs Labor	Equipment	Total	Total Incl O&P
3956	1-1/4" x 1"	Q-15	9.76	1.639	Ea.	39	87.50	5.95	132.45	182
3958	1-1/2" x 3/4"		9.88	1.619		65.50	86.50	5.90	157.90	208
3960	1-1/2" x 1"		9.47	1.690		52.50	90	6.15	148.65	201
3962	2" x 1"		8.65	1.850		30	98.50	6.70	135.20	189
3964	2" x 1-1/2"		8.16	1.961		26.50	104	7.10	137.60	195
3966	2-1/2" x 1"		5.71	2.802		113	149	10.15	272.15	360
3968	2-1/2" x 2"		5.21	3.071		58	164	11.15	233.15	325
3970	3" x 1"		4.88	3.279		78	175	11.90	264.90	365
3972	3" x 1-1/2"		4.72	3.390		40	181	12.30	233.30	330
3974	3" x 2"		4.51	3.548		37.50	189	12.90	239.40	340
3976	4" x 2"		3.69	4.336		53.50	231	15.75	300.25	425
3978	4" x 3"		2.77	5.776		40	310	21	371	530
3980	5" x 3"		2.56	6.250		237	335	22.50	594.50	790
3982	5" x 4"		2.27	7.048		191	375	25.50	591.50	805
3984	6" x 3"	Q-16	3.57	6.723		122	370	16.25	508.25	710
3986	6" x 4"		3.19	7.524		101	415	18.20	534.20	755
3988	8" x 4"		2.44	9.836		315	545	24	884	1,200
3990	8" x 6"		2.22	10.811		236	595	26	857	1,200
3992	10" x 6"		1.91	12.565		440	695	30.50	1,165.50	1,575
3994	10" x 8"		1.61	14.907		365	825	36	1,226	1,700
3995	12" x 6"		1.63	14.724		675	815	35.50	1,525.50	2,000
3996	12" x 8"		1.41	17.021		620	940	41	1,601	2,150
3997	12" x 10"		1.27	18.898		435	1,050	45.50	1,530.50	2,100
4000	Socket weld joint, 3000 lb., type 304									
4100	90° Elbow									
4140	1/4"	Q-15	13.47	1.188	Ea.	44	63.50	4.31	111.81	149
4150	3/8"		12.97	1.234		57	65.50	4.48	126.98	167
4160	1/2"		12.21	1.310		62.50	70	4.76	137.26	179
4170	3/4"		10.68	1.498		72	80	5.45	157.45	205
4180	1"		9.74	1.643		108	87.50	5.95	201.45	258
4190	1-1/4"		9.05	1.768		189	94	6.40	289.40	355
4200	1-1/2"		8.55	1.871		229	99.50	6.80	335.30	410
4210	2"		7.26	2.204		370	117	8	495	590
4300	45° Elbow									
4340	1/4"	Q-15	13.47	1.188	Ea.	83	63.50	4.31	150.81	191
4350	3/8"		12.97	1.234		83	65.50	4.48	152.98	195
4360	1/2"		12.21	1.310		83	70	4.76	157.76	201
4370	3/4"		10.68	1.498		94	80	5.45	179.45	229
4380	1"		9.74	1.643		136	87.50	5.95	229.45	289
4390	1-1/4"		9.05	1.768		224	94	6.40	324.40	395
4400	1-1/2"		8.55	1.871		225	99.50	6.80	331.30	405
4410	2"		7.26	2.204		410	117	8	535	635
4500	Tee									
4540	1/4"	Q-15	8.97	1.784	Ea.	58	95	6.45	159.45	214
4550	3/8"		8.64	1.852		69	98.50	6.70	174.20	232
4560	1/2"		8.13	1.968		85	105	7.15	197.15	259
4570	3/4"		7.12	2.247		98	120	8.15	226.15	298
4580	1"		6.48	2.469		132	132	8.95	272.95	355
4590	1-1/4"		6.03	2.653		234	141	9.65	384.65	480
4600	1-1/2"		5.69	2.812		335	150	10.20	495.20	605
4610	2"		4.83	3.313		510	177	12	699	840
5000	Socket weld joint, 3000 lb., type 316									
5100	90° Elbow									

For customer support on your Mechanical Cost Data, call 877.739.3689.

195

22 11 13.66 Pipe Fittings, Stainless Steel	Crew	Daily Output	Labor-Hours	Unit	Material	2016 Bare Costs Labor	Equipment	Total	Total Incl O&P
5140	1/4"	Q-15	13.47	1.188	Ea.	54	63.50	4.31	121.81	159
5150	3/8"		12.97	1.234		63	65.50	4.48	132.98	173
5160	1/2"		12.21	1.310		76	70	4.76	150.76	194
5170	3/4"		10.68	1.498		100	80	5.45	185.45	236
5180	1"		9.74	1.643		142	87.50	5.95	235.45	295
5190	1-1/4"		9.05	1.768		253	94	6.40	353.40	425
5200	1-1/2"		8.55	1.871		286	99.50	6.80	392.30	470
5210	2"		7.26	2.204		485	117	8	610	720
5300	45° Elbow									
5340	1/4"	Q-15	13.47	1.188	Ea.	104	63.50	4.31	171.81	214
5350	3/8"		12.97	1.234		104	65.50	4.48	173.98	218
5360	1/2"		12.21	1.310		109	70	4.76	183.76	230
5370	3/4"		10.68	1.498		121	80	5.45	206.45	259
5380	1"		9.74	1.643		182	87.50	5.95	275.45	340
5390	1-1/4"		9.05	1.768		260	94	6.40	360.40	435
5400	1-1/2"		8.55	1.871		292	99.50	6.80	398.30	475
5410	2"		7.26	2.204		435	117	8	560	665
5500	Tee									
5540	1/4"	Q-15	8.97	1.784	Ea.	73	95	6.45	174.45	231
5550	3/8"		8.64	1.852		89.50	98.50	6.70	194.70	254
5560	1/2"		8.13	1.968		99.50	105	7.15	211.65	275
5570	3/4"		7.12	2.247		123	120	8.15	251.15	325
5580	1"		6.48	2.469		188	132	8.95	328.95	415
5590	1-1/4"		6.03	2.653		299	141	9.65	449.65	555
5600	1-1/2"		5.69	2.812		420	150	10.20	580.20	700
5610	2"		4.83	3.313		670	177	12	859	1,025
5700	For socket weld joint, 6000 lb., type 304 and 316, add					100%				
6000	Threaded companion flange									
6010	Stainless steel, 150 lb., type 304									
6020	1/2" diam.	1 Plum	30	.267	Ea.	45	15.80		60.80	73.50
6030	3/4" diam.		28	.286		50	16.90		66.90	80.50
6040	1" diam.		27	.296		55	17.55		72.55	87
6050	1-1/4" diam.	Q-1	44	.364		71	19.35		90.35	107
6060	1-1/2" diam.		40	.400		71	21.50		92.50	110
6070	2" diam.		36	.444		93	23.50		116.50	138
6080	2-1/2" diam.		28	.571		130	30.50		160.50	189
6090	3" diam.		20	.800		136	42.50		178.50	215
6110	4" diam.		12	1.333		186	71		257	310
6130	6" diam.	Q-2	14	1.714		335	94.50		429.50	510
6140	8" diam.	"	12	2		645	111		756	875
6150	For type 316 add					40%				
6260	Weld flanges, stainless steel, type 304									
6270	Slip on, 150 lb. (welded, front and back)									
6280	1/2" diam.	Q-15	12.70	1.260	Ea.	40	67	4.57	111.57	150
6290	3/4" diam.		11.11	1.440		41	76.50	5.25	122.75	167
6300	1" diam.		10.13	1.579		45	84	5.75	134.75	183
6310	1-1/4" diam.		9.41	1.700		61	90.50	6.15	157.65	211
6320	1-1/2" diam.		8.89	1.800		61	96	6.55	163.55	219
6330	2" diam.		7.55	2.119		79	113	7.70	199.70	265
6340	2-1/2" diam.		3.98	4.020		111	214	14.60	339.60	465
6350	3" diam.		3.21	4.984		119	266	18.10	403.10	550
6370	4" diam.		2.44	6.557		162	350	24	536	730
6390	6" diam.	Q-16	1.89	12.698		247	700	30.50	977.50	1,350

For customer support on your Mechanical Cost Data, call 877.739.3689.

22 11 13.66 Pipe Fittings, Stainless Steel		Crew	Daily Output	Labor-Hours	Unit	Material	2016 Bare Costs Labor	Equipment	Total	Total Incl O&P
6400	8" diam.	Q-16	1.22	19.672	Ea.	465	1,075	47.50	1,587.50	2,225
6410	For type 316, add					40%				
6530	Weld neck 150 lb.									
6540	1/2" diam.	Q-15	25.40	.630	Ea.	20.50	33.50	2.29	56.29	75.50
6550	3/4" diam.		22.22	.720		25	38.50	2.61	66.11	88
6560	1" diam.		20.25	.790		27	42	2.87	71.87	96
6570	1-1/4" diam.		18.82	.850		43.50	45.50	3.09	92.09	119
6580	1-1/2" diam.		17.78	.900		38	48	3.27	89.27	118
6590	2" diam.		15.09	1.060		40.50	56.50	3.85	100.85	134
6600	2-1/2" diam.		7.96	2.010		65	107	7.30	179.30	242
6610	3" diam.		6.43	2.488		66.50	133	9.05	208.55	283
6630	4" diam.		4.88	3.279		101	175	11.90	287.90	390
6640	5" diam.		4.26	3.756		132	200	13.65	345.65	460
6650	6" diam.	Q-16	5.66	4.240		145	234	10.25	389.25	525
6652	8" diam.		3.65	6.575		261	365	15.90	641.90	855
6654	10" diam.		2.88	8.333		365	460	20	845	1,125
6656	12" diam.		2.28	10.526		705	580	25.50	1,310.50	1,675
6670	For type 316 add					23%				
7000	Threaded joint, 150 lb., type 304									
7030	90° elbow									
7040	1/8"	1 Plum	13	.615	Ea.	27.50	36.50		64	85
7050	1/4"		13	.615		27.50	36.50		64	85
7070	3/8"		13	.615		32	36.50		68.50	90
7080	1/2"		12	.667		28.50	39.50		68	91
7090	3/4"		11	.727		34	43		77	103
7100	1"		10	.800		46.50	47.50		94	123
7110	1-1/4"	Q-1	17	.941		72	50		122	155
7120	1-1/2"		16	1		83.50	53.50		137	172
7130	2"		14	1.143		120	61		181	224
7140	2-1/2"		11	1.455		292	77.50		369.50	435
7150	3"		8	2		425	107		532	630
7160	4"	Q-2	11	2.182		725	121		846	975
7180	45° elbow									
7190	1/8"	1 Plum	13	.615	Ea.	40	36.50		76.50	99
7200	1/4"		13	.615		40	36.50		76.50	99
7210	3/8"		13	.615		40.50	36.50		77	99.50
7220	1/2"		12	.667		41	39.50		80.50	105
7230	3/4"		11	.727		44	43		87	114
7240	1"		10	.800		51	47.50		98.50	128
7250	1-1/4"	Q-1	17	.941		71.50	50		121.50	154
7260	1-1/2"		16	1		93	53.50		146.50	183
7270	2"		14	1.143		131	61		192	236
7280	2-1/2"		11	1.455		415	77.50		492.50	570
7290	3"		8	2		605	107		712	825
7300	4"	Q-2	11	2.182		1,100	121		1,221	1,375
7320	Tee, straight									
7330	1/8"	1 Plum	9	.889	Ea.	42	52.50		94.50	126
7340	1/4"		9	.889		42	52.50		94.50	126
7350	3/8"		9	.889		45	52.50		97.50	129
7360	1/2"		8	1		42.50	59		101.50	137
7370	3/4"		7	1.143		45.50	67.50		113	152
7380	1"		6.50	1.231		58.50	73		131.50	175
7390	1-1/4"	Q-1	11	1.455		98	77.50		175.50	225

For customer support on your Mechanical Cost Data, call 877.739.3689.

197

22 11 13.66 Pipe Fittings, Stainless Steel		Crew	Daily Output	Labor-Hours	Unit	Material	2016 Bare Costs Labor	Equipment	Total	Total Incl O&P
7400	1-1/2"	Q-1	10	1.600	Ea.	126	85.50		211.50	267
7410	2"		9	1.778		156	94.50		250.50	315
7420	2-1/2"		7	2.286		420	122		542	645
7430	3"		5	3.200		640	171		811	960
7440	4"	Q-2	7	3.429		1,575	189		1,764	2,025
7460	Coupling, straight									
7470	1/8"	1 Plum	19	.421	Ea.	10.55	25		35.55	49
7480	1/4"		19	.421		12.50	25		37.50	51.50
7490	3/8"		19	.421		14.95	25		39.95	54
7500	1/2"		19	.421		19.85	25		44.85	59.50
7510	3/4"		18	.444		26.50	26.50		53	68.50
7520	1"		15	.533		42.50	31.50		74	94.50
7530	1-1/4"	Q-1	26	.615		68.50	33		101.50	125
7540	1-1/2"		24	.667		77	35.50		112.50	138
7550	2"		21	.762		127	40.50		167.50	202
7560	2-1/2"		18	.889		296	47.50		343.50	395
7570	3"		14	1.143		400	61		461	535
7580	4"	Q-2	16	1.500		565	83		648	745
7600	Reducer, concentric, 1/2"	1 Plum	12	.667		21.50	39.50		61	83
7610	3/4"		11	.727		27.50	43		70.50	95.50
7612	1"		10	.800		45.50	47.50		93	122
7614	1-1/4"	Q-1	17	.941		97	50		147	183
7616	1-1/2"		16	1		106	53.50		159.50	198
7618	2"		14	1.143		165	61		226	274
7620	2-1/2"		11	1.455		435	77.50		512.50	595
7622	3"		8	2		485	107		592	695
7624	4"	Q-2	11	2.182		805	121		926	1,075
7710	Union									
7720	1/8"	1 Plum	12	.667	Ea.	50	39.50		89.50	115
7730	1/4"		12	.667		50	39.50		89.50	115
7740	3/8"		12	.667		58.50	39.50		98	124
7750	1/2"		11	.727		73	43		116	145
7760	3/4"		10	.800		100	47.50		147.50	182
7770	1"		9	.889		145	52.50		197.50	239
7780	1-1/4"	Q-1	16	1		360	53.50		413.50	475
7790	1-1/2"		15	1.067		390	57		447	515
7800	2"		13	1.231		490	65.50		555.50	640
7810	2-1/2"		10	1.600		925	85.50		1,010.50	1,150
7820	3"		7	2.286		1,225	122		1,347	1,525
7830	4"	Q-2	10	2.400		1,700	133		1,833	2,075
7838	Caps									
7840	1/2"	1 Plum	24	.333	Ea.	13.30	19.75		33.05	44.50
7841	3/4"		22	.364		19.35	21.50		40.85	54
7842	1"		20	.400		31	23.50		54.50	69.50
7843	1-1/2"	Q-1	32	.500		78	26.50		104.50	126
7844	2"	"	28	.571		99	30.50		129.50	155
7845	4"	Q-2	22	1.091		390	60.50		450.50	520
7850	For 150 lb., type 316, add					25%				
8750	Threaded joint, 2000 lb., type 304									
8770	90° Elbow									
8780	1/8"	1 Plum	13	.615	Ea.	32.50	36.50		69	90.50
8790	1/4"		13	.615		32.50	36.50		69	90.50
8800	3/8"		13	.615		40	36.50		76.50	99

For customer support on your Mechanical Cost Data, call 877.739.3689.

22 11 13 – Facility Water Distribution Piping

22 11 13.66 Pipe Fittings, Stainless Steel		Crew	Daily Output	Labor-Hours	Unit	Material	2016 Bare Costs Labor	Equipment	Total	Total Incl O&P
8810	1/2"	1 Plum	12	.667	Ea.	54.50	39.50		94	119
8820	3/4"		11	.727		63	43		106	135
8830	1"		10	.800		83.50	47.50		131	164
8840	1-1/4"	Q-1	17	.941		136	50		186	226
8850	1-1/2"		16	1		218	53.50		271.50	320
8860	2"		14	1.143		305	61		366	425
8880	45° Elbow									
8890	1/8"	1 Plum	13	.615	Ea.	67.50	36.50		104	130
8900	1/4"		13	.615		67.50	36.50		104	130
8910	3/8"		13	.615		84.50	36.50		121	148
8920	1/2"		12	.667		84.50	39.50		124	153
8930	3/4"		11	.727		93	43		136	167
8940	1"		10	.800		111	47.50		158.50	194
8950	1-1/4"	Q-1	17	.941		194	50		244	289
8960	1-1/2"		16	1		260	53.50		313.50	365
8970	2"		14	1.143		299	61		360	420
8990	Tee, straight									
9000	1/8"	1 Plum	9	.889	Ea.	42	52.50		94.50	126
9010	1/4"		9	.889		42	52.50		94.50	126
9020	3/8"		9	.889		54	52.50		106.50	139
9030	1/2"		8	1		69	59		128	166
9040	3/4"		7	1.143		78.50	67.50		146	188
9050	1"		6.50	1.231		108	73		181	229
9060	1-1/4"	Q-1	11	1.455		182	77.50		259.50	320
9070	1-1/2"		10	1.600		305	85.50		390.50	465
9080	2"		9	1.778		420	94.50		514.50	610
9100	For couplings and unions use 3000 lb., type 304									
9120	2000 lb., type 316									
9130	90° Elbow									
9140	1/8"	1 Plum	13	.615	Ea.	40.50	36.50		77	99.50
9150	1/4"		13	.615		40.50	36.50		77	99.50
9160	3/8"		13	.615		47	36.50		83.50	107
9170	1/2"		12	.667		61	39.50		100.50	127
9180	3/4"		11	.727		75	43		118	148
9190	1"		10	.800		111	47.50		158.50	195
9200	1-1/4"	Q-1	17	.941		202	50		252	298
9210	1-1/2"		16	1		233	53.50		286.50	335
9220	2"		14	1.143		395	61		456	525
9240	45° Elbow									
9250	1/8"	1 Plum	13	.615	Ea.	85.50	36.50		122	149
9260	1/4"		13	.615		85.50	36.50		122	149
9270	3/8"		13	.615		85.50	36.50		122	149
9280	1/2"		12	.667		85.50	39.50		125	154
9300	3/4"		11	.727		96.50	43		139.50	171
9310	1"		10	.800		146	47.50		193.50	232
9320	1-1/4"	Q-1	17	.941		208	50		258	305
9330	1-1/2"		16	1		285	53.50		338.50	395
9340	2"		14	1.143		350	61		411	475
9360	Tee, straight									
9370	1/8"	1 Plum	9	.889	Ea.	51.50	52.50		104	136
9380	1/4"		9	.889		51.50	52.50		104	136
9390	3/8"		9	.889		63	52.50		115.50	149
9400	1/2"		8	1		78	59		137	176

22 11 13.66 Pipe Fittings, Stainless Steel		Crew	Daily Output	Labor-Hours	Unit	Material	2016 Bare Costs Labor	Equipment	Total	Total Incl O&P
9410	3/4"	1 Plum	7	1.143	Ea.	96.50	67.50		164	208
9420	1"		6.50	1.231		147	73		220	272
9430	1-1/4"	Q-1	11	1.455		239	77.50		316.50	380
9440	1-1/2"		10	1.600		345	85.50		430.50	505
9450	2"		9	1.778		545	94.50		639.50	745
9470	For couplings and unions use 3000 lb., type 316									
9490	3000 lb., type 304									
9510	Coupling									
9520	1/8"	1 Plum	19	.421	Ea.	12.80	25		37.80	51.50
9530	1/4"		19	.421		13.90	25		38.90	53
9540	3/8"		19	.421		16.20	25		41.20	55.50
9550	1/2"		19	.421		19.50	25		44.50	59
9560	3/4"		18	.444		26	26.50		52.50	68
9570	1"		15	.533		44	31.50		75.50	96
9580	1-1/4"	Q-1	26	.615		107	33		140	167
9590	1-1/2"		24	.667		125	35.50		160.50	192
9600	2"		21	.762		165	40.50		205.50	243
9620	Union									
9630	1/8"	1 Plum	12	.667	Ea.	98	39.50		137.50	168
9640	1/4"		12	.667		98	39.50		137.50	168
9650	3/8"		12	.667		105	39.50		144.50	175
9660	1/2"		11	.727		105	43		148	180
9670	3/4"		10	.800		128	47.50		175.50	213
9680	1"		9	.889		198	52.50		250.50	297
9690	1-1/4"	Q-1	16	1		350	53.50		403.50	465
9700	1-1/2"		15	1.067		410	57		467	540
9710	2"		13	1.231		550	65.50		615.50	705
9730	3000 lb., type 316									
9750	Coupling									
9770	1/8"	1 Plum	19	.421	Ea.	14.40	25		39.40	53.50
9780	1/4"		19	.421		16.25	25		41.25	55.50
9790	3/8"		19	.421		17.40	25		42.40	56.50
9800	1/2"		19	.421		24.50	25		49.50	64.50
9810	3/4"		18	.444		36.50	26.50		63	79.50
9820	1"		15	.533		60.50	31.50		92	115
9830	1-1/4"	Q-1	26	.615		136	33		169	200
9840	1-1/2"		24	.667		159	35.50		194.50	229
9850	2"		21	.762		219	40.50		259.50	305
9870	Union									
9880	1/8"	1 Plum	12	.667	Ea.	108	39.50		147.50	178
9890	1/4"		12	.667		108	39.50		147.50	178
9900	3/8"		12	.667		125	39.50		164.50	197
9910	1/2"		11	.727		125	43		168	202
9920	3/4"		10	.800		163	47.50		210.50	251
9930	1"		9	.889		257	52.50		309.50	360
9940	1-1/4"	Q-1	16	1		450	53.50		503.50	575
9950	1-1/2"		15	1.067		555	57		612	695
9960	2"		13	1.231		705	65.50		770.50	875

22 11 13.74 Pipe, Plastic		Crew	Daily Output	Labor-Hours	Unit	Material	2016 Bare Costs Labor	Equipment	Total	Total Incl O&P
0010	**PIPE, PLASTIC**									
0020	Fiberglass reinforced, couplings 10' O.C., clevis hanger assy's, 3 per 10'									
0080	General service									
0120	2" diameter	Q-1	59	.271	L.F.	12.70	14.45		27.15	36
0140	3" diameter		52	.308		16.30	16.40		32.70	42.50
0150	4" diameter		48	.333		21.50	17.75		39.25	50.50
0160	6" diameter		39	.410		36	22		58	73
0170	8" diameter	Q-2	49	.490		56	27		83	103
0180	10" diameter		41	.585		81	32.50		113.50	138
0190	12" diameter		36	.667		104	37		141	171
0600	PVC, high impact/pressure, cplgs. 10' O.C., clevis hanger assy's, 3 per 10'									
1020	Schedule 80									
1070	1/2" diameter	1 Plum	50	.160	L.F.	6.95	9.45		16.40	22
1080	3/4" diameter		47	.170		8.05	10.10		18.15	24
1090	1" diameter		43	.186		9.95	11		20.95	27.50
1100	1-1/4" diameter		39	.205		12.25	12.15		24.40	32
1110	1-1/2" diameter		34	.235		13.90	13.95		27.85	36.50
1120	2" diameter	Q-1	55	.291		17.30	15.50		32.80	42.50
1140	3" diameter		50	.320		32.50	17.05		49.55	61.50
1150	4" diameter		46	.348		45.50	18.55		64.05	78
1170	6" diameter		38	.421		98.50	22.50		121	142
1730	To delete coupling & hangers, subtract									
1740	1/2" diam.					62%	80%			
1750	3/4" diam. to 1-1/4" diam.					58%	73%			
1760	1-1/2" diam. to 6" diam.					40%	57%			
1800	PVC, couplings 10' O.C., clevis hanger assemblies, 3 per 10'									
1820	Schedule 40									
1860	1/2" diameter	1 Plum	54	.148	L.F.	4.96	8.75		13.71	18.70
1870	3/4" diameter		51	.157		5.30	9.30		14.60	19.85
1880	1" diameter		46	.174		5.95	10.30		16.25	22
1890	1-1/4" diameter		42	.190		6.70	11.30		18	24.50
1900	1-1/2" diameter		36	.222		7	13.15		20.15	27.50
1910	2" diameter	Q-1	59	.271		8.25	14.45		22.70	31
1920	2-1/2" diameter		56	.286		10.65	15.20		25.85	35
1930	3" diameter		53	.302		13	16.10		29.10	39
1940	4" diameter		48	.333		16.60	17.75		34.35	45.50
1950	5" diameter		43	.372		27.50	19.85		47.35	60
1960	6" diameter		39	.410		28.50	22		50.50	64.50
1970	8" diameter	Q-2	48	.500		35.50	27.50		63	80.50
1980	10" diameter		43	.558		78	31		109	133
1990	12" diameter		42	.571		93.50	31.50		125	151
2000	14" diameter		31	.774		149	43		192	229
2010	16" diameter		23	1.043		214	57.50		271.50	320
2340	To delete coupling & hangers, subtract									
2360	1/2" diam. to 1-1/4" diam.					65%	74%			
2370	1-1/2" diam. to 6" diam.					44%	57%			
2380	8" diam. to 12" diam.					41%	53%			
2390	14" diam. to 16" diam.					48%	45%			
2420	Schedule 80									
2440	1/4" diameter	1 Plum	58	.138	L.F.	4.41	8.15		12.56	17.15
2450	3/8" diameter		55	.145		4.41	8.60		13.01	17.85
2460	1/2" diameter		50	.160		5.10	9.45		14.55	19.90

For customer support on your Mechanical Cost Data, call 877.739.3689.

201

22 11 13 – Facility Water Distribution Piping

22 11 13.74 Pipe, Plastic	Crew	Daily Output	Labor-Hours	Unit	Material	2016 Bare Costs Labor	Equipment	Total	Total Incl O&P
2470	3/4" diameter	1 Plum	47	.170	L.F.	5.45	10.10		15.55	21
2480	1" diameter		43	.186		6.20	11		17.20	23.50
2490	1-1/4" diameter		39	.205		7	12.15		19.15	26
2500	1-1/2" diameter		34	.235		7.45	13.95		21.40	29
2510	2" diameter	Q-1	55	.291		8.40	15.50		23.90	33
2520	2-1/2" diameter		52	.308		10.15	16.40		26.55	35.50
2530	3" diameter		50	.320		14.40	17.05		31.45	41.50
2540	4" diameter		46	.348		20	18.55		38.55	50
2550	5" diameter		42	.381		30.50	20.50		51	64
2560	6" diameter		38	.421		37	22.50		59.50	75
2570	8" diameter	Q-2	47	.511		46.50	28		74.50	93.50
2580	10" diameter		42	.571		111	31.50		142.50	170
2590	12" diameter		38	.632		134	35		169	201
2830	To delete coupling & hangers, subtract									
2840	1/4" diam. to 1/2" diam.					66%	80%			
2850	3/4" diam. to 1-1/4" diam.					61%	73%			
2860	1-1/2" diam. to 6" diam.					41%	57%			
2870	8" diam. to 12" diam.					31%	50%			
2900	Schedule 120									
2910	1/2" diameter	1 Plum	50	.160	L.F.	5.60	9.45		15.05	20.50
2950	3/4" diameter		47	.170		6.15	10.10		16.25	22
2960	1" diameter		43	.186		7.25	11		18.25	24.50
2970	1-1/4" diameter		39	.205		8.55	12.15		20.70	28
2980	1-1/2" diameter		33	.242		9.30	14.35		23.65	32
2990	2" diameter	Q-1	54	.296		11.10	15.80		26.90	36
3000	2-1/2" diameter		52	.308		16.45	16.40		32.85	42.50
3010	3" diameter		49	.327		19.90	17.40		37.30	48.50
3020	4" diameter		45	.356		28.50	18.95		47.45	60
3030	6" diameter		37	.432		52	23		75	92
3240	To delete coupling & hangers, subtract									
3250	1/2" diam. to 1-1/4" diam.					52%	74%			
3260	1-1/2" diam. to 4" diam.					30%	57%			
3270	6" diam.					17%	50%			
3300	PVC, pressure, couplings 10' O.C., clevis hanger assy's, 3 per 10'									
3310	SDR 26, 160 psi									
3350	1-1/4" diameter	1 Plum	42	.190	L.F.	6.35	11.30		17.65	24
3360	1-1/2" diameter	"	36	.222		6.55	13.15		19.70	27
3370	2" diameter	Q-1	59	.271		7.50	14.45		21.95	30.50
3380	2-1/2" diameter		56	.286		10.10	15.20		25.30	34
3390	3" diameter		53	.302		12.70	16.10		28.80	38.50
3400	4" diameter		48	.333		17.10	17.75		34.85	46
3420	6" diameter		39	.410		32	22		54	68
3430	8" diameter	Q-2	48	.500		47	27.50		74.50	93
3660	To delete coupling & clevis hanger assy's, subtract									
3670	1-1/4" diam.					63%	68%			
3680	1-1/2" diam. to 4" diam.					48%	57%			
3690	6" diam. to 8" diam.					60%	54%			
3720	SDR 21, 200 psi, 1/2" diameter	1 Plum	54	.148	L.F.	4.83	8.75		13.58	18.55
3740	3/4" diameter		51	.157		5.20	9.30		14.50	19.70
3750	1" diameter		46	.174		5.70	10.30		16	22
3760	1-1/4" diameter		42	.190		6.55	11.30		17.85	24
3770	1-1/2" diameter		36	.222		6.95	13.15		20.10	27.50
3780	2" diameter	Q-1	59	.271		7.90	14.45		22.35	30.50

22 11 13.74 Pipe, Plastic		Crew	Daily Output	Labor-Hours	Unit	Material	2016 Bare Costs Labor	Equipment	Total	Total Incl O&P
3790	2-1/2" diameter	Q-1	56	.286	L.F.	12.30	15.20		27.50	36.50
3800	3" diameter		53	.302		13.60	16.10		29.70	39.50
3810	4" diameter		48	.333		21	17.75		38.75	50
3830	6" diameter		39	.410		35.50	22		57.50	72
3840	8" diameter	Q-2	48	.500		52.50	27.50		80	99
4000	To delete coupling & hangers, subtract									
4010	1/2" diam. to 3/4 diam.					71%	77%			
4020	1" diam. to 1-1/4" diam.					63%	70%			
4030	1-1/2" diam. to 6" diam.					44%	57%			
4040	8" diam.					46%	54%			
4100	DWV type, schedule 40, couplings 10' O.C., clevis hanger assy's, 3 per 10'									
4210	ABS, schedule 40, foam core type									
4212	Plain end black									
4214	1-1/2" diameter	1 Plum	39	.205	L.F.	5.30	12.15		17.45	24
4216	2" diameter	Q-1	62	.258		5.70	13.75		19.45	27.50
4218	3" diameter		56	.286		8.15	15.20		23.35	32
4220	4" diameter		51	.314		10.45	16.70		27.15	36.50
4222	6" diameter		42	.381		18.80	20.50		39.30	51
4240	To delete coupling & hangers, subtract									
4244	1-1/2" diam. to 6" diam.					43%	48%			
4400	PVC									
4410	1-1/4" diameter	1 Plum	42	.190	L.F.	5.50	11.30		16.80	23
4420	1-1/2" diameter	"	36	.222		5.35	13.15		18.50	25.50
4460	2" diameter	Q-1	59	.271		5.70	14.45		20.15	28.50
4470	3" diameter		53	.302		8.20	16.10		24.30	33.50
4480	4" diameter		48	.333		10.25	17.75		28	38.50
4490	6" diameter		39	.410		16.90	22		38.90	51.50
4500	8" diameter	Q-2	48	.500		22.50	27.50		50	66
4510	To delete coupling & hangers, subtract									
4520	1-1/4" diam. to 1-1/2" diam.					48%	60%			
4530	2" diam. to 8" diam.					42%	54%			
4550	PVC, schedule 40, foam core type									
4552	Plain end, white									
4554	1-1/2" diameter	1 Plum	39	.205	L.F.	5.05	12.15		17.20	24
4556	2" diameter	Q-1	62	.258		5.40	13.75		19.15	27
4558	3" diameter		56	.286		7.60	15.20		22.80	31.50
4560	4" diameter		51	.314		9.30	16.70		26	35.50
4562	6" diameter		42	.381		15.05	20.50		35.55	47
4564	8" diameter	Q-2	51	.471		19.40	26		45.40	60.50
4568	10" diameter		48	.500		23.50	27.50		51	67.50
4570	12" diameter		46	.522		28.50	29		57.50	75
4580	To delete coupling & hangers, subtract									
4582	1-1/2" dia to 2" dia					58%	54%			
4584	3" dia to 12" dia					46%	42%			
4800	PVC, clear pipe, cplgs. 10' O.C., clevis hanger assy's 3 per 10', Sched. 40									
4840	1/4" diameter	1 Plum	59	.136	L.F.	5	8.05		13.05	17.60
4850	3/8" diameter		56	.143		5.35	8.45		13.80	18.65
4860	1/2" diameter		54	.148		6.05	8.75		14.80	19.90
4870	3/4" diameter		51	.157		6.75	9.30		16.05	21.50
4880	1" diameter		46	.174		8.40	10.30		18.70	25
4890	1-1/4" diameter		42	.190		9.95	11.30		21.25	28
4900	1-1/2" diameter		36	.222		11.05	13.15		24.20	32
4910	2" diameter	Q-1	59	.271		13.50	14.45		27.95	37

For customer support on your Mechanical Cost Data, call 877.739.3689.

203

22 11 Facility Water Distribution

22 11 13 – Facility Water Distribution Piping

22 11 13.74 Pipe, Plastic		Crew	Daily Output	Labor-Hours	Unit	Material	2016 Bare Costs Labor	Equipment	Total	Total Incl O&P
4920	2-1/2" diameter	Q-1	56	.286	L.F.	19.75	15.20		34.95	45
4930	3" diameter		53	.302		24.50	16.10		40.60	51.50
4940	3-1/2" diameter		50	.320		33.50	17.05		50.55	62.50
4950	4" diameter		48	.333		35	17.75		52.75	65.50
5250	To delete coupling & hangers, subtract									
5260	1/4" diam. to 3/8" diam.					60%	81%			
5270	1/2" diam. to 3/4" diam.					41%	77%			
5280	1" diam. to 1-1/2" diam.					26%	67%			
5290	2" diam. to 4" diam.					16%	58%			
5300	CPVC, socket joint, couplings 10' O.C., clevis hanger assemblies, 3 per 10'									
5302	Schedule 40									
5304	1/2" diameter	1 Plum	54	.148	L.F.	6.05	8.75		14.80	19.90
5305	3/4" diameter		51	.157		6.95	9.30		16.25	21.50
5306	1" diameter		46	.174		8.35	10.30		18.65	25
5307	1-1/4" diameter		42	.190		10	11.30		21.30	28
5308	1-1/2" diameter		36	.222		11.20	13.15		24.35	32
5309	2" diameter	Q-1	59	.271		13.30	14.45		27.75	36.50
5310	2-1/2" diameter		56	.286		20.50	15.20		35.70	46
5311	3" diameter		53	.302		24.50	16.10		40.60	51.50
5312	4" diameter		48	.333		32	17.75		49.75	62.50
5314	6" diameter		43	.372		60	19.85		79.85	96
5318	To delete coupling & hangers, subtract									
5319	1/2" diam. to 3/4" diam.					37%	77%			
5320	1" diam. to 1-1/4" diam.					27%	70%			
5321	1-1/2" diam. to 3" diam.					21%	57%			
5322	4" diam. to 6" diam.					16%	57%			
5324	Schedule 80									
5325	1/2" diameter	1 Plum	50	.160	L.F.	6.75	9.45		16.20	22
5326	3/4" diameter		47	.170		7.20	10.10		17.30	23
5327	1" diameter		43	.186		8.95	11		19.95	26.50
5328	1-1/4" diameter		39	.205		10.95	12.15		23.10	30.50
5329	1-1/2" diameter		34	.235		12.45	13.95		26.40	34.50
5330	2" diameter	Q-1	55	.291		15.25	15.50		30.75	40.50
5331	2-1/2" diameter		52	.308		23.50	16.40		39.90	50
5332	3" diameter		50	.320		27	17.05		44.05	55
5333	4" diameter		46	.348		37	18.55		55.55	68.50
5334	6" diameter		38	.421		70.50	22.50		93	112
5335	8" diameter	Q-2	47	.511		179	28		207	239
5339	To delete couplings & hangers, subtract									
5340	1/2" diam. to 3/4" diam.					44%	77%			
5341	1" diam. to 1-1/4" diam.					32%	71%			
5342	1-1/2" diam. to 4" diam.					25%	58%			
5343	6" diam. to 8" diam.					20%	53%			
5360	CPVC, threaded, couplings 10' O.C., clevis hanger assemblies, 3 per 10'									
5380	Schedule 40									
5460	1/2" diameter	1 Plum	54	.148	L.F.	6.90	8.75		15.65	21
5470	3/4" diameter		51	.157		8.35	9.30		17.65	23
5480	1" diameter		46	.174		9.85	10.30		20.15	26.50
5490	1-1/4" diameter		42	.190		11.15	11.30		22.45	29.50
5500	1-1/2" diameter		36	.222		12.20	13.15		25.35	33.50
5510	2" diameter	Q-1	59	.271		14.55	14.45		29	38
5520	2-1/2" diameter		56	.286		22	15.20		37.20	47
5530	3" diameter		53	.302		26.50	16.10		42.60	53.50

22 11 Facility Water Distribution

22 11 13 – Facility Water Distribution Piping

22 11 13.74 Pipe, Plastic		Crew	Daily Output	Labor-Hours	Unit	Material	2016 Bare Costs Labor	Equipment	Total	Total Incl O&P
5540	4" diameter	Q-1	48	.333	L.F.	39.50	17.75		57.25	70.50
5550	6" diameter	↓	43	.372	↓	63.50	19.85		83.35	100
5730	To delete coupling & hangers, subtract									
5740	1/2" diam. to 3/4" diam.					37%	77%			
5750	1" diam. to 1-1/4" diam.					27%	70%			
5760	1-1/2" diam. to 3" diam.					21%	57%			
5770	4" diam. to 6" diam.					16%	57%			
5800	Schedule 80									
5860	1/2" diameter	1 Plum	50	.160	L.F.	7.60	9.45		17.05	22.50
5870	3/4" diameter		47	.170		8.60	10.10		18.70	24.50
5880	1" diameter		43	.186		10.50	11		21.50	28
5890	1-1/4" diameter		39	.205		12.15	12.15		24.30	31.50
5900	1-1/2" diameter	↓	34	.235		13.45	13.95		27.40	36
5910	2" diameter	Q-1	55	.291		16.50	15.50		32	41.50
5920	2-1/2" diameter		52	.308		24.50	16.40		40.90	51.50
5930	3" diameter		50	.320		29	17.05		46.05	57
5940	4" diameter		46	.348		44.50	18.55		63.05	77
5950	6" diameter	↓	38	.421		74	22.50		96.50	116
5960	8" diameter	Q-2	47	.511	↓	178	28		206	238
6060	To delete couplings & hangers, subtract									
6070	1/2" diam. to 3/4" diam.					44%	77%			
6080	1" diam. to 1-1/4" diam.					32%	71%			
6090	1-1/2" diam. to 4" diam.					25%	58%			
6100	6" diam. to 8" diam.					20%	53%			
6240	CTS, 1/2" diameter	1 Plum	54	.148	L.F.	5.55	8.75		14.30	19.35
6250	3/4" diameter		51	.157		6.80	9.30		16.10	21.50
6260	1" diameter		46	.174		10.20	10.30		20.50	27
6270	1-1/4"		42	.190		14.75	11.30		26.05	33.50
6280	1-1/2" diameter	↓	36	.222		18.40	13.15		31.55	40
6290	2" diameter	Q-1	59	.271	↓	28.50	14.45		42.95	53.50
6370	To delete coupling & hangers, subtract									
6380	1/2" diam.					51%	79%			
6390	3/4" diam.					40%	76%			
6392	1" thru 2" diam.					72%	68%			
7280	PEX, flexible, no couplings or hangers									
7282	Note: For labor costs add 25% to the couplings and fittings labor total.									
7285	For fittings see section 23 83 16.10 7000									
7300	Non-barrier type, hot/cold tubing rolls									
7310	1/4" diameter x 100'				L.F.	.49			.49	.54
7350	3/8" diameter x 100'					.55			.55	.61
7360	1/2" diameter x 100'					.61			.61	.67
7370	1/2" diameter x 500'					.61			.61	.67
7380	1/2" diameter x 1000'					.61			.61	.67
7400	3/4" diameter x 100'					1.11			1.11	1.22
7410	3/4" diameter x 500'					1.11			1.11	1.22
7420	3/4" diameter x 1000'					1.11			1.11	1.22
7460	1" diameter x 100'					1.90			1.90	2.09
7470	1" diameter x 300'					1.90			1.90	2.09
7480	1" diameter x 500'					1.90			1.90	2.09
7500	1-1/4" diameter x 100'					3.23			3.23	3.55
7510	1-1/4" diameter x 300'					3.23			3.23	3.55
7540	1-1/2" diameter x 100'					4.40			4.40	4.84
7550	1-1/2" diameter x 300'					4.40			4.40	4.84

For customer support on your Mechanical Cost Data, call 877.739.3689.

205

22 11 13.74 Pipe, Plastic		Crew	Daily Output	Labor-Hours	Unit	Material	2016 Bare Costs Labor	Equipment	Total	Total Incl O&P
7596	Most sizes available in red or blue									
7700	Non-barrier type, hot/cold tubing straight lengths									
7710	1/2" diameter x 20'				L.F.	.61			.61	.67
7750	3/4" diameter x 20'					1.11			1.11	1.22
7760	1" diameter x 20'					1.90			1.90	2.09
7770	1-1/4" diameter x 20'					3.23			3.23	3.55
7780	1-1/2" diameter x 20'					4.40			4.40	4.84
7790	2" diameter					8.60			8.60	9.45
7796	Most sizes available in red or blue									
9000	Polypropylene pipe									
9002	For fusion weld fittings and accessories see line 22 11 13.76 9400									
9004	Note: sizes 1/2" thru 4" use socket fusion									
9005	Sizes 6" thru 10" use butt fusion									
9010	SDR 7.4, (domestic hot water piping)									
9011	Enhanced to minimize thermal expansion and high temperature life									
9016	13' lengths, size is ID, includes joints 13' O.C. and hangers 3 per 10'									
9020	3/8" diameter	1 Plum	53	.151	L.F.	2.40	8.95		11.35	16.15
9022	1/2" diameter		52	.154		2.85	9.10		11.95	16.90
9024	3/4" diameter		50	.160		3.36	9.45		12.81	18
9026	1" diameter		45	.178		4.39	10.50		14.89	20.50
9028	1-1/4" diameter		40	.200		6.25	11.85		18.10	25
9030	1-1/2" diameter		35	.229		8.85	13.55		22.40	30.50
9032	2" diameter	Q-1	58	.276		12.45	14.70		27.15	35.50
9034	2-1/2" diameter		55	.291		17.10	15.50		32.60	42.50
9036	3" diameter		52	.308		22.50	16.40		38.90	49
9038	3-1/2" diameter		49	.327		31.50	17.40		48.90	61
9040	4" diameter		46	.348		34	18.55		52.55	65.50
9042	6" diameter		39	.410		46	22		68	83.50
9044	8" diameter	Q-2	48	.500		70	27.50		97.50	119
9046	10" diameter	"	43	.558		110	31		141	168
9050	To delete joint & hangers, subtract									
9052	3/8" diam. to 1" diam.					45%	65%			
9054	1-1/4" diam. to 4" diam.					15%	45%			
9056	6" diam. to 10" diam.					5%	24%			
9060	SDR 11, (domestic cold water piping)									
9062	13' lengths, size is ID, includes joints 13' O.C. and hangers 3 per 10'									
9064	1/2" diameter	1 Plum	57	.140	L.F.	2.53	8.30		10.83	15.35
9066	3/4" diameter		54	.148		2.85	8.75		11.60	16.40
9068	1" diameter		49	.163		3.46	9.65		13.11	18.40
9070	1-1/4" diameter		45	.178		4.90	10.50		15.40	21.50
9072	1-1/2" diameter		40	.200		6.45	11.85		18.30	25
9074	2" diameter	Q-1	62	.258		9.10	13.75		22.85	31
9076	2-1/2" diameter		59	.271		11.90	14.45		26.35	35
9078	3" diameter		56	.286		16	15.20		31.20	40.50
9080	3-1/2" diameter		53	.302		23	16.10		39.10	50
9082	4" diameter		50	.320		26	17.05		43.05	54.50
9084	6" diameter		47	.340		31.50	18.15		49.65	62
9086	8" diameter	Q-2	51	.471		48	26		74	91.50
9088	10" diameter	"	46	.522		72.50	29		101.50	124
9090	To delete joint & hangers, subtract									
9092	1/2" diam. to 1" diam.					45%	65%			
9094	1-1/4" diam. to 4" diam.					15%	45%			
9096	6" diam. to 10" diam.					5%	24%			

22 11 Facility Water Distribution

22 11 13 – Facility Water Distribution Piping

22 11 13.76 Pipe Fittings, Plastic	Crew	Daily Output	Labor-Hours	Unit	Material	2016 Bare Costs Labor	Equipment	Total	Total Incl O&P
0010 **PIPE FITTINGS, PLASTIC**									
0030 Epoxy resin, fiberglass reinforced, general service									
0100 3"	Q-1	20.80	.769	Ea.	114	41		155	188
0110 4"		16.50	.970		116	51.50		167.50	206
0120 6"	↓	10.10	1.584		227	84.50		311.50	375
0130 8"	Q-2	9.30	2.581		420	143		563	675
0140 10"		8.50	2.824		525	156		681	815
0150 12"	↓	7.60	3.158	↓	755	174		929	1,100
0170 Elbow, 90°, flanged									
0172 2"	Q-1	23	.696	Ea.	199	37		236	275
0173 3"		16	1		229	53.50		282.50	330
0174 4"		13	1.231		298	65.50		363.50	430
0176 6"	↓	8	2		540	107		647	755
0177 8"	Q-2	9	2.667		980	147		1,127	1,300
0178 10"		7	3.429		1,325	189		1,514	1,750
0179 12"	↓	5	4.800	↓	1,800	265		2,065	2,375
0186 Elbow, 45°, flanged									
0188 2"	Q-1	23	.696	Ea.	199	37		236	275
0189 3"		16	1		229	53.50		282.50	330
0190 4"		13	1.231		298	65.50		363.50	430
0192 6"	↓	8	2		540	107		647	755
0193 8"	Q-2	9	2.667		925	147		1,072	1,250
0194 10"		7	3.429		1,175	189		1,364	1,575
0195 12"	↓	5	4.800	↓	1,500	265		1,765	2,050
0352 Tee, flanged									
0354 2"	Q-1	17	.941	Ea.	270	50		320	375
0355 3"		10	1.600		355	85.50		440.50	525
0356 4"		8	2		400	107		507	600
0358 6"	↓	5	3.200		690	171		861	1,000
0359 8"	Q-2	6	4		1,300	221		1,521	1,750
0360 10"		5	4.800		1,900	265		2,165	2,475
0361 12"	↓	4	6	↓	2,600	330		2,930	3,375
0365 Wye, flanged									
0367 2"	Q-1	17	.941	Ea.	535	50		585	665
0368 3"		10	1.600		735	85.50		820.50	940
0369 4"		8	2		980	107		1,087	1,225
0371 6"	↓	5	3.200		1,375	171		1,546	1,775
0372 8"	Q-2	6	4		2,275	221		2,496	2,825
0373 10"		5	4.800		3,525	265		3,790	4,275
0374 12"	↓	4	6	↓	5,125	330		5,455	6,150
0380 Couplings									
0410 2"	Q-1	33.10	.483	Ea.	23	26		49	64
0420 3"		20.80	.769		25	41		66	89.50
0430 4"		16.50	.970		33.50	51.50		85	115
0440 6"	↓	10.10	1.584		79.50	84.50		164	214
0450 8"	Q-2	9.30	2.581		135	143		278	365
0460 10"		8.50	2.824		198	156		354	450
0470 12"	↓	7.60	3.158	↓	265	174		439	555
0473 High corrosion resistant couplings, add					30%				
0474 Reducer, concentric, flanged									
0475 2" x 1-1/2"	Q-1	30	.533	Ea.	520	28.50		548.50	615
0476 3" x 2"	↓	24	.667	↓	565	35.50		600.50	675

22 11 13.76 Pipe Fittings, Plastic		Crew	Daily Output	Labor-Hours	Unit	Material	2016 Bare Costs Labor	2016 Bare Costs Equipment	Total	Total Incl O&P
0477	4" x 3"	Q-1	19	.842	Ea.	635	45		680	770
0479	6" x 4"	↓	15	1.067		635	57		692	785
0480	8" x 6"	Q-2	16	1.500		1,050	83		1,133	1,275
0481	10" x 8"		13	1.846		965	102		1,067	1,225
0482	12" x 10"	↓	11	2.182	↓	1,500	121		1,621	1,825
0486	Adapter, bell x male or female									
0488	2"	Q-1	28	.571	Ea.	26	30.50		56.50	74.50
0489	3"		20	.800		38.50	42.50		81	107
0491	4"		17	.941		50.50	50		100.50	131
0492	6"	↓	12	1.333		108	71		179	226
0493	8"	Q-2	15	1.600		156	88.50		244.50	305
0494	10"	"	11	2.182	↓	224	121		345	430
0528	Flange									
0532	2"	Q-1	46	.348	Ea.	37	18.55		55.55	68.50
0533	3"		32	.500		47	26.50		73.50	91.50
0534	4"		26	.615		61	33		94	117
0536	6"	↓	16	1		116	53.50		169.50	208
0537	8"	Q-2	18	1.333		177	73.50		250.50	305
0538	10"		14	1.714		251	94.50		345.50	420
0539	12"	↓	10	2.400	↓	315	133		448	545
2100	PVC schedule 80, socket joint									
2110	90° elbow, 1/2"	1 Plum	30.30	.264	Ea.	2.61	15.65		18.26	26.50
2130	3/4"		26	.308		3.32	18.20		21.52	31
2140	1"		22.70	.352		5.20	21		26.20	37
2150	1-1/4"		20.20	.396		7.20	23.50		30.70	43.50
2160	1-1/2"	↓	18.20	.440		7.70	26		33.70	48
2170	2"	Q-1	33.10	.483		9.30	26		35.30	49
2180	3"		20.80	.769		24.50	41		65.50	89
2190	4"	↓	16.50	.970		37	51.50		88.50	119
2200	6"	↓	10.10	1.584		106	84.50		190.50	243
2210	8"	Q-2	9.30	2.581		291	143		434	535
2250	45° elbow, 1/2"	1 Plum	30.30	.264		4.90	15.65		20.55	29
2270	3/4"	↓	26	.308		7.50	18.20		25.70	36
2280	1"		22.70	.352		11.20	21		32.20	44
2290	1-1/4"		20.20	.396		14.30	23.50		37.80	51.50
2300	1-1/2"		18.20	.440		16.90	26		42.90	58
2310	2"	Q-1	33.10	.483		22	26		48	63
2320	3"		20.80	.769		56	41		97	124
2330	4"		16.50	.970		101	51.50		152.50	189
2340	6"	↓	10.10	1.584		127	84.50		211.50	267
2350	8"	Q-2	9.30	2.581		275	143		418	520
2400	Tee, 1/2"	1 Plum	20.20	.396		7.35	23.50		30.85	43.50
2420	3/4"		17.30	.462		7.70	27.50		35.20	50
2430	1"		15.20	.526		9.60	31		40.60	57.50
2440	1-1/4"		13.50	.593		26.50	35		61.50	82
2450	1-1/2"	↓	12.10	.661		26.50	39		65.50	88
2460	2"	Q-1	20	.800		33	42.50		75.50	101
2470	3"		13.90	1.151		45	61.50		106.50	142
2480	4"		11	1.455		52	77.50		129.50	174
2490	6"	↓	6.70	2.388		178	127		305	385
2500	8"	Q-2	6.20	3.871		410	214		624	780
2510	Flange, socket, 150 lb., 1/2"	1 Plum	55.60	.144		14.25	8.50		22.75	28.50
2514	3/4"	↓	47.60	.168	↓	15.25	9.95		25.20	32

22 11 13 – Facility Water Distribution Piping

22 11 13.76 Pipe Fittings, Plastic		Crew	Daily Output	Labor-Hours	Unit	Material	2016 Bare Costs Labor	Equipment	Total	Total Incl O&P
2518	1"	1 Plum	41.70	.192	Ea.	16.95	11.35		28.30	36
2522	1-1/2"		33.30	.240		17.40	14.20		31.60	40.50
2526	2"	Q-1	60.60	.264		24	14.05		38.05	47
2530	4"		30.30	.528		51.50	28		79.50	99
2534	6"		18.50	.865		80.50	46		126.50	159
2538	8"	Q-2	17.10	1.404		144	77.50		221.50	276
2550	Coupling, 1/2"	1 Plum	30.30	.264		4.71	15.65		20.36	28.50
2570	3/4"		26	.308		6.35	18.20		24.55	34.50
2580	1"		22.70	.352		6.55	21		27.55	38.50
2590	1-1/4"		20.20	.396		10	23.50		33.50	46.50
2600	1-1/2"		18.20	.440		10.75	26		36.75	51.50
2610	2"	Q-1	33.10	.483		11.55	26		37.55	51.50
2620	3"		20.80	.769		32.50	41		73.50	98
2630	4"		16.50	.970		41	51.50		92.50	123
2640	6"		10.10	1.584		88	84.50		172.50	224
2650	8"	Q-2	9.30	2.581		120	143		263	345
2660	10"		8.50	2.824		420	156		576	695
2670	12"		7.60	3.158		475	174		649	785
2700	PVC (white), schedule 40, socket joints									
2760	90° elbow, 1/2"	1 Plum	33.30	.240	Ea.	.50	14.20		14.70	22
2770	3/4"		28.60	.280		.57	16.55		17.12	25.50
2780	1"		25	.320		1.01	18.95		19.96	29.50
2790	1-1/4"		22.20	.360		1.79	21.50		23.29	34
2800	1-1/2"		20	.400		1.94	23.50		25.44	37.50
2810	2"	Q-1	36.40	.440		3.03	23.50		26.53	39
2820	2-1/2"		26.70	.599		9.25	32		41.25	58
2830	3"		22.90	.699		11.05	37		48.05	68
2840	4"		18.20	.879		19.75	47		66.75	92
2850	5"		12.10	1.322		51	70.50		121.50	162
2860	6"		11.10	1.441		63	77		140	185
2870	8"	Q-2	10.30	2.330		162	129		291	370
2980	45° elbow, 1/2"	1 Plum	33.30	.240		.83	14.20		15.03	22.50
2990	3/4"		28.60	.280		1.29	16.55		17.84	26.50
3000	1"		25	.320		1.55	18.95		20.50	30
3010	1-1/4"		22.20	.360		2.14	21.50		23.64	34.50
3020	1-1/2"		20	.400		2.72	23.50		26.22	38.50
3030	2"	Q-1	36.40	.440		3.56	23.50		27.06	39.50
3040	2-1/2"		26.70	.599		9.25	32		41.25	58
3050	3"		22.90	.699		14.35	37		51.35	72
3060	4"		18.20	.879		25.50	47		72.50	99
3070	5"		12.10	1.322		51	70.50		121.50	162
3080	6"		11.10	1.441		63.50	77		140.50	186
3090	8"	Q-2	10.30	2.330		153	129		282	360
3180	Tee, 1/2"	1 Plum	22.20	.360		.63	21.50		22.13	32.50
3190	3/4"		19	.421		.73	25		25.73	38.50
3200	1"		16.70	.479		1.36	28.50		29.86	44.50
3210	1-1/4"		14.80	.541		2.11	32		34.11	51
3220	1-1/2"		13.30	.602		2.57	35.50		38.07	56.50
3230	2"	Q-1	24.20	.661		3.74	35		38.74	57
3240	2-1/2"		17.80	.899		12.35	48		60.35	86
3250	3"		15.20	1.053		16.20	56		72.20	102
3260	4"		12.10	1.322		29.50	70.50		100	139
3270	5"		8.10	1.975		71	105		176	237

For customer support on your Mechanical Cost Data, call 877.739.3689.

209

22 11 13 – Facility Water Distribution Piping

22 11 13.76 Pipe Fittings, Plastic		Crew	Daily Output	Labor-Hours	Unit	Material	2016 Bare Costs Labor	Equipment	Total	Total Incl O&P
3280	6"	Q-1	7.40	2.162	Ea.	98.50	115		213.50	283
3290	8"	Q-2	6.80	3.529		229	195		424	545
3380	Coupling, 1/2"	1 Plum	33.30	.240		.33	14.20		14.53	22
3390	3/4"		28.60	.280		.46	16.55		17.01	25.50
3400	1"		25	.320		.80	18.95		19.75	29.50
3410	1-1/4"		22.20	.360		1.11	21.50		22.61	33
3420	1-1/2"	▼	20	.400		1.19	23.50		24.69	37
3430	2"	Q-1	36.40	.440		1.80	23.50		25.30	37.50
3440	2-1/2"		26.70	.599		4	32		36	52.50
3450	3"		22.90	.699		6.25	37		43.25	63
3460	4"		18.20	.879		9.10	47		56.10	80.50
3470	5"		12.10	1.322		16.60	70.50		87.10	124
3480	6"	▼	11.10	1.441		28.50	77		105.50	148
3490	8"	Q-2	10.30	2.330		53.50	129		182.50	253
3600	Cap, schedule 40, PVC socket 1/2"	1 Plum	60.60	.132		.46	7.80		8.26	12.30
3610	3/4"		51.90	.154		.53	9.15		9.68	14.35
3620	1"		45.50	.176		.83	10.40		11.23	16.60
3630	1-1/4"		40.40	.198		1.19	11.70		12.89	19
3640	1-1/2"	▼	36.40	.220		1.29	13		14.29	21
3650	2"	Q-1	66.10	.242		1.56	12.90		14.46	21
3660	2-1/2"		48.50	.330		4.97	17.60		22.57	32
3670	3"		41.60	.385		5.35	20.50		25.85	37
3680	4"		33.10	.483		12.35	26		38.35	52.50
3690	6"	▼	20.20	.792		29.50	42		71.50	96
3700	8"	Q-2	18.60	1.290	▼	74.50	71.50		146	190
3710	Reducing insert, schedule 40, socket weld									
3712	3/4"	1 Plum	31.50	.254	Ea.	.53	15.05		15.58	23
3713	1"		27.50	.291		.96	17.20		18.16	27
3715	1-1/2"	▼	22	.364		1.37	21.50		22.87	34
3716	2"	Q-1	40	.400		2.27	21.50		23.77	34.50
3717	4"		20	.800	·	12	42.50		54.50	77.50
3718	6"	▼	12.20	1.311		29.50	70		99.50	138
3719	8"	Q-2	11.30	2.124	▼	117	117		234	305
3730	Reducing insert, socket weld x female/male thread									
3732	1/2"	1 Plum	38.30	.209	Ea.	2.39	12.35		14.74	21.50
3733	3/4"		32.90	.243		1.48	14.40		15.88	23
3734	1"		28.80	.278		2.08	16.45		18.53	27.50
3736	1-1/2"	▼	23	.348		3.74	20.50		24.24	35
3737	2"	Q-1	41.90	.382		4	20.50		24.50	35
3738	4"	"	20.90	.766	▼	38.50	41		79.50	104
3742	Male adapter, socket weld x male thread									
3744	1/2"	1 Plum	38.30	.209	Ea.	.46	12.35		12.81	19.15
3745	3/4"		32.90	.243		.50	14.40		14.90	22
3746	1"		28.80	.278		.92	16.45		17.37	26
3748	1-1/2"	▼	23	.348		1.48	20.50		21.98	32.50
3749	2"	Q-1	41.90	.382		1.94	20.50		22.44	32.50
3750	4"	"	20.90	.766	▼	10.70	41		51.70	73.50
3754	Female adapter, socket weld x female thread									
3756	1/2"	1 Plum	38.30	.209	Ea.	.57	12.35		12.92	19.30
3757	3/4"		32.90	.243		.73	14.40		15.13	22.50
3758	1"		28.80	.278		.83	16.45		17.28	26
3760	1-1/2"	▼	23	.348		1.48	20.50		21.98	32.50
3761	2"	Q-1	41.90	.382		2	20.50		22.50	32.50

22 11 13.76 Pipe Fittings, Plastic		Crew	Daily Output	Labor-Hours	Unit	Material	2016 Bare Costs Labor	Equipment	Total	Total Incl O&P
3762	4"	Q-1	20.90	.766	Ea.	11.20	41		52.20	74
3800	PVC, schedule 80, socket joints									
3810	Reducing insert									
3812	3/4"	1 Plum	28.60	.280	Ea.	1.52	16.55		18.07	26.50
3813	1"		25	.320		4.35	18.95		23.30	33.50
3815	1-1/2"		20	.400		9.45	23.50		32.95	46
3816	2"	Q-1	36.40	.440		13.25	23.50		36.75	50
3817	4"		18.20	.879		50.50	47		97.50	126
3818	6"		11.10	1.441		70	77		147	193
3819	8"	Q-2	10.20	2.353		405	130		535	640
3830	Reducing insert, socket weld x female/male thread									
3832	1/2"	1 Plum	34.80	.230	Ea.	9.50	13.60		23.10	31
3833	3/4"		29.90	.268		5.85	15.85		21.70	30.50
3834	1"		26.10	.307		9.15	18.15		27.30	37.50
3836	1-1/2"		20.90	.383		11.50	22.50		34	46.50
3837	2"	Q-1	38	.421		16.80	22.50		39.30	52.50
3838	4"	"	19	.842		81.50	45		126.50	157
3844	Adapter, male socket x male thread									
3846	1/2"	1 Plum	34.80	.230	Ea.	3.67	13.60		17.27	24.50
3847	3/4"		29.90	.268		4.05	15.85		19.90	28.50
3848	1"		26.10	.307		7	18.15		25.15	35
3850	1-1/2"		20.90	.383		11.75	22.50		34.25	47
3851	2"	Q-1	38	.421		17	22.50		39.50	52.50
3852	4"	"	19	.842		38	45		83	110
3860	Adapter, female socket x female thread									
3862	1/2"	1 Plum	34.80	.230	Ea.	4.41	13.60		18.01	25.50
3863	3/4"		29.90	.268		6.55	15.85		22.40	31
3864	1"		26.10	.307		9.65	18.15		27.80	38
3866	1-1/2"		20.90	.383		19.20	22.50		41.70	55
3867	2"	Q-1	38	.421		33.50	22.50		56	71
3868	4"	"	19	.842		102	45		147	180
3872	Union, socket joints									
3874	1/2"	1 Plum	25.80	.310	Ea.	9.65	18.35		28	38
3875	3/4"		22.10	.362		12.25	21.50		33.75	46
3876	1"		19.30	.415		14	24.50		38.50	52.50
3878	1-1/2"		15.50	.516		31.50	30.50		62	80.50
3879	2"	Q-1	28.10	.569		42.50	30.50		73	93
3888	Cap									
3890	1/2"	1 Plum	54.50	.147	Ea.	4.61	8.70		13.31	18.15
3891	3/4"		46.70	.171		4.86	10.15		15.01	20.50
3892	1"		41	.195		8.65	11.55		20.20	27
3894	1-1/2"		32.80	.244		10.40	14.45		24.85	33.50
3895	2"	Q-1	59.50	.269		27.50	14.35		41.85	51.50
3896	4"		30	.533		83	28.50		111.50	134
3897	6"		18.20	.879		206	47		253	298
3898	8"	Q-2	16.70	1.437		265	79.50		344.50	410
4500	DWV, ABS, non pressure, socket joints									
4540	1/4 Bend, 1-1/4"	1 Plum	20.20	.396	Ea.	3.17	23.50		26.67	39
4560	1-1/2"	"	18.20	.440		2.36	26		28.36	42
4570	2"	Q-1	33.10	.483		3.75	26		29.75	43
4580	3"		20.80	.769		9.45	41		50.45	72.50
4590	4"		16.50	.970		19.60	51.50		71.10	99.50
4600	6"		10.10	1.584		82.50	84.50		167	218

For customer support on your Mechanical Cost Data, call 877.739.3689.

211

22 11 13 – Facility Water Distribution Piping

22 11 13.76 Pipe Fittings, Plastic		Crew	Daily Output	Labor-Hours	Unit	Material	2016 Bare Costs Labor	Equipment	Total	Total Incl O&P
4650	1/8 Bend, same as 1/4 Bend									
4800	Tee, sanitary									
4820	1-1/4"	1 Plum	13.50	.593	Ea.	4.21	35		39.21	57.50
4830	1-1/2"	"	12.10	.661		3.60	39		42.60	63
4840	2"	Q-1	20	.800		5.55	42.50		48.05	70.50
4850	3"		13.90	1.151		15.25	61.50		76.75	109
4860	4"		11	1.455		27	77.50		104.50	147
4862	Tee, sanitary, reducing, 2" x 1-1/2"		22	.727		4.84	39		43.84	64
4864	3" x 2"		15.30	1.046		9.50	55.50		65	94.50
4868	4" x 3"		12.10	1.322		26	70.50		96.50	135
4870	Combination Y and 1/8 bend									
4872	1-1/2"	1 Plum	12.10	.661	Ea.	8.60	39		47.60	68.50
4874	2"	Q-1	20	.800		9.65	42.50		52.15	75
4876	3"		13.90	1.151		23	61.50		84.50	118
4878	4"		11	1.455		43.50	77.50		121	165
4880	3" x 1-1/2"		15.50	1.032		22	55		77	108
4882	4" x 3"		12.10	1.322		34.50	70.50		105	144
4900	Wye, 1-1/4"	1 Plum	13.50	.593		4.85	35		39.85	58.50
4902	1-1/2"	"	12.10	.661		5.50	39		44.50	65
4904	2"	Q-1	20	.800		7.20	42.50		49.70	72.50
4906	3"		13.90	1.151		16.95	61.50		78.45	111
4908	4"		11	1.455		34.50	77.50		112	155
4910	6"		6.70	2.388		103	127		230	305
4918	3" x 1-1/2"		15.50	1.032		13.60	55		68.60	98
4920	4" x 3"		12.10	1.322		27	70.50		97.50	136
4922	6" x 4"		6.90	2.319		82	124		206	277
4930	Double Wye, 1-1/2"	1 Plum	9.10	.879		17.15	52		69.15	97.50
4932	2"	Q-1	16.60	.964		20.50	51.50		72	100
4934	3"		10.40	1.538		49	82		131	178
4936	4"		8.25	1.939		96.50	103		199.50	262
4940	2" x 1-1/2"		16.80	.952		18.90	50.50		69.40	97.50
4942	3" x 2"		10.60	1.509		34.50	80.50		115	159
4944	4" x 3"		8.45	1.893		76.50	101		177.50	236
4946	6" x 4"		7.25	2.207		111	118		229	299
4950	Reducer bushing, 2" x 1-1/2"		36.40	.440		1.92	23.50		25.42	37.50
4952	3" x 1-1/2"		27.30	.586		8.15	31		39.15	56
4954	4" x 2"		18.20	.879		15.65	47		62.65	87.50
4956	6" x 4"		11.10	1.441		43.50	77		120.50	164
4960	Couplings, 1-1/2"	1 Plum	18.20	.440		1.17	26		27.17	41
4962	2"	Q-1	33.10	.483		1.57	26		27.57	40.50
4963	3"		20.80	.769		4.41	41		45.41	67
4964	4"		16.50	.970		7.95	51.50		59.45	87
4966	6"		10.10	1.584		33	84.50		117.50	164
4970	2" x 1-1/2"		33.30	.480		3.39	25.50		28.89	42
4972	3" x 1-1/2"		21	.762		7.65	40.50		48.15	70
4974	4" x 3"		16.70	.958		14.95	51		65.95	93.50
4978	Closet flange, 4"	1 Plum	32	.250		8.15	14.80		22.95	31.50
4980	4" x 3"	"	34	.235		9.85	13.95		23.80	32
5000	DWV, PVC, schedule 40, socket joints									
5040	1/4 bend, 1-1/4"	1 Plum	20.20	.396	Ea.	6.40	23.50		29.90	42.50
5060	1-1/2"	"	18.20	.440		1.82	26		27.82	41.50
5070	2"	Q-1	33.10	.483		2.87	26		28.87	42
5080	3"		20.80	.769		8.45	41		49.45	71.50

22 11 13 – Facility Water Distribution Piping

22 11 13.76 Pipe Fittings, Plastic	Crew	Daily Output	Labor-Hours	Unit	Material	2016 Bare Costs Labor	Equipment	Total	Total Incl O&P
5090	4"	Q-1	16.50	.970	Ea.	16.60	51.50		68.10	96.50
5100	6"	↓	10.10	1.584		56.50	84.50		141	189
5105	8"	Q-2	9.30	2.581		75.50	143		218.50	298
5106	10"	"	8.50	2.824		305	156		461	570
5110	1/4 bend, long sweep, 1-1/2"	1 Plum	18.20	.440		4.21	26		30.21	44
5112	2"	Q-1	33.10	.483		4.70	26		30.70	44
5114	3"		20.80	.769		10.75	41		51.75	74
5116	4"	↓	16.50	.970		20.50	51.50		72	101
5150	1/8 bend, 1-1/4"	1 Plum	20.20	.396		4.32	23.50		27.82	40.50
5170	1-1/2"	"	18.20	.440		1.78	26		27.78	41.50
5180	2"	Q-1	33.10	.483		2.57	26		28.57	42
5190	3"		20.80	.769		7.55	41		48.55	70.50
5200	4"		16.50	.970		13.25	51.50		64.75	92.50
5210	6"	↓	10.10	1.584		51.50	84.50		136	184
5215	8"	Q-2	9.30	2.581		165	143		308	395
5216	10"		8.50	2.824		217	156		373	475
5217	12"	↓	7.60	3.158		298	174		472	590
5250	Tee, sanitary 1-1/4"	1 Plum	13.50	.593		6.65	35		41.65	60.50
5254	1-1/2"	"	12.10	.661		3.17	39		42.17	62.50
5255	2"	Q-1	20	.800		4.67	42.50		47.17	69.50
5256	3"		13.90	1.151		12.25	61.50		73.75	106
5257	4"		11	1.455		22.50	77.50		100	142
5259	6"	↓	6.70	2.388		91	127		218	292
5261	8"	Q-2	6.20	3.871	↓	199	214		413	545
5276	Tee, sanitary, reducing									
5281	2" x 1-1/2" x 1-1/2"	Q-1	23	.696	Ea.	4.12	37		41.12	60.50
5282	2" x 1-1/2" x 2"		22	.727		4.98	39		43.98	64
5283	2" x 2" x 1-1/2"		22	.727		4.13	39		43.13	63
5284	3" x 3" x 1-1/2"		15.50	1.032		8.95	55		63.95	93
5285	3" x 3" x 2"		15.30	1.046		9.20	55.50		64.70	94
5286	4" x 4" x 1-1/2"		12.30	1.301		23.50	69.50		93	131
5287	4" x 4" x 2"		12.20	1.311		19.50	70		89.50	127
5288	4" x 4" x 3"		12.10	1.322		26.50	70.50		97	135
5291	6" x 6" x 4"	↓	6.90	2.319	↓	88	124		212	283
5294	Tee, double sanitary									
5295	1-1/2"	1 Plum	9.10	.879	Ea.	7.05	52		59.05	86.50
5296	2"	Q-1	16.60	.964		9.55	51.50		61.05	88
5297	3"		10.40	1.538		26.50	82		108.50	154
5298	4"	↓	8.25	1.939	↓	43	103		146	203
5303	Wye, reducing									
5304	2" x 1-1/2" x 1-1/2"	Q-1	23	.696	Ea.	8.05	37		45.05	65
5305	2" x 2" x 1-1/2"		22	.727		7	39		46	66
5306	3" x 3" x 2"		15.30	1.046		22.50	55.50		78	109
5307	4" x 4" x 2"		12.20	1.311		16.80	70		86.80	123
5309	4" x 4" x 3"	↓	12.10	1.322		22.50	70.50		93	131
5314	Combination Y & 1/8 bend, 1-1/2"	1 Plum	12.10	.661		7.70	39		46.70	67.50
5315	2"	Q-1	20	.800		8.20	42.50		50.70	73.50
5317	3"		13.90	1.151		21	61.50		82.50	116
5318	4"		11	1.455		42	77.50		119.50	163
5319	6"	↓	6.70	2.388		162	127		289	370
5320	8"	Q-2	6.20	3.871		233	214		447	580
5321	10"		5.70	4.211		850	233		1,083	1,275
5322	12"	↓	5.10	4.706	↓	1,025	260		1,285	1,525

22 11 13.76 Pipe Fittings, Plastic		Crew	Daily Output	Labor-Hours	Unit	Material	2016 Bare Costs Labor	Equipment	Total	Total Incl O&P
5324	Combination Y & 1/8 bend, reducing									
5325	2" x 2" x 1-1/2"	Q-1	22	.727	Ea.	10.80	39		49.80	70.50
5327	3" x 3" x 1-1/2"		15.50	1.032		19.25	55		74.25	104
5328	3" x 3" x 2"		15.30	1.046		14.40	55.50		69.90	100
5329	4" x 4" x 2"		12.20	1.311		22	70		92	129
5331	Wye, 1-1/4"	1 Plum	13.50	.593		8.50	35		43.50	62.50
5332	1-1/2"	"	12.10	.661		5.80	39		44.80	65.50
5333	2"	Q-1	20	.800		5.70	42.50		48.20	71
5334	3"		13.90	1.151		15.35	61.50		76.85	109
5335	4"		11	1.455		28	77.50		105.50	148
5336	6"		6.70	2.388		81.50	127		208.50	282
5337	8"	Q-2	6.20	3.871		144	214		358	485
5338	10"		5.70	4.211		305	233		538	685
5339	12"		5.10	4.706		500	260		760	940
5341	2" x 1-1/2"	Q-1	22	.727		7	39		46	66
5342	3" x 1-1/2"		15.50	1.032		10.40	55		65.40	94.50
5343	4" x 3"		12.10	1.322		22.50	70.50		93	131
5344	6" x 4"		6.90	2.319		62	124		186	254
5345	8" x 6"	Q-2	6.40	3.750		135	207		342	465
5347	Double wye, 1-1/2"	1 Plum	9.10	.879		13	52		65	93
5348	2"	Q-1	16.60	.964		14.50	51.50		66	93.50
5349	3"		10.40	1.538		30.50	82		112.50	158
5350	4"		8.25	1.939		62	103		165	224
5353	Double wye, reducing									
5354	2" x 2" x 1-1/2" x 1-1/2"	Q-1	16.80	.952	Ea.	13.25	50.50		63.75	91
5355	3" x 3" x 2" x 2"		10.60	1.509		22.50	80.50		103	146
5356	4" x 4" x 3" x 3"		8.45	1.893		49	101		150	206
5357	6" x 6" x 4" x 4"		7.25	2.207		170	118		288	365
5374	Coupling, 1-1/4"	1 Plum	20.20	.396		4	23.50		27.50	40
5376	1-1/2"	"	18.20	.440		.85	26		26.85	40.50
5378	2"	Q-1	33.10	.483		1.17	26		27.17	40.50
5380	3"		20.80	.769		4.05	41		45.05	66.50
5390	4"		16.50	.970		6.95	51.50		58.45	85.50
5400	6"		10.10	1.584		22	84.50		106.50	151
5402	8"	Q-2	9.30	2.581		36.50	143		179.50	255
5404	2" x 1-1/2"	Q-1	33.30	.480		2.61	25.50		28.11	41.50
5406	3" x 1-1/2"		21	.762		7.70	40.50		48.20	70
5408	4" x 3"		16.70	.958		12.45	51		63.45	90.50
5410	Reducer bushing, 2" x 1-1/4"		36.50	.438		3.35	23.50		26.85	38.50
5411	2" x 1-1/2"		36.40	.440		1.49	23.50		24.99	37
5412	3" x 1-1/2"		27.30	.586		7.20	31		38.20	55
5413	3" x 2"		27.10	.590		3.76	31.50		35.26	51.50
5414	4" x 2"		18.20	.879		12.55	47		59.55	84.50
5415	4" x 3"		16.70	.958		6.40	51		57.40	84
5416	6" x 4"		11.10	1.441		32	77		109	152
5418	8" x 6"	Q-2	10.20	2.353		63	130		193	266
5425	Closet flange 4"	Q-1	32	.500		8.55	26.50		35.05	49.50
5426	4" x 3"	"	34	.471		9.70	25		34.70	48.50
5450	Solvent cement for PVC, industrial grade, per quart				Qt.	28.50			28.50	31.50
5500	CPVC, Schedule 80, threaded joints									
5540	90° Elbow, 1/4"	1 Plum	32	.250	Ea.	12.75	14.80		27.55	36.50
5560	1/2"		30.30	.264		7.40	15.65		23.05	31.50
5570	3/4"		26	.308		11.05	18.20		29.25	39.50

22 11 13.76 Pipe Fittings, Plastic		Crew	Daily Output	Labor-Hours	Unit	Material	2016 Bare Costs Labor	Equipment	Total	Total Incl O&P
5580	1"	1 Plum	22.70	.352	Ea.	15.55	21		36.55	48.50
5590	1-1/4"		20.20	.396		30	23.50		53.50	68.50
5600	1-1/2"		18.20	.440		32.50	26		58.50	75
5610	2"	Q-1	33.10	.483		43	26		69	86.50
5620	2-1/2"		24.20	.661		134	35		169	200
5630	3"		20.80	.769		144	41		185	220
5640	4"		16.50	.970		225	51.50		276.50	325
5650	6"		10.10	1.584		265	84.50		349.50	420
5660	45° Elbow same as 90° Elbow									
5700	Tee, 1/4"	1 Plum	22	.364	Ea.	25	21.50		46.50	60
5702	1/2"		20.20	.396		25	23.50		48.50	63
5704	3/4"		17.30	.462		35.50	27.50		63	81
5706	1"		15.20	.526		38.50	31		69.50	89.50
5708	1-1/4"		13.50	.593		38.50	35		73.50	95.50
5710	1-1/2"		12.10	.661		40.50	39		79.50	104
5712	2"	Q-1	20	.800		45	42.50		87.50	114
5714	2-1/2"		16.20	.988		222	52.50		274.50	325
5716	3"		13.90	1.151		257	61.50		318.50	375
5718	4"		11	1.455		605	77.50		682.50	785
5720	6"		6.70	2.388		700	127		827	960
5730	Coupling, 1/4"	1 Plum	32	.250		16.25	14.80		31.05	40.50
5732	1/2"		30.30	.264		13.35	15.65		29	38
5734	3/4"		26	.308		21	18.20		39.20	50.50
5736	1"		22.70	.352		24.50	21		45.50	58.50
5738	1-1/4"		20.20	.396		26	23.50		49.50	64
5740	1-1/2"		18.20	.440		28	26		54	70
5742	2"	Q-1	33.10	.483		33	26		59	75
5744	2-1/2"		24.20	.661		59	35		94	118
5746	3"		20.80	.769		69	41		110	138
5748	4"		16.50	.970		140	51.50		191.50	232
5750	6"		10.10	1.584		191	84.50		275.50	335
5752	8"	Q-2	9.30	2.581		410	143		553	665
5900	CPVC, Schedule 80, socket joints									
5904	90° Elbow, 1/4"	1 Plum	32	.250	Ea.	12.15	14.80		26.95	36
5906	1/2"		30.30	.264		4.76	15.65		20.41	29
5908	3/4"		26	.308		6.10	18.20		24.30	34
5910	1"		22.70	.352		9.65	21		30.65	42
5912	1-1/4"		20.20	.396		21	23.50		44.50	58.50
5914	1-1/2"		18.20	.440		23.50	26		49.50	65
5916	2"	Q-1	33.10	.483		28	26		54	70
5918	2-1/2"		24.20	.661		64.50	35		99.50	124
5920	3"		20.80	.769		73	41		114	143
5922	4"		16.50	.970		132	51.50		183.50	223
5924	6"		10.10	1.584		265	84.50		349.50	420
5926	8"		9.30	1.720		650	91.50		741.50	855
5930	45° Elbow, 1/4"	1 Plum	32	.250		18.05	14.80		32.85	42.50
5932	1/2"		30.30	.264		5.80	15.65		21.45	30
5934	3/4"		26	.308		8.35	18.20		26.55	36.50
5936	1"		22.70	.352		13.35	21		34.35	46
5938	1-1/4"		20.20	.396		26.50	23.50		50	64.50
5940	1-1/2"		18.20	.440		27	26		53	69
5942	2"	Q-1	33.10	.483		30.50	26		56.50	72.50
5944	2-1/2"		24.20	.661		62	35		97	121

For customer support on your Mechanical Cost Data, call 877.739.3689.

215

22 11 13.76 Pipe Fittings, Plastic		Crew	Daily Output	Labor-Hours	Unit	Material	2016 Bare Costs Labor	Equipment	Total	Total Incl O&P
5946	3"	Q-1	20.80	.769	Ea.	79.50	41		120.50	150
5948	4"		16.50	.970		109	51.50		160.50	198
5950	6"		10.10	1.584		320	84.50		404.50	475
5952	8"		9.30	1.720		700	91.50		791.50	910
5960	Tee, 1/4"	1 Plum	22	.364		11.15	21.50		32.65	45
5962	1/2"		20.20	.396		11.15	23.50		34.65	48
5964	3/4"		17.30	.462		11.35	27.50		38.85	54
5966	1"		15.20	.526		13.90	31		44.90	62.50
5968	1-1/4"		13.50	.593		29.50	35		64.50	85.50
5970	1-1/2"		12.10	.661		33.50	39		72.50	96
5972	2"	Q-1	20	.800		38	42.50		80.50	106
5974	2-1/2"		16.20	.988		95	52.50		147.50	185
5976	3"		13.90	1.151		95	61.50		156.50	198
5978	4"		11	1.455		127	77.50		204.50	256
5980	6"		6.70	2.388		330	127		457	555
5982	8"	Q-2	6.20	3.871		945	214		1,159	1,375
5990	Coupling, 1/4"	1 Plum	32	.250		12.90	14.80		27.70	36.50
5992	1/2"		30.30	.264		5.05	15.65		20.70	29
5994	3/4"		26	.308		7.05	18.20		25.25	35.50
5996	1"		22.70	.352		9.45	21		30.45	42
5998	1-1/4"		20.20	.396		14.15	23.50		37.65	51
6000	1-1/2"		18.20	.440		17.85	26		43.85	59
6002	2"	Q-1	33.10	.483		20.50	26		46.50	62
6004	2-1/2"		24.20	.661		46	35		81	104
6006	3"		20.80	.769		50	41		91	117
6008	4"		16.50	.970		65.50	51.50		117	151
6010	6"		10.10	1.584		155	84.50		239.50	297
6012	8"	Q-2	9.30	2.581		420	143		563	675
6200	CTS, 100 psi at 180°F, hot and cold water									
6230	90° Elbow, 1/2"	1 Plum	20	.400	Ea.	.51	23.50		24.01	36
6250	3/4"		19	.421		.84	25		25.84	38.50
6251	1"		16	.500		2.30	29.50		31.80	47
6252	1-1/4"		15	.533		4.88	31.50		36.38	53
6253	1-1/2"		14	.571		8.80	34		42.80	60.50
6254	2"	Q-1	23	.696		16.90	37		53.90	74.50
6260	45° Elbow, 1/2"	1 Plum	20	.400		.65	23.50		24.15	36
6280	3/4"		19	.421		1.12	25		26.12	38.50
6281	1"		16	.500		2.84	29.50		32.34	47.50
6282	1-1/4"		15	.533		6.05	31.50		37.55	54
6283	1-1/2"		14	.571		8.75	34		42.75	60.50
6284	2"	Q-1	23	.696		18.30	37		55.30	76
6290	Tee, 1/2"	1 Plum	13	.615		.66	36.50		37.16	55.50
6310	3/4"		12	.667		1.23	39.50		40.73	61
6311	1"		11	.727		6.45	43		49.45	72
6312	1-1/4"		10	.800		9.85	47.50		57.35	82.50
6313	1-1/2"		10	.800		12.85	47.50		60.35	85.50
6314	2"	Q-1	17	.941		21	50		71	98.50
6320	Coupling, 1/2"	1 Plum	22	.364		.42	21.50		21.92	33
6340	3/4"		21	.381		.56	22.50		23.06	34.50
6341	1"		18	.444		2.50	26.50		29	42.50
6342	1-1/4"		17	.471		3.22	28		31.22	45.50
6343	1-1/2"		16	.500		4.55	29.50		34.05	49.50
6344	2"	Q-1	28	.571		8.65	30.50		39.15	55.50

For customer support on your Mechanical Cost Data, call 877.739.3689.

22 11 Facility Water Distribution

22 11 13 – Facility Water Distribution Piping

22 11 13.76 Pipe Fittings, Plastic		Crew	Daily Output	Labor-Hours	Unit	Material	2016 Bare Costs Labor	Equipment	Total	Total Incl O&P
6360	Solvent cement for CPVC, commercial grade, per quart				Qt.	46.50			46.50	51
7340	PVC flange, slip-on, Sch 80 std., 1/2"	1 Plum	22	.364	Ea.	14.25	21.50		35.75	48
7350	3/4"		21	.381		15.25	22.50		37.75	51
7360	1"		18	.444		16.95	26.50		43.45	58
7370	1-1/4"		17	.471		17.50	28		45.50	61.50
7380	1-1/2"		16	.500		17.85	29.50		47.35	64
7390	2"	Q-1	26	.615		24	33		57	75.50
7400	2-1/2"		24	.667		36.50	35.50		72	94
7410	3"		18	.889		40.50	47.50		88	116
7420	4"		15	1.067		51.50	57		108.50	143
7430	6"		10	1.600		80.50	85.50		166	218
7440	8"	Q-2	11	2.182		144	121		265	340
7550	Union, schedule 40, socket joints, 1/2"	1 Plum	19	.421		5.45	25		30.45	43.50
7560	3/4"		18	.444		5.65	26.50		32.15	45.50
7570	1"		15	.533		5.80	31.50		37.30	54
7580	1-1/4"		14	.571		17.50	34		51.50	70.50
7590	1-1/2"		13	.615		19.50	36.50		56	76.50
7600	2"	Q-1	20	.800		26.50	42.50		69	93.50
7992	Polybutyl/polyethyl pipe, for copper fittings see Line 22 11 13.25 7000									
8000	Compression type, PVC, 160 psi cold water									
8010	Coupling, 3/4" CTS	1 Plum	21	.381	Ea.	4.06	22.50		26.56	38.50
8020	1" CTS		18	.444		5.05	26.50		31.55	45
8030	1-1/4" CTS		17	.471		7.10	28		35.10	50
8040	1-1/2" CTS		16	.500		9.75	29.50		39.25	55
8050	2" CTS		15	.533		13.50	31.50		45	62.50
8060	Female adapter, 3/4" FPT x 3/4" CTS		23	.348		5.20	20.50		25.70	36.50
8070	3/4" FPT x 1" CTS		21	.381		6.05	22.50		28.55	40.50
8080	1" FPT x 1" CTS		20	.400		6.05	23.50		29.55	42
8090	1-1/4" FPT x 1-1/4" CTS		18	.444		7.90	26.50		34.40	48
8100	1-1/2" FPT x 1-1/2" CTS		16	.500		9	29.50		38.50	54.50
8110	2" FPT x 2" CTS		13	.615		13.10	36.50		49.60	69.50
8130	Male adapter, 3/4" MPT x 3/4" CTS		23	.348		4.32	20.50		24.82	36
8140	3/4" MPT x 1" CTS		21	.381		5.15	22.50		27.65	39.50
8150	1" MPT x 1" CTS		20	.400		5.15	23.50		28.65	41
8160	1-1/4" MPT x 1-1/4" CTS		18	.444		6.90	26.50		33.40	47
8170	1-1/2" MPT x 1-1/2" CTS		16	.500		8.30	29.50		37.80	53.50
8180	2" MPT x 2" CTS		13	.615		10.80	36.50		47.30	67
8200	Spigot adapter, 3/4" IPS x 3/4" CTS		23	.348		2.17	20.50		22.67	33.50
8210	3/4" IPS x 1" CTS		21	.381		2.67	22.50		25.17	37
8220	1" IPS x 1" CTS		20	.400		2.66	23.50		26.16	38.50
8230	1-1/4" IPS x 1-1/4" CTS		18	.444		4.01	26.50		30.51	44
8240	1-1/2" IPS x 1-1/2" CTS		16	.500		4.20	29.50		33.70	49
8250	2" IPS x 2" CTS		13	.615		5.20	36.50		41.70	60.50
8270	Price includes insert stiffeners									
8280	250 psi is same price as 160 psi									
8300	Insert type, nylon, 160 & 250 psi, cold water									
8310	Clamp ring stainless steel, 3/4" IPS	1 Plum	115	.070	Ea.	2.69	4.12		6.81	9.15
8320	1" IPS		107	.075		2.75	4.43		7.18	9.75
8330	1-1/4" IPS		101	.079		2.77	4.69		7.46	10.15
8340	1-1/2" IPS		95	.084		3.71	4.99		8.70	11.60
8350	2" IPS		85	.094		4.27	5.55		9.82	13.10
8370	Coupling, 3/4" IPS		22	.364		.93	21.50		22.43	33.50
8380	1" IPS		19	.421		.97	25		25.97	38.50

For customer support on your Mechanical Cost Data, call 877.739.3689.

217

22 11 13 – Facility Water Distribution Piping

22 11 13.76 Pipe Fittings, Plastic		Crew	Daily Output	Labor-Hours	Unit	Material	2016 Bare Costs Labor	Equipment	Total	Total Incl O&P
8390	1-1/4" IPS	1 Plum	18	.444	Ea.	1.43	26.50		27.93	41
8400	1-1/2" IPS		17	.471		1.70	28		29.70	44
8410	2" IPS		16	.500		3.78	29.50		33.28	48.50
8430	Elbow, 90°, 3/4" IPS		22	.364		1.86	21.50		23.36	34.50
8440	1" IPS		19	.421		2.06	25		27.06	40
8450	1-1/4" IPS		18	.444		2.30	26.50		28.80	42
8460	1-1/2" IPS		17	.471		2.71	28		30.71	45
8470	2" IPS		16	.500		3.78	29.50		33.28	48.50
8490	Male adapter, 3/4" IPS x 3/4" MPT		25	.320		.93	18.95		19.88	29.50
8500	1" IPS x 1" MPT		21	.381		.96	22.50		23.46	35
8510	1-1/4" IPS x 1-1/4" MPT		20	.400		1.52	23.50		25.02	37
8520	1-1/2" IPS x 1-1/2" MPT		18	.444		1.70	26.50		28.20	41.50
8530	2" IPS x 2" MPT		15	.533		3.26	31.50		34.76	51
8550	Tee, 3/4" IPS		14	.571		1.79	34		35.79	53
8560	1" IPS		13	.615		2.34	36.50		38.84	57.50
8570	1-1/4" IPS		12	.667		3.65	39.50		43.15	63.50
8580	1-1/2" IPS		11	.727		4.14	43		47.14	69.50
8590	2" IPS		10	.800		8.20	47.50		55.70	80.50
8610	Insert type, PVC, 100 psi @ 180°F, hot & cold water									
8620	Coupler, male, 3/8" CTS x 3/8" MPT	1 Plum	29	.276	Ea.	.76	16.35		17.11	25.50
8630	3/8" CTS x 1/2" MPT		28	.286		.76	16.90		17.66	26.50
8640	1/2" CTS x 1/2" MPT		27	.296		.76	17.55		18.31	27.50
8650	1/2" CTS x 3/4" MPT		26	.308		2.40	18.20		20.60	30
8660	3/4" CTS x 1/2" MPT		25	.320		2.18	18.95		21.13	31
8670	3/4" CTS x 3/4" MPT		25	.320		2.58	18.95		21.53	31.50
8700	Coupling, 3/8" CTS x 1/2" CTS		25	.320		4.58	18.95		23.53	33.50
8710	1/2" CTS		23	.348		5.50	20.50		26	37
8730	3/4" CTS		22	.364		11.60	21.50		33.10	45.50
8750	Elbow 90°, 3/8" CTS		25	.320		4	18.95		22.95	33
8760	1/2" CTS		23	.348		4.73	20.50		25.23	36
8770	3/4" CTS		22	.364		7.25	21.50		28.75	40.50
8800	Rings, crimp, copper, 3/8" CTS		120	.067		.18	3.95		4.13	6.15
8810	1/2" CTS		117	.068		.18	4.05		4.23	6.30
8820	3/4" CTS		115	.070		.23	4.12		4.35	6.45
8850	Reducer tee, bronze, 3/8" x 3/8" x 1/2" CTS		17	.471		5	28		33	47.50
8860	1/2" x 1/2" x 3/4" CTS		15	.533		6.55	31.50		38.05	54.50
8870	3/4" x 1/2" x 1/2" CTS		14	.571		6.55	34		40.55	58
8890	3/4" x 3/4" x 1/2" CTS		14	.571		6.65	34		40.65	58.50
8900	1" x 1/2" x 1/2" CTS		14	.571		11	34		45	63
8930	Tee, 3/8" CTS		17	.471		3.31	28		31.31	45.50
8940	1/2" CTS		15	.533		2.27	31.50		33.77	50
8950	3/4" CTS		14	.571		3.50	34		37.50	55
8960	Copper rings included in fitting price									
9000	Flare type, assembled, acetal, hot & cold water									
9010	Coupling, 1/4" & 3/8" CTS	1 Plum	24	.333	Ea.	3.36	19.75		23.11	33.50
9020	1/2" CTS		22	.364		3.84	21.50		25.34	36.50
9030	3/4" CTS		21	.381		5.70	22.50		28.20	40.50
9040	1" CTS		18	.444		7.25	26.50		33.75	47.50
9050	Elbow 90°, 1/4" CTS		26	.308		3.73	18.20		21.93	31.50
9060	3/8" CTS		24	.333		4	19.75		23.75	34.50
9070	1/2" CTS		22	.364		4.73	21.50		26.23	37.50
9080	3/4" CTS		21	.381		7.25	22.50		29.75	42
9090	1" CTS		18	.444		9.15	26.50		35.65	49.50

22 11 13.76 Pipe Fittings, Plastic		Crew	Daily Output	Labor-Hours	Unit	Material	2016 Bare Costs Labor	Equipment	Total	Total Incl O&P
9110	Tee, 1/4" CTS	1 Plum	16	.500	Ea.	4.09	29.50		33.59	49
9114	3/8" CTS		15	.533		4.13	31.50		35.63	52
9120	1/2" CTS		14	.571		5.25	34		39.25	57
9130	3/4" CTS		13	.615		8.25	36.50		44.75	64
9140	1" CTS		12	.667		11.05	39.50		50.55	71.50
9400	Polypropylene, fittings and accessories									
9404	Fittings fusion welded, sizes are I.D.									
9408	Note: sizes 1/2" thru 4" use socket fusion									
9410	Sizes 6" thru 10" use butt fusion									
9416	Coupling									
9420	3/8"	1 Plum	39	.205	Ea.	.88	12.15		13.03	19.30
9422	1/2"		37.40	.214		1.15	12.65		13.80	20.50
9424	3/4"		35.40	.226		1.28	13.40		14.68	21.50
9426	1"		29.70	.269		1.69	15.95		17.64	26
9428	1-1/4"		27.60	.290		2.02	17.15		19.17	28
9430	1-1/2"		24.80	.323		4.25	19.10		23.35	33.50
9432	2"	Q-1	43	.372		8.50	19.85		28.35	39.50
9434	2-1/2"		35.60	.449		9.50	24		33.50	46.50
9436	3"		30.90	.518		21	27.50		48.50	64.50
9438	3-1/2"		27.80	.576		34	30.50		64.50	84
9440	4"		25	.640		44.50	34		78.50	101
9442	Reducing coupling, female to female									
9446	2" to 1-1/2"	Q-1	49	.327	Ea.	12.25	17.40		29.65	40
9448	2-1/2" to 2"		41.20	.388		13.45	20.50		33.95	46
9450	3" to 2-1/2"		33.10	.483		18.15	26		44.15	59
9470	Reducing bushing, female to female									
9472	1/2" to 3/8"	1 Plum	38.20	.209	Ea.	1.15	12.40		13.55	19.95
9474	3/4" to 3/8" or 1/2"		36.50	.219		1.28	13		14.28	21
9476	1" to 3/4" or 1/2"		33.10	.242		1.71	14.30		16.01	23.50
9478	1-1/4" to 3/4" or 1"		28.70	.279		2.63	16.50		19.13	28
9480	1-1/2" to 1/2" thru 1-1/4"		26.20	.305		4.36	18.10		22.46	32.50
9482	2" to 1/2" thru 1-1/2"	Q-1	43	.372		8.70	19.85		28.55	39.50
9484	2-1/2" to 1/2" thru 2"		41.20	.388		9.75	20.50		30.25	41.50
9486	3" to 1-1/2" thru 2-1/2"		33.10	.483		21.50	26		47.50	63
9488	3-1/2" to 2" thru 3"		29.20	.548		34.50	29		63.50	82
9490	4" to 2-1/2" thru 3-1/2"		26.20	.611		54	32.50		86.50	109
9491	6" to 4" SDR 7.4		16.50	.970		65	51.50		116.50	150
9492	6" to 4" SDR 11		16.50	.970		65	51.50		116.50	150
9493	8" to 6" SDR 7.4		10.10	1.584		95.50	84.50		180	232
9494	8" to 6" SDR 11		10.10	1.584		87.50	84.50		172	224
9495	10" to 8" SDR 7.4		7.80	2.051		131	109		240	310
9496	10" to 8" SDR 11		7.80	2.051		114	109		223	291
9500	90° Elbow									
9504	3/8"	1 Plum	39	.205	Ea.	.97	12.15		13.12	19.40
9506	1/2"		37.40	.214		1.23	12.65		13.88	20.50
9508	3/4"		35.40	.226		1.58	13.40		14.98	21.50
9510	1"		29.70	.269		2.28	15.95		18.23	26.50
9512	1-1/4"		27.60	.290		3.51	17.15		20.66	30
9514	1-1/2"		24.80	.323		7.55	19.10		26.65	37.50
9516	2"	Q-1	43	.372		11.60	19.85		31.45	43
9518	2-1/2"		35.60	.449		26	24		50	64.50
9520	3"		30.90	.518		43	27.50		70.50	88.50
9522	3-1/2"		27.80	.576		61	30.50		91.50	114

22 11 13.76 Pipe Fittings, Plastic		Crew	Daily Output	Labor-Hours	Unit	Material	2016 Bare Costs Labor	Equipment	Total	Total Incl O&P
9524	4"	Q-1	25	.640	Ea.	93.50	34		127.50	155
9526	6" SDR 7.4		5.55	2.883		107	154		261	350
9528	6" SDR 11	▼	5.55	2.883		83.50	154		237.50	325
9530	8" SDR 7.4	Q-2	8.10	2.963		415	164		579	705
9532	8" SDR 11		8.10	2.963		355	164		519	635
9534	10" SDR 7.4		7.50	3.200		635	177		812	965
9536	10" SDR 11	▼	7.50	3.200	▼	575	177		752	900
9551	45° Elbow									
9554	3/8"	1 Plum	39	.205	Ea.	.94	12.15		13.09	19.40
9556	1/2"		37.40	.214		1.23	12.65		13.88	20.50
9558	3/4"		35.40	.226		1.58	13.40		14.98	21.50
9564	1"		29.70	.269		2.28	15.95		18.23	26.50
9566	1-1/4"		27.60	.290		3.51	17.15		20.66	30
9568	1-1/2"	▼	24.80	.323		7.55	19.10		26.65	37.50
9570	2"	Q-1	43	.372		11.45	19.85		31.30	42.50
9572	2-1/2"		35.60	.449		25.50	24		49.50	64
9574	3"		30.90	.518		47	27.50		74.50	93
9576	3-1/2"		27.80	.576		67	30.50		97.50	120
9578	4"		25	.640		103	34		137	165
9580	6" SDR 7.4		5.55	2.883		119	154		273	365
9582	6" SDR 11	▼	5.55	2.883		98	154		252	340
9584	8" SDR 7.4	Q-2	8.10	2.963		325	164		489	605
9586	8" SDR 11		8.10	2.963		284	164		448	555
9588	10" SDR 7.4		7.50	3.200		525	177		702	840
9590	10" SDR 11	▼	7.50	3.200	▼	435	177		612	740
9600	Tee									
9604	3/8"	1 Plum	26	.308	Ea.	1.22	18.20		19.42	29
9606	1/2"		24.90	.321		1.66	19		20.66	30.50
9608	3/4"		23.70	.338		2.28	20		22.28	32.50
9610	1"		19.90	.402		2.90	24		26.90	39
9612	1-1/4"		18.50	.432		4.43	25.50		29.93	43.50
9614	1-1/2"	▼	16.60	.482		12.65	28.50		41.15	57
9616	2"	Q-1	26.80	.597		18.10	32		50.10	68
9618	2-1/2"		23.80	.672		30.50	36		66.50	87.50
9620	3"		20.60	.777		50	41.50		91.50	118
9622	3-1/2"		18.40	.870		78	46.50		124.50	156
9624	4"		16.70	.958		104	51		155	191
9626	6" SDR 7.4		3.70	4.324		129	230		359	490
9628	6" SDR 11	▼	3.70	4.324		98	230		328	460
9630	8" SDR 7.4	Q-2	5.40	4.444		355	246		601	760
9632	8" SDR 11		5.40	4.444		310	246		556	715
9634	10" SDR 7.4		5	4.800		610	265		875	1,075
9636	10" SDR 11	▼	5	4.800	▼	530	265		795	980
9638	For reducing tee use same tee price									
9660	End cap									
9662	3/8"	1 Plum	78	.103	Ea.	1.22	6.05		7.27	10.50
9664	1/2"		74.60	.107		1.80	6.35		8.15	11.60
9666	3/4"		71.40	.112		2.28	6.65		8.93	12.50
9668	1"		59.50	.134		2.76	7.95		10.71	15.05
9670	1-1/4"		55.60	.144		4.36	8.50		12.86	17.65
9672	1-1/2"	▼	49.50	.162		6	9.55		15.55	21
9674	2"	Q-1	80	.200		10.05	10.65		20.70	27
9676	2-1/2"	▼	71.40	.224	▼	14.55	11.95		26.50	34

22 11 13 – Facility Water Distribution Piping

22 11 13.76 Pipe Fittings, Plastic		Crew	Daily Output	Labor-Hours	Unit	Material	2016 Bare Costs Labor	Equipment	Total	Total Incl O&P
9678	3"	Q-1	61.70	.259	Ea.	33	13.80		46.80	57
9680	3-1/2"		55.20	.290		39.50	15.45		54.95	67
9682	4"		50	.320		60.50	17.05		77.55	92
9684	6" SDR 7.4		26.50	.604		83	32		115	140
9686	6" SDR 11		26.50	.604		83	32		115	140
9688	8" SDR 7.4	Q-2	16.30	1.472		83	81.50		164.50	215
9690	8" SDR 11		16.30	1.472		72	81.50		153.50	202
9692	10" SDR 7.4		14.90	1.611		124	89		213	271
9694	10" SDR 11		14.90	1.611		106	89		195	250
9800	Accessories and tools									
9802	Pipe clamps for suspension, not including rod or beam clamp									
9804	3/8"	1 Plum	74	.108	Ea.	2.13	6.40		8.53	12
9805	1/2"		70	.114		2.67	6.75		9.42	13.15
9806	3/4"		68	.118		3.07	6.95		10.02	13.90
9807	1"		66	.121		3.35	7.20		10.55	14.55
9808	1-1/4"		64	.125		3.44	7.40		10.84	14.95
9809	1-1/2"		62	.129		3.76	7.65		11.41	15.70
9810	2"	Q-1	110	.145		4.66	7.75		12.41	16.85
9811	2-1/2"		104	.154		5.95	8.20		14.15	18.90
9812	3"		98	.163		6.45	8.70		15.15	20
9813	3-1/2"		92	.174		7.05	9.25		16.30	22
9814	4"		86	.186		7.90	9.90		17.80	23.50
9815	6"		70	.229		9.65	12.20		21.85	29
9816	8"	Q-2	100	.240		34.50	13.25		47.75	58
9817	10"	"	94	.255		39.50	14.10		53.60	65
9820	Pipe cutter									
9822	For 3/8" thru 1-1/4"				Ea.	107			107	117
9824	For 1-1/2" thru 4"				"	280			280	305
9826	Note: Pipes may be cut with standard									
9827	iron saw with blades for plastic.									
9982	For plastic hangers see Line 22 05 29.10 8000									
9986	For copper/brass fittings see Line 22 11 13.25 7000									

22 11 13.78 Pipe, High Density Polyethylene Plastic (HDPE)

		Crew	Daily Output	Labor-Hours	Unit	Material	2016 Bare Costs Labor	Equipment	Total	Total Incl O&P
0010	**PIPE, HIGH DENSITY POLYETHYLENE PLASTIC (HDPE)**									
0020	Not incl. hangers, trenching, backfill, hoisting or digging equipment.									
0030	Standard length is 40', add a weld for each joint									
0040	Single wall									
0050	Straight									
0054	1" diameter DR 11				L.F.	.79			.79	.87
0058	1-1/2" diameter DR 11					1			1	1.10
0062	2" diameter DR 11					1.67			1.67	1.84
0066	3" diameter DR 11					2.01			2.01	2.21
0070	3" diameter DR 17					1.61			1.61	1.77
0074	4" diameter DR 11					3.36			3.36	3.70
0078	4" diameter DR 17					3.36			3.36	3.70
0082	6" diameter DR 11					8.35			8.35	9.20
0086	6" diameter DR 17					5.55			5.55	6.10
0090	8" diameter DR 11					13.90			13.90	15.30
0094	8" diameter DR 26					6.50			6.50	7.15
0098	10" diameter DR 11					22			22	24
0102	10" diameter DR 26					10.05			10.05	11.05
0106	12" diameter DR 11					32			32	35

For customer support on your Mechanical Cost Data, call 877.739.3689.

221

22 11 13.78 Pipe, High Density Polyethylene Plastic (HDPE)	Crew	Daily Output	Labor-Hours	Unit	Material	2016 Bare Costs Labor	Equipment	Total	Total Incl O&P
0110	12" diameter DR 26				L.F.	15.05			15.05	16.60
0114	16" diameter DR 11					48.50			48.50	53.50
0118	16" diameter DR 26					22			22	24
0122	18" diameter DR 11					62			62	68
0126	18" diameter DR 26					28.50			28.50	31.50
0130	20" diameter DR 11					75.50			75.50	83
0134	20" diameter DR 26					33.50			33.50	37
0138	22" diameter DR 11					92			92	101
0142	22" diameter DR 26					42			42	46
0146	24" diameter DR 11					109			109	120
0150	24" diameter DR 26					48.50			48.50	53.50
0154	28" diameter DR 17					100			100	111
0158	28" diameter DR 26					67			67	73.50
0162	30" diameter DR 21					94			94	103
0166	30" diameter DR 26					77			77	84.50
0170	36" diameter DR 26					77			77	84.50
0174	42" diameter DR 26					149			149	164
0178	48" diameter DR 26					196			196	216
0182	54" diameter DR 26				▼	246			246	271
0300	90° Elbow									
0304	1" diameter DR 11				Ea.	5.60			5.60	6.15
0308	1-1/2" diameter DR 11					7			7	7.70
0312	2" diameter DR 11					7			7	7.70
0316	3" diameter DR 11					14			14	15.40
0320	3" diameter DR 17					14			14	15.40
0324	4" diameter DR 11					19.60			19.60	21.50
0328	4" diameter DR 17					19.60			19.60	21.50
0332	6" diameter DR 11					45			45	49.50
0336	6" diameter DR 17					45			45	49.50
0340	8" diameter DR 11					111			111	122
0344	8" diameter DR 26					98			98	108
0348	10" diameter DR 11					415			415	455
0352	10" diameter DR 26					380			380	420
0356	12" diameter DR 11					435			435	480
0360	12" diameter DR 26					395			395	435
0364	16" diameter DR 11					515			515	570
0368	16" diameter DR 26					510			510	560
0372	18" diameter DR 11					650			650	715
0376	18" diameter DR 26					610			610	670
0380	20" diameter DR 11					765			765	840
0384	20" diameter DR 26					740			740	810
0388	22" diameter DR 11					790			790	870
0392	22" diameter DR 26					765			765	840
0396	24" diameter DR 11					890			890	980
0400	24" diameter DR 26					865			865	950
0404	28" diameter DR 17					1,075			1,075	1,200
0408	28" diameter DR 26					1,025			1,025	1,125
0412	30" diameter DR 17					1,525			1,525	1,675
0416	30" diameter DR 26					1,400			1,400	1,550
0420	36" diameter DR 26					1,775			1,775	1,950
0424	42" diameter DR 26					2,300			2,300	2,525
0428	48" diameter DR 26					2,675			2,675	2,950
0432	54" diameter DR 26				▼	6,375			6,375	7,000

For customer support on your Mechanical Cost Data, call 877.739.3689.

22 11 13 – Facility Water Distribution Piping

22 11 13.78 Pipe, High Density Polyethylene Plastic (HDPE)	Crew	Daily Output	Labor-Hours	Unit	Material	2016 Bare Costs Labor	Equipment	Total	Total Incl O&P
0500	45° Elbow									
0512	2" diameter DR 11				Ea.	5.60			5.60	6.15
0516	3" diameter DR 11					14			14	15.40
0520	3" diameter DR 17					14			14	15.40
0524	4" diameter DR 11					19.60			19.60	21.50
0528	4" diameter DR 17					19.60			19.60	21.50
0532	6" diameter DR 11					45			45	49.50
0536	6" diameter DR 17					45			45	49.50
0540	8" diameter DR 11					111			111	122
0544	8" diameter DR 26					64.50			64.50	71
0548	10" diameter DR 11					415			415	455
0552	10" diameter DR 26					380			380	420
0556	12" diameter DR 11					435			435	480
0560	12" diameter DR 26					395			395	435
0564	16" diameter DR 11					229			229	252
0568	16" diameter DR 26					216			216	238
0572	18" diameter DR 11					243			243	268
0576	18" diameter DR 26					223			223	245
0580	20" diameter DR 11					370			370	405
0584	20" diameter DR 26					350			350	385
0588	22" diameter DR 11					510			510	560
0592	22" diameter DR 26					485			485	530
0596	24" diameter DR 11					625			625	685
0600	24" diameter DR 26					600			600	660
0604	28" diameter DR 17					710			710	780
0608	28" diameter DR 26					690			690	755
0612	30" diameter DR 17					880			880	965
0616	30" diameter DR 26					855			855	940
0620	36" diameter DR 26					1,075			1,075	1,200
0624	42" diameter DR 26					1,400			1,400	1,525
0628	48" diameter DR 26					1,525			1,525	1,675
0632	54" diameter DR 26					2,025			2,025	2,225
0700	Tee									
0704	1" diameter DR 11				Ea.	7.30			7.30	8.05
0708	1-1/2" diameter DR 11					10.25			10.25	11.30
0712	2" diameter DR 11					8.80			8.80	9.65
0716	3" diameter DR 11					16.10			16.10	17.70
0720	3" diameter DR 17					16.10			16.10	17.70
0724	4" diameter DR 11					23.50			23.50	26
0728	4" diameter DR 17					23.50			23.50	26
0732	6" diameter DR 11					58.50			58.50	64.50
0736	6" diameter DR 17					58.50			58.50	64.50
0740	8" diameter DR 11					145			145	159
0744	8" diameter DR 17					145			145	159
0748	10" diameter DR 11					430			430	475
0752	10" diameter DR 17					430			430	475
0756	12" diameter DR 11					575			575	635
0760	12" diameter DR 17					575			575	635
0764	16" diameter DR 11					305			305	335
0768	16" diameter DR 17					251			251	276
0772	18" diameter DR 11					430			430	470
0776	18" diameter DR 17					355			355	390
0780	20" diameter DR 11					525			525	575

22 11 13 – Facility Water Distribution Piping

22 11 13.78 Pipe, High Density Polyethylene Plastic (HDPE)		Crew	Daily Output	Labor-Hours	Unit	Material	2016 Bare Costs Labor	Equipment	Total	Total Incl O&P
0784	20" diameter DR 17				Ea.	430			430	470
0788	22" diameter DR 11					670			670	740
0792	22" diameter DR 17					530			530	585
0796	24" diameter DR 11					835			835	915
0800	24" diameter DR 17					705			705	775
0804	28" diameter DR 17					1,350			1,350	1,500
0812	30" diameter DR 17					1,550			1,550	1,725
0820	36" diameter DR 17					2,575			2,575	2,825
0824	42" diameter DR 26					2,875			2,875	3,150
0828	48" diameter DR 26					3,075			3,075	3,375
1000	Flange adptr, w/back-up ring and 1/2 cost of plated bolt set									
1004	1" diameter DR 11				Ea.	28			28	30.50
1008	1-1/2" diameter DR 11					28			28	30.50
1012	2" diameter DR 11					17.55			17.55	19.35
1016	3" diameter DR 11					20.50			20.50	22.50
1020	3" diameter DR 17					20.50			20.50	22.50
1024	4" diameter DR 11					28			28	30.50
1028	4" diameter DR 17					28			28	30.50
1032	6" diameter DR 11					39.50			39.50	43.50
1036	6" diameter DR 17					39.50			39.50	43.50
1040	8" diameter DR 11					57			57	63
1044	8" diameter DR 26					57			57	63
1048	10" diameter DR 11					91			91	100
1052	10" diameter DR 26					91			91	100
1056	12" diameter DR 11					133			133	147
1060	12" diameter DR 26					133			133	147
1064	16" diameter DR 11					286			286	315
1068	16" diameter DR 26					286			286	315
1072	18" diameter DR 11					370			370	410
1076	18" diameter DR 26					370			370	410
1080	20" diameter DR 11					515			515	565
1084	20" diameter DR 26					515			515	565
1088	22" diameter DR 11					560			560	615
1092	22" diameter DR 26					555			555	610
1096	24" diameter DR 17					605			605	665
1100	24" diameter DR 32.5					605			605	665
1104	28" diameter DR 15.5					820			820	900
1108	28" diameter DR 32.5					820			820	900
1112	30" diameter DR 11					950			950	1,050
1116	30" diameter DR 21					950			950	1,050
1120	36" diameter DR 26					1,050			1,050	1,150
1124	42" diameter DR 26					1,175			1,175	1,300
1128	48" diameter DR 26					1,450			1,450	1,600
1132	54" diameter DR 26					1,750			1,750	1,925
1200	Reducer									
1208	2" x 1-1/2" diameter DR 11				Ea.	8.40			8.40	9.25
1212	3" x 2" diameter DR 11					8.40			8.40	9.25
1216	4" x 2" diameter DR 11					9.80			9.80	10.80
1220	4" x 3" diameter DR 11					12.60			12.60	13.85
1224	6" x 4" diameter DR 11					29.50			29.50	32.50
1228	8" x 6" diameter DR 11					45			45	49.50
1232	10" x 8" diameter DR 11					77			77	85
1236	12" x 8" diameter DR 11					126			126	139

22 11 13 – Facility Water Distribution Piping

22 11 13.78 Pipe, High Density Polyethylene Plastic (HDPE)	Crew	Daily Output	Labor-Hours	Unit	Material	2016 Bare Costs Labor	Equipment	Total	Total Incl O&P
1240	12" x 10" diameter DR 11				Ea.	101			101	111
1244	14" x 12" diameter DR 11					112			112	123
1248	16" x 14" diameter DR 11					143			143	157
1252	18" x 16" diameter DR 11					174			174	191
1256	20" x 18" diameter DR 11					345			345	380
1260	22" x 20" diameter DR 11					425			425	465
1264	24" x 22" diameter DR 11					475			475	525
1268	26" x 24" diameter DR 11					560			560	615
1272	28" x 24" diameter DR 11					715			715	785
1276	32" x 28" diameter DR 17					925			925	1,025
1280	36" x 32" diameter DR 17				▼	1,250			1,250	1,375
4000	Welding labor per joint, not including welding machine									
4010	Pipe joint size (cost based on thickest wall for each diam.)									
4030	1" pipe size	4 Skwk	273	.117	Ea.		5.85		5.85	9
4040	1-1/2" pipe size		175	.183			9.10		9.10	14.05
4050	2" pipe size		128	.250			12.50		12.50	19.20
4060	3" pipe size		100	.320			15.95		15.95	24.50
4070	4" pipe size	▼	77	.416			20.50		20.50	32
4080	6" pipe size	5 Skwk	63	.635			31.50		31.50	49
4090	8" pipe size		48	.833			41.50		41.50	64
4100	10" pipe size	▼	40	1			50		50	77
4110	12" pipe size	6 Skwk	41	1.171			58.50		58.50	90
4120	16" pipe size		34	1.412			70.50		70.50	108
4130	18" pipe size	▼	32	1.500			75		75	115
4140	20" pipe size	8 Skwk	37	1.730			86.50		86.50	133
4150	22" pipe size		35	1.829			91.50		91.50	141
4160	24" pipe size		34	1.882			94		94	145
4170	28" pipe size		33	1.939			97		97	149
4180	30" pipe size		32	2			100		100	154
4190	36" pipe size		31	2.065			103		103	159
4200	42" pipe size	▼	30	2.133			106		106	164
4210	48" pipe size	9 Skwk	33	2.182			109		109	168
4220	54" pipe size	"	31	2.323	▼		116		116	178
4300	Note: Cost for set up each time welder is moved.									
4301	Add 50% of a weld cost									
4310	Welder usually remains stationary with pipe moved through it.									
4340	Weld machine, rental per day based on diam. capacity									
4350	1" thru 2" diameter				Ea.			40.50	40.50	44.50
4360	3" thru 4" diameter							46	46	50.50
4370	6" thru 8" diameter							103	103	113
4380	10" thru 12" diameter							178	178	196
4390	16" thru 18" diameter							259	259	285
4400	20" thru 24" diameter							500	500	550
4410	28" thru 32" diameter							545	545	600
4420	36" diameter							570	570	625
4430	42" thru 54" diameter				▼			890	890	980
5000	Dual wall contained pipe									
5040	Straight									
5054	1" DR 11 x 3" DR 11				L.F.	7.45			7.45	8.15
5058	1" DR 11 x 4" DR 11					7.90			7.90	8.70
5062	1-1/2" DR 11 x 4" DR 17					8.15			8.15	8.95
5066	2" DR 11 x 4" DR 17					8.80			8.80	9.65
5070	2" DR 11 x 6" DR 17				▼	13.15			13.15	14.45

22 11 13.78 Pipe, High Density Polyethylene Plastic (HDPE)	Crew	Daily Output	Labor-Hours	Unit	Material	2016 Bare Costs Labor	Equipment	Total	Total Incl O&P
5074	3" DR 11 x 6" DR 17				L.F.	14.65			14.65	16.10
5078	3" DR 11 x 6" DR 26					12.60			12.60	13.85
5086	3" DR 17 x 8" DR 17					18.60			18.60	20.50
5090	4" DR 11 x 8" DR 17					22			22	24
5094	4" DR 17 x 8" DR 26					17.20			17.20	18.95
5098	6" DR 11 x 10" DR 17					33.50			33.50	37
5102	6" DR 17 x 10" DR 26					25.50			25.50	28
5106	6" DR 26 x 10" DR 26					24			24	26.50
5110	8" DR 17 x 12" DR 26					36			36	39.50
5114	8" DR 26 x 12" DR 32.5					30.50			30.50	33.50
5118	10" DR 17 x 14" DR 26					50			50	55
5122	10" DR 17 x 16" DR 26					55			55	60.50
5126	10" DR 26 x 16" DR 26					49.50			49.50	54.50
5130	12" DR 26 x 16" DR 26					56.50			56.50	62
5134	12" DR 17 x 18" DR 26					72			72	79
5138	12" DR 26 x 18" DR 26					64.50			64.50	71
5142	14" DR 26 x 20" DR 32.5					71			71	78
5146	16" DR 26 x 22" DR 32.5					107			107	118
5150	18" DR 26 x 24" DR 32.5					98			98	108
5154	20" DR 32.5 x 28" DR 32.5					127			127	140
5158	22" DR 32.5 x 30" DR 32.5					127			127	140
5162	24" DR 32.5 x 32" DR 32.5					162			162	178
5166	36" DR 32.5 x 42" DR 32.5				↓	274			274	300
5300	Force transfer coupling									
5354	1" DR 11 x 3" DR 11				Ea.	267			267	294
5358	1" DR 11 x 4" DR 17					281			281	310
5362	1-1/2" DR 11 x 4" DR 17					294			294	325
5366	2" DR 11 x 4" DR 17					310			310	340
5370	2" DR 11 x 6" DR 17					445			445	490
5374	3" DR 11 x 6" DR 17					445			445	490
5378	3" DR 11 x 6" DR 26					445			445	490
5382	3" DR 11 x 8" DR 11					470			470	520
5386	3" DR 11 x 8" DR 17					470			470	520
5390	4" DR 11 x 8" DR 17					470			470	520
5394	4" DR 17 x 8" DR 26					415			415	455
5398	6" DR 11 x 10" DR 17					600			600	660
5402	6" DR 17 x 10" DR 26					495			495	540
5406	6" DR 26 x 10" DR 26					495			495	540
5410	8" DR 17 x 12" DR 26					655			655	720
5414	8" DR 26 x 12" DR 32.5					530			530	580
5418	10" DR 17 x 14" DR 26					860			860	945
5422	10" DR 17 x 16" DR 26					860			860	945
5426	10" DR 26 x 16" DR 26					860			860	945
5430	12" DR 26 x 16" DR 26					1,075			1,075	1,175
5434	12" DR 17 x 18" DR 26					1,100			1,100	1,225
5438	12" DR 26 x 18" DR 26					1,100			1,100	1,225
5442	14" DR 26 x 20" DR 32.5					1,175			1,175	1,275
5446	16" DR 26 x 22" DR 32.5					1,200			1,200	1,325
5450	18" DR 26 x 24" DR 32.5					1,300			1,300	1,425
5454	20" DR 32.5 x 28" DR 32.5					1,575			1,575	1,750
5458	22" DR 32.5 x 30" DR 32.5					1,750			1,750	1,925
5462	24" DR 32.5 x 32" DR 32.5					1,900			1,900	2,075
5466	36" DR 32.5 x 42" DR 32.5				↓	3,150			3,150	3,450

For customer support on your Mechanical Cost Data, call 877.739.3689.

22 11 13 – Facility Water Distribution Piping

22 11 13.78 Pipe, High Density Polyethylene Plastic (HDPE)	Crew	Daily Output	Labor-Hours	Unit	Material	2016 Bare Costs Labor	Equipment	Total	Total Incl O&P
5600	90° Elbow								
5654	1" DR 11 x 3" DR 11				Ea.	221			221	243
5658	1" DR 11 x 4" DR 17					212			212	233
5662	1-1/2" DR 11 x 4" DR 17					231			231	254
5666	2" DR 11 x 4" DR 17					245			245	270
5670	2" DR 11 x 6" DR 17					310			310	345
5674	3" DR 11 x 6" DR 17					355			355	395
5678	3" DR 17 x 6" DR 26					285			285	315
5682	3" DR 17 x 8" DR 11					575			575	635
5686	3" DR 17 x 8" DR 17					450			450	495
5690	4" DR 11 x 8" DR 17					525			525	580
5694	4" DR 17 x 8" DR 26					420			420	460
5698	6" DR 11 x 10" DR 17					700			700	770
5702	6" DR 17 x 10" DR 26					535			535	590
5706	6" DR 26 x 10" DR 26					505			505	555
5710	8" DR 17 x 12" DR 26					800			800	880
5714	8" DR 26 x 12" DR 32.5					740			740	815
5718	10" DR 17 x 14" DR 26					1,250			1,250	1,375
5722	10" DR 17 x 16" DR 26					1,200			1,200	1,325
5726	10" DR 26 x 16" DR 26					1,125			1,125	1,225
5730	12" DR 26 x 16" DR 26					1,225			1,225	1,350
5734	12" DR 17 x 18" DR 26					1,475			1,475	1,625
5738	12" DR 26 x 18" DR 26					1,350			1,350	1,500
5742	14" DR 26 x 20" DR 32.5					1,775			1,775	1,975
5746	16" DR 26 x 22" DR 32.5					1,800			1,800	1,975
5750	18" DR 26 x 24" DR 32.5					2,200			2,200	2,425
5754	20" DR 32.5 x 28" DR 32.5					2,550			2,550	2,800
5758	22" DR 32.5 x 30" DR 32.5					3,450			3,450	3,800
5762	24" DR 32.5 x 32" DR 32.5					3,375			3,375	3,700
5766	36" DR 32.5 x 42" DR 32.5					5,475			5,475	6,025
5800	45° Elbow									
5804	1" DR 11 x 3" DR 11				Ea.	132			132	146
5808	1" DR 11 x 4" DR 17					131			131	144
5812	1-1/2" DR 11 x 4" DR 17					141			141	155
5816	2" DR 11 x 4" DR 17					154			154	169
5820	2" DR 11 x 6" DR 17					193			193	213
5824	3" DR 11 x 6" DR 17					215			215	236
5828	3" DR 17 x 6" DR 26					178			178	195
5832	3" DR 17 x 8" DR 11					330			330	360
5836	3" DR 17 x 8" DR 17					262			262	289
5840	4" DR 11 x 8" DR 17					305			305	335
5844	4" DR 17 x 8" DR 26					250			250	275
5848	6" DR 11 x 10" DR 17					405			405	445
5852	6" DR 17 x 10" DR 26					320			320	350
5856	6" DR 26 x 10" DR 26					299			299	330
5860	8" DR 17 x 12" DR 26					485			485	530
5864	8" DR 26 x 12" DR 32.5					450			450	495
5868	10" DR 17 x 14" DR 26					755			755	830
5872	10" DR 17 x 16" DR 26					740			740	815
5876	10" DR 26 x 16" DR 26					695			695	765
5880	12" DR 26 x 16" DR 26					780			780	860
5884	12" DR 17 x 18" DR 26					920			920	1,000
5888	12" DR 26 x 18" DR 26					860			860	945

For customer support on your Mechanical Cost Data, call 877.739.3689.

227

22 11 13.78 Pipe, High Density Polyethylene Plastic (HDPE)	Crew	Daily Output	Labor-Hours	Unit	Material	2016 Bare Costs Labor	Equipment	Total	Total Incl O&P
5892	14" DR 26 x 20" DR 32.5				Ea.	1,100			1,100	1,225
5896	16" DR 26 x 22" DR 32.5					1,150			1,150	1,275
5900	18" DR 26 x 24" DR 32.5					1,375			1,375	1,500
5904	20" DR 32.5 x 28" DR 32.5					1,675			1,675	1,850
5908	22" DR 32.5 x 30" DR 32.5					2,125			2,125	2,350
5912	24" DR 32.5 x 32" DR 32.5					2,125			2,125	2,325
5916	36" DR 32.5 x 42" DR 32.5				▼	3,300			3,300	3,625
6000	Access port with 4" riser									
6050	1" DR 11 x 4" DR 17				Ea.	289			289	320
6054	1-1/2" DR 11 x 4" DR 17					295			295	325
6058	2" DR 11 x 6" DR 17					360			360	395
6062	3" DR 11 x 6" DR 17					360			360	395
6066	3" DR 17 x 6" DR 26					355			355	390
6070	3" DR 17 x 8" DR 11					405			405	445
6074	3" DR 17 x 8" DR 17					390			390	425
6078	4" DR 11 x 8" DR 17					400			400	440
6082	4" DR 17 x 8" DR 26					385			385	420
6086	6" DR 11 x 10" DR 17					500			500	550
6090	6" DR 17 x 10" DR 26					460			460	510
6094	6" DR 26 x 10" DR 26					450			450	495
6098	8" DR 17 x 12" DR 26					485			485	535
6102	8" DR 26 x 12" DR 32.5				▼	460			460	510
6200	End termination with vent plug									
6204	1" DR 11 x 3" DR 11				Ea.	239			239	263
6208	1" DR 11 x 4" DR 17					256			256	281
6212	1-1/2" DR 11 x 4" DR 17					256			256	281
6216	2" DR 11 x 4" DR 17					278			278	305
6220	2" DR 11 x 6" DR 17					315			315	350
6224	3" DR 11 x 6" DR 17					315			315	350
6228	3" DR 17 x 6" DR 26					315			315	350
6232	3" DR 17 x 8" DR 11					440			440	480
6236	3" DR 17 x 8" DR 17					440			440	480
6240	4" DR 11 x 8" DR 17					440			440	480
6244	4" DR 17 x 8" DR 26					380			380	420
6248	6" DR 11 x 10" DR 17					520			520	570
6252	6" DR 17 x 10" DR 26					430			430	470
6256	6" DR 26 x 10" DR 26					430			430	470
6260	8" DR 17 x 12" DR 26					635			635	695
6264	8" DR 26 x 12" DR 32.5					505			505	555
6268	10" DR 17 x 14" DR 26					660			660	725
6272	10" DR 17 x 16" DR 26					660			660	725
6276	10" DR 26 x 16" DR 26					660			660	725
6280	12" DR 26 x 16" DR 26					660			660	725
6284	12" DR 17 x 18" DR 26					865			865	950
6288	12" DR 26 x 18" DR 26					865			865	950
6292	14" DR 26 x 20" DR 32.5					880			880	970
6296	16" DR 26 x 22" DR 32.5					1,025			1,025	1,125
6300	18" DR 26 x 24" DR 32.5					1,050			1,050	1,150
6304	20" DR 32.5 x 28" DR 32.5					1,350			1,350	1,475
6308	22" DR 32.5 x 30" DR 32.5					1,475			1,475	1,625
6312	24" DR 32.5 x 32" DR 32.5					1,650			1,650	1,800
6316	36" DR 32.5 x 42" DR 32.5				▼	2,400			2,400	2,625
6600	Tee									

22 11 13 – Facility Water Distribution Piping

22 11 13.78 Pipe, High Density Polyethylene Plastic (HDPE)	Crew	Daily Output	Labor-Hours	Unit	Material	2016 Bare Costs Labor	Equipment	Total	Total Incl O&P
6604	1" DR 11 x 3" DR 11				Ea.	243			243	268
6608	1" DR 11 x 4" DR 17					300			300	330
6612	1-1/2" DR 11 x 4" DR 17					335			335	370
6616	2" DR 11 x 4" DR 17					365			365	405
6620	2" DR 11 x 6" DR 17					445			445	490
6624	3" DR 11 x 6" DR 17					515			515	565
6628	3" DR 17 x 6" DR 26					500			500	550
6632	3" DR 17 x 8" DR 11					605			605	665
6636	3" DR 17 x 8" DR 17					575			575	635
6640	4" DR 11 x 8" DR 17					625			625	690
6644	4" DR 17 x 8" DR 26					600			600	660
6648	6" DR 11 x 10" DR 17					785			785	865
6652	6" DR 17 x 10" DR 26					740			740	815
6656	6" DR 26 x 10" DR 26					735			735	810
6660	8" DR 17 x 12" DR 26					735			735	810
6664	8" DR 26 x 12" DR 32.5					960			960	1,050
6668	10" DR 17 x 14" DR 26					1,100			1,100	1,225
6672	10" DR 17 x 16" DR 26					1,175			1,175	1,275
6676	10" DR 26 x 16" DR 26					1,175			1,175	1,275
6680	12" DR 26 x 16" DR 26					1,250			1,250	1,375
6684	12" DR 17 x 18" DR 26					1,250			1,250	1,375
6688	12" DR 26 x 18" DR 26					1,400			1,400	1,525
6692	14" DR 26 x 20" DR 32.5					1,850			1,850	2,025
6696	16" DR 26 x 22" DR 32.5					2,250			2,250	2,475
6700	18" DR 26 x 24" DR 32.5					2,450			2,450	2,700
6704	20" DR 32.5 x 28" DR 32.5					3,100			3,100	3,400
6708	22" DR 32.5 x 30" DR 32.5					3,675			3,675	4,050
6712	24" DR 32.5 x 32" DR 32.5					4,400			4,400	4,825
6716	36" DR 32.5 x 42" DR 32.5					8,975			8,975	9,850
6800	Wye									
6816	2" DR 11 x 4" DR 17				Ea.	465			465	510
6820	2" DR 11 x 6" DR 17					525			525	580
6824	3" DR 11 x 6" DR 17					530			530	585
6828	3" DR 17 x 6" DR 26					515			515	565
6832	3" DR 17 x 8" DR 11					645			645	710
6836	3" DR 17 x 8" DR 17					605			605	665
6840	4" DR 11 x 8" DR 17					660			660	725
6844	4" DR 17 x 8" DR 26					625			625	690
6848	6" DR 11 x 10" DR 17					815			815	895
6852	6" DR 17 x 10" DR 26					755			755	835
6856	6" DR 26 x 10" DR 26					755			755	830
6860	8" DR 17 x 12" DR 26					1,075			1,075	1,175
6864	8" DR 26 x 12" DR 32.5					1,050			1,050	1,150
6868	10" DR 17 x 14" DR 26					1,300			1,300	1,425
6872	10" DR 17 x 16" DR 26					1,375			1,375	1,525
6876	10" DR 26 x 16" DR 26					1,350			1,350	1,500
6880	12" DR 26 x 16" DR 26					1,525			1,525	1,675
6884	12" DR 17 x 18" DR 26					1,725			1,725	1,900
6888	12" DR 26 x 18" DR 26					1,750			1,750	1,925
6892	14" DR 26 x 20" DR 32.5					2,125			2,125	2,325
6896	16" DR 26 x 22" DR 32.5					3,050			3,050	3,350
6900	18" DR 26 x 24" DR 32.5					3,600			3,600	3,950
6904	20" DR 32.5 x 28" DR 32.5					4,500			4,500	4,950

For customer support on your Mechanical Cost Data, call 877.739.3689.

229

22 11 13 – Facility Water Distribution Piping

22 11 13.78 Pipe, High Density Polyethylene Plastic (HDPE)	Crew	Daily Output	Labor-Hours	Unit	Material	2016 Bare Costs Labor	Equipment	Total	Total Incl O&P
6908	22" DR 32.5 x 30" DR 32.5				Ea.	5,125			5,125	5,625
6912	24" DR 32.5 x 32" DR 32.5				↓	5,825			5,825	6,425
9000	Welding labor per joint, not including welding machine									
9010	Pipe joint size, outer pipe (cost based on the thickest walls)									
9020	Straight pipe									
9050	3" pipe size	4 Skwk	96	.333	Ea.		16.65		16.65	25.50
9060	4" pipe size	"	77	.416			20.50		20.50	32
9070	6" pipe size	5 Skwk	60	.667			33.50		33.50	51
9080	8" pipe size	"	40	1			50		50	77
9090	10" pipe size	6 Skwk	41	1.171			58.50		58.50	90
9100	12" pipe size		39	1.231			61.50		61.50	94.50
9110	14" pipe size		38	1.263			63		63	97
9120	16" pipe size	↓	35	1.371			68.50		68.50	105
9130	18" pipe size	8 Skwk	45	1.422			71		71	109
9140	20" pipe size		42	1.524			76		76	117
9150	22" pipe size		40	1.600			80		80	123
9160	24" pipe size		38	1.684			84		84	129
9170	28" pipe size		37	1.730			86.50		86.50	133
9180	30" pipe size		36	1.778			88.50		88.50	137
9190	32" pipe size		35	1.829			91.50		91.50	141
9200	42" pipe size	↓	32	2	↓		100		100	154
9300	Note: Cost for set up each time welder is moved.									
9301	Add 100% of weld labor cost									
9310	For handling between fitting welds add 50% of weld labor cost									
9320	Welder usually remains stationary with pipe moved through it									
9360	Weld machine, rental per day based on diam. capacity									
9380	3" thru 4" diameter				Ea.			63.50	63.50	70
9390	6" thru 8" diameter							207	207	228
9400	10" thru 12" diameter							310	310	340
9410	14" thru 18" diameter							420	420	460
9420	20" thru 24" diameter							750	750	825
9430	28" thru 32" diameter							880	880	970
9440	42" thru 58" diameter				↓			910	910	1,000

22 11 19 – Domestic Water Piping Specialties

22 11 19.10 Flexible Connectors

		Crew	Daily Output	Labor-Hours	Unit	Material	2016 Bare Costs Labor	Equipment	Total	Total Incl O&P
0010	**FLEXIBLE CONNECTORS**, Corrugated, 7/8" O.D., 1/2" I.D.									
0050	Gas, seamless brass, steel fittings									
0200	12" long	1 Plum	36	.222	Ea.	6.20	13.15		19.35	26.50
0220	18" long		36	.222		7.70	13.15		20.85	28.50
0240	24" long		34	.235		9.15	13.95		23.10	31
0260	30" long		34	.235		9.85	13.95		23.80	32
0280	36" long		32	.250		10.90	14.80		25.70	34.50
0320	48" long		30	.267		13.85	15.80		29.65	39
0340	60" long		30	.267		16.45	15.80		32.25	42
0360	72" long	↓	30	.267	↓	19	15.80		34.80	45
2000	Water, copper tubing, dielectric separators									
2100	12" long	1 Plum	36	.222	Ea.	6.20	13.15		19.35	26.50
2220	15" long		36	.222		6.90	13.15		20.05	27.50
2240	18" long		36	.222		7.50	13.15		20.65	28
2260	24" long	↓	34	.235	↓	9.20	13.95		23.15	31

22 11 19.14 Flexible Metal Hose	Crew	Daily Output	Labor-Hours	Unit	Material	2016 Bare Costs Labor	Equipment	Total	Total Incl O&P
0010 **FLEXIBLE METAL HOSE**, Connectors, standard lengths									
0100 Bronze braided, bronze ends									
0120 3/8" diameter x 12"	1 Stpi	26	.308	Ea.	21	18.70		39.70	51
0140 1/2" diameter x 12"		24	.333		22	20		42	54.50
0160 3/4" diameter x 12"		20	.400		32	24.50		56.50	71.50
0180 1" diameter x 18"		19	.421		38	25.50		63.50	80.50
0200 1-1/2" diameter x 18"		13	.615		52	37.50		89.50	114
0220 2" diameter x 18"		11	.727		68	44		112	142
1000 Carbon steel ends									
1020 1/4" diameter x 12"	1 Stpi	28	.286	Ea.	25	17.35		42.35	53.50
1040 3/8" diameter x 12"		26	.308		25.50	18.70		44.20	56
1060 1/2" diameter x 12"		24	.333		28	20		48	61.50
1080 1/2" diameter x 24"		24	.333		49.50	20		69.50	85
1120 3/4" diameter x 12"		20	.400		45.50	24.50		70	86.50
1140 3/4" diameter x 24"		20	.400		61.50	24.50		86	104
1160 3/4" diameter x 36"		20	.400		73	24.50		97.50	117
1180 1" diameter x 18"		19	.421		57	25.50		82.50	101
1200 1" diameter x 30"		19	.421		71	25.50		96.50	117
1220 1" diameter x 36"		19	.421		85.50	25.50		111	133
1240 1-1/4" diameter x 18"		15	.533		79	32.50		111.50	136
1260 1-1/4" diameter x 36"		15	.533		114	32.50		146.50	175
1280 1-1/2" diameter x 18"		13	.615		111	37.50		148.50	179
1300 1-1/2" diameter x 36"		13	.615		124	37.50		161.50	194
1320 2" diameter x 24"		11	.727		150	44		194	232
1340 2" diameter x 36"		11	.727		154	44		198	236
1360 2-1/2" diameter x 24"		9	.889		355	54		409	470
1380 2-1/2" diameter x 36"		9	.889		560	54		614	695
1400 3" diameter x 24"		7	1.143		485	69.50		554.50	640
1420 3" diameter x 36"		7	1.143		670	69.50		739.50	840
2000 Carbon steel braid, carbon steel solid ends									
2100 1/2" diameter x 12"	1 Stpi	24	.333	Ea.	44.50	20		64.50	79.50
2120 3/4" diameter x 12"		20	.400		67.50	24.50		92	111
2140 1" diameter x 12"		19	.421		96.50	25.50		122	145
2160 1-1/4" diameter x 12"		15	.533		28	32.50		60.50	80
2180 1-1/2" diameter x 12"		13	.615		30	37.50		67.50	89.50
3000 Stainless steel braid, welded on carbon steel ends									
3100 1/2" diameter x 12"	1 Stpi	24	.333	Ea.	61	20		81	97.50
3120 3/4" diameter x 12"		20	.400		75	24.50		99.50	119
3140 3/4" diameter x 24"		20	.400		86	24.50		110.50	131
3160 3/4" diameter x 36"		20	.400		99	24.50		123.50	146
3180 1" diameter x 12"		19	.421		89	25.50		114.50	136
3200 1" diameter x 24"		19	.421		107	25.50		132.50	157
3220 1" diameter x 36"		19	.421		123	25.50		148.50	175
3240 1-1/4" diameter x 12"		15	.533		134	32.50		166.50	196
3260 1-1/4" diameter x 24"		15	.533		146	32.50		178.50	210
3280 1-1/4" diameter x 36"		15	.533		166	32.50		198.50	231
3300 1-1/2" diameter x 12"		13	.615		44	37.50		81.50	105
3320 1-1/2" diameter x 24"		13	.615		160	37.50		197.50	233
3340 1-1/2" diameter x 36"		13	.615		187	37.50		224.50	263
3400 Metal stainless steel braid, over corrugated stainless steel, flanged ends									
3410 150 PSI									
3420 1/2" diameter x 12"	1 Stpi	24	.333	Ea.	120	20		140	163

For customer support on your Mechanical Cost Data, call 877.739.3689.

231

22 11 19 – Domestic Water Piping Specialties

22 11 19.14 Flexible Metal Hose

		Crew	Daily Output	Labor-Hours	Unit	Material	2016 Bare Costs Labor	Equipment	Total	Total Incl O&P
3430	1" diameter x 12"	1 Stpi	20	.400	Ea.	180	24.50		204.50	235
3440	1-1/2" diameter x 12"		15	.533		204	32.50		236.50	273
3450	2-1/2" diameter x 9"		12	.667		108	40.50		148.50	180
3460	3" diameter x 9"		9	.889		98	54		152	190
3470	4" diameter x 9"		7	1.143		126	69.50		195.50	244
3480	4" diameter x 30"		5	1.600		560	97		657	760
3490	4" diameter x 36"		4.80	1.667		580	101		681	795
3500	6" diameter x 11"		5	1.600		225	97		322	395
3510	6" diameter x 36"		3.80	2.105		685	128		813	945
3520	8" diameter x 12"		4	2		410	121		531	635
3530	10" diameter x 13"		3	2.667		625	162		787	935
3540	12" diameter x 14"	Q-5	4	4		950	219		1,169	1,375
6000	Molded rubber with helical wire reinforcement									
6010	150 PSI									
6020	1-1/2" diameter x 12"	1 Stpi	15	.533	Ea.	90	32.50		122.50	148
6030	2" diameter x 12"		12	.667		240	40.50		280.50	325
6040	3" diameter x 12"		8	1		251	60.50		311.50	370
6050	4" diameter x 12"		6	1.333		320	81		401	470
6060	6" diameter x 18"		4	2		475	121		596	710
6070	8" diameter x 24"		3	2.667		665	162		827	980
6080	10" diameter x 24"		2	4		810	243		1,053	1,250
6090	12" diameter x 24"	Q-5	3	5.333		925	291		1,216	1,475
7000	Molded teflon with stainless steel flanges									
7010	150 PSI									
7020	2-1/2" diameter x 3-3/16"	Q-1	7.80	2.051	Ea.	2,325	109		2,434	2,725
7030	3" diameter x 3-5/8"		6.50	2.462		1,625	131		1,756	1,975
7040	4" diameter x 3-5/8"		5	3.200		2,100	171		2,271	2,575
7050	6" diameter x 4"		4.30	3.721		2,950	198		3,148	3,550
7060	8" diameter x 6"		3.80	4.211		4,700	224		4,924	5,525

22 11 19.18 Mixing Valve

		Crew	Daily Output	Labor-Hours	Unit	Material	2016 Bare Costs Labor	Equipment	Total	Total Incl O&P
0010	**MIXING VALVE**, Automatic, water tempering.									
0040	1/2" size	1 Stpi	19	.421	Ea.	555	25.50		580.50	650
0050	3/4" size		18	.444		555	27		582	650
0100	1" size		16	.500		845	30.50		875.50	975
0120	1-1/4" size		13	.615		1,150	37.50		1,187.50	1,325
0140	1-1/2" size		10	.800		1,400	48.50		1,448.50	1,600
0160	2" size		8	1		1,750	60.50		1,810.50	2,025
0170	2-1/2" size		6	1.333		1,750	81		1,831	2,050
0180	3" size		4	2		4,125	121		4,246	4,725
0190	4" size		3	2.667		4,125	162		4,287	4,800

22 11 19.22 Pressure Reducing Valve

		Crew	Daily Output	Labor-Hours	Unit	Material	2016 Bare Costs Labor	Equipment	Total	Total Incl O&P
0010	**PRESSURE REDUCING VALVE**, Steam, pilot operated.									
0100	Threaded, iron body									
0200	1-1/2" size	1 Stpi	8	1	Ea.	2,000	60.50		2,060.50	2,300
0220	2" size	"	5	1.600	"	2,300	97		2,397	2,675
1000	Flanged, iron body, 125 lb. flanges									
1020	2" size	1 Stpi	8	1	Ea.	2,325	60.50		2,385.50	2,650
1040	2-1/2" size	"	4	2		2,675	121		2,796	3,100
1060	3" size	Q-5	4.50	3.556		3,000	194		3,194	3,600
1080	4" size	"	3	5.333		4,375	291		4,666	5,250
1500	For 250 lb. flanges, add					5%				

22 11 19.26 Pressure Regulators		Crew	Daily Output	Labor-Hours	Unit	Material	2016 Bare Costs Labor	Equipment	Total	Total Incl O&P
0010	**PRESSURE REGULATORS**									
0100	Gas appliance regulators									
0106	Main burner and pilot applications									
0108	Rubber seat poppet type									
0109	1/8" pipe size	1 Stpi	24	.333	Ea.	16.50	20		36.50	48.50
0110	1/4" pipe size		24	.333		18.95	20		38.95	51.50
0112	3/8" pipe size		24	.333		20	20		40	52.50
0113	1/2" pipe size		24	.333		22.50	20		42.50	55
0114	3/4" pipe size		20	.400		26	24.50		50.50	65.50
0122	Lever action type									
0123	3/8" pipe size	1 Stpi	24	.333	Ea.	24.50	20		44.50	57.50
0124	1/2" pipe size		24	.333		24.50	20		44.50	57.50
0125	3/4" pipe size		20	.400		48	24.50		72.50	89.50
0126	1" pipe size		19	.421		48	25.50		73.50	91.50
0132	Double diaphragm type									
0133	3/8" pipe size	1 Stpi	24	.333	Ea.	37	20		57	71
0134	1/2" pipe size		24	.333		50.50	20		70.50	86.50
0135	3/4" pipe size		20	.400		84.50	24.50		109	130
0136	1" pipe size		19	.421		103	25.50		128.50	153
0137	1-1/4" pipe size		15	.533		360	32.50		392.50	445
0138	1-1/2" pipe size		13	.615		660	37.50		697.50	780
0139	2" pipe size		11	.727		660	44		704	790
0140	2-1/2" pipe size	Q-5	15	1.067		1,275	58.50		1,333.50	1,500
0141	3" pipe size		13	1.231		1,275	67		1,342	1,500
0142	4" pipe size (flanged)		8	2		2,275	109		2,384	2,675
0160	Main burner only									
0162	Straight-thru-flow design									
0163	1/2" pipe size	1 Stpi	24	.333	Ea.	41.50	20		61.50	76
0164	3/4" pipe size		20	.400		56.50	24.50		81	98.50
0165	1" pipe size		19	.421		80	25.50		105.50	127
0166	1-1/4" pipe size		15	.533		80	32.50		112.50	137
0200	Oil, light, hot water, ordinary steam, threaded									
0220	Bronze body, 1/4" size	1 Stpi	24	.333	Ea.	122	20		142	166
0230	3/8" size		24	.333		144	20		164	190
0240	1/2" size		24	.333		153	20		173	199
0250	3/4" size		20	.400		185	24.50		209.50	241
0260	1" size		19	.421		242	25.50		267.50	305
0270	1-1/4" size		15	.533		335	32.50		367.50	420
0280	1-1/2" size		13	.615		375	37.50		412.50	465
0290	2" size		11	.727		540	44		584	660
0320	Iron body, 1/4" size		24	.333		157	20		177	203
0330	3/8" size		24	.333		169	20		189	217
0340	1/2" size		24	.333		210	20		230	262
0350	3/4" size		20	.400		251	24.50		275.50	315
0360	1" size		19	.421		380	25.50		405.50	460
0370	1-1/4" size		15	.533		515	32.50		547.50	615
0380	1-1/2" size		13	.615		555	37.50		592.50	665
0390	2" size		11	.727		1,000	44		1,044	1,175
0500	Oil, heavy, viscous fluids, threaded									
0520	Bronze body, 3/8" size	1 Stpi	24	.333	Ea.	269	20		289	325
0530	1/2" size		24	.333		345	20		365	410
0540	3/4" size		20	.400		385	24.50		409.50	460

For customer support on your Mechanical Cost Data, call 877.739.3689.

233

22 11 Facility Water Distribution

22 11 19 – Domestic Water Piping Specialties

22 11 19.26 Pressure Regulators		Crew	Daily Output	Labor-Hours	Unit	Material	2016 Bare Costs Labor	Equipment	Total	Total Incl O&P
0550	1" size	1 Stpi	19	.421	Ea.	480	25.50		505.50	570
0560	1-1/4" size		15	.533		680	32.50		712.50	795
0570	1-1/2" size		13	.615		780	37.50		817.50	910
0600	Iron body, 3/8" size		24	.333		214	20		234	267
0620	1/2" size		24	.333		260	20		280	315
0630	3/4" size		20	.400		294	24.50		318.50	360
0640	1" size		19	.421		350	25.50		375.50	425
0650	1-1/4" size		15	.533		505	32.50		537.50	605
0660	1-1/2" size	↓	13	.615	↓	540	37.50		577.50	650
0800	Process steam, wet or super heated, monel trim, threaded									
0820	Bronze body, 1/4" size	1 Stpi	24	.333	Ea.	615	20		635	705
0830	3/8" size		24	.333		615	20		635	705
0840	1/2" size		24	.333		705	20		725	805
0850	3/4" size		20	.400		830	24.50		854.50	945
0860	1" size		19	.421		1,050	25.50		1,075.50	1,200
0870	1-1/4" size		15	.533		1,300	32.50		1,332.50	1,475
0880	1-1/2" size		13	.615		1,550	37.50		1,587.50	1,750
0920	Iron body, max 125 PSIG press out, 1/4" size		24	.333		815	20		835	925
0930	3/8" size		24	.333		815	20		835	925
0940	1/2" size		24	.333		1,000	20		1,020	1,150
0950	3/4" size		20	.400		1,225	24.50		1,249.50	1,375
0960	1" size		19	.421		1,575	25.50		1,600.50	1,775
0970	1-1/4" size		15	.533		1,875	32.50		1,907.50	2,125
0980	1-1/2" size	↓	13	.615	↓	2,250	37.50		2,287.50	2,525
3000	Steam, high capacity, bronze body, stainless steel trim									
3020	Threaded, 1/2" diameter	1 Stpi	24	.333	Ea.	2,350	20		2,370	2,600
3030	3/4" diameter		24	.333		2,350	20		2,370	2,600
3040	1" diameter		19	.421		2,625	25.50		2,650.50	2,925
3060	1-1/4" diameter		15	.533		2,900	32.50		2,932.50	3,225
3080	1-1/2" diameter		13	.615		3,300	37.50		3,337.50	3,700
3100	2" diameter	↓	11	.727		4,050	44		4,094	4,525
3120	2-1/2" diameter	Q-5	12	1.333		5,075	73		5,148	5,700
3140	3" diameter	"	11	1.455	↓	5,800	79.50		5,879.50	6,500
3500	Flanged connection, iron body, 125 lb. W.S.P.									
3520	3" diameter	Q-5	11	1.455	Ea.	6,350	79.50		6,429.50	7,100
3540	4" diameter	"	5	3.200	"	8,000	175		8,175	9,075
9002	For water pressure regulators, see Section 22 05 23.20									

22 11 19.30 Pressure and Temperature Safety Plug

	PRESSURE & TEMPERATURE SAFETY PLUG	Crew	Daily Output	Labor-Hours	Unit	Material	Labor	Equipment	Total	Total Incl O&P
0010	**PRESSURE & TEMPERATURE SAFETY PLUG**									
1000	3/4" external thread, 3/8" diam. element									
1020	Carbon steel									
1050	7-1/2" insertion	1 Stpi	32	.250	Ea.	45	15.20		60.20	72.50
1120	304 stainless steel									
1150	7-1/2" insertion	1 Stpi	32	.250	Ea.	45	15.20		60.20	72.50
1220	316 Stainless steel									
1250	7-1/2" insertion	1 Stpi	32	.250	Ea.	49	15.20		64.20	77

22 11 19.32 Pressure and Temperature Measurement Plug

	PRESSURE & TEMPERATURE MEASUREMENT PLUG	Crew	Daily Output	Labor-Hours	Unit	Material	Labor	Equipment	Total	Total Incl O&P
0010	**PRESSURE & TEMPERATURE MEASUREMENT PLUG**									
0020	A permanent access port for insertion of									
0030	a pressure or temperature measuring probe									
0100	Plug, brass									
0110	1/4" MNPT, 1-1/2" long	1 Stpi	32	.250	Ea.	5.05	15.20		20.25	28.50

22 11 Facility Water Distribution

22 11 19 – Domestic Water Piping Specialties

22 11 19.32 Pressure and Temperature Measurement Plug		Crew	Daily Output	Labor-Hours	Unit	Material	2016 Bare Costs Labor	Equipment	Total	Total Incl O&P
0120	3" long	1 Stpi	31	.258	Ea.	9.70	15.65		25.35	34
0140	1/2" MNPT, 1-1/2" long		30	.267		6.75	16.20		22.95	32
0150	3" long	↓	29	.276	↓	11.95	16.75		28.70	38.50
0200	Pressure gauge probe adapter									
0210	1/8" diameter, 1-1/2" probe				Ea.	16.55			16.55	18.20
0220	3" probe				"	28			28	30.50
0300	Temperature gauge, 5" stem									
0310	Analog				Ea.	10.40			10.40	11.45
0330	Digital				"	28.50			28.50	31.50
0400	Pressure gauge, compound									
0410	1/4" MNPT				Ea.	27.50			27.50	30.50
0500	Pressure and temperature test kit									
0510	Contains 2 thermometers and 2 pressure gauges									
0520	Kit				Ea.	330			330	365

22 11 19.34 Sleeves and Escutcheons

		Crew	Daily Output	Labor-Hours	Unit	Material	2016 Bare Costs Labor	Equipment	Total	Total Incl O&P
0010	**SLEEVES & ESCUTCHEONS**									
0100	Pipe sleeve									
0110	Steel, w/water stop, 12" long, with link seal									
0120	2" diam. for 1/2" carrier pipe	1 Plum	8.40	.952	Ea.	60.50	56.50		117	152
0130	2-1/2" diam. for 3/4" carrier pipe		8	1		68.50	59		127.50	165
0140	2-1/2" diam. for 1" carrier pipe		8	1		64.50	59		123.50	161
0150	3" diam. for 1-1/4" carrier pipe		7.20	1.111		83.50	66		149.50	191
0160	3-1/2" diam. for 1-1/2" carrier pipe		6.80	1.176		84	69.50		153.50	198
0170	4" diam. for 2" carrier pipe		6	1.333		90	79		169	218
0180	4" diam. for 2-1/2" carrier pipe		6	1.333		91.50	79		170.50	220
0190	5" diam. for 3" carrier pipe		5.40	1.481		104	87.50		191.50	246
0200	6" diam. for 4" carrier pipe	↓	4.80	1.667		125	98.50		223.50	287
0210	10" diam. for 6" carrier pipe	Q-1	8	2		249	107		356	435
0220	12" diam. for 8" carrier pipe		7.20	2.222		335	118		453	550
0230	14" diam. for 10" carrier pipe		6.40	2.500		350	133		483	585
0240	16" diam. for 12" carrier pipe		5.80	2.759		390	147		537	650
0250	18" diam. for 14" carrier pipe		5.20	3.077		720	164		884	1,025
0260	24" diam. for 18" carrier pipe		4	4		1,075	213		1,288	1,500
0270	24" diam. for 20" carrier pipe		4	4		900	213		1,113	1,300
0280	30" diam. for 24" carrier pipe	↓	3.20	5	↓	1,325	266		1,591	1,850
0500	Wall sleeve									
0510	Ductile iron with rubber gasket seal									
0520	3"	1 Plum	8.40	.952	Ea.	700	56.50		756.50	855
0530	4"		7.20	1.111		750	66		816	925
0540	6"		6	1.333		920	79		999	1,150
0550	8"		4	2		1,125	118		1,243	1,425
0560	10"		3	2.667		1,375	158		1,533	1,775
0570	12"	↓	2.40	3.333	↓	1,625	197		1,822	2,100
5000	Escutcheon									
5100	Split ring, pipe									
5110	Chrome plated									
5120	1/2"	1 Plum	160	.050	Ea.	1.12	2.96		4.08	5.70
5130	3/4"		160	.050		1.25	2.96		4.21	5.85
5140	1"		135	.059		1.34	3.51		4.85	6.75
5150	1-1/2"		115	.070		1.88	4.12		6	8.25
5160	2"		100	.080		2.09	4.74		6.83	9.45
5170	4"	↓	80	.100		4.55	5.90		10.45	13.95

For customer support on your Mechanical Cost Data, call 877.739.3689.

235

22 11 Facility Water Distribution

22 11 19 – Domestic Water Piping Specialties

22 11 19.34 Sleeves and Escutcheons	Crew	Daily Output	Labor-Hours	Unit	Material	2016 Bare Costs Labor	Equipment	Total	Total Incl O&P
5180	6"	1 Plum	68	.118	Ea.	7.55	6.95		14.50	18.80
5400	Shallow flange type									
5410	Chrome plated steel									
5420	1/2" CTS	1 Plum	180	.044	Ea.	.33	2.63		2.96	4.33
5430	3/4" CTS		180	.044		.35	2.63		2.98	4.36
5440	1/2" IPS		180	.044		.37	2.63		3	4.38
5450	3/4" IPS		180	.044		.41	2.63		3.04	4.42
5460	1" IPS		175	.046		.47	2.71		3.18	4.60
5470	1-1/2" IPS		170	.047		.78	2.79		3.57	5.05
5480	2" IPS		160	.050		.94	2.96		3.90	5.50

22 11 19.38 Water Supply Meters

		Crew	Daily Output	Labor-Hours	Unit	Material	2016 Bare Costs Labor	Equipment	Total	Total Incl O&P
0010	**WATER SUPPLY METERS**									
1000	Detector, serves dual systems such as fire and domestic or									
1020	process water, wide range cap., UL and FM approved									
1100	3" mainline x 2" by-pass, 400 GPM	Q-1	3.60	4.444	Ea.	7,450	237		7,687	8,525
1140	4" mainline x 2" by-pass, 700 GPM	"	2.50	6.400		7,450	340		7,790	8,700
1180	6" mainline x 3" by-pass, 1600 GPM	Q-2	2.60	9.231		11,400	510		11,910	13,300
1220	8" mainline x 4" by-pass, 2800 GPM		2.10	11.429		16,900	630		17,530	19,500
1260	10" mainline x 6" by-pass, 4400 GPM		2	12		24,100	665		24,765	27,500
1300	10" x 12" mainlines x 6" by-pass, 5400 GPM		1.70	14.118		32,700	780		33,480	37,200
2000	Domestic/commercial, bronze									
2020	Threaded									
2060	5/8" diameter, to 20 GPM	1 Plum	16	.500	Ea.	50	29.50		79.50	99.50
2080	3/4" diameter, to 30 GPM		14	.571		91	34		125	151
2100	1" diameter, to 50 GPM		12	.667		138	39.50		177.50	212
2300	Threaded/flanged									
2340	1-1/2" diameter, to 100 GPM	1 Plum	8	1	Ea.	340	59		399	460
2360	2" diameter, to 160 GPM	"	6	1.333	"	460	79		539	625
2600	Flanged, compound									
2640	3" diameter, 320 GPM	Q-1	3	5.333	Ea.	2,150	284		2,434	2,800
2660	4" diameter, to 500 GPM		1.50	10.667		3,450	570		4,020	4,650
2680	6" diameter, to 1,000 GPM		1	16		5,650	850		6,500	7,475
2700	8" diameter, to 1,800 GPM		.80	20		8,775	1,075		9,850	11,300
7000	Turbine									
7260	Flanged									
7300	2" diameter, to 160 GPM	1 Plum	7	1.143	Ea.	580	67.50		647.50	740
7320	3" diameter, to 450 GPM	Q-1	3.60	4.444		920	237		1,157	1,350
7340	4" diameter, to 650 GPM	"	2.50	6.400		1,525	340		1,865	2,200
7360	6" diameter, to 1800 GPM	Q-2	2.60	9.231		2,775	510		3,285	3,825
7380	8" diameter, to 2500 GPM		2.10	11.429		4,250	630		4,880	5,625
7400	10" diameter, to 5500 GPM		1.70	14.118		5,700	780		6,480	7,450

22 11 19.42 Backflow Preventers

		Crew	Daily Output	Labor-Hours	Unit	Material	2016 Bare Costs Labor	Equipment	Total	Total Incl O&P
0010	**BACKFLOW PREVENTERS**, Includes valves									
0020	and four test cocks, corrosion resistant, automatic operation									
1000	Double check principle									
1010	Threaded, with ball valves									
1020	3/4" pipe size	1 Plum	16	.500	Ea.	205	29.50		234.50	270
1030	1" pipe size		14	.571		234	34		268	310
1040	1-1/2" pipe size		10	.800		500	47.50		547.50	620
1050	2" pipe size		7	1.143		580	67.50		647.50	740
1080	Threaded, with gate valves									
1100	3/4" pipe size	1 Plum	16	.500	Ea.	960	29.50		989.50	1,100

22 11 19.42 Backflow Preventers

		Crew	Daily Output	Labor-Hours	Unit	Material	2016 Bare Costs Labor	Equipment	Total	Total Incl O&P
1120	1" pipe size	1 Plum	14	.571	Ea.	970	34		1,004	1,125
1140	1-1/2" pipe size		10	.800		1,250	47.50		1,297.50	1,450
1160	2" pipe size	↓	7	1.143	↓	1,525	67.50		1,592.50	1,775
1200	Flanged, valves are gate									
1210	3" pipe size	Q-1	4.50	3.556	Ea.	2,350	189		2,539	2,850
1220	4" pipe size	"	3	5.333		2,425	284		2,709	3,100
1230	6" pipe size	Q-2	3	8		3,600	440		4,040	4,625
1240	8" pipe size		2	12		6,525	665		7,190	8,175
1250	10" pipe size	↓	1	24	↓	9,675	1,325		11,000	12,600
1300	Flanged, valves are OS&Y									
1370	1" pipe size	1 Plum	5	1.600	Ea.	1,075	94.50		1,169.50	1,325
1374	1-1/2" pipe size		5	1.600		1,350	94.50		1,444.50	1,650
1378	2" pipe size	↓	4.80	1.667		1,650	98.50		1,748.50	1,950
1380	3" pipe size	Q-1	4.50	3.556		2,850	189		3,039	3,400
1400	4" pipe size	"	3	5.333		3,600	284		3,884	4,375
1420	6" pipe size	Q-2	3	8		5,925	440		6,365	7,200
1430	8" pipe size	"	2	12	↓	12,600	665		13,265	14,900
4000	Reduced pressure principle									
4100	Threaded, bronze, valves are ball									
4120	3/4" pipe size	1 Plum	16	.500	Ea.	450	29.50		479.50	540
4140	1" pipe size		14	.571		485	34		519	585
4150	1-1/4" pipe size		12	.667		890	39.50		929.50	1,050
4160	1-1/2" pipe size		10	.800		975	47.50		1,022.50	1,150
4180	2" pipe size	↓	7	1.143	↓	1,100	67.50		1,167.50	1,300
5000	Flanged, bronze, valves are OS&Y									
5060	2-1/2" pipe size	Q-1	5	3.200	Ea.	4,950	171		5,121	5,700
5080	3" pipe size		4.50	3.556		5,625	189		5,814	6,475
5100	4" pipe size	↓	3	5.333		7,050	284		7,334	8,175
5120	6" pipe size	Q-2	3	8	↓	10,200	440		10,640	12,000
5200	Flanged, iron, valves are gate									
5210	2-1/2" pipe size	Q-1	5	3.200	Ea.	2,225	171		2,396	2,700
5220	3" pipe size		4.50	3.556		2,325	189		2,514	2,825
5230	4" pipe size	↓	3	5.333		3,150	284		3,434	3,875
5240	6" pipe size	Q-2	3	8		4,425	440		4,865	5,550
5250	8" pipe size		2	12		7,950	665		8,615	9,750
5260	10" pipe size	↓	1	24	↓	11,200	1,325		12,525	14,300
5600	Flanged, iron, valves are OS&Y									
5660	2-1/2" pipe size	Q-1	5	3.200	Ea.	2,525	171		2,696	3,025
5680	3" pipe size		4.50	3.556		2,650	189		2,839	3,200
5700	4" pipe size	↓	3	5.333		3,350	284		3,634	4,100
5720	6" pipe size	Q-2	3	8		4,825	440		5,265	6,000
5740	8" pipe size		2	12		8,500	665		9,165	10,400
5760	10" pipe size	↓	1	24	↓	11,400	1,325		12,725	14,500

22 11 19.50 Vacuum Breakers

		Crew	Daily Output	Labor-Hours	Unit	Material	2016 Bare Costs Labor	Equipment	Total	Total Incl O&P
0010	**VACUUM BREAKERS**									
0013	See also backflow preventers Section 22 11 19.42									
1000	Anti-siphon continuous pressure type									
1010	Max. 150 PSI - 210°F									
1020	Bronze body									
1030	1/2" size	1 Stpi	24	.333	Ea.	161	20		181	208
1040	3/4" size		20	.400	↓	161	24.50		185.50	214
1050	1" size	↓	19	.421		166	25.50		191.50	222

For customer support on your Mechanical Cost Data, call 877.739.3689.

237

22 11 19.50 Vacuum Breakers

22 11 19.50 Vacuum Breakers		Crew	Daily Output	Labor-Hours	Unit	Material	2016 Bare Costs Labor	Equipment	Total	Total Incl O&P
1060	1-1/4" size	1 Stpi	15	.533	Ea.	325	32.50		357.50	410
1070	1-1/2" size		13	.615		405	37.50		442.50	500
1080	2" size		11	.727		415	44		459	520
1200	Max. 125 PSI with atmospheric vent									
1210	Brass, in-line construction									
1220	1/4" size	1 Stpi	24	.333	Ea.	126	20		146	170
1230	3/8" size	"	24	.333		126	20		146	170
1260	For polished chrome finish, add					13%				
2000	Anti-siphon, non-continuous pressure type									
2010	Hot or cold water 125 PSI - 210°F									
2020	Bronze body									
2030	1/4" size	1 Stpi	24	.333	Ea.	69	20		89	107
2040	3/8" size		24	.333		69	20		89	107
2050	1/2" size		24	.333		78	20		98	116
2060	3/4" size		20	.400		93	24.50		117.50	139
2070	1" size		19	.421		144	25.50		169.50	197
2080	1-1/4" size		15	.533		252	32.50		284.50	325
2090	1-1/2" size		13	.615		296	37.50		333.50	380
2100	2" size		11	.727		460	44		504	570
2110	2-1/2" size		8	1		1,325	60.50		1,385.50	1,550
2120	3" size		6	1.333		1,750	81		1,831	2,050
2150	For polished chrome finish, add					50%				
3000	Air gap fitting									
3020	1/2" NPT size	1 Plum	19	.421	Ea.	54	25		79	96.50
3030	1" NPT size	"	15	.533		62	31.50		93.50	116
3040	2" NPT size	Q-1	21	.762		121	40.50		161.50	195
3050	3" NPT size		14	1.143		239	61		300	355
3060	4" NPT size		10	1.600		239	85.50		324.50	390

22 11 19.54 Water Hammer Arresters/Shock Absorbers

22 11 19.54 Water Hammer Arresters/Shock Absorbers		Crew	Daily Output	Labor-Hours	Unit	Material	2016 Bare Costs Labor	Equipment	Total	Total Incl O&P
0010	**WATER HAMMER ARRESTERS/SHOCK ABSORBERS**									
0490	Copper									
0500	3/4" male I.P.S. For 1 to 11 fixtures	1 Plum	12	.667	Ea.	28	39.50		67.50	90.50
0600	1" male I.P.S. For 12 to 32 fixtures		8	1		47.50	59		106.50	142
0700	1-1/4" male I.P.S. For 33 to 60 fixtures		8	1		48	59		107	143
0800	1-1/2" male I.P.S. For 61 to 113 fixtures		8	1		69	59		128	166
0900	2" male I.P.S. For 114 to 154 fixtures		8	1		101	59		160	201
1000	2-1/2" male I.P.S. For 155 to 330 fixtures		4	2		305	118		423	515
4000	Bellows type									
4010	3/4" FNPT, to 11 fixture units	1 Plum	10	.800	Ea.	273	47.50		320.50	370
4020	1" FNPT, to 32 fixture units		8.80	.909		550	54		604	685
4030	1" FNPT, to 60 fixture units		8.80	.909		830	54		884	995
4040	1" FNPT, to 113 fixture units		8.80	.909		2,075	54		2,129	2,350
4050	1" FNPT, to 154 fixture units		6.60	1.212		2,350	72		2,422	2,675
4060	1-1/2" FNPT, to 300 fixture units		5	1.600		2,875	94.50		2,969.50	3,325

22 11 19.64 Hydrants

22 11 19.64 Hydrants		Crew	Daily Output	Labor-Hours	Unit	Material	2016 Bare Costs Labor	Equipment	Total	Total Incl O&P
0010	**HYDRANTS**									
0050	Wall type, moderate climate, bronze, encased									
0200	3/4" IPS connection	1 Plum	16	.500	Ea.	800	29.50		829.50	925
0300	1" IPS connection		14	.571		910	34		944	1,050
0500	Anti-siphon type, 3/4" connection		16	.500		690	29.50		719.50	805
1000	Non-freeze, bronze, exposed									
1100	3/4" IPS connection, 4" to 9" thick wall	1 Plum	14	.571	Ea.	515	34		549	615

22 11 19 – Domestic Water Piping Specialties

22 11 19.64 Hydrants	Crew	Daily Output	Labor-Hours	Unit	Material	2016 Bare Costs Labor	Equipment	Total	Total Incl O&P
1120	10" to 14" thick wall	1 Plum	12	.667	Ea.	560	39.50		599.50	675
1140	15" to 19" thick wall		12	.667		625	39.50		664.50	750
1160	20" to 24" thick wall	↓	10	.800	↓	680	47.50		727.50	820
1200	For 1" IPS connection, add					15%	10%			
1240	For 3/4" adapter type vacuum breaker, add				Ea.	65			65	71.50
1280	For anti-siphon type, add				"	136			136	149
2000	Non-freeze bronze, encased, anti-siphon type									
2100	3/4" IPS connection, 5" to 9" thick wall	1 Plum	14	.571	Ea.	1,175	34		1,209	1,350
2120	10" to 14" thick wall		12	.667		1,400	39.50		1,439.50	1,600
2140	15" to 19" thick wall		12	.667		1,475	39.50		1,514.50	1,675
2160	20" to 24" thick wall	↓	10	.800	↓	1,525	47.50		1,572.50	1,750
2200	For 1" IPS connection, add					10%	10%			
3000	Ground box type, bronze frame, 3/4" IPS connection									
3080	Non-freeze, all bronze, polished face, set flush									
3100	2 feet depth of bury	1 Plum	8	1	Ea.	1,025	59		1,084	1,225
3120	3 feet depth of bury		8	1		1,100	59		1,159	1,300
3140	4 feet depth of bury		8	1		1,175	59		1,234	1,400
3160	5 feet depth of bury		7	1.143		1,275	67.50		1,342.50	1,500
3180	6 feet depth of bury		7	1.143		1,350	67.50		1,417.50	1,575
3200	7 feet depth of bury		6	1.333		1,425	79		1,504	1,675
3220	8 feet depth of bury		5	1.600		1,500	94.50		1,594.50	1,800
3240	9 feet depth of bury		4	2		1,575	118		1,693	1,900
3260	10 feet depth of bury	↓	4	2		1,650	118		1,768	2,000
3400	For 1" IPS connection, add					15%	10%			
3450	For 1-1/4" IPS connection, add					325%	14%			
3500	For 1-1/2" connection, add					370%	18%			
3550	For 2" connection, add					445%	24%			
3600	For tapped drain port in box, add				↓	86			86	94.50
4000	Non-freeze, CI body, bronze frame & scoriated cover									
4010	with hose storage									
4100	2 feet depth of bury	1 Plum	7	1.143	Ea.	1,875	67.50		1,942.50	2,175
4120	3 feet depth of bury		7	1.143		1,975	67.50		2,042.50	2,275
4140	4 feet depth of bury		7	1.143		2,025	67.50		2,092.50	2,325
4160	5 feet depth of bury		6.50	1.231		2,075	73		2,148	2,375
4180	6 feet depth of bury		6	1.333		2,100	79		2,179	2,450
4200	7 feet depth of bury		5.50	1.455		2,175	86		2,261	2,525
4220	8 feet depth of bury		5	1.600		2,200	94.50		2,294.50	2,575
4240	9 feet depth of bury		4.50	1.778		2,250	105		2,355	2,625
4260	10 feet depth of bury	↓	4	2		2,400	118		2,518	2,825
4280	For 1" IPS connection, add					390			390	425
4300	For tapped drain port in box, add				↓	86			86	94.50
5000	Moderate climate, all bronze, polished face									
5020	and scoriated cover, set flush									
5100	3/4" IPS connection	1 Plum	16	.500	Ea.	710	29.50		739.50	825
5120	1" IPS connection	"	14	.571		870	34		904	1,000
5200	For tapped drain port in box, add				↓	86			86	94.50
6000	Ground post type, all non-freeze, all bronze, aluminum casing									
6010	guard, exposed head, 3/4" IPS connection									
6100	2 feet depth of bury	1 Plum	8	1	Ea.	925	59		984	1,125
6120	3 feet depth of bury		8	1		1,000	59		1,059	1,200
6140	4 feet depth of bury		8	1		1,075	59		1,134	1,300
6160	5 feet depth of bury		7	1.143		1,250	67.50		1,317.50	1,475
6180	6 feet depth of bury	↓	7	1.143		1,325	67.50		1,392.50	1,575

22 11 19 – Domestic Water Piping Specialties

22 11 19.64 Hydrants	Crew	Daily Output	Labor-Hours	Unit	Material	2016 Bare Costs Labor	Equipment	Total	Total Incl O&P
6200	7 feet depth of bury	1 Plum	6	1.333	Ea.	1,425	79		1,504	1,700
6220	8 feet depth of bury		5	1.600		1,525	94.50		1,619.50	1,825
6240	9 feet depth of bury		4	2		1,600	118		1,718	1,950
6260	10 feet depth of bury		4	2		1,700	118		1,818	2,050
6300	For 1" IPS connection, add					40%	10%			
6350	For 1-1/4" IPS connection, add					140%	14%			
6400	For 1-1/2" IPS connection, add					225%	18%			
6450	For 2" IPS connection, add					315%	24%			

22 11 23 – Domestic Water Pumps

22 11 23.10 General Utility Pumps

		Crew	Daily Output	Labor-Hours	Unit	Material	2016 Bare Costs Labor	Equipment	Total	Total Incl O&P
0010	**GENERAL UTILITY PUMPS**									
2000	Single stage									
3000	Double suction,									
3140	50 HP, 5"D. x 6"S.	Q-2	.33	72.727	Ea.	9,925	4,025		13,950	17,000
3180	60 HP, 6"D. x 8"S.	Q-3	.30	106		14,500	6,000		20,500	25,000
3190	75 HP, to 2500 GPM		.28	114		22,000	6,450		28,450	33,900
3220	100 HP, to 3000 GPM		.26	123		28,000	6,925		34,925	41,200
3240	150 HP, to 4000 GPM		.24	133		37,300	7,525		44,825	52,500
4000	Centrifugal, end suction, mounted on base									
4010	Horizontal mounted, with drip proof motor, rated @ 100' head									
4020	Vertical split case, single stage									
4040	100 GPM, 5 HP, 1-1/2" discharge	Q-1	1.70	9.412	Ea.	3,975	500		4,475	5,125
4050	200 GPM, 10 HP, 2" discharge		1.30	12.308		4,325	655		4,980	5,750
4060	250 GPM, 10 HP, 3" discharge		1.28	12.500		4,825	665		5,490	6,300
4070	300 GPM, 15 HP, 2" discharge	Q-2	1.56	15.385		5,650	850		6,500	7,500
4080	500 GPM, 20 HP, 4" discharge		1.44	16.667		6,500	920		7,420	8,550
4090	750 GPM, 30 HP, 4" discharge		1.20	20		6,850	1,100		7,950	9,225
4100	1050 GPM, 40 HP, 5" discharge		1	24		9,925	1,325		11,250	12,900
4110	1500 GPM, 60 HP, 6" discharge		.60	40		14,500	2,200		16,700	19,200
4120	2000 GPM, 75 HP, 6" discharge		.50	48		15,800	2,650		18,450	21,400
4130	3000 GPM, 100 HP, 8" discharge		.40	60		20,500	3,325		23,825	27,500
4200	Horizontal split case, single stage									
4210	100 GPM, 7.5 HP, 1-1/2" discharge	Q-1	1.70	9.412	Ea.	3,775	500		4,275	4,900
4220	250 GPM, 15 HP, 2-1/2" discharge	"	1.30	12.308		8,675	655		9,330	10,500
4230	500 GPM, 20 HP, 4" discharge	Q-2	1.60	15		10,600	830		11,430	12,900
4240	750 GPM, 25 HP, 5" discharge		1.54	15.584		11,000	860		11,860	13,400
4250	1000 GPM, 40 HP, 5" discharge		1.20	20		15,300	1,100		16,400	18,500
4260	1500 GPM, 50 HP, 6" discharge	Q-3	1.42	22.535		18,000	1,275		19,275	21,700
4270	2000 GPM, 75 HP, 8" discharge		1.14	28.070		24,400	1,575		25,975	29,300
4280	3000 GPM, 100 HP, 10" discharge		.96	33.333		27,300	1,875		29,175	32,900
4290	3500 GPM, 150 HP, 10" discharge		.86	37.209		35,900	2,100		38,000	42,700
4300	4000 GPM, 200 HP, 10" discharge		.66	48.485		42,900	2,725		45,625	51,000
4330	Horizontal split case, two stage, 500' head									
4340	100 GPM, 40 HP, 1-1/2" discharge	Q-2	1.70	14.118	Ea.	17,200	780		17,980	20,100
4350	200 GPM, 50 HP, 1-1/2" discharge	"	1.44	16.667		18,000	920		18,920	21,200
4360	300 GPM, 75 HP, 2" discharge	Q-3	1.57	20.382		24,700	1,150		25,850	28,900
4370	400 GPM, 100 HP, 3" discharge		1.14	28.070		29,200	1,575		30,775	34,500
4380	800 GPM, 200 HP, 4" discharge		.86	37.209		38,700	2,100		40,800	45,800
5000	Centrifugal, in-line									
5006	Vertical mount, iron body, 125 lb. flgd, 1800 RPM TEFC mtr									
5010	Single stage									
5012	.5 HP, 1-1/2" suction & discharge	Q-1	3.20	5	Ea.	1,375	266		1,641	1,925

22 11 23 – Domestic Water Pumps

22 11 23.10 General Utility Pumps		Crew	Daily Output	Labor-Hours	Unit	Material	2016 Bare Costs Labor	Equipment	Total	Total Incl O&P
5014	.75 HP, 2" suction & discharge	Q-1	2.80	5.714	Ea.	2,625	305		2,930	3,350
5015	1 HP, 3" suction & discharge		2.50	6.400		2,950	340		3,290	3,775
5016	1.5 HP, 3" suction & discharge		2.30	6.957		3,000	370		3,370	3,825
5018	2 HP, 4" suction & discharge		2.10	7.619		3,875	405		4,280	4,900
5020	3 HP, 5" suction & discharge		1.90	8.421		4,550	450		5,000	5,675
5030	5 HP, 6" suction & discharge		1.60	10		5,175	535		5,710	6,475
5040	7.5 HP, 6" suction & discharge		1.30	12.308		5,950	655		6,605	7,525
5050	10 HP, 8" suction & discharge	Q-2	1.80	13.333		7,925	735		8,660	9,800
5060	15 HP, 8" suction & discharge		1.70	14.118		9,275	780		10,055	11,400
5064	20 HP, 8" suction & discharge		1.60	15		10,700	830		11,530	13,100
5070	30 HP, 8" suction & discharge		1.50	16		13,700	885		14,585	16,300
5080	40 HP, 8" suction & discharge		1.40	17.143		16,400	945		17,345	19,500
5090	50 HP, 8" suction & discharge		1.30	18.462		16,600	1,025		17,625	19,900
5094	60 HP, 8" suction & discharge		.90	26.667		17,500	1,475		18,975	21,500
5100	75 HP, 8" suction & discharge	Q-3	.60	53.333		20,000	3,000		23,000	26,500
5110	100 HP, 8" suction & discharge	"	.50	64		22,800	3,600		26,400	30,600

22 11 23.11 Miscellaneous Pumps

		Crew	Daily Output	Labor-Hours	Unit	Material	2016 Bare Costs Labor	Equipment	Total	Total Incl O&P
0010	**MISCELLANEOUS PUMPS**									
0020	Water pump, portable, gasoline powered									
0100	170 GPH, 2" discharge	Q-1	11	1.455	Ea.	685	77.50		762.50	870
0110	343 GPH, 3" discharge		10.50	1.524		805	81		886	1,000
0120	608 GPH, 4" discharge		10	1.600		1,075	85.50		1,160.50	1,325
0500	Pump, propylene body, housing and impeller									
0510	22 GPM, 1/3 HP, 40' HD	1 Plum	5	1.600	Ea.	895	94.50		989.50	1,125
0520	33 GPM, 1/2 HP, 40' HD	Q-1	5	3.200		820	171		991	1,150
0530	53 GPM, 3/4 HP, 40' HD	"	4	4		1,325	213		1,538	1,800
0600	Rotary pump, CI									
0614	282 GPH, 3/4 HP, 1" discharge	Q-1	4.50	3.556	Ea.	1,225	189		1,414	1,625
0618	277 GPH, 1 HP, 1" discharge		4	4		1,225	213		1,438	1,675
0624	1100 GPH, 1.5 HP, 1-1/4" discharge		3.60	4.444		3,225	237		3,462	3,900
0628	1900 GPH, 2 HP, 1-1/4" discharge		3.20	5		4,550	266		4,816	5,425
1000	Turbine pump, CI									
1010	50 GPM, 2 HP, 3" discharge	Q-1	.80	20	Ea.	3,700	1,075		4,775	5,675
1020	100 GPM, 3 HP, 4" discharge	Q-2	.96	25		6,100	1,375		7,475	8,775
1030	250 GPM, 15 HP, 6" discharge	"	.94	25.532		7,575	1,400		8,975	10,500
1040	500 GPM, 25 HP, 6" discharge	Q-3	1.22	26.230		7,625	1,475		9,100	10,600
1050	1000 GPM, 50 HP, 8" discharge		1.14	28.070		10,500	1,575		12,075	14,000
1060	2000 GPM, 100 HP, 10" discharge		1	32		12,500	1,800		14,300	16,400
1070	3000 GPM, 150 HP, 10" discharge		.80	40		20,100	2,250		22,350	25,500
1080	4000 GPM, 200 HP, 12" discharge		.70	45.714		22,000	2,575		24,575	28,100
1090	6000 GPM, 300 HP, 14" discharge		.60	53.333		33,500	3,000		36,500	41,300
1100	10,000 GPM, 300 HP, 18" discharge		.58	55.172		38,700	3,100		41,800	47,300
2000	Centrifugal stainless steel pumps									
2100	100 GPM, 100' TDH, 3 HP	Q-1	1.80	8.889	Ea.	11,000	475		11,475	12,800
2130	250 GPM, 100' TDH, 10 HP	"	1.28	12.500		15,500	665		16,165	18,000
2160	500 GPM, 100' TDH, 20 HP	Q-2	1.44	16.667		17,700	920		18,620	20,800
2200	Vertical turbine stainless steel pumps									
2220	100 GPM, 100' TDH, 7.5 HP	Q-2	.95	25.263	Ea.	62,000	1,400		63,400	70,000
2240	250 GPM, 100' TDH, 15 HP		.94	25.532		68,500	1,400		69,900	77,500
2260	500 GPM, 100' TDH, 20 HP		.93	25.806		75,000	1,425		76,425	84,500
2280	750 GPM, 100' TDH, 30 HP	Q-3	1.19	26.891		81,500	1,525		83,025	92,500
2300	1000 GPM, 100' TDH, 40 HP	"	1.16	27.586		88,500	1,550		90,050	99,500

For customer support on your Mechanical Cost Data, call 877.739.3689.

241

22 11 Facility Water Distribution

22 11 23 – Domestic Water Pumps

22 11 23.11 Miscellaneous Pumps

	Crew	Daily Output	Labor-Hours	Unit	Material	2016 Bare Costs Labor	Equipment	Total	Total Incl O&P
2320	100 GPM, 200' TDH, 15 HP	Q-2	.94	25.532	Ea.	95,000	1,400		96,400	106,500
2340	250 GPM, 200' TDH, 25 HP	Q-3	1.22	26.230		101,500	1,475		102,975	114,000
2360	500 GPM, 200' TDH, 40 HP		1.16	27.586		108,000	1,550		109,550	121,500
2380	750 GPM, 200' TDH, 60 HP		1.13	28.319		109,000	1,600		110,600	122,500
2400	1000 GPM, 200' TDH, 75 HP		1.12	28.571		112,500	1,600		114,100	126,000

22 11 23.13 Domestic-Water Packaged Booster Pumps

		Crew	Daily Output	Labor-Hours	Unit	Material	2016 Bare Costs Labor	Equipment	Total	Total Incl O&P
0010	**DOMESTIC-WATER PACKAGED BOOSTER PUMPS**									
0200	Pump system, with diaphragm tank, control, press. switch									
0300	1 HP pump	Q-1	1.30	12.308	Ea.	6,750	655		7,405	8,425
0400	1-1/2 HP pump		1.25	12.800		6,825	680		7,505	8,525
0420	2 HP pump		1.20	13.333		7,000	710		7,710	8,775
0440	3 HP pump		1.10	14.545		7,100	775		7,875	8,975
0460	5 HP pump	Q-2	1.50	16		7,850	885		8,735	9,950
0480	7-1/2 HP pump		1.42	16.901		8,725	935		9,660	11,000
0500	10 HP pump		1.34	17.910		9,150	990		10,140	11,600
2000	Pump system, variable speed, base, controls, starter									
2010	Duplex, 100' head									
2020	400 GPM, 7-1/2HP, 4" discharge	Q-2	.70	34.286	Ea.	40,100	1,900		42,000	47,000
2025	Triplex, 100' head									
2030	1000 GPM, 15HP, 6" discharge	Q-2	.50	48	Ea.	58,500	2,650		61,150	68,000
2040	1700 GPM, 30HP, 6" discharge	"	.30	80	"	70,500	4,425		74,925	84,000

22 12 Facility Potable-Water Storage Tanks

22 12 21 – Facility Underground Potable-Water Storage Tanks

22 12 21.13 Fiberglass, Undrgrnd Pot.-Water Storage Tanks

		Crew	Daily Output	Labor-Hours	Unit	Material	2016 Bare Costs Labor	Equipment	Total	Total Incl O&P
0010	**FIBERGLASS, UNDERGROUND POTABLE-WATER STORAGE TANKS**									
0020	Excludes excavation, backfill, & piping									
0030	Single wall									
2000	600 gallon capacity	B-21B	3.75	10.667	Ea.	3,475	440	176	4,091	4,675
2010	1,000 gallon capacity		3.50	11.429		4,475	470	188	5,133	5,850
2020	2,000 gallon capacity		3.25	12.308		6,500	505	203	7,208	8,150
2030	4,000 gallon capacity		3	13.333		9,000	550	219	9,769	11,000
2040	6,000 gallon capacity		2.65	15.094		10,100	620	248	10,968	12,300
2050	8,000 gallon capacity		2.30	17.391		12,100	715	286	13,101	14,700
2060	10,000 gallon capacity		2	20		13,900	825	330	15,055	16,900
2070	12,000 gallon capacity		1.50	26.667		19,200	1,100	440	20,740	23,300
2080	15,000 gallon capacity		1	40		22,300	1,650	660	24,610	27,800
2090	20,000 gallon capacity		.75	53.333		28,600	2,200	880	31,680	35,800
2100	25,000 gallon capacity		.50	80		42,800	3,300	1,325	47,425	53,500
2110	30,000 gallon capacity		.35	114		52,000	4,700	1,875	58,575	66,500
2120	40,000 gallon capacity		.30	133		75,000	5,500	2,200	82,700	93,500

22 12 23 – Facility Indoor Potable-Water Storage Tanks

22 12 23.13 Facility Steel, Indoor Pot.-Water Storage Tanks

		Crew	Daily Output	Labor-Hours	Unit	Material	2016 Bare Costs Labor	Equipment	Total	Total Incl O&P
0010	**FACILITY STEEL, INDOOR POT.-WATER STORAGE TANKS**									
2000	Galvanized steel, 15 gal., 14" diam., 26" LOA	1 Plum	12	.667	Ea.	1,475	39.50		1,514.50	1,675
2060	30 gal., 14" diam. x 49" LOA		11	.727		1,750	43		1,793	2,000
2080	80 gal., 20" diam. x 64" LOA		9	.889		2,725	52.50		2,777.50	3,075
2100	135 gal., 24" diam. x 75" LOA		6	1.333		3,975	79		4,054	4,500
2120	240 gal., 30" diam. x 86" LOA		4	2		7,775	118		7,893	8,725
2140	300 gal., 36" diam. x 76" LOA		3	2.667		11,100	158		11,258	12,400

22 12 Facility Potable-Water Storage Tanks

22 12 23 – Facility Indoor Potable-Water Storage Tanks

22 12 23.13 Facility Steel, Indoor Pot.-Water Storage Tanks	Crew	Daily Output	Labor-Hours	Unit	Material	2016 Bare Costs Labor	Equipment	Total	Total Incl O&P
2160	400 gal., 36" diam. x 100" LOA	Q-1	4	4	Ea.	13,600	213		13,813	15,300
2180	500 gal., 36" diam., x 126" LOA	"	3	5.333		16,800	284		17,084	18,900
3000	Glass lined, P.E., 80 gal., 20" diam. x 60" LOA	1 Plum	9	.889		3,950	52.50		4,002.50	4,425
3060	140 gal., 24" diam. x 75" LOA		6	1.333		5,250	79		5,329	5,900
3080	200 gal., 30" diam. x 71" LOA		4	2		7,825	118		7,943	8,775
3100	350 gal., 36" diam. x 86" LOA		3	2.667		10,600	158		10,758	11,900
3120	450 gal., 42" diam. x 79" LOA	Q-1	4	4		12,900	213		13,113	14,400
3140	600 gal., 48" diam. x 81" LOA		3	5.333		15,700	284		15,984	17,700
3160	750 gal., 48" diam. x 105" LOA		3	5.333		17,600	284		17,884	19,800
3180	900 gal.,54" diam. x 95" LOA		2.50	6.400		24,500	340		24,840	27,500
3200	1500 gal., 54" diam. x 153" LOA		2	8		31,500	425		31,925	35,300
3220	1800 gal., 54" diam. x 181" LOA		1.50	10.667		32,300	570		32,870	36,400
3240	2000 gal., 60" diam. x 165" LOA		1	16		40,400	850		41,250	45,800
3260	3500 gal., 72" diam. x 201" LOA	Q-2	1.50	16		51,000	885		51,885	58,000

22 13 Facility Sanitary Sewerage

22 13 16 – Sanitary Waste and Vent Piping

22 13 16.40 Pipe Fittings, Cast Iron for Drainage

		Crew	Daily Output	Labor-Hours	Unit	Material	2016 Bare Costs Labor	Equipment	Total	Total Incl O&P
0010	**PIPE FITTINGS, CAST IRON FOR DRAINAGE**, Special									
1000	Drip pan elbow, (safety valve discharge elbow)									
1010	Cast iron, threaded inlet									
1014	2-1/2"	Q-1	8	2	Ea.	1,400	107		1,507	1,675
1015	3"		6.40	2.500		1,500	133		1,633	1,850
1017	4"		4.80	3.333		2,000	178		2,178	2,475
1018	Cast iron, flanged inlet									
1019	6"	Q-2	3.60	6.667	Ea.	2,075	370		2,445	2,825
1020	8"	"	2.60	9.231	"	2,175	510		2,685	3,175

22 13 16.60 Traps

		Crew	Daily Output	Labor-Hours	Unit	Material	2016 Bare Costs Labor	Equipment	Total	Total Incl O&P
0010	**TRAPS**									
4700	Copper, drainage, drum trap									
4800	3" x 5" solid, 1-1/2" pipe size	1 Plum	16	.500	Ea.	106	29.50		135.50	162
4840	3" x 6" swivel, 1-1/2" pipe size	"	16	.500	"	160	29.50		189.50	221
5100	P trap, standard pattern									
5200	1-1/4" pipe size	1 Plum	18	.444	Ea.	78.50	26.50		105	126
5240	1-1/2" pipe size		17	.471		72	28		100	121
5260	2" pipe size		15	.533		111	31.50		142.50	170
5280	3" pipe size		11	.727		281	43		324	375
5340	With cleanout, swivel joint and slip joint									
5360	1-1/4" pipe size	1 Plum	18	.444	Ea.	99.50	26.50		126	149
5400	1-1/2" pipe size		17	.471		109	28		137	162
5420	2" pipe size		15	.533		177	31.50		208.50	242
6710	ABS DWV P trap, solvent weld joint									
6720	1-1/2" pipe size	1 Plum	18	.444	Ea.	7.75	26.50		34.25	48
6722	2" pipe size		17	.471		10.20	28		38.20	53
6724	3" pipe size		15	.533		40	31.50		71.50	91.50
6726	4" pipe size		14	.571		80	34		114	139
6732	PVC DWV P trap, solvent weld joint									
6733	1-1/2" pipe size	1 Plum	18	.444	Ea.	6.10	26.50		32.60	46
6734	2" pipe size		17	.471		8.20	28		36.20	51
6735	3" pipe size		15	.533		28	31.50		59.50	78
6736	4" pipe size		14	.571		61.50	34		95.50	119

For customer support on your Mechanical Cost Data, call 877.739.3689.

243

22 13 16.60 Traps	Crew	Daily Output	Labor-Hours	Unit	Material	2016 Bare Costs Labor	Equipment	Total	Total Incl O&P
6760	PP DWV, dilution trap, 1-1/2" pipe size	1 Plum	16	.500	Ea.	238	29.50		267.50	305
6770	P trap, 1-1/2" pipe size		17	.471		59	28		87	107
6780	2" pipe size		16	.500		87.50	29.50		117	141
6790	3" pipe size		14	.571		160	34		194	227
6800	4" pipe size		13	.615		270	36.50		306.50	350
6830	S trap, 1-1/2" pipe size		16	.500		49	29.50		78.50	98
6840	2" pipe size		15	.533		73	31.50		104.50	128
6850	Universal trap, 1-1/2" pipe size		14	.571		96.50	34		130.50	157
6860	PVC DWV hub x hub, basin trap, 1-1/4" pipe size		18	.444		41.50	26.50		68	85
6870	Sink P trap, 1-1/2" pipe size		18	.444		11.20	26.50		37.70	52
6880	Tubular S trap, 1-1/2" pipe size	▼	17	.471	▼	26.50	28		54.50	71
6890	PVC sch. 40 DWV, drum trap									
6900	1-1/2" pipe size	1 Plum	16	.500	Ea.	27.50	29.50		57	74.50
6910	P trap, 1-1/2" pipe size		18	.444		6.75	26.50		33.25	47
6920	2" pipe size		17	.471		9.15	28		37.15	52
6930	3" pipe size		15	.533		31	31.50		62.50	81.50
6940	4" pipe size		14	.571		70.50	34		104.50	129
6950	P trap w/clean out, 1-1/2" pipe size		18	.444		11.20	26.50		37.70	52
6960	2" pipe size		17	.471		18.80	28		46.80	62.50
6970	P trap adjustable, 1-1/2" pipe size		17	.471		8.95	28		36.95	52
6980	P trap adj. w/union & cleanout, 1-1/2" pipe size	▼	16	.500	▼	35.50	29.50		65	83.50

22 13 16.80 Vent Flashing and Caps

		Crew	Daily Output	Labor-Hours	Unit	Material	2016 Bare Costs Labor	Equipment	Total	Total Incl O&P
0010	**VENT FLASHING AND CAPS**									
0120	Vent caps									
0140	Cast iron									
0180	2-1/2" - 3-5/8" pipe	1 Plum	21	.381	Ea.	45	22.50		67.50	83.50
0190	4" - 4-1/8" pipe	"	19	.421	"	55	25		80	98
0900	Vent flashing									
1000	Aluminum with lead ring									
1020	1-1/4" pipe	1 Plum	20	.400	Ea.	4.63	23.50		28.13	40.50
1030	1-1/2" pipe		20	.400		4.90	23.50		28.40	41
1040	2" pipe		18	.444		4.97	26.50		31.47	45
1050	3" pipe		17	.471		5.50	28		33.50	48
1060	4" pipe	▼	16	.500	▼	6.65	29.50		36.15	52
1350	Copper with neoprene ring									
1400	1-1/4" pipe	1 Plum	20	.400	Ea.	68	23.50		91.50	110
1430	1-1/2" pipe		20	.400		68	23.50		91.50	110
1440	2" pipe		18	.444		68	26.50		94.50	114
1450	3" pipe		17	.471		82	28		110	132
1460	4" pipe	▼	16	.500	▼	82	29.50		111.50	135
2000	Galvanized with neoprene ring									
2020	1-1/4" pipe	1 Plum	20	.400	Ea.	14.90	23.50		38.40	52
2030	1-1/2" pipe		20	.400		14.90	23.50		38.40	52
2040	2" pipe		18	.444		15.40	26.50		41.90	56.50
2050	3" pipe		17	.471		17.05	28		45.05	61
2060	4" pipe	▼	16	.500	▼	19.75	29.50		49.25	66
2980	Neoprene, one piece									
3000	1-1/4" pipe	1 Plum	24	.333	Ea.	2.93	19.75		22.68	33
3030	1-1/2" pipe		24	.333		2.95	19.75		22.70	33.50
3040	2" pipe		23	.348		2.95	20.50		23.45	34.50
3050	3" pipe		21	.381		3.39	22.50		25.89	37.50
3060	4" pipe	▼	20	.400	▼	5	23.50		28.50	41

244

For customer support on your Mechanical Cost Data, call 877.739.3689.

22 13 Facility Sanitary Sewerage

22 13 16 – Sanitary Waste and Vent Piping

22 13 16.80 Vent Flashing and Caps

		Crew	Daily Output	Labor-Hours	Unit	Material	2016 Bare Costs Labor	Equipment	Total	Total Incl O&P
4000	Lead, 4#, 8" skirt, vent through roof									
4100	2" pipe	1 Plum	18	.444	Ea.	41.50	26.50		68	85.50
4110	3" pipe		17	.471		47.50	28		75.50	94
4120	4" pipe		16	.500		57	29.50		86.50	107
4130	6" pipe		14	.571		81	34		115	140

22 13 26 – Sanitary Waste Separators

22 13 26.10 Separators

		Crew	Daily Output	Labor-Hours	Unit	Material	2016 Bare Costs Labor	Equipment	Total	Total Incl O&P
0010	**SEPARATORS**, Entrainment eliminator, steel body, 150 PSIG									
0100	1/4" size	1 Stpi	24	.333	Ea.	259	20		279	315
0120	1/2" size		24	.333		268	20		288	325
0140	3/4" size		20	.400		280	24.50		304.50	345
0160	1" size		19	.421		289	25.50		314.50	360
0180	1-1/4" size		15	.533		305	32.50		337.50	390
0200	1-1/2" size		13	.615		340	37.50		377.50	425
0220	2" size		11	.727		375	44		419	480
0240	2-1/2" size	Q-5	15	1.067		1,600	58.50		1,658.50	1,875
0260	3" size		13	1.231		1,800	67		1,867	2,100
0280	4" size		10	1.600		2,050	87.50		2,137.50	2,375
0300	5" size		6	2.667		2,425	146		2,571	2,900
0320	6" size		3	5.333		2,650	291		2,941	3,350
0340	8" size	Q-6	4.40	5.455		3,350	310		3,660	4,175
0360	10" size	"	4	6		4,750	340		5,090	5,750
1000	For 300 PSIG, add					15%				

22 13 29 – Sanitary Sewerage Pumps

22 13 29.13 Wet-Pit-Mounted, Vertical Sewerage Pumps

		Crew	Daily Output	Labor-Hours	Unit	Material	2016 Bare Costs Labor	Equipment	Total	Total Incl O&P
0010	**WET-PIT-MOUNTED, VERTICAL SEWERAGE PUMPS**									
0020	Controls incl. alarm/disconnect panel w/wire. Excavation not included									
0260	Simplex, 9 GPM at 60 PSIG, 91 gal. tank				Ea.	3,375			3,375	3,725
0300	Unit with manway, 26" I.D., 18" high					3,925			3,925	4,325
0340	26" I.D., 36" high					3,975			3,975	4,375
0380	43" I.D., 4' high					4,100			4,100	4,500
0600	Simplex, 9 GPM at 60 PSIG, 150 gal. tank, indoor					3,825			3,825	4,200
0700	Unit with manway, 26" I.D., 36" high					4,500			4,500	4,925
0740	26" I.D., 4' high					4,700			4,700	5,175
2000	Duplex, 18 GPM at 60 PSIG, 150 gal. tank, indoor					7,525			7,525	8,275
2060	Unit with manway, 43" I.D., 4' high					8,350			8,350	9,175
2400	For core only					1,950			1,950	2,150
3000	Indoor residential type installation									
3020	Simplex, 9 GPM at 60 PSIG, 91 gal. HDPE tank				Ea.	3,450			3,450	3,800

22 13 29.14 Sewage Ejector Pumps

		Crew	Daily Output	Labor-Hours	Unit	Material	2016 Bare Costs Labor	Equipment	Total	Total Incl O&P
0010	**SEWAGE EJECTOR PUMPS**, With operating and level controls									
0100	Simplex system incl. tank, cover, pump 15' head									
0500	37 gal. PE tank, 12 GPM, 1/2 HP, 2" discharge	Q-1	3.20	5	Ea.	490	266		756	940
0510	3" discharge		3.10	5.161		535	275		810	1,000
0530	87 GPM, .7 HP, 2" discharge		3.20	5		755	266		1,021	1,225
0540	3" discharge		3.10	5.161		820	275		1,095	1,325
0600	45 gal. coated stl. tank, 12 GPM, 1/2 HP, 2" discharge		3	5.333		880	284		1,164	1,400
0610	3" discharge		2.90	5.517		915	294		1,209	1,450
0630	87 GPM, .7 HP, 2" discharge		3	5.333		1,125	284		1,409	1,675
0640	3" discharge		2.90	5.517		1,200	294		1,494	1,750
0660	134 GPM, 1 HP, 2" discharge		2.80	5.714		1,225	305		1,530	1,800

For customer support on your Mechanical Cost Data, call 877.739.3689.

245

22 13 29 – Sanitary Sewerage Pumps

22 13 29.14 Sewage Ejector Pumps		Crew	Daily Output	Labor-Hours	Unit	Material	2016 Bare Costs Labor	Equipment	Total	Total Incl O&P
0680	3" discharge	Q-1	2.70	5.926	Ea.	1,275	315		1,590	1,900
0700	70 gal. PE tank, 12 GPM, 1/2 HP, 2" discharge		2.60	6.154		950	330		1,280	1,550
0710	3" discharge		2.40	6.667		1,000	355		1,355	1,625
0730	87 GPM, 0.7 HP, 2" discharge		2.50	6.400		1,225	340		1,565	1,875
0740	3" discharge		2.30	6.957		1,300	370		1,670	1,975
0760	134 GPM, 1 HP, 2" discharge		2.20	7.273		1,325	385		1,710	2,050
0770	3" discharge		2	8		1,425	425		1,850	2,225
0800	75 gal. coated stl. tank, 12 GPM, 1/2 HP, 2" discharge		2.40	6.667		1,050	355		1,405	1,675
0810	3" discharge		2.20	7.273		1,100	385		1,485	1,775
0830	87 GPM, .7 HP, 2" discharge		2.30	6.957		1,325	370		1,695	2,000
0840	3" discharge		2.10	7.619		1,400	405		1,805	2,150
0860	134 GPM, 1 HP, 2" discharge		2	8		1,425	425		1,850	2,225
0880	3" discharge		1.80	8.889		1,500	475		1,975	2,375
1040	Duplex system incl. tank, covers, pumps									
1060	110 gal. fiberglass tank, 24 GPM, 1/2 HP, 2" discharge	Q-1	1.60	10	Ea.	1,900	535		2,435	2,900
1080	3" discharge		1.40	11.429		2,000	610		2,610	3,125
1100	174 GPM, .7 HP, 2" discharge		1.50	10.667		2,450	570		3,020	3,550
1120	3" discharge		1.30	12.308		2,550	655		3,205	3,825
1140	268 GPM, 1 HP, 2" discharge		1.20	13.333		2,675	710		3,385	4,025
1160	3" discharge		1	16		2,775	850		3,625	4,325
1260	135 gal. coated stl. tank, 24 GPM, 1/2 HP, 2" discharge	Q-2	1.70	14.118		1,975	780		2,755	3,325
2000	3" discharge		1.60	15		2,075	830		2,905	3,550
2640	174 GPM, .7 HP, 2" discharge		1.60	15		2,575	830		3,405	4,075
2660	3" discharge		1.50	16		2,700	885		3,585	4,300
2700	268 GPM, 1 HP, 2" discharge		1.30	18.462		2,800	1,025		3,825	4,625
3040	3" discharge		1.10	21.818		2,975	1,200		4,175	5,100
3060	275 gal. coated stl. tank, 24 GPM, 1/2 HP, 2" discharge		1.50	16		2,450	885		3,335	4,025
3080	3" discharge		1.40	17.143		2,500	945		3,445	4,150
3100	174 GPM, .7 HP, 2" discharge		1.40	17.143		3,175	945		4,120	4,900
3120	3" discharge		1.30	18.462		3,350	1,025		4,375	5,250
3140	268 GPM, 1 HP, 2" discharge		1.10	21.818		3,475	1,200		4,675	5,650
3160	3" discharge		.90	26.667		3,650	1,475		5,125	6,250
3260	Pump system accessories, add									
3300	Alarm horn and lights, 115 V mercury switch	Q-1	8	2	Ea.	100	107		207	271
3340	Switch, mag. contactor, alarm bell, light, 3 level control		5	3.200		485	171		656	790
3380	Alternator, mercury switch activated		4	4		875	213		1,088	1,275

22 14 Facility Storm Drainage

22 14 29 – Sump Pumps

22 14 29.13 Wet-Pit-Mounted, Vertical Sump Pumps

		Crew	Daily Output	Labor-Hours	Unit	Material	Labor	Equipment	Total	Total Incl O&P
0010	**WET-PIT-MOUNTED, VERTICAL SUMP PUMPS**									
0400	Molded PVC base, 21 GPM at 15' head, 1/3 HP	1 Plum	5	1.600	Ea.	140	94.50		234.50	296
0800	Iron base, 21 GPM at 15' head, 1/3 HP		5	1.600		168	94.50		262.50	330
1200	Solid brass, 21 GPM at 15' head, 1/3 HP		5	1.600		297	94.50		391.50	470
2000	Sump pump, single stage									
2010	25 GPM, 1 HP, 1-1/2" discharge	Q-1	1.80	8.889	Ea.	3,825	475		4,300	4,950
2020	75 GPM, 1-1/2 HP, 2" discharge		1.50	10.667		4,050	570		4,620	5,300
2030	100 GPM, 2 HP, 2-1/2" discharge		1.30	12.308		4,125	655		4,780	5,550
2040	150 GPM, 3 HP, 3" discharge		1.10	14.545		4,125	775		4,900	5,725
2050	200 GPM, 3 HP, 3" discharge		1	16		4,375	850		5,225	6,100
2060	300 GPM, 10 HP, 4" discharge	Q-2	1.20	20		4,725	1,100		5,825	6,875

22 14 Facility Storm Drainage

22 14 29 – Sump Pumps

22 14 29.13 Wet-Pit-Mounted, Vertical Sump Pumps		Crew	Daily Output	Labor-Hours	Unit	Material	2016 Bare Costs Labor	Equipment	Total	Total Incl O&P
2070	500 GPM, 15 HP, 5" discharge	Q-2	1.10	21.818	Ea.	5,375	1,200		6,575	7,725
2080	800 GPM, 20 HP, 6" discharge		1	24		6,350	1,325		7,675	8,975
2090	1000 GPM, 30 HP, 6" discharge		.85	28.235		6,975	1,550		8,525	10,000
2100	1600 GPM, 50 HP, 8" discharge		.72	33.333		10,900	1,850		12,750	14,800
2110	2000 GPM, 60 HP, 8" discharge	Q-3	.85	37.647		11,100	2,125		13,225	15,400
2202	For general purpose float switch, copper coated float, add	Q-1	5	3.200		103	171		274	370

22 14 29.16 Submersible Sump Pumps

		Crew	Daily Output	Labor-Hours	Unit	Material	2016 Bare Costs Labor	Equipment	Total	Total Incl O&P
0010	**SUBMERSIBLE SUMP PUMPS**									
1000	Elevator sump pumps, automatic									
1010	Complete systems, pump, oil detector, controls and alarm									
1020	1-1/2" discharge, does not include the sump pit/tank.									
1040	1/3 HP, 115 V	1 Plum	4.40	1.818	Ea.	1,900	108		2,008	2,250
1050	1/2 HP, 115 V		4	2		1,975	118		2,093	2,350
1060	1/2 HP, 230 V		4	2		2,025	118		2,143	2,400
1070	3/4 HP, 115 V		3.60	2.222		2,050	132		2,182	2,475
1080	3/4 HP, 230 V		3.60	2.222		2,100	132		2,232	2,525
1100	Sump pump only									
1110	1/3 HP, 115 V	1 Plum	6.40	1.250	Ea.	174	74		248	305
1120	1/2 HP, 115 V		5.80	1.379		248	81.50		329.50	395
1130	1/2 HP, 230 V		5.80	1.379		293	81.50		374.50	445
1140	3/4 HP, 115 V		5.40	1.481		330	87.50		417.50	495
1150	3/4 HP, 230 V		5.40	1.481		380	87.50		467.50	545
1200	Oil detector, control and alarm only									
1210	115 V	1 Plum	8	1	Ea.	1,725	59		1,784	2,000
1220	230 V	"	8	1	"	1,725	59		1,784	2,000
7000	Sump pump, automatic									
7100	Plastic, 1-1/4" discharge, 1/4 HP	1 Plum	6.40	1.250	Ea.	150	74		224	277
7140	1/3 HP		6	1.333		217	79		296	360
7160	1/2 HP		5.40	1.481		266	87.50		353.50	425
7180	1-1/2" discharge, 1/2 HP		5.20	1.538		305	91		396	470
7500	Cast iron, 1-1/4" discharge, 1/4 HP		6	1.333		211	79		290	350
7540	1/3 HP		6	1.333		248	79		327	390
7560	1/2 HP		5	1.600		300	94.50		394.50	475

22 14 53 – Rainwater Storage Tanks

22 14 53.13 Fiberglass, Rainwater Storage Tank

		Crew	Daily Output	Labor-Hours	Unit	Material	2016 Bare Costs Labor	Equipment	Total	Total Incl O&P
0010	**FIBERGLASS, RAINWATER STORAGE TANK**									
2000	600 gallon	B-21B	3.75	10.667	Ea.	3,475	440	176	4,091	4,675
2010	1,000 gallon		3.50	11.429		4,475	470	188	5,133	5,850
2020	2,000 gallon		3.25	12.308		6,500	505	203	7,208	8,150
2030	4,000 gallon		3	13.333		9,000	550	219	9,769	11,000
2040	6,000 gallon		2.65	15.094		10,100	620	248	10,968	12,300
2050	8,000 gallon		2.30	17.391		12,100	715	286	13,101	14,700
2060	10,000 gallon		2	20		13,900	825	330	15,055	16,900
2070	12,000 gallon		1.50	26.667		19,200	1,100	440	20,740	23,300
2080	15,000 gallon		1	40		22,300	1,650	660	24,610	27,800
2090	20,000 gallon		.75	53.333		28,600	2,200	880	31,680	35,800
2100	25,000 gallon		.50	80		42,800	3,300	1,325	47,425	53,500
2110	30,000 gallon		.35	114		52,000	4,700	1,875	58,575	66,500
2120	40,000 gallon		.30	133		75,000	5,500	2,200	82,700	93,500

For customer support on your Mechanical Cost Data, call 877.739.3689.

247

22 15 13 – General Service Compressed-Air Piping

22 15 13.10 Compressor Accessories		Crew	Daily Output	Labor-Hours	Unit	Material	2016 Bare Costs Labor	Equipment	Total	Total Incl O&P
0010	**COMPRESSOR ACCESSORIES**									
1700	Refrigerated air dryers with ambient air filters									
1710	10 CFM	Q-5	8	2	Ea.	830	109		939	1,075
1720	25 CFM		6.60	2.424		1,050	132		1,182	1,375
1730	50 CFM		6.20	2.581		1,850	141		1,991	2,275
1740	75 CFM		5.80	2.759		2,200	151		2,351	2,650
1750	100 CFM		5.60	2.857		2,650	156		2,806	3,150
3460	Air filter, regulator, lubricator combination									
3470	Flush mount									
3480	Adjustable range 0-140 psi									
3500	1/8" NPT, 34 SCFM	1 Stpi	17	.471	Ea.	108	28.50		136.50	162
3510	1/4" NPT, 61 SCFM		17	.471		179	28.50		207.50	240
3520	3/8" NPT, 85 SCFM		16	.500		184	30.50		214.50	249
3530	1/2" NPT, 150 SCFM		15	.533		179	32.50		211.50	246
3540	3/4" NPT, 171 SCFM		14	.571		237	34.50		271.50	315
3550	1" NPT, 150 SCFM		13	.615		370	37.50		407.50	460
4000	Couplers, air line, sleeve type									
4010	Female, connection size NPT									
4020	1/4"	1 Stpi	38	.211	Ea.	9.95	12.80		22.75	30.50
4030	3/8"		36	.222		8.55	13.50		22.05	30
4040	1/2"		35	.229		21.50	13.85		35.35	45
4050	3/4"		34	.235		22.50	14.30		36.80	46
4100	Male									
4110	1/4"	1 Stpi	38	.211	Ea.	9	12.80		21.80	29
4120	3/8"		36	.222		11.35	13.50		24.85	33
4130	1/2"		35	.229		21	13.85		34.85	44.50
4140	3/4"		34	.235		24	14.30		38.30	48
4150	Coupler, combined male and female halves									
4160	1/2"	1 Stpi	17	.471	Ea.	43	28.50		71.50	90
4170	3/4"	"	15	.533	"	46.50	32.50		79	100

22 15 19 – General Service Packaged Air Compressors and Receivers

22 15 19.10 Air Compressors

22 15 19.10 Air Compressors		Crew	Daily Output	Labor-Hours	Unit	Material	2016 Bare Costs Labor	Equipment	Total	Total Incl O&P
0010	**AIR COMPRESSORS**									
5250	Air, reciprocating air cooled, splash lubricated, tank mounted									
5300	Single stage, 1 phase, 140 psi									
5303	1/2 HP, 17 gal. tank	1 Stpi	3	2.667	Ea.	2,275	162		2,437	2,750
5305	3/4 HP, 30 gal. tank		2.60	3.077		2,450	187		2,637	2,975
5307	1 HP, 30 gal. tank		2.20	3.636		2,550	221		2,771	3,125
5309	2 HP, 30 gal. tank	Q-5	4	4		2,850	219		3,069	3,475
5310	3 HP, 30 gal. tank		3.60	4.444		2,900	243		3,143	3,550
5314	3 HP, 60 gal. tank		3.50	4.571		3,700	250		3,950	4,450
5320	5 HP, 60 gal. tank		3.20	5		3,875	273		4,148	4,675
5330	5 HP, 80 gal. tank		3	5.333		4,175	291		4,466	5,050
5340	7.5 HP, 80 gal. tank		2.60	6.154		5,450	335		5,785	6,500
5600	2 stage pkg., 3 phase									
5650	6 CFM at 125 psi 1-1/2 HP, 60 gal. tank	Q-5	3	5.333	Ea.	3,400	291		3,691	4,200
5670	10.9 CFM at 125 psi, 3 HP, 80 gal. tank		1.50	10.667		3,950	585		4,535	5,225
5680	38.7 CFM at 125 psi, 10 HP, 120 gal. tank		.60	26.667		7,050	1,450		8,500	9,950
5690	105 CFM at 125 psi, 25 HP, 250 gal. tank	Q-6	.60	40		13,800	2,275		16,075	18,600
5800	With single stage pump									
5850	8.3 CFM at 125 psi, 2 HP, 80 gal. tank	Q-6	3.50	6.857	Ea.	3,775	390		4,165	4,725
5860	38.7 CFM at 125 psi, 10 HP, 120 gal. tank	"	.90	26.667	"	7,050	1,500		8,550	10,000

For customer support on your Mechanical Cost Data, call 877.739.3689.

22 15 General Service Compressed-Air Systems

22 15 19 – General Service Packaged Air Compressors and Receivers

22 15 19.10 Air Compressors		Crew	Daily Output	Labor-Hours	Unit	Material	2016 Bare Costs Labor	Equipment	Total	Total Incl O&P
6000	Reciprocating, 2 stage, tank mtd, 3 Ph., Cap. rated @175 PSIG									
6050	Pressure lubricated, hvy. duty, 9.7 CFM, 3 HP, 120 gal. tank	Q-5	1.30	12.308	Ea.	5,825	670		6,495	7,450
6054	5 CFM, 1-1/2 HP, 80 gal. tank		2.80	5.714		3,625	310		3,935	4,475
6056	6.4 CFM, 2 HP, 80 gal. tank		2	8		3,775	435		4,210	4,800
6058	8.1 CFM, 3 HP, 80 gal. tank		1.70	9.412		3,950	515		4,465	5,125
6059	14.8 CFM, 5 HP, 80 gal. tank		1	16		4,150	875		5,025	5,900
6060	16.5 CFM, 5 HP, 120 gal. tank		1	16		6,000	875		6,875	7,950
6063	13 CFM, 6 HP, 80 gal. tank		.90	17.778		6,075	970		7,045	8,150
6066	19.8 CFM, 7.5 HP, 80 gal. tank		.80	20		6,075	1,100		7,175	8,325
6070	25.8 CFM, 7-1/2 HP, 120 gal. tank		.80	20		8,850	1,100		9,950	11,400
6078	34.8 CFM, 10 HP, 80 gal. tank		.70	22.857		8,500	1,250		9,750	11,200
6080	34.8 CFM, 10 HP, 120 gal. tank		.60	26.667		9,025	1,450		10,475	12,100
6090	53.7 CFM, 15 HP, 120 gal. tank	Q-6	.80	30		11,500	1,700		13,200	15,200
6100	76.7 CFM, 20 HP, 120 gal. tank		.70	34.286		14,400	1,950		16,350	18,700
6104	76.7 CFM, 20 HP, 240 gal. tank		.68	35.294		15,200	2,000		17,200	19,800
6110	90.1 CFM, 25 HP, 120 gal. tank		.63	38.095		14,800	2,150		16,950	19,600
6120	101 CFM, 30 HP, 120 gal. tank		.57	42.105		15,100	2,375		17,475	20,200
6130	101 CFM, 30 HP, 250 gal. tank		.52	46.154		15,900	2,625		18,525	21,500
6200	Oil-less, 13.6 CFM, 5 HP, 120 gal. tank	Q-5	.88	18.182		17,200	995		18,195	20,400
6210	13.6 CFM, 5 HP, 250 gal. tank		.80	20		18,400	1,100		19,500	22,000
6220	18.2 CFM, 7.5 HP, 120 gal. tank		.73	21.918		17,200	1,200		18,400	20,700
6230	18.2 CFM, 7.5 HP, 250 gal. tank		.67	23.881		18,400	1,300		19,700	22,300
6250	30.5 CFM, 10 HP, 120 gal. tank		.57	28.070		20,200	1,525		21,725	24,600
6260	30.5 CFM, 10 HP, 250 gal. tank		.53	30.189		21,500	1,650		23,150	26,100
6270	41.3 CFM, 15 HP, 120 gal. tank	Q-6	.70	34.286		22,000	1,950		23,950	27,100
6280	41.3 CFM, 15 HP, 250 gal. tank	"	.67	35.821		23,300	2,025		25,325	28,700

22 31 Domestic Water Softeners

22 31 13 – Residential Domestic Water Softeners

22 31 13.10 Residential Water Softeners

0010	**RESIDENTIAL WATER SOFTENERS**									
7350	Water softener, automatic, to 30 grains per gallon	2 Plum	5	3.200	Ea.	400	189		589	725
7400	To 100 grains per gallon	"	4	4	"	850	237		1,087	1,300

22 35 Domestic Water Heat Exchangers

22 35 30 – Water Heating by Steam

22 35 30.10 Water Heating Transfer Package

0010	**WATER HEATING TRANSFER PACKAGE**, Complete controls,									
0020	expansion tank, converter, air separator									
1000	Hot water, 180°F enter, 200°F leaving, 15# steam									
1010	One pump system, 28 GPM	Q-6	.75	32	Ea.	20,800	1,825		22,625	25,600
1020	35 GPM		.70	34.286		22,700	1,950		24,650	27,800
1040	55 GPM		.65	36.923		26,800	2,100		28,900	32,700
1060	130 GPM		.55	43.636		33,800	2,475		36,275	40,900
1080	255 GPM		.40	60		44,600	3,400		48,000	54,000
1100	550 GPM		.30	80		60,500	4,525		65,025	73,500
1120	800 GPM		.25	96		72,000	5,450		77,450	87,000
1220	Two pump system, 28 GPM		.70	34.286		28,400	1,950		30,350	34,100
1240	35 GPM		.65	36.923		33,200	2,100		35,300	39,700

For customer support on your Mechanical Cost Data, call 877.739.3689.

249

22 35 Domestic Water Heat Exchangers

22 35 30 – Water Heating by Steam

22 35 30.10 Water Heating Transfer Package	Crew	Daily Output	Labor-Hours	Unit	Material	2016 Bare Costs Labor	2016 Bare Costs Equipment	Total	Total Incl O&P
1260	55 GPM	Q-6	.60	40	Ea.	35,800	2,275		38,075	42,800
1280	130 GPM		.50	48		47,100	2,725		49,825	56,000
1300	255 GPM		.35	68.571		60,500	3,875		64,375	72,500
1320	550 GPM		.25	96		76,000	5,450		81,450	91,500
1340	800 GPM		.20	120		97,500	6,800		104,300	118,000

22 41 Residential Plumbing Fixtures

22 41 39 – Residential Faucets, Supplies and Trim

22 41 39.10 Faucets and Fittings

		Crew	Daily Output	Labor-Hours	Unit	Material	2016 Bare Costs Labor	2016 Bare Costs Equipment	Total	Total Incl O&P
0010	**FAUCETS AND FITTINGS**									
5000	Sillcock, compact, brass, IPS or copper to hose	1 Plum	24	.333	Ea.	10.20	19.75		29.95	41.50
6000	Stop and waste valves, bronze									
6100	Angle, solder end 1/2"	1 Plum	24	.333	Ea.	12.65	19.75		32.40	44
6110	3/4"		20	.400		14.85	23.50		38.35	52
6300	Straightway, solder end 3/8"		24	.333		17.50	19.75		37.25	49.50
6310	1/2"		24	.333		17.50	19.75		37.25	49.50
6320	3/4"		20	.400		20	23.50		43.50	57.50
6410	Straightway, threaded 1/2"		24	.333		21.50	19.75		41.25	53.50
6420	3/4"		20	.400		23.50	23.50		47	61
6430	1"		19	.421		27	25		52	67.50

22 51 Swimming Pool Plumbing Systems

22 51 19 – Swimming Pool Water Treatment Equipment

22 51 19.50 Swimming Pool Filtration Equipment

		Crew	Daily Output	Labor-Hours	Unit	Material	2016 Bare Costs Labor	2016 Bare Costs Equipment	Total	Total Incl O&P
0010	**SWIMMING POOL FILTRATION EQUIPMENT**									
0900	Filter system, sand or diatomite type, incl. pump, 6,000 gal./hr.	2 Plum	1.80	8.889	Total	2,025	525		2,550	3,025
1020	Add for chlorination system, 800 S.F. pool		3	5.333	Ea.	185	315		500	680
1040	5,000 S.F. pool		3	5.333	"	1,850	315		2,165	2,500

22 52 Fountain Plumbing Systems

22 52 16 – Fountain Pumps

22 52 16.10 Fountain Water Pumps

		Crew	Daily Output	Labor-Hours	Unit	Material	2016 Bare Costs Labor	2016 Bare Costs Equipment	Total	Total Incl O&P
0010	**FOUNTAIN WATER PUMPS**									
0100	Pump w/controls									
0200	Single phase, 100' cord, 1/2 H.P. pump	2 Skwk	4.40	3.636	Ea.	1,200	181		1,381	1,575
0300	3/4 H.P. pump		4.30	3.721		1,275	186		1,461	1,700
0400	1 H.P. pump		4.20	3.810		1,875	190		2,065	2,375
0500	1-1/2 H.P. pump		4.10	3.902		2,550	195		2,745	3,100
0600	2 H.P. pump		4	4		3,925	200		4,125	4,625
0700	Three phase, 200' cord, 5 H.P. pump		3.90	4.103		5,475	205		5,680	6,350
0800	7-1/2 H.P. pump		3.80	4.211		9,450	210		9,660	10,700
0900	10 H.P. pump		3.70	4.324		14,000	216		14,216	15,700
1000	15 H.P. pump		3.60	4.444		17,800	222		18,022	19,900
2000	DESIGN NOTE: Use two horsepower per surface acre.									

22 52 Fountain Plumbing Systems

22 52 33 – Fountain Ancillary

22 52 33.10 Fountain Miscellaneous	Crew	Daily Output	Labor-Hours	Unit	Material	2016 Bare Costs Labor	Equipment	Total	Total Incl O&P
0010 **FOUNTAIN MISCELLANEOUS**									
1300 Lights w/mounting kits, 200 watt	2 Skwk	18	.889	Ea.	1,175	44.50		1,219.50	1,375
1400 300 watt		18	.889		1,325	44.50		1,369.50	1,550
1500 500 watt		18	.889		1,500	44.50		1,544.50	1,725
1600 Color blender	↓	12	1.333	↓	585	66.50		651.50	740

22 62 Vacuum Systems for Laboratory and Healthcare Facilities

22 62 19 – Vacuum Equipment for Laboratory and Healthcare Facilities

22 62 19.70 Healthcare Vacuum Equipment

22 62 19.70 Healthcare Vacuum Equipment	Crew	Daily Output	Labor-Hours	Unit	Material	2016 Bare Costs Labor	Equipment	Total	Total Incl O&P
0010 **HEALTHCARE VACUUM EQUIPMENT**									
0300 Dental oral									
0310 Duplex									
0330 165 SCFM with 77 gal. separator	Q-2	1.30	18.462	Ea.	52,000	1,025		53,025	58,500
1100 Vacuum system									
1110 Vacuum outlet alarm panel	1 Plum	3.20	2.500	Ea.	1,100	148		1,248	1,450
2000 Medical, with receiver									
2100 Rotary vane type, lubricated, with controls									
2110 Simplex									
2120 1.5 HP, 80 gal tank	Q-1	8	2	Ea.	6,525	107		6,632	7,325
2130 2 HP, 80 gal tank		7	2.286		6,750	122		6,872	7,600
2140 3 HP, 80 gal tank		6.60	2.424		7,125	129		7,254	8,025
2150 5 HP, 80 gal tank	↓	6	2.667	↓	7,775	142		7,917	8,775
2200 Duplex									
2210 1 HP, 80 gal tank	Q-1	8.40	1.905	Ea.	10,400	101		10,501	11,700
2220 1.5 HP, 80 gal tank		7.80	2.051		10,700	109		10,809	12,000
2230 2 HP, 80 gal tank		6.80	2.353		11,600	125		11,725	12,900
2240 3 HP, 120 gal tank		6	2.667		12,600	142		12,742	14,000
2250 5 HP, 120 gal tank		5.40	2.963		14,300	158		14,458	15,900
2260 7.5 HP, 200 gal tank	Q-2	7	3.429		20,400	189		20,589	22,800
2270 10 HP, 200 gal tank		6.40	3.750		23,600	207		23,807	26,300
2280 15 HP, 200 gal tank		5.80	4.138		42,200	229		42,429	46,800
2290 20 HP, 200 gal tank		5	4.800		48,100	265		48,365	53,500
2300 25 HP, 200 gal tank	↓	4	6	↓	56,000	330		56,330	62,000
2400 Triplex									
2410 7.5 HP, 200 gal tank	Q-2	6.40	3.750	Ea.	32,400	207		32,607	36,000
2420 10 HP, 200 gal tank		5.70	4.211		37,800	233		38,033	41,900
2430 15 HP, 200 gal tank		4.90	4.898		64,000	271		64,271	70,500
2440 20 HP, 200 gal tank		4	6		73,000	330		73,330	80,500
2450 25 HP, 200 gal tank	↓	3.70	6.486		84,000	360		84,360	93,000
2500 Quadruplex									
2510 7.5 HP, 200 gal tank	Q-2	5.30	4.528	Ea.	40,400	250		40,650	44,800
2520 10 HP, 200 gal tank		4.70	5.106		46,600	282		46,882	51,500
2530 15 HP, 200 gal tank		4	6		84,000	330		84,330	93,000
2540 20 HP, 200 gal tank		3.60	6.667		96,000	370		96,370	106,000
2550 25 HP, 200 gal tank	↓	3.20	7.500	↓	111,500	415		111,915	123,000
4000 Liquid ring type , water sealed, with controls									
4200 Duplex									
4210 1.5 HP, 120 gal tank	Q-1	6	2.667	Ea.	20,600	142		20,742	22,900
4220 3 HP, 120 gal tank	"	5.50	2.909		21,500	155		21,655	23,900
4230 4 HP, 120 gal tank	Q-2	6.80	3.529		22,900	195		23,095	25,500

22 62 Vacuum Systems for Laboratory and Healthcare Facilities

22 62 19 – Vacuum Equipment for Laboratory and Healthcare Facilities

22 62 19.70 Healthcare Vacuum Equipment	Crew	Daily Output	Labor-Hours	Unit	Material	2016 Bare Costs Labor	Equipment	Total	Total Incl O&P
4240	5 HP, 120 gal tank	Q-2	6.50	3.692	Ea.	24,200	204		24,404	26,900
4250	7.5 HP, 200 gal tank		6.20	3.871		29,400	214		29,614	32,700
4260	10 HP, 200 gal tank		5.80	4.138		36,700	229		36,929	40,700
4270	15 HP, 200 gal tank		5.10	4.706		44,500	260		44,760	49,400
4280	20 HP, 200 gal tank		4.60	5.217		57,500	288		57,788	64,000
4290	30 HP, 200 gal tank		4	6		69,000	330		69,330	76,000

22 63 Gas Systems for Laboratory and Healthcare Facilities

22 63 13 – Gas Piping for Laboratory and Healthcare Facilities

22 63 13.70 Healthcare Gas Piping

		Crew	Daily Output	Labor-Hours	Unit	Material	2016 Bare Costs Labor	Equipment	Total	Total Incl O&P
0010	**HEALTHCARE GAS PIPING**									
1000	Nitrogen or oxygen system									
1010	Cylinder manifold									
1020	5 cylinder	1 Plum	.80	10	Ea.	5,950	590		6,540	7,450
1026	10 cylinder	"	.40	20	"	7,175	1,175		8,350	9,650
3000	Outlets and valves									
3010	Recessed, wall mounted									
3012	Single outlet	1 Plum	3.20	2.500	Ea.	71.50	148		219.50	300
3100	Ceiling outlet									
3190	Zone valve with box									
3210	2" valve size	1 Plum	3.20	2.500	Ea.	420	148		568	685
4000	Alarm panel, medical gases and vacuum									
4010	Alarm panel	1 Plum	3.20	2.500	Ea.	1,100	148		1,248	1,450

Estimating Tips
The labor adjustment factors listed in Subdivision 22 01 02.20 also apply to Division 23.

23 10 00 Facility Fuel Systems
- The prices in this subdivision for above- and below-ground storage tanks do not include foundations or hold-down slabs, unless noted. The estimator should refer to Divisions 3 and 31 for foundation system pricing. In addition to the foundations, required tank accessories, such as tank gauges, leak detection devices, and additional manholes and piping, must be added to the tank prices.

23 50 00 Central Heating Equipment
- When estimating the cost of an HVAC system, check to see who is responsible for providing and installing the temperature control system. It is possible to overlook controls, assuming that they would be included in the electrical estimate.

- When looking up a boiler, be careful on specified capacity. Some manufacturers rate their products on output while others use input.
- Include HVAC insulation for pipe, boiler, and duct (wrap and liner).
- Be careful when looking up mechanical items to get the correct pressure rating and connection type (thread, weld, flange).

23 70 00 Central HVAC Equipment
- Combination heating and cooling units are sized by the air conditioning requirements. (See Reference No. R236000-20 for preliminary sizing guide.)
- A ton of air conditioning is nominally 400 CFM.
- Rectangular duct is taken off by the linear foot for each size, but its cost is usually estimated by the pound. Remember that SMACNA standards now base duct on internal pressure.
- Prefabricated duct is estimated and purchased like pipe: straight sections and fittings.

- Note that cranes or other lifting equipment are not included on any lines in Division 23. For example, if a crane is required to lift a heavy piece of pipe into place high above a gym floor, or to put a rooftop unit on the roof of a four-story building, etc., it must be added. Due to the potential for extreme variation—from nothing additional required to a major crane or helicopter—we feel that including a nominal amount for "lifting contingency" would be useless and detract from the accuracy of the estimate. When using equipment rental cost data from RSMeans, do not forget to include the cost of the operator(s).

Reference Numbers
Reference numbers are shown at the beginning of some major classifications. These numbers refer to related items in the Reference Section. The reference information may be an estimating procedure, an alternate pricing method, or technical information.

Note: Not all subdivisions listed here necessarily appear. ∎

23 05 Common Work Results for HVAC

23 05 02 – HVAC General

23 05 02.10 Air Conditioning, General	Crew	Daily Output	Labor-Hours	Unit	Material	2016 Bare Costs Labor	Equipment	Total	Total Incl O&P
0010	**AIR CONDITIONING, GENERAL** Prices are for standard efficiencies (SEER 13)									
0020	for upgrade to SEER 14 add					10%				

23 05 05 – Selective Demolition for HVAC

23 05 05.10 HVAC Demolition

	23 05 05.10 HVAC Demolition	Crew	Daily Output	Labor-Hours	Unit	Material	Labor	Equipment	Total	Total Incl O&P
0010	**HVAC DEMOLITION** [R220105-10]									
0100	Air conditioner, split unit, 3 ton	Q-5	2	8	Ea.		435		435	660
0150	Package unit, 3 ton [R024119-10]	Q-6	3	8			455		455	685
0190	Rooftop, self contained, up to 5 ton	1 Plum	1.20	6.667			395		395	595
0250	Air curtain	Q-9	20	.800	L.F.		41		41	63
0254	Air filters, up thru 16,000 CFM		20	.800	Ea.		41		41	63
0256	20,000 thru 60,000 CFM		16	1			51.50		51.50	79
0297	Boiler blowdown	Q-5	8	2			109		109	165
0298	Boilers									
0300	Electric, up thru 148 kW	Q-19	2	12	Ea.		655		655	990
0310	150 thru 518 kW	"	1	24			1,325		1,325	1,975
0320	550 thru 2000 kW	Q-21	.40	80			4,500		4,500	6,775
0330	2070 kW and up	"	.30	106			6,000		6,000	9,050
0340	Gas and/or oil, up thru 150 MBH	Q-7	2.20	14.545			840		840	1,275
0350	160 thru 2000 MBH		.80	40			2,300		2,300	3,500
0360	2100 thru 4500 MBH		.50	64			3,700		3,700	5,575
0370	4600 thru 7000 MBH		.30	106			6,175		6,175	9,300
0380	7100 thru 12,000 MBH		.16	200			11,600		11,600	17,400
0390	12,200 thru 25,000 MBH		.12	266			15,400		15,400	23,300
0400	Central station air handler unit, up thru 15 ton	Q-5	1.60	10			545		545	825
0410	17.5 thru 30 ton	"	.80	20			1,100		1,100	1,650
0430	Computer room unit									
0434	Air cooled split, up thru 10 ton	Q-5	.67	23.881	Ea.		1,300		1,300	1,975
0436	12 thru 23 ton		.53	30.189			1,650		1,650	2,500
0440	Chilled water, up thru 10 ton		1.30	12.308			670		670	1,025
0444	12 thru 23 ton		1	16			875		875	1,325
0450	Glycol system, up thru 10 ton		.53	30.189			1,650		1,650	2,500
0454	12 thru 23 ton		.40	40			2,175		2,175	3,300
0460	Water cooled, not including condenser, up thru 10 ton		.80	20			1,100		1,100	1,650
0464	12 thru 23 ton		.60	26.667			1,450		1,450	2,200
0600	Condenser, up thru 50 ton		1	16			875		875	1,325
0610	51 thru 100 ton	Q-6	.90	26.667			1,500		1,500	2,275
0620	101 thru 1000 ton	"	.70	34.286			1,950		1,950	2,925
0660	Condensing unit, up thru 10 ton	Q-5	1.25	12.800			700		700	1,050
0670	11 thru 50 ton	"	.40	40			2,175		2,175	3,300
0680	60 thru 100 ton	Q-6	.30	80			4,525		4,525	6,850
0700	Cooling tower, up thru 400 ton		.80	30			1,700		1,700	2,575
0710	450 thru 600 ton		.53	45.283			2,575		2,575	3,875
0720	700 thru 1300 ton		.40	60			3,400		3,400	5,125
0780	Dehumidifier, up thru 155 lb./hr.	Q-1	8	2			107		107	161
0790	240 lb./hr. and up	"	2	8			425		425	645
1560	Ductwork									
1570	Metal, steel, sst, fabricated	Q-9	1000	.016	Lb.		.82		.82	1.26
1580	Aluminum, fabricated		485	.033	"		1.70		1.70	2.60
1590	Spiral, prefabricated		400	.040	L.F.		2.06		2.06	3.15
1600	Fiberglass, prefabricated		400	.040			2.06		2.06	3.15
1610	Flex, prefabricated		500	.032			1.65		1.65	2.52
1620	Glass fiber reinforced plastic, prefabricated		280	.057			2.94		2.94	4.50

23 05 05.10 HVAC Demolition	Crew	Daily Output	Labor-Hours	Unit	Material	2016 Bare Costs Labor	Equipment	Total	Total Incl O&P
1630	Diffusers, registers or grills, up thru 20" max dimension	1 Shee	50	.160	Ea.		9.15		9.15	14
1640	21 thru 36" max dimension		36	.222			12.70		12.70	19.45
1650	Above 36" max dimension	↓	30	.267			15.25		15.25	23.50
1700	Evaporator, up thru 12,000 BTUH	Q-5	5.30	3.019			165		165	249
1710	12,500 thru 30,000 BTUH	"	2.70	5.926			325		325	490
1720	31,000 BTUH and up	Q-6	1.50	16			905		905	1,375
1730	Evaporative cooler, up thru 5 H.P.	Q-9	2.70	5.926			305		305	465
1740	10 thru 30 H.P.	"	.67	23.881	↓		1,225		1,225	1,875
1750	Exhaust systems									
1760	Exhaust components	1 Shee	8	1	System		57.50		57.50	87.50
1770	Weld fume hoods	"	20	.400	Ea.		23		23	35
2120	Fans, up thru 1 H.P. or 2000 CFM	Q-9	8	2			103		103	158
2124	1-1/2 thru 10 H.P. or 20,000 CFM	↓	5.30	3.019			156		156	238
2128	15 thru 30 H.P. or above 20,000 CFM	↓	4	4			206		206	315
2150	Fan coil air conditioner, chilled water, up thru 7.5 ton	Q-5	14	1.143			62.50		62.50	94
2154	Direct expansion, up thru 10 ton		8	2			109		109	165
2158	11 thru 30 ton	↓	2	8			435		435	660
2170	Flue shutter damper	Q-9	8	2			103		103	158
2200	Furnace, electric	Q-20	2	10			520		520	795
2300	Gas or oil, under 120 MBH	Q-9	4	4			206		206	315
2340	Over 120 MBH	"	3	5.333			275		275	420
2730	Heating and ventilating unit	Q-5	2.70	5.926			325		325	490
2740	Heater, electric, wall, baseboard and quartz	1 Elec	10	.800			44		44	66
2750	Heater, electric, unit, cabinet, fan and convector	"	8	1			55		55	82.50
2760	Heat exchanger, shell and tube type	Q-5	1.60	10			545		545	825
2770	Plate type	Q-6	.60	40	↓		2,275		2,275	3,425
2810	Heat pump									
2820	Air source, split, 4 thru 10 ton	Q-5	.90	17.778	Ea.		970		970	1,475
2830	15 thru 25 ton	Q-6	.80	30			1,700		1,700	2,575
2850	Single package, up thru 12 ton	Q-5	1	16			875		875	1,325
2860	Water source, up thru 15 ton	"	.90	17.778			970		970	1,475
2870	20 thru 50 ton	Q-6	.80	30			1,700		1,700	2,575
2910	Heat recovery package, up thru 20,000 CFM	Q-5	2	8			435		435	660
2920	25,000 CFM and up		1.20	13.333			730		730	1,100
2930	Heat transfer package, up thru 130 GPM		.80	20			1,100		1,100	1,650
2934	255 thru 800 GPM		.42	38.095			2,075		2,075	3,150
2940	Humidifier		10.60	1.509			82.50		82.50	124
2961	Hydronic unit heaters, up thru 200 MBH		14	1.143			62.50		62.50	94
2962	Above 200 MBH		8	2			109		109	165
2964	Valance units	↓	32	.500	↓		27.50		27.50	41
2966	Radiant floor heating									
2967	System valves, controls, manifolds	Q-5	16	1	Ea.		54.50		54.50	82.50
2968	Per room distribution		8	2			109		109	165
2970	Hydronic heating, baseboard radiation		16	1			54.50		54.50	82.50
2976	Convectors and free standing radiators	↓	18	.889			48.50		48.50	73.50
2980	Induced draft fan, up thru 1 H.P.	Q-9	4.60	3.478			179		179	274
2984	1-1/2 H.P. thru 7-1/2 H.P.	"	2.20	7.273			375		375	575
2988	Infrared unit	Q-5	16	1	↓		54.50		54.50	82.50
2992	Louvers	1 Shee	46	.174	S.F.		9.95		9.95	15.20
3000	Mechanical equipment, light items. Unit is weight, not cooling.	Q-5	.90	17.778	Ton		970		970	1,475
3600	Heavy items		1.10	14.545	"		795		795	1,200
3720	Make-up air unit, up thru 6000 CFM		3	5.333	Ea.		291		291	440
3730	6500 thru 30,000 CFM	↓	1.60	10	↓		545		545	825

23 05 05 – Selective Demolition for HVAC

23 05 05.10 HVAC Demolition	Crew	Daily Output	Labor-Hours	Unit	Material	2016 Bare Costs Labor	Equipment	Total	Total Incl O&P
3740	35,000 thru 75,000 CFM	Q-6	1	24	Ea.		1,350		1,350	2,050
3800	Mixing boxes, constant and VAV	Q-9	18	.889			46		46	70
4000	Packaged terminal air conditioner, up thru 18,000 BTUH	Q-5	8	2			109		109	165
4010	24,000 thru 48,000 BTUH	"	2.80	5.714	↓		310		310	470
5000	Refrigerant compressor, reciprocating or scroll									
5010	Up thru 5 ton	1 Stpi	6	1.333	Ea.		81		81	122
5020	5.08 thru 10 ton	Q-5	6	2.667			146		146	220
5030	15 thru 50 ton	"	3	5.333			291		291	440
5040	60 thru 130 ton	Q-6	2.80	8.571	↓		485		485	735
5090	Remove refrigerant from system	1 Stpi	40	.200	Lb.		12.15		12.15	18.30
5100	Rooftop air conditioner, up thru 10 ton	Q-5	1.40	11.429	Ea.		625		625	940
5110	12 thru 40 ton	Q-6	1	24			1,350		1,350	2,050
5120	50 thru 140 ton		.50	48			2,725		2,725	4,100
5130	150 thru 300 ton	↓	.30	80			4,525		4,525	6,850
6000	Self contained single package air conditioner, up thru 10 ton	Q-5	1.60	10			545		545	825
6010	15 thru 60 ton	Q-6	1.20	20			1,125		1,125	1,700
6100	Space heaters, up thru 200 MBH	Q-5	10	1.600			87.50		87.50	132
6110	Over 200 MBH		5	3.200			175		175	264
6200	Split ductless, both sections	↓	8	2			109		109	165
6300	Steam condensate meter	1 Stpi	11	.727			44		44	66.50
6600	Thru-the-wall air conditioner	L-2	8	2	↓		84.50		84.50	130
7000	Vent chimney, prefabricated, up thru 12" diameter	Q-9	94	.170	V.L.F.		8.75		8.75	13.40
7010	14" thru 36" diameter		40	.400			20.50		20.50	31.50
7020	38" thru 48" diameter	↓	32	.500			26		26	39.50
7030	54" thru 60" diameter	Q-10	14	1.714	↓		91.50		91.50	140
7400	Ventilators, up thru 14" neck diameter	Q-9	58	.276	Ea.		14.20		14.20	21.50
7410	16" thru 50" neck diameter		40	.400			20.50		20.50	31.50
7450	Relief vent, up thru 24" x 96"		22	.727			37.50		37.50	57.50
7460	48" x 60" thru 96" x 144"	↓	10	1.600			82.50		82.50	126
8000	Water chiller up thru 10 ton	Q-5	2.50	6.400			350		350	530
8010	15 thru 100 ton	Q-6	.48	50			2,825		2,825	4,275
8020	110 thru 500 ton	Q-7	.29	110			6,375		6,375	9,625
8030	600 thru 1000 ton		.23	139			8,050		8,050	12,100
8040	1100 ton and up	↓	.20	160			9,250		9,250	14,000
8400	Window air conditioner	1 Carp	16	.500	↓		24		24	37

23 05 23 – General-Duty Valves for HVAC Piping

23 05 23.20 Valves, Bronze/Brass

		Crew	Daily Output	Labor-Hours	Unit	Material	2016 Bare Costs Labor	Equipment	Total	Total Incl O&P
0010	**VALVES, BRONZE/BRASS**									
0020	Brass									
1300	Ball combination valves, shut-off and union									
1310	Solder, with strainer, drain and PT ports									
1320	1/2"	1 Stpi	17	.471	Ea.	75	28.50		103.50	126
1330	3/4"		16	.500		78	30.50		108.50	132
1340	1"		14	.571		123	34.50		157.50	188
1350	1-1/4"		12	.667		142	40.50		182.50	217
1360	1-1/2"		10	.800		221	48.50		269.50	315
1370	2"	↓	8	1	↓	260	60.50		320.50	380
1410	Threaded, with strainer, drain and PT ports									
1420	1/2"	1 Stpi	20	.400	Ea.	75	24.50		99.50	119
1430	3/4"		18	.444		78	27		105	127
1440	1"		17	.471		123	28.50		151.50	178
1450	1-1/4"	↓	12	.667		142	40.50		182.50	217

23 05 23 – General-Duty Valves for HVAC Piping

23 05 23.20 Valves, Bronze/Brass		Crew	Daily Output	Labor-Hours	Unit	Material	2016 Bare Costs Labor	2016 Bare Costs Equipment	Total	Total Incl O&P
1460	1-1/2"	1 Stpi	11	.727	Ea.	221	44		265	310
1470	2"		9	.889		260	54		314	370

23 05 23.30 Valves, Iron Body

		Crew	Daily Output	Labor-Hours	Unit	Material	Labor	Equipment	Total	Total Incl O&P
0010	**VALVES, IRON BODY** R220523-90									
0022	For grooved joint, see Section 22 11 13.48									
0100	Angle, 125 lb.									
0110	Flanged									
0116	2"	1 Plum	5	1.600	Ea.	1,275	94.50		1,369.50	1,550
0118	4"	Q-1	3	5.333		2,100	284		2,384	2,750
0120	6"	Q-2	3	8		4,125	440		4,565	5,200
0122	8"	"	2.50	9.600		5,800	530		6,330	7,175
0560	Butterfly, lug type, pneumatic operator, 2" size	1 Stpi	14	.571		495	34.50		529.50	600
0570	3"	Q-1	8	2		535	107		642	745
0580	4"	"	5	3.200		635	171		806	950
0590	6"	Q-2	5	4.800		820	265		1,085	1,300
0600	8"		4.50	5.333		1,075	295		1,370	1,650
0610	10"		4	6		1,350	330		1,680	1,975
0620	12"		3	8		2,000	440		2,440	2,875
0630	14"		2.30	10.435		2,275	575		2,850	3,375
0640	18"		1.50	16		3,850	885		4,735	5,550
0650	20"		1	24		4,925	1,325		6,250	7,400
0790	Butterfly, lug type, gear operated, 2" size, 200 lb. except noted	1 Plum	14	.571		390	34		424	480
0800	2-1/2"	Q-1	9	1.778		395	94.50		489.50	580
0810	3"		8	2		415	107		522	615
0820	4"		5	3.200		470	171		641	770
0830	5"	Q-2	5	4.800		585	265		850	1,050
0840	6"		5	4.800		645	265		910	1,100
0850	8"		4.50	5.333		835	295		1,130	1,375
0860	10"		4	6		1,175	330		1,505	1,775
0870	12"		3	8		1,600	440		2,040	2,450
0880	14", 150 lb.		2.30	10.435		3,225	575		3,800	4,425
0890	16", 150 lb.		1.75	13.714		4,700	760		5,460	6,325
0900	18", 150 lb.		1.50	16		5,900	885		6,785	7,800
0910	20", 150 lb.		1	24		7,875	1,325		9,200	10,700
0930	24", 150 lb.		.75	32		13,400	1,775		15,175	17,400
1020	Butterfly, wafer type, gear actuator, 200 lb.									
1030	2"	1 Plum	14	.571	Ea.	96	34		130	156
1040	2-1/2"	Q-1	9	1.778		97.50	94.50		192	250
1050	3"		8	2		101	107		208	272
1060	4"		5	3.200		113	171		284	380
1070	5"	Q-2	5	4.800		126	265		391	540
1080	6"		5	4.800		143	265		408	560
1090	8"		4.50	5.333		178	295		473	640
1100	10"		4	6		239	330		569	765
1110	12"		3	8		455	440		895	1,175
1200	Wafer type, lever actuator, 200 lb.									
1220	2"	1 Plum	14	.571	Ea.	157	34		191	224
1230	2-1/2"	Q-1	9	1.778		161	94.50		255.50	320
1240	3"		8	2		190	107		297	370
1250	4"		5	3.200		210	171		381	490
1260	5"	Q-2	5	4.800		320	265		585	755
1270	6"		5	4.800		350	265		615	785

For customer support on your Mechanical Cost Data, call 877.739.3689.

257

23 05 23 – General-Duty Valves for HVAC Piping

23 05 23.30 Valves, Iron Body		Crew	Daily Output	Labor-Hours	Unit	Material	2016 Bare Costs Labor	Equipment	Total	Total Incl O&P
1280	8"	Q-2	4.50	5.333	Ea.	540	295		835	1,050
1290	10"		4	6		755	330		1,085	1,325
1300	12"		3	8		1,250	440		1,690	2,050
1650	Gate, 125 lb., N.R.S.									
2150	Flanged									
2200	2"	1 Plum	5	1.600	Ea.	745	94.50		839.50	965
2240	2-1/2"	Q-1	5	3.200		765	171		936	1,100
2260	3"		4.50	3.556		855	189		1,044	1,225
2280	4"		3	5.333		1,225	284		1,509	1,775
2290	5"	Q-2	3.40	7.059		2,100	390		2,490	2,900
2300	6"		3	8		2,100	440		2,540	2,975
2320	8"		2.50	9.600		3,575	530		4,105	4,750
2340	10"		2.20	10.909		6,300	605		6,905	7,850
2360	12"		1.70	14.118		8,675	780		9,455	10,700
2370	14"		1.30	18.462		12,600	1,025		13,625	15,500
2380	16"		1	24		17,600	1,325		18,925	21,400
2420	For 250 lb. flanged, add					200%	10%			
3550	OS&Y, 125 lb., flanged									
3600	2"	1 Plum	5	1.600	Ea.	500	94.50		594.50	695
3640	2-1/2"	Q-1	5	3.200		505	171		676	810
3660	3"		4.50	3.556		555	189		744	895
3670	3-1/2"		3	5.333		760	284		1,044	1,275
3680	4"		3	5.333		810	284		1,094	1,325
3690	5"	Q-2	3.40	7.059		1,300	390		1,690	2,025
3700	6"		3	8		1,300	440		1,740	2,100
3720	8"		2.50	9.600		2,300	530		2,830	3,325
3740	10"		2.20	10.909		4,250	605		4,855	5,575
3760	12"		1.70	14.118		5,775	780		6,555	7,550
3770	14"		1.30	18.462		12,600	1,025		13,625	15,500
3780	16"		1	24		18,000	1,325		19,325	21,800
3790	18"		.80	30		21,800	1,650		23,450	26,500
3800	20"		.60	40		30,400	2,200		32,600	36,700
3830	24"		.50	48		45,100	2,650		47,750	53,500
3900	For 175 lb., flanged, add					200%	10%			
4350	Globe, OS&Y									
4540	Class 125, flanged									
4550	2"	1 Plum	5	1.600	Ea.	985	94.50		1,079.50	1,225
4560	2-1/2"	Q-1	5	3.200		990	171		1,161	1,350
4570	3"		4.50	3.556		1,200	189		1,389	1,600
4580	4"		3	5.333		1,725	284		2,009	2,325
4590	5"	Q-2	3.40	7.059		3,150	390		3,540	4,050
4600	6"		3	8		3,150	440		3,590	4,125
4610	8"		2.50	9.600		6,150	530		6,680	7,550
4612	10"		2.20	10.909		9,600	605		10,205	11,500
4614	12"		1.70	14.118		10,700	780		11,480	13,000
5040	Class 250, flanged									
5050	2"	1 Plum	4.50	1.778	Ea.	1,575	105		1,680	1,875
5060	2-1/2"	Q-1	4.50	3.556		2,050	189		2,239	2,525
5070	3"		4	4		2,125	213		2,338	2,650
5080	4"		2.70	5.926		3,100	315		3,415	3,875
5090	5"	Q-2	3	8		5,750	440		6,190	7,000
5100	6"		2.70	8.889		5,575	490		6,065	6,875
5110	8"		2.20	10.909		9,425	605		10,030	11,300

For customer support on your Mechanical Cost Data, call 877.739.3689.

23 05 23.30 Valves, Iron Body		Crew	Daily Output	Labor-Hours	Unit	Material	2016 Bare Costs Labor	Equipment	Total	Total Incl O&P
5120	10"	Q-2	2	12	Ea.	13,400	665		14,065	15,800
5130	12"	↓	1.60	15		20,200	830		21,030	23,500
5240	Valve sprocket rim w/chain, for 2" valve	1 Stpi	30	.267		127	16.20		143.20	165
5250	2-1/2" valve		27	.296		127	18		145	167
5260	3-1/2" valve		25	.320		127	19.40		146.40	170
5270	6" valve		20	.400		127	24.50		151.50	177
5280	8" valve		18	.444		156	27		183	212
5290	12" valve		16	.500		156	30.50		186.50	217
5300	16" valve		12	.667		215	40.50		255.50	297
5310	20" valve		10	.800		215	48.50		263.50	310
5320	36" valve	↓	8	1	↓	380	60.50		440.50	510
5450	Swing check, 125 lb., threaded									
5500	2"	1 Plum	11	.727	Ea.	450	43		493	560
5540	2-1/2"	Q-1	15	1.067		575	57		632	720
5550	3"		13	1.231		615	65.50		680.50	775
5560	4"	↓	10	1.600	↓	990	85.50		1,075.50	1,225
5950	Flanged									
5994	1"	1 Plum	7	1.143	Ea.	253	67.50		320.50	380
5998	1-1/2"		6	1.333		370	79		449	525
6000	2"	↓	5	1.600		440	94.50		534.50	630
6040	2-1/2"	Q-1	5	3.200		400	171		571	700
6050	3"		4.50	3.556		430	189		619	760
6060	4"	↓	3	5.333		675	284		959	1,175
6070	6"	Q-2	3	8		1,150	440		1,590	1,950
6080	8"		2.50	9.600		2,175	530		2,705	3,200
6090	10"		2.20	10.909		3,700	605		4,305	4,975
6100	12"		1.70	14.118		5,775	780		6,555	7,525
6102	14"		1.55	15.484		11,700	855		12,555	14,200
6104	16"		1.40	17.143		15,600	945		16,545	18,600
6110	18"		1.30	18.462		23,700	1,025		24,725	27,700
6112	20"		1	24		37,800	1,325		39,125	43,600
6114	24"	↓	.75	32	↓	52,000	1,775		53,775	60,000
6160	For 250 lb. flanged, add					200%	20%			
6600	Silent check, bronze trim									
6610	Compact wafer type, for 125 or 150 lb. flanges									
6630	1-1/2"	1 Plum	11	.727	Ea.	136	43		179	214
6640	2"	"	9	.889		165	52.50		217.50	262
6650	2-1/2"	Q-1	9	1.778		180	94.50		274.50	340
6660	3"		8	2		192	107		299	370
6670	4"	↓	5	3.200		245	171		416	525
6680	5"	Q-2	6	4		325	221		546	690
6690	6"		6	4		430	221		651	810
6700	8"		4.50	5.333		885	295		1,180	1,425
6710	10"		4	6		1,525	330		1,855	2,175
6720	12"	↓	3	8	↓	2,900	440		3,340	3,850
6740	For 250 or 300 lb. flanges, thru 6" no change									
6741	For 8" and 10", add				Ea.	11%	10%			
6750	Twin disc									
6752	2"	1 Plum	9	.889	Ea.	340	52.50		392.50	455
6754	4"	Q-1	5	3.200		560	171		731	870
6756	6"	Q-2	5	4.800		830	265		1,095	1,300
6758	8"		4.50	5.333		1,275	295		1,570	1,850
6760	10"		4	6	↓	2,025	330		2,355	2,725

For customer support on your Mechanical Cost Data, call 877.739.3689.

259

23 05 23.30 Valves, Iron Body

		Crew	Daily Output	Labor-Hours	Unit	Material	Labor	Equipment	Total	Total Incl O&P
6762	12"	Q-2	3	8	Ea.	2,625	440		3,065	3,575
6764	18"		1.50	16		11,400	885		12,285	13,800
6766	24"		.75	32		16,000	1,775		17,775	20,300
6800	Full flange type, 150 lb.									
6810	1"	1 Plum	14	.571	Ea.	159	34		193	226
6811	1-1/4"		12	.667		165	39.50		204.50	242
6812	1-1/2"		11	.727		192	43		235	276
6813	2"		9	.889		221	52.50		273.50	325
6814	2-1/2"	Q-1	9	1.778		266	94.50		360.50	435
6815	3"		8	2		277	107		384	465
6816	4"		5	3.200		340	171		511	630
6817	5"	Q-2	6	4		540	221		761	930
6818	6"		5	4.800		710	265		975	1,175
6819	8"		4.50	5.333		1,175	295		1,470	1,750
6820	10"		4	6		1,500	330		1,830	2,150
6840	For 250 lb., add					30%	10%			
6900	Globe type, 125 lb.									
6910	2"	1 Plum	9	.889	Ea.	300	52.50		352.50	410
6911	2-1/2"	Q-1	9	1.778		405	94.50		499.50	590
6912	3"		8	2		475	107		582	685
6913	4"		5	3.200		675	171		846	995
6914	5"	Q-2	6	4		940	221		1,161	1,350
6915	6"		5	4.800		1,175	265		1,440	1,700
6916	8"		4.50	5.333		2,125	295		2,420	2,775
6917	10"		4	6		2,725	330		3,055	3,475
6918	12"		3	8		4,475	440		4,915	5,600
6919	14"		2.30	10.435		6,125	575		6,700	7,625
6920	16"		1.75	13.714		6,775	760		7,535	8,625
6921	18"		1.50	16		9,800	885		10,685	12,100
6922	20"		1	24		11,900	1,325		13,225	15,100
6923	24"		.75	32		14,800	1,775		16,575	18,900
6940	For 250 lb., add					40%	10%			
6980	Screwed end type, 125 lb.									
6981	1"	1 Plum	19	.421	Ea.	74	25		99	119
6982	1-1/4"		15	.533		97.50	31.50		129	155
6983	1-1/2"		13	.615		118	36.50		154.50	185
6984	2"		11	.727		162	43		205	243

23 05 23.50 Valves, Multipurpose

		Crew	Daily Output	Labor-Hours	Unit	Material	Labor	Equipment	Total	Total Incl O&P
0010	**VALVES, MULTIPURPOSE**									
0100	Functions as a shut off, balancing, check & metering valve									
1000	Cast iron body									
1010	Threaded									
1020	1-1/2" size	1 Stpi	11	.727	Ea.	395	44		439	500
1030	2" size	"	8	1		480	60.50		540.50	615
1040	2-1/2" size	Q-5	5	3.200		520	175		695	835
1200	Flanged									
1210	3" size	Q-5	4.20	3.810	Ea.	600	208		808	975
1220	4" size	"	3	5.333		1,200	291		1,491	1,775
1230	5" size	Q-6	3.80	6.316		1,425	360		1,785	2,125
1240	6" size		3	8		1,775	455		2,230	2,650
1250	8" size		2.50	9.600		3,075	545		3,620	4,225
1260	10" size		2.20	10.909		4,425	620		5,045	5,800

23 05 23 – General-Duty Valves for HVAC Piping

23 05 23.50 Valves, Multipurpose	Crew	Daily Output	Labor-Hours	Unit	Material	2016 Bare Costs Labor	Equipment	Total	Total Incl O&P
1270	12" size	Q-6	2.10	11.429	Ea.	7,975	645		8,620	9,750
1280	14" size	↓	2	12	↓	11,600	680		12,280	13,700

23 05 23.70 Valves, Semi-Steel

		Crew	Daily Output	Labor-Hours	Unit	Material	Labor	Equipment	Total	Total Incl O&P
0010	**VALVES, SEMI-STEEL** R220523-90									
1020	Lubricated plug valve, threaded, 200 psi									
1030	1/2"	1 Plum	18	.444	Ea.	99	26.50		125.50	149
1040	3/4"		16	.500		99	29.50		128.50	154
1050	1"		14	.571		126	34		160	190
1060	1-1/4"		12	.667		150	39.50		189.50	225
1070	1-1/2"		11	.727		161	43		204	242
1080	2"	↓	8	1		196	59		255	305
1090	2-1/2"	Q-1	5	3.200		300	171		471	585
1100	3"	"	4.50	3.556	↓	365	189		554	685
6990	Flanged, 200 psi									
7000	2"	1 Plum	8	1	Ea.	239	59		298	350
7010	2-1/2"	Q-1	5	3.200		340	171		511	630
7020	3"		4.50	3.556		415	189		604	740
7030	4"		3	5.333		520	284		804	1,000
7036	5"	↓	2.50	6.400		1,425	340		1,765	2,100
7040	6"	Q-2	3	8		1,725	440		2,165	2,575
7050	8"		2.50	9.600		1,825	530		2,355	2,800
7060	10"		2.20	10.909		2,800	605		3,405	4,000
7070	12"		1.70	14.118	↓	5,400	780		6,180	7,125

23 05 23.80 Valves, Steel

		Crew	Daily Output	Labor-Hours	Unit	Material	Labor	Equipment	Total	Total Incl O&P
0010	**VALVES, STEEL** R220523-90									
0800	Cast									
1350	Check valve, swing type, 150 lb., flanged									
1370	1"	1 Plum	10	.800	Ea.	360	47.50		407.50	465
1400	2"	"	8	1		580	59		639	730
1440	2-1/2"	Q-1	5	3.200		670	171		841	995
1450	3"		4.50	3.556		685	189		874	1,025
1460	4"	↓	3	5.333		1,025	284		1,309	1,550
1470	6"	Q-2	3	8		1,625	440		2,065	2,450
1480	8"		2.50	9.600		2,650	530		3,180	3,700
1490	10"		2.20	10.909		3,900	605		4,505	5,200
1500	12"		1.70	14.118		5,675	780		6,455	7,425
1510	14"		1.30	18.462		8,500	1,025		9,525	10,900
1520	16"	↓	1	24		10,900	1,325		12,225	14,000
1540	For 300 lb., flanged, add					50%	15%			
1548	For 600 lb., flanged, add					110%	20%			
1571	300 lb., 2"	1 Plum	7.40	1.081		620	64		684	775
1572	2-1/2"	Q-1	4.20	3.810		930	203		1,133	1,325
1573	3"		4	4		930	213		1,143	1,350
1574	4"	↓	2.80	5.714		1,275	305		1,580	1,850
1575	6"	Q-2	2.90	8.276		2,450	455		2,905	3,400
1576	8"		2.40	10		3,625	555		4,180	4,800
1577	10"		2.10	11.429		5,450	630		6,080	6,950
1578	12"		1.60	15		7,900	830		8,730	9,925
1579	14"		1.20	20		11,700	1,100		12,800	14,600
1581	16"	↓	.90	26.667	↓	15,000	1,475		16,475	18,600
1950	Gate valve, 150 lb., flanged									
2000	2"	1 Plum	8	1	Ea.	625	59		684	775

For customer support on your Mechanical Cost Data, call 877.739.3689.

261

23 05 23.80 Valves, Steel		Crew	Daily Output	Labor-Hours	Unit	Material	2016 Bare Costs Labor	Equipment	Total	Total Incl O&P
2040	2-1/2"	Q-1	5	3.200	Ea.	885	171		1,056	1,225
2050	3"		4.50	3.556		885	189		1,074	1,250
2060	4"		3	5.333		1,125	284		1,409	1,650
2070	6"	Q-2	3	8		1,800	440		2,240	2,650
2080	8"		2.50	9.600		2,725	530		3,255	3,800
2090	10"		2.20	10.909		4,050	605		4,655	5,350
2100	12"		1.70	14.118		5,525	780		6,305	7,250
2110	14"		1.30	18.462		9,000	1,025		10,025	11,500
2120	16"		1	24		11,700	1,325		13,025	14,800
2130	18"		.80	30		15,200	1,650		16,850	19,300
2140	20"		.60	40		17,600	2,200		19,800	22,700
2650	300 lb., flanged									
2700	2"	1 Plum	7.40	1.081	Ea.	850	64		914	1,025
2740	2-1/2"	Q-1	4.20	3.810		1,175	203		1,378	1,575
2750	3"		4	4		1,175	213		1,388	1,600
2760	4"		2.80	5.714		1,625	305		1,930	2,250
2770	6"	Q-2	2.90	8.276		2,750	455		3,205	3,725
2780	8"		2.40	10		4,325	555		4,880	5,575
2790	10"		2.10	11.429		5,800	630		6,430	7,325
2800	12"		1.60	15		8,350	830		9,180	10,500
2810	14"		1.20	20		16,700	1,100		17,800	20,100
2820	16"		.90	26.667		21,000	1,475		22,475	25,300
2830	18"		.70	34.286		29,900	1,900		31,800	35,800
2840	20"		.50	48		33,300	2,650		35,950	40,600
3650	Globe valve, 150 lb., flanged									
3700	2"	1 Plum	8	1	Ea.	785	59		844	955
3740	2-1/2"	Q-1	5	3.200		995	171		1,166	1,350
3750	3"		4.50	3.556		995	189		1,184	1,375
3760	4"		3	5.333		1,450	284		1,734	2,025
3770	6"	Q-2	3	8		2,275	440		2,715	3,200
3780	8"		2.50	9.600		4,275	530		4,805	5,500
3790	10"		2.20	10.909		7,850	605		8,455	9,550
3800	12"		1.70	14.118		10,400	780		11,180	12,700
4080	300 lb., flanged									
4100	2"	1 Plum	7.40	1.081	Ea.	1,050	64		1,114	1,250
4140	2-1/2"	Q-1	4.20	3.810		1,425	203		1,628	1,875
4150	3"		4	4		1,425	213		1,638	1,900
4160	4"		2.80	5.714		1,975	305		2,280	2,625
4170	6"	Q-2	2.90	8.276		3,575	455		4,030	4,625
4180	8"		2.40	10		6,050	555		6,605	7,500
4190	10"		2.10	11.429		11,600	630		12,230	13,800
4200	12"		1.60	15		14,000	830		14,830	16,700
4680	600 lb., flanged									
4700	2"	1 Plum	7	1.143	Ea.	1,450	67.50		1,517.50	1,700
4740	2-1/2"	Q-1	4	4		2,350	213		2,563	2,900
4750	3"		3.60	4.444		2,350	237		2,587	2,925
4760	4"		2.50	6.400		3,600	340		3,940	4,475
4770	6"	Q-2	2.60	9.231		7,600	510		8,110	9,150
4780	8"	"	2.10	11.429		12,700	630		13,330	15,000
4800	Silent check, 316 S.S. trim									
4810	Full flange type, 150 lb.									
4811	1"	1 Plum	14	.571	Ea.	535	34		569	640
4812	1-1/4"		12	.667		535	39.50		574.50	650

23 05 23 – General-Duty Valves for HVAC Piping

23 05 23.80 Valves, Steel		Crew	Daily Output	Labor-Hours	Unit	Material	2016 Bare Costs Labor	Equipment	Total	Total Incl O&P
4813	1-1/2"	1 Plum	11	.727	Ea.	590	43		633	710
4814	2"		9	.889		710	52.50		762.50	860
4815	2-1/2"	Q-1	9	1.778		855	94.50		949.50	1,075
4816	3"		8	2		995	107		1,102	1,250
4817	4"		5	3.200		1,350	171		1,521	1,750
4818	5"	Q-2	6	4		1,700	221		1,921	2,200
4819	6"		5	4.800		2,200	265		2,465	2,825
4820	8"		4.50	5.333		3,175	295		3,470	3,950
4821	10"		4	6		4,475	330		4,805	5,425
4840	For 300 lb., add					20%	10%			
4860	For 600 lb., add					50%	15%			
4900	Globe type, 150 lb.									
4910	2"	1 Plum	9	.889	Ea.	840	52.50		892.50	1,000
4911	2-1/2"	Q-1	9	1.778		1,000	94.50		1,094.50	1,275
4912	3"		8	2		1,125	107		1,232	1,400
4913	4"		5	3.200		1,500	171		1,671	1,900
4914	5"	Q-2	6	4		1,875	221		2,096	2,400
4915	6"		5	4.800		2,325	265		2,590	2,950
4916	8"		4.50	5.333		3,250	295		3,545	4,025
4917	10"		4	6		5,625	330		5,955	6,700
4918	12"		3	8		6,900	440		7,340	8,275
4919	14"		2.30	10.435		10,500	575		11,075	12,400
4920	16"		1.75	13.714		14,600	760		15,360	17,300
4921	18"		1.50	16		15,400	885		16,285	18,200
4940	For 300 lb., add					30%	10%			
4960	For 600 lb., add					60%	15%			
5150	Forged									
5340	Ball valve, 1500 psi, threaded, 1/4" size	1 Plum	24	.333	Ea.	62.50	19.75		82.25	99
5350	3/8"		24	.333		76	19.75		95.75	114
5360	1/2"		24	.333		76.50	19.75		96.25	115
5370	3/4"		20	.400		102	23.50		125.50	148
5380	1"		19	.421		133	25		158	185
5390	1-1/4"		15	.533		208	31.50		239.50	276
5400	1-1/2"		13	.615		221	36.50		257.50	298
5410	2"		11	.727		305	43		348	405
5460	Ball valve, 800 lb., socket weld, 1/4" size	Q-15	19	.842		76	45	3.06	124.06	154
5470	3/8"		19	.842		76	45	3.06	124.06	154
5480	1/2"		19	.842		94	45	3.06	142.06	174
5490	3/4"		19	.842		125	45	3.06	173.06	208
5500	1"		15	1.067		156	57	3.87	216.87	261
5510	1-1/4"		13	1.231		208	65.50	4.47	277.97	330
5520	1-1/2"		11	1.455		271	77.50	5.30	353.80	420
5530	2"		8.50	1.882		360	100	6.85	466.85	555
5550	Ball valve, 150 lb., flanged									
5560	4"	Q-1	3	5.333	Ea.	1,125	284		1,409	1,650
5570	6"	Q-2	3	8		2,350	440		2,790	3,250
5580	8"	"	2.50	9.600		4,675	530		5,205	5,950
5650	Check valve, class 800, horizontal									
5651	Socket									
5652	1/4"	Q-15	19	.842	Ea.	94.50	45	3.06	142.56	175
5654	3/8"		19	.842		94.50	45	3.06	142.56	175
5656	1/2"		19	.842		94.50	45	3.06	142.56	175
5658	3/4"		19	.842		101	45	3.06	149.06	182

For customer support on your Mechanical Cost Data, call 877.739.3689.

263

23 05 23.80 Valves, Steel		Crew	Daily Output	Labor-Hours	Unit	Material	2016 Bare Costs Labor	Equipment	Total	Total Incl O&P
5660	1"	Q-15	15	1.067	Ea.	119	57	3.87	179.87	221
5662	1-1/4"		13	1.231		233	65.50	4.47	302.97	360
5664	1-1/2"		11	1.455		233	77.50	5.30	315.80	380
5666	2"		8.50	1.882		325	100	6.85	431.85	520
5698	Threaded									
5700	1/4"	1 Plum	24	.333	Ea.	94.50	19.75		114.25	134
5720	3/8"		24	.333		94.50	19.75		114.25	134
5730	1/2"		24	.333		94.50	19.75		114.25	134
5740	3/4"		20	.400		101	23.50		124.50	147
5750	1"		19	.421		119	25		144	169
5760	1-1/4"		15	.533		233	31.50		264.50	305
5770	1-1/2"		13	.615		233	36.50		269.50	310
5780	2"		11	.727		325	43		368	425
5840	For class 150, flanged, add					100%	15%			
5860	For class 300, flanged, add					120%	20%			
6100	Gate, class 800, OS&Y, socket									
6102	3/8"	Q-15	19	.842	Ea.	65.50	45	3.06	113.56	143
6103	1/2"		19	.842		65.50	45	3.06	113.56	143
6104	3/4"		19	.842		72	45	3.06	120.06	150
6105	1"		15	1.067		87	57	3.87	147.87	186
6106	1-1/4"		13	1.231		165	65.50	4.47	234.97	285
6107	1-1/2"		11	1.455		165	77.50	5.30	247.80	305
6108	2"		8.50	1.882		185	100	6.85	291.85	365
6118	Threaded									
6120	3/8"	1 Plum	24	.333	Ea.	65.50	19.75		85.25	102
6130	1/2"		24	.333		65.50	19.75		85.25	102
6140	3/4"		20	.400		72	23.50		95.50	115
6150	1"		19	.421		87	25		112	134
6160	1-1/4"		15	.533		165	31.50		196.50	229
6170	1-1/2"		13	.615		165	36.50		201.50	236
6180	2"		11	.727		185	43		228	269
6260	For OS&Y, flanged, add					100%	20%			
6700	Globe, OS&Y, class 800, socket									
6710	1/4"	Q-15	19	.842	Ea.	101	45	3.06	149.06	182
6720	3/8"		19	.842		101	45	3.06	149.06	182
6730	1/2"		19	.842		101	45	3.06	149.06	182
6740	3/4"		19	.842		116	45	3.06	164.06	199
6750	1"		15	1.067		151	57	3.87	211.87	256
6760	1-1/4"		13	1.231		295	65.50	4.47	364.97	430
6770	1-1/2"		11	1.455		295	77.50	5.30	377.80	450
6780	2"		8.50	1.882		380	100	6.85	486.85	575
6860	For OS&Y, flanged, add					300%	20%			
6880	Threaded									
6882	1/4"	1 Plum	24	.333	Ea.	101	19.75		120.75	141
6884	3/8"		24	.333		101	19.75		120.75	141
6886	1/2"		24	.333		101	19.75		120.75	141
6888	3/4"		20	.400		116	23.50		139.50	164
6890	1"		19	.421		151	25		176	204
6892	1-1/4"		15	.533		295	31.50		326.50	375
6894	1-1/2"		13	.615		295	36.50		331.50	380
6896	2"		11	.727		380	43		423	480

23 05 23.90 Valves, Stainless Steel		Crew	Daily Output	Labor-Hours	Unit	Material	2016 Bare Costs Labor	Equipment	Total	Total Incl O&P
0010	**VALVES, STAINLESS STEEL** R220523-90									
1610	Ball, threaded 1/4"	1 Stpi	24	.333	Ea.	50	20		70	85.50
1620	3/8"		24	.333		50	20		70	85.50
1630	1/2"		22	.364		50	22		72	88.50
1640	3/4"		20	.400		82.50	24.50		107	128
1650	1"		19	.421		101	25.50		126.50	150
1660	1-1/4		15	.533		192	32.50		224.50	260
1670	1-1/2"		13	.615		200	37.50		237.50	277
1680	2"		11	.727		260	44		304	355
1700	Check, 200 lb., threaded									
1710	1/4"	1 Plum	24	.333	Ea.	143	19.75		162.75	188
1720	1/2"		22	.364		143	21.50		164.50	191
1730	3/4"		20	.400		148	23.50		171.50	198
1750	1"		19	.421		192	25		217	249
1760	1-1/2"		13	.615		360	36.50		396.50	450
1770	2"		11	.727		610	43		653	735
1800	150 lb., flanged									
1810	2-1/2"	Q-1	5	3.200	Ea.	1,875	171		2,046	2,300
1820	3"		4.50	3.556		1,875	189		2,064	2,325
1830	4"		3	5.333		2,775	284		3,059	3,500
1840	6"	Q-2	3	8		4,925	440		5,365	6,100
1850	8"	"	2.50	9.600		10,100	530		10,630	11,900
2100	Gate, OS&Y, 150 lb., flanged									
2120	1/2"	1 Plum	18	.444	Ea.	595	26.50		621.50	695
2140	3/4"		16	.500		575	29.50		604.50	680
2150	1"		14	.571		725	34		759	850
2160	1-1/2"		11	.727		1,400	43		1,443	1,625
2170	2"		8	1		1,650	59		1,709	1,925
2180	2-1/2"	Q-1	5	3.200		2,025	171		2,196	2,475
2190	3"		4.50	3.556		2,025	189		2,214	2,500
2200	4"		3	5.333		2,975	284		3,259	3,700
2205	5"		2.80	5.714		4,800	305		5,105	5,725
2210	6"	Q-2	3	8		5,575	440		6,015	6,800
2220	8"		2.50	9.600		9,675	530		10,205	11,400
2230	10"		2.30	10.435		16,800	575		17,375	19,400
2240	12"		1.90	12.632		22,500	700		23,200	25,900
2260	For 300 lb., flanged, add					120%	15%			
2600	600 lb., flanged									
2620	1/2"	1 Plum	16	.500	Ea.	213	29.50		242.50	279
2640	3/4"		14	.571		230	34		264	305
2650	1"		12	.667		277	39.50		316.50	365
2660	1-1/2"		10	.800		445	47.50		492.50	555
2670	2"		7	1.143		610	67.50		677.50	770
2680	2-1/2"	Q-1	4	4		7,375	213		7,588	8,450
2690	3"	"	3.60	4.444		7,375	237		7,612	8,475
3100	Globe, OS&Y, 150 lb., flanged									
3120	1/2"	1 Plum	18	.444	Ea.	570	26.50		596.50	665
3140	3/4"		16	.500		620	29.50		649.50	725
3150	1"		14	.571		810	34		844	940
3160	1-1/2"		11	.727		1,275	43		1,318	1,475
3170	2"		8	1		1,650	59		1,709	1,925
3180	2-1/2"	Q-1	5	3.200		3,425	171		3,596	4,000

For customer support on your Mechanical Cost Data, call 877.739.3689.

265

23 05 23 – General-Duty Valves for HVAC Piping

23 05 23.90 Valves, Stainless Steel		Crew	Daily Output	Labor-Hours	Unit	Material	2016 Bare Costs Labor	Equipment	Total	Total Incl O&P
3190	3"	Q-1	4.50	3.556	Ea.	3,425	189		3,614	4,025
3200	4"	↓	3	5.333		5,450	284		5,734	6,425
3210	6"	Q-2	3	8	↓	9,200	440		9,640	10,800
5000	Silent check, 316 S.S. body and trim									
5010	Compact wafer type, 300 lb.									
5020	1"	1 Plum	14	.571	Ea.	895	34		929	1,025
5021	1-1/4"		12	.667		1,000	39.50		1,039.50	1,150
5022	1-1/2"		11	.727		1,000	43		1,043	1,175
5023	2"	↓	9	.889		1,150	52.50		1,202.50	1,325
5024	2-1/2"	Q-1	9	1.778		1,575	94.50		1,669.50	1,875
5025	3"	"	8	2	↓	1,800	107		1,907	2,150
5100	Full flange wafer type, 150 lb.									
5110	1"	1 Plum	14	.571	Ea.	630	34		664	740
5111	1-1/4"		12	.667		675	39.50		714.50	805
5112	1-1/2"		11	.727		775	43		818	915
5113	2"		9	.889		1,000	52.50		1,052.50	1,175
5114	2-1/2"	Q-1	9	1.778		1,200	94.50		1,294.50	1,450
5115	3"		8	2		1,350	107		1,457	1,650
5116	4"	↓	5	3.200		2,025	171		2,196	2,475
5117	5"	Q-2	6	4		2,350	221		2,571	2,900
5118	6"		5	4.800		2,950	265		3,215	3,625
5119	8"		4.50	5.333		4,900	295		5,195	5,850
5120	10"	↓	4	6	↓	7,475	330		7,805	8,725
5200	Globe type, 300 lb.									
5210	2"	1 Plum	9	.889	Ea.	1,125	52.50		1,177.50	1,300
5211	2-1/2"	Q-1	9	1.778		1,375	94.50		1,469.50	1,675
5212	3"		8	2		1,700	107		1,807	2,025
5213	4"	↓	5	3.200		2,375	171		2,546	2,850
5214	5"	Q-2	6	4		3,300	221		3,521	3,950
5215	6"		5	4.800		3,725	265		3,990	4,500
5216	8"		4.50	5.333		5,275	295		5,570	6,250
5217	10"	↓	4	6	↓	7,475	330		7,805	8,725
5300	Screwed end type, 300 lb.									
5310	1/2"	1 Plum	24	.333	Ea.	79.50	19.75		99.25	117
5311	3/4"		20	.400		94	23.50		117.50	140
5312	1"		19	.421		109	25		134	158
5313	1-1/4"		15	.533		146	31.50		177.50	208
5314	1-1/2"		13	.615		157	36.50		193.50	227
5315	2"	↓	11	.727	↓	194	43		237	278

23 05 93 – Testing, Adjusting, and Balancing for HVAC

23 05 93.10 Balancing, Air

		Crew	Daily Output	Labor-Hours	Unit	Material	Labor	Equipment	Total	Total Incl O&P
0010	**BALANCING, AIR** (Subcontractor's quote incl. material and labor) R230500-10									
0900	Heating and ventilating equipment									
1000	Centrifugal fans, utility sets				Ea.				420	420
1100	Heating and ventilating unit								630	630
1200	In-line fan								630	630
1300	Propeller and wall fan								119	119
1400	Roof exhaust fan								280	280
2000	Air conditioning equipment, central station								910	910
2100	Built-up low pressure unit								840	840
2200	Built-up high pressure unit								980	980
2300	Built-up high pressure dual duct								1,550	1,550

23 05 93 – Testing, Adjusting, and Balancing for HVAC

23 05 93.10 Balancing, Air	Crew	Daily Output	Labor-Hours	Unit	Material	2016 Bare Costs Labor	Equipment	Total	Total Incl O&P
2400	Built-up variable volume				Ea.				1,825	1,825
2500	Multi-zone A.C. and heating unit								630	630
2600	For each zone over one, add								140	140
2700	Package A.C. unit								350	350
2800	Rooftop heating and cooling unit								490	490
3000	Supply, return, exhaust, registers & diffusers, avg. height ceiling								84	84
3100	High ceiling								126	126
3200	Floor height								70	70
3300	Off mixing box								56	56
3500	Induction unit								91	91
3600	Lab fume hood								420	420
3700	Linear supply								210	210
3800	Linear supply high								245	245
4000	Linear return								70	70
4100	Light troffers								84	84
4200	Moduline - master								84	84
4300	Moduline - slaves								42	42
4400	Regenerators								560	560
4500	Taps into ceiling plenums								105	105
4600	Variable volume boxes								84	84

23 05 93.20 Balancing, Water

		Crew	Daily Output	Labor-Hours	Unit	Material	Labor	Equipment	Total	Total Incl O&P
0010	**BALANCING, WATER** (Subcontractor's quote incl. material and labor) R230500-10									
0050	Air cooled condenser				Ea.				256	256
0080	Boiler								515	515
0100	Cabinet unit heater								88	88
0200	Chiller								620	620
0300	Convector								73	73
0400	Converter								365	365
0500	Cooling tower								475	475
0600	Fan coil unit, unit ventilator								132	132
0700	Fin tube and radiant panels								146	146
0800	Main and duct re-heat coils								135	135
0810	Heat exchanger								135	135
0900	Main balancing cocks								110	110
1000	Pumps								320	320
1100	Unit heater								102	102

23 05 93.50 Piping, Testing

		Crew	Daily Output	Labor-Hours	Unit	Material	Labor	Equipment	Total	Total Incl O&P
0010	**PIPING, TESTING**									
0100	Nondestructive testing									
0110	Nondestructive hydraulic pressure test, isolate & 1 hr. hold									
0120	1" - 4" pipe									
0140	0 – 250 L.F.	1 Stpi	1.33	6.015	Ea.		365		365	550
0160	250 – 500 L.F.	"	.80	10			605		605	915
0180	500 – 1000 L.F.	Q-5	1.14	14.035			765		765	1,150
0200	1000 – 2000 L.F.	"	.80	20			1,100		1,100	1,650
0300	6" - 10" pipe									
0320	0 – 250 L.F.	Q-5	1	16	Ea.		875		875	1,325
0340	250 – 500 L.F.		.73	21.918			1,200		1,200	1,800
0360	500 – 1000 L.F.		.53	30.189			1,650		1,650	2,500
0380	1000 – 2000 L.F.		.38	42.105			2,300		2,300	3,475
1000	Pneumatic pressure test, includes soaping joints									
1120	1" - 4" pipe									

For customer support on your Mechanical Cost Data, call 877.739.3689.

267

23 05 Common Work Results for HVAC

23 05 93 – Testing, Adjusting, and Balancing for HVAC

23 05 93.50 Piping, Testing		Crew	Daily Output	Labor-Hours	Unit	Material	2016 Bare Costs Labor	Equipment	Total	Total Incl O&P
1140	0 – 250 L.F.	Q-5	2.67	5.993	Ea.	12.20	325		337.20	510
1160	250 – 500 L.F.		1.33	12.030		24.50	655		679.50	1,025
1180	500 – 1000 L.F.		.80	20		36.50	1,100		1,136.50	1,700
1200	1000 – 2000 L.F.		.50	32		49	1,750		1,799	2,700
1300	6" - 10" pipe									
1320	0 – 250 L.F.	Q-5	1.33	12.030	Ea.	12.20	655		667.20	1,000
1340	250 – 500 L.F.		.67	23.881		24.50	1,300		1,324.50	2,000
1360	500 – 1000 L.F.		.40	40		49	2,175		2,224	3,350
1380	1000 – 2000 L.F.		.25	64		61	3,500		3,561	5,350
2000	X-Ray of welds									
2110	2" diam.	1 Stpi	8	1	Ea.	15.50	60.50		76	109
2120	3" diam.		8	1		15.50	60.50		76	109
2130	4" diam.		8	1		23.50	60.50		84	117
2140	6" diam.		8	1		23.50	60.50		84	117
2150	8" diam.		6.60	1.212		23.50	73.50		97	137
2160	10" diam.		6	1.333		31	81		112	156
3000	Liquid penetration of welds									
3110	2" diam.	1 Stpi	14	.571	Ea.	5	34.50		39.50	58
3120	3" diam.		13.60	.588		5	35.50		40.50	59.50
3130	4" diam.		13.40	.597		5	36		41	60
3140	6" diam.		13.20	.606		5	37		42	61
3150	8" diam.		13	.615		7.50	37.50		45	65
3160	10" diam.		12.80	.625		7.50	38		45.50	66

23 07 HVAC Insulation

23 07 13 – Duct Insulation

23 07 13.10 Duct Thermal Insulation

			Crew	Daily Output	Labor-Hours	Unit	Material	2016 Bare Costs Labor	Equipment	Total	Total Incl O&P
0010	**DUCT THERMAL INSULATION**										
0100	Rule of thumb, as a percentage of total mechanical costs					Job				10%	10%
0110	Insulation req'd. is based on the surface size/area to be covered										
3000	Ductwork										
3020	Blanket type, fiberglass, flexible										
3030	Fire rated for grease and hazardous exhaust ducts										
3060	1-1/2" thick		Q-14	84	.190	S.F.	4.54	9.15		13.69	19.25
3090	Fire rated for plenums										
3100	1/2" x 24" x 25'		Q-14	1.94	8.247	Roll	167	395		562	800
3110	1/2" x 24" x 25'			98	.163	S.F.	3.35	7.85		11.20	15.90
3120	1/2" x 48" x 25'			1.04	15.385	Roll	335	740		1,075	1,525
3126	1/2" x 48" x 25'			104	.154	S.F.	3.35	7.40		10.75	15.20
3140	FSK vapor barrier wrap, .75 lb. density										
3160	1" thick	G	Q-14	350	.046	S.F.	.18	2.20		2.38	3.62
3170	1-1/2" thick	G		320	.050		.22	2.40		2.62	3.98
3180	2" thick	G		300	.053		.26	2.56		2.82	4.28
3190	3" thick	G		260	.062		.36	2.96		3.32	5
3200	4" thick	G		242	.066		.51	3.18		3.69	5.50
3210	Vinyl jacket, same as FSK										
3280	Unfaced, 1 lb. density										
3310	1" thick	G	Q-14	360	.044	S.F.	.19	2.14		2.33	3.53
3320	1-1/2" thick	G		330	.048		.29	2.33		2.62	3.95
3330	2" thick	G		310	.052		.35	2.48		2.83	4.25
3400	FSK facing, 1 lb. density										

23 07 13 – Duct Insulation

23 07 13.10 Duct Thermal Insulation		Crew	Daily Output	Labor-Hours	Unit	Material	2016 Bare Costs Labor	Equipment	Total	Total Incl O&P
3420	1-1/2" thick	G Q-14	310	.052	S.F.	.23	2.48		2.71	4.11
3430	2" thick	G "	300	.053	"	.35	2.56		2.91	4.38
3450	FSK facing, 1.5 lb. density									
3470	1-1/2" thick	G Q-14	300	.053	S.F.	.36	2.56		2.92	4.39
3480	2" thick	G "	290	.055	"	.43	2.65		3.08	4.60
3730	Sheet insulation									
3760	Polyethylene foam, closed cell, UV resistant									
3770	Standard temperature (-90°F to +212°F)									
3771	1/4" thick	G Q-14	450	.036	S.F.	1.74	1.71		3.45	4.57
3772	3/8" thick	G	440	.036		2.48	1.75		4.23	5.45
3773	1/2" thick	G	420	.038		3.05	1.83		4.88	6.20
3774	3/4" thick	G	400	.040		4.36	1.92		6.28	7.80
3775	1" thick	G	380	.042		5.90	2.02		7.92	9.60
3776	1-1/2" thick	G	360	.044		9.30	2.14		11.44	13.55
3777	2" thick	G	340	.047		12.30	2.26		14.56	17.05
3778	2-1/2" thick	G	320	.050		15.80	2.40		18.20	21
3779	Adhesive (see line 7878)									
3780	Foam, rubber									
3782	1" thick	G 1 Stpi	50	.160	S.F.	2.89	9.70		12.59	17.85
3795	Finishes									
3800	Stainless steel woven mesh	Q-14	100	.160	S.F.	.74	7.70		8.44	12.75
3810	For .010" stainless steel, add		160	.100		2.88	4.81		7.69	10.65
3820	18 oz. fiberglass cloth, pasted on		170	.094		.67	4.52		5.19	7.80
3900	8 oz. canvas, pasted on		180	.089		.21	4.27		4.48	6.90
3940	For .016" aluminum jacket, add		200	.080		.92	3.84		4.76	7
7000	Board insulation									
7020	Mineral wool, 1200° F									
7022	6 lb. density, plain									
7024	1" thick	G Q-14	370	.043	S.F.	.33	2.08		2.41	3.59
7026	1-1/2" thick	G	350	.046		.50	2.20		2.70	3.97
7028	2" thick	G	330	.048		.67	2.33		3	4.37
7030	3" thick	G	300	.053		.83	2.56		3.39	4.90
7032	4" thick	G	280	.057		1.33	2.75		4.08	5.75
7038	8 lb. density, plain									
7040	1" thick	G Q-14	360	.044	S.F.	.40	2.14		2.54	3.76
7042	1-1/2" thick	G	340	.047		.61	2.26		2.87	4.19
7044	2" thick	G	320	.050		.81	2.40		3.21	4.63
7046	3" thick	G	290	.055		1.21	2.65		3.86	5.45
7048	4" thick	G	270	.059		1.61	2.85		4.46	6.20
7060	10 lb. density, plain									
7062	1" thick	G Q-14	350	.046	S.F.	.55	2.20		2.75	4.03
7064	1-1/2" thick	G	330	.048		.83	2.33		3.16	4.54
7066	2" thick	G	310	.052		1.11	2.48		3.59	5.10
7068	3" thick	G	280	.057		1.66	2.75		4.41	6.10
7070	4" thick	G	260	.062		2.21	2.96		5.17	7.05
7878	Contact cement, quart can				Ea.	10.50			10.50	11.55

23 07 16 – HVAC Equipment Insulation

23 07 16.10 HVAC Equipment Thermal Insulation

		Crew	Daily Output	Labor-Hours	Unit	Material	Labor	Equipment	Total	Total Incl O&P
0010	**HVAC EQUIPMENT THERMAL INSULATION**									
0100	Rule of thumb, as a percentage of total mechanical costs				Job				10%	10%
0110	Insulation req'd. is based on the surface size/area to be covered									
1000	Boiler, 1-1/2" calcium silicate only	G Q-14	110	.145	S.F.	4.64	7		11.64	16

For customer support on your Mechanical Cost Data, call 877.739.3689.

269

23 07 HVAC Insulation

23 07 16 – HVAC Equipment Insulation

23 07 16.10 HVAC Equipment Thermal Insulation		Crew	Daily Output	Labor-Hours	Unit	Material	2016 Bare Costs Labor	Equipment	Total	Total Incl O&P
1020	Plus 2" fiberglass	G	Q-14	80	.200	S.F.	5.65	9.60		15.25	21
2000	Breeching, 2" calcium silicate										
2020	Rectangular	G	Q-14	42	.381	S.F.	9.10	18.30		27.40	38.50
2040	Round	G	"	38.70	.413	"	9.50	19.85		29.35	41.50
2300	Calcium silicate block, + 200°F to + 1200°F										
2310	On irregular surfaces, valves and fittings										
2340	1" thick	G	Q-14	30	.533	S.F.	4.12	25.50		29.62	44.50
2360	1-1/2" thick	G		25	.640		4.64	31		35.64	53
2380	2" thick	G		22	.727		6.10	35		41.10	61
2400	3" thick	↓		18	.889	↓	9.35	42.50		51.85	77
2410	On plane surfaces										
2420	1" thick	G	Q-14	126	.127	S.F.	4.12	6.10		10.22	14.05
2430	1-1/2" thick	G		120	.133		4.64	6.40		11.04	15.05
2440	2" thick	G		100	.160		6.10	7.70		13.80	18.65
2450	3" thick	G	↓	70	.229	↓	9.35	11		20.35	27.50

23 09 Instrumentation and Control for HVAC

23 09 13 – Instrumentation and Control Devices for HVAC

23 09 13.60 Water Level Controls

		Crew	Daily Output	Labor-Hours	Unit	Material	2016 Bare Costs Labor	Equipment	Total	Total Incl O&P
0010	**WATER LEVEL CONTROLS**										
1000	Electric water feeder	1 Stpi	12	.667	Ea.	300	40.50		340.50	390
2000	Feeder cut-off combination									
2100	Steam system up to 5000 sq. ft.	1 Stpi	12	.667	Ea.	650	40.50		690.50	775
2200	Steam system above 5000 sq. ft.		12	.667		795	40.50		835.50	935
2300	Steam and hot water, high pressure	↓	10	.800	↓	940	48.50		988.50	1,100
3000	Low water cut-off for hot water boiler, 50 psi maximum									
3100	1" top & bottom equalizing pipes, manual reset	1 Stpi	14	.571	Ea.	360	34.50		394.50	450
3200	1" top & bottom equalizing pipes		14	.571		360	34.50		394.50	450
3300	2-1/2" side connection for nipple-to-boiler	↓	14	.571	↓	325	34.50		359.50	415
4000	Low water cut-off for low pressure steam with quick hook-up ftgs.									
4100	For installation in gauge glass tappings	1 Stpi	16	.500	Ea.	263	30.50		293.50	335
4200	Built-in type, 2-1/2" tap - 3-1/8" insertion		16	.500		218	30.50		248.50	285
4300	Built-in type, 2-1/2" tap - 1-3/4" insertion		16	.500		233	30.50		263.50	300
4400	Side connection to 2-1/2" tapping		16	.500		248	30.50		278.50	320
5000	Pump control, low water cut-off and alarm switch	↓	14	.571	↓	695	34.50		729.50	820
9000	Water gauges, complete									
9010	Rough brass, wheel type									
9020	125 PSI at 350°F									
9030	3/8" pipe size	1 Stpi	11	.727	Ea.	65	44		109	138
9040	1/2" pipe size	"	10	.800	"	71	48.50		119.50	152
9060	200 PSI at 400°F									
9070	3/8" pipe size	1 Stpi	11	.727	Ea.	81	44		125	156
9080	1/2" pipe size		10	.800		81	48.50		129.50	163
9090	3/4" pipe size	↓	9	.889		103	54		157	195
9130	Rough brass, chain lever type									
9140	250 PSI at 400°F									
9200	Polished brass, wheel type									
9210	200 PSI at 400°F									
9220	3/8" pipe size	1 Stpi	11	.727	Ea.	152	44		196	234
9230	1/2" pipe size		10	.800		154	48.50		202.50	244
9240	3/4" pipe size	↓	9	.889	↓	163	54		217	262

23 09 13 – Instrumentation and Control Devices for HVAC

23 09 13.60 Water Level Controls		Crew	Daily Output	Labor-Hours	Unit	Material	2016 Bare Costs Labor	Equipment	Total	Total Incl O&P
9260	Polished brass, chain lever type									
9270	250 PSI at 400°F									
9280	1/2" pipe size	1 Stpi	10	.800	Ea.	223	48.50		271.50	320
9290	3/4" pipe size	"	9	.889	"	243	54		297	350
9400	Bronze, high pressure, ASME									
9410	1/2" pipe size	1 Stpi	10	.800	Ea.	263	48.50		311.50	365
9420	3/4" pipe size	"	9	.889	"	297	54		351	405
9460	Chain lever type									
9470	1/2" pipe size	1 Stpi	10	.800	Ea.	355	48.50		403.50	465
9480	3/4" pipe size	"	9	.889	"	395	54		449	515
9500	316 stainless steel, high pressure, ASME									
9510	500 PSI at 450°F									
9520	1/2" pipe size	1 Stpi	10	.800	Ea.	845	48.50		893.50	1,000
9530	3/4" pipe size	"	9	.889	"	820	54		874	980

23 09 23 – Direct-Digital Control System for HVAC

23 09 23.10 Control Components/DDC Systems

23 09 23.10 Control Components/DDC Systems		Crew	Daily Output	Labor-Hours	Unit	Material	2016 Bare Costs Labor	Equipment	Total	Total Incl O&P
0010	CONTROL COMPONENTS/DDC SYSTEMS (Sub's quote incl. M & L)									
0100	Analog inputs									
0110	Sensors (avg. 50' run in 1/2" EMT)									
0120	Duct temperature				Ea.				415	415
0130	Space temperature								665	665
0140	Duct humidity, +/- 3%								695	695
0150	Space humidity, +/- 2%								1,075	1,075
0160	Duct static pressure								565	565
0170	CFM/Transducer								765	765
0172	Water temperature								655	655
0174	Water flow								2,400	2,400
0176	Water pressure differential								975	975
0177	Steam flow								2,400	2,400
0178	Steam pressure								1,025	1,025
0180	KW/Transducer								1,350	1,350
0182	KWH totalization (not incl. elec. meter pulse xmtr.)								625	625
0190	Space static pressure								1,075	1,075
1000	Analog outputs (avg. 50' run in 1/2" EMT)									
1010	P/I Transducer				Ea.				635	635
1020	Analog output, matl. in MUX								305	305
1030	Pneumatic (not incl. control device)								645	645
1040	Electric (not incl. control device)								380	380
2000	Status (Alarms)									
2100	Digital inputs (avg. 50' run in 1/2" EMT)									
2110	Freeze				Ea.				435	435
2120	Fire								395	395
2130	Differential pressure, (air)								595	595
2140	Differential pressure, (water)								975	975
2150	Current sensor								435	435
2160	Duct high temperature thermostat								570	570
2170	Duct smoke detector								705	705
2200	Digital output (avg. 50' run in 1/2" EMT)									
2210	Start/stop				Ea.				340	340
2220	On/off (maintained contact)				"				585	585
3000	Controller MUX panel, incl. function boards									
3100	48 point				Ea.				5,275	5,275

For customer support on your Mechanical Cost Data, call 877.739.3689.

271

23 09 Instrumentation and Control for HVAC

23 09 23 – Direct-Digital Control System for HVAC

23 09 23.10 Control Components/DDC Systems		Crew	Daily Output	Labor-Hours	Unit	Material	2016 Bare Costs Labor	Equipment	Total	Total Incl O&P
3110	128 point				Ea.				7,225	7,225
3200	DDC controller (avg. 50' run in conduit)									
3210	Mechanical room									
3214	16 point controller (incl. 120 volt/1 phase power supply)				Ea.				3,275	3,275
3229	32 point controller (incl. 120 volt/1 phase power supply)				"				5,425	5,425
3230	Includes software programming and checkout									
3260	Space									
3266	VAV terminal box (incl. space temp. sensor)				Ea.				840	840
3280	Host computer (avg. 50' run in conduit)									
3281	Package complete with PC, keyboard,									
3282	printer, monitor, basic software				Ea.				3,150	3,150
4000	Front end costs									
4100	Computer (P.C.) with software program				Ea.				6,350	6,350
4200	Color graphics software								3,925	3,925
4300	Color graphics slides								490	490
4350	Additional printer								980	980
4400	Communications trunk cable				L.F.				3.80	3.80
4500	Engineering labor, (not incl. dftg.)				Point				94	94
4600	Calibration labor								120	120
4700	Start-up, checkout labor								120	120
4800	Programming labor, as req'd									
5000	Communications bus (data transmission cable)									
5010	#18 twisted shielded pair in 1/2" EMT conduit				C.L.F.				380	380
8000	Applications software									
8050	Basic maintenance manager software (not incl. data base entry)				Ea.				1,950	1,950
8100	Time program				Point				6.85	6.85
8120	Duty cycle								13.65	13.65
8140	Optimum start/stop								41.50	41.50
8160	Demand limiting								20.50	20.50
8180	Enthalpy program								41.50	41.50
8200	Boiler optimization				Ea.				1,225	1,225
8220	Chiller optimization				"				1,625	1,625
8240	Custom applications									
8260	Cost varies with complexity									

23 09 33 – Electric and Electronic Control System for HVAC

23 09 33.10 Electronic Control Systems

0010	**ELECTRONIC CONTROL SYSTEMS** R230500-10									
0020	For electronic costs, add to Section 23 09 43.10				Ea.				15%	15%

23 09 43 – Pneumatic Control System for HVAC

23 09 43.10 Pneumatic Control Systems

0010	**PNEUMATIC CONTROL SYSTEMS** R230500-10									
0011	Including a nominal 50 ft. of tubing. Add control panelboard if req'd.									
0100	Heating and ventilating, split system									
0200	Mixed air control, economizer cycle, panel readout, tubing									
0220	Up to 10 tons 〔G〕	Q-19	.68	35.294	Ea.	4,275	1,925		6,200	7,600
0240	For 10 to 20 tons 〔G〕		.63	37.915		4,575	2,075		6,650	8,150
0260	For over 20 tons 〔G〕		.58	41.096		4,950	2,250		7,200	8,850
0270	Enthalpy cycle, up to 10 tons		.50	48.387		4,700	2,650		7,350	9,175
0280	For 10 to 20 tons		.46	52.174		5,075	2,850		7,925	9,900
0290	For over 20 tons		.42	56.604		5,500	3,100		8,600	10,700
0300	Heating coil, hot water, 3 way valve,									
0320	Freezestat, limit control on discharge, readout	Q-5	.69	23.088	Ea.	3,175	1,250		4,425	5,375

272   **For customer support on your Mechanical Cost Data, call 877.739.3689.**

23 09 Instrumentation and Control for HVAC

23 09 43 – Pneumatic Control System for HVAC

23 09 43.10 Pneumatic Control Systems

		Crew	Daily Output	Labor-Hours	Unit	Material	2016 Bare Costs Labor	2016 Bare Costs Equipment	Total	Total Incl O&P
0500	Cooling coil, chilled water, room									
0520	Thermostat, 3 way valve	Q-5	2	8	Ea.	1,400	435		1,835	2,200
0600	Cooling tower, fan cycle, damper control,									
0620	Control system including water readout in/out at panel	Q-19	.67	35.821	Ea.	5,600	1,950		7,550	9,125
1000	Unit ventilator, day/night operation,									
1100	freezestat, ASHRAE, cycle 2	Q-19	.91	26.374	Ea.	3,100	1,450		4,550	5,600
2000	Compensated hot water from boiler, valve control,									
2100	readout and reset at panel, up to 60 GPM	Q-19	.55	43.956	Ea.	5,800	2,400		8,200	10,000
2120	For 120 GPM		.51	47.059		6,225	2,575		8,800	10,700
2140	For 240 GPM		.49	49.180		6,500	2,700		9,200	11,200
3000	Boiler room combustion air, damper to 5 S.F., controls		1.37	17.582		2,800	965		3,765	4,525
3500	Fan coil, heating and cooling valves, 4 pipe control system		3	8		1,275	440		1,715	2,050
3600	Heat exchanger system controls	▼	.86	27.907	▼	2,725	1,525		4,250	5,300
3900	Multizone control (one per zone), includes thermostat, damper									
3910	motor and reset of discharge temperature	Q-5	.51	31.373	Ea.	2,800	1,725		4,525	5,650
4000	Pneumatic thermostat, including controlling room radiator valve	"	2.43	6.593		845	360		1,205	1,475
4040	Program energy saving optimizer [G]	Q-19	1.21	19.786		7,000	1,075		8,075	9,325
4060	Pump control system	"	3	8		1,300	440		1,740	2,075
4080	Reheat coil control system, not incl. coil	Q-5	2.43	6.593	▼	1,100	360		1,460	1,750
4500	Air supply for pneumatic control system									
4600	Tank mounted duplex compressor, starter, alternator,									
4620	piping, dryer, PRV station and filter									
4630	1/2 HP	Q-19	.68	35.139	Ea.	10,400	1,925		12,325	14,400
4640	3/4 HP		.64	37.383		10,900	2,050		12,950	15,100
4650	1 HP		.61	39.539		11,900	2,175		14,075	16,400
4660	1-1/2 HP		.58	41.739		12,700	2,275		14,975	17,500
4680	3 HP		.55	43.956		17,300	2,400		19,700	22,700
4690	5 HP	▼	.42	57.143	▼	30,200	3,125		33,325	38,000
4800	Main air supply, includes 3/8" copper main and labor	Q-5	1.82	8.791	C.L.F.	350	480		830	1,100
4810	If poly tubing used, deduct								30%	30%
7000	Static pressure control for air handling unit, includes pressure									
7010	sensor, receiver controller, readout and damper motors	Q-19	.64	37.383	Ea.	8,300	2,050		10,350	12,200
7020	If return air fan requires control, add								70%	70%
8600	VAV boxes, incl. thermostat, damper motor, reheat coil & tubing	Q-5	1.46	10.989		1,300	600		1,900	2,325
8610	If no reheat coil, deduct				▼				204	204

23 09 53 – Pneumatic and Electric Control System for HVAC

23 09 53.10 Control Components

		Crew	Daily Output	Labor-Hours	Unit	Material	2016 Bare Costs Labor	2016 Bare Costs Equipment	Total	Total Incl O&P
0010	**CONTROL COMPONENTS**									
0500	Aquastats									
0508	Immersion type									
0510	High/low limit, breaks contact w/temp rise	1 Stpi	4	2	Ea.	325	121		446	545
0514	Sequencing, break 2 switches w/temp rise		4	2		305	121		426	520
0518	Circulating, makes contact w/temp rise	▼	4	2	▼	230	121		351	435
0600	Carbon monoxide detector system									
0606	Panel	1 Stpi	4	2	Ea.	980	121		1,101	1,250
0610	Sensor		7.30	1.096		660	66.50		726.50	825
0680	Controller for VAV box, includes actuator	▼	7.30	1.096	▼	278	66.50		344.50	405
0700	Controller, receiver									
0730	Pneumatic, panel mount, single input	1 Plum	8	1	Ea.	570	59		629	715
0740	With conversion mounting bracket		8	1		570	59		629	715
0750	Dual input, with control point adjustment	▼	7	1.143		785	67.50		852.50	965
0850	Electric, single snap switch	1 Elec	4	2	▼	485	110		595	695

For customer support on your Mechanical Cost Data, call 877.739.3689.

273

23 09 53 – Pneumatic and Electric Control System for HVAC

23 09 53.10 Control Components	Crew	Daily Output	Labor-Hours	Unit	Material	2016 Bare Costs Labor	Equipment	Total	Total Incl O&P
0860	Dual snap switches	1 Elec	3	2.667	Ea.	650	147		797	935
0870	Humidity controller		8	1		266	55		321	375
0880	Load limiting controller		8	1		1,075	55		1,130	1,250
0890	Temperature controller	↓	8	1	↓	600	55		655	745
0900	Control panel readout									
0910	Panel up to 12 indicators	1 Stpi	1.20	6.667	Ea.	380	405		785	1,025
0914	Panel up to 24 indicators		.86	9.302		385	565		950	1,275
0918	Panel up to 48 indicators	↓	.50	16	↓	930	970		1,900	2,500
1000	Enthalpy control, boiler water temperature control									
1010	governed by outdoor temperature, with timer	1 Elec	3	2.667	Ea.	440	147		587	705
1300	Energy control/monitor									
1320	BTU computer meter and controller	1 Stpi	1	8	Ea.	1,100	485		1,585	1,950
1600	Flow meters									
1610	Gas	1 Stpi	4	2	Ea.	166	121		287	365
1620	Liquid	"	4	2	"	166	121		287	365
1640	Freezestat									
1644	20' sensing element, adjustable	1 Stpi	3.50	2.286	Ea.	97.50	139		236.50	315
2000	Gauges, pressure or vacuum									
2100	2" diameter dial	1 Stpi	32	.250	Ea.	9.10	15.20		24.30	33
2200	2-1/2" diameter dial		32	.250		10.95	15.20		26.15	35
2300	3-1/2" diameter dial		32	.250		18.70	15.20		33.90	43.50
2400	4-1/2" diameter dial	↓	32	.250	↓	21.50	15.20		36.70	46.50
2700	Flanged iron case, black ring									
2800	3-1/2" diameter dial	1 Stpi	32	.250	Ea.	102	15.20		117.20	135
2900	4-1/2" diameter dial		32	.250		105	15.20		120.20	139
3000	6" diameter dial	↓	32	.250	↓	167	15.20		182.20	206
3010	Steel case, 0 – 300 psi									
3012	2" diam. dial	1 Stpi	16	.500	Ea.	7.60	30.50		38.10	54.50
3014	4" diam. dial	"	16	.500	"	21	30.50		51.50	69
3020	Aluminum case, 0 – 300 psi									
3022	3-1/2" diam. dial	1 Stpi	16	.500	Ea.	10.15	30.50		40.65	57
3024	4-1/2" diam. dial		16	.500		115	30.50		145.50	173
3026	6" diam. dial		16	.500		189	30.50		219.50	253
3028	8-1/2" diam. dial	↓	16	.500	↓	320	30.50		350.50	400
3030	Brass case, 0 – 300 psi									
3032	2" diam. dial	1 Stpi	16	.500	Ea.	44.50	30.50		75	95
3034	4-1/2" diam. dial	"	16	.500	"	110	30.50		140.50	167
3040	Steel case, high pressure, 0 -10,000 psi									
3042	4-1/2" diam. dial	1 Stpi	16	.500	Ea.	299	30.50		329.50	375
3044	6-1/2" diam. dial		16	.500		305	30.50		335.50	380
3046	8-1/2" diam. dial	↓	16	.500	↓	400	30.50		430.50	485
3080	Pressure gauge, differential, magnehelic									
3084	0 – 2" W.C., with air filter kit	1 Stpi	6	1.333	Ea.	72	81		153	201
3300	For compound pressure-vacuum, add					18%				
3350	Humidistat									
3360	Pneumatic operation									
3361	Room humidistat, direct acting	1 Stpi	12	.667	Ea.	395	40.50		435.50	495
3362	Room humidistat, reverse acting		12	.667		395	40.50		435.50	495
3363	Room humidity transmitter		17	.471		425	28.50		453.50	510
3364	Duct mounted controller		12	.667		465	40.50		505.50	570
3365	Duct mounted transmitter		12	.667		420	40.50		460.50	525
3366	Humidity indicator, 3-1/2"	↓	28	.286		163	17.35		180.35	205
3390	Electric operated	1 Shee	8	1	↓	43	57.50		100.50	135

23 09 53 – Pneumatic and Electric Control System for HVAC

23 09 53.10 Control Components	Crew	Daily Output	Labor-Hours	Unit	Material	2016 Bare Costs Labor	Equipment	Total	Total Incl O&P
3400 Relays									
3430 Pneumatic/electric	1 Plum	16	.500	Ea.	365	29.50		394.50	445
3440 Pneumatic proportioning		8	1		285	59		344	405
3450 Pneumatic switching		12	.667		187	39.50		226.50	266
3460 Selector, 3 point		6	1.333		133	79		212	266
3470 Pneumatic time delay	▼	8	1	▼	360	59		419	490
3500 Sensor, air operated									
3520 Humidity	1 Plum	16	.500	Ea.	465	29.50		494.50	555
3540 Pressure		16	.500		68.50	29.50		98	120
3560 Temperature	▼	12	.667	▼	201	39.50		240.50	281
3600 Electric operated									
3620 Humidity	1 Elec	8	1	Ea.	199	55		254	300
3650 Pressure		8	1		258	55		313	365
3680 Temperature	▼	10	.800	▼	118	44		162	195
3700 Switches									
3710 Minimum position									
3720 Electrical	1 Stpi	4	2	Ea.	21	121		142	206
3730 Pneumatic	"	4	2	"	111	121		232	305
4000 Thermometers									
4100 Dial type, 3-1/2" diameter, vapor type, union connection	1 Stpi	32	.250	Ea.	315	15.20		330.20	370
4120 Liquid type, union connection		32	.250		610	15.20		625.20	700
4130 Remote reading, 15' capillary		32	.250		228	15.20		243.20	273
4500 Stem type, 6-1/2" case, 2" stem, 1/2" NPT		32	.250		68	15.20		83.20	97.50
4520 4" stem, 1/2" NPT		32	.250		83	15.20		98.20	115
4600 9" case, 3-1/2" stem, 3/4" NPT		28	.286		115	17.35		132.35	152
4620 6" stem, 3/4" NPT		28	.286		133	17.35		150.35	172
4640 8" stem, 3/4" NPT		28	.286		217	17.35		234.35	265
4660 12" stem, 1" NPT	▼	26	.308	▼	187	18.70		205.70	234
4670 Bi-metal, dial type, steel case brass stem									
4672 2" dial, 4" - 9" stem	1 Stpi	16	.500	Ea.	39	30.50		69.50	88.50
4673 2-1/2" dial, 4" - 9" stem		16	.500		39	30.50		69.50	88.50
4674 3-1/2" dial, 4" - 9" stem	▼	16	.500		48	30.50		78.50	99
4680 Mercury filled, industrial, union connection type									
4682 Angle stem, 7" scale	1 Stpi	16	.500	Ea.	57	30.50		87.50	109
4683 9" scale		16	.500		153	30.50		183.50	214
4684 12" scale		16	.500		197	30.50		227.50	262
4686 Straight stem, 7" scale		16	.500		150	30.50		180.50	211
4687 9" scale		16	.500		167	30.50		197.50	230
4688 12" scale	▼	16	.500	▼	206	30.50		236.50	273
4690 Mercury filled, industrial, separable socket type, with well									
4692 Angle stem, with socket, 7" scale	1 Stpi	16	.500	Ea.	181	30.50		211.50	245
4693 9" scale		16	.500		181	30.50		211.50	245
4694 12" scale		16	.500		214	30.50		244.50	281
4696 Straight stem, with socket, 7" scale		16	.500		141	30.50		171.50	201
4697 9" scale		16	.500		141	30.50		171.50	201
4698 12" scale	▼	16	.500	▼	174	30.50		204.50	237
5000 Thermostats									
5030 Manual	1 Stpi	8	1	Ea.	53	60.50		113.50	150
5040 1 set back, electric, timed **G**		8	1		38	60.50		98.50	133
5050 2 set back, electric, timed **G**		8	1		190	60.50		250.50	300
5100 Locking cover	▼	20	.400		19.55	24.50		44.05	58
5200 24 hour, automatic, clock **G**	1 Shee	8	1		165	57.50		222.50	270
5220 Electric, low voltage, 2 wire	1 Elec	13	.615		48.50	34		82.50	104

23 09 53 – Pneumatic and Electric Control System for HVAC

23 09 53.10 Control Components		Crew	Daily Output	Labor-Hours	Unit	Material	2016 Bare Costs Labor	Equipment	Total	Total Incl O&P
5230	3 wire	1 Elec	10	.800	Ea.	38.50	44		82.50	108
5236	Heating/cooling, low voltage, with clock	↓	8	1	↓	264	55		319	375
5240	Pneumatic									
5250	Single temp., single pressure	1 Stpi	8	1	Ea.	264	60.50		324.50	380
5251	Dual pressure		8	1		400	60.50		460.50	530
5252	Dual temp., dual pressure		8	1		355	60.50		415.50	480
5253	Reverse acting w/averaging element		8	1		274	60.50		334.50	390
5254	Heating-cooling w/deadband		8	1		590	60.50		650.50	740
5255	Integral w/piston top valve actuator		8	1		243	60.50		303.50	360
5256	Dual temp., dual pressure		8	1		248	60.50		308.50	365
5257	Low limit, 8' averaging element		8	1		226	60.50		286.50	340
5258	Room single temp. proportional		8	1		92.50	60.50		153	194
5260	Dual temp, direct acting for VAV fan		8	1		320	60.50		380.50	445
5262	Capillary tube type, 20', 30°F to 100°F	↓	8	1	↓	325	60.50		385.50	450
5300	Transmitter, pneumatic									
5320	Temperature averaging element	Q-1	8	2	Ea.	159	107		266	335
5350	Pressure differential	1 Plum	7	1.143		1,325	67.50		1,392.50	1,575
5370	Humidity, duct		8	1		420	59		479	555
5380	Room		12	.667		425	39.50		464.50	525
5390	Temperature, with averaging element	↓	6	1.333	↓	216	79		295	355
5500	Timer/time clocks									
5510	7 day, 12 hr. battery	1 Stpi	2.60	3.077	Ea.	193	187		380	495
5600	Recorders									
5610	Hydrograph humidity, wall mtd., aluminum case, grad. chart									
5612	8"	1 Stpi	1.60	5	Ea.	925	305		1,230	1,475
5614	10"	"	1.60	5	"	930	305		1,235	1,475
5620	Thermo-hydrograph, wall mtd., grad. chart, 5' SS probe									
5622	8"	1 Stpi	1.60	5	Ea.	795	305		1,100	1,325
5624	10"	"	1.60	5	"	1,450	305		1,755	2,050
5630	Time of operation, wall mtd., 1 pen cap, 24 hr. clock									
5632	6"	1 Stpi	1.60	5	Ea.	495	305		800	1,000
5634	8"	"	1.60	5	"	660	305		965	1,175
5640	Pressure & vacuum, bourdon tube type									
5642	Flush mount, 1 pen									
5644	4"	1 Stpi	1.60	5	Ea.	450	305		755	955
5645	6"		1.60	5		390	305		695	890
5646	8"		1.60	5		630	305		935	1,150
5648	10"	↓	1.60	5	↓	1,050	305		1,355	1,625
6000	Valves, motorized zone									
6100	Sweat connections, 1/2" C x C	1 Stpi	20	.400	Ea.	171	24.50		195.50	225
6110	3/4" C x C		20	.400		172	24.50		196.50	226
6120	1" C x C		19	.421		223	25.50		248.50	284
6140	1/2" C x C, with end switch, 2 wire		20	.400		169	24.50		193.50	223
6150	3/4" C x C, with end switch, 2 wire		20	.400		181	24.50		205.50	236
6160	1" C x C, with end switch, 2 wire	↓	19	.421	↓	206	25.50		231.50	265
7090	Valves, motor controlled, including actuator									
7100	Electric motor actuated									
7200	Brass, two way, screwed									
7210	1/2" pipe size	L-6	36	.333	Ea.	310	19.30		329.30	370
7220	3/4" pipe size		30	.400		545	23		568	635
7230	1" pipe size		28	.429		630	25		655	730
7240	1-1/2" pipe size		19	.632		690	36.50		726.50	815
7250	2" pipe size	↓	16	.750	↓	1,025	43.50		1,068.50	1,200

23 09 53.10 Control Components	Crew	Daily Output	Labor-Hours	Unit	Material	2016 Bare Costs Labor	Equipment	Total	Total Incl O&P
7350	Brass, three way, screwed								
7360	1/2" pipe size	L-6	33	.364	Ea.	390	21		411	455
7370	3/4" pipe size		27	.444		425	25.50		450.50	510
7380	1" pipe size		25.50	.471		485	27		512	575
7384	1-1/4" pipe size		21	.571		555	33		588	660
7390	1-1/2" pipe size		17	.706		610	41		651	730
7400	2" pipe size	↓	14	.857	↓	920	49.50		969.50	1,075
7550	Iron body, two way, flanged								
7560	2-1/2" pipe size	L-6	4	3	Ea.	1,375	173		1,548	1,750
7570	3" pipe size		3	4		1,400	231		1,631	1,900
7580	4" pipe size	↓	2	6	↓	2,675	345		3,020	3,475
7850	Iron body, three way, flanged								
7860	2-1/2" pipe size	L-6	3	4	Ea.	1,400	231		1,631	1,875
7870	3" pipe size		2.50	4.800		1,575	278		1,853	2,150
7880	4" pipe size	↓	2	6	↓	2,025	345		2,370	2,750
8000	Pneumatic, air operated								
8050	Brass, two way, screwed								
8060	1/2" pipe size, class 250	1 Plum	24	.333	Ea.	214	19.75		233.75	266
8070	3/4" pipe size, class 250		20	.400		256	23.50		279.50	320
8080	1" pipe size, class 250		19	.421		299	25		324	370
8090	1-1/4" pipe size, class 125		15	.533		370	31.50		401.50	460
8100	1-1/2" pipe size, class 125		13	.615		475	36.50		511.50	575
8110	2" pipe size, class 125	↓	11	.727	↓	550	43		593	670
8180	Brass, three way, screwed								
8190	1/2" pipe size, class 250	1 Plum	22	.364	Ea.	214	21.50		235.50	269
8200	3/4" pipe size, class 250		18	.444		256	26.50		282.50	320
8210	1" pipe size, class 250		17	.471		299	28		327	370
8214	1-1/4" pipe size, class 250		14	.571		370	34		404	460
8220	1-1/2" pipe size, class 125		11	.727		475	43		518	585
8230	2" pipe size, class 125	↓	9	.889		550	52.50		602.50	685
8450	Iron body, two way, flanged								
8510	2-1/2" pipe size, class 125	Q-1	5	3.200	Ea.	1,100	171		1,271	1,450
8520	3" pipe size, class 125		4.50	3.556		1,200	189		1,389	1,600
8530	4" pipe size, class 125	↓	3	5.333		1,500	284		1,784	2,075
8540	5" pipe size, class 125	Q-2	3.40	7.059		3,300	390		3,690	4,250
8550	6" pipe size, class 125	"	3	8	↓	4,100	440		4,540	5,175
8560	Iron body, three way, flanged								
8570	2-1/2" pipe size, class 125	Q-1	4.50	3.556	Ea.	1,200	189		1,389	1,575
8580	3" pipe size, class 125		4	4		1,475	213		1,688	1,950
8590	4" pipe size, class 125	↓	2.50	6.400		2,775	340		3,115	3,575
8600	6" pipe size, class 125	Q-2	3	8	↓	4,200	440		4,640	5,300
9005	Pneumatic system misc. components								
9010	Adding/subtracting repeater	1 Stpi	20	.400	Ea.	206	24.50		230.50	264
9020	Adjustable ratio network		16	.500		195	30.50		225.50	260
9030	Comparator	↓	20	.400	↓	410	24.50		434.50	485
9040	Cumulator								
9041	Air switching	1 Stpi	20	.400	Ea.	187	24.50		211.50	243
9042	Averaging		20	.400		234	24.50		258.50	294
9043	2:1 ratio		20	.400		395	24.50		419.50	470
9044	Sequencing		20	.400		120	24.50		144.50	168
9045	Two-position		16	.500		234	30.50		264.50	305
9046	Two-position pilot	↓	16	.500	↓	445	30.50		475.50	535
9050	Damper actuator								

For customer support on your Mechanical Cost Data, call 877.739.3689.

277

23 09 53 – Pneumatic and Electric Control System for HVAC

23 09 53.10 Control Components	Crew	Daily Output	Labor-Hours	Unit	Material	2016 Bare Costs Labor	Equipment	Total	Total Incl O&P
9051 For smoke control	1 Stpi	8	1	Ea.	272	60.50		332.50	390
9052 For ventilation	"	8	1	"	292	60.50		352.50	410
9053 Series duplex pedestal mounted									
9054 For inlet vanes on fans, compressors	1 Stpi	6	1.333	Ea.	2,375	81		2,456	2,725
9060 Enthalpy logic center		16	.500		395	30.50		425.50	480
9070 High/low pressure selector		22	.364		75.50	22		97.50	117
9080 Signal limiter		20	.400		120	24.50		144.50	168
9081 Transmitter	▼	28	.286	▼	45.50	17.35		62.85	76
9090 Optimal start									
9092 Mass temperature transmitter	1 Stpi	18	.444	Ea.	179	27		206	238
9100 Step controller with positioner, time delay restrictor									
9110 recycler air valve, switches and cam settings									
9112 With cabinet									
9113 6 points	1 Stpi	4	2	Ea.	1,200	121		1,321	1,500
9114 6 points, 6 manual sequences		2	4		1,050	243		1,293	1,550
9115 8 points	▼	3	2.667	▼	1,325	162		1,487	1,700
9200 Pressure controller and switches									
9210 High static pressure limit	1 Stpi	8	1	Ea.	330	60.50		390.50	455
9220 Pressure transmitter		8	1		282	60.50		342.50	400
9230 Differential pressure transmitter		8	1		1,100	60.50		1,160.50	1,325
9240 Static pressure transmitter		8	1		1,100	60.50		1,160.50	1,325
9250 Proportional-only control, single input		6	1.333		570	81		651	745
9260 Proportional plus integral, single input		6	1.333		805	81		886	1,000
9270 Proportional-only, dual input		6	1.333		785	81		866	985
9280 Proportional plus integral, dual input		6	1.333		985	81		1,066	1,200
9281 Time delay for above proportional units		16	.500		197	30.50		227.50	263
9290 Proportional/2 position controller units		6	1.333		550	81		631	725
9300 Differential pressure control direct/reverse	▼	6	1.333	▼	500	81		581	670
9310 Air pressure reducing valve									
9311 1/8" size	1 Stpi	18	.444	Ea.	46.50	27		73.50	91.50
9315 Precision valve, 1/4" size		17	.471		261	28.50		289.50	330
9320 Air flow controller		12	.667		221	40.50		261.50	305
9330 Booster relay volume amplifier		18	.444		128	27		155	182
9340 Series restrictor, straight		60	.133		6.55	8.10		14.65	19.40
9341 T-fitting		40	.200		9.60	12.15		21.75	29
9342 In-line adjustable		48	.167		42.50	10.10		52.60	62.50
9350 Diode tee		40	.200		22	12.15		34.15	42.50
9351 Restrictor tee		40	.200		22	12.15		34.15	42.50
9360 Pneumatic gradual switch		8	1		370	60.50		430.50	495
9361 Selector switch		8	1		133	60.50		193.50	239
9370 Electro-pneumatic motor driven servo		10	.800		1,200	48.50		1,248.50	1,400
9390 Fan control switch and mounting base		16	.500		134	30.50		164.50	193
9400 Circulating pump sequencer	▼	14	.571	▼	1,550	34.50		1,584.50	1,750
9410 Pneumatic tubing, fittings and accessories									
9414 Tubing, urethane									
9415 1/8" OD x 1/16" ID	1 Stpi	120	.067	L.F.	.86	4.05		4.91	7.05
9416 1/4" OD x 1/8" ID		115	.070		.89	4.22		5.11	7.35
9417 5/32" OD x 3/32" ID	▼	110	.073	▼	.50	4.41		4.91	7.20
9420 Coupling, straight									
9422 Barb x barb									
9423 1/4" x 5/32"	1 Stpi	160	.050	Ea.	.99	3.04		4.03	5.65
9424 1/4" x 1/4"		158	.051		.75	3.07		3.82	5.45
9425 3/8" x 3/8"	▼	154	.052	▼	.71	3.15		3.86	5.55

23 09 53 – Pneumatic and Electric Control System for HVAC

23 09 53.10 Control Components	Crew	Daily Output	Labor-Hours	Unit	Material	2016 Bare Costs Labor	Equipment	Total	Total Incl O&P
9426	3/8" x 1/4"	1 Stpi	150	.053	Ea.	.68	3.24		3.92	5.65
9427	1/2" x 1/4"		148	.054		6.45	3.28		9.73	12.05
9428	1/2" x 3/8"		144	.056		1.30	3.37		4.67	6.55
9429	1/2" x 1/2"		140	.057		.86	3.47		4.33	6.20
9440	Tube x tube									
9441	1/4" x 1/4"	1 Stpi	100	.080	Ea.	2.68	4.86		7.54	10.30
9442	3/8" x 1/4"		96	.083		5.40	5.05		10.45	13.60
9443	3/8" x 3/8"		92	.087		3.33	5.30		8.63	11.60
9444	1/2" x 3/8"		88	.091		5.50	5.50		11	14.40
9445	1/2" x 1/2"		84	.095		6.10	5.80		11.90	15.40
9450	Elbow coupling									
9452	Barb x barb									
9454	1/4" x 1/4"	1 Stpi	158	.051	Ea.	1.52	3.07		4.59	6.30
9455	3/8" x 3/8"		154	.052		4.08	3.15		7.23	9.25
9456	1/2" x 1/2"		140	.057		5.25	3.47		8.72	11
9460	Tube x tube									
9462	1/2" x 1/2"	1 Stpi	84	.095	Ea.	8.80	5.80		14.60	18.40
9470	Tee coupling									
9472	Barb x barb x barb									
9474	1/4" x 1/4" x 5/32"	1 Stpi	108	.074	Ea.	1.77	4.50		6.27	8.75
9475	1/4" x 1/4" x 1/4"		104	.077		1.20	4.67		5.87	8.35
9476	3/8" x 3/8" x 5/32"		102	.078		3.75	4.76		8.51	11.35
9477	3/8" x 3/8" x 1/4"		99	.081		1.77	4.91		6.68	9.35
9478	3/8" x 3/8" x 3/8"		98	.082		4.77	4.95		9.72	12.75
9479	1/2" x 1/2" x 1/4"		96	.083		4.86	5.05		9.91	13
9480	1/2" x 1/2" x 3/8"		95	.084		3.22	5.10		8.32	11.25
9481	1/2" x 1/2" x 1/2"		92	.087		5.45	5.30		10.75	13.95
9484	Tube x tube x tube									
9485	1/4" x 1/4" x 1/4"	1 Stpi	66	.121	Ea.	5.70	7.35		13.05	17.40
9486	3/8" x 1/4" x 1/4"		64	.125		11.70	7.60		19.30	24.50
9487	3/8" x 3/8" x 1/4"		61	.131		7.70	7.95		15.65	20.50
9488	3/8" x 3/8" x 3/8"		60	.133		8.10	8.10		16.20	21
9489	1/2" x 1/2" x 3/8"		58	.138		18.60	8.35		26.95	33
9490	1/2" x 1/2" x 1/2"		56	.143		13	8.65		21.65	27.50
9492	Needle valve									
9494	Tube x tube									
9495	1/4" x 1/4"	1 Stpi	86	.093	Ea.	4.87	5.65		10.52	13.85
9496	3/8" x 3/8"	"	82	.098	"	10.40	5.90		16.30	20.50
9600	Electronic system misc. components									
9610	Electric motor damper actuator	1 Elec	8	1	Ea.	630	55		685	775
9620	Damper position indicator		10	.800		146	44		190	226
9630	Pneumatic-electronic transducer		12	.667		71	36.50		107.50	133
9700	Modulating step controller									
9701	For 2-5 steps	1 Elec	5.30	1.509	Ea.	810	83		893	1,025
9702	6-10 steps		3.20	2.500		221	138		359	450
9703	11-15 steps		2.70	2.963		455	163		618	745
9704	16-20 steps		1.60	5		680	276		956	1,150
9710	Staging network									
9740	Accessory unit regulator									
9750	Accessory power supply	1 Elec	18	.444	Ea.	240	24.50		264.50	300
9760	Step down transformer	"	16	.500	"	220	27.50		247.50	283

For customer support on your Mechanical Cost Data, call 877.739.3689.

279

23 11 13.10 Fuel Oil Specialties		Crew	Daily Output	Labor-Hours	Unit	Material	2016 Bare Costs Labor	Equipment	Total	Total Incl O&P
0010	**FUEL OIL SPECIALTIES**									
0020	Foot valve, single poppet, metal to metal construction									
0040	Bevel seat, 1/2" diameter	1 Stpi	20	.400	Ea.	61.50	24.50		86	105
0060	3/4" diameter		18	.444		71	27		98	119
0080	1" diameter		16	.500		91	30.50		121.50	146
0100	1-1/4" diameter		15	.533		118	32.50		150.50	179
0120	1-1/2" diameter		13	.615		163	37.50		200.50	237
0140	2" diameter		11	.727		169	44		213	253
0160	Foot valve, double poppet, metal to metal construction									
0164	1" diameter	1 Stpi	15	.533	Ea.	196	32.50		228.50	265
0166	1-1/2" diameter	"	12	.667	"	235	40.50		275.50	320
0400	Fuel fill box, flush type									
0408	Nonlocking, watertight									
0410	1-1/2" diameter	1 Stpi	12	.667	Ea.	22	40.50		62.50	85
0440	2" diameter		10	.800		18.75	48.50		67.25	94
0450	2-1/2" diameter		9	.889		62.50	54		116.50	151
0460	3" diameter		7	1.143		55.50	69.50		125	166
0470	4" diameter		5	1.600		76.50	97		173.50	232
0500	Locking inner cover									
0510	2" diameter	1 Stpi	8	1	Ea.	84	60.50		144.50	184
0520	2-1/2" diameter		7	1.143		113	69.50		182.50	229
0530	3" diameter		5	1.600		121	97		218	280
0540	4" diameter		4	2		155	121		276	355
0600	Fuel system components									
0620	Spill container	1 Stpi	4	2	Ea.	840	121		961	1,100
0640	Fill adapter, 4", straight drop		8	1		71	60.50		131.50	170
0680	Fill cap, 4"		30	.267		38	16.20		54.20	66.50
0700	Extractor fitting, 4" x 1-1/2"		8	1		455	60.50		515.50	590
0740	Vapor hose adapter, 4"		8	1		114	60.50		174.50	217
0760	Wood gage stick, 10'					15			15	16.50
1000	Oil filters, 3/8" IPT, 20 gal. per hour	1 Stpi	20	.400		41	24.50		65.50	81.50
1020	32 gal. per hour		18	.444		46.50	27		73.50	92
1040	40 gal. per hour		16	.500		52	30.50		82.50	104
1060	50 gal. per hour		14	.571		60	34.50		94.50	119
2000	Remote tank gauging system, self contained									
2100	Single tank kit/8 sensors inputs w/printer	1 Stpi	2.50	3.200	Ea.	4,025	194		4,219	4,725
2120	Two tank kit/8 sensors inputs w/printer		2	4		5,425	243		5,668	6,325
3000	Valve, ball check, globe type, 3/8" diameter		24	.333		12.80	20		32.80	44.50
3500	Fusible, 3/8" diameter		24	.333		13.10	20		33.10	45
3600	1/2" diameter		24	.333		32	20		52	65.50
3610	3/4" diameter		20	.400		70	24.50		94.50	114
3620	1" diameter		19	.421		205	25.50		230.50	265
4000	Nonfusible, 3/8" diameter		24	.333		22.50	20		42.50	55
4500	Shutoff, gate type, lever handle, spring-fusible kit									
4520	1/4" diameter	1 Stpi	14	.571	Ea.	34.50	34.50		69	90.50
4540	3/8" diameter		12	.667		33.50	40.50		74	98
4560	1/2" diameter		10	.800		49.50	48.50		98	128
4570	3/4" diameter		8	1		86	60.50		146.50	186
4580	Lever handle, requires weight and fusible kit									
4600	1" diameter	1 Stpi	9	.889	Ea.	261	54		315	370
4620	1-1/4" diameter		8	1		320	60.50		380.50	440
4640	1-1/2" diameter		7	1.143		300	69.50		369.50	440
4660	2" diameter		6	1.333		545	81		626	720

23 11 Facility Fuel Piping

23 11 13 – Facility Fuel-Oil Piping

23 11 13.10 Fuel Oil Specialties	Crew	Daily Output	Labor-Hours	Unit	Material	2016 Bare Costs Labor	Equipment	Total	Total Incl O&P
4680	For fusible link, weight and braided wire, add				Ea.	5%				
5000	Vent alarm, whistling signal					26			26	28.50
5500	Vent protector/breather, 1-1/4" diameter	1 Stpi	32	.250		13.60	15.20		28.80	38
5520	1-1/2" diameter		32	.250		15.20	15.20		30.40	39.50
5540	2" diameter		32	.250		30.50	15.20		45.70	56.50
5560	3" diameter		28	.286		46.50	17.35		63.85	77
5580	4" diameter		24	.333		56	20		76	92
5600	Dust cap, breather, 2"	▼	40	.200	▼	10.90	12.15		23.05	30.50
8000	Fuel oil and tank heaters									
8020	Electric, capacity rated at 230 volts									
8040	Immersion element in steel manifold									
8060	96 GPH at 50°F rise	Q-5	6.40	2.500	Ea.	1,175	137		1,312	1,500
8070	128 GPH at 50°F rise		6.20	2.581		1,225	141		1,366	1,575
8080	160 GPH at 50°F rise		5.90	2.712		1,350	148		1,498	1,725
8090	192 GPH at 50°F rise		5.50	2.909		1,475	159		1,634	1,875
8100	240 GPH at 50°F rise		5.10	3.137		1,700	171		1,871	2,125
8110	288 GPH at 50°F rise		4.60	3.478		1,925	190		2,115	2,400
8120	384 GPH at 50°F rise		3.10	5.161		2,250	282		2,532	2,900
8130	480 GPH at 50°F rise		2.30	6.957		2,825	380		3,205	3,675
8140	576 GPH at 50°F rise	▼	2.10	7.619	▼	2,825	415		3,240	3,725
8300	Suction stub, immersion type									
8320	75" long, 750 watts	1 Stpi	14	.571	Ea.	805	34.50		839.50	940
8330	99" long, 2000 watts		12	.667		1,125	40.50		1,165.50	1,300
8340	123" long, 3000 watts	▼	10	.800	▼	1,225	48.50		1,273.50	1,400
8660	Steam, cross flow, rated at 5 PSIG									
8680	42 GPH	Q-5	7	2.286	Ea.	1,850	125		1,975	2,250
8690	73 GPH		6.70	2.388		2,150	130		2,280	2,575
8700	112 GPH		6.20	2.581		2,250	141		2,391	2,700
8710	158 GPH		5.80	2.759		2,450	151		2,601	2,925
8720	187 GPH		4	4		3,350	219		3,569	4,025
8730	270 GPH	▼	3.60	4.444		3,625	243		3,868	4,350
8740	365 GPH	Q-6	4.90	4.898		4,250	277		4,527	5,100
8750	635 GPH		3.70	6.486		5,375	365		5,740	6,450
8760	845 GPH		2.50	9.600		7,850	545		8,395	9,475
8770	1420 GPH		1.60	15		12,800	850		13,650	15,400
8780	2100 GPH	▼	1.10	21.818	▼	17,600	1,225		18,825	21,200

23 11 23 – Facility Natural-Gas Piping

23 11 23.10 Gas Meters

		Crew	Daily Output	Labor-Hours	Unit	Material	2016 Bare Costs Labor	Equipment	Total	Total Incl O&P
0010	**GAS METERS**									
4000	Residential									
4010	Gas meter, residential, 3/4" pipe size	1 Plum	14	.571	Ea.	239	34		273	315
4020	Gas meter, residential, 1" pipe size		12	.667		192	39.50		231.50	271
4030	Gas meter, residential, 1-1/4" pipe size	▼	10	.800	▼	194	47.50		241.50	285

23 11 23.20 Gas Piping, Flexible (Csst)

		Crew	Daily Output	Labor-Hours	Unit	Material	2016 Bare Costs Labor	Equipment	Total	Total Incl O&P
0010	**GAS PIPING, FLEXIBLE (CSST)**									
0100	Tubing with lightning protection									
0110	3/8"	1 Stpi	65	.123	L.F.	2.60	7.45		10.05	14.10
0120	1/2"		62	.129		2.91	7.85		10.76	15
0130	3/4"		60	.133		3.78	8.10		11.88	16.35
0140	1"		55	.145		5.40	8.85		14.25	19.25
0150	1-1/4"		50	.160		6.75	9.70		16.45	22
0160	1-1/2"		45	.178		11.85	10.80		22.65	29.50

For customer support on your Mechanical Cost Data, call 877.739.3689.

281

23 11 23 – Facility Natural-Gas Piping

23 11 23.20 Gas Piping, Flexible (Csst)	Crew	Daily Output	Labor-Hours	Unit	Material	2016 Bare Costs Labor	Equipment	Total	Total Incl O&P
0170	2"	1 Stpi	40	.200	L.F.	17.05	12.15		29.20	37
0200	Tubing for underground/underslab burial									
0210	3/8"	1 Stpi	65	.123	L.F.	3.60	7.45		11.05	15.20
0220	1/2"		62	.129		4.11	7.85		11.96	16.30
0230	3/4"		60	.133		5.20	8.10		13.30	17.95
0240	1"		55	.145		7	8.85		15.85	21
0250	1-1/4"		50	.160		9.40	9.70		19.10	25
0260	1-1/2"		45	.178		17.65	10.80		28.45	35.50
0270	2"	↓	40	.200	↓	21	12.15		33.15	41.50
3000	Fittings									
3010	Straight									
3100	Tube to NPT									
3110	3/8"	1 Stpi	29	.276	Ea.	10.35	16.75		27.10	37
3120	1/2"		27	.296		11.20	18		29.20	39.50
3130	3/4"		25	.320		15.25	19.40		34.65	46.50
3140	1"		23	.348		23.50	21		44.50	58
3150	1-1/4"		20	.400		52.50	24.50		77	94
3160	1-1/2"		17	.471		108	28.50		136.50	161
3170	2"	↓	15	.533	↓	183	32.50		215.50	251
3200	Coupling									
3210	3/8"	1 Stpi	29	.276	Ea.	18.10	16.75		34.85	45.50
3220	1/2"		27	.296		20.50	18		38.50	49.50
3230	3/4"		25	.320		28.50	19.40		47.90	60.50
3240	1"		23	.348		44	21		65	80
3250	1-1/4"		20	.400		97.50	24.50		122	144
3260	1-1/2"		17	.471		204	28.50		232.50	267
3270	2"	↓	15	.533	↓	350	32.50		382.50	435
3300	Flange fitting									
3310	3/8"	1 Stpi	25	.320	Ea.	15.90	19.40		35.30	47
3320	1/2"		22	.364		15.20	22		37.20	50
3330	3/4"		19	.421		19.65	25.50		45.15	60
3340	1"		16	.500		27	30.50		57.50	76
3350	1-1/4"	↓	12	.667	↓	63	40.50		103.50	130
3400	90° Flange valve									
3410	3/8"	1 Stpi	25	.320	Ea.	30.50	19.40		49.90	63
3420	1/2"		22	.364		31	22		53	67.50
3430	3/4"	↓	19	.421	↓	38	25.50		63.50	80
4000	Tee									
4120	1/2"	1 Stpi	20.50	.390	Ea.	32.50	23.50		56	71.50
4130	3/4"		19	.421		43	25.50		68.50	86
4140	1"	↓	17.50	.457		77.50	28		105.50	128
5000	Reducing									
5110	Tube to NPT									
5120	3/4" to 1/2" NPT	1 Stpi	26	.308	Ea.	16.35	18.70		35.05	46
5130	1" to 3/4" NPT	"	24	.333	"	25.50	20		45.50	58.50
5200	Reducing tee									
5210	1/2" x 3/8" x 3/8"	1 Stpi	21	.381	Ea.	43	23		66	82
5220	3/4" x 1/2" x 1/2"		20	.400		40	24.50		64.50	80.50
5230	1" x 3/4" x 1/2"		18	.444		66.50	27		93.50	114
5240	1-1/4" x 1-1/4" x 1"		15.60	.513		178	31		209	242
5250	1-1/2" x 1-1/2" x 1-1/4"		13.30	.602		271	36.50		307.50	355
5260	2" x 2" x 1-1/2"	↓	11.80	.678		405	41		446	510
5300	Manifold with four ports and mounting bracket									

23 11 Facility Fuel Piping

23 11 23 – Facility Natural-Gas Piping

23 11 23.20 Gas Piping, Flexible (Csst)	Crew	Daily Output	Labor-Hours	Unit	Material	2016 Bare Costs Labor	Equipment	Total	Total Incl O&P
5302	Labor to mount manifold does not include making pipe								
5304	connections which are included in fitting labor.								
5310	3/4" x 1/2" x 1/2"(4)	1 Stpi	76	.105	Ea.	35	6.40		41.40	48
5330	1-1/4" x 1" x 3/4"(4)		72	.111		50.50	6.75		57.25	65.50
5350	2" x 1-1/2" x 1"(4)		68	.118		63	7.15		70.15	80.50
5600	Protective striker plate								
5610	Quarter plate, 3" x 2"	1 Stpi	88	.091	Ea.	.81	5.50		6.31	9.25
5620	Half plate, 3" x 7"		82	.098		1.58	5.90		7.48	10.70
5630	Full plate, 3" x 12"		78	.103		3.27	6.25		9.52	13

23 12 Facility Fuel Pumps

23 12 13 – Facility Fuel-Oil Pumps

23 12 13.10 Pump and Motor Sets

		Crew	Daily Output	Labor-Hours	Unit	Material	2016 Bare Costs Labor	Equipment	Total	Total Incl O&P
0010	**PUMP AND MOTOR SETS**									
1810	Light fuel and diesel oils									
1820	20 GPH 1/3 HP	Q-5	6	2.667	Ea.	1,050	146		1,196	1,375
1830	27 GPH, 1/3 HP		6	2.667		1,050	146		1,196	1,375
1840	80 GPH, 1/3 HP		5	3.200		1,050	175		1,225	1,425
1850	145 GPH, 1/2 HP		4	4		1,100	219		1,319	1,550
1860	277 GPH, 1 HP		4	4		1,225	219		1,444	1,675
1870	700 GPH, 1-1/2 HP		3	5.333		2,750	291		3,041	3,475
1880	1000 GPH, 2 HP		3	5.333		3,225	291		3,516	4,000
1890	1800 GPH, 5 HP		1.80	8.889		4,550	485		5,035	5,750

23 13 Facility Fuel-Storage Tanks

23 13 13 – Facility Underground Fuel-Oil, Storage Tanks

23 13 13.09 Single-Wall Steel Fuel-Oil Tanks

		Crew	Daily Output	Labor-Hours	Unit	Material	2016 Bare Costs Labor	Equipment	Total	Total Incl O&P
0010	**SINGLE-WALL STEEL FUEL-OIL TANKS**									
5000	Tanks, steel ugnd., sti-p3, not incl. hold-down bars									
5500	Excavation, pad, pumps and piping not included									
5510	Single wall, 500 gallon capacity, 7 ga. shell	Q-5	2.70	5.926	Ea.	2,050	325		2,375	2,750
5520	1,000 gallon capacity, 7 ga. shell	"	2.50	6.400		3,825	350		4,175	4,725
5530	2,000 gallon capacity, 1/4" thick shell	Q-7	4.60	6.957		6,200	400		6,600	7,425
5535	2,500 gallon capacity, 7 ga. shell	Q-5	3	5.333		6,850	291		7,141	7,975
5540	5,000 gallon capacity, 1/4" thick shell	Q-7	3.20	10		11,500	580		12,080	13,600
5560	10,000 gallon capacity, 1/4" thick shell		2	16		12,300	925		13,225	15,000
5580	15,000 gallon capacity, 5/16" thick shell		1.70	18.824		19,000	1,100		20,100	22,600
5600	20,000 gallon capacity, 5/16" thick shell		1.50	21.333		26,700	1,225		27,925	31,200
5610	25,000 gallon capacity, 3/8" thick shell		1.30	24.615		37,900	1,425		39,325	43,900
5620	30,000 gallon capacity, 3/8" thick shell		1.10	29.091		38,400	1,675		40,075	44,700
5630	40,000 gallon capacity, 3/8" thick shell		.90	35.556		42,000	2,050		44,050	49,300
5640	50,000 gallon capacity, 3/8" thick shell		.80	40		46,600	2,300		48,900	55,000

For customer support on your Mechanical Cost Data, call 877.739.3689.

283

23 13 13 – Facility Underground Fuel-Oil, Storage Tanks

23 13 13.13 Dbl-Wall Steel, Undrgrnd Fuel-Oil, Stor. Tanks		Crew	Daily Output	Labor-Hours	Unit	Material	2016 Bare Costs Labor	Equipment	Total	Total Incl O&P
0010	**DOUBLE-WALL STEEL, UNDERGROUND FUEL-OIL, STORAGE TANKS**									
6200	Steel, underground, 360°, double wall, U.L. listed,									
6210	with sti-P3 corrosion protection,									
6220	(dielectric coating, cathodic protection, electrical									
6230	isolation) 30 year warranty,									
6240	not incl. manholes or hold-downs.									
6250	500 gallon capacity	Q-5	2.40	6.667	Ea.	5,575	365		5,940	6,700
6260	1,000 gallon capactiy	"	2.25	7.111		6,900	390		7,290	8,150
6270	2,000 gallon capacity	Q-7	4.16	7.692		9,050	445		9,495	10,600
6280	3,000 gallon capacity		3.90	8.205		10,900	475		11,375	12,700
6290	4,000 gallon capacity		3.64	8.791		13,800	510		14,310	15,900
6300	5,000 gallon capacity		2.91	10.997		19,900	635		20,535	22,900
6310	6,000 gallon capacity		2.42	13.223		25,600	765		26,365	29,300
6320	8,000 gallon capacity		2.08	15.385		24,600	890		25,490	28,400
6330	10,000 gallon capacity		1.82	17.582		27,700	1,025		28,725	32,000
6340	12,000 gallon capacity		1.70	18.824		30,600	1,100		31,700	35,400
6350	15,000 gallon capacity		1.33	24.060		31,900	1,400		33,300	37,200
6360	20,000 gallon capacity		1.33	24.060		32,700	1,400		34,100	38,000
6370	25,000 gallon capacity		1.16	27.586		72,500	1,600		74,100	82,500
6380	30,000 gallon capacity		1.03	31.068		87,500	1,800		89,300	98,500
6390	40,000 gallon capacity		.80	40		112,500	2,300		114,800	127,000
6395	50,000 gallon capacity		.73	43.836		136,500	2,525		139,025	154,500
6400	For hold-downs 500-2000 gal., add		16	2	Set	183	116		299	375
6410	For hold-downs 3000-6000 gal., add		12	2.667		370	154		524	645
6420	For hold-downs 8000-12,000 gal., add		11	2.909		445	168		613	745
6430	For hold-downs 15,000 gal., add		9	3.556		640	205		845	1,025
6440	For hold-downs 20,000 gal., add		8	4		735	231		966	1,150
6450	For hold-downs 20,000 gal. plus, add		6	5.333		960	310		1,270	1,525
6500	For manways, add				Ea.	1,800			1,800	2,000
6600	In place with hold-downs									
6652	550 gallon capacity	Q-5	1.84	8.696	Ea.	5,775	475		6,250	7,075

23 13 13.23 Glass-Fiber-Reinfcd-Plastic, Fuel-Oil, Storage

		Crew	Daily Output	Labor-Hours	Unit	Material	2016 Bare Costs Labor	Equipment	Total	Total Incl O&P
0010	**GLASS-FIBER-REINFCD-PLASTIC, UNDERGRND. FUEL-OIL, STORAGE**									
0210	Fiberglass, underground, single wall, U.L. listed, not including									
0220	manway or hold-down strap									
0225	550 gallon capacity	Q-5	2.67	5.993	Ea.	3,350	325		3,675	4,175
0230	1,000 gallon capacity	"	2.46	6.504		4,325	355		4,680	5,275
0240	2,000 gallon capacity	Q-7	4.57	7.002		6,250	405		6,655	7,475
0245	3,000 gallon capacity		3.90	8.205		7,475	475		7,950	8,950
0250	4,000 gallon capacity		3.55	9.014		8,675	520		9,195	10,300
0255	5,000 gallon capacity		3.20	10		9,650	580		10,230	11,500
0260	6,000 gallon capacity		2.67	11.985		11,000	695		11,695	13,200
0270	8,000 gallon capacity		2.29	13.974		11,800	810		12,610	14,200
0280	10,000 gallon capacity		2	16		13,400	925		14,325	16,200
0282	12,000 gallon capacity		1.88	17.021		15,300	985		16,285	18,300
0284	15,000 gallon capacity		1.68	19.048		21,500	1,100		22,600	25,300
0290	20,000 gallon capacity		1.45	22.069		27,600	1,275		28,875	32,300
0300	25,000 gallon capacity		1.28	25		41,300	1,450		42,750	47,700
0320	30,000 gallon capacity		1.14	28.070		50,000	1,625		51,625	57,500
0340	40,000 gallon capacity		.89	35.955		90,000	2,075		92,075	102,000
0360	48,000 gallon capacity		.81	39.506		134,000	2,275		136,275	151,000
0500	For manway, fittings and hold-downs, add					20%	15%			

23 13 13.23 Glass-Fiber-Reinfcd-Plastic, Fuel-Oil, Storage	Crew	Daily Output	Labor-Hours	Unit	Material	2016 Bare Costs Labor	Equipment	Total	Total Incl O&P
0600	For manways, add				Ea.	2,100			2,100	2,300
1000	For helical heating coil, add	Q-5	2.50	6.400	↓	5,025	350		5,375	6,050
1020	Fiberglass, underground, double wall, U.L. listed									
1030	includes manways, not incl. hold-down straps									
1040	600 gallon capacity	Q-5	2.42	6.612	Ea.	7,550	360		7,910	8,850
1050	1,000 gallon capacity	"	2.25	7.111		10,300	390		10,690	12,000
1060	2,500 gallon capacity	Q-7	4.16	7.692		15,500	445		15,945	17,800
1070	3,000 gallon capacity		3.90	8.205		16,800	475		17,275	19,200
1080	4,000 gallon capacity		3.64	8.791		17,100	510		17,610	19,600
1090	6,000 gallon capacity		2.42	13.223		22,200	765		22,965	25,600
1100	8,000 gallon capacity		2.08	15.385		24,700	890		25,590	28,600
1110	10,000 gallon capacity		1.82	17.582		28,600	1,025		29,625	33,000
1120	12,000 gallon capacity		1.70	18.824		35,800	1,100		36,900	41,000
1122	15,000 gallon capacity		1.52	21.053		47,500	1,225		48,725	54,500
1124	20,000 gallon capacity		1.33	24.060		58,500	1,400		59,900	66,500
1126	25,000 gallon capacity		1.16	27.586		80,500	1,600		82,100	91,000
1128	30,000 gallon capacity	↓	1.03	31.068		96,500	1,800		98,300	109,000
1140	For hold-down straps, add				↓	2%	10%			
1150	For hold-downs 500-4000 gal., add	Q-7	16	2	Set	470	116		586	690
1160	For hold-downs 5000-15000 gal., add		8	4		940	231		1,171	1,375
1170	For hold-downs 20,000 gal., add		5.33	6.004		1,400	345		1,745	2,075
1180	For hold-downs 25,000 gal., add		4	8		1,875	460		2,335	2,775
1190	For hold-downs 30,000 gal., add	↓	2.60	12.308	↓	2,825	710		3,535	4,175
2210	Fiberglass, underground, single wall, U.L. listed, including									
2220	hold-down straps, no manways									
2225	550 gallon capacity	Q-5	2	8	Ea.	3,825	435		4,260	4,850
2230	1,000 gallon capacity	"	1.88	8.511		4,800	465		5,265	5,975
2240	2,000 gallon capacity	Q-7	3.55	9.014		6,725	520		7,245	8,175
2250	4,000 gallon capacity		2.90	11.034		9,150	640		9,790	11,100
2260	6,000 gallon capacity		2	16		11,900	925		12,825	14,500
2270	8,000 gallon capacity		1.78	17.978		12,700	1,050		13,750	15,600
2280	10,000 gallon capacity		1.60	20		14,400	1,150		15,550	17,600
2282	12,000 gallon capacity		1.52	21.053		16,200	1,225		17,425	19,600
2284	15,000 gallon capacity		1.39	23.022		22,400	1,325		23,725	26,700
2290	20,000 gallon capacity		1.14	28.070		29,100	1,625		30,725	34,500
2300	25,000 gallon capacity		.96	33.333		43,200	1,925		45,125	50,500
2320	30,000 gallon capacity	↓	.80	40	↓	53,000	2,300		55,300	61,500
3020	Fiberglass, underground, double wall, U.L. listed									
3030	includes manways and hold-down straps									
3040	600 gallon capacity	Q-5	1.86	8.602	Ea.	8,025	470		8,495	9,525
3050	1,000 gallon capacity	"	1.70	9.412		10,800	515		11,315	12,700
3060	2,500 gallon capacity	Q-7	3.29	9.726		16,000	560		16,560	18,500
3070	3,000 gallon capacity		3.13	10.224		17,300	590		17,890	19,900
3080	4,000 gallon capacity		2.93	10.922		17,500	630		18,130	20,300
3090	6,000 gallon capacity		1.86	17.204		23,100	995		24,095	27,000
3100	8,000 gallon capacity		1.65	19.394		25,700	1,125		26,825	29,900
3110	10,000 gallon capacity		1.48	21.622		29,600	1,250		30,850	34,400
3120	12,000 gallon capacity		1.40	22.857		36,700	1,325		38,025	42,400
3122	15,000 gallon capacity		1.28	25		48,500	1,450		49,950	55,500
3124	20,000 gallon capacity		1.06	30.189		60,000	1,750		61,750	68,500
3126	25,000 gallon capacity		.90	35.556		82,500	2,050		84,550	93,500
3128	30,000 gallon capacity	↓	.74	43.243	↓	99,500	2,500		102,000	113,500

For customer support on your Mechanical Cost Data, call 877.739.3689.

285

23 13 Facility Fuel-Storage Tanks

23 13 23 – Facility Aboveground Fuel-Oil, Storage Tanks

23 13 23.13 Vertical, Steel, Abvground Fuel-Oil, Stor. Tanks	Crew	Daily Output	Labor-Hours	Unit	Material	2016 Bare Costs Labor	Equipment	Total	Total Incl O&P
0010 **VERTICAL, STEEL, ABOVEGROUND FUEL-OIL, STORAGE TANKS**									
4000 Fixed roof oil storage tanks, steel (1 BBL=42 gal. w/foundation 3'D x 1'W)									
4200 5,000 barrels				Ea.				194,000	213,500
4300 24,000 barrels								333,500	367,000
4500 56,000 barrels								729,000	802,000
4600 110,000 barrels								1,060,000	1,166,000
4800 143,000 barrels								1,250,000	1,375,000
4900 225,000 barrels								1,360,000	1,496,000
5100 Floating roof gasoline tanks, steel, 5,000 barrels (w/foundation 3'D x 1'W)								204,000	225,000
5200 25,000 barrels								381,000	419,000
5400 55,000 barrels								839,000	923,000
5500 100,000 barrels								1,253,000	1,379,000
5700 150,000 barrels								1,532,000	1,685,000
5800 225,000 barrels								2,300,000	2,783,000

23 13 23.16 Horizontal, Stl, Abvgrd Fuel-Oil, Storage Tanks

	Crew	Daily Output	Labor-Hours	Unit	Material	2016 Bare Costs Labor	Equipment	Total	Total Incl O&P
0010 **HORIZONTAL, STEEL, ABOVEGROUND FUEL-OIL, STORAGE TANKS**									
3000 Steel, storage, above ground, including cradles, coating,									
3020 fittings, not including foundation, pumps or piping									
3040 Single wall, 275 gallon	Q-5	5	3.200	Ea.	490	175		665	805
3060 550 gallon	"	2.70	5.926		3,925	325		4,250	4,825
3080 1,000 gallon	Q-7	5	6.400		6,450	370		6,820	7,650
3100 1,500 gallon		4.75	6.737		9,225	390		9,615	10,700
3120 2,000 gallon		4.60	6.957		11,200	400		11,600	12,900
3140 5,000 gallon		3.20	10		20,500	580		21,080	23,400
3150 10,000 gallon		2	16		36,900	925		37,825	41,900
3160 15,000 gallon		1.70	18.824		47,300	1,100		48,400	53,500
3170 20,000 gallon		1.45	22.069		61,500	1,275		62,775	69,500
3180 25,000 gallon		1.30	24.615		71,500	1,425		72,925	80,500
3190 30,000 gallon		1.10	29.091		85,500	1,675		87,175	97,000
3320 Double wall, 500 gallon capacity	Q-5	2.40	6.667		2,625	365		2,990	3,425
3330 2000 gallon capacity	Q-7	4.15	7.711		9,975	445		10,420	11,700
3340 4000 gallon capacity		3.60	8.889		17,800	515		18,315	20,400
3350 6000 gallon capacity		2.40	13.333		21,000	770		21,770	24,300
3360 8000 gallon capacity		2	16		27,000	925		27,925	31,100
3370 10000 gallon capacity		1.80	17.778		30,200	1,025		31,225	34,800
3380 15000 gallon capacity		1.50	21.333		45,900	1,225		47,125	52,500
3390 20000 gallon capacity		1.30	24.615		52,500	1,425		53,925	59,500
3400 25000 gallon capacity		1.15	27.826		63,500	1,600		65,100	72,500
3410 30000 gallon capacity		1	32		69,500	1,850		71,350	79,500

23 13 23.26 Horizontal, Conc., Abvgrd Fuel-Oil, Stor. Tanks

	Crew	Daily Output	Labor-Hours	Unit	Material	2016 Bare Costs Labor	Equipment	Total	Total Incl O&P
0010 **HORIZONTAL, CONCRETE, ABOVEGROUND FUEL-OIL, STORAGE TANKS**									
0050 Concrete, storage, above ground, including pad & pump									
0100 500 gallon	F-3	2	20	Ea.	10,000	985	330	11,315	12,900
0200 1,000 gallon	"	2	20		14,000	985	330	15,315	17,300
0300 2,000 gallon	F-4	2	24		18,000	1,175	560	19,735	22,200
0400 4,000 gallon		2	24		23,000	1,175	560	24,735	27,700
0500 8,000 gallon		2	24		36,000	1,175	560	37,735	42,000
0600 12,000 gallon		2	24		48,000	1,175	560	49,735	55,500

23 21 20 – Hydronic HVAC Piping Specialties

23 21 20.10 Air Control	Crew	Daily Output	Labor-Hours	Unit	Material	2016 Bare Costs Labor	Equipment	Total	Total Incl O&P
0010 AIR CONTROL									
0030 Air separator, with strainer									
0040 2" diameter	Q-5	6	2.667	Ea.	1,225	146		1,371	1,575
0080 2-1/2" diameter		5	3.200		1,375	175		1,550	1,800
0100 3" diameter		4	4		2,125	219		2,344	2,675
0120 4" diameter		3	5.333		3,050	291		3,341	3,825
0130 5" diameter	Q-6	3.60	6.667		3,900	380		4,280	4,875
0140 6" diameter		3.40	7.059		4,675	400		5,075	5,750
0160 8" diameter		3	8		6,975	455		7,430	8,350
0180 10" diameter		2.20	10.909		10,800	620		11,420	12,800
0200 12" diameter		1.70	14.118		18,000	800		18,800	21,000
0210 14" diameter		1.30	18.462		21,400	1,050		22,450	25,100
0220 16" diameter		1	24		31,900	1,350		33,250	37,200
0230 18" diameter		.80	30		42,000	1,700		43,700	48,800
0240 20" diameter		.60	40		46,900	2,275		49,175	55,000
0300 Without strainer									
0310 2" diameter	Q-5	6	2.667	Ea.	760	146		906	1,050
0320 2-1/2" diameter		5	3.200		910	175		1,085	1,275
0330 3" diameter		4	4		1,275	219		1,494	1,725
0340 4" diameter		3	5.333		2,000	291		2,291	2,650
0350 5" diameter		2.40	6.667		2,725	365		3,090	3,550
0360 6" diameter	Q-6	3.40	7.059		3,125	400		3,525	4,025
0370 8" diameter		3	8		4,275	455		4,730	5,375
0380 10" diameter		2.20	10.909		6,300	620		6,920	7,875
0390 12" diameter		1.70	14.118		9,700	800		10,500	11,900
0400 14" diameter		1.30	18.462		14,500	1,050		15,550	17,600
0410 16" diameter		1	24		20,300	1,350		21,650	24,400
0420 18" diameter		.80	30		26,500	1,700		28,200	31,700
0430 20" diameter		.60	40		31,700	2,275		33,975	38,300
1000 Micro-bubble separator for total air removal, closed loop system									
1010 Requires bladder type tank in system.									
1020 Water (hot or chilled) or glycol system									
1030 Threaded									
1040 3/4" diameter	1 Stpi	20	.400	Ea.	79	24.50		103.50	124
1050 1" diameter		19	.421		89	25.50		114.50	137
1060 1-1/4" diameter		16	.500		120	30.50		150.50	178
1070 1-1/2" diameter		13	.615		160	37.50		197.50	233
1080 2" diameter		11	.727		1,000	44		1,044	1,175
1090 2-1/2" diameter	Q-5	15	1.067		1,125	58.50		1,183.50	1,325
1100 3" diameter		13	1.231		1,600	67		1,667	1,875
1110 4" diameter		10	1.600		1,725	87.50		1,812.50	2,025
1230 Flanged									
1250 2" diameter	1 Stpi	8	1	Ea.	1,525	60.50		1,585.50	1,775
1260 2-1/2" diameter	Q-5	5	3.200		1,500	175		1,675	1,925
1270 3" diameter		4.50	3.556		2,025	194		2,219	2,525
1280 4" diameter		3	5.333		2,250	291		2,541	2,925
1290 5" diameter	Q-6	3.40	7.059		2,975	400		3,375	3,875
1300 6" diameter	"	3	8		3,875	455		4,330	4,925
1400 Larger sizes available									
1590 With extended tank dirt catcher									
1600 Flanged									
1620 2" diameter	1 Stpi	7.40	1.081	Ea.	3,000	65.50		3,065.50	3,400
1630 2-1/2" diameter	Q-5	4.20	3.810		3,200	208		3,408	3,850

23 21 20 – Hydronic HVAC Piping Specialties

23 21 20.10 Air Control		Crew	Daily Output	Labor-Hours	Unit	Material	2016 Bare Costs Labor	Equipment	Total	Total Incl O&P
1640	3" diameter	Q-5	4	4	Ea.	4,300	219		4,519	5,050
1650	4" diameter	↓	2.80	5.714		4,775	310		5,085	5,725
1660	5" diameter	Q-6	3.10	7.742		6,325	440		6,765	7,600
1670	6" diameter	"	2.90	8.276	↓	8,250	470		8,720	9,775
1800	With drain/dismantle/cleaning access flange									
1810	Flanged									
1820	2" diameter	Q-5	4.60	3.478	Ea.	2,575	190		2,765	3,125
1830	2-1/2" diameter	↓	4	4		2,775	219		2,994	3,375
1840	3" diameter		3.60	4.444		3,750	243		3,993	4,500
1850	4" diameter	↓	2.50	6.400		4,150	350		4,500	5,100
1860	5" diameter	Q-6	2.80	8.571		5,500	485		5,985	6,775
1870	6" diameter	"	2.60	9.231	↓	7,175	525		7,700	8,675
2000	Boiler air fitting, separator									
2010	1-1/4"	Q-5	22	.727	Ea.	231	39.50		270.50	315
2020	1-1/2"		20	.800		445	43.50		488.50	555
2021	2"	↓	18	.889	↓	775	48.50		823.50	925
2400	Compression tank air fitting									
2410	For tanks 9" to 24" diameter	1 Stpi	15	.533	Ea.	79.50	32.50		112	137
2420	For tanks 100 gallon or larger	"	12	.667	"	298	40.50		338.50	390

23 21 20.14 Air Purging Scoop

		Crew	Daily Output	Labor-Hours	Unit	Material	Labor	Equipment	Total	Total Incl O&P
0010	**AIR PURGING SCOOP**, With tappings.									
0020	For air vent and expansion tank connection									
0100	1" pipe size, threaded	1 Stpi	19	.421	Ea.	25	25.50		50.50	66
0110	1-1/4" pipe size, threaded		15	.533		25	32.50		57.50	76.50
0120	1-1/2" pipe size, threaded		13	.615		55	37.50		92.50	117
0130	2" pipe size, threaded	↓	11	.727		63	44		107	136
0140	2-1/2" pipe size, threaded	Q-5	15	1.067		134	58.50		192.50	236
0150	3" pipe size, threaded		13	1.231		175	67		242	294
0160	4" 150 lb. flanges	↓	3	5.333	↓	390	291		681	870

23 21 20.18 Automatic Air Vent

		Crew	Daily Output	Labor-Hours	Unit	Material	Labor	Equipment	Total	Total Incl O&P
0010	**AUTOMATIC AIR VENT**									
0020	Cast iron body, stainless steel internals, float type									
0060	1/2" NPT inlet, 300 psi	1 Stpi	12	.667	Ea.	142	40.50		182.50	217
0140	3/4" NPT inlet, 300 psi		12	.667		142	40.50		182.50	217
0180	1/2" NPT inlet, 250 psi		10	.800		360	48.50		408.50	475
0220	3/4" NPT inlet, 250 psi		10	.800		360	48.50		408.50	475
0260	1" NPT inlet, 250 psi	↓	10	.800		530	48.50		578.50	660
0340	1-1/2" NPT inlet, 250 psi	Q-5	12	1.333		1,125	73		1,198	1,325
0380	2" NPT inlet, 250 psi	"	12	1.333	↓	1,125	73		1,198	1,325
0600	Forged steel body, stainless steel internals, float type									
0640	1/2" NPT inlet, 750 psi	1 Stpi	12	.667	Ea.	1,200	40.50		1,240.50	1,375
0680	3/4" NPT inlet, 750 psi		12	.667		1,200	40.50		1,240.50	1,375
0760	3/4" NPT inlet, 1000 psi	↓	10	.800		1,800	48.50		1,848.50	2,075
0800	1" NPT inlet, 1000 psi	Q-5	12	1.333		1,800	73		1,873	2,100
0880	1-1/2" NPT inlet, 1000 psi		10	1.600		5,100	87.50		5,187.50	5,725
0920	2" NPT inlet, 1000 psi	↓	10	1.600	↓	5,100	87.50		5,187.50	5,725
1100	Formed steel body, noncorrosive									
1110	1/8" NPT inlet 150 psi	1 Stpi	32	.250	Ea.	12.85	15.20		28.05	37
1120	1/4" NPT inlet 150 psi		32	.250		43.50	15.20		58.70	70.50
1130	3/4" NPT inlet 150 psi	↓	32	.250		43.50	15.20		58.70	70.50
1300	Chrome plated brass, automatic/manual, for radiators									
1310	1/8" NPT inlet, nickel plated brass	1 Stpi	32	.250	Ea.	7.45	15.20		22.65	31

23 21 Hydronic Piping and Pumps

23 21 20 – Hydronic HVAC Piping Specialties

23 21 20.22 Circuit Sensor

		Crew	Daily Output	Labor-Hours	Unit	Material	2016 Bare Costs Labor	Equipment	Total	Total Incl O&P
0010	**CIRCUIT SENSOR**, Flow meter									
0020	Metering stations									
0040	Wafer orifice insert type									
0060	2-1/2" pipe size	Q-5	12	1.333	Ea.	218	73		291	350
0100	3" pipe size		11	1.455		248	79.50		327.50	390
0140	4" pipe size		8	2		284	109		393	475
0180	5" pipe size		7.30	2.192		360	120		480	575
0220	6" pipe size		6.40	2.500		430	137		567	675
0260	8" pipe size	Q-6	5.30	4.528		595	257		852	1,050
0280	10" pipe size		4.60	5.217		685	296		981	1,200
0360	12" pipe size		4.20	5.714		1,125	325		1,450	1,750

23 21 20.26 Circuit Setter

		Crew	Daily Output	Labor-Hours	Unit	Material	2016 Bare Costs Labor	Equipment	Total	Total Incl O&P
0010	**CIRCUIT SETTER**, Balance valve									
0012	Bronze body, soldered									
0013	1/2" size	1 Stpi	24	.333	Ea.	74	20		94	112
0014	3/4" size	"	20	.400	"	78	24.50		102.50	123
0018	Threaded									
0019	1/2" pipe size	1 Stpi	22	.364	Ea.	76.50	22		98.50	118
0020	3/4" pipe size		20	.400		82	24.50		106.50	127
0040	1" pipe size		18	.444		106	27		133	158
0050	1-1/4" pipe size		15	.533		156	32.50		188.50	220
0060	1-1/2" pipe size		12	.667		183	40.50		223.50	263
0080	2" pipe size		10	.800		263	48.50		311.50	365
0100	2-1/2" pipe size	Q-5	15	1.067		595	58.50		653.50	745
0120	3" pipe size	"	10	1.600		835	87.50		922.50	1,050
0130	Cast iron body, flanged									
0136	3" pipe size, flanged	Q-5	4	4	Ea.	835	219		1,054	1,250
0140	4" pipe size	"	3	5.333		1,250	291		1,541	1,825
0200	For differential meter, accurate to 1%, add					965			965	1,075
0300	For differential meter 400 psi/200°F continuous, add					2,375			2,375	2,625

23 21 20.30 Cocks, Drains and Specialties

		Crew	Daily Output	Labor-Hours	Unit	Material	2016 Bare Costs Labor	Equipment	Total	Total Incl O&P
0010	**COCKS, DRAINS AND SPECIALTIES**									
1000	Boiler drain									
1010	Pipe thread to hose									
1020	Bronze									
1030	1/2" size	1 Stpi	36	.222	Ea.	9.75	13.50		23.25	31
1040	3/4" size	"	34	.235	"	10.20	14.30		24.50	32.50
1100	Solder to hose									
1110	Bronze									
1120	1/2" size	1 Stpi	46	.174	Ea.	12.55	10.55		23.10	30
1130	3/4" size	"	44	.182	"	13.60	11.05		24.65	31.50
1600	With built-in vacuum breaker									
1610	1/2" I.P. or solder	1 Stpi	36	.222	Ea.	42.50	13.50		56	67.50
1630	With tamper proof vacuum breaker									
1640	1/2" I.P. or solder	1 Stpi	36	.222	Ea.	47.50	13.50		61	72.50
1650	3/4" I.P. or solder	"	34	.235	"	48	14.30		62.30	74.50
3000	Cocks									
3010	Air, lever or tee handle									
3020	Bronze, single thread									
3030	1/8" size	1 Stpi	52	.154	Ea.	10.85	9.35		20.20	26
3040	1/4" size		46	.174		11.45	10.55		22	28.50
3050	3/8" size		40	.200		11.60	12.15		23.75	31

23 21 20.30 Cocks, Drains and Specialties

		Crew	Daily Output	Labor-Hours	Unit	Material	2016 Bare Costs Labor	Equipment	Total	Total Incl O&P
3060	1/2" size	1 Stpi	36	.222	Ea.	13.70	13.50		27.20	35.50
3100	Bronze, double thread									
3110	1/8" size	1 Stpi	26	.308	Ea.	13.70	18.70		32.40	43
3120	1/4" size		22	.364		14.20	22		36.20	49
3130	3/8" size		18	.444		14.80	27		41.80	57
3140	1/2" size	↓	15	.533	↓	18.15	32.50		50.65	69
4000	Steam									
4010	Bronze									
4020	1/8" size	1 Stpi	26	.308	Ea.	22	18.70		40.70	52
4030	1/4" size		22	.364		18.95	22		40.95	54.50
4040	3/8" size		18	.444		27.50	27		54.50	70.50
4050	1/2" size		15	.533		27.50	32.50		60	79
4060	3/4" size		14	.571		34	34.50		68.50	90
4070	1" size		13	.615		49.50	37.50		87	111
4080	1-1/4" size		11	.727		67	44		111	141
4090	1-1/2" size		10	.800		104	48.50		152.50	188
4100	2" size	↓	9	.889	↓	148	54		202	244
4300	Bronze, 3 way									
4320	1/4" size	1 Stpi	15	.533	Ea.	47.50	32.50		80	101
4330	3/8" size		12	.667		47.50	40.50		88	113
4340	1/2" size		10	.800		47.50	48.50		96	126
4350	3/4" size		9.50	.842		81	51		132	167
4360	1" size		8.50	.941		98	57		155	194
4370	1-1/4" size		7.50	1.067		147	65		212	260
4380	1-1/2" size		6.50	1.231		170	74.50		244.50	300
4390	2" size	↓	6	1.333	↓	252	81		333	400
4500	Gauge cock, brass									
4510	1/4" FPT	1 Stpi	24	.333	Ea.	11.45	20		31.45	43
4512	1/4" MPT	"	24	.333	"	13.75	20		33.75	45.50
4600	Pigtail, steam syphon									
4604	1/4"	1 Stpi	24	.333	Ea.	20	20		40	52.50
4650	Snubber valve									
4654	1/4"	1 Stpi	22	.364	Ea.	11	22		33	45.50
4660	Nipple, black steel									
4664	1/4" x 3"	1 Stpi	37	.216	Ea.	2.86	13.10		15.96	23

23 21 20.34 Dielectric Unions

		Crew	Daily Output	Labor-Hours	Unit	Material	2016 Bare Costs Labor	Equipment	Total	Total Incl O&P
0010	**DIELECTRIC UNIONS**, Standard gaskets for water and air									
0020	250 psi maximum pressure									
0280	Female IPT to sweat, straight									
0300	1/2" pipe size	1 Plum	24	.333	Ea.	7.45	19.75		27.20	38
0340	3/4" pipe size		20	.400		8.30	23.50		31.80	44.50
0360	1" pipe size		19	.421		11.25	25		36.25	50
0380	1-1/4" pipe size		15	.533		17.45	31.50		48.95	66.50
0400	1-1/2" pipe size		13	.615		26.50	36.50		63	84
0420	2" pipe size	↓	11	.727	↓	35.50	43		78.50	104
0580	Female IPT to brass pipe thread, straight									
0600	1/2" pipe size	1 Plum	24	.333	Ea.	16.10	19.75		35.85	47.50
0640	3/4" pipe size		20	.400		17.75	23.50		41.25	55
0660	1" pipe size		19	.421		33	25		58	73.50
0680	1-1/4" pipe size		15	.533		40	31.50		71.50	92
0700	1-1/2" pipe size		13	.615		58.50	36.50		95	120
0720	2" pipe size	↓	11	.727	↓	114	43		157	191

23 21 20.34 Dielectric Unions		Crew	Daily Output	Labor-Hours	Unit	Material	2016 Bare Costs Labor	Equipment	Total	Total Incl O&P
0780	Female IPT to female IPT, straight									
0800	1/2" pipe size	1 Plum	24	.333	Ea.	15	19.75		34.75	46.50
0840	3/4" pipe size		20	.400		17.05	23.50		40.55	54.50
0860	1" pipe size		19	.421		22.50	25		47.50	62
0880	1-1/4" pipe size		15	.533		30	31.50		61.50	80.50
0900	1-1/2" pipe size		13	.615		46	36.50		82.50	106
0920	2" pipe size		11	.727		68	43		111	140
2000	175 psi maximum pressure									
2180	Female IPT to sweat									
2240	2" pipe size	1 Plum	9	.889	Ea.	154	52.50		206.50	249
2260	2-1/2" pipe size	Q-1	15	1.067		166	57		223	269
2280	3" pipe size		14	1.143		229	61		290	345
2300	4" pipe size		11	1.455		605	77.50		682.50	780
2480	Female IPT to brass pipe									
2500	1-1/2" pipe size	1 Plum	11	.727	Ea.	189	43		232	273
2540	2" pipe size	"	9	.889		222	52.50		274.50	325
2560	2-1/2" pipe size	Q-1	15	1.067		325	57		382	445
2580	3" pipe size		14	1.143		380	61		441	505
2600	4" pipe size		11	1.455		680	77.50		757.50	865

23 21 20.38 Expansion Couplings

		Crew	Daily Output	Labor-Hours	Unit	Material	2016 Bare Costs Labor	Equipment	Total	Total Incl O&P
0010	**EXPANSION COUPLINGS**, Hydronic									
0100	Copper to copper, sweat									
1000	Baseboard riser fitting, 5" stub by coupling 12" long									
1020	1/2" diameter	1 Stpi	24	.333	Ea.	20.50	20		40.50	53
1040	3/4" diameter		20	.400		29	24.50		53.50	68.50
1060	1" diameter		19	.421		40.50	25.50		66	83
1080	1-1/4" diameter		15	.533		52.50	32.50		85	107
1180	9" Stub by tubing 8" long									
1200	1/2" diameter	1 Stpi	24	.333	Ea.	18.30	20		38.30	50.50
1220	3/4" diameter		20	.400		24.50	24.50		49	63
1240	1" diameter		19	.421		33.50	25.50		59	75.50
1260	1-1/4" diameter		15	.533		47	32.50		79.50	101

23 21 20.42 Expansion Joints

		Crew	Daily Output	Labor-Hours	Unit	Material	2016 Bare Costs Labor	Equipment	Total	Total Incl O&P
0010	**EXPANSION JOINTS**									
0100	Bellows type, neoprene cover, flanged spool									
0140	6" face to face, 1-1/4" diameter	1 Stpi	11	.727	Ea.	255	44		299	350
0160	1-1/2" diameter	"	10.60	.755		255	46		301	350
0180	2" diameter	Q-5	13.30	1.203		258	65.50		323.50	385
0190	2-1/2" diameter		12.40	1.290		267	70.50		337.50	400
0200	3" diameter		11.40	1.404		299	76.50		375.50	445
0210	4" diameter		8.40	1.905		325	104		429	510
0220	5" diameter		7.60	2.105		395	115		510	610
0230	6" diameter		6.80	2.353		405	129		534	645
0240	8" diameter		5.40	2.963		475	162		637	765
0250	10" diameter		5	3.200		650	175		825	980
0260	12" diameter		4.60	3.478		745	190		935	1,100
0480	10" face to face, 2" diameter		13	1.231		370	67		437	510
0500	2-1/2" diameter		12	1.333		390	73		463	540
0520	3" diameter		11	1.455		400	79.50		479.50	560
0540	4" diameter		8	2		455	109		564	665
0560	5" diameter		7	2.286		540	125		665	785
0580	6" diameter		6	2.667		560	146		706	835

For customer support on your Mechanical Cost Data, call 877.739.3689.

291

23 21 20.42 Expansion Joints		Crew	Daily Output	Labor-Hours	Unit	Material	2016 Bare Costs Labor	Equipment	Total	Total Incl O&P
0600	8" diameter	Q-5	5	3.200	Ea.	670	175		845	1,000
0620	10" diameter		4.60	3.478		735	190		925	1,100
0640	12" diameter		4	4		910	219		1,129	1,325
0660	14" diameter		3.80	4.211		1,125	230		1,355	1,600
0680	16" diameter		2.90	5.517		1,300	300		1,600	1,875
0700	18" diameter		2.50	6.400		1,475	350		1,825	2,150
0720	20" diameter		2.10	7.619		1,550	415		1,965	2,325
0740	24" diameter		1.80	8.889		1,800	485		2,285	2,700
0760	26" diameter		1.40	11.429		2,025	625		2,650	3,175
0780	30" diameter		1.20	13.333		2,250	730		2,980	3,575
0800	36" diameter		1	16		2,800	875		3,675	4,400
1000	Bellows with internal sleeves and external covers									
1010	Stainless steel, 150 lb.									
1020	With male threads									
1040	3/4" diameter	1 Stpi	20	.400	Ea.	82	24.50		106.50	127
1050	1" diameter		19	.421		89	25.50		114.50	137
1060	1-1/4" diameter		16	.500		96	30.50		126.50	151
1070	1-1/2" diameter		13	.615		106	37.50		143.50	174
1080	2" diameter		11	.727		115	44		159	193
1110	With flanged ends									
1120	1-1/4" diameter	1 Stpi	12	.667	Ea.	106	40.50		146.50	177
1130	1-1/2" diameter		11	.727		121	44		165	200
1140	2" diameter		9	.889		165	54		219	264
1150	3" diameter	Q-5	8	2		230	109		339	420
1160	4" diameter	"	5	3.200		755	175		930	1,100
1170	5" diameter	Q-6	6	4		895	227		1,122	1,325
1180	6" diameter		5	4.800		1,075	272		1,347	1,575
1190	8" diameter		4.50	5.333		1,450	300		1,750	2,025
1200	10" diameter		4	6		1,825	340		2,165	2,525
1210	12" diameter		3.60	6.667		2,850	380		3,230	3,700

23 21 20.46 Expansion Tanks		Crew	Daily Output	Labor-Hours	Unit	Material	2016 Bare Costs Labor	Equipment	Total	Total Incl O&P
0010	**EXPANSION TANKS**									
1400	Plastic, corrosion resistant, see Plumbing Cost Data									
1507	Underground fuel-oil storage tanks, see Section 23 13 13									
1512	Tank leak detection systems, see Section 28 33 33.50									
2000	Steel, liquid expansion, ASME, painted, 15 gallon capacity	Q-5	17	.941	Ea.	665	51.50		716.50	810
2020	24 gallon capacity		14	1.143		740	62.50		802.50	910
2040	30 gallon capacity		12	1.333		740	73		813	925
2060	40 gallon capacity		10	1.600		870	87.50		957.50	1,075
2080	60 gallon capacity		8	2		1,050	109		1,159	1,325
2100	80 gallon capacity		7	2.286		1,125	125		1,250	1,425
2120	100 gallon capacity		6	2.667		1,500	146		1,646	1,900
2130	120 gallon capacity		5	3.200		1,600	175		1,775	2,050
2140	135 gallon capacity		4.50	3.556		1,675	194		1,869	2,150
2150	175 gallon capacity		4	4		2,650	219		2,869	3,225
2160	220 gallon capacity		3.60	4.444		3,000	243		3,243	3,675
2170	240 gallon capacity		3.30	4.848		3,125	265		3,390	3,825
2180	305 gallon capacity		3	5.333		4,400	291		4,691	5,300
2190	400 gallon capacity		2.80	5.714		5,400	310		5,710	6,425
2360	Galvanized									
2370	15 gallon capacity	Q-5	17	.941	Ea.	1,175	51.50		1,226.50	1,350
2380	24 gallon capacity		14	1.143		1,200	62.50		1,262.50	1,425

23 21 20.46 Expansion Tanks		Crew	Daily Output	Labor-Hours	Unit	Material	2016 Bare Costs Labor	Equipment	Total	Total Incl O&P
2390	30 gallon capacity	Q-5	12	1.333	Ea.	1,400	73		1,473	1,625
2400	40 gallon capacity		10	1.600		1,650	87.50		1,737.50	1,925
2410	60 gallon capacity		8	2		1,875	109		1,984	2,250
2420	80 gallon capacity		7	2.286		2,150	125		2,275	2,575
2430	100 gallon capacity		6	2.667		2,825	146		2,971	3,325
2440	120 gallon capacity		5	3.200		3,025	175		3,200	3,600
2450	135 gallon capacity		4.50	3.556		3,175	194		3,369	3,800
2460	175 gallon capacity		4	4		5,225	219		5,444	6,050
2470	220 gallon capacity		3.60	4.444		5,875	243		6,118	6,850
2480	240 gallon capacity		3.30	4.848		6,175	265		6,440	7,175
2490	305 gallon capacity		3	5.333		8,950	291		9,241	10,300
2500	400 gallon capacity		2.80	5.714		11,000	310		11,310	12,600
3000	Steel ASME expansion, rubber diaphragm, 19 gal. cap. accept.		12	1.333		2,525	73		2,598	2,875
3020	31 gallon capacity		8	2		2,800	109		2,909	3,275
3040	61 gallon capacity		6	2.667		3,950	146		4,096	4,575
3060	79 gallon capacity		5	3.200		4,025	175		4,200	4,700
3080	119 gallon capacity		4	4		4,250	219		4,469	5,000
3100	158 gallon capacity		3.80	4.211		5,900	230		6,130	6,850
3120	211 gallon capacity		3.30	4.848		6,825	265		7,090	7,925
3140	317 gallon capacity		2.80	5.714		8,925	310		9,235	10,300
3160	422 gallon capacity		2.60	6.154		13,200	335		13,535	15,000
3180	528 gallon capacity		2.40	6.667		14,500	365		14,865	16,500

23 21 20.50 Float Valves

		Crew	Daily Output	Labor-Hours	Unit	Material	Labor	Equipment	Total	Total Incl O&P
0010	**FLOAT VALVES**									
0020	With ball and bracket									
0030	Single seat, threaded									
0040	Brass body									
0050	1/2"	1 Stpi	11	.727	Ea.	91	44		135	167
0060	3/4"		9	.889		103	54		157	196
0070	1"		7	1.143		137	69.50		206.50	256
0080	1-1/2"		4.50	1.778		208	108		316	390
0090	2"		3.60	2.222		212	135		347	435
0300	For condensate receivers, CI, in-line mount									
0320	1" inlet	1 Stpi	7	1.143	Ea.	137	69.50		206.50	256
0360	For condensate receiver, CI, external float, flanged tank mount									
0370	3/4" inlet	1 Stpi	5	1.600	Ea.	103	97		200	261

23 21 20.54 Flow Check Control

		Crew	Daily Output	Labor-Hours	Unit	Material	Labor	Equipment	Total	Total Incl O&P
0010	**FLOW CHECK CONTROL**									
0100	Bronze body, soldered									
0110	3/4" size	1 Stpi	20	.400	Ea.	66	24.50		90.50	109
0120	1" size	"	19	.421	"	79	25.50		104.50	126
0200	Cast iron body, threaded									
0210	3/4" size	1 Stpi	20	.400	Ea.	51	24.50		75.50	92.50
0220	1" size		19	.421		57.50	25.50		83	102
0230	1-1/4" size		15	.533		69.50	32.50		102	126
0240	1-1/2" size		13	.615		106	37.50		143.50	174
0250	2" size		11	.727		153	44		197	235
0300	Flanged inlet, threaded outlet									
0310	2-1/2" size	1 Stpi	8	1	Ea.	320	60.50		380.50	440
0320	3" size	Q-5	9	1.778		380	97		477	565
0330	4" size	"	7	2.286		740	125		865	1,000

23 21 20.58 Hydronic Heating Control Valves	Crew	Daily Output	Labor-Hours	Unit	Material	2016 Bare Costs Labor	Equipment	Total	Total Incl O&P
0010 **HYDRONIC HEATING CONTROL VALVES**									
0050 Hot water, nonelectric, thermostatic									
0100 Radiator supply, 1/2" diameter	1 Stpi	24	.333	Ea.	68.50	20		88.50	106
0120 3/4" diameter		20	.400		71.50	24.50		96	115
0140 1" diameter		19	.421		88	25.50		113.50	135
0160 1-1/4" diameter		15	.533		125	32.50		157.50	186
0500 For low pressure steam, add					25%				
1000 Manual, radiator supply									
1010 1/2" pipe size, angle union	1 Stpi	24	.333	Ea.	52.50	20		72.50	88.50
1020 3/4" pipe size, angle union		20	.400		66	24.50		90.50	110
1030 1" pipe size, angle union		19	.421		85.50	25.50		111	133
1100 Radiator, balancing, straight, sweat connections									
1110 1/2" pipe size	1 Stpi	24	.333	Ea.	20	20		40	52.50
1120 3/4" pipe size		20	.400		28	24.50		52.50	67.50
1130 1" pipe size		19	.421		45.50	25.50		71	88.50
1200 Steam, radiator, supply									
1210 1/2" pipe size, angle union	1 Stpi	24	.333	Ea.	49	20		69	84.50
1220 3/4" pipe size, angle union		20	.400		55.50	24.50		80	98
1230 1" pipe size, angle union		19	.421		64.50	25.50		90	110
1240 1-1/4" pipe size, angle union		15	.533		82.50	32.50		115	140
8000 System balancing and shut-off									
8020 Butterfly, quarter turn, calibrated, threaded or solder									
8040 Bronze, -30°F to +350°F, pressure to 175 psi									
8060 1/2" size	1 Stpi	22	.364	Ea.	19.40	22		41.40	55
8070 3/4" size		20	.400		31	24.50		55.50	70.50
8080 1" size		19	.421		37.50	25.50		63	80
8090 1-1/4" size		15	.533		60.50	32.50		93	116
8100 1-1/2" size		13	.615		78	37.50		115.50	143
8110 2" size		11	.727		97.50	44		141.50	174

23 21 20.62 Liquid Drainers

23 21 20.62 Liquid Drainers	Crew	Daily Output	Labor-Hours	Unit	Material	2016 Bare Costs Labor	Equipment	Total	Total Incl O&P
0010 **LIQUID DRAINERS**									
0100 Guided lever type									
0110 Cast iron body, threaded									
0120 1/2" pipe size	1 Stpi	22	.364	Ea.	114	22		136	160
0130 3/4" pipe size		20	.400		300	24.50		324.50	365
0140 1" pipe size		19	.421		480	25.50		505.50	570
0150 1-1/2" pipe size		11	.727		1,050	44		1,094	1,225
0160 2" pipe size		8	1		1,050	60.50		1,110.50	1,250
0200 Forged steel body, threaded									
0210 1/2" pipe size	1 Stpi	18	.444	Ea.	1,150	27		1,177	1,300
0220 3/4" pipe size		16	.500		1,150	30.50		1,180.50	1,300
0230 1" pipe size		14	.571		1,725	34.50		1,759.50	1,925
0240 1-1/2" pipe size	Q-5	10	1.600		4,975	87.50		5,062.50	5,600
0250 2" pipe size	"	8	2		4,975	109		5,084	5,650
0300 Socket weld									
0310 1/2" pipe size	Q-17	16	1	Ea.	1,325	54.50	3.63	1,383.13	1,525
0320 3/4" pipe size		14	1.143		1,325	62.50	4.15	1,391.65	1,550
0330 1" pipe size		12	1.333		1,900	73	4.84	1,977.84	2,225
0340 1-1/2" pipe size		9	1.778		5,175	97	6.45	5,278.45	5,850
0350 2" pipe size		7	2.286		5,175	125	8.30	5,308.30	5,900
0400 Flanged									
0410 1/2" pipe size	1 Stpi	12	.667	Ea.	2,725	40.50		2,765.50	3,050

23 21 20.62 Liquid Drainers	Crew	Daily Output	Labor-Hours	Unit	Material	2016 Bare Costs Labor	Equipment	Total	Total Incl O&P
0420	3/4" pipe size	1 Stpi	10	.800	Ea.	2,725	48.50		2,773.50	3,075
0430	1" pipe size		7	1.143		3,425	69.50		3,494.50	3,850
0440	1-1/2" pipe size	Q-5	9	1.778		6,950	97		7,047	7,800
0450	2" pipe size	"	7	2.286		6,950	125		7,075	7,850
1000	Fixed lever and snap action type									
1100	Cast iron body, threaded									
1110	1/2" pipe size	1 Stpi	22	.364	Ea.	177	22		199	229
1120	3/4" pipe size		20	.400		465	24.50		489.50	550
1130	1" pipe size		19	.421		465	25.50		490.50	555
1200	Forged steel body, threaded									
1210	1/2" pipe size	1 Stpi	18	.444	Ea.	1,250	27		1,277	1,400
1220	3/4" pipe size		16	.500		1,250	30.50		1,280.50	1,400
1230	1" pipe size		14	.571		3,050	34.50		3,084.50	3,400
1240	1-1/4" pipe size		12	.667		3,050	40.50		3,090.50	3,400
2000	High pressure high leverage type									
2100	Forged steel body, threaded									
2110	1/2" pipe size	1 Stpi	18	.444	Ea.	2,425	27		2,452	2,725
2120	3/4" pipe size		16	.500		2,425	30.50		2,455.50	2,725
2130	1" pipe size		14	.571		4,250	34.50		4,284.50	4,725
2140	1-1/4" pipe size		12	.667		4,250	40.50		4,290.50	4,750
2150	1-1/2" pipe size	Q-5	10	1.600		6,300	87.50		6,387.50	7,075
2160	2" pipe size	"	8	2		6,300	109		6,409	7,125
2200	Socket weld									
2210	1/2" pipe size	Q-17	16	1	Ea.	2,625	54.50	3.63	2,683.13	2,950
2220	3/4" pipe size		14	1.143		2,625	62.50	4.15	2,691.65	2,975
2230	1" pipe size		12	1.333		4,250	73	4.84	4,327.84	4,825
2240	1-1/4" pipe size		11	1.455		4,250	79.50	5.30	4,334.80	4,825
2250	1-1/2" pipe size		9	1.778		6,300	97	6.45	6,403.45	7,100
2260	2" pipe size		7	2.286		6,300	125	8.30	6,433.30	7,150
2300	Flanged									
2310	1/2" pipe size	1 Stpi	12	.667	Ea.	3,500	40.50		3,540.50	3,925
2320	3/4" pipe size		10	.800		3,500	48.50		3,548.50	3,950
2330	1" pipe size		7	1.143		5,550	69.50		5,619.50	6,200
2340	1-1/4" pipe size		6	1.333		5,550	81		5,631	6,225
2350	1-1/2" pipe size	Q-5	9	1.778		7,575	97		7,672	8,500
2360	2" pipe size	"	7	2.286		7,575	125		7,700	8,550
3000	Guided lever dual gravity type									
3100	Cast iron body, threaded									
3110	1/2" pipe size	1 Stpi	22	.364	Ea.	440	22		462	520
3120	3/4" pipe size		20	.400		440	24.50		464.50	520
3130	1" pipe size		19	.421		645	25.50		670.50	745
3140	1-1/2" pipe size		11	.727		1,250	44		1,294	1,450
3150	2" pipe size		8	1		1,250	60.50		1,310.50	1,475
3200	Forged steel body, threaded									
3210	1/2" pipe size	1 Stpi	18	.444	Ea.	1,325	27		1,352	1,500
3220	3/4" pipe size		16	.500		1,325	30.50		1,355.50	1,500
3230	1" pipe size		14	.571		1,950	34.50		1,984.50	2,200
3240	1-1/2" pipe size	Q-5	10	1.600		5,225	87.50		5,312.50	5,875
3250	2" pipe size	"	8	2		5,225	109		5,334	5,925
3300	Socket weld									
3310	1/2" pipe size	Q-17	16	1	Ea.	1,500	54.50	3.63	1,558.13	1,725
3320	3/4" pipe size		14	1.143		1,500	62.50	4.15	1,566.65	1,750
3330	1" pipe size		12	1.333		2,150	73	4.84	2,227.84	2,500

23 21 Hydronic Piping and Pumps

23 21 20 – Hydronic HVAC Piping Specialties

23 21 20.62 Liquid Drainers	Crew	Daily Output	Labor-Hours	Unit	Material	2016 Bare Costs Labor	Equipment	Total	Total Incl O&P
3340	1-1/2" pipe size	Q-17	9	1.778	Ea.	5,425	97	6.45	5,528.45	6,125
3350	2" pipe size	↓	7	2.286	↓	5,425	125	8.30	5,558.30	6,175
3400	Flanged									
3410	1/2" pipe size	1 Stpi	12	.667	Ea.	2,425	40.50		2,465.50	2,700
3420	3/4" pipe size		10	.800		2,425	48.50		2,473.50	2,725
3430	1" pipe size	↓	7	1.143		3,150	69.50		3,219.50	3,575
3440	1-1/2" pipe size	Q-5	9	1.778		6,700	97		6,797	7,525
3450	2" pipe size	"	7	2.286	↓	6,700	125		6,825	7,575

23 21 20.66 Monoflow Tee Fitting

		Crew	Daily Output	Labor-Hours	Unit	Material	2016 Bare Costs Labor	Equipment	Total	Total Incl O&P
0010	**MONOFLOW TEE FITTING**									
1100	For one pipe hydronic, supply and return									
1110	Copper, soldered									
1120	3/4" x 1/2" size	1 Stpi	13	.615	Ea.	21.50	37.50		59	80
1130	1" x 1/2" size		12	.667		21.50	40.50		62	84.50
1140	1" x 3/4" size		11	.727		21.50	44		65.50	90
1150	1-1/4" x 1/2" size		11	.727		22	44		66	91
1160	1-1/4" x 3/4" size		10	.800		22	48.50		70.50	98
1170	1-1/2" x 3/4" size		10	.800		31.50	48.50		80	108
1180	1-1/2" x 1" size		9	.889		31.50	54		85.50	116
1190	2" x 3/4" size		9	.889		49	54		103	135
1200	2" x 1" size	↓	8	1	↓	49	60.50		109.50	145

23 21 20.70 Steam Traps

		Crew	Daily Output	Labor-Hours	Unit	Material	2016 Bare Costs Labor	Equipment	Total	Total Incl O&P
0010	**STEAM TRAPS**									
0030	Cast iron body, threaded									
0040	Inverted bucket									
0050	1/2" pipe size	1 Stpi	12	.667	Ea.	142	40.50		182.50	217
0070	3/4" pipe size		10	.800		248	48.50		296.50	345
0100	1" pipe size		9	.889		380	54		434	500
0120	1-1/4" pipe size		8	1		575	60.50		635.50	725
0130	1-1/2" pipe size		7	1.143		725	69.50		794.50	905
0140	2" pipe size	↓	6	1.333	↓	1,125	81		1,206	1,375
0200	With thermic vent & check valve									
0202	1/2" pipe size	1 Stpi	12	.667	Ea.	168	40.50		208.50	246
0204	3/4" pipe size		10	.800		285	48.50		333.50	390
0206	1" pipe size		9	.889		415	54		469	535
0208	1-1/4" pipe size		8	1		625	60.50		685.50	780
0210	1-1/2" pipe size		7	1.143		800	69.50		869.50	985
0212	2" pipe size	↓	6	1.333	↓	1,225	81		1,306	1,475
1000	Float & thermostatic, 15 psi									
1010	3/4" pipe size	1 Stpi	16	.500	Ea.	135	30.50		165.50	194
1020	1" pipe size		15	.533		196	32.50		228.50	264
1030	1-1/4" pipe size		13	.615		196	37.50		233.50	272
1040	1-1/2" pipe size		9	.889		285	54		339	395
1060	2" pipe size	↓	6	1.333		525	81		606	695
1100	For vacuum breaker, add				↓	72			72	79.50
1110	Differential condensate controller									
1120	With union and manual metering valve									
1130	1/2" pipe size	1 Stpi	24	.333	Ea.	620	20		640	715
1140	3/4" pipe size		20	.400		720	24.50		744.50	830
1150	1" pipe size		19	.421		970	25.50		995.50	1,125
1160	1-1/4" pipe size		15	.533		1,300	32.50		1,332.50	1,475
1170	1-1/2" pipe size	↓	13	.615	↓	1,575	37.50		1,612.50	1,775

23 21 Hydronic Piping and Pumps

23 21 20 – Hydronic HVAC Piping Specialties

23 21 20.70 Steam Traps	Crew	Daily Output	Labor-Hours	Unit	Material	2016 Bare Costs Labor	Equipment	Total	Total Incl O&P
1180 2" pipe size	1 Stpi	11	.727	Ea.	2,050	44		2,094	2,325
1290 Brass body, threaded									
1300 Thermostatic, angle union, 25 psi									
1310 1/2" pipe size	1 Stpi	24	.333	Ea.	64	20		84	101
1320 3/4" pipe size		20	.400		97.50	24.50		122	144
1330 1" pipe size		19	.421		143	25.50		168.50	196
1990 Carbon steel body, threaded									
2000 Controlled disc, integral strainer & blow down									
2010 Threaded									
2020 3/8" pipe size	1 Stpi	24	.333	Ea.	278	20		298	335
2030 1/2" pipe size		24	.333		278	20		298	335
2040 3/4" pipe size		20	.400		375	24.50		399.50	445
2050 1" pipe size		19	.421		480	25.50		505.50	565
2100 Socket weld									
2110 3/8" pipe size	Q-17	17	.941	Ea.	335	51.50	3.42	389.92	445
2120 1/2" pipe size		16	1		335	54.50	3.63	393.13	450
2130 3/4" pipe size		14	1.143		430	62.50	4.15	496.65	575
2140 1" pipe size		12	1.333		540	73	4.84	617.84	710
2170 Flanged, 600 lb. ASA									
2180 3/8" pipe size	1 Stpi	12	.667	Ea.	750	40.50		790.50	885
2190 1/2" pipe size		12	.667		750	40.50		790.50	885
2200 3/4" pipe size		10	.800		895	48.50		943.50	1,050
2210 1" pipe size		7	1.143		1,075	69.50		1,144.50	1,275
5000 Forged steel body									
5010 Inverted bucket									
5020 Threaded									
5030 1/2" pipe size	1 Stpi	12	.667	Ea.	505	40.50		545.50	615
5040 3/4" pipe size		10	.800		875	48.50		923.50	1,025
5050 1" pipe size		9	.889		1,300	54		1,354	1,525
5060 1-1/4" pipe size		8	1		1,900	60.50		1,960.50	2,200
5070 1-1/2" pipe size		7	1.143		2,700	69.50		2,769.50	3,075
5080 2" pipe size		6	1.333		4,725	81		4,806	5,325
5100 Socket weld									
5110 1/2" pipe size	Q-17	16	1	Ea.	605	54.50	3.63	663.13	755
5120 3/4" pipe size		14	1.143		975	62.50	4.15	1,041.65	1,175
5130 1" pipe size		12	1.333		1,425	73	4.84	1,502.84	1,675
5140 1-1/4" pipe size		10	1.600		2,000	87.50	5.80	2,093.30	2,350
5150 1-1/2" pipe size		9	1.778		2,800	97	6.45	2,903.45	3,225
5160 2" pipe size		7	2.286		4,850	125	8.30	4,983.30	5,550
5200 Flanged									
5210 1/2" pipe size	1 Stpi	12	.667	Ea.	935	40.50		975.50	1,075
5220 3/4" pipe size		10	.800		1,525	48.50		1,573.50	1,750
5230 1" pipe size		7	1.143		2,075	69.50		2,144.50	2,400
5240 1-1/4" pipe size		6	1.333		2,825	81		2,906	3,225
5250 1-1/2" pipe size	Q-5	9	1.778		3,600	97		3,697	4,125
5260 2" pipe size	"	7	2.286		5,675	125		5,800	6,450
6000 Stainless steel body									
6010 Inverted bucket									
6020 Threaded									
6030 1/2" pipe size	1 Stpi	12	.667	Ea.	167	40.50		207.50	244
6040 3/4" pipe size		10	.800		219	48.50		267.50	315
6050 1" pipe size		9	.889		780	54		834	935
6100 Threaded with thermic vent									

23 21 20.70 Steam Traps		Crew	Daily Output	Labor-Hours	Unit	Material	2016 Bare Costs Labor	Equipment	Total	Total Incl O&P
6110	1/2" pipe size	1 Stpi	12	.667	Ea.	167	40.50		207.50	244
6120	3/4" pipe size		10	.800		219	48.50		267.50	315
6130	1" pipe size		9	.889		780	54		834	935
6200	Threaded with check valve									
6210	1/2" pipe size	1 Stpi	12	.667	Ea.	239	40.50		279.50	325
6220	3/4" pipe size	"	10	.800	"	325	48.50		373.50	435
6300	Socket weld									
6310	1/2" pipe size	Q-17	16	1	Ea.	167	54.50	3.63	225.13	269
6320	3/4" pipe size	"	14	1.143	"	219	62.50	4.15	285.65	340

23 21 20.74 Strainers, Basket Type

		Crew	Daily Output	Labor-Hours	Unit	Material	2016 Bare Costs Labor	Equipment	Total	Total Incl O&P
0010	**STRAINERS, BASKET TYPE**, Perforated stainless steel basket									
0100	Brass or monel available									
2000	Simplex style									
2300	Bronze body									
2320	Screwed, 3/8" pipe size	1 Stpi	22	.364	Ea.	172	22		194	223
2340	1/2" pipe size		20	.400		305	24.50		329.50	375
2360	3/4" pipe size		17	.471		410	28.50		438.50	495
2380	1" pipe size		15	.533		410	32.50		442.50	500
2400	1-1/4" pipe size		13	.615		600	37.50		637.50	715
2420	1-1/2" pipe size		12	.667		605	40.50		645.50	725
2440	2" pipe size		10	.800		890	48.50		938.50	1,050
2460	2-1/2" pipe size	Q-5	15	1.067		770	58.50		828.50	940
2480	3" pipe size	"	14	1.143		1,125	62.50		1,187.50	1,325
2600	Flanged, 2" pipe size	1 Stpi	6	1.333		820	81		901	1,025
2620	2-1/2" pipe size	Q-5	4.50	3.556		1,275	194		1,469	1,725
2640	3" pipe size		3.50	4.571		1,450	250		1,700	1,975
2660	4" pipe size		3	5.333		2,450	291		2,741	3,150
2680	5" pipe size	Q-6	3.40	7.059		3,750	400		4,150	4,725
2700	6" pipe size		3	8		4,775	455		5,230	5,925
2710	8" pipe size		2.50	9.600		7,750	545		8,295	9,350
3600	Iron body									
3700	Screwed, 3/8" pipe size	1 Stpi	22	.364	Ea.	98.50	22		120.50	143
3720	1/2" pipe size		20	.400		102	24.50		126.50	149
3740	3/4" pipe size		17	.471		130	28.50		158.50	186
3760	1" pipe size		15	.533		133	32.50		165.50	195
3780	1-1/4" pipe size		13	.615		173	37.50		210.50	247
3800	1-1/2" pipe size		12	.667		191	40.50		231.50	271
3820	2" pipe size		10	.800		228	48.50		276.50	325
3840	2-1/2" pipe size	Q-5	15	1.067		305	58.50		363.50	425
3860	3" pipe size	"	14	1.143		370	62.50		432.50	500
4000	Flanged, 2" pipe size	1 Stpi	6	1.333		355	81		436	510
4020	2-1/2" pipe size	Q-5	4.50	3.556		475	194		669	820
4040	3" pipe size		3.50	4.571		500	250		750	925
4060	4" pipe size		3	5.333		760	291		1,051	1,275
4080	5" pipe size	Q-6	3.40	7.059		1,150	400		1,550	1,875
4100	6" pipe size		3	8		1,475	455		1,930	2,300
4120	8" pipe size		2.50	9.600		2,675	545		3,220	3,775
4140	10" pipe size		2.20	10.909		5,875	620		6,495	7,400
6000	Cast steel body									
6400	Screwed, 1" pipe size	1 Stpi	15	.533	Ea.	224	32.50		256.50	296
6410	1-1/4" pipe size		13	.615		335	37.50		372.50	420
6420	1-1/2" pipe size		12	.667		370	40.50		410.50	470

23 21 20.74 Strainers, Basket Type		Crew	Daily Output	Labor-Hours	Unit	Material	2016 Bare Costs Labor	Equipment	Total	Total Incl O&P
6440	2" pipe size	1 Stpi	10	.800	Ea.	465	48.50		513.50	585
6460	2-1/2" pipe size	Q-5	15	1.067		670	58.50		728.50	825
6480	3" pipe size	"	14	1.143		860	62.50		922.50	1,050
6560	Flanged, 2" pipe size	1 Stpi	6	1.333		840	81		921	1,050
6580	2-1/2" pipe size	Q-5	4.50	3.556		1,300	194		1,494	1,725
6600	3" pipe size		3.50	4.571		1,375	250		1,625	1,900
6620	4" pipe size		3	5.333		1,950	291		2,241	2,600
6640	6" pipe size	Q-6	3	8		3,575	455		4,030	4,600
6660	8" pipe size	"	2.50	9.600		5,725	545		6,270	7,125
7000	Stainless steel body									
7200	Screwed, 1" pipe size	1 Stpi	15	.533	Ea.	283	32.50		315.50	360
7210	1-1/4" pipe size		13	.615		440	37.50		477.50	540
7220	1-1/2" pipe size		12	.667		440	40.50		480.50	545
7240	2" pipe size		10	.800		655	48.50		703.50	795
7260	2-1/2" pipe size	Q-5	15	1.067		930	58.50		988.50	1,125
7280	3" pipe size	"	14	1.143		1,275	62.50		1,337.50	1,500
7400	Flanged, 2" pipe size	1 Stpi	6	1.333		1,125	81		1,206	1,350
7420	2-1/2" pipe size	Q-5	4.50	3.556		2,025	194		2,219	2,550
7440	3" pipe size		3.50	4.571		2,075	250		2,325	2,675
7460	4" pipe size		3	5.333		3,275	291		3,566	4,050
7480	6" pipe size	Q-6	3	8		6,550	455		7,005	7,900
7500	8" pipe size	"	2.50	9.600		9,550	545		10,095	11,300
8100	Duplex style									
8200	Bronze body									
8240	Screwed, 3/4" pipe size	1 Stpi	16	.500	Ea.	1,175	30.50		1,205.50	1,325
8260	1" pipe size		14	.571		1,175	34.50		1,209.50	1,325
8280	1-1/4" pipe size		12	.667		2,825	40.50		2,865.50	3,150
8300	1-1/2" pipe size		11	.727		2,825	44		2,869	3,175
8320	2" pipe size		9	.889		3,700	54		3,754	4,125
8340	2-1/2" pipe size	Q-5	14	1.143		4,775	62.50		4,837.50	5,350
8420	Flanged, 2" pipe size	1 Stpi	6	1.333		3,975	81		4,056	4,500
8440	2-1/2" pipe size	Q-5	4.50	3.556		5,475	194		5,669	6,325
8460	3" pipe size		3.50	4.571		5,975	250		6,225	6,950
8480	4" pipe size		3	5.333		8,575	291		8,866	9,875
8500	5" pipe size	Q-6	3.40	7.059		18,600	400		19,000	21,000
8520	6" pipe size	"	3	8		19,100	455		19,555	21,700
8700	Iron body									
8740	Screwed, 3/4" pipe size	1 Stpi	16	.500	Ea.	1,050	30.50		1,080.50	1,200
8760	1" pipe size		14	.571		1,050	34.50		1,084.50	1,200
8780	1-1/4" pipe size		12	.667		1,175	40.50		1,215.50	1,325
8800	1-1/2" pipe size		11	.727		1,175	44		1,219	1,350
8820	2" pipe size		9	.889		1,975	54		2,029	2,250
8840	2-1/2" pipe size	Q-5	14	1.143		2,175	62.50		2,237.50	2,500
9000	Flanged, 2" pipe size	1 Stpi	6	1.333		2,100	81		2,181	2,450
9020	2-1/2" pipe size	Q-5	4.50	3.556		2,250	194		2,444	2,775
9040	3" pipe size		3.50	4.571		2,450	250		2,700	3,075
9060	4" pipe size		3	5.333		4,150	291		4,441	5,000
9080	5" pipe size	Q-6	3.40	7.059		9,025	400		9,425	10,500
9100	6" pipe size		3	8		9,025	455		9,480	10,600
9120	8" pipe size		2.50	9.600		17,900	545		18,445	20,500
9140	10" pipe size		2.20	10.909		25,500	620		26,120	29,000
9160	12" pipe size		1.70	14.118		28,100	800		28,900	32,200
9170	14" pipe size		1.40	17.143		33,200	970		34,170	38,000

For customer support on your Mechanical Cost Data, call 877.739.3689.

299

23 21 20.74 Strainers, Basket Type

		Crew	Daily Output	Labor-Hours	Unit	Material	2016 Bare Costs Labor	Equipment	Total	Total Incl O&P
9180	16" pipe size	Q-6	1	24	Ea.	40,800	1,350		42,150	47,000
9300	Cast steel body									
9340	Screwed, 1" pipe size	1 Stpi	14	.571	Ea.	1,350	34.50		1,384.50	1,525
9360	1-1/2" pipe size		11	.727		2,175	44		2,219	2,475
9380	2" pipe size		9	.889		2,875	54		2,929	3,250
9460	Flanged, 2" pipe size	↓	6	1.333		3,225	81		3,306	3,675
9480	2-1/2" pipe size	Q-5	4.50	3.556		5,200	194		5,394	6,000
9500	3" pipe size		3.50	4.571		5,650	250		5,900	6,600
9520	4" pipe size	↓	3	5.333		6,975	291		7,266	8,125
9540	6" pipe size	Q-6	3	8		13,100	455		13,555	15,100
9560	8" pipe size	"	2.50	9.600	↓	31,500	545		32,045	35,400
9700	Stainless steel body									
9740	Screwed, 1" pipe size	1 Stpi	14	.571	Ea.	1,900	34.50		1,934.50	2,150
9760	1-1/2" pipe size		11	.727		2,850	44		2,894	3,225
9780	2" pipe size		9	.889		4,000	54		4,054	4,475
9860	Flanged, 2" pipe size	↓	6	1.333		4,350	81		4,431	4,925
9880	2-1/2" pipe size	Q-5	4.50	3.556		7,300	194		7,494	8,325
9900	3" pipe size		3.50	4.571		7,950	250		8,200	9,125
9920	4" pipe size	↓	3	5.333		10,400	291		10,691	11,800
9940	6" pipe size	Q-6	3	8		15,800	455		16,255	18,100
9960	8" pipe size	"	2.50	9.600	↓	44,200	545		44,745	49,400

23 21 20.76 Strainers, Y Type, Bronze Body

		Crew	Daily Output	Labor-Hours	Unit	Material	2016 Bare Costs Labor	Equipment	Total	Total Incl O&P
0010	**STRAINERS, Y TYPE, BRONZE BODY**									
0050	Screwed, 125 lb., 1/4" pipe size	1 Stpi	24	.333	Ea.	27	20		47	60
0070	3/8" pipe size		24	.333		33.50	20		53.50	67.50
0100	1/2" pipe size		20	.400		33.50	24.50		58	73.50
0120	3/4" pipe size		19	.421		41.50	25.50		67	84
0140	1" pipe size		17	.471		49	28.50		77.50	97
0150	1-1/4" pipe size		15	.533		98	32.50		130.50	157
0160	1-1/2" pipe size		14	.571		106	34.50		140.50	169
0180	2" pipe size		13	.615		139	37.50		176.50	210
0182	3" pipe size	↓	12	.667		930	40.50		970.50	1,075
0200	300 lb., 2-1/2" pipe size	Q-5	17	.941		660	51.50		711.50	805
0220	3" pipe size		16	1		1,175	54.50		1,229.50	1,350
0240	4" pipe size	↓	15	1.067	↓	2,300	58.50		2,358.50	2,625
0500	For 300 lb. rating 1/4" thru 2", add					15%				
1000	Flanged, 150 lb., 1-1/2" pipe size	1 Stpi	11	.727	Ea.	475	44		519	585
1020	2" pipe size	"	8	1		640	60.50		700.50	795
1030	2-1/2" pipe size	Q-5	5	3.200		950	175		1,125	1,325
1040	3" pipe size		4.50	3.556		1,175	194		1,369	1,575
1060	4" pipe size		3	5.333		1,775	291		2,066	2,400
1080	5" pipe size	Q-6	3.40	7.059		1,775	400		2,175	2,550
1100	6" pipe size		3	8		3,400	455		3,855	4,400
1106	8" pipe size	↓	2.60	9.231	↓	3,725	525		4,250	4,900
1500	For 300 lb. rating, add					40%				

23 21 20.78 Strainers, Y Type, Iron Body

		Crew	Daily Output	Labor-Hours	Unit	Material	2016 Bare Costs Labor	Equipment	Total	Total Incl O&P
0010	**STRAINERS, Y TYPE, IRON BODY**									
0050	Screwed, 250 lb., 1/4" pipe size	1 Stpi	20	.400	Ea.	11.05	24.50		35.55	48.50
0070	3/8" pipe size		20	.400		11.05	24.50		35.55	48.50
0100	1/2" pipe size		20	.400		11.05	24.50		35.55	48.50
0120	3/4" pipe size		18	.444		12.95	27		39.95	55
0140	1" pipe size	↓	16	.500		16.80	30.50		47.30	64.50

For customer support on your Mechanical Cost Data, call 877.739.3689.

23 21 20.78 Strainers, Y Type, Iron Body

		Crew	Daily Output	Labor-Hours	Unit	Material	2016 Bare Costs Labor	2016 Bare Costs Equipment	Total	Total Incl O&P
0150	1-1/4" pipe size	1 Stpi	15	.533	Ea.	24	32.50		56.50	75.50
0160	1-1/2" pipe size		12	.667		32.50	40.50		73	97
0180	2" pipe size		8	1		49	60.50		109.50	146
0200	2-1/2" pipe size	Q-5	12	1.333		268	73		341	405
0220	3" pipe size		11	1.455		289	79.50		368.50	440
0240	4" pipe size		5	3.200		490	175		665	805
0500	For galvanized body, add					50%				
1000	Flanged, 125 lb., 1-1/2" pipe size	1 Stpi	11	.727	Ea.	104	44		148	181
1020	2" pipe size	"	8	1		109	60.50		169.50	212
1030	2-1/2" pipe size	Q-5	5	3.200		112	175		287	385
1040	3" pipe size		4.50	3.556		179	194		373	490
1060	4" pipe size		3	5.333		296	291		587	765
1080	5" pipe size	Q-6	3.40	7.059		375	400		775	1,025
1100	6" pipe size		3	8		595	455		1,050	1,350
1120	8" pipe size		2.50	9.600		805	545		1,350	1,700
1140	10" pipe size		2	12		1,825	680		2,505	3,025
1160	12" pipe size		1.70	14.118		2,200	800		3,000	3,625
1170	14" pipe size		1.30	18.462		4,050	1,050		5,100	6,050
1180	16" pipe size		1	24		5,750	1,350		7,100	8,375
1500	For 250 lb. rating, add					20%				
2000	For galvanized body, add					50%				
2500	For steel body, add					40%				

23 21 20.80 Suction Diffusers

		Crew	Daily Output	Labor-Hours	Unit	Material	2016 Bare Costs Labor	2016 Bare Costs Equipment	Total	Total Incl O&P
0010	**SUCTION DIFFUSERS**									
0100	Cast iron body with integral straightening vanes, strainer									
1000	Flanged									
1010	2" inlet, 1-1/2" pump side	1 Stpi	6	1.333	Ea.	335	81		416	490
1020	2" pump side	"	5	1.600		350	97		447	530
1030	3" inlet, 2" pump side	Q-5	6.50	2.462		420	134		554	670
1040	2-1/2" pump side		5.30	3.019		605	165		770	915
1050	3" pump side		4.50	3.556		625	194		819	985
1060	4" inlet, 3" pump side		3.50	4.571		710	250		960	1,150
1070	4" pump side		3	5.333		825	291		1,116	1,350
1080	5" inlet, 4" pump side	Q-6	3.80	6.316		940	360		1,300	1,575
1090	5" pump side		3.40	7.059		1,125	400		1,525	1,825
1100	6" inlet, 4" pump side		3.30	7.273		980	410		1,390	1,700
1110	5" pump side		3.10	7.742		1,175	440		1,615	1,950
1120	6" pump side		3	8		1,250	455		1,705	2,050
1130	8" inlet, 6" pump side		2.70	8.889		1,375	505		1,880	2,275
1140	8" pump side		2.50	9.600		2,425	545		2,970	3,475
1150	10" inlet, 6" pump side		2.40	10		2,775	565		3,340	3,925
1160	8" pump side		2.30	10.435		2,875	590		3,465	4,075
1170	10" pump side		2.20	10.909		4,025	620		4,645	5,350
1180	12" inlet, 8" pump side		2	12		4,250	680		4,930	5,700
1190	10" pump side		1.80	13.333		4,700	755		5,455	6,325
1200	12" pump side		1.70	14.118		5,300	800		6,100	7,025
1210	14" inlet, 12" pump side		1.50	16		6,850	905		7,755	8,925
1220	14" pump side		1.30	18.462		7,825	1,050		8,875	10,200

23 21 20.84 Thermoflo Indicator

		Crew	Daily Output	Labor-Hours	Unit	Material	2016 Bare Costs Labor	2016 Bare Costs Equipment	Total	Total Incl O&P
0010	**THERMOFLO INDICATOR**, For balancing									
1000	Sweat connections, 1-1/4" pipe size	1 Stpi	12	.667	Ea.	675	40.50		715.50	805
1020	1-1/2" pipe size		10	.800		690	48.50		738.50	830

For customer support on your Mechanical Cost Data, call 877.739.3689.

301

23 21 20.84 Thermoflo Indicator	Crew	Daily Output	Labor-Hours	Unit	Material	2016 Bare Costs Labor	2016 Bare Costs Equipment	Total	Total Incl O&P
1040	2" pipe size	1 Stpi	8	1	Ea.	720	60.50		780.50	885
1060	2-1/2" pipe size	↓	7	1.143		1,100	69.50		1,169.50	1,300
2000	Flange connections, 3" pipe size	Q-5	5	3.200		1,325	175		1,500	1,725
2020	4" pipe size		4	4		1,575	219		1,794	2,075
2030	5" pipe size		3.50	4.571		1,975	250		2,225	2,550
2040	6" pipe size		3	5.333		2,125	291		2,416	2,775
2060	8" pipe size	↓	2	8	↓	2,575	435		3,010	3,475

23 21 20.88 Venturi Flow	Crew	Daily Output	Labor-Hours	Unit	Material	2016 Bare Costs Labor	2016 Bare Costs Equipment	Total	Total Incl O&P
0010	**VENTURI FLOW**, Measuring device									
0050	1/2" diameter	1 Stpi	24	.333	Ea.	281	20		301	340
0100	3/4" diameter		20	.400		257	24.50		281.50	320
0120	1" diameter		19	.421		276	25.50		301.50	345
0140	1-1/4" diameter		15	.533		340	32.50		372.50	425
0160	1-1/2" diameter		13	.615		355	37.50		392.50	445
0180	2" diameter	↓	11	.727		365	44		409	470
0200	2-1/2" diameter	Q-5	16	1		500	54.50		554.50	635
0220	3" diameter		14	1.143		515	62.50		577.50	665
0240	4" diameter	↓	11	1.455		775	79.50		854.50	970
0260	5" diameter	Q-6	4	6		1,025	340		1,365	1,650
0280	6" diameter		3.50	6.857		1,125	390		1,515	1,825
0300	8" diameter		3	8		1,450	455		1,905	2,275
0320	10" diameter		2	12		3,425	680		4,105	4,800
0330	12" diameter		1.80	13.333		4,950	755		5,705	6,600
0340	14" diameter		1.60	15		5,600	850		6,450	7,450
0350	16" diameter	↓	1.40	17.143		6,250	970		7,220	8,350
0500	For meter, add				↓	2,125			2,125	2,350

23 21 20.94 Weld End Ball Joints	Crew	Daily Output	Labor-Hours	Unit	Material	2016 Bare Costs Labor	2016 Bare Costs Equipment	Total	Total Incl O&P
0010	**WELD END BALL JOINTS**, Steel									
0050	2-1/2" diameter	Q-17	13	1.231	Ea.	615	67	4.47	686.47	785
0100	3" diameter		12	1.333		720	73	4.84	797.84	905
0120	4" diameter	↓	11	1.455		1,175	79.50	5.30	1,259.80	1,400
0140	5" diameter	Q-18	14	1.714		1,350	97	4.15	1,451.15	1,625
0160	6" diameter		12	2		2,000	113	4.84	2,117.84	2,375
0180	8" diameter		9	2.667		3,025	151	6.45	3,182.45	3,550
0200	10" diameter		8	3		4,075	170	7.25	4,252.25	4,750
0220	12" diameter	↓	6	4	↓	6,500	227	9.70	6,736.70	7,500

23 21 20.98 Zone Valves	Crew	Daily Output	Labor-Hours	Unit	Material	2016 Bare Costs Labor	2016 Bare Costs Equipment	Total	Total Incl O&P
0010	**ZONE VALVES**									
1000	Bronze body									
1010	2 way, 125 psi									
1020	Standard, 65' pump head									
1030	1/2" soldered	1 Stpi	24	.333	Ea.	124	20		144	167
1040	3/4" soldered		20	.400		124	24.50		148.50	173
1050	1" soldered		19	.421		133	25.50		158.50	185
1060	1-1/4" soldered	↓	15	.533	↓	155	32.50		187.50	220
1200	High head, 125' pump head									
1210	1/2" soldered	1 Stpi	24	.333	Ea.	174	20		194	222
1220	3/4" soldered		20	.400		180	24.50		204.50	235
1230	1" soldered		19	.421		196	25.50		221.50	255
1300	Geothermal, 150' pump head									
1310	3/4" soldered	1 Stpi	20	.400	Ea.	207	24.50		231.50	264
1320	1" soldered	↓	19	.421	↓	219	25.50		244.50	280

23 21 20 – Hydronic HVAC Piping Specialties

23 21 20.98 Zone Valves		Crew	Daily Output	Labor-Hours	Unit	Material	2016 Bare Costs Labor	Equipment	Total	Total Incl O&P
1330	3/4" threaded	1 Stpi	20	.400	Ea.	207	24.50		231.50	264
3000	3 way, 125 psi									
3200	Bypass, 65' pump head									
3210	1/2" soldered	1 Stpi	24	.333	Ea.	181	20		201	230
3220	3/4" soldered		20	.400		187	24.50		211.50	243
3230	1" soldered		19	.421		213	25.50		238.50	273
9000	Transformer, for up to 3 zone valves	↓	20	.400	↓	26	24.50		50.50	65.50

23 21 23 – Hydronic Pumps

23 21 23.13 In-Line Centrifugal Hydronic Pumps

		Crew	Daily Output	Labor-Hours	Unit	Material	2016 Bare Costs Labor	Equipment	Total	Total Incl O&P
0010	**IN-LINE CENTRIFUGAL HYDRONIC PUMPS**									
0600	Bronze, sweat connections, 1/40 HP, in line									
0640	3/4" size	Q-1	16	1	Ea.	225	53.50		278.50	330
1000	Flange connection, 3/4" to 1-1/2" size									
1040	1/12 HP	Q-1	6	2.667	Ea.	580	142		722	855
1060	1/8 HP		6	2.667		1,000	142		1,142	1,325
1100	1/3 HP		6	2.667		1,125	142		1,267	1,450
1140	2" size, 1/6 HP		5	3.200		1,425	171		1,596	1,825
1180	2-1/2" size, 1/4 HP		5	3.200		1,825	171		1,996	2,275
1220	3" size, 1/4 HP		4	4		1,925	213		2,138	2,425
1260	1/3 HP		4	4		2,350	213		2,563	2,900
1300	1/2 HP		4	4		2,400	213		2,613	2,950
1340	3/4 HP		4	4		2,625	213		2,838	3,200
1380	1 HP	↓	4	4	↓	4,225	213		4,438	4,975
2000	Cast iron, flange connection									
2040	3/4" to 1-1/2" size, in line, 1/12 HP	Q-1	6	2.667	Ea.	375	142		517	630
2060	1/8 HP		6	2.667		625	142		767	905
2100	1/3 HP		6	2.667		700	142		842	985
2140	2" size, 1/6 HP		5	3.200		765	171		936	1,100
2180	2-1/2" size, 1/4 HP		5	3.200		990	171		1,161	1,350
2220	3" size, 1/4 HP		4	4		1,000	213		1,213	1,425
2260	1/3 HP		4	4		1,350	213		1,563	1,825
2300	1/2 HP		4	4		1,400	213		1,613	1,875
2340	3/4 HP		4	4		1,625	213		1,838	2,125
2380	1 HP	↓	4	4	↓	2,350	213		2,563	2,900
2600	For nonferrous impeller, add					3%				
3000	High head, bronze impeller									
3030	1-1/2" size 1/2 HP	Q-1	5	3.200	Ea.	1,200	171		1,371	1,575
3040	1-1/2" size 3/4 HP		5	3.200		1,300	171		1,471	1,700
3050	2" size 1 HP		4	4		1,600	213		1,813	2,075
3090	2" size 1-1/2 HP	↓	4	4	↓	1,975	213		2,188	2,500
4000	Close coupled, end suction, bronze impeller									
4040	1-1/2" size, 1-1/2 HP, to 40 GPM	Q-1	3	5.333	Ea.	2,325	284		2,609	3,000
4090	2" size, 2 HP, to 50 GPM		3	5.333		2,775	284		3,059	3,475
4100	2" size, 3 HP, to 90 GPM		2.30	6.957		2,900	370		3,270	3,725
4190	2-1/2" size, 3 HP, to 150 GPM		2	8		3,125	425		3,550	4,075
4300	3" size, 5 HP, to 225 GPM		1.80	8.889		3,575	475		4,050	4,650
4410	3" size, 10 HP, to 350 GPM		1.60	10		5,225	535		5,760	6,550
4420	4" size, 7-1/2 HP, to 350 GPM	↓	1.60	10		5,275	535		5,810	6,600
4520	4" size, 10 HP, to 600 GPM	Q-2	1.70	14.118		5,325	780		6,105	7,025
4530	5" size, 15 HP, to 1000 GPM		1.70	14.118		5,325	780		6,105	7,050
4610	5" size, 20 HP, to 1350 GPM		1.50	16		5,675	885		6,560	7,575
4620	5" size, 25 HP, to 1550 GPM	↓	1.50	16	↓	7,750	885		8,635	9,850

23 21 23.13 In-Line Centrifugal Hydronic Pumps	Crew	Daily Output	Labor-Hours	Unit	Material	2016 Bare Costs Labor	Equipment	Total	Total Incl O&P
5000	Base mounted, bronze impeller, coupling guard									
5040	1-1/2" size, 1-1/2 HP, to 40 GPM	Q-1	2.30	6.957	Ea.	6,200	370		6,570	7,375
5090	2" size, 2 HP, to 50 GPM		2.30	6.957		6,950	370		7,320	8,200
5100	2" size, 3 HP, to 90 GPM		2	8		7,625	425		8,050	9,050
5190	2-1/2" size, 3 HP, to 150 GPM		1.80	8.889		8,200	475		8,675	9,750
5300	3" size, 5 HP, to 225 GPM		1.60	10		9,425	535		9,960	11,200
5410	4" size, 5 HP, to 350 GPM		1.50	10.667		10,200	570		10,770	12,200
5420	4" size, 7-1/2 HP, to 350 GPM		1.50	10.667		11,600	570		12,170	13,600
5520	5" size, 10 HP, to 600 GPM	Q-2	1.60	15		15,000	830		15,830	17,800
5530	5" size, 15 HP, to 1000 GPM		1.60	15		16,500	830		17,330	19,500
5610	6" size, 20 HP, to 1350 GPM		1.40	17.143		18,300	945		19,245	21,500
5620	6" size, 25 HP, to 1550 GPM		1.40	17.143		20,500	945		21,445	24,000
5800	The above pump capacities are based on 1800 RPM,									
5810	at a 60 foot head. Increasing the RPM									
5820	or decreasing the head will increase the GPM.									

23 21 29 – Automatic Condensate Pump Units

23 21 29.10 Condensate Removal Pump System

		Crew	Daily Output	Labor-Hours	Unit	Material	2016 Bare Costs Labor	Equipment	Total	Total Incl O&P
0010	**CONDENSATE REMOVAL PUMP SYSTEM**									
0020	Pump with 1 gal. ABS tank									
0100	115 V									
0120	1/50 HP, 200 GPH	G 1 Stpi	12	.667	Ea.	197	40.50		237.50	278
0140	1/18 HP, 270 GPH	G	10	.800		210	48.50		258.50	305
0160	1/5 HP, 450 GPH	G	8	1		470	60.50		530.50	605
0200	230 V									
0240	1/18 HP, 270 GPH	1 Stpi	10	.800	Ea.	217	48.50		265.50	315
0260	1/5 HP, 450 GPH	G "	8	1	"	520	60.50		580.50	660

23 22 Steam and Condensate Piping and Pumps
23 22 13 – Steam and Condensate Heating Piping

23 22 13.23 Aboveground Steam and Condensate Piping

		Crew	Daily Output	Labor-Hours	Unit	Material	2016 Bare Costs Labor	Equipment	Total	Total Incl O&P
0010	**ABOVEGROUND STEAM AND CONDENSATE HEATING PIPING**									
0020	Condensate meter									
0100	500 lb. per hour	1 Stpi	14	.571	Ea.	3,475	34.50		3,509.50	3,875
0140	1500 lb. per hour		7	1.143		4,100	69.50		4,169.50	4,600
0160	3000 lb. per hour		5	1.600		4,775	97		4,872	5,425
0200	12,000 lb. per hour	Q-5	3.50	4.571		5,625	250		5,875	6,575

23 22 23 – Steam Condensate Pumps

23 22 23.10 Condensate Return System

		Crew	Daily Output	Labor-Hours	Unit	Material	2016 Bare Costs Labor	Equipment	Total	Total Incl O&P
0010	**CONDENSATE RETURN SYSTEM**									
2000	Simplex									
2010	With pump, motor, CI receiver, float switch									
2020	3/4 HP, 15 GPM	Q-1	1.80	8.889	Ea.	6,475	475		6,950	7,850
2100	Duplex									
2110	With 2 pumps and motors, CI receiver, float switch, alternator									
2120	3/4 HP, 15 GPM, 15 gal. CI rcvr.	Q-1	1.40	11.429	Ea.	7,075	610		7,685	8,700
2130	1 HP, 25 GPM		1.20	13.333		8,500	710		9,210	10,400
2140	1-1/2 HP, 45 GPM		1	16		9,850	850		10,700	12,100
2150	1-1/2 HP, 60 GPM		1	16		11,100	850		11,950	13,500

23 23 Refrigerant Piping

23 23 13 – Refrigerant Piping Valves

23 23 13.10 Valves		Crew	Daily Output	Labor-Hours	Unit	Material	2016 Bare Costs Labor	Equipment	Total	Total Incl O&P
0010	**VALVES**									
8100	Check valve, soldered									
8110	5/8"	1 Stpi	36	.222	Ea.	87	13.50		100.50	117
8114	7/8"		26	.308		93	18.70		111.70	130
8118	1-1/8"		18	.444		124	27		151	177
8122	1-3/8"		14	.571		199	34.50		233.50	272
8126	1-5/8"		13	.615		234	37.50		271.50	315
8130	2-1/8"		12	.667		284	40.50		324.50	370
8134	2-5/8"	Q-5	22	.727		405	39.50		444.50	505
8138	3-1/8"	"	20	.800		540	43.50		583.50	655
8500	Refrigeration valve, packless, soldered									
8510	1/2"	1 Stpi	38	.211	Ea.	48.50	12.80		61.30	72.50
8514	5/8"		36	.222		48.50	13.50		62	73.50
8518	7/8"		26	.308		190	18.70		208.70	237
8520	Packed, soldered									
8522	1-1/8"	1 Stpi	18	.444	Ea.	200	27		227	261
8526	1-3/8"		14	.571		320	34.50		354.50	410
8530	1-5/8"		13	.615		330	37.50		367.50	420
8534	2-1/8"		12	.667		670	40.50		710.50	795
8538	2-5/8"	Q-5	22	.727		975	39.50		1,014.50	1,125
8542	3-1/8"		20	.800		1,150	43.50		1,193.50	1,350
8546	4-1/8"		18	.889		1,800	48.50		1,848.50	2,050
8600	Solenoid valve, flange/solder									
8610	1/2"	1 Stpi	38	.211	Ea.	221	12.80		233.80	262
8614	5/8"		36	.222		286	13.50		299.50	335
8618	3/4"		30	.267		350	16.20		366.20	410
8622	7/8"		26	.308		430	18.70		448.70	505
8626	1-1/8"		18	.444		580	27		607	680
8630	1-3/8"		14	.571		680	34.50		714.50	800
8634	1-5/8"		13	.615		955	37.50		992.50	1,100
8638	2-1/8"		12	.667		1,050	40.50		1,090.50	1,200
8800	Thermostatic valve, flange/solder									
8810	1/2 – 3 ton, 3/8" x 5/8"	1 Stpi	9	.889	Ea.	125	54		179	220
8814	4 – 5 ton, 1/2" x 7/8"		7	1.143		150	69.50		219.50	270
8818	6 – 8 ton, 5/8" x 7/8"		5	1.600		204	97		301	370
8822	7 – 12 ton, 7/8" x 1-1/8"		4	2		241	121		362	450
8826	15 – 20 ton, 7/8" x 1-3/8"		3.20	2.500		241	152		393	495

23 23 16 – Refrigerant Piping Specialties

23 23 16.10 Refrigerant Piping Component Specialties

		Crew	Daily Output	Labor-Hours	Unit	Material	Labor	Equipment	Total	Total Incl O&P
0010	**REFRIGERANT PIPING COMPONENT SPECIALTIES**									
0600	Accumulator									
0610	3/4"	1 Stpi	8.80	.909	Ea.	52.50	55		107.50	141
0614	7/8"		6.40	1.250		67	76		143	189
0618	1-1/8"		4.80	1.667		116	101		217	281
0622	1-3/8"		4	2		122	121		243	315
0626	1-5/8"		3.20	2.500		129	152		281	370
0630	2-1/8"		2.40	3.333		325	202		527	665
0700	Condensate drip/drain pan, electric,10" x 9" x 2-3/4"		24	.333		115	20		135	157
1000	Filter dryer									
1010	Replaceable core type, solder									
1020	1/2"	1 Stpi	20	.400	Ea.	61	24.50		85.50	104
1030	5/8"		19	.421		62.50	25.50		88	107

For customer support on your Mechanical Cost Data, call 877.739.3689.

305

23 23 16.10 Refrigerant Piping Component Specialties		Crew	Daily Output	Labor-Hours	Unit	Material	2016 Bare Costs Labor	Equipment	Total	Total Incl O&P
1040	7/8"	1 Stpi	18	.444	Ea.	162	27		189	220
1050	1-1/8"		15	.533		193	32.50		225.50	261
1060	1-3/8"		14	.571		207	34.50		241.50	281
1070	1-5/8"		12	.667		218	40.50		258.50	300
1080	2-1/8"		10	.800		400	48.50		448.50	515
1090	2-5/8"		9	.889		735	54		789	885
1100	3-1/8"		8	1		790	60.50		850.50	960
1200	Sealed in-line, solder									
1210	1/4"	1 Stpi	22	.364	Ea.	30.50	22		52.50	67
1220	3/8"		21	.381		25.50	23		48.50	63
1230	1/2"		20	.400		30.50	24.50		55	70
1260	5/8"		19	.421		41.50	25.50		67	84
1270	7/8"		18	.444		45	27		72	90
1290	1-1/8"		15	.533		68	32.50		100.50	124
4000	P-Trap, suction line, solder									
4010	5/8"	1 Stpi	19	.421	Ea.	33.50	25.50		59	75
4020	3/4"		19	.421		49.50	25.50		75	93
4030	7/8"		18	.444		37	27		64	81.50
4040	1-1/8"		15	.533		56.50	32.50		89	111
4050	1-3/8"		14	.571		102	34.50		136.50	165
4060	1-5/8"		12	.667		139	40.50		179.50	213
4070	2-1/8"		10	.800		179	48.50		227.50	271
5000	Sightglass									
5010	Moisture and liquid indicator, solder									
5020	1/4"	1 Stpi	22	.364	Ea.	18	22		40	53.50
5030	3/8"		21	.381		18.15	23		41.15	55
5040	1/2"		20	.400		23	24.50		47.50	61.50
5050	5/8"		19	.421		23.50	25.50		49	64.50
5060	7/8"		18	.444		35	27		62	79
5070	1-1/8"		15	.533		38	32.50		70.50	90.50
5080	1-3/8"		14	.571		75.50	34.50		110	136
5090	1-5/8"		12	.667		85	40.50		125.50	155
5100	2-1/8"		10	.800		88.50	48.50	.	137	171
7400	Vacuum pump set									
7410	Two stage, high vacuum continuous duty	1 Stpi	8	1	Ea.	4,150	60.50		4,210.50	4,650

23 23 16.16 Refrigerant Line Sets

		Crew	Daily Output	Labor-Hours	Unit	Material	Labor	Equipment	Total	Total Incl O&P
0010	**REFRIGERANT LINE SETS**									
0100	Copper tube									
0110	1/2" insulation, both tubes									
0120	Combination 1/4" and 1/2" tubes									
0130	10' set	Q-5	42	.381	Ea.	70.50	21		91.50	109
0140	20' set		40	.400		110	22		132	154
0150	30' set		37	.432		148	23.50		171.50	198
0160	40' set		35	.457		193	25		218	251
0170	50' set		32	.500		229	27.50		256.50	293
0180	100' set		22	.727		520	39.50		559.50	635
0300	Combination 1/4" and 3/4" tubes									
0310	10' set	Q-5	40	.400	Ea.	78.50	22		100.50	120
0320	20' set		38	.421		139	23		162	188
0330	30' set		35	.457		208	25		233	267
0340	40' set		33	.485		272	26.50		298.50	340
0350	50' set		30	.533		345	29		374	420

23 23 16 – Refrigerant Piping Specialties

23 23 16.16 Refrigerant Line Sets		Crew	Daily Output	Labor-Hours	Unit	Material	2016 Bare Costs Labor	Equipment	Total	Total Incl O&P
0380	100' set	Q-5	20	.800	Ea.	780	43.50		823.50	925
0500	Combination 3/8" & 3/4" tubes									
0510	10' set	Q-5	28	.571	Ea.	90.50	31		121.50	147
0520	20' set		36	.444		141	24.50		165.50	192
0530	30' set		34	.471		191	25.50		216.50	250
0540	40' set		31	.516		252	28		280	320
0550	50' set		28	.571		296	31		327	370
0580	100' set		18	.889		875	48.50		923.50	1,025
0700	Combination 3/8" & 1-1/8" tubes									
0710	10' set	Q-5	36	.444	Ea.	158	24.50		182.50	211
0720	20' set		33	.485		265	26.50		291.50	330
0730	30' set		31	.516		355	28		383	435
0740	40' set		28	.571		535	31		566	630
0750	50' set		26	.615		615	33.50		648.50	730
0900	Combination 1/2" & 3/4" tubes									
0910	10' set	Q-5	37	.432	Ea.	95.50	23.50		119	141
0920	20' set		35	.457		169	25		194	224
0930	30' set		33	.485		255	26.50		281.50	320
0940	40' set		30	.533		335	29		364	415
0950	50' set		27	.593		420	32.50		452.50	515
0980	100' set		17	.941		950	51.50		1,001.50	1,125
2100	Combination 1/2" & 1-1/8" tubes									
2110	10' set	Q-5	35	.457	Ea.	163	25		188	217
2120	20' set		31	.516		279	28		307	350
2130	30' set		29	.552		420	30		450	505
2140	40' set		25	.640		565	35		600	675
2150	50' set		14	1.143		625	62.50		687.50	780
2300	For 1" thick insulation add					30%	15%			

23 23 23 – Refrigerants

23 23 23.10 Anti-Freeze

		Crew	Daily Output	Labor-Hours	Unit	Material	Labor	Equipment	Total	Total Incl O&P
0010	**ANTI-FREEZE**, Inhibited									
0900	Ethylene glycol concentrated									
1000	55 gallon drums, small quantities					Gal.	11.95			11.95	13.10
1200	Large quantities						9			9	9.90
2000	Propylene glycol, for solar heat, small quantities						15			15	16.50
2100	Large quantities						9.55			9.55	10.50

23 23 23.20 Refrigerant

		Crew	Daily Output	Labor-Hours	Unit	Material	Labor	Equipment	Total	Total Incl O&P
0010	**REFRIGERANT**									
4420	R-22, 30 lb. disposable cylinder					Lb.	26.50			26.50	29
4428	R-134A, 30 lb. disposable cylinder						13.25			13.25	14.60
4434	R-407C, 30 lb. disposable cylinder						9.95			9.95	10.90
4440	R-408A, 25 lb. disposable cylinder						37.50			37.50	41
4450	R-410A, 25 lb. disposable cylinder						9.50			9.50	10.45
4470	R-507, 25 lb. disposable cylinder						10.25			10.25	11.25

For customer support on your Mechanical Cost Data, call 877.739.3689.

307

23 31 13.13 Rectangular Metal Ducts	Crew	Daily Output	Labor-Hours	Unit	Material	2016 Bare Costs Labor	2016 Bare Costs Equipment	Total	Total Incl O&P
0010 **RECTANGULAR METAL DUCTS**									
0020 Fabricated rectangular, includes fittings, joints, supports,									
0021 allowance for flexible connections and field sketches.									
0030 Does not include "as-built dwgs." or insulation. R233100-30									
0031 NOTE: Fabrication and installation are combined									
0040 as LABOR cost. Approx. 25% fittings assumed. R233100-40									
0042 Fabrication/Inst. is to commercial quality standards									
0043 (SMACNA or equiv.) for structure, sealing, leak testing, etc. R233100-50									
0050 Add to labor for elevated installation									
0051 of fabricated ductwork									
0052 10' to 15' high						6%			
0053 15' to 20' high						12%			
0054 20' to 25' high						15%			
0055 25' to 30' high						21%			
0056 30' to 35' high						24%			
0057 35' to 40' high						30%			
0058 Over 40' high						33%			
0072 For duct insulation see Line 23 07 13.10 3000									
0100 Aluminum, alloy 3003-H14, under 100 lb.	Q-10	75	.320	Lb.	3.20	17.10		20.30	29.50
0110 100 to 500 lb.		80	.300		1.88	16.05		17.93	26.50
0120 500 to 1,000 lb.		95	.253		1.82	13.50		15.32	22.50
0140 1,000 to 2,000 lb.		120	.200		1.77	10.70		12.47	18.30
0150 2,000 to 5,000 lb.		130	.185		1.77	9.85		11.62	17.05
0160 Over 5,000 lb.		145	.166		1.77	8.85		10.62	15.45
0500 Galvanized steel, under 200 lb.		235	.102		.57	5.45		6.02	9
0520 200 to 500 lb.		245	.098		.56	5.25		5.81	8.60
0540 500 to 1,000 lb.		255	.094		.55	5.05		5.60	8.30
0560 1,000 to 2,000 lb.		265	.091		.54	4.84		5.38	8
0570 2,000 to 5,000 lb.		275	.087		.53	4.66		5.19	7.75
0580 Over 5,000 lb.		285	.084		.53	4.50		5.03	7.50
0600 For large quantities special prices available from supplier									
1000 Stainless steel, type 304, under 100 lb.	Q-10	165	.145	Lb.	4.20	7.75		11.95	16.50
1020 100 to 500 lb.		175	.137		3.85	7.35		11.20	15.45
1030 500 to 1,000 lb.		190	.126		2.68	6.75		9.43	13.25
1040 1,000 to 2,000 lb.		200	.120		2.39	6.40		8.79	12.45
1050 2,000 to 5,000 lb.		225	.107		2.31	5.70		8.01	11.25
1060 Over 5,000 lb.		235	.102		2.31	5.45		7.76	10.90
1080 Note: Minimum order cost exceeds per lb. cost for min. wt.									
1100 For medium pressure ductwork, add				Lb.		15%			
1200 For high pressure ductwork, add						40%			
1210 For welded ductwork, add						85%			
1220 For 30% fittings, add						11%			
1224 For 40% fittings, add						34%			
1228 For 50% fittings, add						56%			
1232 For 60% fittings, add						79%			
1236 For 70% fittings, add						101%			
1240 For 80% fittings, add						124%			
1244 For 90% fittings, add						147%			
1248 For 100% fittings, add						169%			
1252 Note: Fittings add includes time for detailing and installation.									
1276 Duct tape, 2" x 60 yd roll				Ea.	7.65			7.65	8.45

23 31 HVAC Ducts and Casings

23 31 13 – Metal Ducts

23 31 13.16 Round and Flat-Oval Spiral Ducts	Crew	Daily Output	Labor-Hours	Unit	Material	2016 Bare Costs Labor	Equipment	Total	Total Incl O&P
0010 ROUND AND FLAT-OVAL SPIRAL DUCTS									
0020 Fabricated round and flat oval spiral, includes hangers,									
0021 supports and field sketches.									
1280 Add to labor for elevated installation									
1282 of prefabricated (purchased) ductwork									
1283 10' to 15' high						10%			
1284 15' to 20' high						20%			
1285 20' to 25' high						25%			
1286 25' to 30' high						35%			
1287 30' to 35' high						40%			
1288 35' to 40' high						50%			
1289 Over 40' high						55%			
5400 Spiral preformed, steel, galv., straight lengths, Max 10" spwg.									
5410 4" diameter, 26 ga.	Q-9	360	.044	L.F.	1.86	2.29		4.15	5.55
5416 5" diameter, 26 ga.		320	.050		1.86	2.58		4.44	6
5420 6" diameter, 26 ga.		280	.057		1.86	2.94		4.80	6.55
5425 7" diameter, 26 ga.		240	.067		2.17	3.44		5.61	7.65
5430 8" diameter, 26 ga.		200	.080		2.48	4.12		6.60	9.05
5440 10" diameter, 26 ga.		160	.100		3.10	5.15		8.25	11.30
5450 12" diameter, 26 ga.		120	.133		3.72	6.85		10.57	14.60
5460 14" diameter, 26 ga.		80	.200		4.34	10.30		14.64	20.50
5480 16" diameter, 24 ga.		60	.267		6.15	13.75		19.90	28
5490 18" diameter, 24 ga.		50	.320		6.90	16.50		23.40	32.50
5500 20" diameter, 24 ga.	Q-10	65	.369		7.65	19.75		27.40	38.50
5510 22" diameter, 24 ga.		60	.400		8.45	21.50		29.95	42
5520 24" diameter, 24 ga.		55	.436		9.20	23.50		32.70	45.50
5540 30" diameter, 22 ga.		45	.533		13.65	28.50		42.15	58.50
5600 36" diameter, 22 ga.		40	.600		16.40	32		48.40	67
5800 Connector, 4" diameter	Q-9	100	.160	Ea.	3.20	8.25		11.45	16.10
5810 5" diameter		94	.170		3.20	8.75		11.95	16.90
5820 6" diameter		88	.182		3.20	9.35		12.55	17.80
5840 8" diameter		78	.205		3.60	10.55		14.15	20
5860 10" diameter		70	.229		4.20	11.80		16	22.50
5880 12" diameter		50	.320		4.70	16.50		21.20	30
5900 14" diameter		44	.364		5.05	18.75		23.80	34
5920 16" diameter		40	.400		5.50	20.50		26	37.50
5930 18" diameter		37	.432		5.85	22.50		28.35	40.50
5940 20" diameter		34	.471		6.65	24.50		31.15	44.50
5950 22" diameter		31	.516		7.80	26.50		34.30	49
5960 24" diameter		28	.571		8.25	29.50		37.75	54
5980 30" diameter		22	.727		14.70	37.50		52.20	73.50
6000 36" diameter		18	.889		17.90	46		63.90	89.50
6300 Elbow, 45°, 4" diameter		60	.267		6.05	13.75		19.80	27.50
6310 5" diameter		52	.308		6.70	15.85		22.55	31.50
6320 6" diameter		44	.364		6.70	18.75		25.45	36
6340 8" diameter		28	.571		8.10	29.50		37.60	54
6360 10" diameter		18	.889		8.50	46		54.50	79.50
6380 12" diameter		13	1.231		9.20	63.50		72.70	107
6400 14" diameter		11	1.455		12.75	75		87.75	129
6420 16" diameter		10	1.600		20	82.50		102.50	148
6430 18" diameter		9.60	1.667		50	86		136	186
6440 20" diameter	Q-10	14	1.714		55	91.50		146.50	201

For customer support on your Mechanical Cost Data, call 877.739.3689.

309

23 31 13.16 Round and Flat-Oval Spiral Ducts		Crew	Daily Output	Labor-Hours	Unit	Material	2016 Bare Costs Labor	Equipment	Total	Total Incl O&P
6450	22" diameter	Q-10	13	1.846	Ea.	65	98.50		163.50	223
6460	24" diameter		12	2		75	107		182	246
6480	30" diameter		9	2.667		108	142		250	335
6500	36" diameter		7	3.429		132	183		315	425
6600	Elbow, 90°, 4" diameter	Q-9	60	.267		4.60	13.75		18.35	26
6610	5" diameter		52	.308		4.60	15.85		20.45	29
6620	6" diameter		44	.364		4.60	18.75		23.35	33.50
6625	7" diameter		36	.444		5.15	23		28.15	40.50
6630	8" diameter		28	.571		5.65	29.50		35.15	51
6640	10" diameter		18	.889		7.40	46		53.40	78
6650	12" diameter		13	1.231		9.25	63.50		72.75	107
6660	14" diameter		11	1.455		15.85	75		90.85	132
6670	16" diameter		10	1.600		18	82.50		100.50	146
6676	18" diameter		9.60	1.667		65	86		151	203
6680	20" diameter	Q-10	14	1.714		70	91.50		161.50	217
6684	22" diameter		13	1.846		75	98.50		173.50	234
6690	24" diameter		12	2		80	107		187	251
6700	30" diameter		9	2.667		144	142		286	375
6710	36" diameter		7	3.429		223	183		406	525
6720	End cap									
6722	4"	Q-9	110	.145	Ea.	4.38	7.50		11.88	16.25
6724	6"		94	.170		4.38	8.75		13.13	18.20
6725	8"		88	.182		5.50	9.35		14.85	20.50
6726	10"		78	.205		6.85	10.55		17.40	23.50
6727	12"		70	.229		7.45	11.80		19.25	26
6728	14"		50	.320		10.45	16.50		26.95	36.50
6729	16"		44	.364		11.65	18.75		30.40	41.50
6730	18"		40	.400		16	20.50		36.50	49
6731	20"		37	.432		32.50	22.50		55	70
6732	22"		34	.471		36.50	24.50		61	77.50
6733	24"		30	.533		41	27.50		68.50	87
6800	Reducing coupling, 6" x 4"		46	.348		18.80	17.90		36.70	48
6820	8" x 6"		40	.400		18.80	20.50		39.30	52
6840	10" x 8"		32	.500		20.50	26		46.50	62
6860	12" x 10"		24	.667		23.50	34.50		58	78.50
6880	14" x 12"		20	.800		25	41		66	90.50
6900	16" x 14"		18	.889		32	46		78	106
6920	18" x 16"		16	1		39	51.50		90.50	122
6940	20" x 18"	Q-10	24	1		45.50	53.50		99	132
6950	22" x 20"		23	1.043		45	56		101	135
6960	24" x 22"		22	1.091		58.50	58.50		117	154
6980	30" x 28"		18	1.333		112	71		183	233
7000	36" x 34"		16	1.500		143	80		223	280
7100	Tee, 90°, 4" diameter	Q-9	40	.400		13.55	20.50		34.05	46.50
7110	5" diameter		34.60	.462		14.15	24		38.15	52
7120	6" diameter		30	.533		14.80	27.50		42.30	58.50
7130	8" diameter		19	.842		18.65	43.50		62.15	87
7140	10" diameter		12	1.333		28	68.50		96.50	136
7150	12" diameter		9	1.778		33	91.50		124.50	177
7160	14" diameter		7	2.286		44.50	118		162.50	229
7170	16" diameter		6.70	2.388		72.50	123		195.50	268
7176	18" diameter		6	2.667		105	137		242	325
7180	20" diameter	Q-10	8.70	2.759		125	147		272	360

23 31 13.16 Round and Flat-Oval Spiral Ducts		Crew	Daily Output	Labor-Hours	Unit	Material	2016 Bare Costs		Total	Total Incl O&P
							Labor	Equipment		
7190	24" diameter	Q-10	7	3.429	Ea.	196	183		379	495
7200	30" diameter		6	4		239	214		453	590
7210	36" diameter		5	4.800		305	256		561	725
7250	Saddle Tap									
7254	4" dia. on 6" dia.	Q-9	36	.444	Ea.	25	23		48	62
7256	5" dia on 7" dia		35.80	.447		25	23		48	62
7258	6" dia. on 8" dia.		35.40	.452		22	23.50		45.50	59.50
7260	6" dia. on 10" dia.		35.20	.455		22	23.50		45.50	60
7262	6" dia. on 12" dia.		35	.457		22.50	23.50		46	61
7264	7" dia. on 9" dia.		35.20	.455		25.50	23.50		49	64
7266	8" dia. on 10" dia.		35	.457		25.50	23.50		49	64
7268	8" dia. on 12" dia.		34.80	.460		22.50	23.50		46	61
7270	8" dia. on 14" dia.		34.50	.464		28.50	24		52.50	68
7272	10" dia. on 12" dia.		34.20	.468		26	24		50	65.50
7274	10" dia. on 14" dia.		33.80	.473		28.50	24.50		53	69
7276	10" dia. on 16" dia.		33.40	.479		28.50	24.50		53	69
7278	10" dia. on 18" dia.		33	.485		28.50	25		53.50	69.50
7280	12" dia. on 14" dia.		32.60	.491		33	25.50		58.50	75
7282	12" dia. on 18" dia.		32.20	.497		36.50	25.50		62	79
7284	14" dia. on 16" dia.		31.60	.506		38	26		64	82
7286	14" dia. on 20" dia.		31	.516		38.50	26.50		65	83
7288	16" dia. on 20" dia.		30	.533		45	27.50		72.50	91.50
7290	18" dia. on 20" dia.		28	.571		52.50	29.50		82	103
7292	20" dia. on 22" dia.		26	.615		69.50	31.50		101	125
7294	22" dia. on 24" dia.		24	.667		74.50	34.50		109	134
7400	Steel, PVC coated both sides, straight lengths									
7410	4" diameter, 26 ga.	Q-9	360	.044	L.F.	4.97	2.29		7.26	8.95
7416	5" diameter, 26 ga.		240	.067		3.60	3.44		7.04	9.20
7420	6" diameter, 26 ga.		280	.057		3.60	2.94		6.54	8.45
7440	8" diameter, 26 ga.		200	.080		4.90	4.12		9.02	11.70
7460	10" diameter, 26 ga.		160	.100		6.10	5.15		11.25	14.60
7480	12" diameter, 26 ga.		120	.133		7.30	6.85		14.15	18.55
7500	14" diameter, 26 ga.		80	.200		8.50	10.30		18.80	25
7520	16" diameter, 24 ga.		60	.267		9.70	13.75		23.45	31.50
7540	18" diameter, 24 ga.		45	.356		10.90	18.30		29.20	40
7560	20" diameter, 24 ga.	Q-10	65	.369		12.10	19.75		31.85	43.50
7580	24" diameter, 24 ga.		55	.436		14.50	23.50		38	51.50
7600	30" diameter, 22 ga.		45	.533		22	28.50		50.50	67.50
7620	36" diameter, 22 ga.		40	.600		26	32		58	78
7890	Connector, 4" diameter	Q-9	100	.160	Ea.	2.60	8.25		10.85	15.45
7896	5" diameter		83	.193		2.90	9.95		12.85	18.40
7900	6" diameter		88	.182		3.30	9.35		12.65	17.95
7920	8" diameter		78	.205		3.90	10.55		14.45	20.50
7940	10" diameter		70	.229		4.50	11.80		16.30	23
7960	12" diameter		50	.320		5.30	16.50		21.80	31
7980	14" diameter		44	.364		5.60	18.75		24.35	34.50
8000	16" diameter		40	.400		6.10	20.50		26.60	38
8020	18" diameter		37	.432		6.80	22.50		29.30	41.50
8040	20" diameter		34	.471		8.90	24.50		33.40	47
8060	24" diameter		28	.571		11	29.50		40.50	57
8080	30" diameter		22	.727		14.70	37.50		52.20	73.50
8100	36" diameter		18	.889		17.90	46		63.90	89.50
8390	Elbow, 45°, 4" diameter		60	.267		23	13.75		36.75	46.50

23 31 13.16 Round and Flat-Oval Spiral Ducts	Crew	Daily Output	Labor-Hours	Unit	Material	2016 Bare Costs Labor	Equipment	Total	Total Incl O&P
8396	5" diameter	Q-9	36	.444	Ea.	23	23		46	60.50
8400	6" diameter		44	.364		23	18.75		41.75	54
8420	8" diameter		28	.571		23	29.50		52.50	70.50
8440	10" diameter		18	.889		23	46		69	95.50
8460	12" diameter		13	1.231		23	63.50		86.50	123
8480	14" diameter		11	1.455		23	75		98	141
8500	16" diameter		10	1.600		23	82.50		105.50	152
8520	18" diameter		9	1.778		25	91.50		116.50	168
8540	20" diameter	Q-10	13	1.846		75.50	98.50		174	235
8560	24" diameter		11	2.182		97.50	117		214.50	285
8580	30" diameter		9	2.667		129	142		271	360
8600	36" diameter		7	3.429		157	183		340	455
8800	Elbow, 90°, 4" diameter	Q-9	60	.267		25	13.75		38.75	48.50
8804	5" diameter		52	.308		25	15.85		40.85	51.50
8806	6" diameter		44	.364		25	18.75		43.75	56
8808	8" diameter		28	.571		25	29.50		54.50	72.50
8810	10" diameter		18	.889		25	46		71	97.50
8812	12" diameter		13	1.231		25	63.50		88.50	125
8814	14" diameter		11	1.455		25	75		100	143
8816	16" diameter		10	1.600		29	82.50		111.50	158
8818	18" diameter		9.60	1.667		35.50	86		121.50	170
8820	20" diameter	Q-10	14	1.714		45	91.50		136.50	190
8822	24" diameter		12	2		104	107		211	277
8824	30" diameter		9	2.667		175	142		317	410
8826	36" diameter		7	3.429		285	183		468	595
9000	Reducing coupling									
9040	6" x 4"	Q-9	46	.348	Ea.	19.50	17.90		37.40	49
9060	8" x 6"		40	.400		19.50	20.50		40	53
9080	10" x 8"		32	.500		20.50	26		46.50	62
9100	12" x 10"		24	.667		23.50	34.50		58	78.50
9120	14" x 12"		20	.800		25	41		66	90.50
9140	16" x 14"		18	.889		32	46		78	106
9160	18" x 16"		16	1		39	51.50		90.50	122
9180	20" x 18"	Q-10	24	1		45.50	53.50		99	132
9200	24" x 22"		22	1.091		58.50	58.50		117	154
9220	30" x 28"		18	1.333		104	71		175	223
9240	36" x 34"		16	1.500		143	80		223	280
9800	Steel, stainless, straight lengths									
9810	3" diameter, 26 Ga.	Q-9	400	.040	L.F.	14.20	2.06		16.26	18.75
9820	4" diameter, 26 Ga.		360	.044		16.45	2.29		18.74	21.50
9830	5" diameter, 26 Ga.		320	.050		17.75	2.58		20.33	23.50
9840	6" diameter, 26 Ga.		280	.057		17.75	2.94		20.69	24
9850	7" diameter, 26 Ga.		240	.067		29.50	3.44		32.94	38
9860	8" diameter, 26 Ga.		200	.080		23	4.12		27.12	32
9870	9" diameter, 26 Ga.		180	.089		33.50	4.58		38.08	44
9880	10" diameter, 26 Ga.		160	.100		43.50	5.15		48.65	56
9890	12" diameter, 26 Ga.		120	.133		61.50	6.85		68.35	78.50
9900	14" diameter, 26 Ga.		80	.200		65	10.30		75.30	87.50

23 31 13.17 Round Grease Duct		Crew	Daily Output	Labor-Hours	Unit	Material	2016 Bare Costs Labor	Equipment	Total	Total Incl O&P
0010	**ROUND GREASE DUCT**									
1280	Add to labor for elevated installation									
1282	of prefabricated (purchased) ductwork									
1283	10' to 15' high						10%			
1284	15' to 20' high						20%			
1285	20' to 25' high						25%			
1286	25' to 30' high						35%			
1287	30' to 35' high						40%			
1288	35' to 40' high						50%			
1289	Over 40' high						55%			
4020	Zero clearance, 2 hr. fire rated, UL listed, 3" double wall									
4030	304 interior, aluminized exterior, all sizes are ID									
4032	For 316 interior and aluminized exterior, add						6%			
4034	For 304 interior and 304 exterior, add						21%			
4036	For 316 interior and 316 exterior, add						40%			
4040	Straight									
4052	6"	Q-9	34.20	.468	L.F.	74.50	24		98.50	119
4054	8"		29.64	.540		85.50	28		113.50	137
4056	10"		27.36	.585		95.50	30		125.50	151
4058	12"		25.10	.637		110	33		143	170
4060	14"		23.94	.668		123	34.50		157.50	188
4062	16"		22.80	.702		138	36		174	208
4064	18"		21.66	.739		157	38		195	230
4066	20"	Q-10	20.52	1.170		178	62.50		240.50	292
4068	22"		19.38	1.238		201	66		267	320
4070	24"		18.24	1.316		227	70.50		297.50	355
4072	26"		17.67	1.358		245	72.50		317.50	380
4074	28"		17.10	1.404		261	75		336	405
4076	30"		15.96	1.504		274	80.50		354.50	425
4078	32"		15.39	1.559		295	83.50		378.50	450
4080	36"		14.25	1.684		330	90		420	500
4100	Adjustable section, 30" long									
4104	6"	Q-9	17.10	.936	Ea.	226	48		274	320
4106	8"		14.82	1.080		282	55.50		337.50	395
4108	10"		13.68	1.170		320	60.50		380.50	440
4110	12"		12.54	1.276		360	66		426	495
4112	14"		11.97	1.337		405	69		474	550
4114	16"		11.40	1.404		455	72.50		527.50	610
4116	18"		10.83	1.477		515	76		591	680
4118	20"	Q-10	10.26	2.339		585	125		710	835
4120	22"		9.69	2.477		660	132		792	925
4122	24"		9.12	2.632		750	141		891	1,050
4124	26"		8.83	2.718		805	145		950	1,100
4126	28"		8.55	2.807		860	150		1,010	1,175
4128	30"		8.26	2.906		915	155		1,070	1,225
4130	32"		7.98	3.008		970	161		1,131	1,325
4132	36"		6.84	3.509		1,075	187		1,262	1,475
4140	Tee, 90°, grease									
4144	6"	Q-9	13.68	1.170	Ea.	400	60.50		460.50	530
4146	8"		12.54	1.276		435	66		501	580
4148	10"		11.97	1.337		490	69		559	645
4150	12"		11.40	1.404		570	72.50		642.50	735

For customer support on your Mechanical Cost Data, call 877.739.3689.

313

23 31 13.17 Round Grease Duct		Crew	Daily Output	Labor-Hours	Unit	Material	2016 Bare Costs Labor	Equipment	Total	Total Incl O&P
4152	14"	Q-9	10.26	1.559	Ea.	645	80.50		725.50	835
4154	16"		9.12	1.754		715	90.50		805.50	925
4156	18"		7.98	2.005		840	103		943	1,075
4158	20"		9.69	1.651		970	85		1,055	1,200
4160	22"		8.26	1.937		1,075	100		1,175	1,350
4162	24"		6.84	2.339		1,200	121		1,321	1,500
4164	26"		6.55	2.443		1,325	126		1,451	1,650
4166	28"	Q-10	6.27	3.828		1,450	205		1,655	1,925
4168	30"		5.98	4.013		1,725	214		1,939	2,225
4170	32"		5.70	4.211		2,025	225		2,250	2,575
4172	36"		5.13	4.678		2,250	250		2,500	2,850
4180	Cleanout Tee Cap									
4184	6"	Q-9	21.09	.759	Ea.	81.50	39		120.50	149
4186	8"		19.38	.826		86	42.50		128.50	160
4188	10"		18.24	.877		92.50	45		137.50	171
4190	12"		17.10	.936		99	48		147	183
4192	14"		15.96	1.003		105	51.50		156.50	194
4194	16"		14.25	1.123		116	58		174	216
4196	18"		13.68	1.170		122	60.50		182.50	226
4198	20"	Q-10	15.39	1.559		135	83.50		218.50	276
4200	22"		12.54	1.914		146	102		248	315
4202	24"		11.97	2.005		160	107		267	340
4204	26"		11.40	2.105		174	112		286	365
4206	28"		10.83	2.216		189	118		307	390
4208	30"		10.26	2.339		206	125		331	415
4210	32"		9.69	2.477		224	132		356	450
4212	36"		8.55	2.807		263	150		413	520
4220	Elbow, 45°									
4224	6"	Q-9	17.10	.936	Ea.	282	48		330	385
4226	8"		14.82	1.080		315	55.50		370.50	435
4228	10"		13.68	1.170		360	60.50		420.50	485
4230	12"		12.54	1.276		410	66		476	550
4232	14"		11.97	1.337		460	69		529	610
4234	16"		11.40	1.404		520	72.50		592.50	685
4236	18"		10.83	1.477		590	76		666	765
4238	20"	Q-10	10.26	2.339		670	125		795	925
4240	22"		9.69	2.477		745	132		877	1,025
4242	24"		9.12	2.632		855	141		996	1,150
4244	26"		8.83	2.718		905	145		1,050	1,225
4246	28"		8.55	2.807		1,075	150		1,225	1,400
4248	30"		8.26	2.906		1,075	155		1,230	1,400
4250	32"		7.98	3.008		1,300	161		1,461	1,675
4252	36"		6.84	3.509		1,725	187		1,912	2,175
4260	Elbow, 90°									
4264	6"	Q-9	17.10	.936	Ea.	565	48		613	695
4266	8"		14.82	1.080		635	55.50		690.50	785
4268	10"		13.68	1.170		720	60.50		780.50	885
4270	12"		12.54	1.276		815	66		881	1,000
4272	14"		11.97	1.337		925	69		994	1,125
4274	16"		11.40	1.404		1,050	72.50		1,122.50	1,250
4276	18"		10.83	1.477		1,275	76		1,351	1,525
4278	20"	Q-10	10.26	2.339		1,325	125		1,450	1,675
4280	22"		9.69	2.477		1,500	132		1,632	1,850

23 31 HVAC Ducts and Casings

23 31 13 – Metal Ducts

	23 31 13.17 Round Grease Duct	Crew	Daily Output	Labor-Hours	Unit	Material	2016 Bare Costs Labor	Equipment	Total	Total Incl O&P
4282	24"	Q-10	9.12	2.632	Ea.	1,700	141		1,841	2,100
4284	26"		8.83	2.718		1,825	145		1,970	2,225
4286	28"		8.55	2.807		1,925	150		2,075	2,325
4288	30"		8.26	2.906		2,150	155		2,305	2,575
4290	32"		7.98	3.008		2,575	161		2,736	3,100
4292	36"		6.84	3.509		3,475	187		3,662	4,100
4300	Support strap									
4304	6"	Q-9	25.50	.627	Ea.	99	32.50		131.50	159
4306	8"		22	.727		116	37.50		153.50	185
4308	10"		20.50	.780		119	40		159	193
4310	12"		18.80	.851		128	44		172	207
4312	14"		17.90	.894		138	46		184	222
4314	16"		17	.941		139	48.50		187.50	226
4316	18"		16.20	.988		141	51		192	233
4318	20"	Q-10	15.40	1.558		156	83.50		239.50	299
4320	22"	"	14.50	1.655		173	88.50		261.50	325
4340	Plate Support Assembly									
4344	6"	Q-9	14.82	1.080	Ea.	139	55.50		194.50	237
4346	8"		12.54	1.276		162	66		228	278
4348	10"		11.40	1.404		176	72.50		248.50	305
4350	12"		10.26	1.559		185	80.50		265.50	325
4352	14"		9.69	1.651		219	85		304	370
4354	16"		9.12	1.754		231	90.50		321.50	390
4356	18"		8.55	1.871		243	96.50		339.50	415
4358	20"	Q-10	9.12	2.632		255	141		396	495
4360	22"		8.55	2.807		261	150		411	515
4362	24"		7.98	3.008		265	161		426	540
4364	26"		7.69	3.121		298	167		465	585
4366	28"		7.41	3.239		325	173		498	625
4368	30"		7.12	3.371		365	180		545	675
4370	32"		6.84	3.509		400	187		587	725
4372	36"		5.70	4.211		480	225		705	875
4380	Wall Guide Assembly									
4384	6"	Q-9	19.10	.838	Ea.	184	43		227	268
4386	8"		16.50	.970		208	50		258	305
4388	10"		15.30	1.046		220	54		274	325
4390	12"		14.10	1.135		230	58.50		288.50	345
4392	14"		13.40	1.194		249	61.50		310.50	365
4394	16"		12.75	1.255		252	64.50		316.50	375
4396	18"		12.10	1.322		255	68		323	385
4398	20"	Q-10	11.50	2.087		285	112		397	485
4400	22"		10.90	2.202		305	118		423	520
4402	24"		10.20	2.353		345	126		471	570
4404	26"		9.90	2.424		380	130		510	620
4406	28"		9.60	2.500		395	134		529	640
4408	30"		9.30	2.581		450	138		588	700
4410	32"		9	2.667		475	142		617	745
4412	36"		7.60	3.158		525	169		694	835
4420	Hood Transition, Flanged or Unflanged									
4424	6"	Q-9	15.60	1.026	Ea.	78	53		131	167
4426	8"		13.50	1.185		89	61		150	192
4428	10"		12.50	1.280		99	66		165	210
4430	12"		11.40	1.404		119	72.50		191.50	242

For customer support on your Mechanical Cost Data, call 877.739.3689.

315

23 31 13.17 Round Grease Duct		Crew	Daily Output	Labor-Hours	Unit	Material	2016 Bare Costs Labor	Equipment	Total	Total Incl O&P
4432	14"	Q-9	10.90	1.468	Ea.	128	75.50		203.50	256
4434	16"		10.40	1.538		138	79.50		217.50	272
4436	18"		9.80	1.633		157	84		241	300
4438	20"	Q-10	9.40	2.553		167	136		303	395
4440	22"		8.80	2.727		176	146		322	415
4442	24"		8.30	2.892		197	155		352	455
4444	26"		8.10	2.963		206	158		364	470
4446	28"		7.80	3.077		217	164		381	490
4448	30"		7.50	3.200		235	171		406	520
4450	32"		7.30	3.288		246	176		422	540
4452	36"		6.20	3.871		274	207		481	615
4460	Fan Adapter									
4464	6"	Q-9	15	1.067	Ea.	219	55		274	325
4466	8"		13	1.231		249	63.50		312.50	370
4468	10"		12	1.333		272	68.50		340.50	405
4470	12"		11	1.455		305	75		380	450
4472	14"		10.50	1.524		330	78.50		408.50	485
4474	16"		10	1.600		360	82.50		442.50	520
4476	18"		9.50	1.684		385	87		472	560
4478	20"	Q-10	9	2.667		415	142		557	675
4480	22"		8.50	2.824		445	151		596	720
4482	24"		8	3		470	160		630	760
4484	26"		7.70	3.117		495	167		662	800
4486	28"		7.50	3.200		525	171		696	840
4488	30"		7.20	3.333		555	178		733	880
4490	32"		7	3.429		585	183		768	920
4492	36"		6	4		640	214		854	1,025
4500	Tapered Increaser/Reducer									
4530	6" x 10" dia	Q-9	13.40	1.194	Ea.	282	61.50		343.50	405
4540	6" x 20" dia		10.10	1.584		282	81.50		363.50	435
4550	8" x 10" dia		13.20	1.212		340	62.50		402.50	470
4560	8" x 20" dia		9.90	1.616		340	83.50		423.50	500
4570	10" x 20" dia		10.60	1.509		395	78		473	555
4580	10" x 28" dia	Q-10	8.60	2.791		395	149		544	665
4590	12" x 20" dia		8.80	2.727		455	146		601	725
4600	14" x 28" dia		8.40	2.857		515	153		668	800
4610	16" x 20" dia		8.60	2.791		570	149		719	855
4620	16" x 30" dia		8.30	2.892		570	155		725	860
4630	18" x 24" dia		8.80	2.727		625	146		771	915
4640	18" x 30" dia		8.10	2.963		625	158		783	930
4650	20" x 24" dia		8.30	2.892		680	155		835	985
4660	20" x 32" dia		8	3		680	160		840	995
4670	24" x 26" dia		8	3		800	160		960	1,125
4680	24" x 36" dia		7.30	3.288		800	176		976	1,150
4690	28" x 30" dia		7.90	3.038		915	162		1,077	1,250
4700	28" x 36" dia		7.10	3.380		915	181		1,096	1,275
4710	30" x 36" dia		6.90	3.478		970	186		1,156	1,350

23 31 HVAC Ducts and Casings

23 31 13 – Metal Ducts

23 31 13.17 Round Grease Duct	Crew	Daily Output	Labor-Hours	Unit	Material	2016 Bare Costs Labor	Equipment	Total	Total Incl O&P
4730	Many intermediate standard sizes of tapered fittings are available								

23 31 13.19 Metal Duct Fittings

		Crew	Daily Output	Labor-Hours	Unit	Material	Labor	Equipment	Total	Total Incl O&P
0010	**METAL DUCT FITTINGS**									
0350	Duct collar, spin-in type									
0352	Without damper									
0354	Round									
0356	4"	1 Shee	32	.250	Ea.	2.61	14.30		16.91	25
0357	5"		30	.267		3.16	15.25		18.41	27
0358	6"		28	.286		3.49	16.35		19.84	29
0359	7"		26	.308		3.80	17.60		21.40	31
0360	8"		24	.333		3.80	19.10		22.90	33
0361	9"		22	.364		3.80	21		24.80	36
0362	10"		20	.400		4.18	23		27.18	39.50
0363	12"		17	.471		5.25	27		32.25	47
2000	Fabrics for flexible connections, with metal edge		100	.080	L.F.	3.44	4.58		8.02	10.80
2100	Without metal edge		160	.050	"	2.46	2.86		5.32	7.10

23 31 16 – Nonmetal Ducts

23 31 16.13 Fibrous-Glass Ducts

		Crew	Daily Output	Labor-Hours	Unit	Material	Labor	Equipment	Total	Total Incl O&P
0010	**FIBROUS-GLASS DUCTS**									
1280	Add to labor for elevated installation									
1282	of prefabricated (purchased) ductwork									
1283	10' to 15' high						10%			
1284	15' to 20' high						20%			
1285	20' to 25' high						25%			
1286	25' to 30' high						35%			
1287	30' to 35' high						40%			
1288	35' to 40' high						50%			
1289	Over 40' high						55%			
3490	Rigid fiberglass duct board, foil reinf. kraft facing									
3500	Rectangular, 1" thick, alum. faced, (FRK), std. weight	Q-10	350	.069	SF Surf	.79	3.66		4.45	6.45

23 31 16.16 Thermoset Fiberglass-Reinforced Plastic Ducts

		Crew	Daily Output	Labor-Hours	Unit	Material	Labor	Equipment	Total	Total Incl O&P
0010	**THERMOSET FIBERGLASS-REINFORCED PLASTIC DUCTS**									
1280	Add to labor for elevated installation									
1282	of prefabricated (purchased) ductwork									
1283	10' to 15' high						10%			
1284	15' to 20' high						20%			
1285	20' to 25' high						25%			
1286	25' to 30' high						35%			
1287	30' to 35' high						40%			
1288	35' to 40' high						50%			
1289	Over 40' high						55%			
3550	Rigid fiberglass reinforced plastic, FM approved									
3552	for acid fume and smoke exhaust system, nonflammable									
3554	Straight, 4" diameter	Q-9	200	.080	L.F.	10.35	4.12		14.47	17.70
3555	6" diameter		146	.110		13.55	5.65		19.20	23.50
3556	8" diameter		106	.151		16.45	7.80		24.25	30
3557	10" diameter		85	.188		19.70	9.70		29.40	36.50
3558	12" diameter		68	.235		23	12.10		35.10	44
3559	14" diameter		50	.320		34	16.50		50.50	62.50
3560	16" diameter		40	.400		38	20.50		58.50	73.50
3561	18" diameter		31	.516		42.50	26.50		69	87
3562	20" diameter	Q-10	44	.545		46.50	29		75.50	96

For customer support on your Mechanical Cost Data, call 877.739.3689.

317

23 31 HVAC Ducts and Casings

23 31 16 – Nonmetal Ducts

23 31 16.16 Thermoset Fiberglass-Reinforced Plastic Ducts	Crew	Daily Output	Labor-Hours	Unit	Material	2016 Bare Costs Labor	Equipment	Total	Total Incl O&P
3563	22" diameter	Q-10	40	.600	L.F.	50.50	32		82.50	105
3564	24" diameter		37	.649		54.50	34.50		89	113
3565	26" diameter		34	.706		58.50	37.50		96	122
3566	28" diameter		28	.857		63	46		109	140
3567	30" diameter		24	1		67	53.50		120.50	155
3568	32" diameter		22.60	1.062		71	56.50		127.50	165
3569	34" diameter		21.60	1.111		75	59.50		134.50	173
3570	36" diameter		20.60	1.165		79.50	62.50		142	182
3571	38" diameter		19.80	1.212		122	65		187	233
3572	42" diameter		18.30	1.311		136	70		206	257
3573	46" diameter		17	1.412		151	75.50		226.50	281
3574	50" diameter		15.90	1.509		164	80.50		244.50	305
3575	54" diameter		14.90	1.611		181	86		267	330
3576	60" diameter		13.60	1.765		207	94.50		301.50	370
3577	64" diameter		12.80	1.875		305	100		405	490
3578	68" diameter		12.20	1.967		320	105		425	515
3579	72" diameter		11.90	2.017		345	108		453	545
3584	Note: joints are cemented with									
3586	fiberglass resin, included in material cost.									
3590	Elbow, 90°, 4" diameter	Q-9	20.30	.788	Ea.	50	40.50		90.50	117
3591	6" diameter		13.80	1.159		69.50	59.50		129	168
3592	8" diameter		10	1.600		77	82.50		159.50	211
3593	10" diameter		7.80	2.051		101	106		207	273
3594	12" diameter		6.50	2.462		128	127		255	335
3595	14" diameter		5.50	2.909		157	150		307	400
3596	16" diameter		4.80	3.333		186	172		358	470
3597	18" diameter		4.40	3.636		267	187		454	580
3598	20" diameter	Q-10	5.90	4.068		282	217		499	640
3599	22" diameter		5.40	4.444		345	237		582	745
3600	24" diameter		4.90	1.898		400	262		662	840
3601	26" diameter		4.60	5.217		405	279		684	870
3602	28" diameter		4.20	5.714		505	305		810	1,025
3603	30" diameter		3.90	6.154		560	330		890	1,125
3604	32" diameter		3.70	6.486		625	345		970	1,225
3605	34" diameter		3.45	6.957		735	370		1,105	1,375
3606	36" diameter		3.20	7.500		845	400		1,245	1,550
3607	38" diameter		3.07	7.818		1,200	420		1,620	1,975
3608	42" diameter		2.70	8.889		1,450	475		1,925	2,300
3609	46" diameter		2.54	9.449		1,675	505		2,180	2,625
3610	50" diameter		2.33	10.300		1,950	550		2,500	3,000
3611	54" diameter		2.14	11.215		2,225	600		2,825	3,375
3612	60" diameter		1.94	12.371		2,975	660		3,635	4,275
3613	64" diameter		1.82	13.187		4,325	705		5,030	5,825
3614	68" diameter		1.71	14.035		4,800	750		5,550	6,425
3615	72" diameter		1.62	14.815		5,300	790		6,090	7,050
3626	Elbow, 45°, 4" diameter	Q-9	22.30	.717		48	37		85	110
3627	6" diameter		15.20	1.053		66.50	54		120.50	156
3628	8" diameter		11	1.455		70.50	75		145.50	193
3629	10" diameter		8.60	1.860		74.50	96		170.50	229
3630	12" diameter		7.20	2.222		81.50	115		196.50	265
3631	14" diameter		6.10	2.623		100	135		235	315
3632	16" diameter		5.28	3.030		119	156		275	370
3633	18" diameter		4.84	3.306		139	170		309	415

318

23 31 16.16 Thermoset Fiberglass-Reinforced Plastic Ducts	Crew	Daily Output	Labor-Hours	Unit	Material	2016 Bare Costs Labor	Equipment	Total	Total Incl O&P
3634	20" diameter	Q-10	6.50	3.692	Ea.	161	197		358	480
3635	22" diameter		5.94	4.040		220	216		436	570
3636	24" diameter		5.40	4.444		251	237		488	640
3637	26" diameter		5.06	4.743		285	253		538	700
3638	28" diameter		4.62	5.195		310	278		588	765
3639	30" diameter		4.30	5.581		355	298		653	845
3640	32" diameter		4.07	5.897		395	315		710	915
3641	34" diameter		3.80	6.316		465	335		800	1,025
3642	36" diameter		3.52	6.818		540	365		905	1,150
3643	38" diameter		3.38	7.101		755	380		1,135	1,400
3644	42" diameter		2.97	8.081		905	430		1,335	1,650
3645	46" diameter		2.80	8.571		1,050	460		1,510	1,875
3646	50" diameter		2.56	9.375		1,225	500		1,725	2,125
3647	54" diameter		2.35	10.213		1,400	545		1,945	2,350
3648	60" diameter		2.13	11.268		1,875	600		2,475	2,975
3649	64" diameter		2	12		2,725	640		3,365	3,950
3650	68" diameter		1.88	12.766		3,000	680		3,680	4,350
3651	72" diameter		1.78	13.483		3,325	720		4,045	4,750
3660	Tee, 90°, 4" diameter	Q-9	14.06	1.138		27.50	58.50		86	120
3661	6" diameter		9.52	1.681		42	86.50		128.50	178
3662	8" diameter		6.98	2.292		57	118		175	244
3663	10" diameter		5.45	2.936		74.50	151		225.50	315
3664	12" diameter		4.60	3.478		93	179		272	375
3665	14" diameter		3.86	4.145		136	214		350	475
3666	16" diameter		3.40	4.706		161	242		403	550
3667	18" diameter		3.05	5.246		191	270		461	625
3668	20" diameter	Q-10	4.12	5.825		224	310		534	720
3669	22" diameter		3.78	6.349		256	340		596	800
3670	24" diameter		3.46	6.936		288	370		658	880
3671	26" diameter		3.20	7.500		325	400		725	975
3672	28" diameter		2.96	8.108		365	435		800	1,050
3673	30" diameter		2.75	8.727		410	465		875	1,175
3674	32" diameter		2.54	9.449		450	505		955	1,275
3675	34" diameter		2.40	10		495	535		1,030	1,350
3676	36" diameter		2.27	10.573		545	565		1,110	1,450
3677	38" diameter		2.15	11.163		815	595		1,410	1,800
3678	42" diameter		1.91	12.565		985	670		1,655	2,100
3679	46" diameter		1.78	13.483		1,150	720		1,870	2,375
3680	50" diameter		1.63	14.724		1,350	785		2,135	2,675
3681	54" diameter		1.50	16		1,550	855		2,405	3,025
3682	60" diameter		1.35	17.778		1,950	950		2,900	3,600
3683	64" diameter		1.27	18.898		2,925	1,000		3,925	4,775
3684	68" diameter		1.20	20		3,275	1,075		4,350	5,225
3685	72" diameter		1.13	21.239		3,650	1,125		4,775	5,750
3690	For Y @ 45°, add					22%				
3700	Blast gate, 4" diameter	Q-9	22.30	.717		133	37		170	203
3701	6" diameter		15.20	1.053		150	54		204	248
3702	8" diameter		11	1.455		202	75		277	335
3703	10" diameter		8.60	1.860		231	96		327	400
3704	12" diameter		7.20	2.222		282	115		397	485
3705	14" diameter		6.10	2.623		310	135		445	550
3706	16" diameter		5.28	3.030		345	156		501	620
3707	18" diameter		4.84	3.306		355	170		525	655

For customer support on your Mechanical Cost Data, call 877.739.3689.

319

23 31 16 – Nonmetal Ducts

23 31 16.16 Thermoset Fiberglass-Reinforced Plastic Ducts		Crew	Daily Output	Labor-Hours	Unit	Material	2016 Bare Costs Labor	Equipment	Total	Total Incl O&P
3708	20" diameter	Q-10	6.50	3.692	Ea.	410	197		607	750
3709	22" diameter		5.94	4.040		440	216		656	815
3710	24" diameter		5.40	4.444		475	237		712	890
3711	26" diameter		5.06	4.743		510	253		763	945
3712	28" diameter		4.62	5.195		550	278		828	1,025
3713	30" diameter		4.30	5.581		580	298		878	1,100
3714	32" diameter		4.07	5.897		635	315		950	1,175
3715	34" diameter		3.80	6.316		690	335		1,025	1,275
3716	36" diameter		3.52	6.818		745	365		1,110	1,375
3717	38" diameter		3.38	7.101		1,000	380		1,380	1,675
3718	42" diameter		2.97	8.081		1,150	430		1,580	1,925
3719	46" diameter		2.80	8.571		1,325	460		1,785	2,150
3720	50" diameter		2.56	9.375		1,525	500		2,025	2,450
3721	54" diameter		2.35	10.213		1,700	545		2,245	2,700
3722	60" diameter		2.13	11.268		2,000	600		2,600	3,125
3723	64" diameter		2	12		2,200	640		2,840	3,400
3724	68" diameter		1.88	12.766		2,450	680		3,130	3,750
3725	72" diameter	▼	1.78	13.483		2,700	720		3,420	4,050
3740	Reducers, 4" diameter	Q-9	24.50	.653		33	33.50		66.50	88
3741	6" diameter		16.70	.958		49.50	49.50		99	130
3742	8" diameter		12	1.333		67	68.50		135.50	179
3743	10" diameter		9.50	1.684		82.50	87		169.50	224
3744	12" diameter		7.90	2.025		99	104		203	268
3745	14" diameter		6.70	2.388		115	123		238	315
3746	16" diameter		5.80	2.759		132	142		274	360
3747	18" diameter	▼	5.30	3.019		149	156		305	400
3748	20" diameter	Q-10	7.20	3.333		165	178		343	455
3749	22" diameter		6.50	3.692		266	197		463	595
3750	24" diameter		5.90	4.068		288	217		505	645
3751	26" diameter		5.60	4.286		310	229		539	695
3752	28" diameter		5.10	4.706		340	251		591	760
3753	30" diameter		4.70	5.106		365	273		638	815
3754	32" diameter		4.50	5.333		390	285		675	865
3755	34" diameter		4.20	5.714		415	305		720	920
3756	36" diameter		3.90	6.154		435	330		765	985
3757	38" diameter		3.70	6.486		665	345		1,010	1,250
3758	42" diameter		3.30	7.273		730	390		1,120	1,400
3759	46" diameter		3.10	7.742		805	415		1,220	1,525
3760	50" diameter		2.80	8.571		870	460		1,330	1,650
3761	54" diameter		2.60	9.231		940	495		1,435	1,775
3762	60" diameter		2.34	10.256		1,050	550		1,600	2,025
3763	64" diameter		2.20	10.909		1,625	585		2,210	2,700
3764	68" diameter		2.06	11.650		1,725	620		2,345	2,850
3765	72" diameter	▼	1.96	12.245		1,850	655		2,505	3,025
3780	Flange, per each, 4" diameter	Q-9	40.60	.394		80.50	20.50		101	120
3781	6" diameter		27.70	.578		93.50	30		123.50	149
3782	8" diameter		20.50	.780		117	40		157	191
3783	10" diameter		16.30	.982		140	50.50		190.50	232
3784	12" diameter		13.90	1.151		161	59.50		220.50	269
3785	14" diameter		11.30	1.416		186	73		259	315
3786	16" diameter		9.70	1.649		210	85		295	360
3787	18" diameter	▼	8.30	1.928		233	99.50		332.50	410
3788	20" diameter	Q-10	11.30	2.124		256	113		369	455

23 31 HVAC Ducts and Casings

23 31 16 – Nonmetal Ducts

	23 31 16.16 Thermoset Fiberglass-Reinforced Plastic Ducts	Crew	Daily Output	Labor-Hours	Unit	Material	2016 Bare Costs Labor	Equipment	Total	Total Incl O&P
3789	22" diameter	Q-10	10.40	2.308	Ea.	279	123		402	495
3790	24" diameter		9.50	2.526		305	135		440	540
3791	26" diameter		8.80	2.727		325	146		471	585
3792	28" diameter		8.20	2.927		350	156		506	625
3793	30" diameter		7.60	3.158		370	169		539	665
3794	32" diameter		7.20	3.333		395	178		573	705
3795	34" diameter		6.80	3.529		420	189		609	750
3796	36" diameter		6.40	3.750		440	200		640	790
3797	38" diameter		6.10	3.934		465	210		675	835
3798	42" diameter		5.50	4.364		510	233		743	915
3799	46" diameter		5	4.800		560	256		816	1,000
3800	50" diameter		4.60	5.217		605	279		884	1,100
3801	54" diameter		4.20	5.714		650	305		955	1,175
3802	60" diameter		3.80	6.316		720	335		1,055	1,300
3803	64" diameter		3.60	6.667		765	355		1,120	1,375
3804	68" diameter		3.40	7.059		810	375		1,185	1,475
3805	72" diameter		3.20	7.500		860	400		1,260	1,550

23 31 16.19 PVC Ducts

		Crew	Daily Output	Labor-Hours	Unit	Material	2016 Bare Costs Labor	Equipment	Total	Total Incl O&P
0010	**PVC DUCTS**									
4000	Rigid plastic, corrosive fume resistant PVC									
4020	Straight, 6" diameter	Q-9	220	.073	L.F.	12.30	3.75		16.05	19.25
4040	8" diameter		160	.100		17.80	5.15		22.95	27.50
4060	10" diameter		120	.133		22.50	6.85		29.35	35
4070	12" diameter		100	.160		26	8.25		34.25	41
4080	14" diameter		70	.229		32.50	11.80		44.30	54
4090	16" diameter		62	.258		37.50	13.30		50.80	62
4100	18" diameter		58	.276		51.50	14.20		65.70	78.50
4110	20" diameter	Q-10	75	.320		58.50	17.10		75.60	90
4130	24" diameter	"	55	.436		66	23.50		89.50	108
4250	Coupling, 6" diameter	Q-9	88	.182	Ea.	25	9.35		34.35	42
4270	8" diameter		78	.205		32.50	10.55		43.05	52
4290	10" diameter		70	.229		37.50	11.80		49.30	59.50
4300	12" diameter		55	.291		41	15		56	68
4310	14" diameter		44	.364		53	18.75		71.75	86.50
4320	16" diameter		40	.400		56	20.50		76.50	93
4330	18" diameter		38	.421		72.50	21.50		94	113
4340	20" diameter	Q-10	55	.436		87	23.50		110.50	131
4360	24" diameter	"	45	.533		109	28.50		137.50	164
4470	Elbow, 90°, 6" diameter	Q-9	44	.364		114	18.75		132.75	155
4490	8" diameter		28	.571		128	29.50		157.50	186
4510	10" diameter		18	.889		156	46		202	241
4520	12" diameter		15	1.067		177	55		232	279
4530	14" diameter		11	1.455		205	75		280	340
4540	16" diameter		10	1.600		267	82.50		349.50	420
4550	18" diameter	Q-10	15	1.600		370	85.50		455.50	535
4560	20" diameter		14	1.714		550	91.50		641.50	745
4580	24" diameter		12	2		675	107		782	910
4750	Elbow 45°, use 90° and deduct					25%				

For customer support on your Mechanical Cost Data, call 877.739.3689.

321

23 31 16.21 Polypropylene Ducts	Crew	Daily Output	Labor-Hours	Unit	Material	2016 Bare Costs Labor	2016 Bare Costs Equipment	Total	Total Incl O&P
0010 POLYPROPYLENE DUCTS									
0020 Rigid plastic, corrosive fume resistant,flame retardant, PP									
0110 Straight, 4" diameter	Q-9	200	.080	L.F.	7.55	4.12		11.67	14.60
0120 6" diameter		146	.110		10.75	5.65		16.40	20.50
0140 8" diameter		106	.151		13.75	7.80		21.55	27
0160 10" diameter		85	.188		20.50	9.70		30.20	37.50
0170 12" diameter		68	.235		39	12.10		51.10	61.50
0180 14" diameter		50	.320		44	16.50		60.50	73.50
0190 16" diameter		40	.400		59.50	20.50		80	97
0200 18" diameter		31	.516		97	26.50		123.50	147
0210 20" diameter	Q-10	44	.545		123	29		152	180
0220 22" diameter		40	.600		137	32		169	200
0230 24" diameter		37	.649		186	34.50		220.50	258
0250 28" diameter		28	.857		250	46		296	345
0270 32" diameter		24	1		283	53.50		336.50	390
0400 Coupling									
0410 4" diameter	Q-9	23	.696	Ea.	21	36		57	78
0420 6" diameter		16	1		21.50	51.50		73	103
0430 8" diameter		11.40	1.404		31	72.50		103.50	146
0440 10" diameter		9	1.778		41	91.50		132.50	185
0450 12" diameter		8	2		48	103		151	211
0460 14" diameter		6.60	2.424		93.50	125		218.50	294
0470 16" diameter		5.70	2.807		103	145		248	335
0480 18" diameter		5.20	3.077		123	159		282	380
0490 20" diameter	Q-10	6.80	3.529		134	189		323	435
0500 22" diameter		6.20	3.871		340	207		547	690
0510 24" diameter		5.80	4.138		435	221		656	820
0520 28" diameter		5.50	4.364		440	233		673	840
0530 32" diameter		5.20	4.615		490	247		737	915
0800 Elbow, 90°									
0810 4" diameter	Q-9	22	.727	Ea.	35	37.50		72.50	96
0820 6" diameter		15.40	1.039		49.50	53.50		103	137
0830 8" diameter		11	1.455		60	75		135	181
0840 10" diameter		8.60	1.860		86.50	96		182.50	242
0850 12" diameter		7.70	2.078		225	107		332	410
0860 14" diameter		6.40	2.500		260	129		389	485
0870 16" diameter		5.40	2.963		273	153		426	535
0880 18" diameter	Q-10	7.50	3.200		705	171		876	1,050
0890 20" diameter		6.60	3.636		775	194		969	1,150
0896 22" diameter		6.10	3.934		1,225	210		1,435	1,675
0900 24" diameter		5.60	4.286		1,100	229		1,329	1,575
0910 28" diameter		5.40	4.444		2,200	237		2,437	2,775
0920 32" diameter		5.20	4.615		2,850	247		3,097	3,525
1000 Elbow, 45°									
1020 4" diameter	Q-9	22	.727	Ea.	23	37.50		60.50	83
1030 6" diameter		15.40	1.039		41.50	53.50		95	128
1040 8" diameter		11	1.455		48	75		123	168
1050 10" diameter		8.60	1.860		65	96		161	219
1060 12" diameter		7.70	2.078		174	107		281	355
1070 14" diameter		6.40	2.500		194	129		323	410
1080 16" diameter		5.40	2.963		226	153		379	480
1090 18" diameter	Q-10	7.50	3.200		445	171		616	750

23 31 HVAC Ducts and Casings

23 31 16 – Nonmetal Ducts

23 31 16.21 Polypropylene Ducts

		Crew	Daily Output	Labor-Hours	Unit	Material	2016 Bare Costs Labor	Equipment	Total	Total Incl O&P
1100	20" diameter	Q-10	6.60	3.636	Ea.	575	194		769	930
1106	22" diameter		6.10	3.934		800	210		1,010	1,200
1110	24" diameter		5.60	4.286		760	229		989	1,175
1120	28" diameter		5.40	4.444		1,475	237		1,712	1,975
1130	32" diameter		5.20	4.615		1,850	247		2,097	2,400
1200	Tee									
1210	4" diameter	Q-9	16.50	.970	Ea.	150	50		200	242
1220	6" diameter		11.60	1.379		183	71		254	310
1230	8" diameter		8.30	1.928		228	99.50		327.50	405
1240	10" diameter		6.50	2.462		315	127		442	540
1250	12" diameter		5.80	2.759		415	142		557	670
1260	14" diameter		4.80	3.333		540	172		712	860
1270	16" diameter		4.05	3.951		600	204		804	965
1280	18" diameter	Q-10	5.60	4.286		790	229		1,019	1,225
1290	20" diameter		4.90	4.898		855	262		1,117	1,350
1300	22" diameter		4.40	5.455		1,025	291		1,316	1,575
1310	24" diameter		4.20	5.714		1,300	305		1,605	1,900
1320	28" diameter		4.10	5.854		2,300	315		2,615	3,000
1330	32" diameter		4	6		2,650	320		2,970	3,425

23 33 Air Duct Accessories

23 33 13 – Dampers

23 33 13.13 Volume-Control Dampers

		Crew	Daily Output	Labor-Hours	Unit	Material	2016 Bare Costs Labor	Equipment	Total	Total Incl O&P
0010	**VOLUME-CONTROL DAMPERS**									
5990	Multi-blade dampers, opposed blade, 8" x 6"	1 Shee	24	.333	Ea.	21.50	19.10		40.60	53
5994	8" x 8"		22	.364		22.50	21		43.50	56.50
5996	10" x 10"		21	.381		26.50	22		48.50	63
6000	12" x 12"		21	.381		29.50	22		51.50	66
6020	12" x 18"		18	.444		39.50	25.50		65	82.50
6030	14" x 10"		20	.400		29	23		52	66.50
6031	14" x 14"		17	.471		35.50	27		62.50	80
6032	16" x 10"		18	.444		31	25.50		56.50	73
6033	16" x 12"		17	.471		35.50	27		62.50	80
6035	16" x 16"		16	.500		44	28.50		72.50	92.50
6036	18" x 14"		16	.500		44	28.50		72.50	92.50
6037	18" x 16"		15	.533		48	30.50		78.50	99.50
6038	18" x 18"		15	.533		52	30.50		82.50	104
6040	18" x 24"		12	.667		67	38		105	132
6060	18" x 28"		10	.800		78	46		124	156
6068	20" x 6"		18	.444		28	25.50		53.50	70
6069	20" x 8"		16	.500		32.50	28.50		61	79.50
6070	20" x 16"		14	.571		52	32.50		84.50	107
6071	20" x 18"		13	.615		56	35		91	116
6072	20" x 20"		13	.615		62.50	35		97.50	123
6073	22" x 6"		20	.400		29	23		52	66.50
6074	22" x 18"		14	.571		62.50	32.50		95	119
6075	22" x 20"		12	.667		67.50	38		105.50	133
6076	24" x 16"		11	.727		61	41.50		102.50	131
6077	22" x 22"		10	.800		72.50	46		118.50	150
6078	24" x 20"		8	1		72.50	57.50		130	168
6080	24" x 24"		8	1		85	57.50		142.50	181

For customer support on your Mechanical Cost Data, call 877.739.3689.

323

23 33 13 – Dampers

23 33 13.13 Volume-Control Dampers		Crew	Daily Output	Labor-Hours	Unit	Material	2016 Bare Costs Labor	Equipment	Total	Total Incl O&P
6100	24" x 28"	1 Shee	6	1.333	Ea.	98	76.50		174.50	225
6110	26" x 26"	↓	6	1.333		95	76.50		171.50	222
6120	28" x 28"	Q-9	11	1.455		107	75		182	232
6130	30" x 18"		10	1.600		84	82.50		166.50	219
6132	30" x 24"		7	2.286		107	118		225	297
6133	30" x 30"		6.60	2.424		137	125		262	340
6135	32" x 32"		6.40	2.500		154	129		283	365
6150	34" x 34"		6	2.667		172	137		309	400
6151	36" x 12"		10	1.600		70	82.50		152.50	203
6152	36" x 16"		8	2		96	103		199	263
6156	36" x 32"		6.30	2.540		172	131		303	390
6157	36" x 34"		6.20	2.581		181	133		314	400
6158	36" x 36"		6	2.667		189	137		326	420
6160	44" x 28"		5.80	2.759		194	142		336	430
6180	48" x 36"		5.60	2.857		253	147		400	505
6200	56" x 36"		5.40	2.963		310	153		463	575
6220	60" x 36"	↓	5.20	3.077		330	159		489	605
6240	60" x 44"	↓	5	3.200	↓	390	165		555	680
7500	Variable volume modulating motorized damper, incl. elect. mtr.									
7504	8" x 6"	1 Shee	15	.533	Ea.	118	30.50		148.50	177
7506	10" x 6"		14	.571		118	32.50		150.50	180
7510	10" x 10"		13	.615		122	35		157	188
7520	12" x 12"		12	.667		127	38		165	198
7522	12" x 16"		11	.727		127	41.50		168.50	204
7524	16" x 10"		12	.667		125	38		163	196
7526	16" x 14"		10	.800		130	46		176	213
7528	16" x 18"		9	.889		135	51		186	226
7540	18" x 12"		10	.800		130	46		176	213
7542	18" x 18"		8	1		138	57.50		195.50	239
7544	20" x 14"		8	1		141	57.50		198.50	243
7546	20" x 18"		7	1.143		147	65.50		212.50	262
7560	24" x 12"		8	1		143	57.50		200.50	245
7562	24" x 18"		7	1.143		154	65.50		219.50	269
7568	28" x 10"		7	1.143		143	65.50		208.50	257
7580	28" x 16"		6	1.333		163	76.50		239.50	296
7590	30" x 14"		5	1.600		151	91.50		242.50	305
7600	30" x 18"		4	2		198	115		313	395
7610	30" x 24"	↓	3.80	2.105		264	121		385	475
7690	48" x 48"	Q-9	6	2.667		895	137		1,032	1,200
7694	6' x 14' w/2 motors	"	3	5.333		3,225	275		3,500	3,975
7700	For thermostat, add	1 Shee	8	1		31	57.50		88.50	122
7800	For transformer 40 VA capacity, add	"	16	.500	↓	19.45	28.50		47.95	65.50
8000	Multi-blade dampers, parallel blade									
8100	8" x 8"	1 Shee	24	.333	Ea.	82	19.10		101.10	119
8120	12" x 8"		22	.364		82	21		103	122
8140	16" x 10"		20	.400		103	23		126	149
8160	18" x 12"		18	.444		113	25.50		138.50	163
8180	22" x 12"		15	.533		122	30.50		152.50	181
8200	24" x 16"		11	.727		134	41.50		175.50	212
8220	28" x 16"		10	.800		150	46		196	235
8240	30" x 16"		8	1		152	57.50		209.50	256
8260	30" x 18"	↓	7	1.143	↓	186	65.50		251.50	305
8400	Round damper, butterfly, vol. control w/lever lock reg.									

23 33 13 – Dampers

23 33 13.13 Volume-Control Dampers		Crew	Daily Output	Labor-Hours	Unit	Material	2016 Bare Costs Labor	Equipment	Total	Total Incl O&P
8410	6" diam.	1 Shee	22	.364	Ea.	25	21		46	59.50
8412	7" diam.		21	.381		25.50	22		47.50	62
8414	8" diam.		20	.400		27.50	23		50.50	65
8416	9" diam.		19	.421		28	24		52	68
8418	10" diam.		18	.444		30	25.50		55.50	72.50
8420	12" diam.		16	.500		31	28.50		59.50	78
8422	14" diam.		14	.571		41	32.50		73.50	95
8424	16" diam.		13	.615		44	35		79	103
8426	18" diam.		12	.667		54	38		92	118
8428	20" diam.		11	.727		65.50	41.50		107	136
8430	24" diam.		10	.800		81.50	46		127.50	160
8432	30" diam.		9	.889		114	51		165	204
8434	36" diam.	▼	8	1	▼	152	57.50		209.50	255
8500	Round motor operated damper									
8510	6" diam.	1 Shee	22	.364	Ea.	61	21		82	99.50
8512	7" diam.		21	.381		65	22		87	105
8514	8" diam.		20	.400		67	23		90	109
8516	10" diam.		18	.444		74	25.50		99.50	121
8518	12" diam.		16	.500		79	28.50		107.50	131
8520	14" diam.	▼	14	.571	▼	90	32.50		122.50	149

23 33 13.16 Fire Dampers

		Crew	Daily Output	Labor-Hours	Unit	Material	Labor	Equipment	Total	Total Incl O&P
0010	**FIRE DAMPERS**									
3000	Fire damper, curtain type, 1-1/2 hr. rated, vertical, 6" x 6"	1 Shee	24	.333	Ea.	24	19.10		43.10	55.50
3020	8" x 6"		22	.364		24	21		45	58.50
3021	8" x 8"		22	.364		24	21		45	58.50
3030	10" x 10"		22	.364		24	21		45	58.50
3040	12" x 6"		22	.364		24	21		45	58.50
3044	12" x 12"		20	.400		28.50	23		51.50	66.50
3060	20" x 6"		18	.444		32	25.50		57.50	74
3080	12" x 8"		22	.364		24	21		45	58.50
3100	24" x 8"		16	.500		36.50	28.50		65	84
3120	12" x 10"		21	.381		24	22		46	60
3140	24" x 10"		15	.533		39	30.50		69.50	89.50
3160	36" x 10"		12	.667		50.50	38		88.50	114
3180	16" x 12"		20	.400		33	23		56	71.50
3200	24" x 12"		13	.615		41.50	35		76.50	99.50
3220	48" x 12"		10	.800		64	46		110	140
3238	14" x 14"		19	.421		36	24		60	76.50
3240	16" x 14"		18	.444		44	25.50		69.50	87.50
3260	24" x 14"		12	.667		53	38		91	117
3280	30" x 14"		11	.727		50.50	41.50		92	119
3298	16" x 16"		18	.444		40	25.50		65.50	83
3300	18" x 16"		17	.471		43	27		70	88
3320	24" x 16"		11	.727		46	41.50		87.50	115
3340	36" x 16"		10	.800		58.50	46		104.50	134
3356	18" x 18"		16	.500		44	28.50		72.50	92.50
3360	24" x 18"		10	.800		54	46		100	129
3380	48" x 18"		8	1		77.50	57.50		135	173
3398	20" x 20"		10	.800		47.50	46		93.50	122
3400	24" x 20"		8	1		54	57.50		111.50	147
3420	36" x 20"		7	1.143		65	65.50		130.50	172
3440	24" x 22"	▼	7	1.143	▼	55.50	65.50		121	161

For customer support on your Mechanical Cost Data, call 877.739.3689.

325

23 33 13.16 Fire Dampers		Crew	Daily Output	Labor-Hours	Unit	Material	2016 Bare Costs Labor	Equipment	Total	Total Incl O&P
3460	30" x 22"	1 Shee	6	1.333	Ea.	60.50	76.50		137	184
3478	24" x 24"		8	1		56.50	57.50		114	150
3480	26" x 24"	↓	7	1.143		56.50	65.50		122	162
3484	32" x 24"	Q-9	13	1.231		67	63.50		130.50	171
3500	48" x 24"		12	1.333		87	68.50		155.50	201
3520	28" x 26"		13	1.231		63.50	63.50		127	167
3540	30" x 28"		12	1.333		70	68.50		138.50	182
3560	48" x 30"		11	1.455		102	75		177	227
3580	48" x 36"		10	1.600		136	82.50		218.50	275
3590	32" x 40"		10	1.600		105	82.50		187.50	242
3600	44" x 44"		10	1.600		144	82.50		226.50	285
3620	48" x 48"	↓	8	2	↓	161	103		264	335
3700	U.L. label included in above									
3800	For horizontal operation, add				Ea.	20%				
3900	For cap for blades out of air stream, add					20%				
4000	For 10" 22 ga., U.L. approved sleeve, add					35%				
4100	For 10" 22 ga., U.L. sleeve, 100% free area, add				↓	65%				
4200	For oversize openings group dampers									
4210	Fire damper, round, Type A, 1-1/2 hour									
4214	4" diam.	1 Shee	20	.400	Ea.	61.50	23		84.50	103
4216	5" diam.		16	.500		50	28.50		78.50	99
4218	6" diam.		14.30	.559		50	32		82	104
4220	7" diam.		13.30	.602		50	34.50		84.50	108
4222	8" diam.		11.40	.702		50	40		90	117
4224	9" diam.		10.80	.741		55.50	42.50		98	126
4226	10" diam.		10	.800		55.50	46		101.50	131
4228	12" diam.		9.50	.842		62.50	48		110.50	143
4230	14" diam.		8.20	.976		77.50	56		133.50	171
4232	16" diam.		7.60	1.053		77.50	60.50		138	178
4234	18" diam.		7.20	1.111		94	63.50		157.50	200
4236	20" diam.		6.40	1.250		97	71.50		168.50	215
4238	22" diam.		6.10	1.311		106	75		181	232
4240	24" diam.		5.70	1.404		111	80.50		191.50	245
4242	26" diam.		5.50	1.455		125	83.50		208.50	265
4244	28" diam.		5.40	1.481		131	85		216	274
4246	30" diam.	↓	5.30	1.509	↓	138	86.50		224.50	283
4502	Power open-spring close, sleeve and actuator motor mounted, 1-1/2 hour									
4506	6" x 6"	1 Shee	24	.333	Ea.	60.50	19.10		79.60	95.50
4510	8" x 8"		22	.364		214	21		235	267
4520	16" x 8"		20	.400		340	23		363	405
4540	18" x 8"		18	.444		345	25.50		370.50	420
4560	20" x 8"		16	.500		345	28.50		373.50	425
4580	10" x 10"		21	.381		310	22		332	375
4600	24" x 10"		15	.533		360	30.50		390.50	440
4620	30" x 10"		12	.667		370	38		408	470
4640	12" x 12"		20	.400		340	23		363	410
4660	18" x 12"		18	.444		355	25.50		380.50	430
4680	24" x 12"		13	.615		365	35		400	455
4700	30" x 12"		11	.727		375	41.50		416.50	480
4720	14" x 14"		17	.471		365	27		392	440
4740	16" x 14"		18	.444		365	25.50		390.50	445
4760	20" x 14"		14	.571		370	32.50		402.50	460
4780	24" x 14"		12	.667		375	38		413	475

For customer support on your Mechanical Cost Data, call 877.739.3689.

23 33 13 – Dampers

	23 33 13.16 Fire Dampers	Crew	Daily Output	Labor-Hours	Unit	Material	2016 Bare Costs Labor	2016 Bare Costs Equipment	Total	Total Incl O&P
4800	30" x 14"	1 Shee	10	.800	Ea.	400	46		446	510
4820	16" x 16"		16	.500		375	28.50		403.50	460
4840	20" x 16"		14	.571		380	32.50		412.50	465
4860	24" x 16"		11	.727		385	41.50		426.50	490
4880	30" x 16"		8	1		415	57.50		472.50	545
4900	18" x 18"		15	.533		400	30.50		430.50	480
5000	24" x 18"		10	.800		405	46		451	515
5020	36" x 18"		7	1.143		440	65.50		505.50	585
5040	20" x 20"		13	.615		405	35		440	500
5060	24" x 20"		8	1		410	57.50		467.50	540
5080	30" x 20"		8	1		430	57.50		487.50	560
5100	36" x 20"		7	1.143		445	65.50		510.50	590
5120	24" x 24"		8	1		430	57.50		487.50	560
5130	30" x 24"	▼	7.60	1.053		430	60.50		490.50	565
5140	36" x 24"	Q-9	12	1.333		475	68.50		543.50	625
5141	36" x 30"		10	1.600		540	82.50		622.50	720
5142	40" x 36"		8	2		890	103		993	1,150
5143	48" x 48"	▼	4	4		1,725	206		1,931	2,225
5150	Damper operator motor, 24 or 120 volt	1 Shee	16	.500	▼	540	28.50		568.50	635

23 33 13.25 Exhaust Vent Damper

		Crew	Daily Output	Labor-Hours	Unit	Material	Labor	Equipment	Total	Total Incl O&P
0010	**EXHAUST VENT DAMPER**, Auto, OB with elect. actuator.									
1110	6" x 6"	1 Shee	24	.333	Ea.	64	19.10		83.10	99.50
1114	8" x 6"		23	.348		65.50	19.90		85.40	103
1116	8" x 8"		22	.364		67.50	21		88.50	107
1118	10" x 6"		23	.348		67.50	19.90		87.40	105
1120	10" x 10"		22	.364		72.50	21		93.50	112
1122	12" x 6"		22	.364		68.50	21		89.50	107
1124	12" x 10"		21	.381		76.50	22		98.50	118
1126	12" x 12"		20	.400		78.50	23		101.50	122
1128	14" x 6"		21	.381		70.50	22		92.50	111
1130	14" x 10"		19	.421		80	24		104	125
1132	16" x 6"		21	.381		71.50	22		93.50	112
1134	16" x 8"		20	.400		79	23		102	122
1136	18" x 6"		20	.400		75.50	23		98.50	118
1138	18" x 8"	▼	19	.421	▼	80	24		104	125

23 33 13.28 Splitter Damper Assembly

		Crew	Daily Output	Labor-Hours	Unit	Material	Labor	Equipment	Total	Total Incl O&P
0010	**SPLITTER DAMPER ASSEMBLY**									
7000	Self locking, 1' rod	1 Shee	24	.333	Ea.	21	19.10		40.10	52
7020	3' rod		22	.364		30	21		51	65
7040	4' rod		20	.400		33.50	23		56.50	72
7060	6' rod	▼	18	.444	▼	41	25.50		66.50	84

23 33 13.32 Relief Damper

		Crew	Daily Output	Labor-Hours	Unit	Material	Labor	Equipment	Total	Total Incl O&P
0010	**RELIEF DAMPER**, Electronic bypass with tight seal									
8310	8" x 6"	1 Shee	22	.364	Ea.	176	21		197	226
8314	10" x 6"		22	.364		176	21		197	226
8318	10" x 10"		21	.381		183	22		205	235
8322	12" x 12"		20	.400		187	23		210	241
8326	12" x 16"		19	.421		194	24		218	251
8330	16" x 10"		20	.400		204	23		227	260
8334	16" x 14"		18	.444		209	25.50		234.50	269
8338	16" x 18"		17	.471		228	27		255	291
8342	18" x 12"		18	.444		228	25.50		253.50	289

For customer support on your Mechanical Cost Data, call 877.739.3689.

327

23 33 Air Duct Accessories

23 33 13 – Dampers

23 33 13.32 Relief Damper		Crew	Daily Output	Labor-Hours	Unit	Material	2016 Bare Costs Labor	Equipment	Total	Total Incl O&P
8346	18" x 18"	1 Shee	15	.533	Ea.	209	30.50		239.50	277
8350	20" x 14"		14	.571		212	32.50		244.50	284
8354	20" x 18"		13	.615		220	35		255	296
8358	24" x 12"		13	.615		217	35		252	292
8362	24" x 18"		10	.800		226	46		272	320
8363	24" x 24"		10	.800		235	46		281	330
8364	24" x 36"		9	.889		280	51		331	390
8365	24" x 48"		6	1.333		315	76.50		391.50	460
8366	28" x 10"		13	.615		155	35		190	224
8370	28" x 16"		10	.800		239	46		285	335
8374	30" x 14"		11	.727		163	41.50		204.50	243
8378	30" x 18"		8	1		246	57.50		303.50	360
8382	30" x 24"		6	1.333		261	76.50		337.50	405
8390	46" x 36"		4	2		345	115		460	555
8394	48" x 48"		3	2.667		380	153		533	650
8396	54" x 36"		2	4		400	229		629	790

23 33 19 – Duct Silencers

23 33 19.10 Duct Silencers

		Crew	Daily Output	Labor-Hours	Unit	Material	2016 Bare Costs Labor	Equipment	Total	Total Incl O&P
0010	**DUCT SILENCERS**									
9000	Silencers, noise control for air flow, duct				MCFM	57			57	63
9004	Duct sound trap, packaged									
9010	12" x 18" x 36", 2000 CFM	1 Shee	12	.667	Ea.	470	38		508	575
9011	24" x 18" x 36", 9000 CFM	Q-9	80	.200		540	10.30		550.30	605
9012	24" x 24" x 36", 9000 CFM		60	.267		805	13.75		818.75	910
9013	24" x 30" x 36", 9000 CFM		50	.320		860	16.50		876.50	970
9014	24" x 36" x 36", 9000 CFM		40	.400		1,075	20.50		1,095.50	1,200
9015	24" x 48" x 36", 9000 CFM		28	.571		1,400	29.50		1,429.50	1,600
9018	36" x 18" x 36", 6300 CFM		42	.381		1,075	19.65		1,094.65	1,200
9019	36" x 36" x 36", 6300 CFM		21	.762		2,150	39.50		2,189.50	2,425
9020	36" x 48" x 36", 6300 CFM		14	1.143		2,800	59		2,859	3,175
9021	36" x 60" x 36", 6300 CFM		13	1.231		3,550	63.50		3,613.50	4,000
9022	48" x 48" x 36", 6300 CFM		12	1.333		3,775	68.50		3,843.50	4,250
9023	24" x 24" x 60", 6000 CFM		3.60	4.444		1,250	229		1,479	1,725
9026	24" x 30" x 60", 7000 CFM		3.50	4.571		1,400	236		1,636	1,875
9030	24" x 36" x 60", 8000 CFM		3.40	4.706		1,525	242		1,767	2,050
9034	24" x 48" x 60", 11,000 CFM		3.30	4.848		1,300	250		1,550	1,825
9038	36" x 18" x 60", 6300 CFM		3.50	4.571		1,025	236		1,261	1,475
9042	36" x 36" x 60", 12,000 CFM		3.30	4.848		1,650	250		1,900	2,200
9046	36" x 48" x 60", 17,000 CFM		3.10	5.161		2,175	266		2,441	2,800
9050	36" x 60" x 60", 20,000 CFM		2.80	5.714		2,650	294		2,944	3,375
9054	48" x 48" x 60", 23,000 CFM		2.60	6.154		3,050	315		3,365	3,850
9056	48" x 60" x 60", 23,000 CFM		2.10	7.619		3,825	395		4,220	4,800

23 33 23 – Turning Vanes

23 33 23.13 Air Turning Vanes

		Crew	Daily Output	Labor-Hours	Unit	Material	2016 Bare Costs Labor	Equipment	Total	Total Incl O&P
0010	**AIR TURNING VANES**									
9400	Turning vane components									
9410	Turning vane rail	1 Shee	160	.050	L.F.	.75	2.86		3.61	5.20
9420	Double thick, factory fab. vane		300	.027		1.17	1.53		2.70	3.62
9428	12" high set		170	.047		1.68	2.69		4.37	5.95
9432	14" high set		160	.050		1.96	2.86		4.82	6.55
9434	16" high set		150	.053		2.24	3.05		5.29	7.15
9436	18" high set		144	.056		2.52	3.18		5.70	7.65

23 33 Air Duct Accessories

23 33 23 – Turning Vanes

23 33 23.13 Air Turning Vanes	Crew	Daily Output	Labor-Hours	Unit	Material	2016 Bare Costs Labor	Equipment	Total	Total Incl O&P
9438	20" high set	1 Shee	138	.058	L.F.	2.80	3.32		6.12	8.15
9440	22" high set		130	.062		3.08	3.52		6.60	8.80
9442	24" high set		124	.065		3.36	3.69		7.05	9.35
9444	26" high set		116	.069		3.64	3.95		7.59	10.05
9446	30" high set	↓	112	.071	↓	4.20	4.09		8.29	10.85

23 33 33 – Duct-Mounting Access Doors

23 33 33.13 Duct Access Doors

		Crew	Daily Output	Labor-Hours	Unit	Material	2016 Bare Costs Labor	Equipment	Total	Total Incl O&P
0010	**DUCT ACCESS DOORS**									
1000	Duct access door, insulated, 6" x 6"	1 Shee	14	.571	Ea.	16.05	32.50		48.55	67.50
1020	10" x 10"		11	.727		18.45	41.50		59.95	84
1040	12" x 12"		10	.800		20	46		66	92
1050	12" x 18"		9	.889		37	51		88	119
1060	16" x 12"		9	.889		29.50	51		80.50	111
1070	18" x 18"		8	1		32.50	57.50		90	123
1074	24" x 18"		8	1		44	57.50		101.50	136
1080	24" x 24"	↓	8	1	↓	47	57.50		104.50	140

23 33 46 – Flexible Ducts

23 33 46.10 Flexible Air Ducts

		Crew	Daily Output	Labor-Hours	Unit	Material	2016 Bare Costs Labor	Equipment	Total	Total Incl O&P
0010	**FLEXIBLE AIR DUCTS**									
1280	Add to labor for elevated installation									
1282	of prefabricated (purchased) ductwork									
1283	10' to 15' high						10%			
1284	15' to 20' high						20%			
1285	20' to 25' high						25%			
1286	25' to 30' high						35%			
1287	30' to 35' high						40%			
1288	35' to 40' high						50%			
1289	Over 40' high						55%			
1300	Flexible, coated fiberglass fabric on corr. resist. metal helix									
1400	pressure to 12" (WG) UL-181									
1500	Noninsulated, 3" diameter	Q-9	400	.040	L.F.	1.12	2.06		3.18	4.38
1520	4" diameter		360	.044		1.16	2.29		3.45	4.78
1540	5" diameter		320	.050		1.30	2.58		3.88	5.35
1560	6" diameter		280	.057		1.50	2.94		4.44	6.15
1580	7" diameter		240	.067		1.53	3.44		4.97	6.95
1600	8" diameter		200	.080		1.91	4.12		6.03	8.40
1620	9" diameter		180	.089		2.05	4.58		6.63	9.25
1640	10" diameter		160	.100		2.46	5.15		7.61	10.60
1660	12" diameter		120	.133		2.94	6.85		9.79	13.75
1680	14" diameter		80	.200		3.56	10.30		13.86	19.65
1700	16" diameter	↓	60	.267	↓	5.05	13.75		18.80	26.50
1800	For plastic cable tie, add				Ea.	.77			.77	.85
1900	Insulated, 1" thick, PE jacket, 3" diameter [G]	Q-9	380	.042	L.F.	2.60	2.17		4.77	6.20
1910	4" diameter [G]		340	.047		2.60	2.43		5.03	6.55
1920	5" diameter [G]		300	.053		2.60	2.75		5.35	7.05
1940	6" diameter [G]		260	.062		2.94	3.17		6.11	8.10
1960	7" diameter [G]		220	.073		3.20	3.75		6.95	9.25
1980	8" diameter [G]		180	.089		3.49	4.58		8.07	10.85
2000	9" diameter [G]		160	.100		3.81	5.15		8.96	12.10
2020	10" diameter [G]		140	.114		4.25	5.90		10.15	13.70
2040	12" diameter [G]		100	.160		4.90	8.25		13.15	18
2060	14" diameter [G]	↓	80	.200	↓	6.05	10.30		16.35	22.50

For customer support on your Mechanical Cost Data, call 877.739.3689.

329

23 33 46.10 Flexible Air Ducts		Crew	Daily Output	Labor-Hours	Unit	Material	2016 Bare Costs Labor	Equipment	Total	Total Incl O&P
2080	16" diameter	G Q-9	60	.267	L.F.	8.35	13.75		22.10	30
2100	18" diameter	G	45	.356		9.85	18.30		28.15	39
2120	20" diameter	G Q-10	65	.369		10.85	19.75		30.60	42
2500	Insulated, heavy duty, coated fiberglass fabric									
2520	4" diameter	G Q-9	340	.047	L.F.	3.36	2.43		5.79	7.40
2540	5" diameter	G	300	.053		3.54	2.75		6.29	8.10
2560	6" diameter	G	260	.062		4.20	3.17		7.37	9.45
2580	7" diameter	G	220	.073		4.58	3.75		8.33	10.80
2600	8" diameter	G	180	.089		5.10	4.58		9.68	12.65
2620	9" diameter	G	160	.100		5.55	5.15		10.70	14.05
2640	10" diameter	G	140	.114		6.40	5.90		12.30	16.05
2660	12" diameter	G	100	.160		7.50	8.25		15.75	21
2680	14" diameter	G	80	.200		8.90	10.30		19.20	25.50
2700	16" diameter	G	60	.267		12.20	13.75		25.95	34.50
2720	18" diameter	G	45	.356		14.45	18.30		32.75	44
2800	Flexible, aluminum, pressure to 12" (WG) UL-181									
2820	Non-insulated									
2830	3" diameter	Q-9	400	.040	L.F.	1.16	2.06		3.22	4.43
2831	4" diameter		360	.044		1.27	2.29		3.56	4.90
2832	5" diameter		320	.050		1.60	2.58		4.18	5.70
2833	6" diameter		280	.057		2.04	2.94		4.98	6.75
2834	7" diameter		240	.067		2.42	3.44		5.86	7.90
2835	8" diameter		200	.080		2.75	4.12		6.87	9.35
2836	9" diameter		180	.089		3.25	4.58		7.83	10.60
2837	10" diameter		160	.100		3.85	5.15		9	12.15
2838	12" diameter		120	.133		5.05	6.85		11.90	16.05
2839	14" diameter		80	.200		5.70	10.30		16	22
2840	15" diameter		70	.229		7.70	11.80		19.50	26.50
2841	16" diameter		60	.267		8.90	13.75		22.65	31
2842	18" diameter		45	.356		10.10	18.30		28.40	39
2843	20" diameter	Q-10	65	.369		11.65	19.75		31.40	43
2880	Insulated, 1" thick with 3/4 lb., PE jacket									
2890	3" diameter	G Q-9	380	.042	L.F.	3.14	2.17		5.31	6.75
2891	4" diameter	G	340	.047		3.14	2.43		5.57	7.15
2892	5" diameter	G	300	.053		3.58	2.75		6.33	8.15
2893	6" diameter	G	260	.062		4.24	3.17		7.41	9.50
2894	7" diameter	G	220	.073		5.10	3.75		8.85	11.40
2895	8" diameter	G	180	.089		5.25	4.58		9.83	12.75
2896	9" diameter	G	160	.100		6.05	5.15		11.20	14.55
2897	10" diameter	G	140	.114		7.10	5.90		13	16.80
2898	12" diameter	G	100	.160		8.35	8.25		16.60	22
2899	14" diameter	G	80	.200		9.45	10.30		19.75	26
2900	15" diameter	G	70	.229		11.65	11.80		23.45	31
2901	16" diameter	G	60	.267		13.20	13.75		26.95	35.50
2902	18" diameter	G	45	.356		15.50	18.30		33.80	45
3000	Transitions, 3" deep, square to round									
3010	6" x 6" to round	Q-9	44	.364	Ea.	9.90	18.75		28.65	39.50
3014	6" x 12" to round		42	.381		17.60	19.65		37.25	49.50
3018	8" x 8" to round		40	.400		11.55	20.50		32.05	44
3022	9" x 9" to round		38	.421		11.55	21.50		33.05	45.50
3026	9" x 12" to round		37	.432		22.50	22.50		45	59
3030	9" x 15" to round		36	.444		26.50	23		49.50	64
3034	10" x 10" to round		34	.471		13.75	24.50		38.25	52

23 33 46 – Flexible Ducts

23 33 46.10 Flexible Air Ducts		Crew	Daily Output	Labor-Hours	Unit	Material	2016 Bare Costs Labor	2016 Bare Costs Equipment	Total	Total Incl O&P
3038	10" x 22" to round	Q-9	32	.500	Ea.	32.50	26		58.50	75
3042	12" x 12" to round		30	.533		13.75	27.50		41.25	57
3046	12" x 18" to round		28	.571		32.50	29.50		62	80.50
3050	12" x 24" to round		26	.615		32.50	31.50		64	84
3054	15" x 15" to round		25	.640		18.70	33		51.70	71
3058	15" x 18" to round		24	.667		35	34.50		69.50	91
3062	18" x 18" to round		22	.727		26.50	37.50		64	86.50
3066	21" x 21" to round		21	.762		29	39.50		68.50	92
3070	22" x 22" to round		20	.800		29	41		70	95
3074	24" x 24" to round	↓	18	.889	↓	33	46		79	107
5000	Flexible, aluminum, acoustical, pressure to 2" (wg), NFPA-90A									
5010	Fiberglass insulation 1-1/2" thick, 1/2 lb. density									
5020	Polyethylene jacket, UL approved									
5026	5" diameter [G]	Q-9	300	.053	L.F.	4.29	2.75		7.04	8.90
5030	6" diameter [G]		260	.062		5.05	3.17		8.22	10.40
5034	7" diameter [G]		220	.073		6.05	3.75		9.80	12.40
5038	8" diameter [G]		180	.089		6.15	4.58		10.73	13.80
5042	9" diameter [G]		160	.100		7.25	5.15		12.40	15.90
5046	10" diameter [G]		140	.114		9.45	5.90		15.35	19.40
5050	12" diameter [G]		100	.160		10.45	8.25		18.70	24
5054	14" diameter [G]		80	.200		11.20	10.30		21.50	28
5058	16" diameter [G]	↓	60	.267	↓	15.85	13.75		29.60	38.50
5072	Hospital grade, PE jacket, UL approved									
5076	5" diameter [G]	Q-9	300	.053	L.F.	4.62	2.75		7.37	9.30
5080	6" diameter [G]		260	.062		5.50	3.17		8.67	10.90
5084	7" diameter [G]		220	.073		6.60	3.75		10.35	13
5088	8" diameter [G]		180	.089		6.80	4.58		11.38	14.50
5092	9" diameter [G]		160	.100		7.90	5.15		13.05	16.60
5096	10" diameter [G]		140	.114		10.35	5.90		16.25	20.50
5100	12" diameter [G]		100	.160		11.45	8.25		19.70	25
5104	14" diameter [G]		80	.200		12.30	10.30		22.60	29.50
5108	16" diameter [G]	↓	60	.267	↓	17.25	13.75		31	40
5140	Flexible, aluminum, silencer, pressure to 12" (wg), NFPA-90A									
5144	Fiberglass insulation 1-1/2" thick, 1/2 lb. density									
5148	Aluminum outer shell, UL approved									
5152	5" diameter [G]	Q-9	290	.055	L.F.	11.85	2.84		14.69	17.35
5156	6" diameter [G]		250	.064		13	3.30		16.30	19.35
5160	7" diameter [G]		210	.076		13.75	3.93		17.68	21
5164	8" diameter [G]		170	.094		14.70	4.85		19.55	23.50
5168	9" diameter [G]		150	.107		17.80	5.50		23.30	28
5172	10" diameter [G]		130	.123		20	6.35		26.35	31.50
5176	12" diameter [G]		90	.178		22.50	9.15		31.65	38.50
5180	14" diameter [G]		70	.229		26.50	11.80		38.30	47
5184	16" diameter [G]	↓	50	.320	↓	34	16.50		50.50	62.50
5200	Hospital grade, aluminum shell, UL approved									
5204	5" diameter [G]	Q-9	280	.057	L.F.	12.25	2.94		15.19	18
5208	6" diameter [G]		240	.067		13.50	3.44		16.94	20
5212	7" diameter [G]		200	.080		14.30	4.12		18.42	22
5216	8" diameter [G]		160	.100		15.30	5.15		20.45	24.50
5220	9" diameter [G]		140	.114		18.55	5.90		24.45	29.50
5224	10" diameter [G]		130	.123		21	6.35		27.35	32.50
5228	12" diameter [G]		80	.200		24.50	10.30		34.80	42.50
5232	14" diameter [G]	↓	60	.267		27.50	13.75		41.25	51.50

23 33 46 – Flexible Ducts

23 33 46.10 Flexible Air Ducts		Crew	Daily Output	Labor-Hours	Unit	Material	2016 Bare Costs Labor	Equipment	Total	Total Incl O&P
5236	16" diameter	**G** Q-9	40	.400	L.F.	35.50	20.50		56	70.50

23 33 53 – Duct Liners

23 33 53.10 Duct Liner Board

			Crew	Daily Output	Labor-Hours	Unit	Material	Labor	Equipment	Total	Total Incl O&P
0010	**DUCT LINER BOARD**										
3340	Board type fiberglass liner, FSK, 1-1/2 lb. density										
3344	1" thick	**G**	Q-14	150	.107	S.F.	.62	5.15		5.77	8.70
3345	1-1/2" thick	**G**		130	.123		.68	5.90		6.58	9.95
3346	2" thick	**G**		120	.133		.79	6.40		7.19	10.80
3348	3" thick	**G**		110	.145		1.02	7		8.02	12
3350	4" thick	**G**		100	.160		1.25	7.70		8.95	13.35
3356	3 lb. density, 1" thick	**G**		150	.107		.79	5.15		5.94	8.85
3358	1-1/2" thick	**G**		130	.123		1	5.90		6.90	10.30
3360	2" thick	**G**		120	.133		1.22	6.40		7.62	11.30
3362	2-1/2" thick	**G**		110	.145		1.43	7		8.43	12.45
3364	3" thick	**G**		100	.160		1.64	7.70		9.34	13.75
3366	4" thick	**G**		90	.178		2.06	8.55		10.61	15.55
3370	6 lb. density, 1" thick	**G**		140	.114		1.12	5.50		6.62	9.80
3374	1-1/2" thick	**G**		120	.133		1.50	6.40		7.90	11.60
3378	2" thick	**G**		100	.160		1.88	7.70		9.58	14
3490	Board type, fiberglass liner, 3 lb. density										
3680	No finish										
3700	1" thick	**G**	Q-14	170	.094	S.F.	.44	4.52		4.96	7.55
3710	1-1/2" thick	**G**		140	.114		.66	5.50		6.16	9.30
3720	2" thick	**G**		130	.123		.88	5.90		6.78	10.15
3940	Board type, non-fibrous foam										
3950	Temperature, bacteria and fungi resistant										
3960	1" thick	**G**	Q-14	150	.107	S.F.	2.38	5.15		7.53	10.60
3970	1-1/2" thick	**G**		130	.123		3.28	5.90		9.18	12.80
3980	2" thick	**G**		120	.133		3.98	6.40		10.38	14.35

23 34 HVAC Fans

23 34 13 – Axial HVAC Fans

23 34 13.10 Axial Flow HVAC Fans

		Crew	Daily Output	Labor-Hours	Unit	Material	Labor	Equipment	Total	Total Incl O&P
0010	**AXIAL FLOW HVAC FANS**									
0020	Air conditioning and process air handling									
0500	Axial flow, constant speed									
0505	Direct drive, 1/8" S.P.									
0510	12", 1060 CFM, 1/6 HP	Q-20	3	6.667	Ea.	655	350		1,005	1,250
0514	12", 2095 CFM, 1/2 HP		3	6.667		660	350		1,010	1,250
0518	16", 2490 CFM, 1/3 HP		2.80	7.143		670	375		1,045	1,300
0522	20", 4130 CFM, 3/4 HP		2.60	7.692		915	400		1,315	1,600
0526	22", 4700 CFM, 3/4 HP		2.60	7.692		1,350	400		1,750	2,100
0530	24", 5850 CFM, 1 HP		2.50	8		1,575	420		1,995	2,350
0534	24", 7925 CFM, 1-1/2 HP		2.40	8.333		1,575	435		2,010	2,400
0538	30", 10640 CFM, 2 HP		2.20	9.091		2,225	475		2,700	3,175
0542	30", 14765 CFM, 2-1/2 HP		2.10	9.524		2,575	500		3,075	3,600
0546	36", 16780 CFM, 2 HP		2.10	9.524		2,675	500		3,175	3,675
0550	36", 22920 CFM, 5 HP		1.80	11.111		2,925	580		3,505	4,100
0560	Belt drive, 1/8" S.P.									
0562	15", 2800 CFM, 1/3 HP	Q-20	3.20	6.250	Ea.	1,200	325		1,525	1,825

23 34 13 – Axial HVAC Fans

23 34 13.10 Axial Flow HVAC Fans	Crew	Daily Output	Labor-Hours	Unit	Material	2016 Bare Costs Labor	Equipment	Total	Total Incl O&P
0564	15", 3400 CFM, 1/2 HP	Q-20	3	6.667	Ea.	1,225	350		1,575	1,875
0568	18", 3280 CFM, 1/3 HP		2.80	7.143		1,225	375		1,600	1,925
0572	18", 3900 CFM, 1/2 HP		2.80	7.143		1,300	375		1,675	2,000
0576	18", 5250 CFM, 1 HP		2.70	7.407		1,425	385		1,810	2,175
0584	24", 6430 CFM, 1 HP		2.50	8		1,875	420		2,295	2,675
0588	24", 8860 CFM, 2 HP		2.40	8.333		1,950	435		2,385	2,800
0592	30", 9250 CFM, 1 HP		2.20	9.091		2,400	475		2,875	3,375
0596	30", 16900 CFM, 5 HP		2	10		2,575	520		3,095	3,625
0604	36", 14475 CFM, 2 HP		2.30	8.696		2,950	455		3,405	3,950
0608	36", 20080 CFM, 5 HP		1.80	11.111		3,025	580		3,605	4,200
0616	42", 29000 CFM, 7-1/2 HP		1.40	14.286		4,100	745		4,845	5,625
1500	Vaneaxial, low pressure, 2000 CFM, 1/2 HP		3.60	5.556		2,250	290		2,540	2,925
1520	4,000 CFM, 1 HP		3.20	6.250		2,625	325		2,950	3,375
1540	8,000 CFM, 2 HP		2.80	7.143		3,325	375		3,700	4,225
1560	16,000 CFM, 5 HP		2.40	8.333		4,650	435		5,085	5,775

23 34 14 – Blower HVAC Fans

23 34 14.10 Blower Type HVAC Fans	Crew	Daily Output	Labor-Hours	Unit	Material	2016 Bare Costs Labor	Equipment	Total	Total Incl O&P
0010	**BLOWER TYPE HVAC FANS**									
2000	Blowers, direct drive with motor, complete									
2020	1045 CFM @ .5" S.P., 1/5 HP	Q-20	18	1.111	Ea.	355	58		413	480
2040	1385 CFM @ .5" S.P., 1/4 HP		18	1.111		350	58		408	475
2060	1640 CFM @ .5" S.P., 1/3 HP		18	1.111		320	58		378	440
2080	1760 CFM @ .5" S.P., 1/2 HP		18	1.111		330	58		388	455
2090	4 speed									
2100	1164 to 1739 CFM @ .5" S.P., 1/3 HP	Q-20	16	1.250	Ea.	335	65.50		400.50	470
2120	1467 to 2218 CFM @ 1.0" S.P., 3/4 HP	"	14	1.429	"	385	74.50		459.50	540
2500	Ceiling fan, right angle, extra quiet, 0.10" S.P.									
2520	95 CFM	Q-20	20	1	Ea.	315	52		367	425
2540	210 CFM		19	1.053		370	55		425	490
2560	385 CFM		18	1.111		470	58		528	605
2580	885 CFM		16	1.250		925	65.50		990.50	1,125
2600	1,650 CFM		13	1.538		1,275	80.50		1,355.50	1,525
2620	2,960 CFM		11	1.818		1,700	95		1,795	2,025
2640	For wall or roof cap, add	1 Shee	16	.500		315	28.50		343.50	390
2660	For straight thru fan, add					10%				
2680	For speed control switch, add	1 Elec	16	.500		171	27.50		198.50	229
7500	Utility set, steel construction, pedestal, 1/4" S.P.									
7520	Direct drive, 150 CFM, 1/8 HP	Q-20	6.40	3.125	Ea.	890	163		1,053	1,225
7540	485 CFM, 1/6 HP		5.80	3.448		1,125	180		1,305	1,500
7560	1950 CFM, 1/2 HP		4.80	4.167		1,325	218		1,543	1,775
7580	2410 CFM, 3/4 HP		4.40	4.545		2,425	237		2,662	3,025
7600	3328 CFM, 1-1/2 HP		3	6.667		2,700	350		3,050	3,500
7680	V-belt drive, drive cover, 3 phase									
7700	800 CFM, 1/4 HP	Q-20	6	3.333	Ea.	1,025	174		1,199	1,400
7720	1,300 CFM, 1/3 HP		5	4		1,050	209		1,259	1,500
7740	2,000 CFM, 1 HP		4.60	4.348		1,250	227		1,477	1,725
7760	2,900 CFM, 3/4 HP		4.20	4.762		1,700	249		1,949	2,250
7780	3,600 CFM, 3/4 HP		4	5		2,075	261		2,336	2,700
7800	4,800 CFM, 1 HP		3.50	5.714		2,475	299		2,774	3,150
7820	6,700 CFM, 1-1/2 HP		3	6.667		3,050	350		3,400	3,875
7830	7,500 CFM, 2 HP		2.50	8		4,125	420		4,545	5,175
7840	11,000 CFM, 3 HP		2	10		5,525	520		6,045	6,875

For customer support on your Mechanical Cost Data, call 877.739.3689.

333

23 34 14 – Blower HVAC Fans

23 34 14.10 Blower Type HVAC Fans		Crew	Daily Output	Labor-Hours	Unit	Material	2016 Bare Costs Labor	Equipment	Total	Total Incl O&P
7860	13,000 CFM, 3 HP	Q-20	1.60	12.500	Ea.	5,600	655		6,255	7,150
7880	15,000 CFM, 5 HP		1	20		5,800	1,050		6,850	7,975
7900	17,000 CFM, 7-1/2 HP		.80	25		6,200	1,300		7,500	8,800
7920	20,000 CFM, 7-1/2 HP		.80	25		7,400	1,300		8,700	10,100

23 34 16 – Centrifugal HVAC Fans

23 34 16.10 Centrifugal Type HVAC Fans

		Crew	Daily Output	Labor-Hours	Unit	Material	2016 Bare Costs Labor	Equipment	Total	Total Incl O&P
0010	**CENTRIFUGAL TYPE HVAC FANS**									
0200	In-line centrifugal, supply/exhaust booster									
0220	aluminum wheel/hub, disconnect switch, 1/4" S.P.									
0240	500 CFM, 10" diameter connection	Q-20	3	6.667	Ea.	1,450	350		1,800	2,100
0260	1,380 CFM, 12" diameter connection		2	10		1,450	520		1,970	2,400
0280	1,520 CFM, 16" diameter connection		2	10		1,500	520		2,020	2,450
0300	2,560 CFM, 18" diameter connection		1	20		1,625	1,050		2,675	3,400
0320	3,480 CFM, 20" diameter connection		.80	25		1,950	1,300		3,250	4,125
0326	5,080 CFM, 20" diameter connection		.75	26.667		2,250	1,400		3,650	4,600
0340	7,500 CFM, 22" diameter connection		.70	28.571		2,750	1,500		4,250	5,300
0350	10,000 CFM, 27" diameter connection		.65	30.769		3,250	1,600		4,850	6,025
3500	Centrifugal, airfoil, motor and drive, complete									
3520	1000 CFM, 1/2 HP	Q-20	2.50	8	Ea.	1,925	420		2,345	2,725
3540	2,000 CFM, 1 HP		2	10		2,175	520		2,695	3,175
3560	4,000 CFM, 3 HP		1.80	11.111		2,750	580		3,330	3,900
3580	8,000 CFM, 7-1/2 HP		1.40	14.286		4,125	745		4,870	5,675
3600	12,000 CFM, 10 HP		1	20		5,500	1,050		6,550	7,675
4000	Single width, belt drive, not incl. motor, capacities									
4020	at 2000 fpm, 2.5" S.P. for indicated motor									
4040	6900 CFM, 5 HP	Q-9	2.40	6.667	Ea.	3,875	345		4,220	4,800
4060	10,340 CFM, 7-1/2 HP		2.20	7.273		5,400	375		5,775	6,500
4080	15,320 CFM, 10 HP		2	8		6,350	410		6,760	7,625
4100	22,780 CFM, 15 HP		1.80	8.889		9,450	460		9,910	11,100
4120	33,840 CFM, 20 HP		1.60	10		12,600	515		13,115	14,600
4140	41,400 CFM, 25 HP		1.40	11.429		16,600	590		17,190	19,200
4160	50,100 CFM, 30 HP		.80	20		20,600	1,025		21,625	24,200
4200	Double width wheel, 12,420 CFM, 7.5 HP		2.20	7.273		5,550	375		5,925	6,675
4220	18,620 CFM, 15 HP		2	8		7,925	410		8,335	9,350
4240	27,580 CFM, 20 HP		1.80	8.889		9,700	460		10,160	11,400
4260	40,980 CFM, 25 HP		1.50	10.667		15,200	550		15,750	17,500
4280	60,920 CFM, 40 HP		1	16		20,500	825		21,325	23,800
4300	74,520 CFM, 50 HP		.80	20		24,200	1,025		25,225	28,300
4320	90,160 CFM, 50 HP		.70	22.857		33,200	1,175		34,375	38,300
4340	110,300 CFM, 60 HP		.50	32		44,900	1,650		46,550	52,000
4360	134,960 CFM, 75 HP		.40	40		59,500	2,050		61,550	68,500
5000	Utility set, centrifugal, V belt drive, motor									
5020	1/4" S.P., 1200 CFM, 1/4 HP	Q-20	6	3.333	Ea.	1,750	174		1,924	2,200
5040	1520 CFM, 1/3 HP		5	4		2,300	209		2,509	2,850
5060	1850 CFM, 1/2 HP		4	5		2,275	261		2,536	2,900
5080	2180 CFM, 3/4 HP		3	6.667		2,700	350		3,050	3,475
5100	1/2" S.P., 3600 CFM, 1 HP		2	10		2,775	520		3,295	3,850
5120	4250 CFM, 1-1/2 HP		1.60	12.500		3,400	655		4,055	4,725
5140	4800 CFM, 2 HP		1.40	14.286		4,100	745		4,845	5,650
5160	6920 CFM, 5 HP		1.30	15.385		5,000	805		5,805	6,725
5180	7700 CFM, 7-1/2 HP		1.20	16.667		6,025	870		6,895	7,950
5200	For explosion proof motor, add					15%				

23 34 16.10 Centrifugal Type HVAC Fans		Crew	Daily Output	Labor-Hours	Unit	Material	2016 Bare Costs Labor	Equipment	Total	Total Incl O&P
5300	Fume exhauster without hose and intake nozzle									
5310	630 CFM, 1-1/2 HP	Q-20	2	10	Ea.	900	520		1,420	1,775
5320	Hose extension kit 5'	"	20	1	"	71.50	52		123.50	158
5500	Fans, industrial exhauster, for air which may contain granular matl.									
5520	1000 CFM, 1-1/2 HP	Q-20	2.50	8	Ea.	2,900	420		3,320	3,825
5540	2000 CFM, 3 HP		2	10		3,550	520		4,070	4,700
5560	4000 CFM, 7-1/2 HP		1.80	11.111		4,925	580		5,505	6,275
5580	8000 CFM, 15 HP		1.40	14.286		6,350	745		7,095	8,100
5600	12,000 CFM, 30 HP	↓	1	20	↓	10,300	1,050		11,350	13,000
7000	Roof exhauster, centrifugal, aluminum housing, 12" galvanized									
7020	curb, bird screen, back draft damper, 1/4" S.P.									
7100	Direct drive, 320 CFM, 11" sq. damper	Q-20	7	2.857	Ea.	725	149		874	1,025
7120	600 CFM, 11" sq. damper		6	3.333		925	174		1,099	1,300
7140	815 CFM, 13" sq. damper		5	4		925	209		1,134	1,350
7160	1450 CFM, 13" sq. damper		4.20	4.762		1,450	249		1,699	1,950
7180	2050 CFM, 16" sq. damper		4	5		1,800	261		2,061	2,375
7200	V-belt drive, 1650 CFM, 12" sq. damper		6	3.333		1,350	174		1,524	1,775
7220	2750 CFM, 21" sq. damper		5	4		1,600	209		1,809	2,075
7230	3500 CFM, 21" sq. damper		4.50	4.444		1,800	232		2,032	2,325
7240	4910 CFM, 23" sq. damper		4	5		2,200	261		2,461	2,825
7260	8525 CFM, 28" sq. damper		3	6.667		2,925	350		3,275	3,725
7280	13,760 CFM, 35" sq. damper		2	10		4,075	520		4,595	5,275
7300	20,558 CFM, 43" sq. damper	↓	1	20		8,200	1,050		9,250	10,600
7320	For 2 speed winding, add					15%				
7340	For explosionproof motor, add					600			600	660
7360	For belt driven, top discharge, add				↓	15%				
7400	Roof mounted kitchen exhaust, aluminum, centrifugal									
7410	Direct drive, 1/4" S.P., 2 speed, temp to 200°F									
7414	1/3 HP, 2038 CFM	Q-20	7	2.857	Ea.	565	149		714	845
7416	1/2 HP, 2393 CFM		6	3.333		540	174		714	860
7418	3/4 HP, 4329 CFM	↓	5	4	↓	845	209		1,054	1,250
7424	Belt drive, 1/4" S.P., temp to 250°F									
7426	3/4 HP, 2757 CFM	Q-20	5	4	Ea.	815	209		1,024	1,225
7428	1-1/2 HP, 4913 CFM		4	5		1,150	261		1,411	1,675
7430	1-1/2 HP, 5798 CFM	↓	4	5	↓	1,200	261		1,461	1,725
7450	Upblast, propeller, w/backdraft damper, bird screen & curb									
7454	30,300 CFM @ 3/8" S.P., 5 HP	Q-20	1.60	12.500	Ea.	4,700	655		5,355	6,175
7458	36,000 CFM @ 1/2" S.P., 15 HP	"	1.60	12.500	"	6,700	655		7,355	8,350
8500	Wall exhausters, centrifugal, auto damper, 1/8" S.P.									
8520	Direct drive, 610 CFM, 1/20 HP	Q-20	14	1.429	Ea.	440	74.50		514.50	600
8540	796 CFM, 1/12 HP		13	1.538		905	80.50		985.50	1,125
8560	822 CFM, 1/6 HP		12	1.667		1,125	87		1,212	1,350
8580	1,320 CFM, 1/4 HP		12	1.667		1,300	87		1,387	1,550
8600	1756 CFM, 1/4 HP		11	1.818		1,300	95		1,395	1,600
8620	1983 CFM, 1/4 HP		10	2		1,350	104		1,454	1,625
8640	2900 CFM, 1/2 HP		9	2.222		1,425	116		1,541	1,750
8660	3307 CFM, 3/4 HP		8	2.500		1,550	131		1,681	1,900
8670	5940 CFM, 1 HP	↓	7	2.857	↓	1,600	149		1,749	2,000
9500	V-belt drive, 3 phase									
9520	2,800 CFM, 1/4 HP	Q-20	9	2.222	Ea.	2,000	116		2,116	2,375
9540	3,740 CFM, 1/2 HP		8	2.500		2,075	131		2,206	2,475
9560	4400 CFM, 3/4 HP		7	2.857		2,100	149		2,249	2,525
9580	5700 CFM, 1-1/2 HP	↓	6	3.333	↓	2,175	174		2,349	2,675

For customer support on your Mechanical Cost Data, call 877.739.3689.

335

23 34 HVAC Fans

23 34 23 – HVAC Power Ventilators

23 34 23.10 HVAC Power Circulators and Ventilators		Crew	Daily Output	Labor-Hours	Unit	Material	2016 Bare Costs Labor	Equipment	Total	Total Incl O&P
0010	**HVAC POWER CIRCULATORS AND VENTILATORS**									
3000	Paddle blade air circulator, 3 speed switch									
3020	42", 5,000 CFM high, 3000 CFM low [G]	1 Elec	2.40	3.333	Ea.	169	184		353	460
3040	52", 6,500 CFM high, 4000 CFM low [G]	"	2.20	3.636	"	225	200		425	545
3100	For antique white motor, same cost									
3200	For brass plated motor, same cost									
3300	For light adaptor kit, add [G]				Ea.	41			41	45
3310	Industrial grade, reversible, 4 blade									
3312	5500 CFM	Q-20	5	4	Ea.	215	209		424	555
3314	7600 CFM		5	4		221	209		430	565
3316	21,015 CFM		4	5		269	261		530	690
4000	High volume, low speed (HVLS), paddle blade air circulator									
4010	Variable speed, reversible, 1 HP motor, motor control panel,									
4020	motor drive cable, control cable with remote, and safety cable.									
4140	8' diameter	Q-5	1.50	10.667	Ea.	4,725	585		5,310	6,075
4150	10' diameter		1.45	11.034		4,750	605		5,355	6,125
4160	12' diameter		1.40	11.429		4,975	625		5,600	6,425
4170	14' diameter		1.35	11.852		5,050	645		5,695	6,550
4180	16' diameter		1.30	12.308		5,100	670		5,770	6,625
4190	18' diameter		1.25	12.800		5,150	700		5,850	6,700
4200	20' diameter		1.20	13.333		5,475	730		6,205	7,125
4210	24' diameter		1	16		5,600	875		6,475	7,500
6000	Propeller exhaust, wall shutter									
6020	Direct drive, one speed, .075" S.P.									
6100	653 CFM, 1/30 HP	Q-20	10	2	Ea.	200	104		304	380
6120	1033 CFM, 1/20 HP		9	2.222		298	116		414	505
6140	1323 CFM, 1/15 HP		8	2.500		335	131		466	565
6160	2444 CFM, 1/4 HP		7	2.857		415	149		564	685
6300	V-belt drive, 3 phase									
6320	6175 CFM, 3/4 HP	Q-20	5	4	Ea.	1,225	209		1,434	1,675
6340	7500 CFM, 3/4 HP		5	4		1,300	209		1,509	1,750
6360	10,100 CFM, 1 HP		4.50	4.444		1,475	232		1,707	1,975
6380	14,300 CFM, 1-1/2 HP		4	5		1,775	261		2,036	2,350
6400	19,800 CFM, 2 HP		3	6.667		2,075	350		2,425	2,825
6420	26,250 CFM, 3 HP		2.60	7.692		2,250	400		2,650	3,075
6440	38,500 CFM, 5 HP		2.20	9.091		2,650	475		3,125	3,650
6460	46,000 CFM, 7-1/2 HP		2	10		2,850	520		3,370	3,925
6480	51,500 CFM, 10 HP		1.80	11.111		2,950	580		3,530	4,125
6490	V-belt drive, 115 V, residential, whole house									
6500	Ceiling-wall, 5200 CFM, 1/4 HP, 30" x 30"	1 Shee	6	1.333	Ea.	565	76.50		641.50	735
6510	7500 CFM, 1/3 HP, 36" x 36"		5	1.600		590	91.50		681.50	790
6520	10,500 CFM, 1/3 HP, 42" x 42"		5	1.600		615	91.50		706.50	820
6530	13,200 CFM, 1/3 HP, 48" x 48"		4	2		680	115		795	925
6540	15,445 CFM, 1/2 HP, 48" x 48"		4	2		710	115		825	955
6550	17,025 CFM, 1/2 HP, 54" x 54"		4	2		1,375	115		1,490	1,675
6560	For two speed motor, add					20%				
6570	Shutter, automatic, ceiling/wall									
6580	30" x 30"	1 Shee	8	1	Ea.	184	57.50		241.50	290
6590	36" x 36"		8	1		210	57.50		267.50	320
6600	42" x 42"		8	1		262	57.50		319.50	375
6610	48" x 48"		7	1.143		285	65.50		350.50	415
6620	54" x 54"		6	1.333		375	76.50		451.50	530

336

23 34 23 – HVAC Power Ventilators

23 34 23.10 HVAC Power Circulators and Ventilators		Crew	Daily Output	Labor-Hours	Unit	Material	2016 Bare Costs Labor	Equipment	Total	Total Incl O&P
6630	Timer, shut off, to 12 hour	1 Shee	20	.400	Ea.	60.50	23		83.50	102
6650	Residential, bath exhaust, grille, back draft damper									
6660	50 CFM	Q-20	24	.833	Ea.	65	43.50		108.50	138
6670	110 CFM		22	.909		100	47.50		147.50	183
6672	180 CFM		22	.909		111	47.50		158.50	195
6673	210 CFM		22	.909		145	47.50		192.50	233
6674	260 CFM		22	.909		176	47.50		223.50	267
6675	300 CFM		22	.909		133	47.50		180.50	219
6680	Light combination, squirrel cage, 100 watt, 70 CFM		24	.833		114	43.50		157.50	191
6700	Light/heater combination, ceiling mounted									
6710	70 CFM, 1450 watt	Q-20	24	.833	Ea.	165	43.50		208.50	247
6800	Heater combination, recessed, 70 CFM		24	.833		68.50	43.50		112	142
6820	With 2 infrared bulbs		23	.870		107	45.50		152.50	187
6840	Wall mount, 170 CFM		22	.909		269	47.50		316.50	370
6846	Ceiling mount, 180 CFM		22	.909		272	47.50		319.50	370
6900	Kitchen exhaust, grille, complete, 160 CFM		22	.909		111	47.50		158.50	195
6910	180 CFM		20	1		93.50	52		145.50	183
6920	270 CFM		18	1.111		176	58		234	283
6930	350 CFM		16	1.250		133	65.50		198.50	246
6940	Residential roof jacks and wall caps									
6944	Wall cap with back draft damper									
6946	3" & 4" diam. round duct	1 Shee	11	.727	Ea.	26	41.50		67.50	92
6948	6" diam. round duct	"	11	.727	"	65	41.50		106.50	135
6958	Roof jack with bird screen and back draft damper									
6960	3" & 4" diam. round duct	1 Shee	11	.727	Ea.	26	41.50		67.50	92.50
6962	3-1/4" x 10" rectangular duct	"	10	.800	"	49	46		95	124
6980	Transition									
6982	3-1/4" x 10" to 6" diam. round	1 Shee	20	.400	Ea.	33	23		56	71.50
8020	Attic, roof type									
8030	Aluminum dome, damper & curb									
8080	12" diameter, 1000 CFM (gravity)	1 Elec	10	.800	Ea.	595	44		639	720
8090	16" diameter, 1500 CFM (gravity)		9	.889		720	49		769	865
8100	20" diameter, 2500 CFM (gravity)		8	1		880	55		935	1,050
8110	26" diameter, 4000 CFM (gravity)		7	1.143		1,075	63		1,138	1,275
8120	32" diameter, 6500 CFM (gravity)		6	1.333		1,475	73.50		1,548.50	1,700
8130	38" diameter, 8000 CFM (gravity)		5	1.600		2,175	88		2,263	2,525
8140	50" diameter, 13,000 CFM (gravity)		4	2		3,150	110		3,260	3,650
8160	Plastic, ABS dome									
8180	1050 CFM	1 Elec	14	.571	Ea.	175	31.50		206.50	240
8200	1600 CFM	"	12	.667	"	262	36.50		298.50	345
8240	Attic, wall type, with shutter, one speed									
8250	12" diameter, 1000 CFM	1 Elec	14	.571	Ea.	410	31.50		441.50	495
8260	14" diameter, 1500 CFM		12	.667		445	36.50		481.50	545
8270	16" diameter, 2000 CFM		9	.889		505	49		554	630
8290	Whole house, wall type, with shutter, one speed									
8300	30" diameter, 4800 CFM	1 Elec	7	1.143	Ea.	1,075	63		1,138	1,275
8310	36" diameter, 7000 CFM		6	1.333		1,175	73.50		1,248.50	1,400
8320	42" diameter, 10,000 CFM		5	1.600		1,325	88		1,413	1,575
8330	48" diameter, 16,000 CFM		4	2		1,625	110		1,735	1,975
8340	For two speed, add					98			98	108
8350	Whole house, lay-down type, with shutter, one speed									
8360	30" diameter, 4500 CFM	1 Elec	8	1	Ea.	1,150	55		1,205	1,325
8370	36" diameter, 6500 CFM		7	1.143		1,225	63		1,288	1,450

23 34 HVAC Fans

23 34 23 – HVAC Power Ventilators

23 34 23.10 HVAC Power Circulators and Ventilators	Crew	Daily Output	Labor-Hours	Unit	Material	2016 Bare Costs Labor	2016 Bare Costs Equipment	Total	Total Incl O&P
8380	42" diameter, 9000 CFM	1 Elec	6	1.333	Ea.	1,350	73.50		1,423.50	1,600
8390	48" diameter, 12,000 CFM	↓	5	1.600		1,550	88		1,638	1,825
8440	For two speed, add					74			74	81
8450	For 12 hour timer switch, add	1 Elec	32	.250	↓	74	13.80		87.80	102

23 34 33 – Air Curtains

23 34 33.10 Air Barrier Curtains

		Crew	Daily Output	Labor-Hours	Unit	Material	Labor	Equipment	Total	Total Incl O&P
0010	**AIR BARRIER CURTAINS**, Incl. motor starters, transformers,									
0050	and door switches									
2450	Conveyor openings or service windows									
3000	Service window, 5' high x 25" wide	2 Shee	5	3.200	Ea.	315	183		498	630
3100	Environmental separation									
3110	Door heights up to 8', low profile, super quiet									
3120	Unheated, variable speed									
3130	36" wide	2 Shee	4	4	Ea.	635	229		864	1,050
3134	42" wide		3.80	4.211		665	241		906	1,100
3138	48" wide		3.60	4.444		690	254		944	1,150
3142	60" wide	↓	3.40	4.706		715	269		984	1,200
3146	72" wide	Q-3	4.60	6.957		890	390		1,280	1,575
3150	96" wide		4.40	7.273		1,350	410		1,760	2,125
3154	120" wide		4.20	7.619		1,400	430		1,830	2,175
3158	144" wide	↓	4	8	↓	1,775	450		2,225	2,625
3200	Door heights up to 10'									
3210	Unheated									
3230	36" wide	2 Shee	3.80	4.211	Ea.	675	241		916	1,125
3234	42" wide		3.60	4.444		695	254		949	1,150
3238	48" wide		3.40	4.706		715	269		984	1,200
3242	60" wide	↓	3.20	5		1,075	286		1,361	1,625
3246	72" wide	Q-3	4.40	7.273		1,125	410		1,535	1,875
3250	96" wide		4.20	7.619		1,350	430		1,780	2,125
3254	120" wide		4	8		1,850	450		2,300	2,725
3258	144" wide	↓	3.80	8.421	↓	2,000	475		2,475	2,925
3300	Door heights up to 12'									
3310	Unheated									
3334	42" wide	2 Shee	3.40	4.706	Ea.	1,000	269		1,269	1,500
3338	48" wide		3.20	5		1,025	286		1,311	1,575
3342	60" wide	↓	3	5.333		1,025	305		1,330	1,600
3346	72" wide	Q-3	4.20	7.619		1,825	430		2,255	2,675
3350	96" wide		4	8		2,000	450		2,450	2,875
3354	120" wide		3.80	8.421		2,450	475		2,925	3,425
3358	144" wide	↓	3.60	8.889	↓	2,625	500		3,125	3,625
3400	Door heights up to 16'									
3410	Unheated									
3438	48" wide	2 Shee	3	5.333	Ea.	1,175	305		1,480	1,750
3442	60" wide	"	2.80	5.714		1,225	325		1,550	1,850
3446	72" wide	Q-3	3.80	8.421		2,100	475		2,575	3,025
3450	96" wide		3.60	8.889		2,375	500		2,875	3,375
3454	120" wide		3.40	9.412		3,225	530		3,755	4,350
3458	144" wide	↓	3.20	10	↓	3,350	565		3,915	4,525
3470	Heated, electric									
3474	48" wide	2 Shee	2.90	5.517	Ea.	1,950	315		2,265	2,600
3478	60" wide	"	2.70	5.926		1,975	340		2,315	2,700
3482	72" wide	Q-3	3.70	8.649	↓	3,400	485		3,885	4,475

23 34 HVAC Fans

23 34 33 – Air Curtains

23 34 33.10 Air Barrier Curtains	Crew	Daily Output	Labor-Hours	Unit	Material	2016 Bare Costs Labor	Equipment	Total	Total Incl O&P
3486	96" wide	Q-3	3.50	9.143	Ea.	3,600	515		4,115	4,725
3490	120" wide		3.30	9.697		4,350	545		4,895	5,625
3494	144" wide		3.10	10.323		4,700	580		5,280	6,025

23 35 Special Exhaust Systems

23 35 16 – Engine Exhaust Systems

23 35 16.10 Engine Exhaust Removal Systems

		Crew	Daily Output	Labor-Hours	Unit	Material	2016 Bare Costs Labor	Equipment	Total	Total Incl O&P
0010	**ENGINE EXHAUST REMOVAL SYSTEMS** D3090-320									
0500	Engine exhaust, garage, in-floor system									
0510	Single tube outlet assemblies									
0520	For transite pipe ducting, self-storing tube									
0530	3" tubing adapter plate	1 Shee	16	.500	Ea.	214	28.50		242.50	279
0540	4" tubing adapter plate		16	.500		218	28.50		246.50	283
0550	5" tubing adapter plate		16	.500		218	28.50		246.50	283
0600	For vitrified tile ducting									
0610	3" tubing adapter plate, self-storing tube	1 Shee	16	.500	Ea.	214	28.50		242.50	280
0620	4" tubing adapter plate, self-storing tube		16	.500		218	28.50		246.50	283
0660	5" tubing adapter plate, self-storing tube		16	.500		218	28.50		246.50	283
0800	Two tube outlet assemblies									
0810	For transite pipe ducting, self-storing tube									
0820	3" tubing, dual exhaust adapter plate	1 Shee	16	.500	Ea.	214	28.50		242.50	279
0850	For vitrified tile ducting									
0860	3" tubing, dual exhaust, self-storing tube	1 Shee	16	.500	Ea.	265	28.50		293.50	335
0870	3" tubing, double outlet, non-storing tubes	"	16	.500	"	265	28.50		293.50	335
0900	Accessories for metal tubing, (overhead systems also)									
0910	Adapters, for metal tubing end									
0920	3" tail pipe type				Ea.	44			44	48.50
0930	4" tail pipe type					47			47	52
0940	5" tail pipe type					48			48	53
0990	5" diesel stack type					282			282	310
1000	6" diesel stack type					291			291	320
1100	Bullnose (guide) required for in-floor assemblies									
1110	3" tubing size				Ea.	25			25	27.50
1120	4" tubing size					26.50			26.50	29
1130	5" tubing size					28			28	31
1150	Plain rings, for tubing end									
1160	3" tubing size				Ea.	21.50			21.50	24
1170	4" tubing size				"	37			37	40.50
1200	Tubing, galvanized, flexible, (for overhead systems also)									
1210	3" ID				L.F.	10.30			10.30	11.35
1220	4" ID					12.65			12.65	13.90
1230	5" ID					14.90			14.90	16.35
1240	6" ID					17.20			17.20	18.90
1250	Stainless steel, flexible, (for overhead system, also)									
1260	3" ID				L.F.	23			23	25
1270	4" ID					31			31	34
1280	5" ID					35			35	38.50
1290	6" ID					41			41	45
1500	Engine exhaust, garage, overhead components, for neoprene tubing									
1510	Alternate metal tubing & accessories see above									
1550	Adapters, for neoprene tubing end									

For customer support on your Mechanical Cost Data, call 877.739.3689.

339

23 35 16.10 Engine Exhaust Removal Systems	Crew	Daily Output	Labor-Hours	Unit	Material	2016 Bare Costs Labor	Equipment	Total	Total Incl O&P
1560 3" tail pipe, adjustable, neoprene				Ea.	49.50			49.50	54.50
1570 3" tail pipe, heavy wall neoprene					56			56	61.50
1580 4" tail pipe, heavy wall neoprene					89			89	97.50
1590 5" tail pipe, heavy wall neoprene					93.50			93.50	103
1650 Connectors, tubing									
1660 3" interior, aluminum				Ea.	21.50			21.50	24
1670 4" interior, aluminum					37			37	40.50
1710 5" interior, neoprene					54.50			54.50	60
1750 3" spiralock, neoprene					21.50			21.50	24
1760 4" spiralock, neoprene					37			37	40.50
1780 Y for 3" ID tubing, neoprene, dual exhaust					174			174	191
1790 Y for 4" ID tubing, aluminum, dual exhaust					174			174	191
1850 Elbows, aluminum, splice into tubing for strap									
1860 3" neoprene tubing size				Ea.	43			43	47.50
1870 4" neoprene tubing size					36			36	39.50
1900 Flange assemblies, connect tubing to overhead duct					65			65	71.50
2000 Hardware and accessories									
2020 Cable, galvanized, 1/8" diameter				L.F.	.46			.46	.51
2040 Cleat, tie down cable or rope				Ea.	4.78			4.78	5.25
2060 Pulley					6.60			6.60	7.25
2080 Pulley hook, universal					4.78			4.78	5.25
2100 Rope, nylon, 1/4" diameter				L.F.	.36			.36	.40
2120 Winch, 1" diameter				Ea.	102			102	112
2150 Lifting strap, mounts on neoprene									
2160 3" tubing size				Ea.	24			24	26.50
2170 4" tubing size					24			24	26.50
2180 5" tubing size					24			24	26.50
2190 6" tubing size					24			24	26.50
2200 Tubing, neoprene, 11' lengths									
2210 3" ID				L.F.	9.95			9.95	10.95
2220 4" ID					16.15			16.15	17.75
2230 5" ID					24			24	26.50
2500 Engine exhaust, thru-door outlet									
2510 3" tube size	1 Carp	16	.500	Ea.	53	24		77	95
2530 4" tube size	"	16	.500	"	65.50	24		89.50	109
3000 Tubing, exhaust, flex hose, with									
3010 coupler, damper and tail pipe adapter									
3020 Neoprene									
3040 3" x 20'	1 Shee	6	1.333	Ea.	300	76.50		376.50	445
3050 4" x 15'		5.40	1.481		390	85		475	560
3060 4" x 20'		5	1.600		470	91.50		561.50	660
3070 5" x 15'		4.40	1.818		535	104		639	745
3100 Galvanized									
3110 3" x 20'	1 Shee	6	1.333	Ea.	295	76.50		371.50	440
3120 4" x 17'		5.60	1.429		320	82		402	480
3130 4" x 20'		5	1.600		360	91.50		451.50	535
3140 5" x 17'		4.60	1.739		385	99.50		484.50	570
8000 Blower, for tailpipe exhaust system									
8010 Direct drive									
8012 495 CFM, 1/3 HP	Q-9	5	3.200	Ea.	2,250	165		2,415	2,725
8014 1445 CFM, 3/4 HP		4	4		2,800	206		3,006	3,425
8016 1840 CFM, 1-1/2 HP		3	5.333		3,350	275		3,625	4,125
8030 Beltdrive									

23 35 Special Exhaust Systems

23 35 16 – Engine Exhaust Systems

23 35 16.10 Engine Exhaust Removal Systems

	23 35 16.10 Engine Exhaust Removal Systems	Crew	Daily Output	Labor-Hours	Unit	Material	2016 Bare Costs Labor	Equipment	Total	Total Incl O&P
8032	1400 CFM, 1 HP	Q-9	4	4	Ea.	2,800	206		3,006	3,425
8034	2023 CFM, 1-1/2 HP		3	5.333		3,875	275		4,150	4,700
8036	2750 CFM, 2 HP		2	8		4,600	410		5,010	5,675
8038	4400 CFM, 3 HP		1.80	8.889		6,775	460		7,235	8,150
8040	7060 CFM, 5 HP		1.40	11.429		9,600	590		10,190	11,500

23 35 43 – Welding Fume Elimination Systems

23 35 43.10 Welding Fume Elimination System Components

	23 35 43.10 Welding Fume Elimination System Components	Crew	Daily Output	Labor-Hours	Unit	Material	2016 Bare Costs Labor	Equipment	Total	Total Incl O&P
0010	**WELDING FUME ELIMINATION SYSTEM COMPONENTS**									
7500	Welding fume elimination accessories for garage exhaust systems									
7600	Cut off (blast gate)									
7610	3" tubing size, 3" x 6" opening	1 Shee	24	.333	Ea.	23	19.10		42.10	54.50
7620	4" tubing size, 4" x 8" opening		24	.333		23	19.10		42.10	54.50
7630	5" tubing size, 5" x 10" opening		24	.333		28	19.10		47.10	60
7640	6" tubing size		24	.333		29	19.10		48.10	60.50
7650	8" tubing size		24	.333		40	19.10		59.10	73
7700	Hoods, magnetic, with handle & screen									
7710	3" tubing size, 3" x 6" opening	1 Shee	24	.333	Ea.	95	19.10		114.10	134
7720	4" tubing size, 4" x 8" opening		24	.333		86.50	19.10		105.60	124
7730	5" tubing size, 5" x 10" opening		24	.333		86.50	19.10		105.60	124

23 36 Air Terminal Units

23 36 13 – Constant-Air-Volume Units

23 36 13.10 Constant Volume Mixing Boxes

	23 36 13.10 Constant Volume Mixing Boxes	Crew	Daily Output	Labor-Hours	Unit	Material	2016 Bare Costs Labor	Equipment	Total	Total Incl O&P
0010	**CONSTANT VOLUME MIXING BOXES**									
5180	Mixing box, includes electric or pneumatic motor									
5192	Recommend use with silencer, see Line 23 33 19.10 0010									
5200	Constant volume, 150 to 270 CFM	Q-9	12	1.333	Ea.	715	68.50		783.50	890
5210	270 to 600 CFM		11	1.455		735	75		810	925
5230	550 to 1000 CFM		9	1.778		735	91.50		826.50	950
5240	1000 to 1600 CFM		8	2		755	103		858	990
5250	1300 to 1900 CFM		6	2.667		770	137		907	1,050
5260	550 to 2640 CFM		5.60	2.857		850	147		997	1,150
5270	650 to 3120 CFM		5.20	3.077		905	159		1,064	1,225

23 36 16 – Variable-Air-Volume Units

23 36 16.10 Variable Volume Mixing Boxes

	23 36 16.10 Variable Volume Mixing Boxes	Crew	Daily Output	Labor-Hours	Unit	Material	2016 Bare Costs Labor	Equipment	Total	Total Incl O&P
0010	**VARIABLE VOLUME MIXING BOXES**									
5180	Mixing box, includes electric or pneumatic motor									
5192	Recommend use with attenuator, see Line 23 33 19.10 0010									
5500	VAV Cool only, pneumatic, pressure independent 300 to 600 CFM	Q-9	11	1.455	Ea.	745	75		820	935
5510	500 to 1000 CFM		9	1.778		765	91.50		856.50	980
5520	800 to 1600 CFM		9	1.778		790	91.50		881.50	1,000
5530	1100 to 2000 CFM		8	2		810	103		913	1,050
5540	1500 to 3000 CFM		7	2.286		860	118		978	1,125
5550	2000 to 4000 CFM		6	2.667		870	137		1,007	1,175
5560	For electric, w/thermostat, pressure dependent, add					23			23	25.50
5600	VAV, HW coils, damper, actuator and t'stat									
5610	200 CFM	Q-9	11	1.455	Ea.	805	75		880	1,000
5620	400 CFM		10	1.600		815	82.50		897.50	1,025
5630	600 CFM		10	1.600		815	82.50		897.50	1,025
5640	800 CFM		8	2		840	103		943	1,075

For customer support on your Mechanical Cost Data, call 877.739.3689.

341

23 36 Air Terminal Units

23 36 16 – Variable-Air-Volume Units

23 36 16.10 Variable Volume Mixing Boxes		Crew	Daily Output	Labor-Hours	Unit	Material	2016 Bare Costs Labor	Equipment	Total	Total Incl O&P
5650	1000 CFM	Q-9	8	2	Ea.	840	103		943	1,075
5660	1250 CFM		6	2.667		915	137		1,052	1,200
5670	1500 CFM		6	2.667		915	137		1,052	1,200
5680	2000 CFM		4	4		995	206		1,201	1,425
5684	3000 CFM	↓	3.80	4.211	↓	1,075	217		1,292	1,525
5700	VAV Cool only, fan powered, damper, actuator, thermostat									
5710	200 CFM	Q-9	10	1.600	Ea.	1,100	82.50		1,182.50	1,350
5720	400 CFM		9	1.778		1,175	91.50		1,266.50	1,450
5730	600 CFM		9	1.778		1,175	91.50		1,266.50	1,450
5740	800 CFM		7	2.286		1,250	118		1,368	1,575
5750	1000 CFM		7	2.286		1,250	118		1,368	1,575
5760	1250 CFM		5	3.200		1,350	165		1,515	1,750
5770	1500 CFM		5	3.200		1,350	165		1,515	1,750
5780	2000 CFM	↓	4	4	↓	1,500	206		1,706	1,975
5800	VAV Fan powr'd, with HW coils, dampers, actuators, t'stat									
5810	200 CFM	Q-9	10	1.600	Ea.	1,375	82.50		1,457.50	1,625
5820	400 CFM		9	1.778		1,400	91.50		1,491.50	1,700
5830	600 CFM		9	1.778		1,400	91.50		1,491.50	1,700
5840	800 CFM		7	2.286		1,500	118		1,618	1,825
5850	1000 CFM		7	2.286		1,500	118		1,618	1,825
5860	1250 CFM		5	3.200		1,700	165		1,865	2,125
5870	1500 CFM		5	3.200		1,700	165		1,865	2,125
5880	2000 CFM	↓	3.50	4.571	↓	1,825	236		2,061	2,350

23 37 Air Outlets and Inlets

23 37 13 – Diffusers, Registers, and Grilles

23 37 13.10 Diffusers

			Crew	Daily Output	Labor-Hours	Unit	Material	2016 Bare Costs Labor	Equipment	Total	Total Incl O&P
0010	**DIFFUSERS**, Aluminum, opposed blade damper unless noted										
0100	Ceiling, linear, also for sidewall										
0120	2" wide	R233100-30	1 Shee	32	.250	L.F.	16.80	14.30		31.10	40.50
0140	3" wide			30	.267		19.15	15.25		34.40	44.50
0160	4" wide	R233700-60		26	.308		22.50	17.60		40.10	51.50
0180	6" wide			24	.333		27.50	19.10		46.60	59.50
0200	8" wide			22	.364		32.50	21		53.50	67.50
0220	10" wide			20	.400		37	23		60	75.50
0240	12" wide		↓	18	.444	↓	41.50	25.50		67	84.50
0260	For floor or sill application, add						15%				
0500	Perforated, 24" x 24" lay-in panel size, 6" x 6"		1 Shee	16	.500	Ea.	157	28.50		185.50	216
0520	8" x 8"			15	.533		165	30.50		195.50	228
0530	9" x 9"			14	.571		167	32.50		199.50	234
0540	10" x 10"			14	.571		168	32.50		200.50	234
0560	12" x 12"			12	.667		175	38		213	251
0580	15" x 15"			11	.727		177	41.50		218.50	259
0590	16" x 16"			11	.727		195	41.50		236.50	279
0600	18" x 18"			10	.800		209	46		255	300
0610	20" x 20"			10	.800		226	46		272	320
0620	24" x 24"			9	.889		248	51		299	350
1000	Rectangular, 1 to 4 way blow, 6" x 6"			16	.500		50	28.50		78.50	99
1010	8" x 8"			15	.533		58	30.50		88.50	110
1014	9" x 9"			15	.533		66	30.50		96.50	119
1016	10" x 10"			15	.533		80	30.50		110.50	134

23 37 Air Outlets and Inlets

23 37 13 – Diffusers, Registers, and Grilles

23 37 13.10 Diffusers		Crew	Daily Output	Labor-Hours	Unit	Material	2016 Bare Costs Labor	Equipment	Total	Total Incl O&P
1020	12" x 6"	1 Shee	15	.533	Ea.	71.50	30.50		102	125
1040	12" x 9"		14	.571		75.50	32.50		108	134
1060	12" x 12"		12	.667		84.50	38		122.50	152
1070	14" x 6"		13	.615		77.50	35		112.50	140
1074	14" x 14"		12	.667		128	38		166	199
1080	18" x 12"		11	.727		137	41.50		178.50	215
1120	15" x 15"		10	.800		110	46		156	191
1140	18" x 15"		9	.889		155	51		206	249
1150	18" x 18"		9	.889		138	51		189	229
1160	21" x 21"		8	1		215	57.50		272.50	325
1170	24" x 12"		10	.800		163	46		209	250
1180	24" x 24"		7	1.143		270	65.50		335.50	395
1500	Round, butterfly damper, steel, diffuser size, 6" diameter		18	.444		9.40	25.50		34.90	49.50
1520	8" diameter		16	.500		10	28.50		38.50	55
1540	10" diameter		14	.571		12.40	32.50		44.90	63.50
1560	12" diameter		12	.667		16.35	38		54.35	76.50
1580	14" diameter		10	.800		20.50	46		66.50	92.50
1600	18" diameter	▼	9	.889	▼	49.50	51		100.50	133
1700	Round, steel, adjustable core, annular segmented damper, diffuser size									
1704	6" diameter	1 Shee	18	.444	Ea.	87	25.50		112.50	135
1708	8" diameter		16	.500		103	28.50		131.50	157
1712	10" diameter		14	.571		122	32.50		154.50	184
1716	12" diameter		12	.667		148	38		186	222
1720	14" diameter		10	.800		188	46		234	277
1724	18" diameter		9	.889		263	51		314	365
1728	20" diameter		9	.889		305	51		356	415
1732	24" diameter		8	1		365	57.50		422.50	490
1736	30" diameter		7	1.143		895	65.50		960.50	1,075
1740	36" diameter		6	1.333		1,100	76.50		1,176.50	1,325
2000	T bar mounting, 24" x 24" lay-in frame, 6" x 6"		16	.500		72	28.50		100.50	124
2020	8" x 8"		14	.571		72.50	32.50		105	130
2040	12" x 12"		12	.667		88.50	38		126.50	156
2060	16" x 16"		11	.727		119	41.50		160.50	195
2080	18" x 18"	▼	10	.800	▼	133	46		179	216
2500	Combination supply and return									
2520	21" x 21" supply, 15" x 15" return	Q-9	10	1.600	Ea.	115	82.50		197.50	252
2540	24" x 24" supply, 18" x 18" return		9.50	1.684		145	87		232	292
2560	27" x 27" supply, 18" x 18" return		9	1.778		177	91.50		268.50	335
2580	30" x 30" supply, 21" x 21" return		8.50	1.882		224	97		321	395
2600	33" x 33" supply, 21" x 21" return		8	2		261	103		364	445
2620	36" x 36" supply, 24" x 24" return	▼	7.50	2.133	▼	305	110		415	510
3000	Baseboard, white enameled steel									
3100	18" long	1 Shee	20	.400	Ea.	4.60	23		27.60	40
3120	24" long		18	.444		9	25.50		34.50	49
3140	48" long	▼	16	.500		16.50	28.50		45	62
3400	For matching return, deduct				▼	10%				
4000	Floor, steel, adjustable pattern									
4100	2" x 10"	1 Shee	34	.235	Ea.	6.20	13.45		19.65	27.50
4120	2" x 12"		32	.250		6.80	14.30		21.10	29.50
4140	2" x 14"		30	.267		7.75	15.25		23	32
4200	4" x 10"		28	.286		7.20	16.35		23.55	33
4220	4" x 12"		26	.308		7.95	17.60		25.55	36
4240	4" x 14"	▼	25	.320	▼	8.95	18.30		27.25	38

23 37 13.10 Diffusers

		Crew	Daily Output	Labor-Hours	Unit	Material	2016 Bare Costs Labor	Equipment	Total	Total Incl O&P
4260	6" x 10"	1 Shee	26	.308	Ea.	10.85	17.60		28.45	39
4280	6" x 12"		24	.333		11.30	19.10		30.40	41.50
4300	6" x 14"		22	.364		11.80	21		32.80	45
4500	Linear bar diffuser in wall, sill or ceiling									
4510	2" wide	1 Shee	32	.250	L.F.	55.50	14.30		69.80	83.50
4512	3" wide		30	.267		60.50	15.25		75.75	90
4514	4" wide		26	.308		74.50	17.60		92.10	109
4516	6" wide		24	.333		92	19.10		111.10	130
4520	10" wide		20	.400		123	23		146	170
4522	12" wide		18	.444		138	25.50		163.50	190
5000	Sidewall, aluminum, 3 way dispersion									
5100	8" x 4"	1 Shee	24	.333	Ea.	3.40	19.10		22.50	32.50
5110	8" x 6"		24	.333		3.64	19.10		22.74	33
5120	10" x 4"		22	.364		3.58	21		24.58	36
5130	10" x 6"		22	.364		3.83	21		24.83	36
5140	10" x 8"		20	.400		5.20	23		28.20	41
5160	12" x 4"		18	.444		3.85	25.50		29.35	43
5170	12" x 6"		17	.471		4.26	27		31.26	45.50
5180	12" x 8"		16	.500		5.50	28.50		34	50
5200	14" x 4"		15	.533		4.29	30.50		34.79	51
5220	14" x 6"		14	.571		4.73	32.50		37.23	55
5240	14" x 8"		13	.615		6.15	35		41.15	61
5260	16" x 6"		12.50	.640		6.75	36.50		43.25	63.50
6000	For steel diffusers instead of aluminum, deduct					10%				

23 37 13.30 Grilles

		Crew	Daily Output	Labor-Hours	Unit	Material	2016 Bare Costs Labor	Equipment	Total	Total Incl O&P
0010	**GRILLES**									
0020	Aluminum, unless noted otherwise									
0100	Air supply, single deflection, adjustable									
0120	8" x 4"	1 Shee	30	.267	Ea.	6.95	15.25		22.20	31
0140	8" x 8"		28	.286		8.45	16.35		24.80	34.50
0160	10" x 4"		24	.333		7.70	19.10		26.80	37.50
0180	10" x 10"		23	.348		9.75	19.90		29.65	41.50
0200	12" x 6"		23	.348		9.10	19.90		29	40.50
0220	12" x 12"		22	.364		11.95	21		32.95	45
0230	14" x 6"		23	.348		10.15	19.90		30.05	41.50
0240	14" x 8"		23	.348		10.85	19.90		30.75	42.50
0260	14" x 14"		22	.364		14.65	21		35.65	48
0270	18" x 8"		23	.348		12.50	19.90		32.40	44.50
0280	18" x 10"		23	.348		14.30	19.90		34.20	46.50
0300	18" x 18"		21	.381		22.50	22		44.50	58
0320	20" x 12"		22	.364		16.25	21		37.25	50
0340	20" x 20"		21	.381		27	22		49	63
0350	24" x 8"		20	.400		15.65	23		38.65	52
0360	24" x 14"		18	.444		22	25.50		47.50	63
0380	24" x 24"		15	.533		38	30.50		68.50	88.50
0400	30" x 8"		20	.400		20	23		43	57
0420	30" x 10"		19	.421		22	24		46	61.50
0440	30" x 12"		18	.444		24.50	25.50		50	66
0460	30" x 16"		17	.471		31.50	27		58.50	76
0480	30" x 18"		17	.471		35.50	27		62.50	80
0500	30" x 30"		14	.571		63	32.50		95.50	120
0520	36" x 12"		17	.471		30	27		57	74

23 37 13.30 Grilles		Crew	Daily Output	Labor-Hours	Unit	Material	2016 Bare Costs Labor	2016 Bare Costs Equipment	Total	Total Incl O&P
0540	36" x 16"	1 Shee	16	.500	Ea.	38.50	28.50		67	86.50
0560	36" x 18"		15	.533		42.50	30.50		73	93.50
0570	36" x 20"		15	.533		48.50	30.50		79	99.50
0580	36" x 24"		14	.571		60.50	32.50		93	117
0600	36" x 28"		13	.615		74.50	35		109.50	136
0620	36" x 30"		12	.667		82	38		120	149
0640	36" x 32"		12	.667		90	38		128	158
0660	36" x 34"		11	.727		98	41.50		139.50	172
0680	36" x 36"		11	.727		106	41.50		147.50	181
0700	For double deflecting, add					70%				
1000	Air return, steel, 6" x 6"	1 Shee	26	.308		18.70	17.60		36.30	47.50
1020	10" x 6"		24	.333		18.70	19.10		37.80	49.50
1040	14" x 6"		23	.348		22.50	19.90		42.40	55.50
1060	10" x 8"		23	.348		22.50	19.90		42.40	55.50
1080	16" x 8"		22	.364		26.50	21		47.50	61
1100	12" x 12"		22	.364		26.50	21		47.50	61
1120	24" x 12"		18	.444		36	25.50		61.50	78.50
1140	30" x 12"		16	.500		46	28.50		74.50	95
1160	14" x 14"		22	.364		32	21		53	67
1180	16" x 16"		22	.364		34	21		55	69.50
1200	18" x 18"		21	.381		38	22		60	75.50
1220	24" x 18"		16	.500		43.50	28.50		72	92
1240	36" x 18"		15	.533		68	30.50		98.50	122
1260	24" x 24"		15	.533		53	30.50		83.50	105
1280	36" x 24"		14	.571		75	32.50		107.50	133
1300	48" x 24"		12	.667		128	38		166	199
1320	48" x 30"		11	.727		164	41.50		205.50	245
1340	36" x 36"		13	.615		100	35		135	164
1360	48" x 36"		11	.727		205	41.50		246.50	290
1380	48" x 48"		8	1		250	57.50		307.50	365
2000	Door grilles, 12" x 12"		22	.364		68	21		89	107
2020	18" x 12"		22	.364		78.50	21		99.50	119
2040	24" x 12"		18	.444		88.50	25.50		114	137
2060	18" x 18"		18	.444		109	25.50		134.50	159
2080	24" x 18"		16	.500		124	28.50		152.50	180
2100	24" x 24"		15	.533		175	30.50		205.50	240
3000	Filter grille with filter, 12" x 12"		24	.333		53	19.10		72.10	87
3020	18" x 12"		20	.400		71	23		94	113
3040	24" x 18"		18	.444		83.50	25.50		109	131
3060	24" x 24"		16	.500		98	28.50		126.50	152
3080	30" x 24"		14	.571		119	32.50		151.50	181
3100	30" x 30"		13	.615		141	35		176	209
3950	Eggcrate, framed, 6" x 6" opening		26	.308		15.40	17.60		33	44
3954	8" x 8" opening		24	.333		18.15	19.10		37.25	49
3960	10" x 10" opening		23	.348		23	19.90		42.90	56
3970	12" x 12" opening		22	.364		26	21		47	60.50
3980	14" x 14" opening		22	.364		33.50	21		54.50	69
3984	16" x 16" opening		21	.381		37.50	22		59.50	74.50
3990	18" x 18" opening		21	.381		45	22		67	83
4020	22" x 22" opening		17	.471		53	27		80	99
4040	24" x 24" opening		15	.533		68	30.50		98.50	122
4044	28" x 28" opening		14	.571		102	32.50		134.50	163
4048	36" x 36" opening		12	.667		117	38		155	188

23 37 13.30 Grilles

		Crew	Daily Output	Labor-Hours	Unit	Material	2016 Bare Costs Labor	Equipment	Total	Total Incl O&P
4050	48" x 24" opening	1 Shee	12	.667	Ea.	149	38		187	222
4060	Eggcrate, lay-in, T-bar system									
4070	48" x 24" sheet	1 Shee	40	.200	Ea.	71	11.45		82.45	95.50
5000	Transfer grille, vision proof, 8" x 4"		30	.267		12.85	15.25		28.10	37.50
5020	8" x 6"		28	.286		12.85	16.35		29.20	39
5040	8" x 8"		26	.308		14.35	17.60		31.95	43
5060	10" x 6"		24	.333		13.15	19.10		32.25	43.50
5080	10" x 10"		23	.348		16.90	19.90		36.80	49
5090	12" x 6"		23	.348		14.35	19.90		34.25	46.50
5100	12" x 10"		22	.364		18.45	21		39.45	52.50
5110	12" x 12"		22	.364		20	21		41	54
5120	14" x 10"		22	.364		20	21		41	54
5140	16" x 8"		21	.381		20	22		42	55.50
5160	18" x 12"		20	.400		26	23		49	63.50
5170	18" x 18"		21	.381		37	22		59	74
5180	20" x 12"		20	.400		28	23		51	66
5200	20" x 20"		19	.421		44.50	24		68.50	86
5220	24" x 12"		17	.471		31.50	27		58.50	76
5240	24" x 24"		16	.500		61	28.50		89.50	111
5260	30" x 6"		15	.533		28	30.50		58.50	77.50
5280	30" x 8"		15	.533		33	30.50		63.50	82.50
5300	30" x 12"		14	.571		41	32.50		73.50	95
5320	30" x 16"		13	.615		48.50	35		83.50	108
5340	30" x 20"		12	.667		61.50	38		99.50	126
5360	30" x 24"		11	.727		75.50	41.50		117	147
5380	30" x 30"		10	.800		100	46		146	180
6000	For steel grilles instead of aluminum in above, deduct					10%				
6200	Plastic, eggcrate, lay-in, T-bar system									
6210	48" x 24" sheet	1 Shee	50	.160	Ea.	19.25	9.15		28.40	35
6250	Steel door louver									
6270	With fire link, steel only									
6350	12" x 18"	1 Shee	22	.364	Ea.	208	21		229	261
6410	12" x 24"		21	.381		238	22		260	296
6470	18" x 18"		20	.400		231	23		254	289
6530	18" x 24"		19	.421		266	24		290	330
6660	24" x 24"		15	.533		269	30.50		299.50	345

23 37 13.60 Registers

		Crew	Daily Output	Labor-Hours	Unit	Material	2016 Bare Costs Labor	Equipment	Total	Total Incl O&P
0010	**REGISTERS**									
0980	Air supply									
1000	Ceiling/wall, O.B. damper, anodized aluminum									
1010	One or two way deflection, adj. curved face bars									
1014	6" x 6"	1 Shee	24	.333	Ea.	12	19.10		31.10	42
1020	8" x 4"		26	.308		12	17.60		29.60	40
1040	8" x 8"		24	.333		15	19.10		34.10	45.50
1060	10" x 6"		20	.400		13.80	23		36.80	50
1080	10" x 10"		19	.421		18.60	24		42.60	57.50
1100	12" x 6"		19	.421		13.95	24		37.95	52.50
1120	12" x 12"		18	.444		22	25.50		47.50	63
1140	14" x 8"		17	.471		19.10	27		46.10	62
1160	14" x 14"		18	.444		25.50	25.50		51	67
1170	16" x 16"		17	.471		28.50	27		55.50	72.50
1180	18" x 8"		18	.444		21.50	25.50		47	62.50

23 37 Air Outlets and Inlets

23 37 13 – Diffusers, Registers, and Grilles

23 37 13.60 Registers		Crew	Daily Output	Labor-Hours	Unit	Material	2016 Bare Costs Labor	Equipment	Total	Total Incl O&P
1200	18" x 18"	1 Shee	17	.471	Ea.	39	27		66	84
1220	20" x 4"		19	.421		19.60	24		43.60	58.50
1240	20" x 6"		18	.444		19.60	25.50		45.10	60.50
1260	20" x 8"		18	.444		21.50	25.50		47	62.50
1280	20" x 20"		17	.471		47.50	27		74.50	93
1290	22" x 22"		15	.533		54.50	30.50		85	107
1300	24" x 4"		17	.471		24	27		51	67.50
1320	24" x 6"		16	.500		24	28.50		52.50	70.50
1340	24" x 8"		13	.615		26.50	35		61.50	83
1350	24" x 18"		12	.667		49	38		87	113
1360	24" x 24"		11	.727		67	41.50		108.50	137
1380	30" x 4"		16	.500		30.50	28.50		59	77.50
1400	30" x 6"		15	.533		30.50	30.50		61	80
1420	30" x 8"		14	.571		34.50	32.50		67	88
1440	30" x 24"		12	.667		86.50	38		124.50	154
1460	30" x 30"		10	.800		111	46		157	192
1504	4 way deflection, adjustable curved face bars									
1510	6" x 6"	1 Shee	26	.308	Ea.	14.40	17.60		32	43
1514	8" x 8"		24	.333		18	19.10		37.10	49
1518	10" x 10"		19	.421		22.50	24		46.50	61.50
1522	12" x 6"		19	.421		16.75	24		40.75	55.50
1526	12" x 12"		18	.444		26.50	25.50		52	68
1530	14" x 14"		18	.444		30.50	25.50		56	73
1534	16" x 16"		17	.471		34	27		61	78.50
1538	18" x 18"		17	.471		47	27		74	92.50
1542	22" x 22"		16	.500		65.50	28.50		94	116
1980	One way deflection, adj. vert. or horiz. face bars									
1990	6" x 6"	1 Shee	26	.308	Ea.	13.10	17.60		30.70	41.50
2000	8" x 4"		26	.308		12.30	17.60		29.90	40.50
2020	8" x 8"		24	.333		14.65	19.10		33.75	45
2040	10" x 6"		20	.400		14.75	23		37.75	51
2060	10" x 10"		19	.421		17.20	24		41.20	56
2080	12" x 6"		19	.421		15.95	24		39.95	54.50
2100	12" x 12"		18	.444		20.50	25.50		46	62
2120	14" x 8"		17	.471		19.05	27		46.05	62
2140	14" x 12"		18	.444		23	25.50		48.50	64
2160	14" x 14"		18	.444		25.50	25.50		51	67.50
2180	16" x 6"		18	.444		17.20	25.50		42.70	58
2200	16" x 12"		18	.444		25	25.50		50.50	66.50
2220	16" x 16"		17	.471		33	27		60	77.50
2240	18" x 8"		18	.444		22	25.50		47.50	63.50
2260	18" x 12"		17	.471		27	27		54	70.50
2280	18" x 18"		16	.500		39	28.50		67.50	87
2300	20" x 10"		18	.444		26.50	25.50		52	68
2320	20" x 16"		16	.500		36.50	28.50		65	84
2340	20" x 20"		16	.500		48	28.50		76.50	96.50
2360	24" x 12"		15	.533		33	30.50		63.50	83
2380	24" x 16"		14	.571		44	32.50		76.50	98.50
2400	24" x 20"		12	.667		55	38		93	119
2420	24" x 24"		10	.800		67.50	46		113.50	144
2440	30" x 12"		13	.615		43	35		78	101
2460	30" x 16"		13	.615		56	35		91	116
2480	30" x 24"		11	.727		87	41.50		128.50	159

For customer support on your Mechanical Cost Data, call 877.739.3689.

347

23 37 13.60 Registers		Crew	Daily Output	Labor-Hours	Unit	Material	2016 Bare Costs Labor	Equipment	Total	Total Incl O&P
2500	30" x 30"	1 Shee	9	.889	Ea.	111	51		162	201
2520	36" x 12"		11	.727		53.50	41.50		95	122
2540	36" x 24"		10	.800		107	46		153	188
2560	36" x 36"		8	1		179	57.50		236.50	285
2600	For 2 way deflect., adj. vert. or horiz. face bars, add					40%				
2700	Above registers in steel instead of aluminum, deduct					10%				
3000	Baseboard, hand adj. damper, enameled steel									
3012	8" x 6"	1 Shee	26	.308	Ea.	5.05	17.60		22.65	32.50
3020	10" x 6"		24	.333		5.50	19.10		24.60	35
3040	12" x 5"		23	.348		6	19.90		25.90	37
3060	12" x 6"		23	.348		6.75	19.90		26.65	38
3080	12" x 8"		22	.364		8.70	21		29.70	41.50
3100	14" x 6"		20	.400		6.50	23		29.50	42
4000	Floor, toe operated damper, enameled steel									
4020	4" x 8"	1 Shee	32	.250	Ea.	9.40	14.30		23.70	32.50
4040	4" x 12"		26	.308		11	17.60		28.60	39
4060	6" x 8"		28	.286		10.10	16.35		26.45	36
4080	6" x 14"		22	.364		12.75	21		33.75	46
4100	8" x 10"		22	.364		11.45	21		32.45	44.50
4120	8" x 16"		20	.400		15.40	23		38.40	52
4140	10" x 10"		20	.400		13.75	23		36.75	50
4160	10" x 16"		18	.444		20.50	25.50		46	62
4180	12" x 12"		18	.444		16.85	25.50		42.35	57.50
4200	12" x 24"		16	.500		28	28.50		56.50	74.50
4220	14" x 14"		16	.500		44	28.50		72.50	92.50
4240	14" x 20"		15	.533		51.50	30.50		82	103
4300	Spiral pipe supply register									
4310	Steel, with air scoop									
4320	4" x 12", for 8" thru 13" diameter duct	1 Shee	25	.320	Ea.	74.50	18.30		92.80	110
4330	4" x 18", for 8" thru 13" diameter duct		18	.444		87	25.50		112.50	135
4340	6" x 12", for 14" thru 21" diameter duct		19	.421		78.50	24		102.50	124
4350	6" x 16", for 14" thru 21" diameter duct		18	.444		89	25.50		114.50	137
4360	6" x 20", for 14" thru 21" diameter duct		17	.471		98.50	27		125.50	149
4370	6" x 24", for 14" thru 21" diameter duct		16	.500		116	28.50		144.50	171
4380	8" x 16", for 22" thru 31" diameter duct		19	.421		94	24		118	140
4390	8" x 18", for 22" thru 31" diameter duct		18	.444		98.50	25.50		124	147
4400	8" x 24", for 22" thru 31" diameter duct		15	.533		122	30.50		152.50	181
4980	Air return									
5000	Ceiling or wall, fixed 45° face blades									
5010	Adjustable O.B. damper, anodized aluminum									
5020	4" x 8"	1 Shee	26	.308	Ea.	12.30	17.60		29.90	40.50
5040	6" x 8"		24	.333		13.50	19.10		32.60	44
5060	6" x 10"		19	.421		14.75	24		38.75	53
5070	6" x 12"		19	.421		15.95	24		39.95	54.50
5080	6" x 16"		18	.444		19.05	25.50		44.55	60
5100	8" x 10"		19	.421		16.05	24		40.05	54.50
5120	8" x 12"		16	.500		17.90	28.50		46.40	63.50
5140	10" x 10"		18	.444		17.20	25.50		42.70	58
5160	10" x 16"		17	.471		23	27		50	66
5170	12" x 12"		18	.444		20.50	25.50		46	62
5180	12" x 18"		18	.444		27	25.50		52.50	68.50
5184	12" x 24"		14	.571		33	32.50		65.50	86.50
5200	12" x 30"		12	.667		43	38		81	106

For customer support on your Mechanical Cost Data, call 877.739.3689.

23 37 13 – Diffusers, Registers, and Grilles

23 37 13.60 Registers		Crew	Daily Output	Labor-Hours	Unit	Material	2016 Bare Costs Labor	Equipment	Total	Total Incl O&P
5210	12" x 36"	1 Shee	10	.800	Ea.	53.50	46		99.50	129
5220	16" x 16"		17	.471		33	27		60	77.50
5240	18" x 18"		16	.500		39	28.50		67.50	87
5250	18" x 24"		13	.615		49.50	35		84.50	109
5254	18" x 30"		11	.727		62.50	41.50		104	133
5260	18" x 36"		10	.800		75.50	46		121.50	153
5270	20" x 20"		10	.800		48	46		94	123
5274	20" x 24"		12	.667		55	38		93	119
5280	24" x 24"		11	.727		67.50	41.50		109	138
5290	24" x 30"		10	.800		87	46		133	166
5300	24" x 36"		8	1		107	57.50		164.50	206
5320	24" x 48"		6	1.333		145	76.50		221.50	276
5340	30" x 30"		6.50	1.231		111	70.50		181.50	231
5344	30" x 120"	Q-9	6	2.667		1,000	137		1,137	1,300
5360	36" x 36"	1 Shee	6	1.333		179	76.50		255.50	315
5370	40" x 32"	"	6	1.333		166	76.50		242.50	299
5400	Ceiling or wall, removable-reversible core, single deflection									
5402	Adjustable O.B. damper, radiused frame with border, aluminum									
5410	8" x 4"	1 Shee	26	.308	Ea.	18.70	17.60		36.30	47.50
5414	8" x 6"		24	.333		18.70	19.10		37.80	49.50
5418	10" x 6"		19	.421		24	24		48	63
5422	10" x 10"		18	.444		27	25.50		52.50	69
5426	12" x 6"		19	.421		30.50	24		54.50	70.50
5430	12" x 12"		18	.444		41.50	25.50		67	85
5434	16" x 16"		16	.500		46	28.50		74.50	94.50
5438	18" x 18"		17	.471		61	27		88	108
5442	20" x 20"		16	.500		69.50	28.50		98	121
5446	24" x 12"		15	.533		46	30.50		76.50	97
5450	24" x 18"		13	.615		69.50	35		104.50	131
5454	24" x 24"		11	.727		97	41.50		138.50	171
5458	30" x 12"		13	.615		61	35		96	122
5462	30" x 18"		12.50	.640		88.50	36.50		125	153
5466	30" x 24"		11	.727		208	41.50		249.50	292
5470	30" x 30"		9	.889		219	51		270	320
5474	36" x 12"		11	.727		76.50	41.50		118	148
5478	36" x 18"		10.50	.762		107	43.50		150.50	185
5482	36" x 24"		10	.800		212	46		258	305
5486	36" x 30"		9	.889		229	51		280	330
5490	36" x 36"		8	1		258	57.50		315.50	370
6000	For steel construction instead of aluminum, deduct					10%				

23 37 15 – Louvers

23 37 15.40 HVAC Louvers

		Crew	Daily Output	Labor-Hours	Unit	Material	2016 Bare Costs Labor	Equipment	Total	Total Incl O&P
0010	**HVAC LOUVERS**									
0100	Aluminum, extruded, with screen, mill finish									
1002	Brick vent, see also Section 04 05 23.19									
1100	Standard, 4" deep, 8" wide, 5" high	1 Shee	24	.333	Ea.	35	19.10		54.10	67.50
1200	Modular, 4" deep, 7-3/4" wide, 5" high		24	.333		36.50	19.10		55.60	69
1300	Speed brick, 4" deep, 11-5/8" wide, 3-7/8" high		24	.333		36.50	19.10		55.60	69
1400	Fuel oil brick, 4" deep, 8" wide, 5" high		24	.333		63	19.10		82.10	98
2000	Cooling tower and mechanical equip., screens, light weight		40	.200	S.F.	15.70	11.45		27.15	35
2020	Standard weight		35	.229		41.50	13.10		54.60	66
2500	Dual combination, automatic, intake or exhaust		20	.400		57	23		80	98

For customer support on your Mechanical Cost Data, call 877.739.3689.

349

23 37 15 – Louvers

23 37 15.40 HVAC Louvers

		Crew	Daily Output	Labor-Hours	Unit	Material	2016 Bare Costs Labor	Equipment	Total	Total Incl O&P
2520	Manual operation	1 Shee	20	.400	S.F.	42.50	23		65.50	82
2540	Electric or pneumatic operation		20	.400	↓	42.50	23		65.50	82
2560	Motor, for electric or pneumatic	↓	14	.571	Ea.	490	32.50		522.50	590
3000	Fixed blade, continuous line									
3100	Mullion type, stormproof	1 Shee	28	.286	S.F.	42.50	16.35		58.85	72
3200	Stormproof		28	.286		42.50	16.35		58.85	72
3300	Vertical line	↓	28	.286	↓	50.50	16.35		66.85	80.50
3500	For damper to use with above, add					50%	30%			
3520	Motor, for damper, electric or pneumatic	1 Shee	14	.571	Ea.	490	32.50		522.50	590
4000	Operating, 45°, manual, electric or pneumatic		24	.333	S.F.	51	19.10		70.10	85
4100	Motor, for electric or pneumatic		14	.571	Ea.	490	32.50		522.50	590
4200	Penthouse, roof		56	.143	S.F.	25	8.20		33.20	40
4300	Walls		40	.200		59	11.45		70.45	82.50
5000	Thinline, under 4" thick, fixed blade	↓	40	.200	↓	24.50	11.45		35.95	44.50
5010	Finishes, applied by mfr. at additional cost, available in colors									
5020	Prime coat only, add				S.F.	3.37			3.37	3.71
5040	Baked enamel finish coating, add					6.20			6.20	6.85
5060	Anodized finish, add					6.75			6.75	7.40
5080	Duranodic finish, add					12.25			12.25	13.45
5100	Fluoropolymer finish coating, add				↓	19.30			19.30	21
9000	Stainless steel, fixed blade, continuous line									
9010	Hospital grade									
9110	20 Ga.	1 Shee	22	.364	S.F.	41	21		62	77
9980	For small orders (under 10 pieces), add				"	25%				

23 37 23 – HVAC Gravity Ventilators

23 37 23.10 HVAC Gravity Air Ventilators

		Crew	Daily Output	Labor-Hours	Unit	Material	2016 Bare Costs Labor	Equipment	Total	Total Incl O&P
0010	**HVAC GRAVITY AIR VENTILATORS**, Includes base									
1280	Rotary ventilators, wind driven, galvanized									
1300	4" neck diameter	Q-9	20	.800	Ea.	47.50	41		88.50	116
1320	5" neck diameter		18	.889		47.50	46		93.50	123
1340	6" neck diameter		16	1		48	51.50		99.50	132
1360	8" neck diameter		14	1.143		56	59		115	152
1380	10" neck diameter		12	1.333		61.50	68.50		130	173
1400	12" neck diameter		10	1.600		67	82.50		149.50	200
1420	14" neck diameter		10	1.600		113	82.50		195.50	250
1440	16" neck diameter		9	1.778		132	91.50		223.50	285
1460	18" neck diameter		9	1.778		159	91.50		250.50	315
1480	20" neck diameter		8	2		189	103		292	365
1500	24" neck diameter		8	2		299	103		402	490
1520	30" neck diameter		7	2.286		360	118		478	575
1540	36" neck diameter	↓	6	2.667	↓	575	137		712	840
1600	For aluminum, add					30%				
2000	Stationary, gravity, syphon, galvanized									
2100	3" neck diameter, 40 CFM	Q-9	24	.667	Ea.	28	34.50		62.50	83.50
2120	4" neck diameter, 50 CFM		20	.800		30	41		71	96
2140	5" neck diameter, 58 CFM		18	.889		33	46		79	106
2160	6" neck diameter, 66 CFM		16	1		34	51.50		85.50	116
2180	7" neck diameter, 86 CFM		15	1.067		45	55		100	134
2200	8" neck diameter, 110 CFM		14	1.143		46	59		105	141
2220	10" neck diameter, 140 CFM		12	1.333		63	68.50		131.50	175
2240	12" neck diameter, 160 CFM		10	1.600		78.50	82.50		161	212
2260	14" neck diameter, 250 CFM		10	1.600		124	82.50		206.50	262

23 37 23.10 HVAC Gravity Air Ventilators	Crew	Daily Output	Labor-Hours	Unit	Material	2016 Bare Costs Labor	Equipment	Total	Total Incl O&P
2280 16" neck diameter, 380 CFM	Q-9	9	1.778	Ea.	160	91.50		251.50	315
2300 18" neck diameter, 500 CFM		9	1.778		180	91.50		271.50	340
2320 20" neck diameter, 625 CFM		8	2		214	103		317	395
2340 24" neck diameter, 900 CFM		8	2		266	103		369	450
2360 30" neck diameter, 1375 CFM		7	2.286		330	118		448	545
2380 36" neck diameter, 2,000 CFM		6	2.667		415	137		552	665
2400 42" neck diameter, 3000 CFM		4	4		670	206		876	1,050
2410 48" neck diameter, 4000 CFM		3	5.333		895	275		1,170	1,400
2500 For aluminum, add					30%				
2520 For stainless steel, add					60%				
2530 For copper, add					100%				
3000 Rotating chimney cap, galvanized, 4" neck diameter	Q-9	20	.800		19.40	41		60.40	84.50
3020 5" neck diameter		18	.889		19.40	46		65.40	91.50
3040 6" neck diameter		16	1		19.40	51.50		70.90	101
3060 7" neck diameter		15	1.067		25	55		80	112
3080 8" neck diameter		14	1.143		25	59		84	118
3100 10" neck diameter		12	1.333		33	68.50		101.50	142
3600 Stationary chimney rain cap, galvanized, 3" neck diameter		24	.667		3.19	34.50		37.69	56
3620 4" neck diameter		20	.800		3.49	41		44.49	67
3640 6" neck diameter		16	1		4.15	51.50		55.65	83.50
3680 8" neck diameter		14	1.143		5.20	59		64.20	96
3700 10" neck diameter		12	1.333		6.30	68.50		74.80	112
3720 12" neck diameter		10	1.600		9.25	82.50		91.75	136
3740 14" neck diameter		10	1.600		14.70	82.50		97.20	142
3760 16" neck diameter		9	1.778		15.45	91.50		106.95	157
3770 18" neck diameter		9	1.778		24.50	91.50		116	167
3780 20" neck diameter		8	2		33	103		136	194
3790 24" neck diameter		8	2		41.50	103		144.50	204
4200 Stationary mushroom, aluminum, 16" orifice diameter		10	1.600		645	82.50		727.50	835
4220 26" orifice diameter		6.15	2.602		955	134		1,089	1,250
4230 30" orifice diameter		5.71	2.802		1,400	144		1,544	1,750
4240 38" orifice diameter		5	3.200		2,000	165		2,165	2,450
4250 42" orifice diameter		4.70	3.404		2,650	175		2,825	3,175
4260 50" orifice diameter		4.44	3.604		3,150	186		3,336	3,750
5000 Relief vent									
5500 Rectangular, aluminum, galvanized curb									
5510 intake/exhaust, 0.033" SP									
5580 500 CFM, 12" x 12"	Q-9	8.60	1.860	Ea.	705	96		801	925
5600 600 CFM, 12" x 16"		8	2		795	103		898	1,025
5620 750 CFM, 12" x 20"		7.20	2.222		850	115		965	1,100
5640 1000 CFM, 12" x 24"		6.60	2.424		890	125		1,015	1,175
5660 1500 CFM, 12" x 36"		5.80	2.759		1,200	142		1,342	1,550
5680 3000 CFM, 20" x 42"		4	4		1,550	206		1,756	2,050
5700 5000 CFM, 20" x 72"		3	5.333		2,250	275		2,525	2,900
5720 6000 CFM, 24" x 72"		2.30	6.957		2,425	360		2,785	3,225
5740 10,000 CFM, 48" x 60"		1.80	8.889		3,075	460		3,535	4,075
5760 12,000 CFM, 48" x 72"		1.60	10		3,625	515		4,140	4,800
5780 13,750 CFM, 60" x 66"		1.40	11.429		4,025	590		4,615	5,325
5800 15,000 CFM, 60" x 72"		1.30	12.308		4,300	635		4,935	5,700
5820 18,000 CFM, 72" x 72"		1.20	13.333		4,875	685		5,560	6,400
5880 Size is throat area, volume is at 500 fpm									
7000 Note: sizes based on exhaust. Intake, with 0.125" SP									
7100 loss, approximately twice listed capacity.									

For customer support on your Mechanical Cost Data, call 877.739.3689.

351

23 38 Ventilation Hoods

23 38 13 – Commercial-Kitchen Hoods

23 38 13.10 Hood and Ventilation Equipment

		Crew	Daily Output	Labor-Hours	Unit	Material	2016 Bare Costs Labor	Equipment	Total	Total Incl O&P
0010	**HOOD AND VENTILATION EQUIPMENT**									
2970	Exhaust hood, sst, gutter on all sides, 4' x 4' x 2'	1 Carp	1.80	4.444	Ea.	4,725	215		4,940	5,525
2980	4' x 4' x 7'	"	1.60	5		7,525	242		7,767	8,650
7800	Vent hood, wall canopy with fire protection, 30"	L-3A	9	1.333		425	70.50		495.50	575
7810	Without fire protection, 36"		10	1.200		460	63.50		523.50	600
7820	Island canopy with fire protection, 30"		7	1.714		790	90.50		880.50	1,000
7830	Without fire protection, 36"		8	1.500		800	79		879	1,000
7840	Back shelf with fire protection, 30"		11	1.091		400	57.50		457.50	530
7850	Without fire protection, black, 36"		12	1		400	52.50		452.50	520
7852	Without fire protection, stainless steel, 36"		12	1		500	52.50		552.50	630
7860	Range hood & CO_2 system, 30"	1 Carp	2.50	3.200		4,050	155		4,205	4,700
7950	Hood fire protection system, electric stove	Q-1	3	5.333		1,875	284		2,159	2,500
7952	Hood fire protection system, gas stove	"	3	5.333		2,200	284		2,484	2,850

23 38 13.20 Kitchen Ventilation

		Crew	Daily Output	Labor-Hours	Unit	Material	2016 Bare Costs Labor	Equipment	Total	Total Incl O&P
0010	**KITCHEN VENTILATION**, Commercial									
6000	Exhaust hoods and fans									
6100	Centrifugal grease extraction									
6110	Water wash type									
6120	Per foot of length	Q-10	7.10	3.380	L.F.	1,800	181		1,981	2,275
6200	Non water wash type									
6210	Per foot of length	Q-10	13.40	1.791	L.F.	1,325	95.50		1,420.50	1,600
6400	Island style ventilator									
6410	Water wash type									
6420	Per foot of length	Q-10	7.10	3.380	L.F.	2,400	181		2,581	2,925
6460	Non water wash type									
6470	Per foot of length	Q-10	13.40	1.791	L.F.	830	95.50		925.50	1,050

23 41 Particulate Air Filtration

23 41 13 – Panel Air Filters

23 41 13.10 Panel Type Air Filters

		Crew	Daily Output	Labor-Hours	Unit	Material	2016 Bare Costs Labor	Equipment	Total	Total Incl O&P
0010	**PANEL TYPE AIR FILTERS**									
2950	Mechanical media filtration units									
3000	High efficiency type, with frame, non-supported [G]				MCFM	35			35	38.50
3100	Supported type [G]				"	50			50	55
5500	Throwaway glass or paper media type, 12" x 36" x 1"				Ea.	2.39			2.39	2.63

23 41 16 – Renewable-Media Air Filters

23 41 16.10 Disposable Media Air Filters

		Crew	Daily Output	Labor-Hours	Unit	Material	2016 Bare Costs Labor	Equipment	Total	Total Incl O&P
0010	**DISPOSABLE MEDIA AIR FILTERS**									
5000	Renewable disposable roll				C.S.F.	5.25			5.25	5.75
5800	Filter, bag type									
5810	90-95% efficiency									
5820	24" x 12" x 29", .75-1.25 MCFM	1 Shee	1.90	4.211	Ea.	57.50	241		298.50	435
5830	24" x 24" x 29", 1.5-2.5 MCFM	"	1.90	4.211	"	101	241		342	480
5850	80-85% efficiency									
5860	24" x 12" x 29", .75-1.25 MCFM	1 Shee	1.90	4.211	Ea.	51	241		292	425
5870	24" x 24" x 29", 1.5-2.5 MCFM	"	1.90	4.211	"	94	241		335	475

23 41 Particulate Air Filtration

23 41 19 – Washable Air Filters

23 41 19.10 Permanent Air Filters	Crew	Daily Output	Labor-Hours	Unit	Material	2016 Bare Costs Labor	Equipment	Total	Total Incl O&P
0010 **PERMANENT AIR FILTERS**									
4500 Permanent washable G				MCFM	20			20	22

23 41 23 – Extended Surface Filters

23 41 23.10 Expanded Surface Filters

	Crew	Daily Output	Labor-Hours	Unit	Material	Labor	Equipment	Total	Total Incl O&P
0010 **EXPANDED SURFACE FILTERS**									
4000 Medium efficiency, extended surface G				MCFM	6			6	6.60

23 41 33 – High-Efficiency Particulate Filtration

23 41 33.10 HEPA Filters

	Crew	Daily Output	Labor-Hours	Unit	Material	Labor	Equipment	Total	Total Incl O&P
0010 **HEPA FILTERS**									
6000 HEPA filter complete w/particle board,									
6010 kraft paper frame, separator material									
6020 95% DOP efficiency									
6030 12" x 12" x 6", 150 CFM	1 Shee	3.70	2.162	Ea.	44	124		168	238
6034 24" x 12" x 6", 375 CFM		1.90	4.211		73.50	241		314.50	450
6038 24" x 18" x 6", 450 CFM		1.90	4.211		90	241		331	470
6042 24" x 24" x 6", 700 CFM		1.90	4.211		101	241		342	480
6046 12" x 12" x 12", 250 CFM		3.70	2.162		74.50	124		198.50	271
6050 24" x 12" x 12", 500 CFM		1.90	4.211		105	241		346	485
6054 24" x 18" x 12", 875 CFM		1.90	4.211		146	241		387	530
6058 24" x 24" x 12", 1000 CFM		1.90	4.211		158	241		399	545
6100 99% DOP efficiency									
6110 12" x 12" x 6", 150 CFM	1 Shee	3.70	2.162	Ea.	46	124		170	240
6114 24" x 12" x 6", 325 CFM		1.90	4.211		76	241		317	455
6118 24" x 18" x 6", 550 CFM		1.90	4.211		93	241		334	470
6122 24" x 24" x 6", 775 CFM		1.90	4.211		105	241		346	485
6124 24" x 48" x 6", 775 CFM		1.30	6.154		201	350		551	760
6126 12" x 12" x 12", 250 CFM		3.60	2.222		77	127		204	279
6130 24" x 12" x 12", 500 CFM		1.90	4.211		109	241		350	490
6134 24" x 18" x 12", 775 CFM		1.90	4.211		152	241		393	535
6138 24" x 24" x 12", 1100 CFM		1.90	4.211		164	241		405	550
6500 HEPA filter housing, 14 ga. galv. sheet metal									
6510 12" x 12" x 6"	1 Shee	2.50	3.200	Ea.	660	183		843	1,000
6514 24" x 12" x 6"		2	4		740	229		969	1,150
6518 12" x 12" x 12"		2.40	3.333		660	191		851	1,025
6522 14" x 12" x 12"		2.30	3.478		660	199		859	1,025
6526 24" x 18" x 6"		1.90	4.211		940	241		1,181	1,400
6530 24" x 24" x 6"		1.80	4.444		855	254		1,109	1,325
6534 24" x 48" x 6"		1.70	4.706		1,150	269		1,419	1,650
6538 24" x 72" x 6"		1.60	5		1,475	286		1,761	2,075
6542 24" x 18" x 12"		1.80	4.444		940	254		1,194	1,425
6546 24" x 24" x 12"		1.70	4.706		1,050	269		1,319	1,550
6550 24" x 48" x 12"		1.60	5		1,225	286		1,511	1,800
6554 24" x 72" x 12"		1.50	5.333		1,525	305		1,830	2,175
6558 48" x 48" x 6"	Q-9	2.80	5.714		1,700	294		1,994	2,300
6562 48" x 72" x 6"		2.60	6.154		2,025	315		2,340	2,700
6566 48" x 96" x 6"		2.40	6.667		2,575	345		2,920	3,350
6570 48" x 48" x 12"		2.70	5.926		1,700	305		2,005	2,325
6574 48" x 72" x 12"		2.50	6.400		2,025	330		2,355	2,725
6578 48" x 96" x 12"		2.30	6.957		2,575	360		2,935	3,375
6582 114" x 72" x 12"		2	8		3,550	410		3,960	4,525

For customer support on your Mechanical Cost Data, call 877.739.3689.

353

23 42 Gas-Phase Air Filtration

23 42 13 – Activated-Carbon Air Filtration

23 42 13.10 Charcoal Type Air Filtration

		Crew	Daily Output	Labor-Hours	Unit	Material	2016 Bare Costs Labor	Equipment	Total	Total Incl O&P
0010	**CHARCOAL TYPE AIR FILTRATION**									
0050	Activated charcoal type, full flow				MCFM	600			600	660
0060	Full flow, impregnated media 12" deep					225			225	248
0070	HEPA filter & frame for field erection					400			400	440
0080	HEPA filter-diffuser, ceiling install.				↓	300			300	330

23 43 Electronic Air Cleaners

23 43 13 – Washable Electronic Air Cleaners

23 43 13.10 Electronic Air Cleaners

		Crew	Daily Output	Labor-Hours	Unit	Material	2016 Bare Costs Labor	Equipment	Total	Total Incl O&P
0010	**ELECTRONIC AIR CLEANERS**									
2000	Electronic air cleaner, duct mounted									
2150	1000 CFM	1 Shee	4	2	Ea.	420	115		535	640
2200	1200 CFM		3.80	2.105		485	121		606	720
2250	1400 CFM		3.60	2.222		500	127		627	745
2260	2000 CFM	↓	3.20	2.500	↓	590	143		733	865

23 51 Breechings, Chimneys, and Stacks

23 51 13 – Draft Control Devices

23 51 13.13 Draft-Induction Fans

		Crew	Daily Output	Labor-Hours	Unit	Material	2016 Bare Costs Labor	Equipment	Total	Total Incl O&P
0010	**DRAFT-INDUCTION FANS**									
1000	Breeching installation									
1800	Hot gas, 600°F, variable pitch pulley and motor									
1840	6" diam. inlet, 1/4 H.P., 1 phase, 400 CFM	Q-9	6	2.667	Ea.	1,500	137		1,637	1,850
1860	8" diam. inlet, 1/4 H.P., 1 phase, 1120 CFM		4	4		2,150	206		2,356	2,675
1870	9" diam. inlet, 3/4 H.P., 1 phase, 1440 CFM		3.60	4.444		2,650	229		2,879	3,250
1880	10" diam. inlet, 3/4 H.P., 1 phase, 2000 CFM		3.30	4.848		2,650	250		2,900	3,275
1900	12" diam. inlet, 3/4 H.P., 3 phase, 2960 CFM		3	5.333		2,925	275		3,200	3,625
1910	14" diam. inlet, 1 H.P., 3 phase, 4160 CFM		2.60	6.154		2,950	315		3,265	3,700
1920	16" diam. inlet, 2 H.P., 3 phase, 5500 CFM		2.30	6.957		3,250	360		3,610	4,125
1950	20" diam. inlet, 3 H.P., 3 phase, 9760 CFM		1.50	10.667		4,675	550		5,225	6,000
1960	22" diam. inlet, 5 H.P., 3 phase, 13,360 CFM		1	16		7,625	825		8,450	9,625
1980	24" diam. inlet, 7-1/2 H.P., 3 phase, 17,760 CFM	↓	.80	20	↓	8,375	1,025		9,400	10,800
2300	For multi-blade damper at fan inlet, add				↓	20%				
3600	Chimney-top installation									
3700	6" size	1 Shee	8	1	Ea.	985	57.50		1,042.50	1,175
3740	8" size		7	1.143		990	65.50		1,055.50	1,200
3750	10" size		6.50	1.231		1,300	70.50		1,370.50	1,525
3780	13" size	↓	6	1.333		1,325	76.50		1,401.50	1,600
3880	For speed control switch, add					75.50			75.50	83
3920	For thermal fan control, add				↓	75.50			75.50	83
5500	Flue blade style damper for draft control,									
5510	locking quadrant blade									
5550	8" size	Q-9	8	2	Ea.	405	103		508	605
5560	9" size		7.50	2.133		405	110		515	615
5570	10" size		7	2.286		415	118		533	635
5580	12" size		6.50	2.462		420	127		547	660
5590	14" size		6	2.667		430	137		567	685
5600	16" size		5.50	2.909		445	150		595	720
5610	18" size	↓	5	3.200	↓	465	165		630	760

23 51 13 – Draft Control Devices

23 51 13.13 Draft-Induction Fans

23 51 13.13 Draft-Induction Fans		Crew	Daily Output	Labor-Hours	Unit	Material	2016 Bare Costs Labor	Equipment	Total	Total Incl O&P
5620	20" size	Q-9	4.50	3.556	Ea.	490	183		673	820
5630	22" size		4	4		515	206		721	880
5640	24" size		3.50	4.571		545	236		781	955
5650	27" size		3	5.333		570	275		845	1,050
5660	30" size		2.50	6.400		625	330		955	1,200
5670	32" size		2	8		655	410		1,065	1,350
5680	36" size	↓	1.50	10.667	↓	725	550		1,275	1,625

23 51 13.16 Vent Dampers

		Crew	Daily Output	Labor-Hours	Unit	Material	Labor	Equipment	Total	Total Incl O&P
0010	**VENT DAMPERS**									
5000	Vent damper, bi-metal, gas, 3" diameter	Q-9	24	.667	Ea.	51.50	34.50		86	110
5010	4" diameter		24	.667		51.50	34.50		86	110
5020	5" diameter		23	.696		51.50	36		87.50	112
5030	6" diameter		22	.727		51.50	37.50		89	115
5040	7" diameter		21	.762		53.50	39.50		93	119
5050	8" diameter	↓	20	.800	↓	57	41		98	126

23 51 13.19 Barometric Dampers

		Crew	Daily Output	Labor-Hours	Unit	Material	Labor	Equipment	Total	Total Incl O&P
0010	**BAROMETRIC DAMPERS**									
1000	Barometric, gas fired system only, 6" size for 5" and 6" pipes	1 Shee	20	.400	Ea.	74.50	23		97.50	117
1020	7" size, for 6" and 7" pipes		19	.421		80	24		104	125
1040	8" size, for 7" and 8" pipes		18	.444		103	25.50		128.50	153
1060	9" size, for 8" and 9" pipes	↓	16	.500	↓	116	28.50		144.50	171
2000	All fuel, oil, oil/gas, coal									
2020	10" for 9" and 10" pipes	1 Shee	15	.533	Ea.	173	30.50		203.50	237
2040	12" for 11" and 12" pipes		15	.533		226	30.50		256.50	295
2060	14" for 13" and 14" pipes		14	.571		293	32.50		325.50	370
2080	16" for 15" and 16" pipes		13	.615		410	35		445	505
2100	18" for 17" and 18" pipes		12	.667		545	38		583	660
2120	20" for 19" and 21" pipes	↓	10	.800		650	46		696	785
2140	24" for 22" and 25" pipes	Q-9	12	1.333		800	68.50		868.50	985
2160	28" for 26" and 30" pipes		10	1.600		990	82.50		1,072.50	1,225
2180	32" for 31" and 34" pipes	↓	8	2		1,275	103		1,378	1,550
3260	For thermal switch for above, add	1 Shee	24	.333	↓	71.50	19.10		90.60	108

23 51 23 – Gas Vents

23 51 23.10 Gas Chimney Vents

		Crew	Daily Output	Labor-Hours	Unit	Material	Labor	Equipment	Total	Total Incl O&P
0010	**GAS CHIMNEY VENTS**, Prefab metal, U.L. listed									
0020	Gas, double wall, galvanized steel									
0080	3" diameter	Q-9	72	.222	V.L.F.	6.60	11.45		18.05	25
0100	4" diameter		68	.235		8.15	12.10		20.25	27.50
0120	5" diameter		64	.250		9.25	12.90		22.15	30
0140	6" diameter		60	.267		11.05	13.75		24.80	33
0160	7" diameter		56	.286		21	14.70		35.70	45.50
0180	8" diameter		52	.308		22.50	15.85		38.35	49
0200	10" diameter		48	.333		43	17.20		60.20	74
0220	12" diameter		44	.364		51.50	18.75		70.25	85
0240	14" diameter		42	.381		85.50	19.65		105.15	124
0260	16" diameter		40	.400		123	20.50		143.50	167
0280	18" diameter	↓	38	.421		151	21.50		172.50	199
0300	20" diameter	Q-10	36	.667		179	35.50		214.50	251
0320	22" diameter		34	.706		227	37.50		264.50	310
0340	24" diameter	↓	32	.750	↓	280	40		320	370
0600	For 4", 5" and 6" oval, add					50%				
0650	Gas, double wall, galvanized steel, fittings									

For customer support on your Mechanical Cost Data, call 877.739.3689.

355

23 51 23.10 Gas Chimney Vents		Crew	Daily Output	Labor-Hours	Unit	Material	2016 Bare Costs Labor	Equipment	Total	Total Incl O&P
0660	Elbow 45°, 3" diameter	Q-9	36	.444	Ea.	11.65	23		34.65	48
0670	4" diameter		34	.471		14.05	24.50		38.55	52.50
0680	5" diameter		32	.500		16.45	26		42.45	57.50
0690	6" diameter		30	.533		20.50	27.50		48	64.50
0700	7" diameter		28	.571		32.50	29.50		62	80.50
0710	8" diameter		26	.615		43	31.50		74.50	96
0720	10" diameter		24	.667		92	34.50		126.50	154
0730	12" diameter		22	.727		98	37.50		135.50	166
0740	14" diameter		21	.762		154	39.50		193.50	229
0750	16" diameter		20	.800		200	41		241	283
0760	18" diameter		19	.842		262	43.50		305.50	355
0770	20" diameter	Q-10	18	1.333		293	71		364	430
0780	22" diameter		17	1.412		470	75.50		545.50	630
0790	24" diameter		16	1.500		595	80		675	780
0916	Adjustable length									
0918	3" diameter, to 12"	Q-9	36	.444	Ea.	14.80	23		37.80	51.50
0920	4" diameter, to 12"		34	.471		17.20	24.50		41.70	56
0924	6" diameter, to 12"		30	.533		22	27.50		49.50	66.50
0928	8" diameter, to 12"		26	.615		41.50	31.50		73	94.50
0930	10" diameter, to 18"		24	.667		128	34.50		162.50	194
0932	12" diameter, to 18"		22	.727		130	37.50		167.50	201
0936	16" diameter, to 18"		20	.800		263	41		304	350
0938	18" diameter, to 18"		19	.842		300	43.50		343.50	395
0944	24" diameter, to 18"	Q-10	16	1.500		575	80		655	755
0950	Elbow 90°, adjustable, 3" diameter	Q-9	36	.444		19.80	23		42.80	57
0960	4" diameter		34	.471		23	24.50		47.50	62.50
0970	5" diameter		32	.500		27.50	26		53.50	70
0980	6" diameter		30	.533		33.50	27.50		61	79
0990	7" diameter		28	.571		58	29.50		87.50	109
1010	8" diameter		26	.615		58.50	31.50		90	113
1020	Wall thimble, 4 to 7" adjustable, 3" diameter		36	.444		12.40	23		35.40	48.50
1022	4" diameter		34	.471		13.95	24.50		38.45	52.50
1024	5" diameter		32	.500		16.95	26		42.95	58
1026	6" diameter		30	.533		17.40	27.50		44.90	61
1028	7" diameter		28	.571		36	29.50		65.50	84.50
1030	8" diameter		26	.615		44.50	31.50		76	97.50
1040	Roof flashing, 3" diameter		36	.444		6.80	23		29.80	42.50
1050	4" diameter		34	.471		7.95	24.50		32.45	45.50
1060	5" diameter		32	.500		24	26		50	66
1070	6" diameter		30	.533		18.90	27.50		46.40	63
1080	7" diameter		28	.571		25	29.50		54.50	72.50
1090	8" diameter		26	.615		26.50	31.50		58	78
1100	10" diameter		24	.667		36	34.50		70.50	92
1110	12" diameter		22	.727		50.50	37.50		88	113
1120	14" diameter		20	.800		114	41		155	189
1130	16" diameter		18	.889		133	46		179	216
1140	18" diameter		16	1		189	51.50		240.50	287
1150	20" diameter	Q-10	18	1.333		250	71		321	385
1160	22" diameter		14	1.714		310	91.50		401.50	480
1170	24" diameter		12	2		360	107		467	565
1200	Tee, 3" diameter	Q-9	27	.593		30.50	30.50		61	80
1210	4" diameter		26	.615		33	31.50		64.50	84.50
1220	5" diameter		25	.640		34.50	33		67.50	88.50

23 51 23.10 Gas Chimney Vents		Crew	Daily Output	Labor-Hours	Unit	Material	2016 Bare Costs Labor	Equipment	Total	Total Incl O&P
1230	6" diameter	Q-9	24	.667	Ea.	39.50	34.50		74	95.50
1240	7" diameter		23	.696		56	36		92	117
1250	8" diameter		22	.727		61.50	37.50		99	125
1260	10" diameter		21	.762		164	39.50		203.50	240
1270	12" diameter		20	.800		169	41		210	248
1280	14" diameter		18	.889		300	46		346	400
1290	16" diameter		16	1		445	51.50		496.50	570
1300	18" diameter	▼	14	1.143		535	59		594	680
1310	20" diameter	Q-10	17	1.412		735	75.50		810.50	925
1320	22" diameter	▼	13	1.846		945	98.50		1,043.50	1,200
1330	24" diameter		12	2		1,075	107		1,182	1,375
1460	Tee cap, 3" diameter	Q-9	45	.356		2.03	18.30		20.33	30
1470	4" diameter		42	.381		2.19	19.65		21.84	32.50
1480	5" diameter		40	.400		2.81	20.50		23.31	34.50
1490	6" diameter		37	.432		4.07	22.50		26.57	38.50
1500	7" diameter		35	.457		6.85	23.50		30.35	43.50
1510	8" diameter		34	.471		7.45	24.50		31.95	45
1520	10" diameter		32	.500		60.50	26		86.50	106
1530	12" diameter		30	.533		60	27.50		87.50	108
1540	14" diameter		28	.571		60.50	29.50		90	112
1550	16" diameter		25	.640		64	33		97	121
1560	18" diameter	▼	24	.667		75.50	34.50		110	136
1570	20" diameter	Q-10	27	.889		85.50	47.50		133	167
1580	22" diameter	▼	22	1.091		315	58.50		373.50	440
1590	24" diameter		21	1.143		355	61		416	485
1750	Top, 3" diameter	Q-9	46	.348		15.50	17.90		33.40	44.50
1760	4" diameter		44	.364		16.30	18.75		35.05	46.50
1770	5" diameter		42	.381		18.05	19.65		37.70	50
1780	6" diameter		40	.400		20	20.50		40.50	53.50
1790	7" diameter		38	.421		44.50	21.50		66	82
1800	8" diameter		36	.444		55	23		78	95.50
1810	10" diameter		34	.471		90.50	24.50		115	137
1820	12" diameter		32	.500		110	26		136	161
1830	14" diameter		30	.533		176	27.50		203.50	236
1840	16" diameter		28	.571		227	29.50		256.50	294
1850	18" diameter	▼	26	.615		310	31.50		341.50	395
1860	20" diameter	Q-10	28	.857		475	46		521	590
1870	22" diameter		22	1.091		720	58.50		778.50	885
1880	24" diameter	▼	20	1.200	▼	900	64		964	1,100
1900	Gas, double wall, galvanized steel, oval									
1904	4" x 1'	Q-9	68	.235	V.L.F.	16.65	12.10		28.75	37
1906	5"		64	.250		39.50	12.90		52.40	63
1908	5"/6" x 1'	▼	60	.267	▼	43.50	13.75		57.25	68.50
1910	Oval fittings									
1912	Adjustable length									
1914	4" diameter to 12" lg.	Q-9	34	.471	Ea.	17.20	24.50		41.70	56
1916	5" diameter to 12" lg.		32	.500		17.90	26		43.90	59
1918	5"/6" diameter to 12"	▼	30	.533	▼	22	27.50		49.50	66.50
1920	Elbow 45°									
1922	4"	Q-9	34	.471	Ea.	28	24.50		52.50	68
1924	5"		32	.500		56.50	26		82.50	102
1926	5"/6"	▼	30	.533		58	27.50		85.50	106
1930	Elbow 45°, flat									

23 51 23 – Gas Vents

23 51 23.10 Gas Chimney Vents

		Crew	Daily Output	Labor-Hours	Unit	Material	2016 Bare Costs Labor	Equipment	Total	Total Incl O&P
1932	4"	Q-9	34	.471	Ea.	28	24.50		52.50	68
1934	5"		32	.500		56.50	26		82.50	102
1936	5"/6"		30	.533		58	27.50		85.50	106
1940	Top									
1942	4"	Q-9	44	.364	Ea.	35	18.75		53.75	67
1944	5"		42	.381		35	19.65		54.65	68.50
1946	5"/6"		40	.400		45.50	20.50		66	81.50
1950	Adjustable flashing									
1952	4"	Q-9	34	.471	Ea.	11.90	24.50		36.40	50
1954	5"		32	.500		34.50	26		60.50	77.50
1956	5"/6"		30	.533		37.50	27.50		65	83.50
1960	Tee									
1962	4"	Q-9	26	.615	Ea.	42	31.50		73.50	94.50
1964	5"		25	.640		85	33		118	144
1966	5"/6"		24	.667		87	34.50		121.50	148
1970	Tee with short snout									
1972	4"	Q-9	26	.615	Ea.	43.50	31.50		75	96.50

23 51 26 – All-Fuel Vent Chimneys

23 51 26.10 All-Fuel Vent Chimneys, Press. Tight, Dbl. Wall

		Crew	Daily Output	Labor-Hours	Unit	Material	2016 Bare Costs Labor	Equipment	Total	Total Incl O&P
0010	**ALL-FUEL VENT CHIMNEYS, PRESSURE TIGHT, DOUBLE WALL**									
3200	All fuel, pressure tight, double wall, 1" insulation, U.L. listed, 1400°F.									
3210	304 stainless steel liner, aluminized steel outer jacket									
3220	6" diameter	Q-9	60	.267	L.F.	55.50	13.75		69.25	82
3221	8" diameter		52	.308		62.50	15.85		78.35	92.50
3222	10" diameter		48	.333		69.50	17.20		86.70	103
3223	12" diameter		44	.364		80	18.75		98.75	117
3224	14" diameter		42	.381		89.50	19.65		109.15	129
3225	16" diameter		40	.400		101	20.50		121.50	143
3226	18" diameter		38	.421		115	21.50		136.50	159
3227	20" diameter	Q-10	36	.667		130	35.50		165.50	198
3228	24" diameter		32	.750		166	40		206	245
3229	28" diameter		30	.800		191	42.50		233.50	276
3230	32" diameter		27	.889		216	47.50		263.50	310
3231	36" diameter		25	.960		241	51.50		292.50	345
3232	42" diameter		22	1.091		282	58.50		340.50	400
3233	48" diameter		19	1.263		320	67.50		387.50	455
3260	For 316 stainless steel liner add					30%				
3280	All fuel, pressure tight, double wall fittings									
3284	304 stainless steel inner, aluminized steel jacket									
3288	Adjustable 20"/29" section									
3292	6" diameter	Q-9	30	.533	Ea.	197	27.50		224.50	259
3293	8" diameter		26	.615		205	31.50		236.50	275
3294	10" diameter		24	.667		232	34.50		266.50	310
3295	12" diameter		22	.727		262	37.50		299.50	345
3296	14" diameter		21	.762		296	39.50		335.50	385
3297	16" diameter		20	.800		335	41		376	430
3298	18" diameter		19	.842		375	43.50		418.50	480
3299	20" diameter	Q-10	18	1.333		425	71		496	580
3300	24" diameter		16	1.500		545	80		625	725
3301	28" diameter		15	1.600		590	85.50		675.50	775
3302	32" diameter		14	1.714		710	91.50		801.50	920
3303	36" diameter		12	2		790	107		897	1,025

23 51 26.10 All-Fuel Vent Chimneys, Press. Tight, Dbl. Wall	Crew	Daily Output	Labor-Hours	Unit	Material	2016 Bare Costs Labor	Equipment	Total	Total Incl O&P
3304	42" diameter	Q-10	11	2.182	Ea.	925	117		1,042	1,200
3305	48" diameter	↓	10	2.400	↓	1,050	128		1,178	1,350
3350	Elbow 90° fixed									
3354	6" diameter	Q-9	30	.533	Ea.	410	27.50		437.50	495
3355	8" diameter		26	.615		465	31.50		496.50	560
3356	10" diameter		24	.667		525	34.50		559.50	635
3357	12" diameter		22	.727		595	37.50		632.50	715
3358	14" diameter		21	.762		675	39.50		714.50	800
3359	16" diameter		20	.800		760	41		801	900
3360	18" diameter	↓	19	.842		860	43.50		903.50	1,000
3361	20" diameter	Q-10	18	1.333		975	71		1,046	1,175
3362	24" diameter		16	1.500		1,250	80		1,330	1,500
3363	28" diameter		15	1.600		1,400	85.50		1,485.50	1,675
3364	32" diameter		14	1.714		1,875	91.50		1,966.50	2,225
3365	36" diameter		12	2		2,550	107		2,657	2,975
3366	42" diameter		11	2.182		3,025	117		3,142	3,500
3367	48" diameter	↓	10	2.400		3,375	128		3,503	3,900
3380	For 316 stainless steel liner, add				↓	30%				
3400	Elbow 45°									
3404	6" diameter	Q-9	30	.533	Ea.	205	27.50		232.50	268
3405	8" diameter		26	.615		231	31.50		262.50	305
3406	10" diameter		24	.667		263	34.50		297.50	340
3407	12" diameter		22	.727		300	37.50		337.50	390
3408	14" diameter		21	.762		340	39.50		379.50	430
3409	16" diameter		20	.800		380	41		421	480
3410	18" diameter	↓	19	.842		430	43.50		473.50	540
3411	20" diameter	Q-10	18	1.333		490	71		561	645
3412	24" diameter		16	1.500		625	80		705	810
3413	28" diameter		15	1.600		700	85.50		785.50	900
3414	32" diameter		14	1.714		945	91.50		1,036.50	1,200
3415	36" diameter		12	2		1,275	107		1,382	1,575
3416	42" diameter		11	2.182		1,475	117		1,592	1,800
3417	48" diameter	↓	10	2.400		1,675	128		1,803	2,050
3430	For 316 stainless steel liner, add				↓	30%				
3450	Tee 90°									
3454	6" diameter	Q-9	24	.667	Ea.	242	34.50		276.50	320
3455	8" diameter		22	.727		265	37.50		302.50	350
3456	10" diameter		21	.762		299	39.50		338.50	390
3457	12" diameter		20	.800		345	41		386	445
3458	14" diameter		18	.889		395	46		441	505
3459	16" diameter		16	1		435	51.50		486.50	555
3460	18" diameter	↓	14	1.143		510	59		569	650
3461	20" diameter	Q-10	17	1.412		590	75.50		665.50	765
3462	24" diameter		12	2		735	107		842	975
3463	28" diameter		11	2.182		875	117		992	1,150
3464	32" diameter		10	2.400		1,225	128		1,353	1,525
3465	36" diameter		9	2.667		1,375	142		1,517	1,725
3466	42" diameter		6	4		1,600	214		1,814	2,100
3467	48" diameter	↓	5	4.800		1,825	256		2,081	2,425
3480	For Tee Cap, add					35%	20%			
3500	For 316 stainless steel liner, add				↓	30%				
3520	Plate support, galvanized									
3524	6" diameter	Q-9	26	.615	Ea.	121	31.50		152.50	182

For customer support on your Mechanical Cost Data, call 877.739.3689.

359

23 51 26.10 All-Fuel Vent Chimneys, Press. Tight, Dbl. Wall	Crew	Daily Output	Labor-Hours	Unit	Material	2016 Bare Costs Labor	Equipment	Total	Total Incl O&P
3525	8" diameter	Q-9	22	.727	Ea.	141	37.50		178.50	213
3526	10" diameter		20	.800		154	41		195	232
3527	12" diameter		18	.889		161	46		207	247
3528	14" diameter		17	.941		191	48.50		239.50	284
3529	16" diameter		16	1		202	51.50		253.50	300
3530	18" diameter		15	1.067		212	55		267	315
3531	20" diameter	Q-10	16	1.500		223	80		303	370
3532	24" diameter		14	1.714		231	91.50		322.50	395
3533	28" diameter		13	1.846		284	98.50		382.50	465
3534	32" diameter		12	2		350	107		457	550
3535	36" diameter		10	2.400		420	128		548	655
3536	42" diameter		9	2.667		490	142		632	760
3537	48" diameter		8	3		560	160		720	860
3570	Bellows, lined									
3574	6" diameter	Q-9	30	.533	Ea.	1,325	27.50		1,352.50	1,500
3575	8" diameter		26	.615		1,375	31.50		1,406.50	1,575
3576	10" diameter		24	.667		1,400	34.50		1,434.50	1,600
3577	12" diameter		22	.727		1,450	37.50		1,487.50	1,650
3578	14" diameter		21	.762		1,500	39.50		1,539.50	1,700
3579	16" diameter		20	.800		1,550	41		1,591	1,775
3580	18" diameter		19	.842		1,600	43.50		1,643.50	1,850
3581	20" diameter	Q-10	18	1.333		1,650	71		1,721	1,925
3582	24" diameter		16	1.500		1,925	80		2,005	2,250
3583	28" diameter		15	1.600		2,050	85.50		2,135.50	2,375
3584	32" diameter		14	1.714		2,300	91.50		2,391.50	2,675
3585	36" diameter		12	2		2,500	107		2,607	2,950
3586	42" diameter		11	2.182		2,825	117		2,942	3,275
3587	48" diameter		10	2.400		3,250	128		3,378	3,775
3590	For all 316 stainless steel construction, add					55%				
3600	Ventilated roof thimble, 304 stainless steel									
3620	6" diameter	Q-9	26	.615	Ea.	252	31.50		283.50	325
3624	8" diameter		22	.727		262	37.50		299.50	345
3625	10" diameter		20	.800		270	41		311	360
3626	12" diameter		18	.889		277	46		323	375
3627	14" diameter		17	.941		288	48.50		336.50	390
3628	16" diameter		16	1		310	51.50		361.50	420
3629	18" diameter		15	1.067		335	55		390	455
3630	20" diameter	Q-10	16	1.500		350	80		430	515
3631	24" diameter		14	1.714		385	91.50		476.50	565
3632	28" diameter		13	1.846		420	98.50		518.50	610
3633	32" diameter		12	2		450	107		557	660
3634	36" diameter		10	2.400		485	128		613	730
3635	42" diameter		9	2.667		535	142		677	810
3636	48" diameter		8	3		590	160		750	890
3650	For 316 stainless steel, add					30%				
3670	Exit cone, 316 stainless steel only									
3674	6" diameter	Q-9	46	.348	Ea.	191	17.90		208.90	238
3675	8" diameter		42	.381		197	19.65		216.65	246
3676	10" diameter		40	.400		208	20.50		228.50	261
3677	12" diameter		38	.421		222	21.50		243.50	277
3678	14" diameter		37	.432		237	22.50		259.50	295
3679	16" diameter		36	.444		298	23		321	360
3680	18" diameter		35	.457		320	23.50		343.50	385

23 51 Breechings, Chimneys, and Stacks

23 51 26 – All-Fuel Vent Chimneys

23 51 26.10 All-Fuel Vent Chimneys, Press. Tight, Dbl. Wall	Crew	Daily Output	Labor-Hours	Unit	Material	2016 Bare Costs Labor	Equipment	Total	Total Incl O&P
3681	20" diameter	Q-10	28	.857	Ea.	385	46		431	490
3682	24" diameter		26	.923		505	49.50		554.50	630
3683	28" diameter		25	.960		575	51.50		626.50	710
3684	32" diameter		24	1		710	53.50		763.50	860
3685	36" diameter		22	1.091		860	58.50		918.50	1,050
3686	42" diameter		21	1.143		1,000	61		1,061	1,200
3687	48" diameter	▼	20	1.200	▼	1,150	64		1,214	1,350
3720	Roof guide, 304 stainless steel									
3724	6" diameter	Q-9	25	.640	Ea.	87.50	33		120.50	147
3725	8" diameter		21	.762		102	39.50		141.50	172
3726	10" diameter		19	.842		111	43.50		154.50	190
3727	12" diameter		17	.941		115	48.50		163.50	201
3728	14" diameter		16	1		134	51.50		185.50	227
3729	16" diameter		15	1.067		142	55		197	240
3730	18" diameter	▼	14	1.143		151	59		210	256
3731	20" diameter	Q-10	15	1.600		158	85.50		243.50	305
3732	24" diameter		13	1.846		165	98.50		263.50	335
3733	28" diameter		12	2		202	107		309	385
3734	32" diameter		11	2.182		248	117		365	450
3735	36" diameter		9	2.667		296	142		438	545
3736	42" diameter		8	3		345	160		505	625
3737	48" diameter	▼	7	3.429		395	183		578	710
3750	For 316 stainless steel, add				▼	30%				
3770	Rain cap with bird screen									
3774	6" diameter	Q-9	46	.348	Ea.	296	17.90		313.90	355
3775	8" diameter		42	.381		345	19.65		364.65	405
3776	10" diameter		40	.400		400	20.50		420.50	470
3777	12" diameter		38	.421		460	21.50		481.50	545
3778	14" diameter		37	.432		525	22.50		547.50	610
3779	16" diameter		36	.444		595	23		618	690
3780	18" diameter		35	.457		690	23.50		713.50	795
3781	20" diameter	Q-10	28	.857		780	46		826	925
3782	24" diameter		26	.923		940	49.50		989.50	1,100
3783	28" diameter		25	.960		1,125	51.50		1,176.50	1,300
3784	32" diameter		24	1		1,275	53.50		1,328.50	1,475
3785	36" diameter		22	1.091		1,500	58.50		1,558.50	1,750
3786	42" diameter		21	1.143		1,750	61		1,811	2,025
3787	48" diameter	▼	20	1.200	▼	2,000	64		2,064	2,300

23 51 26.30 All-Fuel Vent Chimneys, Double Wall, St. Stl.

		Crew	Daily Output	Labor-Hours	Unit	Material	2016 Bare Costs Labor	Equipment	Total	Total Incl O&P
0010	**ALL-FUEL VENT CHIMNEYS, DOUBLE WALL, STAINLESS STEEL**									
7780	All fuel, pressure tight, double wall, 4" insulation, U.L. listed, 1400°F.									
7790	304 stainless steel liner, aluminized steel outer jacket									
7800	6" diameter	Q-9	60	.267	V.L.F.	65	13.75		78.75	92.50
7804	8" diameter		52	.308		75	15.85		90.85	106
7806	10" diameter		48	.333		83.50	17.20		100.70	118
7808	12" diameter		44	.364		95.50	18.75		114.25	134
7810	14" diameter	▼	42	.381	▼	107	19.65		126.65	148
7880	For 316 stainless steel liner add				L.F.	30%				
8000	All fuel, double wall, stainless steel fittings									
8010	Roof support 6" diameter	Q-9	30	.533	Ea.	115	27.50		142.50	169
8030	8" diameter		26	.615		134	31.50		165.50	197
8040	10" diameter	▼	24	.667	▼	142	34.50		176.50	209

For customer support on your Mechanical Cost Data, call 877.739.3689.

361

23 51 Breechings, Chimneys, and Stacks

23 51 26 – All-Fuel Vent Chimneys

23 51 26.30 All-Fuel Vent Chimneys, Double Wall, St. Stl.		Crew	Daily Output	Labor-Hours	Unit	Material	2016 Bare Costs Labor	Equipment	Total	Total Incl O&P
8050	12" diameter	Q-9	22	.727	Ea.	151	37.50		188.50	224
8060	14" diameter		21	.762		158	39.50		197.50	234
8100	Elbow 45°, 6" diameter		30	.533		246	27.50		273.50	310
8140	8" diameter		26	.615		276	31.50		307.50	355
8160	10" diameter		24	.667		315	34.50		349.50	400
8180	12" diameter		22	.727		360	37.50		397.50	455
8200	14" diameter		21	.762		400	39.50		439.50	500
8300	Insulated tee, 6" diameter		30	.533		289	27.50		316.50	360
8360	8" diameter		26	.615		320	31.50		351.50	400
8380	10" diameter		24	.667		355	34.50		389.50	445
8400	12" diameter		22	.727		405	37.50		442.50	505
8420	14" diameter		20	.800		470	41		511	585
8500	Boot tee, 6" diameter		28	.571		550	29.50		579.50	650
8520	8" diameter		24	.667		600	34.50		634.50	715
8530	10" diameter		22	.727		700	37.50		737.50	830
8540	12" diameter		20	.800		830	41		871	975
8550	14" diameter		18	.889		945	46		991	1,100
8600	Rain cap with bird screen, 6" diameter		30	.533		296	27.50		323.50	365
8640	8" diameter		26	.615		345	31.50		376.50	425
8660	10" diameter		24	.667		400	34.50		434.50	495
8680	12" diameter		22	.727		460	37.50		497.50	570
8700	14" diameter		21	.762		525	39.50		564.50	635
8800	Flat roof flashing, 6" diameter		30	.533		108	27.50		135.50	161
8840	8" diameter		26	.615		118	31.50		149.50	179
8860	10" diameter		24	.667		127	34.50		161.50	192
8880	12" diameter		22	.727		139	37.50		176.50	211
8900	14" diameter		21	.762		142	39.50		181.50	216

23 52 Heating Boilers

23 52 13 – Electric Boilers

23 52 13.10 Electric Boilers, ASME

			Crew	Daily Output	Labor-Hours	Unit	Material	2016 Bare Costs Labor	Equipment	Total	Total Incl O&P
0010	**ELECTRIC BOILERS, ASME**, Standard controls and trim	R235000-30									
1000	Steam, 6 KW, 20.5 MBH		Q-19	1.20	20	Ea.	4,100	1,100		5,200	6,150
1040	9 KW, 30.7 MBH	D3020–102		1.20	20		4,200	1,100		5,300	6,250
1060	18 KW, 61.4 MBH			1.20	20		4,325	1,100		5,425	6,400
1080	24 KW, 81.8 MBH	D3020–104		1.10	21.818		5,000	1,200		6,200	7,300
1120	36 KW, 123 MBH			1.10	21.818		5,525	1,200		6,725	7,875
1160	60 KW, 205 MBH			1	24		6,925	1,325		8,250	9,575
1220	112 KW, 382 MBH			.75	32		9,475	1,750		11,225	13,100
1240	148 KW, 505 MBH			.65	36.923		9,975	2,025		12,000	14,100
1260	168 KW, 573 MBH			.60	40		20,400	2,200		22,600	25,800
1280	222 KW, 758 MBH			.55	43.636		23,800	2,400		26,200	29,800
1300	296 KW, 1010 MBH			.45	53.333		25,000	2,925		27,925	32,000
1320	300 KW, 1023 MBH			.40	60		26,400	3,275		29,675	34,000
1340	370 KW, 1263 MBH			.35	68.571		29,200	3,750		32,950	37,800
1360	444 KW, 1515 MBH			.30	80		31,400	4,375		35,775	41,100
1380	518 KW, 1768 MBH		Q-21	.36	88.889		32,600	5,000		37,600	43,400
1400	592 KW, 2020 MBH			.34	94.118		35,400	5,300		40,700	46,900
1420	666 KW, 2273 MBH			.32	100		36,800	5,625		42,425	49,000
1460	740 KW, 2526 MBH			.28	114		38,300	6,425		44,725	52,000
1480	814 KW, 2778 MBH			.25	128		40,600	7,200		47,800	55,500

23 52 13 – Electric Boilers

23 52 13.10 Electric Boilers, ASME		Crew	Daily Output	Labor-Hours	Unit	Material	2016 Bare Costs Labor	Equipment	Total	Total Incl O&P
1500	962 KW, 3283 MBH	Q-21	.22	145	Ea.	44,900	8,175		53,075	61,500
1520	1036 KW, 3536 MBH		.20	160		47,600	9,000		56,600	66,000
1540	1110 KW, 3788 MBH		.19	168		49,600	9,475		59,075	69,000
1560	2070 KW, 7063 MBH		.18	177		68,000	10,000		78,000	90,000
1580	2250 KW, 7677 MBH		.17	188		80,000	10,600		90,600	104,000
1600	2,340 KW, 7984 MBH		.16	200		86,500	11,300		97,800	112,000
2000	Hot water, 7.5 KW, 25.6 MBH	Q-19	1.30	18.462		4,950	1,000		5,950	6,975
2020	15 KW, 51.2 MBH		1.30	18.462		5,000	1,000		6,000	7,025
2040	30 KW, 102 MBH		1.20	20		5,325	1,100		6,425	7,500
2060	45 KW, 164 MBH		1.20	20		5,425	1,100		6,525	7,600
2070	60 KW, 205 MBH		1.20	20		5,525	1,100		6,625	7,725
2080	75 KW, 256 MBH		1.10	21.818		5,900	1,200		7,100	8,300
2100	90 KW, 307 MBH		1.10	21.818		6,000	1,200		7,200	8,400
2120	105 KW, 358 MBH		1	24		6,275	1,325		7,600	8,875
2140	120 KW, 410 MBH		.90	26.667		6,375	1,450		7,825	9,225
2160	135 KW, 461 MBH		.75	32		7,400	1,750		9,150	10,800
2180	150 KW, 512 MBH		.65	36.923		7,500	2,025		9,525	11,300
2200	165 KW, 563 MBH		.60	40		7,600	2,200		9,800	11,700
2220	296 KW, 1010 MBH		.55	43.636		16,100	2,400		18,500	21,300
2280	370 KW, 1263 MBH		.40	60		17,900	3,275		21,175	24,700
2300	444 KW, 1515 MBH		.35	68.571		20,800	3,750		24,550	28,600
2340	518 KW, 1768 MBH	Q-21	.44	72.727		23,100	4,100		27,200	31,600
2360	592 KW, 2020 MBH		.43	74.419		26,100	4,175		30,275	35,100
2400	666 KW, 2273 MBH		.40	80		27,300	4,500		31,800	36,800
2420	740 KW, 2526 MBH		.39	82.051		28,500	4,625		33,125	38,400
2440	814 KW, 2778 MBH		.38	84.211		30,500	4,750		35,250	40,600
2460	888 KW, 3031 MBH		.37	86.486		32,500	4,875		37,375	43,100
2480	962 KW, 3283 MBH		.36	88.889		34,000	5,000		39,000	44,900
2500	1036 KW, 3536 MBH		.34	94.118		36,300	5,300		41,600	47,900
2520	1110 KW, 3788 MBH		.33	96.970		38,800	5,450		44,250	51,000
2540	1440 KW, 4915 MBH		.32	100		43,800	5,625		49,425	56,500
2560	1560 KW, 5323 MBH		.31	103		46,800	5,800		52,600	60,500
2580	1680 KW, 5733 MBH		.30	106		50,000	6,000		56,000	64,000
2600	1800 KW, 6143 MBH		.29	110		54,500	6,200		60,700	69,500
2620	1980 KW, 6757 MBH		.28	114		58,000	6,425		64,425	73,500
2640	2100 KW, 7167 MBH		.27	118		63,500	6,675		70,175	80,000
2660	2220 KW, 7576 MBH		.26	123		65,000	6,925		71,925	82,000
2680	2400 KW, 8191 MBH		.25	128		68,000	7,200		75,200	85,500
2700	2610 KW, 8905 MBH		.24	133		72,500	7,500		80,000	91,000
2720	2790 KW, 9519 MBH		.23	139		75,500	7,825		83,325	95,000
2740	2970 KW, 10133 MBH		.21	152		77,000	8,575		85,575	98,000
2760	3150 KW, 10748 MBH		.19	168		82,500	9,475		91,975	105,500
2780	3240 KW, 11055 MBH		.18	177		84,500	10,000		94,500	108,000
2800	3420 KW, 11669 MBH		.17	188		90,000	10,600		100,600	115,000
2820	3600 KW, 12,283 MBH		.16	200		94,000	11,300		105,300	120,500

23 52 16 – Condensing Boilers

23 52 16.24 Condensing Boilers

			Crew	Daily Output	Labor-Hours	Unit	Material	2016 Bare Costs Labor	Equipment	Total	Total Incl O&P
0010	**CONDENSING BOILERS**, Cast iron, high efficiency										
0020	Packaged with standard controls, circulator and trim										
0030	Intermittent (spark) pilot, natural or LP gas										
0040	Hot water, DOE MBH output, (AFUE)										
0100	42 MBH, (84.0%)	G	Q-5	1.80	8.889	Ea.	1,575	485		2,060	2,475

For customer support on your Mechanical Cost Data, call 877.739.3689.

363

23 52 Heating Boilers

23 52 16 – Condensing Boilers

23 52 16.24 Condensing Boilers

		Crew	Daily Output	Labor-Hours	Unit	Material	2016 Bare Costs Labor	Equipment	Total	Total Incl O&P
0120	57 MBH, (84.3%) G	Q-5	1.60	10	Ea.	1,775	545		2,320	2,775
0140	85 MBH, (84.0%) G		1.40	11.429		1,950	625		2,575	3,075
0160	112 MBH, (83.7%) G		1.20	13.333		2,175	730		2,905	3,500
0180	140 MBH, (83.3%) G	Q-6	1.60	15		2,450	850		3,300	3,975
0200	167 MBH, (83.0%) G		1.40	17.143		2,775	970		3,745	4,525
0220	194 MBH, (82.7%) G		1.20	20		3,025	1,125		4,150	5,025

23 52 19 – Pulse Combustion Boilers

23 52 19.20 Pulse Type Combustion Boilers

		Crew	Daily Output	Labor-Hours	Unit	Material	2016 Bare Costs Labor	Equipment	Total	Total Incl O&P
0010	**PULSE TYPE COMBUSTION BOILERS**, High efficiency									
7990	Special feature gas fired boilers									
8000	Pulse combustion, standard controls/trim									
8010	Hot water, DOE MBH output, (AFUE %)									
8030	71 MBH, 95.2%	Q-5	1.60	10	Ea.	3,450	545		3,995	4,625
8050	94 MBH, 95.3% G		1.40	11.429		3,825	625		4,450	5,150
8080	139 MBH, 95.6% G		1.20	13.333		4,300	730		5,030	5,825
8090	207 MBH, 95.4%		1.16	13.793		4,850	755		5,605	6,475
8120	270 MBH, 96.4%		1.12	14.286		6,725	780		7,505	8,575
8130	365 MBH, 91.7%		1.07	14.953		7,600	815		8,415	9,575

23 52 23 – Cast-Iron Boilers

23 52 23.20 Gas-Fired Boilers

		Crew	Daily Output	Labor-Hours	Unit	Material	2016 Bare Costs Labor	Equipment	Total	Total Incl O&P
0010	**GAS-FIRED BOILERS**, Natural or propane, standard controls, packaged R235000-10									
1000	Cast iron, with insulated jacket									
2000	Steam, gross output, 81 MBH R235000-20	Q-7	1.40	22.857	Ea.	2,450	1,325		3,775	4,700
2020	102 MBH		1.30	24.615		2,675	1,425		4,100	5,100
2040	122 MBH R235000-30		1	32		2,925	1,850		4,775	6,025
2060	163 MBH		.90	35.556		3,275	2,050		5,325	6,700
2080	203 MBH		.90	35.556		3,825	2,050		5,875	7,300
2100	240 MBH		.85	37.647		3,925	2,175		6,100	7,575
2120	280 MBH		.80	40		4,700	2,300		7,000	8,675
2140	320 MBH		.70	45.714		5,000	2,650		7,650	9,475
2160	360 MBH		.63	51.200		5,575	2,950		8,525	10,600
2180	400 MBH		.56	56.838		5,900	3,275		9,175	11,500
2200	440 MBH		.51	62.500		6,350	3,600		9,950	12,500
2220	544 MBH		.45	71.588		10,100	4,125		14,225	17,400
2240	765 MBH		.43	74.419		12,200	4,300		16,500	19,900
2260	892 MBH		.38	84.211		14,100	4,875		18,975	22,900
2280	1275 MBH		.34	94.118		17,300	5,450		22,750	27,200
2300	1530 MBH		.32	100		18,500	5,775		24,275	29,100
2320	1,875 MBH		.30	106		25,200	6,175		31,375	37,000
2340	2170 MBH		.26	122		27,000	7,050		34,050	40,300
2360	2675 MBH		.20	163		28,300	9,425		37,725	45,300
2380	3060 MBH		.19	172		29,300	9,950		39,250	47,300
2400	3570 MBH		.18	181		35,800	10,500		46,300	55,500
2420	4207 MBH		.16	205		39,200	11,900		51,100	61,000
2440	4,720 MBH		.15	207		68,000	12,000		80,000	93,000
2460	5660 MBH		.15	220		77,000	12,800		89,800	103,500
2480	6,100 MBH		.13	246		88,000	14,200		102,200	118,500
2500	6390 MBH		.12	266		93,000	15,400		108,400	125,500
2520	6680 MBH		.11	290		94,000	16,800		110,800	129,000
2540	6,970 MBH		.10	320		98,500	18,500		117,000	136,500
3000	Hot water, gross output, 80 MBH		1.46	21.918		1,975	1,275		3,250	4,075
3020	100 MBH		1.35	23.704		2,550	1,375		3,925	4,900

23 52 23 – Cast-Iron Boilers

23 52 23.20 Gas-Fired Boilers		Crew	Daily Output	Labor-Hours	Unit	Material	2016 Bare Costs Labor	Equipment	Total	Total Incl O&P
3040	122 MBH	Q-7	1.10	29.091	Ea.	2,675	1,675		4,350	5,450
3060	163 MBH		1	32		3,150	1,850		5,000	6,275
3080	203 MBH		1	32		3,625	1,850		5,475	6,800
3100	240 MBH		.95	33.684		3,700	1,950		5,650	6,975
3120	280 MBH		.90	35.556		4,325	2,050		6,375	7,850
3140	320 MBH		.80	40		4,675	2,300		6,975	8,650
3160	360 MBH		.71	45.070		5,400	2,600		8,000	9,875
3180	400 MBH		.64	50		5,575	2,900		8,475	10,500
3200	440 MBH		.58	54.983		5,950	3,175		9,125	11,400
3220	544 MBH		.51	62.992		9,800	3,650		13,450	16,300
3240	765 MBH		.46	70.022		11,900	4,050		15,950	19,100
3260	1,088 MBH		.40	80		13,500	4,625		18,125	21,900
3280	1,275 MBH		.36	89.888		17,300	5,200		22,500	26,900
3300	1,530 MBH		.31	104		18,200	6,075		24,275	29,200
3320	2,000 MBH		.26	125		21,400	7,225		28,625	34,400
3340	2,312 MBH		.22	148		24,400	8,550		32,950	39,800
3360	2,856 MBH		.20	160		30,800	9,250		40,050	47,900
3380	3,264 MBH		.18	179		32,700	10,400		43,100	51,500
3400	3,996 MBH		.16	195		35,800	11,300		47,100	56,500
3420	4,488 MBH		.15	210		39,400	12,200		51,600	61,500
3440	4,720 MBH		.15	220		74,000	12,800		86,800	100,500
3460	5,520 MBH		.14	228		95,000	13,200		108,200	124,500
3480	6,100 MBH		.13	250		116,500	14,400		130,900	150,500
3500	6,390 MBH		.11	285		121,500	16,500		138,000	159,000
3520	6,680 MBH		.10	310		126,500	18,000		144,500	166,500
3540	6,970 MBH		.09	359		131,500	20,800		152,300	176,000
7000	For tankless water heater, add					10%				
7050	For additional zone valves up to 312 MBH add					194			194	214

23 52 23.30 Gas/Oil Fired Boilers

		Crew	Daily Output	Labor-Hours	Unit	Material	2016 Bare Costs Labor	Equipment	Total	Total Incl O&P
0010	**GAS/OIL FIRED BOILERS**, Combination with burners and controls, packaged									
1000	Cast iron with insulated jacket									
2000	Steam, gross output, 720 MBH	Q-7	.43	74.074	Ea.	14,700	4,275		18,975	22,700
2020	810 MBH		.38	83.990		14,700	4,850		19,550	23,500
2040	1,084 MBH		.34	93.023		17,000	5,375		22,375	26,800
2060	1,360 MBH		.33	98.160		19,500	5,675		25,175	30,000
2080	1,600 MBH		.30	107		20,900	6,175		27,075	32,300
2100	2,040 MBH		.25	130		24,900	7,525		32,425	38,700
2120	2,450 MBH		.21	156		26,600	9,025		35,625	42,800
2140	2,700 MBH		.19	165		28,100	9,575		37,675	45,400
2160	3,000 MBH		.18	175		30,100	10,200		40,300	48,400
2180	3,270 MBH		.17	183		32,900	10,600		43,500	52,000
2200	3,770 MBH		.17	191		67,500	11,100		78,600	90,500
2220	4,070 MBH		.16	200		71,000	11,600		82,600	95,500
2240	4,650 MBH		.15	210		75,000	12,200		87,200	101,000
2260	5,230 MBH		.14	223		83,000	12,900		95,900	110,500
2280	5,520 MBH		.14	235		90,500	13,600		104,100	120,000
2300	5,810 MBH		.13	248		92,000	14,300		106,300	122,500
2320	6,100 MBH		.12	260		94,000	15,000		109,000	126,000
2340	6,390 MBH		.11	296		97,000	17,100		114,100	132,500
2360	6,680 MBH		.10	320		100,000	18,500		118,500	138,000
2380	6,970 MBH		.09	372		102,500	21,500		124,000	145,500
2900	Hot water, gross output									

23 52 Heating Boilers

23 52 23 – Cast-Iron Boilers

23 52 23.30 Gas/Oil Fired Boilers

		Crew	Daily Output	Labor-Hours	Unit	Material	2016 Bare Costs Labor	Equipment	Total	Total Incl O&P
2910	200 MBH	Q-6	.62	39.024	Ea.	10,600	2,200		12,800	15,000
2920	300 MBH		.49	49.080		10,600	2,775		13,375	15,900
2930	400 MBH		.41	57.971		12,400	3,275		15,675	18,700
2940	500 MBH		.36	67.039		13,400	3,800		17,200	20,400
3000	584 MBH	Q-7	.44	72.072		14,800	4,175		18,975	22,600
3020	876 MBH		.41	79.012		19,600	4,575		24,175	28,400
3040	1,168 MBH		.31	103		28,800	6,000		34,800	40,700
3060	1,460 MBH		.28	113		36,800	6,525		43,325	50,500
3080	2,044 MBH		.26	122		41,700	7,050		48,750	56,500
3100	2,628 MBH		.21	150		45,800	8,675		54,475	63,500
3120	3,210 MBH		.18	174		48,900	10,100		59,000	69,000
3140	3,796 MBH		.17	186		54,500	10,800		65,300	76,000
3160	4,088 MBH		.16	195		62,000	11,300		73,300	85,500
3180	4,672 MBH		.16	203		68,000	11,800		79,800	92,500
3200	5,256 MBH		.15	217		76,000	12,600		88,600	102,500
3220	6,000 MBH, 179 BHP		.13	256		109,000	14,800		123,800	142,500
3240	7,130 MBH, 213 BHP		.08	385		114,000	22,300		136,300	159,000
3260	9,800 MBH, 286 BHP		.06	533		131,000	30,800		161,800	191,000
3280	10,900 MBH, 325.6 BHP		.05	592		159,000	34,200		193,200	226,500
3290	12,200 MBH, 364.5 BHP		.05	666		173,500	38,500		212,000	249,000
3300	13,500 MBH, 403.3 BHP		.04	727		192,500	42,000		234,500	275,000

23 52 23.40 Oil-Fired Boilers

		Crew	Daily Output	Labor-Hours	Unit	Material	2016 Bare Costs Labor	Equipment	Total	Total Incl O&P
0010	**OIL-FIRED BOILERS**, Standard controls, flame retention burner, packaged									
1000	Cast iron, with insulated flush jacket									
2000	Steam, gross output, 109 MBH	Q-7	1.20	26.667	Ea.	2,325	1,550		3,875	4,875
2020	144 MBH		1.10	29.091		2,625	1,675		4,300	5,400
2040	173 MBH		1	32		2,925	1,850		4,775	6,025
2060	207 MBH		.90	35.556		3,175	2,050		5,225	6,600
2080	236 MBH		.85	37.647		3,725	2,175		5,900	7,375
2100	300 MBH		.70	45.714		4,700	2,650		7,350	9,150
2120	480 MBH		.50	64		6,075	3,700		9,775	12,300
2140	665 MBH		.45	71.111		8,625	4,100		12,725	15,700
2160	794 MBH		.41	78.049		9,575	4,500		14,075	17,300
2180	1,084 MBH		.38	85.106		10,900	4,925		15,825	19,400
2200	1,360 MBH		.33	98.160		12,500	5,675		18,175	22,300
2220	1,600 MBH		.26	122		14,000	7,050		21,050	26,100
2240	2,175 MBH		.24	133		18,100	7,750		25,850	31,700
2260	2,480 MBH		.21	156		20,600	9,025		29,625	36,200
2280	3,000 MBH		.19	170		23,700	9,825		33,525	40,900
2300	3,550 MBH		.17	187		27,400	10,800		38,200	46,400
2320	3,820 MBH		.16	200		39,000	11,600		50,600	60,500
2340	4,360 MBH		.15	214		42,800	12,400		55,200	65,500
2360	4,940 MBH		.14	225		67,000	13,000		80,000	93,000
2380	5,520 MBH		.14	235		78,500	13,600		92,100	107,000
2400	6,100 MBH		.13	256		90,000	14,800		104,800	121,500
2420	6,390 MBH		.11	290		95,000	16,800		111,800	130,000
2440	6,680 MBH		.10	313		97,500	18,100		115,600	135,000
2460	6,970 MBH		.09	363		101,000	21,000		122,000	142,500
3000	Hot water, same price as steam									
4000	For tankless coil in smaller sizes, add				Ea.	15%				

23 52 23 – Cast-Iron Boilers

23 52 23.60 Solid-Fuel Boilers		Crew	Daily Output	Labor-Hours	Unit	Material	2016 Bare Costs Labor	Equipment	Total	Total Incl O&P
0010	**SOLID-FUEL BOILERS**									
3000	Stoker fired (coal/wood/biomass) cast iron with flush jacket and									
3400	insulation, steam or water, gross output, 1280 MBH	Q-6	.36	66.667	Ea.	170,000	3,775		173,775	192,500
3420	1460 MBH		.30	80		175,000	4,525		179,525	199,500
3440	1640 MBH		.28	85.714		187,500	4,850		192,350	214,000
3460	1820 MBH		.26	92.308		203,000	5,225		208,225	231,500
3480	2000 MBH		.25	96		219,500	5,450		224,950	249,000
3500	2360 MBH		.23	104		226,000	5,900		231,900	257,500
3540	2725 MBH		.20	120		232,500	6,800		239,300	266,500
3800	2950 MBH	Q-7	.16	200		239,000	11,600		250,600	280,500
3820	3210 MBH		.15	213		252,500	12,300		264,800	296,000
3840	3480 MBH		.14	228		254,000	13,200		267,200	299,000
3860	3745 MBH		.14	228		255,500	13,200		268,700	301,000
3880	4000 MBH		.13	246		257,000	14,200		271,200	304,000
3900	4200 MBH		.13	246		258,500	14,200		272,700	306,000
3920	4400 MBH		.12	266		260,000	15,400		275,400	309,500
3940	4600 MBH		.12	266		263,500	15,400		278,900	313,000

23 52 26 – Steel Boilers

23 52 26.40 Oil-Fired Boilers

		Crew	Daily Output	Labor-Hours	Unit	Material	Labor	Equipment	Total	Total Incl O&P
0010	**OIL-FIRED BOILERS**, Standard controls, flame retention burner									
5000	Steel, with insulated flush jacket									
7000	Hot water, gross output, 103 MBH	Q-6	1.60	15	Ea.	1,825	850		2,675	3,275
7020	122 MBH		1.45	16.506		2,000	935		2,935	3,600
7040	137 MBH		1.36	17.595		2,150	995		3,145	3,850
7060	168 MBH		1.30	18.405		2,250	1,050		3,300	4,050
7080	225 MBH		1.22	19.704		3,250	1,125		4,375	5,250
7100	315 MBH		.96	25.105		7,000	1,425		8,425	9,850
7120	420 MBH		.70	34.483		7,350	1,950		9,300	11,100
7140	525 MBH		.57	42.403		8,700	2,400		11,100	13,200
7180	735 MBH		.48	50.104		10,200	2,850		13,050	15,500
7220	1,050 MBH		.37	65.753		20,300	3,725		24,025	27,900
7280	2,310 MBH		.21	114		26,100	6,500		32,600	38,500
7320	3,150 MBH		.13	184		30,300	10,500		40,800	49,200
7340	For tankless coil in steam or hot water, add					7%				

23 52 26.70 Packaged Water Tube Boilers

		Crew	Daily Output	Labor-Hours	Unit	Material	Labor	Equipment	Total	Total Incl O&P
0010	**PACKAGED WATER TUBE BOILERS**									
2000	Packaged water tube, #2 oil, steam or hot water, gross output									
2010	200 MBH	Q-6	1.50	16	Ea.	4,225	905		5,130	6,025
2014	275 MBH		1.40	17.143		4,825	970		5,795	6,775
2018	360 MBH		1.10	21.818		6,825	1,225		8,050	9,375
2022	520 MBH		.65	36.923		8,200	2,100		10,300	12,200
2026	600 MBH		.60	40		8,375	2,275		10,650	12,700
2030	720 MBH		.55	43.636		9,975	2,475		12,450	14,700
2034	960 MBH		.48	50		12,300	2,825		15,125	17,900
2040	1200 MBH	Q-7	.50	64		22,200	3,700		25,900	30,000
2044	1440 MBH		.45	71.111		29,800	4,100		33,900	39,000
2060	1600 MBH		.40	80		35,100	4,625		39,725	45,600
2068	1920 MBH		.35	91.429		38,000	5,275		43,275	49,800
2072	2160 MBH		.33	96.970		42,200	5,600		47,800	55,000
2080	2400 MBH		.30	106		47,200	6,175		53,375	61,500
2100	3200 MBH		.25	128		54,000	7,400		61,400	70,500

For customer support on your Mechanical Cost Data, call 877.739.3689.

367

23 52 26 — Steel Boilers

23 52 26.70 Packaged Water Tube Boilers		Crew	Daily Output	Labor-Hours	Unit	Material	2016 Bare Costs Labor	Equipment	Total	Total Incl O&P
2120	4800 MBH	Q-7	.20	160	Ea.	70,500	9,250		79,750	91,500
2200	Gas fired									
2204	200 MBH	Q-6	1.50	16	Ea.	6,025	905		6,930	8,000
2208	275 MBH		1.40	17.143		6,350	970		7,320	8,475
2212	360 MBH		1.10	21.818		6,825	1,225		8,050	9,375
2216	520 MBH		.65	36.923		8,300	2,100		10,400	12,300
2220	600 MBH		.60	40		8,375	2,275		10,650	12,700
2224	720 MBH		.55	43.636		9,975	2,475		12,450	14,700
2228	960 MBH	▼	.48	50		12,300	2,825		15,125	17,900
2232	1220 MBH	Q-7	.50	64		18,500	3,700		22,200	25,900
2236	1440 MBH		.45	71.111		19,000	4,100		23,100	27,100
2240	1680 MBH		.40	80		26,800	4,625		31,425	36,500
2244	1920 MBH		.35	91.429		26,900	5,275		32,175	37,600
2248	2160 MBH		.33	96.970		27,300	5,600		32,900	38,500
2252	2400 MBH	▼	.30	106	▼	30,700	6,175		36,875	43,000

23 52 28 — Swimming Pool Boilers

23 52 28.10 Swimming Pool Heaters

		Crew	Daily Output	Labor-Hours	Unit	Material	2016 Bare Costs Labor	Equipment	Total	Total Incl O&P
0010	**SWIMMING POOL HEATERS**, Not including wiring, external									
0020	piping, base or pad,									
0160	Gas fired, input, 155 MBH	Q-6	1.50	16	Ea.	2,100	905		3,005	3,700
0200	199 MBH		1	24		2,225	1,350		3,575	4,500
0220	250 MBH		.70	34.286		2,425	1,950		4,375	5,600
0240	300 MBH		.60	40		2,550	2,275		4,825	6,225
0260	399 MBH		.50	48		2,850	2,725		5,575	7,250
0280	500 MBH		.40	60		9,325	3,400		12,725	15,400
0300	650 MBH		.35	68.571		9,925	3,875		13,800	16,800
0320	750 MBH		.33	72.727		10,800	4,125		14,925	18,100
0360	990 MBH		.22	109		14,600	6,175		20,775	25,300
0370	1,260 MBH		.21	114		17,100	6,475		23,575	28,600
0380	1,440 MBH		.19	126		18,200	7,150		25,350	30,800
0400	1,800 MBH		.14	171		20,300	9,700		30,000	37,000
0410	2,070 MBH	▼	.13	184		23,900	10,500		34,400	42,100
2000	Electric, 12 KW, 4,800 gallon pool	Q-19	3	8		2,100	440		2,540	2,975
2020	15 KW, 7,200 gallon pool		2.80	8.571		2,150	470		2,620	3,050
2040	24 KW, 9,600 gallon pool		2.40	10		2,475	550		3,025	3,550
2060	30 KW, 12,000 gallon pool		2	12		2,525	655		3,180	3,775
2080	36 KW, 14,400 gallon pool		1.60	15		2,900	820		3,720	4,425
2100	57 KW, 24,000 gallon pool	▼	1.20	20	▼	3,650	1,100		4,750	5,650
9000	To select pool heater: 12 BTUH x S.F. pool area									
9010	X temperature differential = required output									
9050	For electric, KW = gallons x 2.5 divided by 1000									
9100	For family home type pool, double the									
9110	Rated gallon capacity = 1/2°F rise per hour									

23 52 39 — Fire-Tube Boilers

23 52 39.13 Scotch Marine Boilers

		Crew	Daily Output	Labor-Hours	Unit	Material	2016 Bare Costs Labor	Equipment	Total	Total Incl O&P
0010	**SCOTCH MARINE BOILERS**									
1000	Packaged fire tube, #2 oil, gross output									
1006	15 PSI steam									
1010	1005 MBH, 30 HP	Q-6	.33	72.072	Ea.	49,300	4,075		53,375	60,000
1014	1675 MBH, 50 HP	"	.26	93.023		62,000	5,275		67,275	76,000
1020	3348 MBH, 100 HP	Q-7	.21	152		83,000	8,800		91,800	105,000
1030	5025 MBH, 150 HP	↓	.17	192	↓	104,500	11,100		115,600	132,000

23 52 39.13 Scotch Marine Boilers		Crew	Daily Output	Labor-Hours	Unit	Material	2016 Bare Costs Labor	Equipment	Total	Total Incl O&P
1040	6696 MBH, 200 HP	Q-7	.14	223	Ea.	124,500	12,900		137,400	156,500
1050	8375 MBH, 250 HP		.14	231		147,000	13,400		160,400	182,000
1060	10,044 MBH, 300 HP		.13	251		158,500	14,600		173,100	196,500
1080	16,740 MBH, 500 HP		.08	380		215,500	22,000		237,500	270,000
1100	23,435 MBH, 700 HP		.07	484		264,000	28,000		292,000	333,000
1102	26,780 MBH, 800 HP		.07	484		264,000	28,000		292,000	333,000
1104	30,125 MBH, 900 HP		.06	500		482,000	28,900		510,900	573,500
1106	33,475 MBH, 1,000 HP		.06	524		510,000	30,300		540,300	606,500
1107	36,825 MBH, 1,100 HP		.06	551		547,500	31,900		579,400	650,000
1108	40,170 MBH, 1,200 HP		.06	581		574,000	33,600		607,600	682,000
1110	46,865 MBH, 1,400 HP		.05	603		604,500	34,900		639,400	717,500
1112	53,560 MBH, 1,600 HP		.05	627		657,000	36,300		693,300	777,500
1114	60,250 MBH, 1,800 HP		.05	666		678,500	38,500		717,000	804,500
1116	66,950 MBH, 2,000 HP		.05	695		718,500	40,200		758,700	851,000
1118	73,650 MBH, 2,200 HP		.04	744		776,500	43,000		819,500	919,000
1120	To fire #6, add		.83	38.554		16,000	2,225		18,225	21,000
1140	To fire #6, and gas, add		.42	76.190		23,500	4,400		27,900	32,500
1160	For high pressure, add					26,100			26,100	28,700
1180	For duplex package feed system									
1200	To 3348 MBH boiler, add	Q-7	.54	59.259	Ea.	12,700	3,425		16,125	19,100
1220	To 6696 MBH boiler, add		.41	78.049		15,000	4,500		19,500	23,300
1240	To 10,044 MBH boiler, add		.38	84.211		17,200	4,875		22,075	26,300
1260	To 16,740 MBH boiler, add		.28	114		19,600	6,600		26,200	31,600
1280	To 23,435 MBH boiler, add		.25	128		23,200	7,400		30,600	36,700
1300	150 PSI steam									
1310	1005 MBH, 30 HP	Q-6	.33	72.072	Ea.	58,000	4,075		62,075	69,500
1320	1675 MBH, 50 HP	"	.26	93.023		70,000	5,275		75,275	84,500
1330	3350 MBH, 100 HP	Q-7	.21	152		100,000	8,800		108,800	123,500
1340	5025 MBH, 150 HP		.17	192		127,000	11,100		138,100	157,000
1350	6700 MBH, 200 HP		.14	223		141,500	12,900		154,400	175,000
1360	10,050 MBH, 300 HP		.13	251		198,500	14,600		213,100	240,500
1400	Packaged scotch marine, #6 oil									
1410	15 PSI steam									
1420	1675 MBH, 50 HP	Q-6	.25	97.166	Ea.	76,000	5,500		81,500	92,000
1430	3350 MBH, 100 HP	Q-7	.20	159		99,000	9,200		108,200	123,000
1440	5025 MBH, 150 HP		.16	202		121,000	11,700		132,700	150,500
1450	6700 MBH, 200 HP		.14	233		138,500	13,500		152,000	172,500
1460	8375 MBH, 250 HP		.13	244		158,500	14,100		172,600	195,500
1470	10,050 MBH, 300 HP		.12	262		178,000	15,200		193,200	219,000
1480	13,400 MBH, 400 HP		.09	351		219,000	20,300		239,300	271,500
1500	150 PSI steam									
1510	1675 MBH, 50 HP	Q-6	.25	97.166	Ea.	90,000	5,500		95,500	107,500
1520	3350 MBH, 100 HP	Q-7	.20	159		113,500	9,200		122,700	139,000
1530	5025 MBH, 150 HP		.16	202		141,000	11,700		152,700	172,500
1540	6700 MBH, 200 HP		.14	233		154,500	13,500		168,000	190,500
1550	8375 MBH, 250 HP		.13	244		187,500	14,100		201,600	228,000
1560	10,050 MBH, 300 HP		.12	262		209,000	15,200		224,200	252,500
1570	13,400 MBH, 400 HP		.09	351		242,000	20,300		262,300	297,000

23 52 Heating Boilers

23 52 84 – Boiler Blowdown

23 52 84.10 Boiler Blowdown Systems	Crew	Daily Output	Labor-Hours	Unit	Material	2016 Bare Costs Labor	Equipment	Total	Total Incl O&P
0010 **BOILER BLOWDOWN SYSTEMS**									
1010 Boiler blowdown, auto/manual to 2000 MBH	Q-5	3.75	4.267	Ea.	5,825	233		6,058	6,750
1020 7300 MBH	"	3	5.333	"	6,650	291		6,941	7,775

23 52 88 – Burners

23 52 88.10 Replacement Type Burners

	Crew	Daily Output	Labor-Hours	Unit	Material	2016 Bare Costs Labor	Equipment	Total	Total Incl O&P
0010 **REPLACEMENT TYPE BURNERS**									
0990 Residential, conversion, gas fired, LP or natural									
1000 Gun type, atmospheric input 50 to 225 MBH	Q-1	2.50	6.400	Ea.	915	340		1,255	1,525
1020 100 to 400 MBH		2	8		1,525	425		1,950	2,325
1040 300 to 1000 MBH	↓	1.70	9.412	↓	4,600	500		5,100	5,825
2000 Commercial and industrial, gas/oil, input									
2050 400 MBH	Q-1	1.50	10.667	Ea.	4,450	570		5,020	5,750
2090 670 MBH		1.40	11.429		4,450	610		5,060	5,825
2140 1155 MBH		1.30	12.308		4,950	655		5,605	6,450
2200 1800 MBH		1.20	13.333		4,975	710		5,685	6,550
2260 3000 MBH		1.10	14.545		5,000	775		5,775	6,675
2320 4100 MBH	↓	1	16	↓	6,800	850		7,650	8,750
3000 Flame retention oil fired assembly, input									
3020 .50 to 2.25 GPH	Q-1	2.40	6.667	Ea.	266	355		621	830
3040 2.0 to 5.0 GPH		2	8		335	425		760	1,025
3060 3.0 to 7.0 GPH		1.80	8.889		575	475		1,050	1,350
3080 6.0 to 12.0 GPH	↓	1.60	10		920	535		1,455	1,800
4600 Gas safety, shut off valve, 3/4" threaded	1 Stpi	20	.400		182	24.50		206.50	237
4610 1" threaded		19	.421		176	25.50		201.50	232
4620 1-1/4" threaded		15	.533		198	32.50		230.50	267
4630 1-1/2" threaded		13	.615		214	37.50		251.50	292
4640 2" threaded	↓	11	.727		240	44		284	330
4650 2-1/2" threaded	Q-1	15	1.067		274	57		331	385
4660 3" threaded		13	1.231		375	65.50		440.50	515
4670 4" flanged	↓	3	5.333		2,700	284		2,984	3,375
4680 6" flanged	Q-2	3	8	↓	6,000	440		6,440	7,275

23 53 Heating Boiler Feedwater Equipment

23 53 24 – Shot Chemical Feeder

23 53 24.10 Chemical Feeder

	Crew	Daily Output	Labor-Hours	Unit	Material	2016 Bare Costs Labor	Equipment	Total	Total Incl O&P
0010 **CHEMICAL FEEDER**									
0200 Shot chem feeder, by pass, in line mount, 300 PSIG, 1.7 gallon	2 Stpi	1.80	8.889	Ea.	272	540		812	1,125
0230 12 gallon		.93	17.204		470	1,050		1,520	2,100
0400 300 lb., ASME, 5 gallon		1.24	12.903		291	785		1,076	1,500
0420 10 gallon	↓	.93	17.204	↓	435	1,050		1,485	2,050

23 54 Furnaces

23 54 13 – Electric-Resistance Furnaces

23 54 13.10 Electric Furnaces

		Crew	Daily Output	Labor-Hours	Unit	Material	2016 Bare Costs Labor	Equipment	Total	Total Incl O&P
0010	**ELECTRIC FURNACES**, Hot air, blowers, std. controls									
0011	not including gas, oil or flue piping									
1000	Electric, UL listed									
1100	34.1 MBH	Q-20	4.40	4.545	Ea.	455	237		692	860
1120	51.6 MBH		4.20	4.762		495	249		744	925
1140	68.3 MBH		4	5		515	261		776	960
1160	85.3 MBH		3.80	5.263		520	275		795	995

23 54 16 – Fuel-Fired Furnaces

23 54 16.13 Gas-Fired Furnaces

		Crew	Daily Output	Labor-Hours	Unit	Material	2016 Bare Costs Labor	Equipment	Total	Total Incl O&P
0010	**GAS-FIRED FURNACES**									
3000	Gas, AGA certified, upflow, direct drive models									
3020	45 MBH input	Q-9	4	4	Ea.	580	206		786	950
3040	60 MBH input		3.80	4.211		595	217		812	985
3060	75 MBH input		3.60	4.444		660	229		889	1,075
3100	100 MBH input		3.20	5		690	258		948	1,150
3120	125 MBH input		3	5.333		725	275		1,000	1,225
3130	150 MBH input		2.80	5.714		1,375	294		1,669	1,950
3140	200 MBH input		2.60	6.154		2,950	315		3,265	3,725
3160	300 MBH input		2.30	6.957		3,250	360		3,610	4,125
3180	400 MBH input		2	8		3,625	410		4,035	4,625
3200	Gas fired wall furnace									
3204	Horizontal flow									
3208	7.7 MBH	Q-9	7	2.286	Ea.	580	118		698	820
3212	14 MBH		6.50	2.462		580	127		707	835
3216	24 MBH		5	3.200		615	165		780	925
3220	49 MBH		4	4		840	206		1,046	1,225
3224	65 MBH		3.60	4.444		1,150	229		1,379	1,625
3240	Up-flow									
3244	7.7 MBH	Q-9	7	2.286	Ea.	580	118		698	820
3248	14 MBH		6.50	2.462		580	127		707	835
3252	24 MBH		5	3.200		600	165		765	910
3254	35 MBH		4.50	3.556		845	183		1,028	1,200
3256	49 MBH		4	4		1,175	206		1,381	1,625
3260	65 MBH		3.60	4.444		1,300	229		1,529	1,775
3500	Gas furnace									
3510	Up-flow									
3514	1200 CFM, 45 MBH	Q-9	4	4	Ea.	1,125	206		1,331	1,575
3518	75 MBH		3.60	4.444		1,250	229		1,479	1,725
3522	100 MBH		3.20	5		1,375	258		1,633	1,900
3526	125 MBH		3	5.333		1,400	275		1,675	1,950
3530	150 MBH		2.80	5.714		1,550	294		1,844	2,175
3540	2000 CFM, 75 MBH		3.60	4.444		1,375	229		1,604	1,850
3544	100 MBH		3.20	5		1,550	258		1,808	2,100
3548	120 MBH		3	5.333		1,625	275		1,900	2,225
3560	Horizontal flow									
3564	1200 CFM, 45 MBH	Q-9	4	4	Ea.	1,300	206		1,506	1,775
3568	60 MBH		3.80	4.211		1,375	217		1,592	1,825
3572	70 MBH		3.60	4.444		1,400	229		1,629	1,900
3576	2000 CFM, 90 MBH		3.30	4.848		1,550	250		1,800	2,075
3580	110 MBH		3.10	5.161		1,800	266		2,066	2,375

23 54 Furnaces

23 54 16 – Fuel-Fired Furnaces

23 54 16.14 Condensing Furnaces

		Crew	Daily Output	Labor-Hours	Unit	Material	2016 Bare Costs Labor	Equipment	Total	Total Incl O&P
0010	**CONDENSING FURNACES**, High efficiency									
0020	Oil fired, packaged, complete									
0030	Upflow									
0040	Output @ 95% A.F.U.E.									
0100	49 MBH @ 1000 CFM	Q-9	3.70	4.324	Ea.	5,500	223		5,723	6,400
0110	73.5 MBH @ 2000 CFM		3.60	4.444		5,750	229		5,979	6,675
0120	96 MBH @ 2000 CFM		3.40	4.706		5,750	242		5,992	6,700
0130	115.6 MBH @ 2000 CFM		3.40	4.706		5,750	242		5,992	6,700
0140	147 MBH @ 2000 CFM		3.30	4.848		15,100	250		15,350	17,100
0150	192 MBH @ 4000 CFM		2.60	6.154		15,700	315		16,015	17,800
0170	231.5 MBH @ 4000 CFM		2.30	6.957		15,700	360		16,060	17,900
0260	For variable speed motor, add					675			675	740
0270	Note: Also available in horizontal, counterflow and lowboy configurations.									

23 54 16.16 Oil-Fired Furnaces

		Crew	Daily Output	Labor-Hours	Unit	Material	2016 Bare Costs Labor	Equipment	Total	Total Incl O&P
0010	**OIL-FIRED FURNACES**									
6000	Oil, UL listed, atomizing gun type burner									
6020	56 MBH output	Q-9	3.60	4.444	Ea.	1,700	229		1,929	2,225
6030	84 MBH output		3.50	4.571		2,500	236		2,736	3,100
6040	95 MBH output		3.40	4.706		2,575	242		2,817	3,200
6060	134 MBH output		3.20	5		2,625	258		2,883	3,275
6080	151 MBH output		3	5.333		2,725	275		3,000	3,425
6100	200 MBH input		2.60	6.154		3,000	315		3,315	3,800
6120	300 MBH input		2.30	6.957		3,775	360		4,135	4,700
6140	400 MBH input		2	8		4,100	410		4,510	5,125
6200	Up flow									
6210	105 MBH	Q-9	3.40	4.706	Ea.	1,825	242		2,067	2,375
6220	140 MBH		3.30	4.848		2,000	250		2,250	2,575
6230	168 MBH		2.90	5.517		3,475	284		3,759	4,250
6260	Lo-boy style									
6270	105 MBH	Q-9	3.40	4.706	Ea.	1,900	242		2,142	2,475
6280	140 MBH		3.30	4.848		2,050	250		2,300	2,650
6290	189 MBH		2.80	5.714		3,300	294		3,594	4,075
6320	Down flow									
6330	105 MBH	Q-9	3.40	4.706	Ea.	1,900	242		2,142	2,475
6340	140 MBH		3.30	4.848		2,050	250		2,300	2,650
6350	168 MBH		2.90	5.517		3,525	284		3,809	4,300
6380	Horizontal									
6390	105 MBH	Q-9	3.40	4.706	Ea.	1,900	242		2,142	2,475
6400	140 MBH		3.30	4.848		2,050	250		2,300	2,650
6410	189 MBH		2.80	5.714		3,300	294		3,594	4,075

23 54 16.21 Solid Fuel-Fired Furnaces

			Crew	Daily Output	Labor-Hours	Unit	Material	2016 Bare Costs Labor	Equipment	Total	Total Incl O&P
0010	**SOLID FUEL-FIRED FURNACES**										
6020	Wood fired furnaces										
6030	Includes hot water coil, thermostat, and auto draft control										
6040	24" long firebox	G	Q-9	4	4	Ea.	4,300	206		4,506	5,050
6050	30" long firebox	G		3.60	4.444		5,025	229		5,254	5,875
6060	With fireplace glass doors	G		3.20	5		6,350	258		6,608	7,400
6200	Wood/oil fired furnaces, includes two thermostats										
6210	Includes hot water coil and auto draft control										
6240	24" long firebox	G	Q-9	3.40	4.706	Ea.	5,150	242		5,392	6,050
6250	30" long firebox	G		3	5.333		5,775	275		6,050	6,775
6260	With fireplace glass doors	G		2.80	5.714		7,075	294		7,369	8,225

23 54 Furnaces

23 54 16 – Fuel-Fired Furnaces

23 54 16.21 Solid Fuel-Fired Furnaces

		Crew	Daily Output	Labor-Hours	Unit	Material	2016 Bare Costs Labor	Equipment	Total	Total Incl O&P
6400	Wood/gas fired furnaces, includes two thermostats									
6410	Includes hot water coil and auto draft control									
6440	24" long firebox	G Q-9	2.80	5.714	Ea.	5,925	294		6,219	6,975
6450	30" long firebox	G	2.40	6.667		6,625	345		6,970	7,825
6460	With fireplace glass doors	G	2	8		7,900	410		8,310	9,325
6600	Wood/oil/gas fired furnaces, optional accessories									
6610	Hot air plenum	Q-9	16	1	Ea.	108	51.50		159.50	198
6620	Safety heat dump		24	.667		94	34.50		128.50	156
6630	Auto air intake		18	.889		148	46		194	233
6640	Cold air return package		14	1.143		159	59		218	265
6650	Wood fork					51			51	56
6700	Wood fired outdoor furnace									
6740	24" long firebox	G Q-9	3.80	4.211	Ea.	4,500	217		4,717	5,275
6760	Wood fired outdoor furnace, optional accessories									
6770	Chimney section, stainless steel, 6" ID x 3' lg.	G Q-9	36	.444	Ea.	165	23		188	217
6780	Chimney cap, stainless steel	G "	40	.400	"	95	20.50		115.50	137
6800	Wood fired hot water furnace									
6820	Includes 200 gal. hot water storage, thermostat, and auto draft control									
6840	30" long firebox	G Q-9	2.10	7.619	Ea.	8,275	395		8,670	9,700
6850	Water to air heat exchanger									
6870	Includes mounting kit and blower relay									
6880	140 MBH, 18.75" W x 18.75" L	Q-9	7.50	2.133	Ea.	370	110		480	575
6890	200 MBH, 24" W x 24" L	"	7	2.286	"	525	118		643	760
6900	Water to water heat exchanger									
6940	100 MBH	Q-9	6.50	2.462	Ea.	370	127		497	605
6960	290 MBH	"	6	2.667	"	535	137		672	800
7000	Optional accessories									
7010	Large volume circulation pump, (2 Included)	Q-9	14	1.143	Ea.	242	59		301	355
7020	Air bleed fittings, (package)		24	.667		50.50	34.50		85	108
7030	Domestic water preheater		6	2.667		219	137		356	450
7040	Smoke pipe kit		4	4		88	206		294	410

23 54 24 – Furnace Components for Cooling

23 54 24.10 Furnace Components and Combinations

		Crew	Daily Output	Labor-Hours	Unit	Material	2016 Bare Costs Labor	Equipment	Total	Total Incl O&P
0010	**FURNACE COMPONENTS AND COMBINATIONS**									
0080	Coils, A.C. evaporator, for gas or oil furnaces									
0090	Add-on, with holding charge									
0100	Upflow									
0120	1-1/2 ton cooling	Q-5	4	4	Ea.	222	219		441	575
0130	2 ton cooling		3.70	4.324		241	236		477	620
0140	3 ton cooling		3.30	4.848		400	265		665	840
0150	4 ton cooling		3	5.333		530	291		821	1,025
0160	5 ton cooling		2.70	5.926		550	325		875	1,100
0300	Downflow									
0330	2-1/2 ton cooling	Q-5	3	5.333	Ea.	330	291		621	805
0340	3-1/2 ton cooling		2.60	6.154		505	335		840	1,050
0350	5 ton cooling		2.20	7.273		550	395		945	1,200
0600	Horizontal									
0630	2 ton cooling	Q-5	3.90	4.103	Ea.	269	224		493	635
0640	3 ton cooling		3.50	4.571		380	250		630	795
0650	4 ton cooling		3.20	5		445	273		718	900
0660	5 ton cooling		2.90	5.517		445	300		745	945
2000	Cased evaporator coils for air handlers									

For customer support on your Mechanical Cost Data, call 877.739.3689.

373

23 54 Furnaces

23 54 24 – Furnace Components for Cooling

23 54 24.10 Furnace Components and Combinations

		Crew	Daily Output	Labor-Hours	Unit	Material	2016 Bare Costs Labor	Equipment	Total	Total Incl O&P
2100	1-1/2 ton cooling	Q-5	4.40	3.636	Ea.	251	199		450	575
2110	2 ton cooling		4.10	3.902		330	213		543	685
2120	2-1/2 ton cooling		3.90	4.103		340	224		564	715
2130	3 ton cooling		3.70	4.324		365	236		601	755
2140	3-1/2 ton cooling		3.50	4.571		455	250		705	880
2150	4 ton cooling		3.20	5		495	273		768	955
2160	5 ton cooling		2.90	5.517		500	300		800	1,000
3010	Air handler, modular									
3100	With cased evaporator cooling coil									
3120	1-1/2 ton cooling	Q-5	3.80	4.211	Ea.	855	230		1,085	1,275
3130	2 ton cooling		3.50	4.571		945	250		1,195	1,425
3140	2-1/2 ton cooling		3.30	4.848		1,000	265		1,265	1,500
3150	3 ton cooling		3.10	5.161		1,150	282		1,432	1,700
3160	3-1/2 ton cooling		2.90	5.517		1,325	300		1,625	1,925
3170	4 ton cooling		2.50	6.400		1,425	350		1,775	2,100
3180	5 ton cooling		2.10	7.619		1,700	415		2,115	2,500
3500	With no cooling coil									
3520	1-1/2 ton coil size	Q-5	12	1.333	Ea.	720	73		793	905
3530	2 ton coil size		10	1.600		720	87.50		807.50	925
3540	2-1/2 ton coil size		10	1.600		730	87.50		817.50	935
3554	3 ton coil size		9	1.778		750	97		847	970
3560	3-1/2 ton coil size		9	1.778		770	97		867	990
3570	4 ton coil size		8.50	1.882		780	103		883	1,025
3580	5 ton coil size		8	2		790	109		899	1,025
4000	With heater									
4120	5 kW, 17.1 MBH	Q-5	16	1	Ea.	455	54.50		509.50	585
4130	7.5 kW, 25.6 MBH		15.60	1.026		1,100	56		1,156	1,275
4140	10 kW, 34.2 MBH		15.20	1.053		1,175	57.50		1,232.50	1,375
4150	12.5 KW, 42.7 MBH		14.80	1.081		1,550	59		1,609	1,800
4160	15 KW, 51.2 MBH		14.40	1.111		1,675	60.50		1,735.50	1,950
4170	25 KW, 85.4 MBH		14	1.143		2,225	62.50		2,287.50	2,550
4180	30 KW, 102 MBH		13	1.231		2,575	67		2,642	2,950

23 55 Fuel-Fired Heaters

23 55 13 – Fuel-Fired Duct Heaters

23 55 13.16 Gas-Fired Duct Heaters

		Crew	Daily Output	Labor-Hours	Unit	Material	2016 Bare Costs Labor	Equipment	Total	Total Incl O&P
0010	**GAS-FIRED DUCT HEATERS**, Includes burner, controls, stainless steel									
0020	heat exchanger. Gas fired, electric ignition									
0030	Indoor installation									
0080	100 MBH output	Q-5	5	3.200	Ea.	2,950	175		3,125	3,525
0100	120 MBH output		4	4		3,225	219		3,444	3,875
0130	200 MBH output		2.70	5.926		4,100	325		4,425	5,000
0140	240 MBH output		2.30	6.957		4,300	380		4,680	5,300
0160	280 MBH output		2	8		4,700	435		5,135	5,800
0180	320 MBH output		1.60	10		5,025	545		5,570	6,350
0300	For powered venter and adapter, add					535			535	590
0502	For required flue pipe, see Section 23 51 23.10									
1000	Outdoor installation, with power venter									
1020	75 MBH output	Q-5	4	4	Ea.	3,600	219		3,819	4,275
1040	94 MBH output		4	4		3,650	219		3,869	4,350
1060	120 MBH output		4	4		3,975	219		4,194	4,700

23 55 Fuel-Fired Heaters

23 55 13 – Fuel-Fired Duct Heaters

23 55 13.16 Gas-Fired Duct Heaters

		Crew	Daily Output	Labor-Hours	Unit	Material	2016 Bare Costs Labor	Equipment	Total	Total Incl O&P
1080	157 MBH output	Q-5	3.50	4.571	Ea.	4,275	250		4,525	5,075
1100	187 MBH output		3	5.333		4,900	291		5,191	5,825
1120	225 MBH output		2.50	6.400		5,100	350		5,450	6,150
1140	300 MBH output		1.80	8.889		7,900	485		8,385	9,400
1160	375 MBH output		1.60	10		8,325	545		8,870	9,975
1180	450 MBH output		1.40	11.429		9,550	625		10,175	11,400
1200	600 MBH output		1	16		10,800	875		11,675	13,200
1300	Aluminized exchanger, subtract					15%				
1500	For two stage gas valve, add					775			775	850

23 55 33 – Fuel-Fired Unit Heaters

23 55 33.13 Oil-Fired Unit Heaters

		Crew	Daily Output	Labor-Hours	Unit	Material	2016 Bare Costs Labor	Equipment	Total	Total Incl O&P
0010	**OIL-FIRED UNIT HEATERS**, Cabinet, grilles, fan, ctrl., burner, no piping									
6000	Oil fired, suspension mounted, 94 MBH output	Q-5	4	4	Ea.	4,925	219		5,144	5,725
6040	140 MBH output		3	5.333		5,175	291		5,466	6,125
6060	184 MBH output		3	5.333		5,500	291		5,791	6,500

23 55 33.16 Gas-Fired Unit Heaters

		Crew	Daily Output	Labor-Hours	Unit	Material	2016 Bare Costs Labor	Equipment	Total	Total Incl O&P
0010	**GAS-FIRED UNIT HEATERS**, Cabinet, grilles, fan, ctrls., burner, no piping									
0022	thermostat, no piping. For flue see Section 23 51 23.10									
1000	Gas fired, floor mounted									
1100	60 MBH output	Q-5	10	1.600	Ea.	925	87.50		1,012.50	1,150
1120	80 MBH output		9	1.778		940	97		1,037	1,175
1140	100 MBH output		8	2		1,025	109		1,134	1,300
1160	120 MBH output		7	2.286		1,225	125		1,350	1,550
1180	180 MBH output		6	2.667		1,475	146		1,621	1,825
1500	Rooftop mounted, power vent, stainless steel exchanger									
1520	75 MBH input	Q-6	4	6	Ea.	5,775	340		6,115	6,875
1540	100 MBH input		3.60	6.667		6,025	380		6,405	7,200
1560	125 MBH input		3.30	7.273		6,300	410		6,710	7,550
1580	150 MBH input		3	8		6,775	455		7,230	8,125
1600	175 MBH input		2.60	9.231		6,950	525		7,475	8,450
1620	225 MBH input		2.30	10.435		7,925	590		8,515	9,600
1640	300 MBH input		1.90	12.632		9,125	715		9,840	11,200
1660	350 MBH input		1.40	17.143		9,750	970		10,720	12,200
1680	450 MBH input		1.20	20		10,500	1,125		11,625	13,200
1720	700 MBH input		.80	30		10,800	1,700		12,500	14,500
1760	1200 MBH input		.30	80		11,300	4,525		15,825	19,300
1900	For aluminized steel exchanger, subtract					10%				
2000	Suspension mounted, propeller fan, 20 MBH output	Q-5	8.50	1.882		1,150	103		1,253	1,400
2020	40 MBH output		7.50	2.133		1,575	117		1,692	1,925
2040	60 MBH output		7	2.286		1,750	125		1,875	2,125
2060	80 MBH output		6	2.667		1,900	146		2,046	2,325
2080	100 MBH output		5.50	2.909		2,050	159		2,209	2,500
2100	130 MBH output		5	3.200		2,250	175		2,425	2,750
2120	140 MBH output		4.50	3.556		2,400	194		2,594	2,925
2140	160 MBH output		4	4		2,450	219		2,669	3,025
2160	180 MBH output		3.50	4.571		2,925	250		3,175	3,575
2180	200 MBH output		3	5.333		2,950	291		3,241	3,700
2200	240 MBH output		2.70	5.926		3,250	325		3,575	4,075
2220	280 MBH output		2.30	6.957		3,575	380		3,955	4,500
2240	320 MBH output		2	8		4,200	435		4,635	5,250
2500	For powered venter and adapter, add					440			440	485
3000	Suspension mounted, blower type, 40 MBH output	Q-5	6.80	2.353		1,000	129		1,129	1,300

For customer support on your Mechanical Cost Data, call 877.739.3689.

375

23 55 Fuel-Fired Heaters

23 55 33 – Fuel-Fired Unit Heaters

23 55 33.16 Gas-Fired Unit Heaters

	23 55 33.16 Gas-Fired Unit Heaters	Crew	Daily Output	Labor-Hours	Unit	Material	2016 Bare Costs Labor	Equipment	Total	Total Incl O&P
3020	60 MBH output	Q-5	6.60	2.424	Ea.	1,025	132		1,157	1,325
3040	84 MBH output		5.80	2.759		1,050	151		1,201	1,400
3060	104 MBH output		5.20	3.077		1,075	168		1,243	1,450
3080	140 MBH output		4.30	3.721		1,100	203		1,303	1,500
3100	180 MBH output		3.30	4.848		1,375	265		1,640	1,925
3120	240 MBH output		2.50	6.400		1,575	350		1,925	2,250
3140	280 MBH output		2	8		1,800	435		2,235	2,625
4000	Suspension mounted, sealed combustion system,									
4020	Aluminized steel exchanger, powered vent									
4040	100 MBH output	Q-5	5	3.200	Ea.	3,050	175		3,225	3,625
4060	120 MBH output		4.70	3.404		3,625	186		3,811	4,275
4080	160 MBH output		3.70	4.324		4,200	236		4,436	4,950
4100	200 MBH output		2.90	5.517		5,125	300		5,425	6,075
4120	240 MBH output		2.50	6.400		5,500	350		5,850	6,575
4140	320 MBH output		1.70	9.412		6,600	515		7,115	8,025
5000	Wall furnace, 17.5 MBH output		6	2.667		745	146		891	1,050
5020	24 MBH output		5	3.200		745	175		920	1,075
5040	35 MBH output		4	4		825	219		1,044	1,225

23 56 Solar Energy Heating Equipment

23 56 16 – Packaged Solar Heating Equipment

23 56 16.40 Solar Heating Systems

	23 56 16.40 Solar Heating Systems		Crew	Daily Output	Labor-Hours	Unit	Material	2016 Bare Costs Labor	Equipment	Total	Total Incl O&P
0010	**SOLAR HEATING SYSTEMS**	R235616-60									
0020	System/Package prices, not including connecting										
0030	pipe, insulation, or special heating/plumbing fixtures										
0152	For solar ultraviolet pipe insulation see Section 22 07 19.10										
0500	Hot water, standard package, low temperature										
0540	1 collector, circulator, fittings, 65 gal. tank	G	Q-1	.50	32	Ea.	3,700	1,700		5,400	6,650
0580	2 collectors, circulator, fittings, 120 gal. tank	G		.40	40		5,075	2,125		7,200	8,825
0620	3 collectors, circulator, fittings, 120 gal. tank	G		.34	47.059		6,925	2,500		9,425	11,400
0700	Medium temperature package										
0720	1 collector, circulator, fittings, 80 gal. tank	G	Q-1	.50	32	Ea.	5,050	1,700		6,750	8,125
0740	2 collectors, circulator, fittings, 120 gal. tank	G		.40	40		6,500	2,125		8,625	10,400
0780	3 collectors, circulator, fittings, 120 gal. tank	G		.30	53.333		7,375	2,850		10,225	12,400
0980	For each additional 120 gal. tank, add	G					1,750			1,750	1,950

23 56 19 – Solar Heating Components

23 56 19.50 Solar Heating Ancillary

	23 56 19.50 Solar Heating Ancillary		Crew	Daily Output	Labor-Hours	Unit	Material	2016 Bare Costs Labor	Equipment	Total	Total Incl O&P
0010	**SOLAR HEATING ANCILLARY**										
2300	Circulators, air										
2310	Blowers										
2330	100-300 S.F. system, 1/10 HP	G	Q-9	16	1	Ea.	249	51.50		300.50	355
2340	300-500 S.F. system, 1/5 HP	G		15	1.067		330	55		385	445
2350	Two speed, 100-300 S.F., 1/10 HP	G		14	1.143		143	59		202	247
2400	Reversible fan, 20" diameter, 2 speed	G		18	.889		113	46		159	194
2550	Booster fan 6" diameter, 120 CFM	G		16	1		36.50	51.50		88	120
2570	6" diameter, 225 CFM	G		16	1		45	51.50		96.50	129
2580	8" diameter, 150 CFM	G		16	1		41	51.50		92.50	125
2590	8" diameter, 310 CFM	G		14	1.143		63.50	59		122.50	160
2600	8" diameter, 425 CFM	G		14	1.143		71	59		130	168
2650	Rheostat	G		32	.500		15.50	26		41.50	56.50

23 56 19 – Solar Heating Components

23 56 19.50 Solar Heating Ancillary		Crew	Daily Output	Labor-Hours	Unit	Material	2016 Bare Costs Labor	Equipment	Total	Total Incl O&P
2660	Shutter/damper	G	Q-9	12	1.333	Ea.	56	68.50		124.50	167
2670	Shutter motor	G	↓	16	1		139	51.50		190.50	232
2800	Circulators, liquid, 1/25 HP, 5.3 GPM	G	Q-1	14	1.143		181	61		242	291
2820	1/20 HP, 17 GPM	G		12	1.333		263	71		334	395
2850	1/20 HP, 17 GPM, stainless steel	G		12	1.333		254	71		325	385
2870	1/12 HP, 30 GPM	G	↓	10	1.600	↓	345	85.50		430.50	510
3000	Collector panels, air with aluminum absorber plate										
3010	Wall or roof mount										
3040	Flat black, plastic glazing										
3080	4' x 8'	G	Q-9	6	2.667	Ea.	660	137		797	940
3100	4' x 10'	G		5	3.200	"	815	165		980	1,150
3200	Flush roof mount, 10' to 16' x 22" wide	G		96	.167	L.F.	133	8.60		141.60	159
3210	Manifold, by L.F. width of collectors	G	↓	160	.100	"	144	5.15		149.15	167
3300	Collector panels, liquid with copper absorber plate										
3320	Black chrome, tempered glass glazing										
3330	Alum. frame, 4' x 8', 5/32" single glazing	G	Q-1	9.50	1.684	Ea.	1,000	89.50		1,089.50	1,225
3390	Alum. frame, 4' x 10', 5/32" single glazing	G		6	2.667		1,150	142		1,292	1,500
3450	Flat black, alum. frame, 3.5' x 7.5'	G		9	1.778		880	94.50		974.50	1,100
3500	4' x 8'	G		5.50	2.909		1,050	155		1,205	1,375
3520	4' x 10'	G		10	1.600		1,250	85.50		1,335.50	1,500
3540	4' x 12.5'	G	↓	5	3.200	↓	1,250	171		1,421	1,625
3550	Liquid with fin tube absorber plate										
3560	Alum. frame 4' x 8' tempered glass	G	Q-1	10	1.600	Ea.	580	85.50		665.50	770
3580	Liquid with vacuum tubes, 4' x 6'-10"	G		9	1.778		905	94.50		999.50	1,150
3600	Liquid, full wetted, plastic, alum. frame, 4' x 10'	G		5	3.200		320	171		491	610
3650	Collector panel mounting, flat roof or ground rack	G		7	2.286	↓	245	122		367	455
3670	Roof clamps	G	↓	70	.229	Set	2.83	12.20		15.03	21.50
3700	Roof strap, teflon	G	1 Plum	205	.039	L.F.	25	2.31		27.31	31
3900	Differential controller with two sensors										
3930	Thermostat, hard wired	G	1 Plum	8	1	Ea.	101	59		160	201
3950	Line cord and receptacle	G		12	.667		130	39.50		169.50	203
4050	Pool valve system	G		2.50	3.200		161	189		350	465
4070	With 12 VAC actuator	G		2	4		325	237		562	710
4080	Pool pump system, 2" pipe size	G		6	1.333		195	79		274	335
4100	Five station with digital read-out	G	↓	3	2.667	↓	263	158		421	525
4150	Sensors										
4200	Brass plug, 1/2" MPT	G	1 Plum	32	.250	Ea.	16	14.80		30.80	40
4210	Brass plug, reversed	G		32	.250		26.50	14.80		41.30	51.50
4220	Freeze prevention	G		32	.250		22.50	14.80		37.30	47.50
4240	Screw attached	G		32	.250		9.85	14.80		24.65	33.50
4250	Brass, immersion	G	↓	32	.250	↓	13.55	14.80		28.35	37.50
4300	Heat exchanger										
4315	includes coil, blower, circulator										
4316	and controller for DHW and space hot air										
4330	Fluid to air coil, up flow, 45 MBH	G	Q-1	4	4	Ea.	315	213		528	670
4380	70 MBH	G		3.50	4.571		360	244		604	765
4400	80 MBH	G	↓	3	5.333	↓	475	284		759	955
4580	Fluid to fluid package includes two circulating pumps										
4590	expansion tank, check valve, relief valve										
4600	controller, high temperature cutoff and sensors	G	Q-1	2.50	6.400	Ea.	800	340		1,140	1,400
4650	Heat transfer fluid										
4700	Propylene glycol, inhibited anti-freeze	G	1 Plum	28	.286	Gal.	15	16.90		31.90	42
4800	Solar storage tanks, knocked down										

For customer support on your Mechanical Cost Data, call 877.739.3689.

377

23 56 19 - Solar Heating Components

23 56 19.50 Solar Heating Ancillary		Crew	Daily Output	Labor-Hours	Unit	Material	2016 Bare Costs Labor	Equipment	Total	Total Incl O&P
4810	Air, galvanized steel clad, double wall, 4" fiberglass insulation									
5120	45 Mil reinforced polypropylene lining,									
5140	4' high, 4' x 4' = 64 C.F./450 gallons	G	Q-9	2	8	Ea.	3,650	410		4,060	4,625
5150	4' x 8' = 128 C.F./900 gallons	G		1.50	10.667		5,450	550		6,000	6,850
5160	4' x 12' = 190 C.F./1300 gallons	G		1.30	12.308		7,275	635		7,910	8,975
5170	8' x 8' = 250 C.F./1700 gallons	G		1	16		7,275	825		8,100	9,250
5190	6'-3" high, 7' x 7' = 306 C.F./2000 gallons	G	Q-10	1.20	20		14,200	1,075		15,275	17,300
5200	7' x 10'-6" = 459 C.F./3000 gallons	G		.80	30		17,800	1,600		19,400	22,100
5210	7' x 14' = 613 C.F./4000 gallons	G		.60	40		21,400	2,125		23,525	26,800
5220	10'-6" x 10'-6" = 689 C.F./4500 gallons	G		.50	48		21,400	2,575		23,975	27,400
5230	10'-6" x 14' = 919 C.F./6000 gallons	G		.40	60		24,900	3,200		28,100	32,300
5240	14' x 14' = 1225 C.F./8000 gallons	G	Q-11	.40	80		28,500	4,350		32,850	38,000
5250	14' x 17'-6" = 1531 C.F./10,000 gallons	G		.30	106		32,100	5,825		37,925	44,200
5260	17'-6" x 17'-6" = 1914 C.F./12,500 gallons	G		.25	128		35,600	6,975		42,575	49,900
5270	17'-6" x 21' = 2297 C.F./15,000 gallons	G		.20	160		39,100	8,725		47,825	56,500
5280	21' x 21' = 2756 C.F./18,000 gallons	G		.18	177		42,700	9,700		52,400	62,000
5290	30 Mil reinforced Hypalon lining, add						.02%				
7000	Solar control valves and vents										
7050	Air purger, 1" pipe size	G	1 Plum	12	.667	Ea.	46.50	39.50		86	111
7070	Air eliminator, automatic 3/4" size	G		32	.250		30	14.80		44.80	55.50
7090	Air vent, automatic, 1/8" fitting	G		32	.250		17.25	14.80		32.05	41.50
7100	Manual, 1/8" NPT	G		32	.250		2.44	14.80		17.24	25
7120	Backflow preventer, 1/2" pipe size	G		16	.500		79	29.50		108.50	131
7130	3/4" pipe size	G		16	.500		80.50	29.50		110	133
7150	Balancing valve, 3/4" pipe size	G		20	.400		67.50	23.50		91	110
7180	Draindown valve, 1/2" copper tube	G		9	.889		212	52.50		264.50	315
7200	Flow control valve, 1/2" pipe size	G		22	.364		130	21.50		151.50	176
7220	Expansion tank, up to 5 gal.	G		32	.250		68	14.80		82.80	97.50
7250	Hydronic controller (aquastat)	G		8	1		186	59		245	295
7400	Pressure gauge, 2" dial	G		32	.250		24	14.80		38.80	49
7450	Relief valve, temp. and pressure 3/4" pipe size	G		30	.267		22.50	15.80		38.30	48.50
7500	Solenoid valve, normally closed										
7520	Brass, 3/4" NPT, 24V	G	1 Plum	9	.889	Ea.	128	52.50		180.50	221
7530	1" NPT, 24V	G		9	.889		207	52.50		259.50	310
7750	Vacuum relief valve, 3/4" pipe size	G		32	.250		29.50	14.80		44.30	55
7800	Thermometers										
7820	Digital temperature monitoring, 4 locations	G	1 Plum	2.50	3.200	Ea.	137	189		326	435
7900	Upright, 1/2" NPT	G		8	1		39.50	59		98.50	133
7970	Remote probe, 2" dial	G		8	1		33	59		92	126
7990	Stem, 2" dial, 9" stem	G		16	.500		21.50	29.50		51	68.50
8250	Water storage tank with heat exchanger and electric element										
8270	66 gal. with 2" x 2 lb. density insulation	G	1 Plum	1.60	5	Ea.	1,600	296		1,896	2,225
8300	80 gal. with 2" x 2 lb. density insulation	G		1.60	5		1,600	296		1,896	2,225
8380	120 gal. with 2" x 2 lb. density insulation	G		1.40	5.714		1,825	340		2,165	2,525
8400	120 gal. with 2" x 2 lb. density insul., 40 S.F. heat coil	G		1.40	5.714		2,325	340		2,665	3,050
8500	Water storage module, plastic										
8600	Tubular, 12" diameter, 4' high	G	1 Carp	48	.167	Ea.	89	8.10		97.10	110
8610	12" diameter, 8' high	G		40	.200		143	9.70		152.70	172
8620	18" diameter, 5' high	G		38	.211		161	10.20		171.20	193
8630	18" diameter, 10' high	G		32	.250		269	12.10		281.10	315
8640	58" diameter, 5' high	G	2 Carp	32	.500		665	24		689	770
8650	Cap, 12" diameter	G					19.80			19.80	22
8660	18" diameter	G					25			25	27.50

23 57 Heat Exchangers for HVAC

23 57 16 – Steam-to-Water Heat Exchangers

23 57 16.10 Shell/Tube Type Steam-to-Water Heat Exch.	Crew	Daily Output	Labor-Hours	Unit	Material	2016 Bare Costs Labor	Equipment	Total	Total Incl O&P
0010 **SHELL AND TUBE TYPE STEAM-TO-WATER HEAT EXCHANGERS**									
0016 Shell & tube type, 2 or 4 pass, 3/4" O.D. copper tubes,									
0020 C.I. heads, C.I. tube sheet, steel shell									
0100 Hot water 40°F to 180°F, by steam at 10 PSI									
0120 8 GPM	Q-5	6	2.667	Ea.	2,250	146		2,396	2,700
0140 10 GPM		5	3.200		3,375	175		3,550	4,000
0160 40 GPM		4	4		5,225	219		5,444	6,075
0180 64 GPM		2	8		8,000	435		8,435	9,450
0200 96 GPM		1	16		10,700	875		11,575	13,100
0220 120 GPM	Q-6	1.50	16		14,000	905		14,905	16,900
0240 168 GPM		1	24		17,200	1,350		18,550	21,100
0260 240 GPM		.80	30		26,900	1,700		28,600	32,200
0300 600 GPM		.70	34.286		58,000	1,950		59,950	66,500
0500 For bronze head and tube sheet, add					50%				

23 57 19 – Liquid-to-Liquid Heat Exchangers

23 57 19.13 Plate-Type, Liquid-to-Liquid Heat Exchangers

	Crew	Daily Output	Labor-Hours	Unit	Material	2016 Bare Costs Labor	Equipment	Total	Total Incl O&P
0010 **PLATE-TYPE, LIQUID-TO-LIQUID HEAT EXCHANGERS**									
3000 Plate type,									
3100 400 GPM	Q-6	.80	30	Ea.	38,200	1,700		39,900	44,600
3120 800 GPM	"	.50	48		66,000	2,725		68,725	76,500
3140 1200 GPM	Q-7	.34	94.118		98,000	5,450		103,450	115,500
3160 1800 GPM	"	.24	133		130,000	7,700		137,700	154,500

23 57 19.16 Shell-Type, Liquid-to-Liquid Heat Exchangers

	Crew	Daily Output	Labor-Hours	Unit	Material	2016 Bare Costs Labor	Equipment	Total	Total Incl O&P
0010 **SHELL-TYPE, LIQUID-TO-LIQUID HEAT EXCHANGERS**									
1000 Hot water 40°F to 140°F, by water at 200°F									
1020 7 GPM	Q-5	6	2.667	Ea.	2,750	146		2,896	3,250
1040 16 GPM		5	3.200		3,900	175		4,075	4,550
1060 34 GPM		4	4		5,900	219		6,119	6,800
1080 55 GPM		3	5.333		8,550	291		8,841	9,850
1100 74 GPM		1.50	10.667		10,700	585		11,285	12,700
1120 86 GPM		1.40	11.429		14,300	625		14,925	16,600
1140 112 GPM	Q-6	2	12		17,800	680		18,480	20,600
1160 126 GPM		1.80	13.333		22,200	755		22,955	25,600
1180 152 GPM		1	24		28,400	1,350		29,750	33,300

23 61 Refrigerant Compressors

23 61 13 – Centrifugal Refrigerant Compressors

23 61 13.10 Centrifugal Compressors

	Crew	Daily Output	Labor-Hours	Unit	Material	2016 Bare Costs Labor	Equipment	Total	Total Incl O&P
0010 **CENTRIFUGAL COMPRESSORS**									
0100 Refrigeration, centrifugal including switches & protective devices									
0210 200 ton	Q-7	.23	137	Ea.	35,000	7,925		42,925	50,500
0220 300 ton		.17	187		38,500	10,800		49,300	58,500
0230 400 ton		.14	226		39,000	13,100		52,100	62,500
0240 500 ton		.14	237		39,200	13,700		52,900	64,000
0250 600 ton		.13	248		47,800	14,300		62,100	74,000
0260 700 ton		.12	260		48,100	15,000		63,100	75,500
0270 800 ton		.12	271		57,000	15,700		72,700	86,000
0280 900 ton		.11	288		65,500	16,700		82,200	97,000
0290 1000 ton		.11	296		74,000	17,100		91,100	107,500
0300 1100 ton		.10	307		76,500	17,800		94,300	111,500

For customer support on your Mechanical Cost Data, call 877.739.3689.

379

23 61 Refrigerant Compressors

23 61 13 – Centrifugal Refrigerant Compressors

23 61 13.10 Centrifugal Compressors	Crew	Daily Output	Labor-Hours	Unit	Material	2016 Bare Costs Labor	Equipment	Total	Total Incl O&P
0310	1200 ton	Q-7	.10	316	Ea.	79,000	18,300		97,300	114,500
0320	1300 ton		.10	326		81,500	18,900		100,400	118,500
0330	1400 ton		.09	340		84,000	19,700		103,700	122,000
0340	1500 ton		.09	351		86,500	20,300		106,800	126,000

23 61 15 – Rotary Refrigerant Compressors

23 61 15.10 Rotary Compressors

		Crew	Daily Output	Labor-Hours	Unit	Material	2016 Bare Costs Labor	Equipment	Total	Total Incl O&P
0010	**ROTARY COMPRESSORS**									
0100	Refrigeration, hermetic, switches and protective devices									
0210	1.25 ton	1 Stpi	3	2.667	Ea.	248	162		410	515
0220	1.42 ton		3	2.667		251	162		413	520
0230	1.68 ton		2.80	2.857		273	173		446	560
0240	2.00 ton		2.60	3.077		281	187		468	590
0250	2.37 ton		2.50	3.200		320	194		514	645
0260	2.67 ton		2.40	3.333		330	202		532	670
0270	3.53 ton		2.30	3.478		365	211		576	720
0280	4.43 ton		2.20	3.636		465	221		686	845
0290	5.08 ton		2.10	3.810		510	231		741	915
0300	5.22 ton		2	4		720	243		963	1,150
0310	7.31 ton	Q-5	3	5.333		1,150	291		1,441	1,725
0320	9.95 ton		2.60	6.154		1,350	335		1,685	2,000
0330	11.6 ton		2.50	6.400		1,750	350		2,100	2,425
0340	15.25 ton		2.30	6.957		2,375	380		2,755	3,200
0350	17.7 ton		2.10	7.619		2,550	415		2,965	3,425

23 61 16 – Reciprocating Refrigerant Compressors

23 61 16.10 Reciprocating Compressors

		Crew	Daily Output	Labor-Hours	Unit	Material	2016 Bare Costs Labor	Equipment	Total	Total Incl O&P
0010	**RECIPROCATING COMPRESSORS**									
0990	Refrigeration, recip. hermetic, switches & protective devices									
1000	10 ton	Q-5	1	16	Ea.	12,800	875		13,675	15,400
1100	20 ton	Q-6	.72	33.333		27,200	1,900		29,100	32,800
1200	30 ton		.64	37.500		28,300	2,125		30,425	34,400
1300	40 ton		.44	54.545		30,500	3,100		33,600	38,200
1400	50 ton		.20	120		32,100	6,800		38,900	45,600
1500	75 ton	Q-7	.27	118		35,500	6,850		42,350	49,300
1600	130 ton	"	.21	152		40,500	8,800		49,300	58,000

23 61 19 – Scroll Refrigerant Compressors

23 61 19.10 Scroll Compressors

		Crew	Daily Output	Labor-Hours	Unit	Material	2016 Bare Costs Labor	Equipment	Total	Total Incl O&P
0010	**SCROLL COMPRESSORS**									
1800	Refrigeration, scroll type									
1810	1.5 ton	1 Stpi	2.70	2.963	Ea.	885	180		1,065	1,250
1820	2.5 ton		2.50	3.200		885	194		1,079	1,275
1830	2.75 ton		2.35	3.404		895	207		1,102	1,300
1840	3 ton		2.30	3.478		940	211		1,151	1,350
1850	3.5 ton		2.25	3.556		965	216		1,181	1,375
1860	4 ton		2.20	3.636		1,200	221		1,421	1,650
1870	5 ton		2.10	3.810		1,250	231		1,481	1,725
1880	6 ton	Q-5	3.70	4.324		1,550	236		1,786	2,050
1900	7.5 ton		3.20	5		2,600	273		2,873	3,250
1910	10 ton		3.10	5.161		2,750	282		3,032	3,450
1920	13 ton		2.90	5.517		2,975	300		3,275	3,725
1960	25 ton		2.20	7.273		5,075	395		5,470	6,200

23 62 Packaged Compressor and Condenser Units

23 62 13 – Packaged Air-Cooled Refrigerant Compressor and Condenser Units

23 62 13.10 Packaged Air-Cooled Refrig. Condensing Units

		Crew	Daily Output	Labor-Hours	Unit	Material	2016 Bare Costs Labor	Equipment	Total	Total Incl O&P
0010	**PACKAGED AIR-COOLED REFRIGERANT CONDENSING UNITS**									
0020	Condensing unit									
0030	Air cooled, compressor, standard controls									
0050	1.5 ton	Q-5	2.50	6.400	Ea.	1,275	350		1,625	1,925
0100	2 ton		2.10	7.619		1,300	415		1,715	2,050
0200	2.5 ton		1.70	9.412		1,425	515		1,940	2,350
0300	3 ton		1.30	12.308		1,475	670		2,145	2,650
0350	3.5 ton		1.10	14.545		1,700	795		2,495	3,075
0400	4 ton		.90	17.778		1,950	970		2,920	3,600
0500	5 ton		.60	26.667		2,325	1,450		3,775	4,775
0550	7.5 ton		.55	29.091		3,275	1,600		4,875	6,000
0560	8.5 ton		.53	30.189		3,450	1,650		5,100	6,300
0600	10 ton		.50	32		3,700	1,750		5,450	6,725
0620	12.5 ton		.48	33.333		4,450	1,825		6,275	7,650
0650	15 ton		.40	40		7,500	2,175		9,675	11,600
0700	20 ton	Q-6	.40	60		10,100	3,400		13,500	16,200
0720	25 ton		.35	68.571		12,700	3,875		16,575	19,900
0750	30 ton		.30	80		14,600	4,525		19,125	23,000
0800	40 ton		.20	120		19,900	6,800		26,700	32,200
0840	50 ton		.18	133		21,600	7,550		29,150	35,100
0860	60 ton		.16	150		25,000	8,500		33,500	40,300
0900	70 ton		.14	171		28,600	9,700		38,300	46,200
1000	80 ton		.12	200		31,800	11,300		43,100	52,000
1010	90 ton		.09	266		35,300	15,100		50,400	61,500
1100	100 ton		.09	266		38,700	15,100		53,800	65,500

23 62 23 – Packaged Water-Cooled Refrigerant Compressor and Condenser Units

23 62 23.10 Packaged Water-Cooled Refrigerant Condensing Units

		Crew	Daily Output	Labor-Hours	Unit	Material	Labor	Equipment	Total	Total Incl O&P
0010	**PACKAGED WATER-COOLED REFRIGERANT CONDENSING UNITS**									
2000	Water cooled, compressor, heat exchanger, controls									
2100	5 ton	Q-5	.70	22.857	Ea.	13,600	1,250		14,850	16,900
2110	10 ton		.60	26.667		15,900	1,450		17,350	19,700
2200	15 ton		.50	32		20,300	1,750		22,050	25,100
2300	20 ton	Q-6	.40	60		25,000	3,400		28,400	32,600
2310	30 ton		.30	80		28,700	4,525		33,225	38,500
2400	40 ton		.20	120		31,200	6,800		38,000	44,600

23 63 Refrigerant Condensers

23 63 13 – Air-Cooled Refrigerant Condensers

23 63 13.10 Air-Cooled Refrig. Condensers

		Crew	Daily Output	Labor-Hours	Unit	Material	Labor	Equipment	Total	Total Incl O&P
0010	**AIR-COOLED REFRIG. CONDENSERS**									
0080	Air cooled, belt drive, propeller fan									
0220	45 ton	Q-6	.70	34.286	Ea.	9,975	1,950		11,925	13,900
0240	50 ton		.69	34.985		10,800	1,975		12,775	14,800
0260	54 ton		.64	37.795		11,800	2,150		13,950	16,100
0280	59 ton		.58	41.308		12,900	2,350		15,250	17,700
0300	65 ton		.53	45.541		15,400	2,575		17,975	20,900
0320	73 ton		.47	51.173		16,900	2,900		19,800	23,000
0340	81 ton		.42	56.738		18,600	3,225		21,825	25,400
0360	86 ton		.40	60.302		19,600	3,425		23,025	26,700
0380	88 ton		.39	61.697		21,000	3,500		24,500	28,400

For customer support on your Mechanical Cost Data, call 877.739.3689.

381

23 63 Refrigerant Condensers

23 63 13 – Air-Cooled Refrigerant Condensers

23 63 13.10 Air-Cooled Refrig. Condensers		Crew	Daily Output	Labor-Hours	Unit	Material	2016 Bare Costs Labor	Equipment	Total	Total Incl O&P
0400	101 ton	Q-7	.45	70.640	Ea.	23,100	4,075		27,175	31,600
0500	159 ton		.31	102		24,500	5,950		30,450	35,900
0600	228 ton	↓	.22	148	↓	52,000	8,550		60,550	70,000
1500	May be specified single or multi-circuit									
1550	Air cooled, direct drive, propeller fan									
1590	1 ton	Q-5	3.80	4.211	Ea.	1,650	230		1,880	2,175
1600	1-1/2 ton		3.60	4.444		1,975	243		2,218	2,550
1620	2 ton		3.20	5		2,175	273		2,448	2,800
1630	3 ton		2.40	6.667		2,450	365		2,815	3,225
1640	5 ton		2	8		5,025	435		5,460	6,175
1650	8 ton		1.80	8.889		5,625	485		6,110	6,900
1660	10 ton		1.40	11.429		6,850	625		7,475	8,500
1670	12 ton		1.30	12.308		7,700	670		8,370	9,500
1680	14 ton		1.20	13.333		8,800	730		9,530	10,800
1690	16 ton		1.10	14.545		10,200	795		10,995	12,400
1700	21 ton		1	16		11,100	875		11,975	13,500
1720	26 ton		.84	19.002		12,500	1,050		13,550	15,400
1740	30 ton	↓	.70	22.792		16,000	1,250		17,250	19,500
1760	41 ton	Q-6	.77	31.008		17,600	1,750		19,350	22,100
1780	52 ton		.66	36.419		25,100	2,075		27,175	30,700
1800	63 ton		.55	44.037		27,700	2,500		30,200	34,300
1820	76 ton		.45	52.980		31,400	3,000		34,400	39,100
1840	86 ton		.40	60		38,900	3,400		42,300	47,800
1860	97 ton	↓	.35	67.989		44,200	3,850		48,050	54,500
1880	105 ton	Q-7	.44	73.563		51,000	4,250		55,250	62,500
1890	118 ton		.39	82.687		55,000	4,775		59,775	67,500
1900	126 ton		.36	88.154		64,000	5,100		69,100	77,500
1910	136 ton		.34	95.238		71,500	5,500		77,000	87,000
1920	142 ton	↓	.32	99.379	↓	73,500	5,750		79,250	89,000

23 63 30 – Lubricants

23 63 30.10 Lubricant Oils

		Crew	Daily Output	Labor-Hours	Unit	Material	2016 Bare Costs Labor	Equipment	Total	Total Incl O&P
0010	**LUBRICANT OILS**									
8000	Oils									
8100	Lubricating									
8120	Oil, lubricating				Oz.	.36			.36	.40
8500	Refrigeration									
8520	Oil, refrigeration				Gal.	41			41	45
8525	Oil, refrigeration				Qt.	10.20			10.20	11.20

23 63 33 – Evaporative Refrigerant Condensers

23 63 33.10 Evaporative Condensers

		Crew	Daily Output	Labor-Hours	Unit	Material	2016 Bare Costs Labor	Equipment	Total	Total Incl O&P
0010	**EVAPORATIVE CONDENSERS**									
3400	Evaporative, copper coil, pump, fan motor									
3440	10 ton	Q-5	.54	29.630	Ea.	8,600	1,625		10,225	11,900
3460	15 ton		.50	32		8,800	1,750		10,550	12,300
3480	20 ton		.47	34.043		9,075	1,850		10,925	12,800
3500	25 ton		.45	35.556		9,275	1,950		11,225	13,100
3520	30 ton	↓	.42	38.095		9,725	2,075		11,800	13,900
3540	40 ton	Q-6	.49	48.980		10,600	2,775		13,375	15,900
3560	50 ton		.39	61.538		12,000	3,475		15,475	18,500
3580	65 ton		.35	68.571		12,900	3,875		16,775	20,000
3600	80 ton		.33	72.727		14,200	4,125		18,325	21,800
3620	90 ton	↓	.29	82.759	↓	15,500	4,700		20,200	24,200

23 63 Refrigerant Condensers

23 63 33 – Evaporative Refrigerant Condensers

23 63 33.10 Evaporative Condensers		Crew	Daily Output	Labor-Hours	Unit	Material	2016 Bare Costs Labor	Equipment	Total	Total Incl O&P
3640	100 ton	Q-7	.36	88.889	Ea.	17,100	5,125		22,225	26,600
3660	110 ton		.33	96.970		18,000	5,600		23,600	28,300
3680	125 ton		.30	106		19,400	6,175		25,575	30,700
3700	135 ton		.28	114		20,500	6,600		27,100	32,500
3720	150 ton		.25	128		22,300	7,400		29,700	35,700
3740	165 ton		.23	139		23,600	8,050		31,650	38,100
3760	185 ton		.22	145		26,700	8,400		35,100	42,000
3860	For fan damper control, add	Q-5	2	8		810	435		1,245	1,550

23 64 Packaged Water Chillers

23 64 13 – Absorption Water Chillers

23 64 13.13 Direct-Fired Absorption Water Chillers

		Crew	Daily Output	Labor-Hours	Unit	Material	Labor	Equipment	Total	Total Incl O&P
0010	**DIRECT-FIRED ABSORPTION WATER CHILLERS**									
3000	Gas fired, air cooled									
3220	5 ton	Q-5	.60	26.667	Ea.	8,975	1,450		10,425	12,100
3270	10 ton	"	.40	40	"	23,100	2,175		25,275	28,700
4000	Water cooled, duplex									
4130	100 ton	Q-7	.13	246	Ea.	136,000	14,200		150,200	171,000
4140	200 ton		.11	283		183,000	16,400		199,400	226,000
4150	300 ton		.11	299		219,500	17,300		236,800	267,500
4160	400 ton		.10	316		285,500	18,300		303,800	341,500
4170	500 ton		.10	329		359,500	19,100		378,600	424,000
4180	600 ton		.09	340		424,500	19,700		444,200	496,000
4190	700 ton		.09	359		470,500	20,800		491,300	549,000
4200	800 ton		.08	380		568,000	22,000		590,000	658,000
4210	900 ton		.08	400		658,000	23,100		681,100	759,000
4220	1000 ton		.08	421		731,500	24,300		755,800	841,000

23 64 13.16 Indirect-Fired Absorption Water Chillers

		Crew	Daily Output	Labor-Hours	Unit	Material	Labor	Equipment	Total	Total Incl O&P
0010	**INDIRECT-FIRED ABSORPTION WATER CHILLERS**									
0020	Steam or hot water, water cooled									
0050	100 ton	Q-7	.13	240	Ea.	125,000	13,900		138,900	158,500
0100	148 ton		.12	258		224,500	14,900		239,400	269,500
0200	200 ton		.12	275		250,000	15,900		265,900	299,000
0240	250 ton		.11	290		275,000	16,800		291,800	328,000
0300	354 ton		.11	304		343,500	17,600		361,100	404,500
0400	420 ton		.10	323		375,500	18,700		394,200	441,000
0500	665 ton		.09	340		527,000	19,700		546,700	609,500
0600	750 ton		.09	363		616,000	21,000		637,000	709,000
0700	850 ton		.08	385		657,500	22,300		679,800	756,500
0800	955 ton		.08	410		703,000	23,700		726,700	809,000
0900	1125 ton		.08	421		811,500	24,300		835,800	929,000
1000	1250 ton		.07	444		872,500	25,700		898,200	999,000
1100	1465 ton		.07	463		1,034,500	26,800		1,061,300	1,178,500
1200	1660 ton		.07	477		1,217,000	27,600		1,244,600	1,380,500
2000	For two stage unit, add					80%	25%			

For customer support on your Mechanical Cost Data, call 877.739.3689.

383

23 64 Packaged Water Chillers

23 64 16 – Centrifugal Water Chillers

23 64 16.10 Centrifugal Type Water Chillers	Crew	Daily Output	Labor-Hours	Unit	Material	2016 Bare Costs Labor	Equipment	Total	Total Incl O&P
0010 **CENTRIFUGAL TYPE WATER CHILLERS**, With standard controls									
0020 Centrifugal liquid chiller, water cooled									
0030 not including water tower									
0100 2000 ton (twin 1000 ton units)	Q-7	.07	477	Ea.	732,500	27,600		760,100	847,000
0274 Centrifugal, packaged unit, water cooled, not incl. tower									
0278 200 ton	Q-7	.19	172	Ea.	122,000	9,950		131,950	149,000
0279 300 ton		.14	233		149,000	13,500		162,500	184,500
0280 400 ton		.11	283		150,500	16,400		166,900	190,500
0282 450 ton		.11	290		165,000	16,800		181,800	206,500
0290 500 ton		.11	296		182,500	17,100		199,600	227,000
0292 550 ton		.11	304		213,000	17,600		230,600	260,500
0300 600 ton		.10	310		227,500	18,000		245,500	277,000
0302 650 ton		.10	320		239,500	18,500		258,000	291,000
0304 700 ton		.10	326		256,500	18,900		275,400	310,500
0306 750 ton		.10	333		274,500	19,300		293,800	331,000
0310 800 ton		.09	340		293,000	19,700		312,700	351,500
0312 850 ton		.09	351		311,000	20,300		331,300	373,000
0316 900 ton		.09	359		329,500	20,800		350,300	394,000
0318 950 ton		.09	363		348,000	21,000		369,000	414,000
0320 1000 ton		.09	372		366,000	21,500		387,500	435,500
0324 1100 ton		.08	385		368,000	22,300		390,300	438,000
0330 1200 ton		.08	395		396,000	22,800		418,800	470,000
0340 1300 ton		.08	410		428,500	23,700		452,200	507,500
0350 1400 ton		.08	421		461,500	24,300		485,800	544,500
0360 1500 ton		.08	426		486,500	24,700		511,200	572,000
0370 1600 ton		.07	438		518,000	25,300		543,300	608,000
0380 1700 ton		.07	450		551,500	26,000		577,500	646,000
0390 1800 ton		.07	457		584,000	26,400		610,400	682,500
0400 1900 ton		.07	470		615,500	27,200		642,700	718,000
0430 2000 ton		.07	492		649,500	28,500		678,000	757,500
0440 2500 ton		.06	524		810,000	30,300		840,300	936,500

23 64 19 – Reciprocating Water Chillers

23 64 19.10 Reciprocating Type Water Chillers

	Crew	Daily Output	Labor-Hours	Unit	Material	Labor	Equipment	Total	Total Incl O&P
0010 **RECIPROCATING TYPE WATER CHILLERS**, With standard controls									
0494 Water chillers, integral air cooled condenser									
0538 60 ton cooling	Q-7	.28	115	Ea.	42,400	6,650		49,050	56,500
0546 70 ton cooling		.27	119		53,000	6,925		59,925	68,500
0552 80 ton cooling		.26	123		56,500	7,125		63,625	73,000
0554 90 ton cooling		.25	125		63,000	7,275		70,275	80,500
0600 100 ton cooling		.25	129		70,500	7,450		77,950	89,500
0620 110 ton cooling		.24	132		76,000	7,650		83,650	95,000
0630 130 ton cooling		.24	135		85,500	7,800		93,300	106,000
0640 150 ton cooling		.23	137		95,500	7,975		103,475	117,000
0650 175 ton cooling		.23	140		107,500	8,150		115,650	131,000
0654 190 ton cooling		.22	144		119,000	8,325		127,325	143,000
0660 210 ton cooling		.22	148		126,000	8,550		134,550	151,500
0662 250 ton cooling		.21	151		146,000	8,775		154,775	173,500
0664 275 ton cooling		.21	156		158,000	9,025		167,025	187,000
0666 300 ton cooling		.20	160		170,500	9,250		179,750	201,500
0668 330 ton cooling		.20	164		192,000	9,475		201,475	226,000
0670 360 ton cooling		.19	168		212,500	9,725		222,225	248,500
0672 390 ton cooling		.19	172		236,500	10,000		246,500	275,000

For customer support on your Mechanical Cost Data, call 877.739.3689.

23 64 19 – Reciprocating Water Chillers

23 64 19.10 Reciprocating Type Water Chillers		Crew	Daily Output	Labor-Hours	Unit	Material	2016 Bare Costs Labor	Equipment	Total	Total Incl O&P
0674	420 ton cooling	Q-7	.18	177	Ea.	261,000	10,300		271,300	303,000
0980	Water cooled, multiple compressor, semi-hermetic, tower not incl.									
1000	15 ton cooling	Q-6	.36	65.934	Ea.	15,300	3,725		19,025	22,400
1020	25 ton cooling	Q-7	.41	78.049		16,200	4,500		20,700	24,600
1040	30 ton cooling		.36	89.888		18,300	5,200		23,500	28,000
1060	35 ton cooling		.31	101		20,200	5,900		26,100	31,100
1080	40 ton cooling		.30	108		21,000	6,250		27,250	32,500
1090	45 ton cooling		.29	111		22,900	6,425		29,325	34,900
1100	50 ton cooling		.28	113		29,700	6,575		36,275	42,600
1120	60 ton cooling		.25	125		30,700	7,275		37,975	44,800
1130	75 ton cooling		.23	139		45,300	8,050		53,350	62,000
1140	85 ton cooling		.21	151		47,700	8,775		56,475	65,500
1150	95 ton cooling		.19	164		51,000	9,525		60,525	71,000
1160	100 ton cooling		.18	179		51,500	10,400		61,900	72,000
1170	115 ton cooling		.17	190		52,000	11,000		63,000	73,500
1180	125 ton cooling		.16	196		54,500	11,300		65,800	76,500
1200	145 ton cooling		.16	202		60,500	11,700		72,200	84,000
1210	155 ton cooling	↓	.15	210	↓	64,000	12,200		76,200	89,000
1300	Water cooled, single compressor, semi-hermetic, tower not incl.									
1320	40 ton cooling,	Q-7	.30	108	Ea.	17,300	6,250		23,550	28,400
1340	50 ton cooling		.28	113		21,500	6,575		28,075	33,500
1360	60 ton cooling	↓	.25	125	↓	23,800	7,275		31,075	37,200
1451	Water cooled, dual compressors, semi-hermetic, tower not incl.									
1500	80 ton cooling	Q-7	.14	222	Ea.	29,600	12,800		42,400	52,000
1520	100 ton cooling		.14	228		40,000	13,200		53,200	64,000
1540	120 ton cooling	↓	.14	231	↓	48,500	13,400		61,900	73,500
4000	Packaged chiller, remote air cooled condensers not incl.									
4020	15 ton cooling	Q-7	.30	108	Ea.	15,500	6,250		21,750	26,500
4030	20 ton cooling		.28	115		16,500	6,700		23,200	28,200
4040	25 ton cooling		.25	125		17,400	7,275		24,675	30,200
4050	35 ton cooling		.24	133		18,400	7,750		26,150	31,900
4060	40 ton cooling		.22	144		19,500	8,325		27,825	34,100
4066	45 ton cooling		.21	149		20,600	8,650		29,250	35,600
4070	50 ton cooling		.21	153		29,600	8,900		38,500	45,900
4080	60 ton cooling		.20	164		32,000	9,475		41,475	49,500
4090	75 ton cooling		.18	173		37,900	10,100		48,000	57,000
4100	85 ton cooling		.17	183		39,700	10,600		50,300	59,500
4110	95 ton cooling		.17	193		40,200	11,200		51,400	61,000
4120	105 ton cooling		.16	203		44,000	11,800		55,800	66,000
4130	115 ton cooling		.15	213		45,400	12,300		57,700	68,500
4140	125 ton cooling		.14	223		45,800	12,900		58,700	70,000
4150	145 ton cooling	↓	.14	233	↓	51,000	13,500		64,500	76,500

23 64 23 – Scroll Water Chillers

23 64 23.10 Scroll Water Chillers		Crew	Daily Output	Labor-Hours	Unit	Material	2016 Bare Costs Labor	Equipment	Total	Total Incl O&P
0010	**SCROLL WATER CHILLERS**, With standard controls									
0480	Packaged w/integral air cooled condenser									
0482	10 ton cooling	Q-7	.34	94.118	Ea.	15,200	5,450		20,650	24,900
0490	15 ton cooling		.37	86.486		18,300	5,000		23,300	27,800
0500	20 ton cooling		.34	94.118		21,700	5,450		27,150	32,000
0510	25 ton cooling		.34	94.118		25,500	5,450		30,950	36,300
0515	30 ton cooling		.31	101		27,400	5,900		33,300	39,000
0517	35 ton cooling	↓	.31	104	↓	30,500	6,075		36,575	42,700

23 64 Packaged Water Chillers

23 64 23 – Scroll Water Chillers

23 64 23.10 Scroll Water Chillers

23 64 23.10 Scroll Water Chillers		Crew	Daily Output	Labor-Hours	Unit	Material	2016 Bare Costs Labor	Equipment	Total	Total Incl O&P
0520	40 ton cooling	Q-7	.30	108	Ea.	33,200	6,250		39,450	45,900
0528	45 ton cooling		.29	109		36,800	6,350		43,150	50,000
0536	50 ton cooling		.28	113		38,600	6,525		45,125	52,500
0680	Scroll water cooled, single compressor, hermetic, tower not incl.									
0700	2 ton cooling	Q-5	.57	28.070	Ea.	3,500	1,525		5,025	6,175
0710	5 ton cooling		.57	28.070		4,175	1,525		5,700	6,925
0720	6 ton cooling		.42	38.005		5,250	2,075		7,325	8,900
0740	8 ton cooling		.31	52.117		5,750	2,850		8,600	10,600
0760	10 ton cooling	Q-6	.36	67.039		6,625	3,800		10,425	13,000
0780	15 ton cooling	"	.33	72.727		10,400	4,125		14,525	17,700
0800	20 ton cooling	Q-7	.38	83.990		11,800	4,850		16,650	20,300
0820	30 ton cooling	"	.33	96.096		13,200	5,550		18,750	23,000

23 64 26 – Rotary-Screw Water Chillers

23 64 26.10 Rotary-Screw Type Water Chillers

23 64 26.10 Rotary-Screw Type Water Chillers		Crew	Daily Output	Labor-Hours	Unit	Material	2016 Bare Costs Labor	Equipment	Total	Total Incl O&P
0010	**ROTARY-SCREW TYPE WATER CHILLERS**, With standard controls									
0110	Screw, liquid chiller, air cooled, insulated evaporator									
0120	130 ton	Q-7	.14	228	Ea.	93,500	13,200		106,700	123,000
0124	160 ton		.13	246		115,000	14,200		129,200	148,000
0128	180 ton		.13	250		129,500	14,400		143,900	164,500
0132	210 ton		.12	258		142,500	14,900		157,400	179,500
0136	270 ton		.12	266		163,000	15,400		178,400	203,000
0140	320 ton		.12	275		204,500	15,900		220,400	249,000
0200	Packaged unit, water cooled, not incl. tower									
0210	80 ton	Q-7	.14	223	Ea.	46,600	12,900		59,500	71,000
0220	100 ton		.14	230		57,500	13,300		70,800	83,500
0230	150 ton		.13	240		75,000	13,900		88,900	103,500
0240	200 ton		.13	251		82,000	14,600		96,600	112,000
0250	250 ton		.12	260		102,000	15,000		117,000	134,500
0260	300 ton		.12	266		124,500	15,400		139,900	160,500
0270	350 ton		.12	275		139,500	15,900		155,400	177,500
1450	Water cooled, tower not included									
1560	135 ton cooling, screw compressors	Q-7	.14	235	Ea.	59,500	13,600		73,100	86,000
1580	150 ton cooling, screw compressors		.13	240		69,000	13,900		82,900	96,500
1620	200 ton cooling, screw compressors		.13	250		95,000	14,400		109,400	126,500
1660	291 ton cooling, screw compressors		.12	260		99,000	15,000		114,000	131,500

23 64 33 – Direct Expansion Water Chillers

23 64 33.10 Direct Expansion Type Water Chillers

23 64 33.10 Direct Expansion Type Water Chillers		Crew	Daily Output	Labor-Hours	Unit	Material	2016 Bare Costs Labor	Equipment	Total	Total Incl O&P
0010	**DIRECT EXPANSION TYPE WATER CHILLERS**, With standard controls									
8000	Direct expansion, shell and tube type, for built up systems									
8020	1 ton	Q-5	2	8	Ea.	7,475	435		7,910	8,875
8030	5 ton		1.90	8.421		12,400	460		12,860	14,400
8040	10 ton		1.70	9.412		15,400	515		15,915	17,700
8050	20 ton		1.50	10.667		17,600	585		18,185	20,200
8060	30 ton		1	16		22,400	875		23,275	25,900
8070	50 ton		.90	17.778		36,000	970		36,970	41,100
8080	100 ton	Q-6	.90	26.667		64,000	1,500		65,500	73,000

23 65 Cooling Towers

23 65 13 – Forced-Draft Cooling Towers

23 65 13.10 Forced-Draft Type Cooling Towers		Crew	Daily Output	Labor-Hours	Unit	Material	2016 Bare Costs Labor	Equipment	Total	Total Incl O&P
0010	**FORCED-DRAFT TYPE COOLING TOWERS**, Packaged units D3030–310									
0070	Galvanized steel									
0080	Induced draft, crossflow									
0100	Vertical, belt drive, 61 tons	Q-6	90	.267	TonAC	215	15.10		230.10	260
0150	100 ton		100	.240		208	13.60		221.60	249
0200	115 ton		109	.220		180	12.45		192.45	218
0250	131 ton		120	.200		217	11.35		228.35	255
0260	162 ton		132	.182		175	10.30		185.30	209
1000	For higher capacities, use multiples									
1500	Induced air, double flow									
1900	Vertical, gear drive, 167 ton	Q-6	126	.190	TonAC	170	10.80		180.80	203
2000	297 ton		129	.186		105	10.55		115.55	131
2100	582 ton		132	.182		58.50	10.30		68.80	79.50
2150	849 ton		142	.169		78.50	9.55		88.05	101
2200	1016 ton		150	.160		79	9.05		88.05	101
2500	Blow through, centrifugal type									
2510	50 ton	Q-6	2.28	10.526	Ea.	12,000	595		12,595	14,100
2520	75 ton		1.52	15.789		15,500	895		16,395	18,400
2524	100 ton		1.22	19.672		18,300	1,125		19,425	21,800
2528	125 ton	Q-7	1.31	24.427		21,400	1,400		22,800	25,600
2532	200 ton		.82	39.216		27,900	2,275		30,175	34,100
2536	250 ton		.65	49.231		33,300	2,850		36,150	41,000
2540	300 ton		.54	59.259		38,300	3,425		41,725	47,400
2544	350 ton		.47	68.085		43,300	3,925		47,225	53,500
2548	400 ton		.41	78.049		45,100	4,500		49,600	56,500
2552	450 ton		.36	88.154		54,500	5,100		59,600	67,000
2556	500 ton		.32	100		58,000	5,775		63,775	72,000
2560	550 ton		.30	107		62,500	6,225		68,725	78,000
2564	600 ton		.27	117		65,500	6,800		72,300	82,500
2568	650 ton		.25	127		72,500	7,375		79,875	91,000
2572	700 ton		.23	137		85,000	7,925		92,925	106,000
2576	750 ton		.22	146		91,500	8,475		99,975	113,500
2580	800 ton		.21	152		99,000	8,800		107,800	122,500
2584	850 ton		.20	156		105,000	9,075		114,075	129,000
2588	900 ton		.20	161		111,500	9,350		120,850	136,500
2592	950 ton		.19	170		117,500	9,825		127,325	144,500
2596	1000 ton		.18	180		121,500	10,400		131,900	149,500
2700	Axial fan, induced draft									
2710	50 ton	Q-6	2.28	10.526	Ea.	10,400	595		10,995	12,400
2720	75 ton		1.52	15.789		12,300	895		13,195	14,900
2724	100 ton		1.22	19.672		13,000	1,125		14,125	16,000
2728	125 ton	Q-7	1.31	24.427		16,000	1,400		17,400	19,700
2732	200 ton		.82	39.216		27,100	2,275		29,375	33,200
2736	250 ton		.65	49.231		30,600	2,850		33,450	37,900
2740	300 ton		.54	59.259		35,700	3,425		39,125	44,500
2744	350 ton		.47	68.085		38,300	3,925		42,225	48,000
2748	400 ton		.41	78.049		43,900	4,500		48,400	55,000
2752	450 ton		.36	88.154		48,700	5,100		53,800	61,000
2756	500 ton		.32	100		51,500	5,775		57,275	65,500
2760	550 ton		.30	107		58,000	6,225		64,225	73,000
2764	600 ton		.27	117		61,000	6,800		67,800	77,500
2768	650 ton		.25	127		66,500	7,375		73,875	84,000
2772	700 ton		.23	137		75,500	7,925		83,425	95,000

23 65 Cooling Towers

23 65 13 – Forced-Draft Cooling Towers

23 65 13.10 Forced-Draft Type Cooling Towers	Crew	Daily Output	Labor-Hours	Unit	Material	2016 Bare Costs Labor	Equipment	Total	Total Incl O&P
2776	750 ton	Q-7	.22	146	Ea.	78,500	8,475		86,975	99,500
2780	800 ton		.21	152		82,000	8,800		90,800	103,500
2784	850 ton		.20	156		82,500	9,075		91,575	104,500
2788	900 ton		.20	161		85,500	9,350		94,850	108,000
2792	950 ton		.19	170		90,500	9,825		100,325	114,500
2796	1000 ton		.18	180		95,500	10,400		105,900	121,000
3000	For higher capacities, use multiples									
3500	For pumps and piping, add	Q-6	38	.632	TonAC	108	36		144	172
4000	For absorption systems, add				"	75%	75%			
4100	Cooling water chemical feeder	Q-5	3	5.333	Ea.	380	291		671	855
5000	Fiberglass tower on galvanized steel support structure									
5010	Draw thru									
5100	100 ton	Q-6	1.40	17.143	Ea.	13,900	970		14,870	16,800
5120	120 ton		1.20	20		16,200	1,125		17,325	19,500
5140	140 ton		1	24		17,500	1,350		18,850	21,300
5160	160 ton		.80	30		19,400	1,700		21,100	24,000
5180	180 ton		.65	36.923		22,100	2,100		24,200	27,500
5200	200 ton		.48	50		25,000	2,825		27,825	31,800
5300	For stainless steel support structure, add					30%				
5360	For higher capacities, use multiples of each size									
6000	Stainless steel									
6010	Induced draft, crossflow, horizontal, belt drive									
6100	57 ton	Q-6	1.50	16	Ea.	27,100	905		28,005	31,200
6120	91 ton		.99	24.242		32,700	1,375		34,075	38,100
6140	111 ton		.43	55.814		43,900	3,150		47,050	53,000
6160	126 ton		.22	109		45,600	6,175		51,775	59,500
6170	Induced draft, crossflow, vertical, gear drive									
6172	167 ton	Q-6	.75	32	Ea.	53,000	1,825		54,825	61,000
6174	297 ton		.43	55.814		59,000	3,150		62,150	70,000
6176	582 ton		.23	104		99,500	5,900		105,400	118,500
6178	849 ton		.17	141		132,000	8,000		140,000	157,500
6180	1016 ton		.15	160		159,000	9,075		168,075	188,000

23 65 33 – Liquid Coolers

23 65 33.10 Liquid Cooler, Closed Circuit Type Cooling Towers

		Crew	Daily Output	Labor-Hours	Unit	Material	2016 Bare Costs Labor	Equipment	Total	Total Incl O&P
0010	**LIQUID COOLER, CLOSED CIRCUIT TYPE COOLING TOWERS**									
0070	Packaged options include: vibration switch, sump sweeper piping,									
0076	grooved coil connections, fan motor inverter capable,									
0080	304 St. St. cold water basin, external platform w/ladder.									
0100	50 ton	Q-6	2	12	Ea.	32,500	680		33,180	36,700
0150	100 ton	"	1.13	21.239		65,000	1,200		66,200	73,500
0200	150 ton	Q-7	.98	32.653		97,500	1,875		99,375	110,000
0250	200 ton		.78	41.026		130,000	2,375		132,375	146,500
0300	250 ton		.63	50.794		162,500	2,925		165,425	183,000

23 72 Air-to-Air Energy Recovery Equipment

23 72 13 – Heat-Wheel Air-to-Air Energy-Recovery Equipment

23 72 13.10 Heat-Wheel Air-to-Air Energy Recovery Equip.

23 72 13.10 Heat-Wheel Air-to-Air Energy Recovery Equip.		Crew	Daily Output	Labor-Hours	Unit	Material	2016 Bare Costs Labor	Equipment	Total	Total Incl O&P
0010	**HEAT-WHEEL AIR-TO-AIR ENERGY RECOVERY EQUIPMENT**									
0100	Air to air									
4000	Enthalpy recovery wheel									
4010	1000 max CFM	G	Q-9	1.20	13.333	Ea.	6,700	685		7,385	8,425
4020	2000 max CFM	G		1	16		7,850	825		8,675	9,875
4030	4000 max CFM	G		.80	20		9,075	1,025		10,100	11,600
4040	6000 max CFM	G		.70	22.857		10,600	1,175		11,775	13,500
4050	8000 max CFM	G	Q-10	1	24		11,700	1,275		12,975	14,900
4060	10,000 max CFM	G		.90	26.667		14,000	1,425		15,425	17,600
4070	20,000 max CFM	G		.80	30		25,400	1,600		27,000	30,500
4080	25,000 max CFM	G		.70	34.286		31,100	1,825		32,925	37,000
4090	30,000 max CFM	G		.50	48		34,500	2,575		37,075	41,800
4100	40,000 max CFM	G		.45	53.333		47,600	2,850		50,450	57,000
4110	50,000 max CFM	G		.40	60		55,500	3,200		58,700	66,000

23 72 16 – Heat-Pipe Air-To-Air Energy-Recovery Equipment

23 72 16.10 Heat Pipes

23 72 16.10 Heat Pipes		Crew	Daily Output	Labor-Hours	Unit	Material	2016 Bare Costs Labor	Equipment	Total	Total Incl O&P
0010	**HEAT PIPES**									
8000	Heat pipe type, glycol, 50% efficient									
8010	100 MBH, 1700 CFM	1 Stpi	.80	10	Ea.	4,300	605		4,905	5,675
8020	160 MBH, 2700 CFM		.60	13.333		5,775	810		6,585	7,575
8030	620 MBH, 4000 CFM		.40	20		8,725	1,225		9,950	11,400

23 73 Indoor Central-Station Air-Handling Units

23 73 13 – Modular Indoor Central-Station Air-Handling Units

23 73 13.10 Air-Handling Units

23 73 13.10 Air-Handling Units		Crew	Daily Output	Labor-Hours	Unit	Material	2016 Bare Costs Labor	Equipment	Total	Total Incl O&P
0010	**AIR-HANDLING UNITS**, Built-Up									
0100	With cooling/heating coil section, filters, mixing box									
0880	Single zone, horizontal/vertical									
0890	Constant volume									
0900	1600 CFM	Q-5	1.20	13.333	Ea.	5,100	730		5,830	6,700
0906	2000 CFM		1.10	14.545		6,125	795		6,920	7,925
0910	3000 CFM		1	16		8,975	875		9,850	11,200
0916	4000 CFM		.95	16.842		11,700	920		12,620	14,300
0920	5000 CFM	Q-6	1.40	17.143		14,300	970		15,270	17,200
0926	6500 CFM		1.30	18.462		17,900	1,050		18,950	21,200
0930	7500 CFM		1.20	20		19,400	1,125		20,525	23,000
0936	9200 CFM		1.10	21.818		22,400	1,225		23,625	26,600
0940	11,500 CFM		1	24		25,500	1,350		26,850	30,200
0946	13,200 CFM		.90	26.667		30,600	1,500		32,100	36,000
0950	16,500 CFM		.80	30		37,700	1,700		39,400	44,100
0960	19,500 CFM		.70	34.286		44,400	1,950		46,350	51,500
0970	22,000 CFM		.60	40		50,000	2,275		52,275	58,500
0980	27,000 CFM		.50	48		60,000	2,725		62,725	70,000
0990	34,000 CFM		.40	60		76,500	3,400		79,900	89,000
1000	40,000 CFM		.30	80		92,000	4,525		96,525	108,000
1010	47,000 CFM		.20	120		100,500	6,800		107,300	121,000
2300	Multi-zone, horizontal or vertical, blow-thru fan									
2310	3000 CFM	Q-5	1	16	Ea.	6,775	875		7,650	8,775
2314	4000 CFM	"	.95	16.842		8,925	920		9,845	11,200
2318	5000 CFM	Q-6	1.40	17.143		11,100	970		12,070	13,700

For customer support on your Mechanical Cost Data, call 877.739.3689.

389

23 73 13 – Modular Indoor Central-Station Air-Handling Units

23 73 13.10 Air-Handling Units	Crew	Daily Output	Labor-Hours	Unit	Material	2016 Bare Costs Labor	Equipment	Total	Total Incl O&P
2322	6500 CFM	Q-6	1.30	18.462	Ea.	14,400	1,050		15,450	17,500
2326	7500 CFM		1.20	20		16,700	1,125		17,825	20,000
2330	9200 CFM		1.10	21.818		20,500	1,225		21,725	24,500
2334	11,500 CFM		1	24		22,600	1,350		23,950	27,000
2336	13,200 CFM		.90	26.667		25,900	1,500		27,400	30,800
2340	16,500 CFM		.80	30		32,300	1,700		34,000	38,100
2344	19,500 CFM		.70	34.286		38,100	1,950		40,050	44,800
2348	22,000 CFM		.60	40		43,000	2,275		45,275	50,500
2352	27,000 CFM		.50	48		48,000	2,725		50,725	57,000
2356	34,000 CFM		.40	60		60,500	3,400		63,900	71,500
2360	40,000 CFM		.30	80		65,500	4,525		70,025	79,500
2364	47,000 CFM		.20	120		83,500	6,800		90,300	102,500

23 73 13.20 Air Handling Units, Packaged Indoor Type

		Crew	Daily Output	Labor-Hours	Unit	Material	Labor	Equipment	Total	Total Incl O&P
0010	**AIR HANDLING UNITS, PACKAGED INDOOR TYPE**									
0040	Cooling coils may be chilled water or DX									
0050	Heating coils may be hot water, steam or electric									
1090	Constant volume									
1200	2000 CFM	Q-5	1.10	14.545	Ea.	11,000	795		11,795	13,300
1400	5000 CFM	Q-6	.80	30		17,600	1,700		19,300	21,900
1550	10,000 CFM		.54	44.444		25,900	2,525		28,425	32,300
1670	15,000 CFM		.38	63.158		53,000	3,575		56,575	64,000
1700	20,000 CFM		.31	77.419		67,500	4,375		71,875	80,500
1710	30,000 CFM		.27	88.889		91,000	5,025		96,025	107,500
2300	Variable air volume									
2330	2000 CFM	Q-5	1	16	Ea.	15,400	875		16,275	18,300
2340	5000 CFM	Q-6	.70	34.286		21,500	1,950		23,450	26,600
2350	10,000 CFM		.50	48		42,000	2,725		44,725	50,500
2360	15,000 CFM		.40	60		61,000	3,400		64,400	72,000
2370	20,000 CFM		.29	82.759		78,000	4,700		82,700	92,500
2380	30,000 CFM		.20	120		94,000	6,800		100,800	114,000

23 73 39 – Indoor, Direct Gas-Fired Heating and Ventilating Units

23 73 39.10 Make-Up Air Unit

		Crew	Daily Output	Labor-Hours	Unit	Material	Labor	Equipment	Total	Total Incl O&P
0010	**MAKE-UP AIR UNIT**									
0020	Indoor suspension, natural/LP gas, direct fired,									
0032	standard control. For flue see Section 23 51 23.10									
0040	70°F temperature rise, MBH is input									
0100	75 MBH input	Q-6	3.60	6.667	Ea.	5,050	380		5,430	6,125
0120	100 MBH input		3.40	7.059		5,250	400		5,650	6,375
0140	125 MBH input		3.20	7.500		5,575	425		6,000	6,775
0160	150 MBH input		3	8		6,125	455		6,580	7,425
0180	175 MBH input		2.80	8.571		6,300	485		6,785	7,650
0200	200 MBH input		2.60	9.231		6,625	525		7,150	8,100
0220	225 MBH input		2.40	10		6,875	565		7,440	8,400
0240	250 MBH input		2	12		7,925	680		8,605	9,725
0260	300 MBH input		1.90	12.632		8,125	715		8,840	10,000
0280	350 MBH input		1.75	13.714		8,975	775		9,750	11,100
0300	400 MBH input		1.60	15		16,200	850		17,050	19,100
0320	840 MBH input		1.20	20		22,200	1,125		23,325	26,100
0340	1200 MBH input		.88	27.273		32,600	1,550		34,150	38,200
0600	For discharge louver assembly, add					5%				
0700	For filters, add					10%				
0800	For air shut-off damper section, add					30%				

23 74 Packaged Outdoor HVAC Equipment

23 74 13 – Packaged, Outdoor, Central-Station Air-Handling Units

23 74 13.10 Packaged, Outdoor Type, Central-Station AHU

		Crew	Daily Output	Labor-Hours	Unit	Material	2016 Bare Costs Labor	Equipment	Total	Total Incl O&P
0010	**PACKAGED, OUTDOOR TYPE, CENTRAL-STATION AIR-HANDLING UNITS**									
2900	Cooling coils may be chilled water or DX									
2910	Heating coils may be hot water, steam or electric									
3000	Weathertight, ground or rooftop									
3010	Constant volume									
3100	2000 CFM	Q-6	1	24	Ea.	23,700	1,350		25,050	28,100
3140	5000 CFM		.79	30.380		28,100	1,725		29,825	33,500
3150	10,000 CFM		.52	46.154		57,500	2,625		60,125	67,500
3160	15,000 CFM		.41	58.537		84,500	3,325		87,825	97,500
3170	20,000 CFM		.30	80		115,500	4,525		120,025	134,000
3180	30,000 CFM		.24	100		147,000	5,675		152,675	170,500
3200	Variable air volume									
3210	2000 CFM	Q-6	.96	25	Ea.	15,200	1,425		16,625	18,800
3240	5000 CFM		.73	32.877		30,800	1,850		32,650	36,600
3250	10,000 CFM		.50	48		66,000	2,725		68,725	77,000
3260	15,000 CFM		.40	60		100,500	3,400		103,900	115,500
3270	20,000 CFM		.29	82.759		130,500	4,700		135,200	150,500
3280	30,000 CFM		.20	120		154,500	6,800		161,300	180,500

23 74 23 – Packaged, Outdoor, Heating-Only Makeup-Air Units

23 74 23.16 Packaged, Ind.-Fired, In/Outdoor

		Crew	Daily Output	Labor-Hours	Unit	Material	2016 Bare Costs Labor	Equipment	Total	Total Incl O&P
0010	**PACKAGED, IND.-FIRED, IN/OUTDOOR**, Heat-Only Makeup-Air Units									
1000	Rooftop unit, natural gas, gravity vent, S.S. exchanger									
1010	70°F temperature rise, MBH is input									
1020	250 MBH	Q-6	4	6	Ea.	15,600	340		15,940	17,700
1040	400 MBH		3.60	6.667		18,100	380		18,480	20,500
1060	550 MBH		3.30	7.273		20,400	410		20,810	23,100
1080	750 MBH		3	8		22,000	455		22,455	24,900
1100	1000 MBH		2.60	9.231		25,500	525		26,025	28,800
1120	1750 MBH		2.30	10.435		30,800	590		31,390	34,800
1140	2500 MBH		1.90	12.632		43,300	715		44,015	48,800
1160	3250 MBH		1.40	17.143		51,000	970		51,970	57,500
1180	4000 MBH		1.20	20		65,000	1,125		66,125	73,500
1200	6000 MBH		1	24		83,000	1,350		84,350	93,500
1600	For cleanable filters, add					5%				
1700	For electric modulating gas control, add					10%				

23 74 33 – Dedicated Outdoor-Air Units

23 74 33.10 Rooftop Air Conditioners

			Crew	Daily Output	Labor-Hours	Unit	Material	2016 Bare Costs Labor	Equipment	Total	Total Incl O&P
0010	**ROOFTOP AIR CONDITIONERS**, Standard controls, curb, economizer										
1000	Single zone, electric cool, gas heat										
1100	3 ton cooling, 60 MBH heating	D3050-150	Q-5	.70	22.857	Ea.	2,800	1,250		4,050	4,950
1120	4 ton cooling, 95 MBH heating	D3050-155		.61	26.403		3,250	1,450		4,700	5,750
1140	5 ton cooling, 112 MBH heating	R236000-10		.56	28.521		4,425	1,550		5,975	7,200
1145	6 ton cooling, 140 MBH heating	R236000-20		.52	30.769		5,050	1,675		6,725	8,100
1150	7.5 ton cooling, 170 MBH heating			.50	32.258		5,875	1,750		7,625	9,125
1156	8.5 ton cooling, 170 MBH heating			.46	34.783		7,100	1,900		9,000	10,700
1160	10 ton cooling, 200 MBH heating		Q-6	.67	35.982		9,100	2,050		11,150	13,100
1170	12.5 ton cooling, 230 MBH heating			.63	37.975		10,300	2,150		12,450	14,600
1180	15 ton cooling, 270 MBH heating			.57	42.032		13,100	2,375		15,475	18,000
1190	17.5 ton cooling, 330 MBH heating			.52	45.889		13,900	2,600		16,500	19,200
1200	20 ton cooling, 360 MBH heating		Q-7	.67	47.976		29,500	2,775		32,275	36,700
1210	25 ton cooling, 450 MBH heating			.56	57.554		32,000	3,325		35,325	40,200
1220	30 ton cooling, 540 MBH heating			.47	68.376		35,600	3,950		39,550	45,200

391

23 74 33.10 Rooftop Air Conditioners		Crew	Daily Output	Labor-Hours	Unit	Material	2016 Bare Costs Labor	Equipment	Total	Total Incl O&P
1240	40 ton cooling, 675 MBH heating	Q-7	.35	91.168	Ea.	43,900	5,275		49,175	56,500
1260	50 ton cooling, 810 MBH heating		.28	113		50,000	6,575		56,575	65,000
1265	60 ton cooling, 900 MBH heating		.23	136		59,500	7,900		67,400	77,500
1270	80 ton cooling, 1000 MBH heating		.20	160		77,500	9,250		86,750	99,000
1275	90 ton cooling, 1200 MBH heating		.17	188		107,000	10,900		117,900	134,500
1280	100 ton cooling, 1350 MBH heating		.14	228		125,500	13,200		138,700	158,000
1300	Electric cool, electric heat									
1310	5 ton	Q-5	.63	25.600	Ea.	4,500	1,400		5,900	7,025
1314	10 ton	Q-6	.74	32.432		8,150	1,825		9,975	11,800
1318	15 ton	"	.64	37.795		12,200	2,150		14,350	16,700
1322	20 ton	Q-7	.74	43.185		16,300	2,500		18,800	21,800
1326	25 ton		.62	51.696		20,400	3,000		23,400	26,900
1330	30 ton		.52	61.657		24,500	3,575		28,075	32,300
1334	40 ton		.39	82.051		32,600	4,750		37,350	43,100
1338	50 ton		.31	102		38,800	5,925		44,725	51,500
1342	60 ton		.26	123		46,500	7,125		53,625	61,500
1346	75 ton		.21	153		58,000	8,900		66,900	77,500
2000	Multizone, electric cool, gas heat, economizer									
2100	15 ton cooling, 360 MBH heating	Q-7	.61	52.545	Ea.	66,000	3,025		69,025	77,000
2120	20 ton cooling, 360 MBH heating		.53	60.038		71,000	3,475		74,475	83,500
2140	25 ton cooling, 450 MBH heating		.45	71.910		85,500	4,150		89,650	100,500
2160	28 ton cooling, 450 MBH heating		.41	79.012		97,000	4,575		101,575	114,000
2180	30 ton cooling, 540 MBH heating		.37	85.562		109,000	4,950		113,950	127,500
2200	40 ton cooling, 540 MBH heating		.28	113		125,000	6,575		131,575	147,500
2210	50 ton cooling, 540 MBH heating		.23	142		155,000	8,225		163,225	183,000
2220	70 ton cooling, 1500 MBH heating		.16	198		168,000	11,500		179,500	202,500
2240	80 ton cooling, 1500 MBH heating		.14	228		192,000	13,200		205,200	231,000
2260	90 ton cooling, 1500 MBH heating		.13	256		201,000	14,800		215,800	244,000
2280	105 ton cooling, 1500 MBH heating		.11	290		222,000	16,800		238,800	269,500
2400	For hot water heat coil, deduct					5%				
2500	For steam heat coil, deduct					2%				
2600	For electric heat, deduct					3%	5%			
4000	Single zone, electric cool, gas heat, variable air volume									
4100	12.5 ton cooling, 230 MBH heating	Q-6	.55	43.716	Ea.	12,800	2,475		15,275	17,700
4120	18 ton cooling, 330 MBH heating	"	.46	52.747		17,100	3,000		20,100	23,300
4140	25 ton cooling, 450 MBH heating	Q-7	.48	66.116		36,300	3,825		40,125	45,800
4160	40 ton cooling, 675 MBH heating		.31	104		50,500	6,075		56,575	65,000
4180	60 ton cooling, 900 MBH heating		.20	157		73,500	9,100		82,600	94,500
4200	80 ton cooling, 1000 MBH heating		.15	209		93,000	12,100		105,100	120,500
5000	Single zone, electric cool only									
5050	3 ton cooling	Q-5	.88	18.203	Ea.	2,700	995		3,695	4,475
5060	4 ton cooling		.76	21.108		3,225	1,150		4,375	5,300
5070	5 ton cooling		.70	22.792		3,575	1,250		4,825	5,825
5080	6 ton cooling		.65	24.502		4,375	1,350		5,725	6,850
5090	7.5 ton cooling		.62	25.806		6,525	1,400		7,925	9,300
5100	8.5 ton cooling		.59	26.981		7,100	1,475		8,575	10,000
5110	10 ton cooling	Q-6	.83	28.812		8,300	1,625		9,925	11,600
5120	12.5 ton cooling		.80	30		11,400	1,700		13,100	15,100
5130	15 ton cooling		.71	33.613		15,400	1,900		17,300	19,800
5140	17.5 ton cooling		.65	36.697		16,300	2,075		18,375	21,100
5150	20 ton cooling	Q-7	.83	38.415		21,400	2,225		23,625	26,900
5160	25 ton cooling		.70	45.977		22,700	2,650		25,350	28,900
5170	30 ton cooling		.59	54.701		24,500	3,150		27,650	31,800

23 74 Packaged Outdoor HVAC Equipment

23 74 33 – Dedicated Outdoor-Air Units

23 74 33.10 Rooftop Air Conditioners	Crew	Daily Output	Labor-Hours	Unit	Material	2016 Bare Costs Labor	Equipment	Total	Total Incl O&P
5180	40 ton cooling	Q-7	.44	73.059	Ea.	32,900	4,225		37,125	42,600
5190	50 ton cooling		.38	84.211		37,800	4,875		42,675	49,000
5200	60 ton cooling	↓	.28	114		46,500	6,600		53,100	61,000
5400	For low heat, add					7%				
5410	For high heat, add				↓	10%				
6000	Single zone electric cooling with variable volume distribution									
6020	20 ton cooling	Q-7	.72	44.199	Ea.	23,500	2,550		26,050	29,700
6030	25 ton cooling		.61	52.893		25,200	3,050		28,250	32,300
6040	30 ton cooling		.51	62.868		29,800	3,625		33,425	38,300
6050	40 ton cooling		.38	83.990		34,300	4,850		39,150	45,000
6060	50 ton cooling		.31	104		34,900	6,075		40,975	47,600
6070	60 ton cooling	↓	.25	125		48,000	7,275		55,275	64,000
6200	For low heat, add					7%				
6210	For high heat, add				↓	10%				
7000	Multizone, cool/heat, variable volume distribution									
7100	50 ton cooling	Q-7	.20	164	Ea.	115,000	9,475		124,475	141,000
7110	70 ton cooling		.14	228		161,000	13,200		174,200	197,500
7120	90 ton cooling		.11	296		185,500	17,100		202,600	230,000
7130	105 ton cooling		.10	333		207,500	19,300		226,800	257,000
7140	120 ton cooling		.08	380		237,000	22,000		259,000	293,500
7150	140 ton cooling	↓	.07	444	↓	253,500	25,700		279,200	318,000
7300	Penthouse unit, cool/heat, variable volume distribution									
7400	150 ton cooling	Q-7	.10	310	Ea.	414,500	18,000		432,500	483,000
7410	170 ton cooling		.10	320		443,000	18,500		461,500	515,500
7420	200 ton cooling		.10	326		521,500	18,900		540,400	602,000
7430	225 ton cooling		.09	340		570,500	19,700		590,200	657,000
7440	250 ton cooling		.09	359		588,500	20,800		609,300	679,000
7450	270 ton cooling		.09	363		635,500	21,000		656,500	730,500
7460	300 ton cooling	↓	.09	372	↓	706,000	21,500		727,500	809,500
9400	Sub assemblies for assembly systems									
9410	Ductwork, per ton, rooftop 1 zone units	Q-9	.96	16.667	Ton	171	860		1,031	1,525
9420	Ductwork, per ton, rooftop multizone units		.50	32		252	1,650		1,902	2,800
9430	Ductwork, VAV cooling/ton, rooftop multizone, 1 zone/ton		.32	50		390	2,575		2,965	4,375
9440	Ductwork per ton, packaged water & air cooled units		1.26	12.698		60.50	655		715.50	1,075
9450	Ductwork per ton, split system remote condensing units		1.32	12.121	↓	57	625		682	1,025
9500	Ductwork package for residential units, 1 ton		.80	20	Ea.	274	1,025		1,299	1,875
9520	2 ton		.40	40		550	2,050		2,600	3,750
9530	3 ton	↓	.27	59.259		820	3,050		3,870	5,575
9540	4 ton	Q-10	.30	80		1,100	4,275		5,375	7,725
9550	5 ton		.23	104		1,375	5,575		6,950	10,000
9560	6 ton		.20	120		1,650	6,400		8,050	11,600
9570	7 ton		.17	141		1,925	7,550		9,475	13,600
9580	8 ton	↓	.15	160	↓	2,200	8,550		10,750	15,500

For customer support on your Mechanical Cost Data, call 877.739.3689.

393

23 76 Evaporative Air-Cooling Equipment

23 76 13 – Direct Evaporative Air Coolers

23 76 13.10 Evaporative Coolers

		Crew	Daily Output	Labor-Hours	Unit	Material	2016 Bare Costs Labor	2016 Bare Costs Equipment	Total	Total Incl O&P
0010	**EVAPORATIVE COOLERS**, (Swamp Coolers) ducted, not incl. duct.									
0100	Side discharge style, capacities at .25" S.P.									
0120	1785 CFM, 1/3 HP, 115 V [G]	Q-9	5	3.200	Ea.	460	165		625	755
0140	2740 CFM, 1/3 HP, 115 V [G]		4.50	3.556		545	183		728	880
0160	3235 CFM, 1/2 HP, 115 V [G]		4	4		550	206		756	920
0180	3615 CFM, 1/2 HP, 230 V [G]		3.60	4.444		680	229		909	1,100
0200	4215 CFM, 3/4 HP, 230 V [G]		3.20	5		720	258		978	1,175
0220	5255 CFM. 1 HP, 115/230 V [G]		3	5.333		1,275	275		1,550	1,825
0240	6090 CFM, 1 HP, 230/460 V [G]		2.80	5.714		1,900	294		2,194	2,550
0260	8300 CFM, 1-1/2 HP, 230/460 V [G]		2.60	6.154		1,900	315		2,215	2,575
0280	8360 CFM, 1-1/2 HP, 230/460 V [G]		2.20	7.273		1,900	375		2,275	2,675
0300	9725 CFM, 2 HP, 230/460 V [G]		1.80	8.889		2,150	460		2,610	3,075
0320	11,715 CFM, 3 HP, 230/460 V [G]		1.40	11.429		2,400	590		2,990	3,550
0340	14,410 CFM, 5 HP, 230/460 V [G]		1	16		2,600	825		3,425	4,100
0400	For two-speed motor, add					5%				
0500	For down discharge style, add					10%				

23 76 26 – Fan Coil Evaporators for Coolers/Freezers

23 76 26.10 Evaporators

		Crew	Daily Output	Labor-Hours	Unit	Material	2016 Bare Costs Labor	2016 Bare Costs Equipment	Total	Total Incl O&P
0010	**EVAPORATORS**, DX coils, remote compressors not included.									
1000	Coolers, reach-in type, above freezing temperatures									
1300	Shallow depth, wall mount, 7 fins per inch, air defrost									
1310	600 BTUH, 8" fan, rust-proof core	Q-5	3.80	4.211	Ea.	690	230		920	1,100
1320	900 BTUH, 8" fan, rust-proof core		3.30	4.848		805	265		1,070	1,275
1330	1200 BTUH, 8" fan, rust-proof core		3	5.333		990	291		1,281	1,550
1340	1800 BTUH, 8" fan, rust-proof core		2.40	6.667		1,025	365		1,390	1,675
1350	2500 BTUH, 10" fan, rust-proof core		2.20	7.273		1,050	395		1,445	1,750
1360	3500 BTUH, 10" fan		2	8		1,125	435		1,560	1,900
1370	4500 BTUH, 10" fan		1.90	8.421		1,200	460		1,660	2,025
1390	For pan drain, add					62			62	68
1600	Undercounter refrigerators, ceiling or wall mount,									
1610	8 fins per inch, air defrost, rust-proof core									
1630	800 BTUH, one 6" fan	Q-5	5	3.200	Ea.	278	175		453	570
1640	1300 BTUH, two 6" fans		4	4		360	219		579	725
1650	1700 BTUH, two 6" fans		3.60	4.444		400	243		643	805
2000	Coolers, reach-in and walk-in types, above freezing temperatures									
2600	Two-way discharge, ceiling mount, 150-4100 CFM,									
2610	above 34°F applications, air defrost									
2630	900 BTUH, 7 fins per inch, 8" fan	Q-5	4	4	Ea.	550	219		769	935
2660	2500 BTUH, 7 fins per inch, 10" fan		2.60	6.154		835	335		1,170	1,425
2690	5500 BTUH, 7 fins per inch, 12" fan		1.70	9.412		1,425	515		1,940	2,350
2720	8500 BTUH, 8 fins per inch, 16" fan		1.20	13.333		1,575	730		2,305	2,825
2750	15,000 BTUH, 7 fins per inch, 18" fan		1.10	14.545		2,550	795		3,345	4,000
2770	24,000 BTUH, 7 fins per inch, two 16" fans		1	16		3,800	875		4,675	5,500
2790	30,000 BTUH, 7 fins per inch, two 18" fans		.90	17.778		4,725	970		5,695	6,675
2850	Two-way discharge, low profile, ceiling mount,									
2860	8 fins per inch, 200-570 CFM, air defrost									
2880	800 BTUH, one 6" fan	Q-5	4	4	Ea.	315	219		534	680
2890	1300 BTUH, two 6" fans		3.50	4.571		380	250		630	795
2900	1800 BTUH, two 6" fans		3.30	4.848		470	265		735	920
2910	2700 BTUH, three 6" fans		2.70	5.926		580	325		905	1,125
3000	Coolers, walk-in type, above freezing temperatures									
3300	General use, ceiling mount, 108-2080 CFM									

23 76 26.10 Evaporators	Crew	Daily Output	Labor-Hours	Unit	Material	2016 Bare Costs Labor	Equipment	Total	Total Incl O&P
3320	600 BTUH, 7 fins per inch, 6" fan	Q-5	5.20	3.077	Ea.	330	168		498	620
3340	1200 BTUH, 7 fins per inch, 8" fan		3.70	4.324		460	236		696	860
3360	1800 BTUH, 7 fins per inch, 10" fan		2.70	5.926		645	325		970	1,200
3380	3500 BTUH, 7 fins per inch, 12" fan		2.40	6.667		945	365		1,310	1,600
3400	5500 BTUH, 8 fins per inch, 12" fan		2	8		1,050	435		1,485	1,800
3430	8500 BTUH, 8 fins per inch, 16" fan		1.40	11.429		1,575	625		2,200	2,675
3460	15,000 BTUH, 7 fins per inch, two 16" fans	▼	1.10	14.545	▼	2,525	795		3,320	3,975
3640	Low velocity, high latent load, ceiling mount,									
3650	1050-2420 CFM, air defrost, 6 fins per inch									
3670	6700 BTUH, two 10" fans	Q-5	1.30	12.308	Ea.	1,725	670		2,395	2,900
3680	10,000 BTUH, three 10" fans		1	16		2,150	875		3,025	3,700
3690	13,500 BTUH, three 10" fans		1	16		2,750	875		3,625	4,350
3700	18,000 BTUH, four 10" fans		1	16		3,425	875		4,300	5,100
3710	26,500 BTUH, four 10" fans	▼	.80	20		4,350	1,100		5,450	6,425
3730	For electric defrost, add				▼	17%				
5000	Freezers and coolers, reach-in type, above 34°F									
5030	to sub-freezing temperature range, low latent load,									
5050	air defrost, 7 fins per inch									
5070	1200 BTUH, 8" fan, rustproof core	Q-5	4.40	3.636	Ea.	585	199		784	945
5080	1500 BTUH, 8" fan, rustproof core		4.20	3.810		605	208		813	980
5090	1800 BTUH, 8" fan, rustproof core		3.70	4.324		620	236		856	1,025
5100	2500 BTUH, 10" fan, rustproof core		2.90	5.517		680	300		980	1,200
5110	3500 BTUH, 10" fan, rustproof core		2.50	6.400		845	350		1,195	1,450
5120	4500 BTUH, 12" fan, rustproof core	▼	2.10	7.619	▼	1,050	415		1,465	1,775
6000	Freezers and coolers, walk-in type									
6050	1960-18,000 CFM, medium profile									
6060	Standard motor, 6 fins per inch, to -30°F, all aluminum									
6080	10,500 BTUH, air defrost, one 18" fan	Q-5	1.40	11.429	Ea.	1,675	625		2,300	2,775
6090	12,500 BTUH, air defrost, one 18" fan		1.40	11.429		1,800	625		2,425	2,925
6100	16,400 BTUH, air defrost, one 18" fan		1.30	12.308		1,825	670		2,495	3,050
6110	20,900 BTUH, air defrost, one 18" fans		1	16		2,125	875		3,000	3,650
6120	27,000 BTUH, air defrost, two 18" fans	▼	1	16		3,075	875		3,950	4,725
6130	32,900 BTUH, air defrost, two 20" fans	Q-6	1.30	18.462		3,425	1,050		4,475	5,350
6140	39,000 BTUH, air defrost, two 20" fans		1.25	19.200		3,950	1,100		5,050	6,000
6150	44,100 BTUH, air defrost, three 20" fans		1	24		4,475	1,350		5,825	6,975
6155	53,000 BTUH, air defrost, three 20" fans		1	24		4,475	1,350		5,825	6,975
6160	66,200 BTUH, air defrost, four 20" fans		1	24		4,950	1,350		6,300	7,500
6170	78,000 BTUH, air defrost, four 20" fans		.80	30		5,700	1,700		7,400	8,850
6180	88,200 BTUH, air defrost, four 24" fans		.60	40		6,150	2,275		8,425	10,200
6190	110,000 BTUH, air defrost, four 24" fans	▼	.50	48		7,425	2,725		10,150	12,300
6330	Hot gas defrost, standard motor units, add					35%				
6370	Electric defrost, 230V, standard motor, add					16%				
6410	460V, standard or high capacity motor, add				▼	25%				
6800	Eight fins per inch, increases BTUH 75%									
6810	Standard capacity, add				Ea.	5%				
6830	Hot gas & electric defrost not recommended									
7000	800-4150 CFM, 12" fans									
7010	Four fins per inch									
7030	3400 BTUH, 1 fan	Q-5	2.80	5.714	Ea.	565	310		875	1,100
7040	4200 BTUH, 1 fan		2.40	6.667		670	365		1,035	1,275
7050	5300 BTUH, 1 fan		1.90	8.421		710	460		1,170	1,475
7060	6800 BTUH, 1 fan		1.70	9.412		800	515		1,315	1,650
7070	8400 BTUH, 2 fans	▼	1.60	10	▼	835	545		1,380	1,750

For customer support on your Mechanical Cost Data, call 877.739.3689.

395

23 76 Evaporative Air-Cooling Equipment

23 76 26 – Fan Coil Evaporators for Coolers/Freezers

23 76 26.10 Evaporators		Crew	Daily Output	Labor-Hours	Unit	Material	2016 Bare Costs Labor	Equipment	Total	Total Incl O&P
7080	10,500 BTUH, 2 fans	Q-5	1.40	11.429	Ea.	970	625		1,595	2,025
7090	13,000 BTUH, 3 fans		1.30	12.308		1,600	670		2,270	2,775
7100	17,000 BTUH, 4 fans		1.20	13.333		1,925	730		2,655	3,225
7110	21,500 BTUH, 5 fans		1	16		2,175	875		3,050	3,725
7160	Six fins per inch increases BTUH 35%, add					5%				
7180	Eight fins per inch increases BTUH 53%,									
7190	not recommended for below 35°F, add				Ea.	8%				
7210	For 12°F temperature differential, add					20%				
7230	For adjustable thermostat control, add					122			122	135
7240	For hot gas defrost, except 8 fin models									
7250	on applications below 35°F, add				Ea.	58%				
7260	For electric defrost, except 8 fin models									
7270	on applications below 35°F, add				Ea.	36%				
8000	Freezers, pass-thru door uprights, walk-in storage									
8300	Low temperature, thin profile, electric defrost									
8310	138-1800 CFM, 5 fins per inch									
8320	900 BTUH, one 6" fan	Q-5	3.30	4.848	Ea.	420	265		685	865
8340	1500 BTUH, two 6" fans		2.20	7.273		635	395		1,030	1,300
8360	2600 BTUH, three 6" fans		1.90	8.421		885	460		1,345	1,675
8370	3300 BTUH, four 6" fans		1.70	9.412		995	515		1,510	1,875
8380	4400 BTUH, five 6" fans		1.40	11.429		1,375	625		2,000	2,475
8400	7000 BTUH, three 10" fans		1.30	12.308		1,725	670		2,395	2,925
8410	8700 BTUH, four 10" fans		1.10	14.545		2,100	795		2,895	3,500
8450	For air defrost, deduct					260			260	286

23 81 Decentralized Unitary HVAC Equipment

23 81 13 – Packaged Terminal Air-Conditioners

23 81 13.10 Packaged Cabinet Type Air-Conditioners

		Crew	Daily Output	Labor-Hours	Unit	Material	Labor	Equipment	Total	Total Incl O&P
0010	**PACKAGED CABINET TYPE AIR-CONDITIONERS**, Cabinet, wall sleeve,									
0100	louver, electric heat, thermostat, manual changeover, 208 V									
0200	6,000 BTUH cooling, 8800 BTU heat	Q-5	6	2.667	Ea.	825	146		971	1,125
0220	9,000 BTUH cooling, 13,900 BTU heat		5	3.200		1,225	175		1,400	1,625
0240	12,000 BTUH cooling, 13,900 BTU heat		4	4		1,350	219		1,569	1,825
0260	15,000 BTUH cooling, 13,900 BTU heat		3	5.333		1,475	291		1,766	2,075
0280	18,000 BTUH cooling, 10 KW heat		2.40	6.667		2,300	365		2,665	3,075
0300	24,000 BTUH cooling, 10 KW heat		1.90	8.421		2,325	460		2,785	3,250
0320	30,000 BTUH cooling, 10 KW heat		1.40	11.429		2,550	625		3,175	3,750
0340	36,000 BTUH cooling, 10 KW heat		1.25	12.800		2,600	700		3,300	3,900
0360	42,000 BTUH cooling, 10 KW heat		1	16		2,975	875		3,850	4,600
0380	48,000 BTUH cooling, 10 KW heat		.90	17.778		3,150	970		4,120	4,950
0500	For hot water coil, increase heat by 10%, add					5%	10%			
1000	For steam, increase heat output by 30%, add					8%	10%			

23 81 19 – Self-Contained Air-Conditioners

23 81 19.10 Window Unit Air Conditioners

		Crew	Daily Output	Labor-Hours	Unit	Material	Labor	Equipment	Total	Total Incl O&P
0010	**WINDOW UNIT AIR CONDITIONERS**									
4000	Portable/window, 15 amp 125 V grounded receptacle required									
4060	5000 BTUH	1 Carp	8	1	Ea.	330	48.50		378.50	435
4340	6000 BTUH		8	1		380	48.50		428.50	495
4480	8000 BTUH		6	1.333		475	64.50		539.50	625
4500	10,000 BTUH		6	1.333		590	64.50		654.50	750

23 81 Decentralized Unitary HVAC Equipment

23 81 19 – Self-Contained Air-Conditioners

23 81 19.10 Window Unit Air Conditioners	Crew	Daily Output	Labor-Hours	Unit	Material	2016 Bare Costs Labor	2016 Bare Costs Equipment	Total	Total Incl O&P
4520	12,000 BTUH	L-2	8	2	Ea.	740	84.50		824.50	945
4600	Window/thru-the-wall, 15 amp 230 V grounded receptacle required									
4780	18,000 BTUH	L-2	6	2.667	Ea.	1,000	113		1,113	1,275
4940	25,000 BTUH		4	4		1,275	169		1,444	1,650
4960	29,000 BTUH		4	4		1,400	169		1,569	1,775

23 81 19.20 Self-Contained Single Package

		Crew	Daily Output	Labor-Hours	Unit	Material	2016 Bare Costs Labor	2016 Bare Costs Equipment	Total	Total Incl O&P
0010	**SELF-CONTAINED SINGLE PACKAGE**									
0100	Air cooled, for free blow or duct, not incl. remote condenser									
0110	Constant volume									
0200	3 ton cooling	Q-5	1	16	Ea.	3,700	875		4,575	5,400
0210	4 ton cooling	"	.80	20		4,025	1,100		5,125	6,075
0220	5 ton cooling	Q-6	1.20	20		4,425	1,125		5,550	6,550
0230	7.5 ton cooling	"	.90	26.667		5,700	1,500		7,200	8,550
0240	10 ton cooling	Q-7	1	32		7,925	1,850		9,775	11,500
0250	15 ton cooling		.95	33.684		10,300	1,950		12,250	14,200
0260	20 ton cooling		.90	35.556		12,800	2,050		14,850	17,200
0270	25 ton cooling		.85	37.647		19,000	2,175		21,175	24,200
0280	30 ton cooling		.80	40		25,300	2,300		27,600	31,300
0300	40 ton cooling		.60	53.333		36,100	3,075		39,175	44,400
0320	50 ton cooling		.50	64		47,200	3,700		50,900	57,500
0340	60 ton cooling	Q-8	.40	80		56,000	4,625	146	60,771	68,500
0380	With hot water heat, variable air volume									
0390	10 ton cooling	Q-7	.96	33.333	Ea.	11,000	1,925		12,925	15,000
0400	20 ton cooling		.86	37.209		18,500	2,150		20,650	23,700
0410	30 ton cooling		.76	42.105		33,600	2,425		36,025	40,700
0420	40 ton cooling		.58	55.172		46,500	3,200		49,700	56,000
0430	50 ton cooling		.48	66.667		62,000	3,850		65,850	74,000
0440	60 ton cooling	Q-8	.38	84.211		75,500	4,875	153	80,528	90,500
0490	For duct mounting no price change									
0500	For steam heating coils, add				Ea.	10%	10%			
0550	Packaged, with electric heat									
0552	Constant volume									
0560	5 ton cooling	Q-6	1.20	20	Ea.	7,925	1,125		9,050	10,400
0570	10 ton cooling	Q-7	1	32		13,300	1,850		15,150	17,400
0580	20 ton cooling		.90	35.556		20,500	2,050		22,550	25,700
0590	30 ton cooling		.80	40		38,100	2,300		40,400	45,400
0600	40 ton cooling		.60	53.333		50,000	3,075		53,075	59,500
0610	50 ton cooling		.50	64		66,500	3,700		70,200	78,500
0700	Variable air volume									
0710	10 ton cooling	Q-7	1	32	Ea.	15,600	1,850		17,450	19,900
0720	20 ton cooling		.90	35.556		24,900	2,050		26,950	30,500
0730	30 ton cooling		.80	40		44,000	2,300		46,300	52,000
0740	40 ton cooling		.60	53.333		56,500	3,075		59,575	67,000
0750	50 ton cooling		.50	64		76,500	3,700		80,200	89,500
0760	60 ton cooling	Q-8	.40	80		87,500	4,625	146	92,271	103,000
1000	Water cooled for free blow or duct, not including tower									
1010	Constant volume									
1100	3 ton cooling	Q-6	1	24	Ea.	3,575	1,350		4,925	5,975
1120	5 ton cooling		1	24		4,650	1,350		6,000	7,150
1130	7.5 ton cooling		.80	30		6,400	1,700		8,100	9,600
1140	10 ton cooling	Q-7	.90	35.556		8,825	2,050		10,875	12,800
1150	15 ton cooling		.85	37.647		12,800	2,175		14,975	17,300

For customer support on your Mechanical Cost Data, call 877.739.3689.

397

23 81 Decentralized Unitary HVAC Equipment

23 81 19 – Self-Contained Air-Conditioners

23 81 19.20 Self-Contained Single Package	Crew	Daily Output	Labor-Hours	Unit	Material	2016 Bare Costs Labor	2016 Bare Costs Equipment	Total	Total Incl O&P
1160	20 ton cooling	Q-7	.80	40	Ea.	28,600	2,300		30,900	34,900
1170	25 ton cooling		.75	42.667		33,700	2,475		36,175	40,800
1180	30 ton cooling		.70	45.714		37,900	2,650		40,550	45,700
1200	40 ton cooling		.40	80		48,000	4,625		52,625	60,000
1220	50 ton cooling		.30	106		59,500	6,175		65,675	75,000
1240	60 ton cooling	Q-8	.30	106		69,500	6,175	194	75,869	86,000
1300	For hot water or steam heat coils, add					12%	10%			
2000	Water cooled with electric heat, not including tower									
2010	Constant volume									
2100	5 ton cooling	Q-6	1.20	20	Ea.	8,150	1,125		9,275	10,700
2120	10 ton cooling	Q-7	1	32		14,200	1,850		16,050	18,400
2130	15 ton cooling	"	.90	35.556		20,400	2,050		22,450	25,600
2400	Variable air volume									
2440	10 ton cooling	Q-7	1	32	Ea.	16,500	1,850		18,350	20,900
2450	15 ton cooling	"	.90	35.556	"	24,800	2,050		26,850	30,400
2600	Water cooled, hot water coils, not including tower									
2610	Variable air volume									
2640	10 ton cooling	Q-7	1	32	Ea.	12,000	1,850		13,850	16,000
2650	15 ton cooling	"	.80	40	"	18,400	2,300		20,700	23,800

23 81 23 – Computer-Room Air-Conditioners

23 81 23.10 Computer Room Units

		Crew	Daily Output	Labor-Hours	Unit	Material	Labor	Equipment	Total	Total Incl O&P
0010	**COMPUTER ROOM UNITS**		D3050-185							
1000	Air cooled, includes remote condenser but not									
1020	interconnecting tubing or refrigerant									
1080	3 ton	Q-5	.50	32	Ea.	19,500	1,750		21,250	24,200
1120	5 ton		.45	35.556		20,900	1,950		22,850	25,800
1160	6 ton		.30	53.333		38,500	2,925		41,425	46,800
1200	8 ton		.27	59.259		38,900	3,225		42,125	47,700
1240	10 ton		.25	64		40,700	3,500		44,200	50,000
1260	12 ton		.24	66.667		42,100	3,650		45,750	52,000
1280	15 ton		.22	72.727		44,800	3,975		48,775	55,500
1290	18 ton		.20	80		51,500	4,375		55,875	63,000
1300	20 ton	Q-6	.26	92.308		53,500	5,225		58,725	67,000
1320	22 ton		.24	100		54,500	5,675		60,175	68,000
1360	30 ton		.21	114		67,000	6,475		73,475	83,500
2200	Chilled water, for connection to									
2220	existing chiller system of adequate capacity									
2260	5 ton	Q-5	.74	21.622	Ea.	14,900	1,175		16,075	18,200
2280	6 ton		.52	30.769		15,000	1,675		16,675	19,000
2300	8 ton		.50	32		15,200	1,750		16,950	19,400
2320	10 ton		.49	32.653		15,300	1,775		17,075	19,600
2330	12 ton		.49	32.990		15,700	1,800		17,500	19,900
2360	15 ton		.48	33.333		16,000	1,825		17,825	20,400
2400	20 ton		.46	34.783		17,000	1,900		18,900	21,600
2440	25 ton	Q-6	.63	38.095		18,100	2,150		20,250	23,200
2460	30 ton		.57	42.105		19,600	2,375		21,975	25,200
2480	35 ton		.52	46.154		19,900	2,625		22,525	25,800
2500	45 ton		.46	52.174		21,000	2,950		23,950	27,600
2520	50 ton		.42	57.143		24,100	3,225		27,325	31,400
2540	60 ton		.38	63.158		25,600	3,575		29,175	33,600
2560	77 ton		.35	68.571		43,300	3,875		47,175	53,500
2580	95 ton		.32	75		50,500	4,250		54,750	62,000

23 81 Decentralized Unitary HVAC Equipment

23 81 23 – Computer-Room Air-Conditioners

23 81 23.10 Computer Room Units	Crew	Daily Output	Labor-Hours	Unit	Material	2016 Bare Costs Labor	Equipment	Total	Total Incl O&P
4000	Glycol system, complete except for interconnecting tubing									
4060	3 ton	Q-5	.40	40	Ea.	24,500	2,175		26,675	30,200
4100	5 ton		.38	42.105		26,800	2,300		29,100	33,000
4120	6 ton		.25	64		39,800	3,500		43,300	49,100
4140	8 ton		.23	69.565		43,000	3,800		46,800	53,000
4160	10 ton		.21	76.190		45,800	4,150		49,950	57,000
4180	12 ton		.19	84.211		49,700	4,600		54,300	61,500
4200	15 ton	Q-6	.26	92.308		56,500	5,225		61,725	70,500
4240	20 ton		.24	100		62,500	5,675		68,175	77,000
4280	22 ton		.23	104		65,500	5,900		71,400	81,000
4430	30 ton		.22	109		81,000	6,175		87,175	99,000
8000	Water cooled system, not including condenser,									
8020	water supply or cooling tower									
8060	3 ton	Q-5	.62	25.806	Ea.	19,900	1,400		21,300	23,900
8100	5 ton		.54	29.630		21,600	1,625		23,225	26,300
8120	6 ton		.35	45.714		30,900	2,500		33,400	37,800
8140	8 ton		.33	48.485		35,400	2,650		38,050	43,000
8160	10 ton		.31	51.613		36,600	2,825		39,425	44,600
8180	12 ton		.29	55.172		39,700	3,025		42,725	48,300
8200	15 ton		.27	59.259		42,900	3,225		46,125	52,000
8240	20 ton	Q-6	.38	63.158		46,100	3,575		49,675	56,000
8280	22 ton		.35	68.571		48,400	3,875		52,275	59,500
8300	30 ton		.30	80		60,000	4,525		64,525	73,500

23 81 26 – Split-System Air-Conditioners

23 81 26.10 Split Ductless Systems	Crew	Daily Output	Labor-Hours	Unit	Material	2016 Bare Costs Labor	Equipment	Total	Total Incl O&P
0010	**SPLIT DUCTLESS SYSTEMS**									
0100	Cooling only, single zone									
0110	Wall mount									
0120	3/4 ton cooling	Q-5	2	8	Ea.	1,100	435		1,535	1,875
0130	1 ton cooling		1.80	8.889		1,225	485		1,710	2,075
0140	1-1/2 ton cooling		1.60	10		1,975	545		2,520	3,000
0150	2 ton cooling		1.40	11.429		2,250	625		2,875	3,425
1000	Ceiling mount									
1020	2 ton cooling	Q-5	1.40	11.429	Ea.	2,000	625		2,625	3,150
1030	3 ton cooling	"	1.20	13.333	"	2,550	730		3,280	3,900
3000	Multizone									
3010	Wall mount									
3020	2 @ 3/4 ton cooling	Q-5	1.80	8.889	Ea.	3,375	485		3,860	4,425
5000	Cooling/Heating									
5010	Wall mount									
5110	1 ton cooling	Q-5	1.70	9.412	Ea.	1,300	515		1,815	2,200
5120	1-1/2 ton cooling	"	1.50	10.667	"	1,975	585		2,560	3,050
7000	Accessories for all split ductless systems									
7010	Add for ambient frost control	Q-5	8	2	Ea.	126	109		235	305
7020	Add for tube/wiring kit, (Line sets)									
7030	15' kit	Q-5	32	.500	Ea.	93	27.50		120.50	143
7036	25' kit		28	.571		186	31		217	251
7040	35' kit		24	.667		196	36.50		232.50	271
7050	50' kit		20	.800		237	43.50		280.50	325

For customer support on your Mechanical Cost Data, call 877.739.3689.

399

23 81 Decentralized Unitary HVAC Equipment

23 81 29 – Variable Refrigerant Flow HVAC Systems

23 81 29.10 Heat Pump, Gas Driven	Crew	Daily Output	Labor-Hours	Unit	Material	2016 Bare Costs Labor	Equipment	Total	Total Incl O&P
0010 **HEAT PUMP, GAS DRIVEN,** Variable refrigerant volume (VRV) type									
0020 Not including interconnecting tubing or multi-zone controls									
1000 For indoor fan VRV type AHU see 23 82 19.40									
1010 Multi-zone split									
1020 Outdoor unit									
1100 8 tons cooling, for up to 17 zones	Q-5	1.30	12.308	Ea.	29,100	670		29,770	33,000
1110 Isolation rails		2.60	6.154	Pair	1,025	335		1,360	1,625
1160 15 tons cooling, for up to 33 zones		1	16	Ea.	38,600	875		39,475	43,700
1170 Isolation rails		2.60	6.154	Pair	1,200	335		1,535	1,825
2000 Packaged unit									
2020 Outdoor unit									
2200 11 tons cooling	Q-5	1.30	12.308	Ea.	37,200	670		37,870	41,900
2210 Roof curb adapter	"	2.60	6.154		645	335		980	1,225
2220 Thermostat	1 Stpi	1.30	6.154		385	375		760	990

23 81 43 – Air-Source Unitary Heat Pumps

23 81 43.10 Air-Source Heat Pumps	Crew	Daily Output	Labor-Hours	Unit	Material	2016 Bare Costs Labor	Equipment	Total	Total Incl O&P
0010 **AIR-SOURCE HEAT PUMPS,** Not including interconnecting tubing									
1000 Air to air, split system, not including curbs, pads, fan coil and ductwork									
1010 For curbs/pads see Section 23 91 00									
1012 Outside condensing unit only, for fan coil see Section 23 82 19.10									
1015 1.5 ton cooling, 7 MBH heat @ 0°F	Q-5	2.40	6.667	Ea.	1,550	365		1,915	2,275
1020 2 ton cooling, 8.5 MBH heat @ 0°F		2	8		1,600	435		2,035	2,400
1030 2.5 ton cooling, 10 MBH heat @ 0°F		1.60	10		1,750	545		2,295	2,750
1040 3 ton cooling, 13 MBH heat @ 0°F		1.20	13.333		2,050	730		2,780	3,350
1050 3.5 ton cooling, 18 MBH heat @ 0°F		1	16		2,175	875		3,050	3,725
1054 4 ton cooling, 24 MBH heat @ 0°F		.80	20		2,375	1,100		3,475	4,275
1060 5 ton cooling, 27 MBH heat @ 0°F		.50	32		2,575	1,750		4,325	5,500
1080 7.5 ton cooling, 33 MBH heat @ 0°F		.45	35.556		4,625	1,950		6,575	8,000
1100 10 ton cooling, 50 MBH heat @ 0°F	Q-6	.64	37.500		6,100	2,125		8,225	9,900
1120 15 ton cooling, 64 MBH heat @ 0°F		.50	48		8,450	2,725		11,175	13,400
1130 20 ton cooling, 85 MBH heat @ 0°F		.35	68.571		17,000	3,875		20,875	24,600
1140 25 ton cooling, 119 MBH heat @ 0°F		.25	96		20,400	5,450		25,850	30,700
1500 Single package, not including curbs, pads, or plenums									
1502 1/2 ton cooling, supplementary heat included	Q-5	8	2	Ea.	2,550	109		2,659	2,975
1504 3/4 ton cooling, supplementary heat supplementary heat included		6	2.667		2,675	146		2,821	3,175
1506 1 ton cooling, supplementary heat included		4	4		2,900	219		3,119	3,525
1510 1.5 ton cooling, 5 MBH heat @ 0°F		1.55	10.323		3,100	565		3,665	4,275
1520 2 ton cooling, 6.5 MBH heat @ 0°F		1.50	10.667		3,100	585		3,685	4,300
1540 2.5 ton cooling, 8 MBH heat @ 0°F		1.40	11.429		3,175	625		3,800	4,450
1560 3 ton cooling, 10 MBH heat @ 0°F		1.20	13.333		3,575	730		4,305	5,025
1570 3.5 ton cooling, 11 MBH heat @ 0°F		1	16		3,900	875		4,775	5,600
1580 4 ton cooling, 13 MBH heat @ 0°F		.96	16.667		4,200	910		5,110	6,000
1620 5 ton cooling, 27 MBH heat @ 0°F		.65	24.615		4,800	1,350		6,150	7,300
1640 7.5 ton cooling, 35 MBH heat @ 0°F		.40	40		7,325	2,175		9,500	11,400
1648 10 ton cooling, 45 MBH heat @ 0°F	Q-6	.40	60		9,925	3,400		13,325	16,000
1652 12 ton cooling, 50 MBH heat @ 0°F	"	.36	66.667		12,200	3,775		15,975	19,100
1696 Supplementary electric heat coil incl., unless noted otherwise									
6000 Air to water, single package, excluding. storage tank and ductwork									
6010 Includes circulating water pump, air duct connections, digital temperature									
6020 controller with remote tank temp. probe and sensor for storage tank.									
6040 Water heating - air cooling capacity									
6110 35.5 MBH heat water, 2.3 Ton cool air	Q-5	1.60	10	Ea.	17,100	545		17,645	19,600

23 81 Decentralized Unitary HVAC Equipment

23 81 43 – Air-Source Unitary Heat Pumps

23 81 43.10 Air-Source Heat Pumps

		Crew	Daily Output	Labor-Hours	Unit	Material	2016 Bare Costs Labor	2016 Bare Costs Equipment	Total	Total Incl O&P
6120	58 MBH heat water, 3.8 Ton cool air	Q-5	1.10	14.545	Ea.	19,500	795		20,295	22,700
6130	76 MBH heat water, 4.9 Ton cool air		.87	18.391		23,300	1,000		24,300	27,100
6140	98 MBH heat water, 6.5 Ton cool air		.62	25.806		30,100	1,400		31,500	35,200
6150	113 MBH heat water, 7.4 Ton cool air		.59	27.119		33,000	1,475		34,475	38,500
6160	142 MBH heat water, 9.2 Ton cool air		.52	30.769		40,000	1,675		41,675	46,500
6170	171 MBH heat water, 11.1 Ton cool air	▼	.49	32.653	▼	47,100	1,775		48,875	54,500

23 81 46 – Water-Source Unitary Heat Pumps

23 81 46.10 Water Source Heat Pumps

		Crew	Daily Output	Labor-Hours	Unit	Material	2016 Bare Costs Labor	2016 Bare Costs Equipment	Total	Total Incl O&P
0010	**WATER SOURCE HEAT PUMPS**, Not incl. connecting tubing or water source									
2000	Water source to air, single package									
2100	1 ton cooling, 13 MBH heat @ 75°F	Q-5	2	8	Ea.	1,775	435		2,210	2,600
2120	1.5 ton cooling, 17 MBH heat @ 75°F		1.80	8.889		1,775	485		2,260	2,700
2140	2 ton cooling, 19 MBH heat @ 75°F		1.70	9.412		2,200	515		2,715	3,200
2160	2.5 ton cooling, 25 MBH heat @ 75°F		1.60	10		2,400	545		2,945	3,475
2180	3 ton cooling, 27 MBH heat @ 75°F		1.40	11.429		2,450	625		3,075	3,650
2190	3.5 ton cooling, 29 MBH heat @ 75°F		1.30	12.308		2,675	670		3,345	3,975
2200	4 ton cooling, 31 MBH heat @ 75°F		1.20	13.333		2,975	730		3,705	4,375
2220	5 ton cooling, 29 MBH heat @ 75°F		.90	17.778		3,275	970		4,245	5,075
2240	7.5 ton cooling, 35 MBH heat @ 75°F		.60	26.667		6,825	1,450		8,275	9,700
2250	8 ton cooling, 40 MBH heat @ 75°F		.58	27.586		7,050	1,500		8,550	10,000
2260	10 ton cooling, 50 MBH heat @ 75°F	▼	.53	30.189		7,700	1,650		9,350	11,000
2280	15 ton cooling, 64 MBH heat @ 75°F	Q-6	.47	51.064		14,000	2,900		16,900	19,800
2300	20 ton cooling, 100 MBH heat @ 75°F		.41	58.537		15,300	3,325		18,625	21,800
2310	25 ton cooling, 100 MBH heat @ 75°F		.32	75		20,800	4,250		25,050	29,300
2320	30 ton cooling, 128 MBH heat @ 75°F		.24	102		22,800	5,775		28,575	33,800
2340	40 ton cooling, 200 MBH heat @ 75°F		.21	117		32,200	6,625		38,825	45,400
2360	50 ton cooling, 200 MBH heat @ 75°F	▼	.15	160		36,400	9,075		45,475	53,500
3960	For supplementary heat coil, add				▼	10%				
4000	For increase in capacity thru use									
4020	of solar collector, size boiler at 60%									

23 82 Convection Heating and Cooling Units

23 82 13 – Valance Heating and Cooling Units

23 82 13.16 Valance Units

		Crew	Daily Output	Labor-Hours	Unit	Material	2016 Bare Costs Labor	2016 Bare Costs Equipment	Total	Total Incl O&P
0010	**VALANCE UNITS**									
6000	Valance units, complete with 1/2" cooling coil, enclosure									
6020	2 tube	Q-5	18	.889	L.F.	33.50	48.50		82	111
6040	3 tube		16	1		40	54.50		94.50	127
6060	4 tube		16	1		48.50	54.50		103	136
6080	5 tube		15	1.067		52	58.50		110.50	145
6100	6 tube		15	1.067		57.50	58.50		116	151
6120	8 tube	▼	14	1.143		83.50	62.50		146	186
6200	For 3/4" cooling coil, add				▼	10%				

For customer support on your Mechanical Cost Data, call 877.739.3689.

401

23 82 16 – Air Coils

23 82 16.10 Flanged Coils	Crew	Daily Output	Labor-Hours	Unit	Material	2016 Bare Costs Labor	Equipment	Total	Total Incl O&P
0010 **FLANGED COILS**									
0100 Basic water, DX, or condenser coils									
0110 Copper tubes, alum. fins, galv. end sheets									
0112 H is finned height, L is finned length									
0120 3/8" x .016 tube, .0065 aluminum fins									
0130 2 row, 8 fins per inch									
0140 4" H x 12" L	Q-5	48	.333	Ea.	1,050	18.20		1,068.20	1,175
0150 4" H x 24" L		38.38	.417		1,100	23		1,123	1,225
0160 4" H x 48" L		19.25	.831		1,150	45.50		1,195.50	1,350
0170 4" H x 72" L		12.80	1.250		1,275	68.50		1,343.50	1,500
0180 6" H x 12" L		48	.333		1,075	18.20		1,093.20	1,200
0190 6" H x 24" L		25.60	.625		1,150	34		1,184	1,300
0200 6" H x 48" L		12.80	1.250		1,200	68.50		1,268.50	1,425
0210 6" H x 72" L		8.53	1.876		1,350	102		1,452	1,650
0220 10" H x 12" L		30.73	.521		1,150	28.50		1,178.50	1,300
0230 10" H x 24" L		15.33	1.044		1,225	57		1,282	1,425
0240 10" H x 48" L		7.69	2.081		1,325	114		1,439	1,625
0250 10" H x 72" L		5.12	3.125		1,500	171		1,671	1,900
0260 12" H x 12" L		25.60	.625		1,175	34		1,209	1,350
0270 12" H x 24" L		12.80	1.250		1,275	68.50		1,343.50	1,500
0280 12" H x 48" L		6.40	2.500		1,375	137		1,512	1,725
0290 12" H x 72" L		4.27	3.747		1,575	205		1,780	2,025
0300 16" H x 16" L		14.38	1.113		1,250	61		1,311	1,475
0310 16" H x 24" L		9.59	1.668		1,350	91		1,441	1,625
0320 16" H x 48" L		4.80	3.333		1,500	182		1,682	1,925
0330 16" H x 72" L		3.20	5		1,725	273		1,998	2,300
0340 20" H x 24" L		7.69	2.081		1,425	114		1,539	1,750
0350 20" H x 48" L		3.84	4.167		1,600	228		1,828	2,125
0360 20" H x 72" L		2.56	6.250		1,875	340		2,215	2,600
0370 24" H x 24" L		6.40	2.500		1,525	137		1,662	1,875
0380 24" H x 48" L		3.20	5		1,725	273		1,998	2,300
0390 24" H x 72" L		2.13	7.512		2,025	410		2,435	2,850
0400 30" H x 28" L		4.39	3.645		1,675	199		1,874	2,150
0410 30" H x 48" L		2.56	6.250		1,925	340		2,265	2,625
0420 30" H x 72" L		1.71	9.357		2,250	510		2,760	3,250
0430 30" H x 84" L		1.46	10.959		2,425	600		3,025	3,575
0440 34" H x 32" L		3.39	4.720		1,825	258		2,083	2,400
0450 34" H x 48" L		2.26	7.080		2,050	385		2,435	2,825
0460 34" H x 72" L		1.51	10.596		2,400	580		2,980	3,525
0470 34" H x 84" L	▼	1.29	12.403	▼	2,600	680		3,280	3,875
0600 For 10 fins per inch, add					1%				
0610 For 12 fins per inch, add					3%				
0620 For 4 row, add				Ea.	30%	50%			
0630 For 6 row, add					60%	100%			
0640 For 8 row, add				▼	80%	150%			
1100 1/2" x .017 tube, .0065 aluminum fins									
1110 2 row, 8 fins per inch									
1120 5" H x 6" L	Q-5	40	.400	Ea.	1,050	22		1,072	1,175
1130 5" H x 20" L		37.10	.431		1,100	23.50		1,123.50	1,225
1140 5" H x 45" L		16.41	.975		1,175	53.50		1,228.50	1,375
1150 5" H x 90" L		8.21	1.949		1,425	106		1,531	1,725
1160 7.5" H x 5" L	▼	38.50	.416	▼	1,075	22.50		1,097.50	1,200

23 82 Convection Heating and Cooling Units

23 82 16 – Air Coils

23 82 16.10 Flanged Coils		Crew	Daily Output	Labor-Hours	Unit	Material	2016 Bare Costs Labor	Equipment	Total	Total Incl O&P
1170	7.5" H x 20" L	Q-5	24.62	.650	Ea.	1,150	35.50		1,185.50	1,300
1180	7.5" H x 45" L		10.94	1.463		1,175	80		1,255	1,425
1190	7.5" H x 90" L		5.46	2.930		1,550	160		1,710	1,950
1200	12.5" H x 10" L		29.49	.543		1,150	29.50		1,179.50	1,325
1210	12.5" H x 20" L		14.75	1.085		1,250	59.50		1,309.50	1,475
1220	12.5" H x 45" L		6.55	2.443		1,375	133		1,508	1,725
1230	12.5" H x 90" L		3.28	4.878		1,775	266		2,041	2,350
1240	15" H x 15" L		16.41	.975		1,225	53.50		1,278.50	1,425
1250	15" H x 30" L		8.21	1.949		1,350	106		1,456	1,650
1260	15" H x 45" L		5.46	2.930		1,450	160		1,610	1,850
1270	15" H x 90" L		2.73	5.861		1,900	320		2,220	2,550
1280	20" H x 20" L		9.22	1.735		1,375	95		1,470	1,675
1290	20" H x 45" L		4.10	3.902		1,600	213		1,813	2,075
1300	20" H x 90" L		2.05	7.805		2,100	425		2,525	2,975
1310	25" H x 25" L		5.90	2.712		1,525	148		1,673	1,925
1320	25" H x 45" L		3.28	4.878		1,750	266		2,016	2,325
1330	25" H x 90" L		1.64	9.756		2,325	535		2,860	3,375
1340	30" H x 30" L		4.10	3.902		1,700	213		1,913	2,200
1350	30" H x 45" L		2.73	5.861		1,900	320		2,220	2,550
1360	30" H x 90" L		1.37	11.679		2,550	640		3,190	3,775
1370	35" H x 35" L		3.01	5.316		2,275	290		2,565	2,950
1380	35" H x 45" L		2.34	6.838		2,025	375		2,400	2,800
1390	35" H x 90" L		1.17	13.675		2,775	745		3,520	4,175
1400	42.5" H x 45" L		1.93	8.290		2,250	455		2,705	3,150
1410	42.5" H x 75" L		1.16	13.793		2,800	755		3,555	4,200
1420	42.5" H x 105" L		.83	19.277		3,350	1,050		4,400	5,275
1600	For 10 fins per inch, add					2%				
1610	For 12 fins per inch, add					5%				
1620	For 4 row, add					35%	50%			
1630	For 6 row, add					65%	100%			
1640	For 8 row, add					90%	150%			
2010	5/8" x .020" tube, .0065 aluminum fins									
2020	2 row, 8 fins per inch									
2030	6" H x 6" L	Q-5	38	.421	Ea.	940	23		963	1,050
2040	6" H x 24" L		24.96	.641		1,000	35		1,035	1,150
2050	6" H x 48" L		12.48	1.282		1,075	70		1,145	1,275
2060	6" H x 96" L		6.24	2.564		1,325	140		1,465	1,650
2070	12" H x 12" L		24.96	.641		1,025	35		1,060	1,175
2080	12" H x 24" L		12.48	1.282		1,100	70		1,170	1,325
2090	12" H x 48" L		6.24	2.564		1,225	140		1,365	1,550
2100	12" H x 96" L		3.12	5.128		2,225	280		2,505	2,875
2110	18" H x 24" L		8.32	1.923		1,225	105		1,330	1,475
2120	18" H x 48" L		4.16	3.846		1,350	210		1,560	1,825
2130	18" H x 96" L		2.08	7.692		1,800	420		2,220	2,600
2140	24" H x 24" L		6.24	2.564		1,325	140		1,465	1,650
2150	24" H x 48" L		3.12	5.128		1,500	280		1,780	2,075
2160	24" H x 96" L		1.56	10.256		2,025	560		2,585	3,075
2170	30" H x 30" L		3.99	4.010		1,475	219		1,694	1,950
2180	30" H x 48" L		2.50	6.400		1,650	350		2,000	2,350
2190	30" H x 96" L		1.25	12.800		2,250	700		2,950	3,500
2200	30" H x 120" L		1	16		2,525	875		3,400	4,100
2210	36" H x 36" L		2.77	5.776		1,675	315		1,990	2,300
2220	36" H x 48" L		2.08	7.692		1,800	420		2,220	2,600

For customer support on your Mechanical Cost Data, call 877.739.3689.

403

23 82 16 – Air Coils

23 82 16.10 Flanged Coils	Crew	Daily Output	Labor-Hours	Unit	Material	2016 Bare Costs Labor	Equipment	Total	Total Incl O&P
2230	36" H x 90" L	Q-5	1.11	14.414	Ea.	2,400	785		3,185	3,825
2240	36" H x 120" L		.83	19.277		2,775	1,050		3,825	4,650
2250	42" H x 48" L		1.78	8.989		1,950	490		2,440	2,900
2260	42" H x 90" L		.95	16.842		2,600	920		3,520	4,275
2270	42" H x 120" L		.71	22.535		3,200	1,225		4,425	5,375
2280	51" H x 48" L		1.47	10.884		2,200	595		2,795	3,325
2290	51" H x 96" L		.73	21.918		3,300	1,200		4,500	5,450
2300	51" H x 120" L	▼	.59	27.119	▼	3,775	1,475		5,250	6,375
2500	For 10 fins per inch, add					3%				
2510	For 12 fins per inch, add					6%				
2520	For 4 row, add				Ea.	40%	50%			
2530	For 6 row, add					80%	100%			
2540	For 8 row, add				▼	110%	150%			
3000	Hot water booster coils									
3010	Copper tubes, alum. fins, galv. end sheets									
3012	H is finned height, L is finned length									
3020	1/2" x .017" tube, .0065 alum. fins									
3030	1 row, 10 fins per inch									
3040	5"H x 10"L	Q-5	40	.400	Ea.	755	22		777	865
3050	5"H x 25"L		33.18	.482		800	26.50		826.50	920
3060	10"H x 10"L		39	.410		785	22.50		807.50	900
3070	10"H x 20"L		20.73	.772		830	42		872	975
3080	10"H x 30"L		13.83	1.157		865	63		928	1,050
3090	15"H x 15"L		18.44	.868		830	47.50		877.50	980
3110	15"H x 20"L		13.83	1.157		870	63		933	1,050
3120	15"H x 30"L		9.22	1.735		915	95		1,010	1,150
3130	20"H x 20"L		10.37	1.543		915	84.50		999.50	1,125
3140	20"H x 30"L		6.91	2.315		970	126		1,096	1,275
3150	20"H x 40"L		5.18	3.089		1,025	169		1,194	1,375
3160	25"H x 25"L		6.64	2.410		990	132		1,122	1,300
3170	25"H x 35"L		4.74	3.376		1,025	184		1,209	1,400
3180	25"H x 45"L		3.69	4.336		1,100	237		1,337	1,550
3190	30"H x 30"L		4.61	3.471		1,075	190		1,265	1,450
3200	30"H x 40"L		3.46	4.624		1,125	253		1,378	1,625
3210	30"H x 50"L	▼	2.76	5.797	▼	1,200	315		1,515	1,800
3300	For 2 row, add				▼	20%	50%			
3400	1/2" x .020" tube, .008 alum. fins									
3410	2 row, 10 fins per inch									
3420	20" H x 27" L	Q-5	7.27	2.201	Ea.	1,125	120		1,245	1,425
3430	20" H x 36" L		5.44	2.941		1,200	161		1,361	1,575
3440	20" H x 45" L		4.35	3.678		1,275	201		1,476	1,700
3450	25" H x 45" L		3.49	4.585		1,400	250		1,650	1,900
3460	25" H x 54" L		2.90	5.517		1,600	300		1,900	2,200
3470	30" H x 54" L		2.42	6.612		1,725	360		2,085	2,450
3480	35" H x 54" L		2.07	7.729		1,875	420		2,295	2,675
3490	35" H x 63" L		1.78	8.989		2,000	490		2,490	2,950
3500	40" H x 72" L		1.36	11.765		2,300	645		2,945	3,500
3510	55" H x 63" L		1.13	14.159		2,700	775		3,475	4,150
3520	55" H x 72" L		.99	16.162		2,875	885		3,760	4,500
3530	55" H x 81" L		.88	18.182		3,125	995		4,120	4,950
3540	75" H x 81" L		.64	25		3,925	1,375		5,300	6,375
3550	75" H x 90" L		.58	27.586		4,175	1,500		5,675	6,875
3560	75" H x 108" L	▼	.48	33.333	▼	4,775	1,825		6,600	8,000

For customer support on your Mechanical Cost Data, call 877.739.3689.

23 82 Convection Heating and Cooling Units

23 82 16 – Air Coils

23 82 16.10 Flanged Coils	Crew	Daily Output	Labor-Hours	Unit	Material	2016 Bare Costs Labor	Equipment	Total	Total Incl O&P
3800	5/8" x .020 tubing, .0065 aluminum fins								
3810	1 row, 10 fins per inch								
3820	6" H x 12" L	Q-5	38	.421	Ea.	680	23		703	785
3830	6" H x 24" L		28.80	.556		720	30.50		750.50	835
3840	6" H x 30" L		23.04	.694		735	38		773	860
3850	12" H x 12" L		28.80	.556		715	30.50		745.50	830
3860	12" H x 24" L		14.40	1.111		765	60.50		825.50	930
3870	12" H x 30" L		11.52	1.389		785	76		861	975
3880	18" H x 18" L		12.80	1.250		760	68.50		828.50	940
3890	18" H x 24" L		9.60	1.667		810	91		901	1,025
3900	18" H x 36" L		6.40	2.500		860	137		997	1,150
3910	24" H x 24" L		7.20	2.222		855	121		976	1,125
3920	24" H x 30" L		4.80	3.333		890	182		1,072	1,250
3930	24" H x 36" L		5.05	3.168		915	173		1,088	1,250
3940	30" H x 30" L		4.61	3.471		940	190		1,130	1,300
3950	30" H x 36" L		3.85	4.159		975	227		1,202	1,425
3960	30" H x 42" L		3.29	4.863		1,000	266		1,266	1,500
3970	36" H x 36" L		3.20	5		1,025	273		1,298	1,525
3980	36" H x 42" L		2.74	5.839		1,075	320		1,395	1,650
4100	For 2 row, add					30%	50%			
4400	Steam and hot water heating coils									
4410	Copper tubes, alum. fins. galv. end sheets									
4412	H is finned height, L is finned length									
4420	5/8" tubing, to 175 PSIG working pressure									
4430	1 row, 8 fins per inch									
4440	6"H x 6"L	Q-5	38	.421	Ea.	1,025	23		1,048	1,150
4450	6"H x 12"L		37	.432		1,050	23.50		1,073.50	1,175
4460	6"H x 24"L		24.48	.654		1,100	35.50		1,135.50	1,275
4470	6"H x 36"L		16.32	.980		1,150	53.50		1,203.50	1,325
4480	6"H x 48"L		12.24	1.307		1,200	71.50		1,271.50	1,400
4490	6"H x 84"L		6.99	2.289		1,575	125		1,700	1,925
4500	6" H x 108" L		5.44	2.941		1,750	161		1,911	2,175
4510	12" H x 12" L		24.48	.654		1,175	35.50		1,210.50	1,325
4520	12" H x 24" L		12.24	1.307		1,250	71.50		1,321.50	1,475
4530	12" H x 36" L		8.16	1.961		1,325	107		1,432	1,600
4540	12" H x 48" L		6.12	2.614		1,400	143		1,543	1,750
4550	12" H x 84" L		3.50	4.571		1,950	250		2,200	2,500
4560	12" H x 108" L		2.72	5.882		2,225	320		2,545	2,925
4570	18" H x 18" L		10.88	1.471		1,325	80.50		1,405.50	1,600
4580	18" H x 24" L		8.16	1.961		1,425	107		1,532	1,725
4590	18" H x 48" L		4.08	3.922		1,625	214		1,839	2,100
4600	18" H x 84" L		2.33	6.867		2,300	375		2,675	3,125
4610	18" H x 108" L		1.81	8.840		2,650	485		3,135	3,650
4620	24" H x 24" L		6.12	2.614		1,650	143		1,793	2,050
4630	24" H x 48" L		3.06	5.229		1,925	286		2,211	2,525
4640	24" H x 84" L		1.75	9.143		2,750	500		3,250	3,775
4650	24" H x 108" L		1.36	11.765		3,200	645		3,845	4,500
4660	36" H x 36" L		2.72	5.882		2,300	320		2,620	3,000
4670	36" H x 48" L		2.04	7.843		2,450	430		2,880	3,350
4680	36" H x 84" L		1.17	13.675		3,625	745		4,370	5,125
4690	36" H x 108" L		.91	17.582		4,250	960		5,210	6,125
4700	48"H x 48"L		1.53	10.458		2,900	570		3,470	4,050
4710	48"H x 84"L		.87	18.391		4,300	1,000		5,300	6,250

For customer support on your Mechanical Cost Data, call 877.739.3689.

405

23 82 Convection Heating and Cooling Units

23 82 16 – Air Coils

23 82 16.10 Flanged Coils	Crew	Daily Output	Labor-Hours	Unit	Material	2016 Bare Costs Labor	Equipment	Total	Total Incl O&P
4720	48"H x 108"L	Q-5	.68	23.529	Ea.	5,100	1,275		6,375	7,575
4730	48"H x 126"L	↓	.58	27.586	↓	5,475	1,500		6,975	8,275
4910	For 10 fins per inch, add					2%				
4920	For 12 fins per inch, add					5%				
4930	For 2 row, add					30%	50%			
4940	For 3 row, add					50%	100%			
4950	For 4 row, add					80%	123%			
4960	For 230 PSIG heavy duty, add					20%	30%			
6000	Steam, heavy duty, 200 PSIG									
6010	Copper tubes, alum. fins, galv. end sheets									
6012	H is finned height, L is finned length									
6020	1" x .035" tubing, .010 aluminum fins									
6030	1 row, 8 fins per inch									
6040	12"H x 12"L	Q-5	24.48	.654	Ea.	1,225	35.50		1,260.50	1,375
6050	12"H x 24"L		12.24	1.307		1,375	71.50		1,446.50	1,625
6060	12"H x 36"L		8.16	1.961		1,500	107		1,607	1,800
6070	12"H x 48"L		6.12	2.614		1,650	143		1,793	2,050
6080	12"H x 84"L		3.50	4.571		2,200	250		2,450	2,800
6090	12"H x 108"L		2.72	5.882		2,550	320		2,870	3,300
6100	18"H x 24"L		8.16	1.961		1,650	107		1,757	1,975
6110	18"H x 36"L		5.44	2.941		1,850	161		2,011	2,275
6120	18"H x 48"L		4.08	3.922		2,050	214		2,264	2,600
6130	18"H x 84"L		2.33	6.867		2,800	375		3,175	3,650
6140	18"H x 108"L		1.81	8.840		3,275	485		3,760	4,325
6150	24"H x 24"L		6.12	2.614		1,850	143		1,993	2,250
6160	24"H x 36"L		4.08	3.922		2,100	214		2,314	2,650
6170	24"H x 48"L		3.06	5.229		2,400	286		2,686	3,050
6180	24"H x 84"L		1.75	9.143		3,300	500		3,800	4,375
6190	24"H x 108"L		1.36	11.765		3,900	645		4,545	5,275
6200	30"H x 36"L		3.26	4.908		2,525	268		2,793	3,175
6210	30"H x 48"L		2.45	6.531		2,850	355		3,205	3,700
6220	30"H x 84"L		1.40	11.429		3,950	625		4,575	5,300
6230	30"H x 108"L		1.09	14.679		4,700	800		5,500	6,375
6240	36"H x 36"L		2.72	5.882		2,800	320		3,120	3,550
6250	36"H x 48"L		2.04	7.843		3,150	430		3,580	4,125
6260	36"H x 84"L		1.17	13.675		4,475	745		5,220	6,050
6270	36"H x 108"L		.91	17.582		5,350	960		6,310	7,350
6280	36"H x 120"L		.82	19.512		5,750	1,075		6,825	7,925
6290	48"H x 48"L		1.53	10.458		4,075	570		4,645	5,350
6300	48"H x 84"L		.87	18.391		5,750	1,000		6,750	7,850
6310	48"H x 108"L		.68	23.529		6,900	1,275		8,175	9,525
6320	48"H x 120"L	↓	.61	26.230		7,400	1,425		8,825	10,300
6400	For 10 fins per inch, add					4%				
6410	For 12 fins per inch, add				↓	10%				

23 82 16.20 Duct Heaters

		Crew	Daily Output	Labor-Hours	Unit	Material	2016 Bare Costs Labor	Equipment	Total	Total Incl O&P
0010	**DUCT HEATERS**, Electric, 480 V, 3 Ph.									
0020	Finned tubular insert, 500°F									
0100	8" wide x 6" high, 4.0 kW	Q-20	16	1.250	Ea.	765	65.50		830.50	945
0120	12" high, 8.0 kW		15	1.333		1,275	69.50		1,344.50	1,500
0140	18" high, 12.0 kW		14	1.429		1,775	74.50		1,849.50	2,075
0160	24" high, 16.0 kW		13	1.538		2,300	80.50		2,380.50	2,650
0180	30" high, 20.0 kW	↓	12	1.667		2,800	87		2,887	3,200

23 82 Convection Heating and Cooling Units

23 82 16 – Air Coils

23 82 16.20 Duct Heaters		Crew	Daily Output	Labor-Hours	Unit	Material	2016 Bare Costs Labor	Equipment	Total	Total Incl O&P
0300	12" wide x 6" high, 6.7 kW	Q-20	15	1.333	Ea.	815	69.50		884.50	1,000
0320	12" high, 13.3 kW		14	1.429		1,325	74.50		1,399.50	1,575
0340	18" high, 20.0 kW		13	1.538		1,850	80.50		1,930.50	2,150
0360	24" high, 26.7 kW		12	1.667		2,375	87		2,462	2,725
0380	30" high, 33.3 kW		11	1.818		2,900	95		2,995	3,325
0500	18" wide x 6" high, 13.3 kW		14	1.429		880	74.50		954.50	1,075
0520	12" high, 26.7 kW		13	1.538		1,500	80.50		1,580.50	1,775
0540	18" high, 40.0 kW		12	1.667		2,000	87		2,087	2,325
0560	24" high, 53.3 kW		11	1.818		2,675	95		2,770	3,075
0580	30" high, 66.7 kW		10	2		3,325	104		3,429	3,825
0700	24" wide x 6" high, 17.8 kW		13	1.538		965	80.50		1,045.50	1,200
0720	12" high, 35.6 kW		12	1.667		1,625	87		1,712	1,925
0740	18" high, 53.3 kW		11	1.818		2,300	95		2,395	2,675
0760	24" high, 71.1 kW		10	2		2,950	104		3,054	3,400
0780	30" high, 88.9 kW		9	2.222		3,625	116		3,741	4,175
0900	30" wide x 6" high, 22.2 kW		12	1.667		1,025	87		1,112	1,250
0920	12" high, 44.4 kW		11	1.818		1,725	95		1,820	2,050
0940	18" high, 66.7 kW		10	2		2,425	104		2,529	2,825
0960	24" high, 88.9 kW		9	2.222		3,125	116		3,241	3,625
0980	30" high, 111.0 kW		8	2.500		3,825	131		3,956	4,425
1400	Note decreased kW available for									
1410	each duct size at same cost									
1420	See line 5000 for modifications and accessories									
2000	Finned tubular flange with insulated									
2020	terminal box, 500°F									
2100	12" wide x 36" high, 54 kW	Q-20	10	2	Ea.	3,775	104		3,879	4,300
2120	40" high, 60 kW		9	2.222		4,350	116		4,466	4,950
2200	24" wide x 36" high, 118.8 kW		9	2.222		4,850	116		4,966	5,500
2220	40" high, 132 kW		8	2.500		4,825	131		4,956	5,500
2400	36" wide x 8" high, 40 kW		11	1.818		1,900	95		1,995	2,250
2420	16" high, 80 kW		10	2		2,575	104		2,679	2,975
2440	24" high, 120 kW		9	2.222		3,350	116		3,466	3,875
2460	32" high, 160 kW		8	2.500		4,300	131		4,431	4,925
2480	36" high, 180 kW		7	2.857		5,300	149		5,449	6,050
2500	40" high, 200 kW		6	3.333		5,875	174		6,049	6,725
2600	40" wide x 8" high, 45 kW		11	1.818		2,000	95		2,095	2,350
2620	16" high, 90 kW		10	2		2,775	104		2,879	3,200
2640	24" high, 135 kW		9	2.222		3,500	116		3,616	4,025
2660	32" high, 180 kW		8	2.500		4,675	131		4,806	5,350
2680	36" high, 202.5 kW		7	2.857		5,375	149		5,524	6,150
2700	40" high, 225 kW		6	3.333		6,125	174		6,299	7,025
2800	48" wide x 8" high, 54.8 kW		10	2		2,100	104		2,204	2,475
2820	16" high, 109.8 kW		9	2.222		2,925	116		3,041	3,375
2840	24" high, 164.4 kW		8	2.500		3,725	131		3,856	4,275
2860	32" high, 219.2 kW		7	2.857		4,950	149		5,099	5,675
2880	36" high, 246.6 kW		6	3.333		5,700	174		5,874	6,550
2900	40" high, 274 kW		5	4		6,450	209		6,659	7,400
3000	56" wide x 8" high, 64 kW		9	2.222		2,400	116		2,516	2,800
3020	16" high, 128 kW		8	2.500		3,325	131		3,456	3,850
3040	24" high, 192 kW		7	2.857		4,025	149		4,174	4,650
3060	32" high, 256 kW		6	3.333		5,625	174		5,799	6,450
3080	36" high, 288kW		5	4		6,450	209		6,659	7,425
3100	40" high, 320kW		4	5		7,125	261		7,386	8,250

For customer support on your Mechanical Cost Data, call 877.739.3689.

407

23 82 Convection Heating and Cooling Units

23 82 16 – Air Coils

23 82 16.20 Duct Heaters

		Crew	Daily Output	Labor-Hours	Unit	Material	2016 Bare Costs Labor	Equipment	Total	Total Incl O&P
3200	64" wide x 8" high, 74kW	Q-20	8	2.500	Ea.	2,475	131		2,606	2,900
3220	16" high, 148kW		7	2.857		3,425	149		3,574	3,975
3240	24" high, 222kW		6	3.333		4,350	174		4,524	5,050
3260	32" high, 296kW		5	4		5,825	209		6,034	6,725
3280	36" high, 333kW		4	5		6,950	261		7,211	8,050
3300	40" high, 370kW		3	6.667		7,675	350		8,025	8,950
3800	Note decreased kW available for									
3820	each duct size at same cost									
5000	Duct heater modifications and accessories									
5120	T.C.O. limit auto or manual reset	Q-20	42	.476	Ea.	126	25		151	176
5140	Thermostat		28	.714		535	37.50		572.50	645
5160	Overheat thermocouple (removable)		7	2.857		760	149		909	1,050
5180	Fan interlock relay		18	1.111		183	58		241	291
5200	Air flow switch		20	1		158	52		210	254
5220	Split terminal box cover		100	.200		52.50	10.45		62.95	73.50
8000	To obtain BTU multiply kW by 3413									

23 82 19 – Fan Coil Units

23 82 19.10 Fan Coil Air Conditioning

		Crew	Daily Output	Labor-Hours	Unit	Material	2016 Bare Costs Labor	Equipment	Total	Total Incl O&P
0010	**FAN COIL AIR CONDITIONING**									
0030	Fan coil AC, cabinet mounted, filters and controls									
0100	Chilled water, 1/2 ton cooling	Q-5	8	2	Ea.	460	109		569	670
0110	3/4 ton cooling		7	2.286		565	125		690	810
0120	1 ton cooling		6	2.667		685	146		831	975
0140	1.5 ton cooling		5.50	2.909		690	159		849	1,000
0150	2 ton cooling		5.25	3.048		995	166		1,161	1,350
0160	2.5 ton cooling		5	3.200		1,450	175		1,625	1,875
0180	3 ton cooling		4	4		1,625	219		1,844	2,100
0262	For hot water coil, add					40%	10%			
0300	Console, 2 pipe with electric heat									
0310	1/2 ton cooling	Q-5	8	2	Ea.	1,325	109		1,434	1,625
0315	3/4 ton cooling		7	2.286		1,450	125		1,575	1,800
0320	1 ton cooling		6	2.667		1,575	146		1,721	1,950
0330	1.5 ton cooling		5.50	2.909		1,925	159		2,084	2,375
0340	2 ton cooling		4.70	3.404		2,550	186		2,736	3,075
0345	2-1/2 ton cooling		4.70	3.404		3,700	186		3,886	4,350
0350	3 ton cooling		4	4		4,975	219		5,194	5,800
0360	4 ton cooling		4	4		5,200	219		5,419	6,025
0940	Direct expansion, for use w/air cooled condensing unit, 1.5 ton cooling		5	3.200		515	175		690	835
0950	2 ton cooling		4.80	3.333		565	182		747	895
0960	2.5 ton cooling		4.40	3.636		590	199		789	950
0970	3 ton cooling		3.80	4.211		730	230		960	1,150
0980	3.5 ton cooling		3.60	4.444		735	243		978	1,175
0990	4 ton cooling		3.40	4.706		870	257		1,127	1,350
1000	5 ton cooling		3	5.333		1,025	291		1,316	1,575
1020	7.5 ton cooling		3	5.333		1,825	291		2,116	2,450
1040	10 ton cooling	Q-6	2.60	9.231		2,075	525		2,600	3,075
1042	12.5 ton cooling		2.30	10.435		2,625	590		3,215	3,800
1050	15 ton cooling		1.60	15		2,925	850		3,775	4,475
1060	20 ton cooling		.70	34.286		4,000	1,950		5,950	7,325
1070	25 ton cooling		.65	36.923		4,800	2,100		6,900	8,425
1080	30 ton cooling		.60	40		5,200	2,275		7,475	9,150
1500	For hot water coil, add					40%	10%			

23 82 19.10 Fan Coil Air Conditioning

		Crew	Daily Output	Labor-Hours	Unit	Material	2016 Bare Costs Labor	Equipment	Total	Total Incl O&P
1512	For condensing unit add see Section 23 62									
3000	Chilled water, horizontal unit, housing, 2 pipe, fan control, no valves									
3100	1/2 ton cooling	Q-5	8	2	Ea.	1,275	109		1,384	1,575
3110	1 ton cooling		6	2.667		1,500	146		1,646	1,875
3120	1.5 ton cooling		5.50	2.909		1,725	159		1,884	2,150
3130	2 ton cooling		5.25	3.048		2,075	166		2,241	2,525
3140	3 ton cooling		4	4		2,475	219		2,694	3,050
3150	3.5 ton cooling		3.80	4.211		2,725	230		2,955	3,350
3160	4 ton cooling		3.80	4.211		2,725	230		2,955	3,350
3170	5 ton cooling		3.40	4.706		3,175	257		3,432	3,900
3180	6 ton cooling		3.40	4.706		3,175	257		3,432	3,900
3190	7 ton cooling		2.90	5.517		3,450	300		3,750	4,250
3200	8 ton cooling		2.90	5.517		3,450	300		3,750	4,250
3210	10 ton cooling	Q-6	2.80	8.571		3,625	485		4,110	4,700
3212	12.5 ton cooling		2.50	9.600		3,200	545		3,745	4,350
3214	15 ton cooling		2.30	10.435		3,475	590		4,065	4,725
3216	20 ton cooling		2.10	11.429		4,000	645		4,645	5,375
3218	25 ton cooling		1.80	13.333		6,400	755		7,155	8,200
3220	30 ton cooling		1.60	15		6,875	850		7,725	8,850
3240	For hot water coil, add					40%	10%			
4000	With electric heat, 2 pipe									
4100	1/2 ton cooling	Q-5	8	2	Ea.	1,500	109		1,609	1,825
4105	3/4 ton cooling		7	2.286		1,675	125		1,800	2,025
4110	1 ton cooling		6	2.667		1,825	146		1,971	2,250
4120	1.5 ton cooling		5.50	2.909		2,025	159		2,184	2,500
4130	2 ton cooling		5.25	3.048		2,575	166		2,741	3,075
4135	2.5 ton cooling		4.80	3.333		3,550	182		3,732	4,175
4140	3 ton cooling		4	4		4,775	219		4,994	5,575
4150	3.5 ton cooling		3.80	4.211		5,700	230		5,930	6,625
4160	4 ton cooling		3.60	4.444		6,575	243		6,818	7,600
4170	5 ton cooling		3.40	4.706		8,025	257		8,282	9,225
4180	6 ton cooling		3.20	5		8,025	273		8,298	9,225
4190	7 ton cooling		2.90	5.517		8,200	300		8,500	9,450
4200	8 ton cooling		2.20	7.273		8,750	395		9,145	10,300
4210	10 ton cooling	Q-6	2.80	8.571		9,500	485		9,985	11,100

23 82 19.20 Heating and Ventilating Units

		Crew	Daily Output	Labor-Hours	Unit	Material	2016 Bare Costs Labor	Equipment	Total	Total Incl O&P
0010	**HEATING AND VENTILATING UNITS**, Classroom units									
0020	Includes filter, heating/cooling coils, standard controls									
0080	750 CFM, 2 tons cooling	Q-6	2	12	Ea.	4,225	680		4,905	5,675
0100	1000 CFM, 2-1/2 tons cooling		1.60	15		4,975	850		5,825	6,750
0120	1250 CFM, 3 tons cooling		1.40	17.143		5,150	970		6,120	7,125
0140	1500 CFM, 4 tons cooling		.80	30		5,500	1,700		7,200	8,625
0160	2000 CFM 5 tons cooling		.50	48		6,475	2,725		9,200	11,200
0500	For electric heat, add					35%				
1000	For no cooling, deduct					25%	10%			

23 82 19.40 Fan Coil Air Conditioning

		Crew	Daily Output	Labor-Hours	Unit	Material	2016 Bare Costs Labor	Equipment	Total	Total Incl O&P
0010	**FAN COIL AIR CONDITIONING**, Variable refrigerant volume (VRV) type									
0020	Not including interconnecting tubing or multi-zone controls									
0030	For VRV condensing unit see section 23 81 29.10									
0050	Indoor type, ducted									
0100	Vertical concealed									
0130	1 ton cooling	[G] Q-5	2.97	5.387	Ea.	2,475	294		2,769	3,175

23 82 19 – Fan Coil Units

23 82 19.40 Fan Coil Air Conditioning		Crew	Daily Output	Labor-Hours	Unit	Material	2016 Bare Costs Labor	Equipment	Total	Total Incl O&P
0140	1.5 ton cooling	G	Q-5	2.77	5.776	Ea.	2,525	315		2,840	3,250
0150	2 ton cooling	G		2.70	5.926		2,725	325		3,050	3,500
0160	2.5 ton cooling	G		2.60	6.154		2,875	335		3,210	3,650
0170	3 ton cooling	G		2.50	6.400		2,925	350		3,275	3,725
0180	3.5 ton cooling	G		2.30	6.957		3,075	380		3,455	3,950
0190	4 ton cooling	G		2.19	7.306		3,075	400		3,475	4,000
0200	4.5 ton cooling	G	↓	2.08	7.692	↓	3,400	420		3,820	4,350
0230	Outside air connection possible										
1100	Ceiling concealed										
1130	.6 ton cooling	G	Q-5	2.60	6.154	Ea.	1,625	335		1,960	2,275
1140	.75 ton cooling	G		2.60	6.154		1,675	335		2,010	2,350
1150	1 ton cooling	G	↓	2.60	6.154		1,775	335		2,110	2,450
1152	Screening door	G	1 Stpi	5.20	1.538		98	93.50		191.50	249
1160	1.5 ton cooling	G	Q-5	2.60	6.154		1,825	335		2,160	2,525
1162	Screening door	G	1 Stpi	5.20	1.538		113	93.50		206.50	266
1170	2 ton cooling	G	Q-5	2.60	6.154		2,150	335		2,485	2,850
1172	Screening door	G	1 Stpi	5.20	1.538		98	93.50		191.50	249
1180	2.5 ton cooling	G	Q-5	2.60	6.154		2,500	335		2,835	3,250
1182	Screening door	G	1 Stpi	5.20	1.538		170	93.50		263.50	330
1190	3 ton cooling	G	Q-5	2.60	6.154		2,650	335		2,985	3,425
1192	Screening door	G	1 Stpi	5.20	1.538		170	93.50		263.50	330
1200	4 ton cooling	G	Q-5	2.30	6.957		2,725	380		3,105	3,550
1202	Screening door	G	1 Stpi	5.20	1.538	↓	170	93.50		263.50	330
1220	6 ton cooling	G	Q-5	2.08	7.692		4,625	420		5,045	5,700
1230	8 ton cooling	G	"	1.89	8.466	↓	5,200	460		5,660	6,425
1260	Outside air connection possible										
4050	Indoor type, duct-free										
4100	Ceiling mounted cassette										
4130	.75 ton cooling	G	Q-5	3.46	4.624	Ea.	1,875	253		2,128	2,450
4140	1 ton cooling	G		2.97	5.387		2,000	294		2,294	2,650
4150	1.5 ton cooling	G		2.77	5.776		2,050	315		2,365	2,725
4160	2 ton cooling	G		2.60	6.154		2,200	335		2,535	2,925
4170	2.5 ton cooling	G		2.50	6.400		2,250	350		2,600	3,000
4180	3 ton cooling	G		2.30	6.957		2,425	380		2,805	3,225
4190	4 ton cooling	G	↓	2.08	7.692		2,700	420		3,120	3,600
4198	For ceiling cassette decoration panel, add	G	1 Stpi	5.20	1.538	↓	294	93.50		387.50	465
4230	Outside air connection possible										
4500	Wall mounted										
4530	.6 ton cooling	G	Q-5	5.20	3.077	Ea.	995	168		1,163	1,350
4540	.75 ton cooling	G		5.20	3.077		1,050	168		1,218	1,400
4550	1 ton cooling	G		4.16	3.846		1,200	210		1,410	1,625
4560	1.5 ton cooling	G		3.77	4.244		1,300	232		1,532	1,800
4570	2 ton cooling	G	↓	3.46	4.624		1,400	253		1,653	1,925
4590	Condensate pump for the above air handlers	G	1 Stpi	6.46	1.238	↓	211	75		286	345
4700	Floor standing unit										
4730	1 ton cooling	G	Q-5	2.60	6.154	Ea.	1,675	335		2,010	2,350
4740	1.5 ton cooling	G		2.60	6.154		1,900	335		2,235	2,600
4750	2 ton cooling	G	↓	2.60	6.154	↓	2,050	335		2,385	2,750
4800	Floor standing unit, concealed										
4830	1 ton cooling	G	Q-5	2.60	6.154	Ea.	1,600	335		1,935	2,275
4840	1.5 ton cooling	G		2.60	6.154		1,825	335		2,160	2,500
4850	2 ton cooling	G	↓	2.60	6.154	↓	1,950	335		2,285	2,650
4880	Outside air connection possible										

23 82 Convection Heating and Cooling Units

23 82 19 – Fan Coil Units

23 82 19.40 Fan Coil Air Conditioning

23 82 19.40 Fan Coil Air Conditioning	Crew	Daily Output	Labor-Hours	Unit	Material	2016 Bare Costs Labor	Equipment	Total	Total Incl O&P
8000	Accessories									
8100	Branch divergence pipe fitting									
8110	Capacity under 76 MBH	1 Stpi	2.50	3.200	Ea.	151	194		345	460
8120	76 MBH to 112 MBH		2.50	3.200		178	194		372	490
8130	112 MBH to 234 MBH	↓	2.50	3.200	↓	505	194		699	850
8200	Header pipe fitting									
8210	Max 4 branches									
8220	Capacity under 76 MBH	1 Stpi	1.70	4.706	Ea.	221	286		507	675
8260	Max 8 branches									
8270	76 MBH to 112 MBH	1 Stpi	1.70	4.706	Ea.	450	286		736	925
8280	112 MBH to 234 MBH	"	1.30	6.154	"	720	375		1,095	1,350

23 82 27 – Infrared Units

23 82 27.10 Infrared Type Heating Units

	23 82 27.10 Infrared Type Heating Units	Crew	Daily Output	Labor-Hours	Unit	Material	Labor	Equipment	Total	Total Incl O&P
0010	**INFRARED TYPE HEATING UNITS**									
0020	Gas fired, unvented, electric ignition, 100% shutoff.									
0030	Piping and wiring not included									
0100	Input, 30 MBH	Q-5	6	2.667	Ea.	1,075	146		1,221	1,425
0120	45 MBH		5	3.200		1,100	175		1,275	1,475
0140	50 MBH		4.50	3.556		1,100	194		1,294	1,500
0160	60 MBH		4	4		1,100	219		1,319	1,525
0180	75 MBH		3	5.333		1,100	291		1,391	1,650
0200	90 MBH		2.50	6.400		1,100	350		1,450	1,725
0220	105 MBH		2	8		1,100	435		1,535	1,850
0240	120 MBH	↓	2	8	↓	1,800	435		2,235	2,625
1000	Gas fired, vented, electric ignition, tubular									
1020	Piping and wiring not included, 20' to 80' lengths									
1030	Single stage, input, 60 MBH	Q-6	4.50	5.333	Ea.	1,450	300		1,750	2,050
1040	80 MBH		3.90	6.154		1,450	350		1,800	2,125
1050	100 MBH		3.40	7.059		1,450	400		1,850	2,200
1060	125 MBH		2.90	8.276		1,450	470		1,920	2,300
1070	150 MBH		2.70	8.889		1,450	505		1,955	2,350
1080	170 MBH		2.50	9.600		1,450	545		1,995	2,425
1090	200 MBH	↓	2.20	10.909	↓	1,650	620		2,270	2,750
1100	Note: Final pricing may vary due to									
1110	tube length and configuration package selected									
1130	Two stage, input, 60 MBH high, 45 MBH low	Q-6	4.50	5.333	Ea.	1,775	300		2,075	2,400
1140	80 MBH high, 60 MBH low		3.90	6.154		1,775	350		2,125	2,475
1150	100 MBH high, 65 MBH low		3.40	7.059		1,775	400		2,175	2,550
1160	125 MBH high, 95 MBH low		2.90	8.276		1,775	470		2,245	2,650
1170	150 MBH high, 100 MBH low		2.70	8.889		1,775	505		2,280	2,700
1180	170 MBH high, 125 MBH low		2.50	9.600		2,000	545		2,545	3,025
1190	200 MBH high, 150 MBH low	↓	2.20	10.909	↓	2,000	620		2,620	3,125
1220	Note: Final pricing may vary due to									
1230	tube length and configuration package selected									
2000	Electric, single or three phase									
2050	6 kW, 20,478 BTU	1 Elec	2.30	3.478	Ea.	605	192		797	955
2100	13.5 KW, 40,956 BTU		2.20	3.636		960	200		1,160	1,350
2150	24 KW, 81,912 BTU	↓	2	4	↓	2,050	220		2,270	2,575
3000	Oil fired, pump, controls, fusible valve, oil supply tank									
3050	91,000 BTU	Q-5	2.50	6.400	Ea.	8,575	350		8,925	9,950
3080	105,000 BTU		2.25	7.111		9,075	390		9,465	10,600
3110	119,000 BTU	↓	2	8	↓	9,950	435		10,385	11,600

23 82 Convection Heating and Cooling Units

23 82 29 – Radiators

23 82 29.10 Hydronic Heating

		Crew	Daily Output	Labor-Hours	Unit	Material	2016 Bare Costs Labor	Equipment	Total	Total Incl O&P
0010	**HYDRONIC HEATING**, Terminal units, not incl. main supply pipe									
1000	Radiation									
1100	Panel, baseboard, C.I., including supports, no covers	Q-5	46	.348	L.F.	38	19		57	70
3000	Radiators, cast iron									
3100	Free standing or wall hung, 6 tube, 25" high	Q-5	96	.167	Section	47.50	9.10		56.60	66
3150	4 tube 25" high		96	.167		34.50	9.10		43.60	52
3200	4 tube, 19" high	↓	96	.167	↓	30.50	9.10		39.60	48
3250	Adj. brackets, 2 per wall radiator up to 30 sections	1 Stpi	32	.250	Ea.	54.50	15.20		69.70	83
3500	Recessed, 20" high x 5" deep, without grille	Q-5	60	.267	Section	40	14.55		54.55	66
3600	For inlet grille, add				"	2.58			2.58	2.84
9500	To convert SFR to BTU rating: Hot water, 150 x SFR									
9510	Forced hot water, 180 x SFR; steam, 240 x SFR									

23 82 33 – Convectors

23 82 33.10 Convector Units

		Crew	Daily Output	Labor-Hours	Unit	Material	2016 Bare Costs Labor	Equipment	Total	Total Incl O&P
0010	**CONVECTOR UNITS**, Terminal units, not incl. main supply pipe									
2204	Convector, multifin, 2 pipe w/cabinet									
2210	17" H x 24" L	Q-5	10	1.600	Ea.	101	87.50		188.50	243
2214	17" H x 36" L		8.60	1.860		151	102		253	320
2218	17" H x 48" L		7.40	2.162		202	118		320	400
2222	21" H x 24" L		9	1.778		101	97		198	258
2226	21" H x 36" L		8.20	1.951		151	107		258	325
2228	21" H x 48" L	↓	6.80	2.353	↓	202	129		331	415
2240	For knob operated damper, add					140%				
2241	For metal trim strips, add	Q-5	64	.250	Ea.	12.90	13.65		26.55	34.50
2243	For snap-on inlet grille, add					10%	10%			
2245	For hinged access door, add	Q-5	64	.250	Ea.	35.50	13.65		49.15	59.50
2246	For air chamber, auto-venting, add	"	58	.276	"	7.85	15.05		22.90	31

23 82 36 – Finned-Tube Radiation Heaters

23 82 36.10 Finned Tube Radiation

		Crew	Daily Output	Labor-Hours	Unit	Material	2016 Bare Costs Labor	Equipment	Total	Total Incl O&P
0010	**FINNED TUBE RADIATION**, Terminal units, not incl. main supply pipe									
1150	Fin tube, wall hung, 14" slope top cover, with damper									
1200	1-1/4" copper tube, 4-1/4" alum. fin	Q-5	38	.421	L.F.	43	23		66	82
1250	1-1/4" steel tube, 4-1/4" steel fin		36	.444		39.50	24.50		64	80
1255	2" steel tube, 4-1/4" steel fin	↓	32	.500	↓	43	27.50		70.50	88
1260	21", two tier, slope top, w/damper									
1262	1-1/4" copper tube, 4-1/4" alum. fin	Q-5	30	.533	L.F.	50.50	29		79.50	99.50
1264	1-1/4" steel tube, 4-1/4" steel fin		28	.571		47	31		78	99
1266	2" steel tube, 4-1/4" steel fin	↓	25	.640	↓	50.50	35		85.50	109
1270	10", flat top, with damper									
1272	1-1/4" copper tube, 4-1/4" alum. fin	Q-5	38	.421	L.F.	21	23		44	57.50
1274	1-1/4" steel tube, 4-1/4" steel fin		36	.444		39.50	24.50		64	80
1276	2" steel tube, 4-1/4" steel fin	↓	32	.500	↓	43	27.50		70.50	88
1280	17", two tier, flat top, w/damper									
1282	1-1/4" copper tube, 4-1/4" alum. fin	Q-5	30	.533	L.F.	50.50	29		79.50	99.50
1284	1-1/4" steel tube, 4-1/4" steel fin		28	.571		47	31		78	99
1286	2" steel tube, 4-1/4" steel fin	↓	25	.640		50.50	35		85.50	109
1301	Rough in wall hung, steel fin w/supply & balance valves	1 Stpi	.91	8.791	Ea.	298	535		833	1,125
1310	Baseboard, pkgd, 1/2" copper tube, alum. fin, 7" high	Q-5	60	.267	L.F.	11.20	14.55		25.75	34.50
1320	3/4" copper tube, alum. fin, 7" high		58	.276		7.35	15.05		22.40	30.50
1340	1" copper tube, alum. fin, 8-7/8" high		56	.286		18.85	15.60		34.45	44
1360	1-1/4" copper tube, alum. fin, 8-7/8" high	↓	54	.296	↓	28	16.20		44.20	55

23 82 Convection Heating and Cooling Units

23 82 36 – Finned-Tube Radiation Heaters

23 82 36.10 Finned Tube Radiation

		Crew	Daily Output	Labor-Hours	Unit	Material	2016 Bare Costs Labor	2016 Bare Costs Equipment	Total	Total Incl O&P
1380	1-1/4" IPS steel tube with steel fins	Q-5	52	.308	L.F.	27.50	16.80		44.30	56
1381	Rough in baseboard panel & fin tube, supply & balance valves	1 Stpi	1.06	7.547	Ea.	241	460		701	955
1500	Note: fin tube may also require corners, caps, etc.									

23 82 39 – Unit Heaters

23 82 39.13 Cabinet Unit Heaters

		Crew	Daily Output	Labor-Hours	Unit	Material	Labor	Equipment	Total	Total Incl O&P
0010	**CABINET UNIT HEATERS**									
5400	Cabinet, horizontal, hot water, blower type									
5420	20 MBH	Q-5	10	1.600	Ea.	950	87.50		1,037.50	1,175
5430	60 MBH		7	2.286		1,650	125		1,775	2,025
5440	100 MBH		5	3.200		1,925	175		2,100	2,400
5450	120 MBH		4	4		2,025	219		2,244	2,550

23 82 39.16 Propeller Unit Heaters

		Crew	Daily Output	Labor-Hours	Unit	Material	Labor	Equipment	Total	Total Incl O&P
0010	**PROPELLER UNIT HEATERS**									
3950	Unit heaters, propeller, 115 V 2 psi steam, 60°F entering air									
4000	Horizontal, 12 MBH	Q-5	12	1.333	Ea.	385	73		458	535
4020	28.2 MBH		10	1.600		500	87.50		587.50	680
4040	36.5 MBH		8	2		585	109		694	805
4060	43.9 MBH		8	2		585	109		694	805
4080	56.5 MBH		7.50	2.133		620	117		737	855
4100	65.6 MBH		7	2.286		640	125		765	890
4120	87.6 MBH		6.50	2.462		690	134		824	965
4140	96.8 MBH		6	2.667		850	146		996	1,150
4160	133.3 MBH		5	3.200		920	175		1,095	1,275
4180	157.6 MBH		4	4		1,125	219		1,344	1,575
4200	197.7 MBH		3	5.333		1,275	291		1,566	1,850
4220	257.2 MBH		2.50	6.400		1,575	350		1,925	2,275
4240	286.9 MBH		2	8		1,775	435		2,210	2,600
4260	364 MBH		1.80	8.889		2,200	485		2,685	3,150
4270	404 MBH		1.60	10		2,300	545		2,845	3,350
4300	Vertical diffuser same price									
4310	Vertical flow, 40 MBH	Q-5	11	1.455	Ea.	575	79.50		654.50	755
4314	58.5 MBH		8	2		595	109		704	820
4318	92 MBH		7	2.286		760	125		885	1,025
4322	109.7 MBH		6	2.667		900	146		1,046	1,200
4326	131.0 MBH		4	4		900	219		1,119	1,325
4330	160.0 MBH		3	5.333		955	291		1,246	1,500
4334	194.0 MBH		2.20	7.273		1,100	395		1,495	1,825
4338	212.0 MBH		2.10	7.619		1,375	415		1,790	2,125
4342	247.0 MBH		1.96	8.163		1,625	445		2,070	2,450
4346	297.0 MBH		1.80	8.889		1,725	485		2,210	2,625
4350	333.0 MBH	Q-6	1.90	12.632		1,875	715		2,590	3,125
4354	420 MBH, (460 V)		1.80	13.333		2,300	755		3,055	3,675
4358	500 MBH, (460 V)		1.71	14.035		3,050	795		3,845	4,550
4362	570 MBH, (460 V)		1.40	17.143		4,175	970		5,145	6,075
4366	620 MBH, (460 V)		1.30	18.462		4,175	1,050		5,225	6,175
4370	960 MBH, (460 V)		1.10	21.818		7,950	1,225		9,175	10,600
4400	Unit heaters, propeller, hot water, 115 V, 60 Deg. F entering air									
4404	Horizontal									
4408	6 MBH	Q-5	11	1.455	Ea.	385	79.50		464.50	545
4412	12 MBH		11	1.455		385	79.50		464.50	545
4416	16 MBH		10	1.600		435	87.50		522.50	605
4420	24 MBH		9	1.778		500	97		597	695

For customer support on your Mechanical Cost Data, call 877.739.3689.

413

23 82 Convection Heating and Cooling Units

23 82 39 – Unit Heaters

23 82 39.16 Propeller Unit Heaters		Crew	Daily Output	Labor-Hours	Unit	Material	2016 Bare Costs Labor	Equipment	Total	Total Incl O&P
4424	29 MBH	Q-5	8	2	Ea.	585	109		694	805
4428	47 MBH		7	2.286		640	125		765	890
4432	63 MBH		6	2.667		690	146		836	980
4436	81 MBH		5.50	2.909		850	159		1,009	1,175
4440	90 MBH		5	3.200		920	175		1,095	1,275
4460	133 MBH		4	4		1,125	219		1,344	1,575
4464	139 MBH		3.50	4.571		1,275	250		1,525	1,775
4468	198 MBH		2.50	6.400		1,575	350		1,925	2,275
4472	224 MBH		2	8		1,775	435		2,210	2,600
4476	273 MBH		1.50	10.667		2,200	585		2,785	3,300
4500	Vertical									
4504	8 MBH	Q-5	13	1.231	Ea.	825	67		892	1,000
4508	12 MBH		13	1.231		825	67		892	1,000
4512	16 MBH		13	1.231		825	67		892	1,000
4516	30 MBH		12	1.333		825	73		898	1,025
4520	43 MBH		11	1.455		855	79.50		934.50	1,050
4524	57 MBH		8	2		1,050	109		1,159	1,325
4528	68 MBH		7	2.286		1,100	125		1,225	1,400
4540	105 MBH		6	2.667		1,275	146		1,421	1,650
4544	123 MBH		5	3.200		1,375	175		1,550	1,775
4550	140 MBH		5	3.200		1,575	175		1,750	2,025
4554	156 MBH		4	4		1,950	219		2,169	2,475
4560	210 MBH		4	4		2,450	219		2,669	3,025
4564	223 MBH		3	5.333		2,675	291		2,966	3,400
4568	257 MBH		2	8		2,675	435		3,110	3,600

23 83 Radiant Heating Units

23 83 16 – Radiant-Heating Hydronic Piping

23 83 16.10 Radiant Floor Heating

		Crew	Daily Output	Labor-Hours	Unit	Material	2016 Bare Costs Labor	Equipment	Total	Total Incl O&P
0010	**RADIANT FLOOR HEATING**									
0100	Tubing, PEX (cross-linked polyethylene)									
0110	Oxygen barrier type for systems with ferrous materials									
0120	1/2"	Q-5	800	.020	L.F.	1.07	1.09		2.16	2.83
0130	3/4"		535	.030		1.51	1.63		3.14	4.13
0140	1"		400	.040		2.36	2.19		4.55	5.90
0200	Non barrier type for ferrous free systems									
0210	1/2"	Q-5	800	.020	L.F.	.61	1.09		1.70	2.32
0220	3/4"		535	.030		1.11	1.63		2.74	3.69
0230	1"		400	.040		1.90	2.19		4.09	5.40
1000	Manifolds									
1110	Brass									
1120	With supply and return valves, flow meter, thermometer,									
1122	auto air vent and drain/fill valve.									
1130	1", 2 circuit	Q-5	14	1.143	Ea.	257	62.50		319.50	375
1140	1", 3 circuit		13.50	1.185		294	65		359	425
1150	1", 4 circuit		13	1.231		320	67		387	450
1154	1", 5 circuit		12.50	1.280		380	70		450	525
1158	1", 6 circuit		12	1.333		415	73		488	565
1162	1", 7 circuit		11.50	1.391		455	76		531	615
1166	1", 8 circuit		11	1.455		500	79.50		579.50	670
1172	1", 9 circuit		10.50	1.524		540	83.50		623.50	720

23 83 Radiant Heating Units

23 83 16 – Radiant-Heating Hydronic Piping

23 83 16.10 Radiant Floor Heating		Crew	Daily Output	Labor-Hours	Unit	Material	2016 Bare Costs Labor	Equipment	Total	Total Incl O&P
1174	1", 10 circuit	Q-5	10	1.600	Ea.	580	87.50		667.50	770
1178	1", 11 circuit		9.50	1.684		605	92		697	805
1182	1", 12 circuit	↓	9	1.778	↓	670	97		767	880
1610	Copper manifold header (cut to size)									
1620	1" header, 12 circuit 1/2" sweat outlets	Q-5	3.33	4.805	Ea.	91	262		353	495
1630	1-1/4" header, 12 circuit 1/2" sweat outlets		3.20	5		106	273		379	525
1640	1-1/4" header, 12 circuit 3/4" sweat outlets		3	5.333		114	291		405	565
1650	1-1/2" header, 12 circuit 3/4" sweat outlets		3.10	5.161		137	282		419	575
1660	2" header, 12 circuit 3/4" sweat outlets	↓	2.90	5.517	↓	201	300		501	675
3000	Valves									
3110	Thermostatic zone valve actuator with end switch	Q-5	40	.400	Ea.	38.50	22		60.50	75.50
3114	Thermostatic zone valve actuator	"	36	.444	"	81	24.50		105.50	126
3120	Motorized straight zone valve with operator complete									
3130	3/4"	Q-5	35	.457	Ea.	130	25		155	181
3140	1"		32	.500		141	27.50		168.50	196
3150	1-1/4"	↓	29.60	.541	↓	179	29.50		208.50	242
3500	4 Way mixing valve, manual, brass									
3530	1"	Q-5	13.30	1.203	Ea.	180	65.50		245.50	297
3540	1-1/4"		11.40	1.404		195	76.50		271.50	330
3550	1-1/2"		11	1.455		249	79.50		328.50	395
3560	2"		10.60	1.509		350	82.50		432.50	515
3800	Mixing valve motor, 4 way for valves, 1" and 1-1/4"		34	.471		355	25.50		380.50	430
3810	Mixing valve motor, 4 way for valves, 1-1/2" and 2"	↓	30	.533	↓	365	29		394	450
5000	Radiant floor heating, zone control panel									
5120	4 Zone actuator valve control, expandable	Q-5	20	.800	Ea.	163	43.50		206.50	245
5130	6 Zone actuator valve control, expandable		18	.889		226	48.50		274.50	320
6070	Thermal track, straight panel for long continuous runs, 5.333 S.F.		40	.400		26.50	22		48.50	62.50
6080	Thermal track, utility panel, for direction reverse at run end, 5.333 S.F.		40	.400		26.50	22		48.50	62.50
6090	Combination panel, for direction reverse plus straight run, 5.333 S.F.	↓	40	.400	↓	26.50	22		48.50	62.50
7000	PEX tubing fittings									
7100	Compression type									
7116	Coupling									
7120	1/2" x 1/2"	1 Stpi	27	.296	Ea.	6.70	18		24.70	34.50
7124	3/4" x 3/4"	"	23	.348	"	10.60	21		31.60	43.50
7130	Adapter									
7132	1/2" x female sweat 1/2"	1 Stpi	27	.296	Ea.	4.36	18		22.36	32
7134	1/2" x female sweat 3/4"		26	.308		4.88	18.70		23.58	33.50
7136	5/8" x female sweat 3/4"	↓	24	.333	↓	7	20		27	38
7140	Elbow									
7142	1/2" x female sweat 1/2"	1 Stpi	27	.296	Ea.	6.65	18		24.65	34.50
7144	1/2" x female sweat 3/4"		26	.308		7.80	18.70		26.50	36.50
7146	5/8" x female sweat 3/4"	↓	24	.333	↓	8.75	20		28.75	40
7200	Insert type									
7206	PEX x male NPT									
7210	1/2" x 1/2"	1 Stpi	29	.276	Ea.	2.59	16.75		19.34	28.50
7220	3/4" x 3/4"		27	.296		3.81	18		21.81	31
7230	1" x 1"	↓	26	.308	↓	6.45	18.70		25.15	35
7300	PEX coupling									
7310	1/2" x 1/2"	1 Stpi	30	.267	Ea.	.98	16.20		17.18	25.50
7320	3/4" x 3/4"		29	.276		1.22	16.75		17.97	27
7330	1" x 1"	↓	28	.286	↓	2.29	17.35		19.64	28.50
7400	PEX stainless crimp ring									
7410	1/2" x 1/2"	1 Stpi	86	.093	Ea.	.37	5.65		6.02	8.90

For customer support on your Mechanical Cost Data, call 877.739.3689.

415

23 83 Radiant Heating Units

23 83 16 – Radiant-Heating Hydronic Piping

23 83 16.10 Radiant Floor Heating	Crew	Daily Output	Labor-Hours	Unit	Material	2016 Bare Costs Labor	Equipment	Total	Total Incl O&P
7420	3/4" x 3/4"	1 Stpi	84	.095	Ea.	.51	5.80		6.31	9.25
7430	1" x 1"	↓	82	.098	↓	.73	5.90		6.63	9.75

23 83 33 – Electric Radiant Heaters

23 83 33.10 Electric Heating

		Crew	Daily Output	Labor-Hours	Unit	Material	Labor	Equipment	Total	Total Incl O&P
0010	**ELECTRIC HEATING**, not incl. conduit or feed wiring									
1100	Rule of thumb: Baseboard units, including control	1 Elec	4.40	1.818	kW	105	100		205	266
1300	Baseboard heaters, 2' long, 350 watt		8	1	Ea.	26	55		81	111
1400	3' long, 750 watt		8	1		31.50	55		86.50	117
1600	4' long, 1000 watt		6.70	1.194		35.50	66		101.50	138
1800	5' long, 935 watt		5.70	1.404		46.50	77.50		124	167
2000	6' long, 1500 watt		5	1.600		51	88		139	188
2200	7' long, 1310 watt		4.40	1.818		53	100		153	209
2400	8' long, 2000 watt		4	2		56	110		166	227
2600	9' long, 1680 watt		3.60	2.222		92	122		214	284
2800	10' long, 1875 watt	↓	3.30	2.424	↓	180	134		314	400
2950	Wall heaters with fan, 120 to 277 volt									
3160	Recessed, residential, 750 watt	1 Elec	6	1.333	Ea.	83.50	73.50		157	202
3170	1000 watt		6	1.333		83.50	73.50		157	202
3180	1250 watt		5	1.600		83.50	88		171.50	224
3190	1500 watt		4	2		83.50	110		193.50	257
3210	2000 watt		4	2		83.50	110		193.50	257
3230	2500 watt		3.50	2.286		209	126		335	420
3240	3000 watt		3	2.667		310	147		457	560
3250	4000 watt		2.70	2.963		310	163		473	585
3260	Commercial, 750 watt		6	1.333		196	73.50		269.50	325
3270	1000 watt		6	1.333		196	73.50		269.50	325
3280	1250 watt		5	1.600		196	88		284	350
3290	1500 watt		4	2		196	110		306	380
3300	2000 watt		4	2		196	110		306	380
3310	2500 watt		3.50	2.286		196	126		322	405
3320	3000 watt		3	2.667		515	147		662	790
3330	4000 watt		2.70	2.963		515	163		678	815
3600	Thermostats, integral		16	.500		26.50	27.50		54	70
3800	Line voltage, 1 pole		8	1		12.85	55		67.85	96.50
3810	2 pole	↓	8	1	↓	18.50	55		73.50	103
4000	Heat trace system, 400 degree									
4020	115 V, 2.5 watts per L.F.	1 Elec	530	.015	L.F.	7.90	.83		8.73	9.95
4030	5 watts per L.F.		530	.015		7.90	.83		8.73	9.95
4050	10 watts per L.F.		530	.015		7.90	.83		8.73	9.95
4060	208 V, 5 watts per L.F.		530	.015		7.90	.83		8.73	9.95
4080	480 V, 8 watts per L.F.	↓	530	.015	↓	7.90	.83		8.73	9.95
4200	Heater raceway									
4260	Heat transfer cement									
4280	1 gallon				Ea.	60			60	66
4300	5 gallon				"	234			234	257
4320	Cable tie									
4340	3/4" pipe size	1 Elec	470	.017	Ea.	.02	.94		.96	1.42
4360	1" pipe size		444	.018		.02	.99		1.01	1.51
4380	1-1/4" pipe size		400	.020		.02	1.10		1.12	1.67
4400	1-1/2" pipe size		355	.023		.02	1.24		1.26	1.88
4420	2" pipe size		320	.025		.02	1.38		1.40	2.08
4440	3" pipe size	↓	160	.050	↓	.02	2.76		2.78	4.14

23 83 Radiant Heating Units

23 83 33 – Electric Radiant Heaters

23 83 33.10 Electric Heating	Crew	Daily Output	Labor-Hours	Unit	Material	2016 Bare Costs Labor	Equipment	Total	Total Incl O&P
4460	4" pipe size	1 Elec	100	.080	Ea.	.02	4.41		4.43	6.60
4480	Thermostat NEMA 3R, 22 amp, 0-150 Deg, 10' cap.		8	1		241	55		296	350
4500	Thermostat NEMA 4X, 25 amp, 40 Deg, 5-1/2' cap.		7	1.143		241	63		304	360
4520	Thermostat NEMA 4X, 22 amp, 25-325 Deg, 10' cap.		7	1.143		650	63		713	810
4540	Thermostat NEMA 4X, 22 amp, 15-140 Deg,		6	1.333		545	73.50		618.50	710
4580	Thermostat NEMA 4,7,9, 22 amp, 25-325 Deg, 10' cap.		3.60	2.222		800	122		922	1,075
4600	Thermostat NEMA 4,7,9, 22 amp, 15-140 Deg,		3	2.667		795	147		942	1,100
4720	Fiberglass application tape, 36 yard roll		11	.727		77.50	40		117.50	146
5000	Radiant heating ceiling panels, 2' x 4', 500 watt		16	.500		410	27.50		437.50	495
5050	750 watt		16	.500		455	27.50		482.50	540
5200	For recessed plaster frame, add		32	.250		119	13.80		132.80	152
5300	Infrared quartz heaters, 120 volts, 1000 watts		6.70	1.194		345	66		411	480
5350	1500 watt		5	1.600		345	88		433	510
5400	240 volts, 1500 watt		5	1.600		345	88		433	510
5450	2000 watt		4	2		345	110		455	545
5500	3000 watt		3	2.667		380	147		527	640
5550	4000 watt		2.60	3.077		380	170		550	675
5570	Modulating control		.80	10		150	550		700	990
5600	Unit heaters, heavy duty, with fan & mounting bracket									
5650	Single phase, 208-240-277 volt, 3 kW	1 Elec	6	1.333	Ea.	735	73.50		808.50	920
5750	5 kW		5.50	1.455		835	80		915	1,050
5800	7 kW		5	1.600		1,000	88		1,088	1,225
5850	10 kW		4	2		1,300	110		1,410	1,600
5950	15 kW		3.80	2.105		2,125	116		2,241	2,525
6000	480 volt, 3 kW		6	1.333		675	73.50		748.50	850
6020	4 kW		5.80	1.379		695	76		771	880
6040	5 kW		5.50	1.455		715	80		795	910
6060	7 kW		5	1.600		1,075	88		1,163	1,300
6080	10 kW		4	2		1,200	110		1,310	1,500
6100	13 kW		3.80	2.105		2,125	116		2,241	2,525
6120	15 kW		3.70	2.162		2,325	119		2,444	2,725
6140	20 kW		3.50	2.286		2,525	126		2,651	2,975
6300	3 phase, 208-240 volt, 5 kW		5.50	1.455		700	80		780	885
6320	7 kW		5	1.600		1,150	88		1,238	1,400
6340	10 kW		4	2		1,275	110		1,385	1,575
6360	15 kW		3.70	2.162		1,900	119		2,019	2,275
6380	20 kW		3.50	2.286		2,500	126		2,626	2,950
6400	25 kW		3.30	2.424		2,950	134		3,084	3,450
6500	480 volt, 5 kW		5.50	1.455		990	80		1,070	1,225
6520	7 kW		5	1.600		1,350	88		1,438	1,625
6540	10 kW		4	2		1,700	110		1,810	2,025
6560	13 kW		3.80	2.105		2,125	116		2,241	2,525
6580	15 kW		3.70	2.162		2,125	119		2,244	2,525
6600	20 kW		3.50	2.286		2,525	126		2,651	2,975
6620	25 kW		3.30	2.424		2,950	134		3,084	3,450
6630	30 kW		3	2.667		3,625	147		3,772	4,200
6640	40 kW		2	4		4,375	220		4,595	5,125
6650	50 kW		1.60	5		5,150	276		5,426	6,050
6800	Vertical discharge heaters, with fan									
6820	Single phase, 208-240-277 volt, 10 kW	1 Elec	4	2	Ea.	1,225	110		1,335	1,525
6840	15 kW		3.70	2.162		1,825	119		1,944	2,175
6900	3 phase, 208-240 volt, 10 kW		4	2		1,275	110		1,385	1,575
6920	15 kW		3.70	2.162		1,900	119		2,019	2,275

For customer support on your Mechanical Cost Data, call 877.739.3689.

417

23 83 Radiant Heating Units

23 83 33 – Electric Radiant Heaters

23 83 33.10 Electric Heating		Crew	Daily Output	Labor-Hours	Unit	Material	2016 Bare Costs Labor	Equipment	Total	Total Incl O&P
6940	20 kW	1 Elec	3.50	2.286	Ea.	2,500	126		2,626	2,950
6960	25 kW		3.30	2.424		2,950	134		3,084	3,450
6980	30 kW		3	2.667		3,325	147		3,472	3,900
7000	40 kW		2	4		4,275	220		4,495	5,025
7020	50 kW		1.60	5		5,150	276		5,426	6,050
7100	480 volt, 10 kW		4	2		1,300	110		1,410	1,600
7120	15 kW		3.70	2.162		2,125	119		2,244	2,525
7140	20 kW		3.50	2.286		2,525	126		2,651	2,975
7160	25 kW		3.30	2.424		2,950	134		3,084	3,450
7180	30 kW		3	2.667		3,625	147		3,772	4,200
7200	40 kW		2	4		4,375	220		4,595	5,125
7220	50 kW	▼	1.60	5	▼	5,150	276		5,426	6,050
7900	Cabinet convector heaters, 240 volt, three phase,									
7920	3' long, 2000 watt	1 Elec	5.30	1.509	Ea.	2,200	83		2,283	2,525
7940	3000 watt		5.30	1.509		2,275	83		2,358	2,650
7960	4000 watt		5.30	1.509		2,350	83		2,433	2,700
7980	6000 watt		4.60	1.739		2,675	96		2,771	3,100
8000	8000 watt		4.60	1.739		2,725	96		2,821	3,150
8020	4' long, 4000 watt		4.60	1.739		2,375	96		2,471	2,775
8040	6000 watt		4	2		2,475	110		2,585	2,900
8060	8000 watt		4	2		2,550	110		2,660	2,975
8080	10,000 watt	▼	4	2	▼	2,575	110		2,685	3,025
8100	Available also in 208 or 277 volt									
8200	Cabinet unit heaters, 120 to 277 volt, 1 pole,									
8220	wall mounted, 2 kW	1 Elec	4.60	1.739	Ea.	2,125	96		2,221	2,500
8230	3 kW		4.60	1.739		2,225	96		2,321	2,600
8240	4 kW		4.40	1.818		2,300	100		2,400	2,675
8250	5 kW		4.40	1.818		2,375	100		2,475	2,775
8260	6 kW		4.20	1.905		2,425	105		2,530	2,825
8270	8 kW		4	2		2,475	110		2,585	2,900
8280	10 kW		3.80	2.105		2,525	116		2,641	2,975
8290	12 kW		3.50	2.286		2,575	126		2,701	3,025
8300	13.5 kW		2.90	2.759		3,100	152		3,252	3,625
8310	16 kW		2.70	2.963		3,125	163		3,288	3,675
8320	20 kW		2.30	3.478		3,975	192		4,167	4,650
8330	24 kW		1.90	4.211		4,025	232		4,257	4,775
8350	Recessed, 2 kW		4.40	1.818		2,150	100		2,250	2,500
8370	3 kW		4.40	1.818		2,225	100		2,325	2,600
8380	4 kW		4.20	1.905		2,300	105		2,405	2,675
8390	5 kW		4.20	1.905		2,375	105		2,480	2,750
8400	6 kW		4	2		2,425	110		2,535	2,825
8410	8 kW		3.80	2.105		2,525	116		2,641	2,950
8420	10 kW		3.50	2.286		3,050	126		3,176	3,575
8430	12 kW		2.90	2.759		3,075	152		3,227	3,600
8440	13.5 kW		2.70	2.963		3,100	163		3,263	3,650
8450	16 kW		2.30	3.478		3,300	192		3,492	3,900
8460	20 kW		1.90	4.211		3,750	232		3,982	4,475
8470	24 kW		1.60	5		3,800	276		4,076	4,600
8490	Ceiling mounted, 2 kW		3.20	2.500		2,125	138		2,263	2,550
8510	3 kW		3.20	2.500		2,225	138		2,363	2,650
8520	4 kW		3	2.667		2,300	147		2,447	2,750
8530	5 kW		3	2.667		2,375	147		2,522	2,825
8540	6 kW		2.80	2.857		2,375	157		2,532	2,850

23 83 33.10 Electric Heating	Crew	Daily Output	Labor-Hours	Unit	Material	2016 Bare Costs Labor	Equipment	Total	Total Incl O&P
8550	8 kW	1 Elec	2.40	3.333	Ea.	2,475	184		2,659	3,000
8560	10 kW		2.20	3.636		2,525	200		2,725	3,075
8570	12 kW		2	4		2,575	220		2,795	3,150
8580	13.5 kW		1.50	5.333		2,625	294		2,919	3,325
8590	16 kW		1.30	6.154		2,700	340		3,040	3,450
8600	20 kW		.90	8.889		3,750	490		4,240	4,850
8610	24 kW	▼	.60	13.333	▼	3,800	735		4,535	5,300
8630	208 to 480 V, 3 pole									
8650	Wall mounted, 2 kW	1 Elec	4.60	1.739	Ea.	2,275	96		2,371	2,650
8670	3 kW		4.60	1.739		2,375	96		2,471	2,750
8680	4 kW		4.40	1.818		2,425	100		2,525	2,825
8690	5 kW		4.40	1.818		2,500	100		2,600	2,900
8700	6 kW		4.20	1.905		2,525	105		2,630	2,925
8710	8 kW		4	2		2,600	110		2,710	3,050
8720	10 kW		3.80	2.105		2,650	116		2,766	3,100
8730	12 kW		3.50	2.286		2,700	126		2,826	3,175
8740	13.5 kW		2.90	2.759		2,750	152		2,902	3,250
8750	16 kW		2.70	2.963		2,800	163		2,963	3,350
8760	20 kW		2.30	3.478		3,875	192		4,067	4,550
8770	24 kW		1.90	4.211		4,125	232		4,357	4,875
8790	Recessed, 2 kW		4.40	1.818		2,275	100		2,375	2,650
8810	3 kW		4.40	1.818		2,300	100		2,400	2,675
8820	4 kW		4.20	1.905		2,400	105		2,505	2,775
8830	5 kW		4.20	1.905		2,425	105		2,530	2,800
8840	6 kW		4	2		2,425	110		2,535	2,850
8850	8 kW		3.80	2.105		2,525	116		2,641	2,950
8860	10 kW		3.50	2.286		2,700	126		2,826	3,150
8870	12 kW		2.90	2.759		2,800	152		2,952	3,325
8880	13.5 kW		2.70	2.963		2,850	163		3,013	3,375
8890	16 kW		2.30	3.478		4,150	192		4,342	4,850
8900	20 kW		1.90	4.211		4,200	232		4,432	4,975
8920	24 kW		1.60	5		4,275	276		4,551	5,100
8940	Ceiling mount, 2 kW		3.20	2.500		2,375	138		2,513	2,800
8950	3 kW		3.20	2.500		2,525	138		2,663	2,975
8960	4 kW		3	2.667		2,525	147		2,672	3,000
8970	5 kW		3	2.667		2,600	147		2,747	3,100
8980	6 kW		2.80	2.857		2,625	157		2,782	3,100
8990	8 kW		2.40	3.333		2,700	184		2,884	3,250
9000	10 kW		2.20	3.636		2,775	200		2,975	3,350
9020	13.5 kW		1.50	5.333		2,850	294		3,144	3,600
9030	16 kW		1.30	6.154		4,150	340		4,490	5,075
9040	20 kW		.90	8.889		4,200	490		4,690	5,350
9060	24 kW	▼	.60	13.333	▼	4,675	735		5,410	6,250

23 84 Humidity Control Equipment

23 84 13 – Humidifiers

23 84 13.10 Humidifier Units

	Crew	Daily Output	Labor-Hours	Unit	Material	2016 Bare Costs Labor	Equipment	Total	Total Incl O&P
0010 **HUMIDIFIER UNITS**									
0520 Steam, room or duct, filter, regulators, auto. controls, 220 V									
0540 11 lb. per hour	Q-5	6	2.667	Ea.	2,700	146		2,846	3,200
0560 22 lb. per hour		5	3.200		2,975	175		3,150	3,550
0580 33 lb. per hour		4	4		3,050	219		3,269	3,700
0600 50 lb. per hour		4	4		3,550	219		3,769	4,225
0620 100 lb. per hour		3	5.333		4,475	291		4,766	5,375
0640 150 lb. per hour		2.50	6.400		6,400	350		6,750	7,575
0660 200 lb. per hour	▼	2	8	▼	8,025	435		8,460	9,475
0700 With blower									
0720 11 lb. per hour	Q-5	5.50	2.909	Ea.	4,300	159		4,459	4,975
0740 22 lb. per hour		4.75	3.368		4,525	184		4,709	5,250
0760 33 lb. per hour		3.75	4.267		4,650	233		4,883	5,475
0780 50 lb. per hour		3.50	4.571		5,325	250		5,575	6,225
0800 100 lb. per hour		2.75	5.818		5,950	320		6,270	7,025
0820 150 lb. per hour		2	8		8,675	435		9,110	10,200
0840 200 lb. per hour	▼	1.50	10.667	▼	9,700	585		10,285	11,600
1000 Steam for duct installation, manifold with pneum. controls									
1010 3 – 191 lb./hr., 24" x 24"	Q-5	5	3.200	Ea.	2,125	175		2,300	2,600
1020 3 – 191 lb./hr., 48" x 48"		4.60	3.478		4,075	190		4,265	4,750
1030 65 – 334 lb./hr., 36" x 36"		4.80	3.333		3,975	182		4,157	4,650
1040 65 – 334 lb./hr., 72" x 72"		4	4		4,300	219		4,519	5,050
1050 100 – 690 lb./hr., 48" x 48"		4	4		5,600	219		5,819	6,475
1060 100 – 690 lb./hr., 84" x 84"		3.60	4.444		5,775	243		6,018	6,725
1070 220 – 2000 lb./hr., 72" x 72"		3.80	4.211		5,850	230		6,080	6,775
1080 220 – 2000 lb./hr., 144" x 144"	▼	3.40	4.706		6,825	257		7,082	7,925
1090 For electrically controlled unit, add				▼	805			805	885
1092 For additional manifold, add					25%	15%			
1200 Electric steam, self generating, with manifold									
1210 1 – 96 lb./hr., 12" x 72"	Q-5	3.60	4.444	Ea.	10,400	243		10,643	11,900
1220 20 – 168 lb./hr., 12" x 72"		3.60	4.444		15,900	243		16,143	17,900
1290 For room mounting, add	▼	16	1	▼	262	54.50		316.50	370
5000 Furnace type, wheel bypass									
5020 10 GPD	1 Stpi	4	2	Ea.	296	121		417	510
5040 14 GPD		3.80	2.105		330	128		458	560
5060 19 GPD	▼	3.60	2.222	▼	565	135		700	830

23 84 16 – Mechanical Dehumidification Units

23 84 16.10 Dehumidifier Units

	Crew	Daily Output	Labor-Hours	Unit	Material	2016 Bare Costs Labor	Equipment	Total	Total Incl O&P
0010 **DEHUMIDIFIER UNITS**									
6000 Self contained with filters and standard controls									
6040 1.5 lb./hr., 50 CFM	1 Plum	8	1	Ea.	8,175	59		8,234	9,100
6060 3 lb./hr., 150 CFM	Q-1	12	1.333		8,550	71		8,621	9,500
6065 6 lb./hr., 150 CFM		9	1.778		13,900	94.50		13,994.50	15,400
6070 16 to 20 lb./hr., 600 CFM		5	3.200		26,500	171		26,671	29,400
6080 30 to 40 lb./hr., 1125 CFM		4	4		41,200	213		41,413	45,600
6090 60 to 75 lb./hr., 2250 CFM		3	5.333		54,000	284		54,284	59,500
6100 120 to 155 lb./hr., 4500 CFM		2	8		95,000	425		95,425	105,000
6110 240 to 310 lb./hr., 9000 CFM	▼	1.50	10.667		136,500	570		137,070	151,000
6120 400 to 515 lb./hr., 15,000 CFM	Q-2	1.60	15		171,500	830		172,330	190,000
6130 530 to 690 lb./hr., 20,000 CFM		1.40	17.143		186,000	945		186,945	206,000
6140 800 to 1030 lb./hr., 30,000 CFM		1.20	20		211,500	1,100		212,600	234,000
6150 1060 to 1375 lb./hr., 40,000 CFM	▼	1	24	▼	294,500	1,325		295,825	326,000

23 91 10 – Prefabricated Curbs, Pads and Stands

23 91 10.10 Prefabricated Pads and Stands	Crew	Daily Output	Labor-Hours	Unit	Material	2016 Bare Costs Labor	Equipment	Total	Total Incl O&P
0010 **PREFABRICATED PADS AND STANDS**									
6000 Pad, fiberglass reinforced concrete with polystyrene foam core									
6050 Condenser, 2" thick, 20" x 38"	1 Shee	8	1	Ea.	28.50	57.50		86	119
6070 24" x 24"		16	.500		22.50	28.50		51	69
6090 24" x 36"		12	.667		36.50	38		74.50	98.50
6110 24" x 42"	↓	8	1		42	57.50		99.50	134
6150 26" x 36"	Q-9	8	2		36.50	103		139.50	198
6170 28" x 38"	1 Shee	8	1		45	57.50		102.50	137
6190 30" x 30"		12	.667		38	38		76	101
6220 30" x 36"	↓	8	1		45	57.50		102.50	137
6240 30" x 40"	Q-9	8	2		49	103		152	212
6260 32" x 32"		7	2.286		42.50	118		160.50	227
6280 36" x 36"		8	2		52.50	103		155.50	216
6300 36" x 40"		7	2.286		59.50	118		177.50	246
6320 36" x 48"		7	2.286		69.50	118		187.50	257
6340 36" x 54"		6	2.667		84.50	137		221.50	305
6360 36" x 60" x 3"		7	2.286		100	118		218	290
6400 41" x 47" x 3"		6	2.667		93	137		230	310
6490 26" round		10	1.600		22.50	82.50		105	151
6550 30" round		9	1.778		29.50	91.50		121	173
6600 36" round	↓	8	2		40.50	103		143.50	203
6700 For 3" thick pad, add				↓	20%	5%			
8800 Stand, corrosion-resistant plastic									
8850 Heat pump, 6" high	1 Shee	32	.250	Ea.	5.70	14.30		20	28.50
8870 12" high	"	32	.250	"	12.90	14.30		27.20	36

Division Notes

	CREW	DAILY OUTPUT	LABOR-HOURS	UNIT	BARE COSTS				TOTAL INCL O&P
					MAT.	LABOR	EQUIP.	TOTAL	

Estimating Tips
26 05 00 Common Work Results for Electrical
- Conduit should be taken off in three main categories—power distribution, branch power, and branch lighting—so the estimator can concentrate on systems and components, therefore making it easier to ensure all items have been accounted for.
- For cost modifications for elevated conduit installation, add the percentages to labor according to the height of installation, and only to the quantities exceeding the different height levels, not to the total conduit quantities.
- Remember that aluminum wiring of equal ampacity is larger in diameter than copper and may require larger conduit.
- If more than three wires at a time are being pulled, deduct percentages from the labor hours of that grouping of wires.
- When taking off grounding systems, identify separately the type and size of wire, and list each unique type of ground connection.
- The estimator should take the weights of materials into consideration when completing a takeoff. Topics to consider include: How will the materials be supported? What methods of support are available? How high will the support structure have to reach? Will the final support structure be able to withstand the total burden? Is the support material included or separate from the fixture, equipment, and material specified?
- Do not overlook the costs for equipment used in the installation. If scaffolding or highlifts are available in the field, contractors may use them in lieu of the proposed ladders and rolling staging.

26 20 00 Low-Voltage Electrical Transmission
- Supports and concrete pads may be shown on drawings for the larger equipment, or the support system may be only a piece of plywood for the back of a panelboard. In either case, it must be included in the costs.

26 40 00 Electrical and Cathodic Protection
- When taking off cathodic protections systems, identify the type and size of cable, and list each unique type of anode connection.

26 50 00 Lighting
- Fixtures should be taken off room by room, using the fixture schedule, specifications, and the ceiling plan. For large concentrations of lighting fixtures in the same area, deduct the percentages from labor hours.

Reference Numbers
Reference numbers are shown at the beginning of some major classifications. These numbers refer to related items in the Reference Section. The reference information may be an estimating procedure, an alternate pricing method, or technical information.

Note: Not all subdivisions listed here necessarily appear. ∎

26 01 Operation and Maintenance of Electrical Systems

26 01 02 – Labor Adjustment

26 01 02.20 Labor Adjustment Factors	Crew	Daily Output	Labor-Hours	Unit	Material	2016 Bare Costs Labor	2016 Bare Costs Equipment	Total	Total Incl O&P
0010 **LABOR ADJUSTMENT FACTORS** (For Div. 26, 27 and 28)									
0100 Subtract from labor for Economy of Scale for Wire									
0110 4-5 wires						25%			
0120 6-10 wires						30%			
0130 11-15 wires						35%			
0140 over 15 wires						40%			

26 05 Common Work Results for Electrical

26 05 33 – Raceway and Boxes for Electrical Systems

26 05 33.95 Cutting and Drilling

		Crew	Daily Output	Labor-Hours	Unit	Material	2016 Bare Costs Labor	2016 Bare Costs Equipment	Total	Total Incl O&P
0010	**CUTTING AND DRILLING**									
0100	Hole drilling to 10' high, concrete wall									
0110	8" thick, 1/2" pipe size	R-31	12	.667	Ea.	.24	36.50	5.35	42.09	61
0120	3/4" pipe size		12	.667		.24	36.50	5.35	42.09	61
0130	1" pipe size		9.50	.842		.38	46.50	6.75	53.63	77.50
0140	1-1/4" pipe size		9.50	.842		.38	46.50	6.75	53.63	77.50
0150	1-1/2" pipe size		9.50	.842		.38	46.50	6.75	53.63	77.50
0160	2" pipe size		4.40	1.818		.51	100	14.60	115.11	167
0170	2-1/2" pipe size		4.40	1.818		.51	100	14.60	115.11	167
0180	3" pipe size		4.40	1.818		.51	100	14.60	115.11	167
0190	3-1/2" pipe size		3.30	2.424		.66	134	19.45	154.11	222
0200	4" pipe size		3.30	2.424		.66	134	19.45	154.11	222
0500	12" thick, 1/2" pipe size		9.40	.851		.36	47	6.85	54.21	78
0520	3/4" pipe size		9.40	.851		.36	47	6.85	54.21	78
0540	1" pipe size		7.30	1.096		.57	60.50	8.80	69.87	101
0560	1-1/4" pipe size		7.30	1.096		.57	60.50	8.80	69.87	101
0570	1-1/2" pipe size		7.30	1.096		.57	60.50	8.80	69.87	101
0580	2" pipe size		3.60	2.222		.77	122	17.85	140.62	203
0590	2-1/2" pipe size		3.60	2.222		.77	122	17.85	140.62	203
0600	3" pipe size		3.60	2.222		.77	122	17.85	140.62	203
0610	3-1/2" pipe size		2.80	2.857		1	157	23	181	262
0630	4" pipe size		2.50	3.200		1	176	25.50	202.50	294
0650	16" thick, 1/2" pipe size		7.60	1.053		.48	58	8.45	66.93	97
0670	3/4" pipe size		7	1.143		.48	63	9.20	72.68	105
0690	1" pipe size		6	1.333		.76	73.50	10.70	84.96	123
0710	1-1/4" pipe size		5.50	1.455		.76	80	11.70	92.46	134
0730	1-1/2" pipe size		5.50	1.455		.76	80	11.70	92.46	134
0750	2" pipe size		3	2.667		1.03	147	21.50	169.53	245
0770	2-1/2" pipe size		2.70	2.963		1.03	163	24	188.03	271
0790	3" pipe size		2.50	3.200		1.03	176	25.50	202.53	294
0810	3-1/2" pipe size		2.30	3.478		1.33	192	28	221.33	320
0830	4" pipe size		2	4		1.33	220	32	253.33	365
0850	20" thick, 1/2" pipe size		6.40	1.250		.59	69	10.05	79.64	115
0870	3/4" pipe size		6	1.333		.59	73.50	10.70	84.79	122
0890	1" pipe size		5	1.600		.95	88	12.85	101.80	147
0910	1-1/4" pipe size		4.80	1.667		.95	92	13.40	106.35	153
0930	1-1/2" pipe size		4.60	1.739		.95	96	13.95	110.90	159
0950	2" pipe size		2.70	2.963		1.28	163	24	188.28	271
0970	2-1/2" pipe size		2.40	3.333		1.28	184	27	212.28	305
0990	3" pipe size		2.20	3.636		1.28	200	29	230.28	335
1010	3-1/2" pipe size		2	4		1.66	220	32	253.66	365

26 05 33.95 Cutting and Drilling	Crew	Daily Output	Labor-Hours	Unit	Material	2016 Bare Costs Labor	Equipment	Total	Total Incl O&P
1030 4" pipe size	R-31	1.70	4.706	Ea.	1.66	259	38	298.66	435
1050 24" thick, 1/2" pipe size		5.50	1.455		.71	80	11.70	92.41	134
1070 3/4" pipe size		5.10	1.569		.71	86.50	12.60	99.81	144
1090 1" pipe size		4.30	1.860		1.14	103	14.95	119.09	171
1110 1-1/4" pipe size		4	2		1.14	110	16.05	127.19	184
1130 1-1/2" pipe size		4	2		1.14	110	16.05	127.19	184
1150 2" pipe size		2.40	3.333		1.54	184	27	212.54	305
1170 2-1/2" pipe size		2.20	3.636		1.54	200	29	230.54	335
1190 3" pipe size		2	4		1.54	220	32	253.54	365
1210 3-1/2" pipe size		1.80	4.444		1.99	245	35.50	282.49	405
1230 4" pipe size		1.50	5.333		1.99	294	43	338.99	490
1500 Brick wall, 8" thick, 1/2" pipe size		18	.444		.24	24.50	3.57	28.31	40.50
1520 3/4" pipe size		18	.444		.24	24.50	3.57	28.31	40.50
1540 1" pipe size		13.30	.602		.38	33	4.83	38.21	55
1560 1-1/4" pipe size		13.30	.602		.38	33	4.83	38.21	55
1580 1-1/2" pipe size		13.30	.602		.38	33	4.83	38.21	55
1600 2" pipe size		5.70	1.404		.51	77.50	11.25	89.26	129
1620 2-1/2" pipe size		5.70	1.404		.51	77.50	11.25	89.26	129
1640 3" pipe size		5.70	1.404		.51	77.50	11.25	89.26	129
1660 3-1/2" pipe size		4.40	1.818		.66	100	14.60	115.26	167
1680 4" pipe size		4	2		.66	110	16.05	126.71	183
1700 12" thick, 1/2" pipe size		14.50	.552		.36	30.50	4.43	35.29	51
1720 3/4" pipe size		14.50	.552		.36	30.50	4.43	35.29	51
1740 1" pipe size		11	.727		.57	40	5.85	46.42	67
1760 1-1/4" pipe size		11	.727		.57	40	5.85	46.42	67
1780 1-1/2" pipe size		11	.727		.57	40	5.85	46.42	67
1800 2" pipe size		5	1.600		.77	88	12.85	101.62	147
1820 2-1/2" pipe size		5	1.600		.77	88	12.85	101.62	147
1840 3" pipe size		5	1.600		.77	88	12.85	101.62	147
1860 3-1/2" pipe size		3.80	2.105		1	116	16.90	133.90	194
1880 4" pipe size		3.30	2.424		1	134	19.45	154.45	223
1900 16" thick, 1/2" pipe size		12.30	.650		.48	36	5.20	41.68	60
1920 3/4" pipe size		12.30	.650		.48	36	5.20	41.68	60
1940 1" pipe size		9.30	.860		.76	47.50	6.90	55.16	79.50
1960 1-1/4" pipe size		9.30	.860		.76	47.50	6.90	55.16	79.50
1980 1-1/2" pipe size		9.30	.860		.76	47.50	6.90	55.16	79.50
2000 2" pipe size		4.40	1.818		1.03	100	14.60	115.63	167
2010 2-1/2" pipe size		4.40	1.818		1.03	100	14.60	115.63	167
2030 3" pipe size		4.40	1.818		1.03	100	14.60	115.63	167
2050 3-1/2" pipe size		3.30	2.424		1.33	134	19.45	154.78	223
2070 4" pipe size		3	2.667		1.33	147	21.50	169.83	245
2090 20" thick, 1/2" pipe size		10.70	.748		.59	41	6	47.59	69
2110 3/4" pipe size		10.70	.748		.59	41	6	47.59	69
2130 1" pipe size		8	1		.95	55	8.05	64	92.50
2150 1-1/4" pipe size		8	1		.95	55	8.05	64	92.50
2170 1-1/2" pipe size		8	1		.95	55	8.05	64	92.50
2190 2" pipe size		4	2		1.28	110	16.05	127.33	184
2210 2-1/2" pipe size		4	2		1.28	110	16.05	127.33	184
2230 3" pipe size		4	2		1.28	110	16.05	127.33	184
2250 3-1/2" pipe size		3	2.667		1.66	147	21.50	170.16	245
2270 4" pipe size		2.70	2.963		1.66	163	24	188.66	272
2290 24" thick, 1/2" pipe size		9.40	.851		.71	47	6.85	54.56	78.50
2310 3/4" pipe size		9.40	.851		.71	47	6.85	54.56	78.50

For customer support on your Mechanical Cost Data, call 877.739.3689.

425

26 05 Common Work Results for Electrical

26 05 33 – Raceway and Boxes for Electrical Systems

26 05 33.95 Cutting and Drilling

	26 05 33.95 Cutting and Drilling	Crew	Daily Output	Labor-Hours	Unit	Material	2016 Bare Costs Labor	Equipment	Total	Total Incl O&P
2330	1" pipe size	R-31	7.10	1.127	Ea.	1.14	62	9.05	72.19	104
2350	1-1/4" pipe size		7.10	1.127		1.14	62	9.05	72.19	104
2370	1-1/2" pipe size		7.10	1.127		1.14	62	9.05	72.19	104
2390	2" pipe size		3.60	2.222		1.54	122	17.85	141.39	204
2410	2-1/2" pipe size		3.60	2.222		1.54	122	17.85	141.39	204
2430	3" pipe size		3.60	2.222		1.54	122	17.85	141.39	204
2450	3-1/2" pipe size		2.80	2.857		1.99	157	23	181.99	263
2470	4" pipe size	▼	2.50	3.200	▼	1.99	176	25.50	203.49	295
3000	Knockouts to 8' high, metal boxes & enclosures									
3020	With hole saw, 1/2" pipe size	1 Elec	53	.151	Ea.		8.30		8.30	12.45
3040	3/4" pipe size		47	.170			9.40		9.40	14.05
3050	1" pipe size		40	.200			11		11	16.50
3060	1-1/4" pipe size		36	.222			12.25		12.25	18.30
3070	1-1/2" pipe size		32	.250			13.80		13.80	20.50
3080	2" pipe size		27	.296			16.35		16.35	24.50
3090	2-1/2" pipe size		20	.400			22		22	33
4010	3" pipe size		16	.500			27.50		27.50	41
4030	3-1/2" pipe size		13	.615			34		34	50.50
4050	4" pipe size	▼	11	.727	▼		40		40	60

26 05 80 – Wiring Connections

26 05 80.10 Motor Connections

	26 05 80.10 Motor Connections	Crew	Daily Output	Labor-Hours	Unit	Material	2016 Bare Costs Labor	Equipment	Total	Total Incl O&P
0010	**MOTOR CONNECTIONS**									
0020	Flexible conduit and fittings, 115 volt, 1 phase, up to 1 HP motor	1 Elec	8	1	Ea.	5.55	55		60.55	88.50
0050	2 HP motor		6.50	1.231		11.35	68		79.35	114
0100	3 HP motor		5.50	1.455		9.85	80		89.85	131
0110	230 volt, 3 phase, 3 HP motor		6.78	1.180		6.85	65		71.85	105
0112	5 HP motor		5.47	1.463		5.70	80.50		86.20	127
0114	7-1/2 HP motor		4.61	1.735		8.70	95.50		104.20	153
0120	10 HP motor		4.20	1.905		18.90	105		123.90	178
0150	15 HP motor		3.30	2.424		18.90	134		152.90	221
0200	25 HP motor		2.70	2.963		25	163		188	272
0400	50 HP motor		2.20	3.636		52.50	200		252.50	360
0600	100 HP motor		1.50	5.333		114	294		408	565
1500	460 volt, 5 HP motor, 3 phase		8	1		5.95	55		60.95	89
1520	10 HP motor		8	1		5.95	55		60.95	89
1530	25 HP motor		6	1.333		12.15	73.50		85.65	123
1540	30 HP motor		6	1.333		12.15	73.50		85.65	123
1550	40 HP motor		5	1.600		17.80	88		105.80	152
1560	50 HP motor		5	1.600		19.80	88		107.80	154
1570	60 HP motor		3.80	2.105		21.50	116		137.50	198
1580	75 HP motor		3.50	2.286		31	126		157	222
1590	100 HP motor		2.50	3.200		46	176		222	315
1600	125 HP motor		2	4		57.50	220		277.50	395
1610	150 HP motor		1.80	4.444		61	245		306	430
1620	200 HP motor		1.50	5.333		95	294		389	545
2005	460 Volt, 5 HP motor, 3 Phase, w/sealtite		8	1		9.50	55		64.50	93
2010	10 HP motor		8	1		9.50	55		64.50	93
2015	25 HP motor		6	1.333		18.55	73.50		92.05	131
2020	30 HP motor		6	1.333		18.55	73.50		92.05	131
2025	40 HP motor		5	1.600		33	88		121	169
2030	50 HP motor		5	1.600		33.50	88		121.50	169
2035	60 HP motor	▼	3.80	2.105	▼	49.50	116		165.50	229

26 05 80 – Wiring Connections

26 05 80.10 Motor Connections		Crew	Daily Output	Labor-Hours	Unit	Material	2016 Bare Costs Labor	Equipment	Total	Total Incl O&P
2040	75 HP motor	1 Elec	3.50	2.286	Ea.	55	126		181	249
2045	100 HP motor		2.50	3.200		74	176		250	345
2055	150 HP motor		1.80	4.444		77.50	245		322.50	450
2060	200 HP motor		1.50	5.333		510	294		804	1,000

26 05 90 – Residential Applications

26 05 90.10 Residential Wiring

		Crew	Daily Output	Labor-Hours	Unit	Material	2016 Bare Costs Labor	Equipment	Total	Total Incl O&P
0010	**RESIDENTIAL WIRING**									
0020	20' avg. runs and #14/2 wiring incl. unless otherwise noted									
1000	Service & panel, includes 24' SE-AL cable, service eye, meter,									
1010	Socket, panel board, main bkr., ground rod, 15 or 20 amp									
1020	1-pole circuit breakers, and misc. hardware									
1100	100 amp, with 10 branch breakers	1 Elec	1.19	6.723	Ea.	485	370		855	1,100
1110	With PVC conduit and wire		.92	8.696		525	480		1,005	1,300
1120	With RGS conduit and wire		.73	10.959		685	605		1,290	1,650
1150	150 amp, with 14 branch breakers		1.03	7.767		705	430		1,135	1,425
1170	With PVC conduit and wire		.82	9.756		785	540		1,325	1,675
1180	With RGS conduit and wire		.67	11.940		1,100	660		1,760	2,175
1200	200 amp, with 18 branch breakers	2 Elec	1.80	8.889		1,050	490		1,540	1,875
1220	With PVC conduit and wire		1.46	10.959		1,125	605		1,730	2,150
1230	With RGS conduit and wire		1.24	12.903		1,525	710		2,235	2,750
1800	Lightning surge suppressor	1 Elec	32	.250		50.50	13.80		64.30	76.50
2000	Switch devices									
2100	Single pole, 15 amp, Ivory, with a 1-gang box, cover plate,									
2110	Type NM (Romex) cable	1 Elec	17.10	.468	Ea.	18.70	26		44.70	59
2120	Type MC (BX) cable		14.30	.559		27.50	31		58.50	76
2130	EMT & wire		5.71	1.401		36	77		113	156
2150	3-way, #14/3, type NM cable		14.55	.550		12.70	30.50		43.20	59.50
2170	Type MC cable		12.31	.650		25	36		61	81
2180	EMT & wire		5	1.600		30	88		118	165
2200	4-way, #14/3, type NM cable		14.55	.550		20.50	30.50		51	68
2220	Type MC cable		12.31	.650		32.50	36		68.50	89.50
2230	EMT & wire		5	1.600		37.50	88		125.50	173
2250	S.P., 20 amp, #12/2, type NM cable		13.33	.600		14.75	33		47.75	65.50
2270	Type MC cable		11.43	.700		21	38.50		59.50	80.50
2280	EMT & wire		4.85	1.649		33.50	91		124.50	173
2290	S.P. rotary dimmer, 600W, no wiring		17	.471		31.50	26		57.50	73.50
2300	S.P. rotary dimmer, 600W, type NM cable		14.55	.550		35.50	30.50		66	84.50
2320	Type MC cable		12.31	.650		44.50	36		80.50	102
2330	EMT & wire		5	1.600		54.50	88		142.50	192
2350	3-way rotary dimmer, type NM cable		13.33	.600		19.85	33		52.85	71.50
2370	Type MC cable		11.43	.700		28.50	38.50		67	89
2380	EMT & wire		4.85	1.649		39	91		130	179
2400	Interval timer wall switch, 20 amp, 1-30 min., #12/2									
2410	Type NM cable	1 Elec	14.55	.550	Ea.	57	30.50		87.50	109
2420	Type MC cable		12.31	.650		62	36		98	122
2430	EMT & wire		5	1.600		76	88		164	216
2500	Decorator style									
2510	S.P., 15 amp, type NM cable	1 Elec	17.10	.468	Ea.	22.50	26		48.50	63.50
2520	Type MC cable		14.30	.559		31.50	31		62.50	80.50
2530	EMT & wire		5.71	1.401		40	77		117	160
2550	3-way, #14/3, type NM cable		14.55	.550		16.60	30.50		47.10	64
2570	Type MC cable		12.31	.650		29	36		65	85.50

For customer support on your Mechanical Cost Data, call 877.739.3689.

427

26 05 90.10 Residential Wiring	Crew	Daily Output	Labor-Hours	Unit	Material	2016 Bare Costs Labor	Equipment	Total	Total Incl O&P
2580	EMT & wire	1 Elec	5	1.600	Ea.	33.50	88		121.50	169
2600	4-way, #14/3, type NM cable		14.55	.550		24.50	30.50		55	72
2620	Type MC cable		12.31	.650		36.50	36		72.50	94
2630	EMT & wire		5	1.600		41.50	88		129.50	178
2650	S.P., 20 amp, #12/2, type NM cable		13.33	.600		18.60	33		51.60	70
2670	Type MC cable		11.43	.700		25	38.50		63.50	85
2680	EMT & wire		4.85	1.649		37.50	91		128.50	177
2700	S.P., slide dimmer, type NM cable		17.10	.468		24.50	26		50.50	65.50
2720	Type MC cable		14.30	.559		33	31		64	82.50
2730	EMT & wire		5.71	1.401		43.50	77		120.50	164
2750	S.P., touch dimmer, type NM cable		17.10	.468		41	26		67	83.50
2770	Type MC cable		14.30	.559		50	31		81	101
2780	EMT & wire		5.71	1.401		60	77		137	182
2800	3-way touch dimmer, type NM cable		13.33	.600		51.50	33		84.50	106
2820	Type MC cable		11.43	.700		60	38.50		98.50	124
2830	EMT & wire	↓	4.85	1.649	↓	70	91		161	213
3000	Combination devices									
3100	S.P. switch/15 amp recpt., Ivory, 1-gang box, plate									
3110	Type NM cable	1 Elec	11.43	.700	Ea.	24	38.50		62.50	84
3120	Type MC cable		10	.800		33	44		77	102
3130	EMT & wire		4.40	1.818		43	100		143	198
3150	S.P. switch/pilot light, type NM cable		11.43	.700		24.50	38.50		63	84.50
3170	Type MC cable		10	.800		33	44		77	103
3180	EMT & wire		4.43	1.806		43	99.50		142.50	197
3190	2-S.P. switches, 2-#14/2, no wiring		14	.571		13.70	31.50		45.20	62
3200	2-S.P. switches, 2-#14/2, type NM cables		10	.800		26.50	44		70.50	95.50
3220	Type MC cable		8.89	.900		40.50	49.50		90	119
3230	EMT & wire		4.10	1.951		45.50	108		153.50	211
3250	3-way switch/15 amp recpt., #14/3, type NM cable		10	.800		31.50	44		75.50	101
3270	Type MC cable		8.89	.900		44	49.50		93.50	123
3280	EMT & wire		4.10	1.951		48.50	108		156.50	215
3300	2-3 way switches, 2-#14/3, type NM cables		8.89	.900		40.50	49.50		90	119
3320	Type MC cable		8	1		61.50	55		116.50	150
3330	EMT & wire		4	2		55.50	110		165.50	226
3350	S.P. switch/20 amp recpt., #12/2, type NM cable		10	.800		31	44		75	101
3370	Type MC cable		8.89	.900		36	49.50		85.50	114
3380	EMT & wire	↓	4.10	1.951	↓	50	108		158	216
3400	Decorator style									
3410	S.P. switch/15 amp recpt., type NM cable	1 Elec	11.43	.700	Ea.	28	38.50		66.50	88.50
3420	Type MC cable		10	.800		36.50	44		80.50	107
3430	EMT & wire		4.40	1.818		47	100		147	202
3450	S.P. switch/pilot light, type NM cable		11.43	.700		28	38.50		66.50	88.50
3470	Type MC cable		10	.800		37	44		81	107
3480	EMT & wire		4.40	1.818		47	100		147	202
3500	2-S.P. switches, 2-#14/2, type NM cables		10	.800		30.50	44		74.50	99.50
3520	Type MC cable		8.89	.900		44	49.50		93.50	123
3530	EMT & wire		4.10	1.951		49.50	108		157.50	216
3550	3-way/15 amp recpt., #14/3, type NM cable		10	.800		35.50	44		79.50	105
3570	Type MC cable		8.89	.900		48	49.50		97.50	127
3580	EMT & wire		4.10	1.951		52.50	108		160.50	219
3650	2-3 way switches, 2-#14/3, type NM cables		8.89	.900		44.50	49.50		94	123
3670	Type MC cable		8	1		65	55		120	154
3680	EMT & wire		4	2		59.50	110		169.50	231

26 05 90.10 Residential Wiring	Crew	Daily Output	Labor-Hours	Unit	Material	2016 Bare Costs Labor	Equipment	Total	Total Incl O&P
3700	S.P. switch/20 amp recpt., #12/2, type NM cable	1 Elec	10	.800	Ea.	35	44		79	105
3720	Type MC cable		8.89	.900		40	49.50		89.50	118
3730	EMT & wire		4.10	1.951		54	108		162	220
4000	Receptacle devices									
4010	Duplex outlet, 15 amp recpt., Ivory, 1-gang box, plate									
4015	Type NM cable	1 Elec	14.55	.550	Ea.	10.30	30.50		40.80	57
4020	Type MC cable		12.31	.650		19	36		55	74.50
4030	EMT & wire		5.33	1.501		28	82.50		110.50	155
4050	With #12/2, type NM cable		12.31	.650		12.65	36		48.65	67.50
4070	Type MC cable		10.67	.750		19	41.50		60.50	83
4080	EMT & wire		4.71	1.699		31.50	93.50		125	175
4100	20 amp recpt., #12/2, type NM cable		12.31	.650		19.30	36		55.30	75
4120	Type MC cable		10.67	.750		25.50	41.50		67	90
4130	EMT & wire		4.71	1.699		38	93.50		131.50	182
4140	For GFI see Section 26 05 90.10 line 4300 below									
4150	Decorator style, 15 amp recpt., type NM cable	1 Elec	14.55	.550	Ea.	14.20	30.50		44.70	61
4170	Type MC cable		12.31	.650		23	36		59	78.50
4180	EMT & wire		5.33	1.501		31.50	82.50		114	159
4200	With #12/2, type NM cable		12.31	.650		16.55	36		52.55	71.50
4220	Type MC cable		10.67	.750		23	41.50		64.50	87
4230	EMT & wire		4.71	1.699		35.50	93.50		129	179
4250	20 amp recpt. #12/2, type NM cable		12.31	.650		23	36		59	79
4270	Type MC cable		10.67	.750		29.50	41.50		71	94.50
4280	EMT & wire		4.71	1.699		42	93.50		135.50	186
4300	GFI, 15 amp recpt., type NM cable		12.31	.650		23	36		59	78.50
4320	Type MC cable		10.67	.750		31.50	41.50		73	96.50
4330	EMT & wire		4.71	1.699		40.50	93.50		134	185
4350	GFI with #12/2, type NM cable		10.67	.750		25	41.50		66.50	89.50
4370	Type MC cable		9.20	.870		31.50	48		79.50	106
4380	EMT & wire		4.21	1.900		44	105		149	206
4400	20 amp recpt., #12/2 type NM cable		10.67	.750		51.50	41.50		93	119
4420	Type MC cable		9.20	.870		58	48		106	135
4430	EMT & wire		4.21	1.900		70.50	105		175.50	235
4500	Weather-proof cover for above receptacles, add		32	.250		2.94	13.80		16.74	23.50
4550	Air conditioner outlet, 20 amp-240 volt recpt.									
4560	30' of #12/2, 2 pole circuit breaker									
4570	Type NM cable	1 Elec	10	.800	Ea.	62	44		106	135
4580	Type MC cable		9	.889		70	49		119	151
4590	EMT & wire		4	2		81	110		191	254
4600	Decorator style, type NM cable		10	.800		67	44		111	140
4620	Type MC cable		9	.889		74.50	49		123.50	156
4630	EMT & wire		4	2		85.50	110		195.50	259
4650	Dryer outlet, 30 amp-240 volt recpt., 20' of #10/3									
4660	2 pole circuit breaker									
4670	Type NM cable	1 Elec	6.41	1.248	Ea.	58	69		127	167
4680	Type MC cable		5.71	1.401		64	77		141	187
4690	EMT & wire		3.48	2.299		75	127		202	273
4700	Range outlet, 50 amp-240 volt recpt., 30' of #8/3									
4710	Type NM cable	1 Elec	4.21	1.900	Ea.	85	105		190	251
4720	Type MC cable		4	2		126	110		236	305
4730	EMT & wire		2.96	2.703		104	149		253	340
4750	Central vacuum outlet, Type NM cable		6.40	1.250		53.50	69		122.50	162
4770	Type MC cable		5.71	1.401		66.50	77		143.50	190

For customer support on your Mechanical Cost Data, call 877.739.3689.

429

26 05 90 – Residential Applications

26 05 90.10 Residential Wiring		Crew	Daily Output	Labor-Hours	Unit	Material	2016 Bare Costs Labor	2016 Bare Costs Equipment	Total	Total Incl O&P
4780	EMT & wire	1 Elec	3.48	2.299	Ea.	81	127		208	279
4800	30 amp-110 volt locking recpt., #10/2 circ. bkr.									
4810	Type NM cable	1 Elec	6.20	1.290	Ea.	61.50	71		132.50	174
4830	EMT & wire	"	3.20	2.500	"	91.50	138		229.50	305
4900	Low voltage outlets									
4910	Telephone recpt., 20' of 4/C phone wire	1 Elec	26	.308	Ea.	8.40	16.95		25.35	34.50
4920	TV recpt., 20' of RG59U coax wire, F type connector	"	16	.500	"	16	27.50		43.50	58.50
4950	Door bell chime, transformer, 2 buttons, 60' of bellwire									
4970	Economy model	1 Elec	11.50	.696	Ea.	50	38.50		88.50	113
4980	Custom model		11.50	.696		99.50	38.50		138	167
4990	Luxury model, 3 buttons	↓	9.50	.842	↓	281	46.50		327.50	380
6000	Lighting outlets									
6050	Wire only (for fixture), type NM cable	1 Elec	32	.250	Ea.	6.35	13.80		20.15	27.50
6070	Type MC cable		24	.333		12.55	18.35		30.90	41.50
6080	EMT & wire		10	.800		20.50	44		64.50	88.50
6100	Box (4"), and wire (for fixture), type NM cable		25	.320		13.90	17.65		31.55	42
6120	Type MC cable		20	.400		20	22		42	55
6130	EMT & wire	↓	11	.727	↓	28	40		68	91
6200	Fixtures (use with lines 6050 or 6100 above)									
6210	Canopy style, economy grade	1 Elec	40	.200	Ea.	27	11		38	46
6220	Custom grade		40	.200		53.50	11		64.50	75.50
6250	Dining room chandelier, economy grade		19	.421		67.50	23		90.50	109
6260	Custom grade		19	.421		315	23		338	380
6270	Luxury grade		15	.533		820	29.50		849.50	950
6310	Kitchen fixture (fluorescent), economy grade		30	.267		71.50	14.70		86.20	101
6320	Custom grade		25	.320		221	17.65		238.65	271
6350	Outdoor, wall mounted, economy grade		30	.267		25.50	14.70		40.20	50
6360	Custom grade		30	.267		120	14.70		134.70	154
6370	Luxury grade		25	.320		248	17.65		265.65	300
6410	Outdoor PAR floodlights, 1 lamp, 150 watt		20	.400		27.50	22		49.50	63
6420	2 lamp, 150 watt each		20	.400		45.50	22		67.50	83
6430	For infrared security sensor, add		32	.250		132	13.80		145.80	166
6450	Outdoor, quartz-halogen, 300 watt flood		20	.400		33	22		55	69.50
6600	Recessed downlight, round, pre-wired, 50 or 75 watt trim		30	.267		68	14.70		82.70	96.50
6610	With shower light trim		30	.267		82.50	14.70		97.20	113
6620	With wall washer trim		28	.286		85	15.75		100.75	117
6630	With eye-ball trim		28	.286		84.50	15.75		100.25	117
6700	Porcelain lamp holder		40	.200		2.52	11		13.52	19.25
6710	With pull switch		40	.200		7.55	11		18.55	25
6750	Fluorescent strip, 2-20 watt tube, wrap around diffuser, 24"		24	.333		42	18.35		60.35	74
6770	2-34 watt tubes, 48"		20	.400		118	22		140	163
6800	Bathroom heat lamp, 1-250 watt		28	.286		37.50	15.75		53.25	64.50
6810	2-250 watt lamps	↓	28	.286	↓	59.50	15.75		75.25	89
6820	For timer switch, see Section 26 05 90.10 line 2400									
6900	Outdoor post lamp, incl. post, fixture, 35' of #14/2									
6910	Type NM cable	1 Elec	3.50	2.286	Ea.	285	126		411	505
6920	Photo-eye, add		27	.296		27.50	16.35		43.85	54.50
6950	Clock dial time switch, 24 hr., w/enclosure, type NM cable		11.43	.700		73	38.50		111.50	138
6970	Type MC cable		11	.727		81.50	40		121.50	150
6980	EMT & wire	↓	4.85	1.649	↓	90.50	91		181.50	236
7000	Alarm systems									
7050	Smoke detectors, box, #14/3, type NM cable	1 Elec	14.55	.550	Ea.	33.50	30.50		64	82.50
7070	Type MC cable	↓	12.31	.650	↓	44.50	36		80.50	103

For customer support on your Mechanical Cost Data, call 877.739.3689.

26 05 90 – Residential Applications

26 05 90.10 Residential Wiring	Crew	Daily Output	Labor-Hours	Unit	Material	2016 Bare Costs Labor	Equipment	Total	Total Incl O&P
7080	EMT & wire	1 Elec	5	1.600	Ea.	49	88		137	186
7090	For relay output to security system, add				↓	10.20			10.20	11.20
8000	Residential equipment									
8050	Disposal hook-up, incl. switch, outlet box, 3' of flex									
8060	20 amp-1 pole circ. bkr., and 25' of #12/2									
8070	Type NM cable	1 Elec	10	.800	Ea.	36.50	44		80.50	106
8080	Type MC cable		8	1		43.50	55		98.50	131
8090	EMT & wire	↓	5	1.600	↓	58.50	88		146.50	196
8100	Trash compactor or dishwasher hook-up, incl. outlet box,									
8110	3' of flex, 15 amp-1 pole circ. bkr., and 25' of #14/2									
8120	Type NM cable	1 Elec	10	.800	Ea.	21	44		65	89.50
8130	Type MC cable		8	1		31	55		86	117
8140	EMT & wire	↓	5	1.600	↓	43.50	88		131.50	180
8150	Hot water sink dispensor hook-up, use line 8100									
8200	Vent/exhaust fan hook-up, type NM cable	1 Elec	32	.250	Ea.	6.35	13.80		20.15	27.50
8220	Type MC cable		24	.333		12.55	18.35		30.90	41.50
8230	EMT & wire	↓	10	.800	↓	20.50	44		64.50	88.50
8250	Bathroom vent fan, 50 CFM (use with above hook-up)									
8260	Economy model	1 Elec	15	.533	Ea.	19.65	29.50		49.15	65.50
8270	Low noise model		15	.533		42	29.50		71.50	90.50
8280	Custom model	↓	12	.667	↓	131	36.50		167.50	199
8300	Bathroom or kitchen vent fan, 110 CFM									
8310	Economy model	1 Elec	15	.533	Ea.	67.50	29.50		97	119
8320	Low noise model	"	15	.533	"	92	29.50		121.50	145
8350	Paddle fan, variable speed (w/o lights)									
8360	Economy model (AC motor)	1 Elec	10	.800	Ea.	109	44		153	186
8362	With light kit		10	.800		150	44		194	231
8370	Custom model (AC motor)		10	.800		262	44		306	355
8372	With light kit		10	.800		300	44		344	400
8380	Luxury model (DC motor)		8	1		288	55		343	400
8382	With light kit		8	1		330	55		385	445
8390	Remote speed switch for above, add	↓	12	.667	↓	37.50	36.50		74	96
8500	Whole house exhaust fan, ceiling mount, 36", variable speed									
8510	Remote switch, incl. shutters, 20 amp-1 pole circ. bkr.									
8520	30' of #12/2, type NM cable	1 Elec	4	2	Ea.	1,325	110		1,435	1,650
8530	Type MC cable		3.50	2.286		1,350	126		1,476	1,675
8540	EMT & wire	↓	3	2.667	↓	1,350	147		1,497	1,725
8600	Whirlpool tub hook-up, incl. timer switch, outlet box									
8610	3' of flex, 20 amp-1 pole GFI circ. bkr.									
8620	30' of #12/2, type NM cable	1 Elec	5	1.600	Ea.	157	88		245	305
8630	Type MC cable		4.20	1.905		160	105		265	335
8640	EMT & wire	↓	3.40	2.353	↓	172	130		302	385
8650	Hot water heater hook-up, incl. 1-2 pole circ. bkr., box;									
8660	3' of flex, 20' of #10/2, type NM cable	1 Elec	5	1.600	Ea.	31.50	88		119.50	167
8670	Type MC cable		4.20	1.905		42.50	105		147.50	204
8680	EMT & wire	↓	3.40	2.353		47	130		177	246
9000	Heating/air conditioning									
9050	Furnace/boiler hook-up, incl. firestat, local on-off switch									
9060	Emergency switch, and 40' of type NM cable	1 Elec	4	2	Ea.	52	110		162	222
9070	Type MC cable		3.50	2.286		65.50	126		191.50	260
9080	EMT & wire	↓	1.50	5.333	↓	83.50	294		377.50	530
9100	Air conditioner hook-up, incl. local 60 amp disc. switch									
9110	3' sealtite, 40 amp, 2 pole circuit breaker									

26 05 Common Work Results for Electrical

26 05 90 – Residential Applications

26 05 90.10 Residential Wiring		Crew	Daily Output	Labor-Hours	Unit	Material	2016 Bare Costs Labor	Equipment	Total	Total Incl O&P
9130	40' of #8/2, type NM cable	1 Elec	3.50	2.286	Ea.	144	126		270	345
9140	Type MC cable	3		2.667		204	147		351	445
9150	EMT & wire	↓	1.30	6.154	↓	185	340		525	710
9200	Heat pump hook-up, 1-40 & 1-100 amp 2 pole circ. bkr.									
9210	Local disconnect switch, 3' sealtite									
9220	40' of #8/2 & 30' of #3/2									
9230	Type NM cable	1 Elec	1.30	6.154	Ea.	460	340		800	1,000
9240	Type MC cable		1.08	7.407		510	410		920	1,175
9250	EMT & wire	↓	.94	8.511	↓	520	470		990	1,275
9500	Thermostat hook-up, using low voltage wire									
9520	Heating only, 25' of #18-3	1 Elec	24	.333	Ea.	8	18.35		26.35	36.50
9530	Heating/cooling, 25' of #18-4	"	20	.400	"	10.50	22		32.50	44.50

26 24 Switchboards and Panelboards

26 24 19 – Motor-Control Centers

26 24 19.40 Motor Starters and Controls

		Crew	Daily Output	Labor-Hours	Unit	Material	2016 Bare Costs Labor	Equipment	Total	Total Incl O&P
0010	**MOTOR STARTERS AND CONTROLS**									
0050	Magnetic, FVNR, with enclosure and heaters, 480 volt									
0080	2 HP, size 00	1 Elec	3.50	2.286	Ea.	197	126		323	405
0100	5 HP, size 0		2.30	3.478		315	192		507	630
0200	10 HP, size 1	↓	1.60	5		267	276		543	705
0300	25 HP, size 2	2 Elec	2.20	7.273		500	400		900	1,150
0400	50 HP, size 3		1.80	8.889		820	490		1,310	1,625
0500	100 HP, size 4		1.20	13.333		1,825	735		2,560	3,100
0600	200 HP, size 5		.90	17.778		4,250	980		5,230	6,150
0610	400 HP, size 6	↓	.80	20		19,000	1,100		20,100	22,600
0620	NEMA 7, 5 HP, size 0	1 Elec	1.60	5		1,400	276		1,676	1,950
0630	10 HP, size 1	"	1.10	7.273		1,475	400		1,875	2,225
0640	25 HP, size 2	2 Elec	1.80	8.889		2,375	490		2,865	3,350
0650	50 HP, size 3		1.20	13.333		3,575	735		4,310	5,025
0660	100 HP, size 4		.90	17.778		5,775	980		6,755	7,825
0670	200 HP, size 5	↓	.50	32		13,800	1,775		15,575	17,900
0700	Combination, with motor circuit protectors, 5 HP, size 0	1 Elec	1.80	4.444		1,000	245		1,245	1,475
0800	10 HP, size 1	"	1.30	6.154		1,050	340		1,390	1,650
0900	25 HP, size 2	2 Elec	2	8		1,475	440		1,915	2,250
1000	50 HP, size 3		1.32	12.121		2,125	670		2,795	3,350
1200	100 HP, size 4	↓	.80	20		4,600	1,100		5,700	6,700
1220	NEMA 7, 5 HP, size 0	1 Elec	1.30	6.154		2,800	340		3,140	3,575
1230	10 HP, size 1	"	1	8		2,875	440		3,315	3,800
1240	25 HP, size 2	2 Elec	1.32	12.121		3,825	670		4,495	5,200
1250	50 HP, size 3		.80	20		6,325	1,100		7,425	8,600
1260	100 HP, size 4		.60	26.667		9,850	1,475		11,325	13,000
1270	200 HP, size 5	↓	.40	40		21,400	2,200		23,600	26,800
1400	Combination, with fused switch, 5 HP, size 0	1 Elec	1.80	4.444		590	245		835	1,025
1600	10 HP, size 1	"	1.30	6.154		630	340		970	1,200
1800	25 HP, size 2	2 Elec	2	8		1,025	440		1,465	1,775
2000	50 HP, size 3		1.32	12.121		1,725	670		2,395	2,900
2200	100 HP, size 4		.80	20	↓	3,025	1,100		4,125	4,975
3500	Magnetic FVNR with NEMA 12, enclosure & heaters, 480 volt									
3600	5 HP, size 0	1 Elec	2.20	3.636	Ea.	224	200		424	545
3700	10 HP, size 1	"	1.50	5.333	↓	335	294		629	810

26 24 19 – Motor-Control Centers

26 24 19.40 Motor Starters and Controls

		Crew	Daily Output	Labor-Hours	Unit	Material	2016 Bare Costs Labor	Equipment	Total	Total Incl O&P
3800	25 HP, size 2	2 Elec	2	8	Ea.	630	440		1,070	1,350
3900	50 HP, size 3		1.60	10		970	550		1,520	1,900
4000	100 HP, size 4		1	16		2,325	880		3,205	3,875
4100	200 HP, size 5		.80	20		5,575	1,100		6,675	7,775
4200	Combination, with motor circuit protectors, 5 HP, size 0	1 Elec	1.70	4.706		740	259		999	1,200
4300	10 HP, size 1	"	1.20	6.667		770	365		1,135	1,400
4400	25 HP, size 2	2 Elec	1.80	8.889		1,150	490		1,640	2,000
4500	50 HP, size 3		1.20	13.333		1,875	735		2,610	3,175
4600	100 HP, size 4		.74	21.622		4,250	1,200		5,450	6,450
4700	Combination, with fused switch, 5 HP, size 0	1 Elec	1.70	4.706		720	259		979	1,175
4800	10 HP, size 1	"	1.20	6.667		750	365		1,115	1,375
4900	25 HP, size 2	2 Elec	1.80	8.889		1,150	490		1,640	1,975
5000	50 HP, size 3		1.20	13.333		1,825	735		2,560	3,100
5100	100 HP, size 4		.74	21.622		3,700	1,200		4,900	5,850
5200	Factory installed controls, adders to size 0 thru 5									
5300	Start-stop push button	1 Elec	32	.250	Ea.	47	13.80		60.80	72
5400	Hand-off-auto-selector switch		32	.250		47	13.80		60.80	72
5500	Pilot light		32	.250		88	13.80		101.80	118
5600	Start-stop-pilot		32	.250		135	13.80		148.80	169
5700	Auxiliary contact, NO or NC		32	.250		64.50	13.80		78.30	91.50
5800	NO-NC		32	.250		129	13.80		142.80	163

26 29 13 – Enclosed Controllers

26 29 13.20 Control Stations

		Crew	Daily Output	Labor-Hours	Unit	Material	2016 Bare Costs Labor	Equipment	Total	Total Incl O&P
0010	**CONTROL STATIONS**									
0050	NEMA 1, heavy duty, stop/start	1 Elec	8	1	Ea.	131	55		186	227
0100	Stop/start, pilot light		6.20	1.290		178	71		249	300
0200	Hand/off/automatic		6.20	1.290		97	71		168	213
0400	Stop/start/reverse		5.30	1.509		177	83		260	320
0500	NEMA 7, heavy duty, stop/start		6	1.333		440	73.50		513.50	590
0600	Stop/start, pilot light		4	2		535	110		645	750

26 29 23 – Variable-Frequency Motor Controllers

26 29 23.10 Variable Frequency Drives/Adj. Frequency Drives

			Crew	Daily Output	Labor-Hours	Unit	Material	2016 Bare Costs Labor	Equipment	Total	Total Incl O&P
0010	**VARIABLE FREQUENCY DRIVES/ADJ. FREQUENCY DRIVES**										
0100	Enclosed (NEMA 1), 460 volt, for 3 HP motor size	G	1 Elec	.80	10	Ea.	1,900	550		2,450	2,925
0110	5 HP motor size	G		.80	10		2,175	550		2,725	3,225
0120	7.5 HP motor size	G		.67	11.940		2,275	660		2,935	3,500
0130	10 HP motor size	G		.67	11.940		2,500	660		3,160	3,725
0140	15 HP motor size	G	2 Elec	.89	17.978		3,225	990		4,215	5,025
0150	20 HP motor size	G		.89	17.978		4,375	990		5,365	6,300
0160	25 HP motor size	G		.67	23.881		5,400	1,325		6,725	7,900
0170	30 HP motor size	G		.67	23.881		5,850	1,325		7,175	8,425
0180	40 HP motor size	G		.67	23.881		6,775	1,325		8,100	9,425
0190	50 HP motor size	G		.53	30.189		9,975	1,675		11,650	13,500
0200	60 HP motor size	G	R-3	.56	35.714		12,100	1,950	251	14,301	16,500
0210	75 HP motor size	G		.56	35.714		13,900	1,950	251	16,101	18,500
0220	100 HP motor size	G		.50	40		15,600	2,200	281	18,081	20,800
0230	125 HP motor size	G		.50	40		17,000	2,200	281	19,481	22,200
0240	150 HP motor size	G		.50	40		20,500	2,200	281	22,981	26,100

For customer support on your Mechanical Cost Data, call 877.739.3689.

433

26 29 23.10 Variable Frequency Drives/Adj. Frequency Drives		Crew	Daily Output	Labor-Hours	Unit	Material	2016 Bare Costs Labor	Equipment	Total	Total Incl O&P
0250	200 HP motor size	G	R-3	.42	47.619	Ea.	26,900	2,600	335	29,835	33,900
1100	Custom-engineered, 460 volt, for 3 HP motor size	G	1 Elec	.56	14.286		3,075	785		3,860	4,550
1110	5 HP motor size	G		.56	14.286		3,075	785		3,860	4,550
1120	7.5 HP motor size	G		.47	17.021		3,225	940		4,165	4,950
1130	10 HP motor size	G		.47	17.021		3,375	940		4,315	5,100
1140	15 HP motor size	G	2 Elec	.62	25.806		4,225	1,425		5,650	6,775
1150	20 HP motor size	G		.62	25.806		4,750	1,425		6,175	7,350
1160	25 HP motor size	G		.47	34.043		5,550	1,875		7,425	8,925
1170	30 HP motor size	G		.47	34.043		6,950	1,875		8,825	10,400
1180	40 HP motor size	G		.47	34.043		8,200	1,875		10,075	11,800
1190	50 HP motor size	G		.37	43.243		9,425	2,375		11,800	14,000
1200	60 HP motor size	G	R-3	.39	51.282		14,200	2,800	360	17,360	20,200
1210	75 HP motor size	G		.39	51.282		15,100	2,800	360	18,260	21,200
1220	100 HP motor size	G		.35	57.143		16,500	3,125	400	20,025	23,300
1230	125 HP motor size	G		.35	57.143		17,700	3,125	400	21,225	24,600
1240	150 HP motor size	G		.35	57.143		20,300	3,125	400	23,825	27,400
1250	200 HP motor size	G		.29	68.966		26,500	3,775	485	30,760	35,300
2000	For complex & special design systems to meet specific										
2010	requirements, obtain quote from vendor.										

Estimating Tips

- When estimating material costs for electronic safety and security systems, it is always prudent to obtain manufacturers' quotations for equipment prices and special installation requirements that affect the total cost.

- Fire alarm systems consist of control panels, annunciator panels, battery with rack, charger, and fire alarm actuating and indicating devices. Some fire alarm systems include speakers, telephone lines, door closer controls, and other components. Be careful not to overlook the costs related to installation for these items. Also be aware of costs for integrated automation instrumentation and terminal devices, control equipment, control wiring, and programming.

- Security equipment includes items such as CCTV, access control, and other detection and identification systems to perform alert and alarm functions. Be sure to consider the costs related to installation for this security equipment, such as for integrated automation instrumentation and terminal devices, control equipment, control wiring, and programming.

Reference Numbers

Reference numbers are shown at the beginning of some major classifications. These numbers refer to related items in the Reference Section. The reference information may be an estimating procedure, an alternate pricing method, or technical information.

Note: Not all subdivisions listed here necessarily appear. ■

Did you know?

RSMeans Online gives you the same access to RSMeans' data with 24/7 access:

- Quickly locate costs in the searchable database.
- Build cost lists, estimates, and reports in minutes.
- Adjust costs to any location in the U.S. and Canada with the click of a button.

Start your free trial today at **www.RSMeansOnline.com**

RSMeansOnline
FROM THE G⊕RDIAN GROUP®

Note: Trade Service, in part, has been used as a reference source for some of the material prices used in Division 28.

28 13 Access Control

28 13 53 – Security Access Detection

28 13 53.13 Security Access Metal Detectors

		Crew	Daily Output	Labor-Hours	Unit	Material	2016 Bare Costs Labor	Equipment	Total	Total Incl O&P
0010	**SECURITY ACCESS METAL DETECTORS**									
0240	Metal detector, hand-held, wand type, unit only				Ea.	129			129	138
0250	Metal detector, walk through portal type, single zone	1 Elec	2	4		4,400	220		4,620	5,175
0260	Multi-zone	"	2	4		5,450	220		5,670	6,325

28 13 53.16 Security Access X-Ray Equipment

		Crew	Daily Output	Labor-Hours	Unit	Material	2016 Bare Costs Labor	Equipment	Total	Total Incl O&P
0010	**SECURITY ACCESS X-RAY EQUIPMENT**									
0290	X-ray machine, desk top, for mail/small packages/letters	1 Elec	4	2	Ea.	3,475	110		3,585	4,000
0300	Conveyor type, incl monitor		2	4		16,000	220		16,220	17,900
0310	Includes additional features		2	4		28,000	220		28,220	31,100
0320	X-ray machine, large unit, for airports, incl monitor	2 Elec	1	16		40,000	880		40,880	45,300
0330	Full console	"	.50	32		68,500	1,775		70,275	78,000

28 13 53.23 Security Access Explosive Detection Equipment

		Crew	Daily Output	Labor-Hours	Unit	Material	2016 Bare Costs Labor	Equipment	Total	Total Incl O&P
0010	**SECURITY ACCESS EXPLOSIVE DETECTION EQUIPMENT**									
0270	Explosives detector, walk through portal type	1 Elec	2	4	Ea.	44,200	220		44,420	48,900
0280	Hand-held, battery operated				"				25,500	28,100

28 16 Intrusion Detection

28 16 16 – Intrusion Detection Systems Infrastructure

28 16 16.50 Intrusion Detection

		Crew	Daily Output	Labor-Hours	Unit	Material	2016 Bare Costs Labor	Equipment	Total	Total Incl O&P
0010	**INTRUSION DETECTION**, not including wires & conduits									
0100	Burglar alarm, battery operated, mechanical trigger	1 Elec	4	2	Ea.	278	110		388	470
0200	Electrical trigger		4	2		330	110		440	530
0400	For outside key control, add		8	1		86	55		141	177
0600	For remote signaling circuitry, add		8	1		137	55		192	233
0800	Card reader, flush type, standard		2.70	2.963		790	163		953	1,125
1000	Multi-code		2.70	2.963		1,200	163		1,363	1,575

28 23 Video Surveillance

28 23 23 – Video Surveillance Systems Infrastructure

28 23 23.50 Video Surveillance Equipments

		Crew	Daily Output	Labor-Hours	Unit	Material	2016 Bare Costs Labor	Equipment	Total	Total Incl O&P
0010	**VIDEO SURVEILLANCE EQUIPMENTS**									
0200	Video cameras, wireless, hidden in exit signs, clocks, etc., incl. receiver	1 Elec	3	2.667	Ea.	152	147		299	385
0210	Accessories for video recorder, single camera		3	2.667		183	147		330	420
0220	For multiple cameras		3	2.667		1,750	147		1,897	2,150
0230	Video cameras, wireless, for under vehicle searching, complete		2	4		9,725	220		9,945	11,000

28 31 Fire Detection and Alarm

28 31 23 – Fire Detection and Alarm Annunciation Panels and Fire Stations

28 31 23.50 Alarm Panels and Devices

		Crew	Daily Output	Labor-Hours	Unit	Material	2016 Bare Costs Labor	Equipment	Total	Total Incl O&P
0010	**ALARM PANELS AND DEVICES**, not including wires & conduits									
3594	Fire, alarm control panel									
3600	4 zone	2 Elec	2	8	Ea.	405	440		845	1,100
3800	8 zone		1	16		690	880		1,570	2,075
4000	12 zone		.67	23.988		2,400	1,325		3,725	4,600
4020	Alarm device	1 Elec	8	1		238	55		293	345
4050	Actuating device		8	1		335	55		390	455
4200	Battery and rack		4	2		410	110		520	620

28 31 Fire Detection and Alarm

28 31 23 – Fire Detection and Alarm Annunciation Panels and Fire Stations

28 31 23.50 Alarm Panels and Devices		Crew	Daily Output	Labor-Hours	Unit	Material	2016 Bare Costs Labor	Equipment	Total	Total Incl O&P
4400	Automatic charger	1 Elec	8	1	Ea.	585	55		640	725
4600	Signal bell		8	1		96.50	55		151.50	189
4800	Trouble buzzer or manual station		8	1		83.50	55		138.50	174
5600	Strobe and horn		5.30	1.509		152	83		235	291
5800	Fire alarm horn		6.70	1.194		61	66		127	166
6000	Door holder, electro-magnetic		4	2		103	110		213	278
6200	Combination holder and closer		3.20	2.500		123	138		261	340
6600	Drill switch		8	1		370	55		425	495
6800	Master box		2.70	2.963		6,400	163		6,563	7,300
7000	Break glass station		8	1		55.50	55		110.50	144
7800	Remote annunciator, 8 zone lamp		1.80	4.444		209	245		454	595
8000	12 zone lamp	2 Elec	2.60	6.154		335	340		675	875
8200	16 zone lamp	"	2.20	7.273		420	400		820	1,075

28 31 43 – Fire Detection Sensors

28 31 43.50 Fire and Heat Detectors

		Crew	Daily Output	Labor-Hours	Unit	Material	Labor	Equipment	Total	Total Incl O&P
0010	**FIRE & HEAT DETECTORS**									
5000	Detector, rate of rise	1 Elec	8	1	Ea.	51	55		106	139
5100	Fixed temperature	"	8	1	"	51	55		106	139

28 31 46 – Smoke Detection Sensors

28 31 46.50 Smoke Detectors

		Crew	Daily Output	Labor-Hours	Unit	Material	Labor	Equipment	Total	Total Incl O&P
0010	**SMOKE DETECTORS**									
5200	Smoke detector, ceiling type	1 Elec	6.20	1.290	Ea.	110	71		181	227
5400	Duct type	"	3.20	2.500	"	325	138		463	565

28 33 Gas Detection and Alarm

28 33 33 – Gas Detection Sensors

28 33 33.50 Tank Leak Detection Systems

		Crew	Daily Output	Labor-Hours	Unit	Material	Labor	Equipment	Total	Total Incl O&P
0010	**TANK LEAK DETECTION SYSTEMS** Liquid and vapor									
0100	For hydrocarbons and hazardous liquids/vapors									
0120	Controller, data acquisition, incl. printer, modem, RS232 port									
0140	24 channel, for use with all probes				Ea.	4,050			4,050	4,475
0160	9 channel, for external monitoring				"	875			875	965
0200	Probes									
0210	Well monitoring									
0220	Liquid phase detection				Ea.	460			460	505
0230	Hydrocarbon vapor, fixed position					480			480	530
0240	Hydrocarbon vapor, float mounted					465			465	510
0250	Both liquid and vapor hydrocarbon					500			500	550
0300	Secondary containment, liquid phase									
0310	Pipe trench/manway sump				Ea.	795			795	875
0320	Double wall pipe and manual sump					795			795	875
0330	Double wall fiberglass annular space					365			365	400
0340	Double wall steel tank annular space					365			365	400

For customer support on your Mechanical Cost Data, call 877.739.3689.

437

28 33 33.50 Tank Leak Detection Systems	Crew	Daily Output	Labor-Hours	Unit	Material	2016 Bare Costs Labor	Equipment	Total	Total Incl O&P
0500 Accessories									
0510 Modem, non-dedicated phone line				Ea.	265			265	292
0600 Monitoring, internal									
0610 Automatic tank gauge, incl. overfill				Ea.	1,400			1,400	1,550
0620 Product line				"	1,400			1,400	1,550
0700 Monitoring, special									
0710 Cathodic protection				Ea.	730			730	805
0720 Annular space chemical monitor				"	995			995	1,100

Estimating Tips

31 05 00 Common Work Results for Earthwork

- Estimating the actual cost of performing earthwork requires careful consideration of the variables involved. This includes items such as type of soil, whether water will be encountered, dewatering, whether banks need bracing, disposal of excavated earth, and length of haul to fill or spoil sites, etc. If the project has large quantities of cut or fill, consider raising or lowering the site to reduce costs, while paying close attention to the effect on site drainage and utilities.

- If the project has large quantities of fill, creating a borrow pit on the site can significantly lower the costs.

- It is very important to consider what time of year the project is scheduled for completion. Bad weather can create large cost overruns from dewatering, site repair, and lost productivity from cold weather.

Reference Numbers

Reference numbers are shown at the beginning of some major classifications. These numbers refer to related items in the Reference Section. The reference information may be an estimating procedure, an alternate pricing method, or technical information.

Note: Not all subdivisions listed here necessarily appear. ■

Did you know?

RSMeans Online gives you the same access to RSMeans' data with 24/7 access:

- Quickly locate costs in the searchable database.
- Build cost lists, estimates, and reports in minutes.
- Adjust costs to any location in the U.S. and Canada with the click of a button.

Start your free trial today at **www.RSMeansOnline.com**

RSMeans Online
FROM THE G⊕RDIAN GROUP®

31 05 Common Work Results for Earthwork

31 05 23 – Cement and Concrete for Earthwork

31 05 23.30 Plant Mixed Bituminous Concrete	Crew	Daily Output	Labor-Hours	Unit	Material	2016 Bare Costs Labor	Equipment	Total	Total Incl O&P
0010 **PLANT MIXED BITUMINOUS CONCRETE**									
0020 Asphaltic concrete plant mix (145 lb. per C.F.)				Ton	68.50			68.50	75.50
0040 Asphaltic concrete less than 300 tons add trucking costs									
0050 See Section 31 23 23.20 for hauling costs									
0200 All weather patching mix, hot				Ton	69			69	76
0250 Cold patch					77.50			77.50	85.50
0300 Berm mix				↓	67.50			67.50	74

31 23 Excavation and Fill

31 23 16 – Excavation

31 23 16.13 Excavating, Trench

		Crew	Daily Output	Labor-Hours	Unit	Material	2016 Bare Costs Labor	Equipment	Total	Total Incl O&P
0010	**EXCAVATING, TRENCH** G1030–805									
0011	Or continuous footing									
0020	Common earth with no sheeting or dewatering included									
0050	1' to 4' deep, 3/8 C.Y. excavator	B-11C	150	.107	B.C.Y.		4.75	2.44	7.19	9.90
0060	1/2 C.Y. excavator	B-11M	200	.080			3.56	1.98	5.54	7.60
0090	4' to 6' deep, 1/2 C.Y. excavator	"	200	.080			3.56	1.98	5.54	7.60
0100	5/8 C.Y. excavator	B-12Q	250	.064			2.89	2.36	5.25	7
0110	3/4 C.Y. excavator	B-12F	300	.053			2.40	2.18	4.58	6.05
0300	1/2 C.Y. excavator, truck mounted	B-12J	200	.080			3.61	4.41	8.02	10.35
0500	6' to 10' deep, 3/4 C.Y. excavator	B-12F	225	.071			3.21	2.91	6.12	8.10
0600	1 C.Y. excavator, truck mounted	B-12K	400	.040			1.80	2.49	4.29	5.50
0900	10' to 14' deep, 3/4 C.Y. excavator	B-12F	200	.080			3.61	3.27	6.88	9.10
1000	1-1/2 C.Y. excavator	B-12B	540	.030			1.34	1.91	3.25	4.13
1300	14' to 20' deep, 1 C.Y. excavator	B-12A	320	.050			2.25	2.54	4.79	6.20
1340	20' to 24' deep, 1 C.Y. excavator	"	288	.056			2.50	2.82	5.32	6.90
1352	4' to 6' deep, 1/2 C.Y. excavator w/trench box	B-13H	188	.085			3.84	5.10	8.94	11.50
1354	5/8 C.Y. excavator	"	235	.068			3.07	4.10	7.17	9.20
1356	3/4 C.Y. excavator	B-13G	282	.057			2.56	2.61	5.17	6.75
1362	6' to 10' deep, 3/4 C.Y. excavator w/trench box		212	.075			3.40	3.47	6.87	9
1374	10' to 14' deep, 3/4 C.Y. excavator w/trench box		188	.085			3.84	3.91	7.75	10.15
1376	1-1/2 C.Y. excavator	B-13E	508	.032			1.42	2.19	3.61	4.56
1381	14' to 20' deep, 1 C.Y. excavator w/trench box	B-13D	301	.053			2.40	2.97	5.37	6.90
1386	20' to 24' deep, 1 C.Y. excavator w/trench box	"	271	.059			2.66	3.30	5.96	7.70
1400	By hand with pick and shovel 2' to 6' deep, light soil	1 Clab	8	1			38		38	58
1500	Heavy soil	"	4	2	↓		76		76	116
1700	For tamping backfilled trenches, air tamp, add	A-1G	100	.080	E.C.Y.		3.03	.57	3.60	5.25
1900	Vibrating plate, add	B-18	180	.133	"		5.15	.26	5.41	8.20
2100	Trim sides and bottom for concrete pours, common earth		1500	.016	S.F.		.62	.03	.65	.98
2300	Hardpan	↓	600	.040	"		1.54	.08	1.62	2.46
5020	Loam & Sandy clay with no sheeting or dewatering included									
5050	1' to 4' deep, 3/8 C.Y. tractor loader/backhoe	B-11C	162	.099	B.C.Y.		4.40	2.26	6.66	9.20
5060	1/2 C.Y. excavator	B-11M	216	.074			3.30	1.83	5.13	7
5080	4' to 6' deep, 1/2 C.Y. excavator	"	216	.074			3.30	1.83	5.13	7
5090	5/8 C.Y. excavator	B-12Q	276	.058			2.61	2.14	4.75	6.35
5100	3/4 C.Y. excavator	B-12F	324	.049			2.23	2.02	4.25	5.60
5130	1/2 C.Y. excavator, truck mounted	B-12J	216	.074			3.34	4.08	7.42	9.60
5140	6' to 10' deep, 3/4 C.Y. excavator	B-12F	243	.066			2.97	2.69	5.66	7.50
5160	1 C.Y. excavator, truck mounted	B-12K	432	.037			1.67	2.31	3.98	5.10
5190	10' to 14' deep, 3/4 C.Y. excavator	B-12F	216	.074			3.34	3.03	6.37	8.45
5210	1-1/2 C.Y. excavator	B-12B	583	.027	↓		1.24	1.76	3	3.82

31 23 16.13 Excavating, Trench

		Crew	Daily Output	Labor-Hours	Unit	Material	2016 Bare Costs Labor	Equipment	Total	Total Incl O&P
5250	14' to 20' deep, 1 C.Y. excavator	B-12A	346	.046	B.C.Y.		2.08	2.35	4.43	5.75
5300	20' to 24' deep, 1 C.Y. excavator	"	311	.051			2.32	2.61	4.93	6.40
5352	4' to 6' deep, 1/2 C.Y. excavator w/trench box	B-13H	205	.078			3.52	4.70	8.22	10.50
5354	5/8 C.Y. excavator	"	257	.062			2.81	3.75	6.56	8.40
5356	3/4 C.Y. excavator	B-13G	308	.052			2.34	2.39	4.73	6.20
5362	6' to 10' deep, 3/4 C.Y. excavator w/trench box		231	.069			3.12	3.18	6.30	8.25
5370	10' to 14' deep, 3/4 C.Y. excavator w/trench box		205	.078			3.52	3.59	7.11	9.30
5374	1-1/2 C.Y. excavator	B-13E	554	.029			1.30	2	3.30	4.18
5382	14' to 20' deep, 1 C.Y. excavator w/trench box	B-13D	329	.049			2.19	2.72	4.91	6.35
5392	20' to 24' deep, 1 C.Y. excavator w/trench box	"	295	.054			2.45	3.03	5.48	7.05
6020	Sand & gravel with no sheeting or dewatering included									
6050	1' to 4' deep, 3/8 C.Y. excavator	B-11C	165	.097	B.C.Y.		4.32	2.21	6.53	9
6060	1/2 C.Y. excavator	B-11M	220	.073			3.24	1.80	5.04	6.90
6080	4' to 6' deep, 1/2 C.Y. excavator	"	220	.073			3.24	1.80	5.04	6.90
6090	5/8 C.Y. excavator	B-12Q	275	.058			2.62	2.14	4.76	6.35
6100	3/4 C.Y. excavator	B-12F	330	.048			2.19	1.98	4.17	5.50
6130	1/2 C.Y. excavator, truck mounted	B-12J	220	.073			3.28	4.01	7.29	9.40
6140	6' to 10' deep, 3/4 C.Y. excavator	B-12F	248	.065			2.91	2.64	5.55	7.35
6160	1 C.Y. excavator, truck mounted	B-12K	440	.036			1.64	2.26	3.90	4.98
6190	10' to 14' deep, 3/4 C.Y. excavator	B-12F	220	.073			3.28	2.97	6.25	8.25
6210	1-1/2 C.Y. excavator	B-12B	594	.027			1.21	1.73	2.94	3.76
6250	14' to 20' deep, 1 C.Y. excavator	B-12A	352	.045			2.05	2.31	4.36	5.65
6300	20' to 24' deep, 1 C.Y. excavator	"	317	.050			2.28	2.56	4.84	6.30
6352	4' to 6' deep, 1/2 C.Y. excavator w/trench box	B-13H	209	.077			3.45	4.61	8.06	10.30
6362	6' to 10' deep, 3/4 C.Y. excavator w/trench box	B-13G	236	.068			3.06	3.11	6.17	8.10
6370	10' to 14' deep, 3/4 C.Y. excavator w/trench box	"	209	.077			3.45	3.52	6.97	9.10
6374	1-1/2 C.Y. excavator	B-13E	564	.028			1.28	1.97	3.25	4.11
6382	14' to 20' deep, 1 C.Y. excavator w/trench box	B-13D	334	.048			2.16	2.68	4.84	6.25
6392	20' to 24' deep, 1 C.Y. excavator w/trench box	"	301	.053			2.40	2.97	5.37	6.90
7020	Dense hard clay with no sheeting or dewatering included									
7050	1' to 4' deep, 3/8 C.Y. excavator	B-11C	132	.121	B.C.Y.		5.40	2.77	8.17	11.25
7060	1/2 C.Y. excavator	B-11M	176	.091			4.05	2.25	6.30	8.65
7080	4' to 6' deep, 1/2 C.Y. excavator	"	176	.091			4.05	2.25	6.30	8.65
7090	5/8 C.Y. excavator	B-12Q	220	.073			3.28	2.68	5.96	7.95
7100	3/4 C.Y. excavator	B-12F	264	.061			2.73	2.48	5.21	6.90
7130	1/2 C.Y. excavator, truck mounted	B-12J	176	.091			4.10	5	9.10	11.75
7140	6' to 10' deep, 3/4 C.Y. excavator	B-12F	198	.081			3.64	3.30	6.94	9.20
7160	1 C.Y. excavator, truck mounted	B-12K	352	.045			2.05	2.83	4.88	6.25
7190	10' to 14' deep, 3/4 C.Y. excavator	B-12F	176	.091			4.10	3.72	7.82	10.35
7210	1-1/2 C.Y. excavator	B-12B	475	.034			1.52	2.17	3.69	4.69
7250	14' to 20' deep, 1 C.Y. excavator	B-12A	282	.057			2.56	2.88	5.44	7.05
7300	20' to 24' deep, 1 C.Y. excavator	"	254	.063			2.84	3.20	6.04	7.85

31 23 16.14 Excavating, Utility Trench

		Crew	Daily Output	Labor-Hours	Unit	Material	2016 Bare Costs Labor	Equipment	Total	Total Incl O&P
0010	**EXCAVATING, UTILITY TRENCH**									
0011	Common earth									
0050	Trenching with chain trencher, 12 H.P., operator walking									
0100	4" wide trench, 12" deep	B-53	800	.010	L.F.		.49	.09	.58	.83
0150	18" deep		750	.011			.52	.09	.61	.89
0200	24" deep		700	.011			.56	.10	.66	.96
0300	6" wide trench, 12" deep		650	.012			.61	.10	.71	1.04
0350	18" deep		600	.013			.66	.11	.77	1.11
0400	24" deep		550	.015			.72	.12	.84	1.22

For customer support on your Mechanical Cost Data, call 877.739.3689.

441

31 23 Excavation and Fill

31 23 16 – Excavation

31 23 16.14 Excavating, Utility Trench

		Crew	Daily Output	Labor-Hours	Unit	Material	2016 Bare Costs Labor	Equipment	Total	Total Incl O&P
0450	36" deep	B-53	450	.018	L.F.		.87	.15	1.02	1.49
0600	8" wide trench, 12" deep		475	.017			.83	.14	.97	1.41
0650	18" deep		400	.020			.98	.17	1.15	1.68
0700	24" deep		350	.023			1.12	.19	1.31	1.91
0750	36" deep		300	.027			1.31	.23	1.54	2.23
1000	Backfill by hand including compaction, add									
1050	4" wide trench, 12" deep	A-1G	800	.010	L.F.		.38	.07	.45	.66
1100	18" deep		530	.015			.57	.11	.68	1
1150	24" deep		400	.020			.76	.14	.90	1.32
1300	6" wide trench, 12" deep		540	.015			.56	.11	.67	.98
1350	18" deep		405	.020			.75	.14	.89	1.30
1400	24" deep		270	.030			1.12	.21	1.33	1.95
1450	36" deep		180	.044			1.68	.32	2	2.93
1600	8" wide trench, 12" deep		400	.020			.76	.14	.90	1.32
1650	18" deep		265	.030			1.14	.21	1.35	2
1700	24" deep		200	.040			1.52	.28	1.80	2.64
1750	36" deep		135	.059			2.25	.42	2.67	3.91
2000	Chain trencher, 40 H.P. operator riding									
2050	6" wide trench and backfill, 12" deep	B-54	1200	.007	L.F.		.33	.28	.61	.81
2100	18" deep		1000	.008			.39	.33	.72	.96
2150	24" deep		975	.008			.40	.34	.74	.99
2200	36" deep		900	.009			.44	.37	.81	1.07
2250	48" deep		750	.011			.52	.45	.97	1.28
2300	60" deep		650	.012			.61	.51	1.12	1.49
2400	8" wide trench and backfill, 12" deep		1000	.008			.39	.33	.72	.96
2450	18" deep		950	.008			.41	.35	.76	1.02
2500	24" deep		900	.009			.44	.37	.81	1.07
2550	36" deep		800	.010			.49	.42	.91	1.20
2600	48" deep		650	.012			.61	.51	1.12	1.49
2700	12" wide trench and backfill, 12" deep		975	.008			.40	.34	.74	.99
2750	18" deep		860	.009			.46	.39	.85	1.12
2800	24" deep		800	.010			.49	.42	.91	1.20
2850	36" deep		725	.011			.54	.46	1	1.33
3000	16" wide trench and backfill, 12" deep		835	.010			.47	.40	.87	1.15
3050	18" deep		750	.011			.52	.45	.97	1.28
3100	24" deep		700	.011			.56	.48	1.04	1.38
3200	Compaction with vibratory plate, add								35%	35%
5100	Hand excavate and trim for pipe bells after trench excavation									
5200	8" pipe	1 Clab	155	.052	L.F.		1.96		1.96	3
5300	18" pipe	"	130	.062	"		2.33		2.33	3.58

31 23 16.16 Structural Excavation for Minor Structures

		Crew	Daily Output	Labor-Hours	Unit	Material	2016 Bare Costs Labor	Equipment	Total	Total Incl O&P
0010	**STRUCTURAL EXCAVATION FOR MINOR STRUCTURES**									
0015	Hand, pits to 6' deep, sandy soil	1 Clab	8	1	B.C.Y.		38		38	58
0100	Heavy soil or clay		4	2			76		76	116
0300	Pits 6' to 12' deep, sandy soil		5	1.600			60.50		60.50	93
0500	Heavy soil or clay		3	2.667			101		101	155
0700	Pits 12' to 18' deep, sandy soil		4	2			76		76	116
0900	Heavy soil or clay		2	4			152		152	233
1500	For wet or muck hand excavation, add to above								50%	50%

31 23 Excavation and Fill

31 23 16 – Excavation

31 23 16.30 Drilling and Blasting Rock

	31 23 16.30 Drilling and Blasting Rock	Crew	Daily Output	Labor-Hours	Unit	Material	2016 Bare Costs Labor	Equipment	Total	Total Incl O&P
0010	**DRILLING AND BLASTING ROCK**									
0020	Rock, open face, under 1500 C.Y.	B-47	225	.107	B.C.Y.	3.82	4.51	7.15	15.48	18.95
0100	Over 1500 C.Y.		300	.080		3.82	3.39	5.35	12.56	15.25
0200	Areas where blasting mats are required, under 1500 C.Y.		175	.137		3.82	5.80	9.20	18.82	23
0250	Over 1500 C.Y.		250	.096		3.82	4.06	6.45	14.33	17.45
2200	Trenches, up to 1500 C.Y.		22	1.091		11.10	46	73	130.10	163
2300	Over 1500 C.Y.		26	.923		11.10	39	62	112.10	140

31 23 19 – Dewatering

31 23 19.20 Dewatering Systems

	31 23 19.20 Dewatering Systems	Crew	Daily Output	Labor-Hours	Unit	Material	2016 Bare Costs Labor	Equipment	Total	Total Incl O&P
0010	**DEWATERING SYSTEMS**									
0020	Excavate drainage trench, 2′ wide, 2′ deep	B-11C	90	.178	C.Y.		7.90	4.06	11.96	16.50
0100	2′ wide, 3′ deep, with backhoe loader	"	135	.119			5.25	2.71	7.96	11.05
0200	Excavate sump pits by hand, light soil	1 Clab	7.10	1.127			42.50		42.50	65.50
0300	Heavy soil	"	3.50	2.286			86.50		86.50	133
0500	Pumping 8 hr., attended 2 hrs. per day, including 20 L.F.									
0550	of suction hose & 100 L.F. discharge hose									
0600	2″ diaphragm pump used for 8 hours	B-10H	4	3	Day		140	18.90	158.90	234
0650	4″ diaphragm pump used for 8 hours	B-10I	4	3			140	31.50	171.50	248
0800	8 hrs. attended, 2″ diaphragm pump	B-10H	1	12			560	75.50	635.50	935
0900	3″ centrifugal pump	B-10J	1	12			560	87	647	945
1000	4″ diaphragm pump	B-10I	1	12			560	126	686	990
1100	6″ centrifugal pump	B-10K	1	12			560	375	935	1,250
1300	CMP, incl. excavation 3′ deep, 12″ diameter	B-6	115	.209	L.F.	11.10	8.70	3.18	22.98	29
1400	18″ diameter		100	.240	"	20	10	3.66	33.66	41.50
1600	Sump hole construction, incl. excavation and gravel, pit		1250	.019	C.F.	1.10	.80	.29	2.19	2.75
1700	With 12″ gravel collar, 12″ pipe, corrugated, 16 ga.		70	.343	L.F.	21.50	14.30	5.20	41	51.50
1800	15″ pipe, corrugated, 16 ga.		55	.436		28	18.15	6.65	52.80	65.50
1900	18″ pipe, corrugated, 16 ga.		50	.480		32.50	20	7.30	59.80	74
2000	24″ pipe, corrugated, 14 ga.		40	.600		38.50	25	9.15	72.65	90.50
2200	Wood lining, up to 4′ x 4′, add		300	.080	SFCA	16.30	3.33	1.22	20.85	24.50
9950	See Section 31 23 19.40 for wellpoints									
9960	See Section 31 23 19.30 for deep well systems									

31 23 19.30 Wells

	31 23 19.30 Wells	Crew	Daily Output	Labor-Hours	Unit	Material	2016 Bare Costs Labor	Equipment	Total	Total Incl O&P
0010	**WELLS**									
0011	For dewatering 10′ to 20′ deep, 2′ diameter									
0020	with steel casing, minimum	B-6	165	.145	V.L.F.	38	6.05	2.22	46.27	53.50
0050	Average		98	.245		43	10.20	3.73	56.93	67
0100	Maximum		49	.490		48	20.50	7.45	75.95	92
0300	For dewatering pumps see 01 54 33 in Reference Section									
0500	For domestic water wells, see Section 33 21 13.10									

31 23 19.40 Wellpoints

	31 23 19.40 Wellpoints		Crew	Daily Output	Labor-Hours	Unit	Material	2016 Bare Costs Labor	Equipment	Total	Total Incl O&P
0010	**WELLPOINTS**	R312319-90									
0011	For equipment rental, see 01 54 33 in Reference Section										
0100	Installation and removal of single stage system										
0110	Labor only, .75 labor-hours per L.F.		1 Clab	10.70	.748	LF Hdr		28.50		28.50	43.50
0200	2.0 labor-hours per L.F.		"	4	2	"		76		76	116
0400	Pump operation, 4 @ 6 hr. shifts										
0410	Per 24 hour day		4 Eqlt	1.27	25.197	Day		1,250		1,250	1,875
0500	Per 168 hour week, 160 hr. straight, 8 hr. double time			.18	177	Week		8,750		8,750	13,200
0550	Per 4.3 week month			.04	800	Month		39,300		39,300	59,500
0600	Complete installation, operation, equipment rental, fuel &										

For customer support on your Mechanical Cost Data, call 877.739.3689.

443

31 23 Excavation and Fill

31 23 19 – Dewatering

31 23 19.40 Wellpoints

		Crew	Daily Output	Labor-Hours	Unit	Material	2016 Bare Costs Labor	Equipment	Total	Total Incl O&P
0610	removal of system with 2" wellpoints 5' O.C.									
0700	100' long header, 6" diameter, first month	4 Eqlt	3.23	9.907	LF Hdr	159	485		644	910
0800	Thereafter, per month		4.13	7.748		127	380		507	715
1000	200' long header, 8" diameter, first month		6	5.333		145	262		407	555
1100	Thereafter, per month		8.39	3.814		71.50	187		258.50	365
1300	500' long header, 8" diameter, first month		10.63	3.010		55.50	148		203.50	286
1400	Thereafter, per month		20.91	1.530		40	75		115	158
1600	1,000' long header, 10" diameter, first month		11.62	2.754		47.50	135		182.50	258
1700	Thereafter, per month		41.81	.765		24	37.50		61.50	83.50
1900	Note: above figures include pumping 168 hrs. per week									
1910	and include the pump operator and one stand-by pump.									

31 23 23 – Fill

31 23 23.13 Backfill

		Crew	Daily Output	Labor-Hours	Unit	Material	2016 Bare Costs Labor	Equipment	Total	Total Incl O&P
0010	**BACKFILL**									
0015	By hand, no compaction, light soil	1 Clab	14	.571	L.C.Y.		21.50		21.50	33
0100	Heavy soil		11	.727	"		27.50		27.50	42.50
0300	Compaction in 6" layers, hand tamp, add to above		20.60	.388	E.C.Y.		14.70		14.70	22.50
0400	Roller compaction operator walking, add	B-10A	100	.120			5.60	1.85	7.45	10.55
0500	Air tamp, add	B-9D	190	.211			8.05	1.35	9.40	13.85
0600	Vibrating plate, add	A-1D	60	.133			5.05	.60	5.65	8.40
0800	Compaction in 12" layers, hand tamp, add to above	1 Clab	34	.235			8.90		8.90	13.70
0900	Roller compaction operator walking, add	B-10A	150	.080			3.74	1.23	4.97	7
1000	Air tamp, add	B-9	285	.140			5.40	.80	6.20	9.15
1100	Vibrating plate, add	A-1E	90	.089			3.37	.52	3.89	5.70
1300	Dozer backfilling, bulk, up to 300' haul, no compaction	B-10B	1200	.010	L.C.Y.		.47	1.16	1.63	1.99
1400	Air tamped, add	B-11B	80	.200	E.C.Y.		8.70	3.66	12.36	17.30
1600	Compacting backfill, 6" to 12" lifts, vibrating roller	B-10C	800	.015			.70	2.25	2.95	3.53
1700	Sheepsfoot roller	B-10D	750	.016			.75	2.43	3.18	3.80
1900	Dozer backfilling, trench, up to 300' haul, no compaction	B-10B	900	.013	L.C.Y.		.62	1.55	2.17	2.66
2000	Air tamped, add	B-11B	80	.200	E.C.Y.		8.70	3.66	12.36	17.30
2200	Compacting backfill, 6" to 12" lifts, vibrating roller	B-10C	700	.017			.80	2.57	3.37	4.05
2300	Sheepsfoot roller	B-10D	650	.018			.86	2.81	3.67	4.40

31 23 23.16 Fill By Borrow and Utility Bedding

		Crew	Daily Output	Labor-Hours	Unit	Material	2016 Bare Costs Labor	Equipment	Total	Total Incl O&P
0010	**FILL BY BORROW AND UTILITY BEDDING**									
0049	Utility bedding, for pipe & conduit, not incl. compaction									
0050	Crushed or screened bank run gravel	B-6	150	.160	L.C.Y.	23	6.65	2.44	32.09	38.50
0100	Crushed stone 3/4" to 1/2"		150	.160		24	6.65	2.44	33.09	39.50
0200	Sand, dead or bank		150	.160		20	6.65	2.44	29.09	35
0500	Compacting bedding in trench	A-1D	90	.089	E.C.Y.		3.37	.40	3.77	5.60
0600	If material source exceeds 2 miles, add for extra mileage.									
0610	See Section 31 23 23.20 for hauling mileage add.									

31 23 23.17 General Fill

		Crew	Daily Output	Labor-Hours	Unit	Material	2016 Bare Costs Labor	Equipment	Total	Total Incl O&P
0010	**GENERAL FILL**									
0011	Spread dumped material, no compaction									
0020	By dozer	B-10B	1000	.012	L.C.Y.		.56	1.40	1.96	2.39
0100	By hand	1 Clab	12	.667	"		25.50		25.50	39
0500	Gravel fill, compacted, under floor slabs, 4" deep	B-37	10000	.005	S.F.	.44	.19	.02	.65	.79
0600	6" deep		8600	.006		.66	.22	.02	.90	1.09
0700	9" deep		7200	.007		1.10	.27	.02	1.39	1.64
0800	12" deep		6000	.008		1.54	.32	.03	1.89	2.21
1000	Alternate pricing method, 4" deep		120	.400	E.C.Y.	33	16.05	1.36	50.41	62.50
1100	6" deep		160	.300		33	12.05	1.02	46.07	56

G1030–805

31 23 Excavation and Fill

31 23 23 – Fill

31 23 23.17 General Fill		Crew	Daily Output	Labor-Hours	Unit	Material	2016 Bare Costs Labor	Equipment	Total	Total Incl O&P
1200	9" deep	B-37	200	.240	E.C.Y.	33	9.65	.81	43.46	52
1300	12" deep	↓	220	.218	↓	33	8.75	.74	42.49	50.50
1400	Granular fill				L.C.Y.	22			22	24

31 23 23.20 Hauling

		Crew	Daily Output	Labor-Hours	Unit	Material	2016 Bare Costs Labor	Equipment	Total	Total Incl O&P
0010	**HAULING**									
0011	Excavated or borrow, loose cubic yards									
0012	no loading equipment, including hauling, waiting, loading/dumping									
0013	time per cycle (wait, load, travel, unload or dump & return)									
0014	8 C.Y. truck, 15 MPH ave, cycle 0.5 miles, 10 min. wait/Ld./Uld.	B-34A	320	.025	L.C.Y.		1.08	1.30	2.38	3.06
0016	cycle 1 mile		272	.029			1.27	1.53	2.80	3.61
0018	cycle 2 miles		208	.038			1.66	2.01	3.67	4.72
0020	cycle 4 miles		144	.056			2.40	2.90	5.30	6.80
0022	cycle 6 miles		112	.071			3.09	3.73	6.82	8.75
0024	cycle 8 miles		88	.091			3.93	4.74	8.67	11.15
0026	20 MPH ave, cycle 0.5 mile		336	.024			1.03	1.24	2.27	2.92
0028	cycle 1 mile		296	.027			1.17	1.41	2.58	3.31
0030	cycle 2 miles		240	.033			1.44	1.74	3.18	4.08
0032	cycle 4 miles		176	.045			1.96	2.37	4.33	5.60
0034	cycle 6 miles		136	.059			2.54	3.07	5.61	7.20
0036	cycle 8 miles		112	.071			3.09	3.73	6.82	8.75
0044	25 MPH ave, cycle 4 miles		192	.042			1.80	2.17	3.97	5.10
0046	cycle 6 miles		160	.050			2.16	2.61	4.77	6.15
0048	cycle 8 miles		128	.063			2.70	3.26	5.96	7.65
0050	30 MPH ave, cycle 4 miles		216	.037			1.60	1.93	3.53	4.55
0052	cycle 6 miles		176	.045			1.96	2.37	4.33	5.60
0054	cycle 8 miles		144	.056			2.40	2.90	5.30	6.80
0114	15 MPH ave, cycle 0.5 mile, 15 min. wait/Ld./Uld.		224	.036			1.54	1.86	3.40	4.38
0116	cycle 1 mile		200	.040			1.73	2.09	3.82	4.91
0118	cycle 2 miles		168	.048			2.06	2.48	4.54	5.85
0120	cycle 4 miles		120	.067			2.88	3.48	6.36	8.20
0122	cycle 6 miles		96	.083			3.60	4.35	7.95	10.25
0124	cycle 8 miles		80	.100			4.32	5.20	9.52	12.30
0126	20 MPH ave, cycle 0.5 mile		232	.034			1.49	1.80	3.29	4.23
0128	cycle 1 mile		208	.038			1.66	2.01	3.67	4.72
0130	cycle 2 miles		184	.043			1.88	2.27	4.15	5.35
0132	cycle 4 miles		144	.056			2.40	2.90	5.30	6.80
0134	cycle 6 miles		112	.071			3.09	3.73	6.82	8.75
0136	cycle 8 miles		96	.083			3.60	4.35	7.95	10.25
0144	25 MPH ave, cycle 4 miles		152	.053			2.27	2.75	5.02	6.45
0146	cycle 6 miles		128	.063			2.70	3.26	5.96	7.65
0148	cycle 8 miles		112	.071			3.09	3.73	6.82	8.75
0150	30 MPH ave, cycle 4 miles		168	.048			2.06	2.48	4.54	5.85
0152	cycle 6 miles		144	.056			2.40	2.90	5.30	6.80
0154	cycle 8 miles		120	.067			2.88	3.48	6.36	8.20
0214	15 MPH ave, cycle 0.5 mile, 20 min wait/Ld./Uld.		176	.045			1.96	2.37	4.33	5.60
0216	cycle 1 mile		160	.050			2.16	2.61	4.77	6.15
0218	cycle 2 miles		136	.059			2.54	3.07	5.61	7.20
0220	cycle 4 miles		104	.077			3.32	4.01	7.33	9.40
0222	cycle 6 miles		88	.091			3.93	4.74	8.67	11.15
0224	cycle 8 miles		72	.111			4.80	5.80	10.60	13.65
0226	20 MPH ave, cycle 0.5 mile		176	.045			1.96	2.37	4.33	5.60
0228	cycle 1 mile	↓	168	.048	↓		2.06	2.48	4.54	5.85

For customer support on your Mechanical Cost Data, call 877.739.3689.

445

31 23 Excavation and Fill

31 23 23 – Fill

31 23 23.20 Hauling		Crew	Daily Output	Labor-Hours	Unit	Material	2016 Bare Costs Labor	Equipment	Total	Total Incl O&P
0230	cycle 2 miles	B-34A	144	.056	L.C.Y.		2.40	2.90	5.30	6.80
0232	cycle 4 miles		120	.067			2.88	3.48	6.36	8.20
0234	cycle 6 miles		96	.083			3.60	4.35	7.95	10.25
0236	cycle 8 miles		88	.091			3.93	4.74	8.67	11.15
0244	25 MPH ave, cycle 4 miles		128	.063			2.70	3.26	5.96	7.65
0246	cycle 6 miles		112	.071			3.09	3.73	6.82	8.75
0248	cycle 8 miles		96	.083			3.60	4.35	7.95	10.25
0250	30 MPH ave, cycle 4 miles		136	.059			2.54	3.07	5.61	7.20
0252	cycle 6 miles		120	.067			2.88	3.48	6.36	8.20
0254	cycle 8 miles		104	.077			3.32	4.01	7.33	9.40
0314	15 MPH ave, cycle 0.5 mile, 25 min wait/Ld./Uld.		144	.056			2.40	2.90	5.30	6.80
0316	cycle 1 mile		128	.063			2.70	3.26	5.96	7.65
0318	cycle 2 miles		112	.071			3.09	3.73	6.82	8.75
0320	cycle 4 miles		96	.083			3.60	4.35	7.95	10.25
0322	cycle 6 miles		80	.100			4.32	5.20	9.52	12.30
0324	cycle 8 miles		64	.125			5.40	6.50	11.90	15.30
0326	20 MPH ave, cycle 0.5 mile		144	.056			2.40	2.90	5.30	6.80
0328	cycle 1 mile		136	.059			2.54	3.07	5.61	7.20
0330	cycle 2 miles		120	.067			2.88	3.48	6.36	8.20
0332	cycle 4 miles		104	.077			3.32	4.01	7.33	9.40
0334	cycle 6 miles		88	.091			3.93	4.74	8.67	11.15
0336	cycle 8 miles		80	.100			4.32	5.20	9.52	12.30
0344	25 MPH ave, cycle 4 miles		112	.071			3.09	3.73	6.82	8.75
0346	cycle 6 miles		96	.083			3.60	4.35	7.95	10.25
0348	cycle 8 miles		88	.091			3.93	4.74	8.67	11.15
0350	30 MPH ave, cycle 4 miles		112	.071			3.09	3.73	6.82	8.75
0352	cycle 6 miles		104	.077			3.32	4.01	7.33	9.40
0354	cycle 8 miles		96	.083			3.60	4.35	7.95	10.25
0414	15 MPH ave, cycle 0.5 mile, 30 min wait/Ld./Uld.		120	.067			2.88	3.48	6.36	8.20
0416	cycle 1 mile		112	.071			3.09	3.73	6.82	8.75
0418	cycle 2 miles		96	.083			3.60	4.35	7.95	10.25
0420	cycle 4 miles		80	.100			4.32	5.20	9.52	12.30
0422	cycle 6 miles		72	.111			4.80	5.80	10.60	13.65
0424	cycle 8 miles		64	.125			5.40	6.50	11.90	15.30
0426	20 MPH ave, cycle 0.5 mile		120	.067			2.88	3.48	6.36	8.20
0428	cycle 1 mile		112	.071			3.09	3.73	6.82	8.75
0430	cycle 2 miles		104	.077			3.32	4.01	7.33	9.40
0432	cycle 4 miles		88	.091			3.93	4.74	8.67	11.15
0434	cycle 6 miles		80	.100			4.32	5.20	9.52	12.30
0436	cycle 8 miles		72	.111			4.80	5.80	10.60	13.65
0444	25 MPH ave, cycle 4 miles		96	.083			3.60	4.35	7.95	10.25
0446	cycle 6 miles		88	.091			3.93	4.74	8.67	11.15
0448	cycle 8 miles		80	.100			4.32	5.20	9.52	12.30
0450	30 MPH ave, cycle 4 miles		96	.083			3.60	4.35	7.95	10.25
0452	cycle 6 miles		88	.091			3.93	4.74	8.67	11.15
0454	cycle 8 miles		80	.100			4.32	5.20	9.52	12.30
0514	15 MPH ave, cycle 0.5 mile, 35 min wait/Ld./Uld.		104	.077			3.32	4.01	7.33	9.40
0516	cycle 1 mile		96	.083			3.60	4.35	7.95	10.25
0518	cycle 2 miles		88	.091			3.93	4.74	8.67	11.15
0520	cycle 4 miles		72	.111			4.80	5.80	10.60	13.65
0522	cycle 6 miles		64	.125			5.40	6.50	11.90	15.30
0524	cycle 8 miles		56	.143			6.15	7.45	13.60	17.50
0526	20 MPH ave, cycle 0.5 mile		104	.077			3.32	4.01	7.33	9.40

31 23 23.20 Hauling		Crew	Daily Output	Labor-Hours	Unit	Material	2016 Bare Costs		Total	Total Incl O&P
							Labor	Equipment		
0528	cycle 1 mile	B-34A	96	.083	L.C.Y.		3.60	4.35	7.95	10.25
0530	cycle 2 miles		96	.083			3.60	4.35	7.95	10.25
0532	cycle 4 miles		80	.100			4.32	5.20	9.52	12.30
0534	cycle 6 miles		72	.111			4.80	5.80	10.60	13.65
0536	cycle 8 miles		64	.125			5.40	6.50	11.90	15.30
0544	25 MPH ave, cycle 4 miles		88	.091			3.93	4.74	8.67	11.15
0546	cycle 6 miles		80	.100			4.32	5.20	9.52	12.30
0548	cycle 8 miles		72	.111			4.80	5.80	10.60	13.65
0550	30 MPH ave, cycle 4 miles		88	.091			3.93	4.74	8.67	11.15
0552	cycle 6 miles		80	.100			4.32	5.20	9.52	12.30
0554	cycle 8 miles		72	.111			4.80	5.80	10.60	13.65
1014	12 C.Y. truck, cycle 0.5 mile, 15 MPH ave, 15 min. wait/Ld./Uld.	B-34B	336	.024			1.03	2.06	3.09	3.81
1016	cycle 1 mile		300	.027			1.15	2.30	3.45	4.27
1018	cycle 2 miles		252	.032			1.37	2.74	4.11	5.10
1020	cycle 4 miles		180	.044			1.92	3.84	5.76	7.10
1022	cycle 6 miles		144	.056			2.40	4.80	7.20	8.95
1024	cycle 8 miles		120	.067			2.88	5.75	8.63	10.70
1025	cycle 10 miles		96	.083			3.60	7.20	10.80	13.35
1026	20 MPH ave, cycle 0.5 mile		348	.023			.99	1.99	2.98	3.68
1028	cycle 1 mile		312	.026			1.11	2.21	3.32	4.11
1030	cycle 2 miles		276	.029			1.25	2.50	3.75	4.64
1032	cycle 4 miles		216	.037			1.60	3.20	4.80	5.95
1034	cycle 6 miles		168	.048			2.06	4.11	6.17	7.65
1036	cycle 8 miles		144	.056			2.40	4.80	7.20	8.95
1038	cycle 10 miles		120	.067			2.88	5.75	8.63	10.70
1040	25 MPH ave, cycle 4 miles		228	.035			1.52	3.03	4.55	5.60
1042	cycle 6 miles		192	.042			1.80	3.60	5.40	6.70
1044	cycle 8 miles		168	.048			2.06	4.11	6.17	7.65
1046	cycle 10 miles		144	.056			2.40	4.80	7.20	8.95
1050	30 MPH ave, cycle 4 miles		252	.032			1.37	2.74	4.11	5.10
1052	cycle 6 miles		216	.037			1.60	3.20	4.80	5.95
1054	cycle 8 miles		180	.044			1.92	3.84	5.76	7.10
1056	cycle 10 miles		156	.051			2.22	4.43	6.65	8.20
1060	35 MPH ave, cycle 4 miles		264	.030			1.31	2.62	3.93	4.86
1062	cycle 6 miles		228	.035			1.52	3.03	4.55	5.60
1064	cycle 8 miles		204	.039			1.69	3.39	5.08	6.30
1066	cycle 10 miles		180	.044			1.92	3.84	5.76	7.10
1068	cycle 20 miles		120	.067			2.88	5.75	8.63	10.70
1069	cycle 30 miles		84	.095			4.11	8.20	12.31	15.25
1070	cycle 40 miles		72	.111			4.80	9.60	14.40	17.80
1072	40 MPH ave, cycle 6 miles		240	.033			1.44	2.88	4.32	5.35
1074	cycle 8 miles		216	.037			1.60	3.20	4.80	5.95
1076	cycle 10 miles		192	.042			1.80	3.60	5.40	6.70
1078	cycle 20 miles		120	.067			2.88	5.75	8.63	10.70
1080	cycle 30 miles		96	.083			3.60	7.20	10.80	13.35
1082	cycle 40 miles		72	.111			4.80	9.60	14.40	17.80
1084	cycle 50 miles		60	.133			5.75	11.50	17.25	21.50
1094	45 MPH ave, cycle 8 miles		216	.037			1.60	3.20	4.80	5.95
1096	cycle 10 miles		204	.039			1.69	3.39	5.08	6.30
1098	cycle 20 miles		132	.061			2.62	5.25	7.87	9.70
1100	cycle 30 miles		108	.074			3.20	6.40	9.60	11.90
1102	cycle 40 miles		84	.095			4.11	8.20	12.31	15.25
1104	cycle 50 miles		72	.111			4.80	9.60	14.40	17.80

For customer support on your Mechanical Cost Data, call 877.739.3689.

447

31 23 23.20 Hauling		Crew	Daily Output	Labor-Hours	Unit	Material	2016 Bare Costs Labor	Equipment	Total	Total Incl O&P
1106	50 MPH ave, cycle 10 miles	B-34B	216	.037	L.C.Y.		1.60	3.20	4.80	5.95
1108	cycle 20 miles		144	.056			2.40	4.80	7.20	8.95
1110	cycle 30 miles		108	.074			3.20	6.40	9.60	11.90
1112	cycle 40 miles		84	.095			4.11	8.20	12.31	15.25
1114	cycle 50 miles		72	.111			4.80	9.60	14.40	17.80
1214	15 MPH ave, cycle 0.5 mile, 20 min. wait/Ld./Uld.		264	.030			1.31	2.62	3.93	4.86
1216	cycle 1 mile		240	.033			1.44	2.88	4.32	5.35
1218	cycle 2 miles		204	.039			1.69	3.39	5.08	6.30
1220	cycle 4 miles		156	.051			2.22	4.43	6.65	8.20
1222	cycle 6 miles		132	.061			2.62	5.25	7.87	9.70
1224	cycle 8 miles		108	.074			3.20	6.40	9.60	11.90
1225	cycle 10 miles		96	.083			3.60	7.20	10.80	13.35
1226	20 MPH ave, cycle 0.5 mile		264	.030			1.31	2.62	3.93	4.86
1228	cycle 1 mile		252	.032			1.37	2.74	4.11	5.10
1230	cycle 2 miles		216	.037			1.60	3.20	4.80	5.95
1232	cycle 4 miles		180	.044			1.92	3.84	5.76	7.10
1234	cycle 6 miles		144	.056			2.40	4.80	7.20	8.95
1236	cycle 8 miles		132	.061			2.62	5.25	7.87	9.70
1238	cycle 10 miles		108	.074			3.20	6.40	9.60	11.90
1240	25 MPH ave, cycle 4 miles		192	.042			1.80	3.60	5.40	6.70
1242	cycle 6 miles		168	.048			2.06	4.11	6.17	7.65
1244	cycle 8 miles		144	.056			2.40	4.80	7.20	8.95
1246	cycle 10 miles		132	.061			2.62	5.25	7.87	9.70
1250	30 MPH ave, cycle 4 miles		204	.039			1.69	3.39	5.08	6.30
1252	cycle 6 miles		180	.044			1.92	3.84	5.76	7.10
1254	cycle 8 miles		156	.051			2.22	4.43	6.65	8.20
1256	cycle 10 miles		144	.056			2.40	4.80	7.20	8.95
1260	35 MPH ave, cycle 4 miles		216	.037			1.60	3.20	4.80	5.95
1262	cycle 6 miles		192	.042			1.80	3.60	5.40	6.70
1264	cycle 8 miles		168	.048			2.06	4.11	6.17	7.65
1266	cycle 10 miles		156	.051			2.22	4.43	6.65	8.20
1268	cycle 20 miles		108	.074			3.20	6.40	9.60	11.90
1269	cycle 30 miles		72	.111			4.80	9.60	14.40	17.80
1270	cycle 40 miles		60	.133			5.75	11.50	17.25	21.50
1272	40 MPH ave, cycle 6 miles		192	.042			1.80	3.60	5.40	6.70
1274	cycle 8 miles		180	.044			1.92	3.84	5.76	7.10
1276	cycle 10 miles		156	.051			2.22	4.43	6.65	8.20
1278	cycle 20 miles		108	.074			3.20	6.40	9.60	11.90
1280	cycle 30 miles		84	.095			4.11	8.20	12.31	15.25
1282	cycle 40 miles		72	.111			4.80	9.60	14.40	17.80
1284	cycle 50 miles		60	.133			5.75	11.50	17.25	21.50
1294	45 MPH ave, cycle 8 miles		180	.044			1.92	3.84	5.76	7.10
1296	cycle 10 miles		168	.048			2.06	4.11	6.17	7.65
1298	cycle 20 miles		120	.067			2.88	5.75	8.63	10.70
1300	cycle 30 miles		96	.083			3.60	7.20	10.80	13.35
1302	cycle 40 miles		72	.111			4.80	9.60	14.40	17.80
1304	cycle 50 miles		60	.133			5.75	11.50	17.25	21.50
1306	50 MPH ave, cycle 10 miles		180	.044			1.92	3.84	5.76	7.10
1308	cycle 20 miles		132	.061			2.62	5.25	7.87	9.70
1310	cycle 30 miles		96	.083			3.60	7.20	10.80	13.35
1312	cycle 40 miles		84	.095			4.11	8.20	12.31	15.25
1314	cycle 50 miles		72	.111			4.80	9.60	14.40	17.80
1414	15 MPH ave, cycle 0.5 mile, 25 min. wait/Ld./Uld.		204	.039			1.69	3.39	5.08	6.30

For customer support on your Mechanical Cost Data, call 877.739.3689.

31 23 23 – Fill

31 23 23.20 Hauling		Crew	Daily Output	Labor-Hours	Unit	Material	2016 Bare Costs Labor	2016 Bare Costs Equipment	Total	Total Incl O&P
1416	cycle 1 mile	B-34B	192	.042	L.C.Y.		1.80	3.60	5.40	6.70
1418	cycle 2 miles		168	.048			2.06	4.11	6.17	7.65
1420	cycle 4 miles		132	.061			2.62	5.25	7.87	9.70
1422	cycle 6 miles		120	.067			2.88	5.75	8.63	10.70
1424	cycle 8 miles		96	.083			3.60	7.20	10.80	13.35
1425	cycle 10 miles		84	.095			4.11	8.20	12.31	15.25
1426	20 MPH ave, cycle 0.5 mile		216	.037			1.60	3.20	4.80	5.95
1428	cycle 1 mile		204	.039			1.69	3.39	5.08	6.30
1430	cycle 2 miles		180	.044			1.92	3.84	5.76	7.10
1432	cycle 4 miles		156	.051			2.22	4.43	6.65	8.20
1434	cycle 6 miles		132	.061			2.62	5.25	7.87	9.70
1436	cycle 8 miles		120	.067			2.88	5.75	8.63	10.70
1438	cycle 10 miles		96	.083			3.60	7.20	10.80	13.35
1440	25 MPH ave, cycle 4 miles		168	.048			2.06	4.11	6.17	7.65
1442	cycle 6 miles		144	.056			2.40	4.80	7.20	8.95
1444	cycle 8 miles		132	.061			2.62	5.25	7.87	9.70
1446	cycle 10 miles		108	.074			3.20	6.40	9.60	11.90
1450	30 MPH ave, cycle 4 miles		168	.048			2.06	4.11	6.17	7.65
1452	cycle 6 miles		156	.051			2.22	4.43	6.65	8.20
1454	cycle 8 miles		132	.061			2.62	5.25	7.87	9.70
1456	cycle 10 miles		120	.067			2.88	5.75	8.63	10.70
1460	35 MPH ave, cycle 4 miles		180	.044			1.92	3.84	5.76	7.10
1462	cycle 6 miles		156	.051			2.22	4.43	6.65	8.20
1464	cycle 8 miles		144	.056			2.40	4.80	7.20	8.95
1466	cycle 10 miles		132	.061			2.62	5.25	7.87	9.70
1468	cycle 20 miles		96	.083			3.60	7.20	10.80	13.35
1469	cycle 30 miles		72	.111			4.80	9.60	14.40	17.80
1470	cycle 40 miles		60	.133			5.75	11.50	17.25	21.50
1472	40 MPH ave, cycle 6 miles		168	.048			2.06	4.11	6.17	7.65
1474	cycle 8 miles		156	.051			2.22	4.43	6.65	8.20
1476	cycle 10 miles		144	.056			2.40	4.80	7.20	8.95
1478	cycle 20 miles		96	.083			3.60	7.20	10.80	13.35
1480	cycle 30 miles		84	.095			4.11	8.20	12.31	15.25
1482	cycle 40 miles		60	.133			5.75	11.50	17.25	21.50
1484	cycle 50 miles		60	.133			5.75	11.50	17.25	21.50
1494	45 MPH ave, cycle 8 miles		156	.051			2.22	4.43	6.65	8.20
1496	cycle 10 miles		144	.056			2.40	4.80	7.20	8.95
1498	cycle 20 miles		108	.074			3.20	6.40	9.60	11.90
1500	cycle 30 miles		84	.095			4.11	8.20	12.31	15.25
1502	cycle 40 miles		72	.111			4.80	9.60	14.40	17.80
1504	cycle 50 miles		60	.133			5.75	11.50	17.25	21.50
1506	50 MPH ave, cycle 10 miles		156	.051			2.22	4.43	6.65	8.20
1508	cycle 20 miles		120	.067			2.88	5.75	8.63	10.70
1510	cycle 30 miles		96	.083			3.60	7.20	10.80	13.35
1512	cycle 40 miles		72	.111			4.80	9.60	14.40	17.80
1514	cycle 50 miles		60	.133			5.75	11.50	17.25	21.50
1614	15 MPH ave, cycle 0.5 mile, 30 min. wait/Ld./Uld.		180	.044			1.92	3.84	5.76	7.10
1616	cycle 1 mile		168	.048			2.06	4.11	6.17	7.65
1618	cycle 2 miles		144	.056			2.40	4.80	7.20	8.95
1620	cycle 4 miles		120	.067			2.88	5.75	8.63	10.70
1622	cycle 6 miles		108	.074			3.20	6.40	9.60	11.90
1624	cycle 8 miles		84	.095			4.11	8.20	12.31	15.25
1625	cycle 10 miles		84	.095			4.11	8.20	12.31	15.25

For customer support on your Mechanical Cost Data, call 877.739.3689.

449

31 23 23.20 Hauling		Crew	Daily Output	Labor-Hours	Unit	Material	2016 Bare Costs Labor	2016 Bare Costs Equipment	Total	Total Incl O&P
1626	20 MPH ave, cycle 0.5 mile	B-34B	180	.044	L.C.Y.		1.92	3.84	5.76	7.10
1628	cycle 1 mile		168	.048			2.06	4.11	6.17	7.65
1630	cycle 2 miles		156	.051			2.22	4.43	6.65	8.20
1632	cycle 4 miles		132	.061			2.62	5.25	7.87	9.70
1634	cycle 6 miles		120	.067			2.88	5.75	8.63	10.70
1636	cycle 8 miles		108	.074			3.20	6.40	9.60	11.90
1638	cycle 10 miles		96	.083			3.60	7.20	10.80	13.35
1640	25 MPH ave, cycle 4 miles		144	.056			2.40	4.80	7.20	8.95
1642	cycle 6 miles		132	.061			2.62	5.25	7.87	9.70
1644	cycle 8 miles		108	.074			3.20	6.40	9.60	11.90
1646	cycle 10 miles		108	.074			3.20	6.40	9.60	11.90
1650	30 MPH ave, cycle 4 miles		144	.056			2.40	4.80	7.20	8.95
1652	cycle 6 miles		132	.061			2.62	5.25	7.87	9.70
1654	cycle 8 miles		120	.067			2.88	5.75	8.63	10.70
1656	cycle 10 miles		108	.074			3.20	6.40	9.60	11.90
1660	35 MPH ave, cycle 4 miles		156	.051			2.22	4.43	6.65	8.20
1662	cycle 6 miles		144	.056			2.40	4.80	7.20	8.95
1664	cycle 8 miles		132	.061			2.62	5.25	7.87	9.70
1666	cycle 10 miles		120	.067			2.88	5.75	8.63	10.70
1668	cycle 20 miles		84	.095			4.11	8.20	12.31	15.25
1669	cycle 30 miles		72	.111			4.80	9.60	14.40	17.80
1670	cycle 40 miles		60	.133			5.75	11.50	17.25	21.50
1672	40 MPH ave, cycle 6 miles		144	.056			2.40	4.80	7.20	8.95
1674	cycle 8 miles		132	.061			2.62	5.25	7.87	9.70
1676	cycle 10 miles		120	.067			2.88	5.75	8.63	10.70
1678	cycle 20 miles		96	.083			3.60	7.20	10.80	13.35
1680	cycle 30 miles		72	.111			4.80	9.60	14.40	17.80
1682	cycle 40 miles		60	.133			5.75	11.50	17.25	21.50
1684	cycle 50 miles		48	.167			7.20	14.40	21.60	27
1694	45 MPH ave, cycle 8 miles		144	.056			2.40	4.80	7.20	8.95
1696	cycle 10 miles		132	.061			2.62	5.25	7.87	9.70
1698	cycle 20 miles		96	.083			3.60	7.20	10.80	13.35
1700	cycle 30 miles		84	.095			4.11	8.20	12.31	15.25
1702	cycle 40 miles		60	.133			5.75	11.50	17.25	21.50
1704	cycle 50 miles		60	.133			5.75	11.50	17.25	21.50
1706	50 MPH ave, cycle 10 miles		132	.061			2.62	5.25	7.87	9.70
1708	cycle 20 miles		108	.074			3.20	6.40	9.60	11.90
1710	cycle 30 miles		84	.095			4.11	8.20	12.31	15.25
1712	cycle 40 miles		72	.111			4.80	9.60	14.40	17.80
1714	cycle 50 miles		60	.133			5.75	11.50	17.25	21.50
2000	Hauling, 8 C.Y. truck, small project cost per hour	B-34A	8	1	Hr.		43	52	95	123
2100	12 C.Y. Truck	B-34B	8	1			43	86.50	129.50	161
2150	16.5 C.Y. Truck	B-34C	8	1			43	90.50	133.50	166
2175	18 C.Y. 8 wheel Truck	B-34I	8	1			43	109	152	186
2200	20 C.Y. Truck	B-34D	8	1			43	92.50	135.50	168
2300	Grading at dump, or embankment if required, by dozer	B-10B	1000	.012	L.C.Y.		.56	1.40	1.96	2.39
2310	Spotter at fill or cut, if required	1 Clab	8	1	Hr.		38		38	58
9014	18 C.Y. truck, 8 wheels,15 min. wait/Ld./Uld.,15 MPH, cycle 0.5 mi.	B-34I	504	.016	L.C.Y.		.69	1.72	2.41	2.94
9016	cycle 1 mile		450	.018			.77	1.93	2.70	3.29
9018	cycle 2 miles		378	.021			.91	2.30	3.21	3.91
9020	cycle 4 miles		270	.030			1.28	3.22	4.50	5.45
9022	cycle 6 miles		216	.037			1.60	4.03	5.63	6.85
9024	cycle 8 miles		180	.044			1.92	4.83	6.75	8.20

31 23 23.20 Hauling		Crew	Daily Output	Labor-Hours	Unit	Material	2016 Bare Costs			Total Incl O&P
							Labor	Equipment	Total	
9025	cycle 10 miles	B-34I	144	.056	L.C.Y.		2.40	6.05	8.45	10.30
9026	20 MPH ave, cycle 0.5 mile		522	.015			.66	1.67	2.33	2.83
9028	cycle 1 mile		468	.017			.74	1.86	2.60	3.16
9030	cycle 2 miles		414	.019			.83	2.10	2.93	3.57
9032	cycle 4 miles		324	.025			1.07	2.68	3.75	4.56
9034	cycle 6 miles		252	.032			1.37	3.45	4.82	5.85
9036	cycle 8 miles		216	.037			1.60	4.03	5.63	6.85
9038	cycle 10 miles		180	.044			1.92	4.83	6.75	8.20
9040	25 MPH ave, cycle 4 miles		342	.023			1.01	2.54	3.55	4.33
9042	cycle 6 miles		288	.028			1.20	3.02	4.22	5.15
9044	cycle 8 miles		252	.032			1.37	3.45	4.82	5.85
9046	cycle 10 miles		216	.037			1.60	4.03	5.63	6.85
9050	30 MPH ave, cycle 4 miles		378	.021			.91	2.30	3.21	3.91
9052	cycle 6 miles		324	.025			1.07	2.68	3.75	4.56
9054	cycle 8 miles		270	.030			1.28	3.22	4.50	5.45
9056	cycle 10 miles		234	.034			1.48	3.72	5.20	6.30
9060	35 MPH ave, cycle 4 miles		396	.020			.87	2.20	3.07	3.73
9062	cycle 6 miles		342	.023			1.01	2.54	3.55	4.33
9064	cycle 8 miles		288	.028			1.20	3.02	4.22	5.15
9066	cycle 10 miles		270	.030			1.28	3.22	4.50	5.45
9068	cycle 20 miles		162	.049			2.13	5.35	7.48	9.10
9070	cycle 30 miles		126	.063			2.74	6.90	9.64	11.75
9072	cycle 40 miles		90	.089			3.84	9.65	13.49	16.45
9074	40 MPH ave, cycle 6 miles		360	.022			.96	2.41	3.37	4.11
9076	cycle 8 miles		324	.025			1.07	2.68	3.75	4.56
9078	cycle 10 miles		288	.028			1.20	3.02	4.22	5.15
9080	cycle 20 miles		180	.044			1.92	4.83	6.75	8.20
9082	cycle 30 miles		144	.056			2.40	6.05	8.45	10.30
9084	cycle 40 miles		108	.074			3.20	8.05	11.25	13.70
9086	cycle 50 miles		90	.089			3.84	9.65	13.49	16.45
9094	45 MPH ave, cycle 8 miles		324	.025			1.07	2.68	3.75	4.56
9096	cycle 10 miles		306	.026			1.13	2.84	3.97	4.83
9098	cycle 20 miles		198	.040			1.75	4.39	6.14	7.45
9100	cycle 30 miles		144	.056			2.40	6.05	8.45	10.30
9102	cycle 40 miles		126	.063			2.74	6.90	9.64	11.75
9104	cycle 50 miles		108	.074			3.20	8.05	11.25	13.70
9106	50 MPH ave, cycle 10 miles		324	.025			1.07	2.68	3.75	4.56
9108	cycle 20 miles		216	.037			1.60	4.03	5.63	6.85
9110	cycle 30 miles		162	.049			2.13	5.35	7.48	9.10
9112	cycle 40 miles		126	.063			2.74	6.90	9.64	11.75
9114	cycle 50 miles		108	.074			3.20	8.05	11.25	13.70
9214	20 min. wait/Ld./Uld.,15 MPH, cycle 0.5 mi.		396	.020			.87	2.20	3.07	3.73
9216	cycle 1 mile		360	.022			.96	2.41	3.37	4.11
9218	cycle 2 miles		306	.026			1.13	2.84	3.97	4.83
9220	cycle 4 miles		234	.034			1.48	3.72	5.20	6.30
9222	cycle 6 miles		198	.040			1.75	4.39	6.14	7.45
9224	cycle 8 miles		162	.049			2.13	5.35	7.48	9.10
9225	cycle 10 miles		144	.056			2.40	6.05	8.45	10.30
9226	20 MPH ave, cycle 0.5 mile		396	.020			.87	2.20	3.07	3.73
9228	cycle 1 mile		378	.021			.91	2.30	3.21	3.91
9230	cycle 2 miles		324	.025			1.07	2.68	3.75	4.56
9232	cycle 4 miles		270	.030			1.28	3.22	4.50	5.45
9234	cycle 6 miles		216	.037			1.60	4.03	5.63	6.85

For customer support on your Mechanical Cost Data, call 877.739.3689.

451

31 23 23.20 Hauling		Crew	Daily Output	Labor-Hours	Unit	Material	2016 Bare Costs Labor	Equipment	Total	Total Incl O&P
9236	cycle 8 miles	B-34I	198	.040	L.C.Y.		1.75	4.39	6.14	7.45
9238	cycle 10 miles		162	.049			2.13	5.35	7.48	9.10
9240	25 MPH ave, cycle 4 miles		288	.028			1.20	3.02	4.22	5.15
9242	cycle 6 miles		252	.032			1.37	3.45	4.82	5.85
9244	cycle 8 miles		216	.037			1.60	4.03	5.63	6.85
9246	cycle 10 miles		198	.040			1.75	4.39	6.14	7.45
9250	30 MPH ave, cycle 4 miles		306	.026			1.13	2.84	3.97	4.83
9252	cycle 6 miles		270	.030			1.28	3.22	4.50	5.45
9254	cycle 8 miles		234	.034			1.48	3.72	5.20	6.30
9256	cycle 10 miles		216	.037			1.60	4.03	5.63	6.85
9260	35 MPH ave, cycle 4 miles		324	.025			1.07	2.68	3.75	4.56
9262	cycle 6 miles		288	.028			1.20	3.02	4.22	5.15
9264	cycle 8 miles		252	.032			1.37	3.45	4.82	5.85
9266	cycle 10 miles		234	.034			1.48	3.72	5.20	6.30
9268	cycle 20 miles		162	.049			2.13	5.35	7.48	9.10
9270	cycle 30 miles		108	.074			3.20	8.05	11.25	13.70
9272	cycle 40 miles		90	.089			3.84	9.65	13.49	16.45
9274	40 MPH ave, cycle 6 miles		288	.028			1.20	3.02	4.22	5.15
9276	cycle 8 miles		270	.030			1.28	3.22	4.50	5.45
9278	cycle 10 miles		234	.034			1.48	3.72	5.20	6.30
9280	cycle 20 miles		162	.049			2.13	5.35	7.48	9.10
9282	cycle 30 miles		126	.063			2.74	6.90	9.64	11.75
9284	cycle 40 miles		108	.074			3.20	8.05	11.25	13.70
9286	cycle 50 miles		90	.089			3.84	9.65	13.49	16.45
9294	45 MPH ave, cycle 8 miles		270	.030			1.28	3.22	4.50	5.45
9296	cycle 10 miles		252	.032			1.37	3.45	4.82	5.85
9298	cycle 20 miles		180	.044			1.92	4.83	6.75	8.20
9300	cycle 30 miles		144	.056			2.40	6.05	8.45	10.30
9302	cycle 40 miles		108	.074			3.20	8.05	11.25	13.70
9304	cycle 50 miles		90	.089			3.84	9.65	13.49	16.45
9306	50 MPH ave, cycle 10 miles		270	.030			1.28	3.22	4.50	5.45
9308	cycle 20 miles		198	.040			1.75	4.39	6.14	7.45
9310	cycle 30 miles		144	.056			2.40	6.05	8.45	10.30
9312	cycle 40 miles		126	.063			2.74	6.90	9.64	11.75
9314	cycle 50 miles		108	.074			3.20	8.05	11.25	13.70
9414	25 min. wait/Ld./Uld.,15 MPH, cycle 0.5 mi.		306	.026			1.13	2.84	3.97	4.83
9416	cycle 1 mile		288	.028			1.20	3.02	4.22	5.15
9418	cycle 2 miles		252	.032			1.37	3.45	4.82	5.85
9420	cycle 4 miles		198	.040			1.75	4.39	6.14	7.45
9422	cycle 6 miles		180	.044			1.92	4.83	6.75	8.20
9424	cycle 8 miles		144	.056			2.40	6.05	8.45	10.30
9425	cycle 10 miles		126	.063			2.74	6.90	9.64	11.75
9426	20 MPH ave, cycle 0.5 mile		324	.025			1.07	2.68	3.75	4.56
9428	cycle 1 mile		306	.026			1.13	2.84	3.97	4.83
9430	cycle 2 miles		270	.030			1.28	3.22	4.50	5.45
9432	cycle 4 miles		234	.034			1.48	3.72	5.20	6.30
9434	cycle 6 miles		198	.040			1.75	4.39	6.14	7.45
9436	cycle 8 miles		180	.044			1.92	4.83	6.75	8.20
9438	cycle 10 miles		144	.056			2.40	6.05	8.45	10.30
9440	25 MPH ave, cycle 4 miles		252	.032			1.37	3.45	4.82	5.85
9442	cycle 6 miles		216	.037			1.60	4.03	5.63	6.85
9444	cycle 8 miles		198	.040			1.75	4.39	6.14	7.45
9446	cycle 10 miles		180	.044			1.92	4.83	6.75	8.20

31 23 Excavation and Fill

31 23 23 – Fill

31 23 23.20 Hauling		Crew	Daily Output	Labor-Hours	Unit	Material	2016 Bare Costs Labor	Equipment	Total	Total Incl O&P
9450	30 MPH ave, cycle 4 miles	B-34I	252	.032	L.C.Y.		1.37	3.45	4.82	5.85
9452	cycle 6 miles		234	.034			1.48	3.72	5.20	6.30
9454	cycle 8 miles		198	.040			1.75	4.39	6.14	7.45
9456	cycle 10 miles		180	.044			1.92	4.83	6.75	8.20
9460	35 MPH ave, cycle 4 miles		270	.030			1.28	3.22	4.50	5.45
9462	cycle 6 miles		234	.034			1.48	3.72	5.20	6.30
9464	cycle 8 miles		216	.037			1.60	4.03	5.63	6.85
9466	cycle 10 miles		198	.040			1.75	4.39	6.14	7.45
9468	cycle 20 miles		144	.056			2.40	6.05	8.45	10.30
9470	cycle 30 miles		108	.074			3.20	8.05	11.25	13.70
9472	cycle 40 miles		90	.089			3.84	9.65	13.49	16.45
9474	40 MPH ave, cycle 6 miles		252	.032			1.37	3.45	4.82	5.85
9476	cycle 8 miles		234	.034			1.48	3.72	5.20	6.30
9478	cycle 10 miles		216	.037			1.60	4.03	5.63	6.85
9480	cycle 20 miles		144	.056			2.40	6.05	8.45	10.30
9482	cycle 30 miles		126	.063			2.74	6.90	9.64	11.75
9484	cycle 40 miles		90	.089			3.84	9.65	13.49	16.45
9486	cycle 50 miles		90	.089			3.84	9.65	13.49	16.45
9494	45 MPH ave, cycle 8 miles		234	.034			1.48	3.72	5.20	6.30
9496	cycle 10 miles		216	.037			1.60	4.03	5.63	6.85
9498	cycle 20 miles		162	.049			2.13	5.35	7.48	9.10
9500	cycle 30 miles		126	.063			2.74	6.90	9.64	11.75
9502	cycle 40 miles		108	.074			3.20	8.05	11.25	13.70
9504	cycle 50 miles		90	.089			3.84	9.65	13.49	16.45
9506	50 MPH ave, cycle 10 miles		234	.034			1.48	3.72	5.20	6.30
9508	cycle 20 miles		180	.044			1.92	4.83	6.75	8.20
9510	cycle 30 miles		144	.056			2.40	6.05	8.45	10.30
9512	cycle 40 miles		108	.074			3.20	8.05	11.25	13.70
9514	cycle 50 miles		90	.089			3.84	9.65	13.49	16.45
9614	30 min. wait/Ld./Uld.,15 MPH, cycle 0.5 mi.		270	.030			1.28	3.22	4.50	5.45
9616	cycle 1 mile		252	.032			1.37	3.45	4.82	5.85
9618	cycle 2 miles		216	.037			1.60	4.03	5.63	6.85
9620	cycle 4 miles		180	.044			1.92	4.83	6.75	8.20
9622	cycle 6 miles		162	.049			2.13	5.35	7.48	9.10
9624	cycle 8 miles		126	.063			2.74	6.90	9.64	11.75
9625	cycle 10 miles		126	.063			2.74	6.90	9.64	11.75
9626	20 MPH ave, cycle 0.5 mile		270	.030			1.28	3.22	4.50	5.45
9628	cycle 1 mile		252	.032			1.37	3.45	4.82	5.85
9630	cycle 2 miles		234	.034			1.48	3.72	5.20	6.30
9632	cycle 4 miles		198	.040			1.75	4.39	6.14	7.45
9634	cycle 6 miles		180	.044			1.92	4.83	6.75	8.20
9636	cycle 8 miles		162	.049			2.13	5.35	7.48	9.10
9638	cycle 10 miles		144	.056			2.40	6.05	8.45	10.30
9640	25 MPH ave, cycle 4 miles		216	.037			1.60	4.03	5.63	6.85
9642	cycle 6 miles		198	.040			1.75	4.39	6.14	7.45
9644	cycle 8 miles		180	.044			1.92	4.83	6.75	8.20
9646	cycle 10 miles		162	.049			2.13	5.35	7.48	9.10
9650	30 MPH ave, cycle 4 miles		216	.037			1.60	4.03	5.63	6.85
9652	cycle 6 miles		198	.040			1.75	4.39	6.14	7.45
9654	cycle 8 miles		180	.044			1.92	4.83	6.75	8.20
9656	cycle 10 miles		162	.049			2.13	5.35	7.48	9.10
9660	35 MPH ave, cycle 4 miles		234	.034			1.48	3.72	5.20	6.30
9662	cycle 6 miles		216	.037			1.60	4.03	5.63	6.85

For customer support on your Mechanical Cost Data, call 877.739.3689.

453

31 23 23.20 Hauling		Crew	Daily Output	Labor-Hours	Unit	Material	2016 Bare Costs Labor	Equipment	Total	Total Incl O&P
9664	cycle 8 miles	B-34I	198	.040	L.C.Y.		1.75	4.39	6.14	7.45
9666	cycle 10 miles		180	.044			1.92	4.83	6.75	8.20
9668	cycle 20 miles		126	.063			2.74	6.90	9.64	11.75
9670	cycle 30 miles		108	.074			3.20	8.05	11.25	13.70
9672	cycle 40 miles		90	.089			3.84	9.65	13.49	16.45
9674	40 MPH ave, cycle 6 miles		216	.037			1.60	4.03	5.63	6.85
9676	cycle 8 miles		198	.040			1.75	4.39	6.14	7.45
9678	cycle 10 miles		180	.044			1.92	4.83	6.75	8.20
9680	cycle 20 miles		144	.056			2.40	6.05	8.45	10.30
9682	cycle 30 miles		108	.074			3.20	8.05	11.25	13.70
9684	cycle 40 miles		90	.089			3.84	9.65	13.49	16.45
9686	cycle 50 miles		72	.111			4.80	12.10	16.90	20.50
9694	45 MPH ave, cycle 8 miles		216	.037			1.60	4.03	5.63	6.85
9696	cycle 10 miles		198	.040			1.75	4.39	6.14	7.45
9698	cycle 20 miles		144	.056			2.40	6.05	8.45	10.30
9700	cycle 30 miles		126	.063			2.74	6.90	9.64	11.75
9702	cycle 40 miles		108	.074			3.20	8.05	11.25	13.70
9704	cycle 50 miles		90	.089			3.84	9.65	13.49	16.45
9706	50 MPH ave, cycle 10 miles		198	.040			1.75	4.39	6.14	7.45
9708	cycle 20 miles		162	.049			2.13	5.35	7.48	9.10
9710	cycle 30 miles		126	.063			2.74	6.90	9.64	11.75
9712	cycle 40 miles		108	.074			3.20	8.05	11.25	13.70
9714	cycle 50 miles		90	.089			3.84	9.65	13.49	16.45

31 41 16.10 Sheet Piling Systems		Crew	Daily Output	Labor-Hours	Unit	Material	2016 Bare Costs Labor	Equipment	Total	Total Incl O&P
0010	**SHEET PILING SYSTEMS**									
0020	Sheet piling, 50,000 psi steel, not incl. wales, 22 psf, left in place	B-40	10.81	5.920	Ton	1,575	292	355	2,222	2,600
0100	Drive, extract & salvage [R314116-45]		6	10.667		515	525	635	1,675	2,100
0300	20' deep excavation, 27 psf, left in place		12.95	4.942		1,575	244	295	2,114	2,450
0400	Drive, extract & salvage		6.55	9.771		515	480	580	1,575	1,950
1200	15' deep excavation, 22 psf, left in place		983	.065	S.F.	18.40	3.21	3.88	25.49	30
1300	Drive, extract & salvage		545	.117		5.80	5.80	7	18.60	23
1500	20' deep excavation, 27 psf, left in place		960	.067		23	3.29	3.97	30.26	35
1600	Drive, extract & salvage		485	.132		7.55	6.50	7.85	21.90	27
2100	Rent steel sheet piling and wales, first month				Ton	310			310	340
2200	Per added month					31			31	34
2300	Rental piling left in place, add to rental					1,175			1,175	1,275
2500	Wales, connections & struts, 2/3 salvage					480			480	530
3900	Wood, solid sheeting, incl. wales, braces and spacers, [R314116-40]									
3910	drive, extract & salvage, 8' deep excavation	B-31	330	.121	S.F.	1.74	4.90	.65	7.29	10.15
4000	10' deep, 50 S.F./hr. in & 150 S.F./hr. out		300	.133		1.79	5.40	.72	7.91	11
4100	12' deep, 45 S.F./hr. in & 135 S.F./hr. out		270	.148		1.84	6	.80	8.64	12.10
4200	14' deep, 42 S.F./hr. in & 126 S.F./hr. out		250	.160		1.89	6.45	.86	9.20	12.90
4300	16' deep, 40 S.F./hr. in & 120 S.F./hr. out		240	.167		1.95	6.75	.90	9.60	13.50
4400	18' deep, 38 S.F./hr. in & 114 S.F./hr. out		230	.174		2.02	7.05	.93	10	14.05
4520	Left in place, 8' deep, 55 S.F./hr.		440	.091		3.13	3.67	.49	7.29	9.65
4540	10' deep, 50 S.F./hr.		400	.100		3.29	4.04	.54	7.87	10.40
4560	12' deep, 45 S.F./hr.		360	.111		3.47	4.49	.60	8.56	11.40
4565	14' deep, 42 S.F./hr.		335	.119		3.68	4.83	.64	9.15	12.15

31 41. Shoring

31 41 16 – Sheet Piling

31 41 16.10 Sheet Piling Systems	Crew	Daily Output	Labor-Hours	Unit	Material	2016 Bare Costs Labor	Equipment	Total	Total Incl O&P
4570	16' deep, 40 S.F./hr.	B-31	320	.125	S.F.	3.91	5.05	.67	9.63	12.80
4700	Alternate pricing, left in place, 8' deep		1.76	22.727	M.B.F.	700	920	122	1,742	2,300
4800	Drive, extract and salvage, 8' deep		1.32	30.303	"	625	1,225	163	2,013	2,750

31 52 Cofferdams

31 52 16 – Timber Cofferdams

31 52 16.10 Cofferdams

		Crew	Daily Output	Labor-Hours	Unit	Material	2016 Bare Costs Labor	Equipment	Total	Total Incl O&P
0010	**COFFERDAMS**									
0011	Incl. mobilization and temporary sheeting									
0080	Soldier beams & lagging H piles with 3" wood sheeting									
0090	horizontal between piles, including removal of wales & braces									
0100	No hydrostatic head, 15' deep, 1 line of braces, minimum	B-50	545	.206	S.F.	8.40	9.60	4.54	22.54	29.50
0200	Maximum		495	.226		9.35	10.60	5	24.95	32.50
1300	No hydrostatic head, left in place, 15' dp., 1 line of braces, min.		635	.176		11.20	8.25	3.90	23.35	29.50
1400	Maximum		575	.195		12	9.10	4.30	25.40	32
2350	Lagging only, 3" thick wood between piles 8' O.C., minimum	B-46	400	.120		1.87	5.20	.11	7.18	10.30
2370	Maximum		250	.192		2.80	8.35	.18	11.33	16.35
2400	Open sheeting no bracing, for trenches to 10' deep, min.		1736	.028		.84	1.20	.03	2.07	2.83
2450	Maximum		1510	.032		.93	1.38	.03	2.34	3.22
2500	Tie-back method, add to open sheeting, add, minimum								20%	20%
2550	Maximum								60%	60%
2700	Tie-backs only, based on tie-backs total length, minimum	B-46	86.80	.553	L.F.	14.65	24	.52	39.17	54
2750	Maximum		38.50	1.247	"	26	54	1.17	81.17	115
3500	Tie-backs only, typical average, 25' long		2	24	Ea.	645	1,050	22.50	1,717.50	2,350
3600	35' long		1.58	30.380	"	860	1,325	28.50	2,213.50	3,050
6000	See also Section 31 41 16.10									

For customer support on your Mechanical Cost Data, call 877.739.3689.

455

Division Notes

	CREW	DAILY OUTPUT	LABOR-HOURS	UNIT	BARE COSTS				TOTAL INCL O&P
					MAT.	LABOR	EQUIP.	TOTAL	

Estimating Tips

32 01 00 Operations and Maintenance of Exterior Improvements

- Recycling of asphalt pavement is becoming very popular and is an alternative to removal and replacement. It can be a good value engineering proposal if removed pavement can be recycled, either at the project site or at another site that is reasonably close to the project site. Sections on repair of flexible and rigid pavement are included.

32 10 00 Bases, Ballasts, and Paving

- When estimating paving, keep in mind the project schedule. Also note that prices for asphalt and concrete are generally higher in the cold seasons. Lines for pavement markings, including tactile warning systems and fence lines, are included.

32 90 00 Planting

- The timing of planting and guarantee specifications often dictate the costs for establishing tree and shrub growth and a stand of grass or ground cover. Establish the work performance schedule to coincide with the local planting season. Maintenance and growth guarantees can add from 20%–100% to the total landscaping cost and can be contractually cumbersome. The cost to replace trees and shrubs can be as high as 5% of the total cost, depending on the planting zone, soil conditions, and time of year.

Reference Numbers

Reference numbers are shown at the beginning of some major classifications. These numbers refer to related items in the Reference Section. The reference information may be an estimating procedure, an alternate pricing method, or technical information.

Note: Not all subdivisions listed here necessarily appear. ■

Did you know?

RSMeans Online gives you the same access to RSMeans' data with 24/7 access:

- Quickly locate costs in the searchable database.
- Build cost lists, estimates, and reports in minutes.
- Adjust costs to any location in the U.S. and Canada with the click of a button.

Start your free trial today at **www.RSMeansOnline.com**

RSMeans Online
FROM THE GORDIAN GROUP®

32 12 Flexible Paving

32 12 16 – Asphalt Paving

32 12 16.13 Plant-Mix Asphalt Paving

		Crew	Daily Output	Labor-Hours	Unit	Material	2016 Bare Costs Labor	Equipment	Total	Total Incl O&P
0010	**PLANT-MIX ASPHALT PAVING**									
0020	And large paved areas with no hauling included									
0025	See Section 31 23 23.20 for hauling costs									
0080	Binder course, 1-1/2" thick	B-25	7725	.011	S.Y.	5.60	.47	.36	6.43	7.30
0120	2" thick		6345	.014		7.45	.58	.44	8.47	9.55
0130	2-1/2" thick		5620	.016		9.30	.65	.50	10.45	11.80
0160	3" thick		4905	.018		11.15	.75	.57	12.47	14.05
0170	3-1/2" thick		4520	.019		13.05	.81	.62	14.48	16.25
0200	4" thick		4140	.021		14.90	.89	.68	16.47	18.50
0300	Wearing course, 1" thick	B-25B	10575	.009		3.70	.39	.29	4.38	4.98
0340	1-1/2" thick		7725	.012		6.20	.53	.39	7.12	8.10
0380	2" thick		6345	.015		8.35	.64	.48	9.47	10.70
0420	2-1/2" thick		5480	.018		10.30	.74	.55	11.59	13.05
0460	3" thick		4900	.020		12.25	.83	.62	13.70	15.45
0470	3-1/2" thick		4520	.021		14.40	.90	.67	15.97	17.95
0480	4" thick		4140	.023		16.45	.98	.73	18.16	20.50
0500	Open graded friction course	B-25C	5000	.010		2.16	.41	.49	3.06	3.54
0800	Alternate method of figuring paving costs									
0810	Binder course, 1-1/2" thick	B-25	630	.140	Ton	68.50	5.80	4.44	78.74	89.50
0811	2" thick		690	.128		68.50	5.30	4.05	77.85	88
0812	3" thick		800	.110		68.50	4.58	3.50	76.58	86.50
0813	4" thick		900	.098		68.50	4.08	3.11	75.69	85
0850	Wearing course, 1" thick	B-25B	575	.167		75	7.10	5.30	87.40	99
0851	1-1/2" thick		630	.152		75	6.45	4.82	86.27	97.50
0852	2" thick		690	.139		75	5.90	4.40	85.30	96.50
0853	2-1/2" thick		765	.125		75	5.35	3.97	84.32	95
0854	3" thick		800	.120		75	5.10	3.80	83.90	94.50
1000	Pavement replacement over trench, 2" thick	B-37	90	.533	S.Y.	7.70	21.50	1.81	31.01	43
1050	4" thick		70	.686		15.20	27.50	2.32	45.02	61.50
1080	6" thick		55	.873		24	35	2.96	61.96	83.50

32 12 16.14 Asphaltic Concrete Paving

		Crew	Daily Output	Labor-Hours	Unit	Material	2016 Bare Costs Labor	Equipment	Total	Total Incl O&P
0011	**ASPHALTIC CONCRETE PAVING**, parking lots & driveways									
0015	No asphalt hauling included									
0018	Use 6.05 C.Y. per inch per M.S.F. for hauling									
0020	6" stone base, 2" binder course, 1" topping	B-25C	9000	.005	S.F.	1.83	.23	.27	2.33	2.66
0025	2" binder course, 2" topping		9000	.005		2.28	.23	.27	2.78	3.16
0030	3" binder course, 2" topping		9000	.005		2.70	.23	.27	3.20	3.62
0035	4" binder course, 2" topping		9000	.005		3.11	.23	.27	3.61	4.07
0040	1.5" binder course, 1" topping		9000	.005		1.62	.23	.27	2.12	2.43
0042	3" binder course, 1" topping		9000	.005		2.24	.23	.27	2.74	3.12
0045	3" binder course, 3" topping		9000	.005		3.15	.23	.27	3.65	4.12
0050	4" binder course, 3" topping		9000	.005		3.57	.23	.27	4.07	4.57
0055	4" binder course, 4" topping		9000	.005		4.02	.23	.27	4.52	5.05
0300	Binder course, 1-1/2" thick		35000	.001		.62	.06	.07	.75	.86
0400	2" thick		25000	.002		.81	.08	.10	.99	1.12
0500	3" thick		15000	.003		1.25	.14	.16	1.55	1.76
0600	4" thick		10800	.004		1.63	.19	.23	2.05	2.33
0800	Sand finish course, 3/4" thick		41000	.001		.33	.05	.06	.44	.51
0900	1" thick		34000	.001		.40	.06	.07	.53	.61
1000	Fill pot holes, hot mix, 2" thick	B-16	4200	.008		.85	.30	.16	1.31	1.58
1100	4" thick		3500	.009		1.25	.36	.20	1.81	2.14
1120	6" thick		3100	.010		1.67	.41	.22	2.30	2.72

32 12 Flexible Paving

32 12 16 – Asphalt Paving

32 12 16.14 Asphaltic Concrete Paving	Crew	Daily Output	Labor-Hours	Unit	Material	2016 Bare Costs Labor	Equipment	Total	Total Incl O&P
1140	Cold patch, 2" thick	B-51	3000	.016	S.F.	.97	.62	.08	1.67	2.11
1160	4" thick		2700	.018		1.85	.69	.09	2.63	3.19
1180	6" thick		1900	.025		2.87	.98	.13	3.98	4.80

32 84 Planting Irrigation

32 84 23 – Underground Sprinklers

32 84 23.10 Sprinkler Irrigation System

		Crew	Daily Output	Labor-Hours	Unit	Material	2016 Bare Costs Labor	Equipment	Total	Total Incl O&P
0010	**SPRINKLER IRRIGATION SYSTEM**									
0011	For lawns									
0800	Residential system, custom, 1" supply	B-20	2000	.012	S.F.	.26	.51		.77	1.07
0900	1-1/2" supply	"	1800	.013	"	.49	.57		1.06	1.41

For customer support on your Mechanical Cost Data, call 877.739.3689.

459

Division Notes

	CREW	DAILY OUTPUT	LABOR-HOURS	UNIT	BARE COSTS				TOTAL INCL O&P
					MAT.	LABOR	EQUIP.	TOTAL	

Estimating Tips

33 10 00 Water Utilities
33 30 00 Sanitary Sewerage Utilities
33 40 00 Storm Drainage Utilities

- Never assume that the water, sewer, and drainage lines will go in at the early stages of the project. Consider the site access needs before dividing the site in half with open trenches, loose pipe, and machinery obstructions. Always inspect the site to establish that the site drawings are complete. Check off all existing utilities on your drawings as you locate them. Be especially careful with underground utilities because appurtenances are sometimes buried during regrading or repaving operations. If you find any discrepancies, mark up the site plan for further research. Differing site conditions can be very costly if discovered later in the project.

- See also Section 33 01 00 for restoration of pipe where removal/ replacement may be undesirable. Use of new types of piping materials can reduce the overall project cost. Owners/design engineers should consider the installing contractor as a valuable source of current information on utility products and local conditions that could lead to significant cost savings.

Reference Numbers

Reference numbers are shown at the beginning of some major classifications. These numbers refer to related items in the Reference Section. The reference information may be an estimating procedure, an alternate pricing method, or technical information.

Note: Not all subdivisions listed here necessarily appear. ■

33 01 Operation and Maintenance of Utilities

33 01 10 – Operation and Maintenance of Water Utilities

33 01 10.10 Corrosion Resistance	Crew	Daily Output	Labor-Hours	Unit	Material	2016 Bare Costs Labor	Equipment	Total	Total Incl O&P
0010 **CORROSION RESISTANCE**									
0012 Wrap & coat, add to pipe, 4" diameter				L.F.	2.20			2.20	2.42
0020 5" diameter					2.65			2.65	2.92
0040 6" diameter					3.25			3.25	3.58
0060 8" diameter					4.03			4.03	4.43
0080 10" diameter					4.89			4.89	5.40
0100 12" diameter					6.20			6.20	6.80
0120 14" diameter					7.25			7.25	8
0140 16" diameter					9.75			9.75	10.75
0160 18" diameter					10			10	11
0180 20" diameter					10.45			10.45	11.50
0200 24" diameter					12.85			12.85	14.10
0220 Small diameter pipe, 1" diameter, add					.95			.95	1.05
0240 2" diameter					1.38			1.38	1.52
0260 2-1/2" diameter					1.63			1.63	1.79
0280 3" diameter					2			2	2.20
0300 Fittings, field covered, add				S.F.	10.25			10.25	11.30
0500 Coating, bituminous, per diameter inch, 1 coat, add				L.F.	.65			.65	.72
0540 3 coat					1.95			1.95	2.15
0560 Coal tar epoxy, per diameter inch, 1 coat, add					.25			.25	.28
0600 3 coat					.70			.70	.77
1000 Polyethylene H.D. extruded, .025" thk., 1/2" diameter add					.10			.10	.11
1020 3/4" diameter					.15			.15	.17
1040 1" diameter					.20			.20	.22
1060 1-1/4" diameter					.25			.25	.28
1080 1-1/2" diameter					.30			.30	.33
1100 .030" thk., 2" diameter					.40			.40	.44
1120 2-1/2" diameter					.50			.50	.55
1140 .035" thk., 3" diameter					.60			.60	.66
1160 3-1/2" diameter					.70			.70	.77
1180 4" diameter					.80			.80	.88
1200 .040" thk, 5" diameter					1			1	1.10
1220 6" diameter					1.20			1.20	1.32
1240 8" diameter					1.55			1.55	1.71
1260 10" diameter					1.95			1.95	2.15
1280 12" diameter					2.35			2.35	2.59
1300 .060" thk., 14" diameter					2.75			2.75	3.03
1320 16" diameter					3.15			3.15	3.47
1340 18" diameter					3.55			3.55	3.91
1360 20" diameter					3.95			3.95	4.35
1380 Fittings, field wrapped, add				S.F.	4.80			4.80	5.30

33 01 10.20 Pipe Repair

	Crew	Daily Output	Labor-Hours	Unit	Material	2016 Bare Costs Labor	Equipment	Total	Total Incl O&P
0010 **PIPE REPAIR**									
0020 Not including excavation or backfill									
0100 Clamp, stainless steel, lightweight, for steel pipe									
0110 3" long, 1/2" diameter pipe	1 Plum	34	.235	Ea.	9.75	13.95		23.70	32
0120 3/4" diameter pipe		32	.250		12.55	14.80		27.35	36.50
0130 1" diameter pipe		30	.267		13.35	15.80		29.15	38.50
0140 1-1/4" diameter pipe		28	.286		13.75	16.90		30.65	40.50
0150 1-1/2" diameter pipe		26	.308		14.55	18.20		32.75	43.50
0160 2" diameter pipe		24	.333		16	19.75		35.75	47.50
0170 2-1/2" diameter pipe		23	.348		17.50	20.50		38	50.50

33 01 10.20 Pipe Repair	Crew	Daily Output	Labor-Hours	Unit	Material	2016 Bare Costs Labor	Equipment	Total	Total Incl O&P
0180	3" diameter pipe	1 Plum	22	.364	Ea.	21	21.50		42.50	55.50
0190	3-1/2" diameter pipe	↓	21	.381		24.50	22.50		47	60.50
0200	4" diameter pipe	B-20	44	.545		25	23		48	62.50
0210	5" diameter pipe		42	.571		29.50	24.50		54	69.50
0220	6" diameter pipe		38	.632		34	27		61	79
0230	8" diameter pipe		30	.800		40.50	34		74.50	97
0240	10" diameter pipe		28	.857		115	36.50		151.50	183
0250	12" diameter pipe		24	1		120	42.50		162.50	198
0260	14" diameter pipe		22	1.091		125	46.50		171.50	210
0270	16" diameter pipe		20	1.200		135	51		186	228
0280	18" diameter pipe		18	1.333		145	57		202	247
0290	20" diameter pipe		16	1.500		150	64		214	263
0300	24" diameter pipe	↓	14	1.714		170	73		243	299
0360	For 6" long, add					100%	40%			
0370	For 9" long, add					200%	100%			
0380	For 12" long, add					300%	150%			
0390	For 18" long, add				↓	500%	200%			
0400	Pipe Freezing for live repairs of systems 3/8 inch to 6 inch									
0410	Note: Pipe Freezing can also be used to install a valve into a live system									
0420	Pipe Freezing each side 3/8 inch	2 Skwk	8	2	Ea.	550	100		650	760
0425	Pipe Freezing each side 3/8 inch, second location same kit		8	2		22.50	100		122.50	179
0430	Pipe Freezing each side 3/4 inch		8	2		535	100		635	740
0435	Pipe Freezing each side 3/4 inch, second location same kit		8	2		22.50	100		122.50	179
0440	Pipe Freezing each side 1-1/2 inch		6	2.667		535	133		668	790
0445	Pipe Freezing each side 1-1/2 inch , second location same kit		6	2.667		22.50	133		155.50	230
0450	Pipe Freezing each side 2 inch		6	2.667		920	133		1,053	1,225
0455	Pipe Freezing each side 2 inch, second location same kit		6	2.667		22.50	133		155.50	230
0460	Pipe Freezing each side 2-1/2 inch-3 inch		6	2.667		935	133		1,068	1,225
0465	Pipe Freeze each side 2-1/2 -3 inch, second location same kit		6	2.667		45.50	133		178.50	255
0470	Pipe Freezing each side 4 inch		4	4		1,575	200		1,775	2,025
0475	Pipe Freezing each side 4 inch, second location same kit		4	4		72	200		272	385
0480	Pipe Freezing each side 5-6 inch		4	4		4,450	200		4,650	5,200
0485	Pipe Freezing each side 5-6 inch, second location same kit	↓	4	4		217	200		417	545
0490	Pipe Freezing extra 20 lb. CO$_2$ cylinders (3/8" to 2" - 1 ea, 3" -2 ea)					232			232	255
0500	Pipe Freezing extra 50 lb. CO$_2$ cylinders (4" - 2 ea, 5"-6" -6 ea)				↓	505			505	555
1000	Clamp, stainless steel, with threaded service tap									
1040	Full seal for iron, steel, PVC pipe									
1100	6" long, 2" diameter pipe	1 Plum	17	.471	Ea.	87	28		115	138
1110	2-1/2" diameter pipe		16	.500		90	29.50		119.50	144
1120	3" diameter pipe		15.60	.513		106	30.50		136.50	162
1130	3-1/2" diameter pipe	↓	15	.533		110	31.50		141.50	169
1140	4" diameter pipe	B-20	32	.750		117	32		149	177
1150	6" diameter pipe		28	.857		146	36.50		182.50	216
1160	8" diameter pipe		21	1.143		171	48.50		219.50	263
1170	10" diameter pipe		20	1.200		224	51		275	325
1180	12" diameter pipe	↓	17	1.412		260	60		320	380
1200	8" long, 2" diameter pipe	1 Plum	11.72	.683		180	40.50		220.50	259
1210	2-1/2" diameter pipe		11	.727		185	43		228	269
1220	3" diameter pipe		10.75	.744		190	44		234	276
1230	3-1/2" diameter pipe		10.34	.774		200	46		246	289
1240	4" diameter pipe	B-20	22	1.091		210	46.50		256.50	305
1250	6" diameter pipe		19.31	1.243		235	53		288	340
1260	8" diameter pipe	↓	14.48	1.657	↓	265	70.50		335.50	400

33 01 10.20 Pipe Repair		Crew	Daily Output	Labor-Hours	Unit	Material	2016 Bare Costs Labor	Equipment	Total	Total Incl O&P
1270	10" diameter pipe	B-20	13.80	1.739	Ea.	335	74		409	485
1280	12" diameter pipe	↓	11.72	2.048		370	87		457	540
1300	12" long, 2" diameter pipe	1 Plum	9.44	.847		257	50		307	360
1310	2-1/2" diameter pipe		8.89	.900		265	53.50		318.50	375
1320	3" diameter pipe		8.67	.923		270	54.50		324.50	380
1330	3-1/2" diameter pipe	↓	8.33	.960		280	57		337	395
1340	4" diameter pipe	B-20	17.78	1.350		310	57.50		367.50	430
1350	6" diameter pipe		15.56	1.542		340	65.50		405.50	475
1360	8" diameter pipe		11.67	2.057		390	87.50		477.50	565
1370	10" diameter pipe		11.11	2.160		485	92		577	675
1380	12" diameter pipe	↓	9.44	2.542		555	108		663	775
1400	20" long, 2" diameter pipe	1 Plum	8.10	.988		300	58.50		358.50	420
1410	2-1/2" diameter pipe		7.62	1.050		325	62		387	455
1420	3" diameter pipe		7.43	1.077		360	63.50		423.50	490
1430	3-1/2" diameter pipe	↓	7.14	1.120		380	66.50		446.50	520
1440	4" diameter pipe	B-20	15.24	1.575		455	67		522	605
1450	6" diameter pipe		13.33	1.800		525	76.50		601.50	695
1460	8" diameter pipe		10	2.400		595	102		697	810
1470	10" diameter pipe		9.52	2.521		720	107		827	955
1480	12" diameter pipe	↓	8.10	2.963	↓	835	126		961	1,125
1600	Clamp, stainless steel, single section									
1640	Full seal for iron, steel, PVC pipe									
1700	6" long, 2" diameter pipe	1 Plum	17	.471	Ea.	87	28		115	138
1710	2-1/2" diameter pipe		16	.500		91	29.50		120.50	145
1720	3" diameter pipe		15.60	.513		105	30.50		135.50	162
1730	3-1/2" diameter pipe	↓	15	.533		110	31.50		141.50	169
1740	4" diameter pipe	B-20	32	.750		118	32		150	179
1750	6" diameter pipe		27	.889		145	38		183	218
1760	8" diameter pipe		21	1.143		170	48.50		218.50	262
1770	10" diameter pipe		20	1.200		225	51		276	325
1780	12" diameter pipe	↓	17	1.412		260	60		320	380
1800	8" long, 2" diameter pipe	1 Plum	11.72	.683		118	40.50		158.50	191
1805	2-1/2" diameter pipe		11.03	.725		123	43		166	200
1810	3" diameter pipe		10.76	.743		129	44		173	209
1815	3-1/2" diameter pipe	↓	10.34	.774		139	46		185	222
1820	4" diameter pipe	B-20	22.07	1.087		146	46.50		192.50	232
1825	6" diameter pipe		19.31	1.243		175	53		228	275
1830	8" diameter pipe		14.48	1.657		205	70.50		275.50	335
1835	10" diameter pipe		13.79	1.740		272	74		346	415
1840	12" diameter pipe	↓	11.72	2.048		310	87		397	475
1850	12" long, 2" diameter pipe	1 Plum	9.44	.847		198	50		248	294
1855	2-1/2" diameter pipe		8.89	.900		202	53.50		255.50	305
1860	3" diameter pipe		8.67	.923		210	54.50		264.50	315
1865	3-1/2" diameter pipe	↓	8.33	.960		220	57		277	330
1870	4" diameter pipe	B-20	17.78	1.350		230	57.50		287.50	340
1875	6" diameter pipe		15.56	1.542		275	65.50		340.50	405
1880	8" diameter pipe		11.67	2.057		320	87.50		407.50	485
1885	10" diameter pipe		11.11	2.160		425	92		517	610
1890	12" diameter pipe	↓	9.44	2.542		495	108		603	710
1900	20" long, 2" diameter pipe	1 Plum	8.10	.988		240	58.50		298.50	355
1905	2-1/2" diameter pipe		7.62	1.050		265	62		327	385
1910	3" diameter pipe		7.43	1.077		305	63.50		368.50	430
1915	3-1/2" diameter pipe	↓	7.14	1.120		320	66.50		386.50	450

33 01 Operation and Maintenance of Utilities

33 01 10 – Operation and Maintenance of Water Utilities

33 01 10.20 Pipe Repair		Crew	Daily Output	Labor-Hours	Unit	Material	2016 Bare Costs Labor	2016 Bare Costs Equipment	Total	Total Incl O&P
1920	4" diameter pipe	B-20	15.24	1.575	Ea.	390	67		457	535
1925	6" diameter pipe		13.33	1.800		465	76.50		541.50	630
1930	8" diameter pipe		10	2.400		530	102		632	740
1935	10" diameter pipe		9.52	2.521		650	107		757	880
1940	12" diameter pipe		8.10	2.963		775	126		901	1,050
2000	Clamp, stainless steel, two section									
2040	Full seal, for iron, steel, PVC pipe									
2100	6" long, 4" diameter pipe	B-20	24	1	Ea.	230	42.50		272.50	320
2110	6" diameter pipe		20	1.200		265	51		316	370
2120	8" diameter pipe		13	1.846		300	78.50		378.50	450
2130	10" diameter pipe		12	2		305	85		390	465
2140	12" diameter pipe		10	2.400		390	102		492	585
2200	9" long, 4" diameter pipe		16	1.500		300	64		364	430
2210	6" diameter pipe		13	1.846		340	78.50		418.50	495
2220	8" diameter pipe		9	2.667		380	114		494	590
2230	10" diameter pipe		8	3		495	128		623	740
2240	12" diameter pipe		7	3.429		565	146		711	850
2250	14" diameter pipe		6.40	3.750		635	160		795	945
2260	16" diameter pipe		6	4		710	170		880	1,050
2270	18" diameter pipe		5	4.800		800	204		1,004	1,200
2280	20" diameter pipe		4.60	5.217		885	222		1,107	1,325
2290	24" diameter pipe		4	6		1,175	255		1,430	1,700
2320	For 12" long, add to 9"					15%	25%			
2330	For 18" long, add to 9"					70%	55%			
8000	For internal cleaning and inspection, see Section 33 01 30.16									
8100	For pipe testing, see Section 23 05 93.50									

33 01 30 – Operation and Maintenance of Sewer Utilities

33 01 30.16 TV Inspection of Sewer Pipelines

0010	**TV INSPECTION OF SEWER PIPELINES**									
0100	Pipe internal cleaning & inspection, cleaning, pressure pipe systems									
0120	Pig method, lengths 1000' to 10,000'									
0140	4" diameter thru 24" diameter, minimum				L.F.				3.60	4.14
0160	Maximum				"				18	21
6000	Sewage/sanitary systems									
6100	Power rodder with header & cutters									
6110	Mobilization charge, minimum				Total				695	800
6120	Mobilization charge, maximum				"				9,125	10,600
6140	Cleaning 4"-12" diameter				L.F.				3.39	3.90
6190	14"-24" diameter								3.98	4.59
6240	30" diameter								5.80	6.65
6250	36" diameter								6.75	7.75
6260	48" diameter								7.70	8.85
6270	60" diameter								8.70	9.95
6280	72" diameter								9.65	11.10
9000	Inspection, television camera with video									
9060	up to 500 linear feet				Total				715	820

33 01 30.71 Rehabilitation of Sewer Utilities

0010	**REHABILITATION OF SEWER UTILITIES**									
0011	300' runs, replace with HDPE pipe									
0020	Not including excavation, backfill, shoring, or dewatering									
0100	6" to 15" diameter, minimum				L.F.				104	114
0200	Maximum								214	236

For customer support on your Mechanical Cost Data, call 877.739.3689.

465

33 01 Operation and Maintenance of Utilities

33 01 30 – Operation and Maintenance of Sewer Utilities

33 01 30.71 Rehabilitation of Sewer Utilities

	33 01 30.71 Rehabilitation of Sewer Utilities	Crew	Daily Output	Labor-Hours	Unit	Material	2016 Bare Costs Labor	Equipment	Total	Total Incl O&P
0300	18" to 36" diameter, minimum				L.F.				208	229
0400	Maximum				↓				435	510
0500	Mobilize and demobilize, minimum				Job				3,000	3,300
0600	Maximum				"				32,100	35,400

33 01 30.74 HDPE Pipe Lining

	33 01 30.74 HDPE Pipe Lining	Crew	Daily Output	Labor-Hours	Unit	Material	2016 Bare Costs Labor	Equipment	Total	Total Incl O&P
0010	**HDPE PIPE LINING**, excludes cleaning and video inspection									
0020	Pipe relined with one pipe size smaller than original (4" for 6")									
0100	6" diameter, original size	B-6B	600	.080	L.F.	2.62	3.09	1.51	7.22	9.25
0150	8" diameter, original size		600	.080		7.30	3.09	1.51	11.90	14.45
0200	10" diameter, original size		600	.080		9.10	3.09	1.51	13.70	16.45
0250	12" diameter, original size		400	.120		15.50	4.63	2.26	22.39	26.50
0300	14" diameter, original size	↓	400	.120		23.50	4.63	2.26	30.39	35
0350	16" diameter, original size	B-6C	300	.160		28.50	6.15	5.30	39.95	47
0400	18" diameter, original size	"	300	.160	↓	34.50	6.15	5.30	45.95	53
1000	Pipe HDPE lining, make service line taps	B-6	4	6	Ea.	105	250	91.50	446.50	595

33 05 Common Work Results for Utilities

33 05 23 – Trenchless Utility Installation

33 05 23.19 Microtunneling

	33 05 23.19 Microtunneling	Crew	Daily Output	Labor-Hours	Unit	Material	2016 Bare Costs Labor	Equipment	Total	Total Incl O&P
0010	**MICROTUNNELING**									
0011	Not including excavation, backfill, shoring,									
0020	or dewatering, average 50'/day, slurry method									
0100	24" to 48" outside diameter, minimum				L.F.				875	965
0110	Adverse conditions, add				%				50%	50%
1000	Rent microtunneling machine, average monthly lease				Month				97,500	107,000
1010	Operating technician				Day				630	705
1100	Mobilization and demobilization, minimum				Job				41,200	45,900
1110	Maximum				"				445,500	490,500

33 05 23.20 Horizontal Boring

	33 05 23.20 Horizontal Boring	Crew	Daily Output	Labor-Hours	Unit	Material	2016 Bare Costs Labor	Equipment	Total	Total Incl O&P
0010	**HORIZONTAL BORING**									
0011	Casing only, 100' minimum,									
0020	not incl. jacking pits or dewatering									
0100	Roadwork, 1/2" thick wall, 24" diameter casing	B-42	20	3.200	L.F.	122	137	68.50	327.50	425
0200	36" diameter		16	4		225	172	86	483	610
0300	48" diameter		15	4.267		310	183	91.50	584.50	725
0500	Railroad work, 24" diameter		15	4.267		122	183	91.50	396.50	520
0600	36" diameter		14	4.571		225	196	98	519	660
0700	48" diameter	↓	12	5.333		310	229	114	653	820
0900	For ledge, add								20%	20%
1000	Small diameter boring, 3", sandy soil	B-82	900	.018		22	.77	.09	22.86	25.50
1040	Rocky soil	"	500	.032	↓	22	1.39	.17	23.56	26.50
1100	Prepare jacking pits, incl. mobilization & demobilization, minimum				Ea.				3,225	3,700
1101	Maximum				"				22,000	25,500

33 05 26 – Utility Identification

33 05 26.05 Utility Connection

	33 05 26.05 Utility Connection	Crew	Daily Output	Labor-Hours	Unit	Material	2016 Bare Costs Labor	Equipment	Total	Total Incl O&P
0010	**UTILITY CONNECTION**									
0020	Water, sanitary, stormwater, gas, single connection	B-14	1	48	Ea.	2,925	1,925	365	5,215	6,575
0030	Telecommunication	"	3	16	"	390	640	122	1,152	1,550

33 05 Common Work Results for Utilities

33 05 26 – Utility Identification

33 05 26.10 Utility Accessories		Crew	Daily Output	Labor-Hours	Unit	Material	2016 Bare Costs Labor	Equipment	Total	Total Incl O&P
0010	**UTILITY ACCESSORIES**	G1030-805								
0400	Underground tape, detectable, reinforced, alum. foil core, 2"	1 Clab	150	.053	C.L.F.	6	2.02		8.02	9.70
0500	6"	"	140	.057	"	25.50	2.17		27.67	31.50

33 11 Water Utility Distribution Piping

33 11 13 – Public Water Utility Distribution Piping

33 11 13.15 Water Supply, Ductile Iron Pipe

		Crew	Daily Output	Labor-Hours	Unit	Material	2016 Bare Costs Labor	Equipment	Total	Total Incl O&P
0010	**WATER SUPPLY, DUCTILE IRON PIPE**	R331113-80								
0020	Not including excavation or backfill									
2000	Pipe, class 50 water piping, 18' lengths									
2020	Mechanical joint, 4" diameter	B-21A	200	.200	L.F.	30.50	9.45	2.40	42.35	50.50
2040	6" diameter		160	.250		32	11.85	3	46.85	56.50
2060	8" diameter		133.33	.300		44.50	14.20	3.59	62.29	74.50
2080	10" diameter		114.29	.350		58.50	16.55	4.19	79.24	93.50
2100	12" diameter		105.26	.380		79	18	4.55	101.55	120
2120	14" diameter		100	.400		93	18.95	4.79	116.74	136
2140	16" diameter		72.73	.550		94.50	26	6.60	127.10	151
2160	18" diameter		68.97	.580		126	27.50	6.95	160.45	188
2170	20" diameter		57.14	.700		127	33	8.40	168.40	200
2180	24" diameter		47.06	.850		141	40	10.20	191.20	227
3000	Push-on joint, 4" diameter		400	.100		15.50	4.73	1.20	21.43	25.50
3020	6" diameter		333.33	.120		18.25	5.70	1.44	25.39	30
3040	8" diameter		200	.200		27.50	9.45	2.40	39.35	47.50
3060	10" diameter		181.82	.220		36.50	10.40	2.64	49.54	58.50
3080	12" diameter		160	.250		38	11.85	3	52.85	63.50
3100	14" diameter		133.33	.300		38	14.20	3.59	55.79	67.50
3120	16" diameter		114.29	.350		41	16.55	4.19	61.74	74.50
3140	18" diameter		100	.400		45.50	18.95	4.79	69.24	84
3160	20" diameter		88.89	.450		47.50	21.50	5.40	74.40	90.50
3180	24" diameter		76.92	.520		59	24.50	6.25	89.75	109
8000	Piping, fittings, mechanical joint, AWWA C110									
8006	90° bend, 4" diameter	B-20A	16	2	Ea.	155	92		247	310
8020	6" diameter		12.80	2.500		229	115		344	425
8040	8" diameter		10.67	2.999		450	138		588	705
8060	10" diameter	B-21A	11.43	3.500		620	166	42	828	975
8080	12" diameter		10.53	3.799		880	180	45.50	1,105.50	1,300
8100	14" diameter		10	4		1,200	189	48	1,437	1,675
8120	16" diameter		7.27	5.502		1,525	260	66	1,851	2,150
8140	18" diameter		6.90	5.797		2,125	274	69.50	2,468.50	2,850
8160	20" diameter		5.71	7.005		2,650	330	84	3,064	3,525
8180	24" diameter		4.70	8.511		4,200	405	102	4,707	5,325
8200	Wye or tee, 4" diameter	B-20A	10.67	2.999		375	138		513	625
8220	6" diameter		8.53	3.751		565	173		738	885
8240	8" diameter		7.11	4.501		895	207		1,102	1,300
8260	10" diameter	B-21A	7.62	5.249		1,300	248	63	1,611	1,875
8280	12" diameter		7.02	5.698		1,700	270	68.50	2,038.50	2,350
8300	14" diameter		6.67	5.997		2,750	284	72	3,106	3,525
8320	16" diameter		4.85	8.247		3,050	390	99	3,539	4,050
8340	18" diameter		4.60	8.696		4,125	410	104	4,639	5,275
8360	20" diameter		3.81	10.499		5,750	495	126	6,371	7,250

For customer support on your Mechanical Cost Data, call 877.739.3689.

467

33 11 13 – Public Water Utility Distribution Piping

33 11 13.15 Water Supply, Ductile Iron Pipe	Crew	Daily Output	Labor-Hours	Unit	Material	2016 Bare Costs Labor	Equipment	Total	Total Incl O&P
8380	24" diameter	B-21A	3.14	12.739	Ea.	9,775	605	153	10,533	11,800
8398	45° bends, 4" diameter	B-20A	16	2		255	92		347	420
8400	6" diameter		12.80	2.500		199	115		314	395
8405	8" diameter	↓	10.67	2.999		291	138		429	530
8410	12" diameter	B-21A	10.53	3.799		730	180	45.50	955.50	1,125
8420	16" diameter		7.27	5.502		1,275	260	66	1,601	1,875
8430	20" diameter		5.71	7.005		2,075	330	84	2,489	2,875
8440	24" diameter		4.70	8.511		2,900	405	102	3,407	3,900
8450	Decreaser, 6" x 4" diameter	B-20A	14.22	2.250		207	104		311	385
8460	8" x 6" diameter	"	11.64	2.749		310	127		437	535
8470	10" x 6" diameter	B-21A	13.33	3.001		390	142	36	568	685
8480	12" x 6" diameter		12.70	3.150		550	149	37.50	736.50	875
8490	16" x 6" diameter		10	4		890	189	48	1,127	1,325
8500	20" x 6" diameter	↓	8.42	4.751	↓	1,625	225	57	1,907	2,175
8552	For water utility valves see Section 33 12 16									
8700	Joint restraint, ductile iron mechanical joints									
8710	4" diameter	B-20A	32	1	Ea.	17.15	46		63.15	89
8720	6" diameter		25.60	1.250		21.50	57.50		79	111
8730	8" diameter		21.33	1.500		30	69		99	138
8740	10" diameter		18.28	1.751		38.50	80.50		119	166
8750	12" diameter		16.84	1.900		50	87.50		137.50	188
8760	14" diameter		16	2		53	92		145	199
8770	16" diameter		11.64	2.749		58	127		185	257
8780	18" diameter		11.03	2.901		73	134		207	283
8785	20" diameter		9.14	3.501		78	161		239	330
8790	24" diameter	↓	7.53	4.250		104	196		300	415
9600	Steel sleeve with tap, 4" diameter	B-20	3	8		405	340		745	970
9620	6" diameter		2	12		470	510		980	1,300
9630	8" diameter	↓	2	12	↓	510	510		1,020	1,350

33 11 13.25 Water Supply, Polyvinyl Chloride Pipe

		Crew	Daily Output	Labor-Hours	Unit	Material	2016 Bare Costs Labor	Equipment	Total	Total Incl O&P
0010	**WATER SUPPLY, POLYVINYL CHLORIDE PIPE** R331113-80									
0020	Not including excavation or backfill, unless specified									
2100	PVC pipe, Class 150, 1-1/2" diameter	Q-1A	750	.013	L.F.	.50	.79		1.29	1.75
2120	2" diameter		686	.015		.73	.87		1.60	2.11
2140	2-1/2" diameter	↓	500	.020		1.25	1.19		2.44	3.18
2160	3" diameter	B-20	430	.056	↓	1.40	2.38		3.78	5.20
3010	AWWA C905, PR 100, DR 25									
3030	14" diameter	B-20A	213	.150	L.F.	13.70	6.90		20.60	25.50
3040	16" diameter		200	.160		19.25	7.35		26.60	32
3050	18" diameter		160	.200		24	9.20		33.20	40.50
3060	20" diameter		133	.241		30	11.10		41.10	50
3070	24" diameter		107	.299		43	13.80		56.80	68.50
3080	30" diameter		80	.400		81	18.45		99.45	117
3090	36" diameter		80	.400		126	18.45		144.45	166
3100	42" diameter		60	.533		168	24.50		192.50	223
3200	48" diameter		60	.533		220	24.50		244.50	280
4520	Pressure pipe Class 150, SDR 18, AWWA C900, 4" diameter		380	.084		2.73	3.88		6.61	8.90
4530	6" diameter		316	.101		5.25	4.67		9.92	12.85
4540	8" diameter		264	.121		8.05	5.60		13.65	17.35
4550	10" diameter		220	.145		12.15	6.70		18.85	23.50
4560	12" diameter	↓	186	.172	↓	16.35	7.95		24.30	30
8000	Fittings with rubber gasket									

33 11 13 – Public Water Utility Distribution Piping

33 11 13.25 Water Supply, Polyvinyl Chloride Pipe	Crew	Daily Output	Labor-Hours	Unit	Material	2016 Bare Costs Labor	Equipment	Total	Total Incl O&P
8003	Class 150, DR 18								
8006	90° Bend , 4" diameter	B-20	100	.240	Ea.	42	10.20		52.20	61.50
8020	6" diameter		90	.267		75.50	11.35		86.85	100
8040	8" diameter		80	.300		144	12.75		156.75	178
8060	10" diameter		50	.480		330	20.50		350.50	390
8080	12" diameter		30	.800		420	34		454	515
8100	Tee, 4" diameter		90	.267		59.50	11.35		70.85	83
8120	6" diameter		80	.300		145	12.75		157.75	180
8140	8" diameter		70	.343		183	14.60		197.60	224
8160	10" diameter		40	.600		585	25.50		610.50	685
8180	12" diameter		20	1.200		765	51		816	920
8200	45° Bend, 4" diameter		100	.240		42.50	10.20		52.70	62
8220	6" diameter		90	.267		74.50	11.35		85.85	99.50
8240	8" diameter		50	.480		140	20.50		160.50	186
8260	10" diameter		50	.480		279	20.50		299.50	335
8280	12" diameter		30	.800		360	34		394	455
8300	Reducing tee 6" x 4"		100	.240		104	10.20		114.20	131
8320	8" x 6"		90	.267		167	11.35		178.35	200
8330	10" x 6"		90	.267		196	11.35		207.35	233
8340	10" x 8"		90	.267		215	11.35		226.35	254
8350	12" x 6"		90	.267		243	11.35		254.35	284
8360	12" x 8"		90	.267		266	11.35		277.35	310
8400	Tapped service tee (threaded type) 6" x 6" x 3/4"		100	.240		94.50	10.20		104.70	120
8430	6" x 6" x 1"		90	.267		94.50	11.35		105.85	121
8440	6" x 6" x 1-1/2"		90	.267		94.50	11.35		105.85	121
8450	6" x 6" x 2"		90	.267		94.50	11.35		105.85	121
8460	8" x 8" x 3/4"		90	.267		140	11.35		151.35	170
8470	8" x 8" x 1"		90	.267		140	11.35		151.35	170
8480	8" x 8" x 1-1/2"		90	.267		140	11.35		151.35	170
8490	8" x 8" x 2"		90	.267		140	11.35		151.35	170
8500	Repair coupling 4"		100	.240		27	10.20		37.20	45
8520	6" diameter		90	.267		40.50	11.35		51.85	62
8540	8" diameter		50	.480		97	20.50		117.50	139
8560	10" diameter		50	.480		203	20.50		223.50	255
8580	12" diameter		50	.480		296	20.50		316.50	355
8600	Plug end 4"		100	.240		23.50	10.20		33.70	41
8620	6" diameter		90	.267		40.50	11.35		51.85	62
8640	8" diameter		50	.480		69.50	20.50		90	108
8660	10" diameter		50	.480		97	20.50		117.50	139
8680	12" diameter		50	.480		119	20.50		139.50	163
8700	PVC pipe, joint restraint									
8710	4" diameter	B-20A	32	1	Ea.	45	46		91	120
8720	6" diameter		25.60	1.250		54.50	57.50		112	148
8730	8" diameter		21.33	1.500		81	69		150	194
8740	10" diameter		18.28	1.751		152	80.50		232.50	290
8750	12" diameter		16.84	1.900		159	87.50		246.50	310
8760	14" diameter		16	2		237	92		329	400
8770	16" diameter		11.64	2.749		320	127		447	545
8780	18" diameter		11.03	2.901		395	134		529	640
8785	20" diameter		9.14	3.501		480	161		641	770
8790	24" diameter		7.53	4.250		565	196		761	920

For customer support on your Mechanical Cost Data, call 877.739.3689.

469

33 11 Water Utility Distribution Piping

33 11 13 – Public Water Utility Distribution Piping

33 11 13.35 Water Supply, HDPE

		Crew	Daily Output	Labor-Hours	Unit	Material	2016 Bare Costs Labor	Equipment	Total	Total Incl O&P
0010	**WATER SUPPLY, HDPE**									
0011	Butt fusion joints, SDR 21 40' lengths not including excavation or backfill									
0100	4" diameter	B-22A	400	.100	L.F.	2.62	4.36	1.64	8.62	11.35
0200	6" diameter		380	.105		7.30	4.59	1.73	13.62	16.95
0300	8" diameter		320	.125		9.10	5.45	2.05	16.60	20.50
0400	10" diameter		300	.133		15.50	5.80	2.19	23.49	28.50
0500	12" diameter		260	.154		23.50	6.70	2.53	32.73	38.50
0600	14" diameter	B-22B	220	.182		28.50	7.90	4.93	41.33	49
0700	16" diameter		180	.222		34.50	9.70	6.05	50.25	59
0800	18" diameter		140	.286		58	12.45	7.75	78.20	91.50
0900	24" diameter		100	.400		63	17.45	10.85	91.30	108
1000	Fittings									
1100	Elbows, 90 degrees									
1200	4" diameter	B-22A	32	1.250	Ea.	18.80	54.50	20.50	93.80	127
1300	6" diameter		28	1.429		48.50	62	23.50	134	175
1400	8" diameter		24	1.667		125	72.50	27.50	225	279
1500	10" diameter		18	2.222		345	97	36.50	478.50	570
1600	12" diameter		12	3.333		360	145	54.50	559.50	675
1700	14" diameter	B-22B	9	4.444		465	194	121	780	945
1800	16" diameter		6	6.667		545	290	181	1,016	1,250
1900	18" diameter		4	10		600	435	271	1,306	1,625
2000	24" diameter		3	13.333		965	580	360	1,905	2,375
2100	Tees									
2200	4" diameter	B-22A	30	1.333	Ea.	25	58	22	105	141
2300	6" diameter		26	1.538		58.50	67	25.50	151	196
2400	8" diameter		22	1.818		149	79	30	258	320
2500	10" diameter		15	2.667		196	116	44	356	440
2600	12" diameter		10	4		520	174	65.50	759.50	915
2700	14" diameter	B-22B	8	5		615	218	136	969	1,150
2800	16" diameter		6	6.667		725	290	181	1,196	1,450
2900	18" diameter		4	10		1,050	435	271	1,756	2,125
3000	24" diameter		2	20		1,600	870	545	3,015	3,700
4100	Caps									
4110	4" diameter	B-22A	34	1.176	Ea.	15.10	51.50	19.30	85.90	117
4120	6" diameter		30	1.333		36.50	58	22	116.50	154
4130	8" diameter		26	1.538		60	67	25.50	152.50	197
4150	10" diameter		20	2		185	87	33	305	375
4160	12" diameter		14	2.857		229	124	47	400	495

33 12 Water Utility Distribution Equipment

33 12 13 – Water Service Connections

33 12 13.15 Tapping, Crosses and Sleeves

		Crew	Daily Output	Labor-Hours	Unit	Material	2016 Bare Costs Labor	Equipment	Total	Total Incl O&P
0010	**TAPPING, CROSSES AND SLEEVES**									
4000	Drill and tap pressurized main (labor only)									
4100	6" main, 1" to 2" service	Q-1	3	5.333	Ea.		284		284	430
4150	8" main, 1" to 2" service	"	2.75	5.818	"		310		310	470
4500	Tap and insert gate valve									
4600	8" main, 4" branch	B-21	3.20	8.750	Ea.		385	44	429	640
4650	6" branch		2.70	10.370			455	52	507	755
4700	10" Main, 4" branch		2.70	10.370			455	52	507	755

33 12 13 – Water Service Connections

33 12 13.15 Tapping, Crosses and Sleeves	Crew	Daily Output	Labor-Hours	Unit	Material	2016 Bare Costs Labor	Equipment	Total	Total Incl O&P
4750	6" branch	B-21	2.35	11.915	Ea.		525	59.50	584.50	865
4800	12" main, 6" branch		2.35	11.915			525	59.50	584.50	865
4850	8" branch		2.35	11.915			525	59.50	584.50	865
7020	Crosses, 4" x 4"		37	.757		1,375	33.50	3.79	1,412.29	1,550
7030	6" x 4"		25	1.120		1,150	49	5.60	1,204.60	1,350
7040	6" x 6"		25	1.120		1,175	49	5.60	1,229.60	1,375
7060	8" x 6"		21	1.333		1,500	58.50	6.70	1,565.20	1,750
7080	8" x 8"		21	1.333		1,800	58.50	6.70	1,865.20	2,075
7100	10" x 6"		21	1.333		2,600	58.50	6.70	2,665.20	2,950
7120	10" x 10"		21	1.333		2,650	58.50	6.70	2,715.20	3,025
7140	12" x 6"		18	1.556		2,600	68.50	7.80	2,676.30	2,975
7160	12" x 12"		18	1.556		3,575	68.50	7.80	3,651.30	4,075
7180	14" x 6"		16	1.750		6,475	77	8.75	6,560.75	7,250
7200	14" x 14"		16	1.750		6,900	77	8.75	6,985.75	7,700
7220	16" x 6"		14	2		6,900	88	10	6,998	7,750
7240	16" x 10"		14	2		7,000	88	10	7,098	7,850
7260	16" x 16"		14	2		7,425	88	10	7,523	8,325
7280	18" x 6"		10	2.800		9,900	123	14.05	10,037.05	11,100
7300	18" x 12"		10	2.800		10,100	123	14.05	10,237.05	11,300
7320	18" x 18"		10	2.800		9,650	123	14.05	9,787.05	10,800
7340	20" x 6"		8	3.500		8,475	154	17.55	8,646.55	9,575
7360	20" x 12"		8	3.500		8,800	154	17.55	8,971.55	9,925
7380	20" x 20"		8	3.500		13,400	154	17.55	13,571.55	15,000
7400	24" x 6"		6	4.667		12,400	205	23.50	12,628.50	13,900
7420	24" x 12"		6	4.667		11,600	205	23.50	11,828.50	13,100
7440	24" x 18"		6	4.667		17,400	205	23.50	17,628.50	19,500
7460	24" x 24"		6	4.667		18,000	205	23.50	18,228.50	20,100
7600	Cut-in sleeves with rubber gaskets, 4"		18	1.556		430	68.50	7.80	506.30	590
7620	6"		12	2.333		505	103	11.70	619.70	725
7640	8"		10	2.800		650	123	14.05	787.05	920
7660	10"		10	2.800		960	123	14.05	1,097.05	1,250
7680	12"		9	3.111		1,225	137	15.60	1,377.60	1,575
7800	Cut-in valves with rubber gaskets, 4"		18	1.556		505	68.50	7.80	581.30	670
7820	6"		12	2.333		675	103	11.70	789.70	910
7840	8"		10	2.800		1,050	123	14.05	1,187.05	1,375
7860	10"		10	2.800		1,150	123	14.05	1,287.05	1,450
7880	12"		9	3.111		1,300	137	15.60	1,452.60	1,650
7900	Tapping Valve 4 inch, MJ, ductile iron		18	1.556		470	68.50	7.80	546.30	630
7920	6 inch, MJ, ductile iron		12	2.333		650	103	11.70	764.70	885
8000	Sleeves with rubber gaskets, 4" x 4"		37	.757		845	33.50	3.79	882.29	985
8010	6" x 4"		25	1.120		900	49	5.60	954.60	1,075
8020	6" x 6"		25	1.120		1,125	49	5.60	1,179.60	1,325
8030	8" x 4"		21	1.333		960	58.50	6.70	1,025.20	1,150
8040	8" x 6"		21	1.333		765	58.50	6.70	830.20	940
8060	8" x 8"		21	1.333		1,250	58.50	6.70	1,315.20	1,475
8070	10" x 4"		21	1.333		1,075	58.50	6.70	1,140.20	1,300
8080	10" x 6"		21	1.333		1,150	58.50	6.70	1,215.20	1,350
8090	10" x 8"		21	1.333		1,325	58.50	6.70	1,390.20	1,550
8100	10" x 10"		21	1.333		1,950	58.50	6.70	2,015.20	2,250
8110	12" x 4"		18	1.556		1,150	68.50	7.80	1,226.30	1,400
8120	12" x 6"		18	1.556		1,200	68.50	7.80	1,276.30	1,425
8130	12" x 8"		18	1.556		1,475	68.50	7.80	1,551.30	1,750
8135	12" x 10"		18	1.556		2,325	68.50	7.80	2,401.30	2,675

For customer support on your Mechanical Cost Data, call 877.739.3689.

471

33 12 Water Utility Distribution Equipment

33 12 13 – Water Service Connections

33 12 13.15 Tapping, Crosses and Sleeves

		Crew	Daily Output	Labor-Hours	Unit	Material	2016 Bare Costs Labor	Equipment	Total	Total Incl O&P
8140	12" x 12"	B-21	18	1.556	Ea.	2,350	68.50	7.80	2,426.30	2,725
8160	14" x 6"		16	1.750		1,900	77	8.75	1,985.75	2,225
8180	14" x 14"		16	1.750		2,500	77	8.75	2,585.75	2,875
8200	16" x 6"		14	2		2,100	88	10	2,198	2,450
8220	16" x 10"		14	2		3,525	88	10	3,623	4,025
8240	16" x 16"		14	2		8,650	88	10	8,748	9,675
8260	18" x 6"		10	2.800		2,250	123	14.05	2,387.05	2,675
8280	18" x 12"		10	2.800		3,625	123	14.05	3,762.05	4,175
8300	18" x 18"		10	2.800		11,600	123	14.05	11,737.05	13,000
8320	20" x 6"		8	3.500		2,550	154	17.55	2,721.55	3,075
8340	20" x 12"		8	3.500		4,400	154	17.55	4,571.55	5,100
8360	20" x 20"		8	3.500		11,000	154	17.55	11,171.55	12,400
8380	24" x 6"		6	4.667		2,700	205	23.50	2,928.50	3,325
8400	24" x 12"		6	4.667		4,400	205	23.50	4,628.50	5,200
8420	24" x 18"		6	4.667		11,700	205	23.50	11,928.50	13,200
8440	24" x 24"	▼	6	4.667		14,800	205	23.50	15,028.50	16,600
8800	Hydrant valve box, 6' long	B-20	20	1.200		280	51		331	390
8820	8' long		18	1.333		355	57		412	475
8830	Valve box w/lid 4' deep		14	1.714		108	73		181	230
8840	Valve box and large base w/lid	▼	14	1.714	▼	325	73		398	470

33 12 16 – Water Utility Distribution Valves

33 12 16.10 Valves

		Crew	Daily Output	Labor-Hours	Unit	Material	2016 Bare Costs Labor	Equipment	Total	Total Incl O&P
0010	**VALVES**, water distribution									
0011	See Sections 22 05 23.20 and 22 05 23.60									
3000	Butterfly valves with boxes, cast iron, mech. jt.									
3100	4" diameter	B-6	6	4	Ea.	575	167	61	803	955
3140	6" diameter		6	4		750	167	61	978	1,150
3180	8" diameter		6	4		940	167	61	1,168	1,350
3300	10" diameter		6	4		1,275	167	61	1,503	1,725
3340	12" diameter		6	4		1,700	167	61	1,928	2,200
3400	14" diameter		4	6		3,550	250	91.50	3,891.50	4,400
3440	16" diameter		4	6		5,025	250	91.50	5,366.50	6,000
3460	18" diameter		4	6		6,225	250	91.50	6,566.50	7,325
3480	20" diameter		4	6		8,200	250	91.50	8,541.50	9,475
3500	24" diameter	▼	4	6	▼	13,700	250	91.50	14,041.50	15,600
3600	With lever operator									
3610	4" diameter	B-6	6	4	Ea.	470	167	61	698	835
3614	6" diameter		6	4		645	167	61	873	1,025
3616	8" diameter		6	4		835	167	61	1,063	1,250
3618	10" diameter		6	4		1,175	167	61	1,403	1,600
3620	12" diameter		6	4		1,600	167	61	1,828	2,100
3622	14" diameter		4	6		3,225	250	91.50	3,566.50	4,025
3624	16" diameter		4	6		4,700	250	91.50	5,041.50	5,650
3626	18" diameter		4	6		5,900	250	91.50	6,241.50	6,950
3628	20" diameter		4	6		7,875	250	91.50	8,216.50	9,125
3630	24" diameter	▼	4	6	▼	13,400	250	91.50	13,741.50	15,200
3700	Check valves, flanged									
3710	4" diameter	B-6	6	4	Ea.	675	167	61	903	1,075
3714	6" diameter		6	4		1,150	167	61	1,378	1,600
3716	8" diameter		6	4		2,175	167	61	2,403	2,725
3718	10" diameter		6	4		3,700	167	61	3,928	4,400
3720	12" diameter		6	4	▼	5,775	167	61	6,003	6,675

33 12 16.10 Valves		Crew	Daily Output	Labor-Hours	Unit	Material	2016 Bare Costs Labor	Equipment	Total	Total Incl O&P
3726	18" diameter	B-6	4	6	Ea.	23,700	250	91.50	24,041.50	26,600
3730	24" diameter	↓	4	6	↓	52,000	250	91.50	52,341.50	58,000
3800	Gate valves, C.I., 125 PSI, mechanical joint, w/boxes									
3810	4" diameter	B-6	6	4	Ea.	915	167	61	1,143	1,325
3814	6" diameter		6	4		1,425	167	61	1,653	1,875
3816	8" diameter		6	4		2,400	167	61	2,628	2,975
3818	10" diameter		6	4		4,350	167	61	4,578	5,100
3820	12" diameter		6	4		5,900	167	61	6,128	6,800
3822	14" diameter		4	6		13,000	250	91.50	13,341.50	14,800
3824	16" diameter		4	6		18,300	250	91.50	18,641.50	20,600
3826	18" diameter		4	6		22,100	250	91.50	22,441.50	24,900
3828	20" diameter		4	6		30,700	250	91.50	31,041.50	34,300
3830	24" diameter		4	6		45,500	250	91.50	45,841.50	50,500
3831	30" diameter		4	6		39,200	250	91.50	39,541.50	43,600
3832	36" diameter	↓	4	6	↓	57,000	250	91.50	57,341.50	63,000

33 12 16.20 Valves		Crew	Daily Output	Labor-Hours	Unit	Material	2016 Bare Costs Labor	Equipment	Total	Total Incl O&P
0010	**VALVES**									
0011	Special trim or use									
9000	Valves, gate valve, N.R.S. PIV with post, 4" diameter	B-6	6	4	Ea.	1,925	167	61	2,153	2,450
9020	6" diameter		6	4		2,425	167	61	2,653	3,000
9040	8" diameter		6	4		3,425	167	61	3,653	4,100
9060	10" diameter		6	4		5,375	167	61	5,603	6,225
9080	12" diameter		6	4		6,900	167	61	7,128	7,925
9100	14" diameter		6	4		13,800	167	61	14,028	15,500
9120	OS&Y, 4" diameter		6	4		750	167	61	978	1,150
9140	6" diameter		6	4		1,000	167	61	1,228	1,425
9160	8" diameter		6	4		1,600	167	61	1,828	2,075
9180	10" diameter		6	4		2,500	167	61	2,728	3,075
9200	12" diameter		6	4		3,725	167	61	3,953	4,425
9220	14" diameter		4	6		5,100	250	91.50	5,441.50	6,075
9400	Check valves, rubber disc, 2-1/2" diameter		6	4		525	167	61	753	900
9420	3" diameter		6	4		540	167	61	768	915
9440	4" diameter		6	4		675	167	61	903	1,075
9480	6" diameter		6	4		1,150	167	61	1,378	1,600
9500	8" diameter		6	4		2,175	167	61	2,403	2,725
9520	10" diameter		6	4		3,700	167	61	3,928	4,400
9540	12" diameter		6	4		5,775	167	61	6,003	6,675
9542	14" diameter		4	6		5,875	250	91.50	6,216.50	6,950
9700	Detector check valves, reducing, 4" diameter		6	4		990	167	61	1,218	1,425
9720	6" diameter		6	4		1,550	167	61	1,778	2,025
9740	8" diameter		6	4		2,300	167	61	2,528	2,875
9760	10" diameter		6	4		4,325	167	61	4,553	5,100
9800	Galvanized, 4" diameter		6	4		1,575	167	61	1,803	2,075
9820	6" diameter		6	4		2,125	167	61	2,353	2,675
9840	8" diameter		6	4		3,350	167	61	3,578	4,000
9860	10" diameter	↓	6	4	↓	5,125	167	61	5,353	5,975

33 16 Water Utility Storage Tanks

33 16 13 – Aboveground Water Utility Storage Tanks

33 16 13.13 Steel Water Storage Tanks

		Crew	Daily Output	Labor-Hours	Unit	Material	2016 Bare Costs Labor	Equipment	Total	Total Incl O&P
0010	**STEEL WATER STORAGE TANKS**									
0910	Steel, ground level, ht./diam. less than 1, not incl. fdn., 100,000 gallons				Ea.				202,000	244,500
1000	250,000 gallons								295,500	324,000
1200	500,000 gallons								417,000	458,500
1250	750,000 gallons								538,000	591,500
1300	1,000,000 gallons								558,000	725,500
1500	2,000,000 gallons								1,043,000	1,148,000
1600	4,000,000 gallons								2,121,000	2,333,000
1800	6,000,000 gallons								3,095,000	3,405,000
1850	8,000,000 gallons								4,068,000	4,475,000
1910	10,000,000 gallons								5,050,000	5,554,500
2100	Steel standpipes, ht./diam. more than 1,100' to overflow, no fdn.									
2200	500,000 gallons				Ea.				546,500	600,500
2400	750,000 gallons								722,500	794,500
2500	1,000,000 gallons								1,060,500	1,167,000
2700	1,500,000 gallons								1,749,000	1,923,000
2800	2,000,000 gallons								2,327,000	2,559,000

33 16 13.16 Prestressed Conc. Water Storage Tanks

		Crew	Daily Output	Labor-Hours	Unit	Material	2016 Bare Costs Labor	Equipment	Total	Total Incl O&P
0010	**PRESTRESSED CONC. WATER STORAGE TANKS**									
0020	Not including fdn., pipe or pumps, 250,000 gallons				Ea.				299,000	329,500
0100	500,000 gallons								487,000	536,000
0300	1,000,000 gallons								707,000	807,500
0400	2,000,000 gallons								1,072,000	1,179,000
0600	4,000,000 gallons								1,706,000	1,877,000
0700	6,000,000 gallons								2,266,000	2,493,000
0750	8,000,000 gallons								2,924,000	3,216,000
0800	10,000,000 gallons								3,533,000	3,886,000

33 16 13.23 Plastic-Coated Fabric Pillow Water Tanks

		Crew	Daily Output	Labor-Hours	Unit	Material	2016 Bare Costs Labor	Equipment	Total	Total Incl O&P
0010	**PLASTIC-COATED FABRIC PILLOW WATER TANKS**									
7000	Water tanks, vinyl coated fabric pillow tanks, freestanding, 5,000 gallons	4 Clab	4	8	Ea.	3,675	305		3,980	4,525
7100	Supporting embankment not included, 25,000 gallons	6 Clab	2	24		13,200	910		14,110	15,900
7200	50,000 gallons	8 Clab	1.50	42.667		18,500	1,625		20,125	22,800
7300	100,000 gallons	9 Clab	.90	80		42,300	3,025		45,325	51,500
7400	150,000 gallons		.50	144		60,500	5,450		65,950	75,500
7500	200,000 gallons		.40	180		75,000	6,825		81,825	93,000
7600	250,000 gallons		.30	240		105,500	9,100		114,600	130,000

33 16 19 – Elevated Water Utility Storage Tanks

33 16 19.50 Elevated Water Storage Tanks

		Crew	Daily Output	Labor-Hours	Unit	Material	2016 Bare Costs Labor	Equipment	Total	Total Incl O&P
0010	**ELEVATED WATER STORAGE TANKS**									
0011	Not incl. pipe, pumps or foundation									
3000	Elevated water tanks, 100' to bottom capacity line, incl. painting									
3010	50,000 gallons				Ea.				185,000	204,000
3300	100,000 gallons								280,000	307,500
3400	250,000 gallons								751,500	826,500
3600	500,000 gallons								1,336,000	1,470,000
3700	750,000 gallons								1,622,000	1,783,500
3900	1,000,000 gallons								2,322,000	2,556,000

33 21 13.10 Wells and Accessories	Crew	Daily Output	Labor-Hours	Unit	Material	2016 Bare Costs Labor	2016 Bare Costs Equipment	Total	Total Incl O&P
0010 **WELLS & ACCESSORIES**									
0011 Domestic									
0100 Drilled, 4" to 6" diameter	B-23	120	.333	L.F.		12.75	24	36.75	45.50
0200 8" diameter	"	95.20	.420	"		16.10	30	46.10	57.50
0400 Gravel pack well, 40' deep, incl. gravel & casing, complete									
0500 24" diameter casing x 18" diameter screen	B-23	.13	307	Total	38,800	11,800	22,000	72,600	85,000
0600 36" diameter casing x 18" diameter screen		.12	333	"	40,000	12,800	23,800	76,600	90,000
0800 Observation wells, 1-1/4" riser pipe		163	.245	V.L.F.	20.50	9.40	17.55	47.45	56
0900 For flush Buffalo roadway box, add	1 Skwk	16.60	.482	Ea.	51	24		75	93
1200 Test well, 2-1/2" diameter, up to 50' deep (15 to 50 GPM)	B-23	1.51	26.490	"	805	1,025	1,900	3,730	4,525
1300 Over 50' deep, add	"	121.80	.328	L.F.	21.50	12.60	23.50	57.60	69
1500 Pumps, installed in wells to 100' deep, 4" submersible									
1510 1/2 H.P.	Q-1	3.22	4.969	Ea.	730	265		995	1,200
1520 3/4 H.P.		2.66	6.015		855	320		1,175	1,425
1600 1 H.P.		2.29	6.987		905	370		1,275	1,550
1700 1-1/2 H.P.	Q-22	1.60	10		1,800	535	410	2,745	3,225
1800 2 H.P.		1.33	12.030		1,725	640	495	2,860	3,400
1900 3 H.P.		1.14	14.035		2,150	750	580	3,480	4,125
2000 5 H.P.		1.14	14.035		2,925	750	580	4,255	4,975
2050 Remove and install motor only, 4 H.P.		1.14	14.035		1,300	750	580	2,630	3,175
3000 Pump, 6" submersible, 25' to 150' deep, 25 H.P., 249 to 297 GPM		.89	17.978		3,400	960	740	5,100	6,025
3100 25' to 500' deep, 30 H.P., 100 to 300 GPM		.73	21.918		4,300	1,175	900	6,375	7,475
8110 Well screen assembly, stainless steel, 2" diameter	B-23A	273	.088	L.F.	81	3.78	9.90	94.68	106
8120 3" diameter		253	.095		145	4.08	10.65	159.73	178
8130 4" diameter		200	.120		181	5.15	13.50	199.65	222
8140 5" diameter		168	.143		190	6.15	16.05	212.20	236
8150 6" diameter		126	.190		215	8.20	21.50	244.70	273
8160 8" diameter		98.50	.244		290	10.45	27.50	327.95	365
8170 10" diameter		73	.329		360	14.15	37	411.15	455
8180 12" diameter		62.50	.384		415	16.50	43	474.50	530
8190 14" diameter		54.30	.442		465	19	49.50	533.50	595
8200 16" diameter		48.30	.497		520	21.50	56	597.50	665
8210 18" diameter		39.20	.612		650	26.50	69	745.50	830
8220 20" diameter		31.20	.769		725	33	86.50	844.50	945
8230 24" diameter		23.80	1.008		875	43.50	113	1,031.50	1,150
8240 26" diameter		21	1.143		960	49	129	1,138	1,275
8244 Well casing or drop pipe, PVC, 0.5" dia		550	.044		1.13	1.88	4.91	7.92	9.50
8245 0.75" diameter		550	.044		1.17	1.88	4.91	7.96	9.55
8246 1" diameter		550	.044		1.35	1.88	4.91	8.14	9.75
8247 1.25" diameter		520	.046		1.55	1.98	5.20	8.73	10.45
8248 1.5" diameter		490	.049		1.62	2.10	5.50	9.22	11.05
8249 1.75" diameter		380	.063		1.67	2.71	7.10	11.48	13.80
8250 2" diameter		280	.086		2.15	3.68	9.65	15.48	18.55
8252 3" diameter		260	.092		3.40	3.97	10.40	17.77	21
8254 4" diameter		205	.117		4.25	5.05	13.15	22.45	27
8255 5" diameter		170	.141		4.41	6.05	15.90	26.36	31.50
8256 6" diameter		130	.185		8.60	7.95	21	37.55	44.50
8258 8" diameter		100	.240		13.05	10.30	27	50.35	59.50
8260 10" diameter		73	.329		13.60	14.15	37	64.75	77
8262 12" diameter		62	.387		16.65	16.65	43.50	76.80	92
8300 Slotted PVC, 1-1/4" diameter		521	.046		2.38	1.98	5.20	9.56	11.35
8310 1-1/2" diameter		488	.049		2.85	2.11	5.55	10.51	12.45
8320 2" diameter		273	.088		3.70	3.78	9.90	17.38	20.50

For customer support on your Mechanical Cost Data, call 877.739.3689.

475

33 21 13.10 Wells and Accessories

		Crew	Daily Output	Labor-Hours	Unit	Material	2016 Bare Costs Labor	2016 Bare Costs Equipment	Total	Total Incl O&P
8330	3" diameter	B-23A	253	.095	L.F.	5.85	4.08	10.65	20.58	24.50
8340	4" diameter		200	.120		6.80	5.15	13.50	25.45	30
8350	5" diameter		168	.143		14.10	6.15	16.05	36.30	42.50
8360	6" diameter		126	.190		16.75	8.20	21.50	46.45	54.50
8370	8" diameter		98.50	.244		26.50	10.45	27.50	64.45	75
8400	Artificial gravel pack, 2" screen, 6" casing	B-23B	174	.138		4.69	5.95	17.55	28.19	33.50
8405	8" casing		111	.216		6	9.30	27.50	42.80	51
8410	10" casing		74.50	.322		8.60	13.85	41	63.45	75.50
8415	12" casing		60	.400		12	17.20	51	80.20	95
8420	14" casing		50.20	.478		14	20.50	61	95.50	114
8425	16" casing		40.70	.590		17.70	25.50	75	118.20	141
8430	18" casing		36	.667		21.50	28.50	84.50	134.50	160
8435	20" casing		29.50	.814		25	35	103	163	195
8440	24" casing		25.70	.934		28.50	40	119	187.50	224
8445	26" casing		24.60	.976		31.50	42	124	197.50	235
8450	30" casing		20	1.200		36.50	51.50	152	240	287
8455	36" casing		16.40	1.463		41	63	186	290	345
8500	Develop well		8	3	Hr.	262	129	380	771	905
8550	Pump test well		8	3		83	129	380	592	710
8560	Standby well	B-23A	8	3		78.50	129	335	542.50	655
8570	Standby, drill rig		8	3			129	335	464	565
8580	Surface seal well, concrete filled		1	24	Ea.	855	1,025	2,700	4,580	5,500
8590	Well test pump, install & remove	B-23	1	40			1,525	2,850	4,375	5,500
8600	Well sterilization, chlorine	2 Clab	1	16		122	605		727	1,075
8610	Well water pressure switch	1 Clab	12	.667		121	25.50		146.50	172
8630	Well, water pressure switch with manual reset	"	12	.667		121	25.50		146.50	172
9950	See Section 31 23 19.40 for wellpoints									
9960	See Section 31 23 19.30 for drainage wells									

33 21 13.20 Water Supply Wells, Pumps

		Crew	Daily Output	Labor-Hours	Unit	Material	2016 Bare Costs Labor	2016 Bare Costs Equipment	Total	Total Incl O&P
0010	**WATER SUPPLY WELLS, PUMPS**									
0011	With pressure control									
1000	Deep well, jet, 42 gal. galvanized tank									
1040	3/4 HP	1 Plum	.80	10	Ea.	1,100	590		1,690	2,125
3000	Shallow well, jet, 30 gal. galvanized tank									
3040	1/2 HP	1 Plum	2	4	Ea.	895	237		1,132	1,350

33 31 Sanitary Utility Sewerage Piping

33 31 13 – Public Sanitary Utility Sewerage Piping

33 31 13.15 Sewage Collection, Concrete Pipe

		Crew	Daily Output	Labor-Hours	Unit	Material	2016 Bare Costs Labor	2016 Bare Costs Equipment	Total	Total Incl O&P
0010	**SEWAGE COLLECTION, CONCRETE PIPE**									
0020	See Section 33 41 13.60 for sewage/drainage collection, concrete pipe									

33 31 13.25 Sewage Collection, Polyvinyl Chloride Pipe

		Crew	Daily Output	Labor-Hours	Unit	Material	2016 Bare Costs Labor	2016 Bare Costs Equipment	Total	Total Incl O&P
0010	**SEWAGE COLLECTION, POLYVINYL CHLORIDE PIPE**									
0020	Not including excavation or backfill									
2000	20' lengths, SDR 35, B&S, 4" diameter	B-20	375	.064	L.F.	1.49	2.72		4.21	5.85
2040	6" diameter		350	.069		3.28	2.92		6.20	8.10
2080	13' lengths , SDR 35, B&S, 8" diameter		335	.072		6.15	3.05		9.20	11.45
2120	10" diameter	B-21	330	.085		11.10	3.73	.43	15.26	18.35
2160	12" diameter		320	.088		12.70	3.85	.44	16.99	20.50
2200	15" diameter		240	.117		13	5.15	.58	18.73	23

33 31 Sanitary Utility Sewerage Piping

33 31 13 – Public Sanitary Utility Sewerage Piping

33 31 13.25 Sewage Collection, Polyvinyl Chloride Pipe	Crew	Daily Output	Labor-Hours	Unit	Material	2016 Bare Costs Labor	Equipment	Total	Total Incl O&P
4000	Piping, DWV PVC, no exc./bkfill., 10' L, Sch 40, 4" diameter	B-20	375	.064	L.F.	3.97	2.72		6.69	8.55
4010	6" diameter		350	.069		8	2.92		10.92	13.30
4020	8" diameter	↓	335	.072	↓	12.50	3.05		15.55	18.45

33 41 Storm Utility Drainage Piping

33 41 13 – Public Storm Utility Drainage Piping

33 41 13.40 Piping, Storm Drainage, Corrugated Metal

		Crew	Daily Output	Labor-Hours	Unit	Material	2016 Bare Costs Labor	Equipment	Total	Total Incl O&P
0010	**PIPING, STORM DRAINAGE, CORRUGATED METAL**									
0020	Not including excavation or backfill									
2000	Corrugated metal pipe, galvanized									
2020	Bituminous coated with paved invert, 20' lengths									
2040	8" diameter, 16 ga.	B-14	330	.145	L.F.	8.75	5.85	1.11	15.71	19.75
2060	10" diameter, 16 ga.		260	.185		9.10	7.40	1.41	17.91	23
2080	12" diameter, 16 ga.		210	.229		11.10	9.15	1.74	21.99	28
2100	15" diameter, 16 ga.		200	.240		15.30	9.65	1.83	26.78	33.50
2120	18" diameter, 16 ga.		190	.253		20	10.15	1.92	32.07	39.50
2140	24" diameter, 14 ga.	↓	160	.300		24.50	12.05	2.28	38.83	48
2160	30" diameter, 14 ga.	B-13	120	.467		29.50	19.30	6.25	55.05	69
2180	36" diameter, 12 ga.		120	.467		34.50	19.30	6.25	60.05	74.50
2200	48" diameter, 12 ga.	↓	100	.560		52	23	7.50	82.50	101
2220	60" diameter, 10 ga.	B-13B	75	.747		79.50	31	14.95	125.45	151
2240	72" diameter, 8 ga.	"	45	1.244	↓	95	51.50	25	171.50	211
2500	Galvanized, uncoated, 20' lengths									
2520	8" diameter, 16 ga.	B-14	355	.135	L.F.	7.85	5.40	1.03	14.28	18.10
2540	10" diameter, 16 ga.		280	.171		9	6.90	1.30	17.20	22
2560	12" diameter, 16 ga.		220	.218		10	8.75	1.66	20.41	26
2580	15" diameter, 16 ga.		220	.218		12.50	8.75	1.66	22.91	29
2600	18" diameter, 16 ga.		205	.234		15.10	9.40	1.78	26.28	33
2620	24" diameter, 14 ga.	↓	175	.274		22	11	2.09	35.09	43
2640	30" diameter, 14 ga.	B-13	130	.431		27	17.85	5.75	50.60	63
2660	36" diameter, 12 ga.		130	.431		31	17.85	5.75	54.60	67.50
2680	48" diameter, 12 ga.	↓	110	.509		46.50	21	6.80	74.30	90.50
2690	60" diameter, 10 ga.	B-13B	78	.718	↓	71.50	29.50	14.40	115.40	140
2780	End sections, 8" diameter	B-14	35	1.371	Ea.	73	55	10.45	138.45	176
2785	10" diameter		35	1.371		77	55	10.45	142.45	180
2790	12" diameter		35	1.371		114	55	10.45	179.45	221
2800	18" diameter	↓	30	1.600		116	64	12.20	192.20	238
2810	24" diameter	B-13	25	2.240		216	92.50	30	338.50	415
2820	30" diameter		25	2.240		330	92.50	30	452.50	540
2825	36" diameter		20	2.800		465	116	37.50	618.50	735
2830	48" diameter	↓	10	5.600		935	232	75	1,242	1,475
2835	60" diameter	B-13B	5	11.200		1,625	465	225	2,315	2,750
2840	72" diameter	"	4	14	↓	1,950	580	281	2,811	3,350

33 41 13.60 Sewage/Drainage Collection, Concrete Pipe

		Crew	Daily Output	Labor-Hours	Unit	Material	2016 Bare Costs Labor	Equipment	Total	Total Incl O&P
0010	**SEWAGE/DRAINAGE COLLECTION, CONCRETE PIPE**									
0020	Not including excavation or backfill									
1000	Non-reinforced pipe, extra strength, B&S or T&G joints									
1010	6" diameter	B-14	265.04	.181	L.F.	7.25	7.25	1.38	15.88	20.50
1020	8" diameter		224	.214		8	8.60	1.63	18.23	23.50
1030	10" diameter		216	.222		8.85	8.90	1.69	19.44	25.50
1040	12" diameter	↓	200	.240	↓	9.95	9.65	1.83	21.43	27.50

For customer support on your Mechanical Cost Data, call 877.739.3689.

477

33 41 13.60 Sewage/Drainage Collection, Concrete Pipe		Crew	Daily Output	Labor-Hours	Unit	Material	2016 Bare Costs Labor	Equipment	Total	Total Incl O&P
1050	15" diameter	B-14	180	.267	L.F.	14	10.70	2.03	26.73	34
1060	18" diameter		144	.333		16.95	13.35	2.54	32.84	42
1070	21" diameter		112	.429		18.05	17.20	3.26	38.51	50
1080	24" diameter		100	.480		19.15	19.25	3.65	42.05	54.50
2000	Reinforced culvert, class 3, no gaskets									
2010	12" diameter	B-14	150	.320	L.F.	12.15	12.85	2.44	27.44	35.50
2020	15" diameter		150	.320		15.70	12.85	2.44	30.99	39.50
2030	18" diameter		132	.364		20	14.60	2.77	37.37	47.50
2035	21" diameter		120	.400		21	16.05	3.04	40.09	51
2040	24" diameter		100	.480		22	19.25	3.65	44.90	57.50
2045	27" diameter	B-13	92	.609		32	25	8.15	65.15	83
2050	30" diameter		88	.636		38	26.50	8.50	73	91.50
2060	36" diameter		72	.778		59.50	32	10.40	101.90	126
2070	42" diameter	B-13B	72	.778		81	32	15.60	128.60	155
2080	48" diameter		64	.875		98	36	17.55	151.55	183
2090	60" diameter		48	1.167		147	48.50	23.50	219	262
2100	72" diameter		40	1.400		221	58	28	307	365
2120	84" diameter		32	1.750		305	72.50	35	412.50	485
2140	96" diameter		24	2.333		365	96.50	47	508.50	600
2200	With gaskets, class 3, 12" diameter	B-21	168	.167		13.35	7.35	.84	21.54	27
2220	15" diameter		160	.175		17.25	7.70	.88	25.83	32
2230	18" diameter		152	.184		22	8.10	.92	31.02	38
2240	24" diameter		136	.206		29	9.05	1.03	39.08	46.50
2260	30" diameter	B-13	88	.636		46	26.50	8.50	81	100
2270	36" diameter	"	72	.778		68.50	32	10.40	110.90	136
2290	48" diameter	B-13B	64	.875		111	36	17.55	164.55	197
2310	72" diameter	"	40	1.400		239	58	28	325	385
2330	Flared ends, 12" diameter	B-21	31	.903	Ea.	225	39.50	4.53	269.03	315
2340	15" diameter		25	1.120		268	49	5.60	322.60	375
2400	18" diameter		20	1.400		310	61.50	7	378.50	440
2420	24" diameter		14	2		370	88	10	468	550
2440	36" diameter	B-13	10	5.600		790	232	75	1,097	1,300
3080	Radius pipe, add to pipe prices, 12" to 60" diameter				L.F.	50%				
3090	Over 60" diameter, add				"	20%				
3500	Reinforced elliptical, 8' lengths, C507 class 3									
3520	14" x 23" inside, round equivalent 18" diameter	B-21	82	.341	L.F.	41	15	1.71	57.71	70
3530	24" x 38" inside, round equivalent 30" diameter	B-13	58	.966		62	40	12.90	114.90	143
3540	29" x 45" inside, round equivalent 36" diameter		52	1.077		78	44.50	14.40	136.90	170
3550	38" x 60" inside, round equivalent 48" diameter		38	1.474		135	61	19.70	215.70	264
3560	48" x 76" inside, round equivalent 60" diameter		26	2.154		185	89	29	303	370
3570	58" x 91" inside, round equivalent 72" diameter		22	2.545		270	105	34	409	495
3780	Concrete slotted pipe, class 4 mortar joint									
3800	12" diameter	B-21	168	.167	L.F.	28	7.35	.84	36.19	43
3840	18" diameter	"	152	.184	"	32	8.10	.92	41.02	48.50
3900	Concrete slotted pipe, Class 4 O-ring joint									
3940	12" diameter	B-21	168	.167	L.F.	28	7.35	.84	36.19	43
3960	18" diameter	"	152	.184	"	32	8.10	.92	41.02	48.50

33 42 16 – Concrete Culverts

33 42 16.15 Oval Arch Culverts		Crew	Daily Output	Labor-Hours	Unit	Material	2016 Bare Costs Labor	Equipment	Total	Total Incl O&P
0010	**OVAL ARCH CULVERTS**									
3000	Corrugated galvanized or aluminum, coated & paved									
3020	17" x 13", 16 ga., 15" equivalent	B-14	200	.240	L.F.	14.50	9.65	1.83	25.98	32.50
3040	21" x 15", 16 ga., 18" equivalent		150	.320		19.25	12.85	2.44	34.54	43.50
3060	28" x 20", 14 ga., 24" equivalent		125	.384		24.50	15.40	2.92	42.82	53.50
3080	35" x 24", 14 ga., 30" equivalent		100	.480		30	19.25	3.65	52.90	66
3100	42" x 29", 12 ga., 36" equivalent	B-13	100	.560		35.50	23	7.50	66	83
3120	49" x 33", 12 ga., 42" equivalent		90	.622		43	26	8.30	77.30	96
3140	57" x 38", 12 ga., 48" equivalent		75	.747		56	31	10	97	120
3160	Steel, plain oval arch culverts, plain									
3180	17" x 13", 16 ga., 15" equivalent	B-14	225	.213	L.F.	13.20	8.55	1.62	23.37	29.50
3200	21" x 15", 16 ga., 18" equivalent		175	.274		16.05	11	2.09	29.14	37
3220	28" x 20", 14 ga., 24" equivalent		150	.320		22.50	12.85	2.44	37.79	47
3240	35" x 24", 14 ga., 30" equivalent	B-13	108	.519		27	21.50	6.95	55.45	70
3260	42" x 29", 12 ga., 36" equivalent		108	.519		32	21.50	6.95	60.45	76
3280	49" x 33", 12 ga., 42" equivalent		92	.609		38	25	8.15	71.15	89.50
3300	57" x 38", 12 ga., 48" equivalent		75	.747		53.50	31	10	94.50	117
3320	End sections, 17" x 13"		22	2.545	Ea.	144	105	34	283	360
3340	42" x 29"		17	3.294	"	395	136	44	575	690
3360	Multi-plate arch, steel	B-20	1690	.014	Lb.	1.30	.60		1.90	2.36

33 44 Storm Utility Water Drains

33 44 13 – Utility Area Drains

33 44 13.13 Catch Basins

		Crew	Daily Output	Labor-Hours	Unit	Material	2016 Bare Costs Labor	Equipment	Total	Total Incl O&P
0010	**CATCH BASINS**									
0011	Not including footing & excavation									
1600	Frames & grates, C.I., 24" square, 500 lb.	B-6	7.80	3.077	Ea.	340	128	47	515	625
1700	26" D shape, 600 lb.		7	3.429		505	143	52	700	830
1800	Light traffic, 18" diameter, 100 lb.		10	2.400		125	100	36.50	261.50	330
1900	24" diameter, 300 lb.		8.70	2.759		197	115	42	354	435
2000	36" diameter, 900 lb.		5.80	4.138		570	172	63	805	965
2100	Heavy traffic, 24" diameter, 400 lb.		7.80	3.077		245	128	47	420	520
2200	36" diameter, 1150 lb.		3	8		795	335	122	1,252	1,525
2300	Mass. State standard, 26" diameter, 475 lb.		7	3.429		266	143	52	461	570
2400	30" diameter, 620 lb.		7	3.429		345	143	52	540	655
2500	Watertight, 24" diameter, 350 lb.		7.80	3.077		320	128	47	495	600
2600	26" diameter, 500 lb.		7	3.429		425	143	52	620	745
2700	32" diameter, 575 lb.		6	4		850	167	61	1,078	1,250
2800	3 piece cover & frame, 10" deep,									
2900	1200 lb., for heavy equipment	B-6	3	8	Ea.	1,050	335	122	1,507	1,825
3000	Raised for paving 1-1/4" to 2" high									
3100	4 piece expansion ring									
3200	20" to 26" diameter	1 Clab	3	2.667	Ea.	158	101		259	330
3300	30" to 36" diameter	"	3	2.667	"	219	101		320	395
3320	Frames and covers, existing, raised for paving, 2", including									
3340	row of brick, concrete collar, up to 12" wide frame	B-6	18	1.333	Ea.	46	55.50	20.50	122	158
3360	20" to 26" wide frame		11	2.182		68.50	91	33	192.50	251
3380	30" to 36" wide frame		9	2.667		84.50	111	40.50	236	305
3400	Inverts, single channel brick	D-1	3	5.333		97	225		322	450
3500	Concrete		5	3.200		107	135		242	325
3600	Triple channel, brick		2	8		148	335		483	680

For customer support on your Mechanical Cost Data, call 877.739.3689.

479

33 44 Storm Utility Water Drains

33 44 13 – Utility Area Drains

33 44 13.13 Catch Basins	Crew	Daily Output	Labor-Hours	Unit	Material	2016 Bare Costs Labor	Equipment	Total	Total Incl O&P
3700 Concrete	D-1	3	5.333	Ea.	141	225		366	500

33 46 Subdrainage

33 46 16 – Subdrainage Piping

33 46 16.25 Piping, Subdrainage, Corrugated Metal

		Crew	Daily Output	Labor-Hours	Unit	Material	2016 Bare Costs Labor	Equipment	Total	Total Incl O&P
0010	**PIPING, SUBDRAINAGE, CORRUGATED METAL**									
0021	Not including excavation and backfill									
2010	Aluminum, perforated									
2020	6" diameter, 18 ga.	B-20	380	.063	L.F.	6.50	2.69		9.19	11.30
2200	8" diameter, 16 ga.	"	370	.065		8.55	2.76		11.31	13.65
2220	10" diameter, 16 ga.	B-21	360	.078		10.70	3.42	.39	14.51	17.45
2240	12" diameter, 16 ga.		285	.098		11.95	4.32	.49	16.76	20.50
2260	18" diameter, 16 ga.		205	.137		17.95	6	.68	24.63	29.50
3000	Uncoated galvanized, perforated									
3020	6" diameter, 18 ga.	B-20	380	.063	L.F.	6.05	2.69		8.74	10.80
3200	8" diameter, 16 ga.	"	370	.065		8.30	2.76		11.06	13.40
3220	10" diameter, 16 ga.	B-21	360	.078		8.80	3.42	.39	12.61	15.40
3240	12" diameter, 16 ga.		285	.098		9.80	4.32	.49	14.61	17.90
3260	18" diameter, 16 ga.		205	.137		15	6	.68	21.68	26.50
4000	Steel, perforated, asphalt coated									
4020	6" diameter, 18 ga.	B-20	380	.063	L.F.	6.55	2.69		9.24	11.35
4030	8" diameter, 18 ga.	"	370	.065		8.60	2.76		11.36	13.70
4040	10" diameter, 16 ga.	B-21	360	.078		10.25	3.42	.39	14.06	17
4050	12" diameter, 16 ga.		285	.098		11.30	4.32	.49	16.11	19.60
4060	18" diameter, 16 ga.		205	.137		17.40	6	.68	24.08	29

33 46 16.30 Piping, Subdrainage, Plastic

		Crew	Daily Output	Labor-Hours	Unit	Material	2016 Bare Costs Labor	Equipment	Total	Total Incl O&P
0010	**PIPING, SUBDRAINAGE, PLASTIC**									
0020	Not including excavation and backfill									
2100	Perforated PVC, 4" diameter	B-14	314	.153	L.F.	1.49	6.15	1.16	8.80	12.30
2110	6" diameter		300	.160		3.28	6.40	1.22	10.90	14.75
2120	8" diameter		290	.166		5.60	6.65	1.26	13.51	17.70
2130	10" diameter		280	.171		8.85	6.90	1.30	17.05	21.50
2140	12" diameter		270	.178		12.70	7.15	1.35	21.20	26.50

33 49 Storm Drainage Structures

33 49 13 – Storm Drainage Manholes, Frames, and Covers

33 49 13.10 Storm Drainage Manholes, Frames and Covers

		Crew	Daily Output	Labor-Hours	Unit	Material	2016 Bare Costs Labor	Equipment	Total	Total Incl O&P
0010	**STORM DRAINAGE MANHOLES, FRAMES & COVERS**									
0020	Excludes footing, excavation, backfill (See line items for frame & cover)									
0050	Brick, 4' inside diameter, 4' deep	D-1	1	16	Ea.	535	675		1,210	1,625
0100	6' deep		.70	22.857		760	965		1,725	2,300
0150	8' deep		.50	32		980	1,350		2,330	3,125
0200	For depths over 8', add		4	4	V.L.F.	83.50	169		252.50	350
0400	Concrete blocks (radial), 4' I.D., 4' deep		1.50	10.667	Ea.	400	450		850	1,125
0500	6' deep		1	16		535	675		1,210	1,600
0600	8' deep		.70	22.857		670	965		1,635	2,200
0700	For depths over 8', add		5.50	2.909	V.L.F.	70	123		193	264
0800	Concrete, cast in place, 4' x 4', 8" thick, 4' deep	C-14H	2	24	Ea.	540	1,125	15.85	1,680.85	2,325
0900	6' deep		1.50	32		780	1,525	21	2,326	3,200

For customer support on your Mechanical Cost Data, call 877.739.3689.

33 49 Storm Drainage Structures

33 49 13 – Storm Drainage Manholes, Frames, and Covers

33 49 13.10 Storm Drainage Manholes, Frames and Covers	Crew	Daily Output	Labor-Hours	Unit	Material	2016 Bare Costs Labor	Equipment	Total	Total Incl O&P
1000	8' deep	C-14H	1	48	Ea.	1,125	2,275	31.50	3,431.50	4,725
1100	For depths over 8', add	↓	8	6	V.L.F.	126	284	3.96	413.96	575
1110	Precast, 4' I.D., 4' deep	B-22	4.10	7.317	Ea.	740	325	51.50	1,116.50	1,375
1120	6' deep		3	10		1,375	445	70	1,890	2,250
1130	8' deep		2	15		1,800	670	105	2,575	3,125
1140	For depths over 8', add	↓	16	1.875	V.L.F.	163	83.50	13.15	259.65	320
1150	5' I.D., 4' deep	B-6	3	8	Ea.	1,575	335	122	2,032	2,400
1160	6' deep		2	12		1,925	500	183	2,608	3,075
1170	8' deep		1.50	16	↓	2,700	665	244	3,609	4,250
1180	For depths over 8', add		12	2	V.L.F.	355	83.50	30.50	469	550
1190	6' I.D., 4' deep		2	12	Ea.	2,225	500	183	2,908	3,425
1200	6' deep		1.50	16		2,850	665	244	3,759	4,450
1210	8' deep		1	24		3,425	1,000	365	4,790	5,675
1220	For depths over 8', add	↓	8	3	V.L.F.	465	125	45.50	635.50	755
1250	Slab tops, precast, 8" thick									
1300	4' diameter manhole	B-6	8	3	Ea.	257	125	45.50	427.50	525
1400	5' diameter manhole		7.50	3.200		420	133	48.50	601.50	715
1500	6' diameter manhole	↓	7	3.429		650	143	52	845	985
3800	Steps, heavyweight cast iron, 7" x 9"	1 Bric	40	.200		19.25	9.25		28.50	35
3900	8" x 9"		40	.200		23	9.25		32.25	39.50
3928	12" x 10-1/2"		40	.200		27	9.25		36.25	43.50
4000	Standard sizes, galvanized steel		40	.200		22	9.25		31.25	38
4100	Aluminum		40	.200		24	9.25		33.25	40.50
4150	Polyethylene	↓	40	.200	↓	26.50	9.25		35.75	43.50

33 51 Natural-Gas Distribution

33 51 13 – Natural-Gas Piping

33 51 13.10 Piping, Gas Service and Distribution, P.E.

		Crew	Daily Output	Labor-Hours	Unit	Material	2016 Bare Costs Labor	Equipment	Total	Total Incl O&P
0010	**PIPING, GAS SERVICE AND DISTRIBUTION, POLYETHYLENE**									
0020	Not including excavation or backfill									
1000	60 psi coils, compression coupling @ 100', 1/2" diameter, SDR 11	B-20A	608	.053	L.F.	.50	2.43		2.93	4.24
1010	1" diameter, SDR 11		544	.059		1.13	2.71		3.84	5.35
1040	1-1/4" diameter, SDR 11		544	.059		1.62	2.71		4.33	5.90
1100	2" diameter, SDR 11		488	.066		2.25	3.02		5.27	7.05
1160	3" diameter, SDR 11	↓	408	.078		4.70	3.61		8.31	10.65
1500	60 PSI 40' joints with coupling, 3" diameter, SDR 11	B-21A	408	.098		9.40	4.64	1.17	15.21	18.65
1540	4" diameter, SDR 11		352	.114		13.50	5.40	1.36	20.26	24.50
1600	6" diameter, SDR 11		328	.122		33.50	5.75	1.46	40.71	47.50
1640	8" diameter, SDR 11	↓	272	.147	↓	46.50	6.95	1.76	55.21	63.50

For customer support on your Mechanical Cost Data, call 877.739.3689.

481

33 52 16 – Gasoline Distribution

33 52 16.13 Gasoline Piping	Crew	Daily Output	Labor-Hours	Unit	Material	2016 Bare Costs Labor	Equipment	Total	Total Incl O&P
0010 **GASOLINE PIPING**									
0020	Primary containment pipe, fiberglass-reinforced								
0030	Plastic pipe 15' & 30' lengths								
0040	2" diameter	Q-6	425	.056	L.F.	6.80	3.20		10	12.35
0050	3" diameter		400	.060		10.60	3.40		14	16.80
0060	4" diameter	▼	375	.064	▼	13.75	3.63		17.38	20.50
0100	Fittings								
0110	Elbows, 90° & 45°, bell-ends, 2"	Q-6	24	1	Ea.	43.50	56.50		100	133
0120	3" diameter		22	1.091		55	62		117	154
0130	4" diameter		20	1.200		70	68		138	180
0200	Tees, bell ends, 2"		21	1.143		60.50	64.50		125	164
0210	3" diameter		18	1.333		64	75.50		139.50	185
0220	4" diameter		15	1.600		84	90.50		174.50	230
0230	Flanges bell ends, 2"		24	1		33.50	56.50		90	123
0240	3" diameter		22	1.091		39	62		101	137
0250	4" diameter		20	1.200		45	68		113	153
0260	Sleeve couplings, 2"		21	1.143		12.40	64.50		76.90	111
0270	3" diameter		18	1.333		17.80	75.50		93.30	134
0280	4" diameter		15	1.600		23	90.50		113.50	163
0290	Threaded adapters, 2"		21	1.143		17.95	64.50		82.45	117
0300	3" diameter		18	1.333		34	75.50		109.50	152
0310	4" diameter		15	1.600		38	90.50		128.50	179
0320	Reducers, 2"		27	.889		27.50	50.50		78	107
0330	3" diameter		22	1.091		27.50	62		89.50	124
0340	4" diameter	▼	20	1.200	▼	36.50	68		104.50	144
1010	Gas station product line for secondary containment (double wall)								
1100	Fiberglass reinforced plastic pipe 25' lengths								
1120	Pipe, plain end, 3" diameter	Q-6	375	.064	L.F.	26.50	3.63		30.13	34.50
1130	4" diameter		350	.069		32	3.88		35.88	41.50
1140	5" diameter		325	.074		36	4.18		40.18	46
1150	6" diameter	▼	300	.080	▼	39	4.53		43.53	50
1200	Fittings								
1230	Elbows, 90° & 45°, 3" diameter	Q-6	18	1.333	Ea.	136	75.50		211.50	264
1240	4" diameter		16	1.500		167	85		252	310
1250	5" diameter		14	1.714		184	97		281	350
1260	6" diameter		12	2		204	113		317	395
1270	Tees, 3" diameter		15	1.600		166	90.50		256.50	320
1280	4" diameter		12	2		202	113		315	395
1290	5" diameter		9	2.667		315	151		466	580
1300	6" diameter		6	4		375	227		602	755
1310	Couplings, 3" diameter		18	1.333		55.50	75.50		131	175
1320	4" diameter		16	1.500		120	85		205	259
1330	5" diameter		14	1.714		215	97		312	385
1340	6" diameter		12	2		315	113		428	520
1350	Cross-over nipples, 3" diameter		18	1.333		10.50	75.50		86	126
1360	4" diameter		16	1.500		12.75	85		97.75	142
1370	5" diameter		14	1.714		15.90	97		112.90	165
1380	6" diameter		12	2		19.10	113		132.10	192
1400	Telescoping, reducers, concentric 4" x 3"		18	1.333		48	75.50		123.50	167
1410	5" x 4"		17	1.412		95.50	80		175.50	226
1420	6" x 5"	▼	16	1.500	▼	234	85		319	385

33 61 Hydronic Energy Distribution

33 61 13 – Underground Hydronic Energy Distribution

33 61 13.20 Pipe Conduit, Prefabricated/Preinsulated	Crew	Daily Output	Labor-Hours	Unit	Material	2016 Bare Costs Labor	Equipment	Total	Total Incl O&P
0010 **PIPE CONDUIT, PREFABRICATED/PREINSULATED**									
0020 Does not include trenching, fittings or crane.									
0300 For cathodic protection, add 12 to 14%									
0310 of total built-up price (casing plus service pipe)									
0580 Polyurethane insulated system, 250°F. max. temp.									
0620 Black steel service pipe, standard wt., 1/2" insulation									
0660 3/4" diam. pipe size	Q-17	54	.296	L.F.	57	16.20	1.08	74.28	88.50
0670 1" diam. pipe size		50	.320		63	17.50	1.16	81.66	97
0680 1-1/4" diam. pipe size		47	.340		70	18.60	1.24	89.84	106
0690 1-1/2" diam. pipe size		45	.356		76	19.40	1.29	96.69	114
0700 2" diam. pipe size		42	.381		79	21	1.38	101.38	120
0710 2-1/2" diam. pipe size		34	.471		80	25.50	1.71	107.21	129
0720 3" diam. pipe size		28	.571		93	31	2.07	126.07	151
0730 4" diam. pipe size		22	.727		117	39.50	2.64	159.14	192
0740 5" diam. pipe size		18	.889		149	48.50	3.23	200.73	241
0750 6" diam. pipe size	Q-18	23	1.043		173	59	2.53	234.53	283
0760 8" diam. pipe size		19	1.263		254	71.50	3.06	328.56	390
0770 10" diam. pipe size		16	1.500		320	85	3.63	408.63	485
0780 12" diam. pipe size		13	1.846		400	105	4.47	509.47	605
0790 14" diam. pipe size		11	2.182		445	124	5.30	574.30	685
0800 16" diam. pipe size		10	2.400		510	136	5.80	651.80	775
0810 18" diam. pipe size		8	3		590	170	7.25	767.25	915
0820 20" diam. pipe size		7	3.429		655	194	8.30	857.30	1,025
0830 24" diam. pipe size		6	4		805	227	9.70	1,041.70	1,225
0900 For 1" thick insulation, add					10%				
0940 For 1-1/2" thick insulation, add					13%				
0980 For 2" thick insulation, add					20%				
1500 Gland seal for system, 3/4" diam. pipe size	Q-17	32	.500	Ea.	775	27.50	1.82	804.32	895
1510 1" diam. pipe size		32	.500		775	27.50	1.82	804.32	895
1540 1-1/4" diam. pipe size		30	.533		825	29	1.94	855.94	955
1550 1-1/2" diam. pipe size		30	.533		825	29	1.94	855.94	955
1560 2" diam. pipe size		28	.571		980	31	2.07	1,013.07	1,125
1570 2-1/2" diam. pipe size		26	.615		1,050	33.50	2.23	1,085.73	1,225
1580 3" diam. pipe size		24	.667		1,125	36.50	2.42	1,163.92	1,300
1590 4" diam. pipe size		22	.727		1,350	39.50	2.64	1,392.14	1,550
1600 5" diam. pipe size		19	.842		1,650	46	3.06	1,699.06	1,900
1610 6" diam. pipe size	Q-18	26	.923		1,750	52.50	2.23	1,804.73	2,000
1620 8" diam. pipe size		25	.960		2,050	54.50	2.32	2,106.82	2,325
1630 10" diam. pipe size		23	1.043		2,375	59	2.53	2,436.53	2,725
1640 12" diam. pipe size		21	1.143		2,625	64.50	2.77	2,692.27	3,000
1650 14" diam. pipe size		19	1.263		2,950	71.50	3.06	3,024.56	3,350
1660 16" diam. pipe size		18	1.333		3,500	75.50	3.23	3,578.73	3,975
1670 18" diam. pipe size		16	1.500		3,700	85	3.63	3,788.63	4,200
1680 20" diam. pipe size		14	1.714		4,200	97	4.15	4,301.15	4,775
1690 24" diam. pipe size		12	2		4,675	113	4.84	4,792.84	5,300
2000 Elbow, 45° for system									
2020 3/4" diam. pipe size	Q-17	14	1.143	Ea.	490	62.50	4.15	556.65	635
2040 1" diam. pipe size		13	1.231		500	67	4.47	571.47	655
2050 1-1/4" diam. pipe size		11	1.455		570	79.50	5.30	654.80	750
2060 1-1/2" diam. pipe size		9	1.778		575	97	6.45	678.45	785
2070 2" diam. pipe size		6	2.667		645	146	9.70	800.70	940
2080 2-1/2" diam. pipe size		4	4		670	219	14.50	903.50	1,075
2090 3" diam. pipe size		3.50	4.571		775	250	16.60	1,041.60	1,250

For customer support on your Mechanical Cost Data, call 877.739.3689.

483

33 61 Hydronic Energy Distribution

33 61 13 – Underground Hydronic Energy Distribution

33 61 13.20 Pipe Conduit, Prefabricated/Preinsulated	Crew	Daily Output	Labor-Hours	Unit	Material	2016 Bare Costs Labor	Equipment	Total	Total Incl O&P
2100	4" diam. pipe size	Q-17	3	5.333	Ea.	905	291	19.35	1,215.35	1,450
2110	5" diam. pipe size		2.80	5.714		1,175	310	20.50	1,505.50	1,775
2120	6" diam. pipe size	Q-18	4	6		1,325	340	14.50	1,679.50	1,975
2130	8" diam. pipe size		3	8		1,925	455	19.35	2,399.35	2,800
2140	10" diam. pipe size		2.40	10		2,350	565	24	2,939	3,450
2150	12" diam. pipe size		2	12		3,100	680	29	3,809	4,450
2160	14" diam. pipe size		1.80	13.333		3,825	755	32.50	4,612.50	5,400
2170	16" diam. pipe size		1.60	15		4,550	850	36.50	5,436.50	6,350
2180	18" diam. pipe size		1.30	18.462		5,725	1,050	44.50	6,819.50	7,925
2190	20" diam. pipe size		1	24		7,200	1,350	58	8,608	10,000
2200	24" diam. pipe size		.70	34.286		9,050	1,950	83	11,083	13,000
2260	For elbow, 90°, add					25%				
2300	For tee, straight, add					85%	30%			
2340	For tee, reducing, add					170%	30%			
2380	For weldolet, straight, add					50%				
2400	Polyurethane insulation, 1"									
2410	FRP carrier and casing									
2420	4"	Q-5	18	.889	L.F.	53.50	48.50		102	132
2422	6"	Q-6	23	1.043		92.50	59		151.50	191
2424	8"		19	1.263		151	71.50		222.50	274
2426	10"		16	1.500		203	85		288	350
2428	12"		13	1.846		265	105		370	450
2430	FRP carrier and PVC casing									
2440	4"	Q-5	18	.889	L.F.	21.50	48.50		70	97
2444	8"	Q-6	19	1.263		45	71.50		116.50	157
2446	10"		16	1.500		64.50	85		149.50	199
2448	12"		13	1.846		82.50	105		187.50	249
2450	PVC carrier and casing									
2460	4"	Q-1	36	.444	L.F.	10.45	23.50		33.95	47
2462	6"	"	29	.552		14.55	29.50		44.05	60.50
2464	8"	Q-2	36	.667		20.50	37		57.50	78
2466	10"		32	.750		28	41.50		69.50	93
2468	12"		31	.774		32.50	43		75.50	100

33 63 Steam Energy Distribution

33 63 13 – Underground Steam and Condensate Distribution Piping

33 63 13.10 Calcium Silicate Insulated System

		Crew	Daily Output	Labor-Hours	Unit	Material	2016 Bare Costs Labor	Equipment	Total	Total Incl O&P
0010	**CALCIUM SILICATE INSULATED SYSTEM**									
0011	High temp. (1200 degrees F)									
2840	Steel casing with protective exterior coating									
2850	6-5/8" diameter	Q-18	52	.462	L.F.	105	26	1.12	132.12	156
2860	8-5/8" diameter		50	.480		114	27	1.16	142.16	167
2870	10-3/4" diameter		47	.511		134	29	1.24	164.24	192
2880	12-3/4" diameter		44	.545		146	31	1.32	178.32	208
2890	14" diameter		41	.585		164	33	1.42	198.42	233
2900	16" diameter		39	.615		176	35	1.49	212.49	248
2910	18" diameter		36	.667		194	38	1.61	233.61	272
2920	20" diameter		34	.706		219	40	1.71	260.71	305
2930	22" diameter		32	.750		300	42.50	1.82	344.32	395
2940	24" diameter		29	.828		345	47	2	394	455
2950	26" diameter		26	.923		395	52.50	2.23	449.73	515

33 63 Steam Energy Distribution

33 63 13 – Underground Steam and Condensate Distribution Piping

33 63 13.10 Calcium Silicate Insulated System		Crew	Daily Output	Labor-Hours	Unit	Material	2016 Bare Costs Labor	Equipment	Total	Total Incl O&P
2960	28" diameter	Q-18	23	1.043	L.F.	490	59	2.53	551.53	630
2970	30" diameter		21	1.143		525	64.50	2.77	592.27	675
2980	32" diameter		19	1.263		590	71.50	3.06	664.56	760
2990	34" diameter		18	1.333		595	75.50	3.23	673.73	775
3000	36" diameter		16	1.500		640	85	3.63	728.63	830
3040	For multi-pipe casings, add					10%				
3060	For oversize casings, add					2%				
3400	Steel casing gland seal, single pipe									
3420	6-5/8" diameter	Q-18	25	.960	Ea.	1,450	54.50	2.32	1,506.82	1,675
3440	8-5/8" diameter		23	1.043		1,700	59	2.53	1,761.53	1,975
3450	10-3/4" diameter		21	1.143		1,900	64.50	2.77	1,967.27	2,200
3460	12-3/4" diameter		19	1.263		2,250	71.50	3.06	2,324.56	2,575
3470	14" diameter		17	1.412		2,500	80	3.42	2,583.42	2,875
3480	16" diameter		16	1.500		2,925	85	3.63	3,013.63	3,350
3490	18" diameter		15	1.600		3,250	90.50	3.87	3,344.37	3,725
3500	20" diameter		13	1.846		3,600	105	4.47	3,709.47	4,125
3510	22" diameter		12	2		4,025	113	4.84	4,142.84	4,600
3520	24" diameter		11	2.182		4,525	124	5.30	4,654.30	5,175
3530	26" diameter		10	2.400		5,175	136	5.80	5,316.80	5,900
3540	28" diameter		9.50	2.526		5,850	143	6.10	5,999.10	6,650
3550	30" diameter		9	2.667		5,950	151	6.45	6,107.45	6,750
3560	32" diameter		8.50	2.824		6,675	160	6.85	6,841.85	7,600
3570	34" diameter		8	3		7,275	170	7.25	7,452.25	8,275
3580	36" diameter		7	3.429		7,750	194	8.30	7,952.30	8,825
3620	For multi-pipe casings, add					5%				
4000	Steel casing anchors, single pipe									
4020	6-5/8" diameter	Q-18	8	3	Ea.	1,300	170	7.25	1,477.25	1,725
4040	8-5/8" diameter		7.50	3.200		1,350	181	7.75	1,538.75	1,775
4050	10-3/4" diameter		7	3.429		1,775	194	8.30	1,977.30	2,250
4060	12-3/4" diameter		6.50	3.692		1,900	209	8.95	2,117.95	2,425
4070	14" diameter		6	4		2,250	227	9.70	2,486.70	2,825
4080	16" diameter		5.50	4.364		2,600	247	10.55	2,857.55	3,225
4090	18" diameter		5	4.800		2,950	272	11.60	3,233.60	3,675
4100	20" diameter		4.50	5.333		3,250	300	12.90	3,562.90	4,050
4110	22" diameter		4	6		3,625	340	14.50	3,979.50	4,500
4120	24" diameter		3.50	6.857		3,925	390	16.60	4,331.60	4,925
4130	26" diameter		3	8		4,450	455	19.35	4,924.35	5,600
4140	28" diameter		2.50	9.600		4,900	545	23	5,468	6,250
4150	30" diameter		2	12		5,200	680	29	5,909	6,775
4160	32" diameter		1.50	16		6,200	905	38.50	7,143.50	8,250
4170	34" diameter		1	24		6,975	1,350	58	8,383	9,800
4180	36" diameter		1	24		7,600	1,350	58	9,008	10,500
4220	For multi-pipe, add					5%	20%			
4800	Steel casing elbow									
4820	6-5/8" diameter	Q-18	15	1.600	Ea.	1,775	90.50	3.87	1,869.37	2,100
4830	8-5/8" diameter		15	1.600		1,900	90.50	3.87	1,994.37	2,250
4850	10-3/4" diameter		14	1.714		2,300	97	4.15	2,401.15	2,675
4860	12-3/4" diameter		13	1.846		2,725	105	4.47	2,834.47	3,175
4870	14" diameter		12	2		2,925	113	4.84	3,042.84	3,400
4880	16" diameter		11	2.182		3,200	124	5.30	3,329.30	3,700
4890	18" diameter		10	2.400		3,650	136	5.80	3,791.80	4,200
4900	20" diameter		9	2.667		3,875	151	6.45	4,032.45	4,500
4910	22" diameter		8	3		4,175	170	7.25	4,352.25	4,875

For customer support on your Mechanical Cost Data, call 877.739.3689.

485

33 63 13 – Underground Steam and Condensate Distribution Piping

33 63 13.10 Calcium Silicate Insulated System		Crew	Daily Output	Labor-Hours	Unit	Material	2016 Bare Costs Labor	Equipment	Total	Total Incl O&P
4920	24" diameter	Q-18	7	3.429	Ea.	4,600	194	8.30	4,802.30	5,375
4930	26" diameter		6	4		5,050	227	9.70	5,286.70	5,900
4940	28" diameter		5	4.800		5,450	272	11.60	5,733.60	6,425
4950	30" diameter		4	6		5,475	340	14.50	5,829.50	6,550
4960	32" diameter		3	8		6,200	455	19.35	6,674.35	7,525
4970	34" diameter		2	12		6,825	680	29	7,534	8,575
4980	36" diameter	▼	2	12	▼	7,275	680	29	7,984	9,050
5500	Black steel service pipe, std. wt., 1" thick insulation									
5510	3/4" diameter pipe size	Q-17	54	.296	L.F.	61.50	16.20	1.08	78.78	93
5540	1" diameter pipe size		50	.320		63.50	17.50	1.16	82.16	97.50
5550	1-1/4" diameter pipe size		47	.340		69.50	18.60	1.24	89.34	105
5560	1-1/2" diameter pipe size		45	.356		73.50	19.40	1.29	94.19	111
5570	2" diameter pipe size		42	.381		80	21	1.38	102.38	121
5580	2-1/2" diameter pipe size		34	.471		84	25.50	1.71	111.21	133
5590	3" diameter pipe size		28	.571		92	31	2.07	125.07	150
5600	4" diameter pipe size		22	.727		97	39.50	2.64	139.14	170
5610	5" diameter pipe size		18	.889		106	48.50	3.23	157.73	194
5620	6" diameter pipe size	Q-18	23	1.043	▼	114	59	2.53	175.53	217
6000	Black steel service pipe, std. wt., 1-1/2" thick insul.									
6010	3/4" diameter pipe size	Q-17	54	.296	L.F.	62.50	16.20	1.08	79.78	94
6040	1" diameter pipe size		50	.320		66.50	17.50	1.16	85.16	101
6050	1-1/4" diameter pipe size		47	.340		72.50	18.60	1.24	92.34	109
6060	1-1/2" diameter pipe size		45	.356		78	19.40	1.29	98.69	117
6070	2" diameter pipe size		42	.381		84	21	1.38	106.38	126
6080	2-1/2" diameter pipe size		34	.471		88	25.50	1.71	115.21	138
6090	3" diameter pipe size		28	.571		98	31	2.07	131.07	157
6100	4" diameter pipe size		22	.727		106	39.50	2.64	148.14	180
6110	5" diameter pipe size	▼	18	.889		114	48.50	3.23	165.73	202
6120	6" diameter pipe size	Q-18	23	1.043		129	59	2.53	190.53	234
6130	8" diameter pipe size		19	1.263		178	71.50	3.06	252.56	305
6140	10" diameter pipe size		16	1.500		238	85	3.63	326.63	395
6150	12" diameter pipe size	▼	13	1.846		297	105	4.47	406.47	490
6190	For 2" thick insulation, add					15%				
6220	For 2-1/2" thick insulation, add					25%				
6260	For 3" thick insulation, add				▼	30%				
6800	Black steel service pipe, ex. hvy. wt., 1" thick insul.									
6820	3/4" diameter pipe size	Q-17	50	.320	L.F.	63.50	17.50	1.16	82.16	97.50
6840	1" diameter pipe size		47	.340		67.50	18.60	1.24	87.34	103
6850	1-1/4" diameter pipe size		44	.364		75.50	19.85	1.32	96.67	114
6860	1-1/2" diameter pipe size		42	.381		77	21	1.38	99.38	118
6870	2" diameter pipe size		40	.400		84	22	1.45	107.45	127
6880	2-1/2" diameter pipe size		31	.516		88	28	1.87	117.87	142
6890	3" diameter pipe size		27	.593		96	32.50	2.15	130.65	157
6900	4" diameter pipe size		21	.762		101	41.50	2.77	145.27	177
6910	5" diameter pipe size	▼	17	.941		109	51.50	3.42	163.92	201
6920	6" diameter pipe size	Q-18	22	1.091	▼	119	62	2.64	183.64	227
7400	Black steel service pipe, ex. hvy. wt., 1-1/2" thick insul.									
7420	3/4" diameter pipe size	Q-17	50	.320	L.F.	54.50	17.50	1.16	73.16	88
7440	1" diameter pipe size		47	.340		59.50	18.60	1.24	79.34	95
7450	1-1/4" diameter pipe size		44	.364		66.50	19.85	1.32	87.67	104
7460	1-1/2" diameter pipe size		42	.381		71.50	21	1.38	93.88	112
7470	2" diameter pipe size		40	.400		76.50	22	1.45	99.95	119
7480	2-1/2" diameter pipe size	▼	31	.516	▼	81	28	1.87	110.87	134

33 63 13 – Underground Steam and Condensate Distribution Piping

33 63 13.10 Calcium Silicate Insulated System	Crew	Daily Output	Labor-Hours	Unit	Material	2016 Bare Costs Labor	Equipment	Total	Total Incl O&P
7490	3" diameter pipe size	Q-17	27	.593	L.F.	90	32.50	2.15	124.65	150
7500	4" diameter pipe size		21	.762		109	41.50	2.77	153.27	186
7510	5" diameter pipe size		17	.941		151	51.50	3.42	205.92	247
7520	6" diameter pipe size	Q-18	22	1.091		158	62	2.64	222.64	270
7530	8" diameter pipe size		18	1.333		213	75.50	3.23	291.73	350
7540	10" diameter pipe size		15	1.600		272	90.50	3.87	366.37	440
7550	12" diameter pipe size		13	1.846		325	105	4.47	434.47	525
7590	For 2" thick insulation, add					13%				
7640	For 2-1/2" thick insulation, add					18%				
7680	For 3" thick insulation, add					24%				

33 63 13.20 Combined Steam Pipe With Condensate Return

		Crew	Daily Output	Labor-Hours	Unit	Material	2016 Bare Costs Labor	Equipment	Total	Total Incl O&P
0010	**COMBINED STEAM PIPE WITH CONDENSATE RETURN**									
9010	8" and 4" in 24" case	Q-18	100	.240	L.F.	390	13.60	.58	404.18	450
9020	6" and 3" in 20" case		104	.231		365	13.05	.56	378.61	420
9030	3" and 1-1/2" in 16" case		110	.218		272	12.35	.53	284.88	320
9040	2" and 1-1/4" in 12-3/4" case		114	.211		233	11.95	.51	245.46	275
9050	1-1/2" and 1-1/4" in 10-3/4" case		116	.207		233	11.70	.50	245.20	274
9100	Steam pipe only (no return)									
9110	6" in 18" case	Q-18	108	.222	L.F.	297	12.60	.54	310.14	345
9120	4" in 14" case		112	.214		233	12.15	.52	245.67	275
9130	3" in 14" case		114	.211		218	11.95	.51	230.46	259
9140	2-1/2" in 12-3/4" case		118	.203		203	11.50	.49	214.99	241
9150	2" in 12-3/4" case		122	.197		193	11.15	.48	204.63	229

For customer support on your Mechanical Cost Data, call 877.739.3689.

487

Division Notes

	CREW	DAILY OUTPUT	LABOR-HOURS	UNIT	BARE COSTS				TOTAL INCL O&P
					MAT.	LABOR	EQUIP.	TOTAL	

Estimating Tips

This section involves equipment and construction costs for air noise and odor pollution control systems. These systems may be interrelated and care must be taken that the complete systems are estimated. For example, air pollution equipment may include dust and air-entrained particles that have to be collected. The vacuum systems could be noisy, requiring silencers to reduce noise pollution, and the collected solids have to be disposed of to prevent solid pollution.

Reference Numbers

Reference numbers are shown at the beginning of some major classifications. These numbers refer to related items in the Reference Section. The reference information may be an estimating procedure, an alternate pricing method, or technical information.

Note: Not all subdivisions listed here necessarily appear. ∎

44 11 16.10 Dust Collection Systems	Crew	Daily Output	Labor-Hours	Unit	Material	2016 Bare Costs Labor	Equipment	Total	Total Incl O&P
0010 **DUST COLLECTION SYSTEMS** Commercial/Industrial									
0120 Central vacuum units									
0130 Includes stand, filters and motorized shaker									
0200 500 CFM, 10" inlet, 2 HP	Q-20	2.40	8.333	Ea.	4,700	435		5,135	5,800
0220 1000 CFM, 10" inlet, 3 HP		2.20	9.091		4,925	475		5,400	6,150
0240 1500 CFM, 10" inlet, 5 HP		2	10		5,200	520		5,720	6,525
0260 3000 CFM, 13" inlet, 10 HP		1.50	13.333		14,600	695		15,295	17,100
0280 5000 CFM, 16" inlet, 2 @ 10 HP	▼	1	20	▼	15,500	1,050		16,550	18,700
1000 Vacuum tubing, galvanized									
1100 2-1/8" OD, 16 ga.	Q-9	440	.036	L.F.	2.95	1.87		4.82	6.10
1110 2-1/2" OD, 16 ga.		420	.038		3.24	1.96		5.20	6.55
1120 3" OD, 16 ga.		400	.040		4.13	2.06		6.19	7.70
1130 3-1/2" OD, 16 ga.		380	.042		5.85	2.17		8.02	9.75
1140 4" OD, 16 ga.		360	.044		6	2.29		8.29	10.10
1150 5" OD, 14 ga.		320	.050		10.80	2.58		13.38	15.80
1160 6" OD, 14 ga.		280	.057		12.10	2.94		15.04	17.80
1170 8" OD, 14 ga.		200	.080		18.30	4.12		22.42	26.50
1180 10" OD, 12 ga.		160	.100		35.50	5.15		40.65	47
1190 12" OD, 12 ga.		120	.133		47	6.85		53.85	62
1200 14" OD, 12 ga.	▼	80	.200	▼	53.50	10.30		63.80	74.50
1940 Hose, flexible wire reinforced rubber									
1956 3" diam.	Q-9	400	.040	L.F.	9.40	2.06		11.46	13.50
1960 4" diam.		360	.044		10.90	2.29		13.19	15.50
1970 5" diam.		320	.050		14.15	2.58		16.73	19.55
1980 6" diam.	▼	280	.057	▼	14.80	2.94		17.74	21
2000 90° Elbow, slip fit									
2110 2-1/8" diam.	Q-9	70	.229	Ea.	10.85	11.80		22.65	30
2120 2-1/2" diam.		65	.246		15.20	12.70		27.90	36
2130 3" diam.		60	.267		20.50	13.75		34.25	43.50
2140 3-1/2" diam.		55	.291		25.50	15		40.50	51
2150 4" diam.		50	.320		31.50	16.50		48	59.50
2160 5" diam.		45	.356		60	18.30		78.30	94
2170 6" diam.		40	.400		82.50	20.50		103	123
2180 8" diam.	▼	30	.533	▼	157	27.50		184.50	214
2400 45° Elbow, slip fit									
2410 2-1/8" diam.	Q-9	70	.229	Ea.	9.50	11.80		21.30	28.50
2420 2-1/2" diam.		65	.246		14	12.70		26.70	35
2430 3" diam.		60	.267		16.90	13.75		30.65	39.50
2440 3-1/2" diam.		55	.291		21.50	15		36.50	46.50
2450 4" diam.		50	.320		27.50	16.50		44	55.50
2460 5" diam.		45	.356		46.50	18.30		64.80	79
2470 6" diam.		40	.400		63	20.50		83.50	101
2480 8" diam.	▼	35	.457	▼	123	23.50		146.50	172
2800 90° TY, slip fit thru 6" diam.									
2810 2-1/8" diam.	Q-9	42	.381	Ea.	17.75	19.65		37.40	49.50
2820 2-1/2" diam.		39	.410		26.50	21		47.50	61.50
2830 3" diam.		36	.444		37	23		60	75.50
2840 3-1/2" diam.		33	.485		39	25		64	80.50
2850 4" diam.		30	.533		41.50	27.50		69	88
2860 5" diam.		27	.593		70.50	30.50		101	124
2870 6" diam.	▼	24	.667		101	34.50		135.50	164
2880 8" diam., butt end					155			155	170
2890 10" diam., butt end			▼		475			475	525

44 11 Particulate Control Equipment

44 11 16 – Fugitive Dust Barrier Systems

44 11 16.10 Dust Collection Systems	Crew	Daily Output	Labor-Hours	Unit	Material	2016 Bare Costs Labor	Equipment	Total	Total Incl O&P
2900	12" diam., butt end				Ea.	475			475	525
2910	14" diam., butt end					855			855	940
2920	6" x 4" diam., butt end					106			106	116
2930	8" x 4" diam., butt end					200			200	220
2940	10" x 4" diam., butt end					263			263	289
2950	12" x 4" diam., butt end					300			300	330
3100	90° Elbow, butt end segmented									
3110	8" diam., butt end, segmented				Ea.	157			157	172
3120	10" diam., butt end, segmented					390			390	430
3130	12" diam., butt end, segmented					495			495	545
3140	14" diam., butt end, segmented					620			620	680
3200	45° Elbow, butt end segmented									
3210	8" diam., butt end, segmented				Ea.	123			123	136
3220	10" diam., butt end, segmented					282			282	310
3230	12" diam., butt end, segmented					288			288	315
3240	14" diam., butt end, segmented					585			585	645
3400	All butt end fittings require one coupling per joint.									
3410	Labor for fitting included with couplings.									
3460	Compression coupling, galvanized, neoprene gasket									
3470	2-1/8" diam.	Q-9	44	.364	Ea.	10.85	18.75		29.60	40.50
3480	2-1/2" diam.		44	.364		10.85	18.75		29.60	40.50
3490	3" diam.		38	.421		15.60	21.50		37.10	50
3500	3-1/2" diam.		35	.457		17.45	23.50		40.95	55
3510	4" diam.		33	.485		18.95	25		43.95	59
3520	5" diam.		29	.552		21.50	28.50		50	67.50
3530	6" diam.		26	.615		25.50	31.50		57	76.50
3540	8" diam.		22	.727		46	37.50		83.50	108
3550	10" diam.		20	.800		67.50	41		108.50	138
3560	12" diam.		18	.889		84	46		130	163
3570	14" diam.		16	1		119	51.50		170.50	210
3800	Air gate valves, galvanized									
3810	2-1/8" diam.	Q-9	30	.533	Ea.	111	27.50		138.50	164
3820	2-1/2" diam.		28	.571		119	29.50		148.50	176
3830	3" diam.		26	.615		127	31.50		158.50	189
3840	4" diam.		23	.696		147	36		183	217
3850	6" diam.		18	.889		205	46		251	295

44 11 96 – Dry Plate Electrostatic Precipitator Equipment

44 11 96.10 Electrostatic Precipitators

		Crew	Daily Output	Labor-Hours	Unit	Material	2016 Bare Costs Labor	Equipment	Total	Total Incl O&P
0010	**ELECTROSTATIC PRECIPITATORS**									
4100	Smoke pollution control									
4110	Exhaust air volume									
4120	1 to 2.4 MCFM	Q-10	2.15	11.163	MCFM	9,075	595		9,670	10,900
4130	2.4 to 4.8 MCFM		2.70	8.889		10,300	475		10,775	12,100
4140	4.8 to 7.2 MCFM		3.60	6.667		10,500	355		10,855	12,100
4150	7.2 to 12 MCFM		5.40	4.444		11,000	237		11,237	12,500

Division Notes

	CREW	DAILY OUTPUT	LABOR-HOURS	UNIT	BARE COSTS				TOTAL INCL O&P
					MAT.	LABOR	EQUIP.	TOTAL	

Estimating Tips

This division contains information about water and wastewater equipment and systems, which was formerly located in Division 44. The main areas of focus are total wastewater treatment plants and components of wastewater treatment plants. Also included in this section are oil/water separators for wastewater treatment.

Reference Numbers

Reference numbers are shown at the beginning of some major classifications. These numbers refer to related items in the Reference Section. The reference information may be an estimating procedure, an alternate pricing method, or technical information.

Note: Not all subdivisions listed here necessarily appear. ■

46 23 Grit Removal And Handling Equipment

46 23 23 – Vortex Grit Removal Equipment

46 23 23.10 Rainwater Filters	Crew	Daily Output	Labor-Hours	Unit	Material	2016 Bare Costs Labor	2016 Bare Costs Equipment	Total	Total Incl O&P
0010 **RAINWATER FILTERS**									
0100　42 gal./min	B-21	3.50	8	Ea.	46,500	350	40	46,890	51,500
0200　65 gal./min		3.50	8		46,500	350	40	46,890	51,500
0300　208 gal./min		3.50	8		46,500	350	40	46,890	51,500

46 25 Oil and Grease Separation and Removal Equipment

46 25 13 – Coalescing Oil-Water Separators

46 25 13.20 Oil/Water Separators

	Crew	Daily Output	Labor-Hours	Unit	Material	2016 Bare Costs Labor	2016 Bare Costs Equipment	Total	Total Incl O&P
0010 **OIL/WATER SEPARATORS**									
0020 Underground, tank only									
0030　Excludes excavation, backfill, & piping									
0100　　200 GPM	B-21	3.50	8	Ea.	46,200	350	40	46,590	51,500
0110　　400 GPM		3.25	8.615		64,500	380	43	64,923	71,000
0120　　600 GPM		2.75	10.182		72,000	445	51	72,496	79,500
0130　　800 GPM		2.50	11.200		81,500	490	56	82,046	90,500
0140　　1000 GPM		2	14		91,000	615	70	91,685	101,000
0150　　1200 GPM		1.50	18.667		104,000	820	93.50	104,913.50	116,000
0160　　1500 GPM		1	28		122,500	1,225	140	123,865	137,000

46 25 16 – API Oil-Water Separators

46 25 16.10 Oil/Water Separators

	Crew	Daily Output	Labor-Hours	Unit	Material	2016 Bare Costs Labor	2016 Bare Costs Equipment	Total	Total Incl O&P
0010 **OIL/WATER SEPARATORS**									
0020 Complete system, not including excavation and backfill									
0100 Treated capacity 0.2 cubic feet per second	B-22	1	30	Ea.	6,500	1,325	211	8,036	9,425
0200　0.5 cubic foot per second	B-13	.75	74.667		14,000	3,100	1,000	18,100	21,200
0300　1.0 cubic foot per second		.60	93.333		23,000	3,875	1,250	28,125	32,600
0400　2.4 – 3 cubic feet per second		.30	186		47,000	7,725	2,500	57,225	66,000
0500　11 cubic feet per second		.17	329		90,000	13,600	4,400	108,000	124,500
0600　22 cubic feet per second		.10	560		128,000	23,200	7,475	158,675	184,000

Assemblies Section

Table of Contents

How RSMeans Assemblies Data Works

Assemblies estimating provides a fast and reasonably accurate way to develop construction costs. An assembly is the grouping of individual work items, with appropriate quantities, to provide a cost for a major construction component in a convenient unit of measure.

An assemblies estimate is often used during early stages of design development to compare the cost impact of various design alternatives on total building cost.

Assemblies estimates are also used as an efficient tool to verify construction estimates.

Assemblies estimates do not require a completed design or detailed drawings. Instead, they are based on the general size of the structure and other known parameters of the project. The degree of accuracy of an assemblies estimate is generally within +/- 15%.

Most RSMeans assemblies consist of three major elements: a graphic, the system components, and the cost data itself. The **Graphic** is a visual representation showing

① Unique 12-character Identifier

RSMeans assemblies are identified by a **unique 12-character identifier**. The assemblies are numbered using UNIFORMAT II, ASTM Standard E1557. The first 5 characters represent this system to Level 3. The last 7 characters represent further breakdown by RSMeans in order to arrange items in understandable groups of similar tasks. Line numbers are consistent across all RSMeans publications, so a line number in any RSMeans assemblies data set will always refer to the same work.

② Reference Box

Information is available in the Reference Section to assist the estimator with estimating procedures, alternate pricing methods, and additional technical information.

The **Reference Box** indicates the exact location of this information in the Reference Section. The "R" stands for "reference," and the remaining characters are the RSMeans line numbers.

③ Narrative Descriptions

RSMeans assemblies descriptions appear in two formats: narrative and table. **Narrative descriptions** are shown in a hierarchical structure to make them readable. In order to read a complete description, read up through the indents to the top of the section. Include everything that is above and to the left that is not contradicted by information below.

D30 HVAC

D3010 Energy Supply

Basis for Heat Loss Estimate, Apartment Type Structures:

1. Masonry walls and flat roof are insulated. U factor is assumed at .08.
2. Window glass area taken as BOCA minimum, 1/10th of floor area. Double insulating glass with 1/4" air space, U = .65.
3. Infiltration = 0.3 C.F. per hour per S.F. of net wall.
4. Concrete floor loss is 2 BTUH per S.F.
5. Temperature difference taken as 70°F.
6. Ventilating or makeup air has not been included and must be added if desired. Air shafts are not used.

System Components	QUANTITY	UNIT	COST EACH		
			MAT.	INST.	TOTAL
① SYSTEM D3010 510 1760					
HEATING SYSTEM, FIN TUBE RADIATION, FORCED HOT WATER					
1,000 S.F. AREA, 10,000 C.F. VOLUME					
Boiler, oil fired, CI, burner, ctrls/insul/breech/pipe/fftng/valves, 109 MBH	1.000	Ea.	4,462.50	4,068.75	8,531.25
Circulating pump, CI flange connection, 1/12 HP	1.000	Ea.	415	214	629
Expansion tank, painted steel, ASME 18 Gal capacity	1.000	Ea.	815	94	909
Storage tank, steel, above ground, 275 Gal capacity w/supports	1.000	Ea.	540	264	804
Copper tubing type L, solder joint, hanger 10' OC, 3/4" diam	100.000	L.F.	575	940	1,515
Radiation, 3/4" copper tube w/alum fin baseboard pkg, 7" high	30.000	L.F.	243	675	918
Pipe covering, calcium silicate w/cover, 1' wall, 3/4' diam	100.000	L.F.	448	705	1,153
TOTAL			7,498.50	6,960.75	14,459.25
COST PER S.F.			7.50	6.96	14.46

D3010 510	Apartment Building Heating - Fin Tube Radiation		COST PER S.F.		
			MAT.	INST.	TOTAL
1740	Heating systems, fin tube radiation, forced hot water				
1760	1,000 S.F. area, 10,000 C.F. volume		7.50	6.95	14.45
1800	10,000 S.F. area, 100,000 C.F. volume	R235000 -10	3.14	4.18	7.32
1840	20,000 S.F. area, 200,000 C.F. volume		3.43	4.69	8.12
1880	30,000 S.F. area, 300,000 C.F. volume	R235000 -20	3.32	4.55	7.87
1890					

③ **②**

For supplemental customizable square foot estimating forms, visit: **www.RSMeans.com/2016extras**

the typical appearance of the assembly in question, frequently accompanied by additional explanatory technical information describing the class of items. The **System Components** is a listing of the individual tasks that make up the assembly, including the quantity and unit of measure for each item, along with the cost of material and installation.

The **Assemblies Data** below lists prices for other similar systems with dimensional and/or size variations.

All RSMeans assemblies costs represent the cost for the installing contractor. An allowance for profit has been added to all material, labor, and equipment rental costs. A markup for labor burdens, including workers'

compensation, fixed overhead, and business overhead, is included with installation costs.

The information in RSMeans cost data represents a "national average" cost. This data should be modified to the project location using the **City Cost Indexes** or **Location Factors** tables found in the Reference Section.

D30 HVAC

D3010 Energy Supply

Basis for Heat Loss Estimate, Apartment Type Structures:

1. Masonry walls and flat roof are insulated. U factor is assumed at .08.
2. Window glass area taken as BOCA minimum, 1/10th of floor area. Double insulating glass with 1/4" air space, U = .65.
3. Infiltration = 0.3 C.F. per hour per S.F. of net wall.
4. Concrete floor loss is 2 BTUH per S.F.
5. Temperature difference taken as 70°F.
6. Ventilating or makeup air has not been included and must be added if desired. Air shafts are not used.

System Components	QUANTITY	UNIT	COST EACH MAT.	COST EACH INST.	COST EACH TOTAL
SYSTEM D3010 510 1760					
HEATING SYSTEM, FIN TUBE RADIATION, FORCED HOT WATER					
1,000 S.F. AREA, 10,000 C.F. VOLUME					
Boiler, oil fired, CI, burner, ctrls/insul/breech/pipe/ftng/valves, 109 MBH	1.000	Ea.	4,462.50	4,068.75	8,531.25
Circulating pump, CI flange connection, 1/12 HP	1.000	Ea.	415	214	629
Expansion tank, painted steel, ASME 18 Gal capacity	1.000	Ea.	815	94	909
Storage tank, steel, above ground, 275 Gal capacity w/supports	1.000	Ea.	540	264	804
Copper tubing type L, solder joint, hanger 10' OC, 3/4" diam	100.000	L.F.	575	940	1,515
Radiation, 3/4" copper tube w/alum fin baseboard pkg, 7" high	30.000	L.F.	243	675	918
Pipe covering, calcium silicate w/cover, 1' wall, 3/4" diam	100.000	L.F.	448	705	1,153
TOTAL			7,498.50	6,960.75	14,459.25
COST PER S.F.			7.50	6.96	14.46

D3010 510	Apartment Building Heating - Fin Tube Radiation		COST PER S.F. MAT.	COST PER S.F. INST.	COST PER S.F. TOTAL
1740	Heating systems, fin tube radiation, forced hot water				
1760	1,000 S.F. area, 10,000 C.F. volume		7.50	6.95	14.45
1800	10,000 S.F. area, 100,000 C.F. volume		3.14	4.18	7.32
1840	20,000 S.F. area, 200,000 C.F. volume	R235000 -10	3.43	4.69	8.12
1880	30,000 S.F. area, 300,000 C.F. volume		3.32	4.55	7.87
1890		R235000 -20			

4 Unit of Measure

All RSMeans assemblies data includes a typical **Unit of Measure** used for estimating that item. For instance, for continuous footings or foundation walls the unit is linear feet (L.F.). For spread footings the unit is each (Ea.). The estimator needs to take special care that the unit in the data matches the unit in the takeoff. Abbreviations and unit conversions can be found in the Reference Section.

5 System Components

System components are listed separately to detail what is included in the development of the total system price.

How RSMeans Assemblies Data Works (Continued)

Sample Estimate

This sample demonstrates the elements of an estimate, including a tally of the RSMeans data lines. Published assemblies costs include all markups for labor burden and profit for the installing contractor. This estimate adds a summary of the markups applied by a general contractor on the installing contractor's work. These figures represent the total cost to the owner. The RSMeans location factor is applied at the bottom of the estimate to adjust the cost of the work to a specific location.

Project Name:	Interior Fit-out, ABC Office				
Location:	**Anywhere, USA**		**Date: 1/1/2016**		**STD**
Assembly Number	**Description**	**Qty.**	**Unit**		**Subtotal**
C1010 ❶ 4 1200	Wood partition, 2 x 4 @ 16" OC w/5/8" FR gypsum board	560.000	S.F.		$2,738.40
C1020 ❶ 4 1800	Metal door & frame, flush hollow core, 3'-0" x 7'-0"	2.000	Ea.		$2,470.00
C3010 230 0080	Painting, brushwork, primer & 2 coats	1,120.000	S.F.		$1,344.00
C3020 410 0140	Carpet, tufted, nylon, roll goods, 12' wide, 26 oz	240.000	S.F.		$813.60
C3030 210 6000	Acoustic ceilings, 24" x 48" tile, tee grid suspension	200.000	S.F.		$1,156.00
D5020 125 0560	Receptacles incl plate, box, conduit, wire, 20 A duplex	8.000	Ea.		$2,168.00
D5020 125 0720	Light switch incl plate, box, conduit, wire, 20 A single pole	2.000	Ea.		$532.00
D5020 210 0560	Fluorescent fixtures, recess mounted, 20 per 1000 SF	200.000	S.F.		$2,186.00
	Assembly Subtotal				**$13,408.00**
	Sales Tax @ ❷		5 %		$ 335.20
	General Requirements @ ❸		7 %		$ 938.56
	Subtotal A				**$14,681.76**
	GC Overhead @ ❹		5 %		$ 734.09
	Subtotal B				**$15,415.85**
	GC Profit @ ❺		5 %		$ 770.79
	Subtotal C				**$16,186.64**
	Adjusted by Location Factor ❻	118.1			$ 19,116.42
	Architects Fee @ ❼		8 %		$ 1,529.31
	Contingency @ ❽		15 %		$ 2,867.46
	Project Total Cost				**$ 23,513.20**

This estimate is based on an interactive spreadsheet.
A copy of this spreadsheet is located on the RSMeans website at
www.RSMeans.com/2016extras.
You are free to download it and adjust it to your methodology.

1 Work Performed

The body of the estimate shows the RSMeans data selected, including line numbers, a brief description of each item, its takeoff quantity and unit, and the total installed cost, including the installing contractor's overhead and profit.

2 Sales Tax

If the work is subject to state or local sales taxes, the amount must be added to the estimate. In a conceptual estimate it can be assumed that one half of the total represents material costs. Therefore, apply the sales tax rate to 50% of the assembly subtotal.

3 General Requirements

This item covers project-wide needs provided by the general contractor. These items vary by project but may include temporary facilities and utilities, security, testing, project cleanup, etc. In assemblies estimates a percentage is used, typically between 5% and 15% of project cost.

4 General Contractor Overhead

This entry represents the general contractor's markup on all work to cover project administration costs.

5 General Contractor Profit

This entry represents the GC's profit on all work performed. The value included here can vary widely by project and is influenced by the GC's perception of the project's financial risk and market conditions.

6 Location Factor

RSMeans published data is based on national average costs. If necessary, adjust the total cost of the project using a location factor from the "Location Factor" table or the "City Cost Indexes" table found in the Reference Section. Use location factors if the work is general, covering the work of multiple trades. If the work is by a single trade (e.g., masonry) use the more specific data found in the City Cost Indexes.

To adjust costs by location factors, multiply the base cost by the factor and divide by 100.

7 Architect's Fee

If appropriate, add the design cost to the project estimate. These fees vary based on project complexity and size. Typical design and engineering fees can be found in the reference section.

8 Contingency

A factor for contingency may be added to any estimate to represent the cost of unknowns that may occur between the time that the estimate is performed and the time the project is constructed. The amount of the allowance will depend on the stage of design at which the estimate is done, and the contractor's assessment of the risk involved.

A20 Basement Construction

A2020 Basement Walls

A2020 220	Subdrainage Piping	COST PER L.F.		
		MAT.	INST.	TOTAL
2000	Piping, excavation & backfill excluded, PVC, perforated			
2110	3" diameter	1.64	4.19	5.83
2130	4" diameter	1.64	4.19	5.83
2140	5" diameter	3.61	4.48	8.09
2150	6" diameter	3.61	4.48	8.09
3000	Metal alum. or steel, perforated asphalt coated			
3150	6" diameter	7.15	4.13	11.28
3160	8" diameter	9.40	4.24	13.64
3170	10" diameter	11.75	5.70	17.45
3180	12" diameter	13.15	7.15	20.30
3220	18" diameter	19.75	9.95	29.70
4000	Corrugated HDPE tubing			
4130	4" diameter	.75	.78	1.53
4150	6" diameter	1.74	1.03	2.77
4160	8" diameter	3.47	1.33	4.80
4180	12" diameter	7.05	4.62	11.67
4200	15" diameter	8.55	5.25	13.80
4220	18" diameter	13.30	7.40	20.70

For customer support on your Mechanical Cost Data, call 877.739.3689.

In this closed-loop indirect collection system, fluid with a low freezing temperature, such as propylene glycol, transports heat from the collectors to water storage. The transfer fluid is contained in a closed-loop consisting of collectors, supply and return piping, and a remote heat exchanger. The heat exchanger transfers heat energy from the

fluid in the collector loop to potable water circulated in a storage loop. A typical two-or-three panel system contains 5 to 6 gallons of heat transfer fluid.

When the collectors become approximately 20°F warmer than the storage temperature, a controller activates the circulator on the collector and storage

loops. The circulators will move the fluid and potable water through the heat exchanger until heat collection no longer occurs. At that point, the system shuts down. Since the heat transfer medium is a fluid with a very low freezing temperature, there is no need for it to be drained from the system between periods of collection.

D20 Plumbing

D2020 Domestic Water Distribution

System Components	QUANTITY	UNIT	COST EACH		
			MAT.	INST.	TOTAL
SYSTEM D2020 265 2760					
SOLAR, CLOSED LOOP, ADD-ON HOT WATER SYS., EXTERNAL HEAT EXCHANGER					
3/4" TUBING, TWO 3'X7' BLACK CHROME COLLECTORS					
A,B,G,L,K,M Heat exchanger fluid-fluid pkg incl 2 circulators, expansion tank,					
Check valve, relief valve, controller, hi temp cutoff, & 2 sensors	1.000	Ea.	880	515	1,395
C Thermometer, 2" dial	3.000	Ea.	72	133.50	205.50
D, T Fill & drain valve, brass, 3/4" connection	1.000	Ea.	11.25	30	41.25
E Air vent, manual, 1/8" fitting	2.000	Ea.	5.36	45	50.36
F Air purger	1.000	Ea.	51.50	59.50	111
H Strainer, Y type, bronze body, 3/4" IPS	1.000	Ea.	45.50	38.50	84
I Valve, gate, bronze, NRS, soldered 3/4" diam	6.000	Ea.	435	213	648
J Neoprene vent flashing	2.000	Ea.	11	71	82
N-1, N Relief valve temp & press, 150 psi 210°F self-closing 3/4" IPS	1.000	Ea.	24.50	24	48.50
O Pipe covering, urethane, ultraviolet cover, 1" wall 3/4" diam	20.000	L.F.	62.20	130	192.20
P Pipe covering, fiberglass, all service jacket, 1" wall, 3/4" diam	50.000	L.F.	49.50	260	309.50
Q Collector panel solar energy blk chrome on copper, 1/8" temp glass 3'x7'	2.000	Ea.	2,200	270	2,470
Roof clamps for solar energy collector panels	2.000	Set	6.22	36.80	43.02
R Valve, swing check, bronze, regrinding disc, 3/4" diam	2.000	Ea.	232	71	303
S Pressure gauge, 60 psi, 2" dial	1.000	Ea.	26.50	22.50	49
U Valve, water tempering, bronze, sweat connections, 3/4" diam	1.000	Ea.	150	35.50	185.50
W-2, V Tank water storage w/heating element, drain, relief valve, existing	1.000	Ea.			
Copper tubing type L, solder joint, hanger 10' OC 3/4" diam	20.000	L.F.	115	188	303
Copper tubing, type M, solder joint, hanger 10' OC 3/4" diam	70.000	L.F.	337.40	640.50	977.90
Sensor wire, #22-2 conductor multistranded	.500	C.L.F.	7.83	33	40.83
Solar energy heat transfer fluid, propylene glycol anti-freeze	6.000	Gal.	99	153	252
Wrought copper fittings & solder, 3/4" diam	76.000	Ea.	211.28	2,850	3,061.28
TOTAL			5,033.04	5,819.80	10,852.84

D2020 265	Solar, Closed Loop, Add-On Hot Water Systems		COST EACH		
			MAT.	INST.	TOTAL
2550	Solar, closed loop, add-on hot water system, external heat exchanger				
2570	3/8" tubing, 3 ea. 4' x 4'-4" vacuum tube collectors		5,575	5,475	11,050
2580	1/2" tubing, 4 ea. 4 x 4'-4" vacuum tube collectors	R235616	6,400	5,875	12,275
2600	2 ea. 3'x7' black chrome collectors	-60	4,600	5,550	10,150
2620	3 ea. 3'x7' black chrome collectors		5,700	5,700	11,400
2640	2 ea. 3'x7' flat black collectors		4,325	5,575	9,900
2660	3 ea. 3'x7' flat black collectors		5,300	5,725	11,025
2700	3/4" tubing, 3 ea. 3'x7' black chrome collectors		6,125	5,975	12,100
2720	3 ea. 3'x7' flat black absorber plate collectors		5,725	6,000	11,725
2740	2 ea. 4'x9' flat black w/plastic glazing collectors		5,125	6,025	11,150
2760	2 ea. 3'x7' black chrome collectors		5,025	5,825	10,850
2780	1" tubing, 4 ea 2'x9' plastic absorber & glazing collectors		6,150	6,775	12,925
2800	4 ea. 3'x7' black chrome absorber collectors		8,000	6,800	14,800
2820	4 ea. 3'x7' flat black absorber collectors		7,450	6,825	14,275

505

In the drainback indirect-collection system, the heat transfer fluid is distilled water contained in a loop consisting of collectors, supply and return piping, and an unpressurized holding tank. A large heat exchanger containing incoming potable water is immersed in the holding tank. When a controller activates solar collection, the distilled water is pumped through the collectors and heated and pumped back down to the holding tank. When the temperature differential between the water in the collectors and water in storage is such that collection no longer occurs, the pump turns off and gravity causes the distilled water in the collector loop to drain back to the holding tank. All the loop piping is pitched so that the water can drain out of the collectors and piping and not freeze there. As hot water is needed in the home, incoming water first flows through the holding tank with the immersed heat exchanger and is warmed and then flows through a conventional heater for any supplemental heating that is necessary.

506

For customer support on your **Mechanical Cost Data**, call 877.739.3689.

D2020 Domestic Water Distribution

System Components	QUANTITY	UNIT	COST EACH		
			MAT.	INST.	TOTAL
SYSTEM D2020 270 2760					
SOLAR, DRAINBACK, ADD ON, HOT WATER, IMMERSED HEAT EXCHANGER					
3/4″ TUBING, THREE EA 3′X7′ BLACK CHROME COLLECTOR					
A, B Differential controller 2 sensors, thermostat, solar energy system	1.000	Ea.	143	59.50	202.50
C Thermometer 2″ dial	3.000	Ea.	72	133.50	205.50
D, T Fill & drain valve, brass, 3/4″ connection	1.000	Ea.	11.25	30	41.25
E-1 Automatic air vent 1/8″ fitting	1.000	Ea.	18.95	22.50	41.45
H Strainer, Y type, bronze body, 3/4″ IPS	1.000	Ea.	45.50	38.50	84
I Valve, gate, bronze, NRS, soldered 3/4″ diam	2.000	Ea.	145	71	216
J Neoprene vent flashing	2.000	Ea.	11	71	82
L Circulator, solar heated liquid, 1/20 HP	1.000	Ea.	289	107	396
N Relief valve temp. & press. 150 psi 210°F self-closing 3/4″ IPS	1.000	Ea.	24.50	24	48.50
O Pipe covering, urethane, ultraviolet cover, 1″ wall, 3/4″ diam	20.000	L.F.	62.20	130	192.20
P Pipe covering, fiberglass, all service jacket, 1″ wall, 3/4″ diam	50.000	L.F.	49.50	260	309.50
Q Collector panel solar energy blk chrome on copper, 1/8″ temp glas 3′x7′	3.000	Ea.	3,300	405	3,705
Roof clamps for solar energy collector panels	3.000	Set	9.33	55.20	64.53
R Valve, swing check, bronze, regrinding disc, 3/4″ diam	1.000	Ea.	116	35.50	151.50
U Valve, water tempering, bronze sweat connections, 3/4″ diam	1.000	Ea.	150	35.50	185.50
V Tank, water storage w/heating element, drain, relief valve, existing	1.000	Ea.			
W Tank, water storage immersed heat exchr elec 2″x1/2# insul 120 gal	1.000	Ea.	2,025	510	2,535
X Valve, globe, bronze, rising stem, 3/4″ diam, soldered	3.000	Ea.	342	106.50	448.50
Y Flow control valve	1.000	Ea.	143	32.50	175.50
Z Valve, ball, bronze, solder 3/4″ diam, solar loop flow control	1.000	Ea.	25.50	35.50	61
Copper tubing, type L, solder joint, hanger 10′ OC 3/4″ diam	20.000	L.F.	115	188	303
Copper tubing, type M, solder joint, hanger 10′ OC 3/4″ diam	70.000	L.F.	337.40	640.50	977.90
Sensor wire, #22-2 conductor, multistranded	.500	C.L.F.	7.83	33	40.83
Wrought copper fittings & solder, 3/4″ diam	76.000	Ea.	211.28	2,850	3,061.28
TOTAL			7,654.24	5,874.20	13,528.44

D2020 270	Solar, Drainback, Hot Water Systems		COST EACH		
			MAT.	INST.	TOTAL
2550	Solar, drainback, hot water, immersed heat exchanger				
2560	3/8″ tubing, 3 ea. 4′ x 4′-4″ vacuum tube collectors	R235616 -60	6,875	5,300	12,175
2580	1/2″ tubing, 4 ea 4′ x 4′-4″ vacuum tube collectors, 80 gal tank		7,725	5,725	13,450
2600	120 gal tank		7,975	5,800	13,775
2640	2 ea. 3′x7′ blk chrome collectors, 80 gal tank		5,925	5,400	11,325
2660	3 ea. 3′x7′ blk chrome collectors, 120 gal tank		7,275	5,600	12,875
2700	2 ea. 3′x7′ flat blk collectors, 120 gal tank		5,900	5,475	11,375
2720	3 ea. 3′x7′ flat blk collectors, 120 gal tank		6,875	5,625	12,500
2760	3/4″ tubing, 3 ea 3′x7′ black chrome collectors, 120 gal tank		7,650	5,875	13,525
2780	3 ea. 3′x7′ flat black absorber collectors, 120 gal tank		7,250	5,900	13,150
2800	2 ea. 4′x9′ flat blk w/plastic glazing collectors 120 gal tank		6,650	5,925	12,575
2840	1″ tubing, 4 ea. 2′x9′ plastic absorber & glazing collectors, 120 gal tank		7,750	6,675	14,425
2860	4 ea. 3′x7′ black chrome absorber collectors, 120 gal tank		9,600	6,700	16,300
2880	4 ea. 3′x7′ flat black absorber collectors, 120 gal tank		9,050	6,725	15,775

For customer support on your Mechanical Cost Data, call 877.739.3689.

507

In the draindown direct-collection system, incoming domestic water is heated in the collectors. When the controller activates solar collection, domestic water is first heated as it flows through the collectors and is then pumped to storage. When conditions are no longer suitable for heat collection, the pump shuts off and the water in the loop drains down and out of the system by means of solenoid valves and properly pitched piping.

D2020 Domestic Water Distribution

System Components	QUANTITY	UNIT	COST EACH		
			MAT.	INST.	TOTAL
SYSTEM D2020 275 2760					
SOLAR, DRAINDOWN, HOT WATER, DIRECT COLLECTION					
3/4″ TUBING, THREE 3′X7′ BLACK CHROME COLLECTORS					
A, B Differential controller, 2 sensors, thermostat, solar energy system	1.000	Ea.	143	59.50	202.50
A-1 Solenoid valve, solar heating loop, brass, 3/4″ diam, 24 volts	3.000	Ea.	423	238.50	661.50
B-1 Solar energy sensor, freeze prevention	1.000	Ea.	25	22.50	47.50
C Thermometer, 2″ dial	3.000	Ea.	72	133.50	205.50
E-1 Vacuum relief valve, 3/4″ diam	1.000	Ea.	32.50	22.50	55
F-1 Air vent, automatic, 1/8″ fitting	1.000	Ea.	18.95	22.50	41.45
H Strainer, Y type, bronze body, 3/4″ IPS	1.000	Ea.	45.50	38.50	84
I Valve, gate, bronze, NRS, soldered, 3/4″ diam	2.000	Ea.	145	71	216
J Vent flashing neoprene	2.000	Ea.	11	71	82
K Circulator, solar heated liquid, 1/25 HP	1.000	Ea.	199	92	291
N Relief valve temp & press 150 psi 210°F self-closing 3/4″ IPS	1.000	Ea.	24.50	24	48.50
O Pipe covering, urethane, ultraviolet cover, 1″ wall, 3/4″ diam	20.000	L.F.	62.20	130	192.20
P Pipe covering, fiberglass, all service jacket, 1″ wall, 3/4″ diam	50.000	L.F.	49.50	260	309.50
Roof clamps for solar energy collector panels	3.000	Set	9.33	55.20	64.53
Q Collector panel solar energy blk chrome on copper, 1/8″ temp glass 3′x7′	3.000	Ea.	3,300	405	3,705
R Valve, swing check, bronze, regrinding disc, 3/4″ diam, soldered	2.000	Ea.	232	71	303
T Drain valve, brass, 3/4″ connection	2.000	Ea.	22.50	60	82.50
U Valve, water tempering, bronze, sweat connections, 3/4″ diam	1.000	Ea.	150	35.50	185.50
W-2, W Tank, water storage elec elem 2″x1/2# insul 120 gal	1.000	Ea.	2,025	510	2,535
X Valve, globe, bronze, rising stem, 3/4″ diam, soldered	1.000	Ea.	114	35.50	149.50
Copper tubing, type L, solder joints, hangers 10′ OC 3/4″ diam	20.000	L.F.	115	188	303
Copper tubing, type M, solder joints, hangers 10′ OC 3/4″ diam	70.000	L.F.	337.40	640.50	977.90
Sensor wire, #22-2 conductor, multistranded	.500	C.L.F.	7.83	33	40.83
Wrought copper fittings & solder, 3/4″ diam	76.000	Ea.	211.28	2,850	3,061.28
TOTAL			7,775.49	6,069.20	13,844.69

D2020 275	Solar, Draindown, Hot Water Systems		COST EACH		
			MAT.	INST.	TOTAL
2550	Solar, draindown, hot water				
2560	3/8″ tubing, 3 ea. 4′ x 4′-4″ vacuum tube collectors, 80 gal tank		6,825	5,775	12,600
2580	1/2″ tubing, 4 ea. 4′ x 4′-4″ vacuum tube collectors, 80 gal tank	R235616 -60	7,825	5,925	13,750
2600	120 gal tank		8,075	6,000	14,075
2640	2 ea. 3′x7′ black chrome collectors, 80 gal tank		6,025	5,600	11,625
2660	3 ea. 3′x7′ black chrome collectors, 120 gal tank		7,400	5,825	13,225
2700	2 ea. 3′x7′ flat black collectors, 120 gal tank		6,025	5,675	11,700
2720	3 ea. 3′x7′ flat black collectors, 120 gal tank		6,900	5,800	12,700
2760	3/4″ tubing, 3 ea. 3′x7′ black chrome collectors, 120 gal tank		7,775	6,075	13,850
2780	3 ea. 3′x7′ flat collectors, 120 gal tank		7,375	6,100	13,475
2800	2 ea. 4′x9′ flat black & plastic glazing collectors, 120 gal tank		6,775	6,125	12,900
2840	1″ tubing, 4 ea. 2′x9′ plastic absorber & glazing collectors, 120 gal tank		11,500	8,075	19,575
2860	4 ea. 3′x7′ black chrome absorber collectors, 120 gal tank		11,000	7,625	18,625
2880	4 ea. 3′x7′ flat black absorber collectors, 120 gal tank		9,350	6,900	16,250

In the recirculation system, (a direct-collection system), incoming domestic water is heated in the collectors. When the controller activates solar collection, domestic water is heated as it flows through the collectors and then it flows back to storage. When conditions are not suitable for heat collection, the pump shuts off and the flow of the water stops. In this type of system, water remains in the collector loop at all times. A "frost sensor" at the collector activates the circulation of warm water from storage through the collectors when protection from freezing is required.

D2020 Domestic Water Distribution

System Components	QUANTITY	UNIT	COST EACH		
			MAT.	INST.	TOTAL
SYSTEM D2020 280 2820					
SOLAR, RECIRCULATION, HOT WATER					
3/4″ TUBING, TWO 3′X7′ BLACK CHROME COLLECTORS					
A, B Differential controller 2 sensors, thermostat, for solar energy system	1.000	Ea.	143	59.50	202.50
A-1 Solenoid valve, solar heating loop, brass, 3/4″ IPS, 24 volts	2.000	Ea.	282	159	441
B-1 Solar energy sensor freeze prevention	1.000	Ea.	25	22.50	47.50
C Thermometer, 2″ dial	3.000	Ea.	72	133.50	205.50
D Drain valve, brass, 3/4″ connection	1.000	Ea.	11.25	30	41.25
F-1 Air vent, automatic, 1/8″ fitting	1.000	Ea.	18.95	22.50	41.45
H Strainer, Y type, bronze body, 3/4″ IPS	1.000	Ea.	45.50	38.50	84
I Valve, gate, bronze, 125 lb, soldered 3/4″ diam	3.000	Ea.	217.50	106.50	324
J Vent flashing, neoprene	2.000	Ea.	11	71	82
L Circulator, solar heated liquid, 1/20 HP	1.000	Ea.	289	107	396
N Relief valve, temp & press 150 psi 210° F self-closing 3/4″ IPS	2.000	Ea.	49	48	97
O Pipe covering, urethane, ultraviolet cover, 1″ wall, 3/4″ diam	20.000	L.F.	62.20	130	192.20
P Pipe covering, fiberglass, all service jacket, 1″ wall 3/4″ diam	50.000	L.F.	49.50	260	309.50
Q Collector panel solar energy blk chrome on copper, 1/8″ temp glass 3′x7′	2.000	Ea.	2,200	270	2,470
Roof clamps for solar energy collector panels	2.000	Set	6.22	36.80	43.02
R Valve, swing check, bronze, 125 lb, regrinding disc, soldered 3/4″ diam	2.000	Ea.	232	71	303
U Valve, water tempering, bronze, sweat connections, 3/4″ diam	1.000	Ea.	150	35.50	185.50
V Tank, water storage, w/heating element, drain, relief valve, existing	1.000	Ea.			
X Valve, globe, bronze, 125 lb, 3/4″ diam	2.000	Ea.	228	71	299
Copper tubing, type L, solder joints, hangers 10′ OC 3/4″ diam	20.000	L.F.	115	188	303
Copper tubing, type M, solder joints, hangers 10′ OC 3/4″ diam	70.000	L.F.	337.40	640.50	977.90
Wrought copper fittings & solder, 3/4″ diam	76.000	Ea.	211.28	2,850	3,061.28
Sensor wire, #22-2 conductor, multistranded	.500	C.L.F.	7.83	33	40.83
TOTAL			**4,763.63**	**5,383.80**	**10,147.43**

D2020 280	Solar Recirculation, Domestic Hot Water Systems		COST EACH		
			MAT.	INST.	TOTAL
2550	Solar recirculation, hot water				
2560	3/8″ tubing, 3 ea. 4′ x 4′-4″ vacuum tube collectors		5,325	5,025	10,350
2580	1/2″ tubing, 4 ea. 4′ x 4′-4″ vacuum tube collectors		6,150	5,450	11,600
2640	2 ea. 3′x7′ black chrome collectors	R235616 -60	4,375	5,125	9,500
2660	3 ea. 3′x7′ black chrome collectors		5,475	5,275	10,750
2700	2 ea. 3′x7′ flat black collectors		4,100	5,125	9,225
2720	3 ea. 3′x7′ flat black collectors		5,050	5,300	10,350
2760	3/4″ tubing, 3 ea. 3′x7′ black chrome collectors		5,875	5,525	11,400
2780	3 ea. 3′x7′ flat black absorber plate collectors		5,450	5,550	11,000
2800	2 ea. 4′x9′ flat black w/plastic glazing collectors		4,875	5,575	10,450
2820	2 ea. 3′x7′ black chrome collectors		4,775	5,375	10,150
2840	1″ tubing, 4 ea. 2′x9′ black plastic absorber & glazing collectors		6,125	6,325	12,450
2860	4 ea. 3′x7′ black chrome absorber collectors		7,975	6,350	14,325
2880	4 ea. 3′x7′ flat black absorber collectors		7,425	6,375	13,800

For customer support on your Mechanical Cost Data, call 877.739.3689.

511

The thermosyphon domestic hot water system, a direct collection system, operates under city water pressure and does not require pumps for system operation. An insulated water storage tank is located above the collectors. As the sun heats the collectors, warm water in them rises by means of natural convection; the colder water in the storage tank flows into the collectors by means of gravity. As long as the sun is shining the water continues to flow through the collectors and to become warmer.

To prevent freezing, the system must be drained or the collectors covered with an insulated lid when the temperature drops below 32°F.

D2020 Domestic Water Distribution

System Components	QUANTITY	UNIT	COST EACH		
			MAT.	INST.	TOTAL
SYSTEM D2020 285 0960					
SOLAR, THERMOSYPHON, WATER HEATER					
3/4″ TUBING, TWO 3′X7′ BLACK CHROME COLLECTORS					
D-1 Framing lumber, fir, 2″ x 6″ x 8′, tank cradle	.008	M.B.F.	5.36	7.60	12.96
F-1 Framing lumber, fir, 2″ x 4″ x 24′, sleepers	.016	M.B.F.	10.48	22.80	33.28
I Valve, gate, bronze, 125 lb, soldered 1/2″ diam	2.000	Ea.	124	60	184
J Vent flashing, neoprene	4.000	Ea.	22	142	164
O Pipe covering, urethane, ultraviolet cover, 1″ wall, 1/2″diam	40.000	L.F.	95.20	256	351.20
P Pipe covering fiberglass all service jacket 1″ wall 1/2″ diam	160.000	L.F.	145.60	798.40	944
Q Collector panel solar, blk chrome on copper, 3/16″ temp glass 3′-6″ x 7.5′	2.000	Ea.	1,930	286	2,216
Y Flow control valve, globe, bronze, 125#, soldered, 1/2″ diam	1.000	Ea.	111	30	141
U Valve, water tempering, bronze,sweat connections, 1/2″ diam	1.000	Ea.	112	30	142
W Tank, water storage, solar energy system, 80 Gal, 2″ x 1/2 lb insul	1.000	Ea.	1,775	445	2,220
Copper tubing type L, solder joints, hangers 10′ OC 1/2″ diam	150.000	L.F.	607.50	1,327.50	1,935
Copper tubing type M, solder joints, hangers 10′ OC 1/2″ diam	50.000	L.F.	176.50	425	601.50
Sensor wire, #22-2 gauge multistranded	.500	C.L.F.	7.83	33	40.83
Wrought copper fittings & solder, 1/2″ diam	75.000	Ea.	93	2,662.50	2,755.50
TOTAL			5,215.47	6,525.80	11,741.27

D2020 285	Thermosyphon, Hot Water		COST EACH		
			MAT.	INST.	TOTAL
0960	Solar, thermosyphon, hot water, two collector system	R235616 -60	5,225	6,525	11,750
0970					

513

For customer support on your Mechanical Cost Data, call 877.739.3689.

ATTIC FLOOR

CWS

HWS

This domestic hot water pre-heat system includes heat exchanger with a circulating pump, blower, air-to-water, coil and controls, mounted in the upper collector manifold. Heat from the hot air coming out of the collectors is transferred through the heat exchanger. For each degree of DHW preheating gained, one degree less heating is needed from the fuel fired water heater. The system is simple, inexpensive to operate and can provide a substantial portion of DHW requirements for modest additional cost.

514

For customer support on your Mechanical Cost Data, call 877.739.3689.

D2020 Domestic Water Distribution

System Components	QUANTITY	UNIT	COST EACH		
			MAT.	INST.	TOTAL
SYSTEM D2020 290 2560					
SOLAR, HOT WATER, AIR TO WATER HEAT EXCHANGE					
THREE COLLECTORS, OPTICAL BLACK ON ALUM., 7.5' x 3.5', 80 GAL TANK					
A, B Differential controller, 2 sensors, thermostat, solar energy system	1.000	Ea.	143	59.50	202.50
C Thermometer, 2" dial	2.000	Ea.	48	89	137
C-1 Heat exchanger, air to fluid, up flow 70 MBH	1.000	Ea.	395	370	765
E Air vent, manual, for solar energy system 1/8" fitting	1.000	Ea.	2.68	22.50	25.18
F Air purger	1.000	Ea.	51.50	59.50	111
G Expansion tank	1.000	Ea.	75	22.50	97.50
H Strainer, Y type, bronze body, 1/2" IPS	1.000	Ea.	37	36.50	73.50
I Valve, gate, bronze, 125 lb, NRS, soldered 1/2" diam	5.000	Ea.	310	150	460
N Relief valve, temp & pressure solar 150 psi 210°F self-closing	1.000	Ea.	24.50	24	48.50
P Pipe covering, fiberglass, all service jacket, 1" wall, 1/2" diam	60.000	L.F.	54.60	299.40	354
Q Collector panel solar energy, air, black on alum plate, 7.5' x 3.5'	3.000	Ea.	2,895	429	3,324
B-1, R-1 Shutter damper for solar heater circulator	1.000	Ea.	61.50	105	166.50
B-1, R-1 Shutter motor for solar heater circulator	1.000	Ea.	153	79	232
R Backflow preventer, 1/2" pipe size	2.000	Ea.	173	89	262
T Drain valve, brass, 3/4" connection	1.000	Ea.	11.25	30	41.25
U Valve, water tempering, bronze, sweat connections, 1/2" diam	1.000	Ea.	112	30	142
W Tank, water storage, solar energy system, 80 Gal, 2" x 2 lb insul	1.000	Ea.	1,775	445	2,220
V Tank, water storage, w/heating element, drain, relief valve, existing	1.000	System			
X Valve, globe, bronze, 125 lb, rising stem, 1/2" diam	1.000	Ea.	111	30	141
Copper tubing type M, solder joints, hangers 10' OC, 1/2" diam	50.000	L.F.	176.50	425	601.50
Copper tubing type L, solder joints, hangers 10' OC 1/2" diam	10.000	L.F.	40.50	88.50	129
Wrought copper fittings & solder, 1/2" diam	10.000	Ea.	12.40	355	367.40
Sensor wire, #22-2 conductor multistranded	.500	C.L.F.	7.83	33	40.83
Q-1, Q-2 Ductwork, fiberglass, aluminized jacket, 1-1/2" thick, 8" diam	32.000	S.F.	180.80	224	404.80
Q-3 Manifold for flush mount solar energy collector panels, air	6.000	L.F.	954	47.40	1,001.40
TOTAL			7,805.06	3,542.80	11,347.86

D2020 290	Solar Hot Water, Air To Water Heat Exchange		COST EACH		
			MAT.	INST.	TOTAL
2550	Solar hot water, air to water heat exchange				
2560	Three collectors, optical black on aluminum, 7.5' x 3.5', 80 Gal tank		7,800	3,550	11,350
2580	Four collectors, optical black on aluminum, 7.5' x 3.5', 80 Gal tank	R235616	9,125	3,750	12,875
2600	Four collectors, optical black on aluminum, 7.5' x 3.5', 120 Gal tank	-60	9,375	3,825	13,200

515

For customer support on your Mechanical Cost Data, call 877.739.3689.

In this closed-loop indirect collection system, fluid with a low freezing temperature, such as propylene glycol, transports heat from the collectors to water storage. The transfer fluid is contained in a closed-loop consisting of collectors, supply and return piping, and a heat exchanger immersed in the storage tank. A typical two-or-three panel system contains 5 to 6 gallons of heat transfer fluid.

When the collectors become approximately 20°F warmer than the storage temperature, a controller activates the circulator. The circulator moves the fluid continuously through the collectors until the temperature difference between the collectors and storage is such that heat collection no longer occurs; at that point, the circulator shuts off. Since the heat transfer fluid has a very low freezing temperature, there is no need for it to be drained from the collectors between periods of collection.

D2020 Domestic Water Distribution

System Components		QUANTITY	UNIT	COST EACH		
				MAT.	INST.	TOTAL
SYSTEM D2020 295 2760						
SOLAR, CLOSED LOOP, HOT WATER SYSTEM, IMMERSED HEAT EXCHANGER						
3/4″ TUBING, THREE 3′ X 7′ BLACK CHROME COLLECTORS						
A, B	Differential controller, 2 sensors, thermostat, solar energy system	1.000	Ea.	143	59.50	202.50
C	Thermometer 2″ dial	3.000	Ea.	72	133.50	205.50
D, T	Fill & drain valves, brass, 3/4″ connection	3.000	Ea.	33.75	90	123.75
E	Air vent, manual, 1/8″ fitting	1.000	Ea.	2.68	22.50	25.18
F	Air purger	1.000	Ea.	51.50	59.50	111
G	Expansion tank	1.000	Ea.	75	22.50	97.50
I	Valve, gate, bronze, NRS, soldered 3/4″ diam	3.000	Ea.	217.50	106.50	324
J	Neoprene vent flashing	2.000	Ea.	11	71	82
K	Circulator, solar heated liquid, 1/25 HP	1.000	Ea.	199	92	291
N-1, N	Relief valve, temp & press 150 psi 210°F self-closing 3/4″ IPS	2.000	Ea.	49	48	97
O	Pipe covering, urethane ultraviolet cover, 1″ wall, 3/4″ diam	20.000	L.F.	62.20	130	192.20
P	Pipe covering, fiberglass, all service jacket, 1″ wall, 3/4″ diam	50.000	L.F.	49.50	260	309.50
	Roof clamps for solar energy collector panel	3.000	Set	9.33	55.20	64.53
Q	Collector panel solar blk chrome on copper, 1/8″ temp glass, 3′x7′	3.000	Ea.	3,300	405	3,705
R-1	Valve, swing check, bronze, regrinding disc, 3/4″ diam, soldered	1.000	Ea.	116	35.50	151.50
S	Pressure gauge, 60 psi, 2-1/2″ dial	1.000	Ea.	26.50	22.50	49
U	Valve, water tempering, bronze, sweat connections, 3/4″ diam	1.000	Ea.	150	35.50	185.50
W-2, W	Tank, water storage immersed heat exchr elec elem 2″x2# insul 120 Gal	1.000	Ea.	2,025	510	2,535
X	Valve, globe, bronze, rising stem, 3/4″ diam, soldered	1.000	Ea.	114	35.50	149.50
	Copper tubing type L, solder joint, hanger 10′ OC 3/4″ diam	20.000	L.F.	115	188	303
	Copper tubing, type M, solder joint, hanger 10′ OC 3/4″ diam	70.000	L.F.	337.40	640.50	977.90
	Sensor wire, #22-2 conductor multistranded	.500	C.L.F.	7.83	33	40.83
	Solar energy heat transfer fluid, propylene glycol, anti-freeze	6.000	Gal.	99	153	252
	Wrought copper fittings & solder, 3/4″ diam	76.000	Ea.	211.28	2,850	3,061.28
	TOTAL			7,477.47	6,058.70	13,536.17

D2020 295	Solar, Closed Loop, Hot Water Systems		COST EACH		
			MAT.	INST.	TOTAL
2550	Solar, closed loop, hot water system, immersed heat exchanger				
2560	3/8″ tubing, 3 ea. 4′ x 4′-4″ vacuum tube collectors, 80 gal. tank		6,725	5,500	12,225
2580	1/2″ tubing, 4 ea. 4′ x 4′-4″ vacuum tube collectors, 80 gal. tank		7,550	5,925	13,475
2600	120 gal. tank		7,800	6,000	13,800
2640	2 ea. 3′x7′ black chrome collectors, 80 gal. tank	R235616 -60	5,825	5,650	11,475
2660	120 gal. tank		6,025	5,650	11,675
2700	2 ea. 3′x7′ flat black collectors, 120 gal. tank		5,750	5,675	11,425
2720	3 ea. 3′x7′ flat black collectors, 120 gal. tank		6,700	5,825	12,525
2760	3/4″ tubing, 3 ea. 3′x7′ black chrome collectors, 120 gal. tank		7,475	6,050	13,525
2780	3 ea. 3′x7′ flat black collectors, 120 gal. tank		7,075	6,075	13,150
2800	2 ea. 4′x9′ flat black w/plastic glazing collectors 120 gal. tank		6,475	6,100	12,575
2840	1″ tubing, 4 ea. 2′x9′ plastic absorber & glazing collectors 120 gal. tank		7,400	6,825	14,225
2860	4 ea. 3′x7′ black chrome collectors, 120 gal. tank		9,250	6,850	16,100
2880	4 ea. 3′x7′ flat black absorber collectors, 120 gal. tank		8,700	6,875	15,575

517

For customer support on your Mechanical Cost Data, call 877.739.3689.

D3010 Energy Supply

**Basis for Heat Loss Estimate,
Apartment Type Structures:**

1. Masonry walls and flat roof are insulated. U factor is assumed at .08.
2. Window glass area taken as BOCA minimum, 1/10th of floor area. Double insulating glass with 1/4″ air space, U = .65.
3. Infiltration = 0.3 C.F. per hour per S.F. of net wall.
4. Concrete floor loss is 2 BTUH per S.F.
5. Temperature difference taken as 70°F.
6. Ventilating or makeup air has not been included and must be added if desired. Air shafts are not used.

System Components	QUANTITY	UNIT	COST EACH		
			MAT.	INST.	TOTAL
SYSTEM D3010 510 1760					
HEATING SYSTEM, FIN TUBE RADIATION, FORCED HOT WATER					
1,000 S.F. AREA, 10,000 C.F. VOLUME					
Boiler, oil fired, CI, burner, ctrls/insul/breech/pipe/ftng/valves, 109 MBH	1.000	Ea.	4,462.50	4,068.75	8,531.25
Circulating pump, CI flange connection, 1/12 HP	1.000	Ea.	415	214	629
Expansion tank, painted steel, ASME 18 Gal capacity	1.000	Ea.	815	94	909
Storage tank, steel, above ground, 275 Gal capacity w/supports	1.000	Ea.	540	264	804
Copper tubing type L, solder joint, hanger 10′ OC, 3/4″ diam	100.000	L.F.	575	940	1,515
Radiation, 3/4″ copper tube w/alum fin baseboard pkg, 7″ high	30.000	L.F.	243	675	918
Pipe covering, calcium silicate w/cover, 1′ wall, 3/4′ diam	100.000	L.F.	448	705	1,153
TOTAL			7,498.50	6,960.75	14,459.25
COST PER S.F.			7.50	6.96	14.46

D3010 510	Apartment Building Heating - Fin Tube Radiation		COST PER S.F.		
			MAT.	INST.	TOTAL
1740	Heating systems, fin tube radiation, forced hot water				
1760	1,000 S.F. area, 10,000 C.F. volume		7.50	6.95	14.45
1800	10,000 S.F. area, 100,000 C.F. volume	R235000 -10	3.14	4.18	7.32
1840	20,000 S.F. area, 200,000 C.F. volume		3.43	4.69	8.12
1880	30,000 S.F. area, 300,000 C.F. volume	R235000 -20	3.32	4.55	7.87
1890					

D30 HVAC

D3010 Energy Supply

Fin Tube Radiator

Basis for Heat Loss Estimate, Factory or Commercial Type Structures:

1. Walls and flat roof are of lightly insulated concrete block or metal. U factor is assumed at .17.
2. Windows and doors are figured at 1-1/2 S.F. per linear foot of building perimeter. The U factor for flat single glass is 1.13.
3. Infiltration is approximately 5 C.F. per hour per S.F. of wall.
4. Concrete floor loss is 2 BTUH per S.F. of floor.
5. Temperature difference is assumed at 70°F.
6. Ventilation or makeup air has not been included and must be added if desired.

System Components	QUANTITY	UNIT	COST EACH		
			MAT.	INST.	TOTAL
SYSTEM D3010 520 1960					
HEATING SYSTEM, FIN TUBE RADIATION, FORCED HOT WATER					
1,000 S.F. BLDG., ONE FLOOR					
Boiler, oil fired, CI, burner/ctrls/insul/breech/pipe/ftngs/valves, 109 MBH	1.000	Ea.	4,462.50	4,068.75	8,531.25
Expansion tank, painted steel, ASME 18 Gal capacity	1.000	Ea.	2,775	110	2,885
Storage tank, steel, above ground, 550 Gal capacity w/supports	1.000	Ea.	4,325	490	4,815
Circulating pump, CI flanged, 1/8 HP	1.000	Ea.	690	214	904
Pipe, steel, black, sch. 40, threaded, cplg & hngr 10'OC, 1-1/2" diam.	260.000	L.F.	1,677	4,186	5,863
Pipe covering, calcium silicate w/cover , 1" wall, 1-1/2" diam	260.000	L.F.	1,180.40	1,885	3,065.40
Radiation, steel 1-1/4" tube & 4-1/4" fin w/cover & damper, wall hung	42.000	L.F.	1,827	1,533	3,360
Rough in, steel fin tube radiation w/supply & balance valves	4.000	Set	1,300	3,220	4,520
TOTAL			18,236.90	15,706.75	33,943.65
COST PER S.F.			18.24	15.71	33.95

D3010 520	Commercial Building Heating - Fin Tube Radiation		COST PER S.F.		
			MAT.	INST.	TOTAL
1940	Heating systems, fin tube radiation, forced hot water				
1960	1,000 S.F. bldg, one floor		18.20	15.70	33.90
2000	10,000 S.F., 100,000 C.F., total two floors		4.70	5.95	10.65
2040	100,000 S.F., 1,000,000 C.F., total three floors	R235000 -10	1.90	2.67	4.57
2080	1,000,000 S.F., 10,000,000 C.F., total five floors		1.01	1.40	2.41
2090		R235000 -20			

D3010 Energy Supply

Unit Heater

Basis for Heat Loss Estimate, Factory or Commercial Type Structures:

1. Walls and flat roof are of lightly insulated concrete block or metal. U factor is assumed at .17.
2. Windows and doors are figured at 1-1/2 S.F. per linear foot of building perimeter. The U factor for flat single glass is 1.13.
3. Infiltration is approximately 5 C.F. per hour per S.F. of wall.
4. Concrete floor loss is 2 BTUH per S.F. of floor.
5. Temperature difference is assumed at 70°F.
6. Ventilation or makeup air has not been included and must be added if desired.

System Components	QUANTITY	UNIT	COST EACH		
			MAT.	INST.	TOTAL
SYSTEM D3010 530 1880					
HEATING SYSTEM, TERMINAL UNIT HEATERS, FORCED HOT WATER					
1,000 S.F. BLDG., ONE FLOOR					
Boiler oil fired, CI, burner/ctrls/insul/breech/pipe/ftngs/valves, 109 MBH	1.000	Ea.	4,462.50	4,068.75	8,531.25
Expansion tank, painted steel, ASME 18 Gal capacity	1.000	Ea.	2,775	110	2,885
Storage tank, steel, above ground, 550 Gal capacity w/supports	1.000	Ea.	4,325	490	4,815
Circulating pump, CI, flanged, 1/8 HP	1.000	Ea.	690	214	904
Pipe, steel, black, schedule 40, threaded, cplg & hngr 10' OC 1-1/2" diam	260.000	L.F.	1,677	4,186	5,863
Pipe covering, calcium silicate w/cover, 1" wall, 1-1/2" diam	260.000	L.F.	1,180.40	1,885	3,065.40
Unit heater, 1 speed propeller, horizontal, 200° EWT, 26.9 MBH	2.000	Ea.	1,280	330	1,610
Unit heater piping hookup with controls	2.000	Set	1,590	3,100	4,690
TOTAL			17,979.90	14,383.75	32,363.65
COST PER S.F.			17.98	14.38	32.36

D3010 530	Commercial Bldg. Heating - Terminal Unit Heaters		COST PER S.F.		
			MAT.	INST.	TOTAL
1860	Heating systems, terminal unit heaters, forced hot water				
1880	1,000 S.F. bldg., one floor		18	14.40	32.40
1920	10,000 S.F. bldg., 100,000 C.F. total two floors		4.25	4.91	9.16
1960	100,000 S.F. bldg., 1,000,000 C.F. total three floors	R235000 -10	1.95	2.43	4.38
2000	1,000,000 S.F. bldg., 10,000,000 C.F. total five floors		1.32	1.57	2.89
2010		R235000 -20			

D3010 Energy Supply

Many styles of active solar energy systems exist. Those shown on the following page represent the majority of systems now being installed in different regions of the country.

The five active domestic hot water (DHW) systems which follow are typically specified as two or three panel systems with additional variations being the type of glazing and size of the storage tanks. Various combinations have been costed for the user's evaluation and comparison. The basic specifications from which the following systems were developed satisfy the construction detail requirements specified in the HUD Intermediate Minimum Property Standards (IMPS). If these standards are not complied with, the renewable energy system's costs could be significantly lower than shown.

To develop the system's specifications and costs it was necessary to make a number of assumptions about the systems. Certain systems are more appropriate to one climatic region than another or the systems may require modifications to be usable in particular locations. Specific instances in which the systems are not appropriate throughout the country as specified, or in which modification will be needed include the following:

- The freeze protection mechanisms provided in the DHW systems vary greatly. In harsh climates, where freeze is a major concern, a closed-loop indirect collection system may be more appropriate than a direct collection system.

- The thermosyphon water heater system described cannot be used when temperatures drop below 32°F.

- In warm climates it may be necessary to modify the systems installed to prevent overheating.

For each renewable resource (solar) system a schematic diagram and descriptive summary of the system is provided along with a list of all the components priced as part of the system. The costs were developed based on these specifications.

Considerations affecting costs which may increase or decrease beyond the estimates presented here include the following:

- Special structural qualities (allowance for earthquake, future expansion, high winds, and unusual spans or shapes);

- Isolated building site or rough terrain that would affect the transportation of personnel, material, or equipment;

- Unusual climatic conditions during the construction process;

- Substitution of other materials or system components for those used in the system specifications.

521

In this closed-loop indirect collection system, fluid with a low freezing temperature, propylene glycol, transports heat from the collectors to water storage. The transfer fluid is contained in a closed loop consisting of collectors, supply and return piping, and a heat exchanger immersed in the storage tank.

When the collectors become approximately 20°F warmer than the storage temperature, the controller activates the circulator. The circulator moves the fluid continuously until the temperature difference between fluid in the collectors and storage is such that the collection will no longer occur and then the circulator turns off. Since the heat transfer fluid has a very low freezing temperature, there is no need for it to be drained from the collectors between periods of collection.

D30 HVAC

D3010 Energy Supply

System Components

System Components	QUANTITY	UNIT	MAT.	INST.	TOTAL
SYSTEM D3010 650 2750					
SOLAR, CLOSED LOOP, SPACE/HOT WATER					
1" TUBING, TEN 3'X7' BLK CHROME ON COPPER ABSORBER COLLECTORS					
A, B Differential controller 2 sensors, thermostat, solar energy system	2.000	Ea.	286	119	405
C Thermometer, 2" dial	10.000	Ea.	240	445	685
C-1 Heat exchanger, solar energy system, fluid to air, up flow, 80 MBH	1.000	Ea.	525	430	955
D, T Fill & drain valves, brass, 3/4" connection	5.000	Ea.	56.25	150	206.25
D-1 Fan center	1.000	Ea.	126	147	273
E Air vent, manual, 1/8" fitting	1.000	Ea.	2.68	22.50	25.18
E-2 Thermostat, 2 stage for sensing room temperature	1.000	Ea.	209	91.50	300.50
F Air purger	2.000	Ea.	103	119	222
F-1 Controller, liquid temperature, solar energy system	1.000	Ea.	119	143	262
G Expansion tank	2.000	Ea.	150	45	195
I Valve, gate, bronze, 125 lb, soldered, 1" diam	5.000	Ea.	450	187.50	637.50
J Vent flashing, neoprene	2.000	Ea.	11	71	82
K Circulator, solar heated liquid, 1/25 HP	2.000	Ea.	398	184	582
L Circulator, solar heated liquid, 1/20 HP	1.000	Ea.	289	107	396
N Relief valve, temp & pressure 150 psi 210°F self-closing	4.000	Ea.	98	96	194
N-1 Relief valve, pressure poppet, bronze, 30 psi, 3/4" IPS	3.000	Ea.	259.50	76.50	336
O Pipe covering, urethane, ultraviolet cover, 1" wall, 1" diam	50.000	L.F.	155.50	325	480.50
P Pipe covering, fiberglass, all service jacket, 1" wall, 1" diam	60.000	L.F.	59.40	312	371.40
Q Collector panel solar energy blk chrome on copper 1/8" temp glass 3'x7'	10.000	Ea.	11,000	1,350	12,350
Roof clamps for solar energy collector panel	10.000	Set	31.10	184	215.10
R Valve, swing check, bronze, 125 lb, regrinding disc, 3/4" & 1" diam	4.000	Ea.	668	150	818
S Pressure gage, 0-60 psi, for solar energy system	2.000	Ea.	53	45	98
U Valve, water tempering, bronze, sweat connections, 3/4" diam	1.000	Ea.	150	35.50	185.50
W-2, W-1, W Tank, water storage immersed heat xchr elec elem 2"x1/2# ins 120 gal	4.000	Ea.	8,100	2,040	10,140
X Valve, globe, bronze, 125 lb, rising stem, 1" diam	3.000	Ea.	531	112.50	643.50
Y Valve, flow control	1.000	Ea.	143	32.50	175.50
Copper tubing, type M, solder joint, hanger 10' OC 1" diam	110.000	L.F.	797.50	1,122	1,919.50
Copper tubing, type L, solder joint, hanger 10' OC 3/4" diam	20.000	L.F.	115	188	303
Wrought copper fittings & solder, 3/4" & 1" diam	121.000	Ea.	828.85	5,384.50	6,213.35
Sensor, wire, #22-2 conductor, multistranded	.700	C.L.F.	10.96	46.20	57.16
Ductwork, galvanized steel, for heat exchanger	8.000	Lb.	5.04	66.80	71.84
Solar energy heat transfer fluid propylene glycol, anti-freeze	25.000	Gal.	412.50	637.50	1,050
TOTAL			26,383.28	14,465.50	40,848.78

D3010 650	Solar, Closed Loop, Space/Hot Water Systems		MAT.	INST.	TOTAL
2540	Solar, closed loop, space/hot water				
2550	1/2" tubing, 12 ea. 4'x4'4" vacuum tube collectors		25,500	13,500	39,000
2600	3/4" tubing, 12 ea. 4'x4'4" vacuum tube collectors		26,600	13,900	40,500
2650	10 ea. 3' x 7' black chrome absorber collectors	R235616-60	25,600	13,500	39,100
2700	10 ea. 3' x 7' flat black absorber collectors		24,300	13,600	37,900
2750	1" tubing, 10 ea. 3' x 7' black chrome absorber collectors		26,400	14,500	40,900
2800	10 ea. 3' x 7' flat black absorber collectors		25,200	14,600	39,800
2850	6 ea. 4' x 9' flat black w/plastic glazing collectors		22,400	14,500	36,900
2900	12 ea. 2' x 9' plastic absorber and glazing collectors		23,200	14,700	37,900

For customer support on your Mechanical Cost Data, call 877.739.3689.

523

This draindown pool system uses a differential thermostat similar to those used in solar domestic hot water and space heating applications. To heat the pool, the pool water passes through the conventional pump-filter loop and then flows through the collectors. When collection is not possible, or when the pool temperature is reached, all water drains from the solar loop back to the pool through the existing piping. The modes are controlled by solenoid valves or other automatic valves in conjunction with a vacuum breaker relief valve, which facilitates draindown.

D3010 Energy Supply

System Components	QUANTITY	UNIT	COST EACH		
			MAT.	INST.	TOTAL
SYSTEM D3010 660 2640					
SOLAR SWIMMING POOL HEATER, ROOF MOUNTED COLLECTORS					
TEN 4' X 10' FULLY WETTED UNGLAZED PLASTIC ABSORBERS					
A Differential thermostat/controller, 110V, adj pool pump system	1.000	Ea.	355	355	710
A-1 Solenoid valve, PVC, normally 1 open 1 closed (included)	2.000	Ea.			
B Sensor, thermistor type (included)	2.000	Ea.			
E-1 Valve, vacuum relief	1.000	Ea.	32.50	22.50	55
Q Collector panel, solar energy, plastic, liquid full wetted, 4' x 10'	10.000	Ea.	3,550	2,570	6,120
R Valve, ball check, PVC, socket, 1-1/2" diam	1.000	Ea.	124	35.50	159.50
Z Valve, ball, PVC, socket, 1-1/2" diam	3.000	Ea.	324	106.50	430.50
Pipe, PVC, sch 40, 1-1/2" diam	80.000	L.F.	616	1,588	2,204
Pipe fittings, PVC sch 40, socket joint, 1-1/2" diam	10.000	Ea.	21.30	355	376.30
Sensor wire, #22-2 conductor, multistranded	.500	C.L.F.	7.83	33	40.83
Roof clamps for solar energy collector panels	10.000	Set	31.10	184	215.10
Roof strap, teflon for solar energy collector panels	26.000	L.F.	715	90.74	805.74
TOTAL			5,776.73	5,340.24	11,116.97

D3010 660	Solar Swimming Pool Heater Systems		COST EACH		
			MAT.	INST.	TOTAL
2530	Solar swimming pool heater systems, roof mounted collectors				
2540	10 ea. 3'x7' black chrome absorber, 1/8" temp. glass		13,200	4,125	17,325
2560	10 ea. 4'x8' black chrome absorber, 3/16" temp. glass	R235616	15,000	4,900	19,900
2580	10 ea. 3'8"x6' flat black absorber, 3/16" temp. glass	-60	11,900	4,200	16,100
2600	10 ea. 4'x9' flat black absorber, plastic glazing		13,700	5,100	18,800
2620	10 ea. 2'x9' rubber absorber, plastic glazing		15,100	5,525	20,625
2640	10 ea. 4'x10' fully wetted unglazed plastic absorber		5,775	5,350	11,125
2660	Ground mounted collectors				
2680	10 ea. 3'x7' black chrome absorber, 1/8" temp. glass		13,200	4,600	17,800
2700	10 ea. 4'x8' black chrome absorber, 3/16" temp. glass		15,000	5,400	20,400
2720	10 ea. 3'8"x6' flat blk absorber, 3/16" temp. glass		11,900	4,700	16,600
2740	10 ea. 4'x9' flat blk absorber, plastic glazing		13,700	5,600	19,300
2760	10 ea. 2'x9' rubber absorber, plastic glazing		15,000	5,850	20,850
2780	10 ea. 4'x10' fully wetted unglazed plastic absorber		5,775	5,825	11,600

The complete Solar Air Heating System provides maximum savings of conventional fuel with both space heating and year-round domestic hot water heating. It allows for the home air conditioning to operate simultaneously and independently from the solar domestic water heating in summer. The system's modes of operation are:

Mode 1: The building is heated directly from the collectors with air circulated by the Solar Air Mover.

Mode 2: When heat is not needed in the building, the dampers change within the air mover to circulate the air from the collectors to the rock storage bin.

Mode 3: When heat is not available from the collector array and is available in rock storage, the air mover draws heated air from rock storage and directs it into the building. When heat is not available from the collectors or the rock storage bin, the auxiliary heating unit will provide heat for the building. The size of the collector array is typically 25% the size of the main floor area.

526

D3010 Energy Supply

System Components	QUANTITY	UNIT	COST EACH		
			MAT.	INST.	TOTAL
SYSTEM D3010 675 1210					
SOLAR, SPACE/HOT WATER, AIR TO WATER HEAT EXCHANGE					
A, B Differential controller 2 sensors thermos., solar energy sys liquid loop	1.000	Ea.	143	59.50	202.50
A-1, B Differential controller 2 sensors 6 station solar energy sys air loop	1.000	Ea.	289	238	527
B-1 Solar energy sensor, freeze prevention	1.000	Ea.	25	22.50	47.50
C Thermometer for solar energy system, 2″ dial	2.000	Ea.	48	89	137
C-1 Heat exchanger, solar energy system, air to fluid, up flow, 70 MBH	1.000	Ea.	395	370	765
D Drain valve, brass, 3/4″ connection	2.000	Ea.	22.50	60	82.50
E Air vent, manual, for solar energy system 1/8″ fitting	1.000	Ea.	2.68	22.50	25.18
E-1 Thermostat, 2 stage for sensing room temperature	1.000	Ea.	209	91.50	300.50
F Air purger	1.000	Ea.	51.50	59.50	111
G Expansion tank, for solar energy system	1.000	Ea.	75	22.50	97.50
I Valve, gate, bronze, 125 lb, NRS, soldered 3/4″ diam	2.000	Ea.	145	71	216
K Circulator, solar heated liquid, 1/25 HP	1.000	Ea.	199	92	291
N Relief valve temp & press 150 psi 210°F self-closing, 3/4″ IPS	1.000	Ea.	24.50	24	48.50
N-1 Relief valve, pressure, poppet, bronze, 30 psi, 3/4″ IPS	1.000	Ea.	86.50	25.50	112
P Pipe covering, fiberglass, all service jacket, 1″ wall, 3/4″ diam	60.000	L.F.	59.40	312	371.40
Q-3 Manifold for flush mount solar energy collector panels	20.000	L.F.	3,180	158	3,338
Q Collector panel solar energy, air, black on alum. plate, 7.5′ x 3.5′	10.000	Ea.	9,650	1,430	11,080
R Valve, swing check, bronze, 125 lb, regrinding disc, 3/4″ diam	2.000	Ea.	232	71	303
S Pressure gage, 2″ dial, for solar energy system	1.000	Ea.	26.50	22.50	49
U Valve, water tempering, bronze, sweat connections, 3/4″ diam	1.000	Ea.	150	35.50	185.50
W-2, W Tank, water storage, solar, elec element 2″x1/2# insul, 80 Gal	1.000	Ea.	1,775	445	2,220
X Valve, globe, bronze, 125 lb, soldered, 3/4″ diam	1.000	Ea.	114	35.50	149.50
Copper tubing type L, solder joints, hangers 10′ OC 3/4″ diam	10.000	L.F.	57.50	94	151.50
Copper tubing type M, solder joints, hangers 10′ OC 3/4″ diam	60.000	L.F.	289.20	549	838.20
Wrought copper fittings & solder, 3/4″ diam	26.000	Ea.	72.28	975	1,047.28
Sensor wire, #22-2 conductor multistranded	1.200	C.L.F.	18.78	79.20	97.98
Q-1 Duct work, rigid fiberglass, rectangular	400.000	S.F.	348	2,240	2,588
Duct work, spiral preformed, steel, PVC coated both sides, 12″ x 10″	8.000	Ea.	208	420	628
R-2 Shutter/damper for solar heater circulator	9.000	Ea.	553.50	945	1,498.50
K-1 Shutter motor for solar heater circulator blower	9.000	Ea.	1,377	711	2,088
K-1 Fan, solar energy heated air circulator, space & DHW system	1.000	Ea.	1,700	2,525	4,225
Z-1 Tank, solar energy air storage, 6′-3″H 7′x7′ = 306 CF/2000 Gal	1.000	Ea.	15,700	1,625	17,325
Z-2 Crushed stone 1-1/2″	11.000	C.Y.	291.50	90.86	382.36
C-2 Thermometer, remote probe, 2″ dial	1.000	Ea.	36.50	89.50	126
R-1 Solenoid valve	1.000	Ea.	141	79.50	220.50
V, R-4, R-3 Furnace, supply diffusers, return grilles, existing	1.000	Ea.	15,700	1,625	17,325
TOTAL			53,395.84	15,805.06	69,200.90

D3010 675	Air To Water Heat Exchange		COST EACH		
			MAT.	INST.	TOTAL
1210 1220	Solar, air to water heat exchange, for space/hot water heating	R235616 -60	53,500	15,800	69,300

527

For customer support on your Mechanical Cost Data, call 877.739.3689.

D30 HVAC

D3020 Heat Generating Systems

Boiler   **Baseboard Radiation**

Small Electric Boiler
System Considerations:
1. Terminal units are fin tube baseboard radiation rated at 720 BTU/hr with 200° water temperature or 820 BTU/hr steam.
2. Primary use being for residential or smaller supplementary areas, the floor levels are based on 7-1/2' ceiling heights.
3. All distribution piping is copper for boilers through 205 MBH. All piping for larger systems is steel pipe.

System Components	QUANTITY	UNIT	MAT.	INST.	TOTAL
SYSTEM D3020 102 1120					
SMALL HEATING SYSTEM, HYDRONIC, ELECTRIC BOILER					
1,480 S.F., 61 MBH, STEAM, 1 FLOOR					
Boiler, electric steam, std cntrls, trim, ftngs and valves, 18 KW, 61.4 MBH	1.000	Ea.	5,225	1,815	7,040
Copper tubing type L, solder joint, hanger 10'OC, 1-1/4" diam	160.000	L.F.	1,768	1,968	3,736
Radiation, 3/4" copper tube w/alum fin baseboard pkg 7" high	60.000	L.F.	486	1,350	1,836
Rough in baseboard panel or fin tube with valves & traps	10.000	Set	2,650	6,900	9,550
Pipe covering, calcium silicate w/cover, 1" wall 1-1/4" diam	160.000	L.F.	716.80	1,160	1,876.80
Low water cut-off, quick hookup, in gage glass tappings	1.000	Ea.	289	46	335
TOTAL			11,134.80	13,239	24,373.80
COST PER S.F.			7.52	8.95	16.47

D3020 102	Small Heating Systems, Hydronic, Electric Boilers		MAT.	INST.	TOTAL
1100	Small heating systems, hydronic, electric boilers				
1120	Steam, 1 floor, 1480 S.F., 61 M.B.H.		7.50	8.92	16.42
1160	3,000 S.F., 123 M.B.H.		5.75	7.85	13.60
1200	5,000 S.F., 205 M.B.H.	R235000 -10	4.98	7.25	12.23
1240	2 floors, 12,400 S.F., 512 M.B.H.		3.88	7.20	11.08
1280	3 floors, 24,800 S.F., 1023 M.B.H.	R235000 -20	4.32	7.10	11.42
1320	34,750 S.F., 1,433 M.B.H.		3.97	6.90	10.87
1360	Hot water, 1 floor, 1,000 S.F., 41 M.B.H.	R235000 -30	12.65	4.95	17.60
1400	2,500 S.F., 103 M.B.H.		8.85	8.95	17.80
1440	2 floors, 4,850 S.F., 205 M.B.H.		8.45	10.75	19.20
1480	3 floors, 9,700 S.F., 410 M.B.H.		8.70	11.10	19.80

528

D3020 Heat Generating Systems

Boiler

Unit Heater

Large Electric Boiler System Considerations:

1. Terminal units are all unit heaters of the same size. Quantities are varied to accommodate total requirements.
2. All air is circulated through the heaters a minimum of three times per hour.
3. As the capacities are adequate for commercial use, floor levels are based on 10' ceiling heights.
4. All distribution piping is black steel pipe.

System Components	QUANTITY	UNIT	COST EACH		
			MAT.	INST.	TOTAL
SYSTEM D3020 104 1240					
LARGE HEATING SYSTEM, HYDRONIC, ELECTRIC BOILER					
9,280 S.F., 135 KW, 461 MBH, 1 FLOOR					
Boiler, electric hot water, std ctrls, trim, ftngs, valves, 135 KW, 461 MBH	1.000	Ea.	12,225	3,975	16,200
Expansion tank, painted steel, 60 Gal capacity ASME	1.000	Ea.	4,350	220	4,570
Circulating pump, CI, close cpld, 50 GPM, 2 HP, 2" pipe conn	1.000	Ea.	3,050	430	3,480
Unit heater, 1 speed propeller, horizontal, 200° EWT, 72.7 MBH	7.000	Ea.	6,545	1,680	8,225
Unit heater piping hookup with controls	7.000	Set	5,565	10,850	16,415
Pipe, steel, black, schedule 40, welded, 2-1/2" diam	380.000	L.F.	4,408	10,966.80	15,374.80
Pipe covering, calcium silicate w/cover, 1" wall, 2-1/2" diam	380.000	L.F.	2,109	2,850	4,959
TOTAL			38,252	30,971.80	69,223.80
COST PER S.F.			4.12	3.34	7.46

D3020 104	Large Heating Systems, Hydronic, Electric Boilers		COST PER S.F.		
			MAT.	INST.	TOTAL
1230	Large heating systems, hydronic, electric boilers				
1240	9,280 S.F., 135 K.W., 461 M.B.H., 1 floor		4.13	3.34	7.47
1280	14,900 S.F., 240 K.W., 820 M.B.H., 2 floors	R235000 -10	5.35	5.40	10.75
1320	18,600 S.F., 296 K.W., 1,010 M.B.H., 3 floors		5.25	5.85	11.10
1360	26,100 S.F., 420 K.W., 1,432 M.B.H., 4 floors	R235000 -20	5.15	5.75	10.90
1400	39,100 S.F., 666 K.W., 2,273 M.B.H., 4 floors		4.43	4.81	9.24
1440	57,700 S.F., 900 K.W., 3,071 M.B.H., 5 floors	R235000 -30	4.24	4.74	8.98
1480	111,700 S.F., 1,800 K.W., 6,148 M.B.H., 6 floors		3.77	4.07	7.84
1520	149,000 S.F., 2,400 K.W., 8,191 M.B.H., 8 floors		3.71	4.05	7.76
1560	223,300 S.F., 3,600 K.W., 12,283 M.B.H., 14 floors		3.91	4.60	8.51

529

For customer support on your Mechanical Cost Data, call 877.739.3689.

D3020 Heat Generating Systems

Unit Heater

Fossil Fuel Boiler System Considerations:

1. Terminal units are horizontal unit heaters. Quantities are varied to accommodate total heat loss per building.
2. Unit heater selection was determined by their capacity to circulate the building volume a minimum of three times per hour in addition to the BTU output.

3. Systems shown are forced hot water. Steam boilers cost slightly more than hot water boilers. However, this is compensated for by the smaller size or fewer terminal units required with steam.
4. Floor levels are based on 10′ story heights.
5. MBH requirements are gross boiler output.

System Components	QUANTITY	UNIT	COST EACH		
			MAT.	INST.	TOTAL
SYSTEM D3020 108 1280					
HEATING SYSTEM, HYDRONIC, FOSSIL FUEL, TERMINAL UNIT HEATERS					
CAST IRON BOILER, GAS, 80 MBH, 1,070 S.F. BUILDING					
Boiler, gas, hot water, CI, burner, controls, insulation, breeching, 80 MBH	1.000	Ea.	2,283.75	1,995	4,278.75
Pipe, steel, black, schedule 40, threaded, cplg & hngr 10′OC, 2″ diam	200.000	L.F.	1,630	4,000	5,630
Unit heater, 1 speed propeller, horizontal, 200° EWT, 72.7 MBH	2.000	Ea.	1,870	480	2,350
Unit heater piping hookup with controls	2.000	Set	1,590	3,100	4,690
Expansion tank, painted steel, ASME, 18 Gal capacity	1.000	Ea.	2,775	110	2,885
Circulating pump, CI, flange connection, 1/12 HP	1.000	Ea.	415	214	629
Pipe covering, calcium silicate w/cover, 1″ wall, 2″ diam	200.000	L.F.	1,040	1,500	2,540
TOTAL			11,603.75	11,399	23,002.75
COST PER S.F.			10.84	10.65	21.49

D3020 108	Heating Systems, Unit Heaters		COST PER S.F.		
			MAT.	INST.	TOTAL
1260	Heating systems, hydronic, fossil fuel, terminal unit heaters,				
1280	Cast iron boiler, gas, 80 M.B.H., 1,070 S.F. bldg.		10.84	10.65	21.49
1320	163 M.B.H., 2,140 S.F. bldg.	R235000 -10	7.30	7.25	14.55
1360	544 M.B.H., 7,250 S.F. bldg.		5.45	5.15	10.60
1400	1,088 M.B.H., 14,500 S.F. bldg.	R235000 -20	4.47	4.85	9.32
1440	3,264 M.B.H., 43,500 S.F. bldg.		3.67	3.66	7.33
1480	5,032 M.B.H., 67,100 S.F. bldg.	R235000 -30	4.21	3.75	7.96
1520	Oil, 109 M.B.H., 1,420 S.F. bldg.		12.10	9.50	21.60
1560	235 M.B.H., 3,150 S.F. bldg.		7.50	6.80	14.30
1600	940 M.B.H., 12,500 S.F. bldg.		5.35	4.43	9.78
1640	1,600 M.B.H., 21,300 S.F. bldg.		4.96	4.24	9.20
1680	2,480 M.B.H., 33,100 S.F. bldg.		4.95	3.82	8.77
1720	3,350 M.B.H., 44,500 S.F. bldg.		4.39	3.91	8.30
1760	Coal, 148 M.B.H., 1,975 S.F. bldg.		94.50	6.20	100.70
1800	300 M.B.H., 4,000 S.F. bldg.		51.50	4.86	56.36
1840	2,360 M.B.H., 31,500 S.F. bldg.		11.35	3.89	15.24

D3020 Heat Generating Systems

Fin Tube Radiator

Fossil Fuel Boiler System Considerations:

1. Terminal units are commercial steel fin tube radiation. Quantities are varied to accommodate total heat loss per building.
2. Systems shown are forced hot water. Steam boilers cost slightly more than hot water boilers. However, this is compensated for by the smaller size or fewer terminal units required with steam.
3. Floor levels are based on 10' story heights.
4. MBH requirements are gross boiler output.

System Components	QUANTITY	UNIT	COST EACH		
			MAT.	INST.	TOTAL
SYSTEM D3020 110 3240					
HEATING SYSTEM, HYDRONIC, FOSSIL FUEL, FIN TUBE RADIATION					
CAST IRON BOILER, GAS, 80 MBH, 1,070 S.F. BLDG.					
Boiler, gas, hot water, CI, burner, controls, insulation, breeching, 80 MBH	1.000	Ea.	2,283.75	1,995	4,278.75
Pipe, steel, black, schedule 40, threaded, 2″ diam	200.000	L.F.	1,630	4,000	5,630
Radiation, steel 1-1/4″ tube & 4-1/4″ fin w/cover & damper, wall hung	80.000	L.F.	3,480	2,920	6,400
Rough-in, wall hung steel fin radiation with supply & balance valves	8.000	Set	2,600	6,440	9,040
Expansion tank, painted steel, ASME, 18 Gal capacity	1.000	Ea.	2,775	110	2,885
Circulating pump, CI flanged, 1/12 HP	1.000	Ea.	415	214	629
Pipe covering, calcium silicate w/cover, 1″ wall, 2″ diam	200.000	L.F.	1,040	1,500	2,540
TOTAL			14,223.75	17,179	31,402.75
COST PER S.F.			13.29	16.06	29.35

D3020 110	Heating System, Fin Tube Radiation		COST PER S.F.		
			MAT.	INST.	TOTAL
3230	Heating systems, hydronic, fossil fuel, fin tube radiation				
3240	Cast iron boiler, gas, 80 MBH, 1,070 S.F. bldg.		13.27	16.07	29.34
3280	169 M.B.H., 2,140 S.F. bldg.		8.40	10.20	18.60
3320	544 M.B.H., 7,250 S.F. bldg.	R235000 -10	7.20	8.75	15.95
3360	1,088 M.B.H., 14,500 S.F. bldg.		6.40	8.55	14.95
3400	3,264 M.B.H., 43,500 S.F. bldg.	R235000 -20	5.75	7.55	13.30
3440	5,032 M.B.H., 67,100 S.F. bldg.		6.30	7.65	13.95
3480	Oil, 109 M.B.H., 1,420 S.F. bldg.	R235000 -30	16.45	17.55	34
3520	235 M.B.H., 3,150 S.F. bldg.		9.25	10.30	19.55
3560	940 M.B.H., 12,500 S.F. bldg.		7.30	8.15	15.45
3600	1,600 M.B.H., 21,300 S.F. bldg.		7.05	8.10	15.15
3640	2,480 M.B.H., 33,100 S.F. bldg.		7.05	7.70	14.75
3680	3,350 M.B.H., 44,500 S.F. bldg.		6.45	7.80	14.25
3720	Coal, 148 M.B.H., 1,975 S.F. bldg.		96	9.70	105.70
3760	300 M.B.H., 4,000 S.F. bldg.		53.50	8.35	61.85
3800	2,360 M.B.H., 31,500 S.F. bldg.		13.35	7.70	21.05
4080	Steel boiler, oil, 97 M.B.H., 1,300 S.F. bldg.		12.25	14.40	26.65
4120	315 M.B.H., 4,550 S.F. bldg.		6.30	7.30	13.60
4160	525 M.B.H., 7,000 S.F. bldg.		9.10	8.45	17.55
4200	1,050 M.B.H., 14,000 S.F. bldg.		8.20	8.45	16.65
4240	2,310 M.B.H., 30,800 S.F. bldg.		7.10	7.75	14.85
4280	3,150 M.B.H., 42,000 S.F. bldg.		6.60	7.85	14.45

531

For customer support on your Mechanical Cost Data, call 877.739.3689.

D3020 Heat Generating Systems

Electric Boiler, Hot Water

System Components	QUANTITY	UNIT	COST EACH		
			MAT.	INST.	TOTAL
SYSTEM D3020 126 1010					
BOILER, ELECTRIC, HOT WATER, 15 kW, 52 MBH					
Boilers, electric, ASME, hot water, 15 KW, 52 MBH	1.000	Ea.	5,500	1,525	7,025
Pipe, black steel, Sch 40, threaded, w/coupling & hangers, 10' OC, 3/4" dia	10.000	L.F.	42.32	134.55	176.87
Pipe, black steel, Sch 40, threaded, w/coupling & hangers, 10' OC, 1" dia	20.000	L.F.	101.70	290.25	391.95
Elbow, 90°, black, straight, 3/4" dia.	6.000	Ea.	27.54	306	333.54
Elbow, 90°, black, straight, 1" dia.	6.000	Ea.	47.70	330	377.70
Tee, black, straight, 3/4" dia.	1.000	Ea.	7.30	79.50	86.80
Tee, black, straight, 1" dia.	2.000	Ea.	25	179	204
Tee, black, reducing, 1" dia.	2.000	Ea.	40	179	219
Union, black with brass seat, 3/4" dia.	1.000	Ea.	20.50	55	75.50
Union, black with brass seat, 1" dia.	2.000	Ea.	53	119	172
Pipe nipples, 3/4" diam.	3.000	Ea.	5.52	17.55	23.07
Pipe nipples, 1" diam.	3.000	Ea.	7.10	20.25	27.35
Valves, bronze, gate, N.R.S., threaded, class 150, 3/4" size	2.000	Ea.	189	71	260
Valves, bronze, gate, N.R.S., threaded, class 150, 1" size	2.000	Ea.	236	75	311
Thermometer, stem type, 9" case, 8" stem, 3/4" NPT	1.000	Ea.	478	52	530
Tank, steel, liquid expansion, ASME, painted, 15 gallon capacity	1.000	Ea.	730	77.50	807.50
Pump, circulating, bronze, flange connection, 3/4" to 1-1/2" size, 1/8 HP	1.000	Ea.	1,100	214	1,314
Insulation, fiberglass pipe covering, 1" wall, 1" IPS	20.000	L.F.	21.40	109	130.40
TOTAL			8,632.08	3,833.60	12,465.68

D3020 126	Electric Boiler, Hot Water	COST EACH		
		MAT.	INST.	TOTAL
1010	Boiler, electric, hot water, 15 kW, 52 MBH	8,625	3,825	12,450
1020	30 kW, 103 MBH	9,300	4,050	13,350
1030	60 kW, 205 MBH	9,500	4,175	13,675
1040	120 kW, 410 MBH	11,300	5,000	16,300
1050	150 kW, 510 MBH	18,400	9,625	28,025
1060	210 kW, 716 MBH	20,700	10,600	31,300
1070	296 kW, 1010 MBH	30,200	10,900	41,100
1080	444 kW, 1515 MBH	36,100	14,700	50,800
1090	666 kW, 2273 MBH	43,600	15,900	59,500
1100	900 kW, 3071 MBH	57,000	18,500	75,500
1110	1320 kW, 4505 MBH	64,000	19,400	83,400
1120	2100 kW, 7167 MBH	93,000	23,600	116,600

D3020 Heat Generating Systems

D3020 126	Electric Boiler, Hot Water	COST EACH		
		MAT.	INST.	TOTAL
1130	2610 kW, 8905 MBH	113,000	28,400	141,400
1140	3600 kW, 12,283 MBH	137,000	34,000	171,000

D3020 128	Electric Boiler, Steam	COST EACH		
		MAT.	INST.	TOTAL
1010	Boiler, electric, steam, 18 kW, 61.4 MBH	6,125	3,675	9,800
1020	36 kW, 123 MBH	7,650	3,900	11,550
1030	60 kW, 205 MBH	17,900	7,925	25,825
1040	120 kW, 409 MBH	20,700	8,725	29,425
1050	150 kW, 512 MBH	24,600	13,100	37,700
1060	210 kW, 716 MBH	40,600	14,100	54,700
1070	300 kW, 1023 MBH	43,400	15,400	58,800
1080	510 kW, 1740 MBH	52,500	20,100	72,600
1090	720 kW, 2456 MBH	57,000	21,100	78,100
1100	1080 kW, 3685 MBH	64,500	25,000	89,500
1110	1260 kW, 4300 MBH	71,500	26,600	98,100
1120	1620 kW, 5527 MBH	75,500	30,200	105,700
1130	2070 kW, 7063 MBH	98,000	31,700	129,700
1140	2340 kW, 7984 MBH	125,500	35,900	161,400

For customer support on your Mechanical Cost Data, call 877.739.3689.

533

D3020 Heat Generating Systems

Expansion Tank

Supply Piping

Gas Supply

Pressure Relief Valve

Return Piping

Circulating Pump

Cast Iron Boiler, Hot Water, Gas Fired

System Components	QUANTITY	UNIT	COST EACH		
			MAT.	INST.	TOTAL
SYSTEM D3020 130 1010					
BOILER, CAST IRON, GAS, HOT WATER, 100 MBH					
Boilers, gas fired, std controls, CI, insulated, HW, gross output 100 MBH	1.000	Ea.	2,825	2,075	4,900
Pipe, black steel, Sch 40, threaded, w/coupling & hangers, 10' OC, 3/4" dia	20.000	L.F.	82.80	263.25	346.05
Pipe, black steel, Sch 40, threaded, w/coupling & hangers, 10' OC, 1" dia.	20.000	L.F.	101.70	290.25	391.95
Elbow, 90°, black, straight, 3/4" dia.	9.000	Ea.	41.31	459	500.31
Elbow, 90°, black, straight, 1" dia.	6.000	Ea.	47.70	330	377.70
Tee, black, straight, 3/4" dia.	2.000	Ea.	14.60	159	173.60
Tee, black, straight, 1" dia.	2.000	Ea.	25	179	204
Tee, black, reducing, 1" dia.	2.000	Ea.	40	179	219
Pipe cap, black, 3/4" dia.	1.000	Ea.	5.20	22.50	27.70
Union, black with brass seat, 3/4" dia.	2.000	Ea.	41	110	151
Union, black with brass seat, 1" dia.	2.000	Ea.	53	119	172
Pipe nipples, black, 3/4" dia.	5.000	Ea.	9.20	29.25	38.45
Pipe nipples, black, 1" dia.	3.000	Ea.	7.10	20.25	27.35
Valves, bronze, gate, N.R.S., threaded, class 150, 3/4" size	2.000	Ea.	189	71	260
Valves, bronze, gate, N.R.S., threaded, class 150, 1" size	2.000	Ea.	236	75	311
Gas cock, brass, 3/4" size	1.000	Ea.	17.05	32.50	49.55
Thermometer, stem type, 9" case, 8" stem, 3/4" NPT	2.000	Ea.	478	52	530
Tank, steel, liquid expansion, ASME, painted, 15 gallon capacity	1.000	Ea.	730	77.50	807.50
Pump, circulating, bronze, flange connection, 3/4" to 1-1/2" size, 1/8 HP	1.000	Ea.	1,100	214	1,314
Vent chimney, all fuel, pressure tight, double wall, SS, 6" dia.	20.000	L.F.	1,220	420	1,640
Vent chimney, elbow, 90° fixed, 6" dia.	2.000	Ea.	910	84	994
Vent chimney, Tee, 6" dia.	2.000	Ea.	532	105	637
Vent chimney, ventilated roof thimble, 6" dia.	1.000	Ea.	278	48.50	326.50
Vent chimney, flat roof flashing, 6" dia.	1.000	Ea.	119	42	161
Vent chimney, stack cap, 6" diameter	1.000	Ea.	325	27.50	352.50
Insulation, fiberglass pipe covering, 1" wall, 1" IPS	20.000	L.F.	21.40	109	130.40
TOTAL			9,449.06	5,593.50	15,042.56

D3020 130	Boiler, Cast Iron, Hot Water, Gas	COST EACH		
		MAT.	INST.	TOTAL
1010	Boiler, cast iron, gas, hot water, 100 MBH	9,450	5,600	15,050
1020	200 MBH	10,800	6,400	17,200
1030	320 MBH	10,900	6,975	17,875
1040	440 MBH	14,800	8,975	23,775
1050	544 MBH	25,100	14,000	39,100
1060	765 MBH	30,100	15,100	45,200

D3020 Heat Generating Systems

D3020 130	Boiler, Cast Iron, Hot Water, Gas	COST EACH		
		MAT.	INST.	TOTAL
1070	1088 MBH	32,100	16,100	48,200
1080	1530 MBH	39,100	21,300	60,400
1090	2312 MBH	46,000	25,000	71,000
1100	2856 MBH	60,500	27,000	87,500
1110	3808 MBH	67,000	31,600	98,600
1120	4720 MBH	110,500	35,500	146,000
1130	6100 MBH	158,500	39,600	198,100
1140	6970 MBH	174,500	48,900	223,400

D3020 134	Boiler, Cast Iron, Steam, Gas	COST EACH		
		MAT.	INST.	TOTAL
1010	Boiler, cast iron, gas, steam, 100 MBH	7,900	5,425	13,325
1020	200 MBH	17,900	10,200	28,100
1030	320 MBH	19,700	11,400	31,100
1040	544 MBH	28,700	17,900	46,600
1050	765 MBH	32,200	18,600	50,800
1060	1275 MBH	37,900	20,400	58,300
1070	2675 MBH	53,500	29,700	83,200
1080	3570 MBH	65,500	33,500	99,000
1090	4720 MBH	101,000	35,700	136,700
1100	6100 MBH	126,500	41,600	168,100
1110	6970 MBH	138,000	48,700	186,700

D3020 136	Boiler, Cast Iron, Hot Water, Gas/Oil	COST EACH		
		MAT.	INST.	TOTAL
1010	Boiler, cast iron, gas & oil, hot water, 584 MBH	30,800	15,600	46,400
1020	876 MBH	38,900	16,700	55,600
1030	1168 MBH	49,000	18,900	67,900
1040	1460 MBH	59,500	22,700	82,200
1050	2044 MBH	65,000	23,500	88,500
1060	2628 MBH	78,500	28,400	106,900
1070	3210 MBH	82,000	30,500	112,500
1080	3796 MBH	88,000	31,500	119,500
1090	4672 MBH	102,500	33,100	135,600
1100	5256 MBH	125,000	40,200	165,200
1110	6000 MBH	150,000	40,800	190,800
1120	9800 MBH	188,000	70,000	258,000
1130	12,200 MBH	234,500	81,500	316,000
1140	13,500 MBH	255,000	87,000	342,000

D3020 138	Boiler, Cast Iron, Steam, Gas/Oil	COST EACH		
		MAT.	INST.	TOTAL
1010	Boiler, cast iron, gas & oil, steam, 810 MBH	35,300	20,200	55,500
1020	1360 MBH	40,600	21,600	62,200
1030	2040 MBH	49,900	27,600	77,500
1040	2700 MBH	53,500	30,800	84,300
1050	3270 MBH	62,500	34,300	96,800
1060	3770 MBH	100,500	35,000	135,500
1070	4650 MBH	112,500	39,900	152,400
1080	5520 MBH	129,500	42,000	171,500
1090	6100 MBH	133,500	44,200	177,700
1100	6970 MBH	143,000	54,000	197,000

D3020 Heat Generating Systems

Base Mounted End—Suction Pump

System Components	QUANTITY	UNIT	COST EACH		
			MAT.	INST.	TOTAL
SYSTEM D3020 330 1010					
PUMP, BASE MTD. WITH MOTOR, END-SUCTION, 2-1/2" SIZE, 3 HP, TO 150 GPM					
Pump, circulating, CI, base mounted, 2-1/2" size, 3 HP, to 150 GPM	1.000	Ea.	9,025	715	9,740
Pipe, black steel, Sch. 40, on yoke & roll hangers, 10' O.C., 2-1/2" dia	12.000	L.F.	139.20	346.32	485.52
Elbow, 90°, weld joint, steel, 2-1/2" pipe size	1.000	Ea.	30.50	169	199.50
Flange, weld neck, 150 LB, 2-1/2" pipe size	8.000	Ea.	280	675.92	955.92
Valve, iron body, gate, 125 lb., N.R.S., flanged, 2-1/2" size	1.000	Ea.	840	257	1,097
Strainer, Y type, iron body, flanged, 125 lb., 2-1/2" pipe size	1.000	Ea.	123	264	387
Multipurpose valve, CI body, 2" size	1.000	Ea.	525	91.50	616.50
Expansion joint, flanged spool, 6" F to F, 2-1/2" dia	2.000	Ea.	588	212	800
T-O-L, weld joint, socket, 1/4" pipe size, nozzle	2.000	Ea.	15.60	117.56	133.16
T-O-L, weld joint, socket, 1/2" pipe size, nozzle	2.000	Ea.	15.60	122.80	138.40
Control gauges, pressure or vacuum, 3-1/2" diameter dial	2.000	Ea.	41	46	87
Pressure/temperature relief plug, 316 SS, 3/4" OD, 7-1/2" insertion	2.000	Ea.	108	46	154
Insulation, fiberglass pipe covering, 1-1/2" wall, 2-1/2" IPS	12.000	L.F.	30	79.80	109.80
Water flow	1.000	Ea.	1,325	1,075	2,400
Pump balancing	1.000	Ea.		320	320
TOTAL			13,085.90	4,537.90	17,623.80

D3020 330	Circulating Pump Systems, End Suction	COST EACH		
		MAT.	INST.	TOTAL
1010	Pump, base mtd with motor, end-suction, 2-1/2" size, 3 HP, to 150 GPM	13,100	4,550	17,650
1020	3" size, 5 HP, to 225 GPM	15,800	5,150	20,950
1030	4" size, 7-1/2 HP, to 350 GPM	18,600	6,100	24,700
1040	5" size, 15 HP, to 1000 GPM	27,400	9,025	36,425
1050	6" size, 25 HP, to 1550 GPM	36,300	10,900	47,200

D3020 340	Circulating Pump Systems, Double Suction	COST EACH		
		MAT.	INST.	TOTAL
1010	Pump, base mtd w/motor, double suction, 6" size, 50 HP, to 1200 GPM	24,600	15,600	40,200
1020	8" size, 75 HP, to 2500 GPM	45,100	20,800	65,900
1030	8" size, 100 HP, to 3000 GPM	60,000	23,200	83,200
1040	10" size, 150 HP, to 4000 GPM	85,000	26,200	111,200

D3030 Cooling Generating Systems

Chilled Water Supply & Return Piping — Air Cooled Water Chiller Unit — Roof — Insulate Return — Fan Coil Unit — Supply — Finish Ceiling

Design Assumptions: The chilled water, air cooled systems priced, utilize reciprocating hermetic compressors and propeller-type condenser fans. Piping with pumps and expansion tanks is included based on a two pipe system. No ducting is included and the fan-coil units are cooling only. Water treatment and balancing are not included. Chilled water piping is insulated. Area distribution is through the use of multiple fan coil units. Fewer but larger fan coil units with duct distribution would be approximately the same S.F. cost.

System Components	QUANTITY	UNIT	COST EACH MAT.	COST EACH INST.	COST EACH TOTAL
SYSTEM D3030 110 1200					
PACKAGED CHILLER, AIR COOLED, WITH FAN COIL UNIT					
APARTMENT CORRIDORS, 3,000 S.F., 5.50 TON					
Fan coil air conditioning unit, cabinet mounted & filters chilled water	1.000	Ea.	3,253.58	604.89	3,858.47
Water chiller, air conditioning unit, air cooled	1.000	Ea.	6,545	2,255	8,800
Chilled water unit coil connections	1.000	Ea.	1,275	1,650	2,925
Chilled water distribution piping	440.000	L.F.	8,426	22,220	30,646
TOTAL			19,499.58	26,729.89	46,229.47
COST PER S.F.			6.50	8.91	15.41

*Cooling requirements would lead to choosing a water cooled unit.

D3030 110	Chilled Water, Air Cooled Condenser Systems	COST PER S.F. MAT.	COST PER S.F. INST.	COST PER S.F. TOTAL
1180	Packaged chiller, air cooled, with fan coil unit			
1200	Apartment corridors, 3,000 S.F., 5.50 ton	6.50	8.90	15.40
1240	6,000 S.F., 11.00 ton	5.40	7.10	12.50
1280	10,000 S.F., 18.33 ton	4.65	5.35	10
1320	20,000 S.F., 36.66 ton	3.63	3.87	7.50
1360	40,000 S.F., 73.33 ton	5.10	3.89	8.99
1440	Banks and libraries, 3,000 S.F., 12.50 ton	9.95	10.35	20.30
1480	6,000 S.F., 25.00 ton	9.10	8.40	17.50
1520	10,000 S.F., 41.66 ton	7.15	5.85	13
1560	20,000 S.F., 83.33 ton	8.50	5.25	13.75
1600	40,000 S.F., 167 ton*			
1680	Bars and taverns, 3,000 S.F., 33.25 ton	17.55	12.10	29.65
1720	6,000 S.F., 66.50 ton	15.75	9.20	24.95
1760	10,000 S.F., 110.83 ton	12.70	2.76	15.46
1800	20,000 S.F., 220 ton*			
1840	40,000 S.F., 440 ton*			
1920	Bowling alleys, 3,000 S.F., 17.00 ton	12.55	11.60	24.15
1960	6,000 S.F., 34.00 ton	9.70	8.10	17.80
2000	10,000 S.F., 56.66 ton	8.40	5.80	14.20
2040	20,000 S.F., 113.33 ton	9.45	5.30	14.75
2080	40,000 S.F., 227 ton*			
2160	Department stores, 3,000 S.F., 8.75 ton	9.15	9.90	19.05
2200	6,000 S.F., 17.50 ton	6.95	7.60	14.55
2240	10,000 S.F., 29.17 ton	5.70	5.45	11.15

D3030 Cooling Generating Systems

D3030 110	Chilled Water, Air Cooled Condenser Systems	COST PER S.F.		
		MAT.	INST.	TOTAL
2280	20,000 S.F., 58.33 ton	4.79	3.99	8.78
2320	40,000 S.F., 116.66 ton	6.20	4	10.20
2400	Drug stores, 3,000 S.F., 20.00 ton	14.25	11.95	26.20
2440	6,000 S.F., 40.00 ton	11.55	8.85	20.40
2480	10,000 S.F., 66.66 ton	12.90	7.50	20.40
2520	20,000 S.F., 133.33 ton	11.10	5.55	16.65
2560	40,000 S.F., 267 ton*			
2640	Factories, 2,000 S.F., 10.00 ton	8.75	9.95	18.70
2680	6,000 S.F., 20.00 ton	7.90	8	15.90
2720	10,000 S.F., 33.33 ton	6.15	5.60	11.75
2760	20,000 S.F., 66.66 ton	7.50	5	12.50
2800	40,000 S.F., 133.33 ton	6.70	4.10	10.80
2880	Food supermarkets, 3,000 S.F., 8.50 ton	9	9.85	18.85
2920	6,000 S.F., 17.00 ton	6.80	7.55	14.35
2960	10,000 S.F., 28.33 ton	5.40	5.30	10.70
3000	20,000 S.F., 56.66 ton	4.59	3.91	8.50
3040	40,000 S.F., 113.33 ton	6.05	4.01	10.06
3120	Medical centers, 3,000 S.F., 7.00 ton	8.10	9.60	17.70
3160	6,000 S.F., 14.00 ton	6.10	7.30	13.40
3200	10,000 S.F., 23.33 ton	5.45	5.60	11.05
3240	20,000 S.F., 46.66 ton	4.35	4.07	8.42
3280	40,000 S.F., 93.33 ton	5.60	3.94	9.54
3360	Offices, 3,000 S.F., 9.50 ton	8.25	9.80	18.05
3400	6,000 S.F., 19.00 ton	7.65	7.95	15.60
3440	10,000 S.F., 31.66 ton	6	5.55	11.55
3480	20,000 S.F., 63.33 ton	7.50	5.10	12.60
3520	40,000 S.F., 126.66 ton	6.60	4.14	10.74
3600	Restaurants, 3,000 S.F., 15.00 ton	11.15	10.75	21.90
3640	6,000 S.F., 30.00 ton	9.25	8.15	17.40
3680	10,000 S.F., 50.00 ton	8.10	5.90	14
3720	20,000 S.F., 100.00 ton	9.65	5.55	15.20
3760	40,000 S.F., 200 ton*			
3840	Schools and colleges, 3,000 S.F., 11.50 ton	9.45	10.20	19.65
3880	6,000 S.F., 23.00 ton	8.60	8.25	16.85
3920	10,000 S.F., 38.33 ton	6.75	5.75	12.50
3960	20,000 S.F., 76.66 ton	8.35	5.30	13.65
4000	40,000 S.F., 153 ton*			

For customer support on your Mechanical Cost Data, call 877.739.3689.

D3030 Cooling Generating Systems

Reciprocating Package Chiller — Condenser Water — Cooling Tower — Cooling Tower Water Makeup — Chilled Water Supply & Return Piping — Roof Structure — Finish Ceiling — Insulate Return — Supply — Fan Coil Unit

General: Water cooled chillers are available in the same sizes as air cooled units. They are also available in larger capacities.

Design Assumptions: The chilled water systems with water cooled condenser include reciprocating hermetic compressors, water cooling tower, pumps, piping and expansion tanks and are based on a two pipe system. Chilled water piping is insulated. No ducts are included and fan-coil units are cooling only. Area distribution is through use of multiple fan coil units. Fewer but larger fan coil units with duct distribution would be approximately the same S.F. cost. Water treatment and balancing are not included.

System Components			COST EACH		
	QUANTITY	UNIT	MAT.	INST.	TOTAL
SYSTEM D3030 115 1320					
PACKAGED CHILLER, WATER COOLED, WITH FAN COIL UNIT					
APARTMENT CORRIDORS, 4,000 S.F., 7.33 TON					
Fan coil air conditioner unit, cabinet mounted & filters, chilled water	2.000	Ea.	4,336.33	806.19	5,142.52
Water chiller, water cooled, 1 compressor, hermetic scroll,	1.000	Ea.	5,793.70	3,938.80	9,732.50
Cooling tower, draw thru single flow, belt drive	1.000	Ea.	1,737.21	168.59	1,905.80
Cooling tower pumps & piping	1.000	System	864.94	395.82	1,260.76
Chilled water unit coil connections	2.000	Ea.	2,550	3,300	5,850
Chilled water distribution piping	520.000	L.F.	9,958	26,260	36,218
TOTAL			25,240.18	34,869.40	60,109.58
COST PER S.F.			6.31	8.72	15.03

*Cooling requirements would lead to choosing a water cooled unit.

D3030 115	Chilled Water, Cooling Tower Systems		COST PER S.F.		
			MAT.	INST.	TOTAL
1300	Packaged chiller, water cooled, with fan coil unit				
1320	Apartment corridors, 4,000 S.F., 7.33 ton		6.30	8.73	15.03
1360	6,000 S.F., 11.00 ton		5.30	7.45	12.75
1400	10,000 S.F., 18.33 ton	R236000 -20	4.96	5.50	10.46
1440	20,000 S.F., 26.66 ton		4	4.10	8.10
1480	40,000 S.F., 73.33 ton		4.99	4.20	9.19
1520	60,000 S.F., 110.00 ton		5.25	4.32	9.57
1600	Banks and libraries, 4,000 S.F., 16.66 ton		10.55	9.45	20
1640	6,000 S.F., 25.00 ton		8.95	8.35	17.30
1680	10,000 S.F., 41.66 ton		7.25	6.15	13.40
1720	20,000 S.F., 83.33 ton		8.10	5.90	14
1760	40,000 S.F., 166.66 ton		8.95	7.20	16.15
1800	60,000 S.F., 250.00 ton		9.20	7.60	16.80
1880	Bars and taverns, 4,000 S.F., 44.33 ton		17	11.75	28.75
1920	6,000 S.F., 66.50 ton		21	12.20	33.20
1960	10,000 S.F., 110.83 ton		17.70	9.60	27.30
2000	20,000 S.F., 221.66 ton		20.50	9.70	30.20
2040	40,000 S.F., 440 ton*				
2080	60,000 S.F., 660 ton*				
2160	Bowling alleys, 4,000 S.F., 22.66 ton		12.35	10.35	22.70
2200	6,000 S.F., 34.00 ton		11.05	9.10	20.15
2240	10,000 S.F., 56.66 ton		9.65	6.65	16.30
2280	20,000 S.F., 113.33 ton		10.05	6.15	16.20
2320	40,000 S.F., 226.66 ton		11.80	7.15	18.95

539

For customer support on your Mechanical Cost Data, call 877.739.3689.

D3030 Cooling Generating Systems

D3030 115	Chilled Water, Cooling Tower Systems	COST PER S.F.		
		MAT.	INST.	TOTAL
2360	60,000 S.F., 340 ton			
2440	Department stores, 4,000 S.F., 11.66 ton	7.20	9.40	16.60
2480	6,000 S.F., 17.50 ton	7.75	7.70	15.45
2520	10,000 S.F., 29.17 ton	6	5.75	11.75
2560	20,000 S.F., 58.33 ton	4.93	4.24	9.17
2600	40,000 S.F., 116.66 ton	6.20	4.40	10.60
2640	60,000 S.F., 175.00 ton	8.10	6.95	15.05
2720	Drug stores, 4,000 S.F., 26.66 ton	13.30	10.65	23.95
2760	6,000 S.F., 40.00 ton	11.45	9.10	20.55
2800	10,000 S.F., 66.66 ton	13.50	8.25	21.75
2840	20,000 S.F., 133.33 ton	11.50	6.40	17.90
2880	40,000 S.F., 266.67 ton	12.10	8.10	20.20
2920	60,000 S.F., 400 ton*			
3000	Factories, 4,000 S.F., 13.33 ton	9.10	9	18.10
3040	6,000 S.F., 20.00 ton	7.85	7.75	15.60
3080	10,000 S.F., 33.33 ton	6.60	5.95	12.55
3120	20,000 S.F., 66.66 ton	7.65	5.30	12.95
3160	40,000 S.F., 133.33 ton	6.85	4.54	11.39
3200	60,000 S.F., 200.00 ton	8.35	7.30	15.65
3280	Food supermarkets, 4,000 S.F., 11.33 ton	7.10	9.30	16.40
3320	6,000 S.F., 17.00 ton	6.95	7.55	14.50
3360	10,000 S.F., 28.33 ton	5.90	5.70	11.60
3400	20,000 S.F., 56.66 ton	5.10	4.26	9.36
3440	40,000 S.F., 113.33 ton	6.15	4.41	10.56
3480	60,000 S.F., 170.00 ton	8.05	6.95	15
3560	Medical centers, 4.000 S.F., 9.33 ton	6.15	8.55	14.70
3600	6,000 S.F., 14.00 ton	6.75	7.40	14.15
3640	10,000 S.F., 23.33 ton	5.40	5.55	10.95
3680	20,000 S.F., 46.66 ton	4.58	4.16	8.74
3720	40,000 S.F., 93.33 ton	5.60	4.32	9.92
3760	60,000 S.F., 140.00 ton	6.95	7.05	14
3840	Offices, 4,000 S.F., 12.66 ton	8.80	8.95	17.75
3880	6,000 S.F., 19.00 ton	7.80	7.95	15.75
3920	10,000 S.F., 31.66 ton	6.55	6	12.55
3960	20,000 S.F., 63.33 ton	7.60	5.30	12.90
4000	40,000 S.F., 126.66 ton	7.50	6.90	14.40
4040	60,000 S.F., 190.00 ton	8.10	7.20	15.30
4120	Restaurants, 4,000 S.F., 20.00 ton	11.05	9.60	20.65
4160	6,000 S.F., 30.00 ton	9.95	8.55	18.50
4200	10,000 S.F., 50.00 ton	8.90	6.40	15.30
4240	20,000 S.F., 100.00 ton	9.45	6.20	15.65
4280	40,000 S.F., 200.00 ton	9.45	6.95	16.40
4320	60,000 S.F., 300.00 ton	10.15	7.85	18
4400	Schools and colleges, 4,000 S.F., 15.33 ton	9.95	9.30	19.25
4440	6,000 S.F., 23.00 ton	8.50	8.15	16.65
4480	10,000 S.F., 38.33 ton	6.85	6.05	12.90
4520	20,000 S.F., 76.66 ton	7.85	5.85	13.70
4560	40,000 S.F., 153.33 ton	8.45	7.05	15.50
4600	60,000 S.F., 230.00 ton	8.55	7.30	15.85

D3030 Cooling Generating Systems

Chilled Water Piping

Reciprocating Chiller, Water Cooled

Condenser Water Piping

System Components	QUANTITY	UNIT	COST EACH		
			MAT.	INST.	TOTAL
SYSTEM D3030 130 1010					
CHILLER, RECIPROCATING, WATER COOLED, STD. CONTROLS, 60 TON					
Water chiller, reciprocating, water cooled, 60 ton cooling	1.000	Ea.	33,800	11,000	44,800
Pipe, black steel, welded, Sch. 40, on yoke & roll hgrs, 10' O.C., 4" dia	40.000	L.F.	770	1,469.20	2,239.20
Elbow, 90°, weld joint, butt, 4" pipe size	8.000	Ea.	424	2,158	2,582
Flange, welding neck, 150 lb., 4" pipe size	16.000	Ea.	744	2,166.40	2,910.40
Valves, iron body, butterfly, lug type, lever actuator, 4" size	4.000	Ea.	924	1,028	1,952
Expansion joints, bellows type, flange spool, 6" F to F, 4" dia	4.000	Ea.	1,420	628	2,048
T-O-L socket, weld joint, 1/4" pipe size, nozzle	4.000	Ea.	31.20	235.12	266.32
T-O-L socket, weld joint, 3/4" pipe size, nozzle	4.000	Ea.	36	258.16	294.16
Thermometers, stem type, 9" case, 8" stem, 3/4" NPT	4.000	Ea.	956	104	1,060
Gauges, pressure or vacuum, 3-1/2" diameter dial	4.000	Ea.	82	92	174
Insulation, fiberglass pipe covering, 1-1/2" wall, 4" IPS	20.000	L.F.	59.40	171	230.40
Chiller balancing	1.000	Ea.		625	625
TOTAL			39,246.60	19,934.88	59,181.48

D3030 130	Reciprocating Chiller, Water Cooled	COST EACH		
		MAT.	INST.	TOTAL
1010	Chiller, reciprocating, water cooled, std. controls, 60 Ton	39,200	19,900	59,100
1020	100 Ton	51,000	29,900	80,900
1030	150 Ton	84,000	33,000	117,000
1040	200 Ton	113,000	34,800	147,800

D3030 135	Reciprocating Chiller, Air Cooled	COST EACH		
		MAT.	INST.	TOTAL
1010	Chiller, reciprocating, air cooled, std. controls, 20 Ton	26,000	11,700	37,700
1020	30 Ton	32,300	12,400	44,700
1030	40 Ton	38,900	13,300	52,200
1040	60 Ton	49,500	14,800	64,300
1050	70 Ton	61,000	15,300	76,300
1060	80 Ton	65,500	15,500	81,000
1070	90 Ton	72,500	15,800	88,300
1080	100 Ton	81,000	16,100	97,100
1090	110 Ton	87,500	17,400	104,900
1100	130 Ton	98,000	17,700	115,700

541

D3030 Cooling Generating Systems

D3030 135	Reciprocating Chiller, Air Cooled	COST EACH		
		MAT.	INST.	TOTAL
1110	150 Ton	109,000	17,900	126,900
1120	175 Ton	122,500	18,200	140,700
1130	210 Ton	143,000	19,700	162,700

D3030 Cooling Generating Systems

Chilled Water Piping

Centrifugal Chiller, Water Cooled, Hermetic

Condenser Water Piping

System Components	QUANTITY	UNIT	COST EACH		
			MAT.	INST.	TOTAL
SYSTEM D3030 140 1010					
CHILLER, CENTRIF., WATER COOLED, PKGD. HERMETIC, STD. CONTROLS, 200 TON					
Water chiller, cntfgl, water cooled, not incl. tower, 200 ton	1.000	Ea.	134,000	15,000	149,000
Pipe, black steel, welded, Sch. 40, on yoke & roll hgrs, 10' OC, 6" dia	40.000	L.F.	2,020	2,290.80	4,310.80
Elbow, 90°, weld joint, 6" pipe size	8.000	Ea.	928	3,302	4,230
Flange, welding neck, 150 lb., 6" pipe size	16.000	Ea.	1,120	3,302.40	4,422.40
Valves, iron body, butterfly, lug type, lever actuator, 6" size	4.000	Ea.	1,540	1,600	3,140
Expansion joints, bellows type, flange spool, 6" F to F, 6" dia	4.000	Ea.	1,800	776	2,576
T-O-L socket, weld jiont, 1/4" pipe size, nozzle	4.000	Ea.	31.20	235.12	266.32
T-O-L socket, weld joint, 3/4" pipe size, nozzle	4.000	Ea.	36	258.16	294.16
Thermometers, stem type, 9" case, 8" stem, 3/4" NPT	4.000	Ea.	956	104	1,060
Gauges, pressure or vacuum, 3-1/2" diameter dial	4.000	Ea.	82	92	174
Insulation, fiberglass pipe covering, 1-1/2" wall, 6" IPS	40.000	L.F.	140.40	436	576.40
Chiller balancing	1.000	Ea.		625	625
TOTAL			142,653.60	28,021.48	170,675.08

D3030 140	Centrifugal Chiller, Water Cooled	COST EACH		
		MAT.	INST.	TOTAL
1010	Chiller, centrif., water cooled, pkgd. hermetic, std controls, 200 Ton	142,500	28,000	170,500
1020	400 Ton	183,000	42,400	225,400
1030	1,000 Ton	445,000	62,000	507,000
1040	1,500 Ton	600,000	77,500	677,500

D3030 Cooling Generating Systems

Chilled Water Piping
(Condenser Water Piping Similar)

Gas Absorption Chiller, Water Cooled

Gas Piping

System Components	QUANTITY	UNIT	COST EACH		
			MAT.	INST.	TOTAL
SYSTEM D3030 145 1010					
CHILLER, GAS FIRED ABSORPTION, WATER COOLED, STD. CONTROLS, 800 TON					
Absorption water chiller, gas fired, water cooled, 800 ton	1.000	Ea.	625,000	33,200	658,200
Pipe, black steel, weld, Sch. 40, on yoke & roll hangers, 10' O.C., 10" dia	20.000	L.F.	2,100	1,723.20	3,823.20
Pipe, black steel, welded, Sch. 40, 14" dia., (two rod roll type hanger)	44.000	L.F.	5,060	6,039.44	11,099.44
Pipe, black steel, weld, Sch. 40, on yoke & roll hangers, 10' O.C., 4" dia	20.000	L.F.	385	734.60	1,119.60
Elbow, 90°, weld joint, 10" pipe size	4.000	Ea.	1,740	2,746	4,486
Elbow, 90°, weld joint, 14" pipe size	6.000	Ea.	6,450	6,192	12,642
Elbow, 90°, weld joint, 4" pipe size	3.000	Ea.	159	809.25	968.25
Flange, welding neck, 150 lb., 10" pipe size	8.000	Ea.	1,592	2,765.20	4,357.20
Flange, welding neck, 150 lb., 14" pipe size	14.000	Ea.	5,810	6,428.80	12,238.80
Flange, welding neck, 150 lb., 4" pipe size	3.000	Ea.	139.50	406.20	545.70
Valves, iron body, butterfly, lug type, gear operated, 10" size	2.000	Ea.	2,550	1,000	3,550
Valves, iron body, butterfly, lug type, gear operated, 14" size	5.000	Ea.	17,750	4,350	22,100
Valves, semi-steel, lubricated plug valve, flanged, 4" pipe size	1.000	Ea.	570	430	1,000
Expansion joints, bellows type, flange spool, 6" F to F, 10" dia	2.000	Ea.	1,430	528	1,958
Expansion joints, bellows type, flange spool, 10" F to F, 14" dia	2.000	Ea.	2,500	690	3,190
T-O-L socket, weld joint, 1/4" pipe size, nozzle	4.000	Ea.	31.20	235.12	266.32
T-O-L socket, weld joint, 3/4" pipe size, nozzle	4.000	Ea.	36	258.16	294.16
Thermometers, stem type, 9" case, 8" stem, 3/4" NPT	4.000	Ea.	956	104	1,060
Gauges, pressure or vacuum, 3-1/2" diameter dial	4.000	Ea.	82	92	174
Insulation, fiberglass pipe covering, 2" wall, 10" IPS	20.000	L.F.	156	342	498
Vent chimney, all fuel, pressure tight, double wall, SS, 12" dia	20.000	L.F.	1,760	570	2,330
Chiller balancing	1.000	Ea.		625	625
TOTAL			676,256.70	70,268.97	746,525.67

D3030 145	Gas Absorption Chiller, Water Cooled	COST EACH		
		MAT.	INST.	TOTAL
1010	Chiller, gas fired absorption, water cooled, std. ctrls, 800 Ton	676,500	70,500	747,000
1020	1000 Ton	889,000	90,000	979,000

D3030 150	Steam Absorption Chiller, Water Cooled	COST EACH		
		MAT.	INST.	TOTAL
1010	Chiller, steam absorption, water cooled, std. ctrls, 750 Ton	748,000	83,500	831,500
1020	955 Ton	872,500	99,000	971,500
1030	1465 Ton	1,264,500	113,500	1,378,000
1040	1660 Ton	1,526,000	129,000	1,655,000

D3030 Cooling Generating Systems

Cold Water Fill

Drain To Waste Piping

Cooling Tower

Condenser Water Piping

System Components	QUANTITY	UNIT	COST EACH		
			MAT.	INST.	TOTAL
SYSTEM D3030 310 1010					
COOLING TOWER, GALVANIZED STEEL, PACKAGED UNIT, DRAW THRU, 60 TON					
Cooling towers, draw thru, single flow, belt drive, 60 tons	1.000	Ea.	14,220	1,380	15,600
Pipe, black steel, welded, Sch. 40, on yoke & roll hangers, 10' OC, 4" dia.	40.000	L.F.	770	1,469.20	2,239.20
Tubing, copper, Type L, couplings & hangers 10' OC, 1-1/2" dia.	20.000	L.F.	277	275	552
Elbow, 90°, copper, cu x cu, 1-1/2" dia.	3.000	Ea.	47.40	165	212.40
Elbow, 90°, weld joint, 4" pipe size	6.000	Ea.	318	1,618.50	1,936.50
Flange, welding neck, 150 lb., 4" pipe size	6.000	Ea.	279	812.40	1,091.40
Valves, bronze, gate, N.R.S., soldered, 125 psi, 1-1/2" size	1.000	Ea.	167	55	222
Valves, iron body, butterfly, lug type, lever actuator, 4" size	2.000	Ea.	462	514	976
Electric heat trace system, 400 degree, 115v, 5 watts per L.F.	160.000	L.F.	1,392	198.40	1,590.40
Insulation, fiberglass pipe covering, , 1-1/2" wall, 4" IPS	40.000	L.F.	118.80	342	460.80
Insulation, fiberglass pipe covering, 1" wall, 1-1/2" IPS	20.000	L.F.	24.80	114	138.80
Cooling tower condenser control system	1.000	Ea.	6,175	2,950	9,125
Cooling tower balancing	1.000	Ea.		475	475
TOTAL			24,251	10,368.50	34,619.50

D3030 310	Galvanized Draw Through Cooling Tower	COST EACH		
		MAT.	INST.	TOTAL
1010	Cooling tower, galvanized steel, packaged unit, draw thru, 60 Ton	24,300	10,400	34,700
1020	110 Ton	33,600	12,200	45,800
1030	300 Ton	49,300	18,400	67,700
1040	600 Ton	61,000	27,500	88,500
1050	1000 Ton	126,500	40,500	167,000

D3030 320	Fiberglass Draw Through Cooling Tower	COST EACH		
		MAT.	INST.	TOTAL
1010	Cooling tower, fiberglass, packaged unit, draw thru, 100 Ton	25,300	10,500	35,800
1020	120 Ton	29,500	11,800	41,300
1030	140 Ton	34,000	15,700	49,700
1040	160 Ton	44,100	20,800	64,900
1050	180 Ton	63,500	30,000	93,500

D3030 330	Stainless Steel Draw Through Cooling Tower	COST EACH		
		MAT.	INST.	TOTAL
1010	Cooling tower, stainless steel, packaged unit, draw thru, 60 Ton	39,800	10,400	50,200
1020	110 Ton	47,700	12,200	59,900
1030	300 Ton	80,000	18,400	98,400
1040	600 Ton	132,000	27,100	159,100
1050	1000 Ton	214,000	40,500	254,500

D3040 Distribution Systems

Two-Way Valve Piping

Three-Way Valve Piping

Built-Up Air Handling Unit

System Components	QUANTITY	UNIT	COST EACH		
			MAT.	INST.	TOTAL
SYSTEM D3040 106 1010					
AHU, FIELD FAB., BLT. UP, COOL/HEAT COILS, FLTRS., CONST. VOL., 40,000 CFM					
AHU, Built-up, cool/heat coils, filters, mix box, constant volume	1.000	Ea.	101,000	6,850	107,850
Pipe, black steel, welded, Sch. 40, on yoke & roll hgrs, 10' OC, 3" dia.	26.000	L.F.	358.80	818.48	1,177.28
Pipe, black steel, welded, Sch. 40, on yoke & roll hgrs, 10' OC, 4" dia	20.000	L.F.	385	734.60	1,119.60
Elbow, 90°, weld joint, 3" pipe size	5.000	Ea.	160	965.50	1,125.50
Elbow, 90°, weld joint, 4" pipe size	4.000	Ea.	212	1,079	1,291
Tee, welded, reducing on outlet, 3" by 2-1/2" pipe	3.000	Ea.	390	1,007.85	1,397.85
Tee, welded, reducing on outlet, 4" by 3" pipe	2.000	Ea.	274	903	1,177
Reducer, eccentric, weld joint, 3" pipe size	4.000	Ea.	260	676	936
Reducer, eccentric, weld joint, 4" pipe size	4.000	Ea.	332	898.60	1,230.60
Flange, weld neck, 150 LB, 2-1/2" pipe size	3.000	Ea.	105	253.47	358.47
Flange, weld neck, 150 lb., 3" pipe size	14.000	Ea.	539	1,351.84	1,890.84
Flange, weld neck, 150 lb., 4" pipe size	10.000	Ea.	465	1,354	1,819
Valves, iron body, butterfly, lug type, lever actuator, 3" size	2.000	Ea.	418	322	740
Valves, iron body, butterfly, lug type, lever actuator, 4" size	2.000	Ea.	462	514	976
Strainer, Y type, iron body, flanged, 125 lb., 3" pipe size	1.000	Ea.	196	293	489
Strainer, Y type, iron body, flanged, 125 lb., 4" pipe size	1.000	Ea.	325	440	765
Valve, electric motor actuated, iron body, 3 way, flanged, 2-1/2" pipe	1.000	Ea.	1,525	350	1,875
Valve, electric motor actuated, iron body, two way, flanged, 3" pipe size	1.000	Ea.	1,550	350	1,900
Expansion joints, bellows type, flange spool, 6" F to F, 3" dia	2.000	Ea.	660	232	892
Expansion joints, bellows type, flange spool, 6" F to F, 4" dia	2.000	Ea.	710	314	1,024
T-O-L socket, weld joint, 1/4" pipe size, nozzle	4.000	Ea.	31.20	235.12	266.32
T-O-L socket, weld joint, 1/2" pipe size, nozzle	12.000	Ea.	93.60	736.80	830.40
Thermometers, stem type, 9" case, 8" stem, 3/4" NPT	4.000	Ea.	956	104	1,060
Pressure/temperature relief plug, 316 SS, 3/4"OD, 7-1/2" insertion	4.000	Ea.	216	92	308
Insulation, fiberglass pipe covering, 1-1/2" wall, 3" IPS	26.000	L.F.	67.86	183.30	251.16
Insulation, fiberglass pipe covering, 1-1/2" wall, 4" IPS	20.000	L.F.	59.40	171	230.40
Fan coil control system	1.000	Ea.	1,400	660	2,060
Re-heat coil balancing	2.000	Ea.		272	272
Air conditioner balancing	1.000	Ea.		840	840
TOTAL			113,150.86	23,001.56	136,152.42

D3040 106	Field Fabricated Air Handling Unit	COST EACH		
		MAT.	INST.	TOTAL
1010	AHU, Field fab, blt up, cool/heat coils, fltrs, const vol, 40,000 CFM	113,000	23,000	136,000
1020	60,000 CFM	129,000	35,700	164,700
1030	75,000 CFM	130,500	37,600	168,100

D3040 Distribution Systems

D3040 108	Field Fabricated VAV Air Handling Unit	COST EACH		
		MAT.	INST.	TOTAL
1010	AHU, Field fab, blt up, cool/heat coils, fltrs, VAV, 75,000 CFM	133,000	43,200	176,200
1020	100,000 CFM	207,000	57,500	264,500
1030	150,000 CFM	339,500	91,500	431,000

For customer support on your Mechanical Cost Data, call 877.739.3689.

547

D3040 Distribution Systems

| Chilled Water Piping | Central Station Air Handling Unit | Hot Water Piping |

System Components	QUANTITY	UNIT	COST EACH		
			MAT.	INST.	TOTAL
SYSTEM D3040 110 1010					
AHU, CENTRAL STATION, COOL/HEAT COILS, CONST. VOL., FLTRS., 2,000 CFM					
Central station air handling unit, chilled water, 2000 CFM	1.000	Ea.	12,100	1,200	13,300
Pipe, black steel, Sch 40, threaded, w/cplgs & hangers, 10' OC, 1" dia.	26.000	L.F.	153.73	438.75	592.48
Pipe, black steel, Sch 40, threaded, w/cplg & hangers, 10' OC, 1-1/4" dia.	20.000	L.F.	148.20	375.70	523.90
Elbow, 90°, black, straight, 1" dia.	4.000	Ea.	31.80	220	251.80
Elbow, 90°, black, straight, 1-1/4" dia.	4.000	Ea.	52.60	234	286.60
Tee, black, reducing, 1" dia.	8.000	Ea.	160	716	876
Tee, black, reducing, 1-1/4" dia.	8.000	Ea.	284	736	1,020
Reducer, concentric, black, 1" dia.	4.000	Ea.	58	190	248
Reducer, concentric, black, 1-1/4" dia.	4.000	Ea.	64.80	198	262.80
Union, black with brass seat, 1" dia.	4.000	Ea.	106	238	344
Union, black with brass seat, 1-1/4" dia.	2.000	Ea.	76	123	199
Pipe nipple, black, 1" dia	13.000	Ea.	30.75	87.75	118.50
Pipe nipple, black, 1-1/4" dia	12.000	Ea.	34.20	86.70	120.90
Valves, bronze, gate, N.R.S., threaded, class 150, 1" size	2.000	Ea.	236	75	311
Valves, bronze, gate, N.R.S., threaded, class 150, 1-1/4" size	2.000	Ea.	320	95	415
Strainer, Y type, bronze body, screwed, 150 lb., 1" pipe size	1.000	Ea.	54	43	97
Strainer, Y type, bronze body, screwed, 150 lb., 1-1/4" pipe size	1.000	Ea.	108	49	157
Valve, electric motor actuated, brass, 3 way, screwed, 3/4" pipe	1.000	Ea.	470	38.50	508.50
Valve, electric motor actuated, brass, 2 way, screwed, 1" pipe size	1.000	Ea.	690	37.50	727.50
Thermometers, stem type, 9" case, 8" stem, 3/4" NPT	4.000	Ea.	956	104	1,060
Pressure/temperature relief plug, 316 SS, 3/4"OD, 7-1/2" insertion	4.000	Ea.	216	92	308
Insulation, fiberglass pipe covering, 1" wall, 1" IPS	26.000	L.F.	27.82	141.70	169.52
Insulation, fiberglass pipe covering, 1" wall, 1-1/4" IPS	20.000	L.F.	23.20	114	137.20
Fan coil control system	1.000	Ea.	1,400	660	2,060
Re-heat coil balancing	2.000	Ea.		272	272
Multizone A/C balancing	1.000	Ea.		630	630
TOTAL			17,801.10	7,195.60	24,996.70

D3040 110	Central Station Air Handling Unit	COST EACH		
		MAT.	INST.	TOTAL
1010	AHU, Central station, cool/heat coils, constant vol, fltrs, 2,000 CFM	17,800	7,200	25,000
1020	5,000 CFM	26,500	9,275	35,775
1030	10,000 CFM	38,800	13,000	51,800
1040	15,000 CFM	71,500	17,500	89,000
1050	20,000 CFM	87,000	20,400	107,400

D30 HVAC

D3040 Distribution Systems

D3040 112	Central Station Air Handling Unit, VAV	COST EACH		
		MAT.	INST.	TOTAL
1010	AHU, Central station, cool/heat coils, VAV, fltrs, 5,000 CFM	30,900	9,625	40,525
1020	10,000 CFM	56,500	13,300	69,800
1030	15,000 CFM	80,000	17,200	97,200
1040	20,000 CFM	98,500	20,900	119,400
1050	30,000 CFM	116,500	24,100	140,600

D30 HVAC

D3040 Distribution Systems

Two-Way Valve Piping

Three-Way Valve Piping

Cooling Coil

Heating Coil

Return Air Fan Section

Exhaust Section
Mixing Box
Filter Section
Coil Section

Roof Mounting Curb

Supply Air Fan Section

Rooftop Air Handling Unit

System Components	QUANTITY	UNIT	MAT.	INST.	TOTAL
SYSTEM D3040 114 1010					
AHU, ROOFTOP, COOL/HEAT COILS, FLTRS., CONST. VOL., 2,000 CFM					
AHU, Built-up, cool/heat coils, filter, mix box, rooftop, constant volume	1.000	Ea.	26,000	2,050	28,050
Pipe, black steel, Sch 40, threaded, w/cplgs & hangers, 10' OC, 1" dia.	26.000	L.F.	153.73	438.75	592.48
Pipe, black steel, Sch 40, threaded, w/cplg & hangers, 10' OC, 1-1/4" dia.	20.000	L.F.	148.20	375.70	523.90
Elbow, 90°, black, straight, 1" dia.	4.000	Ea.	31.80	220	251.80
Elbow, 90°, black, straight, 1-1/4" dia.	4.000	Ea.	52.60	234	286.60
Tee, black, reducing, 1" dia.	8.000	Ea.	160	716	876
Tee, black, reducing, 1-1/4" dia.	8.000	Ea.	284	736	1,020
Reducer, black, concentric, 1" dia.	4.000	Ea.	58	190	248
Reducer, black, concentric, 1-1/4" dia.	4.000	Ea.	64.80	198	262.80
Union, black with brass seat, 1" dia.	4.000	Ea.	106	238	344
Union, black with brass seat, 1-1/4" dia.	2.000	Ea.	76	123	199
Pipe nipples, 1" dia	13.000	Ea.	30.75	87.75	118.50
Pipe nipples, 1-1/4" dia	12.000	Ea.	34.20	86.70	120.90
Valves, bronze, gate, N.R.S., threaded, class 150, 1" size	2.000	Ea.	236	75	311
Valves, bronze, gate, N.R.S., threaded, class 150, 1-1/4" size	2.000	Ea.	320	95	415
Strainer, Y type, bronze body, screwed, 150 lb., 1" pipe size	1.000	Ea.	54	43	97
Strainer, Y type, bronze body, screwed, 150 lb., 1-1/4" pipe size	1.000	Ea.	108	49	157
Valve, electric motor actuated, brass, 3 way, screwed, 3/4" pipe	1.000	Ea.	470	38.50	508.50
Valve, electric motor actuated, brass, 2 way, screwed, 1" pipe size	1.000	Ea.	690	37.50	727.50
Thermometers, stem type, 9" case, 8" stem, 3/4" NPT	4.000	Ea.	956	104	1,060
Pressure/temperature relief plug, 316 SS, 3/4"OD, 7-1/2" insertion	4.000	Ea.	216	92	308
Insulation, fiberglass pipe covering, 1" wall, 1" IPS	26.000	L.F.	27.82	141.70	169.52
Insulation, fiberglass pipe covering, 1" wall, 1-1/4" IPS	20.000	L.F.	23.20	114	137.20
Insulation pipe covering .016" aluminum jacket finish, add	45.000	S.F.	45.45	270	315.45
Fan coil control system	1.000	Ea.	1,400	660	2,060
Re-heat coil balancing	2.000	Ea.		272	272
Rooftop unit heat/cool balancing	1.000	Ea.		490	490
TOTAL			31,746.55	8,175.60	39,922.15

D3040 114	Rooftop Air Handling Unit	MAT.	INST.	TOTAL
1010	AHU, Rooftop, cool/heat coils, constant vol., fltrs, 2,000 CFM	31,700	8,175	39,875
1020	5,000 CFM	38,200	9,500	47,700

550

D30 HVAC

D3040 Distribution Systems

D3040 114	Rooftop Air Handling Unit	COST EACH		
		MAT.	INST.	TOTAL
1030	10,000 CFM	74,000	13,300	87,300
1040	15,000 CFM	106,000	17,400	123,400
1050	20,000 CFM	140,000	21,000	161,000

D3040 116	Rooftop Air Handling Unit, VAV	COST EACH		
		MAT.	INST.	TOTAL
1010	AHU, Rooftop, cool/heat coils, VAV, fltrs, 5,000 CFM	41,100	9,700	50,800
1020	10,000 CFM	83,500	13,400	96,900
1030	15,000 CFM	124,000	17,500	141,500
1040	20,000 CFM	156,500	21,300	177,800
1050	30,000 CFM	183,000	24,500	207,500

D3040 Distribution Systems

Galvanized Steel Duct

Fiberglass Duct Insulation

Flexible Fiberglass Duct

Horizontal Fan Coil Air Conditioning System

System Components	QUANTITY	UNIT	COST EACH MAT.	INST.	TOTAL
SYSTEM D3040 124 1010					
FAN COIL A/C SYSTEM, HORIZONTAL, W/HOUSING, CONTROLS, 2 PIPE, 1/2 TON					
Fan coil A/C, horizontal housing, filters, chilled water, 1/2 ton cooling	1.000	Ea.	1,400	165	1,565
Pipe, black steel, Sch 40, threaded, w/cplgs & hangers, 10' OC, 3/4" dia.	20.000	L.F.	79.12	251.55	330.67
Elbow, 90°, black, straight, 3/4" dia.	6.000	Ea.	27.54	306	333.54
Tee, black, straight, 3/4" dia.	2.000	Ea.	14.60	159	173.60
Union, black with brass seat, 3/4" dia.	2.000	Ea.	41	110	151
Pipe nipples, 3/4" diam	3.000	Ea.	5.52	17.55	23.07
Valves, bronze, gate, N.R.S., threaded, class 150, 3/4" size	2.000	Ea.	189	71	260
Circuit setter, balance valve, bronze body, threaded, 3/4" pipe size	1.000	Ea.	90	36.50	126.50
Insulation, fiberglass pipe covering, 1" wall, 3/4" IPS	20.000	L.F.	19.80	104	123.80
Ductwork, 12" x 8" fabricated, galvanized steel, 12 LF	55.000	Lb.	34.65	459.25	493.90
Insulation, ductwork, blanket type, fiberglass, 1" thk, 1-1/2 LB density	40.000	S.F.	199.60	570	769.60
Diffusers, aluminum, OB damper, ceiling, perf, 24"x24" panel size, 6"x6"	2.000	Ea.	344	88	432
Ductwork, flexible, fiberglass fabric, insulated, 1"thk, PE jacket, 6" dia	16.000	L.F.	51.68	77.60	129.28
Round volume control damper 6" dia.	2.000	Ea.	55	64	119
Fan coil unit balancing	1.000	Ea.		91	91
Re-heat coil balancing	1.000	Ea.		136	136
Diffuser/register, high, balancing	2.000	Ea.		252	252
TOTAL			2,551.51	2,958.45	5,509.96

D3040 118	Fan Coil A/C Unit, Two Pipe	COST EACH MAT.	INST.	TOTAL
1010	Fan coil A/C system, cabinet mounted, controls, 2 pipe, 1/2 Ton	970	1,350	2,320
1020	1 Ton	1,375	1,525	2,900
1030	1-1/2 Ton	1,375	1,525	2,900
1040	2 Ton	2,500	1,875	4,375
1050	3 Ton	3,175	1,950	5,125

D3040 120	Fan Coil A/C Unit, Two Pipe, Electric Heat	COST EACH MAT.	INST.	TOTAL
1010	Fan coil A/C system, cabinet mntd, elect. ht, controls, 2 pipe, 1/2 Ton	1,925	1,350	3,275
1020	1 Ton	2,325	1,525	3,850
1030	1-1/2 Ton	2,725	1,525	4,250
1040	2 Ton	4,200	1,900	6,100
1050	3 Ton	6,875	1,950	8,825

D3040 Distribution Systems

D3040 122	Fan Coil A/C Unit, Four Pipe	COST EACH		
		MAT.	INST.	TOTAL
1010	Fan coil A/C system, cabinet mounted, controls, 4 pipe, 1/2 Ton	1,650	2,625	4,275
1020	1 Ton	2,125	2,800	4,925
1030	1-1/2 Ton	2,150	2,825	4,975
1040	2 Ton	2,850	2,925	5,775
1050	3 Ton	3,925	3,125	7,050

D3040 124	Fan Coil A/C, Horizontal, Duct Mount, 2 Pipe	COST EACH		
		MAT.	INST.	TOTAL
1010	Fan coil A/C system, horizontal w/housing, controls, 2 pipe, 1/2 Ton	2,550	2,950	5,500
1020	1 Ton	3,475	4,700	8,175
1030	1-1/2 Ton	4,325	6,325	10,650
1040	2 Ton	5,775	7,850	13,625
1050	3 Ton	7,325	10,900	18,225
1060	3-1/2 Ton	7,600	11,200	18,800
1070	4 Ton	8,050	13,000	21,050
1080	5 Ton	9,750	16,700	26,450
1090	6 Ton	9,950	17,900	27,850
1100	7 Ton	9,925	18,300	28,225
1110	8 Ton	9,925	18,600	28,525
1120	10 Ton	10,500	19,100	29,600

D3040 126	Fan Coil A/C, Horiz., Duct Mount, 2 Pipe, Elec. Ht.	COST EACH		
		MAT.	INST.	TOTAL
1010	Fan coil A/C system, horiz. hsng, elect. ht, ctrls, 2 pipe, 1/2 Ton	2,800	2,950	5,750
1020	1 Ton	3,850	4,700	8,550
1030	1-1/2 Ton	4,675	6,325	11,000
1040	2 Ton	6,325	7,850	14,175
1050	3 Ton	9,850	11,100	20,950
1060	3-1/2 Ton	10,900	11,200	22,100
1070	4 Ton	12,300	13,000	25,300
1080	5 Ton	15,100	16,700	31,800
1090	6 Ton	15,300	17,900	33,200
1100	7 Ton	15,700	18,600	34,300
1110	8 Ton	15,900	19,100	35,000
1120	10 Ton	16,900	18,300	35,200

D3040 128	Fan Coil A/C, Horizontal, Duct Mount, 4 Pipe	COST EACH		
		MAT.	INST.	TOTAL
1010	Fan coil A/C system, horiz. w/cabinet, controls, 4 pipe, 1/2 Ton	3,575	4,225	7,800
1020	1 Ton	4,600	6,125	10,725
1030	1-1/2 Ton	5,550	7,600	13,150
1040	2 Ton	6,600	8,900	15,500
1050	3 Ton	8,475	12,300	20,775
1060	3.5 Ton	10,000	15,600	25,600
1070	4 Ton	10,000	15,700	25,700
1080	5 Ton	12,500	22,100	34,600
1090	6 Ton	13,900	26,800	40,700
1100	7 Ton	15,900	30,100	46,000
1110	8 Ton	16,500	32,300	48,800
1120	10 Ton	19,500	38,900	58,400

Handwritten annotations in right margin: $/Ton — 7750, 6925, 7314, 6425, 6920, 6783

For customer support on your Mechanical Cost Data, call 877.739.3689.

553

VAV Terminal   Fan Powered VAV Terminal

System Components	QUANTITY	UNIT	COST EACH		
			MAT.	INST.	TOTAL
SYSTEM D3040 132 1010					
VAV TERMINAL, COOLING ONLY, WITH ACTUATOR/CONTROLS, 200 CFM					
Mixing box variable volume 300 to 600 CFM, cool only	1.000	Ea.	902	126.50	1,028.50
Ductwork, 12" x 8" fabricated, galvanized steel, 12 LF	55.000	Lb.	34.65	459.25	493.90
Insulation, ductwork, blanket type, fiberglass, 1" thk, 1-1/2 LB density	40.000	S.F.	199.60	570	769.60
Diffusers, aluminum, OB damper, ceiling, perf, 24"x24" panel size, 6"x6"	2.000	Ea.	344	88	432
Ductwork, flexible, fiberglass fabric, insulated, 1"thk, PE jacket, 6" dia	16.000	L.F.	51.68	77.60	129.28
Round volume control damper 6" dia.	2.000	Ea.	55	64	119
VAV box balancing	1.000	Ea.		84	84
Diffuser/register, high, balancing	2.000	Ea.		252	252
TOTAL			1,586.93	1,721.35	3,308.28

D3040 132	VAV Terminal, Cooling Only	COST EACH		
		MAT.	INST.	TOTAL
1010	VAV Terminal, cooling only, with actuator/controls, 200 CFM	1,575	1,725	3,300
1020	400 CFM	2,100	3,325	5,425
1030	600 CFM	2,725	4,925	7,650
1040	800 CFM	3,025	6,225	9,250
1050	1000 CFM	3,475	7,175	10,650
1060	1250 CFM	4,175	9,350	13,525
1070	1500 CFM	4,675	11,200	15,875
1080	2000 CFM	6,100	15,800	21,900

D3040 134	VAV Terminal, Hot Water Reheat	COST EACH		
		MAT.	INST.	TOTAL
1010	VAV Terminal, cooling, HW reheat, with actuator/controls, 200 CFM	2,675	2,850	5,525
1020	400 CFM	3,225	4,325	7,550
1030	600 CFM	3,825	6,025	9,850
1040	800 CFM	4,150	7,350	11,500
1050	1000 CFM	4,550	8,300	12,850
1060	1250 CFM	5,350	10,500	15,850
1070	1500 CFM	6,075	12,700	18,775
1080	2000 CFM	7,575	17,300	24,875

D3040 136	Fan Powered VAV Terminal, Cooling Only	COST EACH		
		MAT.	INST.	TOTAL
1010	VAV Terminal, cooling, fan powrd, with actuator/controls, 200 CFM	1,900	1,725	3,625
1020	400 CFM	2,500	3,150	5,650

D30 HVAC

D3040   Distribution Systems

D3040 136	Fan Powered VAV Terminal, Cooling Only	COST EACH		
		MAT.	INST.	TOTAL
1030	600 CFM	3,100	4,925	8,025
1040	800 CFM	3,500	6,150	9,650
1050	1000 CFM	3,900	7,075	10,975
1060	1250 CFM	4,725	9,425	14,150
1070	1500 CFM	5,150	11,300	16,450
1080	2000 CFM	6,700	16,100	22,800

D3040 138	Fan Powered VAV Terminal, Hot Water Reheat	COST EACH		
		MAT.	INST.	TOTAL
1010	VAV Terminal, cool, HW reht, fan powrd, with actuator/ctrls, 200 CFM	3,300	2,825	6,125
1020	400 CFM	3,875	4,325	8,200
1030	600 CFM	4,475	6,050	10,525
1040	800 CFM	4,875	7,275	12,150
1050	1000 CFM	5,275	8,225	13,500
1060	1250 CFM	6,225	10,600	16,825
1070	1500 CFM	6,975	12,700	19,675
1080	2000 CFM	8,500	17,600	26,100

For customer support on your Mechanical Cost Data, call 877.739.3689.

555

D3040 Distribution Systems

Axial Flow Centrifugal Fan
Belt Drive

Belt Drive Utility Set

Centrifugal Roof Exhaust Fan
Direct Drive

System Components	QUANTITY	UNIT	COST EACH		
			MAT.	INST.	TOTAL
SYSTEM D3040 220 1010					
FAN SYSTEM, IN-LINE CENTRIFUGAL, 500 CFM					
Fans, in-line centrifugal, supply/exhaust, 500 CFM, 10" dia conn	1.000	Ea.	1,575	530	2,105
Louver, alum., w/screen, damper, mill fin, fxd blade, cont line, stormproof	4.050	S.F.	190.35	101.25	291.60
Motor for louver damper, electric or pneumatic	1.000	Ea.	540	50	590
Ductwork, 12" x 8" fabricated, galvanized steel, 60 LF	200.000	Lb.	161.20	2,080	2,241.20
Grilles, aluminum, air supply, single deflection, adjustable, 12" x 6"	1.000	Ea.	30	91.50	121.50
Fire damper, curtain type, vertical, 12" x 8"	1.000	Ea.	26.50	32	58.50
Duct access door, insulated, 10" x 10"	1.000	Ea.	20.50	63.50	84
Duct access door, insulated, 16" x 12"	1.000	Ea.	32.50	78	110.50
Pressure controller/switch, air flow controller	1.000	Ea.	243	61	304
Fan control switch and mounting base	1.000	Ea.	147	46	193
Fan balancing	1.000	Ea.		420	420
Diffuser/register balancing	3.000	Ea.		252	252
TOTAL			2,966.05	3,805.25	6,771.30

D3040 220	Centrifugal In Line Fan Systems	COST EACH		
		MAT.	INST.	TOTAL
1010	Fan system, in-line centrifugal, 500 CFM	2,975	3,800	6,775
1020	1300 CFM	3,375	6,600	9,975
1030	1500 CFM	3,675	7,625	11,300
1040	2500 CFM	5,425	23,300	28,725
1050	3500 CFM	6,825	31,400	38,225
1060	5000 CFM	8,950	46,600	55,550
1070	7500 CFM	10,500	47,700	58,200
1080	10,000 CFM	12,600	56,000	68,600

D3040 230	Utility Set Fan Systems	COST EACH		
		MAT.	INST.	TOTAL
1010	Utility fan set system, belt drive, 2000 CFM	2,700	12,300	15,000
1020	3500 CFM	4,525	21,800	26,325
1030	5000 CFM	5,625	30,300	35,925
1040	7500 CFM	8,025	35,200	43,225
1050	10,000 CFM	10,700	46,200	56,900
1060	15,000 CFM	10,900	47,200	58,100
1070	20,000 CFM	13,300	53,500	66,800

D30 HVAC

D3040 Distribution Systems

D3040 240	Roof Exhaust Fan Systems	COST EACH		
		MAT.	INST.	TOTAL
1010	Roof vent. system, centrifugal, alum., galv curb, BDD, 500 CFM	1,425	2,500	3,925
1020	800 CFM	1,675	4,825	6,500
1030	1500 CFM	2,425	7,150	9,575
1040	2750 CFM	3,425	16,200	19,625
1050	3500 CFM	4,150	20,900	25,050
1060	5000 CFM	5,725	34,400	40,125
1070	8500 CFM	7,525	44,200	51,725
1080	13,800 CFM	10,700	64,000	74,700

D3040 Distribution Systems

Commercial/Industrial Dust Collection System

System Components	QUANTITY	UNIT	COST EACH MAT.	COST EACH INST.	COST EACH TOTAL
SYSTEM D3040 250 1010					
COMMERCIAL/INDUSTRIAL VACUUM DUST COLLECTION SYSTEM, 500 CFM					
Dust collection central vac unit inc. stand, filter & shaker, 500 CFM 2 HP	1.000	Ea.	5,150	660	5,810
Galvanized tubing, 16 ga., 4″ OD	80.000	L.F.	528	280	808
Galvanized tubing, 90° ell, 4″ dia.	3.000	Ea.	103.50	75	178.50
Galvanized tubing, 45° ell, 4″ dia.	4.000	Ea.	122	100	222
Galvanized tubing, TY slip fit, 4″ dia.	4.000	Ea.	184	168	352
Flexible rubber hose, 4″ dia.	32.000	L.F.	384	112	496
Galvanized air gate valve, 4″ dia.	4.000	Ea.	648	220	868
Galvanized compression coupling, 4″ dia.	36.000	Ea.	756	1,368	2,124
Pipe hangers, steel, 4″ pipe size	16.000	Ea.	416	187.20	603.20
TOTAL			8,291.50	3,170.20	11,461.70

D3040 250	Commercial/Industrial Vacuum Dust Collection	COST EACH MAT.	COST EACH INST.	COST EACH TOTAL
1010	Commercial/industrial, vacuum dust collection system, 500 CFM	8,300	3,175	11,475
1020	1000 CFM	11,400	3,625	15,025
1030	1500 CFM	14,000	4,000	18,000
1040	3000 CFM	31,000	5,600	36,600
1050	5000 CFM	40,900	8,425	49,325

D3040 Distribution Systems

Gas Fired Make-up Air Unit

Roof Fan →

Grease Filters

Make-up Air

Exhaust Air

Removable Grease Trough

Commercial Kitchen Exhaust/Make-up Air Rooftop System

System Components	QUANTITY	UNIT	COST EACH		
			MAT.	INST.	TOTAL
SYSTEM D3040 260 1010					
COMMERCIAL KITCHEN EXHAUST/MAKE-UP AIR SYSTEM, ROOFTOP, GAS, 2000 CFM					
Make-up air rooftop unit, gas, 750 MBH	1.000	Ea.	5,541.80	156.87	5,698.67
Pipe, black steel, Sch 40, thrded, nipples, w/cplngs/hngrs, 10' OC, 1" dia.	20.000	L.F.	104.06	297	401.06
Elbow, 90°, black steel, straight, 3/4" dia.	2.000	Ea.	9.18	102	111.18
Elbow, 90°, black steel, straight, 1" dia.	3.000	Ea.	23.85	165	188.85
Tee, black steel, reducing, 1" dia.	1.000	Ea.	20	89.50	109.50
Union, black with brass seat, 1" dia.	1.000	Ea.	26.50	59.50	86
Pipe nipples, black, 3/4" dia	2.000	Ea.	7.36	23.40	30.76
Pipe cap, malleable iron, black, 1" dia.	1.000	Ea.	6.30	24	30.30
Gas cock, brass, 1" size	1.000	Ea.	34	37.50	71.50
Ductwork, 20" x 14" fabricated, galv. steel, 30 LF	220.000	Lb.	134.20	1,694	1,828.20
Ductwork, 18" x 10" fabricated, galv. steel, 12 LF	72.000	Lb.	43.92	554.40	598.32
Ductwork, 30" x 10" fabricated, galv. steel, 30 LF	260.000	Lb.	158.60	2,002	2,160.60
Ductwork, 16" x 10" fabricated, galv. steel, 12 LF	70.000	Lb.	42.70	539	581.70
Kitchen ventilation, island style, water wash, Stainless Steel	5.000	L.F.	13,250	1,380	14,630
Control system, Make-up Air Unit/Exhauster	1.000	Ea.	3,075	2,575	5,650
Rooftop unit heat/cool balancing	1.000	Ea.		490	490
Roof fan balancing	1.000	Ea.		280	280
TOTAL			22,477.47	10,469.17	32,946.64

D3040 260	Kitchen Exhaust/Make-Up Air	COST EACH		
		MAT.	INST.	TOTAL
1010	Commercial kitchen exhaust/make-up air system, rooftop, gas, 2000 CFM	22,500	10,500	33,000
1020	3000 CFM	34,200	14,600	48,800
1030	5000 CFM	40,100	17,200	57,300
1040	8000 CFM	57,500	25,500	83,000
1050	12,000 CFM	75,500	29,900	105,400
1060	16,000 CFM	93,000	45,900	138,900

559

D3040   Distribution Systems

Plate Heat Exchanger   Shell and Tube Heat Exchanger

System Components	QUANTITY	UNIT	COST EACH		
			MAT.	INST.	TOTAL
SYSTEM D3040 620 1010					
SHELL & TUBE HEAT EXCHANGER, 40 GPM					
Heat exchanger, 4 pass, 3/4" O.D. copper tubes, by steam at 10 psi, 40 GPM	1.000	Ea.	5,750	330	6,080
Pipe, black steel, Sch 40, threaded, w/couplings & hangers, 10' OC, 2" dia.	40.000	L.F.	411.58	1,010	1,421.58
Elbow, 90°, straight, 2" dia.	16.000	Ea.	488	1,144	1,632
Reducer, black steel, concentric, 2" dia.	2.000	Ea.	67	123	190
Valves, bronze, gate, rising stem, threaded, class 150, 2" size	6.000	Ea.	1,578	390	1,968
Valves, bronze, globe, class 150, rising stem, threaded, 2" size	2.000	Ea.	1,030	130	1,160
Strainers, Y type, bronze body, screwed, 125 lb, 2" pipe size	2.000	Ea.	306	113	419
Valve, electric motor actuated, brass, 2 way, screwed, 1-1/2" pipe size	1.000	Ea.	760	55	815
Union, black with brass seat, 2" dia.	8.000	Ea.	444	604	1,048
Tee, black, straight, 2" dia.	9.000	Ea.	337.50	1,053	1,390.50
Tee, black, reducing run and outlet, 2" dia.	4.000	Ea.	214	468	682
Pipe nipple, black, 2" dia.	21.000	Ea.	4.08	10	14.08
Thermometers, stem type, 9" case, 8" stem, 3/4" NPT	2.000	Ea.	478	52	530
Gauges, pressure or vacuum, 3-1/2" diameter dial	2.000	Ea.	41	46	87
Insulation, fiberglass pipe covering, 1" wall, 2" IPS	40.000	L.F.	53.60	240	293.60
Heat exchanger control system	1.000	Ea.	3,000	2,300	5,300
Coil balancing	1.000	Ea.		135	135
TOTAL			14,962.76	8,203	23,165.76

D3040 610	Heat Exchanger, Plate Type	COST EACH		
		MAT.	INST.	TOTAL
1010	Plate heat exchanger, 400 GPM	59,500	16,600	76,100
1020	800 GPM	95,000	21,100	116,100
1030	1200 GPM	142,500	28,600	171,100
1040	1800 GPM	193,000	35,600	228,600

D3040 620	Heat Exchanger, Shell & Tube	COST EACH		
		MAT.	INST.	TOTAL
1010	Shell & tube heat exchanger, 40 GPM	14,962.75	8,203	23,173
1020	96 GPM	27,600	14,500	42,100
1030	240 GPM	50,000	19,800	69,800
1040	600 GPM	98,000	27,800	125,800

D3050   Terminal & Package Units

Cabinet Unit Heater   **Gas Unit Heater**   **Hydronic Unit Heater**

System Components	QUANTITY	UNIT	COST EACH		
			MAT.	INST.	TOTAL
SYSTEM D3050 120 1010					
SPACE HEATER, SUSPENDED, GAS FIRED, PROPELLER FAN, 20 MBH					
Space heater, propeller fan, 20 MBH output	1.000	Ea.	1,250	155	1,405
Pipe, black steel, Sch 40, threaded, w/coupling & hangers, 10' OC, 3/4" dia	20.000	L.F.	77.28	245.70	322.98
Elbow, 90°, black steel, straight, 1/2" dia.	3.000	Ea.	11.40	142.50	153.90
Elbow, 90°, black steel, straight, 3/4" dia.	3.000	Ea.	13.77	153	166.77
Tee, black steel, reducing, 3/4" dia.	1.000	Ea.	13.85	79.50	93.35
Union, black with brass seat, 3/4" dia.	1.000	Ea.	20.50	55	75.50
Pipe nipples, black, 1/2" dia	2.000	Ea.	6.38	22.70	29.08
Pipe nipples, black, 3/4" dia	2.000	Ea.	3.68	11.70	15.38
Pipe cap, black, 3/4" dia.	1.000	Ea.	5.20	22.50	27.70
Gas cock, brass, 3/4" size	1.000	Ea.	17.05	32.50	49.55
Thermostat, 1 set back, electric, timed	1.000	Ea.	41.50	91.50	133
Wiring, thermostat hook-up,25' of #18-3	1.000	Ea.	8.80	27.50	36.30
Vent chimney, prefab metal, U.L. listed, gas, double wall, galv. st, 4" dia	12.000	L.F.	108	222.60	330.60
Vent chimney, gas, double wall, galv. steel, elbow 90°, 4" dia.	2.000	Ea.	51	74	125
Vent chimney, gas, double wall, galv. steel, Tee, 4" dia.	1.000	Ea.	36	48.50	84.50
Vent chimney, gas, double wall, galv. steel, T cap, 4" dia.	1.000	Ea.	2.41	30	32.41
Vent chimney, gas, double wall, galv. steel, roof flashing, 4" dia.	1.000	Ea.	8.70	37	45.70
Vent chimney, gas, double wall, galv. steel, top, 4" dia.	1.000	Ea.	17.95	28.50	46.45
TOTAL			1,693.47	1,479.70	3,173.17

D3050 120	Unit Heaters, Gas	COST EACH		
		MAT.	INST.	TOTAL
1010	Space heater, suspended, gas fired, propeller fan, 20 MBH	1,693.47	1,479.70	3,168.40
1020	60 MBH	2,425	1,550	3,975
1030	100 MBH	2,775	1,625	4,400
1040	160 MBH	3,300	1,800	5,100
1050	200 MBH	4,125	1,975	6,100
1060	280 MBH	4,800	2,125	6,925
1070	320 MBH	6,350	2,400	8,750

D3050 130	Unit Heaters, Hydronic	COST EACH		
		MAT.	INST.	TOTAL
1010	Space heater, suspended, horiz. mount, HW, prop. fan, 20 MBH	2,200	1,575	3,775
1020	60 MBH	2,975	1,775	4,750

561

D3050 Terminal & Package Units

D3050 130	Unit Heaters, Hydronic	COST EACH		
		MAT.	INST.	TOTAL
1030	100 MBH	3,575	1,950	5,525
1040	150 MBH	4,075	2,150	6,225
1050	200 MBH	4,600	2,425	7,025
1060	300 MBH	5,425	2,850	8,275

D3050 140	Cabinet Unit Heaters, Hydronic	COST EACH		
		MAT.	INST.	TOTAL
1010	Unit heater, cabinet type, horizontal blower, hot water, 20 MBH	2,600	1,550	4,150
1020	60 MBH	3,800	1,700	5,500
1030	100 MBH	4,450	1,875	6,325
1040	120 MBH	4,550	1,950	6,500

D3050 Terminal & Package Units

Roof — **Roof Top Unit**

Return Duct

Return Grille

Supply Diff.

Insulated Supply Duct

Finish Ceiling

System Description: Rooftop single zone units are electric cooling and gas heat. Duct systems are low velocity, galvanized steel supply and return. Price variations between sizes are due to several factors. Jumps in the cost of the rooftop unit occur when the manufacturer shifts from the largest capacity unit on a small frame to the smallest capacity on the next larger frame, or changes from one compressor to two. As the unit capacity increases for larger areas the duct distribution grows in proportion. For most applications there is a tradeoff point where it is less expensive and more efficient to utilize smaller units with short simple distribution systems. Larger units also require larger initial supply and return ducts which can create a space problem. Supplemental heat may be desired in colder locations. The table below is based on one unit supplying the area listed. The 10,000 S.F. unit for bars and taverns is not listed because a nominal 110 ton unit would be required and this is above the normal single zone rooftop capacity.

System Components	QUANTITY	UNIT	COST EACH		
			MAT.	INST.	TOTAL
SYSTEM D3050 150 1280					
ROOFTOP, SINGLE ZONE, AIR CONDITIONER					
APARTMENT CORRIDORS, 500 S.F., .92 TON					
Rooftop air conditioner, 1 zone, electric cool, standard controls, curb	1.000	Ea.	1,414.50	862.50	2,277
Ductwork package for rooftop single zone units	1.000	System	172.96	1,219	1,391.96
TOTAL			1,587.46	2,081.50	3,668.96
COST PER S.F.			3.17	4.16	7.33

*Size would suggest multiple units.

D3050 150	Rooftop Single Zone Unit Systems	COST PER S.F.		
		MAT.	INST.	TOTAL
1260	Rooftop, single zone, air conditioner			
1280	Apartment corridors, 500 S.F., .92 ton	3.15	4.15	7.30
1320	1,000 S.F., 1.83 ton	3.15	4.14	7.29
1360	1500 S.F., 2.75 ton	1.98	3.43	5.41
1400	3,000 S.F., 5.50 ton	2.12	3.29	5.41
1440	5,000 S.F., 9.17 ton	2.17	2.99	5.16
1480	10,000 S.F., 18.33 ton	3.32	2.81	6.13
1560	Banks or libraries, 500 S.F., 2.08 ton	7.20	9.40	16.60
1600	1,000 S.F., 4.17 ton	4.51	7.80	12.31
1640	1,500 S.F., 6.25 ton	4.82	7.50	12.32
1680	3,000 S.F., 12.50 ton	4.95	6.80	11.75
1720	5,000 S.F., 20.80 ton	7.55	6.40	13.95
1760	10,000 S.F., 41.67 ton	5.80	6.35	12.15
1840	Bars and taverns, 500 S.F. 5.54 ton	11.80	12.55	24.35
1880	1,000 S.F., 11.08 ton	12.10	10.75	22.85
1920	1,500 S.F., 16.62 ton	11.70	10	21.70
1960	3,000 S.F., 33.25 ton	15.50	9.55	25.05
2000	5,000 S.F., 55.42 ton	13.25	9.55	22.80
2040	10,000 S.F., 110.83 ton*			
2080	Bowling alleys, 500 S.F., 2.83 ton	6.10	10.60	16.70
2120	1,000 S.F., 5.67 ton	6.55	10.15	16.70
2160	1,500 S.F., 8.50 ton	6.75	9.25	16
2200	3,000 S.F., 17.00 ton	6.50	8.85	15.35
2240	5,000 S.F., 28.33 ton	8.45	8.65	17.10
2280	10,000 S.F., 56.67 ton	7.30	8.65	15.95
2360	Department stores, 500 S.F., 1.46 ton	5.05	6.60	11.65

R236000 -20

563

For customer support on your Mechanical Cost Data, call 877.739.3689.

D30 HVAC

D3050 Terminal & Package Units

D3050 150	Rooftop Single Zone Unit Systems	MAT.	INST.	TOTAL
2400	1,000 S.F., 2.92 ton	3.16	5.45	8.61
2440	1,500 S.F., 4.37 ton	3.38	5.25	8.63
2480	3,000 S.F., 8.75 ton	3.47	4.76	8.23
2520	5,000 S.F., 14.58 ton	3.35	4.56	7.91
2560	10,000 S.F., 29.17 ton	4.36	4.45	8.81
2640	Drug stores, 500 S.F., 3.33 ton	7.20	12.45	19.65
2680	1,000 S.F., 6.67 ton	7.70	11.95	19.65
2720	1,500 S.F., 10.00 ton	7.90	10.90	18.80
2760	3,000 S.F., 20.00 ton	12.10	10.20	22.30
2800	5,000 S.F., 33.33 ton	9.95	10.15	20.10
2840	10,000 S.F., 66.67 ton	8.60	10.15	18.75
2920	Factories, 500 S.F., 1.67 ton	5.75	7.55	13.30
2960	1,000 S.F., 3.33 ton	3.61	6.20	9.81
3000	1,500 S.F., 5.00 ton	3.86	6	9.86
3040	3,000 S.F., 10.00 ton	3.96	5.45	9.41
3080	5,000 S.F., 16.67 ton	3.83	5.20	9.03
3120	10,000 S.F., 33.33 ton	4.99	5.10	10.09
3200	Food supermarkets, 500 S.F., 1.42 ton	4.88	6.40	11.28
3240	1,000 S.F., 2.83 ton	3.03	5.25	8.28
3280	1,500 S.F., 4.25 ton	3.28	5.10	8.38
3320	3,000 S.F., 8.50 ton	3.36	4.62	7.98
3360	5,000 S.F., 14.17 ton	3.25	4.44	7.69
3400	10,000 S.F., 28.33 ton	4.23	4.31	8.54
3480	Medical centers, 500 S.F., 1.17 ton	4.03	5.30	9.33
3520	1,000 S.F., 2.33 ton	4.03	5.30	9.33
3560	1,500 S.F., 3.50 ton	2.53	4.36	6.89
3600	3,000 S.F., 7.00 ton	2.70	4.19	6.89
3640	5,000 S.F., 11.67 ton	2.77	3.81	6.58
3680	10,000 S.F., 23.33 ton	4.23	3.58	7.81
3760	Offices, 500 S.F., 1.58 ton	5.45	7.15	12.60
3800	1,000 S.F., 3.17 ton	3.43	5.90	9.33
3840	1,500 S.F., 4.75 ton	3.67	5.70	9.37
3880	3,000 S.F., 9.50 ton	3.77	5.15	8.92
3920	5,000 S.F., 15.83 ton	3.64	4.95	8.59
3960	10,000 S.F., 31.67 ton	4.74	4.83	9.57
4000	Restaurants, 500 S.F., 2.50 ton	8.65	11.30	19.95
4040	1,000 S.F., 5.00 ton	5.80	9	14.80
4080	1,500 S.F., 7.50 ton	5.95	8.15	14.10
4120	3,000 S.F., 15.00 ton	5.75	7.85	13.60
4160	5,000 S.F., 25.00 ton	9.05	7.65	16.70
4200	10,000 S.F., 50.00 ton	6.45	7.60	14.05
4240	Schools and colleges, 500 S.F., 1.92 ton	6.60	8.70	15.30
4280	1,000 S.F., 3.83 ton	4.14	7.15	11.29
4320	1,500 S.F., 5.75 ton	4.44	6.90	11.34
4360	3,000 S.F., 11.50 ton	4.55	6.25	10.80
4400	5,000 S.F., 19.17 ton	6.95	5.90	12.85
4440	10,000 S.F., 38.33 ton	5.35	5.85	11.20
5000				
5100				
5200				
9000	Components of ductwork packages for above systems, per ton of cooling:			
9010	Ductwork; galvanized steel, 120 pounds			
9020	Insulation; fiberglass 2" thick, FRK faced, 52 S.F.			
9030	Diffusers; aluminum, 24" x 12", one			
9040	Registers; aluminum, return with dampers, one			

564

For customer support on your Mechanical Cost Data, call 877.739.3689.

System Description: Rooftop units are multizone with up to 12 zones, and include electric cooling, gas heat, thermostats, filters, supply and return fans complete. Duct systems are low velocity, galvanized steel supply and return with insulated supplies.

Multizone units cost more per ton of cooling than single zone. However, they offer flexibility where load conditions are varied due to heat generating areas or exposure to radiational heating. For example, perimeter offices on the "sunny side" may require cooling at the same

time "shady side" or central offices may require heating. It is possible to accomplish similar results using duct heaters in branches of the single zone unit. However, heater location could be a problem and total system operating energy efficiency could be lower.

System Components	QUANTITY	UNIT	COST EACH		
			MAT.	INST.	TOTAL
SYSTEM D3050 155 1280					
ROOFTOP, MULTIZONE, AIR CONDITIONER					
APARTMENT CORRIDORS, 3,000 S.F., 5.50 TON					
Rooftop multizone unit, standard controls, curb	1.000	Ea.	31,900	2,013	33,913
Ductwork package for rooftop multizone units	1.000	System	1,523.50	13,887.50	15,411
TOTAL			33,423.50	15,900.50	49,324
COST PER S.F.			11.14	5.30	16.44

Note A: Small single zone unit recommended.
Note B: A combination of multizone units recommended.

D3050 155	Rooftop Multizone Unit Systems		COST PER S.F.		
			MAT.	INST.	TOTAL
1240	Rooftop, multizone, air conditioner				
1260	Apartment corridors, 1,500 S.F., 2.75 ton. See Note A.				
1280	3,000 S.F., 5.50 ton	R236000 -20	11.13	5.30	16.43
1320	10,000 S.F., 18.30 ton		7.70	5.10	12.80
1360	15,000 S.F., 27.50 ton		7.50	5.10	12.60
1400	20,000 S.F., 36.70 ton		7.85	5.10	12.95
1440	25,000 S.F., 45.80 ton		7.30	5.10	12.40
1520	Banks or libraries, 1,500 S.F., 6.25 ton		25.50	12.05	37.55
1560	3,000 S.F., 12.50 ton		21.50	11.80	33.30
1600	10,000 S.F., 41.67 ton		16.65	11.65	28.30
1640	15,000 S.F., 62.50 ton		12.15	11.55	23.70
1680	20,000 S.F., 83.33 ton		12.15	11.55	23.70
1720	25,000 S.F., 104.00 ton		10.80	11.50	22.30
1800	Bars and taverns, 1,500 S.F., 16.62 ton		55	17.35	72.35
1840	3,000 S.F., 33.24 ton		46	16.75	62.75
1880	10,000 S.F., 110.83 ton		27.50	16.65	44.15
1920	15,000 S.F., 165 ton, See Note B.				
1960	20,000 S.F., 220 ton, See Note B.				
2000	25,000 S.F., 275 ton, See Note B.				
2080	Bowling alleys, 1,500 S.F., 8.50 ton		34.50	16.40	50.90
2120	3,000 S.F., 17.00 ton		29	16.05	45.05
2160	10,000 S.F., 56.70 ton		22.50	15.85	38.35
2200	15,000 S.F., 85.00 ton		16.50	15.70	32.20
2240	20,000 S.F., 113.00 ton		14.70	15.65	30.35
2280	25,000 S.F., 140.00 ton see Note B.				
2360	Department stores, 1,500 S.F., 4.37 ton, See Note A.				

565

For customer support on your Mechanical Cost Data, call 877.739.3689.

D3050 Terminal & Package Units

D3050 155	Rooftop Multizone Unit Systems	COST PER S.F.		
		MAT.	INST.	TOTAL
2400	3,000 S.F., 8.75 ton	17.75	8.45	26.20
2440	10,000 S.F., 29.17 ton	12.45	8.10	20.55
2480	15,000 S.F., 43.75 ton	11.65	8.15	19.80
2520	20,000 S.F., 58.33 ton	8.50	8.10	16.60
2560	25,000 S.F., 72.92 ton	8.50	8.10	16.60
2640	Drug stores, 1,500 S.F., 10.00 ton	40.50	19.25	59.75
2680	3,000 S.F., 20.00 ton	28	18.55	46.55
2720	10,000 S.F., 66.66 ton	19.45	18.50	37.95
2760	15,000 S.F., 100.00 ton	17.35	18.45	35.80
2800	20,000 S.F., 135 ton, See Note B.			
2840	25,000 S.F., 165 ton, See Note B.			
2920	Factories, 1,500 S.F., 5 ton, See Note A.			
2960	3,000 S.F., 10.00 ton	20.50	9.65	30.15
3000	10,000 S.F., 33.33 ton	14.25	9.25	23.50
3040	15,000 S.F., 50.00 ton	13.30	9.30	22.60
3080	20,000 S.F., 66.66 ton	9.75	9.25	19
3120	25,000 S.F., 83.33 ton	9.70	9.25	18.95
3200	Food supermarkets, 1,500 S.F., 4.25 ton, See Note A.			
3240	3,000 S.F., 8.50 ton	17.20	8.20	25.40
3280	10,000 S.F., 28.33 ton	11.60	7.85	19.45
3320	15,000 S.F., 42.50 ton	11.30	7.90	19.20
3360	20,000 S.F., 56.67 ton	8.25	7.85	16.10
3400	25,000 S.F., 70.83 ton	8.25	7.85	16.10
3480	Medical centers, 1,500 S.F., 3.5 ton, See Note A.			
3520	3,000 S.F., 7.00 ton	14.20	6.75	20.95
3560	10,000 S.F., 23.33 ton	9.40	6.50	15.90
3600	15,000 S.F., 35.00 ton	10	6.45	16.45
3640	20,000 S.F., 46.66 ton	9.30	6.50	15.80
3680	25,000 S.F., 58.33 ton	6.80	6.45	13.25
3760	Offices, 1,500 S.F., 4.75 ton, See Note A.			
3800	3,000 S.F., 9.50 ton	19.25	9.15	28.40
3840	10,000 S.F., 31.66 ton	13.55	8.80	22.35
3880	15,000 S.F., 47.50 ton	12.65	8.85	21.50
3920	20,000 S.F., 63.33 ton	9.25	8.80	18.05
3960	25,000 S.F., 79.16 ton	9.25	8.80	18.05
4000	Restaurants, 1,500 S.F., 7.50 ton	30.50	14.45	44.95
4040	3,000 S.F., 15.00 ton	25.50	14.15	39.65
4080	10,000 S.F., 50.00 ton	19.95	13.95	33.90
4120	15,000 S.F., 75.00 ton	14.60	13.85	28.45
4160	20,000 S.F., 100.00 ton	13	13.85	26.85
4200	25,000 S.F., 125 ton, See Note B.			
4240	Schools and colleges, 1,500 S.F., 5.75 ton	23.50	11.10	34.60
4280	3,000 S.F., 11.50 ton	19.55	10.85	30.40
4320	10,000 S.F., 38.33 ton	15.30	10.70	26
4360	15,000 S.F., 57.50 ton	11.20	10.65	21.85
4400	20,000 S.F., 76.66 ton	11.15	10.65	21.80
4440	25,000 S.F., 95.83 ton	10.50	10.65	21.15
4450				
9000	Components of ductwork packages for above systems, per ton of cooling:			
9010	Ductwork; galvanized steel, 240 pounds			
9020	Insulation; fiberglass, 2" thick, FRK faced, 104 S.F.			
9030	Diffusers; aluminum, 24" x 12", two			
9040	Registers; aluminum, return with dampers, one			

D3050   Terminal & Package Units

System Description: Self-contained, single package water cooled units include cooling tower, pump, piping allowance. Systems for 1000 S.F. and up include duct and diffusers to provide for even distribution of air. Smaller units distribute

air through a supply air plenum, which is integral with the unit.

Returns are not ducted and supplies are not insulated.

Hot water or steam heating coils are included but piping to boiler and the boiler itself is not included.

Where local codes or conditions permit single pass cooling for the smaller units, deduct 10%.

System Components	QUANTITY	UNIT	COST EACH MAT.	COST EACH INST.	COST EACH TOTAL
SYSTEM D3050 160 1300					
SELF-CONTAINED, WATER COOLED UNIT					
APARTMENT CORRIDORS, 500 S.F., .92 TON					
Self contained, water cooled, single package air conditioner unit	1.000	Ea.	1,318.80	688.80	2,007.60
Ductwork package for water or air cooled packaged units	1.000	System	61.18	920	981.18
Cooling tower, draw thru single flow, belt drive	1.000	Ea.	218.04	21.16	239.20
Cooling tower pumps & piping	1.000	System	108.56	49.68	158.24
TOTAL			1,706.58	1,679.64	3,386.22
COST PER S.F.			3.41	3.36	6.77

D3050 160	Self-contained, Water Cooled Unit Systems	COST PER S.F. MAT.	COST PER S.F. INST.	COST PER S.F. TOTAL
1280	Self-contained, water cooled unit	3.42	3.36	6.78
1300	Apartment corridors, 500 S.F., .92 ton	3.40	3.35	6.75
1320	1,000 S.F., 1.83 ton	3.40	3.35	6.75
1360	3,000 S.F., 5.50 ton	2.83	2.80	5.63
1400	5,000 S.F., 9.17 ton	2.73	2.60	5.33
1440	10,000 S.F., 18.33 ton	3.94	2.32	6.26
1520	Banks or libraries, 500 S.F., 2.08 ton	7.45	3.44	10.89
1560	1,000 S.F., 4.17 ton	6.45	6.35	12.80
1600	3,000 S.F., 12.50 ton	6.20	5.90	12.10
1640	5,000 S.F., 20.80 ton	8.95	5.30	14.25
1680	10,000 S.F., 41.66 ton	7.80	5.30	13.10
1760	Bars and taverns, 500 S.F., 5.54 ton	16.30	5.80	22.10
1800	1,000 S.F., 11.08 ton	16.05	10.15	26.20
1840	3,000 S.F., 33.25 ton	21	8	29
1880	5,000 S.F., 55.42 ton	19.85	8.35	28.20
1920	10,000 S.F., 110.00 ton	19.30	8.20	27.50
2000	Bowling alleys, 500 S.F., 2.83 ton	10.15	4.70	14.85
2040	1,000 S.F., 5.66 ton	8.75	8.65	17.40
2080	3,000 S.F., 17.00 ton	12.20	7.20	19.40
2120	5,000 S.F., 28.33 ton	11.05	6.95	18
2160	10,000 S.F., 56.66 ton	10.35	7.10	17.45
2200	Department stores, 500 S.F., 1.46 ton	5.25	2.42	7.67
2240	1,000 S.F., 2.92 ton	4.49	4.47	8.96
2280	3,000 S.F., 8.75 ton	4.33	4.15	8.48
2320	5,000 S.F., 14.58 ton	6.25	3.71	9.96
2360	10,000 S.F., 29.17 ton	5.70	3.57	9.27

Note (at row 1360): R236000 -20

567

For customer support on your Mechanical Cost Data, call 877.739.3689.

D3050 Terminal & Package Units

D3050 160	Self-contained, Water Cooled Unit Systems	COST PER S.F.		
		MAT.	INST.	TOTAL
2440	Drug stores, 500 S.F., 3.33 ton	11.95	5.50	17.45
2480	1,000 S.F., 6.66 ton	10.30	10.20	20.50
2520	3,000 S.F., 20.00 ton	14.30	8.45	22.75
2560	5,000 S.F., 33.33 ton	15.05	8.35	23.40
2600	10,000 S.F., 66.66 ton	12.15	8.35	20.50
2680	Factories, 500 S.F., 1.66 ton	5.95	2.76	8.71
2720	1,000 S.F. 3.37 ton	5.20	5.15	10.35
2760	3,000 S.F., 10.00 ton	4.96	4.73	9.69
2800	5,000 S.F., 16.66 ton	7.15	4.23	11.38
2840	10,000 S.F., 33.33 ton	6.50	4.08	10.58
2920	Food supermarkets, 500 S.F., 1.42 ton	5.05	2.35	7.40
2960	1,000 S.F., 2.83 ton	5.25	5.20	10.45
3000	3,000 S.F., 8.50 ton	4.37	4.33	8.70
3040	5,000 S.F., 14.17 ton	4.21	4.02	8.23
3080	10,000 S.F., 28.33 ton	5.50	3.46	8.96
3160	Medical centers, 500 S.F., 1.17 ton	4.18	1.93	6.11
3200	1,000 S.F., 2.33 ton	4.35	4.26	8.61
3240	3,000 S.F., 7.00 ton	3.61	3.56	7.17
3280	5,000 S.F., 11.66 ton	3.48	3.31	6.79
3320	10,000 S.F., 23.33 ton	5	2.96	7.96
3400	Offices, 500 S.F., 1.58 ton	5.65	2.62	8.27
3440	1,000 S.F., 3.17 ton	5.90	5.80	11.70
3480	3,000 S.F., 9.50 ton	4.71	4.49	9.20
3520	5,000 S.F., 15.83 ton	6.80	4.02	10.82
3560	10,000 S.F., 31.67 ton	6.15	3.87	10.02
3640	Restaurants, 500 S.F., 2.50 ton	8.95	4.15	13.10
3680	1,000 S.F., 5.00 ton	7.70	7.65	15.35
3720	3,000 S.F., 15.00 ton	7.45	7.10	14.55
3760	5,000 S.F., 25.00 ton	10.75	6.35	17.10
3800	10,000 S.F., 50.00 ton	6.70	5.85	12.55
3880	Schools and colleges, 500 S.F., 1.92 ton	6.85	3.18	10.03
3920	1,000 S.F., 3.83 ton	5.90	5.85	11.75
3960	3,000 S.F., 11.50 ton	5.70	5.45	11.15
4000	5,000 S.F., 19.17 ton	8.25	4.87	13.12
4040	10,000 S.F., 38.33 ton	7.20	4.87	12.07
4050				
9000	Components of ductwork packages for above systems, per ton of cooling:			
9010	Ductwork; galvanized steel, 108 pounds			
9020	Diffusers, aluminum, 24" x 12", two			

D3050 Terminal & Package Units

System Description: Self-contained air cooled units with remote air cooled condenser and interconnecting tubing. Systems for 1000 S.F. and up include duct and diffusers. Smaller units distribute air directly.

Returns are not ducted and supplies are not insulated.

Potential savings may be realized by using a single zone rooftop system or through-the-wall unit, especially in the smaller capacities, if the application permits.

Hot water or steam heating coils are included but piping to boiler and the boiler itself is not included.

Condenserless models are available for 15% less where remote refrigerant source is available.

System Components	QUANTITY	UNIT	COST EACH		
			MAT.	INST.	TOTAL
SYSTEM D3050 165 1320					
SELF-CONTAINED, AIR COOLED UNIT					
APARTMENT CORRIDORS, 500 S.F., .92 TON					
Air cooled, package unit	1.000	Ea.	1,385.50	450.50	1,836
Ductwork package for water or air cooled packaged units	1.000	System	61.18	920	981.18
Refrigerant piping	1.000	System	372.60	607.20	979.80
Air cooled condenser, direct drive, propeller fan	1.000	Ea.	829.25	170.50	999.75
TOTAL			2,648.53	2,148.20	4,796.73
COST PER S.F.			5.30	4.30	9.60

D3050 165	Self-contained, Air Cooled Unit Systems		COST PER S.F.		
			MAT.	INST.	TOTAL
1300	Self-contained, air cooled unit				
1320	Apartment corridors, 500 S.F., .92 ton		5.30	4.30	9.60
1360	1,000 S.F., 1.83 ton		5.20	4.27	9.47
1400	3,000 S.F., 5.50 ton	R236000 -20	4.85	3.97	8.82
1440	5,000 S.F., 9.17 ton		4	3.78	7.78
1480	10,000 S.F., 18.33 ton		3.34	3.47	6.81
1560	Banks or libraries, 500 S.F., 2.08 ton		11.50	5.50	17
1600	1,000 S.F., 4.17 ton		10.95	9	19.95
1640	3,000 S.F., 12.50 ton		9.10	8.60	17.70
1680	5,000 S.F., 20.80 ton		7.70	7.90	15.60
1720	10,000 S.F., 41.66 ton		8.50	7.70	16.20
1800	Bars and taverns, 500 S.F., 5.54 ton		26.50	12	38.50
1840	1,000 S.F., 11.08 ton		24	17.30	41.30
1880	3,000 S.F., 33.25 ton		22.50	14.95	37.45
1920	5,000 S.F., 55.42 ton		22.50	15	37.50
1960	10,000 S.F., 110.00 ton		23	14.90	37.90
2040	Bowling alleys, 500 S.F., 2.83 ton		15.80	7.55	23.35
2080	1,000 S.F., 5.66 ton		14.95	12.30	27.25
2120	3,000 S.F., 17.00 ton		10.40	10.75	21.15
2160	5,000 S.F., 28.33 ton		11.75	10.50	22.25
2200	10,000 S.F., 56.66 ton		11.80	10.45	22.25
2240	Department stores, 500 S.F., 1.46 ton		8.15	3.87	12.02
2280	1,000 S.F., 2.92 ton		7.70	6.30	14
2320	3,000 S.F., 8.75 ton		6.35	6	12.35

D3050 Terminal & Package Units

D3050 165	Self-contained, Air Cooled Unit Systems	COST PER S.F.		
		MAT.	INST.	TOTAL
2360	5,000 S.F., 14.58 ton	6.35	6	12.35
2400	10,000 S.F., 29.17 ton	6.05	5.40	11.45
2480	Drug stores, 500 S.F., 3.33 ton	18.60	8.85	27.45
2520	1,000 S.F., 6.66 ton	17.60	14.45	32.05
2560	3,000 S.F., 20.00 ton	12.40	12.65	25.05
2600	5,000 S.F., 33.33 ton	13.85	12.35	26.20
2640	10,000 S.F., 66.66 ton	14.05	12.35	26.40
2720	Factories, 500 S.F., 1.66 ton	9.45	4.46	13.91
2760	1,000 S.F., 3.33 ton	8.85	7.25	16.10
2800	3,000 S.F., 10.00 ton	7.30	6.85	14.15
2840	5,000 S.F., 16.66 ton	6.05	6.30	12.35
2880	10,000 S.F., 33.33 ton	6.90	6.15	13.05
2960	Food supermarkets, 500 S.F., 1.42 ton	7.90	3.77	11.67
3000	1,000 S.F., 2.83 ton	8.05	6.60	14.65
3040	3,000 S.F., 8.50 ton	7.50	6.15	13.65
3080	5,000 S.F., 14.17 ton	6.20	5.85	12.05
3120	10,000 S.F., 28.33 ton	5.90	5.25	11.15
3200	Medical centers, 500 S.F., 1.17 ton	6.55	3.11	9.66
3240	1,000 S.F., 2.33 ton	6.70	5.45	12.15
3280	3,000 S.F., 7.00 ton	6.20	5.05	11.25
3320	5,000 S.F., 16.66 ton	5.10	4.81	9.91
3360	10,000 S.F., 23.33 ton	4.27	4.42	8.69
3440	Offices, 500 S.F., 1.58 ton	8.85	4.21	13.06
3480	1,000 S.F., 3.16 ton	9.10	7.40	16.50
3520	3,000 S.F., 9.50 ton	6.95	6.55	13.50
3560	5,000 S.F., 15.83 ton	5.85	6	11.85
3600	10,000 S.F., 31.66 ton	6.60	5.85	12.45
3680	Restaurants, 500 S.F., 2.50 ton	13.90	6.60	20.50
3720	1,000 S.F., 5.00 ton	13.25	10.85	24.10
3760	3,000 S.F., 15.00 ton	10.95	10.30	21.25
3800	5,000 S.F., 25.00 ton	10.35	9.25	19.60
3840	10,000 S.F., 50.00 ton	10.75	9.20	19.95
3920	Schools and colleges, 500 S.F., 1.92 ton	10.70	5.10	15.80
3960	1,000 S.F., 3.83 ton	10.15	8.30	18.45
4000	3,000 S.F., 11.50 ton	8.35	7.90	16.25
4040	5,000 S.F., 19.17 ton	7	7.25	14.25
4080	10,000 S.F., 38.33 ton	7.80	7.10	14.90
4090				
9000	Components of ductwork packages for above systems, per ton of cooling:			
9010	Ductwork; galvanized steel, 102 pounds			
9020	Diffusers; aluminum, 24" x 12", two			

D3050   Terminal & Package Units

Air Cooled Condensing Unit
Refrigerant Piping
Roof
Supply Duct
Fin. Ceiling
Return Grille
DX Air Handling Unit
Supply Diffuser

General: Split systems offer several important advantages which should be evaluated when a selection is to be made. They provide a greater degree of flexibility in component selection which permits an accurate match-up of the proper equipment size and type with the particular needs of the building. This allows for maximum use of modern

energy saving concepts in heating and cooling. Outdoor installation of the air cooled condensing unit allows space savings in the building and also isolates the equipment operating sounds from building occupants.

Design Assumptions: The systems below are comprised of a direct expansion air handling unit and air cooled condensing

unit with interconnecting copper tubing. Ducts and diffusers are also included for distribution of air. Systems are priced for cooling only. Heat can be added as desired either by putting hot water/steam coils into the air unit or into the duct supplying the particular area of need. Gas fired duct furnaces are also available. Refrigerant liquid line is insulated.

System Components			COST EACH		
	QUANTITY	UNIT	MAT.	INST.	TOTAL
SYSTEM D3050 170 1280					
SPLIT SYSTEM, AIR COOLED CONDENSING UNIT					
APARTMENT CORRIDORS, 1,000 S.F., 1.80 TON					
Fan coil AC unit, cabinet mntd & filters direct expansion air cool	1.000	Ea.	411.75	161.04	572.79
Ductwork package, for split system, remote condensing unit	1.000	System	57.19	873.83	931.02
Refrigeration piping	1.000	System	345.87	1,006.50	1,352.37
Condensing unit, air cooled, incls compressor & standard controls	1.000	Ea.	1,708	646.60	2,354.60
TOTAL			2,522.81	2,687.97	5,210.78
COST PER S.F.			2.52	2.69	5.21

*Cooling requirements would lead to choosing a water cooled unit.

D3050 170	Split Systems With Air Cooled Condensing Units		COST PER S.F.		
			MAT.	INST.	TOTAL
1260	Split system, air cooled condensing unit				
1280	Apartment corridors, 1,000 S.F., 1.83 ton		2.53	2.70	5.23
1320	2,000 S.F., 3.66 ton	R236000	1.79	2.71	4.50
1360	5,000 S.F., 9.17 ton	-20	1.63	3.39	5.02
1400	10,000 S.F., 18.33 ton		2.03	3.70	5.73
1440	20,000 S.F., 36.66 ton		2.15	3.76	5.91
1520	Banks and libraries, 1,000 S.F., 4.17 ton		4.07	6.20	10.27
1560	2,000 S.F., 8.33 ton		3.70	7.70	11.40
1600	5,000 S.F., 20.80 ton		4.63	8.40	13.03
1640	10,000 S.F., 41.66 ton		4.92	8.55	13.47
1680	20,000 S.F., 83.32 ton		5.95	8.90	14.85
1760	Bars and taverns, 1,000 S.F., 11.08 ton		8.45	12.15	20.60
1800	2,000 S.F., 22.16 ton		12.60	14.55	27.15
1840	5,000 S.F., 55.42 ton		10.80	13.60	24.40
1880	10,000 S.F., 110.84 ton		13.10	14.20	27.30
1920	20,000 S.F., 220 ton*				
2000	Bowling alleys, 1,000 S.F., 5.66 ton		5.60	11.50	17.10
2040	2,000 S.F., 11.33 ton		5	10.50	15.50
2080	5,000 S.F., 28.33 ton		6.30	11.45	17.75
2120	10,000 S.F., 56.66 ton		6.70	11.65	18.35
2160	20,000 S.F., 113.32 ton		8.75	12.60	21.35
2320	Department stores, 1,000 S.F., 2.92 ton		2.88	4.25	7.13
2360	2,000 S.F., 5.83 ton		2.89	5.95	8.84
2400	5,000 S.F., 14.58 ton		2.58	5.40	7.98

571

D3050 Terminal & Package Units

D3050 170	Split Systems With Air Cooled Condensing Units	COST PER S.F.		
		MAT.	INST.	TOTAL
2440	10,000 S.F., 29.17 ton	3.24	5.90	9.14
2480	20,000 S.F., 58.33 ton	3.44	6	9.44
2560	Drug stores, 1,000 S.F., 6.66 ton	6.60	13.55	20.15
2600	2,000 S.F., 13.32 ton	5.90	12.30	18.20
2640	5,000 S.F., 33.33 ton	7.85	13.70	21.55
2680	10,000 S.F., 66.66 ton	8.10	14.30	22.40
2720	20,000 S.F., 133.32 ton*			
2800	Factories, 1,000 S.F., 3.33 ton	3.28	4.85	8.13
2840	2,000 S.F., 6.66 ton	3.31	6.75	10.06
2880	5,000 S.F., 16.66 ton	3.70	6.70	10.40
2920	10,000 S.F., 33.33 ton	3.93	6.85	10.78
2960	20,000 S.F., 66.66 ton	4.05	7.15	11.20
3040	Food supermarkets, 1,000 S.F., 2.83 ton	2.80	4.13	6.93
3080	2,000 S.F., 5.66 ton	2.82	5.75	8.57
3120	5,000 S.F., 14.66 ton	2.51	5.25	7.76
3160	10,000 S.F., 28.33 ton	3.15	5.70	8.85
3200	20,000 S.F., 56.66 ton	3.34	5.80	9.14
3280	Medical centers, 1,000 S.F., 2.33 ton	2.51	3.32	5.83
3320	2,000 S.F., 4.66 ton	2.31	4.75	7.06
3360	5,000 S.F., 11.66 ton	2.07	4.31	6.38
3400	10,000 S.F., 23.33 ton	2.60	4.71	7.31
3440	20,000 S.F., 46.66 ton	2.76	4.79	7.55
3520	Offices, 1,000 S.F., 3.17 ton	3.12	4.61	7.73
3560	2,000 S.F., 6.33 ton	3.14	6.45	9.59
3600	5,000 S.F., 15.83 ton	2.81	5.85	8.66
3640	10,000 S.F., 31.66 ton	3.53	6.40	9.93
3680	20,000 S.F., 63.32 ton	3.84	6.80	10.64
3760	Restaurants, 1,000 S.F., 5.00 ton	4.97	10.15	15.12
3800	2,000 S.F., 10.00 ton	4.44	9.25	13.69
3840	5,000 S.F., 25.00 ton	5.55	10.10	15.65
3880	10,000 S.F., 50.00 ton	5.90	10.30	16.20
3920	20,000 S.F., 100.00 ton	7.75	11.15	18.90
4000	Schools and colleges, 1,000 S.F., 3.83 ton	3.74	5.70	9.44
4040	2,000 S.F., 7.66 ton	3.39	7.10	10.49
4080	5,000 S.F., 19.17 ton	4.26	7.75	12.01
4120	10,000 S.F., 38.33 ton	4.52	7.85	12.37
4160	20,000 S.F., 76.66 ton	4.65	8.20	12.85
5000				
5100				
5200				
9000	Components of ductwork packages for above systems, per ton of cooling:			
9010	Ductwork; galvanized steel, 102 pounds			
9020	Diffusers; aluminum, 24" x 12", two			

D3050 Terminal & Package Units

Rooftop Air Conditioner Unit

System Components			COST EACH		
	QUANTITY	UNIT	MAT.	INST.	TOTAL
SYSTEM D3050 175 1010					
A/C, ROOFTOP, DX COOL, GAS HEAT, CURB, ECONOMIZER, FILTERS., 5 TON					
Roof top A/C, curb, economizer, sgl zone, elec cool, gas ht, 5 ton, 112 MBH	1.000	Ea.	4,850	2,350	7,200
Pipe, black steel, Sch 40, threaded, w/cplgs & hangers, 10' OC, 1" dia	20.000	L.F.	99.33	283.50	382.83
Elbow, 90°, black, straight, 3/4" dia.	3.000	Ea.	13.77	153	166.77
Elbow, 90°, black, straight, 1" dia.	3.000	Ea.	23.85	165	188.85
Tee, black, reducing, 1" dia.	1.000	Ea.	20	89.50	109.50
Union, black with brass seat, 1" dia.	1.000	Ea.	26.50	59.50	86
Pipe nipple, black, 3/4" dia	2.000	Ea.	7.36	23.40	30.76
Pipe nipple, black, 1" dia	2.000	Ea.	4.73	13.50	18.23
Cap, black, 1" dia.	1.000	Ea.	6.30	24	30.30
Gas cock, brass, 1" size	1.000	Ea.	34	37.50	71.50
Control system, pneumatic, Rooftop A/C unit	1.000	Ea.	3,075	2,575	5,650
Rooftop unit heat/cool balancing	1.000	Ea.		490	490
TOTAL			8,160.84	6,263.90	14,424.74

D3050 175	Rooftop Air Conditioner, Const. Volume	COST EACH		
		MAT.	INST.	TOTAL
1010	A/C, Rooftop, DX cool, gas heat, curb, economizer, fltrs, 5 Ton	8,160.84	6,263.90	14,425.80
1020	7-1/2 Ton	9,775	6,525	16,300
1030	12-1/2 Ton	14,700	7,175	21,875
1040	18 Ton	18,700	7,850	26,550
1050	25 Ton	38,600	8,950	47,550
1060	40 Ton	51,500	11,900	63,400

D3050 180	Rooftop Air Conditioner, Variable Air Volume	COST EACH		
		MAT.	INST.	TOTAL
1010	A/C, Rooftop, DX cool, gas heat, curb, ecmizr, fltrs, VAV, 12-1/2 Ton	17,400	7,650	25,050
1020	18 Ton	22,200	8,425	30,625
1030	25 Ton	43,400	9,700	53,100
1040	40 Ton	59,500	13,200	72,700
1050	60 Ton	84,500	17,900	102,400
1060	80 Ton	106,000	22,400	128,400

573

For customer support on your Mechanical Cost Data, call 877.739.3689.

D3050 Terminal & Package Units

Computer rooms impose special requirements on air conditioning systems. A prime requirement is reliability, due to the potential monetary loss that could be incurred by a system failure. A second basic requirement is the tolerance of control with which temperature and humidity are regulated, and dust eliminated. As the air conditioning system reliability is so vital, the additional cost of reserve capacity and redundant components is often justified.

System Descriptions: Computer areas may be environmentally controlled by one of three methods as follows:

1. Self-contained Units
These are units built to higher standards of performance and reliability.

They usually contain alarms and controls to indicate component operation failure, filter change, etc. It should be remembered that these units in the room will occupy space that is relatively expensive to build and that all alterations and service of the equipment will also have to be accomplished within the computer area.

2. Decentralized Air Handling Units In operation these are similar to the self-contained units except that their cooling capability comes from remotely located refrigeration equipment as refrigerant or chilled water. As no compressors or refrigerating equipment are required in the air units, they are smaller and require less service than

self-contained units. An added plus for this type of system occurs if some of the computer components themselves also require chilled water for cooling.

3. Central System Supply Cooling is obtained from a central source which, since it is not located within the computer room, may have excess capacity and permit greater flexibility without interfering with the computer components. System performance criteria must still be met.

Note: The costs shown below do not include an allowance for ductwork or piping.

D3050 185	Computer Room Cooling Units	COST EACH		
		MAT.	INST.	TOTAL
0560	Computer room unit, air cooled, includes remote condenser			
0580	3 ton	21,500	2,650	24,150
0600	5 ton	22,900	2,925	25,825
0620	8 ton	42,800	4,875	47,675
0640	10 ton	44,800	5,275	50,075
0660	15 ton	49,300	6,000	55,300
0680	20 ton	59,500	8,550	68,050
0700	23 ton	73,500	9,775	83,275
0800	Chilled water, for connection to existing chiller system			
0820	5 ton	16,400	1,775	18,175
0840	8 ton	16,700	2,650	19,350
0860	10 ton	16,900	2,700	19,600
0880	15 ton	17,600	2,750	20,350
0900	20 ton	18,700	2,875	21,575
0920	23 ton	19,900	3,250	23,150
1000	Glycol system, complete except for interconnecting tubing			
1020	3 ton	26,900	3,300	30,200
1040	5 ton	29,500	3,475	32,975
1060	8 ton	47,300	5,725	53,025
1080	10 ton	50,500	6,275	56,775
1100	15 ton	62,500	7,900	70,400
1120	20 ton	68,500	8,550	77,050
1140	23 ton	72,000	8,925	80,925
1240	Water cooled, not including condenser water supply or cooling tower			
1260	3 ton	21,800	2,125	23,925
1280	5 ton	23,800	2,450	26,250
1300	8 ton	39,000	4,000	43,000
1320	15 ton	47,200	4,875	52,075
1340	20 ton	50,500	5,400	55,900
1360	23 ton	53,500	5,850	59,350

D3050 201	Packaged A/C, Elec. Ht., Const. Volume	COST EACH		
		MAT.	INST.	TOTAL
1010	A/C Packaged, DX, air cooled, electric heat, constant vol., 5 Ton	9,600	6,050	15,650
1020	10 Ton	15,700	8,200	23,900
1030	20 Ton	24,200	10,900	35,100
1040	30 Ton	43,600	12,100	55,700

D3050 Terminal & Package Units

D3050 201	Packaged A/C, Elec. Ht., Const. Volume	COST EACH		
		MAT.	INST.	TOTAL
1050	40 Ton	57,000	13,600	70,600
1060	50 Ton	75,000	15,100	90,100

D3050 202	Packaged Air Conditioner, Elect. Heat, VAV	COST EACH		
		MAT.	INST.	TOTAL
1010	A/C Packaged, DX, air cooled, electric heat, VAV, 10 Ton	17,600	7,675	25,275
1020	20 Ton	28,100	10,100	38,200
1030	30 Ton	49,100	11,000	60,100
1040	40 Ton	63,500	12,600	76,100
1050	50 Ton	85,000	14,100	99,100
1060	60 Ton	97,000	16,500	113,500

D3050 203	Packaged A/C, Hot Wtr. Heat, Const. Volume	COST EACH		
		MAT.	INST.	TOTAL
1010	A/C Packaged, DX, air cooled, H/W heat, constant vol., 5 Ton	8,500	8,175	16,675
1020	10 Ton	12,900	10,700	23,600
1030	20 Ton	20,000	13,600	33,600
1040	30 Ton	36,700	17,500	54,200
1050	40 Ton	50,500	19,500	70,000
1060	50 Ton	64,500	21,100	85,600

D3050 204	Packaged A/C, Hot Wtr. Heat, VAV	COST EACH		
		MAT.	INST.	TOTAL
1010	A/C Packaged, DX, air cooled, H/W heat, VAV, 10 Ton	14,800	9,975	24,775
1020	20 Ton	24,300	13,100	37,400
1030	30 Ton	42,100	16,300	58,400
1040	40 Ton	56,500	17,500	74,000
1050	50 Ton	74,500	20,400	94,900
1060	60 Ton	89,500	22,900	112,400

For customer support on your Mechanical Cost Data, call 877.739.3689.

575

D30 HVAC

D3050 Terminal & Package Units

Condenser Supply Water Piping   Condenser Return Water Piping

Self–Contained Air Conditioner Unit, Water Cooled

System Components	QUANTITY	UNIT	COST EACH		
			MAT.	INST.	TOTAL
SYSTEM D3050 210 1010					
A/C, SELF CONTAINED, SINGLE PKG., WATER COOLED, ELECT. HEAT, 5 TON					
Self-contained, water cooled, elect. heat, not inc tower, 5 ton, const vol	1.000	Ea.	8,975	1,700	10,675
Pipe, black steel, Sch 40, threaded, w/cplg & hangers, 10' OC, 1-1/4" dia.	20.000	L.F.	122.55	310.68	433.23
Elbow, 90°, black, straight, 1-1/4" dia.	6.000	Ea.	78.90	351	429.90
Tee, black, straight, 1-1/4" dia.	2.000	Ea.	40	184	224
Tee, black, reducing, 1-1/4" dia.	2.000	Ea.	71	184	255
Thermometers, stem type, 9" case, 8" stem, 3/4" NPT	2.000	Ea.	478	52	530
Union, black with brass seat, 1-1/4" dia.	2.000	Ea.	76	123	199
Pipe nipple, black, 1-1/4" dia	3.000	Ea.	8.55	21.68	30.23
Valves, bronze, gate, N.R.S., threaded, class 150, 1-1/4" size	2.000	Ea.	320	95	415
Circuit setter, bal valve, bronze body, threaded, 1-1/4" pipe size	1.000	Ea.	171	49	220
Insulation, fiberglass pipe covering, 1" wall, 1-1/4" IPS	20.000	L.F.	23.20	114	137.20
Control system, pneumatic, A/C Unit with heat	1.000	Ea.	3,075	2,575	5,650
Re-heat coil balancing	1.000	Ea.		136	136
Rooftop unit heat/cool balancing	1.000	Ea.		490	490
TOTAL			13,439.20	6,385.36	19,824.56

D3050 210	AC Unit, Package, Elec. Ht., Water Cooled	COST EACH		
		MAT.	INST.	TOTAL
1010	A/C, Self contained, single pkg., water cooled, elect. heat, 5 Ton	13,439.20	6,385.35	19,847
1020	10 Ton	20,700	7,875	28,575
1030	20 Ton	27,800	9,975	37,775
1040	30 Ton	47,400	10,900	58,300
1050	40 Ton	60,500	12,100	72,600
1060	50 Ton	79,000	14,400	93,400

D3050 215	AC Unit, Package, Elec. Ht., Water Cooled, VAV	COST EACH		
		MAT.	INST.	TOTAL
1010	A/C, Self contained, single pkg., water cooled, elect. ht, VAV, 10 Ton	23,200	7,875	31,075
1020	20 Ton	32,600	9,975	42,575
1030	30 Ton	54,000	10,900	64,900
1040	40 Ton	68,000	12,100	80,100
1050	50 Ton	90,000	14,400	104,400
1060	60 Ton	103,500	17,400	120,900

D3050 Terminal & Package Units

D3050 220	AC Unit, Package, Hot Water Coil, Water Cooled	COST EACH		
		MAT.	INST.	TOTAL
1010	A/C, Self contn'd, single pkg., water cool, H/W ht, const. vol, 5 Ton	12,000	8,850	20,850
1020	10 Ton	17,300	10,600	27,900
1030	20 Ton	43,100	13,200	56,300
1040	30 Ton	56,000	16,600	72,600
1050	40 Ton	69,000	20,400	89,400
1060	50 Ton	84,500	24,900	109,400

D3050 225	AC Unit, Package, HW Coil, Water Cooled, VAV	COST EACH		
		MAT.	INST.	TOTAL
1010	A/C, Self contn'd, single pkg., water cool, H/W ht, VAV, 10 Ton	19,800	9,950	29,750
1020	20 Ton	28,900	12,800	41,700
1030	30 Ton	47,200	15,900	63,100
1040	40 Ton	61,500	17,500	79,000
1050	50 Ton	80,500	20,500	101,000
1060	60 Ton	95,500	22,300	117,800

577

For customer support on your Mechanical Cost Data, call 877.739.3689.

D3050   Terminal & Package Units

Note: Water source units do not include an allowance for any type of water source or piping to it.

Heat Pump, Air/Air, Rooftop w/Economizer

System Components	QUANTITY	UNIT	COST EACH		
			MAT.	INST.	TOTAL
SYSTEM D3050 245 1010					
HEAT PUMP, ROOFTOP, AIR/AIR, CURB., ECONOMIZER., SUP. ELECT. HEAT, 10 TON					
Heat pump, air to air, rooftop, curb, economizer, elect ht, 10 ton cooling	1.000	Ea.	67,500	5,125	72,625
Ductwork, 32" x 12" fabricated, galvanized steel, 16 LF	176.000	Lb.	110.88	1,469.60	1,580.48
Insulation, ductwork, blanket type, fiberglass, 1" thk, 1-1/2 LB density	117.000	S.F.	583.83	1,667.25	2,251.08
Rooftop unit heat/cool balancing	1.000	Ea.		490	490
TOTAL			68,194.71	8,751.85	76,946.56

D3050 230	Heat Pump, Water Source, Central Station	COST EACH		
		MAT.	INST.	TOTAL
1010	Heat pump, central station, water source, constant vol., 5 Ton	5,000	3,575	8,575
1020	10 Ton	10,500	5,000	15,500
1030	20 Ton	19,000	9,300	28,300
1040	30 Ton	27,600	13,600	41,200
1050	40 Ton	37,900	14,800	52,700
1060	50 Ton	42,900	19,900	62,800

D3050 235	Heat Pump, Water Source, Console	COST EACH		
		MAT.	INST.	TOTAL
1020	Heat pump, console, water source, 1 Ton	2,550	1,950	4,500
1030	1-1/2 Ton	2,575	2,025	4,600
1040	2 Ton	3,825	2,400	6,225
1050	3 Ton	4,100	2,550	6,650
1060	3-1/2 Ton	4,350	2,650	7,000

D3050 240	Heat Pump, Horizontal, Duct Mounted	COST EACH		
		MAT.	INST.	TOTAL
1020	Heat pump, horizontal, ducted, water source, 1 Ton	3,025	4,175	7,200
1030	1-1/2 Ton	3,250	5,225	8,475
1040	2 Ton	4,500	6,250	10,750
1050	3 Ton	5,150	8,200	13,350
1060	3-1/2 Ton	5,400	8,475	13,875

D30 HVAC

D3050 Terminal & Package Units

D3050 245	Heat Pump, Rooftop, Air/Air	COST EACH		
		MAT.	INST.	TOTAL
1010	Heat pump, rooftop, air/air, curb, economizer, sup. elect heat, 10 Ton	68,000	8,750	76,750
1020	20 Ton	97,500	14,300	111,800
1030	30 Ton	131,000	19,700	150,700
1040	40 Ton	146,000	22,800	168,800
1050	50 Ton	170,000	25,600	195,600
1060	60 Ton	179,500	28,900	208,400

D3050 Terminal & Package Units

Thru-Wall Heat Pump   Thru-Wall Air Conditioner

System Components	QUANTITY	UNIT	COST EACH		
			MAT.	INST.	TOTAL
SYSTEM D3050 255 1010 **A/C UNIT, THRU-THE-WALL, SUPPLEMENTAL ELECT. HEAT, CABINET, LOUVER, 1/2 TON** A/C unit, thru-wall, electric heat., cabinet, louver, 1/2 ton	1.000	Ea.	910	220	1,130
TOTAL			910	220	1,130

D3050 255	**Thru-Wall A/C Unit**	COST EACH		
		MAT.	INST.	TOTAL
1010	A/C Unit, thru-the-wall, supplemental elect. heat, cabinet, louver, 1/2 Ton	910	220	1,130
1020	3/4 Ton	1,350	264	1,614
1030	1 Ton	1,500	330	1,830
1040	1 1/2 Ton	2,525	550	3,075
1050	2 Ton	2,550	695	3,245

D3050 260	**Thru-Wall Heat Pump**	COST EACH		
		MAT.	INST.	TOTAL
1010	Heat pump, thru-the-wall, cabinet, louver, 1/2 Ton	2,800	165	2,965
1020	3/4 Ton	2,950	220	3,170
1030	1 Ton	3,200	330	3,530
1040	Supplemental. elect. heat, 1-1/2 Ton	3,425	850	4,275
1050	2 Ton	3,425	880	4,305

D3090 Other HVAC Systems/Equip

Roof Exhaust Fan

10" Dia. Fiberglass 90° Elbows →

10" Dia. Fiberglass Reinforced Duct

Fume Hood →

Fume Hood Exhaust

System Components	QUANTITY	UNIT	COST EACH		
			MAT.	INST.	TOTAL
SYSTEM D3090 310 1010					
FUME HOOD EXHAUST SYSTEM, 3' LONG, 1000 CFM					
Fume hood	3.000	L.F.	3,225	1,485	4,710
Ductwork, plastic, FM approved for acid fume & smoke, straight, 10" dia	60.000	L.F.	1,290	888	2,178
Ductwork, plastic, FM approved for acid fume & smoke, elbow 90°, 10" dia	3.000	Ea.	333	486	819
Fan, corrosive fume resistant plastic, 1630 CFM, 1/2 HP	1.000	Ea.	6,675	395	7,070
Fan control switch & mounting base	1.000	Ea.	147	46	193
Roof fan balancing	1.000	Ea.		280	280
TOTAL			11,670	3,580	15,250

D3090 310	Fume Hood Exhaust Systems	COST EACH		
		MAT.	INST.	TOTAL
1010	Fume hood exhaust system, 3' long, 1000 CFM	11,700	3,575	15,275
1020	4' long, 2000 CFM	13,400	4,525	17,925
1030	6' long, 3500 CFM	17,500	6,775	24,275
1040	6' long, 5000 CFM	27,000	7,950	34,950
1050	10' long, 8000 CFM	34,100	10,900	45,000

D3090   Other HVAC Systems/Equip

Cast Iron Garage Exhaust System

Dual Exhaust System

System Components	QUANTITY	UNIT	COST EACH		
			MAT.	INST.	TOTAL
SYSTEM D3090 320 1040					
GARAGE, EXHAUST, SINGLE 3″ EXHAUST OUTLET, CARS & LIGHT TRUCKS					
A Outlet top assy, for engine exhaust system with adapters and ftngs, 3″ diam	1.000	Ea.	236	44	280
F Bullnose (guide) for engine exhaust system, 3″ diam	1.000	Ea.	27.50		27.50
G Galvanized flexible tubing for engine exhaust system, 3″ diam	8.000	L.F.	90.80		90.80
H Adapter for metal tubing end of engine exhaust system, 3″ tail pipe	1.000	Ea.	48.50		48.50
J Pipe, sewer, cast iron, push-on joint, 8″ diam.	18.000	L.F.	1,080	792	1,872
Excavating utility trench, chain trencher, 8″ wide, 24″ deep	18.000	L.F.		34.38	34.38
Backfill utility trench by hand, incl. compaction, 8″ wide 24″ deep	18.000	L.F.		47.52	47.52
Stand for blower, concrete over polystyrene core, 6″ high	1.000	Ea.	6.25	22	28.25
AC&V duct spiral reducer 10″x8″	1.000	Ea.	22.50	39.50	62
AC&V duct spiral reducer 12″x10″	1.000	Ea.	26	52.50	78.50
AC&V duct spiral preformed 45° elbow, 8″ diam	1.000	Ea.	8.90	45	53.90
AC&V utility fan, belt drive, 3 phase, 2000 CFM, 1 HP	2.000	Ea.	2,750	690	3,440
Safety switch, heavy duty fused, 240V, 3 pole, 30 amp	1.000	Ea.	110	206	316
TOTAL			4,406.45	1,972.90	6,379.35

D3090 320	Garage Exhaust Systems	COST PER BAY		
		MAT.	INST.	TOTAL
1040	Garage, single 3″ exhaust outlet, cars & light trucks, one bay	4,400	1,975	6,375
1060	Additional bays up to seven bays	1,000	530	1,530
1500	4″ outlet, trucks, one bay	4,425	1,975	6,400
1520	Additional bays up to six bays	1,025	530	1,555
1600	5″ outlet, diesel trucks, one bay	4,725	1,975	6,700
1650	Additional single bays up to six	1,425	625	2,050
1700	Two adjoining bays	4,725	1,975	6,700
2000	Dual exhaust, 3″ outlets, pair of adjoining bays	5,475	2,475	7,950
2100	Additional pairs of adjoining bays	1,600	625	2,225

D3090 Other HVAC Systems/Equip

System Components			COST EACH		
	QUANTITY	UNIT	MAT.	INST.	TOTAL
SYSTEM D3090 750 1000					
SEISMIC PAD FOR 20-38 SELF-CONTAINED UNIT					
Form equipment pad	28.000	L.F.	29.96	123.48	153.44
Reinforcing in hold down pad, #3 to #7	60.000	Lb.	31.80	37.80	69.60
Concrete 3000 PSI	.866	C.Y.	102.19		102.19
Place concrete pad	.866	C.Y.		32.21	32.21
Finish concrete equipment pad	47.000	S.F.		39.48	39.48
Drill for 3/8 inch slab anchors	10.000	Ea.	.60	94.50	95.10
Exisiting slab anchors 3/8 inch	10.000	Ea.	10	42.50	52.50
Equipment anchors 3/8 inch for plate/foot	8.000	Ea.	12.96	36.32	49.28
TOTAL			187.51	406.29	593.80

D3090 750	Seismic Concrete Equipment Pads	COST PER EACH		
		MAT.	INST.	TOTAL
1000	Seismic pad for 20-38 tons Self-contained AC unit without economizer .25g	188	405	593
1010	0.5 g seismic acceleration	188	405	593
1020	0.75 g seismic acceleration	188	405	593
1030	1.0 g seismic acceleration	188	405	593
1040	1.5 g seismic acceleration	192	460	652
1050	2.0 g seismic acceleration	208	635	843
1060	2.5 g seismic acceleration	220	800	1,020
1100	Seismic pad for 40-80 tons Self contained AC unit without economizer .25g	294	590	884
1110	0.5 g seismic acceleration	294	590	884
1120	0.75 g seismic acceleration	294	590	884
1130	1.0 g seismic acceleration	294	590	884
1140	1.5 g seismic acceleration	325	840	1,165
1150	2.0 g seismic acceleration	385	1,300	1,685
1160	2.5 g seismic acceleration	455	1,850	2,305

D3090 Other HVAC Systems/Equip

D3090 750	Seismic Concrete Equipment Pads	COST PER EACH		
		MAT.	INST.	TOTAL
1200	Seismic pad for 80-110 tons Self contained AC unit without economizer .25g	325	635	960
1210	0.5 g seismic acceleration	325	635	960
1220	0.75 g seismic acceleration	325	635	960
1230	1.0 g seismic acceleration	325	635	960
1240	1.5 g seismic acceleration	355	900	1,255
1250	2.0 g seismic acceleration	420	1,400	1,820
1260	2.5 g seismic acceleration	495	2,000	2,495
1990	Seismic pad for self-contained AC units for isolator/snubber bolts			
2000	Seismic pad for 20-38 tons Self contained AC unit without economizer .25g	200	440	640
2010	0.5 g seismic acceleration	200	440	640
2020	0.75 g seismic acceleration	200	440	640
2030	1.0 g seismic acceleration	200	440	640
2040	1.5 g seismic acceleration	205	495	700
2050	2.0 g seismic acceleration	224	680	904
2060	2.5 g seismic acceleration	237	845	1,082
2100	Seismic pad for 40-80 tons Self contained AC unit without economizer .25g	310	635	945
2110	0.5 g seismic acceleration	310	635	945
2120	0.75 g seismic acceleration	310	635	945
2130	1.0 g seismic acceleration	310	635	945
2140	1.5 g seismic acceleration	340	885	1,225
2150	2.0 g seismic acceleration	405	1,350	1,755
2160	2.5 g seismic acceleration	480	1,925	2,405
2200	Seismic pad for 80-110 tons Self contained AC unit without economizer .25g	340	680	1,020
2210	0.5 g seismic acceleration	340	680	1,020
2220	0.75 g seismic acceleration	340	680	1,020
2230	1.0 g seismic acceleration	340	680	1,020
2240	1.5 g seismic acceleration	370	945	1,315
2250	2.0 g seismic acceleration	445	1,475	1,920
2260	2.5 g seismic acceleration	520	2,075	2,595
2990	Seismic pad for self-contained AC economizer units for plate/foot attachment			
3000	Seismic pad for 20-38 tons Self contained AC unit economizer .25 g-2.0 g	125	237	362
3010	2.5 g seismic acceleration	133	315	448
3100	Seismic pad for 40-80 tons Self contained AC unit economizer .25-1.5 g	167	315	482
3110	2.0 g seismic acceleration	173	365	538
3120	2.5 g seismic acceleration	181	430	611
3190	Seismic pad for self-contained AC unit economizer for isolator/snubber bolts			
3200	Seismic pad for 20-38 tons Self contained AC unit economizer .25 g-2.0 g	130	250	380
3210	2.5 g seismic acceleration	141	340	481
3300	Seismic pad for 40-80 tons Self contained AC unit economizer .25-1.5 g	177	340	517
3310	2.0 g seismic acceleration	183	390	573
3320	2.5 g seismic acceleration	191	460	651

G1030 Site Earthwork

Trenching Systems are shown on a cost per linear foot basis. The systems include: excavation; backfill and removal of spoil; and compaction for various depths and trench bottom widths. The backfill has been reduced to accommodate a pipe of suitable diameter and bedding.

The slope for trench sides varies from none to 1:1.

The Expanded System Listing shows Trenching Systems that range from 2' to 12' in width. Depths range from 2' to 25'.

System Components	QUANTITY	UNIT	COST PER L.F. EQUIP.	LABOR	TOTAL
SYSTEM G1030 805 1310					
TRENCHING COMMON EARTH, NO SLOPE, 2' WIDE, 2' DP, 3/8 C.Y. BUCKET					
Excavation, trench, hyd. backhoe, track mtd., 3/8 C.Y. bucket	.148	B.C.Y.	.40	1.07	1.47
Backfill and load spoil, from stockpile	.153	L.C.Y.	.13	.33	.46
Compaction by vibrating plate, 6" lifts, 4 passes	.118	E.C.Y.	.03	.39	.42
Remove excess spoil, 8 C.Y. dump truck, 2 mile roundtrip	.040	L.C.Y.	.15	.17	.32
TOTAL			.71	1.96	2.67

G1030 805	Trenching Common Earth	COST PER L.F. EQUIP.	LABOR	TOTAL
1310	Trenching, common earth, no slope, 2' wide, 2' deep, 3/8 C.Y. bucket	.71	1.96	2.67
1320	3' deep, 3/8 C.Y. bucket	1	2.94	3.94
1330	4' deep, 3/8 C.Y. bucket	1.29	3.92	5.21
1340	6' deep, 3/8 C.Y. bucket	1.68	5.10	6.78
1350	8' deep, 1/2 C.Y. bucket	2.22	6.75	8.97
1360	10' deep, 1 C.Y. bucket	3.46	8.05	11.51
1400	4' wide, 2' deep, 3/8 C.Y. bucket	1.61	3.89	5.50
1410	3' deep, 3/8 C.Y. bucket	2.21	5.85	8.06
1420	4' deep, 1/2 C.Y. bucket	2.57	6.55	9.12
1430	6' deep, 1/2 C.Y. bucket	4.16	10.50	14.66
1440	8' deep, 1/2 C.Y. bucket	6.65	13.60	20.25
1450	10' deep, 1 C.Y. bucket	8	16.85	24.85
1460	12' deep, 1 C.Y. bucket	10.30	21.50	31.80
1470	15' deep, 1-1/2 C.Y. bucket	9.45	19.20	28.65
1480	18' deep, 2-1/2 C.Y. bucket	13.05	27	40.05
1520	6' wide, 6' deep, 5/8 C.Y. bucket w/trench box	8.55	15.70	24.25
1530	8' deep, 3/4 C.Y. bucket	11.30	20.50	31.80
1540	10' deep, 1 C.Y. bucket	11.55	21.50	33.05
1550	12' deep, 1-1/2 C.Y. bucket	12.40	23	35.40
1560	16' deep, 2-1/2 C.Y. bucket	16.85	28.50	45.35
1570	20' deep, 3-1/2 C.Y. bucket	22	34	56
1580	24' deep, 3-1/2 C.Y. bucket	26	41	67
1640	8' wide, 12' deep, 1-1/2 C.Y. bucket w/trench box	17.35	29	46.35
1650	15' deep, 1-1/2 C.Y. bucket	22.50	38	60.50
1660	18' deep, 2-1/2 C.Y. bucket	24.50	38	62.50
1680	24' deep, 3-1/2 C.Y. bucket	35.50	53	88.50
1730	10' wide, 20' deep, 3-1/2 C.Y. bucket w/trench box	29	50.50	79.50
1740	24' deep, 3-1/2 C.Y. bucket	42.50	60.50	103
1780	12' wide, 20' deep, 3-1/2 C.Y. bucket w/trench box	45.50	64	109.50
1790	25' deep, bucket	56	81.50	137.50
1800	1/2 to 1 slope, 2' wide, 2' deep, 3/8 C.Y. bucket	1	2.94	3.94
1810	3' deep, 3/8 C.Y. bucket	1.67	5.15	6.82
1820	4' deep, 3/8 C.Y. bucket	2.50	7.85	10.35
1840	6' deep, 3/8 C.Y. bucket	4.02	12.75	16.77

G1030 Site Earthwork

G1030 805	Trenching Common Earth	COST PER L.F.		
		EQUIP.	LABOR	TOTAL
1860	8' deep, 1/2 C.Y. bucket	6.40	20.50	26.90
1880	10' deep, 1 C.Y. bucket	11.90	28.50	40.40
2300	4' wide, 2' deep, 3/8 C.Y. bucket	1.91	4.87	6.78
2310	3' deep, 3/8 C.Y. bucket	2.89	8.05	10.94
2320	4' deep, 1/2 C.Y. bucket	3.63	9.95	13.58
2340	6' deep, 1/2 C.Y. bucket	6.95	18.65	25.60
2360	8' deep, 1/2 C.Y. bucket	12.80	27.50	40.30
2380	10' deep, 1 C.Y. bucket	17.70	38.50	56.20
2400	12' deep, 1 C.Y. bucket	24.50	52	76.50
2430	15' deep, 1-1/2 C.Y. bucket	26.50	56	82.50
2460	18' deep, 2-1/2 C.Y. bucket	46	87.50	133.50
2840	6' wide, 6' deep, 5/8 C.Y. bucket w/trench box	12.60	23	35.60
2860	8' deep, 3/4 C.Y. bucket	18.25	35	53.25
2880	10' deep, 1 C.Y. bucket	18.65	35.50	54.15
2900	12' deep, 1-1/2 C.Y. bucket	23.50	46.50	70
2940	16' deep, 2-1/2 C.Y. bucket	38.50	68	106.50
2980	20' deep, 3-1/2 C.Y. bucket	55	92	147
3020	24' deep, 3-1/2 C.Y. bucket	78	126	204
3100	8' wide, 12' deep, 1-1/2 C.Y. bucket w/trench box	29	53	82
3120	15' deep, 1-1/2 C.Y. bucket	42.50	76.50	119
3140	18' deep, 2-1/2 C.Y. bucket	53	91	144
3180	24' deep, 3-1/2 C.Y. bucket	87.50	138	225.50
3270	10' wide, 20' deep, 3-1/2 C.Y. bucket w/trench box	56	106	162
3280	24' deep, 3-1/2 C.Y. bucket	96.50	150	246.50
3370	12' wide, 20' deep, 3-1/2 C.Y. bucket w/trench box	81	123	204
3380	25' deep, 3-1/2 C.Y. bucket	113	175	288
3500	1 to 1 slope, 2' wide, 2' deep, 3/8 C.Y. bucket	1.29	3.92	5.21
3520	3' deep, 3/8 C.Y. bucket	3.73	8.95	12.68
3540	4' deep, 3/8 C.Y. bucket	3.69	11.80	15.49
3560	6' deep, 1/2 C.Y. bucket	4.02	12.75	16.77
3580	8' deep, 1/2 C.Y. bucket	7.95	25.50	33.45
3600	10' deep, 1 C.Y. bucket	20.50	48.50	69
3800	4' wide, 2' deep, 3/8 C.Y. bucket	2.21	5.85	8.06
3820	3' deep, 3/8 C.Y. bucket	3.56	10.30	13.86
3840	4' deep, 1/2 C.Y. bucket	4.66	13.35	18.01
3860	6' deep, 1/2 C.Y. bucket	9.80	27	36.80
3880	8' deep, 1/2 C.Y. bucket	19.05	41.50	60.55
3900	10' deep, 1 C.Y. bucket	27.50	60	87.50
3920	12' deep, 1 C.Y. bucket	40.50	87	127.50
3940	15' deep, 1-1/2 C.Y. bucket	44	93	137
3960	18' deep, 2-1/2 C.Y. bucket	63.50	121	184.50
4030	6' wide, 6' deep, 5/8 C.Y. bucket w/trench box	16.50	31.50	48
4040	8' deep, 3/4 C.Y. bucket	23.50	43	66.50
4050	10' deep, 1 C.Y. bucket	27	52	79
4060	12' deep, 1-1/2 C.Y. bucket	36	71	107
4070	16' deep, 2-1/2 C.Y. bucket	60	108	168
4080	20' deep, 3-1/2 C.Y. bucket	89	150	239
4090	24' deep, 3-1/2 C.Y. bucket	130	211	341
4500	8' wide, 12' deep, 1-1/2 C.Y. bucket w/trench box	41	77	118
4550	15' deep, 1-1/2 C.Y. bucket	62	115	177
4600	18' deep, 2-1/2 C.Y. bucket	80.50	141	221.50
4650	24' deep, 3-1/2 C.Y. bucket	139	222	361
4800	10' wide, 20' deep, 3-1/2 C.Y. bucket w/trench box	83.50	161	244.50
4850	24' deep, 3-1/2 C.Y. bucket	148	235	383
4950	12' wide, 20' deep, 3-1/2 C.Y. bucket w/trench box	117	181	298
4980	25' deep, 3-1/2 C.Y. bucket	168	265	433

G1030 Site Earthwork

Trenching Systems are shown on a cost per linear foot basis. The systems include: excavation; backfill and removal of spoil; and compaction for various depths and trench bottom widths. The backfill has been reduced to accommodate a pipe of suitable diameter and bedding.

The slope for trench sides varies from none to 1:1.

The Expanded System Listing shows Trenching Systems that range from 2' to 12' in width. Depths range from 2' to 25'.

System Components	QUANTITY	UNIT	COST PER L.F.		
			EQUIP.	LABOR	TOTAL
SYSTEM G1030 806 1310					
TRENCHING LOAM & SANDY CLAY, NO SLOPE, 2' WIDE, 2' DP, 3/8 C.Y. BUCKET					
Excavation, trench, hyd. backhoe, track mtd., 3/8 C.Y. bucket	.148	B.C.Y.	.37	.99	1.36
Backfill and load spoil, from stockpile	.165	L.C.Y.	.14	.35	.49
Compaction by vibrating plate 18" wide, 6" lifts, 4 passes	.118	E.C.Y.	.03	.39	.42
Remove excess spoil, 8 C.Y. dump truck, 2 mile roundtrip	.042	L.C.Y.	.16	.18	.34
TOTAL			.70	1.91	2.61

G1030 806	Trenching Loam & Sandy Clay	COST PER L.F.		
		EQUIP.	LABOR	TOTAL
1310	Trenching, loam & sandy clay, no slope, 2' wide, 2' deep, 3/8 C.Y. bucket	.70	1.91	2.61
1320	3' deep, 3/8 C.Y. bucket	1.09	3.15	4.24
1330	4' deep, 3/8 C.Y. bucket	1.27	3.83	5.10
1340	6' deep, 3/8 C.Y. bucket	1.80	4.56	6.35
1350	8' deep, 1/2 C.Y. bucket	2.38	6.05	8.45
1360	10' deep, 1 C.Y. bucket	2.69	6.45	9.15
1400	4' wide, 2' deep, 3/8 C.Y. bucket	1.61	3.82	5.45
1410	3' deep, 3/8 C.Y. bucket	2.19	5.75	7.95
1420	4' deep, 1/2 C.Y. bucket	2.56	6.45	9
1430	6' deep, 1/2 C.Y. bucket	4.41	9.45	13.85
1440	8' deep, 1/2 C.Y. bucket	6.45	13.40	19.85
1450	10' deep, 1 C.Y. bucket	6.45	13.70	20
1460	12' deep, 1 C.Y. bucket	8.15	17.05	25
1470	15' deep, 1-1/2 C.Y. bucket	9.85	19.85	29.50
1480	18' deep, 2-1/2 C.Y. bucket	11.50	22	33.50
1520	6' wide, 6' deep, 5/8 C.Y. bucket w/trench box	8.30	15.40	23.50
1530	8' deep, 3/4 C.Y. bucket	10.95	20	31
1540	10' deep, 1 C.Y. bucket	10.65	20.50	31
1550	12' deep, 1-1/2 C.Y. bucket	12.10	23	35
1560	16' deep, 2-1/2 C.Y. bucket	16.45	29	45.50
1570	20' deep, 3-1/2 C.Y. bucket	20.50	34	54.50
1580	24' deep, 3-1/2 C.Y. bucket	25.50	41.50	67
1640	8' wide, 12' deep, 1-1/4 C.Y. bucket w/trench box	17.10	29	46
1650	15' deep, 1-1/2 C.Y. bucket	21	37	58
1660	18' deep, 2-1/2 C.Y. bucket	25	41.50	66.50
1680	24' deep, 3-1/2 C.Y. bucket	34.50	54	88.50
1730	10' wide, 20' deep, 3-1/2 C.Y. bucket w/trench box	35	54	89
1740	24' deep, 3-1/2 C.Y. bucket	43.50	67	111
1780	12' wide, 20' deep, 3-1/2 C.Y. bucket w/trench box	42	64	106
1790	25' deep, 3-1/2 C.Y. bucket	54.50	83	138
1800	1/2:1 slope, 2' wide, 2' deep, 3/8 C.Y. bucket	.98	2.88	3.86
1810	3' deep, 3/8 C.Y. bucket	1.63	5.05	6.70
1820	4' deep, 3/8 C.Y. bucket	2.44	7.70	10.15
1840	6' deep, 3/8 C.Y. bucket	4.31	11.40	15.70

G1030 Site Earthwork

G1030 806	Trenching Loam & Sandy Clay	COST PER L.F.		
		EQUIP.	LABOR	TOTAL
1860	8' deep, 1/2 C.Y. bucket	6.85	18.20	25
1880	10' deep, 1 C.Y. bucket	9.15	23	32
2300	4' wide, 2' deep, 3/8 C.Y. bucket	1.90	4.78	6.70
2310	3' deep, 3/8 C.Y. bucket	2.85	7.90	10.75
2320	4' deep, 1/2 C.Y. bucket	3.59	9.80	13.40
2340	6' deep, 1/2 C.Y. bucket	7.40	16.80	24
2360	8' deep, 1/2 C.Y. bucket	12.50	27	39.50
2380	10' deep, 1 C.Y. bucket	14.20	31.50	45.50
2400	12' deep, 1 C.Y. bucket	24	52	76
2430	15' deep, 1-1/2 C.Y. bucket	28	58	86
2460	18' deep, 2-1/2 C.Y. bucket	45.50	88.50	134
2840	6' wide, 6' deep, 5/8 C.Y. bucket w/trench box	12.05	23.50	35.50
2860	8' deep, 3/4 C.Y. bucket	17.60	34	51.50
2880	10' deep, 1 C.Y. bucket	19	38.50	57.50
2900	12' deep, 1-1/2 C.Y. bucket	23.50	47	70.50
2940	16' deep, 2-1/2 C.Y. bucket	37.50	69	107
2980	20' deep, 3-1/2 C.Y. bucket	53.50	93	147
3020	24' deep, 3-1/2 C.Y. bucket	75.50	128	204
3100	8' wide, 12' deep, 1-1/2 C.Y. bucket w/trench box	28.50	53	81.50
3120	15' deep, 1-1/2 C.Y. bucket	39	74.50	114
3140	18' deep, 2-1/2 C.Y. bucket	52	92.50	145
3180	24' deep, 3-1/2 C.Y. bucket	85	140	225
3270	10' wide, 20' deep, 3-1/2 C.Y. bucket w/trench box	68	113	181
3280	24' deep, 3-1/2 C.Y. bucket	93.50	153	247
3320	12' wide, 20' deep, 3-1/2 C.Y. bucket w/trench box	75	123	198
3380	25' deep, 3-1/2 C.Y. bucket w/trench box	103	165	268
3500	1:1 slope, 2' wide, 2' deep, 3/8 C.Y. bucket	1.27	3.84	5.10
3520	3' deep, 3/8 C.Y. bucket	2.29	7.20	9.50
3540	4' deep, 3/8 C.Y. bucket	3.59	11.55	15.15
3560	6' deep, 1/2 C.Y. bucket	4.31	11.40	15.70
3580	8' deep, 1/2 C.Y. bucket	11.35	30.50	42
3600	10' deep, 1 C.Y. bucket	15.65	39	54.50
3800	4' wide, 2' deep, 3/8 C.Y. bucket	2.19	5.75	7.95
3820	3' deep, 1/2 C.Y. bucket	3.50	10.10	13.60
3840	4' deep, 1/2 C.Y. bucket	4.60	13.15	17.75
3860	6' deep, 1/2 C.Y. bucket	10.35	24	34.50
3880	8' deep, 1/2 C.Y. bucket	18.50	41	59.50
3900	10' deep, 1 C.Y. bucket	22	49	71
3920	12' deep, 1 C.Y. bucket	31.50	69.50	101
3940	15' deep, 1-1/2 C.Y. bucket	46	96	142
3960	18' deep, 2-1/2 C.Y. bucket	62	122	184
4030	6' wide, 6' deep, 5/8 C.Y. bucket w/trench box	15.80	31	47
4040	8' deep, 3/4 C.Y. bucket	24.50	48	72.50
4050	10' deep, 1 C.Y. bucket	27.50	56.50	84
4060	12' deep, 1-1/2 C.Y. bucket	35	71	106
4070	16' deep, 2-1/2 C.Y. bucket	58.50	109	168
4080	20' deep, 3-1/2 C.Y. bucket	86.50	152	239
4090	24' deep, 3-1/2 C.Y. bucket	126	214	340
4500	8' wide, 12' deep, 1-1/4 C.Y. bucket w/trench box	40	77	117
4550	15' deep, 1-1/2 C.Y. bucket	57	112	169
4600	18' deep, 2-1/2 C.Y. bucket	78.50	143	222
4650	24' deep, 3-1/2 C.Y. bucket	135	226	360
4800	10' wide, 20' deep, 3-1/2 C.Y. bucket w/trench box	101	172	273
4850	24' deep, 3-1/2 C.Y. bucket	144	239	385
4950	12' wide, 20' deep, 3-1/2 C.Y. bucket w/trench box	108	182	290
4980	25' deep, 3-1/2 C.Y. bucket	163	269	430

G1030 Site Earthwork

Trenching Systems are shown on a cost per linear foot basis. The systems include: excavation; backfill and removal of spoil; and compaction for various depths and trench bottom widths. The backfill has been reduced to accommodate a pipe of suitable diameter and bedding.

The slope for trench sides varies from none to 1:1.

The Expanded System Listing shows Trenching Systems that range from 2′ to 12′ in width. Depths range from 2′ to 25′.

System Components	QUANTITY	UNIT	COST PER L.F.		
			EQUIP.	LABOR	TOTAL
SYSTEM G1030 807 1310					
TRENCHING SAND & GRAVEL, NO SLOPE, 2′ WIDE, 2′ DEEP, 3/8 C.Y. BUCKET					
Excavation, trench, hyd. backhoe, track mtd., 3/8 C.Y. bucket	.148	B.C.Y.	.36	.97	1.33
Backfill and load spoil, from stockpile	.140	L.C.Y.	.11	.30	.41
Compaction by vibrating plate 18″ wide, 6″ lifts, 4 passes	.118	E.C.Y.	.03	.39	.42
Remove excess spoil, 8 C.Y. dump truck, 2 mile roundtrip	.035	L.C.Y.	.14	.15	.29
TOTAL			.64	1.81	2.45

G1030 807	Trenching Sand & Gravel	COST PER L.F.		
		EQUIP.	LABOR	TOTAL
1310	Trenching, sand & gravel, no slope, 2′ wide, 2′ deep, 3/8 C.Y. bucket	.64	1.81	2.45
1320	3′ deep, 3/8 C.Y. bucket	1.03	3.02	4.05
1330	4′ deep, 3/8 C.Y. bucket	1.19	3.64	4.83
1340	6′ deep, 3/8 C.Y. bucket	1.71	4.33	6.05
1350	8′ deep, 1/2 C.Y. bucket	2.25	5.75	8
1360	10′ deep, 1 C.Y. bucket	2.50	6.05	8.55
1400	4′ wide, 2′ deep, 3/8 C.Y. bucket	1.46	3.59	5.05
1410	3′ deep, 3/8 C.Y. bucket	2.01	5.45	7.45
1420	4′ deep, 1/2 C.Y. bucket	2.34	6.10	8.45
1430	6′ deep, 1/2 C.Y. bucket	4.18	9	13.20
1440	8′ deep, 1/2 C.Y. bucket	6.10	12.70	18.80
1450	10′ deep, 1 C.Y. bucket	6.10	12.95	19.05
1460	12′ deep, 1 C.Y. bucket	7.70	16.15	24
1470	15′ deep, 1-1/2 C.Y. bucket	9.25	18.70	28
1480	18′ deep, 2-1/2 C.Y. bucket	10.85	20.50	31.50
1520	6′ wide, 6′ deep, 5/8 C.Y. bucket w/trench box	7.80	14.55	22.50
1530	8′ deep, 3/4 C.Y. bucket	10.35	19.15	29.50
1540	10′ deep, 1 C.Y. bucket	10	19.35	29.50
1550	12′ deep, 1-1/2 C.Y. bucket	11.45	21.50	33
1560	16′ deep, 2 C.Y. bucket	15.70	27.50	43
1570	20′ deep, 3-1/2 C.Y. bucket	19.25	32	51.50
1580	24′ deep, 3-1/2 C.Y. bucket	24	39	63
1640	8′ wide, 12′ deep, 1-1/2 C.Y. bucket w/trench box	16.10	27.50	43.50
1650	15′ deep, 1-1/2 C.Y. bucket	19.55	35	54.50
1660	18′ deep, 2-1/2 C.Y. bucket	23.50	39	62.50
1680	24′ deep, 3-1/2 C.Y. bucket	32.50	50.50	83
1730	10′ wide, 20′ deep, 3-1/2 C.Y. bucket w/trench box	33	50.50	83.50
1740	24′ deep, 3-1/2 C.Y. bucket	41	62.50	104
1780	12′ wide, 20′ deep, 3-1/2 C.Y. bucket w/trench box	39.50	60	99.50
1790	25′ deep, 3-1/2 C.Y. bucket	51.50	77.50	129
1800	1/2:1 slope, 2′ wide, 2′ deep, 3/8 C.Y. bucket	.92	2.73	3.65
1810	3′ deep, 3/8 C.Y. bucket	1.54	4.79	6.35
1820	4′ deep, 3/8 C.Y. bucket	2.29	7.30	9.60
1840	6′ deep, 3/8 C.Y. bucket	4.11	10.90	15

G10 Site Preparation

G1030 Site Earthwork

G1030 807	Trenching Sand & Gravel	COST PER L.F.		
		EQUIP.	LABOR	TOTAL
1860	8' deep, 1/2 C.Y. bucket	6.55	17.40	24
1880	10' deep, 1 C.Y. bucket	8.55	21.50	30
2300	4' wide, 2' deep, 3/8 C.Y. bucket	1.74	4.51	6.25
2310	3' deep, 3/8 C.Y. bucket	2.63	7.50	10.15
2320	4' deep, 1/2 C.Y. bucket	3.31	9.30	12.60
2340	6' deep, 1/2 C.Y. bucket	7.05	16.05	23
2360	8' deep, 1/2 C.Y. bucket	11.85	25.50	37.50
2380	10' deep, 1 C.Y. bucket	13.45	29.50	43
2400	12' deep, 1 C.Y. bucket	22.50	49.50	72
2430	15' deep, 1-1/2 C.Y. bucket	26.50	55	81.50
2460	18' deep, 2-1/2 C.Y. bucket	43	83.50	127
2840	6' wide, 6' deep, 5/8 C.Y. bucket w/trench box	11.40	22	33.50
2860	8' deep, 3/4 C.Y. bucket	16.70	32.50	49
2880	10' deep, 1 C.Y. bucket	17.95	36.50	54.50
2900	12' deep, 1-1/2 C.Y. bucket	22.50	44.50	67
2940	16' deep, 2 C.Y. bucket	36	65	101
2980	20' deep, 3-1/2 C.Y. bucket	50.50	87.50	138
3020	24' deep, 3-1/2 C.Y. bucket	72	120	192
3100	8' wide, 12' deep, 1-1/4 C.Y. bucket w/trench box	27	50	77
3120	15' deep, 1-1/2 C.Y. bucket	36.50	70.50	107
3140	18' deep, 2-1/2 C.Y. bucket	49	87	136
3180	24' deep, 3-1/2 C.Y. bucket	80.50	131	212
3270	10' wide, 20' deep, 3-1/2 C.Y. bucket w/trench box	64.50	106	171
3280	24' deep, 3-1/2 C.Y. bucket	88.50	143	232
3370	12' wide, 20' deep, 3-1/2 C.Y. bucket w/trench box	72	118	190
3380	25' deep, 3-1/2 C.Y. bucket	103	165	268
3500	1:1 slope, 2' wide, 2' deep, 3/8 C.Y. bucket	2.06	4.55	6.60
3520	3' deep, 3/8 C.Y. bucket	2.16	6.85	9
3540	4' deep, 3/8 C.Y. bucket	3.38	11	14.40
3560	6' deep, 3/8 C.Y. bucket	4.11	10.90	15
3580	8' deep, 1/2 C.Y. bucket	10.80	29	40
3600	10' deep, 1 C.Y. bucket	14.65	37	51.50
3800	4' wide, 2' deep, 3/8 C.Y. bucket	2.01	5.45	7.45
3820	3' deep, 3/8 C.Y. bucket	3.25	9.55	12.80
3840	4' deep, 1/2 C.Y. bucket	4.26	12.50	16.75
3860	6' deep, 1/2 C.Y. bucket	9.90	23	33
3880	8' deep, 1/2 C.Y. bucket	17.55	38.50	56
3900	10' deep, 1 C.Y. bucket	21	46.50	67.50
3920	12' deep, 1 C.Y. bucket	30	66	96
3940	15' deep, 1-1/2 C.Y. bucket	43.50	91	135
3960	18' deep, 2-1/2 C.Y. bucket	59	115	174
4030	6' wide, 6' deep, 5/8 C.Y. bucket w/trench box	15	29.50	44.50
4040	8' deep, 3/4 C.Y. bucket	23	45.50	68.50
4050	10' deep, 1 C.Y. bucket	26	53.50	79.50
4060	12' deep, 1-1/2 C.Y. bucket	33.50	67	101
4070	16' deep, 2 C.Y. bucket	56.50	103	160
4080	20' deep, 3-1/2 C.Y. bucket	82	143	225
4090	24' deep, 3-1/2 C.Y. bucket	119	201	320
4500	8' wide, 12' deep, 1-1/2 C.Y. bucket w/trench box	38	73	111
4550	15' deep, 1-1/2 C.Y. bucket	53.50	106	160
4600	18' deep, 2-1/2 C.Y. bucket	74.50	135	210
4650	24' deep, 3-1/2 C.Y. bucket	128	212	340
4800	10' wide, 20' deep, 3-1/2 C.Y. bucket w/trench box	95.50	161	257
4850	24' deep, 3-1/2 C.Y. bucket	136	224	360
4950	12' wide, 20' deep, 3-1/2 C.Y. bucket w/trench box	103	171	274
4980	25' deep, 3-1/2 C.Y. bucket	155	253	410

G1030 Site Earthwork

The Pipe Bedding System is shown for various pipe diameters. Compacted bank sand is used for pipe bedding and to fill 12″ over the pipe. No backfill is included. Various side slopes are shown to accommodate different soil conditions. Pipe sizes vary from 6″ to 84″ diameter.

System Components	QUANTITY	UNIT	COST PER L.F.		
			MAT.	INST.	TOTAL
SYSTEM G1030 815 1440					
PIPE BEDDING, SIDE SLOPE 0 TO 1, 1′ WIDE, PIPE SIZE 6″ DIAMETER					
Borrow, bank sand, 2 mile haul, machine spread	.086	C.Y.	1.88	.71	2.59
Compaction, vibrating plate	.086	C.Y.		.22	.22
TOTAL			1.88	.93	2.81

G1030 815	Pipe Bedding	COST PER L.F.		
		MAT.	INST.	TOTAL
1440	Pipe bedding, side slope 0 to 1, 1′ wide, pipe size 6″ diameter	1.88	.93	2.81
1460	2′ wide, pipe size 8″ diameter	4.07	1.99	6.06
1480	Pipe size 10″ diameter	4.16	2.04	6.20
1500	Pipe size 12″ diameter	4.25	2.08	6.33
1520	3′ wide, pipe size 14″ diameter	6.90	3.38	10.28
1540	Pipe size 15″ diameter	6.95	3.41	10.36
1560	Pipe size 16″ diameter	7	3.44	10.44
1580	Pipe size 18″ diameter	7.15	3.52	10.67
1600	4′ wide, pipe size 20″ diameter	10.15	4.99	15.14
1620	Pipe size 21″ diameter	10.25	5.05	15.30
1640	Pipe size 24″ diameter	10.45	5.15	15.60
1660	Pipe size 30″ diameter	10.65	5.25	15.90
1680	6′ wide, pipe size 32″ diameter	18.25	8.95	27.20
1700	Pipe size 36″ diameter	18.70	9.15	27.85
1720	7′ wide, pipe size 48″ diameter	29.50	14.55	44.05
1740	8′ wide, pipe size 60″ diameter	36	17.75	53.75
1760	10′ wide, pipe size 72″ diameter	50.50	24.50	75
1780	12′ wide, pipe size 84″ diameter	66.50	32.50	99
2140	Side slope 1/2 to 1, 1′ wide, pipe size 6″ diameter	3.51	1.71	5.22
2160	2′ wide, pipe size 8″ diameter	5.95	2.92	8.87
2180	Pipe size 10″ diameter	6.40	3.14	9.54
2200	Pipe size 12″ diameter	6.80	3.32	10.12
2220	3′ wide, pipe size 14″ diameter	9.80	4.81	14.61
2240	Pipe size 15″ diameter	10	4.91	14.91
2260	Pipe size 16″ diameter	10.25	5.05	15.30
2280	Pipe size 18″ diameter	10.80	5.30	16.10
2300	4′ wide, pipe size 20″ diameter	14.25	7	21.25
2320	Pipe size 21″ diameter	14.55	7.15	21.70
2340	Pipe size 24″ diameter	15.50	7.60	23.10
2360	Pipe size 30″ diameter	17.20	8.45	25.65
2380	6′ wide, pipe size 32″ diameter	25.50	12.40	37.90
2400	Pipe size 36″ diameter	27	13.20	40.20
2420	7′ wide, pipe size 48″ diameter	42	20.50	62.50
2440	8′ wide, pipe size 60″ diameter	53.50	26	79.50
2460	10′ wide, pipe size 72″ diameter	73	36	109
2480	12′ wide, pipe size 84″ diameter	96	47	143
2620	Side slope 1 to 1, 1′ wide, pipe size 6″ diameter	5.15	2.52	7.67
2640	2′ wide, pipe size 8″ diameter	7.90	3.89	11.79

G1030 Site Earthwork

G1030 815	Pipe Bedding	COST PER L.F.		
		MAT.	INST.	TOTAL
2660	Pipe size 10" diameter	8.60	4.22	12.82
2680	Pipe size 12" diameter	9.35	4.60	13.95
2700	3' wide, pipe size 14" diameter	12.70	6.20	18.90
2720	Pipe size 15" diameter	13.10	6.40	19.50
2740	Pipe size 16" diameter	13.55	6.65	20.20
2760	Pipe size 18" diameter	14.45	7.10	21.55
2780	4' wide, pipe size 20" diameter	18.35	9	27.35
2800	Pipe size 21" diameter	18.85	9.25	28.10
2820	Pipe size 24" diameter	20.50	10.05	30.55
2840	Pipe size 30" diameter	23.50	11.60	35.10
2860	6' wide, pipe size 32" diameter	32.50	15.90	48.40
2880	Pipe size 36" diameter	35	17.30	52.30
2900	7' wide, pipe size 48" diameter	54.50	26.50	81
2920	8' wide, pipe size 60" diameter	70.50	34.50	105
2940	10' wide, pipe size 72" diameter	96	47	143
2960	12' wide, pipe size 84" diameter	125	61.50	186.50

For customer support on your Mechanical Cost Data, call 877.739.3689.

593

G3010   Water Supply

G3010 110	Water Distribution Piping	COST PER L.F.		
		MAT.	INST.	TOTAL
2000	Piping, excav. & backfill excl., ductile iron class 250, mech. joint			
2130	4" diameter	33.50	17	50.50
2150	6" diameter	35	21.50	56.50
2160	8" diameter	49	25.50	74.50
2170	10" diameter	64	29.50	93.50
2180	12" diameter	87	32.50	119.50
2210	16" diameter	104	47	151
2220	18" diameter	139	49	188
3000	Tyton joint			
3130	4" diameter	17.05	8.50	25.55
3150	6" diameter	20	10.20	30.20
3160	8" diameter	30.50	17	47.50
3170	10" diameter	40	18.70	58.70
3180	12" diameter	42	21.50	63.50
3210	16" diameter	45	29.50	74.50
3220	18" diameter	50	34	84
3230	20" diameter	52	38.50	90.50
3250	24" diameter	65	44.50	109.50
4000	Copper tubing, type K			
4050	3/4" diameter	7.30	3.22	10.52
4060	1" diameter	9.70	4.02	13.72
4080	1-1/2" diameter	15.50	4.85	20.35
4090	2" diameter	24	5.60	29.60
4110	3" diameter	49	9.60	58.60
4130	4" diameter	74.50	13.55	88.05
4150	6" diameter	161	25	186
5000	Polyvinyl chloride class 160, S.D.R. 26			
5130	1-1/2" diameter	.55	1.20	1.75
5150	2" diameter	.80	1.31	2.11
5160	4" diameter	3	5.90	8.90
5170	6" diameter	5.75	7.10	12.85
5180	8" diameter	8.85	8.50	17.35
6000	Polyethylene 160 psi, S.D.R. 7			
6050	3/4" diameter	.37	1.71	2.08
6060	1" diameter	.58	1.85	2.43
6080	1-1/2" diameter	1.16	2	3.16
6090	2" diameter	1.93	2.46	4.39

G3020 Sanitary Sewer

G3020 110	Drainage & Sewage Piping	COST PER L.F.		
		MAT.	INST.	TOTAL
2000	Piping, excavation & backfill excluded, PVC, plain			
2130	4" diameter	1.64	4.19	5.83
2150	6" diameter	3.61	4.48	8.09
2160	8" diameter	6.75	4.69	11.44
2900	Box culvert, precast, 8' long			
3000	6' x 3'	275	35	310
3020	6' x 7'	345	39	384
3040	8' x 3'	345	43	388
3060	8' x 8'	420	49	469
3080	10' x 3'	480	44.50	524.50
3100	10' x 8'	545	61	606
3120	12' x 3'	710	49	759
3140	12' x 8'	895	73	968
4000	Concrete, nonreinforced			
4150	6" diameter	8	12.60	20.60
4160	8" diameter	8.75	14.95	23.70
4170	10" diameter	9.75	15.50	25.25
4180	12" diameter	10.95	16.75	27.70
4200	15" diameter	15.40	18.60	34
4220	18" diameter	18.65	23.50	42.15
4250	24" diameter	21	33.50	54.50
4400	Reinforced, no gasket			
4580	12" diameter	13.35	22.50	35.85
4600	15" diameter	17.25	22.50	39.75
4620	18" diameter	22	25.50	47.50
4650	24" diameter	24	33.50	57.50
4670	30" diameter	42	49.50	91.50
4680	36" diameter	65.50	60.50	126
4690	42" diameter	89	66	155
4700	48" diameter	108	75	183
4720	60" diameter	162	99.50	261.50
4730	72" diameter	243	120	363
4740	84" diameter	335	150	485
4800	With gasket			
4980	12" diameter	14.70	12.15	26.85
5000	15" diameter	19	12.75	31.75
5020	18" diameter	24.50	13.40	37.90
5050	24" diameter	31.50	15	46.50
5070	30" diameter	50.50	49.50	100
5080	36" diameter	75.50	60.50	136
5090	42" diameter	97.50	60	157.50
5100	48" diameter	122	75	197
5120	60" diameter	194	96	290
5130	72" diameter	263	120	383
5140	84" diameter	400	199	599
5700	Corrugated metal, alum. or galv. bit. coated			
5760	8" diameter	9.65	10.10	19.75
5770	10" diameter	10	12.90	22.90
5780	12" diameter	12.20	15.95	28.15
5800	15" diameter	16.85	16.75	33.60
5820	18" diameter	22	17.60	39.60
5850	24" diameter	27	21	48
5870	30" diameter	32.50	36.50	69
5880	36" diameter	38	36.50	74.50
5900	48" diameter	57	44	101
5920	60" diameter	87.50	63.50	151
5930	72" diameter	105	106	211
6000	Plain			

For customer support on your Mechanical Cost Data, call 877.739.3689.

595

G3020 Sanitary Sewer

G3020 110	Drainage & Sewage Piping	COST PER L.F.		
		MAT.	INST.	TOTAL
6060	8" diameter	8.65	9.45	18.10
6070	10" diameter	9.90	11.95	21.85
6080	12" diameter	11	15.25	26.25
6100	15" diameter	13.75	15.25	29
6120	18" diameter	16.60	16.30	32.90
6140	24" diameter	24	19.15	43.15
6170	30" diameter	29.50	33.50	63
6180	36" diameter	34	33.50	67.50
6200	48" diameter	51	39.50	90.50
6220	60" diameter	78.50	61.50	140
6230	72" diameter	178	156	334
6300	Steel or alum. oval arch, coated & paved invert			
6400	15" equivalent diameter	15.95	16.75	32.70
6420	18" equivalent diameter	21	22.50	43.50
6450	24" equivalent diameter	27	26.50	53.50
6470	30" equivalent diameter	32.50	33.50	66
6480	36" equivalent diameter	39	44	83
6490	42" equivalent diameter	47.50	48.50	96
6500	48" equivalent diameter	61.50	58	119.50
6600	Plain			
6700	15" equivalent diameter	14.50	14.90	29.40
6720	18" equivalent diameter	17.65	19.15	36.80
6750	24" equivalent diameter	24.50	22.50	47
6770	30" equivalent diameter	29.50	40.50	70
6780	36" equivalent diameter	35.50	40.50	76
6790	42" equivalent diameter	42	47.50	89.50
6800	48" equivalent diameter	59	58	117
8000	Polyvinyl chloride SDR 35			
8130	4" diameter	1.64	4.19	5.83
8150	6" diameter	3.61	4.48	8.09
8160	8" diameter	6.75	4.69	11.44
8170	10" diameter	12.20	6.15	18.35
8180	12" diameter	14	6.40	20.40
8200	15" diameter	14.30	8.50	22.80

G3030 Storm Sewer

Manhole   Catch Basin

The Manhole and Catch Basin System includes: excavation with a backhoe; a formed concrete footing; frame and cover; cast iron steps and compacted backfill.

The Expanded System Listing shows manholes that have a 4', 5' and 6' inside diameter riser. Depths range from 4' to 14'. Construction material shown is either concrete, concrete block, precast concrete, or brick.

System Components	QUANTITY	UNIT	COST PER EACH		
			MAT.	INST.	TOTAL
SYSTEM G3030 210 1920					
MANHOLE/CATCH BASIN, BRICK, 4' I.D. RISER, 4' DEEP					
Excavation, hydraulic backhoe, 3/8 C.Y. bucket	14.815	B.C.Y.		119.71	119.71
Trim sides and bottom of excavation	64.000	S.F.		62.72	62.72
Forms in place, manhole base, 4 uses	20.000	SFCA	15.40	109	124.40
Reinforcing in place footings, #4 to #7	.019	Ton	19.95	23.75	43.70
Concrete, 3000 psi	.925	C.Y.	109.15		109.15
Place and vibrate concrete, footing, direct chute	.925	C.Y.		49.73	49.73
Catch basin or MH, brick, 4' ID, 4' deep	1.000	Ea.	590	1,025	1,615
Catch basin or MH steps; heavy galvanized cast iron	1.000	Ea.	21	14.10	35.10
Catch basin or MH frame and cover	1.000	Ea.	270	247.50	517.50
Fill, granular	12.954	L.C.Y.	310.90		310.90
Backfill, spread with wheeled front end loader	12.954	L.C.Y.		34.46	34.46
Backfill compaction, 12" lifts, air tamp	12.954	E.C.Y.		118.27	118.27
TOTAL			1,336.40	1,804.24	3,140.64

G3030 210	Manholes & Catch Basins	COST PER EACH		
		MAT.	INST.	TOTAL
1920	Manhole/catch basin, brick, 4' I.D. riser, 4' deep	1,325	1,800	3,125
1940	6' deep	1,900	2,500	4,400
1960	8' deep	2,525	3,425	5,950
1980	10' deep	2,975	4,225	7,200
3000	12' deep	3,725	4,600	8,325
3020	14' deep	4,575	6,425	11,000
3200	Block, 4' I.D. riser, 4' deep	1,175	1,450	2,625
3220	6' deep	1,650	2,050	3,700
3240	8' deep	2,200	2,850	5,050
3260	10' deep	2,600	3,500	6,100
3280	12' deep	3,325	4,450	7,775
3300	14' deep	4,150	5,450	9,600
4620	Concrete, cast-in-place, 4' I.D. riser, 4' deep	1,325	2,525	3,850
4640	6' deep	1,900	3,375	5,275
4660	8' deep	2,675	4,875	7,550
4680	10' deep	3,225	6,050	9,275
4700	12' deep	4,050	7,500	11,550
4720	14' deep	5,000	8,975	13,975
5820	Concrete, precast, 4' I.D. riser, 4' deep	1,550	1,350	2,900
5840	6' deep	2,550	1,800	4,350

597

G30 Site Mechanical Utilities

G3030 Storm Sewer

G3030 210	Manholes & Catch Basins	COST PER EACH		
		MAT.	INST.	TOTAL
5860	8' deep	3,425	2,500	5,925
5880	10' deep	4,050	3,100	7,150
5900	12' deep	4,975	3,825	8,800
5920	14' deep	6,000	4,850	10,850
6000	5' I.D. riser, 4' deep	2,575	1,525	4,100
6020	6' deep	3,250	2,125	5,375
6040	8' deep	4,500	2,825	7,325
6060	10' deep	5,825	3,575	9,400
6080	12' deep	7,250	4,550	11,800
6100	14' deep	8,800	5,550	14,350
6200	6' I.D. riser, 4' deep	3,425	2,000	5,425
6220	6' deep	4,500	2,650	7,150
6240	8' deep	5,600	3,700	9,300
6260	10' deep	7,225	4,700	11,925
6280	12' deep	8,975	5,875	14,850
6300	14' deep	10,900	7,125	18,025

G3060 Fuel Distribution

G3060 110	Gas Service Piping	COST PER L.F.		
		MAT.	INST.	TOTAL
2000	Piping, excavation & backfill excluded, polyethylene			
2070	1-1/4" diam, SDR 10	1.78	4.12	5.90
2090	2" diam, SDR 11	2.48	4.59	7.07
2110	3" diam, SDR 11.5	5.15	5.50	10.65
2130	4" diam, SDR 11	14.85	9.65	24.50
2150	6" diam, SDR 21	37	10.35	47.35
2160	8" diam, SDR 21	51	12.50	63.50
3000	Steel, schedule 40, plain end, tarred & wrapped			
3060	1" diameter	5.50	9.25	14.75
3090	2" diameter	8.65	9.95	18.60
3110	3" diameter	14.30	10.70	25
3130	4" diameter	18.60	16.65	35.25
3140	5" diameter	27	19.35	46.35
3150	6" diameter	33	23.50	56.50
3160	8" diameter	52.50	30.50	83

G3060 Fuel Distribution

Fiberglass Underground Fuel Storage Tank, Double Wall

System Components	QUANTITY	UNIT	COST EACH		
			MAT.	INST.	TOTAL
SYSTEM G3060 310 1010					
STORAGE TANK, FUEL, UNDERGROUND, DOUBLE WALL FIBERGLASS, 600 GAL.					
Excavating trench, no sheeting or dewatering, 6'-10 ' D, 3/4 CY hyd backhoe	18.000	Ea.		145.44	145.44
Corrosion resistance wrap & coat, small diam. pipe, 1" diam., add	120.000	Ea.	126		126
Fiberglass-reinforced elbows 90 & 45 degree, 2 inch	7.000	Ea.	332.50	598.50	931
Fiberglass-reinforced threaded adapters, 2 inch	2.000	Ea.	39.50	195	234.50
Remote tank gauging system, 30', 5" pointer travel	1.000	Ea.	5,950	365	6,315
Corrosion resistance wrap & coat, small diam. pipe. 4 " diam. add	30.000	Ea.	72.60		72.60
Pea gravel	9.000	Ea.	1,080	454.50	1,534.50
Reinforcing in hold down pad, #3 to #7	120.000	Ea.	63.60	75.60	139.20
Tank leak detector system, 8 channel, external monitoring	1.000	Ea.	965		965
Tubbing copper, type L, 3/4" diam.	120.000	Ea.	690	1,128	1,818
Elbow 90 degree ,wrought cu x cu 3/4 " diam.	4.000	Ea.	11.12	150	161.12
Pipe, black steel welded, sch 40, 3" diam.	30.000	Ea.	414	944.40	1,358.40
Elbow 90 degree, steel welded butt joint, 3 " diam.	3.000	Ea.	96	579.30	675.30
Fiberglass-reinforced plastic pipe, 2 inch	156.000	Ea.	1,170	753.48	1,923.48
Fiberglass-reinforced plastic pipe, 3 inch	36.000	Ea.	1,044	196.20	1,240.20
Fiberglass-reinforced elbows 90 & 45 degree, 3 inch	3.000	Ea.	450	342	792
Tank leak detection system, probes, well monitoring, liquid phase detection	2.000	Ea.	1,010		1,010
Double wall fiberglass annular space	1.000	Ea.	400		400
Flange, weld neck, 150 lb, 3 " pipe size	1.000	Ea.	38.50	96.56	135.06
Vent protector/breather, 2" diam.	1.000	Ea.	33.50	23	56.50
Corrosion resistance wrap & coat, small diam. pipe, 2" diam., add	36.000	Ea.	54.72		54.72
Concrete hold down pad, 6' x 5' x 8" thick	30.000	Ea.	93.60	47.10	140.70
600 gallon capacity, underground storage tank, double wall fiberglass	1.000	Ea.	8,300	545	8,845
Tank, for hold-downs 500-4000 gal., add	1.000	Ea.	515	174	689
Foot valve, single poppet, 3/4" diam.	1.000	Ea.	78.50	40.50	119
Fuel fill box, locking inner cover, 3" diam.	1.000	Ea.	133	147	280
Fuel oil specialties, valve,ball chk, globe type fusible, 3/4" diam.	2.000	Ea.	154	73	227
TOTAL			23,315.14	7,073.58	30,388.72

G3060 310	Fiberglass Fuel Tank, Double Wall	COST EACH		
		MAT.	INST.	TOTAL
1010	Storage tank, fuel, underground, double wall fiberglass, 600 Gal.	23,300	7,075	30,375
1020	2500 Gal.	36,900	9,700	46,600
1030	4000 Gal.	43,100	12,000	55,100
1040	6000 Gal.	50,500	13,500	64,000

G3060 Fuel Distribution

G3060 310	Fiberglass Fuel Tank, Double Wall	COST EACH		
		MAT.	INST.	TOTAL
1050	8000 Gal.	58,000	16,300	74,300
1060	10,000 Gal.	66,000	18,700	84,700
1070	15,000 Gal.	91,000	21,500	112,500
1080	20,000 Gal.	114,000	26,000	140,000
1090	25,000 Gal.	145,500	30,100	175,600
1100	30,000 Gal.	166,000	31,900	197,900

Steel Underground Fuel Storage Tank, Double Wall

System Components	QUANTITY	UNIT	COST EACH		
			MAT.	INST.	TOTAL
SYSTEM G3060 320 1010					
STORAGE TANK, FUEL, UNDERGROUND, DOUBLE WALL STEEL, 500 GAL.					
Fiberglass-reinforced plastic pipe, 2 inch	156.000	Ea.	1,170	753.48	1,923.48
Fiberglass-reinforced plastic pipe, 3 inch	36.000	Ea.	1,044	196.20	1,240.20
Fiberglass-reinforced elbows 90 & 45 degree, 3 inch	3.000	Ea.	450	342	792
Tank leak detection system, probes, well monitoring, liquid phase detection	2.000	Ea.	1,010		1,010
Flange, weld neck, 150 lb, 3" pipe size	1.000	Ea.	38.50	96.56	135.06
Vent protector/breather, 2" diam.	1.000	Ea.	33.50	23	56.50
Concrete hold down pad, 6' x 5' x 8" thick	30.000	Ea.	93.60	47.10	140.70
Tanks, for hold-downs, 500-2000 gal, add	1.000	Ea.	201	174	375
Double wall steel tank annular space	1.000	Ea.	400		400
Foot valve, single poppet, 3/4" diam.	1.000	Ea.	78.50	40.50	119
Fuel fill box, locking inner cover, 3" diam.	1.000	Ea.	133	147	280
Fuel oil specialties, valve, ball chk, globe type fusible, 3/4" diam.	2.000	Ea.	154	73	227
Pea gravel	9.000	Ea.	1,080	454.50	1,534.50
Reinforcing in hold down pad, #3 to #7	120.000	Ea.	63.60	75.60	139.20
Tank, for manways, add	1.000	Ea.	2,000		2,000
Tank leak detector system, 8 channel, external monitoring	1.000	Ea.	965		965
Tubbing copper, type L, 3/4" diam.	120.000	Ea.	690	1,128	1,818
Elbow 90 degree, wrought cu x cu 3/4" diam.	8.000	Ea.	22.24	300	322.24
Pipe, black steel welded, sch 40, 3" diam.	30.000	Ea.	414	944.40	1,358.40
Elbow 90 degree, steel welded butt joint, 3" diam.	3.000	Ea.	96	579.30	675.30
Excavating trench, no sheeting or dewatering, 6'-10' D, 3/4 CY hyd backhoe	18.000	Ea.		145.44	145.44
Fiberglass-reinforced elbows 90 & 45 degree, 2 inch	7.000	Ea.	332.50	598.50	931
Fiberglass-reinforced threaded adapters, 2 inch	2.000	Ea.	39.50	195	234.50
500 gallon capacity, underground storage tank, double wall steel	1.000	Ea.	6,150	550	6,700
Remote tank gauging system, 30', 5" pointer travel	1.000	Ea.	5,950	365	6,315
TOTAL			22,608.94	7,228.58	29,837.52

G3060 320	Steel Fuel Tank, Double Wall	COST EACH		
		MAT.	INST.	TOTAL
1010	Storage tank, fuel, underground, double wall steel, 500 Gal.	22,600	7,225	29,825
1020	2000 Gal.	30,200	9,400	39,600
1030	4000 Gal.	41,000	12,100	53,100
1040	6000 Gal.	55,500	13,400	68,900
1050	8000 Gal.	59,500	16,600	76,100
1060	10,000 Gal.	66,500	19,000	85,500

G3060 Fuel Distribution

G3060 320	Steel Fuel Tank, Double Wall	COST EACH		
		MAT.	INST.	TOTAL
1070	15,000 Gal.	75,500	22,100	97,600
1080	20,000 Gal.	87,500	26,400	113,900
1090	25,000 Gal.	136,000	30,100	166,100
1100	30,000 Gal.	155,000	32,100	187,100
1110	40,000 Gal.	195,000	40,300	235,300
1120	50,000 Gal.	236,000	48,200	284,200

G3060 325	Steel Tank, Above Ground, Single Wall	COST EACH		
		MAT.	INST.	TOTAL
1010	Storage tank, fuel, above ground, single wall steel, 550 Gal.	14,000	5,050	19,050
1020	2000 Gal.	22,000	5,150	27,150
1030	5000 Gal.	33,700	7,925	41,625
1040	10,000 Gal.	51,500	8,450	59,950
1050	15,000 Gal.	63,000	8,700	71,700
1060	20,000 Gal.	78,500	8,975	87,475
1070	25,000 Gal.	89,500	9,200	98,700
1080	30,000 Gal.	105,500	9,575	115,075

G3060 330	Steel Tank, Above Ground, Double Wall	COST EACH		
		MAT.	INST.	TOTAL
1010	Storage tank, fuel, above ground, double wall steel, 500 Gal.	14,000	5,100	19,100
1020	2000 Gal.	22,100	5,225	27,325
1030	4000 Gal.	31,100	5,325	36,425
1040	6000 Gal.	36,000	8,225	44,225
1050	8000 Gal.	43,000	8,450	51,450
1060	10,000 Gal.	46,500	8,600	55,100
1070	15,000 Gal.	64,000	8,900	72,900
1080	20,000 Gal.	71,000	9,200	80,200
1090	25,000 Gal.	83,500	9,475	92,975
1100	30,000 Gal.	90,000	9,850	99,850

For customer support on your Mechanical Cost Data, call 877.739.3689.

603

Reference Section

All the reference information is in one section, making it easy to find what you need to know . . . and easy to use the data set on a daily basis. This section is visually identified by a vertical black bar on the page edges.

In this Reference Section, we've included Equipment Rental Costs, a listing of rental and operating costs; Crew Listings, a full listing of all crews and equipment, and their costs; Historical Cost Indexes for cost comparisons over time; City Cost Indexes and Location Factors for adjusting costs to the region you are in; Reference Tables, where you will find explanations, estimating information and procedures, or technical data; Change Orders, information on pricing changes to contract documents; and an explanation of all the Abbreviations in the data set.

Table of Contents

605

Estimating Tips

- This section contains the average costs to rent and operate hundreds of pieces of construction equipment. This is useful information when estimating the time and material requirements of any particular operation in order to establish a unit or total cost. Equipment costs include not only rental, but also operating costs for equipment under normal use.

Rental Costs

- Equipment rental rates are obtained from the following industry sources throughout North America: contractors, suppliers, dealers, manufacturers, and distributors.

- Rental rates vary throughout the country, with larger cities generally having lower rates. Lease plans for new equipment are available for periods in excess of six months, with a percentage of payments applying toward purchase.

- Monthly rental rates vary from 2% to 5% of the purchase price of the equipment depending on the anticipated life of the equipment and its wearing parts.

- Weekly rental rates are about 1/3 the monthly rates, and daily rental rates are about 1/3 the weekly rate.

- Rental rates can also be treated as reimbursement costs for contractor-owned equipment. Owned equipment costs include depreciation, loan payments, interest, taxes, insurance, storage, and major repairs.

Operating Costs

- The operating costs include parts and labor for routine servicing, such as repair and replacement of pumps, filters and worn lines. Normal operating expendables, such as fuel, lubricants, tires and electricity (where applicable), are also included.

- Extraordinary operating expendables with highly variable wear patterns, such as diamond bits and blades, are excluded. These costs can be found as material costs in the Unit Price section.

- The hourly operating costs listed do not include the operator's wages.

Equipment Cost/Day

- Any power equipment required by a crew is shown in the Crew Listings with a daily cost.

- The daily cost of equipment needed by a crew is based on dividing the weekly rental rate by 5 (number of working days in the week), and then adding the hourly operating cost times 8 (the number of hours in a day). This "Equipment Cost/Day" is shown in the far right column of the Equipment Rental pages.

- If equipment is needed for only one or two days, it is best to develop your own cost by including components for daily rent and hourly operating cost. This is important when the listed Crew for a task does not contain the equipment needed, such as a crane for lifting mechanical heating/cooling equipment up onto a roof.

- If the quantity of work is less than the crew's Daily Output shown for a Unit Price line item that includes a bare unit equipment cost, it is recommended to estimate one day's rental cost and operating cost for equipment shown in the Crew Listing for that line item.

Mobilization/ Demobilization

- The cost to move construction equipment from an equipment yard or rental company to the job site and back again is not included in equipment rental costs listed in the Reference Section, nor in the bare equipment cost of any unit price line item, nor in any equipment costs shown in the Crew listings.

- Mobilization (to the site) and demobilization (from the site) costs can be found in the Unit Price Section.

- If a piece of equipment is already at the job site, it is not appropriate to utilize mobilization/demobilization. costs again in an estimate. ■

01 54 33 | Equipment Rental

		UNIT	HOURLY OPER. COST	RENT PER DAY	RENT PER WEEK	RENT PER MONTH	EQUIPMENT COST/DAY	
10	**0010**	**CONCRETE EQUIPMENT RENTAL** without operators							**10**
	0200	Bucket, concrete lightweight, 1/2 C.Y.	Ea.	.80	23.50	70	210	20.40	
	0300	1 C.Y.		.85	28	84	252	23.60	
	0400	1-1/2 C.Y.		1.10	38.50	115	345	31.80	
	0500	2 C.Y.		1.20	45	135	405	36.60	
	0580	8 C.Y.		6.00	257	770	2,300	202	
	0600	Cart, concrete, self-propelled, operator walking, 10 C.F.		3.25	56.50	170	510	60	
	0700	Operator riding, 18 C.F.		5.35	95	285	855	99.80	
	0800	Conveyer for concrete, portable, gas, 16" wide, 26' long		12.60	123	370	1,100	174.80	
	0900	46' long		13.00	148	445	1,325	193	
	1000	56' long		13.15	157	470	1,400	199.20	
	1100	Core drill, electric, 2-1/2 H.P., 1" to 8" bit diameter		2.15	78.50	235	705	64.20	
	1150	11 H.P., 8" to 18" cores		6.00	115	345	1,025	117	
	1200	Finisher, concrete floor, gas, riding trowel, 96" wide		12.90	145	435	1,300	190.20	
	1300	Gas, walk-behind, 3 blade, 36" trowel		2.15	20	60	180	29.20	
	1400	4 blade, 48" trowel		4.25	27.50	83	249	50.60	
	1500	Float, hand-operated (Bull float) 48" wide		.08	13.35	40	120	8.65	
	1570	Curb builder, 14 H.P., gas, single screw		14.40	248	745	2,225	264.20	
	1590	Double screw		15.10	292	875	2,625	295.80	
	1600	Floor grinder, concrete and terrazzo, electric, 22" path		2.70	167	500	1,500	121.60	
	1700	Edger, concrete, electric, 7" path		1.04	51.50	155	465	39.30	
	1750	Vacuum pick-up system for floor grinders, wet/dry		1.49	81.50	245	735	60.90	
	1800	Mixer, powered, mortar and concrete, gas, 6 C.F., 18 H.P.		8.65	118	355	1,075	140.20	
	1900	10 C.F., 25 H.P.		10.85	143	430	1,300	172.80	
	2000	16 C.F.		11.15	165	495	1,475	188.20	
	2100	Concrete, stationary, tilt drum, 2 C.Y.		6.80	232	695	2,075	193.40	
	2120	Pump, concrete, truck mounted 4" line 80' boom		24.70	885	2,650	7,950	727.60	
	2140	5" line, 110' boom		32.05	1,175	3,510	10,500	958.40	
	2160	Mud jack, 50 C.F. per hr.		7.25	128	385	1,150	135	
	2180	225 C.F. per hr.		9.55	147	440	1,325	164.40	
	2190	Shotcrete pump rig, 12 C.Y./hr.		15.00	223	670	2,000	254	
	2200	35 C.Y./hr.		17.50	240	720	2,150	284	
	2600	Saw, concrete, manual, gas, 18 H.P.		6.80	45	135	405	81.40	
	2650	Self-propelled, gas, 30 H.P.		13.45	102	305	915	168.60	
	2675	V-groove crack chaser, manual, gas, 6 H.P.		2.35	17.35	52	156	29.20	
	2700	Vibrators, concrete, electric, 60 cycle, 2 H.P.		.41	8.65	26	78	8.50	
	2800	3 H.P.		.60	11.65	35	105	11.80	
	2900	Gas engine, 5 H.P.		2.05	16.35	49	147	26.20	
	3000	8 H.P.		2.80	15.35	46	138	31.60	
	3050	Vibrating screed, gas engine, 8 H.P.		2.58	71.50	215	645	63.65	
	3120	Concrete transit mixer, 6 x 4, 250 H.P., 8 C.Y., rear discharge		61.15	570	1,715	5,150	832.20	
	3200	Front discharge		72.05	700	2,100	6,300	996.40	
	3300	6 x 6, 285 H.P., 12 C.Y., rear discharge		71.10	660	1,980	5,950	964.80	
	3400	Front discharge		74.30	710	2,125	6,375	1,019	
20	**0010**	**EARTHWORK EQUIPMENT RENTAL** without operators							**20**
	0040	Aggregate spreader, push type 8' to 12' wide	Ea.	3.05	25.50	76	228	39.60	
	0045	Tailgate type, 8' wide		2.90	32.50	98	294	42.80	
	0055	Earth auger, truck-mounted, for fence & sign posts, utility poles		19.10	445	1,340	4,025	420.80	
	0060	For borings and monitoring wells		46.50	690	2,070	6,200	786	
	0070	Portable, trailer mounted		3.25	33	99	297	45.80	
	0075	Truck-mounted, for caissons, water wells		98.70	2,950	8,815	26,400	2,553	
	0080	Horizontal boring machine, 12" to 36" diameter, 45 H.P.		24.40	195	585	1,750	312.20	
	0090	12" to 48" diameter, 65 H.P.		34.30	340	1,015	3,050	477.40	
	0095	Auger, for fence posts, gas engine, hand held		.60	6	18	54	8.40	
	0100	Excavator, diesel hydraulic, crawler mounted, 1/2 C.Y. cap.		24.45	415	1,250	3,750	445.60	
	0120	5/8 C.Y. capacity		32.30	550	1,655	4,975	589.40	
	0140	3/4 C.Y. capacity		35.25	620	1,860	5,575	654	
	0150	1 C.Y. capacity		49.35	695	2,090	6,275	812.80	

R015433 -10

		UNIT	HOURLY OPER. COST	RENT PER DAY	RENT PER WEEK	RENT PER MONTH	EQUIPMENT COST/DAY
**01 54 33	Equipment Rental**							
20	0200	1-1/2 C.Y. capacity	Ea.	59.40	925	2,770	8,300	1,029
	0300	2 C.Y. capacity		66.90	1,075	3,205	9,625	1,176
	0320	2-1/2 C.Y. capacity		102.00	1,275	3,830	11,500	1,582
	0325	3-1/2 C.Y. capacity		143.70	2,150	6,460	19,400	2,442
	0330	4-1/2 C.Y. capacity		172.45	2,650	7,970	23,900	2,974
	0335	6 C.Y. capacity		218.85	2,925	8,750	26,300	3,501
	0340	7 C.Y. capacity		221.10	3,050	9,140	27,400	3,597
	0342	Excavator attachments, bucket thumbs		3.15	245	735	2,200	172.20
	0345	Grapples		2.70	193	580	1,750	137.60
	0346	Hydraulic hammer for boom mounting, 4000 ft lb.		12.15	350	1,045	3,125	306.20
	0347	5000 ft lb.		14.10	425	1,275	3,825	367.80
	0348	8000 ft lb.		20.90	625	1,870	5,600	541.20
	0349	12,000 ft lb.		22.90	750	2,245	6,725	632.20
	0350	Gradall type, truck mounted, 3 ton @ 15' radius, 5/8 C.Y.		48.75	820	2,460	7,375	882
	0370	1 C.Y. capacity		53.85	940	2,825	8,475	995.80
	0400	Backhoe-loader, 40 to 45 H.P., 5/8 C.Y. capacity		14.85	242	725	2,175	263.80
	0450	45 H.P. to 60 H.P., 3/4 C.Y. capacity		23.30	298	895	2,675	365.40
	0460	80 H.P., 1-1/4 C.Y. capacity		25.90	315	945	2,825	396.20
	0470	112 H.P., 1-1/2 C.Y. capacity		40.25	640	1,915	5,750	705
	0482	Backhoe-loader attachment, compactor, 20,000 lb.		5.70	142	425	1,275	130.60
	0485	Hydraulic hammer, 750 ft lb.		3.20	95	285	855	82.60
	0486	Hydraulic hammer, 1200 ft lb.		6.20	217	650	1,950	179.60
	0500	Brush chipper, gas engine, 6" cutter head, 35 H.P.		11.10	103	310	930	150.80
	0550	Diesel engine, 12" cutter head, 130 H.P.		27.45	282	845	2,525	388.60
	0600	15" cutter head, 165 H.P.		33.00	340	1,020	3,050	468
	0750	Bucket, clamshell, general purpose, 3/8 C.Y.		1.25	38.50	115	345	33
	0800	1/2 C.Y.		1.35	45	135	405	37.80
	0850	3/4 C.Y.		1.50	55	165	495	45
	0900	1 C.Y.		1.55	58.50	175	525	47.40
	0950	1-1/2 C.Y.		2.50	80	240	720	68
	1000	2 C.Y.		2.65	90	270	810	75.20
	1010	Bucket, dragline, medium duty, 1/2 C.Y.		.70	23.50	70	210	19.60
	1020	3/4 C.Y.		.75	24.50	73	219	20.60
	1030	1 C.Y.		.75	26	78	234	21.60
	1040	1-1/2 C.Y.		1.20	40	120	360	33.60
	1050	2 C.Y.		1.25	45	135	405	37
	1070	3 C.Y.		1.95	61.50	185	555	52.60
	1200	Compactor, manually guided 2-drum vibratory smooth roller, 7.5 H.P.		7.35	210	630	1,900	184.80
	1250	Rammer/tamper, gas, 8"		2.75	46.50	140	420	50
	1260	15"		3.10	53.50	160	480	56.80
	1300	Vibratory plate, gas, 18" plate, 3000 lb. blow		2.70	24.50	73	219	36.20
	1350	21" plate, 5000 lb. blow		3.35	33	99	297	46.60
	1370	Curb builder/extruder, 14 H.P., gas, single screw		14.40	248	745	2,225	264.20
	1390	Double screw		15.10	292	875	2,625	295.80
	1500	Disc harrow attachment, for tractor		.45	75	225	675	48.60
	1810	Feller buncher, shearing & accumulating trees, 100 H.P.		47.95	755	2,270	6,800	837.60
	1860	Grader, self-propelled, 25,000 lb.		39.40	665	1,990	5,975	713.20
	1910	30,000 lb.		43.05	615	1,850	5,550	714.40
	1920	40,000 lb.		64.75	1,125	3,385	10,200	1,195
	1930	55,000 lb.		84.00	1,775	5,345	16,000	1,741
	1950	Hammer, pavement breaker, self-propelled, diesel, 1000 to 1250 lb.		29.55	350	1,055	3,175	447.40
	2000	1300 to 1500 lb.		44.33	705	2,110	6,325	776.65
	2050	Pile driving hammer, steam or air, 4150 ft lb. @ 225 bpm		10.25	480	1,435	4,300	369
	2100	8750 ft lb. @ 145 bpm		12.25	665	2,000	6,000	498
	2150	15,000 ft lb. @ 60 bpm		13.80	800	2,405	7,225	591.40
	2200	24,450 ft lb. @ 111 bpm		14.85	890	2,670	8,000	652.80
	2250	Leads, 60' high for pile driving hammers up to 20,000 ft lb.		3.25	80	240	720	74
	2300	90' high for hammers over 20,000 ft lb.		4.95	141	424	1,275	124.40

01 54 33 | Equipment Rental

		UNIT	HOURLY OPER. COST	RENT PER DAY	RENT PER WEEK	RENT PER MONTH	EQUIPMENT COST/DAY	
20	2350	Diesel type hammer, 22,400 ft lb.	Ea.	18.00	390	1,170	3,500	378	20
	2400	41,300 ft lb.		27.45	505	1,515	4,550	522.60	
	2450	141,000 ft lb.		47.40	880	2,640	7,925	907.20	
	2500	Vib. elec. hammer/extractor, 200 kW diesel generator, 34 H.P.		54.65	645	1,930	5,800	823.20	
	2550	80 H.P.		100.35	945	2,830	8,500	1,369	
	2600	150 H.P.		190.65	1,800	5,435	16,300	2,612	
	2800	Log chipper, up to 22" diameter, 600 H.P.		63.25	640	1,920	5,750	890	
	2850	Logger, for skidding & stacking logs, 150 H.P.		52.35	830	2,490	7,475	916.80	
	2860	Mulcher, diesel powered, trailer mounted		23.95	213	640	1,925	319.60	
	2900	Rake, spring tooth, with tractor		18.08	345	1,042	3,125	353.05	
	3000	Roller, vibratory, tandem, smooth drum, 20 H.P.		8.95	152	455	1,375	162.60	
	3050	35 H.P.		11.05	252	755	2,275	239.40	
	3100	Towed type vibratory compactor, smooth drum, 50 H.P.		25.70	325	980	2,950	401.60	
	3150	Sheepsfoot, 50 H.P.		27.05	350	1,050	3,150	426.40	
	3170	Landfill compactor, 220 H.P.		88.90	1,550	4,640	13,900	1,639	
	3200	Pneumatic tire roller, 80 H.P.		16.35	360	1,085	3,250	347.80	
	3250	120 H.P.		24.55	615	1,840	5,525	564.40	
	3300	Sheepsfoot vibratory roller, 240 H.P.		67.15	1,125	3,355	10,100	1,208	
	3320	340 H.P.		97.60	1,650	4,960	14,900	1,773	
	3350	Smooth drum vibratory roller, 75 H.P.		24.75	615	1,840	5,525	566	
	3400	125 H.P.		32.10	725	2,175	6,525	691.80	
	3410	Rotary mower, brush, 60", with tractor		22.30	320	955	2,875	369.40	
	3420	Rototiller, walk-behind, gas, 5 H.P.		1.76	51.50	155	465	45.10	
	3422	8 H.P.		2.75	80	240	720	70	
	3440	Scrapers, towed type, 7 C.Y. capacity		5.65	113	340	1,025	113.20	
	3450	10 C.Y. capacity		6.50	155	465	1,400	145	
	3500	15 C.Y. capacity		6.95	180	540	1,625	163.60	
	3525	Self-propelled, single engine, 14 C.Y. capacity		130.55	1,875	5,640	16,900	2,172	
	3550	Dual engine, 21 C.Y. capacity		173.20	2,225	6,645	19,900	2,715	
	3600	31 C.Y. capacity		230.80	3,150	9,445	28,300	3,735	
	3640	44 C.Y. capacity		286.85	4,075	12,220	36,700	4,739	
	3650	Elevating type, single engine, 11 C.Y. capacity		71.80	1,000	2,995	8,975	1,173	
	3700	22 C.Y. capacity		138.90	2,400	7,185	21,600	2,548	
	3710	Screening plant 110 H.P. w/5' x 10' screen		19.62	375	1,125	3,375	381.95	
	3720	5' x 16' screen		24.98	485	1,450	4,350	489.85	
	3850	Shovel, crawler-mounted, front-loading, 7 C.Y. capacity		257.95	3,575	10,740	32,200	4,212	
	3855	12 C.Y. capacity		419.50	4,950	14,860	44,600	6,328	
	3860	Shovel/backhoe bucket, 1/2 C.Y.		2.40	66.50	200	600	59.20	
	3870	3/4 C.Y.		2.45	73.50	220	660	63.60	
	3880	1 C.Y.		2.55	83.50	250	750	70.40	
	3890	1-1/2 C.Y.		2.70	96.50	290	870	79.60	
	3910	3 C.Y.		3.05	130	390	1,175	102.40	
	3950	Stump chipper, 18" deep, 30 H.P.		6.83	198	593	1,775	173.25	
	4110	Dozer, crawler, torque converter, diesel 80 H.P.		28.80	400	1,205	3,625	471.40	
	4150	105 H.P.		35.30	535	1,605	4,825	603.40	
	4200	140 H.P.		50.95	790	2,365	7,100	880.60	
	4260	200 H.P.		76.95	1,300	3,905	11,700	1,397	
	4310	300 H.P.		100.25	1,800	5,425	16,300	1,887	
	4360	410 H.P.		132.10	2,225	6,675	20,000	2,392	
	4370	500 H.P.		170.65	3,275	9,820	29,500	3,329	
	4380	700 H.P.		280.40	5,175	15,520	46,600	5,347	
	4400	Loader, crawler, torque conv., diesel, 1-1/2 C.Y., 80 H.P.		31.25	460	1,375	4,125	525	
	4450	1-1/2 to 1-3/4 C.Y., 95 H.P.		34.40	600	1,795	5,375	634.20	
	4510	1-3/4 to 2-1/4 C.Y., 130 H.P.		53.50	895	2,680	8,050	964	
	4530	2-1/2 to 3-1/4 C.Y., 190 H.P.		65.50	1,125	3,365	10,100	1,197	
	4560	3-1/2 to 5 C.Y., 275 H.P.		87.45	1,450	4,330	13,000	1,566	
	4610	Front end loader, 4WD, articulated frame, diesel, 1 to 1-1/4 C.Y., 70 H.P.		19.75	235	705	2,125	299	
	4620	1-1/2 to 1-3/4 C.Y., 95 H.P.		24.70	300	905	2,725	378.60	

01 54 33 | Equipment Rental

			UNIT	HOURLY OPER. COST	RENT PER DAY	RENT PER WEEK	RENT PER MONTH	EQUIPMENT COST/DAY	
20	4650	1-3/4 to 2 C.Y., 130 H.P.	Ea.	29.25	365	1,095	3,275	453	20
	4710	2-1/2 to 3-1/2 C.Y., 145 H.P.		33.25	415	1,250	3,750	516	
	4730	3 to 4-1/2 C.Y., 185 H.P.		42.60	535	1,605	4,825	661.80	
	4760	5-1/4 to 5-3/4 C.Y., 270 H.P.		67.20	925	2,770	8,300	1,092	
	4810	7 to 9 C.Y., 475 H.P.		113.75	1,750	5,270	15,800	1,964	
	4870	9 - 11 C.Y., 620 H.P.		160.25	3,000	9,015	27,000	3,085	
	4880	Skid steer loader, wheeled, 10 C.F., 30 H.P. gas		10.40	150	450	1,350	173.20	
	4890	1 C.Y., 78 H.P., diesel		20.20	257	770	2,300	315.60	
	4892	Skid-steer attachment, auger		.47	77.50	233	700	50.35	
	4893	Backhoe		.66	110	329	985	71.10	
	4894	Broom		.70	117	351	1,050	75.80	
	4895	Forks		.22	37	111	335	23.95	
	4896	Grapple		.53	88	264	790	57.05	
	4897	Concrete hammer		1.00	167	502	1,500	108.40	
	4898	Tree spade		.93	156	467	1,400	100.85	
	4899	Trencher		.59	98.50	295	885	63.70	
	4900	Trencher, chain, boom type, gas, operator walking, 12 H.P.		5.00	46.50	140	420	68	
	4910	Operator riding, 40 H.P.		20.05	290	870	2,600	334.40	
	5000	Wheel type, diesel, 4' deep, 12" wide		85.15	835	2,510	7,525	1,183	
	5100	6' deep, 20" wide		91.45	1,925	5,795	17,400	1,891	
	5150	Chain type, diesel, 5' deep, 8" wide		32.25	470	1,405	4,225	539	
	5200	Diesel, 8' deep, 16" wide		133.50	2,975	8,960	26,900	2,860	
	5202	Rock trencher, wheel type, 6" wide x 18" deep		33.65	575	1,730	5,200	615.20	
	5206	Chain type, 18" wide x 7' deep		110.75	2,850	8,525	25,600	2,591	
	5210	Tree spade, self-propelled		12.30	267	800	2,400	258.40	
	5250	Truck, dump, 2-axle, 12 ton, 8 C.Y. payload, 220 H.P.		34.30	238	715	2,150	417.40	
	5300	Three axle dump, 16 ton, 12 C.Y. payload, 400 H.P.		60.85	340	1,020	3,050	690.80	
	5310	Four axle dump, 25 ton, 18 C.Y. payload, 450 H.P.		71.55	495	1,485	4,450	869.40	
	5350	Dump trailer only, rear dump, 16-1/2 C.Y.		5.35	138	415	1,250	125.80	
	5400	20 C.Y.		5.80	157	470	1,400	140.40	
	5450	Flatbed, single axle, 1-1/2 ton rating		25.90	68.50	205	615	248.20	
	5500	3 ton rating		31.05	98.50	295	885	307.40	
	5550	Off highway rear dump, 25 ton capacity		73.60	1,300	3,905	11,700	1,370	
	5600	35 ton capacity		82.20	1,450	4,355	13,100	1,529	
	5610	50 ton capacity		102.35	1,725	5,210	15,600	1,861	
	5620	65 ton capacity		105.35	1,775	5,360	16,100	1,915	
	5630	100 ton capacity		150.95	2,900	8,710	26,100	2,950	
	6000	Vibratory plow, 25 H.P., walking		8.50	61.50	185	555	105	
40	0010	**GENERAL EQUIPMENT RENTAL** without operators	R015433 -10						40
	0150	Aerial lift, scissor type, to 15' high, 1000 lb. cap., electric	Ea.	3.15	51.50	155	465	56.20	
	0160	To 25' high, 2000 lb. capacity		3.60	68.50	205	615	69.80	
	0170	Telescoping boom to 40' high, 500 lb. capacity, diesel		13.50	335	1,005	3,025	309	
	0180	To 45' high, 500 lb. capacity		14.75	365	1,090	3,275	336	
	0190	To 60' high, 600 lb. capacity		17.30	515	1,550	4,650	448.40	
	0195	Air compressor, portable, 6.5 CFM, electric		.79	13	39	117	14.10	
	0196	Gasoline		.70	19.65	59	177	17.40	
	0200	Towed type, gas engine, 60 CFM		14.10	50	150	450	142.80	
	0300	160 CFM		16.40	51.50	155	465	162.20	
	0400	Diesel engine, rotary screw, 250 CFM		16.35	112	335	1,000	197.80	
	0500	365 CFM		22.05	137	410	1,225	258.40	
	0550	450 CFM		27.95	170	510	1,525	325.60	
	0600	600 CFM		49.10	235	705	2,125	533.80	
	0700	750 CFM		49.25	245	735	2,200	541	
	0800	For silenced models, small sizes, add to rent		3%	5%	5%	5%		
	0900	Large sizes, add to rent		5%	7%	7%	7%		
	0930	Air tools, breaker, pavement, 60 lb.	Ea.	.50	10.35	31	93	10.20	
	0940	80 lb.		.50	10.35	31	93	10.20	
	0950	Drills, hand (jackhammer) 65 lb.		.60	17.65	53	159	15.40	

01 54 33 | Equipment Rental

		UNIT	HOURLY OPER. COST	RENT PER DAY	RENT PER WEEK	RENT PER MONTH	EQUIPMENT COST/DAY	
40	0960	Track or wagon, swing boom, 4" drifter	Ea.	62.25	910	2,735	8,200	1,045	**40**
	0970	5" drifter		73.80	1,100	3,290	9,875	1,248	
	0975	Track mounted quarry drill, 6" diameter drill		126.40	1,675	4,990	15,000	2,009	
	0980	Dust control per drill		1.04	25	75	225	23.30	
	0990	Hammer, chipping, 12 lb.		.55	26	78	234	20	
	1000	Hose, air with couplings, 50' long, 3/4" diameter		.03	4.67	14	42	3.05	
	1100	1" diameter		.03	5.65	17	51	3.65	
	1200	1-1/2" diameter		.05	8	24	72	5.20	
	1300	2" diameter		.06	10.65	32	96	6.90	
	1400	2-1/2" diameter		.11	19	57	171	12.30	
	1410	3" diameter		.13	21.50	64	192	13.85	
	1450	Drill, steel, 7/8" x 2'		.06	9.35	28	84	6.10	
	1460	7/8" x 6'		.05	8.65	26	78	5.60	
	1520	Moil points		.02	4	12	36	2.55	
	1525	Pneumatic nailer w/accessories		.55	36.50	109	325	26.20	
	1530	Sheeting driver for 60 lb. breaker		.04	6.65	20	60	4.30	
	1540	For 90 lb. breaker		.14	9	27	81	6.50	
	1550	Spade, 25 lb.		.50	7	21	63	8.20	
	1560	Tamper, single, 35 lb.		.50	33.50	100	300	24	
	1570	Triple, 140 lb.		.75	50	150	450	36	
	1580	Wrenches, impact, air powered, up to 3/4" bolt		.40	12.65	38	114	10.80	
	1590	Up to 1-1/4" bolt		.50	23.50	71	213	18.20	
	1600	Barricades, barrels, reflectorized, 1 to 99 barrels		.03	4.60	13.80	41.50	3	
	1610	100 to 200 barrels		.02	3.53	10.60	32	2.30	
	1620	Barrels with flashers, 1 to 99 barrels		.03	5.25	15.80	47.50	3.40	
	1630	100 to 200 barrels		.03	4.20	12.60	38	2.75	
	1640	Barrels with steady burn type C lights		.04	7	21	63	4.50	
	1650	Illuminated board, trailer mounted, with generator		3.50	130	390	1,175	106	
	1670	Portable barricade, stock, with flashers, 1 to 6 units		.03	5.25	15.80	47.50	3.40	
	1680	25 to 50 units		.03	4.90	14.70	44	3.20	
	1685	Butt fusion machine, wheeled, 1.5 HP electric, 2" - 8" diameter pipe		2.63	167	500	1,500	121.05	
	1690	Tracked, 20 HP diesel, 4"-12" diameter pipe		10.39	490	1,465	4,400	376.10	
	1695	83 HP diesel, 8" - 24" diameter pipe		27.35	975	2,930	8,800	804.80	
	1700	Carts, brick, hand powered, 1000 lb. capacity		.41	68.50	205	615	44.30	
	1800	Gas engine, 1500 lb., 7-1/2' lift		3.61	108	324	970	93.70	
	1822	Dehumidifier, medium, 6 lb./hr., 150 CFM		1.08	67.50	202.50	610	49.15	
	1824	Large, 18 lb./hr., 600 CFM		2.06	129	385.50	1,150	93.60	
	1830	Distributor, asphalt, trailer mounted, 2000 gal., 38 H.P. diesel		10.25	345	1,035	3,100	289	
	1840	3000 gal., 38 H.P. diesel		11.80	375	1,125	3,375	319.40	
	1850	Drill, rotary hammer, electric		1.07	26.50	80	240	24.55	
	1860	Carbide bit, 1-1/2" diameter, add to electric rotary hammer		.03	4.44	13.33	40	2.90	
	1865	Rotary, crawler, 250 H.P.		147.20	2,150	6,480	19,400	2,474	
	1870	Emulsion sprayer, 65 gal., 5 H.P. gas engine		2.79	101	303	910	82.90	
	1880	200 gal., 5 H.P. engine		8.05	168	505	1,525	165.40	
	1900	Floor auto-scrubbing machine, walk-behind, 28" path		5.13	335	1,000	3,000	241.05	
	1930	Floodlight, mercury vapor, or quartz, on tripod, 1000 watt		.44	20.50	62	186	15.90	
	1940	2000 watt		.83	41.50	124	370	31.45	
	1950	Floodlights, trailer mounted with generator, 1 - 300 watt light		3.60	71.50	215	645	71.80	
	1960	2 - 1000 watt lights		4.80	96.50	290	870	96.40	
	2000	4 - 300 watt lights		4.50	91.50	275	825	91	
	2005	Foam spray rig, incl. box trailer, compressor, generator, proportioner		33.75	495	1,485	4,450	567	
	2020	Forklift, straight mast, 12' lift, 5000 lb., 2 wheel drive, gas		26.25	207	620	1,850	334	
	2040	21' lift, 5000 lb., 4 wheel drive, diesel		20.45	250	750	2,250	313.60	
	2050	For rough terrain, 42' lift, 35' reach, 9000 lb., 110 H.P.		29.40	500	1,495	4,475	534.20	
	2060	For plant, 4 ton capacity, 80 H.P., 2 wheel drive, gas		16.05	95	285	855	185.40	
	2080	10 ton capacity, 120 H.P., 2 wheel drive, diesel		23.20	165	495	1,475	284.60	
	2100	Generator, electric, gas engine, 1.5 kW to 3 kW		3.70	11.65	35	105	36.60	
	2200	5 kW		4.80	15.35	46	138	47.60	

01 54 33 | Equipment Rental

			UNIT	HOURLY OPER. COST	RENT PER DAY	RENT PER WEEK	RENT PER MONTH	EQUIPMENT COST/DAY	
40	2300	10 kW	Ea.	9.05	38.50	115	345	95.40	40
	2400	25 kW		10.25	86.50	260	780	134	
	2500	Diesel engine, 20 kW		11.80	70	210	630	136.40	
	2600	50 kW		22.50	105	315	945	243	
	2700	100 kW		41.50	135	405	1,225	413	
	2800	250 kW		81.70	257	770	2,300	807.60	
	2850	Hammer, hydraulic, for mounting on boom, to 500 ft lb.		2.50	75	225	675	65	
	2860	1000 ft lb.		4.25	125	375	1,125	109	
	2900	Heaters, space, oil or electric, 50 MBH		1.59	7.65	23	69	17.30	
	3000	100 MBH		2.93	10.65	32	96	29.85	
	3100	300 MBH		8.54	38.50	115	345	91.30	
	3150	500 MBH		14.00	45	135	405	139	
	3200	Hose, water, suction with coupling, 20' long, 2" diameter		.02	3	9	27	1.95	
	3210	3" diameter		.03	4.33	13	39	2.85	
	3220	4" diameter		.03	5	15	45	3.25	
	3230	6" diameter		.11	17.65	53	159	11.50	
	3240	8" diameter		.32	53.50	161	485	34.75	
	3250	Discharge hose with coupling, 50' long, 2" diameter		.01	1.33	4	12	.90	
	3260	3" diameter		.01	2.33	7	21	1.50	
	3270	4" diameter		.02	3.67	11	33	2.35	
	3280	6" diameter		.06	9.35	28	84	6.10	
	3290	8" diameter		.20	33	99	297	21.40	
	3295	Insulation blower		.78	6	18	54	9.85	
	3300	Ladders, extension type, 16' to 36' long		.14	22.50	68	204	14.70	
	3400	40' to 60' long		.19	31	93	279	20.10	
	3405	Lance for cutting concrete		2.28	64	192	575	56.65	
	3407	Lawn mower, rotary, 22", 5 H.P.		1.55	39.50	118	355	36	
	3408	48" self propelled		3.36	109	327	980	92.30	
	3410	Level, electronic, automatic, with tripod and leveling rod		.91	60.50	182	545	43.70	
	3430	Laser type, for pipe and sewer line and grade		.73	48.50	145	435	34.85	
	3440	Rotating beam for interior control		.78	52	156	470	37.45	
	3460	Builder's optical transit, with tripod and rod		.09	14.65	44	132	9.50	
	3500	Light towers, towable, with diesel generator, 2000 watt		4.50	91.50	275	825	91	
	3600	4000 watt		4.80	96.50	290	870	96.40	
	3700	Mixer, powered, plaster and mortar, 6 C.F., 7 H.P.		2.70	20	60	180	33.60	
	3800	10 C.F., 9 H.P.		2.85	31.50	95	285	41.80	
	3850	Nailer, pneumatic		.55	36.50	109	325	26.20	
	3900	Paint sprayers complete, 8 CFM		1.05	70	210	630	50.40	
	4000	17 CFM		1.98	132	395	1,175	94.85	
	4020	Pavers, bituminous, rubber tires, 8' wide, 50 H.P., diesel		31.90	510	1,535	4,600	562.20	
	4030	10' wide, 150 H.P.		106.55	1,875	5,635	16,900	1,979	
	4050	Crawler, 8' wide, 100 H.P., diesel		90.15	1,900	5,670	17,000	1,855	
	4060	10' wide, 150 H.P.		113.75	2,300	6,890	20,700	2,288	
	4070	Concrete paver, 12' to 24' wide, 250 H.P.		100.45	1,550	4,685	14,100	1,741	
	4080	Placer-spreader-trimmer, 24' wide, 300 H.P.		149.30	2,675	8,010	24,000	2,796	
	4100	Pump, centrifugal gas pump, 1-1/2" diam., 65 GPM		4.05	51.50	155	465	63.40	
	4200	2" diameter, 130 GPM		5.45	56.50	170	510	77.60	
	4300	3" diameter, 250 GPM		5.75	58.50	175	525	81	
	4400	6" diameter, 1500 GPM		30.35	180	540	1,625	350.80	
	4500	Submersible electric pump, 1-1/4" diameter, 55 GPM		.40	17	51	153	13.40	
	4600	1-1/2" diameter, 83 GPM		.44	19.35	58	174	15.10	
	4700	2" diameter, 120 GPM		1.55	24.50	73	219	27	
	4800	3" diameter, 300 GPM		2.79	43.50	130	390	48.30	
	4900	4" diameter, 560 GPM		14.32	162	485	1,450	211.55	
	5000	6" diameter, 1590 GPM		21.51	217	650	1,950	302.10	
	5100	Diaphragm pump, gas, single, 1-1/2" diameter		1.08	49.50	148	445	38.25	
	5200	2" diameter		4.35	61.50	185	555	71.80	
	5300	3" diameter		4.35	65	195	585	73.80	

01 54 33 | Equipment Rental

		UNIT	HOURLY OPER. COST	RENT PER DAY	RENT PER WEEK	RENT PER MONTH	EQUIPMENT COST/DAY
5400	Double, 4" diameter	Ea.	6.50	110	330	990	118
5450	Pressure washer 5 GPM, 3000 psi		4.95	53.50	160	480	71.60
5460	7 GPM, 3000 psi		6.45	61.50	185	555	88.60
5500	Trash pump, self-priming, gas, 2" diameter		4.65	22	66	198	50.40
5600	Diesel, 4" diameter		8.95	91.50	275	825	126.60
5650	Diesel, 6" diameter		24.25	153	460	1,375	286
5655	Grout Pump		26.25	272	815	2,450	373
5700	Salamanders, L.P. gas fired, 100,000 Btu		3.13	13	39	117	32.85
5705	50,000 Btu		1.74	10.35	31	93	20.10
5720	Sandblaster, portable, open top, 3 C.F. capacity		.60	26.50	80	240	20.80
5730	6 C.F. capacity		.95	40	120	360	31.60
5740	Accessories for above		.13	22	66	198	14.25
5750	Sander, floor		.98	27.50	83	249	24.45
5760	Edger		.53	15.65	47	141	13.65
5800	Saw, chain, gas engine, 18" long		2.30	22	66	198	31.60
5900	Hydraulic powered, 36" long		.75	65	195	585	45
5950	60" long		.75	66.50	200	600	46
6000	Masonry, table mounted, 14" diameter, 5 H.P.		1.32	56.50	170	510	44.55
6050	Portable cut-off, 8 H.P.		2.50	33.50	100	300	40
6100	Circular, hand held, electric, 7-1/4" diameter		.23	4.67	14	42	4.65
6200	12" diameter		.23	8	24	72	6.65
6250	Wall saw, w/hydraulic power, 10 H.P.		9.75	61.50	185	555	115
6275	Shot blaster, walk-behind, 20" wide		4.80	295	885	2,650	215.40
6280	Sidewalk broom, walk-behind		2.04	65	195	585	55.30
6300	Steam cleaner, 100 gallons per hour		3.80	78.50	235	705	77.40
6310	200 gallons per hour		5.40	95	285	855	100.20
6340	Tar Kettle/Pot, 400 gallons		16.30	76.50	230	690	176.40
6350	Torch, cutting, acetylene-oxygen, 150' hose, excludes gases		.30	15	45	135	11.40
6360	Hourly operating cost includes tips and gas		21.00				168
6410	Toilet, portable chemical		.13	21.50	64	192	13.85
6420	Recycle flush type		.15	25.50	77	231	16.60
6430	Toilet, fresh water flush, garden hose,		.19	31	93	279	20.10
6440	Hoisted, non-flush, for high rise		.15	25	75	225	16.20
6465	Tractor, farm with attachment		21.35	305	910	2,725	352.80
6480	Trailers, platform, flush deck, 2 axle, 3 ton capacity		1.45	19.65	59	177	23.40
6500	25 ton capacity		5.40	117	350	1,050	113.20
6600	40 ton capacity		6.95	165	495	1,475	154.60
6700	3 axle, 50 ton capacity		7.50	183	550	1,650	170
6800	75 ton capacity		9.30	240	720	2,150	218.40
6810	Trailer mounted cable reel for high voltage line work		5.57	265	795	2,375	203.55
6820	Trailer mounted cable tensioning rig		11.06	525	1,580	4,750	404.50
6830	Cable pulling rig		71.33	2,950	8,850	26,600	2,341
6850	Portable cable/wire puller, 8000 lb max pulling capacity		3.71	167	502	1,500	130.10
6900	Water tank trailer, engine driven discharge, 5000 gallons		6.95	143	430	1,300	141.60
6925	10,000 gallons		9.45	198	595	1,775	194.60
6950	Water truck, off highway, 6000 gallons		87.60	780	2,335	7,000	1,168
7010	Tram car for high voltage line work, powered, 2 conductor		6.66	144	431	1,300	139.50
7020	Transit (builder's level) with tripod		.09	14.65	44	132	9.50
7030	Trench box, 3000 lb., 6' x 8'		.56	93.50	280	840	60.50
7040	7200 lb., 6' x 20'		.75	125	375	1,125	81
7050	8000 lb., 8' x 16'		1.08	180	540	1,625	116.65
7060	9500 lb., 8' x 20'		1.21	201	603	1,800	130.30
7065	11,000 lb., 8' x 24'		1.27	211	633	1,900	136.75
7070	12,000 lb., 10' x 20'		1.50	249	748	2,250	161.60
7100	Truck, pickup, 3/4 ton, 2 wheel drive		13.85	58.50	175	525	145.80
7200	4 wheel drive		14.10	73.50	220	660	156.80
7250	Crew carrier, 9 passenger		19.60	86.50	260	780	208.80
7290	Flat bed truck, 20,000 lb. GVW		21.60	125	375	1,125	247.80

40 ... **40**

01 54 33 | Equipment Rental

		UNIT	HOURLY OPER. COST	RENT PER DAY	RENT PER WEEK	RENT PER MONTH	EQUIPMENT COST/DAY
40								**40**
7300	Tractor, 4 x 2, 220 H.P.	Ea.	30.10	197	590	1,775	358.80
7410	330 H.P.		44.65	272	815	2,450	520.20
7500	6 x 4, 380 H.P.		51.25	315	950	2,850	600
7600	450 H.P.		62.10	385	1,150	3,450	726.80
7610	Tractor, with A frame, boom and winch, 225 H.P.		33.00	273	820	2,450	428
7620	Vacuum truck, hazardous material, 2500 gallons		11.85	292	875	2,625	269.80
7625	5,000 gallons		13.20	410	1,230	3,700	351.60
7650	Vacuum, HEPA, 16 gallon, wet/dry		.85	18	54	162	17.60
7655	55 gallon, wet/dry		.80	27	81	243	22.60
7660	Water tank, portable		.16	26.50	80	240	17.30
7690	Sewer/catch basin vacuum, 14 C.Y., 1500 gallons		17.54	615	1,850	5,550	510.30
7700	Welder, electric, 200 amp		3.88	16.35	49	147	40.85
7800	300 amp		5.76	20	60	180	58.10
7900	Gas engine, 200 amp		14.45	24.50	73	219	130.20
8000	300 amp		16.55	24.50	74	222	147.20
8100	Wheelbarrow, any size		.08	13	39	117	8.45
8200	Wrecking ball, 4000 lb.		2.30	68.50	205	615	59.40
50								**50**
0010	**HIGHWAY EQUIPMENT RENTAL** without operators						
0050	Asphalt batch plant, portable drum mixer, 100 ton/hr.	Ea.	80.81	1,450	4,380	13,100	1,522
0060	200 ton/hr.		92.38	1,550	4,655	14,000	1,670
0070	300 ton/hr.		109.99	1,825	5,465	16,400	1,973
0100	Backhoe attachment, long stick, up to 185 H.P., 10.5' long		.36	23.50	71	213	17.10
0140	Up to 250 H.P., 12' long		.39	25.50	77	231	18.50
0180	Over 250 H.P., 15' long		.54	35.50	107	320	25.70
0200	Special dipper arm, up to 100 H.P., 32' long		1.10	73	219	655	52.60
0240	Over 100 H.P., 33' long		1.37	91	273	820	65.55
0280	Catch basin/sewer cleaning truck, 3 ton, 9 C.Y., 1000 gal.		42.90	390	1,170	3,500	577.20
0300	Concrete batch plant, portable, electric, 200 C.Y./hr.		22.52	515	1,545	4,625	489.15
0520	Grader/dozer attachment, ripper/scarifier, rear mounted, up to 135 H.P.		2.90	58.50	175	525	58.20
0540	Up to 180 H.P.		3.85	88.50	265	795	83.80
0580	Up to 250 H.P.		5.50	142	425	1,275	129
0700	Pvmt. removal bucket, for hyd. excavator, up to 90 H.P.		1.85	51.50	155	465	45.80
0740	Up to 200 H.P.		2.05	68.50	205	615	57.40
0780	Over 200 H.P.		2.20	85	255	765	68.60
0900	Aggregate spreader, self-propelled, 187 H.P.		52.46	685	2,050	6,150	829.70
1000	Chemical spreader, 3 C.Y.		3.25	43.50	130	390	52
1900	Hammermill, traveling, 250 H.P.		70.05	2,125	6,360	19,100	1,832
2000	Horizontal borer, 3" diameter, 13 H.P. gas driven		6.40	55	165	495	84.20
2150	Horizontal directional drill, 20,000 lb. thrust, 78 H.P. diesel		31.00	680	2,045	6,125	657
2160	30,000 lb. thrust, 115 H.P.		38.60	1,050	3,135	9,400	935.80
2170	50,000 lb. thrust, 170 H.P.		55.35	1,325	4,005	12,000	1,244
2190	Mud trailer for HDD, 1500 gallons, 175 H.P., gas		32.25	153	460	1,375	350
2200	Hydromulcher, diesel, 3000 gallon, for truck mounting		23.30	247	740	2,225	334.40
2300	Gas, 600 gallon		8.60	98.50	295	885	127.80
2400	Joint & crack cleaner, walk behind, 25 H.P.		4.00	50	150	450	62
2500	Filler, trailer mounted, 400 gallons, 20 H.P.		9.45	212	635	1,900	202.60
3000	Paint striper, self-propelled, 40 gallon, 22 H.P.		7.25	157	470	1,400	152
3100	120 gallon, 120 H.P.		23.15	400	1,195	3,575	424.20
3200	Post drivers, 6" I-Beam frame, for truck mounting		19.00	385	1,155	3,475	383
3400	Road sweeper, self-propelled, 8' wide, 90 H.P.		39.40	585	1,755	5,275	666.20
3450	Road sweeper, vacuum assisted, 4 C.Y., 220 gallons		75.35	635	1,905	5,725	983.80
4000	Road mixer, self-propelled, 130 H.P.		47.70	785	2,350	7,050	851.60
4100	310 H.P.		83.25	2,125	6,395	19,200	1,945
4220	Cold mix paver, incl. pug mill and bitumen tank, 165 H.P.		98.45	2,400	7,190	21,600	2,226
4240	Pavement brush, towed		3.15	93.50	280	840	81.20
4250	Paver, asphalt, wheel or crawler, 130 H.P., diesel		98.15	2,375	7,125	21,400	2,210
4300	Paver, road widener, gas 1' to 6', 67 H.P.		49.10	915	2,750	8,250	942.80

R015433 -10

01 54 33 | Equipment Rental

			UNIT	HOURLY OPER. COST	RENT PER DAY	RENT PER WEEK	RENT PER MONTH	EQUIPMENT COST/DAY	
50	4400	Diesel, 2' to 14', 88 H.P.	Ea.	62.75	1,075	3,205	9,625	1,143	**50**
	4600	Slipform pavers, curb and gutter, 2 track, 75 H.P.		61.85	1,000	3,035	9,100	1,102	
	4700	4 track, 165 H.P.		41.15	750	2,255	6,775	780.20	
	4800	Median barrier, 215 H.P.		66.45	1,225	3,670	11,000	1,266	
	4901	Trailer, low bed, 75 ton capacity		10.05	238	715	2,150	223.40	
	5000	Road planer, walk behind, 10" cutting width, 10 H.P.		3.70	33.50	100	300	49.60	
	5100	Self-propelled, 12" cutting width, 64 H.P.		10.50	113	340	1,025	152	
	5120	Traffic line remover, metal ball blaster, truck mounted, 115 H.P.		48.45	765	2,295	6,875	846.60	
	5140	Grinder, truck mounted, 115 H.P.		54.85	830	2,490	7,475	936.80	
	5160	Walk-behind, 11 H.P.		4.20	53.50	160	480	65.60	
	5200	Pavement profiler, 4' to 6' wide, 450 H.P.		252.75	3,400	10,165	30,500	4,055	
	5300	8' to 10' wide, 750 H.P.		395.65	4,450	13,350	40,100	5,835	
	5400	Roadway plate, steel, 1" x 8' x 20'		.08	13.65	41	123	8.85	
	5600	Stabilizer, self-propelled, 150 H.P.		48.85	625	1,870	5,600	764.80	
	5700	310 H.P.		96.75	1,775	5,295	15,900	1,833	
	5800	Striper, truck mounted, 120 gallon paint, 460 H.P.		64.90	485	1,455	4,375	810.20	
	5900	Thermal paint heating kettle, 115 gallons		8.29	25.50	77	231	81.70	
	6000	Tar kettle, 330 gallon, trailer mounted		12.54	58.50	175	525	135.30	
	7000	Tunnel locomotive, diesel, 8 to 12 ton		32.15	595	1,780	5,350	613.20	
	7005	Electric, 10 ton		26.90	680	2,035	6,100	622.20	
	7010	Muck cars, 1/2 C.Y. capacity		2.10	24.50	74	222	31.60	
	7020	1 C.Y. capacity		2.30	33	99	297	38.20	
	7030	2 C.Y. capacity		2.45	38.50	115	345	42.60	
	7040	Side dump, 2 C.Y. capacity		2.65	45	135	405	48.20	
	7050	3 C.Y. capacity		3.60	51.50	155	465	59.80	
	7060	5 C.Y. capacity		5.15	65	195	585	80.20	
	7100	Ventilating blower for tunnel, 7-1/2 H.P.		2.10	51.50	155	465	47.80	
	7110	10 H.P.		2.28	53.50	160	480	50.25	
	7120	20 H.P.		3.54	69.50	208	625	69.90	
	7140	40 H.P.		5.81	98.50	295	885	105.50	
	7160	60 H.P.		8.92	153	460	1,375	163.35	
	7175	75 H.P.		11.71	210	630	1,900	219.70	
	7180	200 H.P.		23.75	305	920	2,750	374	
	7800	Windrow loader, elevating		55.85	1,325	4,000	12,000	1,247	
60	0010	**LIFTING AND HOISTING EQUIPMENT RENTAL** without operators							**60**
	0120	Aerial lift truck, 2 person, to 80'	Ea.	26.40	745	2,240	6,725	659.20	
	0140	Boom work platform, 40' snorkel		11.75	285	855	2,575	265	
	0150	Crane, flatbed mounted, 3 ton capacity		16.15	198	595	1,775	248.20	
	0200	Crane, climbing, 106' jib, 6000 lb. capacity, 410 fpm		39.85	1,675	5,060	15,200	1,331	
	0300	101' jib, 10,250 lb. capacity, 270 fpm		46.60	2,125	6,410	19,200	1,655	
	0500	Tower, static, 130' high, 106' jib, 6200 lb. capacity at 400 fpm		43.75	1,950	5,840	17,500	1,518	
	0600	Crawler mounted, lattice boom, 1/2 C.Y., 15 tons at 12' radius		36.96	665	2,000	6,000	695.70	
	0700	3/4 C.Y., 20 tons at 12' radius		49.28	835	2,510	7,525	896.25	
	0800	1 C.Y., 25 tons at 12' radius		65.70	1,125	3,340	10,000	1,194	
	0900	1-1/2 C.Y., 40 tons at 12' radius		65.75	1,125	3,385	10,200	1,203	
	1000	2 C.Y., 50 tons at 12' radius		69.75	1,325	3,940	11,800	1,346	
	1100	3 C.Y., 75 tons at 12' radius		74.30	1,525	4,565	13,700	1,507	
	1200	100 ton capacity, 60' boom		84.05	1,650	4,975	14,900	1,667	
	1300	165 ton capacity, 60' boom		106.80	2,050	6,170	18,500	2,088	
	1400	200 ton capacity, 70' boom		129.75	2,575	7,740	23,200	2,586	
	1500	350 ton capacity, 80' boom		181.25	3,850	11,575	34,700	3,765	
	1600	Truck mounted, lattice boom, 6 x 4, 20 tons at 10' radius		35.42	1,150	3,420	10,300	967.35	
	1700	25 tons at 10' radius		38.28	1,250	3,720	11,200	1,050	
	1800	8 x 4, 30 tons at 10' radius		41.49	1,325	3,960	11,900	1,124	
	1900	40 tons at 12' radius		44.28	1,375	4,140	12,400	1,182	
	2000	60 tons at 15' radius		49.79	1,450	4,380	13,100	1,274	
	2050	82 tons at 15' radius		55.71	1,550	4,680	14,000	1,382	
	2100	90 tons at 15' radius		62.48	1,700	5,100	15,300	1,520	

Reference codes in table: 0010 → R015433-10; 0140 → R015433-15; 0200 → R312316-45

01 54 33 | Equipment Rental

		UNIT	HOURLY OPER. COST	RENT PER DAY	RENT PER WEEK	RENT PER MONTH	EQUIPMENT COST/DAY	
60	2200	115 tons at 15' radius	Ea.	70.50	1,900	5,700	17,100	1,704	60
	2300	150 tons at 18' radius		83.50	2,000	6,000	18,000	1,868	
	2350	165 tons at 18' radius		82.77	2,125	6,360	19,100	1,934	
	2400	Truck mounted, hydraulic, 12 ton capacity		41.55	545	1,630	4,900	658.40	
	2500	25 ton capacity		44.10	660	1,980	5,950	748.80	
	2550	33 ton capacity		44.65	675	2,030	6,100	763.20	
	2560	40 ton capacity		57.95	795	2,390	7,175	941.60	
	2600	55 ton capacity		74.95	870	2,615	7,850	1,123	
	2700	80 ton capacity		98.50	1,400	4,210	12,600	1,630	
	2720	100 ton capacity		92.65	1,475	4,425	13,300	1,626	
	2740	120 ton capacity		107.65	1,625	4,845	14,500	1,830	
	2760	150 ton capacity		126.15	2,100	6,320	19,000	2,273	
	2800	Self-propelled, 4 x 4, with telescoping boom, 5 ton		17.35	237	710	2,125	280.80	
	2900	12-1/2 ton capacity		31.65	375	1,130	3,400	479.20	
	3000	15 ton capacity		32.35	395	1,190	3,575	496.80	
	3050	20 ton capacity		35.10	455	1,370	4,100	554.80	
	3100	25 ton capacity		36.70	520	1,565	4,700	606.60	
	3150	40 ton capacity		45.10	585	1,755	5,275	711.80	
	3200	Derricks, guy, 20 ton capacity, 60' boom, 75' mast		23.97	415	1,238	3,725	439.35	
	3300	100' boom, 115' mast		37.86	710	2,130	6,400	728.90	
	3400	Stiffleg, 20 ton capacity, 70' boom, 37' mast		26.57	535	1,610	4,825	534.55	
	3500	100' boom, 47' mast		41.01	860	2,580	7,750	844.10	
	3550	Helicopter, small, lift to 1250 lb. maximum, w/pilot		96.80	3,325	10,000	30,000	2,774	
	3600	Hoists, chain type, overhead, manual, 3/4 ton		.15	.33	1	3	1.40	
	3900	10 ton		.75	6	18	54	9.60	
	4000	Hoist and tower, 5000 lb. cap., portable electric, 40' high		5.02	238	713	2,150	182.75	
	4100	For each added 10' section, add		.11	18.65	56	168	12.10	
	4200	Hoist and single tubular tower, 5000 lb. electric, 100' high		6.80	330	996	3,000	253.60	
	4300	For each added 6'-6" section, add		.19	32.50	97	291	20.90	
	4400	Hoist and double tubular tower, 5000 lb., 100' high		7.30	365	1,097	3,300	277.80	
	4500	For each added 6'-6" section, add		.21	35.50	107	320	23.10	
	4550	Hoist and tower, mast type, 6000 lb., 100' high		7.87	380	1,137	3,400	290.35	
	4570	For each added 10' section, add		.13	22	66	198	14.25	
	4600	Hoist and tower, personnel, electric, 2000 lb., 100' @ 125 fpm		16.61	1,000	3,030	9,100	738.90	
	4700	3000 lb., 100' @ 200 fpm		18.97	1,150	3,430	10,300	837.75	
	4800	3000 lb., 150' @ 300 fpm		21.02	1,275	3,840	11,500	936.15	
	4900	4000 lb., 100' @ 300 fpm		21.78	1,300	3,920	11,800	958.25	
	5000	6000 lb., 100' @ 275 fpm		23.46	1,375	4,110	12,300	1,010	
	5100	For added heights up to 500', add	L.F.	.01	1.67	5	15	1.10	
	5200	Jacks, hydraulic, 20 ton	Ea.	.05	2	6	18	1.60	
	5500	100 ton		.40	11.65	35	105	10.20	
	6100	Jacks, hydraulic, climbing w/50' jackrods, control console, 30 ton cap.		2.05	137	410	1,225	98.40	
	6150	For each added 10' jackrod section, add		.05	3.33	10	30	2.40	
	6300	50 ton capacity		3.30	220	659	1,975	158.20	
	6350	For each added 10' jackrod section, add		.06	4	12	36	2.90	
	6500	125 ton capacity		8.60	575	1,720	5,150	412.80	
	6550	For each added 10' jackrod section, add		.59	39	117	350	28.10	
	6600	Cable jack, 10 ton capacity with 200' cable		1.72	114	343	1,025	82.35	
	6650	For each added 50' of cable, add		.21	13.65	41	123	9.90	
70	0010	**WELLPOINT EQUIPMENT RENTAL** without operators R015433-10							70
	0020	Based on 2 months rental							
	0100	Combination jetting & wellpoint pump, 60 H.P. diesel	Ea.	16.29	340	1,016	3,050	333.50	
	0200	High pressure gas jet pump, 200 H.P., 300 psi	"	36.68	289	868	2,600	467.05	
	0300	Discharge pipe, 8" diameter	L.F.	.01	.55	1.65	4.95	.40	
	0350	12" diameter		.01	.81	2.43	7.30	.55	
	0400	Header pipe, flows up to 150 GPM, 4" diameter		.01	.50	1.50	4.50	.40	
	0500	400 GPM, 6" diameter		.01	.58	1.75	5.25	.45	

01 54 33 | Equipment Rental

			UNIT	HOURLY OPER. COST	RENT PER DAY	RENT PER WEEK	RENT PER MONTH	EQUIPMENT COST/DAY	
70	0600	800 GPM, 8" diameter	L.F.	.01	.81	2.43	7.30	.55	70
	0700	1500 GPM, 10" diameter		.01	.85	2.56	7.70	.60	
	0800	2500 GPM, 12" diameter		.02	1.61	4.83	14.50	1.15	
	0900	4500 GPM, 16" diameter		.03	2.06	6.19	18.55	1.50	
	0950	For quick coupling aluminum and plastic pipe, add		.03	2.14	6.41	19.25	1.50	
	1100	Wellpoint, 25' long, with fittings & riser pipe, 1-1/2" or 2" diameter	Ea.	.06	4.26	12.79	38.50	3.05	
	1200	Wellpoint pump, diesel powered, 4" suction, 20 H.P.		7.16	195	585	1,750	174.30	
	1300	6" suction, 30 H.P.		9.67	242	726	2,175	222.55	
	1400	8" suction, 40 H.P.		13.09	330	996	3,000	303.90	
	1500	10" suction, 75 H.P.		19.62	390	1,164	3,500	389.75	
	1600	12" suction, 100 H.P.		28.25	615	1,850	5,550	596	
	1700	12" suction, 175 H.P.		41.13	685	2,050	6,150	739.05	
80	0010	**MARINE EQUIPMENT RENTAL** without operators R015433 -10							80
	0200	Barge, 400 Ton, 30' wide x 90' long	Ea.	17.40	1,075	3,200	9,600	779.20	
	0240	800 Ton, 45' wide x 90' long		21.05	1,300	3,880	11,600	944.40	
	2000	Tugboat, diesel, 100 H.P.		37.80	217	650	1,950	432.40	
	2040	250 H.P.		78.95	395	1,185	3,550	868.60	
	2080	380 H.P.		155.50	1,175	3,535	10,600	1,951	
	3000	Small work boat, gas, 16-foot, 50 H.P.		17.35	60	180	540	174.80	
	4000	Large, diesel, 48-foot, 200 H.P.		88.50	1,250	3,735	11,200	1,455	

Crews - Standard

Crew A-1

Crew A-1	\ Bare Costs Hr.	Daily	Incl. Subs O&P Hr.	Daily	Cost/L-H Bare Costs	Incl. O&P
1 Building Laborer	$37.90	$303.20	$58.15	$465.20	$37.90	$58.15
1 Concrete Saw, Gas Manual		81.40		89.54	10.18	11.19
8 L.H., Daily Totals		$384.60		$554.74	$48.08	$69.34

Crew A-1A

Crew A-1A	Hr.	Daily	Hr.	Daily	Bare Costs	Incl. O&P
1 Skilled Worker	$49.90	$399.20	$76.85	$614.80	$49.90	$76.85
1 Shot Blaster, 20"		215.40		236.94	26.93	29.62
8 L.H., Daily Totals		$614.60		$851.74	$76.83	$106.47

Crew A-1B

Crew A-1B	Hr.	Daily	Hr.	Daily	Bare Costs	Incl. O&P
1 Building Laborer	$37.90	$303.20	$58.15	$465.20	$37.90	$58.15
1 Concrete Saw		168.60		185.46	21.07	23.18
8 L.H., Daily Totals		$471.80		$650.66	$58.98	$81.33

Crew A-1C

Crew A-1C	Hr.	Daily	Hr.	Daily	Bare Costs	Incl. O&P
1 Building Laborer	$37.90	$303.20	$58.15	$465.20	$37.90	$58.15
1 Chain Saw, Gas, 18"		31.60		34.76	3.95	4.34
8 L.H., Daily Totals		$334.80		$499.96	$41.85	$62.49

Crew A-1D

Crew A-1D	Hr.	Daily	Hr.	Daily	Bare Costs	Incl. O&P
1 Building Laborer	$37.90	$303.20	$58.15	$465.20	$37.90	$58.15
1 Vibrating Plate, Gas, 18"		36.20		39.82	4.53	4.98
8 L.H., Daily Totals		$339.40		$505.02	$42.42	$63.13

Crew A-1E

Crew A-1E	Hr.	Daily	Hr.	Daily	Bare Costs	Incl. O&P
1 Building Laborer	$37.90	$303.20	$58.15	$465.20	$37.90	$58.15
1 Vibrating Plate, Gas, 21"		46.60		51.26	5.83	6.41
8 L.H., Daily Totals		$349.80		$516.46	$43.73	$64.56

Crew A-1F

Crew A-1F	Hr.	Daily	Hr.	Daily	Bare Costs	Incl. O&P
1 Building Laborer	$37.90	$303.20	$58.15	$465.20	$37.90	$58.15
1 Rammer/Tamper, Gas, 8"		50.00		55.00	6.25	6.88
8 L.H., Daily Totals		$353.20		$520.20	$44.15	$65.03

Crew A-1G

Crew A-1G	Hr.	Daily	Hr.	Daily	Bare Costs	Incl. O&P
1 Building Laborer	$37.90	$303.20	$58.15	$465.20	$37.90	$58.15
1 Rammer/Tamper, Gas, 15"		56.80		62.48	7.10	7.81
8 L.H., Daily Totals		$360.00		$527.68	$45.00	$65.96

Crew A-1H

Crew A-1H	Hr.	Daily	Hr.	Daily	Bare Costs	Incl. O&P
1 Building Laborer	$37.90	$303.20	$58.15	$465.20	$37.90	$58.15
1 Exterior Steam Cleaner		77.40		85.14	9.68	10.64
8 L.H., Daily Totals		$380.60		$550.34	$47.58	$68.79

Crew A-1J

Crew A-1J	Hr.	Daily	Hr.	Daily	Bare Costs	Incl. O&P
1 Building Laborer	$37.90	$303.20	$58.15	$465.20	$37.90	$58.15
1 Cultivator, Walk-Behind, 5 H.P.		45.10		49.61	5.64	6.20
8 L.H., Daily Totals		$348.30		$514.81	$43.54	$64.35

Crew A-1K

Crew A-1K	Hr.	Daily	Hr.	Daily	Bare Costs	Incl. O&P
1 Building Laborer	$37.90	$303.20	$58.15	$465.20	$37.90	$58.15
1 Cultivator, Walk-Behind, 8 H.P.		70.00		77.00	8.75	9.63
8 L.H., Daily Totals		$373.20		$542.20	$46.65	$67.78

Crew A-1M

Crew A-1M	Hr.	Daily	Hr.	Daily	Bare Costs	Incl. O&P
1 Building Laborer	$37.90	$303.20	$58.15	$465.20	$37.90	$58.15
1 Snow Blower, Walk-Behind		55.30		60.83	6.91	7.60
8 L.H., Daily Totals		$358.50		$526.03	$44.81	$65.75

Crew A-2

Crew A-2	Hr.	Daily	Hr.	Daily	Bare Costs	Incl. O&P
2 Laborers	$37.90	$606.40	$58.15	$930.40	$39.25	$59.88
1 Truck Driver (light)	41.95	335.60	63.35	506.80		
1 Flatbed Truck, Gas, 1.5 Ton		248.20		273.02	10.34	11.38
24 L.H., Daily Totals		$1190.20		$1710.22	$49.59	$71.26

Crew A-2A

Crew A-2A	Hr.	Daily	Hr.	Daily	Bare Costs	Incl. O&P
2 Laborers	$37.90	$606.40	$58.15	$930.40	$39.25	$59.88
1 Truck Driver (light)	41.95	335.60	63.35	506.80		
1 Flatbed Truck, Gas, 1.5 Ton		248.20		273.02		
1 Concrete Saw		168.60		185.46	17.37	19.10
24 L.H., Daily Totals		$1358.80		$1895.68	$56.62	$78.99

Crew A-2B

Crew A-2B	Hr.	Daily	Hr.	Daily	Bare Costs	Incl. O&P
1 Truck Driver (light)	$41.95	$335.60	$63.35	$506.80	$41.95	$63.35
1 Flatbed Truck, Gas, 1.5 Ton		248.20		273.02	31.02	34.13
8 L.H., Daily Totals		$583.80		$779.82	$72.97	$97.48

Crew A-3A

Crew A-3A	Hr.	Daily	Hr.	Daily	Bare Costs	Incl. O&P
1 Equip. Oper. (light)	$49.15	$393.20	$74.35	$594.80	$49.15	$74.35
1 Pickup Truck, 4x4, 3/4 Ton		156.80		172.48	19.60	21.56
8 L.H., Daily Totals		$550.00		$767.28	$68.75	$95.91

Crew A-3B

Crew A-3B	Hr.	Daily	Hr.	Daily	Bare Costs	Incl. O&P
1 Equip. Oper. (medium)	$51.10	$408.80	$77.30	$618.40	$47.15	$71.28
1 Truck Driver (heavy)	43.20	345.60	65.25	522.00		
1 Dump Truck, 12 C.Y., 400 H.P.		690.80		759.88		
1 F.E. Loader, W.M.,2.5 C.Y.		516.00		567.60	75.42	82.97
16 L.H., Daily Totals		$1961.20		$2467.88	$122.58	$154.24

Crew A-3C

Crew A-3C	Hr.	Daily	Hr.	Daily	Bare Costs	Incl. O&P
1 Equip. Oper. (light)	$49.15	$393.20	$74.35	$594.80	$49.15	$74.35
1 Loader, Skid Steer, 78 H.P.		315.60		347.16	39.45	43.40
8 L.H., Daily Totals		$708.80		$941.96	$88.60	$117.75

Crew A-3D

Crew A-3D	Hr.	Daily	Hr.	Daily	Bare Costs	Incl. O&P
1 Truck Driver (light)	$41.95	$335.60	$63.35	$506.80	$41.95	$63.35
1 Pickup Truck, 4x4, 3/4 Ton		156.80		172.48		
1 Flatbed Trailer, 25 Ton		113.20		124.52	33.75	37.13
8 L.H., Daily Totals		$605.60		$803.80	$75.70	$100.47

Crew A-3E

Crew A-3E	Hr.	Daily	Hr.	Daily	Bare Costs	Incl. O&P
1 Equip. Oper. (crane)	$52.25	$418.00	$79.05	$632.40	$47.73	$72.15
1 Truck Driver (heavy)	43.20	345.60	65.25	522.00		
1 Pickup Truck, 4x4, 3/4 Ton		156.80		172.48	9.80	10.78
16 L.H., Daily Totals		$920.40		$1326.88	$57.52	$82.93

Crew A-3F

Crew A-3F	Hr.	Daily	Hr.	Daily	Bare Costs	Incl. O&P
1 Equip. Oper. (crane)	$52.25	$418.00	$79.05	$632.40	$47.73	$72.15
1 Truck Driver (heavy)	43.20	345.60	65.25	522.00		
1 Pickup Truck, 4x4, 3/4 Ton		156.80		172.48		
1 Truck Tractor, 6x4, 380 H.P.		600.00		660.00		
1 Lowbed Trailer, 75 Ton		223.40		245.74	61.26	67.39
16 L.H., Daily Totals		$1743.80		$2232.62	$108.99	$139.54

For customer support on your Mechanical Cost Data, call 877.739.3689.

619

Crews - Standard

Crew A-3G

Crew A-3G	Bare Costs Hr.	Daily	Incl. Subs O&P Hr.	Daily	Cost Per Labor-Hour Bare Costs	Incl. O&P
1 Equip. Oper. (crane)	$52.25	$418.00	$79.05	$632.40	$47.73	$72.15
1 Truck Driver (heavy)	43.20	345.60	65.25	522.00		
1 Pickup Truck, 4x4, 3/4 Ton		156.80		172.48		
1 Truck Tractor, 6x4, 450 H.P.		726.80		799.48		
1 Lowbed Trailer, 75 Ton		223.40		245.74	69.19	76.11
16 L.H., Daily Totals		$1870.60		$2372.10	$116.91	$148.26

Crew A-3H

Crew A-3H	Bare Costs Hr.	Daily	Incl. Subs O&P Hr.	Daily	Cost Per Labor-Hour Bare Costs	Incl. O&P
1 Equip. Oper. (crane)	$52.25	$418.00	$79.05	$632.40	$52.25	$79.05
1 Hyd. Crane, 12 Ton (Daily)		877.40		965.14	109.68	120.64
8 L.H., Daily Totals		$1295.40		$1597.54	$161.93	$199.69

Crew A-3I

Crew A-3I	Bare Costs Hr.	Daily	Incl. Subs O&P Hr.	Daily	Cost Per Labor-Hour Bare Costs	Incl. O&P
1 Equip. Oper. (crane)	$52.25	$418.00	$79.05	$632.40	$52.25	$79.05
1 Hyd. Crane, 25 Ton (Daily)		1013.00		1114.30	126.63	139.29
8 L.H., Daily Totals		$1431.00		$1746.70	$178.88	$218.34

Crew A-3J

Crew A-3J	Bare Costs Hr.	Daily	Incl. Subs O&P Hr.	Daily	Cost Per Labor-Hour Bare Costs	Incl. O&P
1 Equip. Oper. (crane)	$52.25	$418.00	$79.05	$632.40	$52.25	$79.05
1 Hyd. Crane, 40 Ton (Daily)		1259.00		1384.90	157.38	173.11
8 L.H., Daily Totals		$1677.00		$2017.30	$209.63	$252.16

Crew A-3K

Crew A-3K	Bare Costs Hr.	Daily	Incl. Subs O&P Hr.	Daily	Cost Per Labor-Hour Bare Costs	Incl. O&P
1 Equip. Oper. (crane)	$52.25	$418.00	$79.05	$632.40	$49.15	$74.35
1 Equip. Oper. (oiler)	46.05	368.40	69.65	557.20		
1 Hyd. Crane, 55 Ton (Daily)		1470.00		1617.00		
1 P/U Truck, 3/4 Ton (Daily)		170.80		187.88	102.55	112.81
16 L.H., Daily Totals		$2427.20		$2994.48	$151.70	$187.16

Crew A-3L

Crew A-3L	Bare Costs Hr.	Daily	Incl. Subs O&P Hr.	Daily	Cost Per Labor-Hour Bare Costs	Incl. O&P
1 Equip. Oper. (crane)	$52.25	$418.00	$79.05	$632.40	$49.15	$74.35
1 Equip. Oper. (oiler)	46.05	368.40	69.65	557.20		
1 Hyd. Crane, 80 Ton (Daily)		2193.00		2412.30		
1 P/U Truck, 3/4 Ton (Daily)		170.80		187.88	147.74	162.51
16 L.H., Daily Totals		$3150.20		$3789.78	$196.89	$236.86

Crew A-3M

Crew A-3M	Bare Costs Hr.	Daily	Incl. Subs O&P Hr.	Daily	Cost Per Labor-Hour Bare Costs	Incl. O&P
1 Equip. Oper. (crane)	$52.25	$418.00	$79.05	$632.40	$49.15	$74.35
1 Equip. Oper. (oiler)	46.05	368.40	69.65	557.20		
1 Hyd. Crane, 100 Ton (Daily)		2216.00		2437.60		
1 P/U Truck, 3/4 Ton (Daily)		170.80		187.88	149.18	164.09
16 L.H., Daily Totals		$3173.20		$3815.08	$198.32	$238.44

Crew A-3N

Crew A-3N	Bare Costs Hr.	Daily	Incl. Subs O&P Hr.	Daily	Cost Per Labor-Hour Bare Costs	Incl. O&P
1 Equip. Oper. (crane)	$52.25	$418.00	$79.05	$632.40	$52.25	$79.05
1 Tower Cane (monthly)		1145.00		1259.50	143.13	157.44
8 L.H., Daily Totals		$1563.00		$1891.90	$195.38	$236.49

Crew A-3P

Crew A-3P	Bare Costs Hr.	Daily	Incl. Subs O&P Hr.	Daily	Cost Per Labor-Hour Bare Costs	Incl. O&P
1 Equip. Oper. (light)	$49.15	$393.20	$74.35	$594.80	$49.15	$74.35
1 A.T. Forklift, 42' lift		534.20		587.62	66.78	73.45
8 L.H., Daily Totals		$927.40		$1182.42	$115.93	$147.80

Crew A-3Q

Crew A-3Q	Bare Costs Hr.	Daily	Incl. Subs O&P Hr.	Daily	Cost Per Labor-Hour Bare Costs	Incl. O&P
1 Equip. Oper. (light)	$49.15	$393.20	$74.35	$594.80	$49.15	$74.35
1 Pickup Truck, 4x4, 3/4 Ton		156.80		172.48		
1 Flatbed trailer, 3 Ton		23.40		25.74	22.52	24.78
8 L.H., Daily Totals		$573.40		$793.02	$71.67	$99.13

Crew A-4

Crew A-4	Bare Costs Hr.	Daily	Incl. Subs O&P Hr.	Daily	Cost Per Labor-Hour Bare Costs	Incl. O&P
2 Carpenters	$48.45	$775.20	$74.30	$1188.80	$45.75	$69.80
1 Painter, Ordinary	40.35	322.80	60.80	486.40		
24 L.H., Daily Totals		$1098.00		$1675.20	$45.75	$69.80

Crew A-5

Crew A-5	Bare Costs Hr.	Daily	Incl. Subs O&P Hr.	Daily	Cost Per Labor-Hour Bare Costs	Incl. O&P
2 Laborers	$37.90	$606.40	$58.15	$930.40	$38.35	$58.73
.25 Truck Driver (light)	41.95	83.90	63.35	126.70		
.25 Flatbed Truck, Gas, 1.5 Ton		62.05		68.25	3.45	3.79
18 L.H., Daily Totals		$752.35		$1125.36	$41.80	$62.52

Crew A-6

Crew A-6	Bare Costs Hr.	Daily	Incl. Subs O&P Hr.	Daily	Cost Per Labor-Hour Bare Costs	Incl. O&P
1 Instrument Man	$49.90	$399.20	$76.85	$614.80	$48.30	$73.95
1 Rodman/Chainman	46.70	373.60	71.05	568.40		
1 Level, Electronic		43.70		48.07	2.73	3.00
16 L.H., Daily Totals		$816.50		$1231.27	$51.03	$76.95

Crew A-7

Crew A-7	Bare Costs Hr.	Daily	Incl. Subs O&P Hr.	Daily	Cost Per Labor-Hour Bare Costs	Incl. O&P
1 Chief of Party	$61.65	$493.20	$93.70	$749.60	$52.75	$80.53
1 Instrument Man	49.90	399.20	76.85	614.80		
1 Rodman/Chainman	46.70	373.60	71.05	568.40		
1 Level, Electronic		43.70		48.07	1.82	2.00
24 L.H., Daily Totals		$1309.70		$1980.87	$54.57	$82.54

Crew A-8

Crew A-8	Bare Costs Hr.	Daily	Incl. Subs O&P Hr.	Daily	Cost Per Labor-Hour Bare Costs	Incl. O&P
1 Chief of Party	$61.65	$493.20	$93.70	$749.60	$51.24	$78.16
1 Instrument Man	49.90	399.20	76.85	614.80		
2 Rodmen/Chainmen	46.70	747.20	71.05	1136.80		
1 Level, Electronic		43.70		48.07	1.37	1.50
32 L.H., Daily Totals		$1683.30		$2549.27	$52.60	$79.66

Crew A-9

Crew A-9	Bare Costs Hr.	Daily	Incl. Subs O&P Hr.	Daily	Cost Per Labor-Hour Bare Costs	Incl. O&P
1 Asbestos Foreman	$53.90	$431.20	$83.85	$670.80	$53.46	$83.19
7 Asbestos Workers	53.40	2990.40	83.10	4653.60		
64 L.H., Daily Totals		$3421.60		$5324.40	$53.46	$83.19

Crew A-10A

Crew A-10A	Bare Costs Hr.	Daily	Incl. Subs O&P Hr.	Daily	Cost Per Labor-Hour Bare Costs	Incl. O&P
1 Asbestos Foreman	$53.90	$431.20	$83.85	$670.80	$53.57	$83.35
2 Asbestos Workers	53.40	854.40	83.10	1329.60		
24 L.H., Daily Totals		$1285.60		$2000.40	$53.57	$83.35

Crew A-10B

Crew A-10B	Bare Costs Hr.	Daily	Incl. Subs O&P Hr.	Daily	Cost Per Labor-Hour Bare Costs	Incl. O&P
1 Asbestos Foreman	$53.90	$431.20	$83.85	$670.80	$53.52	$83.29
3 Asbestos Workers	53.40	1281.60	83.10	1994.40		
32 L.H., Daily Totals		$1712.80		$2665.20	$53.52	$83.29

Crew A-10C

Crew A-10C	Bare Costs Hr.	Daily	Incl. Subs O&P Hr.	Daily	Cost Per Labor-Hour Bare Costs	Incl. O&P
3 Asbestos Workers	$53.40	$1281.60	$83.10	$1994.40	$53.40	$83.10
1 Flatbed Truck, Gas, 1.5 Ton		248.20		273.02	10.34	11.38
24 L.H., Daily Totals		$1529.80		$2267.42	$63.74	$94.48

Crew A-10D

Crew A-10D	Bare Costs Hr.	Daily	Incl. Subs O&P Hr.	Daily	Cost Per Labor-Hour Bare Costs	Incl. O&P
2 Asbestos Workers	$53.40	$854.40	$83.10	$1329.60	$51.27	$78.72
1 Equip. Oper. (crane)	52.25	418.00	79.05	632.40		
1 Equip. Oper. (oiler)	46.05	368.40	69.65	557.20		
1 Hydraulic Crane, 33 Ton		763.20		839.52	23.85	26.23
32 L.H., Daily Totals		$2404.00		$3358.72	$75.13	$104.96

Crew No.	Bare Costs		Incl. Subs O&P		Cost Per Labor-Hour	

Crew A-11	Hr.	Daily	Hr.	Daily	Bare Costs	Incl. O&P
1 Asbestos Foreman	$53.90	$431.20	$83.85	$670.80	$53.46	$83.19
7 Asbestos Workers	53.40	2990.40	83.10	4653.60		
2 Chip. Hammers, 12 Lb., Elec.		40.00		44.00	.63	.69
64 L.H., Daily Totals		$3461.60		$5368.40	$54.09	$83.88

Crew A-12	Hr.	Daily	Hr.	Daily	Bare Costs	Incl. O&P
1 Asbestos Foreman	$53.90	$431.20	$83.85	$670.80	$53.46	$83.19
7 Asbestos Workers	53.40	2990.40	83.10	4653.60		
1 Trk-Mtd Vac, 14 CY, 1500 Gal.		510.30		561.33		
1 Flatbed Truck, 20,000 GVW		247.80		272.58	11.85	13.03
64 L.H., Daily Totals		$4179.70		$6158.31	$65.31	$96.22

Crew A-13	Hr.	Daily	Hr.	Daily	Bare Costs	Incl. O&P
1 Equip. Oper. (light)	$49.15	$393.20	$74.35	$594.80	$49.15	$74.35
1 Trk-Mtd Vac, 14 CY, 1500 Gal.		510.30		561.33		
1 Flatbed Truck, 20,000 GVW		247.80		272.58	94.76	104.24
8 L.H., Daily Totals		$1151.30		$1428.71	$143.91	$178.59

Crew B-1	Hr.	Daily	Hr.	Daily	Bare Costs	Incl. O&P
1 Labor Foreman (outside)	$39.90	$319.20	$61.20	$489.60	$38.57	$59.17
2 Laborers	37.90	606.40	58.15	930.40		
24 L.H., Daily Totals		$925.60		$1420.00	$38.57	$59.17

Crew B-1A	Hr.	Daily	Hr.	Daily	Bare Costs	Incl. O&P
1 Labor Foreman (outside)	$39.90	$319.20	$61.20	$489.60	$38.57	$59.17
2 Laborers	37.90	606.40	58.15	930.40		
2 Cutting Torches		22.80		25.08		
2 Sets of Gases		336.00		369.60	14.95	16.45
24 L.H., Daily Totals		$1284.40		$1814.68	$53.52	$75.61

Crew B-1B	Hr.	Daily	Hr.	Daily	Bare Costs	Incl. O&P
1 Labor Foreman (outside)	$39.90	$319.20	$61.20	$489.60	$41.99	$64.14
2 Laborers	37.90	606.40	58.15	930.40		
1 Equip. Oper. (crane)	52.25	418.00	79.05	632.40		
2 Cutting Torches		22.80		25.08		
2 Sets of Gases		336.00		369.60		
1 Hyd. Crane, 12 Ton		658.40		724.24	31.79	34.97
32 L.H., Daily Totals		$2360.80		$3171.32	$73.78	$99.10

Crew B-1C	Hr.	Daily	Hr.	Daily	Bare Costs	Incl. O&P
1 Labor Foreman (outside)	$39.90	$319.20	$61.20	$489.60	$38.57	$59.17
2 Laborers	37.90	606.40	58.15	930.40		
1 Aerial Lift Truck, 60' Boom		448.40		493.24	18.68	20.55
24 L.H., Daily Totals		$1374.00		$1913.24	$57.25	$79.72

Crew B-1D	Hr.	Daily	Hr.	Daily	Bare Costs	Incl. O&P
2 Laborers	$37.90	$606.40	$58.15	$930.40	$37.90	$58.15
1 Small Work Boat, Gas, 50 H.P.		174.80		192.28		
1 Pressure Washer, 7 GPM		88.60		97.46	16.46	18.11
16 L.H., Daily Totals		$869.80		$1220.14	$54.36	$76.26

Crew B-1E	Hr.	Daily	Hr.	Daily	Bare Costs	Incl. O&P
1 Labor Foreman (outside)	$39.90	$319.20	$61.20	$489.60	$38.40	$58.91
3 Laborers	37.90	909.60	58.15	1395.60		
1 Work Boat, Diesel, 200 H.P.		1455.00		1600.50		
2 Pressure Washers, 7 GPM		177.20		194.92	51.01	56.11
32 L.H., Daily Totals		$2861.00		$3680.62	$89.41	$115.02

Crew B-1F	Hr.	Daily	Hr.	Daily	Bare Costs	Incl. O&P
2 Skilled Workers	$49.90	$798.40	$76.85	$1229.60	$45.90	$70.62
1 Laborer	37.90	303.20	58.15	465.20		
1 Small Work Boat, Gas, 50 H.P.		174.80		192.28		
1 Pressure Washer, 7 GPM		88.60		97.46	10.98	12.07
24 L.H., Daily Totals		$1365.00		$1984.54	$56.88	$82.69

Crew B-1G	Hr.	Daily	Hr.	Daily	Bare Costs	Incl. O&P
2 Laborers	$37.90	$606.40	$58.15	$930.40	$37.90	$58.15
1 Small Work Boat, Gas, 50 H.P.		174.80		192.28	10.93	12.02
16 L.H., Daily Totals		$781.20		$1122.68	$48.83	$70.17

Crew B-1H	Hr.	Daily	Hr.	Daily	Bare Costs	Incl. O&P
2 Skilled Workers	$49.90	$798.40	$76.85	$1229.60	$45.90	$70.62
1 Laborer	37.90	303.20	58.15	465.20		
1 Small Work Boat, Gas, 50 H.P.		174.80		192.28	7.28	8.01
24 L.H., Daily Totals		$1276.40		$1887.08	$53.18	$78.63

Crew B-1J	Hr.	Daily	Hr.	Daily	Bare Costs	Incl. O&P
1 Labor Foreman (inside)	$38.40	$307.20	$58.90	$471.20	$38.15	$58.52
1 Laborer	37.90	303.20	58.15	465.20		
16 L.H., Daily Totals		$610.40		$936.40	$38.15	$58.52

Crew B-1K	Hr.	Daily	Hr.	Daily	Bare Costs	Incl. O&P
1 Carpenter Foreman (inside)	$48.95	$391.60	$75.10	$600.80	$48.70	$74.70
1 Carpenter	48.45	387.60	74.30	594.40		
16 L.H., Daily Totals		$779.20		$1195.20	$48.70	$74.70

Crew B-2	Hr.	Daily	Hr.	Daily	Bare Costs	Incl. O&P
1 Labor Foreman (outside)	$39.90	$319.20	$61.20	$489.60	$38.30	$58.76
4 Laborers	37.90	1212.80	58.15	1860.80		
40 L.H., Daily Totals		$1532.00		$2350.40	$38.30	$58.76

Crew B-2A	Hr.	Daily	Hr.	Daily	Bare Costs	Incl. O&P
1 Labor Foreman (outside)	$39.90	$319.20	$61.20	$489.60	$38.57	$59.17
2 Laborers	37.90	606.40	58.15	930.40		
1 Aerial Lift Truck, 60' Boom		448.40		493.24	18.68	20.55
24 L.H., Daily Totals		$1374.00		$1913.24	$57.25	$79.72

Crew B-3	Hr.	Daily	Hr.	Daily	Bare Costs	Incl. O&P
1 Labor Foreman (outside)	$39.90	$319.20	$61.20	$489.60	$42.20	$64.22
2 Laborers	37.90	606.40	58.15	930.40		
1 Equip. Oper. (medium)	51.10	408.80	77.30	618.40		
2 Truck Drivers (heavy)	43.20	691.20	65.25	1044.00		
1 Crawler Loader, 3 C.Y.		1197.00		1316.70		
2 Dump Trucks, 12 C.Y., 400 H.P.		1381.60		1519.76	53.72	59.09
48 L.H., Daily Totals		$4604.20		$5918.86	$95.92	$123.31

Crew B-3A	Hr.	Daily	Hr.	Daily	Bare Costs	Incl. O&P
4 Laborers	$37.90	$1212.80	$58.15	$1860.80	$40.54	$61.98
1 Equip. Oper. (medium)	51.10	408.80	77.30	618.40		
1 Hyd. Excavator, 1.5 C.Y.		1029.00		1131.90	25.73	28.30
40 L.H., Daily Totals		$2650.60		$3611.10	$66.27	$90.28

Crew B-3B	Hr.	Daily	Hr.	Daily	Bare Costs	Incl. O&P
2 Laborers	$37.90	$606.40	$58.15	$930.40	$42.52	$64.71
1 Equip. Oper. (medium)	51.10	408.80	77.30	618.40		
1 Truck Driver (heavy)	43.20	345.60	65.25	522.00		
1 Backhoe Loader, 80 H.P.		396.20		435.82		
1 Dump Truck, 12 C.Y., 400 H.P.		690.80		759.88	33.97	37.37
32 L.H., Daily Totals		$2447.80		$3266.50	$76.49	$102.08

Crew No.	Bare Costs		Incl. Subs O&P		Cost Per Labor-Hour	
Crew B-3C	**Hr.**	**Daily**	**Hr.**	**Daily**	Bare Costs	Incl. O&P
3 Laborers	$37.90	$909.60	$58.15	$1395.60	$41.20	$62.94
1 Equip. Oper. (medium)	51.10	408.80	77.30	618.40		
1 Crawler Loader, 4 C.Y.		1566.00		1722.60	48.94	53.83
32 L.H., Daily Totals		$2884.40		$3736.60	$90.14	$116.77
Crew B-4	**Hr.**	**Daily**	**Hr.**	**Daily**	Bare Costs	Incl. O&P
1 Labor Foreman (outside)	$39.90	$319.20	$61.20	$489.60	$39.12	$59.84
4 Laborers	37.90	1212.80	58.15	1860.80		
1 Truck Driver (heavy)	43.20	345.60	65.25	522.00		
1 Truck Tractor, 220 H.P.		358.80		394.68		
1 Flatbed Trailer, 40 Ton		154.60		170.06	10.70	11.77
48 L.H., Daily Totals		$2391.00		$3437.14	$49.81	$71.61
Crew B-5	**Hr.**	**Daily**	**Hr.**	**Daily**	Bare Costs	Incl. O&P
1 Labor Foreman (outside)	$39.90	$319.20	$61.20	$489.60	$41.96	$64.06
4 Laborers	37.90	1212.80	58.15	1860.80		
2 Equip. Opers. (medium)	51.10	817.60	77.30	1236.80		
1 Air Compressor, 250 cfm		197.80		217.58		
2 Breakers, Pavement, 60 lb.		20.40		22.44		
2 -50' Air Hoses, 1.5"		10.40		11.44		
1 Crawler Loader, 3 C.Y.		1197.00		1316.70	25.46	28.00
56 L.H., Daily Totals		$3775.20		$5155.36	$67.41	$92.06
Crew B-5A	**Hr.**	**Daily**	**Hr.**	**Daily**	Bare Costs	Incl. O&P
1 Labor Foreman (outside)	$39.90	$319.20	$61.20	$489.60	$42.09	$64.13
6 Laborers	37.90	1819.20	58.15	2791.20		
2 Equip. Opers. (medium)	51.10	817.60	77.30	1236.80		
1 Equip. Oper. (light)	49.15	393.20	74.35	594.80		
2 Truck Drivers (heavy)	43.20	691.20	65.25	1044.00		
1 Air Compressor, 365 cfm		258.40		284.24		
2 Breakers, Pavement, 60 lb.		20.40		22.44		
8 -50' Air Hoses, 1"		29.20		32.12		
2 Dump Trucks, 8 C.Y., 220 H.P.		834.80		918.28	11.90	13.09
96 L.H., Daily Totals		$5183.20		$7413.48	$53.99	$77.22
Crew B-5B	**Hr.**	**Daily**	**Hr.**	**Daily**	Bare Costs	Incl. O&P
1 Powderman	$49.90	$399.20	$76.85	$614.80	$46.95	$71.20
2 Equip. Opers. (medium)	51.10	817.60	77.30	1236.80		
3 Truck Drivers (heavy)	43.20	1036.80	65.25	1566.00		
1 F.E. Loader, W.M., 2.5 C.Y.		516.00		567.60		
3 Dump Trucks, 12 C.Y., 400 H.P.		2072.40		2279.64		
1 Air Compressor, 365 CFM		258.40		284.24	59.31	65.24
48 L.H., Daily Totals		$5100.40		$6549.08	$106.26	$136.44
Crew B-5C	**Hr.**	**Daily**	**Hr.**	**Daily**	Bare Costs	Incl. O&P
3 Laborers	$37.90	$909.60	$58.15	$1395.60	$43.69	$66.37
1 Equip. Oper. (medium)	51.10	408.80	77.30	618.40		
2 Truck Drivers (heavy)	43.20	691.20	65.25	1044.00		
1 Equip. Oper. (crane)	52.25	418.00	79.05	632.40		
1 Equip. Oper. (oiler)	46.05	368.40	69.65	557.20		
2 Dump Trucks, 12 C.Y., 400 H.P.		1381.60		1519.76		
1 Crawler Loader, 4 C.Y.		1566.00		1722.60		
1 S.P. Crane, 4x4, 25 Ton		606.60		667.26	55.53	61.09
64 L.H., Daily Totals		$6350.20		$8157.22	$99.22	$127.46

Crew No.	Bare Costs		Incl. Subs O&P		Cost Per Labor-Hour	
Crew B-5D	**Hr.**	**Daily**	**Hr.**	**Daily**	Bare Costs	Incl. O&P
1 Labor Foreman (outside)	$39.90	$319.20	$61.20	$489.60	$42.11	$64.21
4 Laborers	37.90	1212.80	58.15	1860.80		
2 Equip. Opers. (medium)	51.10	817.60	77.30	1236.80		
1 Truck Driver (heavy)	43.20	345.60	65.25	522.00		
1 Air Compressor, 250 cfm		197.80		217.58		
2 Breakers, Pavement, 60 lb.		20.40		22.44		
2 -50' Air Hoses, 1.5"		10.40		11.44		
1 Crawler Loader, 3 C.Y.		1197.00		1316.70		
1 Dump Truck, 12 C.Y., 400 H.P.		690.80		759.88	33.07	36.38
64 L.H., Daily Totals		$4811.60		$6437.24	$75.18	$100.58
Crew B-6	**Hr.**	**Daily**	**Hr.**	**Daily**	Bare Costs	Incl. O&P
2 Laborers	$37.90	$606.40	$58.15	$930.40	$41.65	$63.55
1 Equip. Oper. (light)	49.15	393.20	74.35	594.80		
1 Backhoe Loader, 48 H.P.		365.40		401.94	15.23	16.75
24 L.H., Daily Totals		$1365.00		$1927.14	$56.88	$80.30
Crew B-6A	**Hr.**	**Daily**	**Hr.**	**Daily**	Bare Costs	Incl. O&P
.5 Labor Foreman (outside)	$39.90	$159.60	$61.20	$244.80	$43.58	$66.42
1 Laborer	37.90	303.20	58.15	465.20		
1 Equip. Oper. (medium)	51.10	408.80	77.30	618.40		
1 Vacuum Truck, 5000 Gal.		351.60		386.76	17.58	19.34
20 L.H., Daily Totals		$1223.20		$1715.16	$61.16	$85.76
Crew B-6B	**Hr.**	**Daily**	**Hr.**	**Daily**	Bare Costs	Incl. O&P
2 Labor Foremen (outside)	$39.90	$638.40	$61.20	$979.20	$38.57	$59.17
4 Laborers	37.90	1212.80	58.15	1860.80		
1 S.P. Crane, 4x4, 5 Ton		280.80		308.88		
1 Flatbed Truck, Gas, 1.5 Ton		248.20		273.02		
1 Butt Fusion Mach., 4"-12" diam.		376.10		413.71	18.86	20.74
48 L.H., Daily Totals		$2756.30		$3835.61	$57.42	$79.91
Crew B-6C	**Hr.**	**Daily**	**Hr.**	**Daily**	Bare Costs	Incl. O&P
2 Labor Foremen (outside)	$39.90	$638.40	$61.20	$979.20	$38.57	$59.17
4 Laborers	37.90	1212.80	58.15	1860.80		
1 S.P. Crane, 4x4, 12 Ton		479.20		527.12		
1 Flatbed Truck, Gas, 3 Ton		307.40		338.14		
1 Butt Fusion Mach., 8"-24" diam.		804.80		885.28	33.15	36.47
48 L.H., Daily Totals		$3442.60		$4590.54	$71.72	$95.64
Crew B-7	**Hr.**	**Daily**	**Hr.**	**Daily**	Bare Costs	Incl. O&P
1 Labor Foreman (outside)	$39.90	$319.20	$61.20	$489.60	$40.43	$61.85
4 Laborers	37.90	1212.80	58.15	1860.80		
1 Equip. Oper. (medium)	51.10	408.80	77.30	618.40		
1 Brush Chipper, 12", 130 H.P.		388.60		427.46		
1 Crawler Loader, 3 C.Y.		1197.00		1316.70		
2 Chain Saws, Gas, 36" Long		90.00		99.00	34.91	38.40
48 L.H., Daily Totals		$3616.40		$4811.96	$75.34	$100.25
Crew B-7A	**Hr.**	**Daily**	**Hr.**	**Daily**	Bare Costs	Incl. O&P
2 Laborers	$37.90	$606.40	$58.15	$930.40	$41.65	$63.55
1 Equip. Oper. (light)	49.15	393.20	74.35	594.80		
1 Rake w/Tractor		353.05		388.36		
2 Chain Saws, Gas, 18"		63.20		69.52	17.34	19.08
24 L.H., Daily Totals		$1415.85		$1983.08	$58.99	$82.63

Crew B-7B

Crew B-7B	Hr.	Daily	Hr.	Daily	Bare Costs	Incl. O&P
1 Labor Foreman (outside)	$39.90	$319.20	$61.20	$489.60	$40.83	$62.34
4 Laborers	37.90	1212.80	58.15	1860.80		
1 Equip. Oper. (medium)	51.10	408.80	77.30	618.40		
1 Truck Driver (heavy)	43.20	345.60	65.25	522.00		
1 Brush Chipper, 12", 130 H.P.		388.60		427.46		
1 Crawler Loader, 3 C.Y.		1197.00		1316.70		
2 Chain Saws, Gas, 36" Long		90.00		99.00		
1 Dump Truck, 8 C.Y., 220 H.P.		417.40		459.14	37.38	41.11
56 L.H., Daily Totals		$4379.40		$5793.10	$78.20	$103.45

Crew B-7C

Crew B-7C	Hr.	Daily	Hr.	Daily	Bare Costs	Incl. O&P
1 Labor Foreman (outside)	$39.90	$319.20	$61.20	$489.60	$40.83	$62.34
4 Laborers	37.90	1212.80	58.15	1860.80		
1 Equip. Oper. (medium)	51.10	408.80	77.30	618.40		
1 Truck Driver (heavy)	43.20	345.60	65.25	522.00		
1 Brush Chipper, 12", 130 H.P.		388.60		427.46		
1 Crawler Loader, 3 C.Y.		1197.00		1316.70		
2 Chain Saws, Gas, 36" Long		90.00		99.00		
1 Dump Truck, 12 C.Y., 400 H.P.		690.80		759.88	42.26	46.48
56 L.H., Daily Totals		$4652.80		$6093.84	$83.09	$108.82

Crew B-8

Crew B-8	Hr.	Daily	Hr.	Daily	Bare Costs	Incl. O&P
1 Labor Foreman (outside)	$39.90	$319.20	$61.20	$489.60	$43.79	$66.53
2 Laborers	37.90	606.40	58.15	930.40		
2 Equip. Opers. (medium)	51.10	817.60	77.30	1236.80		
1 Equip. Oper. (oiler)	46.05	368.40	69.65	557.20		
2 Truck Drivers (heavy)	43.20	691.20	65.25	1044.00		
1 Hyd. Crane, 25 Ton		748.80		823.68		
1 Crawler Loader, 3 C.Y.		1197.00		1316.70		
2 Dump Trucks, 12 C.Y., 400 H.P.		1381.60		1519.76	51.99	57.19
64 L.H., Daily Totals		$6130.20		$7918.14	$95.78	$123.72

Crew B-9

Crew B-9	Hr.	Daily	Hr.	Daily	Bare Costs	Incl. O&P
1 Labor Foreman (outside)	$39.90	$319.20	$61.20	$489.60	$38.30	$58.76
4 Laborers	37.90	1212.80	58.15	1860.80		
1 Air Compressor, 250 cfm		197.80		217.58		
2 Breakers, Pavement, 60 lb.		20.40		22.44		
2 -50' Air Hoses, 1.5"		10.40		11.44	5.71	6.29
40 L.H., Daily Totals		$1760.60		$2601.86	$44.02	$65.05

Crew B-9A

Crew B-9A	Hr.	Daily	Hr.	Daily	Bare Costs	Incl. O&P
2 Laborers	$37.90	$606.40	$58.15	$930.40	$39.67	$60.52
1 Truck Driver (heavy)	43.20	345.60	65.25	522.00		
1 Water Tank Trailer, 5000 Gal.		141.60		155.76		
1 Truck Tractor, 220 H.P.		358.80		394.68		
2 -50' Discharge Hoses, 3"		3.00		3.30	20.98	23.07
24 L.H., Daily Totals		$1455.40		$2006.14	$60.64	$83.59

Crew B-9B

Crew B-9B	Hr.	Daily	Hr.	Daily	Bare Costs	Incl. O&P
2 Laborers	$37.90	$606.40	$58.15	$930.40	$39.67	$60.52
1 Truck Driver (heavy)	43.20	345.60	65.25	522.00		
2 -50' Discharge Hoses, 3"		3.00		3.30		
1 Water Tank Trailer, 5000 Gal.		141.60		155.76		
1 Truck Tractor, 220 H.P.		358.80		394.68		
1 Pressure Washer		71.60		78.76	23.96	26.35
24 L.H., Daily Totals		$1527.00		$2084.90	$63.63	$86.87

Crew B-9D

Crew B-9D	Hr.	Daily	Hr.	Daily	Bare Costs	Incl. O&P
1 Labor Foreman (outside)	$39.90	$319.20	$61.20	$489.60	$38.30	$58.76
4 Common Laborers	37.90	1212.80	58.15	1860.80		
1 Air Compressor, 250 cfm		197.80		217.58		
2 -50' Air Hoses, 1.5"		10.40		11.44		
2 Air Powered Tampers		48.00		52.80	6.41	7.05
40 L.H., Daily Totals		$1788.20		$2632.22	$44.70	$65.81

Crew B-10

Crew B-10	Hr.	Daily	Hr.	Daily	Bare Costs	Incl. O&P
1 Equip. Oper. (medium)	$51.10	$408.80	$77.30	$618.40	$46.70	$70.92
.5 Laborer	37.90	151.60	58.15	232.60		
12 L.H., Daily Totals		$560.40		$851.00	$46.70	$70.92

Crew B-10A

Crew B-10A	Hr.	Daily	Hr.	Daily	Bare Costs	Incl. O&P
1 Equip. Oper. (medium)	$51.10	$408.80	$77.30	$618.40	$46.70	$70.92
.5 Laborer	37.90	151.60	58.15	232.60		
1 Roller, 2-Drum, W.B., 7.5 H.P.		184.80		203.28	15.40	16.94
12 L.H., Daily Totals		$745.20		$1054.28	$62.10	$87.86

Crew B-10B

Crew B-10B	Hr.	Daily	Hr.	Daily	Bare Costs	Incl. O&P
1 Equip. Oper. (medium)	$51.10	$408.80	$77.30	$618.40	$46.70	$70.92
.5 Laborer	37.90	151.60	58.15	232.60		
1 Dozer, 200 H.P.		1397.00		1536.70	116.42	128.06
12 L.H., Daily Totals		$1957.40		$2387.70	$163.12	$198.97

Crew B-10C

Crew B-10C	Hr.	Daily	Hr.	Daily	Bare Costs	Incl. O&P
1 Equip. Oper. (medium)	$51.10	$408.80	$77.30	$618.40	$46.70	$70.92
.5 Laborer	37.90	151.60	58.15	232.60		
1 Dozer, 200 H.P.		1397.00		1536.70		
1 Vibratory Roller, Towed, 23 Ton		401.60		441.76	149.88	164.87
12 L.H., Daily Totals		$2359.00		$2829.46	$196.58	$235.79

Crew B-10D

Crew B-10D	Hr.	Daily	Hr.	Daily	Bare Costs	Incl. O&P
1 Equip. Oper. (medium)	$51.10	$408.80	$77.30	$618.40	$46.70	$70.92
.5 Laborer	37.90	151.60	58.15	232.60		
1 Dozer, 200 H.P.		1397.00		1536.70		
1 Sheepsft. Roller, Towed		426.40		469.04	151.95	167.15
12 L.H., Daily Totals		$2383.80		$2856.74	$198.65	$238.06

Crew B-10E

Crew B-10E	Hr.	Daily	Hr.	Daily	Bare Costs	Incl. O&P
1 Equip. Oper. (medium)	$51.10	$408.80	$77.30	$618.40	$46.70	$70.92
.5 Laborer	37.90	151.60	58.15	232.60		
1 Tandem Roller, 5 Ton		162.60		178.86	13.55	14.90
12 L.H., Daily Totals		$723.00		$1029.86	$60.25	$85.82

Crew B-10F

Crew B-10F	Hr.	Daily	Hr.	Daily	Bare Costs	Incl. O&P
1 Equip. Oper. (medium)	$51.10	$408.80	$77.30	$618.40	$46.70	$70.92
.5 Laborer	37.90	151.60	58.15	232.60		
1 Tandem Roller, 10 Ton		239.40		263.34	19.95	21.95
12 L.H., Daily Totals		$799.80		$1114.34	$66.65	$92.86

Crew B-10G

Crew B-10G	Hr.	Daily	Hr.	Daily	Bare Costs	Incl. O&P
1 Equip. Oper. (medium)	$51.10	$408.80	$77.30	$618.40	$46.70	$70.92
.5 Laborer	37.90	151.60	58.15	232.60		
1 Sheepsfoot Roller, 240 H.P.		1208.00		1328.80	100.67	110.73
12 L.H., Daily Totals		$1768.40		$2179.80	$147.37	$181.65

For customer support on your **Mechanical Cost Data**, call 877.739.3689.

623

Crews - Standard

Crew B-10H

	Hr.	Daily	Hr.	Daily	Bare Costs	Incl. O&P
1 Equip. Oper. (medium)	$51.10	$408.80	$77.30	$618.40	$46.70	$70.92
.5 Laborer	37.90	151.60	58.15	232.60		
1 Diaphragm Water Pump, 2"		71.80		78.98		
1 -20' Suction Hose, 2"		1.95		2.15		
2 -50' Discharge Hoses, 2"		1.80		1.98	6.30	6.93
12 L.H., Daily Totals		$635.95		$934.11	$53.00	$77.84

Crew B-10I

	Hr.	Daily	Hr.	Daily	Bare Costs	Incl. O&P
1 Equip. Oper. (medium)	$51.10	$408.80	$77.30	$618.40	$46.70	$70.92
.5 Laborer	37.90	151.60	58.15	232.60		
1 Diaphragm Water Pump, 4"		118.00		129.80		
1 -20' Suction Hose, 4"		3.25		3.58		
2 -50' Discharge Hoses, 4"		4.70		5.17	10.50	11.55
12 L.H., Daily Totals		$686.35		$989.54	$57.20	$82.46

Crew B-10J

	Hr.	Daily	Hr.	Daily	Bare Costs	Incl. O&P
1 Equip. Oper. (medium)	$51.10	$408.80	$77.30	$618.40	$46.70	$70.92
.5 Laborer	37.90	151.60	58.15	232.60		
1 Centrifugal Water Pump, 3"		81.00		89.10		
1 -20' Suction Hose, 3"		2.85		3.13		
2 -50' Discharge Hoses, 3"		3.00		3.30	7.24	7.96
12 L.H., Daily Totals		$647.25		$946.53	$53.94	$78.88

Crew B-10K

	Hr.	Daily	Hr.	Daily	Bare Costs	Incl. O&P
1 Equip. Oper. (medium)	$51.10	$408.80	$77.30	$618.40	$46.70	$70.92
.5 Laborer	37.90	151.60	58.15	232.60		
1 Centr. Water Pump, 6"		350.80		385.88		
1 -20' Suction Hose, 6"		11.50		12.65		
2 -50' Discharge Hoses, 6"		12.20		13.42	31.21	34.33
12 L.H., Daily Totals		$934.90		$1262.95	$77.91	$105.25

Crew B-10L

	Hr.	Daily	Hr.	Daily	Bare Costs	Incl. O&P
1 Equip. Oper. (medium)	$51.10	$408.80	$77.30	$618.40	$46.70	$70.92
.5 Laborer	37.90	151.60	58.15	232.60		
1 Dozer, 80 H.P.		471.40		518.54	39.28	43.21
12 L.H., Daily Totals		$1031.80		$1369.54	$85.98	$114.13

Crew B-10M

	Hr.	Daily	Hr.	Daily	Bare Costs	Incl. O&P
1 Equip. Oper. (medium)	$51.10	$408.80	$77.30	$618.40	$46.70	$70.92
.5 Laborer	37.90	151.60	58.15	232.60		
1 Dozer, 300 H.P.		1887.00		2075.70	157.25	172.97
12 L.H., Daily Totals		$2447.40		$2926.70	$203.95	$243.89

Crew B-10N

	Hr.	Daily	Hr.	Daily	Bare Costs	Incl. O&P
1 Equip. Oper. (medium)	$51.10	$408.80	$77.30	$618.40	$46.70	$70.92
.5 Laborer	37.90	151.60	58.15	232.60		
1 F.E. Loader, T.M., 1.5 C.Y		525.00		577.50	43.75	48.13
12 L.H., Daily Totals		$1085.40		$1428.50	$90.45	$119.04

Crew B-10O

	Hr.	Daily	Hr.	Daily	Bare Costs	Incl. O&P
1 Equip. Oper. (medium)	$51.10	$408.80	$77.30	$618.40	$46.70	$70.92
.5 Laborer	37.90	151.60	58.15	232.60		
1 F.E. Loader, T.M., 2.25 C.Y.		964.00		1060.40	80.33	88.37
12 L.H., Daily Totals		$1524.40		$1911.40	$127.03	$159.28

Crew B-10P

	Hr.	Daily	Hr.	Daily	Bare Costs	Incl. O&P
1 Equip. Oper. (medium)	$51.10	$408.80	$77.30	$618.40	$46.70	$70.92
.5 Laborer	37.90	151.60	58.15	232.60		
1 Crawler Loader, 3 C.Y.		1197.00		1316.70	99.75	109.72
12 L.H., Daily Totals		$1757.40		$2167.70	$146.45	$180.64

Crew B-10Q

	Hr.	Daily	Hr.	Daily	Bare Costs	Incl. O&P
1 Equip. Oper. (medium)	$51.10	$408.80	$77.30	$618.40	$46.70	$70.92
.5 Laborer	37.90	151.60	58.15	232.60		
1 Crawler Loader, 4 C.Y.		1566.00		1722.60	130.50	143.55
12 L.H., Daily Totals		$2126.40		$2573.60	$177.20	$214.47

Crew B-10R

	Hr.	Daily	Hr.	Daily	Bare Costs	Incl. O&P
1 Equip. Oper. (medium)	$51.10	$408.80	$77.30	$618.40	$46.70	$70.92
.5 Laborer	37.90	151.60	58.15	232.60		
1 F.E. Loader, W.M., 1 C.Y.		299.00		328.90	24.92	27.41
12 L.H., Daily Totals		$859.40		$1179.90	$71.62	$98.33

Crew B-10S

	Hr.	Daily	Hr.	Daily	Bare Costs	Incl. O&P
1 Equip. Oper. (medium)	$51.10	$408.80	$77.30	$618.40	$46.70	$70.92
.5 Laborer	37.90	151.60	58.15	232.60		
1 F.E. Loader, W.M., 1.5 C.Y.		378.60		416.46	31.55	34.70
12 L.H., Daily Totals		$939.00		$1267.46	$78.25	$105.62

Crew B-10T

	Hr.	Daily	Hr.	Daily	Bare Costs	Incl. O&P
1 Equip. Oper. (medium)	$51.10	$408.80	$77.30	$618.40	$46.70	$70.92
.5 Laborer	37.90	151.60	58.15	232.60		
1 F.E. Loader, W.M., 2.5 C.Y.		516.00		567.60	43.00	47.30
12 L.H., Daily Totals		$1076.40		$1418.60	$89.70	$118.22

Crew B-10U

	Hr.	Daily	Hr.	Daily	Bare Costs	Incl. O&P
1 Equip. Oper. (medium)	$51.10	$408.80	$77.30	$618.40	$46.70	$70.92
.5 Laborer	37.90	151.60	58.15	232.60		
1 F.E. Loader, W.M., 5.5 C.Y.		1092.00		1201.20	91.00	100.10
12 L.H., Daily Totals		$1652.40		$2052.20	$137.70	$171.02

Crew B-10V

	Hr.	Daily	Hr.	Daily	Bare Costs	Incl. O&P
1 Equip. Oper. (medium)	$51.10	$408.80	$77.30	$618.40	$46.70	$70.92
.5 Laborer	37.90	151.60	58.15	232.60		
1 Dozer, 700 H.P.		5347.00		5881.70	445.58	490.14
12 L.H., Daily Totals		$5907.40		$6732.70	$492.28	$561.06

Crew B-10W

	Hr.	Daily	Hr.	Daily	Bare Costs	Incl. O&P
1 Equip. Oper. (medium)	$51.10	$408.80	$77.30	$618.40	$46.70	$70.92
.5 Laborer	37.90	151.60	58.15	232.60		
1 Dozer, 105 H.P.		603.40		663.74	50.28	55.31
12 L.H., Daily Totals		$1163.80		$1514.74	$96.98	$126.23

Crew B-10X

	Hr.	Daily	Hr.	Daily	Bare Costs	Incl. O&P
1 Equip. Oper. (medium)	$51.10	$408.80	$77.30	$618.40	$46.70	$70.92
.5 Laborer	37.90	151.60	58.15	232.60		
1 Dozer, 410 H.P.		2392.00		2631.20	199.33	219.27
12 L.H., Daily Totals		$2952.40		$3482.20	$246.03	$290.18

Crew B-10Y

	Hr.	Daily	Hr.	Daily	Bare Costs	Incl. O&P
1 Equip. Oper. (medium)	$51.10	$408.80	$77.30	$618.40	$46.70	$70.92
.5 Laborer	37.90	151.60	58.15	232.60		
1 Vibr. Roller, Towed, 12 Ton		566.00		622.60	47.17	51.88
12 L.H., Daily Totals		$1126.40		$1473.60	$93.87	$122.80

Crew B-11A

	Hr.	Daily	Hr.	Daily	Bare Costs	Incl. O&P
1 Equipment Oper. (med.)	$51.10	$408.80	$77.30	$618.40	$44.50	$67.72
1 Laborer	37.90	303.20	58.15	465.20		
1 Dozer, 200 H.P.		1397.00		1536.70	87.31	96.04
16 L.H., Daily Totals		$2109.00		$2620.30	$131.81	$163.77

For customer support on your Mechanical Cost Data, call 877.739.3689.

Crew No.	Bare Costs		Incl. Subs O&P		Cost Per Labor-Hour	

Crew B-11B

	Hr.	Daily	Hr.	Daily	Bare Costs	Incl. O&P
1 Equipment Oper. (light)	$49.15	$393.20	$74.35	$594.80	$43.52	$66.25
1 Laborer	37.90	303.20	58.15	465.20		
1 Air Powered Tamper		24.00		26.40		
1 Air Compressor, 365 cfm		258.40		284.24		
2 -50' Air Hoses, 1.5"		10.40		11.44	18.30	20.13
16 L.H., Daily Totals		$989.20		$1382.08	$61.83	$86.38

Crew B-11C

	Hr.	Daily	Hr.	Daily	Bare Costs	Incl. O&P
1 Equipment Oper. (med.)	$51.10	$408.80	$77.30	$618.40	$44.50	$67.72
1 Laborer	37.90	303.20	58.15	465.20		
1 Backhoe Loader, 48 H.P.		365.40		401.94	22.84	25.12
16 L.H., Daily Totals		$1077.40		$1485.54	$67.34	$92.85

Crew B-11J

	Hr.	Daily	Hr.	Daily	Bare Costs	Incl. O&P
1 Equipment Oper. (med.)	$51.10	$408.80	$77.30	$618.40	$44.50	$67.72
1 Laborer	37.90	303.20	58.15	465.20		
1 Grader, 30,000 Lbs.		714.40		785.84		
1 Ripper, Beam & 1 Shank		83.80		92.18	49.89	54.88
16 L.H., Daily Totals		$1510.20		$1961.62	$94.39	$122.60

Crew B-11K

	Hr.	Daily	Hr.	Daily	Bare Costs	Incl. O&P
1 Equipment Oper. (med.)	$51.10	$408.80	$77.30	$618.40	$44.50	$67.72
1 Laborer	37.90	303.20	58.15	465.20		
1 Trencher, Chain Type, 8' D		2860.00		3146.00	178.75	196.63
16 L.H., Daily Totals		$3572.00		$4229.60	$223.25	$264.35

Crew B-11L

	Hr.	Daily	Hr.	Daily	Bare Costs	Incl. O&P
1 Equipment Oper. (med.)	$51.10	$408.80	$77.30	$618.40	$44.50	$67.72
1 Laborer	37.90	303.20	58.15	465.20		
1 Grader, 30,000 Lbs.		714.40		785.84	44.65	49.12
16 L.H., Daily Totals		$1426.40		$1869.44	$89.15	$116.84

Crew B-11M

	Hr.	Daily	Hr.	Daily	Bare Costs	Incl. O&P
1 Equipment Oper. (med.)	$51.10	$408.80	$77.30	$618.40	$44.50	$67.72
1 Laborer	37.90	303.20	58.15	465.20		
1 Backhoe Loader, 80 H.P.		396.20		435.82	24.76	27.24
16 L.H., Daily Totals		$1108.20		$1519.42	$69.26	$94.96

Crew B-11N

	Hr.	Daily	Hr.	Daily	Bare Costs	Incl. O&P
1 Labor Foreman (outside)	$39.90	$319.20	$61.20	$489.60	$44.59	$67.48
2 Equipment Operators (med.)	51.10	817.60	77.30	1236.80		
6 Truck Drivers (heavy)	43.20	2073.60	65.25	3132.00		
1 F.E. Loader, W.M., 5.5 C.Y.		1092.00		1201.20		
1 Dozer, 410 H.P.		2392.00		2631.20		
6 Dump Trucks, Off Hwy., 50 Ton		11166.00		12282.60	203.47	223.82
72 L.H., Daily Totals		$17860.40		$20973.40	$248.06	$291.30

Crew B-11Q

	Hr.	Daily	Hr.	Daily	Bare Costs	Incl. O&P
1 Equipment Operator (med.)	$51.10	$408.80	$77.30	$618.40	$46.70	$70.92
.5 Laborer	37.90	151.60	58.15	232.60		
1 Dozer, 140 H.P.		880.60		968.66	73.38	80.72
12 L.H., Daily Totals		$1441.00		$1819.66	$120.08	$151.64

Crew B-11R

	Hr.	Daily	Hr.	Daily	Bare Costs	Incl. O&P
1 Equipment Operator (med.)	$51.10	$408.80	$77.30	$618.40	$46.70	$70.92
.5 Laborer	37.90	151.60	58.15	232.60		
1 Dozer, 200 H.P.		1397.00		1536.70	116.42	128.06
12 L.H., Daily Totals		$1957.40		$2387.70	$163.12	$198.97

Crew B-11S

	Hr.	Daily	Hr.	Daily	Bare Costs	Incl. O&P
1 Equipment Operator (med.)	$51.10	$408.80	$77.30	$618.40	$46.70	$70.92
.5 Laborer	37.90	151.60	58.15	232.60		
1 Dozer, 300 H.P.		1887.00		2075.70		
1 Ripper, Beam & 1 Shank		83.80		92.18	164.23	180.66
12 L.H., Daily Totals		$2531.20		$3018.88	$210.93	$251.57

Crew B-11T

	Hr.	Daily	Hr.	Daily	Bare Costs	Incl. O&P
1 Equipment Operator (med.)	$51.10	$408.80	$77.30	$618.40	$46.70	$70.92
.5 Laborer	37.90	151.60	58.15	232.60		
1 Dozer, 410 H.P.		2392.00		2631.20		
1 Ripper, Beam & 2 Shanks		129.00		141.90	210.08	231.09
12 L.H., Daily Totals		$3081.40		$3624.10	$256.78	$302.01

Crew B-11U

	Hr.	Daily	Hr.	Daily	Bare Costs	Incl. O&P
1 Equipment Operator (med.)	$51.10	$408.80	$77.30	$618.40	$46.70	$70.92
.5 Laborer	37.90	151.60	58.15	232.60		
1 Dozer, 520 H.P.		3329.00		3661.90	277.42	305.16
12 L.H., Daily Totals		$3889.40		$4512.90	$324.12	$376.07

Crew B-11V

	Hr.	Daily	Hr.	Daily	Bare Costs	Incl. O&P
3 Laborers	$37.90	$909.60	$58.15	$1395.60	$37.90	$58.15
1 Roller, 2-Drum, W.B., 7.5 H.P.		184.80		203.28	7.70	8.47
24 L.H., Daily Totals		$1094.40		$1598.88	$45.60	$66.62

Crew B-11W

	Hr.	Daily	Hr.	Daily	Bare Costs	Incl. O&P
1 Equipment Operator (med.)	$51.10	$408.80	$77.30	$618.40	$43.42	$65.66
1 Common Laborer	37.90	303.20	58.15	465.20		
10 Truck Drivers (heavy)	43.20	3456.00	65.25	5220.00		
1 Dozer, 200 H.P.		1397.00		1536.70		
1 Vibratory Roller, Towed, 23 Ton		401.60		441.76		
10 Dump Trucks, 8 C.Y., 220 H.P.		4174.00		4591.40	62.21	68.44
96 L.H., Daily Totals		$10140.60		$12873.46	$105.63	$134.10

Crew B-11Y

	Hr.	Daily	Hr.	Daily	Bare Costs	Incl. O&P
1 Labor Foreman (outside)	$39.90	$319.20	$61.20	$489.60	$42.52	$64.87
5 Common Laborers	37.90	1516.00	58.15	2326.00		
3 Equipment Operators (med.)	51.10	1226.40	77.30	1855.20		
1 Dozer, 80 H.P.		471.40		518.54		
2 Rollers, 2-Drum, W.B., 7.5 H.P.		369.60		406.56		
4 Vibrating Plates, Gas, 21"		186.40		205.04	14.27	15.70
72 L.H., Daily Totals		$4089.00		$5800.94	$56.79	$80.57

Crew B-12A

	Hr.	Daily	Hr.	Daily	Bare Costs	Incl. O&P
1 Equip. Oper. (crane)	$52.25	$418.00	$79.05	$632.40	$45.08	$68.60
1 Laborer	37.90	303.20	58.15	465.20		
1 Hyd. Excavator, 1 C.Y.		812.80		894.08	50.80	55.88
16 L.H., Daily Totals		$1534.00		$1991.68	$95.88	$124.48

Crew B-12B

	Hr.	Daily	Hr.	Daily	Bare Costs	Incl. O&P
1 Equip. Oper. (crane)	$52.25	$418.00	$79.05	$632.40	$45.08	$68.60
1 Laborer	37.90	303.20	58.15	465.20		
1 Hyd. Excavator, 1.5 C.Y.		1029.00		1131.90	64.31	70.74
16 L.H., Daily Totals		$1750.20		$2229.50	$109.39	$139.34

Crew B-12C

	Hr.	Daily	Hr.	Daily	Bare Costs	Incl. O&P
1 Equip. Oper. (crane)	$52.25	$418.00	$79.05	$632.40	$45.08	$68.60
1 Laborer	37.90	303.20	58.15	465.20		
1 Hyd. Excavator, 2 C.Y.		1176.00		1293.60	73.50	80.85
16 L.H., Daily Totals		$1897.20		$2391.20	$118.58	$149.45

Crew B-12D

	Bare Costs Hr.	Daily	Incl. Subs O&P Hr.	Daily	Cost Per Labor-Hour Bare Costs	Incl. O&P
1 Equip. Oper. (crane)	$52.25	$418.00	$79.05	$632.40	$45.08	$68.60
1 Laborer	37.90	303.20	58.15	465.20		
1 Hyd. Excavator, 3.5 C.Y.		2442.00		2686.20	152.63	167.89
16 L.H., Daily Totals		$3163.20		$3783.80	$197.70	$236.49

Crew B-12E

	Bare Costs Hr.	Daily	Incl. Subs O&P Hr.	Daily	Cost Per Labor-Hour Bare Costs	Incl. O&P
1 Equip. Oper. (crane)	$52.25	$418.00	$79.05	$632.40	$45.08	$68.60
1 Laborer	37.90	303.20	58.15	465.20		
1 Hyd. Excavator, .5 C.Y.		445.60		490.16	27.85	30.64
16 L.H., Daily Totals		$1166.80		$1587.76	$72.92	$99.23

Crew B-12F

	Bare Costs Hr.	Daily	Incl. Subs O&P Hr.	Daily	Cost Per Labor-Hour Bare Costs	Incl. O&P
1 Equip. Oper. (crane)	$52.25	$418.00	$79.05	$632.40	$45.08	$68.60
1 Laborer	37.90	303.20	58.15	465.20		
1 Hyd. Excavator, .75 C.Y.		654.00		719.40	40.88	44.96
16 L.H., Daily Totals		$1375.20		$1817.00	$85.95	$113.56

Crew B-12G

	Bare Costs Hr.	Daily	Incl. Subs O&P Hr.	Daily	Cost Per Labor-Hour Bare Costs	Incl. O&P
1 Equip. Oper. (crane)	$52.25	$418.00	$79.05	$632.40	$45.08	$68.60
1 Laborer	37.90	303.20	58.15	465.20		
1 Crawler Crane, 15 Ton		695.70		765.27		
1 Clamshell Bucket, .5 C.Y.		37.80		41.58	45.84	50.43
16 L.H., Daily Totals		$1454.70		$1904.45	$90.92	$119.03

Crew B-12H

	Bare Costs Hr.	Daily	Incl. Subs O&P Hr.	Daily	Cost Per Labor-Hour Bare Costs	Incl. O&P
1 Equip. Oper. (crane)	$52.25	$418.00	$79.05	$632.40	$45.08	$68.60
1 Laborer	37.90	303.20	58.15	465.20		
1 Crawler Crane, 25 Ton		1194.00		1313.40		
1 Clamshell Bucket, 1 C.Y.		47.40		52.14	77.59	85.35
16 L.H., Daily Totals		$1962.60		$2463.14	$122.66	$153.95

Crew B-12I

	Bare Costs Hr.	Daily	Incl. Subs O&P Hr.	Daily	Cost Per Labor-Hour Bare Costs	Incl. O&P
1 Equip. Oper. (crane)	$52.25	$418.00	$79.05	$632.40	$45.08	$68.60
1 Laborer	37.90	303.20	58.15	465.20		
1 Crawler Crane, 20 Ton		896.25		985.88		
1 Dragline Bucket, .75 C.Y.		20.60		22.66	57.30	63.03
16 L.H., Daily Totals		$1638.05		$2106.14	$102.38	$131.63

Crew B-12J

	Bare Costs Hr.	Daily	Incl. Subs O&P Hr.	Daily	Cost Per Labor-Hour Bare Costs	Incl. O&P
1 Equip. Oper. (crane)	$52.25	$418.00	$79.05	$632.40	$45.08	$68.60
1 Laborer	37.90	303.20	58.15	465.20		
1 Gradall, 5/8 C.Y.		882.00		970.20	55.13	60.64
16 L.H., Daily Totals		$1603.20		$2067.80	$100.20	$129.24

Crew B-12K

	Bare Costs Hr.	Daily	Incl. Subs O&P Hr.	Daily	Cost Per Labor-Hour Bare Costs	Incl. O&P
1 Equip. Oper. (crane)	$52.25	$418.00	$79.05	$632.40	$45.08	$68.60
1 Laborer	37.90	303.20	58.15	465.20		
1 Gradall, 3 Ton, 1 C.Y.		995.80		1095.38	62.24	68.46
16 L.H., Daily Totals		$1717.00		$2192.98	$107.31	$137.06

Crew B-12L

	Bare Costs Hr.	Daily	Incl. Subs O&P Hr.	Daily	Cost Per Labor-Hour Bare Costs	Incl. O&P
1 Equip. Oper. (crane)	$52.25	$418.00	$79.05	$632.40	$45.08	$68.60
1 Laborer	37.90	303.20	58.15	465.20		
1 Crawler Crane, 15 Ton		695.70		765.27		
1 F.E. Attachment, .5 C.Y.		59.20		65.12	47.18	51.90
16 L.H., Daily Totals		$1476.10		$1927.99	$92.26	$120.50

Crew B-12M

	Bare Costs Hr.	Daily	Incl. Subs O&P Hr.	Daily	Cost Per Labor-Hour Bare Costs	Incl. O&P
1 Equip. Oper. (crane)	$52.25	$418.00	$79.05	$632.40	$45.08	$68.60
1 Laborer	37.90	303.20	58.15	465.20		
1 Crawler Crane, 20 Ton		896.25		985.88		
1 F.E. Attachment, .75 C.Y.		63.60		69.96	59.99	65.99
16 L.H., Daily Totals		$1681.05		$2153.43	$105.07	$134.59

Crew B-12N

	Bare Costs Hr.	Daily	Incl. Subs O&P Hr.	Daily	Cost Per Labor-Hour Bare Costs	Incl. O&P
1 Equip. Oper. (crane)	$52.25	$418.00	$79.05	$632.40	$45.08	$68.60
1 Laborer	37.90	303.20	58.15	465.20		
1 Crawler Crane, 25 Ton		1194.00		1313.40		
1 F.E. Attachment, 1 C.Y.		70.40		77.44	79.03	86.93
16 L.H., Daily Totals		$1985.60		$2488.44	$124.10	$155.53

Crew B-12O

	Bare Costs Hr.	Daily	Incl. Subs O&P Hr.	Daily	Cost Per Labor-Hour Bare Costs	Incl. O&P
1 Equip. Oper. (crane)	$52.25	$418.00	$79.05	$632.40	$45.08	$68.60
1 Laborer	37.90	303.20	58.15	465.20		
1 Crawler Crane, 40 Ton		1203.00		1323.30		
1 F.E. Attachment, 1.5 C.Y.		79.60		87.56	80.16	88.18
16 L.H., Daily Totals		$2003.80		$2508.46	$125.24	$156.78

Crew B-12P

	Bare Costs Hr.	Daily	Incl. Subs O&P Hr.	Daily	Cost Per Labor-Hour Bare Costs	Incl. O&P
1 Equip. Oper. (crane)	$52.25	$418.00	$79.05	$632.40	$45.08	$68.60
1 Laborer	37.90	303.20	58.15	465.20		
1 Crawler Crane, 40 Ton		1203.00		1323.30		
1 Dragline Bucket, 1.5 C.Y.		33.60		36.96	77.29	85.02
16 L.H., Daily Totals		$1957.80		$2457.86	$122.36	$153.62

Crew B-12Q

	Bare Costs Hr.	Daily	Incl. Subs O&P Hr.	Daily	Cost Per Labor-Hour Bare Costs	Incl. O&P
1 Equip. Oper. (crane)	$52.25	$418.00	$79.05	$632.40	$45.08	$68.60
1 Laborer	37.90	303.20	58.15	465.20		
1 Hyd. Excavator, 5/8 C.Y.		589.40		648.34	36.84	40.52
16 L.H., Daily Totals		$1310.60		$1745.94	$81.91	$109.12

Crew B-12S

	Bare Costs Hr.	Daily	Incl. Subs O&P Hr.	Daily	Cost Per Labor-Hour Bare Costs	Incl. O&P
1 Equip. Oper. (crane)	$52.25	$418.00	$79.05	$632.40	$45.08	$68.60
1 Laborer	37.90	303.20	58.15	465.20		
1 Hyd. Excavator, 2.5 C.Y.		1582.00		1740.20	98.88	108.76
16 L.H., Daily Totals		$2303.20		$2837.80	$143.95	$177.36

Crew B-12T

	Bare Costs Hr.	Daily	Incl. Subs O&P Hr.	Daily	Cost Per Labor-Hour Bare Costs	Incl. O&P
1 Equip. Oper. (crane)	$52.25	$418.00	$79.05	$632.40	$45.08	$68.60
1 Laborer	37.90	303.20	58.15	465.20		
1 Crawler Crane, 75 Ton		1507.00		1657.70		
1 F.E. Attachment, 3 C.Y.		102.40		112.64	100.59	110.65
16 L.H., Daily Totals		$2330.60		$2867.94	$145.66	$179.25

Crew B-12V

	Bare Costs Hr.	Daily	Incl. Subs O&P Hr.	Daily	Cost Per Labor-Hour Bare Costs	Incl. O&P
1 Equip. Oper. (crane)	$52.25	$418.00	$79.05	$632.40	$45.08	$68.60
1 Laborer	37.90	303.20	58.15	465.20		
1 Crawler Crane, 75 Ton		1507.00		1657.70		
1 Dragline Bucket, 3 C.Y.		52.60		57.86	97.47	107.22
16 L.H., Daily Totals		$2280.80		$2813.16	$142.55	$175.82

Crew B-12Y

	Bare Costs Hr.	Daily	Incl. Subs O&P Hr.	Daily	Cost Per Labor-Hour Bare Costs	Incl. O&P
1 Equip. Oper. (crane)	$52.25	$418.00	$79.05	$632.40	$42.68	$65.12
2 Laborers	37.90	606.40	58.15	930.40		
1 Hyd. Excavator, 3.5 C.Y.		2442.00		2686.20	101.75	111.93
24 L.H., Daily Totals		$3466.40		$4249.00	$144.43	$177.04

Crews - Standard

Crew B-12Z

	Hr.	Daily	Hr.	Daily	Bare Costs	Incl. O&P
1 Equip. Oper. (crane)	$52.25	$418.00	$79.05	$632.40	$42.68	$65.12
2 Laborers	37.90	606.40	58.15	930.40		
1 Hyd. Excavator, 2.5 C.Y.		1582.00		1740.20	65.92	72.51
24 L.H., Daily Totals		$2606.40		$3303.00	$108.60	$137.63

Crew B-13

	Hr.	Daily	Hr.	Daily	Bare Costs	Incl. O&P
1 Labor Foreman (outside)	$39.90	$319.20	$61.20	$489.60	$41.40	$63.21
4 Laborers	37.90	1212.80	58.15	1860.80		
1 Equip. Oper. (crane)	52.25	418.00	79.05	632.40		
1 Equip. Oper. (oiler)	46.05	368.40	69.65	557.20		
1 Hyd. Crane, 25 Ton		748.80		823.68	13.37	14.71
56 L.H., Daily Totals		$3067.20		$4363.68	$54.77	$77.92

Crew B-13A

	Hr.	Daily	Hr.	Daily	Bare Costs	Incl. O&P
1 Labor Foreman (outside)	$39.90	$319.20	$61.20	$489.60	$43.47	$66.09
2 Laborers	37.90	606.40	58.15	930.40		
2 Equipment Operators (med.)	51.10	817.60	77.30	1236.80		
2 Truck Drivers (heavy)	43.20	691.20	65.25	1044.00		
1 Crawler Crane, 75 Ton		1507.00		1657.70		
1 Crawler Loader, 4 C.Y.		1566.00		1722.60		
2 Dump Trucks, 8 C.Y., 220 H.P.		834.80		918.28	69.78	76.76
56 L.H., Daily Totals		$6342.20		$7999.38	$113.25	$142.85

Crew B-13B

	Hr.	Daily	Hr.	Daily	Bare Costs	Incl. O&P
1 Labor Foreman (outside)	$39.90	$319.20	$61.20	$489.60	$41.40	$63.21
4 Laborers	37.90	1212.80	58.15	1860.80		
1 Equip. Oper. (crane)	52.25	418.00	79.05	632.40		
1 Equip. Oper. (oiler)	46.05	368.40	69.65	557.20		
1 Hyd. Crane, 55 Ton		1123.00		1235.30	20.05	22.06
56 L.H., Daily Totals		$3441.40		$4775.30	$61.45	$85.27

Crew B-13C

	Hr.	Daily	Hr.	Daily	Bare Costs	Incl. O&P
1 Labor Foreman (outside)	$39.90	$319.20	$61.20	$489.60	$41.40	$63.21
4 Laborers	37.90	1212.80	58.15	1860.80		
1 Equip. Oper. (crane)	52.25	418.00	79.05	632.40		
1 Equip. Oper. (oiler)	46.05	368.40	69.65	557.20		
1 Crawler Crane, 100 Ton		1667.00		1833.70	29.77	32.74
56 L.H., Daily Totals		$3985.40		$5373.70	$71.17	$95.96

Crew B-13D

	Hr.	Daily	Hr.	Daily	Bare Costs	Incl. O&P
1 Laborer	$37.90	$303.20	$58.15	$465.20	$45.08	$68.60
1 Equip. Oper. (crane)	52.25	418.00	79.05	632.40		
1 Hyd. Excavator, 1 C.Y.		812.80		894.08		
1 Trench Box		81.00		89.10	55.86	61.45
16 L.H., Daily Totals		$1615.00		$2080.78	$100.94	$130.05

Crew B-13E

	Hr.	Daily	Hr.	Daily	Bare Costs	Incl. O&P
1 Laborer	$37.90	$303.20	$58.15	$465.20	$45.08	$68.60
1 Equip. Oper. (crane)	52.25	418.00	79.05	632.40		
1 Hyd. Excavator, 1.5 C.Y.		1029.00		1131.90		
1 Trench Box		81.00		89.10	69.38	76.31
16 L.H., Daily Totals		$1831.20		$2318.60	$114.45	$144.91

Crew B-13F

	Hr.	Daily	Hr.	Daily	Bare Costs	Incl. O&P
1 Laborer	$37.90	$303.20	$58.15	$465.20	$45.08	$68.60
1 Equip. Oper. (crane)	52.25	418.00	79.05	632.40		
1 Hyd. Excavator, 3.5 C.Y.		2442.00		2686.20		
1 Trench Box		81.00		89.10	157.69	173.46
16 L.H., Daily Totals		$3244.20		$3872.90	$202.76	$242.06

Crew B-13G

	Hr.	Daily	Hr.	Daily	Bare Costs	Incl. O&P
1 Laborer	$37.90	$303.20	$58.15	$465.20	$45.08	$68.60
1 Equip. Oper. (crane)	52.25	418.00	79.05	632.40		
1 Hyd. Excavator, .75 C.Y.		654.00		719.40		
1 Trench Box		81.00		89.10	45.94	50.53
16 L.H., Daily Totals		$1456.20		$1906.10	$91.01	$119.13

Crew B-13H

	Hr.	Daily	Hr.	Daily	Bare Costs	Incl. O&P
1 Laborer	$37.90	$303.20	$58.15	$465.20	$45.08	$68.60
1 Equip. Oper. (crane)	52.25	418.00	79.05	632.40		
1 Gradall, 5/8 C.Y.		882.00		970.20		
1 Trench Box		81.00		89.10	60.19	66.21
16 L.H., Daily Totals		$1684.20		$2156.90	$105.26	$134.81

Crew B-13I

	Hr.	Daily	Hr.	Daily	Bare Costs	Incl. O&P
1 Laborer	$37.90	$303.20	$58.15	$465.20	$45.08	$68.60
1 Equip. Oper. (crane)	52.25	418.00	79.05	632.40		
1 Gradall, 3 Ton, 1 C.Y.		995.80		1095.38		
1 Trench Box		81.00		89.10	67.30	74.03
16 L.H., Daily Totals		$1798.00		$2282.08	$112.38	$142.63

Crew B-13J

	Hr.	Daily	Hr.	Daily	Bare Costs	Incl. O&P
1 Laborer	$37.90	$303.20	$58.15	$465.20	$45.08	$68.60
1 Equip. Oper. (crane)	52.25	418.00	79.05	632.40		
1 Hyd. Excavator, 2.5 C.Y.		1582.00		1740.20		
1 Trench Box		81.00		89.10	103.94	114.33
16 L.H., Daily Totals		$2384.20		$2926.90	$149.01	$182.93

Crew B-13K

	Hr.	Daily	Hr.	Daily	Bare Costs	Incl. O&P
2 Equip. Opers. (crane)	$52.25	$836.00	$79.05	$1264.80	$52.25	$79.05
1 Hyd. Excavator, .75 C.Y.		654.00		719.40		
1 Hyd. Hammer, 4000 ft-lb		306.20		336.82		
1 Hyd. Excavator, .75 C.Y.		654.00		719.40	100.89	110.98
16 L.H., Daily Totals		$2450.20		$3040.42	$153.14	$190.03

Crew B-13L

	Hr.	Daily	Hr.	Daily	Bare Costs	Incl. O&P
2 Equip. Opers. (crane)	$52.25	$836.00	$79.05	$1264.80	$52.25	$79.05
1 Hyd. Excavator, 1.5 C.Y.		1029.00		1131.90		
1 Hyd. Hammer, 5000 ft-lb		367.80		404.58		
1 Hyd. Excavator, .75 C.Y.		654.00		719.40	128.18	140.99
16 L.H., Daily Totals		$2886.80		$3520.68	$180.43	$220.04

Crew B-13M

	Hr.	Daily	Hr.	Daily	Bare Costs	Incl. O&P
2 Equip. Opers. (crane)	$52.25	$836.00	$79.05	$1264.80	$52.25	$79.05
1 Hyd. Excavator, 2.5 C.Y.		1582.00		1740.20		
1 Hyd. Hammer, 8000 ft-lb		541.20		595.32		
1 Hyd. Excavator, 1.5 C.Y.		1029.00		1131.90	197.01	216.71
16 L.H., Daily Totals		$3988.20		$4732.22	$249.26	$295.76

Crew B-13N

	Hr.	Daily	Hr.	Daily	Bare Costs	Incl. O&P
2 Equip. Opers. (crane)	$52.25	$836.00	$79.05	$1264.80	$52.25	$79.05
1 Hyd. Excavator, 3.5 C.Y.		2442.00		2686.20		
1 Hyd. Hammer, 12,000 ft-lb		632.20		695.42		
1 Hyd. Excavator, 1.5 C.Y.		1029.00		1131.90	256.45	282.10
16 L.H., Daily Totals		$4939.20		$5778.32	$308.70	$361.14

For customer support on your Mechanical Cost Data, call 877.739.3689.

627

Crew B-14

Crew B-14	Bare Costs Hr.	Daily	Incl. Subs O&P Hr.	Daily	Cost Per Labor-Hour Bare Costs	Incl. O&P
1 Labor Foreman (outside)	$39.90	$319.20	$61.20	$489.60	$40.11	$61.36
4 Laborers	37.90	1212.80	58.15	1860.80		
1 Equip. Oper. (light)	49.15	393.20	74.35	594.80		
1 Backhoe Loader, 48 H.P.		365.40		401.94	7.61	8.37
48 L.H., Daily Totals		$2290.60		$3347.14	$47.72	$69.73

Crew B-14A

Crew B-14A	Bare Costs Hr.	Daily	Incl. Subs O&P Hr.	Daily	Cost Per Labor-Hour Bare Costs	Incl. O&P
1 Equip. Oper. (crane)	$52.25	$418.00	$79.05	$632.40	$47.47	$72.08
.5 Laborer	37.90	151.60	58.15	232.60		
1 Hyd. Excavator, 4.5 C.Y.		2974.00		3271.40	247.83	272.62
12 L.H., Daily Totals		$3543.60		$4136.40	$295.30	$344.70

Crew B-14B

Crew B-14B	Bare Costs Hr.	Daily	Incl. Subs O&P Hr.	Daily	Cost Per Labor-Hour Bare Costs	Incl. O&P
1 Equip. Oper. (crane)	$52.25	$418.00	$79.05	$632.40	$47.47	$72.08
.5 Laborer	37.90	151.60	58.15	232.60		
1 Hyd. Excavator, 6 C.Y.		3501.00		3851.10	291.75	320.93
12 L.H., Daily Totals		$4070.60		$4716.10	$339.22	$393.01

Crew B-14C

Crew B-14C	Bare Costs Hr.	Daily	Incl. Subs O&P Hr.	Daily	Cost Per Labor-Hour Bare Costs	Incl. O&P
1 Equip. Oper. (crane)	$52.25	$418.00	$79.05	$632.40	$47.47	$72.08
.5 Laborer	37.90	151.60	58.15	232.60		
1 Hyd. Excavator, 7 C.Y.		3597.00		3956.70	299.75	329.73
12 L.H., Daily Totals		$4166.60		$4821.70	$347.22	$401.81

Crew B-14F

Crew B-14F	Bare Costs Hr.	Daily	Incl. Subs O&P Hr.	Daily	Cost Per Labor-Hour Bare Costs	Incl. O&P
1 Equip. Oper. (crane)	$52.25	$418.00	$79.05	$632.40	$47.47	$72.08
.5 Laborer	37.90	151.60	58.15	232.60		
1 Hyd. Shovel, 7 C.Y.		4212.00		4633.20	351.00	386.10
12 L.H., Daily Totals		$4781.60		$5498.20	$398.47	$458.18

Crew B-14G

Crew B-14G	Bare Costs Hr.	Daily	Incl. Subs O&P Hr.	Daily	Cost Per Labor-Hour Bare Costs	Incl. O&P
1 Equip. Oper. (crane)	$52.25	$418.00	$79.05	$632.40	$47.47	$72.08
.5 Laborer	37.90	151.60	58.15	232.60		
1 Hyd. Shovel, 12 C.Y.		6328.00		6960.80	527.33	580.07
12 L.H., Daily Totals		$6897.60		$7825.80	$574.80	$652.15

Crew B-14J

Crew B-14J	Bare Costs Hr.	Daily	Incl. Subs O&P Hr.	Daily	Cost Per Labor-Hour Bare Costs	Incl. O&P
1 Equip. Oper. (medium)	$51.10	$408.80	$77.30	$618.40	$46.70	$70.92
.5 Laborer	37.90	151.60	58.15	232.60		
1 F.E. Loader, 8 C.Y.		1964.00		2160.40	163.67	180.03
12 L.H., Daily Totals		$2524.40		$3011.40	$210.37	$250.95

Crew B-14K

Crew B-14K	Bare Costs Hr.	Daily	Incl. Subs O&P Hr.	Daily	Cost Per Labor-Hour Bare Costs	Incl. O&P
1 Equip. Oper. (medium)	$51.10	$408.80	$77.30	$618.40	$46.70	$70.92
.5 Laborer	37.90	151.60	58.15	232.60		
1 F.E. Loader, 10 C.Y.		3085.00		3393.50	257.08	282.79
12 L.H., Daily Totals		$3645.40		$4244.50	$303.78	$353.71

Crew B-15

Crew B-15	Bare Costs Hr.	Daily	Incl. Subs O&P Hr.	Daily	Cost Per Labor-Hour Bare Costs	Incl. O&P
1 Equipment Oper. (med.)	$51.10	$408.80	$77.30	$618.40	$44.70	$67.68
.5 Laborer	37.90	151.60	58.15	232.60		
2 Truck Drivers (heavy)	43.20	691.20	65.25	1044.00		
2 Dump Trucks, 12 C.Y., 400 H.P.		1381.60		1519.76		
1 Dozer, 200 H.P.		1397.00		1536.70	99.24	109.16
28 L.H., Daily Totals		$4030.20		$4951.46	$143.94	$176.84

Crew B-16

Crew B-16	Bare Costs Hr.	Daily	Incl. Subs O&P Hr.	Daily	Cost Per Labor-Hour Bare Costs	Incl. O&P
1 Labor Foreman (outside)	$39.90	$319.20	$61.20	$489.60	$39.73	$60.69
2 Laborers	37.90	606.40	58.15	930.40		
1 Truck Driver (heavy)	43.20	345.60	65.25	522.00		
1 Dump Truck, 12 C.Y., 400 H.P.		690.80		759.88	21.59	23.75
32 L.H., Daily Totals		$1962.00		$2701.88	$61.31	$84.43

Crew B-17

Crew B-17	Bare Costs Hr.	Daily	Incl. Subs O&P Hr.	Daily	Cost Per Labor-Hour Bare Costs	Incl. O&P
2 Laborers	$37.90	$606.40	$58.15	$930.40	$42.04	$63.98
1 Equip. Oper. (light)	49.15	393.20	74.35	594.80		
1 Truck Driver (heavy)	43.20	345.60	65.25	522.00		
1 Backhoe Loader, 48 H.P.		365.40		401.94		
1 Dump Truck, 8 C.Y., 220 H.P.		417.40		459.14	24.46	26.91
32 L.H., Daily Totals		$2128.00		$2908.28	$66.50	$90.88

Crew B-17A

Crew B-17A	Bare Costs Hr.	Daily	Incl. Subs O&P Hr.	Daily	Cost Per Labor-Hour Bare Costs	Incl. O&P
2 Labor Foremen (outside)	$39.90	$638.40	$61.20	$979.20	$40.90	$62.81
6 Laborers	37.90	1819.20	58.15	2791.20		
1 Skilled Worker Foreman (out)	51.90	415.20	79.95	639.60		
1 Skilled Worker	49.90	399.20	76.85	614.80		
80 L.H., Daily Totals		$3272.00		$5024.80	$40.90	$62.81

Crew B-17B

Crew B-17B	Bare Costs Hr.	Daily	Incl. Subs O&P Hr.	Daily	Cost Per Labor-Hour Bare Costs	Incl. O&P
2 Laborers	$37.90	$606.40	$58.15	$930.40	$42.04	$63.98
1 Equip. Oper. (light)	49.15	393.20	74.35	594.80		
1 Truck Driver (heavy)	43.20	345.60	65.25	522.00		
1 Backhoe Loader, 48 H.P.		365.40		401.94		
1 Dump Truck, 12 C.Y., 400 H.P.		690.80		759.88	33.01	36.31
32 L.H., Daily Totals		$2401.40		$3209.02	$75.04	$100.28

Crew B-18

Crew B-18	Bare Costs Hr.	Daily	Incl. Subs O&P Hr.	Daily	Cost Per Labor-Hour Bare Costs	Incl. O&P
1 Labor Foreman (outside)	$39.90	$319.20	$61.20	$489.60	$38.57	$59.17
2 Laborers	37.90	606.40	58.15	930.40		
1 Vibrating Plate, Gas, 21"		46.60		51.26	1.94	2.14
24 L.H., Daily Totals		$972.20		$1471.26	$40.51	$61.30

Crew B-19

Crew B-19	Bare Costs Hr.	Daily	Incl. Subs O&P Hr.	Daily	Cost Per Labor-Hour Bare Costs	Incl. O&P
1 Pile Driver Foreman (outside)	$50.40	$403.20	$79.95	$639.60	$49.32	$76.84
4 Pile Drivers	48.40	1548.80	76.75	2456.00		
2 Equip. Opers. (crane)	52.25	836.00	79.05	1264.80		
1 Equip. Oper. (oiler)	46.05	368.40	69.65	557.20		
1 Crawler Crane, 40 Ton		1203.00		1323.30		
1 Lead, 90' High		124.40		136.84		
1 Hammer, Diesel, 22k ft-lb		378.00		415.80	26.65	29.31
64 L.H., Daily Totals		$4861.80		$6793.54	$75.97	$106.15

Crew B-19A

Crew B-19A	Bare Costs Hr.	Daily	Incl. Subs O&P Hr.	Daily	Cost Per Labor-Hour Bare Costs	Incl. O&P
1 Pile Driver Foreman (outside)	$50.40	$403.20	$79.95	$639.60	$49.32	$76.84
4 Pile Drivers	48.40	1548.80	76.75	2456.00		
2 Equip. Opers. (crane)	52.25	836.00	79.05	1264.80		
1 Equip. Oper. (oiler)	46.05	368.40	69.65	557.20		
1 Crawler Crane, 75 Ton		1507.00		1657.70		
1 Lead, 90' high		124.40		136.84		
1 Hammer, Diesel, 41k ft-lb		522.60		574.86	33.66	37.02
64 L.H., Daily Totals		$5310.40		$7287.00	$82.98	$113.86

Crew No.	Bare Costs		Incl. Subs O&P		Cost Per Labor-Hour	
Crew B-19B	Hr.	Daily	Hr.	Daily	Bare Costs	Incl. O&P
1 Pile Driver Foreman (outside)	$50.40	$403.20	$79.95	$639.60	$49.32	$76.84
4 Pile Drivers	48.40	1548.80	76.75	2456.00		
2 Equip. Opers. (crane)	52.25	836.00	79.05	1264.80		
1 Equip. Oper. (oiler)	46.05	368.40	69.65	557.20		
1 Crawler Crane, 40 Ton		1203.00		1323.30		
1 Lead, 90' High		124.40		136.84		
1 Hammer, Diesel, 22k ft-lb		378.00		415.80		
1 Barge, 400 Ton		779.20		857.12	38.82	42.70
64 L.H., Daily Totals		$5641.00		$7650.66	$88.14	$119.54

Crew No.	Bare Costs		Incl. Subs O&P		Cost Per Labor-Hour	
Crew B-19C	Hr.	Daily	Hr.	Daily	Bare Costs	Incl. O&P
1 Pile Driver Foreman (outside)	$50.40	$403.20	$79.95	$639.60	$49.32	$76.84
4 Pile Drivers	48.40	1548.80	76.75	2456.00		
2 Equip. Opers. (crane)	52.25	836.00	79.05	1264.80		
1 Equip. Oper. (oiler)	46.05	368.40	69.65	557.20		
1 Crawler Crane, 75 Ton		1507.00		1657.70		
1 Lead, 90' High		124.40		136.84		
1 Hammer, Diesel, 41k ft-lb		522.60		574.86		
1 Barge, 400 Ton		779.20		857.12	45.83	50.41
64 L.H., Daily Totals		$6089.60		$8144.12	$95.15	$127.25

Crew No.	Bare Costs		Incl. Subs O&P		Cost Per Labor-Hour	
Crew B-20	Hr.	Daily	Hr.	Daily	Bare Costs	Incl. O&P
1 Labor Foreman (outside)	$39.90	$319.20	$61.20	$489.60	$42.57	$65.40
1 Skilled Worker	49.90	399.20	76.85	614.80		
1 Laborer	37.90	303.20	58.15	465.20		
24 L.H., Daily Totals		$1021.60		$1569.60	$42.57	$65.40

Crew No.	Bare Costs		Incl. Subs O&P		Cost Per Labor-Hour	
Crew B-20A	Hr.	Daily	Hr.	Daily	Bare Costs	Incl. O&P
1 Labor Foreman (outside)	$39.90	$319.20	$61.20	$489.60	$46.09	$70.04
1 Laborer	37.90	303.20	58.15	465.20		
1 Plumber	59.20	473.60	89.35	714.80		
1 Plumber Apprentice	47.35	378.80	71.45	571.60		
32 L.H., Daily Totals		$1474.80		$2241.20	$46.09	$70.04

Crew No.	Bare Costs		Incl. Subs O&P		Cost Per Labor-Hour	
Crew B-21	Hr.	Daily	Hr.	Daily	Bare Costs	Incl. O&P
1 Labor Foreman (outside)	$39.90	$319.20	$61.20	$489.60	$43.95	$67.35
1 Skilled Worker	49.90	399.20	76.85	614.80		
1 Laborer	37.90	303.20	58.15	465.20		
.5 Equip. Oper. (crane)	52.25	209.00	79.05	316.20		
.5 S.P. Crane, 4x4, 5 Ton		140.40		154.44	5.01	5.52
28 L.H., Daily Totals		$1371.00		$2040.24	$48.96	$72.87

Crew No.	Bare Costs		Incl. Subs O&P		Cost Per Labor-Hour	
Crew B-21A	Hr.	Daily	Hr.	Daily	Bare Costs	Incl. O&P
1 Labor Foreman (outside)	$39.90	$319.20	$61.20	$489.60	$47.32	$71.84
1 Laborer	37.90	303.20	58.15	465.20		
1 Plumber	59.20	473.60	89.35	714.80		
1 Plumber Apprentice	47.35	378.80	71.45	571.60		
1 Equip. Oper. (crane)	52.25	418.00	79.05	632.40		
1 S.P. Crane, 4x4, 12 Ton		479.20		527.12	11.98	13.18
40 L.H., Daily Totals		$2372.00		$3400.72	$59.30	$85.02

Crew No.	Bare Costs		Incl. Subs O&P		Cost Per Labor-Hour	
Crew B-21B	Hr.	Daily	Hr.	Daily	Bare Costs	Incl. O&P
1 Labor Foreman (outside)	$39.90	$319.20	$61.20	$489.60	$41.17	$62.94
3 Laborers	37.90	909.60	58.15	1395.60		
1 Equip. Oper. (crane)	52.25	418.00	79.05	632.40		
1 Hyd. Crane, 12 Ton		658.40		724.24	16.46	18.11
40 L.H., Daily Totals		$2305.20		$3241.84	$57.63	$81.05

Crew No.	Bare Costs		Incl. Subs O&P		Cost Per Labor-Hour	
Crew B-21C	Hr.	Daily	Hr.	Daily	Bare Costs	Incl. O&P
1 Labor Foreman (outside)	$39.90	$319.20	$61.20	$489.60	$41.40	$63.21
4 Laborers	37.90	1212.80	58.15	1860.80		
1 Equip. Oper. (crane)	52.25	418.00	79.05	632.40		
1 Equip. Oper. (oiler)	46.05	368.40	69.65	557.20		
2 Cutting Torches		22.80		25.08		
2 Sets of Gases		336.00		369.60		
1 Lattice Boom Crane, 90 Ton		1520.00		1672.00	33.55	36.91
56 L.H., Daily Totals		$4197.20		$5606.68	$74.95	$100.12

Crew No.	Bare Costs		Incl. Subs O&P		Cost Per Labor-Hour	
Crew B-22	Hr.	Daily	Hr.	Daily	Bare Costs	Incl. O&P
1 Labor Foreman (outside)	$39.90	$319.20	$61.20	$489.60	$44.50	$68.13
1 Skilled Worker	49.90	399.20	76.85	614.80		
1 Laborer	37.90	303.20	58.15	465.20		
.75 Equip. Oper. (crane)	52.25	313.50	79.05	474.30		
.75 S.P. Crane, 4x4, 5 Ton		210.60		231.66	7.02	7.72
30 L.H., Daily Totals		$1545.70		$2275.56	$51.52	$75.85

Crew No.	Bare Costs		Incl. Subs O&P		Cost Per Labor-Hour	
Crew B-22A	Hr.	Daily	Hr.	Daily	Bare Costs	Incl. O&P
1 Labor Foreman (outside)	$39.90	$319.20	$61.20	$489.60	$43.57	$66.68
1 Skilled Worker	49.90	399.20	76.85	614.80		
2 Laborers	37.90	606.40	58.15	930.40		
1 Equipment Operator, Crane	52.25	418.00	79.05	632.40		
1 S.P. Crane, 4x4, 5 Ton		280.80		308.88		
1 Butt Fusion Mach., 4"-12" diam.		376.10		413.71	16.42	18.06
40 L.H., Daily Totals		$2399.70		$3389.79	$59.99	$84.74

Crew No.	Bare Costs		Incl. Subs O&P		Cost Per Labor-Hour	
Crew B-22B	Hr.	Daily	Hr.	Daily	Bare Costs	Incl. O&P
1 Labor Foreman (outside)	$39.90	$319.20	$61.20	$489.60	$43.57	$66.68
1 Skilled Worker	49.90	399.20	76.85	614.80		
2 Laborers	37.90	606.40	58.15	930.40		
1 Equip. Oper. (crane)	52.25	418.00	79.05	632.40		
1 S.P. Crane, 4x4, 5 Ton		280.80		308.88		
1 Butt Fusion Mach., 8"-24" diam.		804.80		885.28	27.14	29.85
40 L.H., Daily Totals		$2828.40		$3861.36	$70.71	$96.53

Crew No.	Bare Costs		Incl. Subs O&P		Cost Per Labor-Hour	
Crew B-22C	Hr.	Daily	Hr.	Daily	Bare Costs	Incl. O&P
1 Skilled Worker	$49.90	$399.20	$76.85	$614.80	$43.90	$67.50
1 Laborer	37.90	303.20	58.15	465.20		
1 Butt Fusion Mach., 2"-8" diam.		121.05		133.16	7.57	8.32
16 L.H., Daily Totals		$823.45		$1213.16	$51.47	$75.82

Crew No.	Bare Costs		Incl. Subs O&P		Cost Per Labor-Hour	
Crew B-23	Hr.	Daily	Hr.	Daily	Bare Costs	Incl. O&P
1 Labor Foreman (outside)	$39.90	$319.20	$61.20	$489.60	$38.30	$58.76
4 Laborers	37.90	1212.80	58.15	1860.80		
1 Drill Rig, Truck-Mounted		2553.00		2808.30		
1 Flatbed Truck, Gas, 3 Ton		307.40		338.14	71.51	78.66
40 L.H., Daily Totals		$4392.40		$5496.84	$109.81	$137.42

Crew No.	Bare Costs		Incl. Subs O&P		Cost Per Labor-Hour	
Crew B-23A	Hr.	Daily	Hr.	Daily	Bare Costs	Incl. O&P
1 Labor Foreman (outside)	$39.90	$319.20	$61.20	$489.60	$42.97	$65.55
1 Laborer	37.90	303.20	58.15	465.20		
1 Equip. Oper. (medium)	51.10	408.80	77.30	618.40		
1 Drill Rig, Truck-Mounted		2553.00		2808.30		
1 Pickup Truck, 3/4 Ton		145.80		160.38	112.45	123.69
24 L.H., Daily Totals		$3730.00		$4541.88	$155.42	$189.25

For customer support on your Mechanical Cost Data, call 877.739.3689.

629

Crew No.	Bare Costs		Incl. Subs O&P		Cost Per Labor-Hour	

Crew B-23B

	Hr.	Daily	Hr.	Daily	Bare Costs	Incl. O&P
1 Labor Foreman (outside)	$39.90	$319.20	$61.20	$489.60	$42.97	$65.55
1 Laborer	37.90	303.20	58.15	465.20		
1 Equip. Oper. (medium)	51.10	408.80	77.30	618.40		
1 Drill Rig, Truck-Mounted		2553.00		2808.30		
1 Pickup Truck, 3/4 Ton		145.80		160.38		
1 Centr. Water Pump, 6"		350.80		385.88	127.07	139.77
24 L.H., Daily Totals		$4080.80		$4927.76	$170.03	$205.32

Crew B-24

	Hr.	Daily	Hr.	Daily	Bare Costs	Incl. O&P
1 Cement Finisher	$45.65	$365.20	$67.70	$541.60	$44.00	$66.72
1 Laborer	37.90	303.20	58.15	465.20		
1 Carpenter	48.45	387.60	74.30	594.40		
24 L.H., Daily Totals		$1056.00		$1601.20	$44.00	$66.72

Crew B-25

	Hr.	Daily	Hr.	Daily	Bare Costs	Incl. O&P
1 Labor Foreman (outside)	$39.90	$319.20	$61.20	$489.60	$41.68	$63.65
7 Laborers	37.90	2122.40	58.15	3256.40		
3 Equip. Opers. (medium)	51.10	1226.40	77.30	1855.20		
1 Asphalt Paver, 130 H.P.		2210.00		2431.00		
1 Tandem Roller, 10 Ton		239.40		263.34		
1 Roller, Pneum. Whl., 12 Ton		347.80		382.58	31.79	34.97
88 L.H., Daily Totals		$6465.20		$8678.12	$73.47	$98.61

Crew B-25B

	Hr.	Daily	Hr.	Daily	Bare Costs	Incl. O&P
1 Labor Foreman (outside)	$39.90	$319.20	$61.20	$489.60	$42.47	$64.79
7 Laborers	37.90	2122.40	58.15	3256.40		
4 Equip. Opers. (medium)	51.10	1635.20	77.30	2473.60		
1 Asphalt Paver, 130 H.P.		2210.00		2431.00		
2 Tandem Rollers, 10 Ton		478.80		526.68		
1 Roller, Pneum. Whl., 12 Ton		347.80		382.58	31.63	34.79
96 L.H., Daily Totals		$7113.40		$9559.86	$74.10	$99.58

Crew B-25C

	Hr.	Daily	Hr.	Daily	Bare Costs	Incl. O&P
1 Labor Foreman (outside)	$39.90	$319.20	$61.20	$489.60	$42.63	$65.04
3 Laborers	37.90	909.60	58.15	1395.60		
2 Equip. Opers. (medium)	51.10	817.60	77.30	1236.80		
1 Asphalt Paver, 130 H.P.		2210.00		2431.00		
1 Tandem Roller, 10 Ton		239.40		263.34	51.03	56.13
48 L.H., Daily Totals		$4495.80		$5816.34	$93.66	$121.17

Crew B-25D

	Hr.	Daily	Hr.	Daily	Bare Costs	Incl. O&P
1 Labor Foreman (outside)	$39.90	$319.20	$61.20	$489.60	$42.81	$65.29
3 Laborers	37.90	909.60	58.15	1395.60		
2.125 Equip. Opers. (medium)	51.10	868.70	77.30	1314.10		
.125 Truck Driver (heavy)	43.20	43.20	65.25	65.25		
.125 Truck Tractor, 6x4, 380 H.P.		75.00		82.50		
.125 Dist. Tanker, 3000 Gallon		39.92		43.92		
1 Asphalt Paver, 130 H.P.		2210.00		2431.00		
1 Tandem Roller, 10 Ton		239.40		263.34	51.29	56.42
50 L.H., Daily Totals		$4705.02		$6085.31	$94.10	$121.71

Crew B-25E

	Hr.	Daily	Hr.	Daily	Bare Costs	Incl. O&P
1 Labor Foreman (outside)	$39.90	$319.20	$61.20	$489.60	$42.98	$65.52
3 Laborers	37.90	909.60	58.15	1395.60		
2.250 Equip. Opers. (medium)	51.10	919.80	77.30	1391.40		
.25 Truck Driver (heavy)	43.20	86.40	65.25	130.50		
.25 Truck Tractor, 6x4, 380 H.P.		150.00		165.00		
.25 Dist. Tanker, 3000 Gallon		79.85		87.83		
1 Asphalt Paver, 130 H.P.		2210.00		2431.00		
1 Tandem Roller, 10 Ton		239.40		263.34	51.52	56.68
52 L.H., Daily Totals		$4914.25		$6354.27	$94.50	$122.20

Crew B-26

	Hr.	Daily	Hr.	Daily	Bare Costs	Incl. O&P
1 Labor Foreman (outside)	$39.90	$319.20	$61.20	$489.60	$42.56	$64.95
6 Laborers	37.90	1819.20	58.15	2791.20		
2 Equip. Opers. (medium)	51.10	817.60	77.30	1236.80		
1 Rodman (reinf.)	53.00	424.00	82.10	656.80		
1 Cement Finisher	45.65	365.20	67.70	541.60		
1 Grader, 30,000 Lbs.		714.40		785.84		
1 Paving Mach. & Equip.		2796.00		3075.60	39.89	43.88
88 L.H., Daily Totals		$7255.60		$9577.44	$82.45	$108.83

Crew B-26A

	Hr.	Daily	Hr.	Daily	Bare Costs	Incl. O&P
1 Labor Foreman (outside)	$39.90	$319.20	$61.20	$489.60	$42.56	$64.95
6 Laborers	37.90	1819.20	58.15	2791.20		
2 Equip. Opers. (medium)	51.10	817.60	77.30	1236.80		
1 Rodman (reinf.)	53.00	424.00	82.10	656.80		
1 Cement Finisher	45.65	365.20	67.70	541.60		
1 Grader, 30,000 Lbs.		714.40		785.84		
1 Paving Mach. & Equip.		2796.00		3075.60		
1 Concrete Saw		168.60		185.46	41.81	45.99
88 L.H., Daily Totals		$7424.20		$9762.90	$84.37	$110.94

Crew B-26B

	Hr.	Daily	Hr.	Daily	Bare Costs	Incl. O&P
1 Labor Foreman (outside)	$39.90	$319.20	$61.20	$489.60	$43.27	$65.98
6 Laborers	37.90	1819.20	58.15	2791.20		
3 Equip. Opers. (medium)	51.10	1226.40	77.30	1855.20		
1 Rodman (reinf.)	53.00	424.00	82.10	656.80		
1 Cement Finisher	45.65	365.20	67.70	541.60		
1 Grader, 30,000 Lbs.		714.40		785.84		
1 Paving Mach. & Equip.		2796.00		3075.60		
1 Concrete Pump, 110' Boom		958.40		1054.24	46.55	51.20
96 L.H., Daily Totals		$8622.80		$11250.08	$89.82	$117.19

Crew B-26C

	Hr.	Daily	Hr.	Daily	Bare Costs	Incl. O&P
1 Labor Foreman (outside)	$39.90	$319.20	$61.20	$489.60	$41.70	$63.72
6 Laborers	37.90	1819.20	58.15	2791.20		
1 Equip. Oper. (medium)	51.10	408.80	77.30	618.40		
1 Rodman (reinf.)	53.00	424.00	82.10	656.80		
1 Cement Finisher	45.65	365.20	67.70	541.60		
1 Paving Mach. & Equip.		2796.00		3075.60		
1 Concrete Saw		168.60		185.46	37.06	40.76
80 L.H., Daily Totals		$6301.00		$8358.66	$78.76	$104.48

Crew B-27

	Hr.	Daily	Hr.	Daily	Bare Costs	Incl. O&P
1 Labor Foreman (outside)	$39.90	$319.20	$61.20	$489.60	$38.40	$58.91
3 Laborers	37.90	909.60	58.15	1395.60		
1 Berm Machine		295.80		325.38	9.24	10.17
32 L.H., Daily Totals		$1524.60		$2210.58	$47.64	$69.08

Crew B-28

	Hr.	Daily	Hr.	Daily	Bare Costs	Incl. O&P
2 Carpenters	$48.45	$775.20	$74.30	$1188.80	$44.93	$68.92
1 Laborer	37.90	303.20	58.15	465.20		
24 L.H., Daily Totals		$1078.40		$1654.00	$44.93	$68.92

Crew B-29

	Hr.	Daily	Hr.	Daily	Bare Costs	Incl. O&P
1 Labor Foreman (outside)	$39.90	$319.20	$61.20	$489.60	$41.40	$63.21
4 Laborers	37.90	1212.80	58.15	1860.80		
1 Equip. Oper. (crane)	52.25	418.00	79.05	632.40		
1 Equip. Oper. (oiler)	46.05	368.40	69.65	557.20		
1 Gradall, 5/8 C.Y.		882.00		970.20	15.75	17.32
56 L.H., Daily Totals		$3200.40		$4510.20	$57.15	$80.54

For customer support on your Mechanical Cost Data, call 877.739.3689.

Crew No.	Bare Costs		Incl. Subs O&P		Cost Per Labor-Hour	
Crew B-30	Hr.	Daily	Hr.	Daily	Bare Costs	Incl. O&P
1 Equip. Oper. (medium)	$51.10	$408.80	$77.30	$618.40	$45.83	$69.27
2 Truck Drivers (heavy)	43.20	691.20	65.25	1044.00		
1 Hyd. Excavator, 1.5 C.Y.		1029.00		1131.90		
2 Dump Trucks, 12 C.Y., 400 H.P.		1381.60		1519.76	100.44	110.49
24 L.H., Daily Totals		$3510.60		$4314.06	$146.28	$179.75

Crew No.	Bare Costs		Incl. Subs O&P		Cost Per Labor-Hour	
Crew B-31	Hr.	Daily	Hr.	Daily	Bare Costs	Incl. O&P
1 Labor Foreman (outside)	$39.90	$319.20	$61.20	$489.60	$40.41	$61.99
3 Laborers	37.90	909.60	58.15	1395.60		
1 Carpenter	48.45	387.60	74.30	594.40		
1 Air Compressor, 250 cfm		197.80		217.58		
1 Sheeting Driver		6.50		7.15		
2 -50' Air Hoses, 1.5"		10.40		11.44	5.37	5.90
40 L.H., Daily Totals		$1831.10		$2715.77	$45.78	$67.89

Crew B-32	Hr.	Daily	Hr.	Daily	Bare Costs	Incl. O&P
1 Laborer	$37.90	$303.20	$58.15	$465.20	$47.80	$72.51
3 Equip. Opers. (medium)	51.10	1226.40	77.30	1855.20		
1 Grader, 30,000 Lbs.		714.40		785.84		
1 Tandem Roller, 10 Ton		239.40		263.34		
1 Dozer, 200 H.P.		1397.00		1536.70	73.46	80.81
32 L.H., Daily Totals		$3880.40		$4906.28	$121.26	$153.32

Crew B-32A	Hr.	Daily	Hr.	Daily	Bare Costs	Incl. O&P
1 Laborer	$37.90	$303.20	$58.15	$465.20	$46.70	$70.92
2 Equip. Opers. (medium)	51.10	817.60	77.30	1236.80		
1 Grader, 30,000 Lbs.		714.40		785.84		
1 Roller, Vibratory, 25 Ton		691.80		760.98	58.59	64.45
24 L.H., Daily Totals		$2527.00		$3248.82	$105.29	$135.37

Crew B-32B	Hr.	Daily	Hr.	Daily	Bare Costs	Incl. O&P
1 Laborer	$37.90	$303.20	$58.15	$465.20	$46.70	$70.92
2 Equip. Opers. (medium)	51.10	817.60	77.30	1236.80		
1 Dozer, 200 H.P.		1397.00		1536.70		
1 Roller, Vibratory, 25 Ton		691.80		760.98	87.03	95.74
24 L.H., Daily Totals		$3209.60		$3999.68	$133.73	$166.65

Crew B-32C	Hr.	Daily	Hr.	Daily	Bare Costs	Incl. O&P
1 Labor Foreman (outside)	$39.90	$319.20	$61.20	$489.60	$44.83	$68.23
2 Laborers	37.90	606.40	58.15	930.40		
3 Equip. Opers. (medium)	51.10	1226.40	77.30	1855.20		
1 Grader, 30,000 Lbs.		714.40		785.84		
1 Tandem Roller, 10 Ton		239.40		263.34		
1 Dozer, 200 H.P.		1397.00		1536.70	48.98	53.87
48 L.H., Daily Totals		$4502.80		$5861.08	$93.81	$122.11

Crew B-33A	Hr.	Daily	Hr.	Daily	Bare Costs	Incl. O&P
1 Equip. Oper. (medium)	$51.10	$408.80	$77.30	$618.40	$47.33	$71.83
.5 Laborer	37.90	151.60	58.15	232.60		
.25 Equip. Oper. (medium)	51.10	102.20	77.30	154.60		
1 Scraper, Towed, 7 C.Y.		113.20		124.52		
1.25 Dozers, 300 H.P.		2358.75		2594.63	176.57	194.22
14 L.H., Daily Totals		$3134.55		$3724.74	$223.90	$266.05

Crew B-33B	Hr.	Daily	Hr.	Daily	Bare Costs	Incl. O&P
1 Equip. Oper. (medium)	$51.10	$408.80	$77.30	$618.40	$47.33	$71.83
.5 Laborer	37.90	151.60	58.15	232.60		
.25 Equip. Oper. (medium)	51.10	102.20	77.30	154.60		
1 Scraper, Towed, 10 C.Y.		145.00		159.50		
1.25 Dozers, 300 H.P.		2358.75		2594.63	178.84	196.72
14 L.H., Daily Totals		$3166.35		$3759.72	$226.17	$268.55

Crew B-33C	Hr.	Daily	Hr.	Daily	Bare Costs	Incl. O&P
1 Equip. Oper. (medium)	$51.10	$408.80	$77.30	$618.40	$47.33	$71.83
.5 Laborer	37.90	151.60	58.15	232.60		
.25 Equip. Oper. (medium)	51.10	102.20	77.30	154.60		
1 Scraper, Towed, 15 C.Y.		163.60		179.96		
1.25 Dozers, 300 H.P.		2358.75		2594.63	180.17	198.18
14 L.H., Daily Totals		$3184.95		$3780.18	$227.50	$270.01

Crew B-33D	Hr.	Daily	Hr.	Daily	Bare Costs	Incl. O&P
1 Equip. Oper. (medium)	$51.10	$408.80	$77.30	$618.40	$47.33	$71.83
.5 Laborer	37.90	151.60	58.15	232.60		
.25 Equip. Oper. (medium)	51.10	102.20	77.30	154.60		
1 S.P. Scraper, 14 C.Y.		2172.00		2389.20		
.25 Dozer, 300 H.P.		471.75		518.92	188.84	207.72
14 L.H., Daily Totals		$3306.35		$3913.72	$236.17	$279.55

Crew B-33E	Hr.	Daily	Hr.	Daily	Bare Costs	Incl. O&P
1 Equip. Oper. (medium)	$51.10	$408.80	$77.30	$618.40	$47.33	$71.83
.5 Laborer	37.90	151.60	58.15	232.60		
.25 Equip. Oper. (medium)	51.10	102.20	77.30	154.60		
1 S.P. Scraper, 21 C.Y.		2715.00		2986.50		
.25 Dozer, 300 H.P.		471.75		518.92	227.63	250.39
14 L.H., Daily Totals		$3849.35		$4511.02	$274.95	$322.22

Crew B-33F	Hr.	Daily	Hr.	Daily	Bare Costs	Incl. O&P
1 Equip. Oper. (medium)	$51.10	$408.80	$77.30	$618.40	$47.33	$71.83
.5 Laborer	37.90	151.60	58.15	232.60		
.25 Equip. Oper. (medium)	51.10	102.20	77.30	154.60		
1 Elev. Scraper, 11 C.Y.		1173.00		1290.30		
.25 Dozer, 300 H.P.		471.75		518.92	117.48	129.23
14 L.H., Daily Totals		$2307.35		$2814.82	$164.81	$201.06

Crew B-33G	Hr.	Daily	Hr.	Daily	Bare Costs	Incl. O&P
1 Equip. Oper. (medium)	$51.10	$408.80	$77.30	$618.40	$47.33	$71.83
.5 Laborer	37.90	151.60	58.15	232.60		
.25 Equip. Oper. (medium)	51.10	102.20	77.30	154.60		
1 Elev. Scraper, 22 C.Y.		2548.00		2802.80		
.25 Dozer, 300 H.P.		471.75		518.92	215.70	237.27
14 L.H., Daily Totals		$3682.35		$4327.32	$263.02	$309.09

Crew B-33H	Hr.	Daily	Hr.	Daily	Bare Costs	Incl. O&P
.5 Laborer	$37.90	$151.60	$58.15	$232.60	$47.33	$71.83
1 Equipment Operator (med.)	51.10	408.80	77.30	618.40		
.25 Equipment Operator (med.)	51.10	102.20	77.30	154.60		
1 S.P. Scraper, 44 C.Y.		4739.00		5212.90		
.25 Dozer, 410 H.P.		598.00		657.80	381.21	419.34
14 L.H., Daily Totals		$5999.60		$6876.30	$428.54	$491.16

Crew B-33J	Hr.	Daily	Hr.	Daily	Bare Costs	Incl. O&P
1 Equipment Operator (med.)	$51.10	$408.80	$77.30	$618.40	$51.10	$77.30
1 S.P. Scraper, 14 C.Y.		2172.00		2389.20	271.50	298.65
8 L.H., Daily Totals		$2580.80		$3007.60	$322.60	$375.95

Crew B-33K	Hr.	Daily	Hr.	Daily	Bare Costs	Incl. O&P
1 Equipment Operator (med.)	$51.10	$408.80	$77.30	$618.40	$47.33	$71.83
.25 Equipment Operator (med.)	51.10	102.20	77.30	154.60		
.5 Laborer	37.90	151.60	58.15	232.60		
1 S.P. Scraper, 31 C.Y.		3735.00		4108.50		
.25 Dozer, 410 H.P.		598.00		657.80	309.50	340.45
14 L.H., Daily Totals		$4995.60		$5771.90	$356.83	$412.28

For customer support on your Mechanical Cost Data, call 877.739.3689.

631

Crew B-34A

Crew No.	Bare Costs Hr.	Bare Costs Daily	Incl. Subs O&P Hr.	Incl. Subs O&P Daily	Cost Per Labor-Hour Bare Costs	Cost Per Labor-Hour Incl. O&P
1 Truck Driver (heavy)	$43.20	$345.60	$65.25	$522.00	$43.20	$65.25
1 Dump Truck, 8 C.Y., 220 H.P.		417.40		459.14	52.17	57.39
8 L.H., Daily Totals		$763.00		$981.14	$95.38	$122.64

Crew B-34B

Crew No.	Bare Costs Hr.	Bare Costs Daily	Incl. Subs O&P Hr.	Incl. Subs O&P Daily	Bare Costs	Incl. O&P
1 Truck Driver (heavy)	$43.20	$345.60	$65.25	$522.00	$43.20	$65.25
1 Dump Truck, 12 C.Y., 400 H.P.		690.80		759.88	86.35	94.98
8 L.H., Daily Totals		$1036.40		$1281.88	$129.55	$160.24

Crew B-34C

Crew No.	Bare Costs Hr.	Bare Costs Daily	Incl. Subs O&P Hr.	Incl. Subs O&P Daily	Bare Costs	Incl. O&P
1 Truck Driver (heavy)	$43.20	$345.60	$65.25	$522.00	$43.20	$65.25
1 Truck Tractor, 6x4, 380 H.P.		600.00		660.00		
1 Dump Trailer, 16.5 C.Y.		125.80		138.38	90.72	99.80
8 L.H., Daily Totals		$1071.40		$1320.38	$133.93	$165.05

Crew B-34D

Crew No.	Bare Costs Hr.	Bare Costs Daily	Incl. Subs O&P Hr.	Incl. Subs O&P Daily	Bare Costs	Incl. O&P
1 Truck Driver (heavy)	$43.20	$345.60	$65.25	$522.00	$43.20	$65.25
1 Truck Tractor, 6x4, 380 H.P.		600.00		660.00		
1 Dump Trailer, 20 C.Y.		140.40		154.44	92.55	101.81
8 L.H., Daily Totals		$1086.00		$1336.44	$135.75	$167.06

Crew B-34E

Crew No.	Bare Costs Hr.	Bare Costs Daily	Incl. Subs O&P Hr.	Incl. Subs O&P Daily	Bare Costs	Incl. O&P
1 Truck Driver (heavy)	$43.20	$345.60	$65.25	$522.00	$43.20	$65.25
1 Dump Truck, Off Hwy., 25 Ton		1370.00		1507.00	171.25	188.38
8 L.H., Daily Totals		$1715.60		$2029.00	$214.45	$253.63

Crew B-34F

Crew No.	Bare Costs Hr.	Bare Costs Daily	Incl. Subs O&P Hr.	Incl. Subs O&P Daily	Bare Costs	Incl. O&P
1 Truck Driver (heavy)	$43.20	$345.60	$65.25	$522.00	$43.20	$65.25
1 Dump Truck, Off Hwy., 35 Ton		1529.00		1681.90	191.13	210.24
8 L.H., Daily Totals		$1874.60		$2203.90	$234.32	$275.49

Crew B-34G

Crew No.	Bare Costs Hr.	Bare Costs Daily	Incl. Subs O&P Hr.	Incl. Subs O&P Daily	Bare Costs	Incl. O&P
1 Truck Driver (heavy)	$43.20	$345.60	$65.25	$522.00	$43.20	$65.25
1 Dump Truck, Off Hwy., 50 Ton		1861.00		2047.10	232.63	255.89
8 L.H., Daily Totals		$2206.60		$2569.10	$275.82	$321.14

Crew B-34H

Crew No.	Bare Costs Hr.	Bare Costs Daily	Incl. Subs O&P Hr.	Incl. Subs O&P Daily	Bare Costs	Incl. O&P
1 Truck Driver (heavy)	$43.20	$345.60	$65.25	$522.00	$43.20	$65.25
1 Dump Truck, Off Hwy., 65 Ton		1915.00		2106.50	239.38	263.31
8 L.H., Daily Totals		$2260.60		$2628.50	$282.57	$328.56

Crew B-34I

Crew No.	Bare Costs Hr.	Bare Costs Daily	Incl. Subs O&P Hr.	Incl. Subs O&P Daily	Bare Costs	Incl. O&P
1 Truck Driver (heavy)	$43.20	$345.60	$65.25	$522.00	$43.20	$65.25
1 Dump Truck, 18 C.Y., 450 H.P.		869.40		956.34	108.68	119.54
8 L.H., Daily Totals		$1215.00		$1478.34	$151.88	$184.79

Crew B-34J

Crew No.	Bare Costs Hr.	Bare Costs Daily	Incl. Subs O&P Hr.	Incl. Subs O&P Daily	Bare Costs	Incl. O&P
1 Truck Driver (heavy)	$43.20	$345.60	$65.25	$522.00	$43.20	$65.25
1 Dump Truck, Off Hwy., 100 Ton		2950.00		3245.00	368.75	405.63
8 L.H., Daily Totals		$3295.60		$3767.00	$411.95	$470.88

Crew B-34K

Crew No.	Bare Costs Hr.	Bare Costs Daily	Incl. Subs O&P Hr.	Incl. Subs O&P Daily	Bare Costs	Incl. O&P
1 Truck Driver (heavy)	$43.20	$345.60	$65.25	$522.00	$43.20	$65.25
1 Truck Tractor, 6x4, 450 H.P.		726.80		799.48		
1 Lowbed Trailer, 75 Ton		223.40		245.74	118.78	130.65
8 L.H., Daily Totals		$1295.80		$1567.22	$161.97	$195.90

Crew B-34L

Crew No.	Bare Costs Hr.	Bare Costs Daily	Incl. Subs O&P Hr.	Incl. Subs O&P Daily	Bare Costs	Incl. O&P
1 Equip. Oper. (light)	$49.15	$393.20	$74.35	$594.80	$49.15	$74.35
1 Flatbed Truck, Gas, 1.5 Ton		248.20		273.02	31.02	34.13
8 L.H., Daily Totals		$641.40		$867.82	$80.17	$108.48

Crew B-34M

Crew No.	Bare Costs Hr.	Bare Costs Daily	Incl. Subs O&P Hr.	Incl. Subs O&P Daily	Bare Costs	Incl. O&P
1 Equip. Oper. (light)	$49.15	$393.20	$74.35	$594.80	$49.15	$74.35
1 Flatbed Truck, Gas, 3 Ton		307.40		338.14	38.42	42.27
8 L.H., Daily Totals		$700.60		$932.94	$87.58	$116.62

Crew B-34N

Crew No.	Bare Costs Hr.	Bare Costs Daily	Incl. Subs O&P Hr.	Incl. Subs O&P Daily	Bare Costs	Incl. O&P
1 Truck Driver (heavy)	$43.20	$345.60	$65.25	$522.00	$47.15	$71.28
1 Equip. Oper. (medium)	51.10	408.80	77.30	618.40		
1 Truck Tractor, 6x4, 380 H.P.		600.00		660.00		
1 Flatbed Trailer, 40 Ton		154.60		170.06	47.16	51.88
16 L.H., Daily Totals		$1509.00		$1970.46	$94.31	$123.15

Crew B-34P

Crew No.	Bare Costs Hr.	Bare Costs Daily	Incl. Subs O&P Hr.	Incl. Subs O&P Daily	Bare Costs	Incl. O&P
1 Pipe Fitter	$60.70	$485.60	$91.60	$732.80	$51.25	$77.42
1 Truck Driver (light)	41.95	335.60	63.35	506.80		
1 Equip. Oper. (medium)	51.10	408.80	77.30	618.40		
1 Flatbed Truck, Gas, 3 Ton		307.40		338.14		
1 Backhoe Loader, 48 H.P.		365.40		401.94	28.03	30.84
24 L.H., Daily Totals		$1902.80		$2598.08	$79.28	$108.25

Crew B-34Q

Crew No.	Bare Costs Hr.	Bare Costs Daily	Incl. Subs O&P Hr.	Incl. Subs O&P Daily	Bare Costs	Incl. O&P
1 Pipe Fitter	$60.70	$485.60	$91.60	$732.80	$51.63	$78.00
1 Truck Driver (light)	41.95	335.60	63.35	506.80		
1 Equip. Oper. (crane)	52.25	418.00	79.05	632.40		
1 Flatbed Trailer, 25 Ton		113.20		124.52		
1 Dump Truck, 8 C.Y., 220 H.P.		417.40		459.14		
1 Hyd. Crane, 25 Ton		748.80		823.68	53.31	58.64
24 L.H., Daily Totals		$2518.60		$3279.34	$104.94	$136.64

Crew B-34R

Crew No.	Bare Costs Hr.	Bare Costs Daily	Incl. Subs O&P Hr.	Incl. Subs O&P Daily	Bare Costs	Incl. O&P
1 Pipe Fitter	$60.70	$485.60	$91.60	$732.80	$51.63	$78.00
1 Truck Driver (light)	41.95	335.60	63.35	506.80		
1 Equip. Oper. (crane)	52.25	418.00	79.05	632.40		
1 Flatbed Trailer, 25 Ton		113.20		124.52		
1 Dump Truck, 8 C.Y., 220 H.P.		417.40		459.14		
1 Hyd. Crane, 25 Ton		748.80		823.68		
1 Hyd. Excavator, 1 C.Y.		812.80		894.08	87.17	95.89
24 L.H., Daily Totals		$3331.40		$4173.42	$138.81	$173.89

Crew B-34S

Crew No.	Bare Costs Hr.	Bare Costs Daily	Incl. Subs O&P Hr.	Incl. Subs O&P Daily	Bare Costs	Incl. O&P
2 Pipe Fitters	$60.70	$971.20	$91.60	$1465.60	$54.21	$81.88
1 Truck Driver (heavy)	43.20	345.60	65.25	522.00		
1 Equip. Oper. (crane)	52.25	418.00	79.05	632.40		
1 Flatbed Trailer, 40 Ton		154.60		170.06		
1 Truck Tractor, 6x4, 380 H.P.		600.00		660.00		
1 Hyd. Crane, 80 Ton		1630.00		1793.00		
1 Hyd. Excavator, 2 C.Y.		1176.00		1293.60	111.27	122.62
32 L.H., Daily Totals		$5295.40		$6536.66	$165.48	$204.27

Crew B-34T

Crew No.	Bare Costs Hr.	Bare Costs Daily	Incl. Subs O&P Hr.	Incl. Subs O&P Daily	Bare Costs	Incl. O&P
2 Pipe Fitters	$60.70	$971.20	$91.60	$1465.60	$54.21	$81.88
1 Truck Driver (heavy)	43.20	345.60	65.25	522.00		
1 Equip. Oper. (crane)	52.25	418.00	79.05	632.40		
1 Flatbed Trailer, 40 Ton		154.60		170.06		
1 Truck Tractor, 6x4, 380 H.P.		600.00		660.00		
1 Hyd. Crane, 80 Ton		1630.00		1793.00	74.52	81.97
32 L.H., Daily Totals		$4119.40		$5243.06	$128.73	$163.85

Crew B-34U

Crew No.	Bare Costs Hr.	Daily	Incl. Subs O&P Hr.	Daily	Cost Per Labor-Hour Bare Costs	Incl. O&P
1 Truck Driver (heavy)	$43.20	$345.60	$65.25	$522.00	$46.17	$69.80
1 Equip. Oper. (light)	49.15	393.20	74.35	594.80		
1 Truck Tractor, 220 H.P.		358.80		394.68		
1 Flatbed Trailer, 25 Ton		113.20		124.52	29.50	32.45
16 L.H., Daily Totals		$1210.80		$1636.00	$75.67	$102.25

Crew B-34V

Crew No.	Bare Costs Hr.	Daily	Incl. Subs O&P Hr.	Daily	Cost Per Labor-Hour Bare Costs	Incl. O&P
1 Truck Driver (heavy)	$43.20	$345.60	$65.25	$522.00	$48.20	$72.88
1 Equip. Oper. (crane)	52.25	418.00	79.05	632.40		
1 Equip. Oper. (light)	49.15	393.20	74.35	594.80		
1 Truck Tractor, 6x4, 450 H.P.		726.80		799.48		
1 Equipment Trailer, 50 Ton		170.00		187.00		
1 Pickup Truck, 4x4, 3/4 Ton		156.80		172.48	43.90	48.29
24 L.H., Daily Totals		$2210.40		$2908.16	$92.10	$121.17

Crew B-34W

Crew No.	Bare Costs Hr.	Daily	Incl. Subs O&P Hr.	Daily	Cost Per Labor-Hour Bare Costs	Incl. O&P
5 Truck Drivers (heavy)	$43.20	$1728.00	$65.25	$2610.00	$45.66	$69.11
2 Equip. Opers. (crane)	52.25	836.00	79.05	1264.80		
1 Equip. Oper. (mechanic)	52.50	420.00	79.45	635.60		
1 Laborer	37.90	303.20	58.15	465.20		
4 Truck Tractors, 6x4, 380 H.P.		2400.00		2640.00		
2 Equipment Trailers, 50 Ton		340.00		374.00		
2 Flatbed Trailers, 40 Ton		309.20		340.12		
1 Pickup Truck, 4x4, 3/4 Ton		156.80		172.48		
1 S.P. Crane, 4x4, 20 Ton		554.80		610.28	52.23	57.46
72 L.H., Daily Totals		$7048.00		$9112.48	$97.89	$126.56

Crew B-35

Crew No.	Bare Costs Hr.	Daily	Incl. Subs O&P Hr.	Daily	Cost Per Labor-Hour Bare Costs	Incl. O&P
1 Labor Foreman (outside)	$39.90	$319.20	$61.20	$489.60	$47.53	$72.38
1 Skilled Worker	49.90	399.20	76.85	614.80		
1 Welder (plumber)	59.20	473.60	89.35	714.80		
1 Laborer	37.90	303.20	58.15	465.20		
1 Equip. Oper. (crane)	52.25	418.00	79.05	632.40		
1 Equip. Oper. (oiler)	46.05	368.40	69.65	557.20		
1 Welder, Electric, 300 amp		58.10		63.91		
1 Hyd. Excavator, .75 C.Y.		654.00		719.40	14.84	16.32
48 L.H., Daily Totals		$2993.70		$4257.31	$62.37	$88.69

Crew B-35A

Crew No.	Bare Costs Hr.	Daily	Incl. Subs O&P Hr.	Daily	Cost Per Labor-Hour Bare Costs	Incl. O&P
1 Labor Foreman (outside)	$39.90	$319.20	$61.20	$489.60	$46.16	$70.34
2 Laborers	37.90	606.40	58.15	930.40		
1 Skilled Worker	49.90	399.20	76.85	614.80		
1 Welder (plumber)	59.20	473.60	89.35	714.80		
1 Equip. Oper. (crane)	52.25	418.00	79.05	632.40		
1 Equip. Oper. (oiler)	46.05	368.40	69.65	557.20		
1 Welder, Gas Engine, 300 amp		147.20		161.92		
1 Crawler Crane, 75 Ton		1507.00		1657.70	29.54	32.49
56 L.H., Daily Totals		$4239.00		$5758.82	$75.70	$102.84

Crew B-36

Crew No.	Bare Costs Hr.	Daily	Incl. Subs O&P Hr.	Daily	Cost Per Labor-Hour Bare Costs	Incl. O&P
1 Labor Foreman (outside)	$39.90	$319.20	$61.20	$489.60	$43.58	$66.42
2 Laborers	37.90	606.40	58.15	930.40		
2 Equip. Opers. (medium)	51.10	817.60	77.30	1236.80		
1 Dozer, 200 H.P.		1397.00		1536.70		
1 Aggregate Spreader		39.60		43.56		
1 Tandem Roller, 10 Ton		239.40		263.34	41.90	46.09
40 L.H., Daily Totals		$3419.20		$4500.40	$85.48	$112.51

Crew B-36A

Crew No.	Bare Costs Hr.	Daily	Incl. Subs O&P Hr.	Daily	Cost Per Labor-Hour Bare Costs	Incl. O&P
1 Labor Foreman (outside)	$39.90	$319.20	$61.20	$489.60	$45.73	$69.53
2 Laborers	37.90	606.40	58.15	930.40		
4 Equip. Opers. (medium)	51.10	1635.20	77.30	2473.60		
1 Dozer, 200 H.P.		1397.00		1536.70		
1 Aggregate Spreader		39.60		43.56		
1 Tandem Roller, 10 Ton		239.40		263.34		
1 Roller, Pneum. Whl., 12 Ton		347.80		382.58	36.14	39.75
56 L.H., Daily Totals		$4584.60		$6119.78	$81.87	$109.28

Crew B-36B

Crew No.	Bare Costs Hr.	Daily	Incl. Subs O&P Hr.	Daily	Cost Per Labor-Hour Bare Costs	Incl. O&P
1 Labor Foreman (outside)	$39.90	$319.20	$61.20	$489.60	$45.41	$68.99
2 Laborers	37.90	606.40	58.15	930.40		
4 Equip. Opers. (medium)	51.10	1635.20	77.30	2473.60		
1 Truck Driver (heavy)	43.20	345.60	65.25	522.00		
1 Grader, 30,000 Lbs.		714.40		785.84		
1 F.E. Loader, Crl, 1.5 C.Y.		634.20		697.62		
1 Dozer, 300 H.P.		1887.00		2075.70		
1 Roller, Vibratory, 25 Ton		691.80		760.98		
1 Truck Tractor, 6x4, 450 H.P.		726.80		799.48		
1 Water Tank Trailer, 5000 Gal.		141.60		155.76	74.93	82.43
64 L.H., Daily Totals		$7702.20		$9690.98	$120.35	$151.42

Crew B-36C

Crew No.	Bare Costs Hr.	Daily	Incl. Subs O&P Hr.	Daily	Cost Per Labor-Hour Bare Costs	Incl. O&P
1 Labor Foreman (outside)	$39.90	$319.20	$61.20	$489.60	$47.28	$71.67
3 Equip. Opers. (medium)	51.10	1226.40	77.30	1855.20		
1 Truck Driver (heavy)	43.20	345.60	65.25	522.00		
1 Grader, 30,000 Lbs.		714.40		785.84		
1 Dozer, 300 H.P.		1887.00		2075.70		
1 Roller, Vibratory, 25 Ton		691.80		760.98		
1 Truck Tractor, 6x4, 450 H.P.		726.80		799.48		
1 Water Tank Trailer, 5000 Gal.		141.60		155.76	104.04	114.44
40 L.H., Daily Totals		$6052.80		$7444.56	$151.32	$186.11

Crew B-36D

Crew No.	Bare Costs Hr.	Daily	Incl. Subs O&P Hr.	Daily	Cost Per Labor-Hour Bare Costs	Incl. O&P
1 Labor Foreman (outside)	$39.90	$319.20	$61.20	$489.60	$48.30	$73.28
3 Equip. Opers. (medium)	51.10	1226.40	77.30	1855.20		
1 Grader, 30,000 Lbs.		714.40		785.84		
1 Dozer, 300 H.P.		1887.00		2075.70		
1 Roller, Vibratory, 25 Ton		691.80		760.98	102.91	113.20
32 L.H., Daily Totals		$4838.80		$5967.32	$151.21	$186.48

Crew B-37

Crew No.	Bare Costs Hr.	Daily	Incl. Subs O&P Hr.	Daily	Cost Per Labor-Hour Bare Costs	Incl. O&P
1 Labor Foreman (outside)	$39.90	$319.20	$61.20	$489.60	$40.11	$61.36
4 Laborers	37.90	1212.80	58.15	1860.80		
1 Equip. Oper. (light)	49.15	393.20	74.35	594.80		
1 Tandem Roller, 5 Ton		162.60		178.86	3.39	3.73
48 L.H., Daily Totals		$2087.80		$3124.06	$43.50	$65.08

Crew B-37A

Crew No.	Bare Costs Hr.	Daily	Incl. Subs O&P Hr.	Daily	Cost Per Labor-Hour Bare Costs	Incl. O&P
2 Laborers	$37.90	$606.40	$58.15	$930.40	$39.25	$59.88
1 Truck Driver (light)	41.95	335.60	63.35	506.80		
1 Flatbed Truck, Gas, 1.5 Ton		248.20		273.02		
1 Tar Kettle, T.M.		135.30		148.83	15.98	17.58
24 L.H., Daily Totals		$1325.50		$1859.05	$55.23	$77.46

Crew B-37B

Crew No.	Bare Costs Hr.	Daily	Incl. Subs O&P Hr.	Daily	Cost Per Labor-Hour Bare Costs	Incl. O&P
3 Laborers	$37.90	$909.60	$58.15	$1395.60	$38.91	$59.45
1 Truck Driver (light)	41.95	335.60	63.35	506.80		
1 Flatbed Truck, Gas, 1.5 Ton		248.20		273.02		
1 Tar Kettle, T.M.		135.30		148.83	11.98	13.18
32 L.H., Daily Totals		$1628.70		$2324.25	$50.90	$72.63

Crew No.	Bare Costs		Incl. Subs O&P		Cost Per Labor-Hour	

Crew B-37C	Hr.	Daily	Hr.	Daily	Bare Costs	Incl. O&P
2 Laborers	$37.90	$606.40	$58.15	$930.40	$39.92	$60.75
2 Truck Drivers (light)	41.95	671.20	63.35	1013.60		
2 Flatbed Trucks, Gas, 1.5 Ton		496.40		546.04		
1 Tar Kettle, T.M.		135.30		148.83	19.74	21.71
32 L.H., Daily Totals		$1909.30		$2638.87	$59.67	$82.46

Crew B-37D	Hr.	Daily	Hr.	Daily	Bare Costs	Incl. O&P
1 Laborer	$37.90	$303.20	$58.15	$465.20	$39.92	$60.75
1 Truck Driver (light)	41.95	335.60	63.35	506.80		
1 Pickup Truck, 3/4 Ton		145.80		160.38	9.11	10.02
16 L.H., Daily Totals		$784.60		$1132.38	$49.04	$70.77

Crew B-37E	Hr.	Daily	Hr.	Daily	Bare Costs	Incl. O&P
3 Laborers	$37.90	$909.60	$58.15	$1395.60	$42.55	$64.69
1 Equip. Oper. (light)	49.15	393.20	74.35	594.80		
1 Equip. Oper. (medium)	51.10	408.80	77.30	618.40		
2 Truck Drivers (light)	41.95	671.20	63.35	1013.60		
4 Barrels w/ Flasher		13.60		14.96		
1 Concrete Saw		168.60		185.46		
1 Rotary Hammer Drill		24.55		27.00		
1 Hammer Drill Bit		2.90		3.19		
1 Loader, Skid Steer, 30 H.P.		173.20		190.52		
1 Conc. Hammer Attach.		108.40		119.24		
1 Vibrating Plate, Gas, 18"		36.20		39.82		
2 Flatbed Trucks, Gas, 1.5 Ton		496.40		546.04	18.28	20.11
56 L.H., Daily Totals		$3406.65		$4748.64	$60.83	$84.80

Crew B-37F	Hr.	Daily	Hr.	Daily	Bare Costs	Incl. O&P
3 Laborers	$37.90	$909.60	$58.15	$1395.60	$38.91	$59.45
1 Truck Driver (light)	41.95	335.60	63.35	506.80		
4 Barrels w/ Flasher		13.60		14.96		
1 Concrete Mixer, 10 C.F.		172.80		190.08		
1 Air Compressor, 60 cfm		142.80		157.08		
1 -50' Air Hose, 3/4"		3.05		3.36		
1 Spade (Chipper)		8.20		9.02		
1 Flatbed Truck, Gas, 1.5 Ton		248.20		273.02	18.40	20.23
32 L.H., Daily Totals		$1833.85		$2549.92	$57.31	$79.68

Crew B-37G	Hr.	Daily	Hr.	Daily	Bare Costs	Incl. O&P
1 Labor Foreman (outside)	$39.90	$319.20	$61.20	$489.60	$40.11	$61.36
4 Laborers	37.90	1212.80	58.15	1860.80		
1 Equip. Oper. (light)	49.15	393.20	74.35	594.80		
1 Berm Machine		295.80		325.38		
1 Tandem Roller, 5 Ton		162.60		178.86	9.55	10.51
48 L.H., Daily Totals		$2383.60		$3449.44	$49.66	$71.86

Crew B-37H	Hr.	Daily	Hr.	Daily	Bare Costs	Incl. O&P
1 Labor Foreman (outside)	$39.90	$319.20	$61.20	$489.60	$40.11	$61.36
4 Laborers	37.90	1212.80	58.15	1860.80		
1 Equip. Oper. (light)	49.15	393.20	74.35	594.80		
1 Tandem Roller, 5 Ton		162.60		178.86		
1 Flatbed Trucks, Gas, 1.5 Ton		248.20		273.02		
1 Tar Kettle, T.M.		135.30		148.83	11.38	12.51
48 L.H., Daily Totals		$2471.30		$3545.91	$51.49	$73.87

Crew B-37I	Hr.	Daily	Hr.	Daily	Bare Costs	Incl. O&P
3 Laborers	$37.90	$909.60	$58.15	$1395.60	$42.55	$64.69
1 Equip. Oper. (light)	49.15	393.20	74.35	594.80		
1 Equip. Oper. (medium)	51.10	408.80	77.30	618.40		
2 Truck Drivers (light)	41.95	671.20	63.35	1013.60		
4 Barrels w/ Flasher		13.60		14.96		
1 Concrete Saw		168.60		185.46		
1 Rotary Hammer Drill		24.55		27.00		
1 Hammer Drill Bit		2.90		3.19		
1 Air Compressor, 60 cfm		142.80		157.08		
1 -50' Air Hose, 3/4"		3.05		3.36		
1 Spade (Chipper)		8.20		9.02		
1 Loader, Skid Steer, 30 H.P.		173.20		190.52		
1 Conc. Hammer Attach.		108.40		119.24		
1 Concrete Mixer, 10 C.F.		172.80		190.08		
1 Vibrating Plate, Gas, 18"		36.20		39.82		
2 Flatbed Trucks, Gas, 1.5 Ton		496.40		546.04	24.12	26.53
56 L.H., Daily Totals		$3733.50		$5108.17	$66.67	$91.22

Crew B-37J	Hr.	Daily	Hr.	Daily	Bare Costs	Incl. O&P
1 Labor Foreman (outside)	$39.90	$319.20	$61.20	$489.60	$40.11	$61.36
4 Laborers	37.90	1212.80	58.15	1860.80		
1 Equip. Oper. (light)	49.15	393.20	74.35	594.80		
1 Air Compressor, 60 cfm		142.80		157.08		
1 -50' Air Hose, 3/4"		3.05		3.36		
2 Concrete Mixers, 10 C.F.		345.60		380.16		
2 Flatbed Trucks, Gas, 1.5 Ton		496.40		546.04		
1 Shot Blaster, 20"		215.40		236.94	25.07	27.57
48 L.H., Daily Totals		$3128.45		$4268.77	$65.18	$88.93

Crew B-37K	Hr.	Daily	Hr.	Daily	Bare Costs	Incl. O&P
1 Labor Foreman (outside)	$39.90	$319.20	$61.20	$489.60	$40.11	$61.36
4 Laborers	37.90	1212.80	58.15	1860.80		
1 Equip. Oper. (light)	49.15	393.20	74.35	594.80		
1 Air Compressor, 60 cfm		142.80		157.08		
1 -50' Air Hose, 3/4"		3.05		3.36		
2 Flatbed Trucks, Gas, 1.5 Ton		496.40		546.04		
1 Shot Blaster, 20"		215.40		236.94	17.87	19.65
48 L.H., Daily Totals		$2782.85		$3888.61	$57.98	$81.01

Crew B-38	Hr.	Daily	Hr.	Daily	Bare Costs	Incl. O&P
1 Labor Foreman (outside)	$39.90	$319.20	$61.20	$489.60	$43.19	$65.83
2 Laborers	37.90	606.40	58.15	930.40		
1 Equip. Oper. (light)	49.15	393.20	74.35	594.80		
1 Equip. Oper. (medium)	51.10	408.80	77.30	618.40		
1 Backhoe Loader, 48 H.P.		365.40		401.94		
1 Hyd. Hammer, (1200 lb.)		179.60		197.56		
1 F.E. Loader, W.M., 4 C.Y.		661.80		727.98		
1 Pvmt. Rem. Bucket		57.40		63.14	31.61	34.77
40 L.H., Daily Totals		$2991.80		$4023.82	$74.80	$100.60

Crew B-39	Hr.	Daily	Hr.	Daily	Bare Costs	Incl. O&P
1 Labor Foreman (outside)	$39.90	$319.20	$61.20	$489.60	$40.11	$61.36
4 Laborers	37.90	1212.80	58.15	1860.80		
1 Equip. Oper. (light)	49.15	393.20	74.35	594.80		
1 Air Compressor, 250 cfm		197.80		217.58		
2 Breakers, Pavement, 60 lb.		20.40		22.44		
2 -50' Air Hoses, 1.5"		10.40		11.44	4.76	5.24
48 L.H., Daily Totals		$2153.80		$3196.66	$44.87	$66.60

Crew No.	Bare Costs Hr.	Daily	Incl. Subs O&P Hr.	Daily	Cost Per Labor-Hour Bare Costs	Incl. O&P
Crew B-40	Hr.	Daily	Hr.	Daily	Bare Costs	Incl. O&P
1 Pile Driver Foreman (outside)	$50.40	$403.20	$79.95	$639.60	$49.32	$76.84
4 Pile Drivers	48.40	1548.80	76.75	2456.00		
2 Equip. Opers. (crane)	52.25	836.00	79.05	1264.80		
1 Equip. Oper. (oiler)	46.05	368.40	69.65	557.20		
1 Crawler Crane, 40 Ton		1203.00		1323.30		
1 Vibratory Hammer & Gen.		2612.00		2873.20	59.61	65.57
64 L.H., Daily Totals		$6971.40		$9114.10	$108.93	$142.41
Crew B-40B	Hr.	Daily	Hr.	Daily	Bare Costs	Incl. O&P
1 Labor Foreman (outside)	$39.90	$319.20	$61.20	$489.60	$41.98	$64.06
3 Laborers	37.90	909.60	58.15	1395.60		
1 Equip. Oper. (crane)	52.25	418.00	79.05	632.40		
1 Equip. Oper. (oiler)	46.05	368.40	69.65	557.20		
1 Lattice Boom Crane, 40 Ton		1182.00		1300.20	24.63	27.09
48 L.H., Daily Totals		$3197.20		$4375.00	$66.61	$91.15
Crew B-41	Hr.	Daily	Hr.	Daily	Bare Costs	Incl. O&P
1 Labor Foreman (outside)	$39.90	$319.20	$61.20	$489.60	$39.29	$60.18
4 Laborers	37.90	1212.80	58.15	1860.80		
.25 Equip. Oper. (crane)	52.25	104.50	79.05	158.10		
.25 Equip. Oper. (oiler)	46.05	92.10	69.65	139.30		
.25 Crawler Crane, 40 Ton		300.75		330.82	6.84	7.52
44 L.H., Daily Totals		$2029.35		$2978.63	$46.12	$67.70
Crew B-42	Hr.	Daily	Hr.	Daily	Bare Costs	Incl. O&P
1 Labor Foreman (outside)	$39.90	$319.20	$61.20	$489.60	$42.88	$66.58
4 Laborers	37.90	1212.80	58.15	1860.80		
1 Equip. Oper. (crane)	52.25	418.00	79.05	632.40		
1 Equip. Oper. (oiler)	46.05	368.40	69.65	557.20		
1 Welder	53.20	425.60	90.10	720.80		
1 Hyd. Crane, 25 Ton		748.80		823.68		
1 Welder, Gas Engine, 300 amp		147.20		161.92		
1 Horz. Boring Csg. Mch.		477.40		525.14	21.46	23.61
64 L.H., Daily Totals		$4117.40		$5771.54	$64.33	$90.18
Crew B-43	Hr.	Daily	Hr.	Daily	Bare Costs	Incl. O&P
1 Labor Foreman (outside)	$39.90	$319.20	$61.20	$489.60	$41.98	$64.06
3 Laborers	37.90	909.60	58.15	1395.60		
1 Equip. Oper. (crane)	52.25	418.00	79.05	632.40		
1 Equip. Oper. (oiler)	46.05	368.40	69.65	557.20		
1 Drill Rig, Truck-Mounted		2553.00		2808.30	53.19	58.51
48 L.H., Daily Totals		$4568.20		$5883.10	$95.17	$122.56
Crew B-44	Hr.	Daily	Hr.	Daily	Bare Costs	Incl. O&P
1 Pile Driver Foreman (outside)	$50.40	$403.20	$79.95	$639.60	$48.30	$75.40
4 Pile Drivers	48.40	1548.80	76.75	2456.00		
2 Equip. Opers. (crane)	52.25	836.00	79.05	1264.80		
1 Laborer	37.90	303.20	58.15	465.20		
1 Crawler Crane, 40 Ton		1203.00		1323.30		
1 Lead, 60' High		74.00		81.40		
1 Hammer, Diesel, 15K ft.-lbs.		591.40		650.54	29.19	32.11
64 L.H., Daily Totals		$4959.60		$6880.84	$77.49	$107.51
Crew B-45	Hr.	Daily	Hr.	Daily	Bare Costs	Incl. O&P
1 Equip. Oper. (medium)	$51.10	$408.80	$77.30	$618.40	$47.15	$71.28
1 Truck Driver (heavy)	43.20	345.60	65.25	522.00		
1 Dist. Tanker, 3000 Gallon		319.40		351.34		
1 Truck Tractor, 6x4, 380 H.P.		600.00		660.00	57.46	63.21
16 L.H., Daily Totals		$1673.80		$2151.74	$104.61	$134.48

Crew No.	Bare Costs Hr.	Daily	Incl. Subs O&P Hr.	Daily	Cost Per Labor-Hour Bare Costs	Incl. O&P
Crew B-46	Hr.	Daily	Hr.	Daily	Bare Costs	Incl. O&P
1 Pile Driver Foreman (outside)	$50.40	$403.20	$79.95	$639.60	$43.48	$67.98
2 Pile Drivers	48.40	774.40	76.75	1228.00		
3 Laborers	37.90	909.60	58.15	1395.60		
1 Chain Saw, Gas, 36" Long		45.00		49.50	.94	1.03
48 L.H., Daily Totals		$2132.20		$3312.70	$44.42	$69.01
Crew B-47	Hr.	Daily	Hr.	Daily	Bare Costs	Incl. O&P
1 Blast Foreman (outside)	$39.90	$319.20	$61.20	$489.60	$42.32	$64.57
1 Driller	37.90	303.20	58.15	465.20		
1 Equip. Oper. (light)	49.15	393.20	74.35	594.80		
1 Air Track Drill, 4"		1045.00		1149.50		
1 Air Compressor, 600 cfm		533.80		587.18		
2 -50' Air Hoses, 3"		27.70		30.47	66.94	73.63
24 L.H., Daily Totals		$2622.10		$3316.75	$109.25	$138.20
Crew B-47A	Hr.	Daily	Hr.	Daily	Bare Costs	Incl. O&P
1 Drilling Foreman (outside)	$39.90	$319.20	$61.20	$489.60	$46.07	$69.97
1 Equip. Oper. (heavy)	52.25	418.00	79.05	632.40		
1 Equip. Oper. (oiler)	46.05	368.40	69.65	557.20		
1 Air Track Drill, 5"		1248.00		1372.80	52.00	57.20
24 L.H., Daily Totals		$2353.60		$3052.00	$98.07	$127.17
Crew B-47C	Hr.	Daily	Hr.	Daily	Bare Costs	Incl. O&P
1 Laborer	$37.90	$303.20	$58.15	$465.20	$43.52	$66.25
1 Equip. Oper. (light)	49.15	393.20	74.35	594.80		
1 Air Compressor, 750 cfm		541.00		595.10		
2 -50' Air Hoses, 3"		27.70		30.47		
1 Air Track Drill, 4"		1045.00		1149.50	100.86	110.94
16 L.H., Daily Totals		$2310.10		$2835.07	$144.38	$177.19
Crew B-47E	Hr.	Daily	Hr.	Daily	Bare Costs	Incl. O&P
1 Labor Foreman (outside)	$39.90	$319.20	$61.20	$489.60	$38.40	$58.91
3 Laborers	37.90	909.60	58.15	1395.60		
1 Flatbed Truck, Gas, 3 Ton		307.40		338.14	9.61	10.57
32 L.H., Daily Totals		$1536.20		$2223.34	$48.01	$69.48
Crew B-47G	Hr.	Daily	Hr.	Daily	Bare Costs	Incl. O&P
1 Labor Foreman (outside)	$39.90	$319.20	$61.20	$489.60	$41.21	$62.96
2 Laborers	37.90	606.40	58.15	930.40		
1 Equip. Oper. (light)	49.15	393.20	74.35	594.80		
1 Air Track Drill, 4"		1045.00		1149.50		
1 Air Compressor, 600 cfm		533.80		587.18		
2 -50' Air Hoses, 3"		27.70		30.47		
1 Gunite Pump Rig		373.00		410.30	61.86	68.05
32 L.H., Daily Totals		$3298.30		$4192.25	$103.07	$131.01
Crew B-47H	Hr.	Daily	Hr.	Daily	Bare Costs	Incl. O&P
1 Skilled Worker Foreman (out)	$51.90	$415.20	$79.95	$639.60	$50.40	$77.63
3 Skilled Workers	49.90	1197.60	76.85	1844.40		
1 Flatbed Truck, Gas, 3 Ton		307.40		338.14	9.61	10.57
32 L.H., Daily Totals		$1920.20		$2822.14	$60.01	$88.19

For customer support on your Mechanical Cost Data, call 877.739.3689.

635

Crew No.	Bare Costs		Incl. Subs O&P		Cost Per Labor-Hour	
Crew B-48	Hr.	Daily	Hr.	Daily	Bare Costs	Incl. O&P
1 Labor Foreman (outside)	$39.90	$319.20	$61.20	$489.60	$43.01	$65.53
3 Laborers	37.90	909.60	58.15	1395.60		
1 Equip. Oper. (crane)	52.25	418.00	79.05	632.40		
1 Equip. Oper. (oiler)	46.05	368.40	69.65	557.20		
1 Equip. Oper. (light)	49.15	393.20	74.35	594.80		
1 Centr. Water Pump, 6"		350.80		385.88		
1 -20' Suction Hose, 6"		11.50		12.65		
1 -50' Discharge Hose, 6"		6.10		6.71		
1 Drill Rig, Truck-Mounted		2553.00		2808.30	52.17	57.38
56 L.H., Daily Totals		$5329.80		$6883.14	$95.17	$122.91
Crew B-49	Hr.	Daily	Hr.	Daily	Bare Costs	Incl. O&P
1 Labor Foreman (outside)	$39.90	$319.20	$61.20	$489.60	$45.10	$69.17
3 Laborers	37.90	909.60	58.15	1395.60		
2 Equip. Opers. (crane)	52.25	836.00	79.05	1264.80		
2 Equip. Opers. (oilers)	46.05	736.80	69.65	1114.40		
1 Equip. Oper. (light)	49.15	393.20	74.35	594.80		
2 Pile Drivers	48.40	774.40	76.75	1228.00		
1 Hyd. Crane, 25 Ton		748.80		823.68		
1 Centr. Water Pump, 6"		350.80		385.88		
1 -20' Suction Hose, 6"		11.50		12.65		
1 -50' Discharge Hose, 6"		6.10		6.71		
1 Drill Rig, Truck-Mounted		2553.00		2808.30	41.71	45.88
88 L.H., Daily Totals		$7639.40		$10124.42	$86.81	$115.05
Crew B-50	Hr.	Daily	Hr.	Daily	Bare Costs	Incl. O&P
2 Pile Driver Foremen (outside)	$50.40	$806.40	$79.95	$1279.20	$46.82	$73.04
6 Pile Drivers	48.40	2323.20	76.75	3684.00		
2 Equip. Opers. (crane)	52.25	836.00	79.05	1264.80		
1 Equip. Oper. (oiler)	46.05	368.40	69.65	557.20		
3 Laborers	37.90	909.60	58.15	1395.60		
1 Crawler Crane, 40 Ton		1203.00		1323.30		
1 Lead, 60' High		74.00		81.40		
1 Hammer, Diesel, 15K ft.-lbs.		591.40		650.54		
1 Air Compressor, 600 cfm		533.80		587.18		
2 -50' Air Hoses, 3"		27.70		30.47		
1 Chain Saw, Gas, 36" Long		45.00		49.50	22.10	24.31
112 L.H., Daily Totals		$7718.50		$10903.19	$68.92	$97.35
Crew B-51	Hr.	Daily	Hr.	Daily	Bare Costs	Incl. O&P
1 Labor Foreman (outside)	$39.90	$319.20	$61.20	$489.60	$38.91	$59.52
4 Laborers	37.90	1212.80	58.15	1860.80		
1 Truck Driver (light)	41.95	335.60	63.35	506.80		
1 Flatbed Truck, Gas, 1.5 Ton		248.20		273.02	5.17	5.69
48 L.H., Daily Totals		$2115.80		$3130.22	$44.08	$65.21
Crew B-52	Hr.	Daily	Hr.	Daily	Bare Costs	Incl. O&P
1 Carpenter Foreman (outside)	$50.45	$403.60	$77.40	$619.20	$44.33	$67.65
1 Carpenter	48.45	387.60	74.30	594.40		
3 Laborers	37.90	909.60	58.15	1395.60		
1 Cement Finisher	45.65	365.20	67.70	541.60		
.5 Rodman (reinf.)	53.00	212.00	82.10	328.40		
.5 Equip. Oper. (medium)	51.10	204.40	77.30	309.20		
.5 Crawler Loader, 3 C.Y.		598.50		658.35	10.69	11.76
56 L.H., Daily Totals		$3080.90		$4446.75	$55.02	$79.41
Crew B-53	Hr.	Daily	Hr.	Daily	Bare Costs	Incl. O&P
1 Equip. Oper. (light)	$49.15	$393.20	$74.35	$594.80	$49.15	$74.35
1 Trencher, Chain, 12 H.P.		68.00		74.80	8.50	9.35
8 L.H., Daily Totals		$461.20		$669.60	$57.65	$83.70

Crew No.	Bare Costs		Incl. Subs O&P		Cost Per Labor-Hour	
Crew B-54	Hr.	Daily	Hr.	Daily	Bare Costs	Incl. O&P
1 Equip. Oper. (light)	$49.15	$393.20	$74.35	$594.80	$49.15	$74.35
1 Trencher, Chain, 40 H.P.		334.40		367.84	41.80	45.98
8 L.H., Daily Totals		$727.60		$962.64	$90.95	$120.33
Crew B-54A	Hr.	Daily	Hr.	Daily	Bare Costs	Incl. O&P
.17 Labor Foreman (outside)	$39.90	$54.26	$61.20	$83.23	$49.47	$74.96
1 Equipment Operator (med.)	51.10	408.80	77.30	618.40		
1 Wheel Trencher, 67 H.P.		1183.00		1301.30	126.39	139.03
9.36 L.H., Daily Totals		$1646.06		$2002.93	$175.86	$213.99
Crew B-54B	Hr.	Daily	Hr.	Daily	Bare Costs	Incl. O&P
.25 Labor Foreman (outside)	$39.90	$79.80	$61.20	$122.40	$48.86	$74.08
1 Equipment Operator (med.)	51.10	408.80	77.30	618.40		
1 Wheel Trencher, 150 H.P.		1891.00		2080.10	189.10	208.01
10 L.H., Daily Totals		$2379.60		$2820.90	$237.96	$282.09
Crew B-54C	Hr.	Daily	Hr.	Daily	Bare Costs	Incl. O&P
1 Laborer	$37.90	$303.20	$58.15	$465.20	$44.50	$67.72
1 Equipment Operator (med.)	51.10	408.80	77.30	618.40		
1 Wheel Trencher, 67 H.P.		1183.00		1301.30	73.94	81.33
16 L.H., Daily Totals		$1895.00		$2384.90	$118.44	$149.06
Crew B-54D	Hr.	Daily	Hr.	Daily	Bare Costs	Incl. O&P
1 Laborer	$37.90	$303.20	$58.15	$465.20	$44.50	$67.72
1 Equipment Operator (med.)	51.10	408.80	77.30	618.40		
1 Rock Trencher, 6" Width		615.20		676.72	38.45	42.30
16 L.H., Daily Totals		$1327.20		$1760.32	$82.95	$110.02
Crew B-54E	Hr.	Daily	Hr.	Daily	Bare Costs	Incl. O&P
1 Laborer	$37.90	$303.20	$58.15	$465.20	$44.50	$67.72
1 Equipment Operator (med.)	51.10	408.80	77.30	618.40		
1 Rock Trencher, 18" Width		2591.00		2850.10	161.94	178.13
16 L.H., Daily Totals		$3303.00		$3933.70	$206.44	$245.86
Crew B-55	Hr.	Daily	Hr.	Daily	Bare Costs	Incl. O&P
2 Laborers	$37.90	$606.40	$58.15	$930.40	$39.25	$59.88
1 Truck Driver (light)	41.95	335.60	63.35	506.80		
1 Truck-Mounted Earth Auger		786.00		864.60		
1 Flatbed Truck, Gas, 3 Ton		307.40		338.14	45.56	50.11
24 L.H., Daily Totals		$2035.40		$2639.94	$84.81	$110.00
Crew B-56	Hr.	Daily	Hr.	Daily	Bare Costs	Incl. O&P
1 Laborer	$37.90	$303.20	$58.15	$465.20	$43.52	$66.25
1 Equip. Oper. (light)	49.15	393.20	74.35	594.80		
1 Air Track Drill, 4"		1045.00		1149.50		
1 Air Compressor, 600 cfm		533.80		587.18		
1 -50' Air Hose, 3"		13.85		15.23	99.54	109.49
16 L.H., Daily Totals		$2289.05		$2811.92	$143.07	$175.74

Crews - Standard

Crew B-57

	Hr.	Daily	Hr.	Daily	Bare Costs	Incl. O&P
1 Labor Foreman (outside)	$39.90	$319.20	$61.20	$489.60	$43.86	$66.76
2 Laborers	37.90	606.40	58.15	930.40		
1 Equip. Oper. (crane)	52.25	418.00	79.05	632.40		
1 Equip. Oper. (light)	49.15	393.20	74.35	594.80		
1 Equip. Oper. (oiler)	46.05	368.40	69.65	557.20		
1 Crawler Crane, 25 Ton		1194.00		1313.40		
1 Clamshell Bucket, 1 C.Y.		47.40		52.14		
1 Centr. Water Pump, 6"		350.80		385.88		
1 -20' Suction Hose, 6"		11.50		12.65		
20 -50' Discharge Hoses, 6"		122.00		134.20	35.95	39.55
48 L.H., Daily Totals		$3830.90		$5102.67	$79.81	$106.31

Crew B-58

	Hr.	Daily	Hr.	Daily	Bare Costs	Incl. O&P
2 Laborers	$37.90	$606.40	$58.15	$930.40	$41.65	$63.55
1 Equip. Oper. (light)	49.15	393.20	74.35	594.80		
1 Backhoe Loader, 48 H.P.		365.40		401.94		
1 Small Helicopter, w/ Pilot		2774.00		3051.40	130.81	143.89
24 L.H., Daily Totals		$4139.00		$4978.54	$172.46	$207.44

Crew B-59

	Hr.	Daily	Hr.	Daily	Bare Costs	Incl. O&P
1 Truck Driver (heavy)	$43.20	$345.60	$65.25	$522.00	$43.20	$65.25
1 Truck Tractor, 220 H.P.		358.80		394.68		
1 Water Tank Trailer, 5000 Gal.		141.60		155.76	62.55	68.81
8 L.H., Daily Totals		$846.00		$1072.44	$105.75	$134.06

Crew B-59A

	Hr.	Daily	Hr.	Daily	Bare Costs	Incl. O&P
2 Laborers	$37.90	$606.40	$58.15	$930.40	$39.67	$60.52
1 Truck Driver (heavy)	43.20	345.60	65.25	522.00		
1 Water Tank Trailer, 5000 Gal.		141.60		155.76		
1 Truck Tractor, 220 H.P.		358.80		394.68	20.85	22.93
24 L.H., Daily Totals		$1452.40		$2002.84	$60.52	$83.45

Crew B-60

	Hr.	Daily	Hr.	Daily	Bare Costs	Incl. O&P
1 Labor Foreman (outside)	$39.90	$319.20	$61.20	$489.60	$44.61	$67.84
2 Laborers	37.90	606.40	58.15	930.40		
1 Equip. Oper. (crane)	52.25	418.00	79.05	632.40		
2 Equip. Opers. (light)	49.15	786.40	74.35	1189.60		
1 Equip. Oper. (oiler)	46.05	368.40	69.65	557.20		
1 Crawler Crane, 40 Ton		1203.00		1323.30		
1 Lead, 60' High		74.00		81.40		
1 Hammer, Diesel, 15K ft.-lbs.		591.40		650.54		
1 Backhoe Loader, 48 H.P.		365.40		401.94	39.89	43.88
56 L.H., Daily Totals		$4732.20		$6256.38	$84.50	$111.72

Crew B-61

	Hr.	Daily	Hr.	Daily	Bare Costs	Incl. O&P
1 Labor Foreman (outside)	$39.90	$319.20	$61.20	$489.60	$40.55	$62.00
3 Laborers	37.90	909.60	58.15	1395.60		
1 Equip. Oper. (light)	49.15	393.20	74.35	594.80		
1 Cement Mixer, 2 C.Y.		193.40		212.74		
1 Air Compressor, 160 cfm		162.20		178.42	8.89	9.78
40 L.H., Daily Totals		$1977.60		$2871.16	$49.44	$71.78

Crew B-62

	Hr.	Daily	Hr.	Daily	Bare Costs	Incl. O&P
2 Laborers	$37.90	$606.40	$58.15	$930.40	$41.65	$63.55
1 Equip. Oper. (light)	49.15	393.20	74.35	594.80		
1 Loader, Skid Steer, 30 H.P.		173.20		190.52	7.22	7.94
24 L.H., Daily Totals		$1172.80		$1715.72	$48.87	$71.49

Crew B-62A

	Hr.	Daily	Hr.	Daily	Bare Costs	Incl. O&P
2 Laborers	$37.90	$606.40	$58.15	$930.40	$41.65	$63.55
1 Equip. Oper. (light)	49.15	393.20	74.35	594.80		
1 Loader, Skid Steer, 30 H.P.		173.20		190.52		
1 Trencher Attachment		63.70		70.07	9.87	10.86
24 L.H., Daily Totals		$1236.50		$1785.79	$51.52	$74.41

Crew B-63

	Hr.	Daily	Hr.	Daily	Bare Costs	Incl. O&P
4 Laborers	$37.90	$1212.80	$58.15	$1860.80	$40.15	$61.39
1 Equip. Oper. (light)	49.15	393.20	74.35	594.80		
1 Loader, Skid Steer, 30 H.P.		173.20		190.52	4.33	4.76
40 L.H., Daily Totals		$1779.20		$2646.12	$44.48	$66.15

Crew B-63B

	Hr.	Daily	Hr.	Daily	Bare Costs	Incl. O&P
1 Labor Foreman (inside)	$38.40	$307.20	$58.90	$471.20	$40.84	$62.39
2 Laborers	37.90	606.40	58.15	930.40		
1 Equip. Oper. (light)	49.15	393.20	74.35	594.80		
1 Loader, Skid Steer, 78 H.P.		315.60		347.16	9.86	10.85
32 L.H., Daily Totals		$1622.40		$2343.56	$50.70	$73.24

Crew B-64

	Hr.	Daily	Hr.	Daily	Bare Costs	Incl. O&P
1 Laborer	$37.90	$303.20	$58.15	$465.20	$39.92	$60.75
1 Truck Driver (light)	41.95	335.60	63.35	506.80		
1 Power Mulcher (small)		150.80		165.88		
1 Flatbed Truck, Gas, 1.5 Ton		248.20		273.02	24.94	27.43
16 L.H., Daily Totals		$1037.80		$1410.90	$64.86	$88.18

Crew B-65

	Hr.	Daily	Hr.	Daily	Bare Costs	Incl. O&P
1 Laborer	$37.90	$303.20	$58.15	$465.20	$39.92	$60.75
1 Truck Driver (light)	41.95	335.60	63.35	506.80		
1 Power Mulcher (Large)		319.60		351.56		
1 Flatbed Truck, Gas, 1.5 Ton		248.20		273.02	35.49	39.04
16 L.H., Daily Totals		$1206.60		$1596.58	$75.41	$99.79

Crew B-66

	Hr.	Daily	Hr.	Daily	Bare Costs	Incl. O&P
1 Equip. Oper. (light)	$49.15	$393.20	$74.35	$594.80	$49.15	$74.35
1 Loader-Backhoe, 40 H.P.		263.80		290.18	32.98	36.27
8 L.H., Daily Totals		$657.00		$884.98	$82.13	$110.62

Crew B-67

	Hr.	Daily	Hr.	Daily	Bare Costs	Incl. O&P
1 Millwright	$51.10	$408.80	$74.90	$599.20	$50.13	$74.63
1 Equip. Oper. (light)	49.15	393.20	74.35	594.80		
1 Forklift, R/T, 4,000 Lb.		313.60		344.96	19.60	21.56
16 L.H., Daily Totals		$1115.60		$1538.96	$69.72	$96.19

Crew B-67B

	Hr.	Daily	Hr.	Daily	Bare Costs	Incl. O&P
1 Millwright Foreman (inside)	$51.60	$412.80	$75.65	$605.20	$51.35	$75.28
1 Millwright	51.10	408.80	74.90	599.20		
16 L.H., Daily Totals		$821.60		$1204.40	$51.35	$75.28

Crew B-68

	Hr.	Daily	Hr.	Daily	Bare Costs	Incl. O&P
2 Millwrights	$51.10	$817.60	$74.90	$1198.40	$50.45	$74.72
1 Equip. Oper. (light)	49.15	393.20	74.35	594.80		
1 Forklift, R/T, 4,000 Lb.		313.60		344.96	13.07	14.37
24 L.H., Daily Totals		$1524.40		$2138.16	$63.52	$89.09

Crew B-68A

	Hr.	Daily	Hr.	Daily	Bare Costs	Incl. O&P
1 Millwright Foreman (inside)	$51.60	$412.80	$75.65	$605.20	$51.27	$75.15
2 Millwrights	51.10	817.60	74.90	1198.40		
1 Forklift, 8,000 Lb.		185.40		203.94	7.72	8.50
24 L.H., Daily Totals		$1415.80		$2007.54	$58.99	$83.65

For customer support on your Mechanical Cost Data, call 877.739.3689.

637

Crew No.	Bare Costs Hr.	Daily	Incl. Subs O&P Hr.	Daily	Cost Per Labor-Hour Bare Costs	Incl. O&P
Crew B-68B	Hr.	Daily	Hr.	Daily	Bare Costs	Incl. O&P
1 Millwright Foreman (inside)	$51.60	$412.80	$75.65	$605.20	$54.63	$81.29
2 Millwrights	51.10	817.60	74.90	1198.40		
2 Electricians	55.10	881.60	82.45	1319.20		
2 Plumbers	59.20	947.20	89.35	1429.60		
1 Forklift, 5,000 Lb.		334.00		367.40	5.96	6.56
56 L.H., Daily Totals		$3393.20		$4919.80	$60.59	$87.85

Crew No.	Hr.	Daily	Hr.	Daily	Bare Costs	Incl. O&P
Crew B-68C	Hr.	Daily	Hr.	Daily	Bare Costs	Incl. O&P
1 Millwright Foreman (inside)	$51.60	$412.80	$75.65	$605.20	$54.25	$80.59
1 Millwright	51.10	408.80	74.90	599.20		
1 Electrician	55.10	440.80	82.45	659.60		
1 Plumber	59.20	473.60	89.35	714.80		
1 Forklift, 5,000 Lb.		334.00		367.40	10.44	11.48
32 L.H., Daily Totals		$2070.00		$2946.00	$64.69	$92.07

Crew No.	Hr.	Daily	Hr.	Daily	Bare Costs	Incl. O&P
Crew B-68D	Hr.	Daily	Hr.	Daily	Bare Costs	Incl. O&P
1 Labor Foreman (inside)	$38.40	$307.20	$58.90	$471.20	$41.82	$63.80
1 Laborer	37.90	303.20	58.15	465.20		
1 Equip. Oper. (light)	49.15	393.20	74.35	594.80		
1 Forklift, 5,000 Lb.		334.00		367.40	13.92	15.31
24 L.H., Daily Totals		$1337.60		$1898.60	$55.73	$79.11

Crew No.	Hr.	Daily	Hr.	Daily	Bare Costs	Incl. O&P
Crew B-68E	Hr.	Daily	Hr.	Daily	Bare Costs	Incl. O&P
1 Struc. Steel Foreman (inside)	$53.70	$429.60	$90.95	$727.60	$53.30	$90.27
3 Struc. Steel Workers	53.20	1276.80	90.10	2162.40		
1 Welder	53.20	425.60	90.10	720.80		
1 Forklift, 8,000 Lb.		185.40		203.94	4.63	5.10
40 L.H., Daily Totals		$2317.40		$3814.74	$57.94	$95.37

Crew No.	Hr.	Daily	Hr.	Daily	Bare Costs	Incl. O&P
Crew B-68F	Hr.	Daily	Hr.	Daily	Bare Costs	Incl. O&P
1 Skilled Worker Foreman (out)	$51.90	$415.20	$79.95	$639.60	$50.57	$77.88
2 Skilled Workers	49.90	798.40	76.85	1229.60		
1 Forklift, 5,000 Lb.		334.00		367.40	13.92	15.31
24 L.H., Daily Totals		$1547.60		$2236.60	$64.48	$93.19

Crew No.	Hr.	Daily	Hr.	Daily	Bare Costs	Incl. O&P
Crew B-68G	Hr.	Daily	Hr.	Daily	Bare Costs	Incl. O&P
2 Structural Steel Workers	$53.20	$851.20	$90.10	$1441.60	$53.20	$90.10
1 Forklift, 5,000 Lb.		334.00		367.40	20.88	22.96
16 L.H., Daily Totals		$1185.20		$1809.00	$74.08	$113.06

Crew No.	Hr.	Daily	Hr.	Daily	Bare Costs	Incl. O&P
Crew B-69	Hr.	Daily	Hr.	Daily	Bare Costs	Incl. O&P
1 Labor Foreman (outside)	$39.90	$319.20	$61.20	$489.60	$41.98	$64.06
3 Laborers	37.90	909.60	58.15	1395.60		
1 Equip. Oper. (crane)	52.25	418.00	79.05	632.40		
1 Equip. Oper. (oiler)	46.05	368.40	69.65	557.20		
1 Hyd. Crane, 80 Ton		1630.00		1793.00	33.96	37.35
48 L.H., Daily Totals		$3645.20		$4867.80	$75.94	$101.41

Crew No.	Hr.	Daily	Hr.	Daily	Bare Costs	Incl. O&P
Crew B-69A	Hr.	Daily	Hr.	Daily	Bare Costs	Incl. O&P
1 Labor Foreman (outside)	$39.90	$319.20	$61.20	$489.60	$41.73	$63.44
3 Laborers	37.90	909.60	58.15	1395.60		
1 Equip. Oper. (medium)	51.10	408.80	77.30	618.40		
1 Concrete Finisher	45.65	365.20	67.70	541.60		
1 Curb/Gutter Paver, 2-Track		1102.00		1212.20	22.96	25.25
48 L.H., Daily Totals		$3104.80		$4257.40	$64.68	$88.70

Crew No.	Hr.	Daily	Hr.	Daily	Bare Costs	Incl. O&P
Crew B-69B	Hr.	Daily	Hr.	Daily	Bare Costs	Incl. O&P
1 Labor Foreman (outside)	$39.90	$319.20	$61.20	$489.60	$41.73	$63.44
3 Laborers	37.90	909.60	58.15	1395.60		
1 Equip. Oper. (medium)	51.10	408.80	77.30	618.40		
1 Cement Finisher	45.65	365.20	67.70	541.60		
1 Curb/Gutter Paver, 4-Track		780.20		858.22	16.25	17.88
48 L.H., Daily Totals		$2783.00		$3903.42	$57.98	$81.32

Crew No.	Hr.	Daily	Hr.	Daily	Bare Costs	Incl. O&P
Crew B-70	Hr.	Daily	Hr.	Daily	Bare Costs	Incl. O&P
1 Labor Foreman (outside)	$39.90	$319.20	$61.20	$489.60	$43.84	$66.79
3 Laborers	37.90	909.60	58.15	1395.60		
3 Equip. Opers. (medium)	51.10	1226.40	77.30	1855.20		
1 Grader, 30,000 Lbs.		714.40		785.84		
1 Ripper, Beam & 1 Shank		83.80		92.18		
1 Road Sweeper, S.P., 8' wide		666.20		732.82		
1 F.E. Loader, W.M., 1.5 C.Y.		378.60		416.46	32.91	36.20
56 L.H., Daily Totals		$4298.20		$5767.70	$76.75	$102.99

Crew No.	Hr.	Daily	Hr.	Daily	Bare Costs	Incl. O&P
Crew B-70A	Hr.	Daily	Hr.	Daily	Bare Costs	Incl. O&P
1 Laborer	$37.90	$303.20	$58.15	$465.20	$48.46	$73.47
4 Equip. Opers. (medium)	51.10	1635.20	77.30	2473.60		
1 Grader, 40,000 Lbs.		1195.00		1314.50		
1 F.E. Loader, W.M., 2.5 C.Y.		516.00		567.60		
1 Dozer, 80 H.P.		471.40		518.54		
1 Roller, Pneum. Whl., 12 Ton		347.80		382.58	63.26	69.58
40 L.H., Daily Totals		$4468.60		$5722.02	$111.72	$143.05

Crew No.	Hr.	Daily	Hr.	Daily	Bare Costs	Incl. O&P
Crew B-71	Hr.	Daily	Hr.	Daily	Bare Costs	Incl. O&P
1 Labor Foreman (outside)	$39.90	$319.20	$61.20	$489.60	$43.84	$66.79
3 Laborers	37.90	909.60	58.15	1395.60		
3 Equip. Opers. (medium)	51.10	1226.40	77.30	1855.20		
1 Pvmt. Profiler, 750 H.P.		5835.00		6418.50		
1 Road Sweeper, S.P., 8' wide		666.20		732.82		
1 F.E. Loader, W.M., 1.5 C.Y.		378.60		416.46	122.85	135.14
56 L.H., Daily Totals		$9335.00		$11308.18	$166.70	$201.93

Crew No.	Hr.	Daily	Hr.	Daily	Bare Costs	Incl. O&P
Crew B-72	Hr.	Daily	Hr.	Daily	Bare Costs	Incl. O&P
1 Labor Foreman (outside)	$39.90	$319.20	$61.20	$489.60	$44.75	$68.11
3 Laborers	37.90	909.60	58.15	1395.60		
4 Equip. Opers. (medium)	51.10	1635.20	77.30	2473.60		
1 Pvmt. Profiler, 750 H.P.		5835.00		6418.50		
1 Hammermill, 250 H.P.		1832.00		2015.20		
1 Windrow Loader		1247.00		1371.70		
1 Mix Paver 165 H.P.		2226.00		2448.60		
1 Roller, Pneum. Whl., 12 Ton		347.80		382.58	179.50	197.45
64 L.H., Daily Totals		$14351.80		$16995.38	$224.25	$265.55

Crew No.	Hr.	Daily	Hr.	Daily	Bare Costs	Incl. O&P
Crew B-73	Hr.	Daily	Hr.	Daily	Bare Costs	Incl. O&P
1 Labor Foreman (outside)	$39.90	$319.20	$61.20	$489.60	$46.40	$70.50
2 Laborers	37.90	606.40	58.15	930.40		
5 Equip. Opers. (medium)	51.10	2044.00	77.30	3092.00		
1 Road Mixer, 310 H.P.		1945.00		2139.50		
1 Tandem Roller, 10 Ton		239.40		263.34		
1 Hammermill, 250 H.P.		1832.00		2015.20		
1 Grader, 30,000 Lbs.		714.40		785.84		
.5 F.E. Loader, W.M., 1.5 C.Y.		189.30		208.23		
.5 Truck Tractor, 220 H.P.		179.40		197.34		
.5 Water Tank Trailer, 5000 Gal.		70.80		77.88	80.79	88.86
64 L.H., Daily Totals		$8139.90		$10199.33	$127.19	$159.36

Crew B-74	Hr.	Daily	Hr.	Daily	Bare Costs	Incl. O&P
1 Labor Foreman (outside)	$39.90	$319.20	$61.20	$489.60	$46.08	$69.88
1 Laborer	37.90	303.20	58.15	465.20		
4 Equip. Opers. (medium)	51.10	1635.20	77.30	2473.60		
2 Truck Drivers (heavy)	43.20	691.20	65.25	1044.00		
1 Grader, 30,000 Lbs.		714.40		785.84		
1 Ripper, Beam & 1 Shank		83.80		92.18		
2 Stabilizers, 310 H.P.		3666.00		4032.60		
1 Flatbed Truck, Gas, 3 Ton		307.40		338.14		
1 Chem. Spreader, Towed		52.00		57.20		
1 Roller, Vibratory, 25 Ton		691.80		760.98		
1 Water Tank Trailer, 5000 Gal.		141.60		155.76		
1 Truck Tractor, 220 H.P.		358.80		394.68	94.00	103.40
64 L.H., Daily Totals		$8964.60		$11089.78	$140.07	$173.28

Crew B-75	Hr.	Daily	Hr.	Daily	Bare Costs	Incl. O&P
1 Labor Foreman (outside)	$39.90	$319.20	$61.20	$489.60	$46.49	$70.54
1 Laborer	37.90	303.20	58.15	465.20		
4 Equip. Opers. (medium)	51.10	1635.20	77.30	2473.60		
1 Truck Driver (heavy)	43.20	345.60	65.25	522.00		
1 Grader, 30,000 Lbs.		714.40		785.84		
1 Ripper, Beam & 1 Shank		83.80		92.18		
2 Stabilizers, 310 H.P.		3666.00		4032.60		
1 Dist. Tanker, 3000 Gallon		319.40		351.34		
1 Truck Tractor, 6x4, 380 H.P.		600.00		660.00		
1 Roller, Vibratory, 25 Ton		691.80		760.98	108.49	119.34
56 L.H., Daily Totals		$8678.60		$10633.34	$154.97	$189.88

Crew B-76	Hr.	Daily	Hr.	Daily	Bare Costs	Incl. O&P
1 Dock Builder Foreman (outside)	$50.40	$403.20	$79.95	$639.60	$49.22	$76.83
5 Dock Builders	48.40	1936.00	76.75	3070.00		
2 Equip. Opers. (crane)	52.25	836.00	79.05	1264.80		
1 Equip. Oper. (oiler)	46.05	368.40	69.65	557.20		
1 Crawler Crane, 50 Ton		1346.00		1480.60		
1 Barge, 400 Ton		779.20		857.12		
1 Hammer, Diesel, 15K ft.-lbs.		591.40		650.54		
1 Lead, 60' High		74.00		81.40		
1 Air Compressor, 600 cfm		533.80		587.18		
2 -50' Air Hoses, 3"		27.70		30.47	46.56	51.21
72 L.H., Daily Totals		$6895.70		$9218.91	$95.77	$128.04

Crew B-76A	Hr.	Daily	Hr.	Daily	Bare Costs	Incl. O&P
1 Labor Foreman (outside)	$39.90	$319.20	$61.20	$489.60	$40.96	$62.58
5 Laborers	37.90	1516.00	58.15	2326.00		
1 Equip. Oper. (crane)	52.25	418.00	79.05	632.40		
1 Equip. Oper. (oiler)	46.05	368.40	69.65	557.20		
1 Crawler Crane, 50 Ton		1346.00		1480.60		
1 Barge, 400 Ton		779.20		857.12	33.21	36.53
64 L.H., Daily Totals		$4746.80		$6342.92	$74.17	$99.11

Crew B-77	Hr.	Daily	Hr.	Daily	Bare Costs	Incl. O&P
1 Labor Foreman (outside)	$39.90	$319.20	$61.20	$489.60	$39.11	$59.80
3 Laborers	37.90	909.60	58.15	1395.60		
1 Truck Driver (light)	41.95	335.60	63.35	506.80		
1 Crack Cleaner, 25 H.P.		62.00		68.20		
1 Crack Filler, Trailer Mtd.		202.60		222.86		
1 Flatbed Truck, Gas, 3 Ton		307.40		338.14	14.30	15.73
40 L.H., Daily Totals		$2136.40		$3021.20	$53.41	$75.53

Crew B-78	Hr.	Daily	Hr.	Daily	Bare Costs	Incl. O&P
1 Labor Foreman (outside)	$39.90	$319.20	$61.20	$489.60	$38.91	$59.52
4 Laborers	37.90	1212.80	58.15	1860.80		
1 Truck Driver (light)	41.95	335.60	63.35	506.80		
1 Paint Striper, S.P., 40 Gallon		152.00		167.20		
1 Flatbed Truck, Gas, 3 Ton		307.40		338.14		
1 Pickup Truck, 3/4 Ton		145.80		160.38	12.61	13.87
48 L.H., Daily Totals		$2472.80		$3522.92	$51.52	$73.39

Crew B-78A	Hr.	Daily	Hr.	Daily	Bare Costs	Incl. O&P
1 Equip. Oper. (light)	$49.15	$393.20	$74.35	$594.80	$49.15	$74.35
1 Line Rem. (Metal Balls) 115 H.P.		846.60		931.26	105.83	116.41
8 L.H., Daily Totals		$1239.80		$1526.06	$154.97	$190.76

Crew B-78B	Hr.	Daily	Hr.	Daily	Bare Costs	Incl. O&P
2 Laborers	$37.90	$606.40	$58.15	$930.40	$39.15	$59.95
.25 Equip. Oper. (light)	49.15	98.30	74.35	148.70		
1 Pickup Truck, 3/4 Ton		145.80		160.38		
1 Line Rem.,11 H.P.,Walk Behind		65.60		72.16		
.25 Road Sweeper, S.P., 8' wide		166.55		183.21	21.00	23.10
18 L.H., Daily Totals		$1082.65		$1494.85	$60.15	$83.05

Crew B-78C	Hr.	Daily	Hr.	Daily	Bare Costs	Incl. O&P
1 Labor Foreman (outside)	$39.90	$319.20	$61.20	$489.60	$38.91	$59.52
4 Laborers	37.90	1212.80	58.15	1860.80		
1 Truck Driver (light)	41.95	335.60	63.35	506.80		
1 Paint Striper, T.M., 120 Gal.		810.20		891.22		
1 Flatbed Truck, Gas, 3 Ton		307.40		338.14		
1 Pickup Truck, 3/4 Ton		145.80		160.38	26.32	28.95
48 L.H., Daily Totals		$3131.00		$4246.94	$65.23	$88.48

Crew B-78D	Hr.	Daily	Hr.	Daily	Bare Costs	Incl. O&P
2 Labor Foremen (outside)	$39.90	$638.40	$61.20	$979.20	$38.70	$59.28
7 Laborers	37.90	2122.40	58.15	3256.40		
1 Truck Driver (light)	41.95	335.60	63.35	506.80		
1 Paint Striper, T.M., 120 Gal.		810.20		891.22		
1 Flatbed Truck, Gas, 3 Ton		307.40		338.14		
3 Pickup Trucks, 3/4 Ton		437.40		481.14		
1 Air Compressor, 60 cfm		142.80		157.08		
1 -50' Air Hose, 3/4"		3.05		3.36		
1 Breakers, Pavement, 60 lb.		10.20		11.22	21.39	23.53
80 L.H., Daily Totals		$4807.45		$6624.56	$60.09	$82.81

Crew B-78E	Hr.	Daily	Hr.	Daily	Bare Costs	Incl. O&P
2 Labor Foremen (outside)	$39.90	$638.40	$61.20	$979.20	$38.57	$59.09
9 Laborers	37.90	2728.80	58.15	4186.80		
1 Truck Driver (light)	41.95	335.60	63.35	506.80		
1 Paint Striper, T.M., 120 Gal.		810.20		891.22		
1 Flatbed Truck, Gas, 3 Ton		307.40		338.14		
4 Pickup Trucks, 3/4 Ton		583.20		641.52		
2 Air Compressors, 60 cfm		285.60		314.16		
2 -50' Air Hoses, 3/4"		6.10		6.71		
2 Breakers, Pavement, 60 lb.		20.40		22.44	20.97	23.06
96 L.H., Daily Totals		$5715.70		$7886.99	$59.54	$82.16

For customer support on your Mechanical Cost Data, call 877.739.3689.

639

Crew No.	Bare Costs Hr.	Bare Costs Daily	Incl. Subs O&P Hr.	Incl. Subs O&P Daily	Cost Per Labor-Hour Bare Costs	Cost Per Labor-Hour Incl. O&P
Crew B-78F	Hr.	Daily	Hr.	Daily	Bare Costs	Incl. O&P
2 Labor Foremen (outside)	$39.90	$638.40	$61.20	$979.20	$38.48	$58.96
11 Laborers	37.90	3335.20	58.15	5117.20		
1 Truck Driver (light)	41.95	335.60	63.35	506.80		
1 Paint Striper, T.M., 120 Gal.		810.20		891.22		
1 Flatbed Truck, Gas, 3 Ton		307.40		338.14		
7 Pickup Trucks, 3/4 Ton		1020.60		1122.66		
3 Air Compressors, 60 cfm		428.40		471.24		
3 -50' Air Hose, 3/4"		9.15		10.07		
3 Breakers, Pavement, 60 lb.		30.60		33.66	23.27	25.60
112 L.H., Daily Totals		$6915.55		$9470.18	$61.75	$84.56
Crew B-79	Hr.	Daily	Hr.	Daily	Bare Costs	Incl. O&P
1 Labor Foreman (outside)	$39.90	$319.20	$61.20	$489.60	$39.11	$59.80
3 Laborers	37.90	909.60	58.15	1395.60		
1 Truck Driver (light)	41.95	335.60	63.35	506.80		
1 Paint Striper, T.M., 120 Gal.		810.20		891.22		
1 Heating Kettle, 115 Gallon		81.70		89.87		
1 Flatbed Truck, Gas, 3 Ton		307.40		338.14		
2 Pickup Trucks, 3/4 Ton		291.60		320.76	37.27	41.00
40 L.H., Daily Totals		$3055.30		$4031.99	$76.38	$100.80
Crew B-79A	Hr.	Daily	Hr.	Daily	Bare Costs	Incl. O&P
1.5 Equip. Oper. (light)	$49.15	$589.80	$74.35	$892.20	$49.15	$74.35
.5 Line Remov. (Grinder) 115 H.P.		468.40		515.24		
1 Line Rem. (Metal Balls) 115 H.P.		846.60		931.26	109.58	120.54
12 L.H., Daily Totals		$1904.80		$2338.70	$158.73	$194.89
Crew B-79B	Hr.	Daily	Hr.	Daily	Bare Costs	Incl. O&P
1 Laborer	$37.90	$303.20	$58.15	$465.20	$37.90	$58.15
1 Set of Gases		168.00		184.80	21.00	23.10
8 L.H., Daily Totals		$471.20		$650.00	$58.90	$81.25
Crew B-79C	Hr.	Daily	Hr.	Daily	Bare Costs	Incl. O&P
1 Labor Foreman (outside)	$39.90	$319.20	$61.20	$489.60	$38.76	$59.33
5 Laborers	37.90	1516.00	58.15	2326.00		
1 Truck Driver (light)	41.95	335.60	63.35	506.80		
1 Paint Striper, T.M., 120 Gal.		810.20		891.22		
1 Heating Kettle, 115 Gallon		81.70		89.87		
1 Flatbed Truck, Gas, 3 Ton		307.40		338.14		
3 Pickup Trucks, 3/4 Ton		437.40		481.14		
1 Air Compressor, 60 cfm		142.80		157.08		
1 -50' Air Hose, 3/4"		3.05		3.36		
1 Breakers, Pavement, 60 lb.		10.20		11.22	32.01	35.21
56 L.H., Daily Totals		$3963.55		$5294.43	$70.78	$94.54
Crew B-79D	Hr.	Daily	Hr.	Daily	Bare Costs	Incl. O&P
2 Labor Foremen (outside)	$39.90	$638.40	$61.20	$979.20	$38.91	$59.56
5 Laborers	37.90	1516.00	58.15	2326.00		
1 Truck Driver (light)	41.95	335.60	63.35	506.80		
1 Paint Striper, T.M., 120 Gal.		810.20		891.22		
1 Heating Kettle, 115 Gallon		81.70		89.87		
1 Flatbed Truck, Gas, 3 Ton		307.40		338.14		
4 Pickup Trucks, 3/4 Ton		583.20		641.52		
1 Air Compressor, 60 cfm		142.80		157.08		
1 -50' Air Hose, 3/4"		3.05		3.36		
1 Breakers, Pavement, 60 lb.		10.20		11.22	30.29	33.32
64 L.H., Daily Totals		$4428.55		$5944.40	$69.20	$92.88

Crew No.	Bare Costs Hr.	Bare Costs Daily	Incl. Subs O&P Hr.	Incl. Subs O&P Daily	Cost Per Labor-Hour Bare Costs	Cost Per Labor-Hour Incl. O&P
Crew B-79E	Hr.	Daily	Hr.	Daily	Bare Costs	Incl. O&P
2 Labor Foremen (outside)	$39.90	$638.40	$61.20	$979.20	$38.70	$59.28
7 Laborers	37.90	2122.40	58.15	3256.40		
1 Truck Driver (light)	41.95	335.60	63.35	506.80		
1 Paint Striper, T.M., 120 Gal.		810.20		891.22		
1 Heating Kettle, 115 Gallon		81.70		89.87		
1 Flatbed Truck, Gas, 3 Ton		307.40		338.14		
5 Pickup Trucks, 3/4 Ton		729.00		801.90		
2 Air Compressors, 60 cfm		285.60		314.16		
2 -50' Air Hoses, 3/4"		6.10		6.71		
2 Breakers, Pavement, 60 lb.		20.40		22.44	28.00	30.81
80 L.H., Daily Totals		$5336.80		$7206.84	$66.71	$90.09
Crew B-80	Hr.	Daily	Hr.	Daily	Bare Costs	Incl. O&P
1 Labor Foreman (outside)	$39.90	$319.20	$61.20	$489.60	$42.23	$64.26
1 Laborer	37.90	303.20	58.15	465.20		
1 Truck Driver (light)	41.95	335.60	63.35	506.80		
1 Equip. Oper. (light)	49.15	393.20	74.35	594.80		
1 Flatbed Truck, Gas, 3 Ton		307.40		338.14		
1 Earth Auger, Truck-Mtd.		420.80		462.88	22.76	25.03
32 L.H., Daily Totals		$2079.40		$2857.42	$64.98	$89.29
Crew B-80A	Hr.	Daily	Hr.	Daily	Bare Costs	Incl. O&P
3 Laborers	$37.90	$909.60	$58.15	$1395.60	$37.90	$58.15
1 Flatbed Truck, Gas, 3 Ton		307.40		338.14	12.81	14.09
24 L.H., Daily Totals		$1217.00		$1733.74	$50.71	$72.24
Crew B-80B	Hr.	Daily	Hr.	Daily	Bare Costs	Incl. O&P
3 Laborers	$37.90	$909.60	$58.15	$1395.60	$40.71	$62.20
1 Equip. Oper. (light)	49.15	393.20	74.35	594.80		
1 Crane, Flatbed Mounted, 3 Ton		248.20		273.02	7.76	8.53
32 L.H., Daily Totals		$1551.00		$2263.42	$48.47	$70.73
Crew B-80C	Hr.	Daily	Hr.	Daily	Bare Costs	Incl. O&P
2 Laborers	$37.90	$606.40	$58.15	$930.40	$39.25	$59.88
1 Truck Driver (light)	41.95	335.60	63.35	506.80		
1 Flatbed Truck, Gas, 1.5 Ton		248.20		273.02		
1 Manual Fence Post Auger, Gas		8.40		9.24	10.69	11.76
24 L.H., Daily Totals		$1198.60		$1719.46	$49.94	$71.64
Crew B-81	Hr.	Daily	Hr.	Daily	Bare Costs	Incl. O&P
1 Laborer	$37.90	$303.20	$58.15	$465.20	$44.07	$66.90
1 Equip. Oper. (medium)	51.10	408.80	77.30	618.40		
1 Truck Driver (heavy)	43.20	345.60	65.25	522.00		
1 Hydromulcher, T.M., 3000 Gal.		334.40		367.84		
1 Truck Tractor, 220 H.P.		358.80		394.68	28.88	31.77
24 L.H., Daily Totals		$1750.80		$2368.12	$72.95	$98.67
Crew B-81A	Hr.	Daily	Hr.	Daily	Bare Costs	Incl. O&P
1 Laborer	$37.90	$303.20	$58.15	$465.20	$39.92	$60.75
1 Truck Driver (light)	41.95	335.60	63.35	506.80		
1 Hydromulcher, T.M., 600 Gal.		127.80		140.58		
1 Flatbed Truck, Gas, 3 Ton		307.40		338.14	27.20	29.92
16 L.H., Daily Totals		$1074.00		$1450.72	$67.13	$90.67
Crew B-82	Hr.	Daily	Hr.	Daily	Bare Costs	Incl. O&P
1 Laborer	$37.90	$303.20	$58.15	$465.20	$43.52	$66.25
1 Equip. Oper. (light)	49.15	393.20	74.35	594.80		
1 Horiz. Borer, 6 H.P.		84.20		92.62	5.26	5.79
16 L.H., Daily Totals		$780.60		$1152.62	$48.79	$72.04

Crew B-82A

Crew No.	Bare Costs Hr.	Bare Costs Daily	Incl. Subs O&P Hr.	Incl. Subs O&P Daily	Cost Per Labor-Hour Bare Costs	Cost Per Labor-Hour Incl. O&P
2 Laborers	$37.90	$606.40	$58.15	$930.40	$43.52	$66.25
2 Equip. Opers. (light)	49.15	786.40	74.35	1189.60		
2 Dump Trucks, 8 C.Y., 220 H.P.		834.80		918.28		
1 Flatbed Trailer, 25 Ton		113.20		124.52		
1 Horiz. Dir. Drill, 20k lb. Thrust		657.00		722.70		
1 Mud Trailer for HDD, 1500 Gal.		350.00		385.00		
1 Pickup Truck, 4x4, 3/4 Ton		156.80		172.48		
1 Flatbed trailer, 3 Ton		23.40		25.74		
1 Loader, Skid Steer, 78 H.P.		315.60		347.16	76.59	84.25
32 L.H., Daily Totals		$3843.60		$4815.88	$120.11	$150.50

Crew B-82B

Crew No.	Bare Costs Hr.	Bare Costs Daily	Incl. Subs O&P Hr.	Incl. Subs O&P Daily	Bare Costs	Incl. O&P
2 Laborers	$37.90	$606.40	$58.15	$930.40	$43.52	$66.25
2 Equip. Opers. (light)	49.15	786.40	74.35	1189.60		
2 Dump Truck, 8 C.Y., 220 H.P.		834.80		918.28		
1 Flatbed Trailer, 25 Ton		113.20		124.52		
1 Horiz. Dir. Drill, 30k lb. Thrust		935.80		1029.38		
1 Mud Trailer for HDD, 1500 Gal.		350.00		385.00		
1 Pickup Truck, 4x4, 3/4 Ton		156.80		172.48		
1 Flatbed trailer, 3 Ton		23.40		25.74		
1 Loader, Skid Steer, 78 H.P.		315.60		347.16	85.30	93.83
32 L.H., Daily Totals		$4122.40		$5122.56	$128.82	$160.08

Crew B-82C

Crew No.	Bare Costs Hr.	Bare Costs Daily	Incl. Subs O&P Hr.	Incl. Subs O&P Daily	Bare Costs	Incl. O&P
2 Laborers	$37.90	$606.40	$58.15	$930.40	$43.52	$66.25
2 Equip. Opers. (light)	49.15	786.40	74.35	1189.60		
2 Dump Trucks, 8 C.Y., 220 H.P.		834.80		918.28		
1 Flatbed Trailer, 25 Ton		113.20		124.52		
1 Horiz. Dir. Drill, 50k lb. Thrust		1244.00		1368.40		
1 Mud Trailer for HDD, 1500 Gal.		350.00		385.00		
1 Pickup Truck, 4x4, 3/4 Ton		156.80		172.48		
1 Flatbed trailer, 3 Ton		23.40		25.74		
1 Loader, Skid Steer, 78 H.P.		315.60		347.16	94.93	104.42
32 L.H., Daily Totals		$4430.60		$5461.58	$138.46	$170.67

Crew B-82D

Crew No.	Bare Costs Hr.	Bare Costs Daily	Incl. Subs O&P Hr.	Incl. Subs O&P Daily	Bare Costs	Incl. O&P
1 Equip. Oper. (light)	$49.15	$393.20	$74.35	$594.80	$49.15	$74.35
1 Mud Trailer for HDD, 1500 Gal.		350.00		385.00	43.75	48.13
8 L.H., Daily Totals		$743.20		$979.80	$92.90	$122.47

Crew B-83

Crew No.	Bare Costs Hr.	Bare Costs Daily	Incl. Subs O&P Hr.	Incl. Subs O&P Daily	Bare Costs	Incl. O&P
1 Tugboat Captain	$51.10	$408.80	$77.30	$618.40	$44.50	$67.72
1 Tugboat Hand	37.90	303.20	58.15	465.20		
1 Tugboat, 250 H.P.		868.60		955.46	54.29	59.72
16 L.H., Daily Totals		$1580.60		$2039.06	$98.79	$127.44

Crew B-84

Crew No.	Bare Costs Hr.	Bare Costs Daily	Incl. Subs O&P Hr.	Incl. Subs O&P Daily	Bare Costs	Incl. O&P
1 Equip. Oper. (medium)	$51.10	$408.80	$77.30	$618.40	$51.10	$77.30
1 Rotary Mower/Tractor		369.40		406.34	46.17	50.79
8 L.H., Daily Totals		$778.20		$1024.74	$97.28	$128.09

Crew B-85

Crew No.	Bare Costs Hr.	Bare Costs Daily	Incl. Subs O&P Hr.	Incl. Subs O&P Daily	Bare Costs	Incl. O&P
3 Laborers	$37.90	$909.60	$58.15	$1395.60	$41.60	$63.40
1 Equip. Oper. (medium)	51.10	408.80	77.30	618.40		
1 Truck Driver (heavy)	43.20	345.60	65.25	522.00		
1 Aerial Lift Truck, 80'		659.20		725.12		
1 Brush Chipper, 12", 130 H.P.		388.60		427.46		
1 Pruning Saw, Rotary		6.65		7.32	26.36	29.00
40 L.H., Daily Totals		$2718.45		$3695.90	$67.96	$92.40

Crew B-86

Crew No.	Bare Costs Hr.	Bare Costs Daily	Incl. Subs O&P Hr.	Incl. Subs O&P Daily	Bare Costs	Incl. O&P
1 Equip. Oper. (medium)	$51.10	$408.80	$77.30	$618.40	$51.10	$77.30
1 Stump Chipper, S.P.		173.25		190.57	21.66	23.82
8 L.H., Daily Totals		$582.05		$808.98	$72.76	$101.12

Crew B-86A

Crew No.	Bare Costs Hr.	Bare Costs Daily	Incl. Subs O&P Hr.	Incl. Subs O&P Daily	Bare Costs	Incl. O&P
1 Equip. Oper. (medium)	$51.10	$408.80	$77.30	$618.40	$51.10	$77.30
1 Grader, 30,000 Lbs.		714.40		785.84	89.30	98.23
8 L.H., Daily Totals		$1123.20		$1404.24	$140.40	$175.53

Crew B-86B

Crew No.	Bare Costs Hr.	Bare Costs Daily	Incl. Subs O&P Hr.	Incl. Subs O&P Daily	Bare Costs	Incl. O&P
1 Equip. Oper. (medium)	$51.10	$408.80	$77.30	$618.40	$51.10	$77.30
1 Dozer, 200 H.P.		1397.00		1536.70	174.63	192.09
8 L.H., Daily Totals		$1805.80		$2155.10	$225.72	$269.39

Crew B-87

Crew No.	Bare Costs Hr.	Bare Costs Daily	Incl. Subs O&P Hr.	Incl. Subs O&P Daily	Bare Costs	Incl. O&P
1 Laborer	$37.90	$303.20	$58.15	$465.20	$48.46	$73.47
4 Equip. Opers. (medium)	51.10	1635.20	77.30	2473.60		
2 Feller Bunchers, 100 H.P.		1675.20		1842.72		
1 Log Chipper, 22" Tree		890.00		979.00		
1 Dozer, 105 H.P.		603.40		663.74		
1 Chain Saw, Gas, 36" Long		45.00		49.50	80.34	88.37
40 L.H., Daily Totals		$5152.00		$6473.76	$128.80	$161.84

Crew B-88

Crew No.	Bare Costs Hr.	Bare Costs Daily	Incl. Subs O&P Hr.	Incl. Subs O&P Daily	Bare Costs	Incl. O&P
1 Laborer	$37.90	$303.20	$58.15	$465.20	$49.21	$74.56
6 Equip. Opers. (medium)	51.10	2452.80	77.30	3710.40		
2 Feller Bunchers, 100 H.P.		1675.20		1842.72		
1 Log Chipper, 22" Tree		890.00		979.00		
2 Log Skidders, 50 H.P.		1833.60		2016.96		
1 Dozer, 105 H.P.		603.40		663.74		
1 Chain Saw, Gas, 36" Long		45.00		49.50	90.13	99.14
56 L.H., Daily Totals		$7803.20		$9727.52	$139.34	$173.71

Crew B-89

Crew No.	Bare Costs Hr.	Bare Costs Daily	Incl. Subs O&P Hr.	Incl. Subs O&P Daily	Bare Costs	Incl. O&P
1 Equip. Oper. (light)	$49.15	$393.20	$74.35	$594.80	$45.55	$68.85
1 Truck Driver (light)	41.95	335.60	63.35	506.80		
1 Flatbed Truck, Gas, 3 Ton		307.40		338.14		
1 Concrete Saw		168.60		185.46		
1 Water Tank, 65 Gal.		17.30		19.03	30.83	33.91
16 L.H., Daily Totals		$1222.10		$1644.23	$76.38	$102.76

Crew B-89A

Crew No.	Bare Costs Hr.	Bare Costs Daily	Incl. Subs O&P Hr.	Incl. Subs O&P Daily	Bare Costs	Incl. O&P
1 Skilled Worker	$49.90	$399.20	$76.85	$614.80	$43.90	$67.50
1 Laborer	37.90	303.20	58.15	465.20		
1 Core Drill (Large)		117.00		128.70	7.31	8.04
16 L.H., Daily Totals		$819.40		$1208.70	$51.21	$75.54

Crew B-89B

Crew No.	Bare Costs Hr.	Bare Costs Daily	Incl. Subs O&P Hr.	Incl. Subs O&P Daily	Bare Costs	Incl. O&P
1 Equip. Oper. (light)	$49.15	$393.20	$74.35	$594.80	$45.55	$68.85
1 Truck Driver (light)	41.95	335.60	63.35	506.80		
1 Wall Saw, Hydraulic, 10 H.P.		115.00		126.50		
1 Generator, Diesel, 100 kW		413.00		454.30		
1 Water Tank, 65 Gal.		17.30		19.03		
1 Flatbed Truck, Gas, 3 Ton		307.40		338.14	53.29	58.62
16 L.H., Daily Totals		$1581.50		$2039.57	$98.84	$127.47

Crew B-90	Hr.	Daily	Hr.	Daily	Bare Costs	Incl. O&P
1 Labor Foreman (outside)	$39.90	$319.20	$61.20	$489.60	$42.29	$64.36
3 Laborers	37.90	909.60	58.15	1395.60		
2 Equip. Opers. (light)	49.15	786.40	74.35	1189.60		
2 Truck Drivers (heavy)	43.20	691.20	65.25	1044.00		
1 Road Mixer, 310 H.P.		1945.00		2139.50		
1 Dist. Truck, 2000 Gal.		289.00		317.90	34.91	38.40
64 L.H., Daily Totals		$4940.40		$6576.20	$77.19	$102.75

Crew B-90A	Hr.	Daily	Hr.	Daily	Bare Costs	Incl. O&P
1 Labor Foreman (outside)	$39.90	$319.20	$61.20	$489.60	$45.73	$69.53
2 Laborers	37.90	606.40	58.15	930.40		
4 Equip. Opers. (medium)	51.10	1635.20	77.30	2473.60		
2 Graders, 30,000 Lbs.		1428.80		1571.68		
1 Tandem Roller, 10 Ton		239.40		263.34		
1 Roller, Pneum. Whl., 12 Ton		347.80		382.58	36.00	39.60
56 L.H., Daily Totals		$4576.80		$6111.20	$81.73	$109.13

Crew B-90B	Hr.	Daily	Hr.	Daily	Bare Costs	Incl. O&P
1 Labor Foreman (outside)	$39.90	$319.20	$61.20	$489.60	$44.83	$68.23
2 Laborers	37.90	606.40	58.15	930.40		
3 Equip. Opers. (medium)	51.10	1226.40	77.30	1855.20		
1 Roller, Pneum. Whl., 12 Ton		347.80		382.58		
1 Road Mixer, 310 H.P.		1945.00		2139.50	47.77	52.54
48 L.H., Daily Totals		$4444.80		$5797.28	$92.60	$120.78

Crew B-90C	Hr.	Daily	Hr.	Daily	Bare Costs	Incl. O&P
1 Labor Foreman (outside)	$39.90	$319.20	$61.20	$489.60	$43.13	$65.59
4 Laborers	37.90	1212.80	58.15	1860.80		
3 Equip. Opers. (medium)	51.10	1226.40	77.30	1855.20		
3 Truck Drivers (heavy)	43.20	1036.80	65.25	1566.00		
3 Road Mixers, 310 H.P.		5835.00		6418.50	66.31	72.94
88 L.H., Daily Totals		$9630.20		$12190.10	$109.43	$138.52

Crew B-90D	Hr.	Daily	Hr.	Daily	Bare Costs	Incl. O&P
1 Labor Foreman (outside)	$39.90	$319.20	$61.20	$489.60	$42.32	$64.44
6 Laborers	37.90	1819.20	58.15	2791.20		
3 Equip. Opers. (medium)	51.10	1226.40	77.30	1855.20		
3 Truck Drivers (heavy)	43.20	1036.80	65.25	1566.00		
3 Road Mixers, 310 H.P.		5835.00		6418.50	56.11	61.72
104 L.H., Daily Totals		$10236.60		$13120.50	$98.43	$126.16

Crew B-90E	Hr.	Daily	Hr.	Daily	Bare Costs	Incl. O&P
1 Labor Foreman (outside)	$39.90	$319.20	$61.20	$489.60	$43.11	$65.66
4 Laborers	37.90	1212.80	58.15	1860.80		
3 Equip. Opers. (medium)	51.10	1226.40	77.30	1855.20		
1 Truck Driver (heavy)	43.20	345.60	65.25	522.00		
1 Road Mixers, 310 H.P.		1945.00		2139.50	27.01	29.72
72 L.H., Daily Totals		$5049.00		$6867.10	$70.13	$95.38

Crew B-91	Hr.	Daily	Hr.	Daily	Bare Costs	Incl. O&P
1 Labor Foreman (outside)	$39.90	$319.20	$61.20	$489.60	$45.41	$68.99
2 Laborers	37.90	606.40	58.15	930.40		
4 Equip. Opers. (medium)	51.10	1635.20	77.30	2473.60		
1 Truck Driver (heavy)	43.20	345.60	65.25	522.00		
1 Dist. Tanker, 3000 Gallon		319.40		351.34		
1 Truck Tractor, 6x4, 380 H.P.		600.00		660.00		
1 Aggreg. Spreader, S.P.		829.70		912.67		
1 Roller, Pneum. Whl., 12 Ton		347.80		382.58		
1 Tandem Roller, 10 Ton		239.40		263.34	36.50	40.16
64 L.H., Daily Totals		$5242.70		$6985.53	$81.92	$109.15

Crew B-91B	Hr.	Daily	Hr.	Daily	Bare Costs	Incl. O&P
1 Laborer	$37.90	$303.20	$58.15	$465.20	$44.50	$67.72
1 Equipment Oper. (med.)	51.10	408.80	77.30	618.40		
1 Road Sweeper, Vac. Assist.		983.80		1082.18	61.49	67.64
16 L.H., Daily Totals		$1695.80		$2165.78	$105.99	$135.36

Crew B-91C	Hr.	Daily	Hr.	Daily	Bare Costs	Incl. O&P
1 Laborer	$37.90	$303.20	$58.15	$465.20	$39.92	$60.75
1 Truck Driver (light)	41.95	335.60	63.35	506.80		
1 Catch Basin Cleaning Truck		577.20		634.92	36.08	39.68
16 L.H., Daily Totals		$1216.00		$1606.92	$76.00	$100.43

Crew B-91D	Hr.	Daily	Hr.	Daily	Bare Costs	Incl. O&P
1 Labor Foreman (outside)	$39.90	$319.20	$61.20	$489.60	$43.95	$66.84
5 Laborers	37.90	1516.00	58.15	2326.00		
5 Equip. Opers. (medium)	51.10	2044.00	77.30	3092.00		
2 Truck Drivers (heavy)	43.20	691.20	65.25	1044.00		
1 Aggreg. Spreader, S.P.		829.70		912.67		
2 Trucks, Tractor, 6x4, 380 H.P.		1200.00		1320.00		
2 Dist. Tankers, 3000 Gallon		638.80		702.68		
2 Pavement Brushes, Towed		162.40		178.64		
2 Rollers, Pneum. Whl., 12 Ton		695.60		765.16	33.91	37.30
104 L.H., Daily Totals		$8096.90		$10830.75	$77.85	$104.14

Crew B-92	Hr.	Daily	Hr.	Daily	Bare Costs	Incl. O&P
1 Labor Foreman (outside)	$39.90	$319.20	$61.20	$489.60	$38.40	$58.91
3 Laborers	37.90	909.60	58.15	1395.60		
1 Crack Cleaner, 25 H.P.		62.00		68.20		
1 Air Compressor, 60 cfm		142.80		157.08		
1 Tar Kettle, T.M.		135.30		148.83		
1 Flatbed Truck, Gas, 3 Ton		307.40		338.14	20.23	22.26
32 L.H., Daily Totals		$1876.30		$2597.45	$58.63	$81.17

Crew B-93	Hr.	Daily	Hr.	Daily	Bare Costs	Incl. O&P
1 Equip. Oper. (medium)	$51.10	$408.80	$77.30	$618.40	$51.10	$77.30
1 Feller Buncher, 100 H.P.		837.60		921.36	104.70	115.17
8 L.H., Daily Totals		$1246.40		$1539.76	$155.80	$192.47

Crew B-94A	Hr.	Daily	Hr.	Daily	Bare Costs	Incl. O&P
1 Laborer	$37.90	$303.20	$58.15	$465.20	$37.90	$58.15
1 Diaphragm Water Pump, 2"		71.80		78.98		
1 -20' Suction Hose, 2"		1.95		2.15		
2 -50' Discharge Hoses, 2"		1.80		1.98	9.44	10.39
8 L.H., Daily Totals		$378.75		$548.30	$47.34	$68.54

Crew B-94B	Hr.	Daily	Hr.	Daily	Bare Costs	Incl. O&P
1 Laborer	$37.90	$303.20	$58.15	$465.20	$37.90	$58.15
1 Diaphragm Water Pump, 4"		118.00		129.80		
1 -20' Suction Hose, 4"		3.25		3.58		
2 -50' Discharge Hoses, 4"		4.70		5.17	15.74	17.32
8 L.H., Daily Totals		$429.15		$603.75	$53.64	$75.47

Crew B-94C	Hr.	Daily	Hr.	Daily	Bare Costs	Incl. O&P
1 Laborer	$37.90	$303.20	$58.15	$465.20	$37.90	$58.15
1 Centrifugal Water Pump, 3"		81.00		89.10		
1 -20' Suction Hose, 3"		2.85		3.13		
2 -50' Discharge Hoses, 3"		3.00		3.30	10.86	11.94
8 L.H., Daily Totals		$390.05		$560.74	$48.76	$70.09

Crew No.	Bare Costs Hr.	Bare Costs Daily	Incl. Subs O&P Hr.	Incl. Subs O&P Daily	Cost Per Labor-Hour Bare Costs	Cost Per Labor-Hour Incl. O&P
Crew B-94D	Hr.	Daily	Hr.	Daily	Bare Costs	Incl. O&P
1 Laborer	$37.90	$303.20	$58.15	$465.20	$37.90	$58.15
1 Centr. Water Pump, 6"		350.80		385.88		
1 -20' Suction Hose, 6"		11.50		12.65		
2 -50' Discharge Hoses, 6"		12.20		13.42	46.81	51.49
8 L.H., Daily Totals		$677.70		$877.15	$84.71	$109.64
Crew C-1	Hr.	Daily	Hr.	Daily	Bare Costs	Incl. O&P
3 Carpenters	$48.45	$1162.80	$74.30	$1783.20	$45.81	$70.26
1 Laborer	37.90	303.20	58.15	465.20		
32 L.H., Daily Totals		$1466.00		$2248.40	$45.81	$70.26
Crew C-2	Hr.	Daily	Hr.	Daily	Bare Costs	Incl. O&P
1 Carpenter Foreman (outside)	$50.45	$403.60	$77.40	$619.20	$47.02	$72.13
4 Carpenters	48.45	1550.40	74.30	2377.60		
1 Laborer	37.90	303.20	58.15	465.20		
48 L.H., Daily Totals		$2257.20		$3462.00	$47.02	$72.13
Crew C-2A	Hr.	Daily	Hr.	Daily	Bare Costs	Incl. O&P
1 Carpenter Foreman (outside)	$50.45	$403.60	$77.40	$619.20	$46.56	$71.03
3 Carpenters	48.45	1162.80	74.30	1783.20		
1 Cement Finisher	45.65	365.20	67.70	541.60		
1 Laborer	37.90	303.20	58.15	465.20		
48 L.H., Daily Totals		$2234.80		$3409.20	$46.56	$71.03
Crew C-3	Hr.	Daily	Hr.	Daily	Bare Costs	Incl. O&P
1 Rodman Foreman (outside)	$55.00	$440.00	$85.20	$681.60	$48.99	$75.53
4 Rodmen (reinf.)	53.00	1696.00	82.10	2627.20		
1 Equip. Oper. (light)	49.15	393.20	74.35	594.80		
2 Laborers	37.90	606.40	58.15	930.40		
3 Stressing Equipment		30.60		33.66		
.5 Grouting Equipment		82.20		90.42	1.76	1.94
64 L.H., Daily Totals		$3248.40		$4958.08	$50.76	$77.47
Crew C-4	Hr.	Daily	Hr.	Daily	Bare Costs	Incl. O&P
1 Rodman Foreman (outside)	$55.00	$440.00	$85.20	$681.60	$53.50	$82.88
3 Rodmen (reinf.)	53.00	1272.00	82.10	1970.40		
3 Stressing Equipment		30.60		33.66	.96	1.05
32 L.H., Daily Totals		$1742.60		$2685.66	$54.46	$83.93
Crew C-4A	Hr.	Daily	Hr.	Daily	Bare Costs	Incl. O&P
2 Rodmen (reinf.)	$53.00	$848.00	$82.10	$1313.60	$53.00	$82.10
4 Stressing Equipment		40.80		44.88	2.55	2.81
16 L.H., Daily Totals		$888.80		$1358.48	$55.55	$84.91
Crew C-5	Hr.	Daily	Hr.	Daily	Bare Costs	Incl. O&P
1 Rodman Foreman (outside)	$55.00	$440.00	$85.20	$681.60	$52.19	$80.33
4 Rodmen (reinf.)	53.00	1696.00	82.10	2627.20		
1 Equip. Oper. (crane)	52.25	418.00	79.05	632.40		
1 Equip. Oper. (oiler)	46.05	368.40	69.65	557.20		
1 Hyd. Crane, 25 Ton		748.80		823.68	13.37	14.71
56 L.H., Daily Totals		$3671.20		$5322.08	$65.56	$95.04
Crew C-6	Hr.	Daily	Hr.	Daily	Bare Costs	Incl. O&P
1 Labor Foreman (outside)	$39.90	$319.20	$61.20	$489.60	$39.52	$60.25
4 Laborers	37.90	1212.80	58.15	1860.80		
1 Cement Finisher	45.65	365.20	67.70	541.60		
2 Gas Engine Vibrators		63.20		69.52	1.32	1.45
48 L.H., Daily Totals		$1960.40		$2961.52	$40.84	$61.70

Crew No.	Bare Costs Hr.	Bare Costs Daily	Incl. Subs O&P Hr.	Incl. Subs O&P Daily	Cost Per Labor-Hour Bare Costs	Cost Per Labor-Hour Incl. O&P
Crew C-7	Hr.	Daily	Hr.	Daily	Bare Costs	Incl. O&P
1 Labor Foreman (outside)	$39.90	$319.20	$61.20	$489.60	$41.36	$62.96
5 Laborers	37.90	1516.00	58.15	2326.00		
1 Cement Finisher	45.65	365.20	67.70	541.60		
1 Equip. Oper. (medium)	51.10	408.80	77.30	618.40		
1 Equip. Oper. (oiler)	46.05	368.40	69.65	557.20		
2 Gas Engine Vibrators		63.20		69.52		
1 Concrete Bucket, 1 C.Y.		23.60		25.96		
1 Hyd. Crane, 55 Ton		1123.00		1235.30	16.80	18.48
72 L.H., Daily Totals		$4187.40		$5863.58	$58.16	$81.44
Crew C-7A	Hr.	Daily	Hr.	Daily	Bare Costs	Incl. O&P
1 Labor Foreman (outside)	$39.90	$319.20	$61.20	$489.60	$39.48	$60.31
5 Laborers	37.90	1516.00	58.15	2326.00		
2 Truck Drivers (heavy)	43.20	691.20	65.25	1044.00		
2 Conc. Transit Mixers		2038.00		2241.80	31.84	35.03
64 L.H., Daily Totals		$4564.40		$6101.40	$71.32	$95.33
Crew C-7B	Hr.	Daily	Hr.	Daily	Bare Costs	Incl. O&P
1 Labor Foreman (outside)	$39.90	$319.20	$61.20	$489.60	$40.96	$62.58
5 Laborers	37.90	1516.00	58.15	2326.00		
1 Equipment Operator, Crane	52.25	418.00	79.05	632.40		
1 Equipment Oiler	46.05	368.40	69.65	557.20		
1 Conc. Bucket, 2 C.Y.		36.60		40.26		
1 Lattice Boom Crane, 165 Ton		1934.00		2127.40	30.79	33.87
64 L.H., Daily Totals		$4592.20		$6172.86	$71.75	$96.45
Crew C-7C	Hr.	Daily	Hr.	Daily	Bare Costs	Incl. O&P
1 Labor Foreman (outside)	$39.90	$319.20	$61.20	$489.60	$41.45	$63.32
5 Laborers	37.90	1516.00	58.15	2326.00		
2 Equipment Operators (med.)	51.10	817.60	77.30	1236.80		
2 F.E. Loaders, W.M., 4 C.Y.		1323.60		1455.96	20.68	22.75
64 L.H., Daily Totals		$3976.40		$5508.36	$62.13	$86.07
Crew C-7D	Hr.	Daily	Hr.	Daily	Bare Costs	Incl. O&P
1 Labor Foreman (outside)	$39.90	$319.20	$61.20	$489.60	$40.07	$61.32
5 Laborers	37.90	1516.00	58.15	2326.00		
1 Equip. Oper. (medium)	51.10	408.80	77.30	618.40		
1 Concrete Conveyer		199.20		219.12	3.56	3.91
56 L.H., Daily Totals		$2443.20		$3653.12	$43.63	$65.23
Crew C-8	Hr.	Daily	Hr.	Daily	Bare Costs	Incl. O&P
1 Labor Foreman (outside)	$39.90	$319.20	$61.20	$489.60	$42.29	$64.05
3 Laborers	37.90	909.60	58.15	1395.60		
2 Cement Finishers	45.65	730.40	67.70	1083.20		
1 Equip. Oper. (medium)	51.10	408.80	77.30	618.40		
1 Concrete Pump (Small)		727.60		800.36	12.99	14.29
56 L.H., Daily Totals		$3095.60		$4387.16	$55.28	$78.34
Crew C-8A	Hr.	Daily	Hr.	Daily	Bare Costs	Incl. O&P
1 Labor Foreman (outside)	$39.90	$319.20	$61.20	$489.60	$40.82	$61.84
3 Laborers	37.90	909.60	58.15	1395.60		
2 Cement Finishers	45.65	730.40	67.70	1083.20		
48 L.H., Daily Totals		$1959.20		$2968.40	$40.82	$61.84

Crews - Standard

Crew C-8B

Crew C-8B	Hr.	Daily	Hr.	Daily	Bare Costs	Incl. O&P
1 Labor Foreman (outside)	$39.90	$319.20	$61.20	$489.60	$40.94	$62.59
3 Laborers	37.90	909.60	58.15	1395.60		
1 Equip. Oper. (medium)	51.10	408.80	77.30	618.40		
1 Vibrating Power Screed		63.65		70.02		
1 Roller, Vibratory, 25 Ton		691.80		760.98		
1 Dozer, 200 H.P.		1397.00		1536.70	53.81	59.19
40 L.H., Daily Totals		$3790.05		$4871.30	$94.75	$121.78

Crew C-8C

Crew C-8C	Hr.	Daily	Hr.	Daily	Bare Costs	Incl. O&P
1 Labor Foreman (outside)	$39.90	$319.20	$61.20	$489.60	$41.73	$63.44
3 Laborers	37.90	909.60	58.15	1395.60		
1 Cement Finisher	45.65	365.20	67.70	541.60		
1 Equip. Oper. (medium)	51.10	408.80	77.30	618.40		
1 Shotcrete Rig, 12 C.Y./hr		254.00		279.40		
1 Air Compressor, 160 cfm		162.20		178.42		
4 -50' Air Hoses, 1"		14.60		16.06		
4 -50' Air Hoses, 2"		27.60		30.36	9.55	10.51
48 L.H., Daily Totals		$2461.20		$3549.44	$51.27	$73.95

Crew C-8D

Crew C-8D	Hr.	Daily	Hr.	Daily	Bare Costs	Incl. O&P
1 Labor Foreman (outside)	$39.90	$319.20	$61.20	$489.60	$43.15	$65.35
1 Laborer	37.90	303.20	58.15	465.20		
1 Cement Finisher	45.65	365.20	67.70	541.60		
1 Equipment Oper. (light)	49.15	393.20	74.35	594.80		
1 Air Compressor, 250 cfm		197.80		217.58		
2 -50' Air Hoses, 1"		7.30		8.03	6.41	7.05
32 L.H., Daily Totals		$1585.90		$2316.81	$49.56	$72.40

Crew C-8E

Crew C-8E	Hr.	Daily	Hr.	Daily	Bare Costs	Incl. O&P
1 Labor Foreman (outside)	$39.90	$319.20	$61.20	$489.60	$41.40	$62.95
3 Laborers	37.90	909.60	58.15	1395.60		
1 Cement Finisher	45.65	365.20	67.70	541.60		
1 Equipment Oper. (light)	49.15	393.20	74.35	594.80		
1 Shotcrete Rig, 35 C.Y./hr		284.00		312.40		
1 Air Compressor, 250 cfm		197.80		217.58		
4 -50' Air Hoses, 1"		14.60		16.06		
4 -50' Air Hoses, 2"		27.60		30.36	10.92	12.01
48 L.H., Daily Totals		$2511.20		$3598.00	$52.32	$74.96

Crew C-10

Crew C-10	Hr.	Daily	Hr.	Daily	Bare Costs	Incl. O&P
1 Laborer	$37.90	$303.20	$58.15	$465.20	$43.07	$64.52
2 Cement Finishers	45.65	730.40	67.70	1083.20		
24 L.H., Daily Totals		$1033.60		$1548.40	$43.07	$64.52

Crew C-10B

Crew C-10B	Hr.	Daily	Hr.	Daily	Bare Costs	Incl. O&P
3 Laborers	$37.90	$909.60	$58.15	$1395.60	$41.00	$61.97
2 Cement Finishers	45.65	730.40	67.70	1083.20		
1 Concrete Mixer, 10 C.F.		172.80		190.08		
2 Trowels, 48" Walk-Behind		101.20		111.32	6.85	7.54
40 L.H., Daily Totals		$1914.00		$2780.20	$47.85	$69.50

Crew C-10C

Crew C-10C	Hr.	Daily	Hr.	Daily	Bare Costs	Incl. O&P
1 Laborer	$37.90	$303.20	$58.15	$465.20	$43.07	$64.52
2 Cement Finishers	45.65	730.40	67.70	1083.20		
1 Trowel, 48" Walk-Behind		50.60		55.66	2.11	2.32
24 L.H., Daily Totals		$1084.20		$1604.06	$45.17	$66.84

Crew C-10D

Crew C-10D	Hr.	Daily	Hr.	Daily	Bare Costs	Incl. O&P
1 Laborer	$37.90	$303.20	$58.15	$465.20	$43.07	$64.52
2 Cement Finishers	45.65	730.40	67.70	1083.20		
1 Vibrating Power Screed		63.65		70.02		
1 Trowel, 48" Walk-Behind		50.60		55.66	4.76	5.24
24 L.H., Daily Totals		$1147.85		$1674.08	$47.83	$69.75

Crew C-10E

Crew C-10E	Hr.	Daily	Hr.	Daily	Bare Costs	Incl. O&P
1 Laborer	$37.90	$303.20	$58.15	$465.20	$43.07	$64.52
2 Cement Finishers	45.65	730.40	67.70	1083.20		
1 Vibrating Power Screed		63.65		70.02		
1 Cement Trowel, 96" Ride-On		190.20		209.22	10.58	11.63
24 L.H., Daily Totals		$1287.45		$1827.64	$53.64	$76.15

Crew C-10F

Crew C-10F	Hr.	Daily	Hr.	Daily	Bare Costs	Incl. O&P
1 Laborer	$37.90	$303.20	$58.15	$465.20	$43.07	$64.52
2 Cement Finishers	45.65	730.40	67.70	1083.20		
1 Aerial Lift Truck, 60' Boom		448.40		493.24	18.68	20.55
24 L.H., Daily Totals		$1482.00		$2041.64	$61.75	$85.07

Crew C-11

Crew C-11	Hr.	Daily	Hr.	Daily	Bare Costs	Incl. O&P
1 Struc. Steel Foreman (outside)	$55.20	$441.60	$93.50	$748.00	$52.52	$86.98
6 Struc. Steel Workers	53.20	2553.60	90.10	4324.80		
1 Equip. Oper. (crane)	52.25	418.00	79.05	632.40		
1 Equip. Oper. (oiler)	46.05	368.40	69.65	557.20		
1 Lattice Boom Crane, 150 Ton		1868.00		2054.80	25.94	28.54
72 L.H., Daily Totals		$5649.60		$8317.20	$78.47	$115.52

Crew C-12

Crew C-12	Hr.	Daily	Hr.	Daily	Bare Costs	Incl. O&P
1 Carpenter Foreman (outside)	$50.45	$403.60	$77.40	$619.20	$47.66	$72.92
3 Carpenters	48.45	1162.80	74.30	1783.20		
1 Laborer	37.90	303.20	58.15	465.20		
1 Equip. Oper. (crane)	52.25	418.00	79.05	632.40		
1 Hyd. Crane, 12 Ton		658.40		724.24	13.72	15.09
48 L.H., Daily Totals		$2946.00		$4224.24	$61.38	$88.00

Crew C-13

Crew C-13	Hr.	Daily	Hr.	Daily	Bare Costs	Incl. O&P
1 Struc. Steel Worker	$53.20	$425.60	$90.10	$720.80	$51.62	$84.83
1 Welder	53.20	425.60	90.10	720.80		
1 Carpenter	48.45	387.60	74.30	594.40		
1 Welder, Gas Engine, 300 amp		147.20		161.92	6.13	6.75
24 L.H., Daily Totals		$1386.00		$2197.92	$57.75	$91.58

Crew C-14

Crew C-14	Hr.	Daily	Hr.	Daily	Bare Costs	Incl. O&P
1 Carpenter Foreman (outside)	$50.45	$403.60	$77.40	$619.20	$46.99	$71.89
5 Carpenters	48.45	1938.00	74.30	2972.00		
4 Laborers	37.90	1212.80	58.15	1860.80		
4 Rodmen (reinf.)	53.00	1696.00	82.10	2627.20		
2 Cement Finishers	45.65	730.40	67.70	1083.20		
1 Equip. Oper. (crane)	52.25	418.00	79.05	632.40		
1 Equip. Oper. (oiler)	46.05	368.40	69.65	557.20		
1 Hyd. Crane, 80 Ton		1630.00		1793.00	11.32	12.45
144 L.H., Daily Totals		$8397.20		$12145.00	$58.31	$84.34

Crews - Standard

Crew No.	Bare Costs		Incl. Subs O&P		Cost Per Labor-Hour	
Crew C-14A	Hr.	Daily	Hr.	Daily	Bare Costs	Incl. O&P
1 Carpenter Foreman (outside)	$50.45	$403.60	$77.40	$619.20	$48.41	$74.24
16 Carpenters	48.45	6201.60	74.30	9510.40		
4 Rodmen (reinf.)	53.00	1696.00	82.10	2627.20		
2 Laborers	37.90	606.40	58.15	930.40		
1 Cement Finisher	45.65	365.20	67.70	541.60		
1 Equip. Oper. (medium)	51.10	408.80	77.30	618.40		
1 Gas Engine Vibrator		31.60		34.76		
1 Concrete Pump (Small)		727.60		800.36	3.80	4.18
200 L.H., Daily Totals		$10440.80		$15682.32	$52.20	$78.41
Crew C-14B	Hr.	Daily	Hr.	Daily	Bare Costs	Incl. O&P
1 Carpenter Foreman (outside)	$50.45	$403.60	$77.40	$619.20	$48.30	$73.98
16 Carpenters	48.45	6201.60	74.30	9510.40		
4 Rodmen (reinf.)	53.00	1696.00	82.10	2627.20		
2 Laborers	37.90	606.40	58.15	930.40		
2 Cement Finishers	45.65	730.40	67.70	1083.20		
1 Equip. Oper. (medium)	51.10	408.80	77.30	618.40		
1 Gas Engine Vibrator		31.60		34.76		
1 Concrete Pump (Small)		727.60		800.36	3.65	4.01
208 L.H., Daily Totals		$10806.00		$16223.92	$51.95	$78.00
Crew C-14C	Hr.	Daily	Hr.	Daily	Bare Costs	Incl. O&P
1 Carpenter Foreman (outside)	$50.45	$403.60	$77.40	$619.20	$46.03	$70.55
6 Carpenters	48.45	2325.60	74.30	3566.40		
2 Rodmen (reinf.)	53.00	848.00	82.10	1313.60		
4 Laborers	37.90	1212.80	58.15	1860.80		
1 Cement Finisher	45.65	365.20	67.70	541.60		
1 Gas Engine Vibrator		31.60		34.76	.28	.31
112 L.H., Daily Totals		$5186.80		$7936.36	$46.31	$70.86
Crew C-14D	Hr.	Daily	Hr.	Daily	Bare Costs	Incl. O&P
1 Carpenter Foreman (outside)	$50.45	$403.60	$77.40	$619.20	$48.04	$73.61
18 Carpenters	48.45	6976.80	74.30	10699.20		
2 Rodmen (reinf.)	53.00	848.00	82.10	1313.60		
2 Laborers	37.90	606.40	58.15	930.40		
1 Cement Finisher	45.65	365.20	67.70	541.60		
1 Equip. Oper. (medium)	51.10	408.80	77.30	618.40		
1 Gas Engine Vibrator		31.60		34.76		
1 Concrete Pump (Small)		727.60		800.36	3.80	4.18
200 L.H., Daily Totals		$10368.00		$15557.52	$51.84	$77.79
Crew C-14E	Hr.	Daily	Hr.	Daily	Bare Costs	Incl. O&P
1 Carpenter Foreman (outside)	$50.45	$403.60	$77.40	$619.20	$47.15	$72.41
2 Carpenters	48.45	775.20	74.30	1188.80		
4 Rodmen (reinf.)	53.00	1696.00	82.10	2627.20		
3 Laborers	37.90	909.60	58.15	1395.60		
1 Cement Finisher	45.65	365.20	67.70	541.60		
1 Gas Engine Vibrator		31.60		34.76	.36	.40
88 L.H., Daily Totals		$4181.20		$6407.16	$47.51	$72.81
Crew C-14F	Hr.	Daily	Hr.	Daily	Bare Costs	Incl. O&P
1 Labor Foreman (outside)	$39.90	$319.20	$61.20	$489.60	$43.29	$64.86
2 Laborers	37.90	606.40	58.15	930.40		
6 Cement Finishers	45.65	2191.20	67.70	3249.60		
1 Gas Engine Vibrator		31.60		34.76	.44	.48
72 L.H., Daily Totals		$3148.40		$4704.36	$43.73	$65.34

Crew No.	Bare Costs		Incl. Subs O&P		Cost Per Labor-Hour	
Crew C-14G	Hr.	Daily	Hr.	Daily	Bare Costs	Incl. O&P
1 Labor Foreman (outside)	$39.90	$319.20	$61.20	$489.60	$42.61	$64.04
2 Laborers	37.90	606.40	58.15	930.40		
4 Cement Finishers	45.65	1460.80	67.70	2166.40		
1 Gas Engine Vibrator		31.60		34.76	.56	.62
56 L.H., Daily Totals		$2418.00		$3621.16	$43.18	$64.66
Crew C-14H	Hr.	Daily	Hr.	Daily	Bare Costs	Incl. O&P
1 Carpenter Foreman (outside)	$50.45	$403.60	$77.40	$619.20	$47.32	$72.33
2 Carpenters	48.45	775.20	74.30	1188.80		
1 Rodman (reinf.)	53.00	424.00	82.10	656.80		
1 Laborer	37.90	303.20	58.15	465.20		
1 Cement Finisher	45.65	365.20	67.70	541.60		
1 Gas Engine Vibrator		31.60		34.76	.66	.72
48 L.H., Daily Totals		$2302.80		$3506.36	$47.98	$73.05
Crew C-14L	Hr.	Daily	Hr.	Daily	Bare Costs	Incl. O&P
1 Carpenter Foreman (outside)	$50.45	$403.60	$77.40	$619.20	$44.87	$68.63
6 Carpenters	48.45	2325.60	74.30	3566.40		
4 Laborers	37.90	1212.80	58.15	1860.80		
1 Cement Finisher	45.65	365.20	67.70	541.60		
1 Gas Engine Vibrator		31.60		34.76	.33	.36
96 L.H., Daily Totals		$4338.80		$6622.76	$45.20	$68.99
Crew C-14M	Hr.	Daily	Hr.	Daily	Bare Costs	Incl. O&P
1 Carpenter Foreman (outside)	$50.45	$403.60	$77.40	$619.20	$46.61	$71.17
2 Carpenters	48.45	775.20	74.30	1188.80		
1 Rodman (reinf.)	53.00	424.00	82.10	656.80		
2 Laborers	37.90	606.40	58.15	930.40		
1 Cement Finisher	45.65	365.20	67.70	541.60		
1 Equip. Oper. (medium)	51.10	408.80	77.30	618.40		
1 Gas Engine Vibrator		31.60		34.76		
1 Concrete Pump (Small)		727.60		800.36	11.86	13.05
64 L.H., Daily Totals		$3742.40		$5390.32	$58.48	$84.22
Crew C-15	Hr.	Daily	Hr.	Daily	Bare Costs	Incl. O&P
1 Carpenter Foreman (outside)	$50.45	$403.60	$77.40	$619.20	$45.04	$68.66
2 Carpenters	48.45	775.20	74.30	1188.80		
3 Laborers	37.90	909.60	58.15	1395.60		
2 Cement Finishers	45.65	730.40	67.70	1083.20		
1 Rodman (reinf.)	53.00	424.00	82.10	656.80		
72 L.H., Daily Totals		$3242.80		$4943.60	$45.04	$68.66
Crew C-16	Hr.	Daily	Hr.	Daily	Bare Costs	Incl. O&P
1 Labor Foreman (outside)	$39.90	$319.20	$61.20	$489.60	$42.29	$64.05
3 Laborers	37.90	909.60	58.15	1395.60		
2 Cement Finishers	45.65	730.40	67.70	1083.20		
1 Equip. Oper. (medium)	51.10	408.80	77.30	618.40		
1 Gunite Pump Rig		373.00		410.30		
2 -50' Air Hoses, 3/4"		6.10		6.71		
2 -50' Air Hoses, 2"		13.80		15.18	7.02	7.72
56 L.H., Daily Totals		$2760.90		$4018.99	$49.30	$71.77
Crew C-16A	Hr.	Daily	Hr.	Daily	Bare Costs	Incl. O&P
1 Laborer	$37.90	$303.20	$58.15	$465.20	$45.08	$67.71
2 Cement Finishers	45.65	730.40	67.70	1083.20		
1 Equip. Oper. (medium)	51.10	408.80	77.30	618.40		
1 Gunite Pump Rig		373.00		410.30		
2 -50' Air Hoses, 3/4"		6.10		6.71		
2 -50' Air Hoses, 2"		13.80		15.18		
1 Aerial Lift Truck, 60' Boom		448.40		493.24	26.29	28.92
32 L.H., Daily Totals		$2283.70		$3092.23	$71.37	$96.63

For customer support on your Mechanical Cost Data, call 877.739.3689.

645

Crews - Standard

(See table above.)

For customer support on your Mechanical Cost Data, call 877.739.3689.

645

Crew No.	Bare Costs		Incl. Subs O&P		Cost Per Labor-Hour	
Crew C-17	Hr.	Daily	Hr.	Daily	Bare Costs	Incl. O&P
2 Skilled Worker Foremen (out)	$51.90	$830.40	$79.95	$1279.20	$50.30	$77.47
8 Skilled Workers	49.90	3193.60	76.85	4918.40		
80 L.H., Daily Totals		$4024.00		$6197.60	$50.30	$77.47
Crew C-17A	Hr.	Daily	Hr.	Daily	Bare Costs	Incl. O&P
2 Skilled Worker Foremen (out)	$51.90	$830.40	$79.95	$1279.20	$50.32	$77.49
8 Skilled Workers	49.90	3193.60	76.85	4918.40		
.125 Equip. Oper. (crane)	52.25	52.25	79.05	79.05		
.125 Hyd. Crane, 80 Ton		203.75		224.13	2.52	2.77
81 L.H., Daily Totals		$4280.00		$6500.77	$52.84	$80.26
Crew C-17B	Hr.	Daily	Hr.	Daily	Bare Costs	Incl. O&P
2 Skilled Worker Foremen (out)	$51.90	$830.40	$79.95	$1279.20	$50.35	$77.51
8 Skilled Workers	49.90	3193.60	76.85	4918.40		
.25 Equip. Oper. (crane)	52.25	104.50	79.05	158.10		
.25 Hyd. Crane, 80 Ton		407.50		448.25		
.25 Trowel, 48" Walk-Behind		12.65		13.91	5.12	5.64
82 L.H., Daily Totals		$4548.65		$6817.86	$55.47	$83.14
Crew C-17C	Hr.	Daily	Hr.	Daily	Bare Costs	Incl. O&P
2 Skilled Worker Foremen (out)	$51.90	$830.40	$79.95	$1279.20	$50.37	$77.53
8 Skilled Workers	49.90	3193.60	76.85	4918.40		
.375 Equip. Oper. (crane)	52.25	156.75	79.05	237.15		
.375 Hyd. Crane, 80 Ton		611.25		672.38	7.36	8.10
83 L.H., Daily Totals		$4792.00		$7107.13	$57.73	$85.63
Crew C-17D	Hr.	Daily	Hr.	Daily	Bare Costs	Incl. O&P
2 Skilled Worker Foremen (out)	$51.90	$830.40	$79.95	$1279.20	$50.39	$77.55
8 Skilled Workers	49.90	3193.60	76.85	4918.40		
.5 Equip. Oper. (crane)	52.25	209.00	79.05	316.20		
.5 Hyd. Crane, 80 Ton		815.00		896.50	9.70	10.67
84 L.H., Daily Totals		$5048.00		$7410.30	$60.10	$88.22
Crew C-17E	Hr.	Daily	Hr.	Daily	Bare Costs	Incl. O&P
2 Skilled Worker Foremen (out)	$51.90	$830.40	$79.95	$1279.20	$50.30	$77.47
8 Skilled Workers	49.90	3193.60	76.85	4918.40		
1 Hyd. Jack with Rods		98.40		108.24	1.23	1.35
80 L.H., Daily Totals		$4122.40		$6305.84	$51.53	$78.82
Crew C-18	Hr.	Daily	Hr.	Daily	Bare Costs	Incl. O&P
.125 Labor Foreman (outside)	$39.90	$39.90	$61.20	$61.20	$38.12	$58.49
1 Laborer	37.90	303.20	58.15	465.20		
1 Concrete Cart, 10 C.F.		60.00		66.00	6.67	7.33
9 L.H., Daily Totals		$403.10		$592.40	$44.79	$65.82
Crew C-19	Hr.	Daily	Hr.	Daily	Bare Costs	Incl. O&P
.125 Labor Foreman (outside)	$39.90	$39.90	$61.20	$61.20	$38.12	$58.49
1 Laborer	37.90	303.20	58.15	465.20		
1 Concrete Cart, 18 C.F.		99.80		109.78	11.09	12.20
9 L.H., Daily Totals		$442.90		$636.18	$49.21	$70.69
Crew C-20	Hr.	Daily	Hr.	Daily	Bare Costs	Incl. O&P
1 Labor Foreman (outside)	$39.90	$319.20	$61.20	$489.60	$40.77	$62.12
5 Laborers	37.90	1516.00	58.15	2326.00		
1 Cement Finisher	45.65	365.20	67.70	541.60		
1 Equip. Oper. (medium)	51.10	408.80	77.30	618.40		
2 Gas Engine Vibrators		63.20		69.52		
1 Concrete Pump (Small)		727.60		800.36	12.36	13.59
64 L.H., Daily Totals		$3400.00		$4845.48	$53.13	$75.71

Crew No.	Bare Costs		Incl. Subs O&P		Cost Per Labor-Hour	
Crew C-21	Hr.	Daily	Hr.	Daily	Bare Costs	Incl. O&P
1 Labor Foreman (outside)	$39.90	$319.20	$61.20	$489.60	$40.77	$62.12
5 Laborers	37.90	1516.00	58.15	2326.00		
1 Cement Finisher	45.65	365.20	67.70	541.60		
1 Equip. Oper. (medium)	51.10	408.80	77.30	618.40		
2 Gas Engine Vibrators		63.20		69.52		
1 Concrete Conveyer		199.20		219.12	4.10	4.51
64 L.H., Daily Totals		$2871.60		$4264.24	$44.87	$66.63
Crew C-22	Hr.	Daily	Hr.	Daily	Bare Costs	Incl. O&P
1 Rodman Foreman (outside)	$55.00	$440.00	$85.20	$681.60	$53.20	$82.32
4 Rodmen (reinf.)	53.00	1696.00	82.10	2627.20		
.125 Equip. Oper. (crane)	52.25	52.25	79.05	79.05		
.125 Equip. Oper. (oiler)	46.05	46.05	69.65	69.65		
.125 Hyd. Crane, 25 Ton		93.60		102.96	2.23	2.45
42 L.H., Daily Totals		$2327.90		$3560.46	$55.43	$84.77
Crew C-23	Hr.	Daily	Hr.	Daily	Bare Costs	Incl. O&P
2 Skilled Worker Foremen (out)	$51.90	$830.40	$79.95	$1279.20	$50.15	$76.97
6 Skilled Workers	49.90	2395.20	76.85	3688.80		
1 Equip. Oper. (crane)	52.25	418.00	79.05	632.40		
1 Equip. Oper. (oiler)	46.05	368.40	69.65	557.20		
1 Lattice Boom Crane, 90 Ton		1520.00		1672.00	19.00	20.90
80 L.H., Daily Totals		$5532.00		$7829.60	$69.15	$97.87
Crew C-23A	Hr.	Daily	Hr.	Daily	Bare Costs	Incl. O&P
1 Labor Foreman (outside)	$39.90	$319.20	$61.20	$489.60	$42.80	$65.24
2 Laborers	37.90	606.40	58.15	930.40		
1 Equip. Oper. (crane)	52.25	418.00	79.05	632.40		
1 Equip. Oper. (oiler)	46.05	368.40	69.65	557.20		
1 Crawler Crane, 100 Ton		1667.00		1833.70		
3 Conc. Buckets, 8 C.Y.		606.00		666.60	56.83	62.51
40 L.H., Daily Totals		$3985.00		$5109.90	$99.63	$127.75
Crew C-24	Hr.	Daily	Hr.	Daily	Bare Costs	Incl. O&P
2 Skilled Worker Foremen (out)	$51.90	$830.40	$79.95	$1279.20	$50.15	$76.97
6 Skilled Workers	49.90	2395.20	76.85	3688.80		
1 Equip. Oper. (crane)	52.25	418.00	79.05	632.40		
1 Equip. Oper. (oiler)	46.05	368.40	69.65	557.20		
1 Lattice Boom Crane, 150 Ton		1868.00		2054.80	23.35	25.68
80 L.H., Daily Totals		$5880.00		$8212.40	$73.50	$102.66
Crew C-25	Hr.	Daily	Hr.	Daily	Bare Costs	Incl. O&P
2 Rodmen (reinf.)	$53.00	$848.00	$82.10	$1313.60	$42.17	$67.75
2 Rodmen Helpers	31.35	501.60	53.40	854.40		
32 L.H., Daily Totals		$1349.60		$2168.00	$42.17	$67.75
Crew C-27	Hr.	Daily	Hr.	Daily	Bare Costs	Incl. O&P
2 Cement Finishers	$45.65	$730.40	$67.70	$1083.20	$45.65	$67.70
1 Concrete Saw		168.60		185.46	10.54	11.59
16 L.H., Daily Totals		$899.00		$1268.66	$56.19	$79.29
Crew C-28	Hr.	Daily	Hr.	Daily	Bare Costs	Incl. O&P
1 Cement Finisher	$45.65	$365.20	$67.70	$541.60	$45.65	$67.70
1 Portable Air Compressor, Gas		17.40		19.14	2.17	2.39
8 L.H., Daily Totals		$382.60		$560.74	$47.83	$70.09
Crew C-29	Hr.	Daily	Hr.	Daily	Bare Costs	Incl. O&P
1 Laborer	$37.90	$303.20	$58.15	$465.20	$37.90	$58.15
1 Pressure Washer		71.60		78.76	8.95	9.85
8 L.H., Daily Totals		$374.80		$543.96	$46.85	$68.00

Crew No.	Bare Costs		Incl. Subs O&P		Cost Per Labor-Hour	
Crew C-30	Hr.	Daily	Hr.	Daily	Bare Costs	Incl. O&P
1 Laborer	$37.90	$303.20	$58.15	$465.20	$37.90	$58.15
1 Concrete Mixer, 10 C.F.		172.80		190.08	21.60	23.76
8 L.H., Daily Totals		$476.00		$655.28	$59.50	$81.91
Crew C-31	Hr.	Daily	Hr.	Daily	Bare Costs	Incl. O&P
1 Cement Finisher	$45.65	$365.20	$67.70	$541.60	$45.65	$67.70
1 Grout Pump		373.00		410.30	46.63	51.29
8 L.H., Daily Totals		$738.20		$951.90	$92.28	$118.99
Crew C-32	Hr.	Daily	Hr.	Daily	Bare Costs	Incl. O&P
1 Cement Finisher	$45.65	$365.20	$67.70	$541.60	$41.77	$62.92
1 Laborer	37.90	303.20	58.15	465.20		
1 Crack Chaser Saw, Gas, 6 H.P.		29.20		32.12		
1 Vacuum Pick-Up System		60.90		66.99	5.63	6.19
16 L.H., Daily Totals		$758.50		$1105.91	$47.41	$69.12
Crew D-1	Hr.	Daily	Hr.	Daily	Bare Costs	Incl. O&P
1 Bricklayer	$46.25	$370.00	$70.60	$564.80	$42.13	$64.33
1 Bricklayer Helper	38.00	304.00	58.05	464.40		
16 L.H., Daily Totals		$674.00		$1029.20	$42.13	$64.33
Crew D-2	Hr.	Daily	Hr.	Daily	Bare Costs	Incl. O&P
3 Bricklayers	$46.25	$1110.00	$70.60	$1694.40	$43.45	$66.37
2 Bricklayer Helpers	38.00	608.00	58.05	928.80		
.5 Carpenter	48.45	193.80	74.30	297.20		
44 L.H., Daily Totals		$1911.80		$2920.40	$43.45	$66.37
Crew D-3	Hr.	Daily	Hr.	Daily	Bare Costs	Incl. O&P
3 Bricklayers	$46.25	$1110.00	$70.60	$1694.40	$43.21	$66.00
2 Bricklayer Helpers	38.00	608.00	58.05	928.80		
.25 Carpenter	48.45	96.90	74.30	148.60		
42 L.H., Daily Totals		$1814.90		$2771.80	$43.21	$66.00
Crew D-4	Hr.	Daily	Hr.	Daily	Bare Costs	Incl. O&P
1 Bricklayer	$46.25	$370.00	$70.60	$564.80	$42.85	$65.26
2 Bricklayer Helpers	38.00	608.00	58.05	928.80		
1 Equip. Oper. (light)	49.15	393.20	74.35	594.80		
1 Grout Pump, 50 C.F./hr.		135.00		148.50	4.22	4.64
32 L.H., Daily Totals		$1506.20		$2236.90	$47.07	$69.90
Crew D-5	Hr.	Daily	Hr.	Daily	Bare Costs	Incl. O&P
1 Bricklayer	46.25	370.00	70.60	564.80	46.25	70.60
8 L.H., Daily Totals		$370.00		$564.80	$46.25	$70.60
Crew D-6	Hr.	Daily	Hr.	Daily	Bare Costs	Incl. O&P
3 Bricklayers	$46.25	$1110.00	$70.60	$1694.40	$42.38	$64.72
3 Bricklayer Helpers	38.00	912.00	58.05	1393.20		
.25 Carpenter	48.45	96.90	74.30	148.60		
50 L.H., Daily Totals		$2118.90		$3236.20	$42.38	$64.72
Crew D-7	Hr.	Daily	Hr.	Daily	Bare Costs	Incl. O&P
1 Tile Layer	$44.50	$356.00	$65.75	$526.00	$39.65	$58.60
1 Tile Layer Helper	34.80	278.40	51.45	411.60		
16 L.H., Daily Totals		$634.40		$937.60	$39.65	$58.60
Crew D-8	Hr.	Daily	Hr.	Daily	Bare Costs	Incl. O&P
3 Bricklayers	$46.25	$1110.00	$70.60	$1694.40	$42.95	$65.58
2 Bricklayer Helpers	38.00	608.00	58.05	928.80		
40 L.H., Daily Totals		$1718.00		$2623.20	$42.95	$65.58
Crew D-9	Hr.	Daily	Hr.	Daily	Bare Costs	Incl. O&P
3 Bricklayers	$46.25	$1110.00	$70.60	$1694.40	$42.13	$64.33
3 Bricklayer Helpers	38.00	912.00	58.05	1393.20		
48 L.H., Daily Totals		$2022.00		$3087.60	$42.13	$64.33
Crew D-10	Hr.	Daily	Hr.	Daily	Bare Costs	Incl. O&P
1 Bricklayer Foreman (outside)	$48.25	$386.00	$73.70	$589.60	$46.19	$70.35
1 Bricklayer	46.25	370.00	70.60	564.80		
1 Bricklayer Helper	38.00	304.00	58.05	464.40		
1 Equip. Oper. (crane)	52.25	418.00	79.05	632.40		
1 S.P. Crane, 4x4, 12 Ton		479.20		527.12	14.98	16.47
32 L.H., Daily Totals		$1957.20		$2778.32	$61.16	$86.82
Crew D-11	Hr.	Daily	Hr.	Daily	Bare Costs	Incl. O&P
1 Bricklayer Foreman (outside)	$48.25	$386.00	$73.70	$589.60	$44.17	$67.45
1 Bricklayer	46.25	370.00	70.60	564.80		
1 Bricklayer Helper	38.00	304.00	58.05	464.40		
24 L.H., Daily Totals		$1060.00		$1618.80	$44.17	$67.45
Crew D-12	Hr.	Daily	Hr.	Daily	Bare Costs	Incl. O&P
1 Bricklayer Foreman (outside)	$48.25	$386.00	$73.70	$589.60	$42.63	$65.10
1 Bricklayer	46.25	370.00	70.60	564.80		
2 Bricklayer Helpers	38.00	608.00	58.05	928.80		
32 L.H., Daily Totals		$1364.00		$2083.20	$42.63	$65.10
Crew D-13	Hr.	Daily	Hr.	Daily	Bare Costs	Incl. O&P
1 Bricklayer Foreman (outside)	$48.25	$386.00	$73.70	$589.60	$45.20	$68.96
1 Bricklayer	46.25	370.00	70.60	564.80		
2 Bricklayer Helpers	38.00	608.00	58.05	928.80		
1 Carpenter	48.45	387.60	74.30	594.40		
1 Equip. Oper. (crane)	52.25	418.00	79.05	632.40		
1 S.P. Crane, 4x4, 12 Ton		479.20		527.12	9.98	10.98
48 L.H., Daily Totals		$2648.80		$3837.12	$55.18	$79.94
Crew E-1	Hr.	Daily	Hr.	Daily	Bare Costs	Incl. O&P
1 Welder Foreman (outside)	$55.20	$441.60	$93.50	$748.00	$52.52	$85.98
1 Welder	53.20	425.60	90.10	720.80		
1 Equip. Oper. (light)	49.15	393.20	74.35	594.80		
1 Welder, Gas Engine, 300 amp		147.20		161.92	6.13	6.75
24 L.H., Daily Totals		$1407.60		$2225.52	$58.65	$92.73
Crew E-2	Hr.	Daily	Hr.	Daily	Bare Costs	Incl. O&P
1 Struc. Steel Foreman (outside)	$55.20	$441.60	$93.50	$748.00	$52.33	$86.09
4 Struc. Steel Workers	53.20	1702.40	90.10	2883.20		
1 Equip. Oper. (crane)	52.25	418.00	79.05	632.40		
1 Equip. Oper. (oiler)	46.05	368.40	69.65	557.20		
1 Lattice Boom Crane, 90 Ton		1520.00		1672.00	27.14	29.86
56 L.H., Daily Totals		$4450.40		$6492.80	$79.47	$115.94
Crew E-3	Hr.	Daily	Hr.	Daily	Bare Costs	Incl. O&P
1 Struc. Steel Foreman (outside)	$55.20	$441.60	$93.50	$748.00	$53.87	$91.23
1 Struc. Steel Worker	53.20	425.60	90.10	720.80		
1 Welder	53.20	425.60	90.10	720.80		
1 Welder, Gas Engine, 300 amp		147.20		161.92	6.13	6.75
24 L.H., Daily Totals		$1440.00		$2351.52	$60.00	$97.98

Crews - Standard

Crew E-3A	Hr.	Daily	Hr.	Daily	Bare Costs	Incl. O&P
1 Struc. Steel Foreman (outside)	$55.20	$441.60	$93.50	$748.00	$53.87	$91.23
1 Struc. Steel Worker	53.20	425.60	90.10	720.80		
1 Welder	53.20	425.60	90.10	720.80		
1 Welder, Gas Engine, 300 amp		147.20		161.92		
1 Aerial Lift Truck, 40' Boom		309.00		339.90	19.01	20.91
24 L.H., Daily Totals		$1749.00		$2691.42	$72.88	$112.14

Crew E-4	Hr.	Daily	Hr.	Daily	Bare Costs	Incl. O&P
1 Struc. Steel Foreman (outside)	$55.20	$441.60	$93.50	$748.00	$53.70	$90.95
3 Struc. Steel Workers	53.20	1276.80	90.10	2162.40		
1 Welder, Gas Engine, 300 amp		147.20		161.92	4.60	5.06
32 L.H., Daily Totals		$1865.60		$3072.32	$58.30	$96.01

Crew E-5	Hr.	Daily	Hr.	Daily	Bare Costs	Incl. O&P
2 Struc. Steel Foremen (outside)	$55.20	$883.20	$93.50	$1496.00	$52.79	$87.63
5 Struc. Steel Workers	53.20	2128.00	90.10	3604.00		
1 Equip. Oper. (crane)	52.25	418.00	79.05	632.40		
1 Welder	53.20	425.60	90.10	720.80		
1 Equip. Oper. (oiler)	46.05	368.40	69.65	557.20		
1 Lattice Boom Crane, 90 Ton		1520.00		1672.00		
1 Welder, Gas Engine, 300 amp		147.20		161.92	20.84	22.92
80 L.H., Daily Totals		$5890.40		$8844.32	$73.63	$110.55

Crew E-6	Hr.	Daily	Hr.	Daily	Bare Costs	Incl. O&P
3 Struc. Steel Foremen (outside)	$55.20	$1324.80	$93.50	$2244.00	$52.82	$87.78
9 Struc. Steel Workers	53.20	3830.40	90.10	6487.20		
1 Equip. Oper. (crane)	52.25	418.00	79.05	632.40		
1 Welder	53.20	425.60	90.10	720.80		
1 Equip. Oper. (oiler)	46.05	368.40	69.65	557.20		
1 Equip. Oper. (light)	49.15	393.20	74.35	594.80		
1 Lattice Boom Crane, 90 Ton		1520.00		1672.00		
1 Welder, Gas Engine, 300 amp		147.20		161.92		
1 Air Compressor, 160 cfm		162.20		178.42		
2 Impact Wrenches		36.40		40.04	14.58	16.03
128 L.H., Daily Totals		$8626.20		$13288.78	$67.39	$103.82

Crew E-7	Hr.	Daily	Hr.	Daily	Bare Costs	Incl. O&P
1 Struc. Steel Foreman (outside)	$55.20	$441.60	$93.50	$748.00	$52.79	$87.63
4 Struc. Steel Workers	53.20	1702.40	90.10	2883.20		
1 Equip. Oper. (crane)	52.25	418.00	79.05	632.40		
1 Equip. Oper. (oiler)	46.05	368.40	69.65	557.20		
1 Welder Foreman (outside)	55.20	441.60	93.50	748.00		
2 Welders	53.20	851.20	90.10	1441.60		
1 Lattice Boom Crane, 90 Ton		1520.00		1672.00		
2 Welders, Gas Engine, 300 amp		294.40		323.84	22.68	24.95
80 L.H., Daily Totals		$6037.60		$9006.24	$75.47	$112.58

Crew E-8	Hr.	Daily	Hr.	Daily	Bare Costs	Incl. O&P
1 Struc. Steel Foreman (outside)	$55.20	$441.60	$93.50	$748.00	$52.57	$86.99
4 Struc. Steel Workers	53.20	1702.40	90.10	2883.20		
1 Welder Foreman (outside)	55.20	441.60	93.50	748.00		
4 Welders	53.20	1702.40	90.10	2883.20		
1 Equip. Oper. (crane)	52.25	418.00	79.05	632.40		
1 Equip. Oper. (oiler)	46.05	368.40	69.65	557.20		
1 Equip. Oper. (light)	49.15	393.20	74.35	594.80		
1 Lattice Boom Crane, 90 Ton		1520.00		1672.00		
4 Welders, Gas Engine, 300 amp		588.80		647.68	20.28	22.30
104 L.H., Daily Totals		$7576.40		$11366.48	$72.85	$109.29

Crew E-9	Hr.	Daily	Hr.	Daily	Bare Costs	Incl. O&P
2 Struc. Steel Foremen (outside)	$55.20	$883.20	$93.50	$1496.00	$52.82	$87.78
5 Struc. Steel Workers	53.20	2128.00	90.10	3604.00		
1 Welder Foreman (outside)	55.20	441.60	93.50	748.00		
5 Welders	53.20	2128.00	90.10	3604.00		
1 Equip. Oper. (crane)	52.25	418.00	79.05	632.40		
1 Equip. Oper. (oiler)	46.05	368.40	69.65	557.20		
1 Equip. Oper. (light)	49.15	393.20	74.35	594.80		
1 Lattice Boom Crane, 90 Ton		1520.00		1672.00		
5 Welders, Gas Engine, 300 amp		736.00		809.60	17.63	19.39
128 L.H., Daily Totals		$9016.40		$13718.00	$70.44	$107.17

Crew E-10	Hr.	Daily	Hr.	Daily	Bare Costs	Incl. O&P
1 Welder Foreman (outside)	$55.20	$441.60	$93.50	$748.00	$54.20	$91.80
1 Welder	53.20	425.60	90.10	720.80		
1 Welder, Gas Engine, 300 amp		147.20		161.92		
1 Flatbed Truck, Gas, 3 Ton		307.40		338.14	28.41	31.25
16 L.H., Daily Totals		$1321.80		$1968.86	$82.61	$123.05

Crew E-11	Hr.	Daily	Hr.	Daily	Bare Costs	Incl. O&P
2 Painters, Struc. Steel	$41.30	$660.80	$70.90	$1134.40	$42.41	$68.58
1 Building Laborer	37.90	303.20	58.15	465.20		
1 Equip. Oper. (light)	49.15	393.20	74.35	594.80		
1 Air Compressor, 250 cfm		197.80		217.58		
1 Sandblaster, Portable, 3 C.F.		20.80		22.88		
1 Set Sand Blasting Accessories		14.25		15.68	7.28	8.00
32 L.H., Daily Totals		$1590.05		$2450.53	$49.69	$76.58

Crew E-11A	Hr.	Daily	Hr.	Daily	Bare Costs	Incl. O&P
2 Painters, Struc. Steel	$41.30	$660.80	$70.90	$1134.40	$42.41	$68.58
1 Building Laborer	37.90	303.20	58.15	465.20		
1 Equip. Oper. (light)	49.15	393.20	74.35	594.80		
1 Air Compressor, 250 cfm		197.80		217.58		
1 Sandblaster, Portable, 3 C.F.		20.80		22.88		
1 Set Sand Blasting Accessories		14.25		15.68		
1 Aerial Lift Truck, 60' Boom		448.40		493.24	21.29	23.42
32 L.H., Daily Totals		$2038.45		$2943.78	$63.70	$91.99

Crew E-11B	Hr.	Daily	Hr.	Daily	Bare Costs	Incl. O&P
2 Painters, Struc. Steel	$41.30	$660.80	$70.90	$1134.40	$40.17	$66.65
1 Building Laborer	37.90	303.20	58.15	465.20		
2 Paint Sprayers, 8 C.F.M.		100.80		110.88		
1 Aerial Lift Truck, 60' Boom		448.40		493.24	22.88	25.17
24 L.H., Daily Totals		$1513.20		$2203.72	$63.05	$91.82

Crew E-12	Hr.	Daily	Hr.	Daily	Bare Costs	Incl. O&P
1 Welder Foreman (outside)	$55.20	$441.60	$93.50	$748.00	$52.17	$83.92
1 Equip. Oper. (light)	49.15	393.20	74.35	594.80		
1 Welder, Gas Engine, 300 amp		147.20		161.92	9.20	10.12
16 L.H., Daily Totals		$982.00		$1504.72	$61.38	$94.05

Crew E-13	Hr.	Daily	Hr.	Daily	Bare Costs	Incl. O&P
1 Welder Foreman (outside)	$55.20	$441.60	$93.50	$748.00	$53.18	$87.12
.5 Equip. Opers. (light)	49.15	196.60	74.35	297.40		
1 Welder, Gas Engine, 300 amp		147.20		161.92	12.27	13.49
12 L.H., Daily Totals		$785.40		$1207.32	$65.45	$100.61

Crew E-14	Hr.	Daily	Hr.	Daily	Bare Costs	Incl. O&P
1 Welder Foreman (outside)	$55.20	$441.60	$93.50	$748.00	$55.20	$93.50
1 Welder, Gas Engine, 300 amp		147.20		161.92	18.40	20.24
8 L.H., Daily Totals		$588.80		$909.92	$73.60	$113.74

Crew No.	Bare Costs		Incl. Subs O&P		Cost Per Labor-Hour	

Crew E-16	Hr.	Daily	Hr.	Daily	Bare Costs	Incl. O&P
1 Welder Foreman (outside)	$55.20	$441.60	$93.50	$748.00	$54.20	$91.80
1 Welder	53.20	425.60	90.10	720.80		
1 Welder, Gas Engine, 300 amp		147.20		161.92	9.20	10.12
16 L.H., Daily Totals		$1014.40		$1630.72	$63.40	$101.92

Crew E-17	Hr.	Daily	Hr.	Daily	Bare Costs	Incl. O&P
1 Struc. Steel Foreman (outside)	$55.20	$441.60	$93.50	$748.00	$54.20	$91.80
1 Structural Steel Worker	53.20	425.60	90.10	720.80		
16 L.H., Daily Totals		$867.20		$1468.80	$54.20	$91.80

Crew E-18	Hr.	Daily	Hr.	Daily	Bare Costs	Incl. O&P
1 Struc. Steel Foreman (outside)	$55.20	$441.60	$93.50	$748.00	$53.18	$88.22
3 Structural Steel Workers	53.20	1276.80	90.10	2162.40		
1 Equipment Operator (med.)	51.10	408.80	77.30	618.40		
1 Lattice Boom Crane, 20 Ton		967.35		1064.09	24.18	26.60
40 L.H., Daily Totals		$3094.55		$4592.89	$77.36	$114.82

Crew E-19	Hr.	Daily	Hr.	Daily	Bare Costs	Incl. O&P
1 Struc. Steel Foreman (outside)	$55.20	$441.60	$93.50	$748.00	$52.52	$85.98
1 Structural Steel Worker	53.20	425.60	90.10	720.80		
1 Equip. Oper. (light)	49.15	393.20	74.35	594.80		
1 Lattice Boom Crane, 20 Ton		967.35		1064.09	40.31	44.34
24 L.H., Daily Totals		$2227.75		$3127.68	$92.82	$130.32

Crew E-20	Hr.	Daily	Hr.	Daily	Bare Costs	Incl. O&P
1 Struc. Steel Foreman (outside)	$55.20	$441.60	$93.50	$748.00	$52.44	$86.59
5 Structural Steel Workers	53.20	2128.00	90.10	3604.00		
1 Equip. Oper. (crane)	52.25	418.00	79.05	632.40		
1 Equip. Oper. (oiler)	46.05	368.40	69.65	557.20		
1 Lattice Boom Crane, 40 Ton		1182.00		1300.20	18.47	20.32
64 L.H., Daily Totals		$4538.00		$6841.80	$70.91	$106.90

Crew E-22	Hr.	Daily	Hr.	Daily	Bare Costs	Incl. O&P
1 Skilled Worker Foreman (out)	$51.90	$415.20	$79.95	$639.60	$50.57	$77.88
2 Skilled Workers	49.90	798.40	76.85	1229.60		
24 L.H., Daily Totals		$1213.60		$1869.20	$50.57	$77.88

Crew E-24	Hr.	Daily	Hr.	Daily	Bare Costs	Incl. O&P
3 Structural Steel Workers	$53.20	$1276.80	$90.10	$2162.40	$52.67	$86.90
1 Equipment Operator (med.)	51.10	408.80	77.30	618.40		
1 Hyd. Crane, 25 Ton		748.80		823.68	23.40	25.74
32 L.H., Daily Totals		$2434.40		$3604.48	$76.08	$112.64

Crew E-25	Hr.	Daily	Hr.	Daily	Bare Costs	Incl. O&P
1 Welder Foreman (outside)	$55.20	$441.60	$93.50	$748.00	$55.20	$93.50
1 Cutting Torch		11.40		12.54	1.43	1.57
8 L.H., Daily Totals		$453.00		$760.54	$56.63	$95.07

Crew E-26	Hr.	Daily	Hr.	Daily	Bare Costs	Incl. O&P
1 Struc. Steel Foreman (outside)	$55.20	$441.60	$93.50	$748.00	$54.34	$90.47
1 Struc. Steel Worker	53.20	425.60	90.10	720.80		
1 Welder	53.20	425.60	90.10	720.80		
.25 Electrician	55.10	110.20	82.45	164.90		
.25 Plumber	59.20	118.40	89.35	178.70		
1 Welder, Gas Engine, 300 amp		147.20		161.92	5.26	5.78
28 L.H., Daily Totals		$1668.60		$2695.12	$59.59	$96.25

Crew F-3	Hr.	Daily	Hr.	Daily	Bare Costs	Incl. O&P
4 Carpenters	$48.45	$1550.40	$74.30	$2377.60	$49.21	$75.25
1 Equip. Oper. (crane)	52.25	418.00	79.05	632.40		
1 Hyd. Crane, 12 Ton		658.40		724.24	16.46	18.11
40 L.H., Daily Totals		$2626.80		$3734.24	$65.67	$93.36

Crew F-4	Hr.	Daily	Hr.	Daily	Bare Costs	Incl. O&P
4 Carpenters	$48.45	$1550.40	$74.30	$2377.60	$48.68	$74.32
1 Equip. Oper. (crane)	52.25	418.00	79.05	632.40		
1 Equip. Oper. (oiler)	46.05	368.40	69.65	557.20		
1 Hyd. Crane, 55 Ton		1123.00		1235.30	23.40	25.74
48 L.H., Daily Totals		$3459.80		$4802.50	$72.08	$100.05

Crew F-5	Hr.	Daily	Hr.	Daily	Bare Costs	Incl. O&P
1 Carpenter Foreman (outside)	$50.45	$403.60	$77.40	$619.20	$48.95	$75.08
3 Carpenters	48.45	1162.80	74.30	1783.20		
32 L.H., Daily Totals		$1566.40		$2402.40	$48.95	$75.08

Crew F-6	Hr.	Daily	Hr.	Daily	Bare Costs	Incl. O&P
2 Carpenters	$48.45	$775.20	$74.30	$1188.80	$44.99	$68.79
2 Building Laborers	37.90	606.40	58.15	930.40		
1 Equip. Oper. (crane)	52.25	418.00	79.05	632.40		
1 Hyd. Crane, 12 Ton		658.40		724.24	16.46	18.11
40 L.H., Daily Totals		$2458.00		$3475.84	$61.45	$86.90

Crew F-7	Hr.	Daily	Hr.	Daily	Bare Costs	Incl. O&P
2 Carpenters	$48.45	$775.20	$74.30	$1188.80	$43.17	$66.22
2 Building Laborers	37.90	606.40	58.15	930.40		
32 L.H., Daily Totals		$1381.60		$2119.20	$43.17	$66.22

Crew G-1	Hr.	Daily	Hr.	Daily	Bare Costs	Incl. O&P
1 Roofer Foreman (outside)	$43.70	$349.60	$74.45	$595.60	$39.03	$66.49
4 Roofers Composition	41.70	1334.40	71.05	2273.60		
2 Roofer Helpers	31.35	501.60	53.40	854.40		
1 Application Equipment		193.00		212.30		
1 Tar Kettle/Pot		176.40		194.04		
1 Crew Truck		208.80		229.68	10.32	11.36
56 L.H., Daily Totals		$2763.80		$4359.62	$49.35	$77.85

Crew G-2	Hr.	Daily	Hr.	Daily	Bare Costs	Incl. O&P
1 Plasterer	$44.90	$359.20	$67.15	$537.20	$40.28	$60.73
1 Plasterer Helper	38.05	304.40	56.90	455.20		
1 Building Laborer	37.90	303.20	58.15	465.20		
1 Grout Pump, 50 C.F./hr.		135.00		148.50	5.63	6.19
24 L.H., Daily Totals		$1101.80		$1606.10	$45.91	$66.92

Crew G-2A	Hr.	Daily	Hr.	Daily	Bare Costs	Incl. O&P
1 Roofer Composition	$41.70	$333.60	$71.05	$568.40	$36.98	$60.87
1 Roofer Helper	31.35	250.80	53.40	427.20		
1 Building Laborer	37.90	303.20	58.15	465.20		
1 Foam Spray Rig, Trailer-Mtd.		567.00		623.70		
1 Pickup Truck, 3/4 Ton		145.80		160.38	29.70	32.67
24 L.H., Daily Totals		$1600.40		$2244.88	$66.68	$93.54

Crew G-3	Hr.	Daily	Hr.	Daily	Bare Costs	Incl. O&P
2 Sheet Metal Workers	$57.25	$916.00	$87.50	$1400.00	$47.58	$72.83
2 Building Laborers	37.90	606.40	58.15	930.40		
32 L.H., Daily Totals		$1522.40		$2330.40	$47.58	$72.83

Crew G-4

Crew G-4	Hr.	Daily	Hr.	Daily	Bare Costs	Incl. O&P
1 Labor Foreman (outside)	$39.90	$319.20	$61.20	$489.60	$38.57	$59.17
2 Building Laborers	37.90	606.40	58.15	930.40		
1 Flatbed Truck, Gas, 1.5 Ton		248.20		273.02		
1 Air Compressor, 160 cfm		162.20		178.42	17.10	18.81
24 L.H., Daily Totals		$1336.00		$1871.44	$55.67	$77.98

Crew G-5

Crew G-5	Hr.	Daily	Hr.	Daily	Bare Costs	Incl. O&P
1 Roofer Foreman (outside)	$43.70	$349.60	$74.45	$595.60	$37.96	$64.67
2 Roofers Composition	41.70	667.20	71.05	1136.80		
2 Roofer Helpers	31.35	501.60	53.40	854.40		
1 Application Equipment		193.00		212.30	4.83	5.31
40 L.H., Daily Totals		$1711.40		$2799.10	$42.78	$69.98

Crew G-6A

Crew G-6A	Hr.	Daily	Hr.	Daily	Bare Costs	Incl. O&P
2 Roofers Composition	$41.70	$667.20	$71.05	$1136.80	$41.70	$71.05
1 Small Compressor, Electric		14.10		15.51		
2 Pneumatic Nailers		52.40		57.64	4.16	4.57
16 L.H., Daily Totals		$733.70		$1209.95	$45.86	$75.62

Crew G-7

Crew G-7	Hr.	Daily	Hr.	Daily	Bare Costs	Incl. O&P
1 Carpenter	$48.45	$387.60	$74.30	$594.40	$48.45	$74.30
1 Small Compressor, Electric		14.10		15.51		
1 Pneumatic Nailer		26.20		28.82	5.04	5.54
8 L.H., Daily Totals		$427.90		$638.73	$53.49	$79.84

Crew H-1

Crew H-1	Hr.	Daily	Hr.	Daily	Bare Costs	Incl. O&P
2 Glaziers	$46.70	$747.20	$71.05	$1136.80	$49.95	$80.58
2 Struc. Steel Workers	53.20	851.20	90.10	1441.60		
32 L.H., Daily Totals		$1598.40		$2578.40	$49.95	$80.58

Crew H-2

Crew H-2	Hr.	Daily	Hr.	Daily	Bare Costs	Incl. O&P
2 Glaziers	$46.70	$747.20	$71.05	$1136.80	$43.77	$66.75
1 Building Laborer	37.90	303.20	58.15	465.20		
24 L.H., Daily Totals		$1050.40		$1602.00	$43.77	$66.75

Crew H-3

Crew H-3	Hr.	Daily	Hr.	Daily	Bare Costs	Incl. O&P
1 Glazier	$46.70	$373.60	$71.05	$568.40	$41.35	$63.40
1 Helper	36.00	288.00	55.75	446.00		
16 L.H., Daily Totals		$661.60		$1014.40	$41.35	$63.40

Crew H-4

Crew H-4	Hr.	Daily	Hr.	Daily	Bare Costs	Incl. O&P
1 Carpenter	$48.45	$387.60	$74.30	$594.40	$44.80	$68.51
1 Carpenter Helper	36.00	288.00	55.75	446.00		
.5 Electrician	55.10	220.40	82.45	329.80		
20 L.H., Daily Totals		$896.00		$1370.20	$44.80	$68.51

Crew J-1

Crew J-1	Hr.	Daily	Hr.	Daily	Bare Costs	Incl. O&P
3 Plasterers	$44.90	$1077.60	$67.15	$1611.60	$42.16	$63.05
2 Plasterer Helpers	38.05	608.80	56.90	910.40		
1 Mixing Machine, 6 C.F.		140.20		154.22	3.50	3.86
40 L.H., Daily Totals		$1826.60		$2676.22	$45.66	$66.91

Crew J-2

Crew J-2	Hr.	Daily	Hr.	Daily	Bare Costs	Incl. O&P
3 Plasterers	$44.90	$1077.60	$67.15	$1611.60	$43.02	$64.17
2 Plasterer Helpers	38.05	608.80	56.90	910.40		
1 Lather	47.30	378.40	69.80	558.40		
1 Mixing Machine, 6 C.F.		140.20		154.22	2.92	3.21
48 L.H., Daily Totals		$2205.00		$3234.62	$45.94	$67.39

Crew J-3

Crew J-3	Hr.	Daily	Hr.	Daily	Bare Costs	Incl. O&P
1 Terrazzo Worker	$44.60	$356.80	$65.90	$527.20	$41.00	$60.60
1 Terrazzo Helper	37.40	299.20	55.30	442.40		
1 Floor Grinder, 22" Path		121.60		133.76		
1 Terrazzo Mixer		188.20		207.02	19.36	21.30
16 L.H., Daily Totals		$965.80		$1310.38	$60.36	$81.90

Crew J-4

Crew J-4	Hr.	Daily	Hr.	Daily	Bare Costs	Incl. O&P
2 Cement Finishers	$45.65	$730.40	$67.70	$1083.20	$43.07	$64.52
1 Laborer	37.90	303.20	58.15	465.20		
1 Floor Grinder, 22" Path		121.60		133.76		
1 Floor Edger, 7" Path		39.30		43.23		
1 Vacuum Pick-Up System		60.90		66.99	9.24	10.17
24 L.H., Daily Totals		$1255.40		$1792.38	$52.31	$74.68

Crew J-4A

Crew J-4A	Hr.	Daily	Hr.	Daily	Bare Costs	Incl. O&P
2 Cement Finishers	$45.65	$730.40	$67.70	$1083.20	$41.77	$62.92
2 Laborers	37.90	606.40	58.15	930.40		
1 Floor Grinder, 22" Path		121.60		133.76		
1 Floor Edger, 7" Path		39.30		43.23		
1 Vacuum Pick-Up System		60.90		66.99		
1 Floor Auto Scrubber		241.05		265.15	14.46	15.91
32 L.H., Daily Totals		$1799.65		$2522.74	$56.24	$78.84

Crew J-4B

Crew J-4B	Hr.	Daily	Hr.	Daily	Bare Costs	Incl. O&P
1 Laborer	$37.90	$303.20	$58.15	$465.20	$37.90	$58.15
1 Floor Auto Scrubber		241.05		265.15	30.13	33.14
8 L.H., Daily Totals		$544.25		$730.36	$68.03	$91.29

Crew J-6

Crew J-6	Hr.	Daily	Hr.	Daily	Bare Costs	Incl. O&P
2 Painters	$40.35	$645.60	$60.80	$972.80	$41.94	$63.52
1 Building Laborer	37.90	303.20	58.15	465.20		
1 Equip. Oper. (light)	49.15	393.20	74.35	594.80		
1 Air Compressor, 250 cfm		197.80		217.58		
1 Sandblaster, Portable, 3 C.F.		20.80		22.88		
1 Set Sand Blasting Accessories		14.25		15.68	7.28	8.00
32 L.H., Daily Totals		$1574.85		$2288.93	$49.21	$71.53

Crew J-7

Crew J-7	Hr.	Daily	Hr.	Daily	Bare Costs	Incl. O&P
2 Painters	$40.35	$645.60	$60.80	$972.80	$40.35	$60.80
1 Floor Belt Sander		24.45		26.90		
1 Floor Sanding Edger		13.65		15.02	2.38	2.62
16 L.H., Daily Totals		$683.70		$1014.71	$42.73	$63.42

Crew K-1

Crew K-1	Hr.	Daily	Hr.	Daily	Bare Costs	Incl. O&P
1 Carpenter	$48.45	$387.60	$74.30	$594.40	$45.20	$68.83
1 Truck Driver (light)	41.95	335.60	63.35	506.80		
1 Flatbed Truck, Gas, 3 Ton		307.40		338.14	19.21	21.13
16 L.H., Daily Totals		$1030.60		$1439.34	$64.41	$89.96

Crew K-2

Crew K-2	Hr.	Daily	Hr.	Daily	Bare Costs	Incl. O&P
1 Struc. Steel Foreman (outside)	$55.20	$441.60	$93.50	$748.00	$50.12	$82.32
1 Struc. Steel Worker	53.20	425.60	90.10	720.80		
1 Truck Driver (light)	41.95	335.60	63.35	506.80		
1 Flatbed Truck, Gas, 3 Ton		307.40		338.14	12.81	14.09
24 L.H., Daily Totals		$1510.20		$2313.74	$62.92	$96.41

Crew L-1

Crew L-1	Hr.	Daily	Hr.	Daily	Bare Costs	Incl. O&P
1 Electrician	$55.10	$440.80	$82.45	$659.60	$57.15	$85.90
1 Plumber	59.20	473.60	89.35	714.80		
16 L.H., Daily Totals		$914.40		$1374.40	$57.15	$85.90

Crew No.	Bare Costs		Incl. Subs O&P		Cost Per Labor-Hour	
Crew L-2	Hr.	Daily	Hr.	Daily	Bare Costs	Incl. O&P
1 Carpenter	$48.45	$387.60	$74.30	$594.40	$42.23	$65.03
1 Carpenter Helper	36.00	288.00	55.75	446.00		
16 L.H., Daily Totals		$675.60		$1040.40	$42.23	$65.03
Crew L-3	Hr.	Daily	Hr.	Daily	Bare Costs	Incl. O&P
1 Carpenter	$48.45	$387.60	$74.30	$594.40	$52.31	$79.64
.5 Electrician	55.10	220.40	82.45	329.80		
.5 Sheet Metal Worker	57.25	229.00	87.50	350.00		
16 L.H., Daily Totals		$837.00		$1274.20	$52.31	$79.64
Crew L-3A	Hr.	Daily	Hr.	Daily	Bare Costs	Incl. O&P
1 Carpenter Foreman (outside)	$50.45	$403.60	$77.40	$619.20	$52.72	$80.77
.5 Sheet Metal Worker	57.25	229.00	87.50	350.00		
12 L.H., Daily Totals		$632.60		$969.20	$52.72	$80.77
Crew L-4	Hr.	Daily	Hr.	Daily	Bare Costs	Incl. O&P
2 Skilled Workers	$49.90	$798.40	$76.85	$1229.60	$45.27	$69.82
1 Helper	36.00	288.00	55.75	446.00		
24 L.H., Daily Totals		$1086.40		$1675.60	$45.27	$69.82
Crew L-5	Hr.	Daily	Hr.	Daily	Bare Costs	Incl. O&P
1 Struc. Steel Foreman (outside)	$55.20	$441.60	$93.50	$748.00	$53.35	$89.01
5 Struc. Steel Workers	53.20	2128.00	90.10	3604.00		
1 Equip. Oper. (crane)	52.25	418.00	79.05	632.40		
1 Hyd. Crane, 25 Ton		748.80		823.68	13.37	14.71
56 L.H., Daily Totals		$3736.40		$5808.08	$66.72	$103.72
Crew L-5A	Hr.	Daily	Hr.	Daily	Bare Costs	Incl. O&P
1 Struc. Steel Foreman (outside)	$55.20	$441.60	$93.50	$748.00	$53.46	$88.19
2 Structural Steel Workers	53.20	851.20	90.10	1441.60		
1 Equip. Oper. (crane)	52.25	418.00	79.05	632.40		
1 S.P. Crane, 4x4, 25 Ton		606.60		667.26	18.96	20.85
32 L.H., Daily Totals		$2317.40		$3489.26	$72.42	$109.04
Crew L-5B	Hr.	Daily	Hr.	Daily	Bare Costs	Incl. O&P
1 Struc. Steel Foreman (outside)	$55.20	$441.60	$93.50	$748.00	$54.61	$85.61
2 Structural Steel Workers	53.20	851.20	90.10	1441.60		
2 Electricians	55.10	881.60	82.45	1319.20		
2 Steamfitters/Pipefitters	60.70	971.20	91.60	1465.60		
1 Equip. Oper. (crane)	52.25	418.00	79.05	632.40		
1 Equip. Oper. (oiler)	46.05	368.40	69.65	557.20		
1 Hyd. Crane, 80 Ton		1630.00		1793.00	22.64	24.90
72 L.H., Daily Totals		$5562.00		$7957.00	$77.25	$110.51
Crew L-6	Hr.	Daily	Hr.	Daily	Bare Costs	Incl. O&P
1 Plumber	$59.20	$473.60	$89.35	$714.80	$57.83	$87.05
.5 Electrician	55.10	220.40	82.45	329.80		
12 L.H., Daily Totals		$694.00		$1044.60	$57.83	$87.05
Crew L-7	Hr.	Daily	Hr.	Daily	Bare Costs	Incl. O&P
2 Carpenters	$48.45	$775.20	$74.30	$1188.80	$46.39	$70.85
1 Building Laborer	37.90	303.20	58.15	465.20		
.5 Electrician	55.10	220.40	82.45	329.80		
28 L.H., Daily Totals		$1298.80		$1983.80	$46.39	$70.85
Crew L-8	Hr.	Daily	Hr.	Daily	Bare Costs	Incl. O&P
2 Carpenters	$48.45	$775.20	$74.30	$1188.80	$50.60	$77.31
.5 Plumber	59.20	236.80	89.35	357.40		
20 L.H., Daily Totals		$1012.00		$1546.20	$50.60	$77.31

Crew No.	Bare Costs		Incl. Subs O&P		Cost Per Labor-Hour	
Crew L-9	Hr.	Daily	Hr.	Daily	Bare Costs	Incl. O&P
1 Labor Foreman (inside)	$38.40	$307.20	$58.90	$471.20	$43.32	$68.12
2 Building Laborers	37.90	606.40	58.15	930.40		
1 Struc. Steel Worker	53.20	425.60	90.10	720.80		
.5 Electrician	55.10	220.40	82.45	329.80		
36 L.H., Daily Totals		$1559.60		$2452.20	$43.32	$68.12
Crew L-10	Hr.	Daily	Hr.	Daily	Bare Costs	Incl. O&P
1 Struc. Steel Foreman (outside)	$55.20	$441.60	$93.50	$748.00	$53.55	$87.55
1 Structural Steel Worker	53.20	425.60	90.10	720.80		
1 Equip. Oper. (crane)	52.25	418.00	79.05	632.40		
1 Hyd. Crane, 12 Ton		658.40		724.24	27.43	30.18
24 L.H., Daily Totals		$1943.60		$2825.44	$80.98	$117.73
Crew L-11	Hr.	Daily	Hr.	Daily	Bare Costs	Incl. O&P
2 Wreckers	$37.90	$606.40	$61.35	$981.60	$44.30	$69.03
1 Equip. Oper. (crane)	52.25	418.00	79.05	632.40		
1 Equip. Oper. (light)	49.15	393.20	74.35	594.80		
1 Hyd. Excavator, 2.5 C.Y.		1582.00		1740.20		
1 Loader, Skid Steer, 78 H.P.		315.60		347.16	59.30	65.23
32 L.H., Daily Totals		$3315.20		$4296.16	$103.60	$134.26
Crew M-1	Hr.	Daily	Hr.	Daily	Bare Costs	Incl. O&P
3 Elevator Constructors	$79.10	$1898.40	$118.00	$2832.00	$75.15	$112.11
1 Elevator Apprentice	63.30	506.40	94.45	755.60		
5 Hand Tools		48.00		52.80	1.50	1.65
32 L.H., Daily Totals		$2452.80		$3640.40	$76.65	$113.76
Crew M-3	Hr.	Daily	Hr.	Daily	Bare Costs	Incl. O&P
1 Electrician Foreman (outside)	$57.10	$456.80	$85.40	$683.20	$58.86	$88.31
1 Common Laborer	37.90	303.20	58.15	465.20		
.25 Equipment Operator (med.)	51.10	102.20	77.30	154.60		
1 Elevator Constructor	79.10	632.80	118.00	944.00		
1 Elevator Apprentice	63.30	506.40	94.45	755.60		
.25 S.P. Crane, 4x4, 20 Ton		138.70		152.57	4.08	4.49
34 L.H., Daily Totals		$2140.10		$3155.17	$62.94	$92.80
Crew M-4	Hr.	Daily	Hr.	Daily	Bare Costs	Incl. O&P
1 Electrician Foreman (outside)	$57.10	$456.80	$85.40	$683.20	$58.22	$87.37
1 Common Laborer	37.90	303.20	58.15	465.20		
.25 Equipment Operator, Crane	52.25	104.50	79.05	158.10		
.25 Equip. Oper. (oiler)	46.05	92.10	69.65	139.30		
1 Elevator Constructor	79.10	632.80	118.00	944.00		
1 Elevator Apprentice	63.30	506.40	94.45	755.60		
.25 S.P. Crane, 4x4, 40 Ton		177.95		195.75	4.94	5.44
36 L.H., Daily Totals		$2273.75		$3341.15	$63.16	$92.81
Crew Q-1	Hr.	Daily	Hr.	Daily	Bare Costs	Incl. O&P
1 Plumber	$59.20	$473.60	$89.35	$714.80	$53.27	$80.40
1 Plumber Apprentice	47.35	378.80	71.45	571.60		
16 L.H., Daily Totals		$852.40		$1286.40	$53.27	$80.40
Crew Q-1A	Hr.	Daily	Hr.	Daily	Bare Costs	Incl. O&P
.25 Plumber Foreman (outside)	$61.20	$122.40	$92.35	$184.70	$59.60	$89.95
1 Plumber	59.20	473.60	89.35	714.80		
10 L.H., Daily Totals		$596.00		$899.50	$59.60	$89.95

For customer support on your **Mechanical Cost Data**, call 877.739.3689.

651

Crew No.	Bare Costs		Incl. Subs O&P		Cost Per Labor-Hour	

Crew Q-1C	Hr.	Daily	Hr.	Daily	Bare Costs	Incl. O&P
1 Plumber	$59.20	$473.60	$89.35	$714.80	$52.55	$79.37
1 Plumber Apprentice	47.35	378.80	71.45	571.60		
1 Equip. Oper. (medium)	51.10	408.80	77.30	618.40		
1 Trencher, Chain Type, 8' D		2860.00		3146.00	119.17	131.08
24 L.H., Daily Totals		$4121.20		$5050.80	$171.72	$210.45

Crew Q-2	Hr.	Daily	Hr.	Daily	Bare Costs	Incl. O&P
2 Plumbers	$59.20	$947.20	$89.35	$1429.60	$55.25	$83.38
1 Plumber Apprentice	47.35	378.80	71.45	571.60		
24 L.H., Daily Totals		$1326.00		$2001.20	$55.25	$83.38

Crew Q-3	Hr.	Daily	Hr.	Daily	Bare Costs	Incl. O&P
1 Plumber Foreman (inside)	$59.70	$477.60	$90.10	$720.80	$56.36	$85.06
2 Plumbers	59.20	947.20	89.35	1429.60		
1 Plumber Apprentice	47.35	378.80	71.45	571.60		
32 L.H., Daily Totals		$1803.60		$2722.00	$56.36	$85.06

Crew Q-4	Hr.	Daily	Hr.	Daily	Bare Costs	Incl. O&P
1 Plumber Foreman (inside)	$59.70	$477.60	$90.10	$720.80	$56.36	$85.06
1 Plumber	59.20	473.60	89.35	714.80		
1 Welder (plumber)	59.20	473.60	89.35	714.80		
1 Plumber Apprentice	47.35	378.80	71.45	571.60		
1 Welder, Electric, 300 amp		58.10		63.91	1.82	2.00
32 L.H., Daily Totals		$1861.70		$2785.91	$58.18	$87.06

Crew Q-5	Hr.	Daily	Hr.	Daily	Bare Costs	Incl. O&P
1 Steamfitter	$60.70	$485.60	$91.60	$732.80	$54.63	$82.42
1 Steamfitter Apprentice	48.55	388.40	73.25	586.00		
16 L.H., Daily Totals		$874.00		$1318.80	$54.63	$82.42

Crew Q-6	Hr.	Daily	Hr.	Daily	Bare Costs	Incl. O&P
2 Steamfitters	$60.70	$971.20	$91.60	$1465.60	$56.65	$85.48
1 Steamfitter Apprentice	48.55	388.40	73.25	586.00		
24 L.H., Daily Totals		$1359.60		$2051.60	$56.65	$85.48

Crew Q-7	Hr.	Daily	Hr.	Daily	Bare Costs	Incl. O&P
1 Steamfitter Foreman (inside)	$61.20	$489.60	$92.35	$738.80	$57.79	$87.20
2 Steamfitters	60.70	971.20	91.60	1465.60		
1 Steamfitter Apprentice	48.55	388.40	73.25	586.00		
32 L.H., Daily Totals		$1849.20		$2790.40	$57.79	$87.20

Crew Q-8	Hr.	Daily	Hr.	Daily	Bare Costs	Incl. O&P
1 Steamfitter Foreman (inside)	$61.20	$489.60	$92.35	$738.80	$57.79	$87.20
1 Steamfitter	60.70	485.60	91.60	732.80		
1 Welder (steamfitter)	60.70	485.60	91.60	732.80		
1 Steamfitter Apprentice	48.55	388.40	73.25	586.00		
1 Welder, Electric, 300 amp		58.10		63.91	1.82	2.00
32 L.H., Daily Totals		$1907.30		$2854.31	$59.60	$89.20

Crew Q-9	Hr.	Daily	Hr.	Daily	Bare Costs	Incl. O&P
1 Sheet Metal Worker	$57.25	$458.00	$87.50	$700.00	$51.52	$78.75
1 Sheet Metal Apprentice	45.80	366.40	70.00	560.00		
16 L.H., Daily Totals		$824.40		$1260.00	$51.52	$78.75

Crew Q-10	Hr.	Daily	Hr.	Daily	Bare Costs	Incl. O&P
2 Sheet Metal Workers	$57.25	$916.00	$87.50	$1400.00	$53.43	$81.67
1 Sheet Metal Apprentice	45.80	366.40	70.00	560.00		
24 L.H., Daily Totals		$1282.40		$1960.00	$53.43	$81.67

Crew No.	Bare Costs		Incl. Subs O&P		Cost Per Labor-Hour	

Crew Q-11	Hr.	Daily	Hr.	Daily	Bare Costs	Incl. O&P
1 Sheet Metal Foreman (inside)	$57.75	$462.00	$88.25	$706.00	$54.51	$83.31
2 Sheet Metal Workers	57.25	916.00	87.50	1400.00		
1 Sheet Metal Apprentice	45.80	366.40	70.00	560.00		
32 L.H., Daily Totals		$1744.40		$2666.00	$54.51	$83.31

Crew Q-12	Hr.	Daily	Hr.	Daily	Bare Costs	Incl. O&P
1 Sprinkler Installer	$57.60	$460.80	$87.30	$698.40	$51.85	$78.60
1 Sprinkler Apprentice	46.10	368.80	69.90	559.20		
16 L.H., Daily Totals		$829.60		$1257.60	$51.85	$78.60

Crew Q-13	Hr.	Daily	Hr.	Daily	Bare Costs	Incl. O&P
1 Sprinkler Foreman (inside)	$58.10	$464.80	$88.10	$704.80	$54.85	$83.15
2 Sprinkler Installers	57.60	921.60	87.30	1396.80		
1 Sprinkler Apprentice	46.10	368.80	69.90	559.20		
32 L.H., Daily Totals		$1755.20		$2660.80	$54.85	$83.15

Crew Q-14	Hr.	Daily	Hr.	Daily	Bare Costs	Incl. O&P
1 Asbestos Worker	$53.40	$427.20	$83.10	$664.80	$48.05	$74.78
1 Asbestos Apprentice	42.70	341.60	66.45	531.60		
16 L.H., Daily Totals		$768.80		$1196.40	$48.05	$74.78

Crew Q-15	Hr.	Daily	Hr.	Daily	Bare Costs	Incl. O&P
1 Plumber	$59.20	$473.60	$89.35	$714.80	$53.27	$80.40
1 Plumber Apprentice	47.35	378.80	71.45	571.60		
1 Welder, Electric, 300 amp		58.10		63.91	3.63	3.99
16 L.H., Daily Totals		$910.50		$1350.31	$56.91	$84.39

Crew Q-16	Hr.	Daily	Hr.	Daily	Bare Costs	Incl. O&P
2 Plumbers	$59.20	$947.20	$89.35	$1429.60	$55.25	$83.38
1 Plumber Apprentice	47.35	378.80	71.45	571.60		
1 Welder, Electric, 300 amp		58.10		63.91	2.42	2.66
24 L.H., Daily Totals		$1384.10		$2065.11	$57.67	$86.05

Crew Q-17	Hr.	Daily	Hr.	Daily	Bare Costs	Incl. O&P
1 Steamfitter	$60.70	$485.60	$91.60	$732.80	$54.63	$82.42
1 Steamfitter Apprentice	48.55	388.40	73.25	586.00		
1 Welder, Electric, 300 amp		58.10		63.91	3.63	3.99
16 L.H., Daily Totals		$932.10		$1382.71	$58.26	$86.42

Crew Q-17A	Hr.	Daily	Hr.	Daily	Bare Costs	Incl. O&P
1 Steamfitter	$60.70	$485.60	$91.60	$732.80	$53.83	$81.30
1 Steamfitter Apprentice	48.55	388.40	73.25	586.00		
1 Equip. Oper. (crane)	52.25	418.00	79.05	632.40		
1 Hyd. Crane, 12 Ton		658.40		724.24		
1 Welder, Electric, 300 amp		58.10		63.91	29.85	32.84
24 L.H., Daily Totals		$2008.50		$2739.35	$83.69	$114.14

Crew Q-18	Hr.	Daily	Hr.	Daily	Bare Costs	Incl. O&P
2 Steamfitters	$60.70	$971.20	$91.60	$1465.60	$56.65	$85.48
1 Steamfitter Apprentice	48.55	388.40	73.25	586.00		
1 Welder, Electric, 300 amp		58.10		63.91	2.42	2.66
24 L.H., Daily Totals		$1417.70		$2115.51	$59.07	$88.15

Crew Q-19	Hr.	Daily	Hr.	Daily	Bare Costs	Incl. O&P
1 Steamfitter	$60.70	$485.60	$91.60	$732.80	$54.78	$82.43
1 Steamfitter Apprentice	48.55	388.40	73.25	586.00		
1 Electrician	55.10	440.80	82.45	659.60		
24 L.H., Daily Totals		$1314.80		$1978.40	$54.78	$82.43

For customer support on your Mechanical Cost Data, call 877.739.3689.

Crew No.	Bare Costs		Incl. Subs O&P		Cost Per Labor-Hour	

Crew Q-20	Hr.	Daily	Hr.	Daily	Bare Costs	Incl. O&P
1 Sheet Metal Worker	$57.25	$458.00	$87.50	$700.00	$52.24	$79.49
1 Sheet Metal Apprentice	45.80	366.40	70.00	560.00		
.5 Electrician	55.10	220.40	82.45	329.80		
20 L.H., Daily Totals		$1044.80		$1589.80	$52.24	$79.49

Crew Q-21	Hr.	Daily	Hr.	Daily	Bare Costs	Incl. O&P
2 Steamfitters	$60.70	$971.20	$91.60	$1465.60	$56.26	$84.72
1 Steamfitter Apprentice	48.55	388.40	73.25	586.00		
1 Electrician	55.10	440.80	82.45	659.60		
32 L.H., Daily Totals		$1800.40		$2711.20	$56.26	$84.72

Crew Q-22	Hr.	Daily	Hr.	Daily	Bare Costs	Incl. O&P
1 Plumber	$59.20	$473.60	$89.35	$714.80	$53.27	$80.40
1 Plumber Apprentice	47.35	378.80	71.45	571.60		
1 Hyd. Crane, 12 Ton		658.40		724.24	41.15	45.27
16 L.H., Daily Totals		$1510.80		$2010.64	$94.42	$125.67

Crew Q-22A	Hr.	Daily	Hr.	Daily	Bare Costs	Incl. O&P
1 Plumber	$59.20	$473.60	$89.35	$714.80	$49.17	$74.50
1 Plumber Apprentice	47.35	378.80	71.45	571.60		
1 Laborer	37.90	303.20	58.15	465.20		
1 Equip. Oper. (crane)	52.25	418.00	79.05	632.40		
1 Hyd. Crane, 12 Ton		658.40		724.24	20.57	22.63
32 L.H., Daily Totals		$2232.00		$3108.24	$69.75	$97.13

Crew Q-23	Hr.	Daily	Hr.	Daily	Bare Costs	Incl. O&P
1 Plumber Foreman (outside)	$61.20	$489.60	$92.35	$738.80	$57.17	$86.33
1 Plumber	59.20	473.60	89.35	714.80		
1 Equip. Oper. (medium)	51.10	408.80	77.30	618.40		
1 Lattice Boom Crane, 20 Ton		967.35		1064.09	40.31	44.34
24 L.H., Daily Totals		$2339.35		$3136.09	$97.47	$130.67

Crew R-1	Hr.	Daily	Hr.	Daily	Bare Costs	Incl. O&P
1 Electrician Foreman	$55.60	$444.80	$83.20	$665.60	$51.52	$77.08
3 Electricians	55.10	1322.40	82.45	1978.80		
2 Electrician Apprentices	44.10	705.60	65.95	1055.20		
48 L.H., Daily Totals		$2472.80		$3699.60	$51.52	$77.08

Crew R-1A	Hr.	Daily	Hr.	Daily	Bare Costs	Incl. O&P
1 Electrician	$55.10	$440.80	$82.45	$659.60	$49.60	$74.20
1 Electrician Apprentice	44.10	352.80	65.95	527.60		
16 L.H., Daily Totals		$793.60		$1187.20	$49.60	$74.20

Crew R-1B	Hr.	Daily	Hr.	Daily	Bare Costs	Incl. O&P
1 Electrician	$55.10	$440.80	$82.45	$659.60	$47.77	$71.45
2 Electrician Apprentices	44.10	705.60	65.95	1055.20		
24 L.H., Daily Totals		$1146.40		$1714.80	$47.77	$71.45

Crew R-1C	Hr.	Daily	Hr.	Daily	Bare Costs	Incl. O&P
2 Electricians	$55.10	$881.60	$82.45	$1319.20	$49.60	$74.20
2 Electrician Apprentices	44.10	705.60	65.95	1055.20		
1 Portable cable puller, 8000 lb.		130.10		143.11	4.07	4.47
32 L.H., Daily Totals		$1717.30		$2517.51	$53.67	$78.67

Crew R-2	Hr.	Daily	Hr.	Daily	Bare Costs	Incl. O&P
1 Electrician Foreman	$55.60	$444.80	$83.20	$665.60	$51.62	$77.36
3 Electricians	55.10	1322.40	82.45	1978.80		
2 Electrician Apprentices	44.10	705.60	65.95	1055.20		
1 Equip. Oper. (crane)	52.25	418.00	79.05	632.40		
1 S.P. Crane, 4x4, 5 Ton		280.80		308.88	5.01	5.52
56 L.H., Daily Totals		$3171.60		$4640.88	$56.64	$82.87

Crew R-3	Hr.	Daily	Hr.	Daily	Bare Costs	Incl. O&P
1 Electrician Foreman	$55.60	$444.80	$83.20	$665.60	$54.73	$82.07
1 Electrician	55.10	440.80	82.45	659.60		
.5 Equip. Oper. (crane)	52.25	209.00	79.05	316.20		
.5 S.P. Crane, 4x4, 5 Ton		140.40		154.44	7.02	7.72
20 L.H., Daily Totals		$1235.00		$1795.84	$61.75	$89.79

Crew R-4	Hr.	Daily	Hr.	Daily	Bare Costs	Incl. O&P
1 Struc. Steel Foreman (outside)	$55.20	$441.60	$93.50	$748.00	$53.98	$89.25
3 Struc. Steel Workers	53.20	1276.80	90.10	2162.40		
1 Electrician	55.10	440.80	82.45	659.60		
1 Welder, Gas Engine, 300 amp		147.20		161.92	3.68	4.05
40 L.H., Daily Totals		$2306.40		$3731.92	$57.66	$93.30

Crew R-5	Hr.	Daily	Hr.	Daily	Bare Costs	Incl. O&P
1 Electrician Foreman	$55.60	$444.80	$83.20	$665.60	$48.20	$72.81
4 Electrician Linemen	55.10	1763.20	82.45	2638.40		
2 Electrician Operators	55.10	881.60	82.45	1319.20		
4 Electrician Groundmen	36.00	1152.00	55.75	1784.00		
1 Crew Truck		208.80		229.68		
1 Flatbed Truck, 20,000 GVW		247.80		272.58		
1 Pickup Truck, 3/4 Ton		145.80		160.38		
.2 Hyd. Crane, 55 Ton		224.60		247.06		
.2 Hyd. Crane, 12 Ton		131.68		144.85		
.2 Earth Auger, Truck-Mtd.		84.16		92.58		
1 Tractor w/Winch		428.00		470.80	16.71	18.39
88 L.H., Daily Totals		$5712.44		$8025.12	$64.91	$91.19

Crew R-6	Hr.	Daily	Hr.	Daily	Bare Costs	Incl. O&P
1 Electrician Foreman	$55.60	$444.80	$83.20	$665.60	$48.20	$72.81
4 Electrician Linemen	55.10	1763.20	82.45	2638.40		
2 Electrician Operators	55.10	881.60	82.45	1319.20		
4 Electrician Groundmen	36.00	1152.00	55.75	1784.00		
1 Crew Truck		208.80		229.68		
1 Flatbed Truck, 20,000 GVW		247.80		272.58		
1 Pickup Truck, 3/4 Ton		145.80		160.38		
.2 Hyd. Crane, 55 Ton		224.60		247.06		
.2 Hyd. Crane, 12 Ton		131.68		144.85		
.2 Earth Auger, Truck-Mtd.		84.16		92.58		
1 Tractor w/Winch		428.00		470.80		
3 Cable Trailers		610.65		671.72		
.5 Tensioning Rig		202.25		222.47		
.5 Cable Pulling Rig		1170.50		1287.55	39.25	43.18
88 L.H., Daily Totals		$7695.84		$10206.86	$87.45	$115.99

Crew R-7	Hr.	Daily	Hr.	Daily	Bare Costs	Incl. O&P
1 Electrician Foreman	$55.60	$444.80	$83.20	$665.60	$39.27	$60.33
5 Electrician Groundmen	36.00	1440.00	55.75	2230.00		
1 Crew Truck		208.80		229.68	4.35	4.79
48 L.H., Daily Totals		$2093.60		$3125.28	$43.62	$65.11

For customer support on your **Mechanical Cost Data**, call 877.739.3689.

653

Crew No.	Bare Costs				Incl. Subs O&P				Cost Per Labor-Hour	
Crew R-8	Hr.	Daily	Hr.	Daily					Bare Costs	Incl. O&P
1 Electrician Foreman	$55.60	$444.80	$83.20	$665.60					$48.82	$73.67
3 Electrician Linemen	55.10	1322.40	82.45	1978.80						
2 Electrician Groundmen	36.00	576.00	55.75	892.00						
1 Pickup Truck, 3/4 Ton		145.80		160.38						
1 Crew Truck		208.80		229.68					7.39	8.13
48 L.H., Daily Totals		$2697.80		$3926.46					$56.20	$81.80
Crew R-9	Hr.	Daily	Hr.	Daily					Bare Costs	Incl. O&P
1 Electrician Foreman	$55.60	$444.80	$83.20	$665.60					$45.61	$69.19
1 Electrician Lineman	55.10	440.80	82.45	659.60						
2 Electrician Operators	55.10	881.60	82.45	1319.20						
4 Electrician Groundmen	36.00	1152.00	55.75	1784.00						
1 Pickup Truck, 3/4 Ton		145.80		160.38						
1 Crew Truck		208.80		229.68					5.54	6.09
64 L.H., Daily Totals		$3273.80		$4818.46					$51.15	$75.29
Crew R-10	Hr.	Daily	Hr.	Daily					Bare Costs	Incl. O&P
1 Electrician Foreman	$55.60	$444.80	$83.20	$665.60					$52.00	$78.13
4 Electrician Linemen	55.10	1763.20	82.45	2638.40						
1 Electrician Groundman	36.00	288.00	55.75	446.00						
1 Crew Truck		208.80		229.68						
3 Tram Cars		418.50		460.35					13.07	14.38
48 L.H., Daily Totals		$3123.30		$4440.03					$65.07	$92.50
Crew R-11	Hr.	Daily	Hr.	Daily					Bare Costs	Incl. O&P
1 Electrician Foreman	$55.60	$444.80	$83.20	$665.60					$52.31	$78.60
4 Electricians	55.10	1763.20	82.45	2638.40						
1 Equip. Oper. (crane)	52.25	418.00	79.05	632.40						
1 Common Laborer	37.90	303.20	58.15	465.20						
1 Crew Truck		208.80		229.68						
1 Hyd. Crane, 12 Ton		658.40		724.24					15.49	17.03
56 L.H., Daily Totals		$3796.40		$5355.52					$67.79	$95.63
Crew R-12	Hr.	Daily	Hr.	Daily					Bare Costs	Incl. O&P
1 Carpenter Foreman (inside)	$48.95	$391.60	$75.10	$600.80					$45.33	$70.21
4 Carpenters	48.45	1550.40	74.30	2377.60						
4 Common Laborers	37.90	1212.80	58.15	1860.80						
1 Equip. Oper. (medium)	51.10	408.80	77.30	618.40						
1 Steel Worker	53.20	425.60	90.10	720.80						
1 Dozer, 200 H.P.		1397.00		1536.70						
1 Pickup Truck, 3/4 Ton		145.80		160.38					17.53	19.29
88 L.H., Daily Totals		$5532.00		$7875.48					$62.86	$89.49
Crew R-13	Hr.	Daily	Hr.	Daily					Bare Costs	Incl. O&P
1 Electrician Foreman	$55.60	$444.80	$83.20	$665.60					$53.34	$79.99
3 Electricians	55.10	1322.40	82.45	1978.80						
.25 Equip. Oper. (crane)	52.25	104.50	79.05	158.10						
1 Equipment Oiler	46.05	368.40	69.65	557.20						
.25 Hydraulic Crane, 33 Ton		190.80		209.88					4.54	5.00
42 L.H., Daily Totals		$2430.90		$3569.58					$57.88	$84.99
Crew R-15	Hr.	Daily	Hr.	Daily					Bare Costs	Incl. O&P
1 Electrician Foreman	$55.60	$444.80	$83.20	$665.60					$54.19	$81.22
4 Electricians	55.10	1763.20	82.45	2638.40						
1 Equipment Oper. (light)	49.15	393.20	74.35	594.80						
1 Aerial Lift Truck, 40' Boom		309.00		339.90					6.44	7.08
48 L.H., Daily Totals		$2910.20		$4238.70					$60.63	$88.31

Crew No.	Bare Costs				Incl. Subs O&P				Cost Per Labor-Hour	
Crew R-15A	Hr.	Daily	Hr.	Daily					Bare Costs	Incl. O&P
1 Electrician Foreman	$55.60	$444.80	$83.20	$665.60					$48.46	$73.13
2 Electricians	55.10	881.60	82.45	1319.20						
2 Common Laborers	37.90	606.40	58.15	930.40						
1 Equip. Oper. (light)	49.15	393.20	74.35	594.80						
1 Aerial Lift Truck, 40' Boom		309.00		339.90					6.44	7.08
48 L.H., Daily Totals		$2635.00		$3849.90					$54.90	$80.21
Crew R-18	Hr.	Daily	Hr.	Daily					Bare Costs	Incl. O&P
.25 Electrician Foreman	$55.60	$111.20	$83.20	$166.40					$48.37	$72.35
1 Electrician	55.10	440.80	82.45	659.60						
2 Electrician Apprentices	44.10	705.60	65.95	1055.20						
26 L.H., Daily Totals		$1257.60		$1881.20					$48.37	$72.35
Crew R-19	Hr.	Daily	Hr.	Daily					Bare Costs	Incl. O&P
.5 Electrician Foreman	$55.60	$222.40	$83.20	$332.80					$55.20	$82.60
2 Electricians	55.10	881.60	82.45	1319.20						
20 L.H., Daily Totals		$1104.00		$1652.00					$55.20	$82.60
Crew R-21	Hr.	Daily	Hr.	Daily					Bare Costs	Incl. O&P
1 Electrician Foreman	$55.60	$444.80	$83.20	$665.60					$55.12	$82.51
3 Electricians	55.10	1322.40	82.45	1978.80						
.1 Equip. Oper. (medium)	51.10	40.88	77.30	61.84						
.1 S.P. Crane, 4x4, 25 Ton		60.66		66.73					1.85	2.03
32.8 L.H., Daily Totals		$1868.74	*	$2772.97					$56.97	$84.54
Crew R-22	Hr.	Daily	Hr.	Daily					Bare Costs	Incl. O&P
.66 Electrician Foreman	$55.60	$293.57	$83.20	$439.30					$50.45	$75.47
2 Electricians	55.10	881.60	82.45	1319.20						
2 Electrician Apprentices	44.10	705.60	65.95	1055.20						
37.28 L.H., Daily Totals		$1880.77		$2813.70					$50.45	$75.47
Crew R-30	Hr.	Daily	Hr.	Daily					Bare Costs	Incl. O&P
.25 Electrician Foreman (outside)	$57.10	$114.20	$85.40	$170.80					$44.67	$67.72
1 Electrician	55.10	440.80	82.45	659.60						
2 Laborers, (Semi-Skilled)	37.90	606.40	58.15	930.40						
26 L.H., Daily Totals		$1161.40		$1760.80					$44.67	$67.72
Crew R-31	Hr.	Daily	Hr.	Daily					Bare Costs	Incl. O&P
1 Electrician	$55.10	$440.80	$82.45	$659.60					$55.10	$82.45
1 Core Drill, Electric, 2.5 H.P.		64.20		70.62					8.03	8.83
8 L.H., Daily Totals		$505.00		$730.22					$63.13	$91.28
Crew W-41E	Hr.	Daily	Hr.	Daily					Bare Costs	Incl. O&P
.5 Plumber Foreman (outside)	$61.20	$244.80	$92.35	$369.40					$51.08	$77.47
1 Plumber	59.20	473.60	89.35	714.80						
1 Laborer	37.90	303.20	58.15	465.20						
20 L.H., Daily Totals		$1021.60		$1549.40					$51.08	$77.47

Historical Cost Indexes

The table below lists both the RSMeans® historical cost index based on Jan. 1, 1993 = 100 as well as the computed value of an index based on Jan. 1, 2016 costs. Since the Jan. 1, 2016 figure is estimated, space is left to write in the actual index figures as they become available through the quarterly *RSMeans Construction Cost Indexes*.

To compute the actual index based on Jan. 1, 2016 = 100, divide the historical cost index for a particular year by the actual Jan. 1, 2016 construction cost index. Space has been left to advance the index figures as the year progresses.

Year	Historical Cost Index Jan. 1, 1993 = 100		Current Index Based on Jan. 1, 2016 = 100		Year	Historical Cost Index Jan. 1, 1993 = 100	Current Index Based on Jan. 1, 2016 = 100		Year	Historical Cost Index Jan. 1, 1993 = 100	Current Index Based on Jan. 1, 2016 = 100	
	Est.	Actual	Est.	Actual		Actual	Est.	Actual		Actual	Est.	Actual
Oct 2016*					July 2001	125.1	60.4		July 1983	80.2	38.7	
July 2016*					2000	120.9	58.3		1982	76.1	36.8	
April 2016*					1999	117.6	56.8		1981	70.0	33.8	
Jan 2016*	207.2		100.0	100.0	1998	115.1	55.6		1980	62.9	30.4	
July 2015		206.2			1997	112.8	54.4		1979	57.8	27.9	
2014		204.9	98.9		1996	110.2	53.2		1978	53.5	25.8	
2013		201.2	97.1		1995	107.6	51.9		1977	49.5	23.9	
2012		194.6	93.9		1994	104.4	50.4		1976	46.9	22.6	
2011		191.2	92.3		1993	101.7	49.1		1975	44.8	21.6	
2010		183.5	88.6		1992	99.4	48.0		1974	41.4	20.0	
2009		180.1	86.9		1991	96.8	46.7		1973	37.7	18.2	
2008		180.4	87.1		1990	94.3	45.5		1972	34.8	16.8	
2007		169.4	81.8		1989	92.1	44.5		1971	32.1	15.5	
2006		162.0	78.2		1988	89.9	43.4		1970	28.7	13.9	
2005		151.6	73.2		1987	87.7	42.3		1969	26.9	13.0	
2004		143.7	69.4		1986	84.2	40.7		1968	24.9	12.0	
2003		132.0	63.7		1985	82.6	39.9		1967	23.5	11.3	
2002		128.7	62.1		1984	82.0	39.6		1966	22.7	11.0	

Adjustments to Costs

The "Historical Cost Index" can be used to convert national average building costs at a particular time to the approximate building costs for some other time.

Example:

Estimate and compare construction costs for different years in the same city.

To estimate the national average construction cost of a building in 1970, knowing that it cost $900,000 in 2016:

INDEX in 1970 = 28.7

INDEX in 2016 = 207.2

Note: The city cost indexes for Canada can be used to convert U.S. national averages to local costs in Canadian dollars.

Example:

To estimate and compare the cost of a building in Toronto, ON in 2016 with the known cost of $600,000 (US$) in New York, NY in 2016:

INDEX Toronto = 109.9

INDEX New York = 131.1

$$\frac{\text{INDEX Toronto}}{\text{INDEX New York}} \times \text{Cost New York} = \text{Cost Toronto}$$

$$\frac{109.9}{131.1} \times \$600,000 = .841 \times \$600,000 = \$502,975$$

The construction cost of the building in Toronto is $502,975 (CN$).

Time Adjustment Using the Historical Cost Indexes:

$$\frac{\text{Index for Year A}}{\text{Index for Year B}} \times \text{Cost in Year B} = \text{Cost in Year A}$$

$$\frac{\text{INDEX 1970}}{\text{INDEX 2016}} \times \text{Cost 2016} = \text{Cost 1970}$$

$$\frac{28.7}{207.2} \times \$900,000 = .139 \times \$900,000 = \$124,662$$

The construction cost of the building in 1970 was $124,662.

*Historical Cost Index updates and other resources are provided on the following website:
http://info.thegordiangroup.com/RSMeans.html

655

For customer support on your Mechanical Cost Data, call 877.739.3689.

How to Use the City Cost Indexes

What you should know before you begin

RSMeans City Cost Indexes (CCI) are an extremely useful tool for when you want to compare costs from city to city and region to region.

This publication contains average construction cost indexes for 731 U.S. and Canadian cities covering over 930 three-digit zip code locations, as listed directly under each city.

Keep in mind that a City Cost Index number is a percentage ratio of a specific city's cost to the national average cost of the same item at a stated time period.

In other words, these index figures represent relative construction factors (or, if you prefer, multipliers) for material and installation costs, as well as the weighted average for Total In Place costs for each CSI MasterFormat division. Installation costs include both labor and equipment rental costs. When estimating equipment rental rates only for a specific location, use 01 54 33 EQUIPMENT RENTAL COSTS in the Reference Section.

The 30 City Average Index is the average of 30 major U.S. cities and serves as a national average.

Index figures for both material and installation are based on the 30 major city average of 100 and represent the cost relationship as of July 1, 2015. The index for each division is computed from representative material and labor quantities for that division. The weighted average for each city is a weighted total of the components listed above it. It does not include relative productivity between trades or cities.

As changes occur in local material prices, labor rates, and equipment rental rates (including fuel costs), the impact of these changes should be accurately measured by the change in the City Cost Index for each particular city (as compared to the 30 city average).

Therefore, if you know (or have estimated) building costs in one city today, you can easily convert those costs to expected building costs in another city.

In addition, by using the Historical Cost Index, you can easily convert national average building costs at a particular time to the approximate building costs for some other time. The City Cost Indexes can then be applied to calculate the costs for a particular city.

Quick calculations

Location Adjustment Using the City Cost Indexes:

$$\frac{\text{Index for City A}}{\text{Index for City B}} \times \text{Cost in City B} = \text{Cost in City A}$$

Time Adjustment for the National Average Using the Historical Cost Index:

$$\frac{\text{Index for Year A}}{\text{Index for Year B}} \times \text{Cost in Year B} = \text{Cost in Year A}$$

Adjustment from the National Average:

$$\frac{\text{Index for City A}}{100} \times \text{National Average Cost} = \text{Cost in City A}$$

Since each of the other RSMeans data sets contain many different items, any *one* item multiplied by the particular city index may give incorrect results. However, the larger the number of items compiled, the closer the results should be to actual costs for that particular city.

The City Cost Indexes for Canadian cities are calculated using Canadian material and equipment prices and labor rates in Canadian dollars. Therefore, indexes for Canadian cities can be used to convert U.S. national average prices to local costs in Canadian dollars.

How to use this section

1. Compare costs from city to city.

In using the RSMeans Indexes, remember that an index number is not a fixed number but a ratio: It's a percentage ratio of a building component's cost at any stated time to the national average cost of that same component at the same time period. Put in the form of an equation:

$$\frac{\text{Specific City Cost}}{\text{National Average Cost}} \times 100 = \text{City Index Number}$$

Therefore, when making cost comparisons between cities, do not subtract one city's index number from the index number of another city and read the result as a percentage difference. Instead, divide one city's index number by that of the other city. The resulting number may then be used as a multiplier to calculate cost differences from city to city.

The formula used to find cost differences between cities for the purpose of comparison is as follows:

$$\frac{\text{City A Index}}{\text{City B Index}} \times \text{City B Cost (Known)} = \text{City A Cost (Unknown)}$$

In addition, you can use RSMeans CCI to calculate and compare costs division by division between cities using the same basic formula. (Just be sure that you're comparing similar divisions.)

2. Compare a specific city's construction costs with the national average.

When you're studying construction location feasibility, it's advisable to compare a prospective project's cost index with an index of the national average cost.

For example, divide the weighted average index of construction costs of a specific city by that of the 30 City Average, which = 100.

$$\frac{\text{City Index}}{100} = \text{\% of National Average}$$

As a result, you get a ratio that indicates the relative cost of construction in that city in comparison with the national average.

3. Convert U.S. national average to actual costs in Canadian City.

$$\frac{\text{Index for Canadian City}}{100} \times \text{National Average Cost} = \text{Cost in Canadian City in \$ CAN}$$

4. Adjust construction cost data based on a national average.

When you use a source of construction cost data which is based on a national average (such as RSMeans cost data), it is necessary to adjust those costs to a specific location.

$$\frac{\text{City Index}}{100} \times \frac{\text{Cost Based on}}{\text{National Average Costs}} = \frac{\text{City Cost}}{\text{(Unknown)}}$$

5. When applying the City Cost Indexes to demolition projects, use the appropriate division installation index. For example, for removal of existing doors and windows, use the Division 8 (Openings) index.

What you might like to know about how we developed the Indexes

The information presented in the CCI is organized according to the Construction Specifications Institute (CSI) MasterFormat 2014 classification system.

To create a reliable index, RSMeans researched the building type most often constructed in the United States and Canada. Because it was concluded that no one type of building completely represented the building construction industry, nine different types of buildings were combined to create a composite model.

The exact material, labor, and equipment quantities are based on detailed analyses of these nine building types, and then each quantity is weighted in proportion to expected usage. These various material items, labor hours, and equipment rental rates are thus combined to form a composite building representing as closely as possible the actual usage of materials, labor, and equipment in the North American building construction industry.

The following structures were chosen to make up that composite model:

1. Factory, 1 story
2. Office, 2–4 stories
3. Store, Retail
4. Town Hall, 2–3 stories
5. High School, 2–3 stories
6. Hospital, 4–8 stories
7. Garage, Parking
8. Apartment, 1–3 stories
9. Hotel/Motel, 2–3 stories

For the purposes of ensuring the timeliness of the data, the components of the index for the composite model have been streamlined. They currently consist of:

- specific quantities of 66 commonly used construction materials;
- specific labor-hours for 21 building construction trades; and
- specific days of equipment rental for 6 types of construction equipment (normally used to install the 66 material items by the 21 trades.) Fuel costs and routine maintenance costs are included in the equipment cost.

A sophisticated computer program handles the updating of all costs for each city on a quarterly basis. Material and equipment price quotations are gathered quarterly from cities in the United States and Canada. These prices and the latest negotiated labor wage rates for 21 different building trades are used to compile the quarterly update of the City Cost Index.

The 30 major U.S. cities used to calculate the national average are:

Atlanta, GA	Memphis, TN
Baltimore, MD	Milwaukee, WI
Boston, MA	Minneapolis, MN
Buffalo, NY	Nashville, TN
Chicago, IL	New Orleans, LA
Cincinnati, OH	New York, NY
Cleveland, OH	Philadelphia, PA
Columbus, OH	Phoenix, AZ
Dallas, TX	Pittsburgh, PA
Denver, CO	St. Louis, MO
Detroit, MI	San Antonio, TX
Houston, TX	San Diego, CA
Indianapolis, IN	San Francisco, CA
Kansas City, MO	Seattle, WA
Los Angeles, CA	Washington, DC

What the CCI does not indicate

The weighted average for each city is a total of the divisional components weighted to reflect typical usage. It does not include the productivity variations between trades or cities.

In addition, the CCI does not take into consideration factors such as the following:

- managerial efficiency
- competitive conditions
- automation
- restrictive union practices
- unique local requirements
- regional variations due to specific building codes

City Cost Indexes

ALABAMA (and United States Average)

DIVISION		UNITED STATES 30 CITY AVERAGE			ANNISTON 362			BIRMINGHAM 350 - 352			BUTLER 369			DECATUR 356			DOTHAN 363		
		MAT.	INST.	TOTAL	MAT.	INST.	TOTAL	MAT.	INST.	TOTAL	MAT.	INST.	TOTAL	MAT.	INST.	TOTAL	MAT.	INST.	TOTAL
015433	CONTRACTOR EQUIPMENT		100.0	100.0		101.9	101.9		102.0	102.0		99.3	99.3		101.9	101.9		99.3	99.3
0241, 31 - 34	SITE & INFRASTRUCTURE, DEMOLITION	100.0	100.0	100.0	89.0	92.6	91.6	94.6	92.9	93.4	103.2	88.6	92.9	87.5	92.5	91.1	100.8	88.6	92.2
0310	Concrete Forming & Accessories	100.0	100.0	100.0	90.2	78.0	79.6	94.5	78.3	80.5	86.5	77.6	78.8	97.0	72.2	75.5	95.6	75.9	78.5
0320	Concrete Reinforcing	100.0	100.0	100.0	87.9	88.3	88.1	97.0	88.3	92.5	92.9	88.2	90.5	90.7	81.5	85.9	92.9	88.3	90.5
0330	Cast-in-Place Concrete	100.0	100.0	100.0	100.5	73.7	89.9	102.9	74.4	91.7	98.0	73.3	88.2	101.1	73.0	90.0	98.0	73.2	88.2
03	CONCRETE	100.0	100.0	100.0	104.1	79.4	92.1	101.8	79.8	91.1	101.5	79.0	92.4	100.0	75.3	88.0	104.3	78.2	91.6
04	MASONRY	100.0	100.0	100.0	99.7	77.7	86.2	96.3	79.1	85.7	104.9	74.7	86.3	94.3	76.6	83.4	106.4	76.7	88.1
05	METALS	100.0	100.0	100.0	95.1	96.0	95.4	95.3	95.9	95.5	94.1	94.1	94.1	97.2	93.4	96.0	94.1	94.0	94.1
06	WOOD, PLASTICS & COMPOSITES	100.0	100.0	100.0	91.9	78.9	84.7	98.0	78.9	87.4	86.6	78.9	82.3	105.1	71.8	86.6	98.5	76.8	86.4
07	THERMAL & MOISTURE PROTECTION	100.0	100.0	100.0	97.2	85.3	92.2	102.2	84.5	94.7	97.2	83.2	91.4	100.6	83.0	93.2	97.2	84.6	91.9
08	OPENINGS	100.0	100.0	100.0	98.4	77.1	93.5	107.4	78.6	100.7	98.5	75.5	93.1	110.2	71.7	101.3	98.5	74.4	92.9
0920	Plaster & Gypsum Board	100.0	100.0	100.0	83.5	78.7	80.3	89.8	78.7	82.4	80.8	78.7	79.4	90.8	71.5	77.9	90.5	76.6	81.2
0950, 0980	Ceilings & Acoustic Treatment	100.0	100.0	100.0	77.8	78.7	78.4	85.5	78.7	81.0	77.8	78.7	78.4	83.0	71.5	75.3	77.8	76.6	77.0
0960	Flooring	100.0	100.0	100.0	92.2	76.7	87.8	98.7	76.7	92.4	97.4	76.7	91.5	95.7	76.7	90.2	103.2	76.7	95.6
0970, 0990	Wall Finishes & Painting/Coating	100.0	100.0	100.0	90.4	69.1	78.0	88.1	55.9	69.4	90.4	69.1	78.0	84.3	63.7	72.3	90.4	69.1	78.0
09	FINISHES	100.0	100.0	100.0	85.2	76.8	80.6	91.9	75.7	82.9	88.4	76.6	81.9	88.5	71.8	79.2	91.3	74.1	81.8
COVERS	DIVS. 10 - 14, 25, 28, 41, 43, 44, 46	100.0	100.0	100.0	100.0	87.3	97.2	100.0	89.2	97.8	100.0	86.9	97.2	100.0	86.2	97.0	100.0	86.7	97.1
21, 22, 23	FIRE SUPPRESSION, PLUMBING & HVAC	100.0	100.0	100.0	100.9	69.8	87.7	100.0	67.3	86.2	98.1	70.7	86.5	100.0	69.2	87.0	98.1	69.2	85.9
26, 27, 3370	ELECTRICAL, COMMUNICATIONS & UTIL.	100.0	100.0	100.0	92.2	62.5	77.2	99.1	63.4	81.1	93.9	62.5	78.0	94.9	67.0	80.8	92.9	62.7	77.7
MF2014	WEIGHTED AVERAGE	100.0	100.0	100.0	97.3	77.5	88.6	99.3	77.3	89.6	97.6	76.7	88.4	98.7	76.0	88.7	97.7	76.1	88.2

ALABAMA

DIVISION		EVERGREEN 364			GADSDEN 359			HUNTSVILLE 357 - 358			JASPER 355			MOBILE 365 - 366			MONTGOMERY 360 - 361		
		MAT.	INST.	TOTAL	MAT.	INST.	TOTAL	MAT.	INST.	TOTAL	MAT.	INST.	TOTAL	MAT.	INST.	TOTAL	MAT.	INST.	TOTAL
015433	CONTRACTOR EQUIPMENT		99.3	99.3		101.9	101.9		101.9	101.9		101.9	101.9		99.3	99.3		99.3	99.3
0241, 31 - 34	SITE & INFRASTRUCTURE, DEMOLITION	103.7	88.5	93.0	93.6	92.6	92.9	87.2	92.5	90.9	93.3	92.6	92.8	95.8	89.8	91.6	93.9	88.4	90.0
0310	Concrete Forming & Accessories	83.1	75.7	76.7	88.9	51.2	56.3	97.0	71.7	75.1	94.3	76.7	79.0	94.6	78.9	81.1	93.7	77.9	80.0
0320	Concrete Reinforcing	93.0	81.4	87.0	96.2	88.2	92.0	90.7	85.0	87.8	90.7	87.0	88.8	90.7	88.2	89.4	98.9	88.2	93.4
0330	Cast-in-Place Concrete	98.0	73.2	88.2	101.1	73.4	90.2	98.4	72.2	88.1	112.0	73.7	96.9	102.7	73.4	91.1	99.3	73.7	89.2
03	CONCRETE	105.4	77.0	91.6	104.8	67.3	86.6	98.7	75.5	87.4	108.5	78.6	94.0	100.6	79.8	90.5	100.3	79.3	90.1
04	MASONRY	104.9	76.7	87.5	92.7	77.7	83.5	95.7	75.1	83.0	90.2	59.0	71.0	103.1	74.7	85.6	100.1	75.6	85.0
05	METALS	94.1	93.9	94.0	95.1	95.3	95.2	97.2	94.7	96.4	95.1	92.8	94.3	96.1	96.7	96.3	95.3	95.8	95.5
06	WOOD, PLASTICS & COMPOSITES	83.1	76.8	79.6	95.7	43.7	66.7	105.1	71.8	86.6	102.2	78.9	89.2	97.2	80.5	87.9	95.1	78.9	86.1
07	THERMAL & MOISTURE PROTECTION	97.2	81.9	90.8	100.7	80.5	92.2	100.5	82.4	92.9	100.7	74.4	89.7	96.9	83.4	91.2	95.6	83.7	90.6
08	OPENINGS	98.5	74.4	92.9	106.5	52.6	94.0	109.9	72.5	101.2	106.5	73.6	98.8	101.5	78.0	96.0	101.4	77.1	95.7
0920	Plaster & Gypsum Board	80.2	76.6	77.8	83.5	42.5	56.1	90.8	71.5	77.9	87.5	78.7	81.6	87.6	80.4	82.8	87.0	78.7	81.5
0950, 0980	Ceilings & Acoustic Treatment	77.8	76.6	77.0	80.5	42.5	55.3	84.8	71.5	76.0	80.5	78.7	79.3	83.0	80.4	81.2	85.3	78.7	80.9
0960	Flooring	95.1	76.7	89.8	91.5	76.7	87.2	95.7	76.7	90.2	93.8	38.7	78.0	102.7	73.4	94.3	99.5	73.4	92.0
0970, 0990	Wall Finishes & Painting/Coating	90.4	69.1	78.0	84.3	71.7	77.0	84.3	69.1	75.5	84.3	38.6	57.7	93.7	69.1	79.4	90.1	71.7	79.4
09	FINISHES	87.7	75.3	80.9	86.1	56.1	69.5	88.9	72.0	79.6	87.3	66.6	75.9	91.4	76.9	83.4	91.2	76.5	83.0
COVERS	DIVS. 10 - 14, 25, 28, 41, 43, 44, 46	100.0	86.7	97.1	100.0	85.5	96.9	100.0	87.8	97.4	100.0	49.8	89.1	100.0	90.1	97.9	100.0	89.4	97.7
21, 22, 23	FIRE SUPPRESSION, PLUMBING & HVAC	98.1	58.2	81.2	102.1	48.5	79.4	100.0	65.6	85.5	102.1	66.6	87.1	99.9	69.2	86.9	100.0	68.8	86.8
26, 27, 3370	ELECTRICAL, COMMUNICATIONS & UTIL.	91.6	62.5	76.9	94.9	63.4	79.0	95.8	67.0	81.3	94.6	63.4	78.8	94.6	59.1	76.6	95.1	64.6	79.6
MF2014	WEIGHTED AVERAGE	97.3	73.6	86.9	98.8	67.1	84.8	98.7	75.3	88.4	99.2	71.8	87.1	98.3	76.6	88.7	97.9	77.0	88.7

ALABAMA / ALASKA

DIVISION		PHENIX CITY 368			SELMA 367			TUSCALOOSA 354			ANCHORAGE 995 - 996			FAIRBANKS 997			JUNEAU 998		
		MAT.	INST.	TOTAL	MAT.	INST.	TOTAL	MAT.	INST.	TOTAL	MAT.	INST.	TOTAL	MAT.	INST.	TOTAL	MAT.	INST.	TOTAL
015433	CONTRACTOR EQUIPMENT		99.3	99.3		99.3	99.3		101.9	101.9		114.7	114.7		114.7	114.7		114.7	114.7
0241, 31 - 34	SITE & INFRASTRUCTURE, DEMOLITION	107.6	88.4	94.0	100.7	88.4	92.0	87.8	92.6	91.2	129.5	130.0	129.9	123.3	130.1	128.1	139.9	130.0	132.9
0310	Concrete Forming & Accessories	90.2	76.3	78.2	87.7	77.3	78.7	96.9	78.0	80.6	119.9	119.2	119.2	128.7	119.5	120.7	130.4	119.2	120.7
0320	Concrete Reinforcing	92.9	81.5	87.0	92.9	88.2	90.5	90.7	88.3	89.5	148.9	116.0	131.8	149.8	116.0	132.3	145.4	116.0	130.2
0330	Cast-in-Place Concrete	98.0	73.6	88.4	98.0	73.4	88.2	102.6	73.7	91.2	120.0	117.7	119.1	123.9	118.4	121.7	129.7	117.7	125.0
03	CONCRETE	108.4	77.4	93.3	103.8	79.0	91.7	100.7	79.4	90.4	130.5	117.3	124.1	120.9	117.7	119.3	134.1	117.3	125.9
04	MASONRY	104.9	77.7	88.1	108.4	77.7	89.5	94.6	77.7	84.2	187.9	125.8	149.7	197.2	126.5	153.6	182.4	125.8	147.6
05	METALS	94.1	93.5	93.9	94.1	95.4	94.5	96.4	96.1	96.3	129.0	104.9	121.4	133.5	104.9	124.4	132.1	104.9	123.4
06	WOOD, PLASTICS & COMPOSITES	91.7	76.8	83.4	88.6	78.9	83.2	105.1	78.9	90.5	117.7	117.7	117.7	129.9	117.7	123.1	126.2	117.7	121.5
07	THERMAL & MOISTURE PROTECTION	97.6	84.0	91.9	97.1	84.2	91.7	100.6	84.2	93.7	156.3	118.3	140.4	171.8	117.9	149.3	177.7	118.3	152.9
08	OPENINGS	98.4	74.4	92.8	98.4	77.1	93.5	109.9	78.6	102.6	130.0	115.8	126.7	138.1	116.1	133.0	130.1	115.8	126.8
0920	Plaster & Gypsum Board	84.8	76.6	79.3	82.8	78.7	80.1	90.8	78.7	82.7	137.8	117.9	124.5	164.8	117.9	133.5	151.3	117.9	129.0
0950, 0980	Ceilings & Acoustic Treatment	77.8	76.6	77.0	77.8	78.7	78.4	84.8	78.7	80.8	121.0	117.9	119.0	124.7	117.9	120.2	133.8	117.9	123.3
0960	Flooring	99.6	73.4	92.0	97.9	28.7	78.0	95.7	76.7	90.2	131.4	128.1	130.4	125.6	128.1	126.3	131.5	128.1	130.5
0970, 0990	Wall Finishes & Painting/Coating	90.4	71.7	79.5	90.4	69.1	78.0	84.3	71.7	77.0	110.5	116.5	114.0	108.6	121.3	116.0	108.2	116.5	113.0
09	FINISHES	90.1	75.2	81.9	88.6	67.5	76.9	88.9	77.1	82.4	130.8	120.8	125.3	131.2	121.5	125.8	134.6	120.8	127.0
COVERS	DIVS. 10 - 14, 25, 28, 41, 43, 44, 46	100.0	89.2	97.7	100.0	87.3	97.2	100.0	89.4	97.7	100.0	112.3	102.7	100.0	112.5	102.7	100.0	112.3	102.7
21, 22, 23	FIRE SUPPRESSION, PLUMBING & HVAC	98.1	68.7	85.7	98.1	65.4	84.3	100.0	71.3	87.9	100.5	105.8	102.7	100.5	109.7	104.4	100.6	105.8	102.8
26, 27, 3370	ELECTRICAL, COMMUNICATIONS & UTIL.	93.4	70.7	81.9	92.7	62.5	77.4	95.4	63.4	79.2	107.1	117.3	112.3	117.5	117.3	117.4	109.5	117.3	113.4
MF2014	WEIGHTED AVERAGE	98.2	77.2	89.0	97.4	74.9	87.5	98.7	78.1	89.6	121.7	115.8	119.1	124.2	116.9	120.9	123.9	115.8	120.3

Table 1

	DIVISION	ALASKA			ARIZONA													
		KETCHIKAN			CHAMBERS			FLAGSTAFF			GLOBE			KINGMAN			MESA/TEMPE		
		999			865			860			855			864			852		
		MAT.	INST.	TOTAL	MAT.	INST.	TOTAL	MAT.	INST.	TOTAL	MAT.	INST.	TOTAL	MAT.	INST.	TOTAL	MAT.	INST.	TOTAL
015433	CONTRACTOR EQUIPMENT		114.7	114.7		91.0	91.0		91.0	91.0		92.0	92.0		91.0	91.0		92.0	92.0
0241, 31 - 34	SITE & INFRASTRUCTURE, DEMOLITION	181.7	130.0	145.2	72.3	95.6	88.8	91.5	95.7	94.5	106.0	96.4	99.2	72.2	95.7	88.8	96.0	96.5	96.4
0310	Concrete Forming & Accessories	120.2	119.2	119.3	99.1	56.7	62.4	104.8	68.7	73.6	96.9	56.8	62.2	97.3	63.3	67.9	100.1	67.5	71.9
0320	Concrete Reinforcing	113.6	116.0	114.8	103.0	79.1	90.6	102.8	79.1	90.5	112.2	79.1	95.0	103.1	79.1	90.7	112.9	79.1	95.4
0330	Cast-in-Place Concrete	250.7	117.7	198.2	87.9	72.0	81.6	87.9	72.1	81.7	93.7	71.6	85.0	87.6	72.2	81.5	94.5	71.8	85.5
03	CONCRETE	205.3	117.3	162.5	98.1	66.3	82.6	119.4	71.7	96.2	115.7	66.2	91.6	97.8	69.3	83.9	106.3	71.1	89.2
04	MASONRY	205.4	125.8	156.4	98.7	63.3	76.9	98.9	63.4	77.0	116.6	63.2	83.7	98.7	63.4	76.9	116.8	63.3	83.8
05	METALS	133.7	104.9	124.5	101.1	72.1	91.9	101.6	72.2	92.2	99.0	72.8	90.7	101.8	72.2	92.4	99.3	73.0	90.9
06	WOOD, PLASTICS & COMPOSITES	120.7	117.7	119.0	97.5	53.1	72.7	103.4	69.4	84.4	90.0	53.2	69.5	92.5	62.0	75.5	93.3	67.5	79.0
07	THERMAL & MOISTURE PROTECTION	176.5	118.3	152.2	94.6	63.7	81.6	96.3	67.7	84.3	99.8	64.2	84.9	94.5	64.2	81.9	99.1	64.2	84.5
08	OPENINGS	133.6	115.8	129.4	105.1	58.6	94.3	105.3	71.5	97.4	93.3	58.7	85.2	105.3	65.7	96.1	93.3	68.8	87.6
0920	Plaster & Gypsum Board	153.5	117.9	129.7	92.2	51.9	65.2	95.3	68.6	77.5	95.2	51.9	66.2	84.7	61.0	68.9	97.2	66.6	76.7
0950, 0980	Ceilings & Acoustic Treatment	118.2	117.9	118.0	99.4	51.9	67.9	100.3	68.6	79.3	93.5	51.9	65.9	100.3	61.0	74.3	93.5	66.6	75.7
0960	Flooring	125.6	128.1	126.3	96.2	61.2	86.1	98.7	34.5	80.2	103.3	61.2	91.2	94.9	66.2	86.6	105.0	47.6	88.5
0970, 0990	Wall Finishes & Painting/Coating	108.6	116.5	113.2	89.5	57.5	70.9	89.5	57.5	70.9	94.1	57.5	72.8	89.5	57.5	70.9	94.1	57.5	72.8
09	FINISHES	132.2	120.8	125.9	94.5	56.2	73.3	97.7	61.0	77.4	98.8	56.4	75.3	93.2	62.4	76.2	98.5	62.7	78.7
COVERS	DIVS. 10 - 14, 25, 28, 41, 43, 44, 46	100.0	112.3	102.7	100.0	82.1	96.1	100.0	83.8	96.5	100.0	82.3	96.2	100.0	83.0	96.3	100.0	83.9	96.5
21, 22, 23	FIRE SUPPRESSION, PLUMBING & HVAC	98.6	105.8	101.6	97.8	79.0	89.9	100.2	79.1	91.3	96.7	79.1	89.3	97.8	79.0	89.9	100.1	79.1	91.2
26, 27, 3370	ELECTRICAL, COMMUNICATIONS & UTIL.	117.4	117.3	117.3	103.0	90.0	96.4	102.0	61.7	81.6	100.2	61.7	80.7	103.0	61.7	82.1	97.1	61.7	79.2
MF2014	WEIGHTED AVERAGE	134.9	115.8	126.5	98.9	73.5	87.7	102.7	71.9	89.2	100.9	69.8	87.2	98.9	71.3	86.7	100.1	72.0	87.7

Table 2

	DIVISION	ARIZONA												ARKANSAS					
		PHOENIX			PRESCOTT			SHOW LOW			TUCSON			BATESVILLE			CAMDEN		
		850,853			863			859			856 - 857			725			717		
		MAT.	INST.	TOTAL	MAT.	INST.	TOTAL	MAT.	INST.	TOTAL	MAT.	INST.	TOTAL	MAT.	INST.	TOTAL	MAT.	INST.	TOTAL
015433	CONTRACTOR EQUIPMENT		92.6	92.6		91.0	91.0		92.0	92.0		92.0	92.0		91.9	91.9		91.9	91.9
0241, 31 - 34	SITE & INFRASTRUCTURE, DEMOLITION	96.5	96.7	96.7	79.1	95.5	90.7	108.3	96.4	99.9	91.5	96.5	95.0	76.0	88.0	84.5	78.4	87.9	85.1
0310	Concrete Forming & Accessories	101.0	67.7	72.2	100.9	51.1	57.8	104.3	67.5	72.5	100.5	67.3	71.8	84.7	46.4	51.6	82.6	35.0	41.4
0320	Concrete Reinforcing	111.1	79.1	94.6	102.8	79.0	90.5	112.9	79.1	95.4	93.4	79.1	86.0	97.6	67.0	81.7	103.9	66.8	84.7
0330	Cast-in-Place Concrete	94.5	71.9	85.6	87.9	71.9	81.6	93.7	71.7	85.0	97.2	71.7	87.2	74.3	75.3	74.7	79.4	75.2	77.7
03	CONCRETE	105.9	71.2	89.0	103.9	63.7	84.4	118.2	71.0	95.3	104.6	71.0	88.2	82.3	60.9	71.9	86.1	55.6	71.3
04	MASONRY	102.9	63.3	78.5	98.8	63.3	77.0	116.7	63.2	83.7	100.9	63.3	77.7	97.8	45.3	65.5	115.2	42.1	70.1
05	METALS	100.7	73.8	92.2	101.6	71.7	92.1	98.8	72.9	90.5	100.0	72.8	91.4	99.3	68.7	89.6	103.6	68.1	92.3
06	WOOD, PLASTICS & COMPOSITES	94.3	67.5	79.4	98.8	45.7	69.3	97.8	67.5	81.0	93.6	67.5	79.1	96.6	44.1	67.4	96.1	28.8	58.6
07	THERMAL & MOISTURE PROTECTION	98.7	66.0	85.0	95.0	62.8	81.6	100.0	71.4	88.0	100.2	63.5	84.8	98.8	52.8	79.5	95.6	50.3	76.7
08	OPENINGS	95.1	70.5	89.4	105.3	54.6	93.5	92.5	66.5	86.5	89.9	70.5	85.4	104.6	44.9	90.7	107.6	39.4	91.8
0920	Plaster & Gypsum Board	99.2	66.6	77.4	92.3	44.3	60.3	99.1	66.6	77.4	102.0	66.6	78.3	82.2	42.8	55.9	82.6	27.1	45.5
0950, 0980	Ceilings & Acoustic Treatment	100.9	66.6	78.2	98.6	44.3	62.7	93.5	66.6	75.7	94.3	66.6	75.9	78.6	42.8	54.9	79.6	27.1	44.8
0960	Flooring	105.3	66.2	94.1	97.2	61.2	86.8	106.9	43.3	88.6	96.1	42.5	80.7	92.5	56.8	82.2	103.0	38.1	84.3
0970, 0990	Wall Finishes & Painting/Coating	94.1	61.8	75.3	89.5	57.5	70.9	94.1	57.5	72.8	95.1	57.5	73.2	91.6	41.2	62.3	96.4	50.4	69.6
09	FINISHES	100.6	66.3	81.6	95.1	51.9	71.2	100.6	61.9	79.2	96.7	61.2	77.1	82.4	46.7	62.6	86.5	35.2	58.1
COVERS	DIVS. 10 - 14, 25, 28, 41, 43, 44, 46	100.0	83.9	96.5	100.0	81.2	95.9	100.0	83.9	96.5	100.0	83.9	96.5	100.0	68.0	93.1	100.0	66.3	92.7
21, 22, 23	FIRE SUPPRESSION, PLUMBING & HVAC	100.0	79.2	91.2	100.2	79.0	91.3	96.7	79.1	89.3	100.1	74.4	89.2	96.7	53.6	78.5	96.7	58.1	80.4
26, 27, 3370	ELECTRICAL, COMMUNICATIONS & UTIL.	102.5	64.1	83.1	101.7	61.7	81.5	97.5	61.7	79.4	99.3	61.7	80.3	98.9	63.0	80.8	96.5	61.7	78.9
MF2014	WEIGHTED AVERAGE	100.5	73.0	88.4	100.4	68.4	86.3	101.1	71.9	88.3	98.9	70.8	86.5	95.3	58.3	79.0	97.6	56.0	79.3

Table 3

	DIVISION	ARKANSAS																	
		FAYETTEVILLE			FORT SMITH			HARRISON			HOT SPRINGS			JONESBORO			LITTLE ROCK		
		727			729			726			719			724			720 - 722		
		MAT.	INST.	TOTAL	MAT.	INST.	TOTAL	MAT.	INST.	TOTAL	MAT.	INST.	TOTAL	MAT.	INST.	TOTAL	MAT.	INST.	TOTAL
015433	CONTRACTOR EQUIPMENT		91.9	91.9		91.9	91.9		91.9	91.9		91.9	91.9		114.2	114.2		91.9	91.9
0241, 31 - 34	SITE & INFRASTRUCTURE, DEMOLITION	75.4	89.7	85.5	80.7	89.8	87.1	81.2	88.1	86.0	81.8	89.2	87.0	101.8	107.2	105.6	90.2	90.7	90.6
0310	Concrete Forming & Accessories	80.3	39.3	44.8	99.4	60.3	65.6	89.2	46.4	52.2	80.1	39.2	44.7	88.3	49.0	54.3	94.1	61.2	65.6
0320	Concrete Reinforcing	97.6	70.7	83.7	98.8	71.5	84.6	97.2	67.1	81.6	101.9	66.9	83.8	94.5	75.9	84.9	104.7	68.6	86.0
0330	Cast-in-Place Concrete	74.3	75.3	74.7	84.9	75.7	81.3	82.4	75.2	79.6	81.2	75.1	78.8	80.9	76.9	79.3	82.3	76.9	80.2
03	CONCRETE	82.0	58.3	70.5	89.9	68.0	79.3	89.4	60.8	75.5	89.5	57.5	74.0	86.8	65.0	76.2	91.0	68.5	80.0
04	MASONRY	88.8	45.3	62.0	96.2	52.4	69.2	98.2	46.9	66.6	86.4	43.5	59.9	90.2	44.7	62.1	94.8	71.2	80.2
05	METALS	99.3	70.0	90.0	101.6	71.5	92.1	100.4	68.6	90.3	103.5	68.1	92.3	95.9	85.4	92.6	101.8	74.9	93.2
06	WOOD, PLASTICS & COMPOSITES	93.1	34.4	60.4	113.4	61.8	84.7	102.5	44.1	70.0	93.4	34.6	60.7	100.8	46.8	70.8	100.9	61.8	79.1
07	THERMAL & MOISTURE PROTECTION	99.6	51.7	79.6	99.9	56.5	81.8	99.1	53.2	79.9	95.9	51.3	77.2	104.0	56.7	84.2	94.7	61.3	80.7
08	OPENINGS	104.6	43.9	90.4	106.7	59.3	95.6	105.4	46.1	91.6	107.6	44.2	92.9	110.6	54.0	97.4	102.5	60.0	92.6
0920	Plaster & Gypsum Board	81.6	32.9	49.0	88.2	61.1	70.1	87.3	42.8	57.5	81.2	33.1	49.1	96.0	45.3	62.1	93.1	61.1	71.7
0950, 0980	Ceilings & Acoustic Treatment	78.6	32.9	48.3	80.3	61.1	67.6	80.3	42.8	55.5	79.6	33.1	48.8	82.5	45.3	57.9	83.4	61.1	68.6
0960	Flooring	89.4	56.8	80.0	99.4	57.3	87.3	94.8	56.8	83.9	101.8	56.8	88.8	66.4	51.4	62.1	97.5	75.2	91.0
0970, 0990	Wall Finishes & Painting/Coating	91.6	34.1	58.2	91.6	53.8	69.6	91.6	41.2	62.3	96.4	55.2	72.4	80.9	48.4	62.0	92.7	54.8	70.7
09	FINISHES	81.4	40.0	58.5	85.9	59.1	71.1	84.5	46.5	63.5	86.3	42.4	62.0	80.0	48.0	62.3	88.3	62.8	74.2
COVERS	DIVS. 10 - 14, 25, 28, 41, 43, 44, 46	100.0	48.4	88.8	100.0	80.7	95.8	100.0	71.2	93.8	100.0	39.1	86.8	100.0	69.2	93.3	100.0	80.8	95.8
21, 22, 23	FIRE SUPPRESSION, PLUMBING & HVAC	96.7	51.4	77.6	100.1	51.5	79.5	96.7	51.4	77.5	96.7	53.3	78.4	100.4	53.5	80.6	99.9	53.9	80.4
26, 27, 3370	ELECTRICAL, COMMUNICATIONS & UTIL.	92.9	59.1	75.8	96.3	59.2	77.5	97.5	59.1	78.1	98.5	69.7	83.9	102.1	63.0	82.3	103.0	71.1	86.8
MF2014	WEIGHTED AVERAGE	94.1	55.6	77.1	97.8	62.3	82.2	96.6	57.6	79.4	96.9	56.9	79.3	97.4	62.6	82.1	98.2	67.4	84.6

For customer support on your Mechanical Cost Data, call 877.739.3689.

659

ARKANSAS / CALIFORNIA

DIVISION		PINE BLUFF 716 MAT.	INST.	TOTAL	RUSSELLVILLE 728 MAT.	INST.	TOTAL	TEXARKANA 718 MAT.	INST.	TOTAL	WEST MEMPHIS 723 MAT.	INST.	TOTAL	ALHAMBRA 917-918 MAT.	INST.	TOTAL	ANAHEIM 928 MAT.	INST.	TOTAL
015433	CONTRACTOR EQUIPMENT		91.9	91.9		91.9	91.9		93.0	93.0		114.2	114.2		95.5	95.5		100.4	100.4
0241, 31 - 34	SITE & INFRASTRUCTURE, DEMOLITION	83.9	90.8	88.8	77.3	88.1	85.0	95.3	91.6	92.6	109.5	107.1	107.8	101.6	106.9	105.3	100.9	107.8	105.8
0310	Concrete Forming & Accessories	79.7	60.9	63.4	85.4	56.0	60.0	85.8	44.1	49.7	93.9	49.1	55.1	115.7	120.5	119.8	103.6	120.9	118.5
0320	Concrete Reinforcing	103.8	67.9	85.2	98.2	66.9	82.0	103.4	67.0	84.5	94.5	72.1	82.9	106.5	115.2	111.0	96.3	115.1	106.0
0330	Cast-in-Place Concrete	81.2	76.9	79.5	77.8	75.2	76.8	88.6	75.4	83.3	84.8	76.9	81.7	89.4	118.0	100.7	90.9	121.8	103.1
03	CONCRETE	90.3	68.2	79.6	85.4	65.1	75.5	88.4	59.8	74.5	93.9	64.4	79.6	99.4	117.6	108.2	102.0	119.1	110.4
04	MASONRY	122.1	58.5	82.9	95.5	44.6	64.2	101.5	44.7	66.5	78.5	44.6	57.6	114.4	120.0	117.9	81.6	117.9	104.0
05	METALS	104.3	74.2	94.8	99.3	68.5	89.5	96.4	68.8	87.6	95.0	85.5	92.0	89.3	99.5	92.5	108.0	101.1	105.8
06	WOOD, PLASTICS & COMPOSITES	93.0	61.8	75.6	97.9	57.2	75.2	100.7	40.7	67.3	106.7	46.8	73.4	95.5	118.3	108.2	97.5	118.6	109.3
07	THERMAL & MOISTURE PROTECTION	96.0	58.0	80.1	99.8	54.0	80.6	96.6	52.2	78.1	104.5	58.4	85.2	95.5	116.1	104.1	101.8	119.9	109.4
08	OPENINGS	108.9	57.5	96.9	104.6	54.2	92.8	114.3	45.2	98.2	108.0	54.1	95.4	90.9	117.3	97.0	108.1	117.9	110.4
0920	Plaster & Gypsum Board	80.9	61.1	67.6	82.2	56.3	64.9	84.0	39.3	54.1	98.4	45.3	62.9	99.2	119.0	112.4	104.5	119.1	114.2
0950, 0980	Ceilings & Acoustic Treatment	79.6	61.1	67.3	78.6	56.3	63.8	82.9	39.3	54.0	80.7	45.3	57.2	101.9	119.0	113.2	105.6	119.1	114.5
0960	Flooring	101.4	57.3	88.8	92.0	56.8	81.9	103.8	49.0	88.0	68.9	51.4	63.9	97.9	107.7	100.7	101.5	108.2	103.4
0970, 0990	Wall Finishes & Painting/Coating	96.4	50.4	69.6	91.6	35.7	59.1	96.4	30.3	57.9	80.9	50.4	63.1	110.9	111.5	111.2	92.9	109.4	102.5
09	FINISHES	86.2	59.1	71.2	82.5	53.9	66.7	88.7	42.2	62.9	81.4	48.4	63.1	101.4	116.9	110.0	99.8	117.2	109.4
COVERS	DIVS. 10 - 14, 25, 28, 41, 43, 44, 46	100.0	80.8	95.8	100.0	69.5	93.4	100.0	65.0	92.4	100.0	53.0	89.8	100.0	110.6	102.3	100.0	111.3	102.5
21, 22, 23	FIRE SUPPRESSION, PLUMBING & HVAC	100.2	53.8	80.6	96.7	53.2	78.3	100.2	59.5	83.0	97.0	66.7	84.2	96.7	115.1	104.5	100.0	119.4	108.2
26, 27, 3370	ELECTRICAL, COMMUNICATIONS & UTIL.	96.7	69.7	83.1	96.3	59.1	77.5	98.4	61.0	79.5	103.7	65.2	84.2	123.0	120.3	121.6	90.6	105.3	98.0
MF2014	WEIGHTED AVERAGE	99.6	65.2	84.5	95.3	59.7	79.7	98.5	58.7	80.9	97.0	65.3	83.0	99.5	114.8	106.2	100.6	114.0	106.5

CALIFORNIA

DIVISION		BAKERSFIELD 932-933 MAT.	INST.	TOTAL	BERKELEY 947 MAT.	INST.	TOTAL	EUREKA 955 MAT.	INST.	TOTAL	FRESNO 936-938 MAT.	INST.	TOTAL	INGLEWOOD 903-905 MAT.	INST.	TOTAL	LONG BEACH 906-908 MAT.	INST.	TOTAL
015433	CONTRACTOR EQUIPMENT		98.4	98.4		101.8	101.8		98.2	98.2		98.4	98.4		97.2	97.2		97.2	97.2
0241, 31 - 34	SITE & INFRASTRUCTURE, DEMOLITION	99.7	105.6	103.9	106.9	108.0	107.7	112.5	104.7	107.0	103.8	105.3	104.9	92.8	103.6	100.5	99.9	103.6	102.5
0310	Concrete Forming & Accessories	101.9	122.9	120.1	114.3	146.6	142.3	112.4	132.4	129.7	101.0	132.0	127.8	105.6	120.9	118.8	100.4	120.9	118.1
0320	Concrete Reinforcing	99.5	115.1	107.6	95.6	116.6	106.5	105.0	116.3	110.9	83.1	116.0	100.1	97.8	115.3	106.9	97.0	115.3	106.5
0330	Cast-in-Place Concrete	89.3	120.9	101.7	120.4	126.4	122.8	98.6	123.0	108.2	96.9	122.6	107.1	84.6	120.8	98.9	96.2	120.8	105.9
03	CONCRETE	95.5	119.7	107.2	108.8	132.3	120.2	113.0	124.9	118.8	99.3	124.5	111.6	92.1	118.8	105.1	101.5	118.8	109.9
04	MASONRY	98.1	117.6	110.1	117.0	137.6	129.7	109.3	136.6	126.1	102.8	131.5	120.5	79.4	120.2	104.5	88.4	120.2	108.0
05	METALS	108.1	100.5	105.7	102.7	102.9	102.8	107.8	102.8	106.2	108.4	101.3	106.1	99.2	101.2	99.8	99.0	101.2	99.7
06	WOOD, PLASTICS & COMPOSITES	90.4	121.8	107.9	100.8	150.5	131.7	111.3	133.6	123.7	100.4	133.6	118.8	95.6	118.7	108.5	89.7	118.7	105.8
07	THERMAL & MOISTURE PROTECTION	96.6	114.9	104.3	100.7	137.5	116.1	100.7	133.3	117.2	88.4	118.6	101.1	101.9	117.3	108.3	102.2	117.8	108.7
08	OPENINGS	98.1	116.9	102.5	93.0	138.4	103.5	107.3	122.9	110.9	100.2	123.4	105.6	89.4	117.5	96.0	89.4	117.5	95.9
0920	Plaster & Gypsum Board	96.5	122.1	113.6	104.1	151.3	135.6	109.9	134.3	126.2	95.3	134.3	121.3	99.8	119.0	112.6	96.1	119.0	111.4
0950, 0980	Ceilings & Acoustic Treatment	94.5	122.1	112.8	104.0	151.3	135.3	110.6	134.3	126.3	91.1	134.3	119.7	101.2	119.0	113.0	101.2	119.0	113.0
0960	Flooring	102.4	108.2	104.1	115.3	126.5	118.5	105.4	77.1	97.3	108.5	102.0	106.6	104.7	107.7	105.6	101.9	107.7	103.6
0970, 0990	Wall Finishes & Painting/Coating	92.3	98.1	95.7	111.7	144.3	130.7	94.4	120.0	109.3	107.6	109.8	108.9	110.4	111.5	111.0	110.4	111.5	111.0
09	FINISHES	96.1	117.7	108.1	106.4	144.1	127.3	105.4	123.1	115.1	98.3	126.2	113.7	103.1	117.3	110.9	102.1	117.3	110.5
COVERS	DIVS. 10 - 14, 25, 28, 41, 43, 44, 46	100.0	109.3	102.0	100.0	125.8	105.6	100.0	122.9	105.0	100.0	122.8	104.9	100.0	111.6	102.5	100.0	111.6	102.5
21, 22, 23	FIRE SUPPRESSION, PLUMBING & HVAC	100.2	109.9	104.3	96.7	149.6	119.0	96.6	111.8	103.0	100.2	116.3	107.0	96.3	115.2	104.3	96.3	115.2	104.3
26, 27, 3370	ELECTRICAL, COMMUNICATIONS & UTIL.	103.4	102.8	103.1	104.5	143.9	124.4	96.7	115.8	106.4	94.3	100.7	97.5	99.5	120.3	110.0	99.3	120.3	109.9
MF2014	WEIGHTED AVERAGE	100.4	111.4	105.2	102.0	135.3	116.7	103.9	118.3	110.2	100.5	116.5	107.5	96.0	115.0	104.4	97.5	115.0	105.2

CALIFORNIA

DIVISION		LOS ANGELES 900-902 MAT.	INST.	TOTAL	MARYSVILLE 959 MAT.	INST.	TOTAL	MODESTO 953 MAT.	INST.	TOTAL	MOJAVE 935 MAT.	INST.	TOTAL	OAKLAND 946 MAT.	INST.	TOTAL	OXNARD 930 MAT.	INST.	TOTAL
015433	CONTRACTOR EQUIPMENT		100.3	100.3		98.2	98.2		98.2	98.2		98.4	98.4		101.8	101.8		97.3	97.3
0241, 31 - 34	SITE & INFRASTRUCTURE, DEMOLITION	100.4	107.0	105.0	108.8	104.7	105.9	103.5	104.9	104.5	96.9	105.6	103.0	112.2	108.0	109.3	104.1	103.6	103.8
0310	Concrete Forming & Accessories	102.5	121.2	118.7	102.3	132.0	128.0	98.7	132.1	127.6	112.7	122.8	121.5	103.0	146.6	140.7	104.2	121.0	118.7
0320	Concrete Reinforcing	98.6	115.5	107.3	105.0	115.7	110.6	108.8	115.7	112.4	100.1	115.1	107.9	97.8	116.6	107.5	98.3	115.0	106.9
0330	Cast-in-Place Concrete	91.3	121.4	103.1	110.2	122.7	115.1	98.6	122.8	108.2	83.8	120.9	98.4	114.2	126.4	119.0	97.6	121.1	106.9
03	CONCRETE	97.4	119.2	108.0	114.1	124.5	119.1	104.8	124.6	114.4	91.3	119.6	105.1	108.6	132.3	120.1	99.8	118.9	109.1
04	MASONRY	95.1	123.0	112.3	110.3	126.8	120.4	106.1	126.8	118.8	100.8	117.4	111.0	123.1	137.6	132.0	103.9	115.6	111.1
05	METALS	105.9	102.4	104.8	107.3	101.4	105.4	104.1	101.6	103.3	105.5	100.4	103.9	98.5	102.8	99.9	103.4	100.7	102.5
06	WOOD, PLASTICS & COMPOSITES	95.2	118.8	108.4	98.8	133.6	118.1	94.7	133.6	116.3	99.6	121.8	112.0	95.8	150.5	126.2	94.9	118.7	108.2
07	THERMAL & MOISTURE PROTECTION	101.3	119.5	108.9	105.2	120.7	111.7	104.8	121.2	111.7	94.0	110.7	101.0	98.9	137.5	115.1	97.0	118.2	105.9
08	OPENINGS	96.0	118.2	101.2	106.5	122.9	110.3	105.4	124.1	109.7	94.7	116.9	99.9	93.0	138.4	103.6	97.5	118.0	102.3
0920	Plaster & Gypsum Board	99.6	119.1	112.6	102.5	134.3	123.7	104.8	134.3	124.5	104.7	122.1	116.4	98.7	151.3	133.8	99.1	119.1	112.4
0950, 0980	Ceilings & Acoustic Treatment	111.2	119.1	116.4	109.8	134.3	126.0	105.6	134.3	124.6	92.5	122.1	112.1	106.6	151.3	136.2	94.8	119.1	110.9
0960	Flooring	102.3	108.2	104.0	101.2	113.0	104.6	101.5	110.1	104.0	109.2	105.9	108.3	108.9	126.5	114.0	101.1	108.2	103.2
0970, 0990	Wall Finishes & Painting/Coating	109.3	111.5	110.6	94.4	120.0	109.3	94.4	119.9	109.3	92.0	130.7	114.5	111.7	144.3	130.7	92.0	104.3	99.1
09	FINISHES	104.6	117.5	111.7	102.3	129.1	117.1	101.4	128.6	116.4	98.5	120.8	110.8	104.7	144.1	126.5	96.0	116.7	107.5
COVERS	DIVS. 10 - 14, 25, 28, 41, 43, 44, 46	100.0	111.8	102.5	100.0	122.8	104.9	100.0	122.8	104.9	100.0	107.0	101.5	100.0	125.7	105.6	100.0	111.5	102.5
21, 22, 23	FIRE SUPPRESSION, PLUMBING & HVAC	100.0	119.5	108.2	96.6	110.0	102.3	100.0	115.0	106.3	96.7	113.9	104.0	100.1	149.6	121.0	100.1	119.4	108.3
26, 27, 3370	ELECTRICAL, COMMUNICATIONS & UTIL.	98.0	123.5	110.9	93.8	106.5	100.3	96.0	103.9	100.0	92.7	99.1	96.0	103.8	138.2	121.2	98.1	106.5	102.3
MF2014	WEIGHTED AVERAGE	100.2	117.2	107.7	103.1	115.9	108.8	102.0	116.6	108.4	97.4	111.9	103.8	102.1	134.5	116.4	99.9	113.5	105.9

CALIFORNIA

	DIVISION	PALM SPRINGS 922			PALO ALTO 943			PASADENA 910-912			REDDING 960			RICHMOND 948			RIVERSIDE 925		
		MAT.	INST.	TOTAL	MAT.	INST.	TOTAL	MAT.	INST.	TOTAL	MAT.	INST.	TOTAL	MAT.	INST.	TOTAL	MAT.	INST.	TOTAL
015433	CONTRACTOR EQUIPMENT		99.2	99.2		101.8	101.8		95.5	95.5		98.2	98.2		101.8	101.8		99.2	99.2
0241, 31 - 34	SITE & INFRASTRUCTURE, DEMOLITION	92.2	105.8	101.8	103.2	108.0	106.6	98.3	106.9	104.4	134.7	104.7	113.5	111.4	108.0	109.0	99.4	105.8	103.9
0310	Concrete Forming & Accessories	100.2	120.8	118.0	101.1	142.5	136.9	104.3	120.5	118.3	106.4	140.5	135.9	117.2	146.4	142.4	104.0	120.9	118.6
0320	Concrete Reinforcing	110.5	115.1	112.9	95.6	116.4	106.3	107.4	115.2	111.5	132.3	115.7	123.7	95.6	116.3	106.3	107.3	115.1	111.3
0330	Cast-in-Place Concrete	86.6	121.8	100.5	101.9	126.3	111.5	84.9	118.0	97.9	119.6	122.8	120.8	117.1	126.3	120.7	94.2	121.8	105.1
03	CONCRETE	96.6	119.1	107.5	97.9	130.4	113.7	94.9	117.6	105.9	128.2	128.3	128.2	110.6	132.1	121.1	102.8	119.1	110.7
04	MASONRY	79.8	117.6	103.1	97.9	133.8	120.0	98.2	120.0	111.7	142.0	126.8	132.6	116.8	132.2	126.3	80.8	117.6	103.5
05	METALS	108.7	101.0	106.3	96.1	102.3	98.1	89.3	99.5	92.6	113.3	101.5	109.5	96.1	102.2	98.1	108.1	101.1	105.9
06	WOOD, PLASTICS & COMPOSITES	92.5	118.6	107.0	93.2	145.3	122.2	82.3	118.3	102.3	107.4	144.9	128.3	111.8	150.5	133.3	97.5	118.6	109.3
07	THERMAL & MOISTURE PROTECTION	101.5	117.4	108.1	98.5	134.4	113.5	95.2	116.1	104.0	126.9	122.5	125.1	99.1	135.5	114.3	102.0	119.8	109.5
08	OPENINGS	103.8	117.9	107.1	93.0	133.8	102.5	90.9	117.3	97.0	121.0	129.0	122.8	93.0	137.1	103.3	106.7	117.9	109.3
0920	Plaster & Gypsum Board	99.7	119.0	112.6	97.0	146.0	129.7	92.6	119.0	110.2	103.6	146.0	131.9	106.0	151.3	136.2	103.7	119.1	114.0
0950, 0980	Ceilings & Acoustic Treatment	102.3	119.0	113.4	104.8	146.0	132.1	101.9	119.0	113.2	132.6	146.0	141.5	104.8	151.3	135.6	109.7	119.1	115.9
0960	Flooring	103.8	118.4	108.0	107.6	126.5	113.0	92.2	107.7	96.7	101.1	113.0	104.5	117.6	126.5	120.1	105.3	108.2	106.2
0970, 0990	Wall Finishes & Painting/Coating	91.4	111.3	102.9	111.7	144.3	130.7	110.9	111.5	111.2	104.0	120.0	113.3	111.7	144.3	130.7	91.4	109.4	101.8
09	FINISHES	98.4	118.9	109.7	103.1	141.1	124.1	98.5	116.9	108.7	109.5	135.9	124.1	108.0	144.1	128.0	101.4	117.2	110.1
COVERS	DIVS. 10 - 14, 25, 28, 41, 43, 44, 46	100.0	111.3	102.5	100.0	125.2	105.5	100.0	110.6	102.3	100.0	124.1	105.2	100.0	125.7	105.6	100.0	111.3	102.5
21, 22, 23	FIRE SUPPRESSION, PLUMBING & HVAC	96.6	114.6	104.2	96.7	144.1	116.8	96.7	115.1	104.5	100.4	110.0	104.5	96.7	146.5	117.7	100.0	119.4	108.2
26, 27, 3370	ELECTRICAL, COMMUNICATIONS & UTIL.	93.3	104.6	99.0	103.7	149.1	126.6	119.8	120.3	120.1	99.7	106.5	103.1	104.2	131.7	118.1	90.4	105.6	98.1
MF2014	WEIGHTED AVERAGE	98.6	112.9	104.9	98.1	133.4	113.7	97.5	114.8	105.1	112.3	117.8	114.7	101.3	132.2	114.9	100.5	113.9	106.4

CALIFORNIA

	DIVISION	SACRAMENTO 942,956 - 958			SALINAS 939			SAN BERNARDINO 923 - 924			SAN DIEGO 919 - 921			SAN FRANCISCO 940 - 941			SAN JOSE 951		
		MAT.	INST.	TOTAL	MAT.	INST.	TOTAL	MAT.	INST.	TOTAL	MAT.	INST.	TOTAL	MAT.	INST.	TOTAL	MAT.	INST.	TOTAL
015433	CONTRACTOR EQUIPMENT		101.0	101.0		98.4	98.4		99.2	99.2		95.7	95.7		111.0	111.0		99.4	99.4
0241, 31 - 34	SITE & INFRASTRUCTURE, DEMOLITION	92.0	114.1	107.6	118.2	105.5	109.2	78.4	105.8	97.7	105.2	100.6	101.9	114.4	114.0	114.1	135.3	99.8	110.2
0310	Concrete Forming & Accessories	101.4	134.8	130.3	107.4	135.1	131.4	107.9	120.6	118.9	105.3	113.9	112.8	102.7	147.6	141.6	104.8	146.4	140.8
0320	Concrete Reinforcing	90.5	116.0	103.7	98.8	116.1	107.8	107.3	115.2	111.4	104.0	115.1	109.7	111.4	117.0	114.3	95.9	116.4	106.5
0330	Cast-in-Place Concrete	96.6	123.9	107.4	96.4	123.1	107.0	65.1	121.7	87.5	93.0	107.5	98.7	117.1	127.8	121.3	114.6	125.5	118.9
03	CONCRETE	98.8	125.9	111.9	109.5	126.1	117.5	76.5	119.0	97.2	100.1	111.2	105.5	111.9	133.3	122.5	111.1	132.3	121.4
04	MASONRY	98.9	131.0	118.7	100.9	131.7	119.9	86.5	119.6	106.9	90.8	116.4	106.6	123.6	140.8	134.2	136.8	133.8	134.9
05	METALS	94.4	97.6	95.4	108.3	102.6	106.5	108.1	100.9	105.8	105.9	100.7	104.3	103.7	110.6	105.9	102.3	107.7	104.1
06	WOOD, PLASTICS & COMPOSITES	90.1	136.7	116.1	99.3	136.5	120.0	101.1	118.6	110.8	95.8	110.8	104.1	95.8	150.6	126.3	106.9	150.2	131.0
07	THERMAL & MOISTURE PROTECTION	107.3	124.3	114.4	94.6	127.7	108.5	100.7	118.1	108.0	100.3	106.0	102.7	100.5	140.9	117.4	101.1	137.5	116.3
08	OPENINGS	106.0	125.9	110.7	98.7	130.8	106.2	103.9	117.9	107.2	98.1	112.2	101.4	97.2	138.5	106.8	96.4	138.3	106.1
0920	Plaster & Gypsum Board	94.3	137.3	123.0	100.1	137.3	124.9	105.5	119.0	114.5	99.6	111.0	107.2	100.6	151.3	134.5	100.5	151.3	134.4
0950, 0980	Ceilings & Acoustic Treatment	104.8	137.3	126.3	92.5	137.3	122.1	105.6	119.0	114.5	109.3	111.0	110.4	113.0	151.3	138.3	104.8	151.3	135.6
0960	Flooring	107.5	113.0	109.1	103.2	119.4	107.8	107.3	105.9	106.9	100.4	109.8	103.1	108.9	126.5	114.0	94.8	126.5	103.9
0970, 0990	Wall Finishes & Painting/Coating	108.6	132.3	122.4	93.0	144.3	122.8	91.4	104.0	98.7	107.3	114.0	111.2	111.7	158.2	138.8	94.8	143.3	123.6
09	FINISHES	101.8	132.4	118.7	98.0	134.5	118.2	99.8	116.0	108.8	104.2	112.9	109.0	106.5	145.8	128.2	100.1	143.9	124.4
COVERS	DIVS. 10 - 14, 25, 28, 41, 43, 44, 46	100.0	123.8	105.2	100.0	123.2	105.0	100.0	108.6	101.9	100.0	109.8	102.1	100.0	126.3	105.7	100.0	125.2	105.5
21, 22, 23	FIRE SUPPRESSION, PLUMBING & HVAC	99.9	120.1	108.4	96.7	121.3	107.1	96.6	114.3	104.1	100.0	118.6	107.8	100.0	178.0	133.0	100.0	149.8	121.1
26, 27, 3370	ELECTRICAL, COMMUNICATIONS & UTIL.	99.4	109.3	104.4	93.7	120.4	107.2	93.3	102.7	98.0	105.8	98.8	102.3	103.8	163.2	133.8	98.6	160.2	129.8
MF2014	WEIGHTED AVERAGE	99.6	120.1	108.6	101.0	122.3	110.4	96.4	112.3	103.4	101.4	109.8	105.1	104.1	146.0	122.5	103.9	136.9	118.4

CALIFORNIA

	DIVISION	SAN LUIS OBISPO 934			SAN MATEO 944			SAN RAFAEL 949			SANTA ANA 926 - 927			SANTA BARBARA 931			SANTA CRUZ 950		
		MAT.	INST.	TOTAL	MAT.	INST.	TOTAL	MAT.	INST.	TOTAL	MAT.	INST.	TOTAL	MAT.	INST.	TOTAL	MAT.	INST.	TOTAL
015433	CONTRACTOR EQUIPMENT		98.4	98.4		101.8	101.8		101.4	101.4		99.2	99.2		98.4	98.4		99.4	99.4
0241, 31 - 34	SITE & INFRASTRUCTURE, DEMOLITION	109.8	105.6	106.8	109.2	108.0	108.4	104.1	113.7	110.9	90.6	105.8	101.3	104.0	105.6	105.1	135.0	99.7	110.0
0310	Concrete Forming & Accessories	114.4	120.8	119.9	107.2	146.4	141.1	112.2	146.4	141.8	108.0	120.6	118.9	105.0	121.0	118.8	104.8	135.2	131.1
0320	Concrete Reinforcing	100.1	115.1	107.9	95.6	116.6	106.4	96.2	116.5	106.8	111.0	115.1	113.1	98.3	115.1	107.0	118.6	116.1	117.3
0330	Cast-in-Place Concrete	103.6	121.0	110.5	113.4	126.3	118.5	131.9	125.3	129.3	83.2	121.7	98.4	97.3	121.0	106.6	113.8	125.0	118.2
03	CONCRETE	107.6	118.8	113.0	107.1	132.2	119.3	128.7	131.7	130.1	94.1	119.0	106.2	99.6	118.9	109.0	114.1	127.0	120.4
04	MASONRY	102.4	117.4	111.6	116.5	136.7	128.9	96.8	136.7	121.4	76.4	117.9	102.0	101.1	117.4	111.1	140.4	131.9	135.3
05	METALS	106.3	100.7	104.5	96.0	102.6	98.1	97.3	99.6	98.0	108.2	100.9	105.9	103.9	100.8	102.9	109.8	106.4	108.7
06	WOOD, PLASTICS & COMPOSITES	101.7	118.7	111.2	100.9	150.5	128.5	98.8	150.2	127.4	103.0	118.6	111.7	94.9	118.7	108.2	106.9	136.6	123.4
07	THERMAL & MOISTURE PROTECTION	94.8	117.3	104.2	98.9	135.5	114.2	103.1	135.6	116.7	101.8	117.5	108.4	94.3	118.4	104.3	101.0	129.7	113.0
08	OPENINGS	96.8	115.3	101.1	93.0	137.1	103.2	103.6	136.5	111.3	103.1	117.9	106.6	98.5	118.0	103.1	97.6	130.8	105.3
0920	Plaster & Gypsum Board	105.7	119.0	114.6	102.0	151.3	134.9	103.7	151.3	135.5	106.8	119.0	114.9	99.1	119.1	112.4	109.4	137.3	128.0
0950, 0980	Ceilings & Acoustic Treatment	92.5	119.0	110.0	104.8	151.3	135.6	111.4	151.3	137.8	105.6	119.0	114.5	94.8	119.1	110.9	107.3	137.3	127.1
0960	Flooring	110.2	108.2	109.6	111.2	126.5	115.6	122.3	116.4	120.6	107.8	105.9	107.3	102.3	108.2	104.0	98.9	119.4	104.8
0970, 0990	Wall Finishes & Painting/Coating	92.0	104.3	99.1	111.7	144.3	130.7	107.7	143.9	128.8	91.4	106.8	100.3	94.4	144.3	123.6	94.3	144.3	123.6
09	FINISHES	99.9	116.6	109.1	105.3	144.1	126.8	108.5	142.2	127.1	101.3	116.3	109.6	96.5	116.7	107.7	103.2	134.6	120.6
COVERS	DIVS. 10 - 14, 25, 28, 41, 43, 44, 46	100.0	118.6	104.0	100.0	125.8	105.6	100.0	125.1	105.4	100.0	111.3	102.5	100.0	111.5	102.5	100.0	123.5	105.1
21, 22, 23	FIRE SUPPRESSION, PLUMBING & HVAC	96.7	116.0	104.9	96.7	144.3	116.8	96.7	167.9	126.8	96.6	114.3	104.1	100.1	119.4	108.3	100.0	121.4	109.0
26, 27, 3370	ELECTRICAL, COMMUNICATIONS & UTIL.	92.7	105.5	99.2	103.7	146.9	125.5	100.6	119.0	109.9	93.3	106.8	100.1	91.8	109.7	100.8	97.9	120.4	109.3
MF2014	WEIGHTED AVERAGE	100.2	113.0	105.9	100.5	134.3	115.4	103.2	135.4	117.4	98.3	112.8	104.7	99.2	114.3	105.9	105.9	122.5	113.2

City Cost Indexes

		CALIFORNIA														COLORADO		
		SANTA ROSA			STOCKTON			SUSANVILLE			VALLEJO			VAN NUYS			ALAMOSA		
	DIVISION	954			952			961			945			913 - 916			811		
		MAT.	INST.	TOTAL	MAT.	INST.	TOTAL	MAT.	INST.	TOTAL	MAT.	INST.	TOTAL	MAT.	INST.	TOTAL	MAT.	INST.	TOTAL
015433	CONTRACTOR EQUIPMENT		98.7	98.7		98.2	98.2		98.2	98.2		101.4	101.4		95.5	95.5		92.0	92.0
0241, 31 - 34	SITE & INFRASTRUCTURE, DEMOLITION	104.9	104.8	104.8	103.2	104.9	104.4	142.7	104.7	115.9	92.5	113.9	107.6	116.2	106.9	109.6	141.6	87.8	103.6
0310	Concrete Forming & Accessories	101.2	146.3	140.2	102.7	134.3	130.0	107.6	132.4	129.1	102.3	145.9	140.0	111.1	120.5	119.2	105.0	65.8	71.1
0320	Concrete Reinforcing	106.0	116.8	111.6	108.8	115.7	112.4	132.3	115.7	123.7	97.4	116.5	107.3	107.4	115.2	111.5	111.5	75.6	92.9
0330	Cast-in-Place Concrete	108.2	124.4	114.6	96.1	122.8	106.6	108.8	122.8	114.3	105.1	125.1	113.0	89.5	118.0	100.7	98.0	77.1	89.7
03	CONCRETE	114.0	131.7	122.6	103.8	125.5	114.4	130.7	124.7	127.8	104.0	131.4	117.3	108.7	117.6	113.0	118.1	71.9	95.6
04	MASONRY	107.6	140.8	128.0	106.0	126.8	118.8	140.1	126.8	131.9	71.6	136.7	111.7	114.4	120.0	117.9	134.3	71.0	95.3
05	METALS	108.5	104.7	107.3	104.3	101.6	103.4	111.9	101.5	108.6	97.2	99.1	97.8	88.4	99.5	92.0	103.4	79.7	95.9
06	WOOD, PLASTICS & COMPOSITES	93.9	150.0	125.1	100.0	136.5	120.3	109.1	134.0	123.0	87.9	150.2	122.6	90.6	118.3	106.0	94.4	65.2	78.1
07	THERMAL & MOISTURE PROTECTION	102.2	137.0	116.8	104.9	120.9	111.6	128.6	121.1	125.5	100.9	135.9	115.5	96.1	116.1	104.5	104.8	77.6	93.4
08	OPENINGS	104.8	138.2	112.5	105.4	124.4	109.8	123.0	123.1	123.0	105.4	137.0	112.7	90.7	117.3	96.9	95.7	72.4	90.3
0920	Plaster & Gypsum Board	101.5	151.3	134.7	104.8	137.3	126.5	104.4	134.7	124.7	98.4	151.3	133.7	97.3	119.0	111.8	81.1	64.0	69.7
0950, 0980	Ceilings & Acoustic Treatment	105.6	151.3	135.8	113.0	137.3	129.1	125.3	134.7	131.5	113.2	151.3	138.4	99.4	119.0	112.4	96.1	64.0	74.9
0960	Flooring	104.3	112.0	106.5	101.5	110.1	104.0	101.5	117.4	106.1	116.4	126.5	119.3	95.2	107.7	98.8	115.9	52.7	97.7
0970, 0990	Wall Finishes & Painting/Coating	91.4	143.9	122.0	94.4	119.9	109.3	104.0	120.0	113.3	108.7	143.9	129.2	110.9	111.5	111.2	103.9	55.3	75.7
09	FINISHES	104.0	141.3	123.0	103.0	130.3	118.1	109.3	130.1	120.8	105.2	143.9	126.6	100.8	116.9	109.7	103.5	62.1	80.6
COVERS	DIVS. 10 - 14, 25, 28, 41, 43, 44, 46	100.0	124.5	105.3	100.0	123.2	105.0	100.0	122.9	105.0	100.0	125.1	105.4	100.0	110.6	102.3	100.0	88.0	97.4
21, 22, 23	FIRE SUPPRESSION, PLUMBING & HVAC	96.6	171.7	128.3	100.0	115.0	106.3	97.0	110.0	102.5	100.0	133.6	114.2	96.7	115.1	104.5	96.6	71.0	85.8
26, 27, 3370	ELECTRICAL, COMMUNICATIONS & UTIL.	93.6	119.1	106.5	96.0	110.2	103.2	100.0	114.9	107.5	96.9	124.3	110.8	119.8	120.3	120.1	98.2	72.4	85.1
MF2014	WEIGHTED AVERAGE	102.6	136.3	117.4	102.1	117.9	109.0	112.0	117.2	114.3	99.1	128.9	112.2	100.4	114.8	106.7	104.3	73.1	90.5

		COLORADO																
		BOULDER			COLORADO SPRINGS			DENVER			DURANGO			FORT COLLINS			FORT MORGAN		
	DIVISION	803			808 - 809			800 - 802			813			805			807		
		MAT.	INST.	TOTAL	MAT.	INST.	TOTAL	MAT.	INST.	TOTAL	MAT.	INST.	TOTAL	MAT.	INST.	TOTAL	MAT.	INST.	TOTAL
015433	CONTRACTOR EQUIPMENT		94.4	94.4		92.2	92.2		97.2	97.2		92.0	92.0		94.4	94.4		94.4	94.4
0241, 31 - 34	SITE & INFRASTRUCTURE, DEMOLITION	95.3	96.3	96.0	97.2	91.7	93.3	96.3	101.0	99.6	134.7	87.8	101.5	107.5	95.4	98.9	97.7	93.7	94.9
0310	Concrete Forming & Accessories	104.3	71.7	76.1	94.6	66.1	69.9	101.2	69.1	73.4	111.6	65.9	72.1	101.8	72.6	76.5	104.7	72.8	77.1
0320	Concrete Reinforcing	96.9	75.8	85.9	96.1	79.5	87.5	96.1	79.5	87.5	111.5	75.7	92.9	97.0	75.7	86.0	97.1	75.6	86.0
0330	Cast-in-Place Concrete	105.5	78.1	94.7	108.4	78.0	96.4	102.6	78.1	92.9	112.6	77.1	98.6	119.1	76.6	102.4	103.5	76.7	92.9
03	CONCRETE	107.0	77.8	92.8	110.5	73.2	92.4	104.5	74.6	90.0	120.1	72.0	96.7	118.1	74.8	97.1	105.3	74.9	90.5
04	MASONRY	94.6	72.6	81.1	95.5	72.4	81.2	97.4	72.9	82.3	120.7	71.7	90.5	112.7	74.7	89.3	109.5	72.0	86.4
05	METALS	98.9	82.5	93.7	102.1	84.7	96.5	104.6	84.6	98.2	103.4	79.9	95.9	100.2	79.4	93.5	98.7	79.3	92.5
06	WOOD, PLASTICS & COMPOSITES	104.7	80.1	91.0	93.9	64.3	77.4	102.3	67.8	83.1	103.8	65.2	82.3	102.0	74.5	86.7	104.7	74.5	87.9
07	THERMAL & MOISTURE PROTECTION	96.9	81.1	90.3	97.8	79.4	90.1	96.2	81.6	90.1	104.8	77.8	93.5	97.3	71.0	86.3	96.9	80.3	89.9
08	OPENINGS	96.3	80.5	92.6	100.4	72.9	94.0	100.9	74.9	94.9	102.8	72.4	95.7	96.3	77.5	91.9	96.2	77.5	91.9
0920	Plaster & Gypsum Board	117.7	79.8	92.4	100.4	63.4	75.7	113.0	67.2	82.4	94.4	64.0	74.1	111.4	74.1	86.5	117.7	74.1	88.6
0950, 0980	Ceilings & Acoustic Treatment	86.8	79.8	82.2	94.2	63.4	73.8	97.1	67.2	77.3	96.1	64.0	74.9	86.8	74.1	78.4	86.8	74.1	78.4
0960	Flooring	113.7	81.8	104.5	104.7	76.7	96.7	109.9	81.8	101.8	121.3	52.7	101.6	109.8	52.7	93.4	114.1	75.5	103.0
0970, 0990	Wall Finishes & Painting/Coating	102.2	66.6	81.5	101.9	55.9	75.2	102.2	69.1	83.0	103.9	55.3	75.7	102.2	40.2	66.1	102.2	53.6	73.9
09	FINISHES	105.9	77.3	90.0	102.3	65.7	82.1	106.2	70.4	86.4	106.2	61.4	81.4	104.3	65.2	82.7	105.9	71.2	86.7
COVERS	DIVS. 10 - 14, 25, 28, 41, 43, 44, 46	100.0	88.9	97.6	100.0	87.3	97.2	100.0	87.6	97.3	100.0	88.0	97.4	100.0	88.1	97.4	100.0	88.1	97.4
21, 22, 23	FIRE SUPPRESSION, PLUMBING & HVAC	96.7	76.5	88.2	100.2	83.8	93.3	100.0	82.5	92.6	96.6	81.9	90.4	100.1	76.1	89.9	96.7	75.0	87.5
26, 27, 3370	ELECTRICAL, COMMUNICATIONS & UTIL.	96.2	83.9	90.0	99.4	81.8	90.5	101.0	83.9	92.3	97.6	68.9	83.1	96.2	83.8	89.9	96.5	83.8	90.1
MF2014	WEIGHTED AVERAGE	99.1	80.3	90.8	101.3	78.7	91.3	101.6	80.4	92.3	104.7	74.9	91.6	102.4	77.6	91.5	99.7	78.0	90.1

		COLORADO																
		GLENWOOD SPRINGS			GOLDEN			GRAND JUNCTION			GREELEY			MONTROSE			PUEBLO		
	DIVISION	816			804			815			806			814			810		
		MAT.	INST.	TOTAL	MAT.	INST.	TOTAL	MAT.	INST.	TOTAL	MAT.	INST.	TOTAL	MAT.	INST.	TOTAL	MAT.	INST.	TOTAL
015433	CONTRACTOR EQUIPMENT		95.3	95.3		94.4	94.4		95.3	95.3		94.4	94.4		93.6	93.6		92.0	92.0
0241, 31 - 34	SITE & INFRASTRUCTURE, DEMOLITION	151.0	96.2	112.2	108.2	95.5	99.2	134.3	95.7	107.0	94.2	95.3	95.0	144.3	91.7	107.1	125.9	89.2	99.9
0310	Concrete Forming & Accessories	101.8	73.2	77.0	97.1	72.8	76.1	110.2	72.7	77.7	99.7	76.2	79.4	101.2	73.2	77.0	107.3	66.1	71.7
0320	Concrete Reinforcing	110.4	75.7	92.4	97.1	75.6	86.0	110.8	75.5	92.5	96.9	74.3	85.2	110.3	75.7	92.4	106.7	79.5	92.6
0330	Cast-in-Place Concrete	97.9	76.5	89.5	103.6	76.4	92.9	108.4	75.5	95.4	99.5	57.0	82.8	97.9	76.6	89.5	97.2	78.2	89.7
03	CONCRETE	123.6	75.1	100.0	116.3	74.8	96.1	116.3	74.4	96.0	101.9	69.5	86.2	113.9	75.1	95.0	106.1	73.3	90.2
04	MASONRY	105.7	71.9	84.9	112.2	72.3	87.6	141.9	71.1	98.3	106.5	48.1	70.5	113.3	71.7	87.7	101.9	72.3	83.6
05	METALS	103.0	80.5	95.9	98.8	79.3	92.6	104.7	79.0	96.5	100.1	76.6	92.6	102.1	79.9	95.1	106.3	85.5	99.7
06	WOOD, PLASTICS & COMPOSITES	89.8	74.6	81.4	96.6	74.5	84.3	101.6	74.6	86.6	99.1	80.1	88.5	90.9	74.8	81.9	97.2	64.6	79.0
07	THERMAL & MOISTURE PROTECTION	104.7	78.8	93.9	97.8	73.3	87.5	103.8	66.8	88.3	96.8	64.1	83.1	104.9	78.8	94.0	103.3	78.0	92.7
08	OPENINGS	101.8	77.6	96.2	96.3	77.5	91.9	102.5	77.6	96.7	96.2	80.4	92.6	102.9	77.6	97.0	97.5	73.1	91.8
0920	Plaster & Gypsum Board	123.8	74.1	90.6	109.1	74.1	85.7	136.8	74.1	94.9	109.8	79.8	89.8	80.4	74.1	76.2	85.5	63.4	70.8
0950, 0980	Ceilings & Acoustic Treatment	95.3	74.1	81.3	86.8	74.1	78.4	95.3	74.1	81.3	86.8	79.8	82.2	96.1	74.1	81.5	103.6	63.4	77.0
0960	Flooring	115.0	70.6	102.3	107.4	52.7	91.6	120.6	52.7	101.1	108.6	52.7	92.5	118.6	70.6	104.8	117.2	81.8	107.0
0970, 0990	Wall Finishes & Painting/Coating	103.9	66.5	82.2	102.2	66.6	81.5	103.9	66.6	82.2	102.2	26.5	58.2	103.9	55.3	75.7	103.9	51.8	73.6
09	FINISHES	109.7	71.9	88.8	104.0	68.9	84.6	111.3	68.3	87.5	103.0	67.1	83.1	104.2	70.7	85.7	104.3	66.5	83.4
COVERS	DIVS. 10 - 14, 25, 28, 41, 43, 44, 46	100.0	88.4	97.5	100.0	88.3	97.5	100.0	88.5	97.5	100.0	88.7	97.6	100.0	88.7	97.6	100.0	88.0	97.4
21, 22, 23	FIRE SUPPRESSION, PLUMBING & HVAC	96.6	75.2	87.5	96.7	75.9	87.9	99.9	86.6	94.3	100.1	74.9	89.4	96.6	86.9	92.5	99.9	71.1	87.8
26, 27, 3370	ELECTRICAL, COMMUNICATIONS & UTIL.	95.1	68.9	81.8	96.5	83.8	90.1	97.3	53.4	75.1	96.2	83.8	89.9	97.3	55.9	76.4	98.2	72.5	85.2
MF2014	WEIGHTED AVERAGE	104.5	76.4	92.1	101.2	77.8	90.9	106.5	75.6	92.9	99.7	74.0	88.4	103.4	76.5	91.6	102.4	74.7	90.2

City Cost Indexes

Table 1

DIVISION		COLORADO — SALIDA 812 MAT.	INST.	TOTAL	CONNECTICUT — BRIDGEPORT 066 MAT.	INST.	TOTAL	BRISTOL 060 MAT.	INST.	TOTAL	HARTFORD 061 MAT.	INST.	TOTAL	MERIDEN 064 MAT.	INST.	TOTAL	NEW BRITAIN 060 MAT.	INST.	TOTAL
015433	CONTRACTOR EQUIPMENT		93.6	93.6		98.1	98.1		98.1	98.1		98.1	98.1		98.6	98.6		98.1	98.1
0241, 31 - 34	SITE & INFRASTRUCTURE, DEMOLITION	134.4	91.7	104.2	108.6	103.1	104.7	107.8	103.1	104.5	103.0	103.1	103.1	105.6	103.8	104.4	108.0	103.1	104.5
0310	Concrete Forming & Accessories	110.1	73.0	78.0	102.8	124.8	121.8	102.8	124.7	121.7	100.4	124.7	121.4	102.4	124.7	121.7	103.2	124.6	121.8
0320	Concrete Reinforcing	110.1	75.7	92.2	105.7	129.4	117.9	105.7	129.4	117.9	101.3	129.4	115.8	105.7	129.4	117.9	105.7	129.3	117.9
0330	Cast-in-Place Concrete	112.2	76.5	98.1	97.9	130.2	110.6	91.8	130.2	106.9	96.1	130.2	109.5	88.3	130.2	104.8	93.2	130.1	107.8
03	CONCRETE	114.7	75.0	95.4	102.0	127.1	114.2	98.9	127.1	112.6	99.3	127.1	112.8	97.2	127.1	111.7	99.7	127.1	113.0
04	MASONRY	143.0	71.7	99.1	120.3	139.8	132.3	110.7	139.8	128.6	118.7	139.8	131.7	110.3	139.8	128.5	113.2	139.8	129.6
05	METALS	101.8	79.7	94.8	95.2	125.6	104.9	95.2	125.5	104.8	99.7	125.5	107.9	92.1	125.5	102.7	91.3	125.4	102.1
06	WOOD, PLASTICS & COMPOSITES	98.1	74.8	85.1	104.0	122.4	114.2	104.0	122.4	114.2	89.9	122.4	108.0	104.0	122.4	114.2	104.0	122.4	114.2
07	THERMAL & MOISTURE PROTECTION	103.8	78.8	93.4	98.4	130.8	112.0	98.5	129.4	111.5	103.9	129.4	114.6	98.5	129.4	111.4	98.5	126.6	110.3
08	OPENINGS	95.8	77.6	91.6	97.8	132.5	105.8	97.8	132.5	105.8	97.3	132.5	105.4	100.1	132.5	107.6	97.8	132.5	105.8
0920	Plaster & Gypsum Board	80.8	74.1	76.3	100.8	122.5	115.3	100.8	122.5	115.3	96.6	122.5	113.9	102.0	122.5	115.7	100.8	122.5	115.3
0950, 0980	Ceilings & Acoustic Treatment	96.1	74.1	81.5	86.2	122.5	110.2	86.2	122.5	110.2	88.5	122.5	111.0	89.3	122.5	111.3	86.2	122.5	110.2
0960	Flooring	124.2	70.6	108.8	98.2	133.3	108.3	98.2	129.6	107.2	96.5	129.6	106.0	98.2	129.6	107.2	98.2	129.6	107.2
0970, 0990	Wall Finishes & Painting/Coating	103.9	55.3	75.7	93.0	126.0	112.2	93.0	126.0	112.2	93.9	126.0	112.5	93.0	126.0	112.2	93.0	126.0	112.2
09	FINISHES	104.8	70.7	85.9	94.6	125.8	111.9	94.7	125.2	111.6	93.4	125.2	111.0	95.5	125.2	112.0	94.7	125.2	111.6
COVERS	DIVS. 10 - 14, 25, 28, 41, 43, 44, 46	100.0	88.8	97.6	100.0	113.0	102.8	100.0	113.0	102.8	100.0	113.0	102.8	100.0	113.0	102.8	100.0	113.0	102.8
21, 22, 23	FIRE SUPPRESSION, PLUMBING & HVAC	96.6	71.0	85.8	99.9	118.8	107.9	99.9	118.8	107.9	99.9	118.8	107.9	96.6	118.8	105.9	99.9	118.7	107.9
26, 27, 3370	ELECTRICAL, COMMUNICATIONS & UTIL.	97.5	72.4	84.8	95.9	111.4	103.7	95.9	113.4	104.8	95.4	113.8	104.7	95.8	112.8	104.4	95.9	113.4	104.8
MF2014	WEIGHTED AVERAGE	103.9	75.4	91.3	99.5	122.0	109.4	98.7	122.2	109.0	99.6	122.2	109.6	97.5	122.2	108.4	98.3	122.1	108.8

Table 2 — CONNECTICUT

DIVISION		NEW HAVEN 065 MAT.	INST.	TOTAL	NEW LONDON 063 MAT.	INST.	TOTAL	NORWALK 068 MAT.	INST.	TOTAL	STAMFORD 069 MAT.	INST.	TOTAL	WATERBURY 067 MAT.	INST.	TOTAL	WILLIMANTIC 062 MAT.	INST.	TOTAL
015433	CONTRACTOR EQUIPMENT		98.6	98.6		98.6	98.6		98.1	98.1		98.1	98.1		98.1	98.1		98.1	98.1
0241, 31 - 34	SITE & INFRASTRUCTURE, DEMOLITION	107.8	103.8	105.0	100.2	103.8	102.8	108.4	103.1	104.7	109.0	103.1	104.9	108.4	103.1	104.7	108.4	103.1	104.6
0310	Concrete Forming & Accessories	102.5	124.7	121.7	102.4	124.8	121.7	102.8	125.2	122.1	102.7	125.2	122.1	102.8	124.7	121.7	102.7	124.7	121.8
0320	Concrete Reinforcing	105.7	129.4	117.9	82.8	129.4	106.9	105.7	129.5	118.0	105.7	129.5	118.0	105.7	129.3	117.9	105.7	129.4	117.9
0330	Cast-in-Place Concrete	94.8	131.0	109.1	80.9	131.0	100.6	96.3	131.6	110.2	97.9	131.6	111.2	97.9	130.1	110.6	91.5	130.2	106.7
03	CONCRETE	113.8	127.4	120.4	87.5	127.4	106.9	101.2	127.8	114.1	102.0	127.8	114.5	102.0	127.1	114.2	98.8	127.1	112.6
04	MASONRY	111.0	139.8	128.7	109.4	139.8	128.1	110.4	141.3	129.4	111.2	141.3	129.7	111.2	139.8	128.8	110.6	139.8	128.6
05	METALS	91.5	125.5	102.3	91.2	125.6	102.1	95.2	126.1	105.0	95.2	126.1	105.0	95.2	125.5	104.8	94.9	125.5	104.7
06	WOOD, PLASTICS & COMPOSITES	104.0	122.4	114.2	104.0	122.4	114.2	104.0	122.4	114.2	104.0	122.4	114.2	104.0	122.4	114.2	104.0	122.4	114.2
07	THERMAL & MOISTURE PROTECTION	98.6	128.8	111.2	98.4	129.6	111.5	98.6	131.4	112.3	98.5	131.4	112.3	98.5	128.6	111.1	98.7	129.6	111.6
08	OPENINGS	97.8	132.5	105.8	100.2	132.5	107.7	97.8	132.5	105.8	97.8	132.5	105.8	97.8	132.5	105.8	100.2	132.5	107.7
0920	Plaster & Gypsum Board	100.8	122.5	115.3	100.8	122.5	115.3	100.8	122.5	115.3	100.8	122.5	115.3	100.8	122.5	115.3	100.8	122.5	115.3
0950, 0980	Ceilings & Acoustic Treatment	86.2	122.5	110.2	84.3	122.5	109.6	86.2	122.5	110.2	86.2	122.5	110.2	86.2	122.5	110.2	84.3	122.5	109.6
0960	Flooring	98.2	133.3	108.3	98.2	133.3	108.3	98.2	129.6	107.2	98.2	129.6	107.2	98.2	129.6	107.2	98.2	133.3	108.3
0970, 0990	Wall Finishes & Painting/Coating	93.0	126.0	112.2	93.0	130.3	114.7	93.0	126.0	112.2	93.0	126.0	112.2	93.0	126.0	112.2	93.0	126.0	112.2
09	FINISHES	94.7	125.8	111.9	93.8	126.3	111.8	94.7	125.2	111.6	94.8	125.2	111.6	94.6	125.2	111.5	94.4	125.8	111.8
COVERS	DIVS. 10 - 14, 25, 28, 41, 43, 44, 46	100.0	113.0	102.8	100.0	113.0	102.8	100.0	113.2	102.9	100.0	113.2	102.9	100.0	113.0	102.8	100.0	113.0	102.8
21, 22, 23	FIRE SUPPRESSION, PLUMBING & HVAC	99.9	118.8	107.9	96.6	118.8	106.0	99.9	118.8	107.9	99.9	118.8	107.9	99.9	118.7	107.9	99.9	118.8	107.9
26, 27, 3370	ELECTRICAL, COMMUNICATIONS & UTIL.	95.8	112.8	104.4	93.0	113.4	103.4	95.9	165.9	131.3	95.9	165.9	131.3	95.4	111.4	103.5	95.9	113.4	104.8
MF2014	WEIGHTED AVERAGE	99.8	122.3	109.7	95.6	122.4	107.4	99.0	129.7	112.5	99.1	129.7	112.6	99.0	121.9	109.1	98.9	122.3	109.2

Table 3

DIVISION		D.C. — WASHINGTON 200 - 205 MAT.	INST.	TOTAL	DELAWARE — DOVER 199 MAT.	INST.	TOTAL	NEWARK 197 MAT.	INST.	TOTAL	WILMINGTON 198 MAT.	INST.	TOTAL	FLORIDA — DAYTONA BEACH 321 MAT.	INST.	TOTAL	FORT LAUDERDALE 333 MAT.	INST.	TOTAL
015433	CONTRACTOR EQUIPMENT		108.3	108.3		117.5	117.5		117.5	117.5		117.7	117.7		99.3	99.3		92.1	92.1
0241, 31 - 34	SITE & INFRASTRUCTURE, DEMOLITION	102.8	97.6	99.1	104.3	111.7	109.6	103.3	111.8	109.3	103.5	112.2	109.6	103.1	89.8	93.7	95.1	77.3	82.5
0310	Concrete Forming & Accessories	99.3	93.4	94.2	98.6	101.4	101.0	99.8	101.4	101.2	99.0	101.4	101.1	101.0	70.2	74.4	97.5	66.7	70.8
0320	Concrete Reinforcing	101.8	93.0	97.3	101.0	101.8	101.4	97.6	101.8	99.8	101.0	101.8	101.4	94.4	74.4	84.0	90.8	68.5	79.2
0330	Cast-in-Place Concrete	112.7	87.2	102.6	98.9	103.8	100.8	86.0	103.8	93.0	93.7	103.8	97.7	90.1	72.8	83.2	95.4	70.3	85.5
03	CONCRETE	106.6	92.2	99.6	101.4	103.3	102.3	94.0	103.3	98.5	98.9	103.3	101.0	94.3	73.1	84.0	97.5	69.6	83.9
04	MASONRY	96.6	85.3	89.6	113.3	100.7	105.5	108.0	100.7	103.5	112.4	100.7	105.2	96.1	65.5	77.2	102.0	69.9	82.2
05	METALS	101.2	109.0	103.7	99.5	117.0	105.1	101.6	117.1	106.5	99.5	117.1	105.1	98.9	92.0	96.7	97.4	89.2	94.8
06	WOOD, PLASTICS & COMPOSITES	95.2	96.6	96.0	92.1	100.2	96.6	100.1	100.2	100.2	87.4	100.2	94.5	100.8	69.9	83.6	87.9	67.4	76.5
07	THERMAL & MOISTURE PROTECTION	104.4	89.0	97.9	105.6	114.3	109.2	107.9	114.3	110.6	104.6	114.3	108.6	98.0	73.8	87.9	104.4	71.1	90.4
08	OPENINGS	98.7	98.5	98.7	89.1	109.4	93.9	93.1	109.4	96.9	87.0	109.4	92.2	95.7	69.9	89.7	95.4	67.0	88.8
0920	Plaster & Gypsum Board	103.2	96.5	98.7	102.8	100.2	101.1	104.6	100.2	101.6	102.0	100.2	100.8	85.8	69.4	74.9	95.8	66.8	76.5
0950, 0980	Ceilings & Acoustic Treatment	104.3	96.5	99.1	93.0	100.2	97.7	90.5	100.2	96.9	92.4	100.2	97.6	83.3	69.4	74.1	88.7	66.8	74.2
0960	Flooring	94.5	93.7	94.2	95.2	108.6	99.0	92.2	108.6	96.9	94.7	108.6	98.7	108.7	63.8	95.8	104.6	77.3	96.8
0970, 0990	Wall Finishes & Painting/Coating	102.9	80.9	90.1	90.7	109.5	101.7	89.3	109.5	101.1	85.4	109.5	99.5	97.4	70.6	81.8	89.2	70.6	78.4
09	FINISHES	96.2	92.6	94.2	94.1	102.9	99.0	93.1	102.9	98.5	93.9	102.9	98.9	94.4	68.4	80.0	94.7	68.7	80.3
COVERS	DIVS. 10 - 14, 25, 28, 41, 43, 44, 46	100.0	101.0	100.2	100.0	101.2	100.3	100.0	101.2	100.3	100.0	101.2	100.3	100.0	86.6	97.1	100.0	86.6	97.1
21, 22, 23	FIRE SUPPRESSION, PLUMBING & HVAC	100.0	92.6	96.9	100.0	116.9	107.2	100.3	116.9	107.3	100.2	116.9	107.3	100.0	78.0	90.7	100.0	62.4	84.1
26, 27, 3370	ELECTRICAL, COMMUNICATIONS & UTIL.	100.5	105.0	102.8	97.1	111.1	104.1	98.5	111.1	104.8	100.2	111.1	105.7	94.2	68.1	81.0	94.8	72.2	83.3
MF2014	WEIGHTED AVERAGE	100.6	95.9	98.5	99.0	109.4	103.6	98.9	109.4	103.5	98.7	109.4	103.4	97.5	75.5	87.8	97.8	71.2	86.1

For customer support on your Mechanical Cost Data, call 877.739.3689.

663

FLORIDA

DIVISION		FORT MYERS 339,341			GAINESVILLE 326,344			JACKSONVILLE 320,322			LAKELAND 338			MELBOURNE 329			MIAMI 330 - 332,340		
		MAT.	INST.	TOTAL	MAT.	INST.	TOTAL	MAT.	INST.	TOTAL	MAT.	INST.	TOTAL	MAT.	INST.	TOTAL	MAT.	INST.	TOTAL
015433	CONTRACTOR EQUIPMENT		99.3	99.3		99.3	99.3		99.3	99.3		99.3	99.3		99.3	99.3		92.1	92.1
0241, 31 - 34	SITE & INFRASTRUCTURE, DEMOLITION	106.0	89.3	94.2	110.9	89.6	95.8	103.1	89.6	93.6	107.9	89.6	95.0	109.9	89.9	95.8	96.4	77.8	83.3
0310	Concrete Forming & Accessories	93.4	66.2	69.9	95.8	60.5	65.3	100.8	60.5	65.9	89.8	68.8	71.6	97.1	71.0	74.5	102.2	69.8	74.2
0320	Concrete Reinforcing	91.8	67.8	79.4	100.2	67.9	83.4	94.4	67.8	80.7	94.1	89.3	91.6	95.5	75.5	85.1	97.5	68.2	82.3
0330	Cast-in-Place Concrete	99.5	70.1	87.9	103.4	70.7	90.5	91.0	70.6	82.9	101.8	72.0	90.0	108.6	73.8	94.9	92.1	75.3	85.5
03	CONCRETE	98.2	69.2	84.1	105.7	66.9	86.8	94.7	66.9	81.2	100.0	74.9	87.8	106.1	74.0	90.5	97.4	72.6	85.3
04	MASONRY	95.2	58.6	72.7	110.2	60.9	79.8	95.8	60.9	74.3	112.5	75.4	89.7	93.8	67.4	77.5	102.5	74.7	85.4
05	METALS	99.5	88.0	95.8	97.8	89.1	95.1	97.5	89.0	94.8	99.4	97.4	98.8	107.1	92.4	102.4	97.9	88.7	95.0
06	WOOD, PLASTICS & COMPOSITES	84.7	67.4	75.1	94.3	59.0	74.7	100.8	59.0	77.5	80.1	68.9	73.8	96.1	69.9	81.5	93.7	67.4	79.0
07	THERMAL & MOISTURE PROTECTION	104.2	68.1	89.1	98.3	68.2	85.7	98.2	68.2	85.7	104.1	73.1	91.1	98.4	72.3	87.5	104.0	73.8	91.4
08	OPENINGS	96.7	67.0	89.7	95.3	62.4	87.7	95.7	62.4	87.9	96.6	72.9	91.1	94.9	70.2	89.2	98.9	67.0	91.5
0920	Plaster & Gypsum Board	91.9	66.8	75.2	82.3	58.3	66.2	85.8	58.3	67.4	88.9	68.4	75.2	82.3	69.4	73.7	93.8	66.8	75.8
0950, 0980	Ceilings & Acoustic Treatment	84.4	66.8	72.8	77.8	58.3	64.9	83.3	58.3	66.7	84.4	68.4	73.8	82.4	69.4	73.8	93.0	66.8	75.7
0960	Flooring	101.4	53.0	87.5	105.8	70.8	95.7	108.7	62.0	95.2	99.1	54.3	86.2	106.0	63.8	93.9	106.5	70.2	96.0
0970, 0990	Wall Finishes & Painting/Coating	93.3	70.6	80.1	97.4	70.6	81.8	97.4	70.6	81.8	93.3	70.6	80.1	97.4	93.2	95.0	86.6	70.6	77.3
09	FINISHES	94.2	63.8	77.4	92.7	62.5	76.0	94.5	60.8	75.8	93.2	65.7	78.0	93.6	71.3	81.2	96.5	69.5	81.6
COVERS	DIVS. 10 - 14, 25, 28, 41, 43, 44, 46	100.0	81.6	96.0	100.0	84.3	96.6	100.0	82.4	96.2	100.0	82.7	96.3	100.0	87.2	97.2	100.0	89.6	97.8
21, 22, 23	FIRE SUPPRESSION, PLUMBING & HVAC	98.2	62.0	82.9	98.7	65.5	84.7	100.0	65.5	85.4	98.2	63.5	83.5	100.0	80.4	91.7	100.0	60.0	83.1
26, 27, 3370	ELECTRICAL, COMMUNICATIONS & UTIL.	96.6	63.0	79.6	94.5	60.9	77.5	93.9	68.7	81.2	95.0	60.4	77.5	95.2	67.3	81.1	98.3	75.3	86.6
MF2014	WEIGHTED AVERAGE	98.1	68.8	85.1	99.0	68.8	85.7	97.3	69.6	85.1	98.8	72.7	87.3	100.1	76.6	89.8	98.9	72.3	87.2

FLORIDA

DIVISION		ORLANDO 327 - 328,347			PANAMA CITY 324			PENSACOLA 325			SARASOTA 342			ST. PETERSBURG 337			TALLAHASSEE 323		
		MAT.	INST.	TOTAL	MAT.	INST.	TOTAL	MAT.	INST.	TOTAL	MAT.	INST.	TOTAL	MAT.	INST.	TOTAL	MAT.	INST.	TOTAL
015433	CONTRACTOR EQUIPMENT		99.3	99.3		99.3	99.3		99.3	99.3		99.3	99.3		99.3	99.3		99.3	99.3
0241, 31 - 34	SITE & INFRASTRUCTURE, DEMOLITION	102.4	89.8	93.5	114.4	89.1	96.5	114.7	89.5	96.9	113.8	89.6	96.7	109.5	89.4	95.3	100.6	89.2	92.5
0310	Concrete Forming & Accessories	104.1	69.9	74.6	100.1	48.9	55.8	98.0	76.3	79.3	97.6	68.8	72.7	97.0	66.1	70.3	102.7	58.1	64.1
0320	Concrete Reinforcing	100.2	75.9	87.6	98.6	81.8	89.9	101.1	82.2	91.3	92.6	89.3	90.9	94.1	89.0	91.5	99.8	67.7	83.2
0330	Cast-in-Place Concrete	100.9	72.6	89.7	95.6	67.1	84.4	118.1	73.0	100.3	106.8	72.0	93.1	102.9	67.9	89.1	99.2	67.3	86.6
03	CONCRETE	101.1	73.3	87.6	103.8	63.1	84.0	114.1	77.4	96.2	102.7	75.0	89.2	101.7	72.4	87.4	101.0	64.6	83.3
04	MASONRY	99.2	65.5	78.4	100.7	47.1	67.6	121.9	64.4	86.4	98.2	75.4	84.2	154.9	62.9	98.2	99.3	53.6	71.1
05	METALS	97.9	92.3	96.1	98.7	93.1	96.9	99.8	94.2	98.0	100.3	97.5	99.4	100.3	97.7	99.5	95.5	88.7	93.3
06	WOOD, PLASTICS & COMPOSITES	93.5	69.9	80.4	99.6	46.9	70.3	97.4	78.4	86.8	99.2	68.9	82.3	89.3	68.9	77.9	100.6	59.0	77.4
07	THERMAL & MOISTURE PROTECTION	98.9	73.8	88.4	98.5	61.1	82.9	98.4	72.5	87.6	99.8	73.1	88.6	104.3	68.4	89.3	99.8	65.0	85.3
08	OPENINGS	101.9	70.3	94.6	93.7	59.4	85.7	93.7	76.6	89.7	99.2	72.9	93.1	95.4	72.9	90.1	100.1	62.4	91.3
0920	Plaster & Gypsum Board	89.4	69.4	76.0	85.0	45.9	58.8	88.0	78.3	81.5	91.5	68.4	76.1	94.3	68.4	77.0	98.0	58.3	71.5
0950, 0980	Ceilings & Acoustic Treatment	90.4	69.4	76.5	82.4	45.9	58.2	82.4	78.3	79.7	85.3	68.4	74.1	86.3	68.4	74.4	90.6	58.3	69.2
0960	Flooring	105.1	63.8	93.2	108.2	41.9	89.1	103.5	60.5	91.1	110.1	55.6	94.4	103.4	68.2	93.3	106.9	59.6	93.3
0970, 0990	Wall Finishes & Painting/Coating	96.9	70.6	81.6	97.4	70.6	81.8	97.4	70.6	81.8	98.3	70.6	82.2	93.3	73.4	81.7	92.1	70.6	79.6
09	FINISHES	96.5	68.4	81.0	95.2	48.2	69.2	93.9	73.1	82.4	98.7	65.9	80.6	95.8	66.8	79.8	97.0	58.8	75.9
COVERS	DIVS. 10 - 14, 25, 28, 41, 43, 44, 46	100.0	86.6	97.1	100.0	75.1	94.6	100.0	81.9	96.1	100.0	82.7	96.3	100.0	80.1	95.7	100.0	78.3	95.3
21, 22, 23	FIRE SUPPRESSION, PLUMBING & HVAC	100.0	61.2	83.6	100.0	52.9	80.1	100.0	58.1	82.3	99.9	63.3	84.4	100.0	59.5	82.9	100.1	56.3	81.6
26, 27, 3370	ELECTRICAL, COMMUNICATIONS & UTIL.	98.6	67.8	83.0	93.0	60.6	76.6	96.5	55.4	75.7	96.4	68.7	82.4	95.0	60.4	77.5	97.9	60.9	79.1
MF2014	WEIGHTED AVERAGE	99.5	72.0	87.4	98.8	62.0	82.6	101.4	71.0	88.0	100.0	73.8	88.5	101.8	70.2	87.9	98.9	65.0	84.0

DIVISION		FLORIDA TAMPA 335 - 336,346			FLORIDA WEST PALM BEACH 334,349			GEORGIA ALBANY 317,398			GEORGIA ATHENS 306			GEORGIA ATLANTA 300 - 303,399			GEORGIA AUGUSTA 308 - 309		
		MAT.	INST.	TOTAL	MAT.	INST.	TOTAL	MAT.	INST.	TOTAL	MAT.	INST.	TOTAL	MAT.	INST.	TOTAL	MAT.	INST.	TOTAL
015433	CONTRACTOR EQUIPMENT		99.3	99.3		92.1	92.1		93.1	93.1		94.5	94.5		95.0	95.0		94.5	94.5
0241, 31 - 34	SITE & INFRASTRUCTURE, DEMOLITION	110.0	89.6	95.6	92.0	77.3	81.6	100.3	78.6	84.9	102.0	94.2	96.5	98.7	95.8	96.7	95.5	94.4	94.7
0310	Concrete Forming & Accessories	100.0	68.8	73.0	101.0	66.5	71.1	94.4	42.5	49.5	94.5	45.7	52.3	98.6	81.2	83.6	95.8	77.4	79.9
0320	Concrete Reinforcing	90.8	89.3	90.0	93.4	67.0	79.7	90.3	79.4	84.6	95.4	77.1	85.9	94.7	80.3	87.2	95.8	73.4	84.2
0330	Cast-in-Place Concrete	100.5	72.0	89.3	90.7	70.2	82.6	91.2	70.4	83.0	108.1	71.5	93.7	108.1	71.9	93.8	102.1	71.7	90.1
03	CONCRETE	100.2	75.0	87.9	94.1	69.2	82.0	94.9	60.6	78.2	102.7	61.2	82.5	100.4	77.8	89.4	95.7	74.8	85.5
04	MASONRY	103.0	75.4	86.0	101.5	61.7	77.0	98.8	51.6	69.7	80.0	55.1	64.6	93.6	66.4	76.8	93.9	54.4	69.5
05	METALS	99.4	97.4	98.7	96.5	88.1	93.8	98.4	87.4	95.0	94.0	73.7	87.5	94.9	76.2	89.0	93.7	73.1	87.1
06	WOOD, PLASTICS & COMPOSITES	93.3	68.9	79.7	92.8	67.4	78.6	88.8	34.8	58.7	98.0	39.5	65.4	102.3	86.2	93.3	99.6	82.4	90.0
07	THERMAL & MOISTURE PROTECTION	104.6	73.1	91.4	104.1	69.6	89.7	98.6	59.7	82.3	95.3	55.0	78.5	95.1	70.7	84.9	94.9	66.0	82.8
08	OPENINGS	96.6	72.9	91.1	94.9	66.7	88.4	90.1	49.5	80.6	88.7	51.6	80.0	93.8	77.6	90.0	88.7	74.3	85.4
0920	Plaster & Gypsum Board	96.5	68.4	77.7	99.9	66.8	77.8	96.6	33.4	54.4	94.9	38.1	57.0	97.3	86.1	89.8	96.3	82.2	86.9
0950, 0980	Ceilings & Acoustic Treatment	88.7	68.4	75.3	84.4	66.8	72.8	82.7	33.4	49.9	96.1	38.1	57.7	96.1	86.1	89.5	97.0	82.2	87.2
0960	Flooring	104.6	54.3	92.4	106.6	63.9	94.3	107.1	45.9	89.5	98.6	51.8	85.1	100.2	73.2	92.4	98.8	50.5	84.9
0970, 0990	Wall Finishes & Painting/Coating	93.3	70.6	80.1	89.2	70.6	78.4	90.1	68.3	77.4	96.8	67.8	79.9	96.8	68.3	80.2	96.8	68.3	80.2
09	FINISHES	97.1	65.7	79.7	94.6	66.0	78.8	96.8	43.2	67.1	99.2	47.1	70.4	99.6	79.3	88.4	99.0	72.5	84.3
COVERS	DIVS. 10 - 14, 25, 28, 41, 43, 44, 46	100.0	84.8	96.7	100.0	86.6	97.1	100.0	80.8	95.8	100.0	81.2	95.9	100.0	86.8	97.1	100.0	86.0	97.0
21, 22, 23	FIRE SUPPRESSION, PLUMBING & HVAC	100.0	63.5	84.6	98.2	61.9	82.9	100.0	69.3	87.0	96.7	69.2	85.1	100.0	73.6	88.9	100.1	61.4	83.7
26, 27, 3370	ELECTRICAL, COMMUNICATIONS & UTIL.	94.7	60.4	77.4	95.8	73.3	84.4	93.3	60.4	76.6	100.9	69.0	84.8	100.2	70.9	85.4	101.5	65.4	83.2
MF2014	WEIGHTED AVERAGE	99.3	72.8	87.6	96.9	69.9	85.0	96.9	63.1	82.0	96.3	65.2	82.6	98.1	76.5	88.6	96.8	69.9	85.0

GEORGIA

	DIVISION	COLUMBUS 318 - 319			DALTON 307			GAINESVILLE 305			MACON 310 - 312			SAVANNAH 313 - 314			STATESBORO 304		
		MAT.	INST.	TOTAL	MAT.	INST.	TOTAL	MAT.	INST.	TOTAL	MAT.	INST.	TOTAL	MAT.	INST.	TOTAL	MAT.	INST.	TOTAL
015433	CONTRACTOR EQUIPMENT		93.1	93.1		108.2	108.2		94.5	94.5		103.3	103.3		94.0	94.0		95.3	95.3
0241, 31 - 34	SITE & INFRASTRUCTURE, DEMOLITION	100.2	78.4	84.8	102.0	99.5	100.2	102.0	94.1	96.4	101.9	93.6	96.0	100.4	79.7	85.8	103.0	78.4	85.6
0310	Concrete Forming & Accessories	94.4	49.9	55.9	87.1	49.4	54.5	98.3	43.0	50.5	93.9	48.0	54.2	95.6	59.2	64.1	81.6	54.1	57.8
0320	Concrete Reinforcing	90.3	80.0	85.0	94.9	73.2	83.6	95.2	77.1	85.8	91.5	79.4	85.3	97.5	73.9	85.3	94.5	41.5	67.0
0330	Cast-in-Place Concrete	90.9	70.2	82.7	105.0	70.5	91.4	113.6	71.3	96.9	89.6	71.5	82.5	94.7	70.4	85.1	107.9	70.1	93.0
03	CONCRETE	94.7	64.0	79.8	101.6	62.6	82.7	104.7	59.8	82.9	94.3	63.5	79.3	97.9	67.1	82.9	102.0	58.9	81.0
04	MASONRY	98.9	62.9	76.7	81.1	46.1	59.5	88.6	55.5	68.2	112.0	51.6	74.8	95.6	62.6	75.3	82.9	49.8	62.5
05	METALS	98.0	87.7	94.7	94.9	85.1	91.8	93.2	73.1	86.8	93.8	86.9	91.6	94.9	86.4	92.2	98.4	75.7	91.2
06	WOOD, PLASTICS & COMPOSITES	88.8	44.8	64.3	80.6	45.8	61.2	102.1	36.6	65.6	95.9	42.3	66.1	95.4	57.3	74.2	73.8	52.0	61.7
07	THERMAL & MOISTURE PROTECTION	98.5	63.6	83.9	97.3	60.0	81.7	95.3	62.2	81.4	97.0	62.3	82.5	96.3	64.0	82.8	95.9	60.5	81.1
08	OPENINGS	90.1	55.0	81.9	89.5	54.4	81.3	88.7	50.0	79.7	89.8	53.7	81.4	97.7	60.6	89.1	90.4	50.7	81.1
0920	Plaster & Gypsum Board	96.6	43.7	61.2	83.6	44.6	57.6	97.3	35.1	55.7	100.7	41.1	60.9	94.8	56.5	69.2	82.9	51.0	61.6
0950, 0980	Ceilings & Acoustic Treatment	82.2	43.7	56.7	108.1	44.6	66.1	96.1	35.1	55.7	77.5	41.1	53.4	89.7	56.5	67.7	104.5	51.0	69.1
0960	Flooring	107.1	45.9	89.5	99.6	45.1	83.9	100.2	44.8	84.2	84.3	45.9	73.2	105.0	64.6	93.4	117.3	43.1	96.0
0970, 0990	Wall Finishes & Painting/Coating	90.1	68.3	77.4	87.5	68.3	76.3	96.8	68.3	80.2	92.2	68.3	78.3	88.8	68.3	76.9	94.5	68.3	79.3
09	FINISHES	96.7	49.3	70.5	108.3	48.9	75.5	99.8	43.4	68.6	85.4	47.7	64.6	96.7	60.3	76.6	111.5	52.9	79.1
COVERS	DIVS. 10 - 14, 25, 28, 41, 43, 44, 46	100.0	82.0	96.1	100.0	27.7	84.3	100.0	38.7	86.7	100.0	81.8	96.1	100.0	82.9	96.3	100.0	45.2	88.1
21, 22, 23	FIRE SUPPRESSION, PLUMBING & HVAC	100.0	64.1	84.8	96.7	61.7	81.9	96.7	69.6	85.3	100.0	70.3	87.5	100.1	64.0	84.8	97.3	66.8	84.4
26, 27, 3370	ELECTRICAL, COMMUNICATIONS & UTIL.	93.4	69.6	81.4	109.4	67.2	88.1	100.9	70.7	85.6	92.5	63.0	77.6	97.2	64.2	80.5	101.0	56.9	78.7
MF2014	WEIGHTED AVERAGE	96.8	66.1	83.3	98.0	63.0	82.6	96.9	63.6	82.2	95.8	66.2	82.7	97.8	67.6	84.5	98.2	61.0	81.8

	DIVISION	GEORGIA VALDOSTA 316			WAYCROSS 315			HAWAII HILO 967			HONOLULU 968			STATES & POSS., GUAM 969			IDAHO BOISE 836 - 837		
		MAT.	INST.	TOTAL	MAT.	INST.	TOTAL	MAT.	INST.	TOTAL	MAT.	INST.	TOTAL	MAT.	INST.	TOTAL	MAT.	INST.	TOTAL
015433	CONTRACTOR EQUIPMENT		93.1	93.1		93.1	93.1		98.9	98.9		98.9	98.9		165.6	165.6		96.7	96.7
0241, 31 - 34	SITE & INFRASTRUCTURE, DEMOLITION	110.0	78.5	87.8	106.7	77.3	85.9	156.6	105.9	120.8	166.0	105.9	123.5	201.7	103.2	132.0	88.3	95.3	93.3
0310	Concrete Forming & Accessories	84.3	43.3	48.9	86.3	67.4	70.0	109.6	135.3	131.8	122.2	135.3	133.5	112.1	62.1	68.9	100.9	75.6	79.0
0320	Concrete Reinforcing	92.3	75.6	83.6	92.3	71.9	81.8	137.7	116.2	126.6	160.1	116.2	137.4	241.9	29.9	132.1	105.3	79.9	92.1
0330	Cast-in-Place Concrete	89.3	70.3	81.8	100.9	70.0	88.7	201.0	126.7	171.7	161.7	126.7	147.9	174.3	106.8	147.6	88.7	88.4	88.6
03	CONCRETE	100.1	60.3	80.8	103.2	70.1	87.1	165.3	127.6	146.9	156.0	127.6	142.2	167.5	71.9	121.1	102.7	80.9	92.1
04	MASONRY	104.6	50.4	71.2	105.4	49.7	71.1	162.6	129.3	142.0	160.0	129.3	141.1	229.1	43.4	114.6	130.2	87.3	103.8
05	METALS	97.6	85.9	93.9	96.6	80.3	91.4	120.1	105.4	115.5	134.0	105.4	125.0	154.9	76.7	130.0	107.6	80.4	98.9
06	WOOD, PLASTICS & COMPOSITES	76.6	36.3	54.2	78.3	70.0	73.6	111.1	138.5	126.3	127.4	138.5	133.6	123.4	65.5	91.1	89.6	72.5	80.1
07	THERMAL & MOISTURE PROTECTION	98.8	58.6	82.0	98.6	62.0	83.3	127.2	123.6	125.7	144.7	123.6	135.9	149.2	65.2	114.0	94.7	80.2	88.6
08	OPENINGS	86.5	49.6	77.9	86.8	64.6	81.6	120.2	131.1	122.7	133.9	131.1	133.2	127.0	52.7	109.7	96.4	66.2	89.4
0920	Plaster & Gypsum Board	89.0	35.0	52.9	89.0	69.5	76.0	106.8	139.3	128.5	149.8	139.3	142.8	214.0	53.6	106.8	95.6	71.7	79.6
0950, 0980	Ceilings & Acoustic Treatment	79.8	35.0	50.1	77.9	69.5	72.4	122.9	139.3	133.7	132.3	139.3	136.9	232.9	53.6	114.2	96.2	71.7	80.0
0960	Flooring	100.2	45.9	84.6	101.5	28.7	80.6	118.5	134.5	123.1	135.8	134.5	135.4	139.7	44.6	112.3	99.1	80.5	93.7
0970, 0990	Wall Finishes & Painting/Coating	90.1	68.3	77.4	90.1	68.3	77.4	99.0	143.8	125.1	107.2	143.8	128.5	104.0	35.9	64.4	94.0	32.6	58.3
09	FINISHES	94.0	44.1	66.4	93.6	60.7	75.4	114.0	137.4	126.9	129.4	137.4	133.8	187.5	57.5	115.6	96.4	72.0	82.9
COVERS	DIVS. 10 - 14, 25, 28, 41, 43, 44, 46	100.0	80.6	95.8	100.0	53.4	89.9	100.0	116.5	103.6	100.0	116.5	103.6	100.0	74.0	94.4	100.0	87.5	97.3
21, 22, 23	FIRE SUPPRESSION, PLUMBING & HVAC	100.0	69.4	87.1	97.9	64.1	83.6	100.4	107.4	103.3	100.5	107.4	103.4	103.3	37.1	75.3	100.0	72.7	88.5
26, 27, 3370	ELECTRICAL, COMMUNICATIONS & UTIL.	91.7	55.8	73.6	95.5	56.9	76.0	106.2	120.2	113.3	107.7	120.2	114.0	149.9	40.9	94.8	97.6	70.8	84.0
MF2014	WEIGHTED AVERAGE	97.0	62.3	81.7	97.1	64.4	82.7	120.2	119.7	120.0	125.0	119.7	122.7	143.0	57.6	105.4	101.5	77.8	91.1

	DIVISION	IDAHO COEUR D'ALENE 838			IDAHO FALLS 834			LEWISTON 835			POCATELLO 832			TWIN FALLS 833			ILLINOIS BLOOMINGTON 617		
		MAT.	INST.	TOTAL	MAT.	INST.	TOTAL	MAT.	INST.	TOTAL	MAT.	INST.	TOTAL	MAT.	INST.	TOTAL	MAT.	INST.	TOTAL
015433	CONTRACTOR EQUIPMENT		91.5	91.5		96.7	96.7		91.5	91.5		96.7	96.7		96.7	96.7		102.9	102.9
0241, 31 - 34	SITE & INFRASTRUCTURE, DEMOLITION	87.0	90.9	89.7	85.9	95.0	92.3	94.0	91.7	92.4	89.1	95.3	93.5	96.0	96.1	96.1	96.6	99.5	98.7
0310	Concrete Forming & Accessories	111.9	79.3	83.7	94.6	78.8	80.9	117.2	80.5	85.5	101.1	75.4	78.9	102.2	53.1	59.7	84.8	120.0	115.2
0320	Concrete Reinforcing	112.6	95.5	103.7	107.2	79.4	92.8	112.6	95.8	103.9	105.7	79.6	92.2	107.5	79.0	92.7	95.8	109.6	103.0
0330	Cast-in-Place Concrete	95.9	86.1	92.0	84.4	73.1	80.0	99.6	86.9	94.6	91.2	88.3	90.1	93.5	62.7	81.4	98.6	114.4	104.8
03	CONCRETE	109.3	84.7	97.3	94.3	77.0	85.9	113.0	85.5	99.7	101.7	80.7	91.5	109.9	62.0	86.6	99.4	116.1	107.5
04	MASONRY	131.7	83.5	102.0	125.1	80.7	97.7	132.1	86.8	104.2	127.5	82.9	100.0	130.3	80.7	99.7	119.3	119.5	119.4
05	METALS	101.1	86.2	96.4	115.9	78.4	104.0	100.5	87.4	96.4	116.0	79.9	104.5	116.0	78.2	104.0	95.3	113.8	101.2
06	WOOD, PLASTICS & COMPOSITES	93.3	78.3	84.9	83.7	80.1	81.7	98.7	78.3	87.3	89.6	72.5	80.1	90.7	45.6	65.6	83.5	119.3	103.4
07	THERMAL & MOISTURE PROTECTION	148.2	82.1	120.5	94.1	70.1	84.1	148.4	86.1	122.3	94.6	72.5	85.3	95.3	72.6	85.8	98.0	112.5	104.1
08	OPENINGS	113.8	70.9	103.8	99.4	68.3	92.2	107.3	78.5	100.6	97.1	65.5	89.8	100.2	47.8	88.0	97.7	117.6	102.3
0920	Plaster & Gypsum Board	172.6	77.7	109.2	80.6	79.5	79.9	174.0	77.7	109.7	82.5	71.7	75.3	84.2	44.0	57.4	92.2	119.9	110.7
0950, 0980	Ceilings & Acoustic Treatment	127.5	77.7	94.5	97.0	79.5	85.4	127.5	77.7	94.5	103.6	71.7	82.5	99.4	44.0	62.7	87.0	119.9	108.8
0960	Flooring	140.8	81.3	123.7	98.8	41.4	82.3	144.3	92.2	129.3	102.5	80.5	96.2	103.8	41.4	85.9	91.9	114.1	98.2
0970, 0990	Wall Finishes & Painting/Coating	112.8	74.1	90.3	94.0	40.3	62.8	112.8	74.1	90.3	93.9	40.3	62.7	94.0	37.3	61.0	84.4	107.9	98.1
09	FINISHES	161.9	79.3	116.2	93.9	69.1	80.2	163.3	81.6	118.1	97.2	72.8	83.7	97.6	48.2	70.3	91.4	118.1	106.2
COVERS	DIVS. 10 - 14, 25, 28, 41, 43, 44, 46	100.0	86.9	97.2	100.0	46.7	88.4	100.0	87.1	97.2	100.0	87.5	97.3	100.0	42.8	87.6	100.0	105.6	101.2
21, 22, 23	FIRE SUPPRESSION, PLUMBING & HVAC	99.6	81.9	92.1	101.0	70.9	88.2	100.9	85.3	94.3	99.9	72.7	88.4	99.9	68.5	86.7	96.5	107.7	101.2
26, 27, 3370	ELECTRICAL, COMMUNICATIONS & UTIL.	90.9	77.0	83.9	90.8	70.5	80.5	89.0	81.9	85.4	95.5	70.5	82.8	92.1	59.7	75.7	96.3	93.9	95.1
MF2014	WEIGHTED AVERAGE	109.3	82.2	97.4	101.1	74.3	89.3	109.4	85.0	98.6	102.5	77.2	91.3	103.8	66.3	87.3	97.7	110.0	103.1

For customer support on your Mechanical Cost Data, call 877.739.3689.

665

City Cost Indexes

ILLINOIS

DIVISION		CARBONDALE 629 MAT.	INST.	TOTAL	CENTRALIA 628 MAT.	INST.	TOTAL	CHAMPAIGN 618-619 MAT.	INST.	TOTAL	CHICAGO 606-608 MAT.	INST.	TOTAL	DECATUR 625 MAT.	INST.	TOTAL	EAST ST. LOUIS 620-622 MAT.	INST.	TOTAL
015433	CONTRACTOR EQUIPMENT		110.4	110.4		110.4	110.4		103.8	103.8		98.3	98.3		103.8	103.8		110.4	110.4
0241, 31 - 34	SITE & INFRASTRUCTURE, DEMOLITION	97.6	100.7	99.8	97.9	101.4	100.4	105.8	100.4	102.0	101.2	102.7	102.2	93.9	100.5	98.6	100.0	101.4	101.0
0310	Concrete Forming & Accessories	92.1	105.7	103.9	93.8	113.3	110.6	91.4	116.9	113.5	98.9	156.0	148.3	94.3	116.5	113.5	89.5	113.6	110.3
0320	Concrete Reinforcing	92.5	110.3	101.7	92.5	110.6	101.8	95.8	105.2	100.7	99.2	164.2	132.9	90.6	101.7	96.3	92.4	110.6	101.8
0330	Cast-in-Place Concrete	90.9	101.3	95.0	91.4	119.0	102.2	114.3	111.8	113.3	105.0	152.2	123.6	99.3	113.2	104.8	92.9	116.1	102.1
03	CONCRETE	88.5	105.9	97.0	89.0	115.4	101.8	112.4	112.9	112.6	100.3	154.8	126.8	99.7	112.6	106.0	90.0	114.6	102.0
04	MASONRY	83.3	108.1	98.6	83.4	116.2	103.6	144.0	120.1	129.3	98.9	160.9	137.1	78.8	118.2	103.1	83.6	116.2	103.7
05	METALS	93.3	120.1	101.8	93.4	121.5	102.3	95.3	109.1	99.7	95.8	135.5	108.4	96.9	108.3	100.6	94.5	121.0	102.3
06	WOOD, PLASTICS & COMPOSITES	90.7	102.5	97.3	93.1	110.9	103.0	90.4	115.7	104.5	98.3	154.3	129.5	91.4	115.7	104.9	88.0	110.9	100.8
07	THERMAL & MOISTURE PROTECTION	97.8	99.9	98.6	97.8	110.1	103.0	98.7	113.7	105.0	100.0	144.5	118.6	103.8	111.2	106.9	97.8	108.2	102.2
08	OPENINGS	91.6	110.9	96.1	91.6	115.5	97.1	98.3	112.0	101.5	112.0	159.2	123.0	103.2	111.0	105.0	91.6	114.9	97.1
0920	Plaster & Gypsum Board	99.9	102.6	101.7	100.9	111.2	107.8	94.5	116.1	108.9	94.5	155.7	135.4	102.5	116.1	111.6	98.9	111.2	107.1
0950, 0980	Ceilings & Acoustic Treatment	87.2	102.6	97.4	87.2	111.2	103.1	87.0	116.1	106.3	96.8	155.7	135.8	93.8	116.1	108.6	87.2	111.2	103.1
0960	Flooring	122.1	115.5	120.2	123.0	111.3	119.6	95.2	104.4	97.8	96.4	148.7	111.4	109.2	116.8	111.4	121.0	111.3	118.2
0970, 0990	Wall Finishes & Painting/Coating	98.4	102.1	100.5	98.4	109.0	104.6	84.4	112.4	100.7	82.2	146.4	119.6	89.7	107.2	99.9	98.4	109.0	104.6
09	FINISHES	100.0	106.0	103.3	100.5	112.5	107.1	93.4	115.1	105.4	95.7	154.5	128.2	99.9	116.4	109.0	99.7	112.6	106.8
COVERS	DIVS. 10 - 14, 25, 28, 41, 43, 44, 46	100.0	99.7	99.9	100.0	101.7	100.4	100.0	105.0	101.1	100.0	124.7	105.4	100.0	104.6	101.0	100.0	101.7	100.4
21, 22, 23	FIRE SUPPRESSION, PLUMBING & HVAC	96.5	104.5	99.9	96.5	94.2	95.6	96.5	106.8	100.9	100.0	134.8	114.7	99.9	98.1	99.2	99.9	98.5	99.3
26, 27, 3370	ELECTRICAL, COMMUNICATIONS & UTIL.	97.2	106.9	102.1	98.5	106.9	102.8	99.4	94.2	96.8	96.5	128.1	112.5	101.0	89.8	95.3	98.1	102.7	100.4
MF2014	WEIGHTED AVERAGE	94.6	106.7	99.9	94.9	108.3	100.8	101.2	108.4	104.4	99.9	140.4	117.8	98.8	105.8	101.9	95.9	108.4	101.4

ILLINOIS

DIVISION		EFFINGHAM 624 MAT.	INST.	TOTAL	GALESBURG 614 MAT.	INST.	TOTAL	JOLIET 604 MAT.	INST.	TOTAL	KANKAKEE 609 MAT.	INST.	TOTAL	LA SALLE 613 MAT.	INST.	TOTAL	NORTH SUBURBAN 600-603 MAT.	INST.	TOTAL
015433	CONTRACTOR EQUIPMENT		103.8	103.8		102.9	102.9		96.2	96.2		96.2	96.2		102.9	102.9		96.2	96.2
0241, 31 - 34	SITE & INFRASTRUCTURE, DEMOLITION	97.9	100.2	99.5	99.1	99.5	99.4	101.2	101.6	101.5	95.1	101.2	99.4	98.5	100.3	99.8	100.5	101.9	101.4
0310	Concrete Forming & Accessories	99.0	115.1	113.0	91.3	118.2	114.6	100.8	158.1	150.4	94.0	142.7	136.1	105.5	123.7	121.3	100.0	154.9	147.5
0320	Concrete Reinforcing	93.5	101.7	97.7	95.3	112.2	104.1	99.2	148.5	124.8	100.1	145.7	123.7	95.5	143.0	120.1	99.2	154.8	128.0
0330	Cast-in-Place Concrete	98.9	108.8	102.8	101.5	108.1	104.1	105.0	145.5	121.0	97.9	131.4	111.1	101.4	121.4	109.3	105.0	148.5	122.2
03	CONCRETE	100.6	110.4	105.4	102.4	113.7	107.9	100.4	150.5	124.8	94.5	138.3	115.8	103.3	126.5	114.6	100.4	151.3	125.1
04	MASONRY	87.7	111.6	102.4	119.5	118.9	119.1	102.3	152.6	133.3	98.4	140.6	124.4	119.5	124.6	122.6	98.9	149.7	130.2
05	METALS	94.1	105.8	97.8	95.3	114.1	101.3	93.8	127.1	104.4	93.8	124.4	103.5	95.4	130.8	106.7	94.9	129.1	105.8
06	WOOD, PLASTICS & COMPOSITES	93.7	115.7	106.0	90.2	117.3	105.3	99.7	159.1	132.8	91.9	142.0	119.8	105.5	122.0	114.7	98.3	154.3	129.5
07	THERMAL & MOISTURE PROTECTION	103.3	106.1	104.5	98.1	110.1	103.1	99.8	142.6	117.7	99.0	134.8	114.0	98.3	119.9	107.4	100.3	140.1	116.9
08	OPENINGS	97.5	110.1	100.6	97.7	112.0	101.0	109.2	157.7	120.5	101.2	147.6	112.0	97.7	127.9	104.7	109.3	156.6	120.3
0920	Plaster & Gypsum Board	102.3	116.1	111.5	94.5	117.7	110.0	91.5	160.7	137.8	88.5	143.1	125.0	101.1	122.6	115.5	94.5	155.7	135.4
0950, 0980	Ceilings & Acoustic Treatment	87.2	116.1	106.3	87.0	117.7	107.3	96.8	160.7	139.1	96.8	143.1	127.5	87.0	122.6	110.6	96.8	155.7	135.8
0960	Flooring	110.3	115.5	111.8	95.0	114.1	100.5	96.0	159.0	114.1	92.8	126.7	102.6	102.0	117.8	106.6	96.4	134.9	107.4
0970, 0990	Wall Finishes & Painting/Coating	80.7	106.6	96.6	84.4	97.0	91.7	80.7	152.6	122.5	80.7	107.9	96.5	84.4	107.9	98.1	82.2	152.5	123.1
09	FINISHES	98.8	115.5	108.1	92.8	116.0	105.6	95.1	158.9	130.4	93.4	134.8	116.3	95.8	120.2	109.3	95.7	152.1	126.9
COVERS	DIVS. 10 - 14, 25, 28, 41, 43, 44, 46	100.0	79.7	95.6	100.0	103.2	100.7	100.0	124.1	105.2	100.0	118.2	103.9	100.0	103.8	100.8	100.0	121.6	104.7
21, 22, 23	FIRE SUPPRESSION, PLUMBING & HVAC	96.6	101.7	98.7	96.5	104.9	100.1	100.0	133.5	114.2	96.7	130.9	111.1	96.5	122.5	107.5	99.9	131.8	113.4
26, 27, 3370	ELECTRICAL, COMMUNICATIONS & UTIL.	99.0	106.8	102.9	97.2	87.4	92.2	95.8	133.3	114.7	91.2	136.1	113.9	94.5	136.1	115.5	95.6	127.5	111.7
MF2014	WEIGHTED AVERAGE	97.3	106.6	101.4	98.4	107.5	102.4	99.4	139.0	116.8	96.0	131.6	111.7	98.6	123.4	109.5	99.4	136.8	115.9

ILLINOIS

DIVISION		PEORIA 615-616 MAT.	INST.	TOTAL	QUINCY 623 MAT.	INST.	TOTAL	ROCK ISLAND 612 MAT.	INST.	TOTAL	ROCKFORD 610-611 MAT.	INST.	TOTAL	SOUTH SUBURBAN 605 MAT.	INST.	TOTAL	SPRINGFIELD 626-627 MAT.	INST.	TOTAL
015433	CONTRACTOR EQUIPMENT		102.9	102.9		103.8	103.8		102.9	102.9		102.9	102.9		96.2	96.2		103.8	103.8
0241, 31 - 34	SITE & INFRASTRUCTURE, DEMOLITION	99.5	99.5	99.5	96.8	100.1	99.1	97.2	98.5	98.1	99.0	100.7	100.2	100.5	101.9	101.4	98.4	100.5	99.9
0310	Concrete Forming & Accessories	94.3	118.8	115.5	96.8	113.6	111.3	92.9	104.0	102.5	98.8	127.6	123.7	100.0	154.9	147.5	95.1	116.3	113.4
0320	Concrete Reinforcing	92.9	112.3	102.9	93.1	104.1	98.8	95.3	106.0	100.9	87.8	135.3	112.4	99.2	154.8	128.0	93.7	103.9	99.0
0330	Cast-in-Place Concrete	98.5	116.8	105.7	99.1	102.6	100.5	99.3	100.7	99.9	100.8	125.1	110.4	105.0	148.5	122.2	94.1	108.1	99.6
03	CONCRETE	99.5	116.0	108.0	100.2	108.2	104.1	100.3	103.6	101.9	100.2	128.0	113.7	100.4	151.3	125.1	98.1	111.2	104.5
04	MASONRY	118.9	118.8	118.9	111.7	104.8	107.4	119.3	100.5	107.7	92.5	134.5	118.4	98.9	149.7	130.2	89.3	118.5	107.3
05	METALS	98.1	114.7	103.4	94.1	108.2	98.6	95.4	109.9	100.0	98.1	127.0	107.3	94.9	129.1	105.8	94.5	109.3	99.2
06	WOOD, PLASTICS & COMPOSITES	98.0	117.3	108.7	91.3	115.7	104.9	91.8	103.9	98.5	98.0	124.0	112.5	98.3	154.3	129.5	92.3	115.7	105.3
07	THERMAL & MOISTURE PROTECTION	98.9	112.0	104.4	103.3	107.0	104.8	98.1	99.4	98.6	101.4	127.3	112.2	100.3	140.1	116.9	106.0	110.9	108.1
08	OPENINGS	104.0	117.2	107.1	98.3	111.8	101.5	97.7	103.2	99.0	104.0	129.0	109.8	109.3	156.6	120.3	102.3	111.7	104.5
0920	Plaster & Gypsum Board	98.0	117.7	111.2	100.9	116.1	111.1	94.5	104.0	100.8	98.0	124.7	115.8	94.5	155.7	135.4	101.2	116.1	111.2
0950, 0980	Ceilings & Acoustic Treatment	91.9	117.7	109.0	87.2	116.1	106.3	87.0	104.0	98.2	91.9	124.7	113.6	96.8	155.7	135.8	96.9	116.1	109.6
0960	Flooring	98.8	114.1	103.2	109.2	103.9	107.6	96.3	101.5	97.8	98.8	117.8	104.3	96.4	134.9	107.4	113.6	103.9	110.6
0970, 0990	Wall Finishes & Painting/Coating	84.4	107.9	98.1	89.7	108.9	100.9	84.4	95.1	90.6	84.4	121.0	105.7	82.2	152.5	123.1	88.5	109.4	100.6
09	FINISHES	95.5	117.2	107.5	98.2	112.5	106.2	93.1	102.8	98.4	95.5	125.0	111.8	95.7	152.1	126.9	102.5	113.9	108.8
COVERS	DIVS. 10 - 14, 25, 28, 41, 43, 44, 46	100.0	105.8	101.3	100.0	81.5	96.0	100.0	99.0	99.8	100.0	113.5	102.9	100.0	121.5	104.7	100.0	104.4	101.0
21, 22, 23	FIRE SUPPRESSION, PLUMBING & HVAC	99.9	104.4	101.8	96.6	100.5	98.2	96.5	99.4	97.7	100.0	115.4	106.5	99.9	131.8	113.4	99.9	103.7	101.5
26, 27, 3370	ELECTRICAL, COMMUNICATIONS & UTIL.	98.2	97.6	97.9	96.5	81.6	89.0	89.9	99.0	94.5	98.5	132.9	115.9	95.6	128.9	112.4	103.6	92.0	97.7
MF2014	WEIGHTED AVERAGE	100.3	109.8	104.5	98.1	101.9	99.8	97.4	101.6	99.2	99.3	123.5	109.9	99.4	137.0	115.9	99.3	106.9	102.6

City Cost Indexes

INDIANA

DIVISION		ANDERSON 460 MAT.	INST.	TOTAL	BLOOMINGTON 474 MAT.	INST.	TOTAL	COLUMBUS 472 MAT.	INST.	TOTAL	EVANSVILLE 476-477 MAT.	INST.	TOTAL	FORT WAYNE 467-468 MAT.	INST.	TOTAL	GARY 463-464 MAT.	INST.	TOTAL
015433	CONTRACTOR EQUIPMENT		95.2	95.2		83.7	83.7		83.7	83.7		113.5	113.5		95.2	95.2		95.2	95.2
0241, 31 - 34	SITE & INFRASTRUCTURE, DEMOLITION	99.2	93.1	94.9	87.2	91.9	90.5	83.7	91.8	89.4	92.6	121.5	113.0	100.1	93.0	95.1	99.9	96.6	97.6
0310	Concrete Forming & Accessories	97.2	81.2	83.3	100.4	79.9	82.6	94.5	78.0	80.3	93.9	81.5	83.2	95.4	75.1	77.8	97.3	116.0	113.5
0320	Concrete Reinforcing	97.5	81.9	89.4	89.4	83.8	86.5	89.8	85.1	87.4	97.9	78.1	87.6	97.5	75.1	85.9	97.5	112.1	105.0
0330	Cast-in-Place Concrete	105.8	79.5	95.4	99.5	77.0	90.6	99.0	77.5	90.5	95.1	87.4	92.0	112.4	76.9	98.4	110.5	113.2	111.6
03	CONCRETE	99.5	81.3	90.6	103.3	79.4	91.7	102.5	79.0	91.1	103.5	83.1	93.6	102.6	76.4	89.9	101.8	114.1	107.8
04	MASONRY	91.5	81.4	85.3	95.9	76.1	83.7	95.7	76.8	84.0	91.1	84.6	87.1	95.1	80.4	86.0	92.9	115.5	106.8
05	METALS	94.5	90.0	93.1	96.8	77.4	90.7	96.9	77.7	90.9	90.4	84.8	88.6	94.5	86.1	91.8	94.5	109.2	99.2
06	WOOD, PLASTICS & COMPOSITES	100.0	81.1	89.5	110.4	80.1	93.5	105.3	77.4	89.8	91.4	80.2	85.2	99.8	73.7	85.3	97.3	114.9	107.1
07	THERMAL & MOISTURE PROTECTION	108.8	75.6	94.9	97.1	76.8	88.6	96.7	80.1	89.7	101.4	82.3	93.4	108.6	73.1	93.8	107.3	107.7	107.5
08	OPENINGS	98.5	81.0	94.4	101.7	81.0	96.9	97.8	79.9	93.6	95.6	79.6	91.9	98.5	73.2	92.6	98.5	120.1	103.5
0920	Plaster & Gypsum Board	105.6	80.8	89.1	98.2	80.2	86.2	95.5	77.5	83.5	93.9	79.1	84.0	105.0	73.2	83.7	98.7	115.5	109.9
0950, 0980	Ceilings & Acoustic Treatment	90.3	80.8	84.0	79.8	80.2	80.1	79.8	77.5	78.2	83.5	79.1	80.6	90.3	73.2	79.0	90.3	115.5	107.0
0960	Flooring	96.6	83.9	93.0	102.8	72.1	94.0	97.4	83.9	93.5	97.2	78.8	91.9	96.6	79.1	91.6	96.6	118.6	102.9
0970, 0990	Wall Finishes & Painting/Coating	95.4	71.0	81.2	89.9	82.4	85.5	85.9	84.0	86.5	95.3	87.9	91.0	95.4	73.9	82.9	95.4	126.0	113.2
09	FINISHES	94.5	80.6	86.8	93.3	78.7	85.2	91.3	79.6	84.8	91.9	81.9	86.4	94.3	75.7	84.0	93.4	117.7	106.9
COVERS	DIVS. 10 - 14, 25, 28, 41, 43, 44, 46	100.0	92.3	98.3	100.0	88.5	97.5	100.0	88.2	97.4	100.0	96.2	99.2	100.0	90.3	97.9	100.0	106.3	101.4
21, 22, 23	FIRE SUPPRESSION, PLUMBING & HVAC	99.9	78.5	90.9	99.6	78.8	90.8	96.3	78.1	88.6	99.9	80.6	91.7	99.9	72.6	88.4	99.9	106.1	102.5
26, 27, 3370	ELECTRICAL, COMMUNICATIONS & UTIL.	88.3	89.2	88.8	99.0	88.3	93.6	98.3	89.2	93.7	95.5	87.6	91.5	88.9	78.6	83.7	99.1	106.8	103.0
MF2014	WEIGHTED AVERAGE	97.1	83.6	91.1	98.8	81.2	91.0	97.1	81.3	90.1	96.6	86.6	92.2	97.6	78.6	89.3	98.4	110.0	103.5

INDIANA

DIVISION		INDIANAPOLIS 461-462 MAT.	INST.	TOTAL	KOKOMO 469 MAT.	INST.	TOTAL	LAFAYETTE 479 MAT.	INST.	TOTAL	LAWRENCEBURG 470 MAT.	INST.	TOTAL	MUNCIE 473 MAT.	INST.	TOTAL	NEW ALBANY 471 MAT.	INST.	TOTAL
015433	CONTRACTOR EQUIPMENT		90.4	90.4		95.2	95.2		83.7	83.7		102.4	102.4		93.7	93.7		92.2	92.2
0241, 31 - 34	SITE & INFRASTRUCTURE, DEMOLITION	99.0	96.2	97.0	95.5	92.9	93.6	84.6	91.8	89.7	82.5	107.3	100.0	87.0	92.3	90.8	79.4	93.8	89.6
0310	Concrete Forming & Accessories	97.6	85.1	86.8	100.5	76.8	80.0	92.1	80.6	82.2	90.9	76.3	78.3	92.0	80.7	82.3	89.1	71.2	73.6
0320	Concrete Reinforcing	97.7	85.6	91.4	88.2	85.3	86.7	89.4	84.1	86.6	88.7	73.4	80.8	98.9	81.9	90.1	90.0	76.2	82.9
0330	Cast-in-Place Concrete	99.8	86.1	94.4	104.7	81.4	95.5	99.6	82.0	92.6	93.1	74.7	85.8	104.5	82.6	95.9	96.1	72.7	86.9
03	CONCRETE	99.6	85.3	92.6	96.3	80.5	88.6	102.8	81.4	92.4	95.7	75.8	86.0	101.8	82.2	92.2	101.1	73.1	87.5
04	MASONRY	92.8	81.8	86.0	91.1	79.1	83.7	101.3	81.6	89.2	79.6	76.6	77.8	97.8	81.5	87.7	86.6	67.9	75.1
05	METALS	95.0	81.4	90.7	91.1	90.5	90.9	95.3	77.6	89.6	92.1	84.3	89.6	98.7	90.0	95.9	93.9	80.4	89.6
06	WOOD, PLASTICS & COMPOSITES	99.8	85.3	91.8	103.3	75.5	87.8	102.5	80.3	90.1	89.7	76.4	82.3	104.1	80.7	91.1	91.7	71.7	80.6
07	THERMAL & MOISTURE PROTECTION	102.2	80.2	93.0	108.4	75.9	94.8	96.7	79.6	89.5	102.4	76.1	91.4	99.7	77.0	90.2	88.6	67.8	79.9
08	OPENINGS	104.4	84.3	99.7	93.6	78.7	90.1	96.2	81.1	92.7	97.6	75.7	92.5	94.9	80.8	91.6	95.1	74.1	90.2
0920	Plaster & Gypsum Board	95.7	85.0	88.5	110.6	75.0	86.8	92.9	80.4	84.6	72.8	76.3	75.2	93.9	80.8	85.2	91.5	71.3	78.0
0950, 0980	Ceilings & Acoustic Treatment	94.4	85.0	88.2	90.3	75.0	80.2	75.7	80.4	78.8	87.6	76.3	80.1	79.8	80.8	80.5	83.5	71.3	75.4
0960	Flooring	97.0	83.9	93.2	101.0	91.4	98.2	96.3	85.3	93.1	71.8	83.9	75.3	96.3	83.9	92.7	94.6	60.4	84.8
0970, 0990	Wall Finishes & Painting/Coating	93.9	84.0	88.1	95.4	89.1	91.7	89.9	85.8	87.5	90.7	75.5	81.9	89.9	71.0	78.9	95.3	82.7	88.0
09	FINISHES	93.8	84.9	88.9	96.3	80.5	87.6	89.7	82.0	85.4	82.0	78.1	79.8	90.4	80.3	84.8	91.1	70.6	79.7
COVERS	DIVS. 10 - 14, 25, 28, 41, 43, 44, 46	100.0	93.6	98.6	100.0	88.9	97.6	100.0	91.3	98.1	100.0	42.7	87.6	100.0	91.2	98.1	100.0	41.6	87.4
21, 22, 23	FIRE SUPPRESSION, PLUMBING & HVAC	99.8	81.7	92.1	96.5	78.2	88.8	96.3	79.0	89.0	97.3	74.8	87.8	99.6	78.4	90.7	96.5	74.4	87.2
26, 27, 3370	ELECTRICAL, COMMUNICATIONS & UTIL.	102.2	89.2	95.6	92.5	78.1	85.2	97.8	82.7	90.2	93.6	75.4	84.4	91.6	87.8	89.7	94.1	75.2	84.5
MF2014	WEIGHTED AVERAGE	99.0	85.2	93.0	95.3	81.5	89.2	96.8	81.9	90.2	93.8	78.2	86.9	97.3	83.4	91.2	94.9	74.1	85.7

INDIANA / IOWA

DIVISION		SOUTH BEND 465-466 MAT.	INST.	TOTAL	TERRE HAUTE 478 MAT.	INST.	TOTAL	WASHINGTON 475 MAT.	INST.	TOTAL	BURLINGTON 526 MAT.	INST.	TOTAL	CARROLL 514 MAT.	INST.	TOTAL	CEDAR RAPIDS 522-524 MAT.	INST.	TOTAL
015433	CONTRACTOR EQUIPMENT		105.1	105.1		113.5	113.5		113.5	113.5		100.0	100.0		100.0	100.0		96.6	96.6
0241, 31 - 34	SITE & INFRASTRUCTURE, DEMOLITION	99.6	93.2	95.1	94.5	121.7	113.7	94.2	118.9	111.6	100.0	97.1	97.9	88.6	97.0	94.5	101.4	95.6	97.3
0310	Concrete Forming & Accessories	99.7	79.7	82.4	94.8	82.9	84.5	95.6	79.9	82.0	95.2	76.8	79.3	82.6	48.8	53.4	101.1	80.0	82.9
0320	Concrete Reinforcing	97.6	79.8	88.4	97.9	81.9	89.6	90.5	48.2	68.6	93.8	85.9	89.7	94.5	81.4	87.7	94.5	83.7	88.9
0330	Cast-in-Place Concrete	100.7	77.6	91.6	92.0	83.1	88.5	100.1	84.8	94.1	109.6	54.3	87.1	109.5	58.6	89.4	109.8	79.6	97.9
03	CONCRETE	96.6	80.4	88.7	106.5	83.0	95.0	112.3	75.7	94.5	101.5	71.8	87.1	100.3	59.7	80.5	101.6	81.2	91.7
04	MASONRY	100.5	78.5	87.0	99.3	80.7	87.8	91.3	82.2	85.7	106.4	64.6	80.6	108.4	73.7	87.1	112.8	77.7	91.1
05	METALS	94.7	102.4	97.1	91.1	86.7	89.7	85.7	68.2	80.1	85.5	95.0	88.5	85.5	90.2	87.0	87.8	92.4	89.3
06	WOOD, PLASTICS & COMPOSITES	95.5	78.9	86.2	93.5	83.0	87.7	93.9	79.9	86.1	92.2	78.4	84.5	79.0	43.3	59.1	98.9	79.8	88.2
07	THERMAL & MOISTURE PROTECTION	102.1	82.5	93.9	101.5	79.8	92.4	101.6	81.0	93.0	104.8	73.4	91.7	105.1	65.2	88.4	105.8	80.0	95.0
08	OPENINGS	96.2	78.7	92.2	96.1	82.1	92.8	92.9	67.7	87.0	98.7	71.9	92.5	103.5	52.8	91.7	103.9	79.9	98.4
0920	Plaster & Gypsum Board	96.8	78.5	84.6	93.9	82.1	86.0	94.0	78.9	83.9	104.0	77.8	86.5	99.3	41.8	60.9	108.5	79.5	89.2
0950, 0980	Ceilings & Acoustic Treatment	93.8	78.5	83.7	83.5	82.1	82.6	78.5	78.9	78.8	95.8	77.8	83.9	95.8	41.8	60.0	98.3	79.5	85.9
0960	Flooring	96.3	87.6	93.8	97.2	81.7	92.7	98.1	73.5	91.0	103.9	36.3	84.5	97.4	31.9	78.6	119.0	81.1	108.1
0970, 0990	Wall Finishes & Painting/Coating	90.4	87.1	88.5	95.3	83.1	88.2	95.3	87.9	91.0	99.3	88.6	93.1	99.3	78.7	87.3	100.6	73.4	84.8
09	FINISHES	95.3	81.8	87.9	91.9	82.8	86.9	91.4	80.3	85.3	100.1	70.5	83.8	96.0	46.9	68.8	105.8	79.5	91.2
COVERS	DIVS. 10 - 14, 25, 28, 41, 43, 44, 46	100.0	93.4	98.6	100.0	94.1	98.7	100.0	96.2	99.2	100.0	88.7	97.6	100.0	63.1	92.0	100.0	91.4	98.1
21, 22, 23	FIRE SUPPRESSION, PLUMBING & HVAC	99.9	76.8	90.1	99.9	79.2	91.2	96.5	79.5	89.4	96.7	76.0	87.9	96.7	74.1	87.1	100.1	79.7	91.4
26, 27, 3370	ELECTRICAL, COMMUNICATIONS & UTIL.	98.5	88.1	93.2	93.9	90.2	92.0	94.3	87.6	90.9	101.7	73.0	87.2	102.4	80.7	91.4	99.3	80.0	89.5
MF2014	WEIGHTED AVERAGE	97.8	84.1	91.8	97.4	86.6	92.6	95.7	82.6	89.9	97.5	76.9	88.4	97.4	71.0	85.8	99.9	82.6	92.3

For customer support on your Mechanical Cost Data, call 877.739.3689.

667

IOWA

DIVISION		COUNCIL BLUFFS 515			CRESTON 508			DAVENPORT 527 - 528			DECORAH 521			DES MOINES 500 - 503,509			DUBUQUE 520		
		MAT.	INST.	TOTAL	MAT.	INST.	TOTAL	MAT.	INST.	TOTAL	MAT.	INST.	TOTAL	MAT.	INST.	TOTAL	MAT.	INST.	TOTAL
015433	CONTRACTOR EQUIPMENT		95.9	95.9		100.0	100.0		100.0	100.0		100.0	100.0		101.7	101.7		95.4	95.4
0241, 31 - 34	SITE & INFRASTRUCTURE, DEMOLITION	105.1	91.7	95.6	94.5	96.1	95.6	100.3	99.2	99.5	98.6	96.1	96.8	100.9	99.5	99.9	99.2	92.9	94.7
0310	Concrete Forming & Accessories	82.0	71.1	72.6	79.1	64.6	66.6	100.6	91.0	92.3	92.7	45.1	51.5	96.9	79.4	81.7	83.4	75.1	76.2
0320	Concrete Reinforcing	96.4	78.1	86.9	89.9	81.6	85.6	94.5	101.6	98.2	93.8	76.1	84.6	94.3	85.7	89.8	93.2	83.1	88.0
0330	Cast-in-Place Concrete	114.2	79.5	100.5	112.6	63.3	93.2	105.8	91.7	100.2	106.6	79.3	95.8	95.7	80.2	89.6	107.6	97.4	103.6
03	CONCRETE	104.0	76.1	90.4	101.8	68.4	85.6	99.6	93.8	96.8	99.4	64.0	82.2	97.2	81.6	89.6	98.4	84.9	91.8
04	MASONRY	113.9	74.7	89.7	109.1	82.5	92.7	109.3	86.4	95.2	129.7	77.3	97.4	100.9	81.0	88.7	113.8	73.5	89.0
05	METALS	92.6	89.7	91.7	88.6	91.4	89.5	87.8	105.1	93.3	85.6	87.7	86.3	94.3	95.1	94.6	86.5	92.3	88.3
06	WOOD, PLASTICS & COMPOSITES	77.8	70.7	73.9	73.1	59.5	65.5	98.9	90.6	94.3	89.2	35.7	59.4	89.4	78.4	83.2	79.4	73.6	76.2
07	THERMAL & MOISTURE PROTECTION	105.2	68.4	89.8	106.1	76.5	93.7	105.2	89.2	98.5	105.0	59.7	86.0	97.4	78.5	89.5	105.4	78.8	94.3
08	OPENINGS	103.0	75.5	96.6	111.0	71.0	101.7	103.9	94.8	101.8	101.9	47.2	89.2	99.4	84.5	96.0	103.0	79.4	97.5
0920	Plaster & Gypsum Board	99.3	70.2	79.8	93.8	58.4	70.2	108.5	90.4	96.4	102.6	34.0	56.8	89.4	77.8	81.7	99.3	73.1	81.8
0950, 0980	Ceilings & Acoustic Treatment	95.8	70.2	78.8	87.2	58.4	68.2	98.3	90.4	93.1	95.8	34.0	54.9	89.7	77.8	81.8	95.8	73.1	80.8
0960	Flooring	95.9	84.1	92.5	86.9	81.7	85.4	106.6	93.2	102.7	103.3	77.8	95.9	93.4	86.1	91.3	109.0	74.0	98.9
0970, 0990	Wall Finishes & Painting/Coating	95.2	65.8	78.1	89.5	80.3	84.1	99.3	95.5	97.1	99.3	32.9	60.7	85.2	88.6	87.2	99.8	89.7	93.9
09	FINISHES	96.6	72.7	83.4	88.6	68.5	77.5	102.1	91.7	96.3	99.7	47.8	71.0	91.3	81.4	85.8	100.7	75.7	86.9
COVERS	DIVS. 10 - 14, 25, 28, 41, 43, 44, 46	100.0	89.0	97.6	100.0	67.4	92.9	100.0	94.4	98.8	100.0	80.5	95.8	100.0	91.8	98.2	100.0	90.1	97.9
21, 22, 23	FIRE SUPPRESSION, PLUMBING & HVAC	100.1	77.2	90.4	96.5	78.9	89.1	100.1	91.9	96.6	96.7	73.4	86.8	99.7	79.0	91.0	100.1	74.5	89.3
26, 27, 3370	ELECTRICAL, COMMUNICATIONS & UTIL.	104.5	80.9	92.6	94.1	80.7	87.3	97.3	92.4	94.8	99.3	44.0	71.4	104.8	80.7	92.6	103.1	79.1	91.0
MF2014	WEIGHTED AVERAGE	100.6	79.0	91.1	97.5	78.3	89.1	99.0	93.6	96.6	98.5	66.6	84.4	98.4	83.9	92.0	99.0	80.6	90.9

IOWA

DIVISION		FORT DODGE 505			MASON CITY 504			OTTUMWA 525			SHENANDOAH 516			SIBLEY 512			SIOUX CITY 510 - 511		
		MAT.	INST.	TOTAL	MAT.	INST.	TOTAL	MAT.	INST.	TOTAL	MAT.	INST.	TOTAL	MAT.	INST.	TOTAL	MAT.	INST.	TOTAL
015433	CONTRACTOR EQUIPMENT		100.0	100.0		100.0	100.0		95.4	95.4		95.9	95.9		100.0	100.0		100.0	100.0
0241, 31 - 34	SITE & INFRASTRUCTURE, DEMOLITION	102.8	94.9	97.2	102.9	96.0	98.0	99.6	91.0	94.5	103.5	90.8	94.5	110.2	94.8	99.3	112.0	95.5	100.3
0310	Concrete Forming & Accessories	79.7	43.7	48.6	83.9	44.4	49.7	90.6	72.2	74.7	83.7	55.1	59.0	84.1	36.7	43.1	101.1	64.7	69.6
0320	Concrete Reinforcing	89.9	66.4	77.8	89.8	81.0	85.2	93.8	86.3	89.9	96.4	67.6	81.5	96.4	65.2	80.2	94.5	77.1	85.5
0330	Cast-in-Place Concrete	105.7	43.2	81.1	105.7	56.7	86.4	110.3	66.9	93.2	110.4	58.9	90.1	108.2	43.8	82.8	108.9	54.6	87.5
03	CONCRETE	97.0	49.4	73.9	97.3	57.0	77.7	101.0	73.9	87.9	101.3	60.1	81.2	100.2	46.2	74.0	101.0	64.6	83.3
04	MASONRY	107.9	37.7	64.7	122.2	70.3	90.2	110.1	58.4	78.2	113.5	74.7	89.6	133.7	38.0	74.7	106.4	56.1	75.4
05	METALS	88.7	81.7	86.5	88.8	89.2	88.9	85.4	92.9	87.8	91.6	84.0	89.2	85.7	80.6	84.1	87.8	87.8	87.8
06	WOOD, PLASTICS & COMPOSITES	73.5	43.6	56.8	77.4	35.7	54.2	86.5	78.1	81.8	79.4	50.9	63.6	80.2	34.5	54.8	98.9	64.3	79.6
07	THERMAL & MOISTURE PROTECTION	105.4	56.6	85.0	104.9	66.4	88.8	105.6	67.6	89.7	104.4	64.6	87.7	104.8	47.0	80.6	105.2	63.7	87.8
08	OPENINGS	104.5	48.0	91.4	96.1	48.6	85.1	103.4	75.8	97.0	93.9	53.1	84.4	99.9	42.7	86.6	103.9	68.4	95.7
0920	Plaster & Gypsum Board	93.8	42.0	59.2	93.8	34.0	53.8	100.3	77.8	85.3	99.3	49.8	66.3	99.3	32.8	54.9	108.5	63.4	78.4
0950, 0980	Ceilings & Acoustic Treatment	87.2	42.0	57.3	87.2	34.0	52.0	95.8	77.8	83.9	95.8	49.8	65.4	95.8	32.8	54.1	98.3	63.4	75.2
0960	Flooring	88.3	45.3	75.9	90.5	68.3	84.1	112.4	70.7	100.4	96.7	32.9	78.4	98.4	32.7	79.5	106.6	54.7	91.7
0970, 0990	Wall Finishes & Painting/Coating	89.5	61.8	73.4	89.5	30.1	54.9	99.8	83.4	90.3	95.2	65.8	78.1	99.3	61.8	77.5	99.3	65.3	79.5
09	FINISHES	90.5	44.7	65.1	91.1	45.3	65.8	102.0	73.5	86.2	96.7	50.7	71.3	99.3	36.9	64.8	103.6	62.5	80.9
COVERS	DIVS. 10 - 14, 25, 28, 41, 43, 44, 46	100.0	79.3	95.5	100.0	84.1	96.5	100.0	84.0	96.5	100.0	64.1	92.2	100.0	78.4	95.3	100.0	87.8	97.4
21, 22, 23	FIRE SUPPRESSION, PLUMBING & HVAC	96.5	64.8	83.1	96.5	62.3	82.1	96.7	70.2	85.5	96.7	87.4	92.8	96.7	67.6	84.4	100.1	75.0	89.5
26, 27, 3370	ELECTRICAL, COMMUNICATIONS & UTIL.	100.2	41.6	70.5	99.3	55.2	77.0	101.5	71.5	86.3	99.3	80.9	90.0	99.3	41.1	69.9	99.3	72.4	85.7
MF2014	WEIGHTED AVERAGE	97.2	57.6	79.8	97.0	64.4	82.6	98.3	74.7	87.9	97.8	73.5	87.1	98.7	56.0	79.9	99.6	72.3	87.6

IOWA / KANSAS

DIVISION		SPENCER 513			WATERLOO 506 - 507			BELLEVILLE 669			COLBY 677			DODGE CITY 678			EMPORIA 668		
		MAT.	INST.	TOTAL	MAT.	INST.	TOTAL	MAT.	INST.	TOTAL	MAT.	INST.	TOTAL	MAT.	INST.	TOTAL	MAT.	INST.	TOTAL
015433	CONTRACTOR EQUIPMENT		100.0	100.0		100.0	100.0		104.8	104.8		104.8	104.8		104.8	104.8		102.9	102.9
0241, 31 - 34	SITE & INFRASTRUCTURE, DEMOLITION	110.3	94.8	99.3	108.1	95.8	99.4	107.7	96.4	99.7	112.6	96.5	101.3	115.2	96.3	101.8	99.7	93.6	95.4
0310	Concrete Forming & Accessories	90.5	36.7	44.0	95.5	54.1	59.7	98.2	53.2	59.3	103.9	57.6	63.8	96.8	57.2	62.6	88.8	64.2	67.6
0320	Concrete Reinforcing	96.4	66.4	80.8	90.5	83.0	86.6	104.0	55.2	78.7	103.3	55.3	78.4	100.7	55.1	77.1	102.6	55.8	78.3
0330	Cast-in-Place Concrete	108.2	43.8	82.8	113.2	76.4	98.7	118.7	55.7	93.9	120.3	67.6	99.5	122.4	55.6	96.1	114.8	53.7	90.7
03	CONCRETE	100.7	46.4	74.3	103.2	68.4	86.3	119.8	55.9	88.8	119.4	62.0	91.5	120.8	57.6	90.1	111.8	60.3	86.8
04	MASONRY	133.7	38.0	74.7	108.7	73.6	87.1	97.7	59.8	74.3	109.3	61.2	79.7	119.2	59.1	82.2	103.8	69.5	82.7
05	METALS	85.7	81.0	84.2	91.1	91.6	91.2	94.6	78.8	89.6	95.0	79.4	90.1	96.5	78.1	90.6	94.3	80.4	89.9
06	WOOD, PLASTICS & COMPOSITES	86.3	34.5	57.5	91.0	46.3	66.1	97.0	50.3	71.0	105.4	55.8	77.8	97.1	55.8	74.1	88.0	64.3	74.8
07	THERMAL & MOISTURE PROTECTION	105.7	47.0	81.2	105.1	71.3	91.0	96.9	60.3	81.6	98.9	61.7	83.3	98.8	60.0	82.6	95.1	75.6	87.0
08	OPENINGS	112.1	43.1	96.0	96.5	62.3	88.5	99.4	47.1	87.2	108.9	50.1	95.2	108.8	55.8	96.5	97.2	63.2	89.3
0920	Plaster & Gypsum Board	100.3	32.8	55.2	102.7	44.9	64.1	95.5	48.9	64.3	102.0	54.6	70.3	96.4	54.6	68.5	92.8	63.3	73.1
0950, 0980	Ceilings & Acoustic Treatment	95.8	32.8	54.1	89.7	44.9	60.0	83.0	48.9	60.4	80.6	54.6	63.4	80.6	54.6	63.4	83.0	63.3	69.9
0960	Flooring	101.5	32.7	81.7	96.3	74.0	89.9	99.9	37.7	82.0	96.5	57.7	85.4	92.7	37.7	76.8	94.7	57.7	84.1
0970, 0990	Wall Finishes & Painting/Coating	99.3	61.8	77.5	89.5	77.8	82.7	85.5	38.9	58.4	93.8	38.9	61.9	93.8	38.9	61.9	85.5	38.9	58.4
09	FINISHES	100.4	36.9	65.2	94.9	58.2	74.6	93.3	48.1	68.3	93.1	55.4	72.3	91.3	51.4	69.2	90.3	60.5	73.8
COVERS	DIVS. 10 - 14, 25, 28, 41, 43, 44, 46	100.0	78.4	95.3	100.0	87.5	97.3	100.0	39.8	86.9	100.0	40.4	87.1	100.0	40.4	87.1	100.0	39.5	86.9
21, 22, 23	FIRE SUPPRESSION, PLUMBING & HVAC	96.7	67.7	84.4	99.9	79.0	91.1	96.5	79.6	89.4	96.6	69.4	85.1	99.9	69.3	87.0	96.5	73.3	86.7
26, 27, 3370	ELECTRICAL, COMMUNICATIONS & UTIL.	101.0	41.1	70.7	95.9	62.9	79.3	100.8	67.2	83.8	100.1	78.0	88.9	97.3	73.2	85.1	98.2	72.2	85.0
MF2014	WEIGHTED AVERAGE	100.5	56.0	80.9	98.4	73.7	87.5	100.1	66.5	85.3	101.8	68.0	86.9	103.0	66.1	86.8	98.3	70.1	85.9

For customer support on your Mechanical Cost Data, call 877.739.3689.

KANSAS

DIVISION		FORT SCOTT 667			HAYS 676			HUTCHINSON 675			INDEPENDENCE 673			KANSAS CITY 660 - 662			LIBERAL 679		
		MAT.	INST.	TOTAL	MAT.	INST.	TOTAL	MAT.	INST.	TOTAL	MAT.	INST.	TOTAL	MAT.	INST.	TOTAL	MAT.	INST.	TOTAL
015433	CONTRACTOR EQUIPMENT		103.8	103.8		104.8	104.8		104.8	104.8		104.8	104.8		101.1	101.1		104.8	104.8
0241, 31 - 34	SITE & INFRASTRUCTURE, DEMOLITION	96.5	94.4	95.0	118.2	96.6	102.9	96.1	96.4	96.3	116.4	96.4	102.3	90.9	94.2	93.2	117.7	96.4	102.6
0310	Concrete Forming & Accessories	106.7	80.6	84.1	101.3	57.7	63.6	91.2	57.7	61.8	112.9	69.1	75.0	103.0	96.5	97.4	97.2	57.1	62.6
0320	Concrete Reinforcing	101.9	92.3	96.9	100.7	55.3	77.2	100.7	55.2	77.1	100.1	62.0	80.4	98.8	98.6	98.7	102.1	55.2	77.8
0330	Cast-in-Place Concrete	106.4	53.7	85.6	94.9	67.7	84.2	87.9	53.1	74.2	123.0	53.4	95.5	91.2	97.3	93.6	94.9	53.1	78.4
03	CONCRETE	106.9	74.3	91.1	110.2	62.0	86.8	91.9	56.8	74.8	122.1	63.5	93.6	98.7	97.5	98.1	112.3	56.8	85.3
04	MASONRY	105.1	60.1	77.4	118.8	61.2	83.3	109.3	59.7	78.7	106.2	65.6	81.2	106.5	100.2	102.6	116.9	59.7	81.7
05	METALS	94.3	93.1	93.9	94.6	79.7	89.8	94.4	78.1	89.2	94.3	81.5	90.2	102.7	101.5	102.3	94.9	78.1	89.5
06	WOOD, PLASTICS & COMPOSITES	107.5	88.9	97.1	102.2	55.8	76.4	91.9	55.8	71.8	116.1	71.1	91.1	103.3	96.9	99.7	97.7	55.8	74.4
07	THERMAL & MOISTURE PROTECTION	96.1	73.3	86.5	99.2	61.2	83.3	97.7	60.2	82.0	98.9	75.4	89.1	95.9	99.1	97.2	99.3	60.2	83.0
08	OPENINGS	97.2	85.7	94.5	108.8	55.8	96.4	108.7	55.8	96.4	106.3	65.7	96.9	98.5	90.1	96.5	108.8	55.8	96.5
0920	Plaster & Gypsum Board	97.8	88.5	91.6	99.4	54.6	69.4	95.4	54.6	68.1	109.0	70.3	83.1	91.4	96.8	95.0	97.0	54.6	68.7
0950, 0980	Ceilings & Acoustic Treatment	83.0	88.5	86.7	80.6	54.6	63.4	80.6	54.6	63.4	80.6	70.3	73.8	83.0	96.8	92.1	80.6	54.6	63.4
0960	Flooring	110.5	37.9	89.6	95.3	61.7	85.6	89.6	37.7	74.7	101.0	37.7	82.8	88.7	101.0	92.2	92.9	37.7	77.0
0970, 0990	Wall Finishes & Painting/Coating	87.2	38.9	59.1	93.8	38.9	61.9	93.8	38.9	61.9	93.8	38.9	61.9	92.7	67.7	78.2	93.8	38.9	61.9
09	FINISHES	90.6	69.8	81.5	92.9	56.1	72.5	88.6	51.4	68.0	95.6	60.5	76.2	90.6	93.1	92.0	92.1	51.4	69.6
COVERS	DIVS. 10 - 14, 25, 28, 41, 43, 44, 46	100.0	45.5	88.2	100.0	40.4	87.1	100.0	40.4	87.1	100.0	42.1	87.5	100.0	78.9	95.4	100.0	40.4	87.1
21, 22, 23	FIRE SUPPRESSION, PLUMBING & HVAC	96.5	67.5	84.3	96.6	69.4	85.1	96.6	69.3	85.1	96.6	70.4	85.5	99.8	95.4	98.0	96.6	67.4	84.3
26, 27, 3370	ELECTRICAL, COMMUNICATIONS & UTIL.	97.5	72.2	84.7	99.1	67.2	83.0	94.8	67.2	80.8	96.7	72.2	84.3	102.9	96.4	99.6	97.3	73.2	85.1
MF2014	WEIGHTED AVERAGE	98.3	73.7	87.5	101.2	66.9	86.1	97.1	65.2	83.1	101.6	70.2	87.8	99.6	95.9	98.0	101.1	65.6	85.5

KANSAS / KENTUCKY

DIVISION		SALINA 674			TOPEKA 664 - 666			WICHITA 670 - 672			ASHLAND 411 - 412			BOWLING GREEN 421 - 422			CAMPTON 413 - 414		
		MAT.	INST.	TOTAL	MAT.	INST.	TOTAL	MAT.	INST.	TOTAL	MAT.	INST.	TOTAL	MAT.	INST.	TOTAL	MAT.	INST.	TOTAL
015433	CONTRACTOR EQUIPMENT		104.8	104.8		102.9	102.9		104.8	104.8		97.2	97.2		92.2	92.2		98.8	98.8
0241, 31 - 34	SITE & INFRASTRUCTURE, DEMOLITION	105.2	96.7	99.2	94.5	92.8	93.3	101.3	95.1	96.9	114.7	83.9	92.9	79.7	94.2	90.0	88.4	95.7	93.6
0310	Concrete Forming & Accessories	93.1	54.4	59.7	100.5	47.3	54.5	99.4	50.1	56.8	87.2	101.7	99.7	85.3	84.5	84.6	88.8	80.4	81.6
0320	Concrete Reinforcing	100.1	75.4	87.3	98.3	100.6	99.5	98.2	77.1	87.3	91.4	111.6	101.9	88.7	90.5	89.6	89.6	100.6	95.3
0330	Cast-in-Place Concrete	106.3	53.9	85.6	95.7	54.9	79.6	99.4	51.6	80.6	87.5	102.9	93.6	86.8	94.2	89.7	96.8	72.3	87.2
03	CONCRETE	106.8	59.6	83.9	99.9	61.4	81.2	101.6	57.1	80.0	96.9	104.5	100.6	95.2	88.9	92.2	99.0	81.6	90.5
04	MASONRY	135.4	61.2	89.7	99.3	60.6	75.5	105.4	49.8	71.1	98.3	108.5	104.6	100.6	84.0	90.3	97.0	68.7	79.5
05	METALS	96.3	87.6	93.5	98.6	98.5	98.6	98.6	85.2	94.3	93.1	113.0	99.4	94.6	87.1	92.2	93.9	91.1	93.0
06	WOOD, PLASTICS & COMPOSITES	93.5	50.7	69.7	100.4	43.8	68.9	101.4	49.5	72.5	74.4	99.9	88.6	86.7	85.3	85.9	85.2	86.0	85.6
07	THERMAL & MOISTURE PROTECTION	98.3	61.5	82.9	99.7	70.4	87.4	97.9	54.8	79.8	92.8	94.3	93.4	88.4	81.5	85.5	101.8	76.2	91.1
08	OPENINGS	108.8	51.1	95.3	104.1	63.0	94.5	111.9	57.7	99.3	94.4	96.6	94.9	95.1	79.7	91.5	96.5	85.1	93.8
0920	Plaster & Gypsum Board	95.4	49.4	64.6	94.8	42.2	59.7	94.1	48.1	63.3	63.3	100.0	87.8	87.9	85.2	86.1	87.9	85.1	86.0
0950, 0980	Ceilings & Acoustic Treatment	80.6	49.4	59.9	87.7	42.2	57.6	84.5	48.1	60.4	78.9	100.0	92.9	83.5	85.2	84.6	83.5	85.1	84.5
0960	Flooring	91.1	61.7	82.6	100.3	76.9	93.6	98.0	58.2	86.6	77.6	104.3	85.2	92.3	83.2	89.7	94.4	39.6	78.7
0970, 0990	Wall Finishes & Painting/Coating	93.8	38.9	61.9	87.1	76.2	80.7	92.8	71.9	80.7	97.3	104.6	101.5	95.3	72.7	82.2	95.3	65.4	77.9
09	FINISHES	89.9	53.2	69.6	94.5	53.9	72.1	93.5	53.1	71.2	79.5	102.7	92.4	89.7	83.5	86.3	90.5	71.8	80.2
COVERS	DIVS. 10 - 14, 25, 28, 41, 43, 44, 46	100.0	86.7	97.1	100.0	42.6	87.6	100.0	83.6	96.4	100.0	91.8	98.2	100.0	60.9	91.5	100.0	54.0	90.0
21, 22, 23	FIRE SUPPRESSION, PLUMBING & HVAC	99.9	69.4	87.0	99.9	71.5	87.9	99.8	64.9	85.0	96.3	93.7	95.2	99.9	83.0	92.8	96.6	78.9	89.1
26, 27, 3370	ELECTRICAL, COMMUNICATIONS & UTIL.	97.1	73.3	85.0	101.7	74.1	87.7	100.7	73.2	86.8	92.0	99.8	96.0	94.4	80.9	87.6	92.0	65.6	78.7
MF2014	WEIGHTED AVERAGE	101.8	68.9	87.3	99.8	69.8	86.6	101.0	66.1	85.6	94.5	99.8	96.8	95.7	84.1	90.6	95.6	77.6	87.7

KENTUCKY

DIVISION		CORBIN 407 - 409			COVINGTON 410			ELIZABETHTOWN 427			FRANKFORT 406			HAZARD 417 - 418			HENDERSON 424		
		MAT.	INST.	TOTAL	MAT.	INST.	TOTAL	MAT.	INST.	TOTAL	MAT.	INST.	TOTAL	MAT.	INST.	TOTAL	MAT.	INST.	TOTAL
015433	CONTRACTOR EQUIPMENT		98.8	98.8		102.4	102.4		92.2	92.2		98.8	98.8		98.8	98.8		113.5	113.5
0241, 31 - 34	SITE & INFRASTRUCTURE, DEMOLITION	88.0	95.5	93.3	84.0	109.4	102.0	74.2	94.3	88.4	85.6	96.8	93.5	86.2	97.1	93.9	82.5	121.7	110.2
0310	Concrete Forming & Accessories	83.8	72.1	73.7	84.4	84.1	84.2	80.2	80.7	80.6	93.3	70.5	73.6	85.4	81.4	82.0	92.0	81.3	82.7
0320	Concrete Reinforcing	90.7	69.2	79.5	88.2	91.8	90.1	89.2	93.3	91.3	95.2	95.4	95.3	90.0	110.4	100.6	88.8	89.4	89.1
0330	Cast-in-Place Concrete	88.2	57.6	76.2	92.7	96.5	94.2	78.5	72.4	76.1	85.2	68.9	78.8	93.1	81.3	88.5	76.7	90.1	82.0
03	CONCRETE	92.0	67.1	79.9	97.3	90.3	93.9	87.1	80.4	83.8	92.5	75.0	84.0	95.7	86.9	91.4	92.2	85.9	89.2
04	MASONRY	95.0	61.6	74.5	112.3	101.1	105.4	83.8	76.9	79.6	90.4	80.8	84.5	95.6	71.2	80.6	104.4	94.0	98.0
05	METALS	89.0	78.4	85.6	92.0	97.0	93.6	93.8	88.2	92.0	91.4	89.6	90.8	93.9	95.2	94.3	85.4	88.5	86.4
06	WOOD, PLASTICS & COMPOSITES	72.8	77.7	75.5	83.2	75.2	78.8	82.0	82.4	82.2	87.6	66.3	75.7	82.3	86.0	84.3	89.1	78.4	83.1
07	THERMAL & MOISTURE PROTECTION	106.4	72.9	92.4	102.6	87.2	96.2	88.1	77.6	83.7	104.8	79.6	94.3	101.7	78.1	91.8	101.0	85.5	94.5
08	OPENINGS	92.9	66.1	86.7	98.5	82.9	94.9	95.1	83.5	92.4	97.1	76.2	92.2	96.8	78.6	92.6	93.3	80.1	90.2
0920	Plaster & Gypsum Board	95.3	76.6	82.8	70.6	75.1	73.6	87.2	82.3	83.9	95.7	64.9	75.1	87.2	85.1	85.8	90.4	77.3	81.6
0950, 0980	Ceilings & Acoustic Treatment	78.5	76.6	77.2	86.8	75.1	79.0	83.5	82.3	82.7	86.8	64.9	72.3	83.5	85.1	84.5	78.5	77.3	77.7
0960	Flooring	92.4	39.6	77.2	69.2	86.0	74.1	89.5	82.7	87.6	100.2	52.3	86.4	92.6	40.3	77.6	96.2	83.5	92.5
0970, 0990	Wall Finishes & Painting/Coating	96.1	49.4	68.9	90.7	83.2	86.3	95.3	78.7	85.6	96.9	82.8	88.7	95.3	65.4	77.9	95.3	96.1	95.7
09	FINISHES	88.6	64.3	75.2	80.9	83.4	82.3	88.4	80.6	84.1	93.9	67.5	79.3	89.7	72.6	80.3	89.6	83.1	86.0
COVERS	DIVS. 10 - 14, 25, 28, 41, 43, 44, 46	100.0	47.3	88.6	100.0	99.2	99.8	100.0	82.2	96.1	100.0	64.0	92.2	100.0	54.8	90.2	100.0	64.0	92.2
21, 22, 23	FIRE SUPPRESSION, PLUMBING & HVAC	96.6	72.0	86.2	97.3	92.1	95.1	96.7	80.6	89.9	100.0	83.2	92.9	96.6	82.5	90.6	96.7	79.1	89.3
26, 27, 3370	ELECTRICAL, COMMUNICATIONS & UTIL.	90.6	79.5	85.0	95.5	80.1	87.7	91.8	80.9	86.3	99.2	82.4	90.7	92.0	57.9	74.8	93.9	80.8	87.2
MF2014	WEIGHTED AVERAGE	93.3	71.9	83.9	95.8	91.3	93.8	92.5	82.2	87.9	96.1	80.2	89.1	95.1	78.6	87.8	93.5	86.3	90.3

For customer support on your Mechanical Cost Data, call 877.739.3689.

669

KENTUCKY

DIVISION		LEXINGTON 403 - 405			LOUISVILLE 400 - 402			OWENSBORO 423			PADUCAH 420			PIKEVILLE 415 - 416			SOMERSET 425 - 426		
		MAT.	INST.	TOTAL	MAT.	INST.	TOTAL	MAT.	INST.	TOTAL	MAT.	INST.	TOTAL	MAT.	INST.	TOTAL	MAT.	INST.	TOTAL
015433	CONTRACTOR EQUIPMENT		98.8	98.8		92.2	92.2		113.5	113.5		113.5	113.5		97.2	97.2		98.8	98.8
0241, 31 - 34	SITE & INFRASTRUCTURE, DEMOLITION	90.1	98.1	95.8	85.3	94.6	91.9	92.6	122.1	113.5	85.3	121.1	110.6	126.0	82.9	95.5	79.1	96.3	91.3
0310	Concrete Forming & Accessories	95.2	71.1	74.4	94.2	84.0	85.4	90.4	84.4	85.2	88.3	82.7	83.5	96.1	86.8	88.1	86.4	79.4	80.4
0320	Concrete Reinforcing	99.5	95.6	97.5	95.7	95.6	95.7	88.8	91.5	90.2	89.4	89.1	89.2	91.9	111.4	102.0	89.2	95.4	92.4
0330	Cast-in-Place Concrete	90.3	91.4	90.7	89.4	74.7	83.6	89.5	90.2	89.8	81.8	86.0	83.5	96.2	97.7	96.8	76.7	97.6	85.0
03	CONCRETE	95.1	82.9	89.2	94.7	83.1	89.1	104.4	87.8	96.3	96.9	85.1	91.2	110.7	96.0	103.6	82.2	88.7	85.4
04	MASONRY	93.2	80.8	85.5	90.6	79.6	83.8	96.3	88.4	91.4	99.1	81.9	88.5	95.3	99.9	98.1	90.4	77.2	82.3
05	METALS	91.4	90.6	91.1	92.3	89.4	91.3	86.9	89.9	87.8	83.9	87.6	85.1	93.0	112.0	99.0	93.8	88.8	92.2
06	WOOD, PLASTICS & COMPOSITES	87.0	66.3	75.5	87.3	85.3	86.2	87.1	82.5	84.5	84.8	82.3	83.4	83.3	85.3	84.4	82.8	80.1	81.3
07	THERMAL & MOISTURE PROTECTION	106.6	80.5	95.7	103.2	79.4	93.2	101.5	85.9	95.0	101.1	85.9	94.8	93.5	89.0	91.6	101.0	79.9	92.2
08	OPENINGS	93.2	72.8	88.4	87.4	85.6	87.0	93.3	85.0	91.4	92.6	79.6	89.5	95.0	86.2	93.0	95.9	80.0	92.2
0920	Plaster & Gypsum Board	104.9	64.9	78.2	98.0	85.2	89.5	89.0	81.6	84.1	88.1	81.3	83.5	65.6	85.1	78.6	87.2	79.1	81.8
0950, 0980	Ceilings & Acoustic Treatment	82.7	64.9	70.9	87.1	85.2	85.9	78.5	81.6	80.5	78.5	81.3	80.3	78.9	85.1	83.0	83.5	79.1	80.6
0960	Flooring	97.6	64.4	88.1	94.8	82.7	91.3	95.5	83.2	92.0	94.3	57.1	83.6	82.0	104.3	88.4	92.9	39.6	77.6
0970, 0990	Wall Finishes & Painting/Coating	96.1	81.0	87.3	94.9	78.7	85.5	95.3	99.4	97.7	95.3	85.0	89.3	97.3	81.2	88.0	95.3	78.7	85.6
09	FINISHES	92.4	69.5	79.8	92.4	83.4	87.4	89.7	85.4	87.3	88.9	78.3	83.0	82.0	89.4	86.1	89.1	71.9	79.6
COVERS	DIVS. 10 - 14, 25, 28, 41, 43, 44, 46	100.0	96.4	99.2	100.0	95.7	99.1	100.0	103.5	100.8	100.0	60.9	91.5	100.0	54.5	90.1	100.0	57.7	90.8
21, 22, 23	FIRE SUPPRESSION, PLUMBING & HVAC	100.0	81.2	92.0	100.0	81.8	92.3	99.9	78.2	90.7	96.7	83.4	91.1	96.3	88.9	93.2	96.7	80.3	89.8
26, 27, 3370	ELECTRICAL, COMMUNICATIONS & UTIL.	93.0	82.4	87.7	97.8	82.4	90.0	93.9	82.8	88.3	96.1	83.6	89.8	94.8	72.9	83.7	92.4	82.4	87.3
MF2014	WEIGHTED AVERAGE	95.5	82.2	89.7	95.1	84.3	90.4	95.8	88.0	92.4	93.7	85.5	90.1	96.8	89.3	93.5	93.0	81.7	88.0

LOUISIANA

DIVISION		ALEXANDRIA 713 - 714			BATON ROUGE 707 - 708			HAMMOND 704			LAFAYETTE 705			LAKE CHARLES 706			MONROE 712		
		MAT.	INST.	TOTAL	MAT.	INST.	TOTAL	MAT.	INST.	TOTAL	MAT.	INST.	TOTAL	MAT.	INST.	TOTAL	MAT.	INST.	TOTAL
015433	CONTRACTOR EQUIPMENT		93.0	93.0		90.7	90.7		91.2	91.2		91.2	91.2		90.7	90.7		93.0	93.0
0241, 31 - 34	SITE & INFRASTRUCTURE, DEMOLITION	101.7	92.9	95.5	103.2	91.3	94.8	100.0	89.6	92.6	101.1	91.9	94.6	101.8	91.3	94.4	101.7	92.3	95.1
0310	Concrete Forming & Accessories	81.7	46.7	51.4	95.0	67.6	71.3	78.1	54.4	57.6	95.9	58.6	63.6	96.7	67.0	71.0	81.2	43.3	48.4
0320	Concrete Reinforcing	105.4	71.4	87.8	98.5	75.3	86.5	97.9	58.7	77.6	99.4	75.4	86.9	99.4	58.4	78.1	104.2	71.1	87.0
0330	Cast-in-Place Concrete	92.5	68.7	83.1	99.1	78.7	91.1	94.2	62.6	81.7	93.7	63.7	81.9	98.6	70.3	87.5	92.5	63.7	81.2
03	CONCRETE	93.9	59.9	77.4	98.6	73.3	86.3	93.0	58.8	76.4	94.1	64.2	79.6	96.6	67.1	82.2	93.7	56.7	75.7
04	MASONRY	119.9	63.8	85.3	92.6	65.1	75.7	93.7	50.0	66.8	93.7	53.4	68.9	93.0	66.4	76.6	114.1	47.8	73.2
05	METALS	95.1	79.3	90.1	99.3	80.0	93.1	91.1	71.9	85.0	90.4	80.2	87.1	90.4	72.6	84.7	95.0	78.8	89.9
06	WOOD, PLASTICS & COMPOSITES	96.1	40.4	65.1	102.2	66.7	82.4	88.4	57.2	71.0	109.3	60.8	82.3	107.5	66.7	84.8	95.5	40.9	65.1
07	THERMAL & MOISTURE PROTECTION	97.1	67.0	84.5	96.4	71.0	85.8	98.0	61.9	82.9	98.5	65.2	84.6	97.6	71.0	86.5	97.1	61.0	82.0
08	OPENINGS	116.0	48.4	100.3	105.7	64.9	96.2	97.7	56.6	88.2	101.3	60.3	91.8	101.3	59.3	91.5	116.0	51.7	100.9
0920	Plaster & Gypsum Board	81.4	39.0	53.1	96.9	66.1	76.3	96.0	56.3	69.5	104.2	60.0	74.7	104.2	66.1	78.7	81.1	39.5	53.3
0950, 0980	Ceilings & Acoustic Treatment	80.4	39.0	53.0	98.6	66.1	77.1	96.3	56.3	69.8	94.6	60.0	71.7	95.5	66.1	76.0	80.4	39.5	53.3
0960	Flooring	101.8	65.3	91.3	101.6	63.0	90.5	96.0	61.4	86.1	105.8	65.3	94.2	105.8	70.1	95.6	101.3	54.9	87.9
0970, 0990	Wall Finishes & Painting/Coating	96.4	64.4	77.8	94.7	46.3	66.6	99.8	48.3	69.9	99.8	57.9	75.5	99.8	50.0	70.8	96.4	48.1	68.3
09	FINISHES	87.7	50.3	67.0	94.8	64.2	77.9	94.2	55.0	72.5	97.9	59.2	76.5	98.2	65.8	80.3	87.5	44.3	63.6
COVERS	DIVS. 10 - 14, 25, 28, 41, 43, 44, 46	100.0	79.1	95.5	100.0	84.9	96.7	100.0	45.3	88.1	100.0	80.0	95.7	100.0	84.7	96.7	100.0	75.8	94.8
21, 22, 23	FIRE SUPPRESSION, PLUMBING & HVAC	100.2	59.7	83.1	100.0	64.8	85.2	96.7	61.5	81.8	100.1	62.7	84.3	100.1	68.3	86.7	100.2	54.2	80.7
26, 27, 3370	ELECTRICAL, COMMUNICATIONS & UTIL.	95.4	57.5	76.2	99.4	61.1	80.0	93.8	56.8	75.1	94.8	66.9	80.7	94.4	66.8	80.5	97.1	59.2	77.9
MF2014	WEIGHTED AVERAGE	99.7	63.3	83.7	99.5	69.9	86.4	95.2	61.0	80.1	97.0	66.6	83.6	97.2	69.9	85.2	99.6	59.3	81.8

LOUISIANA / MAINE

DIVISION		NEW ORLEANS 700 - 701			SHREVEPORT 710 - 711			THIBODAUX 703			AUGUSTA 043			BANGOR 044			BATH 045		
		MAT.	INST.	TOTAL	MAT.	INST.	TOTAL	MAT.	INST.	TOTAL	MAT.	INST.	TOTAL	MAT.	INST.	TOTAL	MAT.	INST.	TOTAL
015433	CONTRACTOR EQUIPMENT		91.5	91.5		93.0	93.0		91.2	91.2		98.1	98.1		98.1	98.1		98.1	98.1
0241, 31 - 34	SITE & INFRASTRUCTURE, DEMOLITION	99.3	92.9	94.8	104.3	92.3	95.8	102.3	89.9	93.5	91.3	98.6	96.5	94.5	98.8	97.5	92.3	98.6	96.8
0310	Concrete Forming & Accessories	96.4	67.7	71.6	96.1	43.8	50.9	90.1	64.4	67.8	101.8	93.2	94.3	96.6	94.2	94.5	92.0	93.3	93.1
0320	Concrete Reinforcing	99.5	59.2	78.6	106.0	60.0	82.2	97.9	58.5	77.5	103.0	102.8	102.9	92.7	104.1	98.6	91.7	104.0	98.1
0330	Cast-in-Place Concrete	93.8	69.5	84.2	95.5	63.9	83.1	101.2	51.1	81.4	95.8	108.1	100.7	75.6	118.2	92.4	75.6	108.2	88.5
03	CONCRETE	95.4	67.3	81.7	97.6	54.9	76.8	98.2	52.9	79.3	101.5	99.4	100.5	91.9	103.6	97.6	91.9	99.8	95.7
04	MASONRY	94.2	61.4	74.0	104.8	50.3	71.2	116.9	48.6	74.8	115.6	63.3	83.4	124.8	103.4	111.6	132.5	85.7	103.7
05	METALS	102.1	73.4	93.0	99.2	74.2	91.2	91.2	71.9	85.0	100.2	85.7	95.6	90.4	87.5	89.5	88.8	87.4	88.4
06	WOOD, PLASTICS & COMPOSITES	103.2	69.1	84.2	104.0	41.4	69.2	96.1	69.1	81.1	95.7	101.7	99.1	98.3	93.1	95.4	92.3	101.7	97.5
07	THERMAL & MOISTURE PROTECTION	97.1	69.7	85.6	95.2	61.8	81.2	97.8	63.3	83.4	99.5	70.6	87.4	97.9	83.3	91.8	97.8	77.2	89.2
08	OPENINGS	101.0	66.7	93.1	112.9	45.8	97.3	102.2	63.2	93.1	107.3	91.9	103.7	101.0	87.2	97.8	100.9	91.9	98.8
0920	Plaster & Gypsum Board	98.4	68.5	78.5	90.9	40.0	56.9	97.6	68.5	78.2	104.7	101.3	102.4	102.7	92.4	95.9	97.8	101.3	100.1
0950, 0980	Ceilings & Acoustic Treatment	95.2	68.5	77.6	84.3	40.0	55.0	96.3	68.5	77.9	92.7	101.3	98.4	79.9	92.4	88.2	79.0	101.3	93.7
0960	Flooring	107.5	65.3	95.4	105.9	58.9	92.4	102.8	42.4	85.4	97.5	53.1	84.7	94.6	109.9	99.0	92.6	53.1	81.2
0970, 0990	Wall Finishes & Painting/Coating	102.8	68.7	83.0	92.7	44.6	64.7	101.1	49.6	71.1	95.7	63.3	76.9	94.7	44.2	65.3	94.7	44.2	65.3
09	FINISHES	100.3	67.3	82.0	92.3	45.1	66.2	96.7	59.0	75.9	95.0	83.3	88.5	91.8	92.1	92.0	90.3	81.2	85.3
COVERS	DIVS. 10 - 14, 25, 28, 41, 43, 44, 46	100.0	83.6	96.4	100.0	79.0	95.4	100.0	81.4	96.0	100.0	97.6	99.5	100.0	109.2	102.0	100.0	97.6	99.5
21, 22, 23	FIRE SUPPRESSION, PLUMBING & HVAC	100.0	66.2	85.7	99.9	58.4	82.4	96.7	62.3	82.2	99.9	63.6	84.6	100.1	73.5	88.9	96.7	63.6	82.8
26, 27, 3370	ELECTRICAL, COMMUNICATIONS & UTIL.	96.5	74.6	85.4	102.5	66.1	84.1	92.6	73.7	83.0	101.0	79.5	90.1	99.5	76.1	87.7	97.6	79.5	88.5
MF2014	WEIGHTED AVERAGE	99.2	70.7	86.6	100.9	60.7	83.2	97.6	65.6	83.5	101.2	81.0	92.3	97.9	88.8	93.9	96.9	83.3	90.9

City Cost Indexes

MAINE

DIVISION		HOULTON 047			KITTERY 039			LEWISTON 042			MACHIAS 046			PORTLAND 040 - 041			ROCKLAND 048		
		MAT.	INST.	TOTAL	MAT.	INST.	TOTAL	MAT.	INST.	TOTAL	MAT.	INST.	TOTAL	MAT.	INST.	TOTAL	MAT.	INST.	TOTAL
015433	CONTRACTOR EQUIPMENT		98.1	98.1		98.1	98.1		98.1	98.1		98.1	98.1		98.1	98.1		98.1	98.1
0241, 31 - 34	SITE & INFRASTRUCTURE, DEMOLITION	94.3	99.7	98.1	86.4	98.7	95.1	92.0	98.8	96.8	93.6	99.7	97.9	91.6	98.8	96.7	89.9	99.7	96.8
0310	Concrete Forming & Accessories	100.7	99.9	100.0	91.3	94.3	93.9	102.7	94.2	95.4	97.5	99.8	99.5	102.5	94.2	95.4	98.6	99.8	99.7
0320	Concrete Reinforcing	92.7	102.6	97.8	88.7	104.1	96.7	113.9	104.1	108.9	92.7	102.6	97.8	103.0	104.1	103.6	92.7	102.6	97.8
0330	Cast-in-Place Concrete	75.6	117.9	92.3	73.4	109.0	87.5	77.1	118.3	93.4	75.6	117.9	92.3	91.4	118.3	102.0	77.2	117.9	93.2
03	CONCRETE	93.0	105.7	99.2	86.4	100.5	93.2	92.5	103.6	97.9	92.4	105.7	98.9	98.5	103.6	101.0	89.8	105.7	97.5
04	MASONRY	106.3	75.9	87.5	121.1	86.7	99.9	106.9	103.4	104.7	106.3	75.9	87.5	121.9	103.4	110.5	99.7	75.9	85.0
05	METALS	89.0	85.7	88.0	85.2	88.1	86.1	94.0	87.6	91.9	89.1	85.7	88.0	100.6	87.6	96.4	88.9	85.7	87.9
06	WOOD, PLASTICS & COMPOSITES	102.3	101.7	102.0	93.5	101.7	98.1	104.3	93.1	98.1	99.2	101.7	100.6	98.0	93.1	95.3	100.1	101.7	101.0
07	THERMAL & MOISTURE PROTECTION	98.0	78.2	89.7	101.5	76.5	91.1	97.7	83.3	91.7	97.9	77.2	89.3	103.4	83.3	95.0	97.6	77.2	89.1
08	OPENINGS	101.0	91.9	98.9	100.2	95.7	99.1	104.3	87.2	100.4	101.0	91.9	98.9	106.7	87.2	102.1	101.0	91.9	98.9
0920	Plaster & Gypsum Board	105.0	101.3	102.5	97.1	101.3	99.9	108.0	92.4	97.6	103.4	101.3	102.0	105.2	92.4	96.7	103.4	101.3	102.0
0950, 0980	Ceilings & Acoustic Treatment	79.0	101.3	93.7	89.6	101.3	97.3	88.1	92.4	91.0	79.0	101.3	93.7	90.9	92.4	91.9	79.0	101.3	93.7
0960	Flooring	96.0	49.7	82.7	94.3	55.4	83.1	97.8	109.9	101.2	95.0	49.7	82.0	99.5	109.9	102.5	95.5	49.7	82.3
0970, 0990	Wall Finishes & Painting/Coating	94.7	137.1	119.4	88.6	115.5	104.2	94.7	44.2	65.3	94.7	137.1	119.4	97.9	44.2	66.7	94.7	137.1	119.4
09	FINISHES	92.4	95.0	93.8	91.8	89.5	90.6	94.9	92.1	93.4	91.8	95.0	93.6	96.2	92.1	93.9	91.6	95.0	93.5
COVERS	DIVS. 10 - 14, 25, 28, 41, 43, 44, 46	100.0	103.4	100.7	100.0	97.9	99.5	100.0	109.2	102.0	100.0	103.4	100.7	100.0	109.2	102.0	100.0	103.4	100.7
21, 22, 23	FIRE SUPPRESSION, PLUMBING & HVAC	96.7	72.5	86.5	96.7	75.9	87.9	100.1	73.5	88.9	96.7	72.5	86.5	99.9	73.5	88.8	96.7	72.5	86.5
26, 27, 3370	ELECTRICAL, COMMUNICATIONS & UTIL.	101.4	79.5	90.3	92.3	79.5	85.8	101.5	80.4	90.8	101.4	79.5	90.3	103.5	80.4	91.8	101.4	79.5	90.3
MF2014	WEIGHTED AVERAGE	96.5	86.9	92.3	94.5	87.4	91.4	98.5	89.4	94.5	96.3	86.9	92.2	101.6	89.4	96.3	95.6	86.9	91.7

MAINE / MARYLAND

DIVISION		MAINE WATERVILLE 049			ANNAPOLIS 214			BALTIMORE 210 - 212			COLLEGE PARK 207 - 208			CUMBERLAND 215			EASTON 216		
		MAT.	INST.	TOTAL	MAT.	INST.	TOTAL	MAT.	INST.	TOTAL	MAT.	INST.	TOTAL	MAT.	INST.	TOTAL	MAT.	INST.	TOTAL
015433	CONTRACTOR EQUIPMENT		98.1	98.1		102.8	102.8		106.6	106.6		108.2	108.2		102.8	102.8		102.8	102.8
0241, 31 - 34	SITE & INFRASTRUCTURE, DEMOLITION	94.1	98.6	97.2	104.6	94.2	97.2	101.9	98.5	99.5	99.7	97.1	97.9	95.5	94.5	94.8	102.9	91.1	94.6
0310	Concrete Forming & Accessories	91.5	93.2	92.9	100.1	72.2	76.0	100.3	73.2	76.9	85.9	67.8	70.2	92.8	81.7	83.2	90.6	71.2	73.8
0320	Concrete Reinforcing	92.7	102.8	98.0	101.2	83.3	92.0	102.0	83.4	92.3	101.8	80.6	90.8	90.6	76.7	83.4	89.9	80.1	84.8
0330	Cast-in-Place Concrete	75.6	108.1	88.4	111.7	75.7	97.5	112.3	76.5	98.2	113.8	75.6	98.7	94.4	85.6	90.9	104.8	47.6	82.2
03	CONCRETE	93.4	99.4	96.3	105.1	76.6	91.2	107.0	77.4	92.6	106.4	74.3	90.8	92.6	83.0	87.9	100.7	65.9	83.8
04	MASONRY	117.7	63.3	84.2	106.3	72.7	85.6	102.9	72.7	84.3	110.1	71.7	86.4	102.5	90.0	94.8	117.0	42.7	71.2
05	METALS	89.0	85.7	88.0	99.7	95.8	98.4	100.0	96.4	98.9	87.9	99.6	91.6	96.3	94.3	95.7	96.6	88.6	94.0
06	WOOD, PLASTICS & COMPOSITES	91.8	101.7	97.3	97.0	72.9	83.6	107.3	74.2	88.9	79.9	67.6	73.0	94.7	80.2	86.6	92.6	78.5	84.8
07	THERMAL & MOISTURE PROTECTION	98.0	70.6	86.5	105.5	80.3	94.2	104.0	80.6	94.2	104.6	78.5	93.7	102.9	80.5	93.5	103.0	58.8	84.5
08	OPENINGS	101.0	91.9	98.9	103.7	78.9	98.0	97.6	80.6	93.6	93.2	73.7	88.7	100.8	81.7	96.3	99.0	71.5	92.6
0920	Plaster & Gypsum Board	97.8	101.3	100.1	101.6	72.4	82.1	104.9	73.6	84.0	94.6	66.7	75.9	104.9	79.9	88.2	104.9	78.2	87.0
0950, 0980	Ceilings & Acoustic Treatment	79.0	101.3	93.7	93.1	72.4	79.4	103.9	73.6	83.8	96.0	66.7	76.6	98.3	79.9	86.1	98.3	78.2	85.0
0960	Flooring	92.2	53.1	81.0	92.3	79.4	88.6	94.3	79.4	90.0	87.7	78.3	85.0	88.8	97.2	91.2	87.9	49.7	76.9
0970, 0990	Wall Finishes & Painting/Coating	94.7	63.3	76.5	86.9	78.0	81.7	95.2	78.0	85.2	102.9	78.0	88.4	92.8	95.6	94.4	92.8	78.0	84.2
09	FINISHES	90.4	83.3	86.5	92.1	73.4	81.8	99.8	74.2	85.7	91.1	69.8	79.3	95.1	85.9	90.0	95.2	69.5	81.0
COVERS	DIVS. 10 - 14, 25, 28, 41, 43, 44, 46	100.0	97.6	99.5	100.0	86.5	97.1	100.0	87.0	97.2	100.0	84.5	96.6	100.0	91.0	98.0	100.0	73.7	94.3
21, 22, 23	FIRE SUPPRESSION, PLUMBING & HVAC	96.7	63.6	82.7	100.0	81.8	92.3	100.0	81.8	92.3	96.7	83.4	91.1	96.5	74.2	87.1	96.5	77.2	88.3
26, 27, 3370	ELECTRICAL, COMMUNICATIONS & UTIL.	101.4	79.5	90.3	101.0	91.8	96.4	99.7	91.8	95.7	100.1	102.2	101.1	98.1	82.1	90.0	97.7	66.7	82.0
MF2014	WEIGHTED AVERAGE	96.8	81.0	89.8	101.0	82.7	92.9	100.9	83.4	93.2	97.1	83.8	91.2	97.3	84.1	91.5	98.9	71.2	86.7

MARYLAND / MASSACHUSETTS

DIVISION		ELKTON 219			HAGERSTOWN 217			SALISBURY 218			SILVER SPRING 209			WALDORF 206			BOSTON 020 - 022, 024		
		MAT.	INST.	TOTAL	MAT.	INST.	TOTAL	MAT.	INST.	TOTAL	MAT.	INST.	TOTAL	MAT.	INST.	TOTAL	MAT.	INST.	TOTAL
015433	CONTRACTOR EQUIPMENT		102.8	102.8		102.8	102.8		102.8	102.8		100.1	100.1		100.1	100.1		104.3	104.3
0241, 31 - 34	SITE & INFRASTRUCTURE, DEMOLITION	89.4	92.0	91.2	93.7	95.0	94.6	102.8	91.2	94.6	88.9	89.2	89.2	95.2	88.9	90.8	98.4	106.8	104.4
0310	Concrete Forming & Accessories	96.8	88.9	90.0	91.7	77.1	79.1	105.4	50.3	57.8	94.6	68.5	72.0	102.1	66.8	71.6	103.5	145.6	139.9
0320	Concrete Reinforcing	89.9	104.1	97.3	90.6	76.7	83.4	89.9	62.8	75.9	100.5	80.5	90.2	101.2	80.5	90.5	102.8	152.6	128.6
0330	Cast-in-Place Concrete	84.8	75.1	81.0	89.9	85.6	88.2	104.8	46.2	81.7	116.5	78.4	101.5	130.5	75.7	108.9	99.8	146.2	118.1
03	CONCRETE	85.3	87.7	86.5	89.0	80.9	85.1	101.7	52.9	78.0	104.8	75.5	90.5	115.5	73.8	95.2	101.2	145.9	122.9
04	MASONRY	101.4	60.1	75.9	108.5	90.0	97.1	116.9	47.0	73.8	109.3	76.7	89.2	94.0	71.7	80.2	107.9	165.2	143.2
05	METALS	96.6	101.7	98.2	96.4	94.2	95.7	96.6	82.6	92.1	92.1	95.7	93.3	92.1	95.7	93.3	99.7	130.0	109.3
06	WOOD, PLASTICS & COMPOSITES	99.9	97.0	98.3	93.8	74.1	82.8	110.7	53.2	78.7	86.6	66.8	75.6	94.0	66.8	78.8	106.4	145.5	128.1
07	THERMAL & MOISTURE PROTECTION	102.6	75.1	91.1	102.6	78.7	92.6	103.3	63.0	86.4	107.8	85.5	98.5	108.3	83.4	97.9	97.7	150.5	119.8
08	OPENINGS	99.0	86.1	96.0	99.0	75.2	93.4	99.2	60.5	90.2	85.4	72.8	82.4	86.0	73.3	83.0	99.8	147.8	111.0
0920	Plaster & Gypsum Board	108.2	97.1	100.8	104.9	73.6	84.0	115.5	52.1	73.2	100.5	66.7	77.9	103.5	66.7	78.9	104.6	146.3	132.4
0950, 0980	Ceilings & Acoustic Treatment	98.3	97.1	97.5	99.2	73.6	82.2	98.3	52.1	67.7	104.3	66.7	79.4	104.3	66.7	79.4	98.1	146.3	130.0
0960	Flooring	90.4	56.2	80.6	88.4	97.2	90.9	94.5	62.2	85.2	93.5	78.3	89.2	90.7	78.3	91.7	95.7	178.6	119.5
0970, 0990	Wall Finishes & Painting/Coating	92.8	77.6	84.0	92.8	78.0	84.2	92.8	77.6	84.0	110.9	78.0	91.7	110.9	78.0	91.7	91.3	166.2	134.9
09	FINISHES	95.6	82.8	88.5	94.9	80.3	86.9	98.7	54.7	74.4	92.4	69.7	79.8	94.2	69.1	80.3	99.6	154.6	130.0
COVERS	DIVS. 10 - 14, 25, 28, 41, 43, 44, 46	100.0	60.2	91.4	100.0	90.3	97.9	100.0	37.6	86.5	100.0	84.3	96.6	100.0	82.6	96.2	100.0	119.8	104.3
21, 22, 23	FIRE SUPPRESSION, PLUMBING & HVAC	96.5	81.5	90.2	99.9	85.9	94.0	96.5	59.3	80.8	96.7	85.5	92.0	96.7	83.0	91.0	99.9	131.8	113.4
26, 27, 3370	ELECTRICAL, COMMUNICATIONS & UTIL.	99.3	92.3	95.8	97.9	82.1	89.9	96.6	66.6	81.4	97.5	102.2	99.9	95.2	102.2	98.7	102.1	134.3	118.4
MF2014	WEIGHTED AVERAGE	96.3	84.2	91.0	97.7	85.2	92.2	99.4	61.8	82.8	96.4	84.0	90.9	97.1	82.6	90.7	100.6	139.1	117.5

For customer support on your Mechanical Cost Data, call 877.739.3689.

671

City Cost Indexes

MASSACHUSETTS

DIVISION		BROCKTON 023 MAT.	INST.	TOTAL	BUZZARDS BAY 025 MAT.	INST.	TOTAL	FALL RIVER 027 MAT.	INST.	TOTAL	FITCHBURG 014 MAT.	INST.	TOTAL	FRAMINGHAM 017 MAT.	INST.	TOTAL	GREENFIELD 013 MAT.	INST.	TOTAL
015433	CONTRACTOR EQUIPMENT		100.3	100.3		100.3	100.3		101.2	101.2		98.1	98.1		99.6	99.6		98.1	98.1
0241, 31 - 34	SITE & INFRASTRUCTURE, DEMOLITION	94.8	103.2	100.8	85.1	103.1	97.9	93.8	103.3	100.5	86.7	103.1	98.3	83.4	102.9	97.2	90.4	101.7	98.4
0310	Concrete Forming & Accessories	101.1	134.6	130.1	98.6	134.6	129.7	101.1	134.8	130.3	95.5	127.0	122.8	103.5	134.9	130.7	93.7	110.0	107.8
0320	Concrete Reinforcing	101.8	152.0	127.8	81.7	158.3	121.3	101.8	158.3	131.1	85.7	151.7	119.9	85.7	152.5	120.3	89.2	125.4	108.0
0330	Cast-in-Place Concrete	95.2	142.8	114.0	79.1	142.8	104.2	92.1	143.3	112.3	80.7	142.7	105.2	80.8	146.3	106.6	83.0	124.6	99.4
03	CONCRETE	99.2	139.6	118.8	83.9	140.7	111.5	97.7	141.0	118.7	83.7	135.9	109.1	86.4	141.0	113.0	87.2	117.2	101.8
04	MASONRY	105.3	151.7	133.9	98.0	159.2	135.7	106.1	159.2	138.8	106.8	158.3	138.5	113.8	159.3	141.9	111.7	130.1	123.1
05	METALS	95.5	126.6	105.4	90.1	129.3	102.6	95.5	129.7	106.4	92.4	123.7	102.3	92.4	127.3	103.5	94.7	108.9	99.2
06	WOOD, PLASTICS & COMPOSITES	103.4	133.8	120.3	100.5	133.8	119.0	103.4	134.0	120.5	99.4	124.0	113.1	105.7	133.5	121.2	97.3	107.1	102.8
07	THERMAL & MOISTURE PROTECTION	98.3	143.8	117.4	97.5	143.6	116.8	98.2	143.0	116.9	99.0	137.8	115.2	99.1	146.5	118.9	99.0	116.4	106.3
08	OPENINGS	100.9	138.5	109.7	96.8	136.4	106.0	100.9	136.6	109.2	102.7	133.0	109.8	93.1	138.4	103.7	103.0	113.6	105.5
0920	Plaster & Gypsum Board	90.7	134.2	119.8	87.1	134.2	118.6	90.7	134.2	119.8	99.3	124.2	115.9	101.9	134.2	123.5	99.7	106.9	104.5
0950, 0980	Ceilings & Acoustic Treatment	96.8	134.2	121.6	83.7	134.2	117.2	96.8	134.2	121.6	85.3	124.2	111.1	85.3	134.2	117.7	92.8	106.9	102.1
0960	Flooring	95.3	178.6	119.3	93.1	178.6	117.7	94.3	178.6	118.5	94.8	178.6	118.9	96.7	178.6	120.2	93.9	139.3	106.9
0970, 0990	Wall Finishes & Painting/Coating	87.8	155.2	127.0	87.8	155.2	127.0	87.8	155.2	127.0	89.2	155.2	127.6	90.1	155.2	128.0	89.2	126.6	111.0
09	FINISHES	94.4	145.4	122.6	89.7	145.4	120.6	94.1	145.6	122.6	90.8	139.6	117.8	91.6	145.3	121.3	92.6	116.9	106.0
COVERS	DIVS. 10 - 14, 25, 28, 41, 43, 44, 46	100.0	117.1	103.7	100.0	117.1	103.7	100.0	117.7	103.8	100.0	105.9	101.3	100.0	116.7	103.6	100.0	105.9	101.3
21, 22, 23	FIRE SUPPRESSION, PLUMBING & HVAC	100.1	110.9	104.7	96.7	110.0	102.3	100.1	111.0	104.7	97.1	112.4	103.6	97.1	124.7	108.8	97.1	101.8	99.1
26, 27, 3370	ELECTRICAL, COMMUNICATIONS & UTIL.	98.4	99.3	98.9	95.8	99.3	97.6	98.3	99.3	98.8	98.2	105.5	101.9	95.1	127.9	111.7	98.2	99.0	98.6
MF2014	WEIGHTED AVERAGE	98.8	125.1	110.4	93.7	125.9	107.9	98.6	126.3	110.8	95.6	124.5	108.3	94.9	132.9	111.6	96.9	110.0	102.6

MASSACHUSETTS

DIVISION		HYANNIS 026 MAT.	INST.	TOTAL	LAWRENCE 019 MAT.	INST.	TOTAL	LOWELL 018 MAT.	INST.	TOTAL	NEW BEDFORD 027 MAT.	INST.	TOTAL	PITTSFIELD 012 MAT.	INST.	TOTAL	SPRINGFIELD 010 - 011 MAT.	INST.	TOTAL
015433	CONTRACTOR EQUIPMENT		100.3	100.3		100.3	100.3		98.1	98.1		101.2	101.2		98.1	98.1		98.1	98.1
0241, 31 - 34	SITE & INFRASTRUCTURE, DEMOLITION	91.2	103.1	99.6	95.6	103.2	101.0	94.6	103.2	100.7	92.3	103.3	100.1	95.6	101.6	99.8	95.0	101.8	99.8
0310	Concrete Forming & Accessories	92.7	134.6	128.9	104.8	135.1	131.0	101.0	135.2	130.6	101.1	134.8	130.3	101.0	109.1	108.0	101.3	110.3	109.1
0320	Concrete Reinforcing	81.7	158.3	121.3	106.0	148.7	128.1	106.9	148.7	128.5	101.8	158.3	131.1	88.5	120.1	104.9	106.9	125.4	116.5
0330	Cast-in-Place Concrete	86.8	142.8	108.9	93.4	143.0	113.0	84.9	143.0	107.9	81.1	143.3	105.7	92.6	123.1	104.7	88.4	125.1	102.9
03	CONCRETE	90.1	140.7	114.7	101.0	139.3	119.6	92.6	139.1	115.2	92.3	141.0	116.0	93.7	115.3	104.2	94.3	117.5	105.6
04	MASONRY	104.3	159.2	138.1	119.6	159.9	144.5	105.8	159.2	138.7	104.2	159.2	138.1	106.4	132.5	122.5	106.1	131.0	121.4
05	METALS	91.6	129.3	103.6	95.1	126.1	105.0	95.1	123.4	104.1	95.5	129.7	106.4	94.9	107.1	98.8	98.1	108.9	101.5
06	WOOD, PLASTICS & COMPOSITES	93.8	133.8	116.0	106.0	133.8	121.5	105.5	133.8	121.2	103.4	134.0	120.5	105.5	107.1	106.4	105.5	107.1	106.4
07	THERMAL & MOISTURE PROTECTION	97.8	143.6	117.0	99.6	145.7	118.9	99.3	146.6	119.1	98.2	143.0	116.9	99.4	116.6	106.6	99.3	116.8	106.6
08	OPENINGS	97.4	136.4	106.5	97.1	137.5	106.5	104.2	137.5	111.9	100.9	136.6	109.2	104.2	112.2	106.0	104.2	113.6	106.4
0920	Plaster & Gypsum Board	82.5	134.2	117.1	104.4	134.2	124.3	104.4	134.2	124.3	90.7	134.2	119.8	104.4	106.9	106.0	104.4	106.9	106.0
0950, 0980	Ceilings & Acoustic Treatment	90.2	134.2	119.3	94.6	134.2	120.9	94.6	134.2	120.9	96.8	134.2	121.6	94.6	106.9	102.7	94.6	106.9	102.7
0960	Flooring	90.4	178.6	115.7	97.2	178.6	120.6	97.2	178.6	120.6	94.3	178.6	118.5	97.7	148.1	112.2	96.5	139.3	108.8
0970, 0990	Wall Finishes & Painting/Coating	87.8	155.2	127.0	89.3	155.2	127.6	89.2	155.2	127.6	87.8	155.2	127.0	89.2	126.6	111.0	90.5	126.6	111.5
09	FINISHES	90.0	145.4	120.7	94.9	145.4	122.8	94.8	145.4	122.8	94.0	145.6	122.5	95.0	118.0	107.7	94.7	117.1	107.1
COVERS	DIVS. 10 - 14, 25, 28, 41, 43, 44, 46	100.0	117.1	103.7	100.0	117.3	103.7	100.0	117.3	103.7	100.0	117.7	103.8	100.0	105.0	101.1	100.0	106.2	101.3
21, 22, 23	FIRE SUPPRESSION, PLUMBING & HVAC	100.1	110.9	104.8	97.3	124.4	110.3	100.1	124.4	110.3	100.1	111.0	104.7	100.1	101.0	100.5	100.1	102.3	101.0
26, 27, 3370	ELECTRICAL, COMMUNICATIONS & UTIL.	96.2	99.3	97.8	97.3	127.9	112.8	97.8	127.9	113.0	99.1	99.3	99.2	97.8	99.0	98.4	97.8	99.0	98.4
MF2014	WEIGHTED AVERAGE	95.9	125.9	109.1	99.2	132.6	113.9	98.4	132.3	113.3	97.9	126.3	110.4	98.5	109.7	103.4	99.1	110.3	104.0

MASSACHUSETTS / MICHIGAN

DIVISION		WORCESTER 015 - 016 MAT.	INST.	TOTAL	ANN ARBOR 481 MAT.	INST.	TOTAL	BATTLE CREEK 490 MAT.	INST.	TOTAL	BAY CITY 487 MAT.	INST.	TOTAL	DEARBORN 481 MAT.	INST.	TOTAL	DETROIT 482 MAT.	INST.	TOTAL
015433	CONTRACTOR EQUIPMENT		98.1	98.1		109.7	109.7		99.2	99.2		109.7	109.7		109.7	109.7		95.4	95.4
0241, 31 - 34	SITE & INFRASTRUCTURE, DEMOLITION	94.9	103.1	100.7	82.3	97.3	92.9	94.7	85.1	87.9	73.7	96.2	89.6	82.1	97.4	92.9	95.3	99.0	97.9
0310	Concrete Forming & Accessories	101.6	127.1	123.6	97.0	113.3	111.1	97.8	84.2	86.0	97.1	84.3	86.1	96.9	114.1	111.8	100.7	114.1	112.3
0320	Concrete Reinforcing	106.9	151.7	130.1	96.8	112.5	104.9	94.9	91.8	93.3	96.8	111.6	104.5	96.8	112.7	105.0	98.7	112.6	105.9
0330	Cast-in-Place Concrete	87.9	142.7	109.5	87.2	105.3	94.3	92.8	97.4	94.6	83.4	88.4	85.4	85.2	103.0	92.2	99.0	103.0	100.6
03	CONCRETE	94.1	135.9	114.4	91.2	110.9	100.8	96.2	89.5	93.0	89.4	92.0	90.7	90.3	110.5	100.1	97.6	109.2	103.2
04	MASONRY	105.6	158.3	138.1	99.9	109.7	105.9	103.8	88.2	94.2	99.4	87.5	92.1	99.7	113.7	108.3	92.0	113.7	105.4
05	METALS	98.1	123.7	106.3	96.0	119.3	103.4	99.3	85.1	94.8	96.7	115.7	102.7	96.1	119.7	103.6	96.4	100.0	97.6
06	WOOD, PLASTICS & COMPOSITES	105.9	124.0	116.0	89.4	115.0	103.6	91.5	82.4	86.5	89.4	83.3	86.0	89.4	115.0	103.6	96.2	115.0	106.7
07	THERMAL & MOISTURE PROTECTION	99.3	138.5	115.7	102.8	107.3	104.7	97.6	83.3	91.6	100.5	87.0	94.9	101.2	113.0	106.2	99.9	113.0	105.3
08	OPENINGS	104.2	133.0	110.9	99.4	110.5	102.0	95.1	78.8	91.3	99.4	91.0	97.4	99.4	111.0	102.1	100.5	111.0	103.0
0920	Plaster & Gypsum Board	104.4	124.2	117.6	101.8	114.9	110.6	93.9	78.3	83.5	101.8	82.4	88.8	101.8	114.9	110.6	100.9	114.9	110.3
0950, 0980	Ceilings & Acoustic Treatment	94.6	124.2	114.2	91.2	114.9	106.9	85.8	78.3	80.9	92.1	82.4	85.7	91.2	114.9	106.9	101.2	114.9	110.3
0960	Flooring	97.2	172.8	118.9	96.6	114.5	101.8	97.6	88.3	95.0	96.6	78.4	91.4	96.3	112.3	100.9	98.0	112.3	102.1
0970, 0990	Wall Finishes & Painting/Coating	89.2	155.2	127.6	85.5	98.7	93.2	92.5	81.9	86.4	85.5	84.2	84.7	85.5	104.1	96.3	88.6	104.1	97.6
09	FINISHES	94.8	138.5	119.0	93.2	112.6	104.0	92.2	84.4	87.8	93.0	82.5	87.2	93.1	113.2	104.2	97.2	113.2	106.1
COVERS	DIVS. 10 - 14, 25, 28, 41, 43, 44, 46	100.0	110.2	102.2	100.0	105.8	101.3	100.0	93.8	98.6	100.0	91.0	98.1	100.0	106.3	101.4	100.0	106.3	101.4
21, 22, 23	FIRE SUPPRESSION, PLUMBING & HVAC	100.1	112.5	105.3	100.1	98.2	99.3	100.2	87.2	94.7	100.1	82.8	92.8	100.1	108.1	103.5	99.9	109.1	103.8
26, 27, 3370	ELECTRICAL, COMMUNICATIONS & UTIL.	97.8	105.5	101.7	96.9	108.0	102.5	93.1	83.1	88.0	96.0	89.6	92.8	96.9	106.2	101.6	97.8	106.1	102.0
MF2014	WEIGHTED AVERAGE	99.0	124.5	110.2	97.0	107.4	101.6	97.5	86.0	92.4	96.4	90.3	93.7	96.8	109.9	102.6	98.2	108.2	102.6

MICHIGAN

DIVISION		FLINT 484 - 485			GAYLORD 497			GRAND RAPIDS 493,495			IRON MOUNTAIN 498 - 499			JACKSON 492			KALAMAZOO 491		
		MAT.	INST.	TOTAL	MAT.	INST.	TOTAL	MAT.	INST.	TOTAL	MAT.	INST.	TOTAL	MAT.	INST.	TOTAL	MAT.	INST.	TOTAL
015433	CONTRACTOR EQUIPMENT		109.7	109.7		104.1	104.1		99.2	99.2		92.3	92.3		104.1	104.1		99.2	99.2
0241, 31 - 34	SITE & INFRASTRUCTURE, DEMOLITION	71.7	96.4	89.1	89.5	82.6	84.6	93.4	85.1	87.5	97.7	91.6	93.4	111.8	84.8	92.7	95.0	85.1	88.0
0310	Concrete Forming & Accessories	99.8	89.1	90.5	96.0	76.7	79.3	97.2	83.6	85.4	87.7	82.7	83.4	92.8	85.7	86.7	97.8	83.9	85.8
0320	Concrete Reinforcing	96.8	112.0	104.7	88.5	117.3	103.4	99.6	91.7	95.5	88.3	98.8	93.7	86.0	111.8	99.4	94.9	92.3	93.5
0330	Cast-in-Place Concrete	87.7	92.2	89.5	92.6	85.4	89.7	97.3	96.6	97.0	109.6	84.4	99.6	92.4	91.1	91.9	94.6	97.3	95.7
03	CONCRETE	91.7	95.5	93.5	92.7	88.4	90.6	99.9	89.0	94.6	102.2	86.5	94.5	87.4	93.4	90.3	99.7	89.4	94.7
04	MASONRY	99.9	97.0	98.1	114.8	79.9	93.3	98.0	87.6	91.6	99.5	88.9	93.0	93.4	91.9	92.5	102.2	90.1	94.7
05	METALS	96.1	116.6	102.6	100.6	113.8	104.8	96.3	84.7	92.6	99.9	92.1	97.5	100.8	114.4	105.1	99.3	85.1	94.8
06	WOOD, PLASTICS & COMPOSITES	92.9	87.4	89.9	84.9	75.7	79.8	94.9	82.0	87.7	80.4	82.2	81.4	83.6	83.5	83.6	91.5	82.4	86.5
07	THERMAL & MOISTURE PROTECTION	100.5	94.2	97.9	96.1	73.4	86.6	99.2	75.3	89.2	99.8	80.3	91.6	95.4	91.0	93.6	97.6	83.5	91.7
08	OPENINGS	99.4	94.0	98.1	95.4	75.9	90.9	103.0	83.0	98.3	102.7	75.5	96.4	94.4	91.3	93.7	95.1	78.9	91.3
0920	Plaster & Gypsum Board	104.1	86.6	92.4	93.7	73.9	80.5	95.5	77.9	83.8	52.2	82.2	72.3	92.0	81.9	85.3	93.9	78.3	83.5
0950, 0980	Ceilings & Acoustic Treatment	91.2	86.6	88.2	84.9	73.9	77.6	94.6	77.9	83.6	83.9	82.2	82.8	84.9	81.9	82.9	85.8	78.3	80.9
0960	Flooring	96.6	92.0	95.3	90.5	88.8	90.0	97.3	82.4	93.0	111.2	95.3	106.6	89.1	79.5	86.4	97.6	79.5	92.4
0970, 0990	Wall Finishes & Painting/Coating	85.5	85.4	85.5	88.8	84.2	86.1	93.6	80.8	86.2	107.6	73.6	87.8	88.8	96.8	93.4	92.5	81.9	86.4
09	FINISHES	92.8	88.6	90.5	91.6	78.2	84.2	95.6	82.9	88.6	92.8	83.9	87.9	92.5	84.9	88.3	92.2	82.4	86.8
COVERS	DIVS. 10 - 14, 25, 28, 41, 43, 44, 46	100.0	92.6	98.4	100.0	86.5	97.1	100.0	93.9	98.7	100.0	84.8	96.7	100.0	101.4	100.3	100.0	93.7	98.6
21, 22, 23	FIRE SUPPRESSION, PLUMBING & HVAC	100.1	88.7	95.3	97.0	78.6	89.2	100.1	82.4	92.6	96.9	84.6	91.7	97.0	87.8	93.1	100.2	80.7	91.9
26, 27, 3370	ELECTRICAL, COMMUNICATIONS & UTIL.	96.9	99.0	97.9	91.1	78.5	84.8	97.7	78.5	88.0	96.9	83.6	90.2	94.8	108.0	101.4	93.0	78.2	85.5
MF2014	WEIGHTED AVERAGE	96.7	95.6	96.2	96.7	83.6	90.9	98.8	83.9	92.3	98.7	85.8	93.0	96.0	94.2	95.2	97.8	83.9	91.7

MICHIGAN / MINNESOTA

DIVISION		LANSING 488 - 489			MUSKEGON 494			ROYAL OAK 480,483			SAGINAW 486			TRAVERSE CITY 496			BEMIDJI 566		
		MAT.	INST.	TOTAL	MAT.	INST.	TOTAL	MAT.	INST.	TOTAL	MAT.	INST.	TOTAL	MAT.	INST.	TOTAL	MAT.	INST.	TOTAL
015433	CONTRACTOR EQUIPMENT		109.7	109.7		99.2	99.2		92.5	92.5		109.7	109.7		92.3	92.3		98.8	98.8
0241, 31 - 34	SITE & INFRASTRUCTURE, DEMOLITION	94.7	96.4	95.9	92.6	85.0	87.2	86.5	96.6	93.7	74.7	96.1	89.8	83.4	91.1	88.9	96.7	98.2	97.8
0310	Concrete Forming & Accessories	95.5	89.6	90.4	98.2	81.6	83.8	92.6	113.6	110.7	97.0	86.7	88.1	87.7	77.4	78.8	86.5	87.6	87.5
0320	Concrete Reinforcing	98.4	111.8	105.4	95.6	91.9	93.7	87.7	118.5	103.7	96.8	111.6	104.5	89.6	91.5	90.6	96.5	112.3	104.7
0330	Cast-in-Place Concrete	97.5	91.0	94.9	92.5	93.0	92.7	76.3	103.6	87.1	86.1	88.3	87.0	85.5	80.2	83.4	104.0	102.1	103.2
03	CONCRETE	97.3	95.2	96.3	94.5	86.9	90.8	78.7	110.0	93.9	90.7	93.0	91.9	84.0	81.4	82.7	96.6	98.4	97.5
04	MASONRY	101.3	95.5	97.7	100.8	88.1	92.9	93.6	111.5	104.6	101.3	87.5	92.8	97.6	84.6	89.6	106.4	108.9	107.9
05	METALS	96.0	116.0	102.3	96.9	84.8	93.1	99.1	96.8	98.4	96.1	115.5	102.3	99.9	89.3	96.5	88.6	123.6	99.8
06	WOOD, PLASTICS & COMPOSITES	90.0	88.3	89.1	88.3	80.5	83.9	85.2	115.0	101.8	86.0	86.8	86.5	80.4	76.8	78.4	69.3	82.6	76.7
07	THERMAL & MOISTURE PROTECTION	99.5	88.0	94.7	96.5	77.2	88.4	99.1	108.8	103.1	101.4	87.5	95.6	98.8	72.0	87.6	106.7	101.4	104.5
08	OPENINGS	101.0	93.8	99.3	94.3	81.9	91.4	99.2	110.3	101.8	97.3	93.0	96.2	102.7	71.4	95.4	104.4	109.7	105.6
0920	Plaster & Gypsum Board	94.3	87.5	89.8	74.9	76.3	75.9	99.4	114.9	109.8	101.8	86.0	91.2	52.2	76.6	68.5	103.7	82.4	89.5
0950, 0980	Ceilings & Acoustic Treatment	90.6	87.5	88.5	86.6	76.3	79.8	90.5	114.9	106.7	91.2	86.0	87.8	83.9	76.6	79.1	123.1	82.4	96.2
0960	Flooring	102.2	82.4	96.5	96.3	82.4	92.3	93.3	109.5	98.0	96.6	78.4	91.4	111.2	91.2	105.4	100.0	120.2	105.8
0970, 0990	Wall Finishes & Painting/Coating	93.8	85.8	89.1	90.7	80.8	84.9	86.8	96.8	92.6	85.5	84.2	84.7	107.6	44.7	71.0	95.2	95.9	95.6
09	FINISHES	96.2	87.5	91.4	88.6	80.7	84.2	91.9	111.7	102.9	92.9	84.6	88.3	91.7	76.5	83.3	104.3	93.4	98.2
COVERS	DIVS. 10 - 14, 25, 28, 41, 43, 44, 46	100.0	101.0	100.2	100.0	93.0	98.5	100.0	100.2	100.0	100.0	91.4	98.1	100.0	83.5	96.4	100.0	96.3	99.2
21, 22, 23	FIRE SUPPRESSION, PLUMBING & HVAC	100.0	88.5	95.1	99.9	81.3	92.1	96.9	106.0	100.8	100.1	82.3	92.5	96.9	78.2	89.0	96.9	82.6	90.8
26, 27, 3370	ELECTRICAL, COMMUNICATIONS & UTIL.	99.8	92.6	96.2	93.3	78.5	85.8	98.9	103.6	101.3	94.9	89.8	92.3	92.6	78.5	85.5	105.2	105.7	105.5
MF2014	WEIGHTED AVERAGE	98.7	94.4	96.8	96.3	83.2	90.5	95.1	106.1	99.9	96.3	90.8	93.9	95.6	80.9	89.1	98.6	99.1	98.8

MINNESOTA

DIVISION		BRAINERD 564			DETROIT LAKES 565			DULUTH 556 - 558			MANKATO 560			MINNEAPOLIS 553 - 555			ROCHESTER 559		
		MAT.	INST.	TOTAL	MAT.	INST.	TOTAL	MAT.	INST.	TOTAL	MAT.	INST.	TOTAL	MAT.	INST.	TOTAL	MAT.	INST.	TOTAL
015433	CONTRACTOR EQUIPMENT		101.5	101.5		98.8	98.8		101.4	101.4		101.5	101.5		105.0	105.0		101.4	101.4
0241, 31 - 34	SITE & INFRASTRUCTURE, DEMOLITION	98.3	103.1	101.7	94.9	98.5	97.5	99.5	102.8	101.8	95.0	103.8	101.2	99.0	108.2	105.5	99.1	102.2	101.3
0310	Concrete Forming & Accessories	87.7	89.4	89.2	83.4	89.4	88.6	98.1	112.2	110.3	95.9	102.7	101.8	100.2	128.5	124.7	100.7	108.3	107.3
0320	Concrete Reinforcing	95.3	112.1	104.0	96.5	112.3	104.7	95.5	112.6	104.3	95.2	119.4	107.7	94.2	119.9	107.5	93.0	119.6	106.8
0330	Cast-in-Place Concrete	113.0	105.6	110.1	101.0	105.0	102.6	110.4	107.8	109.4	104.2	107.0	105.3	101.3	119.0	108.3	107.3	102.5	105.4
03	CONCRETE	100.7	100.4	100.5	94.1	100.2	97.1	105.6	111.5	108.4	96.1	108.1	102.0	99.4	124.0	111.3	100.6	109.3	104.8
04	MASONRY	132.6	111.5	119.6	132.1	114.5	121.2	108.3	121.8	116.7	119.9	112.6	115.4	109.9	131.5	123.2	106.6	117.2	113.1
05	METALS	89.7	123.1	100.3	88.6	123.4	99.7	94.9	125.7	104.7	89.5	127.3	101.5	95.8	130.8	106.9	95.0	129.3	105.9
06	WOOD, PLASTICS & COMPOSITES	86.4	82.4	84.2	66.4	82.6	75.4	89.2	110.6	101.1	95.7	100.6	98.4	98.1	126.6	114.0	98.1	106.5	102.8
07	THERMAL & MOISTURE PROTECTION	105.2	108.1	106.4	106.5	110.3	108.1	103.6	115.5	108.6	105.7	100.7	103.6	102.6	129.7	113.9	107.4	104.1	106.0
08	OPENINGS	90.3	109.6	94.8	104.4	109.7	105.6	111.2	118.3	112.8	95.2	117.8	100.5	106.2	136.3	113.2	106.4	125.3	110.8
0920	Plaster & Gypsum Board	89.4	82.4	84.8	102.7	82.4	89.2	93.0	111.3	105.2	94.1	101.1	98.8	92.1	127.7	115.9	100.4	107.1	104.9
0950, 0980	Ceilings & Acoustic Treatment	57.8	82.4	74.1	123.1	82.4	96.2	99.7	111.3	107.4	57.8	101.1	86.5	101.4	127.7	118.8	95.8	107.1	103.3
0960	Flooring	98.8	120.2	105.0	98.7	120.2	104.9	99.8	123.4	106.6	101.0	125.5	108.0	99.0	122.0	105.1	100.4	121.1	106.3
0970, 0990	Wall Finishes & Painting/Coating	89.3	95.9	93.1	95.2	95.9	95.6	89.3	110.3	101.5	101.0	105.1	103.4	106.1	131.3	120.8	90.5	104.9	98.9
09	FINISHES	87.1	94.6	91.2	103.6	94.7	98.7	96.8	114.4	106.5	89.1	107.1	99.0	100.3	128.0	115.6	97.0	110.4	104.4
COVERS	DIVS. 10 - 14, 25, 28, 41, 43, 44, 46	100.0	97.6	99.5	100.0	98.0	99.6	100.0	96.7	99.3	100.0	99.3	99.9	100.0	110.0	102.2	100.0	103.6	100.8
21, 22, 23	FIRE SUPPRESSION, PLUMBING & HVAC	96.1	85.3	91.5	96.9	85.2	91.9	99.8	103.5	101.3	96.1	87.6	92.5	99.8	107.2	102.9	99.8	96.4	98.4
26, 27, 3370	ELECTRICAL, COMMUNICATIONS & UTIL.	102.8	102.4	102.6	105.0	70.6	87.6	100.9	102.4	101.6	109.3	88.5	98.8	102.6	113.2	108.0	101.3	88.5	94.9
MF2014	WEIGHTED AVERAGE	97.3	100.5	98.7	99.4	96.1	98.0	101.3	110.5	105.3	97.5	102.7	99.8	100.7	119.9	109.1	100.3	106.3	103.0

For customer support on your Mechanical Cost Data, call 877.739.3689.

673

City Cost Indexes

MINNESOTA / MISSISSIPPI

DIVISION		SAINT PAUL 550-551 MAT.	INST.	TOTAL	ST. CLOUD 563 MAT.	INST.	TOTAL	THIEF RIVER FALLS 567 MAT.	INST.	TOTAL	WILLMAR 562 MAT.	INST.	TOTAL	WINDOM 561 MAT.	INST.	TOTAL	BILOXI 395 MAT.	INST.	TOTAL
015433	CONTRACTOR EQUIPMENT		101.4	101.4		101.5	101.5		98.8	98.8		101.5	101.5		101.5	101.5		99.9	99.9
0241, 31-34	SITE & INFRASTRUCTURE, DEMOLITION	95.7	103.5	101.2	93.5	105.2	101.8	95.6	98.2	97.4	92.9	102.8	99.9	86.7	101.8	97.4	104.1	88.2	92.9
0310	Concrete Forming & Accessories	97.8	129.3	125.0	84.8	116.5	112.3	87.2	86.8	86.8	84.6	88.9	88.3	89.2	83.3	84.1	94.3	52.2	57.9
0320	Concrete Reinforcing	96.3	119.9	108.5	95.4	119.4	107.8	96.9	112.2	104.8	95.0	119.1	107.5	95.0	118.6	107.2	89.3	58.0	73.1
0330	Cast-in-Place Concrete	115.4	118.4	116.6	99.8	114.3	105.5	103.1	87.2	96.8	101.3	84.3	94.6	87.7	90.3	88.7	115.4	62.4	94.5
03	CONCRETE	108.2	124.1	116.0	91.9	116.8	104.0	95.3	92.9	94.1	91.9	94.3	93.0	82.7	93.6	88.0	106.6	58.6	83.3
04	MASONRY	110.1	131.5	123.3	115.8	118.0	117.2	106.4	108.9	107.9	120.1	111.7	114.9	131.8	93.5	108.2	104.3	53.5	89.9
05	METALS	94.9	130.7	106.3	90.3	128.2	102.4	88.8	122.8	99.6	89.5	126.3	101.2	89.4	124.6	100.6	91.9	85.7	89.9
06	WOOD, PLASTICS & COMPOSITES	94.9	127.6	113.1	83.8	114.5	100.9	70.2	82.6	77.1	83.5	82.6	83.0	87.7	80.0	83.4	100.2	54.2	74.6
07	THERMAL & MOISTURE PROTECTION	102.6	129.4	113.8	105.5	116.0	109.8	107.6	98.3	103.7	105.2	106.0	105.5	105.2	90.4	99.0	97.2	60.9	82.0
08	OPENINGS	103.9	136.9	111.6	95.7	129.7	103.6	104.4	109.7	105.6	92.5	91.5	92.3	96.4	89.7	94.8	97.8	56.0	88.1
0920	Plaster & Gypsum Board	96.5	128.8	118.1	89.4	115.4	106.8	103.4	82.4	89.4	89.4	82.7	84.9	89.4	80.0	83.1	96.6	53.3	67.7
0950, 0980	Ceilings & Acoustic Treatment	98.3	128.8	118.5	57.8	115.4	95.9	123.1	82.4	96.2	57.8	82.7	74.2	57.8	80.0	72.5	87.2	53.3	64.8
0960	Flooring	96.0	120.2	103.0	95.3	120.2	102.4	99.7	120.2	105.6	96.8	116.1	102.3	99.6	116.1	104.3	98.2	69.2	89.9
0970, 0990	Wall Finishes & Painting/Coating	88.5	126.1	110.4	101.0	131.3	118.7	95.2	95.9	95.6	95.2	95.9	95.6	95.2	105.1	101.0	85.2	43.4	60.9
09	FINISHES	95.0	128.0	113.3	86.6	119.0	104.5	104.1	93.4	98.2	86.6	94.0	90.7	86.9	91.2	89.3	91.3	54.1	70.7
COVERS	DIVS. 10-14, 25, 28, 41, 43, 44, 46	100.0	109.9	102.1	100.0	101.2	100.3	100.0	96.2	99.2	100.0	97.6	99.5	100.0	91.1	98.1	100.0	69.1	93.3
21, 22, 23	FIRE SUPPRESSION, PLUMBING & HVAC	99.7	113.6	105.6	99.4	103.5	101.2	96.9	82.3	90.7	96.1	95.8	96.0	96.1	83.3	90.7	100.0	49.7	78.7
26, 27, 3370	ELECTRICAL, COMMUNICATIONS & UTIL.	101.4	113.2	107.4	102.8	113.2	108.1	102.3	70.6	86.3	102.8	84.6	93.6	109.3	88.5	98.8	100.4	56.5	78.2
MF2014	WEIGHTED AVERAGE	100.7	120.9	109.6	96.8	114.0	104.4	98.2	93.3	96.0	95.7	98.8	97.1	96.2	93.5	95.0	98.8	60.5	81.9

MISSISSIPPI

DIVISION		CLARKSDALE 386 MAT.	INST.	TOTAL	COLUMBUS 397 MAT.	INST.	TOTAL	GREENVILLE 387 MAT.	INST.	TOTAL	GREENWOOD 389 MAT.	INST.	TOTAL	JACKSON 390-392 MAT.	INST.	TOTAL	LAUREL 394 MAT.	INST.	TOTAL
015433	CONTRACTOR EQUIPMENT		99.9	99.9		99.9	99.9		99.9	99.9		99.9	99.9		99.9	99.9		99.9	99.9
0241, 31-34	SITE & INFRASTRUCTURE, DEMOLITION	97.7	87.6	90.6	102.7	87.8	92.2	103.4	88.7	93.0	100.5	87.3	91.2	99.2	88.7	91.8	108.3	87.7	93.7
0310	Concrete Forming & Accessories	86.5	48.4	53.5	83.1	49.7	54.2	83.1	67.0	69.1	96.1	47.9	54.4	92.6	67.5	70.9	83.2	51.3	55.6
0320	Concrete Reinforcing	99.8	76.7	87.9	95.7	42.0	67.9	100.4	60.2	79.5	99.8	51.9	75.0	99.3	56.9	77.3	96.3	38.6	66.4
0330	Cast-in-Place Concrete	105.8	53.1	85.0	117.7	57.8	94.1	109.0	59.4	89.4	113.8	52.6	89.6	100.5	69.2	88.2	115.1	54.5	91.2
03	CONCRETE	101.1	57.1	79.7	108.2	52.9	81.3	106.9	64.6	86.4	107.9	52.0	80.7	100.8	67.5	84.7	111.1	51.9	82.3
04	MASONRY	98.7	44.8	65.5	132.8	51.4	82.6	147.4	65.7	97.0	99.4	44.7	65.7	112.3	65.7	83.6	128.4	47.9	78.8
05	METALS	90.9	90.5	90.8	88.9	76.1	84.8	91.9	86.4	90.2	90.9	76.8	86.4	98.5	85.6	94.4	89.0	75.1	84.6
06	WOOD, PLASTICS & COMPOSITES	85.5	49.5	65.4	86.1	50.4	66.2	82.2	68.5	74.6	98.6	49.5	71.2	95.8	68.5	80.6	87.2	52.7	68.0
07	THERMAL & MOISTURE PROTECTION	97.4	54.2	79.3	97.5	54.2	77.6	97.7	66.5	84.7	97.8	54.2	79.5	95.8	67.9	84.1	97.3	52.1	78.4
08	OPENINGS	96.4	57.7	87.4	97.5	48.0	86.0	96.1	65.0	88.8	96.4	49.2	85.4	100.7	64.1	92.2	94.5	48.8	83.8
0920	Plaster & Gypsum Board	84.0	48.5	60.3	87.2	49.4	62.0	83.7	68.0	73.2	94.9	48.5	63.9	90.3	68.0	75.4	87.2	51.8	63.6
0950, 0980	Ceilings & Acoustic Treatment	82.9	48.5	60.1	82.0	49.4	60.4	85.6	68.0	74.0	82.9	48.5	60.1	89.6	68.0	75.3	82.0	51.8	62.0
0960	Flooring	103.8	50.8	88.6	91.7	57.0	81.7	102.0	50.8	87.3	110.1	50.8	93.0	96.2	69.2	88.4	90.3	50.8	78.9
0970, 0990	Wall Finishes & Painting/Coating	93.9	46.1	66.1	85.2	53.0	66.5	93.9	49.8	68.2	93.9	46.1	66.1	82.5	49.8	63.5	85.2	60.9	71.1
09	FINISHES	92.6	48.4	68.2	87.0	51.7	67.4	93.2	62.6	76.3	96.4	48.4	69.9	91.0	66.3	77.3	87.1	52.7	68.1
COVERS	DIVS. 10-14, 25, 28, 41, 43, 44, 46	100.0	54.1	90.1	100.0	55.2	90.3	100.0	74.6	94.5	100.0	54.1	90.1	100.0	74.6	94.5	100.0	36.3	86.2
21, 22, 23	FIRE SUPPRESSION, PLUMBING & HVAC	98.5	41.2	74.3	98.1	37.1	72.3	100.0	51.0	79.3	98.5	38.9	73.3	100.0	62.8	84.3	98.2	45.3	75.8
26, 27, 3370	ELECTRICAL, COMMUNICATIONS & UTIL.	96.3	46.0	70.9	97.9	48.7	73.0	96.3	58.4	77.1	96.3	43.1	69.4	101.8	58.4	79.9	99.4	57.0	78.0
MF2014	WEIGHTED AVERAGE	96.6	55.3	78.4	98.6	53.4	78.7	100.3	65.1	84.8	97.9	52.0	77.7	99.8	68.4	86.0	98.8	55.4	79.7

MISSISSIPPI / MISSOURI

DIVISION		MCCOMB 396 MAT.	INST.	TOTAL	MERIDIAN 393 MAT.	INST.	TOTAL	TUPELO 388 MAT.	INST.	TOTAL	BOWLING GREEN 633 MAT.	INST.	TOTAL	CAPE GIRARDEAU 637 MAT.	INST.	TOTAL	CHILLICOTHE 646 MAT.	INST.	TOTAL
015433	CONTRACTOR EQUIPMENT		99.9	99.9		99.9	99.9		99.9	99.9		106.9	106.9		106.9	106.9		101.8	101.8
0241, 31-34	SITE & INFRASTRUCTURE, DEMOLITION	95.5	87.4	89.8	99.6	88.5	91.8	95.2	87.5	89.7	92.9	94.0	93.7	94.4	93.8	94.0	106.8	92.5	96.6
0310	Concrete Forming & Accessories	83.1	49.6	54.1	80.4	66.6	68.5	83.6	50.2	54.7	92.9	92.0	92.1	85.7	82.7	83.1	86.1	91.3	90.6
0320	Concrete Reinforcing	96.8	40.5	67.7	95.7	56.9	75.6	97.7	52.0	74.0	102.7	101.8	102.2	104.0	87.7	95.5	98.8	105.4	102.2
0330	Cast-in-Place Concrete	102.2	52.4	82.5	109.3	67.7	92.9	105.8	74.0	93.2	95.5	101.7	98.0	94.5	87.0	91.6	93.1	86.1	90.3
03	CONCRETE	97.0	50.6	74.5	101.8	66.6	84.7	100.6	60.2	81.0	99.1	98.3	98.7	98.2	86.6	92.6	102.7	92.8	97.9
04	MASONRY	134.2	44.2	78.7	104.0	62.8	78.6	135.2	47.7	81.2	115.2	99.5	105.5	112.2	79.7	92.1	99.2	99.6	99.4
05	METALS	89.2	73.7	84.2	90.0	85.8	88.7	90.8	77.3	86.5	94.8	116.0	101.5	95.9	108.6	99.9	89.9	108.9	95.9
06	WOOD, PLASTICS & COMPOSITES	86.1	52.2	67.2	83.6	68.5	75.2	82.8	50.4	64.7	89.7	92.1	91.1	82.7	82.4	82.5	91.5	90.2	90.7
07	THERMAL & MOISTURE PROTECTION	96.7	52.9	78.3	96.8	66.7	84.2	97.4	57.9	80.9	100.0	99.8	99.9	99.9	85.4	93.9	92.4	96.7	94.2
08	OPENINGS	97.5	49.5	86.3	97.2	65.5	89.8	96.3	49.1	85.3	100.0	99.7	99.9	100.0	80.4	95.4	95.2	94.2	95.0
0920	Plaster & Gypsum Board	87.2	51.2	63.2	87.2	68.0	74.4	83.7	49.4	60.8	90.7	92.2	91.7	89.7	82.2	84.7	102.5	89.7	94.0
0950, 0980	Ceilings & Acoustic Treatment	82.0	51.2	61.6	83.9	68.0	73.4	82.9	49.4	60.7	96.4	92.2	93.6	96.4	82.2	87.0	95.7	89.7	91.7
0960	Flooring	91.7	50.8	79.9	90.2	69.2	84.2	102.2	50.8	87.5	97.0	96.0	96.7	93.3	82.7	90.3	97.3	101.5	98.5
0970, 0990	Wall Finishes & Painting/Coating	85.2	42.3	60.3	85.2	76.2	79.9	93.9	58.7	73.4	94.8	102.8	99.4	94.8	73.2	82.3	91.6	106.3	100.2
09	FINISHES	86.3	49.5	65.9	86.6	68.4	76.5	92.1	51.1	69.4	97.7	93.7	95.5	96.4	80.8	87.8	101.3	94.2	97.4
COVERS	DIVS. 10-14, 25, 28, 41, 43, 44, 46	100.0	57.4	90.8	100.0	73.7	94.3	100.0	55.2	90.3	100.0	86.3	97.0	100.0	90.5	97.9	100.0	86.6	97.1
21, 22, 23	FIRE SUPPRESSION, PLUMBING & HVAC	98.1	34.2	71.1	100.0	61.4	83.7	98.6	42.8	75.0	96.5	95.2	96.0	99.9	95.9	98.2	96.7	98.5	97.5
26, 27, 3370	ELECTRICAL, COMMUNICATIONS & UTIL.	96.4	49.1	72.5	99.4	57.4	78.2	96.0	48.9	72.2	104.8	82.5	93.5	104.7	110.1	107.4	95.1	109.4	102.3
MF2014	WEIGHTED AVERAGE	97.1	51.5	77.0	97.1	67.9	84.2	98.2	55.6	79.4	99.1	96.0	97.7	99.6	92.8	96.6	96.8	98.6	97.6

		MISSOURI																
	DIVISION	COLUMBIA			FLAT RIVER			HANNIBAL			HARRISONVILLE			JEFFERSON CITY			JOPLIN		
		652			636			634			647			650 - 651			648		
		MAT.	INST.	TOTAL	MAT.	INST.	TOTAL	MAT.	INST.	TOTAL	MAT.	INST.	TOTAL	MAT.	INST.	TOTAL	MAT.	INST.	TOTAL
015433	CONTRACTOR EQUIPMENT		110.4	110.4		106.9	106.9		106.9	106.9		101.7	101.7		110.4	110.4		105.6	105.6
0241, 31 - 34	SITE & INFRASTRUCTURE, DEMOLITION	99.8	98.7	99.0	95.5	94.0	94.4	90.5	94.0	93.0	97.7	93.5	94.7	99.6	98.6	98.9	107.4	97.8	100.6
0310	Concrete Forming & Accessories	84.6	83.3	83.4	99.0	86.5	88.2	91.1	76.0	78.0	83.2	96.6	94.8	95.3	80.2	82.2	97.6	73.6	76.8
0320	Concrete Reinforcing	96.6	120.3	108.9	104.0	106.5	105.3	102.1	101.7	101.9	98.4	117.0	108.1	104.3	112.9	108.8	102.0	97.4	99.6
0330	Cast-in-Place Concrete	90.4	87.6	89.3	98.6	93.9	96.8	90.4	101.5	94.8	95.4	103.8	98.7	95.9	86.7	92.3	100.8	77.8	91.8
03	CONCRETE	91.5	93.1	92.3	102.3	94.1	98.3	95.2	91.1	93.2	98.6	103.4	100.9	98.6	90.0	94.4	101.6	80.5	91.3
04	MASONRY	153.6	90.5	114.7	112.6	81.1	93.2	106.7	99.5	102.3	93.9	105.6	101.2	108.7	90.5	97.5	93.0	85.7	88.5
05	METALS	97.6	123.4	105.8	94.7	116.9	101.7	94.8	115.6	101.4	90.3	114.4	97.9	97.2	120.4	104.6	92.8	101.1	95.5
06	WOOD, PLASTICS & COMPOSITES	83.8	80.0	81.7	97.8	85.8	91.1	87.9	71.0	78.5	88.0	94.3	91.5	95.1	76.0	84.4	103.3	71.7	85.7
07	THERMAL & MOISTURE PROTECTION	94.6	88.3	92.0	100.2	94.4	97.8	99.9	95.8	98.2	91.6	104.3	97.0	100.9	87.8	95.4	91.7	79.4	86.5
08	OPENINGS	101.7	95.1	100.2	99.9	97.7	99.4	100.0	82.1	95.8	95.2	101.7	96.7	98.0	91.0	96.3	96.3	76.3	91.7
0920	Plaster & Gypsum Board	93.4	79.5	84.1	96.7	85.6	89.3	90.1	70.5	77.0	98.2	93.9	95.3	97.1	75.3	82.6	108.7	70.7	83.3
0950, 0980	Ceilings & Acoustic Treatment	89.7	79.5	82.9	96.4	85.6	89.3	96.4	70.5	79.2	95.7	93.9	94.5	91.9	75.3	80.9	96.6	70.7	79.4
0960	Flooring	97.6	94.6	96.7	100.4	82.7	95.3	96.2	96.1	96.2	92.6	101.5	95.1	104.5	94.6	101.7	123.5	72.0	108.7
0970, 0990	Wall Finishes & Painting/Coating	92.5	82.1	86.4	94.8	78.4	85.2	94.8	102.8	99.4	96.1	106.3	102.0	89.9	82.1	85.3	91.2	74.9	81.7
09	FINISHES	91.8	83.8	87.4	99.7	83.9	91.0	97.1	80.5	87.9	98.9	97.9	98.3	95.7	81.4	87.8	108.4	73.4	89.0
COVERS	DIVS. 10 - 14, 25, 28, 41, 43, 44, 46	100.0	96.6	99.3	100.0	91.3	98.1	100.0	84.0	96.5	100.0	88.6	97.5	100.0	96.2	99.2	100.0	85.7	96.9
21, 22, 23	FIRE SUPPRESSION, PLUMBING & HVAC	99.8	98.9	99.4	96.5	96.8	96.7	96.5	97.3	96.9	96.7	99.0	97.6	99.8	98.9	99.4	100.1	72.2	88.3
26, 27, 3370	ELECTRICAL, COMMUNICATIONS & UTIL.	99.0	119.2	109.2	109.2	100.8	104.9	103.5	82.5	92.9	100.9	98.8	99.8	104.3	119.2	111.9	93.2	76.9	85.0
MF2014	WEIGHTED AVERAGE	100.3	99.7	100.0	100.0	95.1	97.9	97.9	92.5	95.6	96.3	101.0	98.4	99.6	98.4	99.1	98.2	81.0	90.7

		MISSOURI																
	DIVISION	KANSAS CITY			KIRKSVILLE			POPLAR BLUFF			ROLLA			SEDALIA			SIKESTON		
		640 - 641			635			639			654 - 655			653			638		
		MAT.	INST.	TOTAL	MAT.	INST.	TOTAL	MAT.	INST.	TOTAL	MAT.	INST.	TOTAL	MAT.	INST.	TOTAL	MAT.	INST.	TOTAL
015433	CONTRACTOR EQUIPMENT		103.1	103.1		97.6	97.6		100.2	100.2		110.4	110.4		101.1	101.1		100.2	100.2
0241, 31 - 34	SITE & INFRASTRUCTURE, DEMOLITION	99.8	96.1	97.2	94.7	89.1	90.8	81.3	93.3	89.8	98.7	98.5	98.6	97.5	93.4	94.9	84.8	93.4	90.9
0310	Concrete Forming & Accessories	96.7	103.3	102.4	83.7	79.0	79.6	84.0	82.2	82.5	92.0	95.6	95.1	89.7	79.3	80.7	85.0	77.4	78.5
0320	Concrete Reinforcing	96.9	120.9	109.3	102.9	97.7	100.2	106.0	79.4	92.2	97.1	94.1	95.5	95.6	116.5	106.4	105.4	87.6	96.2
0330	Cast-in-Place Concrete	93.8	107.1	99.0	98.6	93.9	96.7	75.9	87.4	80.4	92.5	98.3	94.8	96.7	93.7	95.5	81.0	85.5	82.8
03	CONCRETE	97.9	108.2	102.9	114.6	88.5	101.9	88.4	84.3	86.5	93.4	97.4	95.3	108.4	92.1	100.5	92.4	83.1	87.9
04	MASONRY	96.2	107.7	103.3	119.1	89.6	100.9	110.5	79.7	91.5	125.4	97.4	108.1	132.1	87.9	104.8	110.2	79.7	91.4
05	METALS	100.2	116.6	105.4	94.4	104.3	97.6	95.0	95.6	95.2	97.0	112.7	102.0	95.9	112.8	101.3	95.3	99.5	96.7
06	WOOD, PLASTICS & COMPOSITES	102.5	102.4	102.4	76.1	75.8	75.9	75.3	82.4	79.3	91.5	96.8	94.5	84.7	76.0	79.9	76.8	75.6	76.2
07	THERMAL & MOISTURE PROTECTION	91.8	105.9	97.7	106.4	94.0	101.2	104.9	85.2	96.7	94.8	96.2	95.4	100.8	93.8	97.9	105.1	85.5	96.9
08	OPENINGS	102.5	107.2	103.6	105.5	83.3	100.3	106.4	78.2	99.9	101.7	89.7	98.9	107.1	87.9	102.6	106.4	76.7	99.5
0920	Plaster & Gypsum Board	106.3	102.2	103.6	85.3	75.3	78.7	85.7	82.2	83.3	95.7	96.7	96.4	89.0	75.3	79.9	87.4	75.2	79.2
0950, 0980	Ceilings & Acoustic Treatment	104.8	102.2	103.1	94.7	75.3	81.9	96.4	82.2	87.0	89.7	96.7	94.3	89.7	75.3	80.2	96.4	75.2	82.4
0960	Flooring	99.6	104.7	101.1	75.4	94.8	80.9	88.4	82.7	86.7	101.3	94.8	99.4	78.3	94.8	83.1	88.9	82.7	87.1
0970, 0990	Wall Finishes & Painting/Coating	96.1	112.2	105.5	90.5	82.1	85.6	89.7	73.2	80.1	92.5	100.7	97.2	92.5	106.3	100.5	89.7	73.2	80.1
09	FINISHES	104.1	104.2	104.2	95.9	80.9	87.6	95.9	81.0	87.6	93.4	96.1	94.9	90.4	83.4	86.5	96.6	76.9	85.7
COVERS	DIVS. 10 - 14, 25, 28, 41, 43, 44, 46	100.0	100.5	100.1	100.0	84.5	96.6	100.0	90.5	97.9	100.0	87.9	97.4	100.0	89.3	97.7	100.0	89.8	97.8
21, 22, 23	FIRE SUPPRESSION, PLUMBING & HVAC	100.0	101.8	100.7	96.6	96.8	96.7	96.6	95.0	95.9	96.4	98.3	97.2	96.4	95.6	96.1	96.6	95.1	96.0
26, 27, 3370	ELECTRICAL, COMMUNICATIONS & UTIL.	102.5	101.6	102.0	103.7	81.9	92.7	104.0	100.7	102.3	97.5	85.8	91.6	98.6	98.8	98.7	103.1	110.0	106.6
MF2014	WEIGHTED AVERAGE	100.2	104.8	102.2	101.5	89.7	96.3	97.9	89.7	94.3	98.3	96.7	97.6	100.7	93.9	97.7	98.4	90.5	94.9

		MISSOURI									MONTANA								
	DIVISION	SPRINGFIELD			ST. JOSEPH			ST. LOUIS			BILLINGS			BUTTE			GREAT FALLS		
		656 - 658			644 - 645			630 - 631			590 - 591			597			594		
		MAT.	INST.	TOTAL	MAT.	INST.	TOTAL	MAT.	INST.	TOTAL	MAT.	INST.	TOTAL	MAT.	INST.	TOTAL	MAT.	INST.	TOTAL
015433	CONTRACTOR EQUIPMENT		103.8	103.8		101.8	101.8		107.8	107.8		99.0	99.0		98.8	98.8		98.8	98.8
0241, 31 - 34	SITE & INFRASTRUCTURE, DEMOLITION	100.7	96.1	97.5	101.4	91.6	94.4	94.8	97.2	96.5	97.0	96.9	97.0	103.2	96.7	98.6	107.0	96.8	99.8
0310	Concrete Forming & Accessories	98.1	78.1	80.8	96.5	91.2	92.0	99.1	102.7	102.3	97.2	70.7	74.3	84.5	70.3	72.2	97.1	70.6	74.2
0320	Concrete Reinforcing	93.0	119.8	106.9	95.7	116.8	106.6	95.1	113.6	104.7	101.4	82.6	91.7	110.0	82.6	95.8	101.5	82.7	91.7
0330	Cast-in-Place Concrete	98.2	78.6	90.5	93.7	103.6	97.6	94.5	103.8	98.2	124.9	73.3	104.6	137.5	68.0	110.1	145.3	67.1	114.5
03	CONCRETE	104.2	86.9	95.8	97.6	100.8	99.2	97.8	106.1	101.8	107.4	74.7	91.5	111.8	72.7	92.8	117.1	72.6	95.5
04	MASONRY	101.2	88.8	93.5	96.0	97.4	96.8	92.9	109.9	103.4	137.3	76.2	99.6	133.0	78.0	99.1	137.6	77.6	100.6
05	METALS	102.2	109.5	104.5	96.5	113.7	102.0	100.6	121.7	107.3	106.7	91.6	101.9	100.8	91.3	97.8	104.2	91.6	100.2
06	WOOD, PLASTICS & COMPOSITES	92.6	76.1	83.4	103.3	89.9	95.8	97.6	101.5	99.8	90.9	69.1	78.7	78.7	69.1	73.3	92.0	69.1	79.2
07	THERMAL & MOISTURE PROTECTION	99.0	81.0	91.5	92.2	97.4	94.4	100.0	105.9	102.5	110.9	71.7	94.5	110.7	69.8	93.6	111.3	71.5	94.6
08	OPENINGS	109.5	83.3	103.4	100.6	99.4	100.3	100.1	109.4	102.3	102.0	71.2	94.9	100.2	68.8	92.9	103.3	69.9	95.5
0920	Plaster & Gypsum Board	96.7	75.4	82.5	110.5	89.4	96.4	97.7	101.8	100.5	108.0	68.5	81.6	107.0	68.5	81.3	116.7	68.5	84.5
0950, 0980	Ceilings & Acoustic Treatment	89.7	75.4	80.2	104.0	89.4	94.3	102.1	101.8	101.9	87.2	68.5	74.8	94.6	68.5	77.3	94.6	68.5	77.3
0960	Flooring	99.9	73.3	92.3	102.7	101.5	102.3	100.1	96.0	99.0	105.0	74.3	96.1	102.2	71.3	93.3	109.9	83.4	102.3
0970, 0990	Wall Finishes & Painting/Coating	86.8	74.9	79.9	91.6	83.9	87.1	94.8	108.7	102.9	98.4	96.6	97.4	97.2	51.2	70.5	97.2	86.6	91.0
09	FINISHES	95.7	77.0	85.3	105.1	91.8	97.8	101.0	101.7	101.4	98.1	73.5	84.5	98.1	68.0	81.5	102.4	74.3	86.8
COVERS	DIVS. 10 - 14, 25, 28, 41, 43, 44, 46	100.0	94.5	98.8	100.0	97.3	99.4	100.0	101.0	100.2	100.0	93.7	98.6	100.0	93.7	98.6	100.0	93.7	98.6
21, 22, 23	FIRE SUPPRESSION, PLUMBING & HVAC	99.9	74.0	88.9	100.1	88.1	95.1	99.9	105.5	102.3	100.0	75.4	89.6	100.1	74.9	89.4	100.1	69.3	87.1
26, 27, 3370	ELECTRICAL, COMMUNICATIONS & UTIL.	102.9	69.1	85.9	100.9	100.8	100.9	107.2	90.6	98.8	97.9	76.6	87.1	104.4	72.6	88.3	97.3	72.6	84.8
MF2014	WEIGHTED AVERAGE	101.8	83.3	93.7	99.4	96.7	98.2	100.2	104.3	102.0	103.7	78.8	92.7	103.7	77.1	92.0	105.2	76.8	92.7

For customer support on your Mechanical Cost Data, call 877.739.3689.

675

MONTANA

DIVISION		HAVRE 595 MAT.	INST.	TOTAL	HELENA 596 MAT.	INST.	TOTAL	KALISPELL 599 MAT.	INST.	TOTAL	MILES CITY 593 MAT.	INST.	TOTAL	MISSOULA 598 MAT.	INST.	TOTAL	WOLF POINT 592 MAT.	INST.	TOTAL
015433	CONTRACTOR EQUIPMENT		98.8	98.8		98.8	98.8		98.8	98.8		98.8	98.8		98.8	98.8		98.8	98.8
0241, 31-34	SITE & INFRASTRUCTURE, DEMOLITION	110.8	96.3	100.5	97.1	96.3	96.5	93.1	96.5	95.5	99.5	96.0	97.0	85.5	96.3	93.1	117.3	96.1	102.3
0310	Concrete Forming & Accessories	77.8	67.8	69.1	100.3	63.3	68.3	87.7	70.3	72.6	95.7	66.4	70.4	87.7	70.1	72.5	88.4	66.7	69.6
0320	Concrete Reinforcing	110.9	82.5	96.2	116.9	82.5	99.1	113.0	82.5	97.2	110.5	82.6	96.0	111.9	82.5	96.7	112.1	82.6	96.8
0330	Cast-in-Place Concrete	148.1	59.3	113.1	107.4	61.8	89.4	119.4	61.6	96.6	130.7	59.2	102.5	101.4	63.8	86.5	146.4	59.6	112.2
03	CONCRETE	120.3	68.6	95.2	103.3	67.4	85.9	100.8	70.5	86.1	108.5	68.0	88.8	88.7	71.2	80.2	123.7	68.2	96.7
04	MASONRY	134.0	72.0	95.8	127.8	77.8	97.0	131.8	78.4	98.9	139.5	67.3	95.0	160.3	78.0	109.6	140.8	68.2	96.0
05	METALS	97.0	90.5	95.0	102.9	89.7	98.7	96.9	90.8	95.0	96.2	90.8	94.5	97.4	90.8	95.3	96.3	90.8	94.6
06	WOOD, PLASTICS & COMPOSITES	71.0	69.1	69.9	92.5	60.1	74.5	81.7	69.1	74.6	89.3	69.1	78.0	81.7	69.1	74.6	81.6	69.1	74.6
07	THERMAL & MOISTURE PROTECTION	111.1	67.1	92.7	105.2	69.6	90.3	110.3	78.7	97.1	110.7	64.7	91.4	109.9	77.7	96.5	111.6	65.1	92.1
08	OPENINGS	100.7	67.0	92.9	102.8	63.7	93.7	100.7	67.1	92.9	100.2	67.0	92.5	100.2	66.8	92.4	100.2	66.8	92.5
0920	Plaster & Gypsum Board	102.7	68.5	79.9	105.9	59.3	74.8	107.0	68.5	81.3	115.9	68.5	84.2	107.0	68.5	81.3	111.0	68.5	82.6
0950, 0980	Ceilings & Acoustic Treatment	93.0	68.5	76.8	95.4	59.3	71.5	93.0	68.5	76.8	91.3	68.5	76.2	93.0	68.5	76.8	91.3	68.5	76.2
0960	Flooring	99.4	71.3	91.3	104.6	56.8	90.9	104.1	71.3	94.7	109.7	67.5	97.6	104.1	71.3	94.7	105.7	67.5	94.8
0970, 0990	Wall Finishes & Painting/Coating	97.2	48.7	69.0	93.2	47.2	66.4	97.2	47.2	68.1	97.2	47.2	68.1	97.2	62.0	76.7	97.2	48.7	69.0
09	FINISHES	97.3	66.3	80.1	99.3	59.8	77.5	98.1	67.7	81.3	101.0	64.4	80.8	97.5	69.2	81.8	100.6	64.7	80.7
COVERS	DIVS. 10-14, 25, 28, 41, 43, 44, 46	100.0	68.9	93.3	100.0	92.8	98.4	100.0	93.8	98.7	100.0	90.4	97.9	100.0	93.7	98.6	100.0	90.6	98.0
21, 22, 23	FIRE SUPPRESSION, PLUMBING & HVAC	96.7	66.6	84.0	99.9	69.3	87.0	96.7	68.7	84.9	96.7	68.9	84.9	100.1	68.5	86.7	96.7	69.2	85.1
26, 27, 3370	ELECTRICAL, COMMUNICATIONS & UTIL.	97.3	72.7	84.8	103.5	72.6	87.9	101.3	72.0	86.5	97.3	77.9	87.5	102.3	72.0	87.0	97.3	77.9	87.5
MF2014	WEIGHTED AVERAGE	102.7	73.0	89.6	102.8	73.6	90.0	100.4	75.5	89.5	101.6	74.0	89.4	101.1	75.7	89.9	103.8	74.2	90.8

NEBRASKA

DIVISION		ALLIANCE 693 MAT.	INST.	TOTAL	COLUMBUS 686 MAT.	INST.	TOTAL	GRAND ISLAND 688 MAT.	INST.	TOTAL	HASTINGS 689 MAT.	INST.	TOTAL	LINCOLN 683-685 MAT.	INST.	TOTAL	MCCOOK 690 MAT.	INST.	TOTAL
015433	CONTRACTOR EQUIPMENT		97.7	97.7		102.9	102.9		102.9	102.9		102.9	102.9		102.9	102.9		102.9	102.9
0241, 31-34	SITE & INFRASTRUCTURE, DEMOLITION	99.7	100.0	99.9	100.2	93.2	95.3	105.0	94.0	97.3	103.7	93.4	96.4	91.0	94.1	93.2	103.5	93.2	96.2
0310	Concrete Forming & Accessories	87.9	54.8	59.2	96.3	65.6	69.7	95.8	70.4	73.9	99.2	69.6	73.6	96.3	72.4	75.6	93.6	54.4	59.7
0320	Concrete Reinforcing	110.2	84.2	96.7	95.7	75.7	85.3	95.1	75.6	85.0	95.1	76.2	85.3	94.2	75.4	84.5	103.1	75.3	88.7
0330	Cast-in-Place Concrete	108.9	78.7	97.0	112.0	81.8	100.1	118.6	66.4	98.0	118.6	62.5	96.4	93.1	73.2	85.3	117.6	58.5	94.3
03	CONCRETE	123.3	68.9	96.8	108.2	73.9	91.5	113.2	70.9	92.7	113.5	69.3	92.0	97.6	74.1	86.2	112.0	60.9	87.2
04	MASONRY	114.5	74.7	90.0	123.8	78.4	95.8	116.1	75.7	91.2	125.7	87.6	102.2	103.8	72.4	84.4	110.1	73.6	87.6
05	METALS	98.9	77.8	92.2	90.7	85.4	89.0	92.5	87.0	90.7	93.2	86.4	91.0	94.4	86.9	92.1	94.0	84.2	90.9
06	WOOD, PLASTICS & COMPOSITES	84.3	50.3	65.4	96.4	64.4	78.5	95.7	69.7	81.2	99.2	69.7	82.8	98.3	72.2	83.8	92.6	50.4	69.1
07	THERMAL & MOISTURE PROTECTION	103.9	64.8	87.5	101.1	68.9	87.6	101.3	76.8	91.0	101.3	77.8	91.5	99.8	76.2	89.9	98.5	65.4	84.7
08	OPENINGS	97.3	56.9	87.9	97.8	64.2	90.0	97.8	67.7	90.8	97.8	67.1	90.7	107.0	65.5	97.3	97.8	54.9	87.8
0920	Plaster & Gypsum Board	82.4	49.0	60.1	95.5	63.4	74.0	94.8	68.9	77.5	96.5	68.9	78.1	105.9	71.5	82.9	94.5	49.0	64.1
0950, 0980	Ceilings & Acoustic Treatment	91.8	49.0	63.5	81.9	63.4	69.6	81.9	68.9	73.3	81.9	68.9	73.3	92.8	71.5	78.7	87.0	49.0	61.8
0960	Flooring	97.4	78.9	92.1	91.5	96.5	92.9	91.2	103.1	94.6	92.6	96.5	93.7	106.0	91.5	101.9	96.1	78.9	91.1
0970, 0990	Wall Finishes & Painting/Coating	150.0	54.8	94.6	73.2	63.7	67.7	73.2	67.6	70.0	73.2	63.7	67.7	90.6	82.5	85.9	84.4	47.7	63.1
09	FINISHES	96.3	57.7	74.9	89.3	69.8	78.5	89.4	74.8	81.3	90.0	73.0	80.6	98.6	77.0	86.7	93.8	56.8	73.3
COVERS	DIVS. 10-14, 25, 28, 41, 43, 44, 46	100.0	62.3	92.0	100.0	95.5	99.0	100.0	88.8	97.6	100.0	86.1	97.0	100.0	89.1	97.6	100.0	62.5	91.9
21, 22, 23	FIRE SUPPRESSION, PLUMBING & HVAC	96.7	69.7	85.3	96.6	75.1	87.5	100.0	77.3	90.4	96.6	75.2	87.6	99.8	77.3	90.3	96.5	76.1	87.9
26, 27, 3370	ELECTRICAL, COMMUNICATIONS & UTIL.	94.5	68.8	81.5	93.6	81.7	87.6	92.3	66.5	79.3	91.7	82.9	87.3	103.9	66.5	85.0	96.4	68.9	82.5
MF2014	WEIGHTED AVERAGE	101.2	70.5	87.7	98.1	77.5	89.0	99.3	76.6	89.3	99.1	78.9	90.2	99.8	76.9	89.7	99.0	70.5	86.4

NEBRASKA / NEVADA

DIVISION		NORFOLK 687 MAT.	INST.	TOTAL	NORTH PLATTE 691 MAT.	INST.	TOTAL	OMAHA 680-681 MAT.	INST.	TOTAL	VALENTINE 692 MAT.	INST.	TOTAL	CARSON CITY 897 MAT.	INST.	TOTAL	ELKO 898 MAT.	INST.	TOTAL
015433	CONTRACTOR EQUIPMENT		93.2	93.2		102.9	102.9		93.2	93.2		96.8	96.8		96.7	96.7		96.7	96.7
0241, 31-34	SITE & INFRASTRUCTURE, DEMOLITION	83.7	92.5	89.9	104.8	93.2	96.6	90.3	93.3	92.4	88.0	98.5	95.4	87.4	97.9	94.8	70.2	96.5	88.8
0310	Concrete Forming & Accessories	82.1	69.0	70.7	96.1	69.3	72.9	95.2	72.3	75.4	84.1	54.1	58.2	105.6	81.9	85.1	112.1	78.7	83.2
0320	Concrete Reinforcing	95.8	65.7	80.2	102.5	75.5	88.5	94.8	75.8	85.0	103.1	65.4	83.6	108.4	115.9	112.3	116.6	85.2	100.3
0330	Cast-in-Place Concrete	112.7	61.5	92.5	117.6	63.5	96.3	93.6	82.0	89.0	103.8	56.4	85.1	92.7	85.2	89.8	89.9	78.7	85.5
03	CONCRETE	106.6	66.2	87.0	112.1	69.3	91.3	97.9	76.5	87.5	109.4	57.8	84.3	102.4	89.6	96.2	99.1	80.3	90.0
04	MASONRY	130.6	78.7	98.6	97.5	74.5	83.3	105.5	79.6	89.5	109.7	73.6	87.5	126.7	74.9	94.8	126.1	71.5	92.5
05	METALS	94.0	73.2	87.4	93.3	84.9	90.6	94.4	78.8	89.4	104.8	73.1	94.7	102.8	100.8	102.2	106.1	87.0	100.0
06	WOOD, PLASTICS & COMPOSITES	80.2	69.2	74.1	94.5	69.7	80.7	93.6	71.7	81.4	79.1	49.9	62.8	88.6	80.4	84.0	99.5	79.1	88.1
07	THERMAL & MOISTURE PROTECTION	100.8	72.1	88.8	98.5	77.9	89.9	96.0	81.3	89.8	99.2	65.5	85.1	104.6	82.3	95.3	101.0	73.9	89.7
08	OPENINGS	99.4	64.5	91.3	97.1	67.1	90.1	104.6	71.2	96.8	99.6	53.9	89.0	98.4	101.3	99.1	100.5	78.8	95.4
0920	Plaster & Gypsum Board	95.6	68.9	77.8	94.5	68.9	77.4	99.5	71.5	80.8	96.3	49.0	64.7	101.4	79.9	87.0	108.2	78.4	88.3
0950, 0980	Ceilings & Acoustic Treatment	95.4	68.9	77.8	95.4	68.9	75.0	93.9	71.5	79.0	104.6	49.0	67.8	91.1	79.9	83.7	93.4	78.4	83.5
0960	Flooring	113.7	96.5	108.7	97.2	96.5	97.0	104.4	88.2	99.7	124.3	76.7	110.6	102.1	74.4	94.1	106.8	49.1	90.2
0970, 0990	Wall Finishes & Painting/Coating	128.4	63.7	90.8	84.4	63.7	72.4	104.5	67.2	82.8	149.5	66.1	101.0	92.7	81.0	85.9	94.0	81.0	86.5
09	FINISHES	106.8	73.0	88.1	94.1	73.0	82.5	101.1	74.3	86.3	115.6	58.0	83.7	96.9	79.5	87.3	97.7	73.4	84.2
COVERS	DIVS. 10-14, 25, 28, 41, 43, 44, 46	100.0	84.7	96.7	100.0	64.7	92.4	100.0	88.0	97.4	100.0	60.6	91.5	100.0	86.4	97.1	100.0	61.0	91.5
21, 22, 23	FIRE SUPPRESSION, PLUMBING & HVAC	96.3	76.0	87.7	99.9	75.1	89.4	99.8	78.6	90.8	96.2	76.0	87.7	100.0	75.2	89.5	98.4	78.5	90.0
26, 27, 3370	ELECTRICAL, COMMUNICATIONS & UTIL.	92.6	81.7	87.1	94.8	76.5	85.5	99.0	81.7	90.2	92.1	85.9	89.0	104.3	97.8	101.0	102.0	95.9	98.9
MF2014	WEIGHTED AVERAGE	99.6	76.0	89.2	98.9	75.9	88.7	99.1	79.5	90.5	101.3	71.8	88.3	101.8	86.7	95.1	101.1	81.4	92.4

DIVISION		NEVADA									NEW HAMPSHIRE								
		ELY			LAS VEGAS			RENO			CHARLESTON			CLAREMONT			CONCORD		
		893			889 - 891			894 - 895			036			037			032 - 033		
		MAT.	INST.	TOTAL	MAT.	INST.	TOTAL	MAT.	INST.	TOTAL	MAT.	INST.	TOTAL	MAT.	INST.	TOTAL	MAT.	INST.	TOTAL
015433	CONTRACTOR EQUIPMENT		96.7	96.7		96.7	96.7		96.7	96.7		98.1	98.1		98.1	98.1		98.1	98.1
0241, 31 - 34	SITE & INFRASTRUCTURE, DEMOLITION	76.0	97.9	91.5	79.4	100.0	94.0	75.7	97.9	91.4	88.2	97.1	94.5	82.0	97.1	92.7	95.6	101.4	99.7
0310	Concrete Forming & Accessories	104.9	102.7	103.0	105.5	108.7	108.2	100.8	81.6	84.2	88.7	79.6	80.9	94.7	79.7	81.7	99.3	92.8	93.7
0320	Concrete Reinforcing	115.2	110.3	112.7	106.3	122.1	114.5	108.9	120.8	115.1	88.7	93.6	91.2	88.7	93.6	91.2	103.5	94.2	98.7
0330	Cast-in-Place Concrete	96.5	103.9	99.4	93.4	110.5	100.1	102.0	85.1	95.3	87.9	104.4	94.4	80.7	104.4	90.1	106.0	116.6	110.2
03	CONCRETE	107.7	104.3	106.0	102.7	111.4	106.9	106.9	90.3	98.9	95.3	90.8	93.1	87.6	90.8	89.2	104.1	101.0	102.6
04	MASONRY	131.9	102.0	113.5	118.6	104.9	110.1	125.2	74.9	94.2	101.5	84.2	90.8	101.5	84.2	90.9	118.5	103.6	109.3
05	METALS	106.1	101.4	104.6	114.3	108.3	112.4	107.7	102.4	106.0	91.8	90.0	91.3	91.8	90.0	91.3	98.1	93.8	96.8
06	WOOD, PLASTICS & COMPOSITES	90.5	103.3	97.6	88.2	106.2	98.2	84.6	80.4	82.3	91.9	87.1	89.2	97.8	87.1	91.8	92.4	90.4	91.3
07	THERMAL & MOISTURE PROTECTION	101.5	93.1	98.0	114.9	101.8	109.4	101.0	82.3	93.2	101.1	78.5	91.6	100.9	78.5	91.5	105.6	101.8	104.0
08	OPENINGS	100.4	96.0	99.4	99.4	116.7	103.4	98.3	88.9	96.1	100.1	74.4	94.1	101.3	74.4	95.0	101.0	92.4	99.0
0920	Plaster & Gypsum Board	103.2	103.3	103.3	98.1	106.3	103.6	93.1	79.9	84.3	97.1	86.2	89.8	98.4	86.2	90.3	101.0	89.7	93.4
0950, 0980	Ceilings & Acoustic Treatment	93.4	103.3	100.0	100.8	106.3	104.5	98.3	79.9	86.1	89.6	86.2	87.4	89.6	86.2	87.4	96.1	89.7	91.8
0960	Flooring	104.1	55.0	90.0	95.4	106.9	98.7	100.7	74.4	93.1	92.7	30.8	74.9	95.3	30.8	76.8	98.5	112.4	102.5
0970, 0990	Wall Finishes & Painting/Coating	94.0	121.0	109.7	96.6	121.7	111.2	94.0	81.0	86.5	88.6	45.7	63.6	88.6	46.0	63.8	89.2	116.5	105.1
09	FINISHES	96.8	96.1	96.4	95.2	109.4	103.1	95.2	79.5	86.5	90.4	67.7	77.8	90.9	67.7	78.1	93.3	98.0	95.9
COVERS	DIVS. 10 - 14, 25, 28, 41, 43, 44, 46	100.0	61.7	91.7	100.0	102.2	100.5	100.0	86.4	97.1	100.0	89.5	97.7	100.0	89.5	97.7	100.0	106.1	101.3
21, 22, 23	FIRE SUPPRESSION, PLUMBING & HVAC	98.4	103.3	100.5	100.2	107.7	103.4	100.1	75.2	89.5	96.7	39.2	72.4	96.7	39.3	72.4	99.9	83.6	93.0
26, 27, 3370	ELECTRICAL, COMMUNICATIONS & UTIL.	102.2	108.2	105.3	106.3	120.7	113.6	102.7	97.8	100.2	93.5	52.5	72.7	93.5	52.5	72.7	94.3	82.9	88.5
MF2014	WEIGHTED AVERAGE	102.4	100.6	101.6	103.6	109.4	106.1	102.2	86.5	95.3	95.7	70.3	84.5	94.9	70.4	84.1	100.0	93.8	97.3

DIVISION		NEW HAMPSHIRE															NEW JERSEY		
		KEENE			LITTLETON			MANCHESTER			NASHUA			PORTSMOUTH			ATLANTIC CITY		
		034			035			031			030			038			082,084		
		MAT.	INST.	TOTAL	MAT.	INST.	TOTAL	MAT.	INST.	TOTAL	MAT.	INST.	TOTAL	MAT.	INST.	TOTAL	MAT.	INST.	TOTAL
015433	CONTRACTOR EQUIPMENT		98.1	98.1		98.1	98.1		98.1	98.1		98.1	98.1		98.1	98.1		96.0	96.0
0241, 31 - 34	SITE & INFRASTRUCTURE, DEMOLITION	95.9	97.4	97.0	82.1	97.4	92.9	93.6	101.4	99.1	97.5	101.4	100.2	90.8	101.4	98.3	95.1	102.9	100.6
0310	Concrete Forming & Accessories	93.3	82.1	83.6	106.4	81.8	85.1	99.2	92.8	93.7	101.8	92.8	94.0	90.2	92.7	92.4	115.1	129.3	127.4
0320	Concrete Reinforcing	88.7	93.8	91.3	89.5	93.7	91.7	103.6	94.2	98.7	110.9	94.2	102.2	88.7	94.2	91.5	77.6	118.4	98.7
0330	Cast-in-Place Concrete	88.3	106.9	95.6	79.2	106.1	89.8	101.1	116.6	107.2	83.6	116.6	96.6	79.3	116.6	94.0	83.3	133.0	102.9
03	CONCRETE	94.9	92.7	93.9	87.1	92.4	89.7	101.7	101.0	101.4	95.3	101.0	98.0	87.2	100.9	93.9	92.1	127.2	109.2
04	MASONRY	104.7	87.8	94.3	114.3	87.8	98.0	112.0	103.6	106.9	106.5	103.6	104.7	101.8	103.6	102.9	118.3	138.7	130.8
05	METALS	92.5	91.2	92.1	92.6	90.9	92.0	100.1	93.8	98.1	98.4	93.8	96.9	94.1	94.0	94.1	93.7	104.5	97.2
06	WOOD, PLASTICS & COMPOSITES	96.1	87.1	91.1	107.1	87.1	95.9	94.6	90.4	92.3	105.6	90.4	97.1	93.1	90.4	91.6	119.4	128.3	124.3
07	THERMAL & MOISTURE PROTECTION	101.6	80.0	92.6	101.0	80.0	92.2	105.0	101.8	103.7	101.9	101.8	101.9	101.6	114.2	106.9	102.9	128.2	113.5
08	OPENINGS	98.7	90.2	96.7	102.3	86.7	98.6	102.8	92.4	100.3	103.5	92.4	100.9	103.9	82.5	98.9	99.1	124.1	104.9
0920	Plaster & Gypsum Board	97.4	86.2	89.9	111.3	86.2	94.6	99.6	89.7	93.0	107.2	89.7	95.5	97.1	89.7	92.1	105.3	128.6	120.9
0950, 0980	Ceilings & Acoustic Treatment	89.6	86.2	87.4	89.6	86.2	87.4	97.0	89.7	92.2	101.2	89.7	93.6	90.5	89.7	90.0	78.5	128.6	111.7
0960	Flooring	94.9	50.6	82.2	106.4	30.8	84.7	92.5	112.4	98.2	99.0	112.4	102.8	92.8	112.4	98.4	101.4	153.7	116.4
0970, 0990	Wall Finishes & Painting/Coating	88.6	116.5	104.8	88.6	59.9	71.9	91.4	116.5	106.0	88.6	116.5	104.8	88.6	116.5	104.8	83.4	130.2	110.6
09	FINISHES	92.4	79.9	85.5	96.1	70.1	81.7	93.4	98.0	96.0	97.5	98.0	97.8	91.4	98.0	95.1	92.2	134.8	115.7
COVERS	DIVS. 10 - 14, 25, 28, 41, 43, 44, 46	100.0	90.7	98.0	100.0	95.7	99.1	100.0	106.1	101.3	100.0	106.1	101.3	100.0	106.1	101.3	100.0	112.5	102.7
21, 22, 23	FIRE SUPPRESSION, PLUMBING & HVAC	96.7	43.0	74.0	96.7	63.2	82.6	99.9	83.6	93.0	100.1	83.6	93.1	100.1	83.6	93.1	99.7	124.1	110.0
26, 27, 3370	ELECTRICAL, COMMUNICATIONS & UTIL.	93.5	62.7	77.9	94.4	55.5	74.7	94.3	81.9	88.0	95.6	81.9	88.7	93.9	82.9	88.4	89.4	137.0	113.5
MF2014	WEIGHTED AVERAGE	96.2	75.6	87.1	96.3	77.6	88.0	99.9	93.6	96.8	99.3	93.6	96.8	96.5	93.7	95.3	97.2	125.3	109.6

DIVISION		NEW JERSEY																	
		CAMDEN			DOVER			ELIZABETH			HACKENSACK			JERSEY CITY			LONG BRANCH		
		081			078			072			076			073			077		
		MAT.	INST.	TOTAL	MAT.	INST.	TOTAL	MAT.	INST.	TOTAL	MAT.	INST.	TOTAL	MAT.	INST.	TOTAL	MAT.	INST.	TOTAL
015433	CONTRACTOR EQUIPMENT		96.0	96.0		98.1	98.1		98.1	98.1		98.1	98.1		96.0	96.0		95.6	95.6
0241, 31 - 34	SITE & INFRASTRUCTURE, DEMOLITION	96.4	103.3	101.3	98.7	104.0	102.4	102.6	104.0	103.6	99.6	104.0	102.7	90.3	103.9	99.9	94.2	103.5	100.7
0310	Concrete Forming & Accessories	105.5	125.5	122.8	101.7	133.2	129.0	114.9	133.2	130.8	101.7	133.2	128.9	106.4	133.3	129.7	106.9	124.4	122.0
0320	Concrete Reinforcing	101.8	120.2	111.3	80.8	130.0	106.3	80.8	130.0	106.3	80.8	130.0	106.3	104.8	130.0	117.9	80.8	129.9	106.2
0330	Cast-in-Place Concrete	80.8	133.0	101.4	93.2	127.4	106.7	80.1	131.2	100.2	91.1	131.1	106.9	72.7	127.5	94.3	80.8	132.3	101.1
03	CONCRETE	92.8	125.9	108.9	95.6	129.5	112.1	91.7	130.8	110.7	93.9	130.7	111.8	89.4	129.4	108.8	93.3	127.0	109.7
04	MASONRY	107.4	138.7	126.7	99.5	137.5	122.9	117.4	137.5	129.7	103.7	137.5	124.5	94.2	137.5	120.8	112.3	132.4	123.5
05	METALS	99.6	105.4	101.5	91.7	113.6	98.7	93.2	113.7	99.7	91.7	113.6	98.7	97.7	111.1	101.9	91.8	110.7	97.8
06	WOOD, PLASTICS & COMPOSITES	107.6	122.9	116.1	101.0	132.9	118.8	116.7	132.9	125.7	101.0	132.9	118.8	101.9	132.9	119.2	103.0	122.9	114.1
07	THERMAL & MOISTURE PROTECTION	102.7	129.5	113.9	102.3	132.1	114.8	102.5	132.6	115.1	102.1	126.4	112.3	101.8	132.1	114.5	102.0	121.4	110.1
08	OPENINGS	101.5	121.6	106.2	105.1	129.2	110.7	103.5	129.2	109.5	102.9	129.2	109.0	101.1	129.2	107.6	97.2	123.7	103.4
0920	Plaster & Gypsum Board	100.5	123.1	115.6	98.9	133.3	121.9	106.9	133.3	124.6	98.9	133.3	121.9	102.4	133.3	123.1	100.9	123.1	115.7
0950, 0980	Ceilings & Acoustic Treatment	87.8	123.1	111.2	77.5	133.3	114.5	79.4	133.3	115.1	77.5	133.3	114.5	88.4	133.3	118.2	77.5	123.1	107.7
0960	Flooring	97.1	153.7	113.4	88.8	176.0	113.9	94.3	176.0	117.8	88.8	176.0	113.9	89.9	176.0	114.7	90.1	171.2	113.5
0970, 0990	Wall Finishes & Painting/Coating	83.4	130.2	110.6	84.5	132.5	112.4	84.5	132.5	112.4	84.5	132.5	112.4	84.6	132.5	112.5	84.6	130.2	111.2
09	FINISHES	92.1	131.6	113.9	88.0	140.5	117.0	91.6	140.5	118.6	87.9	140.5	117.0	91.0	141.2	118.8	88.9	133.6	113.6
COVERS	DIVS. 10 - 14, 25, 28, 41, 43, 44, 46	100.0	111.9	102.6	100.0	116.9	103.7	100.0	116.9	103.7	100.0	116.9	103.7	100.0	116.9	103.7	100.0	110.5	102.3
21, 22, 23	FIRE SUPPRESSION, PLUMBING & HVAC	100.0	124.2	110.2	99.7	129.0	112.1	99.9	126.5	111.2	99.7	129.0	112.1	99.9	129.1	112.2	99.7	128.1	111.7
26, 27, 3370	ELECTRICAL, COMMUNICATIONS & UTIL.	93.5	137.0	115.5	90.0	140.1	115.3	90.5	140.1	115.6	90.0	139.1	114.8	94.1	139.0	116.8	89.7	129.8	110.0
MF2014	WEIGHTED AVERAGE	98.4	124.7	110.0	96.7	129.2	111.0	97.8	128.8	111.5	96.5	129.0	110.8	96.7	128.9	110.9	96.0	124.7	108.6

For customer support on your Mechanical Cost Data, call 877.739.3689.

677

NEW JERSEY

DIVISION		NEW BRUNSWICK 088-089			NEWARK 070-071			PATERSON 074-075			POINT PLEASANT 087			SUMMIT 079			TRENTON 085-086		
		MAT.	INST.	TOTAL	MAT.	INST.	TOTAL	MAT.	INST.	TOTAL	MAT.	INST.	TOTAL	MAT.	INST.	TOTAL	MAT.	INST.	TOTAL
015433	CONTRACTOR EQUIPMENT		95.6	95.6		98.1	98.1		98.1	98.1		95.6	95.6		98.1	98.1		95.6	95.6
0241, 31 - 34	SITE & INFRASTRUCTURE, DEMOLITION	107.7	103.7	104.9	103.6	104.0	103.9	101.5	104.0	103.3	109.3	103.5	105.2	100.2	104.0	102.9	94.2	103.4	100.7
0310	Concrete Forming & Accessories	108.9	133.3	130.0	102.4	133.5	129.3	103.9	133.4	129.4	102.8	124.2	121.3	104.7	133.2	129.4	104.5	131.5	127.9
0320	Concrete Reinforcing	78.5	129.9	105.1	103.1	130.1	117.1	104.8	130.0	117.9	78.5	129.8	105.1	80.8	130.0	106.3	103.0	102.8	102.9
0330	Cast-in-Place Concrete	103.0	133.0	114.8	102.8	131.3	114.0	92.7	131.2	107.9	103.0	130.8	114.0	77.4	131.2	98.6	101.3	130.8	112.9
03	CONCRETE	109.3	131.2	120.0	102.7	130.9	116.4	99.0	130.9	114.5	108.9	126.4	117.4	88.7	130.8	109.2	102.1	124.7	113.1
04	MASONRY	116.0	136.1	128.4	109.5	137.5	126.7	100.1	137.5	123.1	103.5	131.4	120.7	102.6	137.5	124.1	113.4	132.4	125.1
05	METALS	93.8	110.9	99.2	99.4	113.9	104.0	93.4	113.8	99.9	93.8	110.4	99.1	91.7	113.7	98.7	99.4	99.9	99.5
06	WOOD, PLASTICS & COMPOSITES	112.6	132.8	123.9	95.4	132.9	116.3	103.7	132.9	120.0	104.9	122.9	114.9	104.9	132.9	120.5	99.7	132.8	118.1
07	THERMAL & MOISTURE PROTECTION	103.1	131.2	114.9	104.1	132.6	116.0	102.4	126.4	112.4	103.2	121.2	110.7	102.7	132.6	115.2	103.6	126.6	113.2
08	OPENINGS	94.0	129.2	102.2	104.7	129.2	110.4	108.0	129.2	113.0	95.9	125.2	102.7	109.1	129.2	113.8	104.3	122.1	108.4
0920	Plaster & Gypsum Board	102.4	133.3	123.1	99.1	133.3	122.0	102.4	133.3	123.1	97.4	123.1	114.6	100.9	133.3	122.6	97.6	133.3	121.5
0950, 0980	Ceilings & Acoustic Treatment	78.5	133.3	114.8	91.1	133.3	119.1	88.4	133.3	118.2	78.5	123.1	108.0	77.5	133.3	114.5	89.4	133.3	118.5
0960	Flooring	98.7	176.0	120.9	90.7	176.0	115.2	89.9	176.0	114.7	95.9	153.7	112.5	90.2	176.0	114.9	97.4	172.2	118.9
0970, 0990	Wall Finishes & Painting/Coating	83.4	132.5	112.0	85.8	132.5	113.0	84.5	132.5	112.4	83.4	130.2	110.6	84.5	132.5	112.4	87.6	130.2	112.4
09	FINISHES	92.1	141.2	119.2	90.0	141.8	118.6	91.1	141.8	119.1	90.6	130.6	112.7	89.0	140.5	117.5	92.0	139.7	118.4
COVERS	DIVS. 10 - 14, 25, 28, 41, 43, 44, 46	100.0	116.8	103.6	100.0	116.9	103.7	100.0	116.9	103.7	100.0	105.7	101.2	100.0	116.9	103.7	100.0	111.7	102.5
21, 22, 23	FIRE SUPPRESSION, PLUMBING & HVAC	99.7	130.0	112.5	100.0	128.4	112.0	99.9	129.1	112.3	99.7	127.9	111.7	99.7	126.5	111.0	100.0	127.9	111.8
26, 27, 3370	ELECTRICAL, COMMUNICATIONS & UTIL.	90.0	135.8	113.1	97.1	139.1	118.3	94.1	140.1	117.4	89.4	129.8	109.8	90.5	140.1	115.6	96.6	134.4	115.7
MF2014	WEIGHTED AVERAGE	98.8	128.7	112.0	100.3	129.3	113.1	98.5	129.4	112.1	98.2	124.0	109.6	96.7	128.8	110.9	100.2	124.9	111.1

DIVISION		NEW JERSEY VINELAND 080,083			NEW MEXICO ALBUQUERQUE 870-872			CARRIZOZO 883			CLOVIS 881			FARMINGTON 874			GALLUP 873		
		MAT.	INST.	TOTAL	MAT.	INST.	TOTAL	MAT.	INST.	TOTAL	MAT.	INST.	TOTAL	MAT.	INST.	TOTAL	MAT.	INST.	TOTAL
015433	CONTRACTOR EQUIPMENT		96.0	96.0		109.4	109.4		109.4	109.4		109.4	109.4		109.4	109.4		109.4	109.4
0241, 31 - 34	SITE & INFRASTRUCTURE, DEMOLITION	99.3	103.1	102.0	91.0	102.3	98.9	110.8	102.3	104.8	98.0	102.3	101.0	97.9	102.3	101.0	107.2	102.3	103.8
0310	Concrete Forming & Accessories	99.8	127.5	123.8	100.9	63.1	68.2	99.1	63.1	67.9	99.1	63.0	67.8	100.9	63.1	68.2	100.9	63.1	68.2
0320	Concrete Reinforcing	77.6	114.8	96.9	110.0	70.3	89.4	113.5	70.3	91.1	114.8	70.3	91.7	120.3	70.3	94.4	115.1	70.3	91.9
0330	Cast-in-Place Concrete	89.8	131.0	106.1	92.1	70.1	83.4	92.8	70.1	83.8	92.7	70.1	83.8	93.0	70.1	84.0	87.5	70.1	80.7
03	CONCRETE	96.8	125.1	110.6	102.4	68.0	85.7	122.2	68.0	95.9	109.7	67.9	89.4	106.4	68.0	87.7	113.5	68.0	91.4
04	MASONRY	104.8	132.4	121.8	111.1	61.2	80.4	111.1	61.2	80.4	111.2	61.2	80.4	120.7	61.2	84.0	105.6	61.2	78.2
05	METALS	93.7	103.3	96.7	108.2	87.8	101.7	104.7	87.8	99.3	104.4	87.6	99.1	105.8	87.8	100.1	104.9	87.8	99.5
06	WOOD, PLASTICS & COMPOSITES	101.7	127.6	116.1	93.4	63.8	76.9	89.6	63.8	75.2	89.6	63.8	75.2	93.5	63.8	76.9	93.5	63.8	76.9
07	THERMAL & MOISTURE PROTECTION	102.7	125.9	112.4	98.6	70.9	87.0	100.2	70.9	87.9	99.0	70.9	87.2	98.8	70.9	87.1	99.9	70.9	87.2
08	OPENINGS	95.4	123.0	101.9	99.3	64.5	91.2	96.0	64.5	88.6	96.1	64.5	88.8	101.8	64.5	93.1	101.9	64.5	93.2
0920	Plaster & Gypsum Board	95.7	127.9	117.2	107.2	62.4	77.3	80.8	62.4	68.5	80.8	62.4	68.5	99.3	62.4	74.7	99.3	62.4	74.7
0950, 0980	Ceilings & Acoustic Treatment	78.5	127.9	111.2	92.5	62.4	72.6	96.1	62.4	73.8	96.1	62.4	73.8	90.4	62.4	71.9	90.4	62.4	71.9
0960	Flooring	94.9	153.7	111.8	98.4	66.6	89.3	104.1	66.6	93.3	104.1	66.6	93.3	100.2	66.6	90.5	100.2	66.6	90.5
0970, 0990	Wall Finishes & Painting/Coating	83.4	130.2	110.6	97.6	52.7	71.5	94.0	52.7	70.0	94.0	52.7	70.0	92.1	52.7	69.2	92.1	52.7	69.2
09	FINISHES	89.3	133.5	113.7	95.8	62.4	77.3	97.5	62.4	78.1	96.2	62.4	77.5	94.6	62.4	76.8	95.9	62.4	77.4
COVERS	DIVS. 10 - 14, 25, 28, 41, 43, 44, 46	100.0	111.2	102.4	100.0	81.7	96.0	100.0	81.7	96.0	100.0	81.7	96.0	100.0	81.7	96.0	100.0	81.7	96.0
21, 22, 23	FIRE SUPPRESSION, PLUMBING & HVAC	99.7	122.3	109.3	100.2	68.4	86.8	98.0	68.4	85.5	98.0	68.1	85.4	100.1	68.4	86.7	98.0	68.4	85.5
26, 27, 3370	ELECTRICAL, COMMUNICATIONS & UTIL.	89.4	137.0	113.5	89.2	71.3	80.2	92.5	71.3	81.8	90.2	71.3	80.6	87.1	71.3	79.1	86.5	71.3	78.8
MF2014	WEIGHTED AVERAGE	96.4	123.6	108.4	100.3	72.1	87.9	102.1	72.1	88.9	99.9	72.0	87.6	101.0	72.1	88.2	100.7	72.1	88.1

NEW MEXICO

DIVISION		LAS CRUCES 880			LAS VEGAS 877			ROSWELL 882			SANTA FE 875			SOCORRO 878			TRUTH/CONSEQUENCES 879		
		MAT.	INST.	TOTAL	MAT.	INST.	TOTAL	MAT.	INST.	TOTAL	MAT.	INST.	TOTAL	MAT.	INST.	TOTAL	MAT.	INST.	TOTAL
015433	CONTRACTOR EQUIPMENT		84.6	84.6		109.4	109.4		109.4	109.4		109.4	109.4		109.4	109.4		84.6	84.6
0241, 31 - 34	SITE & INFRASTRUCTURE, DEMOLITION	98.3	81.2	86.2	97.2	102.3	100.8	100.3	102.3	101.7	100.2	102.3	101.7	93.3	102.3	99.7	113.8	81.2	90.8
0310	Concrete Forming & Accessories	95.6	61.9	66.5	100.9	63.1	68.2	99.1	63.1	67.9	99.9	63.1	68.1	100.9	63.1	68.2	98.6	61.9	66.9
0320	Concrete Reinforcing	111.1	70.1	89.9	117.1	70.3	92.8	114.8	70.3	91.7	109.3	70.3	89.1	119.3	70.3	93.9	113.3	70.2	91.0
0330	Cast-in-Place Concrete	87.6	62.4	77.6	90.4	70.1	82.4	92.7	70.1	83.8	95.9	70.1	85.7	88.6	70.1	81.3	97.1	62.4	80.4
03	CONCRETE	86.9	64.5	76.0	103.7	68.0	86.4	110.6	68.0	89.9	103.8	68.0	86.4	102.6	68.0	85.8	95.4	64.5	80.4
04	MASONRY	106.5	60.9	78.4	105.9	61.2	78.3	122.7	61.2	84.8	107.9	61.2	79.1	105.7	61.2	78.3	102.8	60.9	77.0
05	METALS	103.2	80.6	96.0	104.6	87.8	99.3	105.6	87.8	100.0	101.8	87.8	97.3	104.9	87.8	99.5	104.5	80.7	96.9
06	WOOD, PLASTICS & COMPOSITES	79.4	62.6	70.1	93.5	63.8	76.9	89.6	63.8	75.2	96.4	63.8	78.2	93.5	63.8	76.9	85.1	62.7	72.6
07	THERMAL & MOISTURE PROTECTION	86.3	66.1	77.8	98.4	70.9	86.9	99.1	70.9	87.3	100.8	70.9	88.3	98.3	70.9	86.8	87.0	66.1	78.2
08	OPENINGS	91.8	63.9	85.3	98.0	64.5	90.2	95.9	64.5	88.6	100.8	64.5	92.4	97.9	64.5	90.1	91.6	63.9	85.2
0920	Plaster & Gypsum Board	79.3	62.4	68.1	99.3	62.4	74.7	80.8	62.4	68.5	110.3	62.4	78.3	99.3	62.4	74.7	100.6	62.4	75.1
0950, 0980	Ceilings & Acoustic Treatment	82.3	62.4	69.2	90.4	62.4	71.9	96.1	62.4	73.8	90.8	62.4	72.0	90.4	62.4	71.9	80.5	62.4	68.5
0960	Flooring	136.5	66.6	116.4	100.2	66.6	90.5	104.1	66.6	93.3	106.7	66.6	95.2	100.2	66.6	90.5	132.5	66.6	113.5
0970, 0990	Wall Finishes & Painting/Coating	83.0	52.7	65.4	92.1	52.7	69.2	94.0	52.7	70.0	96.4	52.7	71.0	92.1	52.7	69.2	84.3	52.7	65.9
09	FINISHES	106.1	61.6	81.5	94.5	62.4	76.7	96.3	62.4	77.6	99.3	62.4	78.9	94.4	62.4	76.7	107.4	61.6	82.0
COVERS	DIVS. 10 - 14, 25, 28, 41, 43, 44, 46	100.0	79.0	95.5	100.0	81.7	96.0	100.0	81.7	96.0	100.0	81.7	96.0	100.0	81.7	96.0	100.0	79.0	95.5
21, 22, 23	FIRE SUPPRESSION, PLUMBING & HVAC	100.5	68.1	86.8	98.0	68.4	85.5	99.9	68.4	86.6	100.1	68.4	86.7	98.0	68.4	85.5	98.1	68.1	85.4
26, 27, 3370	ELECTRICAL, COMMUNICATIONS & UTIL.	91.9	66.3	78.9	88.7	71.3	79.9	91.5	71.3	81.3	99.4	71.3	85.2	86.9	71.3	79.0	90.2	71.3	80.6
MF2014	WEIGHTED AVERAGE	97.5	68.1	84.6	99.0	72.1	87.1	101.4	72.1	88.5	101.1	72.1	88.3	98.6	72.1	86.9	98.3	68.8	85.3

City Cost Indexes

		NEW MEXICO			NEW YORK														
		TUCUMCARI			ALBANY			BINGHAMTON			BRONX			BROOKLYN			BUFFALO		
	DIVISION	884			120 - 122			137 - 139			104			112			140 - 142		
		MAT.	INST.	TOTAL	MAT.	INST.	TOTAL	MAT.	INST.	TOTAL	MAT.	INST.	TOTAL	MAT.	INST.	TOTAL	MAT.	INST.	TOTAL
015433	CONTRACTOR EQUIPMENT		109.4	109.4		112.6	112.6		116.0	116.0		106.3	106.3		110.9	110.9		96.0	96.0
0241, 31 - 34	SITE & INFRASTRUCTURE, DEMOLITION	97.6	102.3	100.9	84.9	103.5	98.1	96.1	92.4	93.5	98.5	113.0	108.8	120.4	124.1	123.1	97.4	97.7	97.6
0310	Concrete Forming & Accessories	99.1	63.0	67.8	100.9	100.2	100.3	101.8	88.4	90.2	97.9	171.2	161.3	108.2	178.3	168.8	102.7	116.4	114.5
0320	Concrete Reinforcing	112.5	70.3	90.6	102.4	101.9	102.1	96.9	97.4	97.2	98.1	212.2	157.2	98.3	207.9	155.1	102.4	101.1	101.7
0330	Cast-in-Place Concrete	92.7	70.1	83.8	80.4	110.3	92.2	106.0	127.1	114.3	95.7	172.9	126.1	108.1	170.9	132.9	103.7	120.1	110.2
03	CONCRETE	108.9	67.9	89.0	92.8	104.6	98.6	97.4	104.9	101.1	94.9	177.4	135.0	108.7	179.1	142.9	100.7	113.9	107.1
04	MASONRY	123.2	61.2	85.0	101.2	112.3	108.0	111.1	102.0	105.5	90.6	175.4	142.8	121.7	175.3	154.8	112.8	122.3	118.6
05	METALS	104.4	87.6	99.1	100.1	111.2	103.6	93.8	119.5	102.0	90.5	152.8	110.3	101.0	150.8	116.8	100.7	95.0	98.8
06	WOOD, PLASTICS & COMPOSITES	89.6	63.8	75.2	98.1	97.6	97.8	105.6	84.6	93.9	93.3	169.2	135.6	107.3	179.0	147.2	99.5	116.4	108.9
07	THERMAL & MOISTURE PROTECTION	99.0	70.9	87.2	110.4	105.2	108.2	109.4	96.7	104.1	107.3	163.2	130.7	109.9	162.9	132.1	102.0	111.6	106.0
08	OPENINGS	95.9	64.5	88.6	100.5	93.4	98.9	92.9	85.2	91.1	90.7	172.2	109.7	90.3	176.5	110.4	97.3	104.0	98.9
0920	Plaster & Gypsum Board	80.8	62.4	68.5	102.2	97.3	99.0	108.2	83.8	91.9	98.3	170.8	146.8	104.3	181.2	155.7	105.3	116.6	112.9
0950, 0980	Ceilings & Acoustic Treatment	96.1	62.4	73.8	90.5	97.3	95.0	91.0	83.8	86.3	84.0	170.8	141.5	87.0	181.2	149.4	101.6	116.6	111.5
0960	Flooring	104.1	66.6	93.3	97.2	109.2	100.7	103.9	102.8	103.6	97.5	189.4	124.0	110.9	189.4	133.4	96.9	119.6	103.4
0970, 0990	Wall Finishes & Painting/Coating	94.0	52.7	70.0	94.2	97.4	96.1	88.7	100.3	95.4	109.5	163.3	140.8	116.9	163.3	143.9	101.5	115.6	109.7
09	FINISHES	96.2	62.4	77.5	94.0	101.2	98.0	93.9	91.3	92.5	94.0	173.4	138.0	106.6	179.2	146.7	99.8	117.8	109.8
COVERS	DIVS. 10 - 14, 25, 28, 41, 43, 44, 46	100.0	81.7	96.0	100.0	99.1	99.8	100.0	97.0	99.3	100.0	132.5	107.0	100.0	132.7	107.1	100.0	106.0	101.3
21, 22, 23	FIRE SUPPRESSION, PLUMBING & HVAC	98.0	68.1	85.4	100.0	104.9	102.1	100.7	91.7	96.9	100.2	166.1	128.0	99.7	165.9	127.7	100.0	97.7	99.0
26, 27, 3370	ELECTRICAL, COMMUNICATIONS & UTIL.	92.5	71.3	81.8	96.0	103.3	99.7	98.2	101.5	99.9	91.5	186.8	139.6	97.9	186.8	142.8	98.7	101.9	100.3
MF2014	WEIGHTED AVERAGE	100.6	72.0	88.0	98.3	104.6	101.1	98.2	98.4	98.3	95.2	166.0	126.4	102.3	167.9	131.2	100.3	106.4	103.0

		NEW YORK																	
		ELMIRA			FAR ROCKAWAY			FLUSHING			GLENS FALLS			HICKSVILLE			JAMAICA		
	DIVISION	148 - 149			116			113			128			115,117,118			114		
		MAT.	INST.	TOTAL	MAT.	INST.	TOTAL	MAT.	INST.	TOTAL	MAT.	INST.	TOTAL	MAT.	INST.	TOTAL	MAT.	INST.	TOTAL
015433	CONTRACTOR EQUIPMENT		118.9	118.9		110.9	110.9		110.9	110.9		112.6	112.6		110.9	110.9		110.9	110.9
0241, 31 - 34	SITE & INFRASTRUCTURE, DEMOLITION	99.7	93.4	95.2	123.6	124.1	124.0	123.6	124.1	124.0	75.0	103.2	94.9	113.3	122.9	120.1	117.9	124.1	122.3
0310	Concrete Forming & Accessories	83.5	93.4	92.1	94.3	170.7	160.4	98.2	170.7	161.0	86.6	93.3	92.4	90.8	152.3	144.0	98.2	170.7	161.0
0320	Concrete Reinforcing	101.3	94.7	97.9	98.3	207.9	155.1	100.0	207.9	155.9	98.2	93.4	95.7	98.3	207.7	155.0	98.3	207.9	155.1
0330	Cast-in-Place Concrete	94.9	104.4	98.6	116.9	170.9	138.2	116.9	170.9	138.2	77.7	107.6	89.5	99.3	165.2	125.3	108.1	170.9	132.9
03	CONCRETE	90.4	99.2	94.7	114.9	175.8	144.5	115.4	175.8	144.7	84.4	99.1	91.6	100.5	165.4	132.0	108.0	175.8	141.0
04	MASONRY	110.0	104.9	106.9	126.2	175.3	156.5	120.1	175.3	154.1	102.8	107.7	105.8	115.7	166.4	147.0	124.0	175.3	155.6
05	METALS	94.9	120.9	103.2	101.0	150.8	116.9	101.0	150.8	116.9	94.0	108.2	98.5	102.5	147.6	116.8	101.0	150.8	116.9
06	WOOD, PLASTICS & COMPOSITES	80.7	92.2	87.1	91.0	168.9	134.3	95.5	168.9	136.3	88.7	90.7	89.8	87.6	149.4	122.0	95.5	168.9	136.3
07	THERMAL & MOISTURE PROTECTION	104.8	94.7	100.6	109.8	161.9	131.6	109.8	161.9	131.6	103.5	101.9	102.8	109.4	154.9	128.4	109.6	161.9	131.5
08	OPENINGS	99.3	88.7	96.8	89.0	171.0	108.1	89.0	171.0	108.1	93.7	87.1	92.2	89.4	160.4	105.9	89.0	171.0	108.1
0920	Plaster & Gypsum Board	100.0	91.8	94.5	93.1	170.8	145.0	95.5	170.8	145.8	94.3	90.3	91.6	92.8	150.8	131.5	95.5	170.8	145.8
0950, 0980	Ceilings & Acoustic Treatment	96.0	91.8	93.2	76.3	170.8	138.9	76.3	170.8	138.9	80.4	90.3	87.0	75.3	150.8	125.3	76.3	170.8	138.9
0960	Flooring	90.7	102.8	94.2	105.5	189.4	129.7	107.1	189.4	130.8	85.5	110.0	92.5	104.4	177.9	125.6	107.1	189.4	130.8
0970, 0990	Wall Finishes & Painting/Coating	95.8	91.7	93.5	116.9	163.3	143.9	116.9	163.3	143.9	91.7	97.4	95.0	116.9	163.3	143.9	116.9	163.3	143.9
09	FINISHES	94.1	94.9	94.5	101.5	173.2	141.1	102.3	173.2	141.5	85.2	96.1	91.2	100.0	157.3	131.7	101.8	173.2	141.3
COVERS	DIVS. 10 - 14, 25, 28, 41, 43, 44, 46	100.0	98.4	99.7	100.0	131.6	106.8	100.0	131.6	106.8	100.0	96.4	99.2	100.0	126.4	105.7	100.0	131.6	106.8
21, 22, 23	FIRE SUPPRESSION, PLUMBING & HVAC	96.9	91.1	94.5	96.3	165.9	125.7	96.3	165.9	125.7	96.8	101.0	98.6	99.7	155.2	123.2	96.3	165.9	125.7
26, 27, 3370	ELECTRICAL, COMMUNICATIONS & UTIL.	96.3	98.8	97.6	104.5	186.8	146.1	104.5	186.8	146.1	91.5	103.3	97.5	97.4	146.6	122.2	96.4	186.8	142.1
MF2014	WEIGHTED AVERAGE	96.9	98.3	97.6	102.5	166.2	130.6	102.4	166.2	130.5	93.3	101.1	96.7	100.3	152.7	123.4	100.7	166.2	129.6

		NEW YORK																	
		JAMESTOWN			KINGSTON			LONG ISLAND CITY			MONTICELLO			MOUNT VERNON			NEW ROCHELLE		
	DIVISION	147			124			111			127			105			108		
		MAT.	INST.	TOTAL	MAT.	INST.	TOTAL	MAT.	INST.	TOTAL	MAT.	INST.	TOTAL	MAT.	INST.	TOTAL	MAT.	INST.	TOTAL
015433	CONTRACTOR EQUIPMENT		92.8	92.8		110.9	110.9		110.9	110.9		110.9	110.9		106.3	106.3		106.3	106.3
0241, 31 - 34	SITE & INFRASTRUCTURE, DEMOLITION	101.1	93.6	95.8	145.7	118.8	126.7	121.4	124.1	123.3	140.5	118.6	125.1	103.8	110.5	108.5	103.4	110.5	108.4
0310	Concrete Forming & Accessories	83.5	87.3	86.8	88.1	101.5	99.7	102.7	170.7	161.6	95.6	100.3	99.7	88.2	137.0	130.4	103.1	136.9	132.4
0320	Concrete Reinforcing	101.4	97.9	99.6	98.6	139.2	119.7	98.3	207.9	155.1	97.8	139.2	119.3	97.1	207.7	154.4	97.1	207.7	154.4
0330	Cast-in-Place Concrete	98.4	103.1	100.3	104.9	136.6	117.4	111.6	170.9	135.0	98.2	134.8	112.7	106.8	147.0	122.7	106.8	147.0	122.7
03	CONCRETE	93.3	94.5	93.9	105.8	120.2	112.8	111.2	175.8	142.6	100.8	120.1	109.7	104.4	152.2	127.6	103.9	152.2	127.4
04	MASONRY	119.6	102.5	109.1	118.2	143.9	134.0	118.4	175.3	153.5	110.7	140.7	129.2	96.0	154.9	132.3	96.0	154.3	132.3
05	METALS	92.5	91.2	92.1	101.7	116.4	106.3	101.0	150.8	116.8	101.6	116.3	106.3	90.3	144.0	107.4	90.6	144.0	107.6
06	WOOD, PLASTICS & COMPOSITES	79.3	84.2	82.0	90.0	93.1	91.7	101.5	168.9	139.0	97.2	93.1	94.9	84.0	131.8	110.6	100.5	131.8	117.9
07	THERMAL & MOISTURE PROTECTION	104.2	93.4	99.7	128.4	135.8	131.5	109.7	161.9	131.6	128.0	134.4	130.7	108.2	147.2	124.5	108.2	147.2	124.6
08	OPENINGS	99.1	85.9	96.1	95.8	114.6	100.2	89.0	171.0	108.1	91.3	114.6	96.7	90.7	150.8	104.7	90.8	150.8	104.7
0920	Plaster & Gypsum Board	87.0	83.6	84.7	96.7	93.0	94.2	100.1	170.8	147.3	97.4	93.0	94.4	93.0	132.3	119.2	106.2	132.3	123.6
0950, 0980	Ceilings & Acoustic Treatment	92.7	83.6	86.7	72.1	93.0	85.9	76.3	170.8	138.9	72.1	93.0	85.9	82.4	132.3	115.4	82.4	132.3	115.4
0960	Flooring	93.3	102.8	96.0	102.7	69.7	93.2	108.8	189.4	132.0	105.2	69.7	95.0	88.9	179.6	115.0	97.0	179.6	120.7
0970, 0990	Wall Finishes & Painting/Coating	97.1	97.1	97.1	121.6	116.8	118.8	116.9	163.3	143.9	121.6	116.8	118.8	107.9	163.3	140.1	107.9	163.3	140.1
09	FINISHES	92.4	90.4	91.3	99.7	94.3	96.7	103.2	173.2	141.9	100.1	93.5	96.4	90.8	146.4	121.6	95.0	146.4	123.4
COVERS	DIVS. 10 - 14, 25, 28, 41, 43, 44, 46	100.0	97.6	99.5	100.0	108.8	101.9	100.0	131.6	106.8	100.0	107.8	101.7	100.0	123.0	105.0	100.0	119.8	104.3
21, 22, 23	FIRE SUPPRESSION, PLUMBING & HVAC	96.7	88.2	93.4	96.7	121.0	107.0	99.7	165.9	127.7	96.7	115.6	104.7	96.9	136.4	113.6	96.9	136.4	113.6
26, 27, 3370	ELECTRICAL, COMMUNICATIONS & UTIL.	95.3	96.6	95.9	92.8	110.2	101.6	96.9	186.8	142.3	92.8	110.2	101.6	89.9	154.0	122.3	89.9	154.0	122.3
MF2014	WEIGHTED AVERAGE	97.0	93.0	95.3	101.8	117.1	108.6	101.9	166.2	130.2	100.3	115.3	106.9	95.5	143.1	116.5	95.9	143.0	116.7

For customer support on your Mechanical Cost Data, call 877.739.3689.

679

City Cost Indexes

NEW YORK

DIVISION		NEW YORK 100-102 MAT.	INST.	TOTAL	NIAGARA FALLS 143 MAT.	INST.	TOTAL	PLATTSBURGH 129 MAT.	INST.	TOTAL	POUGHKEEPSIE 125-126 MAT.	INST.	TOTAL	QUEENS 110 MAT.	INST.	TOTAL	RIVERHEAD 119 MAT.	INST.	TOTAL
015433	CONTRACTOR EQUIPMENT		106.9	106.9		92.8	92.8		96.6	96.6		110.9	110.9		110.9	110.9		110.9	110.9
0241, 31 - 34	SITE & INFRASTRUCTURE, DEMOLITION	106.0	114.0	111.7	103.3	95.1	97.5	110.7	100.9	103.7	141.6	119.5	126.0	116.7	124.1	122.0	114.6	122.9	120.4
0310	Concrete Forming & Accessories	101.9	183.8	172.7	83.5	115.2	110.9	92.2	92.9	92.8	88.1	160.5	150.7	91.0	170.7	160.0	95.4	151.9	144.2
0320	Concrete Reinforcing	103.8	212.3	160.0	100.1	101.5	100.8	102.3	101.5	101.9	98.6	139.5	119.8	100.0	207.9	155.9	100.2	207.7	155.9
0330	Cast-in-Place Concrete	107.6	176.1	134.6	101.8	121.5	109.5	95.1	103.2	98.3	101.6	139.5	116.5	102.7	170.9	129.6	101.0	165.0	126.2
03	CONCRETE	105.1	184.1	143.5	95.5	113.8	104.4	97.9	97.5	97.7	103.1	147.7	124.8	103.8	175.8	138.8	101.1	165.1	132.2
04	MASONRY	100.4	179.8	149.3	127.8	130.5	129.4	96.5	101.1	99.4	110.3	147.0	132.9	112.5	175.3	151.2	121.8	166.4	149.3
05	METALS	102.4	152.8	118.4	95.0	93.1	94.4	97.9	89.6	95.2	101.7	119.2	107.3	101.0	150.8	116.8	103.0	147.1	117.0
06	WOOD, PLASTICS & COMPOSITES	97.3	184.0	145.5	79.3	110.8	96.8	95.3	90.3	92.5	90.0	168.9	133.9	87.7	168.9	132.9	92.6	149.4	124.2
07	THERMAL & MOISTURE PROTECTION	107.3	166.8	132.2	104.3	114.2	108.4	122.3	99.5	112.8	128.3	145.1	135.3	109.4	161.9	131.4	110.4	154.9	129.0
08	OPENINGS	96.7	180.2	116.1	99.2	101.1	99.6	101.6	88.2	98.5	95.8	155.0	109.6	89.0	171.0	108.1	89.4	160.4	105.9
0920	Plaster & Gypsum Board	104.4	186.0	158.9	87.0	110.9	103.0	113.9	89.4	97.6	96.7	170.8	146.2	92.8	170.8	144.9	94.0	150.8	131.9
0950, 0980	Ceilings & Acoustic Treatment	99.7	186.0	156.8	92.7	110.9	104.8	97.9	89.4	92.3	72.1	170.8	137.5	76.3	170.8	138.9	76.2	150.8	125.6
0960	Flooring	98.9	189.4	124.9	93.3	119.6	100.9	108.3	109.3	108.6	102.7	169.5	121.9	104.4	189.4	128.9	105.5	137.1	114.6
0970, 0990	Wall Finishes & Painting/Coating	109.5	159.4	138.5	97.1	115.1	107.6	117.6	94.8	104.3	121.6	117.2	119.0	116.9	163.3	143.9	116.9	163.3	143.9
09	FINISHES	99.2	182.9	145.5	92.6	116.5	105.8	97.7	95.5	96.5	99.5	160.1	133.0	100.4	173.2	140.6	100.7	149.2	127.5
COVERS	DIVS. 10 - 14, 25, 28, 41, 43, 44, 46	100.0	135.6	107.7	100.0	108.1	101.8	100.0	97.3	99.4	100.0	120.4	104.4	100.0	131.6	106.8	100.0	126.4	105.7
21, 22, 23	FIRE SUPPRESSION, PLUMBING & HVAC	100.1	168.3	128.9	96.7	101.1	98.6	96.7	101.1	98.6	96.7	125.2	108.8	99.7	165.9	127.7	100.0	152.0	122.0
26, 27, 3370	ELECTRICAL, COMMUNICATIONS & UTIL.	97.6	186.8	142.7	94.0	101.3	97.7	90.0	92.3	91.2	92.8	119.0	106.0	97.4	186.8	142.6	98.8	146.6	123.0
MF2014	WEIGHTED AVERAGE	100.7	169.8	131.1	98.0	107.2	102.1	98.4	96.7	97.7	101.0	135.2	116.1	100.3	166.2	129.3	101.1	150.9	123.0

NEW YORK

DIVISION		ROCHESTER 144-146 MAT.	INST.	TOTAL	SCHENECTADY 123 MAT.	INST.	TOTAL	STATEN ISLAND 103 MAT.	INST.	TOTAL	SUFFERN 109 MAT.	INST.	TOTAL	SYRACUSE 130-132 MAT.	INST.	TOTAL	UTICA 133-135 MAT.	INST.	TOTAL
015433	CONTRACTOR EQUIPMENT		117.6	117.6		112.6	112.6		106.3	106.3		106.3	106.3		112.6	112.6		112.6	112.6
0241, 31 - 34	SITE & INFRASTRUCTURE, DEMOLITION	90.7	109.1	103.7	85.6	103.5	98.3	107.9	113.0	111.5	100.6	108.6	106.3	94.9	102.7	100.4	73.6	101.2	93.1
0310	Concrete Forming & Accessories	98.4	98.7	98.6	103.9	100.2	100.7	87.8	178.7	166.5	96.4	140.5	134.5	100.8	91.6	92.9	102.1	87.6	89.6
0320	Concrete Reinforcing	102.8	94.7	98.6	97.2	101.9	99.6	98.1	208.0	155.0	97.1	143.0	120.9	97.9	97.5	97.7	97.9	104.7	101.4
0330	Cast-in-Place Concrete	96.2	105.4	99.8	93.7	110.3	100.3	106.9	172.9	132.9	103.6	139.5	117.8	98.4	104.3	100.7	90.0	103.7	95.4
03	CONCRETE	99.3	101.1	100.2	98.4	104.6	101.4	106.0	180.1	142.0	101.0	139.4	119.7	99.9	97.9	98.9	97.7	97.2	97.4
04	MASONRY	109.1	107.8	108.3	99.3	112.3	107.3	102.7	175.4	147.5	95.8	149.9	129.2	103.2	106.9	105.5	94.3	103.5	100.0
05	METALS	99.8	109.2	102.8	98.3	111.2	102.4	88.7	152.8	109.1	88.7	120.7	98.8	97.2	107.7	100.5	95.0	110.0	99.8
06	WOOD, PLASTICS & COMPOSITES	95.0	97.5	96.4	107.9	97.6	102.1	82.6	179.4	136.5	93.1	141.7	120.2	101.9	88.8	94.6	101.9	84.6	92.3
07	THERMAL & MOISTURE PROTECTION	104.5	101.7	103.3	105.0	105.2	105.1	107.6	164.3	131.3	108.1	143.4	122.9	104.1	98.2	101.6	92.2	97.4	94.4
08	OPENINGS	108.0	92.3	104.4	99.5	93.4	98.1	90.7	177.7	110.9	90.8	141.9	102.7	95.1	86.2	93.0	98.0	85.0	95.0
0920	Plaster & Gypsum Board	100.2	97.4	98.3	103.7	97.3	99.5	93.0	181.2	151.9	97.6	142.5	127.6	99.5	88.3	92.0	99.5	84.0	89.1
0950, 0980	Ceilings & Acoustic Treatment	100.3	97.4	98.4	86.9	97.3	93.8	84.0	181.2	148.4	82.4	142.5	122.2	91.0	88.3	89.2	91.0	84.0	86.3
0960	Flooring	92.7	111.7	98.2	93.0	109.2	97.7	92.6	189.4	120.4	92.7	177.9	117.2	92.6	98.5	94.3	90.4	98.3	92.6
0970, 0990	Wall Finishes & Painting/Coating	92.2	102.9	98.4	91.7	97.4	95.0	105.9	163.3	140.8	107.9	131.5	121.6	93.2	102.7	98.8	86.5	102.7	95.9
09	FINISHES	98.0	101.7	100.0	90.9	101.2	96.6	92.6	179.5	140.7	92.3	143.6	120.7	92.8	93.6	93.3	91.0	90.5	90.7
COVERS	DIVS. 10 - 14, 25, 28, 41, 43, 44, 46	100.0	100.2	100.1	100.0	99.1	99.8	100.0	133.6	107.3	100.0	122.7	104.9	100.0	97.6	99.5	100.0	95.9	99.1
21, 22, 23	FIRE SUPPRESSION, PLUMBING & HVAC	100.0	90.1	95.9	100.2	104.5	102.0	100.2	166.1	128.0	96.9	128.6	110.3	100.3	92.0	96.8	100.3	93.2	97.3
26, 27, 3370	ELECTRICAL, COMMUNICATIONS & UTIL.	99.3	97.7	98.5	95.5	103.3	99.5	91.5	186.8	139.6	96.0	122.6	109.4	98.3	101.4	99.8	96.6	101.4	99.0
MF2014	WEIGHTED AVERAGE	100.8	100.0	100.5	98.1	104.5	101.0	96.8	167.5	128.0	95.5	131.9	111.5	98.5	98.2	98.3	96.5	97.5	97.0

NEW YORK / NORTH CAROLINA

DIVISION		WATERTOWN 136 MAT.	INST.	TOTAL	WHITE PLAINS 106 MAT.	INST.	TOTAL	YONKERS 107 MAT.	INST.	TOTAL	ASHEVILLE 287-288 MAT.	INST.	TOTAL	CHARLOTTE 281-282 MAT.	INST.	TOTAL	DURHAM 277 MAT.	INST.	TOTAL
015433	CONTRACTOR EQUIPMENT		112.6	112.6		106.3	106.3		106.3	106.3		98.9	98.9		98.9	98.9		104.4	104.4
0241, 31 - 34	SITE & INFRASTRUCTURE, DEMOLITION	81.2	102.7	96.5	98.3	110.5	106.9	105.5	110.5	109.0	99.6	78.9	85.0	101.5	79.1	85.7	99.6	88.1	91.4
0310	Concrete Forming & Accessories	86.8	94.6	93.5	101.4	137.0	132.2	101.6	144.6	138.8	93.8	60.1	64.7	96.4	61.2	65.9	97.1	59.8	64.8
0320	Concrete Reinforcing	98.5	97.5	98.0	97.1	207.7	154.4	101.0	207.7	156.3	99.1	62.6	80.2	103.6	58.1	80.0	101.4	57.5	78.7
0330	Cast-in-Place Concrete	104.8	107.4	105.8	94.9	147.0	115.5	106.1	147.1	123.3	113.7	65.1	94.6	116.1	66.9	96.7	91.0	64.8	80.7
03	CONCRETE	110.2	100.3	105.8	94.5	152.2	122.6	104.0	155.7	129.1	109.1	63.8	87.1	108.8	64.0	87.0	97.7	62.6	80.7
04	MASONRY	95.5	109.2	103.9	95.2	154.9	132.0	98.5	154.9	133.8	93.4	47.8	65.3	100.1	55.5	72.6	89.2	40.8	59.3
05	METALS	95.1	107.7	99.1	90.2	144.0	107.3	98.7	144.3	113.2	94.0	85.3	91.3	94.9	83.6	91.3	109.7	82.8	101.1
06	WOOD, PLASTICS & COMPOSITES	84.1	92.1	88.5	98.2	131.8	116.9	98.1	141.7	122.4	93.9	64.6	77.6	92.4	64.6	76.9	94.5	64.6	77.8
07	THERMAL & MOISTURE PROTECTION	92.5	100.5	95.8	107.9	147.2	124.4	108.3	148.8	125.2	103.2	53.7	82.5	98.1	56.4	80.6	104.6	53.2	83.1
08	OPENINGS	98.0	88.9	95.9	90.8	150.8	104.7	94.0	156.3	108.5	96.2	64.2	88.8	97.1	63.0	89.2	107.1	63.0	96.8
0920	Plaster & Gypsum Board	90.2	91.7	91.2	100.6	132.3	121.8	104.0	142.5	129.7	97.8	63.3	74.8	97.2	63.3	74.6	97.5	63.3	74.7
0950, 0980	Ceilings & Acoustic Treatment	91.0	91.7	91.5	82.4	132.3	115.4	98.1	142.5	127.5	85.0	63.3	70.7	87.3	63.3	71.4	86.6	63.3	71.2
0960	Flooring	83.4	98.3	87.7	95.2	179.6	119.5	94.7	189.4	121.9	96.7	47.2	82.5	96.4	41.7	80.7	104.1	46.4	87.5
0970, 0990	Wall Finishes & Painting/Coating	86.5	99.3	93.9	107.9	163.3	140.1	107.9	163.3	140.1	102.1	62.0	78.8	102.3	62.0	78.9	96.0	62.0	76.2
09	FINISHES	88.2	95.5	92.3	93.1	146.4	122.6	97.4	154.2	128.8	92.8	58.3	73.7	92.3	58.1	73.3	94.1	58.1	74.2
COVERS	DIVS. 10 - 14, 25, 28, 41, 43, 44, 46	100.0	98.5	99.7	100.0	123.0	105.0	100.0	125.4	105.5	100.0	81.0	95.9	100.0	82.1	96.1	100.0	80.8	95.8
21, 22, 23	FIRE SUPPRESSION, PLUMBING & HVAC	100.3	85.8	94.2	100.3	136.4	115.6	100.3	136.5	115.6	100.5	55.6	81.5	99.9	57.1	81.8	100.6	55.2	81.4
26, 27, 3370	ELECTRICAL, COMMUNICATIONS & UTIL.	98.2	92.4	95.3	89.9	154.0	122.3	96.1	163.1	129.9	100.5	57.5	78.7	99.5	60.5	79.8	93.9	56.3	74.9
MF2014	WEIGHTED AVERAGE	98.0	96.7	97.4	95.2	143.1	116.3	99.4	146.3	120.1	99.0	62.4	82.9	99.1	63.8	83.6	100.7	61.7	83.5

For customer support on your Mechanical Cost Data, call 877.739.3689.

NORTH CAROLINA

DIVISION		ELIZABETH CITY 279			FAYETTEVILLE 283			GASTONIA 280			GREENSBORO 270,272-274			HICKORY 286			KINSTON 285		
		MAT.	INST.	TOTAL	MAT.	INST.	TOTAL	MAT.	INST.	TOTAL	MAT.	INST.	TOTAL	MAT.	INST.	TOTAL	MAT.	INST.	TOTAL
015433	CONTRACTOR EQUIPMENT		108.7	108.7		104.4	104.4		98.9	98.9		104.4	104.4		104.4	104.4		104.4	104.4
0241, 31 - 34	SITE & INFRASTRUCTURE, DEMOLITION	103.8	89.7	93.8	99.0	88.0	91.2	99.5	78.9	85.0	99.5	88.1	91.4	98.5	87.8	91.0	97.6	88.7	91.3
0310	Concrete Forming & Accessories	83.4	42.3	47.9	93.4	59.5	64.1	100.4	60.0	65.5	96.9	59.8	64.8	90.2	59.6	63.8	86.6	63.5	66.6
0320	Concrete Reinforcing	99.3	45.5	71.4	103.0	57.5	79.4	99.5	58.0	78.0	100.3	57.5	78.1	99.1	57.6	77.6	98.6	45.4	71.0
0330	Cast-in-Place Concrete	91.2	66.5	81.4	119.3	64.4	97.6	111.2	65.2	93.0	90.4	64.9	80.3	113.7	64.7	94.4	109.8	71.1	94.5
03	CONCRETE	97.9	53.1	76.1	111.3	62.3	87.5	107.5	62.9	85.8	97.2	62.6	80.4	108.9	62.4	86.3	105.5	64.1	85.4
04	MASONRY	101.6	47.8	68.5	96.7	39.3	61.3	97.8	50.6	68.7	84.8	41.2	57.9	82.2	43.5	58.3	89.1	60.5	71.5
05	METALS	96.9	78.8	91.1	112.9	82.6	103.2	94.8	83.4	91.2	102.8	82.6	96.4	94.1	81.4	90.1	93.0	77.7	88.2
06	WOOD, PLASTICS & COMPOSITES	79.7	42.1	58.8	93.2	64.6	77.2	101.9	64.6	81.1	94.3	64.6	77.7	88.9	64.6	75.4	85.9	64.6	74.0
07	THERMAL & MOISTURE PROTECTION	103.9	50.7	81.6	102.7	51.5	81.3	103.5	54.2	82.8	104.4	51.3	82.2	103.5	52.3	82.1	103.4	57.6	84.2
08	OPENINGS	103.5	47.4	90.5	96.3	63.0	88.6	100.2	63.0	91.5	107.1	63.0	96.8	96.2	62.5	88.4	96.3	59.7	87.8
0920	Plaster & Gypsum Board	90.7	39.6	56.6	101.9	63.3	76.1	103.7	63.3	76.7	99.0	63.3	75.2	97.8	63.3	74.8	97.4	63.3	74.6
0950, 0980	Ceilings & Acoustic Treatment	86.6	39.6	55.5	85.8	63.3	70.9	87.5	63.3	71.5	86.6	63.3	71.2	85.0	63.3	70.7	87.5	63.3	71.5
0960	Flooring	95.9	22.3	74.8	96.9	41.8	81.1	100.2	41.0	83.2	104.1	38.2	85.1	96.6	31.6	77.9	93.8	22.0	73.1
0970, 0990	Wall Finishes & Painting/Coating	96.0	62.0	76.2	102.1	62.0	78.8	102.1	61.5	78.5	96.0	62.0	76.2	102.1	62.0	78.8	102.1	62.0	78.8
09	FINISHES	91.1	39.7	62.7	93.5	57.0	73.3	95.1	57.1	74.1	94.4	56.6	73.5	92.9	54.9	71.9	92.4	56.1	72.3
COVERS	DIVS. 10 - 14, 25, 28, 41, 43, 44, 46	100.0	84.7	96.7	100.0	80.6	95.8	100.0	81.1	95.9	100.0	81.0	95.9	100.0	81.0	95.9	100.0	84.7	96.7
21, 22, 23	FIRE SUPPRESSION, PLUMBING & HVAC	97.1	54.3	79.0	100.2	54.9	81.1	100.5	55.5	81.5	100.5	55.3	81.4	97.1	55.4	79.5	97.1	60.7	81.7
26, 27, 3370	ELECTRICAL, COMMUNICATIONS & UTIL.	93.7	34.3	63.7	99.5	51.3	75.1	99.9	59.1	79.3	93.1	57.5	75.1	98.1	59.1	78.4	97.9	45.8	71.5
MF2014	WEIGHTED AVERAGE	97.8	54.5	78.7	102.2	60.6	83.9	99.8	62.4	83.3	99.2	61.7	82.7	97.3	61.8	81.7	97.0	63.0	82.0

NORTH CAROLINA / NORTH DAKOTA

DIVISION		MURPHY 289			RALEIGH 275 - 276			ROCKY MOUNT 278			WILMINGTON 284			WINSTON-SALEM 271			BISMARCK 585		
		MAT.	INST.	TOTAL	MAT.	INST.	TOTAL	MAT.	INST.	TOTAL	MAT.	INST.	TOTAL	MAT.	INST.	TOTAL	MAT.	INST.	TOTAL
015433	CONTRACTOR EQUIPMENT		98.9	98.9		104.4	104.4		104.4	104.4		98.9	98.9		104.4	104.4		98.8	98.8
0241, 31 - 34	SITE & INFRASTRUCTURE, DEMOLITION	100.8	78.7	85.2	99.1	88.1	91.3	102.0	88.7	92.6	100.9	79.1	85.5	99.8	88.1	91.5	100.2	97.6	98.4
0310	Concrete Forming & Accessories	101.0	39.1	47.4	95.0	60.0	64.7	89.4	63.6	67.1	95.3	61.1	65.7	98.8	60.0	65.2	107.8	41.1	50.1
0320	Concrete Reinforcing	98.6	45.0	70.8	102.4	57.5	79.2	99.3	57.3	77.6	99.8	57.5	77.9	100.3	57.5	78.1	98.1	96.2	97.1
0330	Cast-in-Place Concrete	117.6	63.8	96.4	93.7	65.3	82.5	89.2	71.1	82.1	113.3	66.8	95.0	92.6	65.2	81.8	96.3	70.8	86.3
03	CONCRETE	112.4	50.6	82.4	97.5	62.8	80.7	98.9	66.3	83.0	109.0	63.8	87.1	98.4	62.8	81.1	98.7	62.9	81.3
04	MASONRY	85.3	41.1	58.1	87.6	42.0	59.5	78.5	49.6	60.7	82.8	44.6	59.2	85.0	41.5	58.2	116.5	66.9	85.9
05	METALS	92.0	77.0	87.2	95.7	82.7	91.5	96.1	81.1	91.3	93.7	82.7	90.2	100.2	82.7	94.7	96.6	92.0	95.1
06	WOOD, PLASTICS & COMPOSITES	102.5	38.3	66.8	89.1	64.6	75.5	86.2	64.6	74.1	95.8	64.6	78.4	94.3	64.6	77.7	98.9	33.4	62.5
07	THERMAL & MOISTURE PROTECTION	103.4	48.3	80.4	98.9	53.3	79.8	104.3	55.3	83.8	103.2	52.7	82.1	104.4	51.7	82.4	110.5	73.2	94.9
08	OPENINGS	96.2	45.3	84.3	103.6	63.0	94.1	102.7	62.5	93.3	96.3	63.0	88.6	107.1	63.0	96.8	111.3	49.8	97.0
0920	Plaster & Gypsum Board	103.1	36.3	58.5	91.9	63.3	72.8	92.7	63.3	73.1	99.7	63.3	75.4	86.6	63.3	75.2	94.7	31.9	52.7
0950, 0980	Ceilings & Acoustic Treatment	85.0	36.3	52.8	87.3	63.3	71.4	84.9	63.3	70.6	85.8	63.3	70.9	86.6	63.3	71.2	112.3	31.9	59.1
0960	Flooring	100.5	23.1	78.3	101.4	43.9	84.8	99.6	22.0	77.3	97.5	48.3	83.4	104.1	46.4	87.5	94.3	50.6	81.8
0970, 0990	Wall Finishes & Painting/Coating	102.1	62.0	78.8	94.4	62.0	75.6	96.0	62.0	76.2	102.1	62.0	78.8	96.0	62.0	76.2	93.5	63.3	75.9
09	FINISHES	94.8	37.5	63.1	93.3	57.8	73.7	92.2	56.1	72.3	93.4	59.4	74.6	94.4	58.2	74.4	99.0	43.7	68.4
COVERS	DIVS. 10 - 14, 25, 28, 41, 43, 44, 46	100.0	77.4	95.1	100.0	81.1	95.9	100.0	84.7	96.7	100.0	82.1	96.1	100.0	81.1	95.9	100.0	80.6	95.8
21, 22, 23	FIRE SUPPRESSION, PLUMBING & HVAC	97.1	54.1	78.9	100.1	55.6	81.3	97.1	61.1	81.9	100.5	57.1	82.1	100.5	55.5	81.5	99.9	71.7	88.0
26, 27, 3370	ELECTRICAL, COMMUNICATIONS & UTIL.	101.4	29.1	64.9	95.7	40.1	67.6	95.5	39.4	67.1	100.7	51.3	75.7	93.1	58.0	75.3	104.6	76.2	90.3
MF2014	WEIGHTED AVERAGE	98.2	51.0	77.4	97.7	59.7	81.0	96.8	61.8	81.4	98.5	61.5	82.2	98.9	62.1	82.7	102.1	70.0	87.9

NORTH DAKOTA

DIVISION		DEVILS LAKE 583			DICKINSON 586			FARGO 580 - 581			GRAND FORKS 582			JAMESTOWN 584			MINOT 587		
		MAT.	INST.	TOTAL	MAT.	INST.	TOTAL	MAT.	INST.	TOTAL	MAT.	INST.	TOTAL	MAT.	INST.	TOTAL	MAT.	INST.	TOTAL
015433	CONTRACTOR EQUIPMENT		98.8	98.8		98.8	98.8		98.8	98.8		98.8	98.8		98.8	98.8		98.8	98.8
0241, 31 - 34	SITE & INFRASTRUCTURE, DEMOLITION	106.4	95.2	98.5	114.7	95.2	100.9	99.6	97.9	98.4	110.4	95.2	99.6	105.4	95.2	98.2	107.9	98.3	101.1
0310	Concrete Forming & Accessories	101.9	36.4	45.3	91.5	36.4	43.8	98.2	48.8	55.5	95.2	36.4	44.3	93.0	36.7	44.3	91.2	70.4	73.2
0320	Concrete Reinforcing	100.5	96.4	98.4	101.4	96.1	98.7	100.9	96.9	98.9	98.9	96.4	97.6	101.0	96.8	98.8	102.4	96.3	99.2
0330	Cast-in-Place Concrete	118.3	67.4	98.2	107.0	67.4	91.4	99.3	87.5	94.4	107.0	67.4	91.4	116.8	67.5	97.4	107.0	77.0	95.2
03	CONCRETE	106.7	59.7	83.9	105.8	59.6	83.3	100.0	72.1	86.4	103.1	59.7	82.0	101.8	78.1	90.3	101.8	78.1	90.3
04	MASONRY	128.1	66.6	90.2	129.9	61.2	87.6	120.9	83.7	98.0	121.5	66.6	87.7	142.1	73.6	99.9	120.5	79.6	95.3
05	METALS	92.3	91.7	92.1	92.2	91.3	91.9	94.2	93.3	93.9	92.2	91.7	92.1	92.2	92.5	92.3	92.5	92.1	92.4
06	WOOD, PLASTICS & COMPOSITES	95.5	30.8	59.5	83.2	30.8	54.0	91.8	41.2	63.6	87.5	30.8	55.9	85.1	30.8	54.9	82.9	67.7	74.4
07	THERMAL & MOISTURE PROTECTION	109.5	72.0	93.8	110.1	70.4	93.5	107.4	79.6	95.8	109.7	72.0	93.9	109.4	74.2	94.7	109.5	81.9	97.9
08	OPENINGS	105.8	48.4	92.4	105.8	48.4	92.4	103.3	54.0	91.8	104.4	48.4	91.3	105.8	48.4	92.4	104.5	68.5	96.1
0920	Plaster & Gypsum Board	117.1	29.2	58.3	107.4	29.2	55.2	96.8	39.8	58.7	108.8	29.2	55.6	108.4	29.2	55.5	107.4	67.1	80.5
0950, 0980	Ceilings & Acoustic Treatment	111.2	29.2	56.9	111.2	29.2	56.9	105.7	39.8	62.1	111.2	29.2	56.9	111.2	29.2	56.9	111.2	67.1	82.0
0960	Flooring	106.3	50.6	90.3	98.9	50.6	85.0	103.5	50.6	88.3	101.1	50.6	86.6	99.8	50.6	84.8	98.7	50.6	84.8
0970, 0990	Wall Finishes & Painting/Coating	94.1	63.3	76.2	94.1	63.3	76.2	89.6	74.2	80.6	94.1	74.2	82.5	94.1	63.3	76.2	94.1	63.3	76.2
09	FINISHES	105.6	40.8	69.7	102.9	40.8	68.5	100.8	50.6	73.0	103.2	41.9	69.3	102.3	40.8	68.3	102.0	66.8	82.6
COVERS	DIVS. 10 - 14, 25, 28, 41, 43, 44, 46	100.0	46.6	88.4	100.0	46.7	88.5	100.0	82.9	96.3	100.0	46.7	88.4	100.0	78.3	95.3	100.0	88.1	97.4
21, 22, 23	FIRE SUPPRESSION, PLUMBING & HVAC	96.8	73.8	87.1	96.8	66.0	83.8	100.0	79.0	91.1	100.2	64.7	85.2	96.8	64.7	83.3	100.2	65.7	85.6
26, 27, 3370	ELECTRICAL, COMMUNICATIONS & UTIL.	100.6	51.1	75.5	109.5	76.2	92.7	104.5	69.2	86.7	104.2	67.0	85.4	100.6	51.1	75.5	107.5	77.2	92.2
MF2014	WEIGHTED AVERAGE	101.7	64.8	85.5	102.6	65.9	86.4	101.1	75.0	89.6	101.9	65.1	85.7	101.9	64.7	85.5	101.9	76.9	90.9

City Cost Indexes

NORTH DAKOTA / OHIO

DIVISION		WILLISTON 588 MAT.	INST.	TOTAL	AKRON 442-443 MAT.	INST.	TOTAL	ATHENS 457 MAT.	INST.	TOTAL	CANTON 446-447 MAT.	INST.	TOTAL	CHILLICOTHE 456 MAT.	INST.	TOTAL	CINCINNATI 451-452 MAT.	INST.	TOTAL
015433	CONTRACTOR EQUIPMENT		98.8	98.8		91.7	91.7		87.2	87.2		91.7	91.7		97.1	97.1		96.9	96.9
0241, 31 - 34	SITE & INFRASTRUCTURE, DEMOLITION	108.3	95.2	99.0	99.0	99.1	99.0	116.1	89.5	97.3	99.1	98.9	99.0	101.0	99.8	100.2	97.8	99.6	99.1
0310	Concrete Forming & Accessories	97.0	36.4	44.6	100.9	93.6	94.6	94.9	98.4	98.0	100.9	83.5	85.8	97.3	90.9	91.8	99.3	78.2	81.0
0320	Concrete Reinforcing	103.3	96.2	99.6	100.2	92.6	96.3	94.1	87.3	90.6	100.2	76.8	88.1	91.0	80.3	85.5	96.4	79.8	87.8
0330	Cast-in-Place Concrete	107.0	67.4	91.4	91.2	98.1	94.0	111.4	96.3	105.4	92.1	95.3	93.4	101.2	98.4	100.1	93.2	91.0	92.3
03	CONCRETE	103.2	59.6	82.0	94.1	94.1	94.1	107.8	94.7	101.4	94.5	85.7	90.3	101.9	91.4	96.8	96.3	83.1	89.9
04	MASONRY	114.4	62.7	82.5	100.3	96.8	98.1	86.4	92.0	89.8	101.0	85.6	91.5	94.0	101.1	98.4	93.6	84.6	88.0
05	METALS	92.4	91.4	92.1	99.2	80.8	93.3	103.6	78.1	95.5	99.2	74.4	91.3	95.6	84.8	92.1	97.9	83.9	93.4
06	WOOD, PLASTICS & COMPOSITES	88.9	30.8	56.6	101.5	92.5	96.5	87.5	103.3	96.3	101.9	81.8	90.7	100.8	87.7	93.5	103.4	75.9	88.0
07	THERMAL & MOISTURE PROTECTION	109.7	70.8	93.4	111.7	95.8	105	101.9	96.9	99.8	112.8	91.0	103.7	103.9	96.9	101.0	101.7	83.5	94.1
08	OPENINGS	105.9	48.4	92.5	104.9	92.5	102.0	95.7	91.9	94.8	98.9	77.0	93.8	87.0	81.9	85.8	94.5	76.8	90.4
0920	Plaster & Gypsum Board	108.8	29.2	55.6	97.2	92.1	93.8	98.8	103.1	101.7	98.2	81.1	86.8	101.8	87.7	92.4	103.7	75.5	84.9
0950, 0980	Ceilings & Acoustic Treatment	111.2	29.2	56.9	94.4	92.1	92.9	105.9	103.1	104.1	94.4	81.1	85.6	100.6	87.7	92.1	101.4	75.5	84.3
0960	Flooring	102.0	50.6	87.2	90.9	90.9	90.9	124.0	95.1	115.7	91.1	78.5	87.5	101.9	93.9	99.6	103.1	88.1	98.8
0970, 0990	Wall Finishes & Painting/Coating	94.1	63.3	76.2	97.5	106.3	102.6	115.2	99.1	105.8	97.5	83.3	89.2	112.3	94.3	101.8	112.3	85.6	96.4
09	FINISHES	103.4	40.8	68.7	95.8	94.2	94.9	105.0	99.6	102.0	96.0	81.9	88.2	101.7	91.7	96.2	102.2	80.2	90.0
COVERS	DIVS. 10 - 14, 25, 28, 41, 43, 44, 46	100.0	46.7	88.5	100.0	98.3	99.6	100.0	53.2	89.9	100.0	95.5	99.0	100.0	93.3	98.5	100.0	89.2	97.7
21, 22, 23	FIRE SUPPRESSION, PLUMBING & HVAC	96.8	66.0	83.8	100.1	95.6	98.2	96.7	52.0	77.8	100.1	98.7	99.5	97.3	94.7	96.2	100.0	83.1	92.8
26, 27, 3370	ELECTRICAL, COMMUNICATIONS & UTIL.	104.6	77.2	90.7	101.4	93.5	97.4	98.0	97.4	97.7	100.7	94.9	97.7	97.4	84.9	91.1	96.5	80.2	88.3
MF2014	WEIGHTED AVERAGE	101.0	66.2	85.7	99.9	93.9	97.3	100.2	83.1	92.6	99.3	89.3	94.9	97.2	92.0	94.9	98.2	83.7	91.8

OHIO

DIVISION		CLEVELAND 441 MAT.	INST.	TOTAL	COLUMBUS 430-432 MAT.	INST.	TOTAL	DAYTON 453-454 MAT.	INST.	TOTAL	HAMILTON 450 MAT.	INST.	TOTAL	LIMA 458 MAT.	INST.	TOTAL	LORAIN 440 MAT.	INST.	TOTAL
015433	CONTRACTOR EQUIPMENT		92.2	92.2		91.9	91.9		91.2	91.2		97.1	97.1		90.1	90.1		91.7	91.7
0241, 31 - 34	SITE & INFRASTRUCTURE, DEMOLITION	98.9	99.6	99.4	99.0	95.5	96.5	96.6	99.0	98.3	96.5	99.2	98.4	109.4	89.6	95.4	98.4	100.3	99.7
0310	Concrete Forming & Accessories	100.9	98.4	98.7	98.7	81.9	84.2	99.3	77.1	80.1	99.3	78.3	81.1	94.8	85.9	87.1	101.0	85.8	87.8
0320	Concrete Reinforcing	100.7	93.0	96.7	101.2	83.1	91.9	96.4	82.1	89.0	96.4	79.8	87.8	94.1	82.2	87.9	100.2	92.9	96.4
0330	Cast-in-Place Concrete	89.5	106.7	96.3	93.0	89.5	91.6	86.8	86.3	86.6	92.9	91.2	92.3	102.4	95.8	99.8	86.8	104.3	93.7
03	CONCRETE	93.3	99.4	96.3	96.8	84.5	90.8	93.1	81.0	87.2	96.1	83.2	89.8	100.4	88.3	94.5	92.0	92.8	92.4
04	MASONRY	105.4	105.2	105.3	98.9	92.3	94.8	93.1	84.5	87.8	93.4	85.1	88.3	116.8	86.0	97.8	96.5	104.2	101.2
05	METALS	101.0	84.1	95.6	97.9	80.5	92.4	97.1	77.6	90.9	97.2	84.0	93.0	103.6	80.5	96.3	99.9	82.3	94.3
06	WOOD, PLASTICS & COMPOSITES	100.6	95.8	97.9	95.2	79.7	86.5	104.7	74.4	87.8	103.4	75.9	88.0	87.4	85.3	86.2	101.5	79.9	89.5
07	THERMAL & MOISTURE PROTECTION	110.4	107.9	109.3	97.7	93.2	95.8	107.3	86.2	98.4	104.0	83.7	95.5	101.4	93.3	98.0	112.8	102.1	108.3
08	OPENINGS	95.4	94.3	95.1	96.7	78.1	92.4	93.8	76.5	89.8	91.6	76.8	88.1	95.7	79.0	91.8	98.9	85.6	95.8
0920	Plaster & Gypsum Board	96.5	95.5	95.8	99.3	79.0	85.7	103.7	74.0	83.8	103.7	75.5	84.9	98.8	84.7	89.4	97.2	79.2	85.1
0950, 0980	Ceilings & Acoustic Treatment	92.8	95.5	94.6	96.2	79.0	84.8	102.3	74.0	83.6	101.4	75.5	84.3	105.0	84.7	91.6	94.4	79.2	84.3
0960	Flooring	90.7	107.1	95.4	93.0	86.8	91.2	105.6	98.7	103.6	103.1	88.1	98.8	122.8	88.6	113.0	91.1	107.1	95.7
0970, 0990	Wall Finishes & Painting/Coating	97.5	106.3	102.6	101.7	94.3	97.4	112.3	85.0	96.4	112.3	84.7	96.3	115.2	82.8	96.3	97.5	106.3	102.6
09	FINISHES	95.4	100.4	98.1	95.9	83.4	89.0	103.1	81.3	91.1	102.1	80.2	90.0	104.0	85.9	94.0	95.8	90.7	93.0
COVERS	DIVS. 10 - 14, 25, 28, 41, 43, 44, 46	100.0	102.2	100.5	100.0	92.8	98.4	100.0	89.0	97.6	100.0	89.4	97.7	100.0	94.6	98.8	100.0	100.1	100.0
21, 22, 23	FIRE SUPPRESSION, PLUMBING & HVAC	100.0	99.4	99.7	100.0	90.4	95.9	100.9	86.1	94.6	100.6	83.2	93.3	96.7	87.9	93.0	100.1	91.1	96.3
26, 27, 3370	ELECTRICAL, COMMUNICATIONS & UTIL.	101.0	104.4	102.7	98.2	87.1	92.6	95.2	82.3	88.7	95.5	84.8	90.1	98.3	83.3	90.7	100.8	87.3	94.0
MF2014	WEIGHTED AVERAGE	99.2	99.4	99.3	98.2	87.4	93.5	97.9	83.9	91.7	97.8	84.4	91.9	100.5	86.3	94.3	98.9	92.2	96.0

OHIO

DIVISION		MANSFIELD 448-449 MAT.	INST.	TOTAL	MARION 433 MAT.	INST.	TOTAL	SPRINGFIELD 455 MAT.	INST.	TOTAL	STEUBENVILLE 439 MAT.	INST.	TOTAL	TOLEDO 434-436 MAT.	INST.	TOTAL	YOUNGSTOWN 444-445 MAT.	INST.	TOTAL
015433	CONTRACTOR EQUIPMENT		91.7	91.7		91.4	91.4		91.2	91.2		95.9	95.9		94.7	94.7		91.7	91.7
0241, 31 - 34	SITE & INFRASTRUCTURE, DEMOLITION	95.0	98.9	97.7	94.7	94.3	94.4	96.9	97.6	97.4	138.5	104.3	114.3	98.2	96.3	96.8	98.9	99.4	99.3
0310	Concrete Forming & Accessories	90.4	82.7	83.8	95.1	80.7	82.6	99.3	81.5	83.9	96.6	88.6	89.7	98.7	96.4	96.7	100.9	86.5	88.4
0320	Concrete Reinforcing	91.2	77.2	83.9	93.3	83.1	88.0	96.4	82.0	89.0	90.9	86.9	88.8	101.2	84.5	92.6	100.2	86.9	93.3
0330	Cast-in-Place Concrete	84.5	90.9	87.0	84.9	89.4	86.7	89.1	86.2	87.9	92.2	93.8	92.8	93.0	101.1	96.1	90.3	97.1	93.0
03	CONCRETE	86.8	84.0	85.4	88.8	83.8	86.4	94.2	82.9	88.7	92.8	89.3	91.1	96.8	95.4	96.1	93.7	89.6	91.7
04	MASONRY	99.4	96.6	97.7	100.2	95.1	97.0	93.3	84.5	87.8	86.2	93.8	90.9	107.0	100.9	103.3	100.7	93.3	96.1
05	METALS	100.1	74.9	92.1	97.0	78.7	91.1	97.1	77.3	90.8	93.5	79.0	88.9	97.7	85.3	93.7	99.3	78.4	92.6
06	WOOD, PLASTICS & COMPOSITES	89.4	79.9	84.1	91.3	79.1	84.5	106.2	80.7	92.0	86.7	87.5	87.1	95.2	95.9	95.6	101.5	85.1	92.4
07	THERMAL & MOISTURE PROTECTION	112.2	94.2	104.7	97.3	98.5	97.8	107.1	86.8	98.6	109.1	95.0	103.2	99.1	101.3	100.1	113.0	93.2	104.7
08	OPENINGS	99.4	75.2	93.8	91.5	75.7	87.8	92.1	77.0	88.6	91.6	83.0	89.6	94.1	90.1	93.2	98.9	84.6	95.6
0920	Plaster & Gypsum Board	90.8	79.2	83.0	97.3	78.4	84.7	103.7	80.5	88.2	95.7	86.6	89.6	99.3	95.7	96.9	97.2	84.5	88.7
0950, 0980	Ceilings & Acoustic Treatment	95.2	79.2	84.6	96.2	78.4	84.4	102.3	80.5	87.9	93.3	86.6	88.8	96.2	95.7	95.9	94.4	84.5	87.9
0960	Flooring	86.0	103.4	91.0	91.5	103.4	94.9	105.6	98.7	103.6	118.2	96.5	112.0	91.9	105.1	95.7	91.1	90.3	90.9
0970, 0990	Wall Finishes & Painting/Coating	97.5	90.0	93.1	101.7	47.5	70.2	112.3	85.0	96.4	114.1	103.1	107.7	101.7	102.0	101.9	97.5	94.1	95.5
09	FINISHES	93.3	86.9	89.8	94.8	81.4	87.4	103.1	85.1	93.1	111.1	91.3	100.1	95.5	98.6	97.2	95.9	87.5	91.3
COVERS	DIVS. 10 - 14, 25, 28, 41, 43, 44, 46	100.0	95.9	99.1	100.0	53.2	89.9	100.0	89.7	97.8	100.0	93.0	98.5	100.0	99.6	99.9	100.0	96.1	99.2
21, 22, 23	FIRE SUPPRESSION, PLUMBING & HVAC	96.7	87.9	93.0	96.7	90.1	93.9	100.9	80.8	92.4	97.1	91.1	94.5	100.0	99.7	99.9	100.1	86.1	94.2
26, 27, 3370	ELECTRICAL, COMMUNICATIONS & UTIL.	98.5	77.8	88.0	92.9	77.8	85.2	95.2	87.1	91.1	88.2	113.3	100.9	98.3	103.9	101.1	100.8	85.3	93.0
MF2014	WEIGHTED AVERAGE	97.1	86.2	92.3	95.1	84.6	90.5	97.8	84.1	91.8	96.7	93.9	95.5	98.3	97.6	98.0	99.2	88.1	94.4

City Cost Indexes

Table 1 — OHIO / OKLAHOMA

DIVISION		ZANESVILLE 437-438 MAT.	INST.	TOTAL	ARDMORE 734 MAT.	INST.	TOTAL	CLINTON 736 MAT.	INST.	TOTAL	DURANT 747 MAT.	INST.	TOTAL	ENID 737 MAT.	INST.	TOTAL	GUYMON 739 MAT.	INST.	TOTAL
015433	CONTRACTOR EQUIPMENT		91.4	91.4		84.3	84.3		83.4	83.4		83.4	83.4		83.4	83.4		83.4	83.4
0241, 31-34	SITE & INFRASTRUCTURE, DEMOLITION	97.7	96.0	96.5	98.3	96.3	96.9	99.9	94.8	96.2	96.4	94.2	94.8	101.5	94.8	96.7	104.2	94.3	97.2
0310	Concrete Forming & Accessories	92.3	81.4	82.9	93.3	42.5	49.3	91.8	48.5	54.4	85.5	46.3	51.6	95.5	36.9	44.8	98.9	49.2	55.9
0320	Concrete Reinforcing	92.8	86.4	89.5	97.1	79.6	88.0	97.6	79.7	88.3	100.6	79.3	89.6	97.0	79.7	88.1	97.6	80.5	88.8
0330	Cast-in-Place Concrete	89.5	89.5	89.5	101.2	48.1	80.3	97.9	49.8	78.9	94.5	47.9	76.1	97.9	60.1	83.0	97.9	90.0	94.8
03	CONCRETE	92.5	84.8	88.8	97.1	51.7	75.0	96.4	55.0	76.3	92.8	53.3	73.6	96.9	53.3	75.7	99.7	69.0	84.8
04	MASONRY	97.8	87.8	91.6	101.1	71.2	82.7	126.0	71.2	92.2	93.0	70.5	79.1	107.0	71.2	84.9	103.5	91.4	96.0
05	METALS	98.4	80.3	90.9	100.9	69.5	90.9	101.0	69.5	91.0	92.6	69.4	85.3	102.5	70.1	92.2	101.6	71.3	91.9
06	WOOD, PLASTICS & COMPOSITES	87.0	79.1	82.6	100.9	36.7	65.1	99.9	44.8	69.2	92.7	41.6	64.3	103.3	28.4	61.6	106.7	44.5	72.1
07	THERMAL & MOISTURE PROTECTION	97.4	93.3	95.7	108.8	64.8	90.4	108.9	65.6	90.8	99.8	64.5	85.1	109.0	65.3	90.7	109.3	76.7	95.7
08	OPENINGS	91.5	78.6	88.5	107.9	50.2	94.5	107.9	54.6	95.5	100.5	52.8	89.4	109.1	45.7	94.4	108.1	54.4	95.6
0920	Plaster & Gypsum Board	93.3	78.4	83.4	92.9	35.3	54.5	92.6	43.7	59.9	78.3	40.5	53.0	93.3	26.8	48.9	93.3	43.4	60.0
0950, 0980	Ceilings & Acoustic Treatment	96.2	78.4	84.4	81.2	35.3	50.8	81.2	43.7	56.4	78.7	40.5	53.4	81.2	26.8	45.2	81.2	43.4	56.2
0960	Flooring	89.4	86.8	88.7	97.4	41.5	81.3	96.1	39.5	79.8	98.5	59.5	87.3	98.1	81.5	93.3	99.8	23.4	77.9
0970, 0990	Wall Finishes & Painting/Coating	101.7	94.3	97.4	87.0	62.5	72.7	87.0	62.5	72.7	95.4	62.5	76.3	87.0	62.5	72.7	87.0	62.5	72.7
09	FINISHES	93.8	83.0	87.9	87.2	42.0	62.2	86.9	46.5	64.6	86.5	48.6	65.5	87.7	45.1	64.1	88.6	45.8	64.9
COVERS	DIVS. 10 - 14, 25, 28, 41, 43, 44, 46	100.0	90.2	97.9	100.0	81.0	95.9	100.0	81.9	96.1	100.0	81.5	96.0	100.0	80.1	95.7	100.0	81.8	96.1
21, 22, 23	FIRE SUPPRESSION, PLUMBING & HVAC	96.7	90.2	93.9	96.9	69.2	85.2	96.9	69.2	85.2	96.9	69.2	85.2	100.2	69.2	87.1	96.9	68.7	85.0
26, 27, 3370	ELECTRICAL, COMMUNICATIONS & UTIL.	93.1	82.7	87.8	93.5	70.1	81.7	94.4	70.1	82.1	97.4	70.1	83.6	94.4	70.1	82.2	96.0	61.0	78.3
MF2014	WEIGHTED AVERAGE	95.6	86.3	91.5	98.6	64.9	83.7	99.8	66.2	85.0	95.5	65.9	82.5	100.2	65.2	84.8	99.7	69.1	86.2

Table 2 — OKLAHOMA

DIVISION		LAWTON 735 MAT.	INST.	TOTAL	MCALESTER 745 MAT.	INST.	TOTAL	MIAMI 743 MAT.	INST.	TOTAL	MUSKOGEE 744 MAT.	INST.	TOTAL	OKLAHOMA CITY 730-731 MAT.	INST.	TOTAL	PONCA CITY 746 MAT.	INST.	TOTAL
015433	CONTRACTOR EQUIPMENT		84.3	84.3		83.4	83.4		93.0	93.0		93.0	93.0		84.6	84.6		83.4	83.4
0241, 31-34	SITE & INFRASTRUCTURE, DEMOLITION	97.4	96.3	96.7	89.7	94.2	92.9	91.0	91.3	91.2	91.3	91.1	91.2	95.4	96.9	96.5	96.9	94.4	95.1
0310	Concrete Forming & Accessories	98.9	46.9	53.9	83.5	45.8	50.9	97.0	67.7	71.6	101.6	36.9	45.6	96.4	60.6	65.5	92.3	48.8	54.7
0320	Concrete Reinforcing	97.3	79.7	88.2	100.2	79.3	89.4	98.7	79.7	88.8	99.6	78.5	88.7	107.5	79.8	93.2	99.6	79.7	89.3
0330	Cast-in-Place Concrete	94.6	60.1	81.0	82.9	50.1	70.0	86.8	52.7	73.4	87.9	52.3	73.8	94.7	75.1	87.0	97.1	90.0	94.3
03	CONCRETE	93.2	57.8	76.0	83.4	53.8	69.0	88.0	65.4	77.0	89.8	51.3	71.1	96.6	69.0	83.2	95.0	68.7	82.2
04	MASONRY	103.4	71.2	83.5	110.9	71.1	86.4	95.9	71.2	80.7	113.0	55.7	77.7	105.9	71.2	84.5	88.5	71.2	77.8
05	METALS	106.4	70.2	94.9	92.6	68.8	85.0	92.5	83.1	89.5	94.0	80.8	89.8	99.2	70.6	90.1	92.5	70.1	85.4
06	WOOD, PLASTICS & COMPOSITES	105.8	41.8	70.1	90.3	41.6	63.2	105.2	70.6	85.9	109.6	29.7	65.1	95.5	59.8	75.6	100.8	44.5	69.5
07	THERMAL & MOISTURE PROTECTION	108.8	66.7	91.2	99.5	65.1	85.1	99.9	68.8	86.9	99.9	60.3	83.4	99.2	70.9	87.4	100.0	71.5	88.1
08	OPENINGS	111.1	53.0	97.5	100.5	52.8	89.4	100.5	68.7	93.1	101.7	46.3	88.8	107.7	62.8	97.3	100.5	54.5	89.8
0920	Plaster & Gypsum Board	95.0	40.6	58.7	77.3	40.5	52.7	84.0	70.1	74.7	85.9	28.0	47.3	100.4	59.1	72.8	83.0	43.4	56.5
0950, 0980	Ceilings & Acoustic Treatment	88.6	40.6	56.8	78.7	40.5	53.4	78.7	70.1	73.0	87.0	28.0	48.0	90.9	59.1	69.9	78.7	43.4	55.3
0960	Flooring	100.2	81.5	94.8	97.2	39.5	80.6	105.3	59.5	92.2	108.1	38.2	88.0	96.9	81.5	92.5	102.0	39.5	84.0
0970, 0990	Wall Finishes & Painting/Coating	87.0	62.5	72.7	95.4	62.5	76.3	95.4	62.5	76.3	95.4	62.5	76.3	88.2	62.5	73.3	95.4	62.5	76.3
09	FINISHES	89.7	53.0	69.4	85.5	44.6	62.9	88.7	65.9	76.1	91.7	37.8	61.9	91.3	63.8	76.1	88.4	46.4	65.1
COVERS	DIVS. 10 - 14, 25, 28, 41, 43, 44, 46	100.0	81.5	96.0	100.0	81.5	96.0	100.0	85.2	96.8	100.0	80.6	95.8	100.0	83.5	96.4	100.0	81.8	96.1
21, 22, 23	FIRE SUPPRESSION, PLUMBING & HVAC	100.2	69.2	87.1	96.9	65.2	83.5	96.9	65.4	83.6	100.0	65.3	85.5	100.0	69.3	87.0	96.9	65.4	83.6
26, 27, 3370	ELECTRICAL, COMMUNICATIONS & UTIL.	96.0	70.4	83.0	95.9	75.2	85.4	97.2	75.3	86.1	95.5	75.3	85.3	100.9	70.1	85.3	95.5	71.8	83.5
MF2014	WEIGHTED AVERAGE	100.7	67.5	86.1	94.8	65.3	81.8	95.2	72.1	85.0	97.5	62.9	82.3	99.8	71.3	87.3	95.5	67.6	83.2

Table 3 — OKLAHOMA / OREGON

DIVISION		POTEAU 749 MAT.	INST.	TOTAL	SHAWNEE 748 MAT.	INST.	TOTAL	TULSA 740-741 MAT.	INST.	TOTAL	WOODWARD 738 MAT.	INST.	TOTAL	BEND 977 MAT.	INST.	TOTAL	EUGENE 974 MAT.	INST.	TOTAL
015433	CONTRACTOR EQUIPMENT		91.9	91.9		83.4	83.4		93.0	93.0		83.4	83.4		98.4	98.4		98.4	98.4
0241, 31-34	SITE & INFRASTRUCTURE, DEMOLITION	77.7	89.2	85.9	100.1	94.4	96.1	97.7	91.3	93.2	100.2	94.8	96.4	109.2	102.6	104.5	98.6	102.6	101.4
0310	Concrete Forming & Accessories	90.2	43.3	49.6	85.4	46.1	51.4	101.7	41.5	49.6	92.0	49.1	54.9	108.4	99.7	100.9	104.9	99.6	100.3
0320	Concrete Reinforcing	100.7	79.7	89.8	99.6	79.3	89.1	99.7	79.7	89.4	97.0	79.8	88.1	100.8	98.5	99.6	105.1	98.5	101.7
0330	Cast-in-Place Concrete	86.8	50.2	72.4	100.2	48.1	79.6	95.7	49.6	77.5	97.9	90.0	94.8	102.9	102.2	102.6	99.6	102.2	100.6
03	CONCRETE	90.2	53.7	72.4	96.7	53.3	75.6	95.2	52.6	74.5	96.7	68.9	83.2	109.8	100.0	105.0	101.1	100.0	100.5
04	MASONRY	96.2	71.2	80.8	112.3	71.2	86.9	96.8	71.2	80.8	96.1	71.2	80.8	110.2	103.2	105.9	107.1	103.2	104.7
05	METALS	92.6	82.8	89.5	92.5	69.3	85.1	97.2	82.8	92.6	101.1	70.4	91.3	93.3	95.6	94.0	94.0	95.5	94.5
06	WOOD, PLASTICS & COMPOSITES	97.5	37.9	64.3	92.6	41.6	64.2	108.8	35.5	68.0	100.1	44.5	69.1	98.3	99.5	99.0	94.4	99.5	97.2
07	THERMAL & MOISTURE PROTECTION	100.0	65.1	85.4	100.0	65.4	85.5	100.0	66.8	86.1	109.0	71.5	93.3	109.3	97.8	104.5	108.4	96.2	103.3
08	OPENINGS	100.5	50.9	88.9	100.5	52.8	89.4	103.5	49.6	90.9	107.9	54.5	95.5	100.2	102.5	100.7	100.5	102.5	100.9
0920	Plaster & Gypsum Board	81.3	36.5	51.4	78.3	40.5	53.0	85.9	34.0	51.2	92.6	43.4	59.7	112.3	99.3	103.6	110.7	99.3	103.1
0950, 0980	Ceilings & Acoustic Treatment	78.7	36.5	50.8	78.7	40.5	53.4	87.0	34.0	51.9	81.2	43.4	56.2	85.2	99.3	94.5	86.2	99.3	94.8
0960	Flooring	101.2	59.5	89.2	98.5	31.5	79.2	106.9	55.7	92.1	96.1	41.5	80.4	107.0	101.2	105.4	105.2	101.2	104.1
0970, 0990	Wall Finishes & Painting/Coating	95.4	62.5	76.3	95.4	62.5	76.3	95.4	62.5	76.3	87.0	62.5	72.7	94.9	75.8	83.8	94.9	75.8	83.8
09	FINISHES	86.3	46.4	64.2	86.8	43.3	62.7	91.5	44.2	65.3	87.0	46.7	64.7	97.4	96.7	97.1	98.9	97.4	98.1
COVERS	DIVS. 10 - 14, 25, 28, 41, 43, 44, 46	100.0	81.4	96.0	100.0	81.5	96.0	100.0	81.2	95.9	100.0	81.8	96.1	100.0	102.6	100.6	100.0	102.6	100.6
21, 22, 23	FIRE SUPPRESSION, PLUMBING & HVAC	96.9	65.4	83.6	96.9	69.2	85.2	100.2	65.2	85.4	96.9	69.3	85.2	96.6	99.9	98.0	99.9	99.8	99.9
26, 27, 3370	ELECTRICAL, COMMUNICATIONS & UTIL.	95.6	75.3	85.3	97.4	70.1	83.6	97.4	64.1	80.5	95.8	70.1	82.8	99.7	95.4	97.5	98.4	95.4	96.9
MF2014	WEIGHTED AVERAGE	94.7	66.4	82.2	97.0	65.3	83.0	98.3	64.5	83.4	98.6	68.4	85.3	100.3	99.2	99.8	99.5	99.2	99.3

For customer support on your Mechanical Cost Data, call 877.739.3689.

683

City Cost Indexes

OREGON

DIVISION		KLAMATH FALLS 976 MAT.	INST.	TOTAL	MEDFORD 975 MAT.	INST.	TOTAL	PENDLETON 978 MAT.	INST.	TOTAL	PORTLAND 970-972 MAT.	INST.	TOTAL	SALEM 973 MAT.	INST.	TOTAL	VALE 979 MAT.	INST.	TOTAL
015433	CONTRACTOR EQUIPMENT		98.4	98.4		98.4	98.4		95.8	95.8		98.4	98.4		98.4	98.4		95.8	95.8
0241, 31 - 34	SITE & INFRASTRUCTURE, DEMOLITION	113.5	102.5	105.7	106.7	102.5	103.8	104.3	95.7	98.2	100.9	102.6	102.1	93.0	102.6	99.8	91.2	95.6	94.3
0310	Concrete Forming & Accessories	101.3	99.4	99.7	100.3	99.4	99.5	101.7	99.8	100.0	106.1	99.7	100.6	105.2	99.7	100.4	108.2	98.8	100.1
0320	Concrete Reinforcing	100.8	98.5	99.6	102.5	98.5	100.4	100.3	98.6	99.4	105.9	98.6	102.1	113.5	98.5	105.8	97.9	98.4	98.1
0330	Cast-in-Place Concrete	102.9	105.1	103.8	102.9	102.1	102.6	103.6	103.5	103.6	102.3	102.2	102.3	94.1	102.2	97.3	81.6	106.2	91.3
03	CONCRETE	112.6	100.9	106.9	107.2	99.8	103.6	93.4	100.5	96.9	102.6	100.1	101.4	98.7	100.0	99.4	78.4	100.9	89.4
04	MASONRY	125.3	103.2	111.7	104.1	103.2	103.5	114.9	103.3	107.7	109.0	103.2	105.4	114.8	103.2	107.6	113.1	103.3	107.1
05	METALS	93.3	95.3	94.0	93.6	95.3	94.1	100.0	96.1	98.7	95.1	95.6	95.2	100.7	95.6	99.1	99.9	95.0	98.3
06	WOOD, PLASTICS & COMPOSITES	89.5	99.5	95.1	88.5	99.5	94.6	91.3	99.6	95.9	95.2	99.5	97.6	91.4	99.5	95.9	99.5	99.6	99.6
07	THERMAL & MOISTURE PROTECTION	109.5	94.0	103.0	109.1	94.0	102.8	102.4	97.4	100.4	108.4	97.8	103.9	105.1	97.8	102.0	101.8	94.6	98.8
08	OPENINGS	100.2	102.5	100.7	103.3	102.5	103.1	96.9	102.5	98.2	98.2	102.5	99.2	100.1	102.5	100.6	96.8	94.0	96.2
0920	Plaster & Gypsum Board	107.4	99.3	102.0	106.8	99.3	101.8	93.8	99.3	97.5	110.2	99.3	102.9	106.7	99.3	101.7	99.8	99.3	99.4
0950, 0980	Ceilings & Acoustic Treatment	92.7	99.3	97.0	99.1	99.3	99.2	60.3	99.3	86.1	88.0	99.3	95.5	93.1	99.3	97.2	60.3	99.3	86.1
0960	Flooring	103.8	101.2	103.1	103.2	101.2	102.6	72.2	101.2	80.6	102.8	101.2	102.4	103.2	101.2	102.6	74.4	101.2	82.1
0970, 0990	Wall Finishes & Painting/Coating	94.9	70.4	80.6	94.9	70.4	80.6	87.3	72.8	78.9	94.8	72.8	82.0	93.3	75.8	83.1	87.3	75.8	80.6
09	FINISHES	101.1	96.9	98.7	101.4	96.9	98.9	71.4	97.2	85.7	98.6	97.1	97.8	97.4	97.4	97.4	71.9	97.5	86.1
COVERS	DIVS. 10 - 14, 25, 28, 41, 43, 44, 46	100.0	100.6	100.0	100.0	102.6	100.6	100.0	88.4	97.5	100.0	102.7	100.6	100.0	102.6	100.6	100.0	102.8	100.6
21, 22, 23	FIRE SUPPRESSION, PLUMBING & HVAC	96.6	99.8	98.0	99.9	99.9	99.9	98.5	112.1	104.3	99.9	99.9	99.9	100.0	99.9	99.9	98.5	72.4	87.5
26, 27, 3370	ELECTRICAL, COMMUNICATIONS & UTIL.	98.5	80.1	89.2	101.8	80.1	90.8	91.4	98.9	95.2	98.6	109.2	103.9	105.2	95.4	100.2	91.4	70.6	80.9
MF2014	WEIGHTED AVERAGE	101.3	97.1	99.4	101.0	96.9	99.2	96.2	101.4	98.5	99.7	101.1	100.3	100.9	99.2	100.2	94.1	89.1	91.9

PENNSYLVANIA

DIVISION		ALLENTOWN 181 MAT.	INST.	TOTAL	ALTOONA 166 MAT.	INST.	TOTAL	BEDFORD 155 MAT.	INST.	TOTAL	BRADFORD 167 MAT.	INST.	TOTAL	BUTLER 160 MAT.	INST.	TOTAL	CHAMBERSBURG 172 MAT.	INST.	TOTAL
015433	CONTRACTOR EQUIPMENT		112.6	112.6		112.6	112.6		111.0	111.0		112.6	112.6		112.6	112.6		111.7	111.7
0241, 31 - 34	SITE & INFRASTRUCTURE, DEMOLITION	93.6	101.9	99.5	96.7	101.4	100.1	105.3	99.2	101.0	92.5	100.5	98.1	88.0	103.0	98.6	89.7	99.0	96.3
0310	Concrete Forming & Accessories	100.2	113.7	111.8	85.1	79.8	80.5	83.8	82.7	82.9	87.2	83.8	84.3	86.6	95.9	94.6	87.9	82.1	82.9
0320	Concrete Reinforcing	97.9	108.9	103.6	94.9	98.1	96.6	94.9	98.0	96.5	96.9	97.9	97.4	95.6	110.7	103.4	92.2	107.6	100.2
0330	Cast-in-Place Concrete	89.2	105.9	95.8	99.4	82.7	92.8	117.7	87.5	105.8	94.9	89.2	92.7	87.8	99.1	92.3	96.7	70.2	86.3
03	CONCRETE	94.4	111.0	102.5	90.6	85.8	88.3	106.0	88.7	97.6	96.4	89.8	93.2	82.5	101.0	91.5	103.3	84.5	94.2
04	MASONRY	98.4	103.6	101.6	101.8	78.2	87.2	105.1	87.2	94.1	99.0	87.5	91.9	104.0	99.5	101.2	106.9	89.3	96.0
05	METALS	97.1	125.7	106.2	91.4	116.9	99.5	99.7	114.6	104.4	95.0	116.5	101.8	91.1	124.2	101.6	93.3	121.4	102.2
06	WOOD, PLASTICS & COMPOSITES	101.3	115.9	109.4	80.2	82.3	81.4	87.2	82.2	84.4	86.6	82.3	84.2	81.6	94.9	89.0	87.1	82.3	84.4
07	THERMAL & MOISTURE PROTECTION	104.0	117.8	109.8	102.8	90.9	97.9	101.9	93.5	98.4	104.0	94.5	100.0	102.5	100.2	101.6	96.1	80.1	89.4
08	OPENINGS	95.1	114.4	99.6	88.7	89.7	89.0	94.9	89.7	93.7	95.1	89.7	93.9	88.7	104.0	92.3	94.6	84.3	92.2
0920	Plaster & Gypsum Board	97.5	116.2	110.0	89.2	81.6	84.1	94.4	81.6	85.9	90.5	81.6	84.6	89.2	94.5	92.8	101.6	81.6	88.3
0950, 0980	Ceilings & Acoustic Treatment	82.8	116.2	104.9	86.9	81.6	83.4	98.6	81.6	87.4	86.2	81.6	83.2	87.8	94.5	92.3	83.2	81.6	82.1
0960	Flooring	92.6	97.4	94.0	86.3	100.0	90.3	91.6	103.0	94.9	86.8	103.0	91.5	87.2	104.2	92.1	94.9	44.4	80.3
0970, 0990	Wall Finishes & Painting/Coating	93.2	110.5	103.3	89.1	112.0	102.4	98.0	112.0	106.1	93.2	112.0	104.1	89.1	112.0	102.4	91.7	112.0	103.5
09	FINISHES	90.7	110.6	101.7	88.8	86.0	87.3	96.3	87.4	91.4	88.8	88.1	88.4	88.8	97.4	93.6	90.3	77.4	83.2
COVERS	DIVS. 10 - 14, 25, 28, 41, 43, 44, 46	100.0	105.8	101.3	100.0	95.4	99.0	100.0	98.5	99.7	100.0	99.4	99.9	100.0	102.6	100.6	100.0	96.8	99.3
21, 22, 23	FIRE SUPPRESSION, PLUMBING & HVAC	100.3	115.8	106.9	99.7	86.8	94.3	96.7	89.4	93.6	97.0	96.2	96.6	96.4	96.2	96.3	96.8	91.8	94.7
26, 27, 3370	ELECTRICAL, COMMUNICATIONS & UTIL.	97.6	100.0	98.8	88.8	111.3	100.2	94.0	111.2	102.7	91.9	111.2	101.7	89.4	111.3	100.4	88.2	89.9	89.1
MF2014	WEIGHTED AVERAGE	97.3	110.6	103.2	94.1	93.5	93.9	98.7	95.3	97.1	95.6	97.4	96.4	92.3	103.2	97.1	95.8	91.2	93.8

PENNSYLVANIA

DIVISION		DOYLESTOWN 189 MAT.	INST.	TOTAL	DUBOIS 158 MAT.	INST.	TOTAL	ERIE 164-165 MAT.	INST.	TOTAL	GREENSBURG 156 MAT.	INST.	TOTAL	HARRISBURG 170-171 MAT.	INST.	TOTAL	HAZLETON 182 MAT.	INST.	TOTAL
015433	CONTRACTOR EQUIPMENT		92.2	92.2		111.0	111.0		112.6	112.6		111.0	111.0		111.7	111.7		112.6	112.6
0241, 31 - 34	SITE & INFRASTRUCTURE, DEMOLITION	106.8	87.4	93.1	110.1	99.8	102.8	93.6	102.2	99.7	101.5	102.1	102.0	89.7	100.8	97.5	86.9	101.8	97.5
0310	Concrete Forming & Accessories	84.1	130.5	124.3	83.3	83.9	83.8	99.6	87.5	89.2	90.1	95.9	95.1	97.6	88.2	89.4	81.7	89.8	88.7
0320	Concrete Reinforcing	94.7	138.0	117.2	94.3	98.2	96.3	96.9	99.7	98.3	94.3	110.7	102.8	101.1	107.9	104.6	95.0	108.7	102.1
0330	Cast-in-Place Concrete	84.3	130.9	102.7	113.5	95.0	106.2	97.7	90.6	94.9	109.2	98.9	105.1	95.1	98.6	96.5	84.3	100.8	90.8
03	CONCRETE	89.9	131.6	110.2	107.0	92.1	99.8	89.7	92.3	91.0	100.7	100.9	100.8	97.8	96.9	97.4	87.1	98.5	92.7
04	MASONRY	101.8	130.4	119.4	105.4	91.1	96.6	90.4	93.0	92.0	115.3	99.6	105.6	105.3	91.3	96.7	111.6	99.1	103.9
05	METALS	94.8	127.1	105.1	99.7	122.0	106.8	91.6	117.2	99.8	99.6	123.2	107.1	99.7	123.7	107.3	96.8	122.9	105.1
06	WOOD, PLASTICS & COMPOSITES	82.2	131.5	109.7	86.2	82.2	84.0	97.4	85.6	90.9	93.5	94.8	94.2	93.6	87.1	90.0	80.8	87.1	84.3
07	THERMAL & MOISTURE PROTECTION	101.5	132.1	114.3	102.2	95.9	99.6	103.3	94.8	99.8	101.8	100.2	101.1	98.8	109.0	103.0	103.5	109.2	105.9
08	OPENINGS	97.4	139.2	107.1	94.9	93.4	94.5	88.9	90.2	89.2	94.8	103.9	97.0	99.2	93.5	97.9	95.6	92.3	94.8
0920	Plaster & Gypsum Board	87.8	132.3	117.5	93.4	81.6	85.5	97.5	85.1	89.2	95.5	94.6	94.9	105.3	86.6	92.8	88.2	86.6	87.1
0950, 0980	Ceilings & Acoustic Treatment	82.0	132.3	115.3	98.6	81.6	87.4	82.8	85.1	84.3	97.8	94.6	95.7	90.8	86.6	88.0	83.7	86.6	85.6
0960	Flooring	77.0	138.7	94.7	91.4	103.0	94.7	92.9	103.0	95.8	95.5	104.2	98.0	100.1	92.1	97.8	84.0	96.1	87.5
0970, 0990	Wall Finishes & Painting/Coating	92.8	151.1	126.7	98.0	112.0	106.1	98.2	97.5	97.8	98.0	106.6	103.0	92.2	99.2	96.3	93.2	115.7	106.3
09	FINISHES	82.2	132.5	109.0	96.5	89.0	92.3	91.7	90.7	91.1	97.1	97.4	97.3	94.9	89.2	91.7	86.6	91.3	89.2
COVERS	DIVS. 10 - 14, 25, 28, 41, 43, 44, 46	100.0	118.5	104.0	100.0	99.1	99.8	100.0	100.7	100.1	100.0	102.5	100.5	100.0	98.0	99.6	100.0	101.4	100.3
21, 22, 23	FIRE SUPPRESSION, PLUMBING & HVAC	96.4	132.1	111.5	96.7	90.3	94.0	99.7	97.6	98.8	96.7	95.1	96.1	99.9	92.9	97.0	97.0	99.8	98.2
26, 27, 3370	ELECTRICAL, COMMUNICATIONS & UTIL.	91.4	129.7	110.7	94.5	111.3	103.0	90.4	99.2	94.9	94.5	111.3	103.0	94.6	89.9	92.2	92.6	97.5	95.1
MF2014	WEIGHTED AVERAGE	94.8	127.4	109.1	99.1	97.5	98.4	94.0	97.5	95.5	98.7	102.8	100.5	98.5	96.5	97.6	95.2	100.3	97.5

For customer support on your Mechanical Cost Data, call 877.739.3689.

City Cost Indexes

PENNSYLVANIA

DIVISION		INDIANA 157			JOHNSTOWN 159			KITTANNING 162			LANCASTER 175 - 176			LEHIGH VALLEY 180			MONTROSE 188		
		MAT.	INST.	TOTAL	MAT.	INST.	TOTAL	MAT.	INST.	TOTAL	MAT.	INST.	TOTAL	MAT.	INST.	TOTAL	MAT.	INST.	TOTAL
015433	CONTRACTOR EQUIPMENT		111.0	111.0		111.0	111.0		112.6	112.6		111.7	111.7		112.6	112.6		112.6	112.6
0241, 31 - 34	SITE & INFRASTRUCTURE, DEMOLITION	99.5	100.1	99.9	105.9	101.3	102.6	90.6	102.7	99.1	81.8	100.9	95.3	90.8	101.3	98.2	89.5	100.9	97.6
0310	Concrete Forming & Accessories	84.4	93.1	91.9	83.3	84.2	84.1	86.6	95.5	94.3	90.0	89.1	89.2	93.6	109.5	107.3	82.7	87.5	86.9
0320	Concrete Reinforcing	93.6	110.7	102.4	94.9	110.4	103.0	95.6	110.8	103.4	91.9	107.9	100.2	95.0	106.8	101.1	99.6	107.9	103.9
0330	Cast-in-Place Concrete	107.1	97.7	103.4	118.7	89.1	107.1	91.2	98.2	94.0	82.3	99.7	89.2	91.2	103.5	96.1	89.4	92.2	90.5
03	CONCRETE	98.2	99.2	98.7	106.8	92.2	99.7	85.1	100.5	92.6	90.7	97.7	94.1	93.6	107.9	100.5	92.2	94.5	93.3
04	MASONRY	101.6	100.0	100.6	102.5	90.3	95.0	106.8	100.0	102.6	113.4	93.8	101.3	98.4	100.6	99.7	98.4	97.3	97.7
05	METALS	99.7	122.4	106.9	99.7	121.5	106.6	91.2	123.8	101.6	93.3	123.7	102.9	96.7	122.5	104.9	95.1	123.2	104.0
06	WOOD, PLASTICS & COMPOSITES	88.0	92.2	90.3	86.2	82.2	84.0	81.6	94.9	89.0	89.8	87.1	88.3	93.1	111.9	103.6	81.6	85.4	83.7
07	THERMAL & MOISTURE PROTECTION	101.7	100.1	101.0	101.9	94.4	98.8	102.6	100.4	101.7	95.6	96.2	95.9	103.9	100.5	102.5	103.6	90.6	98.2
08	OPENINGS	94.9	98.8	95.8	94.9	93.4	94.5	88.7	104.0	92.3	94.6	97.9	95.3	95.6	111.7	99.3	92.3	91.3	92.1
0920	Plaster & Gypsum Board	94.7	91.9	92.8	93.2	81.6	85.5	89.2	94.5	92.8	103.6	86.6	92.2	91.2	112.1	105.1	88.8	84.8	86.2
0950, 0980	Ceilings & Acoustic Treatment	98.6	91.9	94.2	97.8	81.6	87.1	87.8	94.5	92.3	83.2	86.6	85.4	83.7	112.1	102.5	86.2	84.8	85.3
0960	Flooring	92.4	103.0	95.4	91.4	103.0	94.7	87.2	103.0	91.7	95.9	103.0	97.9	89.5	92.1	90.3	84.7	103.0	90.0
0970, 0990	Wall Finishes & Painting/Coating	98.0	112.0	106.1	98.0	112.0	106.1	89.1	112.0	102.4	91.7	99.2	96.1	93.2	66.4	77.7	93.2	115.7	106.3
09	FINISHES	96.2	95.5	95.8	95.9	89.4	92.3	88.9	97.0	93.4	90.3	91.6	91.0	89.0	101.1	95.7	87.7	92.5	90.4
COVERS	DIVS. 10 - 14, 25, 28, 41, 43, 44, 46	100.0	101.8	100.4	100.0	99.2	99.8	100.0	102.2	100.5	100.0	98.8	99.7	100.0	103.1	100.7	100.0	100.6	100.1
21, 22, 23	FIRE SUPPRESSION, PLUMBING & HVAC	96.7	92.8	95.1	96.7	93.0	95.2	96.4	96.5	96.4	96.8	94.2	95.7	97.0	110.8	102.8	97.0	98.1	97.5
26, 27, 3370	ELECTRICAL, COMMUNICATIONS & UTIL.	94.5	111.3	103.0	94.5	111.2	103.0	88.8	111.3	100.2	89.5	97.1	93.3	92.6	142.3	117.8	91.9	97.4	94.7
MF2014	WEIGHTED AVERAGE	97.6	101.4	99.2	98.7	98.1	98.5	92.8	103.1	97.3	94.6	98.3	96.2	95.7	112.3	103.0	94.6	98.7	96.4

PENNSYLVANIA

DIVISION		NEW CASTLE 161			NORRISTOWN 194			OIL CITY 163			PHILADELPHIA 190 - 191			PITTSBURGH 150 - 152			POTTSVILLE 179		
		MAT.	INST.	TOTAL	MAT.	INST.	TOTAL	MAT.	INST.	TOTAL	MAT.	INST.	TOTAL	MAT.	INST.	TOTAL	MAT.	INST.	TOTAL
015433	CONTRACTOR EQUIPMENT		112.6	112.6		96.7	96.7		112.6	112.6		96.1	96.1		112.1	112.1		111.7	111.7
0241, 31 - 34	SITE & INFRASTRUCTURE, DEMOLITION	88.4	103.0	98.7	96.3	99.0	98.2	87.0	100.8	96.8	99.4	98.6	98.9	105.1	104.1	104.4	84.7	100.5	95.9
0310	Concrete Forming & Accessories	86.6	97.7	96.2	85.3	131.5	125.3	86.6	92.6	91.8	101.8	139.1	134.1	97.5	98.3	98.2	81.5	89.1	88.1
0320	Concrete Reinforcing	94.4	94.1	94.2	96.2	138.2	118.0	95.6	93.9	94.7	101.7	138.2	120.6	95.4	110.7	103.3	91.1	107.9	99.8
0330	Cast-in-Place Concrete	88.6	96.9	91.9	81.2	129.2	100.1	86.1	94.7	89.5	90.1	130.8	106.2	113.5	99.0	107.8	87.6	100.2	92.6
03	CONCRETE	82.9	98.0	90.2	88.6	131.5	109.4	81.3	94.8	87.9	96.1	135.5	115.3	104.2	102.0	103.1	94.5	97.9	96.1
04	MASONRY	103.5	99.6	101.1	115.4	130.9	125.0	103.0	95.8	98.6	100.3	132.2	120.0	96.7	99.6	98.5	106.2	93.8	98.6
05	METALS	91.2	116.5	99.3	97.4	128.2	107.2	91.2	114.5	98.6	101.2	129.0	110.0	101.2	123.6	108.3	93.5	122.9	102.8
06	WOOD, PLASTICS & COMPOSITES	81.6	97.6	90.5	84.7	131.4	110.7	81.6	92.3	87.5	102.3	140.5	123.6	102.7	97.6	99.9	79.8	87.1	83.9
07	THERMAL & MOISTURE PROTECTION	102.5	97.5	100.4	107.5	134.7	118.9	102.4	98.4	100.7	102.3	137.2	116.9	102.0	101.6	101.9	95.7	108.8	101.2
08	OPENINGS	88.7	97.7	90.8	88.3	139.2	100.2	88.7	86.9	88.3	101.0	145.6	111.4	98.6	105.4	100.2	94.6	91.5	93.9
0920	Plaster & Gypsum Board	89.2	97.4	94.7	91.1	132.3	118.6	89.2	91.9	91.0	103.3	141.6	128.9	99.8	97.4	98.2	98.0	86.6	90.4
0950, 0980	Ceilings & Acoustic Treatment	87.8	97.4	94.2	85.5	132.3	116.5	87.8	91.9	90.5	94.8	141.6	125.8	98.6	97.4	97.8	83.2	86.6	85.4
0960	Flooring	87.2	103.0	91.7	87.2	138.7	102.0	87.2	103.0	91.7	97.4	138.7	109.3	99.5	104.2	100.8	91.8	103.0	95.0
0970, 0990	Wall Finishes & Painting/Coating	89.1	112.0	102.4	90.5	151.1	125.8	89.1	112.0	102.4	92.8	154.7	128.8	98.0	114.7	107.7	91.7	115.7	105.7
09	FINISHES	88.8	100.0	95.0	88.2	134.4	113.7	88.7	95.0	92.2	98.4	140.9	121.9	99.3	100.5	99.9	88.5	93.4	91.2
COVERS	DIVS. 10 - 14, 25, 28, 41, 43, 44, 46	100.0	102.9	100.6	100.0	118.4	104.0	100.0	101.9	100.4	100.0	120.4	104.4	100.0	102.8	100.6	100.0	100.4	100.1
21, 22, 23	FIRE SUPPRESSION, PLUMBING & HVAC	96.4	93.2	95.1	96.8	132.5	111.9	96.4	96.1	96.3	100.3	133.2	114.2	100.1	99.7	99.9	96.8	98.3	97.4
26, 27, 3370	ELECTRICAL, COMMUNICATIONS & UTIL.	89.4	102.8	96.2	93.4	147.2	120.6	91.0	111.2	101.2	96.2	156.2	126.5	96.5	111.3	104.0	87.8	94.8	91.3
MF2014	WEIGHTED AVERAGE	92.4	100.3	95.9	95.4	131.2	111.2	92.3	99.7	95.6	99.5	134.7	115.0	100.2	104.6	102.1	94.4	99.1	96.5

PENNSYLVANIA

DIVISION		READING 195 - 196			SCRANTON 184 - 185			STATE COLLEGE 168			STROUDSBURG 183			SUNBURY 178			UNIONTOWN 154		
		MAT.	INST.	TOTAL	MAT.	INST.	TOTAL	MAT.	INST.	TOTAL	MAT.	INST.	TOTAL	MAT.	INST.	TOTAL	MAT.	INST.	TOTAL
015433	CONTRACTOR EQUIPMENT		117.5	117.5		112.6	112.6		111.7	111.7		112.6	112.6		112.6	112.6		111.0	111.0
0241, 31 - 34	SITE & INFRASTRUCTURE, DEMOLITION	100.8	111.6	108.4	94.1	102.1	99.8	84.5	100.0	95.5	88.5	100.9	97.2	96.8	101.7	100.3	100.1	102.1	101.6
0310	Concrete Forming & Accessories	101.7	90.2	91.7	100.4	88.6	90.2	85.2	79.5	80.3	88.1	89.4	89.3	93.9	87.9	88.7	77.6	95.8	93.3
0320	Concrete Reinforcing	97.6	106.4	102.1	97.9	106.0	102.1	96.2	98.2	97.2	98.2	108.4	103.5	93.7	107.8	101.0	94.3	110.7	102.8
0330	Cast-in-Place Concrete	72.8	100.0	83.6	93.1	101.2	96.3	89.8	74.7	83.9	87.8	74.9	82.7	95.8	94.8	95.4	107.1	98.8	103.8
03	CONCRETE	87.7	98.0	92.7	96.4	97.6	97.0	96.3	83.4	89.9	91.0	89.5	90.3	98.3	95.5	97.0	97.8	100.8	99.3
04	MASONRY	104.1	95.7	98.9	98.8	99.2	99.1	104.4	79.5	89.0	96.3	102.4	100.1	106.8	91.4	97.3	116.3	99.5	106.0
05	METALS	97.8	124.0	106.1	99.3	122.0	106.5	94.8	116.7	101.7	96.8	122.1	104.9	93.2	123.3	102.8	99.4	123.1	106.9
06	WOOD, PLASTICS & COMPOSITES	102.7	86.9	93.9	101.3	85.4	92.4	88.5	82.3	85.0	87.5	85.4	86.3	87.6	87.1	87.3	80.6	94.8	88.5
07	THERMAL & MOISTURE PROTECTION	108.0	112.0	109.7	103.9	97.9	101.4	103.2	88.2	96.9	103.7	84.7	95.8	96.8	103.7	99.7	101.6	100.2	101.0
08	OPENINGS	92.8	97.8	94.0	95.1	90.4	94.0	92.2	89.7	91.6	95.6	84.9	93.1	94.7	92.0	94.1	94.8	103.9	96.9
0920	Plaster & Gypsum Board	102.0	86.6	91.7	99.5	84.8	89.7	91.0	81.6	84.7	89.8	84.8	86.5	97.2	86.6	90.1	91.5	94.5	93.6
0950, 0980	Ceilings & Acoustic Treatment	78.8	86.6	84.0	91.0	84.8	86.9	82.9	81.6	82.0	82.0	84.8	83.9	79.8	86.6	84.3	97.8	94.5	95.7
0960	Flooring	92.2	138.7	105.6	92.6	96.1	93.6	90.0	103.0	93.7	87.3	97.4	90.2	92.6	103.0	95.6	88.4	104.2	92.9
0970, 0990	Wall Finishes & Painting/Coating	89.3	109.7	101.1	93.2	115.7	106.3	93.2	112.0	104.1	93.2	115.7	106.3	91.7	115.7	105.7	91.7	112.0	106.1
09	FINISHES	90.1	99.9	95.5	92.8	91.8	92.2	88.1	86.3	87.1	87.6	91.7	89.9	89.4	92.8	91.3	94.3	97.3	96.0
COVERS	DIVS. 10 - 14, 25, 28, 41, 43, 44, 46	100.0	100.2	100.0	100.0	101.2	100.3	100.0	93.6	98.6	100.0	103.0	100.7	100.0	99.6	99.9	100.0	102.5	100.5
21, 22, 23	FIRE SUPPRESSION, PLUMBING & HVAC	100.3	111.1	104.8	100.3	99.8	100.1	97.0	86.7	92.6	97.0	102.1	99.2	96.8	92.8	95.1	96.7	95.1	96.1
26, 27, 3370	ELECTRICAL, COMMUNICATIONS & UTIL.	99.3	97.1	98.2	97.6	97.5	97.5	91.1	111.2	101.3	92.6	142.5	117.8	88.1	94.8	91.5	91.9	111.3	101.7
MF2014	WEIGHTED AVERAGE	97.1	104.5	100.4	98.1	99.8	98.8	95.1	93.0	94.2	95.1	105.0	99.4	95.3	97.3	96.2	97.7	102.8	99.9

PENNSYLVANIA

DIVISION		WASHINGTON 153 MAT.	INST.	TOTAL	WELLSBORO 169 MAT.	INST.	TOTAL	WESTCHESTER 193 MAT.	INST.	TOTAL	WILKES-BARRE 186-187 MAT.	INST.	TOTAL	WILLIAMSPORT 177 MAT.	INST.	TOTAL	YORK 173-174 MAT.	INST.	TOTAL
015433	CONTRACTOR EQUIPMENT		111.0	111.0		112.6	112.6		96.7	96.7		112.6	112.6		112.6	112.6		111.7	111.7
0241, 31-34	SITE & INFRASTRUCTURE, DEMOLITION	100.2	102.1	101.6	96.0	100.7	99.3	102.2	96.4	98.1	86.5	102.1	97.6	87.9	100.7	97.0	85.5	100.9	96.4
0310	Concrete Forming & Accessories	84.5	95.9	94.4	86.6	86.2	86.2	92.2	131.1	125.9	90.7	90.3	90.3	90.5	88.3	88.6	84.9	89.1	88.5
0320	Concrete Reinforcing	94.3	110.8	102.8	96.2	107.8	102.2	95.3	138.0	117.4	96.9	108.7	103.0	93.0	108.4	101.0	93.7	107.9	101.1
0330	Cast-in-Place Concrete	107.1	98.9	103.9	94.1	90.1	92.5	90.0	131.7	106.4	84.3	101.4	91.0	81.1	92.6	85.6	88.1	99.7	92.7
03	CONCRETE	98.3	100.9	99.6	98.7	93.1	96.0	96.3	132.1	113.7	88.0	98.9	93.3	85.8	95.1	90.3	95.7	97.7	96.7
04	MASONRY	101.9	101.2	101.4	105.3	91.9	97.0	109.7	133.3	124.2	112.0	99.2	104.1	97.9	94.0	95.5	108.0	93.8	99.3
05	METALS	99.3	123.4	107.0	94.9	122.9	103.8	97.4	126.9	106.8	95.0	123.8	104.2	93.3	123.9	103.0	94.9	123.7	104.1
06	WOOD, PLASTICS & COMPOSITES	88.0	94.8	91.8	86.1	85.6	85.8	91.6	131.4	113.8	89.9	87.1	88.3	84.3	87.1	85.9	83.3	87.1	85.4
07	THERMAL & MOISTURE PROTECTION	101.7	100.6	101.2	104.2	92.0	99.1	107.9	131.3	117.7	103.5	109.2	105.9	96.3	106.2	100.4	95.8	110.0	101.7
08	OPENINGS	94.8	103.9	96.9	95.1	91.5	94.3	88.3	131.2	98.3	92.3	98.6	93.8	94.7	90.8	93.8	94.6	93.5	94.3
0920	Plaster & Gypsum Board	94.5	94.5	94.5	89.7	85.1	86.6	92.4	132.3	119.0	90.8	86.6	88.0	98.0	86.6	90.4	98.6	86.6	90.6
0950, 0980	Ceilings & Acoustic Treatment	97.8	94.5	95.7	82.9	85.1	84.3	85.5	132.3	116.5	86.2	86.6	86.4	83.2	86.6	85.4	82.2	86.6	85.1
0960	Flooring	92.4	104.2	95.8	86.4	103.0	91.2	90.3	138.7	104.2	88.2	96.1	90.5	91.4	93.4	92.0	93.2	103.0	96.0
0970, 0990	Wall Finishes & Painting/Coating	98.0	112.0	106.1	93.2	115.7	106.3	90.5	151.1	125.8	93.2	115.7	106.3	91.7	115.7	105.7	91.7	91.6	90.4
09	FINISHES	96.0	97.6	96.9	88.2	90.3	89.3	89.8	133.0	113.7	88.8	92.8	91.0	89.1	91.4	90.3	88.8	91.6	90.4
COVERS	DIVS. 10-14, 25, 28, 41, 43, 44, 46	100.0	102.5	100.5	100.0	99.4	99.9	100.0	118.9	104.1	100.0	101.4	100.3	100.0	98.6	99.7	100.0	98.8	99.7
21, 22, 23	FIRE SUPPRESSION, PLUMBING & HVAC	96.7	97.8	97.2	97.0	92.7	95.2	96.8	133.5	112.3	97.0	99.9	98.2	96.8	93.7	95.5	100.1	94.2	97.6
26, 27, 3370	ELECTRICAL, COMMUNICATIONS & UTIL.	94.0	111.3	102.7	91.9	97.4	94.7	93.3	114.0	103.7	92.6	97.5	95.1	88.5	85.1	86.8	89.5	84.6	87.0
MF2014	WEIGHTED AVERAGE	97.5	103.6	100.2	96.2	96.6	96.3	96.3	126.3	109.5	94.9	101.0	97.6	93.2	96.2	94.5	95.9	96.8	96.3

DIVISION		PUERTO RICO SAN JUAN 009 MAT.	INST.	TOTAL	RHODE ISLAND NEWPORT 028 MAT.	INST.	TOTAL	PROVIDENCE 029 MAT.	INST.	TOTAL	SOUTH CAROLINA AIKEN 298 MAT.	INST.	TOTAL	BEAUFORT 299 MAT.	INST.	TOTAL	CHARLESTON 294 MAT.	INST.	TOTAL
015433	CONTRACTOR EQUIPMENT		87.3	87.3		100.0	100.0		100.0	100.0		104.0	104.0		104.0	104.0		104.0	104.0
0241, 31-34	SITE & INFRASTRUCTURE, DEMOLITION	134.0	89.2	102.3	89.5	102.5	98.7	92.2	102.5	99.5	121.0	89.3	98.6	116.3	88.7	96.8	101.4	89.0	92.6
0310	Concrete Forming & Accessories	94.1	17.2	27.6	101.0	120.9	118.3	99.0	120.9	118.0	95.1	63.8	68.0	94.2	40.5	47.7	93.3	63.4	67.5
0320	Concrete Reinforcing	192.3	12.5	99.2	101.8	147.6	125.5	102.6	147.6	125.9	98.1	64.4	80.7	97.1	25.7	60.2	97.0	65.3	80.6
0330	Cast-in-Place Concrete	97.6	29.2	70.6	77.4	122.2	95.1	97.2	122.2	107.1	82.3	68.2	76.7	82.3	67.6	76.5	96.5	68.1	85.3
03	CONCRETE	105.5	21.4	64.6	90.5	125.7	107.6	100.4	125.7	112.7	108.3	66.7	88.1	105.4	49.2	78.1	99.5	66.7	83.6
04	MASONRY	93.8	16.8	46.3	99.3	133.2	120.2	110.9	133.2	124.6	80.2	56.5	65.6	94.7	39.5	60.7	96.0	58.9	73.1
05	METALS	116.1	34.2	90.1	95.5	123.6	104.4	99.7	123.6	107.3	95.8	85.2	92.5	95.9	73.1	88.6	97.8	85.7	93.9
06	WOOD, PLASTICS & COMPOSITES	101.8	16.7	54.4	103.4	119.3	112.2	97.3	119.3	109.5	96.3	66.3	79.6	94.9	36.5	62.4	93.8	66.1	78.4
07	THERMAL & MOISTURE PROTECTION	128.3	20.7	83.3	98.1	120.0	107.2	102.1	120.0	109.6	100.5	63.2	84.9	100.3	49.6	79.1	99.5	61.7	83.3
08	OPENINGS	155.2	14.8	122.5	100.9	127.1	107.0	105.1	127.1	110.2	101.1	63.5	92.4	101.2	40.9	87.1	105.3	63.9	95.7
0920	Plaster & Gypsum Board	156.0	14.4	61.4	89.9	119.3	109.6	97.1	119.3	111.9	97.2	65.1	75.8	100.1	34.5	56.3	101.7	65.0	77.1
0950, 0980	Ceilings & Acoustic Treatment	218.5	14.4	83.4	87.0	119.3	108.4	87.2	119.3	108.5	88.3	65.1	73.0	90.8	34.5	53.5	90.8	65.0	73.7
0960	Flooring	220.6	15.9	161.8	94.3	136.8	106.5	93.6	136.8	106.1	102.7	65.8	92.1	104.3	49.2	88.4	103.9	65.4	92.8
0970, 0990	Wall Finishes & Painting/Coating	191.2	16.8	89.7	87.8	136.7	116.2	86.3	136.7	115.6	92.0	62.7	74.9	92.0	34.7	58.7	92.0	62.7	74.9
09	FINISHES	202.3	17.0	99.8	91.8	125.7	110.6	90.1	125.7	109.8	95.9	64.1	78.3	96.9	40.2	65.5	95.1	64.1	77.9
COVERS	DIVS. 10-14, 25, 28, 41, 43, 44, 46	100.0	16.7	81.9	100.0	108.1	101.8	100.0	108.1	101.8	100.0	72.1	94.0	100.0	76.6	94.9	100.0	72.1	94.0
21, 22, 23	FIRE SUPPRESSION, PLUMBING & HVAC	103.8	13.8	65.7	100.1	110.1	104.3	100.0	110.1	104.3	97.1	57.8	80.5	97.1	57.4	80.3	100.5	57.3	82.2
26, 27, 3370	ELECTRICAL, COMMUNICATIONS & UTIL.	120.4	18.1	68.7	99.1	99.6	99.3	96.6	99.6	98.1	98.9	65.4	81.9	102.6	33.4	67.6	100.9	59.2	79.8
MF2014	WEIGHTED AVERAGE	122.0	24.5	79.0	97.2	116.8	105.8	99.8	116.8	107.3	98.9	66.8	84.7	99.6	52.5	78.8	99.8	66.1	84.1

SOUTH CAROLINA / SOUTH DAKOTA

DIVISION		COLUMBIA 290-292 MAT.	INST.	TOTAL	FLORENCE 295 MAT.	INST.	TOTAL	GREENVILLE 296 MAT.	INST.	TOTAL	ROCK HILL 297 MAT.	INST.	TOTAL	SPARTANBURG 293 MAT.	INST.	TOTAL	ABERDEEN 574 MAT.	INST.	TOTAL
015433	CONTRACTOR EQUIPMENT		104.0	104.0		104.0	104.0		104.0	104.0		104.0	104.0		104.0	104.0		98.8	98.8
0241, 31-34	SITE & INFRASTRUCTURE, DEMOLITION	100.1	89.0	92.3	110.6	89.0	95.3	105.8	88.4	93.5	103.5	88.2	92.7	105.6	88.4	93.4	101.8	94.2	96.4
0310	Concrete Forming & Accessories	91.4	71.8	74.4	81.6	71.5	72.8	92.7	71.2	74.1	91.0	41.1	47.9	95.5	71.4	74.6	96.0	36.1	44.2
0320	Concrete Reinforcing	102.4	65.4	83.2	96.6	65.3	80.4	96.5	44.9	69.8	97.3	43.8	69.6	96.5	59.2	77.2	104.7	37.7	70.0
0330	Cast-in-Place Concrete	97.6	68.2	86.0	82.2	68.1	76.6	82.2	67.9	76.6	82.2	68.2	76.7	82.2	68.0	76.6	109.7	43.8	83.7
03	CONCRETE	99.2	70.5	85.2	99.4	70.3	85.3	97.9	66.4	82.6	95.7	52.8	74.8	98.1	69.2	84.0	105.8	40.6	74.1
04	MASONRY	91.2	58.9	71.3	80.2	58.9	67.1	77.7	58.9	66.1	102.2	41.6	64.8	80.2	58.9	67.1	121.0	55.1	80.4
05	METALS	95.0	86.0	92.2	96.7	85.7	93.2	96.7	78.4	90.8	95.9	75.1	89.3	96.7	85.3	93.1	92.8	65.3	84.0
06	WOOD, PLASTICS & COMPOSITES	91.6	77.1	83.5	81.4	77.1	79.0	93.5	77.1	84.3	92.1	37.6	61.7	97.3	77.1	86.0	96.1	34.7	61.9
07	THERMAL & MOISTURE PROTECTION	95.7	62.9	82.0	99.7	62.9	84.3	99.7	62.9	84.3	99.5	49.9	78.8	99.7	62.9	84.3	100.6	49.8	79.4
08	OPENINGS	101.9	69.8	94.4	101.2	69.8	93.9	101.1	65.1	92.8	101.2	43.5	87.8	101.1	69.6	93.8	101.7	35.6	86.3
0920	Plaster & Gypsum Board	95.4	76.2	82.6	91.5	76.2	81.3	95.9	76.2	82.7	95.6	35.6	55.5	98.6	76.2	83.6	107.1	33.1	57.7
0950, 0980	Ceilings & Acoustic Treatment	92.3	76.2	81.6	89.1	76.2	80.6	88.3	76.2	80.3	88.3	35.6	53.4	88.3	76.2	80.3	93.0	33.1	53.4
0960	Flooring	98.0	65.4	88.6	95.0	41.9	79.8	101.5	54.9	88.1	100.4	42.7	83.8	102.9	54.9	89.1	102.7	48.4	87.1
0970, 0990	Wall Finishes & Painting/Coating	89.4	62.7	73.9	92.0	62.7	74.9	92.0	62.7	74.9	92.0	62.7	74.9	92.0	62.7	74.9	94.2	37.1	61.0
09	FINISHES	93.1	70.6	80.6	91.8	66.6	77.9	93.8	68.8	80.0	93.2	42.4	65.1	94.6	68.8	80.3	98.1	37.6	64.7
COVERS	DIVS. 10-14, 25, 28, 41, 43, 44, 46	100.0	73.3	94.2	100.0	73.3	94.2	100.0	73.3	94.2	100.0	68.9	93.3	100.0	73.3	94.2	100.0	39.3	86.8
21, 22, 23	FIRE SUPPRESSION, PLUMBING & HVAC	99.9	57.8	82.1	100.5	57.4	82.2	100.5	57.1	82.1	97.1	57.0	80.2	100.5	57.3	82.2	100.0	43.4	76.1
26, 27, 3370	ELECTRICAL, COMMUNICATIONS & UTIL.	100.6	61.8	81.0	98.8	61.8	80.1	101.0	58.4	79.4	101.0	56.2	78.4	101.0	58.4	79.4	103.5	76.4	89.8
MF2014	WEIGHTED AVERAGE	98.2	68.3	85.0	98.0	67.7	84.7	98.1	66.0	83.9	98.0	56.5	79.7	98.3	67.3	84.6	101.0	53.7	80.1

For customer support on your Mechanical Cost Data, call 877.739.3689.

City Cost Indexes

SOUTH DAKOTA

DIVISION	MITCHELL 573 MAT.	INST.	TOTAL	MOBRIDGE 576 MAT.	INST.	TOTAL	PIERRE 575 MAT.	INST.	TOTAL	RAPID CITY 577 MAT.	INST.	TOTAL	SIOUX FALLS 570-571 MAT.	INST.	TOTAL	WATERTOWN 572 MAT.	INST.	TOTAL
015433 CONTRACTOR EQUIPMENT		98.8	98.8		98.8	98.8		98.8	98.8		98.8	98.8		99.8	99.8		98.8	98.8
0241,31-34 SITE & INFRASTRUCTURE, DEMOLITION	98.5	94.2	95.4	98.4	94.2	95.4	101.4	94.4	96.5	100.1	94.3	96.0	93.0	96.1	95.2	98.3	94.2	95.4
0310 Concrete Forming & Accessories	95.1	36.0	44.0	85.7	35.9	42.6	98.4	38.4	46.5	103.9	35.7	44.9	100.9	45.8	53.3	82.3	35.7	42.0
0320 Concrete Reinforcing	104.1	45.8	73.9	106.8	37.7	71.0	105.5	70.3	87.3	97.9	70.5	83.7	105.5	70.5	87.4	101.2	37.8	68.4
0330 Cast-in-Place Concrete	106.6	41.4	80.9	106.6	43.6	81.8	100.6	42.8	77.8	105.8	41.9	80.6	91.0	44.9	72.8	106.6	44.9	82.3
03 CONCRETE	103.4	41.1	73.1	103.2	40.4	72.7	101.3	47.4	75.1	102.7	46.0	75.1	96.7	51.5	74.7	102.1	40.8	72.3
04 MASONRY	108.2	53.0	74.2	117.3	55.1	79.0	121.5	55.8	81.0	117.5	56.1	79.6	106.3	57.7	76.3	146.6	57.2	91.5
05 METALS	91.8	65.0	83.3	91.8	64.8	83.3	95.8	79.8	90.7	94.5	80.4	90.0	94.2	80.2	89.8	91.8	65.3	83.4
06 WOOD, PLASTICS & COMPOSITES	95.0	35.2	61.7	84.1	34.9	56.7	100.7	35.4	64.4	99.9	32.1	62.2	95.7	45.1	67.5	80.4	34.7	55.0
07 THERMAL & MOISTURE PROTECTION	100.4	47.7	78.3	100.5	49.8	79.3	102.3	49.2	80.1	101.1	49.4	79.4	98.6	52.1	79.3	100.2	50.1	79.3
08 OPENINGS	100.7	36.1	85.7	103.5	35.2	87.6	106.1	45.3	92.0	105.8	43.5	91.3	106.6	50.6	93.5	100.7	35.3	85.5
0920 Plaster & Gypsum Board	105.5	33.7	57.5	99.8	33.4	55.4	99.1	33.9	55.6	106.7	30.5	55.8	97.3	43.9	61.6	97.5	33.1	54.5
0950,0980 Ceilings & Acoustic Treatment	88.8	33.7	52.3	93.0	33.4	53.5	93.8	33.9	54.1	95.4	30.5	52.4	88.8	43.9	59.1	88.8	33.1	52.0
0960 Flooring	102.2	48.4	86.7	97.5	51.3	84.2	101.5	34.5	82.2	102.0	75.6	94.4	97.9	75.2	91.3	95.9	48.4	82.3
0970,0990 Wall Finishes & Painting/Coating	94.2	40.6	63.0	94.2	41.7	63.6	95.4	44.6	65.9	94.2	44.6	65.3	94.6	44.6	65.5	94.2	37.1	61.0
09 FINISHES	96.7	38.3	64.4	95.4	38.8	64.1	97.8	36.8	64.0	98.3	42.5	67.5	94.6	50.6	70.3	93.7	37.6	62.7
COVERS DIVS. 10-14, 25, 28, 41, 43, 44, 46	100.0	36.3	86.2	100.0	43.8	87.8	100.0	74.5	94.5	100.0	73.5	94.3	100.0	75.6	94.7	100.0	39.2	86.8
21,22,23 FIRE SUPPRESSION, PLUMBING & HVAC	96.7	39.2	72.4	96.7	39.4	72.5	99.9	68.2	86.5	100.0	67.3	86.2	99.9	42.7	75.7	96.7	42.2	73.6
26,27,3370 ELECTRICAL, COMMUNICATIONS & UTIL.	101.8	43.6	72.4	103.5	44.5	73.6	110.2	51.4	80.5	99.9	51.4	75.4	99.3	76.7	87.9	100.9	48.9	74.6
MF2014 WEIGHTED AVERAGE	98.6	48.1	76.4	99.4	48.7	77.0	102.2	59.3	83.3	100.8	59.6	82.7	99.0	60.5	82.0	99.9	49.9	77.9

TENNESSEE

DIVISION	CHATTANOOGA 373-374 MAT.	INST.	TOTAL	COLUMBIA 384 MAT.	INST.	TOTAL	COOKEVILLE 385 MAT.	INST.	TOTAL	JACKSON 383 MAT.	INST.	TOTAL	JOHNSON CITY 376 MAT.	INST.	TOTAL	KNOXVILLE 377-379 MAT.	INST.	TOTAL
015433 CONTRACTOR EQUIPMENT		104.7	104.7		99.3	99.3		99.3	99.3		105.8	105.8		98.3	98.3		98.3	98.3
0241,31-34 SITE & INFRASTRUCTURE, DEMOLITION	103.1	98.1	99.5	88.5	88.5	88.5	94.0	85.7	88.1	97.1	98.6	98.1	110.0	85.1	92.4	89.7	87.8	88.4
0310 Concrete Forming & Accessories	97.8	57.8	63.2	83.4	61.0	64.0	83.6	34.7	41.3	90.6	44.6	50.8	83.9	37.8	44.0	96.1	61.4	66.1
0320 Concrete Reinforcing	89.2	66.6	77.5	87.5	62.4	74.5	87.5	62.5	75.1	87.5	69.2	78.0	89.8	61.5	75.1	89.2	61.6	74.9
0330 Cast-in-Place Concrete	101.5	63.2	86.4	94.3	51.3	77.4	106.9	43.1	81.8	104.4	72.5	91.8	81.6	59.5	72.9	95.4	66.3	83.9
03 CONCRETE	100.1	63.0	82.1	99.3	59.7	80.1	110.1	45.2	78.5	100.4	60.8	81.1	109.5	51.8	81.5	97.4	64.8	81.5
04 MASONRY	108.0	52.2	73.6	124.3	54.4	81.2	118.8	43.6	72.4	124.5	47.0	76.7	123.4	44.6	74.8	84.8	54.3	66.0
05 METALS	95.7	89.9	93.8	91.9	89.1	91.0	92.0	88.5	90.9	94.2	91.8	93.4	93.0	86.8	91.0	96.2	87.8	93.5
06 WOOD, PLASTICS & COMPOSITES	106.2	59.1	80.0	72.8	62.2	66.9	73.0	32.5	50.4	89.0	43.8	63.8	78.4	35.6	54.6	92.4	60.9	74.9
07 THERMAL & MOISTURE PROTECTION	100.5	61.7	84.3	95.2	63.3	81.9	95.7	53.2	77.9	97.6	59.8	81.8	95.3	53.6	77.9	93.4	63.7	81.0
08 OPENINGS	99.2	58.0	89.6	92.2	56.8	83.9	92.2	41.5	80.4	99.2	52.4	88.3	95.6	44.3	83.6	93.3	58.2	85.1
0920 Plaster & Gypsum Board	80.5	58.4	65.7	81.8	61.5	68.3	81.8	31.0	47.9	83.8	42.6	56.3	99.2	34.3	55.8	106.4	60.3	75.6
0950,0980 Ceilings & Acoustic Treatment	99.9	58.4	72.4	77.9	61.5	67.1	77.9	31.0	46.8	86.1	42.6	57.3	96.3	34.3	55.2	97.2	60.3	72.7
0960 Flooring	101.2	57.3	88.6	86.7	22.0	68.1	86.8	56.3	78.0	86.2	39.2	72.7	95.4	39.1	79.2	100.9	51.7	86.8
0970,0990 Wall Finishes & Painting/Coating	98.2	63.4	78.0	84.5	63.4	72.2	84.5	63.4	72.2	86.1	63.4	72.9	95.6	63.4	76.9	95.6	63.4	76.9
09 FINISHES	96.8	58.2	75.4	88.2	54.1	69.3	88.7	40.2	61.8	88.5	43.9	63.9	100.1	39.2	66.4	94.1	59.4	74.9
COVERS DIVS. 10-14, 25, 28, 41, 43, 44, 46	100.0	41.1	87.2	100.0	54.4	90.1	100.0	39.7	86.9	100.0	49.6	89.1	100.0	76.0	94.8	100.0	82.5	96.2
21,22,23 FIRE SUPPRESSION, PLUMBING & HVAC	100.2	62.0	84.1	97.9	79.5	90.1	97.9	74.0	87.8	100.2	68.1	86.6	99.9	58.5	82.4	99.9	67.8	86.3
26,27,3370 ELECTRICAL, COMMUNICATIONS & UTIL.	103.1	71.2	87.0	93.6	57.0	75.1	95.1	62.6	78.7	99.5	58.1	78.6	94.1	46.0	69.8	99.7	57.0	78.1
MF2014 WEIGHTED AVERAGE	99.9	66.6	85.2	96.2	67.2	83.4	97.5	60.0	81.0	99.0	63.5	83.4	99.9	56.4	80.7	96.5	66.8	83.5

TENNESSEE / TEXAS

DIVISION	MCKENZIE 382 MAT.	INST.	TOTAL	MEMPHIS 375,380-381 MAT.	INST.	TOTAL	NASHVILLE 370-372 MAT.	INST.	TOTAL	ABILENE 795-796 MAT.	INST.	TOTAL	AMARILLO 790-791 MAT.	INST.	TOTAL	AUSTIN 786-787 MAT.	INST.	TOTAL
015433 CONTRACTOR EQUIPMENT		99.3	99.3		102.9	102.9		104.9	104.9		93.0	93.0		93.0	93.0		92.1	92.1
0241,31-34 SITE & INFRASTRUCTURE, DEMOLITION	93.7	85.9	88.2	93.0	94.6	94.1	97.5	97.4	97.4	101.9	93.0	95.6	97.8	92.0	93.7	100.8	91.0	93.9
0310 Concrete Forming & Accessories	91.6	37.1	44.4	97.4	66.2	70.4	96.9	65.8	70.0	99.6	64.0	68.8	97.5	55.6	61.3	98.2	56.9	62.5
0320 Concrete Reinforcing	87.7	69.0	78.0	97.1	69.4	82.7	94.1	66.9	80.0	102.1	51.9	76.1	111.9	50.4	80.0	100.8	51.2	75.1
0330 Cast-in-Place Concrete	104.6	56.8	85.8	92.7	77.6	86.8	91.9	66.1	81.7	96.5	64.8	84.0	88.8	64.7	79.3	90.1	66.7	80.9
03 CONCRETE	108.6	52.1	81.1	98.7	72.2	85.8	95.7	67.7	82.1	94.7	62.7	79.2	94.3	58.6	77.0	93.2	59.9	77.0
04 MASONRY	123.2	46.0	75.6	99.4	56.2	72.8	91.2	57.7	70.6	101.2	63.4	77.9	104.2	57.3	75.3	107.4	57.1	76.4
05 METALS	92.0	91.2	91.7	97.9	93.1	96.4	96.5	91.9	95.0	102.4	70.3	92.2	97.3	69.1	88.4	100.5	68.0	90.2
06 WOOD, PLASTICS & COMPOSITES	81.8	35.0	55.7	98.3	67.6	81.2	102.4	66.3	82.3	104.8	66.3	83.4	103.2	55.4	76.5	95.9	57.0	74.2
07 THERMAL & MOISTURE PROTECTION	95.7	47.8	75.7	98.2	67.4	85.3	95.0	65.5	82.6	100.2	67.0	86.3	96.3	63.0	82.4	99.9	64.4	85.1
08 OPENINGS	92.2	45.1	81.2	99.2	67.5	91.8	98.9	67.5	91.6	105.6	61.6	95.4	109.0	55.3	96.5	109.8	54.5	96.9
0920 Plaster & Gypsum Board	84.8	33.6	50.6	93.8	67.0	75.9	92.7	65.7	74.7	86.3	65.7	72.5	96.6	54.4	68.4	87.6	56.1	66.6
0950,0980 Ceilings & Acoustic Treatment	77.9	33.6	48.6	96.2	67.0	76.9	97.1	65.7	76.3	87.8	65.7	73.2	94.3	54.4	67.9	84.1	56.1	65.6
0960 Flooring	89.8	37.8	74.9	99.7	55.0	86.8	98.9	59.8	87.7	104.5	73.6	95.6	103.1	63.0	91.6	99.7	63.0	89.2
0970,0990 Wall Finishes & Painting/Coating	84.5	49.6	64.2	91.4	63.4	75.1	97.4	63.4	77.6	92.9	56.1	71.5	85.3	56.1	68.3	91.6	48.8	66.7
09 FINISHES	89.9	37.1	60.7	95.0	63.4	77.5	100.1	64.0	80.1	89.9	65.2	76.3	93.0	56.6	72.9	90.0	56.8	71.6
COVERS DIVS. 10-14, 25, 28, 41, 43, 44, 46	100.0	26.5	84.1	100.0	82.5	96.2	100.0	83.0	96.3	100.0	82.4	96.2	100.0	79.1	95.5	100.0	81.1	95.8
21,22,23 FIRE SUPPRESSION, PLUMBING & HVAC	97.9	68.3	85.4	100.0	73.4	88.8	99.9	83.2	92.9	100.2	49.9	78.9	100.0	54.3	80.7	99.9	59.5	82.8
26,27,3370 ELECTRICAL, COMMUNICATIONS & UTIL.	94.9	62.4	78.5	101.5	64.8	82.9	100.1	64.3	82.0	101.7	48.3	74.7	101.4	63.7	82.3	100.7	63.4	81.9
MF2014 WEIGHTED AVERAGE	97.7	59.4	80.8	98.9	72.5	87.3	98.2	74.1	87.6	100.0	62.3	83.4	99.6	62.3	83.1	100.0	63.4	83.9

For customer support on your Mechanical Cost Data, call 877.739.3689.

687

TEXAS

DIVISION		BEAUMONT 776 - 777			BROWNWOOD 768			BRYAN 778			CHILDRESS 792			CORPUS CHRISTI 783 - 784			DALLAS 752 - 753		
		MAT.	INST.	TOTAL	MAT.	INST.	TOTAL	MAT.	INST.	TOTAL	MAT.	INST.	TOTAL	MAT.	INST.	TOTAL	MAT.	INST.	TOTAL
015433	CONTRACTOR EQUIPMENT		97.5	97.5		93.0	93.0		97.5	97.5		93.0	93.0		101.2	101.2		104.0	104.0
0241, 31 - 34	SITE & INFRASTRUCTURE, DEMOLITION	87.8	97.4	94.6	106.4	93.0	96.9	79.3	97.5	92.1	113.2	91.4	97.8	145.7	87.1	104.3	107.1	93.1	97.2
0310	Concrete Forming & Accessories	105.1	60.4	66.4	101.6	59.0	64.8	83.2	64.9	67.4	97.8	63.8	68.4	102.7	57.3	63.4	98.5	64.5	69.1
0320	Concrete Reinforcing	99.1	64.5	81.2	99.3	51.5	74.5	101.6	51.3	75.5	102.2	51.5	76.0	87.8	47.9	67.1	100.2	53.7	76.1
0330	Cast-in-Place Concrete	94.4	65.9	83.1	105.4	64.7	89.4	75.6	64.5	71.2	99.0	64.7	85.5	116.4	65.6	96.4	112.6	66.6	94.5
03	CONCRETE	97.0	63.8	80.9	102.4	60.4	82.0	81.4	62.9	72.4	103.1	62.5	83.4	102.2	60.0	81.7	103.9	64.6	84.8
04	MASONRY	104.3	64.6	79.8	134.7	56.5	86.5	143.7	59.5	91.8	105.3	56.5	75.2	92.0	56.4	70.1	97.7	56.5	72.3
05	METALS	97.1	76.1	90.4	98.9	69.6	89.6	96.8	70.5	88.4	99.9	69.6	90.3	96.7	80.0	91.4	101.2	82.9	95.4
06	WOOD, PLASTICS & COMPOSITES	119.2	59.9	86.2	103.9	59.9	79.4	83.0	66.3	73.7	104.2	66.3	83.1	120.9	57.5	85.6	100.3	66.6	81.6
07	THERMAL & MOISTURE PROTECTION	102.1	67.6	87.7	95.7	64.5	82.6	94.1	66.9	82.7	100.8	64.1	85.4	105.1	65.9	88.7	92.2	66.1	81.3
08	OPENINGS	101.7	61.0	92.2	101.0	58.1	91.0	103.0	61.1	93.3	100.6	61.6	91.6	110.6	53.8	97.3	96.4	62.3	88.4
0920	Plaster & Gypsum Board	103.9	59.1	74.0	86.2	59.1	68.1	92.6	65.7	74.6	85.9	65.7	72.4	97.4	56.5	70.1	93.2	65.7	74.8
0950, 0980	Ceilings & Acoustic Treatment	100.7	59.1	73.1	75.3	59.1	64.6	93.8	65.7	75.2	86.2	65.7	72.6	86.2	56.5	66.5	90.8	65.7	74.2
0960	Flooring	117.8	71.8	104.6	87.9	63.0	80.8	87.8	63.0	80.7	102.7	63.0	91.3	115.0	63.0	100.1	99.0	63.0	88.6
0970, 0990	Wall Finishes & Painting/Coating	93.9	59.4	73.8	92.9	56.1	71.5	91.5	62.2	74.5	92.9	56.1	71.5	111.7	48.8	75.1	105.2	56.1	76.6
09	FINISHES	96.9	62.0	77.6	82.4	59.3	69.6	84.5	64.4	73.4	90.2	63.1	75.2	102.6	57.2	77.5	96.3	63.4	78.1
COVERS	DIVS. 10 - 14, 25, 28, 41, 43, 44, 46	100.0	83.6	96.5	100.0	79.6	95.6	100.0	83.0	96.3	100.0	80.3	95.7	100.0	82.9	96.3	100.0	83.2	96.4
21, 22, 23	FIRE SUPPRESSION, PLUMBING & HVAC	100.2	63.5	84.7	96.8	49.5	76.8	96.8	65.3	83.5	96.8	54.3	78.9	100.2	58.3	82.5	99.9	62.7	84.2
26, 27, 3370	ELECTRICAL, COMMUNICATIONS & UTIL.	96.6	70.1	83.2	93.4	52.6	72.8	94.8	69.7	82.1	101.7	63.7	82.5	97.7	57.2	77.2	94.4	65.5	79.8
MF2014	WEIGHTED AVERAGE	98.9	68.9	85.7	99.1	60.6	82.1	96.4	68.4	84.1	99.7	64.0	84.0	102.0	63.2	84.9	99.2	67.9	85.4

TEXAS

DIVISION		DEL RIO 788			DENTON 762			EASTLAND 764			EL PASO 798 - 799,885			FORT WORTH 760 - 761			GALVESTON 775		
		MAT.	INST.	TOTAL	MAT.	INST.	TOTAL	MAT.	INST.	TOTAL	MAT.	INST.	TOTAL	MAT.	INST.	TOTAL	MAT.	INST.	TOTAL
015433	CONTRACTOR EQUIPMENT		92.1	92.1		100.7	100.7		93.0	93.0		93.1	93.1		93.0	93.0		108.6	108.6
0241, 31 - 34	SITE & INFRASTRUCTURE, DEMOLITION	126.0	91.0	101.2	105.9	84.6	90.8	109.4	91.0	96.4	102.9	91.0	94.5	101.4	93.0	95.5	103.4	95.7	98.0
0310	Concrete Forming & Accessories	99.0	57.0	62.7	109.1	64.0	70.1	102.4	63.7	68.9	98.3	61.9	66.8	100.0	64.1	69.0	92.7	62.3	66.4
0320	Concrete Reinforcing	88.4	47.9	67.4	100.7	51.6	75.3	99.5	51.1	74.4	106.7	51.5	78.1	106.8	53.6	79.2	101.1	64.3	82.0
0330	Cast-in-Place Concrete	125.3	64.3	101.2	81.5	65.9	75.4	111.4	64.7	93.0	90.3	64.9	80.3	97.0	64.9	84.3	100.4	65.6	86.7
03	CONCRETE	122.5	58.5	91.4	81.2	63.9	72.8	107.2	62.3	85.4	94.4	61.6	78.4	97.8	63.1	80.9	98.6	65.5	82.5
04	MASONRY	107.8	56.3	76.1	144.4	56.6	90.3	102.1	65.3	79.4	95.4	58.6	72.7	98.9	56.5	72.8	100.3	59.5	75.2
05	METALS	96.4	66.5	86.9	98.5	83.1	93.6	98.7	69.8	89.5	97.6	67.9	88.2	100.5	70.7	91.0	98.3	90.0	95.7
06	WOOD, PLASTICS & COMPOSITES	99.9	57.4	76.2	115.7	66.5	88.3	111.0	66.3	86.1	89.3	63.6	75.0	99.1	66.3	80.8	101.2	62.7	79.8
07	THERMAL & MOISTURE PROTECTION	101.9	64.9	86.4	93.7	66.1	82.2	96.1	67.4	84.1	98.7	64.1	84.2	92.3	65.4	81.0	93.2	67.0	82.2
08	OPENINGS	105.7	53.7	93.6	119.2	61.7	105.8	72.5	61.4	69.9	99.2	56.5	89.3	104.3	62.1	94.4	107.3	62.5	96.9
0920	Plaster & Gypsum Board	93.1	56.5	68.7	90.4	65.7	73.9	86.2	65.7	72.5	93.4	62.9	73.0	90.3	65.7	73.9	98.7	61.9	74.1
0950, 0980	Ceilings & Acoustic Treatment	82.7	56.5	65.4	80.3	65.7	70.6	75.3	65.7	69.0	89.4	62.9	71.8	85.1	65.7	72.3	97.1	61.9	73.8
0960	Flooring	99.2	63.0	88.8	82.9	63.0	77.2	111.4	63.0	97.5	101.0	63.0	90.1	108.2	63.0	95.2	103.1	63.0	91.5
0970, 0990	Wall Finishes & Painting/Coating	99.1	48.8	69.8	102.9	56.1	75.6	94.3	56.1	72.1	94.6	53.7	70.8	92.0	56.1	71.1	102.7	60.0	77.9
09	FINISHES	94.6	57.0	73.8	81.0	63.3	71.2	89.7	63.1	75.0	91.7	61.6	75.1	92.0	63.1	76.0	94.0	62.0	76.3
COVERS	DIVS. 10 - 14, 25, 28, 41, 43, 44, 46	100.0	78.8	95.4	100.0	80.7	95.8	100.0	80.3	95.7	100.0	81.5	96.0	100.0	82.5	96.2	100.0	84.2	96.6
21, 22, 23	FIRE SUPPRESSION, PLUMBING & HVAC	96.7	56.7	79.8	96.8	60.2	81.3	96.8	49.4	76.8	99.9	49.9	78.8	99.9	59.2	82.7	96.7	65.5	83.5
26, 27, 3370	ELECTRICAL, COMMUNICATIONS & UTIL.	99.8	68.3	83.9	95.6	65.5	80.4	93.3	65.5	79.2	103.7	56.9	80.1	94.7	65.5	79.9	96.3	57.2	76.5
MF2014	WEIGHTED AVERAGE	102.5	63.0	85.1	99.2	66.5	84.8	95.6	64.1	81.8	98.3	61.6	82.2	98.8	65.7	84.2	98.7	68.5	85.4

TEXAS

DIVISION		GIDDINGS 789			GREENVILLE 754			HOUSTON 770 - 772			HUNTSVILLE 773			LAREDO 780			LONGVIEW 756		
		MAT.	INST.	TOTAL	MAT.	INST.	TOTAL	MAT.	INST.	TOTAL	MAT.	INST.	TOTAL	MAT.	INST.	TOTAL	MAT.	INST.	TOTAL
015433	CONTRACTOR EQUIPMENT		92.1	92.1		101.4	101.4		108.5	108.5		97.5	97.5		92.1	92.1		94.3	94.3
0241, 31 - 34	SITE & INFRASTRUCTURE, DEMOLITION	110.8	91.0	96.8	100.4	87.5	91.3	102.2	95.6	97.5	94.5	97.4	96.6	104.3	91.0	94.9	99.8	94.5	96.0
0310	Concrete Forming & Accessories	96.2	57.1	62.4	89.5	64.1	67.5	94.8	62.3	66.7	90.5	60.0	64.2	99.1	57.1	62.8	85.4	63.9	66.8
0320	Concrete Reinforcing	89.0	45.5	66.4	100.7	51.6	75.3	100.8	54.3	76.7	101.9	51.3	75.7	88.4	47.9	67.4	99.6	51.6	74.7
0330	Cast-in-Place Concrete	106.1	64.3	89.7	102.6	65.9	88.1	97.3	65.6	84.8	103.9	65.7	88.9	89.6	64.3	79.7	119.2	64.7	97.7
03	CONCRETE	99.1	58.2	79.2	95.9	63.8	80.3	96.2	63.6	80.4	104.8	61.2	83.6	93.1	58.6	76.3	112.4	62.4	88.1
04	MASONRY	116.2	57.1	79.8	146.8	56.5	91.1	100.1	62.1	76.7	142.3	59.5	91.2	100.8	57.1	73.9	143.4	57.2	90.3
05	METALS	95.9	65.7	86.3	98.6	81.7	93.2	101.2	86.2	96.4	96.7	70.4	88.3	99.2	66.7	88.8	91.8	68.7	84.5
06	WOOD, PLASTICS & COMPOSITES	99.0	57.4	75.8	90.4	66.6	77.1	103.6	62.7	80.8	91.8	59.9	74.0	99.9	57.4	76.2	84.3	66.4	74.4
07	THERMAL & MOISTURE PROTECTION	102.5	65.0	86.8	92.2	65.4	81.0	92.8	67.4	82.2	95.1	66.2	83.1	100.8	64.5	85.6	93.8	64.8	81.7
08	OPENINGS	104.8	53.0	92.7	94.6	61.7	87.0	110.4	59.9	98.6	103.0	57.6	92.5	105.8	53.7	93.7	85.6	61.6	80.0
0920	Plaster & Gypsum Board	92.1	56.5	68.3	87.2	65.7	72.8	100.9	61.9	74.8	96.3	59.1	71.4	93.8	56.5	68.9	85.6	65.7	72.3
0950, 0980	Ceilings & Acoustic Treatment	82.7	56.5	65.4	86.7	65.7	72.8	102.0	61.9	75.4	93.8	59.1	70.8	86.0	56.5	66.5	82.6	65.7	71.4
0960	Flooring	98.6	63.0	88.3	93.4	63.0	84.6	104.6	63.0	92.6	92.5	63.0	84.0	99.0	63.0	88.7	99.5	63.0	89.0
0970, 0990	Wall Finishes & Painting/Coating	99.1	48.8	69.8	105.2	56.1	76.6	102.7	62.2	79.2	91.5	59.4	72.8	99.1	48.8	69.8	95.6	56.1	72.6
09	FINISHES	93.2	57.0	73.2	92.4	63.3	76.3	101.3	62.2	79.7	87.3	60.2	72.3	93.7	57.0	73.4	97.2	63.2	78.4
COVERS	DIVS. 10 - 14, 25, 28, 41, 43, 44, 46	100.0	79.1	95.5	100.0	80.9	95.9	100.0	84.3	96.6	100.0	79.1	95.5	100.0	80.8	95.8	100.0	80.5	95.8
21, 22, 23	FIRE SUPPRESSION, PLUMBING & HVAC	96.8	64.0	82.9	96.7	61.9	82.0	100.0	65.4	85.4	96.8	65.3	83.5	100.1	63.5	84.6	96.6	59.9	81.1
26, 27, 3370	ELECTRICAL, COMMUNICATIONS & UTIL.	96.8	63.4	79.9	91.5	65.5	78.4	98.1	69.7	83.8	94.8	67.5	81.0	100.0	60.5	80.0	91.8	55.2	73.3
MF2014	WEIGHTED AVERAGE	99.3	63.8	83.7	98.4	67.0	84.6	100.7	69.8	87.1	99.7	66.9	85.3	99.4	63.5	83.6	98.4	64.3	83.4

TEXAS

DIVISION		LUBBOCK 793-794 MAT.	INST.	TOTAL	LUFKIN 759 MAT.	INST.	TOTAL	MCALLEN 785 MAT.	INST.	TOTAL	MCKINNEY 750 MAT.	INST.	TOTAL	MIDLAND 797 MAT.	INST.	TOTAL	ODESSA 797 MAT.	INST.	TOTAL
015433	CONTRACTOR EQUIPMENT		103.6	103.6		94.3	94.3		101.3	101.3		101.4	101.4		103.6	103.6		93.0	93.0
0241, 31-34	SITE & INFRASTRUCTURE, DEMOLITION	127.3	89.9	100.8	94.8	96.0	95.7	150.4	87.1	105.7	96.8	87.5	90.3	130.6	89.8	101.8	102.1	91.4	94.5
0310	Concrete Forming & Accessories	98.2	55.7	61.5	88.7	60.4	64.2	103.8	56.8	63.1	88.5	64.1	67.4	102.2	63.8	69.0	99.5	63.7	68.6
0320	Concrete Reinforcing	103.3	52.0	76.7	101.4	64.2	82.1	87.9	47.8	67.2	100.7	50.7	74.8	104.3	51.6	77.0	102.1	51.6	75.9
0330	Cast-in-Place Concrete	96.7	65.9	84.5	106.7	65.6	90.5	126.4	65.3	102.3	96.2	65.8	84.2	102.8	65.9	88.2	96.5	64.7	84.0
03	CONCRETE	93.5	60.2	77.3	104.0	63.5	84.3	110.2	59.6	85.6	91.0	63.6	77.7	98.0	63.8	81.4	94.7	62.5	79.0
04	MASONRY	100.5	63.8	77.9	112.1	59.4	79.6	107.7	56.2	76.0	157.0	56.5	95.1	117.8	57.3	80.5	101.2	57.3	74.1
05	METALS	106.0	83.5	98.9	98.7	72.6	90.4	96.4	79.6	91.0	98.5	81.2	93.0	104.3	83.0	97.5	101.8	69.6	91.5
06	WOOD, PLASTICS & COMPOSITES	103.9	55.5	77.0	92.3	60.0	74.3	119.5	57.5	85.0	89.1	66.6	76.6	108.5	66.5	85.1	104.8	66.3	83.4
07	THERMAL & MOISTURE PROTECTION	89.6	65.6	79.6	93.6	65.1	81.7	105.4	59.6	86.3	92.1	65.4	80.9	89.9	65.0	79.5	100.2	65.4	85.6
08	OPENINGS	113.4	55.6	100.0	65.8	61.4	64.7	109.5	53.8	96.5	94.6	61.4	86.9	114.9	61.7	102.5	105.6	61.6	95.4
0920	Plaster & Gypsum Board	86.5	54.4	65.0	85.1	59.1	67.7	98.2	56.5	70.3	86.6	65.7	72.6	88.1	65.7	73.1	86.3	65.7	72.5
0950, 0980	Ceilings & Acoustic Treatment	88.6	54.4	66.0	77.6	59.1	65.4	86.0	56.5	66.5	86.7	65.7	72.8	87.0	65.7	72.9	87.8	65.7	73.2
0960	Flooring	97.0	63.0	87.3	132.8	63.0	112.7	114.3	61.3	99.1	92.9	63.0	84.3	98.5	63.0	88.3	104.5	63.0	92.6
0970, 0990	Wall Finishes & Painting/Coating	103.4	56.1	75.9	95.6	56.1	72.6	111.7	48.8	75.1	105.2	56.1	76.6	103.4	56.1	75.9	92.9	56.1	71.5
09	FINISHES	92.1	56.7	72.5	105.6	60.4	80.6	102.9	56.8	77.4	92.0	63.3	76.1	92.8	63.3	76.4	89.9	63.1	75.1
COVERS	DIVS. 10-14, 25, 28, 41, 43, 44, 46	100.0	81.5	96.0	100.0	79.4	95.5	100.0	81.0	95.9	100.0	80.9	95.9	100.0	81.7	96.0	100.0	81.3	96.0
21, 22, 23	FIRE SUPPRESSION, PLUMBING & HVAC	99.7	50.2	78.8	96.6	64.6	83.1	96.8	58.2	80.5	96.7	61.8	81.9	96.3	50.0	76.7	100.2	49.9	78.9
26, 27, 3370	ELECTRICAL, COMMUNICATIONS & UTIL.	100.5	63.7	81.9	92.9	70.0	81.4	97.5	36.1	66.4	91.6	65.5	78.4	100.4	63.7	81.9	101.8	63.7	82.5
MF2014	WEIGHTED AVERAGE	101.5	63.6	84.8	95.6	67.7	83.3	102.9	59.9	83.9	98.2	66.9	84.4	102.1	64.6	85.6	99.9	63.2	83.7

TEXAS

DIVISION		PALESTINE 758 MAT.	INST.	TOTAL	SAN ANGELO 769 MAT.	INST.	TOTAL	SAN ANTONIO 781-782 MAT.	INST.	TOTAL	TEMPLE 765 MAT.	INST.	TOTAL	TEXARKANA 755 MAT.	INST.	TOTAL	TYLER 757 MAT.	INST.	TOTAL
015433	CONTRACTOR EQUIPMENT		94.3	94.3		93.0	93.0		94.7	94.7		93.0	93.0		94.3	94.3		94.3	94.3
0241, 31-34	SITE & INFRASTRUCTURE, DEMOLITION	100.4	94.8	96.4	102.5	91.9	95.9	103.8	94.3	97.1	90.3	92.5	91.8	89.1	94.1	92.7	99.2	94.6	95.9
0310	Concrete Forming & Accessories	79.5	63.9	66.0	101.9	57.9	63.9	99.2	52.2	62.9	105.5	57.0	63.6	96.0	57.2	62.5	90.6	64.0	67.6
0320	Concrete Reinforcing	98.9	51.6	74.4	99.1	51.6	74.5	93.2	51.2	71.5	99.3	48.6	73.0	98.7	53.6	75.3	99.6	53.6	75.8
0330	Cast-in-Place Concrete	97.5	64.7	84.6	99.5	65.9	86.2	94.8	65.0	83.1	81.5	64.2	74.6	98.3	64.7	85.0	117.1	64.7	96.4
03	CONCRETE	104.5	62.5	84.0	97.8	60.3	79.6	94.2	59.5	77.3	83.9	58.6	71.6	96.6	59.8	78.7	111.8	62.9	88.0
04	MASONRY	106.3	57.2	76.0	131.0	59.3	86.8	97.7	57.1	72.7	143.5	57.1	90.2	160.8	57.2	97.0	151.9	57.2	93.5
05	METALS	98.5	68.7	89.0	99.1	69.7	89.7	101.2	68.2	90.7	98.8	67.0	88.7	91.7	69.5	84.6	98.3	69.6	89.2
06	WOOD, PLASTICS & COMPOSITES	82.8	66.4	73.7	104.2	57.4	78.2	98.4	57.5	75.6	113.8	57.4	82.4	96.1	57.5	74.6	93.9	66.4	78.6
07	THERMAL & MOISTURE PROTECTION	94.1	64.8	81.8	95.6	65.6	83.0	98.1	65.6	84.5	95.2	64.4	82.3	93.4	63.9	81.0	93.9	64.8	81.7
08	OPENINGS	65.7	61.6	64.8	101.0	56.7	90.7	104.7	54.6	93.0	69.0	55.9	65.9	85.6	57.3	79.0	65.7	62.1	64.8
0920	Plaster & Gypsum Board	82.1	65.7	71.1	86.2	56.5	66.4	94.8	56.5	69.2	86.2	56.5	66.4	89.9	56.5	67.6	85.1	65.7	72.1
0950, 0980	Ceilings & Acoustic Treatment	77.6	65.7	69.7	75.3	56.5	62.9	94.3	56.5	69.3	75.3	56.5	62.9	82.6	56.5	65.3	77.6	65.7	69.7
0960	Flooring	122.1	63.0	105.1	88.1	63.0	80.9	106.4	63.0	93.9	113.3	63.0	98.8	109.3	63.0	96.0	135.4	63.0	114.6
0970, 0990	Wall Finishes & Painting/Coating	95.6	56.1	72.6	92.9	56.1	71.5	104.6	48.8	72.1	94.3	48.8	67.8	95.6	56.1	72.6	95.6	56.1	72.6
09	FINISHES	102.6	63.2	80.8	82.1	58.3	69.0	101.3	57.1	76.8	89.0	57.0	71.3	100.1	57.9	76.8	106.7	63.2	82.7
COVERS	DIVS. 10-14, 25, 28, 41, 43, 44, 46	100.0	80.5	95.8	100.0	81.8	96.1	100.0	81.0	95.9	100.0	79.4	95.5	100.0	81.6	96.0	100.0	82.6	96.2
21, 22, 23	FIRE SUPPRESSION, PLUMBING & HVAC	96.6	60.4	81.3	96.8	50.9	77.4	100.1	63.4	84.6	96.8	54.1	78.8	96.6	58.7	80.6	96.6	62.6	82.2
26, 27, 3370	ELECTRICAL, COMMUNICATIONS & UTIL.	89.6	55.2	72.2	97.0	44.5	70.4	102.0	60.5	81.0	94.4	59.3	76.7	92.8	61.1	76.8	91.8	55.7	73.5
MF2014	WEIGHTED AVERAGE	94.8	64.4	81.4	98.7	59.9	81.6	100.3	64.1	84.4	94.1	61.6	79.8	97.6	63.6	82.6	98.4	63.8	83.8

TEXAS / UTAH

DIVISION		VICTORIA 779 MAT.	INST.	TOTAL	WACO 766-767 MAT.	INST.	TOTAL	WAXAHACHIE 751 MAT.	INST.	TOTAL	WHARTON 774 MAT.	INST.	TOTAL	WICHITA FALLS 763 MAT.	INST.	TOTAL	LOGAN 843 MAT.	INST.	TOTAL
015433	CONTRACTOR EQUIPMENT		107.2	107.2		93.0	93.0		101.4	101.4		108.6	108.6		93.0	93.0		96.0	96.0
0241, 31-34	SITE & INFRASTRUCTURE, DEMOLITION	108.3	93.2	97.6	99.2	93.0	94.8	98.5	87.5	90.7	113.3	95.6	100.8	100.0	93.0	95.1	100.4	94.6	96.3
0310	Concrete Forming & Accessories	92.6	58.2	62.8	103.9	63.9	69.3	88.5	63.9	67.2	87.3	60.3	63.9	103.9	64.1	69.5	104.7	60.7	66.6
0320	Concrete Reinforcing	97.0	48.0	71.6	99.0	51.3	74.3	100.7	50.7	74.8	101.0	51.3	75.2	99.0	51.1	74.1	105.7	80.4	92.6
0330	Cast-in-Place Concrete	112.7	66.9	94.6	88.2	67.1	79.8	101.5	65.8	87.4	115.8	66.8	96.5	94.2	64.9	82.6	84.6	72.5	79.8
03	CONCRETE	106.3	61.1	84.3	90.5	63.3	77.3	94.9	63.5	79.6	110.4	62.6	87.2	93.5	62.6	78.5	110.5	68.9	90.3
04	MASONRY	117.8	59.6	81.9	100.3	57.2	73.8	147.3	56.5	91.3	101.6	59.5	75.7	100.8	63.4	77.8	112.9	63.0	82.1
05	METALS	96.9	83.4	92.6	101.2	69.6	91.1	98.6	81.1	93.0	98.3	84.8	94.0	101.1	70.4	91.4	107.5	79.2	98.5
06	WOOD, PLASTICS & COMPOSITES	105.0	57.5	78.6	112.0	66.3	86.6	89.1	66.6	76.6	94.8	60.0	75.4	112.0	66.3	86.6	82.1	58.1	68.8
07	THERMAL & MOISTURE PROTECTION	97.1	66.7	84.4	96.0	65.3	83.2	92.2	65.3	80.9	93.6	67.0	82.5	96.0	67.0	83.9	97.2	66.2	84.2
08	OPENINGS	107.1	55.4	95.1	81.1	61.5	76.5	94.6	61.4	86.9	107.3	57.7	95.8	81.1	61.6	76.5	91.8	60.4	84.5
0920	Plaster & Gypsum Board	95.9	56.5	69.6	86.8	65.7	72.7	87.0	65.7	72.8	94.7	59.1	70.9	86.8	65.7	72.7	81.0	56.9	64.9
0950, 0980	Ceilings & Acoustic Treatment	97.9	56.5	70.5	77.8	65.7	69.8	88.3	65.7	73.4	97.1	59.1	71.9	77.8	65.7	69.8	97.0	56.9	70.4
0960	Flooring	101.2	63.0	90.2	112.5	63.0	98.3	92.9	63.0	84.3	100.0	76.9	93.4	113.2	73.6	101.8	104.1	63.1	92.3
0970, 0990	Wall Finishes & Painting/Coating	102.9	59.4	77.6	94.3	48.8	67.8	105.2	56.1	76.6	102.7	59.4	77.5	96.4	56.1	72.9	94.0	62.6	75.7
09	FINISHES	91.7	58.8	73.5	89.9	62.4	74.7	92.6	63.3	76.4	93.3	62.7	76.3	90.3	65.2	76.4	97.0	60.3	76.7
COVERS	DIVS. 10-14, 25, 28, 41, 43, 44, 46	100.0	79.1	95.5	100.0	82.4	96.2	100.0	80.9	95.9	100.0	79.4	95.5	100.0	80.4	95.7	100.0	86.0	97.0
21, 22, 23	FIRE SUPPRESSION, PLUMBING & HVAC	96.8	65.3	83.5	100.2	54.2	80.7	96.7	55.0	79.1	96.7	65.2	83.4	100.2	53.7	80.5	99.9	69.0	86.9
26, 27, 3370	ELECTRICAL, COMMUNICATIONS & UTIL.	100.8	57.2	78.7	97.2	59.3	78.0	91.6	65.5	78.4	99.7	67.5	83.5	98.8	60.0	79.2	96.7	75.7	86.1
MF2014	WEIGHTED AVERAGE	100.7	66.1	85.5	95.9	63.7	81.7	98.3	65.4	83.8	100.6	68.7	86.5	96.5	64.6	82.5	101.3	71.2	88.0

For customer support on your Mechanical Cost Data, call 877.739.3689.

689

UTAH / VERMONT

DIVISION		OGDEN 842,844 MAT.	INST.	TOTAL	PRICE 845 MAT.	INST.	TOTAL	PROVO 846-847 MAT.	INST.	TOTAL	SALT LAKE CITY 840-841 MAT.	INST.	TOTAL	BELLOWS FALLS 051 MAT.	INST.	TOTAL	BENNINGTON 052 MAT.	INST.	TOTAL
015433	CONTRACTOR EQUIPMENT		96.0	96.0		95.0	95.0		95.0	95.0		96.0	96.0		98.1	98.1		98.1	98.1
0241, 31 - 34	SITE & INFRASTRUCTURE, DEMOLITION	88.1	94.6	92.7	97.7	92.6	94.1	96.4	93.0	94.0	87.7	94.6	92.5	90.3	97.9	95.7	89.6	97.9	95.5
0310	Concrete Forming & Accessories	104.7	60.7	66.6	107.2	47.6	55.7	106.4	60.7	66.9	107.1	60.7	67.0	97.3	83.3	85.1	94.9	103.7	102.5
0320	Concrete Reinforcing	105.3	80.4	92.4	113.3	80.3	96.2	114.3	80.4	96.8	107.7	80.4	93.6	84.8	84.8	84.8	84.8	84.7	84.8
0330	Cast-in-Place Concrete	85.9	72.5	80.6	84.7	60.4	75.1	84.7	72.5	79.9	93.8	72.5	85.4	85.2	114.0	96.6	85.2	114.0	96.6
03	CONCRETE	99.3	68.9	84.5	111.9	58.9	86.1	110.2	68.9	90.1	119.8	68.9	95.0	90.2	94.0	92.0	90.0	103.1	96.4
04	MASONRY	106.9	63.0	79.8	118.5	61.0	83.0	118.7	63.0	84.3	121.3	63.0	85.3	110.3	94.7	100.7	121.4	94.7	104.9
05	METALS	108.0	79.2	98.8	104.6	77.7	96.0	105.6	79.2	97.2	111.5	79.2	101.3	91.9	87.2	90.4	91.8	87.1	90.3
06	WOOD, PLASTICS & COMPOSITES	82.1	58.1	68.8	85.2	42.7	61.6	83.5	58.1	69.4	83.9	58.1	69.6	109.3	80.6	93.3	106.4	108.6	107.6
07	THERMAL & MOISTURE PROTECTION	96.1	66.2	83.6	99.0	61.0	83.1	99.0	66.2	85.2	103.4	66.2	87.9	97.0	81.0	90.3	97.0	79.4	89.6
08	OPENINGS	91.8	60.4	84.5	95.8	48.2	84.7	95.8	60.4	87.6	93.7	60.4	85.9	103.6	85.0	99.2	103.6	100.3	102.8
0920	Plaster & Gypsum Board	81.0	56.9	64.9	83.6	41.1	55.2	81.3	56.9	65.0	94.0	56.9	69.2	98.3	79.6	85.8	96.7	108.4	104.5
0950, 0980	Ceilings & Acoustic Treatment	97.0	56.9	70.4	97.0	41.1	60.0	97.0	56.9	70.4	89.6	56.9	67.9	86.1	79.6	81.8	86.1	108.4	100.8
0960	Flooring	101.8	63.1	90.7	105.3	47.2	88.6	104.9	63.1	92.9	106.2	63.1	93.8	100.9	100.6	100.8	100.0	100.6	100.2
0970, 0990	Wall Finishes & Painting/Coating	94.0	62.6	75.7	94.0	37.5	61.2	94.0	62.6	75.7	97.2	62.6	77.1	88.7	111.0	101.7	88.7	111.0	101.7
09	FINISHES	95.1	60.3	75.9	98.2	44.8	68.8	95.8	60.3	77.0	97.4	60.3	76.9	92.3	88.6	90.3	91.8	105.3	99.3
COVERS	DIVS. 10 - 14, 25, 28, 41, 43, 44, 46	100.0	86.0	97.0	100.0	58.7	91.1	100.0	86.0	97.0	100.0	86.0	97.0	100.0	94.8	98.9	100.0	97.7	99.5
21, 22, 23	FIRE SUPPRESSION, PLUMBING & HVAC	99.9	69.0	86.9	98.2	67.6	85.3	99.9	69.0	86.9	100.1	69.0	87.0	97.0	95.4	96.3	97.0	95.4	96.3
26, 27, 3370	ELECTRICAL, COMMUNICATIONS & UTIL.	97.0	75.7	86.2	102.1	75.7	88.7	97.4	75.7	86.5	99.4	75.7	87.4	100.0	88.7	94.3	100.0	60.7	80.1
MF2014	WEIGHTED AVERAGE	99.3	71.2	86.9	102.0	65.3	85.8	101.8	71.1	88.2	103.8	71.2	89.4	96.9	91.7	94.6	97.3	92.3	95.1

VERMONT

DIVISION		BRATTLEBORO 053 MAT.	INST.	TOTAL	BURLINGTON 054 MAT.	INST.	TOTAL	GUILDHALL 059 MAT.	INST.	TOTAL	MONTPELIER 056 MAT.	INST.	TOTAL	RUTLAND 057 MAT.	INST.	TOTAL	ST. JOHNSBURY 058 MAT.	INST.	TOTAL
015433	CONTRACTOR EQUIPMENT		98.1	98.1		98.1	98.1		98.1	98.1		98.1	98.1		98.1	98.1		98.1	98.1
0241, 31 - 34	SITE & INFRASTRUCTURE, DEMOLITION	91.2	97.9	95.9	96.1	97.8	97.3	89.3	96.5	94.4	92.1	97.5	95.9	93.8	97.6	96.5	89.4	96.5	94.4
0310	Concrete Forming & Accessories	97.6	83.1	85.0	98.6	85.2	87.0	95.3	76.1	78.7	97.0	81.8	83.8	97.9	85.1	86.9	93.7	76.1	78.5
0320	Concrete Reinforcing	83.9	84.7	84.4	104.7	84.8	94.4	85.6	84.6	85.1	103.7	84.8	93.9	106.0	84.8	95.1	83.9	84.6	84.3
0330	Cast-in-Place Concrete	87.9	114.0	98.2	101.7	112.9	106.1	82.6	104.4	91.2	101.6	112.9	106.0	83.5	112.9	95.1	82.6	104.4	91.2
03	CONCRETE	92.2	93.9	93.0	101.1	94.5	97.9	87.7	87.5	87.6	100.8	92.9	97.0	93.2	94.5	93.8	87.4	87.5	87.4
04	MASONRY	120.8	94.7	104.7	127.2	93.3	106.3	121.2	77.8	94.4	121.9	93.3	104.2	100.1	93.3	95.9	151.0	77.8	105.9
05	METALS	91.8	87.1	90.3	99.7	86.4	95.5	91.9	86.2	90.1	97.8	86.4	94.1	97.9	86.4	94.2	91.9	86.2	90.1
06	WOOD, PLASTICS & COMPOSITES	109.6	80.6	93.5	103.6	85.2	93.3	105.9	80.6	91.8	96.8	80.6	87.8	109.8	85.2	96.1	101.0	80.6	89.6
07	THERMAL & MOISTURE PROTECTION	97.1	76.5	88.5	104.3	78.8	93.7	96.9	69.3	85.3	104.5	78.3	93.6	97.2	78.8	89.5	96.8	69.3	85.3
08	OPENINGS	103.6	85.0	99.2	110.2	83.5	104.0	103.6	81.0	98.3	110.5	81.1	103.7	106.9	83.5	101.4	103.6	81.0	98.3
0920	Plaster & Gypsum Board	98.3	79.6	85.8	101.5	84.3	90.0	104.6	79.6	87.9	100.9	79.6	86.7	98.7	84.3	89.1	106.0	79.6	88.4
0950, 0980	Ceilings & Acoustic Treatment	86.1	79.6	81.8	92.9	84.3	87.2	86.1	79.6	81.8	91.3	79.6	83.6	89.6	84.3	86.1	86.1	79.6	81.8
0960	Flooring	101.1	100.6	100.9	100.7	100.6	100.7	104.1	100.6	103.1	104.7	100.6	103.5	100.9	100.6	100.8	107.8	100.6	105.7
0970, 0990	Wall Finishes & Painting/Coating	88.7	111.0	101.7	93.7	90.4	91.8	88.7	90.4	89.7	92.7	90.4	91.3	88.7	90.4	89.7	88.7	90.4	89.7
09	FINISHES	92.5	88.6	90.3	94.7	88.7	91.4	94.1	82.3	87.5	95.3	86.0	90.2	93.3	88.7	90.7	95.4	82.3	88.1
COVERS	DIVS. 10 - 14, 25, 28, 41, 43, 44, 46	100.0	94.8	98.9	100.0	94.7	98.8	100.0	89.1	97.6	100.0	94.2	98.7	100.0	94.7	98.8	100.0	89.1	97.6
21, 22, 23	FIRE SUPPRESSION, PLUMBING & HVAC	97.0	95.4	96.3	100.1	71.1	87.9	97.0	63.3	82.7	96.7	71.1	85.9	100.3	71.1	88.0	97.0	63.3	82.7
26, 27, 3370	ELECTRICAL, COMMUNICATIONS & UTIL.	100.0	88.7	94.3	100.4	60.7	80.3	100.0	60.7	80.1	98.6	60.7	79.4	100.0	60.7	80.1	100.0	60.7	80.1
MF2014	WEIGHTED AVERAGE	97.7	91.6	95.0	102.2	82.5	93.5	97.2	76.8	88.2	100.6	81.8	92.3	99.0	82.5	91.7	98.7	76.8	89.1

VERMONT / VIRGINIA

DIVISION		WHITE RIVER JCT. 050 MAT.	INST.	TOTAL	ALEXANDRIA 223 MAT.	INST.	TOTAL	ARLINGTON 222 MAT.	INST.	TOTAL	BRISTOL 242 MAT.	INST.	TOTAL	CHARLOTTESVILLE 229 MAT.	INST.	TOTAL	CULPEPER 227 MAT.	INST.	TOTAL
015433	CONTRACTOR EQUIPMENT		98.1	98.1		105.8	105.8		104.4	104.4		104.4	104.4		108.8	108.8		104.4	104.4
0241, 31 - 34	SITE & INFRASTRUCTURE, DEMOLITION	93.8	96.6	95.8	114.5	92.5	98.9	124.8	90.2	100.3	109.2	88.3	94.4	113.7	90.4	97.2	112.0	90.0	96.5
0310	Concrete Forming & Accessories	92.2	76.7	78.8	91.4	86.9	87.5	90.5	87.1	87.5	86.6	41.3	47.4	85.0	48.3	53.2	82.1	86.5	85.9
0320	Concrete Reinforcing	84.8	84.6	84.7	86.0	89.0	87.5	97.1	89.0	92.9	97.0	70.4	83.2	96.4	74.0	84.8	97.1	88.9	92.8
0330	Cast-in-Place Concrete	87.9	105.2	94.8	104.7	81.1	95.4	101.9	81.2	93.7	101.4	47.9	80.3	105.5	56.1	86.0	104.3	70.2	90.9
03	CONCRETE	93.9	88.0	91.0	103.8	86.0	95.2	108.6	86.3	97.7	104.7	51.0	78.6	105.4	57.6	82.1	102.3	82.2	92.5
04	MASONRY	135.5	79.3	100.9	92.7	77.5	83.4	108.4	78.0	89.6	97.3	51.2	68.9	122.2	55.1	80.8	109.0	77.9	89.8
05	METALS	91.9	86.2	90.1	101.7	97.8	100.5	100.3	99.6	100.1	99.1	88.0	95.6	99.4	92.2	97.1	99.5	97.9	99.0
06	WOOD, PLASTICS & COMPOSITES	103.3	80.6	90.7	95.9	89.9	92.6	92.6	89.9	91.1	85.5	38.6	59.4	84.1	44.1	61.8	82.9	89.9	86.8
07	THERMAL & MOISTURE PROTECTION	97.2	70.0	85.8	102.4	84.3	94.9	104.4	84.9	96.3	104.0	59.3	85.3	103.6	67.3	88.4	103.8	84.9	95.9
08	OPENINGS	103.6	81.0	98.3	102.0	85.2	98.1	100.0	85.9	96.7	103.3	45.8	89.9	101.3	51.0	89.6	101.7	85.9	98.0
0920	Plaster & Gypsum Board	95.4	79.6	84.8	102.2	89.4	93.7	98.8	89.4	92.5	95.0	36.7	56.0	95.0	41.6	59.3	95.2	89.4	91.3
0950, 0980	Ceilings & Acoustic Treatment	86.1	79.6	81.8	91.5	89.4	90.1	89.9	89.4	89.6	89.0	36.7	54.4	89.0	41.6	57.6	89.9	89.4	89.6
0960	Flooring	98.8	100.6	99.3	100.7	83.1	95.7	99.1	83.1	94.5	95.4	63.9	86.4	94.0	63.9	85.3	94.0	83.1	90.8
0970, 0990	Wall Finishes & Painting/Coating	88.7	90.4	89.7	111.5	80.3	93.4	111.5	80.3	93.4	98.4	54.3	72.7	98.4	64.0	78.4	111.5	64.0	83.9
09	FINISHES	91.6	82.6	86.7	97.1	86.0	91.0	97.0	85.9	90.8	93.5	45.8	67.1	93.1	51.2	69.9	93.8	84.0	88.4
COVERS	DIVS. 10 - 14, 25, 28, 41, 43, 44, 46	100.0	89.6	97.7	100.0	90.9	98.0	100.0	88.6	97.5	100.0	65.9	92.6	100.0	85.3	96.8	100.0	90.6	98.0
21, 22, 23	FIRE SUPPRESSION, PLUMBING & HVAC	97.0	64.1	83.1	100.4	87.4	94.9	100.4	87.6	95.0	97.0	50.0	77.1	97.0	69.9	85.6	97.0	73.7	87.1
26, 27, 3370	ELECTRICAL, COMMUNICATIONS & UTIL.	100.0	60.7	80.1	97.6	101.6	99.6	95.4	101.6	98.5	97.3	35.1	65.9	97.2	73.1	85.0	99.6	101.6	100.6
MF2014	WEIGHTED AVERAGE	98.5	77.3	89.2	100.6	89.4	95.6	101.5	89.4	96.2	99.4	54.8	79.7	100.5	67.9	86.1	99.9	85.5	93.6

VIRGINIA

DIVISION		FAIRFAX 220-221			FARMVILLE 239			FREDERICKSBURG 224-225			GRUNDY 246			HARRISONBURG 228			LYNCHBURG 245		
		MAT.	INST.	TOTAL	MAT.	INST.	TOTAL	MAT.	INST.	TOTAL	MAT.	INST.	TOTAL	MAT.	INST.	TOTAL	MAT.	INST.	TOTAL
015433	CONTRACTOR EQUIPMENT		104.4	104.4		108.8	108.8		104.4	104.4		104.4	104.4		104.4	104.4		104.4	104.4
0241, 31 - 34	SITE & INFRASTRUCTURE, DEMOLITION	123.4	90.1	99.9	110.7	89.8	95.9	111.6	90.0	96.3	107.0	87.0	92.9	120.3	88.5	97.9	107.9	88.7	94.3
0310	Concrete Forming & Accessories	85.0	86.2	86.0	94.8	42.4	49.5	85.0	85.9	85.7	89.6	35.3	42.7	81.0	41.4	46.7	86.6	59.5	63.2
0320	Concrete Reinforcing	97.1	89.0	92.9	93.6	49.3	70.7	97.8	88.9	93.2	95.7	49.8	71.9	97.1	64.1	80.0	96.4	70.5	83.0
0330	Cast-in-Place Concrete	101.9	93.4	98.5	103.6	52.9	83.6	103.4	69.0	89.8	101.4	46.8	79.9	101.9	57.3	84.3	101.4	61.1	85.5
03	CONCRETE	108.2	90.0	99.3	102.1	49.2	76.4	102.1	81.5	92.1	103.4	43.7	74.4	105.9	53.0	80.2	103.3	63.6	84.0
04	MASONRY	108.2	75.7	88.2	107.4	44.6	68.7	108.1	75.6	88.1	97.7	47.1	66.5	106.0	50.1	71.5	113.6	55.0	77.5
05	METALS	99.5	99.7	99.6	97.5	78.6	91.5	99.5	98.9	99.3	99.1	71.7	90.4	99.4	85.2	94.9	99.3	88.7	95.9
06	WOOD, PLASTICS & COMPOSITES	85.5	89.9	88.0	93.8	42.6	65.3	85.5	89.9	88.0	88.2	33.9	57.9	81.9	37.7	57.3	85.5	60.5	71.6
07	THERMAL & MOISTURE PROTECTION	104.3	75.6	92.3	102.0	48.2	79.5	103.8	83.8	95.4	104.0	44.3	79.0	104.2	62.6	86.8	103.8	65.9	87.9
08	OPENINGS	100.0	85.9	96.7	103.5	41.8	89.1	101.3	85.9	97.7	103.3	33.4	87.0	101.7	48.0	89.2	101.6	57.8	91.4
0920	Plaster & Gypsum Board	95.2	89.4	91.3	102.7	40.0	60.8	95.2	89.4	91.3	95.0	31.8	52.8	95.0	35.7	55.4	95.0	59.2	71.1
0950, 0980	Ceilings & Acoustic Treatment	89.9	89.4	89.6	86.8	40.0	55.8	89.9	89.4	89.6	89.0	31.8	51.1	89.0	35.7	53.7	89.0	59.2	69.3
0960	Flooring	95.8	83.1	92.1	100.4	57.9	88.2	95.8	83.1	92.1	96.8	32.4	78.3	93.7	71.1	87.2	95.4	63.9	86.4
0970, 0990	Wall Finishes & Painting/Coating	111.5	80.3	93.4	98.1	38.8	63.6	111.5	64.0	83.9	98.4	35.1	61.6	111.5	54.3	78.2	98.4	54.3	72.7
09	FINISHES	95.5	85.3	89.8	95.2	44.7	67.3	94.3	83.4	88.3	93.7	34.1	60.7	94.2	46.5	67.8	93.3	59.0	74.3
COVERS	DIVS. 10 - 14, 25, 28, 41, 43, 44, 46	100.0	89.8	97.8	100.0	44.4	88.0	100.0	85.9	96.9	100.0	42.4	87.5	100.0	65.8	92.6	100.0	69.7	93.4
21, 22, 23	FIRE SUPPRESSION, PLUMBING & HVAC	97.0	86.5	92.6	97.1	42.9	74.2	97.0	86.1	92.4	97.0	63.5	82.8	97.0	66.3	84.0	97.0	69.3	85.3
26, 27, 3370	ELECTRICAL, COMMUNICATIONS & UTIL.	98.3	101.6	100.0	91.8	47.9	69.6	95.5	101.6	98.6	97.3	44.3	70.5	97.4	96.9	97.2	98.3	52.3	75.0
MF2014	WEIGHTED AVERAGE	100.7	89.2	95.6	99.0	52.1	78.3	99.4	87.7	94.2	99.2	52.6	78.7	100.2	66.9	85.5	99.9	66.2	85.1

VIRGINIA

DIVISION		NEWPORT NEWS 236			NORFOLK 233-235			PETERSBURG 238			PORTSMOUTH 237			PULASKI 243			RICHMOND 230-232		
		MAT.	INST.	TOTAL	MAT.	INST.	TOTAL	MAT.	INST.	TOTAL	MAT.	INST.	TOTAL	MAT.	INST.	TOTAL	MAT.	INST.	TOTAL
015433	CONTRACTOR EQUIPMENT		108.8	108.8		109.4	109.4		108.8	108.8		108.7	108.7		104.4	104.4		108.8	108.8
0241, 31 - 34	SITE & INFRASTRUCTURE, DEMOLITION	109.6	91.0	96.5	108.0	92.0	96.7	113.3	91.2	97.7	108.2	90.9	95.9	106.2	87.7	93.2	104.1	91.2	95.0
0310	Concrete Forming & Accessories	94.1	62.4	66.7	94.4	62.5	66.8	88.1	55.8	60.2	84.4	51.6	56.1	89.6	37.6	44.6	93.3	55.8	60.9
0320	Concrete Reinforcing	93.4	71.2	81.9	102.2	71.2	86.1	93.0	74.2	83.3	93.0	71.2	81.7	95.7	63.9	79.2	102.5	74.2	87.9
0330	Cast-in-Place Concrete	100.7	65.4	86.8	103.1	65.4	88.3	106.9	53.1	85.7	99.7	65.3	86.2	101.4	47.6	80.2	97.2	53.1	79.8
03	CONCRETE	99.3	66.5	83.4	100.4	66.6	84.0	104.7	60.0	83.0	98.1	61.7	80.4	103.4	47.9	76.4	97.5	60.0	79.3
04	MASONRY	101.5	52.5	71.3	106.5	52.5	73.2	116.3	52.4	76.9	107.8	53.7	74.5	92.9	48.0	65.2	105.6	52.4	72.8
05	METALS	99.9	90.8	97.0	101.5	90.9	98.2	97.6	92.7	96.1	98.8	90.9	96.3	99.2	83.9	94.3	101.6	92.7	98.8
06	WOOD, PLASTICS & COMPOSITES	92.8	64.0	76.8	89.8	64.0	75.4	85.2	56.6	69.3	82.0	49.7	64.0	88.2	35.5	58.9	89.4	56.6	71.1
07	THERMAL & MOISTURE PROTECTION	102.0	67.0	87.4	98.9	67.0	85.6	102.0	67.0	87.3	102.0	65.2	86.6	104.0	49.9	81.3	100.6	67.0	86.5
08	OPENINGS	103.8	60.4	93.7	100.5	60.4	91.2	103.1	57.8	92.6	103.9	52.6	91.9	103.3	41.0	88.8	102.4	57.8	92.1
0920	Plaster & Gypsum Board	103.4	62.1	75.8	97.8	62.1	73.9	96.9	54.5	68.5	97.1	47.4	63.9	95.0	33.5	53.9	96.8	54.5	68.5
0950, 0980	Ceilings & Acoustic Treatment	90.1	62.1	71.5	91.6	62.1	72.1	87.6	54.5	65.7	90.1	47.4	61.8	89.0	33.5	52.3	90.7	54.5	66.7
0960	Flooring	100.4	63.9	89.9	97.8	63.9	88.1	96.0	68.4	88.1	92.8	63.9	84.5	96.8	63.9	87.3	98.2	68.4	89.6
0970, 0990	Wall Finishes & Painting/Coating	98.1	42.0	65.5	97.7	64.0	78.1	98.1	64.0	78.3	98.1	64.0	78.3	98.4	54.3	72.7	96.1	64.0	77.5
09	FINISHES	95.8	60.4	76.3	94.3	62.8	76.9	93.5	58.2	74.0	92.7	53.8	71.2	93.7	42.5	65.4	95.0	58.2	74.6
COVERS	DIVS. 10 - 14, 25, 28, 41, 43, 44, 46	100.0	87.2	97.2	100.0	87.2	97.2	100.0	84.9	96.7	100.0	70.1	93.5	100.0	42.7	87.6	100.0	84.9	96.7
21, 22, 23	FIRE SUPPRESSION, PLUMBING & HVAC	100.5	63.9	85.0	100.2	64.3	85.0	97.1	67.1	84.4	100.5	64.1	85.1	97.0	63.9	83.0	99.9	67.1	86.0
26, 27, 3370	ELECTRICAL, COMMUNICATIONS & UTIL.	94.3	65.3	79.6	97.7	61.9	79.6	94.4	73.1	83.7	92.8	61.9	77.2	97.3	54.0	75.4	97.1	73.1	85.0
MF2014	WEIGHTED AVERAGE	99.8	68.3	85.9	100.1	68.3	86.1	99.8	68.7	86.1	99.3	65.4	84.4	99.0	57.5	80.7	99.8	68.7	86.1

DIVISION		VIRGINIA									WASHINGTON								
		ROANOKE 240-241			STAUNTON 244			WINCHESTER 226			CLARKSTON 994			EVERETT 982			OLYMPIA 985		
		MAT.	INST.	TOTAL	MAT.	INST.	TOTAL	MAT.	INST.	TOTAL	MAT.	INST.	TOTAL	MAT.	INST.	TOTAL	MAT.	INST.	TOTAL
015433	CONTRACTOR EQUIPMENT		104.4	104.4		108.8	108.8		104.4	104.4		90.9	90.9		102.2	102.2		102.2	102.2
0241, 31 - 34	SITE & INFRASTRUCTURE, DEMOLITION	106.8	88.7	94.0	110.1	89.9	95.8	119.0	90.1	98.6	103.2	89.7	93.6	91.8	110.4	104.9	90.9	110.4	104.7
0310	Concrete Forming & Accessories	95.8	59.8	64.7	89.3	48.6	54.1	83.4	68.9	70.8	110.3	67.0	72.9	113.0	97.7	99.8	102.0	97.9	98.4
0320	Concrete Reinforcing	96.7	70.6	83.2	96.4	48.4	71.5	96.4	88.9	92.5	110.5	86.4	98.0	108.9	97.7	103.1	114.3	97.8	105.7
0330	Cast-in-Place Concrete	115.2	61.2	93.9	105.5	50.9	83.9	101.9	70.3	89.4	93.0	89.5	91.6	101.0	105.6	102.8	88.8	105.7	95.5
03	CONCRETE	107.8	63.8	86.4	104.7	51.2	78.7	105.4	74.3	90.3	104.6	78.5	91.9	96.1	100.0	98.0	91.9	100.1	95.9
04	MASONRY	98.9	57.8	73.6	108.6	51.1	73.2	103.6	78.4	88.0	106.9	84.8	93.3	111.8	100.4	104.8	114.1	101.9	106.6
05	METALS	101.5	89.1	97.6	99.4	82.0	93.8	99.5	98.4	99.2	100.3	81.7	94.4	101.6	91.0	98.2	102.2	91.1	98.7
06	WOOD, PLASTICS & COMPOSITES	96.8	60.5	76.6	88.2	48.2	65.9	84.1	65.8	73.9	108.4	61.7	82.4	109.9	96.6	102.5	92.8	96.6	94.9
07	THERMAL & MOISTURE PROTECTION	103.6	69.4	89.3	103.6	50.8	81.5	104.3	80.7	94.4	150.2	81.8	121.6	108.7	100.7	105.4	104.8	100.6	103.0
08	OPENINGS	102.0	57.8	91.7	101.6	46.7	88.9	103.3	72.2	96.1	121.1	66.3	108.4	109.7	97.4	106.8	112.3	97.4	108.8
0920	Plaster & Gypsum Board	102.2	59.2	73.5	95.0	45.8	62.1	95.2	64.6	74.8	146.8	60.4	89.0	110.5	96.5	101.2	103.3	96.5	98.8
0950, 0980	Ceilings & Acoustic Treatment	91.5	59.2	70.1	89.0	45.8	60.4	89.9	64.6	73.2	102.5	60.4	74.6	102.0	96.5	98.4	97.8	96.5	97.0
0960	Flooring	100.7	63.9	90.2	96.4	38.0	79.6	95.2	83.1	91.7	92.9	77.0	88.3	109.8	109.5	109.7	97.9	114.6	102.7
0970, 0990	Wall Finishes & Painting/Coating	98.4	54.3	72.7	98.4	34.5	61.2	111.5	87.2	97.3	81.3	73.9	77.0	93.3	91.5	92.2	84.2	91.5	88.4
09	FINISHES	96.0	59.8	76.0	93.6	45.3	66.8	94.8	72.1	82.2	109.8	68.1	86.8	105.2	99.0	101.8	97.0	100.0	98.7
COVERS	DIVS. 10 - 14, 25, 28, 41, 43, 44, 46	100.0	69.7	93.4	100.0	69.0	93.3	100.0	87.9	97.4	100.0	74.2	94.4	100.0	97.6	99.5	100.0	99.9	100.0
21, 22, 23	FIRE SUPPRESSION, PLUMBING & HVAC	100.4	67.4	86.4	97.0	58.2	80.6	97.0	87.6	93.0	97.3	83.9	91.6	100.2	98.1	99.3	99.9	98.2	99.2
26, 27, 3370	ELECTRICAL, COMMUNICATIONS & UTIL.	97.3	54.0	75.4	96.2	76.7	86.4	95.9	103.9	99.9	87.1	96.7	92.0	101.8	97.7	99.7	99.5	100.2	99.9
MF2014	WEIGHTED AVERAGE	101.0	66.6	85.8	99.7	61.7	83.0	100.0	85.2	93.5	103.8	81.9	94.1	102.2	99.0	100.8	101.0	99.7	100.4

For customer support on your Mechanical Cost Data, call 877.739.3689.

691

WASHINGTON

DIVISION		RICHLAND 993 MAT.	INST.	TOTAL	SEATTLE 980-981,987 MAT.	INST.	TOTAL	SPOKANE 990-992 MAT.	INST.	TOTAL	TACOMA 983-984 MAT.	INST.	TOTAL	VANCOUVER 986 MAT.	INST.	TOTAL	WENATCHEE 988 MAT.	INST.	TOTAL
015433	CONTRACTOR EQUIPMENT		90.9	90.9		102.3	102.3		90.9	90.9		102.2	102.2		97.8	97.8		102.2	102.2
0241, 31-34	SITE & INFRASTRUCTURE, DEMOLITION	105.6	90.5	94.9	96.0	108.6	104.9	104.9	90.4	94.7	94.9	110.4	105.9	104.6	97.8	99.8	103.9	109.0	107.5
0310	Concrete Forming & Accessories	110.4	78.2	82.6	104.4	98.1	98.9	115.5	77.7	82.8	103.4	97.8	98.6	103.8	91.6	93.2	105.2	75.7	79.6
0320	Concrete Reinforcing	105.9	86.5	95.9	114.3	97.8	105.7	106.6	86.4	96.2	107.7	97.8	102.6	108.6	97.4	102.8	108.5	86.7	97.2
0330	Cast-in-Place Concrete	93.2	85.2	90.0	103.8	105.8	104.6	96.8	85.0	92.2	103.9	105.7	104.6	115.9	98.6	109.1	106.0	77.0	94.6
03	CONCRETE	104.2	82.2	93.5	99.8	100.3	100.1	106.4	81.9	94.5	97.8	100.1	98.9	107.6	94.9	101.4	105.1	78.4	92.1
04	MASONRY	108.5	85.3	94.2	109.4	101.9	104.8	109.0	85.3	94.4	110.9	101.9	105.4	110.3	95.5	101.2	114.0	91.0	99.9
05	METALS	100.7	84.7	95.6	101.1	92.8	98.5	103.3	84.2	97.2	103.2	91.1	99.4	100.8	91.6	97.9	100.8	85.1	95.8
06	WOOD, PLASTICS & COMPOSITES	108.5	75.4	90.0	102.5	96.6	99.2	117.0	75.4	93.9	99.7	96.6	98.0	91.2	91.1	91.1	101.2	73.2	85.6
07	THERMAL & MOISTURE PROTECTION	151.4	81.5	122.2	106.6	101.1	104.3	147.7	83.2	120.7	108.5	100.6	105.2	108.6	93.9	102.4	109.1	87.8	100.2
08	OPENINGS	119.0	72.7	108.2	107.2	97.4	104.9	119.6	70.7	108.2	110.5	97.4	107.4	105.7	92.2	102.6	110.0	72.3	101.2
0920	Plaster & Gypsum Board	146.8	74.5	98.5	107.0	96.5	100.0	137.1	74.5	95.3	111.2	96.5	101.4	109.3	91.1	97.1	113.8	72.4	86.1
0950, 0980	Ceilings & Acoustic Treatment	109.0	74.5	86.2	114.9	96.5	102.7	104.4	74.5	84.6	105.3	96.5	99.5	102.4	91.1	94.9	97.7	72.4	80.9
0960	Flooring	93.3	84.7	90.8	104.2	102.6	103.8	92.2	94.8	92.9	102.2	112.2	105.1	107.9	102.1	106.2	105.3	60.2	92.4
0970, 0990	Wall Finishes & Painting/Coating	81.3	73.9	77.0	106.4	91.5	97.7	81.4	70.9	75.3	93.3	91.5	92.2	95.4	71.7	81.6	93.3	72.6	81.2
09	FINISHES	111.6	77.9	93.0	107.2	97.7	101.9	108.9	79.6	92.7	103.9	99.6	101.5	102.0	91.3	96.1	104.4	71.3	86.1
COVERS	DIVS. 10-14, 25, 28, 41, 43, 44, 46	100.0	85.2	96.8	100.0	99.9	100.0	100.0	85.2	96.8	100.0	99.9	100.0	100.0	96.4	99.2	100.0	74.9	94.6
21, 22, 23	FIRE SUPPRESSION, PLUMBING & HVAC	100.7	111.7	105.4	100.1	114.2	106.0	100.6	84.0	93.6	100.2	98.2	99.4	100.3	92.1	96.8	96.9	85.9	92.2
26, 27, 3370	ELECTRICAL, COMMUNICATIONS & UTIL.	84.9	96.8	90.9	101.8	107.1	104.5	83.7	80.1	81.9	101.6	100.2	100.9	106.4	100.2	103.3	102.4	97.7	100.0
MF2014	WEIGHTED AVERAGE	104.4	90.7	98.4	102.3	103.8	103.0	104.8	82.6	95.0	102.6	99.6	101.3	103.4	94.4	99.4	102.8	85.9	95.3

WASHINGTON / WEST VIRGINIA

DIVISION		YAKIMA 989 MAT.	INST.	TOTAL	BECKLEY 258-259 MAT.	INST.	TOTAL	BLUEFIELD 247-248 MAT.	INST.	TOTAL	BUCKHANNON 262 MAT.	INST.	TOTAL	CHARLESTON 250-253 MAT.	INST.	TOTAL	CLARKSBURG 263-264 MAT.	INST.	TOTAL
015433	CONTRACTOR EQUIPMENT		102.2	102.2		104.4	104.4		104.4	104.4		104.4	104.4		104.4	104.4		104.4	104.4
0241, 31-34	SITE & INFRASTRUCTURE, DEMOLITION	97.3	109.8	106.1	100.5	92.1	94.6	100.9	92.1	94.7	107.2	92.0	96.5	99.9	93.2	95.2	107.8	92.0	96.7
0310	Concrete Forming & Accessories	103.8	93.4	94.8	84.7	91.1	90.2	86.5	90.9	90.3	85.9	91.3	90.6	96.9	92.1	92.8	83.4	91.3	90.2
0320	Concrete Reinforcing	108.2	86.5	97.9	93.5	89.7	91.5	95.3	84.9	89.9	96.0	84.7	90.1	101.2	89.8	95.3	96.0	85.2	90.4
0330	Cast-in-Place Concrete	110.9	84.8	100.6	99.5	101.3	100.2	99.2	101.2	100.0	98.9	99.2	99.0	96.2	130.8	109.9	108.4	97.4	104.1
03	CONCRETE	102.6	88.9	95.9	100.0	94.9	97.5	99.1	93.9	96.6	102.2	93.4	97.9	98.8	105.4	102.0	106.3	92.9	99.8
04	MASONRY	103.9	84.0	91.6	96.2	96.5	96.4	94.7	96.5	95.8	106.5	90.9	96.9	92.8	98.0	96.0	110.7	90.9	98.5
05	METALS	101.5	85.3	96.3	101.5	100.3	101.1	99.4	98.5	99.1	99.5	98.4	99.2	100.1	100.8	100.3	99.5	98.5	99.2
06	WOOD, PLASTICS & COMPOSITES	100.1	96.6	98.2	88.2	90.4	89.4	87.4	90.4	89.0	86.7	91.2	89.2	97.0	90.4	93.3	83.5	91.2	87.8
07	THERMAL & MOISTURE PROTECTION	108.6	83.8	98.2	103.2	89.0	97.3	103.7	89.0	97.5	104.0	88.6	97.5	100.4	89.3	95.8	103.9	88.3	97.4
08	OPENINGS	110.0	78.6	102.7	102.2	83.9	98.0	103.7	82.8	98.8	103.7	83.3	99.0	100.8	83.9	96.9	103.7	83.4	99.0
0920	Plaster & Gypsum Board	110.8	96.5	101.3	94.6	89.9	91.4	94.2	89.9	91.3	94.6	90.7	92.0	98.0	89.9	92.6	92.6	90.7	91.4
0950, 0980	Ceilings & Acoustic Treatment	99.5	96.5	97.5	83.3	89.9	87.7	87.4	89.9	89.0	89.0	90.7	90.2	91.5	89.9	90.4	89.0	90.7	90.2
0960	Flooring	103.2	61.9	91.3	98.6	113.4	102.9	93.1	113.4	99.0	92.9	104.7	96.3	103.0	113.4	106.0	91.8	104.7	95.5
0970, 0990	Wall Finishes & Painting/Coating	93.3	70.9	80.2	89.4	94.7	92.5	98.4	94.7	96.2	98.4	94.3	96.0	88.5	100.8	95.7	98.4	94.3	96.0
09	FINISHES	103.0	85.3	93.2	91.8	95.7	93.9	91.7	95.7	93.9	92.6	94.2	93.5	96.1	96.7	96.4	91.9	94.2	93.2
COVERS	DIVS. 10-14, 25, 28, 41, 43, 44, 46	100.0	97.1	99.4	100.0	101.2	100.3	100.0	101.2	100.3	100.0	100.4	100.1	100.0	101.1	100.2	100.0	100.4	100.1
21, 22, 23	FIRE SUPPRESSION, PLUMBING & HVAC	100.2	111.1	104.9	97.2	98.3	97.7	97.0	91.8	94.8	97.0	97.9	97.4	100.2	99.3	99.8	97.0	97.9	97.4
26, 27, 3370	ELECTRICAL, COMMUNICATIONS & UTIL.	104.3	96.8	100.5	98.1	90.4	94.2	96.3	92.2	94.2	97.6	97.2	97.4	103.2	92.6	97.8	97.6	97.2	97.4
MF2014	WEIGHTED AVERAGE	102.8	94.9	99.3	98.8	95.1	97.2	98.2	93.6	96.2	99.6	94.7	97.4	99.7	97.4	98.7	100.2	94.7	97.7

WEST VIRGINIA

DIVISION		GASSAWAY 266 MAT.	INST.	TOTAL	HUNTINGTON 255-257 MAT.	INST.	TOTAL	LEWISBURG 249 MAT.	INST.	TOTAL	MARTINSBURG 254 MAT.	INST.	TOTAL	MORGANTOWN 265 MAT.	INST.	TOTAL	PARKERSBURG 261 MAT.	INST.	TOTAL
015433	CONTRACTOR EQUIPMENT		104.4	104.4		104.4	104.4		104.4	104.4		104.4	104.4		104.4	104.4		104.4	104.4
0241, 31-34	SITE & INFRASTRUCTURE, DEMOLITION	104.6	92.1	95.7	105.1	93.3	96.8	116.8	92.1	99.3	104.3	92.5	96.0	101.9	92.7	95.4	110.1	93.2	98.1
0310	Concrete Forming & Accessories	85.4	89.0	88.5	96.4	92.9	93.3	83.7	90.4	89.5	84.8	82.0	82.3	83.7	91.4	90.4	88.0	88.3	88.3
0320	Concrete Reinforcing	96.0	90.7	93.2	94.9	95.2	95.1	96.0	84.7	90.1	93.5	90.5	91.9	96.0	85.2	90.4	95.3	84.7	89.8
0330	Cast-in-Place Concrete	103.5	129.3	113.7	108.5	102.5	106.1	99.3	101.0	99.9	104.2	95.9	100.9	98.9	99.2	99.0	101.1	96.0	99.1
03	CONCRETE	102.5	103.6	103.1	105.4	97.1	101.4	109.0	93.6	101.5	103.8	89.0	96.6	98.8	93.6	96.3	104.2	91.0	97.8
04	MASONRY	111.4	92.5	99.8	94.6	101.9	99.1	97.1	93.6	94.9	97.5	90.6	93.3	128.4	90.9	105.5	84.0	93.3	89.7
05	METALS	99.5	100.7	99.9	103.9	102.6	103.5	99.4	97.8	98.9	101.9	96.3	100.1	99.6	98.6	99.3	100.2	98.4	99.6
06	WOOD, PLASTICS & COMPOSITES	85.9	87.7	86.9	98.8	90.6	94.2	83.8	90.4	87.5	88.2	80.2	83.8	83.8	91.2	87.9	86.7	86.1	86.4
07	THERMAL & MOISTURE PROTECTION	103.7	88.3	97.3	103.4	90.3	97.9	104.6	88.2	97.7	103.5	85.2	95.8	103.8	88.6	97.4	103.6	89.0	97.5
08	OPENINGS	101.7	82.7	97.3	101.5	85.2	97.7	103.7	82.8	98.8	104.3	72.4	96.9	105.0	83.4	100.0	102.6	80.5	97.4
0920	Plaster & Gypsum Board	94.0	87.1	89.4	102.1	90.1	94.1	92.6	89.9	90.8	95.2	79.4	84.7	92.6	90.7	91.4	95.0	85.5	88.6
0950, 0980	Ceilings & Acoustic Treatment	89.0	87.1	87.8	85.8	90.1	88.7	89.0	89.9	89.6	85.8	79.4	81.6	89.0	90.7	90.2	89.0	85.5	86.7
0960	Flooring	92.6	113.4	98.6	106.5	117.1	109.6	91.9	113.4	98.1	98.6	113.4	102.9	91.9	104.7	95.9	95.9	113.4	100.9
0970, 0990	Wall Finishes & Painting/Coating	98.4	100.8	99.8	89.4	94.7	92.5	98.4	75.4	85.0	89.4	41.4	61.5	98.4	94.3	96.0	98.4	94.7	96.2
09	FINISHES	92.1	94.5	93.5	95.7	97.4	96.7	93.0	93.6	93.4	92.6	82.9	87.3	91.5	94.2	93.0	93.6	93.4	93.5
COVERS	DIVS. 10-14, 25, 28, 41, 43, 44, 46	100.0	100.0	100.0	100.0	101.8	100.4	100.0	54.7	90.2	100.0	95.3	99.0	100.0	98.6	99.7	100.0	100.5	100.1
21, 22, 23	FIRE SUPPRESSION, PLUMBING & HVAC	97.0	91.4	94.6	100.6	100.3	100.5	97.0	98.1	97.5	97.2	85.5	92.3	97.0	97.9	97.4	100.3	90.8	96.3
26, 27, 3370	ELECTRICAL, COMMUNICATIONS & UTIL.	97.6	92.6	95.1	102.1	96.3	99.2	93.8	92.2	93.0	104.3	82.1	93.1	97.8	97.2	97.5	97.7	93.9	95.8
MF2014	WEIGHTED AVERAGE	99.5	94.4	97.3	101.4	97.8	99.8	99.8	92.8	96.7	100.4	87.0	94.5	100.2	94.8	97.8	99.6	92.5	96.5

City Cost Indexes

WEST VIRGINIA / WISCONSIN

DIVISION		PETERSBURG 268 MAT.	INST.	TOTAL	ROMNEY 267 MAT.	INST.	TOTAL	WHEELING 260 MAT.	INST.	TOTAL	BELOIT 535 MAT.	INST.	TOTAL	EAU CLAIRE 547 MAT.	INST.	TOTAL	GREEN BAY 541-543 MAT.	INST.	TOTAL
015433	CONTRACTOR EQUIPMENT		104.4	104.4		104.4	104.4		104.4	104.4		100.9	100.9		101.1	101.1		98.8	98.8
0241, 31 - 34	SITE & INFRASTRUCTURE, DEMOLITION	101.1	92.6	95.1	104.0	92.7	96.0	110.9	92.8	98.1	96.7	106.4	103.5	97.6	103.9	102.1	101.0	100.1	100.4
0310	Concrete Forming & Accessories	86.9	82.3	83.0	82.9	83.0	83.0	89.6	91.4	91.2	99.1	97.5	97.8	97.4	98.5	98.4	106.3	101.1	101.8
0320	Concrete Reinforcing	95.3	81.7	88.3	96.0	90.6	93.2	94.7	85.2	89.8	91.7	114.8	103.7	93.2	98.2	95.8	91.4	91.9	91.7
0330	Cast-in-Place Concrete	98.9	96.4	97.9	103.5	96.7	100.8	101.1	94.2	98.4	102.5	99.4	101.3	102.0	98.0	100.5	105.5	109.6	107.1
03	CONCRETE	99.0	87.9	93.6	102.4	89.9	96.3	104.3	91.9	98.2	98.0	101.5	99.7	98.5	98.5	98.5	101.7	102.4	102.0
04	MASONRY	101.4	90.9	94.9	97.4	91.6	93.8	109.7	91.5	98.5	104.2	105.4	105.0	98.0	102.8	101.0	130.9	118.9	123.5
05	METALS	99.6	96.5	98.6	99.7	100.1	99.8	100.4	98.4	99.7	93.7	106.8	97.9	91.4	101.4	94.6	93.7	99.1	95.4
06	WOOD, PLASTICS & COMPOSITES	87.6	80.2	83.5	83.0	80.2	81.5	88.2	91.2	89.9	98.1	95.9	96.9	101.7	98.0	99.6	107.0	98.0	102.0
07	THERMAL & MOISTURE PROTECTION	103.8	81.3	94.4	103.9	86.2	96.5	104.0	88.0	97.3	102.6	94.6	99.2	104.2	99.9	102.4	106.2	91.4	100.0
08	OPENINGS	105.1	76.6	98.4	105.0	78.6	98.8	103.5	83.4	98.8	101.0	106.3	102.3	107.7	94.5	104.7	103.8	99.6	102.8
0920	Plaster & Gypsum Board	94.6	79.4	84.5	92.3	79.4	83.7	95.0	90.7	92.1	93.1	96.2	95.2	104.9	98.3	100.5	101.6	98.3	99.4
0950, 0980	Ceilings & Acoustic Treatment	89.0	79.4	82.7	89.0	79.4	82.7	89.0	90.7	90.2	81.8	96.2	91.3	89.2	98.3	95.2	82.8	98.3	93.0
0960	Flooring	93.7	104.7	96.9	91.7	106.0	95.8	96.8	104.7	99.1	98.0	116.6	103.3	88.9	109.7	94.9	106.5	126.8	112.3
0970, 0990	Wall Finishes & Painting/Coating	98.4	94.3	96.0	98.4	94.3	96.0	98.4	94.3	96.0	98.4	103.4	101.3	88.6	86.5	87.4	98.5	83.2	89.6
09	FINISHES	92.3	87.6	89.7	91.6	87.9	89.6	93.9	94.3	94.1	94.5	101.3	98.3	93.1	99.6	96.7	97.7	104.5	101.5
COVERS	DIVS. 10 - 14, 25, 28, 41, 43, 44, 46	100.0	61.9	91.7	100.0	99.1	99.8	100.0	93.1	98.5	100.0	97.2	99.4	100.0	97.6	99.5	100.0	100.1	100.0
21, 22, 23	FIRE SUPPRESSION, PLUMBING & HVAC	97.0	91.4	94.6	97.0	91.2	94.5	100.4	98.2	99.5	100.0	97.5	99.0	100.0	86.7	94.4	100.2	88.5	95.3
26, 27, 3370	ELECTRICAL, COMMUNICATIONS & UTIL.	100.7	82.1	91.3	100.0	82.1	91.0	95.1	97.2	96.1	99.5	87.3	93.3	104.9	86.8	95.7	99.4	81.8	90.5
MF2014	WEIGHTED AVERAGE	99.3	87.7	94.2	99.4	89.8	95.1	100.8	94.5	98.0	98.6	99.8	99.1	99.2	95.5	97.6	101.2	97.5	99.5

WISCONSIN

DIVISION		KENOSHA 531 MAT.	INST.	TOTAL	LA CROSSE 546 MAT.	INST.	TOTAL	LANCASTER 538 MAT.	INST.	TOTAL	MADISON 537 MAT.	INST.	TOTAL	MILWAUKEE 530,532 MAT.	INST.	TOTAL	NEW RICHMOND 540 MAT.	INST.	TOTAL
015433	CONTRACTOR EQUIPMENT		98.8	98.8		101.1	101.1		100.9	100.9		100.9	100.9		88.1	88.1		101.5	101.5
0241, 31 - 34	SITE & INFRASTRUCTURE, DEMOLITION	102.2	103.3	103.0	91.3	103.7	100.1	95.9	106.4	103.3	94.8	106.4	103.0	93.6	96.9	95.9	95.7	104.1	101.6
0310	Concrete Forming & Accessories	107.5	98.3	99.5	84.0	98.3	96.4	98.5	96.2	96.5	102.7	98.8	99.4	101.9	119.4	117.0	92.6	98.2	97.4
0320	Concrete Reinforcing	91.6	96.9	94.3	92.9	90.6	91.7	93.0	90.4	91.7	96.0	90.6	93.2	94.8	97.1	95.9	90.5	97.9	94.3
0330	Cast-in-Place Concrete	111.5	110.1	110.9	91.8	94.8	93.0	101.9	99.8	101.1	92.8	101.9	96.4	97.5	111.1	102.9	106.2	81.4	96.4
03	CONCRETE	102.9	102.1	102.5	89.6	95.9	92.7	97.7	96.5	97.1	95.5	98.4	96.9	96.2	111.4	103.6	96.5	92.6	94.6
04	MASONRY	101.5	117.1	111.1	97.1	105.5	102.3	104.3	103.8	104.0	107.1	104.8	105.7	108.7	119.5	115.3	125.5	101.8	110.9
05	METALS	94.6	100.6	96.5	91.3	98.3	93.5	91.2	95.5	92.6	94.6	96.8	95.3	96.0	92.8	95.0	91.5	99.4	94.0
06	WOOD, PLASTICS & COMPOSITES	102.7	93.2	97.4	87.0	98.0	93.2	97.4	95.9	96.6	95.6	97.8	96.8	103.2	120.8	113.0	90.8	99.0	95.4
07	THERMAL & MOISTURE PROTECTION	102.9	106.9	104.6	103.6	102.0	102.9	102.4	81.9	93.8	100.4	102.8	101.4	102.0	111.7	106.0	105.5	90.8	99.3
08	OPENINGS	95.3	98.2	96.0	107.7	89.3	103.4	96.8	83.3	93.6	107.1	99.1	105.2	98.9	113.3	102.3	92.6	94.7	93.1
0920	Plaster & Gypsum Board	84.9	93.4	90.5	99.7	98.3	98.8	92.4	96.2	94.9	103.5	98.1	99.9	97.7	121.6	113.7	90.2	99.5	96.4
0950, 0980	Ceilings & Acoustic Treatment	81.8	93.4	89.5	88.4	98.3	94.9	78.5	96.2	90.2	86.9	98.1	94.3	89.5	121.6	110.7	56.9	99.5	85.1
0960	Flooring	115.8	108.3	113.7	82.3	109.7	90.1	97.5	106.3	100.0	99.4	116.9	104.4	98.6	111.1	102.2	98.9	109.7	102.0
0970, 0990	Wall Finishes & Painting/Coating	109.1	117.9	114.2	88.6	77.9	82.4	98.4	103.4	101.3	94.4	103.4	99.6	103.6	120.4	113.4	101.0	86.5	92.5
09	FINISHES	99.7	101.7	100.8	89.7	98.6	94.7	93.5	99.6	96.9	95.1	102.6	99.3	99.1	119.0	110.1	87.8	98.4	93.7
COVERS	DIVS. 10 - 14, 25, 28, 41, 43, 44, 46	100.0	99.7	99.9	100.0	97.6	99.5	100.0	90.7	98.0	100.0	100.0	100.0	100.0	103.3	100.7	100.0	93.5	98.6
21, 22, 23	FIRE SUPPRESSION, PLUMBING & HVAC	100.2	96.6	98.7	100.0	86.6	94.3	96.7	86.5	92.4	99.9	94.5	97.6	99.9	102.4	100.9	96.1	86.2	91.9
26, 27, 3370	ELECTRICAL, COMMUNICATIONS & UTIL.	99.9	99.2	99.5	105.1	86.8	95.9	99.3	86.8	93.0	101.1	96.3	98.7	98.6	101.0	99.8	103.0	86.8	94.8
MF2014	WEIGHTED AVERAGE	99.2	101.7	100.3	97.6	94.8	96.4	96.8	93.7	95.4	99.3	99.1	99.2	98.9	106.9	102.4	97.1	93.8	95.6

WISCONSIN

DIVISION		OSHKOSH 549 MAT.	INST.	TOTAL	PORTAGE 539 MAT.	INST.	TOTAL	RACINE 534 MAT.	INST.	TOTAL	RHINELANDER 545 MAT.	INST.	TOTAL	SUPERIOR 548 MAT.	INST.	TOTAL	WAUSAU 544 MAT.	INST.	TOTAL
015433	CONTRACTOR EQUIPMENT		98.8	98.8		100.9	100.9		100.9	100.9		98.8	98.8		101.5	101.5		98.8	98.8
0241, 31 - 34	SITE & INFRASTRUCTURE, DEMOLITION	92.6	99.4	97.4	86.7	106.2	100.5	96.5	107.2	104.1	104.9	99.4	101.0	92.3	104.0	100.6	88.4	101.1	97.4
0310	Concrete Forming & Accessories	88.9	95.5	94.6	90.7	96.5	95.7	99.5	98.3	98.5	86.4	95.6	94.4	90.6	89.6	89.7	88.4	100.7	99.0
0320	Concrete Reinforcing	91.5	91.4	91.4	93.1	90.5	91.8	91.7	96.9	94.4	91.7	90.1	90.9	90.5	96.2	93.4	91.7	90.4	91.0
0330	Cast-in-Place Concrete	97.9	98.1	98.0	87.5	102.0	93.2	100.5	109.8	104.2	110.9	99.7	106.5	100.1	102.2	100.9	91.2	103.1	95.9
03	CONCRETE	91.4	95.8	93.6	85.6	97.4	91.3	97.0	102.1	99.5	103.5	96.2	100.0	90.9	95.5	93.2	86.4	99.7	92.9
04	MASONRY	112.2	102.5	106.2	103.2	105.4	104.6	104.2	117.1	112.1	131.3	102.5	113.6	124.7	106.7	113.6	111.6	111.1	111.3
05	METALS	91.7	96.7	93.3	91.9	95.9	93.2	95.4	100.7	97.0	91.6	97.1	93.3	92.4	100.7	95.1	91.5	97.7	93.4
06	WOOD, PLASTICS & COMPOSITES	86.9	96.0	92.0	87.9	95.9	92.4	98.4	93.2	95.5	84.6	96.0	90.9	89.2	87.0	88.0	86.4	98.0	92.9
07	THERMAL & MOISTURE PROTECTION	105.3	84.1	96.4	101.9	91.4	97.5	102.6	107.6	104.7	106.2	84.1	96.9	105.1	91.1	99.3	105.2	86.6	97.4
08	OPENINGS	99.9	93.5	98.4	96.9	98.1	97.2	101.0	98.2	100.4	99.9	84.6	96.4	92.1	90.0	91.6	100.1	94.1	98.7
0920	Plaster & Gypsum Board	89.0	96.2	93.8	86.0	96.2	92.8	93.1	93.4	93.3	89.0	96.2	93.8	90.1	87.2	88.1	89.0	98.3	95.2
0950, 0980	Ceilings & Acoustic Treatment	82.8	96.2	91.6	81.0	96.2	91.0	81.8	93.4	89.5	82.8	96.2	91.6	57.8	87.2	77.2	82.8	98.3	93.0
0960	Flooring	96.8	109.7	100.5	93.6	116.6	100.2	98.0	108.3	100.9	96.1	112.7	100.9	100.1	122.6	106.6	96.7	112.7	101.3
0970, 0990	Wall Finishes & Painting/Coating	95.2	101.4	98.8	98.4	103.4	101.3	98.3	117.9	109.7	95.2	60.4	74.9	89.3	102.8	97.2	95.2	83.2	88.2
09	FINISHES	92.2	98.1	95.4	91.4	100.3	96.3	94.5	101.7	98.5	93.0	94.7	94.0	87.1	96.7	92.4	91.9	101.7	97.3
COVERS	DIVS. 10 - 14, 25, 28, 41, 43, 44, 46	100.0	62.7	91.9	100.0	61.5	91.7	100.0	99.7	99.9	100.0	92.5	98.4	100.0	91.5	98.1	100.0	100.1	100.0
21, 22, 23	FIRE SUPPRESSION, PLUMBING & HVAC	96.9	84.5	91.7	96.7	94.4	95.7	100.0	96.6	98.6	96.9	85.8	92.2	96.1	94.4	95.3	96.9	89.7	93.9
26, 27, 3370	ELECTRICAL, COMMUNICATIONS & UTIL.	103.5	79.8	91.6	102.9	91.4	97.1	99.3	98.8	99.1	102.9	81.2	91.9	107.8	100.2	104.0	104.7	83.9	94.2
MF2014	WEIGHTED AVERAGE	97.2	91.1	94.5	95.3	96.4	95.8	98.7	102.0	100.2	99.8	91.8	96.3	96.8	97.7	97.2	96.5	96.1	96.3

For customer support on your Mechanical Cost Data, call 877.739.3689.

693

City Cost Indexes

WYOMING

DIVISION		CASPER 826 MAT.	INST.	TOTAL	CHEYENNE 820 MAT.	INST.	TOTAL	NEWCASTLE 827 MAT.	INST.	TOTAL	RAWLINS 823 MAT.	INST.	TOTAL	RIVERTON 825 MAT.	INST.	TOTAL	ROCK SPRINGS 829-831 MAT.	INST.	TOTAL
015433	CONTRACTOR EQUIPMENT		96.7	96.7		96.7	96.7		96.7	96.7		96.7	96.7		96.7	96.7		96.7	96.7
0241, 31-34	SITE & INFRASTRUCTURE, DEMOLITION	99.9	93.2	95.1	95.1	93.2	93.7	87.5	93.1	91.5	101.8	93.1	95.7	95.0	93.2	93.7	91.8	92.8	92.5
0310	Concrete Forming & Accessories	103.0	48.8	56.1	104.6	60.1	66.1	95.2	72.1	75.2	99.5	72.4	76.0	94.1	60.1	64.7	101.9	60.5	66.1
0320	Concrete Reinforcing	106.5	70.2	87.7	102.0	70.2	85.5	109.7	70.4	89.3	109.4	70.4	89.2	110.4	70.2	89.5	110.4	69.8	89.4
0330	Cast-in-Place Concrete	94.5	59.8	80.8	89.3	59.9	77.7	90.2	58.9	77.9	90.3	66.6	80.9	90.2	60.1	78.3	90.2	58.9	77.8
03	CONCRETE	102.1	57.2	80.3	101.1	62.4	82.3	101.4	67.4	84.9	116.0	70.1	93.7	110.2	62.4	87.0	101.9	62.1	82.5
04	MASONRY	113.5	52.3	75.8	111.0	52.3	74.8	107.5	58.4	77.3	107.5	58.4	77.3	107.6	52.3	73.5	173.9	50.1	97.6
05	METALS	104.0	71.6	93.7	106.5	71.7	95.4	102.5	71.6	92.7	102.5	71.9	92.8	102.6	72.0	92.9	103.4	71.2	93.1
06	WOOD, PLASTICS & COMPOSITES	91.5	48.0	67.3	89.4	63.3	74.8	80.4	78.3	79.2	84.0	78.3	80.8	79.4	63.3	70.4	88.8	64.6	75.3
07	THERMAL & MOISTURE PROTECTION	108.9	52.7	85.4	103.1	54.3	82.7	104.8	57.0	84.8	106.3	69.3	90.8	105.7	55.1	84.5	104.9	58.5	85.5
08	OPENINGS	106.9	51.4	94.0	106.1	59.7	95.3	110.5	67.4	100.5	110.2	67.4	100.2	110.4	59.7	98.6	110.9	59.9	99.1
0920	Plaster & Gypsum Board	98.1	46.5	63.6	90.2	62.2	71.5	86.7	77.7	80.7	87.1	77.7	80.8	86.7	62.2	70.3	98.3	63.5	75.1
0950, 0980	Ceilings & Acoustic Treatment	98.8	46.5	64.2	92.6	62.2	72.5	94.5	77.7	83.3	94.5	77.7	83.3	94.5	62.2	73.1	94.5	63.5	74.0
0960	Flooring	116.1	67.2	102.0	114.0	67.2	100.5	107.3	49.7	90.7	110.2	49.7	92.8	106.8	67.2	95.4	112.6	49.7	94.5
0970, 0990	Wall Finishes & Painting/Coating	90.7	38.5	60.4	96.4	38.5	62.7	93.0	57.4	72.3	93.0	57.4	72.3	93.0	38.5	61.3	93.0	55.4	71.1
09	FINISHES	103.0	50.3	73.9	100.2	59.4	77.6	95.2	67.5	79.9	97.6	67.5	80.9	96.0	59.4	75.7	98.5	58.1	76.1
COVERS	DIVS. 10 - 14, 25, 28, 41, 43, 44, 46	100.0	85.3	96.8	100.0	87.0	97.2	100.0	89.5	97.7	100.0	89.4	97.7	100.0	81.0	95.9	100.0	78.7	95.4
21, 22, 23	FIRE SUPPRESSION, PLUMBING & HVAC	99.9	68.1	86.5	99.9	68.1	86.5	98.1	69.3	85.9	98.1	69.3	86.0	98.1	68.1	85.4	99.9	67.3	86.1
26, 27, 3370	ELECTRICAL, COMMUNICATIONS & UTIL.	98.2	64.4	81.1	100.0	65.9	82.8	98.9	85.4	92.1	98.9	71.2	84.9	98.9	65.9	82.2	97.2	85.4	91.2
MF2014	WEIGHTED AVERAGE	102.5	63.8	85.5	102.3	66.5	86.5	100.8	72.4	88.3	103.1	71.2	89.0	102.1	66.4	86.4	104.9	68.4	88.8

WYOMING / CANADA

DIVISION		SHERIDAN 828 MAT.	INST.	TOTAL	WHEATLAND 822 MAT.	INST.	TOTAL	WORLAND 824 MAT.	INST.	TOTAL	YELLOWSTONE NAT'L PA 821 MAT.	INST.	TOTAL	BARRIE, ONTARIO MAT.	INST.	TOTAL	BATHURST, NEW BRUNSWICK MAT.	INST.	TOTAL
015433	CONTRACTOR EQUIPMENT		96.7	96.7		96.7	96.7		96.7	96.7		96.7	96.7		101.8	101.8		101.6	101.6
0241, 31-34	SITE & INFRASTRUCTURE, DEMOLITION	95.3	93.2	93.8	92.3	92.9	92.7	89.7	92.8	91.9	89.8	93.1	92.1	119.8	99.9	105.7	105.0	95.8	98.5
0310	Concrete Forming & Accessories	102.2	60.1	65.8	97.2	46.7	53.5	97.3	60.3	65.3	97.3	61.9	66.7	125.0	86.9	92.0	103.7	61.8	67.4
0320	Concrete Reinforcing	110.4	70.2	89.5	109.7	70.3	89.3	110.4	69.7	89.3	112.3	70.4	90.6	175.6	82.8	127.5	136.4	55.9	94.7
0330	Cast-in-Place Concrete	93.3	59.9	80.1	94.1	56.6	79.3	90.2	58.8	77.8	90.2	61.1	78.7	163.9	88.0	134.0	130.5	61.0	103.1
03	CONCRETE	110.1	62.4	86.9	106.5	55.2	81.6	101.6	62.0	82.4	101.9	63.6	83.3	150.4	86.8	119.5	123.6	61.2	93.3
04	MASONRY	107.8	52.3	73.6	107.9	54.4	74.9	107.6	42.3	67.3	107.6	58.7	77.5	173.7	97.6	126.8	161.5	64.7	101.8
05	METALS	106.4	71.7	95.4	102.4	71.3	92.5	102.6	70.6	92.5	103.3	71.6	93.2	109.3	89.7	103.0	109.9	71.6	97.7
06	WOOD, PLASTICS & COMPOSITES	90.7	63.3	75.4	82.0	46.3	62.1	82.1	64.6	72.3	82.1	64.6	72.3	118.8	85.2	100.1	100.0	61.8	78.7
07	THERMAL & MOISTURE PROTECTION	106.0	54.3	84.4	105.1	52.2	82.9	104.9	53.8	83.5	104.3	57.5	84.7	113.6	87.2	102.6	109.4	60.9	89.1
08	OPENINGS	111.1	59.7	99.2	109.0	50.7	95.5	110.8	59.9	98.9	103.3	59.9	93.2	94.7	84.0	92.2	89.3	54.3	81.2
0920	Plaster & Gypsum Board	111.2	62.2	78.4	86.7	44.7	58.7	86.7	63.5	71.2	86.9	63.5	71.3	148.8	84.7	106.0	121.2	60.6	80.7
0950, 0980	Ceilings & Acoustic Treatment	97.3	62.2	74.0	94.5	44.7	61.5	94.5	63.5	74.0	95.3	63.5	74.3	92.2	84.7	87.3	108.0	60.6	76.6
0960	Flooring	111.4	67.2	98.7	108.9	48.8	91.6	108.9	49.7	91.9	108.9	49.7	91.9	126.6	95.0	117.5	107.8	45.3	89.9
0970, 0990	Wall Finishes & Painting/Coating	95.2	38.5	62.3	93.0	45.0	65.1	93.0	55.4	71.1	93.0	55.4	71.1	107.4	89.9	97.2	113.2	51.6	77.4
09	FINISHES	103.3	59.4	79.0	96.0	45.9	68.3	95.7	58.1	74.9	95.9	59.2	75.6	114.4	88.9	100.3	109.7	57.9	81.1
COVERS	DIVS. 10 - 14, 25, 28, 41, 43, 44, 46	100.0	87.0	97.2	100.0	77.4	95.1	100.0	79.2	95.5	100.0	80.5	95.8	139.2	72.3	124.7	131.1	63.4	116.5
21, 22, 23	FIRE SUPPRESSION, PLUMBING & HVAC	98.1	68.1	85.4	98.1	67.2	85.1	98.1	67.2	85.1	98.1	69.3	85.9	103.5	98.2	101.3	103.6	69.0	89.0
26, 27, 3370	ELECTRICAL, COMMUNICATIONS & UTIL.	101.2	65.9	83.4	98.9	85.4	92.1	98.9	85.4	92.1	97.8	96.0	96.9	114.6	89.8	102.0	109.3	61.4	85.1
MF2014	WEIGHTED AVERAGE	103.7	66.5	87.3	101.4	65.5	85.6	101.0	67.4	86.2	100.2	71.5	87.6	118.3	91.7	106.6	111.8	66.3	91.8

CANADA

DIVISION		BRANDON, MANITOBA MAT.	INST.	TOTAL	BRANTFORD, ONTARIO MAT.	INST.	TOTAL	BRIDGEWATER, NOVA SCOTIA MAT.	INST.	TOTAL	CALGARY, ALBERTA MAT.	INST.	TOTAL	CAP-DE-LA-MADELEINE, QUEBEC MAT.	INST.	TOTAL	CHARLESBOURG, QUEBEC MAT.	INST.	TOTAL
015433	CONTRACTOR EQUIPMENT		104.1	104.1		101.8	101.8		101.4	101.4		106.9	106.9		102.2	102.2		102.2	102.2
0241, 31-34	SITE & INFRASTRUCTURE, DEMOLITION	138.3	98.8	110.4	119.2	100.2	105.8	103.9	97.5	99.4	123.0	104.8	110.2	99.4	99.1	99.2	99.4	99.1	99.2
0310	Concrete Forming & Accessories	146.7	70.4	80.7	124.1	94.1	98.2	96.7	69.5	73.2	123.4	93.3	97.4	130.0	85.4	91.4	130.0	85.4	91.4
0320	Concrete Reinforcing	173.8	52.6	111.0	160.7	81.6	119.7	136.7	46.2	89.8	128.3	73.5	99.9	136.7	77.9	106.3	136.7	77.9	106.3
0330	Cast-in-Place Concrete	131.5	75.8	109.5	150.0	109.2	133.9	154.1	71.3	121.5	177.2	104.2	148.4	121.1	96.7	111.4	121.1	96.7	111.4
03	CONCRETE	144.3	69.7	108.0	139.3	97.0	118.7	137.4	66.6	103.0	151.7	93.5	123.4	122.0	88.1	105.5	122.0	88.1	105.5
04	MASONRY	242.4	67.3	134.5	169.6	102.1	128.0	165.0	71.6	107.5	197.5	87.1	129.5	165.5	86.7	116.9	165.5	86.7	116.9
05	METALS	125.4	76.7	109.9	110.2	90.3	103.9	109.3	74.1	98.2	140.6	89.1	124.2	108.6	86.9	101.7	108.6	86.9	101.7
06	WOOD, PLASTICS & COMPOSITES	151.0	71.0	106.5	121.6	92.5	105.4	92.3	68.6	79.1	98.5	94.0	96.0	132.5	85.0	106.0	132.5	85.0	106.0
07	THERMAL & MOISTURE PROTECTION	141.4	70.3	111.7	117.1	92.6	106.9	111.8	68.5	93.7	122.5	90.9	109.3	110.6	87.9	101.1	110.6	87.9	101.1
08	OPENINGS	105.5	62.9	95.6	93.6	90.0	92.8	87.5	63.0	81.8	87.7	84.1	86.9	94.8	78.7	91.1	94.8	78.7	91.1
0920	Plaster & Gypsum Board	111.0	69.7	83.4	117.6	92.2	100.7	119.5	67.6	84.8	123.5	93.3	103.3	142.0	84.4	103.5	142.0	84.4	103.5
0950, 0980	Ceilings & Acoustic Treatment	108.2	69.7	82.7	98.9	92.2	94.5	98.9	67.6	78.2	141.8	93.3	109.7	98.9	84.4	89.3	98.9	84.4	89.3
0960	Flooring	143.1	67.5	121.4	122.5	95.0	114.6	103.5	64.6	92.3	129.7	91.2	118.6	122.5	95.7	114.8	122.5	95.7	114.8
0970, 0990	Wall Finishes & Painting/Coating	120.2	58.7	84.4	112.3	98.7	104.3	112.3	64.1	84.2	125.5	112.4	117.9	112.3	91.7	100.3	112.3	91.7	100.3
09	FINISHES	123.4	69.0	93.3	112.3	95.1	102.8	106.3	68.2	85.2	130.2	95.3	110.9	115.1	88.2	100.2	115.1	88.2	100.2
COVERS	DIVS. 10 - 14, 25, 28, 41, 43, 44, 46	131.1	65.6	116.9	131.1	74.4	118.8	131.1	64.6	116.7	131.1	94.9	123.3	131.1	82.8	120.7	131.1	82.8	120.7
21, 22, 23	FIRE SUPPRESSION, PLUMBING & HVAC	103.7	83.2	95.0	103.6	101.2	102.6	103.6	83.5	95.1	105.5	90.0	98.9	103.9	90.0	98.1	103.9	90.0	98.1
26, 27, 3370	ELECTRICAL, COMMUNICATIONS & UTIL.	116.2	69.1	92.4	108.9	89.3	99.0	113.6	63.9	88.5	103.7	100.6	102.1	108.4	72.1	90.0	108.4	72.1	90.0
MF2014	WEIGHTED AVERAGE	126.4	74.8	103.7	115.5	95.5	106.7	113.4	73.6	95.9	123.8	93.4	110.4	112.8	86.4	101.1	112.8	86.4	101.1

For customer support on your Mechanical Cost Data, call 877.739.3689.

CANADA

DIVISION		CHARLOTTETOWN, PRINCE EDWARD ISLAND			CHICOUTIMI, QUEBEC			CORNER BROOK, NEWFOUNDLAND			CORNWALL, ONTARIO			DALHOUSIE, NEW BRUNSWICK			DARTMOUTH, NOVA SCOTIA		
		MAT.	INST.	TOTAL	MAT.	INST.	TOTAL	MAT.	INST.	TOTAL	MAT.	INST.	TOTAL	MAT.	INST.	TOTAL	MAT.	INST.	TOTAL
015433	CONTRACTOR EQUIPMENT		101.3	101.3		102.2	102.2		102.5	102.5		101.8	101.8		101.6	101.6		101.4	101.4
0241, 31 - 34	SITE & INFRASTRUCTURE, DEMOLITION	125.1	95.0	103.8	102.0	98.4	99.5	143.8	96.2	110.2	117.2	99.7	104.8	102.3	95.8	97.7	129.5	97.5	106.9
0310	Concrete Forming & Accessories	115.8	56.5	64.5	132.5	90.5	96.2	123.6	59.8	68.4	121.9	87.0	91.7	102.6	62.0	67.5	113.2	69.5	75.4
0320	Concrete Reinforcing	143.0	46.2	92.9	102.9	94.8	98.7	159.4	48.0	101.7	160.7	81.3	119.6	138.2	55.9	95.6	166.6	46.2	104.2
0330	Cast-in-Place Concrete	157.7	60.3	119.3	122.9	98.8	113.4	153.0	69.6	120.1	134.9	99.5	121.0	127.2	61.1	101.1	147.7	71.3	117.6
03	CONCRETE	143.3	56.8	101.3	113.9	94.1	104.3	182.4	62.0	123.9	131.7	90.4	111.7	129.7	61.3	96.4	161.7	66.6	115.4
04	MASONRY	179.0	62.3	107.1	163.6	94.6	121.1	238.0	63.9	130.7	168.4	93.6	122.3	162.9	64.7	102.3	253.5	71.6	141.4
05	METALS	134.3	67.8	113.2	109.7	91.6	103.9	125.8	72.8	108.9	110.2	89.0	103.5	106.6	71.8	95.5	125.9	74.1	109.5
06	WOOD, PLASTICS & COMPOSITES	96.0	55.8	73.6	133.0	90.5	109.3	129.6	58.7	90.1	120.0	86.0	101.1	99.3	61.8	78.4	117.9	68.6	90.5
07	THERMAL & MOISTURE PROTECTION	127.9	59.2	99.2	109.8	94.2	103.3	146.4	60.5	110.5	117.0	87.3	104.6	113.7	60.9	91.6	143.4	68.5	112.0
08	OPENINGS	85.7	48.5	77.0	93.5	78.2	90.0	112.1	55.4	98.9	94.8	83.7	92.3	90.4	54.3	82.0	95.6	63.0	88.0
0920	Plaster & Gypsum Board	118.5	54.4	75.7	139.5	90.0	106.4	141.5	57.3	85.3	170.7	85.6	113.8	129.8	60.0	83.6	137.5	67.6	90.8
0950, 0980	Ceilings & Acoustic Treatment	121.7	54.4	77.2	107.1	90.0	95.8	109.0	57.3	74.8	101.4	85.6	90.9	102.1	60.6	74.6	115.6	67.6	83.8
0960	Flooring	116.4	60.5	100.3	124.9	95.7	116.5	123.5	55.0	103.8	122.5	93.7	114.2	109.5	69.3	98.0	118.0	64.6	102.7
0970, 0990	Wall Finishes & Painting/Coating	124.1	43.0	76.9	113.2	106.4	109.2	120.1	61.8	86.2	112.3	91.9	100.4	115.5	51.6	78.3	120.1	64.1	87.5
09	FINISHES	121.5	56.1	85.3	117.0	93.8	104.1	122.6	59.1	87.5	120.4	89.0	103.0	111.6	62.6	84.5	120.4	68.2	91.6
COVERS	DIVS. 10 - 14, 25, 28, 41, 43, 44, 46	131.1	62.9	116.4	131.1	85.8	121.3	131.1	63.6	116.5	131.1	71.9	118.3	131.1	63.4	116.5	131.1	64.6	116.7
21, 22, 23	FIRE SUPPRESSION, PLUMBING & HVAC	104.1	63.1	86.8	103.6	93.5	99.3	103.7	71.0	89.9	103.9	99.1	101.9	103.6	69.0	89.0	103.7	83.5	95.1
26, 27, 3370	ELECTRICAL, COMMUNICATIONS & UTIL.	107.8	51.9	79.5	106.5	86.1	96.2	113.8	57.8	85.5	109.5	90.2	99.8	111.4	58.0	84.4	118.3	63.9	90.8
MF2014	WEIGHTED AVERAGE	120.4	61.9	94.6	111.7	92.0	103.0	131.2	66.6	102.7	115.4	92.0	105.1	112.6	66.5	92.3	127.5	73.6	103.7

CANADA

DIVISION		EDMONTON, ALBERTA			FORT MCMURRAY, ALBERTA			FREDERICTON, NEW BRUNSWICK			GATINEAU, QUEBEC			GRANBY, QUEBEC			HALIFAX, NOVA SCOTIA		
		MAT.	INST.	TOTAL	MAT.	INST.	TOTAL	MAT.	INST.	TOTAL	MAT.	INST.	TOTAL	MAT.	INST.	TOTAL	MAT.	INST.	TOTAL
015433	CONTRACTOR EQUIPMENT		106.9	106.9		104.2	104.2		101.6	101.6		102.2	102.2		102.2	102.2		101.4	101.4
0241, 31 - 34	SITE & INFRASTRUCTURE, DEMOLITION	139.3	104.8	114.9	124.4	101.0	107.8	106.8	95.9	99.1	99.2	99.1	99.1	99.7	99.1	99.3	108.2	97.7	100.7
0310	Concrete Forming & Accessories	119.2	93.3	96.8	121.9	89.8	94.1	122.5	62.3	70.4	130.0	85.3	91.3	130.0	85.2	91.3	113.2	76.6	81.5
0320	Concrete Reinforcing	128.2	73.5	99.9	148.7	73.5	109.7	133.9	56.0	93.6	144.6	77.9	110.1	144.6	77.9	110.0	142.3	61.9	100.7
0330	Cast-in-Place Concrete	183.2	104.2	152.0	199.2	104.1	161.7	124.5	61.1	99.5	119.4	96.6	110.4	123.4	96.6	112.8	128.0	80.9	109.4
03	CONCRETE	154.4	93.5	124.8	161.5	91.9	127.7	126.2	61.5	94.7	122.3	88.0	105.7	124.3	88.0	106.6	127.8	75.9	102.6
04	MASONRY	190.2	87.1	126.7	210.0	88.7	135.3	187.2	66.2	112.6	165.4	86.7	116.9	165.7	86.7	117.0	184.5	84.3	122.8
05	METALS	140.5	89.1	124.2	136.9	88.7	121.5	130.4	72.3	111.9	108.6	86.8	101.7	108.6	86.8	101.6	132.4	81.4	116.2
06	WOOD, PLASTICS & COMPOSITES	98.3	94.0	95.9	115.9	88.8	100.8	110.8	61.8	83.5	132.5	85.0	106.0	132.5	85.0	106.0	96.3	75.2	84.5
07	THERMAL & MOISTURE PROTECTION	125.4	90.9	111.0	125.7	90.0	110.7	123.6	61.9	97.8	110.6	87.9	101.1	110.6	86.4	100.5	120.6	77.7	102.7
08	OPENINGS	85.5	84.1	85.2	94.8	81.2	91.7	89.5	53.2	81.0	94.8	74.1	90.0	94.8	74.1	90.0	93.9	70.3	88.4
0920	Plaster & Gypsum Board	116.1	93.3	100.8	116.3	88.0	97.4	119.7	60.6	80.2	115.9	84.4	94.8	115.9	84.4	94.8	114.9	74.3	87.8
0950, 0980	Ceilings & Acoustic Treatment	134.6	93.3	107.2	106.3	88.0	94.2	121.9	60.6	81.3	98.9	84.4	89.3	98.9	84.4	89.3	120.2	74.3	89.9
0960	Flooring	125.6	91.2	115.7	122.5	91.2	113.5	125.2	72.4	110.0	122.5	95.7	114.8	122.5	95.7	114.8	106.8	87.0	101.1
0970, 0990	Wall Finishes & Painting/Coating	116.7	112.4	114.2	112.4	95.5	102.5	121.3	65.8	89.0	112.3	91.7	100.3	112.3	91.7	100.3	115.1	87.4	99.0
09	FINISHES	125.5	95.3	108.8	115.0	90.7	101.6	123.3	64.7	90.9	111.3	88.2	98.5	111.3	88.2	98.5	115.1	79.9	95.6
COVERS	DIVS. 10 - 14, 25, 28, 41, 43, 44, 46	131.1	94.9	123.3	131.1	94.6	123.2	131.1	63.4	116.5	131.1	82.8	120.7	131.1	82.8	120.7	131.1	67.8	117.4
21, 22, 23	FIRE SUPPRESSION, PLUMBING & HVAC	105.4	90.0	98.9	104.1	97.9	101.5	104.0	78.5	93.2	103.9	90.0	98.1	103.6	90.0	97.8	104.9	76.8	93.0
26, 27, 3370	ELECTRICAL, COMMUNICATIONS & UTIL.	106.4	100.6	103.5	103.7	84.8	94.2	112.8	76.5	94.4	108.4	72.1	90.0	108.9	72.1	90.3	105.2	80.6	92.8
MF2014	WEIGHTED AVERAGE	123.9	93.4	110.5	124.5	91.7	110.0	118.8	71.5	97.9	112.5	86.1	100.9	112.7	86.1	101.0	118.2	79.9	101.3

CANADA

DIVISION		HAMILTON, ONTARIO			HULL, QUEBEC			JOLIETTE, QUEBEC			KAMLOOPS, BRITISH COLUMBIA			KINGSTON, ONTARIO			KITCHENER, ONTARIO		
		MAT.	INST.	TOTAL	MAT.	INST.	TOTAL	MAT.	INST.	TOTAL	MAT.	INST.	TOTAL	MAT.	INST.	TOTAL	MAT.	INST.	TOTAL
015433	CONTRACTOR EQUIPMENT		108.6	108.6		102.2	102.2		102.2	102.2		105.5	105.5		104.1	104.1		104.0	104.0
0241, 31 - 34	SITE & INFRASTRUCTURE, DEMOLITION	112.3	112.4	112.4	99.2	99.1	99.1	99.9	99.1	99.3	121.4	102.6	108.1	117.2	103.6	107.6	99.4	104.5	103.0
0310	Concrete Forming & Accessories	120.6	92.0	95.9	130.0	85.3	91.3	130.0	85.4	91.4	122.4	88.2	92.8	122.0	87.1	91.8	112.8	84.8	88.6
0320	Concrete Reinforcing	137.3	91.0	113.3	144.6	77.9	110.1	136.7	77.9	106.3	107.2	75.9	91.0	160.7	81.3	119.6	96.8	90.9	93.7
0330	Cast-in-Place Concrete	135.9	100.0	121.7	119.4	96.6	110.4	124.4	96.7	113.5	107.1	100.5	104.8	134.9	99.5	121.0	126.0	92.6	112.9
03	CONCRETE	126.7	94.6	111.1	122.3	88.0	105.7	123.6	88.1	106.3	130.5	90.3	110.9	133.6	90.4	112.6	108.4	88.8	99.0
04	MASONRY	178.7	100.7	130.7	165.4	86.7	116.9	165.8	86.7	117.0	172.6	94.5	124.5	175.3	93.6	125.0	148.4	97.2	116.8
05	METALS	126.6	92.1	115.6	108.6	86.8	101.7	108.6	86.9	101.7	110.8	86.9	103.2	111.8	89.0	104.5	117.4	91.8	109.2
06	WOOD, PLASTICS & COMPOSITES	97.8	90.8	93.9	132.5	85.0	106.0	132.5	85.0	106.0	104.3	86.4	94.3	120.0	86.2	101.2	109.2	82.6	94.4
07	THERMAL & MOISTURE PROTECTION	118.6	93.5	108.1	110.6	87.9	101.1	110.6	87.9	101.1	127.0	85.0	109.4	117.0	88.4	105.0	111.2	90.8	102.7
08	OPENINGS	90.4	89.5	90.2	94.8	74.1	90.0	94.8	78.7	91.1	91.5	83.2	89.6	94.8	83.4	92.2	86.4	83.3	85.7
0920	Plaster & Gypsum Board	131.9	90.4	104.2	115.9	84.4	94.8	142.0	84.4	103.5	103.5	85.5	91.5	173.8	85.7	114.9	106.5	82.0	90.2
0950, 0980	Ceilings & Acoustic Treatment	127.6	90.4	103.0	98.9	84.4	89.3	98.9	84.4	89.3	98.9	85.5	90.0	114.6	85.7	95.5	101.7	82.0	88.7
0960	Flooring	120.0	100.7	114.5	122.5	95.7	114.8	122.5	95.7	114.8	121.1	53.5	101.7	122.5	93.7	114.2	106.9	100.7	107.1
0970, 0990	Wall Finishes & Painting/Coating	109.4	102.7	105.5	112.3	91.7	100.3	112.3	91.7	100.3	112.3	84.0	95.8	112.3	84.9	96.4	106.2	92.6	98.3
09	FINISHES	122.0	94.7	106.9	111.3	88.2	98.5	115.1	88.2	100.2	111.7	81.8	95.2	123.7	88.3	104.1	104.8	88.1	95.6
COVERS	DIVS. 10 - 14, 25, 28, 41, 43, 44, 46	131.1	98.2	124.0	131.1	82.8	120.7	131.1	82.8	120.7	131.1	92.9	122.9	131.1	71.9	118.3	131.1	96.4	123.6
21, 22, 23	FIRE SUPPRESSION, PLUMBING & HVAC	105.7	88.2	98.3	103.6	90.0	97.8	103.6	90.0	97.9	103.6	93.4	99.3	103.9	99.2	101.9	104.7	86.7	97.1
26, 27, 3370	ELECTRICAL, COMMUNICATIONS & UTIL.	101.9	99.7	100.8	110.0	72.1	90.8	108.9	72.1	90.3	111.9	81.3	96.4	109.5	88.9	99.1	106.3	97.3	101.7
MF2014	WEIGHTED AVERAGE	116.9	95.6	107.5	112.6	86.1	100.9	113.0	86.4	101.3	115.0	89.3	103.7	116.5	92.1	105.7	109.9	91.8	101.9

For customer support on your Mechanical Cost Data, call 877.739.3689.

695

City Cost Indexes

CANADA

DIVISION		LAVAL, QUEBEC			LETHBRIDGE, ALBERTA			LLOYDMINSTER, ALBERTA			LONDON, ONTARIO			MEDICINE HAT, ALBERTA			MONCTON, NEW BRUNSWICK		
		MAT.	INST.	TOTAL	MAT.	INST.	TOTAL	MAT.	INST.	TOTAL	MAT.	INST.	TOTAL	MAT.	INST.	TOTAL	MAT.	INST.	TOTAL
015433	CONTRACTOR EQUIPMENT		102.2	102.2		104.2	104.2		104.2	104.2		104.2	104.2		104.2	104.2		101.6	101.6
0241, 31 - 34	SITE & INFRASTRUCTURE, DEMOLITION	99.7	99.1	99.3	116.6	101.6	106.0	116.4	101.0	105.5	108.2	104.6	105.7	115.2	101.0	105.2	104.3	96.2	98.6
0310	Concrete Forming & Accessories	130.2	85.3	91.4	123.4	89.8	94.4	121.5	80.3	85.9	122.1	86.1	90.9	123.3	80.3	86.1	103.7	64.4	69.7
0320	Concrete Reinforcing	144.6	77.9	110.1	148.7	73.5	109.7	148.7	73.5	109.7	136.3	90.9	112.8	148.7	73.4	109.7	136.4	60.9	97.3
0330	Cast-in-Place Concrete	123.4	96.6	112.8	149.4	104.2	131.6	138.6	100.5	123.6	139.6	97.6	123.0	138.6	100.5	123.6	125.7	79.2	107.3
03	CONCRETE	124.3	88.0	106.7	137.2	91.9	115.2	131.8	86.4	109.7	128.5	91.0	110.3	131.9	86.4	109.6	121.2	69.8	96.2
04	MASONRY	165.7	86.7	117.0	183.5	88.7	125.1	165.0	82.1	113.9	178.4	98.3	129.0	165.0	82.1	113.9	161.1	66.3	102.7
05	METALS	108.7	86.8	101.7	130.7	88.8	117.4	111.0	88.6	103.9	126.6	91.4	115.4	111.0	88.5	103.8	109.9	81.1	100.7
06	WOOD, PLASTICS & COMPOSITES	132.6	85.0	106.1	119.3	88.8	102.3	115.9	79.3	95.5	107.2	83.7	94.1	119.3	79.3	97.0	100.0	63.8	79.8
07	THERMAL & MOISTURE PROTECTION	111.1	87.9	101.4	122.9	90.0	109.1	119.5	85.7	105.3	120.2	91.6	108.3	125.9	85.7	109.1	113.7	65.0	93.3
08	OPENINGS	94.8	74.1	90.0	94.8	81.2	91.7	94.8	76.0	90.5	84.0	84.6	84.1	94.8	76.0	90.5	89.3	60.9	82.7
0920	Plaster & Gypsum Board	116.2	84.4	94.9	107.7	88.0	94.5	103.6	78.3	86.7	138.0	83.2	101.4	105.9	78.3	87.4	121.2	62.6	82.0
0950, 0980	Ceilings & Acoustic Treatment	98.9	84.4	89.3	106.3	88.0	94.2	98.9	78.3	85.2	125.9	83.2	97.6	98.9	78.3	85.2	108.0	62.6	77.9
0960	Flooring	122.5	95.7	114.8	122.5	91.2	113.5	122.5	91.2	113.5	118.7	100.7	113.5	122.5	91.2	113.5	107.8	71.0	97.3
0970, 0990	Wall Finishes & Painting/Coating	112.3	91.7	100.3	112.2	104.3	107.6	112.4	81.2	94.3	114.9	99.4	105.9	112.2	81.2	94.2	113.2	52.9	78.1
09	FINISHES	111.3	88.2	98.6	112.9	91.7	101.1	110.8	82.0	94.8	123.2	89.8	104.7	110.9	82.0	94.9	109.7	64.4	84.7
COVERS	DIVS. 10 - 14, 25, 28, 41, 43, 44, 46	131.1	82.8	120.7	131.1	94.6	123.2	131.1	91.4	122.5	131.1	96.9	123.7	131.1	91.4	122.5	131.1	65.1	116.8
21, 22, 23	FIRE SUPPRESSION, PLUMBING & HVAC	104.7	90.0	98.5	103.9	94.6	100.0	103.9	94.6	100.0	105.6	85.7	97.2	103.6	91.3	98.4	103.6	71.4	90.0
26, 27, 3370	ELECTRICAL, COMMUNICATIONS & UTIL.	107.0	72.1	89.3	104.9	84.8	94.8	102.9	84.8	93.8	100.2	96.8	98.5	102.9	84.8	93.8	113.0	63.3	87.9
MF2014	WEIGHTED AVERAGE	112.8	86.1	101.1	119.1	91.2	106.8	113.9	87.9	102.5	116.3	92.2	105.7	114.1	87.2	102.2	112.0	70.7	93.8

CANADA

DIVISION		MONTREAL, QUEBEC			MOOSE JAW, SASKATCHEWAN			NEW GLASGOW, NOVA SCOTIA			NEWCASTLE, NEW BRUNSWICK			NORTH BAY, ONTARIO			OSHAWA, ONTARIO		
		MAT.	INST.	TOTAL	MAT.	INST.	TOTAL	MAT.	INST.	TOTAL	MAT.	INST.	TOTAL	MAT.	INST.	TOTAL	MAT.	INST.	TOTAL
015433	CONTRACTOR EQUIPMENT		103.8	103.8		100.3	100.3		101.4	101.4		101.6	101.6		101.8	101.8		104.0	104.0
0241, 31 - 34	SITE & INFRASTRUCTURE, DEMOLITION	111.1	98.7	102.4	116.7	95.3	101.6	123.3	97.5	105.0	105.0	95.8	98.5	138.8	99.3	110.9	110.7	103.8	105.8
0310	Concrete Forming & Accessories	118.4	90.8	94.6	106.5	58.7	65.2	113.2	69.5	75.4	103.7	62.0	67.6	147.6	84.5	93.0	118.3	89.2	93.1
0320	Concrete Reinforcing	131.7	94.9	112.6	104.9	61.2	82.2	159.4	46.2	100.8	136.4	55.9	94.7	188.3	80.9	132.6	153.3	86.4	118.7
0330	Cast-in-Place Concrete	150.2	100.1	130.4	134.5	70.2	109.1	147.7	71.3	117.6	130.5	61.1	103.1	142.3	86.1	120.1	145.7	88.7	123.2
03	CONCRETE	132.8	94.7	114.3	116.3	63.9	90.8	160.6	66.6	114.9	123.6	61.3	93.3	162.1	84.7	124.5	132.1	88.7	111.0
04	MASONRY	168.7	94.6	123.1	163.6	64.1	102.3	237.5	71.6	135.3	161.5	64.7	101.8	244.1	89.5	148.8	151.4	96.3	117.4
05	METALS	130.5	92.1	118.3	107.5	74.0	96.9	123.6	74.1	107.9	109.9	71.8	97.8	124.5	88.7	113.1	108.5	92.1	103.3
06	WOOD, PLASTICS & COMPOSITES	102.9	90.7	96.1	101.7	57.1	76.9	117.9	68.6	90.5	100.0	61.8	78.7	153.9	84.7	115.3	115.8	88.0	100.3
07	THERMAL & MOISTURE PROTECTION	119.9	94.7	109.4	110.0	62.8	90.2	143.4	68.5	112.0	113.7	60.9	91.6	150.4	83.7	122.5	112.1	87.9	102.0
08	OPENINGS	89.6	80.3	87.5	90.7	55.1	82.4	95.6	63.0	88.0	89.3	54.3	81.2	103.9	81.3	98.6	91.9	87.8	91.0
0920	Plaster & Gypsum Board	120.3	90.0	100.0	100.5	55.8	70.6	135.7	67.6	90.2	121.2	60.6	80.7	130.3	84.1	99.5	109.7	87.5	94.9
0950, 0980	Ceilings & Acoustic Treatment	130.9	90.0	103.8	98.9	55.8	70.4	108.2	67.6	81.3	108.0	60.6	76.6	108.2	84.1	92.3	98.4	87.5	91.2
0960	Flooring	122.9	98.0	115.7	111.9	60.8	97.2	118.0	64.6	102.7	107.8	69.3	96.8	143.1	93.7	128.9	112.8	103.3	110.0
0970, 0990	Wall Finishes & Painting/Coating	110.8	106.4	108.2	112.3	66.8	85.8	120.1	64.1	87.5	113.2	51.6	77.4	120.1	91.2	103.3	106.2	106.7	106.5
09	FINISHES	120.5	94.4	106.0	107.0	59.6	80.8	118.5	68.2	90.7	109.7	62.6	83.7	125.9	87.3	104.5	106.1	93.5	99.2
COVERS	DIVS. 10 - 14, 25, 28, 41, 43, 44, 46	131.1	86.4	121.4	131.1	62.7	116.3	131.1	64.6	116.7	131.1	63.4	116.5	131.1	70.7	118.0	131.1	97.3	123.8
21, 22, 23	FIRE SUPPRESSION, PLUMBING & HVAC	105.5	93.6	100.4	103.9	75.3	91.8	103.7	83.5	95.1	103.6	69.0	89.0	103.7	97.1	100.9	104.7	102.6	103.8
26, 27, 3370	ELECTRICAL, COMMUNICATIONS & UTIL.	105.7	86.1	95.8	110.9	62.3	86.3	114.1	63.9	88.7	108.7	61.4	84.8	114.4	90.1	102.1	107.4	92.6	99.9
MF2014	WEIGHTED AVERAGE	117.9	92.4	106.7	111.2	68.5	92.4	125.5	73.6	102.6	111.9	67.0	92.1	128.6	89.8	111.5	112.5	95.3	104.9

CANADA

DIVISION		OTTAWA, ONTARIO			OWEN SOUND, ONTARIO			PETERBOROUGH, ONTARIO			PORTAGE LA PRAIRIE, MANITOBA			PRINCE ALBERT, SASKATCHEWAN			PRINCE GEORGE, BRITISH COLUMBIA		
		MAT.	INST.	TOTAL	MAT.	INST.	TOTAL	MAT.	INST.	TOTAL	MAT.	INST.	TOTAL	MAT.	INST.	TOTAL	MAT.	INST.	TOTAL
015433	CONTRACTOR EQUIPMENT		104.0	104.0		101.8	101.8		101.8	101.8		104.1	104.1		100.3	100.3		105.5	105.5
0241, 31 - 34	SITE & INFRASTRUCTURE, DEMOLITION	105.9	104.3	104.8	119.8	99.7	105.6	119.2	99.6	105.4	117.9	98.7	104.4	112.1	95.5	100.4	125.0	102.6	109.2
0310	Concrete Forming & Accessories	122.1	88.9	93.4	125.0	83.1	88.7	124.1	85.6	90.8	123.5	69.9	77.1	106.5	58.5	65.0	111.9	83.1	87.0
0320	Concrete Reinforcing	129.0	90.9	109.3	175.6	82.8	127.5	160.7	81.3	119.6	148.7	52.6	98.9	109.6	61.1	84.5	107.2	75.9	91.0
0330	Cast-in-Place Concrete	133.4	98.9	119.8	163.9	82.1	131.7	150.0	87.7	125.4	138.6	75.2	113.6	121.9	70.1	101.5	134.9	100.5	121.4
03	CONCRETE	124.4	92.7	109.0	150.4	83.1	117.7	139.3	85.8	113.3	125.6	69.3	98.2	110.8	63.8	87.9	143.1	88.0	116.3
04	MASONRY	169.5	98.4	125.7	173.7	95.2	125.3	169.6	96.2	124.4	168.1	66.3	105.3	162.7	64.1	101.9	174.8	94.5	125.3
05	METALS	127.9	91.3	116.3	109.3	89.5	103.0	110.2	89.2	103.5	111.0	76.7	100.1	107.5	73.8	96.8	110.8	87.0	103.2
06	WOOD, PLASTICS & COMPOSITES	108.2	87.6	96.7	118.8	81.5	98.0	121.6	83.4	100.3	119.3	71.0	92.4	101.7	57.1	76.9	104.3	79.4	90.4
07	THERMAL & MOISTURE PROTECTION	126.1	91.7	111.7	113.6	84.5	101.4	117.1	89.2	105.4	110.4	69.8	93.4	109.9	61.7	89.7	120.9	84.2	105.6
08	OPENINGS	94.8	86.6	92.9	94.7	80.7	91.4	93.6	83.0	91.2	94.8	62.9	87.4	89.6	55.1	81.5	91.5	79.4	88.7
0920	Plaster & Gypsum Board	147.1	87.2	107.0	148.8	80.9	103.4	117.6	82.8	94.4	105.6	69.7	81.6	100.5	55.8	70.6	103.5	78.3	86.6
0950, 0980	Ceilings & Acoustic Treatment	129.0	87.2	101.3	92.2	80.9	84.7	98.9	82.8	88.3	98.9	69.7	79.6	98.9	55.8	70.4	98.9	78.3	85.2
0960	Flooring	111.4	96.0	107.0	126.6	95.0	117.5	122.5	93.7	114.2	122.5	67.5	106.7	111.9	60.8	97.2	117.7	73.3	104.9
0970, 0990	Wall Finishes & Painting/Coating	115.3	94.2	103.0	107.4	89.9	97.2	112.3	93.5	101.3	112.4	58.7	81.2	112.3	57.0	80.1	112.3	84.0	95.8
09	FINISHES	124.1	90.7	105.7	114.4	86.1	98.8	112.3	87.9	98.8	110.9	68.7	87.6	107.0	58.6	80.2	110.7	81.0	94.2
COVERS	DIVS. 10 - 14, 25, 28, 41, 43, 44, 46	131.1	95.1	123.3	139.2	71.1	124.5	131.1	72.1	118.4	131.1	65.2	116.9	131.1	62.7	116.3	131.1	92.1	122.7
21, 22, 23	FIRE SUPPRESSION, PLUMBING & HVAC	105.7	86.2	97.5	103.5	97.0	100.8	103.6	100.6	102.3	103.6	82.6	94.7	103.9	67.8	88.7	103.6	93.4	99.3
26, 27, 3370	ELECTRICAL, COMMUNICATIONS & UTIL.	105.1	98.0	101.5	115.9	88.9	102.2	108.9	89.8	99.2	110.6	59.6	84.8	110.9	62.3	86.3	109.0	81.3	95.0
MF2014	WEIGHTED AVERAGE	117.6	92.9	106.7	118.4	89.9	105.8	115.5	91.7	105.0	113.9	73.1	95.9	110.3	66.8	91.1	116.0	88.6	104.0

City Cost Indexes

CANADA

DIVISION		QUEBEC CITY, QUEBEC MAT.	INST.	TOTAL	RED DEER, ALBERTA MAT.	INST.	TOTAL	REGINA, SASKATCHEWAN MAT.	INST.	TOTAL	RIMOUSKI, QUEBEC MAT.	INST.	TOTAL	ROUYN-NORANDA, QUEBEC MAT.	INST.	TOTAL	SAINT-HYACINTHE, QUEBEC MAT.	INST.	TOTAL
015433	CONTRACTOR EQUIPMENT		104.2	104.2		104.2	104.2		100.3	100.3		102.2	102.2		102.2	102.2		102.2	102.2
0241, 31 - 34	SITE & INFRASTRUCTURE, DEMOLITION	113.3	98.8	103.0	115.2	101.0	105.2	121.5	97.5	104.5	99.7	98.4	98.8	99.2	99.1	99.1	99.7	99.1	99.3
0310	Concrete Forming & Accessories	120.1	91.1	95.0	138.6	80.3	88.2	118.3	88.7	92.7	130.0	90.5	95.9	130.0	85.3	91.3	130.0	85.3	91.3
0320	Concrete Reinforcing	138.1	94.9	115.7	148.7	73.4	109.7	133.4	83.9	107.8	103.1	94.8	98.8	144.6	77.9	110.1	144.6	77.9	110.1
0330	Cast-in-Place Concrete	138.0	100.5	123.2	138.6	100.5	123.6	150.3	94.9	128.5	125.6	98.8	115.0	119.4	96.6	110.4	123.4	96.6	112.8
03	CONCRETE	127.9	95.0	111.9	132.9	86.4	110.3	138.9	89.9	115.1	119.3	94.1	107.1	122.3	88.0	105.7	123.4	88.0	106.7
04	MASONRY	166.4	94.7	122.2	165.0	82.1	113.9	183.8	96.2	129.8	165.2	94.6	121.7	165.4	86.7	116.9	165.6	86.7	117.0
05	METALS	130.5	92.4	118.4	111.0	88.5	103.8	140.1	86.6	123.1	108.1	91.6	102.9	108.6	86.8	101.7	108.6	86.8	101.7
06	WOOD, PLASTICS & COMPOSITES	108.1	90.8	98.5	119.3	79.3	97.0	104.7	88.8	95.9	132.5	90.5	109.1	132.5	85.0	106.0	132.5	85.0	106.0
07	THERMAL & MOISTURE PROTECTION	120.5	94.9	109.8	135.9	85.7	114.9	126.1	84.1	108.5	110.6	94.2	103.7	110.6	87.9	101.1	110.9	87.9	101.3
08	OPENINGS	92.6	88.0	91.5	94.8	76.0	90.5	91.3	78.7	88.3	94.4	78.2	90.6	94.8	74.1	90.0	94.8	74.1	90.0
0920	Plaster & Gypsum Board	120.7	90.0	100.2	105.9	78.3	87.4	132.2	88.4	103.0	141.8	90.0	107.2	115.7	84.4	94.8	115.7	84.4	94.8
0950, 0980	Ceilings & Acoustic Treatment	132.6	90.0	104.4	98.9	78.3	85.2	137.7	88.4	105.1	98.1	90.0	92.7	98.1	84.4	89.0	98.1	84.4	89.0
0960	Flooring	122.7	98.0	115.6	124.9	91.2	115.2	121.9	104.9	117.0	122.5	95.7	114.8	122.5	95.7	114.8	122.5	95.7	114.8
0970, 0990	Wall Finishes & Painting/Coating	126.8	106.4	114.9	112.2	81.2	94.2	118.6	92.9	103.6	112.3	106.4	108.8	112.3	91.7	100.3	112.3	91.7	100.3
09	FINISHES	122.3	94.4	106.9	111.7	82.0	95.2	130.4	93.0	109.7	114.9	93.8	103.2	111.1	88.2	98.4	111.1	88.2	98.4
COVERS	DIVS. 10 - 14, 25, 28, 41, 43, 44, 46	131.1	86.5	121.5	131.1	91.4	122.5	131.1	71.6	118.2	131.1	85.8	121.3	131.1	82.8	120.7	131.1	82.8	120.7
21, 22, 23	FIRE SUPPRESSION, PLUMBING & HVAC	105.3	93.6	100.4	103.6	91.3	98.4	105.0	84.0	96.1	103.6	93.5	99.3	103.6	90.0	97.8	100.3	90.0	95.9
26, 27, 3370	ELECTRICAL, COMMUNICATIONS & UTIL.	106.1	86.1	96.0	102.9	84.8	93.8	112.7	91.7	102.1	108.9	86.1	97.4	108.9	72.1	90.3	109.5	72.1	90.6
MF2014	WEIGHTED AVERAGE	117.9	92.8	106.8	114.6	87.2	102.5	123.0	89.0	108.0	112.3	92.0	103.4	112.5	86.1	100.9	112.0	86.1	100.6

CANADA

DIVISION		SAINT JOHN, NEW BRUNSWICK MAT.	INST.	TOTAL	SARNIA, ONTARIO MAT.	INST.	TOTAL	SASKATOON, SASKATCHEWAN MAT.	INST.	TOTAL	SAULT STE. MARIE, ONTARIO MAT.	INST.	TOTAL	SHERBROOKE, QUEBEC MAT.	INST.	TOTAL	SOREL, QUEBEC MAT.	INST.	TOTAL
015433	CONTRACTOR EQUIPMENT		101.6	101.6		101.8	101.8		100.3	100.3		101.8	101.8		102.2	102.2		102.2	102.2
0241, 31 - 34	SITE & INFRASTRUCTURE, DEMOLITION	105.1	97.3	99.6	117.7	99.7	105.0	112.8	97.5	102.0	107.5	99.3	101.7	99.7	99.1	99.3	99.9	99.1	99.3
0310	Concrete Forming & Accessories	123.9	65.2	73.1	122.8	92.5	96.6	106.3	88.8	91.1	112.1	85.4	89.0	130.0	85.3	91.3	130.0	85.4	91.4
0320	Concrete Reinforcing	136.4	61.3	97.5	114.3	82.6	97.9	110.8	83.9	96.9	103.2	81.2	91.8	144.6	77.9	110.1	136.7	77.9	106.3
0330	Cast-in-Place Concrete	128.6	80.2	109.5	138.4	101.0	123.7	130.6	94.9	116.5	124.4	86.3	109.3	123.4	96.6	112.8	124.4	96.7	113.5
03	CONCRETE	124.1	70.6	98.1	126.7	93.6	110.6	117.3	90.0	104.0	111.4	85.2	98.7	124.3	88.0	106.7	123.6	88.1	106.3
04	MASONRY	182.3	73.4	115.2	181.0	99.0	130.5	171.8	96.2	125.2	166.2	94.0	121.7	165.7	86.7	117.0	165.8	86.7	117.0
05	METALS	109.8	82.2	101.0	110.1	89.5	103.6	106.9	86.6	100.4	109.4	89.7	103.1	108.6	86.8	101.7	108.6	86.9	101.7
06	WOOD, PLASTICS & COMPOSITES	122.2	63.1	89.3	120.5	91.5	104.3	99.4	88.8	93.5	108.3	85.3	95.5	132.5	85.0	106.0	132.5	85.0	106.0
07	THERMAL & MOISTURE PROTECTION	114.0	69.4	95.3	117.2	92.6	106.9	110.1	84.5	99.3	116.0	86.5	103.6	110.6	87.9	101.1	110.6	87.9	101.1
08	OPENINGS	89.3	59.5	82.3	96.0	86.8	93.8	90.2	78.7	87.5	87.8	83.1	86.7	94.8	74.1	90.0	94.8	78.7	91.1
0920	Plaster & Gypsum Board	133.9	62.0	85.9	136.7	91.2	106.3	115.0	88.4	97.2	108.8	84.8	92.8	115.7	84.4	94.8	141.8	84.4	103.4
0950, 0980	Ceilings & Acoustic Treatment	112.9	62.0	79.2	103.0	91.2	95.2	119.5	88.4	98.9	98.9	84.8	89.6	98.1	84.4	89.0	98.1	84.4	89.0
0960	Flooring	120.3	71.0	106.2	122.5	102.4	116.7	114.4	104.9	111.7	114.9	99.3	110.4	122.5	95.7	114.8	122.5	95.7	114.8
0970, 0990	Wall Finishes & Painting/Coating	113.2	80.2	94.0	112.3	105.7	108.4	115.5	92.9	102.3	112.3	97.9	103.9	112.3	91.7	100.3	112.3	91.7	100.3
09	FINISHES	116.4	67.3	89.3	116.0	96.1	105.0	115.4	93.0	103.0	108.0	89.4	97.7	111.1	88.2	98.4	114.9	88.2	100.1
COVERS	DIVS. 10 - 14, 25, 28, 41, 43, 44, 46	131.1	65.6	116.9	131.1	73.5	118.7	131.1	72.0	118.3	131.1	95.8	123.5	131.1	82.8	120.7	131.1	82.8	120.7
21, 22, 23	FIRE SUPPRESSION, PLUMBING & HVAC	103.6	80.9	94.0	103.6	106.6	104.9	104.9	84.0	96.0	103.6	94.6	99.8	103.9	90.0	98.1	103.6	90.0	97.9
26, 27, 3370	ELECTRICAL, COMMUNICATIONS & UTIL.	115.6	90.2	102.7	111.4	92.0	101.6	110.1	91.7	100.8	110.2	90.1	100.1	108.9	72.1	90.3	108.9	72.1	90.3
MF2014	WEIGHTED AVERAGE	114.4	77.8	98.3	115.3	96.1	106.9	112.3	89.0	102.0	110.7	91.1	102.1	112.8	86.1	101.1	113.0	86.4	101.2

CANADA

DIVISION		ST. CATHARINES, ONTARIO MAT.	INST.	TOTAL	SAINT-JEROME, QUEBEC MAT.	INST.	TOTAL	ST. JOHN'S, NEWFOUNDLAND MAT.	INST.	TOTAL	SUDBURY, ONTARIO MAT.	INST.	TOTAL	SUMMERSIDE, PRINCE EDWARD ISLAND MAT.	INST.	TOTAL	SYDNEY, NOVA SCOTIA MAT.	INST.	TOTAL
015433	CONTRACTOR EQUIPMENT		101.8	101.8		102.2	102.2		102.5	102.5		101.8	101.8		101.3	101.3		101.4	101.4
0241, 31 - 34	SITE & INFRASTRUCTURE, DEMOLITION	99.7	100.9	100.6	99.2	99.1	99.1	122.6	99.0	105.9	99.8	100.5	100.3	134.1	95.0	106.5	118.8	97.5	103.7
0310	Concrete Forming & Accessories	110.9	91.6	94.2	130.0	85.3	91.3	125.6	82.3	88.1	106.7	86.5	89.3	113.5	56.6	64.2	113.2	69.5	75.4
0320	Concrete Reinforcing	97.6	90.9	94.2	144.6	77.9	110.1	159.3	74.6	115.4	98.4	90.3	94.2	157.2	46.2	99.7	159.4	46.2	100.8
0330	Cast-in-Place Concrete	120.4	99.6	112.2	119.4	96.6	110.4	158.3	95.2	133.4	121.4	96.0	111.4	140.2	60.3	108.7	114.0	71.3	97.1
03	CONCRETE	105.6	94.2	100.1	122.3	88.0	105.7	147.2	85.5	117.2	105.9	90.6	98.5	170.1	56.8	115.1	144.0	66.6	106.4
04	MASONRY	148.0	100.3	118.6	165.4	86.7	116.9	189.1	92.0	129.3	148.0	96.1	116.0	236.7	62.3	129.2	234.8	71.6	134.2
05	METALS	107.6	91.8	102.6	108.6	86.8	101.7	138.5	83.7	121.1	107.6	91.2	102.4	123.6	67.8	105.9	123.6	74.1	107.9
06	WOOD, PLASTICS & COMPOSITES	106.9	90.7	97.9	132.5	85.0	106.0	109.2	80.0	92.9	103.1	85.3	93.2	118.2	55.8	83.5	117.9	68.6	90.5
07	THERMAL & MOISTURE PROTECTION	111.2	94.3	104.2	110.6	87.9	101.1	137.4	89.4	117.3	110.7	89.6	101.9	142.6	59.9	108.0	143.4	68.5	112.0
08	OPENINGS	85.9	88.1	86.4	94.8	74.1	90.0	90.6	72.4	86.4	86.6	83.5	85.9	107.9	48.5	94.0	95.6	63.0	88.0
0920	Plaster & Gypsum Board	97.9	90.4	92.9	115.7	84.4	94.8	133.7	79.3	97.3	99.8	84.8	89.8	136.0	54.4	81.5	135.7	67.6	90.2
0950, 0980	Ceilings & Acoustic Treatment	98.4	90.4	93.1	98.1	84.4	89.0	119.4	79.3	92.8	93.5	84.8	87.7	108.2	54.4	72.6	108.2	67.6	81.3
0960	Flooring	108.4	97.3	105.2	122.5	95.7	114.8	119.8	57.0	101.7	106.4	99.3	104.3	118.1	60.5	101.5	118.1	64.6	102.7
0970, 0990	Wall Finishes & Painting/Coating	106.2	116.0	111.9	112.3	91.7	100.3	117.5	95.9	104.9	106.2	93.6	98.9	120.1	43.0	75.2	120.1	64.1	87.5
09	FINISHES	102.5	95.4	98.6	111.1	88.2	98.4	125.8	79.0	99.9	101.1	89.4	94.6	119.6	56.1	84.5	118.5	68.2	90.7
COVERS	DIVS. 10 - 14, 25, 28, 41, 43, 44, 46	131.1	74.4	118.8	131.1	82.8	120.7	131.1	70.4	118.0	131.1	96.6	123.7	131.1	62.9	116.4	131.1	64.6	116.7
21, 22, 23	FIRE SUPPRESSION, PLUMBING & HVAC	104.7	86.9	97.2	103.6	90.0	97.8	105.2	81.2	95.0	103.5	87.6	96.8	103.7	63.1	86.5	103.7	83.5	95.1
26, 27, 3370	ELECTRICAL, COMMUNICATIONS & UTIL.	107.7	98.1	102.8	109.5	72.1	90.6	111.9	82.5	97.0	108.2	98.4	103.2	113.3	51.9	82.2	114.1	63.9	88.7
MF2014	WEIGHTED AVERAGE	107.8	93.3	101.5	112.5	86.1	100.9	123.9	83.9	106.3	107.6	92.1	100.8	128.1	62.0	99.0	123.4	73.6	101.4

For customer support on your Mechanical Cost Data, call 877.739.3689.

697

City Cost Indexes

		THUNDER BAY, ONTARIO			TIMMINS, ONTARIO			TORONTO, ONTARIO			TROIS-RIVIERES, QUEBEC			TRURO, NOVA SCOTIA			VANCOUVER, BRITISH COLUMBIA		
DIVISION		MAT.	INST.	TOTAL	MAT.	INST.	TOTAL	MAT.	INST.	TOTAL	MAT.	INST.	TOTAL	MAT.	INST.	TOTAL	MAT.	INST.	TOTAL
015433	CONTRACTOR EQUIPMENT		101.8	101.8		101.8	101.8		104.2	104.2		102.2	102.2		101.4	101.4		112.6	112.6
0241, 31 - 34	SITE & INFRASTRUCTURE, DEMOLITION	104.6	100.8	101.9	119.2	99.3	105.1	128.3	105.3	112.1	119.2	99.1	105.0	104.2	97.5	99.4	114.4	107.3	109.4
0310	Concrete Forming & Accessories	118.3	89.6	93.5	124.1	84.5	89.8	119.1	99.2	101.9	155.3	85.4	94.9	96.7	69.5	73.2	120.1	88.9	93.1
0320	Concrete Reinforcing	87.3	89.8	88.6	160.7	80.9	119.4	137.5	95.7	115.9	159.4	77.9	117.2	136.7	46.2	89.8	124.2	81.1	101.9
0330	Cast-in-Place Concrete	132.5	98.7	119.1	150.0	86.1	124.8	124.2	111.3	119.1	118.0	96.7	109.6	155.8	71.3	122.4	130.7	98.7	118.1
03	CONCRETE	113.4	92.8	103.4	139.3	84.7	112.7	120.9	102.5	112.0	146.5	88.1	118.1	138.2	66.6	103.4	129.2	91.1	110.7
04	MASONRY	148.6	100.1	118.8	169.6	89.5	120.2	177.0	113.3	137.7	239.2	86.7	145.2	165.2	71.6	107.5	166.0	93.2	121.1
05	METALS	107.4	90.8	102.1	110.2	88.7	103.4	128.5	94.4	117.7	122.8	86.9	111.4	109.3	74.2	98.2	130.8	91.3	118.2
06	WOOD, PLASTICS & COMPOSITES	115.8	87.8	100.2	121.6	84.7	101.0	104.2	96.7	100.0	168.0	85.0	121.8	92.3	68.6	79.1	103.9	88.5	95.3
07	THERMAL & MOISTURE PROTECTION	111.5	91.8	103.3	117.1	83.7	103.2	127.9	101.6	116.9	141.6	87.9	119.2	111.8	68.5	93.7	127.7	86.7	110.5
08	OPENINGS	85.1	85.9	85.3	93.6	81.3	90.8	88.8	95.7	90.4	105.5	78.7	99.2	87.5	63.0	81.8	87.3	85.3	86.8
0920	Plaster & Gypsum Board	123.8	87.4	99.5	117.6	84.1	95.3	122.9	96.5	105.2	162.7	84.4	110.4	119.5	67.6	84.8	112.8	87.5	95.9
0950, 0980	Ceilings & Acoustic Treatment	93.5	87.4	89.4	98.9	84.1	89.1	120.6	96.5	104.6	107.3	84.4	92.1	98.9	67.6	78.2	129.6	87.5	101.7
0960	Flooring	112.8	104.3	110.3	122.5	93.7	114.2	114.5	106.7	112.3	143.1	95.7	129.5	103.5	64.6	92.3	129.8	94.5	119.6
0970, 0990	Wall Finishes & Painting/Coating	106.2	94.6	99.4	112.3	91.2	100.0	107.3	106.7	107.0	120.1	91.7	103.6	112.3	64.1	84.2	118.4	98.0	106.6
09	FINISHES	106.8	92.8	99.1	112.3	87.3	98.4	117.1	101.4	108.4	129.6	88.2	106.7	106.3	68.2	85.2	124.4	90.9	105.9
COVERS	DIVS. 10 - 14, 25, 28, 41, 43, 44, 46	131.1	74.4	118.9	131.1	70.7	118.0	131.1	101.0	124.6	131.1	82.8	120.7	131.1	64.6	116.7	131.1	94.5	123.2
21, 22, 23	FIRE SUPPRESSION, PLUMBING & HVAC	104.7	87.0	97.2	103.6	97.1	100.8	105.3	96.8	101.7	103.7	90.0	97.9	103.6	83.5	95.1	105.5	82.0	95.6
26, 27, 3370	ELECTRICAL, COMMUNICATIONS & UTIL.	106.3	96.3	101.2	110.2	90.1	100.1	105.1	100.3	102.7	114.4	72.1	93.0	108.7	63.9	86.0	103.8	82.6	93.1
MF2014	WEIGHTED AVERAGE	109.1	92.3	101.7	115.6	89.8	104.2	116.9	100.9	109.9	126.1	86.4	108.6	113.0	73.6	95.6	117.7	89.2	105.1

		VICTORIA, BRITISH COLUMBIA			WHITEHORSE, YUKON			WINDSOR, ONTARIO			WINNIPEG, MANITOBA			YARMOUTH, NOVA SCOTIA			YELLOWKNIFE, NWT		
DIVISION		MAT.	INST.	TOTAL	MAT.	INST.	TOTAL	MAT.	INST.	TOTAL	MAT.	INST.	TOTAL	MAT.	INST.	TOTAL	MAT.	INST.	TOTAL
015433	CONTRACTOR EQUIPMENT		108.7	108.7		101.6	101.6		101.8	101.8		106.2	106.2		101.4	101.4		101.4	101.4
0241, 31 - 34	SITE & INFRASTRUCTURE, DEMOLITION	123.8	106.7	111.7	142.7	96.0	109.7	95.8	100.8	99.4	114.7	100.4	104.6	123.1	97.5	105.0	159.6	100.3	117.7
0310	Concrete Forming & Accessories	111.3	88.2	91.3	131.5	57.8	67.7	118.3	88.0	92.1	126.3	67.3	75.2	113.2	69.5	75.4	135.0	77.9	85.6
0320	Concrete Reinforcing	108.4	81.0	94.2	156.7	59.3	106.3	95.7	90.8	93.2	132.4	57.2	93.5	159.4	46.2	100.8	141.6	61.9	100.4
0330	Cast-in-Place Concrete	135.3	98.0	120.6	186.2	69.4	140.1	107.3	100.9	114.5	146.8	74.1	118.2	146.2	71.3	116.6	197.0	78.5	129.1
03	CONCRETE	147.5	90.3	119.7	173.7	62.9	119.8	107.3	93.0	100.3	137.6	68.4	103.9	159.8	66.6	114.5	177.0	78.5	129.1
04	MASONRY	174.8	93.2	124.5	273.0	62.5	143.3	148.1	99.5	118.1	177.0	70.5	111.4	237.4	71.6	135.2	269.2	74.5	149.2
05	METALS	109.1	86.7	102.0	140.1	73.6	118.9	107.5	91.3	102.3	149.0	75.1	125.5	123.6	74.1	107.9	138.0	78.5	119.1
06	WOOD, PLASTICS & COMPOSITES	103.3	88.4	95.0	120.8	56.3	84.9	115.8	86.0	99.2	112.1	67.8	87.4	117.9	68.6	90.5	125.9	79.1	99.9
07	THERMAL & MOISTURE PROTECTION	120.1	85.8	105.8	145.5	60.8	110.1	111.3	91.8	103.1	125.9	70.4	102.7	143.4	68.5	112.0	155.5	76.7	122.8
08	OPENINGS	91.7	81.2	89.3	110.5	53.9	97.3	84.9	85.4	85.0	85.1	60.7	79.5	95.6	63.0	88.0	102.3	66.9	94.1
0920	Plaster & Gypsum Board	108.3	87.5	94.4	154.7	54.9	88.0	110.7	85.6	93.9	119.2	66.3	83.8	135.7	67.6	90.2	155.3	78.4	103.9
0950, 0980	Ceilings & Acoustic Treatment	101.3	87.5	92.2	140.7	54.9	84.1	93.5	85.6	88.2	139.9	66.3	91.2	108.2	67.6	81.3	138.8	78.4	98.8
0960	Flooring	120.2	73.3	106.7	140.7	59.0	117.2	112.8	101.5	109.5	119.4	72.7	105.9	118.0	64.6	102.7	138.5	89.0	124.2
0970, 0990	Wall Finishes & Painting/Coating	115.5	95.6	103.9	130.7	56.0	87.3	106.2	95.5	100.0	117.3	56.5	81.9	120.1	64.1	87.5	128.8	80.3	100.6
09	FINISHES	113.7	87.0	99.0	147.7	57.6	97.8	104.6	91.2	97.2	126.0	67.5	93.7	118.5	68.2	90.7	142.2	79.6	107.6
COVERS	DIVS. 10 - 14, 25, 28, 41, 43, 44, 46	131.1	69.7	117.8	131.1	62.0	116.2	131.1	73.9	118.7	131.1	66.8	117.2	131.1	64.6	116.7	131.1	65.1	116.8
21, 22, 23	FIRE SUPPRESSION, PLUMBING & HVAC	103.6	80.1	93.7	104.2	74.0	91.4	104.7	87.0	97.2	105.4	66.2	88.8	103.7	83.5	95.1	104.3	92.9	99.5
26, 27, 3370	ELECTRICAL, COMMUNICATIONS & UTIL.	110.3	82.1	96.0	124.4	61.8	92.8	110.4	98.2	104.3	109.9	67.5	88.5	114.1	63.9	88.7	126.9	84.6	105.5
MF2014	WEIGHTED AVERAGE	116.6	86.7	103.4	137.1	67.5	106.4	108.3	92.3	101.3	122.6	70.8	99.8	125.4	73.6	102.6	136.7	82.9	113.0

Location Factors - Commercial

Costs shown in RSMeans cost data publications are based on national averages for materials and installation. To adjust these costs to a specific location, simply multiply the base cost by the factor and divide by 100 for that city. The data is arranged alphabetically by state and postal zip code numbers. For a city not listed, use the factor for a nearby city with similar economic characteristics.

STATE/ZIP	CITY	MAT.	INST.	TOTAL
ALABAMA				
350-352	Birmingham	99.3	77.3	89.6
354	Tuscaloosa	98.7	78.1	89.6
355	Jasper	99.2	71.8	87.1
356	Decatur	98.7	76.0	88.7
357-358	Huntsville	98.7	75.3	88.4
359	Gadsden	98.8	67.1	84.8
360-361	Montgomery	97.9	77.0	88.7
362	Anniston	97.3	77.5	88.6
363	Dothan	97.7	76.1	88.2
364	Evergreen	97.3	73.6	86.9
365-366	Mobile	98.3	76.6	88.7
367	Selma	97.4	74.9	87.5
368	Phenix City	98.2	77.2	89.0
369	Butler	97.6	76.7	88.4
ALASKA				
995-996	Anchorage	121.7	115.8	119.1
997	Fairbanks	124.2	116.9	120.9
998	Juneau	123.9	115.8	120.3
999	Ketchikan	134.9	115.8	126.5
ARIZONA				
850,853	Phoenix	100.5	73.0	88.4
851,852	Mesa/Tempe	100.1	72.0	87.7
855	Globe	100.9	69.8	87.2
856-857	Tucson	98.9	70.8	86.5
859	Show Low	101.1	71.9	88.3
860	Flagstaff	102.7	71.9	89.2
863	Prescott	100.4	68.4	86.3
864	Kingman	98.9	71.3	86.7
865	Chambers	98.9	73.5	87.7
ARKANSAS				
716	Pine Bluff	99.6	65.2	84.5
717	Camden	97.6	56.0	79.3
718	Texarkana	98.5	58.7	80.9
719	Hot Springs	96.9	56.9	79.3
720-722	Little Rock	98.2	67.4	84.6
723	West Memphis	97.0	65.3	83.0
724	Jonesboro	97.4	62.6	82.1
725	Batesville	95.3	58.3	79.0
726	Harrison	96.6	57.6	79.4
727	Fayetteville	94.1	55.6	77.1
728	Russellville	95.3	59.7	79.7
729	Fort Smith	97.8	62.3	82.2
CALIFORNIA				
900-902	Los Angeles	100.2	117.2	107.7
903-905	Inglewood	96.0	115.0	104.4
906-908	Long Beach	97.5	115.0	105.2
910-912	Pasadena	97.5	114.8	105.1
913-916	Van Nuys	100.4	114.8	106.7
917-918	Alhambra	99.5	114.8	106.2
919-921	San Diego	101.4	109.8	105.1
922	Palm Springs	98.6	112.9	104.9
923-924	San Bernardino	96.4	112.3	103.4
925	Riverside	100.5	113.9	106.4
926-927	Santa Ana	98.3	112.8	104.7
928	Anaheim	100.6	114.0	106.5
930	Oxnard	99.9	113.5	105.9
931	Santa Barbara	99.2	114.3	105.9
932-933	Bakersfield	100.4	111.4	105.2
934	San Luis Obispo	100.2	113.0	105.9
935	Mojave	97.4	111.9	103.8
936-938	Fresno	100.5	116.5	107.5
939	Salinas	101.0	122.3	110.4
940-941	San Francisco	104.1	146.0	122.5
942,956-958	Sacramento	99.6	120.1	108.6
943	Palo Alto	98.1	133.4	113.7
944	San Mateo	100.5	134.3	115.4
945	Vallejo	99.1	128.9	112.2
946	Oakland	102.1	134.5	116.4
947	Berkeley	102.0	135.3	116.7
948	Richmond	101.3	132.2	114.9
949	San Rafael	103.2	135.4	117.4
950	Santa Cruz	105.9	122.5	113.2

STATE/ZIP	CITY	MAT.	INST.	TOTAL
CALIFORNIA (CONT'D)				
951	San Jose	103.9	136.9	118.4
952	Stockton	102.1	117.9	109.0
953	Modesto	102.0	116.6	108.4
954	Santa Rosa	102.6	136.3	117.4
955	Eureka	103.9	118.3	110.2
959	Marysville	103.1	115.9	108.8
960	Redding	112.3	117.8	114.7
961	Susanville	112.0	117.2	114.3
COLORADO				
800-802	Denver	101.6	80.4	92.3
803	Boulder	99.1	80.3	90.8
804	Golden	101.2	77.8	90.9
805	Fort Collins	102.4	77.6	91.5
806	Greeley	99.7	74.0	88.4
807	Fort Morgan	99.7	78.0	90.1
808-809	Colorado Springs	101.3	78.7	91.3
810	Pueblo	102.4	74.7	90.2
811	Alamosa	104.3	73.1	90.5
812	Salida	103.9	75.4	91.3
813	Durango	104.7	74.9	91.6
814	Montrose	103.4	76.5	91.6
815	Grand Junction	106.5	75.6	92.9
816	Glenwood Springs	104.5	76.4	92.1
CONNECTICUT				
060	New Britain	98.3	122.1	108.8
061	Hartford	99.6	122.2	109.6
062	Willimantic	98.9	122.3	109.2
063	New London	95.6	122.4	107.4
064	Meriden	97.5	122.2	108.4
065	New Haven	99.8	122.3	109.7
066	Bridgeport	99.5	122.0	109.4
067	Waterbury	99.0	121.9	109.1
068	Norwalk	99.0	129.7	112.5
069	Stamford	99.1	129.7	112.6
D.C.				
200-205	Washington	100.6	95.9	98.5
DELAWARE				
197	Newark	98.9	109.4	103.5
198	Wilmington	98.7	109.4	103.4
199	Dover	99.0	109.4	103.6
FLORIDA				
320,322	Jacksonville	97.3	69.6	85.1
321	Daytona Beach	97.5	75.5	87.8
323	Tallahassee	98.9	65.0	84.0
324	Panama City	98.8	62.0	82.6
325	Pensacola	101.4	71.0	88.0
326,344	Gainesville	99.0	68.8	85.7
327-328,347	Orlando	99.5	72.0	87.4
329	Melbourne	100.1	76.6	89.8
330-332,340	Miami	98.9	72.3	87.2
333	Fort Lauderdale	97.8	71.2	86.1
334,349	West Palm Beach	96.9	69.9	85.0
335-336,346	Tampa	99.3	72.8	87.6
337	St. Petersburg	101.8	70.2	87.9
338	Lakeland	98.8	72.7	87.3
339,341	Fort Myers	98.1	68.8	85.1
342	Sarasota	100.0	73.8	88.5
GEORGIA				
300-303,399	Atlanta	98.1	76.5	88.6
304	Statesboro	98.2	61.0	81.8
305	Gainesville	96.9	63.6	82.2
306	Athens	96.3	65.2	82.6
307	Dalton	98.0	63.0	82.6
308-309	Augusta	96.8	69.9	85.0
310-312	Macon	95.8	66.2	82.7
313-314	Savannah	97.8	67.6	84.5
315	Waycross	97.1	64.4	82.7
316	Valdosta	97.0	62.3	81.7
317,398	Albany	96.9	63.1	82.0
318-319	Columbus	96.8	66.1	83.3

Location Factors - Commercial

STATE/ZIP	CITY	MAT.	INST.	TOTAL
HAWAII				
967	Hilo	120.2	119.7	120.0
968	Honolulu	125.0	119.7	122.7
STATES & POSS.				
969	Guam	143.0	57.6	105.4
IDAHO				
832	Pocatello	102.5	77.2	91.3
833	Twin Falls	103.8	66.3	87.3
834	Idaho Falls	101.1	74.3	89.3
835	Lewiston	109.4	85.0	98.6
836-837	Boise	101.5	77.8	91.1
838	Coeur d'Alene	109.3	82.2	97.4
ILLINOIS				
600-603	North Suburban	99.4	136.8	115.9
604	Joliet	99.4	139.0	116.8
605	South Suburban	99.4	137.0	115.9
606-608	Chicago	99.9	140.4	117.8
609	Kankakee	96.0	131.6	111.7
610-611	Rockford	99.3	123.5	109.9
612	Rock Island	97.4	101.6	99.2
613	La Salle	98.6	123.4	109.5
614	Galesburg	98.4	107.5	102.4
615-616	Peoria	100.3	109.8	104.5
617	Bloomington	97.7	110.0	103.1
618-619	Champaign	101.2	108.4	104.4
620-622	East St. Louis	95.9	108.4	101.4
623	Quincy	98.1	101.9	99.8
624	Effingham	97.3	106.6	101.4
625	Decatur	98.8	105.8	101.9
626-627	Springfield	99.3	106.9	102.6
628	Centralia	94.9	108.3	100.8
629	Carbondale	94.6	106.7	99.9
INDIANA				
460	Anderson	97.1	83.6	91.1
461-462	Indianapolis	99.0	85.2	93.0
463-464	Gary	98.4	110.0	103.5
465-466	South Bend	97.8	84.1	91.8
467-468	Fort Wayne	97.6	78.6	89.3
469	Kokomo	95.3	81.5	89.2
470	Lawrenceburg	93.8	78.2	86.9
471	New Albany	94.9	74.1	85.7
472	Columbus	97.1	81.3	90.1
473	Muncie	97.3	83.4	91.2
474	Bloomington	98.8	81.2	91.0
475	Washington	95.7	82.6	89.9
476-477	Evansville	96.6	86.6	92.2
478	Terre Haute	97.4	86.6	92.6
479	Lafayette	96.8	81.9	90.2
IOWA				
500-503,509	Des Moines	98.4	83.9	92.0
504	Mason City	97.0	64.4	82.6
505	Fort Dodge	97.2	57.6	79.8
506-507	Waterloo	98.4	73.7	87.5
508	Creston	97.5	78.3	89.1
510-511	Sioux City	99.6	72.3	87.6
512	Sibley	98.7	56.0	79.9
513	Spencer	100.5	56.0	80.9
514	Carroll	97.4	71.0	85.8
515	Council Bluffs	100.6	79.0	91.1
516	Shenandoah	97.8	73.5	87.1
520	Dubuque	99.0	80.2	90.9
521	Decorah	98.5	66.6	84.4
522-524	Cedar Rapids	99.9	82.6	92.3
525	Ottumwa	98.3	74.7	87.9
526	Burlington	97.5	76.9	88.4
527-528	Davenport	99.0	93.6	96.6
KANSAS				
660-662	Kansas City	99.6	95.9	98.0
664-666	Topeka	99.8	69.8	86.6
667	Fort Scott	98.3	73.7	87.5
668	Emporia	98.3	70.1	85.9
669	Belleville	100.1	66.5	85.3
670-672	Wichita	101.0	66.1	85.6
673	Independence	101.6	70.2	87.8
674	Salina	101.8	68.9	87.3
675	Hutchinson	97.1	65.2	83.1
676	Hays	101.2	66.9	86.1
677	Colby	101.8	68.0	86.9

STATE/ZIP	CITY	MAT.	INST.	TOTAL
KANSAS (CONT'D)				
678	Dodge City	103.0	66.1	86.8
679	Liberal	101.1	65.6	85.5
KENTUCKY				
400-402	Louisville	95.1	84.3	90.4
403-405	Lexington	95.5	82.2	89.7
406	Frankfort	96.1	80.2	89.1
407-409	Corbin	93.3	71.9	83.9
410	Covington	95.8	91.3	93.8
411-412	Ashland	94.5	99.8	96.8
413-414	Campton	95.6	77.6	87.7
415-416	Pikeville	96.8	89.3	93.5
417-418	Hazard	95.1	78.6	87.8
420	Paducah	93.7	85.5	90.1
421-422	Bowling Green	95.7	84.1	90.6
423	Owensboro	95.8	88.0	92.4
424	Henderson	93.5	86.3	90.3
425-426	Somerset	93.0	81.7	88.0
427	Elizabethtown	92.5	82.2	87.9
LOUISIANA				
700-701	New Orleans	99.2	70.7	86.6
703	Thibodaux	97.6	65.6	83.5
704	Hammond	95.2	61.0	80.1
705	Lafayette	97.0	66.6	83.6
706	Lake Charles	97.2	69.9	85.2
707-708	Baton Rouge	99.5	69.8	86.4
710-711	Shreveport	100.9	60.7	83.2
712	Monroe	99.6	59.3	81.8
713-714	Alexandria	99.7	63.3	83.7
MAINE				
039	Kittery	94.5	87.4	91.4
040-041	Portland	101.6	89.4	96.3
042	Lewiston	98.5	89.4	94.5
043	Augusta	101.2	81.0	92.3
044	Bangor	97.9	88.8	93.9
045	Bath	96.9	83.3	90.9
046	Machias	96.3	86.9	92.2
047	Houlton	96.5	86.9	92.3
048	Rockland	95.6	86.9	91.7
049	Waterville	96.8	81.0	89.8
MARYLAND				
206	Waldorf	97.1	82.6	90.7
207-208	College Park	97.1	83.8	91.2
209	Silver Spring	96.4	84.0	90.9
210-212	Baltimore	100.9	83.4	93.2
214	Annapolis	101.0	82.7	92.9
215	Cumberland	97.3	84.1	91.5
216	Easton	98.9	71.2	86.7
217	Hagerstown	97.7	85.2	92.2
218	Salisbury	99.4	61.8	82.8
219	Elkton	96.3	84.2	91.0
MASSACHUSETTS				
010-011	Springfield	99.1	110.3	104.0
012	Pittsfield	98.5	109.7	103.4
013	Greenfield	96.9	110.0	102.6
014	Fitchburg	95.6	124.5	108.3
015-016	Worcester	99.0	124.5	110.2
017	Framingham	94.9	132.9	111.6
018	Lowell	98.4	132.3	113.3
019	Lawrence	99.2	132.6	113.9
020-022, 024	Boston	100.6	139.1	117.5
023	Brockton	98.8	125.1	110.4
025	Buzzards Bay	93.7	125.9	107.9
026	Hyannis	95.9	125.9	109.1
027	New Bedford	97.9	126.3	110.4
MICHIGAN				
480,483	Royal Oak	95.1	106.1	99.9
481	Ann Arbor	97.0	107.4	101.6
482	Detroit	98.2	108.2	102.6
484-485	Flint	96.7	95.6	96.2
486	Saginaw	96.3	90.8	93.9
487	Bay City	96.4	90.3	93.7
488-489	Lansing	98.7	94.4	96.8
490	Battle Creek	97.5	86.0	92.4
491	Kalamazoo	97.8	83.9	91.7
492	Jackson	96.0	94.2	95.2
493,495	Grand Rapids	98.8	83.9	92.3
494	Muskegon	96.3	83.2	90.5

Location Factors - Commercial

STATE/ZIP	CITY	MAT.	INST.	TOTAL	STATE/ZIP	CITY	MAT.	INST.	TOTAL
MICHIGAN (CONT'D)					**NEW HAMPSHIRE (CONT'D)**				
496	Traverse City	95.6	80.9	89.1	032-033	Concord	100.0	93.8	97.3
497	Gaylord	96.7	83.6	90.9	034	Keene	96.2	75.6	87.1
498-499	Iron Mountain	98.7	85.8	93.0	035	Littleton	96.3	77.6	88.0
					036	Charleston	95.7	70.3	84.5
MINNESOTA					037	Claremont	94.9	70.4	84.1
550-551	Saint Paul	100.7	120.9	109.6	038	Portsmouth	96.5	93.7	95.3
553-555	Minneapolis	100.7	119.9	109.1					
556-558	Duluth	101.3	110.5	105.3	**NEW JERSEY**				
559	Rochester	100.3	106.3	103.0	070-071	Newark	100.3	129.3	113.1
560	Mankato	97.5	102.7	99.8	072	Elizabeth	97.8	128.8	111.5
561	Windom	96.2	93.5	95.0	073	Jersey City	96.7	128.9	110.9
562	Willmar	95.7	98.8	97.1	074-075	Paterson	98.5	129.4	112.1
563	St. Cloud	96.8	114.0	104.4	076	Hackensack	96.5	129.0	110.8
564	Brainerd	97.3	100.5	98.7	077	Long Branch	96.0	124.7	108.6
565	Detroit Lakes	99.4	96.1	98.0	078	Dover	96.7	129.2	111.0
566	Bemidji	98.6	99.1	98.8	079	Summit	96.7	128.8	110.9
567	Thief River Falls	98.2	93.3	96.0	080,083	Vineland	96.4	123.6	108.4
					081	Camden	98.4	124.7	110.0
MISSISSIPPI					082,084	Atlantic City	97.2	125.3	109.6
386	Clarksdale	96.6	55.3	78.4	085-086	Trenton	100.2	124.9	111.1
387	Greenville	100.3	65.1	84.8	087	Point Pleasant	98.2	124.0	109.6
388	Tupelo	98.2	55.6	79.4	088-089	New Brunswick	98.8	128.7	112.0
389	Greenwood	97.9	52.0	77.7					
390-392	Jackson	99.8	68.4	86.0	**NEW MEXICO**				
393	Meridian	97.1	67.9	84.2	870-872	Albuquerque	100.3	72.1	87.9
394	Laurel	98.8	55.4	79.7	873	Gallup	100.7	72.1	88.1
395	Biloxi	98.8	60.5	81.9	874	Farmington	101.0	72.1	88.2
396	McComb	97.1	51.5	77.0	875	Santa Fe	101.1	72.1	88.3
397	Columbus	98.6	53.4	78.7	877	Las Vegas	99.0	72.1	87.1
					878	Socorro	98.6	72.1	86.9
MISSOURI					879	Truth/Consequences	98.3	68.8	85.3
630-631	St. Louis	100.2	104.3	102.0	880	Las Cruces	97.5	68.1	84.6
633	Bowling Green	99.1	96.0	97.7	881	Clovis	99.9	72.0	87.6
634	Hannibal	97.9	92.5	95.6	882	Roswell	101.4	72.1	88.5
635	Kirksville	101.5	89.7	96.3	883	Carrizozo	102.1	72.1	88.9
636	Flat River	100.0	95.1	97.9	884	Tucumcari	100.6	72.0	88.0
637	Cape Girardeau	99.6	92.8	96.6					
638	Sikeston	98.4	90.5	94.9	**NEW YORK**				
639	Poplar Bluff	97.9	89.7	94.3	100-102	New York	100.7	169.8	131.1
640-641	Kansas City	100.2	104.8	102.2	103	Staten Island	96.8	167.5	128.0
644-645	St. Joseph	99.4	96.7	98.2	104	Bronx	95.2	166.0	126.4
646	Chillicothe	96.8	98.6	97.6	105	Mount Vernon	95.5	143.1	116.5
647	Harrisonville	96.3	101.0	98.4	106	White Plains	95.2	143.1	116.3
648	Joplin	98.2	81.0	90.7	107	Yonkers	99.4	146.3	120.1
650-651	Jefferson City	99.6	98.4	99.1	108	New Rochelle	95.9	143.0	116.7
652	Columbia	100.3	99.7	100.0	109	Suffern	95.5	131.9	111.5
653	Sedalia	100.7	93.9	97.7	110	Queens	100.3	166.2	129.3
654-655	Rolla	98.3	96.7	97.6	111	Long Island City	101.9	166.2	130.2
656-658	Springfield	101.8	83.3	93.7	112	Brooklyn	102.3	167.9	131.2
					113	Flushing	102.4	166.2	130.5
MONTANA					114	Jamaica	100.7	166.2	129.6
590-591	Billings	103.7	78.8	92.7	115,117,118	Hicksville	100.3	152.7	123.4
592	Wolf Point	103.8	74.2	90.8	116	Far Rockaway	102.5	166.2	130.6
593	Miles City	101.6	74.0	89.4	119	Riverhead	101.1	150.9	123.0
594	Great Falls	105.2	76.8	92.7	120-122	Albany	98.3	104.6	101.1
595	Havre	102.7	73.0	89.6	123	Schenectady	98.1	104.5	101.0
596	Helena	102.8	73.6	90.0	124	Kingston	101.8	117.1	108.6
597	Butte	103.7	77.1	92.0	125-126	Poughkeepsie	101.0	135.2	116.1
598	Missoula	101.1	75.7	89.9	127	Monticello	100.3	115.3	106.9
599	Kalispell	100.4	75.5	89.5	128	Glens Falls	93.3	101.1	96.7
					129	Plattsburgh	98.4	96.7	97.7
NEBRASKA					130-132	Syracuse	98.5	98.2	98.3
680-681	Omaha	99.1	79.5	90.5	133-135	Utica	96.5	97.5	97.0
683-685	Lincoln	99.8	76.9	89.7	136	Watertown	98.0	96.7	97.4
686	Columbus	98.1	77.5	89.0	137-139	Binghamton	98.2	98.4	98.3
687	Norfolk	99.6	76.0	89.2	140-142	Buffalo	100.3	106.4	103.0
688	Grand Island	99.3	76.6	89.3	143	Niagara Falls	98.0	107.2	102.1
689	Hastings	99.1	78.9	90.2	144-146	Rochester	100.8	100.0	100.5
690	McCook	99.0	70.5	86.4	147	Jamestown	97.0	93.0	95.3
691	North Platte	98.9	75.9	88.7	148-149	Elmira	96.9	98.3	97.6
692	Valentine	101.3	71.8	88.3					
693	Alliance	101.2	70.5	87.7	**NORTH CAROLINA**				
					270,272-274	Greensboro	99.2	61.7	82.7
NEVADA					271	Winston-Salem	98.9	62.1	82.7
889-891	Las Vegas	103.6	109.4	106.1	275-276	Raleigh	97.7	59.7	81.0
893	Ely	102.4	100.6	101.6	277	Durham	100.7	61.7	83.5
894-895	Reno	102.2	86.5	95.3	278	Rocky Mount	96.8	61.8	81.4
897	Carson City	101.8	86.7	95.1	279	Elizabeth City	97.8	54.5	78.7
898	Elko	101.1	81.4	92.4	280	Gastonia	99.8	62.4	83.3
					281-282	Charlotte	99.1	63.8	83.6
NEW HAMPSHIRE					283	Fayetteville	102.2	60.6	83.9
030	Nashua	99.3	93.6	96.8	284	Wilmington	98.5	61.5	82.2
031	Manchester	99.9	93.6	97.1	285	Kinston	97.0	63.0	82.0

For customer support on your Mechanical Cost Data, call 877.739.3689.

701

Location Factors - Commercial

STATE/ZIP	CITY	MAT.	INST.	TOTAL
NORTH CAROLINA (CONT'D)				
286	Hickory	97.3	61.8	81.7
287-288	Asheville	99.0	62.4	82.9
289	Murphy	98.2	51.0	77.4
NORTH DAKOTA				
580-581	Fargo	101.1	75.0	89.6
582	Grand Forks	101.9	65.1	85.7
583	Devils Lake	101.7	64.8	85.5
584	Jamestown	101.9	64.7	85.5
585	Bismarck	102.1	70.0	87.9
586	Dickinson	102.6	65.9	86.4
587	Minot	101.9	76.9	90.9
588	Williston	101.0	66.2	85.7
OHIO				
430-432	Columbus	98.2	87.4	93.5
433	Marion	95.1	84.6	90.5
434-436	Toledo	98.3	97.6	98.0
437-438	Zanesville	95.6	86.3	91.5
439	Steubenville	96.7	93.9	95.5
440	Lorain	98.9	92.2	96.0
441	Cleveland	99.2	99.4	99.3
442-443	Akron	99.9	93.9	97.3
444-445	Youngstown	99.2	88.1	94.4
446-447	Canton	99.3	89.3	94.9
448-449	Mansfield	97.1	86.2	92.3
450	Hamilton	97.8	84.4	91.9
451-452	Cincinnati	98.2	83.7	91.8
453-454	Dayton	97.9	83.9	91.7
455	Springfield	97.8	84.1	91.8
456	Chillicothe	97.2	92.0	94.9
457	Athens	100.2	83.1	92.6
458	Lima	100.5	86.3	94.3
OKLAHOMA				
730-731	Oklahoma City	99.8	71.3	87.3
734	Ardmore	98.6	64.9	83.7
735	Lawton	100.7	67.5	86.1
736	Clinton	99.8	66.2	85.0
737	Enid	100.2	65.2	84.8
738	Woodward	98.6	68.4	85.3
739	Guymon	99.7	69.1	86.2
740-741	Tulsa	98.3	64.5	83.4
743	Miami	95.2	72.1	85.0
744	Muskogee	97.5	62.9	82.3
745	McAlester	94.8	65.3	81.8
746	Ponca City	95.5	67.6	83.2
747	Durant	95.5	65.9	82.5
748	Shawnee	97.0	65.3	83.0
749	Poteau	94.7	66.4	82.2
OREGON				
970-972	Portland	99.7	101.1	100.3
973	Salem	100.9	99.2	100.2
974	Eugene	99.5	99.2	99.3
975	Medford	101.0	96.9	99.2
976	Klamath Falls	101.3	97.1	99.4
977	Bend	100.3	99.2	99.8
978	Pendleton	96.2	101.4	98.5
979	Vale	94.1	89.1	91.9
PENNSYLVANIA				
150-152	Pittsburgh	100.2	104.6	102.1
153	Washington	97.5	103.6	100.2
154	Uniontown	97.7	102.8	99.9
155	Bedford	98.7	95.3	97.2
156	Greensburg	98.7	102.8	100.5
157	Indiana	97.6	101.4	99.2
158	Dubois	99.1	97.5	98.4
159	Johnstown	98.7	98.1	98.5
160	Butler	92.3	103.2	97.1
161	New Castle	92.4	100.3	95.9
162	Kittanning	92.8	103.1	97.3
163	Oil City	92.3	99.7	95.6
164-165	Erie	94.0	97.5	95.5
166	Altoona	94.1	93.5	93.9
167	Bradford	95.6	97.4	96.4
168	State College	95.1	93.0	94.2
169	Wellsboro	96.2	96.6	96.3
170-171	Harrisburg	98.5	96.5	97.6
172	Chambersburg	95.8	91.2	93.8
173-174	York	95.9	96.8	96.3
175-176	Lancaster	94.6	98.3	96.2

STATE/ZIP	CITY	MAT.	INST.	TOTAL
PENNSYLVANIA (CONT'D)				
177	Williamsport	93.2	96.2	94.5
178	Sunbury	95.3	97.3	96.2
179	Pottsville	94.4	99.1	96.5
180	Lehigh Valley	95.7	112.3	103.0
181	Allentown	97.3	110.6	103.2
182	Hazleton	95.2	100.3	97.5
183	Stroudsburg	95.1	105.0	99.4
184-185	Scranton	98.1	99.8	98.8
186-187	Wilkes-Barre	94.9	101.0	97.6
188	Montrose	94.6	98.7	96.4
189	Doylestown	94.8	127.4	109.1
190-191	Philadelphia	99.5	134.7	115.0
193	Westchester	96.3	126.3	109.5
194	Norristown	95.4	131.2	111.2
195-196	Reading	97.1	104.5	100.4
PUERTO RICO				
009	San Juan	122.0	24.5	79.0
RHODE ISLAND				
028	Newport	97.2	116.8	105.8
029	Providence	99.8	116.8	107.3
SOUTH CAROLINA				
290-292	Columbia	98.2	68.3	85.0
293	Spartanburg	98.3	67.3	84.6
294	Charleston	99.8	66.1	84.9
295	Florence	98.0	67.7	84.7
296	Greenville	98.1	66.0	83.9
297	Rock Hill	98.0	56.5	79.7
298	Aiken	98.9	66.8	84.7
299	Beaufort	99.6	52.5	78.8
SOUTH DAKOTA				
570-571	Sioux Falls	99.0	60.5	82.0
572	Watertown	99.9	49.9	77.9
573	Mitchell	98.6	48.1	76.4
574	Aberdeen	101.0	53.7	80.1
575	Pierre	102.2	59.3	83.3
576	Mobridge	99.4	48.7	77.0
577	Rapid City	100.8	59.6	82.7
TENNESSEE				
370-372	Nashville	98.2	74.1	87.6
373-374	Chattanooga	99.9	66.6	85.2
375,380-381	Memphis	98.9	72.5	87.3
376	Johnson City	99.9	56.4	80.7
377-379	Knoxville	96.5	66.8	83.5
382	McKenzie	97.7	59.4	80.8
383	Jackson	99.0	63.5	83.4
384	Columbia	96.2	67.2	83.4
385	Cookeville	97.5	60.0	81.0
TEXAS				
750	McKinney	98.2	66.9	84.4
751	Waxahackie	98.3	65.4	83.8
752-753	Dallas	99.2	67.9	85.4
754	Greenville	98.4	67.0	84.6
755	Texarkana	97.6	63.6	82.6
756	Longview	98.4	64.3	83.4
757	Tyler	98.4	65.2	83.8
758	Palestine	94.8	64.4	81.4
759	Lufkin	95.6	67.7	83.3
760-761	Fort Worth	98.8	65.7	84.2
762	Denton	99.2	66.5	84.8
763	Wichita Falls	96.5	64.6	82.5
764	Eastland	95.6	64.1	81.8
765	Temple	94.1	61.6	79.8
766-767	Waco	95.9	63.7	81.7
768	Brownwood	99.1	60.6	82.1
769	San Angelo	98.7	59.9	81.6
770-772	Houston	100.7	69.8	87.1
773	Huntsville	99.7	66.9	85.3
774	Wharton	100.6	68.7	86.5
775	Galveston	98.7	68.5	85.4
776-777	Beaumont	98.9	68.9	85.7
778	Bryan	96.4	68.4	84.1
779	Victoria	100.7	66.1	85.5
780	Laredo	99.4	63.5	83.6
781-782	San Antonio	100.3	64.1	84.4
783-784	Corpus Christi	102.0	63.2	84.9
785	McAllen	102.9	59.9	83.9
786-787	Austin	100.0	63.4	83.9

Location Factors - Commercial

STATE/ZIP	CITY	MAT.	INST.	TOTAL
TEXAS (CONT'D)				
788	Del Rio	102.5	63.0	85.1
789	Giddings	99.3	63.8	83.7
790-791	Amarillo	99.6	62.3	83.1
792	Childress	99.7	64.0	84.0
793-794	Lubbock	101.5	63.6	84.8
795-796	Abilene	100.0	62.3	83.4
797	Midland	102.1	64.6	85.6
798-799,885	El Paso	98.3	61.6	82.2
UTAH				
840-841	Salt Lake City	103.8	71.2	89.4
842,844	Ogden	99.3	71.2	86.9
843	Logan	101.3	71.2	88.0
845	Price	102.0	65.3	85.8
846-847	Provo	101.8	71.1	88.2
VERMONT				
050	White River Jct.	98.5	77.3	89.2
051	Bellows Falls	96.9	91.7	94.6
052	Bennington	97.3	92.3	95.1
053	Brattleboro	97.7	91.6	95.0
054	Burlington	102.2	82.5	93.5
056	Montpelier	100.6	81.8	92.3
057	Rutland	99.0	82.5	91.7
058	St. Johnsbury	98.7	76.8	89.1
059	Guildhall	97.2	76.8	88.2
VIRGINIA				
220-221	Fairfax	100.7	89.2	95.6
222	Arlington	101.5	89.4	96.2
223	Alexandria	100.6	89.4	95.6
224-225	Fredericksburg	99.4	87.7	94.2
226	Winchester	100.0	85.2	93.5
227	Culpeper	99.9	85.5	93.6
228	Harrisonburg	100.2	66.9	85.5
229	Charlottesville	100.5	67.9	86.1
230-232	Richmond	99.8	68.7	86.1
233-235	Norfolk	100.1	68.3	86.1
236	Newport News	99.8	68.3	85.9
237	Portsmouth	99.3	65.4	84.4
238	Petersburg	99.8	68.7	86.1
239	Farmville	99.0	52.1	78.3
240-241	Roanoke	101.0	66.6	85.8
242	Bristol	99.4	54.8	79.7
243	Pulaski	99.0	57.5	80.7
244	Staunton	99.7	61.7	83.0
245	Lynchburg	99.9	66.2	85.1
246	Grundy	99.2	52.6	78.7
WASHINGTON				
980-981,987	Seattle	102.3	103.8	103.0
982	Everett	102.2	99.0	100.8
983-984	Tacoma	102.6	99.6	101.3
985	Olympia	101.0	99.7	100.4
986	Vancouver	103.4	94.4	99.4
988	Wenatchee	102.8	85.9	95.3
989	Yakima	102.8	94.9	99.3
990-992	Spokane	104.8	82.6	95.0
993	Richland	104.4	90.7	98.4
994	Clarkston	103.8	81.9	94.1
WEST VIRGINIA				
247-248	Bluefield	98.2	93.6	96.2
249	Lewisburg	99.8	92.8	96.7
250-253	Charleston	99.7	97.4	98.7
254	Martinsburg	100.4	87.0	94.5
255-257	Huntington	101.4	97.8	99.8
258-259	Beckley	98.8	95.1	97.2
260	Wheeling	100.8	94.5	98.0
261	Parkersburg	99.6	92.5	96.5
262	Buckhannon	99.6	94.7	97.4
263-264	Clarksburg	100.2	94.7	97.7
265	Morgantown	100.2	94.8	97.8
266	Gassaway	99.5	94.4	97.3
267	Romney	99.4	89.8	95.1
268	Petersburg	99.3	87.7	94.2
WISCONSIN				
530,532	Milwaukee	98.9	106.9	102.4
531	Kenosha	99.2	101.7	100.3
534	Racine	98.7	102.0	100.2
535	Beloit	98.6	99.8	99.1
537	Madison	99.3	99.1	99.2

STATE/ZIP	CITY	MAT.	INST.	TOTAL
WISCONSIN (CONT'D)				
538	Lancaster	96.8	93.7	95.4
539	Portage	95.3	96.4	95.8
540	New Richmond	97.1	93.8	95.6
541-543	Green Bay	101.2	97.5	99.5
544	Wausau	96.5	96.1	96.3
545	Rhinelander	99.8	91.8	96.3
546	La Crosse	97.6	94.8	96.4
547	Eau Claire	99.2	95.5	97.6
548	Superior	96.8	97.7	97.2
549	Oshkosh	97.2	91.1	94.5
WYOMING				
820	Cheyenne	102.3	66.5	86.5
821	Yellowstone Nat'l Park	100.2	71.5	87.6
822	Wheatland	101.4	65.5	85.6
823	Rawlins	103.1	71.2	89.0
824	Worland	101.0	67.4	86.2
825	Riverton	102.1	66.4	86.4
826	Casper	102.5	63.8	85.5
827	Newcastle	100.8	72.4	88.3
828	Sheridan	103.7	66.5	87.3
829-831	Rock Springs	104.9	68.4	88.8
CANADIAN FACTORS (reflect Canadian currency)				
ALBERTA				
	Calgary	123.8	93.4	110.4
	Edmonton	123.9	93.4	110.5
	Fort McMurray	124.5	91.7	110.0
	Lethbridge	119.1	91.2	106.8
	Lloydminster	113.9	87.9	102.5
	Medicine Hat	114.1	87.2	102.2
	Red Deer	114.6	87.2	102.5
BRITISH COLUMBIA				
	Kamloops	115.0	89.3	103.7
	Prince George	116.0	88.6	104.0
	Vancouver	117.7	89.2	105.1
	Victoria	116.6	86.7	103.4
MANITOBA				
	Brandon	126.4	74.8	103.7
	Portage la Prairie	113.9	73.1	95.9
	Winnipeg	122.6	70.8	99.8
NEW BRUNSWICK				
	Bathurst	111.8	66.3	91.8
	Dalhousie	112.6	66.5	92.3
	Fredericton	118.8	71.5	97.9
	Moncton	112.0	70.7	93.8
	Newcastle	111.9	67.0	92.1
	St. John	114.4	77.8	98.3
NEWFOUNDLAND				
	Corner Brook	131.2	66.6	102.7
	St. John's	123.9	83.9	106.3
NORTHWEST TERRITORIES				
	Yellowknife	136.7	82.9	113.0
NOVA SCOTIA				
	Bridgewater	113.4	73.6	95.9
	Dartmouth	127.5	73.6	103.7
	Halifax	118.2	79.9	101.3
	New Glasgow	125.5	73.6	102.6
	Sydney	123.4	73.6	101.4
	Truro	113.0	73.6	95.6
	Yarmouth	125.4	73.6	102.6
ONTARIO				
	Barrie	118.3	91.7	106.6
	Brantford	115.5	95.5	106.7
	Cornwall	115.4	92.0	105.1
	Hamilton	116.9	95.6	107.5
	Kingston	116.5	92.1	105.7
	Kitchener	109.9	91.8	101.9
	London	116.3	92.2	105.7
	North Bay	128.6	89.8	111.5
	Oshawa	112.5	95.3	104.9
	Ottawa	117.6	92.9	106.7
	Owen Sound	118.4	89.9	105.8
	Peterborough	115.5	91.7	105.0
	Sarnia	115.3	96.1	106.9

For customer support on your Mechanical Cost Data, call 877.739.3689.

Location Factors - Commercial

STATE/ZIP	CITY	MAT.	INST.	TOTAL
ONTARIO (CONT'D)				
	Sault Ste. Marie	110.7	91.1	102.1
	St. Catharines	107.8	93.3	101.5
	Sudbury	107.6	92.1	100.8
	Thunder Bay	109.1	92.3	101.7
	Timmins	115.6	89.8	104.2
	Toronto	116.9	100.9	109.9
	Windsor	108.3	92.3	101.3
PRINCE EDWARD ISLAND				
	Charlottetown	120.4	61.9	94.6
	Summerside	128.1	62.0	99.0
QUEBEC				
	Cap-de-la-Madeleine	112.8	86.4	101.1
	Charlesbourg	112.8	86.4	101.1
	Chicoutimi	111.7	92.0	103.0
	Gatineau	112.5	86.1	100.9
	Granby	112.7	86.1	101.0
	Hull	112.6	86.1	100.9
	Joliette	113.0	86.4	101.3
	Laval	112.8	86.1	101.1
	Montreal	117.9	92.4	106.7
	Quebec City	117.9	92.8	106.8
	Rimouski	112.3	92.0	103.4
	Rouyn-Noranda	112.5	86.1	100.9
	Saint-Hyacinthe	112.0	86.1	100.6
	Sherbrooke	112.8	86.1	101.1
	Sorel	113.0	86.4	101.2
	Saint-Jerome	112.5	86.1	100.9
	Trois-Rivieres	126.1	86.4	108.6
SASKATCHEWAN				
	Moose Jaw	111.2	68.5	92.4
	Prince Albert	110.3	66.8	91.1
	Regina	123.0	89.0	108.0
	Saskatoon	112.3	89.0	102.0
YUKON				
	Whitehorse	137.1	67.5	106.4

R011105-05   Tips for Accurate Estimating

1. Use pre-printed or columnar forms for orderly sequence of dimensions and locations and for recording telephone quotations.

2. Use only the front side of each paper or form except for certain pre-printed summary forms.

3. Be consistent in listing dimensions: For example, length x width x height. This helps in rechecking to ensure that, the total length of partitions is appropriate for the building area.

4. Use printed (rather than measured) dimensions where given.

5. Add up multiple printed dimensions for a single entry where possible.

6. Measure all other dimensions carefully.

7. Use each set of dimensions to calculate multiple related quantities.

8. Convert foot and inch measurements to decimal feet when listing. Memorize decimal equivalents to .01 parts of a foot (1/8″ equals approximately .01′).

9. Do not "round off" quantities until the final summary.

10. Mark drawings with different colors as items are taken off.

11. Keep similar items together, different items separate.

12. Identify location and drawing numbers to aid in future checking for completeness.

13. Measure or list everything on the drawings or mentioned in the specifications.

14. It may be necessary to list items not called for to make the job complete.

15. Be alert for: Notes on plans such as N.T.S. (not to scale); changes in scale throughout the drawings; reduced size drawings; discrepancies between the specifications and the drawings.

16. Develop a consistent pattern of performing an estimate. For example:
    a. Start the quantity takeoff at the lower floor and move to the next higher floor.
    b. Proceed from the main section of the building to the wings.
    c. Proceed from south to north or vice versa, clockwise or counterclockwise.
    d. Take off floor plan quantities first, elevations next, then detail drawings.

17. List all gross dimensions that can be either used again for different quantities, or used as a rough check of other quantities for verification (exterior perimeter, gross floor area, individual floor areas, etc.).

18. Utilize design symmetry or repetition (repetitive floors, repetitive wings, symmetrical design around a center line, similar room layouts, etc.). Note: Extreme caution is needed here so as not to omit or duplicate an area.

19. Do not convert units until the final total is obtained. For instance, when estimating concrete work, keep all units to the nearest cubic foot, then summarize and convert to cubic yards.

20. When figuring alternatives, it is best to total all items involved in the basic system, then total all items involved in the alternates. Therefore you work with positive numbers in all cases. When adds and deducts are used, it is often confusing whether to add or subtract a portion of an item; especially on a complicated or involved alternate.

R011105-10   Unit Gross Area Requirements

The figures in the table below indicate typical ranges in square feet as a function of the "occupant" unit. This table is best used in the preliminary design stages to help determine the probable size requirement for the total project.

Building Type	Unit	Gross Area in S.F.		
		1/4	Median	3/4
Apartments	Unit	660	860	1,100
Auditorium & Play Theaters	Seat	18	25	38
Bowling Alleys	Lane		940	
Churches & Synagogues	Seat	20	28	39
Dormitories	Bed	200	230	275
Fraternity & Sorority Houses	Bed	220	315	370
Garages, Parking	Car	325	355	385
Hospitals	Bed	685	850	1,075
Hotels	Rental Unit	475	600	710
Housing for the elderly	Unit	515	635	755
Housing, Public	Unit	700	875	1,030
Ice Skating Rinks	Total	27,000	30,000	36,000
Motels	Rental Unit	360	465	620
Nursing Homes	Bed	290	350	450
Restaurants	Seat	23	29	39
Schools, Elementary	Pupil	65	77	90
Junior High & Middle		85	110	129
Senior High		102	130	145
Vocational		110	135	195
Shooting Ranges	Point		450	
Theaters & Movies	Seat		15	

R011105-20   Floor Area Ratios

Table below lists commonly used gross to net area and net to gross area ratios expressed in % for various building types.

Building Type	Gross to Net Ratio	Net to Gross Ratio	Building Type	Gross to Net Ratio	Net to Gross Ratio
Apartment	156	64	School Buildings (campus type)		
Bank	140	72	Administrative	150	67
Church	142	70	Auditorium	142	70
Courthouse	162	61	Biology	161	62
Department Store	123	81	Chemistry	170	59
Garage	118	85	Classroom	152	66
Hospital	183	55	Dining Hall	138	72
Hotel	158	63	Dormitory	154	65
Laboratory	171	58	Engineering	164	61
Library	132	76	Fraternity	160	63
Office	135	75	Gymnasium	142	70
Restaurant	141	70	Science	167	60
Warehouse	108	93	Service	120	83
			Student Union	172	59

The gross area of a building is the total floor area based on outside dimensions.

The net area of a building is the usable floor area for the function intended and excludes such items as stairways, corridors and mechanical rooms. In the case of a commercial building, it might be considered as the "leasable area."

R011105-30   Occupancy Determinations

Function of Space	SF/Person Required
Accessory storage areas, mechanical equipment rooms	300
Agriculture Building	300
Aircraft Hangars	500
Airport Terminal	
Baggage claim	20
Baggage handling	300
Concourse	100
Waiting areas	15
Assembly	
Gaming floors (keno, slots, etc.)	11
Exhibit Gallery and Museum	30
Assembly w/ fixed seats	load determined by seat number
Assembly w/o fixed seats	
Concentrated (chairs only-not fixed)	7
Standing space	5
Unconcentrated (tables and chairs)	15
Bowling centers, allow 5 persons for each lane including 15 feet of runway, and for additional areas	7
Business areas	100
Courtrooms-other than fixed seating areas	40
Day care	35
Dormitories	50
Educational	
Classroom areas	20
Shops and other vocational room areas	50
Exercise rooms	50
Fabrication and Manufacturing areas where hazardous materials are used	200
Industrial areas	100
Institutional areas	
Inpatient treatment areas	240
Outpatient areas	100
Sleeping areas	120
Kitchens commercial	200
Library	
Reading rooms	50
Stack area	100
Mercantile	
Areas on other floors	60
Basement and grade floor areas	30
Storage, stock, shipping areas	300
Parking garages	200
Residential	200
Skating rinks, swimming pools	
Rink and pool	50
Decks	15
Stages and platforms	15
Warehouses	500

Excerpted from the 2012 *International Building Code*, Copyright 2011. Washington, D.C.: International Code Council. Reproduced with permission. All rights reserved. www.ICCSAFE.org

707

For customer support on your Mechanical Cost Data, call 877.739.3689.

R011105-50   Metric Conversion Factors

Description: This table is primarily for converting customary U.S. units in the left hand column to SI metric units in the right hand column. In addition, conversion factors for some commonly encountered Canadian and non-SI metric units are included.

	If You Know		Multiply By		To Find
Length	Inches	x	25.4[a]	=	Millimeters
	Feet	x	0.3048[a]	=	Meters
	Yards	x	0.9144[a]	=	Meters
	Miles (statute)	x	1.609	=	Kilometers
Area	Square inches	x	645.2	=	Square millimeters
	Square feet	x	0.0929	=	Square meters
	Square yards	x	0.8361	=	Square meters
Volume (Capacity)	Cubic inches	x	16,387	=	Cubic millimeters
	Cubic feet	x	0.02832	=	Cubic meters
	Cubic yards	x	0.7646	=	Cubic meters
	Gallons (U.S. liquids)[b]	x	0.003785	=	Cubic meters[c]
	Gallons (Canadian liquid)[b]	x	0.004546	=	Cubic meters[c]
	Ounces (U.S. liquid)[b]	x	29.57	=	Milliliters[c, d]
	Quarts (U.S. liquid)[b]	x	0.9464	=	Liters[c, d]
	Gallons (U.S. liquid)[b]	x	3.785	=	Liters[c, d]
Force	Kilograms force[d]	x	9.807	=	Newtons
	Pounds force	x	4.448	=	Newtons
	Pounds force	x	0.4536	=	Kilograms force[d]
	Kips	x	4448	=	Newtons
	Kips	x	453.6	=	Kilograms force[d]
Pressure, Stress, Strength (Force per unit area)	Kilograms force per square centimeter[d]	x	0.09807	=	Megapascals
	Pounds force per square inch (psi)	x	0.006895	=	Megapascals
	Kips per square inch	x	6.895	=	Megapascals
	Pounds force per square inch (psi)	x	0.07031	=	Kilograms force per square centimeter[d]
	Pounds force per square foot	x	47.88	=	Pascals
	Pounds force per square foot	x	4.882	=	Kilograms force per square meter[d]
Flow	Cubic feet per minute	x	0.4719	=	Liters per second
	Gallons per minute	x	0.0631	=	Liters per second
	Gallons per hour	x	1.05	=	Milliliters per second
Bending Moment Or Torque	Inch-pounds force	x	0.01152	=	Meter-kilograms force[d]
	Inch-pounds force	x	0.1130	=	Newton meters
	Foot-pounds force	x	0.1383	=	Meter-kilograms force[d]
	Foot-pounds force	x	1.356	=	Newton-meters
	Meter-kilograms force[d]	x	9.807	=	Newton-meters
Mass	Ounces (avoirdupois)	x	28.35	=	Grams
	Pounds (avoirdupois)	x	0.4536	=	Kilograms
	Tons (metric)	x	1000	=	Kilograms
	Tons, short (2000 pounds)	x	907.2	=	Kilograms
	Tons, short (2000 pounds)	x	0.9072	=	Megagrams[e]
Mass per Unit Volume	Pounds mass per cubic foot	x	16.02	=	Kilograms per cubic meter
	Pounds mass per cubic yard	x	0.5933	=	Kilograms per cubic meter
	Pounds mass per gallon (U.S. liquid)[b]	x	119.8	=	Kilograms per cubic meter
	Pounds mass per gallon (Canadian liquid)[b]	x	99.78	=	Kilograms per cubic meter
Temperature	Degrees Fahrenheit	(F-32)/1.8		=	Degrees Celsius
	Degrees Fahrenheit	(F+459.67)/1.8		=	Degrees Kelvin
	Degrees Celsius	C+273.15		=	Degrees Kelvin

[a]The factor given is exact
[b]One U.S. gallon = 0.8327 Canadian gallon
[c]1 liter = 1000 milliliters = 1000 cubic centimeters
1 cubic decimeter = 0.001 cubic meter

[d]Metric but not SI unit
[e]Called "tonne" in England and "metric ton" in other metric countries

For customer support on your **Mechanical Cost Data**, call 877.739.3689.

Reference Tables

R011105-60 Weights and Measures

Measures of Length
1 Mile = 1760 Yards = 5280 Feet
1 Yard = 3 Feet = 36 inches
1 Foot = 12 Inches
1 Mil = 0.001 Inch
1 Fathom = 2 Yards = 6 Feet
1 Rod = 5.5 Yards = 16.5 Feet
1 Hand = 4 Inches
1 Span = 9 Inches
1 Micro-inch = One Millionth Inch or 0.000001 Inch
1 Micron = One Millionth Meter + 0.00003937 Inch

Surveyor's Measure
1 Mile = 8 Furlongs = 80 Chains
1 Furlong = 10 Chains = 220 Yards
1 Chain = 4 Rods = 22 Yards = 66 Feet = 100 Links
1 Link = 7.92 Inches

Square Measure
1 Square Mile = 640 Acres = 6400 Square Chains
1 Acre = 10 Square Chains = 4840 Square Yards =
   43,560 Sq. Ft.
1 Square Chain = 16 Square Rods = 484 Square Yards =
   4356 Sq. Ft.
1 Square Rod = 30.25 Square Yards = 272.25 Square Feet = 625 Square
   Lines
1 Square Yard = 9 Square Feet
1 Square Foot = 144 Square Inches
An Acre equals a Square 208.7 Feet per Side

Cubic Measure
1 Cubic Yard = 27 Cubic Feet
1 Cubic Foot = 1728 Cubic Inches
1 Cord of Wood = 4 x 4 x 8 Feet = 128 Cubic Feet
1 Perch of Masonry = 16½ x 1½ x 1 Foot = 24.75 Cubic Feet

Avoirdupois or Commercial Weight
1 Gross or Long Ton = 2240 Pounds
1 Net or Short Ton = 2000 Pounds
1 Pound = 16 Ounces = 7000 Grains
1 Ounce = 16 Drachms = 437.5 Grains
1 Stone = 14 Pounds

Power
1 British Thermal Unit per Hour = 0.2931 Watts
1 Ton (Refrigeration) = 3.517 Kilowatts
1 Horsepower (Boiler) = 9.81 Kilowatts
1 Horsepower (550 ft-lb/s) = 0.746 Kilowatts

Shipping Measure
For Measuring Internal Capacity of a Vessel:
   1 Register Ton = 100 Cubic Feet

For Measurement of Cargo:
   Approximately 40 Cubic Feet of Merchandise is considered a Shipping
   Ton, unless that bulk would weigh more than 2000 Pounds, in which case
   Freight Charge may be based upon weight.

40 Cubic Feet = 32.143 U.S. Bushels = 31.16 Imp. Bushels

Liquid Measure
1 Imperial Gallon = 1.2009 U.S. Gallon = 277.42 Cu. In.
1 Cubic Foot = 7.48 U.S. Gallons

R011110-10   Architectural Fees

Tabulated below are typical percentage fees by project size, for good professional architectural service. Fees may vary from those listed depending upon degree of design difficulty and economic conditions in any particular area.

Rates can be interpolated horizontally and vertically. Various portions of the same project requiring different rates should be adjusted proportionately. For alterations, add 50% to the fee for the first $500,000 of project cost and add 25% to the fee for project cost over $500,000.

Architectural fees tabulated below include Structural, Mechanical and Electrical Engineering Fees. They do not include the fees for special consultants such as kitchen planning, security, acoustical, interior design, etc.

Civil Engineering fees are included in the Architectural fee for project sites requiring minimal design such as city sites. However, separate Civil Engineering fees must be added when utility connections require design, drainage calculations are needed, stepped foundations are required, or provisions are required to protect adjacent wetlands.

Building Types	Total Project Size in Thousands of Dollars						
	100	250	500	1,000	5,000	10,000	50,000
Factories, garages, warehouses, repetitive housing	9.0%	8.0%	7.0%	6.2%	5.3%	4.9%	4.5%
Apartments, banks, schools, libraries, offices, municipal buildings	12.2	12.3	9.2	8.0	7.0	6.6	6.2
Churches, hospitals, homes, laboratories, museums, research	15.0	13.6	12.7	11.9	9.5	8.8	8.0
Memorials, monumental work, decorative furnishings	—	16.0	14.5	13.1	10.0	9.0	8.3

R011110-30   Engineering Fees

Typical **Structural Engineering Fees** based on type of construction and total project size. These fees are included in Architectural Fees.

Type of Construction	Total Project Size (in thousands of dollars)			
	$500	$500-$1,000	$1,000-$5,000	Over $5000
Industrial buildings, factories & warehouses	Technical payroll times 2.0 to 2.5	1.60%	1.25%	1.00%
Hotels, apartments, offices, dormitories, hospitals, public buildings, food stores		2.00%	1.70%	1.20%
Museums, banks, churches and cathedrals		2.00%	1.75%	1.25%
Thin shells, prestressed concrete, earthquake resistive		2.00%	1.75%	1.50%
Parking ramps, auditoriums, stadiums, convention halls, hangars & boiler houses		2.50%	2.00%	1.75%
Special buildings, major alterations, underpinning & future expansion		Add to above 0.5%	Add to above 0.5%	Add to above 0.5%

For complex reinforced concrete or unusually complicated structures, add 20% to 50%.

Typical **Mechanical and Electrical Engineering Fees** are based on the size of the subcontract. The fee structure for both is shown below. These fees are included in Architectural Fees.

Type of Construction	Subcontract Size							
	$25,000	$50,000	$100,000	$225,000	$350,000	$500,000	$750,000	$1,000,000
Simple structures	6.4%	5.7%	4.8%	4.5%	4.4%	4.3%	4.2%	4.1%
Intermediate structures	8.0	7.3	6.5	5.6	5.1	5.0	4.9	4.8
Complex structures	10.1	9.0	9.0	8.0	7.5	7.5	7.0	7.0

For renovations, add 15% to 25% to applicable fee.

R012153-10   Repair and Remodeling

Cost figures are based on new construction utilizing the most cost-effective combination of labor, equipment and material with the work scheduled in proper sequence to allow the various trades to accomplish their work in an efficient manner.

The costs for repair and remodeling work must be modified due to the following factors that may be present in any given repair and remodeling project.

1. Equipment usage curtailment due to the physical limitations of the project, with only hand-operated equipment being used.

2. Increased requirement for shoring and bracing to hold up the building while structural changes are being made and to allow for temporary storage of construction materials on above-grade floors.

3. Material handling becomes more costly due to having to move within the confines of an enclosed building. For multi-story construction, low capacity elevators and stairwells may be the only access to the upper floors.

4. Large amount of cutting and patching and attempting to match the existing construction is required. It is often more economical to remove entire walls rather than create many new door and window openings. This sort of trade-off has to be carefully analyzed.

5. Cost of protection of completed work is increased since the usual sequence of construction usually cannot be accomplished.

6. Economies of scale usually associated with new construction may not be present. If small quantities of components must be custom fabricated due to job requirements, unit costs will naturally increase. Also, if only small work areas are available at a given time, job scheduling between trades becomes difficult and subcontractor quotations may reflect the excessive start-up and shut-down phases of the job.

7. Work may have to be done on other than normal shifts and may have to be done around an existing production facility which has to stay in production during the course of the repair and remodeling.

8. Dust and noise protection of adjoining non-construction areas can involve substantial special protection and alter usual construction methods.

9. Job may be delayed due to unexpected conditions discovered during demolition or removal. These delays ultimately increase construction costs.

10. Piping and ductwork runs may not be as simple as for new construction. Wiring may have to be snaked through walls and floors.

11. Matching "existing construction" may be impossible because materials may no longer be manufactured. Substitutions may be expensive.

12. Weather protection of existing structure requires additional temporary structures to protect building at openings.

13. On small projects, because of local conditions, it may be necessary to pay a tradesman for a minimum of four hours for a task that is completed in one hour.

All of the above areas can contribute to increased costs for a repair and remodeling project. Each of the above factors should be considered in the planning, bidding and construction stage in order to minimize the increased costs associated with repair and remodeling jobs.

R012909-80   Sales Tax by State

State sales tax on materials is tabulated below (5 states have no sales tax). Many states allow local jurisdictions, such as a county or city, to levy additional sales tax.

Some projects may be sales tax exempt, particularly those constructed with public funds.

State	Tax (%)	State	Tax (%)	State	Tax (%)	State	Tax (%)
Alabama	4	Illinois	6.25	Montana	0	Rhode Island	7
Alaska	0	Indiana	7	Nebraska	5.5	South Carolina	6
Arizona	5.6	Iowa	6	Nevada	6.85	South Dakota	4
Arkansas	6	Kansas	6.15	New Hampshire	0	Tennessee	7
California	7.5	Kentucky	6	New Jersey	7	Texas	6.25
Colorado	2.9	Louisiana	4	New Mexico	5.125	Utah	5.95
Connecticut	6.35	Maine	5.5	New York	4	Vermont	6
Delaware	0	Maryland	6	North Carolina	4.75	Virginia	5.3
District of Columbia	5.75	Massachusetts	6.25	North Dakota	5	Washington	6.5
Florida	6	Michigan	6	Ohio	5.75	West Virginia	6
Georgia	4	Minnesota	6.875	Oklahoma	4.5	Wisconsin	5
Hawaii	4	Mississippi	7	Oregon	0	Wyoming	4
Idaho	6	Missouri	4.225	Pennsylvania	6	Average	5.08%

Sales Tax by Province (Canada)

GST - a value-added tax, which the government imposes on most goods and services provided in or imported into Canada. PST - a retail sales tax, which three of the provinces impose on the prices of most goods and some

services. QST - a value-added tax, similar to the federal GST, which Quebec imposes. HST - Five provinces have combined their retail sales taxes with the federal GST into one harmonized tax.

Province	PST (%)	QST (%)	GST(%)	HST(%)
Alberta	0	0	5	0
British Columbia	7	0	5	0
Manitoba	8	0	5	0
New Brunswick	0	0	0	13
Newfoundland	0	0	0	13
Northwest Territories	0	0	5	0
Nova Scotia	0	0	0	15
Ontario	0	0	0	13
Prince Edward Island	0	0	0	14
Quebec	0	9.975	5	0
Saskatchewan	5	0	5	0
Yukon	0	0	5	0

R012909-85   Unemployment Taxes and Social Security Taxes

State unemployment tax rates vary not only from state to state, but also with the experience rating of the contractor. The federal unemployment tax rate is 6.0% of the first $7,000 of wages. This is reduced by a credit of up to 5.4% for timely payment to the state. The minimum federal unemployment tax is 0.6% after all credits.

Social security (FICA) for 2016 is estimated at time of publication to be 7.65% of wages up to $118,500.

R012909-90   Overtime

One way to improve the completion date of a project or eliminate negative float from a schedule is to compress activity duration times. This can be achieved by increasing the crew size or working overtime with the proposed crew.

To determine the costs of working overtime to compress activity duration times, consider the following examples. Below is an overtime efficiency and

cost chart based on a five, six, or seven day week with an eight through twelve hour day. Payroll percentage increases for time and one half and double times are shown for the various working days.

Days per Week	Hours per Day	Production Efficiency					Payroll Cost Factors	
		1st Week	2nd Week	3rd Week	4th Week	Average 4 Weeks	@ 1-1/2 Times	@ 2 Times
5	8	100%	100%	100%	100%	100%	1.000	1.000
	9	100	100	95	90	96	1.056	1.111
	10	100	95	90	85	93	1.100	1.200
	11	95	90	75	65	81	1.136	1.273
	12	90	85	70	60	76	1.167	1.333
6	8	100	100	95	90	96	1.083	1.167
	9	100	95	90	85	93	1.130	1.259
	10	95	90	85	80	88	1.167	1.333
	11	95	85	70	65	79	1.197	1.394
	12	90	80	65	60	74	1.222	1.444
7	8	100	95	85	75	89	1.143	1.286
	9	95	90	80	70	84	1.183	1.365
	10	90	85	75	65	79	1.214	1.429
	11	85	80	65	60	73	1.240	1.481
	12	85	75	60	55	69	1.262	1.524

R013113-40   Builder's Risk Insurance

Builder's Risk Insurance is insurance on a building during construction. Premiums are paid by the owner or the contractor. Blasting, collapse and underground insurance would raise total insurance costs above those listed. Floater policy for materials delivered to the job runs $.75 to $1.25 per $100 value. Contractor equipment insurance runs $.50 to $1.50 per $100 value. Insurance for miscellaneous tools to $1,500 value runs from $3.00 to $7.50 per $100 value.

Tabulated below are New England Builder's Risk insurance rates in dollars per $100 value for $1,000 deductible. For $25,000 deductible, rates can be reduced 13% to 34%. On contracts over $1,000,000, rates may be lower than those tabulated. Policies are written annually for the total completed value in place. For "all risk" insurance (excluding flood, earthquake and certain other perils) add $.025 to total rates below.

Coverage	Frame Construction (Class 1)		Average	Brick Construction (Class 4)		Average	Fire Resistive (Class 6)		Average
	Range			Range			Range		
Fire Insurance	$.350 to	$ .850	$.600	$.158 to	$.189	$.174	$.052 to	$.080	$.070
Extended Coverage	.115 to	.200	.158	.080 to	.105	.101	.081 to	.105	.100
Vandalism	.012 to	.016	.014	.008 to	.011	.011	.008 to	.011	.010
Total Annual Rate	$.477 to	$1.066	$.772	$.246 to	$.305	$.286	$.141 to	$.196	$.180

R013113-50   General Contractor's Overhead

There are two distinct types of overhead on a construction project: Project Overhead and Main Office Overhead. Project Overhead includes those costs at a construction site not directly associated with the installation of construction materials. Examples of Project Overhead costs include the following:

1. Superintendent
2. Construction office and storage trailers
3. Temporary sanitary facilities
4. Temporary utilities
5. Security fencing
6. Photographs
7. Clean up
8. Performance and payment bonds

The above Project Overhead items are also referred to as General Requirements and therefore are estimated in Division 1. Division 1 is the first division listed in the CSI MasterFormat but it is usually the last division estimated. The sum of the costs in Divisions 1 through 49 is referred to as the sum of the direct costs.

All construction projects also include indirect costs. The primary components of indirect costs are the contractor's Main Office Overhead and profit. The amount of the Main Office Overhead expense varies depending on the following:

1. Owner's compensation
2. Project managers and estimator's wages
3. Clerical support wages
4. Office rent and utilities
5. Corporate legal and accounting costs
6. Advertising
7. Automobile expenses
8. Association dues
9. Travel and entertainment expenses

These costs are usually calculated as a percentage of annual sales volume. This percentage can range from 35% for a small contractor doing less than $500,000 to 5% for a large contractor with sales in excess of $100 million.

714

For customer support on your Mechanical Cost Data, call 877.739.3689.

R013113-60   Workers' Compensation Insurance Rates by Trade

The table below tabulates the national averages for workers' compensation insurance rates by trade and type of building. The average "Insurance Rate" is multiplied by the "% of Building Cost" for each trade. This produces

the "Workers' Compensation" cost by % of total labor cost, to be added for each trade by building type to determine the weighted average workers' compensation rate for the building types analyzed.

Trade	Insurance Rate (% Labor Cost)			% of Building Cost			Workers' Compensation		
	Range		Average	Office Bldgs.	Schools & Apts.	Mfg.	Office Bldgs.	Schools & Apts.	Mfg.
Excavation, Grading, etc.	3.1 % to	21.5%	9.3%	4.8%	4.9%	4.5%	0.45%	0.46%	0.42%
Piles & Foundations	6.4 to	32.8	14.6	7.1	5.2	8.7	1.04	0.76	1.27
Concrete	4.3 to	32.5	12.6	5.0	14.8	3.7	0.63	1.86	0.47
Masonry	4.6 to	36.8	13.7	6.9	7.5	1.9	0.95	1.03	0.26
Structural Steel	6.5 to	95.8	27.5	10.7	3.9	17.6	2.94	1.07	4.84
Miscellaneous & Ornamental Metals	4.1 to	28.9	11.7	2.8	4.0	3.6	0.33	0.47	0.42
Carpentry & Millwork	5.0 to	42.8	14.4	3.7	4.0	0.5	0.53	0.58	0.07
Metal or Composition Siding	6.5 to	75.9	18.7	2.3	0.3	4.3	0.43	0.06	0.80
Roofing	6.5 to	107.1	31.4	2.3	2.6	3.1	0.72	0.82	0.97
Doors & Hardware	4.1 to	42.8	11.4	0.9	1.4	0.4	0.10	0.16	0.05
Sash & Glazing	4.8 to	28.0	13.1	3.5	4.0	1.0	0.46	0.52	0.13
Lath & Plaster	2.8 to	41.3	10.5	3.3	6.9	0.8	0.35	0.72	0.08
Tile, Marble & Floors	2.9 to	22.4	8.8	2.6	3.0	0.5	0.23	0.26	0.04
Acoustical Ceilings	3.1 to	33.8	8.6	2.4	0.2	0.3	0.21	0.02	0.03
Painting	4.1 to	34.3	11.7	1.5	1.6	1.6	0.18	0.19	0.19
Interior Partitions	5.0 to	42.8	14.4	3.9	4.3	4.4	0.56	0.62	0.63
Miscellaneous Items	2.5 to	105.9	12.6	5.2	3.7	9.7	0.65	0.47	1.22
Elevators	1.4 to	13.6	5.2	2.1	1.1	2.2	0.11	0.06	0.11
Sprinklers	2.8 to	25.9	7.6	0.5	—	2.0	0.04	—	0.15
Plumbing	1.9 to	19.5	6.9	4.9	7.2	5.2	0.34	0.50	0.36
Heat., Vent., Air Conditioning	3.8 to	18.4	8.8	13.5	11.0	12.9	1.19	0.97	1.14
Electrical	2.3 to	14.4	5.6	10.1	8.4	11.1	0.57	0.47	0.62
Total	1.4 % to	107.1%	—	100.0%	100.0%	100.0%	13.01%	12.07%	14.27%

Overall Weighted Average 13.12%

Workers' Compensation Insurance Rates by States

The table below lists the weighted average workers' compensation base rate for each state with a factor comparing this with the national average of 13.0%.

State	Weighted Average	Factor	State	Weighted Average	Factor	State	Weighted Average	Factor
Alabama	19.5%	150	Kentucky	14.3%	110	North Dakota	9.2%	71
Alaska	13.7	105	Louisiana	18.6	143	Ohio	8.2	63
Arizona	12.6	97	Maine	12.2	94	Oklahoma	12.5	96
Arkansas	8.3	64	Maryland	14.1	108	Oregon	11.8	91
California	25.8	198	Massachusetts	11.6	89	Pennsylvania	17.4	134
Colorado	7.2	55	Michigan	16.3	125	Rhode Island	12.1	93
Connecticut	23.7	182	Minnesota	22.6	174	South Carolina	18.9	145
Delaware	11.9	92	Mississippi	13.2	102	South Dakota	14.9	115
District of Columbia	11.0	85	Missouri	14.8	114	Tennessee	12.0	92
Florida	10.7	82	Montana	8.5	65	Texas	9.1	70
Georgia	28.1	216	Nebraska	17.7	136	Utah	8.8	68
Hawaii	8.1	62	Nevada	8.7	67	Vermont	13.1	101
Idaho	10.3	79	New Hampshire	20.0	154	Virginia	9.5	73
Illinois	27.1	208	New Jersey	14.4	111	Washington	10.5	81
Indiana	5.5	42	New Mexico	14.4	111	West Virginia	8.0	62
Iowa	14.6	112	New York	17.2	132	Wisconsin	13.1	101
Kansas	8.4	65	North Carolina	16.7	128	Wyoming	6.1	47

Weighted Average for U.S. is   13.7% of payroll = 100%

The weighted average skilled worker rate for 35 trades is 13.0%. For bidding purposes, apply the full value of workers' compensation directly to total labor costs, or if labor is 38%, materials 42% and overhead and profit 20% of total cost, carry 38/80 x 13.0% = 6.2% of cost (before overhead and profit)

into overhead. Rates vary not only from state to state but also with the experience rating of the contractor.

Rates are the most current available at the time of publication.

715

General Requirements   R0131 Project Management & Coordination

R013113-80   Performance Bond

This table shows the cost of a Performance Bond for a construction job scheduled to be completed in 12 months. Add 1% of the premium cost per month for jobs requiring more than 12 months to complete. The rates are "standard" rates offered to contractors that the bonding company considers financially sound and capable of doing the work. Preferred rates are offered by some bonding companies based upon financial strength of the contractor. Actual rates vary from contractor to contractor and from bonding company to bonding company. Contractors should prequalify through a bonding agency before submitting a bid on a contract that requires a bond.

Contract Amount			Building Construction Class B Projects			Highways & Bridges					
						Class A New Construction			Class A-1 Highway Resurfacing		
First	$ 100,000	bid	$25.00 per M			$15.00 per M			$9.40 per M		
Next	400,000	bid	$ 2,500	plus	$15.00 per M	$ 1,500	plus	$10.00 per M	$ 940	plus	$7.20 per M
Next	2,000,000	bid	8,500	plus	10.00 per M	5,500	plus	7.00 per M	3,820	plus	5.00 per M
Next	2,500,000	bid	28,500	plus	7.50 per M	19,500	plus	5.50 per M	15,820	plus	4.50 per M
Next	2,500,000	bid	47,250	plus	7.00 per M	33,250	plus	5.00 per M	28,320	plus	4.50 per M
Over	7,500,000	bid	64,750	plus	6.00 per M	45,750	plus	4.50 per M	39,570	plus	4.00 per M

General Requirements   R0154 Construction Aids

R015423-10   Steel Tubular Scaffolding

On new construction, tubular scaffolding is efficient up to 60' high or five stories. Above this it is usually better to use a hung scaffolding if construction permits. Swing scaffolding operations may interfere with tenants. In this case, the tubular is more practical at all heights.

In repairing or cleaning the front of an existing building the cost of tubular scaffolding per S.F. of building front increases as the height increases above the first tier. The first tier cost is relatively high due to leveling and alignment.

The minimum efficient crew for erecting and dismantling is three workers. They can set up and remove 18 frame sections per day up to 5 stories high. For 6 to 12 stories high, a crew of four is most efficient. Use two or more on top and two on the bottom for handing up or hoisting. They can also set up and remove 18 frame sections per day. At 7' horizontal spacing, this will run about 800 S.F. per day of erecting and dismantling. Time for placing and removing planks must be added to the above. A crew of three can place and remove 72 planks per day up to 5 stories. For over 5 stories, a crew of four can place and remove 80 planks per day.

The table below shows the number of pieces required to erect tubular steel scaffolding for 1000 S.F. of building frontage. This area is made up of a scaffolding system that is 12 frames (11 bays) long by 2 frames high.

For jobs under twenty-five frames, add 50% to rental cost. Rental rates will be lower for jobs over three months duration. Large quantities for long periods can reduce rental rates by 20%.

Description of Component	Number of Pieces for 1000 S.F. of Building Front	Unit
5' Wide Standard Frame, 6'-4" High	24	Ea.
Leveling Jack & Plate	24	
Cross Brace	44	
Side Arm Bracket, 21"	12	
Guardrail Post	12	
Guardrail, 7' section	22	
Stairway Section	2	
Stairway Starter Bar	1	
Stairway Inside Handrail	2	
Stairway Outside Handrail	2	
Walk-Thru Frame Guardrail	2	

Scaffolding is often used as falsework over 15' high during construction of cast-in-place concrete beams and slabs. Two foot wide scaffolding is generally used for heavy beam construction. The span between frames depends upon the load to be carried with a maximum span of 5'.

Heavy duty shoring frames with a capacity of 10,000#/leg can be spaced up to 10' O.C. depending upon form support design and loading.

Scaffolding used as horizontal shoring requires less than half the material required with conventional shoring.

On new construction, erection is done by carpenters.

Rolling towers supporting horizontal shores can reduce labor and speed the job. For maintenance work, catwalks with spans up to 70' can be supported by the rolling towers.

R015433-10   Contractor Equipment

Rental Rates shown elsewhere in the data set pertain to late model high quality machines in excellent working condition, rented from equipment dealers. Rental rates from contractors may be substantially lower than the rental rates from equipment dealers depending upon economic conditions; for older, less productive machines, reduce rates by a maximum of 15%. Any overtime must be added to the base rates. For shift work, rates are lower. Usual rule of thumb is 150% of one shift rate for two shifts; 200% for three shifts.

For periods of less than one week, operated equipment is usually more economical to rent than renting bare equipment and hiring an operator.

Costs to move equipment to a job site (mobilization) or from a job site (demobilization) are not included in rental rates, nor in any Equipment costs on any Unit Price line items or crew listings. These costs can be found elsewhere. If a piece of equipment is already at a job site, it is not appropriate to utilize mob/demob costs in an estimate again.

Rental rates vary throughout the country with larger cities generally having lower rates. Lease plans for new equipment are available for periods in excess of six months with a percentage of payments applying toward purchase.

Rental rates can also be treated as reimbursement costs for contractor-owned equipment. Owned equipment costs include depreciation, loan payments, interest, taxes, insurance, storage, and major repairs.

Monthly rental rates vary from 2% to 5% of the cost of the equipment depending on the anticipated life of the equipment and its wearing parts. Weekly rates are about 1/3 the monthly rates and daily rental rates are about 1/3 the weekly rates.

The hourly operating costs for each piece of equipment include costs to the user such as fuel, oil, lubrication, normal expendables for the equipment, and a percentage of the mechanic's wages chargeable to maintenance. The hourly operating costs listed do not include the operator's wages.

The daily cost for equipment used in the standard crews is figured by dividing the weekly rate by five, then adding eight times the hourly operating cost to give the total daily equipment cost, not including the operator. This figure is in the right hand column of the Equipment listings under Equipment Cost/Day.

Pile Driving rates shown for the pile hammer and extractor do not include leads, cranes, boilers or compressors. Vibratory pile driving requires an added field specialist during set-up and pile driving operation for the electric model. The hydraulic model requires a field specialist for set-up only. Up to 125 reuses of sheet piling are possible using vibratory drivers. For normal conditions, crane capacity for hammer type and size is as follows.

Crane Capacity	Hammer Type and Size		
	Air or Steam	Diesel	Vibratory
25 ton	to 8,750 ft.-lb.		70 H.P.
40 ton	15,000 ft.-lb.	to 32,000 ft.-lb.	170 H.P.
60 ton	25,000 ft.-lb.		300 H.P.
100 ton		112,000 ft.-lb.	

Cranes should be specified for the job by size, building and site characteristics, availability, performance characteristics, and duration of time required.

Backhoes & Shovels rent for about the same as equivalent size cranes but maintenance and operating expenses are higher. The crane operator's rate must be adjusted for high boom heights. Average adjustments: for 150' boom add 2% per hour; over 185', add 4% per hour; over 210', add 6% per hour; over 250', add 8% per hour and over 295', add 12% per hour.

Tower Cranes of the climbing or static type have jibs from 50' to 200' and capacities at maximum reach range from 4,000 to 14,000 pounds. Lifting capacities increase up to maximum load as the hook radius decreases.

Typical rental rates, based on purchase price, are about 2% to 3% per month.

Erection and dismantling run between 500 and 2000 labor hours. Climbing operation takes 10 labor hours per 20' climb. Crane dead time is about 5 hours per 40' climb. If crane is bolted to side of the building add cost of ties and extra mast sections. Climbing cranes have from 80' to 180' of mast while static cranes have 80' to 800' of mast.

Truck Cranes can be converted to tower cranes by using tower attachments. Mast heights over 400' have been used.

A single 100' high material **Hoist and Tower** can be erected and dismantled in about 400 labor hours; a double 100' high hoist and tower in about 600 labor hours. Erection times for additional heights are 3 and 4 labor hours per vertical foot respectively up to 150', and 4 to 5 labor hours per vertical foot over 150' high. A 40' high portable Buck hoist takes about 160 labor hours to erect and dismantle. Additional heights take 2 labor hours per vertical foot to 80' and 3 labor hours per vertical foot for the next 100'. Most material hoists do not meet local code requirements for carrying personnel.

A 150' high **Personnel Hoist** requires about 500 to 800 labor hours to erect and dismantle. Budget erection time at 5 labor hours per vertical foot for all trades. Local code requirements or labor scarcity requiring overtime can add up to 50% to any of the above erection costs.

Earthmoving Equipment: The selection of earthmoving equipment depends upon the type and quantity of material, moisture content, haul distance, haul road, time available, and equipment available. Short haul cut and fill operations may require dozers only, while another operation may require excavators, a fleet of trucks, and spreading and compaction equipment. Stockpiled material and granular material are easily excavated with front end loaders. Scrapers are most economically used with hauls between 300' and 1-1/2 miles if adequate haul roads can be maintained. Shovels are often used for blasted rock and any material where a vertical face of 8' or more can be excavated. Special conditions may dictate the use of draglines, clamshells, or backhoes. Spreading and compaction equipment must be matched to the soil characteristics, the compaction required and the rate the fill is being supplied.

R015436-50   Mobilization

Costs to move rented construction equipment to a job site from an equipment dealer's or contractor's yard (mobilization) or off the job site (demobilization) are not included in the rental or operating rates, nor in the equipment cost on a unit price line or in a crew listing. These costs can be found consolidated in the Mobilization section of the data and elsewhere in particular site work sections. If a piece of equipment is already on the job site, it is not appropriate to include mob/demob costs in a new estimate that requires use of that equipment. The following table identifies approximate sizes of rented construction equipment that would be hauled on a towed trailer. Because this listing is not all-encompassing, the user can infer as to what size trailer might be required for a piece of equipment not listed.

3-ton Trailer	20-ton Trailer	40-ton Trailer	50-ton Trailer
20 H.P. Excavator	110 H.P. Excavator	200 H.P. Excavator	270 H.P. Excavator
50 H.P. Skid Steer	165 H.P. Dozer	300 H.P. Dozer	Small Crawler Crane
35 H.P. Roller	150 H.P. Roller	400 H.P. Scraper	500 H.P. Scraper
40 H.P. Trencher	Backhoe	450 H.P. Art. Dump Truck	500 H.P. Art. Dump Truck

For customer support on your Mechanical Cost Data, call 877.739.3689.

Existing Conditions   R0241 Demolition

R024119-10   Demolition Defined

Whole Building Demolition - Demolition of the whole building with no concern for any particular building element, component, or material type being demolished. This type of demolition is accomplished with large pieces of construction equipment that break up the structure, load it into trucks and haul it to a disposal site, but disposal or dump fees are not included. Demolition of below-grade foundation elements, such as footings, foundation walls, grade beams, slabs on grade, etc., is not included. Certain mechanical equipment containing flammable liquids or ozone-depleting refrigerants, electric lighting elements, communication equipment components, and other building elements may contain hazardous waste, and must be removed, either selectively or carefully, as hazardous waste before the building can be demolished.

Foundation Demolition - Demolition of below-grade foundation footings, foundation walls, grade beams, and slabs on grade. This type of demolition is accomplished by hand or pneumatic hand tools, and does not include saw cutting, or handling, loading, hauling, or disposal of the debris.

Gutting - Removal of building interior finishes and electrical/mechanical systems down to the load-bearing and sub-floor elements of the rough building frame, with no concern for any particular building element, component, or material type being demolished. This type of demolition is accomplished by hand or pneumatic hand tools, and includes loading into trucks, but not hauling, disposal or dump fees, scaffolding, or shoring. Certain mechanical equipment containing flammable liquids or ozone-depleting refrigerants, electric lighting elements, communication equipment components, and other building elements may contain hazardous waste, and must be removed, either selectively or carefully, as hazardous waste, before the building is gutted.

Selective Demolition - Demolition of a selected building element, component, or finish, with some concern for surrounding or adjacent elements, components, or finishes (see the first Subdivision (s) at the beginning of appropriate Divisions). This type of demolition is accomplished by hand or pneumatic hand tools, and does not include handling, loading,

storing, hauling, or disposal of the debris, scaffolding, or shoring. "Gutting" methods may be used in order to save time, but damage that is caused to surrounding or adjacent elements, components, or finishes may have to be repaired at a later time.

Careful Removal - Removal of a piece of service equipment, building element or component, or material type, with great concern for both the removed item and surrounding or adjacent elements, components or finishes. The purpose of careful removal may be to protect the removed item for later re-use, preserve a higher salvage value of the removed item, or replace an item while taking care to protect surrounding or adjacent elements, components, connections, or finishes from cosmetic and/or structural damage. An approximation of the time required to perform this type of removal is 1/3 to 1/2 the time it would take to install a new item of like kind (see Reference Number R220105-10). This type of removal is accomplished by hand or pneumatic hand tools, and does not include loading, hauling, or storing the removed item, scaffolding, shoring, or lifting equipment.

Cutout Demolition - Demolition of a small quantity of floor, wall, roof, or other assembly, with concern for the appearance and structural integrity of the surrounding materials. This type of demolition is accomplished by hand or pneumatic hand tools, and does not include saw cutting, handling, loading, hauling, or disposal of debris, scaffolding, or shoring.

Rubbish Handling - Work activities that involve handling, loading or hauling of debris. Generally, the cost of rubbish handling must be added to the cost of all types of demolition, with the exception of whole building demolition.

Minor Site Demolition - Demolition of site elements outside the footprint of a building. This type of demolition is accomplished by hand or pneumatic hand tools, or with larger pieces of construction equipment, and may include loading a removed item onto a truck (check the Crew for equipment used). It does not include saw cutting, hauling or disposal of debris, and, sometimes, handling or loading.

R024119-20   Dumpsters

Dumpster rental costs on construction sites are presented in two ways.

The cost per week rental includes the delivery of the dumpster; its pulling or emptying once per week, and its final removal. The assumption is made that the dumpster contractor could choose to empty a dumpster by simply bringing in an empty unit and removing the full one. These costs also include the disposal of the materials in the dumpster.

The Alternate Pricing can be used when actual planned conditions are not approximated by the weekly numbers. For example, these lines can be used when a dumpster is needed for 4 weeks and will need to be emptied 2 or 3 times per week. Conversely the Alternate Pricing lines can be used when a dumpster will be rented for several weeks or months but needs to be emptied only a few times over this period.

Existing Conditions   R0265 Underground Storage Tank Removal

R026510-20   Underground Storage Tank Removal

Underground Storage Tank Removal can be divided into two categories: Non-Leaking and Leaking. Prior to removing an underground storage tank, tests should be made, with the proper authorities present, to determine whether a tank has been leaking or the surrounding soil has been contaminated.

To safely remove Liquid Underground Storage Tanks:
1. Excavate to the top of the tank.
2. Disconnect all piping.
3. Open all tank vents and access ports.

4. Remove all liquids and/or sludge.
5. Purge the tank with an inert gas.
6. Provide access to the inside of the tank and clean out the interior using proper personal protective equipment (PPE).
7. Excavate soil surrounding the tank using proper PPE for on-site personnel.
8. Pull and properly dispose of the tank.
9. Clean up the site of all contaminated material.
10. Install new tanks or close the excavation.

719

Existing Conditions — R0282 Asbestos Remediation

R028213-20 Asbestos Removal Process

Asbestos removal is accomplished by a specialty contractor who understands the federal and state regulations regarding the handling and disposal of the material. The process of asbestos removal is divided into many individual steps. An accurate estimate can be calculated only after all the steps have been priced.

The steps are generally as follows:

1. Obtain an asbestos abatement plan from an industrial hygienist.
2. Monitor the air quality in and around the removal area and along the path of travel between the removal area and transport area. This establishes the background contamination.
3. Construct a two part decontamination chamber at entrance to removal area.
4. Install a HEPA filter to create a negative pressure in the removal area.
5. Install wall, floor and ceiling protection as required by the plan, usually 2 layers of fireproof 6 mil polyethylene.
6. Industrial hygienist visually inspects work area to verify compliance with plan.
7. Provide temporary supports for conduit and piping affected by the removal process.
8. Proceed with asbestos removal and bagging process. Monitor air quality as described in Step #2. Discontinue operations when contaminate levels exceed applicable standards.
9. Document the legal disposal of materials in accordance with EPA standards.
10. Thoroughly clean removal area including all ledges, crevices and surfaces.
11. Post abatement inspection by industrial hygienist to verify plan compliance.
12. Provide a certificate from a licensed industrial hygienist attesting that contaminate levels are within acceptable standards before returning area to regular use.

Thermal & Moist. Protec. — R0784 Firestopping

R078413-30 Firestopping

Firestopping is the sealing of structural, mechanical, electrical and other penetrations through fire-rated assemblies. The basic components of firestop systems are safing insulation and firestop sealant on both sides of wall penetrations and the top side of floor penetrations.

Pipe penetrations are assumed to be through concrete, grout, or joint compound and can be sleeved or unsleeved. Costs for the penetrations and sleeves are not included. An annular space of 1″ is assumed. Escutcheons are not included.

A metallic pipe is assumed to be copper, aluminum, cast iron or similar metallic material. An insulated metallic pipe is assumed to be covered with a thermal insulating jacket of varying thickness and materials.

A non-metallic pipe is assumed to be PVC, CPVC, FR Polypropylene or similar plastic piping material. Intumescent firestop sealants or wrap strips are included. Collars on both sides of wall penetrations and a sheet metal plate on the underside of floor penetrations are included.

Ductwork is assumed to be sheet metal, stainless steel or similar metallic material. Duct penetrations are assumed to be through concrete, grout or joint compound. Costs for penetrations and sleeves are not included. An annular space of 1/2″ is assumed.

Multi-trade openings include costs for sheet metal forms, firestop mortar, wrap strips, collars and sealants as necessary.

Structural penetrations joints are assumed to be 1/2″ or less. CMU walls are assumed to be within 1-1/2″ of the metal deck. Drywall walls are assumed to be tight to the underside of metal decking.

Metal panel, glass or curtain wall systems include a spandrel area of 5′ filled with mineral wool foil-faced insulation. Fasteners and stiffeners are included.

720

Reference Tables

For customer support on your Mechanical Cost Data, call 877.739.3689.

Finishes | R0991 Painting

R099100-10 Painting Estimating Techniques

Proper estimating methodology is needed to obtain an accurate painting estimate. There is no known reliable shortcut or square foot method. The following steps should be followed:

- List all surfaces to be painted, with an accurate quantity (area) of each. Items having similar surface condition, finish, application method and accessibility may be grouped together.

- List all the tasks required for each surface to be painted, including surface preparation, masking, and protection of adjacent surfaces. Surface preparation may include minor repairs, washing, sanding and puttying.

- Select the proper Means line for each task. Review and consider all adjustments to labor and materials for type of paint and location of work. Apply the height adjustment carefully. For instance, when applying the adjustment for work over 8' high to a wall that is 12' high, apply the adjustment only to the area between 8' and 12' high, and not to the entire wall.

When applying more than one percent (%) adjustment, apply each to the base cost of the data, rather than applying one percentage adjustment on top of the other.

When estimating the cost of painting walls and ceilings remember to add the brushwork for all cut-ins at inside corners and around windows and doors as a LF measure. One linear foot of cut-in with a brush equals one square foot of painting.

All items for spray painting include the labor for roll-back.

Deduct for openings greater than 100 SF or openings that extend from floor to ceiling and are greater than 5' wide. Do not deduct small openings.

The cost of brushes, rollers, ladders and spray equipment are considered part of a painting contractor's overhead, and should not be added to the estimate. The cost of rented equipment such as scaffolding and swing staging should be added to the estimate.

Plumbing | R2201 Operation & Maintenance of Plumbing

R220102-20 Labor Adjustment Factors

Labor Adjustment Factors are provided for Divisions 21, 22, and 23 to assist the mechanical estimator account for the various complexities and special conditions of any particular project. While a single percentage has been entered on each line of Division 22 01 02.20, it should be understood that these are just suggested midpoints of ranges of values commonly used by mechanical estimators. They may be increased or decreased depending on the severity of the special conditions.

The group for "existing occupied buildings" has been the subject of requests for explanation. Actually there are two stages to this group: buildings that are existing and "finished" but unoccupied, and those that also are occupied. Buildings that are "finished" may result in higher labor costs due to the

workers having to be more careful not to damage finished walls, ceilings, floors, etc. and may necessitate special protective coverings and barriers. Also corridor bends and doorways may not accommodate long pieces of pipe or larger pieces of equipment. Work above an already hung ceiling can be very time consuming. The addition of occupants may force the work to be done on premium time (nights and/or weekends), eliminate the possible use of some preferred tools such as pneumatic drivers, powder charged drivers, etc. The estimator should evaluate the access to the work area and just how the work is going to be accomplished to arrive at an increase in labor costs over "normal" new construction productivity.

R220105-10 Demolition (Selective vs. Removal for Replacement)

Demolition can be divided into two basic categories.

One type of demolition involves the removal of material with no concern for its replacement. The labor-hours to estimate this work are found under "Selective Demolition" in the Fire Protection, Plumbing and HVAC Divisions. It is selective in that individual items or all the material installed as a system or trade grouping such as plumbing or heating systems are removed. This may be accomplished by the easiest way possible, such as sawing, torch cutting, or sledge hammering as well as simple unbolting.

The second type of demolition is the removal of some items for repair or replacement. This removal may involve careful draining, opening of unions, disconnecting and tagging electrical connections, capping pipes/ducts to prevent entry of debris or leakage of the material contained as well as transporting the item away from its in-place location to a truck/dumpster. An approximation of the time required to accomplish this type of demolition is to use half of the time indicated as necessary to install a new unit. For example: installation of a new pump might be listed as requiring 6 labor-hours so if we had to estimate the removal of the old pump we would allow an additional 3 hours for a total of 9 hours. That is, the complete replacement of a defective pump with a new pump would be estimated to take 9 labor-hours.

R220523-80 Valve Materials

VALVE MATERIALS

Bronze:
Bronze is one of the oldest materials used to make valves. It is most commonly used in hot and cold water systems and other non-corrosive services. It is often used as a seating surface in larger iron body valves to ensure tight closure.

Carbon Steel:
Carbon steel is a high strength material. Therefore, valves made from this metal are used in higher pressure services, such as steam lines up to 600 psi at 850° F. Many steel valves are available with butt-weld ends for economy and are generally used in high pressure steam service as well as other higher pressure non-corrosive services.

Forged Steel:
Valves from tough carbon steel are used in service up to 2000 psi and temperatures up to 1000° F in Gate, Globe and Check valves.

Iron:
Valves are normally used in medium to large pipe lines to control non-corrosive fluid and gases, where pressures do not exceed 250 psi at 450° F or 500 psi cold water, oil or gas.

Stainless Steel:
Developed steel alloys can be used in over 90% corrosive services.

Plastic PVC:
This is used in a great variety of valves generally in high corrosive service with lower temperatures and pressures.

VALVE SERVICE PRESSURES

Pressure ratings on valves provide an indication of the safe operating pressure for a valve at some elevated temperature. This temperature is dependent upon the materials used and the fabrication of the valve. When specific data is not available, a good "rule-of-thumb" to follow is the temperature of saturated steam on the primary rating indicated on the valve body. Example: The valve has the number 150S printed on the side indicating 150 psi and hence, a maximum operating temperature of 367° F (temperature of saturated steam and 150 psi).

DEFINITIONS
1. "WOG" – Water, oil, gas (cold working pressures).
2. "SWP" – Steam working pressure.
3. 100% area (full port) – means the area through the valve is equal to or greater than the area of standard pipe.
4. "Standard Opening" – means that the area through the valve is less than the area of standard pipe and therefore these valves should be used only where restriction of flow is unimportant.
5. "Round Port" – means the valve has a full round opening through the plug and body, of the same size and area as standard pipe.
6. "Rectangular Port" – valves have rectangular shaped ports through the plug body. The area of the port is either equal to 100% of the area of standard pipe, or restricted (standard opening). In either case it is clearly marked.
7. "ANSI" – American National Standards Institute.

R220523-90 Valve Selection Considerations

INTRODUCTION: In any piping application, valve performance is critical. Valves should be selected to give the best performance at the lowest cost.

The following is a list of performance characteristics generally expected of valves.
1. Stopping flow or starting it.
2. Throttling flow (Modulation).
3. Flow direction changing.
4. Checking backflow (Permitting flow in only one direction).
5. Relieving or regulating pressure.

In order to properly select the right valve, some facts must be determined.

A. What liquid or gas will flow through the valve?
B. Does the fluid contain suspended particles?
C. Does the fluid remain in liquid form at all times?
D. Which metals does fluid corrode?
E. What are the pressure and temperature limits? (As temperature and pressure rise, so will the price of the valve.)
F. Is there constant line pressure?
G. Is the valve merely an on-off valve?
H. Will checking of backflow be required?
I. Will the valve operate frequently or infrequently?

Valves are classified by design type into such classifications as Gate, Globe, Angle, Check, Ball, Butterfly and Plug. They are also classified by end connection, stem, pressure restrictions and material such as bronze, cast iron, etc. Each valve has a specific use. A quality valve used correctly will provide a lifetime of trouble-free service, but a high quality valve installed in the wrong service may require frequent attention.

STEM TYPES
(OS & Y)—Rising Stem-Outside Screw and Yoke

Offers a visual indication of whether the valve is open or closed. Recommended where high temperatures, corrosives, and solids in the line might cause damage to inside-valve stem threads. The stem threads are engaged by the yoke bushing so the stem rises through the hand wheel as it is turned.

(R.S.)—Rising Stem-Inside Screw

Adequate clearance for operation must be provided because both the hand wheel and the stem rise.
The valve wedge position is indicated by the position of the stem and hand wheel.

(N.R.S.)—Non-Rising Stem-Inside Screw

A minimum clearance is required for operating this type of valve. Excessive wear or damage to stem threads inside the valve may be caused by heat, corrosion, and solids. Because the hand wheel and stem do not rise, wedge position cannot be visually determined.

VALVE TYPES
Gate Valves

Provide full flow, minute pressure drop, minimum turbulence and minimum fluid trapped in the line.
They are normally used where operation is infrequent.

Globe Valves

Globe valves are designed for throttling and/or frequent operation with positive shut-off. Particular attention must be paid to the several types of seating materials available to avoid unnecessary wear. The seats must be compatible with the fluid in service and may be composition or metal. The configuration of the Globe valve opening causes turbulence which results in increased resistance. Most bronze Globe valves are rising stem-inside screw, but they are also available on O.S. & Y.

Angle Valves

The fundamental difference between the Angle valve and the Globe valve is the fluid flow through the Angle valve. It makes a 90° turn and offers less resistance to flow than the Globe valve while replacing an elbow. An Angle valve thus reduces the number of joints and installation time.

Check Valves

Check valves are designed to prevent backflow by automatically seating when the direction of fluid is reversed.

Swing Check valves are generally installed with Gate-valves, as they provide comparable full flow. Usually recommended for lines where flow velocities are low and should not be used on lines with pulsating flow. Recommended for horizontal installation, or in vertical lines only where flow is upward.

Lift Check Valves

These are commonly used with Globe and Angle valves since they have similar diaphragm seating arrangements and are recommended for preventing backflow of steam, air, gas and water, and on vapor lines with high flow velocities. For horizontal lines, horizontal lift checks should be used and vertical lift checks for vertical lines.

Ball Valves

Ball valves are light and easily installed, yet because of modern elastomeric seats, provide tight closure. Flow is controlled by rotating up to 90° a drilled ball which fits tightly against resilient seals. This ball seats with flow in either direction, and valve handle indicates the degree of opening. Recommended for frequent operation readily adaptable to automation, ideal for installation where space is limited.

Butterfly Valves

Butterfly valves provide bubble-tight closure with excellent throttling characteristics. They can be used for full-open, closed and for throttling applications.

The Butterfly valve consists of a disc within the valve body which is controlled by a shaft. In its closed position, the valve disc seals against a resilient seat. The disc position throughout the full 90° rotation is visually indicated by the position of the operator.

A Butterfly valve is only a fraction of the weight of a Gate valve and requires no gaskets between flanges in most cases. Recommended for frequent operation and adaptable to automation where space is limited.

Wafer and Lug type bodies when installed between two pipe flanges, can be easily removed from the line. The pressure of the bolted flanges holds the valve in place.

Locating lugs makes installation easier.

Plug Valves

Lubricated plug valves, because of the wide range of service to which they are adapted, may be classified as all purpose valves. They can be safely used at all pressure and vacuums, and at all temperatures up to the limits of available lubricants. They are the most satisfactory valves for the handling of gritty suspensions and many other destructive, erosive, corrosive and chemical solutions.

R221113-50 Pipe Material Considerations

1. Malleable fittings should be used for gas service.
2. Malleable fittings are used where there are stresses/strains due to expansion and vibration.
3. Cast fittings may be broken as an aid to disassembling heating lines frozen by long use, temperature and minerals.
4. A cast iron pipe is extensively used for underground and submerged service.
5. Type M (light wall) copper tubing is available in hard temper only and is used for nonpressure and less severe applications than K and L.

6. Type L (medium wall) copper tubing, available hard or soft for interior service.
7. Type K (heavy wall) copper tubing, available in hard or soft temper for use where conditions are severe. For underground and interior service.
8. Hard drawn tubing requires fewer hangers or supports but should not be bent. Silver brazed fittings are recommended, but soft solder is normally used.
9. Type DMV (very light wall) copper tubing designed for drainage, waste and vent plus other non-critical pressure services.

Domestic/Imported Pipe and Fittings Costs

The prices shown in this publication for steel/cast iron pipe and steel, cast iron, and malleable iron fittings are based on domestic production sold at the normal trade discounts. The above listed items of foreign manufacture may be available at prices 1/3 to 1/2 of those shown. Some imported items after minor machining or finishing operations are being sold as domestic to further complicate the system.

Caution: Most pipe prices in this data set also include a coupling and pipe hangers which for the larger sizes can add significantly to the per foot cost and should be taken into account when comparing "book cost" with the quoted supplier's cost.

725

For customer support on your Mechanical Cost Data, call 877.739.3689.

R221113-70   Piping to 10' High

When taking off pipe, it is important to identify the different material types and joining procedures, as well as distances between supports and components required for proper support.

During the takeoff, measure through all fittings. Do not subtract the lengths of the fittings, valves, or strainers, etc. This added length plus the final rounding of the totals will compensate for nipples and waste.

When rounding off totals always increase the actual amount to correspond with manufacturer's shipping lengths.

   A. Both red brass and yellow brass pipe are normally furnished in 12' lengths, plain end. The Unit Price section includes in the linear foot costs two field threads and one coupling per 10' length. A carbon steel clevis type hanger assembly every 10' is also prorated into the linear foot costs, including both material and labor.

   B. Cast iron soil pipe is furnished in either 5' or 10' lengths. For pricing purposes, the Unit Price section features 10' lengths with a joint and a carbon steel clevis hanger assembly every 5' prorated into the per foot costs of both material and labor.

       Three methods of joining are considered: lead and oakum poured joints or push-on gasket type joints for the bell and spigot pipe and a joint clamp for the no-hub soil pipe. The labor and material costs for each of these individual joining procedures are also prorated into the linear costs per foot.

   C. Copper tubing covers types K, L, M, and DWV which are furnished in 20' lengths. Means pricing data is based on a tubing cut each length and a coupling and two soft soldered joints every 10'. A carbon steel, clevis type hanger assembly every 10' is also prorated into the per foot costs. The prices for refrigeration tubing are for materials only. Labor for full lengths may be based on the type L labor but short cut measures in tight areas can increase the installation labor-hours from 20 to 40%.

   D. Corrosion-resistant piping does not lend itself to one particular standard of hanging or support assembly due to its diversity of application and placement. The several varieties of corrosion-resistant piping do not include any material or labor costs for hanger assemblies (See the Unit Price section for appropriate selection).

   E. Glass pipe is furnished in standard lengths either 5' or 10' long, beaded on one end. Special orders for diverse lengths beaded on both ends are also available. For pricing purposes, R.S. Means features 10' lengths with a coupling and a carbon steel band hanger assembly every 10' prorated into the per foot linear costs.

       Glass pipe is also available with conical ends and standard lengths ranging from 6" through 3' in 6" increments, then up to 10' in 12" increments. Special lengths can be customized for particular installation requirements.

       For pricing purposes, Means has based the labor and material pricing on 10' lengths. Included in these costs per linear foot are the prorated costs for a flanged assembly every 10' consisting of two flanges, a gasket, two insertable seals, and the required number of bolts and nuts. A carbon steel band hanger assembly based on 10' center lines has also been prorated into the costs per foot for labor and materials.

   F. Plastic pipe of several compositions and joining methods are considered. Fiberglass reinforced pipe (FRP) is priced, based on 10' lengths (20' lengths are also available), with coupling and epoxy joints every 10'. FRP is furnished in both "General Service" and "High Strength." A carbon steel clevis hanger assembly, 3 for every 10', is built into the prorated labor and material costs on a per foot basis.

       The PVC and CPVC pipe schedules 40, 80 and 120, plus SDR ratings are all based on 20' lengths with a coupling installed every 10', as well as a carbon steel clevis hanger assembly every 3'. The PVC and

ABS type DWV piping is based on 10' lengths with solvent weld couplings every 10', and with carbon steel clevis hanger assemblies, 3 for every 10'. The rest of the plastic piping in this section is based on flexible 100' coils and does not include any coupling or supports.

This section ends with PVC drain and sewer piping based on 10' lengths with bell and spigot ends and 0-ring type, push-on joints.

   G. Stainless steel piping includes both weld end and threaded piping, both in the type 304 and 316 specification and in the following schedules, 5, 10, 40, 80, and 160. Although this piping is usually furnished in 20' lengths, this cost grouping has a joint (either heli-arc butt-welded or threads and coupling) every 10'. A carbon steel clevis type hanger assembly is also included at 10' intervals and prorated into the linear foot costs.

   H. Carbon steel pipe includes both black and galvanized. This section encompasses schedules 40 (standard) and 80 (extra heavy).

       Several common methods of joining steel pipe — such as thread and coupled, butt welded, and flanged (150 lb. weld neck flanges) are also included.

       For estimating purposes, it is assumed that the piping is purchased in 20' lengths and that a compatible joint is made up every 10'. These joints are prorated into the labor and material costs per linear foot. The following hanger and support assemblies every 10' are also included: carbon steel clevis for the T & C pipe, and single rod roll type for both the welded and flanged piping. All of these hangers are oversized to accommodate pipe insulation 3/4" thick through 5" pipe size and 1-1/2" thick from 6" through 12" pipe size.

   I. Grooved joint steel pipe is priced both black and galvanized, in schedules 10, 40, and 80, furnished in 20' lengths. This section describes two joining methods: cut groove and roll groove. The schedule 10 piping is roll-grooved, while the heavier schedules are cut-grooved. The labor and material costs are prorated into per linear foot prices, including a coupled joint every 10', as well as a carbon steel clevis hanger assembly.

Notes:

The pipe hanger assemblies mentioned in the preceding paragraphs include the described hanger; appropriately sized steel, box-type insert and nut; plus 18" of threaded hanger rod.

C clamps are used when the pipe is to be supported from steel shapes rather than anchored in the slab. C clamps are slightly less costly than inserts. However, to save time in estimating, it is advisable to use the given line number cost, rather than substituting a C clamp for the insert.

Add to piping labor for elevated installation:

10' to 14.5' high	10%	30' to 34.5' high	40%
15' to 19.5' high	20%	35' to 39.5' high	50%
20' to 24.5' high	25%	Over 40' and higher	55%
25' to 29.5' high	35%		

When using the percentage adds for elevated piping installations as shown above, bear in mind that the given heights are for the pipe supports, even though the insert, anchor, or clamp may be several feet higher than the pipe itself.

An allowance has been included in the piping installation time for testing and minor tightening of leaking joints, fittings, stuffing boxes, packing glands, etc. For extraordinary test requirements such as x-rays, prolonged pressure or demonstration tests, a percentage of the piping labor, based on the estimator's experience, must be added to the labor total. A testing service specializing in weld x-rays should be consulted for pricing if it is an estimate requirement. Equipment installation time includes start-up with associated adjustments.

Heating, Ventilating & A.C. R2305 Common Work Results for HVAC

R230500-10 Subcontractors

On the unit cost section of the R.S. Means Cost Data, the last column is entitled "Total Incl. O&P". This is normally the cost of the installing contractor. In the HVAC Division, this is the cost of the mechanical contractor. If the particular work being estimated is to be performed by a sub to the mechanical contractor, the mechanical's profit and handling charge (usually 10%) is added to the total of the last column.

Heating, Ventilating & A.C. R2331 HVAC Ducts & Casings

R233100-30 Duct Fabrication/Installation

The labor cost for fabricated sheet metal duct includes both the cost of fabrication and installation of the duct. The split is approximately 40% for fabrication, 60% for installation. It is for this reason that the percentage add for elevated installation of fabricated duct is less than the percentage add for prefabricated/preformed duct.

Example: assume a piece of duct cost $100 installed (labor only)

$$\text{Sheet Metal Fabrication} = 40\% = \$40$$
$$\text{Installation} = 60\% = \$60$$

The add for elevated installation is:

$$\frac{\text{Based on total labor}}{\text{(fabrication \& installation)}} = \$100 \text{ x } 6\% = \$6.00$$

$$\frac{\text{Based on installation cost only}}{\text{(Material purchased prefabricated)}} = \$60 \text{ x } 10\% = \$6.00$$

The $6.00 markup (10' to 15' high) is the same.

727

Heating, Ventilating & A.C. R2331 HVAC Ducts & Casings

R233100-40 Steel Sheet Metal Calculator (Weight in Lb./Ft. of Length)

Gauge	26	24	22	20	18	16
Wt.-Lb./S.F.	.906	1.156	1.406	1.656	2.156	2.656
SMACNA Max. Dimension – Long Side	*	30″	54″	84″	85″ Up	*
Sum-2 sides						
2	.3	.40	.50	.60	.80	.90
3	.5	.65	.80	.90	1.1	1.4
4	.7	.85	1.0	1.2	1.5	1.8
5	.8	1.1	1.3	1.5	1.9	2.3
6	1.0	1.3	1.5	1.7	2.3	2.7
7	1.2	1.5	1.8	2.0	2.7	3.2
8	1.3	1.7	2.0	2.3	3.0	3.6
9	1.5	1.9	2.3	2.6	3.4	4.1
10	1.7	2.2	2.5	2.9	3.8	4.5
11	1.8	2.4	2.8	3.2	4.2	5.0
12	2.0	2.6	3.0	3.5	4.6	5.4
13	2.2	2.8	3.3	3.8	4.9	5.9
14	2.3	3.0	3.5	4.1	5.3	6.3
15	2.5	3.2	3.8	4.4	5.7	6.8
16	2.7	3.4	4.0	4.6	6.1	7.2
17	2.8	3.7	4.3	4.9	6.5	7.7
18	3.0	3.9	4.5	5.2	6.8	8.1
19	3.2	4.1	4.8	5.5	7.2	8.6
20	3.3	4.3	5.0	5.8	7.6	9.0
21	3.5	4.5	5.3	6.1	8.0	9.5
22	3.7	4.7	5.5	6.4	8.4	9.9
23	3.8	5.0	5.8	6.7	8.7	10.4
24	4.0	5.2	6.0	7.0	9.1	10.8
25	4.2	5.4	6.3	7.3	9.5	11.3
26	4.3	5.6	6.5	7.5	9.9	11.7
27	4.5	5.8	6.8	7.8	10.3	12.2
28	4.7	6.0	7.0	8.1	10.6	12.6
29	4.8	6.2	7.3	8.4	11.0	13.1
30	5.0	6.5	7.5	8.7	11.4	13.5
31	5.2	6.7	7.8	9.0	11.8	14.0
32	5.3	6.9	8.0	9.3	12.2	14.4
33	5.5	7.1	8.3	9.6	12.5	14.9
34	5.7	7.3	8.5	9.9	12.9	15.3
35	5.8	7.5	8.8	10.2	13.3	15.8
36	6.0	7.8	9.0	10.4	13.7	16.2
37	6.2	8.0	9.3	10.7	14.1	16.7
38	6.3	8.2	9.5	11.0	14.4	17.1
39	6.5	8.4	9.8	11.3	14.8	17.6
40	6.7	8.6	10.0	11.6	15.2	18.0
41	6.8	8.8	10.3	11.9	15.6	18.5
42	7.0	9.0	10.5	12.2	16.0	18.9
43	7.2	9.2	10.8	12.5	16.3	19.4
44	7.3	9.5	11.0	12.8	16.7	19.8
45	7.5	9.7	11.3	13.1	17.1	20.3
46	7.7	9.9	11.5	13.3	17.5	20.7
47	7.8	10.1	11.8	13.6	17.9	21.2
48	8.0	10.3	12.0	13.9	18.2	21.6
49	8.2	10.5	12.3	14.2	18.6	22.1
50	8.3	10.7	12.5	14.5	19.0	22.5
51	8.5	11.0	12.8	14.8	19.4	23.0
52	8.7	11.2	13.0	15.1	19.8	23.4
53	8.8	11.4	13.3	15.4	20.1	23.9
54	9.0	11.6	13.5	15.7	20.5	24.3
55	9.2	11.8	13.8	16.0	20.9	24.8

Gauge	26	24	22	20	18	16
Wt.-Lb./S.F.	.906	1.156	1.406	1.656	2.156	2.656
SMACNA Max. Dimension – Long Side	*	30″	54″	84″	85″ Up	*
Sum-2 Sides						
56	9.3	12.0	14.0	16.2	21.3	25.2
57	9.5	12.3	14.3	16.5	21.7	25.7
58	9.7	12.5	14.5	16.8	22.0	26.1
59	9.8	12.7	14.8	17.1	22.4	26.6
60	10.0	12.9	15.0	17.4	22.8	27.0
61	10.2	13.1	15.3	17.7	23.2	27.5
62	10.3	13.3	15.5	18.0	23.6	27.9
63	10.5	13.5	15.8	18.3	24.0	28.4
64	10.7	13.7	16.0	18.6	24.3	28.8
65	10.8	13.9	16.3	18.9	24.7	29.3
66	11.0	14.1	16.5	19.1	25.1	29.7
67	11.2	14.3	16.8	19.4	25.5	30.2
68	11.3	14.6	17.0	19.7	25.8	30.6
69	11.5	14.8	17.3	20.0	26.2	31.1
70	11.7	15.0	17.5	20.3	26.6	31.5
71	11.8	15.2	17.8	20.6	27.0	32.0
72	12.0	15.4	18.0	20.9	27.4	32.4
73	12.2	15.6	18.3	21.2	27.7	32.9
74	12.3	15.8	18.5	21.5	28.1	33.3
75	12.5	16.1	18.8	21.8	28.5	33.8
76	12.7	16.3	19.0	22.0	28.9	34.2
77	12.8	16.5	19.3	22.3	29.3	34.7
78	13.0	16.7	19.5	22.6	29.6	35.1
79	13.2	16.9	19.8	22.9	30.0	35.6
80	13.3	17.1	20.0	23.2	30.4	36.0
81	13.5	17.3	20.3	23.5	30.8	36.5
82	13.7	17.5	20.5	23.8	31.2	36.9
83	13.8	17.8	20.8	24.1	31.5	37.4
84	14.0	18.0	21.0	24.4	31.9	37.8
85	14.2	18.2	21.3	24.7	32.3	38.3
86	14.3	18.4	21.5	24.9	32.7	38.7
87	14.5	18.6	21.8	25.2	33.1	39.2
88	14.7	18.8	22.0	25.5	33.4	39.6
89	14.8	19.0	22.3	25.8	33.8	40.1
90	15.0	19.3	22.5	26.1	34.2	40.5
91	15.2	19.5	22.8	26.4	34.6	41.0
92	15.3	19.7	23.0	26.7	35.0	41.4
93	15.5	19.9	23.3	27.0	35.3	41.9
94	15.7	20.1	23.5	27.3	35.7	42.3
95	15.8	20.3	23.8	27.6	36.1	42.8
96	16.0	20.5	24.0	27.8	36.5	43.2
97	16.2	20.8	24.3	28.1	36.9	43.7
98	16.3	21.0	24.5	28.4	37.2	44.1
99	16.5	21.2	24.8	28.7	37.6	44.6
100	16.7	21.4	25.0	29.0	38.0	45.0
101	16.8	21.6	25.3	29.3	38.4	45.5
102	17.0	21.8	25.5	29.6	38.8	45.9
103	17.2	22.0	25.8	29.9	39.1	46.4
104	17.3	22.3	26.0	30.2	39.5	46.8
105	17.5	22.5	26.3	30.5	39.9	47.3
106	17.7	22.7	26.5	30.7	40.3	47.7
107	17.8	22.9	26.8	31.0	40.7	48.2
108	18.0	23.1	27.0	31.3	41.0	48.6
109	18.2	23.3	27.3	31.6	41.4	49.1
110	18.3	23.5	27.5	31.9	41.8	49.5

Example: If duct is 34″ x 20″ x 15′ long, 34″ is greater than 30″ maximum, for 24 ga. so must be 22 ga. 34″ + 20″ = 54″ going across from 54″ find 13.5 lb. per foot. 13.5 x 15′ = 202.5 lbs. For S.F. of surface area 202.5 ÷ 1.406 = 144 S.F.

Note: Figures include an allowance for scrap.
*Do Not use unless engineer specified. 26GA is very light and sometimes used for toilet exhaust. 16 GA is heavy plate and mostly specified for plenums and hoods.

R233100-50 Ductwork Packages (per Ton of Cooling)

System	Sheet Metal	Insulation	Diffusers	Return Register
Roof Top Unit Single Zone	120 Lbs.	52 S.F.	1	1
Roof Top Unit Multizone	240 Lbs.	104 S.F.	2	1
Self-contained Air or Water Cooled	108 Lbs.	—	2	—
Split System Air Cooled	102 Lbs.	—	2	—

Systems reflect most common usage.
Refer to system graphics for duct layout.

Heating, Ventilating & A.C. R2337 Air Outlets & Inlets

R233700-60   Diffuser Evaluation

CFM = V × An × K where V = Outlet velocity in feet per minute. An = Neck area in square feet and K = Diffuser delivery factor. An undersized diffuser for a desired CFM will produce a high velocity and noise level. When air moves past people at a velocity in excess of 25 FPM, an annoying draft is felt. An oversized diffuser will result in low velocity with poor mixing. Consideration must be given to avoid vertical stratification or horizontal areas of stagnation.

Reference Tables

R235000-10 Heating Systems

Heating Systems

The basic function of a heating system is to bring an enclosed volume up to a desired temperature and then maintain that temperature within a reasonable range. To accomplish this, the selected system must have sufficient capacity to offset transmission losses resulting from the temperature difference on the interior and exterior of the enclosing walls in addition to losses due to cold air infiltration through cracks, crevices and around doors and windows. The amount of heat to be furnished is dependent upon the building size, construction, temperature difference, air leakage, use, shape, orientation and exposure. Air circulation is also an important consideration. Circulation will prevent stratification which could result in heat losses through uneven temperatures at various levels. For example, the most

efficient use of unit heaters can usually be achieved by circulating the space volume through the total number of units once every 20 minutes or 3 times an hour. This general rule must, of course, be adapted for special cases such as large buildings with low ratios of heat transmitting surface to cubical volume. The type of occupancy of a building will have considerable bearing on the number of heat transmitting units and the location selected. It is axiomatic, however, that the basis of any successful heating system is to provide the maximum amount of heat at the points of maximum heat loss such as exposed walls, windows, and doors. Large roof areas, wind direction, and wide doorways create problems of excessive heat loss and require special consideration and treatment.

Heat Transmission

Heat transfer is an important parameter to be considered during selection of the exterior wall style, material and window area. A high rate of transfer will permit greater heat loss during the wintertime with the resultant increase in heating energy costs and a greater rate of heat gain in the summer with proportionally greater cooling cost. Several terms are used to describe various aspects of heat transfer. However, for general estimating purposes this data set lists U values for systems of construction materials. U is the "overall heat transfer coefficient." It is defined as the heat flow per hour through one square foot when the temperature difference in the air on either side of the structure wall, roof, ceiling or floor is one degree Fahrenheit. The structural segment may be a single homogeneous material or a composite.

Total heat transfer is found using the following equation:

$Q = AU(T_2 - T_1)$ where

- Q = Heat flow, BTU per hour
- A = Area, square feet
- U = Overall heat transfer coefficient
- $(T_2 - T_1)$ = Difference in temperature of air on each side of the construction component. (Also abbreviated TD)

Note that heat can flow through all surfaces of any building and this flow is in addition to heat gain or loss due to ventilation, infiltration and generation (appliances, machinery, people).

R235000-20 Heating Approximations for Quick Estimating

Oil Piping & Boiler Room Piping

Small System . . . . 20 to 30% of Boiler

Complex System
with Pumps, Headers, Etc. . . . . 80 to 110% of Boiler

Breeching With Insulation:

Small . . . . 10 to 15% of Boiler

Large . . . . 15 to 25% of Boiler

Coils: . . . . 15 to 30% of Containing Unit

Balancing (Independent) . . . . 1/2% of H.V.A.C. Estimate

Quality/Complexity Adjustment: For all heating installations add these adjustments to the estimate to more closely allow for the equipment and conditions of the particular job under consideration.

Economy installation, add . . . . 0 to 5% of System

Good quality, medium complexity, add . . . . 5 to 15% of System

Above average quality and complexity, add . . . . 15 to 25% of System

730

R235000-30 The Basics of a Heating System

The function of a heating system is to achieve and maintain a desired temperature in a room or building by replacing the amount of heat being dissipated. There are four kinds of heating systems: hot-water, steam, warm-air and electric resistance. Each has certain essential and similar elements with the exception of electric resistance heating.

The basic elements of a heating system are:

A. A **combustion chamber** in which fuel is burned and heat transferred to a conveying medium.

B. The **"fluid"** used for conveying the heat (water, steam or air).

C. **Conductors** or pipes for transporting the fluid to specific desired locations.

D. A means of disseminating the heat, sometimes called **terminal units.**

A. The **combustion chamber** in a furnace heats air which is then distributed. This is called a warm-air system.

The combustion chamber in a boiler heats water which is either distributed as hot water or steam and this is termed a hydronic system.

The maximum allowable working pressures are limited by ASME "Code for Heating Boilers" to 15 PSI for steam and 160 PSI for hot-water heating boilers, with a maximum temperature limitation of 250° F. Hot water boilers are generally rated for a working pressure of 30 PSI. High pressure boilers are governed by the ASME "Code for Power Boilers" which is used almost universally for boilers operating over 15 PSIG. High pressure boilers used for a combination of heating/process loads are usually designed for 150 PSIG.

Boiler ratings are usually indicated as either Gross or Net Output. The Gross Load is equal to the Net Load plus a piping and pickup allowance. When this allowance cannot be determined, divide the gross output rating by 1.25 for a value equal to or greater than the net heat loss requirement of the building.

B. Of the three **fluids** used, steam carries the greatest amount of heat per unit volume. This is due to the fact that it gives up its latent heat of vaporization at a temperature considerably above room temperature. Another advantage is that the pressure to produce a positive circulation is readily available. Piping conducts the steam to terminal units and returns condensate to the boiler.

The **steam system** is well adapted to large buildings because of its positive circulation, its comparatively economical installation and its ability to deliver large quantities of heat. Nearly all large office buildings, stores, hotels, and industrial buildings are so heated, in addition to many residences.

Hot water, when used as the heat carrying fluid, gives up a portion of its sensible heat and then returns to the boiler or heating apparatus for reheating. As the heat conveyed by each pound of water is about one-fiftieth of the heat conveyed by a pound of steam, it is necessary to circulate about fifty times as much water as steam by weight (although only one-thirtieth as much by volume). The hot water system is usually, although not necessarily, designed to operate at temperatures below that of the ordinary steam system and so the amount of heat transfer surface must be correspondingly greater. A temperature of 190° F to 200° F is normally the maximum. Circulation in small buildings may depend on the difference in density between hot water and the cool water returning to the boiler; circulating pumps are normally used to maintain a desired rate of flow. Pumps permit a greater degree of flexibility and better control.

In **warm-air** furnace systems, cool air is taken from one or more points in the building, passed over the combustion chamber and flue gas passages and then distributed through a duct system. A disadvantage of this system is that the ducts take up much more building volume than steam or hot water pipes. Advantages of this system are the relative ease with which humidification can be accomplished by the evaporation of water as the air circulates through the heater, and the lack of need for expensive disseminating units as the warm air simply becomes part of the interior atmosphere of the building.

C. Conductors (pipes and ducts) have been lightly treated in the discussion of conveying fluids. For more detailed information such as sizing and distribution methods, the reader is referred to technical publications such as the American Society of Heating, Refrigerating and Air-Conditioning Engineers "Handbook of Fundamentals."

D. Terminal units come in an almost infinite variety of sizes and styles, but the basic principles of operation are very limited. As previously mentioned, warm-air systems require only a simple register or diffuser to mix heated air with that present in the room. Special application items such as radiant coils and infrared heaters are available to meet particular conditions but are not usually considered for general heating needs. Most heating is accomplished by having air flow over coils or pipes containing the heat transporting medium (steam, hot-water, electricity). These units, while varied, may be separated into two general types, (1) radiator/convectors and (2) unit heaters.

Radiator/convectors may be cast, fin-tube or pipe assemblies. They may be direct, indirect, exposed, concealed or mounted within a cabinet enclosure, upright or baseboard style. These units are often collectively referred to as "radiatiors" or "radiation" although none gives off heat either entirely by radiation or by convection but rather a combination of both. The air flows over the units as a gravity "current." It is necessary to have one or more heat-emitting units in each room. The most efficient placement is low along an outside wall or under a window to counteract the cold coming into the room and achieve an even distribution.

In contrast to radiator/convectors which operate most effectively against the walls of smaller rooms, **unit heaters** utilize a fan to move air over heating coils and are very effective in locations of relatively large volume. Unit heaters, while usually suspended overhead, may be floor mounted. They also may take in fresh outside air for ventilation. The heat distributed by unit heaters may be from a remote source and conveyed by a fluid or it may be from the combustion of fuel in each individual heater. In the latter case the only piping required would be for fuel, however, a vent for the products of combustion would be necessary.

The following list gives may of the advantages of unit heaters for applications other than office or residential:

a. Large capacity so smaller number of units are required,
b. Piping system simplified, **c.** Space saved where they are located overhead out of the way, **d.** Rapid heating directed where needed with effective wide distribution, **e.** Difference between floor and ceiling temperature reduced, **f.** Circulation of air obtained, and ventilation with introduction of fresh air possible, **g.** Heat output flexible and easily controlled.

731

R235616-60 Solar Heating (Space and Hot Water)

Collectors should face as close to due South as possible, but variations of up to 20 degrees on either side of true South are acceptable. Local climate and collector type may influence the choice between east or west deviations. Obviously they should be located so they are not shaded from the sun's rays. Incline collectors at a slope of latitude minus 5 degrees for domestic hot water and latitude plus 15 degrees for space heating.

Flat plate collectors consist of a number of components as follows: Insulation to reduce heat loss through the bottom and sides of the collector. The enclosure which contains all the components in this assembly is usually weatherproof and prevents dust, wind and water from coming in contact with the absorber plate. The cover plate usually consists of one or more layers of a variety of glass or plastic and reduces the reradiation by creating an air space which traps the heat between the cover and the absorber plates.

The absorber plate must have a good thermal bond with the fluid passages. The absorber plate is usually metallic and treated with a surface coating which improves absorptivity. Black or dark paints or selective coatings are used for this purpose, and the design of this passage and plate combination helps determine a solar system's effectiveness.

Heat transfer fluid passage tubes are attached above and below or integral with an absorber plate for the purpose of transferring thermal energy from the absorber plate to a heat transfer medium. The heat exchanger is a device for transferring thermal energy from one fluid to another.

Piping and storage tanks should be well insulated to minimize heat losses.

Size domestic water heating storage tanks to hold 20 gallons of water per user, minimum, plus 10 gallons per dishwasher or washing machine. For domestic water heating an optimum collector size is approximately 3/4 square foot of area per gallon of water storage. For space heating of residences and small commercial applications the collector is commonly sized between 30% and 50% of the internal floor area. For space heating of large commercial applications, collector areas less than 30% of the internal floor area can still provide significant heat reductions.

A supplementary heat source is recommended for Northern states for December through February.

The solar energy transmission per square foot of collector surface varies greatly with the material used. Initial cost, heat transmittance and useful life are obviously interrelated.

R236000-10 Air Conditioning

General: The purpose of air conditioning is to control the environment of a space so that comfort is provided for the occupants and/or conditions are suitable for the processes or equipment contained therein. The several items which should be evaluated to define system objectives are:

Temperature Control
Humidity Control
Cleanliness
Odor, smoke and fumes
Ventilation

Efforts to control the above parameters must also include consideration of the degree or tolerance of variation, the noise level introduced, the velocity of air motion and the energy requirements to accomplish the desired results.

The variation in **temperature** and **humidity** is a function of the sensor and the controller. The controller reacts to a signal from the sensor and produces the appropriate suitable response in either the terminal unit, the conductor of the transporting medium (air, steam, chilled water, etc.), or the source (boiler, evaporating coils, etc.).

The **noise level** is a by-product of the energy supplied to moving components of the system. Those items which usually contribute the most noise are pumps, blowers, fans, compressors and diffusers. The level of noise can be partially controlled through use of vibration pads, isolators, proper sizing, shields, baffles and sound absorbing liners.

Some **air motion** is necessary to prevent stagnation and stratification. The maximum acceptable velocity varies with the degree of heating or cooling

which is taking place. Most people feel air moving past them at velocities in excess of 25 FPM as an annoying draft. However, velocities up to 45 FPM may be acceptable in certain cases. Ventilation, expressed as air changes per hour and percentage of fresh air, is usually an item regulated by local codes.

Selection of the system to be used for a particular application is usually a trade-off. In some cases the building size, style, or room available for mechanical use limits the range of possibilities. Prime factors influencing the decision are first cost and total life (operating, maintenance and replacement costs). The accuracy with which each parameter is determined will be an important measure of the reliability of the decision and subsequent satisfactory operation of the installed system.

Heat delivery may be desired from an air conditioning system. Heating capability usually is added as follows: A gas fired burner or hot water/steam/electric coils may be added to the air handling unit directly and heat all air equally. For limited or localized heat requirements the water/steam/electric coils may be inserted into the duct branch supplying the cold areas. Gas fired duct furnaces are also available.

Note: When water or steam coils are used the cost of the piping and boiler must also be added. For a rough estimate use the cost per square foot of the appropriate sized hydronic system with unit heaters. This will provide a cost for the boiler and piping, and the unit heaters of the system would equate to the approximate cost of the heating coils.

R236000-20   Air Conditioning Requirements

BTUs per hour per S.F. of floor area and S.F. per ton of air conditioning.

Type of Building	BTU/Hr per S.F.	S.F. per Ton	Type of Building	BTU/Hr per S.F.	S.F. per Ton	Type of Building	BTU/Hr per S.F.	S.F. per Ton
Apartments, Individual	26	450	Dormitory, Rooms	40	300	Libraries	50	240
Corridors	22	550	Corridors	30	400	Low Rise Office, Exterior	38	320
Auditoriums & Theaters	40	300/18*	Dress Shops	43	280	Interior	33	360
Banks	50	240	Drug Stores	80	150	Medical Centers	28	425
Barber Shops	48	250	Factories	40	300	Motels	28	425
Bars & Taverns	133	90	High Rise Office—Ext. Rms.	46	263	Office (small suite)	43	280
Beauty Parlors	66	180	Interior Rooms	37	325	Post Office, Individual Office	42	285
Bowling Alleys	68	175	Hospitals, Core	43	280	Central Area	46	260
Churches	36	330/20*	Perimeter	46	260	Residences	20	600
Cocktail Lounges	68	175	Hotel, Guest Rooms	44	275	Restaurants	60	200
Computer Rooms	141	85	Corridors	30	400	Schools & Colleges	46	260
Dental Offices	52	230	Public Spaces	55	220	Shoe Stores	55	220
Dept. Stores, Basement	34	350	Industrial Plants, Offices	38	320	Shop'g. Ctrs., Supermarkets	34	350
Main Floor	40	300	General Offices	34	350	Retail Stores	48	250
Upper Floor	30	400	Plant Areas	40	300	Specialty	60	200

*Persons per ton
12,000 BTU = 1 ton of air conditioning

R236000-30 Psychrometric Table

Dewpoint or Saturation Temperature (F)

Relative humidity (%)	32	35	40	45	50	55	60	65	70	75	80	85	90	95	100
100	32	35	40	45	50	55	60	65	70	75	80	85	90	95	100
90	30	33	37	42	47	52	57	62	67	72	77	82	87	92	97
80	27	30	34	39	44	49	54	58	64	68	73	78	83	88	93
70	24	27	31	36	40	45	50	55	60	64	69	74	79	84	88
60	20	24	28	32	36	41	46	51	55	60	65	69	74	79	83
50	16	20	24	28	33	36	41	46	50	55	60	64	69	73	78
40	12	15	18	23	27	31	35	40	45	49	53	58	62	67	71
30	8	10	14	18	21	25	29	33	37	42	46	50	54	59	62
20	6	7	8	9	13	16	20	24	28	31	35	40	43	48	52
10	4	4	5	5	6	8	9	10	13	17	20	24	27	30	34
	32	35	40	45	50	55	60	65	70	75	80	85	90	95	100

Dry bulb temperature (F)

This table shows the relationship between RELATIVE HUMIDITY, DRY BULB TEMPERATURE AND DEWPOINT.

As an example, assume that the thermometer in a room reads 75° F, and we know that the relative humidity is 50%. The chart shows the dewpoint temperature to be 55° F. That is, any surface colder than 55° F will "sweat" or collect condensing moisture. This surface could be the outside of an uninsulated chilled water pipe in the summertime, or the inside surface of a wall or deck in the wintertime. After determining the extreme ambient parameters, the table at the left is useful in determining which surfaces need insulation or vapor barrier protection.

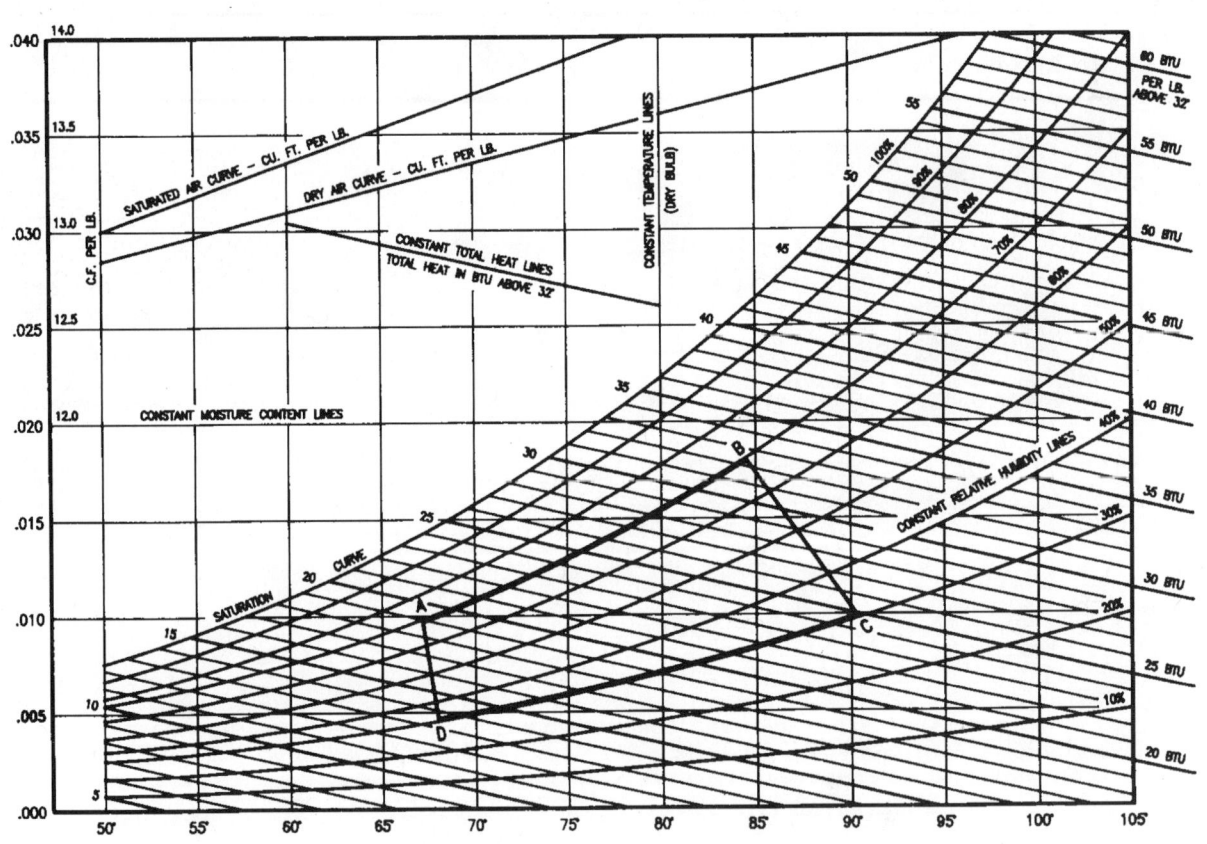

TEMPERATURE DEGREES FAHRENHEIT
TOTAL PRESSURE = 14.696 LB. PER SQ. IN. ABS.

Psychrometric chart showing different variables based on one pound of dry air. Space marked A B C D is temperature-humidity range which is most comfortable for majority of people.

R236000-90   Quality/Complexity Adjustment for Air Conditioning Systems

Economy installation, add . . . . 0 to 5%

Good quality, medium complexity, add . . . . 5 to 15%

Above average quality and complexity, add . . . . 15 to 25%

Add the above adjustments to the estimate to more closely allow for the equipment and conditions of the particular job under consideration.

735

Fig. R238313-11

R238313-10   Heat Trace Systems

Before you can determine the cost of a HEAT TRACE installation, the method of attachment must be established. There are (4) common methods:

1. Cable is simply attached to the pipe with polyester tape every 12'.
2. Cable is attached with a continuous cover of 2" wide aluminum tape.
3. Cable is attached with factory extruded heat transfer cement and covered with metallic raceway with clips every 10'.
4. Cable is attached between layers of pipe insulation using either clips or polyester tape.

In all of the above methods each component of the system must be priced individually.

Example: Components for method 3 must include:
A. Heat trace cable by voltage and watts per linear foot.
B. Heat transfer cement, 1 gallon per 60 linear feet of cover.
C. Metallic raceway by size and type.
D. Raceway clips by size of pipe.

When taking off linear foot lengths of cable add the following for each valve in the system. (E)

SCREWED OR WELDED VALVE:			FLANGED VALVE:			BUTTERFLY VALVES:		
1/2"	=	6"	1/2"	=	1' -0"	1/2"	=	0'
3/4"	=	9"	3/4"	=	1' -6"	3/4"	=	0'
1"	=	1' -0"	1"	=	2' -0"	1"	=	1' -0"
1-1/2"	=	1' -6"	1-1/2"	=	2' -6"	1-1/2"	=	1' -6"
2"	=	2'	2"	=	2' -6"	2"	=	2' -0"
2-1/2"	=	2' -6"	2-1/2"	=	3' -0"	2-1/2"	=	2' -6"
3"	=	2' -6"	3"	=	3' -6"	3"	=	2' -6"
4"	=	4' -0"	4"	=	4' -0"	4"	=	3' -0"
6"	=	7' -0"	6"	=	8' -0"	6"	=	3' -6"
8"	=	9' -6"	8"	=	11' -0"	8"	=	4' -0"
10"	=	12' -6"	10"	=	14' -0"	10"	=	4' -0"
12"	=	15' -0"	12"	=	16' -6"	12"	=	5' -0"
14"	=	18' -0"	14"	=	19' -6"	14"	=	5' -6"
16"	=	21' -6"	16"	=	23' -0"	16"	=	6' -0"
18"	=	25' -6"	18"	=	27' -0"	18"	=	6' -6"
20"	=	28' -6"	20"	=	30' -0"	20"	=	7' -0"
24"	=	34' -0"	24"	=	36' -0"	24"	=	8' -0"
30"	=	40' -0"	30"	=	42' -0"	30"	=	10' -0"

736

For customer support on your Mechanical Cost Data, call 877.739.3689.

R238313-10   Heat Trace Systems (cont.)

Add the following quantities of heat transfer cement to linear foot totals for each valve:

Nominal Valve Size	Gallons of Cement per Valve
1/2"	0.14
3/4"	0.21
1"	0.29
1-1/2"	0.36
2"	0.43
2-1/2"	0.70
3"	0.71
4"	1.00
6"	1.43
8"	1.48
10"	1.50
12"	1.60
14"	1.75
16"	2.00
18"	2.25
20"	2.50
24"	3.00
30"	3.75

The following must be added to the list of components to accurately price HEAT TRACE systems:

1. Expediter fitting and clamp fasteners (F)
2. Junction box and nipple connected to expediter fitting (G)
3. Field installed terminal blocks within junction box
4. Ground lugs
5. Piping from power source to expediter fitting
6. Controls
7. Thermostats
8. Branch wiring
9. Cable splices
10. End of cable terminations
11. Branch piping fittings and boxes

Deduct the following percentages from labor if cable lengths in the same area exceed:

150' to 250'	10%	351' to 500'	20%
251' to 350'	15%	Over 500'	25%

Add the following percentages to labor for elevated installations:

15' to 20' high	10%	31' to 35' high	40%
21' to 25' high	20%	36' to 40' high	50%
26' to 30' high	30%	Over 40' high	60%

R238313-20   Spiral-Wrapped Heat Trace Cable (Pitch Table)

In order to increase the amount of heat, occasionally heat trace cable is wrapped in a spiral fashion around a pipe; increasing the number of feet of heater cable per linear foot of pipe.

Engineers first determine the heat loss per foot of pipe (based on the insulating material, its thickness, and the temperature differential across it). A ratio is then calculated by the formula:

$$\text{Feet of Heat Trace per Foot of Pipe} = \frac{\text{Watts/Foot of Heat Loss}}{\text{Watts/Foot of the Cable}}$$

The linear distance between wraps (pitch) is then taken from a chart or table. Generally, the pitch is listed on a drawing leaving the estimator to calculate the total length of heat tape required. An approximation may be taken from this table.

Feet of Heat Trace Per Foot of Pipe																	
Nominal Pipe Size in Inches																	
Pitch In Inches	1	1¼	1½	2	2½	3	4	6	8	10	12	14	16	18	20	24
3.5	1.80															
4	1.65															
5	1.46	1.60	1.80													
6	1.34	1.45	1.55	1.75												
7	1.25	1.35	1.43	1.57	1.75											
8	1.20	1.28	1.34	1.45	1.60	1.80										
9	1.16	1.23	1.28	1.37	1.51	1.68										
10	1.13	1.19	1.24	1.32	1.44	1.57	1.82									
15	1.06	1.08	1.10	1.15	1.21	1.29	1.42	1.78								
20	1.04	1.05	1.06	1.08	1.13	1.17	1.25	1.49	1.73							
25		1.04	1.04	1.06	1.08	1.11	1.17	1.33	1.51	1.72						
30				1.04	1.05	1.07	1.12	1.24	1.37	1.54	1.70	1.80				
35					1.06	1.06	1.09	1.17	1.28	1.42	1.54	1.64	1.78			
40						1.05	1.07	1.14	1.22	1.33	1.44	1.52	1.64	1.75		
50							1.05	1.09	1.15	1.22	1.29	1.35	1.44	1.53	1.64	1.83
60								1.06	1.11	1.16	1.21	1.25	1.31	1.39	1.46	1.62
70								1.05	1.08	1.12	1.17	1.19	1.24	1.30	1.35	1.47
80									1.06	1.09	1.13	1.15	1.19	1.24	1.28	1.38
90									1.04	1.06	1.10	1.13	1.16	1.19	1.23	1.32
100										1.05	1.08	1.10	1.13	1.15	1.19	1.23

Note: Common practice would normally limit the lower end of the table to 5% of additional heat and above 80% an engineer would likely opt for two (2) parallel cables.

R312319-90   Wellpoints

A single stage wellpoint system is usually limited to dewatering an average 15' depth below normal ground water level. Multi-stage systems are employed for greater depth with the pumping equipment installed only at the lowest header level. Ejectors with unlimited lift capacity can be economical when two or more stages of wellpoints can be replaced or when horizontal clearance is restricted, such as in deep trenches or tunneling projects, and where low water flows are expected. Wellpoints are usually spaced on 2-1/2' to 10' centers along a header pipe. Wellpoint spacing, header size, and pump size are all determined by the expected flow as dictated by soil conditions.

In almost all soils encountered in wellpoint dewatering, the wellpoints may be jetted into place. Cemented soils and stiff clays may require sand wicks about 12" in diameter around each wellpoint to increase efficiency and eliminate weeping into the excavation. These sand wicks require 1/2 to 3 C.Y. of washed filter sand and are installed by using a 12" diameter steel casing and hole puncher jetted into the ground 2' deeper than the wellpoint. Rock may require predrilled holes.

Labor required for the complete installation and removal of a single stage wellpoint system is in the range of 3/4 to 2 labor-hours per linear foot of header, depending upon jetting conditions, wellpoint spacing, etc.

Continuous pumping is necessary except in some free draining soil where temporary flooding is permissible (as in trenches which are backfilled after each day's work). Good practice requires provision of a stand-by pump during the continuous pumping operation.

Systems for continuous trenching below the water table should be installed three to four times the length of expected daily progress to ensure uninterrupted digging, and header pipe size should not be changed during the job.

For pervious free draining soils, deep wells in place of wellpoints may be economical because of lower installation and maintenance costs. Daily production ranges between two to three wells per day, for 25' to 40' depths, to one well per day for depths over 50'.

Detailed analysis and estimating for any dewatering problem is available at no cost from wellpoint manufacturers. Major firms will quote "sufficient equipment" quotes or their affiliates will offer lump sum proposals to cover complete dewatering responsibility.

Description for 200' System with 8" Header		Quantities
Equipment & Material	Wellpoints 25' long, 2" diameter @ 5' O.C.	40 Each
	Header pipe, 8" diameter	200 L.F.
	Discharge pipe, 8" diameter	100 L.F.
	8" valves	3 Each
	Combination jetting & wellpoint pump (standby)	1 Each
	Wellpoint pump, 8" diameter	1 Each
	Transportation to and from site	1 Day
	Fuel for 30 days x 60 gal./day	1800 Gallons
	Lubricants for 30 days x 16 lbs./day	480 Lbs.
	Sand for points	40 C.Y.
Labor	Technician to supervise installation	1 Week
	Labor for installation and removal of system	300 Labor-hours
	4 Operators straight time 40 hrs./wk. for 4.33 wks.	693 Hrs.
	4 Operators overtime 2 hrs./wk. for 4.33 wks.	35 Hrs.

R314116-40   Wood Sheet Piling

Wood sheet piling may be used for depths to 20' where there is no ground water. If moderate ground water is encountered Tongue & Groove

sheeting will help to keep it out. When considerable ground water is present, steel sheeting must be used.

For estimating purposes on trench excavation, sizes are as follows:

Depth	Sheeting	Wales	Braces	B.F. per S.F.
To 8'	3 x 12's	6 x 8's, 2 line	6 x 8's, @ 10'	4.0 @ 8'
8' x 12'	3 x 12's	10 x 10's, 2 line	10 x 10's, @ 9'	5.0 average
12' to 20'	3 x 12's	12 x 12's, 3 line	12 x 12's, @ 8'	7.0 average

Sheeting to be toed in at least 2' depending upon soil conditions. A five person crew with an air compressor and sheeting driver can drive and brace 440 SF/day at 8' deep, 360 SF/day at 12' deep, and 320 SF/day at 16' deep.

For normal soils, piling can be pulled in 1/3 the time to install. Pulling difficulty increases with the time in the ground. Production can be increased by high pressure jetting.

R314116-45   Steel Sheet Piling

Limiting weights are 22 to 38#/S.F. of wall surface with 27#/S.F. average for usual types and sizes. (Weights of piles themselves are from 30.7#/L.F. to 57#/L.F. but they are 15" to 21" wide.) Lightweight sections 12" to 28" wide from 3 ga. to 12 ga. thick are also available for shallow excavations. Piles may be driven two at a time with an impact or vibratory hammer (use vibratory to pull) hung from a crane without leads. A reasonable estimate of the life of steel sheet piling is 10 uses with up to 125 uses possible if a vibratory hammer is used. Used piling costs from 50% to 80% of new piling depending on location and market conditions. Sheet piling and H piles

can be rented for about 30% of the delivered mill price for the first month and 5% per month thereafter. Allow 1 labor-hour per pile for cleaning and trimming after driving. These costs increase with depth and hydrostatic head. Vibratory drivers are faster in wet granular soils and are excellent for pile extraction. Pulling difficulty increases with the time in the ground and may cost more than driving. It is often economical to abandon the sheet piling, especially if it can be used as the outer wall form. Allow about 1/3 additional length or more for toeing into ground. Add bracing, waler and strut costs. Waler costs can equal the cost per ton of sheeting.

R331113-80   Piping Designations

There are several systems currently in use to describe pipe and fittings. The following paragraphs will help to identify and clarify classifications of piping systems used for water distribution.

Piping may be classified by schedule. Piping schedules include 5S, 10S, 10, 20, 30, Standard, 40, 60, Extra Strong, 80, 100, 120, 140, 160 and Double Extra Strong. These schedules are dependent upon the pipe wall thickness. The wall thickness of a particular schedule may vary with pipe size.

Ductile iron pipe for water distribution is classified by Pressure Classes such as Class 150, 200, 250, 300 and 350. These classes are actually the rated water working pressure of the pipe in pounds per square inch (psi). The pipe in these pressure classes is designed to withstand the rated water working pressure plus a surge allowance of 100 psi.

The American Water Works Association (AWWA) provides standards for various types of **plastic pipe**. C-900 is the specification for polyvinyl chloride (PVC) piping used for water distribution in sizes ranging from 4" through 12". C-901 is the specification for polyethylene (PE) pressure pipe, tubing and fittings used for water distribution in sizes ranging from 1/2" through 3". C-905 is the specification for PVC piping sizes 14" and greater.

PVC pressure-rated pipe is identified using the standard dimensional ratio (SDR) method. This method is defined by the American Society for Testing and Materials (ASTM) Standard D 2241. This pipe is available in SDR numbers 64, 41, 32.5, 26, 21, 17, and 13.5. A pipe with an SDR of 64 will have the thinnest wall while a pipe with an SDR of 13.5 will have the thickest wall. When the pressure rating (PR) of a pipe is given in psi, it is based on a line supplying water at 73 degrees F.

The National Sanitation Foundation (NSF) seal of approval is applied to products that can be used with potable water. These products have been tested to ANSI/NSF Standard 14.

Valves and strainers are classified by American National Standards Institute (ANSI) Classes. These Classes are 125, 150, 200, 250, 300, 400, 600, 900, 1500 and 2500. Within each class there is an operating pressure range dependent upon temperature. Design parameters should be compared to the appropriate material dependent, pressure-temperature rating chart for accurate valve selection.

739

For customer support on your Mechanical Cost Data, call 877.739.3689.

Change Orders

Change Order Considerations

A change order is a written document, usually prepared by the design professional, and signed by the owner, the architect/engineer, and the contractor. A change order states the agreement of the parties to: an addition, deletion, or revision in the work; an adjustment in the contract sum, if any; or an adjustment in the contract time, if any. Change orders, or "extras" in the construction process occur after execution of the construction contract and impact architects/engineers, contractors, and owners.

Change orders that are properly recognized and managed can ensure orderly, professional, and profitable progress for all who are involved in the project. There are many causes for change orders and change order requests. In all cases, change orders or change order requests should be addressed promptly and in a precise and prescribed manner. The following paragraphs include information regarding change order pricing and procedures.

The Causes of Change Orders

Reasons for issuing change orders include:

- Unforeseen field conditions that require a change in the work
- Correction of design discrepancies, errors, or omissions in the contract documents
- Owner-requested changes, either by design criteria, scope of work, or project objectives
- Completion date changes for reasons unrelated to the construction process
- Changes in building code interpretations, or other public authority requirements that require a change in the work
- Changes in availability of existing or new materials and products

Procedures

Properly written contract documents must include the correct change order procedures for all parties—owners, design professionals and contractors—to follow in order to avoid costly delays and litigation.

Being "in the right" is not always a sufficient or acceptable defense. The contract provisions requiring notification and documentation must be adhered to within a defined or reasonable time frame.

The appropriate method of handling change orders is by a written proposal and acceptance by all parties involved. Prior to starting work on a project, all parties should identify their

authorized agents who may sign and accept change orders, as well as any limits placed on their authority.

Time may be a critical factor when the need for a change arises. For such cases, the contractor might be directed to proceed on a "time and materials" basis, rather than wait for all paperwork to be processed—a delay that could impede progress. In this situation, the contractor must still follow the prescribed change order procedures including, but not limited to, notification and documentation.

Lack of documentation can be very costly, especially if legal judgments are to be made, and if certain field personnel are no longer available. For time and material change orders, the contractor should keep accurate daily records of all labor and material allocated to the change.

Owners or awarding authorities who do considerable and continual building construction (such as the federal government) realize the inevitability of change orders for numerous reasons, both predictable and unpredictable. As a result, the federal government, the American Institute of Architects (AIA), the Engineers Joint Contract Documents Committee (EJCDC) and other contractor, legal, and technical organizations have developed standards and procedures to be followed by all parties to achieve contract continuance and timely completion, while being financially fair to all concerned.

Pricing Change Orders

When pricing change orders, regardless of their cause, the most significant factor is when the change occurs. The need for a change may be perceived in the field or requested by the architect/engineer *before* any of the actual installation has begun, or may evolve or appear *during* construction when the item of work in question is partially installed. In the latter cases, the original sequence of construction is disrupted, along with all contiguous and supporting systems. Change orders cause the greatest impact when they occur *after* the installation has been completed and must be uncovered, or even replaced. Post-completion changes may be caused by necessary design changes, product failure, or changes in the owner's requirements that are not discovered until the building or the systems begin to function.

Specified procedures of notification and record keeping must be adhered to and enforced regardless of the stage of construction: *before, during,* or *after* installation. Some bidding documents anticipate change orders by requiring that unit prices including overhead and profit percentages—for additional as well as deductible changes—be listed. Generally these unit prices do not fully take into account the ripple effect, or impact on other trades, and should be used for general guidance only.

When pricing change orders, it is important to classify the time frame in which the change occurs. There are two basic time frames for change orders: *pre-installation change orders*, which occur before the start of construction, and *post-installation change orders*, which involve reworking after the original installation. Change orders that occur between these stages may be priced according to the extent of work completed using a combination of techniques developed for pricing *pre-* and *post-installation* changes.

Factors To Consider When Pricing Change Orders

As an estimator begins to prepare a change order, the following questions should be reviewed to determine their impact on the final price.

General

- *Is the change order work* pre-installation *or* post-installation?

  Change order work costs vary according to how much of the installation has been completed. Once workers have the project scoped in their minds, even though they have not started, it can be difficult to refocus. Consequently they may spend more than the normal amount of time understanding the change. Also, modifications to work in place, such as trimming or refitting, usually take more time than was initially estimated. The greater the amount of work in place, the more reluctant workers are to change it. Psychologically they may resent the change and as a result the rework takes longer than normal. Post-installation change order estimates must include demolition of existing work as required to accomplish the change. If the work is performed at a later time, additional obstacles, such as building finishes, may be present which must be protected. Regardless of whether the change occurs

    For customer support on your Mechanical Cost Data, call 877.739.3689.

pre-installation or post-installation, attempt to isolate the identifiable factors and price them separately. For example, add shipping costs that may be required pre-installation or any demolition required post-installation. Then analyze the potential impact on productivity of psychological and/or learning curve factors and adjust the output rates accordingly. One approach is to break down the typical workday into segments and quantify the impact on each segment.

Change Order Installation Efficiency

The labor-hours expressed (for new construction) are based on average installation time, using an efficiency level. For change order situations, adjustments to this efficiency level should reflect the daily labor-hour allocation for that particular occurrence.

- *Will the change substantially delay the original completion date?*

  A significant change in the project may cause the original completion date to be extended. The extended schedule may subject the contractor to new wage rates dictated by relevant labor contracts. Project supervision and other project overhead must also be extended beyond the original completion date. The schedule extension may also put installation into a new weather season. For example, underground piping scheduled for October installation was delayed until January. As a result, frost penetrated the trench area, thereby changing the degree of difficulty of the task. Changes and delays may have a ripple effect throughout the project. This effect must be analyzed and negotiated with the owner.

- *What is the net effect of a deduct change order?*

  In most cases, change orders resulting in a deduction or credit reflect only bare costs. The contractor may retain the overhead and profit based on the original bid.

Materials

- *Will you have to pay more or less for the new material, required by the change order, than you paid for the original purchase?*

  The same material prices or discounts will usually apply to materials purchased for change orders as new construction. In some

instances, however, the contractor may forfeit the advantages of competitive pricing for change orders. Consider the following example:

A contractor purchased over $20,000 worth of fan coil units for an installation, and obtained the maximum discount. Some time later it was determined the project required an additional matching unit. The contractor has to purchase this unit from the original supplier to ensure a match. The supplier at this time may not discount the unit because of the small quantity, and the fact that he is no longer in a competitive situation. The impact of quantity on purchase can add between 0% and 25% to material prices and/or subcontractor quotes.

- *If materials have been ordered or delivered to the job site, will they be subject to a cancellation charge or restocking fee?*

  Check with the supplier to determine if ordered materials are subject to a cancellation charge. Delivered materials not used as result of a change order may be subject to a restocking fee if returned to the supplier. Common restocking charges run between 20% and 40%. Also, delivery charges to return the goods to the supplier must be added.

Labor

- *How efficient is the existing crew at the actual installation?*

  Is the same crew that performed the initial work going to do the change order? Possibly the change consists of the installation of a unit identical to one already installed; therefore, the change should take less time. Be sure to consider this potential productivity increase and modify the productivity rates accordingly.

- *If the crew size is increased, what impact will that have on supervision requirements?*

  Under most bargaining agreements or management practices, there is a point at which a working foreman is replaced by a nonworking foreman. This replacement increases project overhead by adding a nonproductive worker. If additional workers are added to accelerate the project or to perform changes while maintaining the schedule, be sure to add additional supervision time if warranted. Calculate the

hours involved and the additional cost directly if possible.

- *What are the other impacts of increased crew size?*

  The larger the crew, the greater the potential for productivity to decrease. Some of the factors that cause this productivity loss are: overcrowding (producing restrictive conditions in the working space), and possibly a shortage of any special tools and equipment required. Such factors affect not only the crew working on the elements directly involved in the change order, but other crews whose movement may also be hampered. As the crew increases, check its basic composition for changes by the addition or deletion of apprentices or nonworking foreman, and quantify the potential effects of equipment shortages or other logistical factors.

- *As new crews, unfamiliar with the project, are brought onto the site, how long will it take them to become oriented to the project requirements?*

  The orientation time for a new crew to become 100% effective varies with the site and type of project. Orientation is easiest at a new construction site, and most difficult at existing, very restrictive renovation sites. The type of work also affects orientation time. When all elements of the work are exposed, such as concrete or masonry work, orientation is decreased. When the work is concealed or less visible, such as existing electrical systems, orientation takes longer. Usually orientation can be accomplished in one day or less. Costs for added orientation should be itemized and added to the total estimated cost.

- *How much actual production can be gained by working overtime?*

  Short term overtime can be used effectively to accomplish more work in a day. However, as overtime is scheduled to run beyond several weeks, studies have shown marked decreases in output. The following chart shows the effect of long term overtime on worker efficiency. If the anticipated change requires extended overtime to keep the job on schedule, these factors can be used as a guide to predict the impact on time and cost. Add project overhead, particularly supervision, that may also be incurred.

For customer support on your **Mechanical Cost Data**, call 877.739.3689.

741

Days per Week	Hours per Day	Production Efficiency					Payroll Cost Factors	
		1st Week	2nd Week	3rd Week	4th Week	Average 4 Weeks	@ 1-1/2 Times	@ 2 Times
5	8	100%	100%	100%	100%	100%	100%	100%
	9	100	100	95	90	96.25	105.6	111.1
	10	100	95	90	85	91.25	110.0	120.0
	11	95	90	75	65	81.25	113.6	127.3
	12	90	85	70	60	76.25	116.7	133.3
6	8	100	100	95	90	96.25	108.3	116.7
	9	100	95	90	85	92.50	113.0	125.9
	10	95	90	85	80	87.50	116.7	133.3
	11	95	85	70	65	78.75	119.7	139.4
	12	90	80	65	60	73.75	122.2	144.4
7	8	100	95	85	75	88.75	114.3	128.6
	9	95	90	80	70	83.75	118.3	136.5
	10	90	85	75	65	78.75	121.4	142.9
	11	85	80	65	60	72.50	124.0	148.1
	12	85	75	60	55	68.75	126.2	152.4

Effects of Overtime

Caution: Under many labor agreements, Sundays and holidays are paid at a higher premium than the normal overtime rate.

The use of long-term overtime is counterproductive on almost any construction job; that is, the longer the period of overtime, the lower the actual production rate. Numerous studies have been conducted, and while they have resulted in slightly different numbers, all reach the same conclusion. The figure above tabulates the effects of overtime work on efficiency.

As illustrated, there can be a difference between the *actual* payroll cost per hour and the *effective* cost per hour for overtime work. This is due to the reduced production efficiency with the increase in weekly hours beyond 40. This difference between actual and effective cost results from overtime work over a prolonged period. Short-term overtime work does not result in as great a reduction in efficiency and, in such cases, effective cost may not vary significantly from the actual payroll cost. As the total hours per week are increased on a regular basis, more time is lost due to fatigue, lowered morale, and an increased accident rate.

As an example, assume a project where workers are working 6 days a week, 10 hours per day. From the figure above (based on productivity studies), the average effective productive hours over a 4-week period are:

$$0.875 \times 60 = 52.5$$

Depending upon the locale and day of week, overtime hours may be paid at time and a half or double time. For time and a half, the overall (average) *actual* payroll cost (including regular and overtime hours) is determined as follows:

$$\frac{40 \text{ reg. hrs.} + (20 \text{ overtime hrs.} \times 1.5)}{60 \text{ hrs.}} = 1.167$$

Based on 60 hours, the payroll cost per hour will be 116.7% of the normal rate at 40 hours per week. However, because the effective production (efficiency) for 60 hours is reduced to the equivalent of 52.5 hours, the effective cost of overtime is calculated as follows:

For time and a half:

$$\frac{40 \text{ reg. hrs.} + (20 \text{ overtime hrs.} \times 1.5)}{52.5 \text{ hrs.}} = 1.33$$

Installed cost will be 133% of the normal rate (for labor).

Thus, when figuring overtime, the actual cost per unit of work will be higher than the apparent overtime payroll dollar increase, due to the reduced productivity of the longer workweek. These efficiency calculations are true only for those cost factors determined by hours worked. Costs that are applied weekly or monthly, such as equipment rentals, will not be similarly affected.

Equipment

- *What equipment is required to complete the change order?*

Change orders may require extending the rental period of equipment already on the job site, or the addition of special equipment brought in to accomplish the change work. In either case, the additional rental charges and operator labor charges must be added.

Summary

The preceding considerations and others you deem appropriate should be analyzed and applied to a change order estimate. The impact of each should be quantified and listed on the estimate to form an audit trail.

Change orders that are properly identified, documented, and managed help to ensure the orderly, professional and profitable progress of the work. They also minimize potential claims or disputes at the end of the project.

Abbreviations

A	Area Square Feet; Ampere	Brk., brk	Brick	Csc	Cosecant
AAFES	Army and Air Force Exchange Service	brkt	Bracket	C.S.F.	Hundred Square Feet
ABS	Acrylonitrile Butadiene Stryrene; Asbestos Bonded Steel	Brs.	Brass	CSI	Construction Specifications Institute
		Brz.	Bronze	CT	Current Transformer
A.C., AC	Alternating Current; Air-Conditioning; Asbestos Cement; Plywood Grade A & C	Bsn.	Basin	CTS	Copper Tube Size
		Btr.	Better	Cu	Copper, Cubic
		BTU	British Thermal Unit	Cu. Ft.	Cubic Foot
		BTUH	BTU per Hour	cw	Continuous Wave
ACI	American Concrete Institute	Bu.	Bushels	C.W.	Cool White; Cold Water
ACR	Air Conditioning Refrigeration	BUR	Built-up Roofing	Cwt.	100 Pounds
ADA	Americans with Disabilities Act	BX	Interlocked Armored Cable	C.W.X.	Cool White Deluxe
AD	Plywood, Grade A & D	°C	Degree Centigrade	C.Y.	Cubic Yard (27 cubic feet)
Addit.	Additional	c	Conductivity, Copper Sweat	C.Y./Hr.	Cubic Yard per Hour
Adj.	Adjustable	C	Hundred; Centigrade	Cyl.	Cylinder
af	Audio-frequency	C/C	Center to Center, Cedar on Cedar	d	Penny (nail size)
AFUE	Annual Fuel Utilization Efficiency	C-C	Center to Center	D	Deep; Depth; Discharge
AGA	American Gas Association	Cab	Cabinet	Dis., Disch.	Discharge
Agg.	Aggregate	Cair.	Air Tool Laborer	Db	Decibel
A.H., Ah	Ampere Hours	Cal.	Caliper	Dbl.	Double
A hr.	Ampere-hour	Calc	Calculated	DC	Direct Current
A.H.U., AHU	Air Handling Unit	Cap.	Capacity	DDC	Direct Digital Control
A.I.A.	American Institute of Architects	Carp.	Carpenter	Demob.	Demobilization
AIC	Ampere Interrupting Capacity	C.B.	Circuit Breaker	d.f.t.	Dry Film Thickness
Allow.	Allowance	C.C.A.	Chromate Copper Arsenate	d.f.u.	Drainage Fixture Units
alt., alt	Alternate	C.C.F.	Hundred Cubic Feet	D.H.	Double Hung
Alum.	Aluminum	cd	Candela	DHW	Domestic Hot Water
a.m.	Ante Meridiem	cd/sf	Candela per Square Foot	DI	Ductile Iron
Amp.	Ampere	CD	Grade of Plywood Face & Back	Diag.	Diagonal
Anod.	Anodized	CDX	Plywood, Grade C & D, exterior glue	Diam., Dia	Diameter
ANSI	American National Standards Institute			Distrib.	Distribution
		Cefi.	Cement Finisher	Div.	Division
APA	American Plywood Association	Cem.	Cement	Dk.	Deck
Approx.	Approximate	CF	Hundred Feet	D.L.	Dead Load; Diesel
Apt.	Apartment	C.F.	Cubic Feet	DLH	Deep Long Span Bar Joist
Asb.	Asbestos	CFM	Cubic Feet per Minute	dlx	Deluxe
A.S.B.C.	American Standard Building Code	CFRP	Carbon Fiber Reinforced Plastic	Do.	Ditto
Asbe.	Asbestos Worker	c.g.	Center of Gravity	DOP	Dioctyl Phthalate Penetration Test (Air Filters)
ASCE	American Society of Civil Engineers	CHW	Chilled Water; Commercial Hot Water		
A.S.H.R.A.E.	American Society of Heating, Refrig. & AC Engineers			Dp., dp	Depth
		C.I., CI	Cast Iron	D.P.S.T.	Double Pole, Single Throw
ASME	American Society of Mechanical Engineers	C.I.P., CIP	Cast in Place	Dr.	Drive
		Circ.	Circuit	DR	Dimension Ratio
ASTM	American Society for Testing and Materials	C.L.	Carload Lot	Drink.	Drinking
		CL	Chain Link	D.S.	Double Strength
Attchmt.	Attachment	Clab.	Common Laborer	D.S.A.	Double Strength A Grade
Avg., Ave.	Average	Clam	Common Maintenance Laborer	D.S.B.	Double Strength B Grade
AWG	American Wire Gauge	C.L.F.	Hundred Linear Feet	Dty.	Duty
AWWA	American Water Works Assoc.	CLF	Current Limiting Fuse	DWV	Drain Waste Vent
Bbl.	Barrel	CLP	Cross Linked Polyethylene	DX	Deluxe White, Direct Expansion
B&B, BB	Grade B and Better; Balled & Burlapped	cm	Centimeter	dyn	Dyne
		CMP	Corr. Metal Pipe	e	Eccentricity
B&S	Bell and Spigot	CMU	Concrete Masonry Unit	E	Equipment Only; East; Emissivity
B.&W.	Black and White	CN	Change Notice	Ea.	Each
b.c.c.	Body-centered Cubic	Col.	Column	EB	Encased Burial
B.C.Y.	Bank Cubic Yards	CO₂	Carbon Dioxide	Econ.	Economy
BE	Bevel End	Comb.	Combination	E.C.Y	Embankment Cubic Yards
B.F.	Board Feet	comm.	Commercial, Communication	EDP	Electronic Data Processing
Bg. cem.	Bag of Cement	Compr.	Compressor	EIFS	Exterior Insulation Finish System
BHP	Boiler Horsepower; Brake Horsepower	Conc.	Concrete	E.D.R.	Equiv. Direct Radiation
		Cont., cont	Continuous; Continued, Container	Eq.	Equation
B.I.	Black Iron	Corr.	Corrugated	EL	Elevation
bidir.	bidirectional	Cos	Cosine	Elec.	Electrician; Electrical
Bit., Bitum.	Bituminous	Cot	Cotangent	Elev.	Elevator; Elevating
Bit., Conc.	Bituminous Concrete	Cov.	Cover	EMT	Electrical Metallic Conduit; Thin Wall Conduit
Bk.	Backed	C/P	Cedar on Paneling		
Bkrs.	Breakers	CPA	Control Point Adjustment	Eng.	Engine, Engineered
Bldg., bldg	Building	Cplg.	Coupling	EPDM	Ethylene Propylene Diene Monomer
Blk.	Block	CPM	Critical Path Method		
Bm.	Beam	CPVC	Chlorinated Polyvinyl Chloride	EPS	Expanded Polystyrene
Boil.	Boilermaker	C.Pr.	Hundred Pair	Eqhv.	Equip. Oper., Heavy
bpm	Blows per Minute	CRC	Cold Rolled Channel	Eqlt.	Equip. Oper., Light
BR	Bedroom	Creos.	Creosote	Eqmd.	Equip. Oper., Medium
Brg., brng.	Bearing	Crpt.	Carpet & Linoleum Layer	Eqmm.	Equip. Oper., Master Mechanic
Brhe.	Bricklayer Helper	CRT	Cathode-ray Tube	Eqol.	Equip. Oper., Oilers
Bric.	Bricklayer	CS	Carbon Steel, Constant Shear Bar Joist	Equip.	Equipment
				ERW	Electric Resistance Welded

Abbreviations

E.S.	Energy Saver	H	High Henry	Lath.	Lather
Est.	Estimated	HC	High Capacity	Lav.	Lavatory
esu	Electrostatic Units	H.D., HD	Heavy Duty; High Density	lb.; #	Pound
E.W.	Each Way	H.D.O.	High Density Overlaid	L.B., LB	Load Bearing; L Conduit Body
EWT	Entering Water Temperature	HDPE	High Density Polyethylene Plastic	L. & E.	Labor & Equipment
Excav.	Excavation	Hdr.	Header	lb./hr.	Pounds per Hour
excl	Excluding	Hdwe.	Hardware	lb./L.F.	Pounds per Linear Foot
Exp., exp	Expansion, Exposure	H.I.D., HID	High Intensity Discharge	lbf/sq.in.	Pound-force per Square Inch
Ext., ext	Exterior; Extension	Help.	Helper Average	L.C.L.	Less than Carload Lot
Extru.	Extrusion	HEPA	High Efficiency Particulate Air	L.C.Y.	Loose Cubic Yard
f.	Fiber Stress		Filter	Ld.	Load
F	Fahrenheit; Female; Fill	Hg	Mercury	LE	Lead Equivalent
Fab., fab	Fabricated; Fabric	HIC	High Interrupting Capacity	LED	Light Emitting Diode
FBGS	Fiberglass	HM	Hollow Metal	L.F.	Linear Foot
F.C.	Footcandles	HMWPE	High Molecular Weight	L.F. Nose	Linear Foot of Stair Nosing
f.c.c.	Face-centered Cubic		Polyethylene	L.F. Rsr	Linear Foot of Stair Riser
f'c.	Compressive Stress in Concrete;	HO	High Output	Lg.	Long; Length; Large
	Extreme Compressive Stress	Horiz.	Horizontal	L & H	Light and Heat
F.E.	Front End	H.P., HP	Horsepower; High Pressure	LH	Long Span Bar Joist
FEP	Fluorinated Ethylene Propylene	H.P.F.	High Power Factor	L.H.	Labor Hours
	(Teflon)	Hr.	Hour	L.L., LL	Live Load
F.G.	Flat Grain	Hrs./Day	Hours per Day	L.L.D.	Lamp Lumen Depreciation
F.H.A.	Federal Housing Administration	HSC	High Short Circuit	lm	Lumen
Fig.	Figure	Ht.	Height	lm/sf	Lumen per Square Foot
Fin.	Finished	Htg.	Heating	lm/W	Lumen per Watt
FIPS	Female Iron Pipe Size	Htrs.	Heaters	LOA	Length Over All
Fixt.	Fixture	HVAC	Heating, Ventilation & Air-	log	Logarithm
FJP	Finger jointed and primed		Conditioning	L-O-L	Lateralolet
Fl. Oz.	Fluid Ounces	Hvy.	Heavy	long.	Longitude
Flr.	Floor	HW	Hot Water	L.P., LP	Liquefied Petroleum; Low Pressure
FM	Frequency Modulation;	Hyd.; Hydr.	Hydraulic	L.P.F.	Low Power Factor
	Factory Mutual	Hz	Hertz (cycles)	LR	Long Radius
Fmg.	Framing	I.	Moment of Inertia	L.S.	Lump Sum
FM/UL	Factory Mutual/Underwriters Labs	IBC	International Building Code	Lt.	Light
Fdn.	Foundation	I.C.	Interrupting Capacity	Lt. Ga.	Light Gauge
FNPT	Female National Pipe Thread	ID	Inside Diameter	L.T.L.	Less than Truckload Lot
Fori.	Foreman, Inside	I.D.	Inside Dimension; Identification	Lt. Wt.	Lightweight
Foro.	Foreman, Outside	I.F.	Inside Frosted	L.V.	Low Voltage
Fount.	Fountain	I.M.C.	Intermediate Metal Conduit	M	Thousand; Material; Male;
fpm	Feet per Minute	In.	Inch		Light Wall Copper Tubing
FPT	Female Pipe Thread	Incan.	Incandescent	M²CA	Meters Squared Contact Area
Fr	Frame	Incl.	Included; Including	m/hr.; M.H.	Man-hour
F.R.	Fire Rating	Int.	Interior	mA	Milliampere
FRK	Foil Reinforced Kraft	Inst.	Installation	Mach.	Machine
FSK	Foil/Scrim/Kraft	Insul., insul	Insulation/Insulated	Mag. Str.	Magnetic Starter
FRP	Fiberglass Reinforced Plastic	I.P.	Iron Pipe	Maint.	Maintenance
FS	Forged Steel	I.P.S., IPS	Iron Pipe Size	Marb.	Marble Setter
FSC	Cast Body; Cast Switch Box	IPT	Iron Pipe Threaded	Mat; Mat'l.	Material
Ft., ft	Foot; Feet	I.W.	Indirect Waste	Max.	Maximum
Ftng.	Fitting	J	Joule	MBF	Thousand Board Feet
Ftg.	Footing	J.I.C.	Joint Industrial Council	MBH	Thousand BTU's per hr.
Ft lb.	Foot Pound	K	Thousand; Thousand Pounds;	MC	Metal Clad Cable
Furn.	Furniture		Heavy Wall Copper Tubing, Kelvin	MCC	Motor Control Center
FVNR	Full Voltage Non-Reversing	K.A.H.	Thousand Amp. Hours	M.C.F.	Thousand Cubic Feet
FVR	Full Voltage Reversing	kcmil	Thousand Circular Mils	MCFM	Thousand Cubic Feet per Minute
FXM	Female by Male	KD	Knock Down	M.C.M.	Thousand Circular Mils
Fy.	Minimum Yield Stress of Steel	K.D.A.T.	Kiln Dried After Treatment	MCP	Motor Circuit Protector
g	Gram	kg	Kilogram	MD	Medium Duty
G	Gauss	kG	Kilogauss	MDF	Medium-density fibreboard
Ga.	Gauge	kgf	Kilogram Force	M.D.O.	Medium Density Overlaid
Gal., gal.	Gallon	kHz	Kilohertz	Med.	Medium
Galv., galv	Galvanized	Kip	1000 Pounds	MF	Thousand Feet
GC/MS	Gas Chromatograph/Mass	KJ	Kilojoule	M.F.B.M.	Thousand Feet Board Measure
	Spectrometer	K.L.	Effective Length Factor	Mfg.	Manufacturing
Gen.	General	K.L.F.	Kips per Linear Foot	Mfrs.	Manufacturers
GFI	Ground Fault Interrupter	Km	Kilometer	mg	Milligram
GFRC	Glass Fiber Reinforced Concrete	KO	Knock Out	MGD	Million Gallons per Day
Glaz.	Glazier	K.S.F.	Kips per Square Foot	MGPH	Million Gallons per Hour
GPD	Gallons per Day	K.S.I.	Kips per Square Inch	MH, M.H.	Manhole; Metal Halide; Man-Hour
gpf	Gallon per Flush	kV	Kilovolt	MHz	Megahertz
GPH	Gallons per Hour	kVA	Kilovolt Ampere	Mi.	Mile
gpm, GPM	Gallons per Minute	kVAR	Kilovar (Reactance)	MI	Malleable Iron; Mineral Insulated
GR	Grade	KW	Kilowatt	MIPS	Male Iron Pipe Size
Gran.	Granular	KWh	Kilowatt-hour	mj	Mechanical Joint
Grnd.	Ground	L	Labor Only; Length; Long;	m	Meter
GVW	Gross Vehicle Weight		Medium Wall Copper Tubing	mm	Millimeter
GWB	Gypsum Wall Board	Lab.	Labor	Mill.	Millwright
		lat	Latitude	Min., min.	Minimum, Minute

744

Misc.	Miscellaneous	PDCA	Painting and Decorating	SC	Screw Cover
ml	Milliliter, Mainline		Contractors of America	SCFM	Standard Cubic Feet per Minute
M.L.F.	Thousand Linear Feet	P.E., PE	Professional Engineer;	Scaf.	Scaffold
Mo.	Month		Porcelain Enamel;	Sch., Sched.	Schedule
Mobil.	Mobilization		Polyethylene; Plain End	S.C.R.	Modular Brick
Mog.	Mogul Base	P.E.C.I.	Porcelain Enamel on Cast Iron	S.D.	Sound Deadening
MPH	Miles per Hour	Perf.	Perforated	SDR	Standard Dimension Ratio
MPT	Male Pipe Thread	PEX	Cross Linked Polyethylene	S.E.	Surfaced Edge
MRGWB	Moisture Resistant Gypsum	Ph.	Phase	Sel.	Select
	Wallboard	P.I.	Pressure Injected	SER, SEU	Service Entrance Cable
MRT	Mile Round Trip	Pile.	Pile Driver	S.F.	Square Foot
ms	Millisecond	Pkg.	Package	S.F.C.A.	Square Foot Contact Area
M.S.F.	Thousand Square Feet	Pl.	Plate	S.F. Flr.	Square Foot of Floor
Mstz.	Mosaic & Terrazzo Worker	Plah.	Plasterer Helper	S.F.G.	Square Foot of Ground
M.S.Y.	Thousand Square Yards	Plas.	Plasterer	S.F. Hor.	Square Foot Horizontal
Mtd., mtd., mtd	Mounted	plf	Pounds Per Linear Foot	SFR	Square Feet of Radiation
Mthe.	Mosaic & Terrazzo Helper	Pluh.	Plumber Helper	S.F. Shlf.	Square Foot of Shelf
Mtng.	Mounting	Plum.	Plumber	S4S	Surface 4 Sides
Mult.	Multi; Multiply	Ply.	Plywood	Shee.	Sheet Metal Worker
M.V.A.	Million Volt Amperes	p.m.	Post Meridiem	Sin.	Sine
M.V.A.R.	Million Volt Amperes Reactance	Pntd.	Painted	Skwk.	Skilled Worker
MV	Megavolt	Pord.	Painter, Ordinary	SL	Saran Lined
MW	Megawatt	pp	Pages	S.L.	Slimline
MXM	Male by Male	PP, PPL	Polypropylene	Sldr.	Solder
MYD	Thousand Yards	P.P.M.	Parts per Million	SLH	Super Long Span Bar Joist
N	Natural; North	Pr.	Pair	S.N.	Solid Neutral
nA	Nanoampere	P.E.S.B.	Pre-engineered Steel Building	SO	Stranded with oil resistant inside
NA	Not Available; Not Applicable	Prefab.	Prefabricated		insulation
N.B.C.	National Building Code	Prefin.	Prefinished	S-O-L	Socketolet
NC	Normally Closed	Prop.	Propelled	sp	Standpipe
NEMA	National Electrical Manufacturers	PSF, psf	Pounds per Square Foot	S.P.	Static Pressure; Single Pole; Self-
	Assoc.	PSI, psi	Pounds per Square Inch		Propelled
NEHB	Bolted Circuit Breaker to 600V.	PSIG	Pounds per Square Inch Gauge	Spri.	Sprinkler Installer
NFPA	National Fire Protection Association	PSP	Plastic Sewer Pipe	spwg	Static Pressure Water Gauge
NLB	Non-Load-Bearing	Pspr.	Painter, Spray	S.P.D.T.	Single Pole, Double Throw
NM	Non-Metallic Cable	Psst.	Painter, Structural Steel	SPF	Spruce Pine Fir; Sprayed
nm	Nanometer	P.T.	Potential Transformer		Polyurethane Foam
No.	Number	P. & T.	Pressure & Temperature	S.P.S.T.	Single Pole, Single Throw
NO	Normally Open	Ptd.	Painted	SPT	Standard Pipe Thread
N.O.C.	Not Otherwise Classified	Ptns.	Partitions	Sq.	Square; 100 Square Feet
Nose.	Nosing	Pu	Ultimate Load	Sq. Hd.	Square Head
NPT	National Pipe Thread	PVC	Polyvinyl Chloride	Sq. In.	Square Inch
NQOD	Combination Plug-on/Bolt on	Pvmt.	Pavement	S.S.	Single Strength; Stainless Steel
	Circuit Breaker to 240V.	PRV	Pressure Relief Valve	S.S.B.	Single Strength B Grade
N.R.C., NRC	Noise Reduction Coefficient/	Pwr.	Power	sst, ss	Stainless Steel
	Nuclear Regulator Commission	Q	Quantity Heat Flow	Sswk.	Structural Steel Worker
N.R.S.	Non Rising Stem	Qt.	Quart	Sswl.	Structural Steel Welder
ns	Nanosecond	Quan., Qty.	Quantity	St.; Stl.	Steel
nW	Nanowatt	Q.C.	Quick Coupling	STC	Sound Transmission Coefficient
OB	Opposing Blade	r	Radius of Gyration	Std.	Standard
OC	On Center	R	Resistance	Stg.	Staging
OD	Outside Diameter	R.C.P.	Reinforced Concrete Pipe	STK	Select Tight Knot
O.D.	Outside Dimension	Rect.	Rectangle	STP	Standard Temperature & Pressure
ODS	Overhead Distribution System	recpt.	Receptacle	Stpi.	Steamfitter, Pipefitter
O.G.	Ogee	Reg.	Regular	Str.	Strength; Starter; Straight
O.H.	Overhead	Reinf.	Reinforced	Strd.	Stranded
O&P	Overhead and Profit	Req'd.	Required	Struct.	Structural
Oper.	Operator	Res.	Resistant	Sty.	Story
Opng.	Opening	Resi.	Residential	Subj.	Subject
Orna.	Ornamental	RF	Radio Frequency	Subs.	Subcontractors
OSB	Oriented Strand Board	RFID	Radio-frequency Identification	Surf.	Surface
OS&Y	Outside Screw and Yoke	Rgh.	Rough	Sw.	Switch
OSHA	Occupational Safety and Health	RGS	Rigid Galvanized Steel	Swbd.	Switchboard
	Act	RHW	Rubber, Heat & Water Resistant;	S.Y.	Square Yard
Ovhd.	Overhead		Residential Hot Water	Syn.	Synthetic
OWG	Oil, Water or Gas	rms	Root Mean Square	S.Y.P.	Southern Yellow Pine
Oz.	Ounce	Rnd.	Round	Sys.	System
P.	Pole; Applied Load; Projection	Rodm.	Rodman	t.	Thickness
p.	Page	Rofc.	Roofer, Composition	T	Temperature; Ton
Pape.	Paperhanger	Rofp.	Roofer, Precast	Tan	Tangent
P.A.P.R.	Powered Air Purifying Respirator	Rohe.	Roofer Helpers (Composition)	T.C.	Terra Cotta
PAR	Parabolic Reflector	Rots.	Roofer, Tile & Slate	T & C	Threaded and Coupled
P.B., PB	Push Button	R.O.W.	Right of Way	T.D.	Temperature Difference
Pc., Pcs.	Piece, Pieces	RPM	Revolutions per Minute	TDD	Telecommunications Device for
P.C.	Portland Cement; Power Connector	R.S.	Rapid Start		the Deaf
P.C.F.	Pounds per Cubic Foot	Rsr	Riser	T.E.M.	Transmission Electron Microscopy
PCM	Phase Contrast Microscopy	RT	Round Trip	temp	Temperature, Tempered, Temporary
		S.	Suction; Single Entrance; South	TFFN	Nylon Jacketed Wire
		SBS	Styrene Butadiere Styrene		

745

TFE	Tetrafluoroethylene (Teflon)	U.L., UL	Underwriters Laboratory	w/	With
T. & G.	Tongue & Groove;	Uld.	Unloading	W.C., WC	Water Column; Water Closet
	Tar & Gravel	Unfin.	Unfinished	W.F.	Wide Flange
Th., Thk.	Thick	UPS	Uninterruptible Power Supply	W.G.	Water Gauge
Thn.	Thin	URD	Underground Residential	Wldg.	Welding
Thrded	Threaded		Distribution	W. Mile	Wire Mile
Tilf.	Tile Layer, Floor	US	United States	W-O-L	Weldolet
Tilh.	Tile Layer, Helper	USGBC	U.S. Green Building Council	W.R.	Water Resistant
THHN	Nylon Jacketed Wire	USP	United States Primed	Wrck.	Wrecker
THW.	Insulated Strand Wire	UTMCD	Uniform Traffic Manual For Control	WSFU	Water Supply Fixture Unit
THWN	Nylon Jacketed Wire		Devices	W.S.P.	Water, Steam, Petroleum
T.L., TL	Truckload	UTP	Unshielded Twisted Pair	WT., Wt.	Weight
T.M.	Track Mounted	V	Volt	WWF	Welded Wire Fabric
Tot.	Total	VA	Volt Amperes	XFER	Transfer
T-O-L	Threadolet	VAT	Vinyl Asbestos Tile	XFMR	Transformer
tmpd	Tempered	V.C.T.	Vinyl Composition Tile	XHD	Extra Heavy Duty
TPO	Thermoplastic Polyolefin	VAV	Variable Air Volume	XHHW	Cross-Linked Polyethylene Wire
T.S.	Trigger Start	VC	Veneer Core	XLPE	Insulation
Tr.	Trade	VDC	Volts Direct Current	XLP	Cross-linked Polyethylene
Transf.	Transformer	Vent.	Ventilation	Xport	Transport
Trhv.	Truck Driver, Heavy	Vert.	Vertical	Y	Wye
Trlr	Trailer	V.F.	Vinyl Faced	yd	Yard
Trlt.	Truck Driver, Light	V.G.	Vertical Grain	yr	Year
TTY	Teletypewriter	VHF	Very High Frequency	Δ	Delta
TV	Television	VHO	Very High Output	%	Percent
T.W.	Thermoplastic Water Resistant	Vib.	Vibrating	~	Approximately
	Wire	VLF	Vertical Linear Foot	Ø	Phase; diameter
UCI	Uniform Construction Index	VOC	Volatile Organic Compound	@	At
UF	Underground Feeder	Vol.	Volume	#	Pound; Number
UGND	Underground Feeder	VRP	Vinyl Reinforced Polyester	<	Less Than
UHF	Ultra High Frequency	W	Wire; Watt; Wide; West	>	Greater Than
U.I.	United Inch			Z	Zone

Index

749

For customer support on your Mechanical Cost Data, call 877.739.3689.

For customer support on your Mechanical Cost Data, call 877.739.3689.

Index

751

For customer support on your **Mechanical Cost Data**, call 877.739.3689.

Index

756

Index

For customer support on your Mechanical Cost Data, call 877.739.3689.

759

Index

761

Division Notes

	CREW	DAILY OUTPUT	LABOR-HOURS	UNIT	BARE COSTS				TOTAL INCL O&P
					MAT.	LABOR	EQUIP.	TOTAL	

Division Notes

	CREW	DAILY OUTPUT	LABOR-HOURS	UNIT	BARE COSTS				TOTAL INCL O&P
					MAT.	LABOR	EQUIP.	TOTAL	

Division Notes

	CREW	DAILY OUTPUT	LABOR-HOURS	UNIT	BARE COSTS				TOTAL INCL O&P
					MAT.	LABOR	EQUIP.	TOTAL	

Other RSMeans Products & Services

RSMeans—a tradition of excellence in construction cost information and services since 1942

Table of Contents
Annual Cost Guides
RSMeans Online
Seminars

For more information visit the RSMeans website at www.RSMeans.com | Unit prices according to the latest MasterFormat

Cost Data Selection Guide

The following table provides definitive information on the content of each cost data publication. The number of lines of data provided in each unit price or assemblies division, as well as the number of crews, is listed for each data set. The presence of other elements such as reference tables, square foot models, equipment rental costs, historical cost indexes, and city cost indexes, is also indicated. You can use the table to help select the RSMeans data set that has the quantity and type of information you most need in your work.

Unit Cost Divisions	Building Construction	Mechanical	Electrical	Commercial Renovation	Square Foot	Site Work Landsc.	Green Building	Interior	Concrete Masonry	Open Shop	Heavy Construction	Light Commercial	Facilities Construction	Plumbing	Residential
1	562	383	400	504		498	205	302	445	561	496	245	1040	393	177
2	776	279	85	732		991	206	399	213	775	733	480	1220	286	273
3	1691	340	230	1084		1483	986	354	2037	1691	1693	482	1791	316	389
4	960	21	0	920		725	180	615	1158	928	615	533	1175	0	445
5	1893	158	155	1090		844	1799	1098	720	1893	1037	979	1909	204	746
6	2453	18	18	2111		110	589	1528	281	2449	123	2141	2125	22	2661
7	1594	215	128	1633		580	763	531	523	1591	26	1328	1695	227	1048
8	2111	81	44	2645		257	1136	1757	105	2096	0	2219	2890	0	1534
9	2004	86	45	1828		310	451	2088	391	1944	15	1668	2252	54	1437
10	1049	17	10	645		216	27	864	157	1049	29	539	1141	233	224
11	1096	206	166	551		129	54	933	28	1071	0	231	1116	169	110
12	530	0	2	285		205	133	1533	14	500	0	259	1554	23	217
13	744	149	158	253		367	122	252	78	720	244	108	759	115	103
14	273	36	0	221		0	0	257	0	273	0	12	293	16	6
21	91	0	16	37		0	0	250	0	91	0	73	535	540	220
22	1168	7548	160	1201		1572	1066	848	20	1157	1681	874	7450	9370	718
23	1178	6975	580	929		157	903	778	38	1171	110	879	5213	1922	469
26	1380	458	10150	1033		793	630	1144	55	1372	562	1310	10099	399	619
27	72	0	339	34		13	0	71	0	72	39	52	321	0	22
28	96	58	136	71		0	21	78	0	99	0	40	151	44	25
31	1510	733	610	806		3263	288	7	1217	1455	3277	604	1569	660	613
32	813	49	8	881		4409	353	405	291	784	1826	421	1687	133	468
33	529	1076	538	252		2173	41	0	236	522	2157	128	1697	1283	154
34	107	0	47	4		190	0	0	31	62	214	0	136	0	0
35	18	0	0	0		327	0	0	0	18	442	0	84	0	0
41	60	0	0	33		8	0	22	0	61	31	0	68	14	0
44	75	79	0	0		0	0	0	0	0	0	0	75	75	0
46	23	16	0	0		274	261	0	0	23	264	0	33	33	0
48	12	0	25	0		0	25	0	0	12	17	12	12	0	12
Totals	24868	18981	14050	19783		19894	10239	16114	8038	24440	15631	15617	50090	16531	12690

Assem Div	Building Construction	Mechanical	Electrical	Commercial Renovation	Square Foot	Site Work Landscape	Assemblies	Green Building	Interior	Concrete Masonry	Heavy Construction	Light Commercial	Facilities Construction	Plumbing	Asm Div	Residential
A		15	0	188	150	577	598	0	0	536	571	154	24	0	1	378
B		0	0	848	2506	0	5666	56	329	1975	368	2098	174	0	2	211
C		0	0	647	928	0	1309	0	1628	146	0	818	249	0	3	588
D		1067	941	712	1859	72	2538	330	825	0	0	1345	1105	1088	4	851
E		0	0	86	261	0	301	0	5	0	0	258	5	0	5	390
F		0	0	0	114	0	114	0	0	0	0	114	3	0	6	357
G		527	447	318	312	3364	792	0	0	534	1349	205	293	677	7	307
															8	760
															9	80
															10	0
															11	0
															12	0
Totals		1609	1388	2799	6130	4013	11318	386	2787	3191	2288	4992	1853	1765		3922

Reference Section	Building Construction Costs	Mechanical	Electrical	Commercial Renovation	Square Foot	Site Work Landscape	Assem.	Green Building	Interior	Concrete Masonry	Open Shop	Heavy Construction	Light Commercial	Facilities Construction	Plumbing	Resi.
Reference Tables	yes	yes	yes	yes	no	yes	yes	yes	yes	yes	yes	yes	yes	yes	yes	yes
Models					111			25					50			28
Crews	575	575	575	553		575		575	575	575	551	575	551	553	575	551
Equipment Rental Costs	yes	yes	yes	yes		yes		yes	yes	yes	yes	yes	yes	yes	yes	yes
Historical Cost Indexes	yes	yes	yes	yes	yes	yes	yes	yes	yes	yes	yes	yes	yes	yes	yes	no
City Cost Indexes	yes	yes	yes	yes	yes	yes	yes	yes	yes	yes	yes	yes	yes	yes	yes	yes

Online   Book   CD   eBook   Online add-on available
                                   for additional fee

For more information visit our website at www.RSMeans.com

(left margin, vertical text) Unit prices according to the latest MasterFormat

(left margin, vertical text) Annual Cost Guides

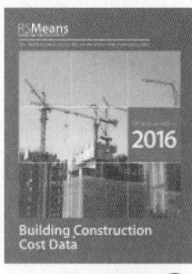

RSMeans Building Construction Cost Data 2016

Offers you unchallenged unit price reliability in an easy-to-use format. Whether used for verifying complete, finished estimates or for periodic checks, it supplies more cost facts better and faster than any comparable source. More than 24,700 unit prices have been updated for 2016. The City Cost Indexes and Location Factors cover more than 930 areas for indexing to any project location in North America. Order and get *RSMeans Quarterly Update Service* FREE.

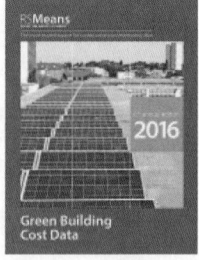

RSMeans Green Building Cost Data 2016

Estimate, plan, and budget the costs of green building for both new commercial construction and renovation work with this sixth edition of *RSMeans Green Building Cost Data*. More than 10,000 unit costs for a wide array of green building products plus assemblies costs. Easily identified cross references to LEED and Green Globes building rating systems criteria.

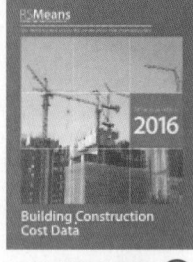

RSMeans Mechanical Cost Data 2016

Total unit and systems price guidance for mechanical construction—materials, parts, fittings, and complete labor cost information. Includes prices for piping, heating, air conditioning, ventilation, and all related construction.

Plus new 2016 unit costs for:

- Thousands of installed HVAC/controls, sub-assemblies and assemblies
- "On-site" Location Factors for more than 930 cities and towns in the U.S. and Canada
- Crews, labor, and equipment

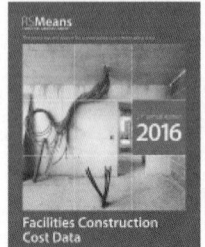

RSMeans Facilities Construction Cost Data 2016

For the maintenance and construction of commercial, industrial, municipal, and institutional properties. Costs are shown for new and remodeling construction and are broken down into materials, labor, equipment, and overhead and profit. Special emphasis is given to sections on mechanical, electrical, furnishings, site work, building maintenance, finish work, and demolition.

More than 49,800 unit costs, plus assemblies costs and a comprehensive Reference Section are included.

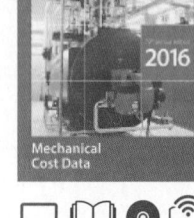

RSMeans Square Foot Costs 2016

Accurate and Easy-to-Use

- Updated price information based on nationwide figures from suppliers, estimators, labor experts, and contractors.
- Green building models
- Realistic graphics, offering true-to-life illustrations of building projects
- Extensive information on using square foot cost data, including sample estimates and alternate pricing methods

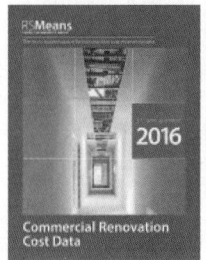

RSMeans Commercial Renovation Cost Data 2016

Commercial/Multi-family Residential

Use this valuable tool to estimate commercial and multi-family residential renovation and remodeling.

Includes: Updated costs for hundreds of unique methods, materials, and conditions that only come up in repair and remodeling, PLUS:

- Unit costs for more than 19,600 construction components
- Installed costs for more than 2,700 assemblies
- More than 930 "on-site" localization factors for the U.S. and Canada

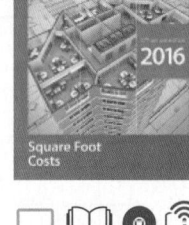

RSMeans Electrical Cost Data 2016

Pricing information for every part of electrical cost planning. More than 13,800 unit and systems costs with design tables; clear specifications and drawings; engineering guides; illustrated estimating procedures; complete labor-hour and materials costs for better scheduling and procurement; and the latest electrical products and construction methods.

- A variety of special electrical systems, including cathodic protection
- Costs for maintenance, demolition, HVAC/mechanical, specialties, equipment, and more

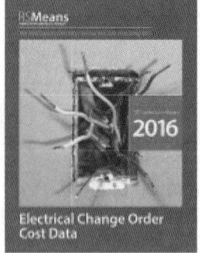

RSMeans Electrical Change Order Cost Data 2016

RSMeans Electrical Change Order Cost Data provides you with electrical unit prices exclusively for pricing change orders. Analyze and check your own change order estimates against those prepared when using RSMeans cost data. It also covers productivity analysis and change order cost justifications. With useful information for calculating the effects of change orders and dealing with their administration.

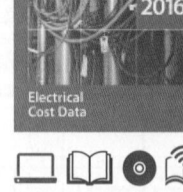

RSMeans Assemblies Cost Data 2016

RSMeans Assemblies Cost Data takes the guesswork out of preliminary or conceptual estimates. Now you don't have to try to calculate the assembled cost by working up individual component costs. We've done all the work for you.

Presents detailed illustrations, descriptions, specifications, and costs for every conceivable building assembly—over 350 types in all—arranged in the easy-to-use UNIFORMAT II system. Each illustrated "assembled" cost includes a complete grouping of materials and associated installation costs, including the installing contractor's overhead and profit.

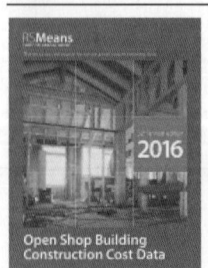

RSMeans Open Shop Building Construction Cost Data 2016

The latest costs for accurate budgeting and estimating of new commercial and residential construction, renovation work, change orders, and cost engineering.

RSMeans Open Shop "BCCD" will assist you to:

- Develop benchmark prices for change orders.
- Plug gaps in preliminary estimates and budgets.
- Estimate complex projects.
- Substantiate invoices on contracts.
- Price ADA-related renovations.

For more information visit our website at www.RSMeans.com

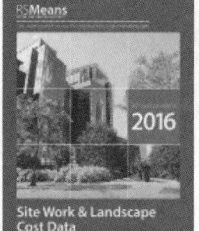

RSMeans Residential Cost Data 2016

Contains square foot costs for 28 basic home models with the look of today, plus hundreds of custom additions and modifications you can quote right off the page. Includes more than 3,900 costs for 89 residential systems. Complete with blank estimating forms, sample estimates, and step-by-step instructions.

Contains line items for cultured stone and brick, PVC trim, lumber, and TPO roofing.

RSMeans Site Work & Landscape Cost Data 2016

Includes unit and assemblies costs for earthwork, sewerage, piped utilities, site improvements, drainage, paving, trees and shrubs, street openings/repairs, underground tanks, and more. Contains more than 60 types of assemblies costs for accurate conceptual estimates.

Includes:

- Estimating for infrastructure improvements
- Environmentally-oriented construction
- ADA-mandated handicapped access
- Hazardous waste line items

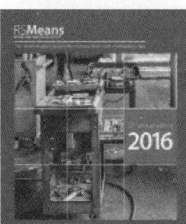

RSMeans Facilities Maintenance & Repair Cost Data 2016

RSMeans Facilities Maintenance & Repair Cost Data gives you a complete system to manage and plan your facility repair and maintenance costs and budget efficiently. Guidelines for auditing a facility and developing an annual maintenance plan. Budgeting is included, along with reference tables on cost and management, and information on frequency and productivity of maintenance operations.

The only nationally recognized source of maintenance and repair costs. Developed in cooperation with the Civil Engineering Research Laboratory (CERL) of the Army Corps of Engineers.

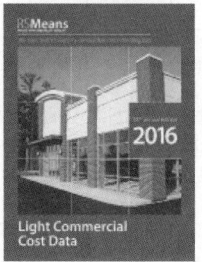

RSMeans Light Commercial Cost Data 2016

Specifically addresses the light commercial market, which is a specialized niche in the construction industry. Aids you, the owner/designer/contractor, in preparing all types of estimates—from budgets to detailed bids. Includes new advances in methods and materials.

The Assemblies section allows you to evaluate alternatives in the early stages of design/planning.

More than 15,500 unit costs ensure that you have the prices you need, when you need them.

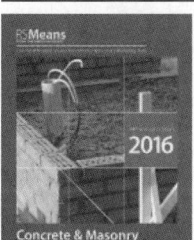

RSMeans Concrete & Masonry Cost Data 2016

Provides you with cost facts for virtually all concrete/masonry estimating needs, from complicated formwork to various sizes and face finishes of brick and block—all in great detail. The comprehensive Unit Price Section contains more than 8,000 selected entries. Also contains an Assemblies [Cost] Section, and a detailed Reference Section that supplements the cost data.

RSMeans Labor Rates for the Construction Industry 2016

Complete information for estimating labor costs, making comparisons, and negotiating wage rates by trade for more than 300 U.S. and Canadian cities. With 46 construction trades in each city, and historical wage rates included for comparison. Each city chart lists the county and is alphabetically arranged with handy visual flip tabs for quick reference.

RSMeans Construction Cost Indexes 2016

What materials and labor costs will change unexpectedly this year? By how much?

- Breakdowns for 318 major cities
- National averages for 30 key cities
- Expanded five major city indexes
- Historical construction cost indexes

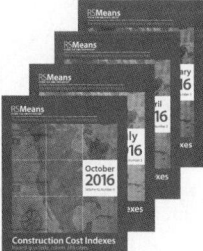

RSMeans Interior Cost Data 2016

Provides you with prices and guidance needed to make accurate interior work estimates. Contains costs on materials, equipment, hardware, custom installations, furnishings, and labor for new and remodel commercial and industrial interior construction, including updated information on office furnishings, as well as reference information.

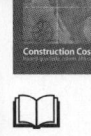

RSMeans Heavy Construction Cost Data 2016

A comprehensive guide to heavy construction costs. Includes costs for highly specialized projects such as tunnels, dams, highways, airports, and waterways. Information on labor rates, equipment, and materials costs is included. Features unit price costs, systems costs, and numerous reference tables for costs and design.

RSMeans Plumbing Cost Data 2016

Comprehensive unit prices and assemblies for plumbing, irrigation systems, commercial and residential fire protection, point-of-use water heaters, and the latest approved materials. This publication and its companion, *RSMeans Mechanical Cost Data*, provide full-range cost estimating coverage for all the mechanical trades.

Contains updated costs for potable water, radiant heat systems, and high efficiency fixtures.

Annual Cost Guides

Unit prices according to the latest MasterFormat

767

RSMeans Online

Competitive Cost Estimates Made Easy

RSMeans Online is a web-based service that provides accurate and up-to-date cost information to help you build competitive estimates or budgets in less time.

Quick, intuitive, easy to use, and automatically updated, RSMeans Online gives you instant access to hundreds of thousands of material, labor, and equipment costs from RSMeans' comprehensive database, delivering the information you need to build budgets and competitive estimates every time.

With RSMeans Online, you can perform quick searches to locate specific costs and adjust costs to reflect prices in your geographic area. Tag and store your favorites for fast access to your frequently used line items and assemblies and clone estimates to save time. System notifications will alert you as updated data becomes available. RSMeans Online is automatically updated throughout the year.

RSMeans Online's visual, interactive estimating features help you create, manage, save and share estimates with ease! It provides increased flexibility with customizable advanced reports. Easily edit custom report templates and import your company logo onto your estimates.

	Standard	Advanced	Professional
Unit Prices	✓	✓	✓
Assemblies	X	✓	✓
Sq Ft Models	X	X	✓
Editable Sq Foot Models	X	X	✓
Editable Assembly Components	X	✓	✓
Custom Cost Data	X	✓	✓
User Defined Components	X	✓	✓
Advanced Reporting & Customization	X	✓	✓
Union Labor Type	✓	✓	✓
Open Shop*	Add on for additional fee	Add on for additional fee	✓
History (3-year lookback)**	Add on for additional fee	Add on for additional fee	Add on for additional fee
Full History (2007)	Add on for additional fee	Add on for additional fee	Add on for additional fee

*Open shop included in Complete Library & Bundles
** 3-year history lookback included with complete library

RSMeans Online

Unit prices according to the latest MasterFormat

Estimate with Precision

Find everything you need to develop complete, accurate estimates.

- Verified costs for construction materials
- Equipment rental costs
- Crew sizing, labor hours and labor rates
- Localized costs for U.S. and Canada

Save Time & Increase Efficiency

Make cost estimating and calculating faster and easier than ever with secure, online estimating tools.

- Quickly locate costs in the searchable database
- Create estimates in minutes with RSMeans cost lines
- Tag and store favorites for fast access to frequently used items

Improve Planning & Decision-Making

Back your estimates with complete, accurate and up-to-date cost data for informed business decisions.

- Verify construction costs from third parties
- Check validity of subcontractor proposals
- Evaluate material and assembly alternatives

Increase Profits

Use RSMeans Online to estimate projects quickly and accurately, so you can gain an edge over your competition.

- Create accurate and competitive bids
- Minimize the risk of cost overruns
- Reduce variability
- Gain control over costs

30 Day Free Trial

Register for a free trial at www.RSMeansOnline.com

- Access to unit prices, building assemblies and facilities repair and remodeling items covering every category of construction
- Powerful tools to customize, save and share cost lists, estimates and reports with ease
- Access to square foot models for quick conceptual estimates
- Allows up to 10 users to evaluate the full range of collaborative functionality used items.

2016 RSMeans Seminar Schedule

Note: call for exact dates and details.

Location	Dates	Location	Dates
Seattle, WA	January and August	Bethesda, MD	June
Dallas/Ft. Worth, TX	January	El Segundo, CA	August
Austin, TX	February	Jacksonville, FL	September
Anchorage, AK	March and September	Dallas, TX	September
Las Vegas, NV	March	Philadelphia, PA	October
New Orleans, LA	March	Houston, TX	October
Washington, DC	April and September	Salt Lake City, UT	November
Phoenix, AZ	April	Baltimore, MD	November
Kansas City, MO	April	Orlando, FL	November
Toronto	May	San Diego, CA	December
Denver, CO	May	San Antonio, TX	December
San Francisco, CA	June	Raleigh, NC	December

☎ 877-620-6245

RSMeans beginning early 2016 will be offering a suite of online self paced offerings. Check our website at www.RSMeans.com for more information.

Professional Development

eLearning Training Sessions

Learn how to use *RSMeans Online*® or the *RSMeans CostWorks*® CD from the convenience of your home or office. Our eLearning training sessions let you join a training conference call and share the instructors' desktops, so you can view the presentation and step-by-step instructions on your own computer screen. The live webinars are held from 9 a.m. to 4 p.m. or 11 a.m. to 6 p.m. eastern standard time, with a one-hour break for lunch.

For these sessions, you must have a computer with high speed Internet access and a compatible Web browser. Learn more at www.RSMeansOnline.com or call for a schedule: 781-422-5115. Webinars are generally held on selected Wednesdays each month.

RSMeans Online	RSMeans CostWorks CD
$299 per person	$299 per person

Professional Development

RSMeans Online Training

Construction estimating is vital to the decision-making process at each state of every project. RSMeansOnline works the way you do. It's systematic, flexible and intuitive. In this one day class you will see how you can estimate any phase of any project faster and better.

Some of what you'll learn:
- Customizing RSMeansOnline
- Making the most of RSMeans "Circle Reference" numbers
- How to integrate your cost data
- Generate reports, exporting estimates to MS Excel, sharing, collaborating and more

Also available: RSMeans Online training webinar

Facilities Construction Estimating

In this *two-day* course, professionals working in facilities management can get help with their daily challenges to establish budgets for all phases of a project.

Some of what you'll learn:
- Determining the full scope of a project
- Identifying the scope of risks and opportunities
- Creative solutions to estimating issues
- Organizing estimates for presentation and discussion
- Special techniques for repair/remodel and maintenance projects
- Negotiating project change orders

Who should attend: facility managers, engineers, contractors, facility tradespeople, planners, and project managers.

Construction Cost Estimating: Concepts and Practice

This introductory course to improve estimating skills and effectiveness starts with the details of interpreting bid documents and ends with the summary of the estimate and bid submission.

Some of what you'll learn:
- Using the plans and specifications for creating estimates
- The takeoff process—deriving all tasks with correct quantities
- Developing pricing using various sources; how subcontractor pricing fits in
- Summarizing the estimate to arrive at the final number
- Formulas for area and cubic measure, adding waste and adjusting productivity to specific projects
- Evaluating subcontractors' proposals and prices
- Adding insurance and bonds
- Understanding how labor costs are calculated
- Submitting bids and proposals

Who should attend: project managers, architects, engineers, owners' representatives, contractors, and anyone who's responsible for budgeting or estimating construction projects.

Maintenance & Repair Estimating for Facilities

This *two-day* course teaches attendees how to plan, budget, and estimate the cost of ongoing and preventive maintenance and repair for existing buildings and grounds.

Some of what you'll learn:
- The most financially favorable maintenance, repair, and replacement scheduling and estimating
- Auditing and value engineering facilities
- Preventive planning and facilities upgrading
- Determining both in-house and contract-out service costs
- Annual, asset-protecting M&R plan

Who should attend: facility managers, maintenance supervisors, buildings and grounds superintendents, plant managers, planners, estimators, and others involved in facilities planning and budgeting.

Practical Project Management for Construction Professionals

In this *two-day* course, acquire the essential knowledge and develop the skills to effectively and efficiently execute the day-to-day responsibilities of the construction project manager.

Some of what you'll learn:
- General conditions of the construction contract
- Contract modifications: change orders and construction change directives
- Negotiations with subcontractors and vendors
- Effective writing: notification and communications
- Dispute resolution: claims and liens

Who should attend: architects, engineers, owners' representatives, and project managers.

Mechanical & Electrical Estimating

This *two-day* course teaches attendees how to prepare more accurate and complete mechanical/electrical estimates, avoid the pitfalls of omission and double-counting, and understand the composition and rationale within the RSMeans mechanical/electrical database.

Some of what you'll learn:
- The unique way mechanical and electrical systems are interrelated
- M&E estimates—conceptual, planning, budgeting, and bidding stages
- Order of magnitude, square foot, assemblies, and unit price estimating
- Comparative cost analysis of equipment and design alternatives

Who should attend: architects, engineers, facilities managers, mechanical and electrical contractors, and others who need a highly reliable method for developing, understanding, and evaluating mechanical and electrical contracts.

Unit Price Estimating

This interactive *two-day* seminar teaches attendees how to interpret project information and process it into final, detailed estimates with the greatest accuracy level.

The most important credential an estimator can take to the job is the ability to visualize construction and estimate accurately.

Some of what you'll learn:
- Interpreting the design in terms of cost
- The most detailed, time-tested methodology for accurate pricing
- Key cost drivers—material, labor, equipment, staging, and subcontracts
- Understanding direct and indirect costs for accurate job cost accounting and change order management

Who should attend: corporate and government estimators and purchasers, architects, engineers, and others who need to produce accurate project estimates.

RSMeans CostWorks CD Training

This *one-day* course helps users become more familiar with the functionality of the *RSMeans CostWorks* program. Each menu, icon, screen, and function found in the program is explained in depth. Time is devoted to hands-on estimating exercises.

Some of what you'll learn:
- Searching the database using all navigation methods
- Exporting RSMeans data to your preferred spreadsheet format
- Viewing crews, assembly components, and much more
- Automatically regionalizing the database

This training session requires you to bring a laptop computer to class.

When you register for this course you will receive an outline for your laptop requirements.

Also offering web training for the RSMeans CostWorks CD!

Facilities Estimating Using the RSMeans CostWorks CD

Combines hands-on skill building with best estimating practices and real-life problems. Brings you up-to-date with key concepts and provides tips, pointers, and guidelines to save time and avoid cost oversights and errors.

Some of what you'll learn:
- Estimating process concepts
- Customizing and adapting RSMeans cost data
- Establishing scope of work to account for all known variables
- Budget estimating: when, why, and how
- Site visits: what to look for—what you can't afford to overlook
- How to estimate repair and remodeling variables

This training session requires you to bring a laptop computer to class.

Who should attend: facility managers, architects, engineers, contractors, facility tradespeople, planners, project managers and anyone involved with JOC, SABRE, or IDIQ.

Conceptual Estimating Using the RSMeans CostWorks CD

This *two-day* class uses the leading industry data and a powerful software package to develop highly accurate conceptual estimates for your construction projects. All attendees must bring a laptop computer loaded with the current year *Square Foot Costs* and the *Assemblies Cost Data* CostWorks titles.

Some of what you'll learn:
- Introduction to conceptual estimating
- Types of conceptual estimates
- Helpful hints
- Order of magnitude estimating
- Square foot estimating
- Assemblies estimating

Who should attend: architects, engineers, contractors, construction estimators, owners' representatives, and anyone looking for an electronic method for performing square foot estimating.

Assessing Scope of Work for Facility Construction Estimating

This *two-day* practical training program addresses the vital importance of understanding the scope of projects in order to produce accurate cost estimates for facility repair and remodeling.

Some of what you'll learn:
- Discussions of site visits, plans/specs, record drawings of facilities, and site-specific lists
- Review of CSI divisions, including means, methods, materials, and the challenges of scoping each topic
- Exercises in scope identification and scope writing for accurate estimating of projects
- Hands-on exercises that require scope, take-off, and pricing

Who should attend: corporate and government estimators, planners, facility managers, and others who need to produce accurate project estimates.

Unit Price Estimating Using the RSMeans CostWorks CD

Step-by-step instructions and practice problems to identify and track key cost drivers—material, labor, equipment, staging, and subcontractors—for each specific task. Learn the most detailed, time-tested methodology for accurately "pricing" these variables, their impact on each other and on total cost.

Some of what you'll learn:
- Unit price cost estimating
- Order of magnitude, square foot, and assemblies estimating
- Quantity takeoff
- Direct and indirect construction costs
- Development of contractors' bill rates
- How to use *RSMeans Building Construction Cost Data*

This training session requires you to bring a laptop computer to class.

Who should attend: architects, engineers, corporate and government estimators, facility managers, and government procurement staff.

Professional Development

771

For more information visit our website at www.RSMeans.com

Registration Information

Register early and save up to $100!
Register 30 days before the start date of a seminar and save $100 off your total fee. Note: This discount can be applied only once per order. It cannot be applied to team discount registrations or any other special offer.

How to register
Register by phone today! The RSMeans toll-free number for making reservations is 781-422-5115.

Two-day seminar registration fee - $935. One-day RSMeans CostWorks® or RSMeans Online training registration fee - $375.
To register by mail, complete the registration form and return, with your full fee, to: RSMeans Seminars, 1099 Hingham Street, Suite 201, Rockland, MA 02370.

One-day Construction Cost Estimating - $575.

Government pricing
All federal government employees save off the regular seminar price. Other promotional discounts cannot be combined with the government discount.

Team discount program
For over five attendee registrations. Call for pricing: 781-422-5115

Multiple course discounts
When signing up for two or more courses, call for pricing.

Refund policy
Cancellations will be accepted up to ten business days prior to the seminar start. There are no refunds for cancellations received later than ten working days prior to the first day of the seminar. A $150 processing fee will be applied for all cancellations. Written notice of the cancellation is required. Substitutions can be made at any time before the session starts. No-shows are subject to the full seminar fee.

AACE approved courses
Many seminars described and offered here have been approved for 14 hours (1.4 recertification credits) of credit by the AACE International Certification Board toward meeting the continuing education requirements for recertification as a Certified Cost Engineer/Certified Cost Consultant.

AIA Continuing Education
We are registered with the AIA Continuing Education System (AIA/CES) and are committed to developing quality learning activities in accordance with the CES criteria. Many seminars meet the AIA/CES criteria for Quality Level 2. AIA members may receive 14 learning units (LUs) for each two-day RSMeans course.

Daily course schedule
The first day of each seminar session begins at 8:30 a.m. and ends at 4:30 p.m. The second day begins at 8:00 a.m. and ends at 4:00 p.m. Participants are urged to bring a hand-held calculator since many actual problems will be worked out in each session.

Continental breakfast
Your registration includes the cost of a continental breakfast and a morning and afternoon refreshment break. These informal segments allow you to discuss topics of mutual interest with other seminar attendees. (You are free to make your own lunch and dinner arrangements.)

Hotel/transportation arrangements
RSMeans arranges to hold a block of rooms at most host hotels. To take advantage of special group rates when making your reservation, be sure to mention that you are attending the RSMeans seminar. You are, of course, free to stay at the lodging place of your choice. (Hotel reservations and transportation arrangements should be made directly by seminar attendees.)

Important
Class sizes are limited, so please register as soon as possible.

Note: Pricing subject to change.

Registration Form   ADDS-1000

Call 781-422-5115 to register or fax this form to 800-632-6732. Visit our website: www.RSMeans.com

Please register the following people for the RSMeans construction seminars as shown here. We understand that we must make our own hotel reservations if overnight stays are necessary.

❑ Full payment of $_____ enclosed.

❑ Bill me.

Please print name of registrant(s).

(To appear on certificate of completion)

_____

_____

_____

_____

P.O. #: _____
GOVERNMENT AGENCIES MUST SUPPLY PURCHASE ORDER NUMBER OR TRAINING FORM.

Firm name_____

Address_____

City/State/Zip_____

Telephone no._____ Fax no._____

E-mail address_____

Charge registration(s) to: ❑ MasterCard ❑ VISA ❑ American Express

Account no._____ Exp. date_____

Cardholder's signature_____

Seminar name_____

Seminar City_____

Please mail check to: 1099 Hingham Street, Suite 201, Rockland, MA 02370 USA